CHILTON'S
ELECTRONIC ENGINE CONTROLS MANUAL

Publisher and Editor-In-Chief Kerry A. Freeman, S.A.E.
Managing Editors Peter M. Conti, Jr. □ W. Calvin Settle, Jr., S.A.E
Assistant Managing Editor Nick D'Andrea
Senior Editors Richard J. Rivele, S.A.E. □ Ron Webb
Project Managers Ken Grabowski, A.S.E., S.A.E. □ Michael L. Grady
Martin J. Gunther □ Steven Morgan □ Richard T. Smith □ Jim Taylor
Service Editors Lawrence C. Braun, S.A.E., A.S.C. □ Robert E. Doughten
Jeff H. Fisher, A.S.E. □ Jacques Gordon □ Ben Greisler, S.A.E.
Jeffrey M. Hoffman □ Steve Horner □ Joseph T. Lafferty
Neil Leonard, A.S.E. □ Kevin Maher □ James R. Marotta
Robert McAnally □ Michael W. Parks □ John H. Rutter
Don Schnell, A.S.E., S.A.E. □ James B. Steele
Larry E. Stiles □ Anthony Tortorici, A.S.E., S.A.E.

Director of Manufacturing Mike D'Imperio
Manager of Manufacturing John F. Butler
Assistant Production Manager Andrea Steiger
Production Assistants Marsha Park Herman □ Monica Santa Maria □ Margaret Stoner
Mechanical Artists Lisa Gressen □ Lorraine Martinelli □ Kim Tansey

National Sales Manager Lawrence R. Rufo
National Sales Administrator Benjamin K. Tatta
National Account Sales Coordinator Jacquelyn T. Powers
Regional Sales Managers Herbert Marshall □ Larry W. Marshall □ Bruce McCorkle

OFFICERS
Senior Vice President Ronald A. Hoxter

AF882571

CHILTON PROFESSIONAL AUTOMOTIVE
*ONE OF THE **DIVERSIFIED PUBLISHING COMPANIES,***
*A PART OF **CAPITAL CITIES/ABC, INC.***
Manufactured in USA
©1992 Chilton Book Company
● Chilton Way, Radnor, Pa. 19089
ISBN 0-8019-8275-8 1234567890 1098765432 ISSN 1050-1134

HOW TO USE THIS MANUAL

For ease of use, this manual is divided into sections:

SECTION 1 Basic Electricity
SECTION 2 Diagnostic Equipment
SECTION 3 Self-Diagnostic Systems
SECTION 4 Electronic Ignition Systems
SECTION 5 Alfa Romeo Fuel Injection Systems
SECTION 6 Audi Fuel Injection Systems
SECTION 7 BMW Fuel Injection Systems
SECTION 8 Jaguar Fuel Injection Systems
SECTION 9 Mercedes-Benz Fuel Injection Systems
SECTION 10 Merkur Fuel Injection Systems
SECTION 11 Peugeot Fuel Injection Systems
SECTION 12 Porsche Fuel Injection Systems
SECTION 13 Saab Fuel Injection Systems
SECTION 14 Sterling Fuel Injection Systems
SECTION 15 Volkswagen Fuel Injection Systems
SECTION 16 Volvo Fuel Injection Systems
SECTION 17 Yugo Fuel Injection Systems
SECTION 18 Feedback Carburetor Systems
SECTION 19 Glossary

The **CONTENTS**, inside the front cover, summarizes the subjects covered in each section

To quickly locate the proper service section, use the chart on the following pages. It references applicable **MODELS** and **SERVICE SECTIONS** for major engine performance control systems.

It is recommended that the service technician be familiar with the applicable **GENERAL INFORMATION** and **SERVICE PRECAUTIONS** (if any) before testing or servicing the system. Major service sections are grouped by individual vehicle or component. Each sub-section contains:

- **GENERAL INFORMATION** Information pertaining to the operation of the system, individual components and the overall logic by which components work together.

- **SERVICE PRECAUTIONS (if any)** Precautions of which the service technician should be aware to prevent injury or damage to the vehicle or components.

- **TESTS, ADJUSTMENTS AND COMPONENT REMOVAL AND INSTALLATION** Performance tests and specifications, adjustments and component replacement procedures.

- **FAULT DIAGNOSIS** Complete troubleshooting for the entire system.

SAFETY NOTICE

Proper service and repair procedures are vital to the safe, reliable operation of all motor vehicles, as well as the personal safety of those performing repairs. This manual outlines procedures for servicing and repairing vehicles using safe, effective methods. The procedures contain many NOTES and CAUTIONS which should be followed along with standard safety procedures to eliminate the possibilty of personal injury or improper service which could damage the vehicle or compromise its safety.

It is important to note that the repair procedures and techniques, tools and parts for servicing motor vehicles, as well as the skill and experience of the individual performing the work vary widely. It is not possible to anticipate all of the conceivable ways or conditions under which vehicles may be serviced, or to provide cautions as to all of the possible hazards that may result. Standard and accepted safety precautions and equipment should be used when handling toxic or flammable fluids, and safety glasses or other protection should be used during cutting, grinding, chiseling, prying, or any other process that can cause material removal or projectiles.

Some procedures require the use of tools specially designed for a specific purpose. Before substituting another tool or procedure, you must be completely satisfied that neither your personal safety, nor the performance of the vehicle will be endangered

PART NUMBERS

Part numbers listed in this reference are not recommendations by Chilton Book Company for any product by brand name. They are references that can be used with interchange manuals and aftermarket supplier catalogs to locate each brand supplier's discrete part number.

Although information in this manual is based on industry sources and is as complete as possible at the time of publication, the possibilty exists that the manufacturer made later changes which could not be included here. While striving for total accuracy, Chilton Book Company cannot assume responsibility for any errors, changes or omissions that may occur in the compilation of this data.

No part of this publication may be reproduced, transmitted or stored in any form or by any means, electronic or mechanical, including photocopy, recording, or by information storage or retrieval system without prior written permission from the publisher.

QUICK LOCATOR INDEX

ALFA ROMEO

Year	Model	Body VIN	Engine Displacement Liters (cc)	Engine ID	Ignition System	Fuel System
1990	Spider	①	2.0 (1962)	AR01544	Electronic	EFI
	Spider Veloce	①	2.0 (1962)	AR01544	Electronic	EFI
	Graduate	①	2.0 (1962)	AR01544	Electronic	EFI
1991	Spider	①	2.0 (1962)	01588	Electronic	EFI
	Spider Veloce	①	2.0 (1962)	01588	Electronic	EFI
	164L	①	3.0 (2959)	G412T1	Electronic	EFI
	164S	①	3.0 (2959)	G412T2	Electronic	EFI
1992	Spider	①	2.0 (1962)	01588	Electronic	EFI
	Spider Veloce	①	2.0 (1962)	01588	Electronic	EFI
	164L	①	3.0 (2959)	G412T1	Electronic	EFI
	164S	①	3.0 (2959)	G412T2	Electronic	EFI

① VIN does not identify the body type
EFI—Electronic Fuel Injection

AUDI

Year	Model	Body VIN	Engine Displacement Liters (cc)	Engine ID (VIN)	Ignition System	Fuel System
1990	80	8A	2.0 (1983)	3A	Motronic	CIS-Motronic
	80 Quattro	8A	2.3 (2309)	NG	CIS-E III	CIS-E III
	90	8A	2.3 (2309)	NG	CIS-E III	CIS-E III
	90 Quattro	8A	2.3 (2309)	7A	MPI	MPI
	Coupé Quattro	8B	2.3 (2309)	7A	MPI	MPI
	100	44	2.3 (2309)	NF	CIS-E III	CIS-E III
	100 Quattro	44	2.3 (2309)	NF	CIS-E III	CIS-E III
	200	44	2.22 (2226)	MC	Electronic	CIS
	200 Quattro	44	2.22 (2226)	MC	Electronic	CIS
	200 Quattro Wagon	44	2.22 (2226)	MC	Electronic	CIS
	V8 Quattro	44	3.6 (3562)	PT	Motronic	Motronic
1991	80	8A	2.3 (2309)	NG	CIS-E III	CIS-E III
	80 Quattro	8A	2.3 (2309)	NG	CIS-E III	CIS-E III
	90	8A	2.3 (2309)	NG	CIS-E III	CIS-E III
	90 Quattro	8A	2.3 (2309)	7A	MPI	MPI
	Coupé Quattro	8B	2.3 (2309)	7A	MPI	MPI
	100	44	2.3 (2309)	NF	CIS-E III	CIS-E III
	100 Quattro	44	2.3 (2309)	NF	CIS-E III	CIS-E III
	200	44	2.22 (2226)	MC	Electronic	CIS
	200 Quattro	44	2.22 (2226)	3B	Motronic	Motronic
	200 Quattro Wagon	44	2.22 (2226)	3B	Motronic	Motronic
	V8 Quattro	4C	3.6 (3562)	PT	Motronic	Motronic
1992	80	8A	2.3 (2309)		CIS-E III	CIS-E III
	80 Quattro	8A	2.3 (2309)		CIS-E III	CIS-E III
	100	44	2.8 (2271)		MPI	MPI
	100S	44	2.8 (2271)		MPI	MPI
	100CS	44	2.8 (2271)		MPI	MPI
	100CS Quattro	44	2.8 (2271)		MPI	MPI
	100CS Quattro Wagon	44	2.8 (2271)		MPI	MPI
	S4	44	2.22 (2226)		Motronic	Motronic
	V8 Quattro	4C	4.2 (4172)	AB	Motronic	Motronic

MPI—Multi-Point Injection

BMW

Year	Model	Body ① VIN	Engine Displacement Liters (cc)	Engine ID	Ignition System	Fuel System
1990	325i	AA13 ②	2.5 (2494)	M20B25	DME	M1.3
	325iX	AB93 ③	2.5 (2494)	M20B25	DME	M1.3
	325iC	BB13	2.5 (2494)	M20B25	DME	M1.3
	525iC	HC13	2.5 (2494)	M20B25	DME	M1.3
	535i	HD13	3.5 (3430)	M30B35M	DME	M1.3
	735i	GB33	3.5 (3430)	M30B35M	DME	M1.3
	735iL	GC43	3.5 (3430)	M30B35M	DME	M1.3
	750iL	GC83	5.0 (4988)	M70B50	DME	M1.2
	M3	AK03	2.3 (2302)	S14B23	DME	M1.2 ④
1991	318i	AJ93	1.8 (1796)	M42B18	DME	M1.7
	318iS	AF93	1.8 (1796)	M42B18	DME	M1.7
	318iC	BA73	1.8 (1796)	M42B18	DME	M1.7
	325i	AA13 ②	2.5 (2494)	M20B25	DME	M1.3
	325iX	AB93 ③	2.5 (2494)	M20B25	DME	M1.3
	325iC	BB13	2.5 (2494)	M20B25	DME	M1.3
	525i	HD53	2.5 (2494)	M50B25	DME	M3.1
	535i	HD13	3.5 (3430)	M30B35M	DME	M1.3
	735i	GB33	3.5 (3430)	M30B35M	DME	M1.3
	735iL	GC43	3.5 (3430)	M30B35M	DME	M1.3
	750iL	GC83	5.0 (4988)	M70B50	DME	M1.7
	850i	EG13	5.0 (4988)	M70B50	DME	M1.7
	M3	AK03	2.3 (2302)	S14B23	DME	M1.2 ④
	M5	HD93	3.6 (34535	S38B36	DME	M1.2 ④
1992	318iC	BA73	1.8 (1796)	M42B18	DME	M1.7
	325iC	BB13	2.5 (2494)	M20B25	DME	M1.3
	325i	CB33	2.5 (2494)	M50B25	DME	M3.1
	525i	HD53	2.5 (2494)	M50B25	DME	M3.1
	535i	HD13	3.5 (3430)	M30B35M	DME	M1.3
	735i	GB43	3.5 (3430)	M30B35M	DME	M1.3
	735iL	GC43	3.5 (3430)	M30B35M	DME	M1.3
	750iL	GC83	5.0 (4988)	M70B50	DME	M1.7
	850i	EG13	5.0 (4988)	M70B50	DME	M1.7
	M5	HD93	3.6 (3453)	S38B36	DME	M1.2 ④

① Digits 4 through 7
② 325i 4 door—AD13, 325iS—AA13
③ 325iX 4 door—AE93
④ Motorsport Motronic injection system

JAGUAR

Year	Model	Body VIN	Engine Displacement Liters (cc)	Engine VIN	Ignition System ①	Fuel System
1990	XJ6	F, H, K	4.0 (3980)	D	Lucas	Lucas
	XJS	N	5.3 (5343)	W	Marelli	Lucas
1991	XJ6	F, H, K	4.0 (3980)	D	Lucas	Lucas
	XJS	N	5.3 (5343)	W	Marelli	Lucas
1992	XJ6	F, H, K	4.0 (3980)	D	Lucas	Lucas
	XJS	N	5.3 (5343)	W	Marelli	Lucas

① Lucas Digital Ignition on XJ6; Marelli-Digital Ignition of XJS

MERCEDES-BENZ

Year	Model	Body ID	Engine Displacement Liter (cc)	Engine ID	Ignition System	Fuel System
1990	190E 2.6	201.029	2.6 (2599)	103.942	EZL	CIS-E
	300D 2.5 Turbo	124.128	2.5 (2497)	602.962	—	EDS
	300E 2.6	124.026	2.6 (2599)	103.940	EZL	CIS-E
	300E	124.030	3.0 (2962)	103.983	EZL	CIS-E
	300CE	124.051	3.0 (2962)	104.980	EZL	CIS-E
	300TE	124.090	3.0 (2962)	103.983	EZL	CIS-E
	300E 4MATIC	124.230	3.0 (2962)	103.985	EZL	CIS-E
	300TE 4MATIC	124.290	3.0 (2962)	103.985	EZL	CIS-E
	350SDL Turbo	126.135	3.5 (3449)	603.970	—	EDS
	300SE	126.024	3.0 (2962)	103.981	EZL	CIS-E
	300SEL	126.025	3.0 (2962)	103.981	EZL	CIS-E
	420SEL	126.035	4.2 (4196)	116.965	EZL	CIS-E
	560SEL	126.039	5.6 (5547)	117.968	EZL	CIS-E
	560SEC	126.045	5.6 (5547)	117.968	EZL	CIS-E
	300SL	129.061	3.0 (2962)	104.981	EZL/AKR	CIS-E
	500SL	129.066	5.0 (4973)	119.960	EZL/AKR	CIS-E
1991	190E 2.3	201.028	2.3 (2299)	102.985	EZL	CIS-E
	190E 2.6	201.029	2.6 (2599)	103.942	EZL	CIS-E
	300D 2.5 Turbo	124.128	2.5 (2497)	602.962	—	EDS
	300E	124.030	3.0 (2960)	103.983	EZL	CIS-E
	300E 2.6	124.026	2.6 (2599)	103.940	EZL	CIS-E
	300CE	124.051	3.0 (2960)	104.980	EZL	CIS-E
	300TE	124.090	3.0 (2960)	103.983	EZL	CIS-E
	300E 4MATIC	124.230	3.0 (2960)	103.985	EZL	CIS-E
	300TE 4MATIC	124.290	3.0 (2960)	103.985	EZL	CIS-E
	350SD Turbo	126.134	3.5 (3449)	603.970	—	EDS
	350SDL Turbo	126.135	3.5 (3449)	603.970	—	EDS

MERCEDES-BENZ

Year	Model	Body ID	Engine Displacement Liter (cc)	Engine ID	Ignition System	Fuel System
1991	300SE	126.024	3.0 (2960)	103.981	EZL	CIS-E
	300SEL	126.025	3.0 (2960)	103.981	EZL	CIS-E
	420SEL	126.035	4.2 (4196)	116.965	EZL	CIS-E
	560SEL	126.039	5.6 (5547)	117.968	EZL	CIS-E
	560SEC	126.045	5.6 (5547)	117.968	EZL	CIS-E
	300SL	129.061	3.0 (2960)	104.981	EZL/AKR	CIS-E
	500SL	129.066	5.0 (4973)	119.960	EZL/AKR	CIS-E
1992	190E 2.3	201.028	2.3 (2299)	102.985	EZL	CIS-E
	190E 2.6	201.029	2.6 (2559)	103.942	EZL	CIS-E
	300D 2.5 Turbo	124.128	2.5 (2497)	602.962	—	EDS
	300E 2.6	124.026	2.6 (2559)	103.940	EZL	CIS-E
	300E	124.030	3.0 (2960)	103.983	EZL	CIS-E
	400E	124.034	4.2 (4196)	119.975	EZL/AKR	LH
	500E	124.036	5.0 (4973)	119.974	EZL/AKR	LH
	300CE	124.051	3.0 (2960)	104.980	EZL	CIS-E
	300TE	124.090	3.0 (2960)	103.983	EZL	CIS-E
	300E 4MATIC	124.230	3.0 (2960)	103.985	EZL	CIS-E
	300TE 4MATIC	124.290	3.0 (2960)	103.985	EZL	CIS-E
	300SD Turbo	140.134	3.5 (3449)	603.971	—	EDS
	300SE	140.032	3.2 (3199)	104.990	EZL/AKR	LH
	400SE	140.042	4.2 (4196)	119.971	EZL/AKR	LH
	500SEL	140.051	5.0 (4973)	119.970	EZL/AKR	LH
	600SEL	140.057	6.0 (5987)	120.980	EZL/AKR	LH
	300SL	129.061	3.0 (2960)	104.981	EZL/AKR	CIS-E
	500SL	129.066	5.0 (4973)	119.960	EZL/AKR	CIS-E

EZL—Electronic Ignition System
EZL/AKR—Electronic Ignition System with Knock Sensor
CIS-E—Continuous Injection System Electronic Control
LH—Mass Air Hot Wire Sensor Fuel Injection
EDS—Electronic Diesel System

MERKUR

Year	Model	Body VIN	Engine Displacement Liters (cc)	Engine VIN	Ignition System	Fuel System
1990	Scorpio	8	2.92 (2900)	Y	TFI	EEC-IV

TFI—Thick-Film Integrated
EEC-IV—Electronic Engine Control

PEUGEOT

Year	Model	Body VIN	Engine Displacement Liters (cc)	Engine ID	Ignition System	Fuel System
1990	405	DA	1.9 (1905)	XU9J2 ②	Motronic M1.3	Motronic M1.3
	405	DA	1.9 (1905)	XU9J4	Motronic M1.3	Motronic M1.3
	505	CF	2.2 (2165)	ZDJL	Bosch Electroinc	Jetronic LU-2
	505	CF	2.8 (2849)	ZN3J	Bosch Digital ①	Jetronic LH 2.2
	505	CE	2.2 (2155)	N9TEA	Bosch Digital ②	Jetronic LU
1991	405	DA	1.9 (1905)	XU9J2	Motronic M1.3	Motronic M1.3
	405	DA	1.9 (1905)	XU9J4	Motronic M1.3	Motronic M1.3
	505	CF	2.2 (2165)	ZDJL	Bosch Electronic	Jetronic LU-2
	505	CE	2.2 (2155)	N9TEA	Bosch Digital ②	Jetronic LU

① EZ 200K system with knock sensor
② EZ 200K system with knock sensor and wastegate control
③ Peugeot did not import cars for 1992.
④ Mi16

PORSCHE

Year	Model	Body④ VIN	Engine Displacement Liters (cc)	Engine ID	Ignition System	Fuel System
1990	911	9, 6, 4	3.6 (3600)	M64/01①	Twin spark	DME
	928	9, 2, 8	5.0 (4957)	M28/47②	EZK	LH-Jetronic
	944	9, 4, 4	3.0 (2990)	M44/41	DME	DME
1991	911	9, 6, 4	3.6 (3600)	M64/01①	Twin spark	DME
	911 Turbo	9, 2, 4	3.3 (3299)	M30/69	Digital spark	CIS
	928	9, 2, 8	5.0 (4957)	M28/47②	EZK	LH-Jetronic
	944	9, 4, 4	3.0 (2990)	M44/41	DME	DME
1992	911	9, 6, 4	3.6 (3600)	M64/01①	Twin spark	DME
	911 Turbo	9, 2, 4	3.3 (3299)	M30/69	Digital spark	CIS
	968	9, 6, 8	3.0 (2990)	M44/43③	DME	DME

CIS—Continuous Injection System
DME—Digital Motor Electronics
① M64/02 with automatic
② M28/42 with automatic
③ M44/44 with automatic
④ 7th, 8th and 12th digits

SAAB

Year	Model	Body① VIN	Engine Displacement Liters (cc)	Engine② ID (VIN)	Ignition System	Fuel System
1990	900	A	2.0 (1985)	B202 (D)	EZK	LH 2.4
	900S	A	2.0 (1985)	B202 (D)	EZK	LH 2.4
	900 Turbo	A	2.0 (1985)	B202 (L)	APC⑤	LH 2.4
	900 Turbo SPG	A	2.0 (1985)	B202 (L)	APC⑤	LH 2.4
	9000S	C	2.0 (1985)	B202 (D)	EZK	LH 2.4
	9000S	C	2.3 (2290)	B234 (B)	DI③	LH 2.4.1
	9000 Turbo	C	2.0 (1985)	B202 (L)	DI-APC④	LH 2.4
	9000 CD Turbo	C	2.0 (1985)	B202 (L)	DI-APC④	LH 2.4
1991	900	A	2.0 (1985)	B202 (D)	EZK	LH 2.4
	900	A	2.1 (2119)	B212 (E)	EZK	LH 2.4.2
	900S	A	2.1 (2119)	B212 (E)	EZK	LH 2.4.2
	900 Turbo	A	2.0 (1985)	B202 (L)	APC⑤	LH 2.4.2
	900 Turbo SPG	A	2.0 (1985)	B202 (L)	APC⑤	LH 2.4.2
	9000	C	2.3 (2290)	B234 (B)	DI③	LH 2.4.2
	9000S	C	2.3 (2290)	B234 (B)	DI③	LH 2.4.2
	9000CD	C	2.3 (2290)	B234 (B)	DI③	LH 2.4.2
	9000 Turbo	C	2.3 (2290)	B234 (M)	DI-APC④	LH 2.4
	9000 Turbo	C	2.3 (2290)	B234 (M)	DI-APC④	LH 2.4.2
	9000 CD Turbo	C	2.3 (2290)	B234 (M)	DI-APC④	LH 2.4.2

SAAB

Year	Model	Body ① VIN	Engine Displacement Liters (cc)	Engine ② ID (VIN)	Ignition System	Fuel System
1992	900	A	2.1 (2119)	B212 (E)	EZK	LH 2.4.2
	900S	A	2.1 (2119)	B212 (E)	EZK	LH 2.4.2
	900 Turbo	A	2.0 (1985)	B202 (L)	APC ⑤	LH 2.4.2
	9000	C	2.3 (2290)	B234 (B)	DI ③	LH 2.4.2
	9000S	C	2.3 (2290)	B234 (B)	DI ③	LH 2.4.2
	9000CD	C	2.3 (2290)	B234 (B)	DI ③	LH 2.4.2
	9000 Turbo	C	2.3 (2290)	B234 (M)	DI-APC ④	LH 2.4-ETS ⑥
	9000 Turbo	C	2.3 (2290)	B234 (M)	DI-APC ④	LH 2.4-ETS ⑥
	9000 Turbo	C	2.3 (2290)	B234 (M)	DI-APC ④	LH 2.4.2
	9000 CD Turbo	C	2.3 (2290)	B234 (M)	DI-APC ④	LH 2.4.2

① 4th VIN digit
② 8th VIN digit
③ Direct ignition (Distributorless; 9000 only)
④ Direct ignition—Automatic performance control (9000 Turbo only)
⑤ Automatic performance control (900 only)
⑥ Electronic throttle control system (with traction control)

STERLING

Year	Model	Body VIN	Engine Displacement Liters (cc)	Engine VIN	Ignition System	Fuel System
1990	827S	4	2.7 (2675)	G, H, J, K	PGM-IG	PGM-FI
	827SL	8	2.7 (2675)	G, H, J, K	PGM-IG	PGM-FI
1991	827S	4	2.7 (2675)	G, H, J, K	PGM-IG	PGM-FI
	827SL	8	2.7 (2675)	G, H, J, K	PGM-IG	PGM-FI

PGM-IG—Programmed Ignition System
PGM-FI—Programmed Fuel Injection System

VOLKSWAGEN

Year	Model	Body ① VIN	Engine Displacement Liters (cc)	Engine ID ②	Ignition System	Fuel System
1990	Fox	30	1.8 (1780)	UM/JN		CIS-E KE Jetronic
	Golf GL	17	1.8 (1780)	RV	Digifant	Digifant
	GTI	1G	1.8 (1780)	RV/PF	Digifant	Digifant
	GTI 16V	1G	2.0 (1984)	9A	Motronic	CIS-E Motronic
	Jetta GL	1G	1.8 (1780)	RV/PF	Digifant	Digifant
	Jetta Carat	1G	1.8 (1780)	RV/PF	Digifant	Digifant
	Jetta GLI 16V	1G	2.0 (1984)	9A	Motronic	CIS-E Motronic
	Cabriolet	15	1.8 (1780)	2M	Digifant	Digifant
	Passat GL	31	2.0 (1984)	9A	Motronic	CIS-E Motronic
	Corrado	50	1.8 (1780)	PG	Digifant	Digifant
	Vanagon	24	2.1 (2109)	MV	Digifant	Digifant
1991	Fox	30	1.8 (1780)	ABG	Digifant	Digifant
	Golf GL	17	1.8 (1780)	RV	Digifant	Digifant
	GTI	1G	1.8 (1780)	RV	Digifant	Digifant
	GTI 16V	1G	2.0 (1984)	9A	Motronic	CIS-E Motronic
	Jetta GL	1G	1.8 (1780)	RV	Digifant	Digifant
	Jetta Carat	1G	1.8 (1780)	RV	Digifant	Digifant
	Jetta GLI 16V	1G	2.0 (1984)	9A	Motronic	CIS-E Motronic
	Cabriolet	15	1.8 (1780)	2M	Digifant	Digifant
	Passat	31	2.0 (1984)	9A	Motronic	CIS-E Motronic
	Corrado	50	1.8 (1780)	PG	Digifant	Digifant
	Vanagon	24	2.1 (2109)	MV	Digifant	Digifant
1992	Fox	30	1.8 (1780)	ABG	Digifant	Digifant
	Golf	17	1.8 (1780)	RV	Digifant	Digifant
	GTI	1G	1.8 (1780)	RV/PF	Digifant	Digifant
	GTI 16V	1G	2.0 (1984)	9A	Motronic	CIS-E Motronic
	Jetta GL	1G	1.8 (1780)	RV	Digifant	Digifant
	Jetta Carat	1G	1.8 (1780)	RV/PF	Digifant	Digifant
	Jetta GLI 16V	1G	2.0 (1984)	9A	Motronic	CIS-E Motronic
	Cabriolet	15	1.8 (1780)	2M	Digifant	Digifant
	Passat CL	31	2.0 (1984)	9A	Motronic	CIS-E Motronic
	Passat GL	31	2.0 (1984)	9A	Motronic	CIS-E Motronic
	Passat GLS	31	2.8 (2861)	AAA	Motronic	Motronic
	Corrado G60	50	1.8 (1780)	PG	Digifant	Digifant
	Corrado SLC	50	2.8 (2861)	AAA	Motronic	Motronic
	Vanagon	24	2.1 (2109)	MV	Digifant	Digifant

① Body VIN is 7th and 8th digits
② Engine code is stamped on the cylinder block.

VOLVO

Year	Model	Body ① VIN	Engine Displacement Liters (cc)	Engine ID (VIN) ②	Ignition System	Fuel System
1990	240	A	2.3 (2316)	B-230F (88)	EZ116K	LH2.4
	240	A	2.3 (2316)	B-230F (88)	EZ116K	LH3.1
	740	F	2.3 (2316)	B-230F (88)	EZ116K	LH2.4
	740	F	2.3 (2316)	B-230F (88)	EZK-S	LH2.4
	740	F	2.3 (2316)	B-230F (88)	REX-1	Regina
	740 GLE	F	2.3 (2316)	B-234F (89)	EZ116K	LH2.4
	740 Turbo	F	2.3 (2316)	B-230F-Turbo (87)	EZ116K	LH2.4
	740 GLE	G	2.8 (2849)	B-280F (69)	EZ115K	LH2.2
	740 GLE Turbo	G	2.3 (2316)	B-230F-Turbo (87)	EZ116K	LH2.4
	780 Coupe	H	2.8 (2849)	B-280F (69)	EZ115K	LH2.2
	780 Turbo	H	2.3 (2316)	B-230F-Turbo (87)	EZ116K	LH2.4
1991	240	A	2.3 (2316)	B-230F (88)	EZ116K	LH2.4
	240	A	2.3 (2316)	B-230F (88)	EZ116K	LH3.1
	740 ③	F	2.3 (2316)	B-230F (88)	REX-1	Regina
	740 ④	F	2.3 (2316)	B-230F (88)	EZ116K	LH2.4
	740 Turbo	F	2.3 (2316)	B-230F-Turbo (87)	EZ116K	LH2.4
	780 Coupe	H	2.3 (2316)	B-230F-Turbo (87)	EZ116K	LH2.4
	940 GLE	J	2.3 (2316)	B-234F (89)	EZ116K	LH2.4
	940 Turbo	J	2.3 (2316)	B-230F-Turbo (87)	EZ116K	LH2.4
	940 SE	J	2.3 (2316)	B-230F-Turbo (87)	EZ116K	LH2.4
1992	240	A	2.3 (2316)	B-230F (82)	EZ116K	LH3.1
	240	A	2.3 (2316)	B-230F (88)	EZ116K	LH2.4
	740 ③	F	2.3 (2316)	B-230F (88)	REX-1	Regina
	740 ④	F	2.3 (2316)	B-230F (88)	EZ116K	LH2.4
	740 Turbo Wagon	F	2.3 (2316)	B-230F-Turbo (87)	EZ116K	LH2.4
	940 GL	J	2.3 (2316)	B-230F (88)	REX-1	Regina
	940 GLE	J	2.3 (2316)	B-234F (89)	EZ116K	LH2.4
	940 Turbo	J	2.3 (2316)	B-230F-Turbo (87)	EZ116K	LH2.4
	960	K	2.9 (2922)	B-6304F (95)	Motronic 1.8 ⑤	Motronic 1.8

① 4th VIN digit
② 6th and 7th VIN digits
③ Except California
④ California
⑤ Distributorless ignition system

YUGO

Year	Model	Body VIN	Engine Displacement Liters (cc)	Engine VIN	Ignition System	Fuel System
1990	GV	BD	1.1 (1116)	2	Electronic	FBC
			1.1 (1116)	6	Electronic	Motronic
1991	GV	BD	1.1 (1116)	6	Electronic	Motronic
1992	GV	BD	1.1 (1116)	6	Electronic	Motronic

FBC—Feedback Carburetor
Motronic—Bosch Motronic MP3.1 fuel system

Basic Electricity

SECTION 1

FUNDAMENTALS OF ELECTRICITY

In order to be able to successfully troubleshoot and repair any electrical problem, one must first understand the theory behind how electricity works. A clear understanding of how electrical current behaves in a circuit and how components such as solenoids and relays work is essential for correct interpretation of test results. The following section covers basic knowledge necessary to perform all service and testing procedures outlined in this manual. This section should be read before attempting any repair or diagnosis procedures.

All matter is made up of tiny particlės called molecules. Each molecule is made up of two or more atoms. Atoms may be divided into even smaller particles called protons, neutrons and electrons. These particles are the same in all matter and differences in materials (hard or soft, conductive or non-conductive) occur only because of the number and arrangement of these particles. In other words, the protons, neutrons and electrons in a drop of water are the same as those in an ounce of lead, there are just more of them (arranged differently) in a lead molecule than in a water molecule. Protons and neutrons packed together form the nucleus of the atom, while electrons orbit around the nucleus much the same way as the planets of the solar system orbit around the sun.

The proton is a small positive natural charge of electricity, while the neutron has no electrical charge. The electron carries a negative charge equal to the positive charge of the proton. Every electrically neutral atom contains the same number of protons and electrons, the exact number of which determines the element. The only difference between a conductor and an insulator is that a conductor possesses free electrons in large quantities, while an insulator has only a few. An element must have very few free electrons to be a good insulator, and vice-versa. When we speak of electricity, we're talking about these free electrons.

In a conductor, the movement of the free electrons is hindered by collisions with the adjoining atoms of the element (matter). This hindrance to movement is called RESISTANCE and it varies with different materials and temperatures. As temperature increases, the movement of the free electrons increases, causing more frequent collisions and therefore increasing resistance to the movement of the electrons. The number of collisions (resistance) also increases with the number of electrons flowing (current). Current is defined as the movement of electrons through a conductor such as a wire. In a conductor (such as copper) electrons can be caused to leave their atoms and move to other atoms. This flow is continuous in that every time an atom gives up an electron, it collects another one to take its place. This movement of electrons is called electric current and is measured in amperes. When 6.28 billion, billion electrons pass a certain point in the circuit in one second, the amount of current flow is called one ampere.

The force or pressure which causes electrons to flow in any conductor (such as a wire) is called voltage. It is measured in volts and is similar to the pressure that causes water to flow in a pipe. Voltage is the difference in electrical pressure measured between two different points in a circuit. In a 12 volt system, for example, the force measured between the two battery posts is 12 volts. Two important concepts are voltage potential and polarity. Voltage potential is the amount of voltage or electrical pressure at a certain point in the circuit with respect to another point. For example, if the voltage potential at one post of the 12 volt battery is zero, the voltage potential at the other post is 12 volts with respect to the first post. One post of the battery is said to be positive (+); the other post is negative (-) and the conventional direction of current flow is from positive to negative in an electrical circuit. It should be noted that the electron flow in the wire is opposite the current flow. In other words, when the circuit is energized, the current flows from positive to negative, but the electrons actually move from negative to positive. The voltage or pressure needed to produce a current flow in a circuit must be greater than the resistance present in the circuit. In other words, if the voltage drop across the resistance is greater than or equal to the voltage input, the voltage potential will be zero—no voltage will flow through the circuit. Resistance to the flow of electrons is measured in ohms. One volt will cause one ampere to flow through a resistance of one ohm.

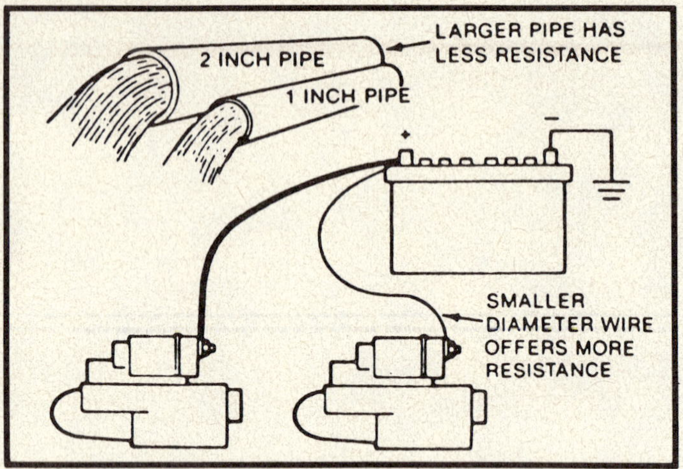

Electrical resistance can be compared to water flow through a pipe. The smaller the wire (pipe), the more resistance to the flow of electrons (water)

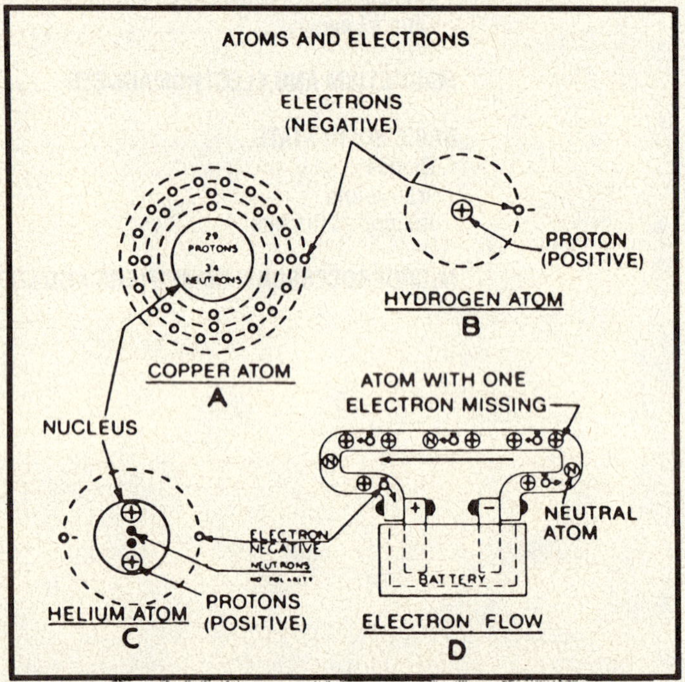

Typical atoms of copper (A), hydrogen (B) and helium (C). Electron flow in battery circuit (D)

Units Of Electrical Measurement

There are three fundamental characteristics of a direct-current electrical circuit: volts, amperes and ohms.

VOLTAGE

The voltage in a circuit controls the intensity with which the loads in the circuit operate. The brightness of a lamp, the heat of an electrical defroster, the speed of a motor are all directly proportional to the voltage, if the resistance in the circuit and/or mechanical load on electric motors remains constant. Voltage available from the battery is constant (normally 12 volts), but as it operates the various loads in the circuit, voltage decreases (drops).

AMPERE

An ampere is the unit of measurement of current in an electrical circuit. One ampere is the quantity of current that will flow through a resistance of one ohm at a pressure of one volt. The amount of current that flows in a circuit is controlled by the voltage and the resistance in the circuit. Current flow is directly proportional to resistance. Thus, as voltage is increased or decreased, current is increased or decreased accordingly. Current is decreased as resistance is increased, however, and is increased as resistance is decreased. With little or no resistance in a circuit, current is high.

OHM

The ohm is the unit of measurement of resistance, represented by the Greek letter Omega (Ω). One ohm is the resistance of a conductor through which a current of one ampere will flow at a pressure of one volt. Electrical resistance can be measured on an instrument called an ohmmeter. The loads (electrical devices) are the primary resistances in a circuit. Loads such as lamps, solenoids, and electric heaters have a resistance that is essentially fixed; at a normal fixed voltage, they will draw a fixed current. Motors, on the other hand, do not have a fixed resistance. Increasing the mechanical load on a motor (such as might be caused by a misadjusted track in a power window system) will decrease the motor speed. The drop in motor rpm has the effect of reducing the internal resistance of the motor because the current draw of the motor varies directly with the mechanical load on the motor, although its actual resistance is unchanged. Thus, as the motor load increases, the current draw of the motor increases, and may increase up to the point where the motor stalls (cannot move the mechanical load).

Circuits are designed with the total resistance of the circuit taken into account. Troubles can arise when unwanted resistances enter into a circuit. If corrosion, dirt, grease, or any other contaminant occurs in places like switches, connectors, and grounds, or if loose connections occur, resistances will develop in these areas. These resistances act like additional loads in the circuit and cause problems.

OHMS LAW

Ohms law is a statement of the relationship between the three fundamental characteristics of an electrical circuit. These rules apply to direct current (DC) only.

Ohms law provides a means to make an accurate circuit analysis without actually seeing the circuit. If, for example, one wanted to check the condition of the rotor winding in a alternator whose specifications indicate that the field (rotor) current draw is normally 2.5 amperes at 12 volts, simply connect the rotor to a 12 volt battery and measure the current with an ammeter. If it measures about 2.5 amperes, the rotor winding can be assumed good.

An ohmmeter can be used to test components that have been removed from the vehicle in much the same manner as an ammeter. Since the voltage and the current of the rotor windings used as an earlier example are known, the resistance can be calculated using Ohms law. The formula would be:

If the rotor resistance measures about 4.8 ohms when checked with an ohmmeter, the winding can be assumed good. By plugging in different specifications, additional circuit information can be determined such as current draw, etc.

$$I = \frac{E}{R} \quad \text{or} \quad \text{AMPERES} = \frac{\text{VOLTS}}{\text{OHMS}}$$

$$R = \frac{E}{I} \quad \text{or} \quad \text{OHMS} = \frac{\text{VOLTS}}{\text{AMPERES}}$$

$$E = I \times R \quad \text{or} \quad \text{VOLTS} = \text{AMPERES} \times \text{OHMS}$$

$$R = \frac{E}{I} \qquad \text{Where: } E = 12 \text{ volts}, \quad I = 2.5 \text{ amperes}$$

$$R = \frac{12 \text{ volts}}{2.5 \text{ amps}} = 4.8 \text{ ohms}$$

Electrical Circuits

An electrical circuit must start from a source of electrical supply and return to that source through a continuous path. Circuits are designed to handle a certain maximum current flow. The maximum allowable current flow is designed higher than the normal current requirements of all the loads in the circuit. Wire size, connections, insulation, etc., are designed to prevent undesirable voltage drop, overheating of conductors, arcing of contacts, and other adverse effects. If the safe maximum current flow level is exceeded, damage to the circuit components will result; it is this condition that circuit protection devices are designed to prevent.

Protection devices are fuses, fusible links or circuit breakers designed to open or break the circuit quickly whenever an overload, such as a short circuit, occurs. By opening the circuit quickly, the circuit protection device prevents damage to the wiring, battery, and other circuit components. Fuses and fusible links are designed to carry a preset maximum amount of current and to melt when that maximum is exceeded, while circuit breakers merely break the connection and may be manually reset. The maximum amperage rating of each fuse is marked on the fuse body and all contain a see-through portion that shows the break in the fuse element when blown. Fusible link maximum amperage rating is indicated by gauge or thickness of the wire. Never replace a blown fuse or fusible link with one of a higher amperage rating.

CAUTION

Resistance wires, like fusible links, are also spliced into conductors in some areas. Do not make the mistake of replacing a fusible link with a resistance wire. Resistance wires are longer than fusible links and are stamped "RESISTOR-DO NOT CUT OR SPLICE."

Circuit breakers consist of two strips of metal which have different coefficients of expansion. As an overload or current flows through the bimetallic strip, the high-expansion metal will elongate due to heat and break the contact. With the circuit open, the bimetal strip cools and shrinks, drawing the strip down until contact is re-established and current flows once again. In actual operation, the contact is broken very quickly if the overload is

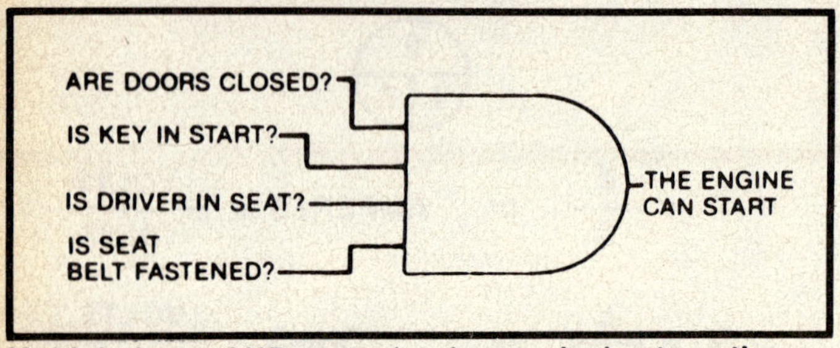

Multiple input AND operation in a typical automotive starting circuit

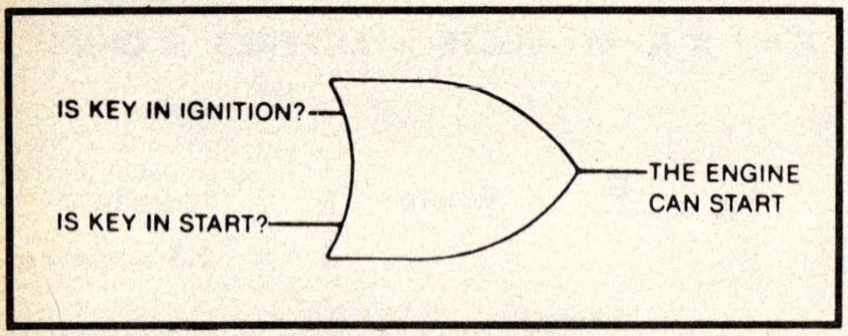

Typical two-input OR circuit operation

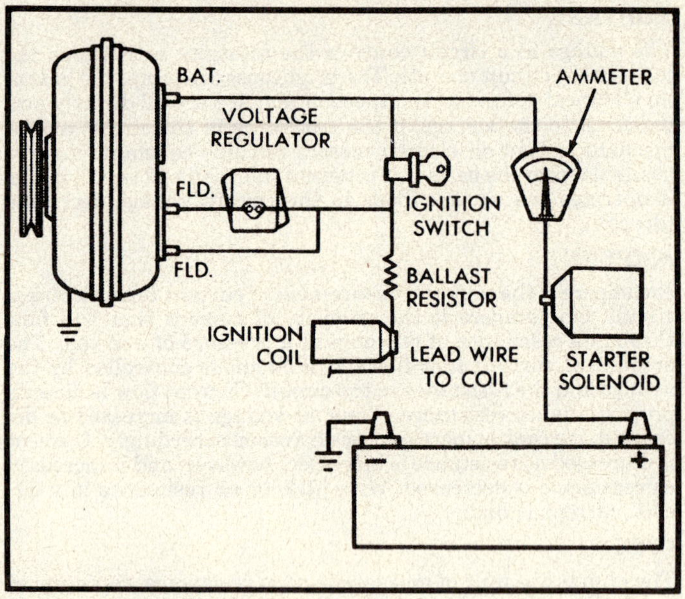

Typical alternator circuit

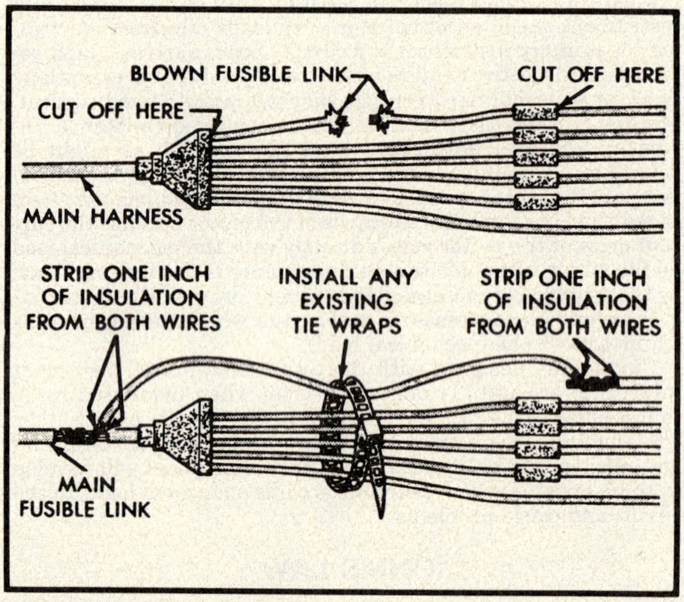

Typical fusible link repair

continuous and the circuit will be repeatedly broken and remade until the source of the overload is corrected.

The self-resetting type of circuit breaker is the one most generally used in automotive electrical systems. On manually reset circuit breakers, a button will pop up on the circuit breaker case. This button must be pushed in to reset the circuit breaker and restore power to the circuit. Always repair the source of the overload before resetting a circuit breaker or replacing a fuse or fusible link. When searching for overloads, keep in mind that the circuit protection devices protect only against overloads between the protection device and ground.

There are two basic types of circuit; Series and Parallel. In a series circuit, all of the elements are connected in chain fashion with the same amount of current passing through each element or load. No matter where an ammeter is connected in a series circuit, it will always read the same. The most important fact to remember about a series circuit is that the sum of the voltages across each element equals the source voltage. The total resistance of a series circuit is equal to the sum of the individual resistances within each element of the circuit. Using ohms law, one can determine the voltage drop across each element in the circuit. If the total resistance and source voltage is known, the amount of current can be calculated. Once the amount of current (amperes) is known, values can be substituted in the Ohms law formula to calculate the voltage drop across each individual element in the series circuit. The individual voltage drops must add up to the same value as the source voltage.

A parallel circuit, unlike a series circuit, contains two or more branches, each branch a separate path independent of the others. The total current draw from the voltage source is the sum of all the currents drawn by each branch. Each branch of a parallel circuit can be analyzed separately. The individual branches can be either simple circuits, series circuits or combinations of series-parallel circuits. Ohms law applies to parallel circuits just as it applies to series circuits, by considering each branch independently of the others. The most important thing to remember is that the voltage across each branch is the same as the source voltage. The current in any branch is that voltage divided by the resistance of the branch. A practical method of determining the resistance of a parallel circuit is to divide the product of the two resistances by the sum of two resistances at a time. Amperes through a parallel circuit is the sum of the amperes through the separate branches. Voltage across a parallel circuit is the same as the voltage across each branch.

By measuring the voltage drops, you are in effect measuring the resistance of each element within the circuit. The greater the voltage drop, the greater the resistance. Voltage drop measurements are a common way of checking circuit resistances in automotive electrical systems. When part of a circuit developes excessive resistance (due to a bad connection) the element will show a higher than normal voltage drop. Normally, automotive wiring is selected to limit voltage drops to a few tenths of a volt. In parallel circuits, the total resistance is less than the sum of the individual resistances; because the current has two paths to take, the total resistance is lower.

Magnetism and Electromagnets

Electricity and magnetism are very closely associated because when electric current passes through a wire, a magnetic field is

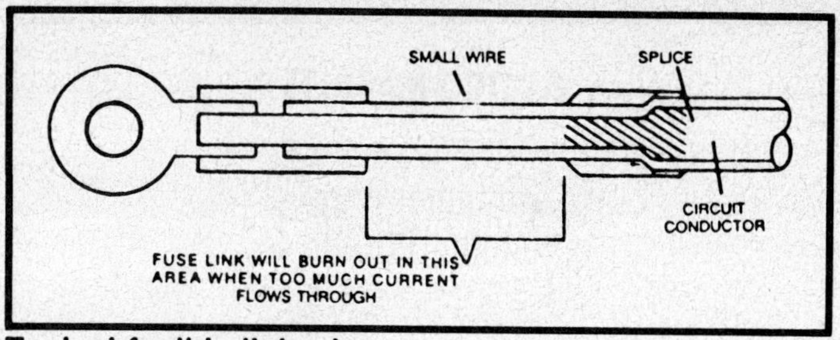

Typical fusible link wire

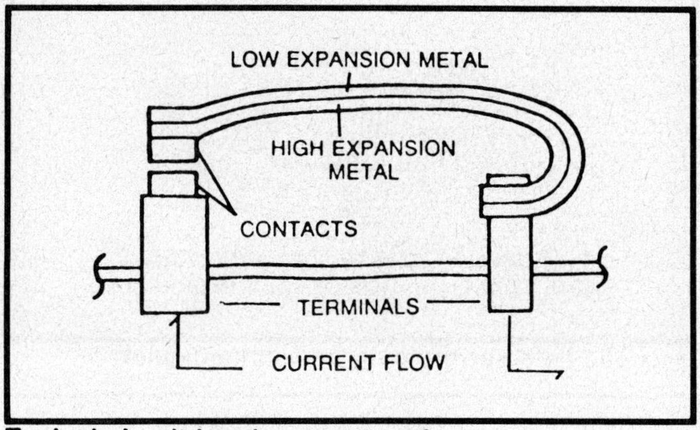

Typical circuit breaker construction

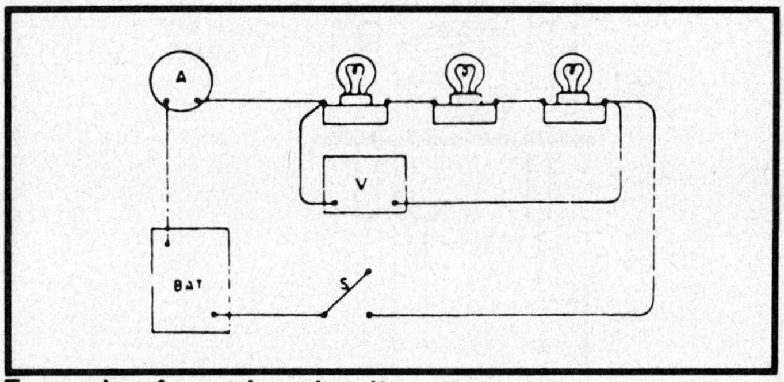

Example of a series circuit

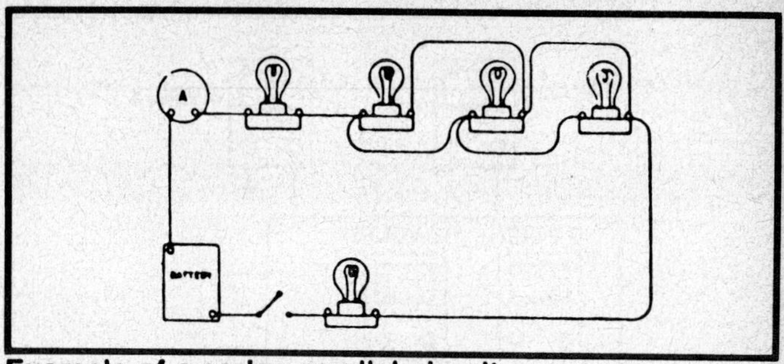

Example of a series-parallel circuit

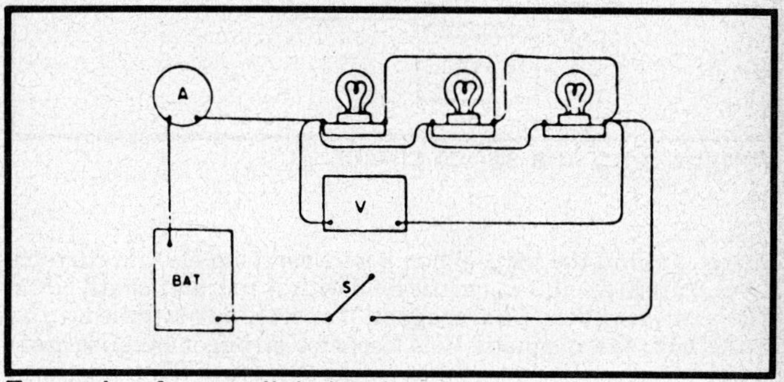

Example of a parallel circuit

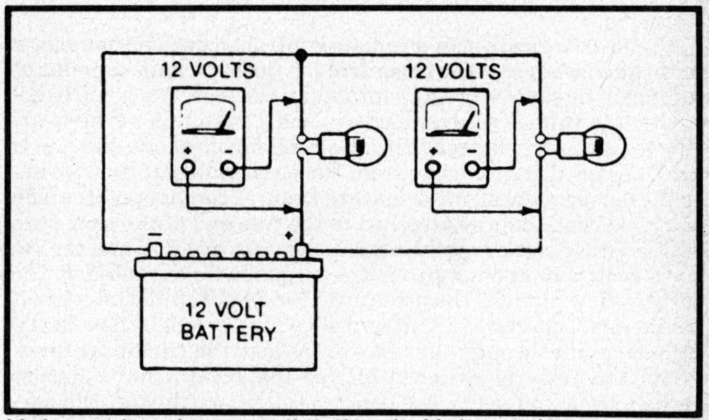

Voltage drop in a parallel circuit. Voltage drop across each lamp is 12 volts

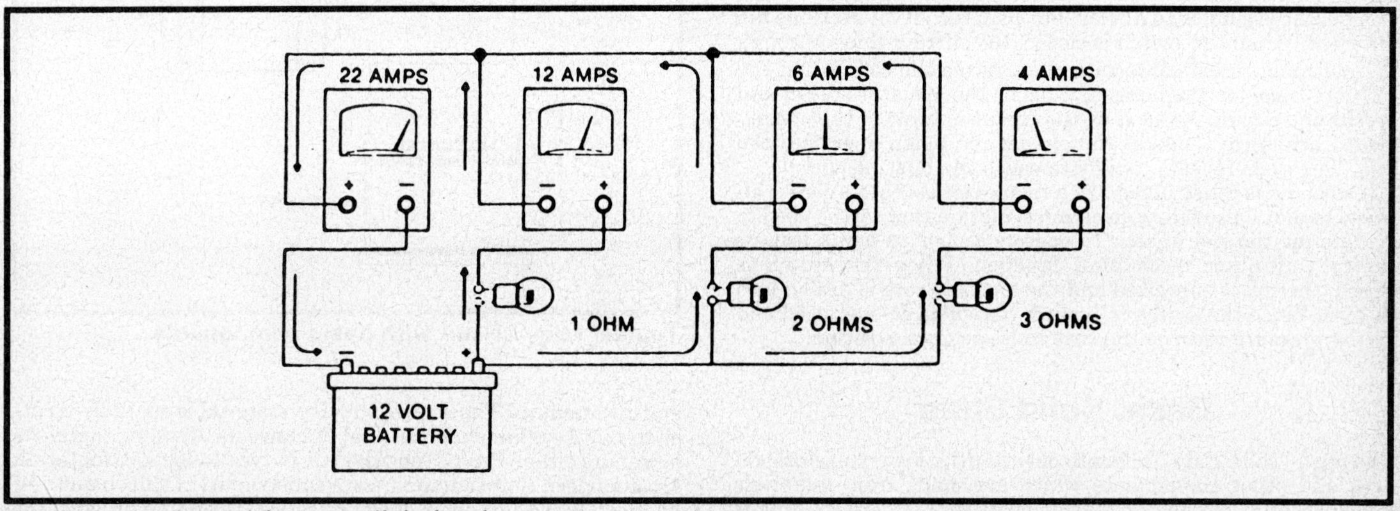

Total current (amps) in a parallel circuit. 4 + 6 + 12 = 22 amps

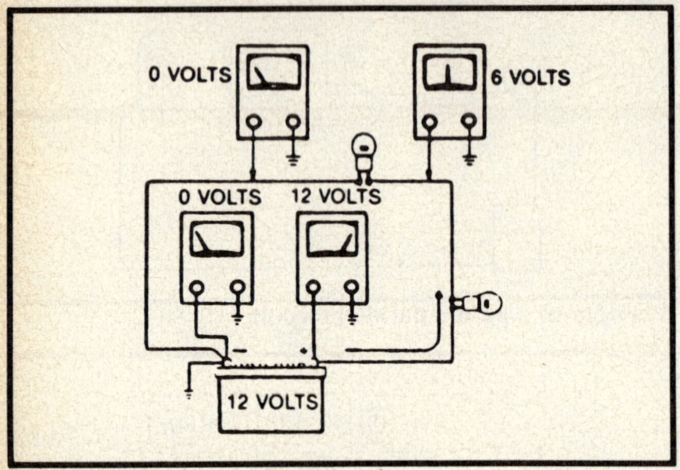

Voltage drop in a series circuit

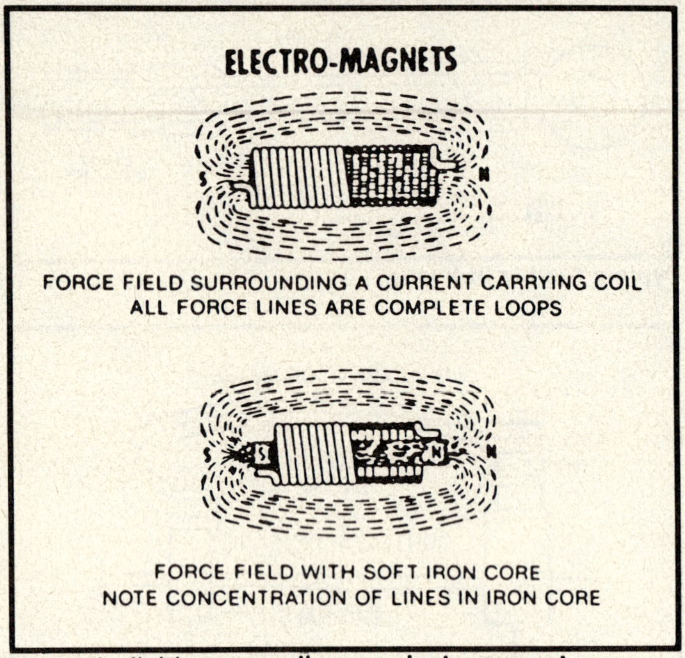

Magnetic field surrounding an electromagnet

created around the wire. When a wire carrying electric current is wound into a coil, a magnetic field with North and South poles is created just like in a bar magnet. If an iron core is placed within the coil, the magnetic field becomes stronger because iron conducts magnetic lines much easier than air. This arrangement is called an electromagnet and is the basic principle behind the operation of such components as relays, buzzers and solenoids.

A relay is basically just a remote-controlled switch that uses a small amount of current to control the flow of a large amount of current. The simplest relay contains an electromagetic coil in series with a voltage source (battery) and a switch. A movable armature made of some magnetic material pivots at one end and is held a small distance away from the electromagnet by a spring or the spring steel of the armature itself. A contact point, made of a good conductor, is attached to the free end of the armature with another contact point a small distance away. When the relay is switched on (energized), the magnetic field created by the current flow attracts the armature, bending it until the contact points meet, closing a circuit and allowing current to flow in the second circuit through the relay to the load the circuit operates. When the relay is switched off (de-energized), the armature springs back and opens the contact points, cutting off the current flow in the secondary, or controlled, circuit. Relays can be designed to be either open or closed when energized, depending on the type of circuit control a manufacturer requires.

A buzzer is similar to a relay, but its internal connections are different. When the switch is closed, the current flows through the normally closed contacts and energizes the coil. When the coil core becomes magnetized, it bends the armature down and breaks the circuit. As soon as the circuit is broken, the spring-loaded armature remakes the circuit and again energizes the coil. This cycle repeats rapidly to cause the buzzing sound.

A solenoid is constructed like a relay, except that its core is allowed to move, providing mechanical motion that can be used to actuate mechanical linkage to operate a door or trunk lock or control any other mechanical function. When the switch is closed, the coil is energized and the movable core is drawn into the coil. When the switch is opened, the coil is de-energized and spring pressure returns the core to its original position.

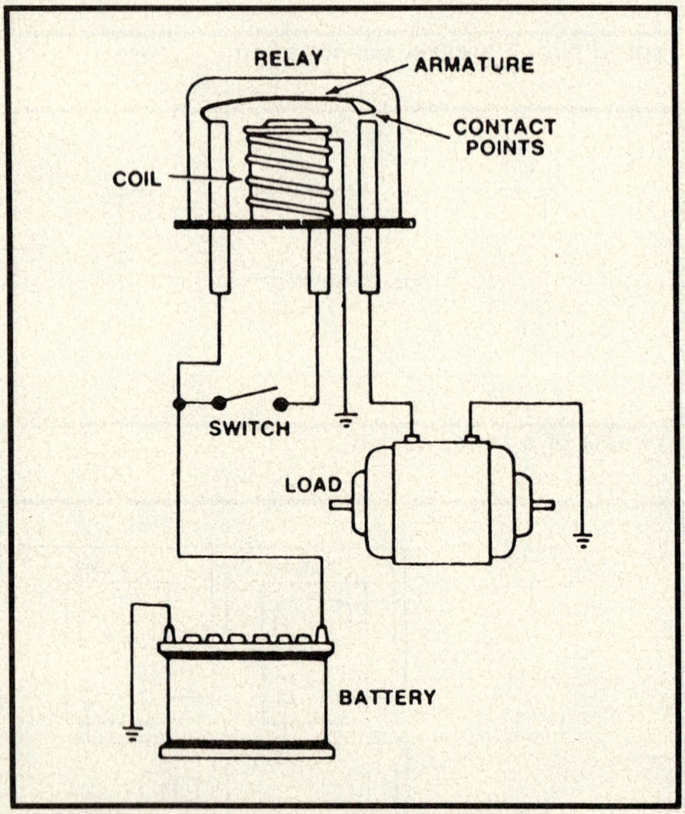

Typical relay circuit with basic components

Basic Solid State

The term "solid state" refers to devices utilizing transistors, diodes and other components which are made from materials known as semiconductors. A semiconductor is a material that is neither a good insulator or a good conductor; principally silicon and germanium. The semiconductor material is specially treated to give it certain qualities that enhance its function, therefore becoming either P-type (positive) or N-type (negative) material. Most modern semiconductors are constructed of silicon and can be made to change their characteristics depending on whether its function calls for an insulator or conductor.

DIODES

The simplest semiconductor function is that of the diode or rectifier (the two terms mean the same thing). A diode will pass current in one direction only, like a one-way valve, because it has low resistance in one direction and high resistance on the other. Whether the diode conducts or not depends on the polarity of the voltage applied to it. A diode has two electrodes, an anode and a cathode. When the anode receives positive (+) voltage and the cathode receives negative (–) voltage, current can flow easily through the diode. When the voltage is reversed, the diode becomes non-conducting and only allows a very slight amount of current to flow in the circuit. Because the semiconductor is not a perfect insulator, a small amount of reverse current leakage will occur, but the amount is usually too small to consider. The application of voltage to maintain the current flow described is called "forward bias."

A light-emitting diode (LED) is made of a particular type of crystal that glows when current is passed through it. LED's are used in display faces of many digital or electronic instrument clusters. LED's are usually arranged to display numbers (digital readout), but can be used to illuminate a variety of electronic graphic displays.

Like any other electrical device, diodes have certain ratings that must be observed and should not be exceeded. The forward current rating (or bias) indicates how much current can safely pass through the diode without causing damage or destroying it. Forward current rating is usually given in either amperes or milliamperes. The voltage drop across a diode remains constant regardless of the current flowing through it. Small diodes designed to carry low amounts of current need no special provision for dissipating the heat generated in any electrical device, but large current carrying diodes are usually mounted on heat sinks to keep the internal temperature from rising to the point where the silicon will melt and destroy the diode. When diodes are operated in a high ambient temperature environment, they must be de-rated to prevent failure.

Another diode specification is its peak inverse voltage rating. This value is the maximum amount of voltage the diode can safely handle when operating in the blocking mode. This value can be anywhere from 50–1000 volts, depending on the diode, and if exceeded can damage the diode just as too much forward current will. Most semiconductor failures are caused by excessive voltage or internal heat.

One can test a diode with a small battery and a lamp with the same voltage rating. With this arrangement one can find a bad diode and determine the polarity of a good one. A diode can fail and cause either a short or open circuit, but in either case it fails to function as a diode. Testing is simply a matter of connecting the test bulb first in one direction and then the other and making sure that current flows in one direction only. If the diode is shorted, the test bulb will remain on no matter how the light is connected.

TRANSISTORS

The transistor is an electrical device used to control voltage within a circuit. A transistor can be considered a "controllable diode" in that, in addition to passing or blocking current, the transistor can control the amount of current passing through it. Simple transistors are composed of three pieces of semiconductor material, P and N type, joined together and enclosed in a container. If two sections of P material and one section of N material are used, it is known as a PNP transistor; if the reverse is true, then it is known as an NPN transistor. The two types cannot be interchanged.

Most modern transistors are made from silicon (earlier transistors were made from germanium) and contain three elements; the emitter, the collector and the base. In addition to passing or blocking current, the transistor can control the

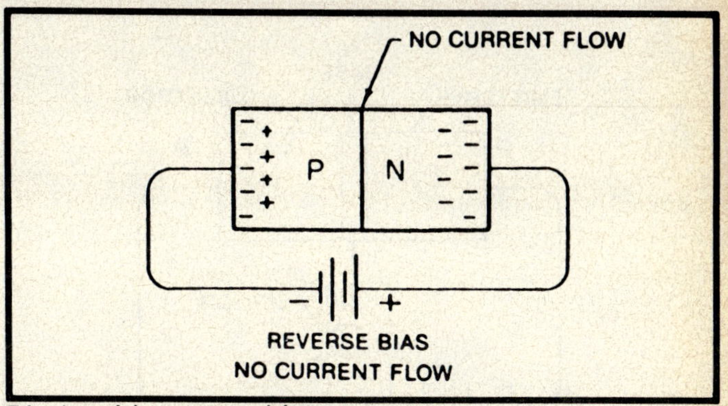

Diode with reverse bias

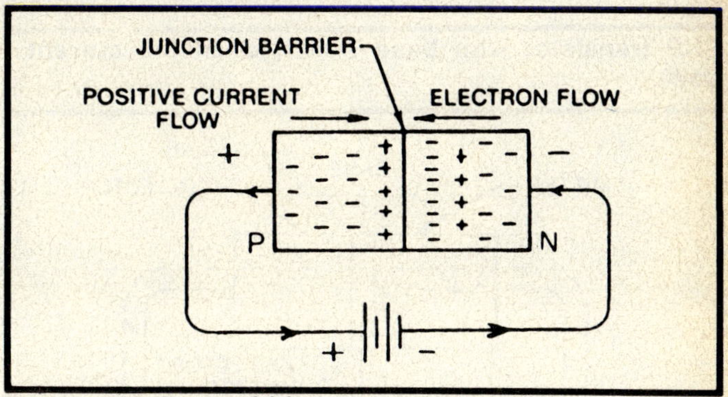

Diode with forward bias

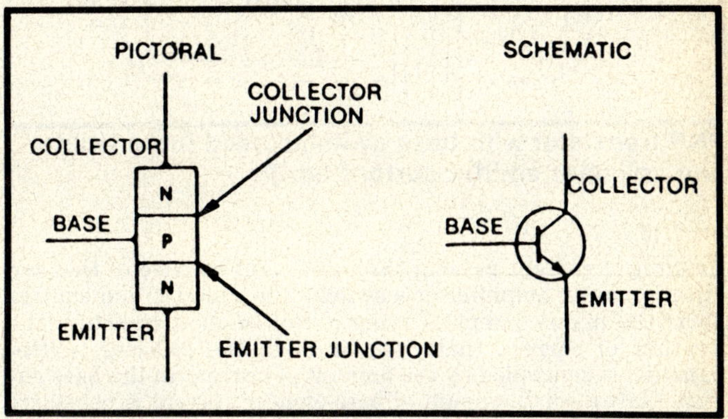

NPN transistor showing both pictorial and schematic illustrations

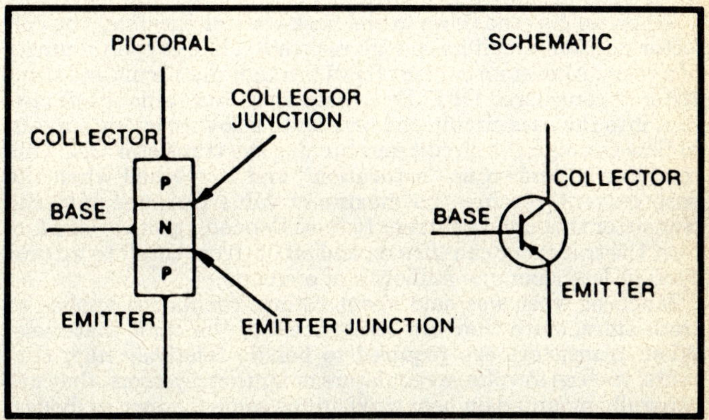

PNP transistor showing both pictorial and schematic illustrations

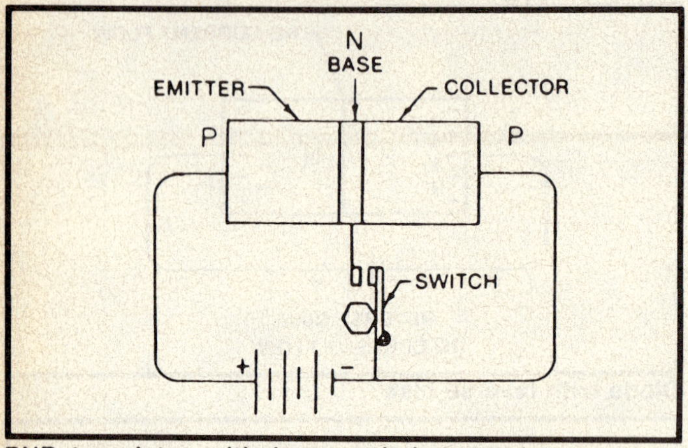

PNP transistor with base switch open (no current flow)

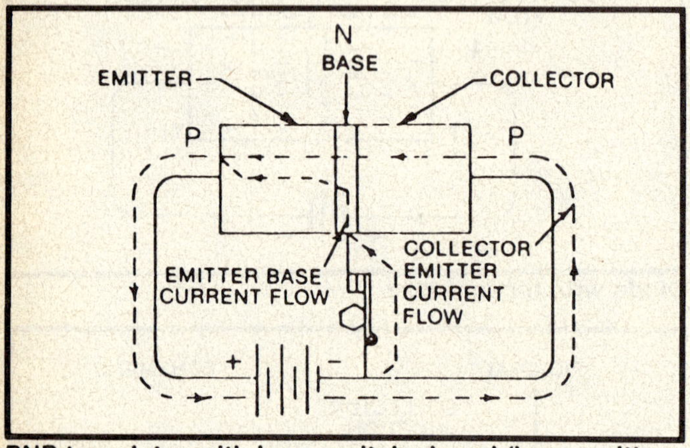

PNP transistor with base switch closed (base emitter and collector emitter current flow)

amount of current passing through it and because of this can function as an amplifier or a switch. The collector and emitter form the main current-carrying circuit of the transistor. The amount of current that flows through the collector-emitter junction is controlled by the amount of current in the base circuit. Only a small amount of base-emitter current is necessary to control a large amount of collector-emitter current (the amplifier effect). In automotive applications, however, the transistor is used primarily as a switch.

When no current flows in the base-emitter junction, the collector-emitter circuit has a high resistance, like to open contacts of a relay. Almost no current flows through the circuit and transistor is considered OFF. By bypassing a small amount of current into the base circuit, the resistance is low, allowing current to flow through the circuit and turning the transistor ON. This condition is known as "saturation" and is reached when the base current reaches the maximum value designed into the transistor that allows current to flow. Depending on various factors, the transistor can turn on and off (go from cutoff to saturation) in less than one millionth of a second.

Much of what was said about ratings for diodes applies to transistors, since they are constructed of the same materials. When transistors are required to handle relatively high currents, such as in voltage regulators or ignition systems, they are generally mounted on heat sinks in the same manner as diodes. They can be damaged or destroyed in the same manner if their voltage ratings are exceeded. A transistor can be checked for proper operation by measuring the resistance with an ohmmeter between the base-emitter terminals and then between the base-collector terminals. The forward resistance should be small, while the reverse resistance should be large. Compare the readings with those from a known good transistor. As a final check, measure the forward and reverse resistance between the collector and emitter terminals.

INTEGRATED CIRCUITS

The integrated circuit (IC) is an extremely sophisticated solid state device that consists of a silicone wafer (or chip) which has been doped, insulated and etched many times so that it contains an entire electrical circuit with transistors, diodes, conductors and capacitors miniaturized within each tiny chip. Integrated circuits are often referred to as "computers on a chip" and are largely responsible for the current boom in electronic control technology.

Microprocessors, Computers and Logic Systems

Mechanical or electromechanical control devices lack the precision necessary to meet the requirements of modern control standards. Since they generally function in an ON/OFF cycle, they do not have the ability to respond to the variety of input conditions common to antilock brakes, climate control and electronic suspension operation. To meet these requirements, manufacturers have gone to solid state logic systems and microprocessors to control the basic functions of suspension, brake and temperature control, as well as other systems and accessories.

One of the vital roles of microprocessor-based systems is their ability to perform logic functions and make decisions. Logic designers use a shorthand notation to indicate whether a voltage is present in a circuit (the number 1) or not present (the number 0). Their systems are designed to respond in different ways depending on the output signal (or the lack of it) from various control devices.

There are 3 basic logic functions or "gates" used to construct a microprocessor control system: the AND gate, the OR gate or the NOT gate. Stated simply, the AND gate works when voltage is present in 2 or more circuits which then energize a third (A and B energize C). The OR gate works when voltage is present at either circuit A or circuit B which then energizes circuit C. The NOT function is performed by a solid state device called an "inverter" which reverses the input from a circuit so that, if voltage is going in, no voltage comes out and vice versa. With these three basic building blocks, a logic designer can create complex systems easily. In actual use, a logic or decision making system may employ many logic gates and receive inputs from a number of sources (sensors), but for the most part, all utilize the basic logic gates discussed above.

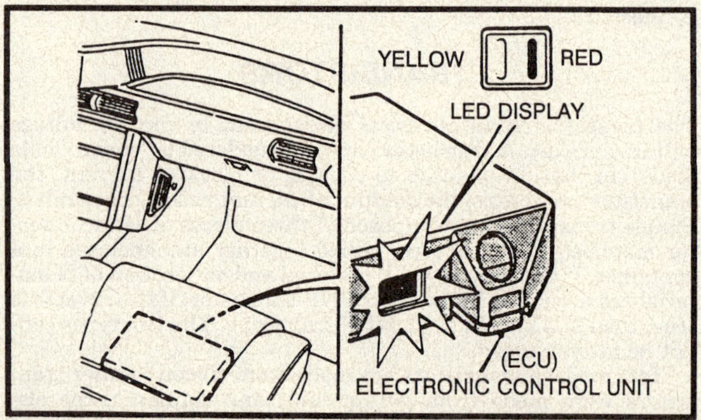

Location of the Electronic Control Unit (ECU)—Sterling

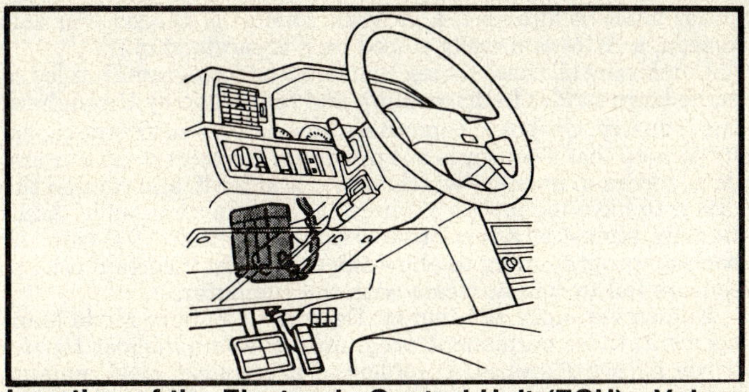

Location of the Electronic Control Unit (ECU) – Volvo 960

Stripped to its bare essentials, a computerized decision-making system is made up of three subsystems:

a. Input devices (sensors or switches)
b. Logic circuits (computer control unit)
c. Output devices (actuators or controls)

The input devices are usually nothing more than switches or sensors that provide a voltage signal to the control unit logic circuits that is read as a 1 or 0 (on or off) by the logic circuits. The output devices are anything from a warning light to solenoid-operated valves, motors, linkage, etc. In most cases, the logic circuits themselves lack sufficient output power to operate these devices directly. Instead, they operate some intermediate device such as a relay or power transistor which in turn operates the appropriate device or control. Many problems diagnosed as computer failures are really the result of a malfunctioning intermediate device like a relay. This must be kept in mind whenever troubleshooting any microprocessor-based control system.

As computer capacity is improved by the manufacturers, so does sensor technology. A few years ago, the on-board computer would receive a message from an engine sensor in a "go or no-go" form; for example, engine coolant temperature is either below or above 150°F. The sensor would or would not pass electric current depending on the temperature. Today's systems allow the same sensor to be much more communicative; it will pass progressively more voltage as the engine warms up. The computer can react to the changing voltage instantly and control other engine functions during the warm-up period.

The logic systems discussed above are called "hardware" systems, because they consist only of the physical electronic components (gates, resistors, transistors, etc.). Hardware systems do not contain a program and are designed to perform specific or "dedicated" functions which cannot readily be changed. For many simple automotive control requirements, such dedicated logic systems are perfectly adequate. When more complex logic functions are required, or where it may be desirable to alter these functions (e.g. from one model vehicle to another) a true computer system is used. A computer can be programmed through its software to perform many different functions and, if that program is stored on a separate integrated circuit chip called a ROM (Read Only Memory), it can be easily changed simply by plugging in a different ROM with the desired program. Most on-board automotive computers are designed with this capability. The on-board computer method of engine control offers the manufacturer a flexible method of responding to data from a variety of input devices and of controlling an equally large variety of output controls. The computer response can be changed quickly and easily by simply modifying its software program.

The microprocessor is the heart of the microcomputer. It is the thinking part of the computer system through which all the data from the various sensors passes. Within the microprocessor, data is acted upon, compared, manipulated or stored for future use. A microprocessor is not necessarily a microcomputer, but the differences between the 2 are becoming very minor. Originally, a microprocessor was a major part of a microcomputer, but today microprocessors are being called "single-chip microcomputers". They contain all the essential elements to make them behave as a computer, including the most important ingredient–the program.

All computers require a program. In a general purpose computer, the program can be easily changed to allow different tasks to be performed. In a "dedicated" computer, such as most on-board automotive computers, the program isn't quite so easily altered. These automotive computers are designed to perform one or more specific tasks, such as maintaining the passenger compartment temperature at a specific, predetermined level. A program is what makes a computer smart; without a program a computer can do absolutely nothing. The term "software" refers to the computer's program that makes the hardware perform the functions needed.

The software program is simply a listing in sequential order of the steps or commands necessary to make a computer perform the desired task. Before the computer can do anything at all, the program must be fed into it by one of several possible methods. A computer can never be "smarter" than the person programming it, but it is a lot faster. Although it cannot perform any calculation or operation that the programmer himself cannot perform, its processing time is measured in millionths of a second.

Because a computer is limited to performing only those operations (instructions) programmed into its memory, the program must be broken down into a large number of very simple steps. Two programmers can come up with 2 different programs, since there is usually more than one way to perform any task or solve a problem. In any computer, however, there is only so much memory space available, so an overly long or inefficient program may not fit into the memory. In addition to performing arithmetic functions (such as with a trip computer), a computer can also store data, look up data in a table and perform the logic functions previously discussed. A Random Access Memory (RAM) al-

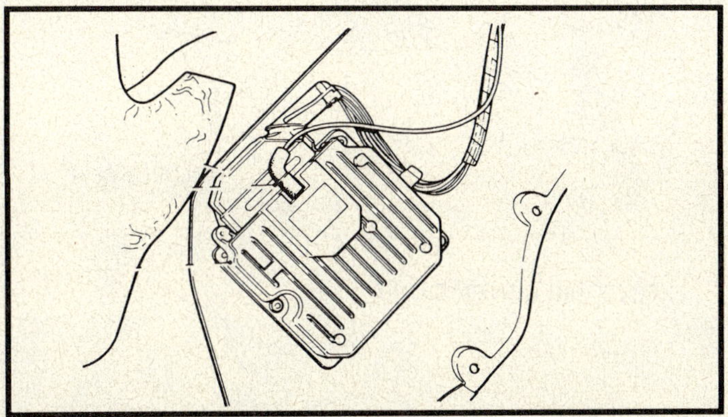

Location of the Electronic Control Unit (ECU) – Jaguar XJS

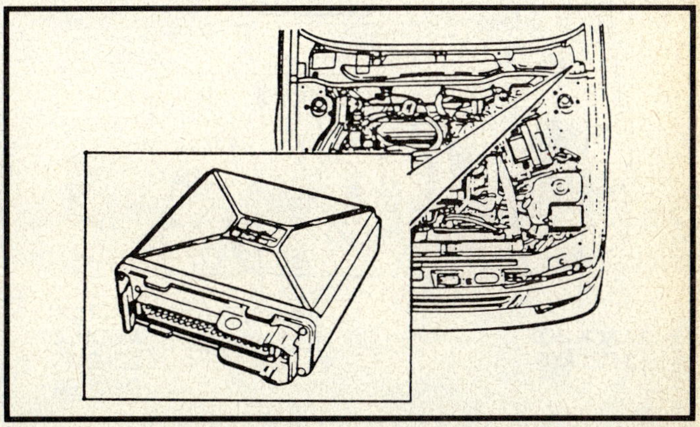

Location of the Electronic Control Unit (ECU) – Saab 9000

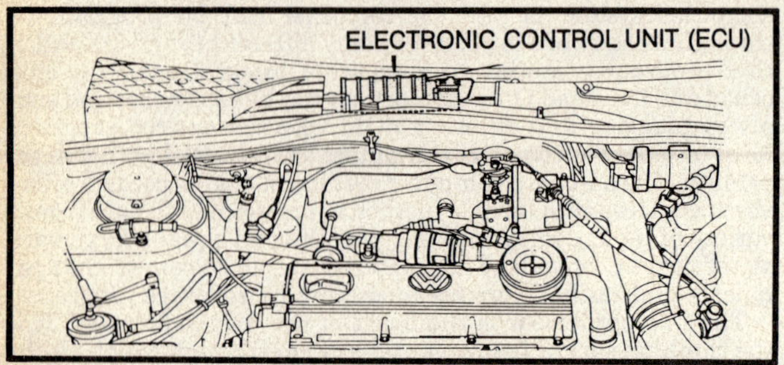

Location of the Electronic Control Unit (ECU) – VW Corrado

lows the computer to store bits of data temporarily while waiting to be acted upon by the program. It may also be used to store output data that is to be sent to an output device. Whatever data is stored in a RAM is lost when power is removed from the system by turning **OFF** the ignition key, for example.

Computers have another type of memory called a Read Only Memory (ROM) which is permanent. This memory is not lost when the power is removed from the system. Most programs for automotive computers are stored on a ROM memory chip. Data is usually in the form of a look-up table that saves computing time and program steps. For example, a computer designed to control the amount of distributor advance can have this information stored in a table. The information that determines distributor advance (engine rpm, manifold vacuum and temperature) is coded to produce the correct amount of distributor advance over a wide range of engine operating conditions. Instead of the computer computing the required advance, it simply looks it up in a pre-programmed table.

Not all electronic control functions can be handled in this manner; some must be computed. On an antilock brake system, for example, the computer must measure the rotation of each separate wheel and then calculate how much brake pressure to apply in order to prevent one wheel from locking up and causing a loss of control.

There are several ways of programming a ROM, but once programmed the ROM cannot be changed. If the ROM is made on the same chip that contains the microprocessor, the whole computer must be altered if a program change is needed. For this reason, a ROM is usually placed on a separate chip.

Combining all these tables, memories and programs has led to some large strides in driveability and fuel economy throughout the industry. On-board computers now possess sufficient capacity to note that a sensor is not providing the correct data or any data, record in memory which sensor is at fault, and remind the driver to have the matter attended to as soon as possible. Additionally, almost all automotive computers contain "fail-safe" or back-up programming to allow the car to run if certain critical sensors fail to communicate with the computer.

Known variously as Limp-in, Default, or Failure Mode Management, these permanent programs provide a lifeboat for the driver in case of on-board problems. For example, most computers monitor the intake air flow and adjust the fuel to the injectors accordingly. The air flow varies drastically between conditions of idle, hard acceleration and highway cruising. If, for some reason–perhaps a loose wiring connector–the sensor message no longer reaches the computer, the computer notes the failure. Instead of shutting the engine off due to lack of necessary data, the computer simply "looks" to its back-up programming which provides it with a fixed value for the air flow. The computer adjusts the fuel flow to meet this assumed value. Engine performance will be compromised (because the metering is fixed, not changing with the conditions) but the engine keeps running and the driver can "limp home" instead of being stranded.

Development is also leading to the use of more than one computer or microprocessor within the vehicle. It is now quite possible to have separate control units for engine function, emission control components, climate control, cruise control, anti-lock brakes and/or passive restraints and seat belts.

As complicated as all these systems may appear, keep in mind that any computer simply processes information. In the automotive world, the computer itself is rarely at fault in a problem car. The network of sensors and wires sending messages to the computer or the network of components and wires receiving the output messages from the computer are generally much more prone to failure or impaired function. Too many people are intimidated by the presence of a computer and immediately blame it instead of hunting down the true cause. Careful, thorough investigation coupled with a firm understanding of how a system functions will lead to the cause of almost any problem.

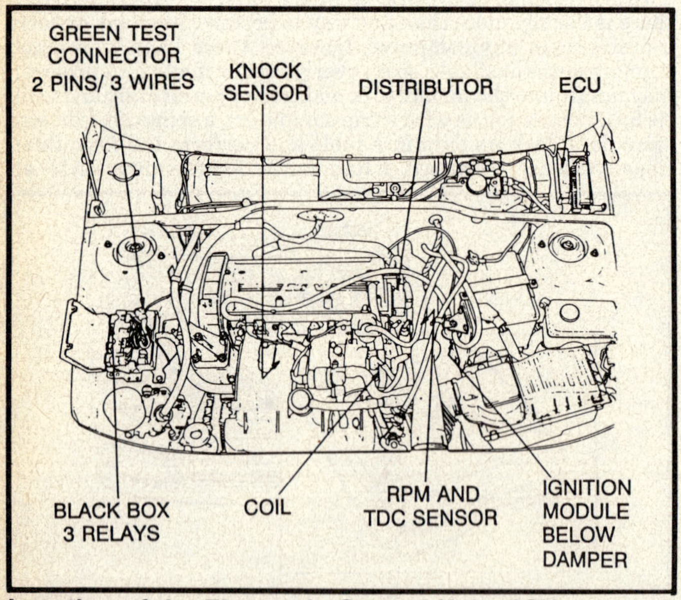

Location of the Electronic Control Unit (ECU) – Peugeot 405

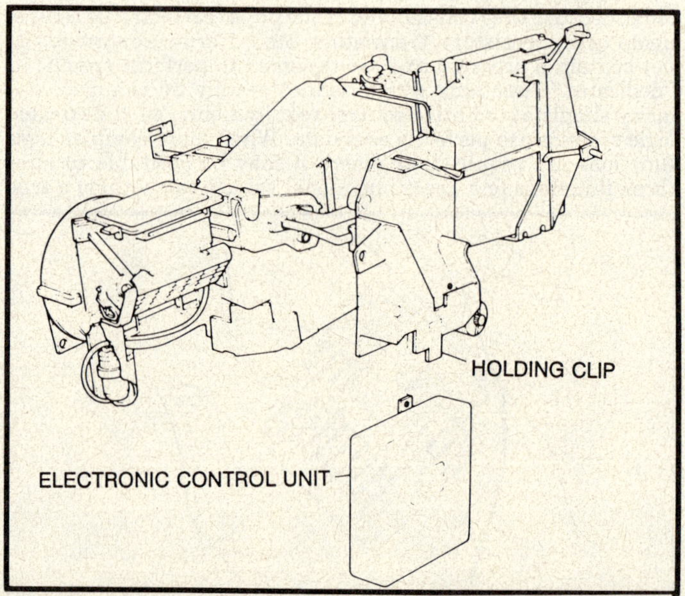

Location of the Electronic Control Unit (ECU) – Audi 80, 90, Coupe Quattro

Diagnostic Equipment and Special Tools

2 SECTION

DIAGNOSTIC EQUIPMENT AND SPECIAL TOOLS

General Information

This section outlines basic diagnostic techniques for dealing with computerized control systems. Along with a general explanation of the various types of test equipment available to aid in servicing modern electronic automotive systems, basic repair techniques for wiring harnesses and connectors are given. Read the basic information before attempting any repairs or testing on any computerized system. This will provide the background information necessary to avoid the most common and obvious mistakes. Likewise, the individual system sections should be read from the beginning to the end before any repair or diagnosis is attempted.

Although the replacement and testing procedures are simple, the systems are not. Unless one has a thorough understanding of all components and their function, the logical test sequences cannot be followed. Minor malfunctions can make a big difference, so it is important to know how each component affects the operation of the system to find the cause of a problem. It is not enough to use the correct test equipment; the test equipment must be used correctly.

SERVICE PRECAUTIONS

CAUTION

Whenever working on or around any computer-based microprocessor control system, always observe these general precautions to prevent the possibility of personal injury or damage to electronic components:

- Never install or remove battery cables with the ignition key **ON** or the engine running. Jumper cables should be connected with the ignition key **OFF** to avoid power surges that can damage electronic control units.
- Always remove the battery cables before charging the battery. Never use a high-output charger on an installed battery. Never attempt to use any type of "hot shot" (24 volt) starting aid.
- Always use a breakout or pin-out box if possible. If probing a connector cannot be avoided, exercise great care when inserting test probes to insure good connections without damaging the connector or spreading the pins. Avoid accidental shorting of terminals during test procedures by probing from the rear or wire side of the connector. Always remove or loosen the backing or rubber housing on the connector before probing at the rear of the connector.
- Never remove or attach wiring harness connectors with the ignition switch **ON**. The resulting voltage spike may destroy an electronic control unit.
- Avoid dropping a component or subjecting it to rough handling. Never apply 12 volts directly to any component unless instructed specifically to do so. Some components are designed to safely handle only 4–5 volts and can be destroyed in seconds if 12 volts are applied directly to the connector.
- Remove the electronic control unit if the vehicle is to be placed in an environment where temperatures exceed approximately 176°F (80°C), such as a paint spray booth.
- If welding is to be done on the vehicle, remove or disconnect all control units on the vehicle including the radio. Connect the welder ground as close as possible to the work area.
- Avoid touching connector pins with the fingers.
- Leave new components and modules in the shipping package until ready to install them.
- Avoid static discharge at or near any solid-state component. Always touch a vehicle ground after sliding across the seat or walking over carpeted or vinyl floors. If possible, wear a grounded wrist strap during all testing and component replacement.
- Never allow welding cables to lie on, near or across any vehicle wiring.
- Avoid draping extension cords or droplight cables on or around vehicle wiring.
- If the vehicle is equipped with air bag or inflatable restraints, always properly disable the system before working on or around system components.
- Always use a digital, high-impedance volt-ohmmeter (DVOM) for testing unless specified otherwise. Minimum impedance should be 10 kilo-ohms per volt.

Organized Troubleshooting

When diagnosing a specific problem, organized troubleshooting is a must. The complexity of modern vehicle systems demands that the approach to any problem be logical and organized. There are certain troubleshooting techniques that are standard:

1. Establish when the problem occurs. Does the problem appear only under certain conditions? Were there any noises, odors, or other unusual symptoms? Make notes on any symptoms found, including warning lights and trouble codes, if applicable.
2. Isolate the problem area. To do this, make some simple tests and observations; then eliminate the systems that are working properly. Experience has shown that many problems in computer controlled systems are the result of fairly simple and obvious causes such as loose or corroded connectors or air leaks in the intake system. Always check the obvious before assuming something complicated is the cause.
3. Once the problem is isolated, test for problems systematically. Are all the components functioning properly? Is there power going to electrical switches and motors? Is there vacuum at vacuum switches and/or actuators? Is there a mechanical problem such as bent linkage or loose screws? Doing careful, systematic checks will often turn up causes on the first inspection without wasting time checking components that have little or no relationship to the problem.
4. Test all repairs after the work is done to make sure that the problem is fixed. Some causes can be traced to more than one component. A careful verification of repair work is important to pick up additional or remaining faults. For example, a blown fuse is a simple problem that may require more than just replacing a fuse. If the problem that caused a fuse to blow isn't searched for, a shorted wire may go undetected.

The diagnostic tree charts are designed to help solve problems by leading the user through closely defined conditions and tests; only the most likely components, vacuum and electrical circuits are checked for proper operation when troubleshooting a particular malfunction. By using the trouble trees to eliminate those items which normally will not cause the fault, a problem can be isolated within one system or circuit without wasting time on unnecessary testing.

Frequent references to special test equipment will be found in the text and in the diagnosis charts. In many cases, a specific testing tool will be necessary to test a system completely. These devices or their equivalents are necessary to perform some of the more complicated test procedures. However, many components can be functionally tested with common shop tools such as volt or ohm meters, test lights, etc. Special test equipment may be purchased from either an aftermarket supplier or the vehicle manufacturer. Care must be taken to insure that the equipment is correct for the vehicle or system on which it will be used.

NOTE: Pinpointing the exact cause of trouble in an electrical system can sometimes only be performed with special equipment. The following describes commonly used test equipment and explains how to put it to best use in diagnosis. Additionally, the tool manufacturer's instruction booklet should be read and clearly understood before attempting any test procedures.

Equipment and Tools

Almost every vehicle manufactured or sold today contains at least one solid-state processor or on-board computer. These units can control many systems and subsystems efficiently, leading to improved driveability, lowered emissions and safer operation. Control modules are currently used to oversee such diverse applications as fuel and ignition management, cruise control, anti-lock brakes, security systems, lighting and suspension.

It is easy to think that a small box containing no moving parts will last the life of the vehicle. In the real world, malfunctions occur due to either electrical problems or external causes. Any computer-based system is extremely sensitive to incorrect voltage; they cannot tolerate careless procedures or improper testing. An inexperienced person can cause extensive and expensive damage in pursuit of a minor problem. Since electricity travels at just under the speed of light, this damage occurs instantly and permanently.

Before beginning any testing or diagnosis on a vehicle, review the safety precautions as well as the procedures to be followed. Make certain that the test equipment to be used is compatible with the vehicle. Read all instructions carefully and double-check test points before connecting probes or other test equipment.

The following section outlines basic diagnostic techniques for dealing with computerized engine control systems. Along with a general explanation of the various types of test equipment available, basic repair techniques for wiring harnesses and connectors are given. Read the basic information before attempting any repairs or testing on any computerized system. This will provide the background information necessary to avoid the most common and obvious mistakes.

The individual system sections for engine controls or fuel injection should be read from the beginning to the end before any repairs or diagnosis is attempted. Although the replacement and testing procedures are simple in themselves, the systems are not. Unless the technician has a thorough understanding of all components and their function within a particular system, the logical test sequence cannot be followed. A minor problem, such as a loose connector, can affect an entire system or several components; it is important to know how each component relates to the operation of the complete system. It is not enough to use the correct test equipment; the test equipment must be used correctly.

JUMPER WIRES

Jumper wires are simple, yet extremely valuable pieces of test equipment. Jumper wires are merely wires of assorted lengths used to bypass sections of a circuit. The simplest type of jumper wire is a length of multi-strand wire with an alligator clip at each end. Jumper wires are usually fabricated from lengths of standard automotive wire and whatever type of connector (alligator clip, spade connector or pin connector) is required for the particular circuit being tested. The well-equipped tool box will have several different styles of jumper wires in several different lengths. Some jumper wires are made with 3 or more terminals coming from a common splice for special-purpose testing. In cramped, hard-to-reach areas it is advisable to have insulated boots over the jumper wire terminals to prevent accidental grounding or sparks, especially when testing fuel system components.

Jumper wires are used primarily to locate open electrical circuits, on either the ground (−) side of the circuit or on the hot (+) side. If an electrical component fails to operate, connect the jumper wire between the component and a good ground. If the component operates only with the jumper installed, the normal ground path is open. If the ground circuit is known to be good but the component does not operate, the circuit between the power feed and component is open.

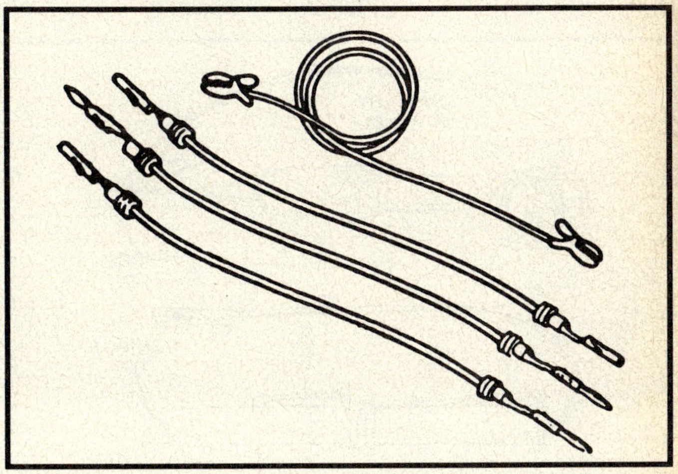

Jumper wires should have assorted connectors

Sometimes a jumper wire connected directly from the battery to the hot terminal of the component can be used, but first make sure the component uses 12 volts in operation. Many electrical components, such as some fuel injectors, are designed to operate on low voltage; running 12 volts to the terminals can burn out the wiring.

NOTE: Inserting an in-line fuseholder into a set of test leads provides a fused jumper, highly recommended for use in testing. This will protect both the component and wiring harnesses in the event of a connection error. A 5 amp fuse will provide protection for most circuits.

Never use jumpers made from wire of lighter gauge (thickness) than used in the circuit being tested. If the jumper wire is of too small gauge, it may overheat or melt; in extreme cases, the heat may cause fire. Never use jumpers to bypass high-resistance loads (such as motors) in a circuit. Bypassing resistances in effect creates a short circuit which may cause damage or fire. Never use a jumper for anything other than temporary bypassing of components in a circuit.

12 VOLT TEST LIGHT

The 12 volt test light is used to check circuits and components while electrical current is flowing through them. It is used for voltage and ground tests. Twelve volt test lights come in different styles but all have 3 main parts; a ground clip, a probe, and a light. The most commonly used 12 volt test lights have pick-type probes. To use a 12 volt test light, connect the ground clip to a good ground and probe wherever necessary with the pick. The pick should be sharp so that it can penetrate wire insulation to make contact with the wire, without making a large hole in the insulation. The wrap-around light is handy in hard to reach areas or where it is difficult to support a wire to push a probe pick into it. To use the wrap around light, hook the wire to be probed with the hook and pull the trigger. A small pick will be forced through the wire insulation into the wire core.

NOTE: Do not use a test light to probe electronic ignition, spark plug or coil wires. Never use a pick-type test light to probe wiring on computer controlled systems unless specifically instructed to do so. Never pierce any shielded cable, such as that used on anti-lock brake systems or data lines between computers. Any wire insulation that is pierced by the test light probe should be taped and sealed with silicone after testing.

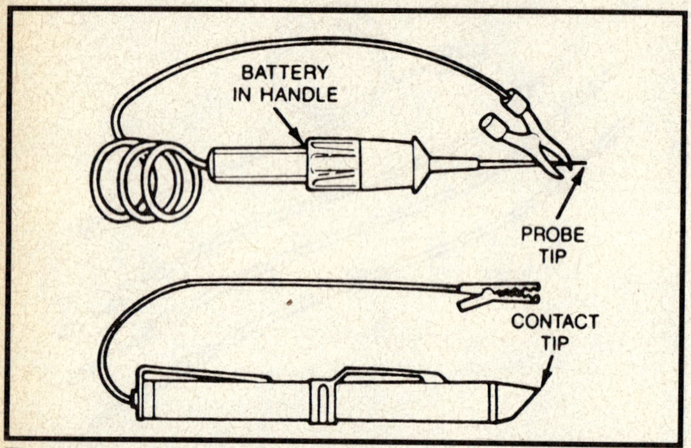

Examples of 12 volt test lights

Like the jumper wire, the 12 volt test light is used to isolate breaks in circuits. While the jumper wire is used to bypass the open to operate the load, the 12 volt test light is used to locate the presence of voltage in a circuit. If the test light glows, there is power up to that point; if the 12 volt test light does not glow when its probe is inserted into the wire or connector, there is an open circuit (no power). Move the test light in successive steps back toward the power source until the light in the handle illuminates. When it lights, the open is between the probe location and the point previously probed.

NOTE: The test light does not detect that 12 volts (or any particular voltage) is present; it only indicates that some voltage is present. It is advisable before using the test light to touch its terminals across the battery posts to make sure the light is operating properly. Do not attempt to determine voltage by the brightness of the lamp; a voltmeter must be used to determine voltage.

Self-Powered Test Light

The self-powered test light usually contains a 1.5 volt penlight battery. One type of self-powered test light is similar in design to the 12 volt test light. This type has both the battery and the light in the handle and pick-type probe tip. The second type has the light toward the open tip, so that the light illuminates the contact point.

The self-powered test light is dual-purpose piece of test equipment. It can be used to test for either open or short circuits when power is isolated from the circuit (continuity test). Do not use a self-powered test light on any computer controlled system or component unless specifically instructed to do so. Many engine sensors can be destroyed by even this small amount of voltage.

OPEN CIRCUIT TESTING

To use the self-powered test light to check for open circuits, first isolate the circuit from the power supply by removing a fuse, disconnecting a wiring harness connector or even turning a dashboard switch **OFF**. Avoid disconnecting the battery cables if at all possible; doing so will erase the electronic memories in the control modules. Besides the obvious effects on the clock and radio, driveability may be affected until the vehicle systems relearn the settings.

Connect the test light ground clip to a good ground and probe sections of the circuit sequentially with the test light, starting from either end of the circuit. If the light is out, the break is between the probe and the circuit ground. If the light is on, the break is between the probe and end of the circuit toward the power source.

SHORT CIRCUIT TESTING

After isolating the circuit both from power and from ground, a self-powered test light, may be used to find shorts within a circuit. Connect the test light ground clip to a good ground and probe any easy-to-reach test point in the circuit. If the light comes on, there is a short somewhere in the circuit. To isolate the short, probe a test point at either end of the isolated circuit (the light should be on). Leave the test light probe connected and open connectors, switches, remove parts, etc., sequentially, until the light goes out. When the light goes out, the short is between the last circuit component opened and the previous circuit opened.

NOTE: The 1.5 volt battery in the test light does not provide much current. A weak battery may not provide enough power to illuminate the test light even when a complete circuit is made. Always make sure that the test battery is strong. To check the battery, briefly touch the ground clip to the probe; if the light glows brightly the battery is strong enough for testing.

Never use a self-powered test light to perform checks for opens or shorts when power is applied to the electrical system under test. The 12-volt vehicle power will quickly burn out the 1.5 volt light bulb in the test light.

VOLTMETER

A voltmeter is used to measure voltage at any point in a circuit, or to measure the voltage drop across any part of a circuit. Voltmeters usually have various scales on the meter dial and a selector switch to allow the selection of different test ranges. The voltmeter has a positive and a negative lead. To avoid damage to the meter, connect the negative lead, usually black, to the negative (−) side of circuit or to ground. Connect the positive lead, usually red, to the positive (+) or power side of the circuit.

A voltmeter can be connected either in parallel or in series with a circuit and has a very high resistance to current flow. When connected in parallel, only a small amount of current will flow through the voltmeter current path; the rest will flow through the normal current path and the circuit will work normally. When the voltmeter is connected in series with a circuit, only a small amount of current can flow through the circuit. The circuit will not work properly, but the voltmeter reading will show if the circuit is complete or not.

Available Voltage Measurement

Set the voltmeter selector switch to the 20V position and connect the meter negative lead to the negative post of the battery and connect the positive meter lead to the positive post of the battery. Read the voltage on the meter or digital display. A well-charged battery should register over 12 volts. If the meter reads below 11.5 volts, the battery power may be insufficient to operate the electrical system properly. This test determines voltage available from the battery and should be the first step in any electrical trouble diagnosis procedure. Many electrical problems, especially on computer controlled systems, can be caused by a low state of charge in the battery. Excessive corrosion at the battery cable terminals can cause a poor contact that will prevent proper charging and full battery current flow.

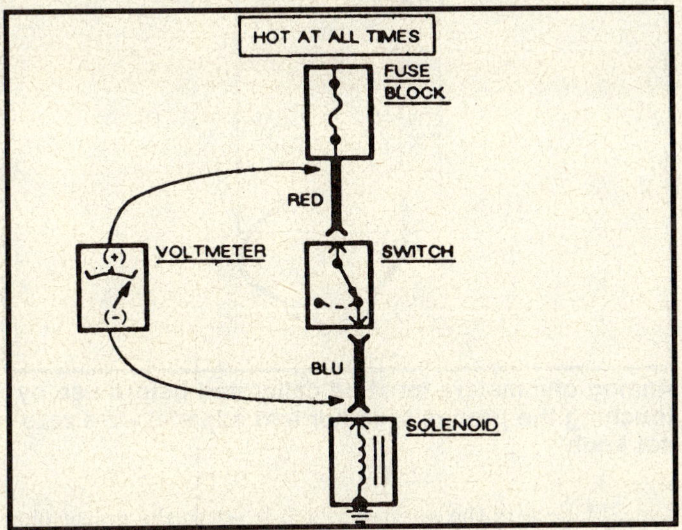

Measuring available voltage in a circuit

Nominal battery voltage is 12 volts but, when fully charged, should be about 13.2 volts. When the battery is supplying current to 1 or more circuits it is said to be under load. When everything is **OFF** the electrical system is under a no-load condition. A fully charged battery showing about 12.5 volts at no load may drop to 12 volts under medium load and will drop even lower under heavy load. If the battery is partially discharged, the voltage decrease under heavy load may be excessive, even though the battery shows 12 volts or more at no load. For this reason, it is important that the battery be fully charged during all testing procedures to avoid errors in diagnosis and incorrect test results.

Voltage Drop

When current flows through a resistance, the voltage beyond the resistance is reduced. The larger the current, the greater the voltage reduction. When the circuit is off, there is no voltage drop because there is no current. All points in the circuit which are connected to the power source are at the same voltage as the power source. In a long circuit with many connectors, a series of small, unwanted voltage drops due to corrosion at the connectors can add up to a total loss of voltage which impairs the operation of the loads in the circuit.

INDIRECT COMPUTATION OF VOLTAGE DROPS

1. Set the voltmeter selector switch to the 20 volt position.
2. Connect the meter negative lead to a good ground.
3. Probe all resistances in the circuit with the positive meter lead.
4. Operate the circuit in all modes and observe the voltage readings.

DIRECT MEASUREMENT OF VOLTAGE DROPS

1. Set the voltmeter switch to the 20 volt position.
2. Connect the voltmeter negative lead to the ground side of the resistance load to be measured.
3. Connect the positive lead to the positive side of the resistance or load to be measured.
4. Read the voltage drop directly on the 20 volt scale.

Too high a voltage indicates too high a resistance. If, for example, a blower motor runs too slowly, there may be too high a resistance in the resistor pack. By taking voltage drop readings

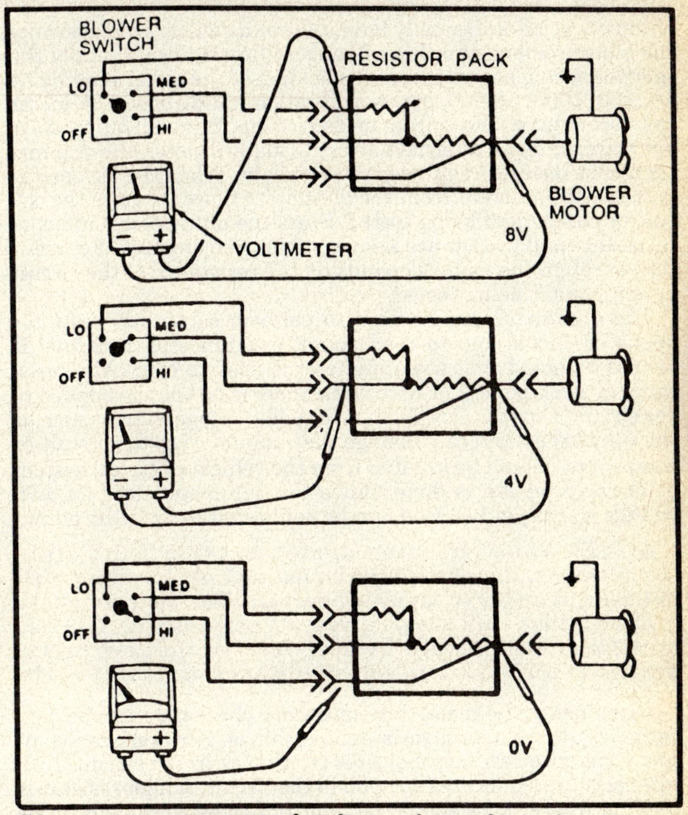

Direct measurement of voltage drops in a circuit

in all parts of the circuit, the problem can be isolated. Too low a voltage drop indicates too low a resistance. If, for example, a blower motor runs too fast in the **MED** and/or **LOW** position, the problem can be isolated to the resistor pack by taking voltage drop readings in all parts of the circuit to locate a possibly shorted resistor. The maximum allowable voltage drop under load is critical, especially if there is more than one high resistance problem in a circuit; all voltage drops are cumulative. A small drop is normal due to the resistance of the conductors.

High Resistance Testing

1. Set the voltmeter selector switch to the 2 volt position.
2. Connect the voltmeter positive lead to the positive post of the battery.
3. Turn **ON** the headlights and heater blower to provide a load.
4. Probe various points in the circuit with the negative voltmeter lead.
5. Read the voltage drop. Some average maximum allowable voltage drops are:
   - Fuse panel – 0.7 volts
   - Ignition switch – 0.5volts
   - Headlight switch – 0.7 volts
   - Ignition coil (+) – 0.5 volts
   - Any other load – 0.5–1.3 volts

NOTE: Voltage drops are all measured while a load is operating; without current flow, there will be no voltage drop.

OHMMETER

The ohmmeter is designed to read resistance (ohms) in a circuit or component. Although there are several different styles of ohmmeters, all will usually have a selector switch which permits the measurement of different ranges of resistance. Usually the selector switch allows the multiplication of the meter reading by 10, 100, 1000, 10,000, etc. A calibration knob allows the meter to be set at zero for accurate measurement. Since all ohmmeters are powered by an internal battery (usually 9 volts), the ohmmeter can be used as a self-powered test light. When the ohmmeter is connected, current from the ohmmeter flows through the circuit or component being tested. Since the ohmmeter's internal resistance and voltage are known values, the amount of current flow through the meter depends on the resistance of the circuit or component being tested.

The ohmmeter can be used to perform continuity tests for opens or shorts and to read actual resistance in a circuit. It should be noted that the ohmmeter is used to check the resistance of a component or wire while there is no voltage applied to the circuit. Current flow from an outside voltage source (such as the vehicle battery) can damage the ohmmeter, so the circuit or component should be isolated from the vehicle electrical system before any testing is done. Since the ohmmeter uses its own voltage source, either lead can be connected to any test point.

NOTE: When checking diodes or other solid state components, the ohmmeter leads can only be connected one way in order to measure current flow in a single direction. Make sure the positive (+) and negative (−) terminal connections are as described in the test procedures to verify the one-way diode operation.

When using the meter for continuity checks, do not be concerned with the actual resistance readings. Zero resistance, or any resistance reading, indicates continuity in the circuit. Infinite resistance indicates an open in the circuit. A high resistance reading where there should be none indicates a problem in the circuit. Checks for short circuits are made in the same manner as checks for open circuits except that the circuit must be isolated from both power and normal ground. Infinite resistance indicates no continuity to ground, while zero resistance indicates a dead short to ground.

Resistance Measurement

The batteries in an ohmmeter may be affected by temperature and will weaken with age. The ohmmeter must be calibrated or "zeroed" before taking measurements. To zero the meter, place the selector switch in its lowest range and touch the 2 leads together. Turn the calibration knob until the meter needle is exactly on zero.

NOTE: All analog (needle) type ohmmeters must be zeroed before use, but some digital ohmmeter models are automatically calibrated when the switch is turned ON. Self-calibrating digital ohmmeters do not have an adjusting knob, but it's a good idea to check for a zero readout before use by touching the leads together. All computer controlled systems require the use of a digital ohmmeter with at least 10 megohms impedance for testing. Before any test procedures are attempted, make sure the ohmmeter used is compatible with the electrical system, or damage to the on-board computer could result.

To measure resistance, first isolate the circuit from the vehicle power source by disconnecting the battery cables or the harness connector. Make sure the key is **OFF** when disconnecting any components or the battery. Where necessary, also isolate at least one side of the circuit to be checked to avoid reading parallel resistances. Parallel circuit resistances will always give a lower reading than the actual resistance of either of the branches. When measuring the resistance of parallel circuits, the total resistance will always be lower than the smallest resistance in the circuit. Connect the meter leads to both sides of the circuit (wire or component) and read the actual measured ohms on the meter

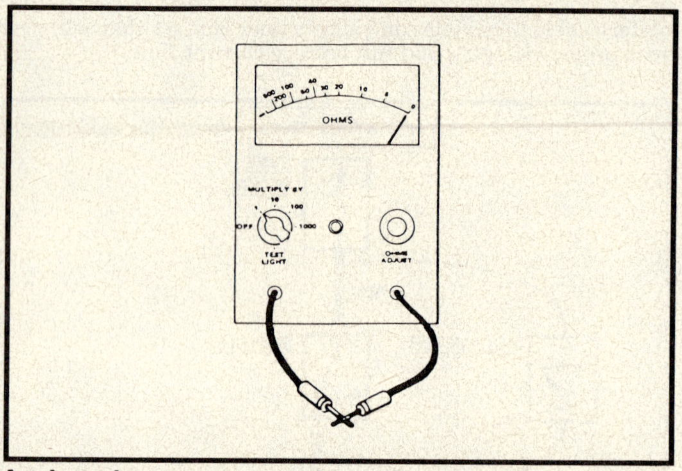

Analog ohmmeters must be calibrated before use by touching the probes together and adjusting the zero-set knob

scale. Make sure the selector switch is set to the proper ohm scale for the circuit being tested to avoid misreading the ohmmeter test value.

NOTE: Never use an ohmmeter with power applied to the circuit. Like the self-powered test light, the ohmmeter is designed to operate on its own power supply. The normal 12 volt automotive system could damage the meter.

AMMETERS

An ammeter measures the amount of current flowing through a circuit in units called amperes or amps. Amperes are units of electron flow which indicate how fast the electrons are flowing through the circuit. Since Ohm's Law dictates that current flow in a circuit is equal to the circuit voltage divided by the total circuit resistance, increasing voltage also increases the current level (amps). Likewise, any decrease in resistance will increase the amount of amps in a circuit. At normal operating voltage, most circuits have a characteristic amount of amperes, called "current draw" which can be measured using an ammeter. By referring to a specified current draw rating, measuring the amperes, and comparing the 2 values, one can determine what is happening within the circuit to aid in diagnosis. An open circuit, for ex-

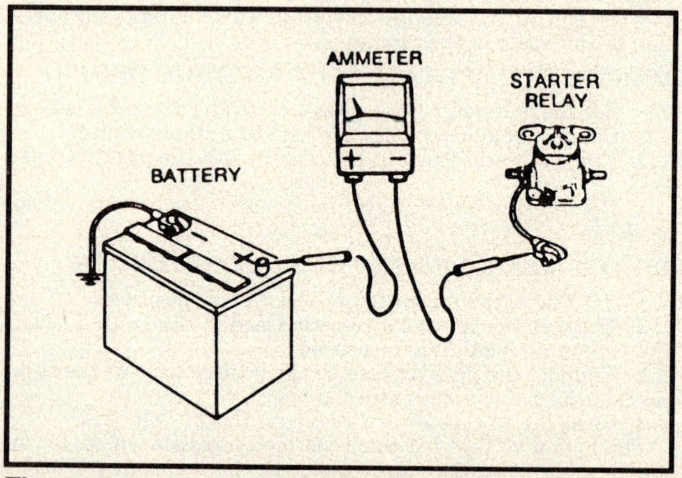

The ammeter is useful in tracking current drain

ample, will not allow any current to flow so the ammeter reading will be zero. More current flows through a heavily loaded circuit or when the charging system is operating.

An ammeter is always connected in series with the circuit being tested. All of the current that normally flows through the circuit must also flow through the ammeter; if there is any other path for the current to follow, the ammeter reading will not be accurate. The ammeter itself has very little resistance to current flow and therefore will not affect the circuit, but it will measure current draw only when the circuit is closed and electricity is flowing. Excessive current draw can blow fuses and/or drain the battery; a reduced current draw can cause motors to run slowly, lights to dim and other components to operate improperly.

NOTE: Even with the ignition OFF, a very small amount of current flows to many of the solid state components in a vehicle. This current, called parasitic draw, represents the power used to maintain memories in such items as the radio, clock, ECU and other components. The current flow is so small that the battery will not be drained if the car is in normal use every few days.

MULTIMETERS

Different combinations of test meters can be built into a single unit designed for specific tests. Some of the more common combination test devices are known as Volt-Amp testers, Tach-Dwell meters, or Digital Multimeters. The Volt-Amp tester is used for charging system, starting system or battery tests and consists of a voltmeter, an ammeter and a variable resistance carbon pile. The voltmeter will usually have at least 2 ranges for use with 6, 12 and 24 volt systems. The ammeter also has more than 1 range for testing various levels of battery loads and starter current draw and the carbon pile can be adjusted to offer different amounts of resistance. The Volt-Amp tester has heavy leads to carry large amounts of current and many later models have an inductive ammeter pickup that clamps around the wire to simplify test connections. On some models, the ammeter also has a zero-center scale to allow testing of charging and starting systems without switching leads or polarity.

A digital multimeter is a voltmeter, ammeter and ohmmeter combined in an instrument which gives a digital readout. These are often used when testing solid state circuits because of their high input impedance (usually 10 megohms or more). It is important to note that the ammeter portion of this meter may only be useful on extremely low current circuits. Refer to the instructions provided with the meter before attempting to use the ammeter on automotive circuits.

The tach-dwell meter combines a tachometer with a dwell (cam angle) meter and is a specialized kind of voltmeter. The tachometer scale is marked to show engine speed in rpm and the dwell scale is marked to show degrees of distributor shaft rotation. In most electronic ignition systems, dwell is determined by the control unit, but the dwell meter can also be used to check the duty cycle (operation) of some electronic engine control systems.

VACUUM LEAK DETECTOR

Although electronically or computer controlled engine management systems have greatly reduced the number of vacuum circuits within the engine compartment, the remaining circuits have become more important. In many cases, the vacuum signal is translated into an electrical signal sent to the engine controller. For this reason, the integrity of the vacuum system must be maintained at all times.

A hand-held electronic vacuum leak detector can be of great value in tracing leaks affecting the system but not easily seen or heard. These battery powered tools employ a microphone and beeper system. The unit is calibrated to respond to ultrasonic signals, well above the level of human hearing. Leaking vacuum will generate sound waves in the 40 kilo-Hz. range and the detector will sound the beeper when this signal is found. Because of the high frequency sensed by the detector, it is not generally affected by normal engine or surrounding shop noises.

Special Test Equipment

A variety of diagnostic tools are available to help troubleshoot and repair computerized engine control systems. Although the most sophisticated units may occupy an entire garage service bay, there are several types of aftermarket electronic testers available. These units allow quick circuit tests of the engine control system by plugging directly into a special connector located in the engine compartment, under the dashboard or in the fusebox. Several tool and equipment manufacturers offer simple, hand-held testers that measure various circuit voltage levels on command. Although these testers usually cost several hundred dollars, consider that the average computer control unit (or ECM) can cost just as much; the money saved by not replacing good sensors or components could justify the purchase price after the first use.

These computerized testers can allow quick and easy test measurements while the engine is operating. In addition, the on-board computer memory can be read, giving access to any stored trouble codes. In the same manner, repairs can be tested to make sure the problem has been corrected. The biggest advantage of these special testers is their relatively easy hookup, minimizing chances of making the wrong connection, getting false voltage readings or damaging the computer.

NOTE: Remember that any meter checks only the presence of electrical signals in a circuit. No meter can detect a faulty component if it is binding or disconnected internally. The presence of voltage does not necessarily mean that the related component is working properly.

SCAN TOOLS

As computers are used to control more circuits on vehicles, the need to communicate with the controllers increases. The hand-held scanner, or scan tool, is capable of reading stored fault codes as well as allowing the operator to see the electrical signals to and from the controller. Depending on the vehicle's diagnostic system, the scan tool may also be used to override certain operating parameters as well.

There are many manufacturers of these tools; a purchaser must be certain that the tool is proper for the intended use. Most manufacturers employ some form of scan or diagnostic tool which may be connected into the engine management system, although the scan tool for one vehicle almost certainly will not work on a different make. Scan tools from outside suppliers can be configured to various makes by inserting the appropriate data module or cartridge.

An example of the usefulness of the scan tool may be seen in the case of a temperature sensor which has changed its electrical characteristics. The engine controller is reacting to an apparently warmer engine that is actually still cold (causing a driveability problem), but the sensor's voltage has not changed enough to set a fault code. Connecting the scan tool, the voltage signal being sent to the ECM may be viewed; comparison to either normal values or a known good vehicle reveals the problem quickly. Additionally, some scan tools may be set for use as a volt- or ohmmeter with the push of a button. The built-in test lead allows certain electrical tests to be performed without swapping tools.

In some cases, the hand scanner may be used to test other computer-controlled vehicle systems such as Antilock Brakes (ABS) or cruise control, but many vehicle manufacturers re-

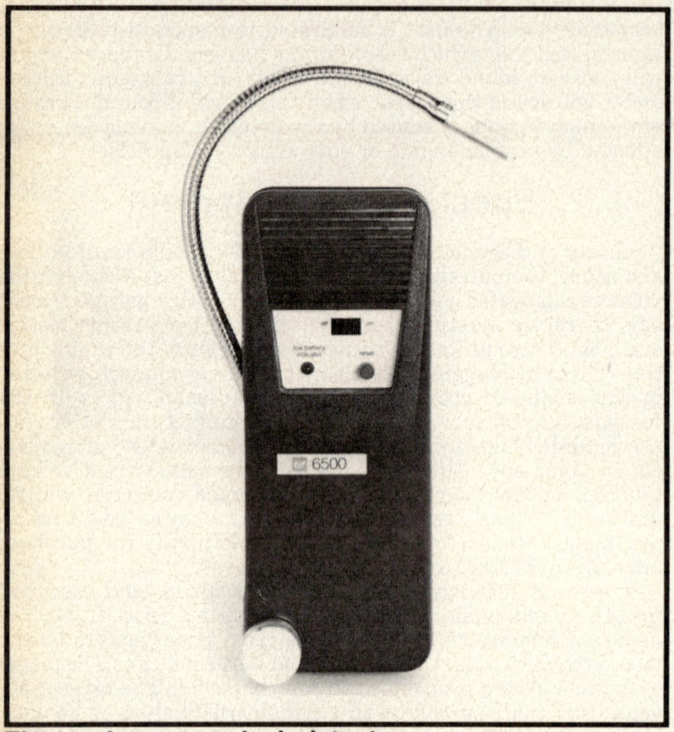

Electronic vacuum leak detector

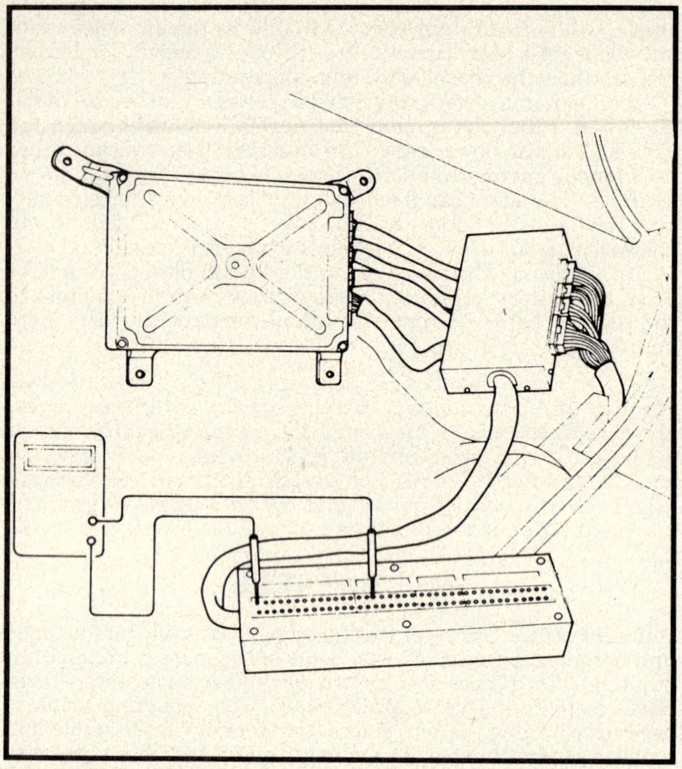

Example of terminal or pin-out box being used for system testing

quire individual tools for the other systems. This is particularly true of ABS and/or air bag systems.

Depending on the vehicle and the amount of computer control used, systems which may be viewed or investigated with a scan tool include:

1. Engine controllers/ECUs
2. Fuel/ignition systems
3. Electronic transmission control
4. Charging system
5. Suspension control functions
6. Anti-Lock brake system.
7. Passive restraint system
8. Anti-theft system
9. Climate Control Systems
10. Body electrical systems including power locks, entertainment systems, sunroofs, and defoggers.

PIN-OUT OR TEST BOXES

Sometimes called a test or diagnostic harness, this test tool connects between the control unit and the wiring harness. A pin-out box routes each of the many wire harness conductors to a discreet, insulated terminal on the box. All testing with the volt-ohmmeter can be done safely at these terminals, eliminating the risk of damage due to backprobing at the control unit.

Additionally, some manufacturers provide plastic overlays for the pinout box. This allows the box to be used on a variety of models; different overlays identify the changes in wire use or labeling. Use of a pin-out box, if available, is highly recommended during diagnosis and circuit checking.

Self-Diagnostics

The control units found on most systems are capable of providing diagnostic help when a malfunction occurs. The units are programmed to monitor certain input and output circuits and react to a major change in the circuit. When a fault is detected, the dashboard warning lamp is usually activated. A numbered code is assigned to the fault and stored in the memory portion of the unit. The stored code(s) become a guide to the faulty system(s).

The stored codes will usually only be output when the system is placed in the diagnostic mode. On most vehicles, this requires a specific action or series of actions, such as installing a jumper across diagnostic terminals or cycling the key ON–OFF–ON–OFF a certain number of times. In many cases, the codes may be retrieved by connecting a voltmeter to the system.

Codes will be output as a series of voltage pulses. Depending on the system, the pulses may be seen as needle sweeps on an analog voltmeter, a flashing of the dashboard warning lamp or the flashing of an LED on the ECU. In some cases, the code appear as an English message on the screen of the scan tool. Remember that these codes indicate failed circuits, not necessarily a failed component. The scan tool message screen may show FAULT 14: THROTTLE POSITION SENSOR, but the problem most likely lies in the wiring and connectors rather than in the sensor itself.

Wiring and Circuits

ELECTRICAL SCHEMATICS

An electrical schematic shows the current paths when a circuit is operating properly. It is essential to understand how a circuit works before trying to figure out why it doesn't. Schematics break the entire electrical system into individual circuits and usually show only one path. In a schematic, no attempt is made to represent wiring and components as they physically appear on the vehicle; switches and other components are shown as simply as possible, usually with standardized symbols. Face views of harness connectors show the cavity or terminal locations in all multi-pin connectors to help locate test points and connector or pin numbers are usually provided.

Schematics are not to drawn to scale; a very short wire will be

Horn

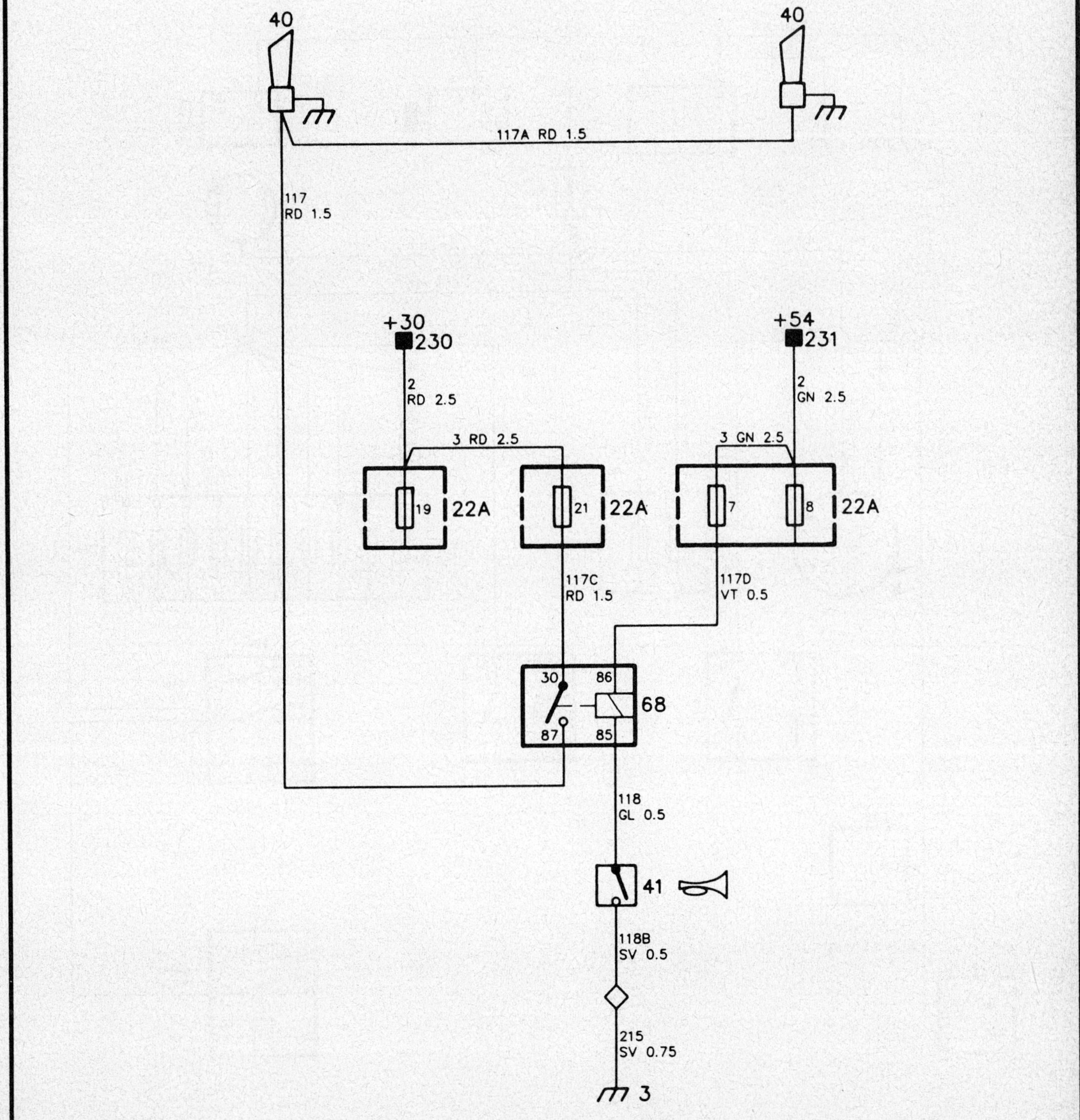

Example of single circuit wiring schematic, also called a current flow diagram

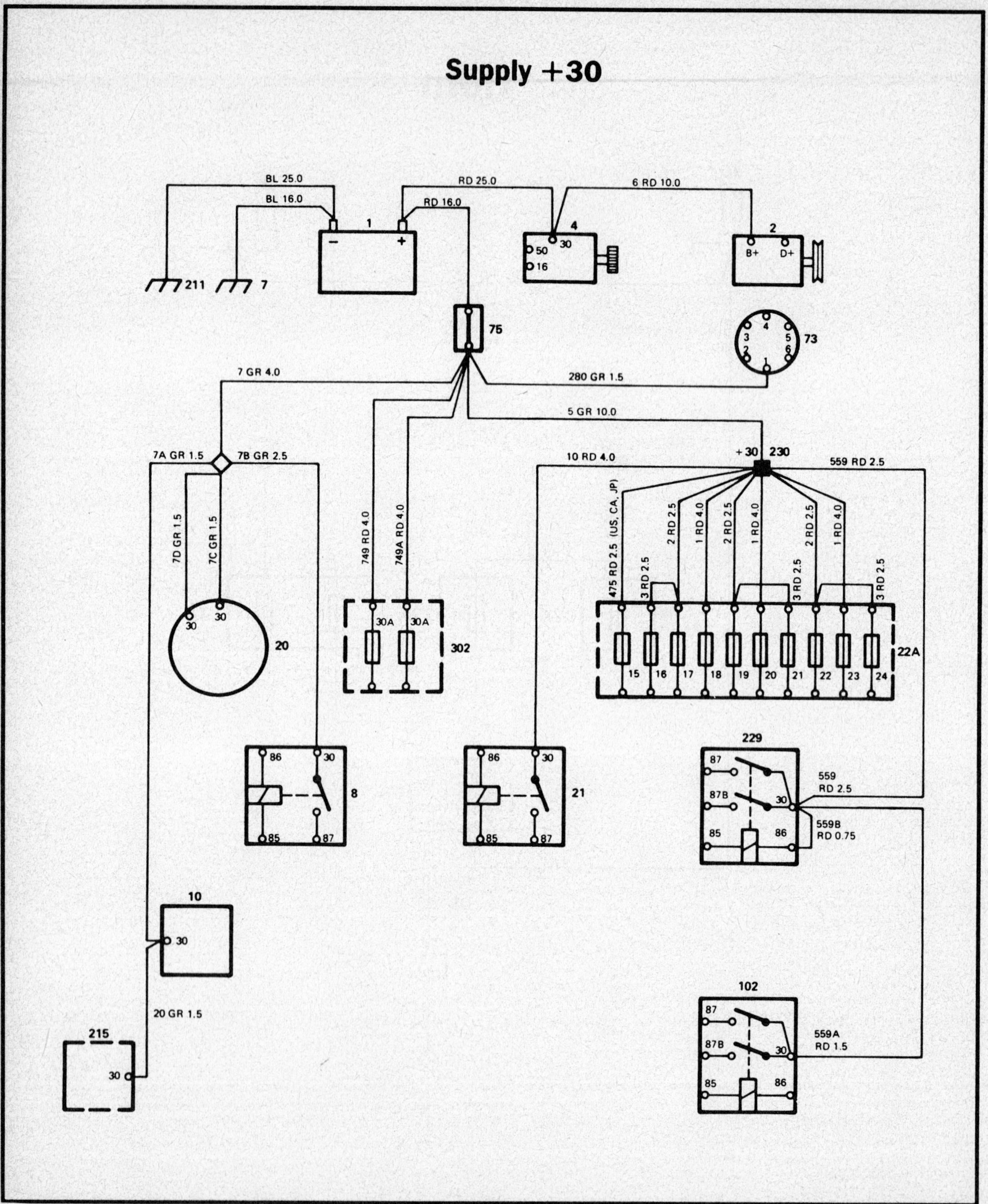

Example of a multi-circuit schematic for power distribution

shown in identical fashion to one running the length of the vehicle. Generally, schematics are drawn to be read either from top-to-bottom or left-to-right. This allows the current flow to be presented in logical order from the power source through all circuit components to ground.

The circuit schematic must be considered an accurate map of the circuit. It should be used to determine the best test points and, in some cases, the test method. Attempting to diagnose any electrical circuit without use of the schematic may lead to wasted time and unnecessary replacement of components.

EUROPEAN WIRING STANDARDS

Most of the vehicles produced in Europe use a standard wiring and terminal identification system. The system is referred to as the DIN standard, comparable to (but different from) the SAE system in the United States.

In wiring diagrams, circuit numbers and terminal identification numbers will always have the same meaning. For example, the battery power circuit is numbered 30. Any relay or component terminal connecting to this circuit will be identified as terminal 30. The starter solenoid circuit is circuit 50, power controlled by the ignition switch is circuit 15 and so on. This system allows a general base of knowledge to be applied to cars of many manufacturers.

Terminal identification is similar. On a typical relay, terminal 87 is the input signal. Terminal 85 is the output end of the winding leading to ground and terminal 87a is the output leg of the circuit. Terminal 1 on an ignition coil is always the primary circuit, terminal 4 is the secondary and terminal 15 is ignition ON battery voltage.

By understanding the European standards, terminals may be easily identified by their number during testing. If, for example, a horn relay has terminals 30, 85.86 and 87 numbered on the case, the technician knows that 30 is battery power, 86 is the start of the winding and 85 is the ground for the winding. Terminal 87 is the output to the horn.

Components also have a standard identification within this system. Relays are marked K, motors with M, controls with N, switches with S and so on. Each manufacturer will usually standardize the number used for a component within their line. If N16 is a fuel pump relay on one wiring schematic, the manufacturer will most likely use N16 to represent a fuel pump relay or controller on other schematics. The number designation may change slightly to indicate a change in circuitry, for example N16/4.

Although the DIN system is commonly used in labeling fuel and emission components, particularly on Bosch systems, a manufacturer may continue to use a unique system of identification of the remainder of the car. Some may add prefixes or suffixes to the standard labels and still others may change a letter because of language differences. Alfa Romeo, for example, has used H as a switch designator instead of S.

WIRE HARNESSES

The average automobile contains about ½ mile of wiring, with hundreds of individual connections. To protect the many wires from damage and to keep them from becoming a confusing tangle, they are organized into bundles, enclosed in plastic or taped together and called wire harnesses. Different wiring harnesses serve different parts of the vehicle. Individual wires are color-coded to help trace them through a harness where sections are hidden from view.

A loose, corroded connection or a replacement wire that is too small for the circuit will add extra resistance and an additional voltage drop to the circuit. A 10% voltage drop can result in slow or erratic motor operation, even though the circuit is complete. Automotive wiring or circuit conductors can be in any of 3 forms:

1. Single strand (solid) wire
2. Multi-strand wire
3. Printed circuitry

Single strand wire has a solid metal core and is usually used inside such components as alternators, motors, relays and other devices. Multi-strand wire has a core made of many small strands of wire twisted together into a single conductor. Most of the wiring in an automotive electrical system is made up of multi-strand wire, either as a single conductor or grouped together in a harness. All wiring is color-coded on the insulator, either as a solid color or as a colored wire with an identification stripe.

A printed circuit is a thin film of copper or other conductor that is printed on an insulator backing. Occasionally, a printed circuit is sandwiched between 2 sheets of plastic for more protection and flexibility. A complete printed circuit, consisting of conductors, insulating material and connectors for lamps or other components is called a printed circuit board. Printed circuitry is used in place of individual wires or harnesses in places where space is limited, such as behind instrument panels.

Wire Gauge

Since computer-controlled automotive electrical systems are very sensitive to changes in resistance, the selection of properly sized wires is critical when systems are repaired. The wire gauge number is an expression of the cross section area of the conductor. The most common system for expressing wire size is the American Wire Gauge (AWG) system.

Wire cross section area is measured in circular mils. A mil is $\frac{1}{1000}$ of an in. (0.001 in); a circular mil is the area of a circle one mil in diameter. For example, a conductor ¼ in. in diameter is 0.250 in. or 250 mils. The circular mil cross section area of the wire is 250 squared or 62,500 circular mils.

Gauge numbers are assigned to conductors of various cross section areas. As gauge number increases, area decreases and the conductor becomes smaller. A 5 gauge conductor is smaller than a 1 gauge conductor and a 10 gauge is smaller than a 5 gauge. As the cross section area of a conductor decreases, resistance increases and so does the gauge number. A conductor with a higher gauge number will carry less current than a conductor with a lower gauge number.

NOTE: Gauge wire size refers to the size of the conductor, not the size of the complete wire. It is possible to have 2 wires of the same gauge with different outer diameters because one may have thicker insulation than the other.

METRIC WIRE SIZES	AWG SIZES
.22	24
.35	22
.5	20
.8	18
1.0	16
2.0	14
3.0	12
5.0	10
8.0	8
13.0	6
19.0	4
32.0	2
50.0	1/0

Wire gauge conversion table

EUROPEAN STANDARD TERMINAL DESIGNATIONS

DEFINITION	TERMINAL	CIRCUIT
Ignition coil, ignition distributor	1	Low voltage
Ignition distributor with two separate electrical circuits	1 a 1 b 2	To ignition contact breaker I To ignition contact breaker II Short-circuit terminal (magneto ignition)
Ignition coil, ignition distributor	4	high voltage
Ignition distributor with two separate electrical circuits	4 a 4 b 15 15 a	From ignition coil I, terminal 4 From ignition coil II, terminal 4 Plus (+) downstream of battery [output of ignition/driving switch] Output at dropping resistor to ignition coil and starter
Glow plug and starter switch	17 19	Start Preheat
Battery	30	Input from + battery terminal, direct
12/24 V series-parallel battery switch	30 a	Input from + terminal of battery II
Return line to battery	31	– battery terminal or ground, direct
Return line to negative battery	31 b	terminal or ground, via switch or relay (switched negative)
12/24 V series-parallel battery switch	31 a 31 c	Return line to – terminal of battery II Return line to – terminal of battery I
Electric Motors	32 33 33 a 33 b 33 f 33 g 33 h 33 L 33 R	Return line ① Main terminal connection ① Self-parking switch-off Shunt field For second lower-speed range For third lower-speed range For fourth lower-speed range Counterclockwise rotation Clockwise rotation
Starter	45	Separate starter relay, output; starter, input (main current)
Two-starter parallel operation Starting relay for engagement current	45 a 45 b 48	Output, starter I Input, starters I and II Output, starter II Terminal on starter and on start-repeating relay for monitoring starting procedure

DEFINITION	TERMINAL	CIRCUIT
Turn-signal Flasher (pulse generator)	49 49 a 49 b 49 c	Input Output Output, second turn-signal circuit Output, third turn-signal circuit
Starter	50	Starter control (direct)
Series-parallel battery switch	50 a	Output for starter control
Starter control	50 b	with parallel operation of two starters with sequential control
Starting relay for sequential control of the engagement current during parallel operation of two starters	50 c 50 d	Input at starting relay for starter I Input at starting relay for starter II
Start-locking relay	50 e 50 f	Input Output
Start-repeating relay	50 g 50 h	Input Output
Alternator	51 51 e	DC voltage at rectifier DC voltage at rectifier with choke coil for daytime driving
Trailer signals	52	Signals from trailer to towing vehicle, general
Wiper motor	53 53 a 53 b 53 c 53 e 53 i	Input (+) Wiper (+), self-parking switch-off Wiper (shunt winding) Electric windshield washer pump Wiper (brake winding) Wiper motor with permanent magnet and third brush (for higher speed)
Stop lamp	54	For lamp combinations and trailer plug connections
Trailer signal	54 g	Pneum. valve for additional retarding brake, electromagnetically actuated
Fog lamp	55	Fog lamp
Headlamp	56 56 a 56 b 56 d	Headlamp High beam, high-beam indicator lamp Low beam Headlamp-flasher contact

EUROPEAN STANDARD TERMINAL DESIGNATIONS

DEFINITION	TERMINAL	CIRCUIT
Lights	57	Side-marker lamp: m-cycles, mopeds. Abroad also cars, trucks, etc.
	57 a	Parking lamp
	57 L	Parking lamp, left
	57 R	Parking lamp, right
	58	Side-marker lamps, tall lamps, license-plate lamps and instrument-panel lamps
	58 b	Tail-lamp changeover for single-axle tractors
	58 c	Trailer plug-and-receptacle assembly for single-conductor tail-lamp cable with fuse in trailer
	58 d	Variable-intensity instrument-panel lamp, tail-lamp and side-marker lamp
	58 L	Left
	58 R	Right, license-plate lamp
Alternator (magneto generator)	59	AC voltage, output Rectifier, input
	59 a	Charging armature, output
	59 b	Tail-light armature, output
	59 c	Stop-lamp armature, output
	61	Alternator charge-Indicator lamp
Tone-sequence control device	71	Input
	71 a	Output to horns 1 & 2, low
	71 b	Output to horns 1 & 2, high
Accessories	72	Alarm switch (rotating beacon)
	75	Radio, cigarette lighter
	75	Speaker
	77	Door-valve control
Switch Break-contact and changeover switches	81	Input
	81 a	1st output, break side
	81 b	2nd output, break side
Make-contact switch	82	Input
	82 a	1st output
	82 b	2nd output
	82 z	1st input
	82 y	2nd input
Multiple-position switch	83	Input
	83 a	Output, position 1
	83 b	Output, position 2
	83 L	Output, left-hand position
	83 R	Output, right-hand position
Current relay	84	Input, actuator and relay contact
	84 a	Output, actuator
	84 b	Output, relay contact
Switching relay	85	Output, actuator (end of winding to ground or negative) Input, actuator
	86	Start of winding
	86 a	Start of winding or 1st winding
	86 b	Winding tap or 2nd winding
Relay contact for break and changeover contacts	87	Input
	87 a	1st output (break side)
	87 b	2nd output
	87 c	3rd output
	87 z	1st input
	87 y	2nd input
	87 x	3rd input
Relay contact for make contact	88	Input
Relay contact for make and changeover contacts (make side)	88 a	1st output
	88 b	2nd output
	88 c	3rd output
Relay contact for make contact	88 z	1st input
	88 y	2nd input
	88 x	3rd input

① Polarity reversal possible at terminals 32–33.

EUROPEAN COMMON DEVICE IDENTIFICATION

LETTER	DEVICE	EXAMPLES
A	System, subassembly, parts group	Car radio, combination unit, trigger box, control unit
B	Transducer for conversion of non-electrical variables to electrical variables, or vice versa	Sensor, probe, pick-up, fanfare horn, horn, microphone, speaker, air-flow sensor
C	Condensers and capacitors of all types	—
D	Binary devices, memory devices	Digital device, integrated circuit (IC), memory, time delay device, timer
E	Devices not listed elsewhere	All types of lighting, heater, air conditioner, spark plug, ignition distributor
F	Protection devices	Polarity protection device, fuse, current protection circuit, overvoltage protection device
G	Power supply	Battery, alternator (generator), rectifying device, battery charger, magneto generator, transformer
H	Indicators, alarms, signalling devices	Acoustic alarm, indicator lamp, turn-signal indicator lamp, turn-signal lamp, stop lamp, high-beam indicator lamp, alternator indicator lamp, indicator lamp, signal lamp, warning buzzer
K	Relays	Battery relay, turn-signal flasher, turn-signal relay, engagement solenoid, starting solenoid, hazard-warning-signal flasher
L	Inductors	Coil, winding
M	Motors	Blower motor, fan motor, windshield-washer motor, windshield-wiper motor, starter motor
N	Regulators, amplifiers	Regulator (electronic or electro-mechanical), voltage stabilizer
P	Measurement, display, test devices	Display and measuring instrument, diagnostic connection, test point, measuring device, testing device, tachometer, speedometer, clock
R	Resistors	Flame glow plug, glow plug, glow-plug indicator resistor, heating resistor, potentiometer, variable resistor (rheostat), dropping resistor, resistance cable, cigarette lighter
S	Switches	Switches/pushbuttons of all types, pressure switch, ignition breaker
T	Transformers	Current transformer, ignition coil
U	Converters of electrical variables into other electrical variables, modulators	DC transformer, converters of all types
V	Semiconductors	Diode, rectifier, semiconductors of all types, transistor, thyristor, Z-diode
W	Transmission paths, conductors, antennas	Car antenna, shielding component, shielded conductor, wiring harness, conductor, common ground line
X	Terminals, plugs, plug-and-socket connections	Electrical connection, terminal, female

Twelve volt automotive electrical systems generally use 10, 12, 14, 16 and 18 gauge wire. Main power distribution circuits and larger accessories usually use 10 and 12 gauge wire. Battery cables are usually 4 or 6 gauge, although 1 and 2 gauge wires are occasionally used. Wire length must also be considered when making repairs to a circuit. As conductor length increases, so does resistance. An 18 gauge wire, for example, can carry a 10 amp load for 10 feet without excessive voltage drop; however if a 15 foot wire is required for the same 10 amp load, a 16 gauge wire must be used.

When backprobing a connector while it is engaged, the order of the terminals must be mentally reversed. The wire color code can help in this situation, as well as any keyway, lock tab or other reference mark.

WIRING REPAIRS

Soldering is a quick, efficient method of joining metals permanently. Everyone who has to make wiring repairs should know how to solder. Electrical connections that are soldered are far less likely to come apart and will conduct electricity much better than connections that are only "pig-tailed" together. The most popular (and preferred) method of soldering is with an electric soldering tool. Soldering irons or guns are available in many sizes and wattage ratings. Irons with higher wattage ratings deliver higher temperatures and recover lost heat faster. A small soldering iron rated for no more than 50 watts is recommended, especially on electrical systems where excess heat can damage the components being soldered.

There are 3 ingredients necessary for successful soldering; proper flux, good solder and sufficient heat. A soldering flux is necessary to clean the metal of tarnish, prepare it for soldering and to enable the solder to spread into tiny crevices. When soldering, always use a resin flux or resin core solder which is non-corrosive and will not attract moisture once the job is finished. Other types of flux (acid core) will leave a residue that will attract moisture and cause the wires to corrode.

Tin is a unique metal with a low melting point. In a molten state, it dissolves and alloys easily with many metals. Solder is made by mixing tin with lead. The most common proportions are 40/60, 50/50 and 60/40, with the percentage of tin listed first. Low priced solders usually contain less tin, making them very difficult for a beginner to use because more heat is required to melt the solder. A common solder is 40/60 which is well suited for general use, but 60/40 melts easier, has more tin for a better joint and is preferred for electrical work.

Soldering Techniques

Successful soldering requires that the metals to be joined be heated to a temperature that will melt the solder (usually 360–460°F). Contrary to popular belief, the purpose of the soldering iron is not to melt the solder itself, but to heat the parts being soldered to a temperature high enough to melt the solder when it touches the work. Melting flux-cored solder on the soldering iron will usually destroy the effectiveness of the flux.

NOTE: Soldering tips are made of copper for good heat conductivity, but must be "tinned" regularly for quick transfer of heat to the project and to prevent the solder from sticking to the iron. To "tin" the iron, simply heat it and touch the flux-cored solder to the tip; the solder will flow over the hot tip. Wipe the excess off with a clean rag, but be careful as the iron will be hot.

After some use, the tip may become pitted. If so, simply dress the tip smooth with a smooth file and "tin" the tip again. An old saying that metals well cleaned are half soldered has much truth; rust, bits of insulation and oil or grease must be removed with a wire brush or emery cloth. For maximum strength in soldered parts, the joint must start off clean and mechanically tight. Weak joints will result if there are gaps.

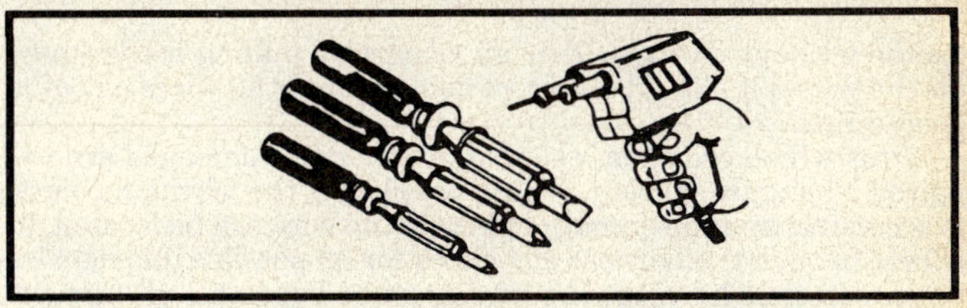

Various types of soldering tools. High wattage tools are not suitable for working on or near solid-state equipment

If a separate soldering flux is used, it should be brushed or swabbed only on areas that are to be soldered. Most solders contain a core of flux and separate fluxing is unnecessary. Hold the work to be soldered firmly. It is best to solder on a wooden board, because a metal vise will only rob heat from the piece to be soldered. Hold the soldering tip with the broadest face against the work to be soldered. Apply solder under the tip close to the work, using enough solder to give a heavy film between the iron and the piece being soldered, while moving slowly and making sure the solder melts properly. Keep the work level or the solder will run to the lowest part. If the soldering tip overheats (the solder coating on the face of the tip burns), it should be retinned.

Once the soldering is completed, let the soldered joint stand until cool. Do not attempt to cool the joint with water or other artificial means. Tape and seal all soldered wire splices after the repair has cooled. The external insulation and weatherproofing must equal that of the undamaged wire.

If soldering a joint is not possible, a crimp connector may be used. Properly installed, a crimp connector will be as reliable as a soldered joint. The connector—usually called a butt connector—must be the correct diameter or gauge for the wires. The wire ends must be clean and square cut. Fit each wire end into the connector and use a crimping tool to close the connector. The most common errors in this type of repair are either not using the correct tool to get an even crimp around the connector or over-tightening the crimp to the point of distorting the connector. Once the connector is distorted, it loses clamping force and the connection becomes loose. In this case, tighter is not better. A proper terminal crimping tool has sized and color-coded notches for delivering the correct amount of pressure to the con-

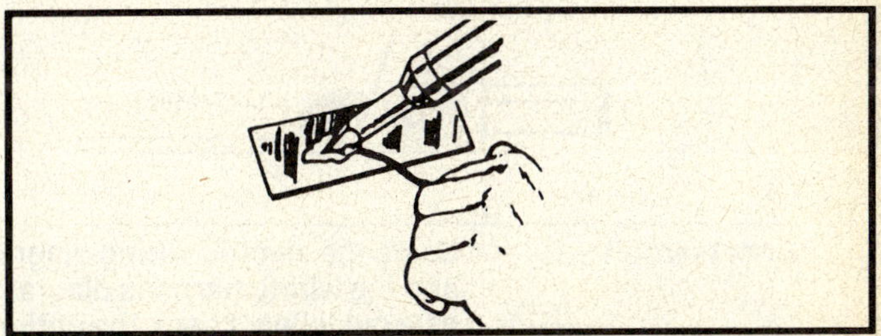

Clean and tin the soldering tool before each use

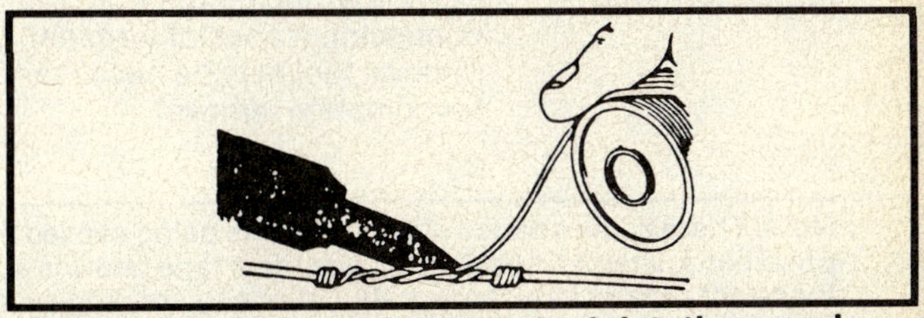

Use the soldering tool to heat the joint, then apply the solder

nector without damage. Test all crimps by pulling moderately on the wires; it should not be possible to pull the wires out of a good crimp.

Wires which are open, exposed or otherwise damaged are repaired by simple splicing. Where possible, if the wiring harness is accessible and the damaged place in the wire can be located, it is best to open the harness and check for all possible damage. In an inaccessible harness, the wire must be bypassed with a new insert, usually taped to the outside of the old harness.

FUSIBLE LINKS AND MAIN FUSES

The main fuses and links used on these vehicles are easily replaced but not repairable. Depending on the vehicle, these high-current fuses may be found either in the underdash fusebox or in the fuse or relay boxes in the engine compartment. Some may be mounted at the battery. The links are simply removed from their mounting and a new one installed. In most cases, the link may be visually checked for integrity.

Wire Harness Connectors

The computer wire harnesses connect the control unit to the various solenoids, switches and sensors used by the control system. Most connectors in the engine compartment or otherwise exposed to the elements are protected against moisture and dirt which could create oxidation and deposits on the terminals. This protection is important because of the very low voltage and current levels used by the computer and sensors. All connectors have a lock which secures the male and female terminals together; some use a secondary lock as well. Connectors attaching a harness to a component are also held by a retaining lock; use care not to damage the component when disconnecting the wiring.

Weatherproof and/or molded-on connectors require complete replacement of the connection. This means splicing a new connector assembly into the harness. All splices in on-board computer systems should be soldered to insure reliable contact. Use care when probing the connections or replacing terminals in them as it is possible to short between adjacent terminals. If this happens to the wrong terminal pair, it is possible to damage certain components. Always use jumper wires between connectors for circuit checking and never probe through weatherproof seals.

Open circuits are often difficult to locate by sight because corrosion or terminal misalignment are hidden by the connectors. Merely wiggling a connector on a sensor or in the wiring harness may correct the open circuit. This should always be considered when an open circuit or a failed sensor is indicated. Intermittent problems may also be caused by oxidized or loose connections.

WIRE HARNESS REPAIR PROCEDURES

Condition	Location	Correction
Non-continuity	Using the electric wiring diagram and the wiring harness diagram as a guideline, check the continuity of the circuit in question by using a tester, and check for breaks, loose connector couplings, or loose terminal crimp contacts. 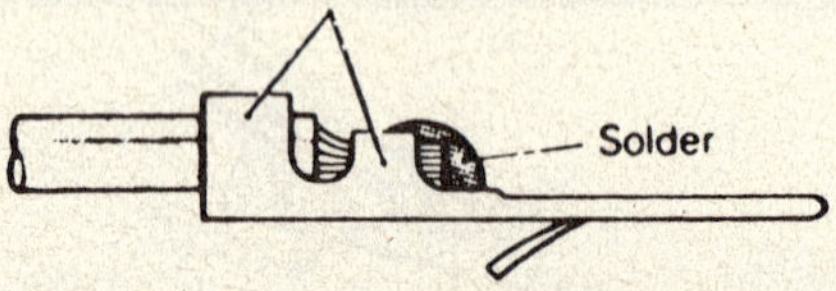	**Breaks**—Reconnect the point of the break by using solder. If the wire is too short and the connection is impossible, extend it by using a wire of the same or larger size.  **Loose couplings**—Hold the connector securely, and insert it until there is a definite joining of the coupling. If the connector is equipped with a locking mechanism, insert the connector until it is locked securely. **Loose terminal crimp contacts**—Remove approximately 2 in. (5mm) of the insulation covering from the end of the wire, crimp the terminal contact by using a pair of pliers, and then, in addition, complete the repair by soldering.
Short-circuit	Using the electric wiring diagram and the wiring harness diagram as a guideline, check the entire circuit for pinched wires.	Remove the pinched portion, and then repair any breaks in the insulation covering with tape. Repair breaks of the wire by soldering.
Loose terminal	Pull the wiring lightly from the connector. A special terminal removal tool may be necessary for complete removal.	Raise the terminal catch pin, and then insert it until a definite clicking sound is heard. 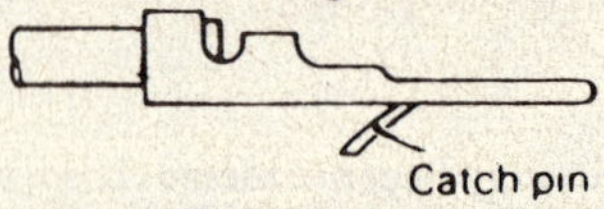

Note: There is the chance of short circuits being caused by insulation damage at soldered points. To avoid this possibility, wrap all splices with electrical tape and use a layer of silicone to seal the connection against moisture. Incorrect repairs can cause malfunctions by creating excessive resistance in a circuit.

Wire harness repair procedures

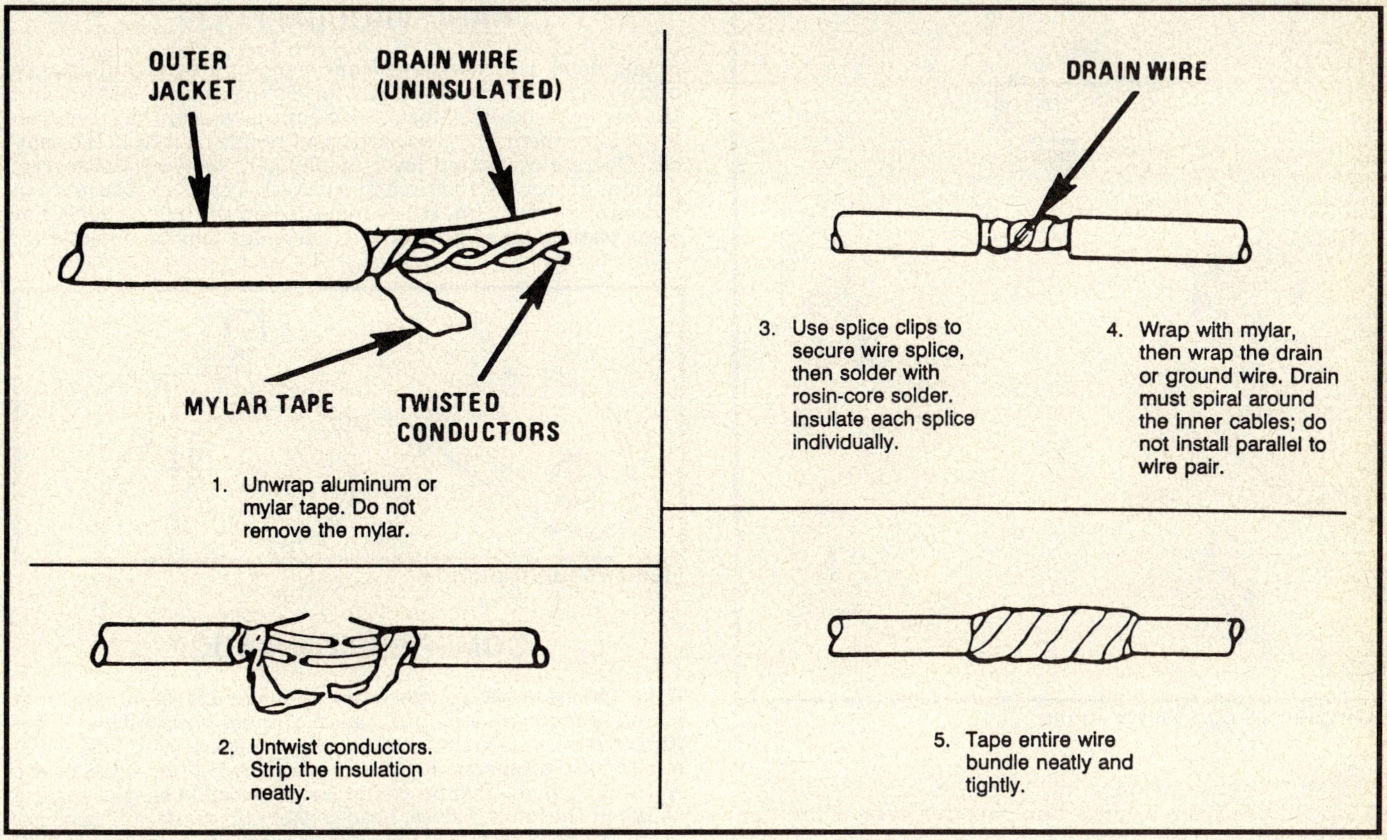

Repair procedures for twisted or shielded cable

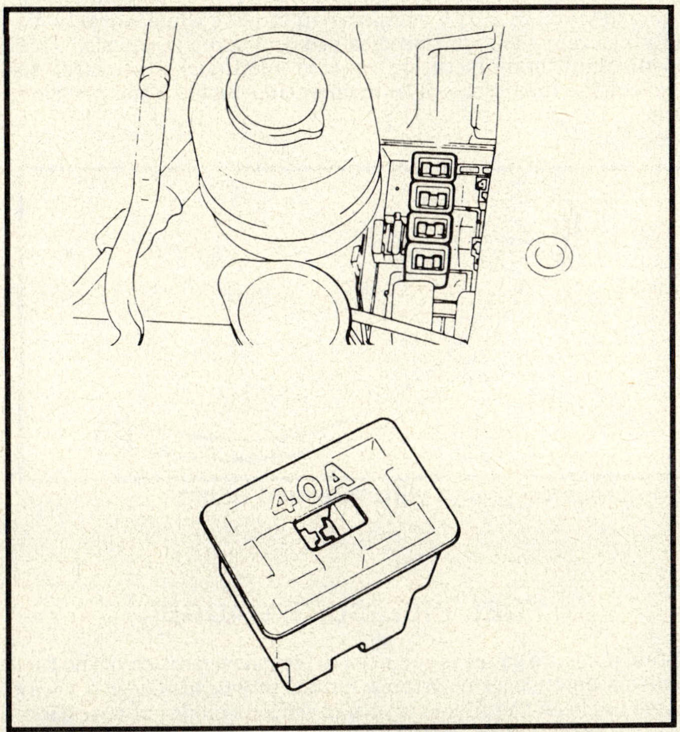

Example of main link location in the engine compartment. This type of link may be inspected visually

When using a circuit tester for diagnosis, always probe connections from the wire side. Be careful not to damage sealed connectors with test probes; never backprobe a sealed or weatherproof connector.

NOTE: Many of the problems in the wiring harness are due to bad ground connections. Always check all vehicle ground connections for corrosion or looseness before performing any other circuit tests. This will eliminate the chance of a bad ground affecting circuit operation.

All wiring harnesses should be replaced with identical parts, using the same gauge wire and connectors. When signal or low-voltage wires are spliced into a harness, choose wire with high temperature insulation only. The low voltage and current levels found in the system require the best possible connection at all wire splices; joints must be mechanically tight and cleanly soldered.

It is seldom necessary to replace a complete harness. If replacement is necessary, pay close attention to insure proper harness routing. Secure the harness with suitable plastic wire clamps to prevent vibrations from causing the harness to wear in spots or contact any hot components.

Repairing Hard Shell Connectors

Unlike molded connectors, the terminal contacts in hard shell connectors can be replaced. Replacement usually involves the use of a special terminal removal tool that moves the locking tangs (barbs) on the connector terminal and allows the connector to be removed from the rear of the shell. The connector shell should be replaced if it shows any evidence of burning, melting, cracks, or breaks. Replace individual terminals that are burnt, corroded, distorted or loose.

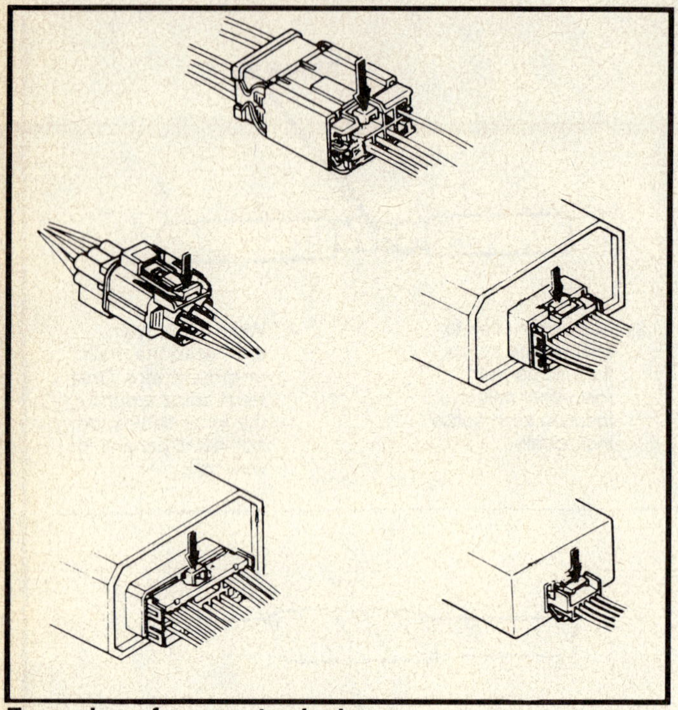

Examples of connector locks

NOTE: The insulation crimp must be tight to prevent the insulation from sliding back on the wire when the wire is pulled. The insulation must be visibly compressed under the crimp tabs, and the ends of the crimp should be turned in for a firm grip on the insulation.

The wire crimp must be made with all wire strands inside the crimp. The terminal must be fully compressed on the wire strands with the ends of the crimp tabs turned in to make a firm grip on the wire. Check all connections with an ohmmeter to insure a good contact. There should be no measurable resistance between the wire and the terminal when connected.

Mechanical Test Equipment

VACUUM GAUGE

Intake manifold vacuum is used to operate various systems and devices on all cars. To correctly diagnose and solve problems in vacuum control systems, a vacuum source is necessary for testing. In some cases, vacuum can be taken from the intake manifold when the engine is running, but vacuum is normally provided by a hand vacuum pump.

Most gauges are graduated in inches of Mercury (in. Hg), although a device called a manometer reads vacuum in inches of water (in. H_2O). The vacuum reading usually varies between 18 and 22 in. Hg at sea level. To test engine vacuum, the vacuum gauge must be connected to a source of manifold vacuum. Many engines have a plug in the intake manifold which can be removed and replaced with an adapter fitting. Connect the vacuum gauge to the fitting with a suitable rubber hose or, if no manifold plug is available, connect the vacuum gauge to any device using manifold vacuum, such as EGR valves, etc. The vacuum gauge can be used to determine the amount of vacuum reaching a component.

HAND VACUUM PUMP

Small, hand-held vacuum pumps come in a variety of designs and provide a source of vacuum for testing components without the engine operating. Most have a built-in vacuum gauge and allow a component to be tested without removing it from the vehicle. Operate the pump lever or plunger, applying the correct amount of vacuum required for the test. The level of vacuum in inches of Mercury (in. Hg) is indicated on the pump gauge. For some testing, an additional vacuum gauge may be necessary.

Hand vacuum pumps

COMPRESSION GAUGE

A compression gauge measures the amount of pressure in pounds per square inch (psi) that a cylinder is producing. Some gauges have a hose that screws into the spark plug hole while others have a tapered rubber tip which is held by hand in the spark plug hole. Engine compression depends on the sealing ability of the rings, valves, head gasket and spark plug gaskets. If any of these parts are not sealing properly, compression will be lost and the power output of the engine will be reduced. The compression in each cylinder should be measured and the variation between cylinders should be noted. The engine should be cranked through 5 or 6 compression strokes while warm, with all plugs removed, ignition disabled and throttle valve(s) wide open. Many manufacturers require additional procedures to prevent the fuel injectors from operating during a compression test.

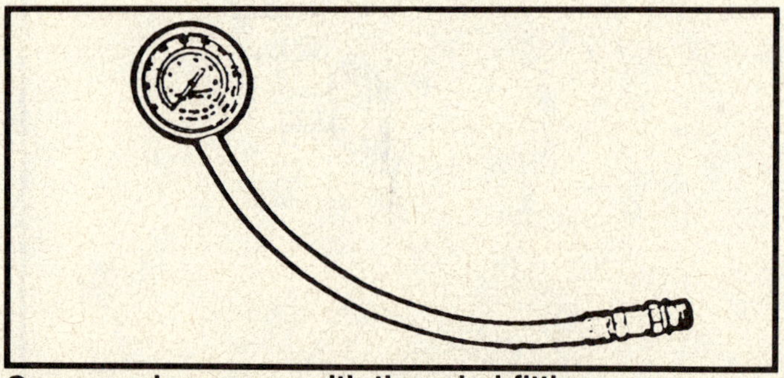

Compression gauge with threaded fitting

FUEL PRESSURE GAUGE

A fuel pressure gauge is required to test the operation of the fuel delivery and injection systems. Some systems also need a 3-way valve to check the fuel pressure in various modes of operation. Gauges may require special adapters for making fuel connections. Always observe the cautions outlined in the Fuel System Service section when working around any pressurized fuel system.

FUEL LINES

CAUTION

Do not attempt to repair high pressure fuel lines. Any sign of scratching, kinking or deformation requires replacement of the line. Replacement lines must always be of the correct material and pressure rating. When working with or around fuel lines, observe the following precautions.

• Treat all drained fuel and fuel-soaked rags as a flammable hazard. Keep containers covered and store rags in an airtight container. Contain spillage; keep work area away from open flame, spark sources including switches and electric motors; observe No Smoking precautions.

• Treat any fuel leak or spillage as a serious hazard. Clean up the spill or repair the leak immediately.

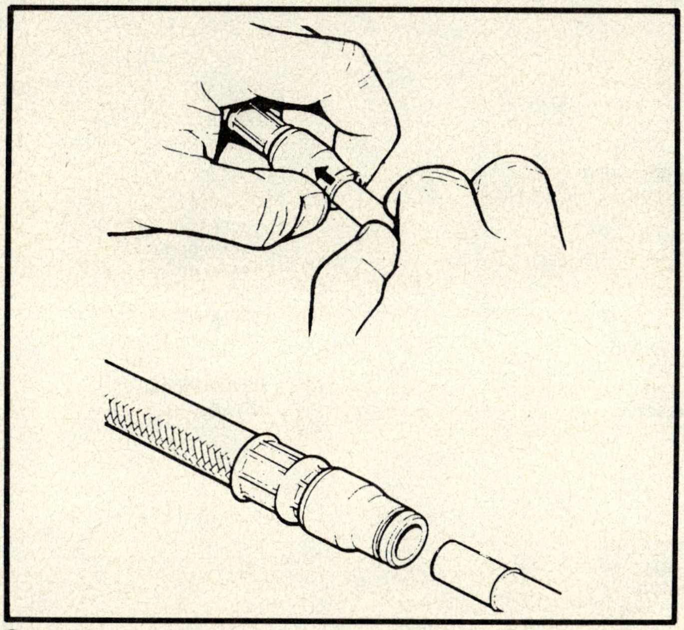

Some fuel systems use quick-connect fittings. Make certain the fitting is correctly seated after assembly

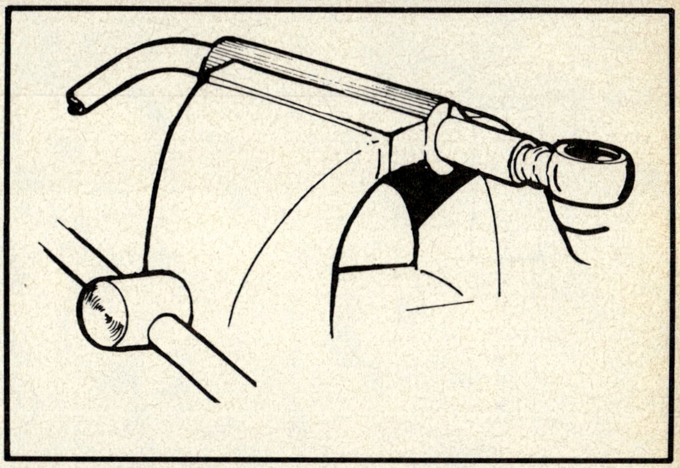

Always protect fuel lines and hoses during repairs

• Keep a dry chemical fire extinguisher (type B-C) within arm's reach in the work area and know how to use it.

• Always relieve fuel system pressure before loosening or disconnecting any lines.

• If repairs involving sparks or high heat are anticipated, identify and protect all fuel lines before work begins. When repairing floorpans, rocker panels, dash or firewalls, carefully check the area on the other side of the panel before beginning work. Removing plastic or nylon fuel lines prior to repair is highly recommended.

• When installing new fuel lines, or performing repairs in the surrounding area, do not hammer directly on the body clips holding the fuel lines. The fuel line may become damaged, resulting in a system leak.

• Inspect connectors for burrs, dirt, etc. Clean connectors with a soft, lint-free cloth.

• Cover all fittings to be disconnected with a rag or towel to prevent fuel spray. Some pressure may remain within the system even after the pump pressure is relieved.

• Always wear eye or full-face protection when working on any part of the fuel system.

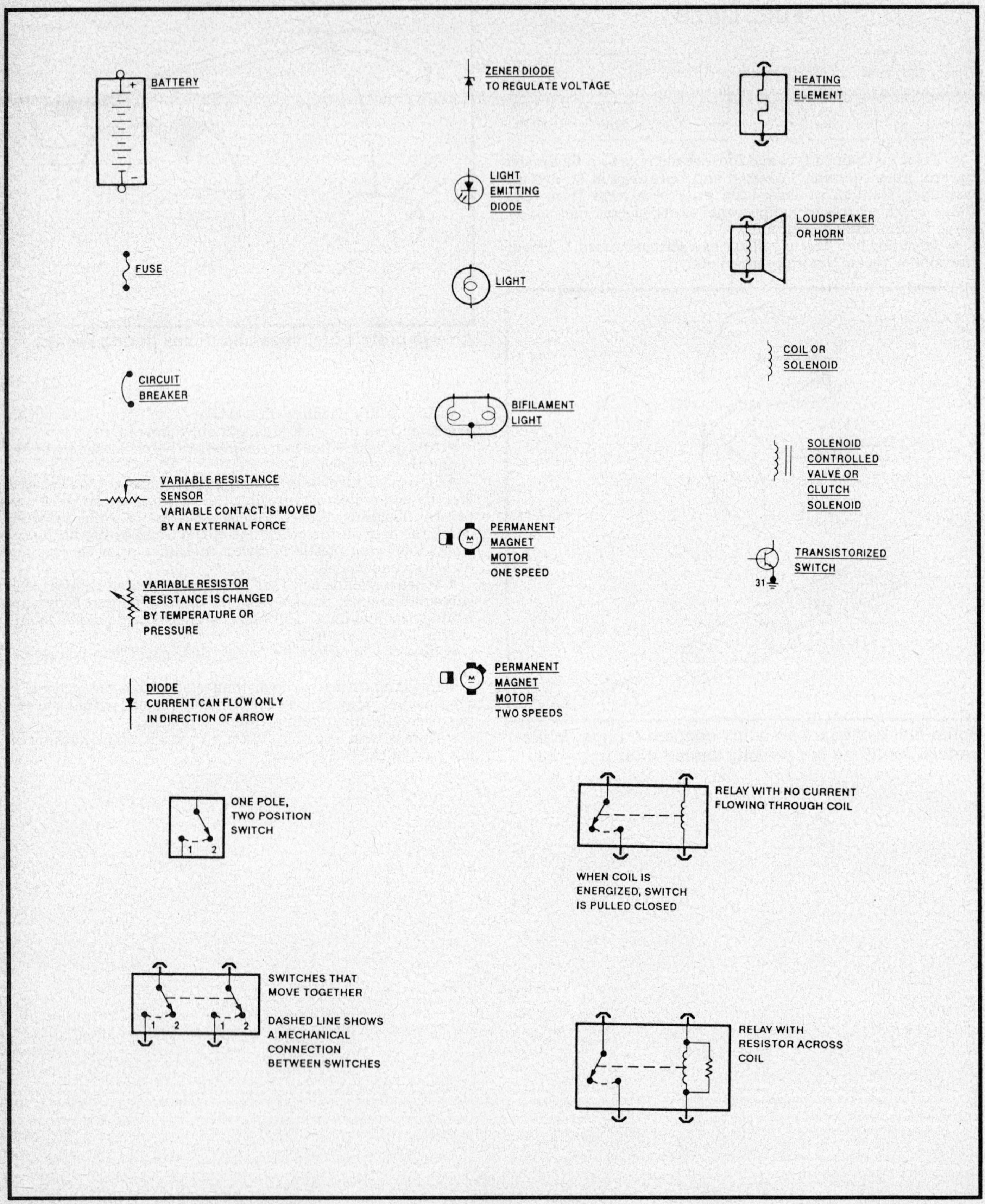

Commonly used electrical symbols

Self-Diagnostic Systems

3 SECTION

ALFA ROMEO

General Information

The Motronic ML-4.1 fuel injection and ignition control system is used on all vehicles. It is a conventional electronically controlled engine management system. The system uses signals sent by various engine sensors to maintain the best engine operation under all driving conditions. The Electronic Control Unit (ECU) also controls the ignition spark timing and advance. Mechanical advance mechanisms are not used.

All engines are equipped with oxygen sensors in the exhaust system. The oxygen sensor determines the oxygen content in the exhaust stream to the ECU and generates a small electrical signal. This signal is sent to the ECU allowing the controller to make fuel and spark corrections when operating in the closed loop mode.

The Motronic ECU control unit is equipped with a self diagnostic feature that provides useful indications about operational malfunctions in the system.

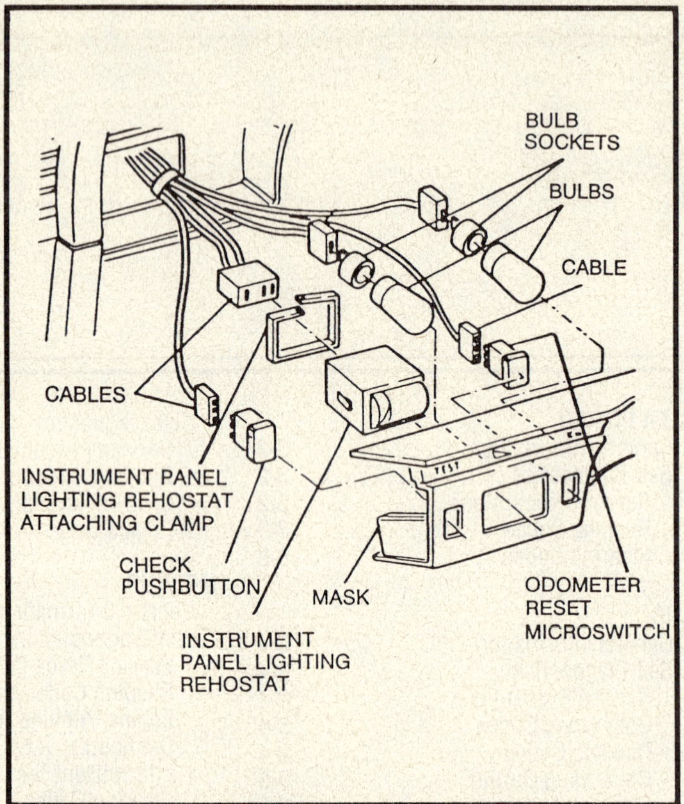

Self-diagnostic test button location at instrument panel—Alfa Romeo 164

Self-Diagnostics

The Motronic control unit is equipped with a self-diagnostic system to help the technician detect problem areas in the engine management system related to the emission, fuel or ignition system.

The CHECK ENGINE light, located on the instrument cluster, will only illuminate when a malfunction is emission related.

Trouble Codes are composed of 4 flashing groups, each of which is a code digit. The first is the ECU identification: the first position signifies Motronic, the second position signifies the method of diagnosis, the third position signifies malfunction signal and the fourth position signifies active diagnosis of actuators. The third and fourth are the specific number of the malfunction.

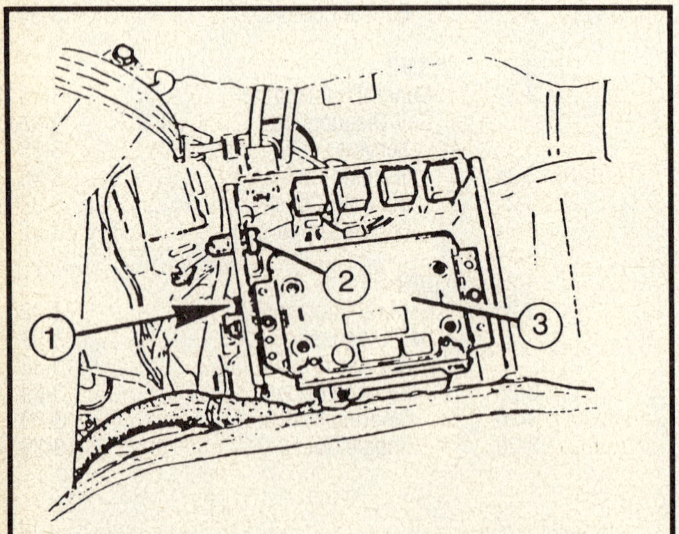

1. CHECK ENGINE warning light
2. Alfa Romeo tester connector
3. Self-diagnostic test button
4. ML4.1 MOTRONIC electronic control unit

Self-diagnostic test button location—Alfa Romeo Spider

SERVICE PRECAUTIONS

- Keep all ECU parts and harnesses dry during service. Protect the ECU and all solid-state components from rough handling or extremes of temperature.
- Before attempting to remove any parts, turn the ignition switch **OFF** and disconnect the battery ground cable.
- Make sure all harness connectors are fastened securely. A poor connection can cause an extremely high voltage surge resulting in damage to integrated circuits.
- Always use a 12 volt battery as a power source.
- Do not disconnect the battery cables with the engine running; never run the engine with the battery cables loose or disconnected.
- Do not attempt to disassemble the ECU unit under any circumstances.
- When connecting or disconnecting pin connectors from the ECU, take care not to bend or break any pin terminals. Check that there are no bends or breaks on the ECU pin terminals before attempting any connections.
- When measuring supply voltage of ECU-controlled components, keep the tester probes separated from each other and from accidental grounding. If the tester probes accidentally make contact with each other during measurement, a short circuit will damage the ECU.
- Remove the ECU if the vehicle is to be exposed to high heat such as a bake oven to cure paint after body refinishing or when welding is be done on the vehicle.
- Use great care when working on or around air bag systems.

Due to back-up circuitry the system may stay armed for a period of time without battery power. When working on any electrical system is the air bag deployment zone (driver's seat area), disconnect the battery and remove the air bag module from the steering wheel. Then reconnect the battery, if necessary, for electrical diagnosis work.

- Don't be hasty in condemning the ECU when problems arise. On most vehicles, regardless of make, the ECU is located in a relatively protected area (under the seat, behind the dashboard, etc.). But the sensors that feed it information exist in very harsh underhood environments such as immersed in scalding coolant or in the hot blast of exhaust gases. In addition, since the signal to/from these sensors is a very small voltage or resistance, every electrical connection for these sensors is a potential trouble area. Look for problems in these areas and especially any and all ground connections before blaming the ECU.

READING CODES

The CHECK ENGINE light, located on the instrument cluster, will only illuminate when a malfunction is detected.

Trouble Codes are composed of 4 flashing groups, each of which is a code digit. The first is the ECU identification: the first position signifies Motronic, the second position signifies the method of diagnosis, the third position signifies malfunction signal and the fourth position signifies active diagnosis of actuators. The third and fourth are the specific number of the malfunction.

Engine Codes

Code 1211—Battery voltage problem. ECU reads under 9 volts or over 16 volts from battery. Problem is outside ECU.
Code 1212—Idle throttle switch problem. ECU not receiving plausible signal from MIN switch. Look for problem in MIN switch and/or wiring or ECU.
Code 1213—Full throttle switch problem. ECU not receiving plausible signal from MAX switch. Look for problem in MAX switch and/or wiring or ECU.
Code 1214—Engine temperature sensor problem. ECU not receiving plausible signal from temperature sensor and is using substitute value. Look for problem in temperature sensor and/ or wiring or ECU.
Code 1221—Airflow Meter problem. ECU receiving signal from flowmeter outside plausible range value and is using substitute value. Look for problem in airflow meter and/or wiring or ECU.
Code 1222—Output stage of ECU for idle speed actuator problem. No plausible signal to actuator. Look for problem in idle air actuator and/or wiring or ECU.
Code 1223—Oxygen sensor regulation to limit stop. ECU receiving signal at minimum or maximum value of correction limit. Look for problem in air/fuel circuit.
Code 1224—Oxygen sensor problem. Look for problem in oxygen sensor and/or wiring or ECU.
Code 1225—Air temperature sensor problem. ECU not receiving plausible signal from airflow meter temperature sensor and is using substitute value. Look for problem in airflow meter temperature sensor and/or wiring or ECU.
Code 1243—Output stage of ECU for fuel pump relay problem. Engine will have stopped. Look for problem in relay and/or wiring or ECU.
Code 1244—Output stage of ECU for emission purge canister valve. Look for problem in purge valve and/or wiring or ECU.
Code 1245—Output stage of ECU for variable valve timing, on engines so equipped. Look for problem in variable valve timing relay and/or wiring or ECU.
Code 1251—Malfunction in ECU. ECU recognizes problem in RAM/ROM memory. Problem is in ECU.
Code 1265—Output stage of ECU not sending signal to CHECK ENGINE warning light.

Spider

The diagnosis procedure can occur only with the ignition **ON** and engine speed up to 2000 rpm. The malfunction signal will continue through the entire rpm range once the diagnosis has started.

To initiate the diagnosis process, press the push button located under the rear parcel shelf, near the electronic control unit. The CHECK ENGINE light will provide the malfunction code number by emitting a series of flashes. Each code is endlessly repeated until the next code is requested. Code 4444 is used to signify no malfunctions.

164

The diagnosis procedure is divided into 2 groups—display of errors stored by the control unit and monitoring of engine actuators by the control unit.

DISPLAY OF ERRORS STORED BY THE CONTROL UNIT

To initiate the diagnosis process on the 164, turn the ignition to key to the **ON** position, then press the push button located at the bottom of the dash near and to the left of the steering column for 2.5–5.0 seconds. The CHECK ENGINE light will provide the malfunction code number by emitting a series of flashes. Each code is repeated until the next code is requested. Code 4444 is used to signify no malfunctions.

When all malfunctions have been displayed, the CHECK ENGINE light will display a series of long flashes.

By pressing the push button for 5–10 seconds, the series of codes will be repeated.

MONITORING OF ENGINE ACTUATORS AND DEVICES

To initiate the monitoring process on the 164, press and hold the push button, located at the bottom of the dash near and to the left of the steering column, while turning the ignition **ON**. Hold the push button for 2.5–5.0 seconds after the ignition key is turned **ON**, then release.

A few seconds after the push button is released, the fuel injectors will begin to operate. This will be followed by the idle adjustment actuator and the evaporation solenoid valve.

By pressing the push button for 5–10 seconds, the process will be repeated. The following codes, flashed on the CHECK ENGINE light, will allow the user to determine which test is in progress:

1-4-1-1: Electro-injectors actuation test activated. Continuous control at a frequency of 10 hertz.

1-4-1-2: Idle adjustment actuator activation test. Continuous control for 1 second at idle position and 2 second at maximum speed.

1-4-1-3: Evaporation solenoid valve actuation test. Continuous control at a frequency of 0.5 hertz.

CLEARING CODES

Trouble codes will be cleared if the battery negative terminal is disconnected for 10 seconds or more. In addition, on the 164, codes may be cleared by pressing the push button for 10 seconds or more.

AUDI

General Information

Four engine management systems are used on Audi vehicles; Bosch Continous Injection System (CIS-E) III on NG, NF and MC engines, Bosch CIS Motronic on 3A and 3B engines, Bosch Motronic on PT engines and Audi MPI on the 7A engine. The CIS-E III system uses separate control units for fuel injection and ignition system control. The Motronic and MPI systems use a single control unit to operate all fuel injection, ignition and emission control functions. All systems are capable of self diagnostics and storing fault codes in memory, but the CIS-E III fault codes are stored in a temporary memory that is erased when the ignition switch is turned **OFF**.

CIS-E III System

Two separate control units are used in this system. The control units operate independently but they do communicate with each other. The fuel injection control unit manages air/fuel mixture and idle control functions. It accepts most of the engine sensor data as well as information from the automatic transmission. The ignition control unit manages all ignition timing functions, including knock regulation.

Test connectors are provided to connect a flashing LED tester or the VAG 1551 diagnostic tester or equivalent. The fuel injection control unit will generate output signals to test 4 of the output devices. On California models, fault codes are stored in permanent memory until intentionally erased. If in doubt, these models can be identified by the part number on the control unit:

Fuel injection control unit—443 906 264B
Ignition system control unit—443 907 397E

On Federal models, the codes are stored in temporary memory, which is erased when the ignition switch is **OFF**. Control units are not interchangable between Federal and California models.

CIS Motronic System

The control unit operates all engine management functions. A manifold pressure sensor is mounted inside the control unit housing with a hose and water separator connecting it to the intake manifold. The hose length is critical and must not be altered when replacing the hose.

The self diagnostic system can differentiate between intermittent and permanent faults and all fault codes are stored in permanent memory until intentionally erased. If a fault has not occurred within the last 50 engine starts, it is erased automatically. Codes can be retrieved using a flash tester or the VAG 1551 diagnostic tester or equivalent. With the Diagnostic Tester, signals can be generated to test certain output devices. Faults that effect exhaust emissions will cause the engine warning light to turn **ON** when the engine is running for as long as the fault exists.

Motronic and MPI Systems

The Motronic system is fully electronic. The single control unit manages all engine functions and stores fault codes in permanent memory. The self diagnostic program is capable of generating signals to test all output devices including the injectors. Fault codes can be retrieved through the engine warning light, an LED test light or with the VAG diagnostic tester or equivalent. Even though there is a permanent fault code memory, the self diagnostic program is designed to be used with the engine running. The output checks can be activated with the Diagnostic Tester or the LED test light when the engine is not running.

Self-Diagnostics

SERVICE PRECAUTIONS

- Do not disconnect the battery or power to the control unit before reading the fault codes. On the Motronic and MPI systems, fault code memory is erased when power is interrupted.
- Make sure the ignition switch is **OFF** before disconnecting any wiring.
- Before removing or installing a control unit, disconnect the negative battery cable. The unit receives power through the main connector at all times and will be permanently damaged if improperly powered up or down.
- Keep all parts and harnesses dry during service. Protect the control unit and all solid-state components from rough handling or extremes of temperature.

GENERATING CODES

CIS Systems

1. The engine must be running to generate fault codes. Read all the Steps in this procedure before starting the engine.
2. Turn the ignition switch **ON** without starting the engine to make sure the engine warning light works, if equipped. If it does not light with the ignition **ON**, engine not running, but does light when attempting to retrieve fault codes, either the wiring between the control units is faulty or the ignition control unit is faulty.
3. The fuel pump relay and fuses 13, 19, 24 and 28 must be good and all ground connections in the engine compartment must be good.
4. To generate fault codes, the vehicle must be driven for at least 5 minutes at normal operating temperature, air conditioner **OFF**. There must be at least one application of full throttle with engine speed above 3000 rpm. On turbocharged engines, full boost should be reached during the full throttle acceleration. After the test drive, allow the engine to idle for at least 2 minutes before retrieving codes.
5. Do not turn the ignition switch **OFF** or the temporary memory will be erased. If the engine stalls, do not restart it. The codes will still be in memory.
6. If the engine will not run, operate the starter for at least 5 seconds and leave the ignition switch **ON**.

READING CODES

With Flash Tester

1. On California models equipped with an engine warning light, an LED tester is not required. On models not equipped with the engine warning light, connect an LED tester to the test connector above the pedals or in the passenger side foot well as shown. Do not connect the jumper wire yet.
2. On all models, codes should be retrieved with the engine running at idle. If the engine will not start, operate the starter for at least 5 seconds and leave the ignition switch **ON**. On CIS systems, the engine should be left running after the procedure for generating codes.
3. When the jumper wire is connected for at least 4 seconds, then disconnected, the program is activated.
4. The engine warning light on the instrument panel or the LED tester will begin to flash the first code. It will continue to flash this code until the jumper wire is connected again.
5. Repeat the 4 second jumper wire connection to proceed to the next code. Do not leave the jumper connected for 10 seconds or the memory will be erased. Engine idle speed may increase slightly when reading injection control unit codes.

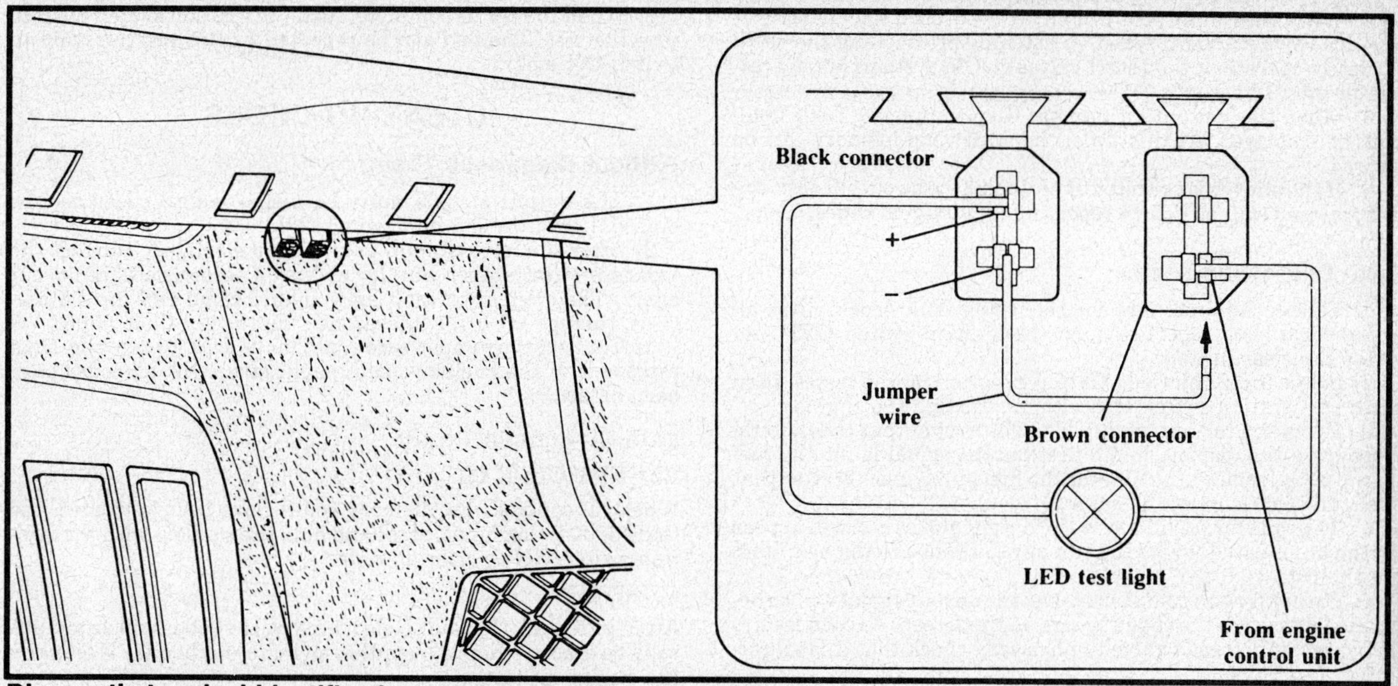

Diagnostic terminal identification used on the Audi V8 Quattro

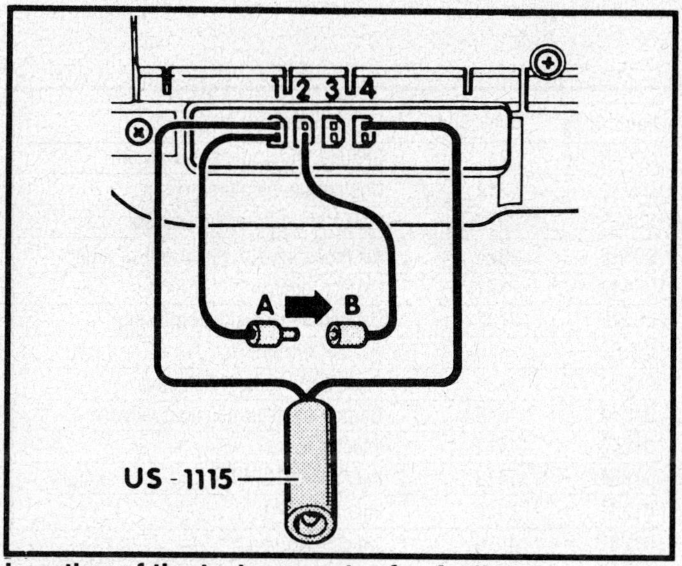

Location of the test connector for Audi self-diagnosis—Except V8 Quattro

6. All codes are 4 digits, with about 2.5 seconds between digits. Codes are displayed in order of importance, usually beginning with ignition system codes. Count the flashes and write down the code, then call the next code.

7. If Code 4444 is the first code, no faults are stored in memory. When the light flasher **ON** for 2.5 seconds with 2.5 second intervals between, this is Code 0000. All fault codes have been reported. On CIS systems, the codes are erased when the ignition switch is turned **OFF**.

With Diagnostic Tester

Connect the VAG 1551 or equivalent diagnostic tester to the diagnostic terminals above the pedals or in the passenger side foot well. The terminals are shaped so they cannot be connected incorrectly. Power for the tester is supplied through fuse 21 or 27. Codes for most on-board control units with memory can be accessed through these connectors.

1. On all models, codes should be retrieved with the engine running at idle. If the engine will not start, operate the starter for at least 5 seconds and leave the ignition switch **ON**. On CIS systems, the engine should be left running after the procedure for generating codes.

2. Operate the tester to select menu option 2, Blink Code Output or Fault Memory Recall. An asterisk (*) will appear and flash the codes, which the tester will count and report on the screen as numbers. If Code 4444 is displayed, no faults are found in memory.

3. If the engine is not running, some codes will be displayed. These can be ignored if the engine has been intentionally stalled but should be investigated if the engine will not start.

4. Press the run key to advance to the next code. Read all the fault codes before starting repairs.

5. When the End of Report Code 0000 is displayed, pressing the run key again will proceed to another control unit or return to the main menu. To stop the program without erasing the codes, turn the ignition switch **OFF** and press the clear button once.

GENERATING OUTPUT SIGNALS

Without Diagnostic Tester

1. With the ignition switch **OFF**, connect the LED test light and the jumper wire to the test connectors.

2. Turn the ignition switch **ON**. The first code will be displayed. If the first code is the fuel pump relay and the pump begins to run, remove the jumper wire quickly to prevent flooding the engine.

3. To go to the next output signal, momentarily disconnect the jumper wire. The next item on the output code list will be activated when the full throttle switch is closed. Be sure to use the correct output code list, the sequence is not the same on all engines.

4. When the full throttle switch is closed, each solenoid or frequency valve can be checked by listening or touching the valve to detect operation. Cold start valves on CIS systems are operated for only 10 seconds.
5. When the last output test has been completed, Code 0000 will be displayed. At this time, the fault code memory can be erased.
6. If the starter is operated at any point in the output test, the control unit will switch to reporting input signal codes.

With Diagnostic Tester

1. Follow the procedure for retrieving fault codes. After all codes have been reported, turn the ignition switch **OFF** and press the clear button.
2. Select the Blink Code Output program on the tester, then turn the ignition switch **ON** without starting the engine.
3. Press and release the run key. The first output code should appear on the display. If the first output signal is for the fuel pump relay, remove the fuse for the fuel pump quickly after that test to avoid flooding the engine.
4. The code for each item in the output signal test will appear in the order listed. Press the run key to change to the next item on the list.
5. Except for the cold start valve, the output signal will be activated as long as the code appears on the screen. As each item is activated, touch each valve to physically check that it is vibrating or humming.
6. When the test is complete, turn the ignition switch **OFF** to stop the test. The test can be repeated by turning the ignition switch **ON** again.

CLEARING CODES

Without Diagnostic Tester

1. The output signals must be tested before codes can be erased. Leave the ignition switch **ON**.
2. After the last output signal code is displayed, connect the test connector jumper wire for at least 4 seconds and remove it again. The engine warning light should come **ON** for 2.5 seconds, then go **OFF** for 2.5 seconds, displaying Code 0000.
3. Connect the jumper wire again for at least 10 seconds, then remove it. If the engine warning light stays **ON**, all codes have been erased.

With Diagnostic Tester

EXCEPT CIS-E III

When all control unit memory codes have been retrieved and Code 0000 is displayed, press and hold the run key with the ignition switch **ON** to clear all codes.

CIE-E III

After activating the fault code memory, the codes will automatically be erased when the ignition switch is turned **OFF** or when the engine is started.

Fault Code	Blink Code	Explanation
00000	4444	No faults in memory
65535	1111	Control unit defective
00513	2111	Engine rpm signal missing
00514	2112	Ignition timing point sender signal
00515	2113	Hall sender signal missing
00516	2121	Idle switch
00535	2141	First knock control program
00524	2142	Knock sensor 1 signal missing
00540	2144	Knock sensor 2 signal missing
00518	2212	Throttle position sensor
00543	2214	Upper rpm limit exceeded
00575	2221	Manifold pressure signal out of limit
00519	2222	Manifold pressure sensor signal out of limit
00528	2223	Altitude sensor signal missing
00544	2224	Boost pressure limit exceeded
00533	2231	Idle speed control out of limit
00532	2234	Control unit supply voltage high/low
00538	2241	Second knock control program
00522	2312	Coolant temperature sensor
00523	2322	Intake air temperature sensor
00553	2324	Air mass sensor signal out of limit
00537	2341	Oxygen control out of limit
00525	2342	Oxygen sensor signal missing
00561	2413	Air/fuel mixture rich
00824	3424	Fault warning light defective
01247	4343	Carbon canister solenoid valve
01249	4411	Injector No. 1
01250	4412	Injector No. 2
01251	4413	Injector No. 3
01252	4414	Injector No. 4
01253	4421	Injector No. 5
01257	4431	Idle stabilizer valve
01262	4442	Wastegate frequency valve
00000	0000	End of output

CIS-Motronic engine fault and blink codes—typical Audi

BMW

General Information

All models use a single control unit for fuel injection and ignition system control. Several other control units are used on various models: automatic transmission, electronic throttle, airbag, anti-lock brakes, anti-theft system, climate control, cruise control, traction control, central heated locks and power windows, seat position memory, driver's trip computer, variable shock absorbers, even the head light washers on some models are electroncially controlled.

Not all control units have their own fault memory, however each control unit can be accessed by the Diagnostic Tester in real time. All units connect to the Diagnostic Connector on the same 2 data link wires in a parallel multi-plex circuit. The TXD wire transmits commands from the tester to each control unit. The RXD wire transmits data from the control unit to the tester. If a control unit has a fault that would cause it to keep the RXD line "busy", the tester would not be able to read any other control unit.

Digital Motor Electronic (DME) Unit

All models use the Bosch DME engine management system. Fuel injection and ignition functions are operated by a single Motronic control unit. Several versions of this unit have been used, depending on the engine series and model year. Although all control units are similar, they are coded to operate in the specific engine/vehicle combination. For example, a 1988 model with the M30 engine operates on the DME 1.1 version, while the same vehicle in 1989 uses DME 1.3. The difference may be due to equipment changes to that model, such as a new cruise control or a different transmission. These and other control units on board communicate with the DME control unit and the DME unit must be appropriately programmed even if engine control functions are unchanged.

Self-Diagnostics

SERVICE PRECAUTIONS

- Do not disconnect the battery or power to the control unit before reading the fault codes. Fault code memroy is erased when power is interrupted.
- Make surc thc ignition switch is **OFF** before disconnecting any wiring.
- Before removing or installing the DME control unit, disconnect the negative battery cable. The unit receives power through the main connector at all times and will be permanently damaged if improperly powered up or down.
- Keep ECU all parts and harnesses dry during service. Protect the ECU and all solid-state components from rough handling or extremes of temperature.
- All airbag system wiring is in a yellow harness. Use extreme care when working around this wiring. Do test these circuits without first disconnecting the airbag units.

READING CODES

With Diagnostic Tester

When the tester is attached to the diagnostic connector, all con-

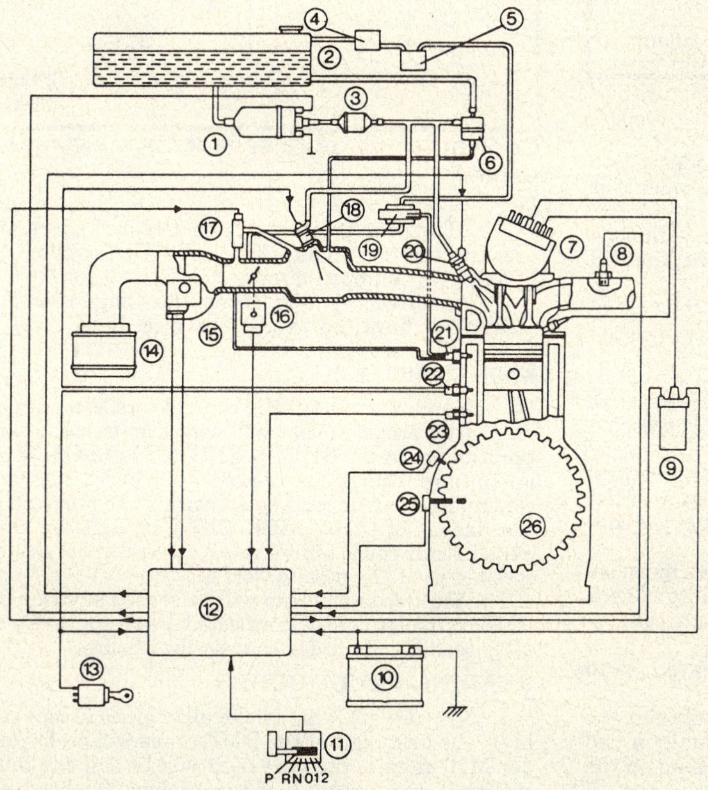

Digital Motor Electronic—typical system

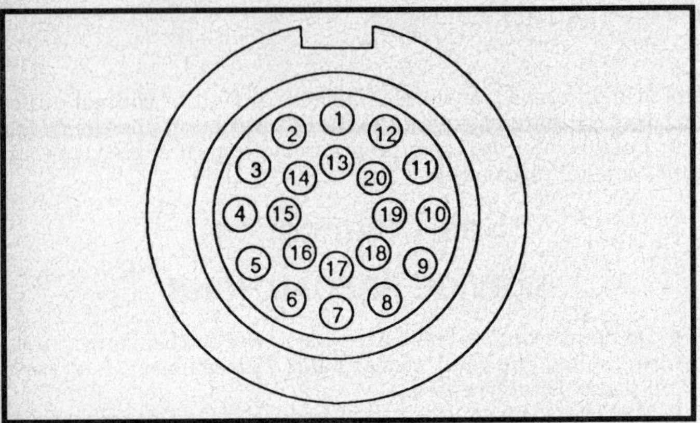

Diagnostic connector; pin 15 is RxD terminal and pin 20 is TxD terminal—BWM

Code	Fault Branch
x211	Control unit (self-test)
x215	Air flow sensor
x221	Oxygen sensor
x222	Oxygen sensor regulating stop
x223	NTC – water
x224	NTC – air
x231	Battery voltage monitor
x232	Idle speed switch
x233	Full load switch
x251	Fuel injector final stage 1
x252	Fuel injector final stage 2
x261	Electric fuel pump relay final stage
x262	Idle speed control final stages
x263	Tank venting final stage
x264	Relay 3 final stage (oxygen sensor heating)

Fault codes—12 cylinder M70 engine

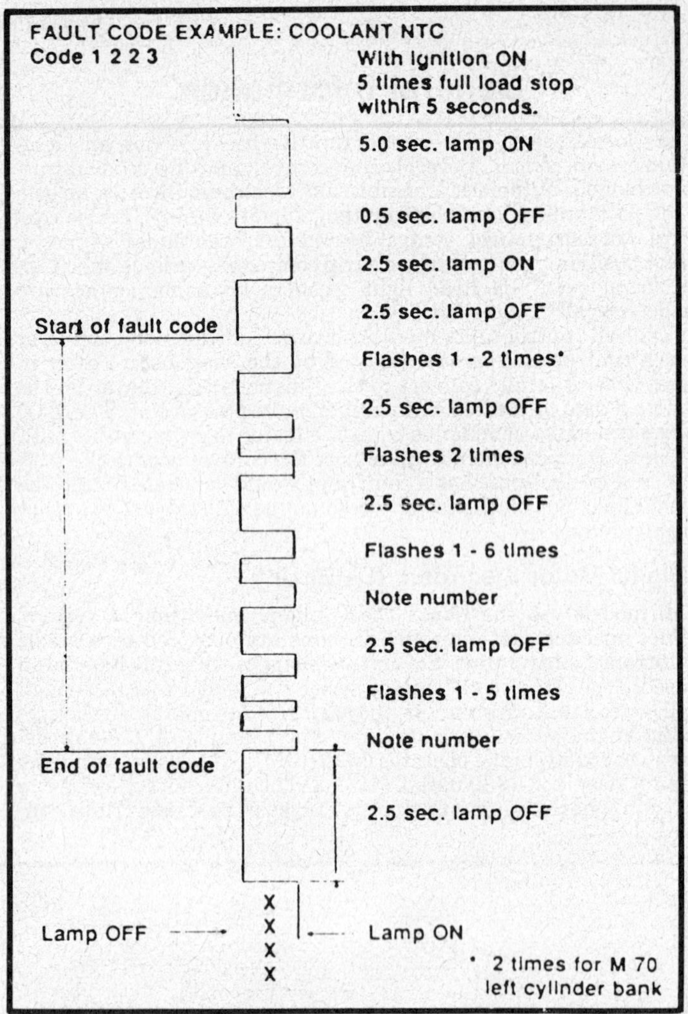

Coolant temperature sensor code example for M70 engine

trol units on the vehicle will present identification code data to the tester for display on the screen. Two options are available at this point:

1. Select a single system for testing or reading the fault code memory. If fault codes are stored in memory, the data line will be highlighted and marked "F".
2. Select the Self Test function to survey the fault code memory in all control units. The test will begin automatically.

ERROR MESSAGE "NOT FOUND"

If the message "NOT FOUND" is displayed on the screen for all control units, one or both of the following conditions is present:

1. One or more control units is faulty.
   a. With the ignition switch **OFF**, disconnect any control unit, turn the ignition **ON** and select Self Test again. If the "NOT FOUND" message is still displayed, reconnect the control unit and disconnect another.
   b. When the "NOT FOUND" message is replaced by an asterisk (*), the last control unit disconnected is faulty. Check the wiring between the control unit and the dignostic connector and power to the control unit.
   c. If the "NOT FOUND" message is still displayed, go to Step 2.
2. The RXD, pin 15 or TXD, pin 20, data link between the control unit and diagnostic connector is faulty, usually a bad connection. The diagnistic connector pins are numbered on the connector.
   a. Turn the ignition switch **OFF** and disconnect all of the control units and the Diagnostic Tester.
   b. Turn the ignition switch **ON** and check the diagnostic connector terminals RXD and TXD for short to ground or voltage or an open circuit.
   c. If all wiring is in order, the Diagnostic Tester may be faulty or have the wrong data disc.

Blink Codes

On some models, the DME control unit is equipped with a self diagnostic program that will detect emissions related malfunctions and turn the CHECK ENGINE light **ON** while the engine is running. Emissions related fault codes stored in the DME control unit can be read by activating the program and counting the flashes of the CHECK ENGINE light on the instrument panel. Fault codes stored in all other control units are only accessible with Diagnostic Tester.

The fault code memory will be erased anytime the battery or DME control unit is disconnected. Fault codes can also be erased after reading the codes as described below.

READING FAULT CODES

1. All codes are 4 digits. On all engines except the 12 cylinder M70, the first digit of the DME codes will be **1**. On the 12 cylinder M70 engine, the first digit will be **2** if the fault is in DME system 2, the left bank of the engine.
2. With the ignition switch **ON** and the engine not running, press the accelerator pedal to the floor 5 times in less than 5 sec-

onds. This will close the full load switch to activate the diagnostic program. If the program will not activate, the full load switch on the throttle body or accelerator pedal sensor is not working.

3. The CHECK ENGINE light will stay **ON** for 5 seconds, then flash once, then begin flashing the first fault code. Each digit will be flashed with about 2.5 seconds between digits. When the whole code has been displayed, the light will stay **ON**. Count the flashes and write the numbers down.

4. The control unit will display only 1 code at a time. To display the next code, repeat the activation sequence described in Step 2. If the first code is **1444**, **2444** or **4444**, no faults are stored in memory.

5. When code **1000** or **2000** is displayed, one or two flashes followed by a long pause before the light stays **ON**, the control unit has reported all fault codes.

CLEARING CODES

With Diagnostic Tester

This function is a choice on the Diagnostic Tester menu, available when the code is displayed. Codes will also be cleared any time power to the control unit is unterrupted.

Without Diagnostic Tester

To erase the fault code memory, first enter the fault code retreival mode described above. When all codes have been read, with the ignition switch **ON** hold the accelerator pedal down to close the full load switch for 10 seconds. Only the codes in the DME control unit memory will be erased. The Diagnostic Tester is required to clear codes from any other control unit.

ENGINE FLASH CODES

Code 1221–Control unit self test
Code 1215–Air flow sensor
Code 1221–Oxygen sensor
Code 1222–Oxygen regulation
Code 1223–Coolant temperature sensor
Code 1224–Air temperature sensor
Code 1231–Battery voltage
Code 1232–Throttle idle switch
Code 1233–Throttle full load switch
Code 1251–Fuel injector group 1, even numbers
Code 1252–Fuel injector group 2, odd numbers
Code 1261–Fuel pump relay
Code 1262–Idle speed control actuator
Code 1263–Carbon canister purge valve
Code 1264–Oxygen sensor heater relay

NOTE: On vehicles equipped with the 12 cylinder M70 engine, if the first digit of the fault code flash is 2, the fault is in DME system 2, cylinders 7–12.

DME SYSTEM COMPARISON

SYSTEM	DME 3.1	DME 1.1–1.3
Ignition	Direct, solid state	Rotary Distributor
Cylinder Identification	Camshaft sensor	Sensor on cylinder 6 ignition lead for semi-sequential fuel injection
Load sensing	Hot wire air mass metering	Volumetric air flow metering
Type of fuel injection	Fully sequential	Parallel or semi-sequential
Throttle position sensing	Throttle butterfly potentiometer; adaptive idle setting; part and full load by way of resistance value	Throttle butterfly switch, idle and full load by mechanical switch contact
Starting	Advance fuel pump run and injection; double spark (max. 250 ms); fully sequential injection	Parallel injection
Full load ignition timing	Altitude compensation; dependent on engine speed and cylinder charge	Dependent on engine speed
Idling control	Adaptive value with/without air conditioning	Single adaptive value
Model application identification	By mapped characteristic	By variant coding
Self-diagnostics	Extended	
Emergency running	Improved from DME 1.1–1.3	
Memory board	40 Kb	32 Kb
Plug pins	88	55
Electromagnetic compatibility	4X multilayer technique	Additional suppression

JAGUAR

General Information

The fuel injection used on all XJS and XJ6 vehicles, is operated by the digital Engine Control Unit (ECU), a solid state computer. Coupled with the Marelli solid-state ignition system, the ECU controls ignition timing and advance as well as fuel metering to the engine.

The systems used on both the 4.0L 6-cylinder engine and the 5.3L V-12 engine provide fuel metering under all operating conditions. The ECU is programmed or mapped with up to 256 combinations of engine speed and load. This allows complete engine control in all situations including cold starting, acceleration, idle and hot restart. The ECU also controls the ignition spark timing and advance; mechanical advance mechanisms are not used.

All engines are equipped with oxygen (Lambda) sensors in the exhaust system. These sensors communicate the amount of oxygen content in the exhaust stream to the ECU, allowing the controller to make fuel and spark corrections when operating in closed loop mode.

ELECTRONIC CONTROL UNIT (ECU)

Fuel metering is controlled by regulating the amount of time the injectors are open during each engine cycle. The ECU contains a microprocessor and discrete electronic circuits connected to sensors and controlled devices on and around the engine. The engine management system maintains ideal engine performance throughout the entire operating range by metering the fuel and adjusting the ignition timing according to the immediate operating condition.

Additionally, the ECU operates the fuel pump, controls the idle speed, provides fuel cut-off during either closed throttle or high rpm operation and provides data to the trip computer re-

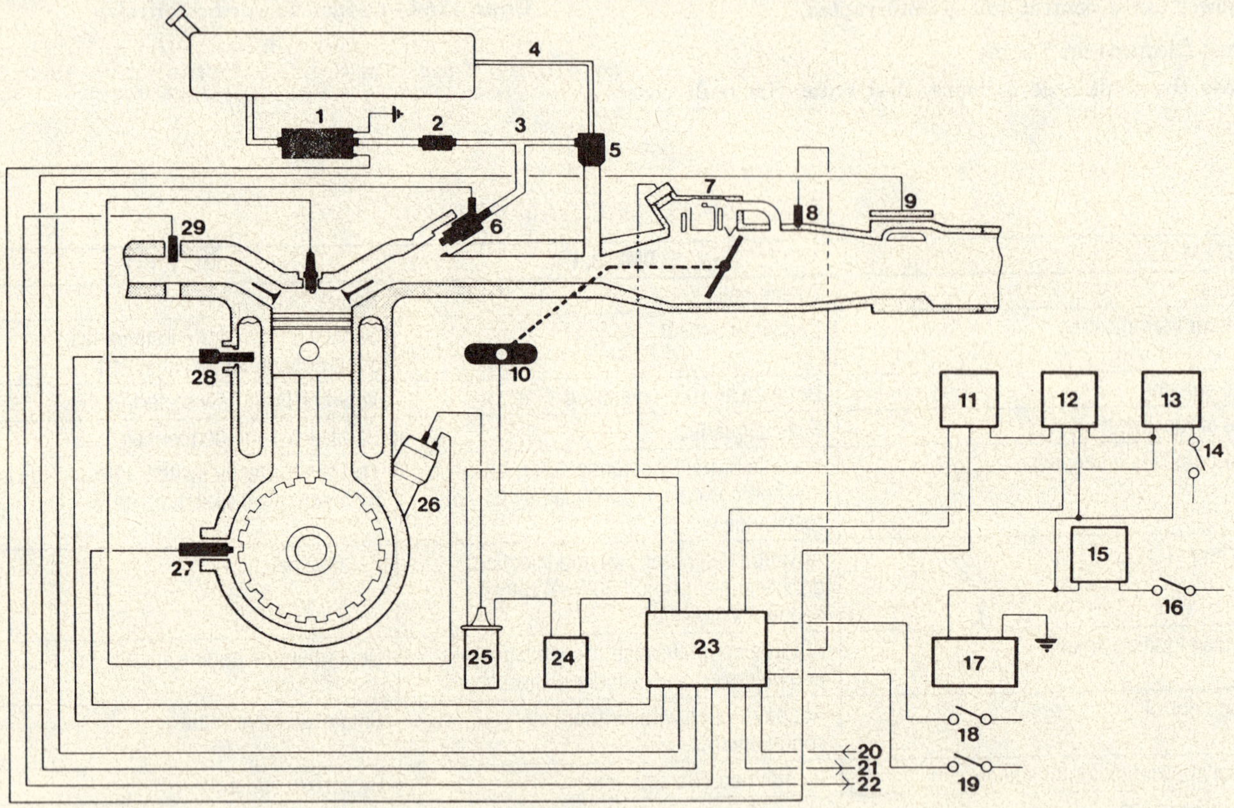

1. Pump
2. Filter
3. Fuel rail
4. Fuel return
5. Pressure regulator
6. Injector
7. Idle speed control valve
8. Air thermistor
9. Airflow meter
10. Throttle position sensor
11. Fuel pump relay
12. Main relay
13. Ignition ON relay
14. Inertia switch
15. Start relay
16. Start switch
17. Battery
18. Gearbox sensor
19. Air conditioning clutch switch
20. Idle override
21. Vehicle condition monitor VCM 1
22. Vehicle condition monitor VCM 2
23. ECU
24. Ignition amplifier
25. Ignition coil
26. High tension distributor
27. Crankshaft sensor
28. Coolant temperature sensor
29. Oxygen (Lambda) sensor

Engine management system—Jaguar 4.0L engine

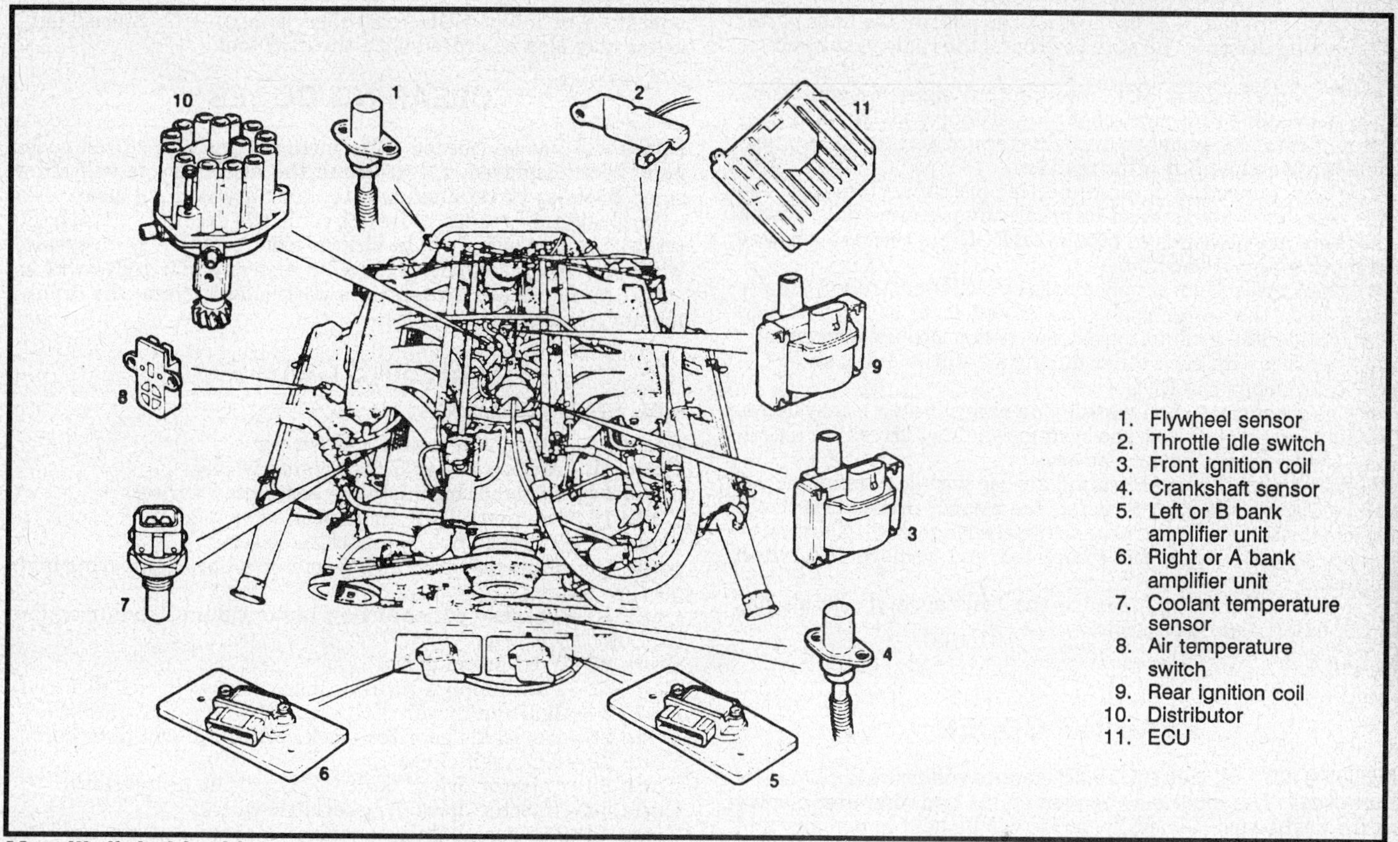

Marelli digital ignition components—Jaguar 5.3L engine

garding fuel consumption and engine condition. The fuel pump is energized only when the ECU detects a cranking or rpm signal. Idle speed control compensates for changing engine loads such as automatic transmission in-gear or air conditioning compressor operating.

Ignition timing is controlled by the ECU, based primarily on information supplied by the airflow meter and the crankshaft sensor. The Marelli Digital Ignition System is a digital microprocessor-controlled control system. The controller, located within the ECU, controls the timing of the low tension voltage, through the amplifiers, to the ignition coils. Primary engine control is derived from the intake manifold vacuum and engine rpm signals; secondary or modifying control comes from the throttle position sensor, coolant temperature sensor and intake air temperature signal.

Back-Up System

The ECU used to control the fuel injection allows the internal self-diagnostic system to evaluate the input and output signals during operation. If the ECU notes the loss or change out-of-range of a sensor signal, the computer will ignore the faulty signal and substitute a fixed value in its place. This value may not necessarily be correct for the immediate driving situation, but will allow the engine to operate. For example, many ECU decisions are based on the engine coolant temperature. Should the coolant temperature sensor signal be lost just after a cold start, the ECU will substitute the default value. This substitution value is based on an engine at normal temperature so vehicle performance will be affected until fully warmed up.

The CHECK ENGINE warning lamp on the dashboard will be lit by the ECU when a fault is detected. If the fault involved loss of a critical signal, the ECU will enter the default value for the signal. This automatic substitution is sometimes referred to as the limp-in mode of operation.

Self-Diagnostics

As the engine control computers became capable of more functions, self-diagnostic and memory circuits were added. These systems allow the controller to note a fault, assign an identity code and store the code in memory for later retrieval. The number, type and meaning of engine codes vary by year and model.

On the 5.3L V-12 engines, the ECU used on pre-1992 engines does not generate or store codes. The 4.0L engines will generate and store fault codes relating to sensor and circuit function.

While most fault codes are held in an electronic memory and are retained even after the ignition is switched **OFF**, certain codes are only held or displayed as long as the ignition is **ON**. If the fault is present at the next restart, the code will reset.

SERVICE PRECAUTIONS

- Keep all ECU parts and harnesses dry during service. Protect the ECU and all solid-state components from rough handling or extremes of temperature.
- Before attempting to remove any parts, turn the ignition switch **OFF** and disconnect the battery ground cable.
- Make sure all harness connectors are fastened securely. A poor connection can cause an extremely high voltage surge, resulting in damage to integrated circuits.
- Always use a 12 volt battery as a power source.
- Do not disconnect the battery cables with the engine running; never run the engine with battery cables loose or disconnected.
- Do not attempt to disassemble the ECU unit under any circumstances.
- If installing a 2-way or CB radio, mobile phone or other radio equipment, keep the antenna as far as possible away from the electronic control unit. Keep the antenna feeder line at least

8 in. away from the ECU harness and do not run the lines parallel for a long distance. Be sure to ground the radio to the vehicle body.

- When performing ECU input/output signal diagnosis, remove the water-proofing rubber plug, if equipped, from the connectors to make it easier to insert tester probes into the connector. Always reinstall it after testing.
- When connecting or disconnecting pin connectors from the ECU, take care not to bend or break any pin terminals. Check that there are no bends or breaks on ECU pin terminals before attempting any connections.
- When measuring supply voltage of ECU-controlled components, keep the tester probes separated from each other and from accidental grounding. If the tester probes accidentally make contact with each other during measurement, a short circuit will damage the ECU.
- Use great care when working on or around air bag systems. Due to back-up circuitry, the system may stay armed for a period of time without battery power.
- When working on or around air bag wiring or components, always disarm the system using the correct procedures. Once disarmed, attach a flag or note to the steering wheel. Re-arm the system when repairs are completed and remove the wheel marker.
- Never attempt to measure the resistance of the air bag squib; detonation may occur.

READING CODES

The 1990–92 4.0L and 1992 5.3L engines will provide diagnostic fault codes. The codes may be read on the trip computer display on the dashboard. Generally, codes will be held in memory and the dash warning lamp will be lit when a fault is sensed by the ECU.

NOTE: Not every stored code will cause the warning lamp to light after the first occurrence. Always check for codes during diagnosis, including cases where the dash warning lamp was not reported lit.

To read the stored codes, bring the vehicle to a complete stop. Switch the ignition **OFF** and wait at least 10 seconds. Turn the ignition to the **II** or **ON** position but do not start or crank the engine. Press the trip computer button **DISP** or **VCM**. After a short period of time, the stored codes will be displayed graphically on the trip computer panel. The display may include the designation FF, an abbreviation for FUEL system FAILURE.

The fault codes are displayed in order of priority, one at a time. The next is only displayed when the preceding one is repaired or cleared.

A computerized, hand-held diagnostic tool is available through Jaguar Cars Ltd. The Jaguar Diagnostic System (JDS) tool, or its equivalent, allows rapid testing and diagnosis of electrical circuits. The tool can be adapted to each motor through the use of specific software. The tool allows testing and limited operation of many ECU controlled components. Stored fault codes may also be erased with the JDS tool.

CLEARING CODES

Codes will not be cleared from memory until all stored codes have been displayed. Interrupting the display cycle will allow some codes to be retained and the memory will not clear.

To clear stored codes, drive the vehicle in excess of 19 mph (30 km/h); the memories will be cleared electronically. Each system will perform its self-check function and any new codes will be set. If no faults are present in a particular system, the failure memory will remain cleared.

ENGINE CODES

4.0L ENGINE

Code 11—Idle pot or associated wiring
Code 12—Airflow meter or associated wiring
Code 14—Coolant thermistor or associated wiring
Code 16—Air thermistor or associated wiring
Code 17—Throttle pot or associated wiring
Code 18—Throttle pot or airflow meter calibration during high air flow
Code 19—Throttle pot or airflow meter calibration during low air flow
Code 22—Fuel pump drive
Code 23—Fuel supply—poor feedback control in rich direction
Code 24—Ignition driver (ECU output to ignition amplifier)
Code 26—Air leak—poor feedback control in lean direction
Code 29—ECU self-check
Code 33—Injector driver fault (ECU output to injectors)
Code 34—Injector drive or associated wiring
Code 44—Oxygen (Lambda) sensor or associated wiring
Code 46—Idle speed control coil 1 driver—(ECU output to idle speed control stepper motor)
Code 47—Idle speed control coil 1 driver—(ECU output to idle speed control stepper motor)
Code 48—Idle speed control motor or valve
Code 68—Road speed sensor or associated wiring
Code 69—Drive/neutral switch or associated wiring

5.3L ENGINE

Code 11—Idle potentiometer
Code 13—ECU pressure sensor
Code 14—Coolant thermistor or associated wiring
Code 16—Air thermistor or associated wiring
Code 17—Throttle potentiometer or associated wiring
Code 18—Throttle potentiometer calibration 1
Code 19—Throttle potentiometer calibration 2
Code 23—Fuel supply
Code 34—Injector fault, A or right bank
Code 36—Injector fault, B or left bank
Code 44—Oxygen (Lambda) sensor fault, A or right bank
Code 45—Oxygen (Lambda) sensor fault, B or left bank
Code 49—Fuel injection ballast resistor or associated wiring

MERCEDES-BENZ

General Information

Mercedes-Benz vehicles utilize Electronic Continuous Injection (CIS-E) and LH Fuel Injection for gasoline powered vehicles, and Electronic Diesel System (EDS) with automatic idle speed control for diesel powered vehicles. These systems monitor sensor output and make corrections to air/fuel mixture and idle speed as necessary to adapt to current operating conditions. Each system is equipped with a self diagnosis system which enables malfunctioning system components to be traced with relative ease.

CIS-E Fuel Injection

The CIS-E fuel injection system is used on all gasoline fueled engines except the 1992 3.2L, 4.2L, 5.0L and 6.0L engines. The basic components of the system are the mechanical airflow sensor and the mixture control unit. Fuel is metered to the individual cylinders dependent on air intake quantity.

The system is a non-engine driven, electronically controlled fuel injection which eliminates mechanical components. The system takes a direct measurement of the inducted air mass by means of a airflow meter and calculates the proper amount of fuel to be injected into the cylinders. The system also uses input signal from the Lambda control (oxygen sensor) to determine injection quantity.

Fuel is drawn by the electric fuel pump through a damper out of the fuel tank and delivered to the fuel distributor on the mixture control unit. A pressure regulator on the mixture control unit keeps fuel pressure at a constant by allowing excess fuel to return to the tank.

The mixture control unit houses the airflow sensor and fuel distributor. The air quantity drawn by the engine is measured in the airflow sensor by means of a plate positioned in the intake tract. For a given quantity of air, the plate assumes a position along the intake tract. This position is recorded by the airflow sensor position indicator and transmitted to the control unit.

The fuel distributor contains an Electro-Hydraulic Actuator (EHA)which acts on a plate valve to regulate fuel pressure. With the engine running, a constant system pressure of approximately 78 psi is applied at the fuel inlet. The plate valve is adjusted by the control unit to control flow rate, in combination with the fixed orifice at the fuel distributor outlet. The corresponding pressure change in the lower chamber of the fuel distributor causes movement of the diaphragm and influences the fuel volume flowing to the injection valve.

The CIS-E fuel injection system has several modes of operation to match current engine operating conditions. The modes are: cold start and warm-up, acceleration enrichment, full load enrichment and deceleration shutoff.

Inputs from various sensors allow the CIS-E control unit to calculate current engine operating conditions. these sensor inputs are: coolant temperature, airflow sensor position indicator, full load contact, idle speed contact, engine speed, oxygen sensor, barometric pressure and shift lever position.

LH Fuel Injection

In order to obtain a more accurate and efficient fuel delivery system under all engine operating conditions, the 1992 3.2L, 4.2L, 5.0L and 6.0L Mercedes-Benz engines use an electronic gasoline injection system with heated wire air mass measurement. This system is commonly referred to as LH Fuel Injection. The L stands for Luftmassenmessung which means air mass measurement, and the H stands for Hitzdraht or heated wire.

The system electronically controls fuel injection which eliminates mechanical components. The system takes a direct measurement of the inducted air mass by means of a hot wire air mass meter and calculates the proper amount of fuel to be injected into the cylinders. The system also uses input signal from the Lambda control (oxygen sensor) to determine injection quantity.

The air mass sensor works with a thin heated platnum resistance wire in the air intake stream. The wire is a temperature dependent resistor which increases resistance with temperature. The wire receives as much heating current as is required to heat it to 356°F (180°C) above the ambient air temperature. The temperature difference is always kept constant. The amount of current required to keep the wire at a constant temperature is used to calculate air mass by the control unit. Dirt deposited on the wire is cleaned by heating the wire to 1832°F (1000°C) after the engine has been stopped.

If a malfunction should occur in the LH fuel injection air mass sensor, the system will revert to a Fixed Operating Mode (FOM). The FOM uses absolute intake manifold pressure, taken from the EZL/AKR ignition control unit, to calculate injection time.

Fuel is supplied by twin electric fuel pumps mounted on the frame rail. The fuel flows to a distribution manifold which ensures equal fuel pressure at all injectors. A pressure regulator, mounted at the end of the distribution manifold, regulates system pressure to 45–55 psi by returning a regulated amount of fuel to the fuel tank. The injection valves, one for each cylinder, meter a precise quantity of fuel into the intake manifold just prior to the intake valve.

The LH fuel injection system has several modes of operation to match current engine operating conditions. The modes are: starting control, after-start enrichment, warmup enrichment, acceleration enrichment, deceleration shutoff, maximum vehicle speed control, engine rpm limitation and fuel safety shutoff.

Inputs from various sensors allow the LH control unit to calculate current engine operating conditions. These sensor inputs are: engine speed, camshaft revolutions, coolant temperature, intake air temperature, idle speed and full load recognition, intake manifold absolute pressure, battery voltage and voltage supply.

Functions performed by the LH control unit are: lambda control, oxygen sensor heater, purging of charcoal canister, air injection, raised transmission shift point, exhaust gas recirculation, camshaft timing adjustment, engine idle speed, fuel shutoff and self diagnostics.

EDS Diesel Fuel Injection

The EDS is used on all Mercedes-Benz diesel engines. The electronic control unit regulates idle speed, EGR, recirculating air and system diagnostic functions. It processes incoming signals and delivers a milliamp current to the vacuum transducer for the EGR and air recirculating valve, as well as to the magnetic actuator of the injection pump.

Idle speed adjustment is accomplished through an rpm sensor which transfers engine speed signals in the form of AC current to the control unit. The control unit processes the rpm signal and performs a nominal-actual value comparison. The idle speed is held constant by the magnetic actuator independent of engine load. At a coolant temperature below 140°F (60°C), the coolant temperature sensor provides a variable signal, increasing the idle speed.

The EDS control unit calculates the appropriate amount of exhaust gas to be recirculated using signals from the fuel rack position sensor, engine speed sensor and others. The vacuum transducer for the EGR valve will then receive a control current input signal and deliver the correct amount of vacuum to the EGR valve. An increase in control current means an increase in EGR quantity and vise-versa.

Recirculation air control is used to improve particulate burnup in the trap oxidizer (thermal reactor). The air recirculating valve is opened or closed depending on various EDS control unit inputs. After the air recirculating valve closes, a residual vacuum of 9 in. Hg remains on the valve. The system is functional only when engine speed is between 1200–3600 rpm or when fuel rack travel is less than 9mm.

Self-Diagnostics

Continuous Injection System (CIS-E)

The malfunction memory of the CIS-E fuel injection system is divided into 2 sections. The Lambda control monitors the input signals, amplifies them, and calculates the output signals for the Electro-Hydraulic Actuator (EHA).

Trouble codes are accessed using a Fixed On/Off Ratio meter at diagnostic test connector X11. Once the malfunctioning component has been repaired, the trouble code will automatically clear.

The second malfunction memory, on-board diagnostics, is used exclusively on California vehicles. This system monitors emission control components that either provide input signals to or receive output signals from the ECU. Malfunctions resulting from interruptions or failure of any component are indicated by the CHECK ENGINE light and stored in the malfunction memory.

Trouble codes are accessed either though an LED on the firewall or through the use of an impulse tester. Each code is called up and cleared individually. Once the on-board diagnostics memory is cleared, the system automatically switches to the lambda control.

LH Fuel Injection

The malfunction memory of the LH fuel injection system combines three different diagnostic functions: diagnostic module, malfunction memory and Lambda control.

All California vehicles are equipped with a diagnostic module. The module is located in the module box and recognizes emission related signals and systems. In case of a malfunction, the system triggers the CHECK ENGINE light to illuminate. Simultaneously, a malfunction code is stored in the diagnostic module memory.

The malfunction memory detects failures which occur while the engine is running. The failures are counted by a malfunction counter. Only if a failure is still present after 4 engine startswill it be stored in memory. Failures which prevent the starting of the engine or severely influence the engine running are stored immediately. The stored malfunctions can be read with the impulse counter at the (X11/4) test connection for diagnosis.

Lambda control trouble codes are accessed using a Fixed On/Off Ratio meter at diagnostic test connector X11. Once the malfunctioning component has been repaired, the trouble code will automatically clear.

Engine Systems (MAS) Control Unit

The MAS control unit is used to combine the functions of the fuel pump relay, ignition system, kickdown circuit and other systems. The RPM signal from the speed sensor is prepared by a dedicated output circuit for distribution through the TNA output terminal.

Stored malfunctions in the system can be accessed with an impulse counter at test connection X11/4 in the engine compartment. The malfunctions remain stored even if the battery or control unit is disconnected.

Electronic Diesel System (EDS)

The EDS control unit regulates idle speed, EGR and system diagnosis functions. As incoming signals arrive, they are processed by the control unit and a milliamp current is sent to the vacuum transducer for the EGR, as well as to the magnetic actuator of the injection pump.

The entire system can be diagnosed by using an impulse tester at diagnostic terminal X11/4 in the engine compartment.

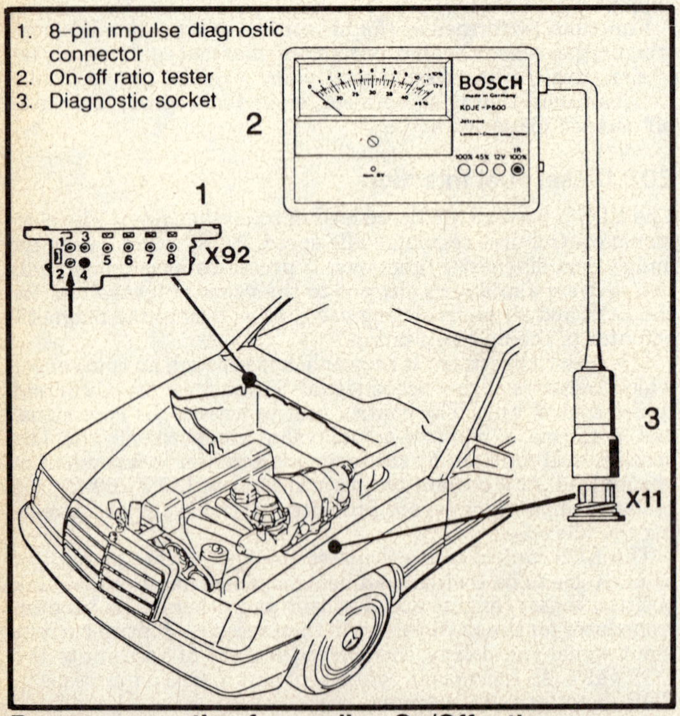

Proper connection for reading On/Off ratio percentages

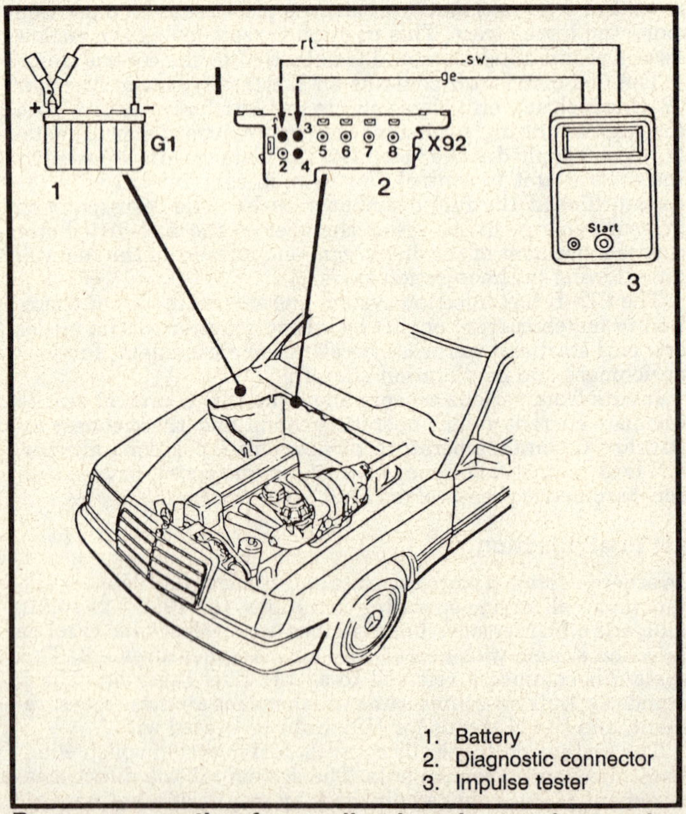

Proper connection for reading impulse readout codes

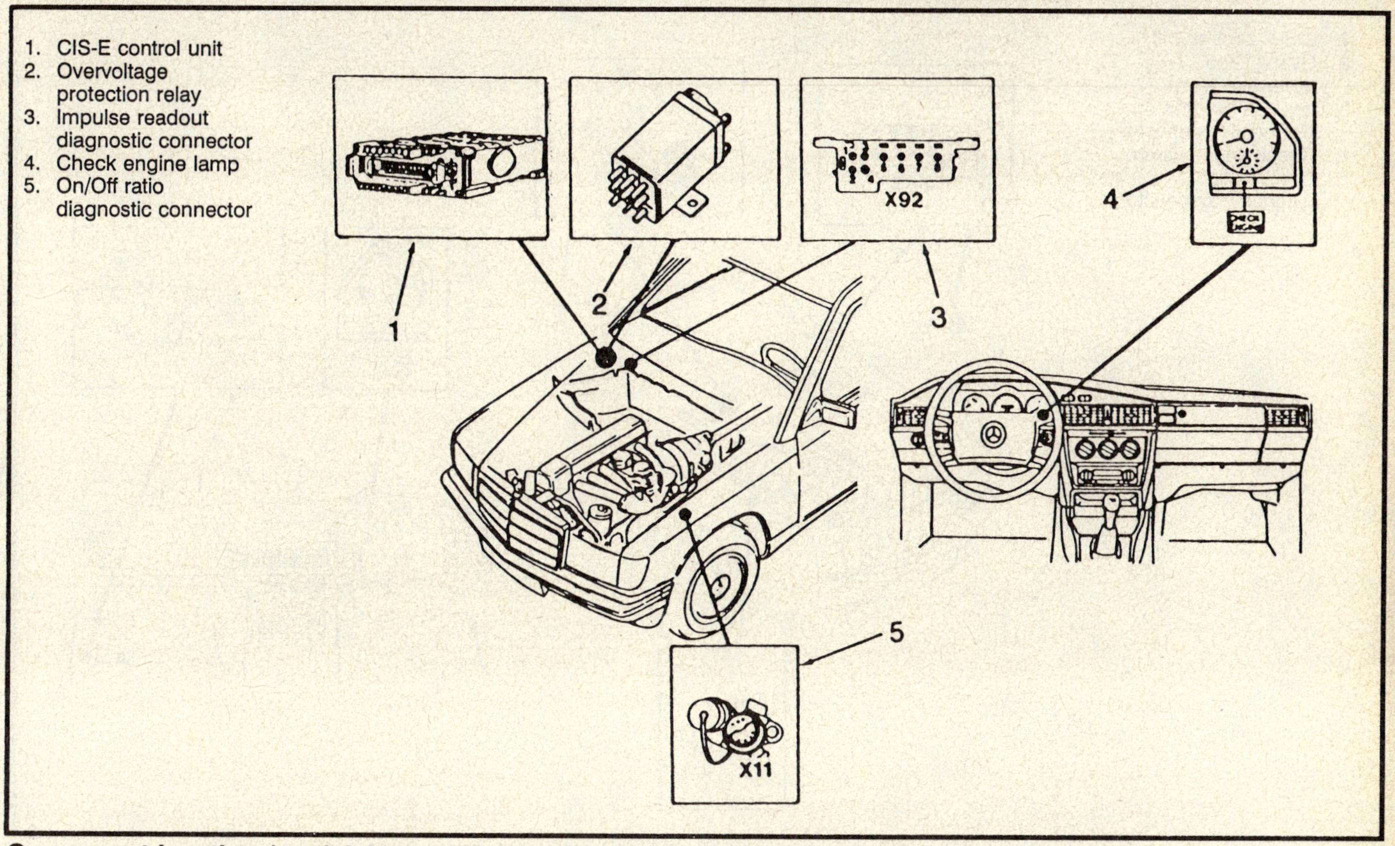

Component location (model 124 and 201)—190 series, 260E and 300 series (except SE/SEL)

Diagnostic connnector—X92

Electronic Idle Speed Control (ELR)

The ELR control unit receives signals from the ring gear speed sensor and performs nominal-actual comparisons to determine engine rpm. The idle speed is held constant by the electromagnetic actuator.

Malfunctions resulting from interruptions or failures are detected, analyzed and transmitted to the test connection (X92). Access to trouble codes is gained via an impulse counter.

SERVICE PRECAUTIONS

- Keep all ECU parts and harnesses dry during service. Protect the ECU and all solid-state components from rough handling or extremes of temperature.
- Use extreme care when working around the ECU or other components; the air bag or SRS wiring may be in the vicinity. On these vehicles the SRS wiring and connectors are red; do not cut or test these circuits.
- Before attempting to remove any parts, turn the ignition switch **OFF** and disconnect the negative battery cable.

READING CODES

Continuous Injection System (CIS-E)

ON-OFF RATIO TESTER (LAMBDA CONTROL)

1. Start the engine and allow it to reach operating temperature. Ensure that all accessories are **OFF**, the selector lever is in **P**, the fuse of the overvoltage protection relay is good and the battery is fully charged.
2. Connect the On-Off Ratio tester to the diagnostic connector terminal (X11) in the engine compartment.
3. Turn the ignition key **ON** and read the tester.
4. Repair or replace the component and repeat the test procedure until no faults are present.

IMPULSE TESTER (ON-BOARD DIAGNOSTICS)

NOTE: Disconnecting the CIS-E control unit, battery or overvoltage protection relay will erase all codes stored in memory.

1. Connect the impulse tester to terminals 1 and 3 of the diagnostic connector (X11/4). The connector is located on the firewall.
2. Turn the ignition **ON** with the selector lever in **P**.

1. CIS-E control unit
2. Overvoltage protection relay
3. Impulse readout diagnostic connector
4. Check engine lamp
5. On/Off ratio diagnostic connector

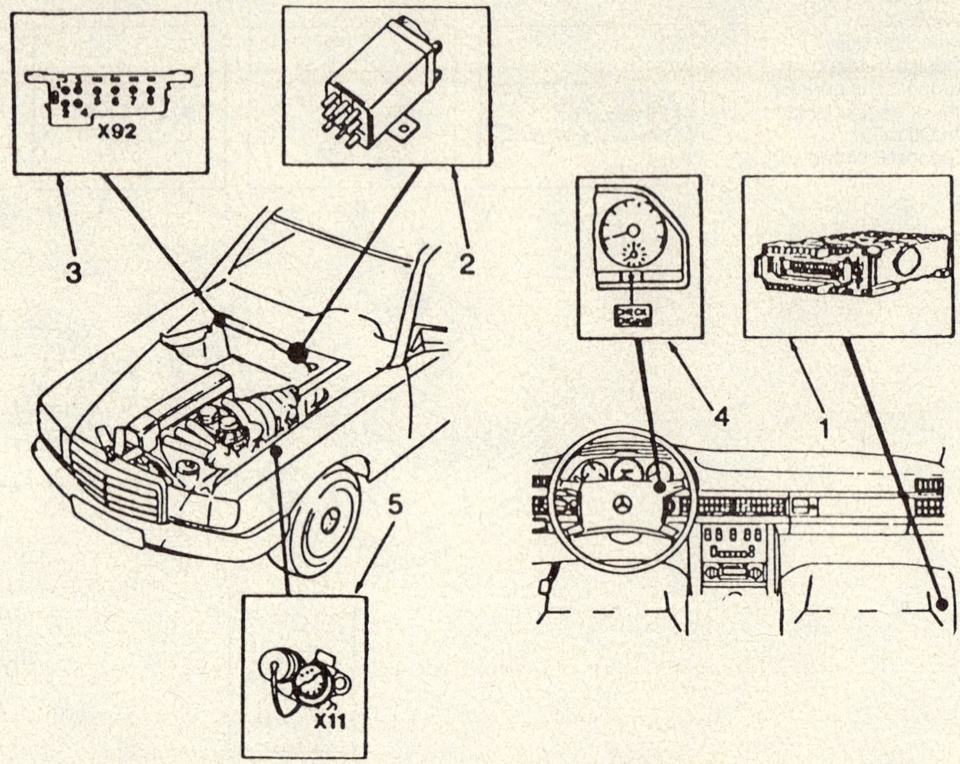

Component location (model 126)—300SE, 300SEL, 300SDL, 420SEL, 560SEL, 560SEC

1. CIS-E control unit
2. Overvoltage protection relay
3. Impulse readout diagnostic connector
4. Check engine lamp
5. On/Off ratio diagnostic connector

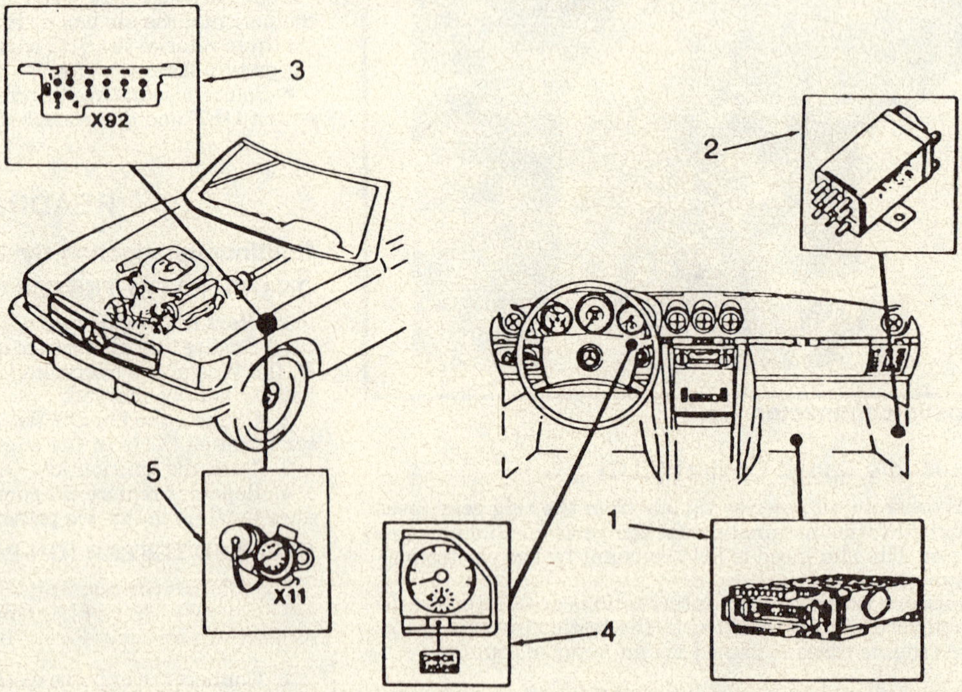

Component location (model 107)—560SL

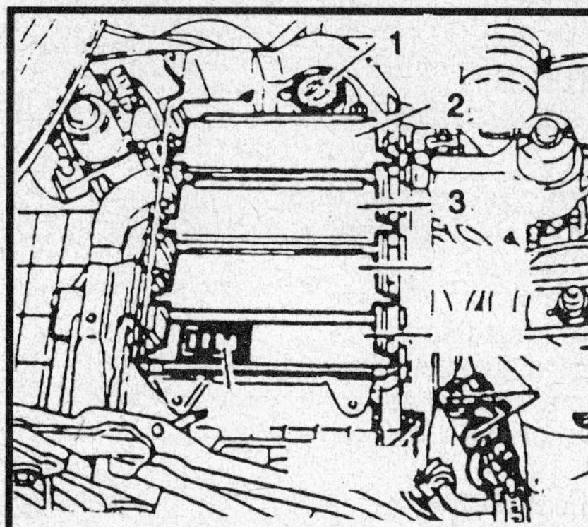

1. Test connection for diagnosis (X11/4)
2. LH control unit
3. Electronic accelerator control unit

Diagnostic module box (model 124 and 140)—1992 300SE, 400E, 400SE, 500E, 500SEL and 600SEL

3. Press the start button for no less than 2, but no more than 4 seconds and observe the indicated impulse readout.
4. Continue to press the start button to obtain additional fault codes. All codes must be read prior to servicing.
5. Repair or replace the component and repeat the test procedure to ensure no faults are present.

LH Fuel Injection

IMPULSE TESTER

1. Connect the impulse tester to the terminals the diagnostic connector (X11/4). The connector is located on the firewall.
2. Turn the ignition **ON** with the selector lever in **P**.
3. Press the start button for no less than 2, but no more than 4 seconds and observe the indicated impulse readout.
4. Continue to press the start button to obtain additional fault codes. All codes must be read prior to servicing.
5. Repair or replace the component and repeat the test procedure to ensure no faults are present.

Electronic Diesel System (EDS)

IMPULSE TESTER

1. Connect the impulse tester to terminals 1 and 4 of the diagnostic connector (X11/4).
2. If the U-Batt LED does not illuminate, check the impulse counter fuse. Ensure that there are 11–14 volts from the diagnostic connector socket 1 to the positive battery cable. Insure that there are 6–12 volts from the diagnostic test connector socket 4 to socket 1.
3. Start the engine and allow it to idle.
4. Push the start button for 2–4 seconds and scan the impulse readout. Record the impulse readout.
5. Push the start button again. If there are no more malfunctions in the system the previous display will be indicated. If there are additional malfunctions, they will be indicated. Repeat the scan test until the first malfunction is displayed again.
6. Run the engine at 900 rpm for at least 5 seconds and push the start button again for 2–4 seconds while maintaining engine speed. Record any malfunctions.
7. Correct the malfunctions.

CLEARING CODES

Electronic Diesel System (EDS)

IMPULSE TESTER

Trouble codes must be erased individually. To erase codes, push the start button and read out the malfunction that has been repaired. Then, hold the start button for 6–8 seconds to erase the code. Perform this procedure until only Code 1 (no malfunctions) remains.

Electronic Idle Speed Adjustment (ELR)

IMPULSE TESTER

Fault codes for the ELR system are not stored in memory. When the faulty component is repaired or replaced, the fault code will automatically erase.

Continuous Injection System (CIS-E)

ON-OFF RATIO TESTER (LAMBDA CONTROL)

Fault codes obtained using an On-Off ratio tester are only present during a malfunction. When the component is repaired or replaced, the fault code will automatically be erased.

IMPULSE TESTER (ON-BOARD DIAGNOSTICS)

1. With the impulse tester connected and the fault code displayed, press the start button for 6 seconds or more.
2. Each impulse code must be erased individually.
3. If there is no number displayed, then the stored fault has been erased.
4. If the display shows the number 1, then there are more system faults in memory.

Fault codes obtained using an impulse tester will be erased if the control unit, overprotection relay or negative battery terminal are disconnected. However, this is not the recommended procedure for erasing the stored codes.

ENGINE CODES

CIS-E Fuel Injection

LAMBDA CONTROL PERCENTAGE CODES

0%—CIS-E control unit receives no voltage or is defective. Circuit to diagnostic socket open or On/Off Ratio tester defective. Air/fuel mixture too rich. Oxygen sensor signal voltage is +12 volts

10%—Airflow sensor position indicator defective or polarity reversed.

20%—Wires in plug of throttle valve switch incorrectly connected connected or short circuit (full load contact closed before engine reaches full load)

30%—Short or open circuit between CIS-E control unit and coolant temperature sensor or coolant temperature sensor defective (idle speed too high)

40%—Short or open circuit to airflow sensor position indicator or airflow sensor position indicator defective.

50%—Oxygen sensor not at operating temperature, open circuit or defective

60%—Not used (4.2L and 5.6L engines)

60%—Speed signal at CIS-E control unit out of range (2.6L, 3.0L and 5.0L engines)

70%—No RPM signal (open circuit) at CIS-E control unit

80%—Short or open circuit to altitude correction capsule or capsule defective (3.0L California, 4.2L and 5.6L engines)

80%—Interrupted data exchange between EZL ignition and CIS-E control units (3.0L and 5.0L 49- state engines)

90%—Current reading out of range at electrohydraulic valve

95%—Deceleration fuel shut-off is currently in operation

100%—Overvoltage protection relay defective (idle speed too high). Air/fuel mixture too lean, oxygen sensor defective (short circuit to ground). CIS-E control unit or On/Off ratio tester defective. No power to electrohydraulic actuator (EHA).

Reading Oscillates—No malfunction within the circuit of the monitored signals. On/Off ratio reading indicated.

ON BOARD DIAGNOSTIC CODES

Code 1—No malfunctions in system
Code 2—Throttle switch full load contact
Code 3—Coolant temperature sensor
Code 4—Airflow sensor position indicator
Code 5—Oxygen sensor
Code 6—Not used
Code 7—RPM signal
Code 8—Altitude correction sensor
Code 9—Electrohydraulic actuator
Code 10—Throttle switch idle contact
Code 11—Air injection system
Code 12—EGR temperature (2.3L engine, California)
Code 12—Absolute pressure values from EZL control unit out of range (2.6L and 3.0L engines)
Code 13—Intake air temperature reading out of range
Code 14—Speed signal out of range
Code 15—Not used
Code 16—EGR
Code 17—Oxygen sensor line shorted
Code 18—Current to idle speed air valve out of range
Code 19—Not used
Code 20—Not used
Code 21—Not used
Code 22—Oxygen sensor heating current out of range
Code 23—Short to positive in purge switchover valve circuit
Code 24—Not used
Code 25—Short circuit to positive in start valve circuit
Code 26—Short circuit to positive in shift point delay circuit
Code 27—Data exchange CIS-E control unit to EZL control unit interrupted
Code 28—Intermittent contact in coolant temperature sensor circuit
Code 29—Difference in coolant temperatures read by CIS-E control unit and EZL control unit
Code 30—Not used
Code 31—Intermittent contact in intake air temperature sensor
Code 32—Not used
Code 33—Not used
Code 34—Coolant temperature from EZL control unit out of range

ENGINE SYSTEMS (MAS) CONTROL UNIT CODES

Code 1—No system malfunctions
Code 2—Fuel pump relay not operating
Code 3—RPM signal interrupted
Code 4—Oxygen sensor heater output failure
Code 5—Air pump output failure
Code 6—Kickdown switch output failure
Code 8—Coolant temperature out of range
Code 9—Circuit 50 failure
Code 10—Start valve output failure
Code 11—A/C compressor engagement signal for idle speed regulation missing
Code 12—A/C compressor output failure
Code 13—Excessive A/C clutch slip
Code 15—Detection of a short in the fuel pump circuit

LH Fuel Injection

ON-BOARD DIAGNOSTICS

Code 1—No malfunctions in system
Code 2—Oxygen sensor
Code 3—Lambda control system function
Code 4—Air injection system function
Code 5—EGR system function
Code 6—Idle speed control
Code 7—Ignition map monitoring
Code 8—Coolant temperature sensor
Code 9—Intake air temperature sensor
Code 10—Air mass sensor with hot wire
Code 11—RPM signal
Code 12—Oxygens sensor heater
Code 13—Signal preparation for camshaft position sensor
Code 14—Intake manifold pressure during starting
Code 15—Full load operation
Code 16—Idle speed operation
Code 17—Data transfer
Code 18—Camshaft position signal
Code 20—Vehicle speed
Code 21—Regeneration switchover valve
Code 22—Camshaft position sensor
Code 23—Intake manifold ressure; engine running
Code 24—Flywheel segments (crankshaft position sensor)
Code 25—Knock sensors
Code 26—Delayed upshift switchover valve
Code 27—Coolant temperature circuits outside range
Code 28—Coolant temperature monitoring

EDS Diesel Fuel Injection

ON-BOARD DIAGNOSTIC CODES

Code 1—All functions in order. If there are driveability complaints, perform the function tests for electronic idle speed and EGR system
Code 2—Fuel rack position sensor
Code 3—Airflow sensor
Code 4—EDS control unit atmospheric pressure sensor
Code 5—EGR valve vacuum transducer or a malfunction in the EGR control unit. Displayed only after 900 rpm for at least 5 seconds. This malfunction is not stored
Code 6—EDS control unit internal voltage supply
Code 7—Starter ring gear speed sensor
Code 8—Coolant temperature sensor
Code 9—Intake air temperature sensor
Code 10—Voltage supply
Code 11—Idle speed control actuator or EGR valve vacuum transducer. Displayed only if there is a short circuit
Code 12—Not used
Code 13—EDS control unit defective (internal malfunction memory)
Code 14—Air pressure sensor defective
Code 15—Intake manifold air pressure control valve vacuum transducer, wastegate vacuum transducer, or malfunction in the intake manifold air pressure control unit.

MERKUR

General Information

The Merkur self-diagnostic system is divided into specialized tests areas; (Key On Engine Off), (Engine Running) and (Continuous Memory). The self test is not a conclusive test by it self, but is used as a part of the functional quick test diagnostic procedure. When the self test mode is activated, it checks the EEC-IV system by testing its memory integrity and processing capability. It also verifies that various sensors and actuators are connected and operating properly.

The Key On Engine Off and Engine Running tests are functional tests which only detect faults present at the time of the self test. Continuous testing is an on going test that stores fault information for retrieval at a later time, during the self test.

Self-Diagnostics

SERVICE PRECAUTIONS

- Before attempting to remove any parts, turn the ignition **OFF** and disconnect the negative battery cable.
- Disconnect solenoids and switches from the harness before measuring for continuity, resistance or energizing by way of 12-volt source.
- When servicing the anti-lock brake system, do not allow brake fluid to come in contact with any of the electrical connectors.
- Do not operate the automatic temperature control system with the blower motor disconnected. Damage may occur to the electronic blower speed controller if cooling air is not provided by the blower motor.
- Probing directly into the vehicle's wiring harness connectors can damage the connector terminals and cause intermittent system malfunctions.
- When lifting, jacking or towing is required (vehicles equipped with air suspension system), power to the air suspension must be shut OFF. Either the negative battery cable must be disconnected or the power switch, located in the luggage compartment, be turned OFF.
- When servicing the air suspension system, be aware that the compressor vent solenoid and all air spring solenoids have internal diodes for electrical noise suppression and are polarized. Care must be taken not to switch battery feed and ground circuits or damage will result.
- All vehicles equipped with air suspension system are fitted with gas pressurized shock absorbers which will extend un-assisted. Do not apply heat or flame to the shock absorber tube.
- When servicing components on or around the steering column (vehicle equipped with air bag system), always deactivate the air bag system.
- Never probe the connectors on the air bag system. Doing so may result in air bag deployment which could result in personal injury.
- All component replacements and wiring repairs (vehicle equipped with air bag system) must be made with the battery ground and backup power supply disconnected.

READING CODES

This first test is entering Key On Engine Off test.

Testing with Voltmeter

NOTE: Voltmeter testing may not be applicable on some Merkur vehicles. To obtain codes for the Scorpio the Star tester or an equivalent testing device is advised.

1. Turn the ignition switch **OFF**.
2. Connect a jumper wire from the Self Test Input (STI) to pin No.2, Signal Return at Self Test connector.
3. Set the analog voltmeter on a DC voltage range to read from 0 to 15 volts.
4. Connect the positive lead of the voltmeter to pin No. 4, Self Test Output (STO) of the Self test connector.
5. Connect a timing light accordingly.
6. Activate Self Test by turning the ignition switch **ON**.

NOTE: Do not depress the throttle during Self Test.

Codes are represented by a pulsing or sweeping movement of the voltmeter's needle across the dial face of the voltmeter. A service code is represented by a 2 digit number. For example:

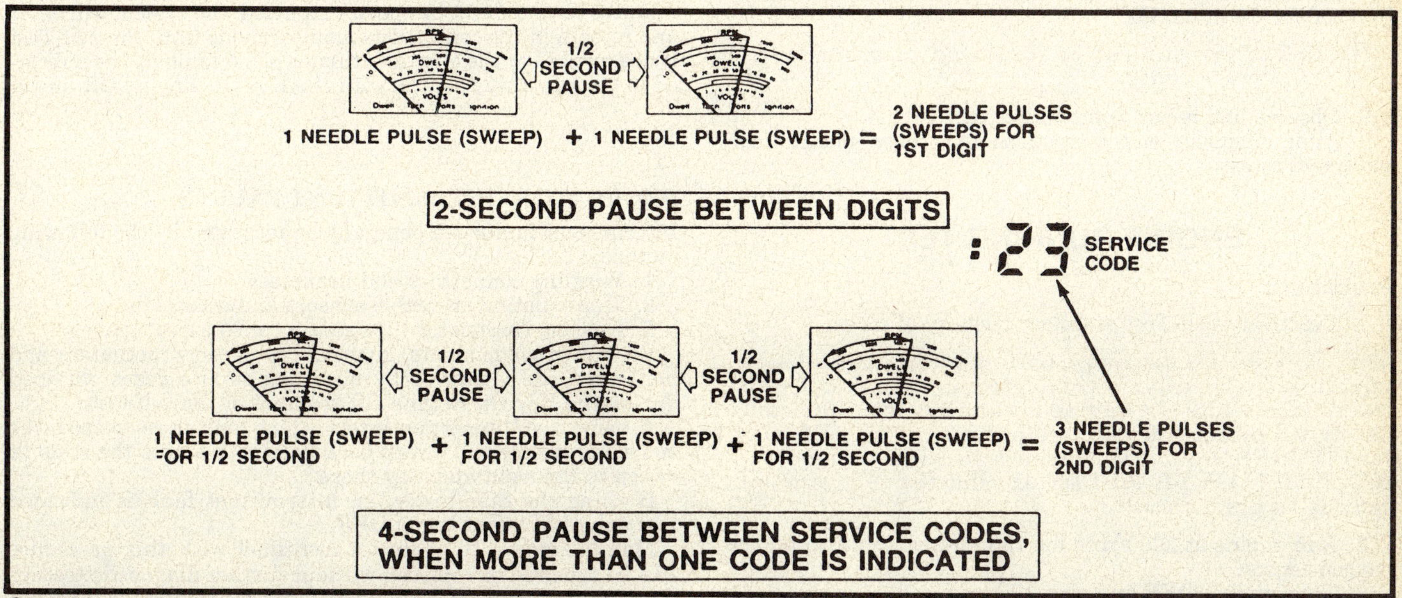

Checking service codes with an Analog voltmeter

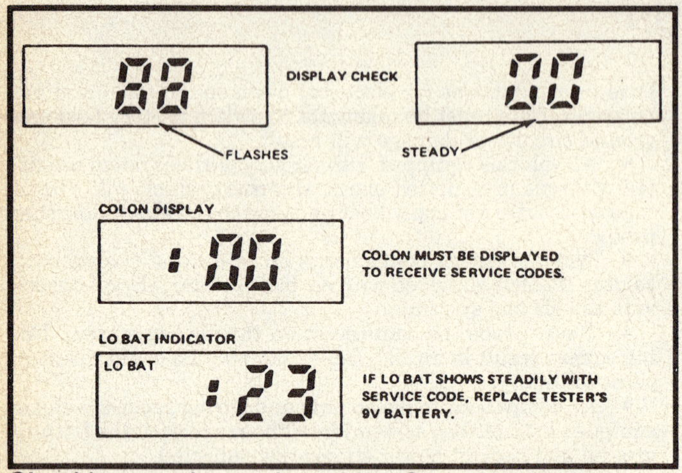

Checking service codes with a Star digital tester

(code 23). The service code 23 will be represented as 2 pulse (sweeps) of the voltmeter's needle, followed by a 2 second pause in between and 3 pulse (sweeps) indicating 23.

7. Observe and record codes indicated by the voltmeter.
8. After diagnosis and corrections have been made, clear codes and retest.

Testing with Star tester

1. Turn the ignition switch **OFF**.
2. Connect the color coded adapter cable leads to the Star tester.
3. Connect the adapter cables service connectors to the vehicles appropriate Self test connectors.
4. Connect a timing light accordingly.
5. Using an adapter cable, connect tester to the vehicles self test connecters.
6. Turn the testers power switch **ON**. The tester will run a display check. The numerals **88** will begin to flash in the display window. A steady 00 will then appear to signify that the tester is prepared to begin the self test and receive the tests service codes.
7. Press the push button at the front of the tester, the button will latch down and a colon will appear in the display window in front of the **00** numerals.

NOTE: The colon must be displayed in order to receive the service codes.

8. Observe and record codes indicated.
9. After diagnosis and corrections have been made, clear codes and retest.

ENGINE RUNNING TEST

Procedure

1. Deactivate Self-Test and clear memory of codes.

NOTE: Vehicles equipped with manual transmission only. If the Star tester is connected and the Self Test is deactivated while the engine is running, after approximately 4–5 minutes from engine start up, the LED may flash ON/OFF. Do not be alarmed, this will occur if the ACT, ECT, MAP, TP OR EDT are shorted to ground or there is an open circuit.

2. Run engine at 2000 rpm for 15 minutes, to warm up the oxygen sensor.
3. Turn engine **OFF** and wait 10 seconds.
4. Activate the Self test.

NOTE: Do not depress the throttle during the Self Test unless a dynamic response code occurs.

5. Start the engine. The Engine Running Self Test will progress in the following sequence:
   a. Engine code will flash.
   b. Run engine at idle.
   c. When the dynamic response code occurs, perform a brief WOT.
   d. Engine Running Service codes will flash.

NOTE: The Engine ID code is equal to ½ the numbers of cylinders. (A code 2 equals a 4 cylinder engine). The Star tester adds a 0 to all single digit readings, so a code 20 indicates a 4 cylinder engine.

The Star tester will also indicate a 10 in the place of 1 for single digit separator code, therefore, the dynamic response code is expressed in the results column as 1(0). Some vehicles do not require a brief WOT and therefore will not display a 1(0).

6. Observe and record codes indicated.

CONTINUOUS TEST

Intermittent Fault Confirmation Check (Wiggle Test)

The ECA is continually looking for shorts, open circuits and other problems within the EEC–IV system and when noted, stores them in the memory, when they occur. Memory codes obtained during the Key On Engine Off segment of the self test, recalls intermittent faults from the ECA memory.

Procedure

1. Repeat the self-test segment when any code other than "System Pass", is generated by the ECA.
2. Attempt to re-create intermittent faults while the test equipment is still connected to the system. This is called an intermittent fault confirmation check.
3. During the Key On Engine Off segment, perform this check with self-test sequence deactivated (ECA not in the self-test mode). A fault is indicated when the VOM deflection is 10.5 volts or greater, or the Star LED turns **OFF**.
4. Activate the Self Test and perform the Engine Running Quick Test.

After the service code output has been completed, do not turn the engine **OFF** or deactivate the Self Test. Approximately 2 minutes after code 11 has been outputted, the system will enter and remain in the continuous monitor mode until the Self Test is deactivated or the engine is turned off. A fault will be indicated in the same way as it is with the Key On Engine Off mode.

RECREATING INTERMITTENT FAULTS

Intermittent faults can generally be recreated by the following methods:

1. Wiggling connectors and harnesses
2. Manipulating moveable sensors and actuators
3. Heating thermistor type sensors with a heat gun

Suspected components, such as the sensors, actuators and harnesses, are identified by matching service codes obtained during the Key On Engine Off segment of the self-test.

If using a voltmeter; an intermittent fault is re-created, the voltmeter needle will sweep back and forth across the scale or sweep to the right and stay there.

If using the Star tester, an intermittent fault is indicated when the LED light turns **OFF**.

Malfunctioning components identified with this procedure can be repaired or replaced without further diagnostic testing. Further testing must be done for hard faults and for intermittent faults that can not be re-created by the above method.

ENGINE CODES

Key On Engine Off

CODE 15—ECA Test Failure
CODE 19—Power to Processor
CODE 21—ECT Out of range
CODE 22—MAP Out of range
CODE 23—TPS Out of range
CODE 24—ACT Out of range
CODE 31—EVR/PRE Out of range
CODE 34—No EGR Flow in test
CODE 35—RPM too low for EGR test
CODE 51—ECT too high
CODE 53—TPS input too high
CODE 54—ACT input to high
CODE 61—ECT input too low
CODE 63—TPS Input too low
CODE 64—ACTA input too low
CODE 67—Neutral drive switch
CODE 84—EGR circuit fault
CODE 85—Canister purge circuit fault
CODE 87—Fuel pump circuit fault
No codes or codes not listed, refer to pinpoint test Q

Engine Running

CODE 12—RPM out of specification
CODE 13—RPM out of specification
CODE 16—RPM too low
CODE 21—ECT out of range
CODE 22—MAP out of range
CODE 23—TPS out of range
CODE 24—ACT out of range
CODE 31—EVR/PFE of limits
CODE 32—EGR not controlling
CODE 33—EVR/PFE out of limits
CODE 34—No EGR flow in test
CODE 35—RPM tool low for EGR test
CODE 41—System always lean
CODE 42—System always rich
CODE 72—No MAP change in (goose) test
CODE 73—No TPS change in (goose) test
CODE 77—Operator did not do (GOOSE) test
CODE 99—ISC has not learned yet
No codes or codes not listed, refer to pinpoint test Q

Continuous Memory

CODE 14—Pip was erratic
CODE 15—ECA Test failure
CODE 18—No tachometer signal
CODE 22—MAP out of range
CODE 31—EVR/PFE out of limit
CODE 32—EGR not controlling
CODE 34—No EGR flow in test
CODE 35—RPM too low for EGR
CODE 51—ECT input too high
CODE 53—TPS input too high
CODE 54—ACT input to high
CODE 61—ECT input too low
CODE 63—TPS input too low
CODE 64—ACT input to low
No codes or codes not listed, refer to pinpoint test Q

CLEARING CODES

1. Perform the Key On Engine Off test.
2. As soon as the first codes are received, disconnect the STI to the Signal Return or unlatch the Star tester connection from the vehicles test connector.

NOTE: Do not disconnect the battery cables to clear code memory.

3. Disconnect the negative battery cable.
4. To clear Continuous Memory codes, exit quick test during code output.
5. Perform the Key On Engine OFF test to confirm the codes have cleared.

NOTE: Some vehicle computers have a learning ability. If a change is noted in vehicle performance after clearing codes, it may be due to the computers learning ability. To restore performance, warm vehicle to normal operating temperature and drive at part throttle with moderate acceleration, before performing any additional diagnosis.

PEUGEOT

General Information

Various electronically controlled Peugeot systems are capable of self-diagnostics. Peugeot uses three fuel injection systems, the LU Jetronic, LH2.2 Jetronic and the M1.3 Motronic systems.

Although all 3 are controlled by an Electronic Control Unit (ECU), only the M1.3 Motronic system contains a self-diagnostic function for the fuel injection system. These fuel injection systems are coupled with either the EZ 200K or the EZ 115K ignition systems, both of which are capable of self-diagnostics.

SYSTEM OPERATION

The Bosch systems used on these vehicles are really Engine Management Systems since fuel and ignition functions are often combined and controlled by 1 Electronic Control Unit (ECU).

The ignition ECU incorporates a microprocessor with memory. Several ignition advance points are programmed into the ECU to provide the optimum engine performance at the limit of engine knock (detonation). Each Electronic Contol Unit (ECU) is unique to its system. The controllers cannot be interchanged.

A self-diagnostic feature is incorporated into the ignition ECU. This enables it to identify some of the faults which may occur in the system during operation. An indicator LED in the lower left corner of the instrument cluster alerts the driver of faults.

Note that the Motronic systems used on XU9J2 and XU9J4 engines are very similar. The main difference is that the XU9J2 engine uses a so-called "simplified" system and does not use a knock (detonation) sensor. The XU9J4 engine uses a knock sensor.

In addition, the LU and LU-2 Jetronic systems are similar. The main difference is that the newer LU-2 system has the cold start enrichment controlled by the ECU, eliminating the cold start injector and thermo-time switch. The LU-2 has its rpm input supplied the coil and uses a different throttle position sensor that the earlier LU Jetronic.

The LH2.2 Jetronic system is similar the the LU series. The main difference is that the LH Jetronic system uses a hot wire air flow sensor to measure the air drawn into the engine. The LH Jetronic system is also equipped with an electronic idle regulator which maintains a constant idle speed regardless of the engine load.

When the engine is cranking, a signal is produced and amplified by the amplifier module. The amplifier module controls a power transistor which makes and breaks the primary coil circuit.

A signal emitted by the Hall sensor provides the ignition ECU with engine speed information. The duration of injection signal emitted by the injection ECU allows selecting an ideal advance setting from the computer's memory.

When detonation exists, the detonation sensor signals the ignition ECU by generating a small electric current. The ECU recognizes this signal and adjusts the spark timing. The spark advance is retarded by a few degrees, the detonation stops and the computer returns the spark to its origninal value. Detonation sensing is done cylinder by cylinder.

If detonation persists, the computer will continue to reduce the advance in 3 degree increments unit the maximum correcting span of 22 degrees has been reached.

An indicator LED in the instrument cluster monitors the proper operation of the ignition system. Under normal operation, with the key **ON** and the engine **OFF**, the LED will be **ON**. Under normal operation, with the engine running, the LED should be **OFF**.

NOTE: Occasional sporadic flashes of the LED with no discernable engine problems may be noticed. This condition is normal.

If problems develop when the vehicle is in operation the LED will blink constantly. The number of blinks, from 1–6, help determine the origin of the fault. The frequency of these blinks will be proportional to engine rpm.

As a safety measure, as soon as a fault in the system is detected, the system will automatically retard the spark advance (20–22 degrees), noticeable by a lack of power and performance of the engine.

Engine Control Units

FUEL INJECTION CONTROL UNIT

The M1.3 Motronic fuel injection system contains a single ECU which controls both the fuel injection and ignition systems. The ECU performs a self-diagnostic check upon initial engine start-up and continues to monitor the performance of its various components while the engine is running. Faults are divided into 2 categories, minor and major. The diagnostic memory will record the occurrence of either type, but will only illuminate the CHECK ENGINE warning lamp when a major fault is found.

IGNITION CONTROL UNIT

On Peugeots whose fuel injection systems do not contain self-diagnostic capabilities, the EZK ignition system is controlled by a separate ECU. The control unit monitors the ignition system for faults and will automatically retard the spark advance to a safety setting (−15 to −24 degrees) when a fault is identified.

An LED indicator is located in the lower left corner of the instrument cluster to warn the driver that a fault has been found. During normal operation, with the ignition **ON** and the engine OFF, the LED will remain ON. During normal operation with the engine running, the LED will remain OFF. In the event that a fault has been detected, the LED will flash a number of times from 1–6 indicating the corresponding fault code.

Self-Diagnostics

SERVICE PRECAUTIONS

- Do not disconnect the battery while the engine is running, or allow the engine to run if the battery terminals are not properly tightened.
- Always disconnect the battery for recharging.
- Do not use an auxiliary starting device which has a voltage rating above 12 volts.
- Before conducting any electrical welding on the vehicle be sure to disconnect all ECU's.Also, do not expose ECU's to temperatures in excess of 176°F (80°C) such as in an infra-red spray booth.
- Do not use a test light or create an arc to check circuit continuity.
- Do not insert the ends of a voltmeter test lead into the receiving end of the component connectors. It is necessary to use the pin-out box or to remove the connector cover and take the test reading where the wires enter the connector.
- Before connecting or disconnecting ECU harness connectors, make sure the ignition switch is **OFF** and the negative battery cable is disconnected to avoid the possibility of damage to the control unit.
- When connecting or disconnecting pin connectors from the ECU, take care not to bend or break any pin terminals. Check that there are no bends or breaks on ECU pin terminals before attempting any connections.
- Before replacing any ECU, perform ECU input/output signal diagnosis to determine if the ECU is functioning properly.
- When measuring supply voltage of ECU controlled components with a circuit tester, separate 1 tester probe from another. If the 2 tester probes accidentally make contact with each other

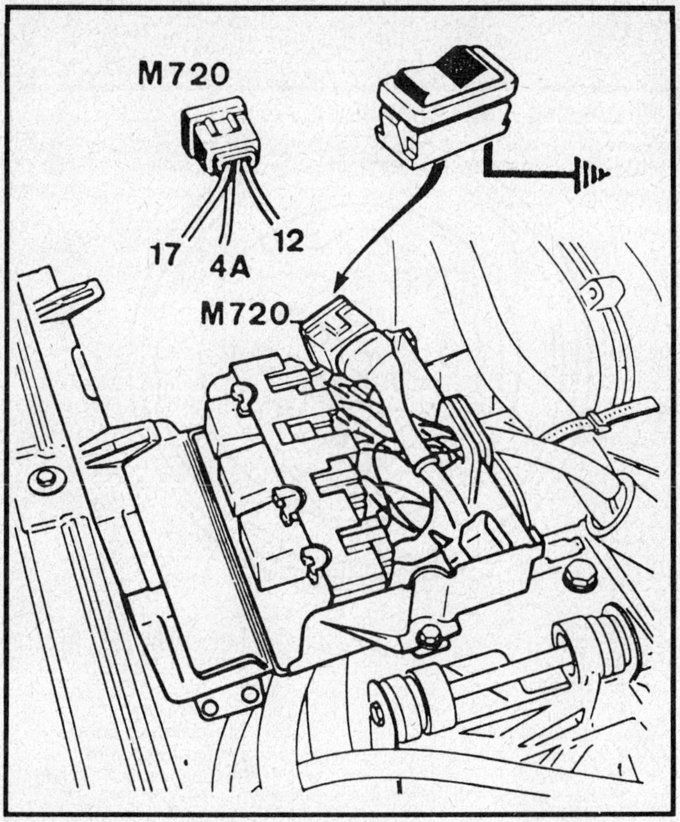

Motronic M1.3 test connector location—Peugeot

during measurement, a short circuit may result and damage the ECU.

READING CODES

Engine Codes

FUEL INJECTION CODES

Codes for the M1.3 Motronic system can be read with the aid of a switched jumper, the TAD 99 tester or equivalent. When using a switched jumper, codes are indicated through flashing of CHECK ENGINE light. A code will display by flashing a number of times for the tens digit and then for the units digit. The two sets of flashes will be separated by a pause. The TAD 99 tester is equipped with a display screen to indicate fault codes numerically.

Without Tester

1. Connect jumper switch (momentary contact type) to pin 1 of the 2-pin green connector (wires 4A and 17) found in the underhood relay box on the passenger side and to ground.
2. Turn the ignition **ON**, then depress jumper switch for 3 seconds, CHECK ENGINE indicator will come ON.
3. Release the switch and watch the CHECK ENGINE light. The light will flash 1 time, remain OFF for 1-5 seconds and flash 2 times. This represents Code 12 and indicates the beginning of the test sequence.
4. Wait for the CHECK ENGINE indicator to come ON.
5. Depress the jumper switch for 3 seconds, CHECK ENGINE indicator will come ON.
6. Release the switch and watch the CHECK ENGINE light, a fault code, if present, will display. The tens digit will flash followed by a 1-5 seconds pause and the units digit.
7. Repeat procedure from Step 4 until Code 11 displays indicating the end of the test sequence.

With TAD 99 Tester

1. Connect the tester, or equivalent, to the green 2-pin connector found in the under-hood relay box on the passenger side and to the battery.
2. Set test selector on position 1, then turn the ignition **ON**.
3. Depress the green test button for at least 3 seconds and release. Code 12 will display to identify the start of the test.
4. Wait a few seconds and depress the green test button for at least 3 seconds and release.
5. The next fault code, if present will display.
6. Repeat procedure from Step 4 until Code 11 displays indicating the end of the test sequence.

IGNITION CODES

No diagnostic procedure or scan tool is necessary to extract fault codes from a vehicle with a separate ECU for the ignition. In the event of a fault, the LED indicator in the instrument display will automatically flash the fault code in cycles. The code is a number from 1–6 and will flash constantly until the fault is corrected.

The frequency of the LED flashing is directly proportional to engine rpm, so it is essential that the code is determined by counting the number of blinks at idle. During normal operation the LED may exhibit occasional sporadic flashes; this does not mean that a fault has been detected.

CLEARING CODES

Engine Codes

FUEL INJECTION CODES

After repairs have been completed, be sure to clear the diagnostic memory.

Without Tester

1. Perform test up to Code 11.
2. Depress the jumper switch for more than 10 seconds, the CHECK ENGINE indicator should be ON.

With TAD 99 Tester

1. Perform test up to Code 11.
2. Depress the RED memory erase button until EF appears on the display and then wait for 00 to display.

IGNITION CODES

After repairs are made the ignition should be switched **OFF** to erase fault codes from memory.

PORSCHE

General Information

Both 911 Carrera models use the Bosch DME engine management system. The ignition and fuel injection systems are operated by a single control unit, which can detect faults in the input and output devices and store the fault code in memory.

The 928 GT and S4 models use the Bosch LH Jetronic engine management system, which includes a separate control unit for fuel injection and ignition system functions. Some sensors connect directly to the LH injection control unit, some connect to the EZK ignition control unit. Raw sensor data or a calculated value is transmitted between control units via the data link circuit. Both control units are capable of detecting, storing and reporting fault codes, beginning at the end of the 1988 model year.

ELECTRONIC CONTROL UNIT

DME System

The control unit in the 911 Carrera is mounted under the driver's seat. The 19 pin diagnostic connector is behind the passenger side kick panel. The memory for fault codes and adaptive control functions is maintained by a constant voltage supply to the control unit. If power is interrupted by disconnecting the battery or the control unit, the fault code memory will be lost. Fault codes can be retrieved using either the Porsche Diagnostic tester 9288 or the flash code tester 9268 and adapter plug. With the 9288 tester, most of the sensors and some of the calculated data can be read directly in real time. The Actual Value menu on that tester will provide the necessary instructions.

LH/EZK System

On all 928 models, both control units are mounted low in the passenger side of the center console. The LH unit has the black connector. Power is supplied to each unit at all times to maintain adaptive control functions and the fault memory. If power is interrupted by disconnecting the battery, both fault code memories will be lost. If one control unit is disconnected, only that memory will be lost.

The 19 pin diagnostic connector is under the sound system booster cover, behind the passenger side kick panel. Fault codes can be retrieved using either the Porsche Diagnostic tester 9288 or the flash code tester 9268 and adapter plug. With the 9288 tester, most of the sensors and some of the calculated data can be read directly in real time. The Actual Value menu on that tester will provide the necessary instructions.

Self-Diagnostics

SERVICE PRECAUTIONS

- Do not disconnect the battery or power to the control unit before reading the fault codes. Fault code memory is erased when power is interrupted.
- Make sure the ignition switch is **OFF** before disconnecting any wiring.
- Before removing or installing a control unit, disconnect the negative battery cable. The unit receives power through the main connector at all times and will be permanently damaged if improperly powered up or down.
- Keep all parts and harnesses dry during service. Protect the control unit and all solid-state components from rough handling or extremes of temperature.
- All airbag system wiring is in a yellow harness. Use extreme care when working around this wiring. Do not test these circuits without first disconnecting the airbag units.

Diagnosing Plug Connections

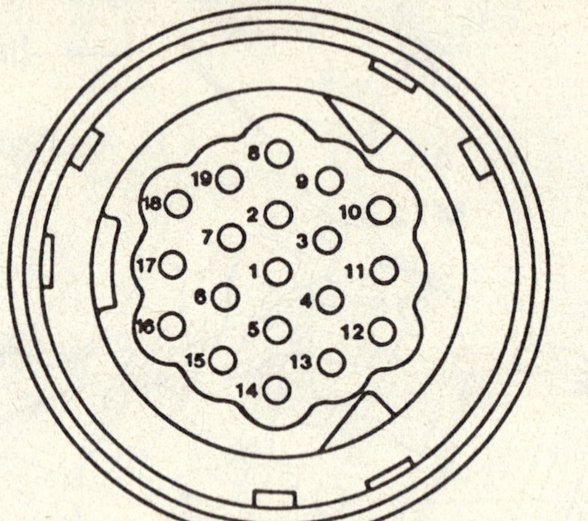

1. –
2. –
3. –
4. –
5. On/Off board
6. Reserved for L and K wire shielding
7. L wire
8. K wire
9. –
10. Ground
11. –
12. Terminal 30
13. Terminal 15
14. ti
15. LH, EZK, lamp wire
16. Tire pressure control system lamp wire
17. Reserve for ti shielding
18. –
19. Knock

Diagnostic plug connector—928 S4 shown

READING CODES

With 9288 Diagnostic Tester

When the tester is attached to the diagnostic connector, the control units will present country and application codes to the tester for display on the screen. The tester will then provide a menu with instructions for retrieving fault codes from the control unit memory. If the tester display shows fault not present, this indicates the fault is intermittent or the conditions under which the fault occurs do not exist at this time. The necessary conditions will be displayed on the screen. If the display shows signal not plausible, the input or output signal does exist but is out of the correct operating range.

With 9268 Diagnostic Tester

The control unit is equipped with a self diagnostic program that will detect emissions related malfunctions and turn the CHECK ENGINE light **ON** while the engine is running. Emissions related fault codes stored in the control unit can be read with the 9268 flash tester. On 1991 and later models, the flash codes can be read on the CHECK ENGINE light on the instrument panel. Only faults that may effect exhaust emissions are reported as flash codes. All other codes are only accessible with the 9288 Diagnostic Tester.

1. If required, connect the flash tester to the diagnostic connector using the adapter connector.
2. Turn the ignition switch **ON** without starting the engine.
3. Fully press the accelerator pedal to close the full load switch. After about 3 seconds the CHECK ENGINE light or tester will flash.
4. When the pedal is released, flash codes will be reported. All codes are 4 digits. Each digit will be flashed with about 2.5 sec-

onds between digits. When the whole code has been displayed, the light will stay **ON** or **OFF**. Count the flashes and write the numbers down. If Code 1500 or 2500 is displayed, no codes are stored in memory.

5. Repeat Steps 3 and 4 until Code 1000 appears, indicating all codes have been reported. On 928 models, the EZK unit will display Code 2000 when all codes have been reported.

NOTE: On all 928 models, if the first digit is 2, the fault is in the EZK ignition system control unit.

If the second digit is 1, the detected fault is current. If the second digit is 2, the detected fault has not occurred during the last running of the vehicle but did occur within the last 50 engine starts. The remaining 2 digits indicate which component or circuit is at fault.

CLEARING CODES

The fault code memory should be cleared before returning the vehicle to service. All Codes in memory and the idle control adaptation are lost when the control unit or the battery is disconnected. To avoid the loss of the learned idle program, use the instructions on the 9288 tester to clear the memory. If this tester is not available, disconnect the battery or control unit to clear the memory. It will be necessary to drive the vehicle for at least 6 minutes and run the engine at idle for about 10 minutes so the control unit can learn idle speed, timing and mixture parameters. Make sure the engine is at operating temperature and that all accessories are **OFF**. The throttle-at-idle position switch must be closed and functioning or system adaptation will not take place.

DME ENGINE FAULT CODES—911, 944 AND 968

Fault Code	Explanation
1000	End of report
1111	Control unit supply voltage out of limit
1112	Throttle idle position switch grounded
1113	Throttle full load position switch grounded
1114	Engine temperature sensor
1115	Throttle idle position switch open
1121	Air flow sensor
1122	Idle speed control circuit
1123	Oxygen regulation out of limit
1124	Oxygen sensor signal
1125	Inlet air temperature sensor signal
1131	Knock sensor 1
1132	Knock sensor 2
1133	Knock regulation in control unit
1134	Hall sender signal missing
1141	Control unit faulty
1143	Carbon canister purge valve
1144	Resonance plate solenoid valve circuit
1145	CHECK ENGINE warning light
1151	Injector, cylinder 1 (911 only)
1152	Injector, cylinder 2 (911 only)
1153	Injector, cylinder 3 (911 only)
1154	Injector, cylinder 4 (911 only)
1155	Injector, cylinder 5 (911 only)
1156	Injector, cylinder 6 (911 only)
1500	No faults in memory

NOTE: If the second digit of the code is 2, the fault is not present at the time of testing. It may be an intermittent fault.

LH ENGINE FAULT CODES—928

Fault Code	Explanation
1000	End of report
1111	Control unit supply voltage out of limit
1112	Throttle idle position switch grounded
1113	Throttle full load position switch grounded
1114	Coolant temperature sensor
1115	Throttle idle position switch open
1121	Air flow sensor
1123	Air/fuel mixture rich
1124	Air/fuel mixture lean
1125	Oxygen sensor signal missing
1141	LH control unit faulty
1500	No faults in memory

NOTE: If the second digit of the code is 2, the fault is not present at the time of testing. It may be an intermittent fault.

SAAB

General Information

The Saab 900 and 9000 are both equipped with Bosch LH-Jetronic fuel injection systems. The 1990 models and in some markets 1991–92 models are equipped with an LH 2.4 fuel system electronic control unit (ECU). Most 1991–92 models are equipped with an LH 2.4.2 fuel system ECU, except the 1991 9000 with the B234 engine, which has an LH 2.4.1 fuel system ECU. Turbocharged 900 models are also equipped with an automatic performance control (APC) ECU and 9000 models are equipped with an integrated Direct Ignition-Automatic Performance Control (DI/APC) control unit, which control ignition functions and turbocharger-related functions.

The LH-system ECU has the particular fuel system identification marked on it for identification. Visually, the LH 2.4 fuel system can be differentiated from the LH 2.4.2 fuel system by the pins on the automatic idle control (AIC) valve. The LH 2.4 fuel system AIC valve has 3 pins and the LH 2.4.2 fuel system AIC valve has 2 pins.

The LH 2.4.1 fuel system differs from the LH 2.4 fuel system in that the cold start injector has been discontinued. The direct ignition system has taken over this function. Also, the vehicle speed sensor is used in the control of the AIC function to tell the fuel system ECU whether the car is moving or at a standstill.

The LH 2.4 (some markets) and LH 2.4.2 fuel systems with the electronic throttle control system (ETS), used on the 1992 9000 with traction control, differ from the systems without ETS in the following ways: The electronically controlled throttle eliminates the need for automatic idling and load control throughout the load range, the throttle angle transmitter is located in the actuator motor, which is integrated with the throttle housing and the electronically controlled throttle system carries out compensation for the air conditioning.

Visually, the throttle housing on engines with ETS is larger, is vacuum operated and has an emergency cable, but in other respects this system operates in the same way as the LH 2.4 and 2.4.2 systems without ETS.

The central component of the LH-Jetronic fuel injection system is the air mass meter that measures the mass of air flow instead of the volume. The microprocessor measures how much electrical energy is used when air flow pass an electrically heated platinum wire in the air mass meter. The higher the rate of air flow, the higher the energy necessary to keep the temperature of the wire constant. At the same time, the microprocessor monitors the engine speed and temperature, calculating the exact amount of fuel needed for optimum performance. The microprocessor also incorporates an rpm limiter that ensures that no opening signals will be transmitted to the injectors at engine speeds above 6000 rpm.

The APC system on turbocharged vehicles enables the engine to achieve optimum performance and good fuel economy, regardless of the grade of fuel being used. A knock sensor, in-conjunction with the pressure transducer and ignition system information, detects knocking in the engine and sends an electrical signal to the microprocessor inside the ECU. The ECU processes these signals and sends electrical pulses to a solenoid valve that controls the charging pressure in the intake manifold. The turbocharger is designed to come into operation at fairly low engine speeds, thereby providing a high torque within the speed range of normal driving. It is water cooled and the coolant for the bearing housing is supplied by a pipe connected to the cooling system.

The DI/APC system was updated in 1991 to include an air temperature sensor, located upstream of the throttle housing. The boost pressure is governed by the position of the throttle valve (butterfly), subject to temperature compensation based on information supplied by the new air temperature sensor. The separate pressure-switch function is discontinued, with the pressure-sensing function now being regulated by the DI/APC-system ECU. The load signal provided by the air mass meter is sent to the LH-system ECU, which will keep the DI/APC system ECU informed of boost status.

Starting in 1991 there is also a spark plug burn-off function that occurs when the engine is stopped. The burn-off function, which operates in all cylinders simultaneously, lasts for 5 seconds at a frequency corresponding to 6000 rpm.

The LH Jetronic system also incorporates an emergency system known as a limp home function. If a malfunction is detected, the limp home feature of the ECU is actuated, enabling the vehicle to continue its journey, but with somewhat diminished performance. If the vehicle is operated in this mode, the check engine light (CEL) on the display panel will be illuminated. An integrated fault-storing capability enables diagnosis to be carried out efficiently.

Self-Diagnostics

SERVICE PRECAUTIONS

- Always disconnect the negative battery terminal to prevent sparks caused by short-circuiting, circuit-breaking, etc.
- Always keep a CO_2 fire extinguisher close on hand.
- Dry sand must be available to soak up any spillage.
- If using a battery charger, make sure that the battery leads are disconnected.
- Never disconnect the battery while the engine is running.
- Never disconnect or connect an ECU connector with the ignition switch in the **ON** position.
- Always remove the control units from the vehicle before carrying out any electrical welding.
- Before conducting a compression test, unplug the connector on the output module.
- Never subject the vehicle to temperatures exceeding 176°F (80°C), such as in a paint booth, unless the control units have first been removed from the vehicle.

READING CODES

Code retrieval is important and necessary for efficient repair of the vehicle. Diagnosis based on symptoms should be viewed as supplementary and not as the primary basis for a course of action.

There are 3 methods for reading codes on the Saab. Two are carried out with scanner-like instruments, and the third uses the CEL. These methods yield 3 different classes of codes. For example, ISAT type codes cannot be retrieved by using the CEL, nor can CEL type codes be retrieved with the ISAT.

With LH System Tester

The Saab LH System Tester 8394223 was developed to simplify service and fault diagnosis work on the LH fuel injection system. The ISAT is now most widely used, but the System Tester can still be used. The tester consists of a test unit, power supply lead, test lead incorporating a 2-way 35-pin connector and a pressure sensor with magnetic base.

The tester is equipped with an automatic program for diagnosing faults, both permanent and intermittent, while the vehicle is operating. Faults detected are then stored in memory for recall after the vehicle is road-tested.

Connect the diagnostic tester, or equivalent, as follows:

1. Insert the power supply lead between the door and body where there is a break in the seal, then run it under the back of the hood on the left hand side.
2. Clean the battery terminals to ensure proper contact with lead clips.

3. Connect the power supply lead to the tester first, then connect the lead clips to the battery.
4. Remove the cover on the left side over the space behind the false bulkhead panel.
5. Remove the antilock brake system (ABS) ECU and bracket, if equipped.
6. Remove the LH system ECU connector. Interface the test lead between the LH system ECU and the connector. Fit a couple of wire ties around the connector and ECU to hold them tightly together.

The tester, designed to perform 3 basic functions, monitor mode, test mode and fuel mode, is now ready to start. If at any time the operation in progress must be interrupted, simultaneously press all 3 control buttons and the tester will revert to the starting point.

MONITOR MODE

The monitor mode can be selected either by switching the ignition **ON**, or if the ignition is **OFF** by pressing the start test button when MON (monitor) appears on the tester display.

When in monitor mode, the tester is used to manually control parameter and functional checks.

TEST MODE

Test mode can only be selected from the monitor mode. Press the start test button to select the test mode. TEST will now appear on the display. In the test mode, the program instructs by way of prompts in the upper part of the display.

FUEL MODE

To select the fuel mode, the ignition switch must be **OFF**. Initially, MON will be displayed on the tester for approximately 5 seconds. During this time the monitor mode can be selected. If none of the tester buttons is activated, FUEL will appear on the display for approximately 2 seconds. To activate the fuel mode, press the start button when FUEL is displayed.

When the tester is in fuel mode the following checks can be performed:

Fuel pump delivery flow.
Fuel pressure and fuel-pressure regulator.
Residual pressure.
Fuel pump delivery pressure.
Delivery flow from injectors.

With SAAB ISAT Tester

The Saab ISAT tester is also available starting in 1990 to extract hard or intermittent fault codes or to issue command codes. This makes fault tracing much easier and enables diagnosis to be carried out more efficiently. ISAT Mode 1 is for fuel system testing Mode 2 is for DI/APC testing.

The diagnostic socket is a black 10-pin connector located under the RH front seat. It is protected by a plastic cover held in

ISAT tester—Saab

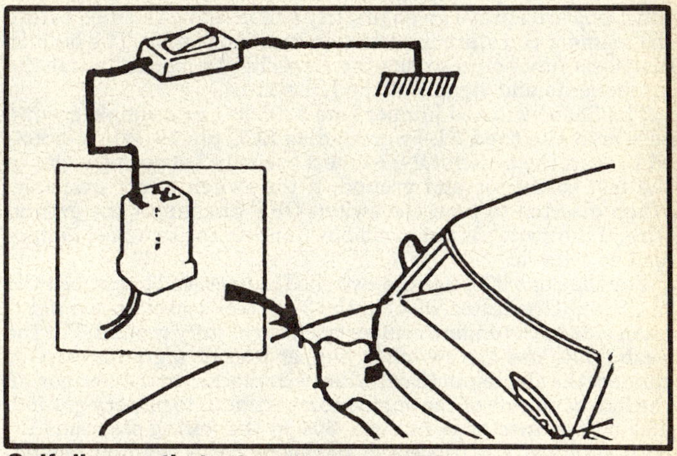

Self-diagnostic test connector—900

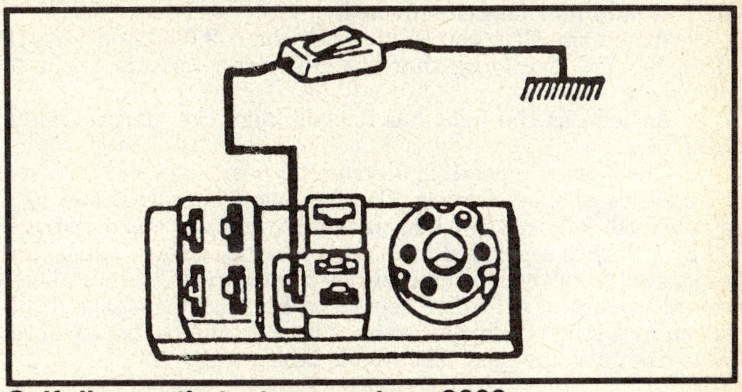

Self-diagnostic test connector—9000

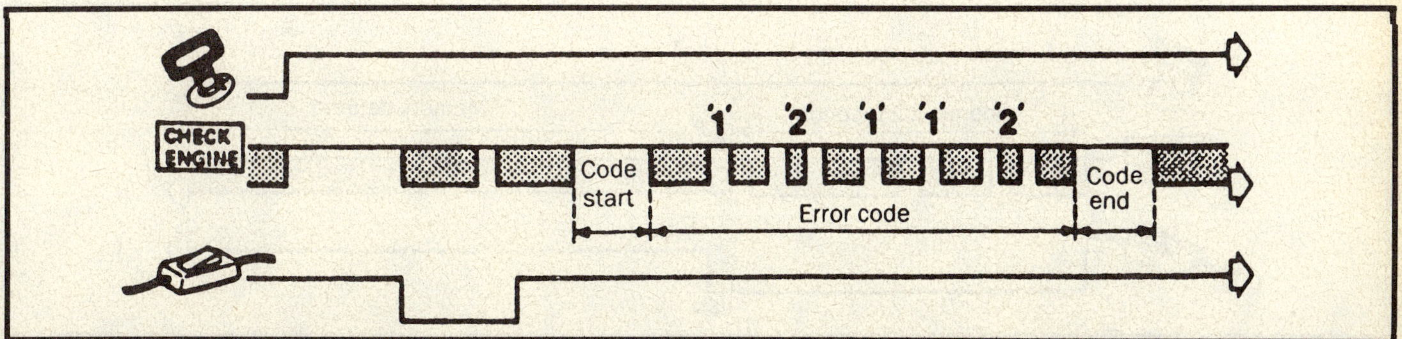

Self-diagnostic Code 12112 example—Saab

place by a screw. Read codes with the ISAT tester, or equivalent, as follows:

1. Connect the tester to the diagnostic socket.
2. Turn the ignition to the **ON** position.
3. Press No.1 on the ISAT tester to identify that the LH system is being investigated.
4. Continue by reading fault codes or issuing command codes.

Without a Diagnostic Tester

The Saab LH2.4, LH2.4.1 and LH2.4.2 fuel injection systems are capable of a self-diagnostic check by way of the flashing of the vehicle's CEL if a scanner is not available. Using such a method of diagnosis on the Saab is of limited value, however, since external monitoring and manipulation of engine parameters is not possible. Furthermore, temporary and intermittent faults are more difficult to trace.

These systems are designed with a built in memory and the ability to store up to 3 intermittent faults at a time. Serious malfunctions are always given priority and must be rectified before the memory can store information on minor faults. The built in diagnosis function also has the capability to manually test the components and signals of the LH system.

The Saab switched jumper lead 8393886, or a simple ground wire activates the CEL by grounding ECU pin 16. When directed to turn the switch **ON**, connect the wire between the pin in the test connector and ground, if the switch is not available. When directed to turn the switch **OFF** disconnect the ground wire, if the switch is not available. Connect the switched jumper lead as follows:

For the Saab 900, use the switched jumper lead to connect the No. 3 pin (Green/Red wire) in the 3-pin test socket on the right-hand side in the engine compartment to battery ground. For the Saab 9000, use the switched jumper lead or a ground wire to connect the single-pin (Green/Red wire) socket in the test box on the left-hand side of the engine compartment to battery ground. The test connector is number 204 in the wiring diagrams for both vehicles. The process of retrieving codes can now begin.

1. Switch the ignition **ON**. The CEL should now illuminate.
2. Set the jumper switch to **ON** (grounding ECU pin 16). The CEL should now be extinguished.
3. Watch the CEL carefully. After about 2.5 seconds, it will flash briefly, signifying that the first error code is about to display.
4. As soon as the light has flashed, turn the jumper switch **OFF**.
5. The first of a possible 3 error codes will now be displayed by a series of short flashes. The number 1 is represented by a single flash followed by a long pause. The number 2 is represented by 2 flashes separated from each other by a short pause, but separated from the next number by a long pause. The number 3 would consist of 3 flashes separated by short pauses and followed by a long pause, and so on. The code will be displayed repeatedly until the next test step is taken.
6. To check for any additional error codes, turn the jumper switch **ON**.
7. Watch the CEL carefully. After a short flash, turn the jumper switch **OFF**.
8. If present, the next error code will now display.
9. Follow the same procedure until all faults have been identified and corrected.
10. To restart the test procedure (return to the first fault), set the jumper switch to **ON**.
11. After 2 short flashes, turn the jumper switch **OFF** or disconnect the ground wire. The fault code for the first fault should now be displayed.
12. Proceed with the test from Step 5.

TESTING COMPONENTS AND SIGNALS

1. Connect the jumper lead in the same manner as for reading fault codes.
2. Set the jumper switch to **ON**.
3. Turn the ignition switch **ON** and wait for a short flash of the CEL.
4. Immediately following the flash, turn the jumper switch **OFF**.
5. The moment the CEL begins flashing, the fuel pump should begin running
6. To move on to the next test, set the jumper switch to **ON**.
7. After a short flash, set the jumper switch to **OFF**. A test code (not a fault code) will be displayed and the corresponding component will activate.
8. Continue through the remaining items in the test sequence in the same method (set switch to **ON**, wait for a short flash, set the switch to **OFF**).

Components and signals are checked in the following order:
Fuel pump (no code displayed),
Injection valves (1.5 ms–10 Hz),
AIC valve (switches between open and closed positions),
ELCD valve (switches between closed and open) – CEL flashing stops,
Drive signal (changes when shifting from **D** to **N**) – CEL flashing stops,
Air conditioning signal (cannot be tested),
Idling signal (changes as accelerator is depressed) – CEL flashing stops,
Full-throttle signal (changes as accelerator is pressed down to the floor) – CEL flashing stops.

CLEARING CODES

With LH System Tester

After completing repairs, reset the tester to the start mode by pressing all 3 control buttons simultaneously. If fault codes are still found, disconnect the battery for at least 10 minutes or road

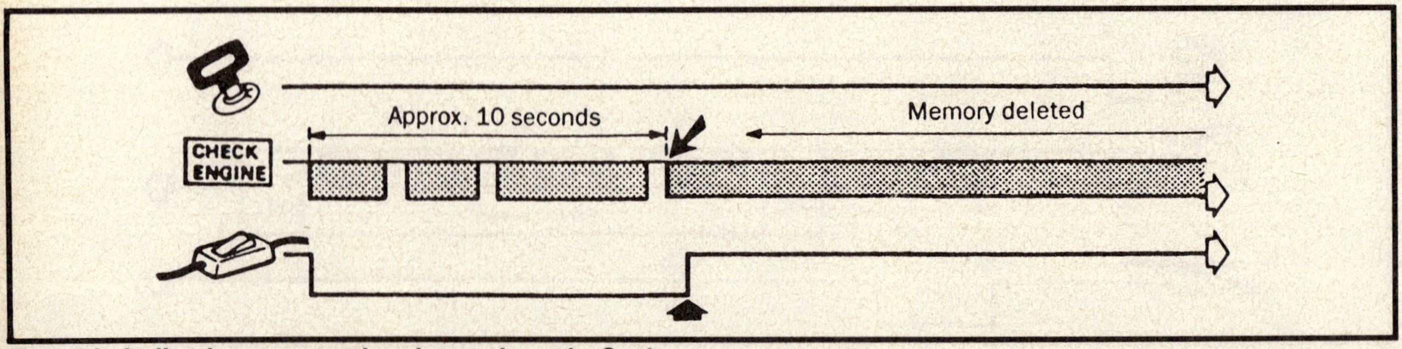

Example indicating memory has been cleared – Saab

test vehicle for 10 minutes or until the CEL extinguishes. The fault code should erase itself after extended operation with the repaired component.

With SAAB ISAT Tester

After repairs are completed, the ISAT command code 900 is used to clear diagnostic memory.

Without LH System Tester

1. Set the jumper switch to **ON**.
2. After 3 short flashes, turn the jumper switch **OFF**.
3. The CEL will now either flash in a continuous series of long flashes (this represents Code 00000) or display the Code 12444, indicating that the contents of the memory have been erased.

ENGINE CODES FOR THE LH SYSTEM TESTER—SAAB

Code	Explanation
E001	No ignition pulse
E002	Throttle Position Sensor (TPS)—idling contacts not closing on idling
E003	Throttle Position Sensor (TPS)—full load contacts constantly open
E004	Battery voltage to Electronic Control Unit (ECU) memory—missing
E005	Electronic Control Unit (ECU) pin 5 not grounding
E006	Air Mass Meter (AMM) not grounding
E007	No signal from Air Mass Meter (AMM)
E008	Air Mass Meter (AMM)—no filament burn-off function
E009	No power to system relay
E013	No signal from temperature sensor
E017	Break in earth circuit continuity
E020	Fuel pump relay—faulty control circuit
E021	System relay—faulty control circuit
E024	Lambda sensor—faulty signal
E033	Signal to AIC valve from Electronic Control Unit (ECU)—missing
E035	No power at +15 supply terminal
E101	Starter motor revolutions too low
E102	Throttle Position Sensor (TPS)—idling contacts not opening on increase from idling to 2500 rpm
E103	Throttle Position Sensor (TPS)—full load contacts constantly closed
E107	Low signal from Air Mass Meter (AMM)
E108	Air Mass Meter (AMM)—filament burn-off function constantly actuated
E109	Low voltage from system relay
E113	Shorting in temperature sensor circuit
E207	High signal from Air Mass Meter (AMM)
E218	Continuous pulses from injectors
E328	Pre-ignition signal constantly grounded

EZK IGNITION CODES FOR THE LH SYSTEM TESTER—SAAB

Number of Flashes	Explanation
1G	Not Applicable
2G	Not Applicable
3G	Not Applicable
4G	A: Knock sensor and cable
4G	B: Internal control unit fault
5G	Incorrect input signal from LH system
R (Sporadic)	Indicates knocking—normal
R (Continuous)	Maximum retardation—check engine for extreme vibrations

R = Red G = Green

ENGINE CODES FOR THE ISAT SYSTEM TESTER—SAAB

Permanent	Intermittent	Explanation
42241	22241	Voltage high 16V
42251	22251	Voltage at ECU pin 4 too low (1V)
42252	22252	Signal low 10V
42291	22291	Voltage at ECU pin 4 outside 10V–16V range
42440	22440	Mixture too rich—no control from Lambda sensor
42441	22441	Mixture rich on idling
42442	22442	Mixture rich on driving
42450	22450	Mixture too lean—no control from Lambda sensor
42451	22451	Mixture lean on idling
42452	22452	Mixture lean on driving
42460	22460	Lambda sensor—signal faulty
42491	22491	Mixture incorrect on idling
42492	22492	Mixture incorrect on driving
44221	24261	Speed signal—missing
44261	24261	Road speed sensor—signal faulty
44671	24671	PRE-IGNition signal—longer than 20 seconds
45641	25641	Air Mass Meter—signal high
45651	25651	Air Mass Meter—signal low
45691	25691	Air Mass Meter—signal faulty (too high/low)
45723	25723	DRIVE signal—faulty
45771	25771	Throttle potentiometer—indicating idling during driving (constantly grounded)
45772	25772	Throttle potentiometer—indicating full throttle and idling simultaneously
46221	26221	Temperature sensor—indicating temperature -90°C during driving (break in circuit)
46271	26271	Temperature sensor—indicating temperature 160°C (constantly grounded)
46391	26391	EGR system malfunction—temperature too low
58121	38121	Air Mass Meter—no filament burn-off function
58321	38321	AIC valve—faulty
58322	38322	ELCD valve—faulty
58371	38371	Injector—faulty
58372	38372	ELCD valve—broken circuit/grounded
58382	38382	ELCD valve—shorting to positive
67192		ROM chip—faulty

DI/APC CODES FOR THE ISAT SYSTEM TESTER—SAAB

Permanent	Intermittent	Explanation
429B0	229B0	Coding fault—engine starts but only baseline charging pressure available
44261	24261	Road speed signal—absent (Turbo)
44360	24360	Crankshaft sensor—signal faulty
44460	24460	Load signal—Tq (faulty)
44461	24461	Adaptation fault—outside limits
44660	24660	Pre-ignition fault
44661	24661	Knock sensor—signal faulty
44662	24662	Pre-ignition synchronization fault
45260	25260	Throttle potentiometer—signal faulty
45360	25360	Brake signal—faulty
46391	26391	Temperature sensor upstream of throttle body—signal faulty
60000	60000	Watchdog—internal monitoring
60001	60001	ECU—ROM
60002	60002	ECU—RAM
11111	11111	Acknowledgement—OK

FLASHING ENGINE CODES—SAAB

Code	Explanation
00000	No more faults or faults not detected
12111	Lambda adaptation fault—fuel-air mixture with throttle open
12112	Lambda adaptation fault—fuel-air mixture with engine idling
12113	Idling control (AIC) adaptation fault—pulse ratio too low
12114	Idling control (AIC) adaptation fault—pulse ratio too high
12211	Incorrect battery voltage with engine running (10V or 16V)
12212	Throttle Position Sensor—faulty idling contacts (grounding when throttle open)
12213	Throttle Position Sensor—faulty full-throttle contacts (grounding when engine idling)
12214	Temperature sensor signal—faulty (signal -90°C or 160°C)
12221	Air Mass Meter signal—missing (engine in limp-home mode)
12222	Idling control system (AIC)—faulty
12223	Fuel-air mixture—lean
12224	Fuel-air mixture—rich
12225	Lambda sensor—faulty or preheating defective (engine temperature must be 80°C)
12231	No ignition signal—always occurs with the engine switched off
12232	Memory voltage 1V

STERLING

General Information

Sterling vehicles utilize the Honda type Programmed Fuel Injection (PGM-FI) System. This system is a fully electronic microprocessor based engine management system. The Electronic Control Unit (ECU) is given responsibility for control of injector timing and duration, intake air control, ignition timing, cold start enrichment, fuel pump control, fuel cut-off, A/C compressor operation, alternator control as well as EGR function and canister purge cycles.

The ECU receives electric signals from many sensors and sources on and around the engine. The signals are processed against pre-programmed values; correct output signals from the ECU are determined by these calculations. The ECU contains additional memories, back-up and fail-safe functions as well as self-diagnostic capabilities.

Diagnosis is generally based on symptom diagnosis and stored fault codes, if any. Testing always requires the use of the diagnostic charts in conjunction with the Acura test harness (pin-out box), a voltmeter and an ohmmeter.

Self-Diagnostics

SERVICE PRECAUTIONS

- Make sure all ECU harness connectors are fastened securely. A poor connection can cause an extremely high voltage surge and result in damage to integrated circuits.
- Keep ECU all parts and harnesses dry during service. Protect the ECU and all solid-state components from rough handling or extremes of temperature.
- Use extreme care when working around the ECU or other components; the airbag or SRS wiring may be in the vicinity. On these vehicles, the SRS wiring and connectors are yellow; do not cut or test these circuits.
- Before attempting to remove any parts, turn the ignition switch **OFF** and disconnect the battery ground cable.
- Always use a 12 volt battery as a power source for the engine, never a booster or high-voltage charging unit.
- Do not disconnect the battery cables with the engine running.
- Do not disconnect any wiring connector with the engine running or the ignition **ON** unless specifically instructed to do so.
- Do not apply battery power directly to injectors.
- Whenever possible, use a flashlight instead of a drop light.
- Relieve fuel system pressure before servicing any fuel system component.
- Always use eye or full-face protection when working around fuel lines, fittings or components.

READING CODES

When a fault is noted, the ECU stores an identifying code and illuminates the CHECK ENGINE light. The code will remain in memory until cleared; the dashboard warning lamp may not illuminate during the next ignition cycle if the fault is no longer present. Not all faults noted by the ECU will trigger the dashboard warning lamp although the fault code will be set in memory. For this reason, troubleshooting should be based on the presence of stored codes, not the illumination of the warning lamp while the car is operating.

Stored codes are displayed by a flashing LED on the ECU. Turn the ignition switch **ON**; the LED will display any stored codes by rhythmic flashing. Some models may have two LEDs on the controller; one is red and one is amber. The red one will flash the fault codes; the amber one is used during idle adjustment and is not related to this procedure.

Codes 1–9 are indicated by a series of short flashes; two-digit codes use a number of long flashes for the first digit followed by the appropriate number of short flashes. For example, Code 43 would be indicated by 4 long flashes followed by 3 short flashes. Codes are separated by a longer pause between transmissions. The position of the codes during output can be helpful in diagnostic work. Multiple codes transmitted in isolated order indicate unique occurrences; a display of showing 1-1-1-pause-9-9-9 indicates two problems or problems occurring at different times. An alternating display, such as 1-9-1-9-1, indicates simultaneous occurrences of the faults.

When counting flashes to determine codes, a code not valid for the vehicle may be found. In this case, first recount the flashes to confirm an accurate count. If necessary, turn the ignition switch **OFF**, then recycle the system and begin the count again. If the Code is not valid for the vehicle, the ECU must be replaced.

Engine Codes

Code LED ON, But No Blinks—disconnected ECU ground wire.
Code LED OFF—ECU ground wire, short circuit in warning light wire or faulty ECU assembly.
Code 1—front oxygen sensor.
Code 2—rear oxygen sensor.
Code 3—MAP sensor or circuit.
Code 4—crank angle sensor or circuit.
Code 5—low vacuum or disconnected MAP sensor hose.
Code 6—coolant temperature sensor or circuit.
Code 7—throttle sensor or circuit.
Code 8—crank angle sensor or circuit (TDC).
Code 9—crank angle sensor or circuit (cylinder).
Code 10—air intake temperature sensor or circuit.
Code 11—improper idle CO; idle mixture adjuster.
Code 12—EGR control system, shorted or disconnected EGR control wire, faulty EGR control system.
Code 13—atmospheric pressure (PA) sensor or circuit.
Code 14—Electronic Idle Control Valve (EICV) or circuit.
Code 15—ignition circuit or faulty igniter unit.
Code 17—road speed transducer/speedo circuit faulty.
Code 18—faulty ignition timing adjuster or circuit.
Code 43—faulty ignition timing adjuster or circuit.
Code 44—faulty ignition timing adjuster or circuit.
Code 45—faulty ignition timing adjuster or circuit.
Code 46—faulty ignition timing adjuster or circuit.

CLEARING CODES

Stored codes are removed from memory by removing power to the ECU. Disconnecting the power may also clear the memories used for other solid-state equipment such as the clock and radio. For this reason, always make note of the radio presets before clearing the system. Additionally, some radios contain anti-theft programming; obtain the owner's code number before clearing the codes.

While disconnecting the battery will clear the memory, this is not the recommended procedure. The memory should be cleared after the ignition is switched **OFF** by removing the appropriate fuse for at least 10 seconds.

VOLKSWAGEN

General Information

There are two basic engine management systems used, CIS-E Motronic and Digifant. The Motronic system, used on some Golf/Jetta models and all Passat models with the 16 valve 9A engine, is equipped with a self diagnostic program that will detect emissions related malfunctions and store fault codes in permanent memory. Certain output tests are also available in this program.

The newer fully electronic Digifant systems are used on all other engines. The Digifant I system, used on California models, is equipped with a self diagnostic program with a permanent memory for storing emissions related fault codes. Digifant II is basically the same system but is not equipped with a self diagnostic program. No fault codes are available on Digifant II systems, even with a diagnostic tester.

Motronic Control Unit

The control unit operates all air/fuel mixture, idle speed and ignition timing functions. All the engine sensors and output devices are monitored by the self diagnostic program built into the control unit. The self diagnostic system can differentiate between intermittent and permanent faults and all fault codes are stored in permanent memory until intentionally erased. If an engine fault has not occurred within the last 50 engine starts, it is erased automatically. Codes can only be retrieved using the VAG 1551 Diagnostic Tester or equivalent. With the Diagnostic Tester, output signals can be generated to test certain output devices.

Digifant Control Unit

The Digifant I system is used on California vehicles. The control unit operates the fuel injectors, idle speed control actuator and ignition timing functions. All the engine sensors and output devices are monitored by the self diagnostic program built into the control unit. The self diagnostic system can differentiate between intermittent and permanent faults and all fault codes are stored in permanent memory until intentionally erased. If an engine fault has not occurred within the last 50 engine starts, it is erased automatically. When a permanent fault is detected, the ENGINE WARNING light on the instrument panel is turned ONuntil the fault is repaired and the memory cleared. Some codes can be retrieved by installing a jumper wire on the test connector and reading the flashes of the ENGINE WARNING light. The remaining codes can only be retrieved with the VAG 1551 Diagnostic Tester or equivalent. With the Diagnostic Tester, signals can be generated to test certain output devices.

Self-Diagnostics

SERVICE PRECAUTIONS

- Do not disconnect the battery or the control unit before reading the fault codes. On the Motronic system, fault code memory is erased when power is interrupted.
- Make sure the ignition switch is **OFF** before disconnecting any wiring.
- Before removing or installing a control unit, disconnect the negative battery cable. The unit receives power through the main connector at all times and will be permanently damaged if improperly powered up or down.
- Keep all parts and harnesses dry during service. Protect the control unit and all solid-state components from rough handling or extremes of temperature.

READING CODES

1. Make sure the ignition switch is **OFF** and that all fuses are good. Make sure all ground connections under the hood are good, especially those for the control unit.
2. Remove the shifter knob and boot and connect the diagnos-

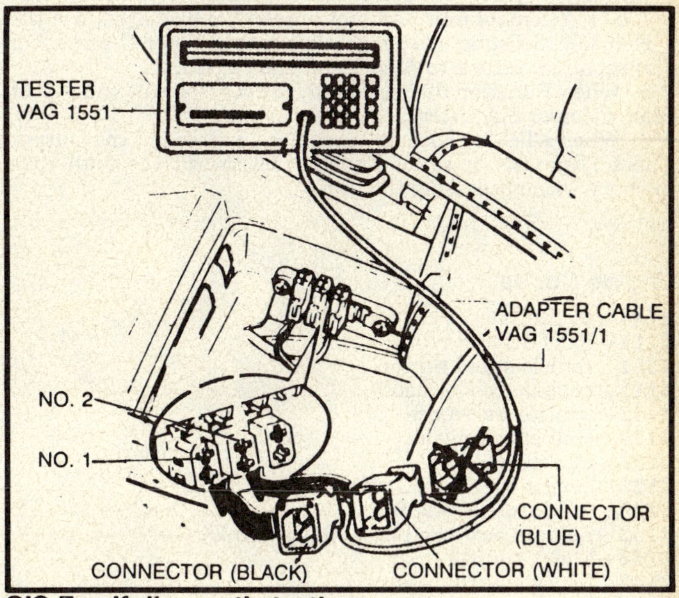

CIS-E self-diagnostic testing

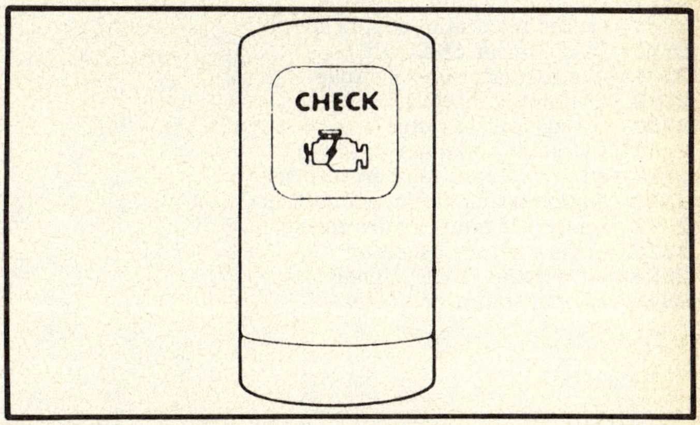

View of the diagnostic fault indicator lamp – Digifant

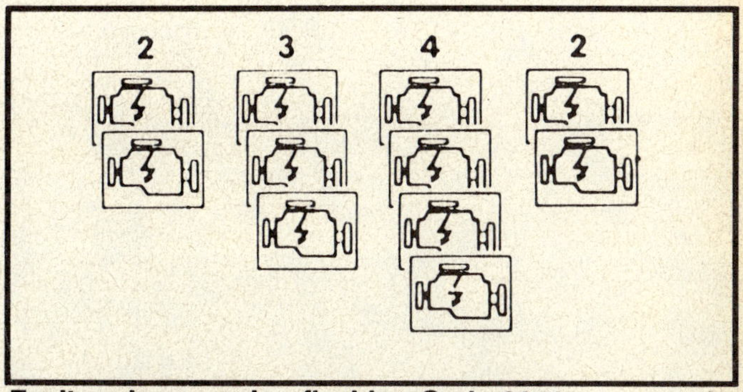

Fault code example – flashing Code 2342

tic tester to the black and white connectors towards the left or front of the vehicle.

3. Turn the tester **ON** to make sure it is receiving power. The screen will display 2 menu options; Rapid Data Transfer and Blink Code Output. Turn the ignition switch **ON**.

4. Select Rapid Data Transfer and Address Word 01, then press the Q button to enter the selection. The tester will display a control unit part number, the system it controls and an application (country) code.

a. If the information is displayed and is correct, press the Run key to proceed: "Select function XX" will be displayed.

b. If "Control unit does not answer" is displayed, use the Help key to display a list of possible causes. When the problem is repaired, return to Step 1 and start again.

5. When Function 02 is selected, the control unit will report fault codes to the tester.

6. When all codes have been reported, proceed to the Output Check Diagnosis or select Function 06 to exit the fault code memory without erasing the codes.

Engine Codes

4444—no faults
1111—ECU defective
1231—vehicle speed sensor
2111—engine speed sensor
2112—ignition reference
2113—Hall effect signal
2121—idle switch
2122—engine speed; Hall effect signal
2123—full throttle switch
2132—no data between fuel and ignition ECUs
2141—knock control
2142—knock control (sensor 1)
2144—knock control (sensor 2)
2212—throttle position sensor
2214—exceed maximum engine speed
2222—MAP sensor or circuit
2231—idle regulator out of range
2232—airflow signal missing
2234—voltage out of range
2242—CO potentionmeter problem
2312—engine coolant temperature sensor
2314—engine/transmission connection
2322—intake air temperature sensor
2323—airflow sensor or circuit
2324—air volume (MAF) signal
2341—oxygen sensor or heater
2342—oxygen sensor signal missing
2411—EGR temperature sensor
2413—mixture problem
4312—EGR frequency valve
4332—ECU output stage
4343—canister solenoid
4431—idle stablizer circuit
4433—fuel pump relay
4411 to 4422—injectors
0000—end of codes

OUTPUT CHECK DIAGNOSIS

Motronic System

Only the Motronic system is equipped with this program. It allows testing most of the engine output devices without running the engine. The program cannot be run without the VAG 1551 Diagnostic Tester or equivalent. During the test, 4 output devices are activated in the following order:

Differential pressure regulator
Carbon canister frequency valve
Idle stabilizer valve
Cold start valve

Testing the differential pressure regulator requires a multi-meter that will read milliamps. The other items can be checked with a voltmeter, test light or by listening and feeling for valve activation. The cold start valve is activated for a limited time to avoid flooding the engine.

1. Connect the diagnostic tester, turn the ignition switch **ON** and confirm that the tester will communicate with the control unit. See the procedure for retrieving fault codes.

2. Select Rapid Data Transfer and Function 03. When the test is started by pressing the Q button (enter), the first output signal is generated.

3. Each time the Run button is pressed, the tester will send an output signal to the next device on the list.

4. When the last item has been tested, select Function 06 to exit the program. To repeat the test, turn the ignition switch **OFF** and **ON** again.

CLEARING FAULT CODES

After all fault codes have been retrieved, select Function 05 and press the Q button to enter the selection. The memory will be erased only if all fault codes have been retrieved. Test drive the vehicle for at least 10 minutes, including at least 1 full throttle application above 3000 rpm. Check the fault code memory again to make sure all faults have been repaired.

VOLVO

General Information

The fuel, ignition and California EGR systems incorporate a built-in self-diagnostic feature. The fuel and ignition systems also have 2 continuous functions testing features. The self-diagnostic function, Control Function 1, is used to retrieve fault codes caused by a malfunctioning sensor or sensor circuit.

In Control Function 1, many different fault codes may be retrieved, but the system can only store up to 3 system faults in each system at any one time. When the engine is running, the control units continuously check for faults related to the following items, for example:

- Control units internal functions
- Coolant temperature sensor
- Air mass meter
- Air temperature sensor (Regina)
- Battery voltage
- Pressure sensor (Regina)
- Ignition settings and engine rpm (through the ignition system control unit).
- Speedometer and knock indicator
- Idle speed air valve and injectors
- Camshaft sensor (Motronic)
- EGR temperature sender (California)

The fuel and ignition diagnostic systems also incorporate 2 other built-in fault tracing functions: Control Functions 2 and 3. Thus, the control units are able to perform 3 different control functions. One is to read fault codes stored in the memory and 2 are for continuous testing of the components in the system.

Control Function 2 is used for testing the various fuel system circuit breakers. As each is activated, functions information is provided by the diagnostic unit diode in the form of series flashes. This function test is generally used after repairs have been made, to ensure that certain controls are connected and operating correctly.

Control Function 3 tests the output functions of the control system. This test consists of initiating a functions cycle, whereby the diagnostic system activates certain control components. It can be determined if the component is operating correctly by placing a hand on the component or by listening for a noise that occurs when the component is activated.

The diagnostic unit is located behind the left spring strut tower in the engine compartment. The diagnostic unit consists of a button, a diode light and a selector plug. When fault tracing the fuel system, the selector plug is placed in socket No. 2. When diagnosing the ignition system or EGR system, socket No. 6 is used, but the Motronic system uses socket No. 2 for the ignition system. Pressing the button once, twice or 3 times chooses the desired control function.

Faults stored in the memory are read through a series of flashes from the diagnostic socket diode light. All fault codes have 3 digits, each capable of ranging from 1 through 9. The fault codes are read from the series of flashes delivered by the diode. Each code requires 3 series of uninterrupted flashes, with a 3 second interval between each series of flashes to make the codes easy to read.

Self-Diagnostics

SERVICE PRECAUTIONS

- Use caution when working around ignition system components. the ignition system operates at high power with dangerous voltages in both the low-tension and high-tension circuits, including connectors and similar fittings.
- Do not operate the fuel pump when the fuel lines are empty.
- Do not arc weld on the vehicle unless the control unit has been removed.
- The control unit must not be exposed to temperatures above 175°F (80°C), such as in a paint booth.
- Make sure that the ignition switch is **OFF** before connecting or disconnecting the ignition coil, spark plug wiring or test instruments.
- Always use new gaskets and seals when remaking a fuel connection.
- When testing the compression on an LH-Jetronic-equipped engine, remove the negative primary lead on the ignition coil.
- When testing the compression on a Regina-equipped engine, remove the primary electrical connector on the ignition coil.
- Do not test the compression on the 960 until the ignition coils are removed, the ignition coil relay is disconnected and fuse No. 31 (to disarm fuel pumps) is removed.
- Do not disconnect the control unit harness connector before all power to it has been removed.
- Make sure all control unit connectors are fastened securely.

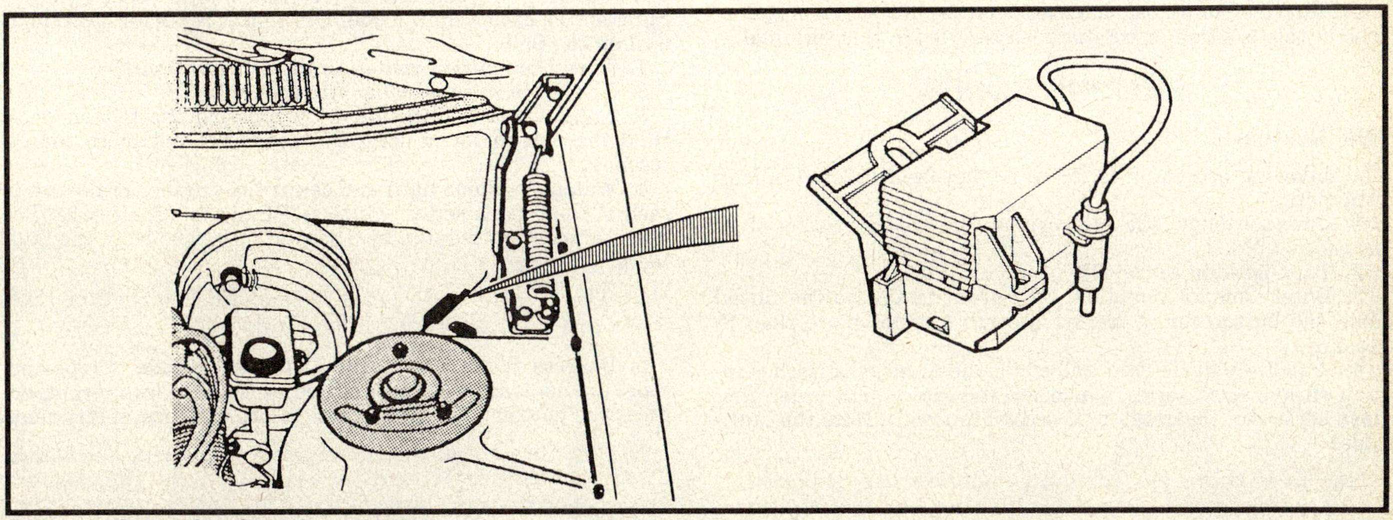

Diagnostic socket location—Volvo

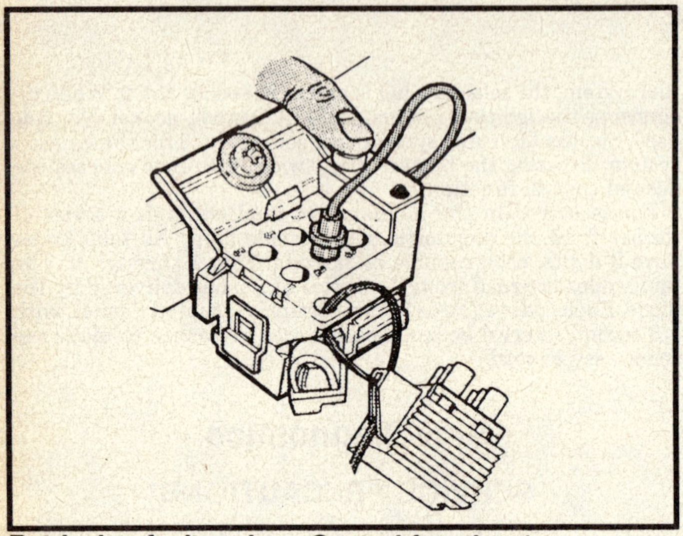

Retrieving fault codes—Control function 1

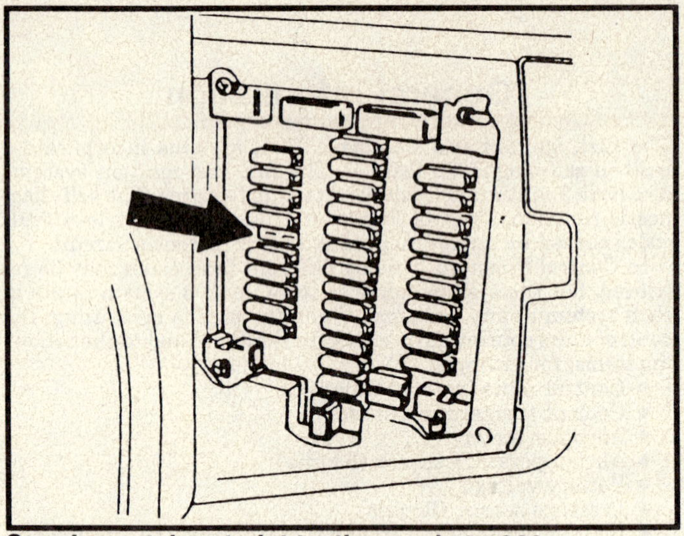

Supplemental restraint testing socket—960

A poor connection can cause an extremely high surge voltage, resulting in damage to integrated circuits.

- Keep any control unit harness at least 4 in. away from adjacent harnesses to prevent a system malfunction due to external electronic "noise."
- Keep all parts and harnesses dry during service.
- Before attempting to remove any parts, turn off the ignition switch and disconnect the battery ground cable.
- Always use a 12 volt battery as a power source.
- Do not attempt to disconnect the battery cables with the engine running.
- Do not attempt to disassemble a control unit under any circumstances.
- If a battery cable is disconnected, the memory will be erased.
- Before connecting or disconnecting control unit ECU harness connectors, make sure the ignition switch is **OFF** and the negative battery cable is disconnected to avoid the possibility of damage to the control unit.
- Always disconnect the yellow air bag connector near the gas pedal before performing any diagnostic, troubleshooting or service procedure that is not even associated with the air bag system. Failure to do so can cause damage to the crash sensor.
- A spare location in the electrical distribution unit is used as a test terminal for air bag diagnostics on the 960. Never install a fuse in this position or connnect accessories to this terminal.

READING CODES

Fuel System

The following procedure is for retrieving fault codes (Control Function 1).

1. Open the diagnostic unit cover and install the selector plug into socket No. 2.
2. Turn ignition to the **ON** position.
3. Enter Control Function 1 by pressing the button once. Hold the button for at least 1 second, but not more than 3 seconds.
4. Watch the diode light and count the number of flashes in each ofthe 3 flash series, which will indicate a fault code. The flash series are separated by 3 second intervals. Note the fault codes.

NOTE: If there are no fault codes in the diagnostic unit, the diode will flash 1-1-1 and the fuel system is operating correctly.

Fault code 213 example—Volvo

5. Depress the button again. If the same code is repeated, there are no additional codes stored. If the code is different, depress the button a third time and record the code, if different.

NOTE: The diagnostic memory is full when it contains 3 fault codes. Until those 3 codes are rectified and the memory erased, the system cannot give information on any other problems.

Ignition System

The following procedure is for retrieving fault codes (Control Function 1), except on the 960. Use socket No. 2 for ignition as well on the 960.

1. Turn the ignition switch to the **ON** position.
2. Insert the selector cable into socket No. 6.
3. Enter Control Function 1 by pressing the button once. Hold the button for at least 1 second, but not more than 3 seconds.
4. Watch the diode light and count the number of flashes in each of the 3 flash series, which will indicate a fault code. The flash series are separated by 3 second intervals. Note the fault codes.

NOTE: If fault code 1-1-1 is flashed, it indicates that there are no fault codes in the memory.

5. Depress the button again. If the same code is repeated, there are no additional codes stored. If the code is different, depress the button a third time and record the code, if different.

NOTE: The diagnostic memory is full when it contains 3 fault codes. Until those 3 are rectified and the memory erased, the system cannot give information on any other problems.

CLEARING CODES

When all the fault codes have been read and corrected, the diagnostic system memory can be erased as follows:

1. Turn ignition switch to the **ON** position.
2. Read fault codes again.
3. Press and hold diagnostic unit button for approximately 5 seconds. Release button. After 3 seconds, the diode light should light.
4. While the diode light is lit, press button again for approximately 5 seconds. Release button. The diode light should go out. Memory is cleared.

NOTE: To ensure that the memory has been erased, press button again for 1 second but not more than 3 seconds. The diode light should flash 1-1-1, indicating that there are no other fault codes.

ENGINE DIAGNOSTIC CODES—VOLVO

Code	Fault
1-1-1	No faults
1-1-2	Fault in control unit
1-1-3	Fault in injector
1-2-1	Signal to/from air mass meter faulty
1-2-1	Signal missing or faulty to/from pressure sensor
1-2-2	Signal missing or faulty to/from air temperature sensor
1-2-3	Signal missing to/from CTS, possible grounding short
1-3-1	Ignition system RPM signal missing
1-3-2	Battery voltage too high/low
1-3-3	Throttle-Switch; idle setting faulty, possible grounding short
1-4-3	Front knock sensor signal absent or faulty
2-1-2	Oxygen sensor (Lambda-Sond) signal missing or is faulty
2-1-3	Throttle-Switch; full load setting faulty, possible grounding short
2-1-4	Timing pick-up signal absent intermittently
2-2-1	Lambda-Sond not operating
2-2-2	Fault in system relay
2-2-3	Signal missing to/from idle valve
2-3-1	Self-Adjusting Lambda-Sond not operating
2-3-2	Self-Adjusting Lambda-Sond not oeprating
2-3-3	Idle valve closed, possibly leaking air
2-4-1	EGR system malfunction
2-4-3	Signal from throttle switch absent or faulty
3-1-1	Signal missing from speedometer
3-1-2	Signal missing for knock related fuel enrichment
3-1-4	Camshaft sensor signal absent
3-2-1	Signal missing or faulty to/from cold start valve
3-2-2	Burn-off cleaning of hot-wire in air mass meter not operating
4-1-3	EGR temperature sensor signal absent or faulty
4-3-3	Rear knock sensor signal absent or faulty

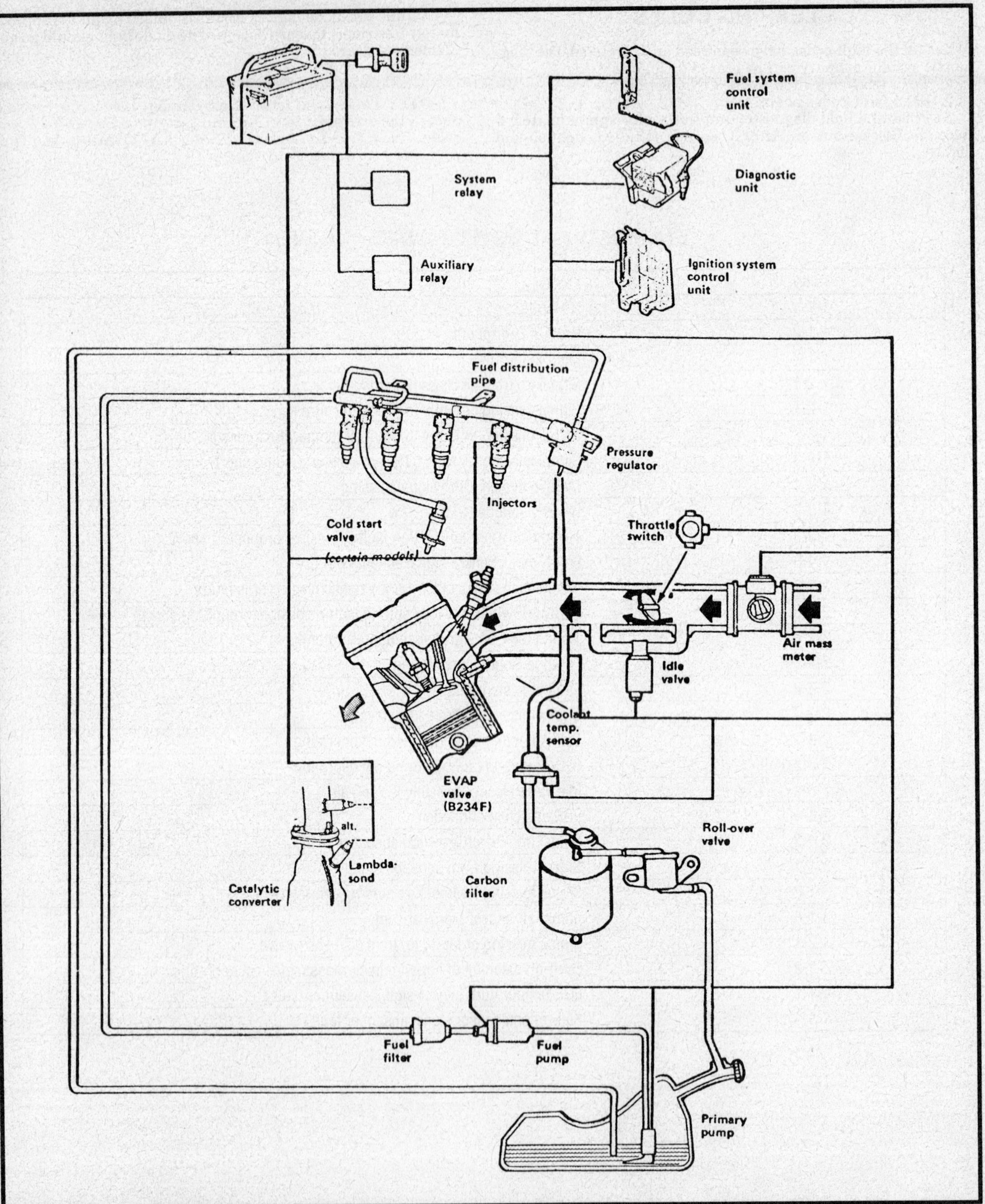

Exploded view of LH-Jetronic fuel injection system– 700 models

YUGO

General Information

The Yugo has been upgraded to use the Bosch Motronic MP 3.1 fuel injection system. This is an improvement over the older design Carter/Weber carburetor. The new fuel injection system allows for improved performance, drivability, fuel economy and the ability to self-diagnose.

NOTE: The Yugo carburetor system did not have the ability to self-diagnose.

The Motronic unit controls both the fuel injection and the electronic ignition. This reduces the total cost of the 2 separate systems since the same sensors are shared by a common control unit.

The Motronic control unit receives data from sensors, interprets the information, compares it to the preprogrammed parameter map and outputs signals to the various output devices such as the injectors. The unit controls fuel to the engine and electronically meters it. It regulates the fuel enrichment for cold starts, warm-up and acceleration. The unit can compensate for additional engine loads, such as the air conditioner compressor.

Self-Diagnostics

SERVICE PRECAUTIONS

- Always disconnect the battery, the Motronic control unit and the ignition coils when using an electric welder on the vehicle or when charging the battery or when the vehicle is placed in a paint drying oven.
- Remove the main Motronic relay to disable both the fuel and ignition systems when checking the compression. Never crank the engine after removing the distributor cap or disconnecting the high tension wire on the ignition coil.
- Never disconnect the battery or wires on the alternator, starter or spark plugs when the engine is running.
- Always work in a well ventilated area to prevent the build up of hazardous fumes.
- Always keep a fire extinguisher within reach when working of the fuel system.
- Wipe up any spilled fuel to prevent fire hazards.
- Relieve the fuel pressure before disconnecting any fuel system components.

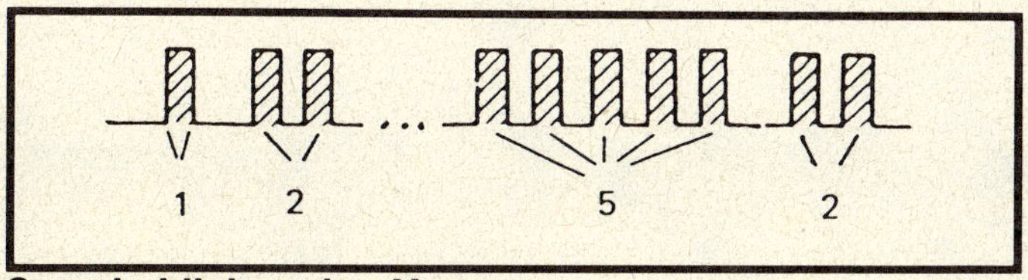

Sample blink code—Yugo

READING CODES

The Motronic control unit will keep in memory any fault codes that have occurred during running. There are 2 methods of retrieving the fault codes from the control unit. The codes can be read using the CHECK ENGINE light flash code or the codes can be retrieved via the Bosch KTS 300 Diagnostic Tester or equivalent.

The blink code numbers do not match the code numbers retrieved by the diagnostic tester. When using the code charts be sure that the correct code form is being used.

Instruction for use of the KTS 300 Diagnostic Tester are included with the device. Equivalent devices will also have instructions for use.

To activate the flash codes, a switch is connected between pin 1 and pin 4 of the diagnostic connector. The 4-pin diagnostic connector is located in the engine compartment, on the left side near the fire wall.

1. Turn the ignition switch to the **START** position. The CHECK ENGINE light should light.
2. Start the engine. If no faults are stored, the light should extinguish during cranking or immediately after starting.
3. If the light extinguishes 3 seconds after the start, there is

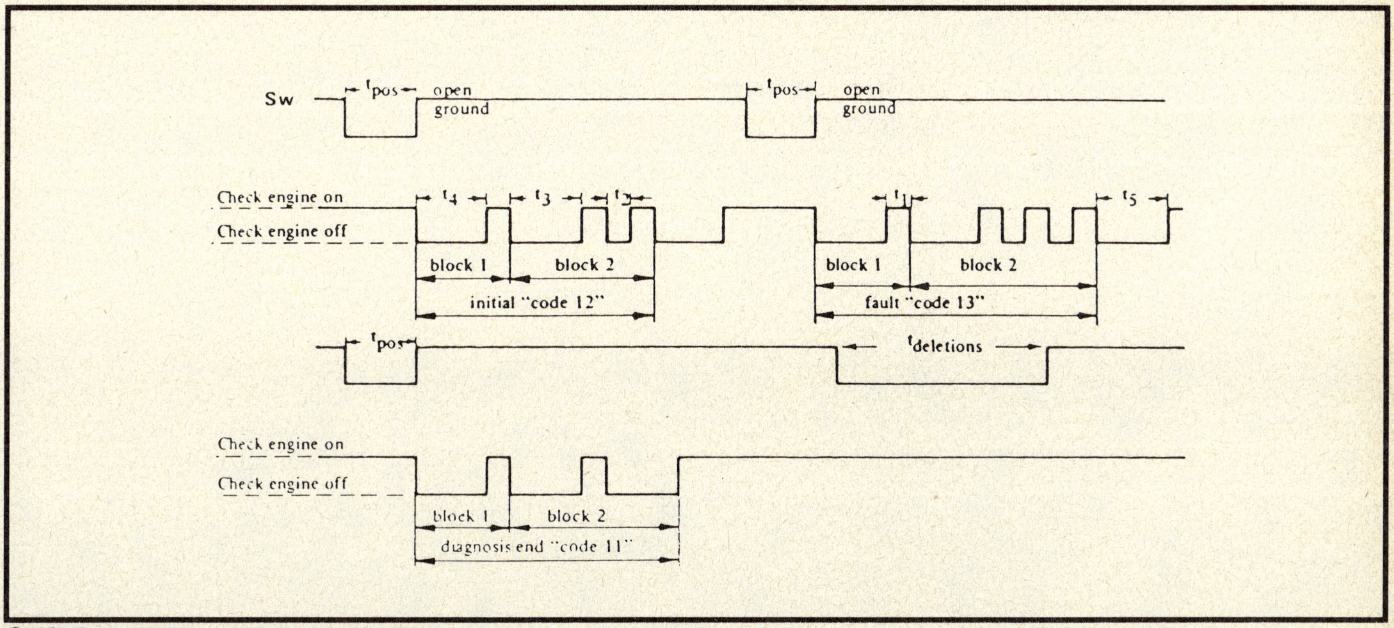

Self-diagnostics sequence—Yugo

at least 1 fault in memory. This is a secondary fault indicating a problem, but it does not affect emission quality.

4. If the light stays on with the engine running, there is a fault that affects emission quality.

5. Turn off the engine. Turn the ignition switch to the **START** position.

6. With the self-diagnostic switch connected between pins 1 and 4 of the selfidiagnostic connector, press the switch for 2.5–10 seconds.

7. The CHECK ENGINE light will extinguish and after 3.6 seconds the CHECK ENGINE light will flash once, then twice. This indicates the Code 12 start code.

8. Press the self diagnostic switch again for 2.5–10 seconds. The next code will flash. Each additional code will need to be activated by pressing the self diagnostic switch. Code 11 is the end code.

CLEARING CODES

1. After end Code 11 is flashed, press the self diagnostic switch for more than 10 seconds. This erases the code memory.

2. The memory can also be erased by disconnecting the negative battery terminal. Any other battery dependant memories, such as radio or alarm system, will be erased, also.

3. When using the KTS 300 Tester, using the ERASE command will erase the code memory.

OUTPUT PERIPHERY TESTING

This test allows the control unit to self-actuate the outputs. Check the outputs by audible means; each output will make a distinctive sound.

1. To activate the output periphery test, a switch is connected between pin 1 and pin 4 of the diagnostic connector. The 4-pin diagnostic connector is located in the engine compartment, on the left side near the fire wall.

2. Press and hold the self diagnostic switch and turn the ignition switch to the **START** position.

3. Keep the diagnostic switch depressed for 3–5 seconds after the ignition switch is turned to the **START** position.

4. The CHECK ENGINE light will extinguish after 2.5 seconds. Code 91 will flash and the injectors will start to operate.

5. Press the self diagnostic switch again for 3–5 seconds. Code 92 will flash and the idle speed regulator will operate.

6. Press the self diagnostic switch again for 3–5 seconds. Code 94 will flash and the tank emission valve will operate.

7. Turn the ignition switch to the **OFF** position.

ENGINE CODES

Code 13—Air temperature sensor—disconnected or short circuit
Code 14—Engine temperature sensor—disconnected or short circuit
Code 21—Throttle sensor—signal too high or too low
Code 22—Idle speed control (first winding)—disconnected or short circuit
Code 22—Idle speed control (second winding)—disconnected or short circuit
Code 33—Vacuum sensor in the intake manifold—vacuum out of specification
Code 41—RPM and TDC voltage signal out of specification
Code 51—Lambda sensor—disconnected or short circuit
Code 52—Lambda sensor—voltage out of specification
Code 53—Battery voltage—too low or too high
Code 54—Defective electronic control unit

Electronic Ignition Systems

4 SECTION

ALFA ROMEO

General Information

Alfa Romeo uses the Bosch Motronic ML 4.1 engine management system which combines electronic ignition and fuel injection functions under 1 Electronic Control Module (ECM). The system depends on signals from sensors located in the most critical points within the supply system in an effort to provide the best engine operation throughout all driving conditions while keeping exhaust emissions low.

SYSTEM OPERATION

Alfa Romeo electronic ignitions are conventional in design. Timing advance functions on both the 4 cylinder and V6 Alfa Romeo engines are controlled by the pre-programmed information in the computer memory in the ECM. The distributor merely distributes the spark through conventional rotor and cap. On 4 cylinder engines, the distributor is mounted to the front cover assembly on the right side of the engine just behind the water

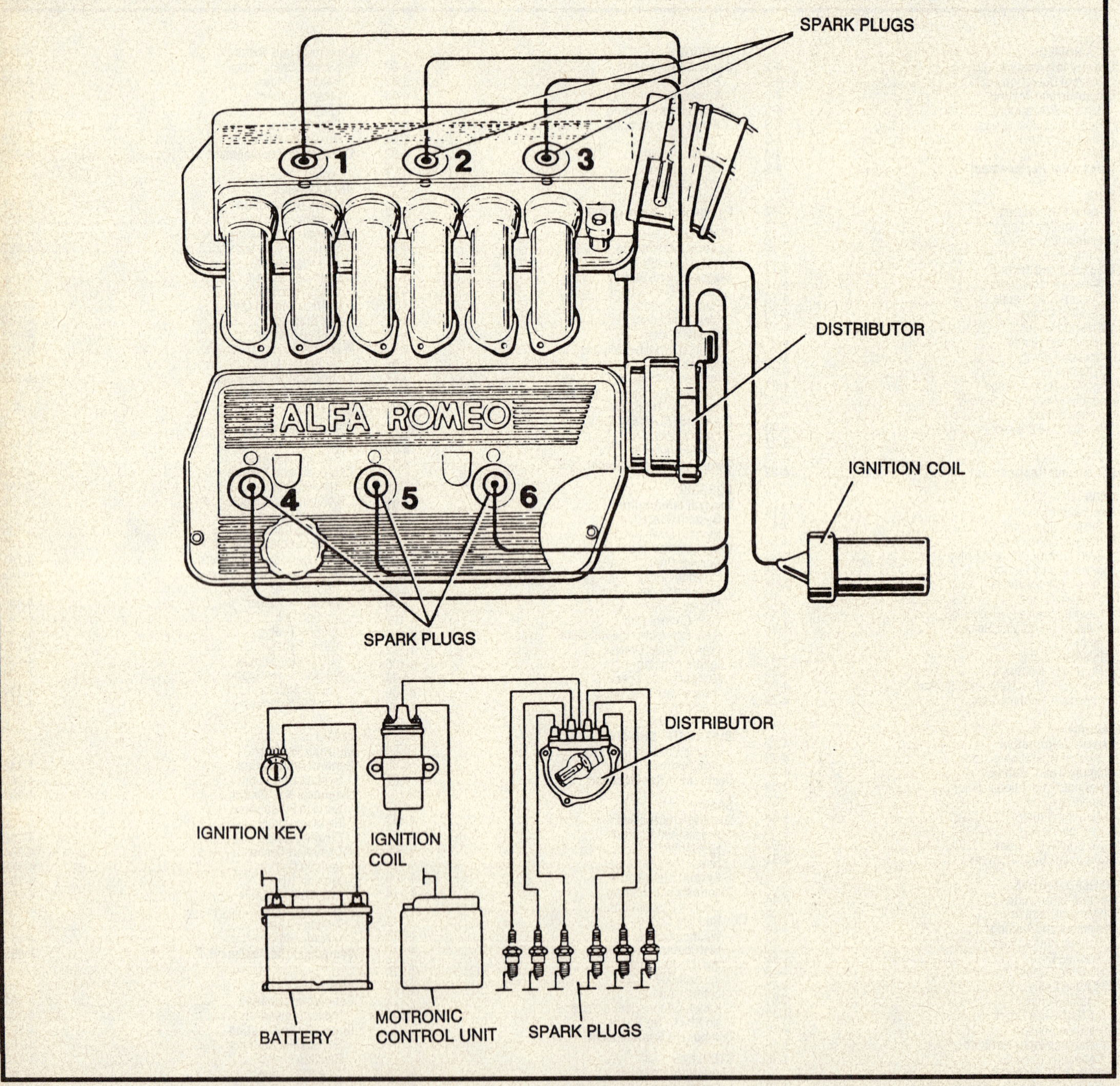

Basic ignition layout. Model 164 V6 engine shown, typical—Alfa Romeo

pump. On V6 engines, the distributor is driven off the rear of the left side camshaft.

The ECM determines engine speed and crankshaft stroke position by means of a crankshaft sensor. The crankshaft sensor is a Hall-effect type switch that is mounted on the front of the engine down near the crankshaft harmonic balancer. A magnetic field is generated and a toothed wheel (trigger wheel) passes through this field. A current is induced which varies with engine speed. One wide gap between the teeth is used to reference the engine stroke. The ECM uses this information to compute ignition timing, firing a conventional design coil.

Diagnosis and Testing

SERVICE PRECAUTIONS

- Use care when working around high-energy electronic ignition systems. High voltage could cause injury to the technician and equipment.
- Before performing any service on the ignition system, make sure the ignition is turned **OFF** and/or the battery negative cable is disconnected.
- Before attempting to remove any parts, turn the ignition switch **OFF** and disconnect the battery ground cable.
- Never connect test equipment such as timing lights, etc., directly to the ignition system for a power source. Always use the vehicle battery for the power source.
- Use only the correct replacement parts. Use of spark plugs, for example, having specifications and sizes different from those specified could cause severe damage and alter emissions levels.
- Use care when working with light alloy parts. Always lubricate the spark plug threads with a suitable lubricant to avoid seizing the plug threads to the cylinder head.
- Keep all ECU parts and harnesses dry during service. Protect the ECU and all solid-state components from rough handling or extremes of temperature.
- Make sure all harness connectors are fastened securely. A poor connection can cause an extremely high voltage surge resulting in damage to integrated circuits.
- Always use a 12 volt battery as a power source.
- Do not disconnect the battery cables with the engine running; never run the engine with the battery cables loose or disconnected.
- Do not attempt to disassemble the ECU unit under any circumstances.
- When connecting or disconnecting pin connectors from the ECU, take care not to bend or break any pin terminals. Check that there are no bends or breaks on the ECU pin terminals before attempting any connections.
- When measuring supply voltage of ECU-controlled components, keep the tester probes separated from each other and from accidental grounding. If the tester probes accidentally make contact with each other during measurement, a short circuit will damage the ECU.
- It is recommended that the ECU control unit be removed if the vehicle is to be exposed to high heat such as a bake oven to cure paint after body refinishing or when welding is be done on the vehicle.
- Use great care when working on or around air bag systems. The system may stay armed for a period of time without battery power. It is recommended that while working on any electrical system in the air bag deployment zone (driver's seat area), disconnect the battery and remove the air bag module from the steering wheel. Then reconnect the battery, if necessary, for electrical diagnosis work.
- Don't be hasty in condemning the ECU when problems arise. On most vehicles, regardless of make, the ECU is located in a relatively protected area (under the seat, behind the dashboard, etc.). But the sensors that feed it information are subjected to vibration, fumes and exist in very harsh underhood environments such as immersed in scalding coolant or in the hot blast of exhaust gases. In addition, since the signal to/from these sensors is a very small voltage or resistance, every electrical connection for these sensors is a potential trouble area. Look for problems in these areas and especially any and all ground connections before blaming the ECU.

READING CODES

The Motronic control unit is equipped with self-diagnostic system allowing a prompt detection of malfunctions related to the emission, fuel or ignition system.

The CHECK ENGINE light, located on the instrument cluster, will only illuminate when a malfunction is emission related.

Trouble Codes are composed of 4 flashing groups, each of which is a code digit. The first is the ECU identification: 1 = Motronic. The second is the method of diagnosis: 2 = malfunction signal, 4 = active diagnosis of actuators. The third and fourth are the specific number of the malfunction.

Engine Codes

Code 1211—Battery voltage problem. ECU reads under 9 volts or over 16 volts from battery. Problem is outside ECU. Perform Battery Failure Test.

Code 1212—Idle throttle switch problem. ECU not receiving plausible signal from MINIMUM switch. Look for problem in MINIMUM switch and/or wiring or ECU. Perform Throttle Valve Switch (Throttle Position Sensor) Test: Minimum Opening.

Code 1213—Full throttle switch problem. ECU not receiving plausible signal from MAXIMUM switch. Look for problem in MAXIMUM switch and/or wiring or ECU. Perform Throttle Valve Switch (Throttle Position Sensor) Test: Maximum Opening.

Code 1214—Engine temperature sensor problem. ECU not receiving plausible signal from temperature sensor and is using substitute value. Look for problem in temperature sensor and/or wiring or ECU. Perform Engine Temperature Sensor Test.

Code 1221—Airflow Meter problem. ECU receiving signal from flowmeter outside plausible range value and is using substitute value. Look for problem in airflow meter and/or wiring or ECU. Perform Airflow Meter Test.

Code 1222—Output stage of ECU for idle speed actuator problem. No plausible signal to actuator. Look for problem in idle air actuator and/or wiring or ECU. Perform Idle Actuator Test.

Code 1223—Oxygen sensor regulation to limit stop. ECU receiving signal at minimum or maximum value of correction limit. Look for problem in air/fuel circuit. Perform Oxygen Sensor Test A.

Code 1224—Oxygen sensor problem. Look for problem in oxygen sensor and/or wiring or ECU. Perform Oxygen Sensor Test B.

Code 1225—Air temperature sensor problem. ECU not receiving plausible signal from airflow meter temperature sensor and is using substitute value. Look for problem in airflow meter temperature sensor and/or wiring or ECU. Perform Air Temperature Sensor Test.

Code 1243—Output stage of ECU for fuel pump relay problem. Engine will have stopped. Look for problem in relay and/or wiring or ECU. Perform Fuel Pump Relay Test.

Code 1244—Output stage of ECU for emission purge canister valve. Look for problem in purge valve and/or wiring or ECU. Perform Evaporative Emission Control Solenoid Test.

Code 1245—Output stage of ECU for variable valve timing, on engines so equipped. Look for problem in variable valve timing relay and/or wiring or ECU. Perform Variable Valve Timing Control Test.

Code 1251—Malfunction in ECU. ECU recognizes problem in RAM/ROM memory. Problem is in ECU.

Code 1265—Output stage of ECU not sending signal to CHECK ENGINE warning light. Look for problem in warning light and/or wiring or ECU.

SPIDER

The diagnosis procedure can occur only with the ignition **ON** and engine speed up to 2000 rpm. The malfunction signal will continue through the entire rpm range once the diagnosis has started.

To initiate the diagnosis process, press the push button located under the rear parcel shelf, near the electronic control unit. The CHECK ENGINE light will provide the malfunction code number by emitting a series of flashes. Each code is endlessly repeated until the next code is requested. Code 4444 is used to signify a no malfunction condition.

164

The diagnosis procedure is divided into 2 groups—display of errors stored by the control unit and monitoring of engine actuators by the control unit.

Display Of Errors Stored By the Control Unit

To initiate the diagnosis process, press the push button located at the bottom of the dash near the steering column for 2.5–5.0 seconds. The CHECK ENGINE light will provide the malfunction code number by emitting a series of flashes. Each code is endlessly repeated until the next code is requested. Code 4444 is used to signify a no malfunction condition.

When all malfunctions have been displayed, the CHECK ENGINE light will display a series of long flashes.

By pressing the push button for 5–10 seconds, the series of codes will be repeated.

MONITORING OF ENGINE ACTUATORS AND DEVICES

To initiate the monitoring process, press and hold the push button, located at the bottom of the dash near the steering column, while turning the ignition **ON**. Hold the push button for 2.5–5.0 seconds after the ignition key is turned **ON**, then release.

A few seconds after the push button is released, the fuel injectors will begin to operate. This will be followed by the idle adjustment actuator and the evaporation solenoid valve.

By pressing the push button for 5–10 seconds, the process will be repeated. The following codes, flashed on the CHECK ENGINE light, will allow the user to determine which test is in progress:

Code 1–4–1–1: Electro-injectors actuation test activated. Continuous control at a frequency of 10 hertz.

Code 1–4–1–2: Idle adjustment actuator activation test. Continuous control for 1 second at idle position and 2 second at maximum speed.

Code 1–4–1–3: Evaporation solenoid valve actuation test. Continuous control at a frequency of 0.5 hertz.

CLEARING CODES

Trouble codes will be cleared if the battery negative terminal is disconnected for 10 seconds or more. In addition, on the 164, codes may be cleared by pressing the push button for 10 seconds or more.

IGNITION TIMING

The Motronic electronic control unit does all the timing changes electronically. Timing is optimized according to battery voltage and engine temperature. The ECM also makes adjustments to avoid knocking and to compensate for the decrease in air quantity at high altitude.

Timing cannot be adjusted manually, nor should timing need to be adjusted. Do not attempt to rotate the distributor or change the firing order of the spark plug wires in the distributor cap.

NO START—NO SPARK

Basic Check

1. Turn ignition key to **ON** or **START**. Check for the following.
2. If nothing happens, look for a problem in the ignition switch.
3. If instrument panel lights illumiate normally, but no noise from starter, check starter motor and starter wiring.
4. If instrument panel lights illuminate normally and starter motor runs normally but the engine does not start, look for an ignition or fuel problem.

Battery Failure Check

If there is a battery failure or a problem in the basic electrical system Code 1–2–1–1 may be set in the on-board diagnostic system.

1. Perform voltage check on battery. Battery should have 11.5–13.0 volts at the terminals. If not, recharge the battery and/or check for possible power leakage such as an automatic switch that won't turn OFF.
2. If battery is okay, check for voltage of between 11.5–13.0 volts between pins 18 (red wire) and pin 5 (black wire, a ground) of the Motronic control unit. If voltage is not okay, repair or replace the wires to the Motronic ECM as necessary.
3. If voltage check in Step 2 is okay, check for 11.5–13.0 volts at pin 35 (red wire) of the ECM. If okay, replace the ECM.
4. If voltage is Step 3 is not okay, check for 11.5–13.0 volts at pin 87 (red wire, also orange and heavy red wire) of the Motronic relay with the diode. If okay, repair the wiring between pin 87 of the relay and pin 35 of the ECM.
5. If the voltage check in Step 4 was not okay, check for 11.5–13.0 volts at pin 86 (pink and blue/red wires) of the Motronic relay.
   a. If okay, go on to the ground check in Step 6.
   b. If not okay, go on to the voltage check Step 7.
6. Perform a ground check by checking for the presence of voltage at pin 85 (black wire, a ground) of the Motronic relay. There should be 0 volts (zero) at pin 85 of the relay since it is a ground. If okay, replace the relay. If not okay, repair the wiring between pin 85 of the relay and ground point.

NOTE: When checking for grounding problems, any resistance reading other than zero ohms is not acceptable. Disassemble the connection to clean any corrosion from the terminals and/or connecting parts. The ground for the Motronic control unit is underhood. It is a multi-wire connection and also provides the ground for the ignition and fuel injection circuits. All ground wires to it are black.

7. If there was no voltage at the relay check in Step 5, perform voltage check by checking for the presence of 11.5–13.0 volts at pin 1 (pink wire) of the round twist-lock multi-wire engine wiring connector. If there is battery voltage here, repair the wiring between pin 1 (pink wire) of the round engine wiring connector and pin 86 (pink wire) of the Motronic relay with diode, through pin 30 (pink wire) of the fuel pump relay.
8. If there was no voltage found in Step 7, check the fuses in the fusebox. Fuse 22 feeds the system. If it is blown, replace it (15 amp). If the fuse is good, check the wiring between pin 1 (pink wire) of the round engine wiring connector and pin 1 (pink wire) of the fuse box.
9. If all of the above steps do not turn up the problem, check for correct operation of the anti-theft system. Activation of the anti-theft system locks the engine ignition circuit.

Ignition Components Check

1. Check that the spark plugs are not fouled. Clean and replace as required.

2. Remove and ground a spark plug, keep connected to high voltage spark plug cable and crank engine. Check for spark. Inspect the high voltage spark plug cables including from coil to distributor and from distributor to spark plugs. Check signs of damage or burn-through. Check the spark plug cable resistance. It should be approximately 2500 ohms. Check the engine ground straps and clean any corrosion that's found.

3. Inspect the inside of the distributor cap and rotor. Clean and replace parts as required. The distributor cap center carbon brush should have approximately 1100 ohms resistance. Clean or replace affected components.

4. Check the ignition coil power supply and primary and secondary resistance. Primary resistance should be 0.45–0.55 ohms. Secondary resistance should be 5400–6600 ohms. Replace coil if faulty.

5. Check the rpm and stroke sensor. Check the connection between the sensor and pin 23 (black wire) of the ECU and between the sensor and pin 25 (yellow wire) of the ECU. Check the air gap between the sensor and the crankshaft mounted trigger wheel. The air gap should be 0.020–0.040 in. Repair wiring and set sensor gap as required.

NOTE: If the engine still will not start after performing these checks and services, a fuel injection problem should be suspected. Troubleshoot fuel system.

Component Replacement

IGNITION COIL

The ignition coil is of conventional design and produces a high voltage capable of providing a spark of duration and intensity required to fire the spark plugs. It is controlled by the ECU.

On most vehicles, the coil is located inside the engine compartment on a bracket near the radiator. Loosen the coil bracket to remove the coil.

When inspecting the coil, pay attention to the coil tower. Carbon tracking and flashover can destroy the coil tower's insulating ability, sending the high voltage right to ground, yet the damage can be hidden under the high tension wire's boot. This can be the cause of a no-start.

CRANKSHAFT SENSOR

Information on engine speed is supplied to the ECU by the Engine RPM and Stroke Sensor. It is a Hall-effect type switch that detects the engine revolutions and stroke. The sensor is mounted on the front of the engine down near the crankshaft harmonic balancer. A magnetic field is generated and a toothed wheel (trigger wheel) passes through this field. A current is induced which varies with engine speed. One wide gap between the teeth is used to reference the engine stroke.

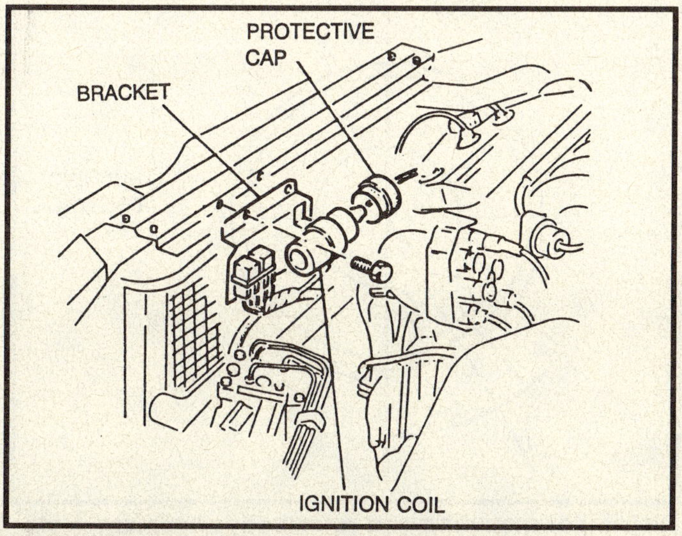

Typical ignition coil mounting near radiator mounting—Alfa Romeo

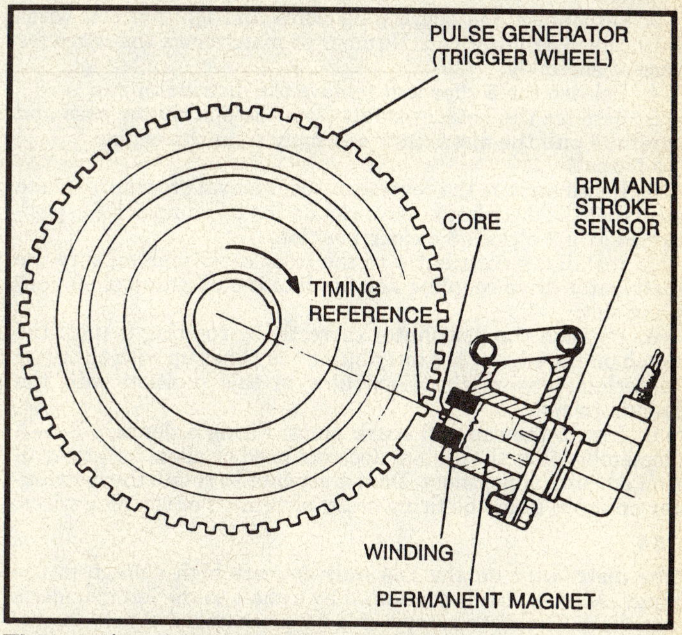

The rpm/stroke sensor is mounted near the crankshaft trigger wheel—Alfa Romeo

When checking the crankshaft sensor, check the gap between the sensor and top of the teeth of the pulley.

Removal and Installation

1. Raise and safely support vehicle.
2. On 164 models, remove the right front wheel.
3. On 164 models, remove the right front inner fender panel, if necessary.
4. Remove the protective covers. On the lower cover on the 164 model, loosen the belt tension pulley of the water pump and air conditioning compressor belt.
5. Remove the crankshaft sensor retaining bolt, disconnect the wire connector and remove from vehicle. Unless necessary, do not remove the sensor mounting bracket.

To install:

6. Installation is the reverse of the removal procedure. Use care when handling the sensor. Remove any protective covering on the end of the sensor.
7. The gap between the trigger wheel and crankshaft sensor must be set. Use a feeler gauge to set the gap to 0.020–0.060 in. on Spiders and 0.020–0.040 in. on 164 models.
8. Reinstall the protective covers, inner fender panel and right front wheel as necessary.

DISTRIBUTOR ASSEMBLY

Removal and Installation

SPIDER

1. Make sure that No. 1 cylinder is on the compression stroke and both valves are closed. Do this by rotating the crankshaft pulley until the notch on the pulley lines up with the reference pointer (pin) mounted on the water pump.
2. Remove the protective covers from the distributor.

3. Disconnect the spark plug wires and ignition coil wires from the distributor cap. Number or matchmark the wires for ease of assembly.
4. Release the 2 clips and remove the distributor cap.
5. Remove the holddown nut from the distributor base and carefully pull the distributor assembly from the engine.

To install:

6. Installation is the reverse of the removal procedure. Make sure that the distributor drive slot on the oil pump drive is positioned at a 2 o'clock/8 o'clock position.
7. Install the distributor to the front cover making sure the distributor drive coupling key fits into the oil pump shaft coupling slot.
8. Position the distributor correctly by rotating it until the notch on the edge of the distributor body lines up with the rotor centerline. Fasten the distributor in this position with the holddown nut.
9. Install the cap and spark plugs. Firing order is 1–3–4–2. Remember that the system does not need or allow for any ignition advance adjustment. Do not attempt to rotate the distributor or try to alter the firing order placement of the plug wires.

164

The distributor on the 164 only delivers high voltage to the plugs. A rotor arm is connected by a shaft to the left cylinder's camshaft. The distributor cap is a 2-piece unit designed to protect the rotor arm and help isolate the radio from ignition electrical noise.

1. Remove the distributor cap cover.
2. Remove the rotor and the bowl shaped lower distributor cap.
3. Remove the 2 screws and remove the distributor body from the cylinder head.
4. If necessary, the distributor body may be disassembled by removing the pin from the distributor drive and removing the components from the distributor stub shaft.

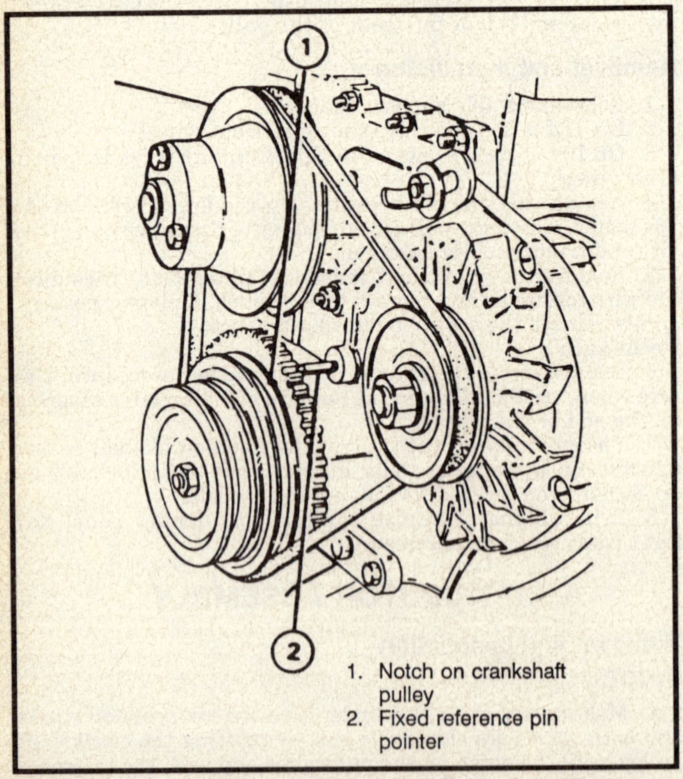

Setting up 4-cylinder engine to TDC No. 1 cylinder—Alfa Romeo Spider

To install:

5. Installation is the reverse of the removal procedure. Assemble any removed parts onto the stub shaft, paying attention to proper assembly of the seals. Lubricate the seals with a suitable grease.
6. Before installing the distributor body to the cylinder head, take note of the slot in the drive. It should be horizontal in the lower half of the drive.

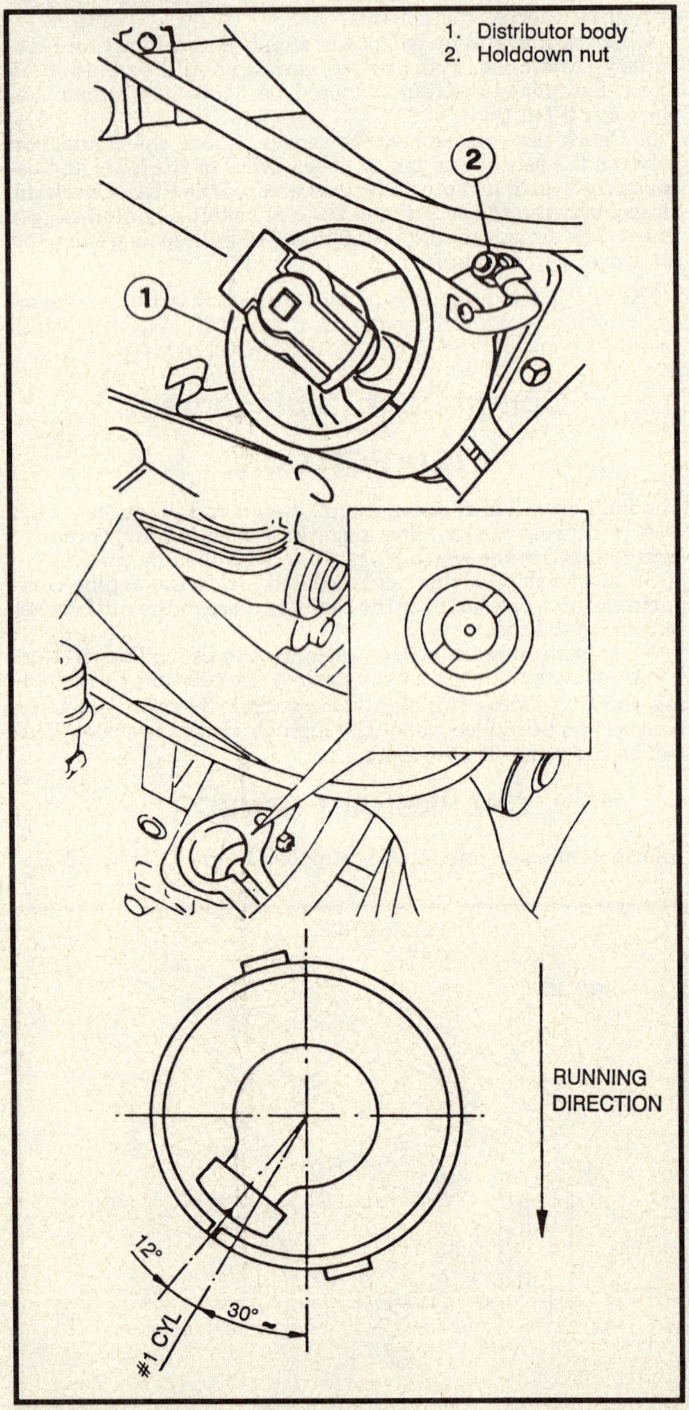

Distributor mounting and rotor/body alignment. Note distributor drive notch alignment at 2 o'clock/8 o'clock—Alfa Romeo Spider

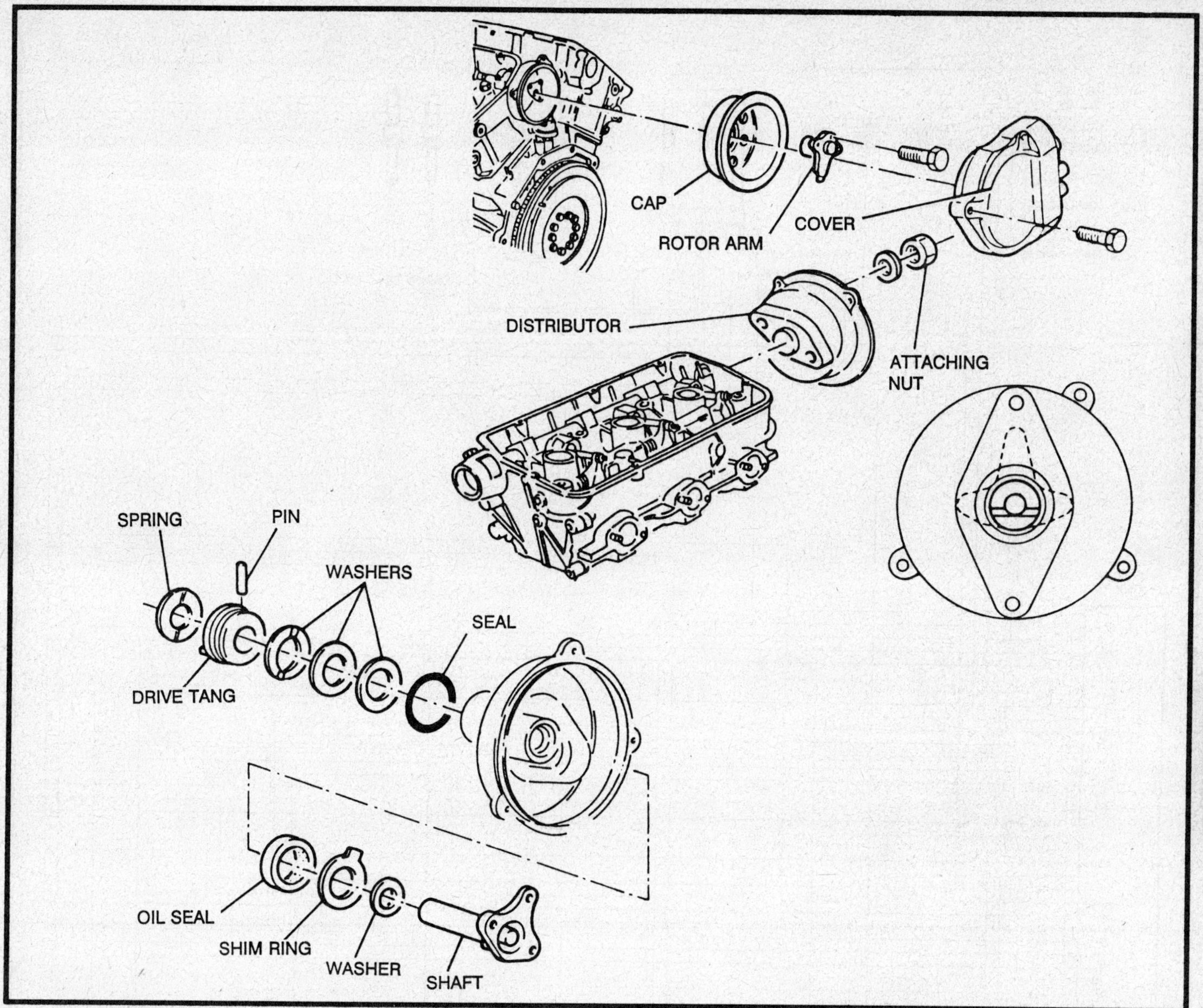

Distributor mounting, components and alignment. Note distributor drive notch alignment: horizontal in the lower half of the drive—Alfa Romeo 164

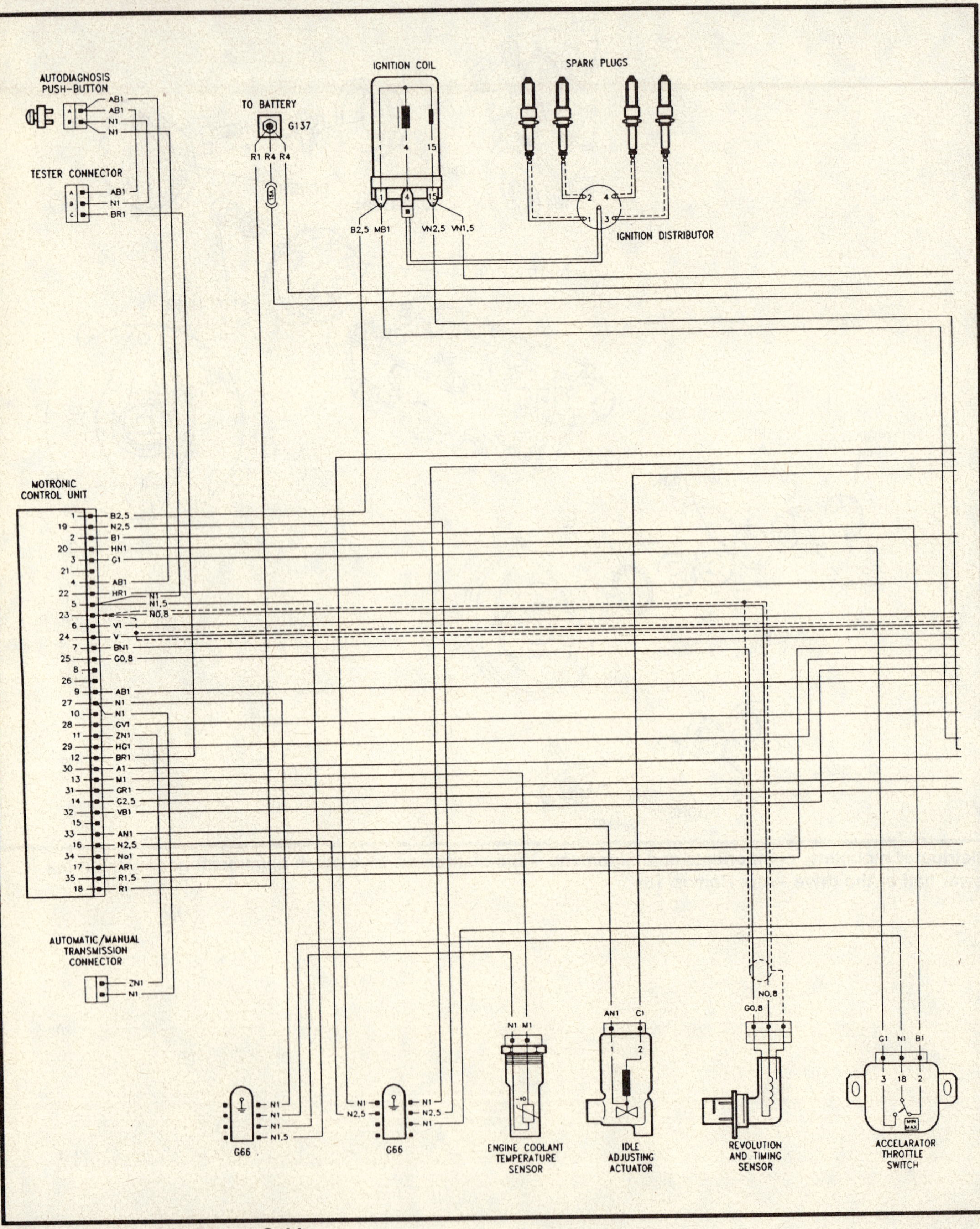

Wiring diagram—Alfa Romeo Spider

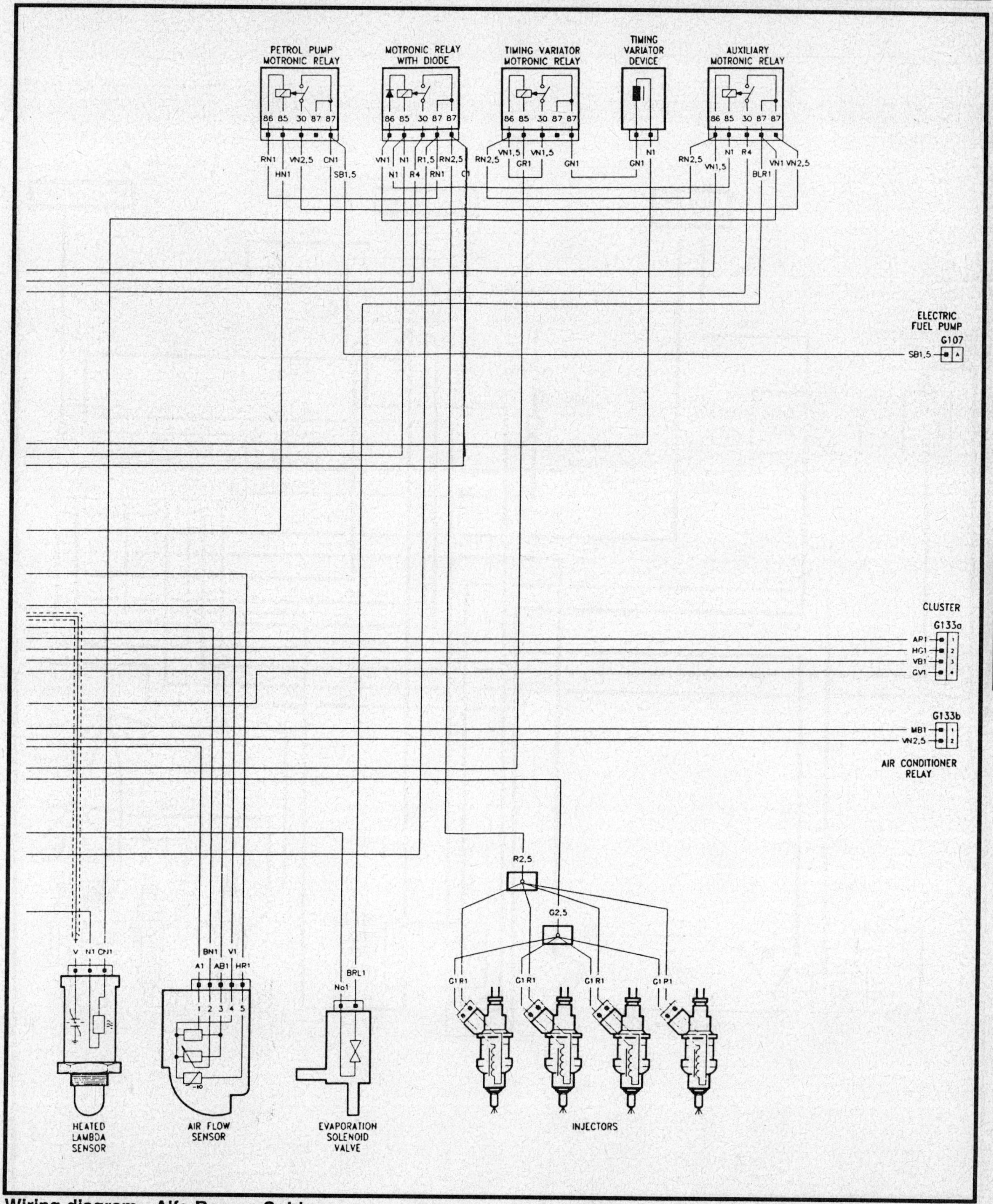

Wiring diagram—Alfa Romeo Spider

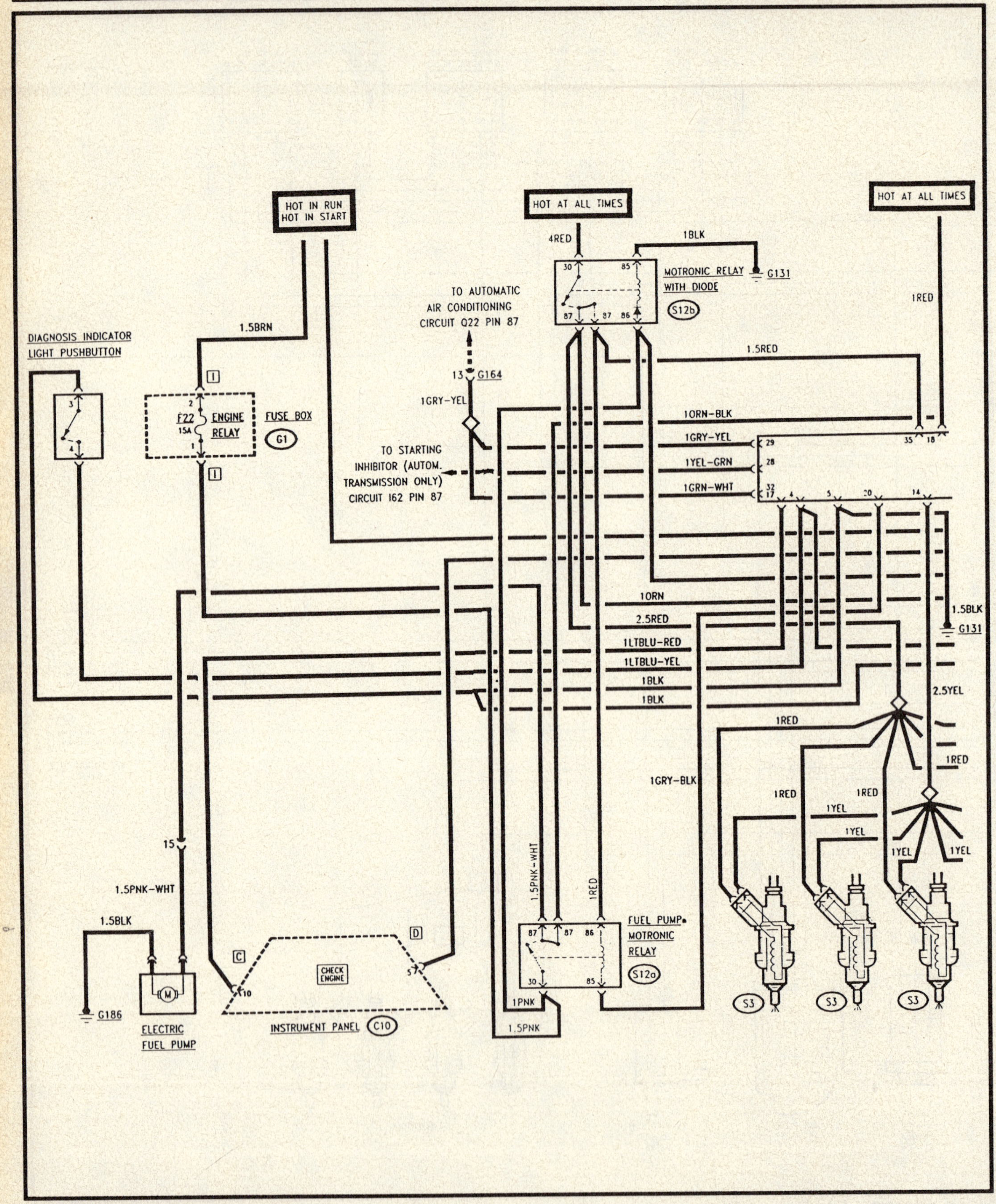

Wiring diagram—Alfa Romeo 164

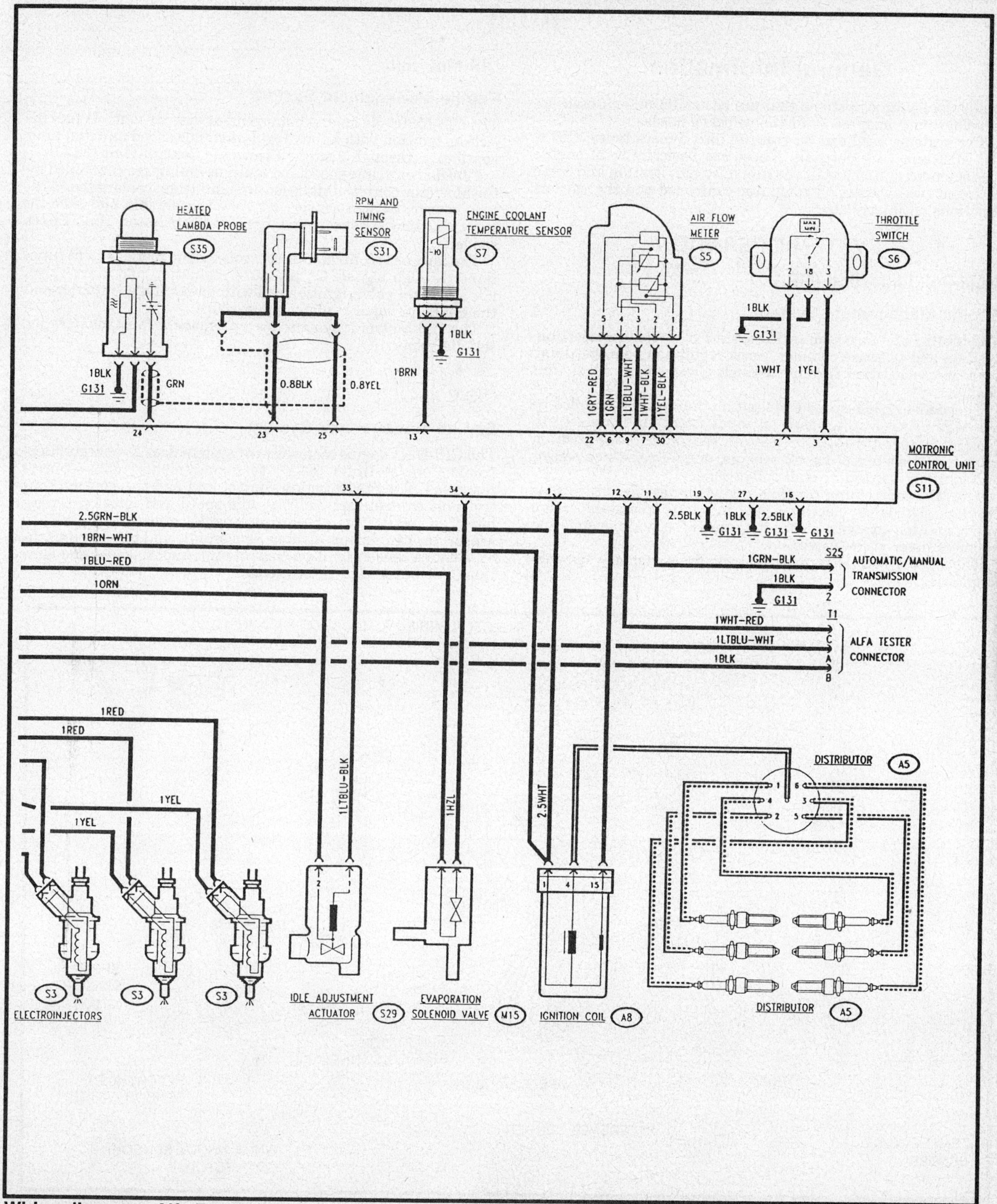

Wiring diagram—Alfa Romeo 164

AUDI

General Information

Audis use ignition systems that are either Bosch or Bosch derived systems, adapted to fit the vehicle's needs.

The systems used can be grouped into 3 main types: CIS-E III, MPI and CIS-Motronic. These are basically total engine management systems with the fuel injection, ignition and some emission sub-systems all computer controlled and the various functions all interrelated.

SYSTEM OPERATION

Multi-Point Injection (MPI)

Engine Management System

The Multi-Point Injection (MPI) system controls fuel injection, ignition and selective cylinder knock regulation, idle speed and fuel tank ventilation through a single Electronic Control Unit (ECU).

Fuel delivery, idle speed and ignition timing are controlled by the MPI control unit, based on input from: air mass sensor, coolant temperature sensor, oxygen sensor, throttle position sensor, idle switch, 2 knock sensors, engine reference sensor and ignition distributor Hall Effect sender.

The MPI control unit controls engine management functions through signals to: electronic fuel injectors, idle stabilization valve, ignition coil with power stage amplifier and the emission vapor canister purge valves.

The MPI system has a permanent fault memory for self-diagnosis.

CIS-Motronic

Engine Management System

The CIS-Motronic engine management system controls fuel injection, ignition with knock regulation, idle speed and fuel tank ventilation through a single Electronic Control Unit (ECU).

Fuel delivery, idle speed and ignition timing are controlled by the Motronic control unit based on input from: coolant temperature sensor, oxygen sensor, air flow sensor, idle and throttle switches, knock sensor and ignition distributor Hall Effect sender.

The Motronic control unit controls engine management functions through signals to: pressure regulator, cold start valve, idle stabilizer valve, ignition coil with power stage amplifier and the emission vapor canister purge valve.

The CIS-Motronic system has a permanent fault memory for self-diagnosis.

CIS-E III

Engine Management System

The CIS-E III engine management system uses 2 separate Electronic Control Units (ECU) to management engine control functions. The fuel injection control unit and the ignition control unit control fuel delivery, idle speed and ignition timing bases on input from these components: coolant temperature sensor, oxygen sensor, air flow sensor, idle and throttle switches, altitude sensor, ignition control unit, knock sensor and ignition distributor Hall Effect sender.

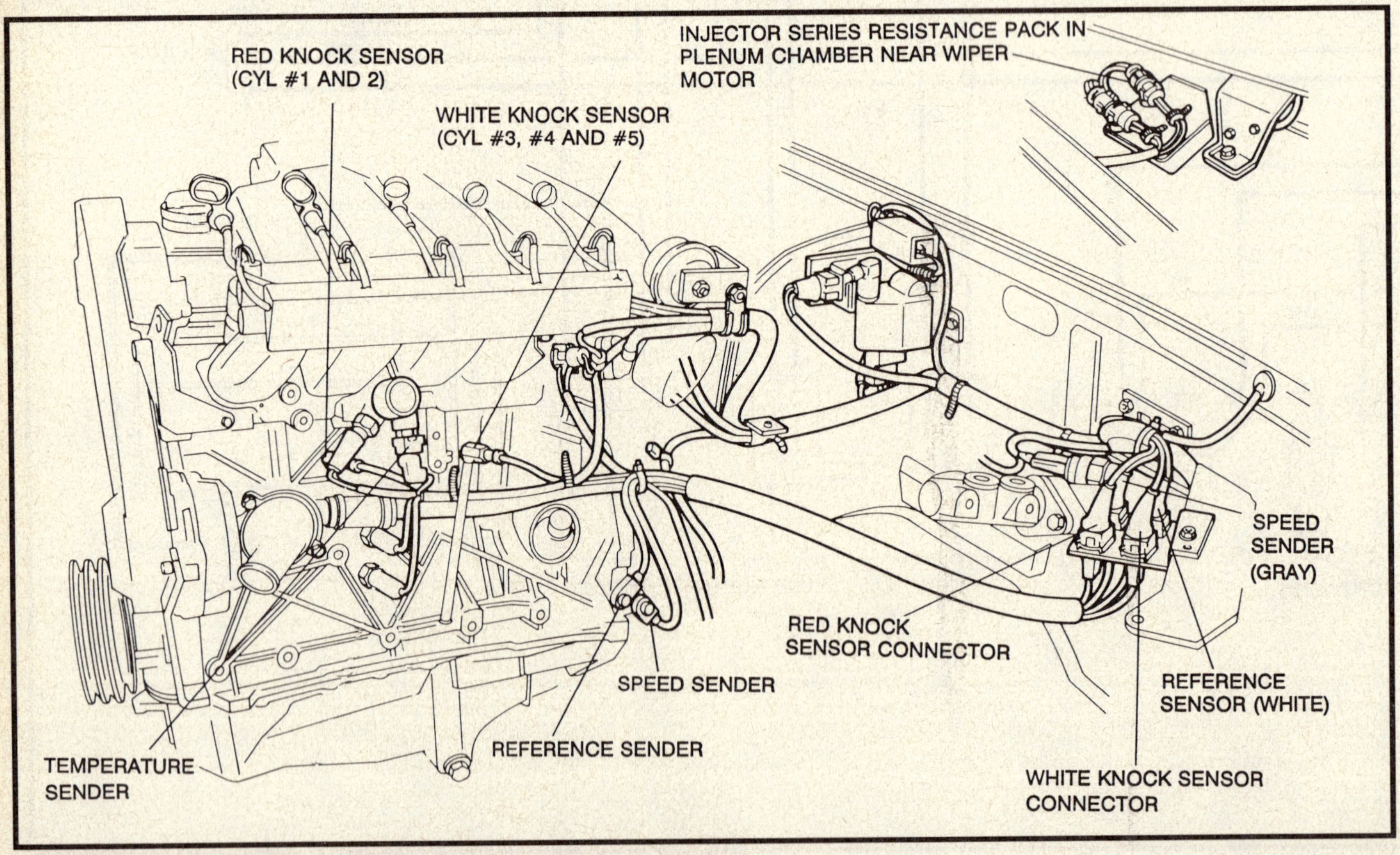

MPI component system locations—Audi 5 cylinder engine

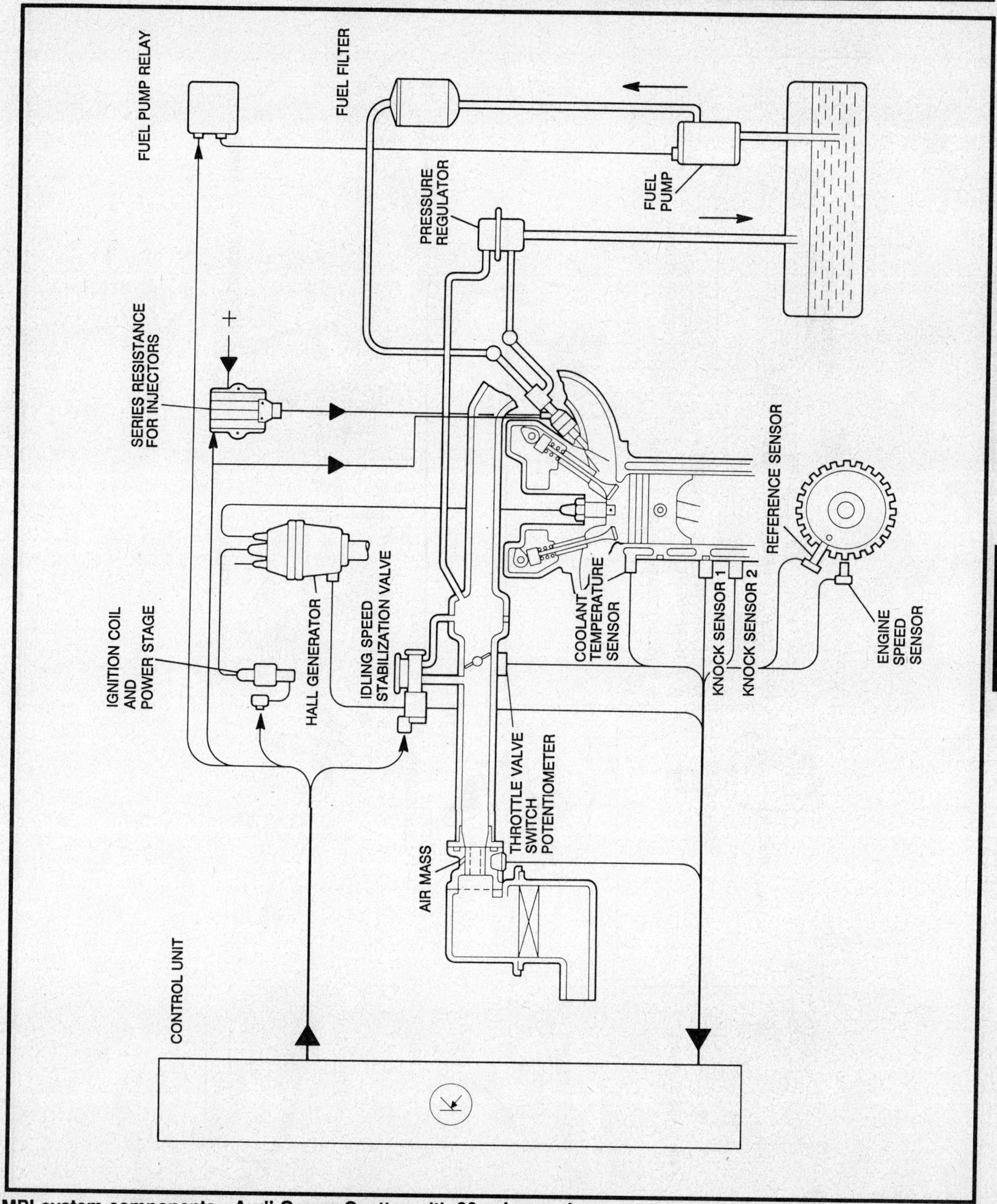

MPI system components—Audi Coupe Quattro with 20 valve engine

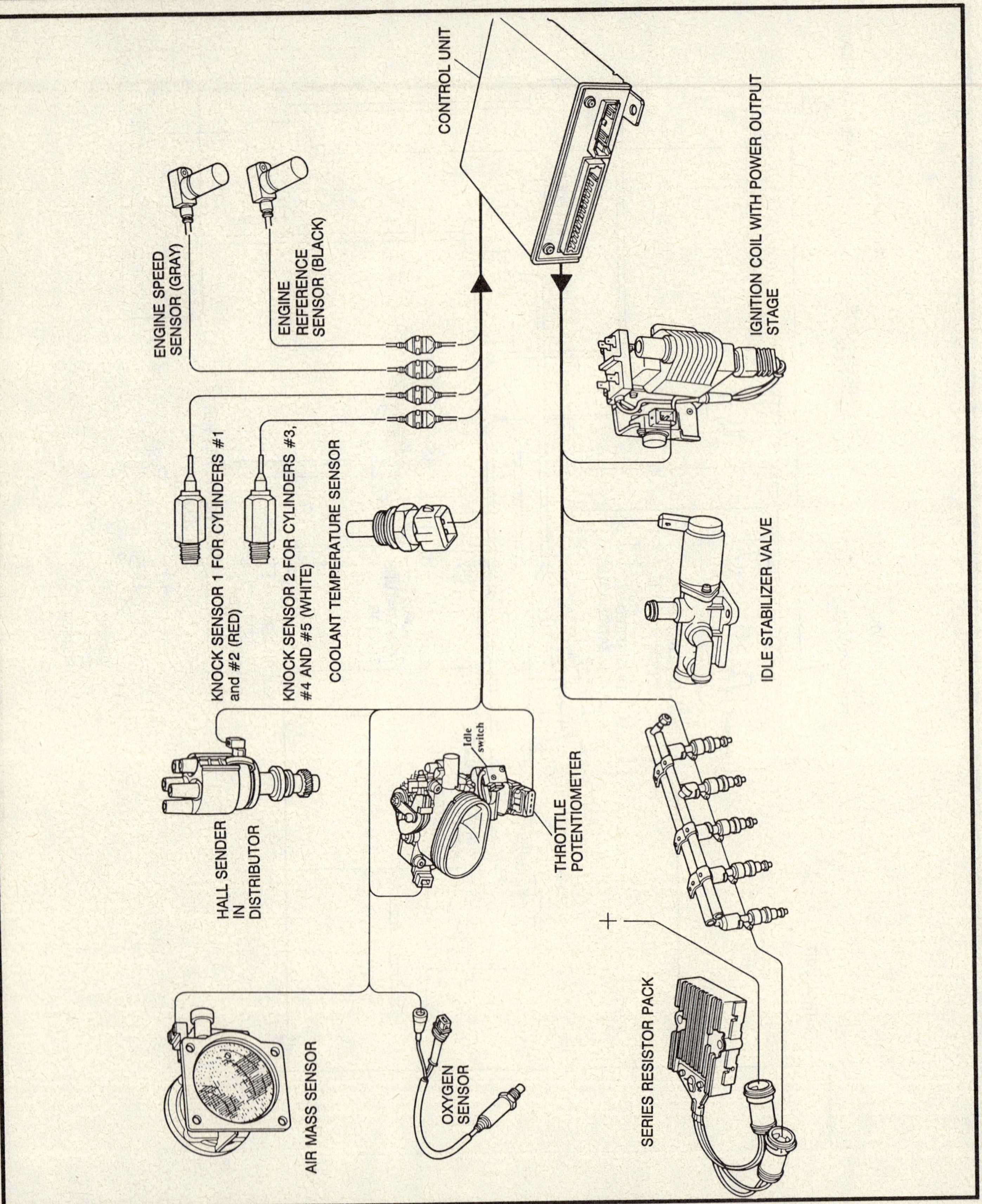

MPI system pictorial schematic—Audi Coupe Quattro with 20 valve engine shown

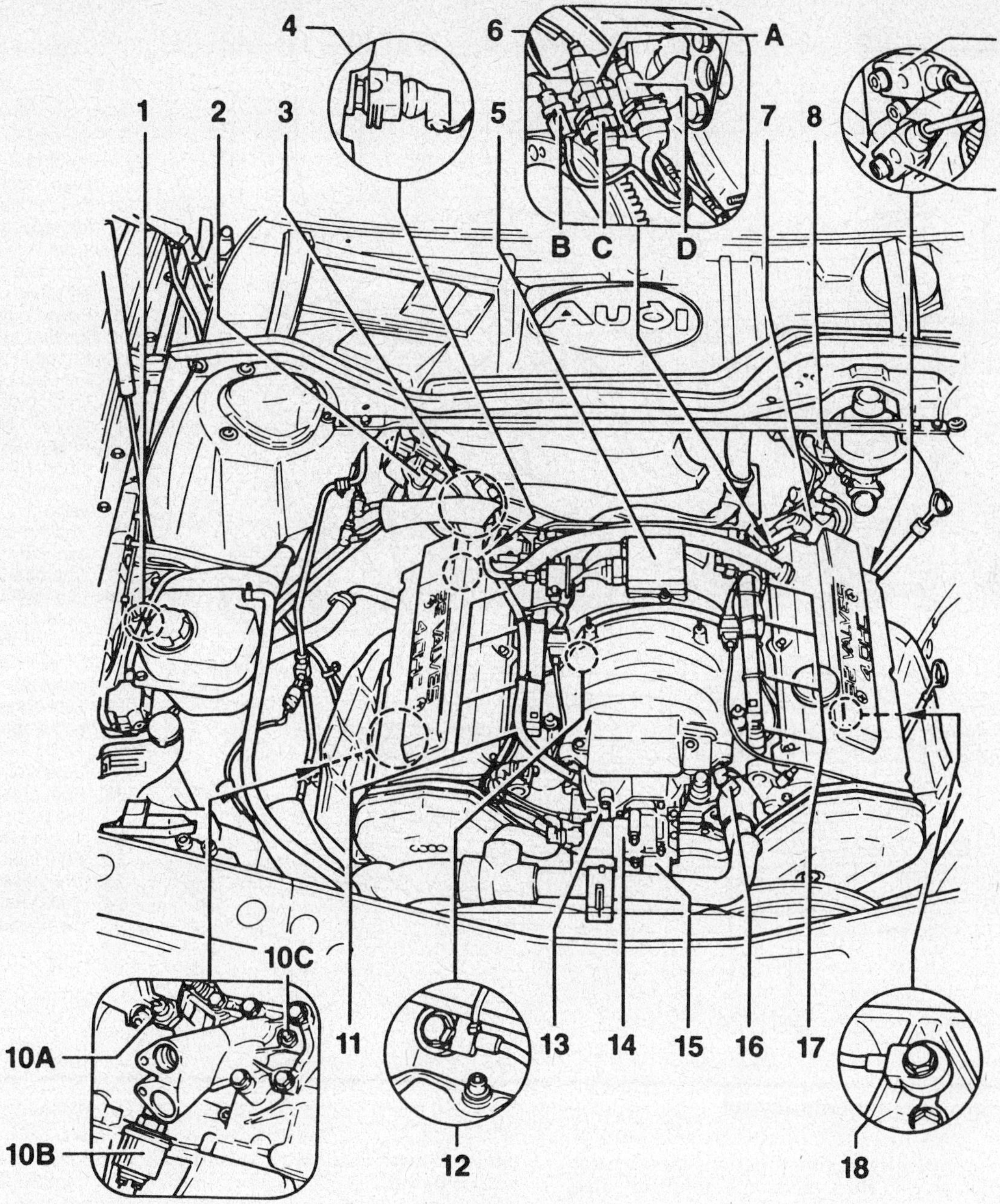

1. Carbon canister solenoid valve
2. Ignition coil (right)
3. Distributor (right)
4. Coolant temperature sender
5. Air mass sensor
6. Harness connector for:
6a. Engine speed sender
6b. Oxygen sensor
6c. Oxygen sensor heater
6d. Ignition reference sender
7. Distributor (left)
8. Ignition coil (left)
9a. Ignition reference sender
9b. Engine speed sender
10a. Oil pressure switch
10b. Dynamic oil pressure switch and sender for oil pressure gauge
10c. Oil temperature sender
11. Fuel injectors, cylinders 1–4
12. Knock sensor (right)
13. Intake air temperature sender
14. Idle stabilizer valve
15. Throttle valve potentiometer
16. Coolant temperature display sender
17. Fuel injectors, cylinders 5–8
18. Knock sensor (left)

CIS—Motronic ignition system—V8 Quattro components layout

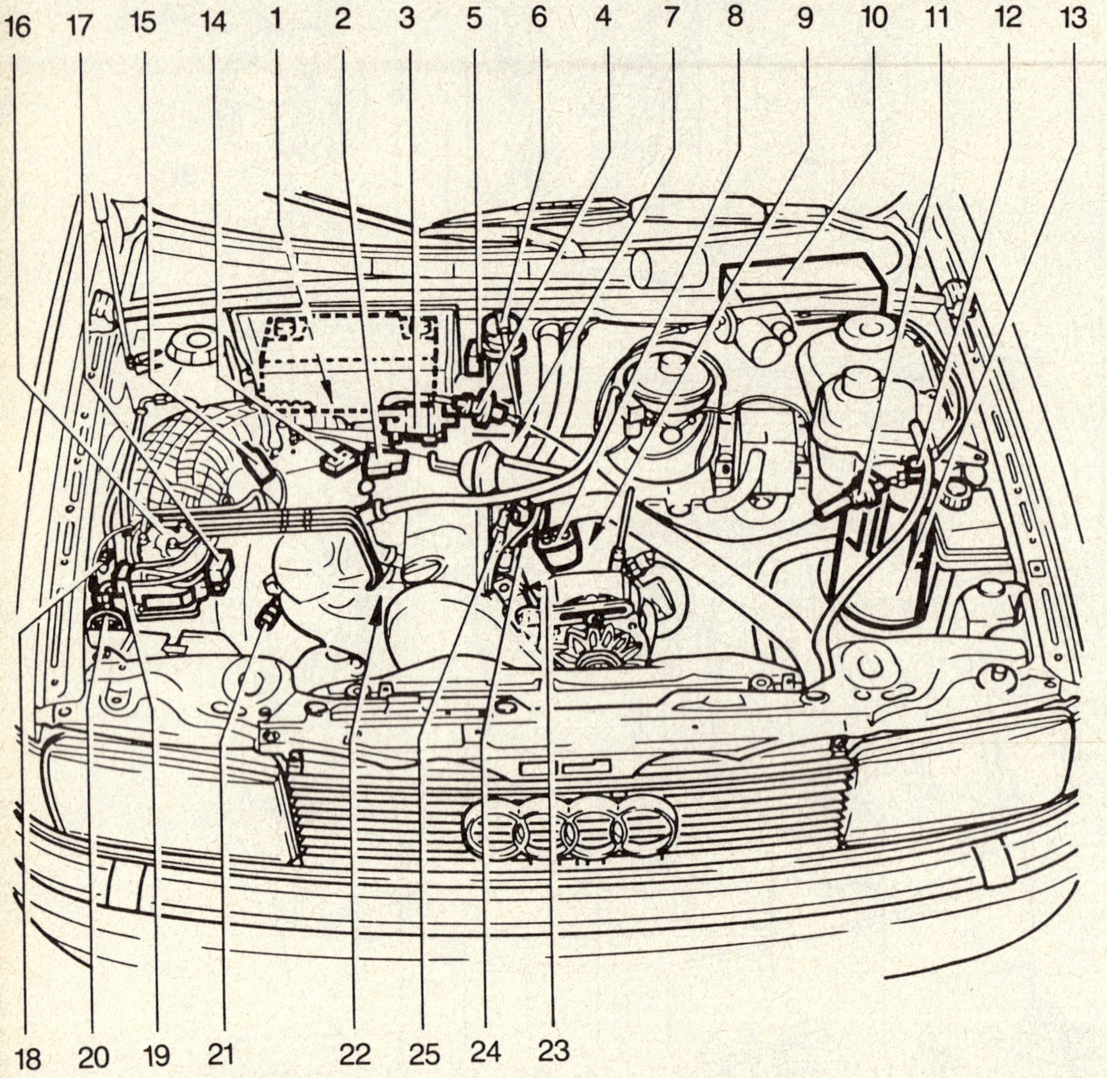

1. Motronic control unit (behind A/C evaporator)
2. Full throttle switch
3. Idle speed stabilizer valve
4. Charcoal canister on/off valve
5. Power output stage
6. Ignition coil
7. Knock sensor harness connector
8. Ignition distributor
9. Oxygen sensor
10. Fuse/relay panel
11. Charcoal canister solenoid (frequency) valve
12. Sealing cap on canister
13. Charcoal canister
14. Idle switch
15. Cold start valve
16. Fuel distributor
17. Differential pressure regulator
18. Test connector for self-diagnosis
19. Air flow sensor potentiometer
20. Fuel pressure regulator
21. CO measuring tube
22. Fuel injection valve
23. Knock sensor
24. Temperature sensor
25. Spark plugs

CIS-Motronic system components layout

The CIS-E III system manages engine functions through signals to: pressure regulator, cold start valve, idle stabilizer valve, ignition coil with power stage amplifier and the emission vapor canister purge valve.

The CIS-E III system has an expanded fault memory for self-diagnosis.

IGNITION CONTROL

On Motronic and MPI systems, ignition is controlled by the engine management system Electronic Control Unit (ECU). Ignition timing is not adjustable. A TDC mark on the crankshaft pulley or on the flywheel is provided for setting valve timing, but no other timing marks are provided. While these ignition systems may share some common parts, there are significant differences.

On the 2.3L engine with the CIS-E III system, the ignition system is operated with a separate control unit. The system includes a distributor with a Hall Effect sender, an ignition coil with a power output stage or amplifier, a coolant temperature sensor and a single knock sensor. The control unit also operates the engine warning light on the instrument panel. Ignition timing is adjustable on this system but the control functions must be by-passed first.

The 2.2L turbocharged engine also uses a distributor with a Hall Effect sender inside. The 2.8L engine, introduced in 1992, is equipped with distributorless ignition, using 2 crankshaft position sensors and a Hall Effect sender. The 4.2L engine, also introduced in 1992, is a derivative of the dual distributor 3.6L V8 used in previous years. It also uses 2 distributors, one for each bank. A single Hall Effect sender is inside the right bank distributor, but the ECU also reads a crankshaft reference sensor to calculate ignition functions.

All models are equipped with engine knock sensors. If knock is detected, the ECU will retard ignition timing to selected cylinders, instead of retarding timing to all cylinders as in previous systems. On turbocharged engines, engine knock will also cause the ECU to reduce boost pressure.

COMPONENT DESCRIPTION

Hall Effect Sender

All models use a Hall Effect sender that consists of a permanent magnet, a semi-conductor chip and rotating vane. When the air gap between the magnet and the chip is open, the magnet causes a voltage to be generated in the chip. When the rotor vane moves into the gap, it changes the magnetic field and the voltage signal is no longer generated. On earlier models, the shape and size of the rotor vanes and windows determined the dwell angle and to some extent, ignition timing. On this version of the system, dwell and timing are fully controlled by the ECU. When the engine is being started, the Hall Effect signal is used to establish a reference point for start of ignition and fuel injection. After the engine is running, other sensors are also used to provide engine speed and crankshaft position signals.

On the 2.2L and 2.3L engines, the Hall Effect sender is in the distributor. On the 4.2L engine, the sender is in the right bank distributor. On the 2.8L engine which uses a distributorless ignition system, the sender is in a housing on the rear of the left cylinder head.

Ignition Reference Sensor

MOTRONIC AND MPI SYSTEMS

This sensor is a magnetic pick-up that senses the position of a pin on the flywheel. As the pin passes the sensor, a magnetic pulse is generated and sent to the ECU. This reference point is used to begin the ignition and fuel injection sequence for that revolution of the crankshaft. If the reference sensor fails while the engine is running, a preset reference point in the ECU memory is used and the ECU will shift into limp-home mode. The engine will not start with a faulty reference sensor.

RPM Sensor

MOTRONIC AND MPI SYSTEMS

This sensor is a magnetic pick-up that reads the passing of the ring gear teeth on the flywheel. As a tooth passes the sensor, a magnetic pulse is generated and sent to the ECU. The time between pulses is used to calculate engine speed.

Knock Sensor

All models use engine knock sensors to calculate ignition timing. Vibrations in the engine cause the quartz crystal in the sensor to generate a voltage. The ECU monitors this signal and can determine if it indicates detonation. Ignition timing can then be adjusted for each cylinder individually. Placement and bolt torque of the knock sensor is critical to proper operation of the sensor. If replacing the knock sensor, be sure to use the correct part, install it in exactly the same location and route the wiring the same way. Always torque the bolt to the proper specification; 15 ft. lbs. (20 Nm) on 2.2L and 4.2L engines or 7 ft. lbs. (10 Nm) on 2.8L engine.

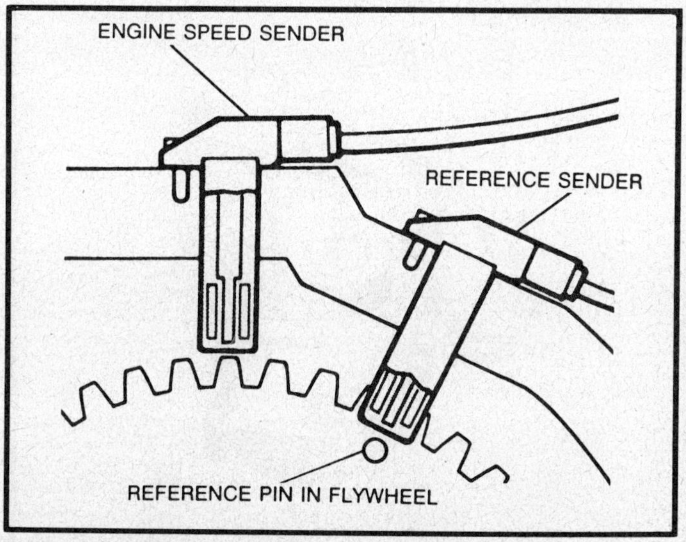

Reference and engine speed senders—typical

Distributorless Ignition

This system is used on the 2.8L engine. It uses most of the same components as the other engines except for the distributor itself. A Hall Effect sender is mounted in a housing on the rear of the left cylinder head. The reference sensor, rpm sensor and knock sensor are the same as used on other models. Two knock sensors are used, one on each side of the engine block. The major difference in the ignition system on this engine is the ignition coil pack in front of the intake manifold and the ignition coil power output stage (amplifier) on the firewall.

Ignition is controlled by the engine management system ECU, which is under the carpet in the passenger side foot well. Ignition timing is not adjustable. A TDC mark on the crankshaft pulley or on the flywheel is provided for setting valve timing, but no other timing marks are provided.

COMPONENT DESCRIPTION

Ignition Coil Power Output Stage

This unit is a solid state switch used to turn the coil primary circuit on and off. The ECU operates the output stage, which then operates the coils so the ECU does not have to handle the high voltages and currents required for the coils. It is not possible to directly test the power output stage, but the signals and wiring to and from the output stage can be tested. If the engine will not start because it has no ignition, start with Ignition Coil Power Output Test.

Ignition Coils

The pack consists of 3 coils in front of the intake manifold. Each coil fires 2 spark plugs at the same time, with the "waste" spark occurring during the companion cylinder's exhaust stroke. Each coil is supplied with 12 volts any time the ignition is ON. The power output stage turns the coil primary circuit on and off by completing the circuit.

Hall Effect Sender

The Hall sender is on the back of the left cylinder head and driven by the camshaft. Its function is the same as the unit on the 4.2L engine.

Diagnosis and Testing

SERVICE PRECAUTIONS

- Always turn the ignition **OFF** when connecting or disconnecting tester leads to the ignition system
- Most of these vehicles are equipped with anti-theft radios that will no longer work if the power to them is ever cut off. Special codes must be used to reset the radio. Always obtain these codes from the customer before disconnecting the battery cables
- Never disconnect the battery with the engine running
- Do not fast charge the battery for more than 15 seconds, during emergency starting
- Do not exceed 16.5 volts charging rate
- Always disconnect the battery when using arc, spot or electrical welding equipment
- Never apply voltage to the control unit to simulate the output signals

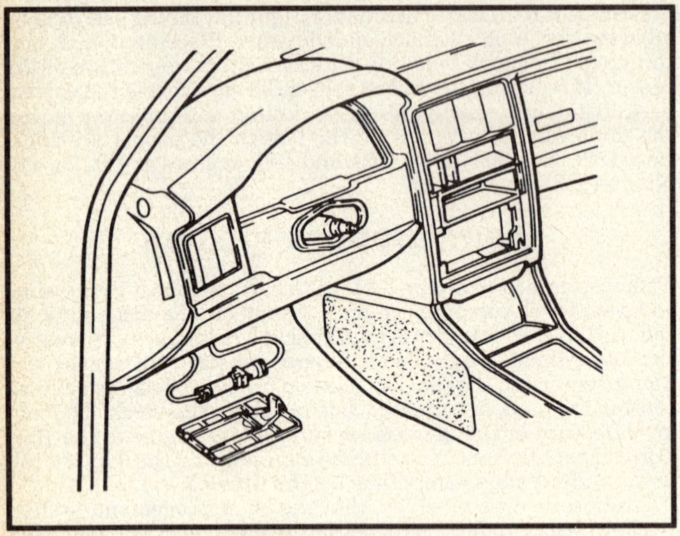

Red single pole air bag connector under instrument panel—Audi 80, 90, Quattro shown

- Always disconnect the distributor high tension lead and jumper it to ground, when service on the ignition system requires cranking the engine
- If the engine must be cranked without spark, disconnect both ignition coil power output stage connectors (V8 Quattro)
- Use care when working around vehicles equipped with a Supplementary Restraint System (SRS), or air bag. On Audi vehicles, look for a connector usually in either the instrument panel harness or air bag harness. This connector should be either red or yellow. This is the power connector to the air bag and should be disconnected when performing service in the air bag deployment area.

Ignition Service Notes

Audi service bulletins provide the following recommendations:

Long cranking times, hot or cold, intermittent stumble and misfiring could be caused by a high resistance at the ground stud. This is caused by the sealing compound (Loctite) used at assembly on the screw-in stud. Remove the ground stud from the intake manifold, clean well, including the internal and external threads. Reinstall stud and ground wire. Always check for poor ground connections. Be especially careful of the ground cable connection at the intake manifold when troubleshooting for intermittent stumble and misfiring.

Fault Code 2132 either with or without a driveabilty complaint could be the result of an incorrect fuel and/or ignition control unit inadvertently installed in the vehicle. Check with the dealer for the correct ignition and fuel injection ECU part numbers.

Squealing/whining noise from the engine upon initial cold start could be caused by insufficient lubrication of the distributor bearing. The factory suggests replacing the entire distributor assembly.

Vehicle no-start, hot or cold, when the weather is below about 45°F could be a problem with the signal from the engine management ECU to the cold start valve. Check for signal to the cold start valve. If no signal, the ECU will have to be replaced. Note that when working on the ECU do not disconnect the ECU from the vehicle wiring harness within 30 seconds of turning the ignition switch **OFF**. This may damage the ECU.

On 5 cylinder engines with knock sensor ignition: if the CHECK ENGINE light comes on and the vehicle looses power, first check for excessive oil in the crankcase. Too much oil could cause enough noise and/or vibration that the knock sensor detects it, misinterprets it as detonation vibration and the ECU will try to correct by retarding the ignition timing. Check oil level and adjust as necessary.

SYSTEM DIAGNOSTICS

All Audi vehicles are equipped with self-diagnostic capabilities. When there is a malfunction in the fuel injection or ignition system, the information is stored in the vehicle's fault memory in the form of a 4 digit code. Each fault code is a 4 digit number that is transmitted by flashing the LED tester or, on California models, the engine warning light on the instrument panel. First the LED will flash for about 2.5 seconds, followed by 2.5 seconds of no flash. This means get ready to read the code. Then the first digit will be shown, wait 2.5 seconds, then the second digit, etc. There is always 2.5 seconds between digits. All digits are between 1 and 4.

It is important that only a suitable tester equipped with a Light Emitting Diode (LED) be used when testing for blink codes.

NOTE: Using a regular test light will draw too much current and will destroy delicate electronic components. Use an LED tester only.

The diagnostic connectors for fault code activation and display are installed in the driver's side footwell. It is not possible to activate the fault memory by means of the fuel pump relay on these models.

Instead, a jumper wire at the diagnostic connectors is used. Locate the diagnostic connector for the vehicle being serviced. Make the correct connections for reading the fault memory of the specific model. In general, a dark colored connector contains 2 terminals. One is for power and is protected by a fuse, usually No. 21. The other terminal is ground. The light colored connector contains a terminal from the engine control unit. Use these connectors to display the contents of the fault memory using an LED tester.

The following is a general outline of diagnostic hook-ups. Specific systems instructions are given further below. Use the diagnostic connectors by connecting the positive terminal of the LED tester to the positive terminal in the dark connector. Connect the negative terminal of the LED tester to the only terminal in the light connector. Connect 1 end of a jumper wire to the negative terminal in the dark connector and touch the other end to the terminal in the light connector for at least 4 seconds. The

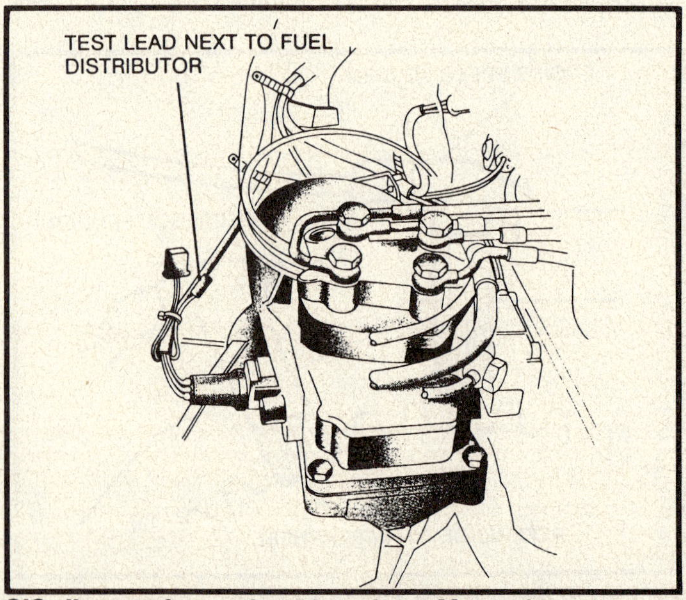

CIS diagnostic test lead location—Motronic

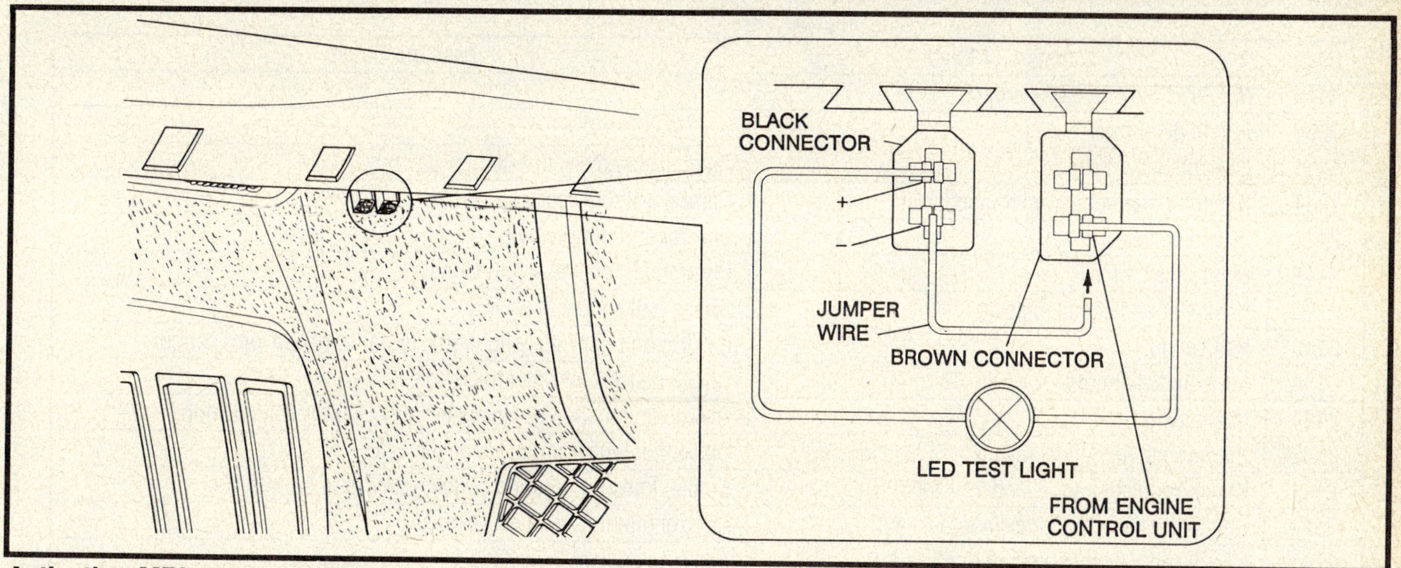

Activating MPI system fault memory using jumper wire and LED test light on diagnostic connectors

fault codes should now be displayed by flashes on the LED tester.

For any fault code displayed, always check for a bad connection first, especially ground connections. Test each component for which a fault code has been displayed before repairing or replacing anything. A fault will remain in memory until erased, even if it has been repaired.

Before performing the input checks, road-test the vehicle for 5 minutes to be certain all necessary information is stored in the memories. If the vehicle cannot be started, operate the starter for approximately 6 seconds to store the necessary inputs.

NOTE: On CIS-E system, do not turn the ignition OFF after road-testing the vehicle or operating the starter. This will erase the fault memory. On CIS-Motronic system. However, the fault memory must be manually cleared.

READING CODES

1. Verify that all fuses are good. Locate the diagnostic connectors in the footwell under the instrument panel.
2. Connect the LED tester and the jumper wires but do not connect the jumper wires together yet. On California vehicles, the engine warning light can be used instead of the LED tester.
3. Start the engine. If the engine will not start, crank the starter for 5–6 seconds and leave the ignition switch **ON**. Connect the jumper wires.
4. The LED tester should now be **ON**. If not, turn the ignition switch **OFF** and check all connections, especially the ECU grounds, and try again.
5. Disconnect the jumper wires. The LED should now start to flash the first fault code. Each code consists of 4 blink pulse groups with a maximum of 4 blink pulses each. There is a pause (light off) of about 2.5 seconds between blink pulse groups. On California models, the instrument cluster engine warning light will also flash the fault codes.
6. The LED will continue to repeat the fault code until the jumpers are connected. Each time the jumper wires are connected for more than 4 seconds, the next 4 digit fault code will be displayed. When Code 0000 is displayed, all faults have been reported. Code 0000 is indicated by flashes in intervals of 2.5 seconds. If the ignition is switched **OFF**, the procedure is interrupted and can be repeated by going back to Step 3. The fault codes are not yet erased.
7. If the ignition is **ON** and the engine is not running, Code 2111 will always be displayed. This is the engine rpm sensor code and can be ignored when the engine is not running.
8. If Code 4444 is displayed, no faults are stored in memory.

CLEARING CODES

The fault memory of the engine control unit can be erased following the display of fault Code 0000. Join the connectors used for reading codes for at least 4 seconds, then separate. The memory is erased when there are no additional flash codes displayed on the LED tester or on the instrument cluster warning lamp.

On the Motronic system as used on the V8 Quattro, if the fault is an intermittent malfunction and does not re-occur during the next 50 engine starts, it will be automatically erased from the diagnostic trouble code memory.

Some models require extracting all codes as well as running an ECU output check before fault memory can be cleared.

If, after attempting to clear the codes from memory, the system is rechecked and codes still appear, then that vehicle's fault memory cannot be canceled without performing the ECU output check.

The during and ECU output check, the ECU output signals power various relays and the electric valves, such as fuel injectors, idle stabilizer valve, charcoal canister solenoid valve and, if equipped, the waste gate frequency valve and the EGR solenoid valve. The ECU outputs can be checked using the LED tester and a jumper wire. When the output check mode is activated, each valve will be operated one at a time. It is necessary to touch and listen at each valve to see that it is operating. This check must be done after reading the fault codes. The output check is interrupted if the engine is started or if a vehicle speed pulse is detected.

1. Perform the diagnostic fault code retrieval as described above and record the fault codes. Turn the ignition switch **OFF**.
2. Connect the jumper wires together, then turn the ignition switch **ON**. Do not start the engine. All of the positions on the gear selector display and the LED tester should be lit. If not, turn the ignition switch **OFF**, check all ECU ground connections and try again.
3. After about 10 seconds, disconnect the jumper wires. The test LED should start to flash and only 1 position on the gear selector display will be lit.

AUDI FAULT CODES

Code	Location of Fault	Possible Cause
4444	No faults have been recorded	
0000	End of diagnosis	
1111	Control unit	Defective memory in ECU
1231	Speedometer sender on transmission	Control unit not receiving signal
2111	Speed sender	Control unit not receiving signal
2112	Reference sender	Control unit not receiving signal
2122	Engine speed signal missing	Check hall sender & wiring
2113	Hall sender	Control unit not receiving signal or out of phase with sender
2114	Ignition distributor	Check distributor adjustment
2121	Idling switch	Switch stuck open or closed or faulty wiring from switch
2132	No data between Fuel Injection ECU to Ign. ECU	Disconnected wiring or defective ECU
2141	Knock regulation for cylinders #1 & 2	Engine knock or pinging causing timing to be retarded
2142	Knock sensor (red) for cylinders #1 & 2	Control unit not receiving signals
2143	Knock regulation for cylinders #3, 4 & 5	Same as above
2144	Knock sensor (white) for cylinders #3, 4 & 5	Same as above
2212	Throttle valve potentiometer	Output signal voltage too high or low
2232	Air mass sensor	Output signal voltage too high or too low compared to engine speed
2233	Air mass sensor	Reference voltage too high
2234	Control unit supply voltage	Voltage supply for control unit too low
2242	CO% potentiometer in air mass sensor	Output signal voltage too high
2312	Coolant temperature sensor	Control unit not receiving signal
2342	Oxygen sensor	Control unit not receiving signal
4431	Idle stabilizer	Open or short circuit to idle stabilizer valve

4. Under the hood, manually move the throttle off of idle. When the throttle is returned and the idle switch is closed, the ECU will begin the output check and the LED will flash the code for the item being operated. Touch the item and listen to make sure it is operating. The first item tested is the fuel injectors, which will all be operated in sequence starting with injector for cylinder number 1. On the 4.2L engine, the fuel injectors will be activated in pairs in the following sequence; 1 and 5, 2 and 6, 3 and 7, 4 and 8.

5. When the last injector has been operated, the ECU automatically moves on to operate the next item. The LED will flash the code for that item and either stop at that item or move on to the next automatically. If it stops, connect the jumper wires for more than 4 seconds to move on to the next output check.

NOTE: The output check can be interrupted and restarted at any time by turning the ignition switch off and starting the procedure again.

6. When the LED flashes Code 0000, all output check items have been tested. This is the only time the fault code memory can be cleared. When the jumper wires are connected for more than 4 seconds, the fault codes will be erased. If a fault still exists, a code will be set when the engine is started.

NO START—NO SPARK

Testing

1. If engine will crank, remove 1 spark plug lead and using a suitable spark plug tester, check for spark.
2. Tester should show a spark when cranking. If there is no spark, remove the distributor cap and crank engine. Check that the rotor turns, to rule out broken distributor cam belt/drive.
3. If the rotor turns and there is no spark, recheck the ignition secondary for signs of damaged or worn parts.
4. Use care when handling parts. For example, the 5 cylinder 20 valve engine has the ignition rotor glued to the distributor shaft and cannot be removed without destroying it. Although the factory recommends replacing the entire distributor, the rotor can be removed by crushing it with a pair of pliers but the replacement must be glued into place. Since the timing mark for #1 cylinder on this engine is hard to see when the distributor is installed, a special tool is used to line up the distributor rotor.
5. If there is no spark and there is no obvious fault in the ignition secondary, check for codes.
   a. Code 2122 means the engine speed signal is missing. Check the Hall sender and wiring.
   b. Code 2132 means there is no data transfer between the fuel injection ECU to the ignition ECU. Check for disconnected or broken wiring or defective ECU.
6. Check the wiring to the speed and the reference sender at the flywheel. Without these signals, the engine will not start.
7. Check all fuses.

IGNITION TIMING

Adjustment

Most all Audi vehicles have the ignition timing controlled by the ECU. Ignition timing is not adjustable. A timing mark is on the crankshaft pulley damper or on the flywheel and may be used as a base setting if the engine was disassembled and the distributor removed.

Some 4 cylinder model 80 and 90 Audi vehicles can have the ignition timing adjusted, but the distributor clamp bolt head is covered with a tamper-proof seal which must be replaced if the distributor is loosened for any reason. For those vehicles where the timing can be adjusted, use the following procedure:

Ignition timing, idle speed and CO are inter-related and must be checked together and adjusted together. Timing adjustment can be done only after the idle speed and CO adjustments are

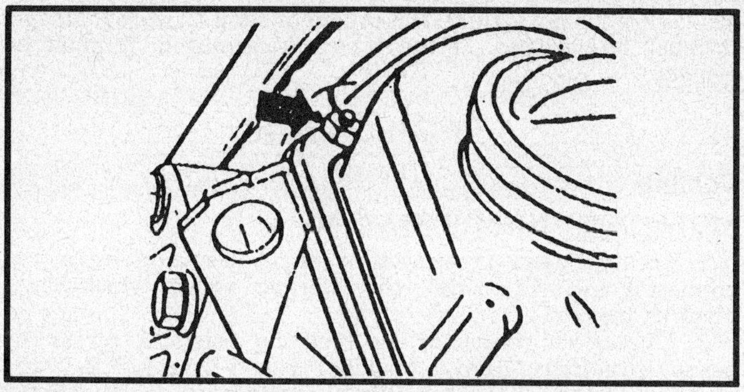

Ignition timing mark location—100/200 vehicles

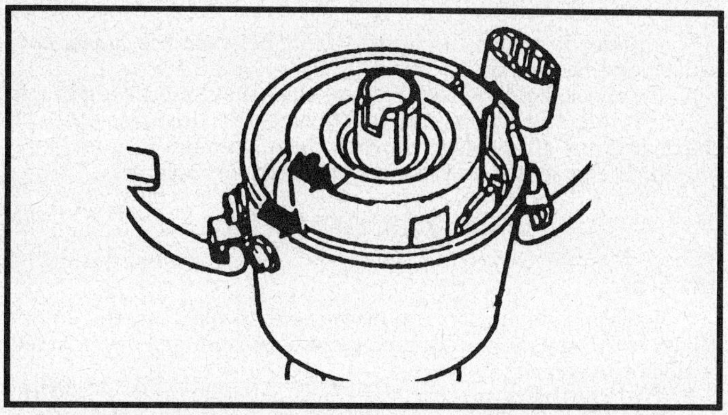

Ignition distributor and rotor timing marks location

correct. A VAG 1551 diagnostic tester or equivalent can be used to display the timing directly or a conventional strobe light can be used. Make sure the fuel pressure test gauges are not installed when checking timing.

1. The engine must be at operating temperature and all accessories must be turned **OFF**.
2. Locate the diagnostic connectors near the accelerator pedal. With the engine at idle, install the jumper wires as shown.
3. On the diagnostic tester or with a strobe light, ignition timing should be:
   a. 4 cylinder: 4–8 degrees BTDC. If timing is not correct, loosen the distributor clamp bolt and turn the distributor as required to set timing to approximately 6 (±) 1 degrees.
   b. 5 cylinder: 13–17 degrees BTDC. If timing is not correct, loosen the distributor clamp bolt and turn the distributor as required to set timing to approximately 15 (±) 1 degrees.
4. Disconnect jumper wires **C** and **D** and rev the engine above 2500 rpm. With the engine at idle, return to these approximate settings.
5. Check idle speed and CO again, adjust as required.

HALL EFFECT SENDER

Testing

2.2L TURBOCHARGED AND 2.3L ENGINES

1. Unplug the connector from the distributor. Connect a voltmeter to the end terminals on the connector and turn the ignition switch **ON**. On the 2.2L engine, there should be 5 volts. On the 2.3L engine, there should be at least 9 volts. The center terminal is the signal return to the ECU.
2. Disconnect and ground the coil wire from the distributor cap. Pierce the insulation or push the connector boot back to connect a test LED to the Hall sender signal wire and reconnect the distributor wiring.
3. Ground the other terminal of the LED and operate the starter. If the LED flashes, the Hall sender is working. If not, the Hall sender or control unit is defective.
4. Check the continuity of the wiring between the distributor and the ECU. If the voltage is reaching the distributor and the wiring is good but the LED does not flash in Step 3, the Hall effect sender is faulty and must be replaced.

4.2L ENGINE

1. The Hall sender is in the right bank distributor. Push the connector boot back so a test LED can be connected to terminals without disconnecting the wiring to the distributor. Connect the LED to terminal 1 and terminal 2.
2. Disconnect and ground both distributor coil wires. When the starter is operated, the LED should flash about once per second.
3. Turn the ignition switch **OFF** and unplug the distributor wiring. Connect a voltmeter between terminal 1 and terminal 3 and turn the ignition **ON**. There should be 4.5–5.5 volts.
4. Turn the ignition switch **OFF** and connect a voltmeter between terminal 2 and terminal 3. Turn the ignition **ON** and check for 4.3–5.2 volts.
5. If the voltages are not correct, check the continuity of the wiring between the distributor and the ECU and check for the proper voltages at the ECU. If the voltages are correct but the LED did not flash, the Hall sender is faulty.

DISTRIBUTORLESS IGNITION SYSTEM

1. Push the connector boot back so a test LED can be connected to terminals without disconnecting the wiring to the sender. Connect the LED to terminal 1 and terminal 2.
2. Unplug connector 1B at the connector station, the white coil primary circuit connector, to disable the ignition system. When the starter is operated, the LED should flash about about once per second.
3. Turn the ignition switch **OFF** and unplug the sender wiring. Turn the ignition switch **ON** and check for voltage between terminal 1 and ground. This is the supply voltage to the sender, terminal 2 is the signal return wire. If voltage is reaching the sender but no signal is returned to the ECU, either the wiring or the Hall effect sender is faulty.

IGNITION COIL

Testing

1. Unplug the connector to the coil and connect a voltmeter to terminal 1 and ground and terminal 1 and terminal 3. With the ignition switch **ON**, there should be 12 volts.
2. Connect a test LED between terminal 2 and terminal 3 and operate the starter. The LED should flash.
3. Turn the ignition switch **OFF**, disconnect the coil output wire and remove the terminal cover. The resistance between terminal 1 and terminal 4 should be 5000–9000 ohms. The resistance between terminal 1 and terminal 15 should be 0.5–1.5 ohms.

DISTRIBUTORLESS IGNITION SYSTEM

1. At the connector station, unplug the white coil power connector 1B. Use a test LED or voltmeter with the negative lead connected to engine ground. With the ignition switch **ON**, there should be 12 volts at each terminal of the connector.
2. Power for the coils comes from the yellow connector in the electronic control box in the passenger side foot well, through thermo-fuse 4. This is a circuit breaker that will reset itself after it had cooled. To test the thermo-fuse, remove it from the socket, allow it to cool and check for continuity across the terminals.
3. To test the secondary side of the coils, turn the ignition **OFF** and disconnect the spark plug wires from the coils. Connect an ohmmeter between terminal 1 and terminal 6, terminal 2 and terminal 4, terminal 3 and terminal 5. The resistance should be 9000–14,000 ohms.

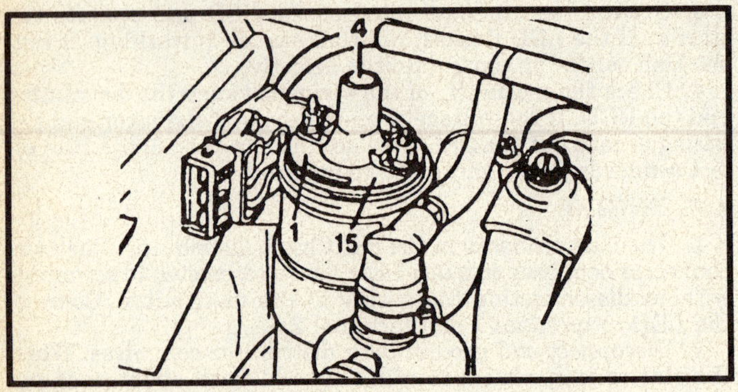

Testing the ignition coil

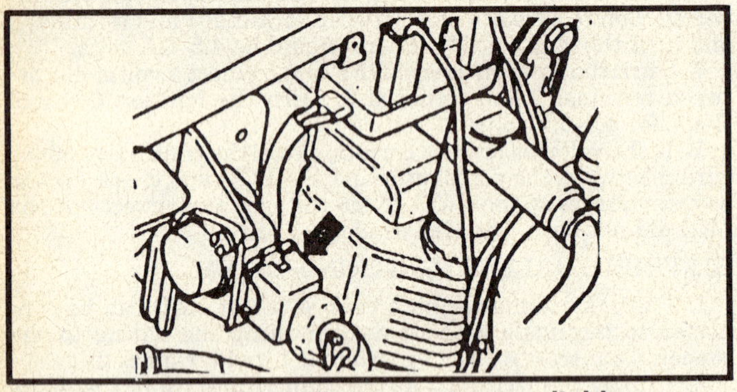

Testing the ignition coil power stage switching module—100/200 vehicles

4. To test the primary side, with the ignition **OFF**, check the resistance between one terminal of the connector in the connector station and a terminal in the 3 pin connector on the power output stage. The resistance should be 0.5–1.0 ohms.

ROTOR

The rotor on all models is a resistor type. Resistance between the center and end terminals of the rotor should be 600–1400 ohms.

IGNITION REFERENCE SENSOR

Testing

MOTRONIC AND MPI SYSTEMS

1. Locate the sensor connector near the firewall. It has a black connector and will be near the rpm sensor which has a grey connector.
2. Unplug the connector and read the resistance across the sensor terminals. Across terminal 1 and terminal 2, there should be about 1000 ohms. Between terminal 1 and terminal 3 and terminal 2 and terminal 3, resistance should be infinite. Terminal 3 is the shield wire and should havecontinuity to ground.
3. Check the continuity of the wiring between the sensor connector and the ECU.
4. To check the pin on the flywheel, remove the starter and rotate the crankshaft till the pin appears in the starter opening. Make sure the pin is tight. If it is bent or shows signs of damage, it must be replaced. If the sensor is damaged, it must be replaced.

RPM SENSOR

Testing

MOTRONIC AND MPI SYSTEMS

1. Locate the sensor connector near the firewall. It has a grey connector and will be near the reference sensor which has a black connector.
2. Unplug the connector and read the resistance across the sensor terminals. Across terminal 1 and terminal 2, there should be about 1000 ohms. Between terminal 1 and terminal 3 and terminal 2 and terminal 3, resistance should be infinite. Terminal 3 is the shield wire and should have continuity to ground.
3. Check the continuity of the wiring between the sensor connector and the ECU.
4. To check the flywheel, remove the sensor mounting bracket and rotate the crankshaft to examine the ring gear teeth. If the gear is out of round or shows signs of damage, it must be replaced. If the sensor is damaged, it must be replaced.

KNOCK SENSOR

Testing

1. Locate each sensor and examine it in place on the engine block. If the sensor or the wiring appears damaged or corroded, replace the sensor.
2. Unplug the connector and check the resistance across the sensor terminals. There should be infinite resistance between all terminals. Terminal 3 is the shield wire and should have continuity to ground.
3. Perform the ECU self diagnostic test.

IGNITION COIL POWER OUTPUT STAGE

Testing

DISTRIBUTORLESS IGNITION SYSTEM

1. Unplug the 4 pin connector and check for continuity to ground on terminal 2 with the ignition switch **OFF**. If there is no ground at terminal 2, check the wiring between the connector and the ECU.
2. If the wiring is good, unplug all the fuel injector connectors. While operating the starter, check for voltage at the connectors for injectors 1, 3 and 4.
3. If the signal is reaching the injectors but there is no ignition, the power output stage or the coil pack may be faulty. If the signal does not reach the injectors and there is no ignition, the ECU or the wiring harness is faulty.

NOTE: If the starter is operated with the power output stage disconnected, a fault code will be stored in the ECU memory. Remember to erase this code when repairs are complete.

Component Replacement

Note: Ignition system components are interrelated with fuel system components. For most components the removal is straight forward, just the removal of the retaining screws. For additional testing or replacement check the fuel system section for this manufacturer.

BMW

General Information

All vehicles covered in this section are equipped with the Bosch Digital Motor Electronic (DME) system. The DME system operates the ignition and fuel injection systems with a single control unit. The DME system combines the L-Jetronic or LH-Jetronic fuel injection with a fully electronic coil ignition.

In 3 Series vehicles, the Electronic Control Unit (ECU) is accessed through the glove compartment opening and is located above the glove compartment under a trim piece. In the 5, 7 and 8 Series vehicles the ECU is mounted under the hood in a covered box on the right side of the firewall. Vehicles with the 12 cylinder M70 use 2 complete DME systems and control units, mounted in the same location.

SYSTEM OPERATION

The DME control unit accepts input signals from the intake air flow or air mass sensor, crankshaft pulse sensor, cylinder identifying sensor, coolant and air temperature sensors an exhaust sensor, a throttle position switch and, if so equipped, the automatic transmission control unit. Optimum fuel quantity and spark timing requirements are calculated based on sensor inputs and internal programs and are adjusted as required for each engine revolution. Output signals operate the idle control valve, fuel pump relay, fuel injectors and the ignition system. Fuel injectors are operated sequentially and air/fuel ratio, ignition timing and idle speed are not adjustable.

Most items of the DME system can be tested with a volt/ohm

DME SYSTEM COMPARISON

SYSTEM	DME 3.1	DME 1.1–1.3
Ignition	Direct, solid state	Rotary Distributor
Cylinder Identification	Camshaft sensor	Sensor on cylinder 6 ignition lead for semi-sequential fuel injection
Load sensing	Hot wire air mass metering	Volumetric air flow metering
Type of fuel injection	Fully sequential	Parallel or semi-sequential
Throttle position sensing	Throttle butterfly potentiometer; adaptive idle setting; part and full load by way of resistance value	Throttle butterfly switch, idle and full load by mechanical switch contact
Starting	Advance fuel pump run and injection; double spark (max. 250 ms); fully sequential injection	Parallel injection
Full load ignition timing	Altitude compensation; dependent on engine speed and cylinder charge	Dependent on engine speed
Idling control	Adaptive value with/without air conditioning	Single adaptive value
Model application identification	By mapped characteristic	By variant coding
Self-diagnostics	Extended	
Emergency running	Improved from DME 1.1–1.3	
Memory board	40 Kb	32 Kb
Plug pins	88	55
Electromagnetic compatibility	4X multilayer technique	Additional suppression

meter although an oscilloscope is required for testing some control functions. A variety of test connectors are also required but some of these can be easily made up as needed. The control unit includes a self diagnostic program that will store fault codes.

The first step in repair or service to any engine management system is to gain as much information as possible from the vehicle owner, especially regarding intermittent problems. Also check any service records which may be available. Before checking any fault codes, it is absolutely essential to check for any obvious mechanical faults or failures. Remember, a trouble code only indicates which sensor or circuit is affected by the problem. Simple mechanical faults such as a vacuum leak or poor electrical connection can cause a fault code.

NOTE: Never disconnect any wiring with the ignition switch ON. Always turn the ignition switch OFF when using an ohmmeter. Any time the battery or main power terminal is disconnected, the control unit fault memory will be erased. Make sure all codes have been retrieved before disconnecting the battery or removing the control unit connector.

The DME system uses a Hall Effect sender, which is required for identification of cylinder No. 1 (ignition TDC). This is installed next to the flywheel, vibration damper or ignition wire. This system also has an inductive sender for speed reference. This sender is located at similar locations and functions the same way. The vibration damper or the ring gear of the flywheel is specially machined for this purpose.

SYSTEM COMPONENTS

The DME ignition system consists of the following components:
- Control unit—digital motor electronics
- Idle speed charging regulator
- Air circulation valve
- Oxygen sensor
- Throttle valve potentiometer
- Cycling valve for charging air pressure regulation

Speed/Position Reference Sensor

Engine rpm and crankshaft position are detected with a sensor mounted on the front of the engine. If this sensor fails, the engine will not run. A special wheel is mounted on the harmonic balancer on the front of the crankshaft. As the teeth in the wheel pass the sensor, a magnetic pulse is generated and sent to the control unit, which reads the pulse frequency as engine rpm. A wide gap in the teeth indicates TDC of number 1 cylinder but the gap will not be at the same place on the wheel as the 0/T timing mark stamped on the wheel used to manually set the engine to TDC. On 12 cylinder engines, 2 sensors are positioned to read the same wheel and each reports to a DME control unit.

The sensor is a magnetic pick-up mounted in a bracket that allows adjustment of the gap between the sensor and the wheel. Power to the sensor and the sensor resistance can be checked with a volt/ohm meter. Sensor function can only be checked with an oscilloscope.

Cylinder Identifying Sensor

This is an inductive pulse sensor similar to the crankshaft sensor. It consists of a coil attached to an ignition wire, supplied with 5 volts from the control unit. When that spark plug is fired, the fluctuation in the return signal is used as a real time reference for sequential fuel injection and ignition control functions. If the sender fails while the engine is running, the engine will be operated in parallel fuel injection control instead of sequential.

There are 2 types of sensors. One type of sensor is attached to spark plug wire No. 6 or 12 near the distributor. The sender can be replaced by cutting the distributor end off the ignition wire and sliding the sender off the end. The other type of sensor reads off the camshaft and is similar to the speed reference sensor. It is bolted into the cam cover and is used on the M42 and M50 engines.

Distributor

The distributor is only required for high secondary voltage distribution. It does not contain any serviceable parts other than the distributor cap and rotor.

DME CONTROL UNIT AND RELAY

3 Series

The DME control unit is located underneath the trim in the glove compartment. The relays are located in the engine compartment. The relays on the M20 equipped 3 series are located just ahead of the drivers side strut tower; the closest relay to the strut tower is the DME relay, the middle being the fuel pump relay and the furthest being the oxygen sensor relay. The relays on the M3 are located at the back edge of the engine compartment on the passenger side; the relay closest to the engine is the fuel pump relay, the middle being the master relay and the furthest being the tank venting relay. The relays on M42 equipped 3 series are located at the back edge of the engine compartment on the drivers side; the relay closest to the engine is the DME relay, the middle being the fuel pump relay and the furthest being the oxygen sensor relay.

5 and 7 Series

M30 AND S38 ENGINES

The DME control unit is mounted in the electronic control box under the hood, on the right side of the firewall. With the cover removed, the DME control unit is closest to the firewall. The box also contains the cruise control module, the automatic transmission control module, the main DME relay, the fuel pump relay, a main power connection and with the M30 engine, the oxygen sensor heater relay. Some models have a fan below the DME control unit that runs whenever the ignition switch is **ON**.

The main DME relay coil receives 12 volts at all times on terminal 86 and the control unit supplies ground to terminal 85 to operate the relay. When the relay is closed, voltage is supplied to the fuel injectors, idle speed actuator, carbon canister purge valve, air mass sensor, ABS control unit, ignition system and the fuel pump relay. On vehicles equipped with the factory installed anti-theft system, the control unit will not operate the relay when the system is armed.

7 and 8 Series

M70 ENGINE

The management system for this engine is two complete and separate fuel and ignition systems, from the fuel pumps all the way to the injectors. The engine is also operated with a "drive by wire" system, meaning the throttles are operated by electric motors instead of being mechanically linked to the pedal. An additional electronic control unit, called the EML, is used to synchronize and operate the throttle motors. If the battery is disconnected or if a control unit is disconnected or replaced, the 3 control units must be synchronized. Other than these features, the control strategy and DME system components are basically the same as other models.

Diagnosis and Testing

SERVICE PRECAUTIONS

BMW offers the following cautions that should be observed to protect the DME system and components:
- Always disconnect the battery, the DME control unit and the ignition coils when using an electric welder on the vehicle or

when charging the battery or when the vehicle is placed in a paint drying oven.

- Remove the main DME relay to disable both the fuel and ignition systems when checking the compression. Never crank the engine after removing the distributor cap or disconnecting the high tension wire on the ignition coil.
- Never disconnect the battery or wires on the alternator, starter or spark plugs when the engine is running.
- Never connect a test lamp on terminal 1 of the ignition coil.
- Never connect terminal 1 of the ignition coil with ground or B+. This means if installing a burglar alarm, terminal 1 should not be used for starter interlocking.
- Whenever performing work on the EML throttle by wire system, a check of the external safety path must be made. This is to check the proper operation of the EML safeguards.
- When working with the EML throttle by wire systems do not place hands or fingers in or near the throttle plates. The throttle motors can close the throttle plates with considerable force.

READING CODES

The DME control unit is equipped with a self diagnostic program that will detect emissions related malfunctions and turn the CHECK ENGINE light **ON** while the engine is running. Emissions related fault codes stored in the DME control unit can be read by activating the program and counting the flashes of the CHECK ENGINE light on the instrument panel. Fault codes stored in all other control units are only accessible with the BMW Service Tester that connects to the diagnostic connector in the engine compartment. The tester is menu driven and will provide operation instructions on the screen.

In the DME control unit, fault codes provide information about which circuit or component is malfunctioning at the time of the test. Intermittent malfunctions are also stored as an existing fault. As useful as these codes are, the first step in repair or service to engine management systems is to gain as much information as possible from the vehicle owner, especially regarding intermittent problems. Also check any service records which may be available. Before repairing any reported faults, it is absolutely essential to check for any obvious mechanical faults or failures. Remember, a fault code only indicates which sensor or circuit is effected by the problem. Simple mechanical faults such as a vacuum leak, bad battery or poor electrical connection can cause a fault code.

The fault code memory will be erased anytime the battery or

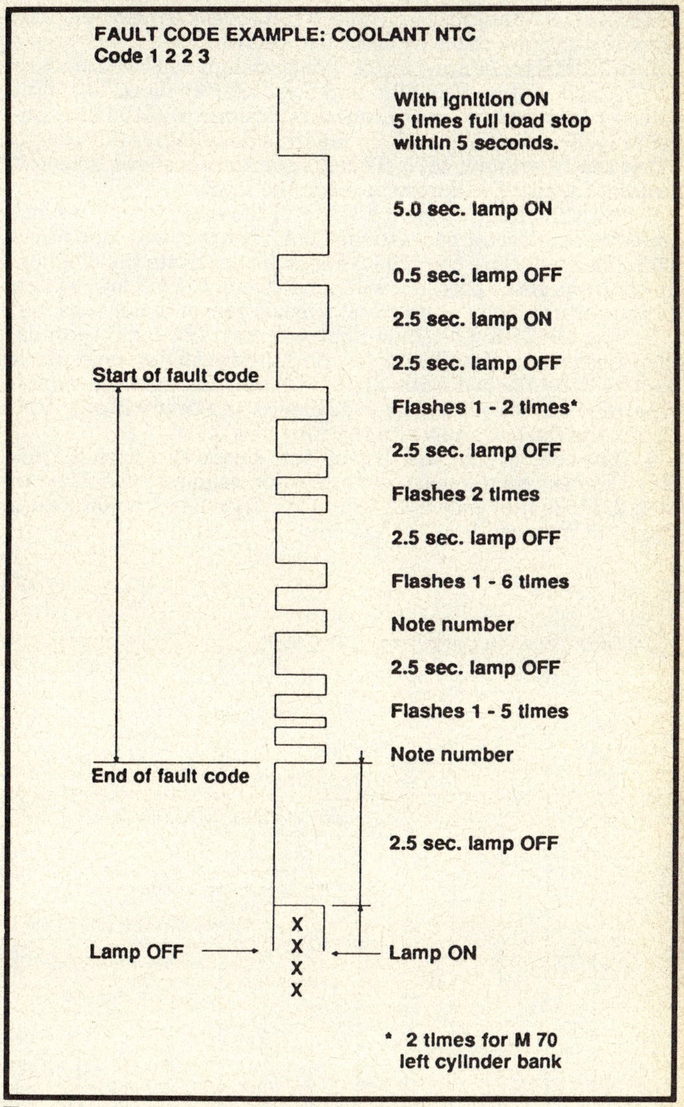

Trouble code output format—BMW

BMW FLASH CODES

x111—Ignition	x262—Idle speed control
x211—Control unit (self test)	x263—Tank venting
x215—Air mass sensor	x264—Oxygen sensor heating relay
x221—Oxygen sensor	x268—Idle speed, CO potentiometer
x222—Oxygen control	x278—Ignition timing correction
x223—Coolant temperature sensor	x512—AC compressor shutoff
x224—Intake air temperature sensor	x513—ECU power input
x231—Control unit voltage monitor	x515—Converter lockup clutch
x232—Idle speed switch	x516—Speed signal
x233—Full load switch	x517—MSR signal (engine drag torque control, part of ASC)
x251—Injector group 1 (cylinders 2, 4, 6)	x518—ASC signal (Automatic Stability Control)
x252—Injector group 2 (cylinders 7, 9, 11)	x818—Final stage
x261—Electric fuel pump relay	x=1 or 2

NOTE: The x represents the numerals 1 or 2. This designates on which DME unit the fault occurred. The numeral 1 indicates the DME unit for the right side of the M70 engine (cylinders 1–6) and it will be the numeral used on all non-M70 engines. The numeral 2 indicates the DME unit for the left side of the M70 engine (cylinders 7–12).

DME control unit is disconnected. Fault codes can also be erased after reading the codes as described below.

1. All DME codes are 4 digits. The first digit will be either **1** or **2**. The digit will be 1 on all 4 and 6 cylinder engines. The digit will be **1** for the right side cylinders (1 through 6) of the 12 cylinder M70. The digit will be **2** for the left side cylinders (7 through 12) of the 12 cylinder M70. This differentiation allows determination of which DME system holds the fault.

2. With the ignition switch **ON** and the engine not running, press the accelerator pedal to the floor 5 times in less than 5 seconds. This will close the full load switch to activate the diagnostic program. If the program will not activate, the full load switch on the throttle body or accelerator pedal sensor is not working.

3. The CHECK ENGINE light will stay **ON** for 5 seconds, then flash once, then begin flashing the first fault code. Each digit will be flashed with about 2.5 seconds between digits. When the whole code has been displayed, the light will stay **ON**. Count the flashes and write the numbers down.

4. The control unit will display only 1 code at a time. To display the next code, repeat the activation sequence described in Step 2. If the first code is Code 1444 or Code 2444, no faults are stored in memory.

5. When Code 1000 or 2000 is displayed, 1 or 2 flashes followed by a long pause before the light stays **ON**, the control unit has reported all fault codes.

CLEARING CODES

1. To erase the fault code memory, hold the accelerator pedal down to close the full load switch for 10 seconds with the ignition **ON** and when Code 1000 or 2000 is displayed.

2. Alternately the DME unit can be unplugged to remove power from the memory circuits. If this method is used, make sure the ignition is **OFF**.

NO START—NO SPARK

Use care when troubleshooting these systems. If the engine is cranked to check for spark with spark plug wires disconnected, it means the ECM doesn't see the load it expects. The result could be a voltage spike that damages the ECM.

DME IGNITION TROUBLESHOOTING

1) Cold engine will not start (oil temp. ≤ 20°C/68°F)
2) Engine starts, but stops again
3) Cold engine hard to start (oil temp. ≤ 20°C/68°F)
4) Warm engine will not start
5) Warm engine hard to start
6) Erratic idling during warm-up phase
7) Idle speed not correct
8) Splashing in intake
9) Hesitation while accelerating
10) Knock when accelerating
11) Hesitation while coasting
12) Misfiring at constant speed
13) Poor acceleration/final output
14) Fuel consumption too high
15) CO/HC not correct

1	2	3	4	5	6	7	8	9	10	11	12	13	14	15	TEST POSITION	
X			X								X				1	Speed sensor/reference mark sensor
X		X	X	X			X				X				2	Ignition coil
		X		X	X	X	X	X	X		X	X		X	3	Spark plugs
		X		X	X		X	X			X				4	High voltage distributor
		X		X	X		X	X			X				5	Ignition lead connectors and leads
X	X	X	X	X					X	X		X	X		6	DME control unit/ignition timing
															7	
															8	
															9	
													X	X	10	Oxygen sensor/emission control and EGR
														X	11	Active carbon filter/purge valve
															12	
	X			X				X				X			13	Tank vent system
															14	
X		X			X	X						X			15	Fuel supply
		X		X	X	X		X		X	X	X		X	16	Intake system
														X	17	Catalytic converter
														X	18	Exhaust system
		X		X	X	X						X		X	19	Crankcase vent/air hose
					X	X									20	Cooling system

1. If testing for spark, use only an approved spark tester, properly grounded, that duplicates the operating load the ECM is designed to handle.
2. If there is no spark when the engine is cranked, the speed sensor/reference sensor should be checked.
3. If there is no spark when the engine is cranked, the ignition coil should be checked.
4. If there is no obvious fault and the speed sensor and coil appear to be in working order, the DME control unit and ignition timing are suspect.

SPEED/POSITION REFERENCE SENSOR

Testing

1. Locate the sensor and disconnect the wiring. With the ignition switch **ON**, there should be 5 volts across the connector terminals from the control unit. If not, check for 5 volts between terminal 1 and ground.
2. Turn the ignition switch **OFF** and connect the ohmmeter to the sensor connector terminals. Check the resistance between terminals 1 and 2. The result should be within 10% of the specified resistance. The sensor should have a resistance more than 100,000 ohms at terminal 3, the shield wire.
3. Connect an oscilloscope to the sensor terminals. While operating the starter, the sensor should produce about a 3 volt peak-to-peak sine wave with a wider wave indicating the reference mark. If there is no signal, check the gap or remove the sen-

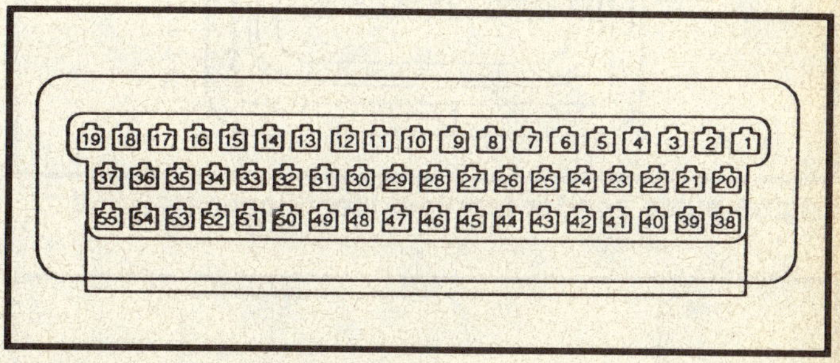

55 pin DME connector – BMW

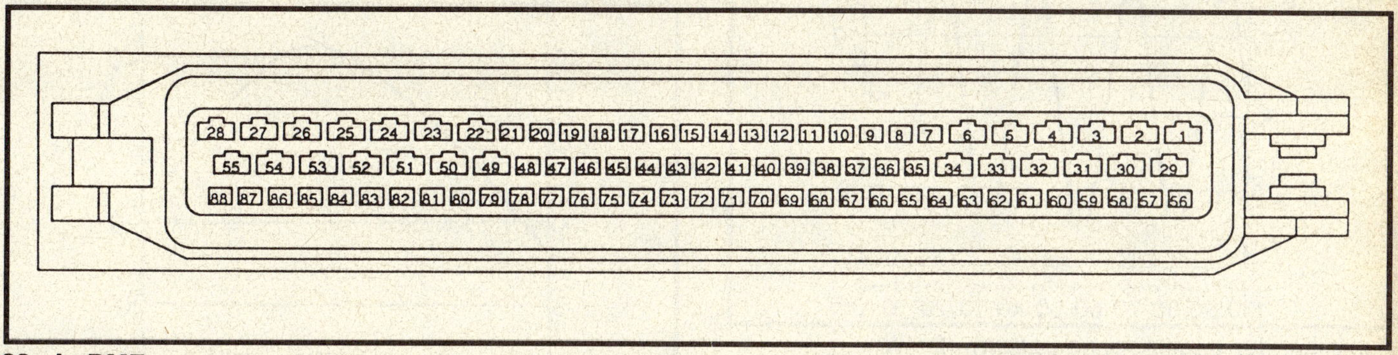

88 pin DME connector – BMW

BMW PULSE SENDER TEST SPECIFICATIONS

Model	Pulse Sender Resistance Ohms @ 68°F (20°C)	Pulse Sender Spacing mm from tooth wheel
318i	1280 ± 10%	1.0 ± 0.3
318is	1280 ± 10%	1.0 ± 0.3
318iC	1280 ± 10%	1.0 ± 0.3
325i (1999–91)	540 ± 10%	1.0 ± 0.3
325i (1992)	1280 ± 10%	1.0 ± 0.3
325iC	540 ± 10%	1.0 ± 0.3
525i (1990–91)	540 ± 10%	1.0 ± 0.3
525i (1992)	1280 ± 10%	1.0 ± 0.3
535i	540 ± 10%	1.0 ± 0.3
735i	540 ± 10%	1.0 ± 0.3
735iL	540 ± 10%	1.0 ± 0.3
750iL	540 ± 10%	1.0 ± 0.3
850i	540 ± 10%	1.0 ± 0.3
M3	960 ± 10%	1.0 ± 0.3
M5	960 ± 10%	1.0 ± 0.3

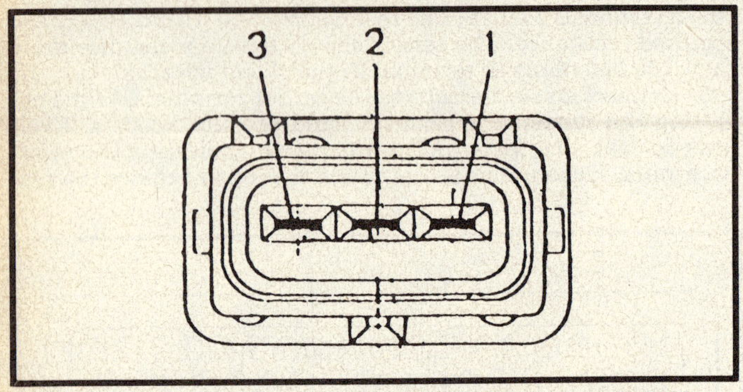

Engine pulse sender terminal identification—BMW

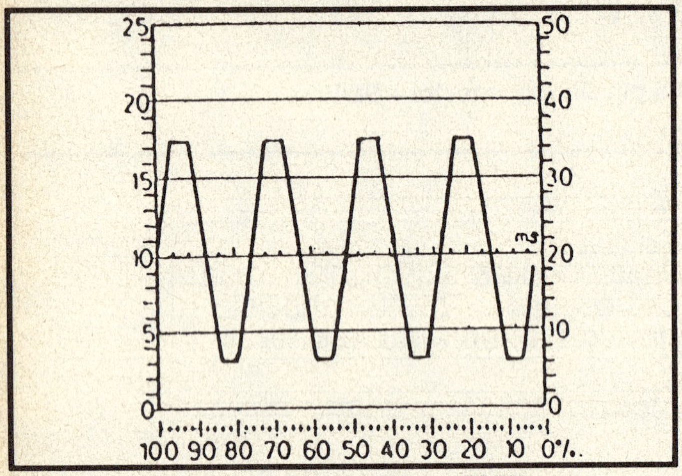

Pulse sender oscilloscope pattern—BMW

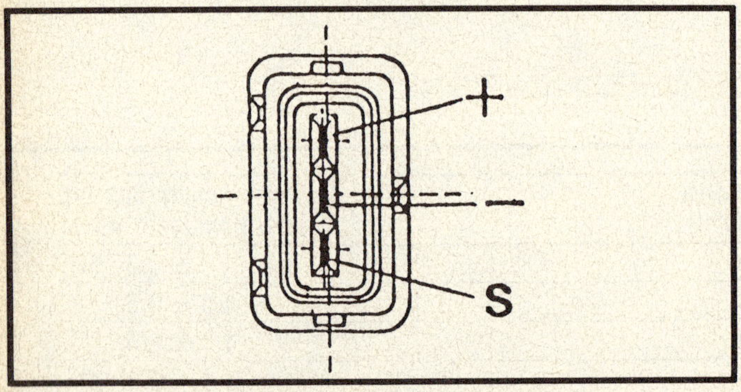

Engine speed sender terminal identification—flywheel type—BMW

sor and check for damage. The gap between the sensor and the wheel should be 0.040 in. (1.0mm).

CYLINDER IDENTIFYING SENDER

Testing

1. With the ignition switch **OFF**, unplug the connector and check the resistance across terminals of the sender connector. Across terminals 1 and 2, resistance should be less than 1 ohm.
2. Terminal 3 is the shield wire, resistance between it and terminal 2 should be more than 10 mega-ohms.

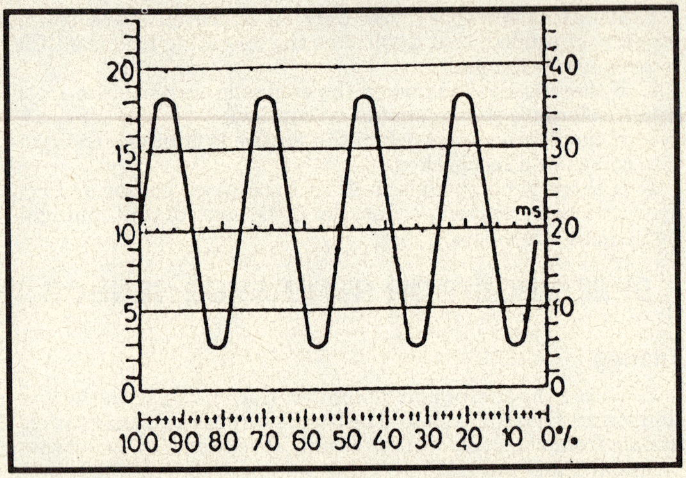

Engine speed sender oscilloscope pattern—flywheel pattern—BMW

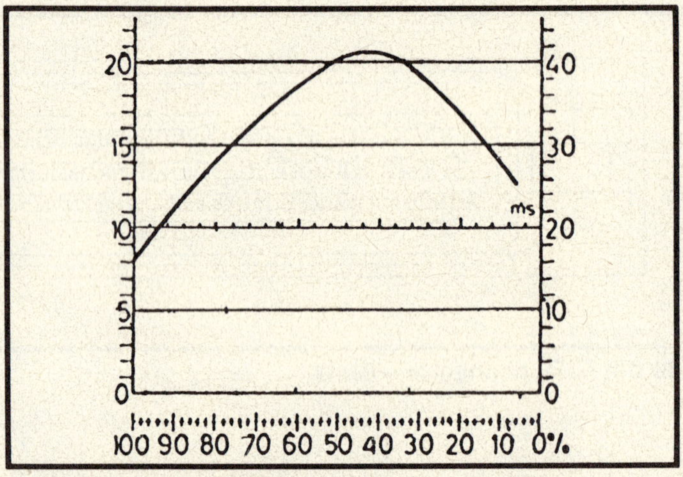

Engine reference sender oscilloscope pattern—flywheel type—BMW

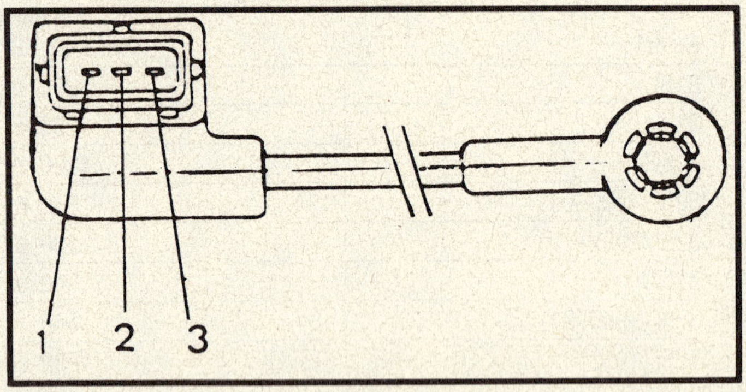

Cylinder identification sender—ignition wire type—BMW

IGNITION COIL

Testing

3 SERIES

1. Measure the resistance of the primary coil terminals.
2. The resistance should be 0.8 ohms ±10%.

3. Measure the resistance of the secondary coil. Between terminal 15 and secondary tower.
4. The resistance should be 8250 ohms +10%.
5. Inspect the connection plate for hair line cranks.
6. Check the tightness plug, between the terminal, if plug is pressed up or out, the coil is defective.
7. Using an oscilloscope, check that the ignition voltage is 6–14 Kilo-volts.

EXCEPT 3 SERIES

1. Measure the resistance of the primary coil terminals.
2. The resistance should be 0.5 ohms ±10%.
3. Measure the resistance of the secondary coil between terminal 15 and secondary tower.
4. The resistance should be 6000 ohms +10%.
5. Inspect the connection plate for hair line cranks.
6. Check the tightness plug, between the terminal, if plug is pressed up or out, the coil is defective.
7. Using an oscilloscope, check that the ignition voltage is 6–14 Kilo-volts.

Component Replacement

SERVICE PRECAUTIONS

NOTE: BMW recommends that all proper cautions be observed to protect the DME system and components and for personal safety.

DME CONTROL UNIT

Removal and Installation

3 SERIES

1. Before removing the control units retrieve any fault codes that may be resident in the units. When power has been removed from the units all fault codes and adaptive memories will be lost.
2. Turn the ignition switch **OFF**.

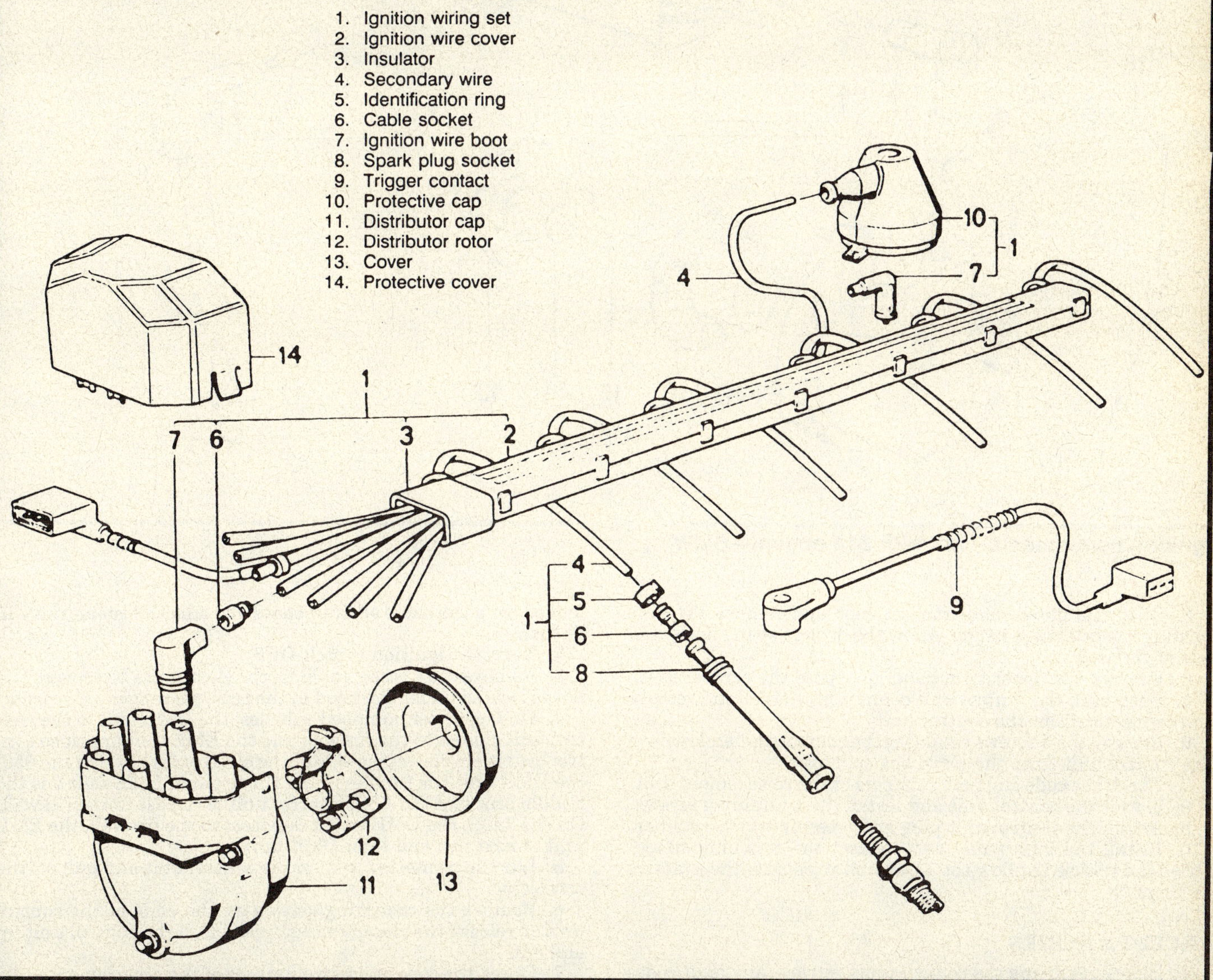

Ignition components—typical 6 cylinder—BMW

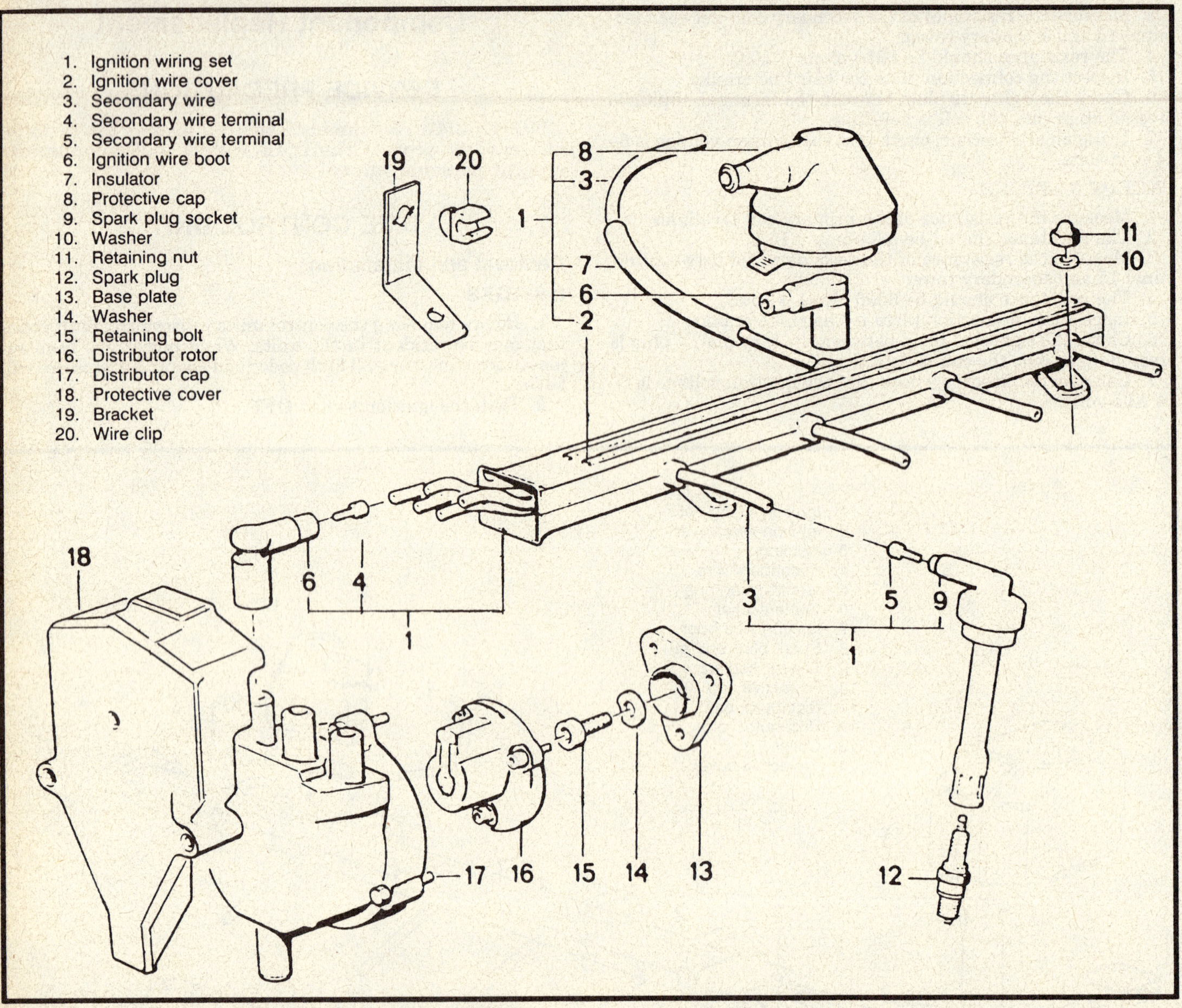

Ignition components—M3 with S14 engine—BMW

3. Open the glove compartment door and remove the pins holding the retaining straps. Allow the door to swing down and out of the way.
4. Remove the upper trim panel to expose the control unit.
5. Push back the control unit connector retainer and remove the connector from the control unit.
6. Remove the 4 screws retaining the control unit and remove the control unit from the glove box opening.
7. Check the code and part numbers of the replacement unit.
8. Mount the control unit and install the 4 retaining screws.
9. Attach the control unit plug and press down the retainer.
10. Install the trim panel and connect the retaining straps. Drive the vehicle to allow the DME unit to relearn the adaptive settings.

EXCEPT 3 SERIES

1. Before removing the control units retrieve any fault codes that may be resident in the units. When power has been removed from the units all fault codes and adaptive memories will be lost.
2. Turn the ignition switch **OFF**.
3. Remove the 4 screws holding the electronics box cover. The box is located under the hood in the rear, passenger side corner.
4. On 6 cylinder equipped vehicles, the DME unit is the control unit closest to the firewall and the EML unit or cruise control unit is on the engine side of the electronics box. On the 850i the DME unit 1 is furthest from the firewall, DME unit 2 is the middle unit and the EML control unit is closest to the firewall. On the 750iL the DME unit 1 is closest to the firewall, the EML unit is next out and then DME unit 2.
5. Lift the connector lock on the control unit and pull off the connector.
6. Remove the mounting screws on the ends of the control unit or release the clamps as equipped. Pull the control unit up and out.
7. Check the code and part numbers of the replacement units and install the units into the correct locations.

8. Replace the mounting screws or attach the clamps. Carefully replace the connector on the control unit and press down the connector lock.

9. Replace the electronics box cover with the 4 screws. On EML equipped vehicles perform the external safety path check. On the M70 equipped vehicles perform the throttle synchronization procedure. Drive the vehicle to allow the DME units to relearn the adaptive settings.

CYLINDER IDENTIFICATION SENDER

Removal and Installation

318i, 318iS, 1992 325i AND 1991-92 525i

1. Turn the ignition switch **OFF**.
2. Unplug the upper of the 2 connectors on the electric lead duct.
3. Unscrew the bolt holding the sender to the engine. Withdraw the sender from the head.
4. Unclip the lead from the lead guide and remove sender from the vehicle.
5. Replace the sender and check that the O–ring is properly seated. Tighten the bolt.
6. Replace the lead in the lead guide and plug in the connector.

EXCEPT 318i, 318iS, 1992 325i AND 1991-92 525i

1. Turn the ignition switch **OFF**.
2. If necessary on the 750iL or 850i, remove the air mass meters and air cleaner assemblies to access the distributors. On the M5 it may be necessary to remove the fan and fan shroud.
3. Remove the covers over the distributor. The cylinder identification sender is on ignition lead 6 on all engines and additionally on ignition lead 12 on the M70 engine.
4. Disconnect the ignition lead and remove the connector from the lead. Pull the sender from the lead.
5. Install the new sender on the lead and install a new connector on the ignition lead.
6. Replace the covers on the distributors and install the air cleaners.

SPEED/POSITION PULSE SENDER

Removal and Installation

750iL AND 850i

Left Cylinder Bank

1. Turn the ignition switch **OFF**.
2. Remove the fan from the engine.
3. Disconnect the sensor plug on the upper timing case cover.
4. Remove the splash guards to gain access to the sensor area.
5. Unscrew the lead holder on the upper timing case cover.
6. Unbolt the pulley for the power steering pump. Unbolt the cooling air duct for the alternator. Unbolt the holders for the hydraulic pipes. Unbolt the hydraulic pump holder from the engine block.
7. Loosen the power steering pump console bolts. Unbolt the pipe holder on the lower timing case cover.
8. Turn the engine over so the sender bolt is accessible through the gap on the increment wheel. Unbolt the sender and push aside the hydraulic pump to remove the sender.
9. Bolt the sender back into position. Bolt the pipe holder on the lower timing case cover. Tighten the power steering console bolts.
10. Tighten the bolts holding the hydraulic pump and the hydraulic pipes. Bolt in the alternator cooling air duct and the power steering pump pulley.
11. Install the lead holder on the upper timing case cover. Replace the splash shields. Connect the sensor plug. Install the fan.

Right Cylinder Bank

1. Turn the ignition switch **OFF**.
2. Remove the fan and belt tensioner from the engine.
3. Disconnect the sensor plug on the upper timing case cover.
4. Remove the splash guards to gain access to the sensor area.
5. Unscrew the lead holder on the upper timing case cover.
6. Unbolt the sender from the engine.
7. Replace the sender and tighten the bolt. Bolt the lead holder on the upper timing case.
8. Replace the splash guards and plug in the sender connector. Replace the belt tensioner and the fan.

EXCEPT 750iL AND 850i

1. Turn the ignition switch **OFF**.
2. Unplug the connector.
3. Unscrew the bolt holding the sender to the engine. Withdraw the sender from the holder.
4. Unclip the lead from the lead guide and remove sender from the vehicle.
5. Replace the sender and check that it is properly seated. Tighten the bolt.
6. Replace the lead in the lead guide and plug in the connector. Make sure that the lead cannot rub on the drive belt.

IGNITION TIMING

Ignition timing is not adjustable on vehicles using the Digital Motor Electronic (DME) control system. If timing is not within specification, check for engine mechanical problem or DME control unit malfunction.

Always check the firing order and secondary wire routing if driveablity problems are reported. The 6 cylinder firing order is 1–5–3–6–2–4 and the 12 cylinder firing order is 1–7–5–11–3–9–6–12–2–8–4–10. If the secondary wiring is routed incorrectly driveablity problems or computer damage could result.

JAGUAR

General Information

Jaguar uses conventional electronic ignition systems controlled by an Electronic Control Unit (ECU). The ECU controls the ignition function and timing on both the 6 and 12 cylinder engines. Low voltage signals are sent from the ECU to the ignition amplifier(s) which are used to trigger the ignition coil(s). The coil(s) then discharge high voltage to the spark plugs via the distributor. No mechanical or vacuum advance mechanisms are used. The distributor only directs the voltage to the appropriate spark plug.

SYSTEM OPERATION

The Lucas ignition system used on the 4.0L engine relies on signals from the airflow meter and crankshaft position sensor. The Marelli system used on the 5.3L engine uses the intake manifold vacuum and engine speed signals as the primary ignition control factors.

An ignition **ON** input signal, taken from the ignition relay, applies power to the ignition coil and energizes the main and fuel pump relays through the ECU. A timer and fuel pump relay permits the fuel pump to work for about 0.5 seconds to pressurize the fuel rail prior to cranking.

When the ignition is turned to the **ON** position, power is applied to the various devices through the main relay. Information from various sensors is then evaluated to determine engine conditions. The computer's memory is accessed to obtain appropriate ignition timing and fuel metering parameters.

The ignition system incorporates an inertia switch as a safety measure. It switches the engine management system (ignition and fuel) OFF in the event of a sudden deceleration.

Diagnosis and Testing

SERVICE PRECAUTIONS

- Use care when working around high-energy electronic ignition systems. High voltage could cause injury to the technician and equipment.
- Before performing any service on the ignition system, make sure the ignition is turned **OFF** and/or the battery negative cable is disconnected.
- Before attempting to remove any parts, turn the ignition switch **OFF** and disconnect the battery ground cable.
- Never connect test equipment such as timing lights, etc., directly to the ignition system for a power source. Always use the vehicle battery for the power source.
- Use only the correct replacement parts. Use of spark plugs, for example, having specifications and sizes different from those specified could cause severe damage and alter emissions levels.
- Use care when working with light alloy parts. Always lubricate the spark plug threads with a suitable lubricant to avoid seizing the plug threads to the cylinder head.
- Keep all ECU parts and harnesses dry during service. Protect the ECU and all solid-state components from rough handling or extremes of temperature.
- Make sure all harness connectors are fastened securely. A poor connection can cause an extremely high voltage surge resulting in damage to integrated circuits.
- Always use a 12 volt battery as a power source.
- Do not disconnect the battery cables with the engine running. Never run the engine with the battery cables loose or disconnected.
- Do not attempt to disassemble the ECU unit under any circumstances.
- When connecting or disconnecting pin connectors from the ECU, take care not to bend or break any pin terminals. Check that there are no bends or breaks on the ECU pin terminals before attempting any connections.
- When measuring supply voltage of ECU-controlled components, keep the tester probes separated from each other and from accidental grounding. If the tester probes accidentally make contact with each other during measurement, a short circuit will damage the ECU.
- The ECU control unit should be removed if the vehicle is to be exposed to high heat such as a bake oven to cure paint after body refinishing or when welding is be done on the vehicle.
- Do not be hasty to condemn the ECU. This system uses voltages and resistances that are very small. Examine the sensors, wiring and connections carefully. The sensors operate in more harsh conditions than the ECU which is protected in the footwell.
- Check all ground connections before condemning the ECU.

READING CODES

Some Jaguar engine management systems have some self-diagnostic capability. The ECU can identify a fault, assign an identity code and store the code in memory for later retrieval. The number, type and meaning of engine codes vary by year and model.

On the 5.3L V12 engines, the ECUs used on 1990 and 1991 vehicles do not generate or store codes. Stored diagnostic or fault codes may be retrieved from 1992 V12 engines. All 1990–92 4.0L engines will generate and store fault codes relating to sensor and circuit function. 1990–92 4.0L and 1992 5.3L engines will also provide diagnostic fault codes. The codes may be read on the trip computer display on the dashboard. Generally, codes will be held in memory and the dash warning lamp will be lit when a fault is sensed by the ECU.

NOTE: Not every stored code will cause the warning lamp to light after the first occurrence. Always check for codes during diagnosis, including cases where the dash warning lamp was not reported lit.

While most fault codes are held in an electronic memory and are retained even after the ignition is switched **OFF**, certain codes are only held or displayed as long as the ignition is **ON**. If the fault is present at the next restart, the code will reset.

To read the stored codes, bring the vehicle to a complete stop. Switch the ignition **OFF** and wait at least 10 seconds. Turn the ignition to the **II** or **ON** position but do not start or crank the engine. Press the trip computer button **DISP** or **VCM**. After a short period of time, the stored codes will be displayed on the trip computer panel. The display may include the designation FF, an abbreviation for FUEL system FAILURE.

The fault codes are displayed in order of priority, one at a time. The next is only displayed when the preceding one is repaired or cleared.

Engine Codes

4.0L ENGINE

Code 11—Idle potentiometer or associated wiring; idle trim pot out of normal operating range.
Code 12—Airflow meter or associated wiring; airflow meter signal out of normal operating range.
Code 14—Coolant temperature sensor (thermistor) or associated wiring; faulty signal during warm-up.
Code 16—Air intake temperature sensor (thermistor) or associated wiring; resistance out of range.
Code 17—Throttle potentiometer or associated wiring; resistance out of range.

Code 18—Throttle potentiometer or airflow meter calibration; throttle pot signal low during high air flow.
Code 19—Throttle potentiometer or airflow meter calibration; throttle pot signal high during low air flow.
Code 22—Fuel pump driver; faulty ECU output to fuel pump relay.
Code 23—Fuel supply—poor feedback control in rich direction
Code 24—Ignition driver; ECU output to ignition amplifier
Code 26—Air leak; poor feedback control in lean direction
Code 29—ECU self-check; internal function.
Code 33—Injector driver fault; ECU output to injectors
Code 34—Injector driver or associated wiring; injector dribble or staying open.
Code 44—Oxygen (Lambda) sensor or associated wiring; feedback signal out of range either rich or lean.
Code 46—Idle speed control, coil 1 driver; ECU output to idle speed control stepper motor
Code 47—Idle speed control, coil 2 driver; ECU output to idle speed control stepper motor.
Code 48—Idle speed control motor; significantly out of position with engine hot.
Code 68—Road speed sensor or associated wiring; indicated speed less than 3 mph (5 kph) at high air flow.
Code 69—Drive/neutral switch or associated wiring; shows cranking in **D** or high airflow in **N**

5.3L ENGINE

Code 11—Idle potentiometer
Code 13—ECU pressure sensor
Code 14—Coolant thermistor or associated wiring
Code 16—Air thermistor or associated wiring
Code 17—Throttle potentiometer or associated wiring
Code 18—Throttle potentiometer calibration 1
Code 19—Throttle potentiometer calibration 2
Code 23—Fuel supply
Code 34—Injector fault, A or right bank
Code 36—Injector fault, B or left bank
Code 44—Oxygen (Lambda) sensor fault, A or right bank
Code 45—Oxygen (Lambda) sensor fault, B or left bank
Code 49—Fuel injection ballast resistor or associated wiring

Jaguar Diagnostic System (JDS)

A computerized, hand-held diagnostic tool is available through Jaguar Cars Ltd. The JDS tool allows rapid testing and diagnosis of electrical circuits. The tool can be adapted to each motor through the use of specific software. The tool allows testing and limited operation of many ECU controlled components. Stored fault codes may also be erased with the JDS tool.

CLEARING CODES

Codes will not be cleared from memory until all stored codes have been displayed. Interrupting the display cycle will allow some codes to be retained and the memory will not clear.

If the JDS tool or its equivalent is not available, clear stored codes by driving the vehicle in excess of 19 mph (30 km/h). The memories will be cleared electronically. Each system will perform its self-check function and any new codes will be set. If no faults are present in a particular system, the failure memory will remain cleared.

IGNITION TIMING

The engine management computer has taken over the timing adjustment functions. The distributor in this system only distributes high voltage to the spark plug cables. There are no auto advance mechanisms within the distributor. Note that if removing the distributor, some distributors may have a small notch in the distributor body which should be lined up with the rotor tip to make distributor service easier.

NO START-NO SPARK

Ignition signals are sent, via a separate ignition amplifier, to the ignition coil which generates high energy pulses for the spark plugs. All of the timing and dwell functions are controlled by the ECU.

If the engine will crank but not start, the traditional methods of identifying problem areas still work. Removing a spark plug wire to check for spark while cranking is an appropriate method of checking the ignition primary and secondary systems. A good, blue spark should jump at least ½ in. to ground while the engine is cranking.

If there is no spark, check the distributor cap and rotor for physical damage. A broken rotor or damaged distributor cap could cause a no-start condition. Inspect the coil tower for burns or signs of flashover. Check the coil wire for damage or burns. Crowded engine compartments make it all the more likely that some electrical component will come into contact with high temperatures and spark plug wires may be damaged. Often these conditions will be hidden by the rubber boots so check carefully, especially if the customer complaint is a no-start in wet weather.

With the distributor cap off, crank the engine. Distributors are driven by an auxiliary shaft. If the rotor does not turn, there is a mechanical problem with the distributor drive.

A more likely problem area is the crankshaft sensor. This is a device which generates engine speed and crankshaft position information. The ECU needs this signal to trigger the ignition and fuel injection system. The sensor system consists of the sensor itself which produces a small magnetic field and a toothed trigger wheel attached to the crankshaft pulley.

Note that 5.3L (V12) engines will use a TDC sensor and a flywheel sensor. These are primary inputs and the engine will not start without these inputs. The TDC sensor is mounted so that it is triggered by a toothed wheel attached to the crankshaft pulley. This gives an 8 degree ATDC reference for each cylinder on **A** bank. The correct sensor air gap is 0.018–0.042 in. The sensor signal can only be checked using an oscilloscope. The flywheel sensor is mounted so that it is triggered by each flywheel ring gear tooth to give an engine speed (rpm) signal. The correct gap is 0.018–0.042 in. The sensor signal can only be checked using an oscilloscope.

Another possibility is that the inertia switch is triggered. An inertia switch is usually secured to the firewall, forward of the A-post fuseblock on the driver's side, under the fuseblock cover. On XJS, look for the inertia switch on the passenger's door pillar next to the footwell. It is a gravity actuated safety device which de-energizes the ignition ON relay, inhibits fuel pump operation and activates the central locking system to open all door locks and close both the luggage compartment and fuel filler locks should the vehicle be subjected to heavy impact force. The inertia switch may be reset by depressing the plunger located on the upper surface of the switch housing.

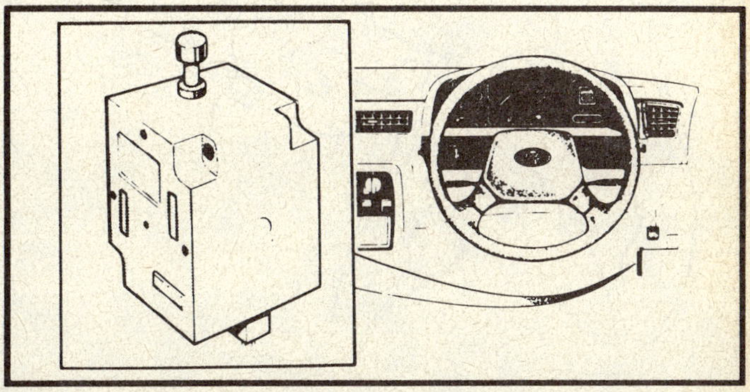

Inertia switch and location. Note pushbutton reset on top

Component Replacement

IGNITION COIL

The ignition coil is a standard ignition coil used on constant energy systems, featuring a low impedance primary winding by a high secondary output.

Removal and Installation

5.3L ENGINE

Front Coil

1. Disconnect the negative battery cable.
2. Install appropriate fender cover to protect the finish.
3. Remove the bolts securing the front high tension distributor coil to the mounting bracket. Remove the coil from the bracket.
4. Disconnect the multi-wire plug connector.
5. Remove the high tension lead and lift out the coil.

To install:

6. Installation is the reverse of the removal procedure. Make sure all wire connections are secure.
7. Install the hold-down screws.
8. Reconnect negative battery cable.

Rear Coil

1. Disconnect the negative battery cable.
2. Under the hood, locate and remove the bolts securing the rear high tension distributor coil mounting bracket.
3. Remove the cable harness P-clips from the coil mounting bracket.
4. Disconnect the multi-wire plug connector.
5. Remove the high tension lead and lift out the coil.

To install:

6. Installation is the reverse of the removal procedure. Install the coil and make sure all connections are secure. Install the cable harness P-clips.
7. Bolt the coil bracket in place and reconnect the battery negative cable.

4.0L ENGINE

1. The ignition coil is a Bosch type and is secured to the right inner fender panel. Disconnect the negative battery cable.
2. Remove the rubber cover from the top of the coil.

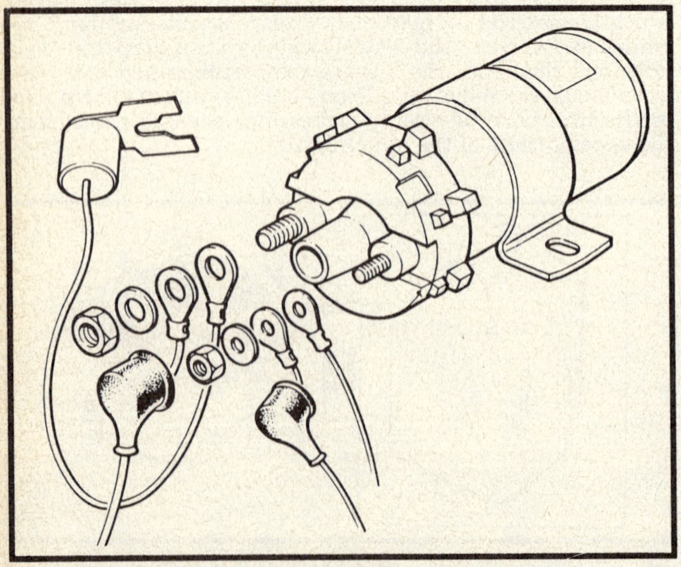

Bosch style ignition coil. Keep low tension connections clean and tight

3. Remove the rubber covers from the small low tension terminals and remove the wire connections.
4. Remove the coil retaining bracket bolts and move the wires including the radio suppression capacitor out of the way.
5. Remove the high tension lead from the center of the coil and lift the coil from the vehicle.

To install:

6. Installation is the reverse of the removal procedure. Install the bracket on the replacement coil and position the coil so that the radio suppression capacitor bracket and its ground lead are below the inner bolt.
7. Tighten the bracket securing bolts, connect the low tension leads and fit the rubber covers.
8. Make sure all electrical connections are secure and reconnect the negative battery cable.

DISTRIBUTOR

Since the engine management computer has taken over the timing adjustment functions, the distributor in this system only distributes high voltage to the spark plug cables. The distributor consists of a body and shaft assembly, a rotor and distributor cap. There are no auto-advance mechanisms within the distributor.

Removal and Installation

5.3L ENGINE

1. Disconnect the negative battery cable.
2. Raise and safely support the vehicle.
3. Remove the flywheel sensor heat shield.
4. Remove the bolts holding the flywheel sensor and remove the sensor from its position. All of this is to for access to allow turning the flywheel to set up the engine to TDC.
5. Rotate the flywheel to obtain the TDC position on the crankshaft pulley. This should align the distributor rotor arm arrow with the nick in the distributor body.

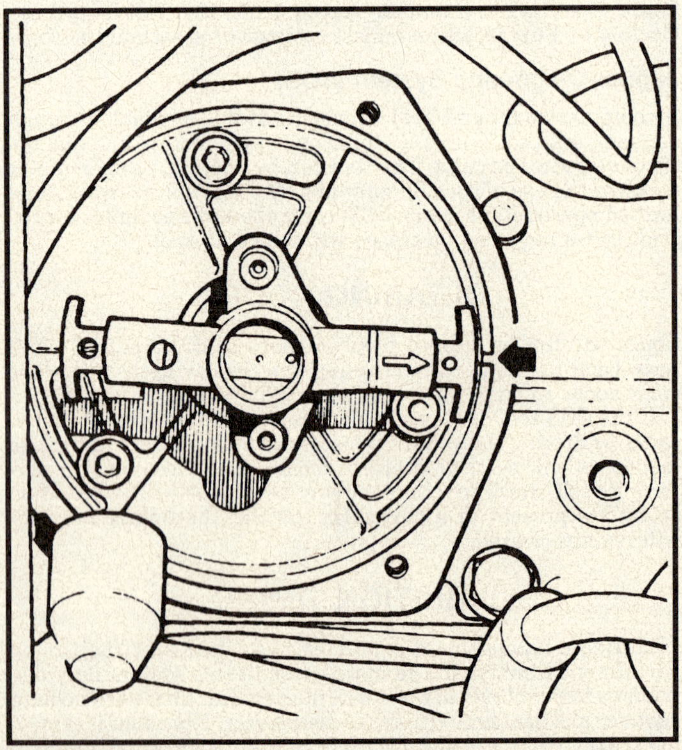

Distributor rotor tip-to-body alignment—Jaguar 5.3L V12 engine

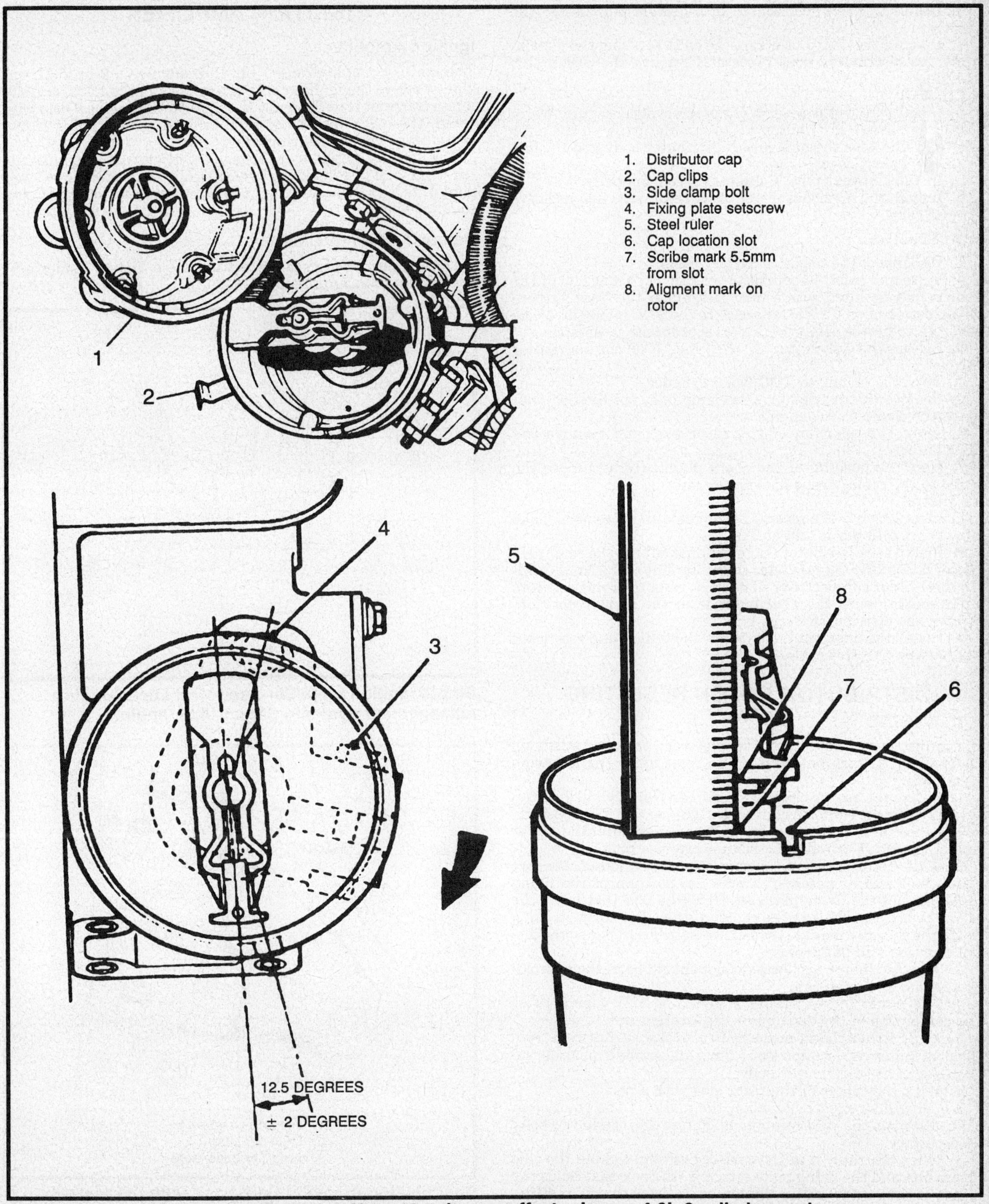

Distributor rotor tip-to-body alignment. Note 12.5 degree offset—Jaguar 4.0L 6 cylinder engine

6. Install the flywheel sensor and heat shield. Lower the vehicle.
7. Remove the distributor cap, the bolts securing the distributor, the distributor vent pipes and remove the distributor assembly.

To install:

8. Installation is the reverse of the removal procedure. When installing the new distributor, make sure the arrow on the rotor arm is in line with the nick in the distributor body and the line on the jackshaft cover.
9. Install the vent pipe, tighten the securing bolts.
10. Install the distributor cap and reconnect the negative ground cable.

4.0L ENGINE

1. Disconnect the negative battery cable.
2. Raise and safely support the vehicle. The front end of the vehicle may be lifted with a floor jack under the front suspension crossmember. This is for access to the TDC indicator. Make sure the vehicle is safely secure on appropriate jackstands.
3. Release the distributor cap clips and lift off the distributor cap.
4. Turn the engine to TDC No. 1 cylinder.
5. Remove the distributor side clamp bolt, not the bolt that holds the clamp to the engine.
6. Note the position of the rotor arm and remove the distributor.
7. Note the position of the spark plug wires and move the wires to the replacement distributor cap.

To install:

8. Installation is the reverse of the removal procedure. Make sure No. 1 cylinder is still at TDC.
9. Install the distributor and position it so that the rotor arm center line is 12½ degrees after the center line of the No. 1 cylinder distributor cap electrode. If it is not, reset the rotor position. After making sure the distributor is in the correct position, tighten the distributor clamp bolt.
10. Install the distributor cap, lower the vehicle and reconnect the battery negative cable.

DISTRIBUTOR ROTOR RESETTING

4.0L ENGINE

If the distributor rotor is in an incorrect position, misfiring could occur. It may be checked and reset using the following procedure:

1. Rotate the engine clockwise until the No. 1 cylinder piston is at TDC on the compression stroke. The correct rotor position is when the rotor arm center line is 12½ degrees after the center line of the No. 1 cylinder distributor cap electrode.
2. If the rotor is not in the correct position, loosen the side clamp bolt and, if necessary, reset the position of the fixing plate, so that the fixing plate setscrew (the bolt that holds the clamp to the engine) is in the center of the adjusting slot. Snug down the setscrew enough to hold the distributor, but so the distributor can still be turned.
3. The distributor cap location slot should be in the approximately 5 o'clock position.
4. The center line of the rotor should be 12½ degrees after the center line of the distributor cap location slot.
5. As a further check, measure in a clockwise direction from the distributor cap location slot 5.5mm and scribe a mark on the shoulder of the distributor body.
6. Mark the center of the rotor arm with a pencil.
7. Using a straight edge, align the mark on the rotor with that scribed on the distributor body. Rotate the distributor body as necessary.
8. When the rotor is in the correct position, tighten the side clamp bolt and the fixing plate setscrew (the bolt that holds the clamp to the engine). After tightening, recheck that the setting has not altered.

IGNITION AMPLIFIER

Ignition Amplifier

The ignition amplifier(s)—1 on 4.0L engine or 2 on 5.3L engine—receive the ignition output signal from the ECU, amplify it and transmit it to the ignition coil(s) on the engine. Once triggered, the coil releases the high tension energy pulse to the spark plugs via the distributor.

The ignition amplifier consists of an encapsulated module fitted to a metal heat sink and is electrically connected by a 7-way plug. It is usually mounted at the right hand side of the engine compartment.

Removal and Installation

1. Disconnect the negative battery cable. Install fender protective covers as required.

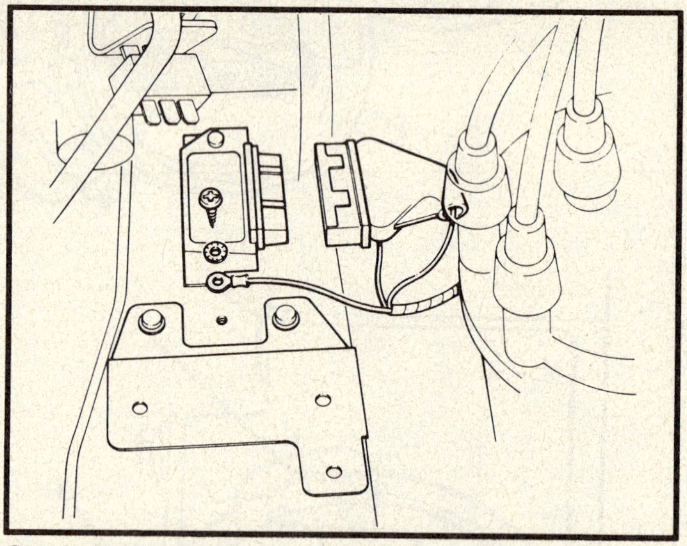

Single ignition amplifier used with Lucas engine management system—Jaguar 4.0L engine

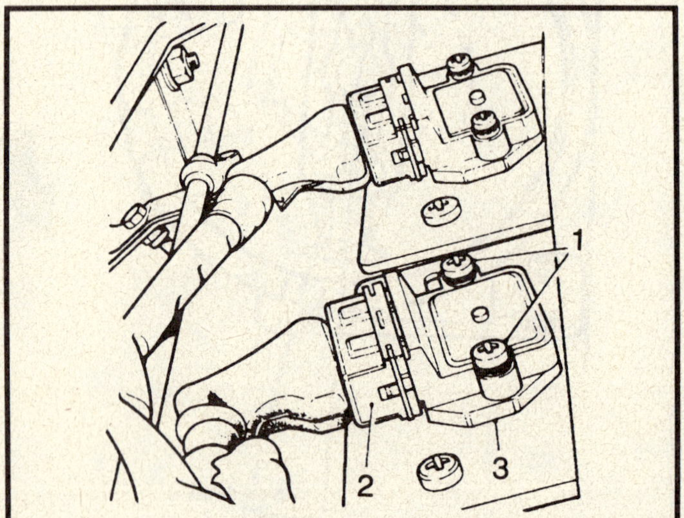

Dual ignition amplifiers are used with the Marelli engine management system—Jaguar 5.3L V12 engine

2. Remove the mounting screws holding the amplifier to its mounting plate. On XJS 4.0L models, it may be necessary to remove the coil bracket for access.
3. Disconnect the amplifier connector (usually black) and remove the amplifier.

To install:

4. Installation is the reverse of the removal procedure. Make sure the electrical connector is secure.
5. Install the amplifier to its mounting plate.
6. Reconnect the negative battery cable.

INERTIA SWITCH

Removal and Installation

XJ-6

1. Disconnect the negative battery cable.
2. Open the driver's door. Remove the screws securing the A-post fuseblock cover and remove the cover.
3. Disconnect the inertia switch electrical connector and remove the switch assembly retaining screws.
4. Remove the switch and bracket as an assembly.
5. Remove the switch to bracket retaining screws and remove the switch from the bracket.

To install:

6. Install the new switch to the bracket.
7. Install the switch and bracket assembly into position.
8. Reconnect the switch electrical connector and secure the fuseblock cover.
9. Reset the inertia switch by depressing the plunger located on the upper surface of the switch housing.
10. Reconnect the negative battery cable.

XJ-S

1. Disconnect the negative battery cable.
2. Open the driver's door remove the driver side dash liner.
3. Remove the switch cover and switch retaining screws.
4. Disconnect the multi-plug cable harness connector and remove the switch.

To install:

5. Install the new switch and connect the multi-plug cable harness connector.
6. Install the switch retaining screws and switch cover.
7. Install the driver side dash liner and reconnect the negative battery cable.

START RELAY

Removal and Installation

1. Disconnect the negative battery cable.
2. Open the passenger front door and remove the passenger side dash liner.
3. Locate and remove the the relay from the main relay connector.

To install:

4. Install the new relay in the main relay connector.
5. Install the passenger side dash liner.
6. Reconnect the negative battery cable.

IGNITION RELAY

Removal and Installation

1. Disconnect the negative battery cable.
2. Open the passenger front door remove the passenger side dash liner.
3. Locate and remove the the relay from the ignition relay connector.

To install:

4. Install the new relay in the ignition relay connector.
5. Install the passenger side dash liner.
6. Reconnect the negative battery cable.

CRANKSHAFT SENSOR

The sensor provides information on both engine speed and crankshaft position to the ECU. The sensor is mounted above a 57-tooth gear on the front of the crankshaft. The teeth passing in front of the sensor generate a small electrical voltage in the sensor. Changes in the voltage signal are proportional to engine speed. The toothed wheel has 3 omitted teeth. As this space passes the sensor, the signal is interrupted and a positive indication of engine position is transmitted.

Removal and Installation

4.0L ENGINE

1. Make certain the ignition is switched **OFF**. Disconnect the negative battery cable.
2. Remove the clips holding the sensor cable.
3. Disconnect the sensor cable at the connector.

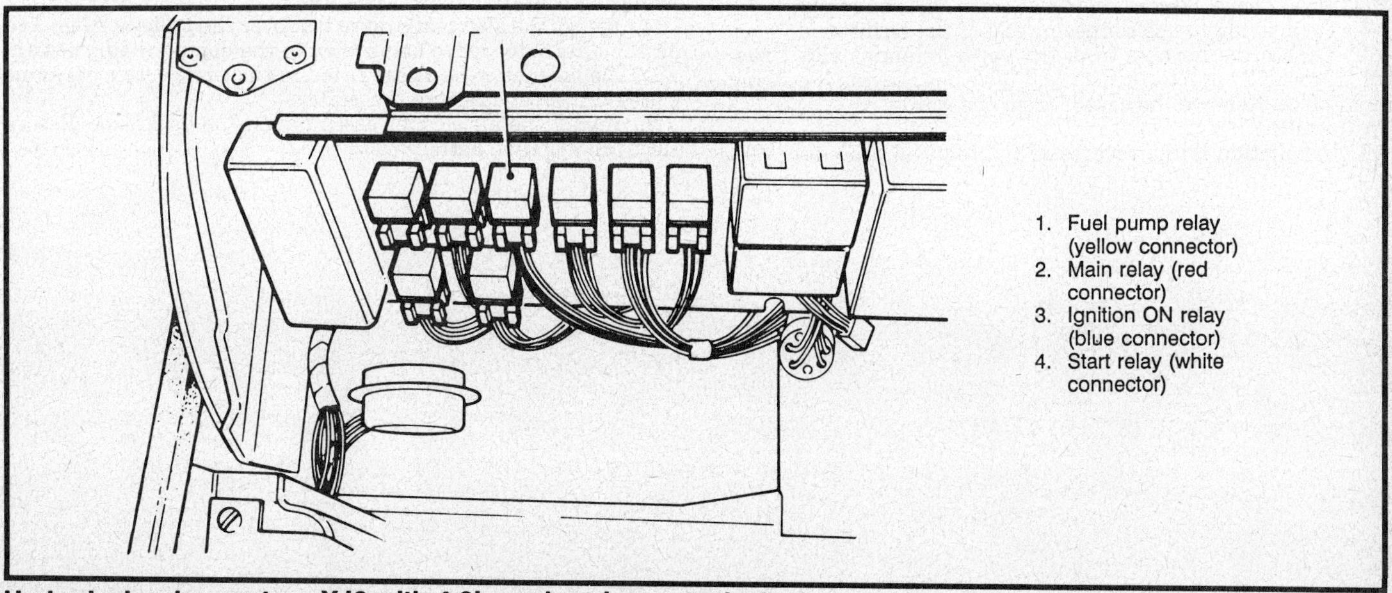

Underdash relay center—XJ6 with 4.0L engine shown, typical

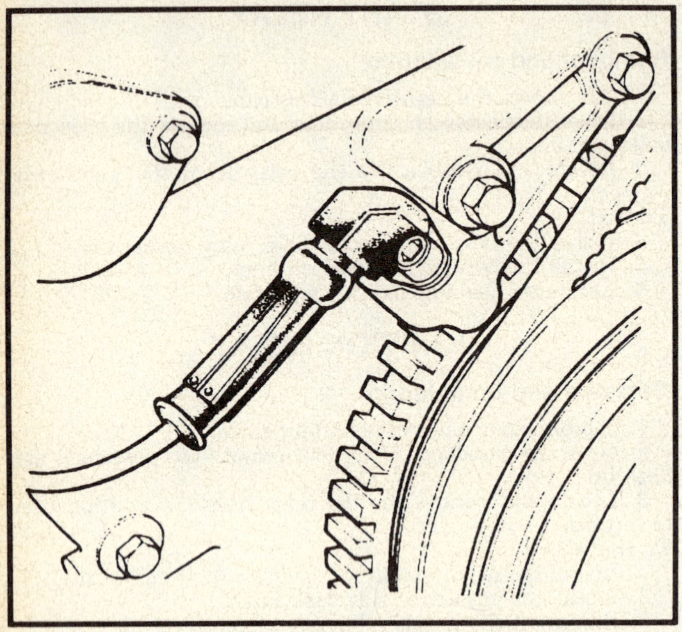

Crankshaft sensor mounting—XJS with 4.0L engine shown, typical

4. Remove the 2 bolts holding the sensor bracket. Remove the bracket with the sensor attached.
5. Remove the sensor from the mount.

To install:

6. Installation is the reverse of the removal procedure. Tighten the mounting bolts to 17–20 ft. lbs. (23–27 Nm). Make certain the sensor is squarely and securely mounted in the holder. Route the wiring correctly and make certain the harness is retained by the clips.

5.3L ENGINE

1. Raise and safely support the vehicle.
2. Remove the bolts holding the sensor. Remove the timing bracket.
3. Remove the sensor.
4. Remove the sensor cable from the lower clips.
5. From above, disconnect the sensor cable connector.
6. Tie a string to the connector end of the harness.
7. Release the harness from the upper retaining clip. From below, carefully pull the sensor and cable out, leaving the string in its place. Remove the sensor from the string.

To install:

8. Installation is the reverse of the removal procedure. Tie the drawstring to the connector of the new sensor. Pull the sensor wiring harness into position from above.
9. Install the new sensor; tighten the bolts to 16–21 ft. lbs. (22–28 Nm).
10. Secure the wiring within the retaining clips.
11. Remove the drawstring. Connect the sensor connector to the wiring harness.
12. Connect the negative battery cable.

CRANKSHAFT SENSOR TOOTHED GEAR

The toothed gear is attached to the crankshaft and has 57 symmetrically spaced rectangular segments around the outside. Three omitted segments, which would give a total segment complement of 60, interrupt the induced signal to provide crankshaft position references.

Removal and Installation

1. Disconnect the negative battery cable.
2. Remove the air cleaner assembly from the backing plate. Clean the cover, box and mating faces.
3. Remove the air conditioning drive pulley. Slacken the alternator drive belt adjusting nut, the link arm adjusting nut and the alternator pivot bolt.
4. Pivot the alternator towards the engine and remove the drive belt.
5. Note the position of the pulley relative to the damper and remove the bolts securing the pulley to the damper.
6. Note the position of the balance weights and remove it from the pulley.
7. Install the front pulley lock (tool 18G1437 or equivalent) to the crankshaft damper and remove the damper retaining bolt.
8. Install the crank pulley remover (tool 18G1436 or equivalent) to the damper and remove the pulley from the crankshaft.
9. Remove the bolts securing the toothed gear to the damper and remove the gear from the damper.
10. Remove the roll pin from the damper and discard the damper.

To install:

11. Fit the roll pin to a new damper.
12. Install the toothed gear to the damper and tighten the retaining bolts.
13. Install the damper to the engine and start the damper retaining bolt. Install the front pulley locking tool and torque the retaining bolt 150 ft. lbs. (204 Nm).
14. Install the alternator drive belt over the pulleys. Align and fit the alternator drive belt pulley to the damper being certain that the balance weights are fitted in their respective positions noted previously. Secure the pulley.
15. Install the air cleaner assembly on the backplate. Reconnect the negative battery cable.

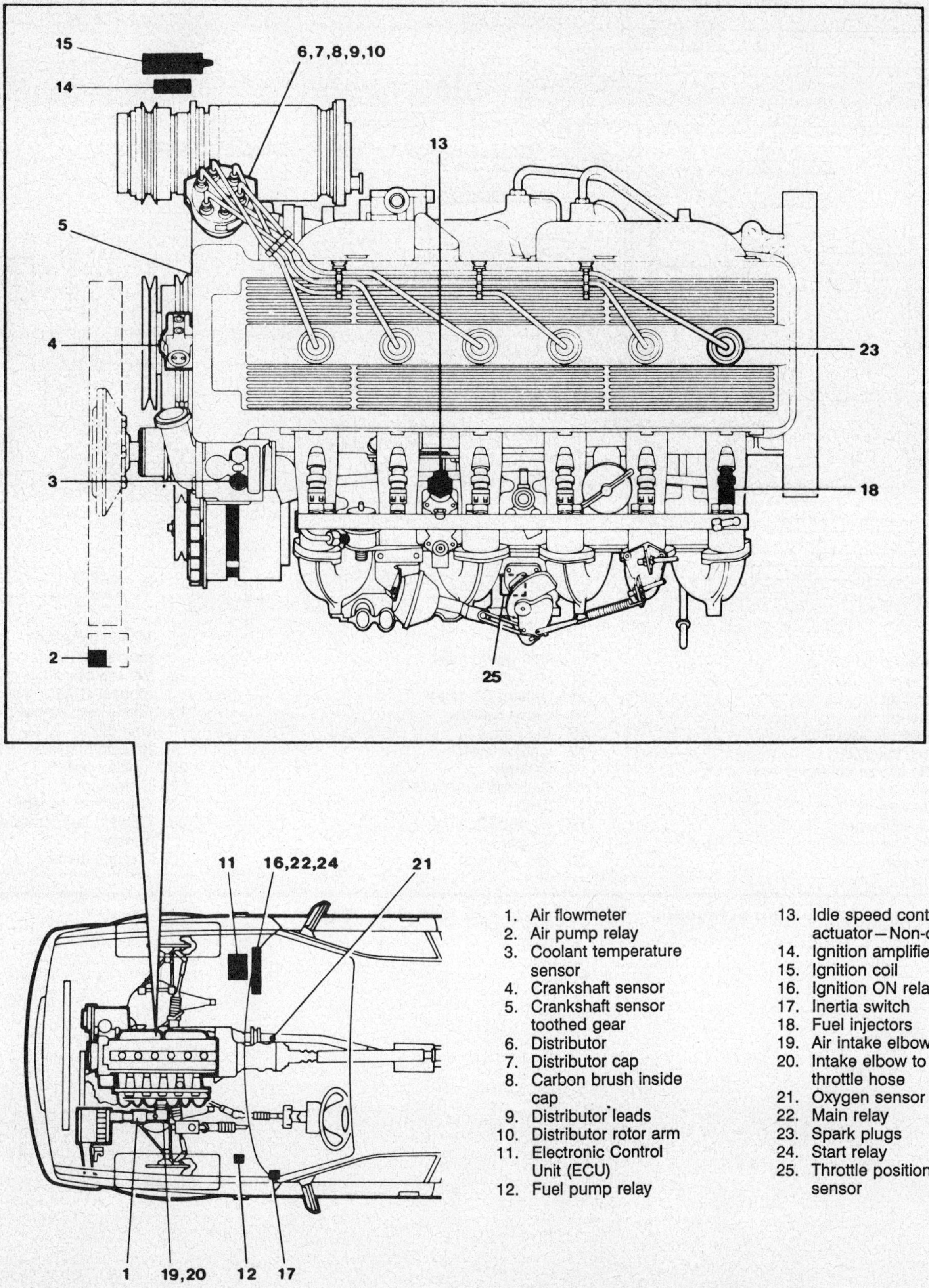

1. Air flowmeter
2. Air pump relay
3. Coolant temperature sensor
4. Crankshaft sensor
5. Crankshaft sensor toothed gear
6. Distributor
7. Distributor cap
8. Carbon brush inside cap
9. Distributor leads
10. Distributor rotor arm
11. Electronic Control Unit (ECU)
12. Fuel pump relay
13. Idle speed control actuator—Non-cruise
14. Ignition amplifier
15. Ignition coil
16. Ignition ON relay
17. Inertia switch
18. Fuel injectors
19. Air intake elbow
20. Intake elbow to throttle hose
21. Oxygen sensor
22. Main relay
23. Spark plugs
24. Start relay
25. Throttle position sensor

Engine management system: component layout— 1991 Jaguar XJ6 with 4.0L engine

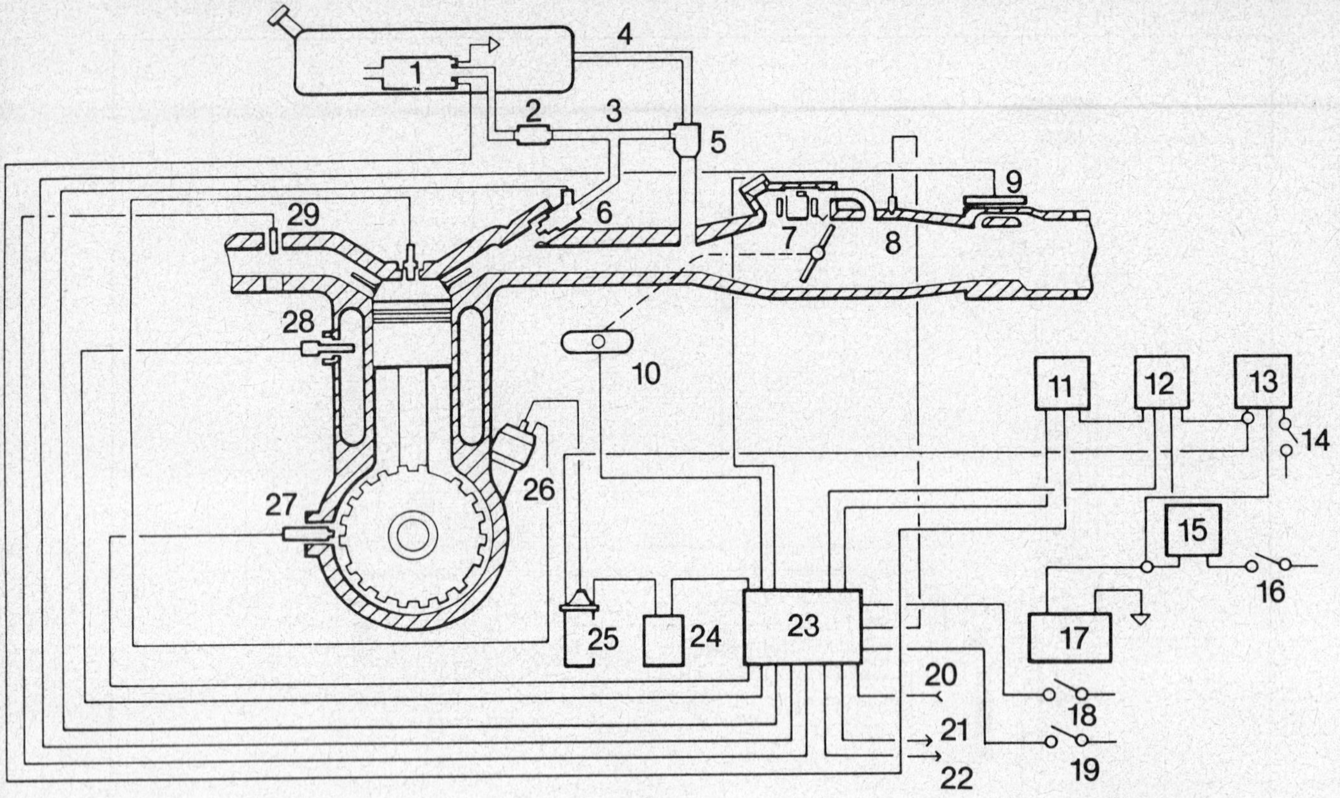

1. In-tank fuel pump
2. Fuel filter
3. Fuel rail
4. Fuel return
5. Pressure regulator
6. Fuel injector
7. Idle speed control actuator
8. Air thermistor
9. Air flowmeter
10. Throttle position sensor
11. Fuel pump relay
12. Main relay
13. Ignition ON relay
14. Inertia switch
15. Start relay
16. Ignition switch
17. Battery
18. Gearbox sensor (A/T only)
19. Air conditioning sensor
20. Idle override
21. Vehicle condition monitor VCM1
22. Vehicle condition monitor CM2
23. Electronic Control Unit (ECU)
24. Ignition amplifier
25. Ignition coil
26. Distributor
27. Crankshaft sensor
28. Coolant temperature sensor
29. Oxygen sensor

Engine management system schematic—1991 Jaguar XJ6 with 4.0L engine

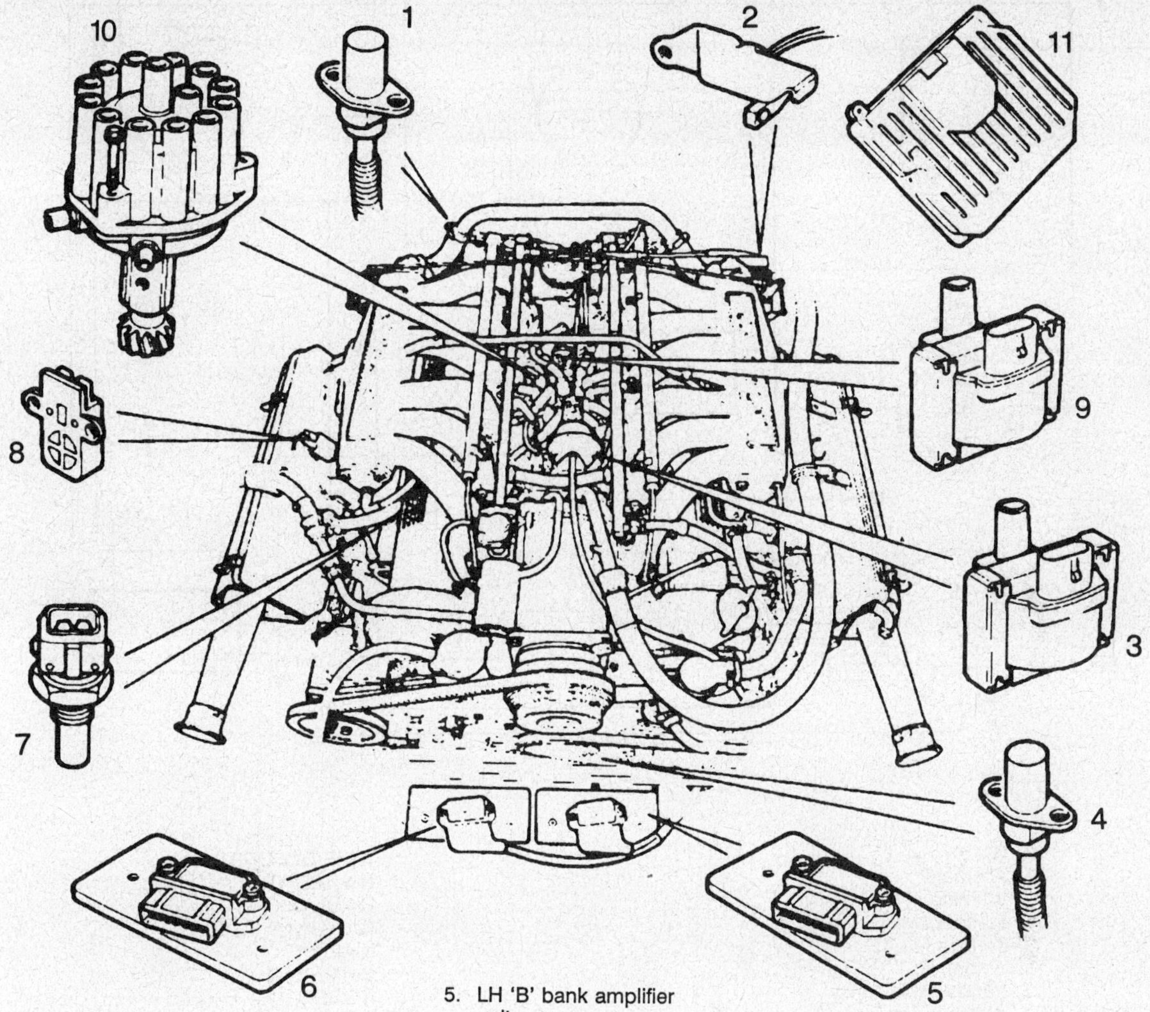

1. Flywheel (rpm) sensor
2. Throttle idle switch
3. Front ignition coil
4. Crankshaft sensor
5. LH 'B' bank amplifier unit
6. RH 'A' bank amplifier unit
7. Coolant temperature sensor
8. Air temperature switch
9. Rear ignition coil
10. Distributor
11. Electronic Control Unit (front passenger footwell)

Engine management system: component layout—1991 Jaguar XJS with 5.3L V12 engine

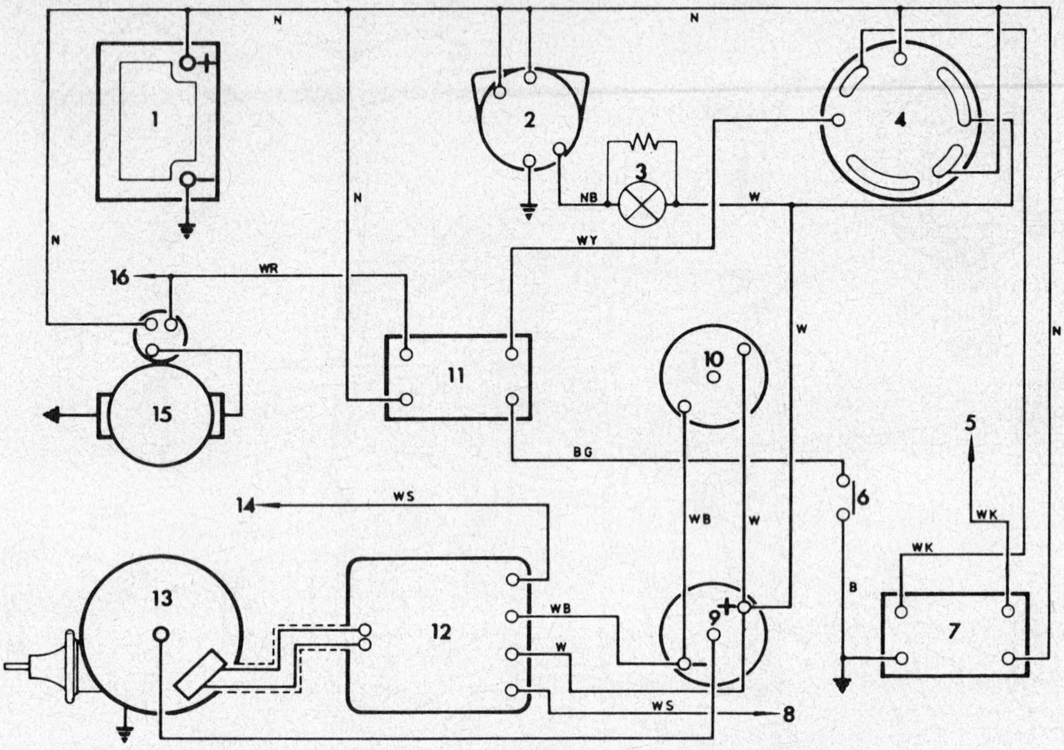

1. Battery
2. Alternator
3. Warning light
4. Ignition switch
5. To fuse No. 12
6. Automatic transmission safety switch
7. Ignition protection relay
8. To tachometer
9. Main coil
10. Auxiliary coil
11. Starter relay
12. Amplifier
13. Distributor
14. To ECU
15. Starter motor
16. To ECU

WIRE COLORS
N = BROWN
B = BLACK
W = WHITE
K = PINK
G = GREEN
R = RED
Y = YELLOW
O = ORANGE
S = SLATE (GRAY)
L = LIGHT
P = PURPLE
2-COLOR, FIRST = MAIN COLOR
SECOND = TRACER

Ignition, alternator and starter schematic— —XJS with 5.3L V12 engine shown

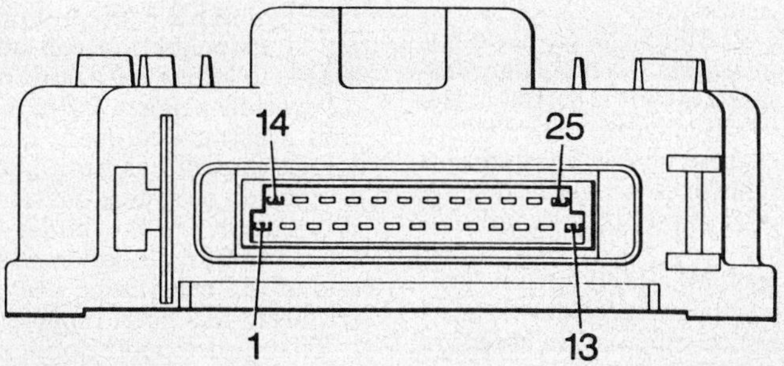

PIN No	CONNECTION		
1	TDC sensor	14	RH amplifier unit (6)
2	TDC sensor earth	15	RH amplifier unit (3)
3	Flywheel sensor earth	16	Flywheel sensor
4	Earth	17	Map selection
5	Switch earths	18	Idle switch
6	Water temperature sensor earth	19	Water temperature sensor
7	Not conected	20	Diagnostic request input
8	Not connected	21	Serial output (diagnostics)
9	LH amplifier unit (6)	22	Gearbox control 1
10	LH amplifier unit (3)	23	Air temperature switch
11	Earth	24	Engine speed output
12	Earth	25	Gearbox control 2
13	Ignition switch		

Typical ECU pin identification—XJS with 5.3L V12 engine shown

MERCEDES-BENZ

General Information

The electronic ignition system used on Mercedes gasoline engines is referred to as the EZL ignition system. The ignition timing and advance for typical load and speed ranges is stored in the ignition Electronic Control Unit (ECU). By means of various input signals, the ECU recognizes the current operating condition of the engine. The ECU uses all available data to match the current operating conditions and arrive at the optimum ignition timing for the respective conditions.

Major components of the system include a high output ignition coil, high voltage distributor, air temperature sensor, position indicator (crank sensor), interference suppression resistors on the distributor cap and an ignition control module.

The electronic ignition system with electronic timing adjustment and anti-knock control (EZL/AKR) is similar to the familiar EZL ignition system. It is used on the 2.6L, 3.0L, 3.2L, 4.2L, 5.0L and 6.0L engines.

The EZL/AKR control unit features a single output stage with an 11 amp ignition coil on 6 cylinder engines or dual output stages to twin 8 amp ignition coils on 8 cylinder engines. On the 12 cylinder engine, twin, mutually independent EZL/AKR ignition systems are used. Each has an 11 amp ignition coil.

SYSTEM OPERATION

EZL IGNITION SYSTEM

The position sensor is mounted at the flywheel and receives signals from segments on the flywheel. On 4 and 6 cylinder engines the flywheel has 2 segments 180 degrees apart. On 8 cylinder engines the flywheel uses 4 segments on the flexplate at 90 degrees apart. The sensor provides information on engine speed and crankshaft location.

Timing is not adjustable, but is based on the resistance of the (R16) reference resistor. If the resistor is not within specification or if increased resistance occurs in this circuit, ignition timing will be affected.

When starting the engine the ignition timing is controlled by the position sensor at the flywheel. When the engine has reached a certain speed the Electronic Control Unit (ECU) no longer uses the fixed ignition timing, but calculates the timing based on current operating conditions. During this warm-up phase certain ignition characteristics are blocked, depending on coolant temperature, to arrive at operating temperature more quickly.

At idle speed and during deceleration, when the idle speed contact of the throttle valve switch is closed, a fixed timing is used. This timing is not influenced by the coolant temperature sensor, intake manifold vacuum or the EZL reference resistor plug. Should the coolant temperature sensor fail, the fixed timing curve on the EZL control unit is no longer used and a certain ignition curve selected by the control unit is used.

At full load, the ignition timing is determined by the full load timing curve in the control unit. The control unit will consider the octane requirement of the engine and then select an advanced ignition curve for peak performance.

The full load condition is recognized by the ignition control unit from input signals sent by the intake manifold vacuum sensor and engine speed sensor. When the coolant temperature exceeds 203°F (95°C), the ignition timing is retarded to avoid a further temperature rise.

Depending on the operating condition, the ignition timing of certain curves can be adjusted by means of the EZL reference resistor plugs. Two types of plugs are used: an adjustable reference resistor plug and a fixed resistance individual reference resistor plug. If a resistor plug is not within specifications or if increased resistance occurs in this circuit, ignition timing will be affected.

EZL/AKR IGNITION SYSTEM

The EZL/AKR control unit features a single output stage with an 11 amp ignition coil on 6 cylinder engines or dual output stages to twin 8 amp ignition coils on 8 cylinder engines. On the 12 cylinder engine, twin, mutually independent EZL/AKR ignition systems are used. Each has an 11 amp ignition coil and knock control has been added to the system.

Ignition coils feed voltage through high tension wires to the distributor. On 6 cylinder engines a single distributor is used. On 8 and 12 cylinder engines, twin distributors, mounted to the end of the camshafts, are used. The moving carbon in the distributors has been replaced by a fixed central electrode.

To improve operational reliability, a data circuit has been installed between the EZL/AKR control unit and either the CIS-E or LH fuel injection control unit. Data concerning air temperature, vacuum, altitude and vehicle speed is sent to the EZL/AKR control unit from the fuel injection control unit. Data concerning coolant temperature, system malfunctions and fixed operating mode conditions is compared by both control units prior to action being taken. The EZL/AKR control unit and either the CIS-E or LH fuel injection control units are assembled in matched sets. Improper matching is recorded as a malfunction.

The ignition timing curves of the EZL/AKR system are designed for optimal output. Should combustion knock (detonation) occur, the position of the affected cylinder is detected by a combination of knock sensors and camshaft position sensors. A signal is sent back to the the anti-knock control in the EZL/AKR control unit, which responds by retarding the ignition point accordingly. Should the cylinder continue to knock, the ignition point will be retarded further until a maximum point, related to coolant temperature, is reached. If no further knocking occurs, the ignition point of the cylinder is advanced by 0.35 degrees per crankshaft revolution.

In the event a malfunction occurs in a knock sensor, the EZL/AKR control unit or a camshaft position sensor, a coolant temperature related retardation of the ignition point is activated for all cylinders.

A transmission overload protection is integrated into the EZL/AKR control unit in order to protect the shift elements of the automatic transmission from thermal overload during gearshifts under load in the upper RPM range. Provided certain conditions are met simultaneously, the ignition point is retarded to 5 degrees BTDC during the 1–2 and 2–3 upshifts. This measure is also used for the 3–2 full load downshifts.

Diagnosis and Testing

SERVICE PRECAUTIONS

- Never disconnect the control unit with the ignition switch ON.
- Do not connect any test light to terminal 1 of the coil. Do not short across terminal 1 and 15 of the ignition coil.
- Take note the coil terminal have different diameter terminal to avoid wrong polarity connections.
- Do not operate the ignition system if any component is disconnected.
- The resistance rating of the secondary wires, distributor cap and rotor are very specific. Make certain that replacement parts are correct for vehicle being serviced.
- Do not pull the coil wire or spark plug wires off while the engine is running.
- When using a cylinder balance tester, which shorts out cylinders, if the engine stalls the tester can not be used with this ignition system.
- If testing for spark is necessary at the secondary wires, it

must be done using a suitable spark tester securely attached to the wire and properly grounded.

- The control unit is mounted with heat conductive paste. Do not remove the protective foil as it does not impair heat dissipation.
- During ignition coil testing, do not load the coil with more than 28 kilo-ohms, to avoid damaging the coil.
- The German wiring color codes differ from the U.S. color codes. Take note of correct color codes before performing tests.

READING CODES

Mercedes Benz uses an electronically controlled Bosch K-Jetronic style Continues Injection System (CIS-E) on all 2.3L, 2.6L and 3.0L engines. On 3.2L, 4.2L, 5.0L and 6.0L engines, Mercedes Benz uses the Bosch LH-Jetronic (LH) style fuel injection. Input signals from various sensors and switches are continuously monitored by the system control units. If the control unit detects a signal out of the normal range, it will automatical-

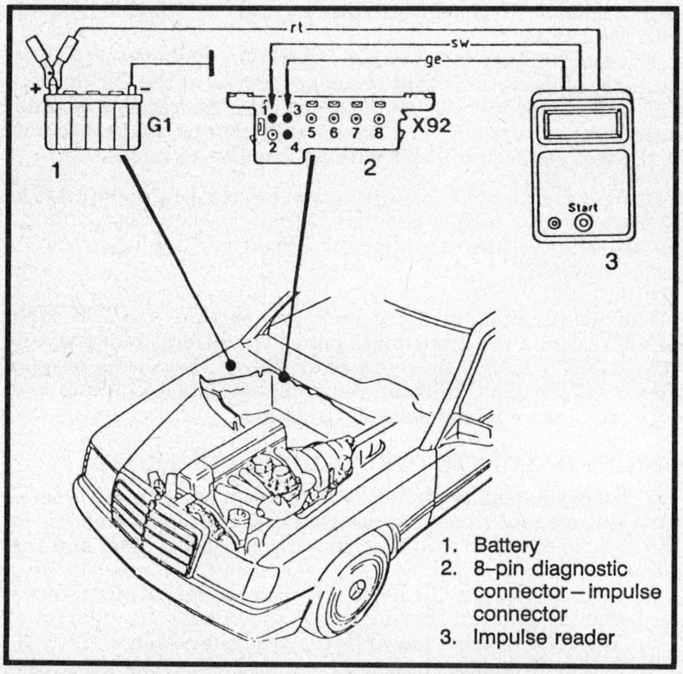

Impulse reader connection – Mercedes-Benz

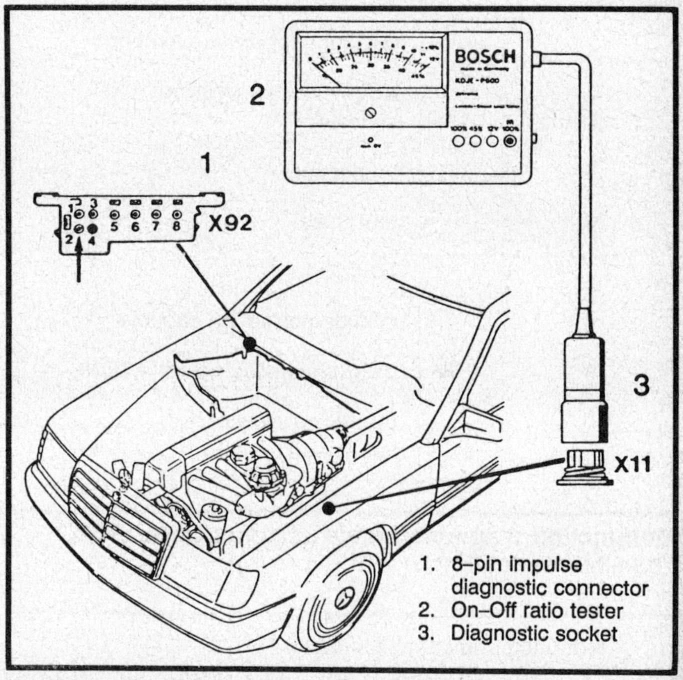

On-Off ratio connection – Mercedes-Benz

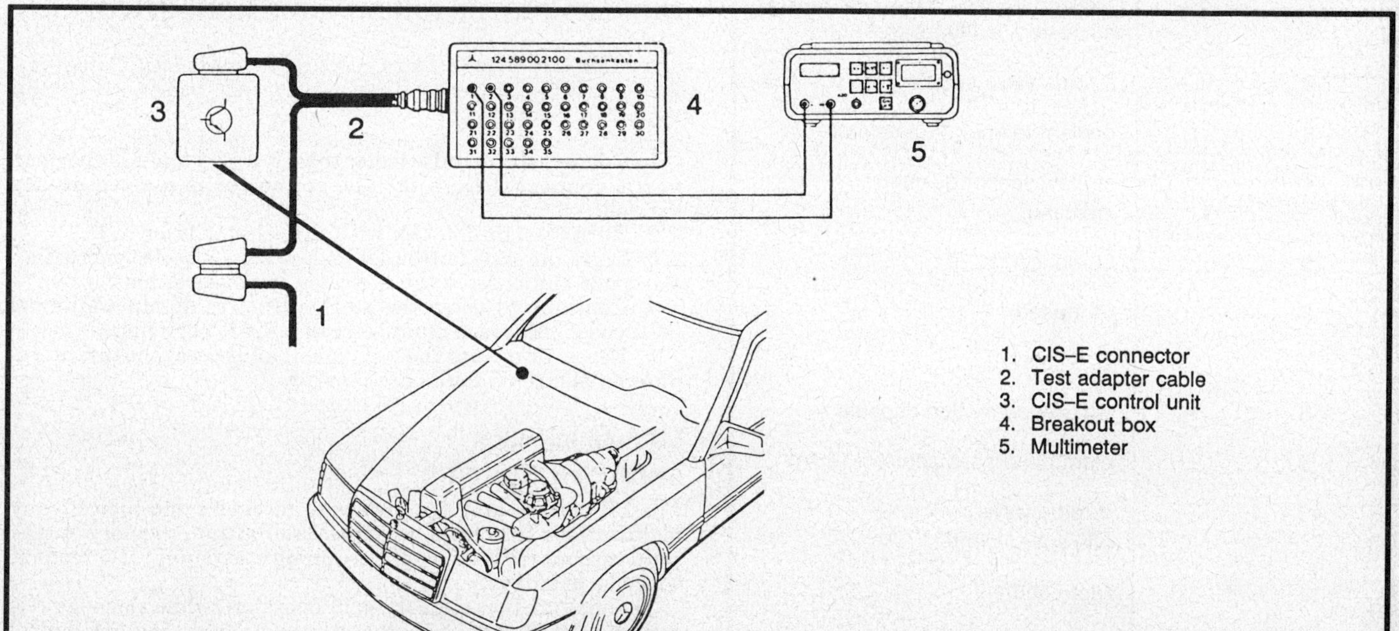

Breakout box connection – Mercedes-Benz

Malfunction diagnostics table

No. of blink impulses	Malfunctioning component
1	no system malfunction
2	throttle valve switch
3	coolant temperature sensor
4	air flow sensor position indicator
5	O_2-sensor
6	(not used)
7	TD-signal
8	altitude correction capsule
9	electrohydraulic actuator (EHA)
10	throttle valve switch idle speed contact

Malfunction diagnostic table—Engine codes 102, 103—Mercedes-Benz

Malfunction diagnostics table

No. of blink impulses	Malfunctioning component
1	no system malfunction
2	throttle valve switch
3	coolant temperature sensor
4	air flow sensor position indicator
5	O_2-sensor
6	(not used)
7	TD-signal
8	altitude correction capsule
9	electrohydraulic actuator (EHA)
10	throttle valve switch idle speed contact
11	(not used)
12	exhaust gas recirculation

Malfunction diagnostic table—Engine codes 116, 117—Mercedes-Benz

ly store a fault code and on some models the "CHECK ENGINE" lamp will illuminate. If the fault is severe enough to affect the driveability of the vehicle, the system will revert to a Fix Operating Mode (FOM).

CIS-E Fuel Injection

SELF DIAGNOSTICS

The CIS-E control unit has 3 ways to recognize and indicate any malfunctions that have occurred: malfunction memory, On-Off ratio readout with the engine running and On-Off ratio readout with the engine off.

The malfunction memory is in operation while the engine is running. Malfunctions that occur are stored in the CIS-E control unit. Intermittent contacts can also be recognized and stored as malfunctions. The stored malfunctions can be read out at the test connector X11/4 with an impulse counter.

NOTE: On the 2.3L engine, disconnecting the battery or control unit will erase any stored malfunction codes. On all other engines, codes must be cleared using the impulse tester.

The malfunction memory is linked to the "CHECK ENGINE" light on the instrument panel. On Federal versions, the "CHECK ENGINE" light comes on only if there is an oxygen sensor failure. On California versions, the light will illuminate with any failure in the emission system.

ON-OFF RATIO TESTER (LAMBDA CONTROL)

1. Start the engine and allow it to reach operating temperature. Ensure that all accessories are **OFF**, the selector lever is in **P**, the fuse of the overvoltage protection relay is good and the battery is fully charged.
2. Connect the On-Off Ratio tester to the diagnostic connector terminal (X11) in the engine compartment.
3. Turn the ignition key **ON** and read the tester.
4. Repair or replace the component and repeat the test procedure until no faults are present.

IMPULSE TESTER (ON-BOARD DIAGNOSTICS)

NOTE: Disconnecting the CIS-E control unit, battery or overvoltage protection relay will erase all codes stored in memory.

1. Connect the impulse tester to terminals 1 and 3 of the diagnostic connector (X11/4). The connector is located on the firewall.
2. Turn the ignition **ON** with the selector lever in **P**.
3. Press the start button for no less than 2, but no more than 4 seconds and observe the indicated impulse readout.
4. Continue to press the start button to obtain additional fault codes. All codes must be read prior to servicing.
5. Repair or replace the component and repeat the test procedure to ensure no faults are present.

LH Fuel Injection

SELF DIAGNOSTICS

The LH control unit has 3 ways to recognize and indicate any malfunctions that have occurred: malfunction memory, on-off ratio readout with the engine running and on-off ratio readout with the engine off.

The malfunction memory is in operation while the engine is running. Malfunctions that occur are stored in the LH control unit. Intermittent contacts can also be recognized and stored as malfunctions. The stored malfunctions can be read out at the test connector X11/4 with an impulse counter. After servicing, engine codes are cleared using the impulse tester.

Troubleshooting chart

Fixed on-off ratio %	Possible cause of failure	Tests to be performed
0 %	CIS-E control unit receives no voltage or is defective. Circuit to diagnostic socket, terminal 3 open or on-off ratio tester defective. Air/fuel mixture too rich. O_2-sensor signal voltage is +12 V.	Check voltage supply, ground connection and wires to diagnostic socket. Check on-off ratio tester. Check O_2-sensor signal.
10 %	Air flow sensor position indicator defective or polarity reversed. Increased rpm. Wires in plug of throttle valve switch incorrectly connected or short circuit (full load contact closed before engine reaches full load).	Check air flow sensor position indicator signal and routing of wires. Check throttle valve switch and wire connections according to wiring diagram. (Defective idle contact, ignition timing increased approx. 10°.)
20 %	Full load contact polarity reversed or defective. Reading at 20 % only with activated microswitch.	Check full load contact.
30 %	Short or open circuit between CIS-E control unit and coolant temperature sensor or coolant temperature sensor defective.	Check coolant temperature sensor.
40 %	Short or open circuit to air flow sensor position indicator or air flow sensor position indicator defective.	Check air flow sensor position indicator.
50 %	O_2-sensor not at operating temperature, open circuit or defective.	Check engine operating temperature. Check O_2-sensor.
60 %	Not used	–
70 %	No TD signal (open circuit) at CIS-E control unit.	Check TD signal.
80 %	Short or open circuit to altitude correction capsule or capsule defective.	Check altitude correction capsule.
90 %	Not used	–
100 %	Air/fuel mixture too lean, O_2-sensor defective (short circuit to ground). CIS-E control unit or on-off ratio tester defective. Defective overvoltage protection relay, no voltage supply.	Check adjustment of lambda control and O_2-sensor signal. Check power supply.
Reading oscillates	No malfunction within the circuit of the monitored signals. On-off ratio reading indicated.	–

On–Off ratio troubleshooting chart – typical – Mercedes-Benz

IMPULSE TESTER

1. Connect the impulse tester to the terminals the diagnostic connector (X11/4). The connector is located on the firewall.
2. Turn the ignition **ON** with the selector lever in **P**.
3. Press the start button for no less than 2, but no more than 4 seconds and observe the indicated impulse readout.
4. Continue to press the start button to obtain additional fault codes. All codes must be read prior to servicing.
5. Repair or replace the component and repeat the test procedure to ensure no faults are present.

PINPOINT TESTS

An On-Off ratio readout can be performed with the engine either running or stopped. With the engine stopped, the input signals to the CIS-E control unit are tested while the components are in a static condition. Malfunctions are not stored and can only be indicated momentarily. With the engine running, momentary and recognized malfunctions can be indicated while they are occurring. All On-Off ratio readouts are measured at the diagnostic test socket X11 using an On-Off ratio tester.

EZL/AKR Ignition System

SELF DIAGNOSTICS

Faults occurring during engine operation are counted by a fault counter. Only if the fault is still present after 8 starting operations will it be stored by the fault memory. Faults which prevent the starting of the engine or severely influence the engine operation will be immediately stored in memory. The fault memory is not erased if the battery is disconnected.

The recorded faults can be read using an impulse tester at the diagnostic test connector X11/4 (38 pole). Readout of the faults must be done when the engine is not running and the ignition is turned **ON**. Faults in the EZL system are diagnosed via a pinpoint test sequence. If a fault cannot be corrected after performing the tests, continue diagnosis in the CIS-E fuel injection system.

Engine Systems Control Unit

SELF DIAGNOSTICS

On SE and SEL models, the MAS control unit malfunctions must be recalled with a pinpoint box. On all other models, the impulse tester can be used at diagnostic connector X11/4. Connect impulse tester to black socket 4 and yellow cable to socket 14.

CLEARING CODES

Continuous Injection System (CIS-E)

ON-OFF RATIO TESTER (LAMBDA CONTROL)

Fault codes obtained using an On-Off ratio tester are only present during a malfunction. When the component is repaired or replaced, the fault code will automatically be erased.

IMPULSE TESTER (ON-BOARD DIAGNOSTICS)

1. With the impulse tester connected and the fault code displayed, press the start button for 6 seconds or more.
2. Each impulse code must be erased individually.
3. If there is no number displayed, then the stored fault has been erased.
4. If the display shows the number 1, then there are more system faults in memory.

Fault codes obtained using an impulse tester will be erased if the control unit, overprotection relay or negative battery terminal are disconnected. However, this is not the recommended procedure for erasing the stored codes.

NO START—NO SPARK

Testing

1. Using a properly grounded spark plug tester, test for spark at spark plugs, while cranking the engine.
2. Check the connection of the ignition coil and distributor.
3. Check the resistance of the spark plug wires.
4. Check that there is battery voltage to terminal 5 of the diagnostic socket or the coil positive terminal with the ignition switch **ON**.
5. Check that there is 0 volts between terminal 5 and 4 of the diagnostic socket.
6. Check that there is battery voltage between terminals 15 and 31 (ground) of the ignition module.
7. Check that there is battery voltage between terminals 16 and 31 (ground) of the ignition module.

REFERENCE RESISTOR (R16/1)

NOTE: United States vehicles are now equipped with an individual (non-adjustable) type resistor, in place of the variable resistance (R16) type plug.

Testing

This resistor must be of the correct resistance. Check the resistance between terminal 3 of the control unit and ground. The resistance should be approximately 750 ohms. If not, repair the circuit or replace the resistor.

A resistor of higher or lower resistance will effect timing; for example: if the resistor only had 470 ohms the timing could be retarded as much as 3 degrees, and a resistance increased to 1.3 kilo-ohms would increase timing. This resistance signal is not to be used as a timing adjustment.

IGNITION COIL

Resistance Test

1. Using a suitable ohmmeter, measure the resistance between the positive and negative terminals.
2. The primary coil resistance (cold) should be 0.3–0.6 ohms.
3. If the resistance is not within specifications, replace the coil.
4. Measure the resistance between the positive terminal and the high tension terminal.
5. The secondary coil resistance (cold) should be 7–13 kilo-ohms.
6. If not within specifications, replace the coil.

POSITION SENSOR

Resistance Test

1. Using a suitable ohmmeter, measure the resistance between the terminals.
2. The resistance should be approximately 680–1200 ohms.
3. If not within specifications, replace the sensor.

REFERENCE INPUT RESISTOR

Testing

1. Using a suitable ohmmeter, measure the resistance between the terminals.
2. The resistance should be approximately 760 ohms.
3. If not within specifications, replace the resistor.

IGNITION TESTING PROCEDURE—MERCEDES-BENZ

Test sequence - survey

A. Engine not running

Check dwell angle at starting speed.

Rated value: 1-30° or 1-50%

yes → Check ignition timing at starting speed.

no → Check voltage supply.
Check open-circuit current cutoff.
Check position pickup.

→ Recheck dwell angle at starting speed.

yes → Check ignition timing at starting speed.

no → Renew ignition control unit.

Check ignition timing at starting speed.

Rated value: TDC ± 2° crankshaft

yes → End of test

no → Check ignition coil.
Check high-voltage distributor.

→ End of test

Note
When complaint cannot be rectified, continue trouble diagnosis in CIS-E injection system.

B. Engine running

Check ignition timing.

Rated value:

yes → Check dwell ange at 3200 min.

no → Check reference resistor plug.
Check throttle valve switch.
Check pressure sensor.
Check temperature sensor coolant.

→ Check dwell ange at 3200 min.
Check high-voltage distributor.
Check voltage supply.
Check ignition coil.

→ End of test.

Note
When complaint cannot be rectified, continue trouble diagnosis in CIS-E injection system.

IGNITION TESTING PROCEDURE—MERCEDES-BENZ

A. Engine fails to run

Check dwell angle at starting speed
Connect measuring instruments.
Start engine.
Check dwell angle on jack 1 of the diagnostic socket.

Rated value: 1-30° or 1-50%

yes | no

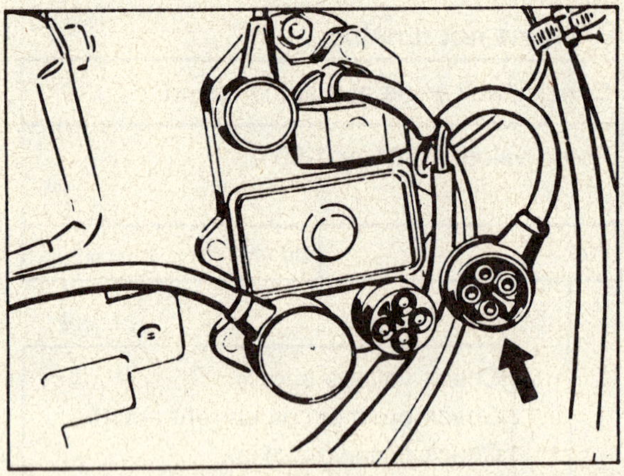

Check voltage supply
Pull off 4-pole round supply plug on ignition control unit.
Switch on ignition.
Measure voltage on plug between terminal 15 (corresponding to jack 5 of the diagnostic socket) and terminal 31.
Rated value: battery voltage.

yes | no

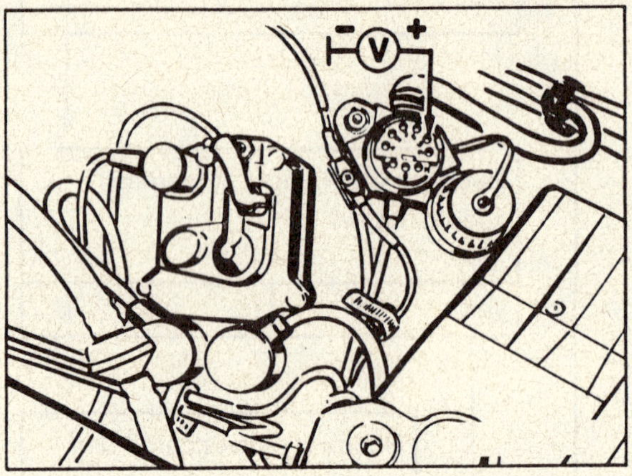

Check voltage supply via ignition lock.
Check ground cable.

Check open-circuit current cutoff
Check voltage differential between jacks 5 and 4 (terminals 15 and 1) of the diagnostic socket.

Rated value: 0 volt

yes | no

Renew ignition control unit and ignition coil.

IGNITION TESTING PROCEDURE—MERCEDES-BENZ

Check position pickup
Switch off ignition.

Pull ignition pickup plug off ignition control unit. Measure pickup resistance between terminals 7 and 31d with the ohmmeter.

Rated value: 680-1200 Ω

yes	no

no → Renew position pickup.

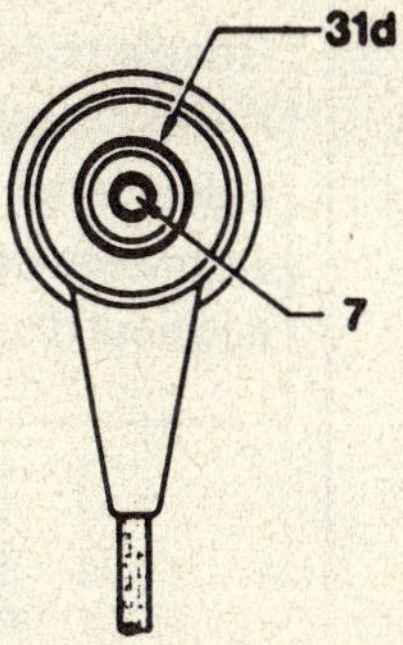

yes → Check position pickup for ground connection.

Check resistance between terminal 7 and vehicle ground.

Rated value: ≥200 kΩ or ∞

yes	no

no → Visual inspection of control cable.
Renew position pickup.

yes → Check alternating voltage from the position pickup with the oscilloscope.
Pull off control cable.
Measure alternating voltage between terminals 7 and 31d at starting speed.

Rated value: $U_s \geq 1$ volt

yes	no

IGNITION TESTING PROCEDURE—MERCEDES-BENZ

Renew position pickup.
If alternating voltage remains below rated value, check segments on the flywheel driven plate.

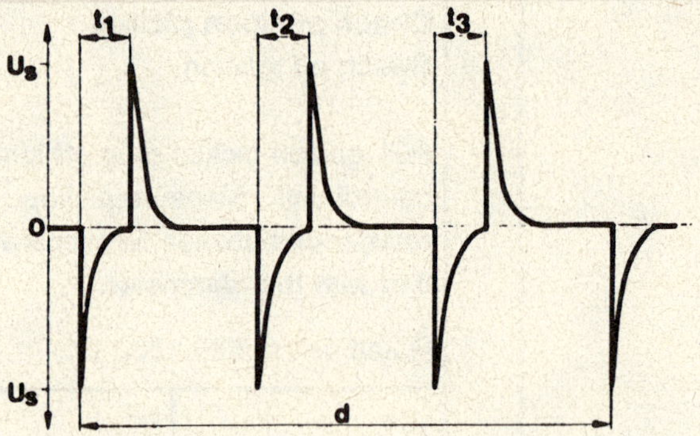

- t1 1st segment
- t2 2nd segment
- t3 3rd segment
- d Period for one crankshaft rotation

Check dwell angle at starting speed
Connect measuring instruments.
Start engine.
Check dwell angle on jack 1 of the diagnostic socket (terminal TD).

Rated value: 1-30° or 1-50%

yes	no

no → Renew ignition control unit.

Check ignition timing at starting speed

Rated value: TDC ± 2°

yes	no

no → **Check ignition coil**
(Unscrew cables from the ignition coil).

Rated values:
Terminals 1 and 15: 0.3-0.6 Ω
Terminals 1 and 4: 8-13 kΩ

yes	no

no → Renew ignition coil.

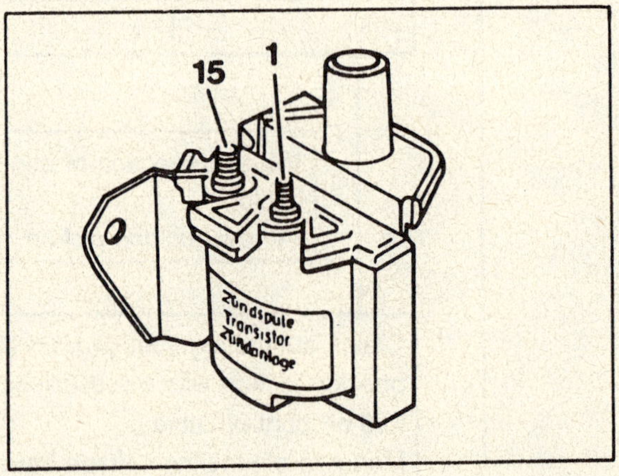

IGNITION TESTING PROCEDURE—MERCEDES-BENZ

Check high-voltage distributor
Visually inspect distributor cap and rotor.
Measure interference suppression resistors.

End of test.

B. Engine running

Check ignition timing
Engine at operating temperature.
Coolant temperature below 95 °C.
Check ignition timing at idle speed.
Check ignition timing at 3200 min with and without vacuum (disconnect vacuum line on control unit).
Observe reference resistor plug position.
Rated values:

yes	no

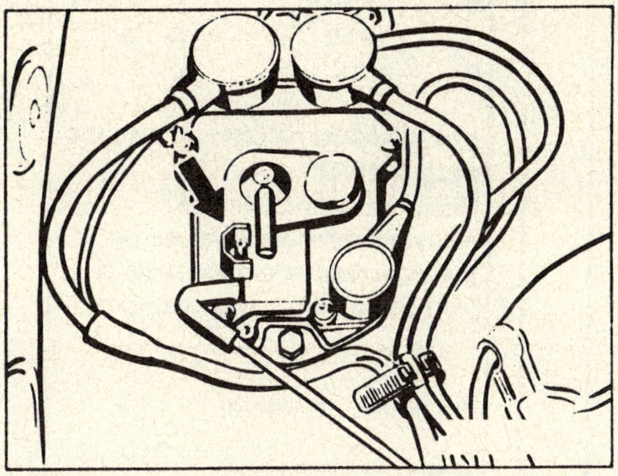

Check reference resistor plug
a) Reference resistor plug (R16)
Compare engine version and reference resistor plug lettering. Pull 4-pole plug-pickup (arrow) off ignition control unit.
Measure resistance value between jack 3 and vehicle ground.

b) Individual reference resistor plug (R16/1)
Compare national version and part number.
Pull 4-pole plug-pickup (arrow) off ignition control unit.
Measure resistance value between jack 3 and vehicle ground.

yes	no

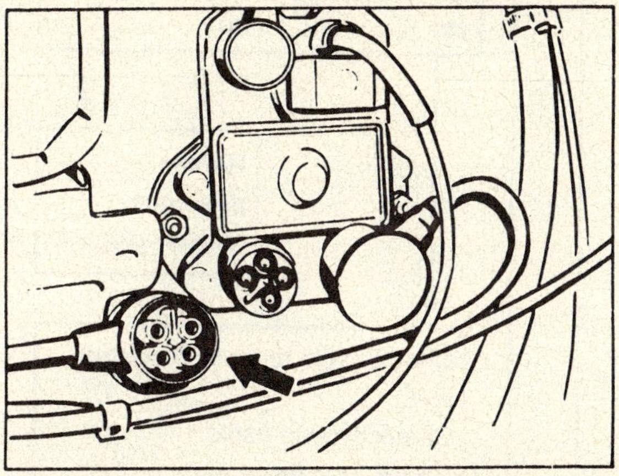

IGNITION TESTING PROCEDURE—MERCEDES-BENZ

a) Reference resistor plug (R16)
Check resistance values in the reference resistor plug (between the outer and center plug pins).

Rated values for all reference resistor plugs:

A		E	470 Ω
B	2.4 kΩ	F	220 Ω
C	1.3 kΩ	G	0 Ω
D	750 Ω		

b) Individual reference resistor plug (R16/1)
Check resistance values measure resistance values on the individual reference resistor plug between the two plug pins.

Part number	Resistance
000 540 24 81	750 Ω

yes | no

Renew reference resistor plug.

Check cable to ignition control unit.
Check ground cable.
Check plug lower part.

yes | no

Renew damaged parts.

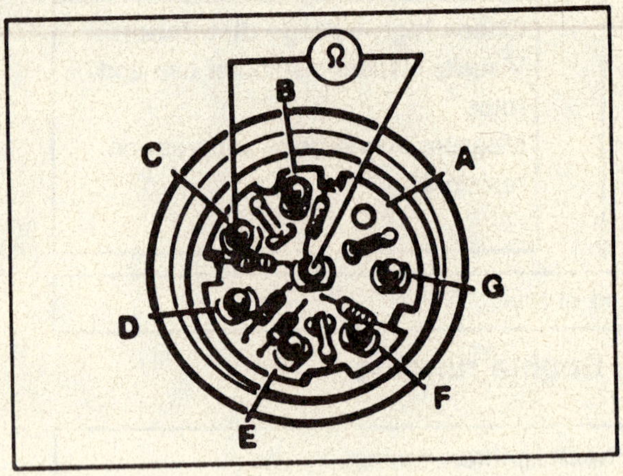

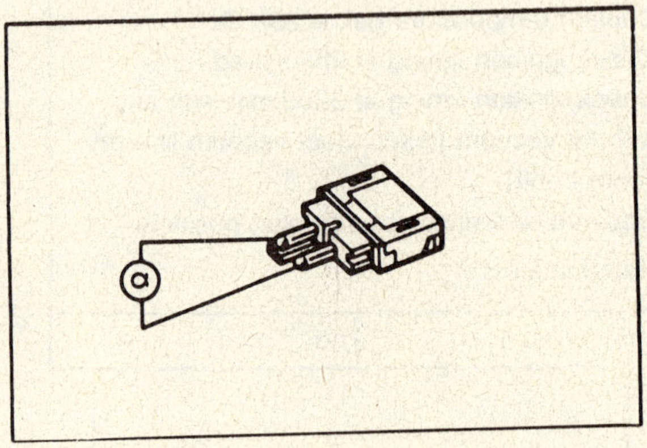

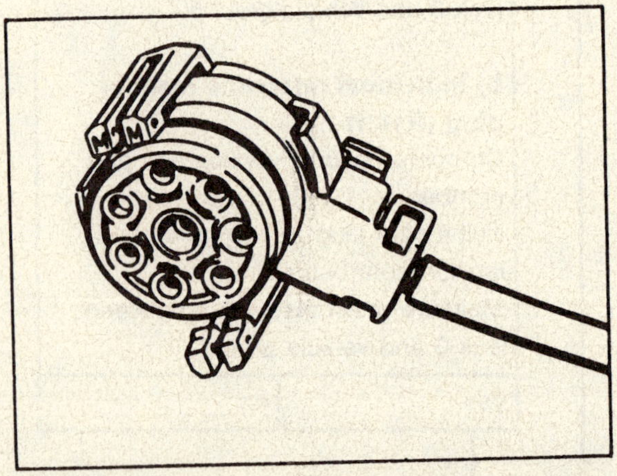

IGNITION TESTING PROCEDURE—MERCEDES-BENZ

Disconnect vacuum line on control unit.
Run engine at approx. 1200 min (not idling).
Ignition timing changes when turning the reference resistor plug.

yes	no

no → Renew ignition control unit.

yes →

Check throttle valve switch
Disconnect coupling on throttle valve switch.
Check resistance of throttle valve switch.

Idle speed position	approx. 0 Ω
Full load	approx. 0 Ω
Part load position	approx. ∞ Ω

yes	no

no → Adjust or renew throttle valve switch.

yes →

Check cable to ignition control unit.
Check ground cables.

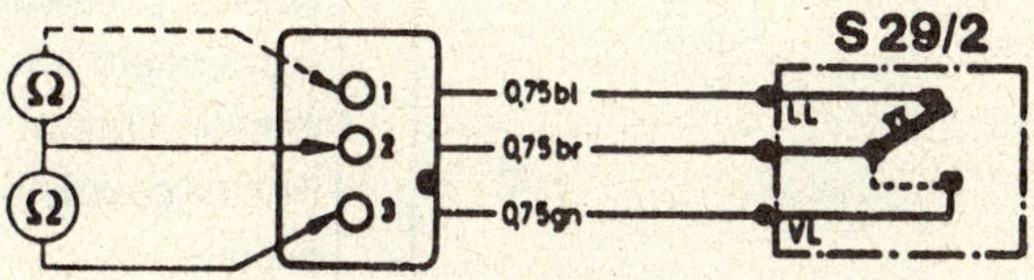

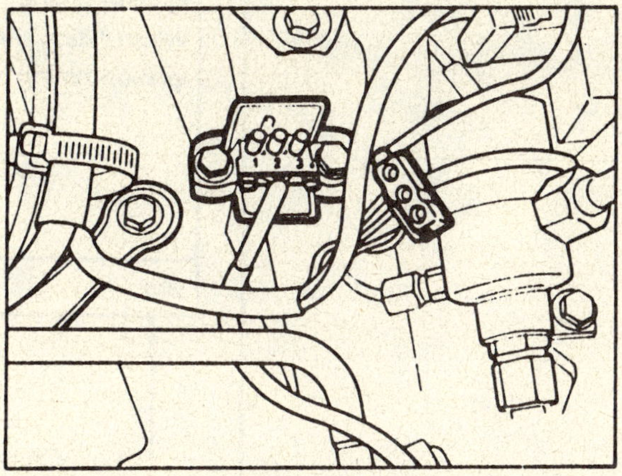

IGNITION TESTING PROCEDURE—MERCEDES-BENZ

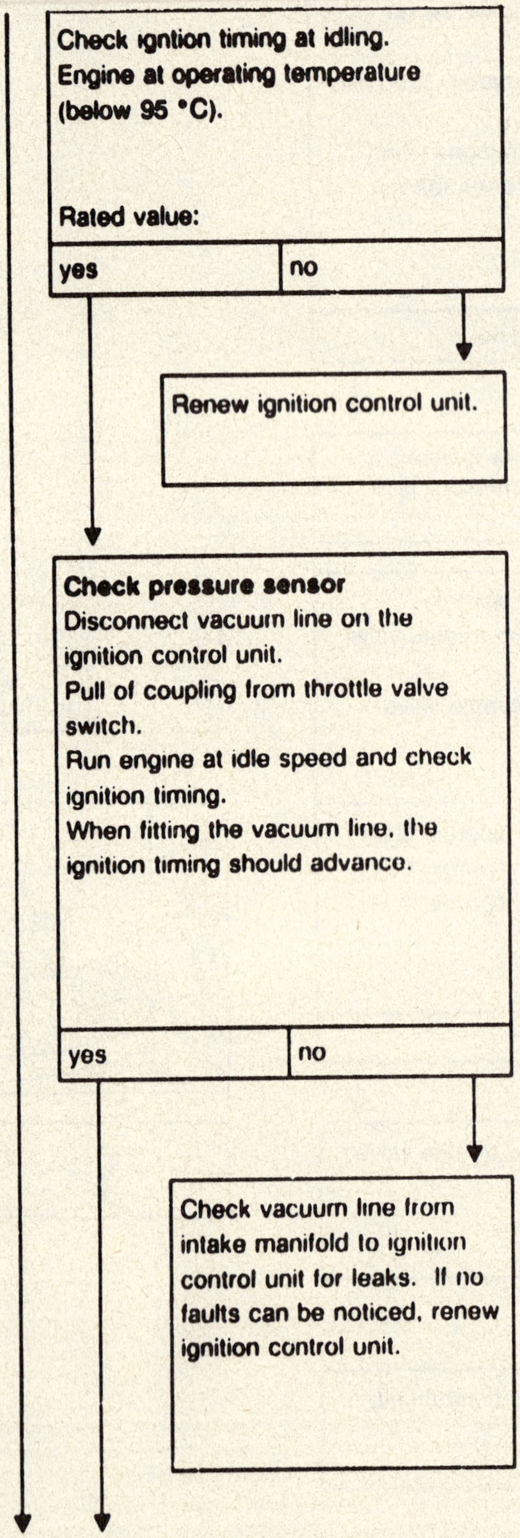

IGNITION TESTING PROCEDURE—MERCEDES-BENZ

Check temperature sensor, coolant
Disconnect 4-pole plug-pickup on control unit.
Measure resistance between jack 1 and ground.
Rated value see diagram.
Check resistance at two temperatures.

Example:
+ 20 °C ⇒ 2.28-2.72 kΩ
+ 80 °C ⇒ 290-364 Ω

yes	no

(no →) Check cable.
Renew temperature sensor, coolant.

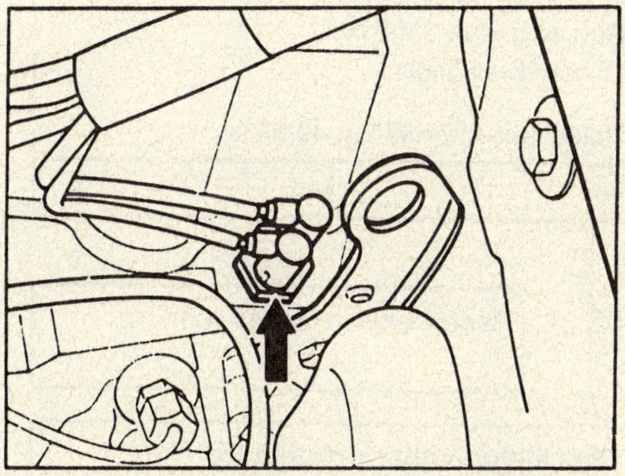

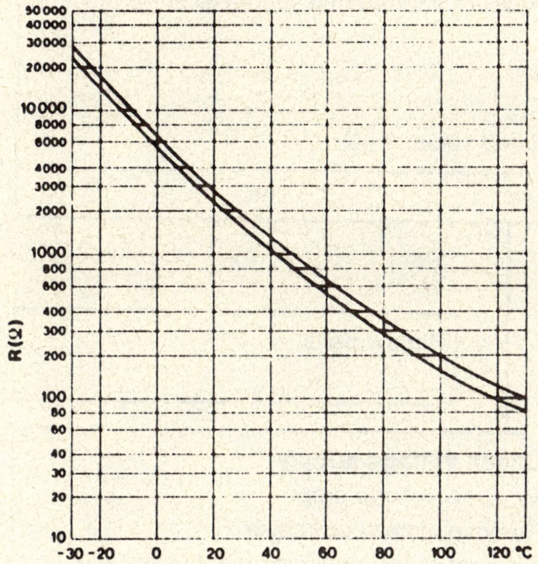

(yes →) Check ignition timing at 3200 min without vacuum.

Rated value:

yes	no

(no →) Renew ignition control unit.

IGNITION TESTING PROCEDURE – MERCEDES-BENZ

Check dwell angle at 3200/min
Vacuum line and 4-pole plug (pickup) on ignition contol unit connected.
Run engine at 3200 min.
Check dwell angle

Rated value: 24-53° or 40-88%

yes	no

no → Renew ignition control unit.

yes ↓

Check high-voltage distributor
Visually inspect distributor cap and rotor.
Measure interference suppression resistors.

Rated value: 1 kΩ

yes	no

no → Renew parts.

yes ↓

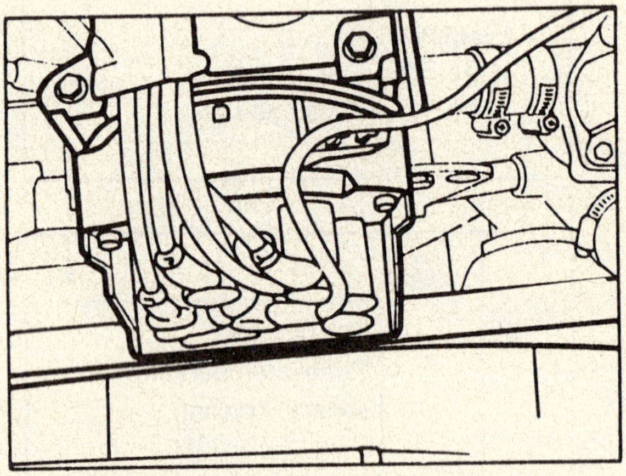

Check voltage supply
Measure voltage drop:
Remove ignition coil cover.
Connect measuring instrument to negative terminal of battery.
Run engine.
Measure voltage on the positive terminal of the battery and subsequently on terminal 15 of the ignition coil.

Rated value: voltage differential < 1 volt

yes	no

IGNITION TESTING PROCEDURE— MERCEDES-BENZ

Check contact resistance:
Loosen battery terminals.
Pull 4-pole plug (supply) off ignition control unit. Switch on ignition.
Measure resistance between positive terminal and terminal 15 on the plug.
Measure resistance between positive terminal and terminal 15 on the ignition coil.
Measure resistance between negative terminal and terminal 31 on the plug.
Measure resistance between terminal 1 on the ignition coil and terminal 16 on the plug.

Rated value: < 1 Ω

yes	no

no → Check plug and screw connections of the voltage supply via ignition lock. Check ground cables.

Check ignition coil
Evaluate by means of oscillogram with running engine.

End of test.

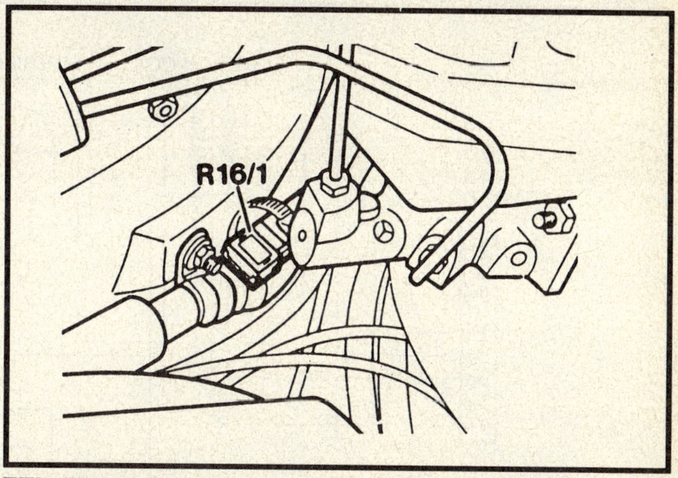

EZL (R16) reference resistor—260E and 300E—Mercedes-Benz

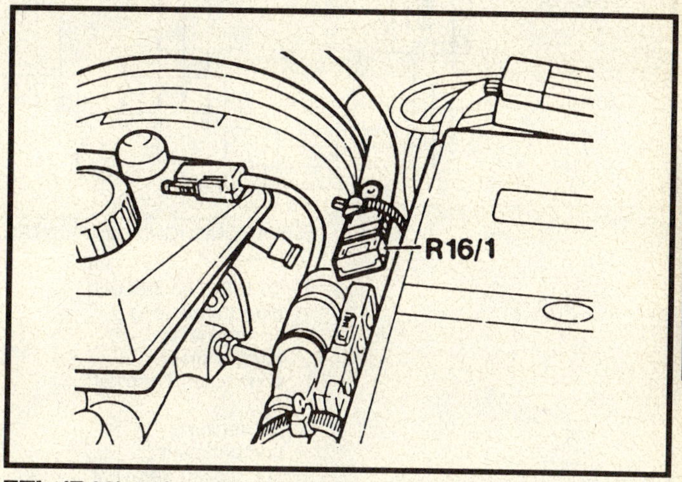

EZL (R16) reference resistor—400 SEL—Mercedes-Benz

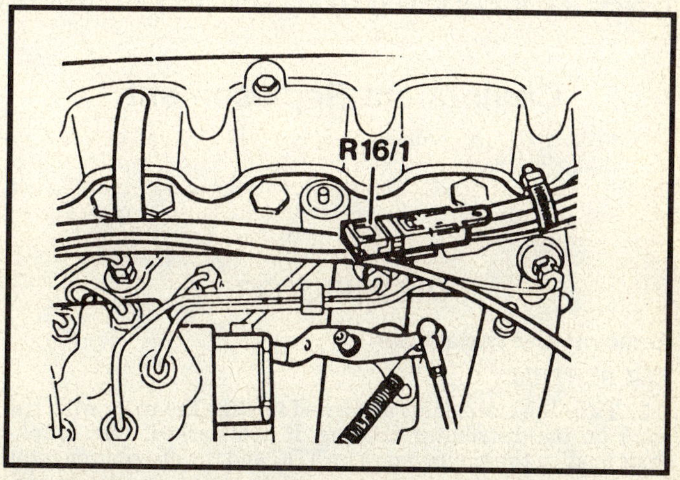

EZL (R16) reference resistor—190E, others similar—Mercedes-Benz

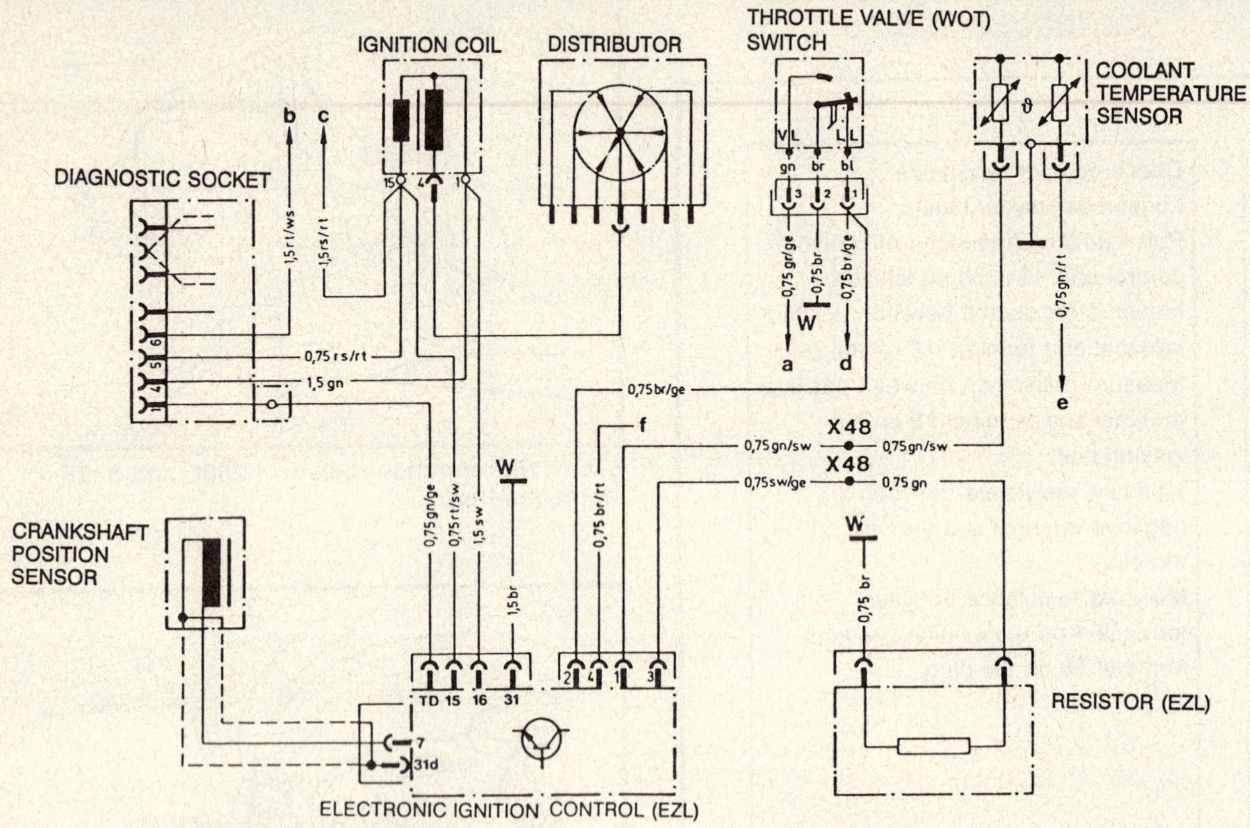

a. To CIS-E control unit plug (terminal 5)
b. To engine compartment connection (terminal 3)
c. To engine compartment connection (terminal 1)
d. To CIS-E control unit (terminal 13)
e. To CIS-E control unit (terminal 21)
f. To electrical center connector U (terminal 6)—when used
W. Ground

Typical electronic ignition (EZL) circuit—Mercedes-Benz

Component Replacement

NOTE: When disconnecting the battery cable, remember that the engine computer, clock, radio and security system memories will be cleared. This can cause the radio not to function due to programmed security codes and temporary driveability problems.

DISTRIBUTOR

Removal and Installation

ELZ SYSTEM

1. Turn the crankshaft as needed to align the rotor with the mark on the distributor housing. If not marked, turn crankshaft to align the timing mark on TDC and the distributor rotor facing No. 1 plug wire.
2. Matchmark the distributor housing to the engine if not equipped with factory markings.

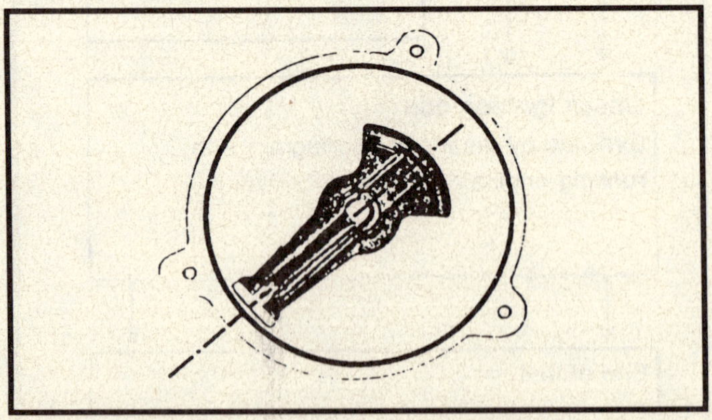

Aligning distributor rotor with mark—Engine codes 116, 117 shown—Mercedes-Benz

3. If necessary, disconnect the negative battery cable.
4. Remove the dust cover, if equipped.
5. Tag and disconnect the coil and spark plug wires.
6. Remove the distributor retaining bolt and remove the distributor.

To install:

7. Install the distributor, taking care to align the distributor housing and rotor with proper matchmarks.
8. Install the distributor cap, spark plug wires and coil wire.
9. Install the dust cover, if necessary.
10. Connect the negative battery cable, reprogram radio security code, as needed.

ELZ/AKR SYSTEM

NOTE: It is not necessary to set the ignition timing after servicing the distributor. The ignition point is fixed by the EZL/AKR control unit.

1. Disconnect the negative battery cable.
2. Remove the cover on the distributor by pulling it straight up.
3. Label and remove the high tension wires.
4. Remove the hex screw and the distributor cap.
5. Remove the rotor by loosening the screws.
6. Remove the follower by loosening the hex screw.
7. Remove the rear cover and O-ring.

To install:

8. Install the rear cover using a new O-ring.
9. Install the follower paying special attention to the locating pin and follower groove.
10. Install the rotor and tighten the screws.
11. Install the distributor cap and tighten the screw.
12. Install the high tension wires and distributor cover.
13. Connect the negative battery cables.

IGNITION COIL

Removal and Installation

1. Disconnect the negative battery cable.
2. Remove the cover over the left headlight.
3. Remove the cover for the ignition coil.
4. Label and remove all electrical connections from the coil.

NOTE: The positive cable has an M5 thread and the negative has an M6 thread.

5. Remove the nuts attaching the coil to the vehicle. Remove the coil.

To install:

6. Install the coil and tighten the nuts.
7. Install all electrical connectors on the coil.
8. Install the coil cover.
9. Install the left headlight cover.
10. Connect the negative battery cable.

EZL/AKR CONTROL UNIT

Removal and Installation

1. Disconnect the negative battery cable.
2. Remove the vacuum connection.
3. Remove the coaxial connector from the crankshaft position sensor.
4. Remove both 8 pin connectors.
5. Remove the 3 pin knock sensor connectors.
6. Remove the trimming plug connector.

NOTE: The control unit is provided with heat conducting paste on the back for improving heat dissipation and is covered with a heat conducting protective foil. The protective foil must not be removed.

7. Remove the control unit attaching nuts. Remove the control unit.

To install:

8. Install the control unit paying special attention to the protective foil. Tighten the attaching nuts.
9. Install the trimming plug connector.
10. Install the 3 pin connector, 8 pin connectors, coaxial connector for the crankshaft position sensor and vacuum connection.
11. Install the negative battery cable.
12. Check the system for faults at diagnostic connector X11/4.

CRANKSHAFT POSITION SENSOR

Removal and Installation

NOTE: Stored faults which may be caused by detaching the cables or simulation during testwork must be erased in the fault memory after servicing.

1. Remove the screw attaching the sensor to the engine.
2. Disconnect the cable leading to the control unit.
3. Remove the crankshaft position sensor.
4. Install the sensor and tighten the screw to 88 inch lbs. (10 Nm).
5. Connect the cable leading to the control unit.
6. Erase fault codes from memory.

Adjustment

1. Rotate the engine until the crankshaft pointer reads 25 degrees BTDC.
2. Measure the distance "A" from the camshaft position sensor to the contact pickup on the camshaft gear.
3. Measure the distance "B" between the contact face and position sensor without the spacer installed.
4. Calculate the actual distance by subtracting distance "B" from distance "A". It should be 0.010–0.020 in. (0.4–0.6mm.). If not, insert an appropriately sized spacer.

KNOCK SENSOR

Removal and Installation

NOTE: On the 4.2L, 5.0L and 6.0L engines, the cables of the knock sensors are combined in a protective tube. They can only be replaced as a unit. All other engines use a single knock sensor.

1. Remove the bolts from the knock sensor.
2. Remove the electrical connector at the control unit.
3. Unclip the protective tube. Remove the knock sensor.
4. Install the knock sensor and clip the protective tube in place.
5. Connect the electrical connector at the control unit.
6. Install the bolts and tighten to 14 ft. lbs. (20 Nm).

TIMING ADJUSTMENT

The distributor used on the EZL system only has the duty of distributing the ignition voltage. Timing is not adjustable. However, timing can be verified using conventional methods.

Checking Ignition Timing

1. Start the engine and allow coolant temperature to reach 167°–194°F (75°–90°C).
2. Disconnect the coolant temperature sensor connector and connect a resistance of 320 ohms between sockets **1** and **3** on the coolant temperature sensor plug.
3. Disconnect the intake air temperature sensor connector.
4. Check ignition timing at idle, it should be 8–12 degrees BTDC with the vacuum hose connected.
5. Check ignition timing at 3200 rpm, it should be 39–43 degrees BTDC with the vacuum hose connected.
6. If the ignition timing is not within specification, perform the ignition test procedure.

MERKUR

General Information

The Merkur Scorpio uses a Thick Film Integrated Ignition System (TFI-IV). The TFI-IV system uses a universal distributor which incorporates a base mounted TFI-IV ignition module self-contained in a molded thermo-plastic unit, a Hall Effect vane switch stator assembly and an octane rod. The TFI-IV system uses a E-core ignition coil and also a separate Electronic Control Assembly (ECA) to collect and transmit all information within the TFI-IV circuitry.

The universal distributor contains a provision to change the basic distributor calibration with the use of a replaceable octane rod, from the standard of 0 degrees to either 3 or 6 degree retard rods. No other calibration is possible.

NOTE: Do not change the ignition timing by the use of a different octane rod without having the proper authority to do so as Federal Emission requirements will be effected.

The TFI distributor contains no vacuum advance components. Initial timing adjustment is not normally required unless the distributor has been removed from the the engine.

The TFI-IV system also incorporates a push start mode which allows push starting, applicable to manual transaxle vehicles only.

SYSTEM OPERATION

System operation is accomplished through the Hall Effect vane switch assembly causing the ignition coil to be switched **OFF** and **ON** by the Electronic Control Assembly (ECA) and TFI-IV modules. The vane switch consist of a Hall Effect sensor on one side and a permanent magnet on the other side.

A rotary vane cup, made of ferrous metal is used to trigger the Hall Effect switch. When the window of the vane cup is in alignment with the magnet and the Hall Effect device, a magnetic flux field is induced from the magnet through the Hall Effect device and back to the magnet at 0 volts.

As the vane passes through this opening, the flux lines are shunted through the vane and back to the magnet. A voltage output is produced as the vane passes through the opening. When the vane clears the opening, the window causes the signal to go to 0 volts. The signal is then used by the EEC-IV system for crankshaft position sensing and the computation of the desired spark advance based on the engine demand an calibration. The conditioned spark advance and voltage distribution is accomplished through a conventional rotor, cap and ignition wires.

In addition to the Hall Effect switch, the ignition module contains the Profile Ignition Pick-up (PIP) sensor, which sends an electronically oriented crankshaft position signal to the ECA and the TFI module circuitry. The ECA, after receiving all sensor information, produces a new signal called the Spout or spark output signal. This Spout signal is then sent back to the TFI module for comparison with the PIP signal. The TFI-IV uses both of these signals to fire the ignition coil at the proper timing interval.

ELECTRONIC CONTROL ASSEMBLY (ECA)

The brain of the EEC-IV system is a micro-processor called the Electronic Control Assembly (ECA). The Electronic Control Assembly (ECA) receives data from a number of sensors, switches and relays. It contains a specific calibration for optimizing emissions, fuel economy and drive ability. Based on information received and programs in memory, the Electronic Control Assembly (ECA) generates output signals to control the various relays, solenoids and actuators.

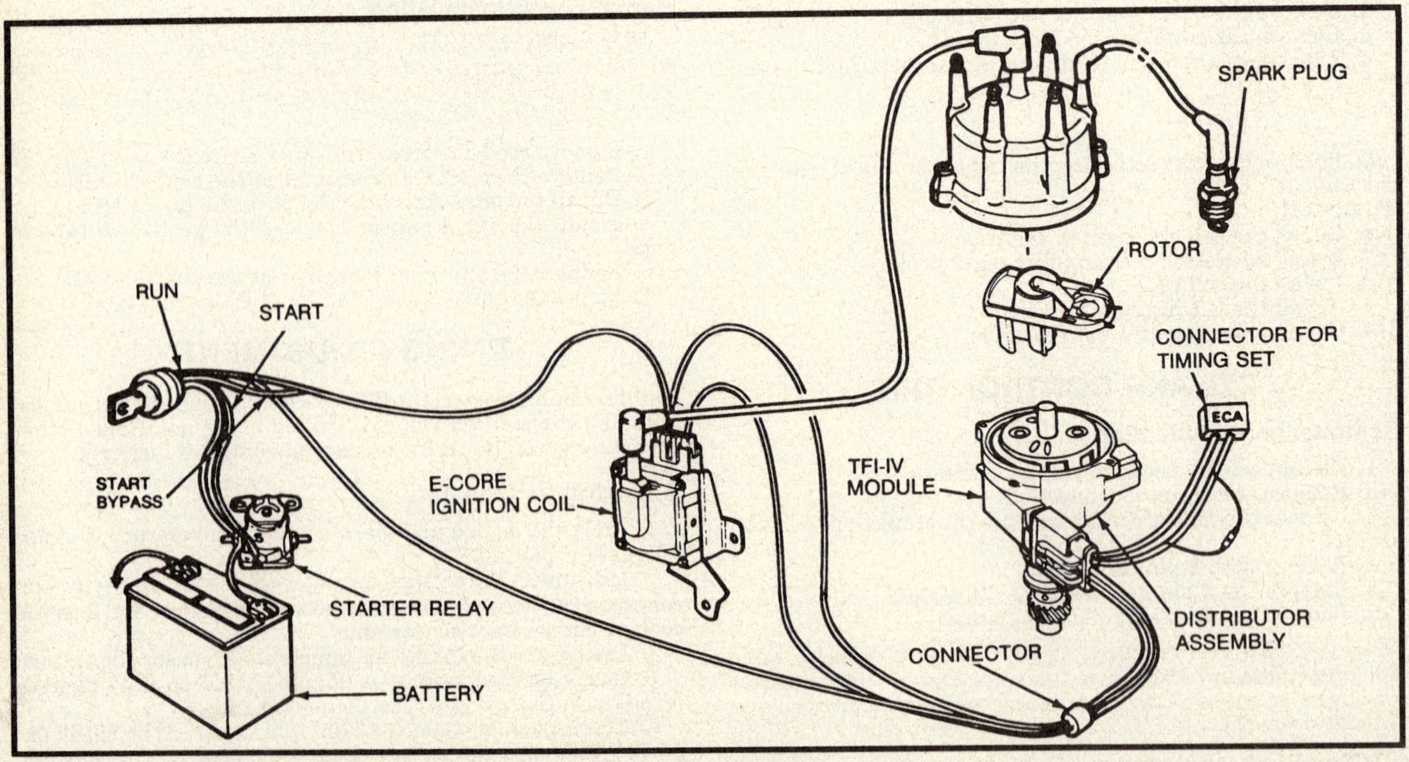

Exploded view of the EEC-IV components—Scorpio, 2.9L engine

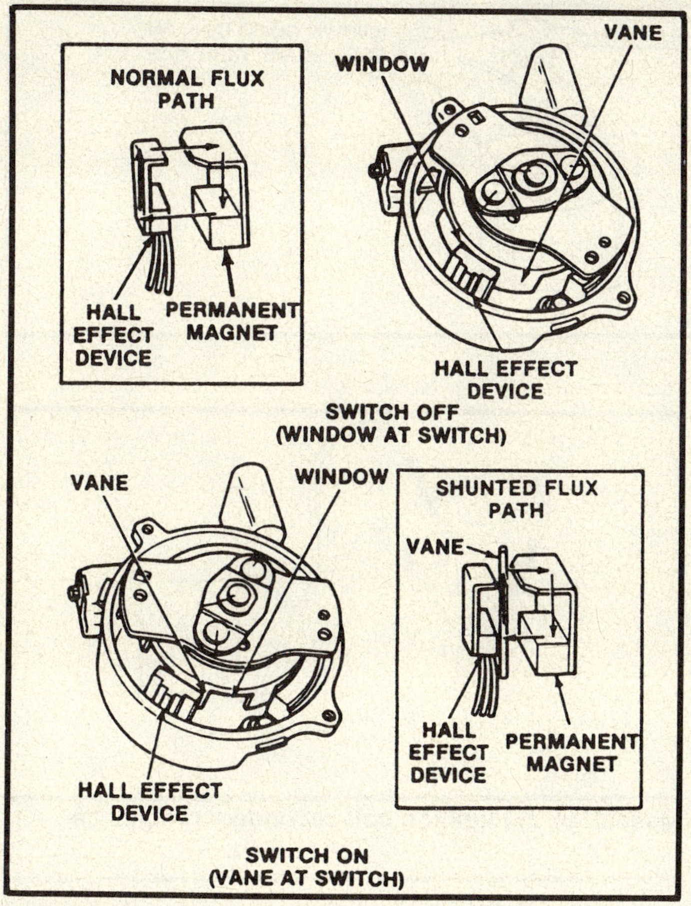

Hall Effect pickup operation—TFI-IV system

COMPONENT DESCRIPTION

DISTRIBUTOR

The distributor is of the universal design. It has a diecast base that incorporates and integrally mounted TFI-IV ignition module and a Hall Effect vane switch stator assembly.

NOTE: Distributor calibration is not require. Initial timing adjustment is not normally required unless the distributor has been removed from the engine.

TFI MODULE

The module is a Thick Film Integrated IV (TFI-IV) design. The TFI-IV module is contained in a molded thermoplastic and is mounted externally on the side of the distributor base. TFI-IV module also incorporates and push start mode, which allows push starting when necessary, (applicable to manual transaxle vehicles only).

IGNITION COIL

The TFI-IV system uses a E-core ignition coil. The E-core ignition coil replaces the conventional oil filled coil used with the older ignition systems. The nomenclature (E-core) is derived from the E-shape laminations of its core.

Diagnosis and Testing

One of the most important checks that must be performed as apart of any diagnostic procedure is a careful visual and physical underhood inspection. This can often lead to correcting a problem without extensive testing.

Visually inspect the engine compartment to ensure all vacuum hoses and spark plug wires are properly routed and securely connected.

Examine all wiring harnesses and connectors for insulation damage and burned, overheated, loose or broken conditions.

Check that the TFI module is securely fastened to the distributor base. Be certain the battery is fully charged. All accessories should be off during diagnosis.

SERVICE PRECAUTIONS

CAUTION

Whenever working on or around any computer base microprocessor control system, always observe these general precautions to prevent the possibility of personal injury or damage to the electronic components.

- Never install or remove battery cables with the key **ON** or the engine running. Jumper cables should be connected with the key **OFF** to avoid both power surges that can damage electronic control units. Engines equipped with computer controlled systems should avoid both giving and getting jump starts due to the possibility of serious damage to components from arcing in the engine compartment when connections are made with the ignition **ON**.
- Always remove the battery cables before charging the battery. Never use a high output charger on and installed battery or attempt to use any type of hot shot, 24 volt starting aid.
- Exerise care when inserting test probes into connecters to insure good connections without damaging the connector or spreading the pins. Always probe connectors from the rear wire side, not the pin side, to avoid accidental shorting of terminals during test procedures.
- Never remove or attach wiring harness connectors with the ignition switch **ON**, especially to an electronic control unit.
- Do not drop any components during service procedures and never apply 12 volts directly to any component unless instructed specifically to do so. Some component electrical windings are designed to safely handle only 4 or 5 volts and can be destroyed in seconds if 12 volts are applied directly to the connector.
- Remove the electronic control unit if the vehicle is to be placed in an environment where temperatures exceed approximately 176°F (80°C), such as a paint spray booth or when arc or gas welding near the control unit location in the car.

IGNITION COIL

Testing

SECONDARY VOLTAGE

1. Disconnect the secondary high tension coil wire from the distributor cap and install a spark tester between the coil wire and ground.
2. Crank the engine. a spark should be noted, but the engine will not start, check the spark plugs, spark plug wiring and fuel system.
3. If there is no spark at the tester:
   a. Check the ignition coil secondary wire resistance; it should be no more than 7000 ohms per foot.
   b. Inspect the ignition coil for damage and/or carbon tracking.
   c. With the distributor cap removed, verify that the distributor shaft turns with the engine; if it does not, repair the engine as required.
   d. Be sure the blade of the rotor signal is coated with silicone compound, approximately $1/32$ in. thick. Do not coat the multi-point rotor.

PRIMARY CIRCUIT SWITCHING

1. Separate the wiring harness connector from the ignition

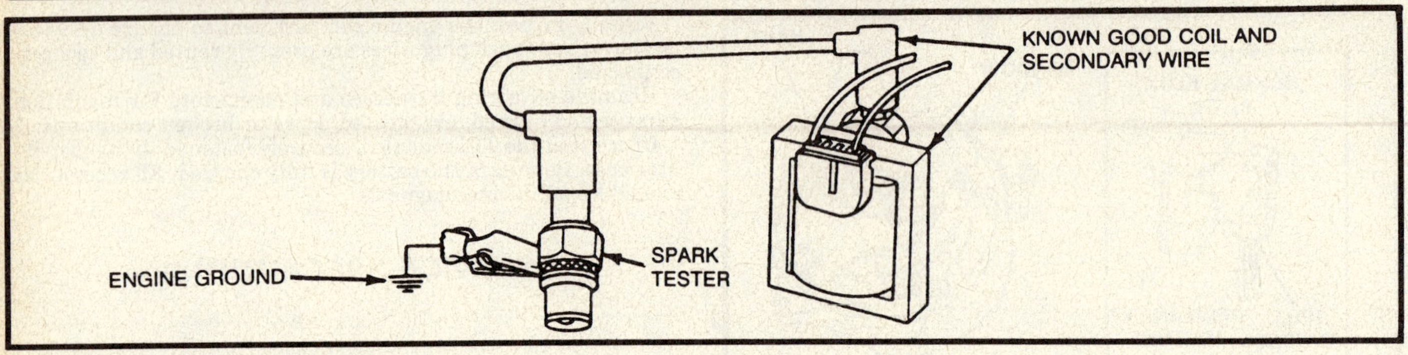

Testing the ignition coil spark

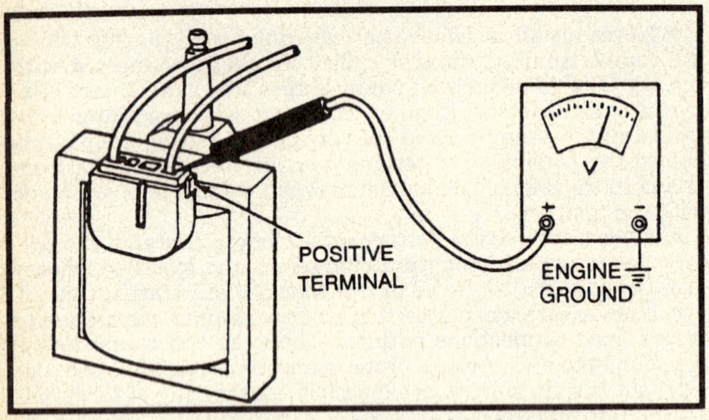

Testing the ignition coil primary circuit voltage

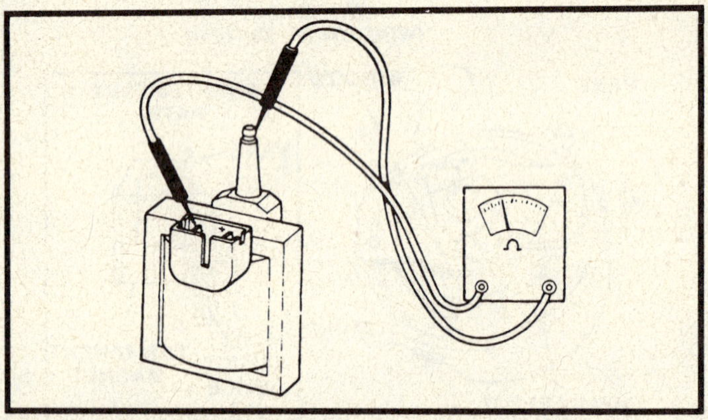

Measuring the ignition coil secondary resistance

module. Inspect for dirt, corrosion and damage. Reconnect the harness connector.

2. Attach a 12 volt test lamp between the ignition coil tach terminal and engine ground.
3. Crank the engine and observe the light.
4. If the lamp flashes or lights but will not flash, proceed to the ignition coil primary resistance test.
5. If the lamp does not light or is very dim, go to the primary circuit continuity test.

PRIMARY RESISTANCE

1. Have the ignition switch in the **OFF** position.
2. Disconnect the ignition coil connector. Inspect it for dirt, corrosion and damage.
3. Using an ohmmeter, measure the resistance from the positive to the negative terminals of the ignition coil.
4. The resistance should be 0.3 ohm; greater than 1.0 ohms, replace the ignition coil.

SECONDARY RESISTANCE

1. With the ignition coil harness connector off, measure the resistance from the negative terminal to the high voltage terminal of the ignition coil connector.
2. The resistance should be 6500–100500 ohms.
3. If the resistance is less than 6500 ohms or greater than 11500 ohms, replace the ignition coil.

PRIMARY VOLTAGE

1. Attach the negative voltmeter lead to the distributor base.
2. Measure the battery voltage.
3. Turn the ignition switch to the **RUN** position.
4. Measure the voltage at the negative terminal of the ignition coil.
5. If the voltage is 90% of battery voltage, inspect the wiring harness between the ignition module and the coil negative terminal.

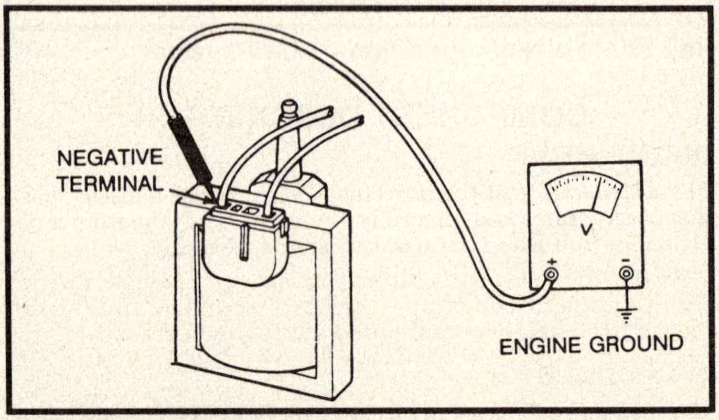

Testing the wiring harness

6. If the voltage is less than 90% of battery voltage, inspect the wiring harness between the ignition module and the coil negative terminal.
7. Turn the ignition switch to the **OFF** position.

SUPPLY VOLTAGE

1. Remove the coil connector.
2. Attach the negative voltmeter lead to the distributor base.
3. Measure the battery voltage.
4. Turn the ignition switch to the **RUN** position.
5. Measure the voltage at the positive terminal of the ignition coil.
6. The voltage should be 90% of battery voltage. Inspect the ignition coil and connector for dirt, corrosion and damage. If required, replace the ignition coil.
7. If the voltage is less than 90% of battery voltage, inspect

and repair, as required, the wiring between the ignition coil and the ignition switch. Check for a worn or damaged ignition switch.

8. Turn the ignition switch to the **OFF** position. Reconnect the ignition module connector.

WIRING HARNESS

Testing

1. Disconnect the the wiring harness connector from the TFI-IV module; the connector tabs must be pushed to disengage the connector. Inspect the connector for damage, dirt and corrosion.
2. Disconnect the wire at the **S** terminal of the starter relay.
3. Attach the negative lead of a voltmeter lead to a small straight pin.
   a. With the ignition switch in the **RUN** position, insert the straight pin into the No. 2 terminal of the TFI-IV module connector. Note the voltage reading, which should be 90 % of battery voltage.
   b. With the ignition switch in the **RUN** position, move the straight pin to the No. 3 connector terminal. Again, note the voltage reading, which should be 90% of battery voltage.
   c. Move the straight pin to the No. 4 connector terminal, turn the ignition switch to **START** position. Note the voltage reading, which should be 90% of the battery voltage.
4. If any reading is less than 90% of battery voltage, inspect the wiring, connectors and/or ignition switch for defects.
5. Turn the ignition switch to the **OFF** position. Remove the straight pin and reconnect the wire at the starter relay.

STATOR

Testing

1. Turn the ignition switch to the **OFF** position.
2. Remove the coil wire and ground it.
3. Attach the negative voltmeter lead to the distributor base.
4. Disconnect the pin-in-line connector near the distributor and attach the positive voltmeter lead to the TFI-IV module side of the connector.
5. Turn the ignition switch to the **ON** position.
6. Bump the starter with the ignition switch and measure the voltage levels with the engine not operating. Record all measurements.
7. If the highest value is less than 90% of battery voltage, replace the stator assembly.
8. If the lowest value is greater than 0.5 volts, remove the distributor from the engine, remove the TFI-IV module from the distributor and inspect the stator connector terminals and the TFI terminals for misalignment. If ok, replace the stator.
9. If the values are between 0.5 volts and 90% of battery voltage, replace the stator assembly.
10. If there are no values between 0.5 volts and 90% of battery voltage, connect a spark tester between the ignition coil wire and the engine ground.
11. Crank the engine. If a spark occurs, check the PIP and ignition ground wires for continuity. Repair as required. If no fault is found, perform an EEC-IV diagnostic check.
12. If no spark occurs, replace the TFI-IV module.

PRIMARY CIRCUIT CONTINUITY

Testing

1. Separate the wiring harness connector from the ignition module. Inspect for dirt, corrosion and damage.
2. Attach the negative voltmeter lead to the distributor base.
3. Measure the battery voltage.
4. Attach the voltmeter lead to a straight pin and insert the pin into the connector terminal No. 2.
5. Turn the ignition switch to the **RUN** position and measure the voltage at terminal No. 2.
6. The voltage should be no less than 90% of battery voltage.
7. If less than 90% of battery voltage, perform the Ignition Coil Primary Voltage test.
8. Turn the ignition switch to the **OFF** position and remove the pin.

Component Replacement

DISTRIBUTOR

Removal and Installation

1. Disconnect the negative battery cable.

NOTE: Prior to removing the distributor cap, mark the position of the No. 1 wire tower on the distributor base for installation purposes.

2. Unscrew the distributor cap retaining bolts. Remove the distributor cap and position it with the attached ignition wires out of the way.
3. Rotate the crankshaft until No. 1 piston is on the compression stroke.
4. Remove the distributor rotor and place it out of the way.
5. Disconnect the TFI harness connector.
6. Remove the distributor hold-down bolt and clamp and remove the distributor from the engine.

To install:

7. Rotate the distributor shaft so that the rotor tip is pointing approximately 20 degrees counterclockwise from the mark previously made on the distributor base.
8. Hold the distributor in correct alignment position (TFI module parallel with the rear face of the engine block). Push the distributor downward into full engagement with the engine block hole. The drive gear helix should rotate the distributor shaft 20 degrees counterclockwise during this operation.

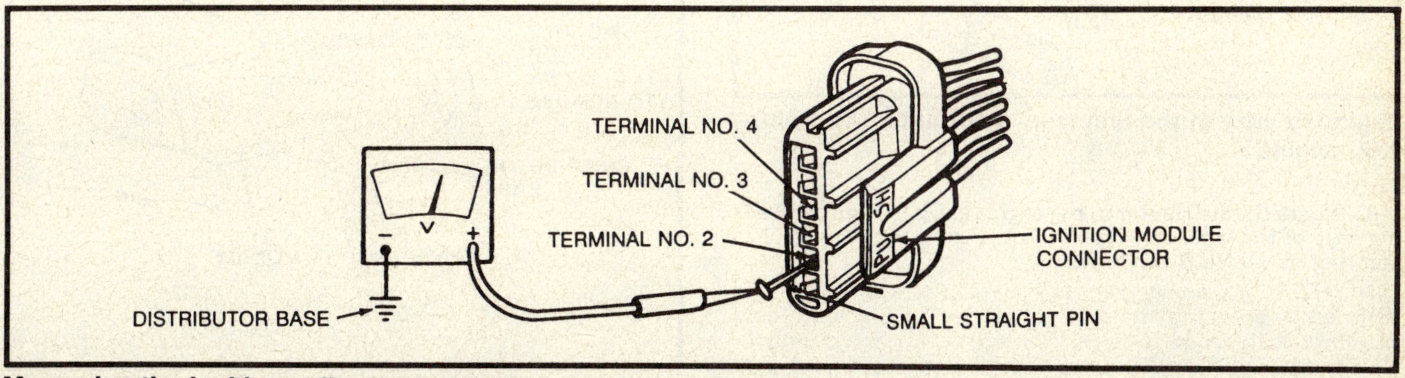

Measuring the ignition coil primary voltage

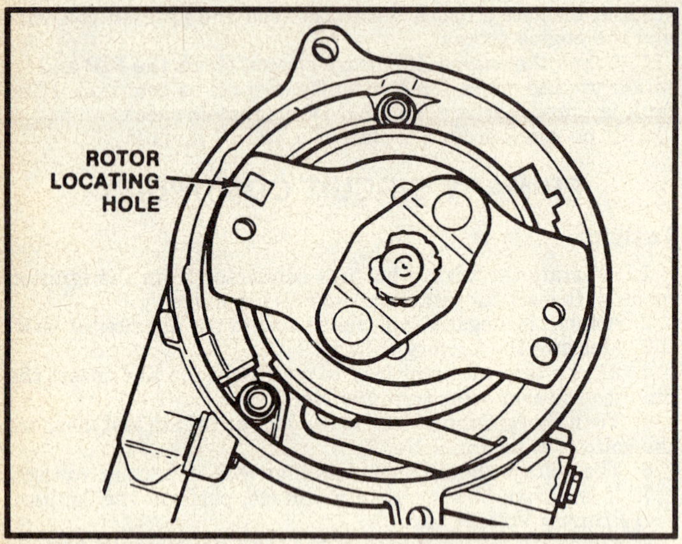

Aligning matchmarks on the distributor base with the rotor locating hole

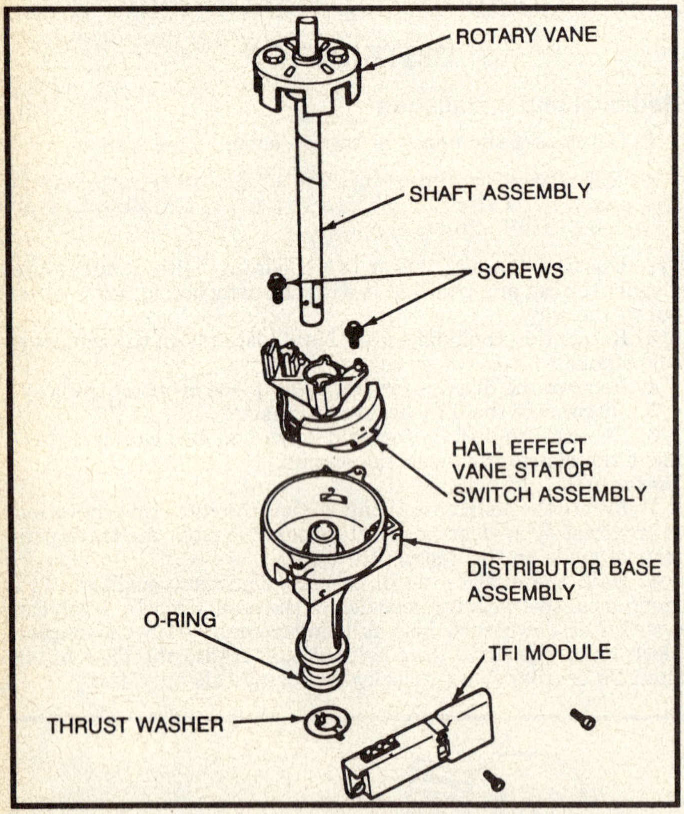

Exploded view of the universal distributor—Scorpio, 2.9L engine

9. Rotate the distributor in the engine block to align the leading edge of the vane and vane switch. Insure that the rotor is pointing to the No. 1 cap tower.

NOTE: If the vane and vane switch stator cannot be aligned by rotating the distributor in the engine block, pull the distributor out of the engine block enough to disengage the distributor and rotate the distributor to engage a different distributor gear tooth.

10. Install the distributor hold-down bolt and clamp. Do not tighten at this time.
11. Connect the TFI harness connector to the TFI module.
12. Install the distributor rotor and tighten the retaining screws.
13. Install the distributor cap and wires. Tighten the distributor cap retaining screws.
14. Insure the ignition wires are securely attached to the distributor cap and spark plugs.
15. Set the initial timing, refer to the Vehicle Emission Control Information Decal.
16. Tighten the distributor hold-down bolt to 12.5–15.5 ft. lbs. (17–21 Nm).

TFI MODULE

Removal and Installation

1. Disconnect the negative battery cable.
2. Remove the distributor cap. Position it and the attached ignition wires out of the way.
3. Remove the distributor rotor.
4. Disconnect the TFI harness connector from the TFI module.
5. Remove the distributor hold-down bolt and clamp. Remove the distributor from the engine.
6. Place the distributor on a work bench. Remove the TFI module retaining screws.
7. Pull the right side of the module down toward the distributor mounting flange and back up to disengage the module terminals from the connector in the distributor base. The module may then be pulled down toward the flange and away from the distributor.

NOTE: Do not attempt to lift the module from the mounting surface prior to moving the entire TFI module toward distributor flange as damage to the pins at the distributor or module connector may occur.

To install:

8. Coat the metal base plate to the TFI module with an approved silicone compound, approximately 1/32 in. thick.
9. Place the TFI module on the distributor base mounting flange.
10. Carefully position the TFI module assembly toward the distributor. Install the distributor connector pins.

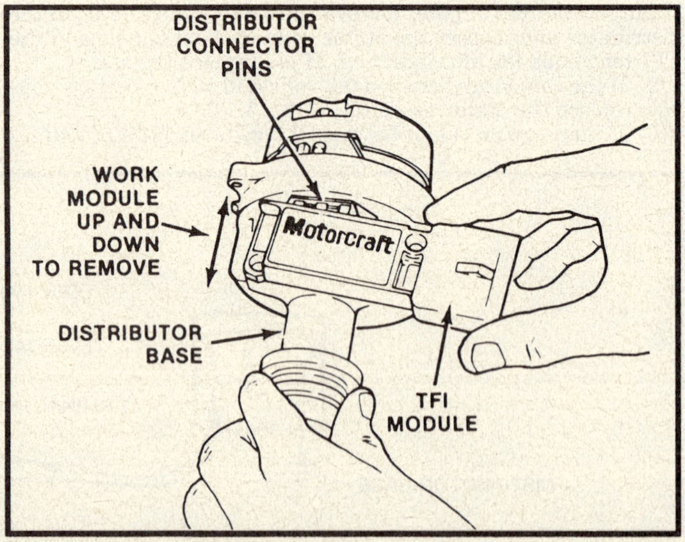

Removing the TFI-IV module

11. Install the TFI module mounting screws. Tighten to 15–35 inch lbs. (1.7–4.0 Nm).
12. Install the distributor into the engine.
13. Install the distributor rotor and cap.
14. Install the TFI harness connector.
15. Set the initial timing, refer to the Vehicle Emission Control Information Decal.

STATOR ASSEMBLY

Removal and Installation

1. Disconnect the negative battery cable.
2. Remove the distributor cap. Position the cap and the attached ignition wires to the side.
3. Remove the TFI module connector.
4. Remove the distributor from the engine and place it on a work bench.
5. Remove the distributor rotor.
6. Remove the TFI module retaining bolts and carefully remove the module. Wipe the silicone lubricant from the distributor base and module, keeping the surfaces free of dirt.

NOTE: To loosen the armature screws, hold the gear in place. Do not hold the armature to loosen the armature screws.

7. Remove the armature retaining screws and remove the armature from the distributor.
8. Matchmark the armature and gear for installation purposes.
9. Using a pin drift tool, drive the roll pin from the distributor shaft gear. Discard the pin.
10. Invert the distributor and place it in a axle bearing seal plate or other suitable support tool. Using a arbor press tool, press the gear off the distributor shaft.
11. Remove the thrust washer from the distributor base.
12. Deburr and polish the shaft with emery paper. Wipe the shaft clean so that it slides freely from the distributor base.
13. Remove the shaft assembly from the distributor.
14. Remove the octane rod retaining screw and octane rod assembly, where applicable.
15. Remove the stator assembly retaining screws, stator retaining ring, where applicable and remove the stator from the distributor housing.

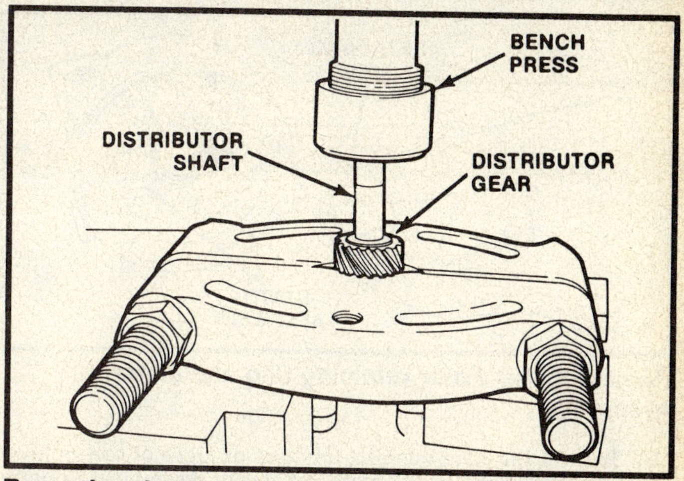

Removing the distributor drive gear from the shaft

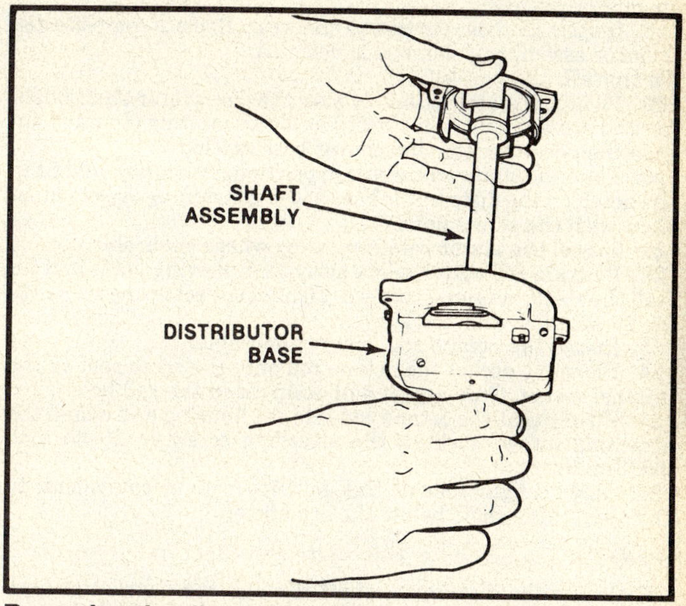

Removing the distributor shaft

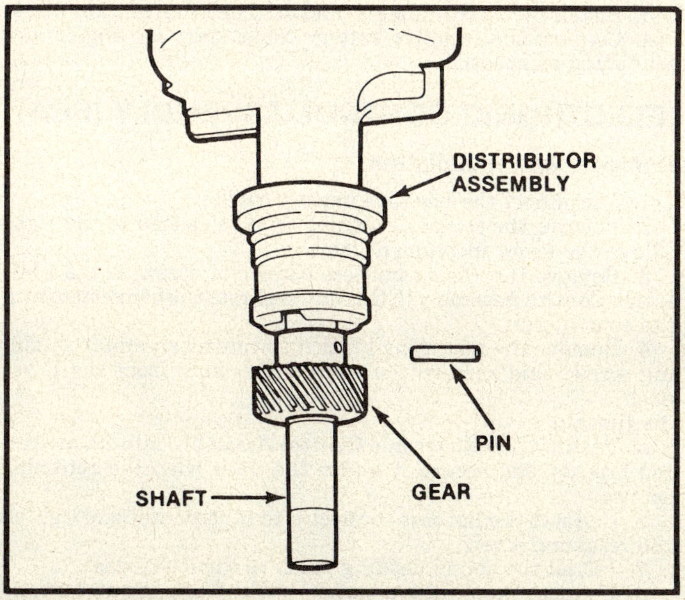

Disassembling the distributor assembly

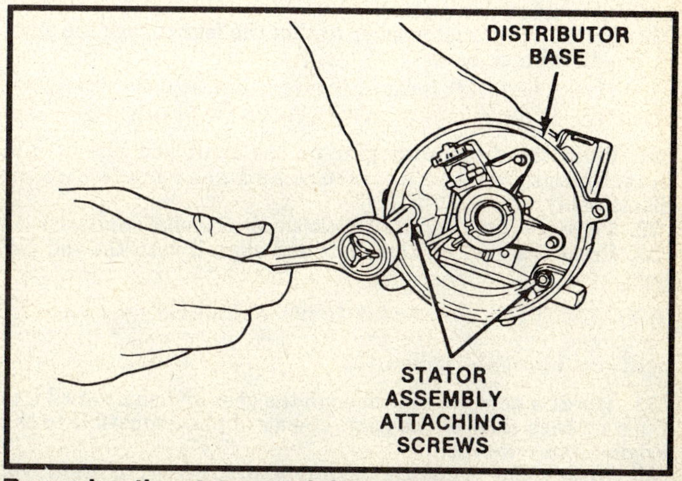

Removing the stator retaining screws

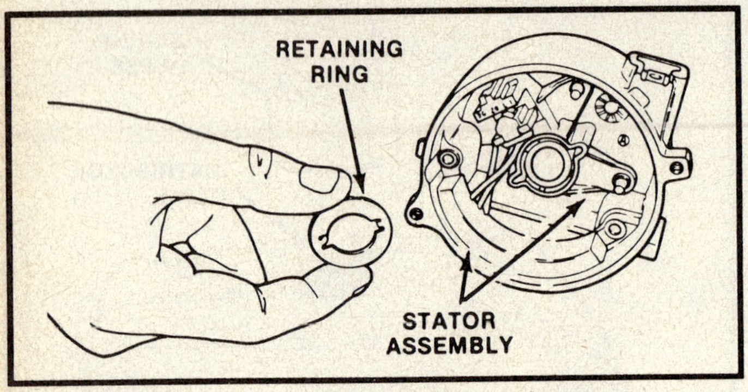

Removing the stator retaining ring and stator assembly

16. Inspect the base bushing for wear or signs of excess heat concentration. Replace complete distributor assembly if damage is detected.
17. Inspect the base O-ring for cuts or damage. Replace O-rings if necessary.
18. Inspect the base for cracks and wear. Replace complete distributor assembly if damage is detected.

To install:

19. Place the stator assembly over the distributor shaft bushing and press to seat it in place. The stator locating pin must engage the pocket in the distributor base casting.
20. Place the stator connector in position. The tab should fit in the notch on the distributor base and the fastening eyelets must align with the screw holes.
21. Install the stator retaining ring, where applicable.
22. Position the stator wires away from moving parts and install the stator retaining screws. Tighten the retaining screws to 25–35 inch lbs. (2.8–3.9 Nm).
23. Install the octane rod, where applicable.
24. Place the end of the octane rod onto the same post as the original stator. Only 1 post will easily fit in the rod hole.
25. Firmly seat the octane rod into the housing and install the attaching screw. Tighten the attaching screw to 25–35 inch lbs.(3–4 Nm).
26. Apply a light coat of ESF-M2C70-A oil or equivalent, to the distributor shaft below the armature.

NOTE: Do not over lubricate the distributor shaft.

27. Insert the distributor shaft through the distributor base bushing.
28. Place a ½ in. deep-well socket over the distributor shaft.
29. Invert the distributor assembly with socket and place it on a suitable arbor press plate.
30. Install the thrust washer so that the legs engage the slots in the distributor base.

NOTE: The shaft hole and gear must be lined up as accurately as possible to ensure ease of roll pin insertion.

31. Place the distributor gear on the shaft and line up the matchmarks on the armature and gear made during disassembly.
32. Place a ⅝ in. deep-well socket over the shaft and gear.
33. Press the gear on the shaft and align it with the roll pin hole.

NOTE: If the gear holes do not align, the gear must be removed, repositioned and pressed on again. A drift pin will not align the holes.

34. Insert a new roll pin through the gear and shaft until it is flush on both sides of the gear. Use a roll pin compatible to the original pin removed.
35. Install the armature and retaining screw, if removed. Tighten the retaining screws to 25–35 inch lbs. (2.8–3.9 Nm).

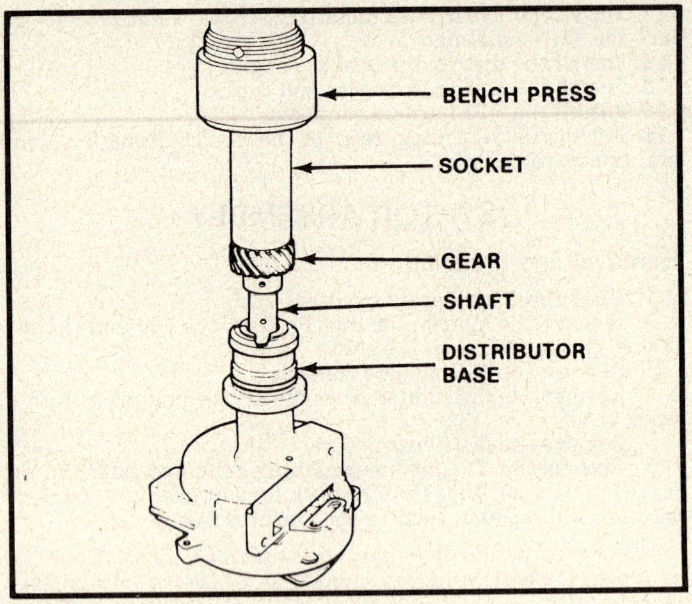

Installing the distributor drive gear

36. Check the distributor for free movement over full rotation of the shaft.

NOTE: If the armature hits the stator, replace the entire distributor assembly.

37. Wipe the back surfaces of the module and distributor mounting face.
35. Apply a thin even coat of an approved silicone dielectric compound to the back of the module.
38. Turn the distributor base upside down so that the stator connector is in full view.
39. Insert the module and ensure that the module pins are inserted into the stator connector. Fully seat the module into the connector and against the distribution base.
40. Install the module retaining screws and tighten to 25–35 inch lbs. (2.8–3.9 Nm).
41. Install the distributor into the engine block.
42. Connect the TFI module harness connector to the module.
43. Install the distributor cap and attached the ignition wires.
44. Connect the negative battery cable, start the engine and set timing as needed.

ELECTRONIC CONTROL ASSEMBLY (ECA)

Removal and Installation

1. Disconnect the negative battery cable.
2. Remove the sound deadening panal attached to the right side of the lower instrument panel.
3. Remove the wiring harness connector screw at the Electronic Control Assembly (ECA) and separate the connector from the control unit.
4. Remove the mounting bracket to control assembly retaining screws and remove the control assembly from the lower dash.

To install:

5. Position the Electronic Control Assembly into its mounting bracket and secure it under the dash with the retaining screws.
6. Connect the harness connector to control unit and tighten the retaining screw.
7. Install the sound deading panal to the lower dash.
8. Connect the negative battery cable. Start the engine and check performance.

IGNITION COIL

Removal and Installation

1. Disconnect the negative battery cable.
2. Carefully remove the high tension coil wire from the E-coil.
3. Disconnect the coil harness connector at the coil.
4. Remove the coil retaining bolts and remove the coil and bracket assembly from the left strut tower.

To install:

5. Position E-coil and bracket on the left shock tower and install the retaining bolts.
6. Connect the coil harness connector to the coil.
7. Plug in the high tension coil wire to the coil.
8. Connect the negative battery cable. Start the engine and check performance.

NO START—NO SPARK

Use care when troubleshooting these systems. If the engine is cranked to check for spark with spark plug wires disconnected, it means the ECM doesn't see the load it expects. The result could be a voltage spike that damages the ECM.

1. If testing for spark, use only an approved spark tester, properly grounded, that duplicates the operating load the ECM is designed to handle.
2. If there is no spark when the engine is cranked, the speed sensor/reference sensor should be checked.
3. If there is no spark when the engine is cranked, the ignition coil should be checked.
4. If there is no obvious fault with the ignition power and coil, then the TFI module should be tested with a TFI module tester or replaced with a known good TFI module.

IGNITION TIMING

ADJUSTMENT

NOTE: Make all adjustments with the engine at normal operating temperature, transmission in neutral (manual) or P (automatic), parking brake applied, wheels blocked and all accessories turned off.

BASE TIMING ADJUSTMENT

NOTE: Make all adjustments with the engine at normal operating temperature, transmission in neutral (manual) or P (automatic), parking brake applied, wheels blocked and all accessories turned off.

1. With engine off, clean and highlight the timing marks on the crankshaft pulley and front cover. Connect a timing light and tachometer to the engine. The ignition coil connector allows a tachometer connection using an alligator clip without removing the coil connector.
2. Disconnect the single wire Spark Output (SPOUT) connector. The connector is located near the distributor about 6 in. from the TFI module. Restart the previously warmed up engine.
3. Check the idle rpm. The engine is equipped with an electronic idle speed control and the idle should be between 825–975 rpm for manual transmission or 925–1075 rpm for automatic transmissions. If idle adjustment is necessary, shut off the engine and disconnect the electrical connector at the idle bypass valve. Restart the engine and run at approximately 2000 rpm, briefly. Allow the engine to return to idle and adjust the idle speed to 725–775 rpm by turning the throttle plate stop screw. Shut off the engine and reconnect the electrical connector to the bypass valve. Restart engine and proceed with ignition timing check.

NOTE: If the underhood calibration/emissions sticker specifications differ from this procedure, follow the specs and directions on the sticker as they will reflect product changes. If the cooling fan comes on during idle speed adjustment, wait until it shuts off or disconnect it temporarily before making adjustment. On Scorpio V6 models, loosen the throttle camplate roller bolt prior to adjustment of the throttle stop adjusting screw. Tighten the camplate roller bolt when adjustment is completed.

4. Point the timing light at the marks. Timing should be 13 degrees Before Top Dead Center (BTDC) for manual transmission models, or 10 degrees BTDC for automatic transmission models. Refer to the Emissions Control Decal, if the information on the decal differs from the specs mention here, set the timing according to the decal. Loosen the distributor hold-down bolt and adjust the timing as required by twisting the distributor assembly clockwise or counterclockwise. Tighten the hold-down bolt and recheck timing.

NOTE: Some distributor hold-down bolts made have a special Torx® head which requires a special wrench.

5. After base timing has been set, shut off the engine and remove all test equipment. Reconnect the single SPOUT wire harness at the distributor.

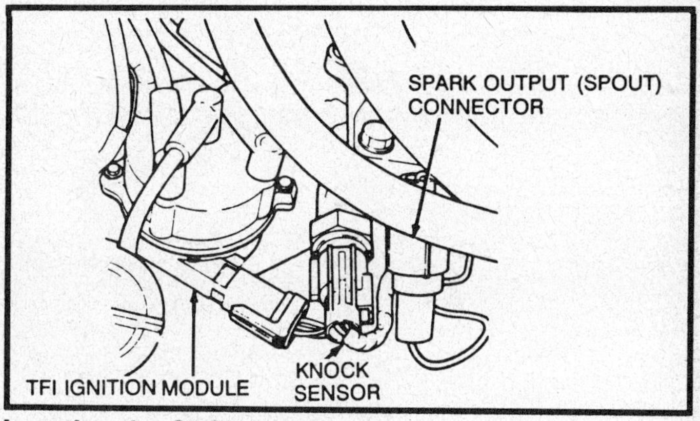

Location the SPOUT connector

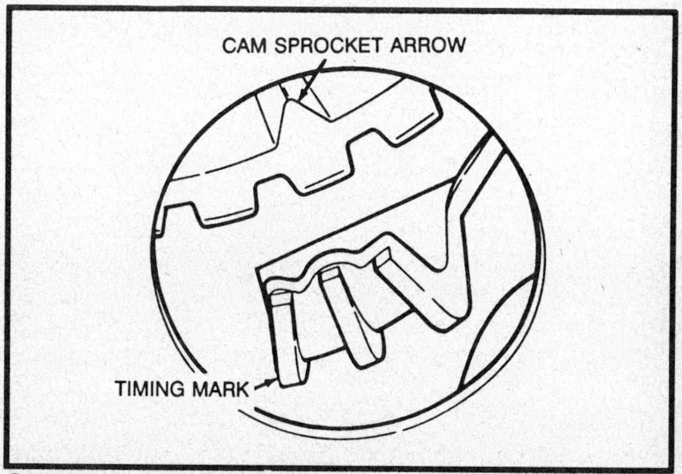

Correct timing mark alignment as viewed from through the access plug on the timing belt outer cover

PEUGEOT

General Information

Peugeot has used a variety of 4 and 6-cylinder engines. Several similar but slightly different Bosch electronic ignition systems have been used depending on the vehicle. Proper engine identification is important. Two identification plates are riveted on the engine block. One has the engine type, the other has the engine number.

Peugeot's gasoline engines include several 4-cylinder models as well as a V-6. The 4-cylinder engines appear in SOHC, DOHC and turbocharged form. The V-6 engine did not appear in the 1991 models; only 4-cylinder gasoline engines were available in the US market. Peugeot withdrew from the US market at the end of 1991. No 1992 cars were imported.

SYSTEM OPERATION

The Bosch systems used on these vehicles are really Engine Management Systems since fuel and ignition functions are often combined and controlled by 1 Electronic Control Unit (ECU).

The ignition ECU incorporates a microprocessor with memory. Several ignition advance points are programmed into the ECU to provide the optimum engine performance at the limit of engine knock (detonation). Each Electronic Contol Unit (ECU) is unique to its system. The controllers cannot be interchanged.

A self-diagnostic feature is incorporated into the ignition ECU. This enables it to identify some of the faults which may occur in the system during operation. An indicator LED in the lower left corner of the instrument cluster alerts the driver of faults.

Note that the Motronic systems used on XU9J2 and XU9J4 engines are very similar. The main difference is that the XU9J2 engine uses a so-called "simplified" system and does not use a knock (detonation) sensor. The XU9J4 engine uses a knock sensor.

In addition, the LU and LU-2 Jetronic systems are similar. The main difference is that the newer LU-2 system has the cold start enrichment controlled by the ECU, eliminating the cold start injector and thermo-time switch. The LU-2 has its rpm input supplied the coil and uses a different throttle position sensor that the earlier LU Jetronic.

The LH2.2 Jetronic system is similar the the LU series. The main difference is that the LH Jetronic system uses a hot wire

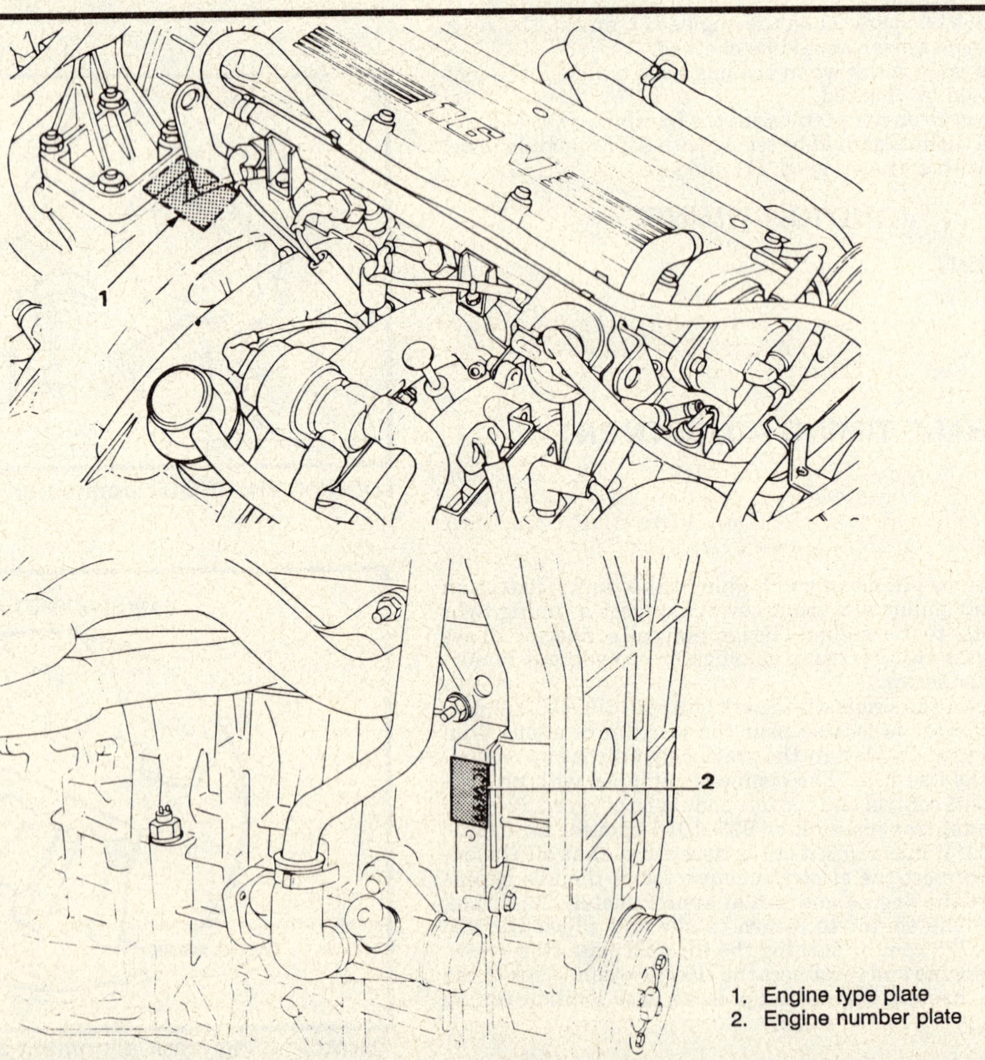

Engine type and engine number plates—Peugeot 405 with XU9J4 engine

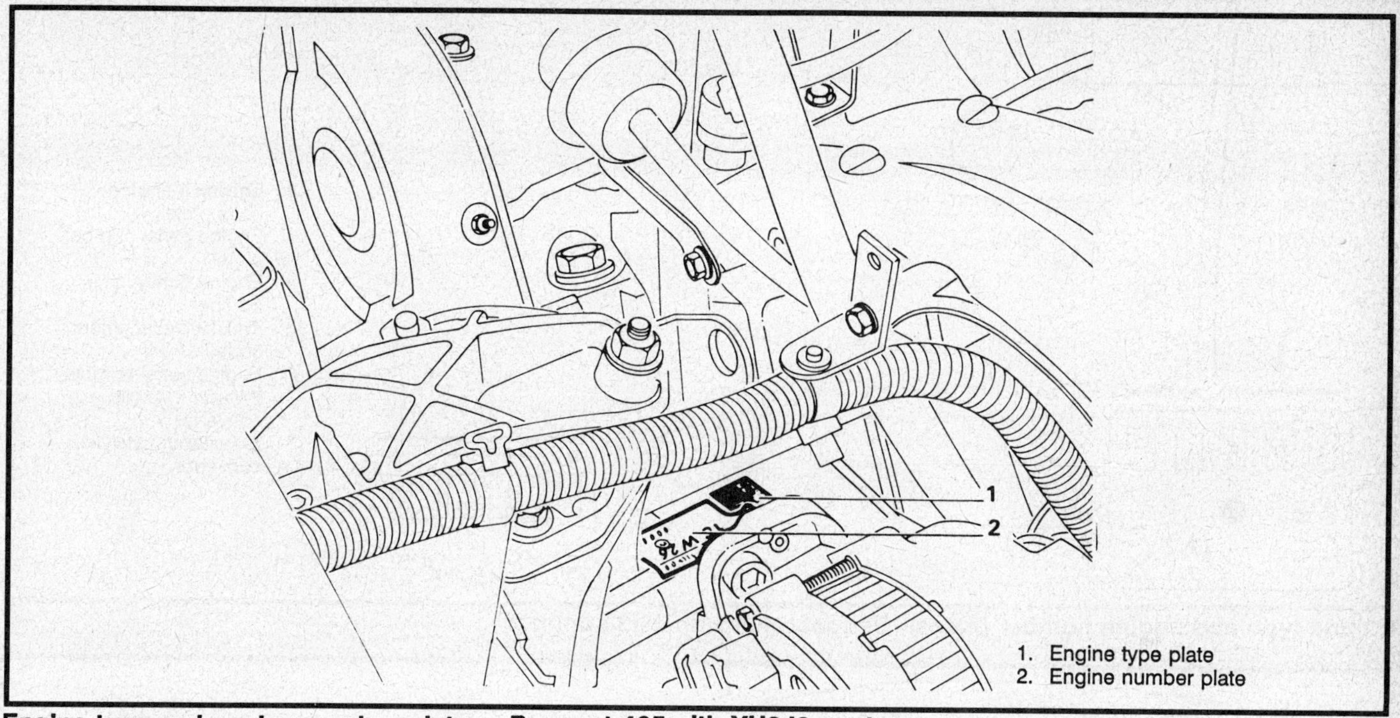

Engine type and engine number plates—Peugeot 405 with XU9J2 engine

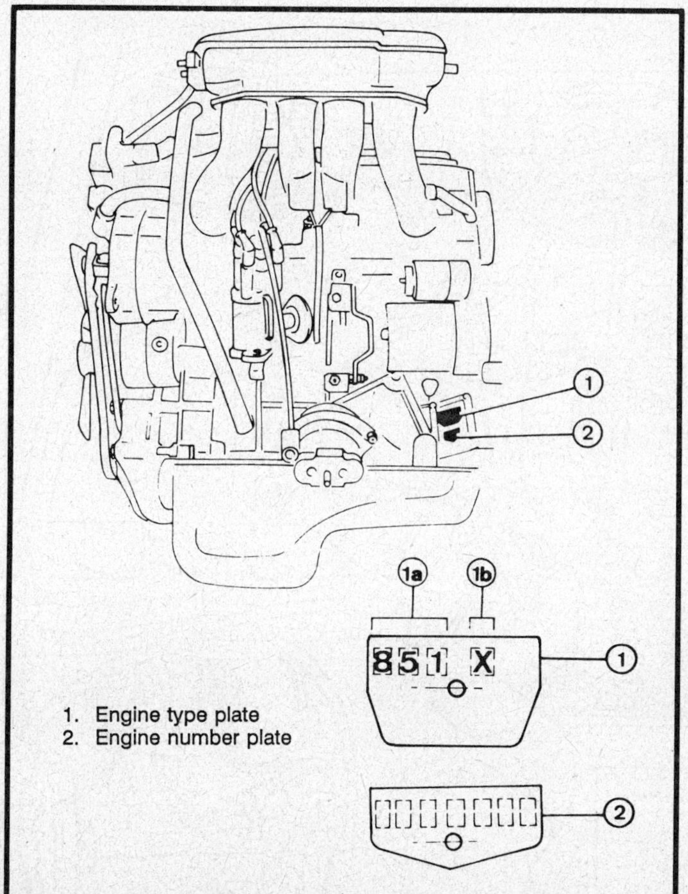

Engine type and engine number plates—Peugeot 505 with ZDJL engine

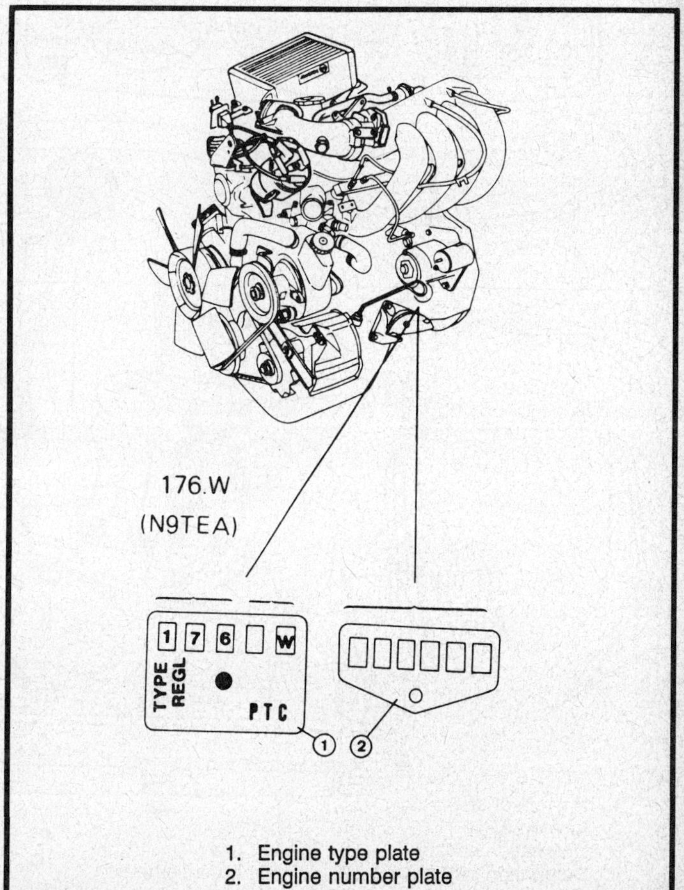

Engine type and engine number plates—Peugeot 505 with N9TEA engine

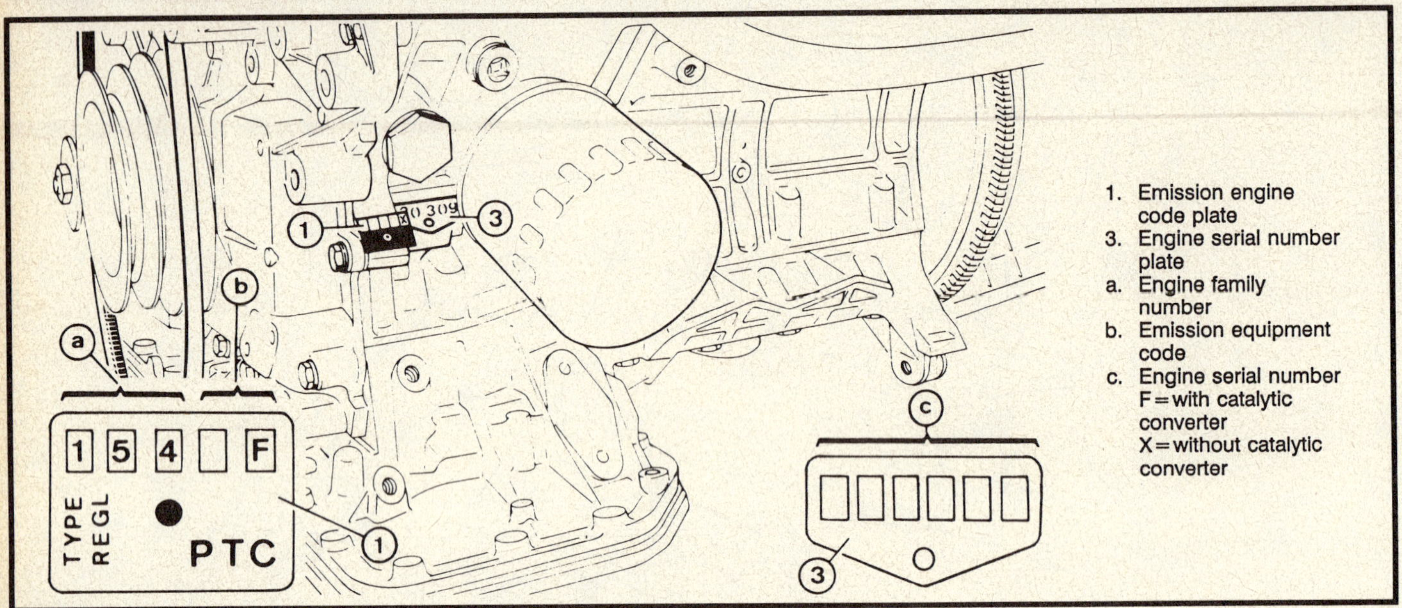

Engine type and engine number plates—Peugeot 505 with ZN3J engine

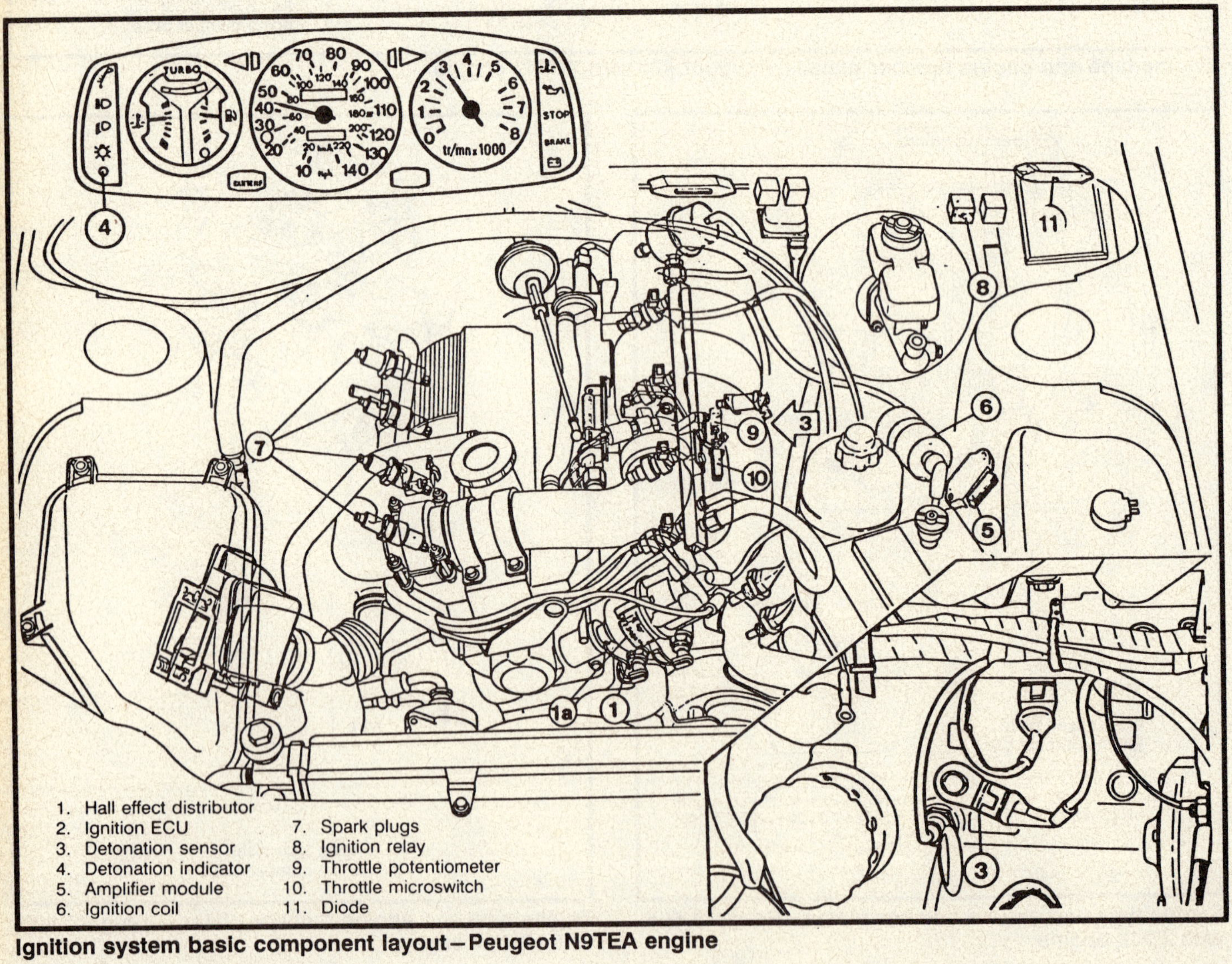

Ignition system basic component layout—Peugeot N9TEA engine

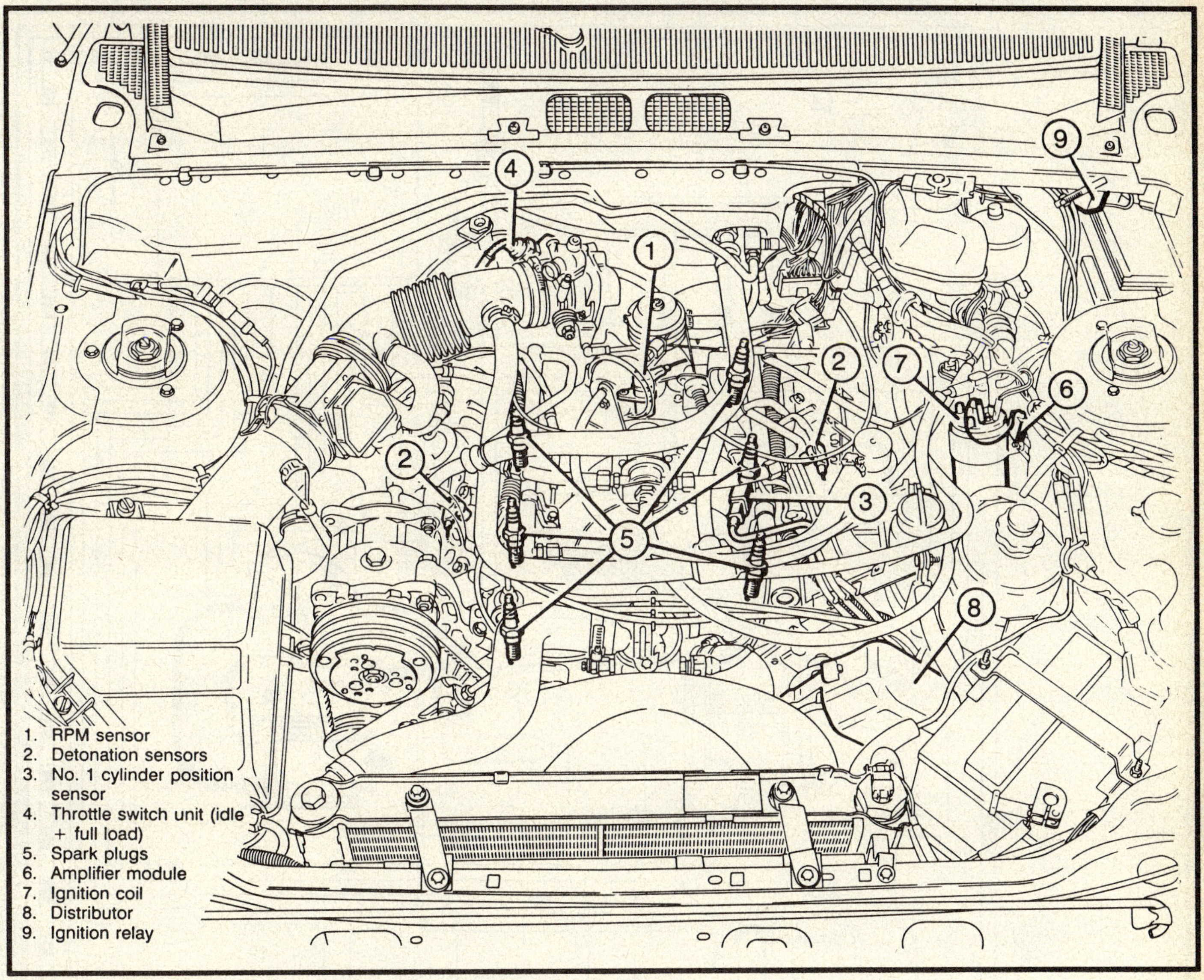

Ignition system basic component layout—Peugeot ZN3J engine

air flow sensor to measure the air drawn into the engine. The LH Jetronic system is also equipped with an electronic idle regulator which maintains a constant idle speed regardless of the engine load.

When the engine is cranking, a signal is produced and amplified by the amplifier module. The amplifier module controls a power transistor which makes and breaks the primary coil circuit.

A signal emitted by the Hall sensor provides the ignition ECU with engine speed information. The duration of injection signal emitted by the injection ECU allows selecting an ideal advance setting from the computer's memory.

When detonation exists, the detonation sensor signals the ignition ECU by generating a small electric current. The ECU recognizes this signal and adjusts the spark timing. The spark advance is retarded by a few degrees, the detonation stops and the computer returns the spark to its orignial value. Detonation sensing is done cylinder by cylinder.

If detonation persists, the computer will continue to reduce the advance in 3 degree increments unit the maximum correcting span of 22 degrees has been reached.

An indicator LED in the instrument cluster monitors the proper operation of the ignition system. Under normal operation, with the key **ON** and the engine **OFF**, the LED will be **ON**. Under normal operation, with the engine running, the LED should be **OFF**.

NOTE: Occasional sporadic flashes of the LED with no discernable engine problems may be noticed. This condition is normal.

If problems develop when the vehicle is in operation the LED will blink constantly. The number of blinks, from 1–6, help determine the origin of the fault. The frequency of these blinks will be proportional to engine rpm.

As a safety measure, as soon as a fault in the system is detected, the system will automatically retard the spark advance (20–22 degrees), noticeable by a lack of power and performance of the engine.

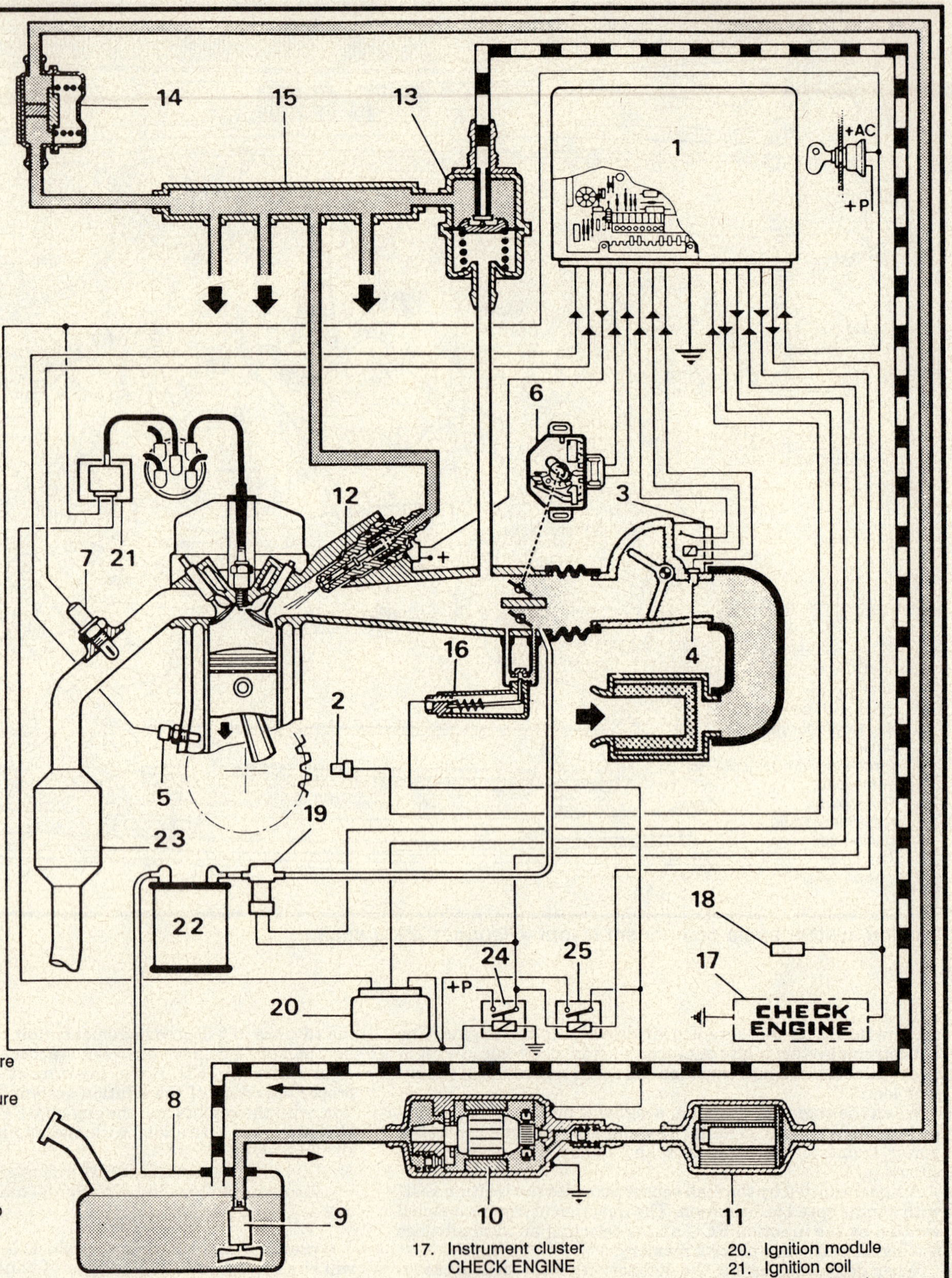

1. ECU
2. RPM and TDC sensor
3. Air flow sensor
4. Airflow temperature sensor
5. Engine temperature sensor
6. Throttle position sensor
7. Oxygen sensor
8. Fuel tank
9. In-tank fuel pump
10. Main fuel pump
11. Fuel filter
12. Fuel injectors
13. Fuel pressure regulator
14. Fuel pulsation damper
15. Fuel rail
16. Auxiliary air valve (cold start)
17. Instrument cluster CHECK ENGINE light
18. Diagnostic connector
19. Charcoal canister purge solenoid
20. Ignition module
21. Ignition coil
22. Charcoal canister
23. Catalytic converter
24. Main relay
25. Fuel pump relay

Engine management system component layout— Peugeot 405 with XU9J2 engine and Motronic M1.3 engine management system, no knock sensor

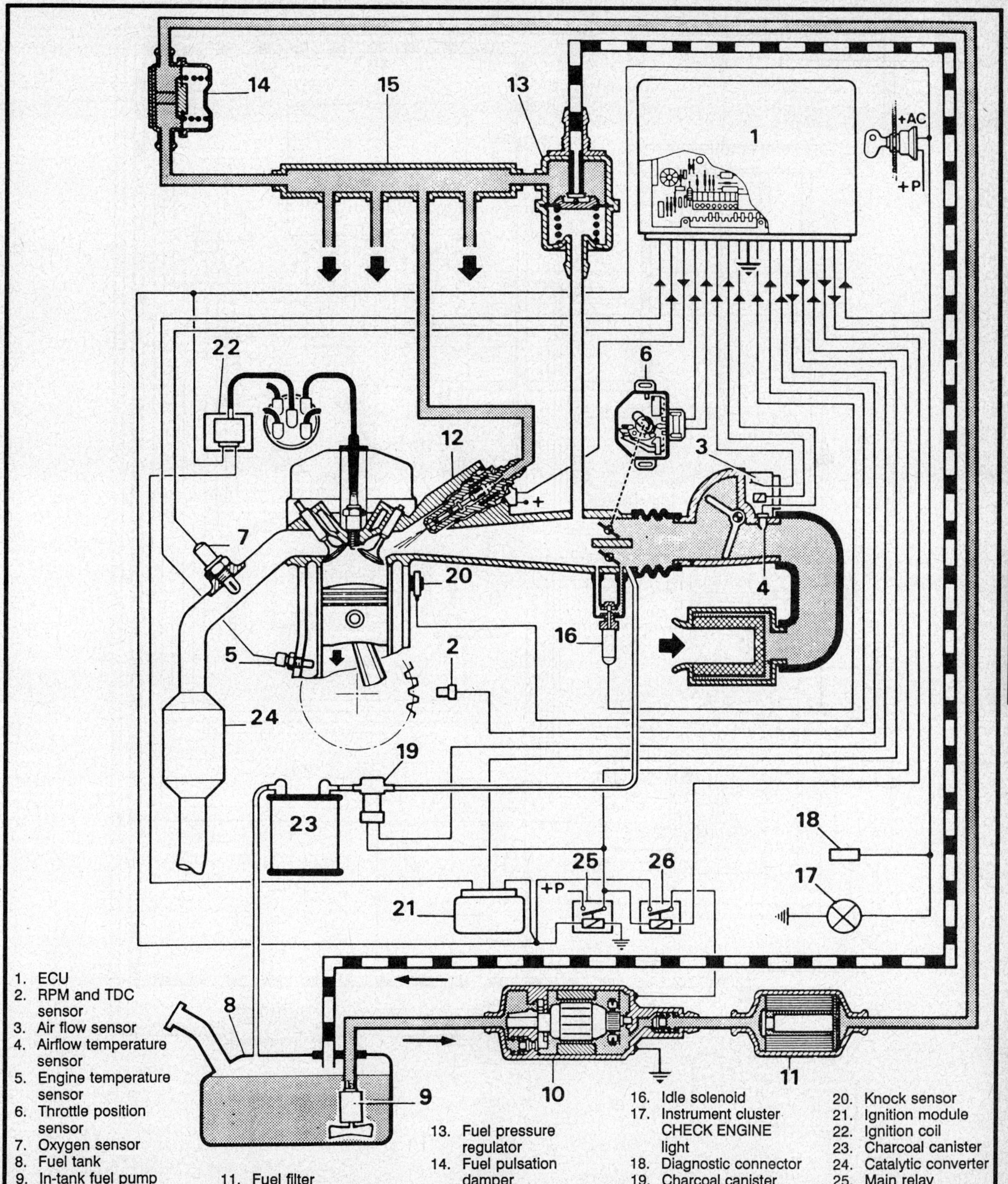

1. ECU
2. RPM and TDC sensor
3. Air flow sensor
4. Airflow temperature sensor
5. Engine temperature sensor
6. Throttle position sensor
7. Oxygen sensor
8. Fuel tank
9. In-tank fuel pump
10. Main fuel pump
11. Fuel filter
12. Fuel injectors
13. Fuel pressure regulator
14. Fuel pulsation damper
15. Fuel rail
16. Idle solenoid
17. Instrument cluster CHECK ENGINE light
18. Diagnostic connector
19. Charcoal canister purge solenoid
20. Knock sensor
21. Ignition module
22. Ignition coil
23. Charcoal canister
24. Catalytic converter
25. Main relay
26. Fuel pump relay

Engine management system component layout – Peugeot 405 with XU9J4 16V engine and Motronic M1.3 engine management system, with knock sensor

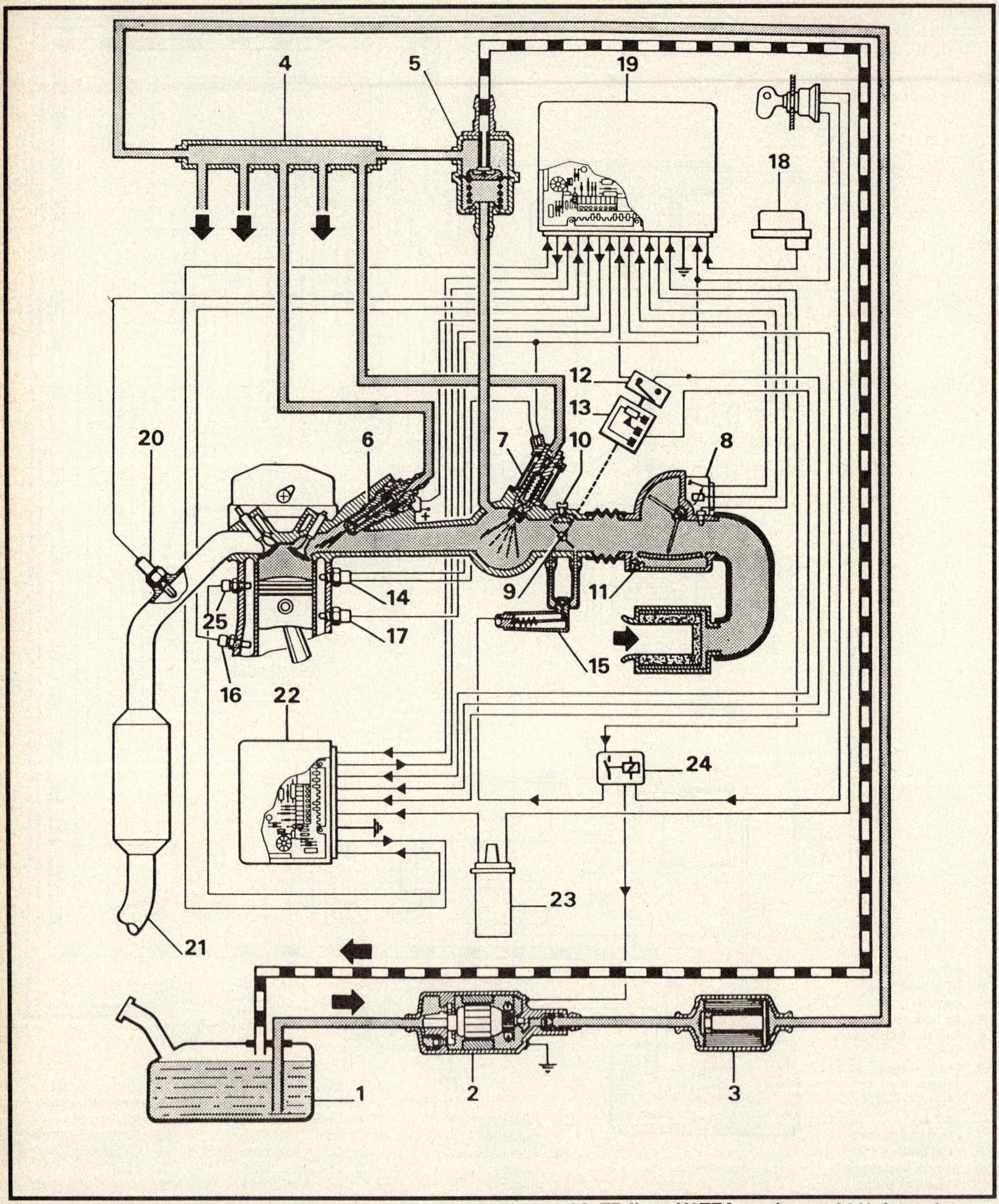

Engine management system component layout— Peugeot 505 with ZDJL or N9TEA engine and LU Jetronic engine management system

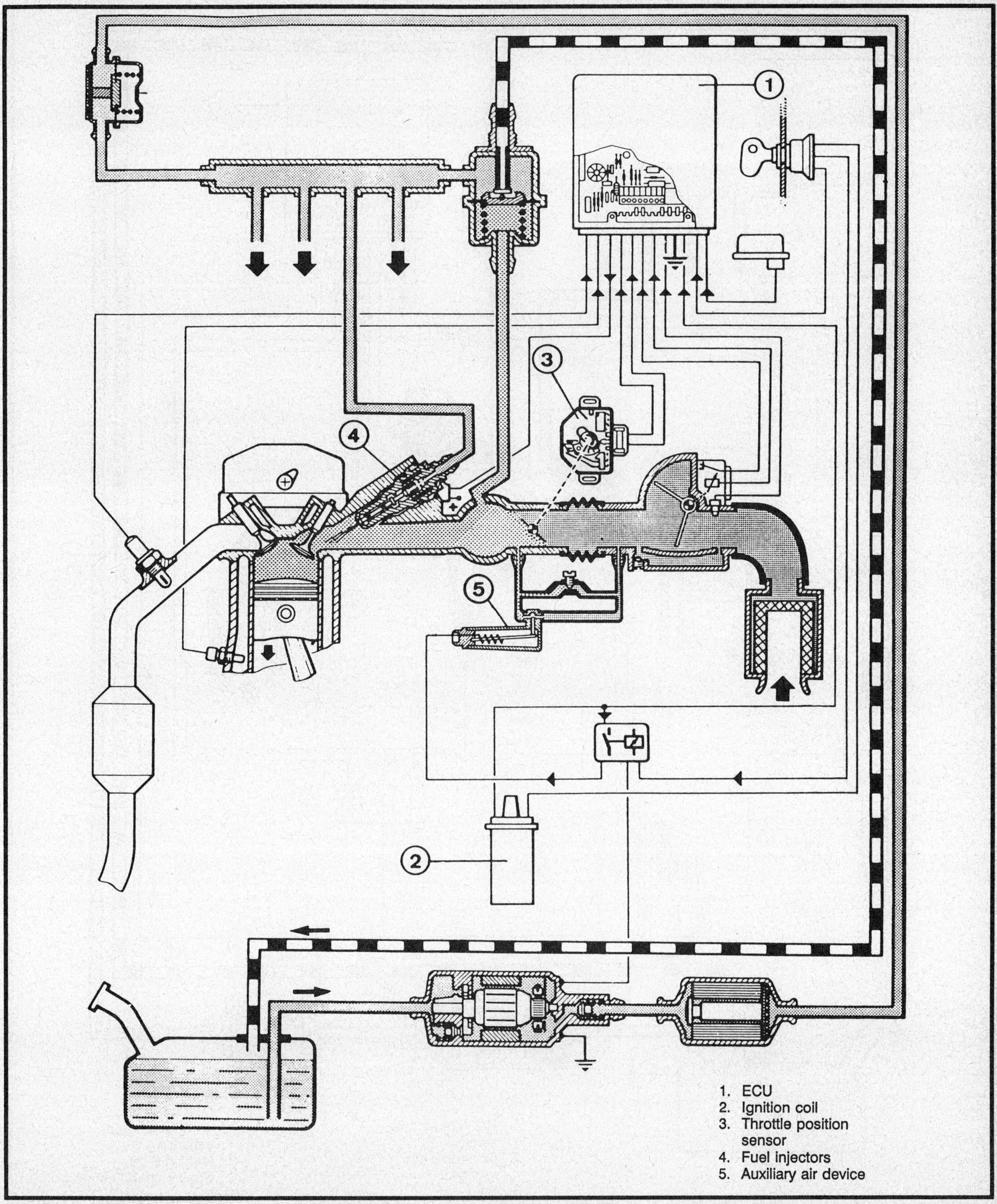

Engine management system component layout—Peugeot 505 with ZDJL or N9TEA engine and LU-2 Jetronic engine management system

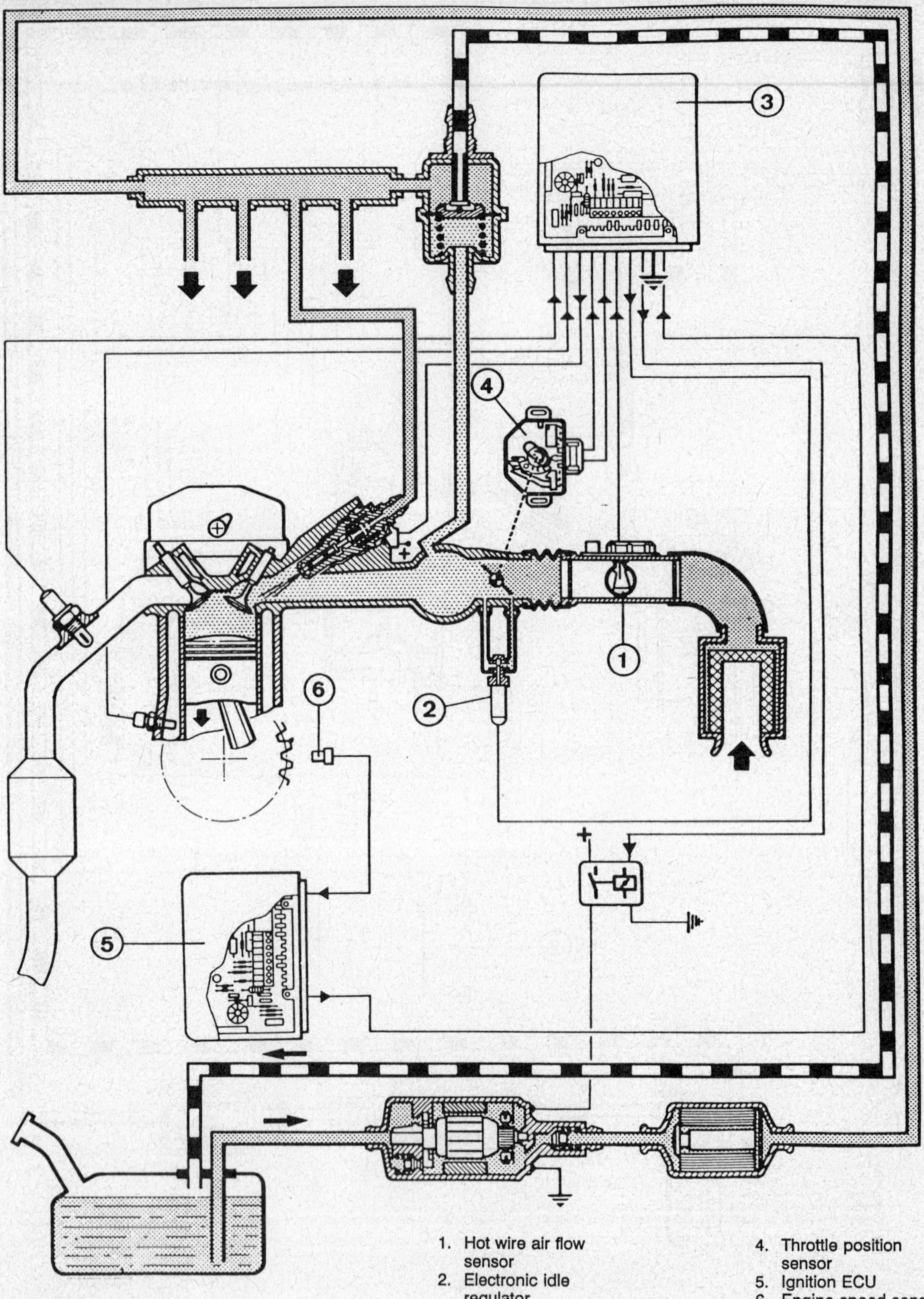

1. Hot wire air flow sensor
2. Electronic idle regulator
3. ECU
4. Throttle position sensor
5. Ignition ECU
6. Engine speed sensor

Engine management system component layout— Peugeot 505 with ZN3J LH2.2 Jetronic engine management system

Diagnosis and Testing

SERVICE PRECAUTIONS

- Use care when working around high-energy electronic ignition systems. High voltage could cause injury to the technician and equipment.
- Before performing any service on the ignition system, make sure the ignition is turned **OFF** and/or the battery negative cable is disconnected.
- Before attempting to remove any parts, turn the ignition switch **OFF** and disconnect the battery ground cable.
- Never connect test equipment such as timing lights, etc., directly to the ignition system for a power source. Always use the vehicle battery for the power source.
- Use only the correct replacement parts. Use of spark plugs, for example, having specifications and sizes different from those specified could cause severe damage and alter emissions levels.
- Use care when working with light alloy parts. Always lubricate the spark plug threads with a suitable lubricant to avoid seizing the plug threads to the cylinder head.
- Keep all ECU parts and harnesses dry during service. Protect the ECU and all solid-state components from rough handling or extremes of temperature.
- Make sure all harness connectors are fastened securely. A poor connection can cause an extremely high voltage surge resulting in damage to integrated circuits.
- Always use a 12 volt battery as a power source.
- Do not disconnect the battery cables with the engine running. Never run the engine with the battery cables loose or disconnected.
- Do not attempt to disassemble the ECU unit under any circumstances.
- When connecting or disconnecting pin connectors from the ECU, take care not to bend or break any pin terminals. Check that there are no bends or breaks on the ECU pin terminals before attempting any connections.
- When measuring supply voltage of ECU-controlled components, keep the tester probes separated from each other and from accidental grounding. If the tester probes accidentally make contact with each other during measurement, a short circuit will damage the ECU.
- It is recommended that the ECU control unit be removed if the vehicle is to be exposed to high heat such as a bake oven to cure paint after body refinishing or when welding is be done on the vehicle.
- Use great care when working on or around air bag systems. The system may stay armed for a period of time without battery power. It is recommended that while working on any electrical system in the air bag deployment zone (driver's seat area), disconnect the battery and remove the air bag module from the steering wheel. Then reconnect the battery, if necessary, for electrical diagnosis work.
- Do not be hasty in condemning the ECU. This system uses voltages and resistances that are very small. Examins the sensors, wiring and connectors carefully. The sensors operate inmore harsh conditions than the ECU which is protected at the rear of the engine compartment.
- Check all ground connections before condemning the ECU.

READING CODES

The Motronic M1.3 fuel system has some self-diagnostic capabilities. The ECU monitors electrical signals during operation. If a signal is interrupted or goes out-of-range when compared to pre-determined values, the ECU will note the fault and store an identifying code in memory. The codes may be read or retrieved during diagnosis.

Once a fault is noted, the ECU will illuminate the CHECK ENGINE warning lamp on the dashboard to advise the operator. Additionally, if the fault involves a signal critical to engine operation, the ECU will immediately subsitute a fixed value for the faulty one. This pre-determined or default value may not be correct for the immediate operating conditions. Engine performance may be affected when the CHECK ENGINE lamp is lit.

The LU and LH fuel systems do not provide fuel system diagnostic codes but will provide a small number of diagnostic codes generated through the ignition ECU. These codes may be helpful in diagnosing certain fuel system components.

Motronic M1.3

The codes may be read by connecting the Peugeot TAD 99 tester, or its equivalent, to the green 2-pin connector and to the battery. Operate the tester according to the manufacturer's instructions.

If the tester is not available, codes may be read through the flashing of the CHECK ENGINE lamp. To read codes:

1. With the ignition **OFF**, connect a switched jumper wire between Terminal 1 of the green, 2-pin diagnostic connector and a good ground.
2. Switch the ignition **ON**. Do not start the engine.
3. Turn the jumper switch ON for 3 seconds; the CHECK ENGINE lamp should be lit.
4. Turn the jumper switch OFF. The CHECK ENGINE warning lamp should blink once, remain off for a short period, then blink twice. This indicates Code 12, beginning of the test. After the last 2 blinks, the lamp should remain lit.
5. Again turn the jumper switch ON for 3 seconds; when the jumper switch is turned OFF the next code, if any, will be transmitted. The first group of flashes represents groups of 10s, the second group of flashes shows the digits. For example, 5 flashes, a pause and 4 flashes represents Code 54.
6. Each time the warning lamp remains lit after a code transmission, repeat Step 5. Consecutively stored codes will be transmitted one at a time.
7. Code 11 indicates the end of the test cycle. Once this code is received, no more faults are stored. If it is necessary to repeat the cycle, the ignition must be switched **OFF**, then **ON** and the procedure repeated from Step 1.

LU and LH Jetronic—Ignition Codes

Vehicles with the EZ 115 K and EZ 200 K ignition systems provide diagnostic codes through the orange LED on the dashboard. Under normal conditions, the LED should be lit when the ignition is **ON**, then go OFF with the engine running. If the light stays on constantly, or never comes on, check the power supply to the lamp and the ground circuit. When a fault is detected, the ignition ECU retards the spark timing. This safety function will be noticed as a lack of power or performance in the engine.

NOTE: An occasional flash or flicker of the LED during engine operation is normal. A fault noted in the system will cause an obvious, rhythmic flashing by the LED.

When a fault is noted by the ignition ECU, the LED will immediately flash in a pattern, representing the single-digit code for the fault. During diagnosis, codes should always be read with the engine idling. On some vehicles, the speed of the flashes is relative to engine speed.

Fault Codes

When using fault codes during diagnosis, remember that the ECU checks electrical signals only. It cannot diagnose mechanical faults such as fouled injectors or vacuum leaks. The code refers to a fault in the entire circuit, not just the named component.

SELF-DIAGNOSTICS—MOTRONIC M1.3 SYSTEM—XU9J2 ENGINE

SELF TEST OF THE MOTRONIC SYSTEM

The different malfunctions which can be detected during the SELF TEST (start mode) or during normal operation, are identified by coded signals during the TEST sequence (see corresponding chapter).

IDENTIFICATION OF THE CODED SIGNALS

* Major fault (causes CHECK ENGINE indicator to come ON)

** Emissions related malfunction only will cause CHECK ENGINE indicator to come ON.

DEFECT CODE →| 0515000	DEFECT CODE |→ 0515001	DEFECTIVE COMPONENT	CHECKS
11	11	End of Test sequence	
12	12	Beginning of Test sequence	
13	13	Air temperature sensor	— Check the defective components and its electrical circuit (see E.C.U. connector terminal checks)
14*	14**	Injection NTC sensor	
21*	21**	Throttle switch (idle position)	
33*	33**	Airflow sensor	
34*	34**	Canister purge electrovalve	
35	35	Throttle switch unit (full load position)	
	41	Engine rpm sensor	
42*	42**	Injectors	
15*	15	Fuel pump(s) relay	— Fuel pump controls
31 32	31 32	Self correction richness regulation function	— Air leak at intake — Exhaust leak
51*	51**		— Check the Lambda sensor electrical circuit — Try again with a new Lambda sensor
52*	52**		— Try again with a new E.C.U.
53	53	Battery voltage	— Charging system
22* 54*	22 54**	E.C.U.	— Try again with a new E.C.U.

SELF-DIAGNOSTICS—MOTRONIC M1.3 SYSTEM—XU9J4 ENGINE

SELF TEST OF THE MOTRONIC SYSTEM

The different malfunctions which can be detected during the SELF TEST (start mode) or during normal operation, are identified by coded signals during the TEST sequence (see corresponding chapter).

IDENTIFICATION OF THE CODED SIGNALS

* Major fault (causes CHECK ENGINE indicator to come ON)

DEFECT CODE	DEFECTIVE COMPONENT	CHECKS
11	End of Test sequence	
12	Beginning of Test sequence	
13	Air temperature sensor	– Check the defective components and its electrical circuit (see E.C.U. connector terminal checks)
14*	Injection NTC sensor	
21*	Throttle switch (idle position)	
22*	Idle regulation electrovalve	
33*	Airflow sensor	
34*	Canister purge electrovalve	
35	Throttle switch unit (full load position)	
42*	Injectors	
15*	Fuel pump(s) relay	– Fuel pump controls
43*	Detonation correction regulation (at max.)	– Detonation sensor – Operation of engine (temperature, spark plugs, ...)
44*	Detonation sensor	– Detonation sensor
31 32	Self correction richness regulation function	– Air leak at intake – Exhaust leak
51*		– Check the Lambda sensor electrical circuit – Try again with a new Lambda sensor
52*		– Try again with a new E.C.U.
53	Battery voltage	– Charging system
54*	E.C.U.	– Try again with a new E.C.U.

MOTRONIC M1.3

Code 11—End of test sequence
Code 12—Beginning of test sequence
Code 13—Air temperature sensor
Code 14—Injection coolant temperture sensor
Code 15—Fuel pump relay
Code 21—Throttle position switch—idle position
Code 22—Idle regulation electro-valve
Code 31—Self-correction richness regulation—air leak at intake
Code 32—Self-correction richness regulation—exhaust leak
Code 33—Airflow sensor
Code 34—Canister purge solenoid
Code 35—Throttle position switch—full throttle position
Code 42—Injector signal or circuit
Code 43—Detonation correction circuit or regulation
Code 44—Detonation sensor
Code 51—Self-correction richness regulation—oxygen (Lambda) sensor or circuit
Code 52—ECU failure
Code 53—Battery voltage
Code 54—ECU failure

NOTE: Codes 13, 31, 32 and 35 will not trigger the CHECK ENGINE warning lamp even though a code is held in memory.

JETRONIC IGNITION CODES

ZN3J V-6 Engine

Code 2—Coolant temperature sensor
Code 4—Detonation sensor or ignition ECU
Code 5—Faulty load signal from injection ECU
Code 6—No. 1 cylinder sensor

N9TEA Turbocharged Engine

Code 1—Continued detonation; maximum correction reached
Code 2—Battery voltage below 10.5 volts
Code 3—Ignition ECU detonation correction circuit faulty; internal failure
Code 4—Faulty signal received from detonation sensor
Code 5—Signal from throttle position sensor (potentiometer) exceeds 4.3 volts
Code 6—Circuit fault between ignition ECU and throttle potentiometer or loss of load signal from injection ECU.

CLEARING CODES

Motronic M1.3

Codes may only be cleared after Code 11, End of Test, has been received. If using the TAD 99 tester, follow the operating instructions. If the tester is not available:

1. Install the switched jumper used for reading codes.
2. Turn the ignition switch **ON**.
3. Retrieve any stored codes. Repeat the procedure until Code 11 is received.
4. After Code 11 is received and the CHECK ENGINE light is lit, turn the jumper wire switch ON for at least 10 seconds.
5. The codes will be cleared during the 10 second period. Turn the jumper wire switch OFF. The dash indicator should remain lit.

Ignition Codes

The ignition codes are only present when the engine is running. Turning the ignition **OFF** automatically clears the codes. Any faults present at the next restart will re-set the codes and cause the dashboard LED to flash.

DETONATION SENSOR

Checking

XU9J4 ENGINE

1. With the sensor connected, warm the engine to normal operating temperature.
2. Pull back the rubber caps from the detonation sensors.
3. Connect a voltmeter (A.C. scale) to the rear of the connector.
4. The voltmeter indicates approximately 0 volt. Rev-up the engine once, the voltmeter should indicate 20 milli-volts.
5. If the voltage readings is incorrect, perform the test using a known good sensor.

N9TEA ENGINE

1. Disconnect the negative battery cable.
2. Remove the ignition ECU connector.
3. Connect an ohmmeter between pin 12 and pin 13 of the ignition ECU connector. The resistance should be between 270–330 kilo-ohms.
4. If out of specification, replace the detonation sensor.

NO START—NO SPARK

Peugeot 405

XU9J4 ENGINE AND

XU9J2 ENGINE

Serial Numbers 0500001–0506989

1. Connect timing light to coil high tension lead and crank starter. If the timing light flashes, check the rotor, cap, spark plugs and wires.
2. If there is no flash from the timing light, leave the ignition key in the **ON** position and check the voltage at pin 2, wire 3A of the coil.
3. If there is battery voltage at wire 3A, check for voltage at the coil tower. If okay, replace the amplifier module. If there is battery voltage at wire 3A but none at the coil tower, replace the coil.
4. If there is no battery voltage, disconnect the amplifier module. If voltage is now available to coil, replace ignition amplifier module.
5. If there is no battery voltage, with the ignition key still **ON**, check for voltage at pin 3, wire 2A of the ignition coil.
6. If there is no battery voltage, look for a continuity problem in this wire since this is the switched battery power **RUN** wire.

Peugeot 405

N9TEA ENGINE, ZDJL ENGINE, AND XU9J2 ENGINE

Serial Numbers from 0506990

1. Connect timing light to coil high tension lead and crank starter. If the timing light flashes, check the rotor, cap, spark plugs and wires.
2. If there is no flash from the timing light, leave the ignition key in the **ON** position and check the voltage at the negative terminal of the coil which is wire 3A.
3. If there is battery voltage at wire 3A, check for voltage on the coil tower. If there is no voltage, replace the coil.
4. If there is voltage at the coil tower, replace the amplifier.
5. If there is no battery voltage at wire 3A in Step 3, disconnect the amplifier module. If voltage is now available to coil, replace ignition amplifier module.
6. If there is no battery voltage, with the ignition key still **ON**, check for voltage at the positive terminal of the coil, wire 2, 2B and 2E.
7. If there is battery voltage, replace the coil. If there is no battery voltage, look for a continuity problem in wire 2B since this is the switched battery power **RUN** wire.

Peugeot 505

ZN3J ENGINE

1. Connect timing light to coil high tension lead and crank starter. If the timing light flashes, check the rotor, cap, spark plugs and wires.
2. If there is no flash from the timing light, leave the ignition key in the **ON** position and check the voltage at the negative terminal, wire 3, at the coil. This wire runs to the ignition amplifier.
3. If there is voltage at wire 3, check for voltage at the coil high tension tower. If okay, check the amplifier module.
4. If there is no voltage at wire 3 on the negative side of the coil, check for voltage at terminal 2 on the positive side of the coil. If there is voltage, replace the coil.
5. If there is no voltage at wire 2 on the positive side of the coil, check for check for voltage at wire 2R of the ignition relay. If there is battery voltage, check the continuity of this wire which runs from the relay to the positive side of the coil.
6. If there is no voltage at wire 2R of the ignition relay, check for voltage at wire 20A of the ignition relay. This wire should be hot right to the battery positive side. If there is no voltage here, check for continuity of this wire.
7. If there is voltage at wire 2R of the ignition relay, with the key still **ON**, check for voltage at wire 32A of the ignition relay. This wire runs to the ignition switch. If there is no voltage, check the continuity of wire 32A between the ignition relay and ignition switch.
8. If there is voltage at wire 32A, check for voltage at wire M1 which is the ground wire of the ignition relay. If there is no voltage, replace the ignition relay. If there is voltage, check the continuity of the M1 ground wire and make sure it is grounded properly.

IGNITION COIL

Testing

If the starter cranks but the engine will not start:

1. Check for 12 volts at the ignition coil connector.
2. Check the ignition coil resistance with an ohmmeter by first disconnecting the wires from the coil. To measure the primary coil, read between the connector at wire 2A which is the positive side and the connector at wire 3A which is the negative side. The reading should be 0.48–0.61 ohms.
3. Measure the resistance of the coil secondary. Read between the connector at wire 3A and the output to the distributor (coil tower). The reading should be 9000–11000 ohms.

IGNITION AMPLIFIER MODULE

The module switches the ignition coil's primary circuit ON and OFF. It incorporates a current limiter to provide a constant current to the primary circuit preventing damage to the coil. The module amplifies the impulses from the ECU to provide proper dwell and spark timing regardless of the engine rpm. Test using the following procedure:

1. To check the ignition amplifier module, disconnect the high tension cable from the distributor cap and connect it to a spark tester.
2. Note that high voltage is present so use insulated or nonmetallic pliers to perform this procedure. Place ignition key in **ON**.
3. Crank engine and look for spark.
4. The ignition amplifier module must be mounted at all times when operating.

IGNITION TIMING

Check and Adjustment

EXCEPT ZN3J V6 ENGINE

1. Warm the engine to normal operating temperature.
2. Make sure the throttle body microswitch is closed. Attach dial-back type timing light, set to 0 degrees, to the No. 1 cylinder spark plug cable.
3. Idle speed should be 900–950 rpm for manual transmission and 950–1000 rpm for automatic.

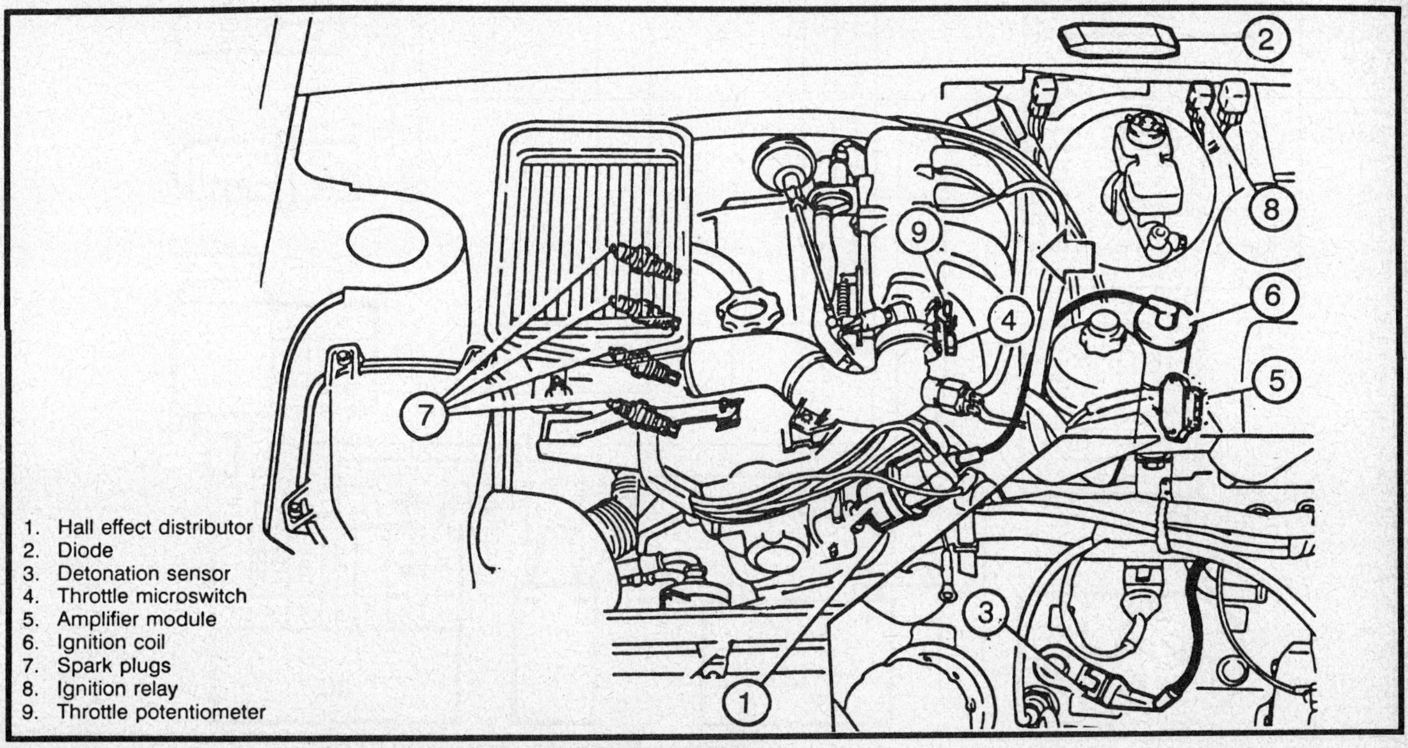

1. Hall effect distributor
2. Diode
3. Detonation sensor
4. Throttle microswitch
5. Amplifier module
6. Ignition coil
7. Spark plugs
8. Ignition relay
9. Throttle potentiometer

Typical ignition system components—Peugeot N9TEA engine

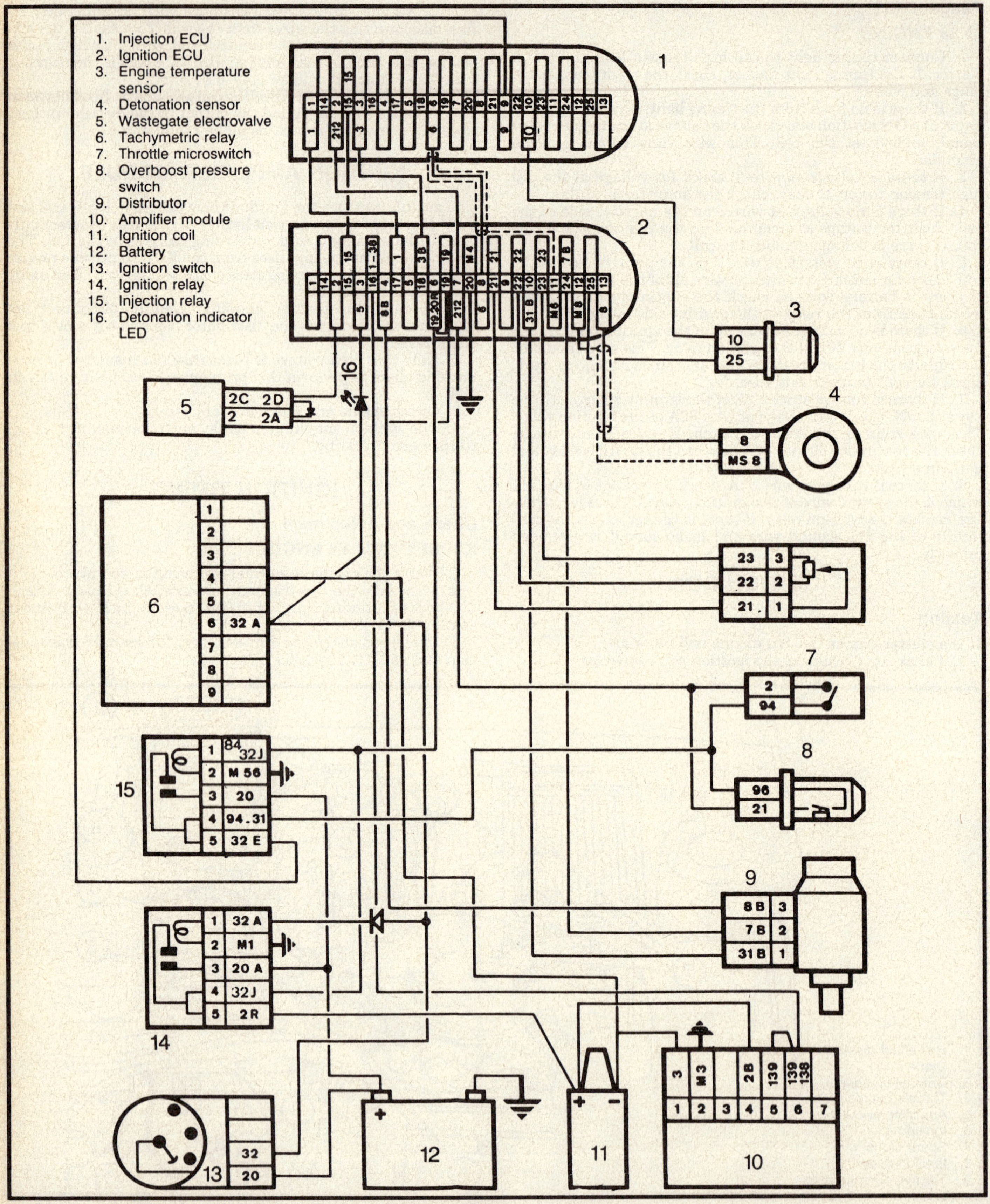

Ignition system basic schematic—Peugeot N9TEA engine

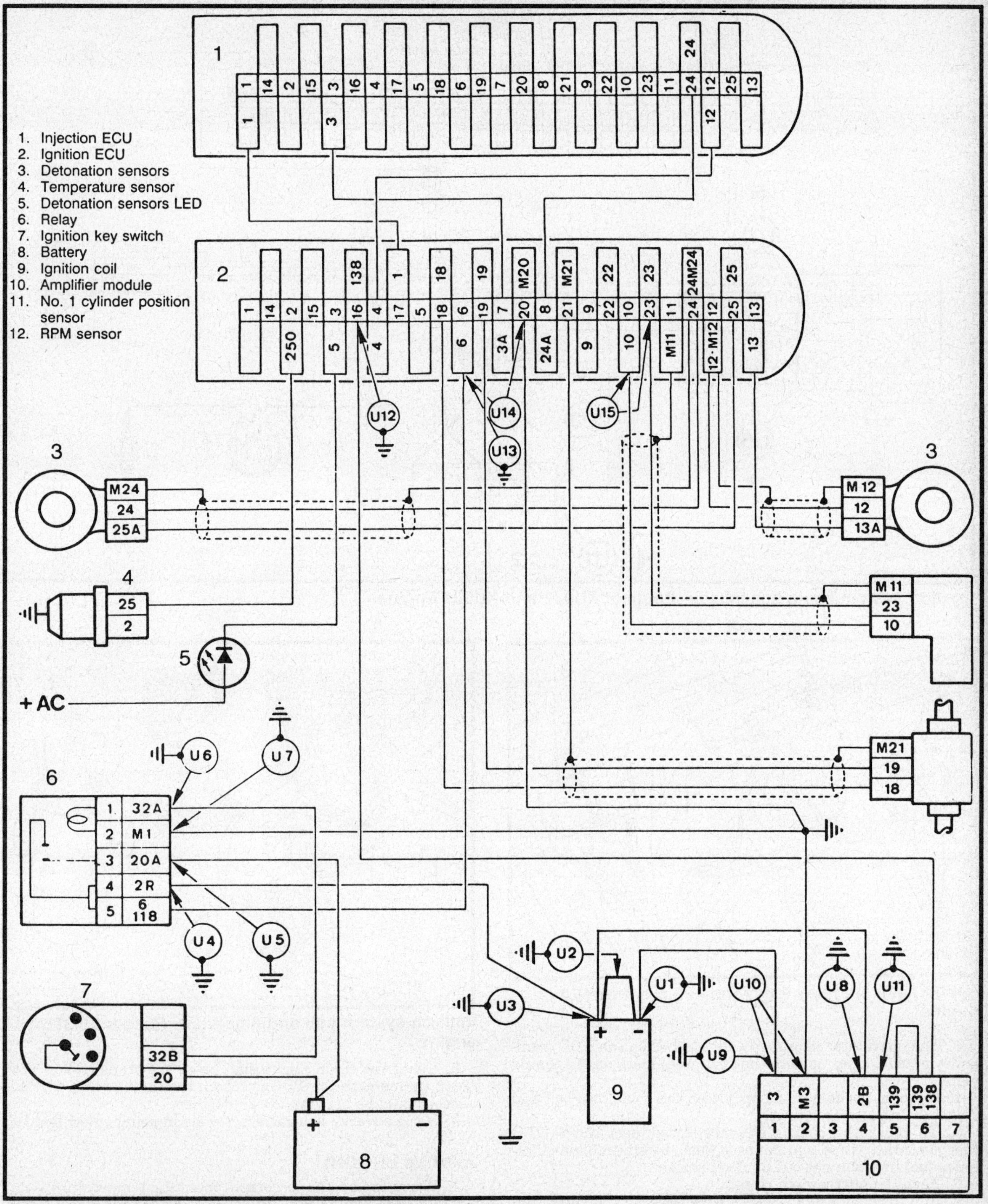

Ignition system basic schematic—Peugeot ZN3J engine

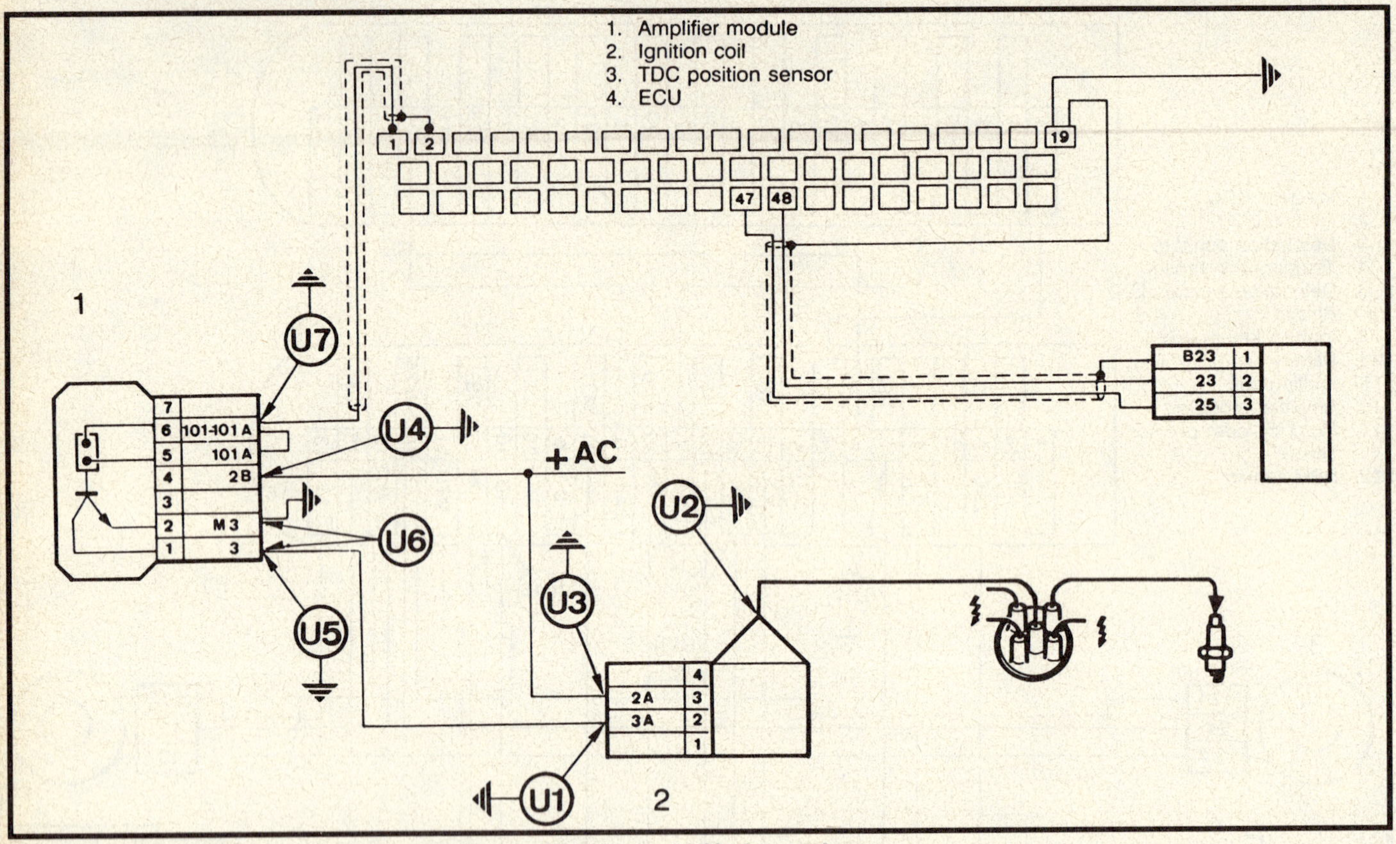

Ignition system basic schematic—Peugeot XU9J2 and XU9J4 engines

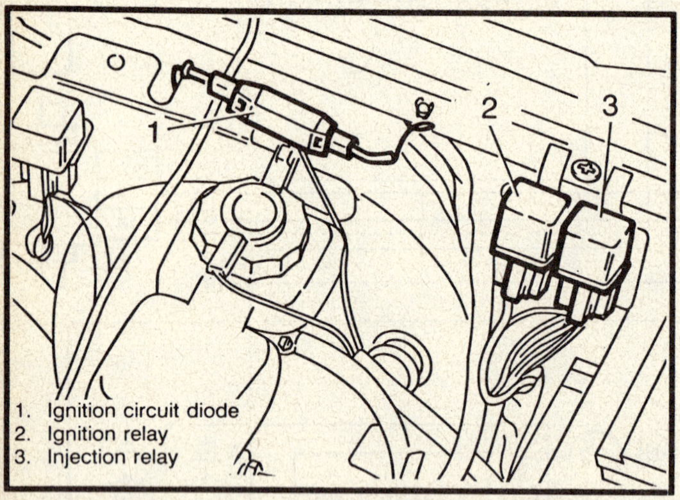

Ignition system relay locations—Peugeot N9TEA engine

4. The distributor is properly adjusted when the TDC crank pulley notch is lines up with the 10 degrees mark on the timing plate.

4. If the marks do not line up, loosen the hold-down bolt and adjust distributor accordingly.

5. To check the timing advance, disconnect the 2–60°C thermocontact wires and connect them together. This eliminates fuel injection cut-out on deceleration.

6. Keep the microswitch closed.

7. Run the engine at a speed ranging between 1600 and 1900 rpm.

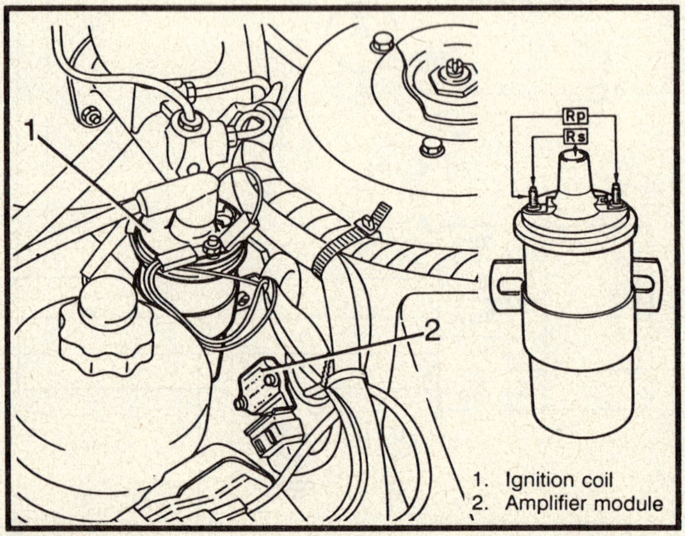

Ignition system coil and amplifier—Peugeot N9TEA engine

8. Using the timing light dial, bring the crank pulley TDC notch in line with the TDC notch on the timing plate. The advance should be between 23–28 degrees.

9. If the advance is incorrect, try again using a new ignition ECU.

ZN3J V6 ENGINE

1. Warm the engine to normal operating temperature.
2. Connect timing light to the No. 1 cylinder high tension lead. A scope may also be connected using the diagnostic plug.

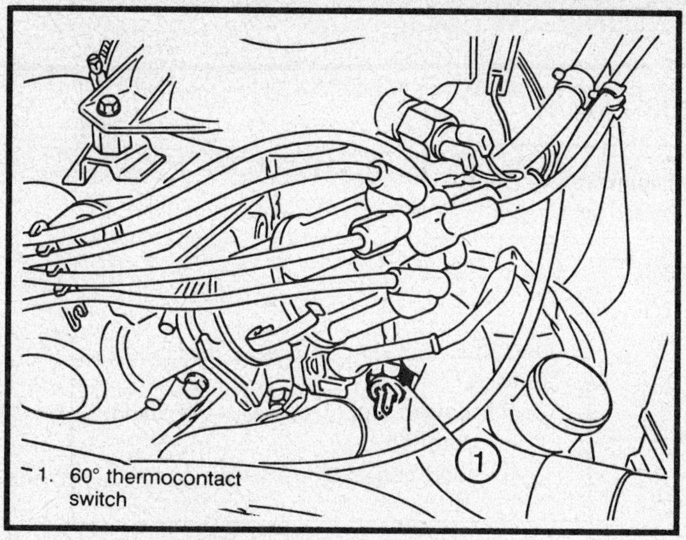

Temperature switch location, typical

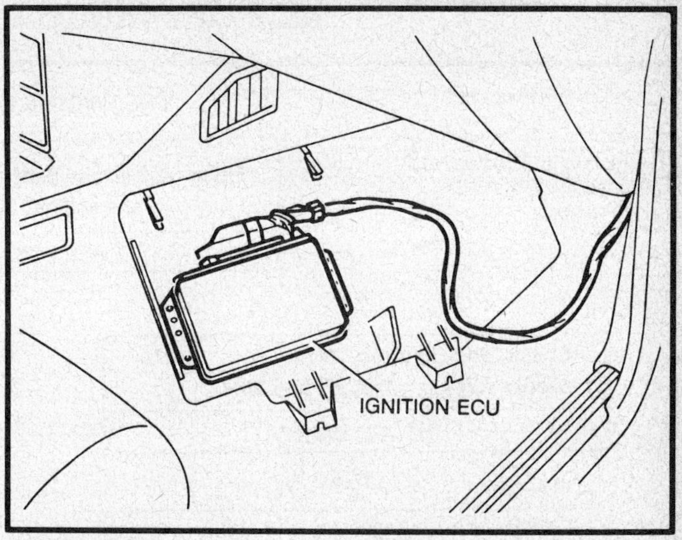

Ignition ECU located under passenger side footrest

3. With the microswitch closed, check to ensure the idle speed is at 910 rpm. If necessary, adjust the idle speed using the air bleed screw on the throttle housing.
4. Check the ignition timing. The ignition timing should be 10 degrees BTDC at 910 rpm.
5. If the ignition timing is not within specification, adjustment is necessary.
6. Loosen the distributor hold-down bolt, turn the distributor until the TDC crank pulley notch is lined up with the 10 degree mark on the timing plate.

Component Removal

DISTRIBUTOR ASSEMBLY

Removal and Installation

EXCEPT ZN3J V6 ENGINE

1. Crank engine to TDC of compression stroke of No. 1 cylinder. This is for reference to when installing distributor. Disconnect negative battery cable.
2. Remove the water shield over the distributor, if used.
3. Remove the distributor cap. If the spark plug cables make this difficult, matchmark the cables for reassembly then disconnect the cables. Make a mark on the distributor body rim next to the rotor tip to help in alignment at installation.
4. Loosen the distributor hold-down bolt, slide the clamp out of the way and lift the distributor up and out of the engine, disconnecting the sensor wires.

To install:

5. Installation is the reverse of the removal procedure. Lubricate the lower part of the distributor with clean engine oil. Lower the distributor body into the engine watching to engage the distributor drive slots.
6. When the distributor is installed, check that the rotor is aligned with the mark on the distributor body rim make at removal.
7. Install the hold-down clamp and bolt. Snug up but do not tighten.
8. Install the rotor, cap and spark plug cables, if removed.
9. Connect negative battery cable.
10. Start and warm engine to operating temperature. Adjust ignition timing. Install distributor cap shielding, if used.

ZN3J V6 ENGINE

1. Disconnect the negative battery cable.
2. Remove the ignition distributor cover and distributor cap.
3. Remove the distributor rotor.
4. Remove the 2 body bolts and the 4 distributor housing bolts.

To install:

5. Installation is the reverse of the removal procedure. Install the 2 body bolts and the 4 distributor housing bolts.
6. Install the rotor, distributor cap and cover.

NO-START TROUBLESHOOTING—N9TEA ENGINE

THE ENGINE DOES NOT START

Hook up a strobe light to the coil wire and crank the engine, using the starter

Does the strobe light flash ?

NO ↓

Check the ignition relay power supply voltage, wires 2R and 20R

U1 = U Battery — NO → Check :
- the power supply to relay and its operation,
- the condition of diode 8DA.

YES ↓

Check the voltage to the coil

U2 = U Battery — NO → Check the wire 2R for continuity

YES ↓

Check the voltage between the (–) terminal of the coil and ground

U3 = U Battery — NO → Check the coil's primary and secondary resistance, coil disconnected.
R primary = 0.85 ohm
R secondary = 6 000 ohms
NOTE - *If U3 is between 0 volts and U battery short in wire 3 or amplifier module.*

YES ↓

Disconnect the amplifier module connector and check the power supply (wire 2B)

U4 = U Battery — NO → Check wire 2B for continuity

YES ↓

Check the voltage of the primary circuit (wire 3)

U5 = U Battery — NO → Check wire 3 for continuity

YES ↓

Remove injection relay Using a test screwdriver with an LED, check for the control signal wire by engaging starter :

* U6 = Presence of a signal — NO → CONTINUE TO TROUBLESHOOT

YES ↓

Replace the amplifier module
Reinstall the injection relay

THE ENGINE STARTS

* **NOTE** - *If no LED tester screwdriver is available, it is still possible to make this test using the AC 20 V scale of a voltmeter :* $\frac{U6}{U7} \simeq 5\ V$

NO-START TROUBLESHOOTING—N9TEA ENGINE

Continued from preceding page. Recap of U6 check

Disconnect the injection relay
- Using a test screwdriver with an LED, check for the control signal (wire 138) by engaging starter.

*U6 = presence of the signal

NO ↓

Disconnect the ignition E.C.U. connector.

Remove the connector's cover

Reconnect the connector

Check the control signal of the amplifier module (wire 138) on pin **16** of connector by engaging starter.

* U7 = presence of the signal — YES → Check continuity of wire 138.

NO ↓

Checking the distributor

Disconnect the connector and check the voltage on pins **(2)** and **(3)**

U8 = 11 volts
U9 = 5 volts

— NO → Check :
- the power supply to ignition E.C.U. pin **(6)** (wire 20R)
- the continuity of wires 8B and 7B.

If power supply and continuity are correct, replace the ignition E.C.U.

YES ↓

Connect the distributor

Place a slot of the rotor with tabs opposite the impulse generator

Check the voltages at pin **(2)**, wire 7B as follows :
: With no feeler gauge in front of HALL generator, U = 0 volt
: With a feeler gauge in front of HALL generator, U = 5 volts

The readings are correct — NO → Replace the HALL generator

YES ↓

Reconnect the amplifier module

Reconnect the injection relay

THE ENGINE STARTS

NO-START TROUBLESHOOTING—N9TEA ENGINE

THE ENGINE DOES NOT START

Hook up a strobe light to the coil wire and crank the engine using the starter.

Does the strobe light flash ?

YES ↓

Disconnect the injection relay

Connect the strobe light to the plug wires and crank engine using starter.

Does the strobe light flash ? — NO → Check the distributor cap, the rotor and the plug wires.

YES ↓

Check the spark plugs

Plugs OK — NO → Clean and gap the plugs or replace them (BOSCH WR 7 DS)
Electrode gap :
$E - 0.9^{+0.1}_{+0}$ mm (0.036"$^{+0.004}_{+0}$)

YES ↓

Recheck the static distributor timing.
Line up :–
– the single notch of the crank pulley with the notch on the timing plate.
– the reference on the distributor housing with the distributor finger

Timing OK — NO → 1) Do the static timing adjustment of the distributor.
2) Do the dynamic timing adjustment of the distributor :
10° BTDC at 910 ± 40 rpm
(see corresponding chapter).

YES ↓

Check for the presence of the signal for timing on pin **(1)** of the injection E.C.U. (see corresponding chapter)

Signal is available — NO → Check wire **(1)** for continuity between ignition E.C.U. and injection E.C.U.
↓
Continuity OK
↓
Make another check using a new ignition E.C.U.

YES ↓

CHECK THE FUEL INJECTION

IGNITION SYSTEM FAULT CODES—N9TEA ENGINE

Fault*	Spark advance correction	Causes	Checks
1	−22°	Continued detonation Maximum correction reached	Cooling system Distributor timing Boost pressure
2	−20°	Battery voltage is under 10.5 volts.	Battery Charging circuit
3	−20° non-revers-ible until ignition switch is turned off	Detonation correction circuit in the ignition E.C.U. defective.	Try with a new ignition E.C.U.
4		Erroneous signal received from detonation sensor (Check for engine speed above 3 200 rpm).	Detonation sensor Electrical circuit continuity.
5		Signal from potentiometer is greater than 4.3 volts.	Try with a new potentiometer
6 A		Connections and wiring between ignition E.C.U. and potentiometer defective.	Continuity of the potentiometer's electrical circuit.
6 B	Mirco-switch / Advance Open / Full Load Closed / Idle	NO LOAD SIGNAL EMANATING FROM THE INJECTION E.C.U.	Electrical circuit continuity between **pin 8** of the ignition E.C.U. and **pin 6** of the injection E.C.U. (wire 6). Check with a new ignition E.C.U. Check with a new injection E.C.U.

*The number of the FAULT represents the number of blinks per cycle of the LED (173) on the instrument cluster.

NOTE - *For the "irreversible" faults (3, 4, 5, 6A), it is necessary to turn ignition key "OFF" to "erase" the fault entered in memory.*

NO-START TROUBLESHOOTING—ZN3J ENGINE

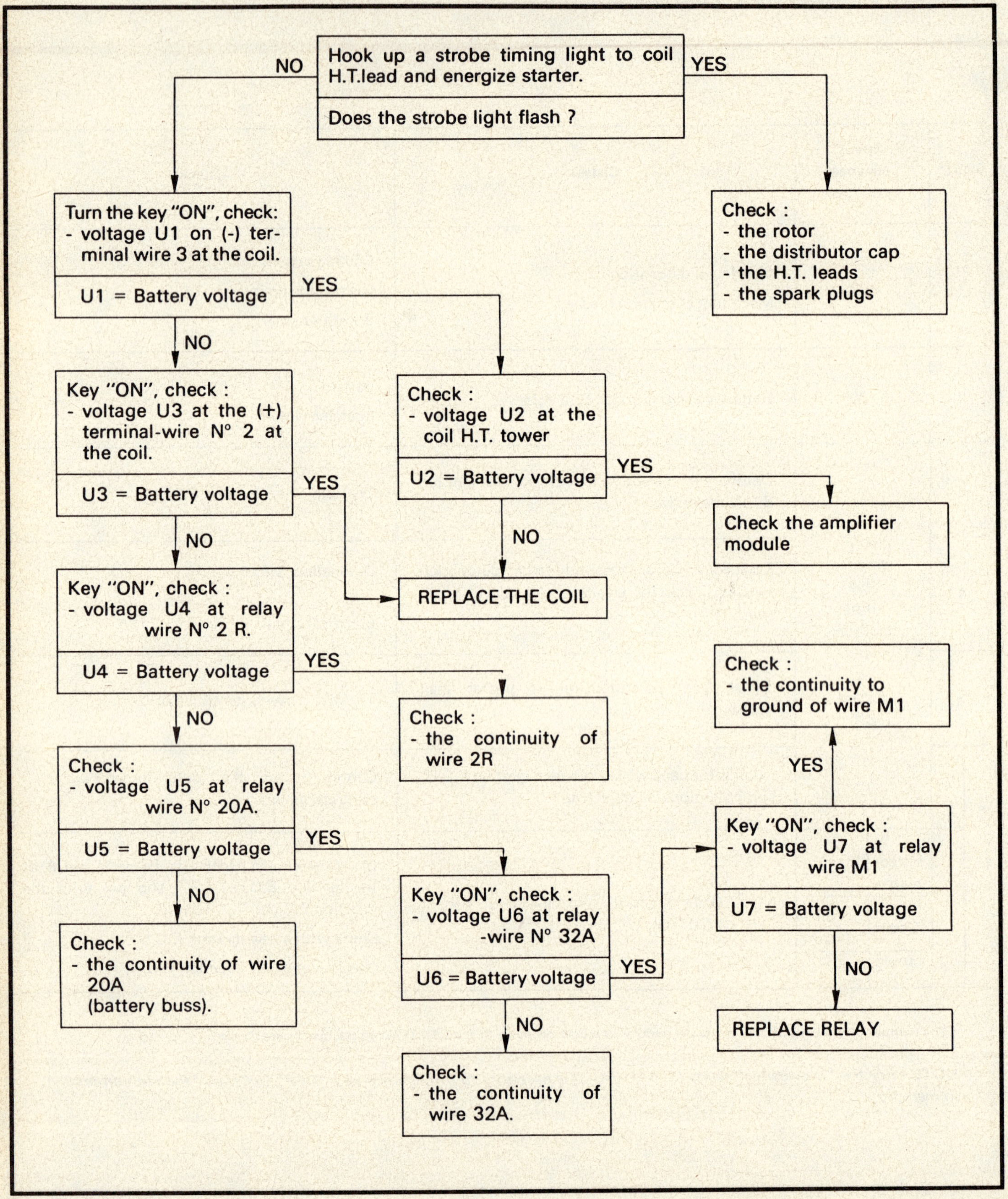

CHECKING THE AMPLIFIER MODULE—ZN3J ENGINE

CHECKING THE AMPLIFIER MODULE.

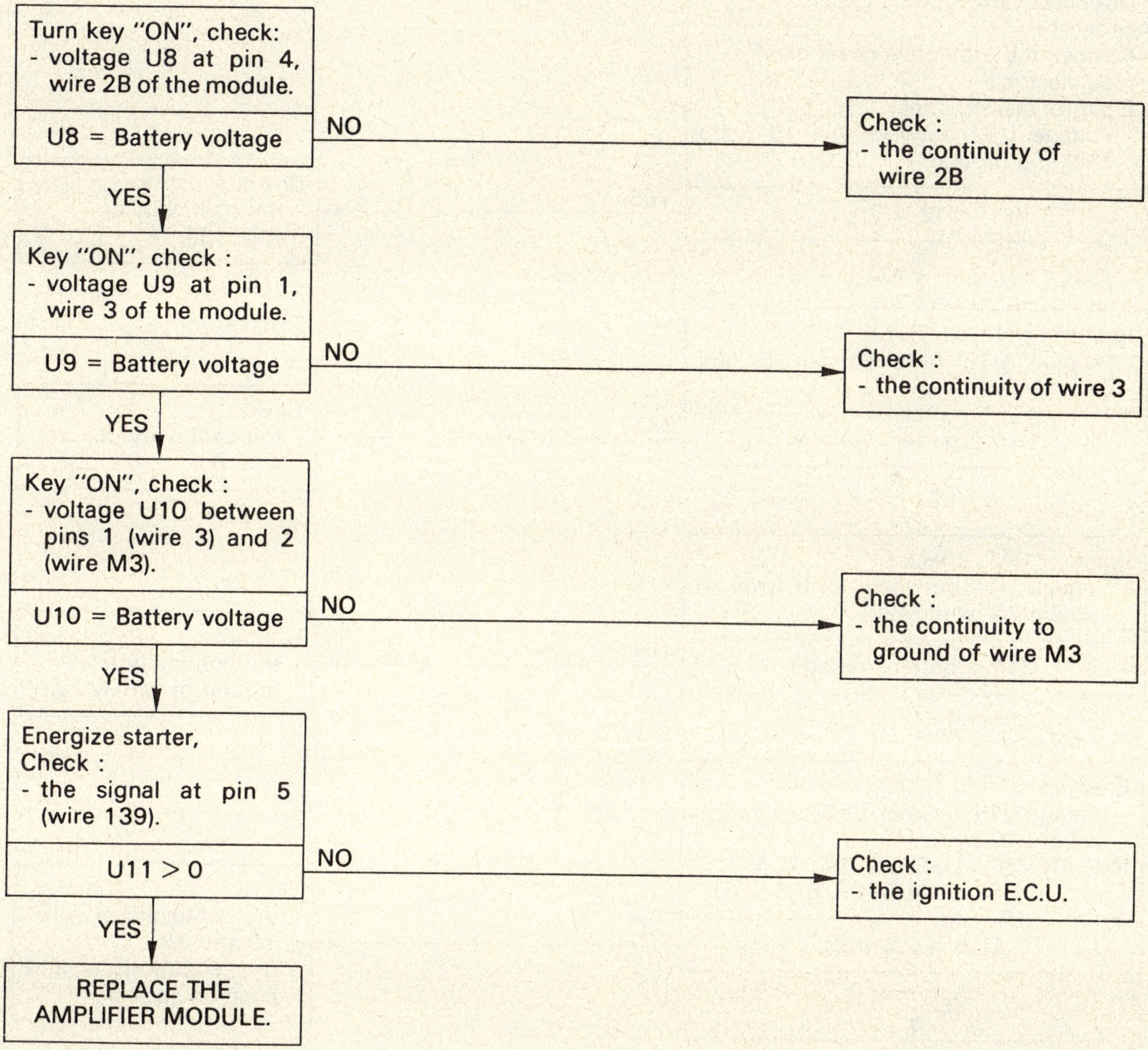

CHECKING THE IGNITION ECU—ZN3J ENGINE

CHECKING THE IGNITION E.C.U.

Disconnect the ignition E.C.U. connector.
Remove the protective cover and reconnect it.
Energize starter, check :
- Voltage U12 between pin 16 (wire 138) and ground.

U12 > 0

YES → Check :
- the continuity of wire 138.

NO ↓

Ignition "ON", check :
- Voltage U13 between pin 6 and ground.

U13 = battery voltage

NO → Check :
- the continuity of wire 6.

YES ↓

Ignition "ON", check :
- Voltage U14 between pin 6 (wire 6) and pin 20 (wire M20).

U14 = battery voltage

NO → Check :
- the continuity to ground of wire M20.

YES ↓

Energize starter, check :
- Voltage U15 between pin 23 (wire 23) and pin 10 (wire 10).

IMPORTANT : USE THE A.C. 20 V SCALE FOR THIS TEST

U15 = 2,5 volts

NO → Check :
- the continuity of wires 10 and 23.
- the engine rpm and position sensor

YES ↓

RUN ANOTHER TEST USING A NEW IGNITION E.C.U.

NO-START TROUBLESHOOTING—XU9J2 AND XU9J4 ENGINES

Preliminary condition : correct power supply to ECU

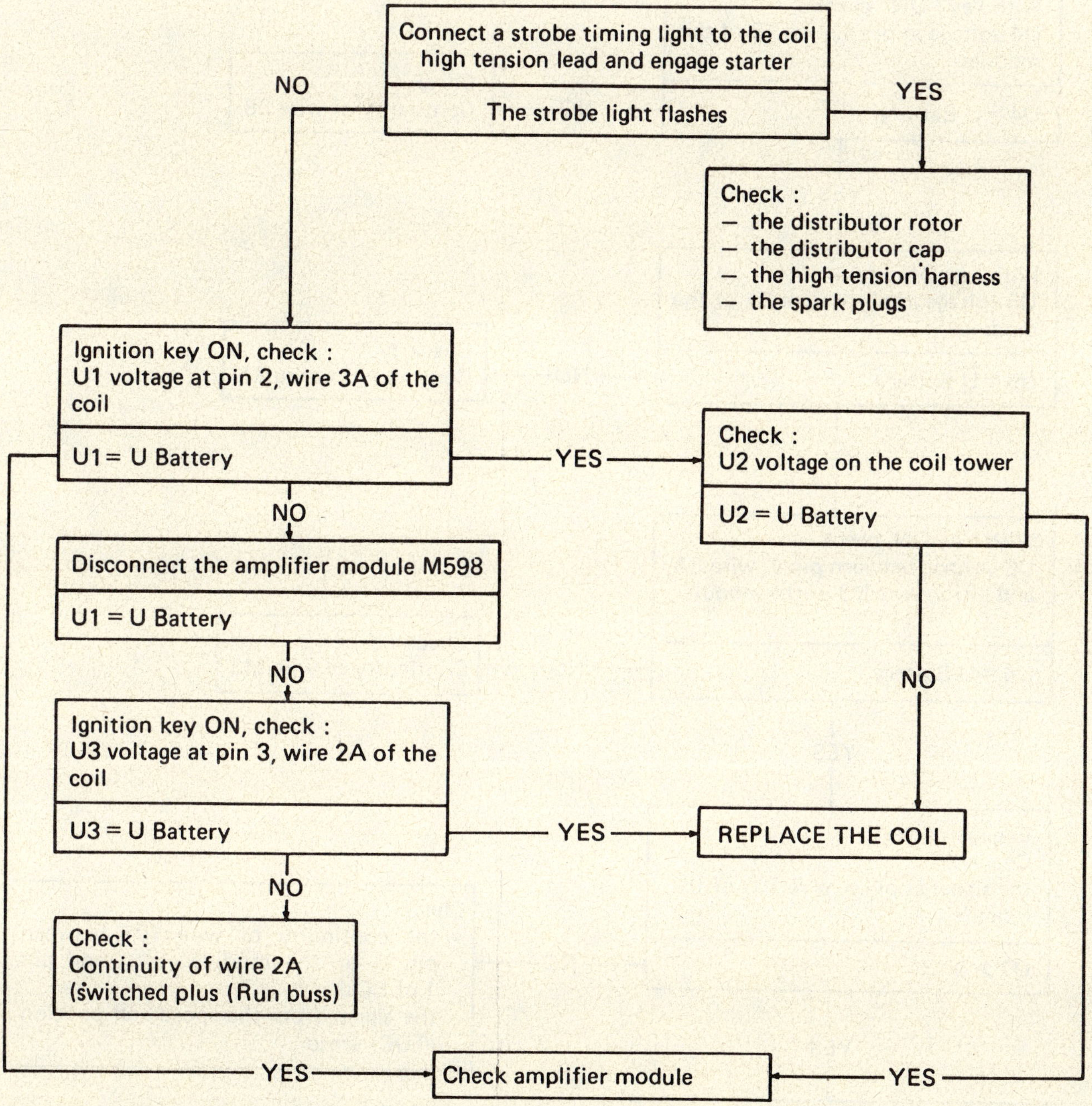

NO-START TROUBLESHOOTING CONTINUED—XU9J2 AND XU9J4 ENGINES

TROUBLESHOOTING TABLE

Checking the amplifier module

With key "ON" check :
U4 voltage at pin 4, wire 2B of the module

U4 = U Battery — NO → Check : Continuity of wire 2B

YES ↓

With key ON, check :
U5 voltage at pin 1, wire 3A of the module

U5 = U Battery — NO → Check : Continuity of wire 3A

YES ↓

With Key ON, check :
U6 voltage between pin 1, wire 3A, and pin 2, wire M3 of the module

U6 = U Battery — NO → Check : Continuity of wire M3

YES ↓

Engage starter
Check :
the signal at pin 6, wire 101 of the module

U7 > 0 — NO → Check :
- the continuity of wire 101 between pin 6 or the module and pin 1 of ECU
- the signal from the speed and position (TDC) sensor

YES ↓

REPLACE THE AMPLIFIER MODULE

If all these checks are correct, REPLACE THE ECU

PORSCHE

General Information

DIGITAL MOTOR ELECTRONIC (DME)

The DME consists of a transistor coil ignition system. In addition, a control unit for knock and charging air pressure (turbocharged) regulation is used. Both electronic components correspond with each other.

With knock and charging air pressure regulation (turbocharged vehicles), it is possible to approach the thermal and mechanical limits of an engine in every operating condition with considerably more sensitivity. This produces higher torque values and noticeably better response of the engine when accelerating, even from low engine speeds.

The DME system of the 911 Carrera and Speedster is equipped with double ignition. A Hall Effect sender, which is required for identification of cylinder No. 1 (ignition TDC), is installed in a double-ignition distributor. This system also has 1 inductive sender for speed and reference mark information. The ring gear of the flywheel is specially machined for this purpose and consists of 60 teeth. Two of the teeth have openings for production of the reference mark signal. The reference mark signal is designed for 84 degree BTDC.

The distributor of the Carrera and Speedster is equipped with a double-ignition and a Hall effect sender. The Hall sender, with which the DME control unit recognizes the ignition timing of cylinder No. 1, is installed in the double-ignition distributor, so that the signals of both knock sensors can be assigned to the firing order appropriately.

The DME ignition system consists of the following components:

- Control unit—digital motor electronics
- Control unit—knock and charging air pressure regulation
- Idle speed charging regulator
- Air circulation valve
- Oxygen sensor
- Throttle valve potentiometer
- Knock sensor
- Cycling valve for charging air pressure regulation

968 MODEL

In 1992, the Porsche 968 was introduced. This vehicle uses a 4 cylinder, 236 hp, 4-valve engine (engine codes M44/43–manual transmission, M44/44–automatic transmission) with variable cam timing. It uses the DME engine management system, level 2.10 with the hot wire LH Jetronic sequential fuel injection with catalytic converter and knock control. The engine speed/reference marker signal is taken from a ring gear milled onto the flywheel. The signal is monitored by an adjustable inductive sensor which must be located 0.8 ± 0.2 mm from the ring gear.

A Hall Effect transmitter is mounted on the exhaust camshaft's drive sprocket. Ignition TDC of cylinder 1 is identified by means of its signal. Ignition firing order is 1–3–4–2. The ECU allocates the 2 knock sensor signals in accordance with the firing order. The Hall Effect signal is also used for sequential fuel injection control. Fuel injection order is 4–2–1–3.

The ECU has diagnostic capability through either the factory scan tool and ocilloscope or the instrument cluster Malfunction Indicator Lamp (MIL). The ECU on this model is located in the front passenger's footwell behind the wooden foot rest board. The Engine Management system is also tied to the 4-speed automatic transmission. To reduce shift impact when changing gears under load, the engine's ignition is briefly retarded according to load and engine speed when the automatic shifts up or down. Engine torque is reduced as a result.

The DME ECU is adaptive. If the multi-pin plug is disconnected from the ECU or if the battery is disconnected, the engine must run for about 10 minutes at operating temperature to adapt. The ECU has a volatile memory. When the ECU is disconnected or the battery cables disconnected, any faults recorded in the fault memory are erased. Test connectors for the diagnostic socket are located on the right side of the passenger footwell.

EZK IGNITION SYSTEM

1990–92 928S4 models are equipped with the Elektrische Zundkontrolle (EZK) breakerless ignition system. This ignition system is controlled by 2 knock sensors, mounted on the engine's upper crankcase section with special screws, but without spring washers. The front knock sensor senses cylinders 1,2,5 and 6, while the rear sensor is for cylinders 3,4,7 and 8.

If knocking occurs, the ignition timing for that cylinder is retarded by 3 degrees. If knocking continues, the ignition timing is retarded in steps of 3 degrees up to a maximum of 9 degrees for each knock. When knocking is no longer recognized, the ignition timing is returned is small steps to its optimal value.

A Hall transmitter, which is mounted behind the camshaft sprocket of cylinder bank 1–4, allows the control unit to register knocking cylinders. The ignition timing of all cylinders will be retarded 6 degrees for safety reasons, if the knock sensors or hall transmitter fails.

Two of the flywheel ring gear teeth are gapped, in order to provide a signal for the reference mark of the speed sensor. In addition, to make the ignition timing visible digitally, 2 pins are pressed into the flywheel for vehicles equipped with automatic transmission or into the pressure plate for manual transmission vehicles. The test connection for making ignition timing visible is located on the left side of the engine in front of the air cleaner.

The LH and EZK control units are capable of self-diagnosis. Tester tool is connected to the diagnosis socket. 1990–92 vehicles are equipped with a 19-pin diagnostic plug connection for performance diagnosing. The diagnosing connection is located underneath the radio booster cover, on passenger side. Tester tool KTS 301 are used to call the fault memories.

An ignition circuit monitoring relay, mounted on the holder of the control unit, monitors both ignition circuits (ignition circuit 1: 1,7,6,4 and ignition circuit 2: 3,2,5,8) and guarantee that the catalytic converter system is not destroyed in case of ignition circuit failure. Thermo-elements is screwed in exhaust ports of cylinders 4 and 8. These thermo-elements monitors the exhaust temperature.

If, for example, an ignition circuit should fail, a relay contact in the relay belonging to the failed ignition circuit is opened and interrupts the supply of power to the fuel injectors of this ignition circuit.

SYSTEM OPERATION

DIGITAL MOTOR ELECTRONIC (DME)

The ignition triggering signal, 1990–92 911 Carrera 4 and Speedster, goes from the DME control unit to 2 final ignition stages. In each of both ignition final stages, a transistor is activated. The secondary circuit of both ignition coils, which are installed on the left hand side in the engine compartment, are then interrupted. In this manner an ignition spark is triggered off parallel by both ignition coils each time. The secondary voltage goes to the double-ignition distributor via the high tension leads to the appropriate spark plug. All ignition leads are marked for better identification of both ignition circuits and cylinders belonging to each circuit.

SYSTEM COMPONENTS

Control Unit

The control unit, mounted on the right side of the passenger's footwell, processes incoming information of air volume, engine speed, intake air temperature, engine temperature, throttle valve position (idle, full throttle), starting and exhaust content (oxygen sensor).

High Altitude Switch

Porsche 944 vehicles are equipped with a high altitude switch. The altitude switch works through a barometric sensor which closes the switch. In areas higher than 3300 ft. (1000m), the engine needs a leaner fuel mixture. A ground signal switches the DME control unit to a leaner fuel mixture (a shorter injection time).

Knock Sensor

The knock sensor consists of a piezo crystal, located between 2 seismic materials. Oscillation, especially engine knocking, produces forces on the crystal and the crystal reacts to these forces with changes in electrical voltage. The knock sensor, 944 Turbo, is bolted on the engine block next to the cylinder. On the 911 Carrera 4, the engine is fitted with 2 knock bridges, on each of which 1 knock sensor is screwed.

Speed and Reference Mark Sensors

The speed and reference mark sensors are inductive sensors. The reference mark sensor detects the position of the crankshaft in relation to TDC. The speed sensor measures engine speed.

Temperature Sensor I

The temperature sensor I, located in the air flow sensor, provides the control unit with information on intake air temperature. The temperature sensor I cannot be replaced. However, its operation can be checked on the plug connection of the air flow sensor.

Temperature Sensor II

The temperature sensor II provides the control unit with information on engine temperature. The control unit uses this information to control the fuel volume during engine warm-up phase.

Distributor

The distributor of the 944 and the 968 is mounted on the end of the camshaft and is driven by the camshaft sprocket. It is only required for high voltage distribution.

The distributor of the 911 Carrera 4 is a double-ignition distributor.

Throttle Switch

The throttle shaft of the throttle housing is connected to a throttle switch. The switch has 2 switching contacts, the idle and full throttle contacts. The idle contact controls the coasting fuel cut-off and supplies the signal for idle stabilization. The full throttle contact reports the full throttle position of the throttle to the control unit. The full throttle contact has a fixed relation to the adjustable idle contact.

Diagnosis and Testing

SERVICE PRECAUTIONS

- Always disconnect the negative battery terminal to prevent sparks cause by short-circuiting, circuit-breaking, etc.
- If using a boost charger, make sure that the battery leads are disconnected.
- Never use an auxiliary starting-assist device to start the engine, which has a voltage rating above 12 volts.
- Never disconnect the battery while the engine is running.
- Only measure resistance values after turning **OFF** the ignition or disconnecting the battery.
- Never replace specified ignition coil with a different ignition coil.
- Never connect ignition coil terminal1 to ground for a burglar alarm or the ignition coil and ECU will be destroyed.
- Never connect the positive battery terminal or a test lamp to ignition coil terminal 1 of the ECU will be destroyed.
- Voltage flashover from ignition coil terminal 4 to coil terminals 1 and 15 must not occur or the ECU will be destroyed.
- Always disconnect the rpm sensor plug for compression tests. Otherwise, hazardous high voltages and insulation damage to the ignition coil, distributor and ignition leads may result.
- Never start the engine as long as the ground lead between body and engine is disconnected or the ECU will be destroyed immediately.

READING CODES

Both the DME and LH/EZK system control units have a fault memory for self-diagnosis of certain faults in the fuel injection and ignition systems. On most vehicles sold outside of California, the fault codes can only be accessed using the Porsche Diagnostic Tester 9268 or 9288. These are hand held units that plug into the diagnostic connector behind the passenger side kick panel. The earlier 9268 tester includes a flashing light to report fault codes. The later 9288 unit will report fault codes, sensor readings and other stored, measured or calculated values. This unit is also menu driven, meaning instructions for test procedures appear right on the screen. The 9288 unit is capable of activating certain output devices for function testing. Instructions for use of the tester is provided with the unit.

Fault codes are set when the control unit is unable to operate the engine according to the normal programs. A Code does not necessarily indicate a faulty component, only the component which will not operate within the normal program. Simple things like a poor connection or a vacuum leak can cause a fault code to be set. Always check for obvious problems such as a bad ground connection or cracked PCV hose before starting other repairs.

On the 928, separate control units are used for fuel and ignition systems. Many of the sensors needed for both systems, report directly to the EZK ignition control unit, so there is a data link between the EZK and the LH control units. Both control units have a self diagnostic program that will store fault codes in memory and the Codes from both units are retrieved at the same time. The first digit of the Code will indicate which control unit (system) has detected the fault. If the first digit is 1, the fault is in the LH fuel injection system. If the first digit is 2, the fault is in the EZK ignition system. Each control unit may report the same fault.

On all 928 and 968 models and on 911 Carrera models sold in California, an on-board Malfunction Indicator Light (MIL) is installed to warn of a failure that might influence emissions. These vehicles are equipped with an on-board diagnostic system that reports fault codes by flashing the warning light. No special tester is needed to access the codes, although the diagnostic tester can still be used as described above. To access the codes on all 928, 968 models and on California 911 Carrera models:

1. Turn the ignition switch **ON** without starting the engine.
2. Fully press the accelerator pedal to close the full load switch. After about 3 seconds the warning light will flash.
3. Release the accelerator to close the idle switch, then count and write down the warning light flashes. Repeat this process until Code 1000 appears.

The control unit will report all Codes stored in memory in numerical order. All Codes are 4 digit numbers. The first digit will be 1 if the fault code is stored in the DME or LH fuel injection control unit. On 928 models, if the first digit is 2, the fault code is in the EZK ignition control unit.

If the second digit is 1, the detected fault is current. If the sec-

DME ENGINE FAULT CODES—911, 944 AND 968

Fault Code	Explanation
1000	End of report
1111	Control unit supply voltage out of limit
1112	Throttle idle position switch grounded
1113	Throttle full load position switch grounded
1114	Engine temperature sensor
1115	Throttle idle position switch open
1121	Air flow sensor
1122	Idle speed control circuit
1123	Oxygen regulation out of limit
1124	Oxygen sensor signal
1125	Inlet air temperature sensor signal
1131	Knock sensor 1
1132	Knock sensor 2
1133	Knock regulation in control unit
1134	Hall sender signal missing
1141	Control unit faulty
1143	Carbon canister purge valve
1144	Resonance plate solenoid valve circuit
1145	CHECK ENGINE warning light
1151	Injector, cylinder 1 (911 only)
1152	Injector, cylinder 2 (911 only)
1153	Injector, cylinder 3 (911 only)
1154	Injector, cylinder 4 (911 only)
1155	Injector, cylinder 5 (911 only)
1156	Injector, cylinder 6 (911 only)
1500	No faults in memory

NOTE: If the second digit of the code is 2, the fault is not present at the time of testing. It may be an intermittent fault.

LH ENGINE FAULT CODES-928

Fault Code	Explanation
1000	End of report
1111	Control unit supply voltage out of limit
1112	Throttle idle position switch grounded
1113	Throttle full load position switch grounded
1114	Coolant temperature sensor
1115	Throttle idle position switch open
1121	Air flow sensor
1123	Air/fuel mixture rich
1124	Air/fuel mixture lean
1125	Oxygen sensor signal missing
1141	LH control unit faulty
1500	No faults in memory

NOTE: If the second digit of the code is 2, the fault is not present at the time of testing. It may be an intermittent fault.

ond digit is 2, the detected fault has not occurred during the last running of the vehicle but did occur within the last 50 engine starts. The remaining 2 digits indicate which component is at fault.

CLEARING CODE MEMORY

The fault code memory should be cleared before returning the vehicle to service. All codes in memory and the idle control adaptation are lost when the control unit or the battery is disconnected. To avoid the loss of the learned idle program, use the instructions on the 9288 tester to clear the memory. If this tester is not available, disconnect the control unit or the battery to clear the memory.

It will be necessary to drive the vehicle for 6–10 minutes at operating temperature and run the engine at idle for about 10 minutes so the control unit can learn idle speed, timing and mixture parameters. Make sure the engine is at operating temperature and that all accessories are **OFF**. The throttle-at-idle position switch must be closed and functioning or system adaptation will not take place.

NO START—NO SPARK

Use care when troubleshooting these systems. If the engine is

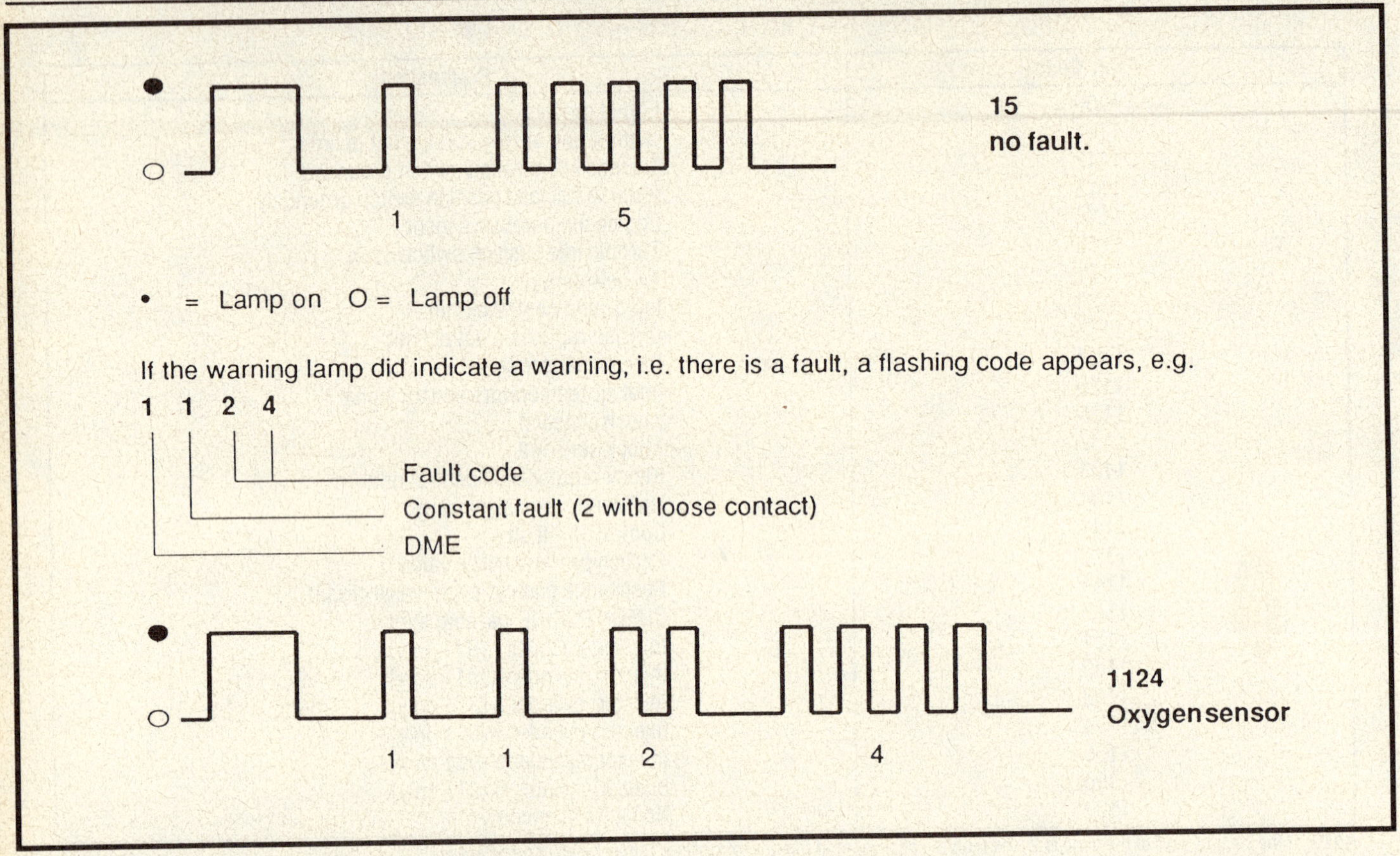

Diagnostic blink code explanation—968 shown

cranked to check for spark with spark plug wires disconnected, it means the ECM doesn't "see" the load it expects. The result could be a voltage spike that damages the ECM.

Testing

1. Using a properly grounded spark plug tester, test for spark at spark plugs, while cranking the engine.
2. Check the connection of the ignition coil and distributor.
3. Check the resistance of the spark plug wires.
4. If there is no spark when the speed sensor/reference sensor should be checked.
5. If there is no spark when the engine is cranked, the ignition coil should be checked.
6. If there is no obvious fault and the speed sensor and coil appear to be in working order, the ECU control unit is suspect.

SPEED SENSOR

Testing

The following procedure requires the use of an oscilloscope.

911, 944, 944 TURBO AND 968 MODELS

1. Hook-up the oscilloscope and adjust it according to the supplier's instruction.
2. Disconnect the DME control unit plug under the driver's seat or passenger footwell in the case of the 968.
3. Connect the oscilloscope tester leads on plug terminals 8 and 27, using the proper test leads. The speed sensor plug on the plug plate in the engine compartment, left hand side, must be connected.
4. Operate the starter. The screen must display sine oscillation, of which the amplitude must be greater than 2.5 volts.
5. If the voltage signal is less than 2.5 volts, repeat the above procedures at the speed sensor plug, connecting the oscilloscope tester leads to the center and bottom pin of the plug.
6. If the voltage signal is still less than 2.5 volts, check for excessive clearance between the sensor and flywheel. The specified clearance is 0.8± 0.3mm.
7. If the clearance is within specification, replace the sensor.

REFERENCE MARK SENSOR

Testing

This following procedure requires the use of an oscilloscope.

911, 944, 944 TURBO AND 968 MODELS

1. Hook-up the oscilloscope and adjust it according to the supplier's instruction.
2. Disconnect the DME control unit plug under the driver's seat.
3. Connect the oscilloscope tester leads on plug terminals 25 and 26, using the proper test leads. The reference mark sensor plug on the plug plate in the engine compartment, left hand side, must be connected.
4. Operate the starter. A single sine oscillation should appear on the screen.

NOTE: It is important that the oscillation begins with a positive flank and the screen scale shows greater than 2.0 volts, with a starter speed of at least 200 rpm. The amount of signal amplitude depends on the starter speed.

5. If the voltage signal is less than 2.0 volts, repeat the above procedures at the reference mark sensor plug, connecting the

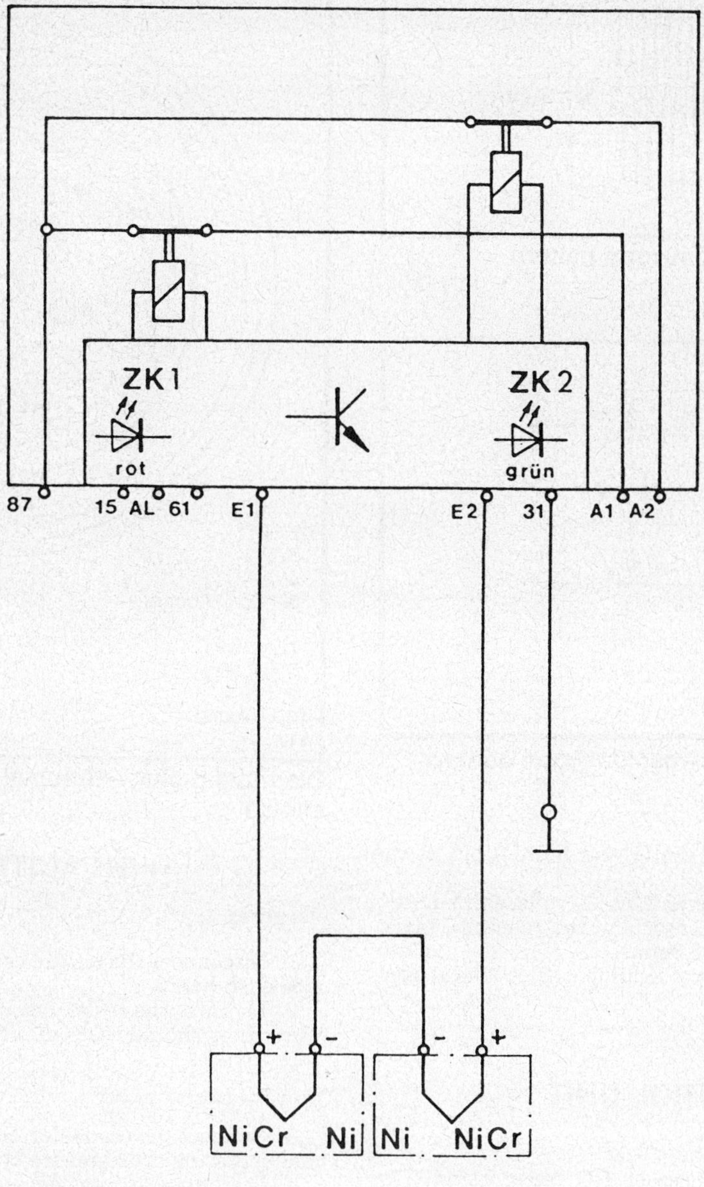

Term. 87	– From LH-Jetronic relay of CEL
Term. 15	– Ignition
Term. AL	– To LH control unit term. 28
Term. 61	– From alternator
Term. E2/E2	– To/from thermo elements
Term. E1	– Ignition circuit I (cyl. 4)
Term. E 2	– Ignition circuit II (cyl. 8)
Term. 31	– Ground
Term. A1	– To fuel injectors for cyl. 1-4-6-7 (ignition circuit I)
Term. A2	– To fuel injectors for cyl. 2-3-5-8 (ignition circuit II)
ZK1/ZK2	– Light emitting diodes for ignition circuit failure indicator (red/green)

Ignition circuit monitoring relay—928S4 wiring diagram

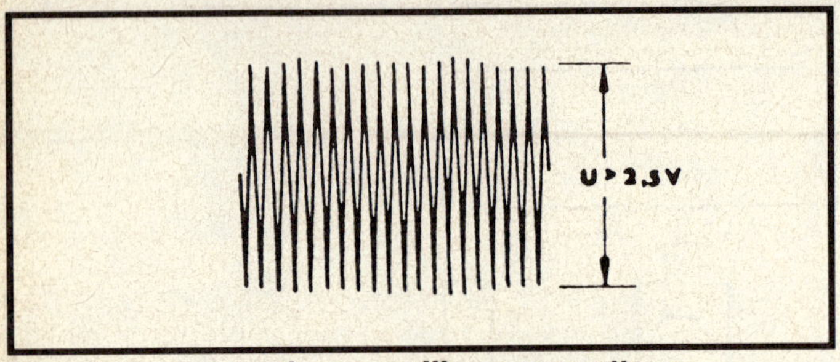

Speed sensor testing—oscilloscope pattern

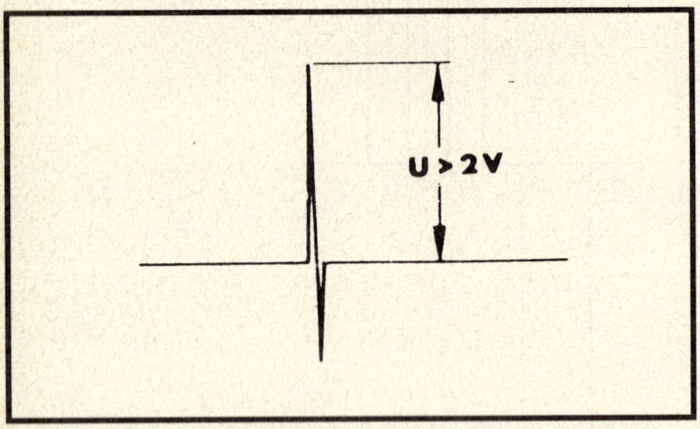

Reference mark sensor test—oscilloscope pattern

Diagnosing Plug Connections

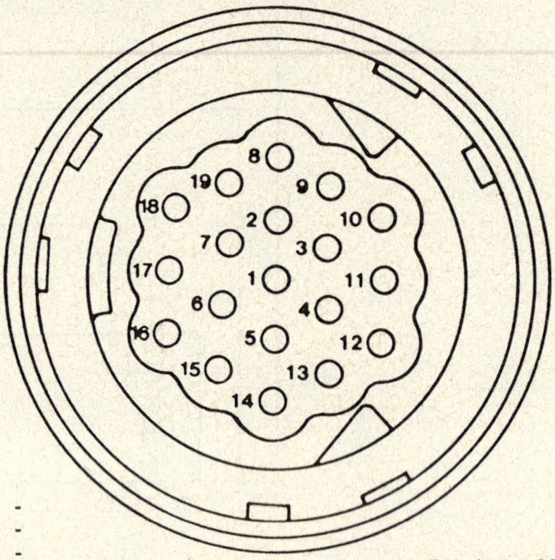

1. -
2. -
3. -
4. -
5. On/Off board
6. L and K wire shield
7. -L wire
8. -K wire
9. -
10. Ground
11. -
12. Bat. (term. 30)
13. Ign. (term. 15)
14. Td
15. LH/EZK, lamp wire
16. Tire press. lamp
17. -Td shielding
18. -
19. Knock

Diagnostic plug—terminal identification—928S4 shown

oscilloscope tester leads to the center and bottom pin of the plug.

6. If the voltage signal is still less than 2.0 volts, check for excessive clearance between the sensor and the reference mark. The specified clearance is 0.8±0.3mm.
7. If the clearance is within specification, replace the sensor.

DME CONTROL UNIT

Testing

1. Connect the CO testing hose or CO tester US4492 or equivalent.
2. Start the engine and run until normal operating temperature is reached.
3. Disconnect the oxygen sensor. Connect the oxygen sensor connector (control unit side) to ground. The CO value should increase.
4. If not, check the wire from the oxygen sensor to the control unit plug terminal.
5. If the wiring is in good condition and contacts were found clean and tight, replacing the DME control unit.
6. If the CO did increase, replace the oxygen sensor.

NOTE: The DME control unit will switch the oxygen sensor off under any of the following conditions:

—Coolant temperature less than 60°F (15°C)
—Coolant temperature less than 113°F (45°C) and engine idling
—Idle switch closed (idle position) and engine above 1000 rpm
—Full throttle switch closed (full throttle)

HIGH ALTITUDE SWITCH

Testing

1. Disconnect the altitude switch plug, located below driver's side dash board.
2. Measure the resistance value across the switch terminals. The meter should read infinity (switch open).

NOTE: The switch is closed in areas above 3280 ft. The meter should read 0 ohm.

3. Connect the CO tester. Start the engine and run until normal operating temperature is reached.
4. Disconnect the oxygen sensor and high altitude switch. Remove the vacuum hose from the fuel pressure regulator and plug the hose.
5. Run the engine at 2000 rpm and note the CO level.
6. Bridge the terminals in the high altitude switch plug. The CO level at 2000 rpm should now be leaner.
7. If not, check the high altitude plug connections. One terminal should go to ground and the other should go to the control unit plug.

THROTTLE SWITCH

Testing

1. Disconnect the negative battery cable. Pull off the control unit plug.
2. Measure the resistance between terminal 2 and ground. The meter should read as follow:
   Throttle switch at idle stop—0 ohm
   Throttle switch open—infinity
   Idle contact must open immediately (maximum 1 degree)

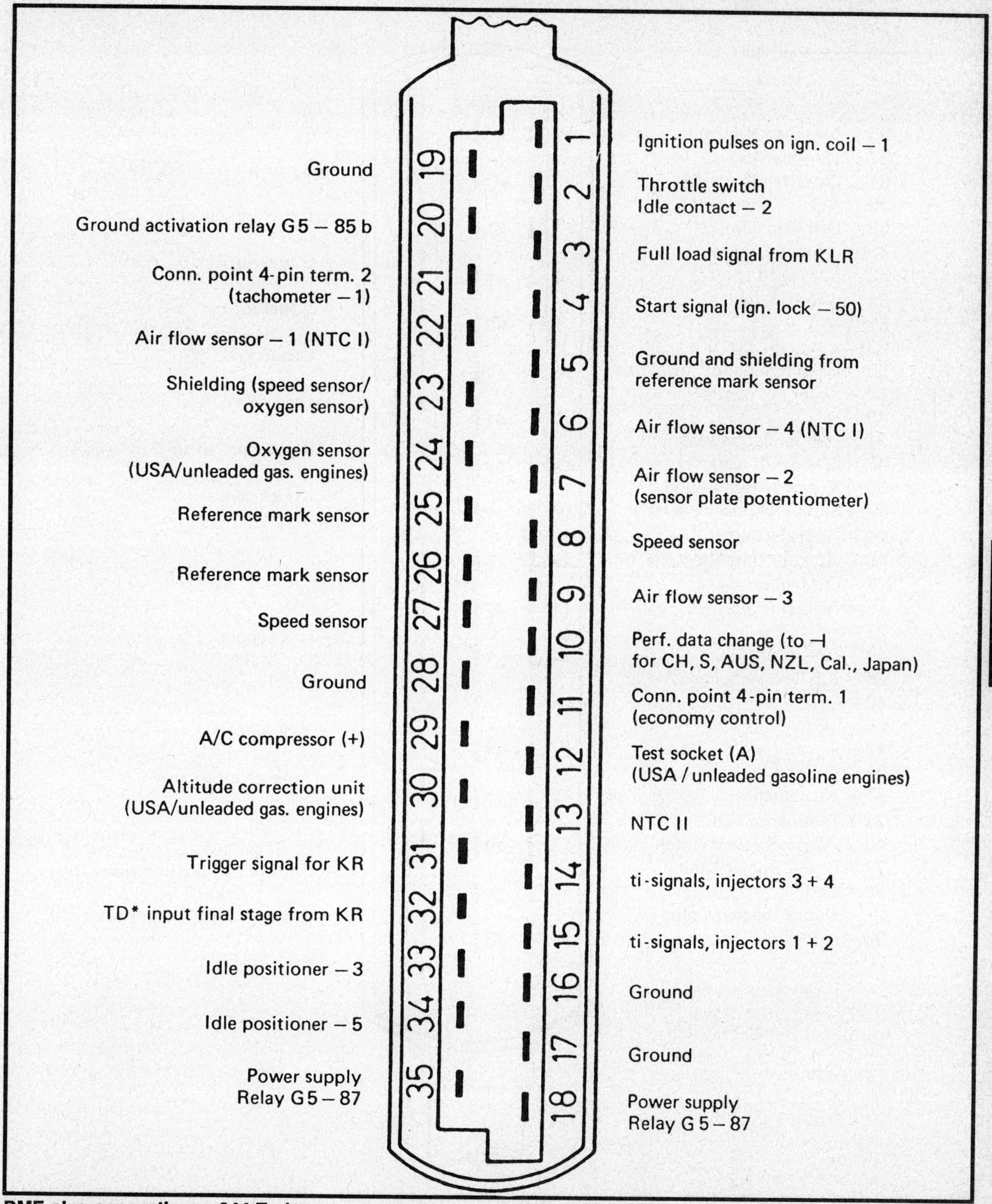

DME plug connections—944 Turbo

55 – Diagnosis K-lead
37 – Terminal 15
54 – Curve family switch
36 – –
53 – Full load contact
35 – t_i fuel inj. cyl. 2
52 – Idle speed contact
34 – t_i fuel inj. cyl. 4
51 – –
33 – t_i fuel inj. cyl. 5
50 – Heating/air cond. regulator
32 – Economy control
49 – –
31 – Hall sender power supply
48 – Speed/ref. mark sender
30 – Knock sensor ground
47 – Speed/ref. mark sender +
29 – Knock sensor II
46 – Altitude potentiometer
28 – Oxygen sensor
45 – NTC II motor
27 – –
44 – Intake air NTC I
26 – Air flow sensor ground
43 – –
25 – –
42 – –
24 – Ground
41 – AC switch
23 – Resonance flap
40 – AC compressor clutch
22 – Check Engine lamp
39 – Coding
21 – Knock diagnosis plug
38 – –
20 – –

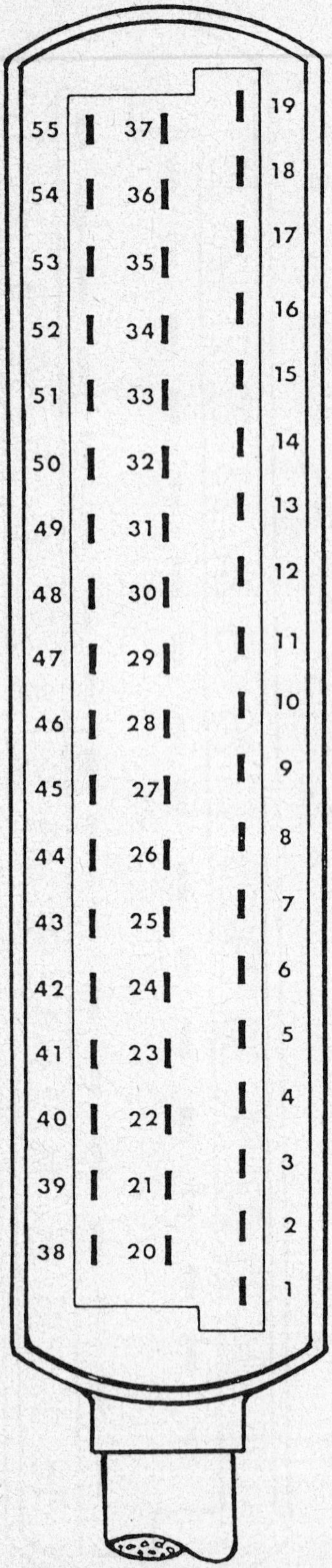

19 – Ground
18 – Battery +
17 – t_i fuel inj. cyl. 1
16 – t_i fuel inj. cyl. 6
15 – t_i fuel inj. cyl. 3
14 – Ground
13 – Diagnosis L-lead
12 – Power sup. to air flow sensor
11 – Knock sensor I
10 – –
9 – –
8 – Hall sender
7 – Air flow sensor
6 – Tachometer
5 – Tank venting valve
4 – Control
3 – DME relay 85 b ground
2 – Ground/shield (final stages)
1 – Final stages

DME plug connections—911 Carrera and Speedster

Pin assignment of DME control unit plug

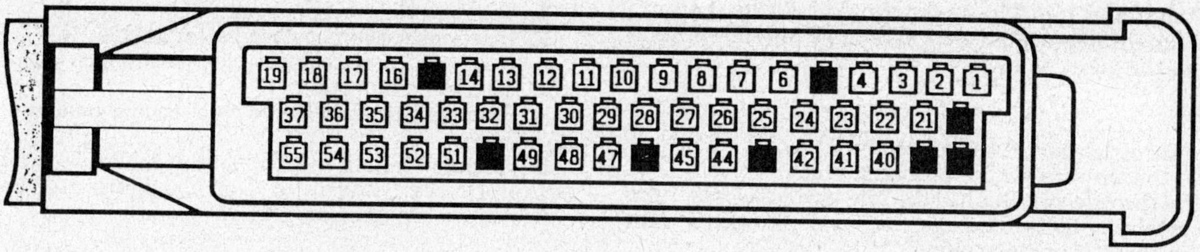

Note

Slots marked black are not used.

1 Ignition coil

2 Ground, ignition output stage (MP IX)

3 Ground (Fuel pump relay)

4 Idle speed positioner, opening winding

5 not used

6 Speed signal at rev counter

7 Air mass sensor signal

8 Hall sender signal

9 Speed signal from speedometer

10 Ground, electronics (MP VIII)

11 Knock sensor 1

12 Positive terminal, Hall sender / throttle

13 Diagnostics L-wire

14 Ground, injection valves (MP IX)

15 not used

16 Injection valve cylinder 3

17 Injection valve cylinder 1

18 Terminal 30

19 Ground, electronics

20 not used

21 Check Engine warning lamp

22 Idle speed positioner, closing winding

23 VarioCam

24 Ground of remaining output stages (MP IX)

25 Hot-wire burnoff

26 Ground, air mass sensor

27 Terminal 15

28 Oxygen sensor/idle CO potentiometer

29 Knock sensor 2

30 Ground NTC/shield knock sensor

31 Knock yes/no

32 Control of AC relay

DME plug connections—968

3. Measure the resistance between terminal 3 and ground. The meter should read as follow:

Throttle switch at idle stop—infinity
Throttle switch open—0 ohm, shortly before full throttle

4. If not, measure the resistance between ground and the throttle switch center pin. The meter should indicate 0 ohm.
5. Repeat the throttle switch tests by connecting the ohmmeter directly to the throttle switch.

Adjustment

1. Set the throttle against the idle stop.
2. Remove the throttle switch retaining bolts.
3. Turn the throttle switch until the idle contact closes, then tighten the retaining bolts. The idle contact must open at less than 1 degree throttle angle opening.

Component Replacement

ELECTRONIC IGNITION CONTROL UNIT

Removal and Installation

1. Disconnect the negative battery cable.
2. Fold back the mat in the passenger's footwell and open the cover for the central electrical board.
3. Disengage the lock on the control unit by pressing toward the housing and pull off the multiple-pin plug. Take off the vacuum connection.
4. Remove the mounting bolts and remove the ECU.

To install:

5. Fit the replacement ECU into the vehicle and secure with the mounting bolts.
6. Install the vacuum connection and multiple-pin plug securely.
7. Secure the mat in the passenger's footwell.
8. Reconnect the negative battery cable.

REFERENCE MARK SENSOR AND SPEED SENSOR

Adjustment

The sensor bracket is mounted on the crankcase flange with 2 bolts. After unscrewing these bolts, the holder can be turned to adjust the sensor clearance.

1. Disconnect the negative battery cable.
2. Remove the speed sensor.
3. Measure the distance from the holder upper surface to the tooth head of the starter ring gear with a depth gauge.
4. Measure the length of the speed sensor.
5. Subtract the distance between the sensor and the reference mark. The difference should be 0.8±0.3mm.
6. If the difference is greater or smaller, turn the holder until the difference between measured distances is correct.
7. Install the sensor. Reconnect the negative battery cable.

911 AND 968 MODELS

The distance of the sender to the flywheel's ring gear is 0.8±0.2mm.

DISTRIBUTOR

Removal and Installation

924 MODELS

1. Set the cylinder No. 1 on TDC.
2. Disconnect the negative battery cable.
3. Remove the distributor cap, disconnect the vacuum hose and electrical connection.
4. Remove the distributor hold—down bolt and remove the distributor.
5. When installing the distributor, make certain the distributor rotor faces cylinder No. 1 on the distributor housing.
6. Adjust the ignition timing.

IGNITION TIMING

Adjustment

911 MODELS

Ignition timing cannot be adjusted. The timing can only be checked at idle speed, with the engine at normal operating temperature. Specifications are as follow:

Ignition timing—3±3 degrees ATDC at 800±40 rpm
Control value—25±3 degrees BTDC at 3800±100 rpm, with bridged test connections B and C (full load contact).

928 MODELS

Whenever engine performance problems or poor power output are experienced, the ignition timing should be checked. The cause for this could be, among others, that the ring gear for the speed/reference mark sensors has moved on the flywheel.

1. Run the engine until operating temperature is reached.
2. Connect an engine tester or timing light and tachometer. The positive terminal for connection of the testing equipment is located above the ignition control unit in the engine compartment.
3. Detach both vacuum hoses at the distributor.
4. Adjust the ignition timing to 31 degrees BTDC, at an engine speed of 3000 rpm. Loosen and turn the distributor to change the ignition timing.
5. Run the engine to a speed of 5000 rpm, at which time the ignition timing must not exceed 36 degree BTDC.
6. Attach both vacuum hoses. Ignition timing must be 0±4 degrees when the engine is at idle speed.

944 TURBO MODELS

Ignition timing cannot be adjusted. The timing can only be checked at idle speed, with the engine at normal operating temperature. Specifications are as follow:

Ignition timing—5±3 degrees BTDC at 840±40 rpm
Control value—35±3 degrees BTDC at 2500 rpm

968 MODELS

Ignition timing is tested with the engine at operating temperature. The factory scan tester is the recommended way to check timing so engine speed and timing can be checked at the same time. Specifications are as follow:

Ignition timing 10±3 degrees BTDC at 840±40 rpm, manual transmission
Ignition timing 10±3 degrees BTDC at 880±40 rpm, automatic transmission

SAAB

General Information

Saab vehicles use the Bosch EZK electronic breakerless ignition system as well as the distributorless Direct Ignition system, some of which are fitted with Automatic Performance Control.

BREAKERLESS IGNITION SYSTEM

The Breakerless Ignition system is fitted on 900 and 9000 vehicles. The distributor uses a Hall Effect sender connected to an output module. The sender consists of a semiconductor element and magnet which acts on the semiconductor. The output module amplifies and transforms the signal from the sender.

Some vehicles (turbocharged engine) have their ignition system fitted with the Automatic Performance Control (APC) system. The APC system enables the engine to achieve optimum performance and good fuel economy, regardless of the grade of fuel being used at any given time. A knock detector, in-conjunction with the pressure transducer and the ignition system information, senses the knocking tendencies in the engine and sends an electrical signal to the microprocessor, inside the ECU. The ECU processes these signals and sends electrical pulses to a solenoid valve which controls the charging pressure in the inlet manifold to matched the grade of fuel being used.

The distributor has a double acting vacuum timing control device which alters the timing to match the load on the engine. In addition to conventional vacuum control, the distributor includes a pressure regulating function which retards the timing under certain conditions. The system is, therefore, able to prevent pinging tendencies when the turbocharger is operating.

EZK IGNITION SYSTEM

The Elektrische Zundkontrolle (EZK) breakerless ignition system is fitted on some vehicles equipped with the 16-valve engine without turbocharger. This ignition system is controlled by a knock sensor to achieve the best possible performance depending on the fuel quality (octane rating) being used. The system is breakerless and is equipped with a crankshaft sensor. The crankshaft signals are applied to the control unit at a rate corresponding to the speed of rotation of the engine. The signals received from the control unit are strengthened by an amplifier and a high tension pulse is generated into the ignition coil secondary winding each time the primary circuit is open.

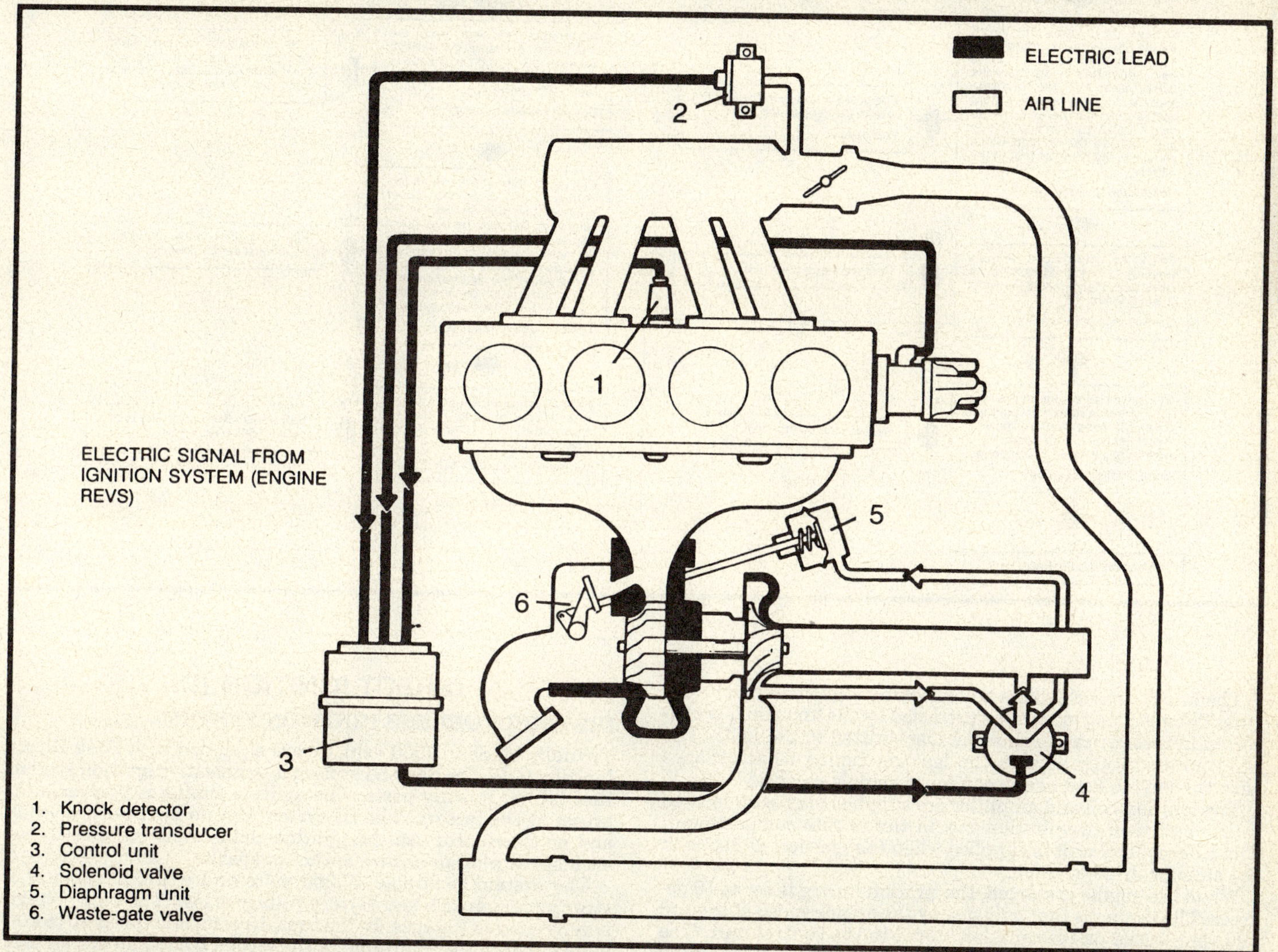

Automatic Performance Control (APC) system—typical

SAAB BREAKERLESS IGNITION SYSTEM TESTING

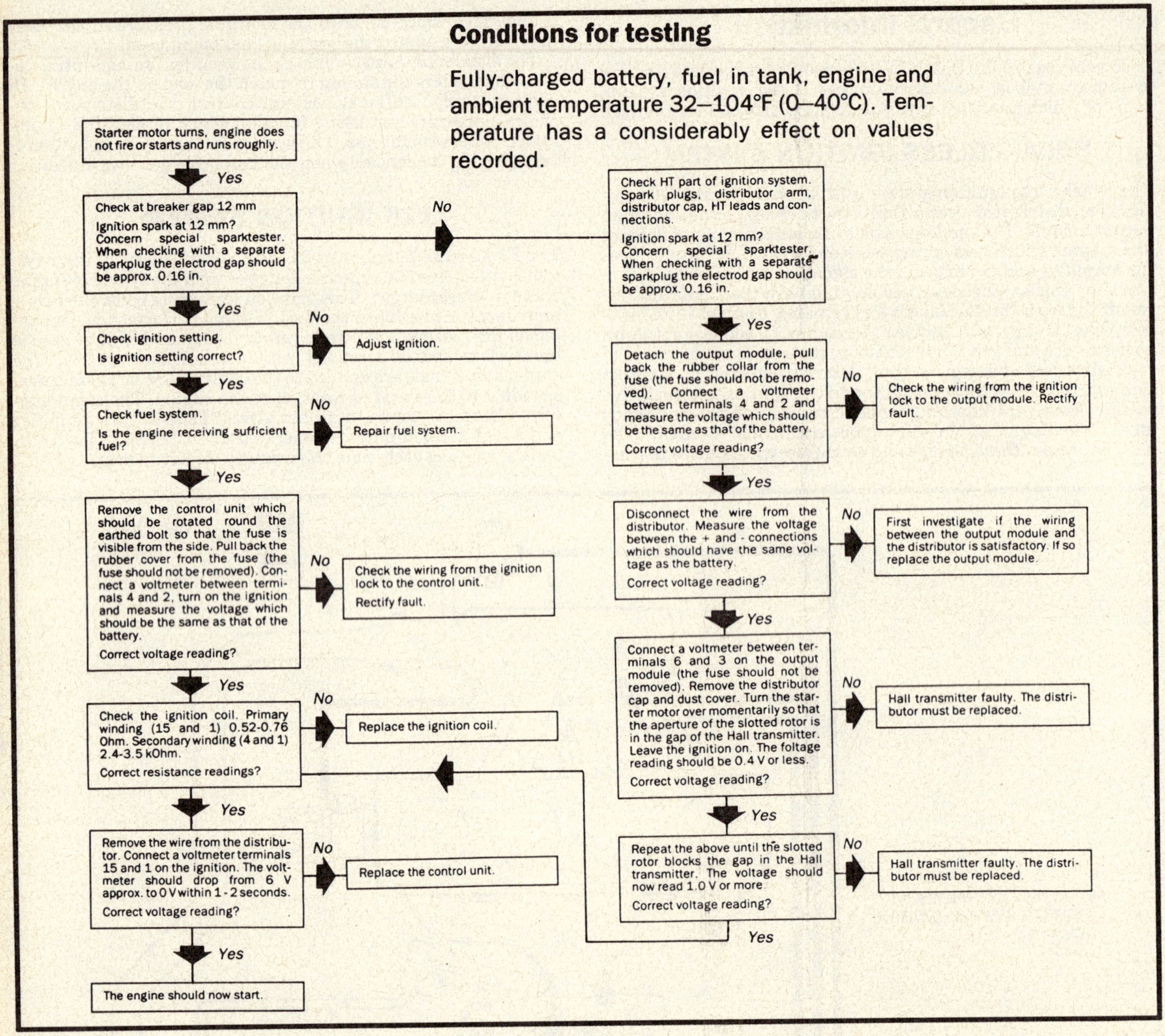

The knock sensor detects any knocking tendencies in the engine. The resulting vibrations produced by the knocking, is converted into electrical signals and transmitted to the ECU. The ECU automatically adjusts the ignition timing in accordance with the engine load and speed input signals received.

The ignition coil and amplifier are supplied via fuse 9 located in the electrical distribution box in the engine compartment. The EZK control unit is supplied when the ignition switch is in the start or drive position.

When the engine is started, the ignition timing is set at 10 degrees. The timing is first adjusted by the engine speed at around 700 rpm, based on information stored in the control unit. The distributor has no centrifugal and vacuum controls. At idling speed of 800 rpm, the ignition timing is set around 14 degrees.

DIRECT IGNITION (DI)

DISTRIBUTORLESS IGNITION SYSTEM

Virtually all Saab 9000 vehicles are equipped with Saab Direct Ignition (DI). The Saab DI system is a capacitive ignition system that has no moving parts. The ignition timing is permanently preset at the factory. The DI system has no high tension leads and no distributor cap. An ignition coil is mounted directly on each spark plug to ensure optimum spark.

The system generates 40,000 volts and will still produce a strong spark even when battery voltage is low. The high voltage is generated in 2 stages. In the first, 12 volts is stepped up to 400 volts by capacitive charging. In the second stage, the 400 volts is stepped up to 40,000 in the 4 ignition coils. The Saab DI coils are

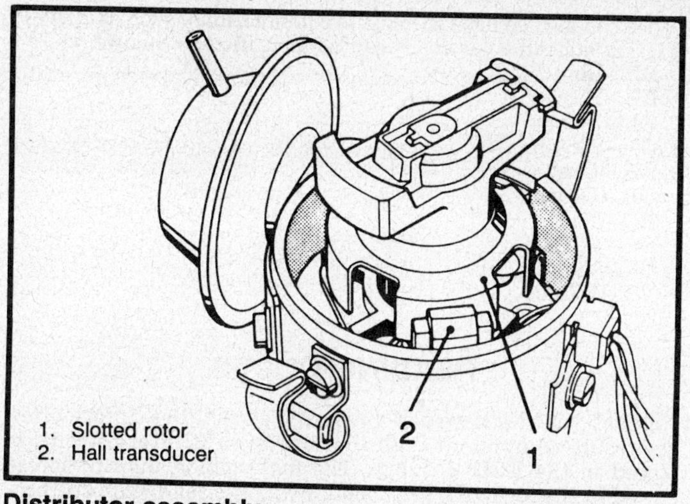

Distributor assembly

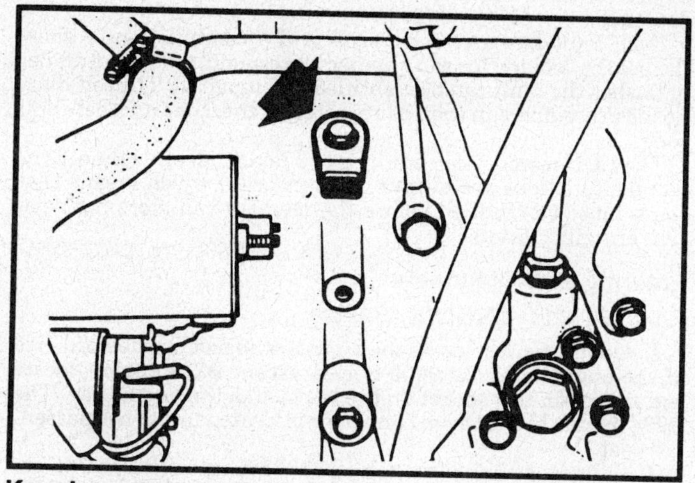

Knock sensor—typical, Saab

very small and are mounted in a cartridge immediately above the spark plugs.

The Saab DI system is designed for very rapid spark generation. Charging is 20 times faster than in a conventional system. This allows whole bursts of sparks to be produced instead of a single spark, helpful in cold weather. This multi-spark burst burns off any carbon deposits or fuel residue on the spark plug electrodes. If, for some reason, the engine fails to start on the first attempt, when the ignition key is released from the starting position, the system produces a massive stream of approximately 1000 sparks on all cylinders at once, to ensure that even the heaviest deposits have been burned off.

The Saab DI has no moving parts and the timing is set at the factory. No adjustments are possible.

SYSTEM OPERATION

Vehicles equipped with Breakerless Ignition and the EZK ignition system have similar functions. When the ignition switch is turn to the **RUN** or **START** positions, voltage will be supplied to the ignition coil and the amplifier. As the engine is cranked and the slot in the rotor is in line with the Hall transducer, the transducer generates a voltage signal at a rate corresponding to the speed of rotation of the engine. The voltage signal is generated by a crankshaft sensor. This signal from the Hall transducer inside the distributor is sent to the amplifier. The signal which opens the primary circuit of the ignition coil is amplified and a high-tension pulse will then be generated to the appropriate spark plug. The output module also monitors and regulates the dwell angle.

An ignition pulse amplifier is included to ensure more reliable control of the ignition pulse-dependent systems. The amplifier also contributes to reducing the level of interference with the radio. The amplifier output leaves a square pulse signal of 12 volts to the timing service instrument socket and to the tachometer.

On vehicles equipped with a knock sensor, the knocking related ignition timing is controlled individually on each cylinders. The ignition timing may thus vary from 1 cylinder to the next. When knocking is detected in a cylinder, the ignition for that particular cylinder will be retarded in steps of approximately 3 degrees until knocking has ceased or until the ignition has been retarded by a total of 13 degrees.

Vehicles equipped with the Direct Ignition (DI) system use a capacitor to fire the ignition spark. When the ignition switch is in the **RUN** position, the capacitor in the ignition cartridge will be charged up to approximately 400 volts. Simultaneously, a voltage of approximately 80 volts is supplied to the spark plug electrodes to determine the firing sequence.

During the starting process, the system first determines which pair of pistons (1–4 or 2–3) is at or approaching TDC. This is accomplished by the crankshaft sensor, which senses the relationship between the field rotor slot for the corresponding cylinder pair and the identification pulse slot.

Under normal conditions, when the cranking speed exceeds 150 rpm, a spark is generated in the cylinder pair whose pistons are in the firing position. The supply voltage to the cylinder, in which combustion is taking place, produces a weak current between the spark plug electrode, which determines the firing sequence.

If the engine fails to start or if it starts and then stops, a burst of sparks (multi-spark) will be fired at all spark plugs to burn off any deposits. This function will be activated only if the ignition key is released and left in the **RUN** position for approximately 5 seconds.

APC System

Turbocharged models are equipped with the Saab DI/APC system. Automatic Performance Control (APC) enables the engine to adjust iteself automatically to the grade of fuel supplied to it at any particular time. The APC function is integrated into the ECU.

SYSTEM COMPONENTS

Amplifier

The amplifier, located on a cooling plate on the left-hand wheel housing, amplifies and converts the pulse from the distributor. The pulse is then used to control the charging and discharging of the ignition coil.

Output Module

The output module is mounted on a cooling fin on the left hand wheel arch. The output module amplifies and transforms the pulse from the distributor.

Ignition Coil

The ignition coil, mounted on the radiator fan housing, is a specially designed coil for the 900 and 9000 vehicles. The ignition coil should not be interchanged with a conventional ignition system coil.

Ignition Cartridge

Saab 9000 vehicles with the Direct Ignition (DI) distributorless ignition uses an ignition cartridge which consists of a cast aluminum housing and a cover in which 4 ignition coils and spark plug connectors are installed. The cartridge also contains a transformer which steps up the voltage to 400 volts and a capacitor which stores electrical energy. Never attempt to start the engine with the ignition cartridge removed and with the spark plugs not grounded or the ignition cartridge will be destroyed. The voltage supply to the DI control unit and the ignition cartridge is taken from the main relay of the LH system. This relay remains energized for 5 seconds after the engine has been switched off by the ignition switch.

The ignition cartridge will be activated when the ignition switch is in the **ON** position. When troubleshooting, check fuse 9 and check that the supply to it is live. Also check the connectors, wiring and ground connections. For addition troubleshooting, Saab recommends their diagnostic equipment be connected to the test socket.

Electronic Control Unit

The LH-system ECU, located adjacent to the bulkhead panel on the left side in the engine compartment, incorporates a microprocessor into which all necessary reference data has been preprogrammed. This computer processes the signals from the various sensors and then computes and controls the opening duration of the injectors. The ECU for the 9000 EZK and DI–APC system is located under the left front seat.

Distributor

The distributor incorporates a slotted rotor, which rotates with the distributor shaft. A hall transducer is fitted to a plate inside the distributor. The slotted rotor alternately interrupts and releases the magnetic field, thereby determining ignition timing.

Knock Sensor

The knock sensor, located on the engine block below the intake manifold, records any knocking which occurs in the engine. The resulting vibrations produce by the knocking, is converted into electrical signals and transmitted to the ECU.

Diagnostic Test Socket

The diagnostic test socket for the ISAT is located under the right side front seat.

Diagnosis and Testing

SERVICE PRECAUTIONS

- Use care when working around high energy ignition systems. These electronic ignition systems have an ignition voltage of over 30,000 volts which may be lethal to sensitive persons, such as those who have a pacemaker.
- Never disconnect the high tension lead from the ignition coil when the engine is running.
- Never attempt to start the engine with the high tension lead from the ignition coil disconnected or with the distributor cap removed.
- Before performing a compression test, always disconnect the amplifier.
- If checking the ignition spark on those engines with conventional distributors, proceed as follows:
  - –Remove the spark plug.
  - –Connect the high tension lead from the ignition coil to the plug or an approved spark plug tester.
  - –Hold the plug against the engine so that it is well grounding and check for a good spark.
- Never disconnect the ECU connector when the control unit alive. After the ignition has been switched off, it takes about 30 seconds for the main relay to trip. So always wait at least 40 seconds before disconnecting the ECU connector.

Color codes are used in the electrical diagrams. The color codes can also be used in various combinations, such as RD/VT, VL/GR, etc. All color codes can be identified as follows:

BL–Blue
BR–Brown
GL–Yellow
GN–Green
GR–Grey
OR–Orange
RD–Red
SV–Black
VL–Violet
VT–White

READING CODES

All Saab vehicles have self-diagnostic capabilities. Information regarding component malfunctions even if intermittent, is stored in the ECU memory. The malfunction information is stored in the form of a fault code. When the self-diagnostic procedure is initiated, the CHECK ENGINE light on the vehicle instrument panel will flash, providing information on any faults detected.

Each fault has a code consisting of a combination of short flashes. By looking for the number corresponding to the number of flashes the fault can be identified and repaired. The self-diagnostic procedure can then be repeated to check for another fault code.

The ECU memory can store up to 3 fault codes at a time. Serious malfunctions are always given priority, which means that these must be rectified before the memory can store information on minor faults.

Entering Self-Diagnosis

EXCEPT 2.3L ENGINE (B234)

1. On 900 Series, locate the 3-pin test socket on the right side of the engine bay. On 9000 Series, except B234 engine, locate the single-pin test socket on the left side in the engine bay. The B234 engine black 10 pin connector is located under the passenger seat.
2. Connect switch jumper lead 8393886 or equivalent, to pin No. 3 and ground. The ignition switch should be in the **OFF** position.
3. Turn the ignition switch **ON**. The CHECK ENGINE light should come ON.
4. Set the switch to the **ON** position, grounding ECU pin No. 16. The CHECK ENGINE light should now go OFF.
5. Observe the CEHCK ENGINE light carefully. After approximately 2½ seconds, it will flash briefly, signifying that the first error code will now be displayed.
6. As soon as the light has flashed, move the switch immediately to the **OFF** position.
7. The first of 3 possible error codes will now be displayed by a series of short flashes of the CHECK ENGINE light.

NOTE: The error code starts and finishes with a long flash of the CHECK ENGINE light. These long flashes are not part of the code itself but serve merely to indicate the beginning and end of the code.

8. When the switch if set to **OFF**, an error code will be flashed repeatedly. The next error code cannot be displayed until the switch has been operated.
9. To check if any additional fault codes have been stored in the memory, proceed as follows:
   a. Set the switch to the **ON** position.
   b. After a short flash of the check engine light, set the switch to the **OFF** position. The next fault code, if any, will now be displayed in the same pattern as the first.
   c. Repeat Steps 1 and 2 to check for additional fault codes. If no faults have been detected or the faults have been recti-

fied, the system will indicate this by a continuous series of long flashes.

2.3L ENGINE (B234)

The B234 engine has enhanced diagnostics. The engine control system is diagnosed using a Saab ISAT diagnostic tester or equivalent. Using the ISAT, a number of input command codes can be generated for checking the function and status of different components. Other LH-Jectronic testers can still be used for fault diagnosis. The ISAT tester is quicker and more accurate.

The black No. 10 pin diagnostic socket is located in the front, underneath the right seat. It is protected by a plastic cover held in place by a quick release screw.

When the ISAT tester is used to simulate signals of functions by communicating with the ECU, the results appear on the display in the form of a 5 character alphanumeric code. For example: to check the position of the throttle butterfly, enter Code 201 via the ISAT keypad. With the engine idling, the code appearing on the display should be 8B101.

CLEARING TROUBLE CODES

1. Set the jumper lead switch to the **ON** position.
2. Wait approximately 2½ seconds.
3. Notice the CHECK ENGINE light. After 3 short flashes, set the switch to the **OFF** position.

NOTE: The contents of the memory cannot be deleted before Code 00000 has been displayed, indicating the end of the error codes.

NO START—NO SPARK

Saab 900

EZK IGNITION SYSTEM

TURBOCHARGED ENGINES

The breakerless ignition system is equipped with a Hall sensor. When the ignition switch is turned to **START** or **ON** position, the ignition coil and amplifier should be energized. The amplifier receives ignition pulses with a 3-core shielded cable from the Hall sensor in the ignition distributor. The pulses are strengthened and adapted in the amplifier. A high-tension pulse is generated in the secondary winding each time a pulse breaks the primary circuit of the ignition coil. The high-tension pulse is then supplied to the corresponding spark plug via the distributor.

The tachometer is supplied from fuse 7. The control pulses required for displaying the engine speed are supplied from the ignition amplifier.

When troubleshooting a No-Start complaint, make sure the ignition is energized with the ignition switch in the **ON** position:

1. Confirm the complaint. Use a suitable and approved spark tester attached to a spark plug wire to check for spark.
2. Check fuse 3.
3. Check that terminal 15 of the ignition coil and terminal 4 on the timing service socket are live.
4. If the tachometer is dead, check fuse 7 and check that there is voltage getting to the fuse. Also check terminal 3 on the circuit board for the tachometer. It too should have power and that ignition pulses are being supplied from the amplifier.
5. Check all connectors, cable harness and ground connections.
6. Saab recommends that further troubleshooting be carried out with their Timing Service Instrument (TSI) at the special diagnostic socket. This socket uses the following terminals:
   a. Terminal 1 is the power supply direct from the battery for the tester tool.
   b. Terminal 2 is ground.
   c. Terminal 3 goes to the solenoid (terminal 50) on the starter motor.
   d. Terminal 4 is the positive power supply from the ignition switch when in **START** or **ON**.
   e. Terminal 5 indicates ignition pulses from the ignition system amplifier.
   f. Terminal 6—Not used.

EZK IGNITION SYSTEM

16 VALVE ENGINE WITHOUT TURBOCHARGER

This version of the EZK ignition system is fitted to 16-valve engines without turbocharger. This breakerless ignition system is equipped with a cranksahft sensor. It is controlled by a knock sensor.

With ignition switch in **START** or **ON**, the ignition coil with integrated amplifier and EZK control unit will be energized. The control unit receives ignition pulses via a 3-core shielded cable from the crankshaft sensor. The frequency of the pulses is proportional to the engine speed. The ECU adjusts timing based on engine speed, load and any knocking.

The ignition pulses supplied by terminal 16 of the control unit are strengthened and adjusted in the ignition coil with integrated amplifier. A high-tension pulse is generated in the secondary winding each time a pulse breaks the primary circuit of the ignition coil. The high-tension pulse is then supplied to the corresponding spark plug via the distributor.

The ignition is set at 10 degrees when the engine started and adjusted as required by the ECU. The distributor has no advance system. At idle, timing should be about 14 degrees.

The tachometer is supplied from fuse 7. The control pulses required for displaying the engine speed are supplied from the ECU.

When troubleshooting a No-Start complaint, make sure the ignition is energized with the ignition switch in the **ON** position:

1. Confirm the complaint. Use a suitable and approved spark tester attached to a spark plug wire to check for spark.
2. Check fuse 3.
3. Check that terminal 15 of the ignition coil and terminal 4 on the timing service socket are live.
4. If the tachometer is dead, check fuse 7 and check that there is voltage getting to the fuse. Also check terminal 3 on the circuit board for the tachometer. It too should have power and that ignition pulses are being supplied from the ECU.
5. Check all connectors, cable harness and ground connections.
6. Activate and read any fault codes before disconnecting the connector from the ECU since the contents of the memory will otherwise be erased when the connector is removed.
7. Saab recommends that further troubleshooting be carried out with their Timing Service Instrument (TSI) at the special diagnostic socket. This socket uses the following terminals:
   a. Terminal 1 is the power supply direct from the battery for the tester tool.
   b. Terminal 2 is ground.
   c. Terminal 3 goes to the solenoid (terminal 50) on the starter motor.
   d. Terminal 4 is the positive power supply from the ignition switch when in **START** or **ON**.
   e. Terminal 5 indicates ignition pulses from the EZK control unit.
   f. Terminal 6—Not used.

As a general rule, the connector to the ECU should always be disconnected when measurements are made on the ignition system cable harness. Use a digital ohmmeter, not a circuit buzzer.

Saab 9000

DIRECT IGNITION (DI)

Distributorless Ignition

Troubleshooting a No-Start complaint on DI equipped vehicles is somewhat different since the traditional method of checking for spark at a plug wire is not possible with the distributorless ignition cartridge used on this model. Attempting to crank the engine with the ignition cartridge removed and with the spark plugs not grounded can serously damage the cartridge which contains the ignition coils.

This vehicle uses a crankshaft sensor located behind the belt pullet on the oil pump casing. A field rotor is installed on the pulley to act on the Hall sensor. The ECU controls the ignition and the crankshaft sensor feeds the ECU. In a no-start situation, check the connections of the crankshaft sensor. In addition:

1. Check fuse 9 and check that the supply to it is alive.
2. Check all connectors, wiring and the system ground point.
3. More extensive troubleshooting should be done with the Saab special diagnostic equipment or equivalent at the test socket. The diagnostic test socket for the ISAT system is located under the right front seat.

KNOCK SENSOR

Testing

1. Disconnect the negative battery cable.
2. Use an ohmmeter to check the shielded cable between the knock sensor and the ECU. Connect the test lead of the ohmmeter to terminals 12 and 13 of the control unit connector.
3. If the circuit is in good condition, install a new cable.
4. If the circuit is not in good condition, connect a jumper across the terminals of the knock sensor. Pull back the rubber cover and connect the jumper on the cable side.
5. Measure between terminals 12 and 13 of the control unit connector.
6. If the circuit is open, install a new cable.
7. If the circuit is not open, install a new knock sensor.

DISTRIBUTOR AND ROTOR ASSEMBLY

Testing

The Hall transducer and slotted rotor cannot be removed. If a fault is developed in the transducer unit, the entire distributor must be replaced.

Measure the resistance of the rotor. The rotor should have a resistance of 1 kilo-ohm. To check the centrifugal control mechanism while in the vehicle, proceed as follows:

1. Connect a tach/dwell meter and a timing lamp to the vehicle. If the ignition service instrument is used, connect it to the socket on the vehicle.
2. Disconnect the vacuum hose.
3. Start the engine and let it run at idle speed.
4. Read the ignition timing.
5. Increase the engine speed and check that the timing changes.

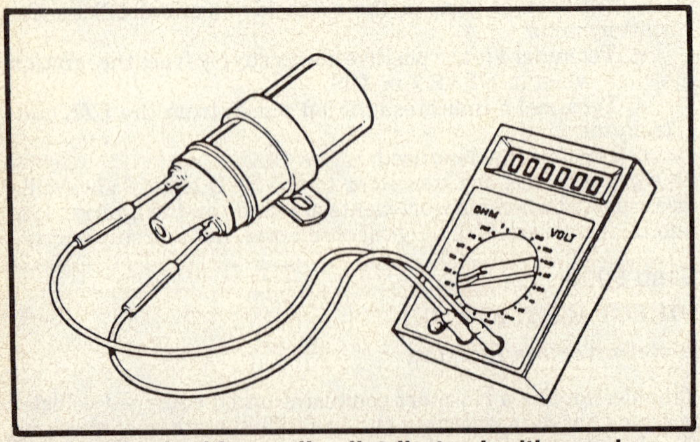

Testing the ignition coil—distributor ignition only

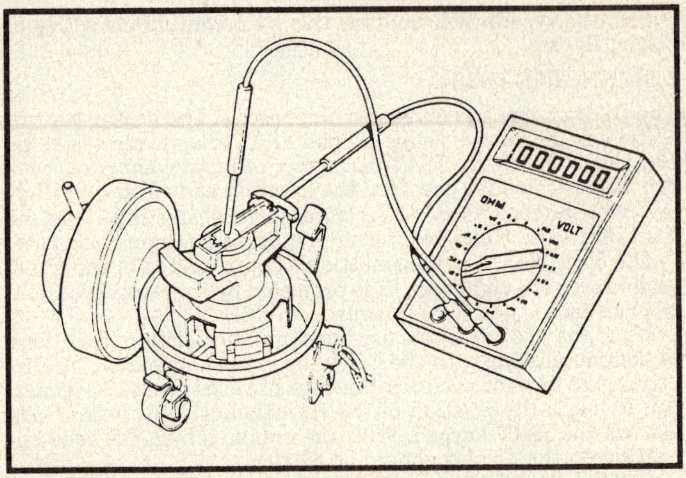

Measuring the rotor resistance—distributor ignition only

IGNITION COIL

Testing

1. Measure the resistance in the primary winding, across terminal 1 (−) and terminal 15 (+). The resistance value should be 0.52–0.76 ohms.
2. Measure the resistance in the secondary winding, across terminal 1 (−) and the high voltage terminal. The resistance value should be 2.4–3.5 ohms.
3. If the resistance value of either measurement is not within specifications, replace the coil.

NOTE: Most faults often occur only when the ignition coil is warm. When in doubt, mount the coil on a test bench and leave it switched on under load for approximately ½ hour. The coil should still performed adequately.

AMPLIFIER

Testing

1. Disconnect the amplifier connector and pull back the rubber cover.
2. Connect a voltmeter to terminal 5 on the rear of the connector.
3. Turn the engine so that the hall sensor is in-line with one of the openings in the slotted rotor.
4. Turn the ignition **ON** and measure the voltage at the instant the ignition is switch **ON**. The voltmeter should read approximately 4–5 volts.
5. Turn the ignition **OFF** and turn the crankshaft so that the hall sensor is in-line with one of the solid segments of the rotor.
6. Turn the ignition **ON** and measure the voltage at the instant the ignition is switch **ON**. The voltmeter should read approximately 0.1 volt.

Component Replacement

IGNITION COIL

Removal and Installation

1. Disconnect the negative battery cable.
2. Roll back the rubber dust cover on the ignition coil.
3. Disconnect the leads from the coil. Remove the coil retaining screw and remove the coil.

4. Fit the replacement coil into position and install the retaining screw.
5. Refit the rubber dust cover to the ignition coil and reconnect the negative battery cable.

OUTPUT MODULE

Removal and Installation

1. Disconnect the negative battery cable.
2. Unplug the connector from the output module.
3. Remove the module retaining screws and remove the output module.
4. Fit the replacement module into position and install the retaining screws.
5. Reconnect the module connector and battery negative cable.

DISTRIBUTOR

Removal and Installation

1. Disconnect the negative battery cable.
2. Remove the distributor cap and disconnect the hall transducer and vacuum hose.
3. Release the clip on the distributor and pull off the distributor.

NOTE: The hall transducer and slotted rotor cannot be removed. If a fault is developed in this unit, the entire distributor must be replaced.

To install:

4. Rotate the distributor shaft until the drive dog engages the slot in the camshaft (the slot is offset). Fit the distributor into position.
5. Install the vacuum hose and reconnect the hall transducer connector.
6. Install the distributor cap and reconnect the negative battery cable.
7. Recheck the ignition timing.

ROTOR

Removal and Installation

1. Disconnect the negative battery cable.
2. Remove the distributor cap.
3. The rotor is glued to the distributor shaft. It must be destroyed to be removed. Break the rotor and crush it with a water pump pliers. Make sure no fragments of the crushed rotor get into the distributor.
4. Coat the distributor shaft with Loctite or equivalent and install the new rotor.
5. Install the distributor cap and reconnect the negative battery cable.

CRANKSHAFT SENSOR

Removal and Installation

900

1. Disconnect the negative battery cable.
2. Remove the engine drive belts and crankshaft pulley.
3. Disconnect the crankshaft sensor electrical connector and remove the crankshaft sensor retaining bolts. Remove the sensor from the vehicle.

To install:

4. Coat the crankshaft sensor retaining bolts with Loctite 270 or equivalent and fit the bolts into the sensor first.
5. Fit the sensor assembly into position and tighten the retaining bolts
6. Install the crankshaft pulley and torque the retaining bolts 104 ft. lbs. (190 Nm).
7. Install the engine drive belts and reconnect the negative battery cable.

9000

1. Disconnect the negative battery cable.
2. Remove the engine drive belts and crankshaft pulley.
3. Cut the strap and separate the connectors. On vehicles equipped with automatic belt tensioner, mark the cables and remove them from the connector.
4. Remove the cable clamp located under the intake manifold.
5. Remove the crankshaft sensor retaining bolts and withdraw the sensor. If the vehicle is equipped with an automatic belt tensioner, remove the upper belt tensioner retaining bolt and move the tensioner to one side.

To install:

6. Coat the crankshaft sensor retaining bolts with Loctite 270 or equivalent and fit the bolts into the sensor first.
7. Fit the sensor assembly into position and tighten the retaining bolts
8. Fit the belt tensioner into position and install the retaining bolt, as required.
9. Install the cable clamp located under the intake manifold. Install the cables in their proper positions and insert them into their connectors. Secure the cables with electrical straps.
10. Install the crankshaft pulley and torque the retaining bolts 104 ft. lbs. (190 Nm).
11. Install the engine drive belts and reconnect the negative battery cable.

FIELD ROTOR

Removal and Installation

1. Disconnect the negative battery cable.
2. Remove the engine drive belts and crankshaft pulley.
3. Remove the 3 field rotor retaining bolts and remove the rotor from the crankshaft pulley.
4. Install the new field rotor to the crankshaft pulley.
5. Install the crankshaft pulley and torque the retaining bolts 104 ft. lbs. (190 Nm).
6. Install the engine drive belts and reconnect the negative battery cable.

ELECTRONIC CONTROL UNIT (ECU)

Removal and Installation

1. Disconnect the negative battery cable.
2. Move the seat to the front position.
3. Remove the control unit cover and disconnect the electrical connector.
4. Remove the scuff plate and protector plate retaining screws, if required.
5. Remove the control unit.
6. Fit the control unit into position.
7. Install the scuff plate and protector plate retaining screws, if required.
8. Connect the electrical connector to the unit.
9. Install the control unit cover. Reconnect the negative battery cable.

IGNITION TIMING

The engine equipped with the EZK ignition has a provision for use of an ignition Timing Service Instrument (TSI). This special socket is intended for use with Saab service tools for checking the timing. Terminals are:

1. Positive supply directly from the battery.
2. Ground
3. Solenoid (terminal 50) of the starter motor.
4. Positive supply across fuse 13 when the ignition switch is in the start of drive position.
5. Ignition pulse from the control unit.
6. Not used.

The factory TSI instrument includes a tachometer, a dwell-angle meter, a stroboscopic lamp, a switch for the starter motor and an ignition timing meter. A special pin is provided on the flywheel and a test socket in incorporated in the flywheel cover. Marks are also provided on the flywheel for checking the timing with a conventional timing lamp.

Saab 9000 Series vehicles are equipped with a distributorless ignition system that is not adjustable. Vehicles without a conventional distributor use Direct Ignition (DI). High voltage is generated in 2 steps. In the first stage, 12 volts is stepped up to 400 volts by capacitive charging. In the second stage, the 400 volts is stepped up to 40,000 volts in the 4 ignition coils, 1 on each spark plug. The DI coils are small enough to be mounted in a cartridge immediately above the spark plugs. The cartridge provides protection against moisture and dirt. The DI system has no moving parts. The timing is factory set and not adjustable.

On vehicles with distributor ignition, a factory Timing Service Instrument (TSI) is intended for use in a special ignition service socket. The ignition is set at 5 degrees when the engine is started. The ignition setting is dependent on engine speed and is adjusted by the ECU. Timing marks are located on the flywheel.

Timing Adjustment

DISTRIBUTOR IGNITION

Do not check or adjust the ignition timing on vehicles equipped with the EZK knock sensor controlled ignition system. Adjust the basic setting only after any repairs are made.

With the flywheel in the 0 degrees position, the mark on the distributor should be in-line with the joint face between the cylinder head and the valve cover.

Ignition timing is set in the conventional manner, using the marks that are located on the flywheel. However, the engine is also equipped for checking the timing using the factory Timing Service Instrument (TSI).

The TSI is connected to the clutch cover by means of a special connector and to the plug lead of No. 1 cylinder by means of a terminal. The TSI is also connected to the ignition service socket at the fuse box and by means of an impulse transmitter at the plug lead for No. 1 cylinder.

The Saab TSI consists of a tachometer, cam angle meter, timing light and switch for operating the starter.

Note: The ignition system components are interrelated with fuel system components. For additional testing, replacement or wiring diagrams see appropriate fuel system section.

SAAB BREAKERLESS IGNITION SYSTEM WIRING DIAGRAM—TYPICAL

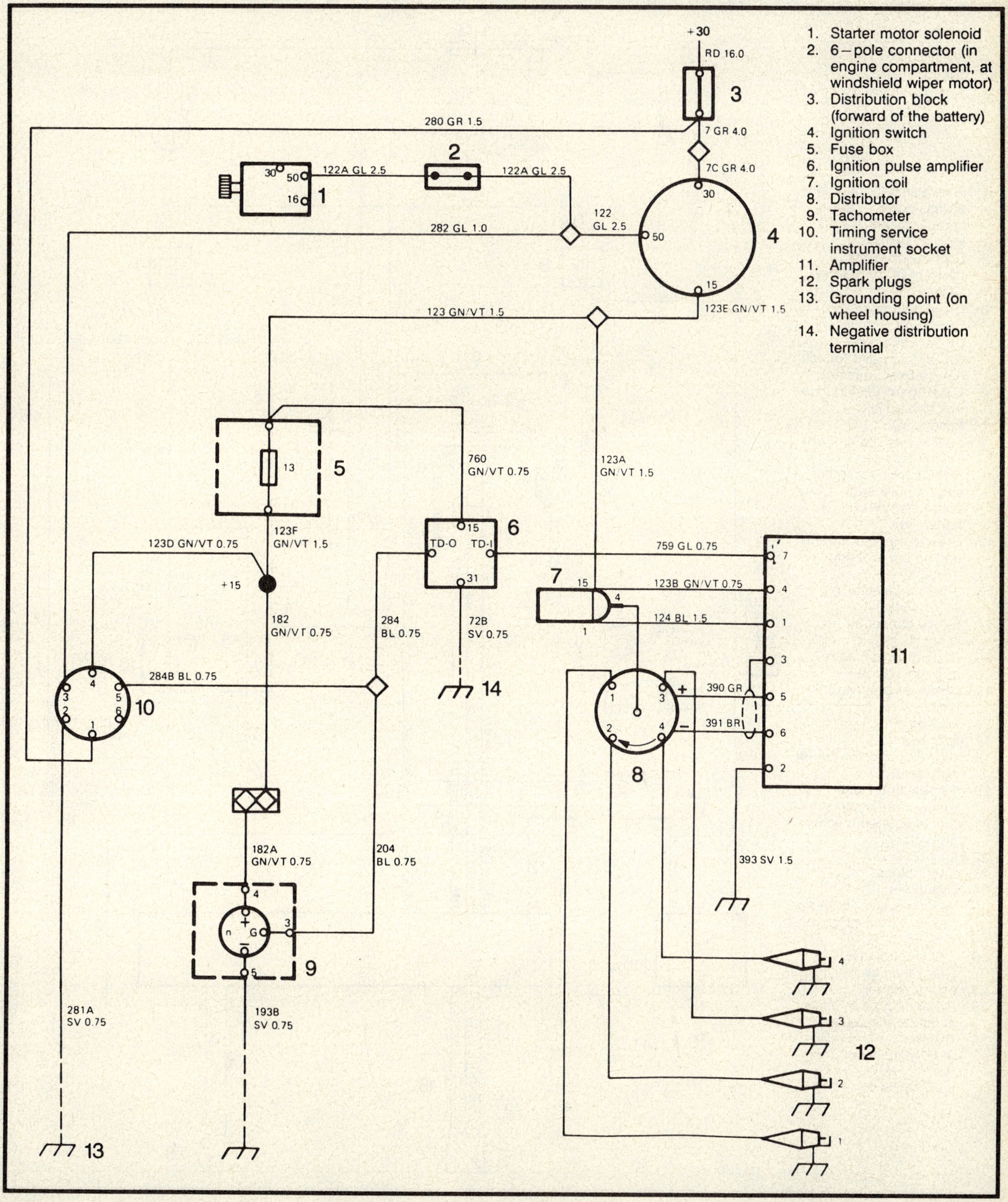

1. Starter motor solenoid
2. 6—pole connector (in engine compartment, at windshield wiper motor)
3. Distribution block (forward of the battery)
4. Ignition switch
5. Fuse box
6. Ignition pulse amplifier
7. Ignition coil
8. Distributor
9. Tachometer
10. Timing service instrument socket
11. Amplifier
12. Spark plugs
13. Grounding point (on wheel housing)
14. Negative distribution terminal

SAAB 9000S VIN D—EZK IGNITION SYSTEM WIRING DIAGRAM

1. Fuse panel (behind the access panel in glove compartment)
2. Ignition switch
3. Distribution block (forward of the battery)
4. Battery
5. Grounding point (on gearbox)
6. Starter solenoid
7. Grounding point (on left hand wheel housing)
8. Fuse board (distribution box behind left headlamp)
9. Fuel system control unit
10. Timing service instrument socket (left side of bulkhead)
11. Throttle angle transmitter
12. Grounding point (rear of engine below intake manifold)
13. 3-pole connector (at wiper motor)
14. 6-pole connector (at wiper motor)
15. Coolant temperature and fuel gauge
16. Check engine warning lamp
17. Tachometer
18. EZK test tap
19. 4-pole connector (under right seat)
20. 10-pole connector (under right seat)
21. Single pole connector (at battery)
22. Ignition coil
23. Spark plugs
24. Distributor
25. Ignition system amplifier (left wheel housing, at battery)
26. EZK control unit
27. Crankshaft sensor
28. Knock sensor
29. Grounding point (facia, at left front speaker)
30. Distribution terminal +15 (distribution box, behind glove compartment)

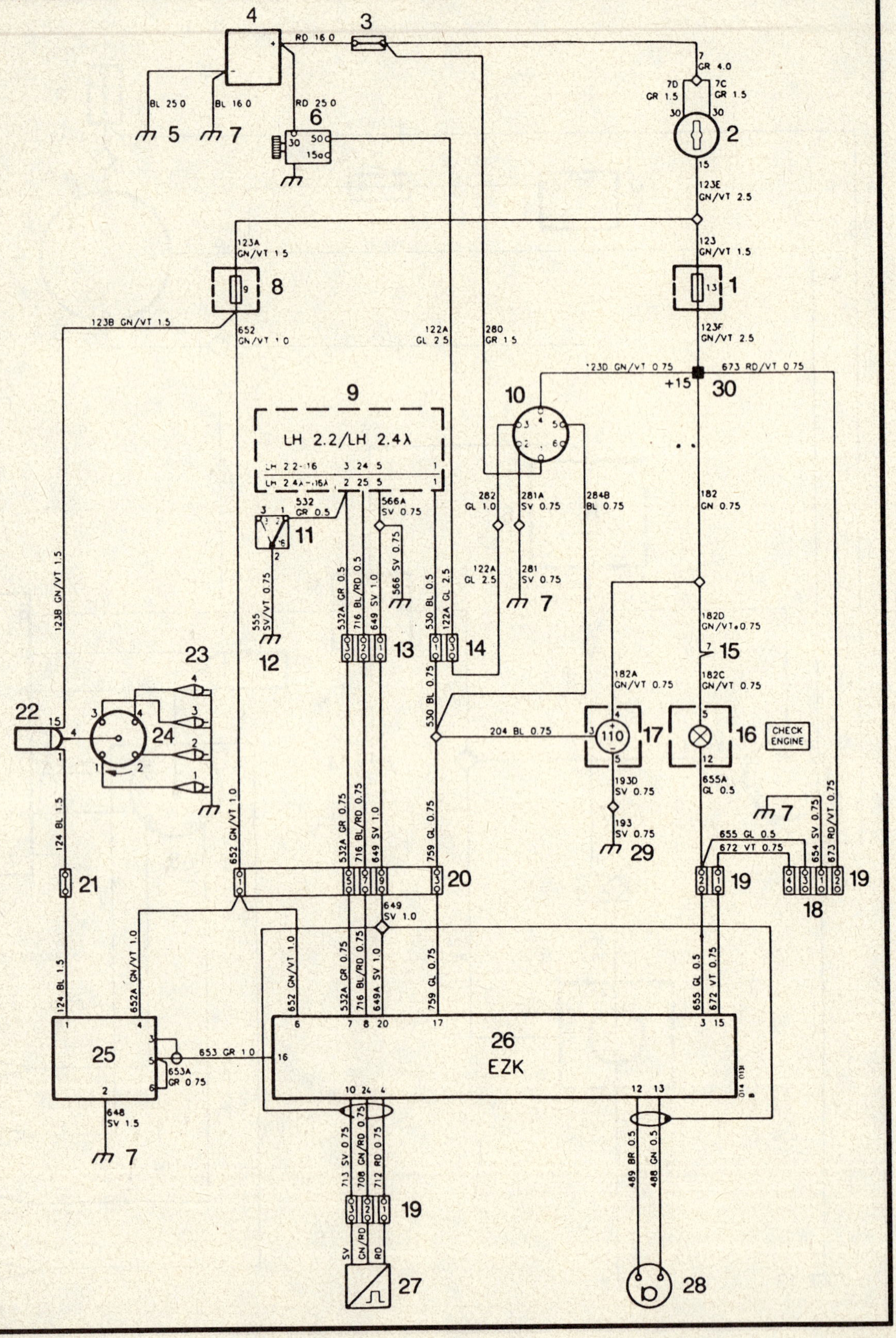

SAAB 9000 VIN L—DI-APC IGNITION SYSTEM WIRING DIAGRAM

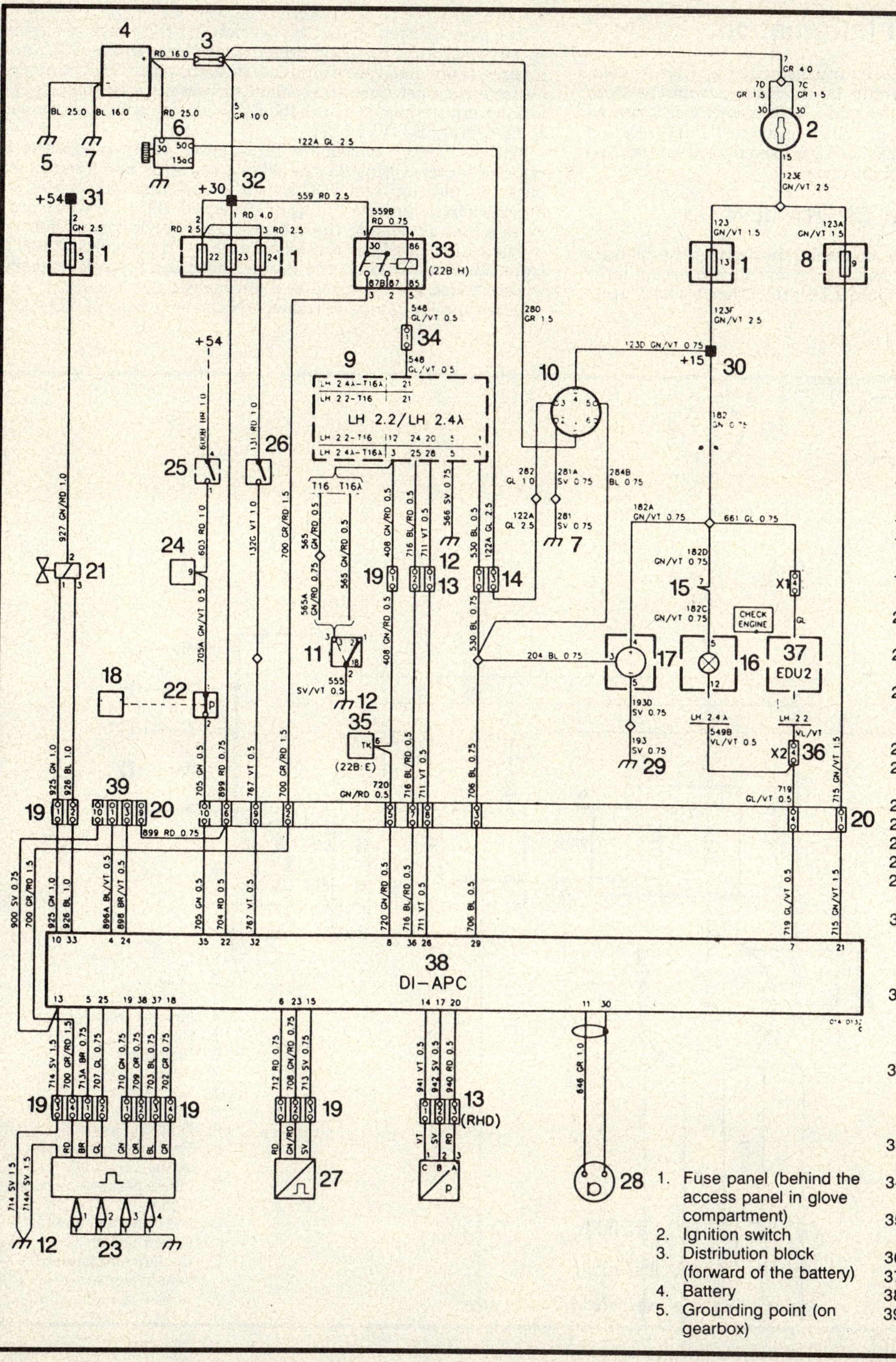

1. Fuse panel (behind the access panel in glove compartment)
2. Ignition switch
3. Distribution block (forward of the battery)
4. Battery
5. Grounding point (on gearbox)
6. Starter solenoid
7. Grounding point (on left hand wheel housing)
8. Fuse board (distribution box behind left headlamp)
9. Fuel system control unit
10. Timing service instrument socket (left side of bulkhead)
11. Throttle angle transmitter
12. Grounding point (rear of engine below intake manifold)
13. 3-pole connector (at wiper motor)
14. 6-pole connector (at wiper motor)
15. Coolant temperature and fuel gauge
16. Check engine warning lamp
17. Tachometer
18. Vacuum pump (cruise control)
19. 4-pole connector (under right seat)
20. 10-pole connector (under right seat)
21. Solenoid valve (on radiator fan casing)
22. Vacuum switch (left of steering column, behind knee gaurd)
23. Spark plugs
24. Cruise control unit (left of steering column)
25. Cruise control selector
26. Brake light switch
27. Crankshaft sensor
28. Knock sensor
29. Grounding point (facia, at left front speaker)
30. Distribution terminal +15 (distribution box, behind glove compartment)
31. Distribution terminal +54 (distribution box, behind glove compartment)
32. Distribution terminal +30 (distribution box, behind glove compartment)
33. Main relay (LH fuel injection system)
34. 6-pole connector (at wiper motor)
35. Time delay relay (for AC and ACC radiator fan)
36. 8-pole connector
37. EDU trip computer
38. DI-APC control unit
39. Diagnostic testing test socket

STERLING

General Information

Sterling 2.7L engines use a fully transistorized ignition system. It is comprised of a distributor driven directly from the front camshaft, an igniter amplifier and a high energy coil mounted on the left side inner fender. The Programmed Fuel Injection (PFI) Electronic Control Unit (ECU) is used to control both the ignition timing and the fuel injection.

SYSTEM OPERATION

The ECU optimizes the ignition timing throughout the engine's operating range. The ECU has 2 independent processing units which check each other. Should 1 fail, the other backs it up to maintain driveability.

The base ignition timing appropriate for the range of engine speeds and intake manifold pressures is stored in the ECU. Adjustments are made according to inputs from the throttle angle, intake air temperature and coolant temperature sensors. An ignition timing adjuster allows the ECU ignition advance to be set at 15 degrees BTDC at idle.

The distributor contains a TDC sensor which consists of a reluctor incorporating 6 poles 60 degrees apart and a magnetic inductive pick-up. The TDC sensor is used during cranking to provide a fixed ignition timing of 10 degrees BTDC and also provides a back-up should the cylinder/crank angle sensor fail.

The igniter amplifier connected directly to the coil contains 2 transistors that switch the ground path for the coil primary windings and are activated at a precise voltage.

When the ignition is switched **ON** the switching of the tran-

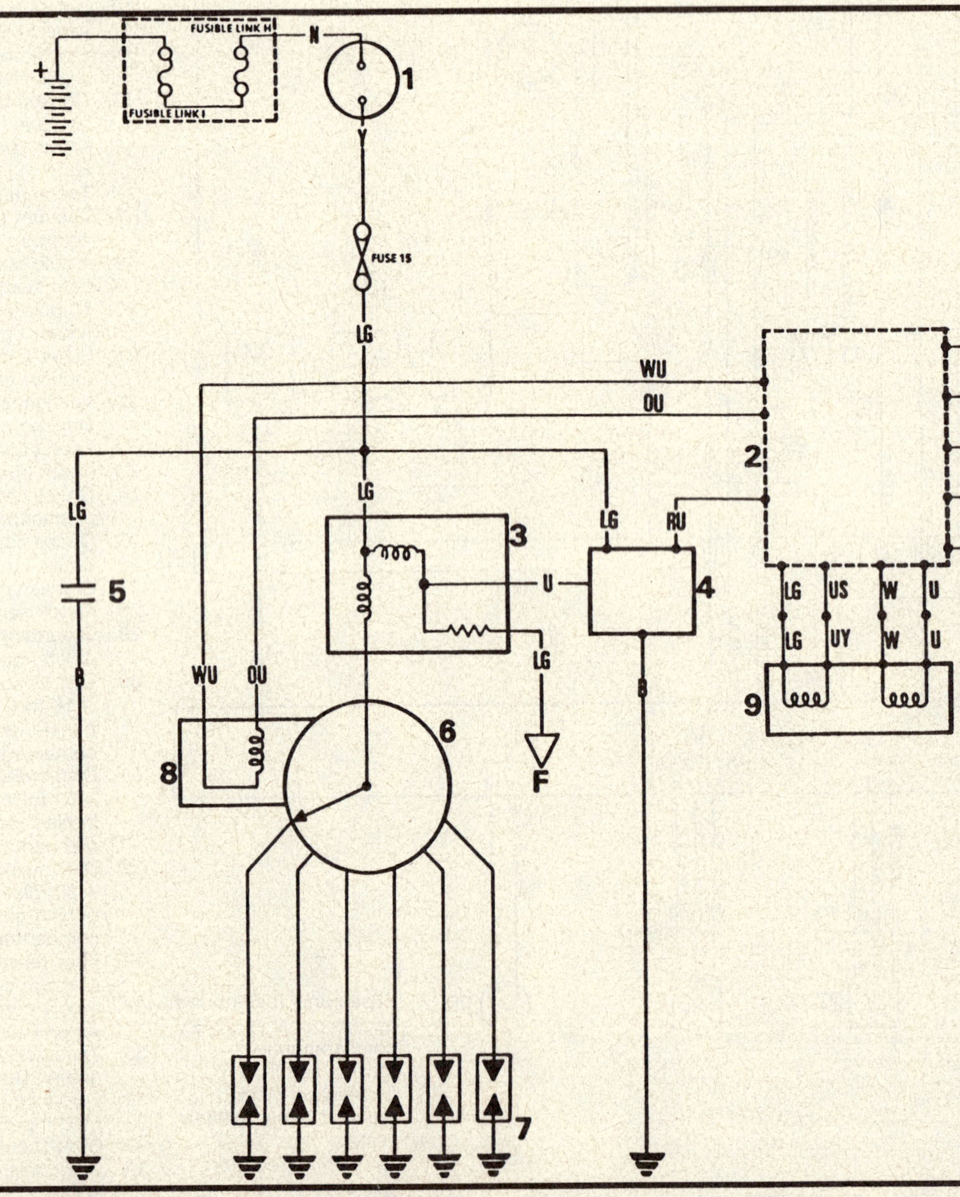

1. Ignition switch
2. ECU
3. Ignition coil
4. Ignition amplifier
5. Condenser
6. Distributor
7. Spark plugs
8. TDC sensor
9. Crank/cylinder sensor
A. Cranking signal
B. Throttle angle sensor signal
C. Intake pressure sensor signal
D. Intake air temperature sensor signal
E. Coolant temperature sensor signal
F. Engine speed signal

Ignition wiring schematic—Sterling 2.7L engine

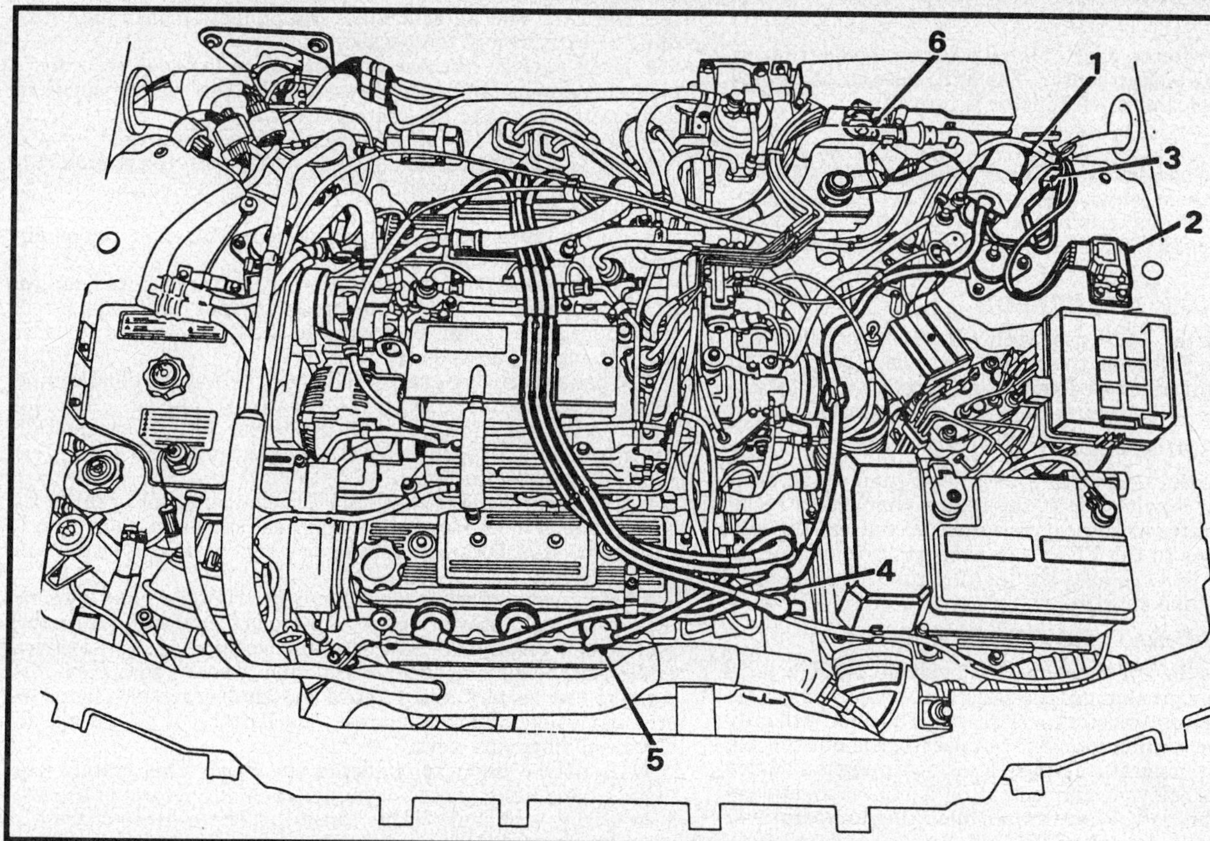

Ignition system basic components location—Sterling 2.7L engine

sistors in the amplifier ensures that no current can flow in the ignition coil primary windings. When the engine is cranked the reluctor moves through the magnetic field created by the stator and when the reluctor teeth are aligned with the stator projections, a small AC voltage is created. The igniter amplifies this voltage and uses it to switch the transistors so that a ground path is provided to the primary windings.

As the reluctor teeth move out of alignment with the stator projections an abrupt change occurs in the AC voltage, the transistors are switched again and the primary circuit ground path is broken. This induces a high voltage in the ignition coil secondary winding.

A time control circuit in the amplifier controls the charging time for the ignition coil according to engine speeds. Low battery voltage is compensated for to prevent secondary voltage drop so that sufficient energy for ignition is obtained.

During cranking, the ignition timing is fixed at 10 degrees BTDC. At idle, the timing is varied to help maintain idle speed. During warmup, the ignition is advanced to maintain driveability. When the engine is hot, the ignition timing is retarded to reduce the possibility of misfire.

IGNITION CONTROL SYSTEM

Ignition timing is provided electronically by the ECU instead of the conventional centrifugal and vacuum advance system used on earlier Sterling applications. The ECU receives input from various sensors and provides precise ignition timing to match engine operating conditions. The electronic engine management system has 2 Central Processing Units (CPU) in the ECU. If 1 CPU fails, the other backs it up to maintain driveability. The self diagnostic function indicates a fault using a Light Emitting Diode (LED) on the ECU case. The LED flashes different fault codes.

The ignition control system consists of the following:

1. Fuel ECU
2. Ignition Coil
3. Igniter unit
4. Distributor
5. Crank/cylinder sensor
6. TDC sensor
7. MAP sensor
8. Throttle angle sensor
9. Intake air temperature sensor
10. Coolant temperature sensor

The ECU receives raw timing signals from the crank/cylinder and TDC sensors. The ECU then uses the sensors to adjust ignition timing to varying speed, pressure, temperature and load conditions. After processing this information, the ECU sends an ignition signal to the igniter unit. The igniter unit alternately grounds and ungrounds the ignition coil primary. This ignition coil sends high voltage ignition pulses to the distributor where it is sent to the spark plugs.

LOW BATTERY VOLTAGE

If the battery voltage drops below 11.8 volts, the ECU compensates by increasing dwell duration to the coil allowing more time for the voltage required to be produced.

CRANK/CYLINDER SENSOR

As the camshaft pulley rotates, a single pole on the pulley passes the crank pickup sensor. As they align, a pulse is created. This is used by the ECU to detect the TDC position for No. 1 cylinder. The camshaft pulley also has 24 poles which pass the cylinder pickup sensor. As each aligns, a pulse is generated and the ECU uses these inputs in establishing engine speed and the position of each piston.

TDC SENSOR

As the distributor rotates, each of the 6 poles on the distributor shaft align with the pickup sensor. The ECU uses these signals to detect TDC when the starter motor is operating.

MAP SENSOR

The ECU provides the sensor with a 5 volt supply. A signal is produced by the sensor which is proportional to manifold absolute pressure. Variations in manifold pressure vary the output voltage. The ECU uses this varying voltage to determine engine load.

THROTTLE ANGLE (TH) SENSOR

The ECU provides the TH sensor with a 5 volt supply. The TH sensor informs the ECU that the engine is at idle. The ECU will now compensate for idle speed variations caused by additional load on the engine using inputs from the other sensors.

INTAKE AIR TEMPERATURE (TA) SENSOR

The ECU provides the TA sensor with a 5 volt supply. The temperature of the air flowing across the sensor changes its resistance. As temperature increases, resistance decreases. The signal from the sensor to the ECU is in proportion to intake air temperature. The ECU retards the ignition timing as the temperature of the intake air rises.

COOLANT TEMPERATURE (TW) SENSOR

The ECU provides the TW sensor with a 5 volt supply. The temperature of the coolant changes the resistance of the sensor. As temperature increases, resistance decreases. The ECU will only change the ignition timing if engine temperature is outside the normal operating temperature range. Ignition timing is advance when the engine is below normal operating temperature and retarded if the engine rises above normal operating temperature.

INERTIA SWITCH

In the event of an accident, the inertia switch is used to cut-off power to the main relay, fuel pump and the injectors. It is located in the RH center console forward of the radio and is reset by pressing the button on top of the switch.

MAIN RELAY

The main relay consists of 2 relays built into 1. One relay operates when the ignition switch is turned to the **ON** position. It supplies battery voltage to the ECU, injectors and the second relay. The second relay operates for approximately 2 seconds when the ignition switch is turned **ON** and by the ECU when the engine is running. It also supplies power to the fuel pump.

Diagnosis and Testing

SERVICE PRECAUTIONS

- Use care when working around high-energy electronic ignition systems. High voltage could cause injury to the technician and equipment.
- Before performing any service on the ignition system, make sure the ignition is turned **OFF** and/or the battery negative cable is disconnected.
- Before attempting to remove any parts, turn the ignition switch **OFF** and disconnect the battery ground cable.
- Never connect test equipment such as timing lights, etc., directly to the ignition system for a power source. Always use the vehicle battery for the power source.
- Use only the correct replacement parts. Use of spark plugs, for example, having specifications and sizes different from those specified could cause severe damage and alter emissions levels.
- Use care when working with light alloy parts. Always lubricate the spark plug threads with a suitable lubricant to avoid seizing the plug threads to the cylinder head.
- Keep all ECU parts and harnesses dry during service. Protect the ECU and all solid-state components from rough handling or extremes of temperature.
- Make sure all harness connectors are fastened securely. A poor connection can cause an extremely high voltage surge resulting in damage to integrated circuits.
- Always use a 12 volt battery as a power source.
- Do not disconnect the battery cables with the engine running; never run the engine with the battery cables loose or disconnected.
- Do not attempt to disassemble the ECU unit under any circumstances.
- When connecting or disconnecting pin connectors from the ECU, take care not to bend or break any pin terminals. Check that there are no bends or breaks on the ECU pin terminals before attempting any connections.
- When measuring supply voltage of ECU-controlled components, keep the tester probes separated from each other and from accidental grounding. If the tester probes accidentally make contact with each other during measurement, a short circuit will damage the ECU.
- It is recommended that the ECU control unit be removed if the vehicle is to be exposed to high heat such as a bake oven to cure paint after body refinishing or when welding is be done on the vehicle.
- Use great care when working on or around air bag systems. The system may stay armed for a period of time without battery power. It is recommended that while working on any electrical system in the air bag deployment zone (driver's seat area), disconnect the battery and remove the air bag module from the steering wheel. Then reconnect the battery, if necessary, for electrical diagnosis work.
- Do not be hasty to condemn the ECU. This system uses voltages and resistances that are very small. Examine the sensors, wiring and connections carefully. The sensors operate in more harsh conditions than the ECU which is in a more protected area.
- Check all ground connections before condemning the ECU.

READING CODES

The ECU incorporates a red LED which blinks between 1 and 18 times providing 16 different fault diagnosis. If the fault occurs in the Programmed Fuel Ignition system, the LED blinks a code then pauses 2 seconds before repeating the sequence. The LED will go out when the ignition is switched **OFF** however, the memory for the fault will not be cancelled. After correcting the fault, disconnect the ECU battery supply by removing the alternator fuse (10A) from the engine compartment fuse carrier for 10 seconds to reset the ECU memory.

When a fault code is displayed on the ECU LED, the CHECK ENGINE warning light on the instrument panel will be continually illuminated to inform the driver of the Fuel Injection system malfunction. The CHECK ENGINE warning light always comes on for 2 seconds after the ignition is switched **ON** as a self-check test (bulb check).

When a fault occurs in a signal from a sensor, the ECU lights the dashboard warning light, stores a failure code in the erasable memory and indicates a code with a red LED light on the ECU box every time the ignition is switched to **ON**.

Problem codes 1–9 are indicated by individual short blinks. Problem codes 10–31 are indicated by a series of long and short blinks. One long blink equals 10 short blinks. Add the long and short blinks together to determine the problem code.

On this generation of self-diagnostics, there is no connector to jumper or buttons to push as on other systems. The blinking diagnostic lights are always available on the ECU unit under the passenger front seat. When the ECU detects a problem with the idle system, the yellow LED lights. If the yellow LED is not illuminated, then no adjustment is required. If the LED blinks, then the Electronic Idle Control Valve (EICV) is beyond value

STERLING 2.7L ENGINE DIAGNOSIS

No. of LED Blinks Between 2 Seconds Pauses		Symptom Indicated	Possible Cause
0	LED On	Engine will not start Engine runs at reduced power	Disconnect ECU ground wire Faulty ECU
	LED Off	Engine will not start No particular symptom shown.	Activated inertia switch. Faulty ECU ground. Disconnected ECU ground wire Short circuit in warning light wire. Faulty ECU—Check system using 2.7 P??? Fast check before changing ECU.
1 2		Incorrect Idle Co	Front oxygen sensor electrical fault. Rear oxygen sensor electrical fault.
3 5		Fuel fouled plug Frequent engine stalling Hesitation	MAP electrical fault. Disconnected MAP pipe.
4		Reduced performance	Crank angle electrical fault.
6		High idle speed Difficult starting at low temperature.	Coolant sensor electrical fault.
7		Hesitation. High idle speed.	Throttle angle sensor incorrectly adjusted or electrical fault.
8		Difficult starting because of no cranking fixed timing.	Distributor TDC sensor electrical fault
9		Reduced performance.	No 1 TDC cylinder sensor electrical fault
10		Slight effect on performance	Intake air temp sensor electrical fault
12		Frequent engine stalling. Erratic running at low speed No fault shown	EGR system electrical fault. Sticking EGR solenoid or blocked pipe
13		Slight effect on performance	Atmospheric pressure sensor electrical fault.
14		Idle speed hunting from 1,200 to 1,500 RPMS. Erratic idle.	EICV electrical fault.
15		Erratic running Engine will not start.	Igniter unit electrical fault
17		No speedometer Stalling when vehicle comes to rest.	Vehicle speed sensor electrical fault
18		Ignition timing or specifications incorrect.	Ignition timing sensor electrical fault.

NOTE: 1. Always check wiring for open or short circuit before changing sensor.
2. If code 16 oo blinks exceed 18, count number of blinks again, if the LED is still showing, these codes, substitute a known good ECU and re-check.

and the adjusting screw of the throttle body must be turned clockwise. If the LED is permanently lit, the EICV is beyond its specified value and the adjusting screw on the throttle body must be turned counterclockwise.

Fault codes are blink codes, counted by the number of flashes of the LED.

Engine Codes

Code LED ON, But No Blinks—Symptom: Engine will not start. Possible cause: Disconnected ECU ground wire.
Code LED OFF–Symptom: Engine will not start, but no particular symptoms. Possible cause: Disconnected ECU ground wire, short circuit in warning light wire or faulty ECU assembly.
Code 1—Symptom: Erratic idle, hesitation, stalling. Possible cause: Faulty or disconnected front oxygen sensor.
Code 2—Symptom: Erratic idle, hesitation, stalling. Possible cause: Faulty or disconnected rear oxygen sensor.
Code 3—Symptom: Fuel fouled spark plugs, frequent stalling, hesitation. Possible cause: Disconnected MAP sensor harness, short, faulty or open in the MAP sensor wire.
Code 4—Symptom: High idle speed, erratic idle. Possible cause: Open or short circuit in crank angle sensor circuit, faulty crank angle sensor.
Code 5—Symptom: Fuel fouled spark plugs, frequent stalling, hesitation, uneven idle. Possible cause: Disconnected MAP sensor hose.
Code 6—Symptom: High idle speed during warmup, high idle speed, difficult starting at low temperatures. Possible cause: Disconnected coolant temperature sensor harness. Short or open circuit in coolant temperature sensor wire. Faulty temperature sensor.
Code 7—Symptom: Poor engine response to rapid throttle opening. High idle speed. Engine does not rev up when cold.

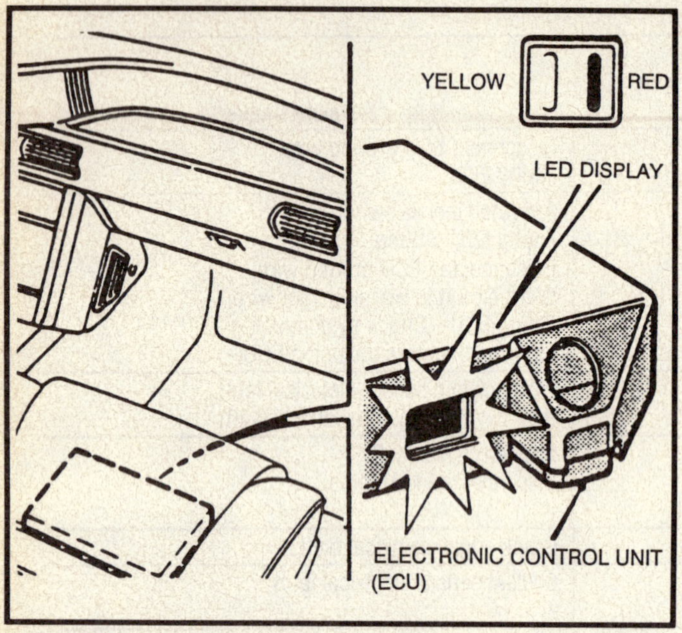

The ECU is located under the passenger side front seat and contains 1 yellow and 1 red LED to aid in diagnostics—Sterling 2.7L engine

Possible cause: Disconnected throttle sensor, open or closed circuit in throttle angle circuit, faulty throttle angle.
Code 8—Symptom: Engine does not rev, high idle speed, erratic idle. Possible cause: Open or closed circuit in crank angle sensor wire, faulty crank angle sensor (TDC).
Code 9—Symptom: Engine does not rev, high idle speed, erratic idle. Possible cause: Open or closed circuit in crank angle sensor wire, faulty crank angle sensor (cylinder).
Code 10—Symptom: High idle speed, erratic idling when very cold. Possible cause: Disconnected intake air temperature sensor. Open or closed circuit in air intake temperature sensor wire.
Code 11—Symptom: High idle speed. Unable to adjust idle CO. Possible cause: Faulty idle mixture adjuster.
Code 12—Symptom: Frequent engine stalling, erratic or unstable running at low speeds, or even no particular symptom. Possible cause: Disconnected EGR control system, shorted or disconnected EGR control wire, faulty EGR control system.
Code 13—Symptom: Poor acceleration at high altitude, difficult starting at high altitude when cold. Possible cause: Disconnected atmospheric pressure (PA) sensor harness, shorted or disconnected PA sensor wire, faulty PA sensor.
Code 14—Symptom: Idle speed "hunting" from 1200–1500 rpm, erratic idle. Possible cause: Shorted or disconnected Electronic Idle Control Valve (EICV) wire, faulty EICV.
Code 15—Symptom: Engine will not start. Possible cause: short or open circuit in ignition circuit, faulty igniter unit.
Code 17—Symptom: No speeedometer, stalling under certain conditions. Possible cause: Road speed transducer/speedo circuit faulty.
Code 18—Symptom: Poor idling and performance. Possible cause: Open or short circuit in ignition timing circuit, faulty ignition timing adjuster.
Code 43—Symptom: Poor idling and performance. Possible cause: Open or short circuit in ignition timing circuit, faulty ignition timing adjuster.
Code 44—Symptom: Poor idling and performance. Possible cause: Open or short circuit in ignition timing circuit, faulty ignition timing adjuster.
Code 45—Symptom: Poor idling and performance. Possible cause: Open or short circuit in ignition timing circuit, faulty ignition timing adjuster.
Code 46—Symptom: Poor idling and performance. Possible cause: Open or short circuit in ignition timing circuit, faulty ignition timing adjuster.

If the ECU control unit blinks codes 11, 16 or exceeds 18 blinks, substitute a known-good ECU and recheck. If the indication goes away, replace the original ECU.

CLEARING CODES

After diagnosis and repairs, the codes may be cleared and the ECU diagnostic system reset. The ECU failure codes are stored in the ECU's erasable memory and will continue to flash its warning codes when the ignition is switched **ON** whether the fault has been corrected or not and will continue to do so until the ECU fault code memory has been reset. This is done by switching the ignition **OFF** and removing the alternator fuse (10A) from the engine compartment fuse carrier for 10 seconds to reset the ECU memory.

IGNITION TIMING

The ECU controls dwell duration according to engine speed, providing reduced dwell at low engine speeds and increased dwell at high engine speeds.

Ignition timing is changed or compensated in the following ways.

CRANKING COMPENSATION

During cranking the ECU receives an engine speed and position signal from the TDC sensor located in the distributor and a cranking input from the ignition switch. The ECU then implements a fixed timing of 10 degrees BTDC.

IDLE COMPENSATION

During idle the ignition is changed to air in the control of the specified idle speed. To achieve this the ECU receives signals from the cyl/crank sensor, MAP sensor, throttle angle sensor and intake air temperature sensor.

WARM UP COMPENSATION

During the warmup period the basic ignition system is modified depending upon intake air temperature.

COOLANT TEMPERATURE COMPENSATION

The ignition timing is advanced during low coolant temperature conditions and retarded when the temperature is high. To achieve this the ECU receives a signal from the coolant temperature sensor.

Checking

Although the ignition timing cannot be changed, it can be checked. If the timing is not correct, specific components of the engine management system should be checked. An ignition timing adjuster allows the ECU ignition advance to be set to 15 degrees BTDC at idle. Check timing using the following procedure:

1. Start the engine and allow to warm to normal operating temperature.
2. With the engine at idle speed, place the transmission in **P** for automatic transmission or **N** for manual transmission.
3. Connect a tachometer according the manufacturer's instructions. Connect a suitable timing light to No. 1 cylinder which is at the camshaft gear end of the rear cylinder head. Aim the timing light at the notches on the crankshaft pulley. The timing should be: 15 degrees ± 2 degrees BTDC at and idle speed of 680 ± 50 rpm.
4. Check the underhood emission label for any factory running changes that might differ from these specifications.
5. If the ignition timing is incorrect, check for faults stored in the ECU using the LED blink codes.

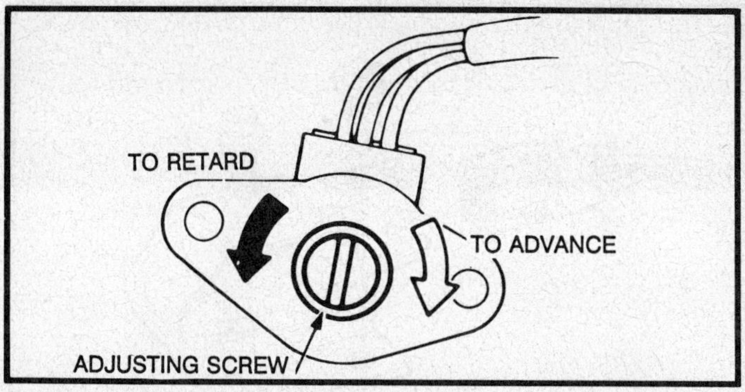

Ignition idle timing adjuster—Sterling 2.7L engine

Adjustment

A problem in the ignition timing adjuster may set Code 18 in the self-diagnostic system.

1. Start the engine and allow it to warm to normal operating temperature.
2. With the engine idling, place the transmission in **P** (automatic) or **N** (manual).
3. Connect a suitable timing light and tachometer to the engine.
4. Point the timing light at the pointer on the timing belt cover.
5. Ignition timing should be 15±2 degrees BTDC at 680±50 rpm. If timing is not within specification, shut down the engine and proceed as follows:
   a. Remove the control box upper cover and remove the timing adjuster from the control box.
   b. Drill the 2 rivets out using $^3/_{16}$ in. drill bit. Remove the cover from the adjuster.
   c. Reconnect adjuster, start the warmed-up engine and adjust the timing as necessary by turning the adjusting screw on the adjuster. Turn the adjuster screw counterclockwise to retard the timing or clockwise to advance timing.
   d. After adjustment, install the cover onto the timing adjuster using new fasteners and assemble control box covers.
6. Recheck timing.

NO START—NO SPARK

Troubleshooting

1. If the engine cranks but there is no spark at the plugs, check fuse 15 in the dashboard fusebox. This fuse connects the **ON** position of the ignition switch to the ignition coil primary.
2. If the fuse is good, remove the distributor cap and check that the rotor turns while the engine is cranked, to rule out cam drive failure. Check the distributor rotor arm for breakage.
3. Check that the connections at the coil, distributor and spark plugs are clean and tight.
4. With the ignition switch in the **ON** position, check for voltage at the coil (light green wire) wire terminal.
5. With the ignition switch in the **ON** position, check for voltage at the igniter (light green wire) wire terminal.
6. Check for signs of high tension carbon tracking on the coil and distributor cap. It the vehicle still fails to start, perform coil resistance check.

Coil Resistance Check

MANUAL TRANSMISSION

1. With the ignition switch **OFF**, disconnect the leads from the coil.
2. Measure the resistance between terminals C and D. It should be approximately 0.351–0.429 ohms.
3. Measure the resistance between terminals B and D. It should be approximately 2090–2310 ohms.
4. Measure the resistance between terminals A and the high tension terminal. It should be approximately 9040–13560 ohms.
5. Check for continuity between terminal A and terminal C. Replace the coil if readings are outside limits.

AUTOMATIC TRANSMISSION

1. With the ignition switch **OFF**, disconnect the leads from the coil.

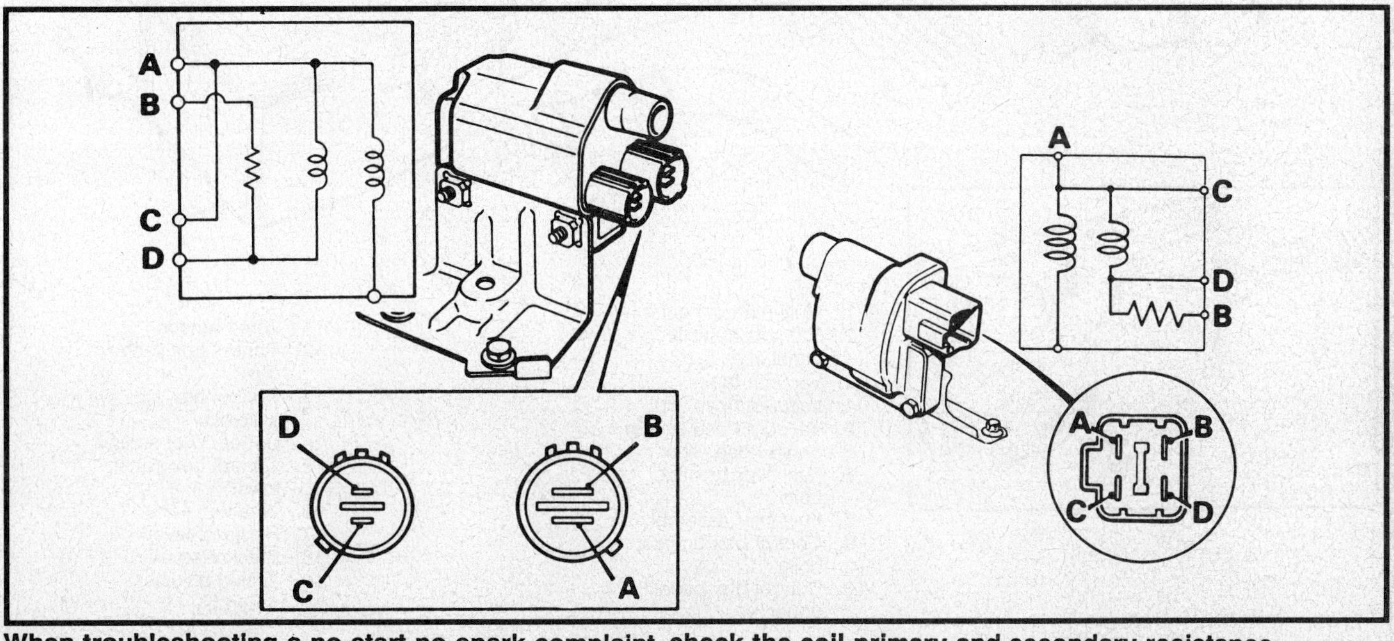

When troubleshooting a no start-no spark complaint, check the coil primary and secondary resistance—Sterling 2.7L engine

2. Measure the resistance between terminals C and D. It should be approximately 0.351–0.429 ohms.
3. Measure the resistance between terminals B and D. It should be approximately 2090–2310 ohms.
4. Measure the resistance between terminals A and the high tension terminal. It should be approximately 14000–21600 ohms.
5. Check for continuity between terminal A and terminal C. Replace the coil if readings are outside limits.

Component Replacement

IGNITION COIL

Removal and Installation

1. Disconnect the negative battery cable.
2. Remove the cover from the coil.
3. Disconnect the high tension lead and the 2 multi-pin plugs from the coil.
4. Remove the coil bolts and lift the coil from the vehicle.

To install:

5. Installation is the reverse of the removal procedure. Check the multi-pin plugs for damage or corrosion.
6. Make sure all electrical connections are secure.

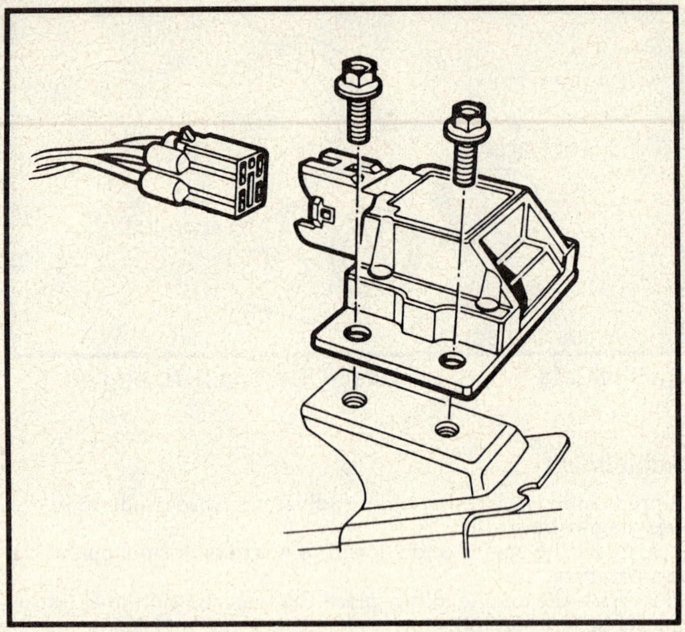

Ignition igniter/amplifier—Sterling 2.7L engine

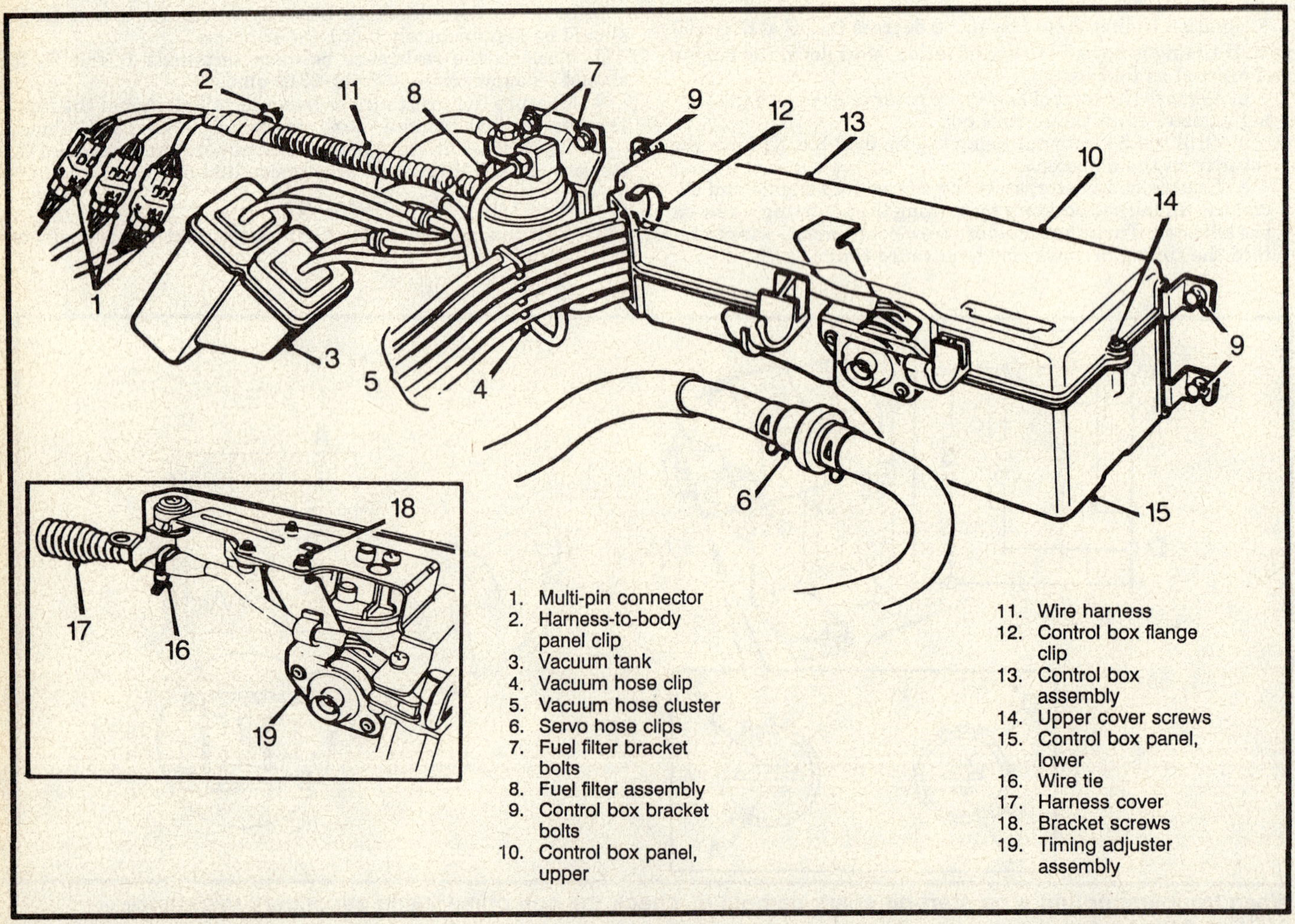

The control box contains a number of components including the idle timing adjuster—Sterling 2.7L engine

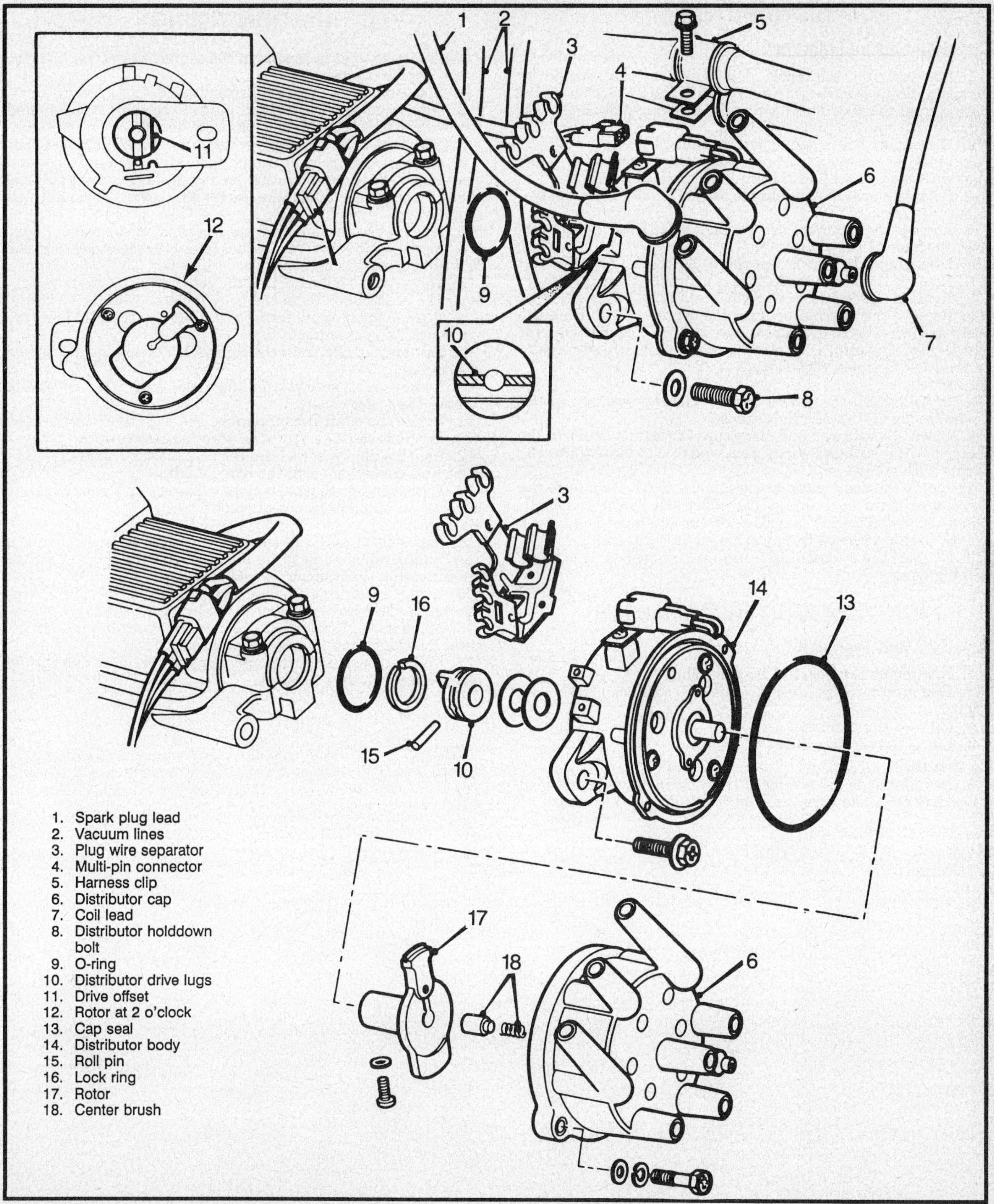

Distributor assembly and related components. Note the offset drive lugs and the rotor at the correct 2 o'clock installation position—Sterling 2.7L engine

DISTRIBUTOR

Removal and Installation

1. For assembly reference, it is suggested that engine be turned until the distributor rotor end is at the 2 o'clock position viewed from the distributor cap end. This should place the distributor drive in the proper position.
2. Disconnect the negative battery cable.
3. Mark for reassembly, if necessary, then remove the spark plug wires and the coil lead from the distributor cap.
4. Remove the multi-pin plug to the EGR valve from the distributor bracket.
5. Remove the vacuum lines the EGR valve from their bracket at the base of the distributor.
6. Disconnect the distributor multi-pin connector.
7. Remove the harness clip that is alongside the distributor.
8. Remove the distributor-to-head retaining bolt.
9. Remove the distributor from the cylinder head. Note the position of the distributor drive lugs and the camshaft slot for reassembly. The rotor is retained by a set screw which needs to be removed for rotor replacement.

To install:

10. Installation is the reverse of the removal procedure. Install a new O-ring on the distributor base.
11. Clean all parts well and make sure the carbon brush in the center of the distributor cap is clean and free to move. Lube the O-ring with engine oil.
12. Install the distributor to the cylinder head. Make sure the distributor drive lugs engage the offset slot in the camshaft. When the distributor is seated, the rotor end must be at the 2 o'clock position viewed from the distributor cap end.
13. Connect the electrical and vacuum connections. Reconnect all plug wires.

IGNITION IGNITER (AMPLIFIER)

Removal and Installation

1. Disconnect the negative battery cable.
2. Disconnect the multi-pin electrical connector from the igniter.
3. Remove the 2 bracket screws.
4. Lift igniter from vehicle.

To install:

5. Installation is the reverse of the removal procedure. Clean all parts well. Make sure the electrical connection is secure.

IGNITION TIMING

The timing adjuster is located in the control box at the left rear of the engine compartment.

1. Disconnect the negative battery cable.
2. Disconnect the 3 multi-pin plugs from their brackets and remove the harness clip from the body panel.
3. Disconnect the rectangular vacuum tank from its bracket.
4. Disconnect the clip holding the multiple vacuum hoses and disconnect the vacuum hoses as required. There should be about 9 hoses so mark the hoses for correct assembly as required.
5. Release the servo hose from the clips on the control box.
6. Remove the bolt from the fuel filter bracket and release the fuel filter from the body panel.
7. Loosen the 3 bolts holding the control box brackets and release the control box from the body panel.
8. Release the harness from behind the fuel filter and remove the control box assembly.
9. Remove the clip from the control box flange near the harness outlet.
10. Remove the 3 screws from the upper control box cover and remove the lower cover.
11. Inside the control box, remove the wire tie from the harness and release the corrugated wire harness cover.
12. Remove the 2 screws from the bracket and remove the timing adjuster/bracket from the control box.
13. If the timing adjuster is to be replaced, drill out the rivets holding the adjuster to the bracket.

To install:

14. Installation is the reverse of the removal procedure. If a new timing adjuster is to be installed, secure to the original bracket using replacement hardware.
15. Install the timing adjuster and bracket to the control box, secure the electrical connector and reassemble the cover pieces.
16. Make sure all electrical and vacuum connections are secure, following matchmarks made at disassembly.
17. Make sure all electrical and vacuum retaining clips are in place and correctly positioned.

Note: The ignition system components are interrelated with fuel system components. For additional testing, replacement or wiring diagrams see appropriate fuel system section.

VOLKSWAGEN

General Information

The Bosch Hall Effect electronic ignition system is used on all Volkswagen models.

SYSTEM OPERATION

There are 2 main system types: Digifant and CIS-E Motronic. Both are Bosch derived systems using electronic ignition control.

These systems consist of a breakerless distributor with a Hall Effect pick-up, an ignition control unit, an ignition coil, an ignition switch and battery. Closely allied with the ignition system is an idle stabilizer, a solid state control unit located between the Hall Effect (ignition) control unit and the distributor's Hall Effect pick-up. It replaces the distributor in sending signals to the ignition control unit when engine speed falls below 940 rpm.

The Hall Effect sending unit (pick-up) is mounted on a switch plate in the distributor. A segmented shutter is attached to the distributor shaft under the rotor. The shutter blades, one for each engine cylinder, pass through the air gap of the Hall Effect sending unit.

This causes a fluctuating voltage signal to be sent from the Hall Effect sending unit, through the idle stabilizer to the ignition control unit.

When engine speed is above 940 rpm, the ignition control unit reads the voltage signal from the Hall Effect unit, and makes and breaks the primary circuit to the ignition coil. Secondary voltage is generated in the coil, and passes through the high tension wire, rotor, distributor cap, and spark plug wires to the spark plugs.

If engine speed drops below 940 rpm, the idle stabilizer (instead of the Hall Effect sending unit) takes over the duty of producing the voltage signal for the ignition control unit. The idle stabilizer, mounted between the distributor and the ignition control unit on the control unit bracket, senses engine speed earlier, causing ignition timing to advance. Advancing ignition timing causes idle speed to increase, and the Hall Effect sending unit to resume its normal operation of sending voltage signals to the ignition control unit.

The 2 connectors at the idle stabilizer unit can be disconnected from the unit and connected to each other to by-pass the idle stabilizer unit for testing. The idle stabilizer is usually mounted on the same bracket as the Hall control unit and is wired between the distributor and the Hall control unit.

Diagnosis and Testing

SERVICE PRECAUTIONS

- Make sure that ignition is in **OFF** position before test equipment or system connectors are either connected or disconnected.
- Some of these vehicles may be equipped with anti-theft radios that will no longer work if the power to them is ever cut off. Special codes must be used to reset the radio. Always obtain these code from the customer before disconnecting the battery cables
- Always disconnect the battery when using arc, spot or electrical welding equipment
- Never disconnect the battery with the engine running
- When cranking engine without starting, as for a compression test, disconnect wire from distributor, and connect to ground. Do not hand hold wire.
- Do not install a standard ignition coil in the system.
- Do not connect a condenser/suppressor or powered test light to negative terminal of ignition coil.
- Do not connect any 12-volt test instruments to positive terminal of ignition coil, as this could damage electronic components.
- Many tests call for a Light Emitting Diode (LED) type tester. This draws very little power. Do not substitute a conventional test light. It draws too much current and will damage delicate components.
- When boost-starting, do not connect a quick-charger for more than 1 minute, nor exceed 16.5 volts with booster.
- Never apply voltage to the control unit to simulate the output signals
- Always disconnect the distributor high tension lead and jumper it to ground, when service on the ignition system requires cranking the engine

SYSTEM DIAGNOSTICS

Some of these vehicles will provide limited ignition diagnostics by means of the Electronic Control Unit. Model year, engine code and Market (49 States or California) all figure in whether a vehicle is equipped with on-board diagnostics or not. In general, if the instrument panel has an On-Board Diagnostic (OBD) lamp, often called a CHECK ENGINE light, the vehicle is likely equipped with some form of diagnostic capability. The OBD lamp in the instrument panel must light up when the ignition is switched to **ON** and then must go out when the engine starts. If there is a problem in the system, the OBD lamp may continue to remain lit even with the engine running.

Note that some models may have different engine management systems, depending on engine and emission certification.

Some models may have diagnostic connectors but the these are for factory testing and no codes are "field accessible." For example, a brown diagnostic connector has been installed in the center console of some vehicles. Do not connect a scan tool to this connector. It is not intended for use outside the factory. If accessed, the codes it provides will indicate faulty components in cases where this is not true.

It should be noted that most fault codes are emission/fuel related, not ignition.

READING CODES

Digifant I System

The Digifant I control unit is used on Cabriolet, Fox and 1991 Golf and Jetta with California certification. This engine management system is a combined ignition and injection system and is equipped with a permanent fault memory. If faults occur in the monitored sensors and/or components, the are stored in the fault memory in code form which can be retrieved to aid troubleshooting. The factory scan tool will be the most useful but blink codes can also be retrieved using a jumper wire and the OBD instrument panel light. Use the following procedure:

1. Turn ignition switch to **ON**.
2. Connect a suitable jumper wire to the diagnostic connectors located in the center console as follows:
   a. Carefully work the shifter boot loose and pull the boot up to access the diagnostic connectors.
   b. Connect 1 end of the jumper wire to the black diagnostic connector No. 1.
   c. Connect the other end of the jumper wire to the white diagnostic connector No. 2.
3. Maintain this connection for approximately 5 seconds. The on-board diagnostic light must begin blinking. Remove jumper wire.
4. Perform this procedure until blink Code 4444 or 0000 appears.
5. To end diagnostic procedure (after 0000 appears), bridge

VOLKSWAGEN FAULT CODES

Code	Explanation or Possible Problem Areas
4444	No faults in memory.
1111	ECU defective.
1231	Vehicle speed sensor signal problem. Check speedometer wiring to transaxle.
2111	Engine speed sensor signal problem. Check wiring or sensor loose/defective.
2112	Ignition reference sensor signal problem. Sensor defective or wire shorted/open.
2113	Hall sender signal missing. Check if rotor out of alignment. Wiring shorted or open.
2121	Idle switch defective or wiring shorted/open.
2122	Engine speed signal missing. Check Hall sender & wiring.
2123	Full throttle switch problem. Defective or check wiring.
2132	No data between F.I. ECU and Ignition ECU. Check for disconnected wiring/defective ECU.
2141	Knock control program wiring. Check timing & fuel quality.
2142	Knock sensor No. 1 signal missing. Defective sensor or open wiring.
2144	Knock sensor No. 2 signal missing. Defective sensor or open wiring.
2212	Throttle position sensor. Open circuit or short to positive or ground.
2214	Maximum engine speed exceeded or sensor faulty.
2222	MAP sensor defective or wire open/shorted.
2231	Idle speed regulator out of limit. Check valve for sticking/dirty.
2232	Airflow sensor signal missing. Check wiring for short/open.
2234	Supply voltage too high or too low. Check alternator.
2242	CO potentiometer problem.
2312	Engine coolant temperature sensor defective. Check wires for short, or replace.
2314	Engine/transmission electrical connector short to ground.
2322	Intake air temperature sensor defective. Check wires for short, or replace.
2323	Airflow sensor defective or wiring open.
2324	Airflow volume meter (MAF) signal defective. Check wiring; air leak to outside air.
2341	Oxygen sensor signal out of limit or heater defective.
2342	Oxygen senosr signal missing. Defective or open wiring.
2411	EGR temperature sensor signal out of limit. Sensor defective or wiring open/shorted.
2413	Mixture problem. Check fuel pressure/air intake/oxygen sensor/exhaust leaks.
4312	EGR frequency valve defective or wiring shorted.
4332	ECU output stage problem. Check for loose terminals.
4343	Emission canister solenoid defective or wire shorted.
4431	Idle stabilizer valve circuit. Check wiring.
4433	Fuel pump relay defective or wire shorted.
4411–4422	Fuel injectors defective or shorted.
0000	End of output.

the diagnostic connectors using jumper as before or switch ignition **OFF.**

Digifant II System

Some Golf/Jetta 1991 California models may have the Digifant II system which can be identified by a rocker switch for accessing the OBD system. The OBD light may light for a fault. Or even if there is no fault in the emission system, but the diagnostic system is too sensitive, which triggers the light.

To cancel the OBD light, use the following procedure:

1. Test drive vehicle for at least 10 minutes.
2. Press and hold the rocker switch.
3. Switch ignition **ON.**
4. Continue pressing switch for about 15 seconds.
5. Turn ignition switch **OFF.**

If the OBD light comes on again after the system has been diagnosed with no faults or the fault has been repaired and emission levels meet specification, the factory recommends that if the customer so desires, the OBD can be disconnected. Use the following procedure:

GOLF

1. Locate and remove the blue/brown wire from the connector near the relay panel.
2. Tape the end of the wire with electrical tape.

JETTA

1. Locate the single connector 20 gauge white/red wire behind the fuse/relay panel.
2. Disconnect this connector. Tape end.

NOTE: The OBD wire originates from the 32 pin harness connector and terminates in the 3 pin connector in the plenum chamber.

Motronic Engine Control System

Corrado, Passat, and Federal certified Golf and Jetta vehicles use the Motronic engine control system. The Motronic ECU module controls the fuel injection and ignition system. It is equipped with a Diagnostic Trouble Code (DTC) memory. If faults occur in the monitored sensors and/or components, the are stored in the fault memory in code form which can be retrieved to aid troubleshooting. The factory scan tool will be the most useful but blink codes can also be retrieved using a jumper wire and OBD instrument panel light. Note that some models may have the adapter cable, or jumper wire stored in the glove compartment. Use the following procedure:

1. Switch the ignition to **ON**.
2. The adapter cable, or jumper wire which should be stored in the glove compartment. Connect this cable to the diagnostic connectors located in the center console as follows:
   a. Carefully work the shifter boot loose and pull the boot up to access the diagnostic connectors.
   b. Bridge the black and white diagnostic connectors using the jumper wire.
3. Maintain this connection for approximately 5 seconds. The on-board diagnostic light must begin blinking.
4. Perform this procedure until blink Code 4444 or 0000 appears.
5. To erase codes after diagnostics use the following procedure:
   a. Switch the ignition **OFF**.
   b. Bridge the diagnostic connectors with the jumper wire.
   c. Switch ignition **ON**. After approximately 5 seconds, remove the jumper wire from the diagnostic connectors.
   d. After blink Code 4444 appears switch ignition **OFF**.
6. Test complete.

CLEARING CODES

Digifant I System

1. Perform diagnostic procedures until blink Code 4444 or 0000 appears.
2. To end diagnostic procedure (after 0000 appears), bridge the diagnostic connectors using jumper as when accessing diagnostic codes or switch ignition **OFF**.

Digifant II System

1. Test drive vehicle for at least 10 minutes.
2. Press and hold the rocker switch.
3. Switch ignition **ON**.
4. Continue pressing switch for about 15 seconds.
5. Turn ignition switch **OFF**.

Motronic Engine Control System

1. To erase codes after diagnostics, switch the ignition **OFF**.
2. Bridge the diagnostic connectors with the jumper wire.
3. Switch ignition **ON**. After approximately 5 seconds, remove the jumper wire from the diagnostic connectors.
4. After blink Code 4444 appears switch ignition **OFF**.
5. Codes are erased.

NO START—NO SPARK

Testing

1. If engine will crank, remove 1 spark plug lead and using a suitable spark plug tester, check for spark.
2. Tester should show a spark when cranking. If there is no spark, remove the distributor cap and crank engine. Check that the rotor turns, to rule out broken distributor cam belt/drive.
3. If the rotor turns and there is no spark, recheck the ignition secondary for signs of damaged or worn parts.
4. Use care when handling parts. For example, some engines such as the 16 valve engine has the ignition rotor glued to the distributor shaft and cannot be removed without destroying it. The rotor can be removed by crushing it with a pair of pliers but the replacement must be glued into place.
5. If there is no spark and there is no obvious fault in the ignition secondary, check for codes.
   a. Note that the Digifant I system, used on California vehicles, has some fault memory in which codes are stored and can be retrieved.
   b. The Digifant II system as used on engine code PG, does not have a self-diagnosis capability.
   c. The Motronic system is equipped with a Diagnostic Trouble Code (DTC) memory and stored codes can be retrieved.
6. Check the wiring to the coil and distributor and all sensors.
7. Check all fuses.

TACHOMETER ADAPTER

1. An adapter is necessary when attaching a conventional tachometer into the Hall Effect ignition system. The Black tachometer lead is attached to an engine ground. The Red tachometer lead is connected to the adapter.
2. The adapter is made by first soldering 2 wires together at one end. One wire, leading to the ignition coil negative terminal, must be equipped with a 1000-ohm, 1-watt resistor. The second wire, leading to engine ground, must be equipped with a 12,000-ohm, 1-watt resistor. Both wires should be soldered to the attaching wires.

DISTRIBUTOR VOLTAGE SUPPLY

Testing

1. Remove the 3-wire connector from the distributor by depressing the retaining clip. Connect a voltmeter positive lead to the Red/Black wire of the harness connector and the negative meter lead to the Brown/White wire (outside terminals of the harness connector). Turn ignition switch to the **ON** position. Voltage reading should be at least 5 volts.
2. If voltage checks okay, proceed to the Hall Effect Sending Unit Check. If there is no voltage, check wiring harness from the distributor to Hall sending unit. Then proceed to Test No. 2, Ignition Control Unit Voltage Check.

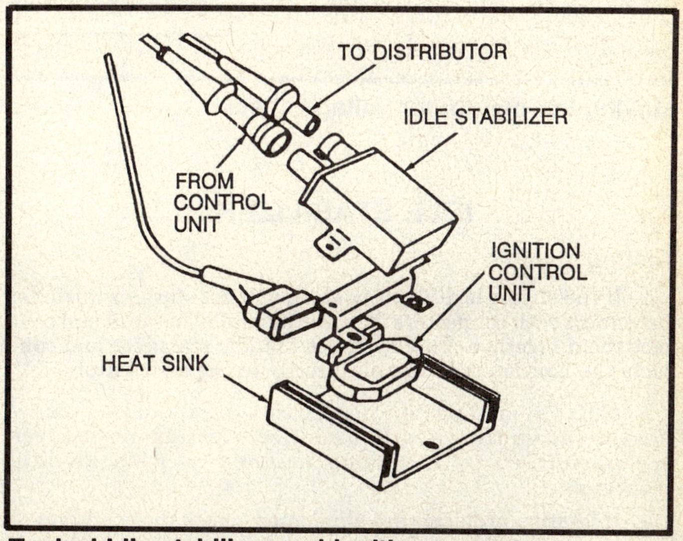

Typical idle stabilizer and ignition control unit

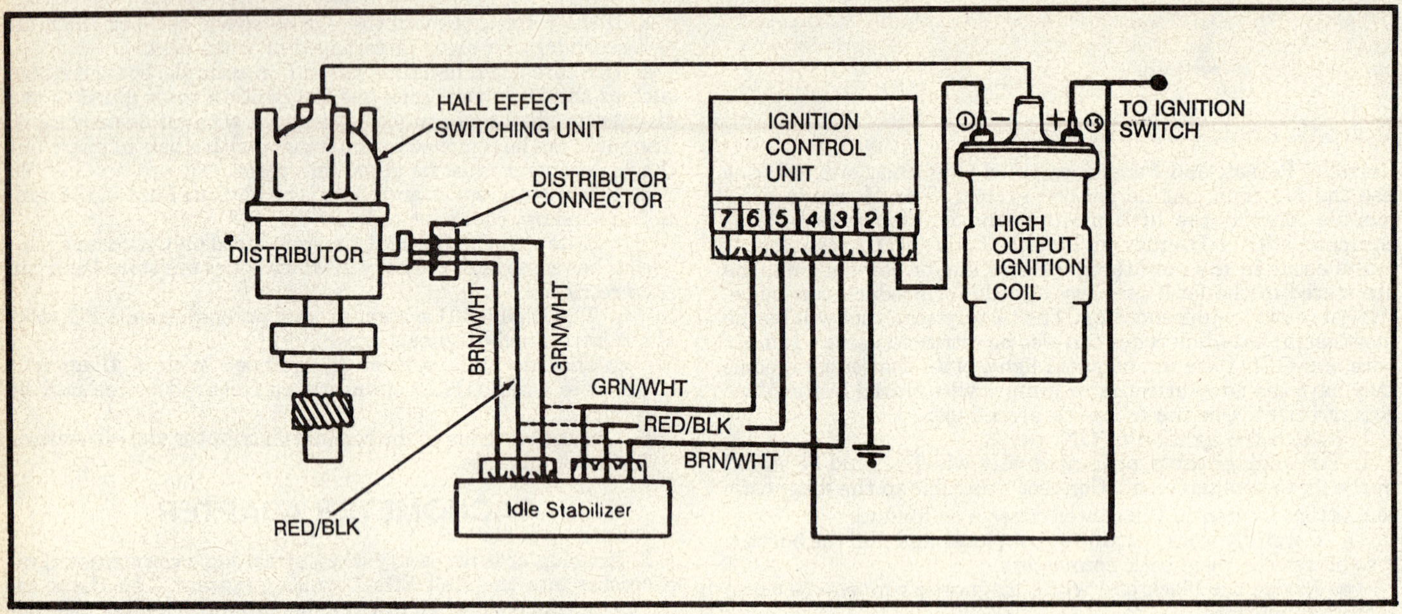

Typical Bosch Hall Effect ignition circuit diagram

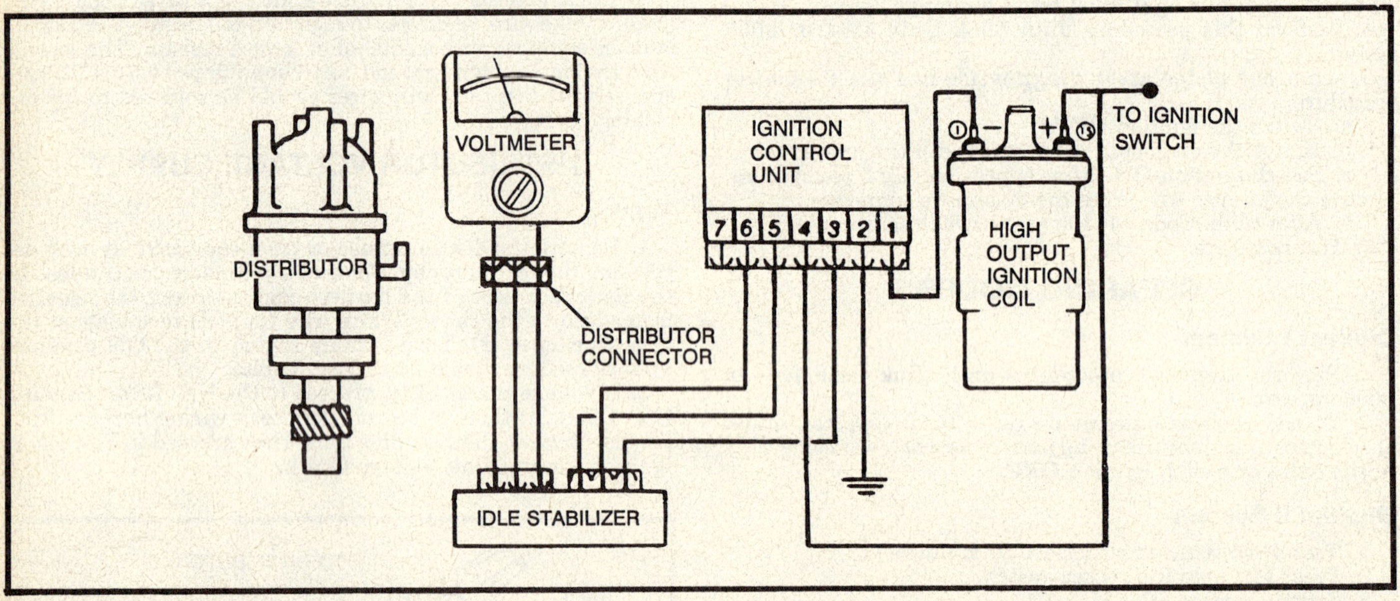

Making the distributor voltage check

IDLE STABILIZER

Testing

1. If the engine is difficult to start, will not start, or misfires, disconnect both connectors from the idle stabilizer unit and connect them together. This bypasses the idle stabilizer and connects the ignition control unit directly to the distributor.

NOTE: On models equipped with an impedance transformer rather than an idle stabilizer, testing procedures are performed in the same manner as with an idle stabilizer.

2. If ignition problems are eliminated, check the condition of the connectors and pins in the idle stabilizer unit, and repair if necessary. Reconnect idle stabilizer.

3. If the system still does not operate properly, replace the idle stabilizer. If the idle stabilizer cannot be pinpointed as the definite cause of the ignition problem, proceed to the next Step.

4. Warm the engine to normal operating temperature and connect a tachometer and a timing light to the engine. Note ignition timing and idle speed. Idle speed must be below 1000 rpm.

5. Turn **ON** all electrical accessories on the vehicle, including the air conditioning. Set the parking brake firmly. On vehicles with an automatic transmission, place the transmission selector in **D**, and slowly increase engine speed to place a load on the engine. On vehicles with a manual transmission, place the transmission in 4th gear, and slip the clutch carefully to place a load on the engine.

6. As engine speed drops, ignition timing should advance. If ignition timing does not advance replace the idle stabilizer unit.

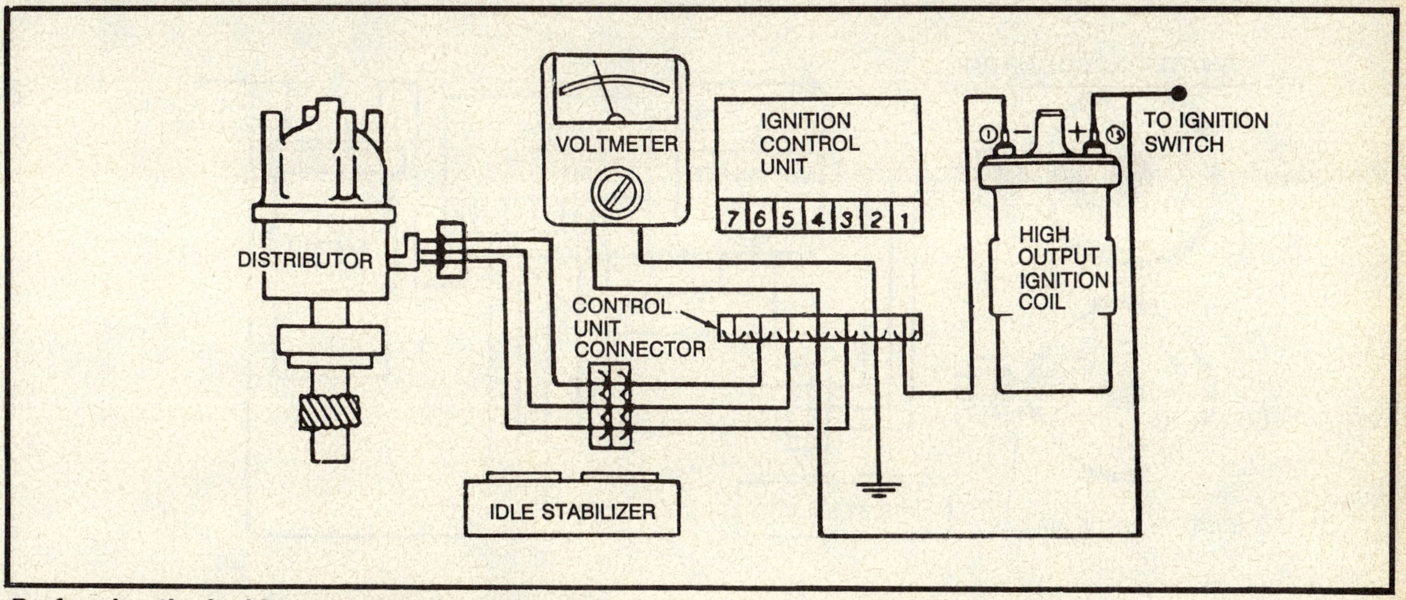

Performing the ignition control unit voltage test

IGNITION CONTROL UNIT

Testing

1. Disconnect both connectors from the idle stabilizer, and connect them together.

NOTE: Conduct ignition coil resistance and idle stabilizer checks before conducting this test.

2. Remove the connector from the ignition control unit. Set a voltmeter to the 16-volt scale. Connect the positive lead to terminal No. 4 and the negative lead to terminal No. 2 in the ignition control unit connector.
3. Turn the ignition switch to the **ON** position. Approximately 12 volts should be present. If not, check wiring from the control unit. Disconnect the Hall Effect connector at the distributor.
4. Turn the ignition **OFF**. Reconnect the connector to the ignition control unit. Disconnect the Hall Effect connector at the distributor.
5. Connect the positive lead of a voltmeter to the coil positive terminal No. 15 and the negative voltmeter lead to the coil negative terminal No. 1. Turn the ignition switch to the **ON** position. At least 2 volts should be present for 1–2 seconds, then voltage should drop to zero. If not, replace the ignition control unit and the ignition coil.
6. Connect a jumper wire to the center terminal in the distributor's Hall Effect harness connector. Momentarily touch

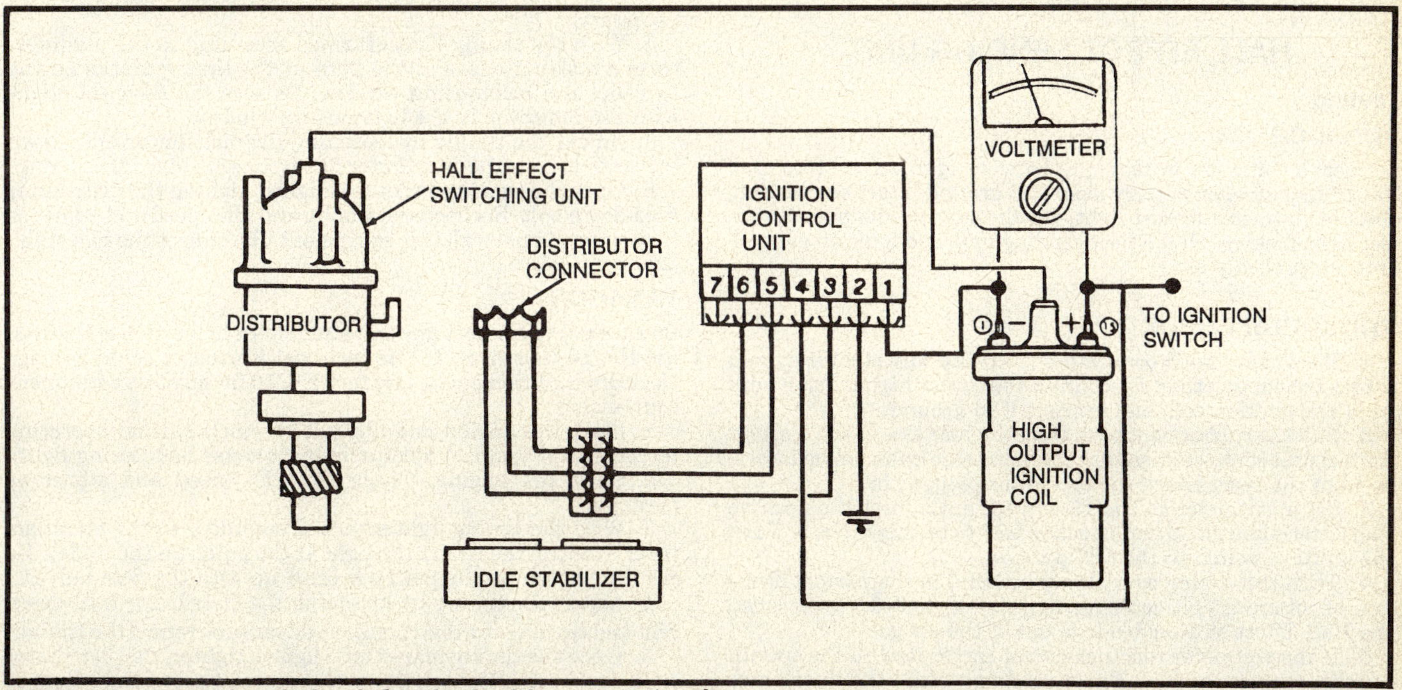

Making the operational check for the ignition control unit

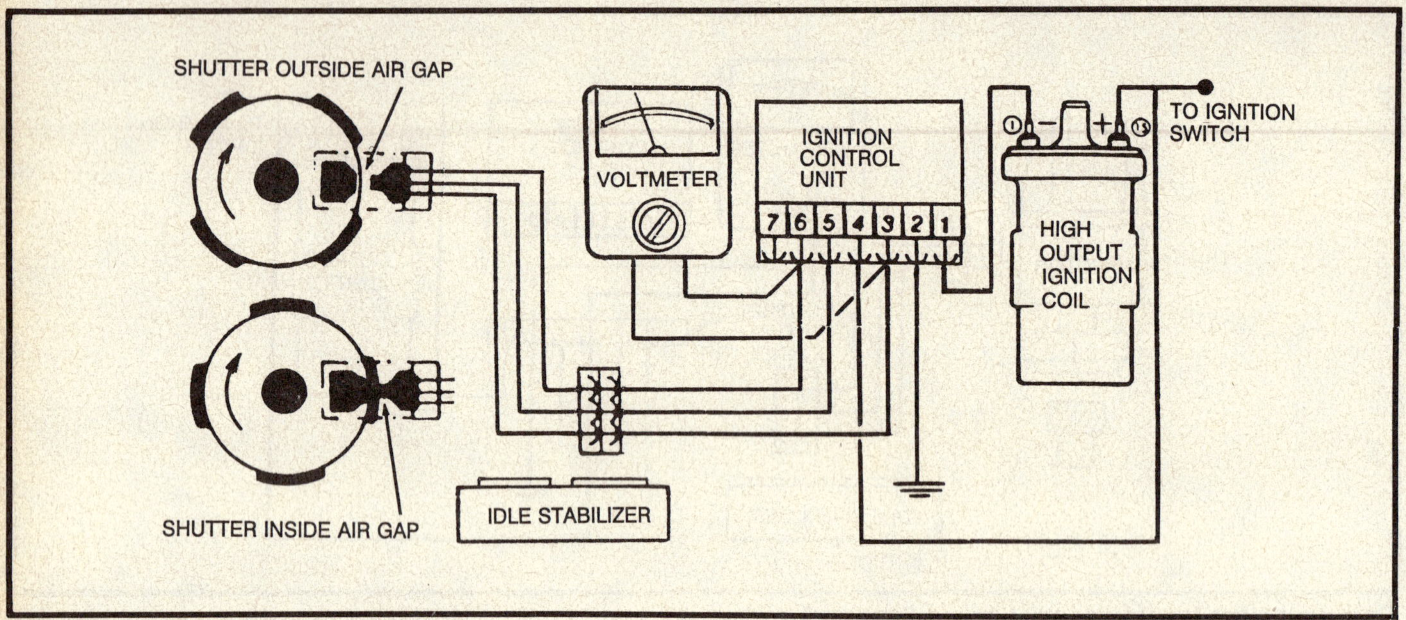

Checking the Hall Effect sending unit

the jumper wire to ground. Voltage should momentarily increase to 2–6 volts. If not, check for continuity between the center terminal of the Hall Effect connector and terminal No. 5 at the ignition control unit.

7. Turn the ignition **OFF**. Connect the voltmeter leads to the outer 2 terminals in the Hall Effect connector. Turn the ignition switch to the **ON** position. At least 5 volts should be present.

8. If voltage is under 5 volts, check wiring from Hall Effect switch connector and ignition control unit. If voltage in Step 7 was correct, but a defect still exists, replace the ignition control unit.

HALL EFFECT SENDING UNIT

Testing

QUICK CHECK

Connect a voltmeter or test light between ignition coil terminal No. 15 (positive, or battery side) and ground. Start engine and observe voltmeter or test light. If voltmeter needle pulsates or test light flickers, Hall Effect sending unit and ignition control unit are probably okay.

THOROUGH CHECK

1. Disconnect both connectors from the idle stabilizer, and connect them together. Disconnect the center high tension wire from the ignition coil, and connect it to ground.
2. Pull the rubber boot back from the rear side of the ignition control unit connector, so that the voltmeter leads can be inserted from the rear. Leave the connector plugged in.
3. Set a voltmeter to the 16-volt scale. Connect the positive lead to terminal No. 6 and negative lead to terminal No. 3. Turn the ignition switch to the **ON** position.
4. Turn the engine over slowly by hand in the normal direction of rotation. Voltmeter reading should be 0–0.7 volts when the Hall Effect shutter blade is out of the air gap.
5. If the voltmeter readings do not correspond to the specifications, replace the Hall Effect sending unit in the distributor (or the distributor).

Component Replacement

IGNITION TIMING

Adjustment

1. Start the engine and allow it to reach normal operating temperature. Shut off all electrical accessories. Be sure the cooling fan is not running during this operation.
2. Stop the engine. On the 4 cylinder engines using a suitable clamp, shut off the supply hose to the idle bypass valve. On the 5 cylinder engines, disconnect the plugs on the idle stabilizer tachometer and timing light.
3. Start the engine. Check the idle speed and adjust as necessary.
4. With the timing light attached according to the manufacturer's instructions, aim the light at the timing marks on the flywheel and bellhousing window. Be sure the flywheel aligns with the indicator in the bellhousing window.
5. Adjust the timing by loosening the distributor hold-down bolt and turning the distributor to advance or retard the timing.
6. Once the timing marks are aligned, tighten the distributor hold-down bolt. Recheck the timing and idle speed and adjust as necessary. Remove all test equipment. Reconnect the idle stabilizer connectors.

VANAGON

Be sure not to connect any test equipment to ignition coil terminal No. 15 (positive). If this terminal is used, it could damage electronic ignition parts. Use fuse No. 10 for any test equipment connections.

1. Start the engine and allow it to reach normal operating temperature. Connect a suitable tachometer and timing light.
2. Start the engine. Check the idle speed and adjust as necessary.
3. With the timing light attached according to the manufacturer's instructions, aim the light at the notch on the pulley. Be sure the notch on the pulley is lined up with the case halves.
4. Adjust the timing by loosening the distributor hold-down bolt and turning the distributor to advance or retard the timing.
5. Once the timing marks are aligned, tighten the distributor hold-down bolt. Recheck the timing and idle speed and adjust as necessary. Remove all test equipment.

VOLVO EZK AND REX 1 IGNITION SYSTEMS

General Information

Volvo uses several different ignition systems. Bosch LH series engine management systems are used on most models. The designation for these ignition systems is: EZ115K and EZ116K. In addition, a few models use the Bendix Regina engine management system and this ignition system is designated REX-1. In 1992, the 960 model was equipped with the Bosch Motronic 1.8 engine management and ignition system.

The EZ115K, EZ116K and REX 1 ignition systems, used on 1990–92 240, 700 and 940 models are like most engine management systems currently in use. The ignition and fuel control functions are closely integrated. Sensors feed information to an on-board computer which makes necessary adjustments. Some self-diagnostic capabilities are built into the computer to help troubleshooting.

SYSTEM OPERATION

Impulse Sensor

The Impulse Sensor, sometimes called an RPM and Crankshaft Position Sensor, is used to determine engine speed and Top Dead Center (TDC). This ensures precise ignition timing. The sensor is located at the rear of the engine block, above the flywheel. Engine speed is transmitted to the fuel control unit. The engine will not start without this signal.

EZ116K

On the flywheel, there are 60 markings for the impulse sensor, 58 which are drilled holes to provide information to the ignition control unit. There are no holes at 2 of the markings. These are 90 degrees Before Top Dead Center (BTDC), for cylinders No. 1 and No. 4. Ignition timing is based on these markings and other information such as engine load and temperature. This means that ignition timing can be controlled and that there is no need for ignition setting to be adjusted.

REX 1

On the flywheel, there are 44 markings for the impulse sensor, 40 which are drilled holes to provide information to the ignition control unit. There are no holes at 2 of the markings. These are 90 degrees Before Top Dead Center (BTDC), for cylinders No. 1 and No. 4. Ignition timing is based on these markings and other information such as engine load and temperature. This means that ignition timing can be controlled and that there is no need for ignition setting to be adjusted.

Knock Sensor

EZ116K

If knocking occurs, the knock sensor corrects ignition timing for each individual cylinder by gradually retarding the ignition for each cylinder that is knocking. If retarding is insufficient, a signal goes to the fuel system control unit, which enriches the air/fuel mixture. This lowers the temperature and stops the knocking.

REX 1

If knocking occurs, the knock sensor corrects ignition timing for each individual cylinder by gradually retarding the ignition for each cylinder that is knocking. The REX 1 incorporates adaptive knock control. Which measures the background noise (due to bearing clearance etc.) in the engine just before firing and adjusts the reference level for knock sensing.

Power Stage

EZ116K

The power stage, also known as an ignition amplifier, receives firing signals from the control unit and acts as an electronic switch. It keeps current through the ignition coil constant regardless of battery voltage and engine speed, so that ignition spark is effective. A current cut-off circuit prevents the ignition

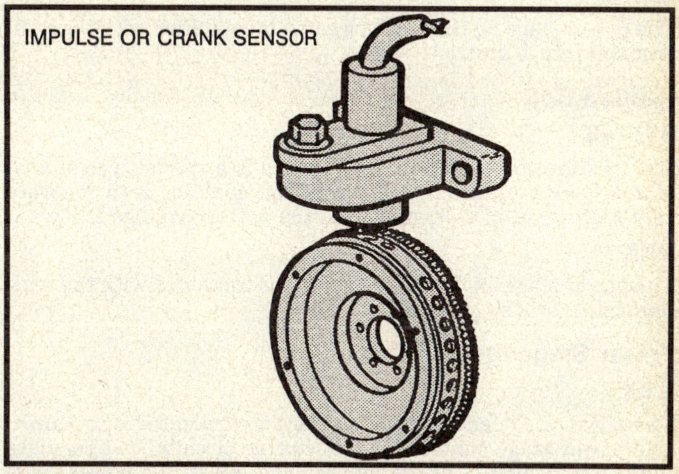

The impulse sender, or rpm/crank sensor, is mounted on the rear of the engine and reads off the flywheel. If there is no signal from this sensor, the engine will not start

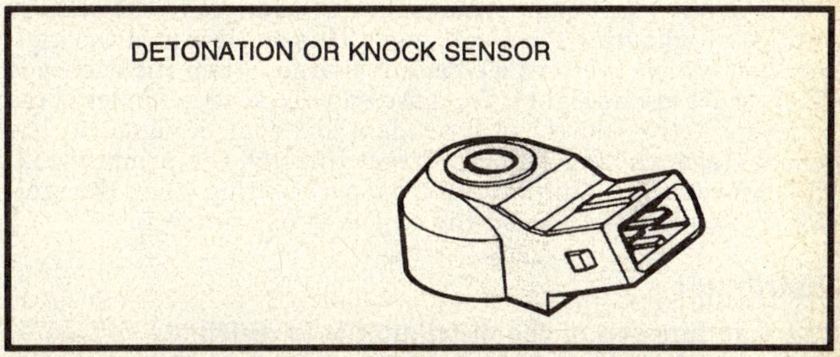

Knock sensors detect detonation and signal the ECU which retards the spark and even richens the fuel mixture, if necessary, to stop engine knock

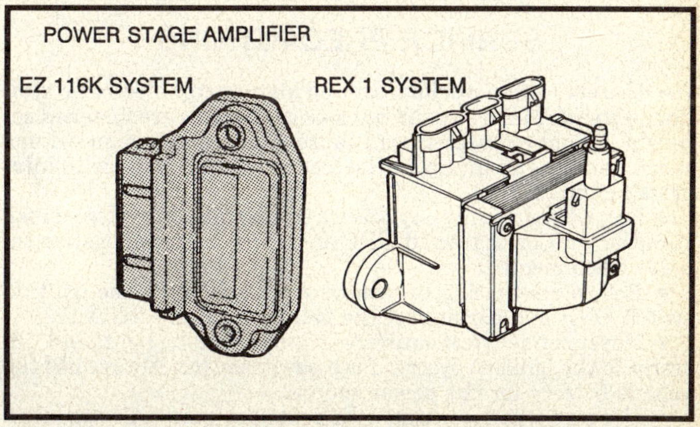

The power stage, or ignition amplifier, differs between the EZ-K and REX1 ignition systems. On REX1 systems, the power stage is built into the coil

coil from overheating if the ignition switch is left in the **ON** with the engine not running.

REX 1

The power stage on the REX 1 system is combined with the ignition coil into 1 unit.

Ignition Coil

EZ116K

The ignition coil used with this system is a special type of coil, with a low resistance primary winding, enabling it to generate very high ignition voltage even if the battery voltage is low.

REX 1

The power stage on the REX 1 system is combined with the ignition coil into a single unit.

Power Stage/Ignition Coil

REX 1

The power stage and ignition coil are incorporated into 1 unit. The power stage contains the electronic circuits that regulate the primary current of the ignition coil. When the power stage receives voltage (high control signal) from the control unit, it sends current through the primary windings of the ignition coil. When the control unit reduces this current (low control signal), the power stage cuts off the primary current of the ignition coil, creating a high voltage in the secondary windings.

The control unit compensates for any changes in the battery voltage by adjusting the dwell angle. This ensures that the ignition coil always receives a constant charge. When the engine is **OFF**, there are no signals from the impulse sensor. Under these conditions, the control unit sends a low control signal to the power stage, cutting off the current through the primary coil. This prevents the ignition coil from overheating when the ignition switch is left **ON** and the engine is not running.

Distributor

The only function of the distributor is to distribute voltage to the spark plugs. There are no advance functions built into the distributor. It is no longer possible to adjust ignition timing through the distributor.

Diagnosis and Testing

SERVICE PRECAUTIONS

- Use care when working around high-energy electronic ignition systems. Make sure all harness connectors are fastened securely. A poor connection can cause an extremely high voltage surge resulting in injury to the technician and damage to integrated circuits.
- Before performing any service on the ignition system, make sure the ignition is turned **OFF** and/or the battery negative cable is disconnected.
- Before attempting to remove any parts, turn the ignition switch **OFF** and disconnect the battery ground cable.
- Never connect test equipment such as timing lights, etc., directly to the ignition system for a power source. Always use the vehicle battery for the power source.
- Use care when working with light alloy parts. Always lubricate the spark plug threads with a suitable lubricant to avoid seizing the plug threads to the cylinder head.
- Keep all ECU parts and harnesses dry during service. Protect the ECU and all solid-state components from rough handling or extremes of temperature.
- Always use a 12 volt battery as a power source.
- Do not disconnect the battery cables with the engine running. Never run the engine with the battery cables loose or disconnected.
- Do not attempt to disassemble the ECU unit under any circumstances.
- When connecting or disconnecting pin connectors from the ECU, take care not to bend or break any pin terminals. Check that there are no bends or breaks on the ECU pin terminals before attempting any connections.
- When measuring supply voltage of ECU-controlled components, keep the tester probes separated from each other and from accidental grounding. If the tester probes accidentally make contact with each other during measurement, a short circuit will damage the ECU.
- It is recommended that the ECU control unit be removed if the vehicle is to be exposed to high heat such as a bake oven to cure paint after body refinishing or when welding is be done on the vehicle.
- Use care when checking for spark to avoid a spark jumping an a component which could be easily damaged. Use an approved spark tester. High-voltage sparks can arc a considerable distance. High voltage arcing through an electronic fuel injector can "spike" and destroy an ECU.
- Do not be hasty in condemning the ECU. This system uses voltages and resistances that are very small. Examine the sensors, wiring and connectors carefully. The sensors operate in more harsh conditions than the ECU which is generally in a more protected location.
- Check all ground connections before condemning the ECU.
- Use care when working around vehicles equipped with Supplementary Restraint System (SRS), often known as "air bags." Volvos equipped with SRS are generally recognized by the letters **SRS** moulded into the steering wheel cover. Follow all precautions to avoid personal injury.
- A spare location in the fuse panel is used as a test terminal for SRS diagnostics. Never install a fuse in this position or connect accessories to this terminal.

SYSTEM DIAGNOSTICS

The EZ115K and EZ116K (called EZ–K) and REX 1 ignition systems, used on 1990–92 240, 700 and 940 models incorporate a diagnostic feature which is built into the system control unit. The system is capable of diagnosing a certain number of faults, but not every possible fault. If power supply to the control unit is lost, control unit connector disconnected or the battery disconnected, all fault codes will be erased.

The system consists of 3 test functions:

1. Test Function 1 (fault code readout): This test gives access to any fault codes which have been stored when the vehicle was being driven. The system is capable of storing up to 3 fault codes.
2. Test Function 2 (input signal check): This test allows a specific component to be checked to determine if the component is operating properly and if it is correctly connected. This test can be used after a repair has been made, to ensure that the component has been correctly installed and is working properly.
3. Test Function 3 (output signal check): This test activates certain control components.

The diagnostic system is controlled by the diagnostic socket, which is located in the engine compartment, on the left shock tower. The diagnostic socket is common to the diagnostic systems of the fuel system and ignition system. Socket No. 6 on the diagnostic socket is used for the diagnosis of faults in the ignition system. The diagnostic socket incorporates a diode light and a button. Test functions are activated by pressing the button 1, 2 or 3 times. Fault codes are determined by the flashing of the diode light, in 3 digit numbers. The fault code digits can range between 1 and 9. The number of flashes indicates the value of the digit, which are separated by 3 second intervals.

When all the fault codes have been read and necessary repairs made, the fault codes must be erased. The fault codes must always be read with the ignition switch in the **ON** position, if not, the codes cannot be erased.

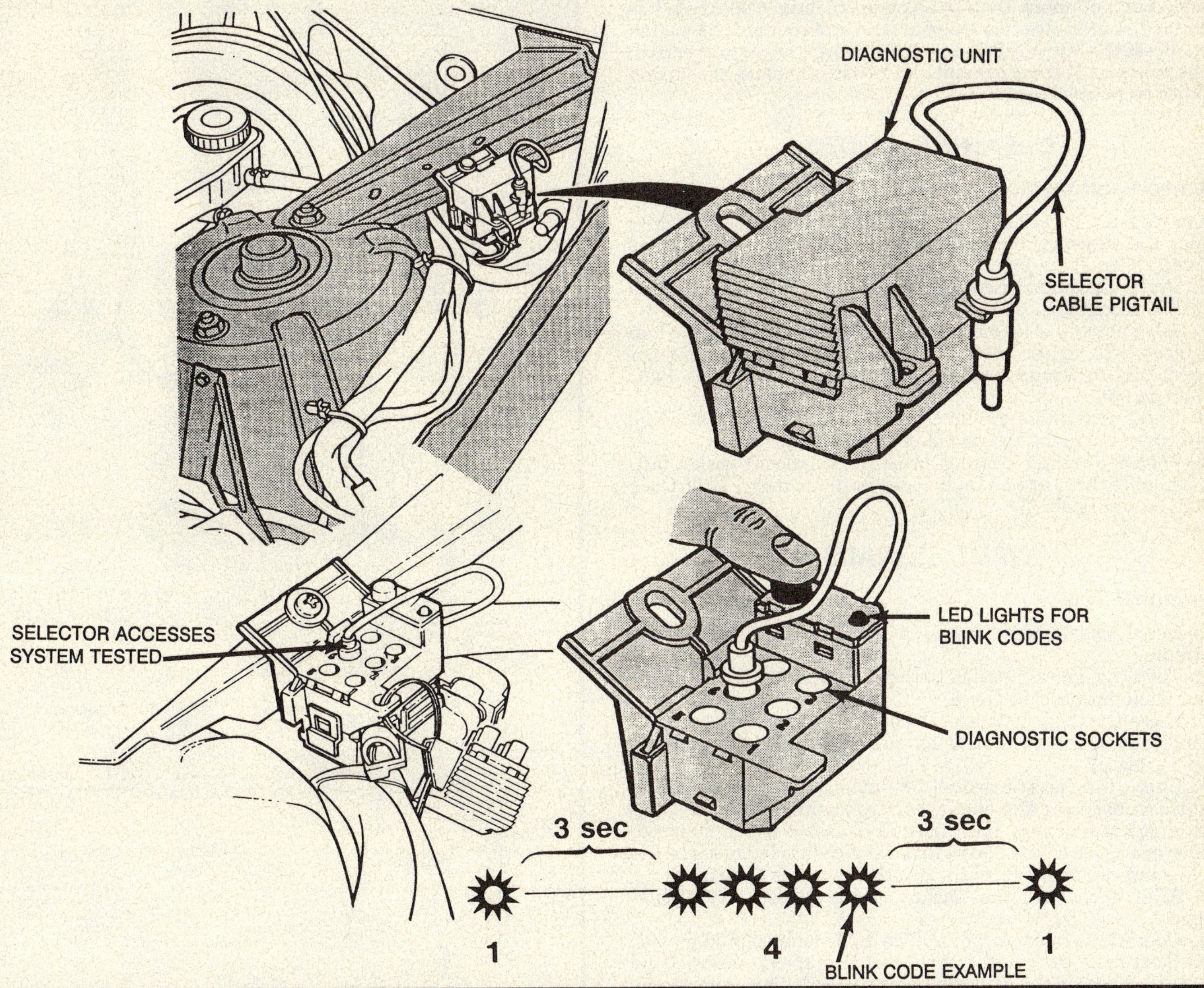

Volvos use a pushbutton diagnostic system. Plugging the selector cable pigtail into different sockets accesses various parts of the system and blink codes aid diagnostics

The EZ115K, EZ116K and REX 1 ignition systems incorporate active anti-knock control and ignition setting control. These 2 controls never need checking or adjustment.

READING CODES

Before any fault codes can be read from the diagnostic socket, the following conditions must be met:

1. The ignition switch must be in the **ON** position.
2. The selector cable must be installed into socket No. 6
3. The diagnostic socket button must be pressed 1 time, for at least 1 second, but not more than 3 seconds.

Every flash and pause between flashes is about half a second. The pauses between flashes is approximately 2.5–3 seconds. The ignition system test function is obtained by pressing the diagnostic button 2 or 3 times, for at least 1 second, but not more than 3 seconds.

The diagnostic unit used on the EZ116K and REX 1 ignition systems has 7 different fault codes. Its memory is capable of storing up to 3 different ignition system faults.

Engine Codes

Code 1-1-1–No fault codes in memory.
Code 1-4-2–Fault in control unit. Engine runs with safety-retarded ignition timing (approximately 10 degrees).
Code 1-4-3–Knock sensor faulty. Engine runs with safety-retarded ignition timing (approximately 10 degrees).
Code 1-4-4–Load signal missing (from fuel system control unit). Control unit selects full-load ignition.
Code 2-1-4–Engine speed sensor faulty.
Code 2-2-4–Coolant temperature sensor inoperative.
Code 2-3-4–Throttle switch for idling faulty. Engine runs with safety-retarded ignition timing (does not apply to REX 1 ignition system).
Code 3-3-4–Throttle switch in idle position–okay (REX 1 ignition only).

Before a fault code can be read, the ignition switch must be in the **ON** position and selector cable installed into socket No. 6. Press the diagnostic socket button 1 time, for at least 1 second, but not more than 3 seconds. If fault Code 1-1-1 is flashed, it indicates that there are no fault codes in the memory. Press the button again to see if there are any more codes in the memory. If there are and a new code is flashed, the button must be pressed again.

CLEARING CODES

Erasing Fault Memory

When all fault codes have been read and the necessary repairs made, the memory must be erased. Erase the memory as follows:

1. Press the diagnostic socket button and hold for approximately 5 seconds.
2. After about 3–4 seconds, the diode light should light. When the diode light lights, press the button again and hold for approximately 5 seconds. Release the button. The diode light should go out.
3. Check that there are no codes in the memory by starting the engine and bring to normal operating temperature.
4. With the engine running, press the diagnostic socket button again. If there are no fault codes in its memory, fault Code 1-1-1 should flash.

INPUT SIGNALS

Functional Test

This test checks the throttle switch and the engine speed sensor as follows:

1. Turn the ignition switch to the **ON** position and install selector cable into socket No. 6.
2. Press the diagnostic socket button 2 times, for at least 1 second, but not more than 3 seconds. The diode light should start to flash.
3. Check the throttle switch by turning the throttle switch slightly from inside the engine compartment. The diode lamp should go out and then flash fault Code 3-3-4 which indicates proper operation of the idle switch. If no code is flashed and the diode lamp continues to flash, throttle switch is faulty.
4. After each test, the flashes of the diode lamp should return.
5. Operate the starter motor. The diode lamp should go out, then flash fault Code 1-4-1 for the engine speed sensor. If no code is flashed and the diode lamp continues to flash, the engine speed sensor is faulty.

If no fault codes are obtained and the diode lamp continues to flash, the switch or the wire are faulty since the control unit did not receive an activation signal. For this test, it does not matter whether a switch contact opens or closes, as long as it is activated.

OUTPUT SIGNALS

Regulation Test

This test applies to the EZ116K ignition system on California models equipped with B230F engines. Press the diagnostic socket button 1 time, for at least 1 second, but not more than 3 seconds. The EGR converter should operate for approximately 10 seconds. If not, the EGR converter is faulty.

NO START—NO SPARK

Testing

1. Attempt to start engine.
2. Check fuse No. 1 (model 740) or fuse No. 31 (model 760) in the fuse box. Note that these fuses must be removed whenever the ECU connector is removed or installed.
3. If fuse No. 1 (or No. 31) is okay, turn ignition switch to **ON**. Open diagnostic socket cover and install selector cable into socket No. 6. Press the diagnostic socket button 2 times, for at least 1 second, but not more than 3 seconds. The LED light

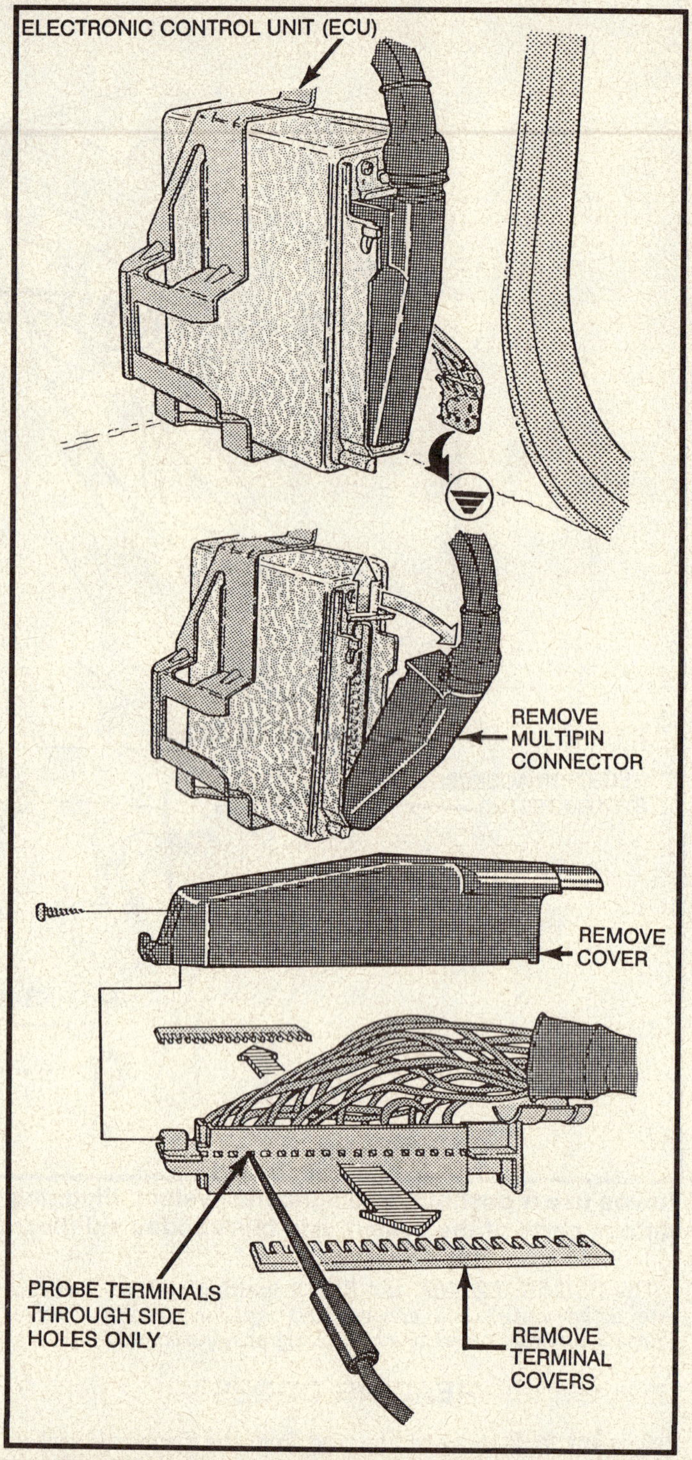

The Electronic Control Unit (ECU) may be accessed for troubleshooting. Use care to probe cable connections from the side openings only

should begin to flash rapidly. This command has the computer start a check of the basic system components. If the LED flashes rapidly, go on to Step 4.

a. If the LED does not blink rapidly, there is a connection problem at the multipin connector at the ECU, which requires separate troubleshooting.

b. To troubleshoot suspected ECU connector problems, turn the ignition **OFF** and locate the ECU so the the ECU multipin connector can be accessed for inspection.

c. On 700 models, if the diode light does not flash, remove panel from under left side of instrument panel, then disconnect connector from control unit. On 240 models, remove panels under instrument panel right side and firewall right side, then disconnect connector from control unit.

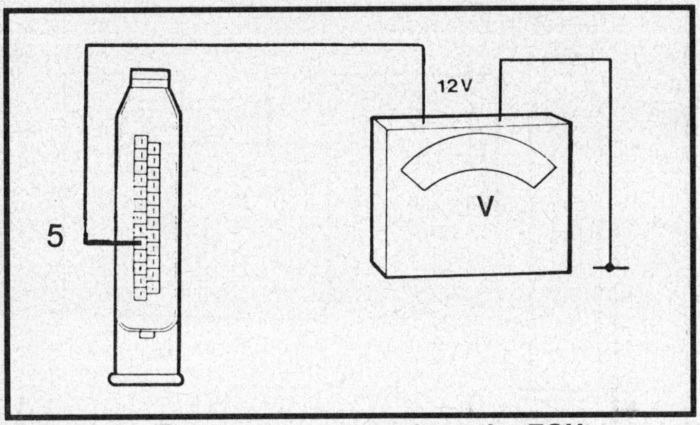

Testing for 12 volts power supply to the ECU

d. Remove protective cover from control unit connector.

e. Check to ensure that no terminal contacts have been pushed down in the connector.

f. Connect a voltmeter between ground and terminal 5 on the control unit. Voltmeter should read approximately 12 volts. If no reading is obtained, check wire from control unit connector to terminal 30 of fuse box.

g. Install selector cable into the underhood diagnostic box socket No. 6. Connect a voltmeter between ground terminal 1 on the control unit. Voltmeter should read approximately 12

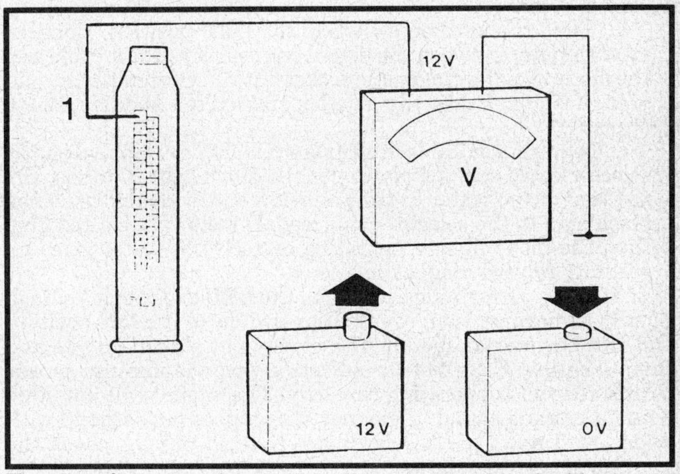

Testing for 12 volts to the diagnostic circuit. Note pushbutton positions

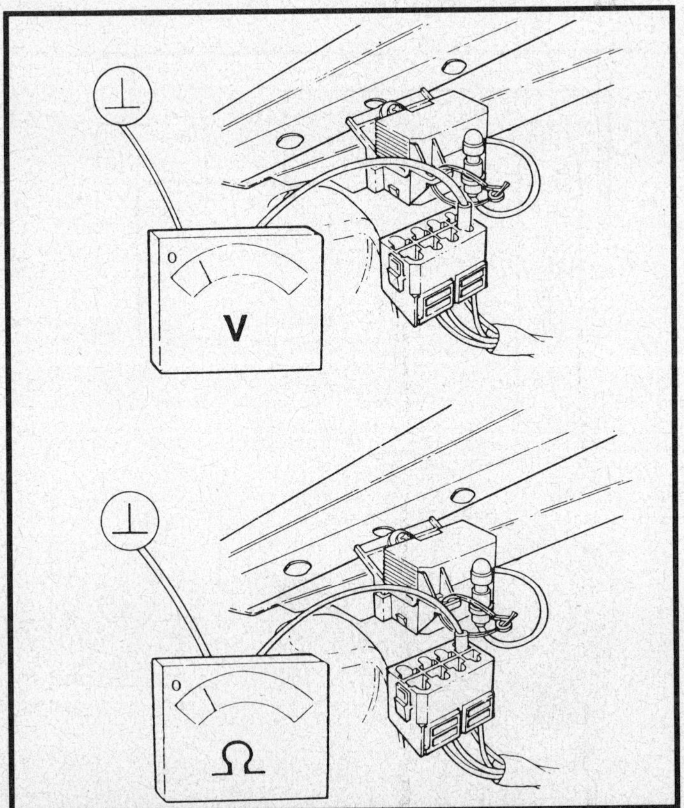

Top: Testing diagnostic connector by checking for 12 volts at blue wire. Bottom: Checking for ground at black wire

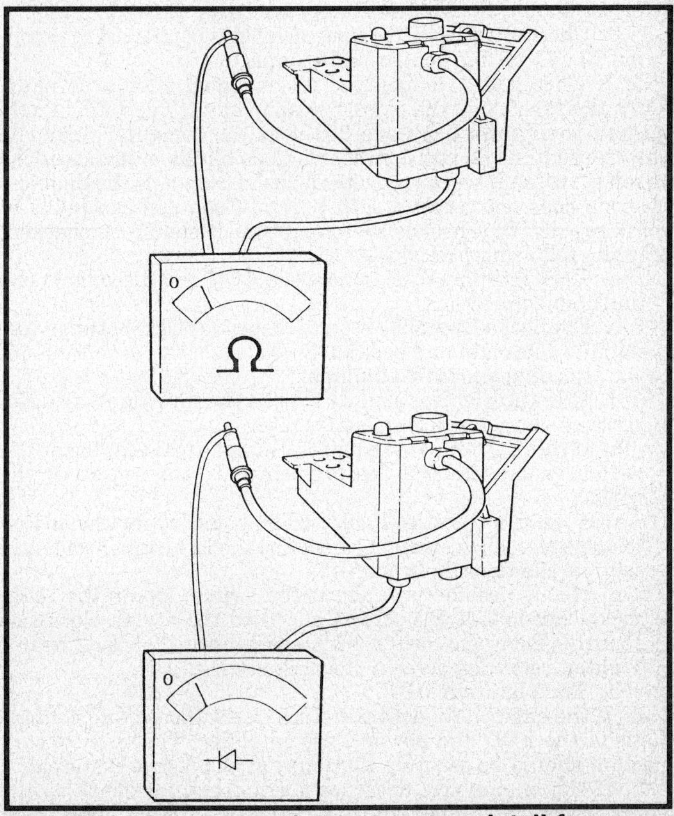

Top: Checking test socket connector pigtail for continuity. Bottom: Checking diagnostic blink code LED light with a diode tester

volts. Press the diagnostic button. Voltmeter should read 0 volts.

h. If there is no voltage at the control unit, check diagnostic socket connector. If voltmeter reads 12 volts when button is pressed, check the test connector.

i. Test the diagnostic socket connector by connecting a voltmeter between ground and the blue wire on the diagnostic socket connector. Voltmeter should read approximately 12 volts.

j. Connect an ohmmeter between ground and the black wire on the diagnostic socket connector. Ohmmeter should read 0 ohms.

k. Turn the ignition switch to the **OFF** position. Connect an ohmmeter between the diagnostic socket selector cable and the pin under the diagnostic socket button. Ohmmeter should read infinity. Press the button, ohmmeter should read 0 ohms.

l. Connect a suitable Multimeter/Diode Tester between the selector cable and the pin under the diode light. Connect the red lead of the tester to the pin below the diode light and the black lead to the selector cable lead. If tester gives a reading, then the diode light is operating correctly. If tester gives no reading, replace diagnostic socket.

4. If, after performing Step 2 and the LED does begin to flash rapidly as normal, turn the ignition switch to the **ON** position and run the starter motor. The diode light should go out and then flash test Code 1–4–1, indicating that the impulse sensor (crank sensor) is operating correctly. The engine will not start without a crank signal. If diagnostic unit does not respond with Code 1–4–1 and the LED continues to flash rapidly, check the impulse sensor as follows:

a. Turn ignition switch **OFF**.

b. Connect an ohmmeter between terminal 10 (red) and terminal 23 (blue) on the ECU control unit connector. Ohmmeter should read between approximately 215–265 ohms.

b. Check that electrical screen/shield is connected to terminal No. 11 on the control unit connector.

5. If Code 1–4–1 blinks, the crank signal should be okay. Turn the throttle in the engine compartment. The LED should go out, then blink out Code 3–3–4 as the computer tests the throttle switch idle position. If this Code blinks, it indicates the throttle switch is working properly; go to Step 5. If the diagnostic code does not respond with normal Code and continues to blink rapidly, the throttle switch resistance should be checked. Use the following procedure:

a. Turn ignition **OFF**, access the ECU and disconnect the multipin connector.

b. Connect a Digital Volt/Ohm Meter (DVOM) between terminal 7 (Orange) and ground.

c. Reading should be 0 ohms.

d. Press accelerator slightly to open the throttle switch. Resistance should now become infinite.

e. If there appears to be a problem, take this reading at the switch to determine if the problem is in the switch or the wiring.

f. Open the throttle slightly while listening for the switch to operate. A click should be heard as the idle contacts open as soon as the throttle is opened.

g. If it is necessary to adjust the switch, loosen the 3MM socket head retaining screws and turn the switch clockwise slightly. Turn the switch back again until a click is heard. Tighten retaining screws. Recheck setting.

h. Turn ignition **OFF**.

6. If the engine still does not start, check the ground connections of the ECU and power stage amplifier. This ground connection should be securely tightened at the intake manifold.

7. If the engine still does not start, check for spark at the spark plugs. Disconnect a spark plug cable from 1 spark plug and connect a spare plug to the spark plug wire. Ground the spark plug and operate the starter motor. If a strong blue/white spark is present, the problem is in the engine or the fuel system.

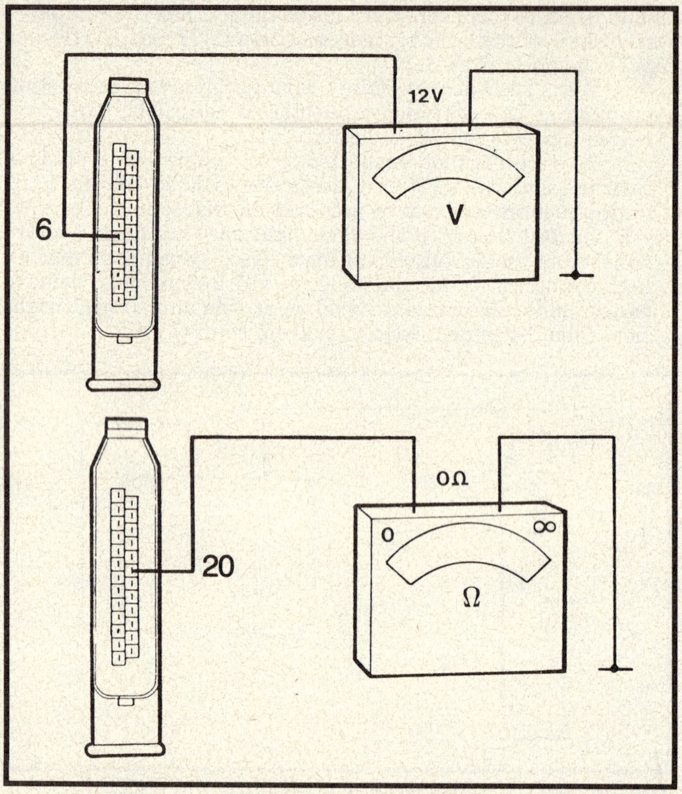

Top: Checking voltage to ECU, ignition ON. Bottom: Checking ECU ground connection

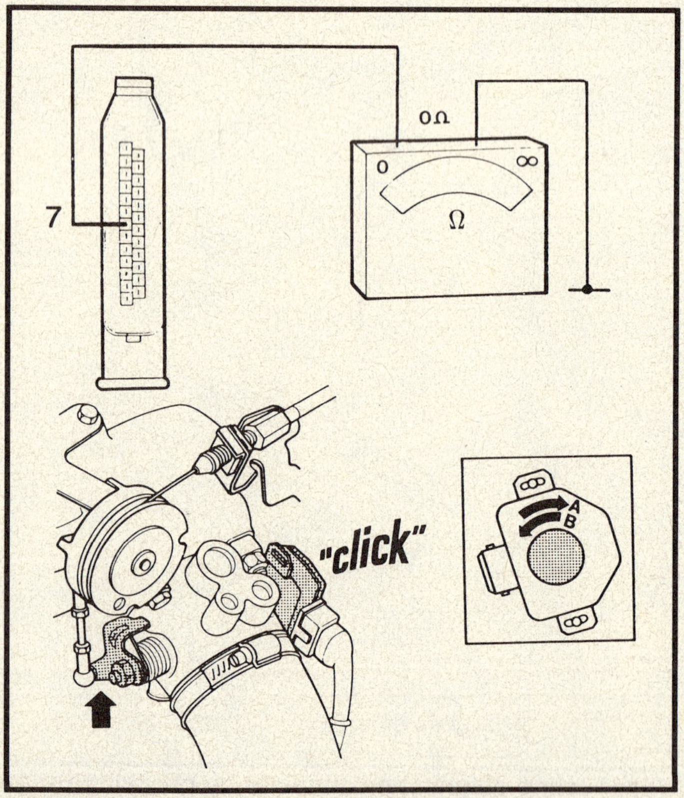

Testing throttle switch at ECU connector and at throttle body

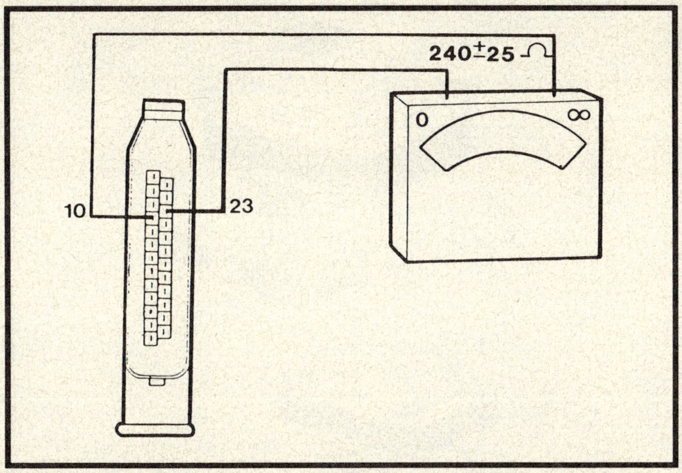

Testing crank sensor resistance at ECU connection

NOTE: Use care when checking for spark. Keep the test spark plug away from the fuel injectors. If a spark happens to flash over to a fuel injector, the voltage spike could feed back and destroy the fuel control unit.

8. If there is a weak spark or no spark, connect a spark plug onto the ignition coil terminal. Ground spark plug and operate starter motor. If a strong blue/white spark is present, check the rotor, distributor housing and spark plug cables and replace if necessary.
9. If there is still a weak spark or no spark, fault is in the ignition coil or low tension section of the ignition system.

If the engine still does not start and tests indicate no spark or a weak spark, it indicates a problem in the coil, the coil control or ignition primary. It could also be a basic problem in the diagnostic system itself which is giving a false problem indication, which should be checked.

Continue troubleshooting by performing checks on the coil and the ignition primary system. Start by checking ECU power supply to see if power is getting to the ECU.

ECU POWER SUPPLY

Testing

Use care when testing around the ECU multipin connector. Never test the terminals from the front. This could result in damage to the terminals and make any faults worse. Note that the multipin connector has a shaped protective cover over the connector end. Remove the protective cover when testing. Terminals should be tested through holes in the side connector. Do not use excessive force. These terminals are very delicate. Terminals are numbered on the side of the connector.

1. After removing the multipin connector, inspect it carefully. Make sure no terminal contacts have been pushed down into the connector. Poor contact may result if any of the contact sleeves have been pushed down.
2. Connect a voltmeter between terminal 5 of the control unit connector and ground. Voltmeter should read 12 volts.
3. If there is no reading, check wire from control unit connector to 30-supply of fusebox.

POWER STAGE (AMPLIFIER) CHECK

Testing

EXCEPT REX 1 SYSTEM

1. Make sure ignition switch is **OFF**. Remove fuse No. 1 (model 740/940) or fuse No. 31 (model 760) from the fusebox.

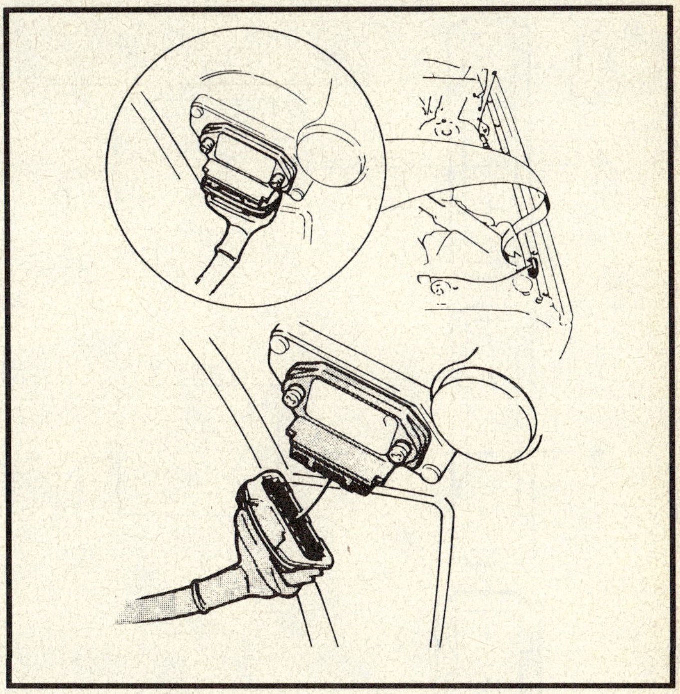

Accessing power stage amplifier connector

Note that these fuses must be removed whenever any connector to or from the ECU is removed or installed.

2. Remove the air filter to gain access to the power stage.
3. Remove the connector from the power stage.
4. Remove the rubber cover from the connector to expose the terminals. Never test the terminals from the front. This could result in damage to the terminals and make any faults worse.
5. Turn the ignition to **ON**.
6. Check the voltage to the power stage and ignition coil by connecting a voltmeter between ground and power stage connector terminal 4. Also test between ground and terminal 15 of the ignition coil. In both cases, voltage should be approximately 12 volts.
7. Crank the starter and check again. Voltage should still be at least 10.5 volts. If voltage is too low, check for a problem in the battery or charging system. If there is no voltage, check the blue wire from the central electrical unit to the ignition coil and power stage. Repair/replace as required.
8. Test that the power stage is receiving signals from the ECU. Connect a voltmeter between terminal 5 and ground. Note voltmeter reading while operating starter motor. Voltmeter should show a pulsating reading between 0–2 volts. If voltmeter showns correct reading, check with a new power stage.
9. To check the power stage ground connection, connect an ohmmeter between connector terminal 2 and ground. Resistance should be 0 ohm.
10. Check that the wire from the ECU to the power stage is properly screened. The screen (shield) should be connected to terminal 3.
11. After testing, turn ignition key **OFF**. Connect power stage connector, air filter, air mass meter and hoses. Check all clips and retainers. Erase fault codes.

REX 1 SYSTEM

1. To check the ignition coil/power stage unit on REX 1 systems, make sure ignition switch is **OFF**. Unplug connector from coil/power stage unit.
2. Remove the ignition coil from the power stage by removing the 2 Torx head screws and lifting off the ignition coil.
3. Measure the resistance between the terminals of the igni-

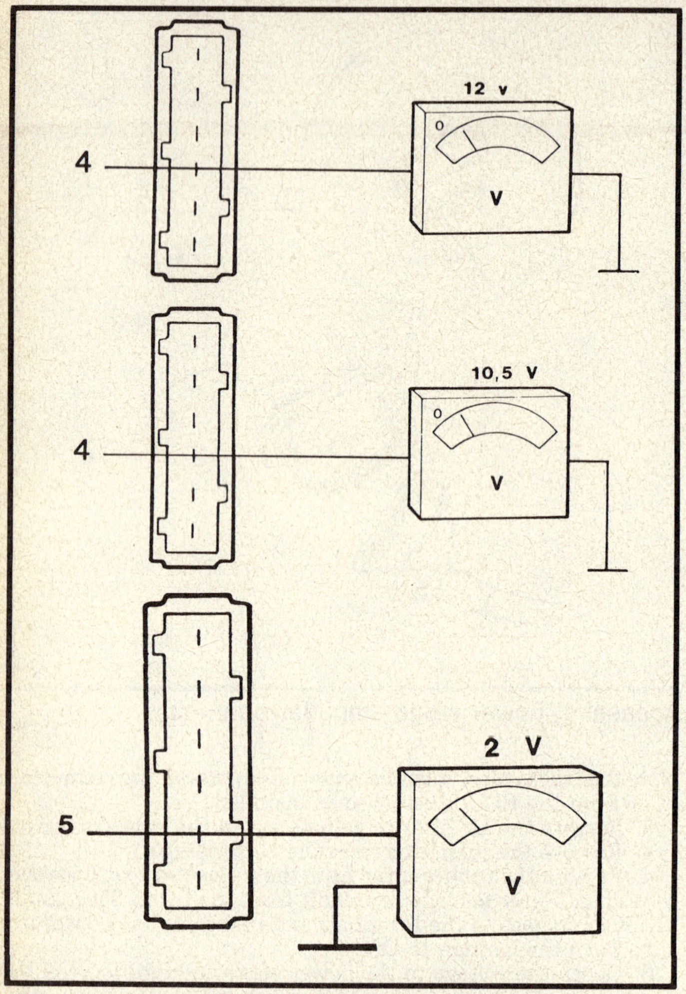

Various voltage checks taken at power stage amplifier connector

tion coil by connecting an ohmmeter between the low voltage terminals (+) and (−). Resistance should be 0.5 ohms.

4. Connect an ohmmeter between the high tension terminal and a low voltage terminal. Resistance should be approximately 5000 ohms.

5. Check that voltage is present at the ignition coil/power stage unit by first turning the ignition **ON**. Connect a voltmeter between ground and terminal A in the 3-way connector. There should be 12 volts. If the voltage is low, or none at all, check the point where the feed wire branches to the ECU and ignition coil/power stage unit.

6. Check that the voltage does not fall below 10.5 volts when the starter motor is operated.

7. Turn ignition key **OFF**.

8. Check the ground connections of the ignition coil/power stage unit by connecting an ohmmeter between ground and terminal B of the 3-way connector. The resistance should not exceed 0.1 ohm. If the resistance is too high, check for bad ground connection.

9. Connect an ohmmeter between ground and terminal A of the 2-way connector. If resistance is more than 0.1 ohm, check for a bad ground connection.

10. Check the signal line between the ignition coil/power stage unit and ECU. Check for power on this circuit. Connecting a shop-made buzzer between terminal B of the 2-way connector and terminal 16 of the ECU may be helpful. If the line is problem free, the buzzer should sound.

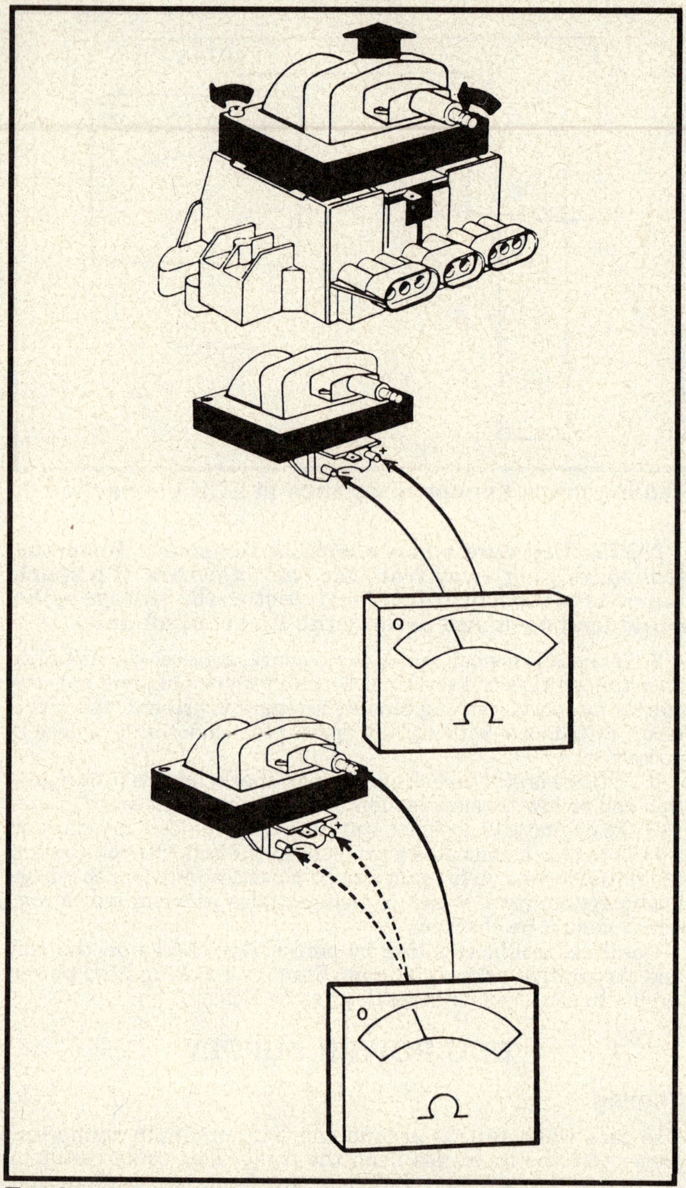

Testing combination power stage amplifier/coil assembly—REX1 ignition system

COIL CHECK

Testing

EXCEPT REX 1

Primary Winding Test

1. Remove fuse No. 1 (model 740/940) or fuse No. 31 (model 760) from the fusebox. Note that these fuses must be removed whenever any connector to or from the ECU is removed or installed.

2. Remove the air filter to gain access to the power stage.

3. Remove the connector from the power stage.

4. Remove the rubber cover from the connector to expose the terminals. Never test the terminals from the front. This could result in damage to the terminals and make any faults worse.

5. Connect an ohmmeter between the power stage amplifier connector terminal 1 and terminal 15 of the ignition coil. Resistance should be 0.6–1.0 ohms.

6. If resistance is low, replace ignition coil.

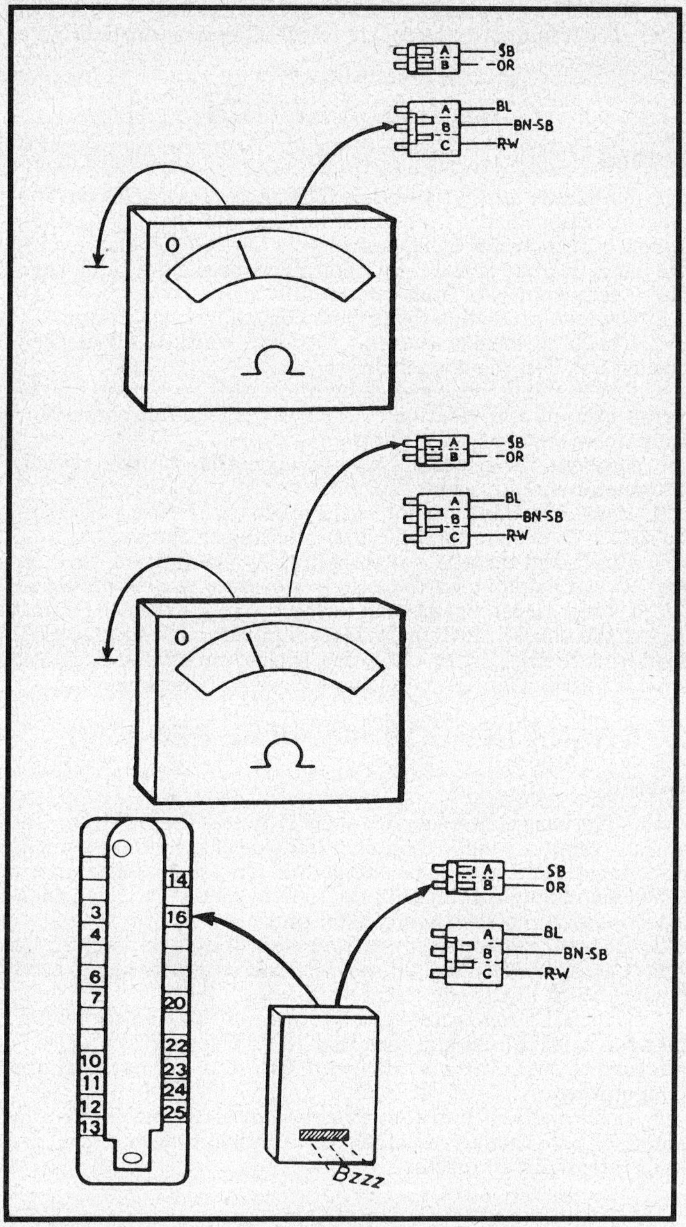

Testing power stage amplifier/coil assembly ground paths and continuity of signal line to ECU—REX1 ignition system

7. If resistance is too high, connect an ohmmeter directly to terminals 1 and 15 of the ignition coil. If resistance is still to high, replace the ignition coil.

8. If resistance is correct (0.6–1.0 ohm), check wire between ignition coil and power stage amplifier connector terminal 1. Replace/repair wire as needed.

Secondary Winding Test

1. Remove fuse No. 1 (model 740) or fuse No. 31 (model 760) from the fusebox. Note that these fuses must be removed whenever any connector to or from the ECU is removed or installed.
2. Remove the air filter to gain access to the power stage.
3. Remove the connector from the power stage.
4. Remove the rubber cover from the connector to expose the terminals. Never test the terminals from the front. This could result in damage to the terminals and make any faults worse.

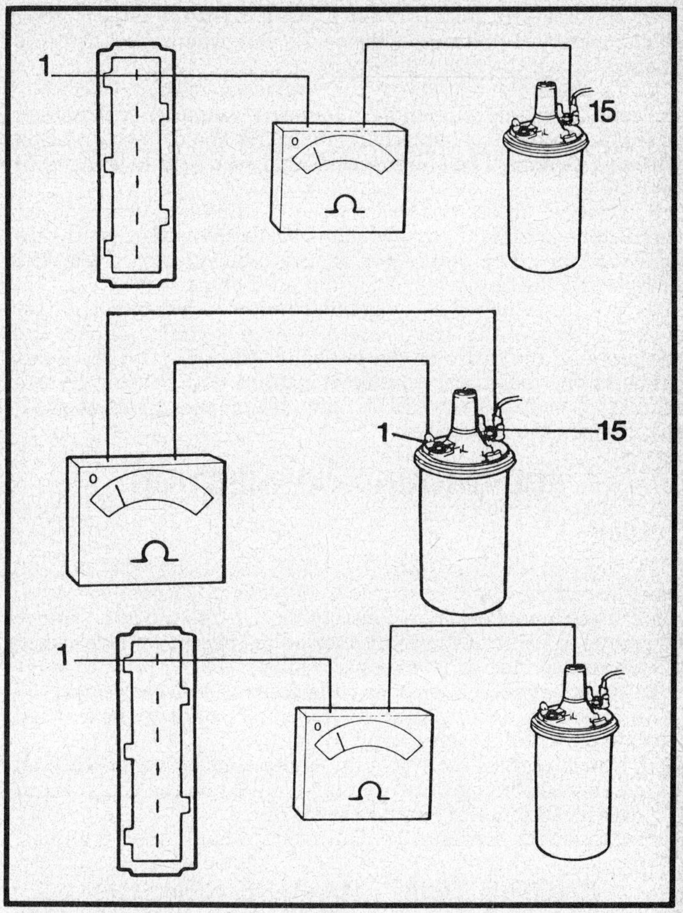

Testing ignition coil primary and secondary—EZ ignition system coil

5. Connect an ohmmeter between the power stage amplifier connector terminal 1 and the ignition coil high tension terminal (coil tower). Resistance should be 6.5–9.0 ohms.
6. If resistance is higher or lower, replace the ignition coil.

REX 1

Since the REX 1 system combines the ignition coil with power stage, testing for the REX 1 coil is contained in the power stage testing procedures.

DIAGNOSTIC SYSTEM CHECK

Testing

These tests are to make sure the diagnostic system is working properly.

1. Make sure ignition switch is **OFF**. Access the ECU. On 700 models, remove panel from under left side of instrument panel, disconnect connector from control unit. On 240 models, remove the panel in front of the right door pillar (passenger side), then disconnect connector from control unit.
2. Remove protective cover from control unit connector.
3. Check to ensure that no terminal contacts have been pushed down in the connector.
4. Insert the free end of the underhood diagnostic connector pigtail into socket No. 6.
5. Connect a voltmeter between ground and terminal 1 of the ECU multipin connector cable. Voltmeter should read 12 volts. Press down the diagnostic button. Voltmeter should now read 0 volts.

6. If there is no voltage at the ECU, measure at the test socket connector. If voltmeter shows 12 volt when the button is pushed, check the test connector.
7. To check the test connector, connect a voltmeter between the connector blue wire and ground. The voltmeter should read 12 volts. Connect an ohmmeter between the connector black wire and ground. The ohmmeter should read 0 ohm. Turn ignition **OFF**.
8. Connect an ohmmeter between the underhood diagnostic connector pigtail and pin 8 which is below the button on the diagnostic unit. The ohmmeter should read infinity. Press the button and the ohmmeter should read 0 ohms.
9. If a diode tester is available, connect it between the test socket LED and the diagnostic connector pigtail. Connect the red probe of the tester to the pin below the LED and the black probe to the diagnostic connector pigtail. If the tester gives a reading, it indicates the LED is okay. If the tester gives no reading, replace the test socket.

ECU GROUND CONNECTION

Testing

1. Make sure ignition switch is **OFF**. Access the ECU. On 700 models, remove panel from under left side of instrument panel, disconnect connector from control unit. On 240 models, remove the panel in front of the right door pillar (passenger side), then disconnect connector from control unit.
2. Remove protective cover from control unit connector.
3. Check to ensure that no terminal contacts have been pushed down in the connector.
4. Check control unit ground connection by connecting an ohmmeter between ground and terminal 20 on the control unit connector. Ohmmeter should read 0 ohms.
5. Also check terminal 14. Ohmmeter should read 0 ohms.

ENGINE TEMPERATURE SENSOR

Testing

1. Make sure ignition switch is **OFF**. Access the ECU. On 700 models, remove panel from under left side of instrument panel, disconnect connector from control unit. On 240 models, remove the panel in front of the right door pillar (passenger side), then disconnect connector from control unit.
2. Remove protective cover from control unit connector.
3. Check to ensure that no terminal contacts have been pushed down in the connector.
4. Check temperature sensor by connecting an ohmmeter between ground and terminal 2 on the control unit connector. At room temperature, the reading should be approximately 3000 ohms. The range is approximately 100 ohms at 250°F, to 5500 ohms at 32°F.

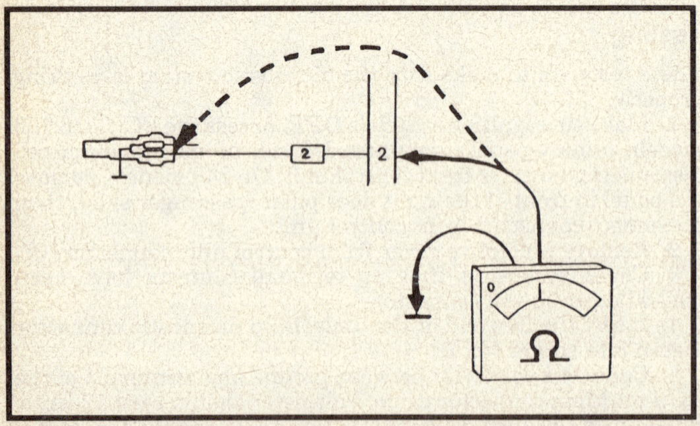

Testing temperature sensor and circuit

5. If reading obtained is incorrect, take reading at sensor to determine if fault is in the sensor or wire. Replace sensor or wire if necessary.

THROTTLE SWITCH

Testing

1. Make sure ignition switch is **OFF**. Access the ECU. On 700 models, remove panel from under left side of instrument panel, disconnect connector from control unit. On 240 models, remove the panel in front of the right door pillar (passenger side), then disconnect connector from control unit.
2. Remove protective cover from control unit connector.
3. Check to ensure that no terminal contacts have been pushed down in the connector.
4. Check the throttle switch by connecting an ohmmeter between ground and terminal 7 on the control unit connector. Ohmmeter should read 0 ohms.
5. Depress gas pedal slightly to open the throttle switch. Ohmmeter should read infinity.
6. If specifications are not as specified, take reading at throttle switch to determine if fault is in wiring or the switch.
7. Check the throttle switch setting by opening the throttle slightly and listening for the switch to operate. A click should be heard (switch opening) as soon as the throttle is opened. If not, adjust the throttle switch by loosening the mounting screws, then turn switch slightly clockwise, then counterclockwise until click is heard. Tighten mounting screws and check setting.

CRANK SENSOR (IMPULSE SENSOR)

Testing

1. Make sure ignition switch is **OFF**. Access the ECU. On 700 models, remove panel from under left side of instrument panel, disconnect connector from control unit. On 240 models, remove the panel in front of the right door pillar (passenger side), then disconnect connector from control unit.
2. Remove protective cover from control unit connector.
3. Check to ensure that no terminal contacts have been pushed down in the connector.
4. Check the engine speed sensor by connecting an ohmmeter between terminal 10 and terminal 23 on the control unit connector. Ohmmeter should read 215–265 ohms at room temperature.
5. If the engine is hot, allowable resistance can reach 280–300 ohms. Ensure that screen shield is connected to terminal 11 on the control unit connector.

KNOCK SENSORS

Testing

1. Make sure ignition switch is **OFF**. Access the ECU. On 700 and on 940 models, remove panel from under left side of instrument panel, disconnect connector from control unit. On 240 models, remove the panel in front of the right door pillar (passenger side), then disconnect connector from control unit.
2. Remove protective cover from control unit connector.
3. Check to ensure that no terminal contacts have been pushed down in the connector.
4. Disconnect electrical connector from knock sensor, then install a jumper wire between terminal 1 and terminal 2.
5. Connect an ohmmeter between terminal 12 and terminal 13 on the control unit connector. Ohmmeter should read 0 ohms. If ohmmeter reads infinity, 1 or both wires are damaged (open circuit).
6. Remove the jumper wire, then check both wires. If wires are not damaged, install new knock sensor. Reconnect connectors and connect the control unit connector.

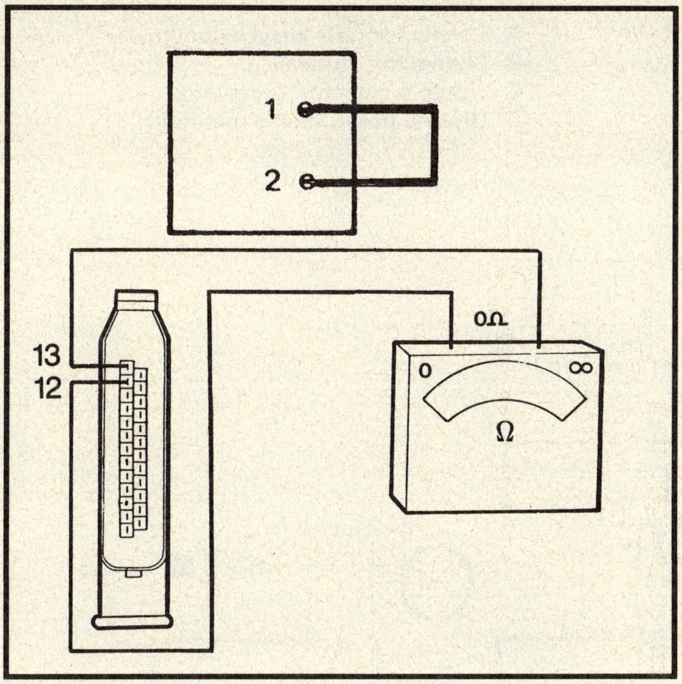

Checking knock sensor and sensor wiring to ECU

IGNITION TIMING

The ignition timing on these vehicles may be checked with a conventional timing light. The timing, however, cannot be adjusted. If the ignition setting is wrong, use the following procedure:

1. Check the throttle switch.
2. Check that the wiring to the crank sensor is correctly connected at the connector in the firewall.
3. Open the cover of the test connector and connect the cable to terminal 6.
4. Turn the ignition **ON**. Select Test Function 1 by pushing button once for more than 1 second and count the number of blinks. Note the number and press again in case there are more fault codes (up to 3). Note the fault codes to begin troubleshooting.
   - **Code 1-1-1** – No fault codes in memory.
   - **Code 1-4-2** – Fault in control unit. Engine runs with safety-retarded ignition timing (approximately 10 degrees).
   - **Code 1-4-3** – Knock sensor faulty. Engine runs with safety-retarded ignition timing (approximately 10 degrees).
   - **Code 1-4-4** – Load signal missing (from fuel system control unit). Control unit selects full-load ignition.
   - **Code 2-1-4** – Engine speed sensor faulty.
   - **Code 2-2-4** – Coolant temperature sensor inoperative.
   - **Code 2-3-4** – Throttle switch for idling faulty. Engine runs with safety-retarded ignition timing (does not apply to REX 1 ignition system).
   - **Code 3-3-4** – Throttle switch in idle position – okay (REX 1 ignition only).
5. If fault Code 1-1-1 (no fault in memory) appears, check the fuel system.
6. If the LED does not light when the button is pressed, or no code is blinked out check the connection at the ECU.

Note: The ignition system components are interrelated with fuel system components. For additional testing, replacement or wiring diagrams see appropriate fuel system section.

Component Replacement

Note: Ignition system components are interrelated with fuel system components. For most components the removal is straight forward, just the removal of the retaining screws. For additional testing or replacement check the fuel system section for this manufacturer.

1. Battery
2. Ignition switch
3. Ignition coil
4. Distributor
5. Spark plug
6. Fusebox
7. Connection box
8. Combined instrument
9. Tachometer
10. Diagnostic unit
11. EGR converter
12. Impulse sensor
13. Ignition control unit
14. In-line fuse
15. Fuel system control unit
16. Power stage
17. Throttle switch
18. Temperature sensor
19. Knock sensor

A **Connector, left suspension tower**
B **Connector, firewall**
C **9-pole connector, glove box**
D **Ground point, intake manifold**

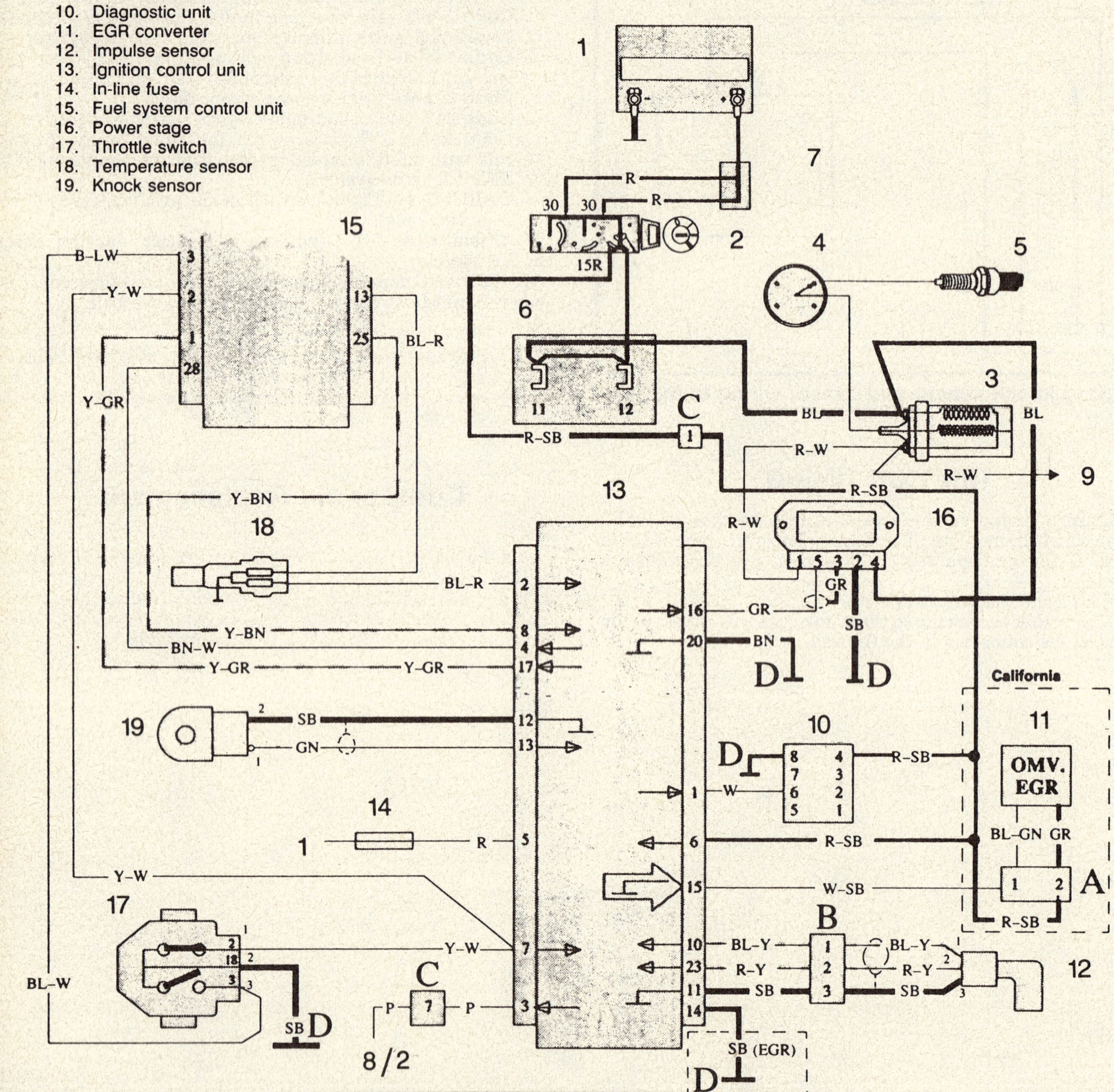

Colour code		
SB = Black	BN = Brown	GN = Green
GR = Grey	Y = Yellow	OR = Orange
W = White	P = Pink	VO = Violet
R = Red	BL = Blue	CU = Copper

EZ116K ignition system wiring–240 series with B230F engine

1. Battery
2. Ignition switch
3. Distributor
4. Spark plug
5. Fusebox
6. Combined instrument (CEL)
7. Positive terminal strip
8. Temperature sensor
9. Rev counter
10. Throttle switch
11. Knock sensor
12. Ignition system control unit
13. Impulse sensor
14. Power stage/ignition coil
15. Fuel system control unit
16. Diagnostic unit

A Connector, RH A-post
B Connector, LH A-post
C Connector, LH wheel housing, single-pole
D Earth terminal, inlet manifold
E Connector, bulkhead
F Connector, LH wheel housing
G Connector, RH wheel housing

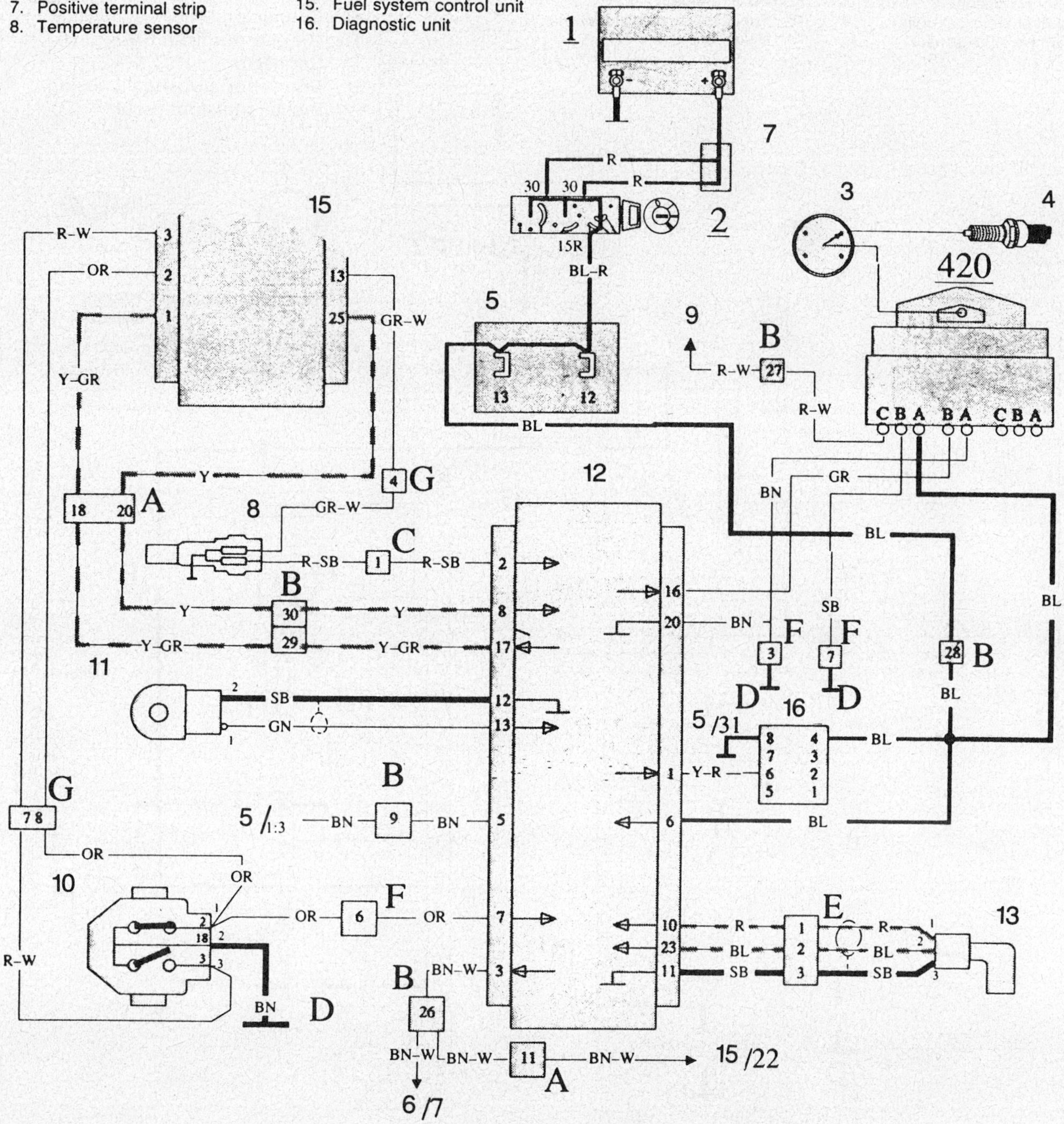

Colour code		
SB = Black	BN = Brown	GN = Green
GR = Grey	Y = Yellow	OR = Orange
W = White	P = Pink	VO = Violet
R = Red	BL = Blue	CU = Copper

REX 1 ignition system wiring–700 series with B230F engine

1. Battery
2. Ignition switch
3. Ignition coil
4. Distributor
5. Spark plugs
6. Fusebox
7. Positive terminal board
8. Ground busbar in central electrical unit
9. Temperature sensor
10. Speedometer
11. Throttle switch
12. Knock sensor
13. Control unit
14. EGR converter
15. EGR temperature sensor
16. Speed/position pick-up
17. Power stage
18. Fuel system control unit
19. Diagnostic unit

A Connector, RH A-post
B Connector, LH A-post
C Connector, LH wheel housing
D Ground terminal, intake manifold
E Connector at bulkhead
F Connector, LH wheel housing
G Connector, RH wheel housing
H Connector, LH wheel housing
I Ground terminal, RH A-post
K Connector
L Connector, LH wheel housing, 2-pole, moisture-proof

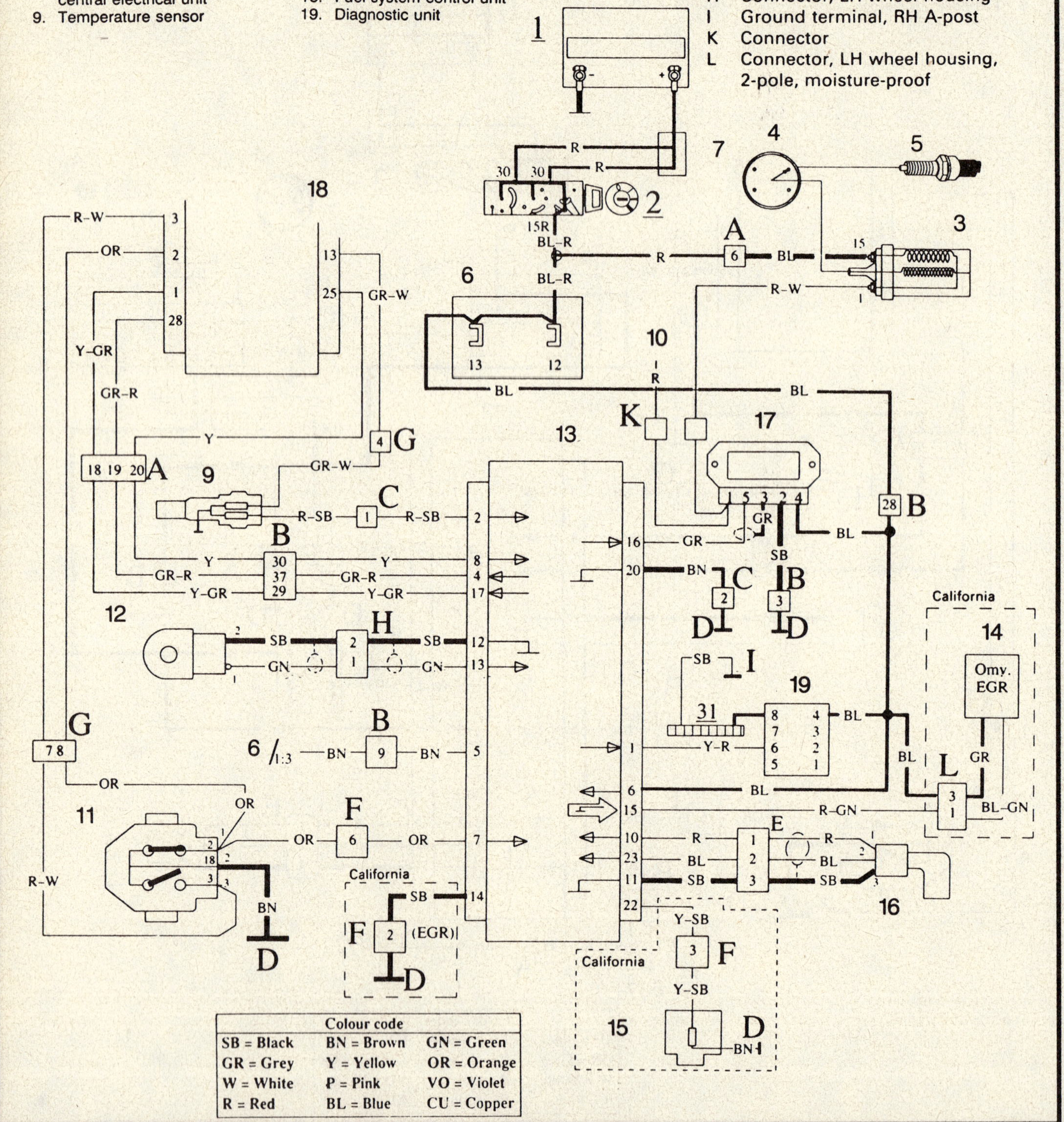

Colour code		
SB = Black	BN = Brown	GN = Green
GR = Grey	Y = Yellow	OR = Orange
W = White	P = Pink	VO = Violet
R = Red	BL = Blue	CU = Copper

EZ116K ignition system wiring–700 series with B234F engine

1. Battery
2. Ignition switch
3. Ignition coil
4. Distributor
5. Spark plugs
6. Fusebox
7. Terminal at combined instrument
8. Positive terminal board
9. Ground busbar in central electrical unit
10. Temperature sensor
11. Throttle switch
12. Knock sensor
13. Control unit
14. EGR converter
15. EGR temperature sensor
16. Speed/position pick-up
17. Power stage
18. Fuel system control unit
19. Diagnostic unit

- A Connector, RH A-post
- B Connector, LH A-post
- C Connector, LH wheel housing
- D Earth terminal, inlet manifold
- E Connector, bulkhead
- F Connector, LH wheel housing
- G Connector, RH wheel housing
- H Connector, LH wheel housing

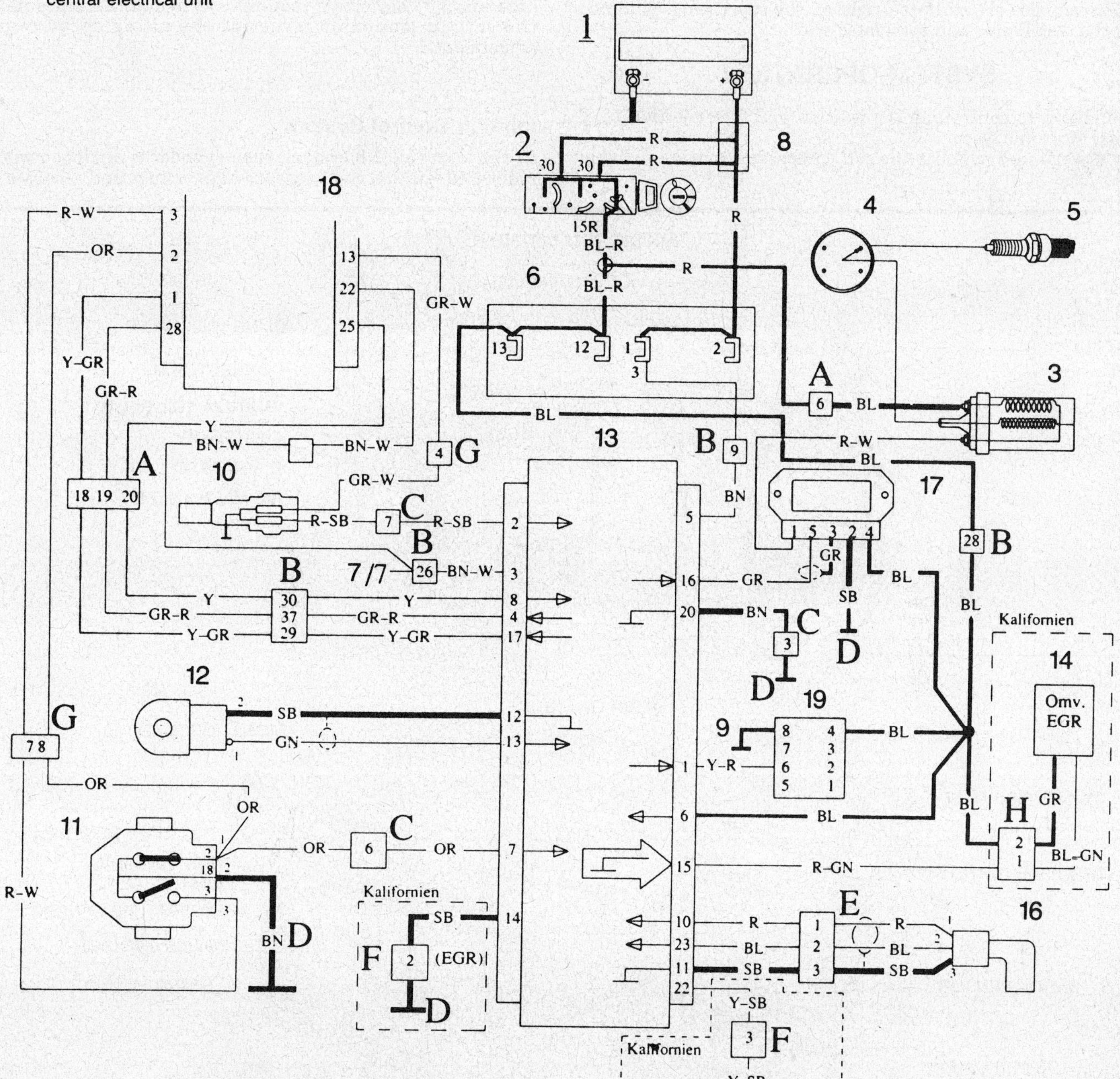

Colour code		
SB = Black	BN = Brown	GN = Green
GR = Grey	Y = Yellow	OR = Orange
W = White	P = Pink	VO = Violet
R = Red	BL = Blue	CU = Copper

EZ116K ignition system wiring–740 and 780 series with B230F turbocharged engine

VOLVO MOTRONIC IGNITION SYSTEM

General Information

The Bosch Motronic 1.8 system was introduced on the 1992 Volvo 960 which is equipped with the 2.9L B-6304F engine. This is a totally integrated engine management system. The fuel and ignition functions are controlled by the same on board computer in the Electronic Control Unit (ECU). This particular system uses a Distributorless Ignition System (DIS) where individual coil assemblies are mounted right on the spark plugs eliminating the distributor and associated wiring.

SYSTEM OPERATION

In addition to controlling the ignition and fuel injection functions, Motronic also:

- Determines whether the A/C compressor may be switched on
- Reduces the engine torque in response to a signal from the automatic transmission control unit, to insure smooth engagement of the different gears and also supplies the transmission control unit with information on engine running conditions for computing gear changes
- Controls the operation of the radiator fan

The control unit is provided with oxygen sensor and idling control functions and will retard the ignition timing to eliminate knock. The service requirement is minimal since neither the carbon monoxide level nor the idling speed require adjustment.

Ignition Control System

In the Motronic 1.8 system, each cylinder is equipped with an individual ignition coil to assure high voltage and accurate con-

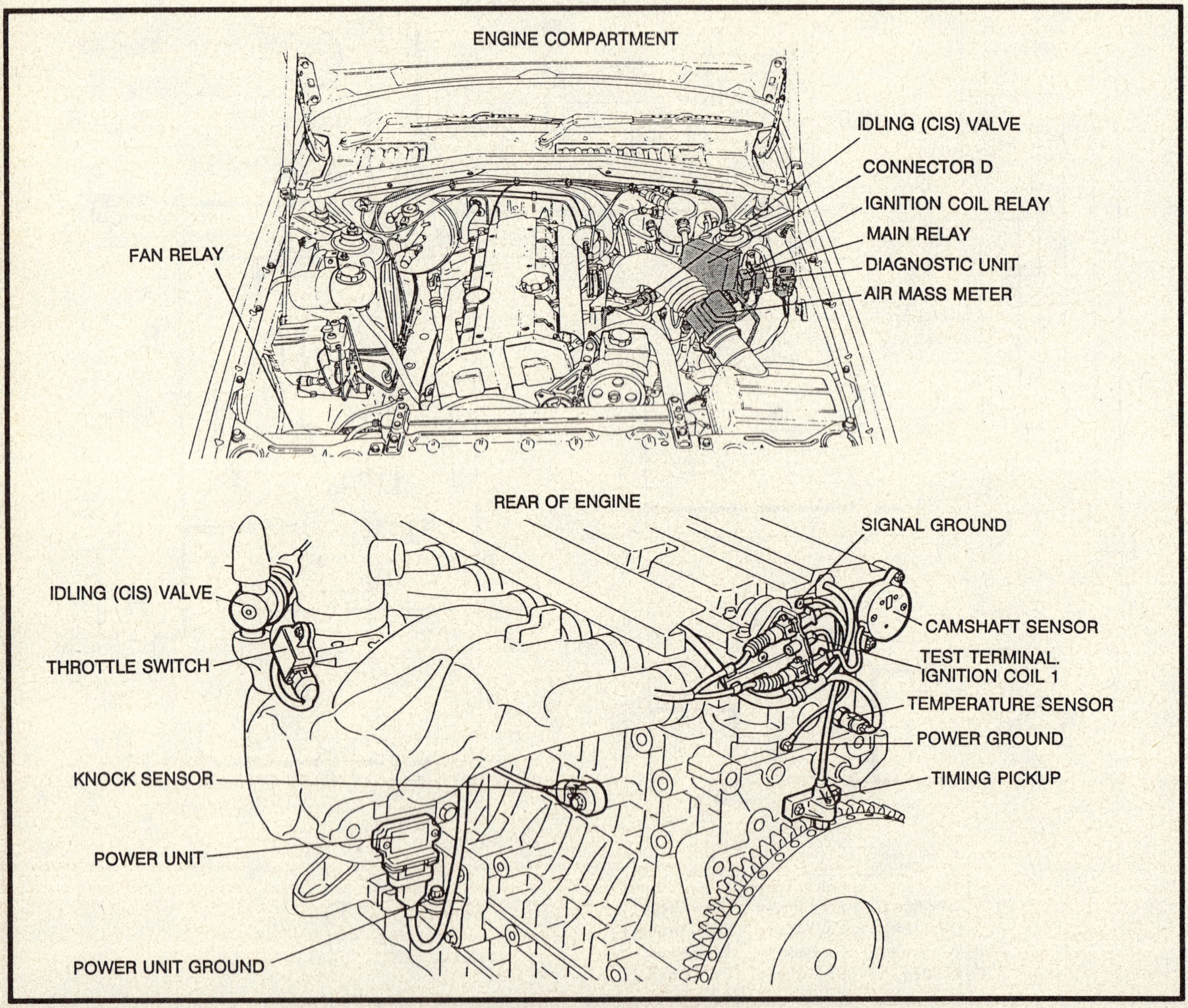

Engine compartment and major component locations—Volvo 960

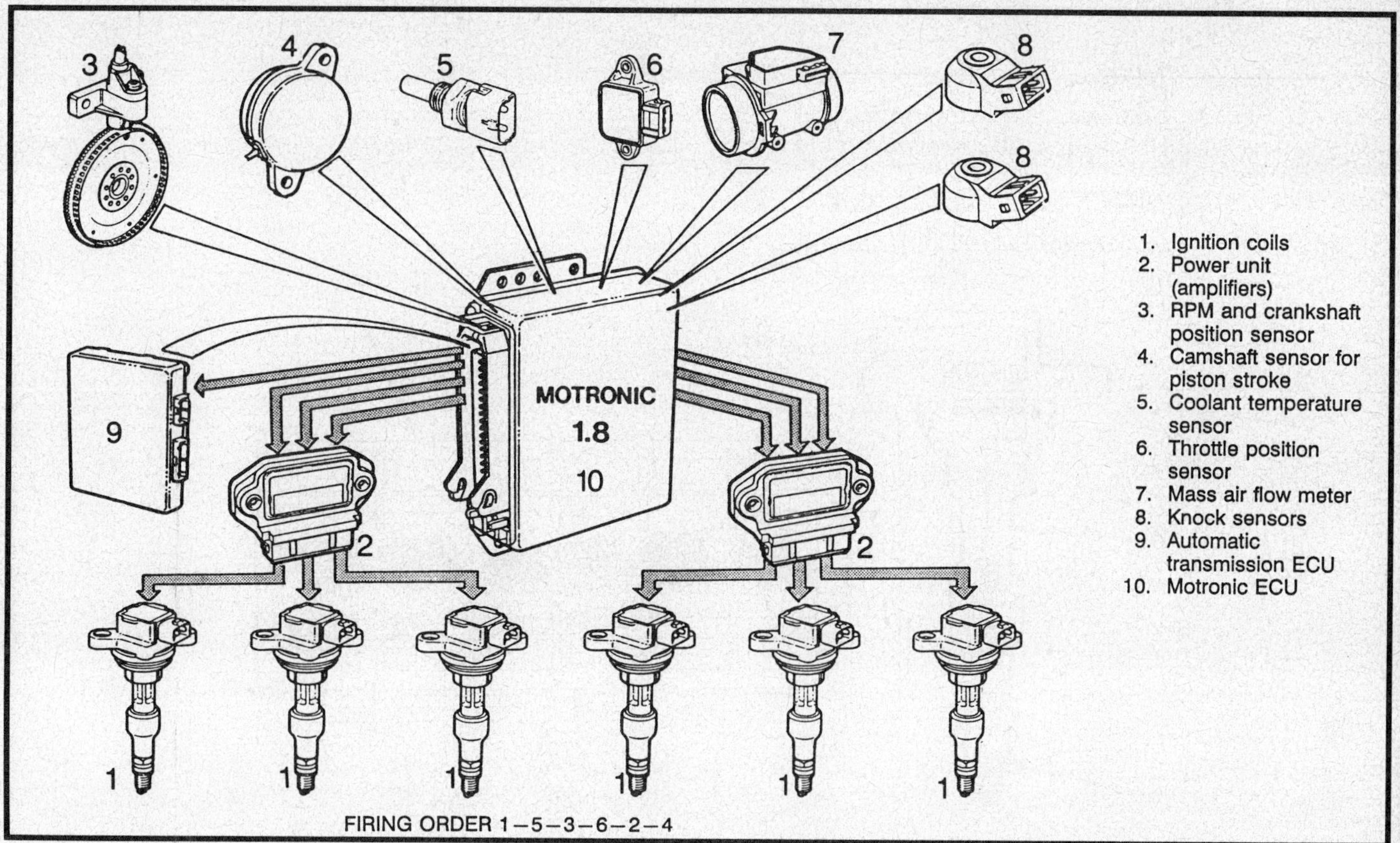

Ignition control major components. Note individual coils installed on the spark plugs—Volvo 960

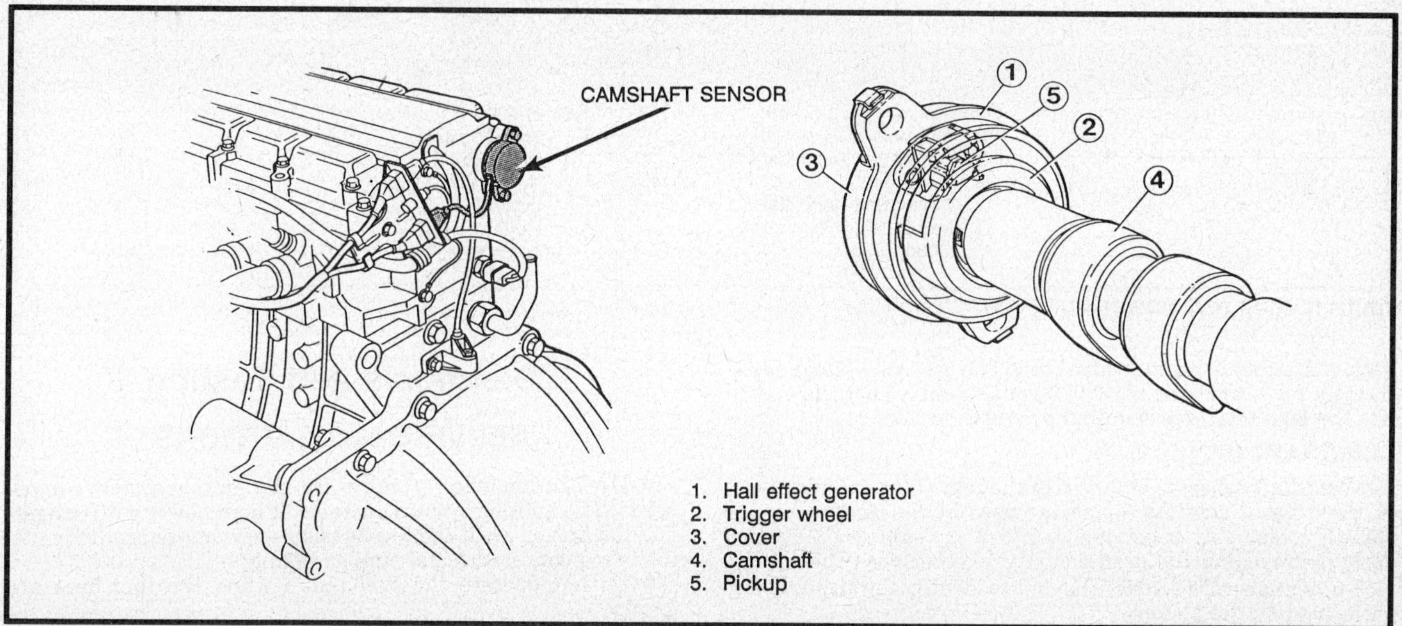

Camshaft sensor is located on the back of the cylinder head driven by the exhaust camshaft—Volvo 960

trol of ignition. There is no distributor or spark plug cables. Firing order is 1-5-3-6-2-4.

The Electronic Control Unit (ECU) computes the instant at which each ignition coil must deliver its pulse. There are 6 coils controlled by 2 power units. The ECU uses information from many sensors to compute the timing.

IGNITION COILS

The ignition coils are mounted directly on the spark plugs. The 2 power units which control the coils are mounted on the intake manifold, 1 on the front, 1 on the rear, for cooling. The front power unit is connected to cylinders 1, 3 and 5 while the rear unit serves cylinders 2, 4 and 6. Each power unit has 3 stages,

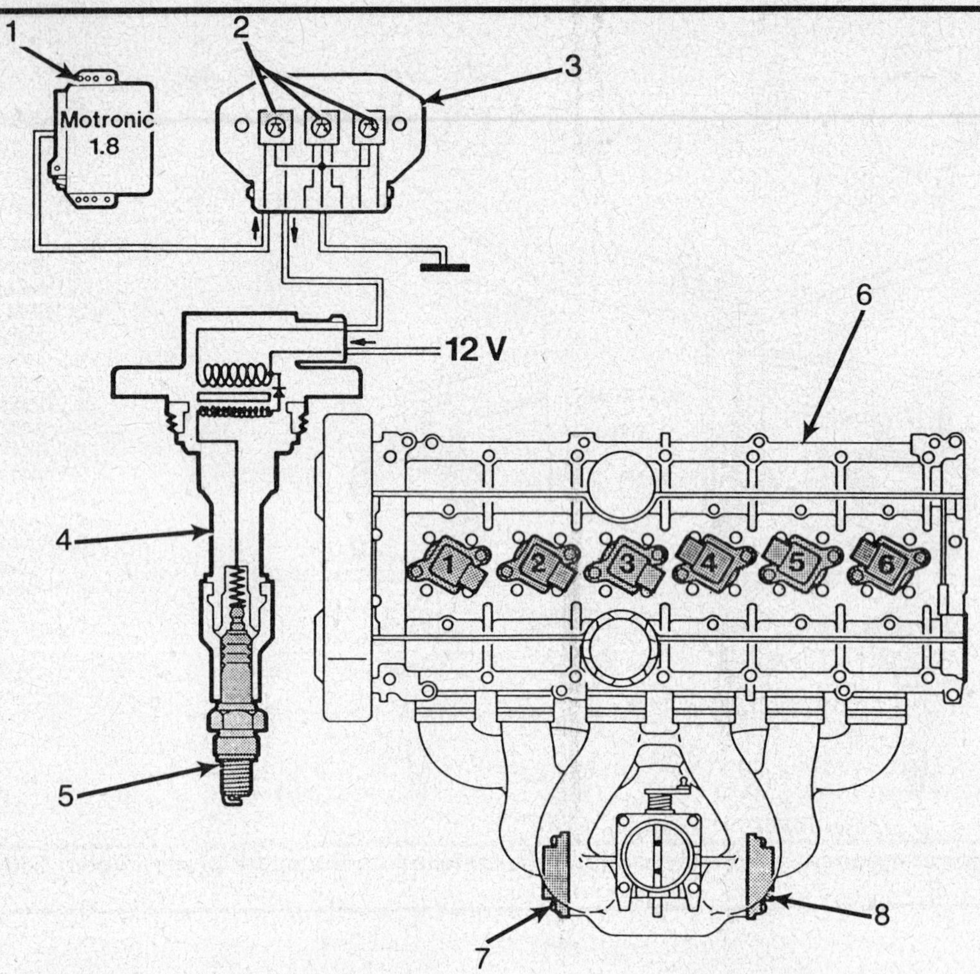

1. Motronic ECU
2. Power unit (amplifier) 3 stages
3. Power unit (amplifier)
4. Individual ignition coil
5. Spark plug
6. B 6304 F engine showing coil locations
7. Front power unit controls 1, 3, 5
8. Rear power unit controls 2, 4, 6

Ignition coils and power units—Volvo 960

each connected to an individual coil. The ignition voltage is extremely high, upwards of 40,000 volts to aid cold starts and to fire the lean mixtures required by emission laws.

CAMSHAFT SENSOR

The camshaft sensor is mounted at the rear of the cylinder head on the exhaust side. Its signals are used by the control unit to identify which pair of pistons (1–6, 5–2 or 3–4) are simultaneously approaching top dead center. Together with the timing pick-up signals, this enables the unit to identify the stroke being performed by the pistons.

The main components of the camshaft sensor are a Hall generator, trigger rotor and cover. As the rotor turns with the camshaft, it alternately shields and exposes the Hall element, generating high 5 volt and low 0 volt signals in turn. Since the vane and opening on the rotor are of the same width, and since the camshaft rotates at only half the crankshaft speed, the output signal changes only after a full crankshaft revolution. After 2 crankshaft revolutions, the signal from the Hall generator has reached 1 maximum and 1 minimum.

Diagnosis and Testing

SERVICE PRECAUTIONS

- Use caution when working around ignition system components. The ignition system operates at high power with dangerous voltages in both the low-tension and high-tension circuits, including connectors and similar fittings.
- Do not operate the fuel pump when the fuel lines are empty.
- Do not arc weld on the vehicle unless all the control units have been removed.
- The control unit must not be exposed to temperatures above 175°F (80°C), such as in a paint booth.
- Do not test the compression on the 960 until the ignition coils are removed, the ignition coil relay is disconnected and fuse No. 31 (to disarm fuel pumps) is removed.
- Do not disconnect the control unit harness connector before fuse No. 24 has been removed.

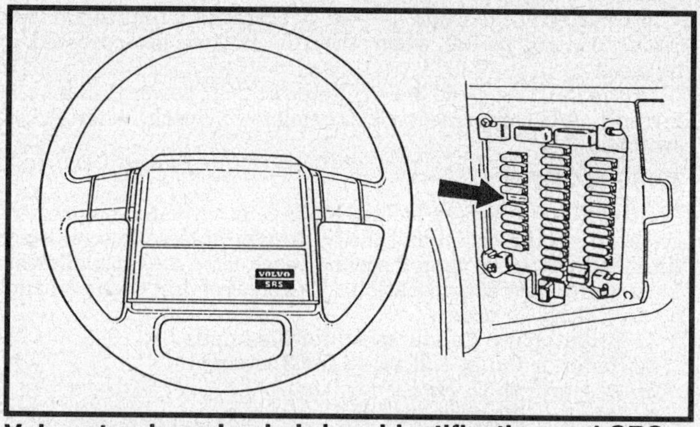

Volvo steering wheel air bag identification and SRS test location. Never install a fuse in this location—Volvo 960

- Make sure all control unit connectors are fastened securely. A poor connection can cause an extremely high surge voltage, resulting in damage to integrated circuits.
- Keep any control unit harness at least 4 in. away from nearby harnesses to prevent a system malfunction due to external electronic interference.
- Keep all parts and harnesses dry during service.
- Before attempting to remove any parts, turn off the ignition switch and disconnect the battery ground cable.
- Always use a 12 volt battery as a power source.
- Do not attempt to disconnect the battery cables with the engine running.
- Do not attempt to disassemble a control unit under any circumstances.
- If a battery cable is disconnected, the memory will be erased.
- Before connecting or disconnecting control unit ECU harness connectors, make sure the ignition switch is **OFF** and the negative battery cable is disconnected to avoid the possibility of damage to the control unit.
- Use care when working around vehicles equipped with Supplementary Restraint System (SRS), often known as "air bags." Volvos equipped with SRS are easily recognized by the letters **SRS** moulded into the steering wheel cover. Follow all precautions
- When working around the instrument panel or steering column areas, take care to make sure that wiring is not pinched, abraded or penetrated by screws when working on soundproofing panels, knee cushion, ignition switch or column covers.
- The SRS crash sensor is located underneath the driver's seat. Never disconnect the sensor connector. Never install accessories near the sensor.
- Never undo the steering gear without first locking the contact reel and removing the steering wheel.
- A spare location in the fuse panel is used as a test terminal for air bag diagnostics on the Volvo 960. Never install a fuse in this position or connnect accessories to this terminal.

READING CODES

To help with troubleshooting, the Motronic control unit is equipped with some diagnostic functions. There are 3 modes:

Self-Diagnosis, which is Test Mode 1
Function Testing, which is Test Mode 2
Control Testing, which is Test Mode 3

Volvos use a small diagnostic unit, or switchbox, usually mounted in the left rear side of the engine compartment. It has a pushbutton, several sockets to access different parts of the system and an LED light which flashes the codes. Access to the diagnostic system is provided by plugging in the diagnostic unit's selector plug, or "pigtail" into socket 2 on the diagnostic unit when the ignition is switched **ON**.

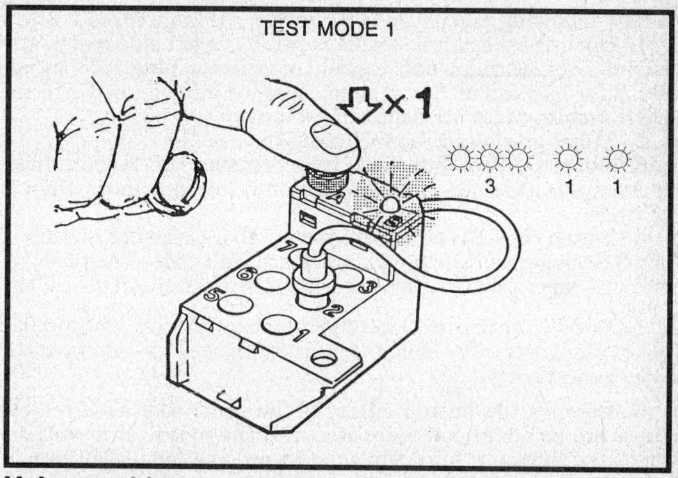

Volvo pushbutton on-board diagnostic unit and example of flash codes. Note different sockets for checking different areas of the vehicle systems—Volvo 960

Self-Diagnostics

TEST MODE 1

The control unit can record and store 18 different fault codes. Fault codes stored in the memory can be displayed by operating the test button. The control unit memory, which stores both the diagnostic codes and adaptive programs, can retain information for periods ranging from 10 minutes to 24 hours after interruption of the power supply.

All self-diagnostic system faults with the exception of a faulty knock sensor, are reversible. This means the ECU will again begin to use a faulty or missing signal immediately after is has been corrected, rather than remaining in the 'limp home' mode. However, the fault code will remain in the memory.

A warning lamp on the instrument cluster indicates emission-related faults which are detectable by the system. All fault codes have 3 digits, each capable of ranging from 1 through 9. The fault codes are read from the series of flashes delivered by the LED. Each code requires 3 series of uninterrupted flashes, with a 3 second interval between each series of flashes to make the codes easy to read.

Flash codes used in the Volvo 960 Motronic system are:

Code 1-1-1—No fault detected by diagnostic system
Code 1-1-2—Control unit problem
Code 1-1-3—Faulty injector
Code 1-2-1—Air mass meter signal absent or faulty
Code 1-2-3—Engine temperature signal absent or faulty
Code 1-3-1—No signal from timing pickup
Code 1-3-2—Battery voltage too high or too low
Code 1-4-3—Front knock sensor signal absent or faulty
Code 2-1-2—Oxygen sensor signal absent or faulty
Code 2-1-4—Timing pickup signal absent intermittently
Code 2-2-3—Idling valve signal absent of faulty
Code 2-3-1—Oxygen sensor lean or rich in part-load range
Code 2-3-2—Oxygen sensor control lean or rich at idling
Code 2-3-3—Adaptive idling control outside control range
Code 2-4-3—Signal from throttle switch absent or faulty
Code 3-1-1—Speedometer signal absent
Code 3-1-4—Camshaft sensor signal absent
Code 3-2-2—Air mass meter burn-off signal absent
Code 4-3-3—Rear knock sensor signal absent or faulty

The following procedure is for retrieving fault codes.

1. Open the diagnostic socket cover in the left side rear of the engine compartment and install the selector plug into socket No. 2 for ignition or fuel system codes, or into socket No. 6 for EGR system codes on California vehicles.
2. Turn ignition to the **ON** position.
3. Enter Control Function 1 by pressing the button once. Hold the button for at least 1 second, but not more than 3 seconds.
4. Watch the LED and count the number of flashes in each of the 3 flash series, which will indicate a fault code. The flash series are separated by 3 second intervals. Note the fault codes.

NOTE: If there are no fault codes in the diagnostic unit, the LED will flash 1-1-1 and the fuel system is operating correctly.

5. Depress the button again. If the same code is repeated, there are no additional codes stored. If the code is different, depress the button a third time and record the code, if different.

NOTE: The diagnostic memory is full when it contains 3 fault codes. Until those 3 are rectified and the memory erased, the system cannot give information on any other problems.

TEST MODE 2

This is function testing, a check to see that various systems are operating and send/receiving signals as required. Function testing is activated by pressing the button twice, holding the button for at least 1 second each time, but not more than 3 seconds each time and also allowing 1–3 seconds between operations, causing the LED to flash. The control unit delivers a diagnostic code (not a fault code) acknowledging the receipt of a signal from each of the following items:

Function Test Code 3–3–2: Throttle switch, when the throttle is moved from the idling position

Function Test Code 3–3–3: Throttle switch, when the throttle is moved from the full-load position

Function Test Code 1–2–4: Start inhibitor switch, via the transmission control unit, when the gear selector lever is moved from 1 of the drive positions to the **N** or **P** position

Function Test Code 1–4–1: Timing pick-up, when the starter motor is running

Function Test Code 1–1–4: Electronic Climate Control (ECC) control panel, when the A/C button is depressed or released

Function Test Code 1–3–4: Relay in ECC power unit and, as a result, A/C compressor electromagnetic clutch, when A/C is switched **ON**

TEST MODE 3

Control Testing, which is Test Mode 3, is activated by pressing the button 3 times, holding the button for at least 1 second each time, but not more than 3 seconds each time and also allowing 1–3 seconds between operations. The control unit then responds by activating, in order:

1. Radiator fan at half speed for 3 seconds
2. Radiator fan at full speed for 3 seconds
3. Injectors at 13 Hz
4. Idling (CIS) valve at 1 Hz
5. Relay in ECC power unit and, as a result, A/C compressor electromagnetic clutch, at 1 Hz.

This activation is repeated twice.

CLEARING CODES

When all the fault codes have been read and corrected, the diagnostic system memory can be erased as follows:

1. Turn ignition switch to the **ON** position.
2. Read fault codes again.
3. Press and hold diagnostic socket button for approximately 5 seconds. Release button. After 3 seconds, the LED should light.
4. While the LED is lit, press button again for approximately 5 seconds. Release button. The LED should go out. Memory is cleared.

NOTE: To ensure that the memory has been erased, press button again for 1 second but not more than 3 seconds. The LED should flash 1-1-1, indicating that the fuel system is operating properly.

NO START—NO SPARK

It is recommended that a breakout box be used when troubleshooting Motronic problems. This is the best way of doing volt-

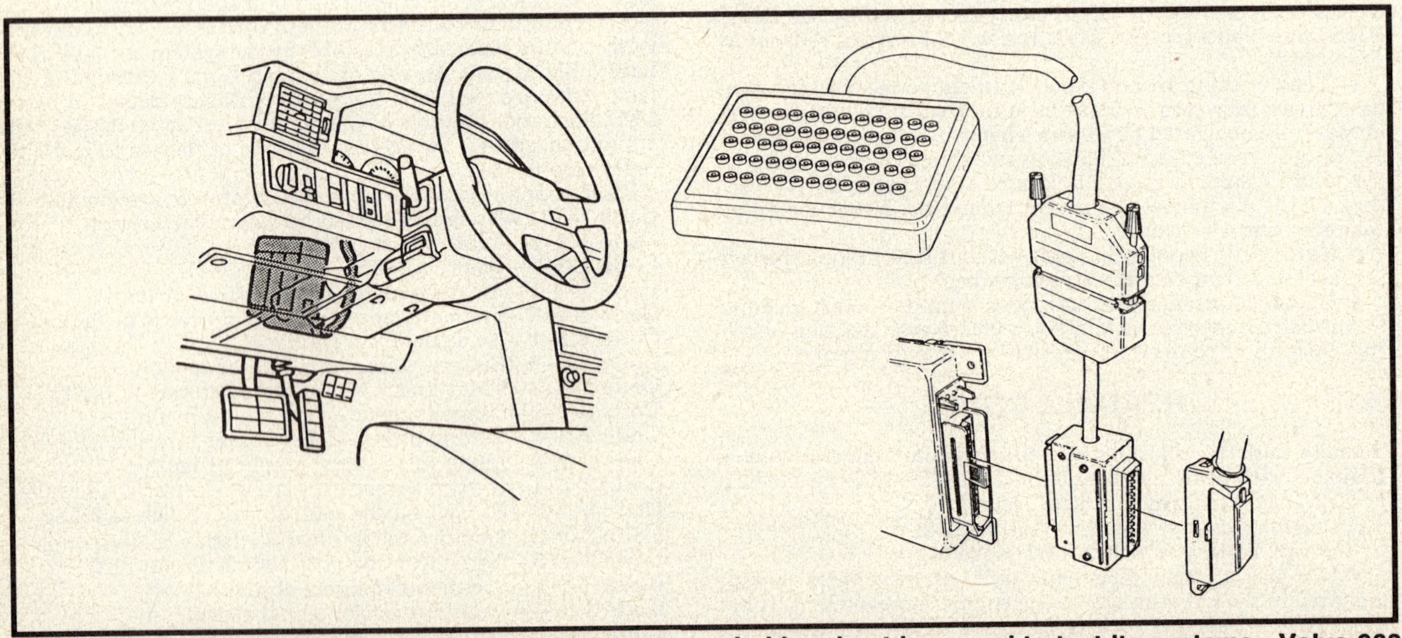

Electronic Control Unit (ECU) location and the recommended breakout box used to test its systems—Volvo 960

age and resistance tests with all connections made and system sensors operating without endangering the ECU or delicate pin connections. All of the following tests and checks are based on using this piece of equipment. If such equipment is available, make sure the ignition switch is **OFF** when connecting or disconnecting the ECU. Remove fuse No. 24 when connecting or disconnecting the ECU.

Ground Connections Check

1. Using a breakout box, with the ignition key **OFF**, connect an ohmmeter between ground and terminals 19, 24, 26, 30 and 48 of the control unit. The meter should indicate approximately 0 ohms in all cases.
2. Check the power unit ground connections under each unit. These 2 components are mounted on the intake manifold. Connect an ohmmeter between ground and terminal 4 on front power unit connector, then between ground and terminal 4 on rear power unit connector. The meter should indicate approximately 0 ohms in both cases.
3. If high resistance is indicated, check the connectors and the wiring.

Voltage Check

1. With the ignition switch **OFF**, check for voltage supply to terminal 18 across fuse No. 24 and to terminal 36 across the main relay.
2. Connect a voltmeter across terminals 18 and 19, then across terminals 36 and 19 of the control unit. The voltmeter should indicate battery voltage in both cases. Check the fuse, wiring and the relay if no reading is obtained.
3. Turn the ignition switch **ON** and check that the voltage falls from battery voltage to approximately 0.1 volt at terminal 36.

Ignition Coil Voltage Supply

Testing

Fault Code 1–3–1 is reported if no signal is received from the camshaft sensor pick-up for 10 engine revolutions. The engine will not start as a result of this.

Fault Code 2–1–4 is reported only if a fault occurs when the engine is running. The code is set if what the control unit computes and what it actually sees are different. A slight roughness may be a symptom felt in this event.

1. Turn the ignition switch **ON**.
2. Connect a voltmeter between ground and the test terminal at the rear of the engine. The reading should be battery voltage.

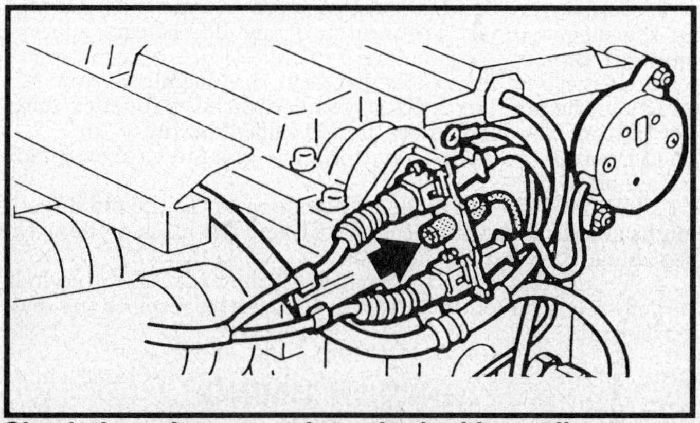

Check the voltage supply to the ignition coils across the coil relay by connecting voltmeter between ground and test terminal at rear of engine (arrow). There should be battery voltage—Volvo 960

3. If the reading is not battery voltage, check the coil relay, wiring and the connectors up to the ignition coils.
4. Attempt to start the engine.
5. If the engine still does not start, remove the ignition coil cover and remove one coil.
6. Fit a spare spark plug in the ignition coil and crank the engine.
7. If no spark is obtained, check the timing pick-up.

Ignition or Fuel System Identification

1. Use a voltmeter to check the injectors.
2. The voltage should be approximately 250–400 millivolts when the engine is hot.
3. If the voltage is correct, either the fault is in the ignition system or the injector is blocked. Check the fuel system for fault codes.

COIL CHECK

Firing Pulse at Coil Test

1. Use test diode 999–5280 or equivalent. Make an adaptor to fit between the ignition coil connector and the test diode.
2. Connect the test diode.
3. Start the engine. The LED should flash when a signal is present.
   a. If there is no signal, check the signal output from the control unit.
   b. If there is a signal present, replace the ignition coil or spark plug.

ECU-TO-COIL SIGNAL

Signal Output Check

1. Connect a voltmeter between ground and the control unit terminal that is connected to the ignition coil.
2. Start the engine.
3. The meter should indicate between 60–100 millivolts if the signal is correct.
4. If the reading is approximately 3.5 volts, there is an open circuit present between the control unit and the ignition coil.
5. If the signal is present, check the power unit and wiring. Stop the engine.

POWER UNIT AND WIRING

Testing

1. With the ignition switch **OFF**, connect the diode tester with the positive side connected to the control unit terminal for cylinder to be checked and to terminal 1 on corresponding ignition coil connector. Also check in reverse direction.
2. Connect the diode tester with the positive side connected to ground and to terminal 1 on the ignition coil connector. Also check in the reverse direction.
3. Check the wiring if a faulty diode is found.

VOLTAGE SIGNAL AND WIRING

Testing

1. Connect a voltmeter across terminals 42 and 19 of the control unit.
2. Turn the ignition switch **ON**.
3. Check the meter reading. It should be approximately 5.5 volts in **P** or **N**.
4. Check the meter reading. It should be approximately 0.5 volts in **D** or **R**.
5. If there is no signal, check the lead between the control unit and the transmission control unit.
6. If the lead is intact, check the signal from the gear selector mechanism to the transmission control unit.

TIMING PICK-UP

Testing

Fault Code 1–3–1 is reported if no signal is received from the camshaft sensor pick-up for 10 engine revolutions. The engine will not start as a result of this.

Fault Code 2–1–4 is reported only if a fault occurs when the engine is running. The code is set if what the control unit computes and what it actually sees are different. A slight roughness may be a symptom felt in this event.

TESTING IN CONTROL FUNCTION 2

1. Enter Control Function 2 and operate the starter motor.
2. Diagnostic Code 1–4–1 should be displayed, indicating a timing pick-up signal.

TIMING PICK-UP SIGNAL CHECK

1. Connect a voltmeter across terminals 47 and 48, and set on the AC millivolt scale.
2. Operate the starter motor and check that the voltmeter displays 300–400 millivolts.

ABSENT OR FAULTY SIGNAL CHECK

1. With the ignition switch **OFF**, remove fuse No. 24 and disconnect the control unit.
2. Connect an ohmmeter between terminals 47 and 48.
3. Check that the ohmmeter reads 200–400 ohms, depending on the temperature.
4. Reconnect the control unit.
5. Measure the resistance between terminals 48 and 19. The resistance should be 0 ohms.
6. Measure the resistance between terminals 48 and 30. The resistance should be 0 ohms.
7. On vehicles with automatic transmissions, check in Control Function 1 if speed signal fault is displayed.

BATTERY VOLTAGE

Testing

Fault Code 1–3–2 will set if the battery voltage falls below 8 volts or exceeds 16 volts for 5 seconds after the engine has started.

VOLTAGE CHECK

1. Connect a voltmeter across terminals 18 and 19 of the control unit.
2. Start the engine and read the battery voltage.
3. Check the battery and charging system if the voltage is in abnormal range.

KNOCK SENSORS

Testing

Fault Codes 1–4–3 and 4–3–3 indicate that the front and rear knock sensor signals, respectively, are absent or faulty. The fault codes will be reported if the following 2 conditions are fulfilled:

1. The engine speed is at least 2970 rpm and a certain engine load has been exceeded.
2. The signal from either of the knock sensors to the control unit falls below 255 millivolts during 96 successive ignitions. The control unit will then safety-retard the timing.

WIRING CHECK

The knock sensors themselves cannot be checked, only the wiring.

1. Disconnect the knock sensor connector and connect a jumper wire between the connector terminals.
2. Remove fuse No. 24 and disconnect the control unit.
3. Connect an ohmmeter between terminal 2 and terminal 11 or 29, depending on which knock sensor is affected. The reading should be 0 ohms.
4. Replace the knock sensor if the reading is correct. Do not replace it if the reading is incorrect. Torque the sensor to 15 ft. lbs. (20 Nm).
5. Reconnect the control unit.

CAMSHAFT SENSOR

Testing

The camshaft sensor is mounted at the end of the camshaft and enables the control unit to determine which cylinder requires fuel and ignition. As the camshaft performs a revolution every 2 crankshaft revolutions, the sensor signal goes high on 1 crankshaft revolution and low on the next. Fault Code 3–1–4 is reported if the signal remains constantly high or low for more than 30 seconds. Under these conditions, the control unit supplies double injection and double ignition while safety-retarding the timing.

CAMSHAFT SENSOR SIGNAL CHECK

1. Remove fuse No. 31 to prevent the engine from starting.
2. Connect a voltmeter across terminals 8 and 19 of the control unit.
3. Operate the starter motor.
4. Check that the reading oscillates between 0.1–0.5 volts.

ABSENT OR FAULTY SIGNAL CHECK

1. Check that the voltage across terminals 10 and 30 of the control unit is approximately 11 volts.
2. Check that terminal 30 is grounded.
3. Connect an ohmmeter between terminals 30 and 19 of the control unit. It should read 0 ohms.

Diagnostic Unit Testing

1. Turn the ignition switch **ON.**
2. Connect the test lead to socket 2 on the diagnostic unit.
3. Connect a voltmeter across terminals 55 and 19 on the control unit connector. The instrument should indicate 12 volts.
4. Press the button on the control unit. The instrument should indicate 0 volt.
5. If there is no voltage at the control unit, measure instead at the diagnostic unit connector.
6. If the voltmeter indicates 12 volts when the button is depressed, check the diagnostic unit.
7. Connect a voltmeter between ground and the blue lead in the diagnostic unit connector. It should read approximately 12 volts.
8. Turn the ignition switch **OFF.**
9. Connect an ohmmeter between ground and the black lead in the diagnostic unit connector. It should indicate approximately 0 ohms.
10. Connect an ohmmeter between the diagnostic unit test plug and the pin that is directly underneath the function selector button. The instrument should indicate infinity.
11. Depress the selector button. The ohmmeter should indicate 0 ohms.
12. Connect the red probe of a diode tester to the pin directly underneath the diagnostic unit LED and the black probe to the test plug, then switch the leads.
13. The LED is operating correctly if the tester shows an indication in one direction only. If it is defective, replace the diagnostic unit.

IGNITION TIMING

Although the timing may be checked, it is not adjustable. All timing functions are carried out by the ECU.The ignition timing may be checked with a onventional inductive timing light. Specification is 5 degrees ± 2 degrees TDC.

Component Replacement

SERVICE PRECAUTIONS

- Use caution when working around ignition system components. The ignition system operates at high power with dangerous voltages in both the low-tension and high-tension circuits, including connectors and similar fittings.
- Do not operate the fuel pump when the fuel lines are empty.
- Do not arc weld on the vehicle unless all the control units have been removed.
- The control units must not be exposed to temperatures above 175°F (80°C), such as in a paint booth.
- Make sure that the ignition switch is **OFF** before connecting or disconnecting the ignition coil, spark plug wiring or test instruments.
- Always use new gaskets and seals when remaking a fuel connection. Torque the fuel filter fittings to 15–26 ft. lbs. (20–35 Nm).
- Do not test the compression on the 960 until the ignition coils are removed, the ignition coil relay is disconnected and fuse No. 31 (to disarm fuel pumps) is removed.
- Do not disconnect the control unit harness connector before fuse No. 24 has been removed.
- Make sure all control unit connectors are fastened securely. A poor connection can cause an extremely high surge voltage, resulting in damage to integrated circuits.
- Keep any control unit harness at least 4 in. away from adjacent harnesses to prevent a system malfunction due to external electronic interference.
- Keep all parts and harnesses dry during service.
- Before attempting to remove any parts, turn off the ignition switch and disconnect the battery ground cable.
- Always use a 12 volt battery as a power source.
- Do not attempt to disconnect the battery cables with the engine running.
- Do not attempt to disassemble a control unit under any circumstances.
- If a battery cable is disconnected, the fault code memory will be erased.
- Before connecting or disconnecting control unit ECU harness connectors, make sure the ignition switch is **OFF** and the negative battery cable is disconnected to avoid the possibility of damage to the control unit.
- A spare location in the fuse box is used as a test terminal for air bag diagnostics on the 960. Never install a fuse in this position or connnect accessories to this terminal.

Note: The ignition system components are interrelated with fuel system components. For additional testing, replacement or wiring diagrams see appropriate fuel system section.

CAMSHAFT SENSOR

Removal and Installation

1. The camshaft sensor is mounted at the rear of the cylinder head on the exhaust side. Disconnect the negative battery cable.
2. Remove any ducting or hoses that may be in the way, matchmarking for assembly.
3. Disconnect the electrical connector taking care not to damage the connector or pins. Ground connections are secured with electrical eyelets and small screws. Use care when disconnecting these pieces.
4. Remove the hold-down bolts for the sensor body.
5. Pull the camshaft sensor away from the cylinder head, noting the drive slots in the camshaft for reassembly.

To install:

6. Installation is the reverse of the removal procedure. Make sure the drive slots in the camshaft properly mesh with the camshaft sensor drive.
7. Install the hold-down bolts and make sure the electrical connection is secure.

IGNITION COILS

Removal and Installation

This system uses 6 individual coils, 1 on each spark plug.

1. Disconnect the negative battery cable.
2. Remove the large coil cover on the top of the engine that is secured to the cam cover. Use care not to drop any of the small retaining screws.
3. Remove any ducting or hoses that may be in the way, matchmarking for assembly.
4. Remove the hold-down capscrews from the coil top.
5. Disconnect the electrical connector from the coil top assembly.
6. Carefully pull the coil from the spark plug, pulling straight up while turning slightly to disengage from the top of the spark plug.

To install:

7. Installation is the reverse of the removal procedure. Press the coil straight down onto the spark plug. There should a slight "click" as the coil seats on the spark plug top.
8. Install the hold-down capscrews, and make sure the electrical connection is secure.
9. Install the coil cover and make sure any hose connections that were removed are now secure.

IGNITION AMPLIFIERS

Removal and Installation

There are 2 ignition amplifiers. They are mounted on the intake manifold. One feeds odd number cylinder coils, the other, even numbers.

1. Remove any ducting or hoses that may be in the way, matchmarking for assembly.
2. Remove the hold-down capscrews from the amplifier being serviced.
3. Carefully disengage the electrical connector and lift the amplifier from the vehicle.

To install:

4. Installation is the reverse of the removal procedure. Use care make sure the hold-down capscrews are firmly in place, but use caution not to over-torque when working around light alloy parts.
5. Make sure the electrical connection is secure.

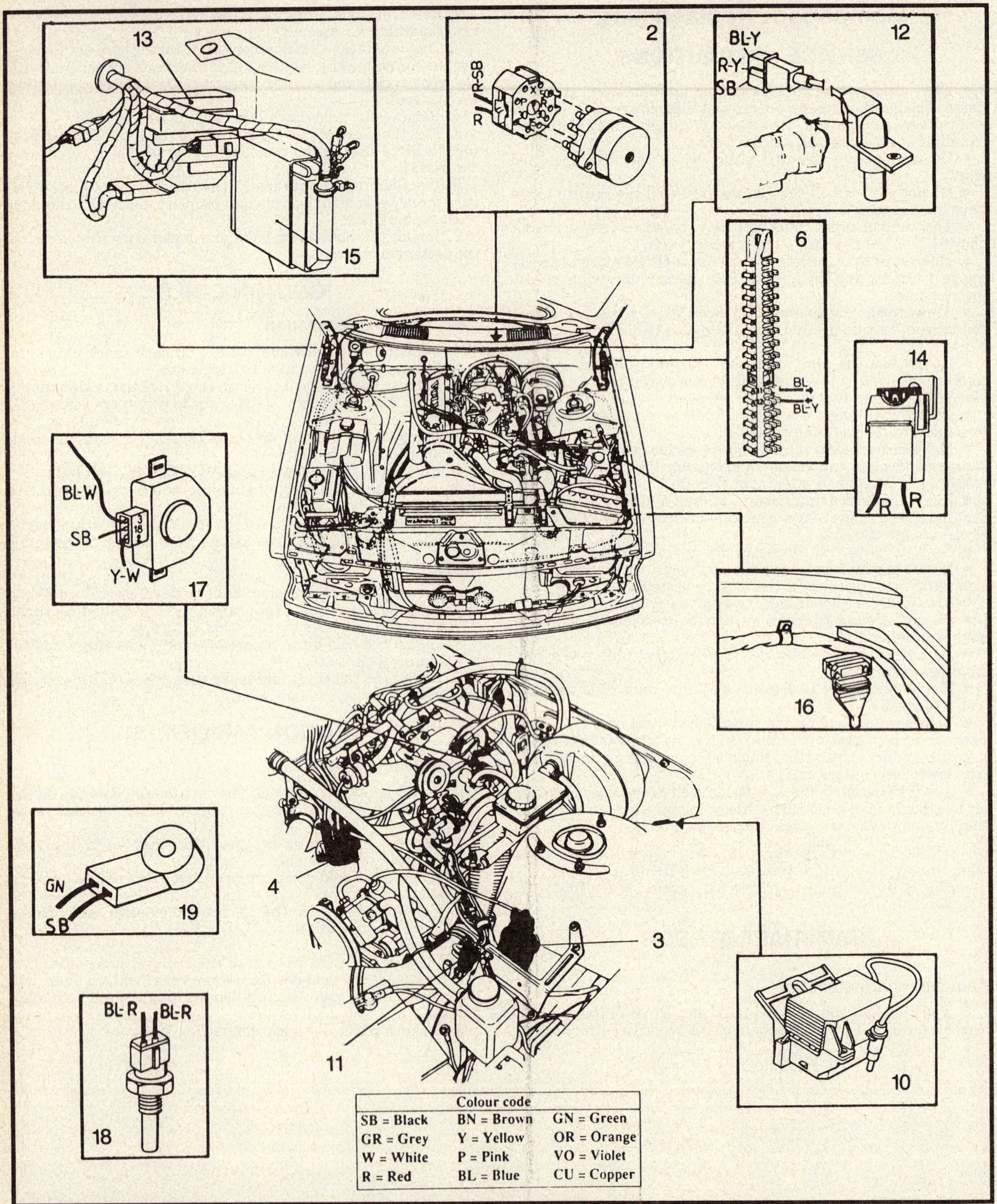

Colour code		
SB = Black	BN = Brown	GN = Green
GR = Grey	Y = Yellow	OR = Orange
W = White	P = Pink	VO = Violet
R = Red	BL = Blue	CU = Copper

EZ116K ignition system components–240 models with B230F engine

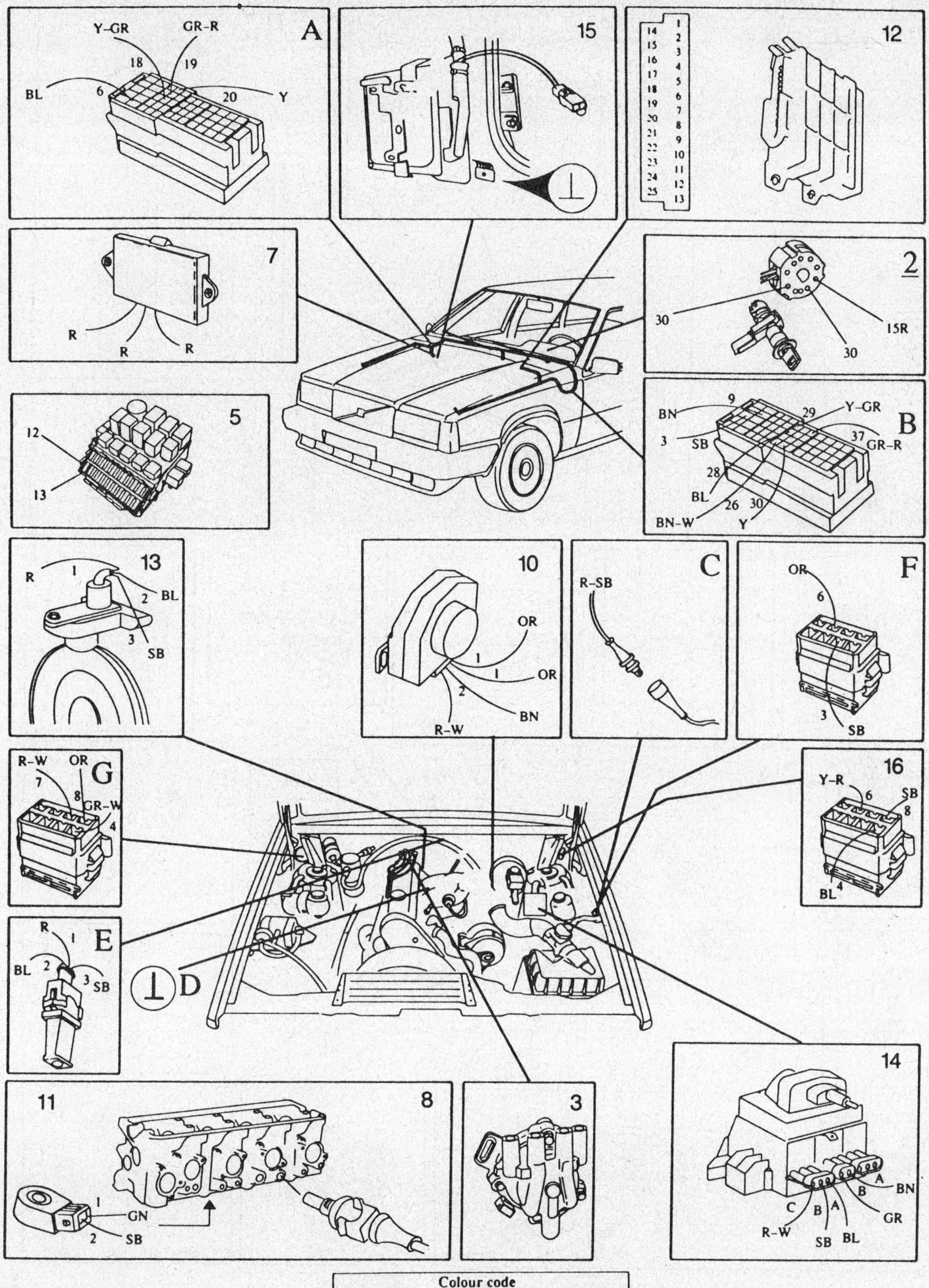

Colour code		
SB = Black	BN = Brown	GN = Green
GR = Grey	Y = Yellow	OR = Orange
W = White	P = Pink	VO = Violet
R = Red	BL = Blue	CU = Copper

REX 1 ignition system components–700 models with B230F engine

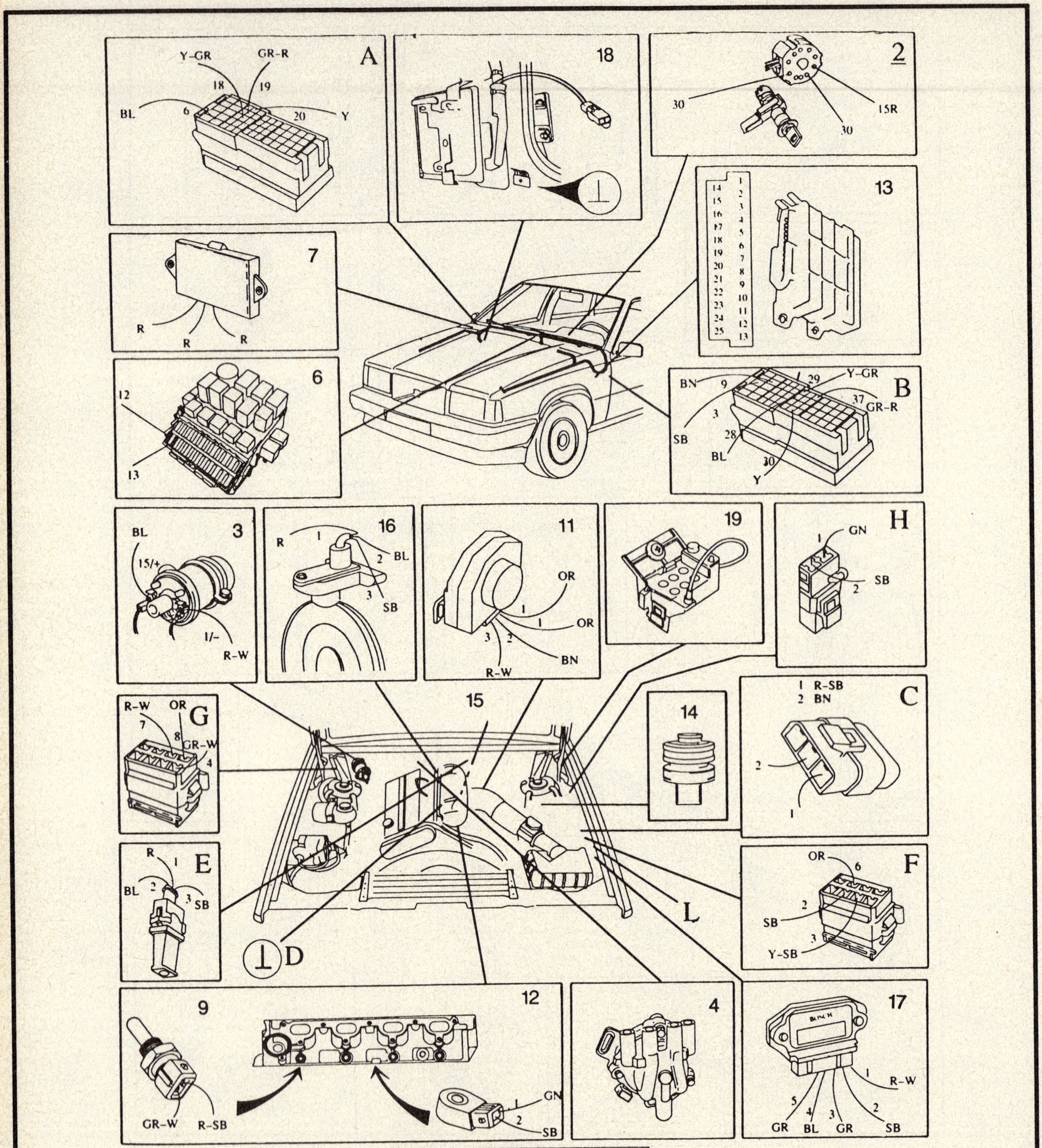

Colour code		
SB = Black	BN = Brown	GN = Green
GR = Grey	Y = Yellow	OR = Orange
W = White	P = Pink	VO = Violet
R = Red	BL = Blue	CU = Copper

EZ116K ignition system components–1990 700 models with B234F engine

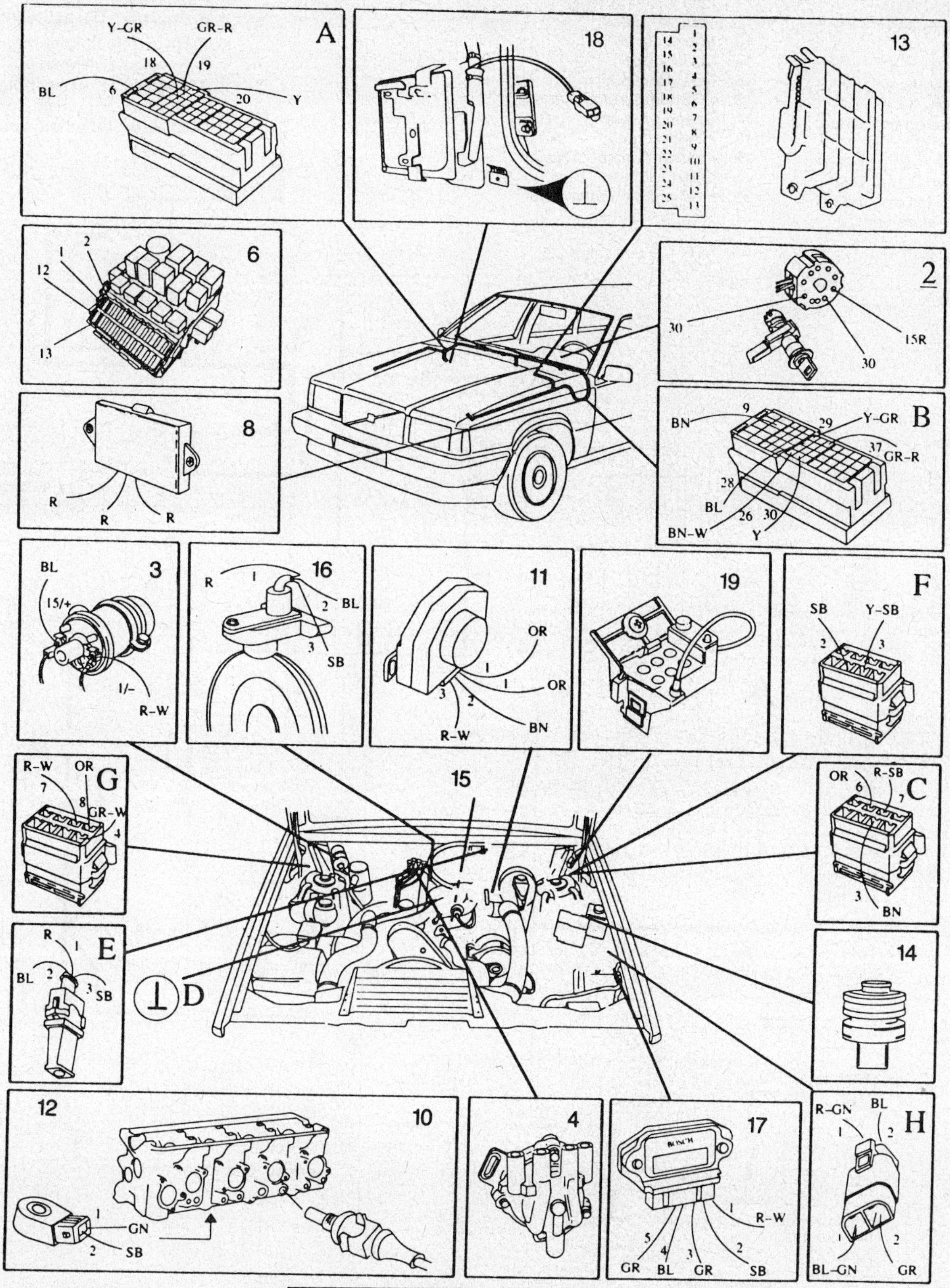

Colour code		
SB = Black	BN = Brown	GN = Green
GR = Grey	Y = Yellow	OR = Orange
W = White	P = Pink	VO = Violet
R = Red	BL = Blue	CU = Copper

EZ116K ignition system components–740 and 780 models with B230F turbocharged engine

1. Battery
2. Ignition switch
3. Ignition coil
4. Distributor
5. Spark plugs
6. Fusebox
7. Terminal on combined instrument
8. Ground busbar in central electrical unit
9. Positive terminal board
10. Temperature sensor
11. Fuel injection relay
12. Throttle switch
13. Knock sensor
14. Control unit
15. EGR converter
16. EGR temperature sensor
17. Speed pick-up
18. Power stage
19. Interference suppression relay
20. Fuel system control unit
21. Diagnostic unit

A Connector, RH A-post
B Connector, LH A-post
C Connector, LH wheel housing
D Earth terminal, inlet manifold
E Connector, bulkhead
F Connector, LH wheel housing
G Connector, RH wheel housing
H Connector, LH wheel housing

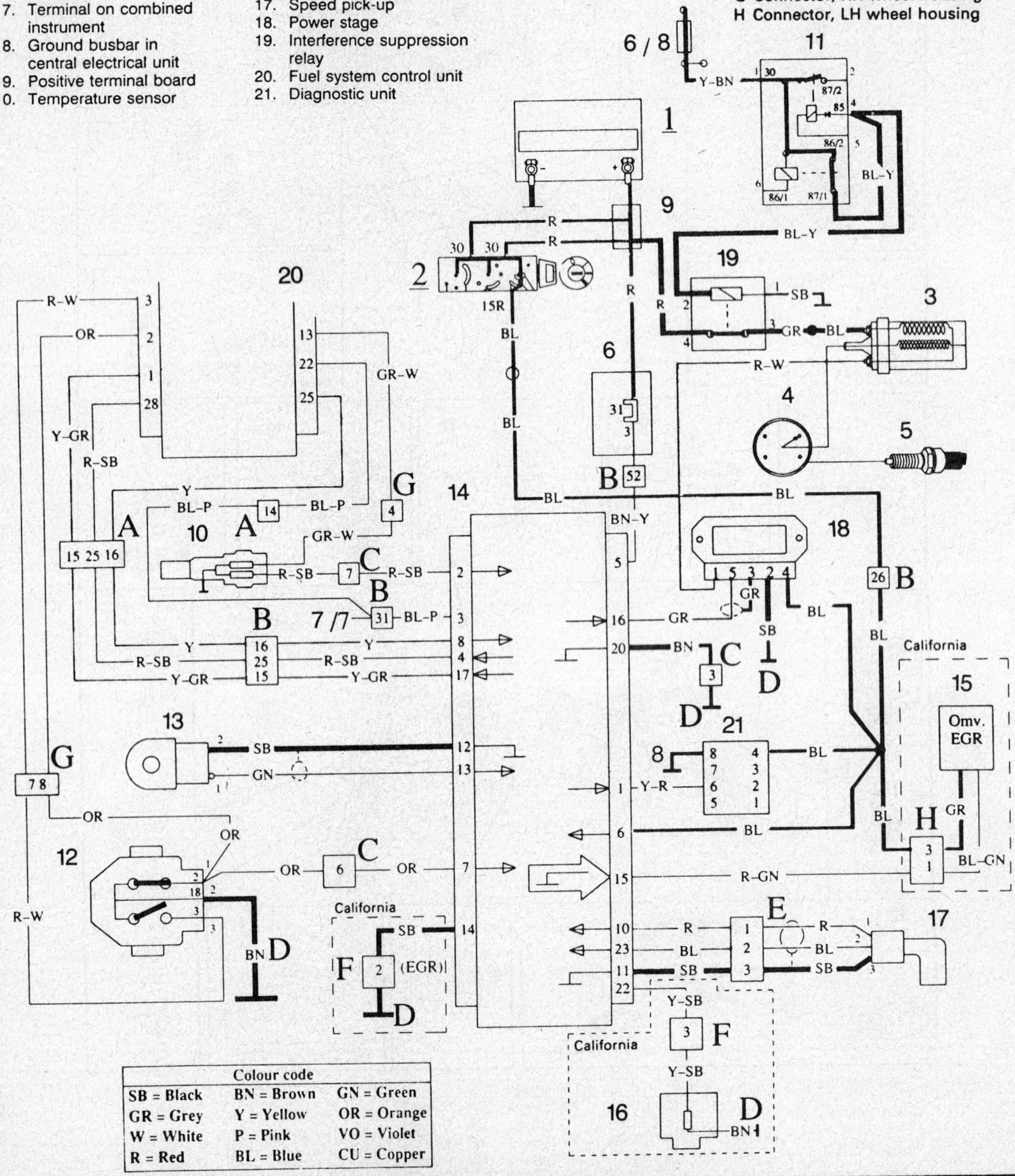

Colour code		
SB = Black	BN = Brown	GN = Green
GR = Grey	Y = Yellow	OR = Orange
W = White	P = Pink	VO = Violet
R = Red	BL = Blue	CU = Copper

EZ116K ignition system wiring–760 models with B230F turbocharged engine

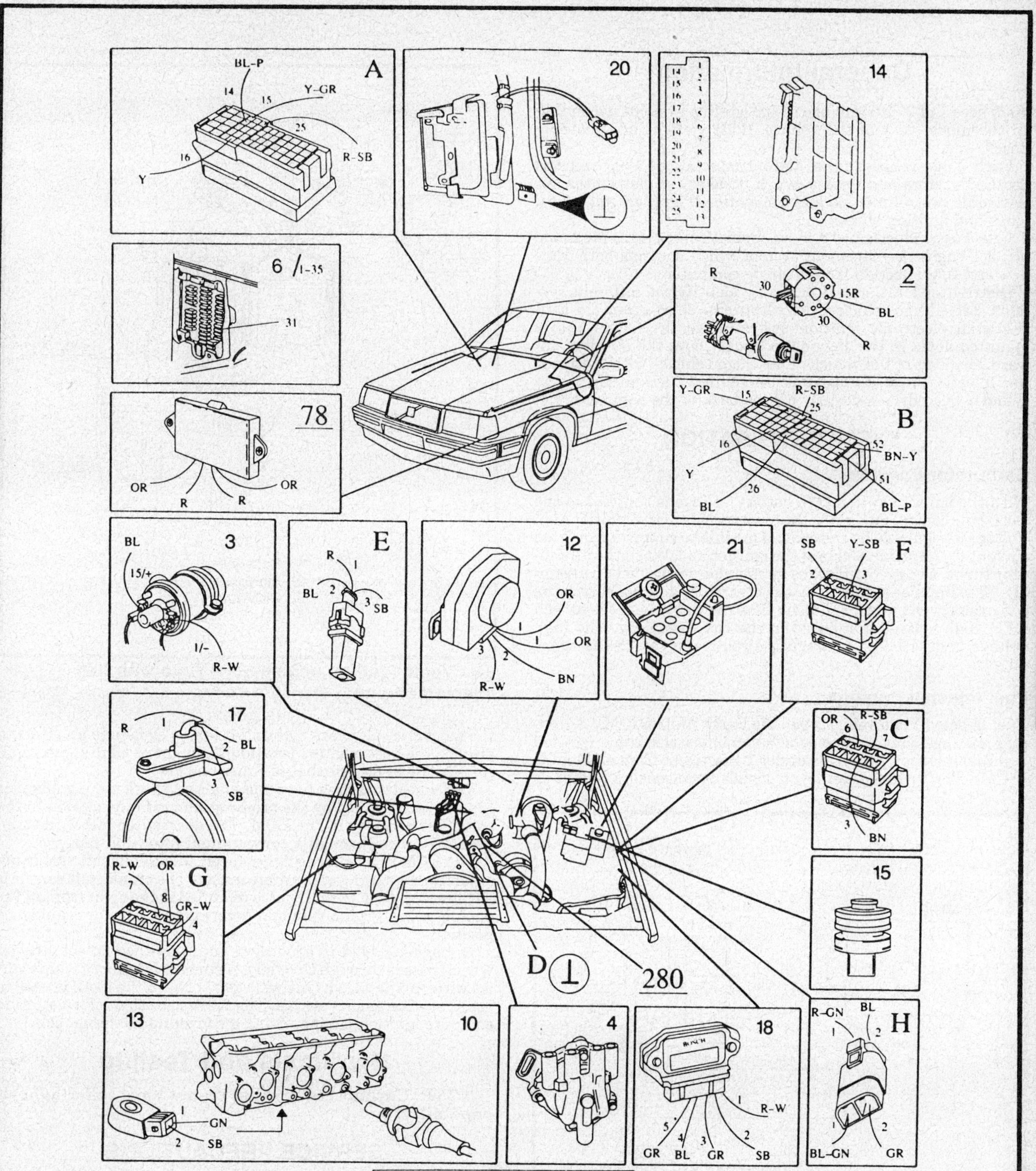

Colour code		
SB = Black	BN = Brown	GN = Green
GR = Grey	Y = Yellow	OR = Orange
W = White	P = Pink	VO = Violet
R = Red	BL = Blue	CU = Copper

EZ116K ignition system components–760 models with B230F turbocharged engine

YUGO

General Information

Yugo uses 1 of 2 different electronic ignition systems depending on the model year of the vehicle. Both systems are made by Bosch.

Early Yugo vehicles up to the introduction of 1990 vehicles had carburetors and used a Bosch breakerless distributor, an electronic control module and conventional ignition coil, spark plugs and ignition wires.

Late Yugo vehicles use a more sophisticated Bosch Motronic MP 3.1 engine management system which controls both ignition and fuel injection from a single computer.

Distributor location is 1 easy way identify the different systems. Early Yugo vehicles with carburetor engines and the conventional electronic ignition system have their distributors mounted down in the engine block just above the oil filter, toward the front of the vehicle. Late Yugo vehicles with Motronic fuel injection have the distributor mounted horizontally on the cylinder head driven directly off the back of the camshaft.

SYSTEM OPERATION

Carburetor Engines

With the ignition switch **ON**, battery voltage is supplied to the electronic control module to the primary side of the ignition coil. Voltage is regulated by the control module to supply a regulated current to the primary side of the ignition coil. When the distributor turns, a trigger generates an impulse on the pickup assembly. This impulse is sensed by the control module and turns the coil primary ON and OFF. Each time the coil primary is turned OFF, high voltage is induced in the coil secondary. The high voltage is distributed through the rotor and cap to the spark plugs.

Fuel Injected Engines

Fuel injected Yugo vehicles use the Bosch Motronic MP 3.1 engine management system which combines electronic ignition and fuel injection functions under 1 Electronic Control Module (ECM). The system depends on signals from sensors located in the most critical points within the supply system in an effort to provide the best engine operation throughout all driving conditions while keeping exhaust emissions low.

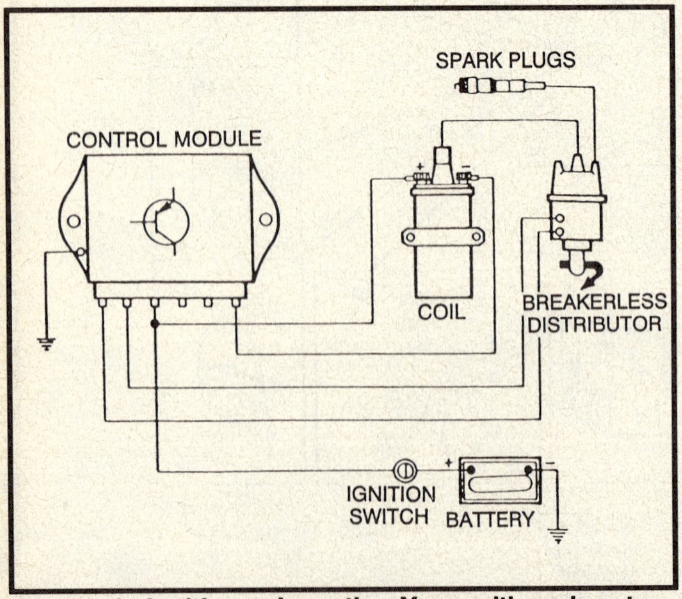

Electronic ignition schematic—Yugo with carburetor engine

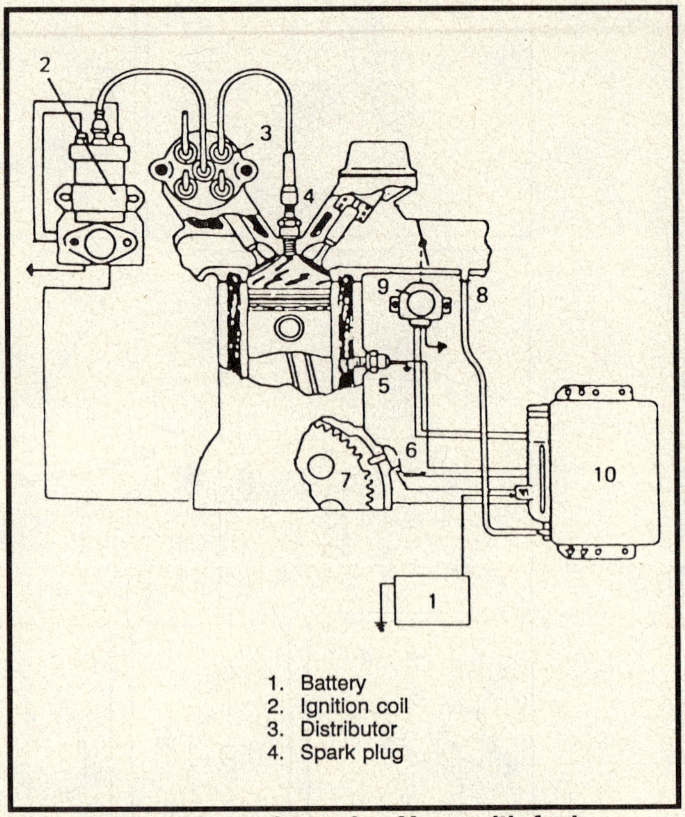

Electronic ignition schematic—Yugo with fuel injected engine

Fuel injected Yugos have the engine's spark timing advance functions controlled by the pre-programmed information in the computer memory in the ECM. The distributor merely distributes the spark through a conventional rotor and cap.

The ECM determines engine speed and crankshaft stroke position by means of a crankshaft sensor. The crankshaft sensor is a Hall-effect type switch that is mounted at the right front of the engine, reads engine rpm and determines stroke or crankshaft position.

A magnetic field is generated and a toothed wheel (trigger wheel) passes through this field. A current is induced which varies with engine speed. One wide gap between the teeth is used to reference the engine stroke. The ECM uses this information to compute ignition timing, firing a conventional design coil.

Diagnosis and Testing

NOTE: Carbureted engines do not have self-diagnosis capability.

SERVICE PRECAUTIONS

- Use care when working around high-energy electronic ignition systems. High voltage could cause injury to the technician and equipment.
- Before performing any service on the ignition system, make sure the ignition is turned **OFF** and/or the battery negative cable is disconnected.

• Before attempting to remove any parts, turn the ignition switch **OFF** and disconnect the battery ground cable.

• Never connect test equipment such as timing lights, etc., directly to the ignition system for a power source. Always use the vehicle battery for the power source.

• Use only the correct replacement parts. Use of spark plugs, for example, having specifications and sizes different from those specified could cause severe damage and alter emissions levels.

• Use care when working with light alloy parts. Always lubricate the spark plug threads with a suitable lubricant to avoid seizing the plug threads to the cylinder head.

• Keep all ECU parts and harnesses dry during service. Protect the ECU and all solid-state components from rough handling or extremes of temperature.

• Make sure all harness connectors are fastened securely. A poor connection can cause an extremely high voltage surge resulting in damage to integrated circuits.

• Always use a 12 volt battery as a power source.

• Do not disconnect the battery cables with the engine running; never run the engine with the battery cables loose or disconnected.

• Do not attempt to disassemble the ECU unit under any circumstances.

• When connecting or disconnecting pin connectors from the ECU, take care not to bend or break any pin terminals. Check that there are no bends or breaks on the ECU pin terminals before attempting any connections.

• When measuring supply voltage of ECU-controlled components, keep the tester probes separated from each other and from accidental grounding. If the tester probes accidentally make contact with each other during measurement, a short circuit will damage the ECU.

• Most vehicle manufacturers recommend that the ECU control unit be removed if the vehicle is to be exposed to high heat such as a bake oven to cure paint after body refinishing or when welding is be done on the vehicle.

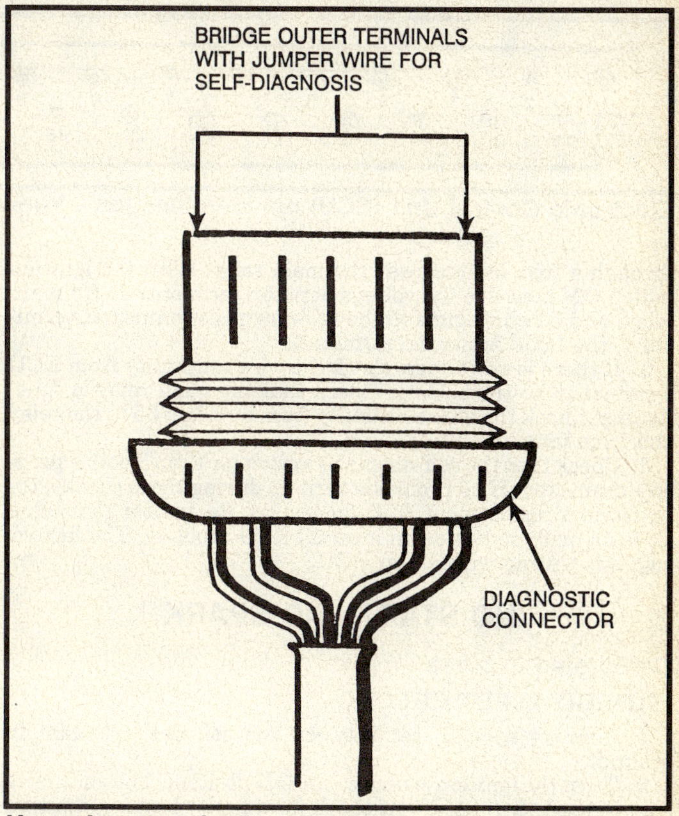

Use a jumper wire to connect the outer terminals together of the self-diagnostic connector to read blink codes—Yugo with fuel injected engine

READING CODES

Carbureted engines do not have self-diagnosis capability. The Motronic control unit used on fuel-injected engines will keep in memory any fault codes that have occurred during running. There are 2 methods of retrieving the fault codes from the control unit. The codes can be read using the CHECK ENGINE light which will flash codes. The CHECK ENGINE light is located on the left side of the instrument panel, just to the left of the Check Transmission indicator. Codes can also be retrieved via the Bosch KTS 300 Diagnostic Tester or equivalent "Scan Tool" if available.

Note that the blink code numbers do not match the code numbers retrieved by the diagnostic tester. When using the code chart be sure that the correct code form is being used.

To activate the flash codes a switch (or jumper wire) is connected between pin 1 and pin 4 (the 2 "outer" terminals of the connector) of the 4-terminal diagnostic connector. The 4 pin diagnostic connector is located in the engine compartment on the left side near the fire wall. These terminals are bridged with a switch or a suitable jumper wire. Enter the self-diagnosis system using the following procedure:

1. Turn the ignition switch to the **ON** position. The Check Engine light should come ON.
2. Start the engine. If no faults are stored, the light should extinguish during cranking or immediately after start.
3. If the light extinguishes 3 seconds after the start there is at least 1 fault in memory. This is a secondary fault indicating a problem, but it does not affect emission quality.
4. If the light stays on with the engine running there is a fault that affects emission quality.
5. Turn the engine OFF. Turn the ignition switch to the **ON** position. Note that faults can be read when the engine is OFF, or when running at less than 2000 rpm.
6. With the self-diagnostic switch (or jumper wire) connected between pins 1 and 4 of the self diagnostic connector, press the switch for 2.5–10 seconds.
7. The Check Engine light will extinguish and after about 3.5 seconds the Check Engine light will flash once, then twice. This indicates the Code 12 start code.
8. Press the self diagnostic switch again for 2.5–10 seconds. The next code will flash. Each additional code will need to be activated by pressing the self diagnostic switch. Code 11 is the end code.

Most Motronic self-diagnostic codes relate to the fuel injection system. Ignition troubleshooting will mostly be confined to blink Code 54 which indicates a defective ECU or blink Code 41 which indicates a problem with the RPM/TDC sensor.

CLEARING TROUBLE CODES

To erase trouble codes from the computer memory, after the End Diagnostics Code 11, input a new call for diagnostics, holding the switch (or jumper wire) for longer than 10 seconds. This should cause a deletion of all stored failures in the ECU.

An alternate method deleting stored Codes is to disconnect the battery terminals for at least 10 seconds.

Testing: Self-Diagnosis Not Working

If the self-diagnosis system doesn't work, perform the following tests:

1. Measure the electronic control unit voltage supply. At the ECU multi-pin connector, there should be battery voltage at the ECU on terminal 18. This should be "hot" right to the battery,

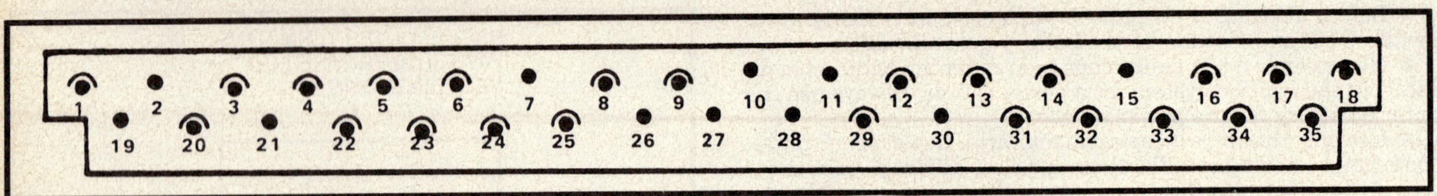

Electronic Control Unit (ECU) pin identification—Yugo with fuel injected engine

through a fuse located near the main relay. With the ignition switch **ON** measure the voltage between terminals 35 (to main relay) and 5 (which runs to the self-diagnosis connector, terminal 1, the Code Actuation terminal).

2. If there is no voltage, check the wire continuity from ECU terminal 18 to the battery. Check that the main relay is functioning. Check the wire continuity from terminal 35 to the relay and from terminal 5 to ground.

3. Check that the self-diagnosis switch (or Scan Tool) is properly connected. If no problems turn up during these checks, try to obtain a replacement ECU for testing. Remember that blink Code 54 or 01 indicates an internal ECU problem. The factory suggests testing with a new ECU.

NO START—NO SPARK

Carbureted Engines

PRIMARY INPUT CHECK

1. Connect a voltmeter between the coil **B+** terminal to ground.
2. Turn the ignition switch to the **ON** position. The voltmeter should read 12 volts. If not check for a faulty battery, ignition switch, wiring or connections.

GROUND TEST

1. Turn the ignition switch **OFF**. Connect an ohmmeter from the control module support mount to the battery ground terminal. The ohmmeter reading should be 0.2 ohms or less.
2. If not, check the module support, mounting and the battery ground connections.
3. Check that the control module casing is clean and that the mounting hardware is clean and tight.

COIL RESISTANCE TEST

1. Disconnect the primary leads from the coil and connect an ohmmeter to the primary terminals of the coil. Note the reading on the ohmmeter. It should be 0.7 ohms.
2. Remove 1 lead from the primary terminal and place it into the high tension tower of the coil. Note the reading; it should be 9500 ohms. Replace the coil if not within specifications.

PICK-UP ASSEMBLY TEST

1. Disconnect the harness from the control module. Connect an ohmmeter to the pick-up assembly connector. The reading should be 990–1200 ohms.
2. Remove 1 ohmmeter lead from the pick-up connector and touch it to the distributor body. The reading should show infinity. Replace the pick-up assembly if not within specifications.
3. Using a nonmagnetic feeler gauge, check the gap between the stator pole and the trigger. Adjust as required. The gap should be 0.011–0.019 in. (0.3–0.5mm).

CONTROL MODULE TEST

1. Reconnect the primary leads to the coil, connect the pick-up assembly to the control module.
2. Disconnect the high tension wire from the distributor cap, but not from the coil.
3. Using an insulated holding tool, hold the high tension wire about 1/4 in. from the ground, crank the engine and check for spark.

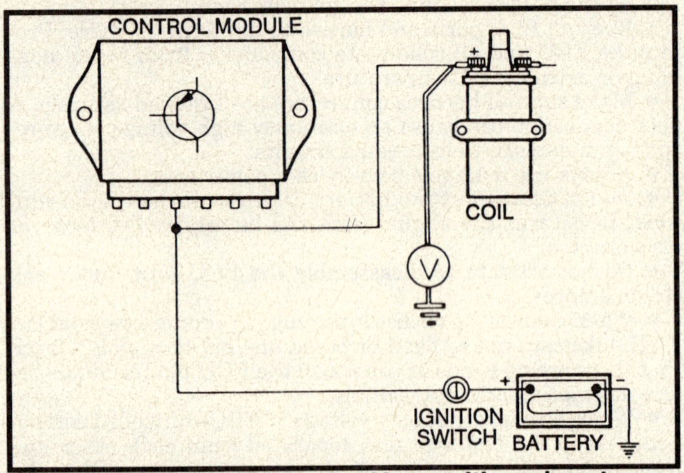

Primary circuit input check—Yugo with carburetor engine

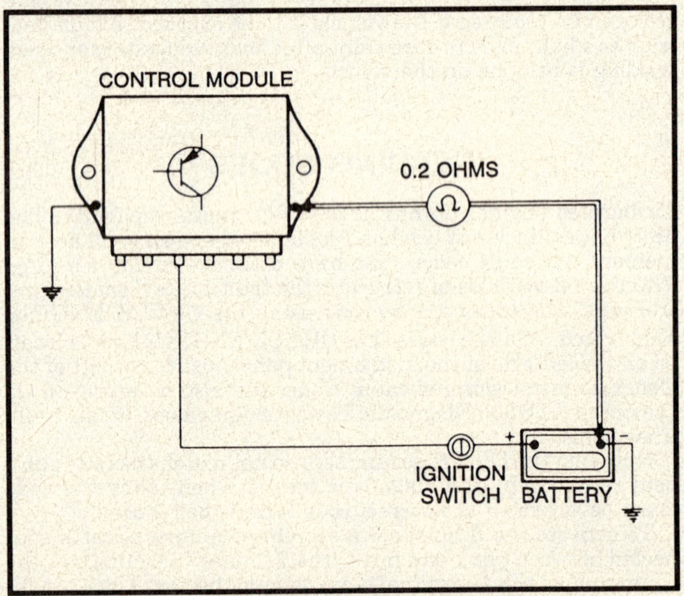

Ground circuit test—Yugo with carburetor engine

NOTE: If no spark is evident and all other tests up to this point check good, replace the control module.

Fuel Injected Engines

IGNITION COIL CHECK

1. Make sure all electrical connections are secure at the coil and control transistor located on the coil mount.
2. Check the coil for corrosion, damage or carbon tracks on the coil tower.
3. Check the ignition coil primary circuit. Place ohmmeter

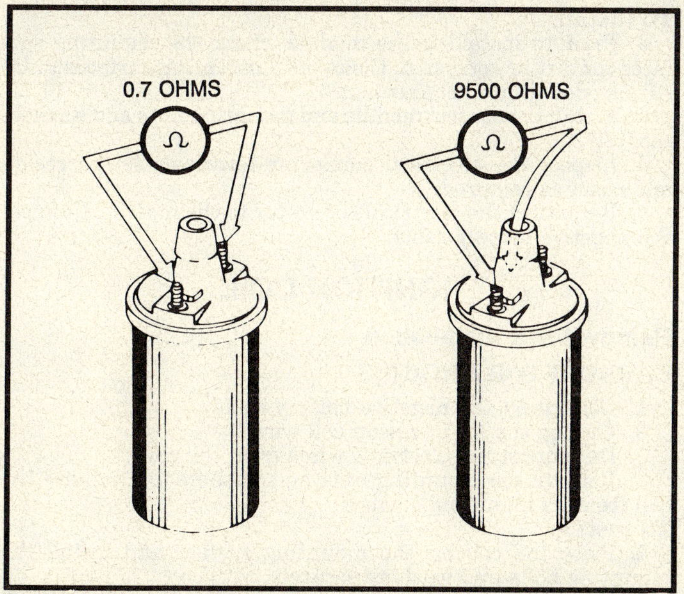

Coil resistance check—Yugo with carburetor engine

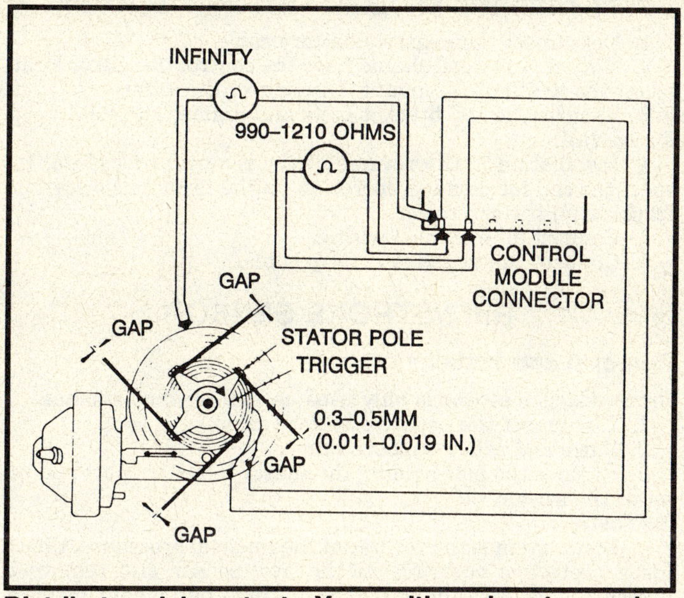

Distributor pick-up test—Yugo with carburetor engine

probes on each of the small coil terminals. Note that coil terminal 15 is positive and coil terminal 1 is negative. Primary resistance should be 0.6–1.0 ohms.

4. Check the ignition coil secondary circuit. Place 1 ohmmeter probe on the terminal 15, next to the control transistor the other probe in the coil center socket. Secondary resistance should be 6400–11000 ohms.

5. If the coil is out of specification, replace it.

6. Check the control transistor. Fold back the connector rubber cover, but do not disconnect. Measure the rear side of the connector with the ignition **ON**. Check for battery voltage between contacts 2–3. If there is no voltage, check if there is 12 volts from the ignition switch contact 15.

7. Check for ground continuity. The control transistor terminal 2 should be ground. Use an ohmmeter to check for ground.

8. If there is still no spark, measure the resistance between contacts 4–2 of the control transistor and contact 2 at the wire end to the negative terminal of the ignition coil.

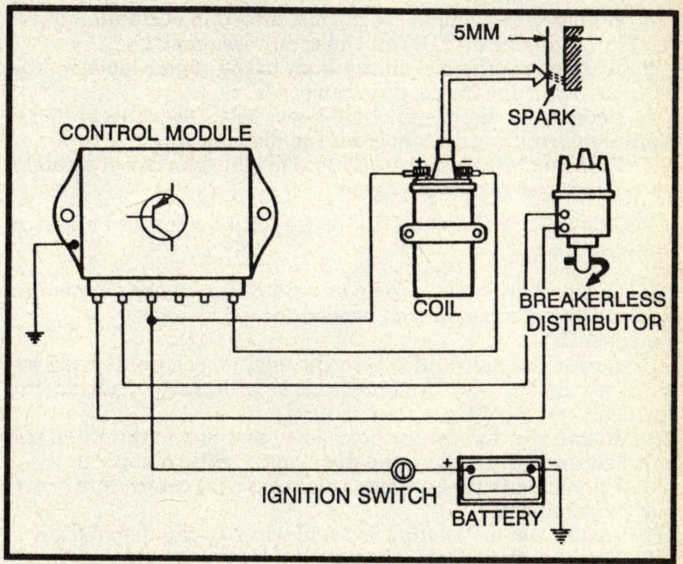

Control module test—Yugo with carburetor engine

a. Resistance between contacts 4–2 should be 740±50 ohms with the positive probe of the ohmmeter on the base of contact 4.

b. Resistance between contacts 4–2 should be 770±50 ohms with the negative probe of the ohmmeter on the base of contact 4.

c. Resistance between contact 4 and the terminal at the end of the wire to ignition coil negative terminal (blue wire) should be 720±50 ohms with the positive probe of the ohmmeter on the base of contact 4.

d. Resistance between contacts 4 and the terminal at the end of the wire to ignition coil negative terminal (blue wire) should be infinite with the negative probe of the ohmmeter on the base of contact 4.

e. Resistance between contact 2 and the terminal at the end of the wire to ignition coil negative terminal (blue wire) should be infinite.

9. If the above resistance test results are not within specifications, replace the transistor.

10. Check the distributor cap and center contact. Measure the resistance of the distributor rotor. It should be 1000–1200 ohms.

RPM/TDC SENSOR CHECK

A defective RPM/TDC sensor should set a blink Code 41.

1. Check wires and connector for damage.
2. Check clearance between sensor and sprocket. Clearance should be 0.020–0.040 in. (0.5–1.0mm).
3. Measure the resistance of the sensor. It should be 540 ohms, plus or minus 10 percent. If out of specification, replace sensor.

Component Replacement

DISTRIBUTOR

Removal and Installation

CARBURETOR ENGINES

1. Disconnect the negative battery cable.
2. Remove the distributor cap from the distributor with the wires attached.
3. Disconnect the vacuum advance line from the distributor vacuum advance diaphragm.

4. Rotate the engine in the normal direction of rotation to set the No.1 cylinder at TDC on the compression stroke.
5. Matchmark the distributor body to the engine block in-line with the rotor for installation purposes.
6. Loosen the distributor hold-down bolt. Disconnect any remaining wiring connectors from the distributor.
7. Remove the hold-down bolt and lift the distributor straight up removing it from the engine.

NOTE: Do not rotate the engine with distributor removed.

8. Inspect the rotor, pick-up and stator for cracks or damage. Replace any components as needed.

To install:

9. Lower the distributor into the engine, positioning the rotor in alignment with the matchmarks previously made during removal.
10. Install the distributor hold-down and bolt and tighten the bolt just enough to allow the distributor to turn slightly.
11. Connect any vacuum lines and electrical connectors previously disconnected.
12. Install the distributor cap and wires to the distributor.
13. Connect the negative battery cable.
14. Start the engine and adjust the timing. Tighten the distributor hold-down bolt.

FUEL INJECTED ENGINES

1. Disconnect the negative battery cable.
2. Remove the distributor cap from the distributor with the wires attached.
3. For references, rotate the engine in the normal direction of rotation to set the No.1 cylinder at TDC on the compression stroke.
4. Remove the distributor holddown fasteners and remove the distributor carefully. Note the relationship between the distributor drive end and the camshaft drive slots for installation purposes.

NOTE: Do not rotate the engine with distributor removed.

5. Inspect the rotor and distributor cap for cracks or damage. Replace any components as needed. There is a replaceable O-ring seal under the rotor mount which may be replaced, if necessary.

To install:

6. Assemble the distributor to the cylinder head, making sure the distributor drive properly engages the end of the camshaft.
7. Install the distributor holddown fasteners.
8. Install the distributor cap and wires to the distributor.
9. Connect the negative battery cable.

CONTROL MODULE

Removal and Installation

1. Disconnect the negative battery cable.
2. Disconnect the electrical connector from the electronic control module.
3. Remove the mounting nuts and washers from the module support and remove the module from the vehicle.

To install:

4. Prior to installing the module, check the mounting surfaces for dirt or corrosion. Clean both mounting surfaces to insure a good ground connection.
5. Install the control module and mounting nuts and washers securely.
6. Inspect the electrical connector terminals for corrosion and clean as required.
7. Reconnect the electrical connector to the module. Connect the negative battery cable.

IGNITION COIL

Removal and Installation

CARBURETOR ENGINES

1. Disconnect the negative battery cable.
2. Unplug the high tension coil wire.
3. Disconnect the electrical wires from the coil.
4. Remove the mounting nuts and washers and remove the coil from its mounting surface.

To install:

5. Place the coil on the mounting surface and install the mounting bolts and washers securely.
6. Connect the electrical wires to the coil and plug in the high tension coil wire.

FUEL INJECTED ENGINES

1. Disconnect the negative battery cable.
2. Note that the coil should have the control transistor located on the ignition coil mount. Disconnect the wires.
3. Remove the coil bracket bolts and remove the coil.

To install:

4. Installation is the reverse of the removal procedure. Inspect the coil for damage, corrosion on the terminals or carbon tracking on the coil tower.
5. Connect all electrical wiring.
6. Connect the negative battery cable.

RPM/STROKE SENSOR

Removal and Installation

An rpm/stroke sensor is only used on fuel injected engines.

1. Disconnect the negative battery cable.
2. Raise and safely support vehicle.
3. Remove the bolt holding the sensor to the support and remove the sensor.

To install:

4. Installation is the reverse of the removal procedure. Check for a protective covering on the new sensor and remove if present. Note that the sensor contains a permanent magnet. Use care not to bump or damage the sensor and keep the tip free of metal chips.
5. Carefully install the sensor making sure it is properly seated. The contacts must be clean.
6. Adjust the clearance between the sensor and sprocket. It should be 0.020–0.040 in (0.5–1mm).

Note: The ignition system components are interrelated with fuel system components. For additional testing, replacement or wiring diagrams see appropriate fuel system section.

Alfa Romeo Fuel Injection

5 SECTION

ALFA ROMEO

General Information

Alfa Romeo uses the Bosch Motronic ML 4.1 electronic ignition and fuel injection system. The system depends on signals from sensors located in the most critical points within the supply system in an effort to provide the best engine operation throughout all driving conditions while keeping exhaust emissions low.

ELECTRONIC CONTROL MODULE

The Electronic Control Unit (ECU) uses sensors to monitor engine rpm, A/C ON or OFF, intake air temperature, engine load, exhaust gas emissions, crankshaft position, intake air quantity, engine temperature and battery voltage. Sensors monitor both air flow and air temperature as well as engine coolant temperature.

The integrated Motronic system controls both ignition and fuel injection. It is equipped for self-diagnosis to help troubleshoot problems.

DATA SENSORS

Data sensors are those which provide input information to the ECU.

Engine Temperature Sensor

The engine temperature sensor decreases resistance as temperature increases. It is powered by a constant voltage from the ECU.

One end is in contact with the engine coolant. The electrical current from the ECU flows through a resistor inside the sensor. It changes in proportion to the engine temperature. On 4-cylinder engines, the sensor is mounted on the side of the block. On 6-cylinder engines, the temperature sensor is mounted on the water pump body.

Airflow Meter Assembly

The airflow meter assembly measures the engine intake air quantity through a throttle valve controlled by the accelerator pedal and sends to the ECU a signal used to determine fuel injection timing.

A spring pushes against the throttle plate while incoming air pressure tends to move the throttle plate. A given quantity of air corresponds to the throttle valve angle, so the incoming air can be measured. An electric signal is generated allowing the ECU to monitor the air flow.

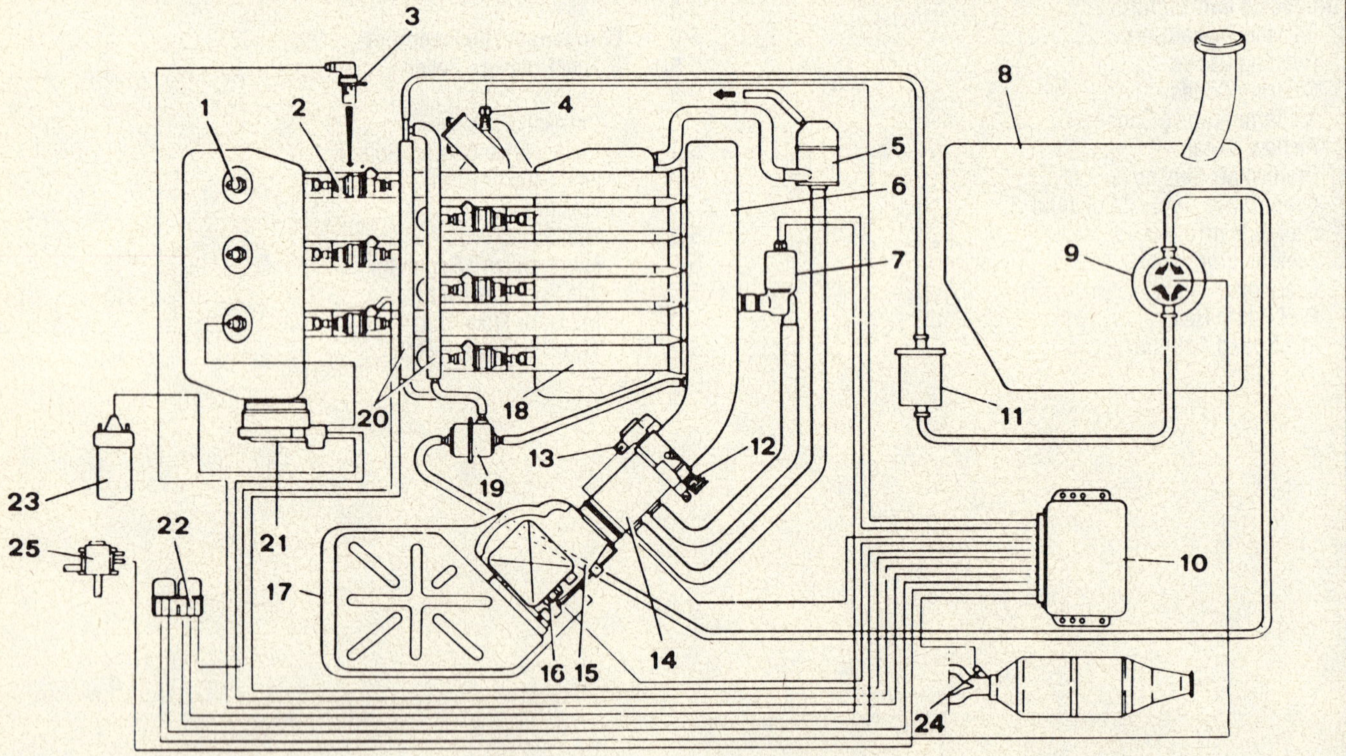

1. Spark plugs
2. Electronic fuel injectors
3. RPM and stroke sensor
4. Engine coolant temperature sensor
5. Oil fume separator
6. Air intake box
7. Idle actuator
8. Fuel tank
9. Fuel pump
10. Motronic ECM
11. Fuel filter
12. Throttle valve
13. Throttle position sensor (min/max switch)
14. Air intake duct
15. Airflow meter
16. Inlet air temperature sensor
17. Air cleaner
18. Intake manifolds
19. Fuel pressure regulator
20. Fuel rails
21. Ignition distributor
22. Relays
23. Ignition coil
24. Oxygen sensor
25. Evaporative emission canister solenoid valve

Basic Motronic engine management system—Alfa Romeo 164 V6 shown, others similar

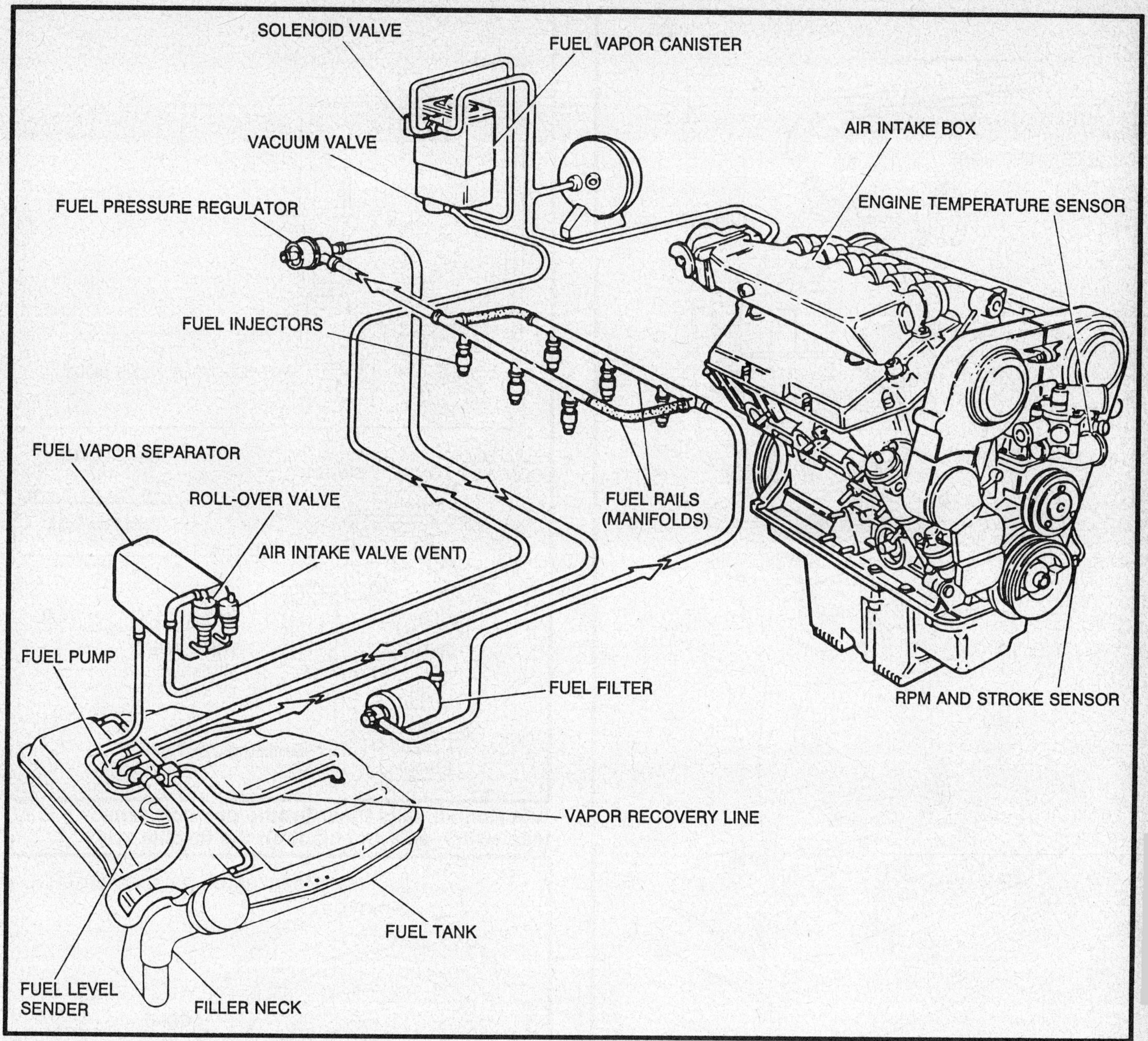

Fuel supply and vapor recovery system—Alfa Romeo 164 V6

Intake Air Temperature Sensor

The intake air temperature sensor is mounted in the air flowmeter assembly. It measures the temperature of the incoming air. The ECU uses this information to determine air density variations. This affects fuel injection timing.

Min/Max Opening Switch

The min/max opening switch is mounted to the throttle valve body. The throttle body controls the quantity of air sent to the manifold according to the accelerator pedal position.

The switch assembly includes 2 microswitches, 1 MIN switch that sends to the ECU a signal proportional to the throttle angle between fully closed and 1 degree. The MAX switch provides a signal proportional to throttle angles higher than 72 degrees.

Both switches provide a signal to block fuel during deceleration (cut-off) and increase fuel quantity as more power is demanded (enrichment).

The throttle body of vehicles equipped with automatic transmission includes a kick-down cam which engages a lower gear when the accelerator pedal is floored. In addition the 164S is equipped with controlled damping suspension which includes a sensor on the throttle which senses throttle openings greater than 20 degrees.

Engine RPM and Stroke Sensor

The engine rpm and stroke sensor is a Hall-effect type switch that detects the engine revolutions and stroke. The sensor is mounted on the front of the engine down near the crankshaft harmonic balancer. A magnetic field is generated and a toothed wheel (trigger wheel) passes through this field. A current is in-

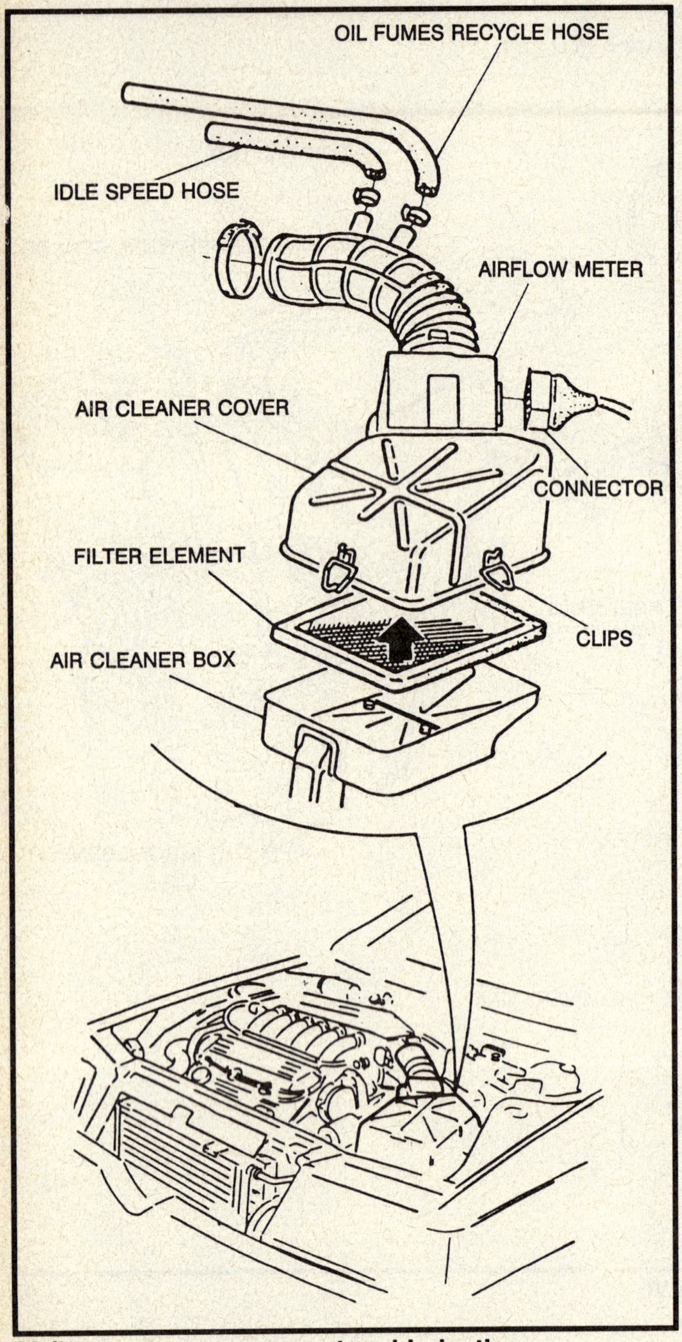

Airflow meter arrangement and induction components—Alfa Romeo

duced which varies with engine speed. One wide gap between the teeth is used to reference the engine stroke.

Oxygen Sensor (Lambda Probe)

The Oxygen Sensor (Lambda Probe) is mounted in the forward part of the catalytic converter. It is in contact with the exhaust gases and generates an electrical signal whose voltage depends on the oxygen concentration in the exhaust gases. The electrical signal allows the oxygen sensor to inform the ECU of the air/fuel mixture. This information allows the ECU to decide which adjustments are necessary to revise the air/fuel for lowest exhaust emissions.

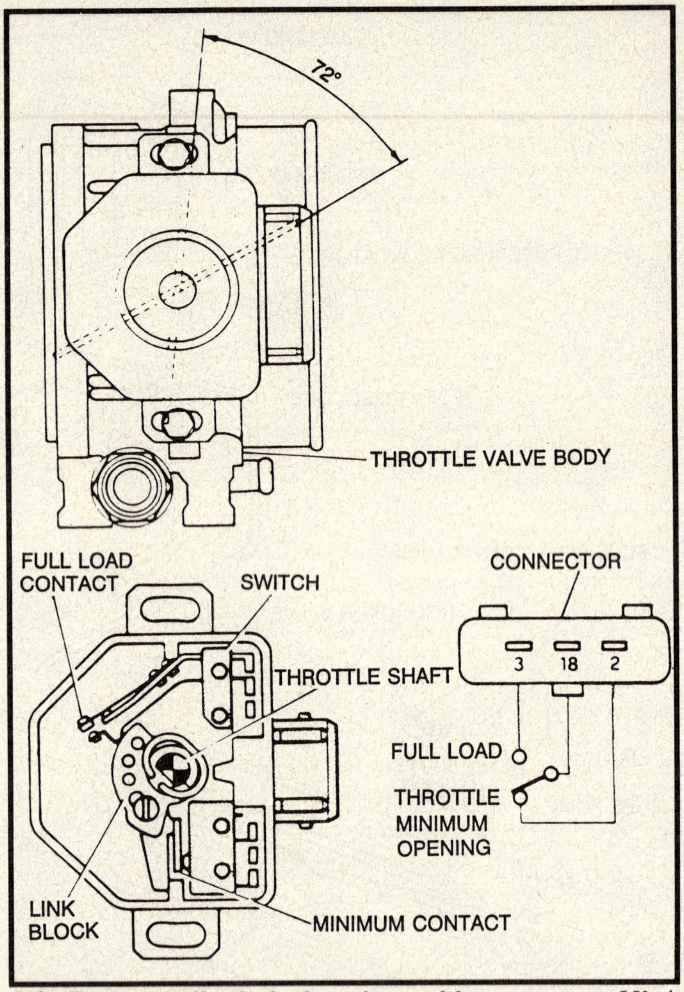

Alfa Romeo calls their throttle position sensor a Min/Max switch and mounts it on the throttle body

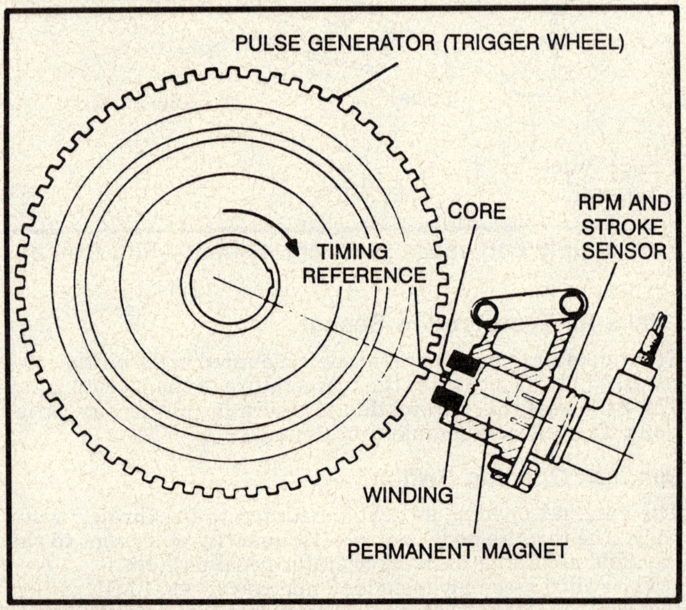

The RPM and stroke sensor mounts near the crankshaft pulley and generates a signal from the toothed trigger wheel

CONTROLLED OUTPUTS

Fuel Pump

The fuel pump(s) are powered only when the ignition is **ON** and the engine is running. This is to avoid fuel spills from ruptured lines or leaks in case of an accident.

On some models, 2 fuel pumps may be used. One pump is in the tank. The other pump is mounted under the vehicle next to the fuel filter.

Ignition Coil

The ignition coil is of conventional design and produces a high voltage capable of providing a spark of duration and intensity required to fire the spark plugs. It is controlled by the ECU.

On most vehicles, the coil is located inside the engine compartment on a bracket near the radiator.

Fuel Injectors

Each cylinder is supplied fuel by 1 electromagnetically operated injector that is connected in parallel with the others. The injector consists mainly of a coil, magnetic core and a needle guide. The magnetic coil moves the needle which is pressed by a spring on the injector sealing seat. The injector needle is actuated by a magnetic field generated by the coil on command from the ECU.

Idle Speed Actuator

The Idle Speed Actuator (IAC) determines how much air is admitted in to the engine during startup and to maintain the engine at idle in any condition.

The actuator is located on the throttle body bypass line. The actuator flow-through section provides at idle an air flow controlled by the accelerator and by the ECU. The actuator is a rotating regulator, working against a closure spring.

Vapor Control Solenoid Valve

The ECU activates the vapor control solenoid to allow fuel vapors stored in the canister to enter the intake manifold. This occurs when the engine is at normal temperature and under load.

FUEL DELIVERY SYSTEM

Fuel is pulled from the tank by an electric fuel pump. Some models have 1 pump in the tank, another on the underbody. A pressure regulator attached to the fuel rail adjusts fuel pressure according to intake vacuum. Fuel in excess of 45 psi on the Spider or 58 psi on the model 164, is returned to the fuel tank.

Fuel Pumps

The primary, or main fuel pump is an electric roller-type pump

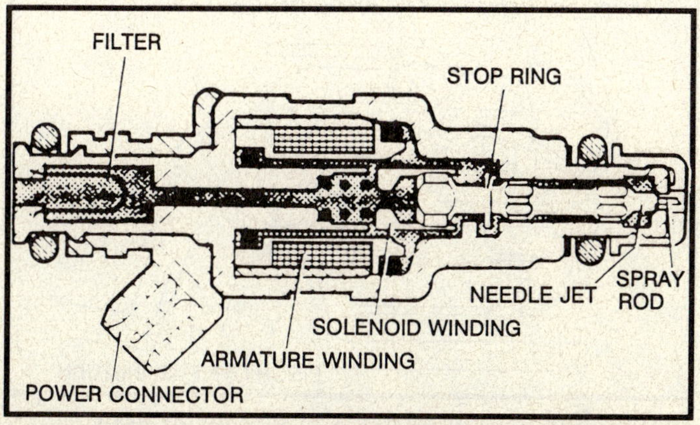

Cutaway of electronically controlled fuel injector

mounted under the vehicle next to the in-line fuel filter. The secondary fuel pump is also an electric pump, mounted inside the fuel tank.

Electric current is supplied to the pumps only when the ignition key is **ON** and the engine running to avoid fuel spills in an accident. The power is controlled by the ECU, through a fuel pump relay.

Fuel Filter

An in-line fuel filter is mounted on the vehicle underbody next to and downstream from the primary fuel pump. The OEM specification is for a 10 micron filter. This is to protect the finely-machined fuel injector nozzles.

Fuel Pressure Regulator

The main purpose of the fuel pressure regulator is to keep a constant fuel pressure based on intake manifold vacuum. In this way, the fuel flow from the injectors to the cylinders is based only on injector opening time, precisely controlled by the ECU. Vacuum from the intake system operates against a diaphragm. When the pressure exceeds the limit, a valve opens and excess fuel is directed back to the fuel tank.

Fuel Injectors

Fuel injectors are conventional high pressure electromagnetic units. They are controlled by the ECU. The injectors fit into the fuel rail, sealed by O-rings. The injectors are not repairable and must be replaced if defective.

Diagnosis and Testing

SERVICE PRECAUTIONS

- Always keep a suitable fire extinguisher close by.
- Always use care when handling fuel. Keep excess fuel in a suitable, approved safety container.
- Fuel system pressure must be relieved before disconnecting any fuel lines or attempting to remove any fuel system components.
- Do not operate the fuel pump when the fuel lines are empty.
- Do not disconnect the injector harness connectors with the engine running.
- Do not apply battery power directly to the injectors.
- Disconnect the ground cable from the battery to avoid accidental short circuits, sparks and fire.
- Always check for leaks after servicing the fuel system.
- Keep all parts and harnesses dry during service.
- Before attempting to remove any parts, turn **OFF** the ignition switch and disconnect the battery ground cable.
- Always use a 12 volt battery as a power source.
- Do not attempt to disconnect the battery cables with the engine running or the ignition key **ON**.
- Never smoke or allow any open flame when servicing the fuel system.
- Be especially careful when working around the fuel system using a drop light. An explosion and fire will occur if cold fuel sprays onto a hot drop light bulb.

READING CODES

The Motronic control unit is equipped with self-diagnostic system allowing a prompt detection of malfunctions related to the emission, fuel or ignition system.

The CHECK ENGINE light, located on the instrument cluster, will only illuminate when a malfunction is emission related.

Trouble codes are composed of 4 flashing groups, each of which is a code digit. The first is the ECU identification: 1 signifies Motronic, 2 signifies the method of diagnosis, 3 signifies malfunction signal, 4 signifies active diagnosis of actuators. The third and fourth are the specific number of the malfunction.

Spider

The diagnosis procedure can occur only with the ignition **ON** and engine speed up to 2000 rpm. The malfunction signal will continue through the entire rpm range once the diagnosis has started.

To initiate the diagnosis process, press the push button located under the rear parcel shelf, near the electronic control unit. The CHECK ENGINE light will provide the malfunction code number by emitting a series of flashes. Each code is endlessly repeated until the next code is requested. Code 4444 is used to signify a no malfunction condition.

164

The diagnosis procedure is divided into 2 groups—display of errors stored by the control unit and monitoring of engine actuators by the control unit.

DISPLAY OF ERRORS STORED BY THE CONTROL UNIT

On the 164, to initiate the diagnosis process, press the push button located at the bottom of the dash near and to the left of the steering column for 2.5–5.0 seconds. The CHECK ENGINE light will provide the malfunction code number by emitting a series of flashes. Each code is repeated until the next code is requested. Code 4444 is used to signify a no malfunction condition.

When all malfunctions have been displayed, the CHECK ENGINE light will display a series of long flashes.

By pressing the push button for 5–10 seconds, the series of codes will be repeated.

MONITORING OF ENGINE ACTUATORS AND DEVICES

On the 164, to initiate the monitoring process, press and hold the push button, located at the bottom of the dash near and to the left of the steering column, while turning the ignition **ON**. Hold the push button for 2.5–5.0 seconds after the ignition key is turned **ON**, then release.

A few seconds after the push button is released, the fuel injectors will begin to operate. This will be followed by the idle adjustment actuator and the evaporation solenoid valve.

By pressing the push button for 5–10 seconds, the process will be repeated. The following codes, flashed on the CHECK ENGINE light, will allow the user to determine which test is in progress:

1–4–1–1: Electro-injectors actuation test activated. Continuous control at a frequency of 10 hertz.

1–4–1–2: Idle adjustment actuator activation test. Continuous control for 1 second at idle position and 2 second at maximum speed.

1–4–1–3: Evaporation solenoid valve actuation test. Continuous control at a frequency of 0.5 hertz.

CLEARING CODES

Trouble codes will be cleared if the battery negative terminal is disconnected for 10 seconds or more. In addition, on the 164, codes may be cleared by pressing the push button for 10 seconds or more.

ENGINE CODES

Code 1211—Battery voltage problem. ECU reads under 9 volts or over 16 volts from battery. Problem is outside ECU. Perform Battery Failure Test.

Code 1212—Idle throttle switch problem. ECU not receiving plausible signal from MIN switch. Look for problem in MIN switch and/or wiring or ECU. Perform Throttle Valve Switch (Throttle Position Sensor) Test: Minimum Opening.

Code 1213—Full throttle switch problem. ECU not receiving plausible signal from MAX switch. Look for problem in MAX switch and/or wiring or ECU. Perform Throttle Valve Switch (Throttle Position Sensor) Test: Maximum Opening.

Code 1214—Engine temperature sensor problem. ECU not receiving plausible signal from temperature sensor and is using substitute value. Look for problem in temperature sensor and/or wiring or ECU. Perform Engine Temperature Sensor Test.

Code 1221—Airflow Meter problem. ECU receiving signal from flowmeter outside plausible range value and is using substitute value. Look for problem in airflow meter and/or wiring or ECU. Perform Airflow Meter Test.

Code 1222—Output stage of ECU for idle speed actuator problem. No plausible signal to actuator. Look for problem in idle air actuator and/or wiring or ECU. Perform Idle Actuator Test.

Code 1223—Oxygen sensor regulation to limit stop. ECU receiving signal at minimum or maximum value of correction limit. Look for problem in air/fuel circuit. Perform Oxygen Sensor Test A.

Code 1224—Oxygen sensor problem. Look for problem in oxygen sensor and/or wiring or ECU. Perform Oxygen Sensor Test B.

Code 1225—Air temperature sensor problem. ECU not receiving plausible signal from airflow meter temperature sensor and is using substitute value. Look for problem in airflow meter temperature sensor and/or wiring or ECU. Perform Air Temperature Sensor Test.

Code 1243—Output stage of ECU for fuel pump relay problem. Engine will have stopped. Look for problem in relay and/or wiring or ECU. Perform Fuel Pump Relay Test.

Code 1244—Output stage of ECU for emission purge canister valve. Look for problem in purge valve and/or wiring or ECU. Perform Evaporative Emission Control Solenoid Test.

Code 1245—Output stage of ECU for variable valve timing, on engines so equipped. Look for problem in variable valve timing relay and/or wiring or ECU. Perform Variable Valve Timing Control Test.

Code 1251—Malfunction in ECU. ECU recognizes problem in RAM/ROM memory. Problem is in ECU.

Code 1265—Output stage of ECU not sending signal to CHECK ENGINE warning light. Look for problem in warning light and/or wiring or ECU.

FUEL SYSTEM PRESSURE

Testing

1. Disconnect the fuel delivery hose from the injector inlet manifold.

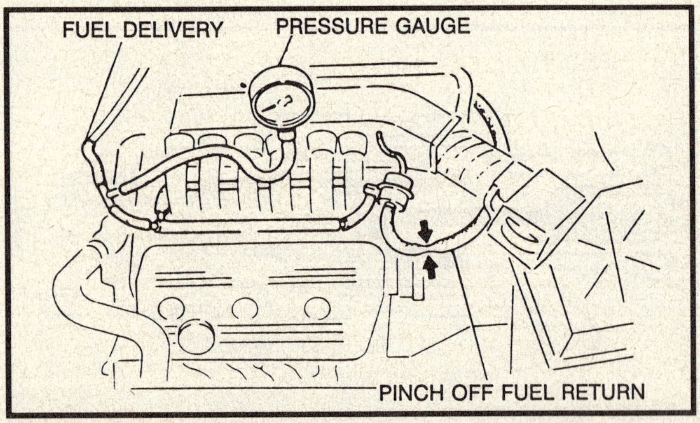

Testing fuel pressure regulator and for system integrity

2. Connect a pressure gauge, through a **T** fitting, at the end of the inlet line previously disconnected.
3. Detach the vacuum hose connecting the pressure regulator to the intake airbox to prevent any uneveness in engine speed which would cause an inaccurate reading.
4. Start the engine and run at idle. Check the fuel pressure. It should be 42–47 psi for 164 models and 41–47 psi for Spider models.
5. Reconnect the hose to the intake air box. When at idle, pressure should now decrease by 7–9 psi and then increase when the throttle opens. If these values are incorrect, check for leaks in the vacuum hose of the fuel pressure regulator.
6. If there are no visible fuel leaks or if there is a persistent fuel smell, check the system for tightness.
7. Keeping the pressure gauge connected and with the engine idling, pinch off the fuel return line from the pressure regulator. The pressure should rise up to about 58 psi. Use care not to exceed this pressure.
8. When this pressure is reached, check for leaks in the fuel lines and connections.
9. If the pressure does not reach the above value and no leaks are detected check the filter and/or pump operation.

FUEL PUMP

Testing

The in-tank pump has a mesh filter to protect the pump. When troubleshooting a fuel problem, look for water in the fuel. Water causes oxidation of the power contacts on the pump, causing a voltage drop. This can reduce fuel pressure and generate air bubbles in the fuel which reduces the quantity of fuel injected. The ECM activates the electric fuel pump through the fuel pump Motronic relay.

Note that replacement OEM pumps may be delivered with a protective oil inside. When installing a new pump, it is not necessary to empty the pump of this oil because the engine will simply burn any residual protective oil along with the fuel.

FUEL INJECTORS

Testing

1. Install an exhaust gas monitor in the tailpipe.
2. Note the percentage of carbon monoxide at the exhaust.
3. One at a time, disconnect the fuel injector electrical connectors. Note the Carbon Monoxide (CO) percentage at the exhaust each time. Check that the percentage keeps constant for each check.
4. If the percentage of CO changes during the test, that injector is likely faulty and should be replaced.
5. Another test is to check the spark plugs. A black color indicates a rich mixture, a white color indicates a lean mixture.
6. To test for injector tightness, remove the injector group and fuel rail(s), but keep these parts connected.
7. Disconnect the injector electrical connectors from the injectors.
8. Crank the engine and check for fuel leaks. If leaks occur, replace faulty injector(s).

CRANKSHAFT SENSOR

Testing

Information on engine speed is supplied to the ECU by the RPM and timing sensor. This sensor is the inductive type, working with a toothed pulley or trigger wheel on the crankshaft.

When checking the crankshaft sensor, check the gap between the sensor and top of the teeth of the pulley. Specification is 0.020–0.040 in.

BATTERY FAILURE

Testing

SPIDER

1. Check for voltage of 11.5–13.0 volts at battery terminals. If not okay, recharge battery, check for battery drain.
2. If voltage at battery is okay, disconnect the ECU cable connector. Place voltmeter positive probe on pin 35 (red wire) of harness connector and black probe to ground. Turn ignition key to **ON**. There should be 11.5–13.0 volts. If not, check for wiring problem.
3. If voltage check at ECU connector is okay, check the wiring between pin 35 (red wire) of the ECU connector and pin 87 (red/black wire and orange wire) of the fuel injector relay (Motronic relay with diode, in wiring diagram) and also between pin 5 (black wire) of the ECU connector and ground.

164

1. Check for voltage of 11.5–13.0 volts at battery terminals. If not okay, recharge battery, check for battery drain.
2. If the battery voltage is okay, check for voltage of 11.5–13.0 volts between pin 18 (red wire) and pin 5 (black wire) of the Motronic ECU. If not okay, repair wires are necessary.
3. If the voltage check at the ECU was okay, check for voltage of 11.5–13.0 volts at pin 35 (red wire, but larger than that at pin 18 in Step 2) of the Motronic ECU. If there is voltage here, replace the ECM.
4. If the was no voltage in Step 3, check for voltage of 11.5–13.0 volts at pin 87 (red wires) of the Motronic relay (called the Motronic relay with diode, in wiring diagram). If there is voltage, repair the wiring between pin 87 of this relay and pin 35 of the ECU.
5. If the was no voltage in Step 4, check for voltage of 11.5–13.0 volts at pin 86 (blue/red wire and pink wire) of the Motronic relay (called Motronic relay with diode, in wiring diagram). If the voltag is okay, go to Step 6. If the voltage is not okay, go to Step 7.
6. If there was voltage in Step 5, perform a ground check. Check for voltage at pin 85 (black wire) of the Motronic relay (called Motronic relay with diode, in wiring diagram). This is a ground wire and there should be zero volts. If okay, replace the relay. If not okay, repair the wiring between pin 85 (black wire) of the relay and the common ground. This is a connection on the passenger side upper fender and should have a number of black wires attached at this point.
7. If the was no voltage in Step 5, check for voltage of 11.5–13.0 volts at pin 1 (pink wire) of the round, twist lock multi-wire connector for the engine wiring. If okay, repair the wiring between pin 1 of this connector and pin 86 (blue/red wire and pink wire) of the Motronic relay (called Motronic relay with diode, in wiring diagram), through pin 30 (pink wire) of the fuel pump relay. If the voltage is not okay, go to Step 8.
8. Perform fuse check. Check fuse F22 in the fuse box. If okay, repair the wiring between pin 1 (pink wire) of the round, twist lock multi-wire connector for the engine wiring and pin 1 (pink wire) of the fuse box. If the fuse is not okay, replace the fuse (15 amp).

THROTTLE VALVE SWITCH MINIMUM OPENING

The minimum/maximum switch supplies the ECU with information regarding the throttle angle and therefore, about engine load.

The switch includes 2 microswitches. One microswitch (idle speed) supplies a signal for throttle angle between 0 degrees (fully closed) and 1 degree. The second microswitch (maximum speed) supplies signals for throttle angles above 60 degrees for the Spider, 72 degrees for model 164.

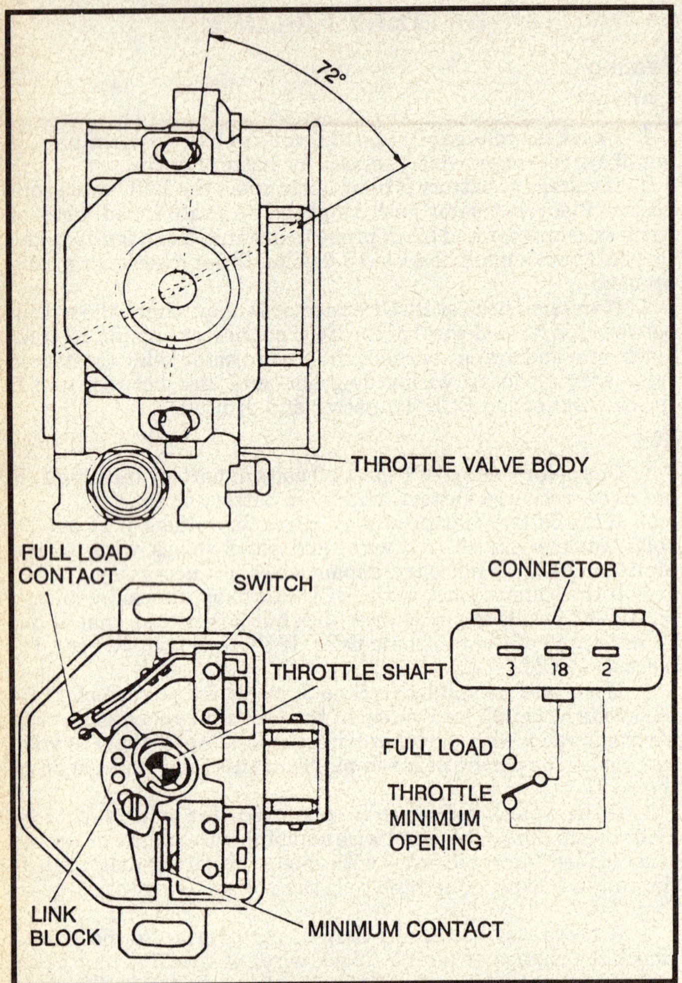

Testing the throttle position sensor or max/min switch.

If the minimum throttle opening switch or its connecting circuit has failed, Code 1–2–1–2 may be set in the self-diagnosic system.

When checking for grounding problems, note that any resistance reading other than zero ohms is not acceptable. Disassemble the connection to clean any corrosion from the terminals and/or connecting parts.

Testing

SPIDER

1. Perform a resistance check. Disconnect the throttle valve switch connector. With an ohmmeter, test between pin 2 (white wire) and pin 18 (black wire, a ground). With the accelerator pedal released, the reading should be zero ohms. Next depress the accelerator and check the make sure the reading is infinite resistance. If not, replace the switch.
2. If the resistance check is okay, disconnect the ECU cable connector. Test between pin 2 (white wire) of the ECU connector and ground. If there is continuity, verify the reading in Step 1 and check the wiring for a loose or intermittent condition between the ECU connector and the throttle switch connector. If there is no continuity between ECU connector pin 2 and ground, check the white wire between the ECU connector and the throttle switch connector.

164

1. With the accelerator pedal released, check for continuity between pin 2 (white wire from ECU) and pin 18 (black wire to ground) of the throttle switch. If not okay, adjust or replace the switch.
2. If the continuity check was okay, perform a grounding check. Check for any voltage at pin 18 (black wire to ground). It should be zero volts. If not okay, repair wiring between pin 18 (black wire to ground) of switch and ground.
3. If the grounding check was okay, perform connections check. With the accelerator pedal released, check for continuity between pin 2 (white wire from ECU) and the ground of the Motronic control unit which is underhood and also provides the ground for the ignition and fuel injection circuits. All ground wires to it are black. If okay, replace the control unit. If not okay, repair wiring between pin 18 (black wire to ground) and pin 2 of the control unit.

THROTTLE VALVE SWITCH MAXIMUM OPENING

The minimum/maximum switch supplies the ECU with information regarding the throttle angle and therefore, about engine load.

The switch includes 2 microswitches. One microswitch (idle speed) supplies a signal for throttle angle between 0 degrees (fully closed) and 1 degree. The second microswitch (maximum speed) supplies signals for throttle angles above 60 degrees for the Spider, 72 degrees for model 164.

When checking for grounding problems, note that any resistance reading other than zero ohms is not acceptable. Disassemble the connection to clean any corrosion from the terminals and/or connecting parts.

If the maximum throttle opening switch or its connecting circuit has failed, Code 1–2–1–3 may be set in the self-diagnosic system.

Testing

SPIDER

1. Perform a resistance check. Disconnect the throttle valve switch connector. With an ohmmeter, test between pin 3 (yellow wire) and pin 18 (black wire, a ground). With the accelerator pedal released, the reading should be infinite. Next depress the accelerator and check the make sure the reading is zero ohms. If not, replace the switch.
2. If the resistance check is okay, disconnect the ECU cable connector. Test between pin 3 (yellow wire) of the ECU connector and ground. Verify the reading in Step 1. With the accelerator pedal released, the reading should be infinite. Next depress the accelerator and check the make sure the reading is zero ohms. If not okay, look for a wiring problem between pin 3 of the ECU wire connector and pin 3 of the throttle switch. If the reading is okay, check the wiring for a loose or intermittent condition between the ECU connector and the throttle switch connector.

164

1. Perform a switch check, using the following procedure. With the accelerator pedal released, check for continuity between pin 3 (yellow wire from ECU) and pin 18 (black wire to ground) of the throttle switch. With the accelerator pedal fully pressed, check for the presence of an open circuit. If not okay, adjust the switch or replace the switch.
2. If the switch check was okay, perform a grounding check. Check for voltage at pin 18 (black wire to ground). It should be zero volts. If not okay, repair wiring between pin 18 (black wire to ground) of switch and ground.
3. If the ground check was okay, perform connections check. With the accelerator pedal released, check for continuity between pin 3 (yellow wire from ECU) and the common ground of the Motronic control unit which is underhood and also provides the ground for the ignition and fuel injection circuits. All ground

wires to it are black. If okay, replace the control unit. If not okay, repair wiring between pin 18 (black wire to ground) and pin 3 of the control unit.

ENGINE TEMPERATURE SENSOR

The temperature sender uses 2 wires. Brown is from the ECU control unit, black is to ground.

If the engine temperature sensor or its connecting circuit has failed, Code 1–2–1–4 may be set in the self-diagnosic system.

Testing

SPIDER

1. Perform a resistance check on the sensor. Disconnect the engine temperature sensor connector. Put the ohmmeter probes on the sensor pins. Read the resistance which should vary according to temperature. If not okay, replace the sensor.
   a. 32°F—approximately 5500 ohms.
   b. 72°F—approximately 1700 ohms.
   c. 212°F—approximately 170 ohms.
2. If the resistance readings are approximately as shown, disconnect the ECU cable connector and test for continuity between pin 13 (brown wire) of the cable connector and ground. The reading should be as in Step 1. If okay, check the wiring between the engine temperature sensor connector, the ECU connector and the sensor ground for a loose or intermittent condition between the ECU connector and the temperature sensor connector. If not okay, look for a wiring problem, break or short, in the brown wire between the sensor and the ECU connector as well as a break or short in the black wire between the sensor connector and its ground connection.

164

1. Perform a resistance check on the sensor. Disconnect the engine temperature sensor connector. Put the ohmmeter probes on the sensor pins. Read the resistance which should vary according to temperature. If not okay, replace the sensor.
   a. 32°F—approximately 7500 ohms.
   b. 68°F—approximately 4500 ohms.
   c. 212°F—approximately 40 ohms.
2. Testing can be done by removing the sender and placing it in a suitable container with ice and/or hot water as required. Use a thermometer to track temperature and an ohmmeter to read resistance. If not okay, replace sensor.
3. If sender checks out okay, perform a continuity check on the circuit. Check for continuity between pin 13 of the Motoronic control unit and the sensor (brown wire). Check for continuity between the ground point and the sensor (black wire). If not okay, repair or replace the wires as necessary. If okay but trouble codes persist, replace the control unit.

AIRFLOW METER

The airflow meter measures air flow and supplies the control unit with a signal that permits the exact metering of fuel. The signal is generated by a potentiometer which sends to the ECU a voltage corresponding to the changing throttle angle. A temperature sensor inside the airflow meter senses intake air temperature. The airflow meter uses 5 wires. All run to the ECU control unit.

If the airflow meter has failed, Code 1–2–2–1 may be set in the self-diagnosic system. Check the airflow meter using the following procedures.

Testing

SPIDER

1. Disconnect the airflow meter connector. Place the probes of an ohmmeter between pin 3 (blue/white wire) and pin 4 (green wire) of the airflow meter sensor. The resistance should be about 400 ohms. Now place the probes on pin 2 (white/black wire) and pin 4 (green wire). The resistance should be about 100 ohms. Check that the resistance varies by operating the throttle valve manually. If not okay, replace the airflow meter.
2. If the airflow meter resistance check is okay, disconnect the ECU connector. Put the ohmmeter probes on pin 9 (blue/white wire) and pin 6 (green wire) of the ECU connector. The resistance should be about 400 ohms. Now place the probes on pin 7 (white/black wire) and pin 6 (green wire) of the ECU connector. The resistance should be about 100 ohms. Check that the resistance varies by operating the throttle valve manually. If not okay, look for a break or short in the wiring between pins 4, 3 and 3 of the airflow meter connector and matching pins 6, 9 and 7 of the ECU connector. If the resistance check was okay, check the wiring between the airflow meter and the ECU connector for a loose or intermittent condition.

164

1. Check for the presence of a resistance of about 400 ohms between pin 3 (light blue/white wire) and pin 4 (green) of the airflow meter. Check for the presence of a resistance of about 5 ohms between pin 2 (white/black wire) and pin 4 (green) of the airflow meter. Operate the airflow meter throttle and check that the resistance varies smoothly. If not okay, replace the airflow meter.
2. If the meter checks out, look for a continuity problem. Check for continuity between pin 2 of the airflow meter and pin 7 of the Motronic ECU control unit (white/black wire). Check for continuity between pin 3 of the airflow meter and pin 9 of the Motronic ECU control unit (light blue/white wire). Check for continuity between pin 4 of the airflow meter and pin 6 of the Motronic ECU control unit (green wire). If all areas check out okay, replace the ECU control unit. If the continuity checks are not okay, repair or replace the wires as necessary.

IDLE SPEED ACTUATOR

When the throttle is closed to slightly open, the ECU controls the airflow through the idle speed actuator which acts as a bypass of the throttle body. The idle adjustment actuator is monitored by the ECU program to determine the exact amount of air necessary to maintain an ideal idle speed.

The idle speed actuator has 2 wires. The orange wire runs to pin 87 of the Motronic relay (called Motronic relay with diode, in wiring diagram) while the blue/black wire runs to pin 33 of the ECU.

If the idle speed actuator has failed, Code 1–2–2–2 may be set in the self-diagnosic system. Test the idle speed actuator using the following procedures.

Testing

SPIDER

1. Test for voltage. Disconnect the idle actuator connector. Place a voltmeter's red probe to the orange wire of the connector and the black probe to ground. There should be battery voltage here only when the ignition key is in the **ON** position. If not okay, look for a problem in the wiring between the idle actuator connector and pin 87 of the Motronic relay (called Motronic relay with diode, in wiring diagram). The relay should be energized when the ignition key is in the **ON** position or the engine would not start.
2. If the voltage test was okay, supply the actuator with 12 volts directly and see if it works. If not, replace the idle actuator.
3. If the idle actuator tested okay, disconnect the ECU connector. With an ohmmeter, check for continuity between pin 33 (blue/black wire) of the ECU and the other end of the blue/black wire at the idle actuator connector. The reading should be zero ohms, indicating continuity. If not, repair the wiring. If the continuity check was okay, check the wiring between the idle actua-

tor connector and the ECU connector for a loose or intermittent condition.

164

1. Perform a voltage check. Check for 12 volts between pin 1 (light blue/black wire) and 2 (orange wire) on the idle speed actuator. If not okay, perform continuity checks.
2. Perform a continuity check. Check for continuity between pin 1 of the actuator and pin 33 of the Motronic ECU control unit (light blue/black wire). If okay, replace the ECU control unit. If not okay, repair or replace the wire as necessary.
3. Perform a continuity check. Check for continuity between pin 87 of the Motronic relay (called Motronic relay with diode, in wiring diagram) and pin 2 of the actuator (orange wire). If okay, replace the idle speed actuator. If not okay, repair or replace the wire as necessary.

OXYGEN SENSOR

The oxygen sensor, called the Lambda Probe by many European auto builders, is one of the main ECU inputs. It generates a small electrical signal. The voltage is dependent on the oxygen concentration in the exhaust. When the probe signal voltage is low, the ECU senses a lean mixture and slightly increases the injected fuel. When the signal voltage is high, the ECU senses a rich mixture and decreases the quantity of injected fuel.

Alfa Romeo uses an oxygen sensor heated by a resistor. It is powered through the Motronic relay with the diode only when the ignition key is in the **RUN** position. At the oxygen sensor, the black wire goes to ground, the orange/black wire is powered by pin 87 of the Motronic relay and the green wire goes to pin 24 of the Motronic ECU.

If the inner part of the oxygen sensor has failed, Code 1–2–2–3 may be set in the self-diagnosic system. Test the oxygen sensor using the following procedures.

Test A

1. Check the air ducts for leaks. If not okay, repair or replace air duct.
2. Check air filter for integrity. If not okay, replace air filter.
3. Check spark plugs and wiring. If not okay, replace defective/broken spark plug(s).
4. Check the fuel injectors and system pressure. Replace defective injector(s) if necessary.
5. On 164 models, perform a continuity check. Check for continuity between pin 14 of the Motronic ECU and the fuel injectors (yellow wire). If okay, replace the ECU. If not okay, repair or replace the wires as necessary.

Test B

If the oxygen sensor has failed, Code 1–2–2–4 may be set in the self-diagnostic system. Check the oxygen sensor using the following procedures.

SPIDER

1. Perform a voltage check. Set a voltmeter to low range. Without disconnecting the connector, place the red tester probe to the green wire to the oxygen sensor and the black probe to ground.
2. Warm the engine to operating temperature. Check the voltage. Normal voltage should be 0.1–1.0 volt.
3. If the voltage is correct, disconnect the ECU connector and check the wiring between the oxygen sensor and the ECU connector for a loose or intermittent condition.
4. If the voltage reading in Step 2 was not okay, check the wiring between pin 24 (green wire) of the ECU connector and the oxygen sensor for continuity. If okay, replace the oxygen sensor. If not okay, repair the wiring.

164

1. Run engine to operating temperature. Check for voltage at the oxygen sensor green wire. Voltage should be 0.1–1.0 volts. If not okay, replace the oxygen sensor.
2. If the voltage check was okay, perform continuity check. Run engine to operating temperature. Check for voltage at pin 24 of the Motronic ECU control unit (green wire). Voltage should be 0.1–1.0 volts. If okay and Code persists, replace the ECU. If not okay, repair or replace the wiring between the oxygen sensor and pin 24 of the ECU and/or between the oxygen sensor and the ground point (black wire).

AIR TEMPERATURE SENSOR

The engine air intake temperature sensor is built into the airflow meter.

If the engine air temperature sensor has failed, Code 1–2–2–5 may be set in the self-diagnosic system. Test the air temperature sensor using the following procedures.

Testing

SPIDER

1. Disconnect the air flow meter connector.
2. Place the test probes of an ohmmeter between pin 5 (gray/red wire) and pin 4 (green wire) of the airflow meter. If resistance is not within specification, replace the airflow meter. It should be:
   a. 32°F – approximately 5500 ohms.
   b. 68°F – approximately 2500 ohms.
   c. 212°F – approximately 180 ohms.
   d. 265°F – approximately 90 ohms.

2. If the resistance readings were within specification, disconnect the ECU cable connector. Place the test probes of an ohmmeter between pin 22 (gray/red wire) and pin 6 (green wire) of the ECU. The resistance readings should be the same as in Step 1. If resistance is not within specification, look for a problem somewhere along the gray/red wire between pin 22 of the ECU and pin 5 of the airflow meter connector.
3. If resistance was within specification yet a Code has been set, check the wiring between the airflow meter connector and the ECU connector for a loose or intermittent condition.

164

1. Perform sensor resistance test. Check for resistance between terminal 5 (gray/red wire) and terminal 4 (green wire) of the airflow meter. If resistance is not within specification, replace the airflow meter. It should be:
   a. 32°F – approximately 6000 ohms.
   b. 72°F – approximately 1500 ohms.
   c. 212°F – approximately 180 ohms.
   d. 265°F – approximately 80 ohms.
2. Perform continuity test. Check for continuity between pin 22 (gray/red wire) and pin 6 (green wire) of the Motronic control unit. If continuity checks out as in Step 1, replace the control unit. If continuity does not check out, repair or replace wires as necessary.

FUEL PUMP RELAY

A failure of the fuel pump relay control final stage may set Code 1–2–4–3. It may be tested using the following procedure.

Testing

SPIDER

1. Disconnect the ECU connector.
2. Disconnect the fuel pump relay.
3. Place the test probes of an ohmmeter between pin 20 (gray/black wire) of the ECU connector and pin 85 (gray/black wire) of the fuel pump relay. There should be good continuity and a reading of about zero ohms. If not, check the wiring between the relay socket and the ECU connector.

4. If the continuity tests okay, switch the meter to volts, put the red probe into pin 86 (red/black wire) of the fuel pump relay and the black probe to ground. Test to see that there is battery voltage only when the ignition key is in the **ON** position. If not, check the wiring to the pump relay socket.
5. If the voltage tests okay, yet a Code has been set, check the wiring between the fuel pump relay socket and the ECU connector for a loose or intermittent condition.

164

1. Perform voltage check. Check for voltage at pin 86 (red wire) of the fuel pump Motronic relay. There should be 12 volts. If not okay, repair wiring between pin 86 (red wire) of the fuel pump relay and pin 87 of the Motronic relay with diode.
2. Perform grounding test. Check for presence of voltage at pin 20 (gray/black wire) of the Motronic ECU control unit. There should be no voltage (0 volts).
   a. If okay, repair wiring between pin 20 (gray/black wire) of the ECU and pin 85 (the other end of the gray/black wire) of the fuel pump relay.
   b. If not okay, replace the ECU.

EVAPORATION EMISSION CONTROL SOLENOID

If the evap solenoid of the emission control system or the ECU output signal to it fails, it may set Code 1–2–4–4. It may be tested using the following procedure.

Testing

SPIDER

1. Disconnect the ECU connector.
2. Disconnect the fuel vapor (sometimes called a canister purge valve) solenoid valve from the vapor canister.
3. Place the test probes of an ohmmeter between pin 34 (light brown wire) of the ECU connector and the black wire of the fuel vapor solenoid valve connector. There should be good continuity and a reading of about zero ohms. If not, check the wiring between the solenoid valve connector and the ECU connector.
4. If the continuity check was okay, apply 12 volts to the solenoid valve and check it for proper function. If it does not work, replace the solenoid valve.
5. If the solenoid valve works when 12 volts is applied, set the meter to volts and connect the red probe to wire feeding the solenoid (usually red/white wire) and connect the black probe to ground. There should be battery voltage here when the ignition key is in the **ON** position. If not okay, check the wiring from the solenoid to its source, 1 of the Motronic relays.
6. If the voltage tests okay, yet a Code has been set, check the wiring between the solenoid valve and the Motronic relay and the ECU connector for a loose or intermittent condition.

164

1. Perform a control unit check. With the engine warm, accelerate with slight pulses and check for the presence of voltage at pin 34 (light brown wire) of the ECU at each pulse of the accelerator pedal. If not okay, replace the ECU.
2. If the control test is okay, perform grounding check. Check for voltage at pin 1 (light brown wire) of the evap solenoid valve. As in Step 1, accelerate with slight pulses and check for the presence of voltage at each pulse of the accelerator pedal. There should be no voltage (0 volts). If not okay, repair wiring between pin 34 (light brown wire) of the ECU and pin 1 (the other end of the light brown wire) of the evap solenoid.
3. If the ground test was okay, perform voltage check. Check for voltage at the evap solenoid (blue/red wire). There should be 12 volts. If okay, replace the evap solenoid valve. If not okay, repair the wiring between the evap solenoid valve and pin 86 (blue/red wire) of Motronic relay with diode.

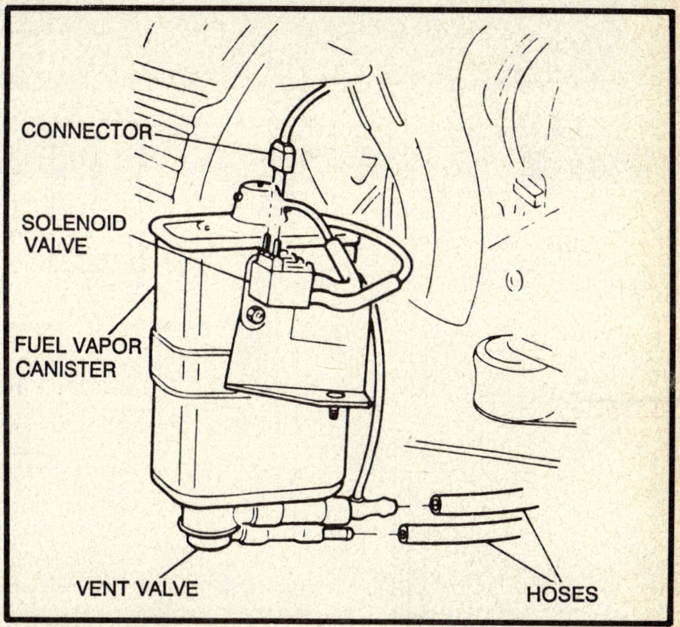

Location of the evaporative emission control solenoid. Model 164 shown, typical

VARIABLE VALVE TIMING CONTROL

If the variable valve timing relay control or its ECU signal fails, it may set Code 1–2–4–5. It may be tested using the following procedure.

Testing

SPIDER

1. Disconnect the ECU connector.
2. Disconnect the valve timing relay.
3. Place the test probes of an ohmmeter between pin 31 (yellow/red wire) of the ECU connector and pin 85 of the valve timing relay socket. There should be good continuity and a reading of about zero ohms. If not, check the wiring between the valve timing relay socket and the ECU connector.
4. If the continuity tests okay, switch the meter to volts, put the red probe into pin 86 (red/black wire) of the valve timing relay socket and the black probe to ground. Test to see that there is battery voltage only when the ignition key is in the **ON** position. If not, check the wiring to the valve timing relay socket.
5. If the voltage tests okay, yet a Code has been set, check the wiring between the valve timing relay socket and the ECU connector for a loose or intermittent condition.

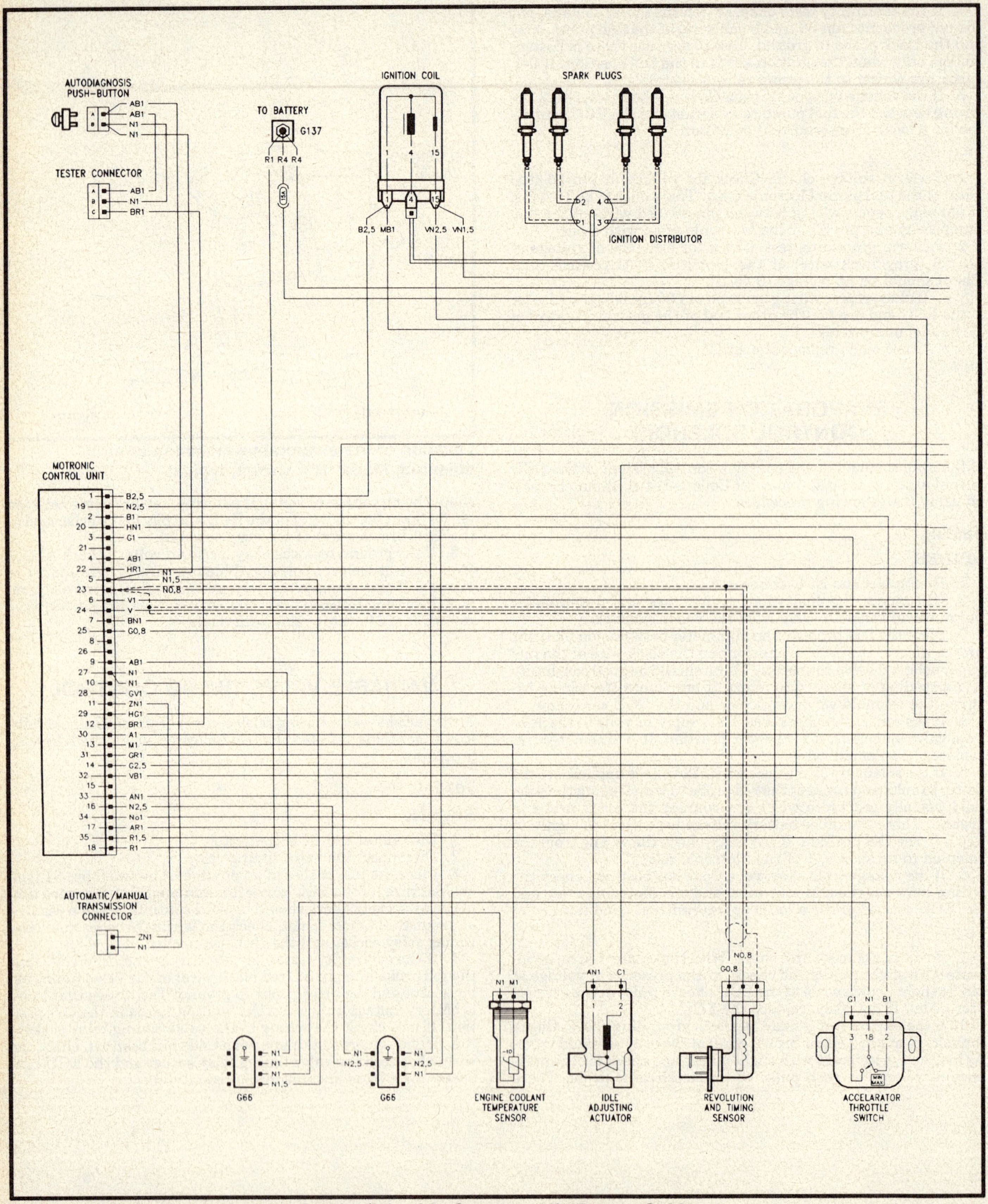

Wiring diagram—Motronic ML4.1 Ignition/Injection System—Alfa Romeo Spider

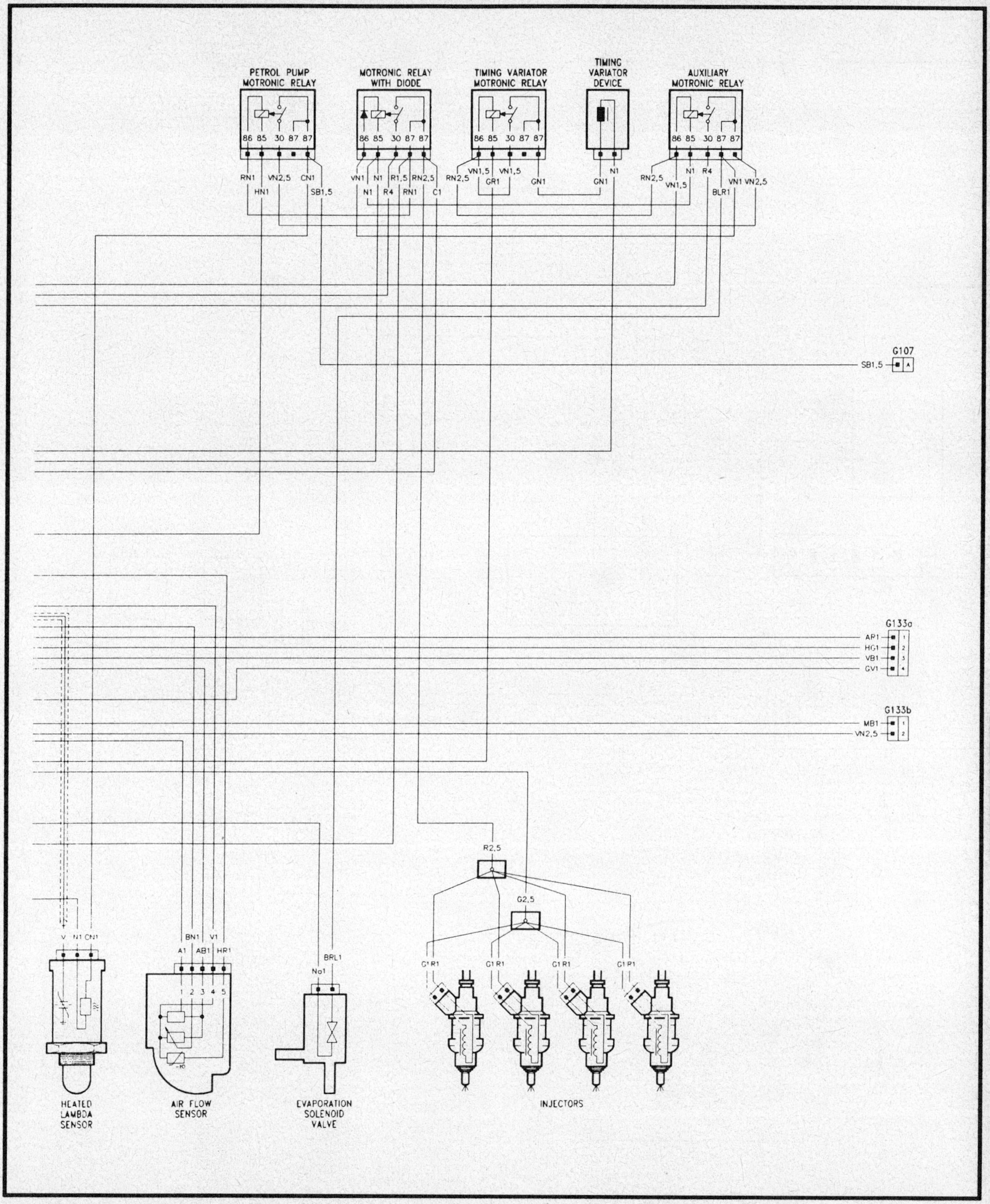

Wiring diagram—Motronic ML4.1 Ignition/Injection System—Alfa Romeo Spider, continued

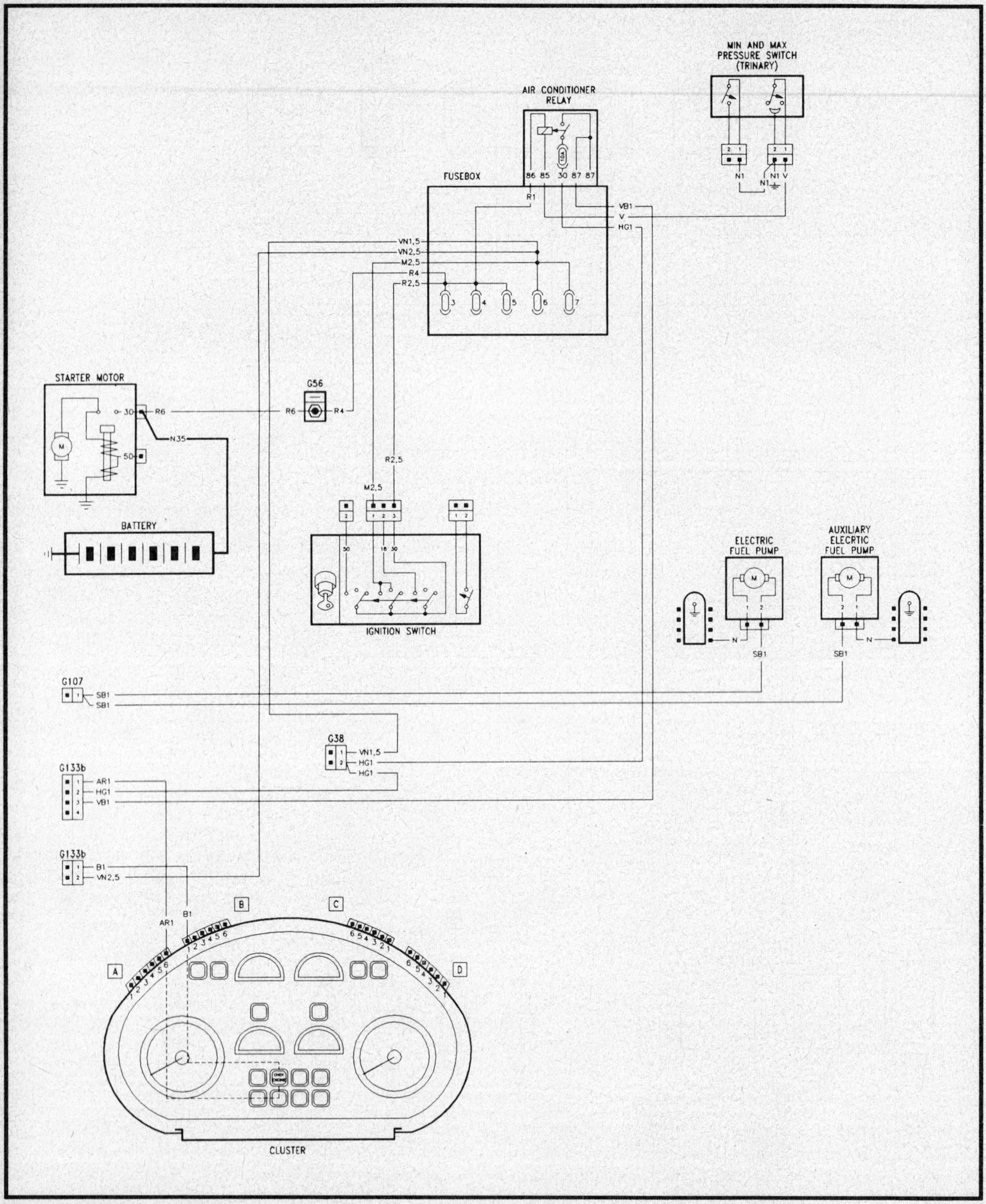

Wiring diagram—Motronic ML4.1 Ignition/Injection System—Alfa Romeo Spider, continued

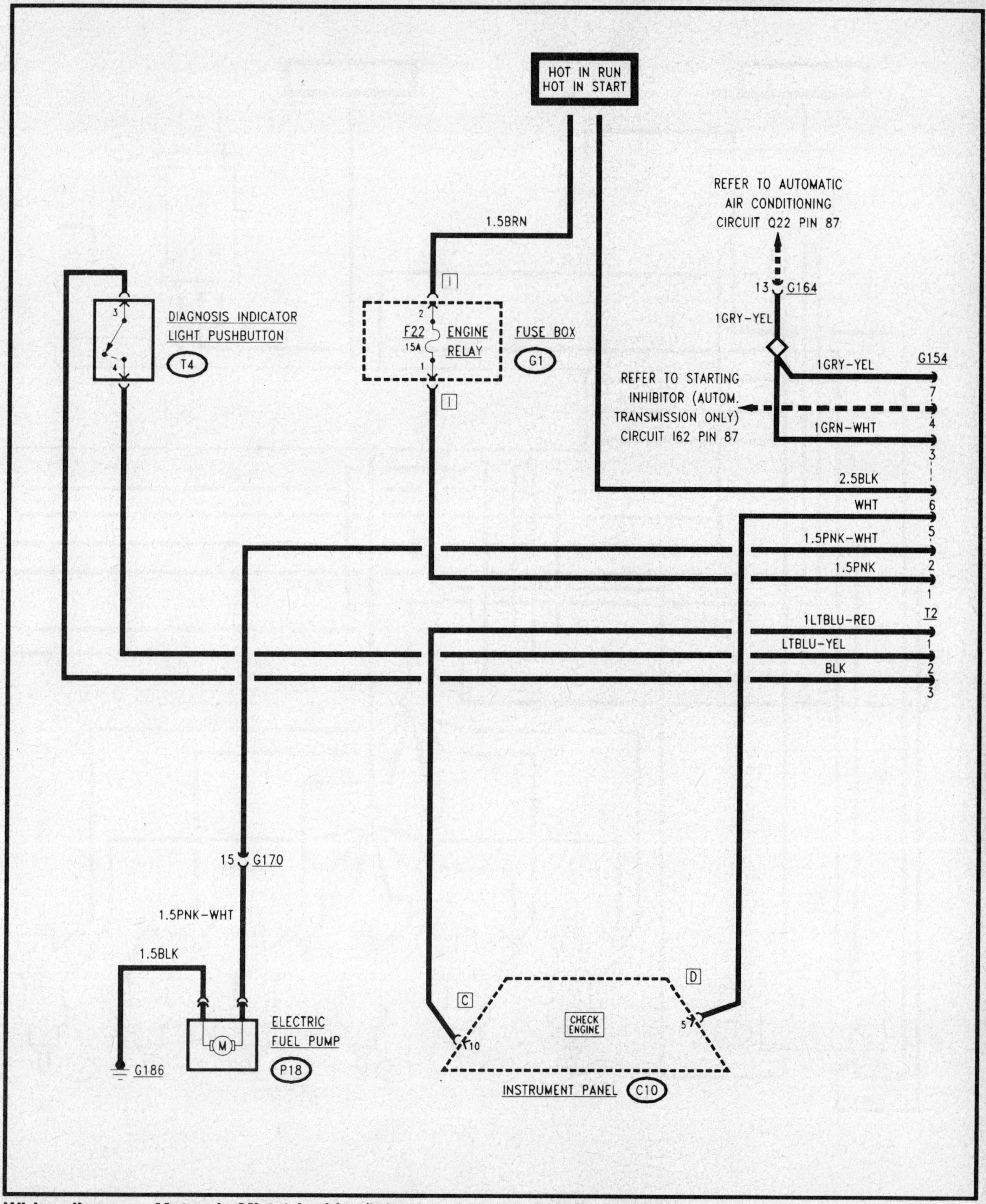

Wiring diagram—Motronic ML4.1 Ignition/Injection System—Alfa Romeo 164

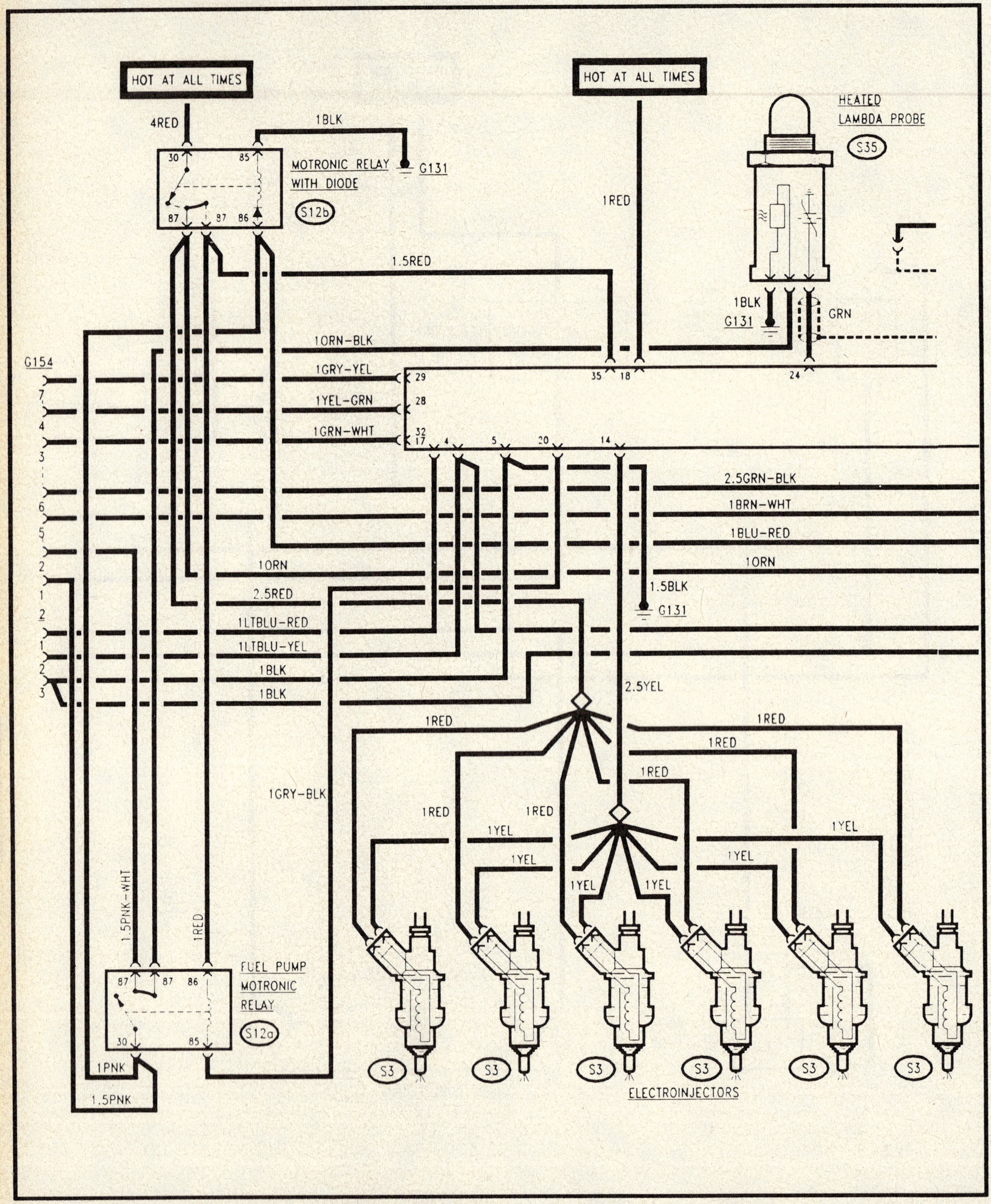

Wiring diagram—Motronic ML4.1 Ignition/Injection System—Alfa Romeo 164, continued

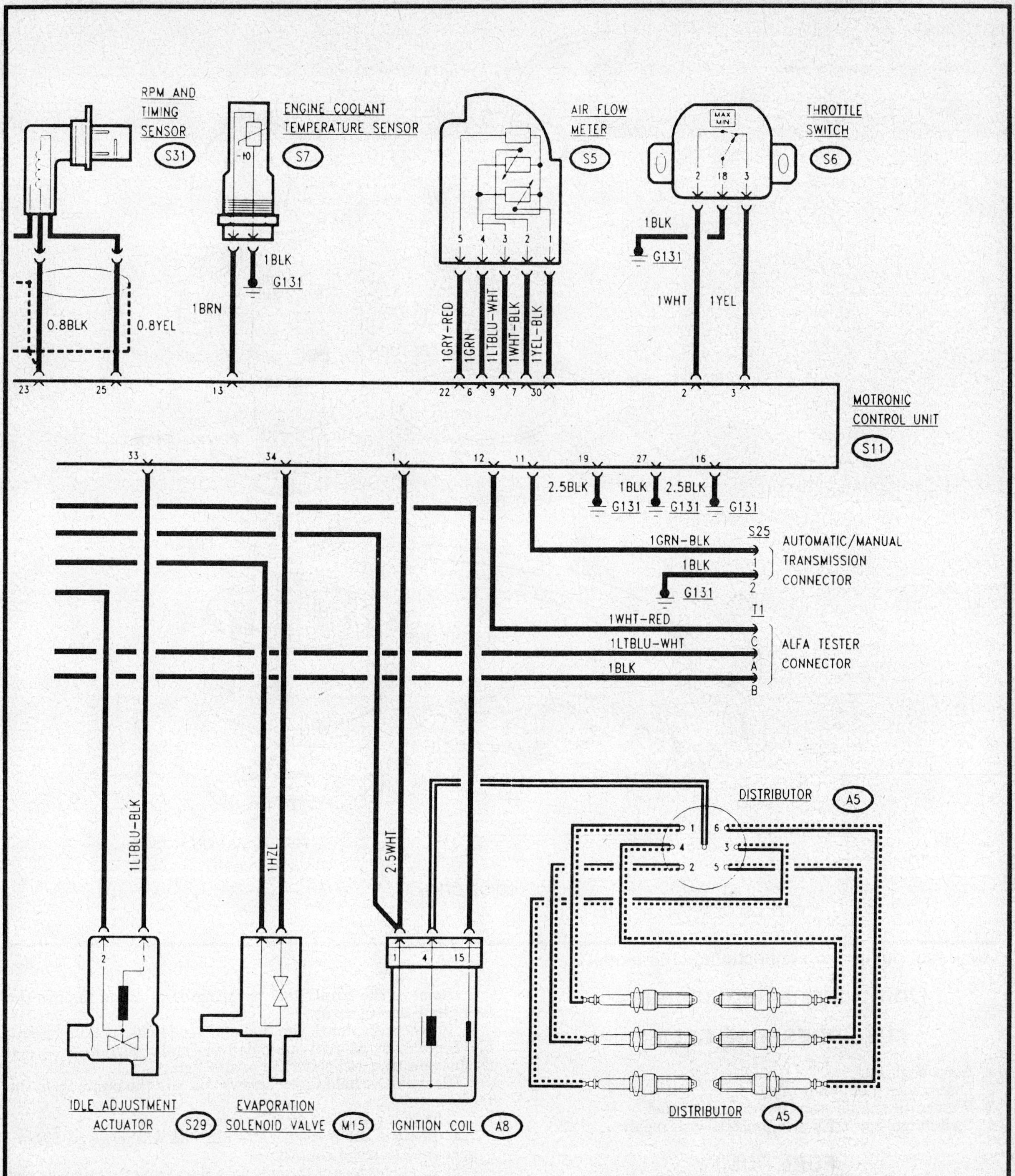

Wiring diagram—Motronic ML4.1 Ignition/Injection System—Alfa Romeo 164, continued

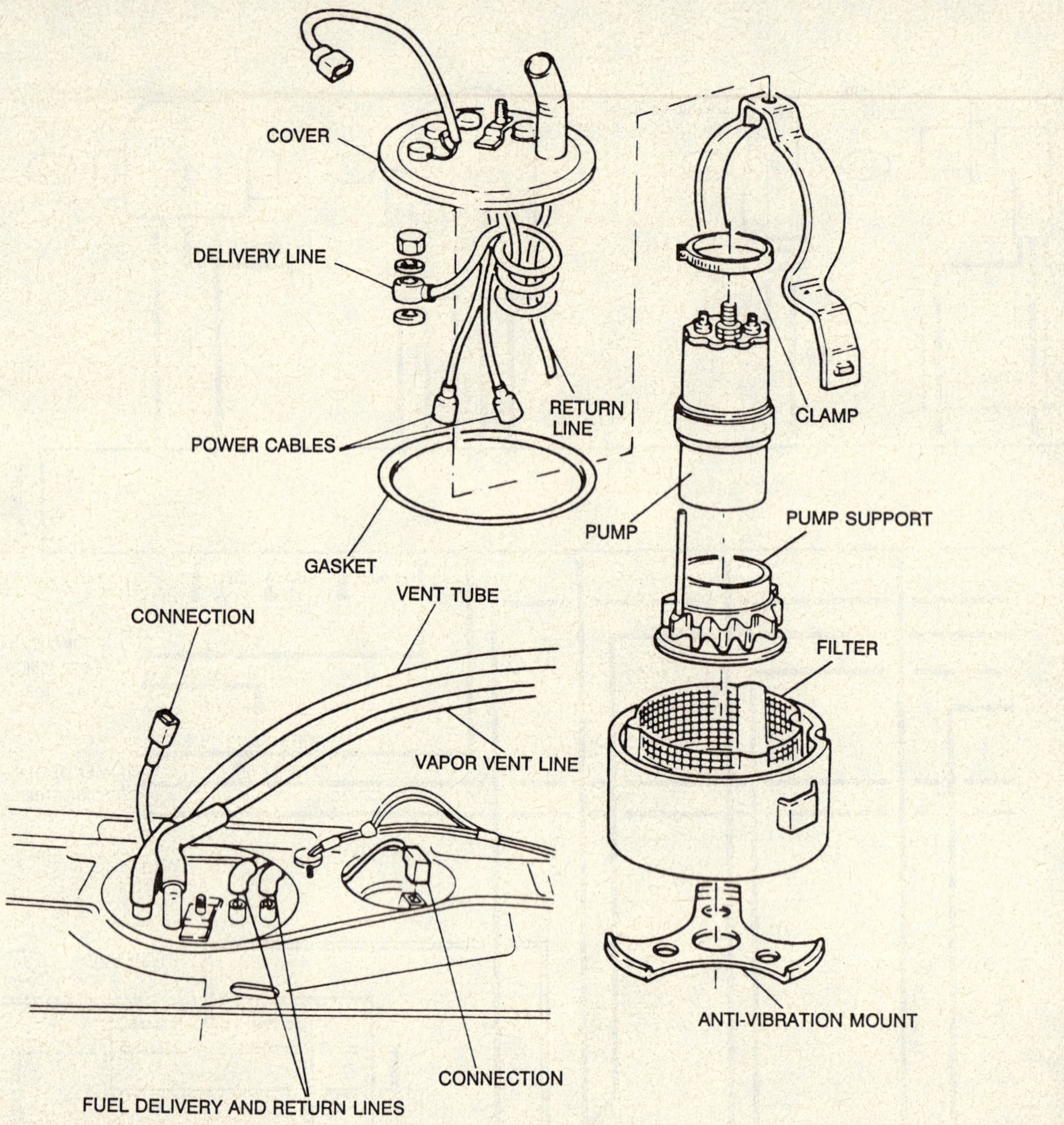

In-tank fuel pump and connections – 164 model

Component Replacement

FUEL PRESSURE RELIEF

1. Start engine.
2. Disconnect fuel pump relay.
3. Wait until engine stops due to lack of fuel.
4. Switch ignition **OFF** and proceed with repair.

FUEL PUMP

Removal and Installation

IN-TANK PUMP

1. Disconnect the negative battery cable.
2. Remove the trunk floor covering. Locate and remove the fuel pump access cover.
3. If necessary, mark the fuel and vent lines to aid reassembly. Disconnect the fuel pump delivery and return lines as well as the vent line and electrical connection.
4. Remove the hold-down screws and pull the pump from the tank.
5. The pump may be disassembled, if necessary.
   a. Remove the pressure line nut and washers and disconnect the electrical connectors.
   b. Remove the nut that secures the spring to the cover and detach the cover together with the lines.
   c. Separate the cover, pump body, gasket and sump filter.

To install:

6. Thoroughly clean the net sump filter. Note that water in

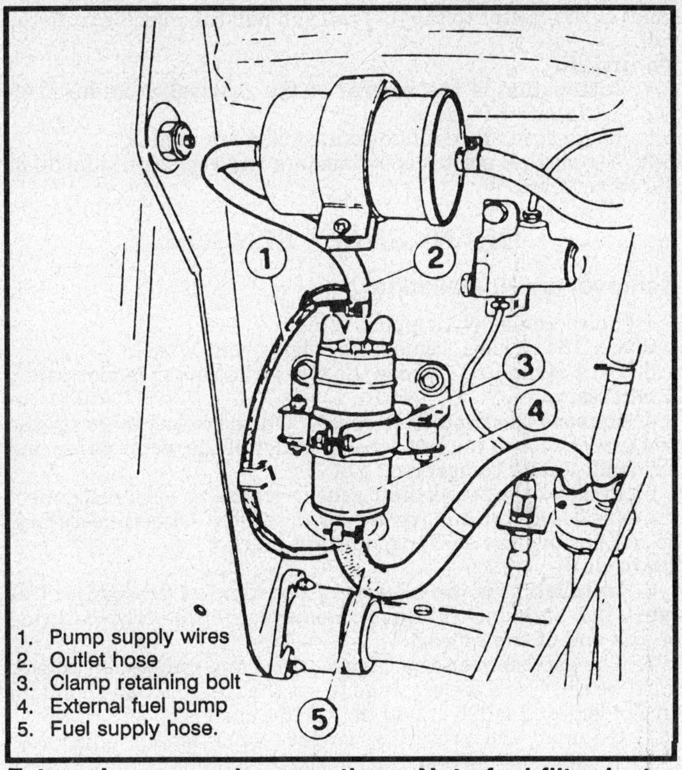

External pump and connections. Note fuel filter just above pump—Spider models

the fuel will cause internal corrosion and could damage the pump. Check the electrical connections carefully.

7. Assemble the pump, if necessary. Use care that all electrical connections are secure.
8. Installation is the reverse of the removal procedure. Always use a new gasket under the pump flange.
9. Make sure all lines are securely connected. Tighten the retaining screws to 7–11 ft. lbs.

EXTERNAL PUMP

1. Disconnect the negative battery cable.
2. Raise and safely support vehicle.
3. Disconnect the electrical supply wires to the pump. Matchmark the wires for reassembly, if necessary.
4. Pinch the hoses to and from the pump, loosen the clamps and disconnect the hoses.
5. Loosen the pump retaining clamp bolt and remove the pump.

To install:

6. Installation is the reverse of the removal procedure. Install the pump, reconnect the hoses and secure with clamps.
7. Reconnect the electrical wires, taking care to properly assemble the connections.
8. Remove the pinch-off pliers from the fuel lines. Reconnect battery.

FUEL FILTER

Removal and Installation

1. Relieve fuel system pressure. Disconnect negative battery cable.
2. Raise and safely support vehicle.
3. Locate the fuel filter under the vehicle mounted to the floor pan. Loosen but do not remove the 2 filter connection bolts. Use care when removing fuel lines since some residual fuel pressure may remain.

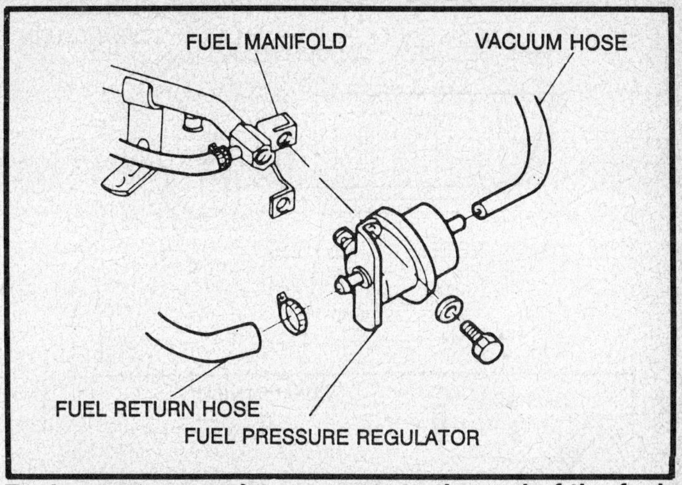

Fuel pressure regulator mounts to the end of the fuel manifold (rail)

4. Remove the fuel filter clamp bolt.
5. Keep a suitable container close to catch and fuel spilled during filter removal. Remove the fittings and the filter.

To install:

6. Installation is the reverse of the removal procedure. Use care not to twist or bend the rigid fuel lines. Torque to 15–17 ft. lbs.
7. Install the new filter taking note of the markings showing fuel flow. Do not overtighten the clamp bolt. Torque to 22–30 ft. lbs.

FUEL PRESSURE REGULATOR

Removal and Installation

1. Relieve fuel system pressure. Disconnect negative battery cable.
2. Use care when removing fuel lines since some residual fuel pressure may remain. Disconnect both the pressure and return fuel lines from the pressure regulator.
3. Remove the retaining bolt from the fuel rail and remove the regulator. Disconnect the vacuum line.

To install:

4. Installation is the reverse of the removal procedure. Use care to make sure the fuel lines are connected properly.
5. Make sure the vacuum line is secure to the pressure regulator.

FUEL INJECTORS

Removal and Installation

1. Relieve fuel system pressure. Disconnect negative battery cable.
2. Remove intake ducting as required for access. Disconnect the vacuum line to the pressure regulator. Remove and mark, if necessary, any wiring in the way such as to the coolant sensor, etc.
3. Disconnect the fuel lines and electrical connectors from the injectors. On the 4 cylinder engine it may be necessary to remove the spark plug wires and distributor cap. Mark wires for reassembly.
4. On V6 engines, remove the stub pipe from the intake manifolds and tilt the intake box backward.
5. Remove the retaining bolts and remove the fuel rails.
6. To remove the individual injectors, on 4 cylinder engines, cut the connecting hose from the injector, saving the bushing on the fuel rail end. On V6 engines, remove the spring clip that

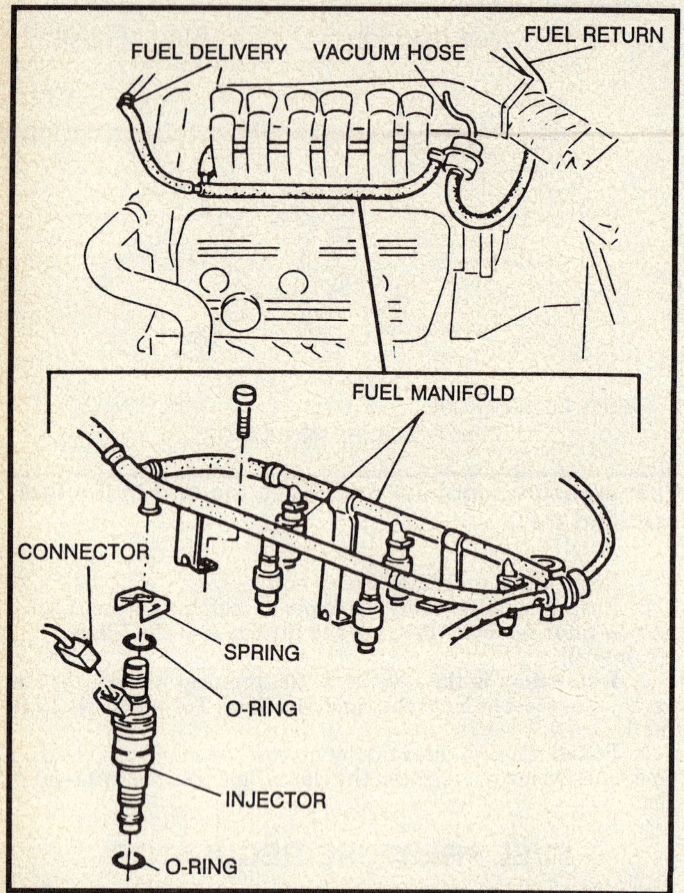

Fuel injector and attaching parts. 164 model with V6 shown

holds each injector to the fuel rail and pull the injector from the rail.

To install:

7. Installation is the reverse of the removal procedure. Always install new O-rings.
8. Make sure all fuel line connections are secure.
9. Connect all electrical connectors and replace intake ducting as required.

CRANKSHAFT SENSOR

Removal and Installation

1. Raise and safely support vehicle.
2. On 164 models, remove the right front wheel.
3. On 164 models, remove the right front inner fender panel, if necessary.
4. Remove the protective covers. On the lower cover on the 164 model, loosen the belt tension pulley of the water pump and air conditioning compressor belt.
5. Remove the crankshaft sensor retaining bolt, disconnect the wire connector and remove from vehicle. Unless necessary, do not remove the sensor mounting bracket.

To install:

6. Installation is the reverse of the removal procedure. Use care when handling the sensor. Remove any protective covering on the end of the sensor.
7. The gap between the trigger wheel and crankshaft sensor must be set. Use a feeler gauge to set the gap to 0.020–0.060 in. on Spiders and 0.020–0.040 in. on 164 models.
8. Reinstall the protective covers, inner fender panel and right front wheel as necessary.

EVAPORATIVE EMISSION CANISTER SOLENOID

Removal and Installation

1. Raise and safely support vehicle.

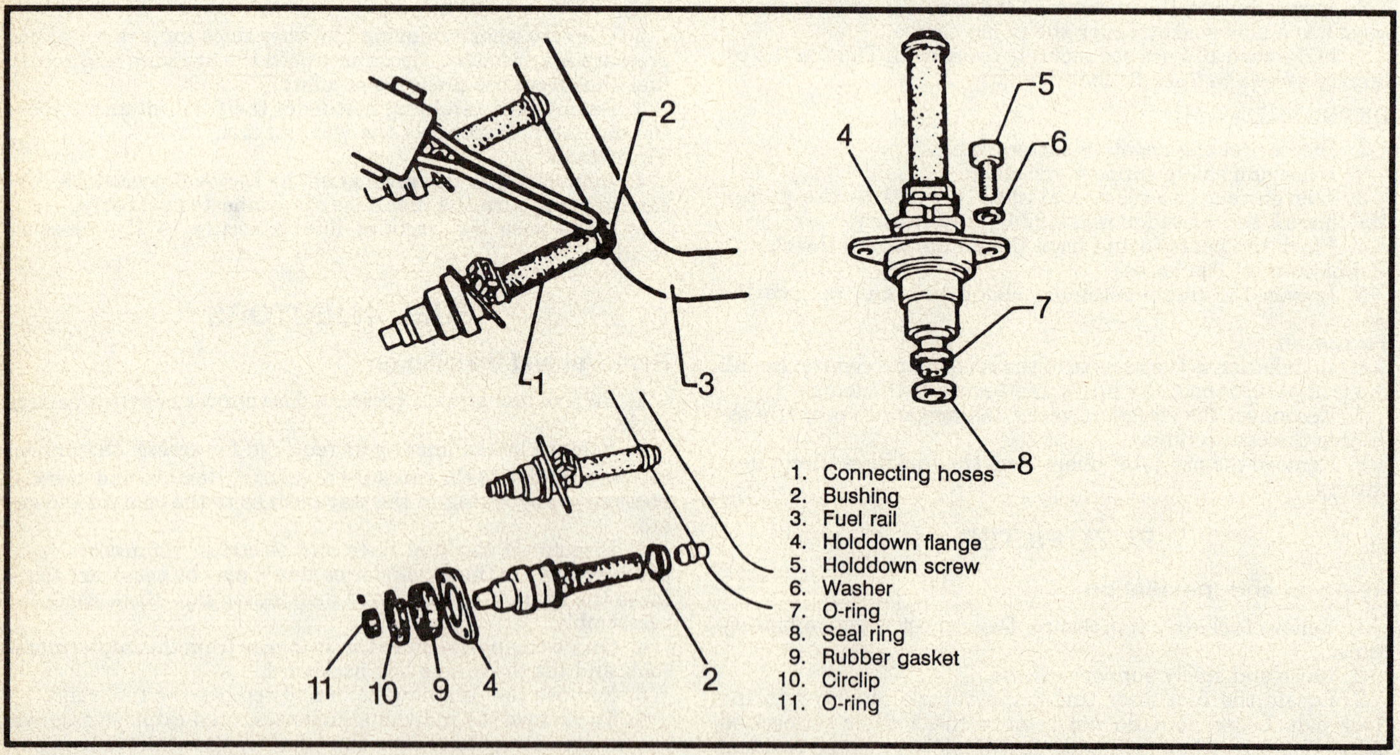

Fuel injector and attaching parts. Spider model with 4 cylinder shown

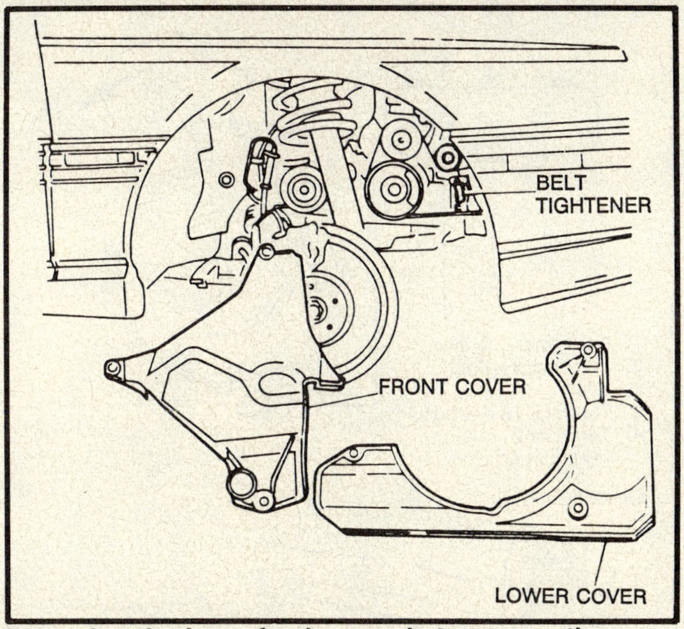

Removing the inner fender panels to access the crankshaft sensor—164 model

2. Remove the front left wheel and fender inner panel.
3. Disconnect the electrical connections from the solenoid valve and the vapor lines from the canister. Matchmark vapor lines, if necessary.
4. Remove the fuel vapor canister.

To install:

5. Installation is the reverse of the removal procedure. Make sure the electrical connections are secure.
6. Make sure the vapor lines are on their proper connectors.
7. Install fender inner panel and front wheel.

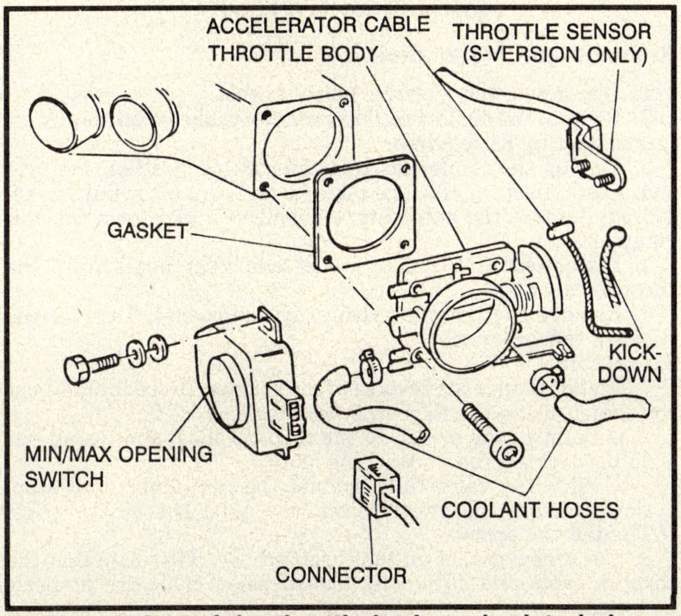

Exploded view of the throttle body and related air intake parts

AIRFLOW METER

Removal and Installation

1. Disconnect the negative battery cable.
2. Remove the flowmeter cover assembly.
3. Separate the airflow meter from the air cleaner cover by removing the attaching screws. Remove the gasket and disconnect the electrical connector.

To install:

4. Installation is the reverse of the removal procedure. Use care handling the electrical connector.
5. Make sure the replacement gasket is in place when tightening the retaining screws.

IDLE SPEED ACTUATOR

Removal and Installation

1. Disconnect negative battery cable.
2. Unplug idle speed actuator connector.
3. Disconnect the bypass hose for the idle speed adjustment.
4. Remove the idle speed actuator from the air intake manifold removing the bolt securing bracket.
5. Remove the gasket.

To install:

6. Installation is the reverse of the removal procedure. Use care when installing the electrical connector.
7. Make sure the idle speed actuator is firmly seated in the gasket.

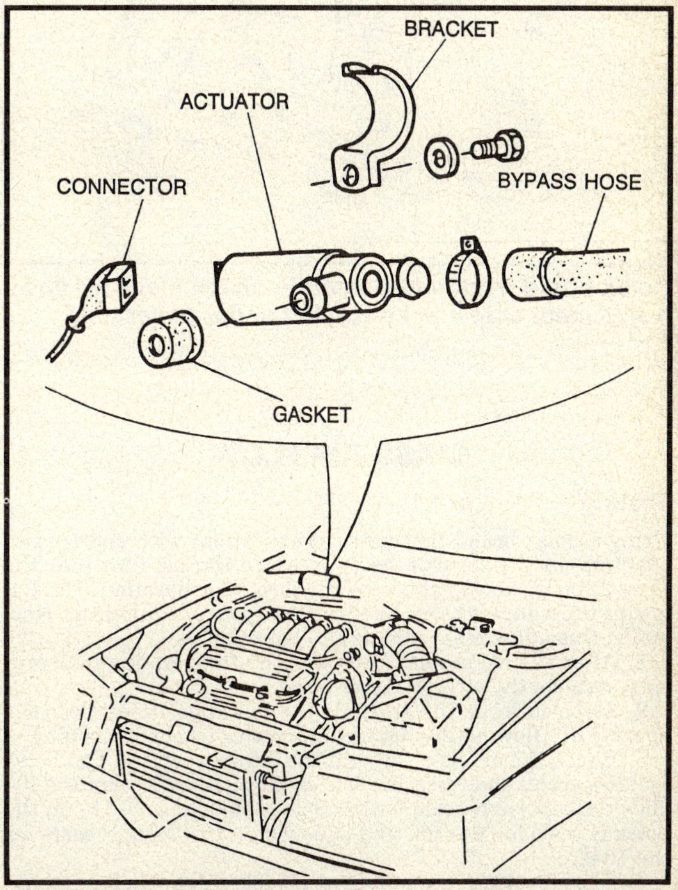

Idle air valve attachment. 164 model shown, typical

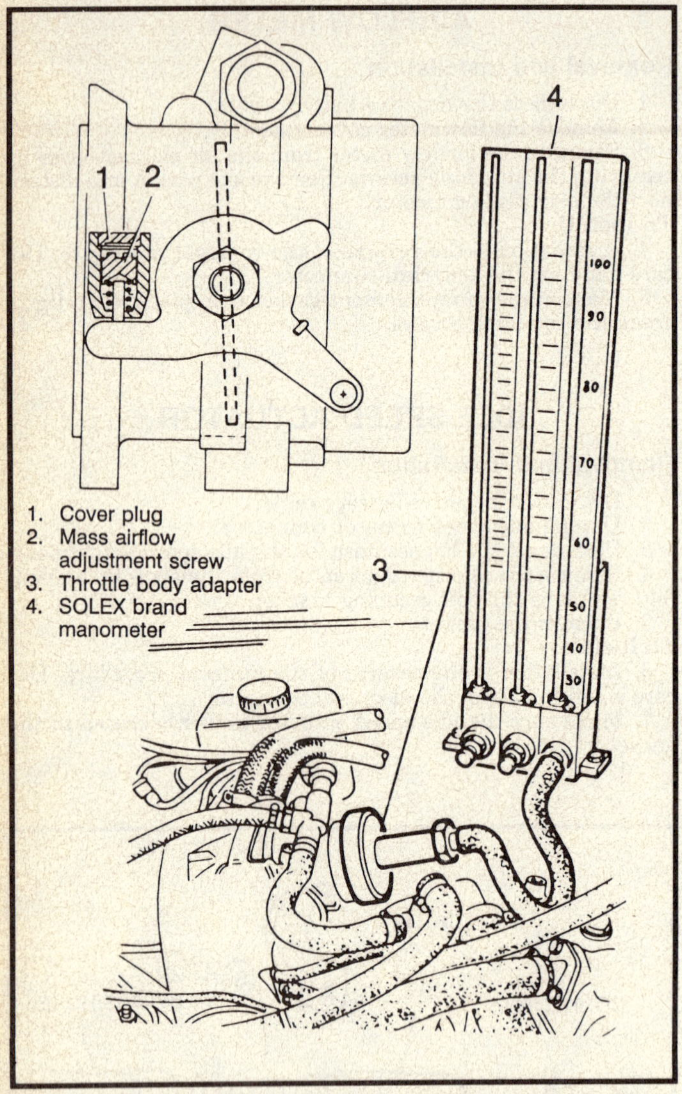

Solex brand manometer used to check mass air flow. Adjustment screw is concealed under cover plug

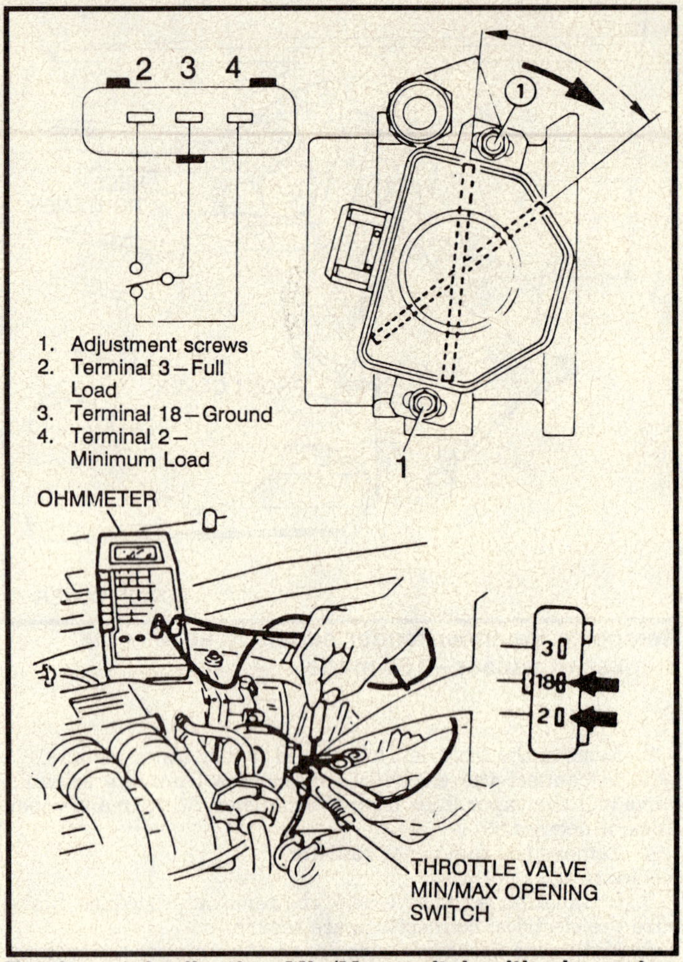

Testing and adjusting Min/Max switch with ohmmeter

MASS AIR FLOW

Testing

Using a Solex brand flowmeter or equivalent with appropriate adapters on the throttle body, measure the air flow into the throttle body thereby checking the throttle calibration. The following procedure is based on using this piece of equipment. Normally, this adjustment should not be required.

1. After the engine has been warmed to operating temperature, remove the airflow meter cover.
2. Make sure the throttle is in the closed position. Place the appropriate flowmeter adapter on the throttle body opening.
3. Check the air flow through the intake opening. The Solex tool has scales that measure the air flow. Specification is 240–260 on the Solex N scale for the Spider models, 290–310 on the Solex N scale for the 164 and 270–290 on the Solex N scale for the 164S.
4. The adjusting screw on most models will be under a cover. The adjusting screw bears against part of the throttle valve arm.

THROTTLE POSITION SENSOR (MIN/MAX SWITCH)

Removal and Installation

1. Disconnect the negative battery cable.
2. Remove the airflow meter/cover assembly as well as any air intake ducting as required.
3. Unplug the connector from the throttle position sensor.
4. Disconnect the throttle cable (s) and connector links as required. Remove the automatic transmission kickdown cable, if equipped.
5. Disconnect the coolant outlet and inlet hoses from the throttle valve body.
6. Remove the throttle valve body and gasket. Take off the position sensor switch.

To install:

7. Installation is the reverse of the removal procedure. Adjust the switch using the following procedure:
   a. Connect the switch to the throttle shaft and install but do not tighten the 2 attaching screws.
   b. With the valve fully against the minimum limit stop, slowly rotate the switch clockwise until the switch trips. Tighten the screws.
8. Use a new gasket on the throttle body. Take care that the throttle cable and automatic transmission cable are properly installed.
9. Install the electrical connector and replace all the intake ducting as required.

Adjustment

The air flow check described above should be performed first before adjusting the throttle position sensor.

1. Unplug the electrical connector from the throttle position sensor and check with an ohmmeter for the following resistance values:

a. With the throttle valve fully closed, between terminal 2 (white wire) and terminal 18 (the ground, black wire) the resistance should be 0 ohms.

b. Slowly rotate the valve. With the meter still connected between terminal 2 (white wire) and terminal 18 (the ground, black wire) the resistance should read infinite before the valve has rotated for less than 1 degree from its previous closed position.

2. If this is not correct, loosen the screws and rotate the switch until the contact is closed (0 ohms). Tighten the screws.

3. Connect the meter between terminal 3 (yellow wire) and terminal 18 (the ground, black wire).

4. Set the throttle valve to the full open position and check the resistance. It should be 0 ohms.

5. If the values are not as listed, adjust the accelerator pedal using a 0.120 (⅛-in) spacer between the bottom of the pedal arm and the limit stop. If this specification is okay and the throttle position sensor readings are still not within specification, replace the switch.

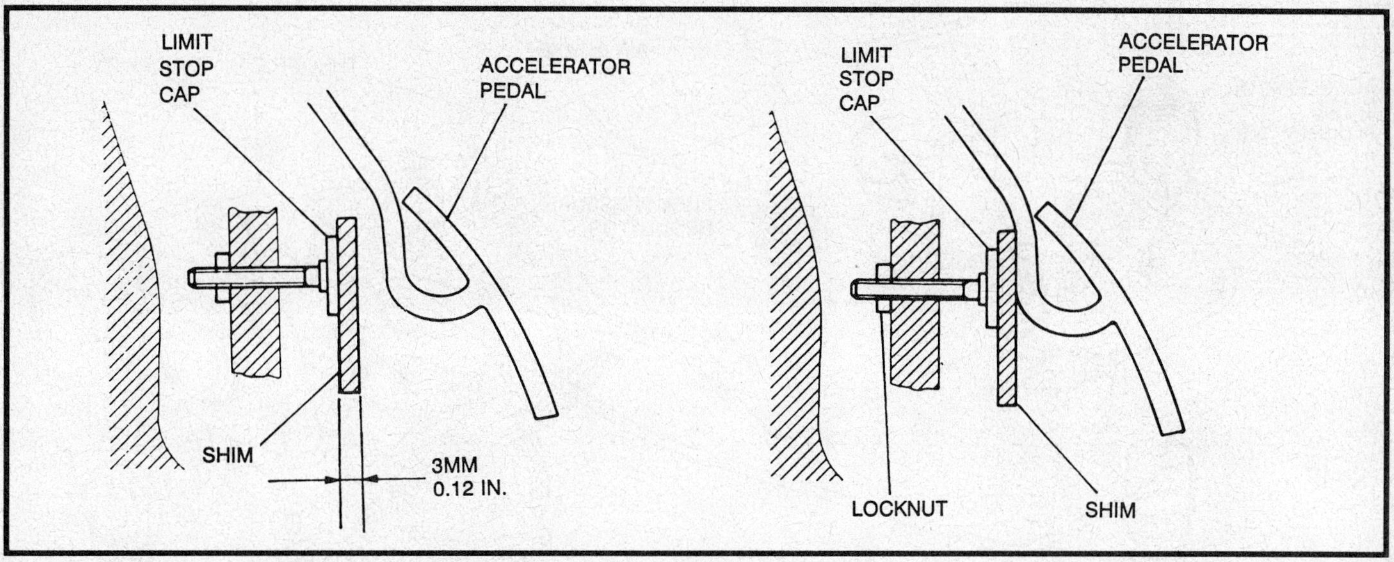

Throttle pedal adjustment should just allow a 0.12 in. (⅛ in.) clearance

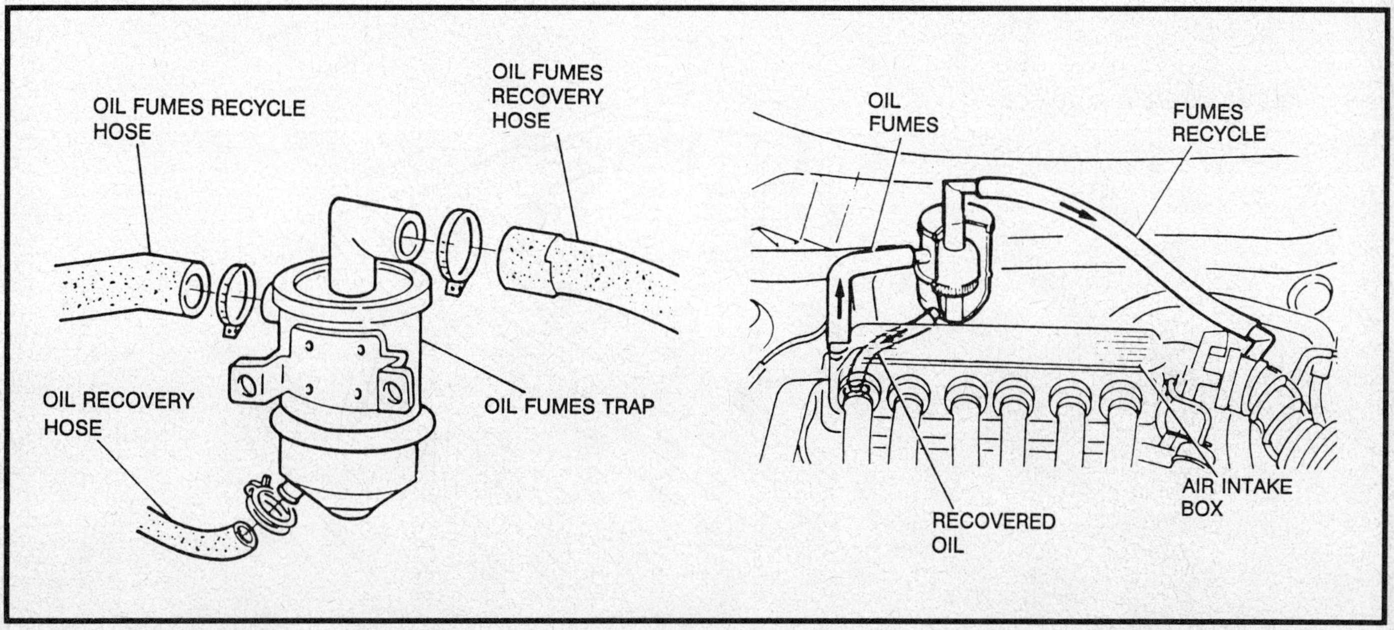

The induction system also separates oil from collected oil vapors, recovers the oil and vents the fumes

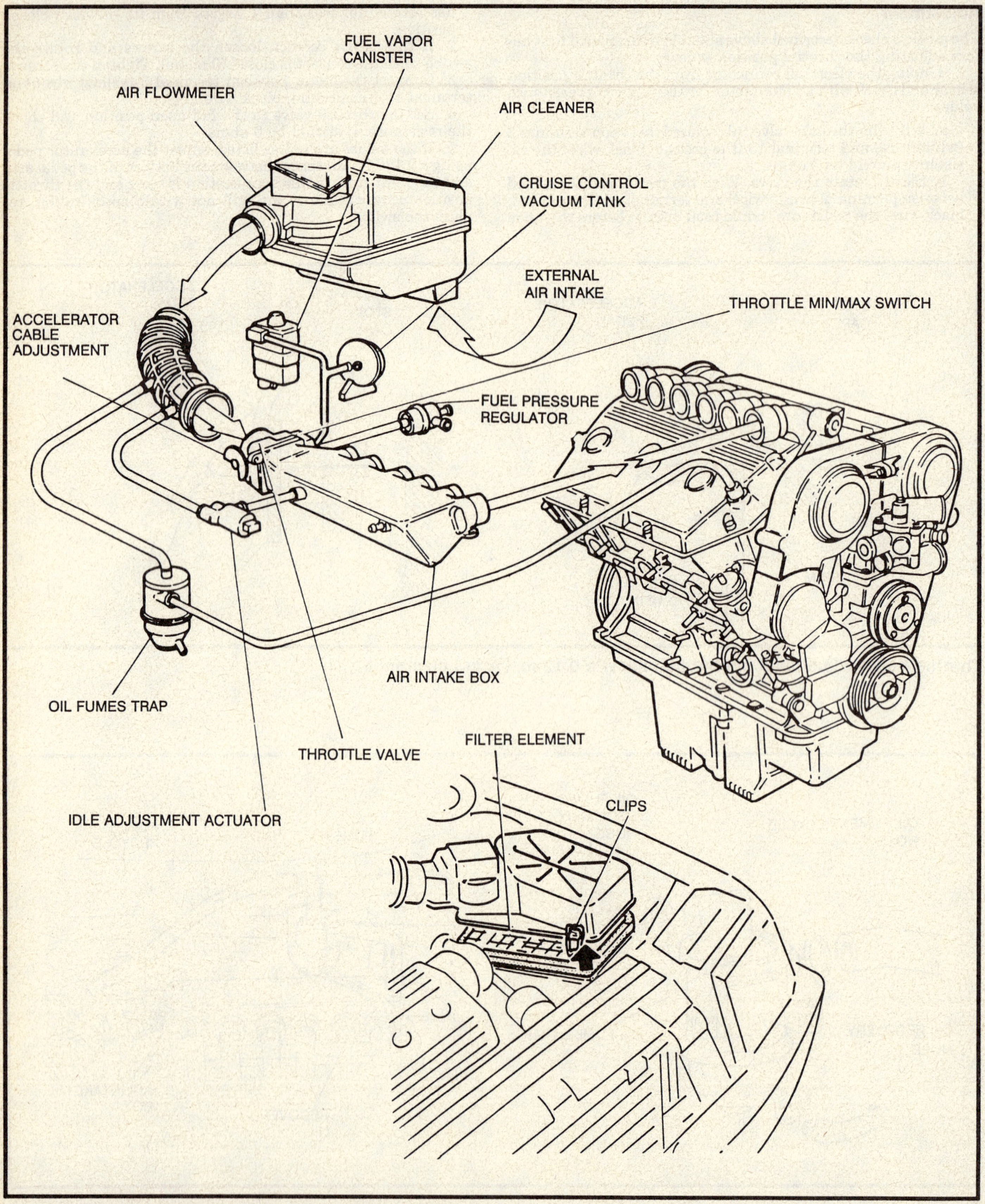

Air induction system—Alfa Romeo 164 shown, typical

Audi Fuel Injection

SECTION 6

AUDI MOTRONIC AND MULTI POINT INJECTION (MPI) FUEL INJECTION SYSTEMS

General Information

The Motronic and MPI fuel injection systems are similar and share most components and modes of operation. Audi uses Motronic to describe the fuel injection system on the V8 Quattro, S4, 200 Quattro and the 200 Quattro Wagon. Audi uses MPI to describe the systems used on the 90 Quattro, Coupe Quattro and the 2.8L V6 equipped 100 series vehicles.

The Motronic and MPI fuel injection systems are self-learning adaptive systems. They continuously learn using a sophisticated feedback system that readjusts various control settings. These new values are then stored in the ECU memory. The adaptive capability allows the systems to compensate for changes in the engine's operating conditions, such as intake leaks, altitude changes or any other system malfunction. If the battery or ECU is disconnected, the vehicle must be driven so ECU can "re-learn" its operating conditions.

Operation of the fuel injection system is based on the infor-

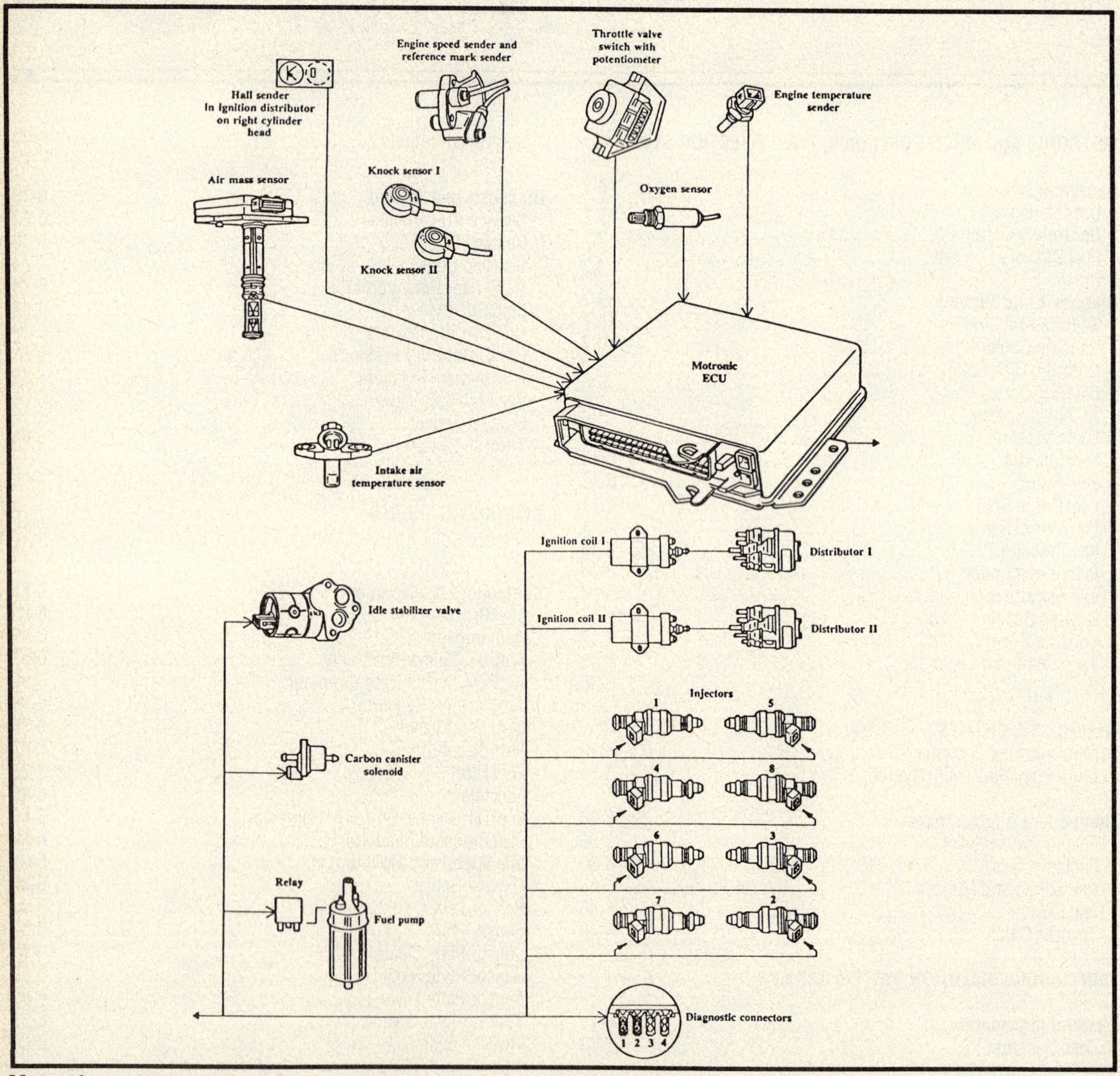

Motronic system components

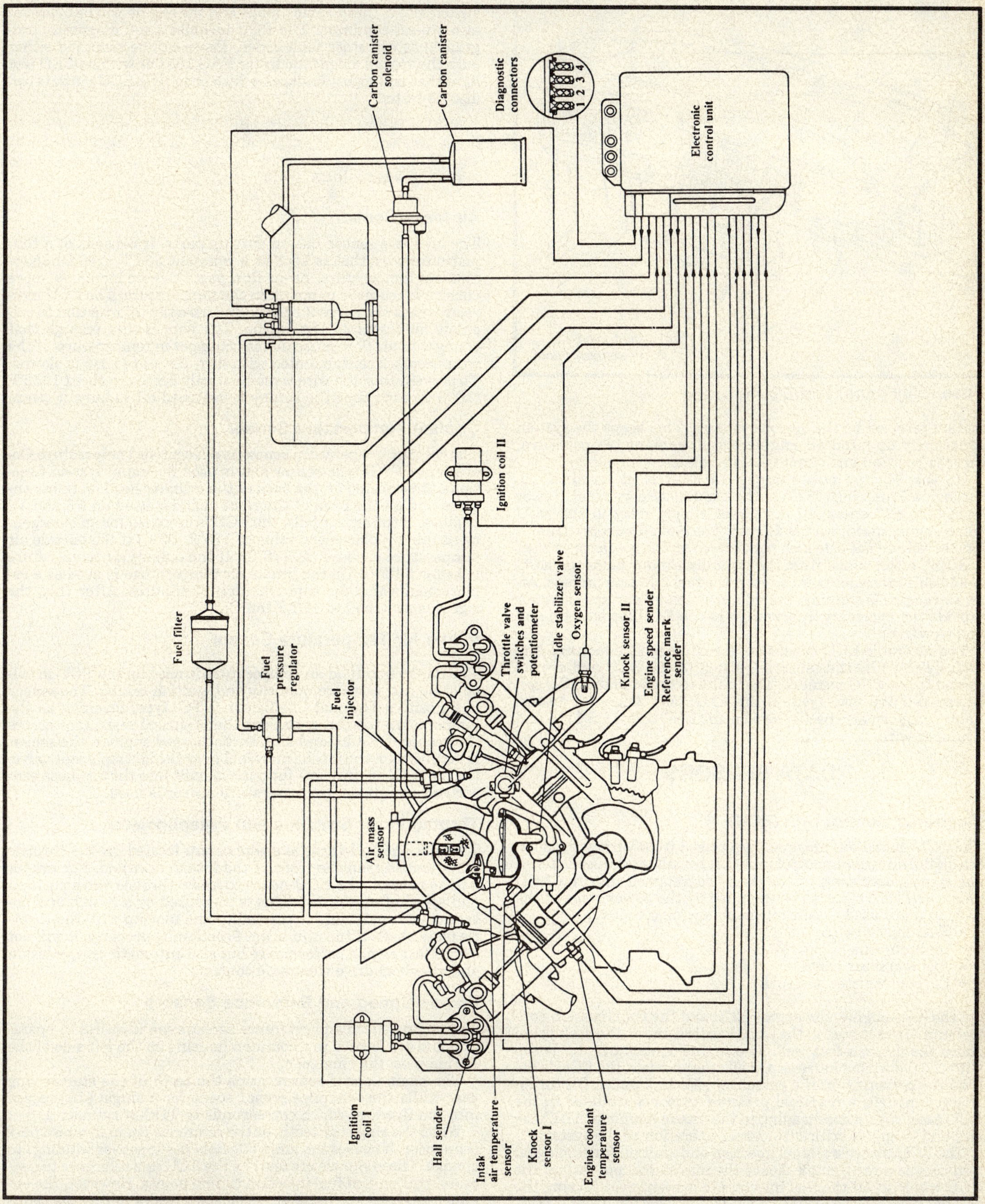

Motronic system schematic

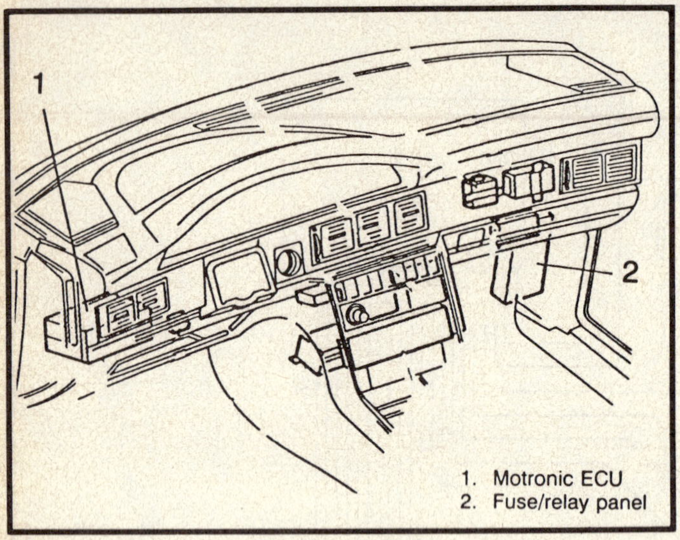

Fuse/relay panel and ECU location

mation received by the various sensors. This keeps the system constantly updated on engine speed, coolant temperature, throttle position and the intake air volume.

On the V8, the power supply to ECU is at terminal 18, through a 5 amp fuse (S27) in the main fuse/relay panel. Power from fuse S27 energizes the power supply relay in the ECU when engine speed reaches 25 rpm. The main fuse/relay panel is located behind the side kick panel cover on the passenger's side.

A Hall effect signal from the right distributor helps the ECU establish a reference point to start the fuel injection process. After the engine is running, the reference sender and speed sensor provide the necessary information to the ECU for ignition and fuel injection.

The Motronic ECU consists of a self-diagnostic feature. Any fault detected by the sensors are sent to the ECU and are recorded in the ECU memory. The faults recorded in the ECU can be recalled and displayed on the VAG 1551 diagnostic tester. Fault codes can also be displayed using LED tester US 1115 and a jumper wire.

DATA SENSORS

Electronic Control Unit (ECU)

The ECU, based on information received from various sensors, is a digital electronic control unit that calculates the correct output signals according to the data programmed into the ECU memory. These signals are amplified by the power stage of the ECU and are used to control the following components:

- Fuel injectors
- Ignition coils
- Idle stabilizer valve
- Carbon canister solenoid

On the 4.2L engine, the engine ECU and the transmission control unit are mounted in the same location, inside the dash panel above the driver's foot well. The engine control unit on is top with the red wiring harness. On all other models, the ECU is behind the carpeting on the passenger side kick panel. On turbocharged models, a manifold pressure sensor is included in the ECU case, with a hose leading to the intake manifold. The hose size and length is critical to proper operation of the system.

The ECU controls all fuel injection and ignition functions and also acts as a rev limiter. Above the maximum rpm rating, the ECU will shut off the fuel injectors to prevent over-revving the engine. If an input or output device fault is detected and it is serious enough to cause engine control problems, the ECU will switch into a "limp home" mode and the engine will be operated at a pre-set condition. The ECU includes a self diagnostic program that will store fault codes. These can be retrieved either with the special diagnostic tester VAG 1551 or with an LED test light that will report faults as a flash code. The ECU outputs can also be tested.

NOTE When removing the ECU connector, wait at least 20 seconds after turning the ignition switch OFF. The ECU must be allowed to complete the air mass sensor wire burn-clean cycle.

Air Mass Sensor

The air mass sensor has no moving parts. It consists of a thin platinum wire that is kept at a constant 356°F (180°C) above ambient temperature. As air flow past the wire increases, more current is required to maintain the temperature. The ECU indirectly reads the current change and uses the information to calculate air flow into the engine. The wire is thin enough that changes in air flow cause instant changes in temperature. If the wire becomes contaminated with dirt, response time is slowed. To prevent this, the wire is momentarily heated to about 1832°F (1000°C) each time the engine is switched off to burn it clean.

Coolant Temperature Sensor

The coolant temperature sensor is a Negative Temperature Coefficient (NTC) type sensor and is used for many system functions. It is located on the back of the cylinder head on inline engines and on the back of the right cylinder head on V6 and V8 engines. If the sensor fails, the ECU is designed for an emergency running replacement value of 175°F (80°C) if the outside air temperature is above 32°F (0°C). If the outside air temperature is below 32°F (0°C), the intake air temperature is used as a replacement value for approximately 3 minutes. After this, the signal is switched to 175°F (80°C).

Intake Air Temperature Sensor

The intake manifold air temperature is used by the ECU in calculating fuel injection quantity and ignition timing. The sensor is a Positive Thermal Coefficient (PTC) type, meaning as the temperature increases, so does the electrical resistance across the sensor. On 4.2L and 2.2L turbocharged engines, the sensor is mounted on the intake manifold near the idle stabilizer valve. On the 2.8L engine, this function is built into the air mass sensor and no intake air temperature sensor is used.

Throttle Valve Switches and Potentiometer

The idle and full throttle switches are located on the throttle valve housing, on the front of the intake manifold. For proper engine control, the ECU needs to know throttle valve position and speed of movement. This is measured by a switch and potentiometer assembly connected to and moving with the throttle plate shaft. If the unit is not functioning properly, it will not only affect engine performance but also automatic transmission and all-wheel drive electronic controls.

Engine Speed and Reference Sensors

The engine speed and reference sensors are identical in operation and are located in a common housing on the left side of the engine near the ring gear.

The engine speed sensor scans the teeth of the starter ring gear while the reference sensor scans for a single pin pressed into the flywheel which corresponds to TDC of cylinder 1.

When the ring gear teeth, or the reference pin in flywheel pass the sensor, it causes the magnetic field in the sensor windings to change. These signals are used by the ECU as a reference to generate ignition and fuel injection timing points. However, the engine cannot be started if either of the sensors fail. If the reference sensor fails while the engine is running, the engine will

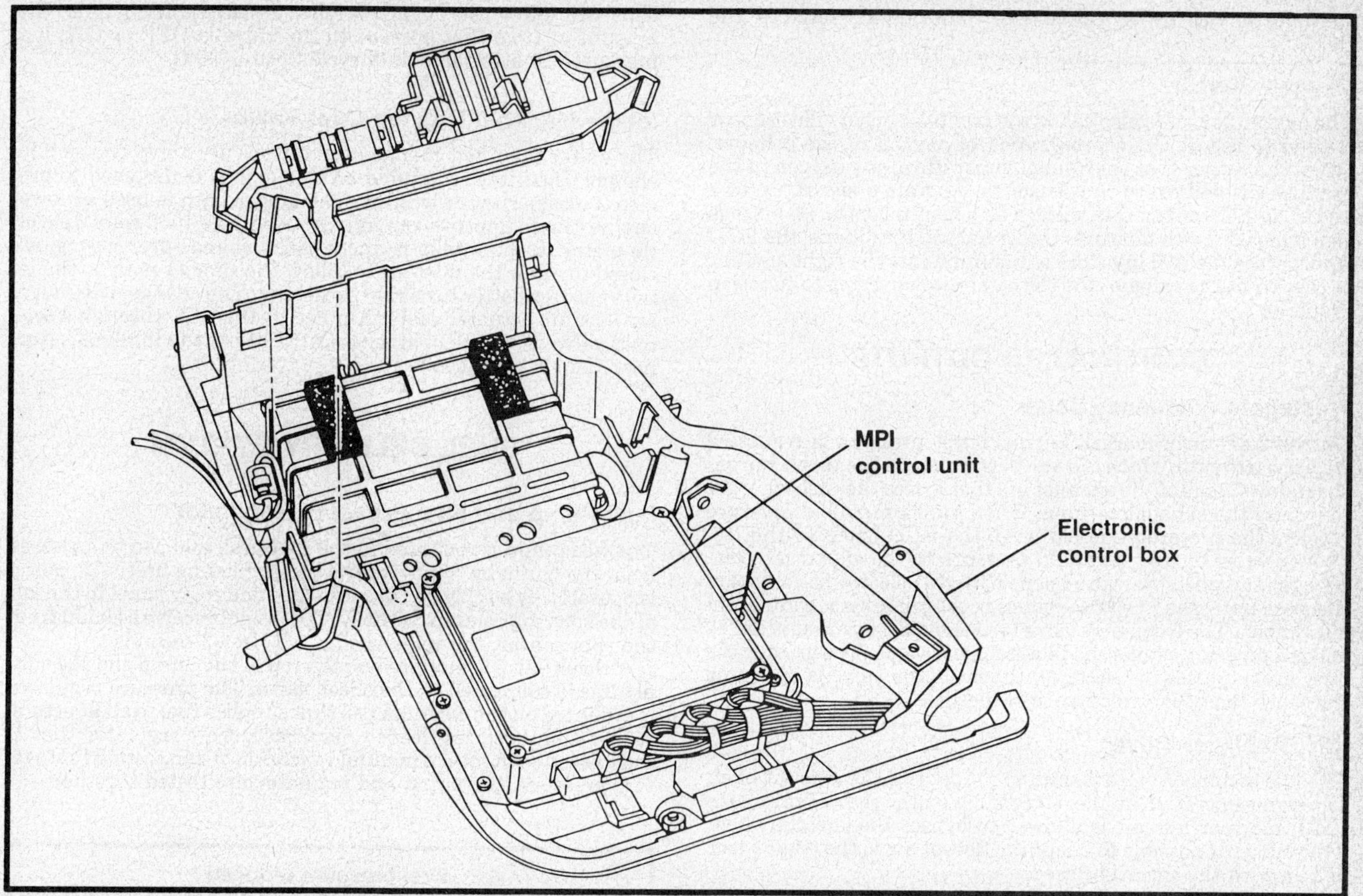

The ECU is mounted in the passenger footwell. Lift the carpet to gain access—1992 Audi 100 vehicles

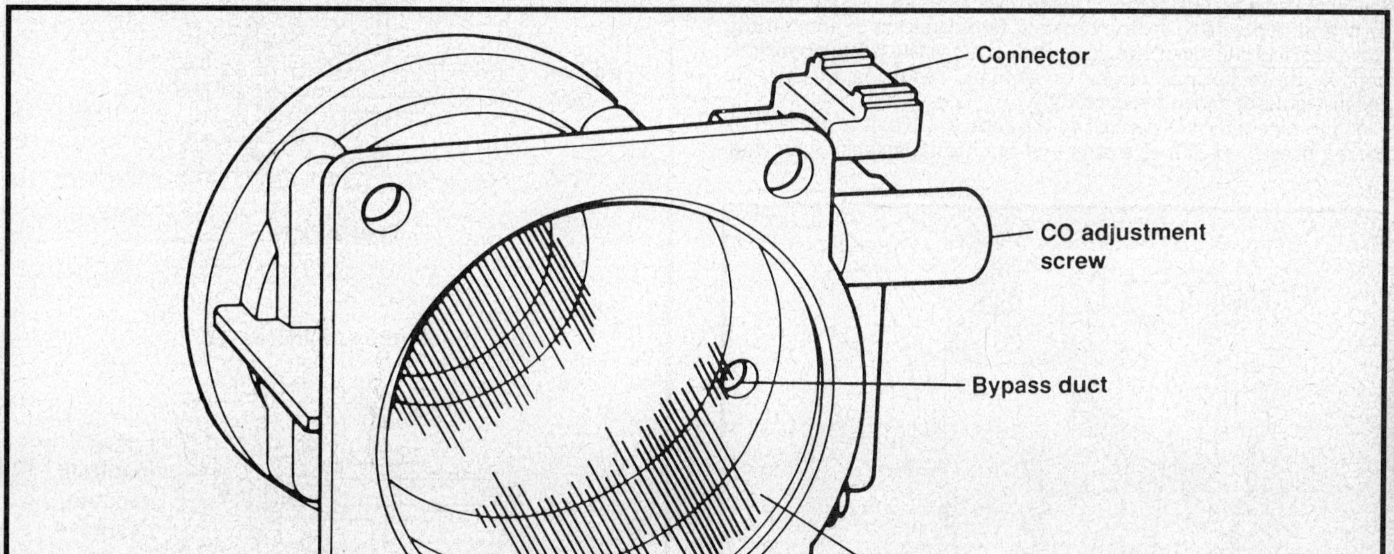

Air mass sensor assembly—1992 Audi 100 vehicles

continue to run using a reference value (data) stored in the ECU.

Oxygen Sensor

The oxygen sensor is also called the Lambda sensor. This sensor is used to report the concentration of oxygen in the exhaust. When the sensor is at operating temperature, any oxygen in the exhaust gas will cause the sensor to generate a signal, up to 1 volt. The ECU reads this voltage as a lean mixture, no voltage signal means a rich mixture. Under normal conditions, the ECU will operate the fuel injectors to maintain just the right amount of oxygen in the exhaust for the catalytic converter to function properly.

CONTROLLED OUTPUTS

Wastegate Frequency Valve

On the 2.2L turbocharged engine, boost pressure is regulated with a waste gate, which is a valve mounted on the turbocharger that allows some of the exhaust gas that drives the turbocharger to bypass the exhaust turbine. When intake manifold pressure reaches the pre-programmed level, the ECU operates the frequency valve to feed manifold pressure to the waste gate diaphragm and push the valve open. With this electronic control of the wastegate, the ECU can reduce boost pressure to control engine knock. The frequency valve is a solenoid valve that is fed 12 volts in pulses, just like the idle stabilizer valve. The greater the duty cycle (pulses per second), the more the valve is open and the more the boost pressure is reduced.

Idle Stabilizer Valve

The idle stabilizer valve is used to the control the amount of air for engine operation at idle. Depending on the signal sent by the ECU, more or less air is allowed to bypass the throttle valve. This makes it possible to adapt the flow of air to the engine load at idle with the throttle closed.

The idle stabilizer consists of a rotary valve connected to an armature. During normal operation, a duty cycle is applied to the armature by the ECU. This causes the valve to rotate in a direction opposite to spring tension. Lengthening or shortening the electrical pulses of the duty cycle causes the opening for the bypass air to become larger or smaller, resulting in the idle speed rising or falling accordingly.

When no current is applied to the rotary valve, it is pressed by spring pressure against a stop and maintains a specific air gap.

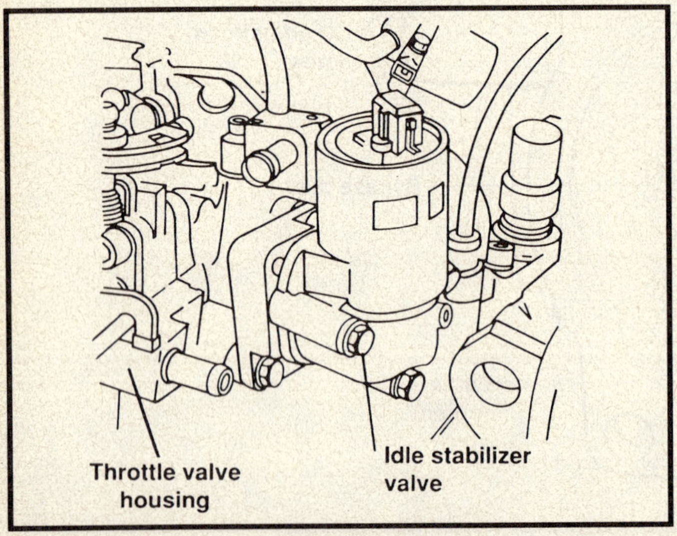

Idle stabilizer mounting—1992 Audi 100 vehicles

This will allow the engine to idle, if the stabilizer fails. The amount of time that current to the valve is OFF or ON, it is measured in a variable duty cycle from 5–95%.

Intake Manifold Change-Over Valve

The intake manifold change-over valve is used only on the 2.8L engine. The intake manifold on this engine is designed to provide 2 intake runner lengths. The longer path is used at lower engine rpm to improve low end torque. Above 4000 rpm, 6 throttle plates in the intake runners called change-over valves, are closed to allow the inlet air to follow the shorter path to the intake port for better high rpm breathing. The change-over valves are vacuum actuated and controlled by the ECU through a solenoid valve. This solenoid is mounted below the intake air duct near the air mass sensor.

FUEL DELIVERY SYSTEM

Fuel Pump and Fuel Pressure Regulator

The fuel pump is mounted in the fuel tank and can be accessed from the trunk by removing the gauge sending unit. The pump is operated by a relay on the main fuse and relay panel to the left of the steering column. If the ECU does not receive a signal from the rpm sensor, the relay is shut off after 2 seconds.

A check valve is mounted on the top of the pump and the supply line is connected to the check valve. The pressure regulator is mounted on the common rail that supplies fuel to all injectors. It is a simple spring loaded diaphragm type regulator that is modulated with intake manifold vacuum. When testing fuel system pressure, the pump and regulator are tested together.

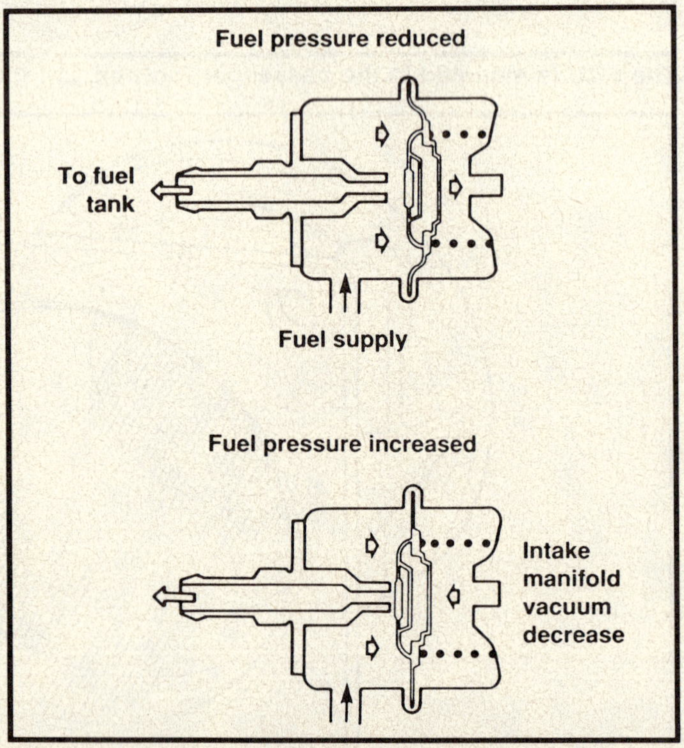

The fuel pressure regulator varies the fuel flow back to the fuel tank in response to manifold vacuum. The lower the vacuum, the less fuel flows back to the fuel tank, increasing the injection pressure

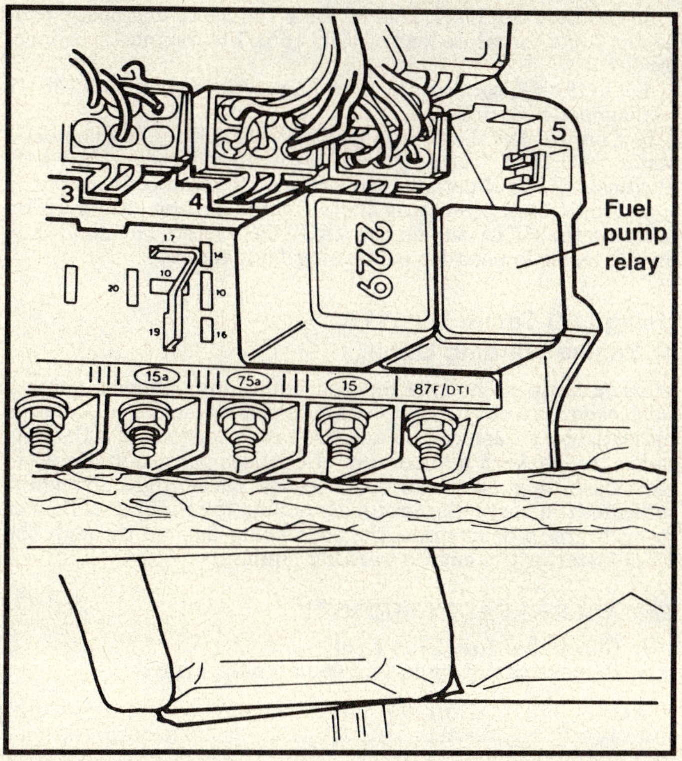

The fuel pump relay is mounted in the relay panel under the dash on the drivers side—1992 Audi 100 vehicles

Fuel Injectors

The fuel injectors are high precision solenoid valves. The opening of the valves is controlled by the ECU. The length of time the injector is held open controls the fuel mixture.

Diagnosis and Testing

SERVICE PRECAUTIONS

To prevent personal injury or damage to the vehicle, follow allservice precautions.

- Always disconnect the negative battery terminal to prevent sparks cause by short-circuiting, circuit-breaking, etc.
- Smoking must not be allowed when working around flammable liquids.
- Always keep a CO_2 fire extinguisher close on hand.
- Absorbant must be available to soak up any spillage.
- Ensure fuel is emptied into a suitable container.
- Always place the transmission gear shift lever in **N** (manual transmission) or **P** (automatic transmission) and set the parking brake and block drive wheels before starting the engine.
- Keep hands, tools and clothing away from engine cooling fan to help prevent personal injury.
- Thoroughly clean all unions and area near connections before disconnecting.

READING CODES AND OUTPUT CHECK

The Motronic control unit (ECU) incorporates a self-diagnosis capability. If a sensor or component malfunctions, they are stored in the ECU memory. Faults that occur occasionally are considered intermittent faults. After the ECU evaluates this information, it differentiates between 31 different faults and stores these faults until the memory is canceled. The ECU is equipped with a diagnosis capability of 10 output checks.

The self-diagnosis function is capable of storing, recalling, canceling output check diagnosis and additional capabilities of basic adjustment, control unit identification and coding. Faults stored by the ECU can be read using the VAG 1551 self-diagnosis tester.

The VAG 1551 self-diagnosis tester consists of a text display, numeric key-pad and a small printer. The unit is connected to the vehicles diagnostic connector, located on the auxiliary relay panel under the front passenger footwell carpeting or in the underhood relay box, through cable VAG 1551/1.

If the VAG 1551 self-diagnosis tester is not available, fault memory recall, output check diagnosis and cancellation of fault memory can be accomplished using LED tester US 1115 or equivalent. When reading a fault code using a LED tester, static and intermittent faults cannot be differentiated. On California models, the flash code can be read off the engine warning lamp on the instrument panel.

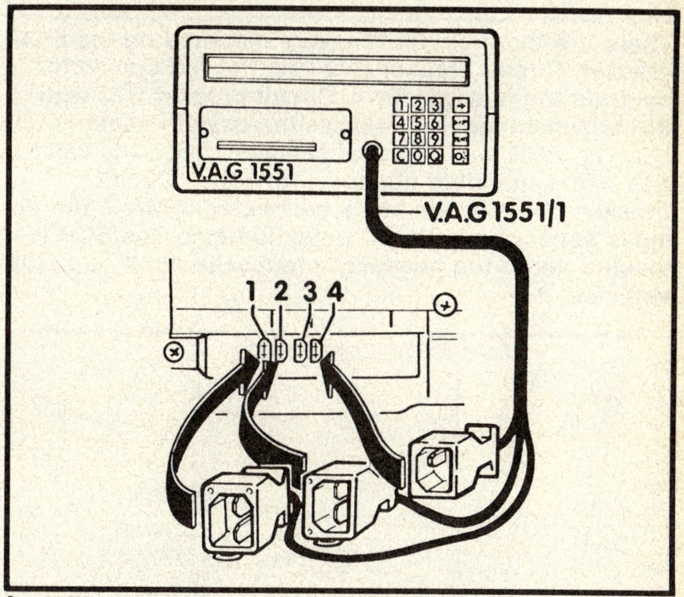

Installing diagnostic tester VAG 1551 and adapter VAG 1551/1

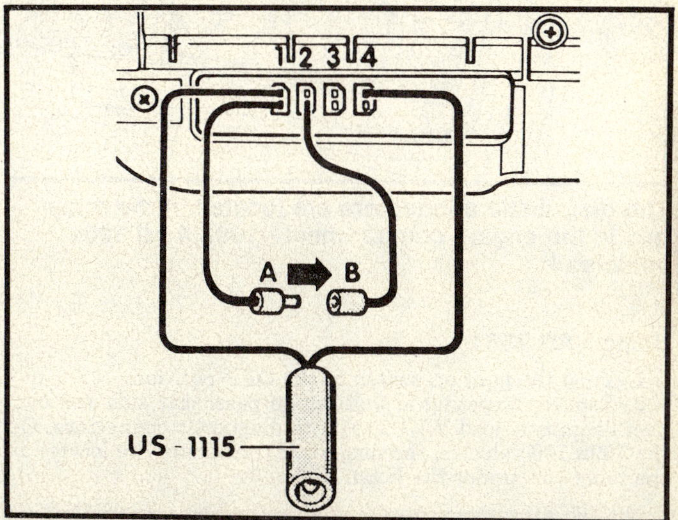

Installing LED tester US 1115 and jumper leads (A/B)

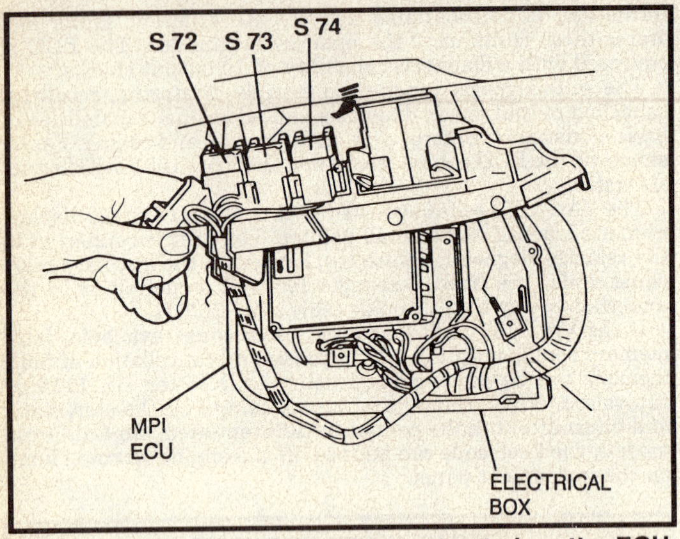

There are three circuit breakers mounted on the ECU bracket. Circuit breaker S72 with the red connector controls the fuel injectors. Circuit breaker S73 with the brown connector controls the oxygen sensors heaters, EGR and charcoal canister frequency valves and intake manifold change-over valve. Circuit breaker S74 with the black connector controls the air mass sensor and the CO potentiometer. The ECU is located under the passenger footwell—1992 Audi 100 vehicles

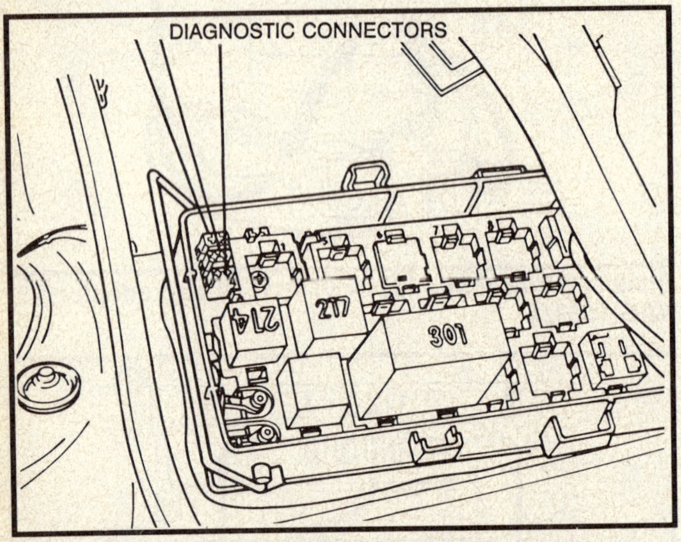

The diagnostic connectors are located in the relay box in the engine compartment—1992 Audi 100 vehicles

Using VAG 1551

1. Turn the ignition switch to the **OFF** position.
2. Remove carpeting in footwell on passenger side and connect diagnostic lead VAG 1151/1 to diagnostic connectors. On the 1992 100 vehicles, the diagnostic connectors are located in the relay box, under the hood.

NOTE: Observe connector shape when connecting diagnostic leads.

3. Connect the black lead of VAG 1551/1 to diagnostic connector 1 and the white lead of VAG 1551/1 to diagnostic connector 2.
4. On the V8, connect the blue diagnostic lead of VAG 1551/1 to diagnostic connector 4.
5. Connect the diagnostic lead of VAG 1551/1 to diagnostic tester VAG 1551. The diagnostic tester should read—VAG self-diagnosis–1 Rapid data transmission or 2 Flash code output.
6. Additional operating instructions can be called up by pressing the HELP key on the VAG 1551 tester. The arrow key on the tester is used for continuing fault tracing.

Using LED Tester US 1115 or Engine Warning Lamp

When a fault memory is recalled, faults may be read using a flash code. In a output check diagnosis, the tested component is indicated by a flash code. Each flash code consists of 4 flashing pulse groups, with a maximum of 4 pulses per group. Between each flashing pulse group, a 2.5 second pause (light off) which indicates the next pulse group. By adding the number of flashes in each group determines the code being flashed through the LED tester or the engine warning lamp.

RECALLING FAULT MEMORY

1. Check that fuse 27 is good.
2. Remove carpeting in footwell on passenger side.

NOTE: On California models, only connect jumper leads (A/B) to lower terminals of diagnostic connectors 1 and 2.

3. Connect the LED tester US 1115 black lead to upper terminal on the yellow diagnostic connector 4 and the red lead of LED tester to the upper terminal of the black diagnostic connector 1.
4. Connect jumper leads to lower terminals of diagnostic connectors 1 and 2. Do not connect leads at this point.
5. Start engine or crank engine for approximately 5–6 seconds. Do not turn ignition switch **OFF**.
6. Connect jumper leads (A/B) for at least 4 seconds. The LED tester and selector lever indicator in dash panel should light. If not, proceed as follows:
   a. Check connections from diagnostic connectors 2 and 4 to the control unit and LED tester connections.
   b. Disconnect jumper leads (A/B). Test lamp should start to flash.
7. Count and note flashes on test lamp or engine warning lamp on California models.

NOTE: At this time, the transmission position indicator shows only the selector lever position or the mechanical limp home mode (3rd gear). Code 2111 (engine speed sender) will be displayed when the ignition switch is ON and the engine not running. This fault display should be ignored when the engine is not running.

8. If the engine does not start, leave ignition switch turned in the **ON** position. Code 2111 should be displayed.
9. Connect jumper leads (A/B) for approximately 4 seconds. Count and note flashes. Repeat Steps 8 and 9 until test lamp shows Code 0000.

OUTPUT CHECK DIAGNOSIS USING LED TESTER US 1115

When performing the output check diagnosis, all injectors, the idle stabilizer valve and the solenoid valve for the carbon canister are checked by sound or by touch. The injectors are activated in pairs; 1 and 5, 2 and 7, 3 and 6, 4 and 8. The output check diagnosis will be interrupted if the engine is started or a speed pulse is detected.

1. Remove carpeting in footwell on passenger side.

NOTE: On California models, only connect jumper leads (A/B) to lower terminals of diagnostic connectors 1 and 2.

2. Connect LED tester US 1115 black lead to the upper terminal on the yellow diagnostic connector 4 and the red lead on the tester to the upper terminal on the black diagnostic connector 1.
3. Connect jumper leads (A/B) to the lower terminals of both diagnostic connectors 1 and 2.
4. Connect jumper leads and turn the ignition switch to the **ON** position. The LED tester and all the display elements on the transmission position indicator must light.
5. If not, check that the ignition switch is turned to the **ON** position, and that test lamp and harness connections to diagnostic connectors 2 and 4 are good.
6. Disconnect jumper leads (A/B). After approximately 10 seconds, the test light should begin to flash. On California models, the engine warning lamp should start to flash.
7. Count and note the number of flashes.
8. Open and release the throttle lever. When throttle lever returns to idle position, injector for cylinder 1 will cycle ON/OFF 5 times. Check operation of injector for cylinder 1 by touching or listening for 5 clicks.

NOTE: After checking the injector for cylinder 1, the ECU will automatically switch to the next injector, which will be displayed on the LED tester or the engine warning lamp, then the remaining injectors can be checked in the same manner.

4411	N 30	Injector cylinder 1
4412	N 31	Injector cylinder 2
4413	N 32	Injector cylinder 3
4414	N 33	Injector cylinder 4
4421	N 83	Injector cylinder 5
4422	N 84	Injector cylinder 6
4423	N 85	Injector cylinder 7
4424	N 86	Injector cylinder 8
4431	N 71	Idle stabilization valve
4343	(N 80)	Carbon canister solenoid valve

Flash code and operation sequence

9. After the ECU checks the injector for cylinder 8, it will automatically switch to the idle stabilizer valve.

NOTE: To switch to the next injector without testing, connect jumper leads (A/B) for approximately 4 seconds, then disconnect leads.

10. Connect jumper leads (A/B) for approximately 4 seconds, then disconnect. Count and note number of flashes on LED tester or engine warning light.

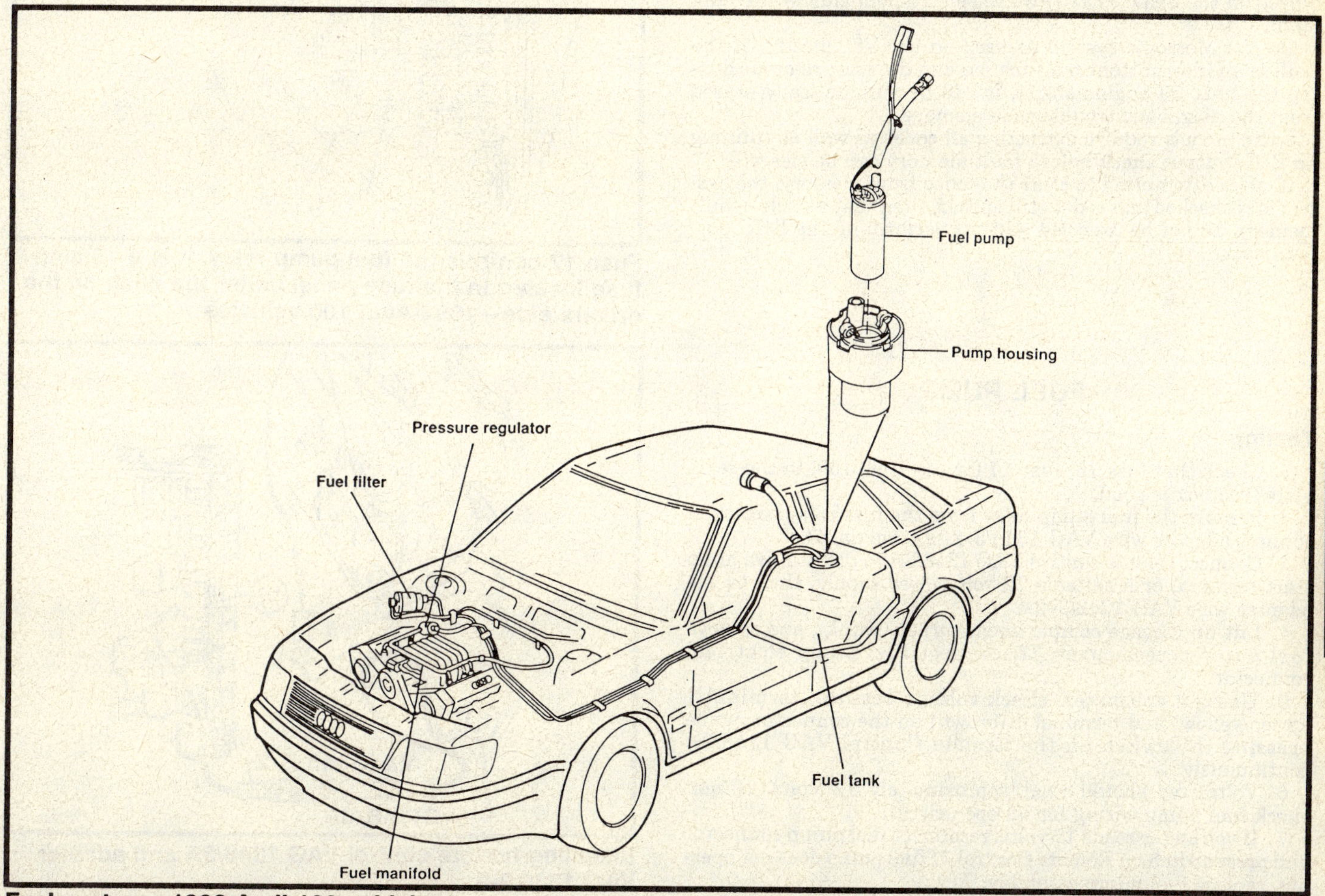

Fuel system—1992 Audi 100 vehicles

NOTE: At this time, the idle stabilizer valve and the solenoid for the carbon canister should operate until jumper leads (A/B) are connected for a minimum of 4 seconds, then disconnected.

11. Repeat Step 10 until LED tester or engine warning light flashes Code 0000. When Code 0000 is obtained, reconnect jumper leads (A/B) for 4 seconds, then disconnect leads. The fault memory in the ECU is then canceled.

Installing Test Box VAG 1598

NOTE:Never connect tester VAG 1598 with tester VAG 1551. Always wait approximately 20 seconds after turning the ignition switch OFF before disconnecting the ECU connector.

1. Make sure ignition switch is in the **OFF** position.
2. Remove the access panel under the steering column.
3. Remove the heating duct, then unclip the footwell light and position aside.
4. Remove the control units attaching bolts. Disconnect the Red connector (upper, Motronic control unit) from the ECU.
5. Connect test box VAG 1598 and adapter cable VAG 1598/5 to the ECU harness connector.

CLEARING CODES

The fault memory of the engine control unit can be erased following the display of fault Code 0000. Join the connectors used for reading codes for at least 4 seconds, then separate. The memory is erased when there are no additional flash codes displayed on the LED tester or on the instrument cluster warning lamp.

On the Motronic system as used on the V8 Quattro, if the fault is an intermittent malfunction and does not re-occur during the next 50 engine starts, it will be automatically erased from the diagnostic trouble code memory.

Some models require extracting all codes as well as running an ECU output check before fault memory can be cleared.

If, after attempting to clear the codes from memory, the system is rechecked and codes still appear, then that vehicle's fault memory cannot be canceled without performing the ECU output check.

FUEL PUMP

Testing

1. Check that fuse 13, fuse 17 for the 1992 100, in the relay plate/fuse box is good.
2. Remove the fuel pump relay from the fuse/relay box, then connect adapter wire VAG 1348/3-2 to terminal 52.
3. Connect remote control VAG 1348/3A to the positive jump start terminal or a suitable 12 volt power supply, then to the adapter wire VAG 1348/3-2.
4. Lift up luggage compartment carpet (trunk), and remove fuel pump access cover. Disconnect fuel pump electrical connector.
5. Using a voltmeter, check voltage between terminal 1 (green/yellow) and terminal 4 (brown) on the connector, while pressing the switch on the Remote Control VAG 1348/3A continuously.
6. Voltmeter should read approximately 12 volts. If not, check fuel pump wiring for an open circuit.
7. If voltmeter reads 12 volts, reconnect fuel pump connector and press switch on Remote Control. If fuel pump does not operate, replace fuel pump assembly.

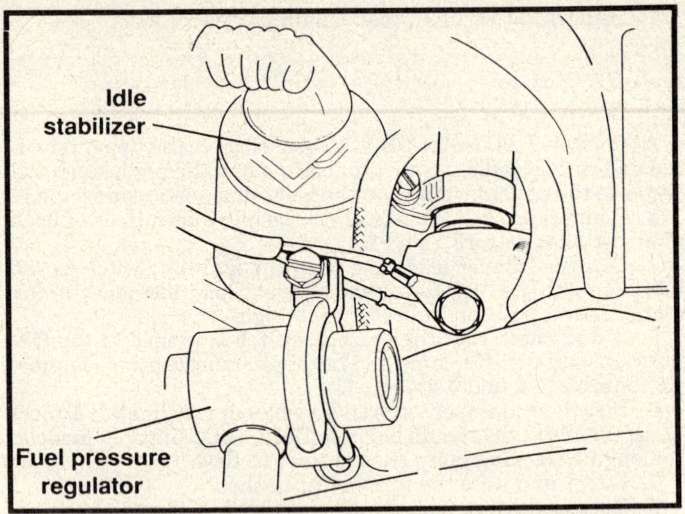

The fuel pressure regulator is mounted next to the idle stabilizer—1992 Audi 100 vehicles

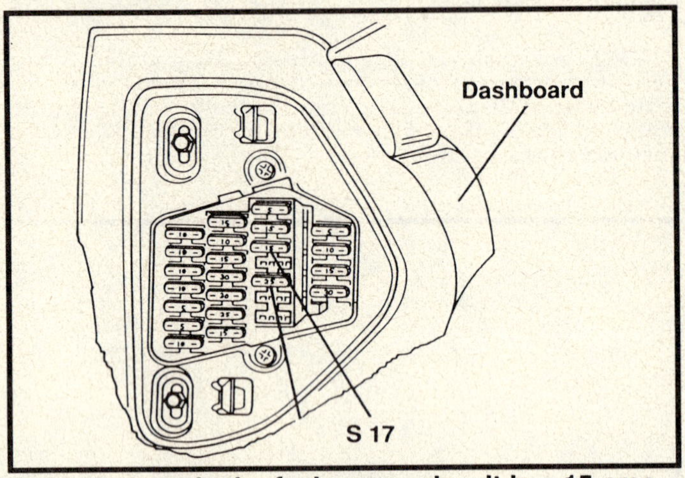

Fuse 17 controls the fuel pump relay. It is a 15 amp fuse located in the fuse panel under the dash on the drivers side—1992 Audi 100 vehicles

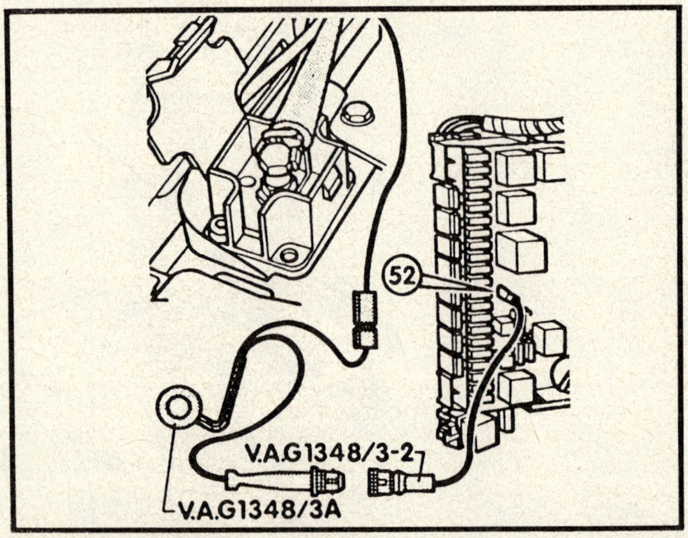

Installing remote control VAG 1348/3A and adapter VAG 1348/3-2

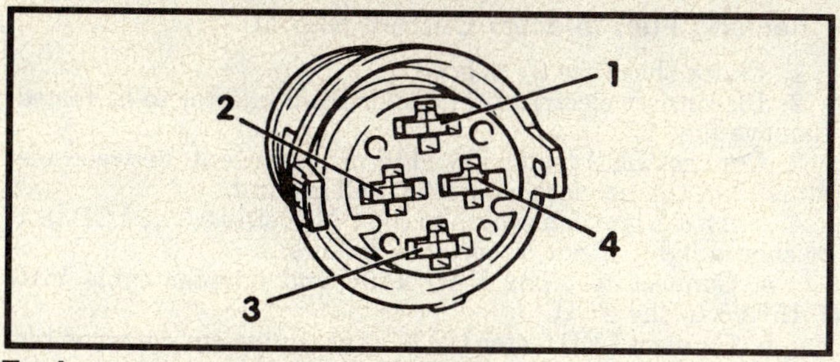

Fuel pump terminal identification

FUEL PUMP CHECK VALVE

Testing

Before testing the fuel pump check valve, check that the fuel pump, pressure regulator and fuel injectors are operating properly. Check that the fuel lines and fittings are tight and not leaking.

1. Disconnect the fuel supply line from the injector manifold.
2. Connect a pressure gauge and adapter assembly between the fuel supply line and the injector manifold as shown. Open pressure gauge lever.
3. Remove the fuel pump relay from the fuse/relay box, then connect adapter wire VAG 1348/3–2 to terminal 52.
4. Connect remote control VAG 1348/3A to the positive jump start terminal or a suitable 12 volt power supply, then to the adapter wire VAG 1348/3–2.
5. Press the switch on the remote control and hold for approximately 30 seconds. Close the pressure gauge lever.
6. Operate the remote control switch so 80 psi, 65 for V8 Quattro, is obtained on the pressure gauge. Wait approximately 10 minutes and recheck gauge. Maximum pressure drop from initial reading is 7 psi.
7. If fuel pump specifications are as indicated, check lines and fittings for leaks and repair as necessary.
8. If fuel pump specifications are low, replace fuel pump check valve.

FUEL PUMP RELAY

Fuel Pump Relay Test

1. Check that fuse 13, 17 for 1992 100, in the fuel/relay box is good.

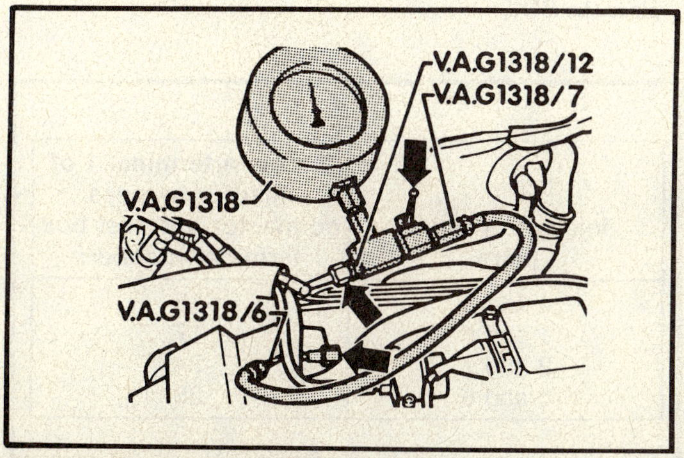

Connecting pressure gauge

2. Connect LED tester US 1115 or equivalent between fuse 13 and ground.
3. Operate starter. LED tester should light and the relay should operate (clicking or vibrating noise).
4. If relay does not operate, check the control signal to the fuel pump relay. If control signal is good, replace the fuel pump relay.

Control Signal Test

1. Remove the fuel pump relay from the fuse/relay panel. Turn the ignition switch to the **ON** position.
2. Connect a multimeter between terminals 46 and 50, then between terminals 48 and 50 on the fuse/relay panel. Multimeter should read approximately 12 volts. Turn ignition switch **OFF**.
3. If specifications are not as indicated, proceed as follows:
   a. Check system wiring for an open circuit. Repair as necessary.
   b. Connect LED tester VAG 1115 or equivalent between terminals 46 and 47. Turn ignition switch to the **ON** position.
   c. The LED tester light should light up (dim). Operate starter motor. LED tester light should light brighter.
   d. If tester light does not become brighter, connect test box VAG 1598 and adapter VAG 1598/5 or equivalent, to the ECU harness.

NOTE: For this test, the ECU harness connector is left disconnected.

   e. Check for continuity between terminal 47 of the fuse/relay panel and terminal 3 on the test box.
   f. If no continuity exists, check for a break in wiring and repair as necessary. If wiring is good, replace ECU.

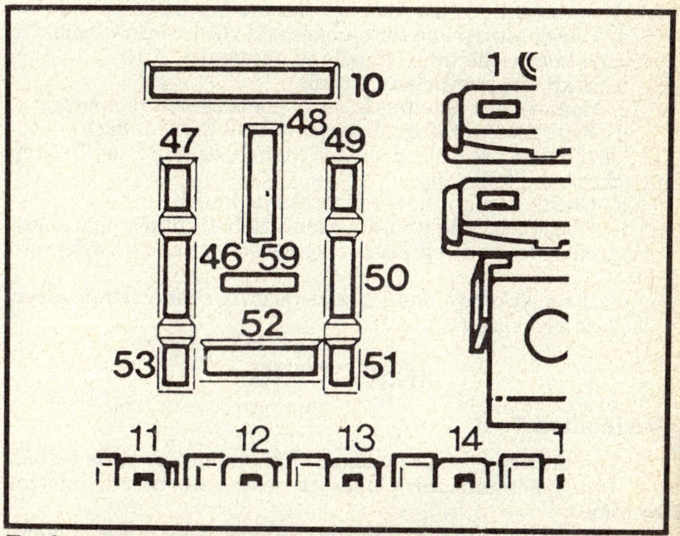

Fuel pump relay terminal identification

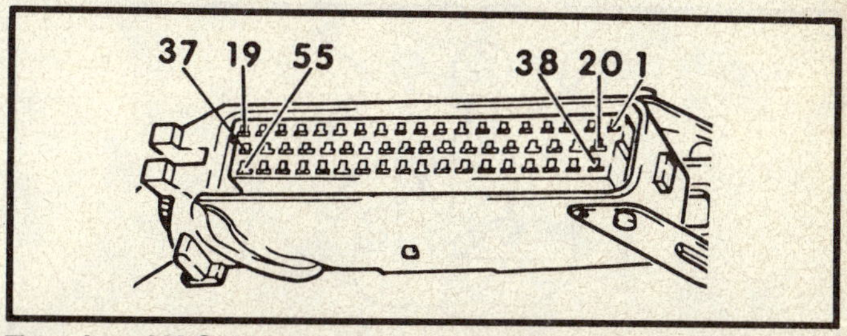

Test box VAG 1598 terminal identification

FUEL SYSTEM

Pressure Test

1. Connect pressure gauge VAG 1318 or equivalent, between the fuel supply line and the fuel rail. Set valve on gauge to the open position.
2. Remove the fuel pump relay from the fuse/relay panel.
3. Connect remote control VAG 1348/3 A with adapter VAG 1348/3-2 or equivalent, to terminal 52 on the fuse/relay panel and to the positive remote connection (12 volt power supply) in the engine compartment.
4. Disconnect an plug vacuum hose from pressure regulator to intake on fuel pressure regulator.

NOTE: If fuel leaks during the following pressure check on the vacuum connection of the pressure regulator, replace pressure regulator.

5. Operate remote control VAG 1348/3A. Fuel pump should operate. If not, check fuel pump and repair as necessary.
6. Operate remote control VAG 1348/3A until fuel pressure no longer increases. Fuel pressure should be 58–62 psi (4.0–4.3 bar) with fuel pump operating, 49–54 psi (3.0–3.8 bar) immediately after fuel pump is switched off.
7. If specifications are not as indicated, check fuel pump and supply line for leaks or damage and repair as necessary. Recheck fuel pressure.
8. If correct specifications are still not obtained, replace pressure regulator and recheck fuel pressure.
9. If fuel pressure exceeds specifications, proceed as follows:
   a. Check fuel rails for fuel flow.
   b. Disconnect return line at strut and place into a suitable container.
   c. Repeat pressure check. If fuel pressure still exceeds specifications, replace fuel rails.
10. If specifications are still not obtained, proceed as follows:
   a. Check return line for damage and the fuel flow from the return line to the tank. Repair as necessary.
   b. Start the engine and let idle.
   c. Connect the vacuum hose to the pressure regulator.
   d. Fuel pressure should decrease by 8 psi (0.6 bar).
11. If the fuel pressure does not change, as indicated in Step 10c, check the following:
   a. Check vacuum hoses for leaks or kinks.
   b. Check that the vacuum connection at intake pipe is not clogged by removing pressure regulator hose and blowing into it.
   c. If no leaks are present and vacuum connection is clear, replace pressure regulator.

INJECTORS

Resistance Test

1. Disconnect electrical connector from injector to be tested.
2. Using a ohmmeter, measure resistance across injector terminals.
3. Ohmmeter should read 15–17 ohms. If specifications are not as indicated, replace injector.

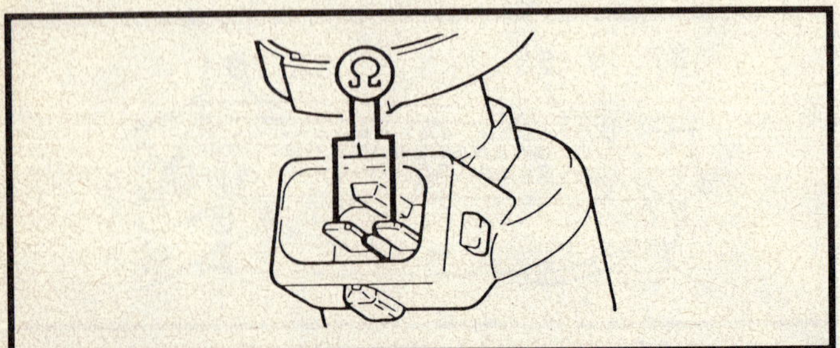

Testing injector

Checking Fuel Injector Current Signal

1. Check that fuse 27 is good.
2. Disconnect electrical connector from injector to be tested. Remove fuse 23.
3. Connect LED tester US 1115 or equivalent, between terminal 2 of the harness connector and ground.
4. Turn the ignition switch to the **ON** position. The LED tester should light. If not, proceed as follows:
   a. Connect test box VAG 1598 and adapter cable VAG 1598/5 to the ECU.
   b. Connect LED tester US 1115 or equivalent, between terminal 19 and terminal 37 on the test box.
   c. Turn the ignition switch to the **ON** position. The LED tester should light.
   d. If LED tester lights, repair open circuit between terminal 37 of the control unit harness connector and terminal 2 of the injector harness being tested.
5. If the LED tester does not light, proceed as follows:
   a. Replace the ECU.
   b. Reinstall fuse 23.
   c. Connect LED tester to terminal 2 of the injector harness connector and ground.
   d. Turn the ignition switch **ON**, then crank engine for a few seconds. LED tester should light when ignition switch is turned ON and during engine cranking.
6. If LED tester light goes out during cranking, proceed as follows:
   a. Check that fuse 23 is good.
   b. Using an ohmmeter, check continuity between terminal 2 of the injector harness connector and fuse 23. Repair as necessary.
   c. Using an ohmmeter, check continuity between fuse 23 and the fuel pump relay terminal 59. Repair as necessary.
   d. Check the fuel pump relay. If necessary, check the current signal to fuel pump relay.
   e. Remove test box and connect LED tester between terminals 1 and 2 of injector harness connector being tested.
   f. Perform the Output Check Diagnosis. While performing the Output Check, when the injector being tested is operated, the LED tester should flash (maximum 5 times).
7. If the LED tester does not flash or light after performing Step 6f, Proceed as follows:
   a. Connect test box VAG 1598 and adapter cable 1598/5 or equivalent, to the ECU harness connector.
   b. Using an ohmmeter, check the following wires for continuity or ground connection, as shown in the following chart.
   c. If an open circuit exists, check circuit between control unit harness connector terminal and terminal 1 of the injector harness connector.
   d. If no open circuits, or short circuits to ground exists, replace the ECU.

Injector/cylinder pairs	Between terminal 1 of injector harness connector and test box terminal number:
4 and 8	16
1 and 5	17
2 and 7	34
3 and 6	35

Checking injector control signals

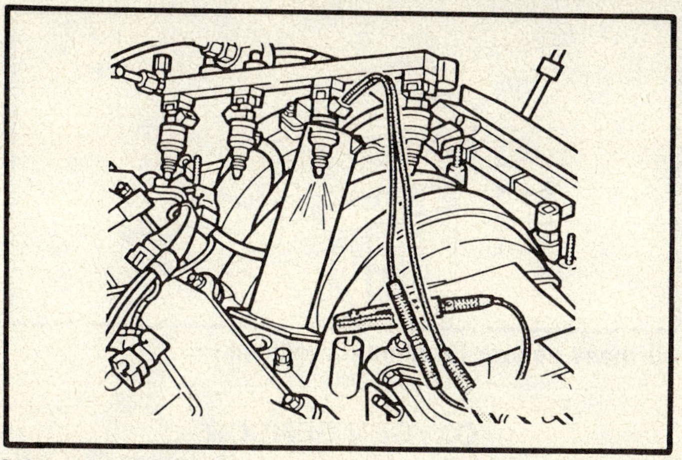

Injector volume test

Injector Leak And Volume Test

NOTE: On some 1990–91 V8 Quattro models leaking injectors could cause starter or flywheel teeth damage due to hydraulic engine lock. If replacing leaking injectors check the starter and flywheel teeth for damage.

1. Connect test box VAG 1598 and adapter cable 1598/5 or equivalent, to the ECU harness connector.
2. Install a jumper wire between terminal 3 and terminal 10 on the test box. Fuel pump should operate.
3. Using a suitable container, check injector for leaks. 1 or 2 drops per minute of fuel is acceptable. If fuel loss exceeds specifications, turn OFF fuel pump and replace injector.
4. Place injector into a suitable measuring container. Operate fuel pump for 30 seconds. Measured volume should be 3–3.4 ounces (90–100 ml).
5. If specifications are not as indicated, replace injector if measured volume from 1 or more injectors is above or below as specified.

AIR MASS SENSOR

Testing

S4, 200 QUATTRO AND WAGON

1. Disconnect the air mass sensor electrical connector.
2. Connect LED tester US 1115 or equivalent to terminal 1 on the harness connector and to the remote positive cable connection (or 12 volt power supply). LED tester should light.
3. If LED tester does not light, proceed as follows:
   a. Check for an open circuit in wiring.
   b. Connect LED tester between terminal 5 on connector and ground.
   c. Turn ignition switch to the **ON** position. LED tester should light.
4. If LED tester does not light after Step 3, proceed as follows:
   a. Connect test box VAG 1598 and adapter VAG 1598/5 or equivalent, to the ECU harness connector.
   b. Connect LED tester between terminal 19 and terminal 37 on the test box. Turn ignition switch to the **ON** position.
   c. LED tester should light. If not, locate and repair if necessary, an open circuit between terminal 37 on the control unit harness connector and terminal 5 on sensor harness connector.
5. Connect LED tester between terminal 5 on the sensor harness connector and ground.
6. Turn the ignition switch **ON** and crank engine for approximately 3–5 seconds. The LED tester should light when the ignition switch is turned ON and when cranking the engine. If not, proceed as follows:
   a. Check that fuse 23 is good.
   b. Using an ohmmeter, check for continuity between terminal 5 on the sensor harness connector and fuse 23.
   c. Using an ohmmeter, check for continuity between fuse 23 and terminal 59 on the fuel pump relay socket.
   d. Check the fuel pump relay, and if necessary, the control signal to the fuel pump relay.
7. Connect test box VAG 1598 and adapter cable VAG 1598/5 to the ECU harness connector.
8. Disconnect air mass sensor electrical connector.
9. Using an ohmmeter, check the following circuits for continuity or shorts:
   a. Air mass sensor connector terminal 2 to test box terminal 26.
   b. Air mass sensor connector terminal 3 to test box terminal 7.
   c. Air mass sensor connector terminal 24 to test box terminal 25.
10. Repair open or short circuits as necessary.
11. Disconnect air mass sensor electrical connector.
12. Connect a multimeter (200 ohm range) to terminals 1 and 2 on the air mass sensor.
13. Disconnect multimeter from harness connector and short leads. Multimeter reading should be the same as reading obtained in Step 2.
14. If reading obtained is greater than 0.1 ohm, proceed as follows:
   a. Replace air mass sensor.
   b. Reconnect air mass sensor connector, then pull back rubber boot on connector to expose terminals.
   c. Connect multimeter (20 volt range) to terminals 1 and 3. Turn ignition switch to the **ON** position. Multimeter should read 1.3–1.6 volts.
   d. Start the engine and let idle. Multimeter should read approximately 2.5 volts.
   e. Increase engine rpm. Voltage should increase to 3–5 volts.
   f. If above voltages are not obtained, replace air mass sensor.
15. Remove fuse 27 from the fuse/relay panel.
16. Connect an ammeter across the fuse terminals. Turn the ignition switch to the **ON** position. Ammeter should read 1.6–1.8 amps.
17. Start the engine and let idle. Ammeter should read 0.2–0.4 amps.
18. Turn the ignition switch **OFF**. Ammeter should read 1.6–1.8 amps. After approximately 3 seconds, amperage should rise above 2 amps for about 1 second.
19. If current does not rise, replace air mass sensor.

90 QUATTRO AND COUPE QUATTRO

1. Check the condition of fuse 27. The fuse must be intact.
2. Disconnect the plug from the air mass sensor. With the notch on the top of the connector, terminal 1 is on the left and terminal 4 on the right.
3. Connect US 1115 LED tester or equivalent between terminal 3 and ground. The LED should glow.
4. If the LED doesn't glow, connect test box 1598 or equivalent with the appropriate adapter to the control unit harness connector A. Keep the other connectors in place. Connector A is the 12 pin connector.
5. Connect the US 1115 tester between terminals 41 and 43 of the test box. With the ignition **ON**, the LED should illuminate. If it doesn't, replace the control unit.
6. Check continuity between pin 1 of the air mass sensor connector and 2 of the control unit connector or terminal 42 of the test box, if used. Check continuity between pin 2 of the air mass sensor connector and terminal 5 of the control unit connector or terminal 45 of the test box, if used. Check continuity between pin 4 of the air mass sensor connector and terminal 1 of the con-

trol unit connector or terminal 41 of the test box, if used. The resistance must not be greater than 0.5 ohms. If resistance is too high or a short exists, repair the wiring.

7. Peel back the rubber covering on the back side of the air mass sensor connector. The terminal numbers are marked on the back side of the connector. Turn **ON** the ignition.

8. The voltage between terminals 2 and 4 must be 1.0–1.5 volts. If the voltage is not correct the CO potentiometer is defective and the air mass sensor must be replaced.

9. Turn **OFF** the ignition. The voltage should be 0.3–1.1 volts.

10. With all electrical accessories turned **OFF** and the engine running check the voltage. The voltage will fluctuate between 0.3 volts and 1.1 volts as a function of the engine speed. If these checks do not work, replace the air mass sensor assembly.

V8 QUATTRO

1. Check the condition of circuit breaker 1 in the electronic control box located in the passenger footwell. The circuit breaker is in the black fuse holder.

2. Disconnect the plug from the air mass sensor. With the notch on the top of the connector, terminal 1 is on the left and terminal 4 on the right.

3. Connect a multimeter between terminal 2 and 3. There should be 12–14 volts.

4. If the voltage is not correct, connect test box 1598 or equivalent with the appropriate adapter to the control unit harness connector A. Keep the other connectors in place. Connector A is the 12 pin connector.

5. Connect the US 1115 tester between terminals 41 and 43 of the test box. With the ignition **ON**, the LED should illuminate. If it doesn't, replace the control unit.

6. Check continuity between pin 1 of the air mass sensor connector and 2 of the control unit connector or terminal 42 of the test box, if used. Check continuity between pin 2 of the air mass sensor connector and terminal 5 of the control unit connector or terminal 45 of the test box, if used. Check continuity between pin 4 of the air mass sensor connector and terminal 1 of the control unit connector or terminal 41 of the test box, if used. The resistance must not be greater than 0.5 ohms. If resistance is too high or a short exists, repair the wiring.

7. Peel back the rubber covering on the back side of the air mass sensor connector. The terminal numbers are marked on the back side of the connector. Turn **ON** the ignition.

8. Turn **ON** the ignition. The voltage should be 1.0–7.5 volts between terminals 2 and 4. Replace the air mass sensor if not correct.

8. The voltage should be 0.3–1.1 volts between terminals 1 and 2.

9. With all electrical accessories turned **OFF** and the engine running check the voltage. The voltage will fluctuate between 1.5 volts and 3.4 volts as a function of the engine speed. If these checks do not work, replace the air mass sensor assembly.

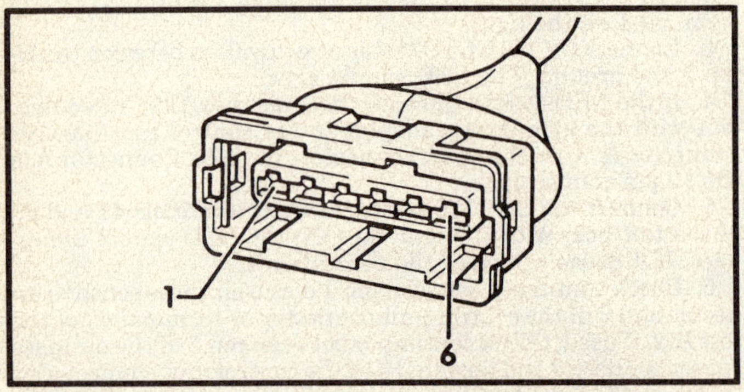

Air mass sensor connector identification

Air mass sensor terminal identification

OXYGEN SENSOR

Testing

OXYGEN SENSOR HEATING

1. On the V8, remove engine compartment support brace attaching bolts. Remove right and left air duct support bolts, then remove air ducts.Remove the 4 bolts attaching the air cleaner, then the air cleaner. Remove the 2 lower air cleaner housing attaching bolts, then remove housing by pressing down and lifting upward. Turn the steering wheel fully to the left.

2. The heater connector is the rectangular connector and the oxygen sensor signal connector is round.

3. On the 90 Quattro and Coupe, the connectors are located next to the intake manifold. On the S4, 200 Quattro and Wagon the connectors are located at the back of the engine compartment at the firewall. On the 100, the connectors are located above the valve covers.

4. Disconnect heater electrical connection from sensor.

5. Connect a multimeter between terminals 1 and 2 on the connector. Turn the ignition switch to the **ON** position. Multimeter should read 12–14 volts.

6. If specifications are not as indicated, perform the following checks as follows:

   a. Check that fuse 5 on the V8, 28 on the 90 Quattro and Coupe, 29 on the S4, 200 Quattro and Wagon, and the circuit breaker in the brown holder on the 100 is good.

   b. Using an ohmmeter, check for continuity between terminal 1 and fuse or circuit breaker.

   c. Check condition of connector pins.

   d. Using an ohmmeter, check for continuity between terminal 2 and ground.

   e. Using an ohmmeter, check for continuity between fuse or circuit breaker and terminal 52, 59 on the 90 Quattro and Coupe, of the fuel pump relay.

7. If specifications are still not as indicated after completing the checks in Step 8, proceed as follows:

   a. Connect VAG 1315 A/1 adapter into oxygen sensor heater plug.

   b. Connect an ammeter to the test adapter.

   c. Start the engine and let idle. Note ammeter reading. Ammeter should read 0.5–3.0 amps (current will decrease as sensor heats up).

8. If above specifications are not as indicated, replace the oxygen sensor.

CONTROL SIGNAL TO OXYGEN SENSOR

1. Disconnect round signal electrical connector from the oxygen sensor.

2. Connect a multimeter (2 volt range) between harness connector and ground. Turn ignition switch to the **ON** position. Multimeter should read 450 ± 50 mV for the V8, S4 and the 200 Quattro and Wagon, 400 ± 50 mV for the 90 Quattro, Coupe and the 100.

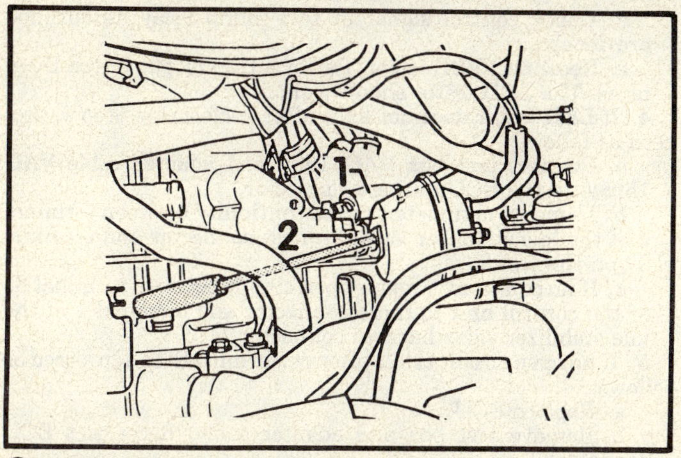

Oxygen sensor connectors (1 and 2)

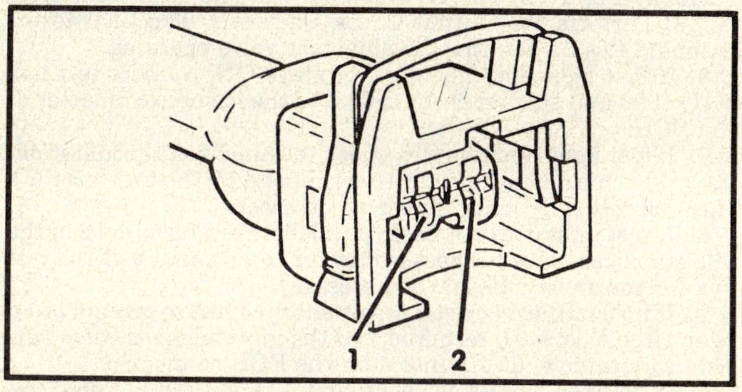

Connector terminal identification (oxygen sensor, idle stabilizer valve, carbon canister solenoid valve and fuel injector)

3. If specifications are not as indicated, proceed as follows:
   a. Connect test box VAG 1598 and adapter cable VAG 1598/5 to the ECU harness connector.

NOTE: Control unit harness connector is not connected.

   b. Using an ohmmeter, check for continuity between the harness connector in the engine compartment and terminal 28 on the test box.
   c. Using an ohmmeter, check for continuity between the engine block and terminal 10 on the test box. Repair wiring as necessary.

CARBON CANISTER SOLENOID VALVE

Testing

1. Remove the coolant expansion tank mounting bolts, but leave hoses connected.
2. Disconnect the solenoid valve electrical connector.
3. Using an ohmmeter, check resistance between solenoid valve terminals. Ohmmeter should read 40–50 ohms.
4. If specifications are not as indicated, replace solenoid valve.

SOLENOID VALVE CONTROL SIGNAL

1. Remove the coolant expansion tank mounting bolts, but leave hoses connected, if obstructing the connector.
2. Disconnect the solenoid valve electrical connector.
3. Connect LED tester US 1115 or equivalent, between terminal 2 of solenoid valve harness connector and ground. Operate starter. LED tester should light.
4. If LED tester does not light, perform the following checks and repair as necessary:
   a. Check fuses 23 and 24.
   b. Using an ohmmeter, check continuity between terminal 2 on the solenoid valve harness connector and fuse 24.
   c. Using an ohmmeter, check continuity between fuse 24 and terminal 59 on the fuel pump relay socket.
   d. Check control signal to fuel pump relay as outlined previously.
   e. Perform Output Check diagnosis. The LED tester should light.
5. If LED tester does not light after performing Step 4, proceed as follows:
   a. Connect test box VAG 1598 and adapter cable VAG 1598/5 to the ECU harness connector.
   b. Using an ohmmeter, check continuity between terminal 2 of the solenoid valve harness connector and terminal 37 on the test box.
   c. If no continuity exists, repair wiring as necessary.
   d. If continuity exists, replace the ECU and remove the test box and adapter cable. Reconnect the ECU harness connector.
   e. Connect LED tester to terminals 1 and 2 on the solenoid valve harness connector. Perform the Output Check diagnosis. When the valve is operated, the LED tester should flash.
6. If LED tester does not flash, or stays on constantly, proceed as follows:
   a. Connect test box VAG 1598 and adapter cable 1598/5 to the ECU harness connector.
   b. Using an ohmmeter, check wiring between terminal 1 of the solenoid valve harness connector and terminal 5 on the test box for a short circuit to ground. Repair as necessary.
   c. If circuit is good, replace the ECU and remove test box and adapter cable. Reconnect ECU connector.

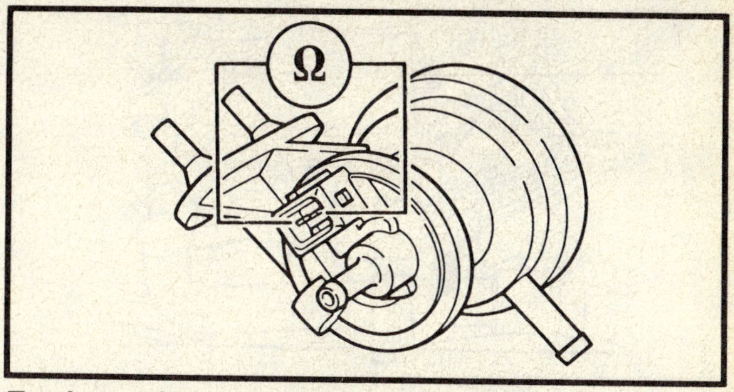

Testing carbon canister solenoid valve

IDLE STABILIZER VALVE

Testing

1. Disconnect idle stabilizer valve electrical connector.
2. Using an ohmmeter, check resistance across idle stabilizer valve terminals. Ohmmeter should read 7.5–8.5 ohms.
3. If specifications are not as indicated, replace idle stabilizer valve.

MECHANICAL CHECK

1. Remove idle stabilizer valve, leaving the electrical connector connected.
2. Visually check the rotary-slide valve for signs of binding or rubbing.

NOTE: When checking the slide valve operation, do not use a screwdriver or any other tool.

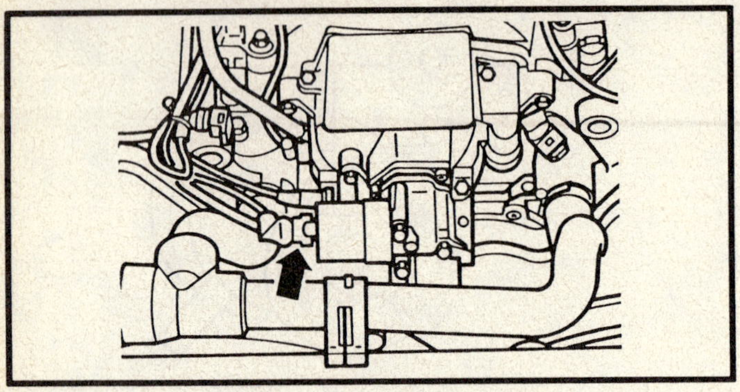

Idle stabilizer valve connector location.

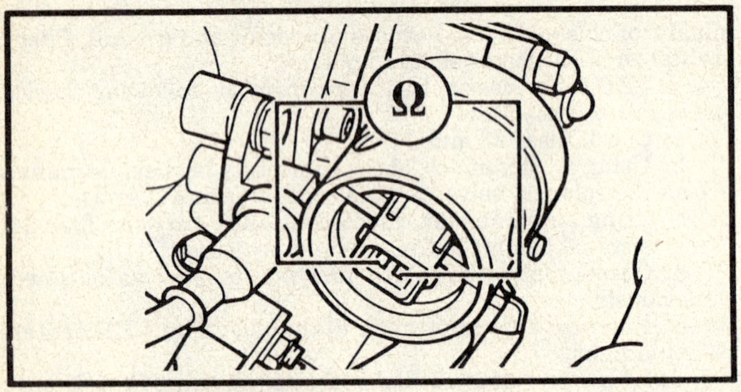

Testing idle stabilizer valve

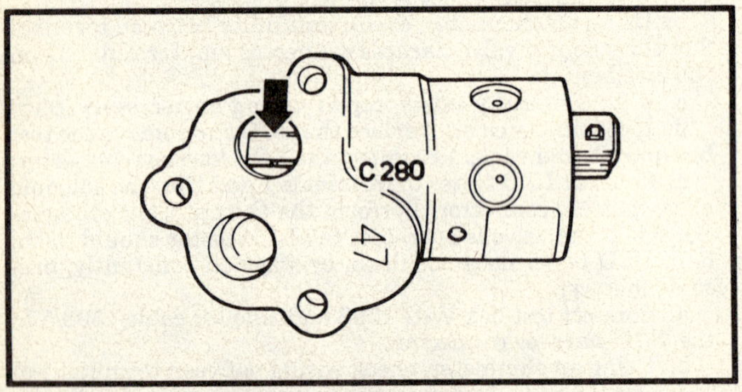

Checking idle stabilizer slide valve

3. Perform the Output Check Diagnosis and send a control signal to the idle stabilizer valve.
4. Check the operation of the slide valve. Replace idle stabilizer valve if their are signs of binding or the rotary-slide does not operate smoothly.

CONTROL SIGNAL CHECK

1. Disconnect electrical connector from idle stabilizer valve. Resistance between the terminals should be 7–11 ohms.
2. Connect LED tester US 1115 and Test Lead VW 1594 or equivalent, between terminal 1 on the harness connector and ground.
3. Operate starter. LED tester should light. If tester does not light, perform the following checks and repair as necessary:
   a. Check fuses 23 and 24.
   b. Using an ohmmeter, check continuity between terminal 1 on the idle stabilizer valve harness connector and fuse 24.
   c. Using an ohmmeter, check continuity between fuse 24 and the fuel pump relay terminal 59.
   d. Check control signal to fuel pump relay as outlined previously.
   e. Reconnect LED tester. Perform the Output Check Diagnosis. The LED tester should light.
4. If LED tester does not light after performing Step 3, proceed as follows:
   a. Connect test box VAG 1598 and adapter cable VAG 1598/5 to the ECU harness connector.
   b. Using an ohmmeter, check continuity between terminal 1 of the idle stabilizer valve harness connector and terminal 37 on the test box.
   c. If no continuity, repair open circuit between terminal 37 of the control unit harness connector and terminal 1 of the idle stabilizer valve harness connector.
5. If no open circuit exists after performing Step 4, proceed as follows:
   a. Replace ECU.
   b. Remove test box and adapter cable. Reconnect ECU connector.
   c. Connect LED tester US 1115 and adapter cable, to terminals 1 and 2 of the idle stabilizer valve harness connector.
   d. Perform the Output Check Diagnosis. The LED tester should flash when the idle stabilizer valve operates.
6. If test light does not flash or stays ON, connect test box VAG 1598 and adapter cable 1598/5 to the harness connector of the ECU.
7. If test light remains ON, check terminal 2 of the idle stabilizer valve harness connector to terminal 4 on the test box for a short circuit to ground. Repair as necessary.
8. It test light does not flash, check between terminal 2 of the idle stabilizer valve harness connector to terminal 4 of the test box for continuity. Repair as necessary.
9. If no continuity exists, repair short circuit to ground or an open circuit between terminal 2 of the idle stabilizer valve harness connector and terminal 4 on the ECU connector.
10. If no open circuit or short circuit to ground is found, replace ECU.

IDLE SPEED

Pre-Check

NOTE: The idle speed is controlled by the ECU and cannot be adjusted.

Before checking idle speed, the following conditions must be met:

1. Start and allow engine to reach normal operating temperature.
2. Check that the throttle valve is in the idle position.
3. All accessories are **OFF**.
4. Fuel pressure gauge not connected.

Check

1. Connect engine tester VAG 1367 as follows:
   a. Negative lead to ground.
   b. Positive lead to remote positive cable connection.
   c. Green lead to terminal 1 on right side of the ignition coil
2. Start the engine, and let idle. Idle speed should be 660–720 rpm.
3. If specifications are not as indicated, recall fault memory as outlined, and if necessary, perform Output Check diagnosis.

CO CONTENT

Pre-Check

NOTE: The CO content is controlled by the ECU and cannot be adjusted.

Before checking CO content, the following conditions must be met:

1. Start and allow engine to reach normal operating temperature.
2. Check that the throttle valve is in the idle position.
3. All accessories are **OFF**.
4. Fuel pressure gauge not connected.
5. Crankcase breather connected.

Check

1. Recall fault memory as outlined previously.
2. Connect CO tester SUN 105, adapter 1318/11 and connecting hose 6006-0019, or equivalent, to the CO tap tube connection (opposite of oxygen sensor).
3. Using a suitable clamp, pinch the upper carbon canister hose.
4. Start the engine and let idle. CO content must be 0.5–0.9%.
5. If specifications are not as indicated, check spark plugs.

NOTE: If one spark plug is not operating properly, the CO content increases approximately 1.5%. If spark plugs are good, but CO content is not within specification, recall fault memory as outlined previously, and if necessary, perform Output Check Diagnosis.

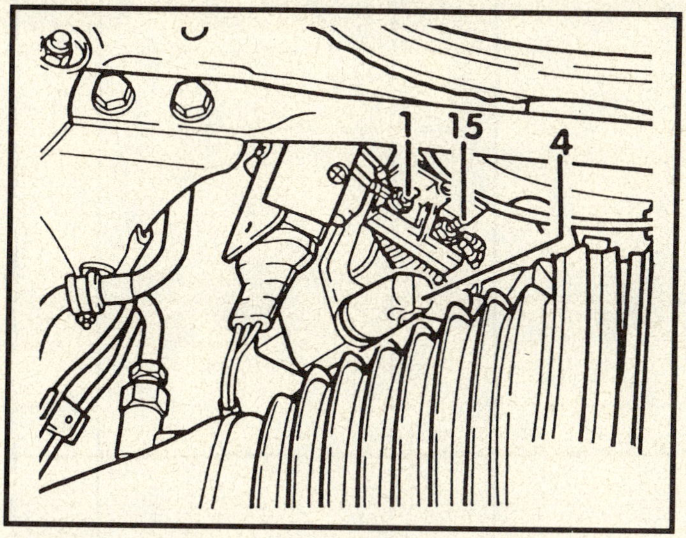

Terminal identification for installing engine tester VAG 1367

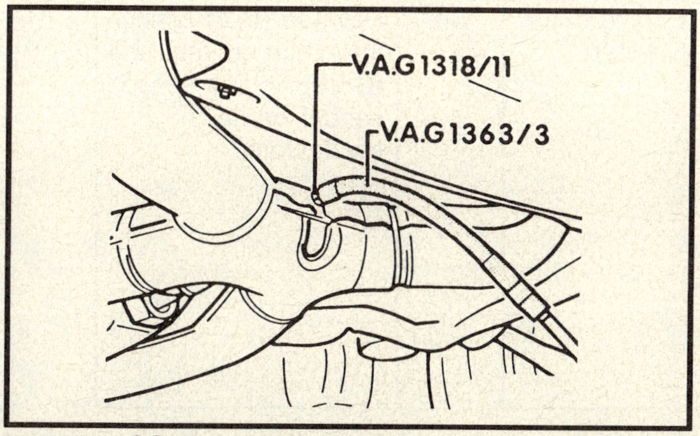

Installing CO tester and adapter

IDLE AND FULL THROTTLE SWITCH

Testing

NOTE: The terminal numbering on the throttle valve connector is reversed when compared to the harness connector.

IDLE SWITCH

1. Disconnect the throttle valve electrical connector.
2. Using an ohmmeter and adapters VW 1594 or equivalent, check resistance across terminals 4 and 6. Ohmmeter should read 0 ohms (continuity).
3. Open throttle approximately 10% of its travel. Ohmmeter should read infinity.
4. If specifications are not as indicated in Steps 2 and 3, or only when the throttle is opened wider than 10% of its travel, adjust the idle switch.
5. After adjustment, if specifications are still not as indicated, replace throttle valve switch assembly.

FULL THROTTLE SWITCH

1. Using an ohmmeter and adapters VW 1594 or equivalent, check resistance between terminals 4 and 5 of the throttle valve connector.
2. Ohmmeter should read infinity. Open the throttle to the wide open position. Ohmmeter should read 0 ohms (continuity).
3. If specifications are not as indicated in Step 2, adjust the idle switch as outlined previously.
4. After adjusting idle switch, if specifications are still not obtained, replace the throttle valve switch assembly.

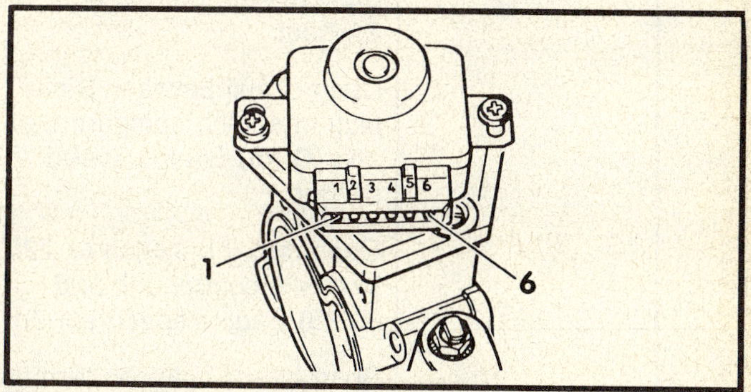

Throttle switch terminal identification

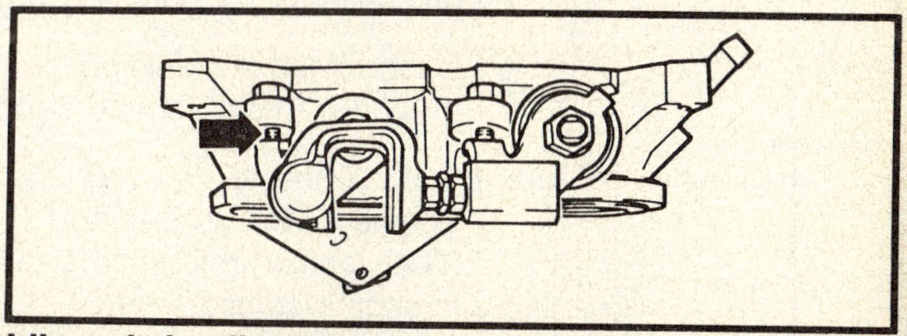

Idle switch adjustment

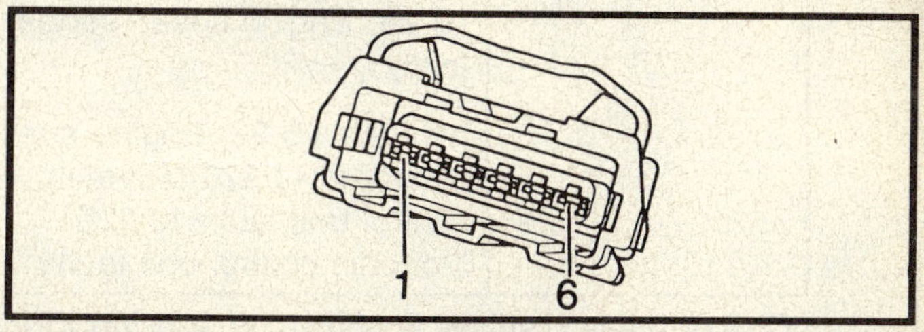

Throttle switch connector identification

MOTRONIC FLASH CODES AND DIAGNOSTIC CHART

Flash Code and Display on VAG 1551	Problem	Indication	Calif. Warn. Light	Repair
1111 Control unit faulty	(**J 17**) Fuel pump relay Ground connections to (**J 220**) Motronic control unit terminals: 10, 14, 19, 24, 26	Engine does not start		Check (**J 17**) fuel pump relay Check ground connection according to wiring diagram
2111 (**G 28**) Engine speed sender *No change in signal	Foreign material on (**G 28**) Engine speed sender Mounting base of (**G 28**) Engine speed sender loose Distance from (**G 28**) Engine speed sender to ring gear of flywheel greater than 1.2 mm (3/64 in.) Broken teeth on ring gear of flywheel Open circuit between 3 pin plug in engine compartment and (**G 28**) Engine speed sender Open circuit between (**J 220**) Motronic control unit and (**G 28**) Engine speed sender Short circuit between terminal 47 and (**J 220**) Motronic control unit shielding or engine ground. Open circuit in shielding of (**G 28**) Engine speed sender	Engine does not start Misfires Engine stalls	On	Check (**G 28**) Engine speed sender Repair short or open circuit according to wiring diagram
*mechanical fault	3 pin connector from (**G 28**) Engine speed sender and (**G 4**) Ignition reference sender in engine compartment reversed (**G 28**) Engine speed sender faulty Input for (**G 28**) Engine speed sender in (**J 220**)Motronic control unit faulty (**J 220**) Motronic control unit faulty	Engine does not start	On	Reverse (**G 4**) Ignition reference sender, and (**G 28**) Engine speed sender harness connectors

*One of these displays is shown, in addition to the component.

MOTRONIC FLASH CODES AND DIAGNOSTIC CHART

Flash Code and Display on VAG 1551	Problem	Indication	Calif. Warn. Light	Repair
2112 (**G 4**) Ignition reference sender *No change in signal *mechanical fault	Mounting base of (**G 4**) Ignition reference sender loose Distance from (**G 4**) Ignition reference sender to pin on ring gear of flywheel greater than 1.2 mm Pin on flywheel bent or broken	Fault before engine start: Engine does not start	On	Check (**G 4**) Ignition reference sender
	Open circuit between 3 pin plug in engine compartment and (**G 4**) Ignition reference sender Open circuit between (**J 220**) Motronic control unit and (**G 4**) Ignition reference sender Short circuit between terminal 49 and (**J 220**) Motronic control unit or shielding, or engine ground Open circuit in shielding of (**G 4**) Ignition reference sender	Fault after engine start: Code (**2112**) is stored		Repair short or open circuit according to wiring diagram
	3 pin plug connection from (**G 4**) Ignition reference sender and (**G 28**) Engine speed sender in the engine compartment reversed (**G 4**) Ignition reference sender faulty Input for (**G 4**) Ignition reference sender in (**J 220**) Motronic control unit faulty (**J 220**) Motronic control unit faulty	Engine does not start	On	

*One of these displays is shown, in addition to the component.

Note on code 2113:

Ignore display if fault is displayed as intermittent.

MOTRONIC FLASH CODES AND DIAGNOSTIC CHART

Flash Code and Display on VAG 1551	Problem	Indication	Calif. Warn. Light	Repair
2113 (**G 40**) Hall sender *No change in signal	No power supply for (**G 40**) Hall sender from (**J 220**) Motronic control unit	Engine lacking power at full load	On	Check (**G 40**) Hall sender
	No ground for (**G 40**) Hall sender	Emission values are out of specification		Repair short or open circuit according to wiring diagram
	Signal lead to (**J 220**) Motronic control unit interrupted or short circuit to ground in signal lead			
	Short circuit between terminal 8 and 12 at (**J 220**) Motronic control unit	High fuel consumption		
	Open circuit between 3 pin plug connection in engine compartment and (**J 220**) Motronic control unit			
	Position of Reference Pin for (**G 4**) Ignition reference sender on flywheel		On	
	(**G 40**) Hall sender faulty			
	Input for (**G 40**) Hall sender in (**J 220**) Motronic control unit faulty (**J 220**) Motronic control unit faulty			

*One of these displays is shown, in addition to the component.

Note on code 2113:
Ignore display if fault is displayed as intermittent.

MOTRONIC FLASH CODES AND DIAGNOSTIC CHART

Flash Code and Display on VAG 1551	Problem	Indication	Calif. Warn. Light	Repair
2114 (**G 40**) Hall sender not at reference mark	Toothed belt skipped tooth or out of position Adjustment of (**G 40**) Hall sender Pin for (**G 4**) Ignition reference sender on flywheel	Same as **2113**	On	Check toothed belt Check (**G 40**) Hall sender, Check pin of (**G 4**) Ignition reference sender
2121 (**F 60**) Idle switch *Signal to positive	Adjust (**F 60**) Idle switch Throttle valve sticking Floor mat presses on accelerator pedal Throttle cable adjustment Open circuit to terminal 52 from (**J 220**) Motronic control unit Input for (**F 60**) Idle switch in (**J 220**) Motronic control unit faulty (**J 220**) Motronic control unit faulty)	Idle stabilization switches to control Idle speed not within specification	On On	Check (**F 60**) Idle switch Adjust throttle cable Repair open or short circuit according to wiring diagram
*Signal to negative	Terminal 52 from (**J 220**) Motronic control unit short circuited to ground Moisture in throttle valve connector			Repair short circuit
2123 (**F 81**) Full throttle switch Signal to negative	Adjustment (**F 81**) Full throttle switch Terminal 53 constantly grounded Moisture in throttle valve connector		Off	Check (**F 81**) Full throttle switch Repair short circuit

*One of these displays is shown, in addition to the component.

MOTRONIC FLASH CODES AND DIAGNOSTIC CHART

Flash Code and Display on VAG 1551	Problem	Indication	Calif. Warn. Light	Repair
2141 First knock regulation (Info) **No change in signal	Knock control module in (**J 220**) Motronic control unit faulty (**J 220**) Motronic control unit faulty	Higher fuel consumption Power loss	On	Replace (**J 220**) Motronic control unit
First knock regulation *Maximum control limit exceeded	Poor fuel quality, less than 95 RON	Limited loss of power	On	Use fuel with at least 95 RON
	Signal wire from (**G 28**) Engine speed sender or (**G 4**) Ignition reference sender interrupted or reversed on (**J 220**) Motronic control unit	Slight increase in fuel consumption Top speed insufficient		Check signal wire from (**G 28**) Engine speed sender and (**G 4**) Ignition reference sender according to wiring diagram
	Abnormal engine noises (auxiliary units loose)	Engine running rough		
	Open circuit in shielding of (**G 61**) Knock sensor I or (**G 66**) Knock sensor II			Repair open circuit
2142 (**G 61**) Knock sensor I *No change in signal	(**G 61**) Knock sensor I loose or plug corroded	High fuel consumption	On	Tightening torque 15-25 Nm (11-18 ft lb)
	Open or short circuit between (**G 61**) Knock sensor I and (**J 220**) Motronic control unit	Power loss		Repair open or short circuit
	Short circuit between (**G 61**) Knock sensor I and ground or to shielding			
	(**G 61**) Knock sensor I faulty			
	Input for (**G 61**) Knock sensor I in (**J 220**) Motronic control unit faulty (**J 220**) Motronic control unit faulty			

*One of these displays is shown, in addition to the component.

- If possible, corresponding cylinder or cylinder pairs are also displayed as additional information to the display ** maximum control limit exceeded.
- Cylinder pairs: 1 and 5, 2 and 7, 3 and 6, and 4 and 8.
- Retard adjustment of ignition timing by knock control is always for corresponding cylinder pairs.

MOTRONIC FLASH CODES AND DIAGNOSTIC CHART

Flash Code and Display on VAG 1551	Problem	Indication	Calif. Warn. Light	Repair
2144 (**G 66**) Knock sensor II *No change in signal	(**G 66**) Knock sensor II loose or plug corroded	High fuel consumption	On	Tightening torque 15-25 Nm (11-18 ft-lb)
		Power loss		Repair open or short circuit
	Open or short circuit between (**G 66**) Knock sensor II and (**J 220**) Motronic control unit			
	Short circuit between (**G 66**) Knock sensor II and ground or to shielding			
	(**G 66**) Knock sensor II faulty			
	Input for (**G 66**) Knock sensor II in (**J 220**) Motronic control unit faulty (**J 220**) Motronic control unit faulty			
2231 Idle speed regulation (Info) *Maximum adaptation limit exceeded	(**N 71**) Idle stabilizer valve binds when closed	Idle speed too low	On	Check (**N 71**) Idle stabilizer valve
	(**G 70**) Air mass sensor map offset			Check (**G 70**) Air mass sensor
	Terminal resistance at connector of (**N 71**) Idle stabilizer valve			
*Minimum adaptation limit not obtained	(**N 71**) Idle stabilizer valve binds when open	Idle speed too high		
	(**G 70**) Air mass sensor map offset			
	Additional air induced between (**G 70**) Air mass sensor and (**N 71**) Idle stabilizer valve			Repair leak

*One of these displays is shown, in addition to the component.

Note flashing code 2232:
Ignore fault display if displayed as an intermittent fault!

MOTRONIC FLASH CODES AND DIAGNOSTIC CHART

Flash Code and Display on VAG 1551	Problem	Indication	Calif. Warn. Light	Repair
2232 (**G 70**) Air mass sensor *Signal too small *Signal too great	Open circuit between (**G 70**) Air mass sensor and (**J 220**) Motronic control unit Voltage supply to (**G 70**) Air mass sensor interrupted or short circuited to ground Short circuit to ground in signal wire to terminal 7 from (**J 220**) Motronic control unit Short circuit in signal wire to terminal 7 from (**J 220**) Motronic control unit (**G 70**) Air mass sensor faulty Input for (**G 70**) Air mass sensor in (**J 220**) Motronic control unit faulty ((**J 220**) Motronic control unit faulty)	Limp home mode Poor fuel intake Reduced power	On	Check (**G 70**) Air mass sensor
2234 Supply voltage (Info) *Signal too high *Signal too low	Supply voltage to control unit greater than 16V Alternator faulty Engine started with two batteries in series Poor ground connection to (**J 220**) Motronic control unit Current drain with ignition OFF Battery discharged	Permanent damage to (**J 220**) Motronic control unit Idle speed out of range Voltage below 6 volts Engine does not start/ crank	Off	Check voltage Check (**J 220**) Motronic control unit

*One of these displays is shown, in addition to the component.

Note flashing code 2234:
Ignore fault if displayed as intermittent.

MOTRONIC FLASH CODES AND DIAGNOSTIC CHART

Flash Code and Display on VAG 1551	Problem	Indication	Calif. Warn. Light	Repair
2312 (**G 62**) Coolant temperature sensor *Signal to negative *Signal to positive	Short circuit to ground Moisture in harness (**G 62**) Coolant temperature sensor connector Open circuit between (**G 62**) Coolant temperature sensor and (**J 220**) Motronic control unit Resistance between (**G 62**) Coolant temperature sensor and (**J 220**) Motronic control unit (**G 62**) Coolant temperature sensor faulty Input from (**G 62**) Coolant temperature sensor in (**J 220**) Motronic control unit faulty ((**J 220**) Motronic control unit faulty)	Cold start difficulties at low temperature Poor idle and acceleration during warm up	On	Check (**G 62**) Coolant temperature sensor
2314 Engine/transmission electrical connection *Signal to negative	Terminal 51 from (**J 220**) Motronic control unit shorted to ground (**J 217**) Automatic transmission control unit faulty	(**J 217**) Automatic transmission control unit goes to limp home mode Possible jerky shifting	On	Check wires according to wiring diagram ((**J 220**) Motronic control unit faulty)

*One of these displays is shown, in addition to the component.

Note flashing code 2234:
Ignore fault if displayed as intermittent.

MOTRONIC FLASH CODES AND DIAGNOSTIC CHART

Flash Code and Display on VAG 1551	Problem	Indication	Calif. Warn. Light	Repair
2322 (**G 42**) Intake air temperature sender *Signal to negative *Signal to positive	Short circuit to ground Open circuit between (**G 42**) Intake air temperature sensor and terminal 44 of (**J 220**) Motronic control unit (**G 42**) Intake air temperature sensor faulty Input of (**G 42**) Intake air temperature sensor faulty, at (**J 220**) Motronic control unit faulty ((**J 220**) Motronic control unit faulty)		Off	Check (**G 42**)
2341 (**G 39**) Oxygen sensor *Control limit exceeded	Fuel tank empty, 10 ltr. (2.5 gal.) minimum resulting in fuel system pressure less than 3.3 bar (48 psi) Ignition system failure	CO before catalyst less than 0.5% Fuel consumption high	On	Fill fuel tank
	Engine intake leak after (**G 70**) Air mass sensor Leak in exhaust system in front of catalyst	Rich exhaust Spark plug fouling CO before catalyst greater than 3%		Repair leak
	Oxygen sensor faulty			Check (**G 39**) Heated oxygen sensor
	Fuel system pressure greater than 4 bar (58 psi)			Check fuel system pressure

*One of these displays is shown, in addition to the component.

MOTRONIC FLASH CODES AND DIAGNOSTIC CHART

Flash Code and Display on VAG 1551	Problem	Indication	Calif. Warn. Light	Repair
2342 (**G 39**) Oxygen sensor *Signal to positive *Signal to negative *No change in signal	Open circuit to (**J 220**) Motronic control unit Sensor heater not functioning Fuse S 5 (sensor heater) faulty Heat resistor in sensor faulty Wiring in sensor heater faulty Short circuit to ground in signal wire Short circuit in signal wire to shielding Fuel tank low, 10 ltr. (2.5 gal) minimum resulting in fuel system pressure less than 3.3 bar (48 psi) Ignition system failure Intake leak after (**G 70**) Air mass sensor Leak in exhaust system before catalyst Oxygen sensor faulty Ground reference wire from terminal 10, of (**J 220**) Motronic control unit to engine block interrupted	Emissions not within specification Increased fuel consumption Rich exhaust Spark plug fouling CO before catalyst above 3% OXS goes to closed loop	On	Repair open circuit Check (**G 39**) Heated oxygen sensor heater
3424 Warning Light (California Engine Warning Light) *Signal on plus(+) *Signal on minus (−)	Disconnected wire between (**J 220**) Motronic control unit and California Engine Warning Light		Off	Eliminate open circuit using wiring diagram
	California Engine Warning Light Faulty			Replace light
	Short circuit between (**J 220**) Motronic control unit and California Engine Warning Light		On	Eliminate short circuit using wiring diagram

*One of these displays is shown, in addition to the component.

MOTRONIC FLASH CODES AND DIAGNOSTIC CHART

Flash Code and Display on VAG 1551	Problem	Indication	Calif. Warn. Light	Repair
4411 (**N 30**) Fuel injector cylinder 1 *Signal to negative	Short circuit to ground Open circuit Fuse 23 faulty	Engine running rough Possible engine stalls	On	Check injectors
*Signal to positive	Short circuit to positive at harness connector or injector	CAT light comes on at instrument cluster with short circuit to ground and high engine load (catalytic converter over-temperature warning) Engine failure		
4412 (**N 31**) Fuel injector cylinder 2 *No change in signal	Short circuit to ground Open circuit Fuse 23 faulty	Engine running rough Possible engine failure	On	Check injectors
	Short circuit to positive at harness connector or injector	CAT light comes on at instrument cluster with short circuit to ground and high engine load (catalytic converter over-temperature warning)		

*One of these displays is shown, in addition to the component.

Note: Injectors 3 and 6,4 and 8, 1 and 5 and 2 and 7 are activated in pairs. The described checks should be carried out on both injectors within a pair.

MOTRONIC FLASH CODES AND DIAGNOSTIC CHART

Flash Code and Display on VAG 1551	Problem	Indication	Calif. Warn. Light	Repair
4413 (**N 32**) Fuel injector cylinder 3	Short circuit to ground Open circuit	Engine running rough	On	Check injectors
*Signal to negative	Fuse 23 faulty	Possible engine stalling		
*Signal to positive	Short circuit to positive at harness connector or injector	CAT light ON, engine failure		
4414 (**N 33**) Fuel injector cylinder 4	Short circuit to ground Open circuit	Engine running rough	On	Check injectors
*Signal to negative	Fuse 23 faulty	Possible engine failure		
*Signal to positive	Short circuit to positive at harness connector or injector	CAT light on, grounded		
4431 (**N 71**) Idle stabilizer valve *Signal to positive *Signal to negative	Short circuit to positive in (**N 71**) Idle stabilizer valve or between (**N 71**) Idle stabilizer valve and terminal 4 from (**J 220**) Motronic control unit	Engine speed out of range at operating temperature	On	Check (**N 71**) Idle stabilizer valve
	Short circuit to ground in (**N 71**) Idle stabilizer valve or between (**N 71**) Idle stabilizer valve and terminal 4 from (**J 220**) Motronic control unit	Possible engine failure with cold engine		
	Fuse 24 faulty Open circuit	Pick-up faults in part load range (lurching)		
		Pick-up faults or idle hunting		

MOTRONIC FLASH CODES AND DIAGNOSTIC CHART

Flash Code and Display on VAG 1551	Problem	Indication	Calif. Warn. Light	Repair
4343 (**N 80**) Carbon canister solenoid *Signal to positive *Signal to negative	Short circuit to positive in (**N 80**) Carbon canister solenoid valve or between (**N 80**) Carbon canister solenoid valve and terminal 5 from (**J 220**) Motronic control unit	Poor response in portion of load range	On	Check (**N 80**) Carbon canister solenoid valve
	Short circuit to ground in (**N 80**) Carbon canister solenoid valve or between (**N 80**) Carbon canister solenoid valve and terminal 5 to (**J 220**) Motronic control unit	Possible fuel odors		
	Fuse 24 faulty			
	Open circuit			

*One of these displays is shown, in addition to the component.

Note: Injectors 3 and 6,4 and 8, 1 and 5 and 2 and 7 are activated in pairs. The described checks should be carried out on both injectors within a pair.

Note

- Before fault elimination procedure or replacing components, check ground connections of (**J 220**) Motronic control unit, terminals 10, 14, 19, 24 and 26 (Specified value: 0-0.2 ohms). Check grounds on engine for corrosion or damage. Check fuel pump relay
- When fault is eliminated, cancel fault memory, conduct road test and recall fault memory again.

MOTRONIC TROUBLESHOOTING CHART

Display breakdown into display sections 1 to 10

Chart of displays in sections 1 to 10

Display section	Display specified values*	Corresponds to measured value	Designation	Remark
1	184 - 217	85° - 105°C (185°-220°F)	Engine temperature	Precondition for all other displays/specified values
2	32 - 42		Engine load	Without auxiliary loads to engine
3	66 - 72	660 - 720/rpm	Engine speed	Without auxiliary loads to engine
4	123 - 140		Operating range of idle speed stabilizer	Change in air mass via (**N 71**) Idle stabilizer valve
5	70 - 126		Idle speed stabilization-curve zero point	Internally calculated value
6	123 - 138		Idle speed stabilization-load adjustment	Change in air mass via (**N 71**) Idle stabilizer valve
7	35 - 44		Idle speed stabilization-curve control	Internally calculated value
8	128 - 133		OXS control	After approximately 1.5 minutes
9	78 - 178		OXS adaptation	If display value low, conduct road test
10	40 - 42		Ignition timing	Ignition timing internally calculated by (**J 220**) Motronic control unit

***Note:** If displayed values do not correspond with specified values, check that air conditioner is switched off. If air conditioner switched off, see **Check table**

MOTRONIC TROUBLESHOOTING CHART CHECK TABLE

Check table

The following table shows the required checks to be carried out if the display values do **NOT** coincide with the specified values.

Note

Before carrying out the checks, ensure that air conditioner and air conditioner compressor are switched **OFF**.

Display section	VAG 1551 Display	Problem	Repair
1	Greater than 217	(**G 62**) Coolant temperature sender	Check (**G 62**) Coolant temperature sender Recall fault memory
1	Less than 184	Thermostat (**G 62**) Coolant temperature sender	Check thermostat Check (**G 62**) Coolant temperature sender Recall fault memory
2	Greater than 42	Air conditioner switched **on** Electrical loads switched **on** OXS control Oxygen sensor Air conditioner compressor in operation although air conditioner switched **off** (**G 70**) Air mass sensor Central hydraulic pump	Switch **off** air conditioner Switch **off** loads OXS control, check display section 8 (of tester VAG 1551) Check Oxygen sensor Check air conditioner Check (**G 19**) Check central hydraulic pump

MOTRONIC TROUBLESHOOTING CHART CHECK TABLE (CONT.)

Display section	VAG 1551 Display	Problem	Repair
2	Less than 32	Excessive air to **(G 19)** Air mass sensor	Determine fault source
		Excessive air to throttle valve	Determine fault source
		Vacuum hose disconnected	Check vacuum system
		Fuel pressure regulator	Check fuel pressure regulator
		Leak in crankcase breather	Check crankcase for air leakage
		Fuel tank ventilation	Check fuel tank ventilation system
		(N 80) Carbon canister solenoid valve 1 sticking	Perform output check diagnosis, **(N 80)** Carbon canister solenoid valve, and **(G 19)**
		(G 70) Air mass sensor	
3	Greater than 72	**(N 71)** Idle speed stabilizer valve at lower control stop	Press accelerator 4 times in intervals of 15 secs.
		Excessive air to throttle valve, injector vacuum hose off	Repair fault source
		Air conditioner	Check air conditioner
		Air conditioner compressor switching signal transmitted, with compressor switched **off**	
		Idle speed stabilizer valve	Perform output check diagnosis,
3	Less than 66	**(N 71)** Idle speed stabilizer valve sticking or faulty	Check **(N 71)** Idle stabilizer valve.
		Air conditioner compressor operates without signal to **(J 220)** Motronic control unit	Check air conditioner
			Perform output check diagnosis
4	123 or less	Excessive air to throttle valve	Repair fault source check **(N 71)** Idle stabilizer valve
		(N 71) Idle speed stabilizer valve binding	

MOTRONIC TROUBLESHOOTING CHART CHECK TABLE (CONT.)

Display section	Display VAG 1551	Problem	Repair
4	140 or greater	Air conditioner compressor operates when switched **off** Battery voltage	Check air conditioner Check battery charge
5	Less than 70	(**N 71**) Idle stabilizer valve	Check (**N 71**) Idle stabilizer valve
5	Greater than 136	Excessive air to (**N 71**) Idle stabilizer valve (**N 71**) Idle stabilizer valve faulty	Repair fault source Check (**N 71**) Idle stabilizer valve
6	Out of 123 - 138 range	Control of (**N 71**) Idle stabilizer valve faulty	Accelerate briefly and observe display, should be between 123 and 138. If no
7	Less than 35 greater than 44	(**N 71**) Idle stabilizer valve faulty	Check (**N 71**) Idle stabilizer valve
8	Constant 128	Fuel level too low Oxygen sensor Coding of (**J 220**) Motronic control unit is zero	Verify 10 ltr. (2.5 gal) minimum fuel in tank Check (**G 39**) oxygen sensor

MOTRONIC TROUBLESHOOTING CHART CHECK TABLE (CONT.)

Display section	Display VAG 1551	Problem	Repair
8	Not between 123 and 133	Oxygen sensor adaptation Oxygen sensor Excessive air to (G 19) Air mass sensor Fuel system pressure Leak in exhaust system	Check display section 9 specified value Check oxygen sensor Repair fault source, check fuel system pressure Check exhaust
8	Inconsistent	Fuel level too low	Verify 10 ltr. (2.5 gal) fuel in tank
9	78 or less	Fuel system pressure Oxygen sensor	Check system pressure Check oxygen sensor
9	178 or greater	Fuel level too low Excessive air intake	Verify 10 ltr. (2.5 gal) minimum fuel in tank
10	43 or greater	(**N 71**) Idle stabilizer valve, incorrect seating position	Perform output check diagnosis Check (**N 71**) Idle stabilizer valve
10	39 or less	(**N 71**) Idle stabilizer valve sticking	Perform output check diagnosis Check (**N 71**) Idle stabilizer valve
10	inconsistent	Excessive air to injectors Spark plugs	Repair fault source Check spark plugs

Component Replacement

FUEL PRESSURE RELIEF

Fuel Pump Relay Method

1. Remove the fuel pump relay.
2. Start the engine and allow to run until it stalls.
3. Switch **OFF** the ignition switch.
4. Replace the fuel pump relay.

Fuel Rail Pressure Tap Method

1. Connect a fuel pressure gauge with bleed down valve to the pressure tap on the fuel rail.
2. Open up the bleed valve on the gauge to relieve the pressure. Catch any escaping fuel.
4. Remove the gauge and cap the tap.

FUEL INJECTORS

Removal and Installation

V8 QUATTRO

NOTE: On some 1990–91 V8 Quattro models leaking injectors could cause starter or flywheel teeth damage due to hydraulic engine lock. If replacing leaking injectors check the starter and flywheel teeth for damage.

1. Disconnect battery ground cable.
2. Remove the left and right support braces, then the intake air ducts.
3. Remove the engine compartment support brace.
4. Remove the upper air cleaner attaching bolts, then the upper air cleaner.
5. Remove the lower air clearer housing attaching bolts, then push housing back and lift outward.

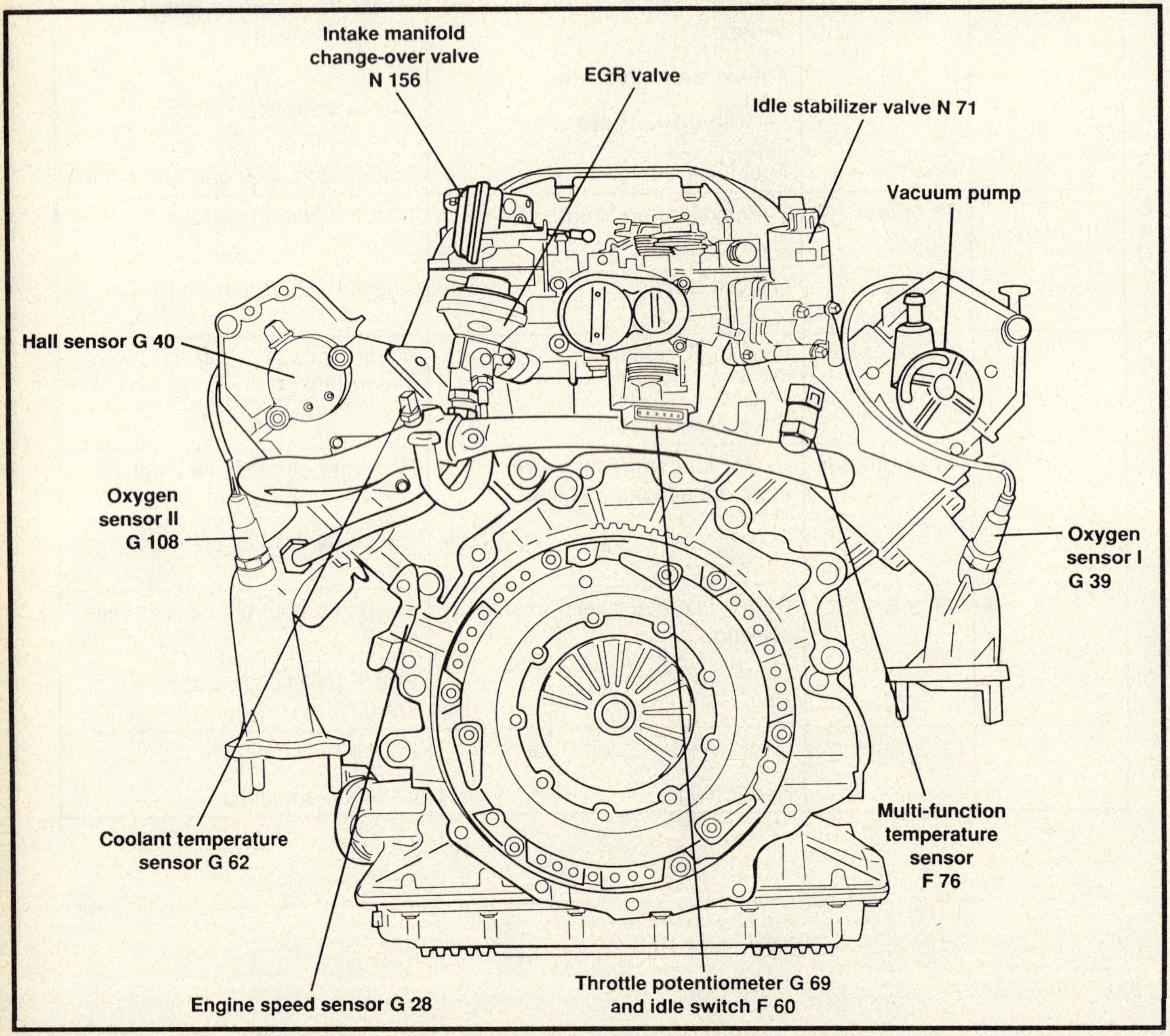

Component location on V6 engine—1992 Audi 100 vehicles

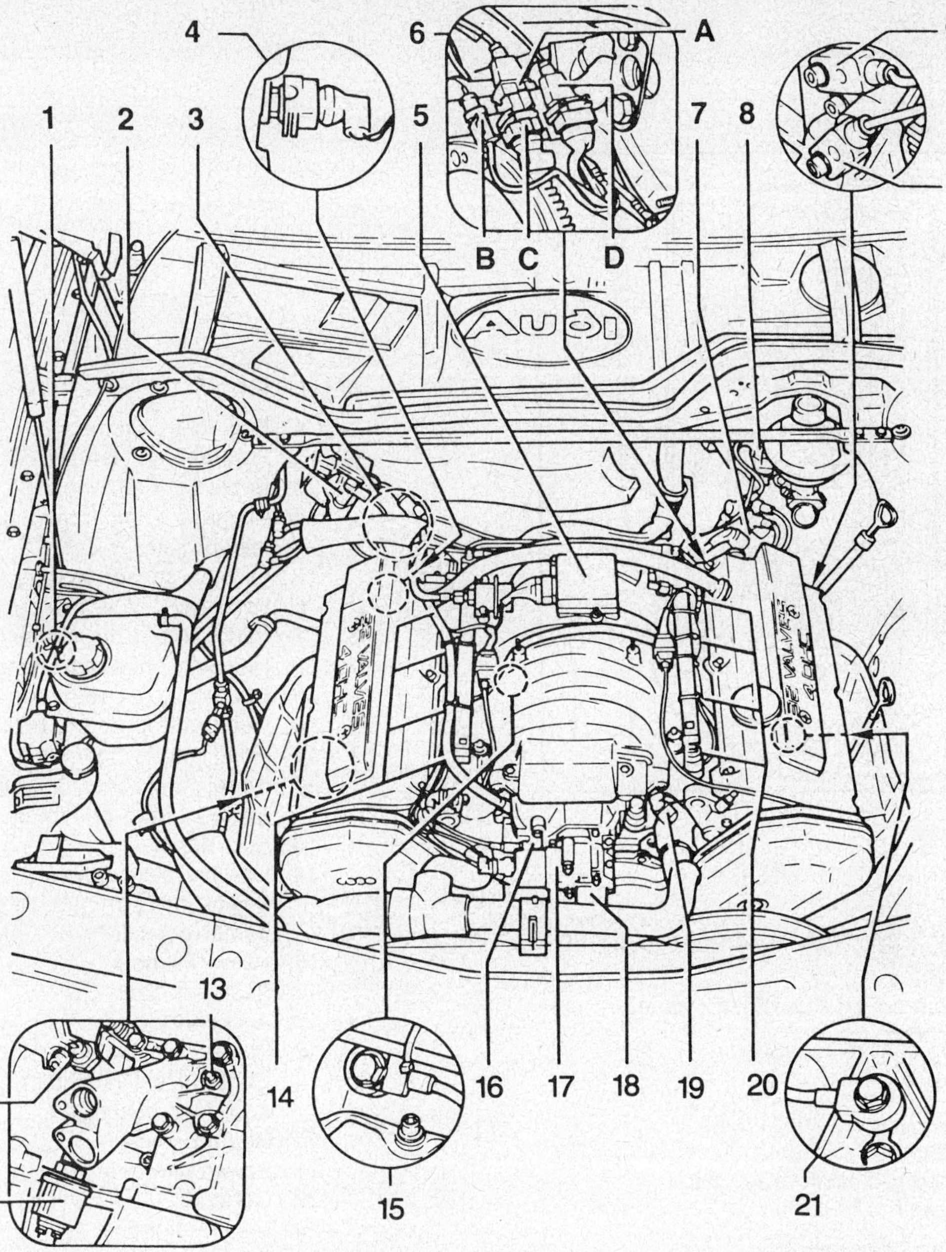

1. Solenoid valve 1 of carbon canister (N 80)
2. Right ignition coil (N)
3. Right distributor (O) with Hall sender (G 40)
4. Coolant temperature sensor (G 62)
5. Air mass sensor (G 19)
6. Harness connectors for:
   A – Engine speed sender (G 28)
   B – Oxygen sensor (signal lead) (G 39)
   C – Oxygen sensor heater (Z 19)
   D – Ignition reference sender (G 4)
7. Left distributor (O 3)
8. Left ignition coil (N 128)
9. Ignition reference sender (G 4)
10. Engine speed sender (G 28)
11. Oil pressure switch
12. Oil pressure sender
13. Oil temperature sensor
14. Fuel injectors, cylinders 1-4
15. Right knock sensor (G 42)
16. Intake air temperature sensor (G 42)
17. Idle stabilizer valve (N 71)
18. Throttle valve potentiometer
19. Coolant temperature display sensor (G 2)
20. Fuel injectors, cylinders 5-8
21. Left knock sensor (G 66)

Motronic components location – Audi V8

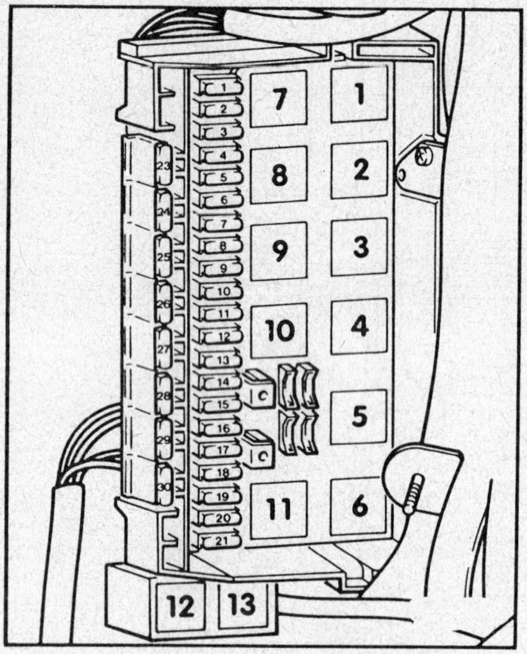

Relay location on Fuse/Relay Panel

1 — Fog light relay, J5
2 — Open
3 — Open
4 — Open
5 — Load reduction relay, J59
6 — Open
7 — Horn relay, J4
8 — Automatic transmission relay, J60
9 — Wash/wipe intermittent relay, J31
10 — Fuel pump relay, J17
11 — Open
12 — Power window, power seat circuit breaker
13 — Radiator cooling fan relay (3rd stage), J135

1 Fog light switches
2 Emergency flash switch
3 Brake light switch
4 Luggage compartment light, interior lights
5 Oxygen sensor heating element
6 Park lights, side marker light, right
7 Park lights, side marker light, left
8 Headlight, hi beam, right
9 Headlight, hi beam, left
10 Headlight, lo beam, right
11 Headlight, lo beam, left
12 Headlight washer, auxilary washer, lamp control unit, rear
13 Fuel pump
14 License plate lights, glove box light, engine compartment light, center console
15 Wiper/washers
16 Rear defogger
17 Air conditioning
18 Power seats with memory
19 Central locking system, door lock heating
20 Heated seats
21 Power windows
22 Open
23 Engine motronic
24 Engine motronic
25 Power seats with memory
26 Automatic transmission
27 Engine motronic
28 Instrument cluster
29 Automatic transmssion
30 Instrument cluster

Fuse/relay panel identification—Audi V8

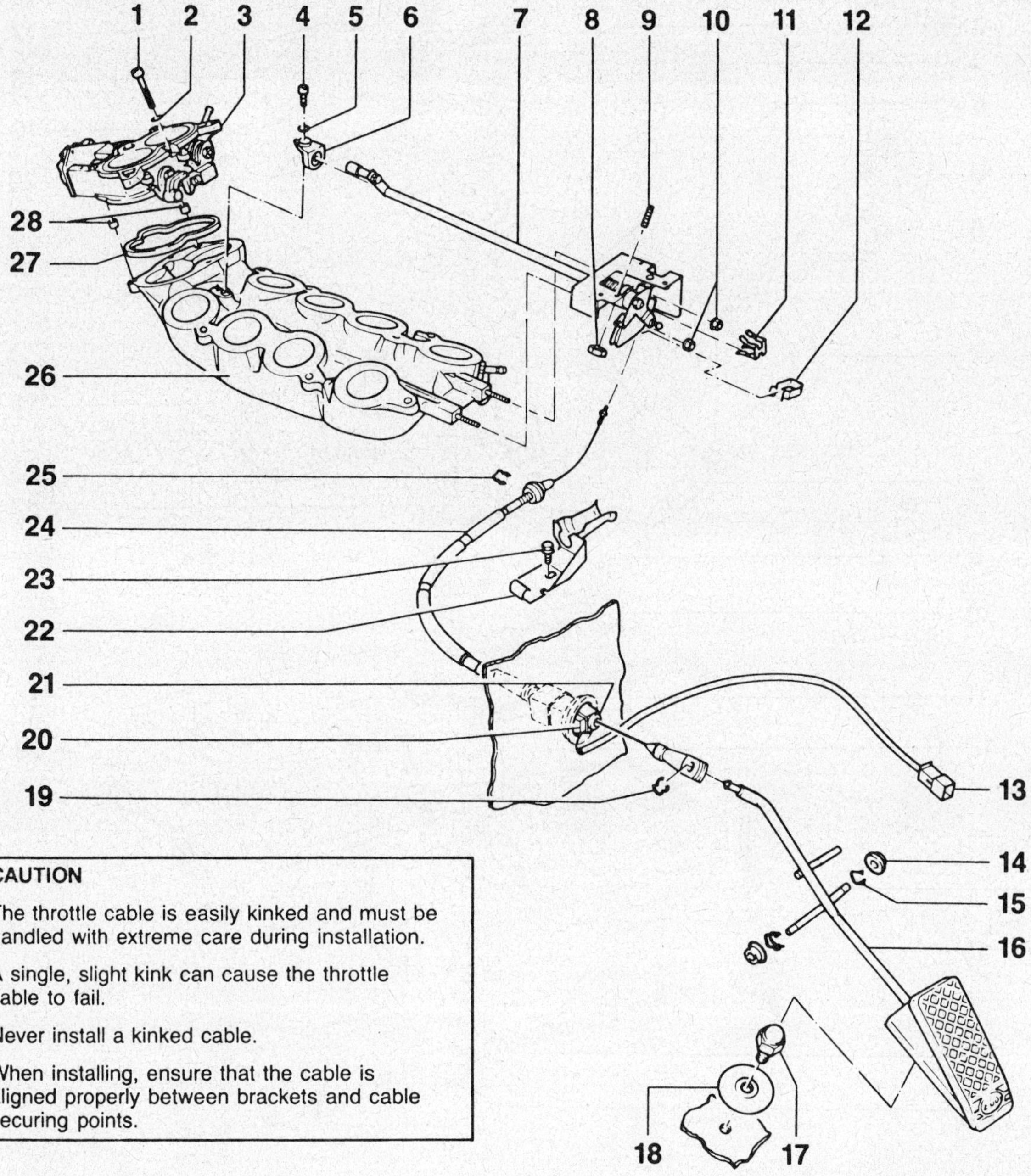

CAUTION

The throttle cable is easily kinked and must be handled with extreme care during installation.

A single, slight kink can cause the throttle cable to fail.

Never install a kinked cable.

When installing, ensure that the cable is aligned properly between brackets and cable securing points.

1. Screw
2. Washer
3. Throttle housing
4. Screw
5. Circlip
6. Throttle linkage mounting
7. Throttle linkage
8. Full throttle stop nut
9. Full throttle stop bolt
10. Nuts
11. Clip
12. Clamp
13. Kick-down switch connector
14. Bearing bushing
15. Clip
16. Accelerator pedal assembly
17. Accelerator pedal stop screw
18. Washer
19. Clip
20. Kick-down switch
21. Square mounting
22. Support bracket
23. Bolt
24. Throttle cable
25. Adjustment clip
26. Intake manifold (lower)
27. Gasket
28. Sleeve

Exploded view of throttle cable—Audi V8

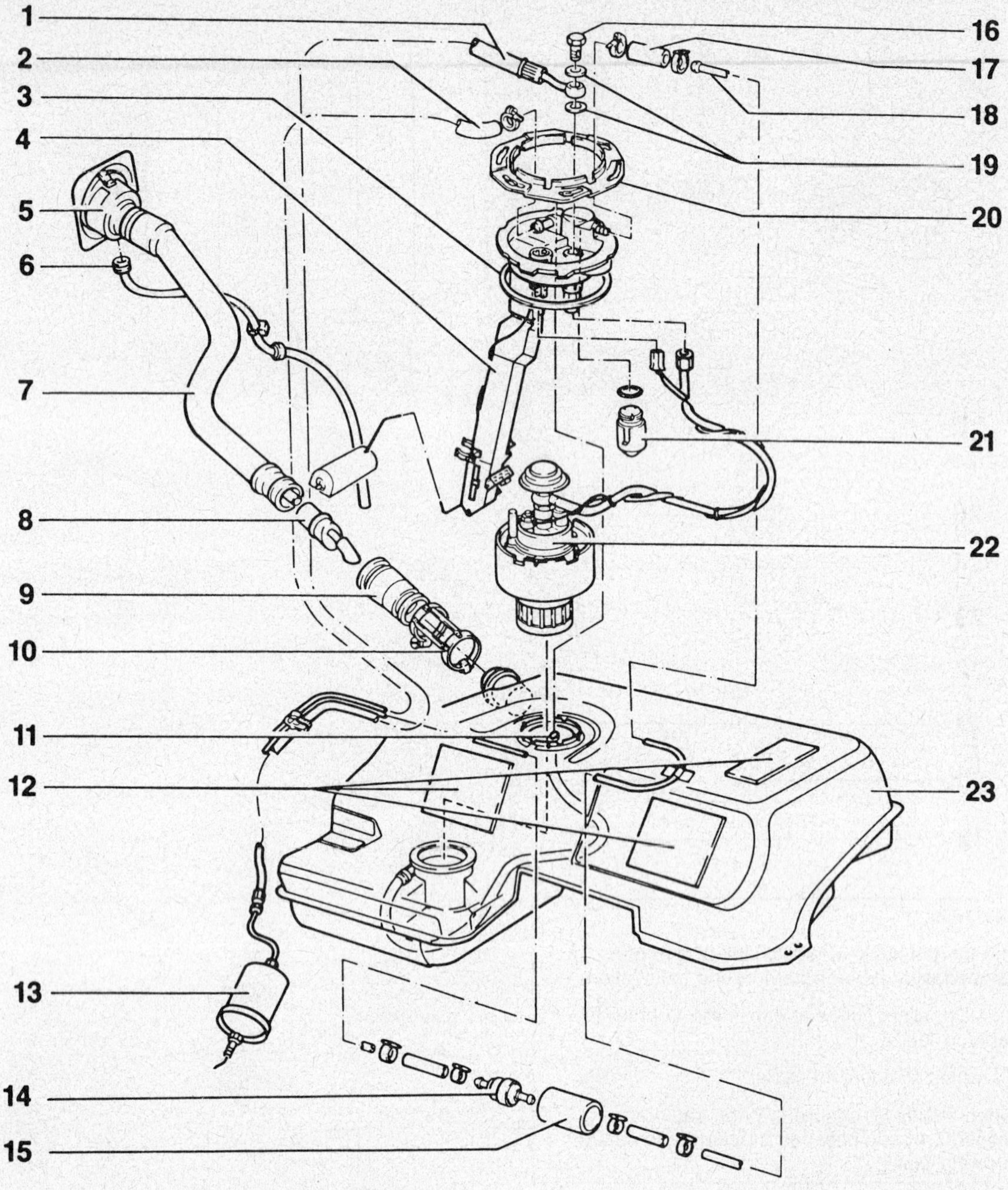

1. Fuel supply line
2. Return line
3. Seal
4. Fuel gauge sending unit
5. Filler neck
6. Overflow hose
7. Filler tube
8. Filler neck, insert
9. Filler tube connecting hose
10. Double clamp
11. Return line
12. Insulating strips
13. Fuel filter
14. Venting and breather valve
15. Support hose
16. Hollow bolt
17. Breather line hose
18. Carbon canister breather hose
19. Seal
20. Cap lock ring
21. Gravity valve
22. Fuel pump
23. Fuel tank

Exploded view of fuel pump assembly—Audi V8

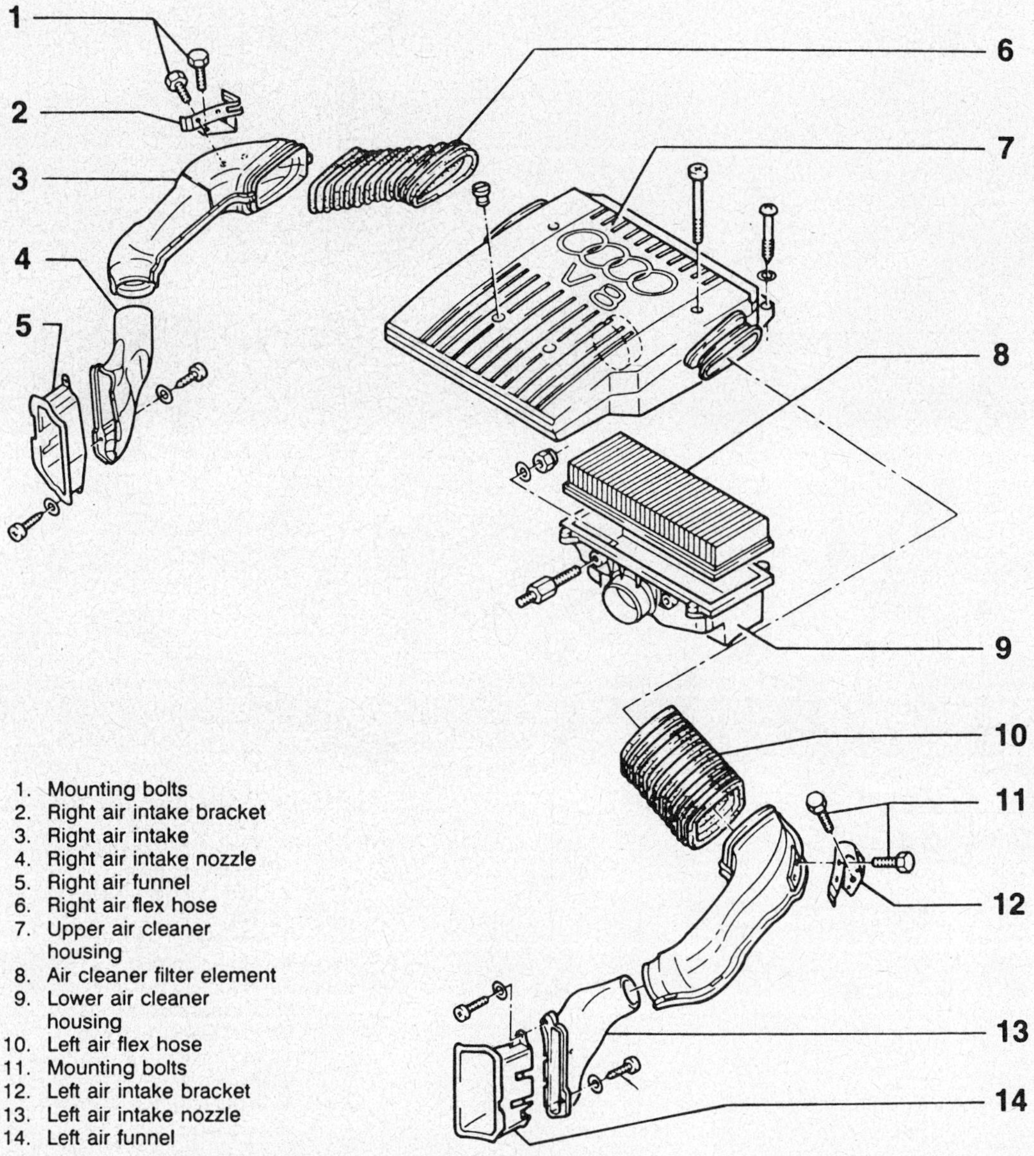

1. Mounting bolts
2. Right air intake bracket
3. Right air intake
4. Right air intake nozzle
5. Right air funnel
6. Right air flex hose
7. Upper air cleaner housing
8. Air cleaner filter element
9. Lower air cleaner housing
10. Left air flex hose
11. Mounting bolts
12. Left air intake bracket
13. Left air intake nozzle
14. Left air funnel

Exploded view of air intake system—Audi V8

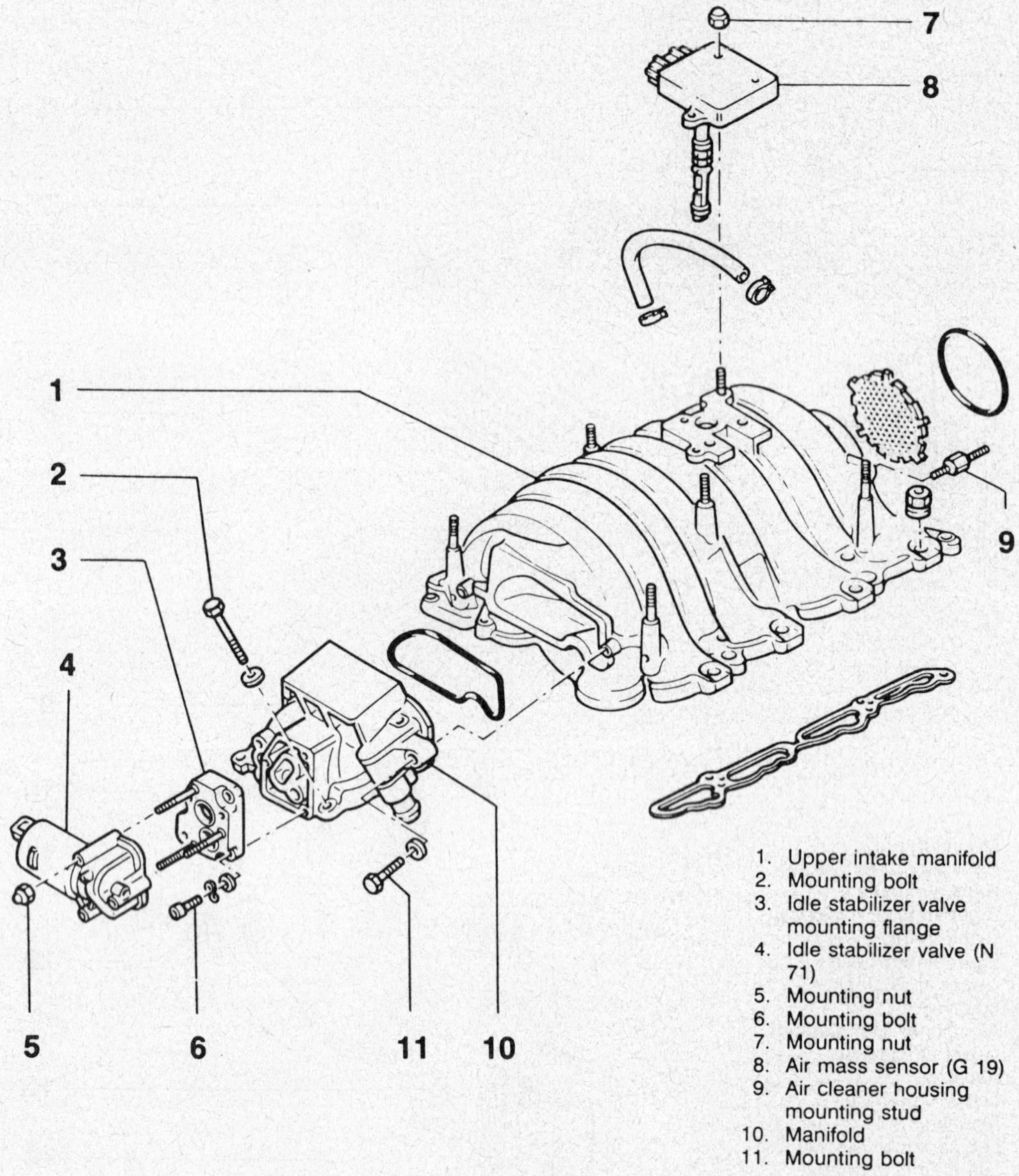

Exploded view of upper intake manifold—Audi V8

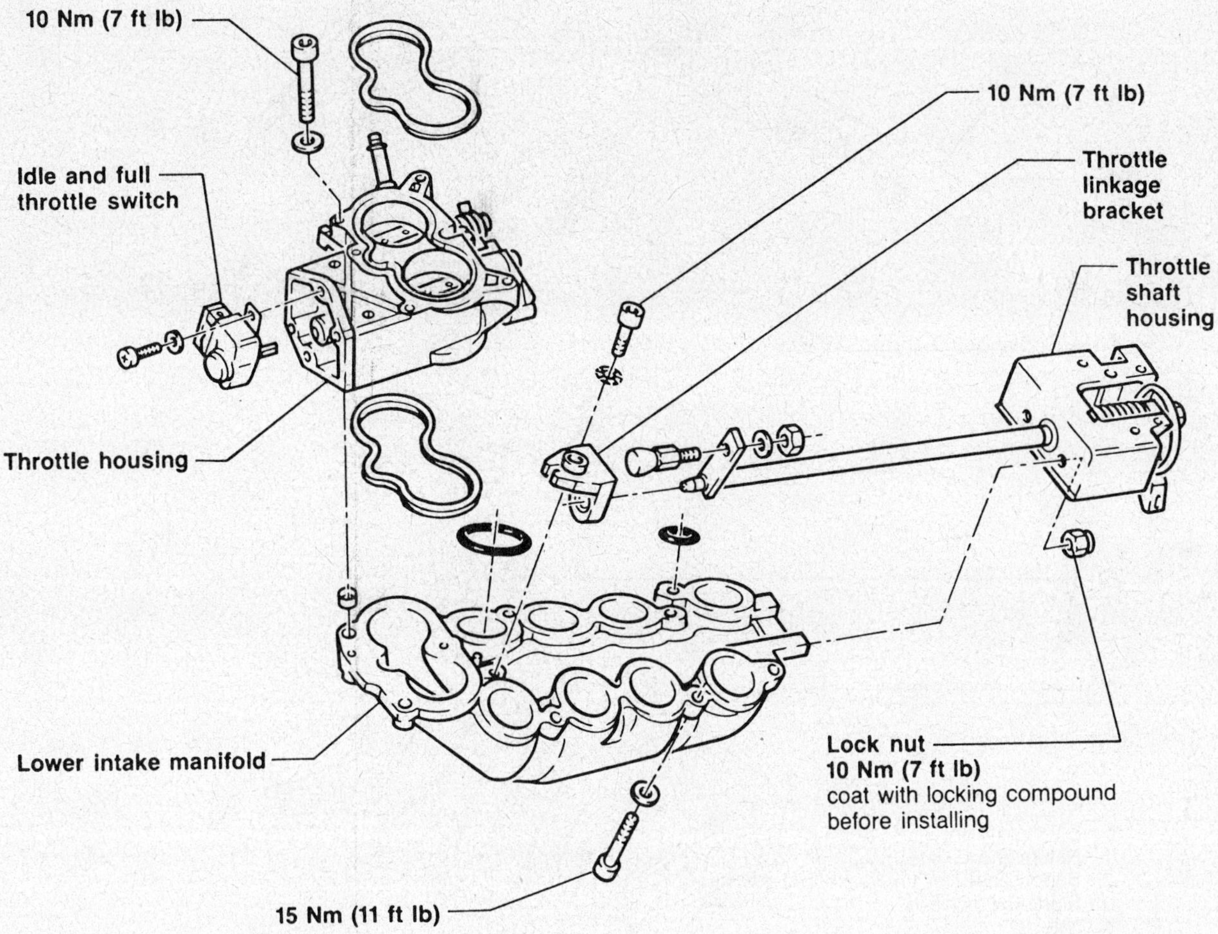

Exploded view of lower intake manifold—Audi V8

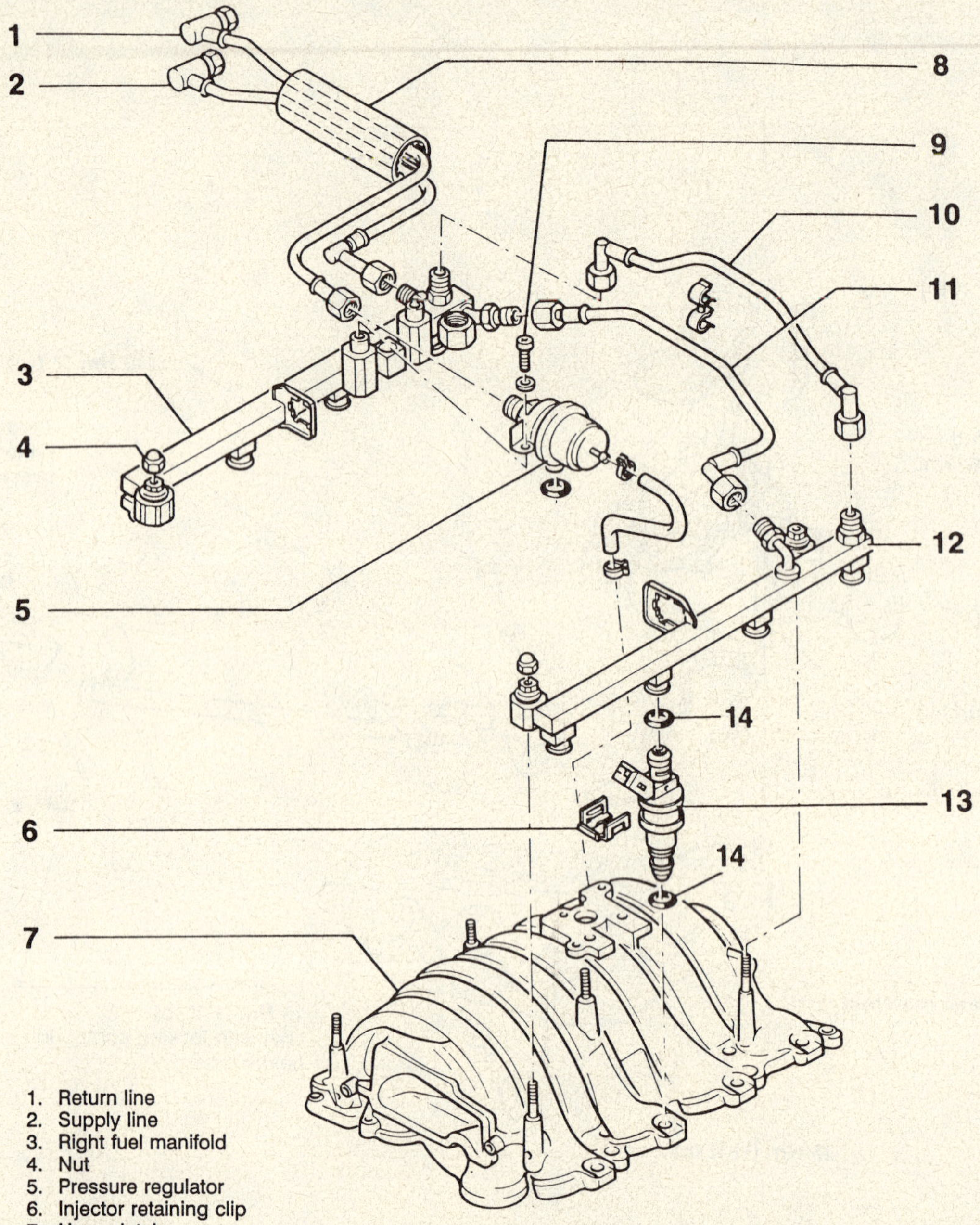

1. Return line
2. Supply line
3. Right fuel manifold
4. Nut
5. Pressure regulator
6. Injector retaining clip
7. Upper intake manifold
8. Insulator
9. Screw
10. Supply line cross-over
11. Return line cross-over
12. Left fuel manifold
13. Fuel injector
14. O-rings

Exploded view of injector assembly—Audi V8

6. Remove upper ventilation hose, then the right fuel rail cable tie.

NOTE: When assembling, install a new cable tie in the same position and location as the original.

7. Remove the engine wiring harness support clip attaching bolts, and position wiring harness aside.
8. Disconnect the throttle valve potentiometer, thermoswitch and idle stabilizer valve electrical connectors.
9. Remove the idle stabilizer valve, then disconnect the vacuum hoses to the carbon canister.
10. Remove the bolts attaching the intake air temperature sensor, then position aside.
11. Disconnect the air mass sensor electrical connectors, then the knock sensor (2 sensors) plug connections.
12. Disconnect the the fuel return and supply lines, then the pressure regulator vacuum hose.
13. Remove the fuel rail attaching bolts, then the fuel rail and injectors as an assembly.
14. Reverse procedure to install. Use new injector O-rings and coat with gasoline before installing.

100

1. Relieve fuel line pressure. Disconnect the injector electrical connectors.
2. Disconnect the fuel feed and fuel return lines.
3. Remove the 4 bolts holding the fuel rail to the intake manifold.
4. Pull up on the fuel rail to remove along with the injectors. Remove the injector clips and injectors from the fuel rail.

To Install:

7. Lubricate the injector seals with clean engine oil. Install the injectors in the fuel rail. Install the clips.
8. Install the fuel rail, fuel lines and torque to 18 ft. lbs (25 Nm).
9. Install the fuel rail mounting bolts and torque to 7 ft. lbs (10 Nm). Check for leaks.

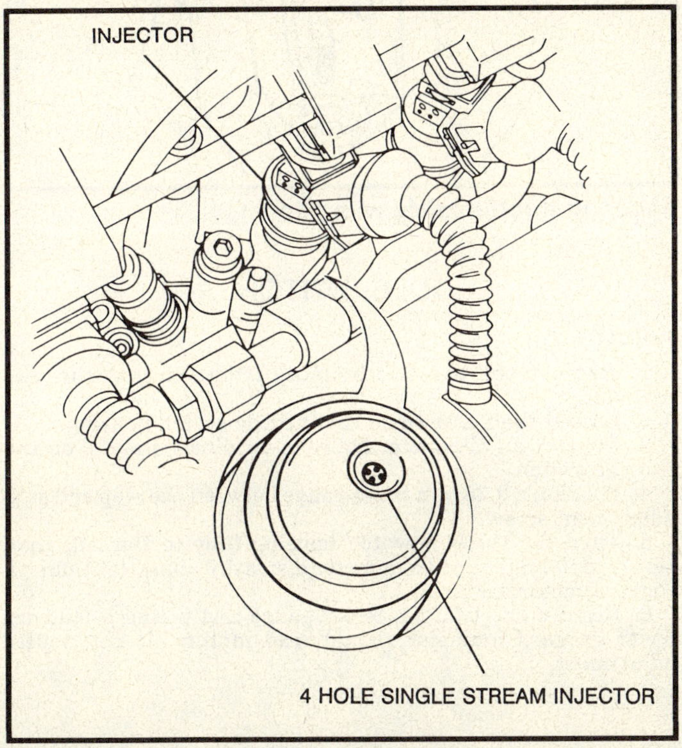

Fuel injectors—1992 Audi 100 vehicles

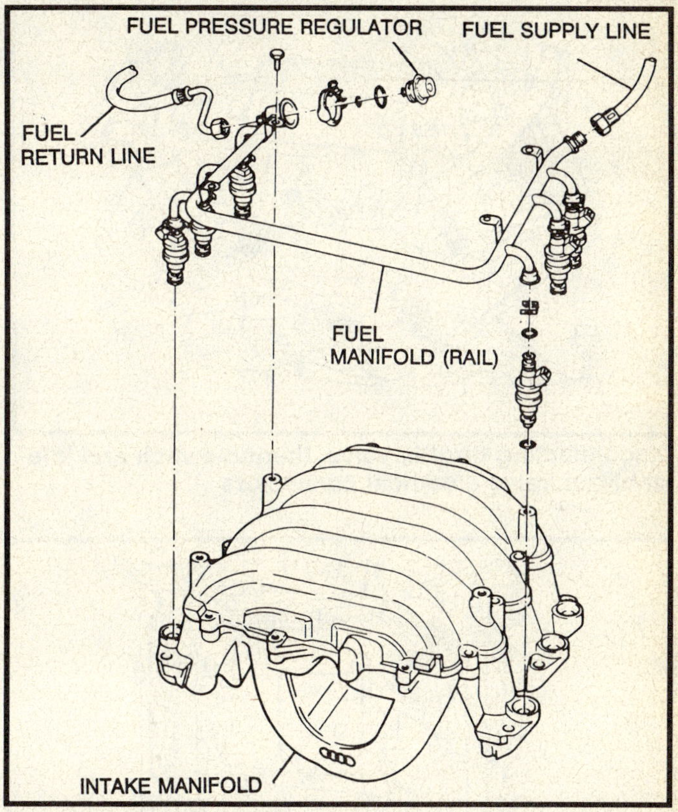

Fuel injectors and fue rail assembly—1992 Audi 100 vehicles

EXCEPT V8 QUATTRO AND 100

1. Relieve the fuel system pressure.
2. Remove the fuel feed line from the fuel rail.
3. Disconnect the fuel return line from the fuel pressure regulator.
4. Remove the fuel pressure regulator mounting hardware and remove the regulator. Disconnect the fuel line from the intake manifold.
5. Pull the fuel rail with the injectors from the intake.
6. Disconnect the electrical connectors and the clips holding the injectors to the fuel rail. Remove the injectors.

To Install:

7. Lubricate the injector seals with clean engine oil. Install the injectors in the fuel rail. Install the clips and electrical connectors.
8. Install the fuel rail, fuel pressure regulator and fuel lines. Torque the fuel line connections to 18 ft. lbs (25 Nm). Check for leaks.

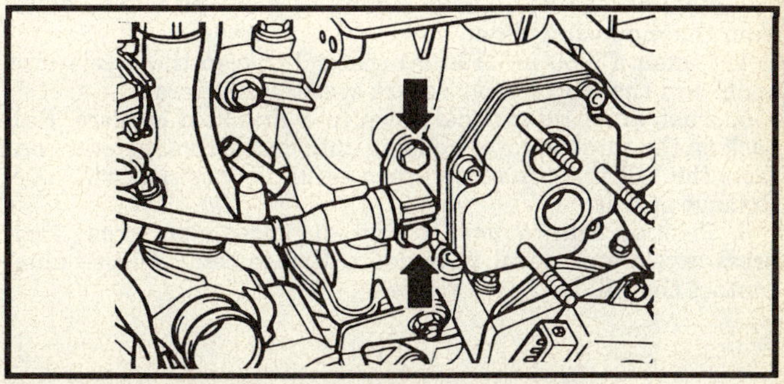

Removing intake air temperature sensor

Disconnecting throttle valve, thermo-switch and idle stabilizer valve electrical connectors

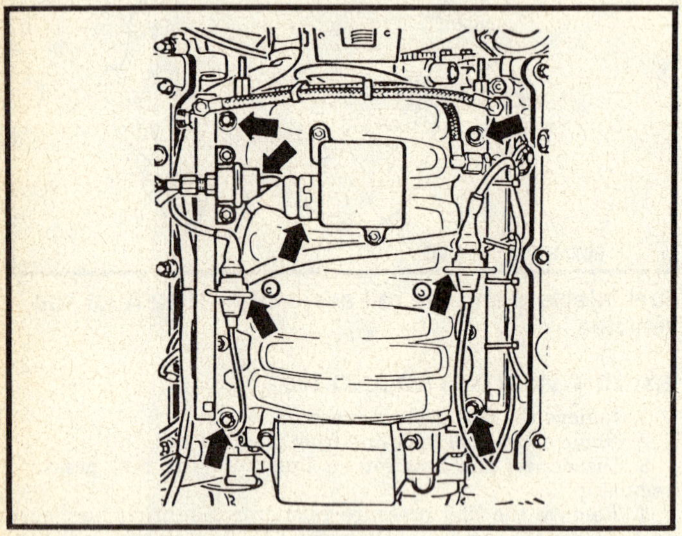

Injector fuel rail bolts location

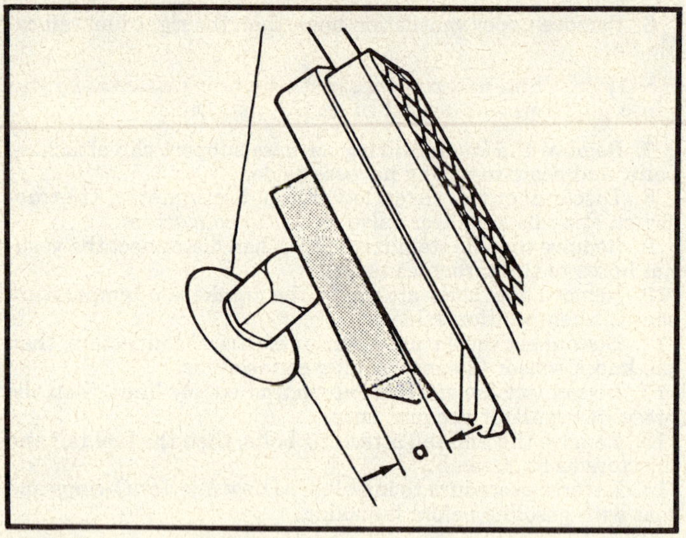

Installing spacer between accelerator pedal and stop screw

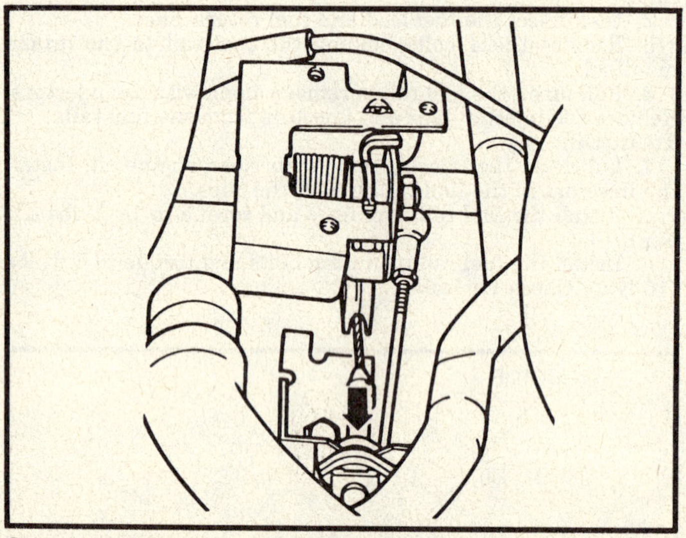

Securing throttle cable with clip

THROTTLE CABLE

Adjust

1. Remove the upper air cleaner housing attaching bolts, then the upper air cleaner housing.
2. Remove the lower air cleaner housing attaching bolts, then push down on housing and lift upward.
3. Remove ignition cable mounting plate attaching bolts, then the mounting plate.
4. Install a 0.60 in. (15mm) spacer between the accelerator pedal and the stop screw. Secure spacer with clamp.
5. Push accelerator pedal down fully and hold in place. Pull back on the throttle cable housing until cruise control lever contacts the full throttle screw, then secure in position with clip. Remove spacer.
6. Position cam by hand to the idle speed stop screw, then press accelerator pedal fully down. Cruise control lever must contact the full load stop screw.

IDLE SWITCH

Adjust

1. Remove the throttle housing from the lower intake manifold.
2. Loosen the 2 screws on the throttle switch housing.
3. Connect an ohmmeter between terminals 4 and 6 on the throttle switch.
4. Position a 0.25 mm feeler gauge between the stop and stop adjustment screw.
5. Turn the throttle switch housing fully to the left, then slowly to the right until ohmmeter displays 0 ohms (continuity). Tighten mounting screws.
6. Remove the 0.25mm feeler gauge and install a 0.30mm feeler gauge. Ohmmeter should read infinity. If not, repeat adjustment.

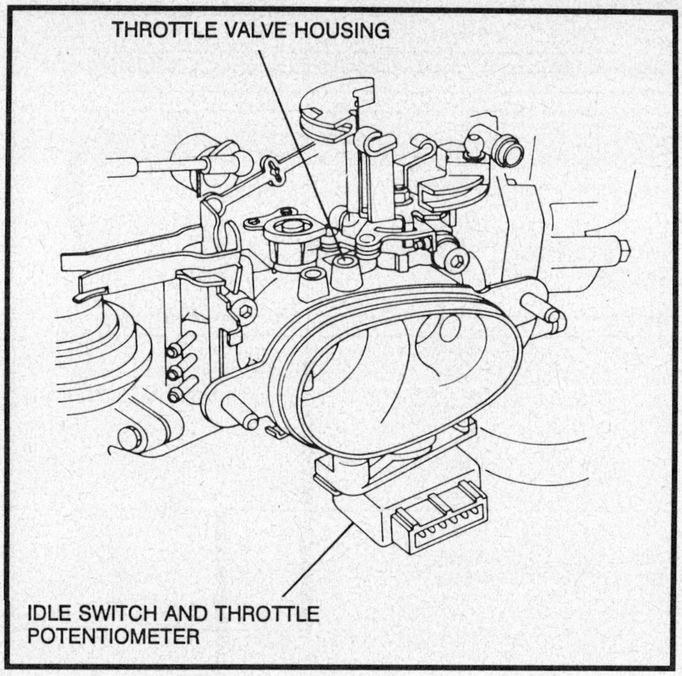

Idle switch assembly mounted on the throttle valve assembly—1992 Audi 100 vehicles

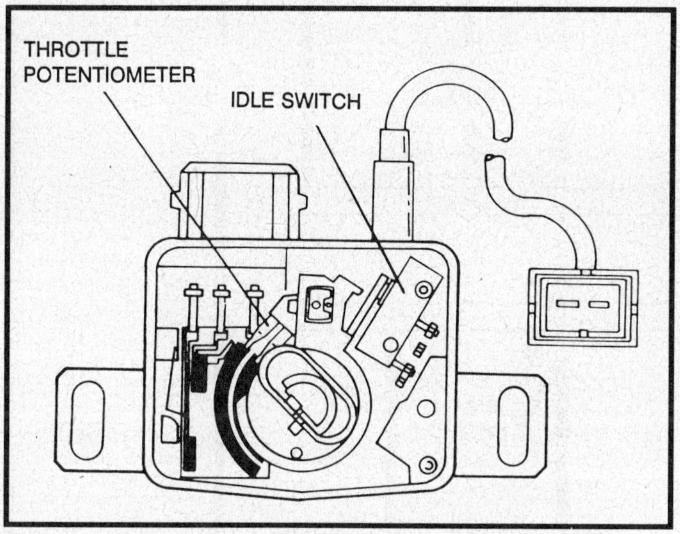

Internal view of the throttle potentiometer and idle switch assembly—1992 Audi 100 vehicles

IDLE SPEED AND MIXTURE (CO)

Adjusting

EXCEPT 90 QUATTRO, COUPE QUATTRO AND 100

The idle speed and CO level are controlled by the engine management system. There are no adjustments possible on these systems. If the CO or idle speed is not within specification, check for vacuum leaks and leaking injectors.

90 QUATTRO AND COUPE QUATTRO

1. Before adjusting the idle and CO, the fault codes should be read, the engine at operating temperature, all electrical accessories turned off and any fuel pressure measuring gauges removed. The engine cooling fan must not run during the test.
2. Disconnect the crankcase ventilation hose and plug. Disconnect the electrical connector from the charcoal canister solenoid valve located next to the coolant expansion tank.
3. Connect a CO tester to the exhaust tap. Start the engine and run at idle speed, 750–850 rpm.
4. If the idle speed is incorrect, adjust the idle. With the engine running below 2000 rpm, insert a fuse in the top of the fuel pump relay. This will set the engine management parameters to predetermined readings. The engine idle speed can be set by turning the idle speed screw on the throttle body.
5. The CO level should be 0.3–1.2%.
6. If the CO level is incorrect, remove the plug over the CO adjusting screw on the air mass sensor. It will be located next to the electrical connection on the sensor. Adjust the screw until the proper level is obtained.
7. Check the idle speed and adjust if necessary. Remove the fuse from the relay and raise the engine speed over 2000 rpm briefly. Check the idle speed and CO. Replace the CO plug and remove the test equipment.

100

The idle speed can not be adjusted on this vehicle. The CO level can be adjusted if necessary, but check for external influences, such as leaking injectors or vacuum leaks, before adjusting the level.

1. Connect a CO tester to one of the exhaust taps. Start the engine and run at idle speed. The CO level should be 0.3–1.2%.
2. If the CO level is incorrect, remove the plug over the CO adjusting screw on the air mass sensor. It will be located next to the electrical connection on the sensor. Adjust the screw until the proper level is obtained.
3. Check the CO at the exhaust tap on the other cylinder bank. Install a new CO plug and remove the test equipment.

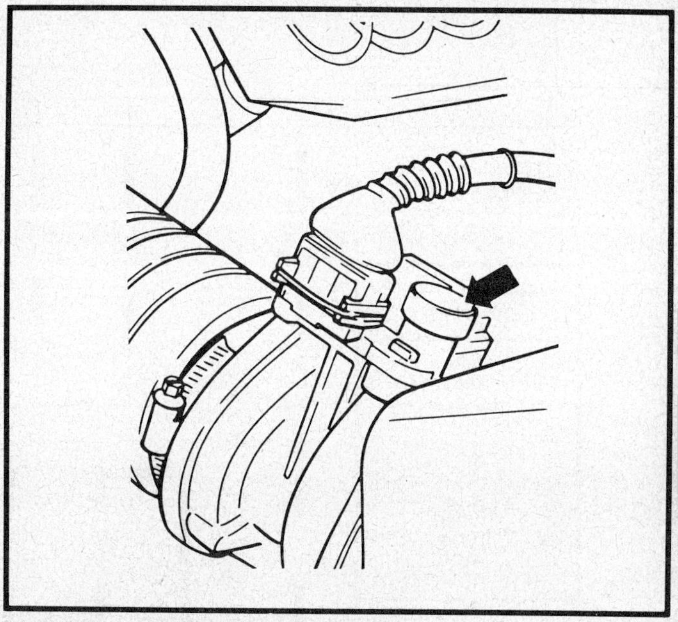

The CO potentiometer is mounted on the air mass sensor—1992 Audi 100 vehicles

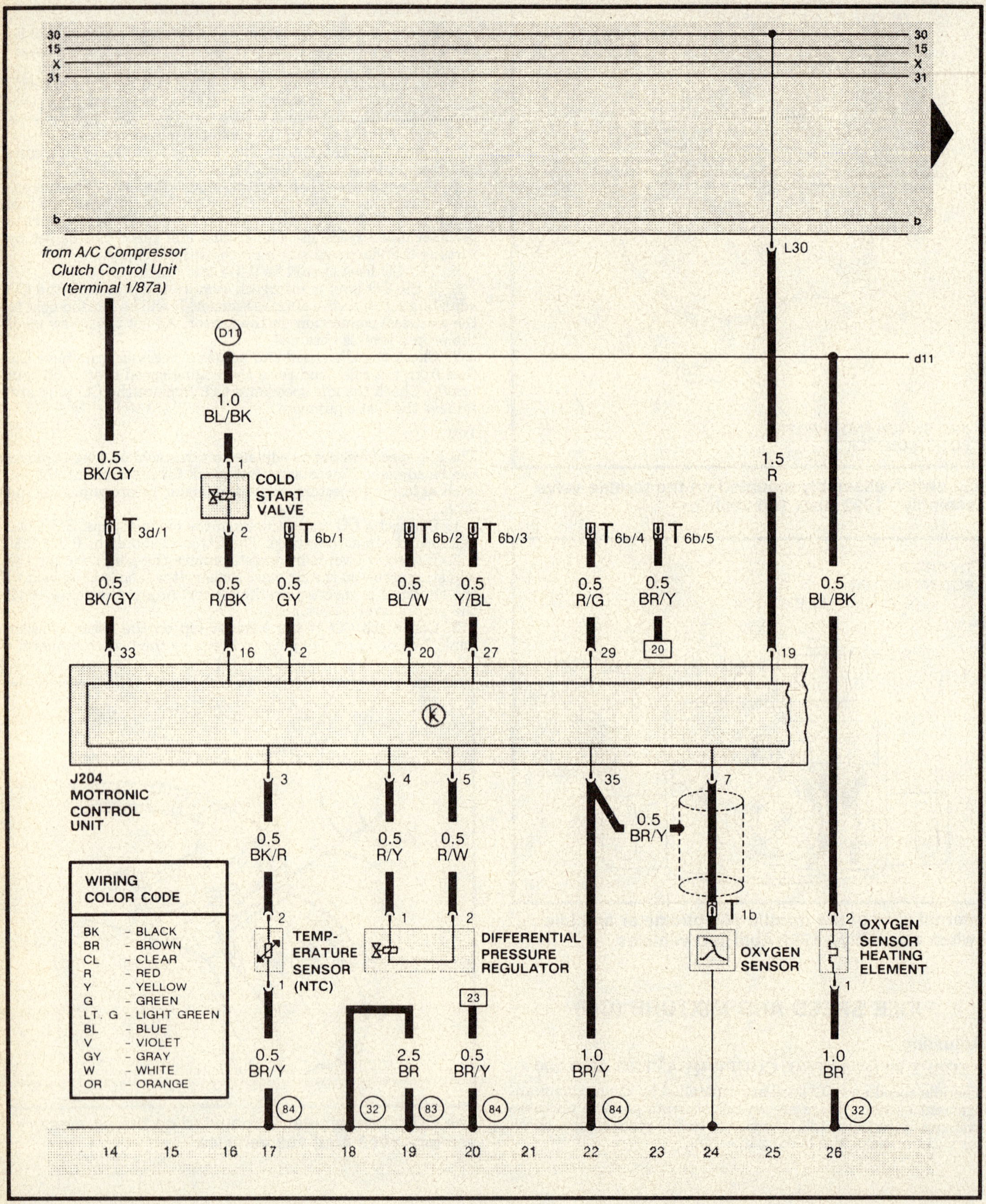

Engine wiring—Audi 80 with Motronic

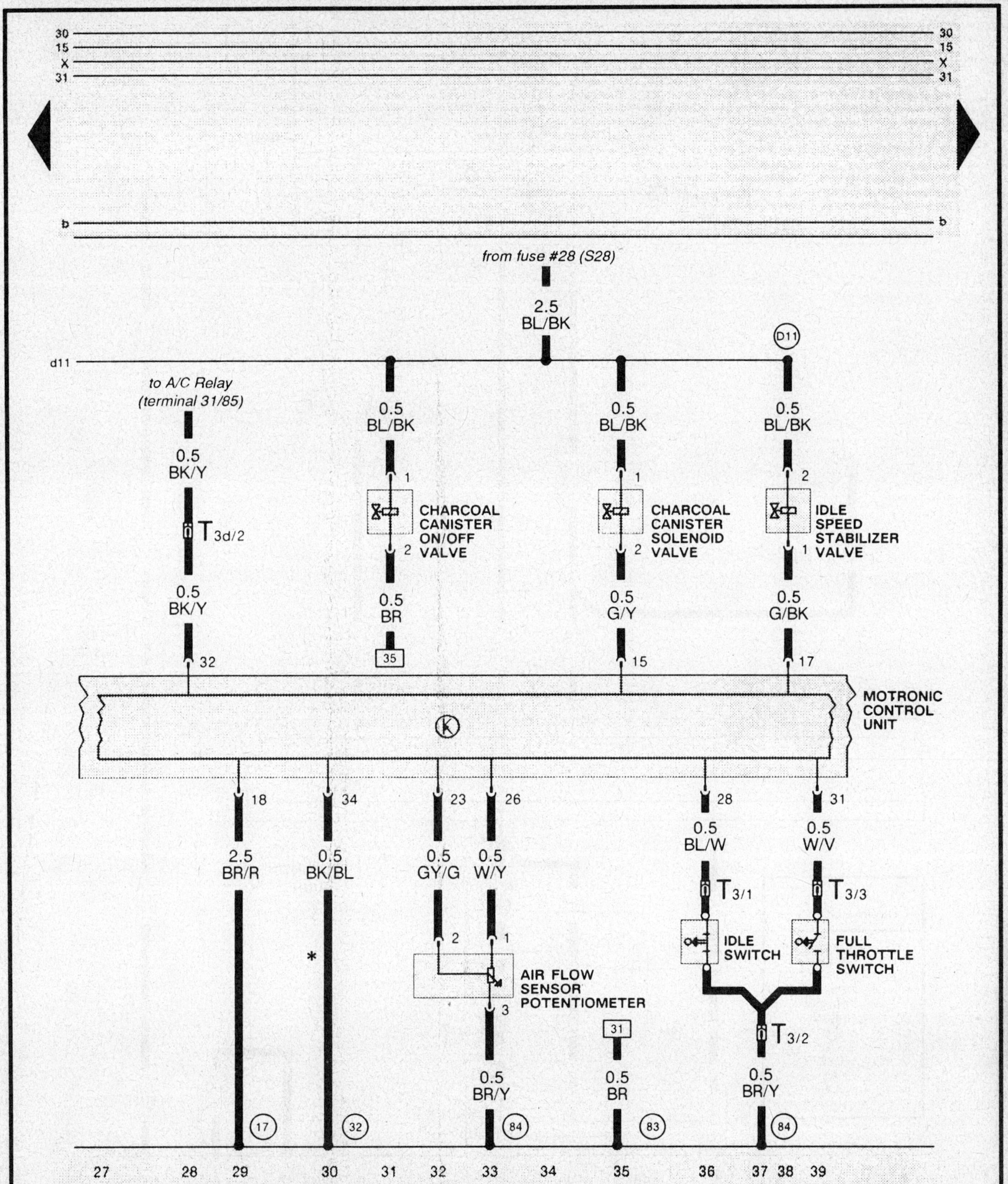

* *Manual trans.: to ground as shown; Auto. trans: to auto. trans. wiring harness*

Engine wiring—Audi 80 with Motronic

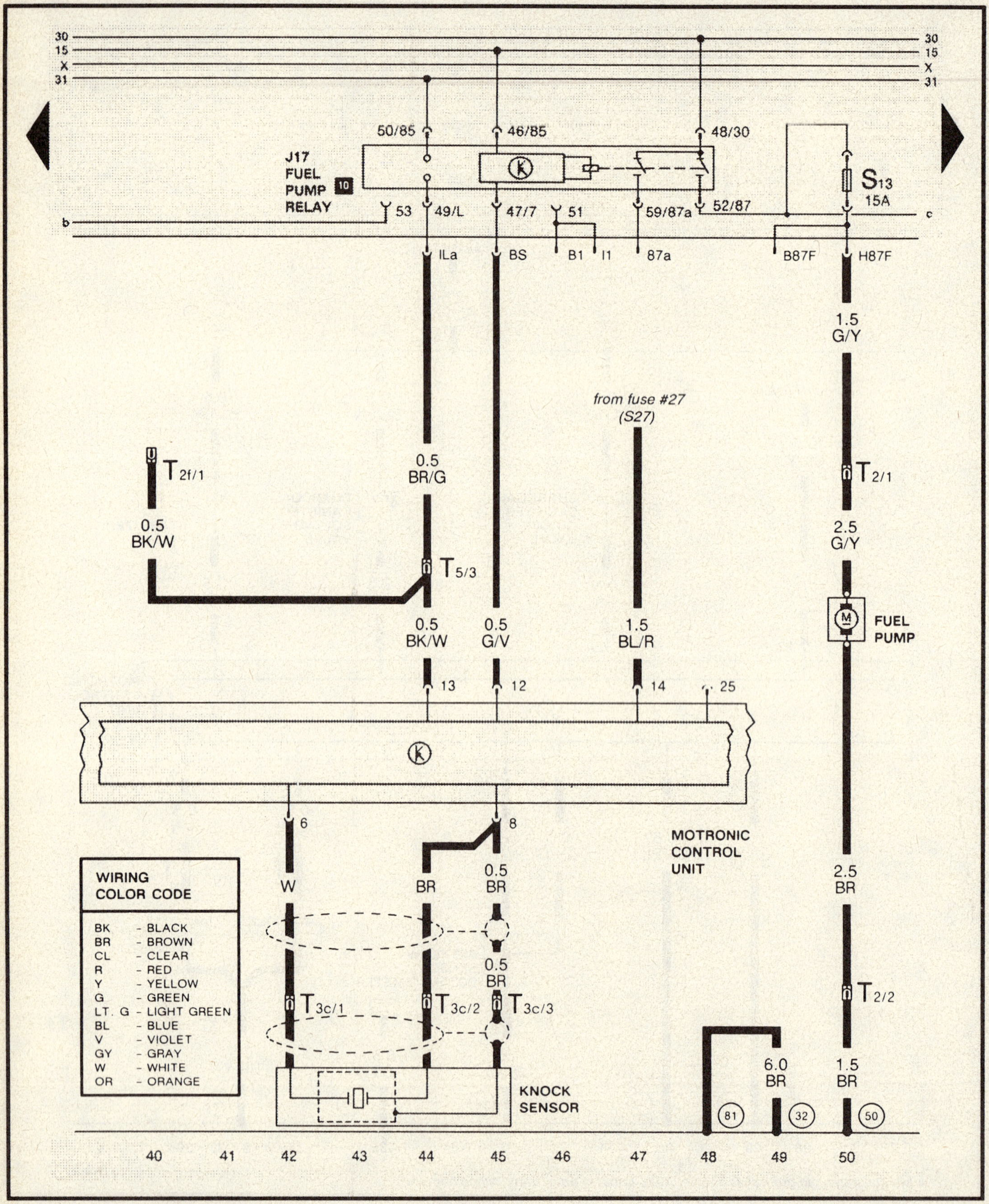

Engine wiring—Audi 80 with Motronic

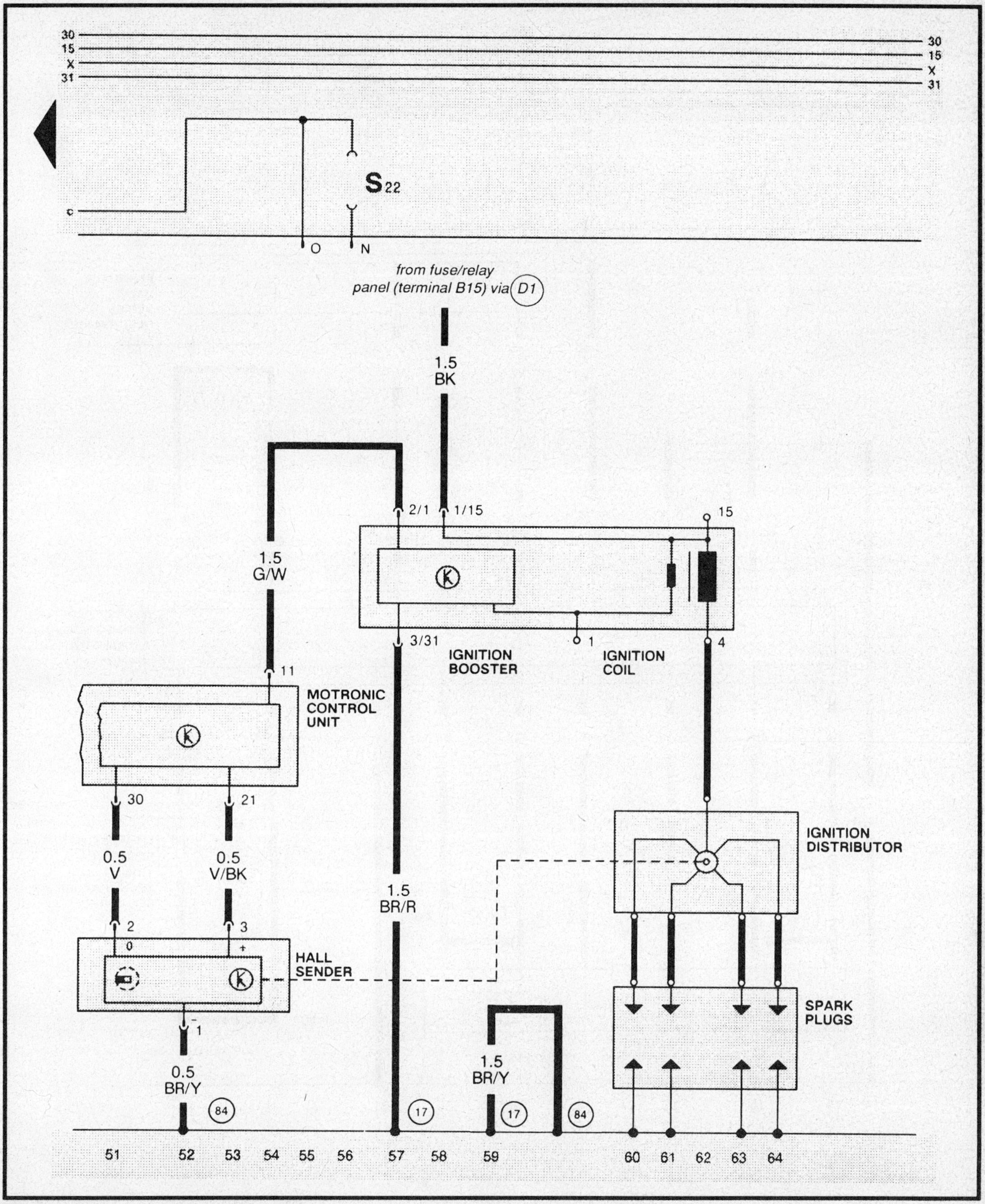

Engine wiring—Audi 80 with Motronic

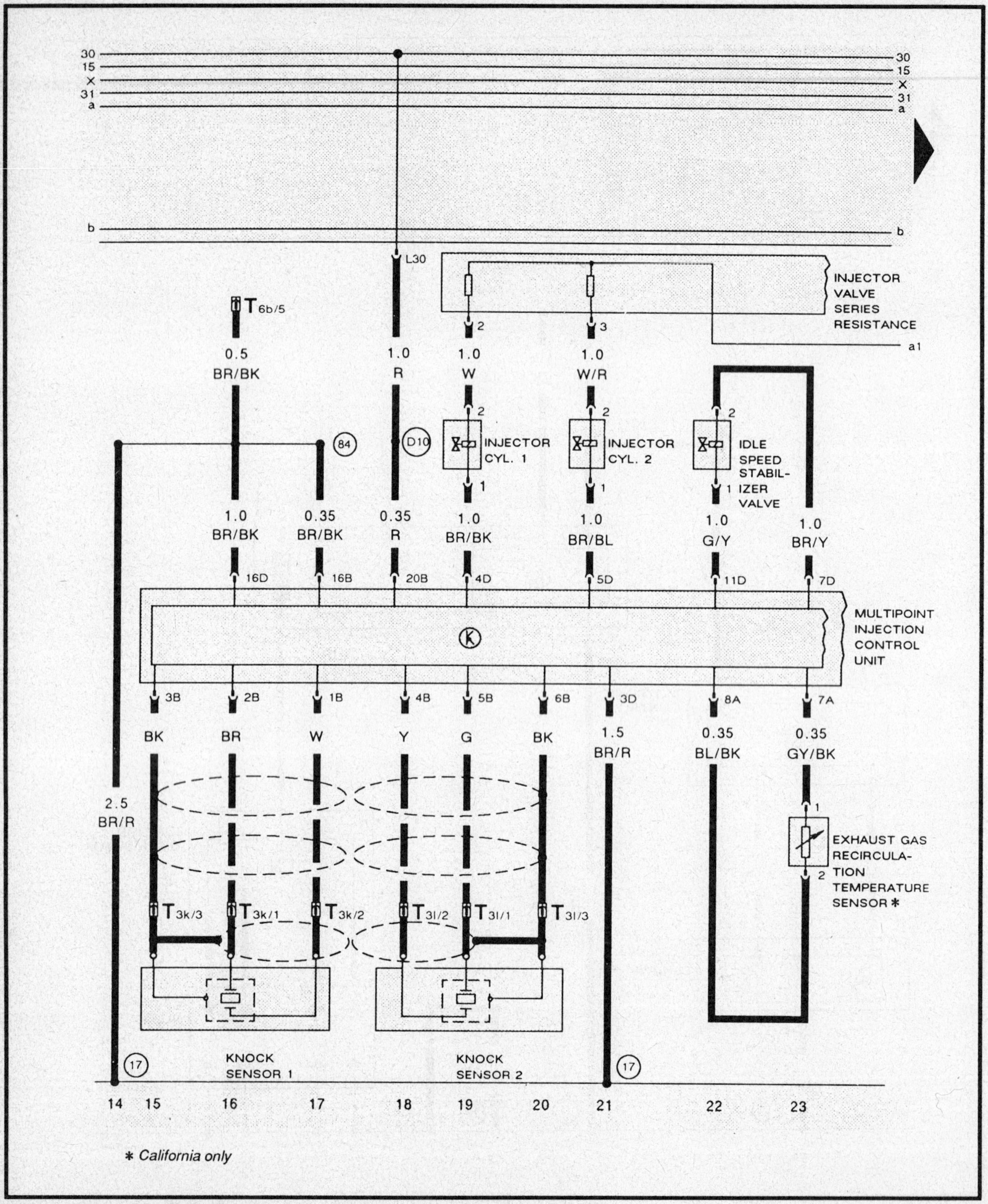

Engine wiring—Audi 90 Quattro 20V and Coupe Quattro 20V with MPI

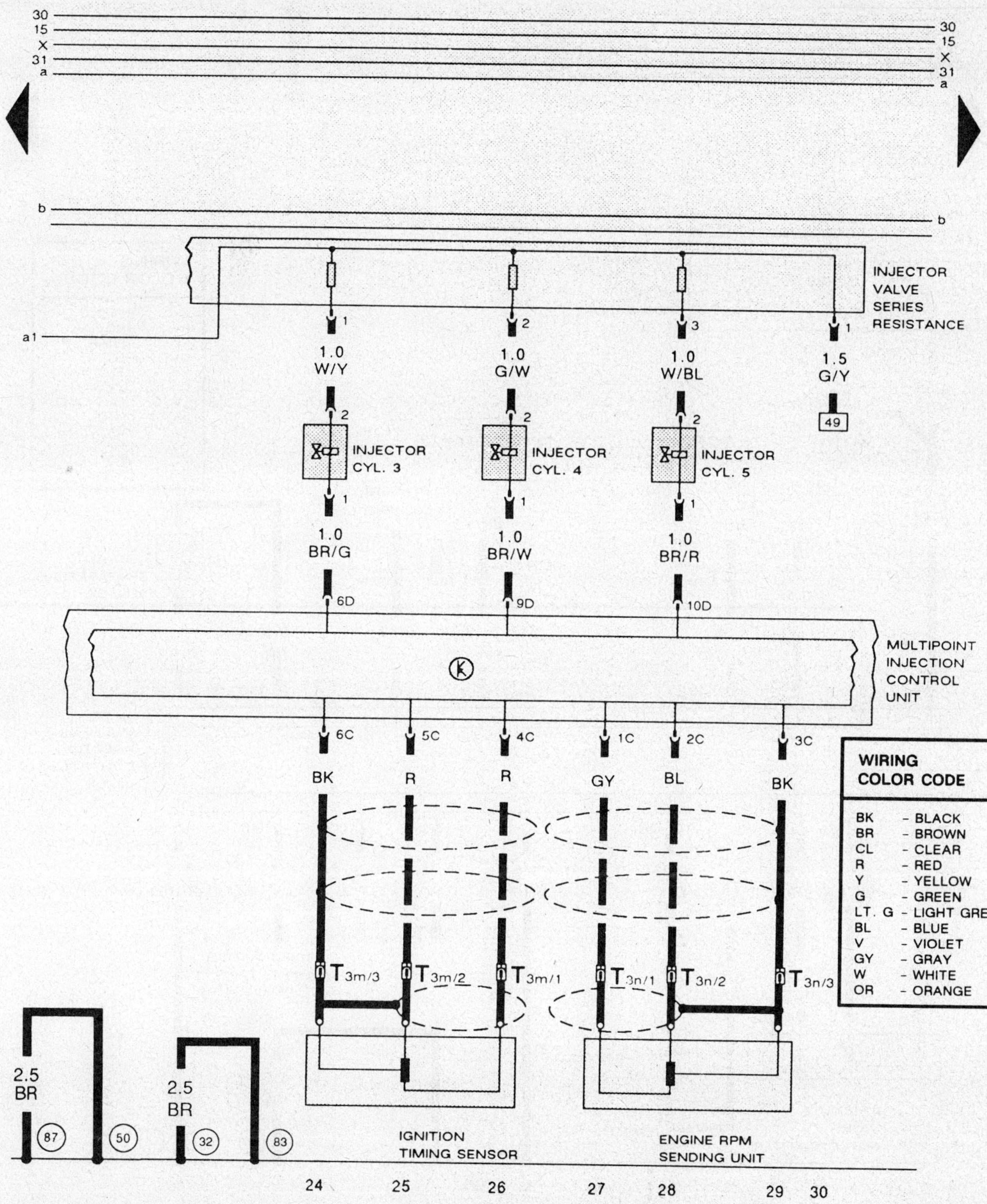

WIRING COLOR CODE

BK	- BLACK
BR	- BROWN
CL	- CLEAR
R	- RED
Y	- YELLOW
G	- GREEN
LT. G	- LIGHT GREEN
BL	- BLUE
V	- VIOLET
GY	- GRAY
W	- WHITE
OR	- ORANGE

Engine wiring—Audi 90 Quattro 20V and Coupe Quattro 20V with MPI

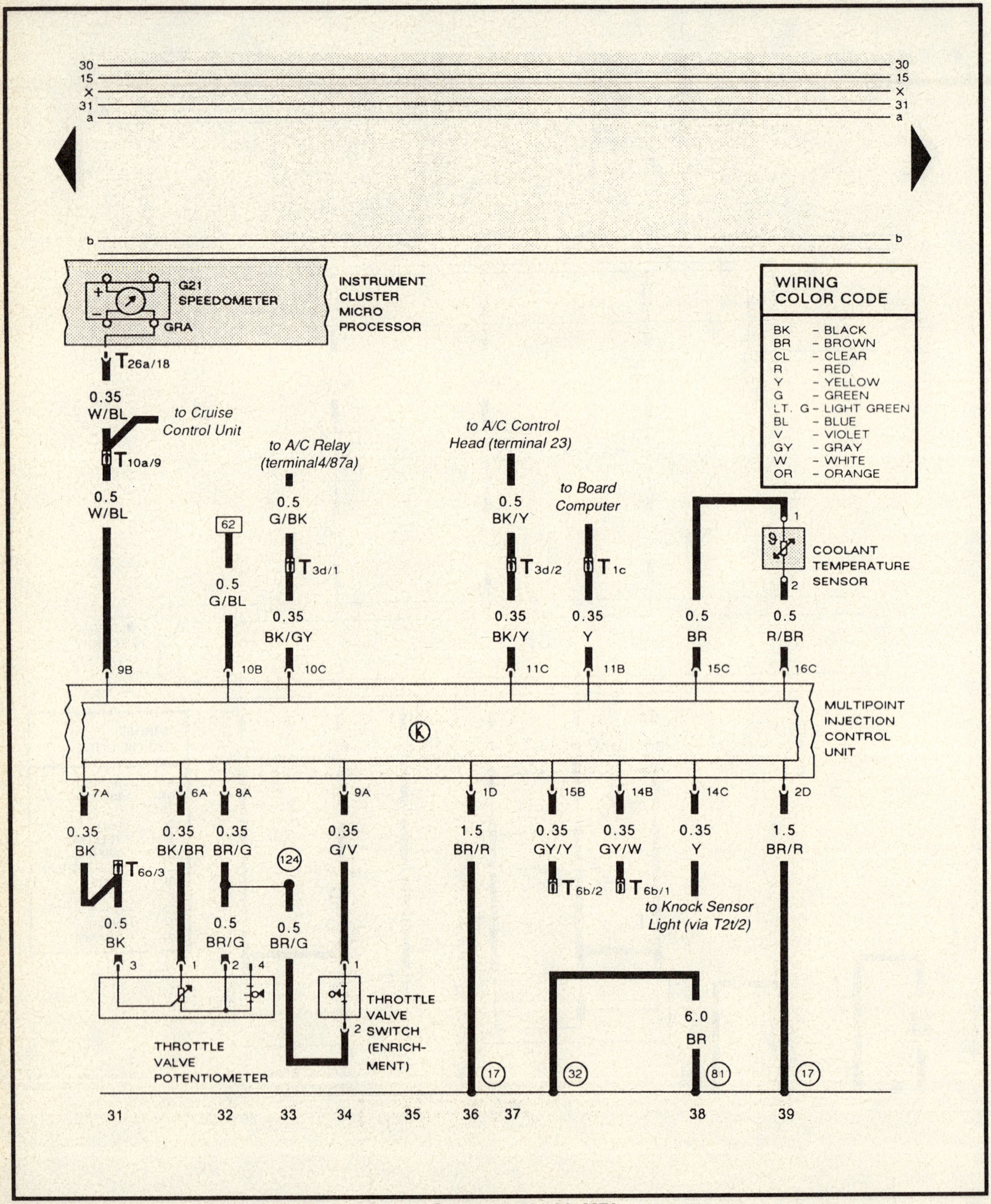

Engine wiring—Audi 90 Quattro 20V and Coupe Quattro 20V with MPI

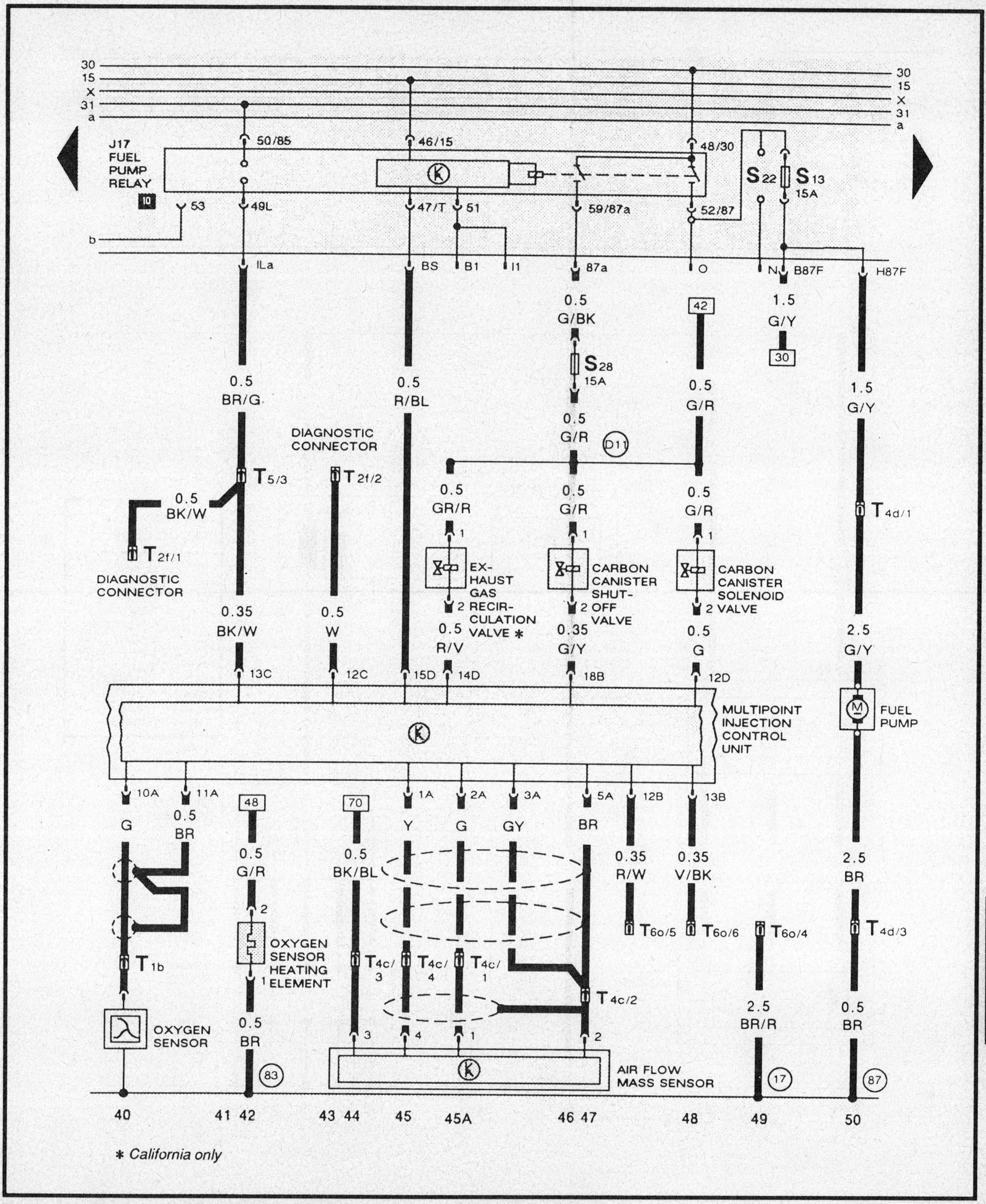

Engine wiring—Audi 90 Quattro 20V and Coupe Quattro 20V with MPI

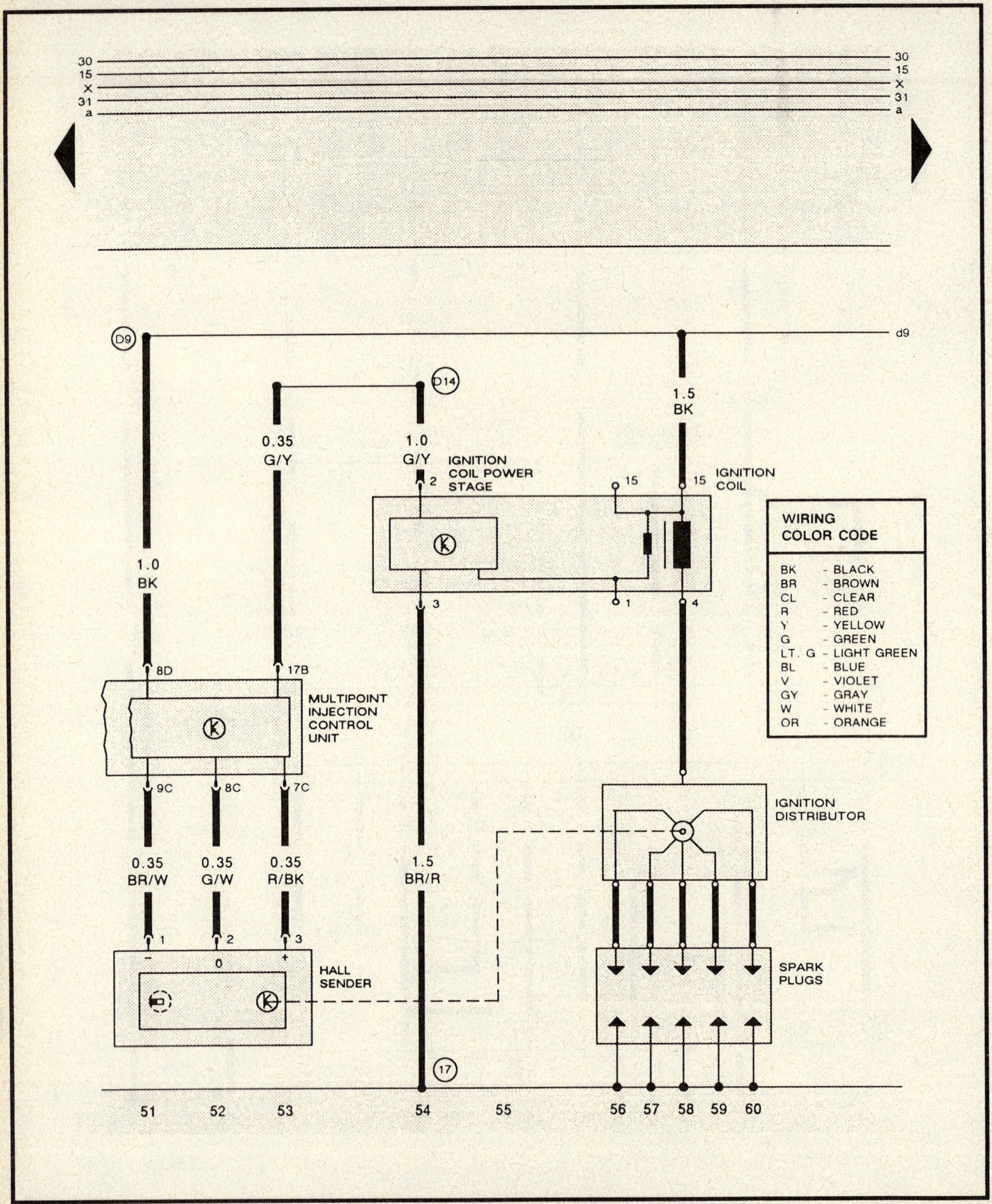

Engine wiring—Audi 90 Quattro 20V and Coupe Quattro 20V with MPI

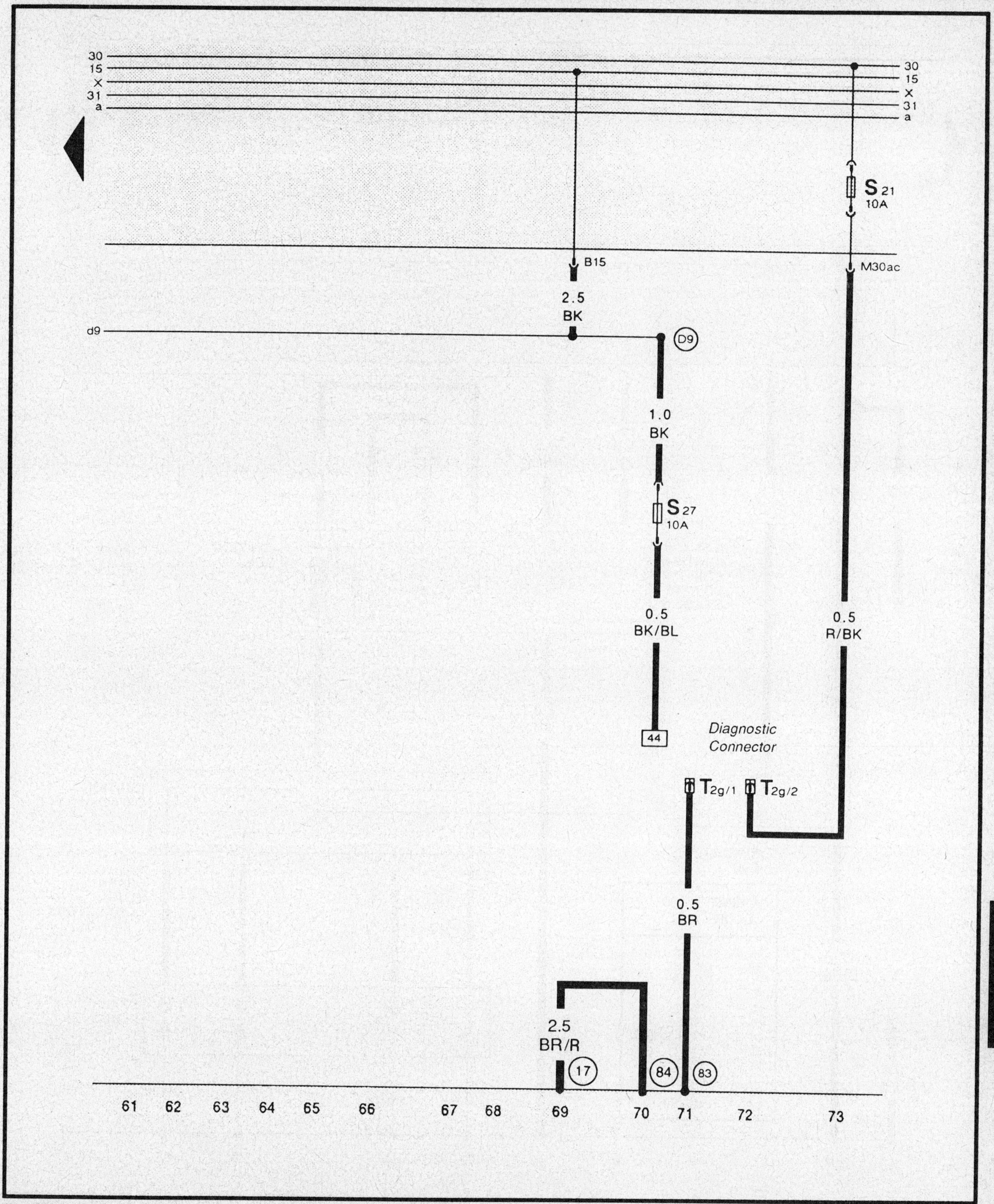

Engine wiring—Audi 90 Quattro 20V and Coupe Quattro 20V with MPI

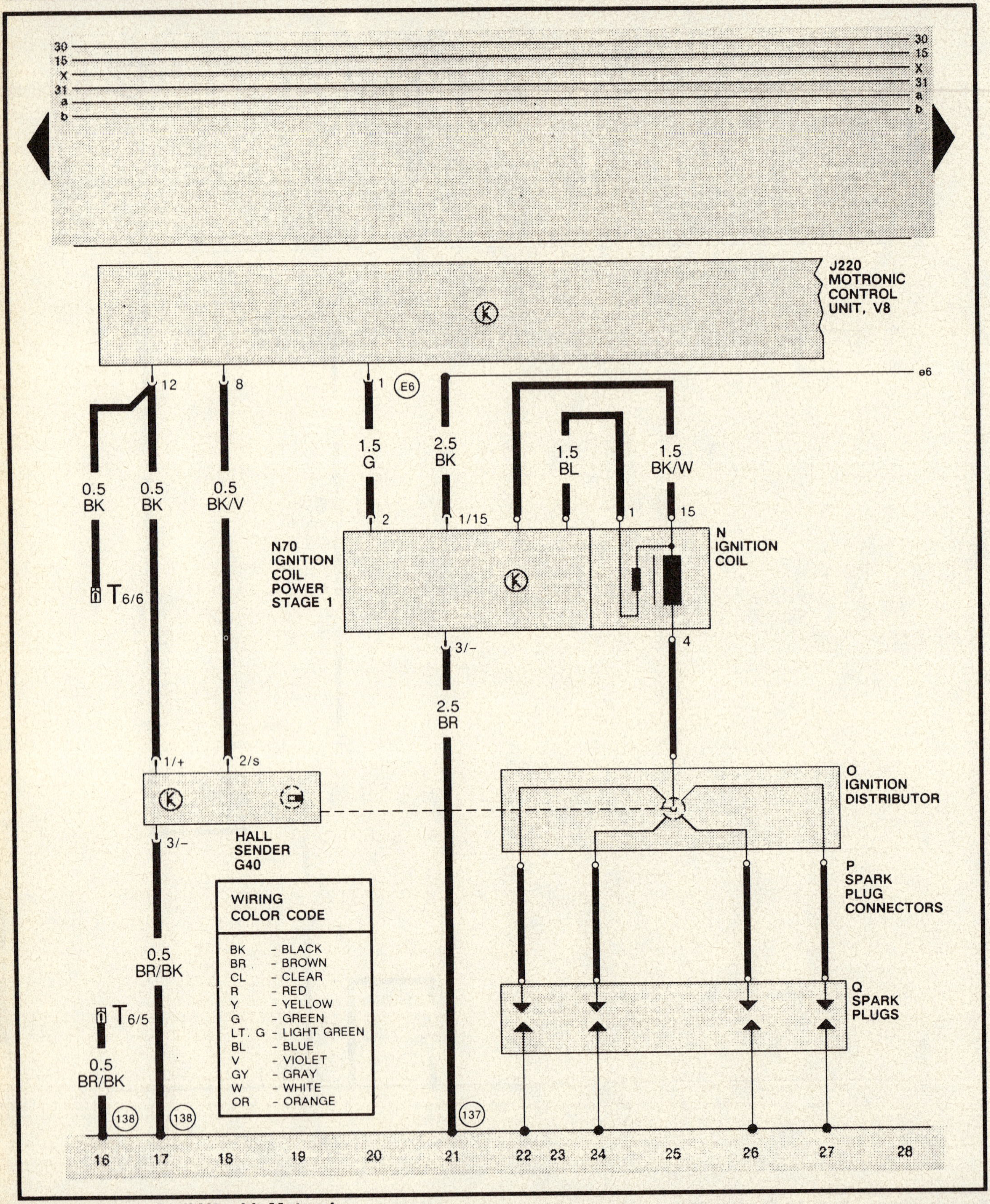

Engine wiring—Audi V8 with Motronic

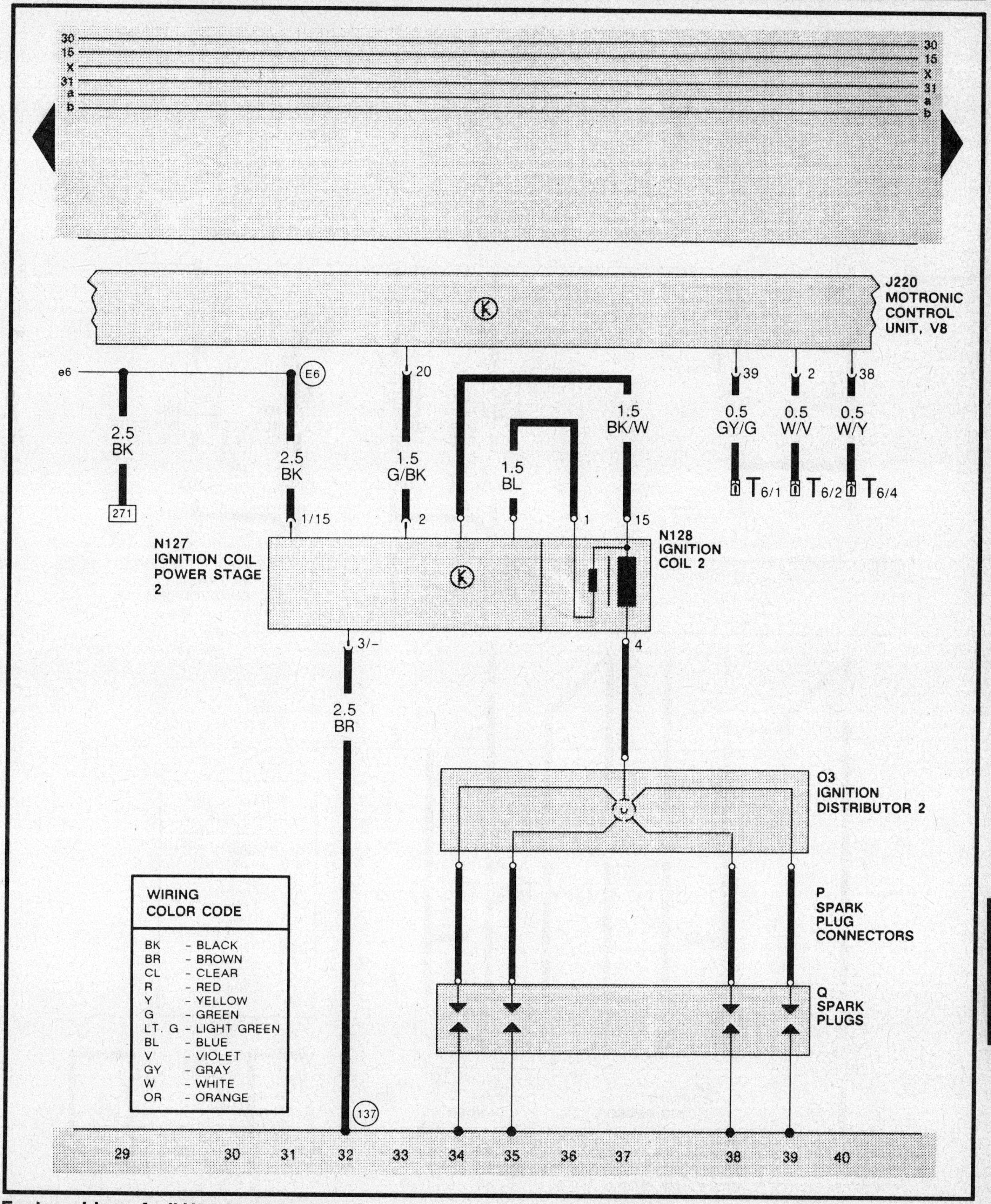

Engine wiring—Audi V8 with Motronic

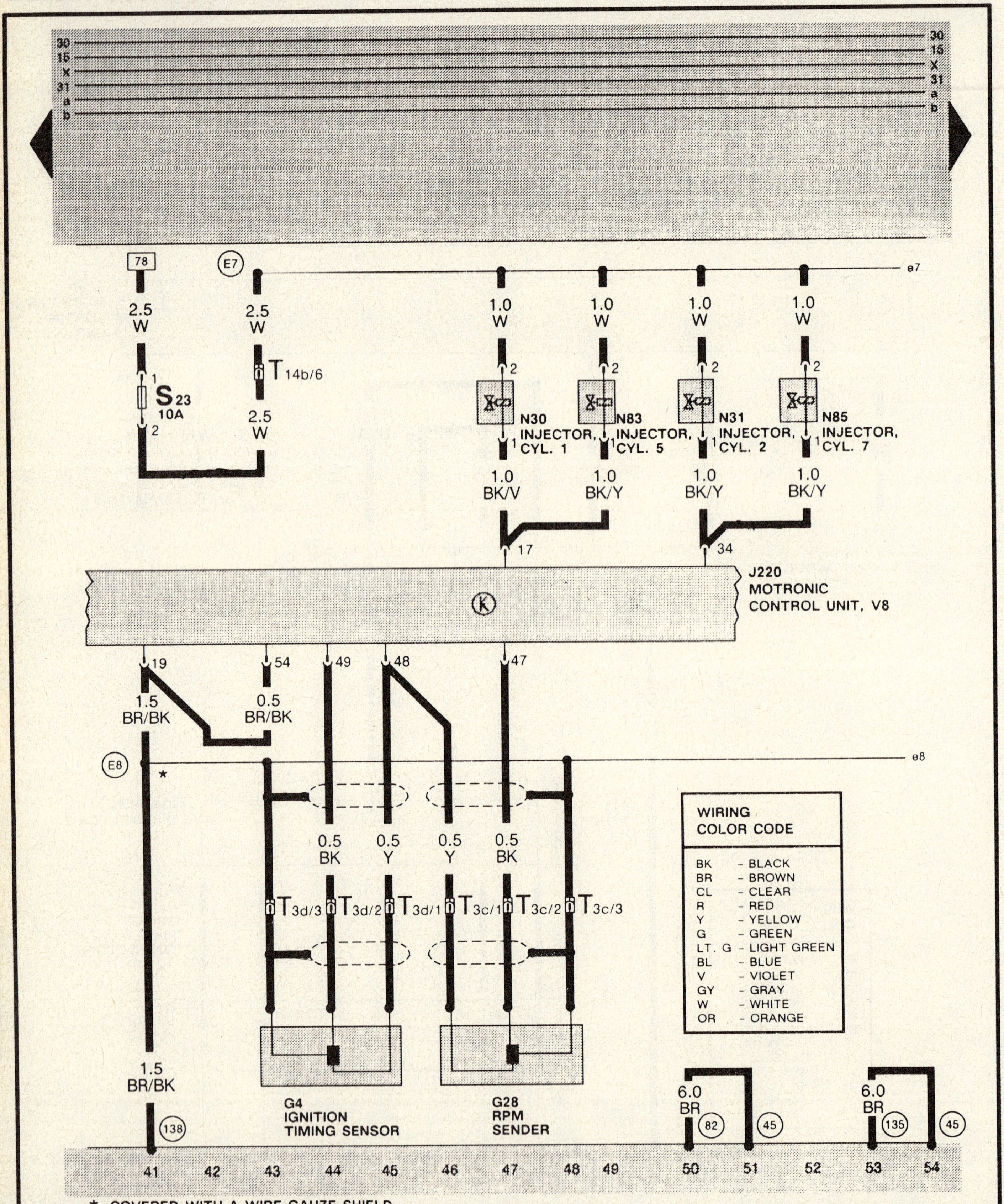

WIRING COLOR CODE	
BK	- BLACK
BR	- BROWN
CL	- CLEAR
R	- RED
Y	- YELLOW
G	- GREEN
LT. G	- LIGHT GREEN
BL	- BLUE
V	- VIOLET
GY	- GRAY
W	- WHITE
OR	- ORANGE

* COVERED WITH A WIRE GAUZE SHIELD

Engine wiring—Audi V8 with Motronic

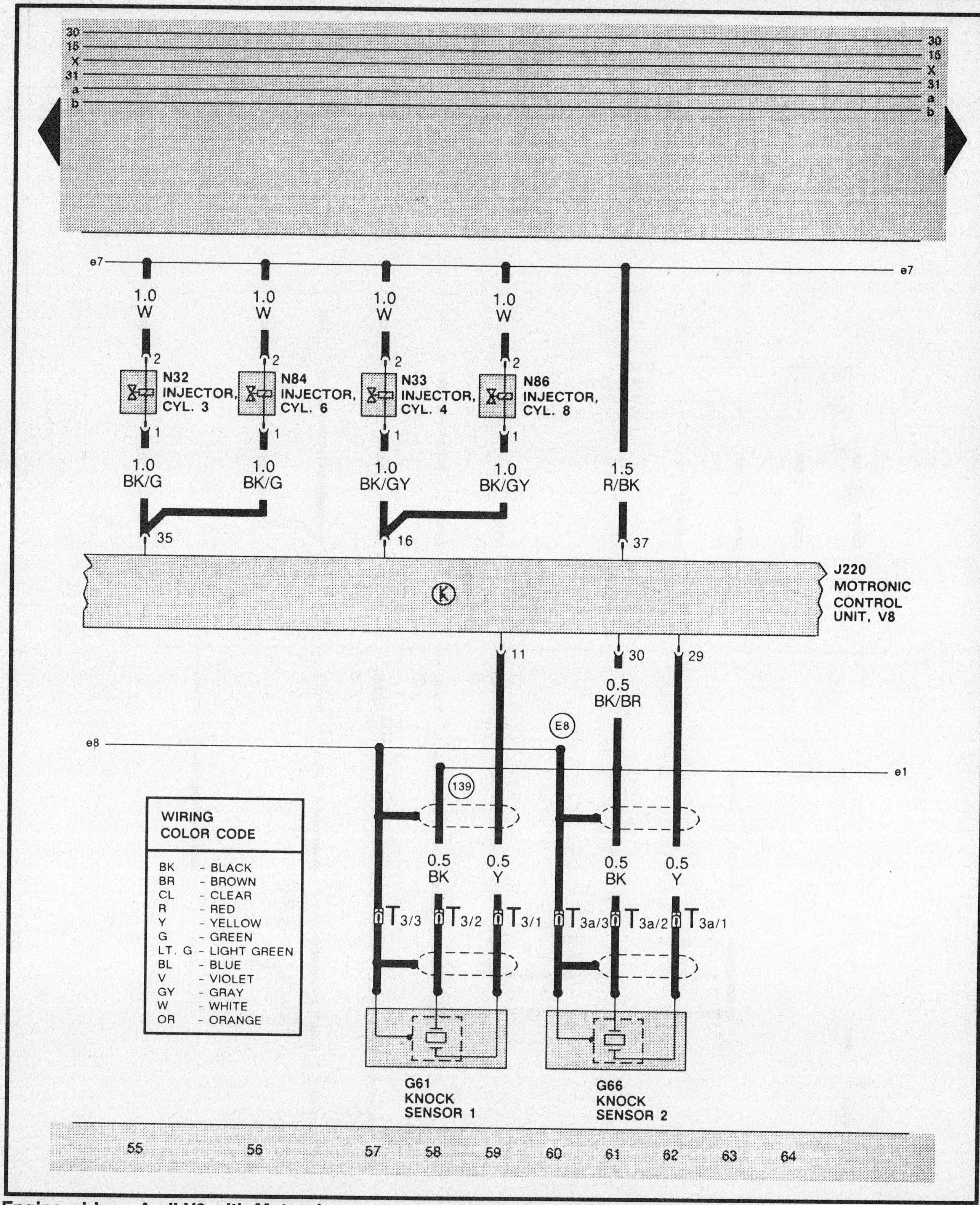

Engine wiring—Audi V8 with Motronic

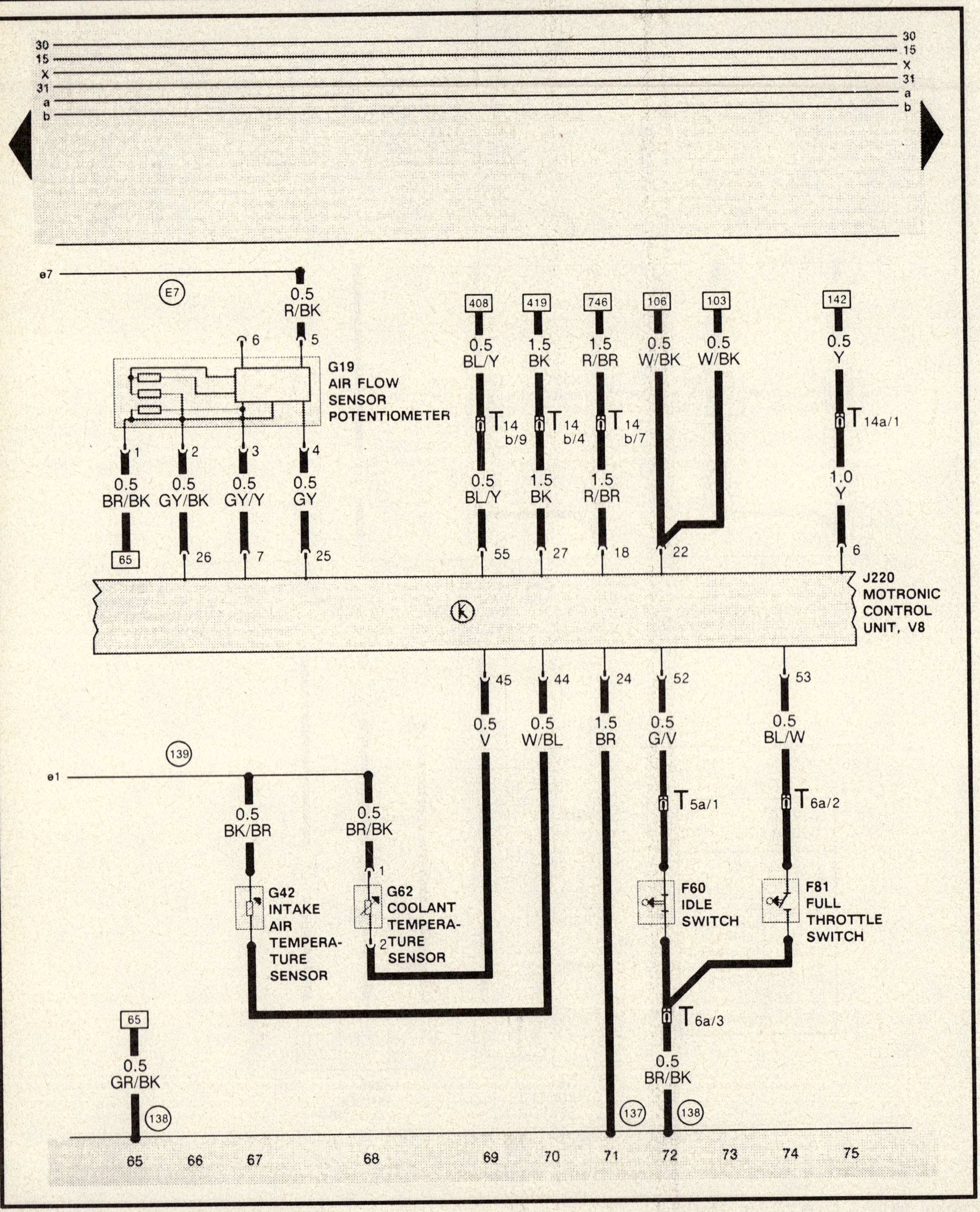

Engine wiring—Audi V8 with Motronic

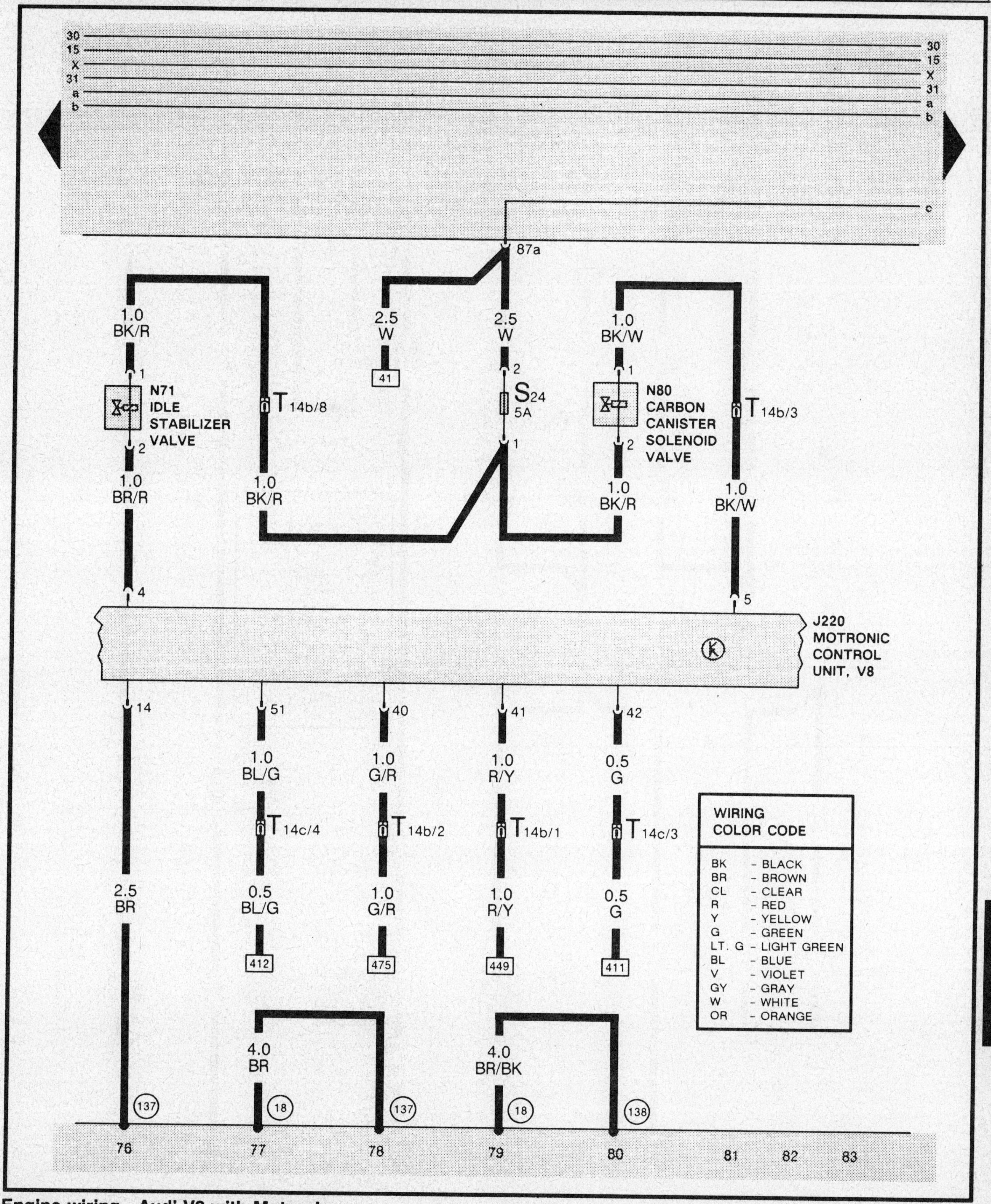

Engine wiring—Audi V8 with Motronic

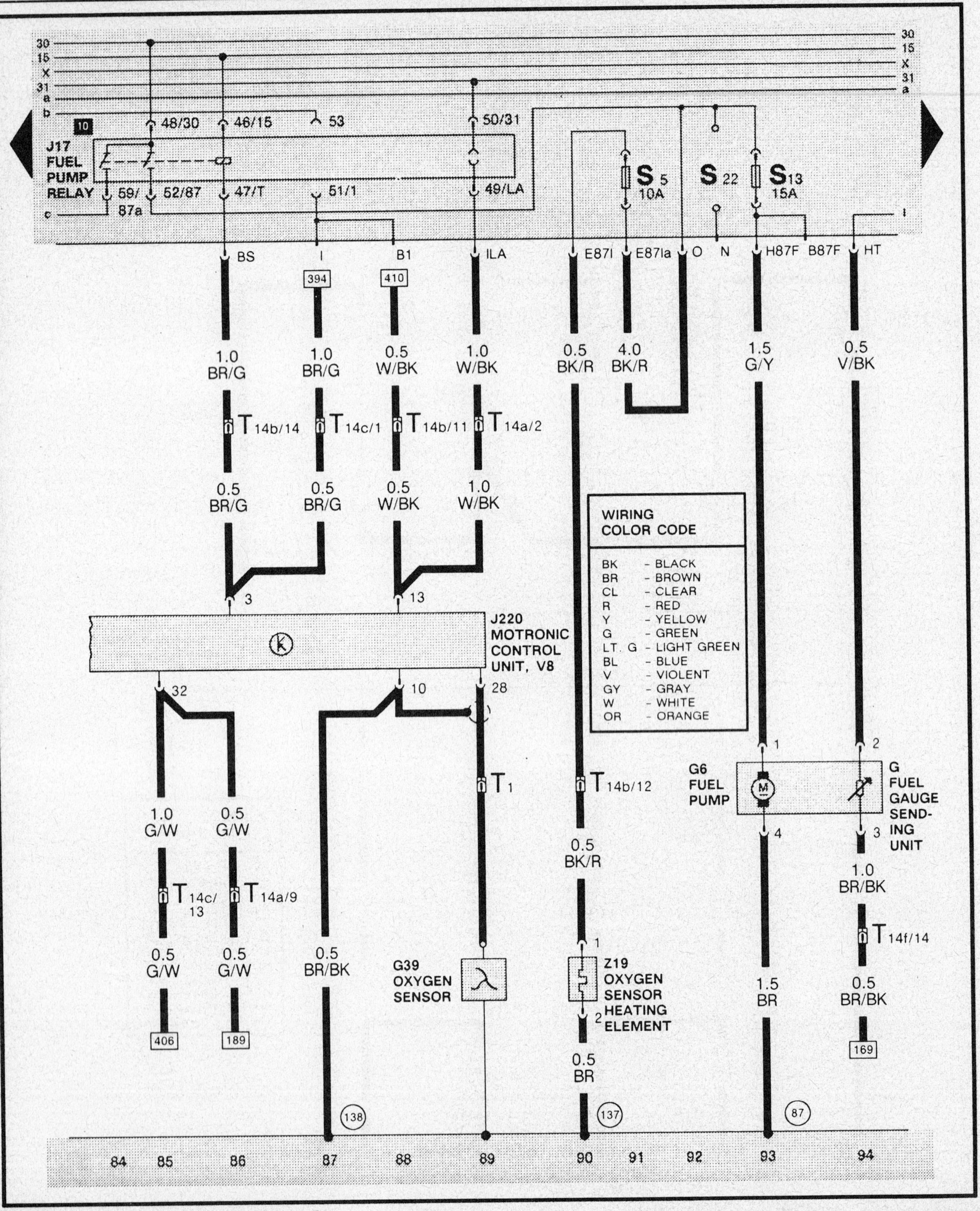

Engine wiring—Audi V8 with Motronic

Color coding

Green = Input signal
Blue = Output signal
Red = Power supply
Brown = Ground

Legend

F 60 Idle switch
F 96 Altitude sensor

G 4 Ignition reference sensor
G 6 Fuel pump
G 28 Engine speed sensor
G 39 Heated oxygen sensor
G 40 Hall sender
G 42 Intake air temperature sensor
G 61 Knock sensor I
G 62 Coolant temperature sensor
G 66 Knock sensor II
G 69 Throttle valve potentiometer
G 70 Hot wire (air mass sensor)

J 17 Fuel pump relay
J 220 Motronic control unit

N 30 Fuel injector, cyl. #1
N 31 Fuel injector, cyl. #2
N 32 Fuel injector, cyl. #3
N 33 Fuel injector, cyl. #4
N 83 Fuel injector, cyl. #5
N 70 Power stage (ignition coil)
N 71 Idle stabilizer valve
N 75 Wastegate frequency valve
N 80 Carbon canister frequency valve

P Spark plug connectors
Q Spark plugs

S 13 Fuse, fuel pump
S 24 Fuse, engine electronics
S 27 Fuse, engine control (Motronic)
S 28 Fuse, engine control (Motronic)
S 29 Fuse, heating element for oxygen sensor

T 6m Coding connector

Z 19 Heating element for oxygen sensor

137 Ground connection at engine block
138 Ground connection at engine block

Color code and legend decoder

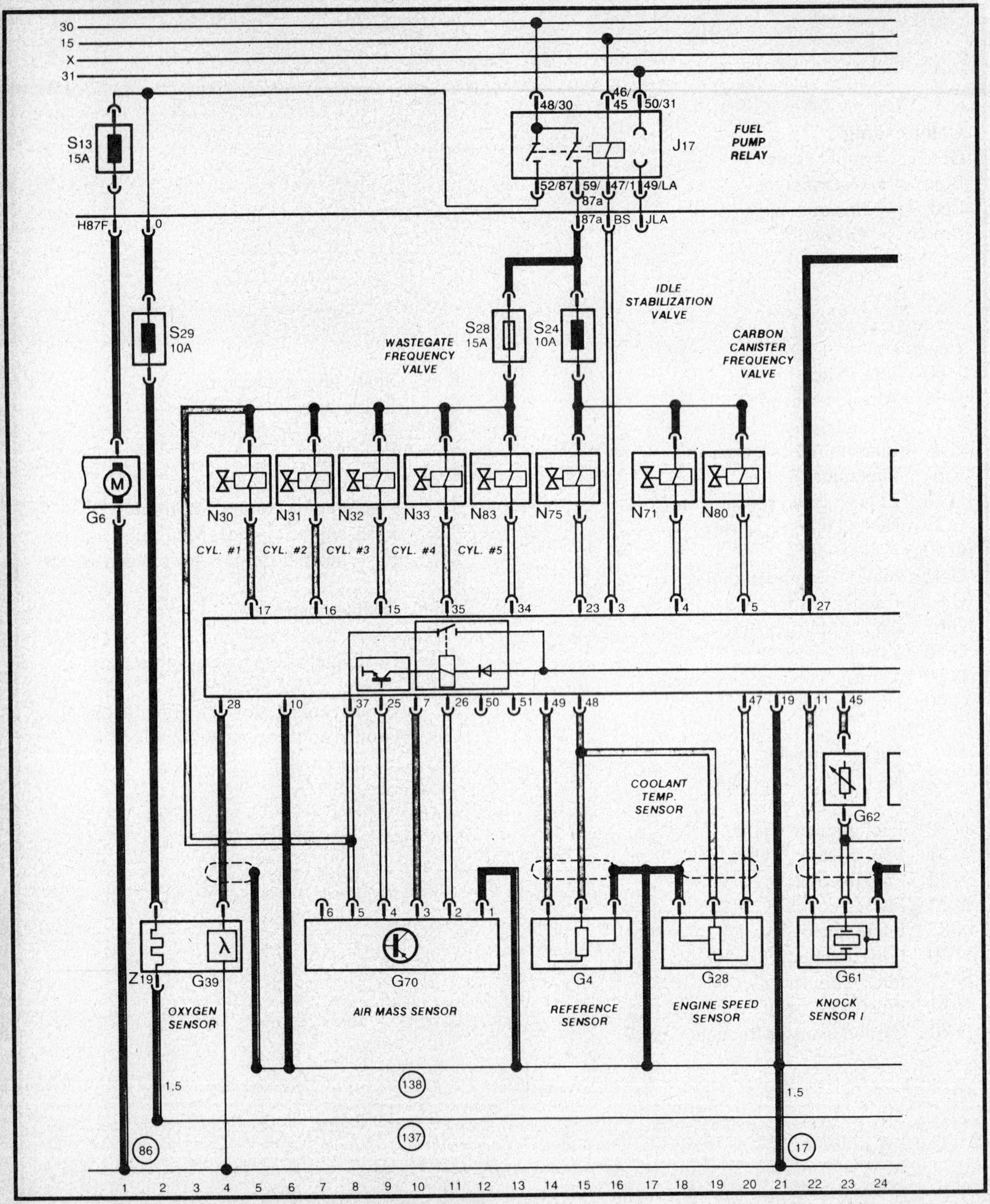

Engine wiring—Audi S4 and 200 Quattro with Motronic

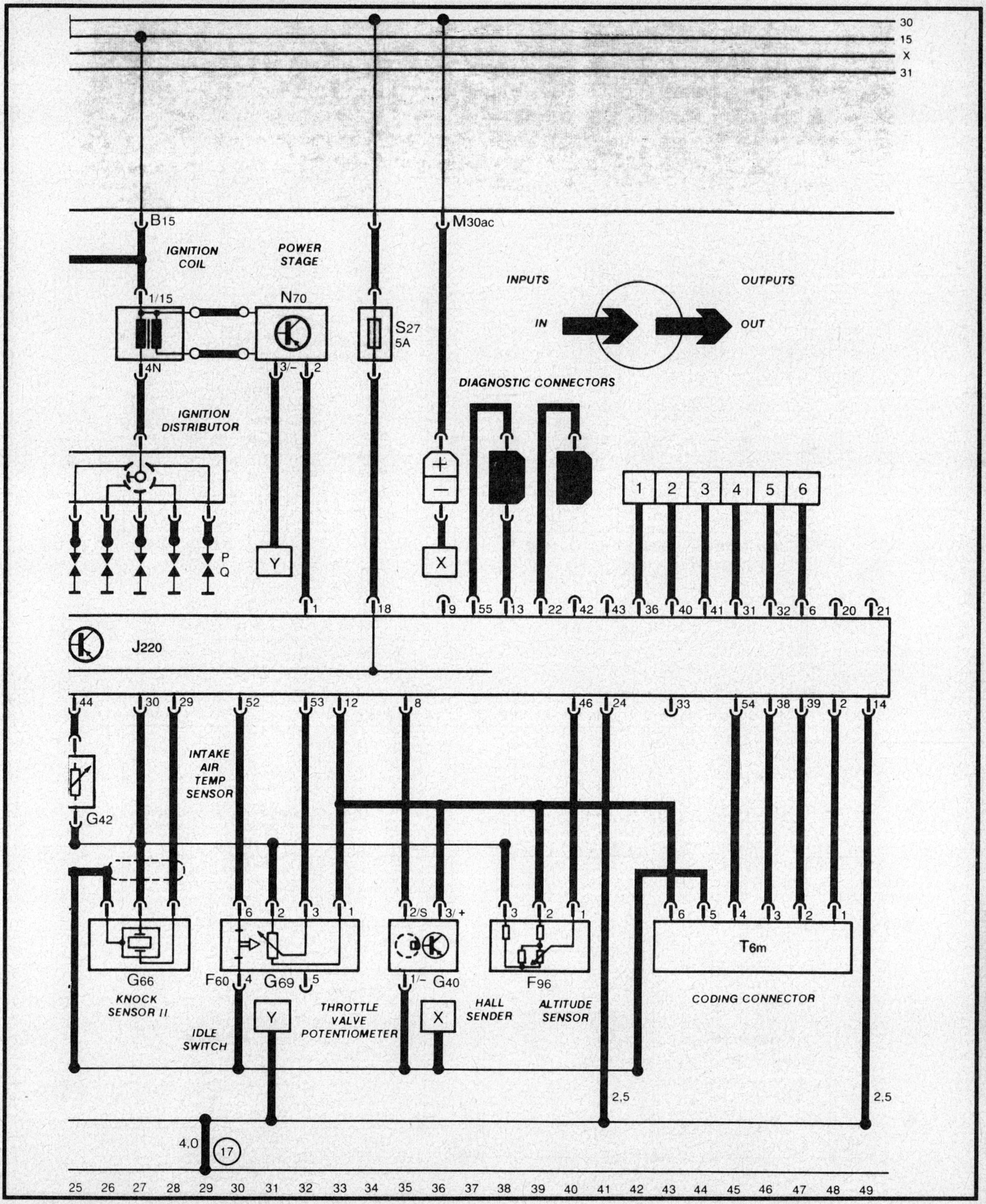

Engine wiring—Audi S4 and 200 Quattro with Motronic

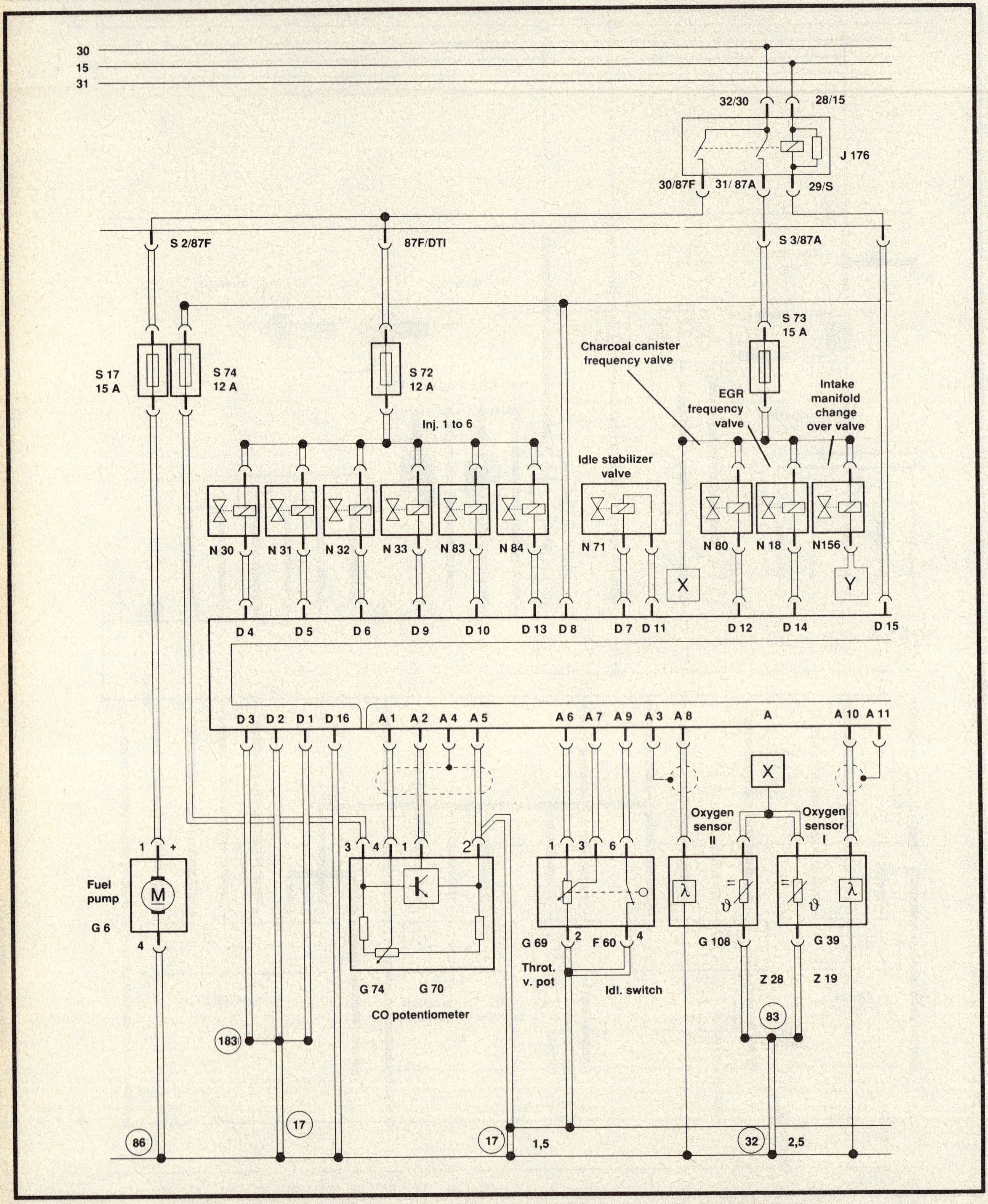

Engine Wiring—Audi 100, 100S, 100CS, 100CS Quattro and 100CS Quattro Wagon with MPI

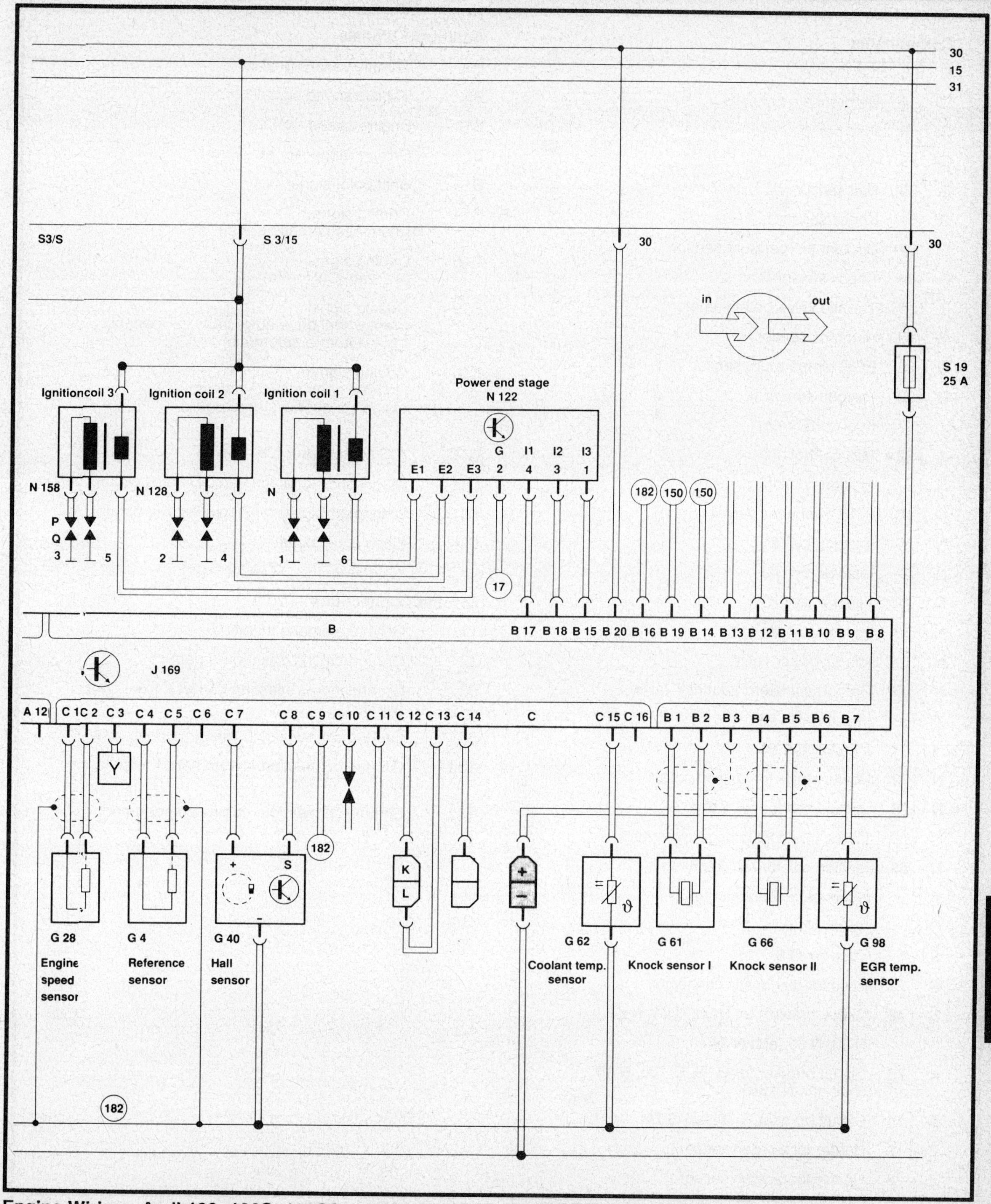

Engine Wiring—Audi 100, 100S, 100CS, 100CS Quattro and 100CS Quattro Wagon with MPI

Components

F	60	Idle switch
G	4	Reference point sensor
G	6	Fuel pump
G	28	Oxygen sensor I
G	39	Hall sensor
G	61	Knock sensor I
G	62	Coolant temperature sensor
G	66	Knock sensor II
G	69	Throttle valve potentiometer
G	70	Air mass sensor
G	98	EGR temperature sensor
G	108	Oxygen sensor II
J	17	Fuel pump relay
J	192	MPI control unit
N		Ignition coil for cyl 1 & 6
N	18	EGR frequency valve
N	30	Injector, cyl. #1
N	31	Injector, cyl. #2
N	32	Injector, cyl. #3
N	33	Injector, cyl. #4
N	71	Idle stabilizer valve
N	80	Carbon canister frequency valve
N	83	Injector, cyl. #5
N	84	Injector, cyl. #6
N	122	Power stage for ignition coils
N	128	Ignition coil for cyl. 2 & 4
N	156	Intake manifold change-over valve
N	158	Ignition coil for cyl. 3 & 5
P		Spark plug connector
Q		Spark plug
S	17	Fuse for G 6
S	19	Fuse for diagnostic connector
S	72	Circuit breaker for N 30, N 31, N 32, N 33, N 83, and N 84
S	73	Circuit breaker for G 39, G 108, N 18, N 80 and N 156
S	74	Circuit breaker G 70 and G 74
Z	19	Heater for oxygen sensor I
Z	28	Heater for oxygen sensor II

Additional Signals

B8	Throttle valve signal
B9	Vehicle speed signal
B10	Engine speed signal
B12	Driving range signal
B13	Shift point signal
B14	Coding signal (manual/automatic trans.)
B16	Coding signal (49 state/California)
B19	Coding signal (front wheel drive auto. trans. or manual trans./Quattro automatic)
C9	Coding signal (oxygen sensor systems/non oxygen sensor systems)
C10	A/C compressor "ON" signal
C12	Diagnostic signal for Rapid Data (K circuit)
C13	Diagnostic signal for Rapid Data (L circuit)
C14	Flash code signal

Ground Connections

17	Ground at intake manifold
32	Ground behind dashboard (left)
83	Ground in fuel injection harness, front right (welded)
86	Ground connection in wiring harness, rear
150	Ground connection in automatic transmission harness (AG4)
182	Ground connection, engine compartment

Legend decoder

AUDI CONTINUOUS INJECTION SYSTEM (CIS-E)

General Information

Audi uses 3 forms of the CIS system; CIS, CIS-E III and CIS Motronic. The CIS E III and the CIS Motronic systems are functional similar and use many of the same components. They will also be considered similar enough to be referred to interchangebly. Both of these ECU controlled versions use a control pressure regulator mounted on the fuel distributor. The CIS system used on the MC 2.22L turbocharged engine uses an externally mounted, thermal controlled warm-up regulator to adjust the control pressure.

The CIS-E system incorporates 2 control units. An Ignition Control Unit (ICU) and a Fuel Injection Control Unit (FICU). The FICU may also be referred to, generically, as the Engine Control Unit (ECU).

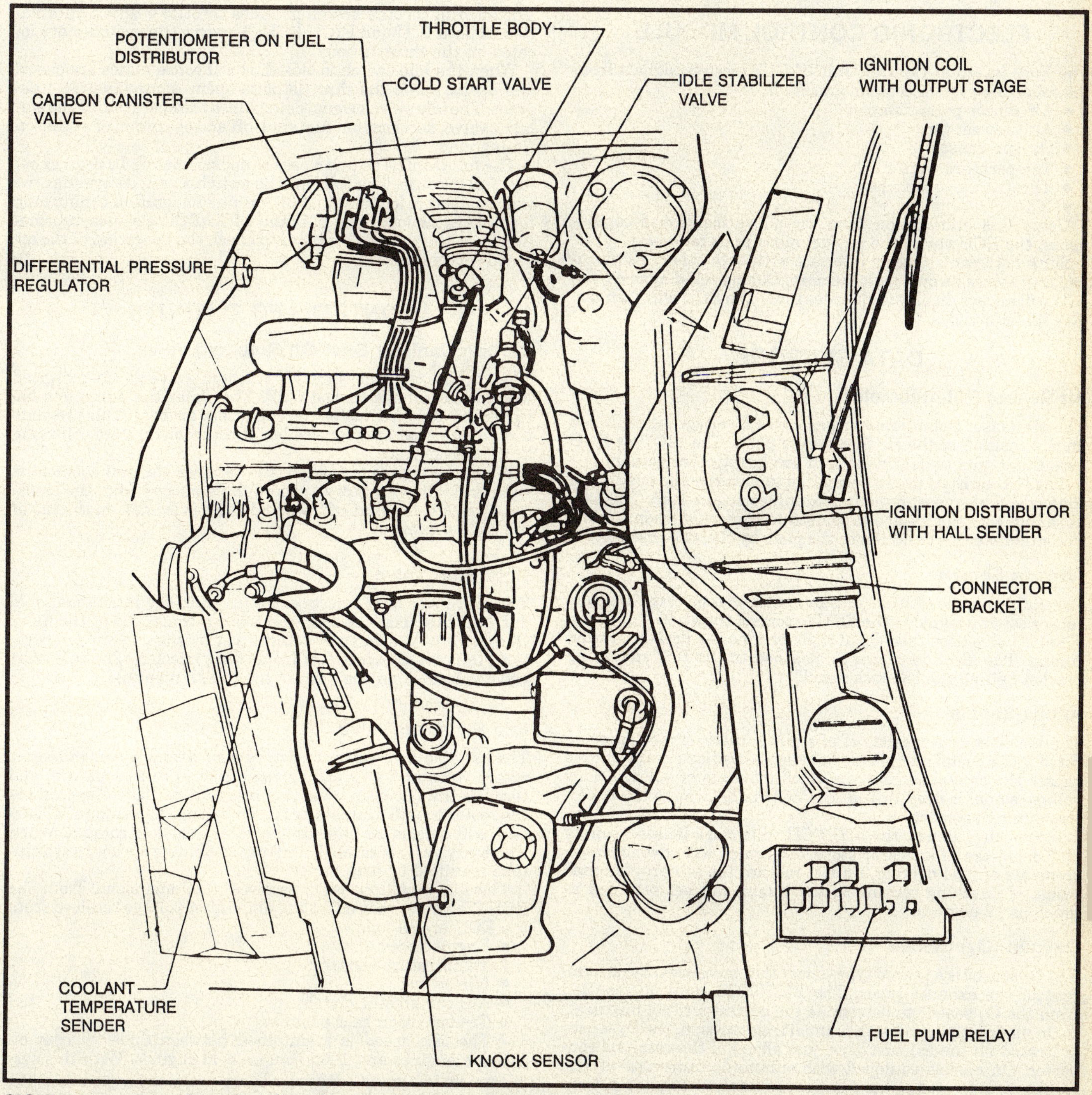

CIS-E components location (5 cylinder shown, 4 cylinder similar)

The CIS-E system has self-diagnosis and troubleshooting capabilities. Input and output signals from various sensors, switches and signaling devices are constantly monitored for faults. These faults are stored in the control unit memory. Faults can be displayed by a flashing 4 digit code sequence from an LED light.

The CIS Motronic system used on Audi 80 and 90 models use a single Electronic Control Unit (ECU), located behind the A/C evaporator assembly. The ECU controls the fuel delivery, ignition system and operation of the emission control components. The CIS Motronic system also incorporates self-diagnostic capabilities.

ELECTRONIC CONTROL MODULE

The Fuel Injection Control Unit (FICU) receives signals from the following system components:

- Air sensor potentiometer
- Altitude sensor
- Oxygen sensor
- Temperature sensor
- Throttle valve switches
- ICU

Using this information from the above mentioned components, the ECU varies the air/fuel mixture as necessary.

Since both control units interface with each other, each control unit stores its own fault memory. During self-diagnosis, the ICU will always display fault codes first. The ECU will then display its fault codes.

DATA SENSORS

Air Sensor Potentiometer

The air sensor potentiometer generates a voltage signal based on the position of the air flow sensor plate. The FICU or ECU uses this signal to determine cold acceleration enrichment.

The ICU or ECU uses information to measure the position of the plate. This measurement determines engine load. This value, along with engine speed, is used to control ignition timing. Potentiometer is connected to the plate by the air sensor lever.

Altitude Sensor

Depending on the altitude or barometric pressure, the altitude sensor sends a signal to the ECU to correct the air fuel mixture. The ICU also uses this signal to determine the proper ignition timing. The altitude sensor is attached to the ICU mounting bracket behind the left kick panel.

Knock Sensor

The knock sensor consists of of a piezoelectric crystal incorporated into a metal and plastic housing. Vibrations in the engine causes the crystal to generate small amounts of voltage. This voltage signal is used by the ICU/ECU to determine the necessary timing retardation.

If knocking occurs, the ICU/ECU will retard ignition timing by 3.4 degrees. If knocking stops, the timing will advance in increments of 0.54 degrees, until it reaches the pre-programmed value. If knocking continues, the timing can bet retarded as much as 12 degrees.

Oxygen (O_2) Sensor

The O_2 sensor is a heated type sensor that measures oxygen content of the exhaust gases. The FICU/ECU uses information from the O_2 sensor to determine the correct air/fuel mixture.

On models equipped with manual transmission, the O_2 sensor is located on the exhaust pipe, just ahead of the catalytic converter. On models equipped with automatic transmission, the O_2 sensor is located at the front of the catalytic converter. Connections for the O_2 sensor and sensor heating element are located on a bracket next to the distributor.

Temperature Sensor

The temperature sensor consists of 2 variable resistors. One resistor determines the coolant temperature, then sends this information to the ICU/ECU. The other resistor, sends the same information to the FICU/ECU.

These signals are used to correct the air/fuel mixture, depending on the coolant temperature. Both resistor have the same resistance value. The temperature sensor is located in the cylinder head coolant outlet flange.

Idle/Full Throttle Switches

The idle/full throttle switch provides throttle position information to both ICU and FICU/ECU. The throttle switches are located on the throttle body.

When the idle switch is closed, the throttle plates are closed and opens, when the throttle plate opens approximately 1 degrees. The idle switch signal determines the operation of the stabilize valve, deceleration fuel shut-off and operation of a ignition timing map.

The full throttle switch closes at approximately 10 degrees before full throttle. The full throttle switch signal determines the full throttle enrichment and operation of a ignition timing map for full engine load. The ICU and FICU/ECU also use information from the throttle switches to check the operation of the air sensor potentiometer.

CONTROLLED OUTPUTS

Carbon Canister Shut-Off Solenoid Valve

The carbon canister shut-off solenoid valve is de-energized when the engine is turned OFF. The closes the purge line between the carbon canister and the intake air boot. This prevents fuel vapors, that may cause a hot start problem, from collecting in the intake tract.

When the engine is started, the canister shut-off solenoid is energized, allowing stored vapors to be purged from the canister. The canister shut-off solenoid valve is located on the top of the carbon canister.

Cold Start Valve

The cold start valve injects extra fuel into the intake manifold during cold starts. The cold start valve is controlled by the ECU. Operation of the valve is determined by the coolant temperature, to prevent excess fuel from being injected. The cold start valve is located on the back of the intake manifold.

Idle Stabilizer Valve

The idle stabilizer valve consists of a small single-wound motor, rotary valve and a return spring attached to a armature, and controls the idle speed. Varying amounts of DC voltage causes the motor to pull against the return spring. The amount of current will determine how far the rotary valve is rotated. When the rotary valve is rotated, air from the intake air boot to the intake manifold by-pass varies.

The idle stabilizer valve receives a voltage signal from the ECU. Variables that effect the amount of voltage received from the ECU are as follows:

- Engine speed
- Temperature sensor
- Idle switch
- A/C compressor clutch
- Transmission gear position

The idle speed is maintained between 650–790 rpm on warm engines and 1000 rpm on cold engines. With this system, no idle speed adjustments are necessary or possible. The idle stabilizer valve is located between the intake air boot and the intake manifold.

Differential Pressure Regulator

The differential pressure regulator controls the fuel flow in the lower chamber of the fuel distributor. This helps to determine the fuel mixture. The regulator designates 0 milli-amps (mA)as a reference point. When the vehicle is at normal operating temperature and all engine controls are operating properly, the regulator operates with current from the ECU between +10 mA (rich) to −10 mA (lean) ranges.

When the starter is operated, current to the differential pressure regulator is increased to enrich the fuel mixture. This will occur whenever the engine is started, and the amount of increase will be regulated by the coolant temperature. When the temperature is extremely cold, the current can be as high as 140 mA.

During cold acceleration, current can raise to approximately 6 mA to enrich the fuel mixture. Enrichment is determined by the engine speed and signals from the temperature sensor.

During full load operation, the full throttle switch closes, sending a signal to the ICU. The ICU sends a signal approximately 3 mA greater than the signal that is currently present. This signal will vary, depending on the engine speed and altitude.

When the vehicle is decelerated, fuel to the injectors is shutoff by reversing the current to −50 mA to −60 mA. The engine speed, at which this reversing will occur, is regulated by the coolant temperature. Current will also be reversed when engine speed reaches 6600 rpm. The differential pressure regulator is attached to the fuel distributor.

Diagnosis and Testing

SERVICE PRECAUTIONS

When testing the CIS-Motronic system, note the following precautions to avoid personal injury and/or damage to the system:

- Do not disconnect the battery when the engine is running.
- Turn ignition switch **OFF**, when connecting or disconnecting tester leads and/or ignition wires.
- Disconnect the battery and the CIS-Motronic control unit when arc, spot or electrical welding.
- If components are heated above 175°F (80°C) from a paint dryer or steam cleaning, wait for components to cool before starting the engine.
- When the high tension ignition wire disconnected and not grounded, do not crank the engine.
- When the coil wire (terminal 4) is disconnected from the distributor, always ground the coil wire using a jumper wire.

READING CODES

The self-diagnostic function of the CIS-Motronic system is used to detect component malfunctions which may occur during vehicle operation. The ECU will store these malfunctions for self-diagnosis. The ECU consists of permanent fault code memory (50 states), as well as a temporary fault code memory (California only). When the ECU detects a temporary fault, the Fault Indicator Light (FIL) on the instrument panel will flash. The FIL will continue to flash until the ignition switch is turned **OFF**. The ECU may clear the fault on the next engine start, and the FIL will not flash unless the fault is detected again. The vehicle must be driven for at least 5 minutes to reset the fault. When the ECU detects a permanent fault, the FIL will flash, but the fault will remain in the ECU memory until erased.

Without VAG 1551

The diagnostic test connectors have been installed in the driver's side footwell. Retrieving fault codes can be accomplished as follows:

1. Connect the positive terminal of the US 1115 LED tester or equivalent, to the positive terminal in connector (A).
2. Connect the negative terminal of LED tester to the terminal in connector (B).
3. Connect a jumper wire to the negative terminal in connector (A), and touch the other end to the terminal in connector (B) for approximately 4 seconds. Fault codes should now display on the LED tester.
4. To retrieve the next fault code, touch the end of the jumper wire to the terminal in connector (B) again for a minimum of 4 seconds.
5. After faults have been corrected, erase the fault memory as outlined.

With VAG 1551

1. Make sure the ignition switch is **OFF** and that all fuses are good. Make sure all ground connections under the hood are good, especially those for the control unit.
2. Connect the diagnostic tester to the black and white connectors in the drivers footwell.
3. Turn the tester **ON** to make sure it is receiving power. The screen will display 2 menu options; Rapid Data Transfer and Blink Code Output. Turn the ignition switch **ON**.
4. Select Rapid Data Transfer and Address Word 01, then press the Q button to enter the selection. The tester will display a control unit part number, the system it controls and an application (country) code.
   a. If the information is displayed and is correct, press the Run key to proceed: "Select function XX" will be displayed.
   b. If "Control unit does not answer" is displayed, use the Help key to display a list of possible causes. When the problem is repaired, return to Step 1 and start again.
5. When Function 02 is selected, the control unit will report fault codes to the tester.
6. When all codes have been reported, select Function 06 to exit the fault code memory without erasing the codes.

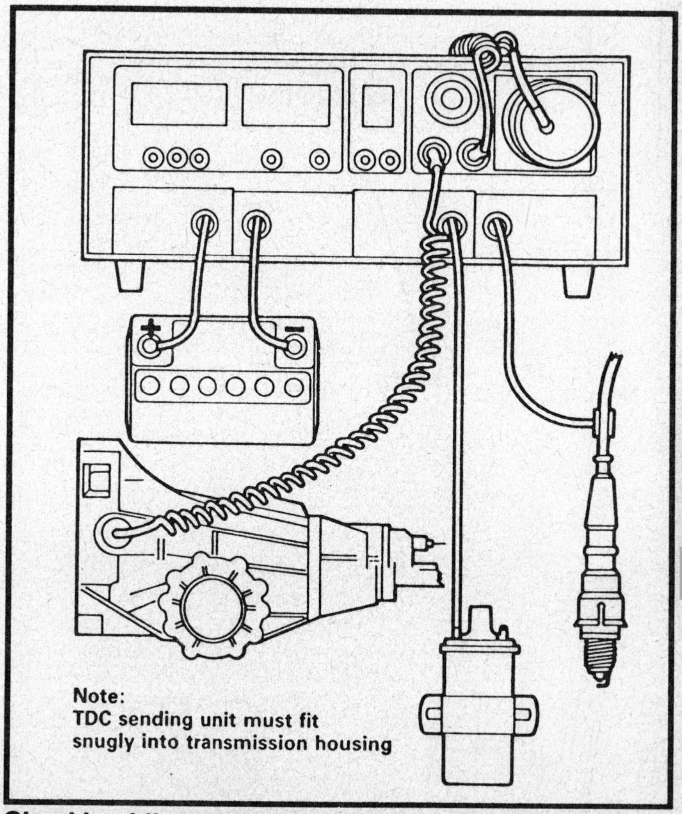

Checking idle stabilization using engine tester VW 1367

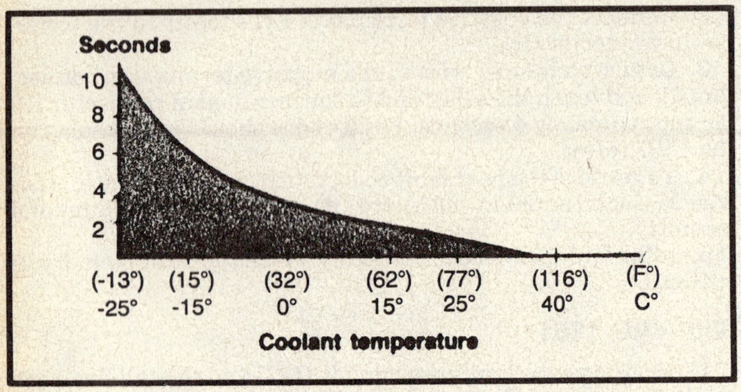

Cold start valve operation graph

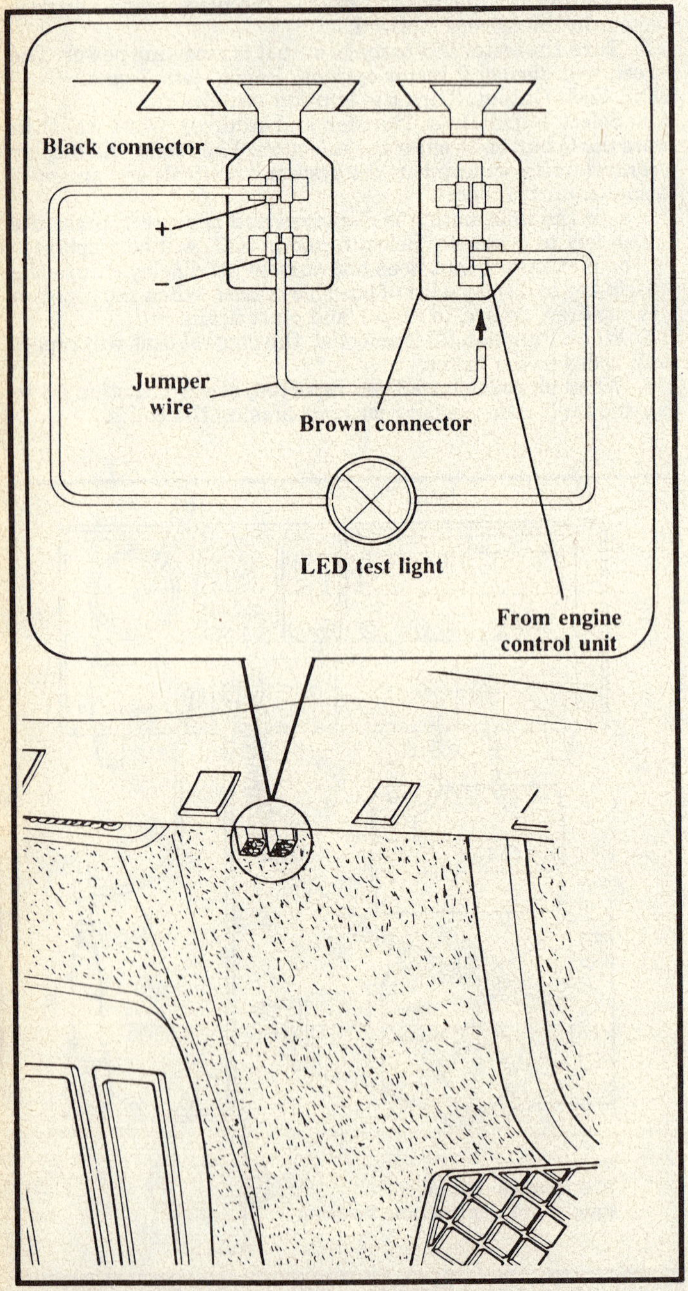

Retrieving fault codes, test connector location

Flash code	Function
1 1 1 1 *	Control unit
2 1 1 3 *	No speed signal from Hall sensor, or, air flow sensor plate not moving freely
2 1 2 1	Idle switch
2 1 2 3 *	Full throttle switch
2 1 4 1 *	Knock control is at control limit
2 1 4 2 *	Knock sensor signal
2 2 3 1	Adjustment limits of idle stabilization exceeded
2 2 3 2 *	Air flow sensor potentiometer
2 3 1 2 *	Coolant temperature sensor
2 3 4 1 *	Oxygen sensor control at control limit
2 3 4 2 *	Oxygen sensor control
2 3 4 3	Fuel mixture adjustment too lean
2 3 4 4	Fuel mixture adjustment too rich
4 4 3 1	Idle stabilizer valve
4 4 4 4	No faults recognized
0 0 0 0	End of fault output display sequence

*Stored in permanent memory unit

Note
Codes not marked with an asterisk are sent to temporary fault storage only.

Fault code chart

CLEARING CODES

Without VAG 1551

1. Ensure that LED tester is connected and ignition switch is in the **OFF** position.
2. Install a spare fuse in the top of the fuel pump relay.
3. Turn ignition switch in the **ON** position. LED tester must light.
4. Remove the fuse after approximately 4 seconds. LED tester must go OFF then begin flashing fault code 0000.
5. Install fuse again for approximately 10 seconds, then remove. LED tester should light and stay on. Fault memory has been erased.

With VAG 1551

1. After all fault codes have been retrieved, select Function 05 and press the Q button to enter the selection. The memory will be erased only if all fault codes have been retrieved.
2. Test drive the vehicle for at least 10 minutes, including at least 1 full throttle application above 3000 rpm.
3. Check the fault code memory again to make sure all faults have been repaired.

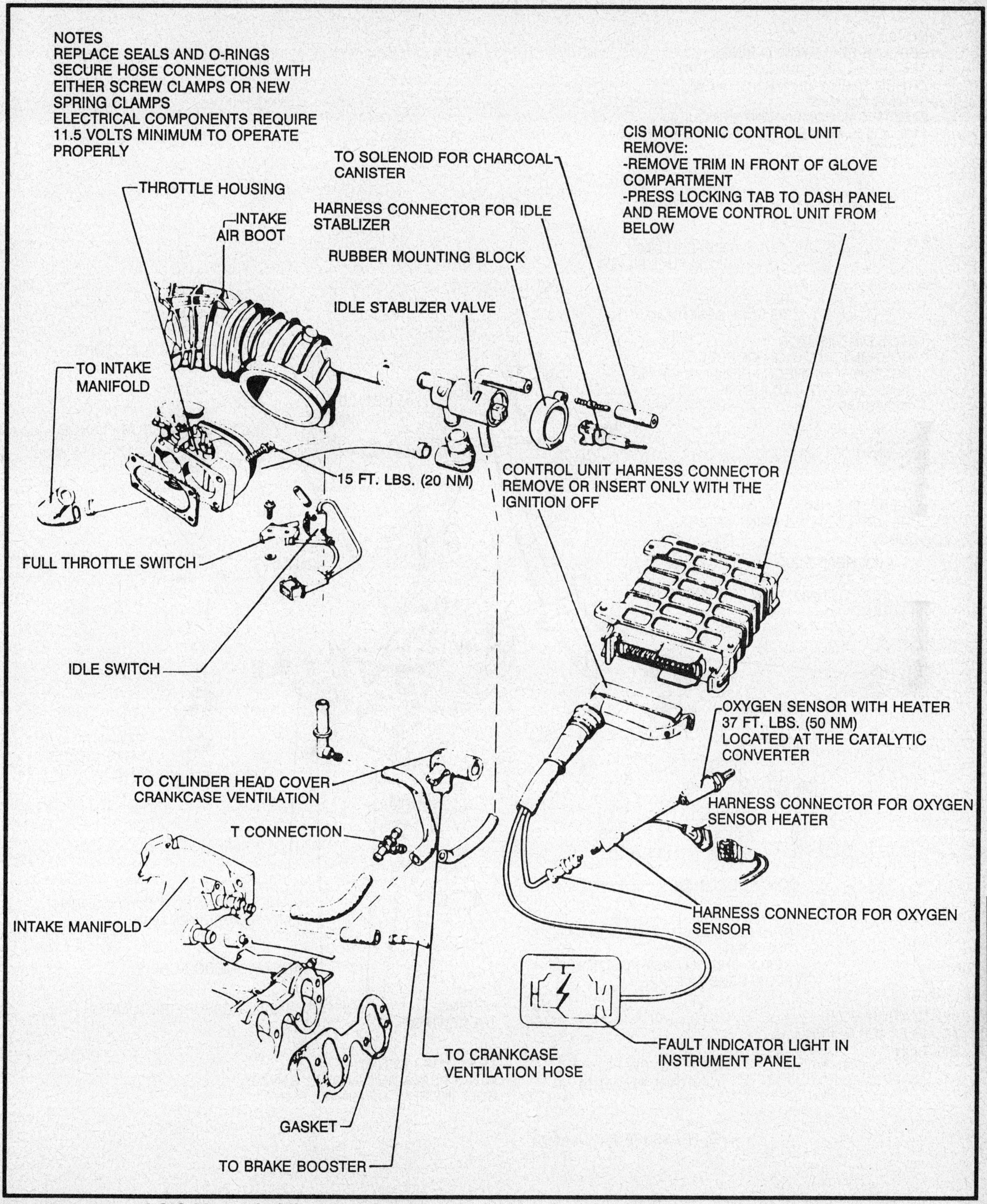

Exploded view of CIS-Motronic idle control system

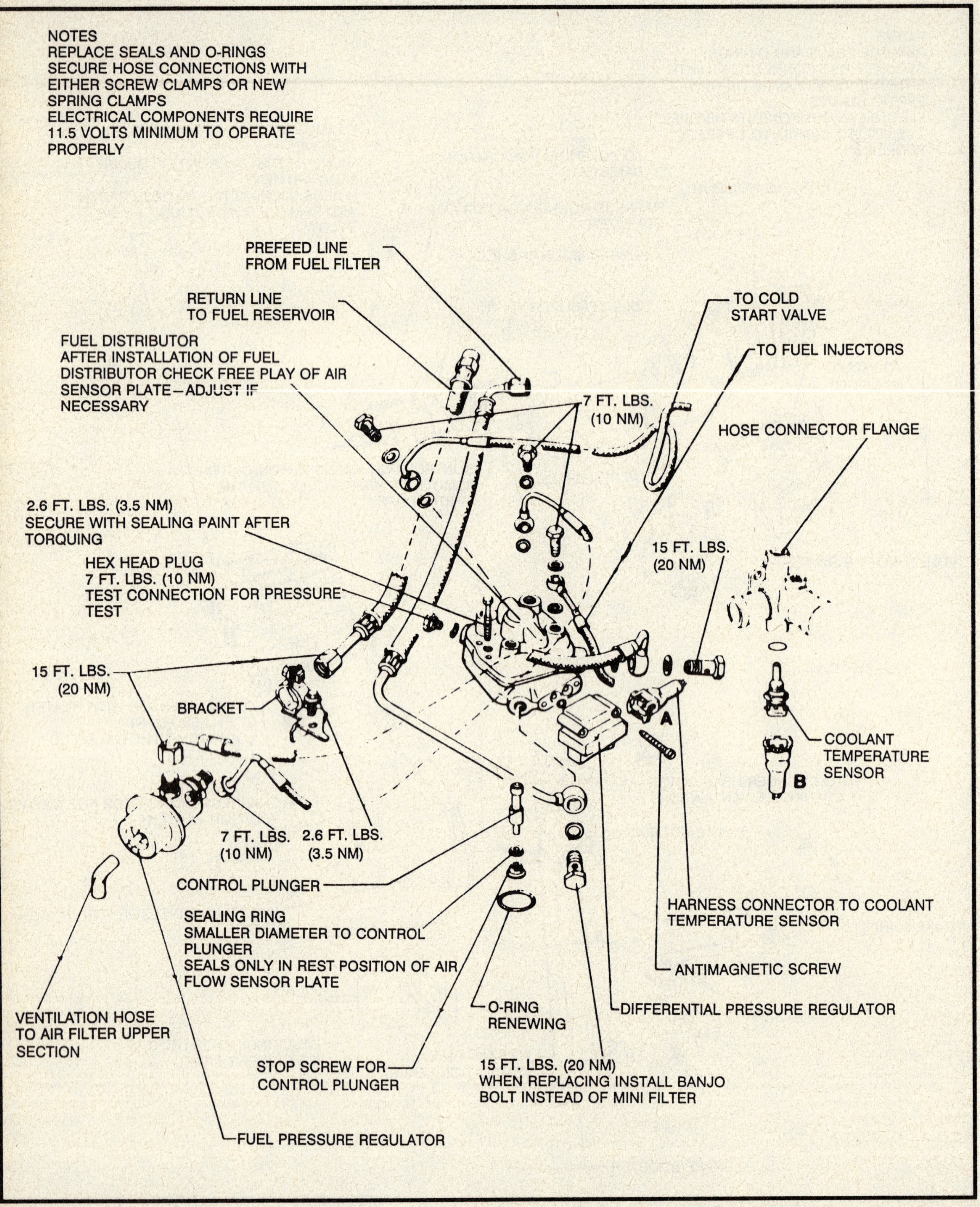

Exploded view of CIS-Motronic fuel system

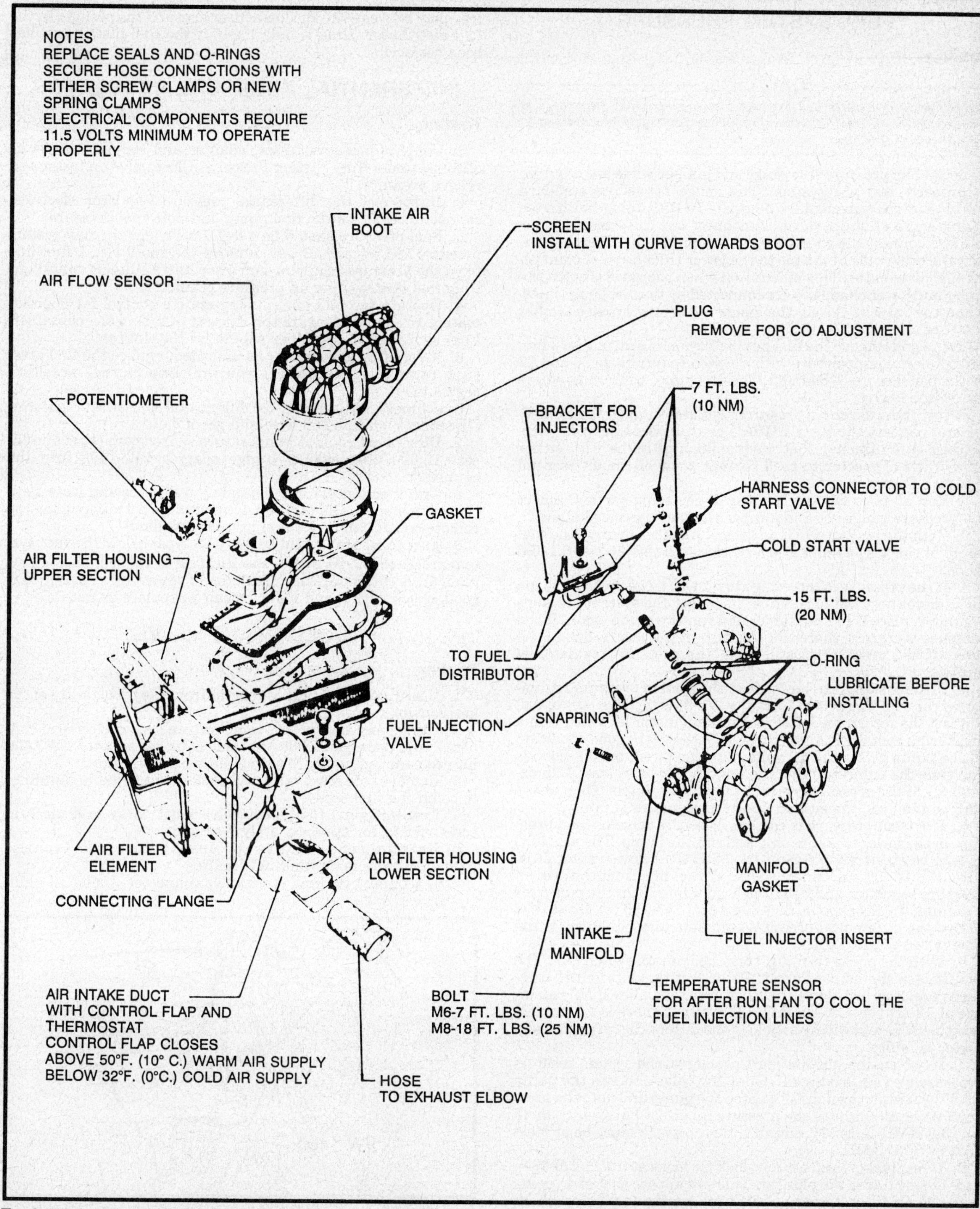

Exploded view of CIS-Motronic air intake system

FUEL SYSTEM PRESSURE

Testing

CAUTION

The following procedure will produce fuel vapors. Make sure there is proper ventilation and take the appropriate fire safety precautions to prevent personal injury.

A special fuel pressure test gauge and jumper wires are required to properly test the system. The gauge set is the standard VW1318 or an equivalent 0–140 psi (0–10 BAR) gauge with a tee fitting and a shut-off valve. Disconnect the cold start injector line and connect the shut-off valve to the fuel distributor. Connect the other side of the tee to the lower chamber test point on the fuel distributor. This means that when the valve is open, the upper and lower chambers are connected by the test gauge lines. When the valve is closed, the gauge reads only lower chamber pressure.

Locate and remove the fuel pump relay on the main fuse panel and connect a jumper wire with a switch to terminals 30 and 87 on the panel or use US4480/2. The fuel pump will run any time the switch is **ON**.

To test the differential pressure regulator, unplug the connector and connect the VAG 1315A/1 or equivalent adapter wire. Connect a multimeter that reads milliamps inline with either wire of the connector to read current draw of the differential pressure regulator.

1. Run all tests before repairing or replacing parts. Connect the pressure gauge set and jumper wires as described above.
2. With the ignition **OFF** and the valve open, operate the fuel pump to check system pressure; it should be 88–95 psi (6.1–6.6 BAR).
3. If the pressure is correct, go to Step 5. If the pressure is too high, disconnect the fuel return line after the system pressure regulator, place it in a container and run the pump again. If the pressure is correct, there is a blockage in the return line to the tank. If the pressure is still high, the pressure regulator is faulty.
4. If the system pressure is low, the system pressure regulator or the fuel pump may be faulty. To flow test the pump, disconnect the return line after pressure regulator and hold it in a measuring container. Make sure the battery is in good condition so the pump gets a full 12 volts. Run the pump for 30 seconds and time the run carefully. The flow should be at least 12.5 oz. (420 cc). If the pressure is low but the flow is higher than about 22.5 oz (760 cc), the system pressure regulator is faulty.
5. If system pressure is correct, close the valve to read pressure in the lower chamber only and run the pump. The pressure should be about 4.5–7.5 psi (0.3–0.5 BAR) below system pressure. If it is not, disconnect the small fuel line from the system pressure regulator and hold it in a container. Run the pump for 1 minute; the flow should be 3.8–4.4 oz. (130–150 cc). If the flow is low, the differential pressure regulator may be faulty. If the flow is correct, procede.
6. With the pump running, turn the ignition switch **ON**. The multimeter should read about 100 mA at the differential pressure regulator. The pressure in the lower chamber should stabilize at 17–21 psi (1.2–1.5 BAR). If either of these is not correct and the ECU and wiring are okay, the differential pressure regulator is faulty.
7. If the engine has hot start problems, the system residual pressure may be leaking off. Open the valve and run the pump with the ignition switch **OFF**. Stop the pump and note the pressure. After 10 minutes the pressure must not be lower than 48 psi (3.3 BAR). After 20 minutes, the pressure must be at least 46.5 psi (3.2 BAR).
8. If residual pressure is low, look for leaks at the fuel distributor O-ring under the plunger. This will appear as fuel dripping onto the air filter. Other possibilities are the fuel pump check valve, a leaking injector, the fuel pressure regulator or improper free-play between the air flow sensor arm and the plunger in the fuel distributor. This is only likely if the fuel distributor has been replaced.

DIFFERENTIAL PRESSURE REGULATOR

Testing

1. Connect pressure tester, adapter and remote control as outlined under Fuel System Pressure. Close valve on tester and relieve pressure.
2. Disconnect the differential pressure regulator electrical connector. Operate the fuel pump and note fuel pressure.
3. Fuel pressure should be 4.3–7.3 psi below the fuel system pressure (89–94 psi). If not, remove the small No. 1 fuel line from the pressure regulator and place into a suitable container. Plug fuel line opening on pressure regulator.
4. Operate the fuel pump using remote control for approximately 1 minute. Measure the amount of fuel in the container. Fuel measured in container should be 130–150cc.
5. If specifications are not as indicated, replace the fuel pressure regulator. Reconnect the fuel line to fuel pressure regulator.
6. Connect multimeter to differential pressure regulator. Disconnect ignition coil wire and ground.
7. Disconnect coolant temperature sensor connector and insert 15,000 ohm side of bridge adapter VW 1490 into the connector.
8. Crank engine for approximately 5 seconds and leave ignition switch in the **ON** position. Operate the fuel pump using the remote control and note fuel pressure reading.
9. Fuel pressure should be 18.9–23.2 psi below the fuel system pressure (89–94 psi). If specification is not as indicated, replace differential pressure regulator. If specification obtained is good, check wiring and ECU. Repair or replace as necessary.

COLD START VALVE

Testing

1. Disconnect electrical connector from the power stage of the ignition coil.
2. Disconnect the coolant temperature sensor electrical connector and install a 15,000 ohm side of bridge adapter VW 1490 into coolant temperature sensor harness connector.
3. Remove cold start valve and place into a suitable container to catch the fuel spray.
4. Crank engine. Cold start valve must spray in a uniform cone pattern for approximately 7 seconds.
5. Dry off tip of cold start valve. Inspect valve after 1 minute to ensure valve is dry and not leaking.
6. Disconnect the cold start valve connector. Using an ohm-

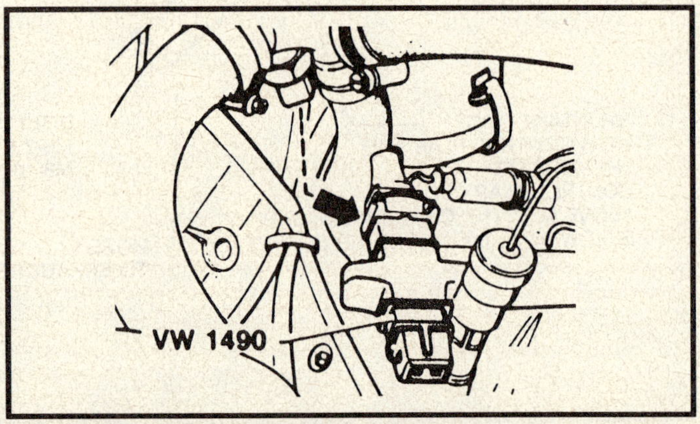

Installing bridge adapter for cold start valve testing

meter, measure resistance across terminal of valve. Ohmmeter should read 10 ohms.

7. If specifications are not as indicated, replace cold start valve. Connect LED tester US 1115 or equivalent to the cold start valve connector. Crank engine. LED tester should light for approximately 7 seconds. If not, replace ECU.

OXYGEN SENSOR

Testing

1. Ensure engine is at operating temperature. Check that the idle speed, voltage to O_2 sensor and coolant temperature sensor are all correct.
2. Connect multimeter US 1119 and adapter 1315A/1 or equivalent, to the pressure regulator.
3. Start engine and let idle for approximately 2 minutes. Note control voltage at differential pressure regulator.
4. Remove a manifold vacuum hose to create a lean mixture. Control voltage must increase. If not, disconnect O_2 sensor harness connector and ground for 20 seconds. If control voltage changes, replace O_2 sensor.
5. If control voltage does not change, check the wire harness for shorts or opens to ground or power. If no shorts are found, replace ECU.

AIR SENSOR POTENTIOMETER

Testing

1. Start and allow engine to reach normal operating temperature.
2. Connect test adapter VW 1501 between potentiometer and harness.
3. Connect multimeter US 1119 or equivalent, to test adapter connectors 1 and 3. Switch multimeter to DC volts and turn the ignition switch **ON**. Multimeter should read 4.35–5.35 volts.
4. Turn ignition switch **OFF** and connect multimeter to test connectors 2 and 3. Start engine and let idle. Multimeter should read 0.4–0.9 volts.

NOTE: The engine cooling fan must not come ON during this test.

5. If specification obtained is slightly outside of tolerance, correct by adjusting the potentiometer using the trim screw. If correct specification is still not obtained, replace potentiometer.

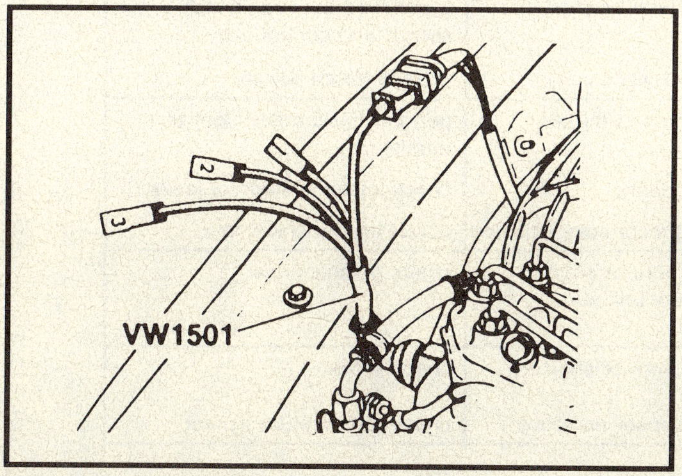

Installing test adapter

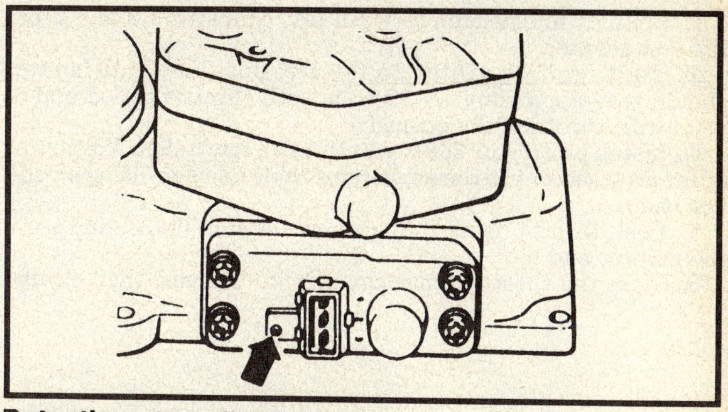

Potentiometer adjusting screw

IDLE STABILIZER VALVE

Testing

1. Close the full throttle switch. Idle stabilizer valve should operate (clicking sound) when the switch is closed.
2. If no sound is heard, disconnect the idle stabilizer valve electrical connector.
3. Connect LED tester US 1115 or equivalent, between terminals 1 and 2 using a jumper wire. If tester lights, replace idle stabilizer valve.
4. If LED tester does not light, connect a multimeter between connector terminal 2 and ground. Multimeter should read 12 volts.
5. If specifications are not as indicated, repair break in wiring. If wiring is good, replace ECU.

COOLANT TEMPERATURE SENSOR (CTS)

Testing

1. Start and allow engine to reach normal operating temperature.
2. Disconnect the CTS electrical connector. Connect voltmeter between CTS terminal 1 and ground.
3. Turn the ignition switch to the **ON** position. Voltmeter should read 4.5–5.5 volts.
4. If no voltage exists, repair break in wiring. If voltage exists and no break in wiring is found, replace the CTS.

IDLE AND FULL THROTTLE SWITCHES

Testing and Adjusting

IDLE THROTTLE SWITCH

1. Connect test adapter VW 1501 between throttle switch and throttle switch harness connector.
2. Connect multimeter to test connectors 1 and 2 on the test adapter. Set multimeter to DC voltage.
3. Turn ignition switch to the **ON** position. Multimeter should read 0 volts at throttle closed and approximately 12 volts with throttle slightly opened
4. Open throttle and close slowly. Using a feeler gauge, check the change-over point of the idle switch between the switch roller and the throttle lever stop.
5. Change over point gap should be 0.0006–0.020 in. (0.15–0.5mm) before idle stop.
6. If specifications are not as indicated, adjust point gap by moving idle switch. Turn ignition switch **OFF**.

FULL THROTTLE SWITCH

1. Connect test adapter VW 1501 between throttle switch and throttle switch harness connector.

2. Connect multimeter between test connectors 2 and 3 on the test adapter.
3. Turn ignition switch to the **ON** position. Multimeter should read approximately 12 volts with throttle closed and 0 volts with throttle fully opened
4. Install protractor 3084 onto throttle shaft, then the pointer for protractor onto the accelerator cable adjuster using a rubber band.
5. Push throttle lever fully opened (at stop limit), and zero the pointer and disc.
6. Close the throttle approximately 20 degrees, then slowly move toward the full throttle position until full throttle switch engages (0 volts on multimeter). Protractor should read 10±2 degrees before contacting the full throttle stop limit.
7. If specifications are not as indicated, adjust by moving the the full throttle switch until the correct specifications are obtained.

NOTE: The roller on the throttle lever must contact the diagonal portion of the lever from the full throttle switch.

CIS-MOTRONIC TROUBLESHOOTING CHART

Code	Location of fault	Problem	Solution
1 1 1 1	control unit	control unit (internal)	replace control unit
2 1 1 3	no speed signal from Hall sensor, or false signal from potentiometer	Hall sensor defective, disconnected wire sensor plate adjustment, air flow sensor lever hard to move hangs up	check Hall sensor check wires check adjustment check adjustment of lever
2 1 2 1	idle switch	idle switch defective, (always closed). wire has a short	check idle switch check wire
2 1 2 3	full throttle switch	full throttle switch defective (always closed). wire has a short	check full throttle switch check wires
2 1 4 1	knock regulation at the control limit *(fault lamp lights up during maximum ignition retard adjustment)	engine vibrates, knocks fuel octane too low ignition timing point adjusted wrong knock sensor wire shielding damaged	compression test, check injection system, change fuel to a higher octane adjust timing, check knock sensor wires
2 1 4 2	knock sensor *(fault lamp lights up continuously until engine is shut off)	disconnected wire or short in sensor wire defective knock sensor	check wire between knock sensor and control unit replace knock sensor
2 2 3 1	idle stabilization control limits exceeded	basic adjustment of throttle intake system leaks ignition timing incorrectly adjusted	perform basic adjustment of throttle, check intake system for leaks adjust timing, check idle,
2 2 3 2	potentiometer on airflow sensor	disconnected wire or short between control unit and potentiometer	check potentiometer,
2 3 1 2	coolant temperature sensor	disconnected wire or short in sensor line, temperature sensor defective	check wires check temperature sensor

CIS-MOTRONIC TROUBLESHOOTING CHART

Code	Location of fault	Problem	Solution
2 3 4 1	oxygen regulation at the control limit (fault lamp lights up, if fault occurs for at least 2 minutes)	CO content not within specifications, oxygen sensor grounded cold start valve leaks charcoal canister open constantly intake system leaks	check CO content check oxygen sensor wire per wiring diagram, check oxygen sensor control, check cold start valve, evaporative system checking check intake system for leaks
2 3 4 2	oxygen sensor does not control *(fault lamp lights up)	disconnected wire to oxygen sensor or defective oxygen sensor	check oxygen sensor control,
2 3 4 3	mixture control, **rich** limit exceeded	idle **NOT OK**	check idle,
2 3 4 4	mixture control, **lean** limit exceeded	intake air system leaks, idle **NOT OK**	check intake system for leaks check idle,
4 4 3 1	idle stabilizer valve	disconnected wire, defective idle stabilizer valve	check wires replace stabilizer valve, replace control unit,
4 4 4 4	no faults stored in memory	—	—

CAUTION

After all test and adjustment work has been performed, erase fault memory

CIS-MOTRONIC VOLTAGE AND RESISTANCE TEST

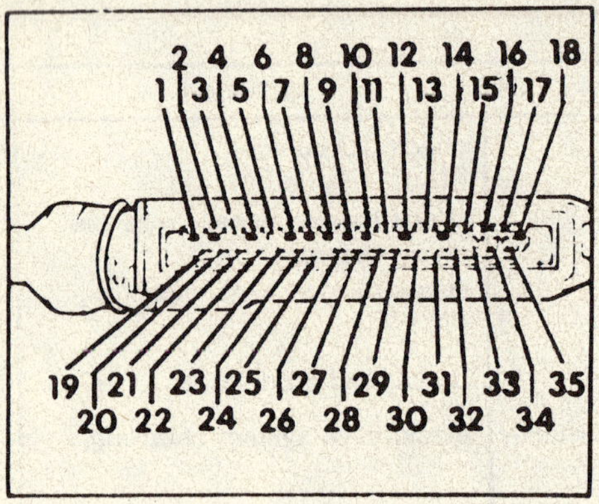

▶ **Terminal numbering for CIS-Motronic control unit harness connector**

Test step	Connection terminal	• Test conditions – Additional work	Specified values
Test range: voltage measurement			
1	14 + 35	• ignition ON	approximately battery voltage
2	17 + 35	• ignition ON	approximately battery voltage
3*	32 + 35	• ignition ON – A C ON	approximately battery voltage
4*	33 + 35	• ignition ON – A C ON	approximately 1 V below battery voltage
5	19 + 35	• ignition ON	approximately battery voltage
6	California version ONLY 13 + 35 connected 13 + 35 connected	• ignition ON • ignition ON – insert fuse in fuel pump relay	fault lamp must light up fault lamp must light up
Test range: resistance measurement			
7	4 + 5		15 to 25 ohms
8	3 + 35		see chart
9	31 + 35	• throttle closed • throttle completely open	open continuity
10	28 + 35	• throttle closed – throttle open	continuity· open
11	49 state version ONLY 18 + 35		continuity
12	13 + 35 13 + 35	– insert fuse in fuel pump relay – ground test connector for vehicle self diagnosis	continuity continuity
13	14 + 16		approximately 10 ohms
14	14 + 15		30 to 60 ohms

CIS-MOTRONIC VOLTAGE AND RESISTANCE TEST

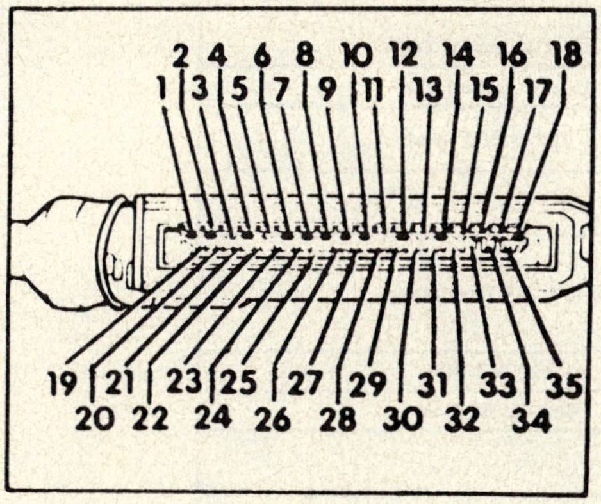

▶ **Terminal numbering for CIS-Motronic control unit harness connector**

Test range: resistance measurement			
Test step	**Connection terminal**	**● Test conditions – Additional work**	**Specified values**
15	 30 + 35 30 + 21	– remove harness connector from Hall sensor (ignition distributor) – connect terminal **1** + **2** to Hall sensor – connect terminal **2** + **3** to Hall sensor	 continuity continuity
16	11 + 35	– remove terminal from power output stage of coil and ground center contact	continuity
17	6 + 8	– separate knock sensor harness connector – connect terminals **1** + **2** – connect terminals **1** + **3**	open continuity continuity
18	7 + 35	– separate harness connection to oxygen sensor and connect wire to control unit to ground	continuity
19	23 + 26	● sensor plate in rest position	approximately 5k ohms
20	26 + 35		approximately 4k ohms
21	vehicles with standard transmission 34 + 35 vehicles with auto transmission 34 + 35	 ● selector lever in **P** or **N** position – insert working points	continuity open
22	12 + 35	– remove fuel pump relay (relay position 10) and ground terminal **47** of fuse/relay panel	continuity

CIS-MOTRONIC SPECIFICATIONS CHART

Engine code	3A
Ignition timing — checking value	4° to 8° Before TDC
Ignition timing — adjusting value	6 ± 1° Before TDC
Idle speed (speed not adjustable, controlled by idle stabilization)	780 to 900 RPM
CO content — checking value (on CO-measuring tube)	0.2 to 1.2 vol. %
CO content — adjusting value (adjustment conducted via differential pressure regulator control current-adjustment)	0 to 5 mA (fluctuating)
CIS-Motronic control unit Part Number	893 907 404
RPM limit	6300 to 6500/min
Fuel deceleration via control current reversal on differential pressure regulator	
Throttle switch, switching points — Switch I — idle switch	0.15 to 0.5 mm (0.006 to 0.020 in.) between point where switch begins operating and throttle lever stops
Throttle switch, switching points — Switch II — full throttle switch	10° ± 2° before full stop
System pressure	6.1 to 6.5 bar (88 to 94 psi) gauge pressure
Differential pressure — 1 — harness connector removed from differential pressure regulator	approximately 0.3 to 0.5 bar (4 to 7 psi) below system pressure
Differential pressure — 2 — harness connector reconnected to differential pressure regulator ■ activate starter for about three seconds	approximately 1.3 to 1.6 bar (19 to 23 psi) below system pressure
Residual pressure — after at least 10 minutes	3.3 bar (47.8 psi) gauge pressure
Residual pressure — after at least 20 minutes	3.2 bar (46.4 psi) gauge pressure
Fuel injection valves — opening pressure	4.3 to 4.6 bar (62 to 67 psi) gauge pressure
Injection quantity tolerance within a set of injectors: idle measurement (position 1 of 1348/1) at 20 ml fuel quantity	maximum 2.0 ml
full throttle measurement (position 2 of 1358/1) at 80 ml fuel quantity	maximum 8.0 ml
Differential pressure regulator resistance	15 to 25 ohms
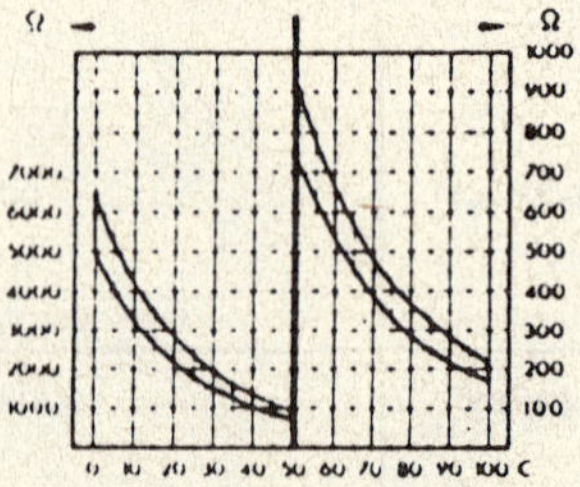	Temperature sensor resistance between terminals Example 1: 2500 ohms @ 20°C (68°F) Example 2: 200 ohms @ 100°C (212°F)
Cold start valve resistance	approximately 10 ohms

CIS-E FAULT CODES (INPUT CHECKS)

Fault codes (input checks)

Code	Location of fault	Problem	Repair procedure
1111	Ignition control unit or fuel injection control unit	Defective memory circuits in control unit	Replace knock regulation control unit
2121	Idle switch	Switch stuck closed or problem in wiring to switch	Check idle switch and wiring
2122	Engine speed signal or Hall sender	No engine speed signal from Terminal 17 of ignition control unit to Terminal 30 of fuel injection control unit	Repair break in wiring using wiring diagram or check Hall sender and electronic control unit power according to the wiring diagram
2123	Full throttle switch	Switch stuck closed or problem in wiring to switch	Check full throttle switch and wiring
2141	Knock regulation	Engine or ignition knock is causing timing to be retarded the maximum amount	Check ignition distributor basic adjustment, compression, and injection system, check fuel octane Check knock sensor wiring using wiring diagram
2142	Knock sensor	Defective sensor or sensor wiring	Check wiring between knock sensor and knock regulation control unit according to wiring diagram
2223	Altitude sensor	No signal from sensor	Check altitude sensor and wiring
2232	Air sensor potentiometer	No signal from potentiometer to fuel injection control unit or break in wire between fuel injection control unit Terminal 21 and ignition control unit Terminal 8	Check potentiometer (on air flow sensor) and wiring Repair break in wiring using wiring diagram
2233	Reference (supply) voltage for air sensor potentiometer and altitude sensor	No reference voltage from Terminal 21 of ignition control unit to Terminal 26 of fuel injection control unit	Check wiring between electronic ignition and fuel injection control unit using wiring diagram
2312	Coolant temperature sensor	No signal from sensor	Check coolant temperature sender and wiring
2341	Oxygen sensor control (at its control limit)	CO content not within specification	Check CO content, adjust if necessary
		Ground short circuit to the oxygen sensor	Check wiring using wiring diagram. Check oxygen sensor control system
		Activated charcoal filter constantly open	Check carbon canister shut-off solenoid
		Air duct shroud or hoses for air intake system are leaking	Check system for leakage, repair as necessary
2342	Oxygen sensor (does not control)	Oxygen sensor lead disconnected or oxygen sensor defective	Check oxygen sensor control system
4431	Idle stabilizer valve	Problem in wiring to idle stabilizer valve	Check wiring for idle stabilizer valve using wiring diagram or perform differential pressure regulator diagnosis
4444	No faults stored in memory	—	—
0000	End of diagnosis	—	—

Note

If fault codes (2121, 2123, 2223) are only displayed during the fault display of a control unit, the first thing to be done is to check the wiring from the control unit to the control unit displaying the fault for a break in wiring (loose terminal).

If the fault lamp blinks while the vehicle is being driven, knock regulation is at the maximum limit of control.

Component Replacement

SERVICE PRECAUTIONS

To prevent personal injury and damage to the vehicle, follow all service precautions.

- Always disconnect the negative battery terminal to prevent sparks cause by short-circuiting, circuit-breaking, etc.
- Smoking must not be allowed when working around flammable liquids.
- Always keep a CO_2 fire extinguisher close on hand.
- Absorbant must be available to soak up any spillage.
- Ensure fuel is emptied into a suitable container.
- Always place the transmission gear shift lever in **N** (manual transmission) or **P** (automatic transmission) and set the parking brake and block drive wheels before starting the engine.
- Keep hands, tools and clothing away from engine cooling fan to help prevent personal injury.
- Thoroughly clean all unions and area near connections before disconnecting.

FUEL PRESSURE RELIEF

1. Follow service precautions to safely perform this procedure to prevent personal injury. Disconnect the negative battery terminal.
2. Wrap a shop rag around the feed to the fuel distributor.
3. Loosen the feed connection to relieve the system pressure. Catch any escaping fuel in the shop rag. Dispose of the rag in a safe manner.
4. Tighten the fuel distributor feed connection.

COLD START VALVE

Removal and Installation

1. Disconnect the negative battery cable.
2. Relieve the fuel system pressure. Wrap a cloth around the connection to catch any escaping fuel.
3. Remove the harness connector and fuel line from the cold start valve.
4. Loosen the cold start valve retaining bolt and remove the cold start valve and gasket from the intake manifold.

To install:

5. Install a new gasket and install the cold start valve to the intake manifold. Install the retaining screws and tighten to 88 inch lbs. (10 Nm).
6. Connect the fuel line and the harness connector to the cold start valve.
7. Turn the ignition switch **ON** and **OFF** several times to pressurize the system. Check for fuel leaks before starting the engine and after starting.

THERMO-TIME SWITCH

Removal and Installation

1. Disconnect the negative battery cable.

CAUTION

Allow the engine to cool sufficiently before attempting any work with the cooling system to avoid personal injury.

2. Drain the engine coolant below the level of the switch. Be careful not to damage the connectors of the switch while removing it from the engine.

To install:

3. Coat the threads of the switch with an approved thread sealant compound.
4. Install the thermo-time switch in the intake manifold. Torque the switch to 22 ft. lbs (30 Nm).
5. Connect the harness connector to the switch.
6. Refill the cooling system and connect the negative battery cable.

FUEL INJECTOR

Removal and Installation

1. Disconnect the negative battery cable.
2. Relieve the fuel system pressure. Wrap a cloth around the connection to catch any escaping fuel.
3. Clean the area around the fuel injectors. Using 2 wrenches, hold the injector with 1 wrench and loosen the fuel line fitting with the other wrench.
4. Remove the fuel line from the injector, do not allow the injector to turn.
5. Carefully pull the injector out from the insert. A injector extractor tool may be required on injectors that are difficult to remove.
6. Using a 13mm hex head socket tool, remove the injector upper insert from the cylinder head.
7. Remove the lower insert and sealing ring from the cylinder head.

To install:

8. Install a new sealing washer and the lower insert into the cylinder head.
9. Coat the upper insert threads with D–6 sealing compound or equivalent. Thread the upper insert into the cylinder head.
10. Using a 13mm hex head socket tool, tighten the the upper insert to 15 ft. lbs. (20 Nm).
11. Make sure all O-rings are in place on the injector. Lubricate the O-rings with fuel prior to installing.
12. Push the injector firmly in position into the upper insert seat.
13. Connect the fuel line to the injector. Using 2 wrenches tighten the fuel line fitting.
14. Turn the ignition switch **ON** and **OFF** several times to pressurize the system. Check for fuel leaks before starting the engine and after starting.

AUXILIARY AIR REGULATOR

The auxiliary air regulator is normally located below the throttle body, mounted to the intake manifold.

Removal and Installation

1. Disconnect the negative battery cable.
2. Remove and plug all vacuum lines to the valve.
3. Disconnect the harness connector at the valve.
4. Remove the auxiliary valve retaining bolts and remove the auxiliary air valve from the intake manifold.

To install:

5. Position the auxiliary air valve to the intake manifold and install the retaining bolts.
6. Connect all vacuum hoses to the auxiliary air valve.
7. Connect the harness connector to the auxiliary air valve.
8. Connect the negative battery cable.

FUEL PUMP

The fuel pump is normally located under the vehicle near the fuel tank or in front of the rear axle on the right side. It is attached to a bracket which also accommodates the fuel filter and the accumulator as an assembly.

Removal and Installation

1. Disconnect the negative battery cable.
2. Relieve the fuel system pressure. Wrap a cloth around the connection to catch any escaping fuel.

NOTE: When replacing the old fuel pump with a replacement pump,the new pump has a positive lock connector that differs from the earlier pumps. The electrical connectors of the fuel pump wiring harness must be changed when installing this newer pump.

3. Raise the vehicle and support it safely.
4. Disconnect the electrical connector from the fuel pump.
5. Remove both wire terminals from the connector housing.
6. Discard the old connector and protective boot.
7. Install the new protective boot over the wires and insert the wire terminals into the new connector.
8. Slide the protective boot over the connector.
9. Disconnect fuel line from the fuel pump.
10. Remove pump mounting bolts and remove the pump.

To install:

11. Position the fuel pump it place and secure it with the mounting bolts.
12. Install new O-ring seals when connecting the fuel lines to the pump.
13. Connect the electrical connector to the fuel pump.
14. Lower the vehicle.
15. Connect the negative battery cable. Turn the ignition switch **ON** and **OFF** several times to pressurize the system.
16. Raise the and support it safely, start the engine and observe for fuel leaks.
17. Shut the engine off and lower the vehicle.

FUEL FILTER

The fuel filter is located under the vehicle, in front of the fuel tank. On some vehicles it is mounted in a bracket assembly with the fuel pump. In other vehicles it is separate.

Removal and Installation

1. Disconnect the negative battery cable.
2. Relieve the fuel system pressure. Wrap a cloth around the connection to catch any escaping fuel.
3. Raise the vehicle and support it safely.
4. Throughly clean all fuel lines connected to the fuel filter and remove them.
5. Unscrew the bracket clamping bolt and remove the fuel filter from the bracket.

To install:

6. Slide the fuel filter in to its mounting bracket, making sure the arrow is pointing to flow towards the fuel distributor.
7. Connect the fuel lines to the filter and tighten the retaining bolts to 18 ft. lbs (25 Nm).

NOTE: Always install new seals and hoses clamps.

8. Connect the negative battery cable. Turn the ignition switch **ON** and **OFF** several times to pressurize the system and check for leaks.

ACCUMULATOR

The accumulator is located under the vehicle, in front of the fuel tank. On some vehicles it is mounted in a bracket assembly with the fuel pump and fuel filter. In other vehicles it is mounted separately.

Removal and Installation

1. Disconnect the negative battery cable.
2. Relieve the fuel system pressure. Wrap a cloth around the connection to catch any escaping fuel.
3. Throughly clean all fuel lines connected to accumulator and remove them.
4. Loosen the bracket clamping bolt and slide the accumulator from the bracket.

To install:

5. Slide the accumulator into its mounting bracket and secure the clamping bolt.
6. Connect the fuel lines and tighten the fittings to 15 ft. lbs. (20 Nm).
7. Connect the negative battery cable. Turn the ignition switch **ON** and **OFF** several times to pressurize the system and check for leaks.

OXYGEN SENSOR

The oxygen sensor is located in the exhaust manifold. The exhaust system should be slightly warm to make removal easier.

Removal and Installation

1. Disconnect the negative battery cable.
2. Disconnect the electrical connector at the oxygen sensor.
3. Unscrew the oxygen sensor and remove it from the exhaust manifold.

To install:

4. Before installing the oxygen sensor, coat the threads with anti-seize compound and install.
5. Connect the electrical sensor to the oxygen sensor. Connect the negative battery cable.

AIR FLOW SENSOR ASSEMBLY

Removal and Installation

1. Disconnect the negative battery cable.
2. Relieve the fuel system pressure. Wrap a cloth around the connection to catch any escaping fuel.
3. Remove rubber air duct from the mixture control unit. If equipped with a venturi air intake screen, remove it from venturi.
4. Throughly clean all fuel lines on the fuel distributor.
5. Disconnect the fuel lines at the fuel distributor. Plug the ends of the lines and the openings of the fuel distributor.
6. Disconnect the harness connectors from the differential pressure regulator and the potentiometer.
7. Remove the pressure regulator and bracket assembly and place it to the side.
8. Remove the screws retaining the air flow sensor assembly to the upper air filter housing and carefully lift the unit out.

NOTE: Prior to installing the mixture control unit, the basic adjustment of the sensor plate must be preformed whenever the fuel distributor or air flow sensor has been replaced.

To install:

9. Inspect the air filter element and replace if needed. Clean any dirt or contamination from the upper and lower air filter housings.
10. Install a new gasket on the air filter upper housing and position the mixture control unit in place.
11. Install the mixture control unit and retaining screws. Tighten the retaining screws to 31 inch lbs. (3.5 Nm) in a diagonal pattern.
12. Install the pressure regulator and bracket assembly. Install the retaining screws and torque it to 31 inch lbs. (3.5 Nm).
13. Connect the harness connectors to the differential pressure regulator and the potentiometer.
14. Connect the fuel injector lines at the top of the fuel distributor. Torque the retaining screws to 7 ft. lbs. (10 Nm).
15. Connect the fuel line from the cold start valve to the fuel distributor. Torque the retaining screw to 15 ft. lbs. (20 Nm).
16. Install the air intake screen, if equipped. Install the rubber air duct on the mixture control unit and tighten the retaining clamp.

NOTE: Improper installation or distortion of the air intake screen may result in interference with the sensor plate. Make sure the screen is installed with the word (open) on the flange facing up.

17. Check the air flow sensor plate for proper position. Adjust if needed.
18. Connect the negative battery cable.
19. Turn the ignition switch **ON** and **OFF** several times to pressurize the system. Check for fuel leaks before starting the engine and after starting.

FUEL DISTRIBUTOR

Removal and Installation

1. Disconnect the negative battery cable.
2. Relieve the fuel system pressure. Wrap a cloth around the connection to catch any escaping fuel.
3. Throughly clean all fuel lines on the fuel distributor and remove them.
4. Disconnect the harness connectors from the differential pressure regulator and potentiometer.
5. Remove the pressure regulator and bracket retaining screws and remove it as an assembly. Place to the side.
6. Remove the air flow sensor retaining screws and remove both the air flow sensor and fuel distributor as an assembly.
7. Remove the fuel distributor retaining screws and carefully lift it from the air flow sensor. Make sure that the control plunger does not fall out of the fuel distributor.

To install:

8. If the control plunger has been removed from the fuel distributor, moisten it with clean gasoline before installing it into the fuel distributor. Be sure to insert the small shoulder on the control plunger first.
9. Install a new O-ring seal before assembling the fuel distributor to the air flow sensor.
10. Assemble the fuel distributor to the air flow sensor and install the retaining screws. Tighten the retaining screws to 31 inch lbs. (3.5 Nm).

NOTE: When ever the air flow sensor or fuel distributor is replaced, the sensor plate basic adjustment must be performed.

11. Install the air flow sensor and fuel distributor as an assembly to the upper air filter housing.
12. Install the air flow sensor retaining screws. Tighten the retaining screws to 31 inch lbs. (3.5 Nm).
13. Install the pressure regulator bracket assembly and retaining screws. Tighten the retaining screws to 31 inch lbs. (3.5 Nm).
14. Connect the fuel injector lines at the top of the fuel distributor. Torque the retaining screws to 7 ft. lbs. (10 Nm).
15. Connect the fuel line from the cold start valve to the fuel distributor. Torque the retaining screw to 15 ft. lbs. (20 Nm).
16. Install the air intake screen if equipped and the air intake hose.
17. Connect the negative battery cable.
18. Turn the ignition switch **ON** and **OFF** several times to pressurize the system. Check for fuel leaks before starting the engine and after starting.

Adjustment

1. Remove the air inlet boot from the sensor plate and operate the starter for about 10 seconds to build fuel system pressure. The engine will not start.
2. Measure the distance from the top of the air flow sensor housing to the sensor plate at the point nearest the fuel distributor.
3. Use a magnet on the sensor plate bolt to lift the plate until resistance is felt. The free-play movement should be 0.019–0.118 in. (0.5–3mm).

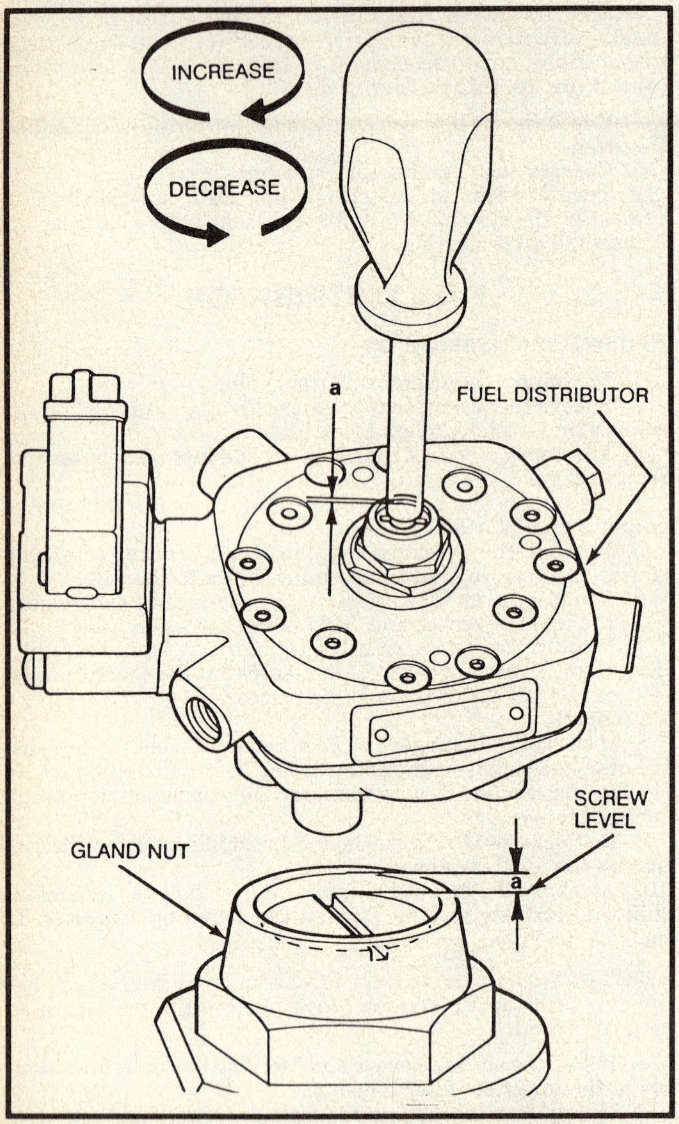

Adjusting the sensor plate with the control plunger stop screw—Audi

4. To adjust, remove the fuel distributor and use a depth caliper to measure the distance from the shoulder of the nut to the stop screw. Distance should be 0.024 in. (0.6mm).
5. Turn the stop screw in or out as required. Turning the screw in ¼ turn will increase the free-play by 0.050 in. (1.3mm).
6. After adjusting plunger free-play, idle speed and CO must be adjusted.

PRESSURE REGULATOR

The pressure regulator is normally located in a bracket, mounted on or near the air flow sensor assembly.

Removal and Installation

1. Disconnect the negative battery cable.
2. Relieve the fuel system pressure. Wrap a cloth around the connection to catch any escaping fuel.
3. Throughly clean the fuel line fittings attached to the pressure regulator and remove them.
4. Unplug the vacuum hose from the pressure regulator.
5. Loosen the bracket clamping bolt and remove the pressure regulator.

To install:

6. Slide the pressure regulator into its mounting bracket and tighten the bracket clamping bolt.
7. Connect the fuel lines to the pressure regulator and plug in the vacuum hose.
8. Connect the negative battery cable.
9. Turn the ignition switch **ON** and **OFF** several times to pressurize the system. Check for fuel leaks before starting the engine and after starting.

DIFFERENTIAL PRESSURE REGULATOR

The differential pressure regulator is attached to the side of the fuel distributor.

Removal and Installation

1. Disconnect the negative battery cable.
2. Relieve the fuel system pressure. Wrap a cloth around the connection to catch any escaping fuel.
3. Disconnect the harness connector from the differential pressure regulator.
4. Before removing, thoroughly clean the exterior and surrounding area of the differential pressure regulator to prevent dirt from entering the fuel distributor or the regulator.
5. Remove the screws retaining the differential pressure regulator to the fuel distributor.
6. Carefully remove the differential pressure regulator from the fuel distributor. Do not lose the O-ring seals as the regulator is being removed.

To install:

7. Replace the O-ring seals before installing the differential pressure regulator.
8. Position the differential pressure regulator to the fuel distributor and install the retaining screws. The retaining screw is nonmagnetic and should not be replaced with a standard screw. Use the original retaining screw or an OEM replacement screw.
9. Connect the negative battery cable.
10. Turn the ignition switch **ON** and **OFF** several times to pressurize the system. Check for fuel leaks before starting the engine and after starting.

SENSOR PLATE HEIGHT

Adjustment

1. Operate the fuel pump for a few seconds to build up system pressure.

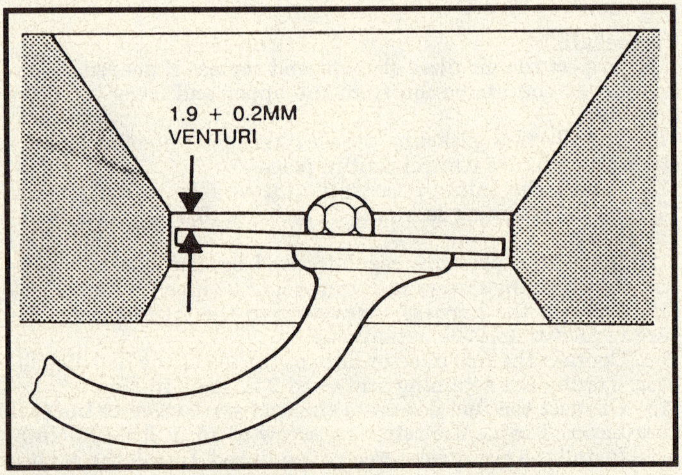

Measuring the sensor plate height—Audi

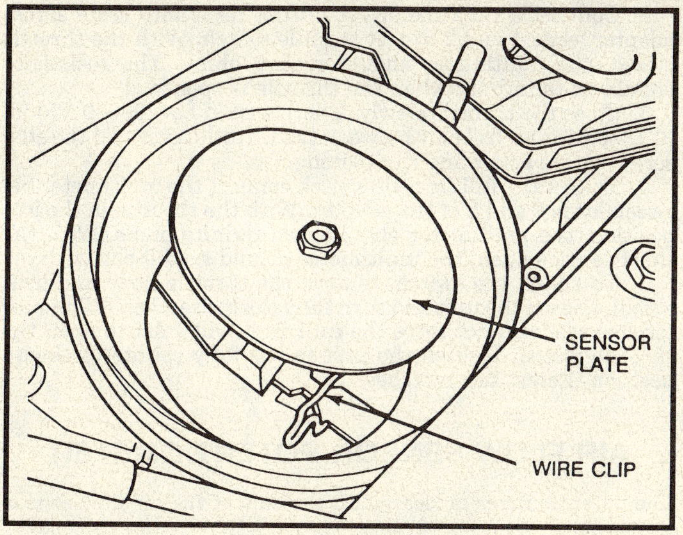

Adjusting the sensor plate height—Audi

2. Remove the rubber connecting boot, from the air sensor housing.
3. Check the position of the sensor plate. The upper edge of the sensor plate must be 1.9 plus (+) 0.2 mm below the cone edge of the air venturi on the side closest to the fuel distributor.
4. Adjust the sensor plate by lifting if fully and bending the wire clip. Be careful not to scratch the venturi of the air flow sensor. Do not bend the leaf spring.

SENSOR PLATE CLEARANCE

Adjustment

1. Lift the sensor plate until even resistance is felt. Check the plated clearance. There should be noticeable clearance at the minimum and a maximum clearance of 1.9 plus (+) 0.2mm. The sensor plate should be even with edge of the venturi cone.
2. If the sensor plate clearances check good, continue to the Sensor Plate and Control Plunger Movement checks.
3. If the sensor plate clearances do not check good, perform a sensor plate lever basic adjustment.

SENSOR PLATE LEVER

Basic Adjustment

NOTE: This adjustment must be made whenever the air flow sensor or the fuel distributor is replaced.

1. Disconnect the negative battery cable.
2. Remove the fuel distributor from the mixture control unit.
3. Check the distance between the seating pads of the removed fuel distributor on the air flow sensor and the roller on the sensor plate lever.
4. The distance should be 0.748±0.004 in. (19.0±0.1mm).
5. If the measurement is incorrect, adjust the CO% adjusting screw to obtain the correct measurement.
6. Reinstall the fuel distributor and recheck the sensor plate clearance.
7. If the sensor plate clearance is correct, readjust the CO%.
8. If it is not correct, remove the fuel distributor and adjust the sensor plate clearance with the control plunger stop screw.
   a. Turn clockwise to increase clearance
   b. Turn counterclockwise to decrease clearance.
   c. Each ¼ turn will change the sensor plate position approximately 0.50 in. (1.3mm).

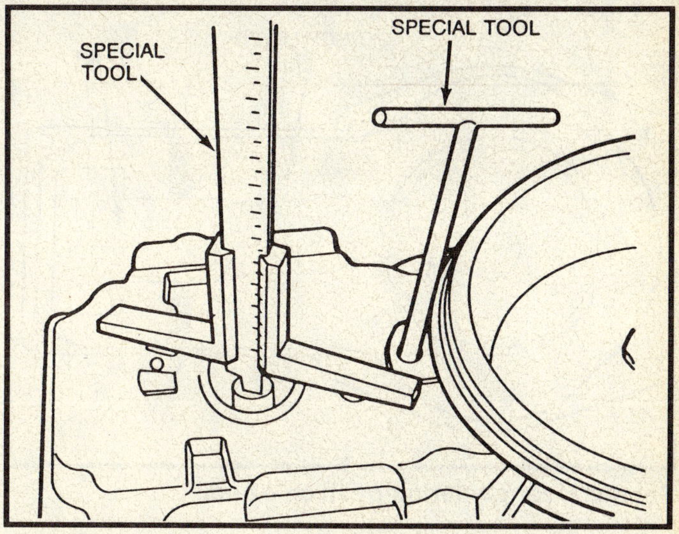

Sensor plate basic adjustment, with fuel distributor removed-measuring distance between seating pads and roller—Audi

9. Reinstall the fuel distributor and check the sensor plate height. The distance should be 0.024 in. (0.6mm) between the stop screw and flange of the hex nut.
10. Connect the negative battery cable.

NOTE: After adjusting the sensor plate position, readjust the idle speed and mA current (CO value) reading.

SENSOR PLATE

Centering Adjustment

1. Disconnect the negative battery cable.
2. Remove the centering plate adjustment bolt from the sensor plate.
3. Coat the bolt with Loctite or an equivalent tread sealant. Reinstall the bolt finger tight.
4. Using a centering tool or a 0.004 in. (10mm) feeler gauge, carefully center the plate.
5. Torque the adjustment screw to 4 ft. lbs. (5–7 Nm).
6. If the plate cannot be centered and the lever appears to be off-center in the air cone, the air sensor housing must be removed to center the lever.

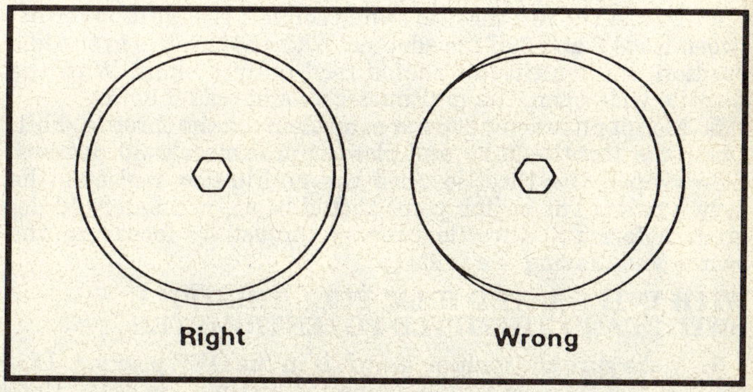

Sensor plate centering position—Audi

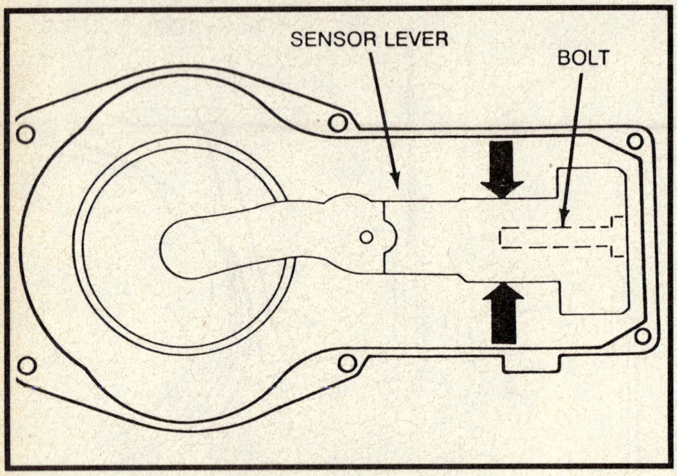

Sensor lever centering position—Audi

SENSOR LEVER

Centering Adjustment

1. Disconnect the negative battery cable.
2. Remove the air sensor housing from the mixture control unit.
3. Remove the clamping bolt on the sensor lever counter weight.
4. Coat the bolt threads with Loctite or an equivalent bolt thread sealant. Install the bolt finger tight.
5. Slide the lever to the center position and tighten the clamping bolt.

THROTTLE SWITCH

Adjustment

WITH SEPARATE IDLE AND FULL THROTTLE SWITCHES

1. Make sure the ignition switch is in the **OFF** position. Disconnect the throttle switch connector and plug in VAG 1501 adapter or equivalent.
2. Connect a multimeter set on the resistance scale across adapter leads 1 and 2 to test the idle switch. With the throttle closed, the multimeter should read 0 ohms. The resistance should go to infinity when the throttle is opened.
3. Close the throttle slowly. There should be 0.006–0.020 in. (0.15–0.50 mm) freeplay between the throttle lever and the stop screw before activating the switch. Adjust the switch by loosen the mounting screw and turning.
4. To test the full throttle switch connect the multimeter between leads 2 and 3 of the adapter. With the throttle in the idle position, the multimeter should read infinite ohms. With the throttle wide open, the multimeter should read 0 ohms.
5. Mount protractor 3084 or equivalent, on the throttle shaft. Open the throttle fully and close approximately 30 degrees. Slowly open the throttle until the multimeter registers the switch point. The switch point should be approximately 10 degrees before full throttle opening. Adjust by loosening the switch and moving the unit.

WITH INTEGRATED IDLE, FULL THROTTLE SWITCH AND THROTTLE POTENTIOMETER

1. Make sure the ignition switch is in the **OFF** position. Disconnect the throttle switch connector and plug in VAG 1501 adapter or equivalent.
2. Connect a multimeter set on the resistance scale across adapter leads 1 and 2 to test the idle switch. With the throttle closed, the multimeter should read 0 ohms. The resistance should go to infinity when the throttle is opened.
3. Close the throttle slowly. There should be 0.006–0.020 in. (0.15–0.50 mm) freeplay between the throttle lever and the stop screw before activating the switch.
4. To test the full throttle switch connect the multimeter between leads 2 and 3 of the adapter. With the throttle in the idle position, the multimeter should read infinite ohms. With the throttle wide open, the multimeter should read 0 ohms.
5. To adjust the switch, remove the throttle body and heat shield. Loosen the unit and turn fully clockwise, then fully counterclockwise. Do not force the unit as it will start to open the throttle plates. Tighten the unit in the fully counterclockwise position. Retest the switches.

AIR FLOW SENSOR POTENTIOMETER

The potentiometer is located on the side of the air flow sensor assembly.

Removal and Installation

1. Disconnect the negative battery cable.
2. Remove the intake air boot from the mixture control unit.
3. Throughly clean all fuel lines on the fuel distributor and remove them.
4. Completely remove the air flow sensor from the air filter housing.
5. Disconnect the harness connector from the potentiometer.
6. Remove the retaining screws from the potentiometer and carefully remove the potentiometer from the air flow sensor body.

NOTE: After the potentiometer has been removed, do not touch the resistance strip or the contact whiskers inside of the potentiometer. This will cause damage to the potentiometer.

To install:

7. Carefully position the potentiometer to the air flow sensor and install the retaining screws.
8. Adjust the potentiometer.
9. Install the mixture control unit to the air filter housing.
10. Connect the fuel lines to the fuel distributor.
11. Connect the negative battery cable.
12. Position the air intake duct to the mixture control unit and tighten the retaining clamp.
13. Turn the ignition switch **ON** and **OFF** several times to pressurize the system. Check for fuel leaks before starting the engine and after starting.

Adjustment

1. Connect adapter VW 1501 between potentiometer and the wiring connector.
2. Connect multimeter between terminal 1 and 3.
3. Set the multimeter to the 20 VDC scale and turn the ignition switch **ON**. The voltage must be between 4.35 volts and 5.35 volts.
4. Turn the ignition switch **OFF**. Connect the multimeter between terminals 1 and 3. Run the engine for a minimum of 2 minutes.
5. Measure the voltage (V1) between terminals 1 and 3. Measure the voltage (V2) between terminals 2 and 3. Compare the voltages to the chart.
6. If the voltage does not fall within the range of the chart, remove the sealing compound over the trim screw and adjust the screw until the range is met.
7. If the range can not be met, replace the air flow sensor. Cover the trim screw with silicone sealer.

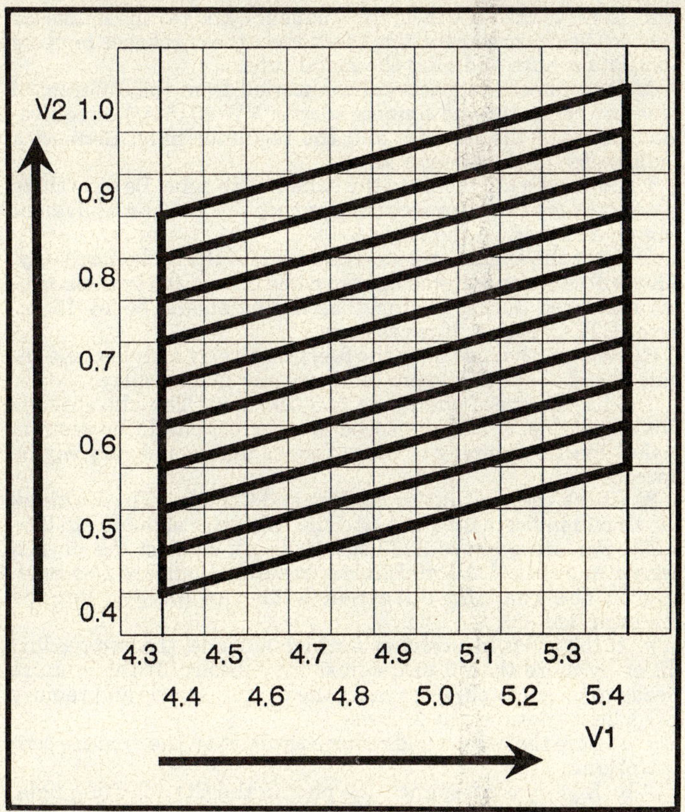

V1 and V2 values for adjusting the air flow sensor potentiometer

IDLE SPEED AND MIXTURE (CO)

Adjustments

Most of these tests require that the engine cooling fan not run during the test. To prevent the engine fan from running, place a large floor fan in front of the vehicle grill to force cool air through the radiator.

NOTE: The following information is being published from the latest information available at the time of publication. If the information differs from the values given on the underhood emission control label, use the data on the label.

1990 80

The idle speed, mixture adjustment and ignition timing must be checked together. Exhaust gas mixture must be checked and adjusted before measuring milliamps (mA), with a special meter and with the oxygen sensor connected. Refer to the underhood vacuum diagram for all hose locations.

1. Start the engine and allow it to reach normal operating temperature. The radiator fan must come on at least once.
2. Switch **OFF** all accessories. The radiator fan and air conditioner must not be running during the test.
3. Pull the crankcase breather hose off the valve cover and allow it to vent to the atmosphere. Pull the crankcase breather hose off the breather housing and allow it to vent to the atmosphere.
4. Remove the sealing plug from the charcoal canister
5. Connect CO tester according to the manufacturer's instructions. Remove the cap from the CO probe receptacle and insert the CO probe. The hose must fit tightly so there is no exhaust leak.
6. Connect adapter VW 1315 A/1 or equivalent to the differential pressure regulator. Connect a multimeter to the adapter and the oxygen sensor test connection. Set the selector switch to DCA 200 mA. The meter must read positive. Reverse the connections if not.
7. Check the idle. The idle should be 780–900 rpm. Check the timing. The timing should be 5–7 degrees BTDC. Unseal the distributor hold-down bolt and adjust as necessary.
8. Briefly accelerate the engine, the meter reading should range between 0–5 mA, fluctuating, with the oxygen sensor connected. The CO value should be 0.2–1.2%.

NOTE: If the engine does not run after the adapter is connected, the connections may be improper. Reverse the plug and try again.

9. If the current reading is not correct, remove CO adjustment plug as follows:
   a. Stop the engine and remove the boot on the mixture control unit.
   b. Lightly centerpunch the plug in the CO adjusting hole.
   c. Drill a $^3/_{32}$ in. (2.5mm) hole in the center of the plug $^9/_{64}$–$^5/_{32}$ (3.5mm) deep.
   d. Carefully clean up any metal shavings and insert a ⅛ in. sheet metal screw into the drilled hole and remove the screw and plug.
10. Start the engine and allow it to idle.
11. Adjust the current reading on the multimeter by turning the CO adjusting screw with tool P377 or equivalent to obtain 0–5mA (reading fluctuates). Turn the adjusting screw clockwise to raise the current reading (CO higher) and screw clockwise to raise the current reading (CO lower).

NOTE: Do not push adjustment tool down when making CO adjustment and do not accelerate the engine with the tool in place. Remove the tool after each adjustment and accelerate the engine briefly before reading the multimeter. Always adjust in direction from high to low.

12. Run the engine and recheck the meter readings. After CO adjustment is complete, recheck the idle speed and adjust if necessary. Reconnect the crankcase breather hose. If, after reconnecting the breather hose, the current reading drops below

specifications, an oil change may be necessary, due to excessive gasoline dilution.

13. Turn the ignition switch **OFF** and drive in a new mixture plug, flush with the mixture control unit. This must be done to meet federal anti-tamper regulations.

14. Remove all test equipment, attach CO probe cap and insert the charcoal canister sealing plug.

1990 80 QUATTRO AND 90, 1991 80, 80 QUATTRO AND 90, 1992 80 AND 80 QUATTRO

1. Warm the engine to operational temperature, oil at 176°F (80°C). All electrical devices must be **OFF**. The engine cooling fan must not be running. Do not connect any fuel pressure gauges. If the injection lines had been disconnected, raise the engine speed past 3000 rpm and allow to idle for 2 minutes.
2. Remove the sealing plug from the charcoal canister. Disconnect the crankcase housing ventilation hose and plug the metal tube.
3. Disconnect the electrical connection from the differential pressure regulator and connect adapter VW 1315 A/1 or equivalent, between the harness and the regulator plug. Connect a multimeter to the adapter leads.
4. Connect a CO tester to the exhaust tap tube. Be sure there is no leakage at the connection.
5. Start the engine and allow to idle. Raise the engine speed to 4000 rpm and allow throttle to snap shut. The reading on the multimeter should indicate a negative 50–60 mA for a moment. If the meter indicates a positive reading, reverse the meter connections. If no reading results, check the idle switch.
6. Measure the CO level. It should be 0.3–1.2% with a reading of 0±3 mA on the multimeter. For areas above 3280 ft. (1000 m), disconnect the oxygen sensor and record the multimeter reading. Compare this reading to the reading obtained with the oxygen sensor connected.
7. If the CO is incorrect or for high altitude, the two reading differ by more than 3 mA, adjust the mixture.
8. If the current reading is not correct, remove CO adjustment plug as follows:
   a. Stop the engine and remove the boot on the mixture control unit.
   b. Lightly centerpunch the plug in the CO adjusting hole.
   c. Drill a 3/32 in. (2.5mm) hole in the center of the plug 9/64–5/32 (3.5mm) deep.
   d. Carefully clean up any metal shavings and insert a 1/8 in. sheet metal screw into the drilled hole and remove the screw and plug.
9. Start the engine and allow it to idle.
10. Adjust the current reading on the multimeter by turning the CO adjusting screw with tool P377 or equivalent to obtain 0±1 mA (reading fluctuates) and on high altitude applications, 0±1 mA of reading taken with the oxygen sensor disconnected. Turn the adjusting screw clockwise to raise the current reading (CO higher) and screw clockwise to raise the current reading (CO lower).

NOTE: Do not push adjustment tool down when making CO adjustment and do not accelerate the engine with the tool in place. Remove the tool after each adjustment and accelerate the engine briefly before reading the multimeter. Always adjust in direction from high to low.

11. Plug the CO adjustment and remove testing equipment. Return the breather hose to its original postion. Connect the differential pressure regulator.

1990–91 100 AND 100 QUATTRO

1. Warm the engine to operational temperature, oil at 176°F (80°C). All electrical devices must be **OFF**. The engine cooling fan must not be running. Do not connect any fuel pressure gauges. If the injection lines had been disconnected, raise the engine speed past 3000 rpm and allow to idle for 2 minutes.
2. Remove the hose from the frequency valve at the intake air boot to the charcoal canister. Disconnect the crankcase housing ventilation hose and plug the metal tube.
3. Disconnect the electrical connection from the differential pressure regulator and connect adapter VW 1315 A/1 or equivalent, between the harness and the regulator plug. Connect a multimeter to the adapter leads.
4. Connect a CO tester to the exhaust tap tube. Be sure there is no leakage at the connection. Connect VW 1367 or equivalent engine analyzer, to the engine.
5. Start the engine and insert a fuse into the fuel pump relay. This will cause the engine light to come on and the ignition timing stabilized during testing. The timing should be 14–16 degrees BTDC. Adjust if neceassary.
6. Remove the fuse from the fuel pump relay and raise the engine speed above 2500 rpm to cancel any fault display.
7. The idle speed should be 650–790 rpm. The idle speed is not adjustable. If the idle speed is incorrect check for vacuum leaks, injector dripping or other factors that would vary engine speed.
8. Check the multimeter reading and CO level. The multimeter should indicate 0±1 mA and the CO tester should read 0.3–1.2%. For areas above 3280 ft. (1000 m), connect the oxygen sensor and record the multimeter reading. Compare this reading to the reading obtained with the oxygen sensor disconnected.
9. If the CO is incorrect or for high altitude, the two reading differ by more than 3 mA, adjust the mixture. If the mixture needs to be adjusted, disconnect the oxygen sensor and remove CO adjustment plug as follows:
   a. Stop the engine and remove the boot on the mixture control unit.
   b. Lightly centerpunch the plug in the CO adjusting hole.
   c. Drill a 3/32 in. (2.5mm) hole in the center of the plug 9/64–5/32 (3.5mm) deep.
   d. Carefully clean up any metal shavings and insert a 1/8 in. sheet metal screw into the drilled hole and remove the screw and plug.
10. Start the engine and allow it to idle.
11. Adjust the current reading on the multimeter by turning the CO adjusting screw with tool P377 or equivalent to obtain 0.6–1.0% CO or on high altitude applications, 0±1 mA of reading taken with the oxygen sensor disconnected. Turn the adjusting screw clockwise to raise the current reading (CO higher) and screw clockwise to raise the current reading (CO lower).

NOTE: Do not push adjustment tool down when making CO adjustment and do not accelerate the engine with the tool in place. Remove the tool after each adjustment and accelerate the engine briefly before reading the multimeter. Always adjust in direction from high to low.

12. Plug the CO adjustment and remove testing equipment. Return the breather hose to its original postion. Connect the differential pressure regulator.

1990 200, 200 QUATTRO AND 200 QUATTRO WAGON, 1991 200

Single Knock Sensor

1. Warm the engine to operational temperature, oil at 176°F (80°C). All electrical devices must be **OFF**. The engine cooling fan must not be running. Do not connect any fuel pressure gauges. If the injection lines had been disconnected, raise the engine speed past 3000 rpm and allow to idle for 2 minutes.
2. Using tester VW 1367 or equivalent, connect the red wire to the coil junction box, the black wire to ground, the green wire to terminal 1 of the coil junction box, and the inductive pickup to cylinder 1 ignition lead.
3. Remove the cap from the charcoal canister tee fitting. Unclamp the breather hose and remove the hose from the fitting. Plug the breather hose with the cap from the tee fitting.

4. Connect a CO tester to the exhaust tap. Be sure there is no leakage from the connection.
5. Connect test adapter VW 1315 A/2 or equivalent, between the harness and the connector on the idle stabilizer valve. Connect a multimeter to the leads of the test adapter. Set the multimeter to the 2A scale.
6. Start the engine and let idle. Manual transmission vehicles should idle at 750–850 rpm and automatic transmission vehicles at 670–770 rpm. Adjust the idle speed by turning the screw on the idle stabilizer valve.
7. Disconnect the green wire from the ignition coil junction block to activate the idle stabilizer preset value. The multimeter will read 410–450 mA. If it does not, check the idle stabilizer system. Connect the green wire to the ignition coil junction block and recheck the multimeter. The multimeter will read 410–450 mA and fluctuate.
8. Remove the inductive pickup from the ignition lead, disconnect the green wire from the ignition coil junction block and connect to test adapter US 1112 or equivalent. Connect US 1112 to the blue and white oxygen sensor test connector located by the throttle body.
9. Select the 4 cylinder range and % button on the analyzer. Start the engine. The duty cycle should be 25–65% and 0.3–1.2% CO.
10. If the CO is greater than 1.2% and the duty cycle is properly 25–65%, check the ignition system, the exhaust system and the fuel distributor for defects.
11. If the % reading is not correct, remove CO adjustment plug as follows:
   a. Stop the engine and remove the boot on the mixture control unit.
   b. Lightly centerpunch the plug in the CO adjusting hole.
   c. Drill a $^{3}/_{32}$ in. (2.5mm) hole in the center of the plug $^{9}/_{64}$–$^{5}/_{32}$ (3.5mm) deep.
   d. Carefully clean up any metal shavings and insert a ⅛ in. sheet metal screw into the drilled hole and remove the screw and plug.
10. Start the engine and allow it to idle.
11. Adjust the duty cycle by turning the CO adjusting screw with tool P377 or equivalent to obtain 42–58%, fluctuating. Turn the adjusting screw clockwise to raise the current reading (CO higher) and screw clockwise to raise the current reading (CO lower).

NOTE: Do not push adjustment tool down when making CO adjustment and do not accelerate the engine with the tool in place. Remove the tool after each adjustment and accelerate the engine briefly before reading the multimeter. Always adjust in direction from high to low.

12. Run the engine and recheck the meter readings. After CO adjustment is complete, recheck the idle speed and adjust if necessary. Reconnect the crankcase breather hose. If, after reconnecting the breather hose, the current reading drops below specifications, an oil change may be necessary, due to excessive gasoline dilution.
13. Turn the ignition switch **OFF** and drive in a new mixture plug, flush with the mixture control unit. This must be done to meet federal anti-tamper regulations.
14. Remove all test equipment, attach CO probe cap and insert the charcoal canister sealing plug.

Dual Knock Sensors

1. Warm the engine to operational temperature, oil at 176°F (80°C). All electrical devices must be **OFF**. The engine cooling fan must not be running. Do not connect any fuel pressure gauges. If the injection lines had been disconnected, raise the engine speed past 3000 rpm and allow to idle for 2 minutes.
2. Disconnect the charcoal canister solenoid valve electrical connectors. One valve is located between the air intake boot and the injection lines and the other is located behind the intake boot by the fuel filter.
3. Remove the cap on the elbow mounted on the air intake boot. The elbow must remain mounted on the air intake boot. Remove the crankcase breather hose from between the air intake boot and the valve cover. Plug the hose.
4. Connect analyzer VW 1367 or equivalent, to terminals 1 and 15 on the ignition coil and as specified by the manufacturer. Connect a CO tester to the exhaust tap. Be sure there is no leakage from the connection.
5. Connect test adapter VW 1315 A/2 or equivalent, between the harness and the connector on the idle stabilizer valve. Connect a multimeter to the leads of the test adapter. Set the multimeter to the 2A acale.
6. Start the engine and let idle at 670–770 rpm. Adjust the idle speed by turning the screw on the idle stabilizer valve.
7. Disconnect the green wire from the ignition coil junction block to activate the idle stabilizer preset value. The multimeter will read 410–450 mA. If it does not, check the idle stabilizer system. Connect the green wire to the ignition coil junction block and recheck the multimeter. The multimeter will read 410–450 mA and fluctuate.
8. Remove the inductive pickup from the ignition lead, disconnect the green wire from the ignition coil junction block and connect to the blue and white oxygen sensor test connector located by the throttle body.
9. Select the 4 cylinder range and % button on the analyzer. Start the engine. The duty cycle should be 25–65% and 0.3–1.2% CO.
10. If the CO is greater than 1.2% and the duty cycle is properly 25–65%, check the ignition system, the exhaust system and the fuel distributor for defects.
11. If the % reading is not correct, remove CO adjustment plug as follows:
   a. Stop the engine and remove the boot on the mixture control unit.
   b. Lightly centerpunch the plug in the CO adjusting hole.
   c. Drill a $^{3}/_{32}$ in. (2.5mm) hole in the center of the plug $^{9}/_{64}$–$^{5}/_{32}$ (3.5mm) deep.
   d. Carefully clean up any metal shavings and insert a ⅛ in. sheet metal screw into the drilled hole and remove the screw and plug.
10. Start the engine and allow it to idle.
11. Adjust the duty cycle by turning the CO adjusting screw with tool P377 or equivalent to obtain 42–58%, fluctuating. Turn the adjusting screw clockwise to raise the current reading (CO higher) and screw clockwise to raise the current reading (CO lower).

NOTE: Do not push adjustment tool down when making CO adjustment and do not accelerate the engine with the tool in place. Remove the tool after each adjustment and accelerate the engine briefly before reading the multimeter. Always adjust in direction from high to low.

12. Run the engine and recheck the meter readings. After CO adjustment is complete, recheck the idle speed and adjust if necessary. Reconnect the crankcase breather hose. If, after reconnecting the breather hose, the current reading drops below specifications, an oil change may be necessary, due to excessive gasoline dilution.
13. Turn the ignition switch **OFF** and drive in a new mixture plug, flush with the mixture control unit. This must be done to meet federal anti-tamper regulations.
14. Remove all test equipment, attach CO probe cap and insert the charcoal canister sealing plug.

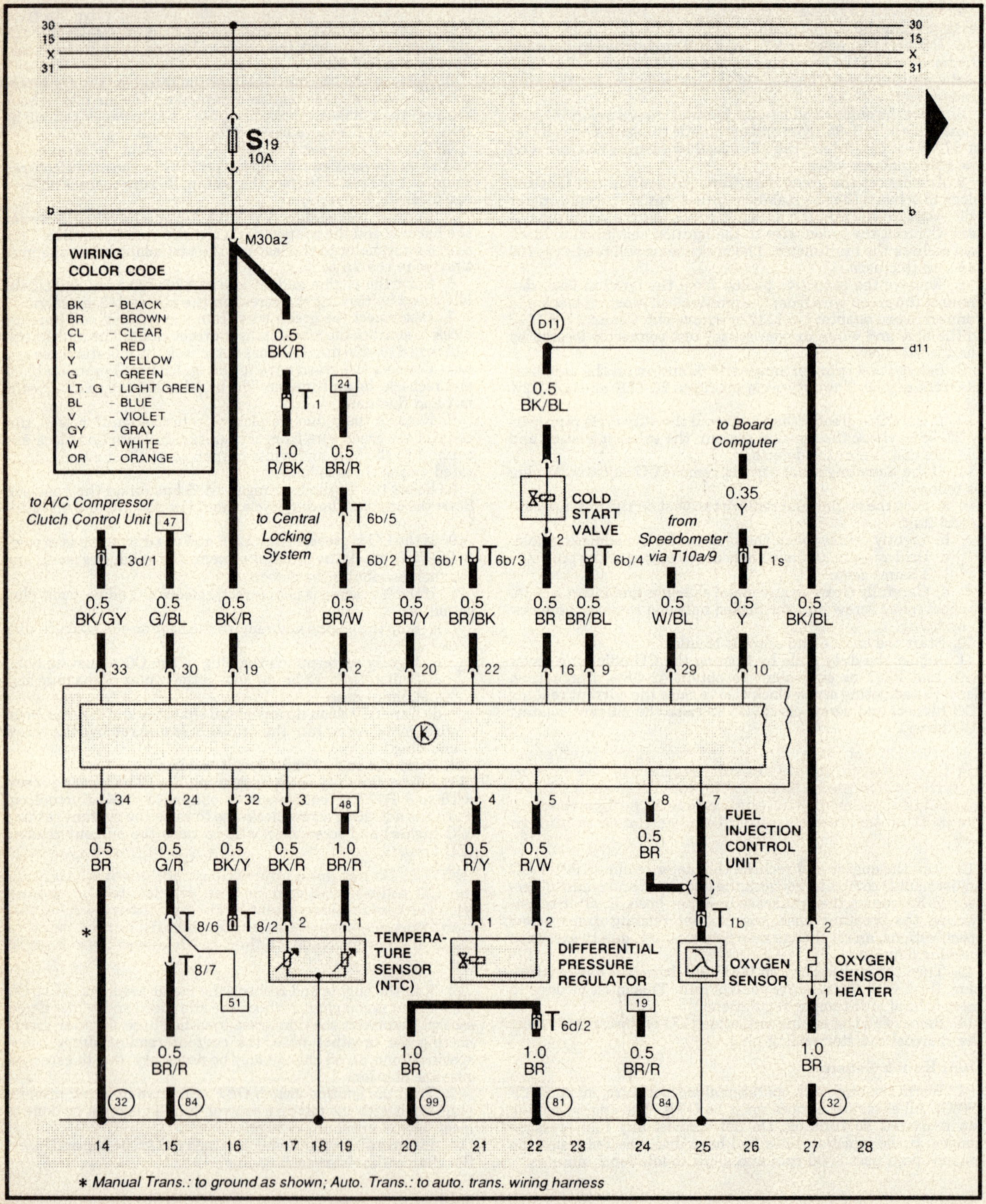

Engine wiring—Audi 80 Quattro and 90 with CIS-E III

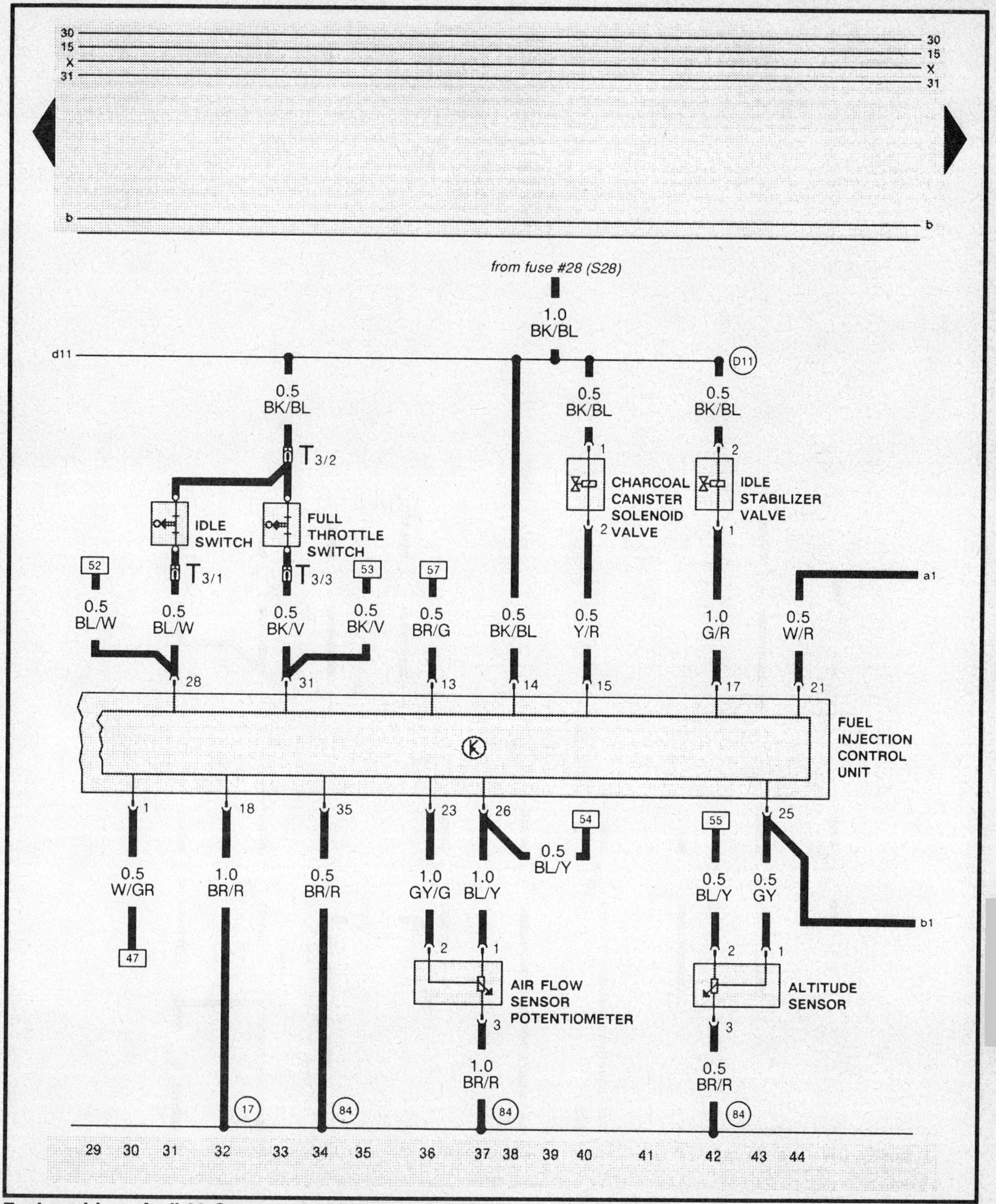

Engine wiring—Audi 80 Quattro and 90 with CIS-E III

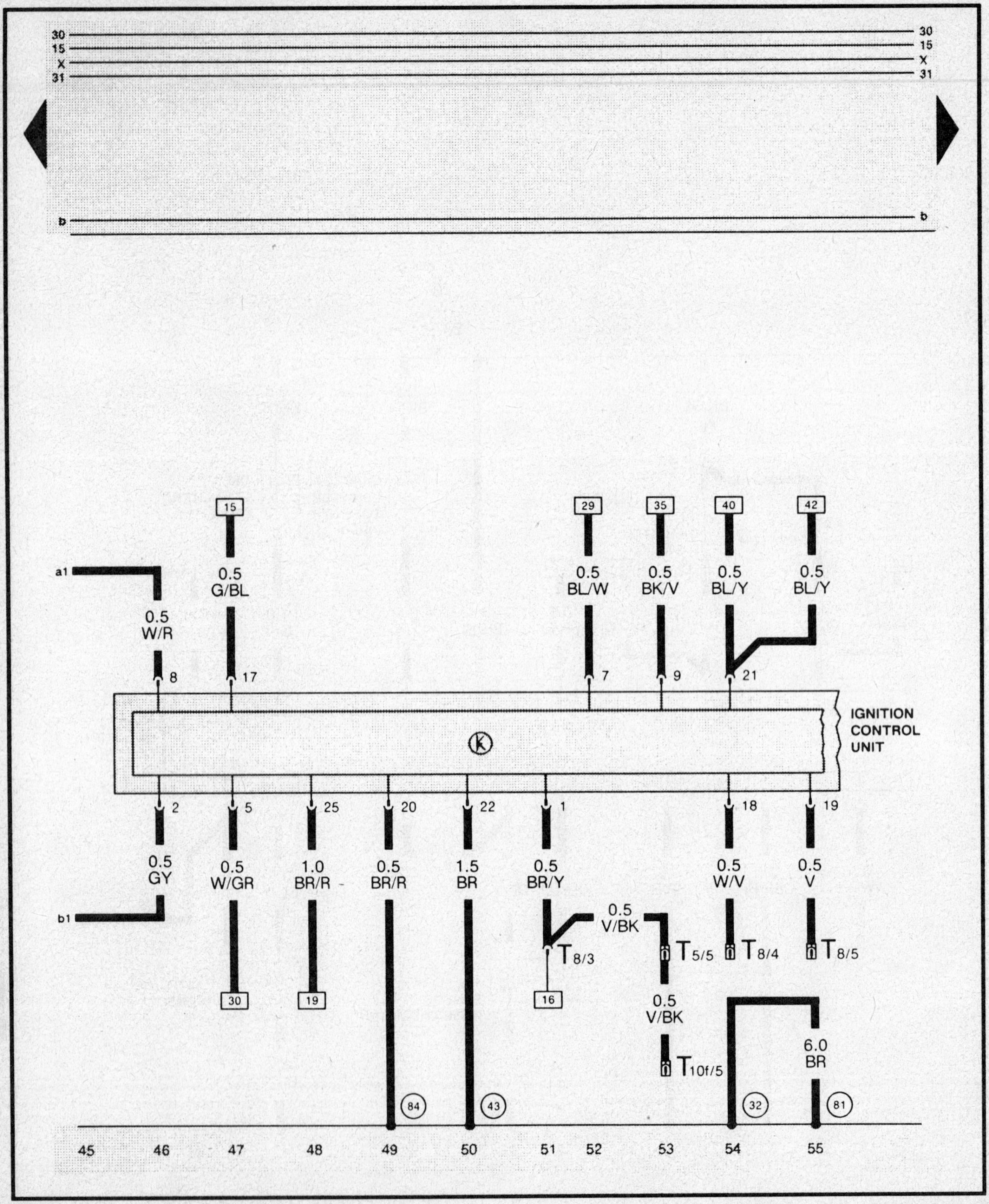

Engine wiring—Audi 80 Quattro and 90 with CIS-E III

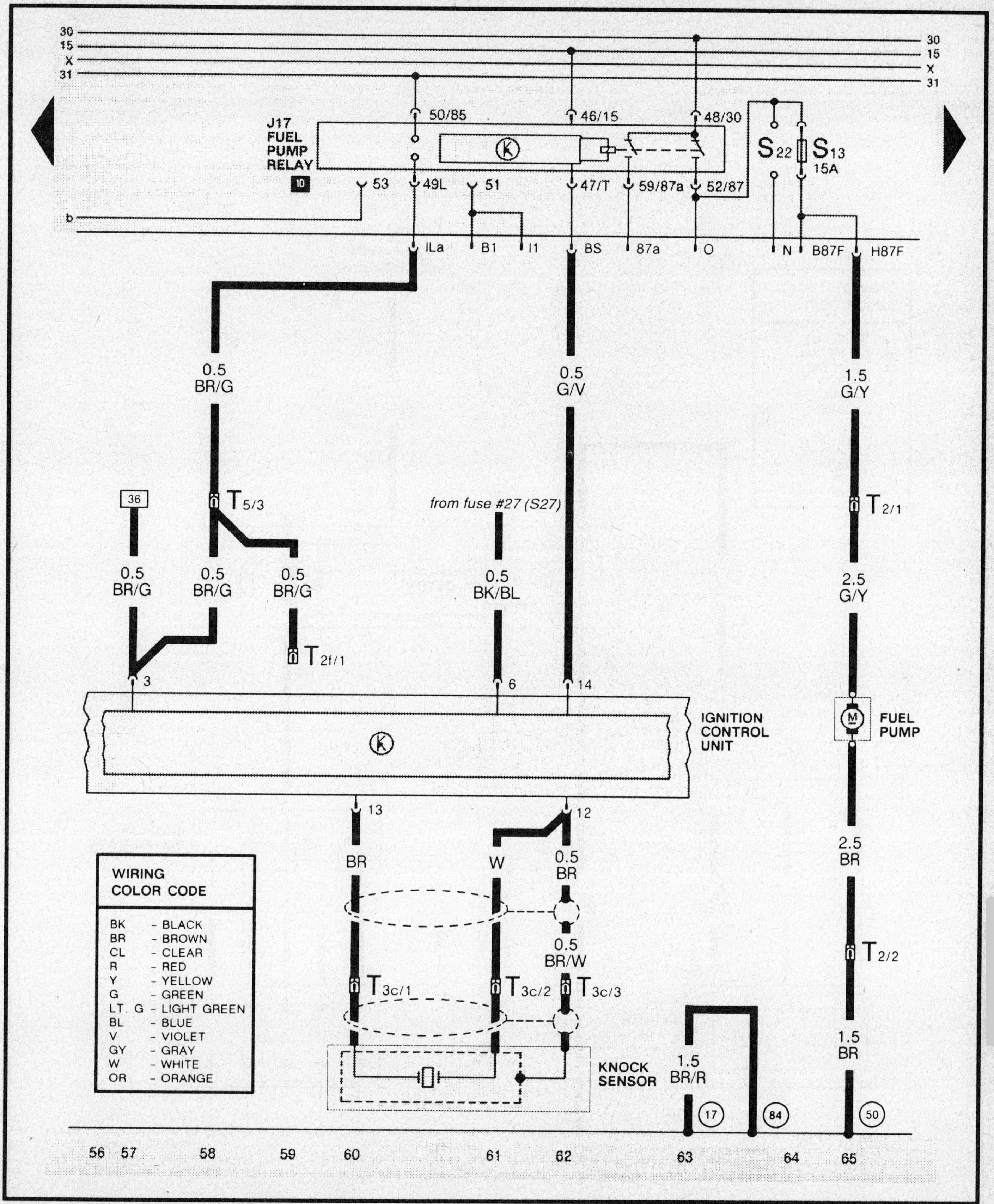

Engine wiring—Audi 80 Quattro and 90 with CIS-E III

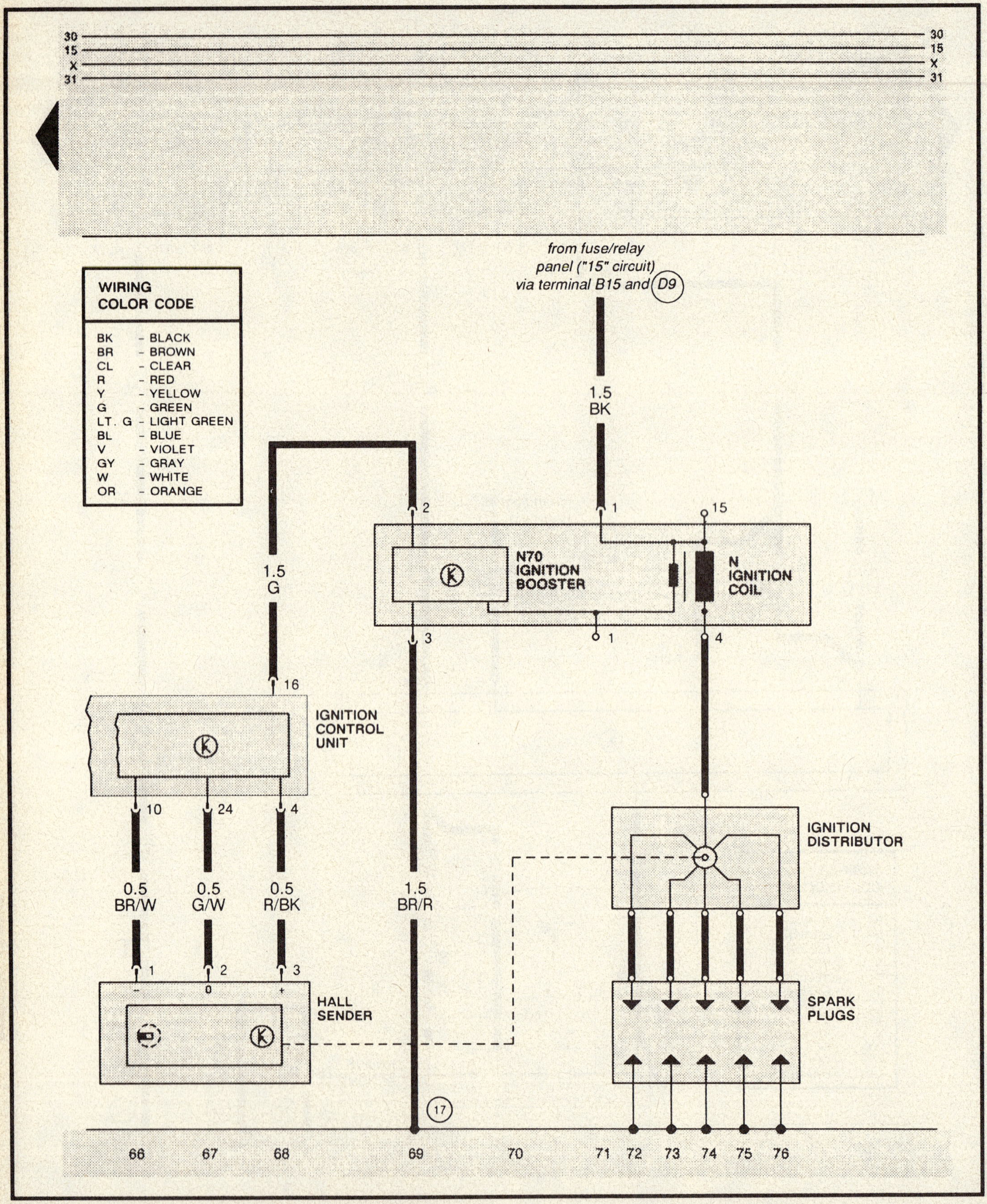

Engine wiring—Audi 80 Quattro and 90 with CIS-E III

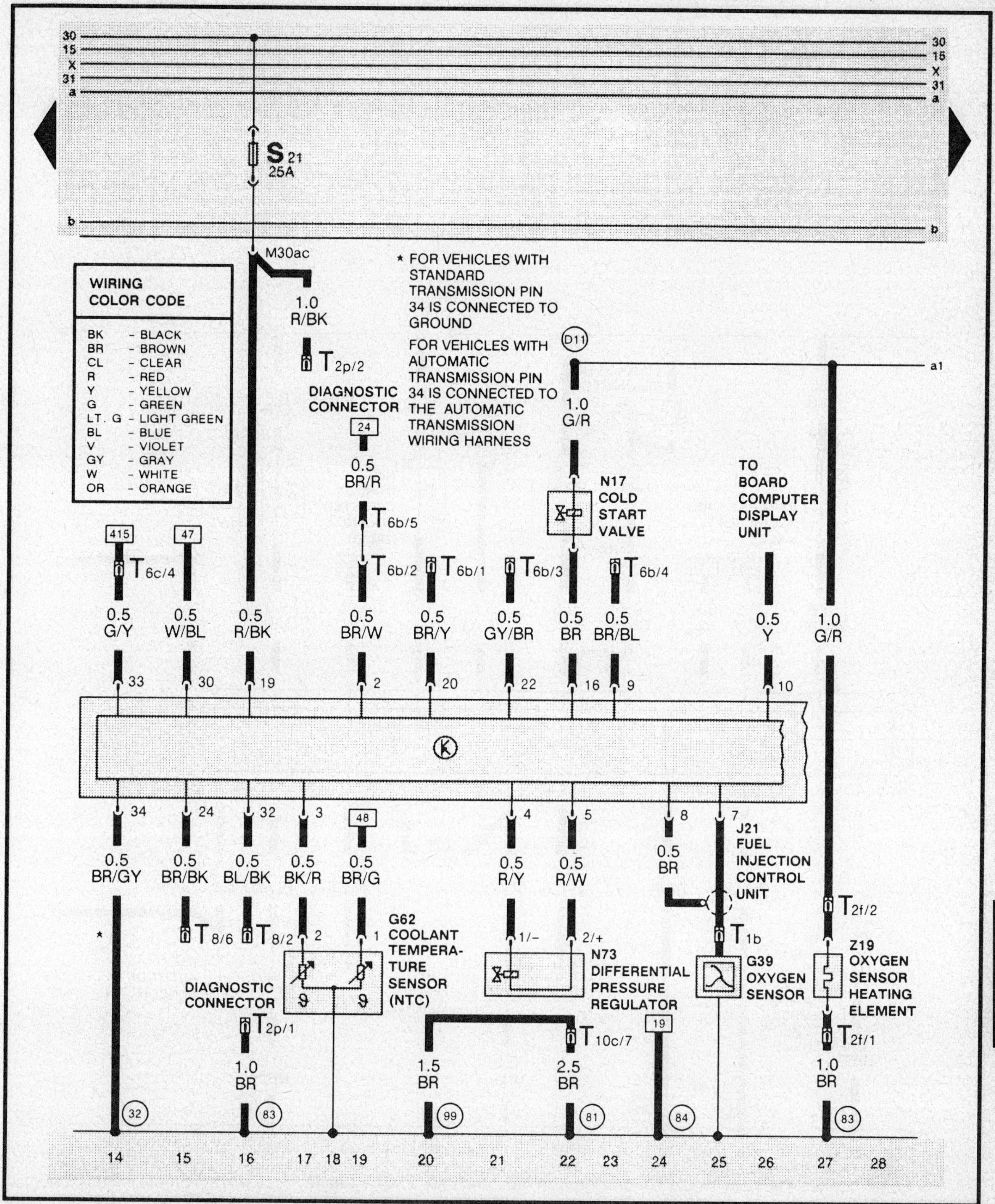

Engine wiring—Audi 100 and 100 Quattro with CIS-E III

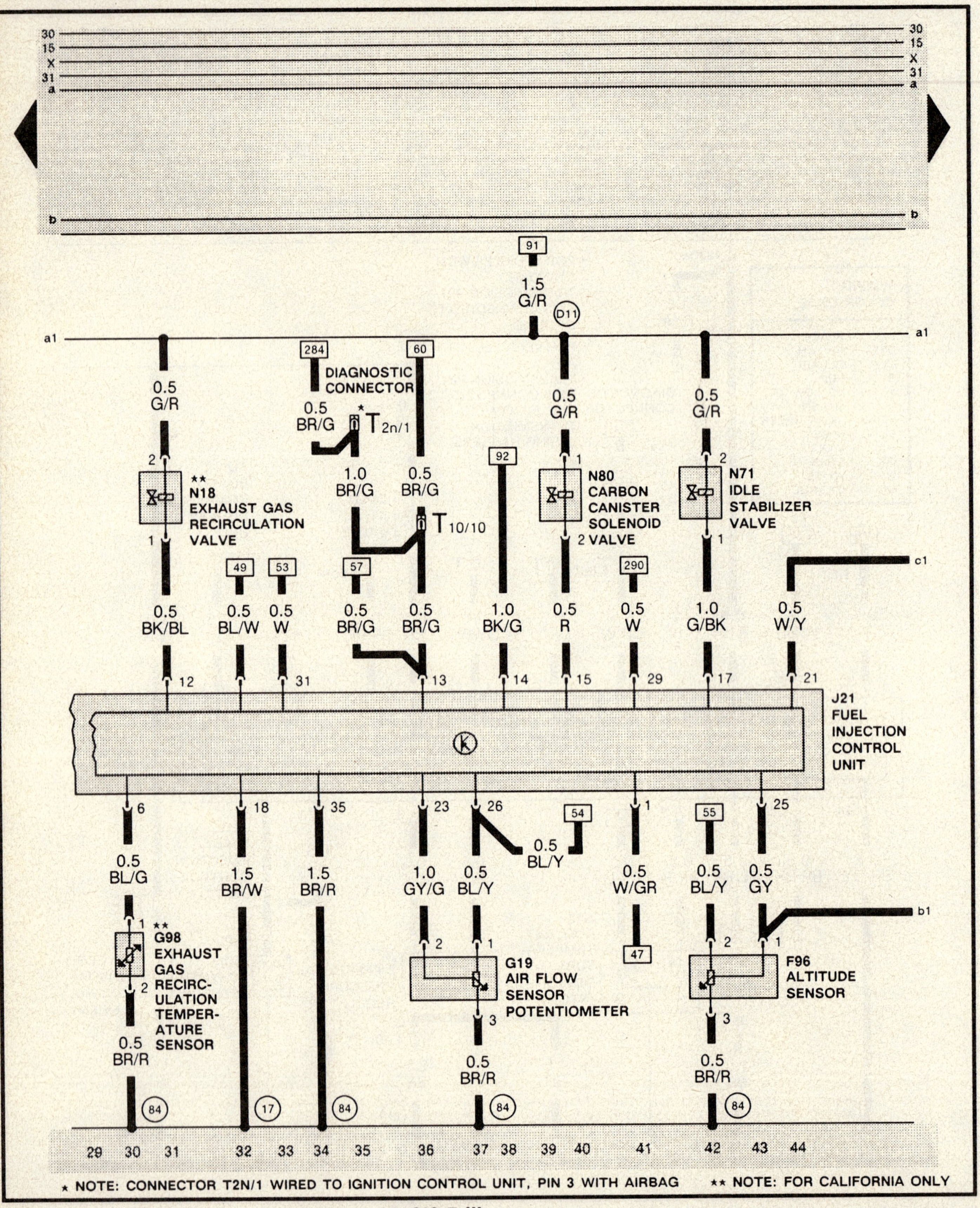

Engine wiring—Audi 100 and 100 Quattro with CIS-E III

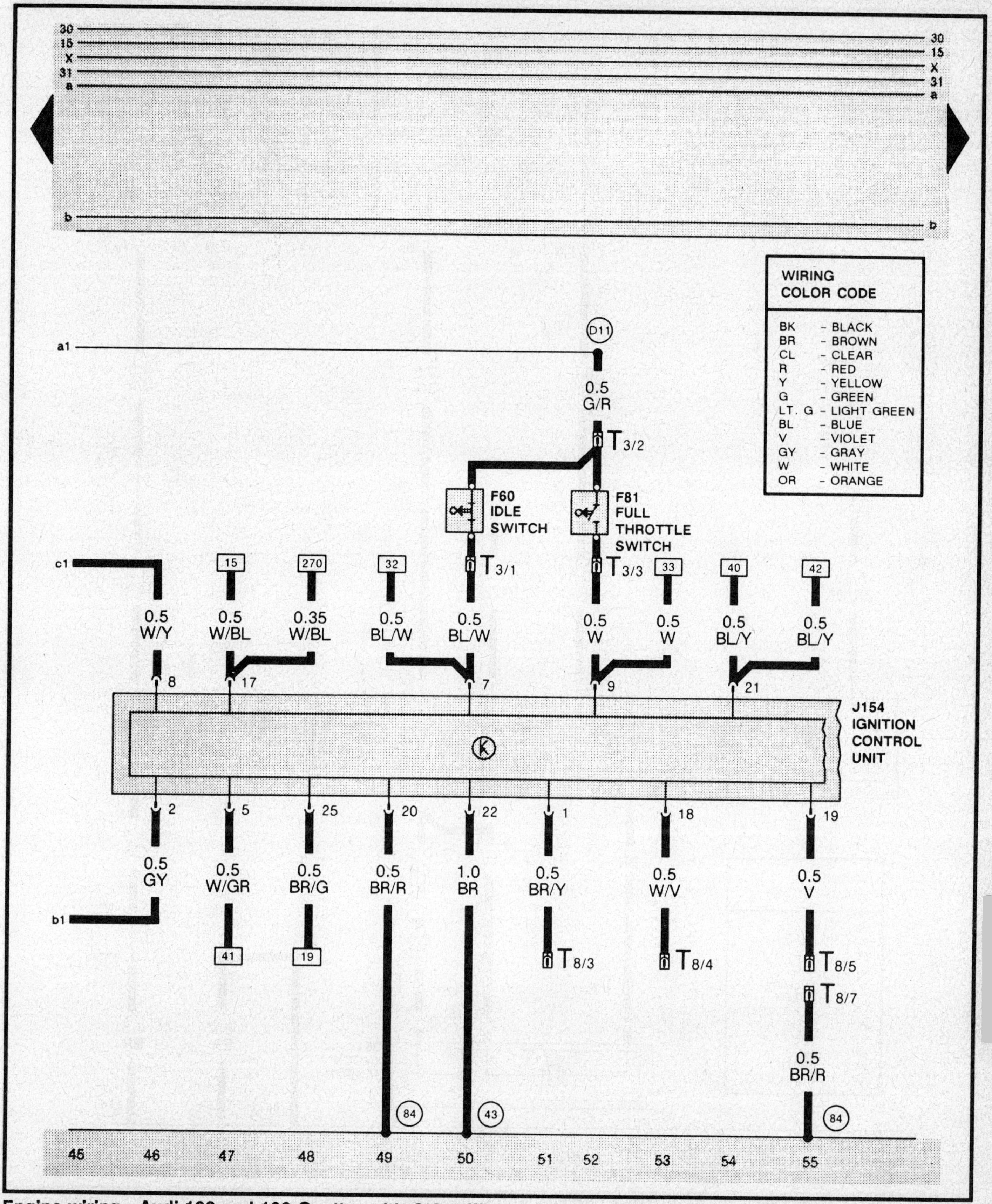

Engine wiring—Audi 100 and 100 Quattro with CIS-E III

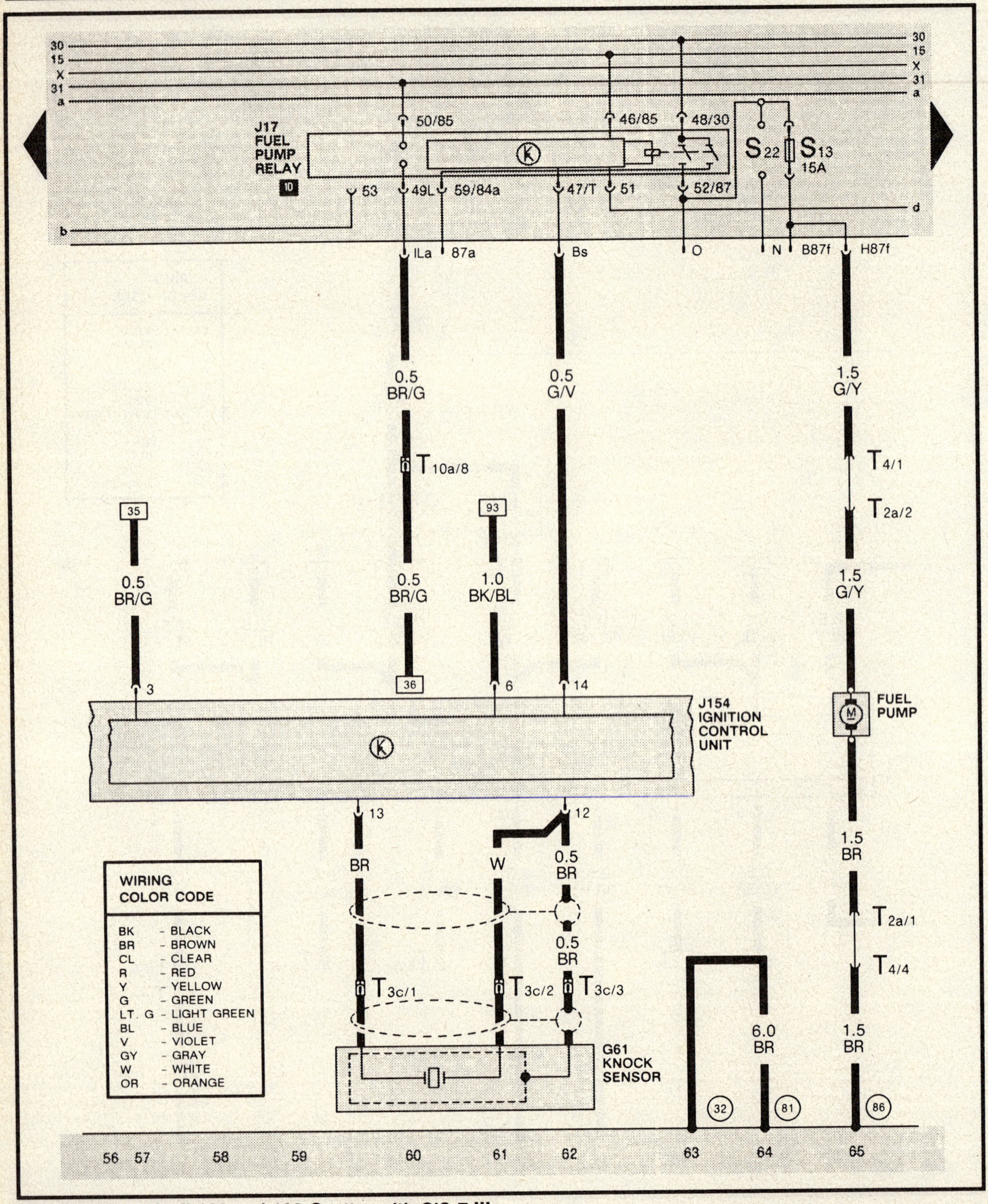

Engine wiring—Audi 100 and 100 Quattro with CIS-E III

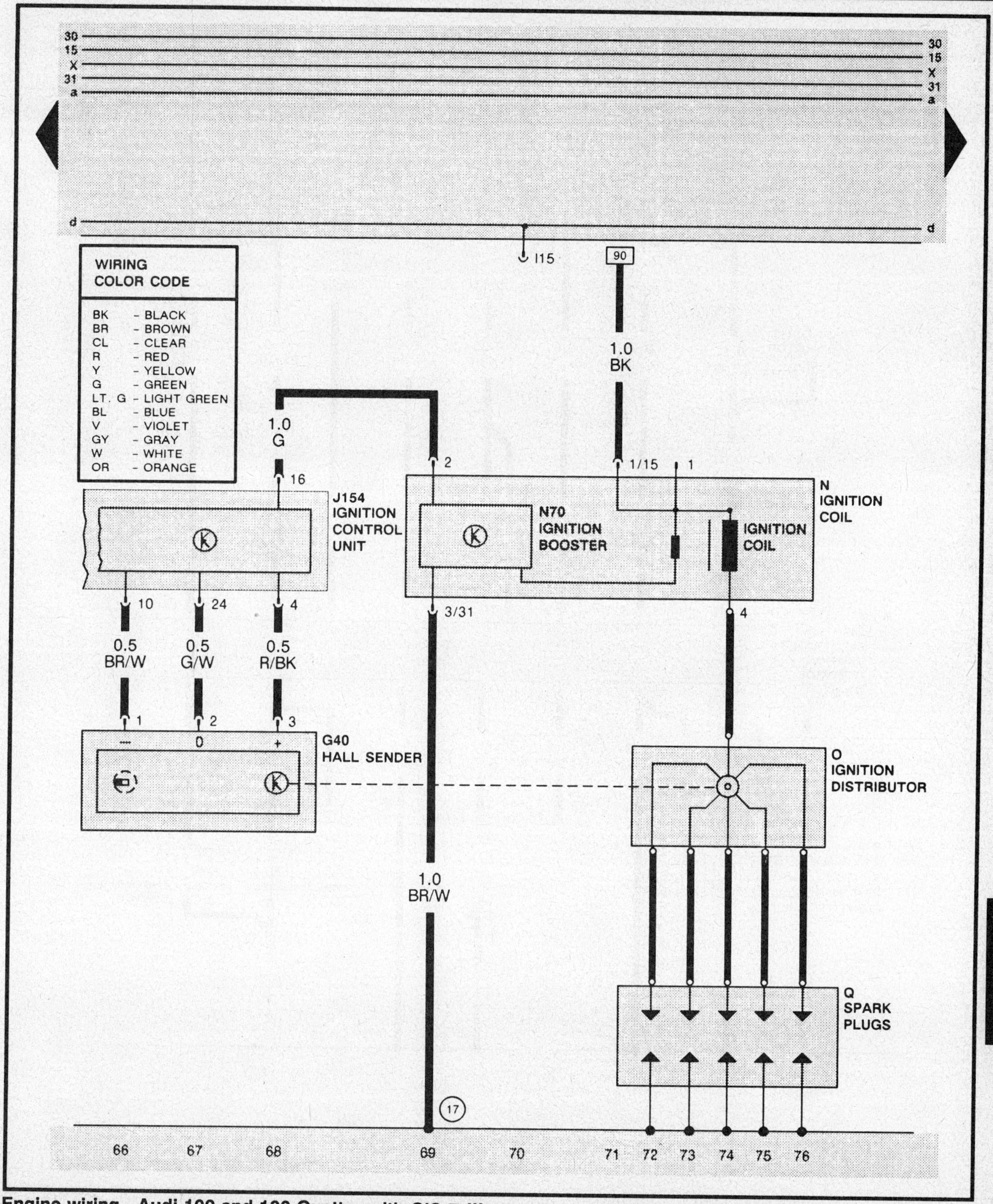

Engine wiring—Audi 100 and 100 Quattro with CIS-E III

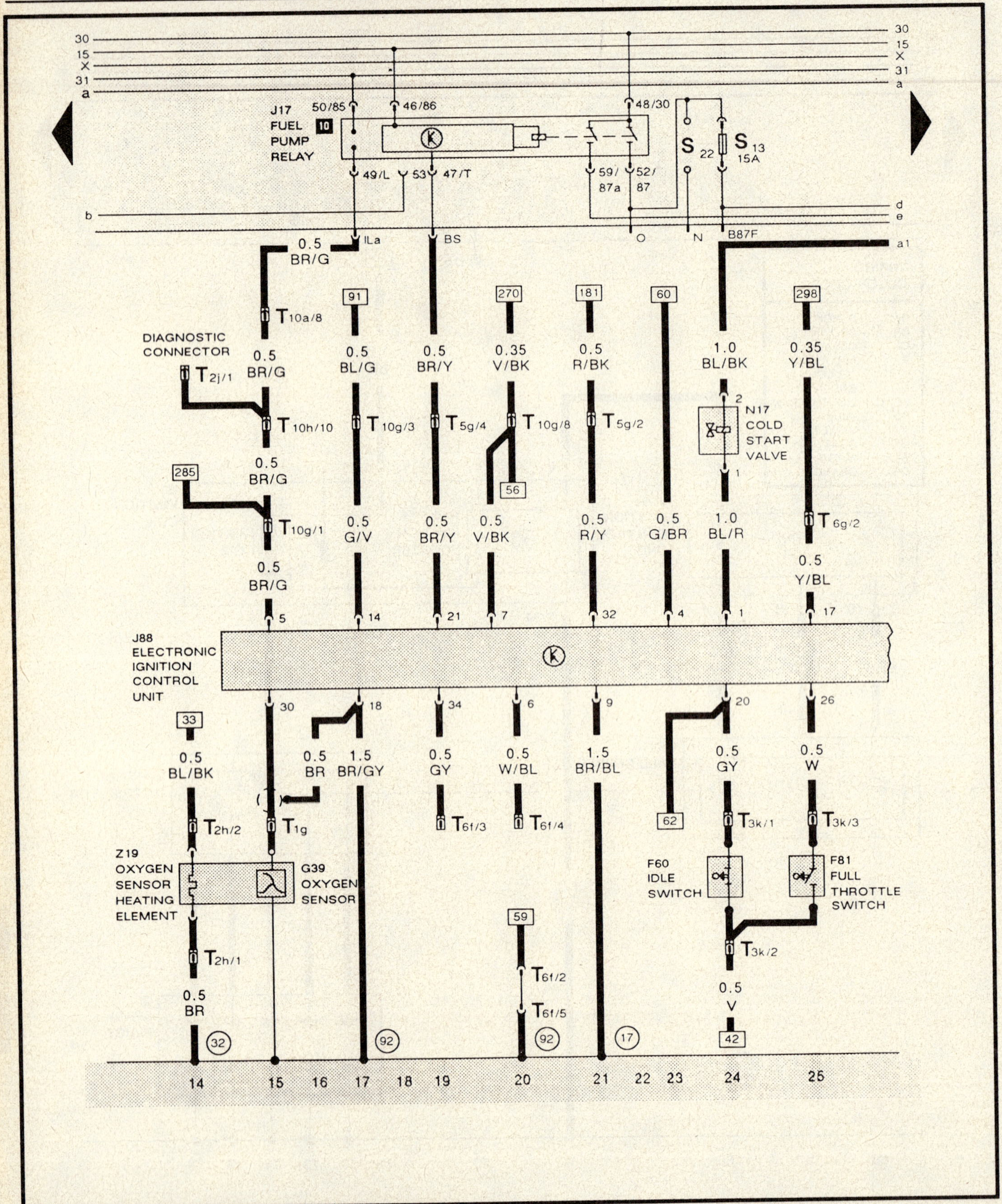

Engine wiring—Audi 200 and 200 Quattro with CIS-E III

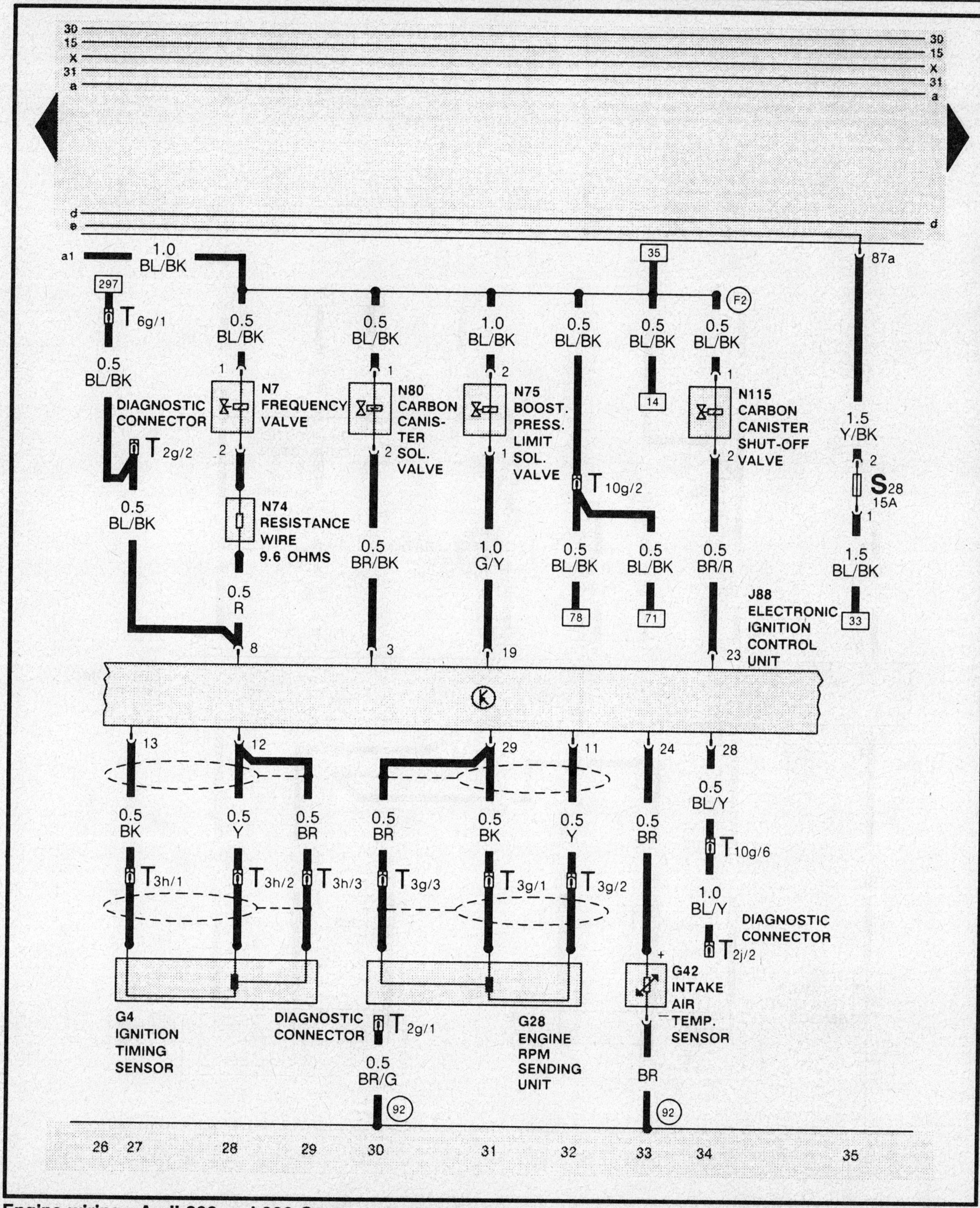

Engine wiring—Audi 200 and 200 Quattro with CIS-E III

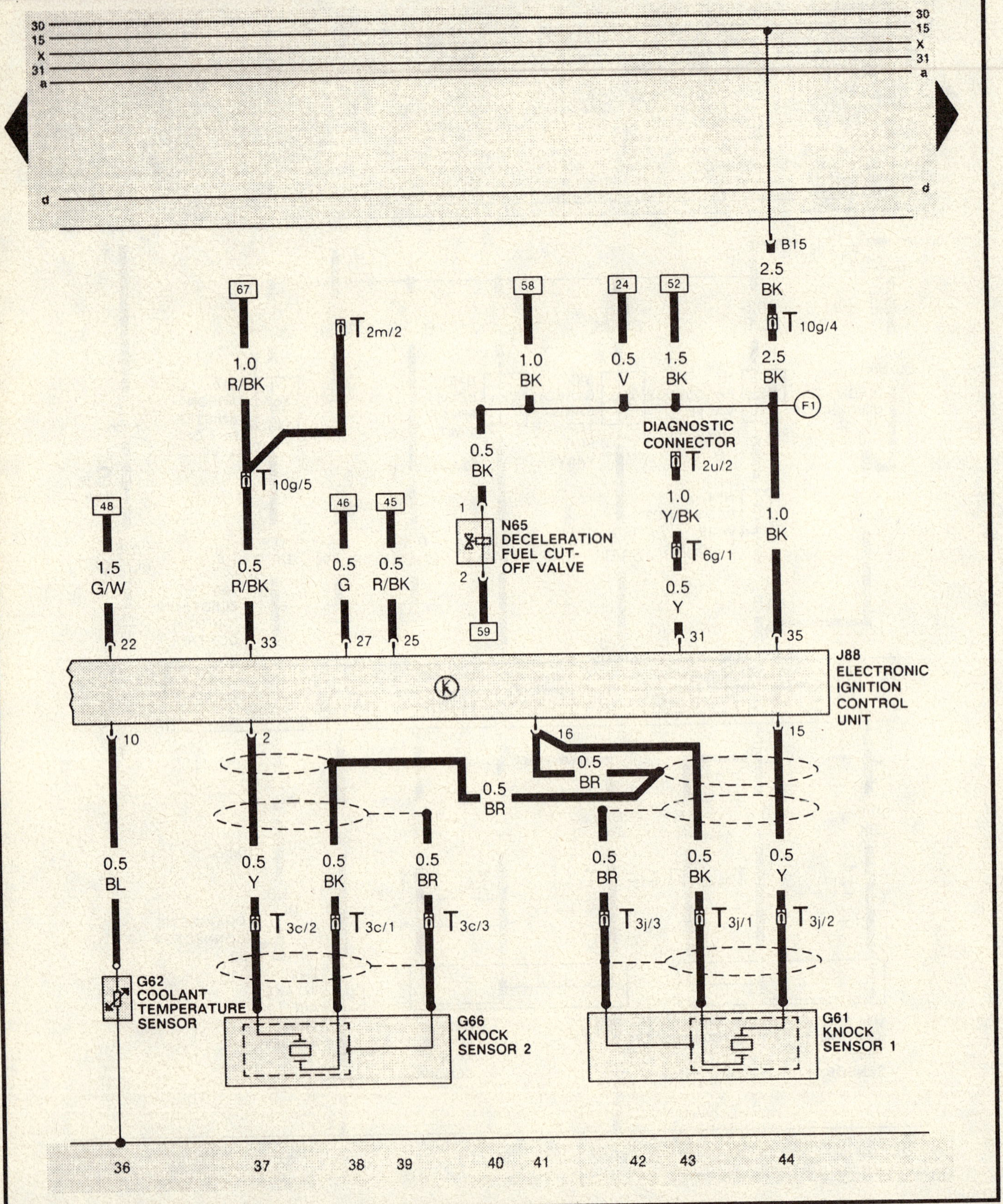

Engine wiring—Audi 200 and 200 Quattro with CIS-E III

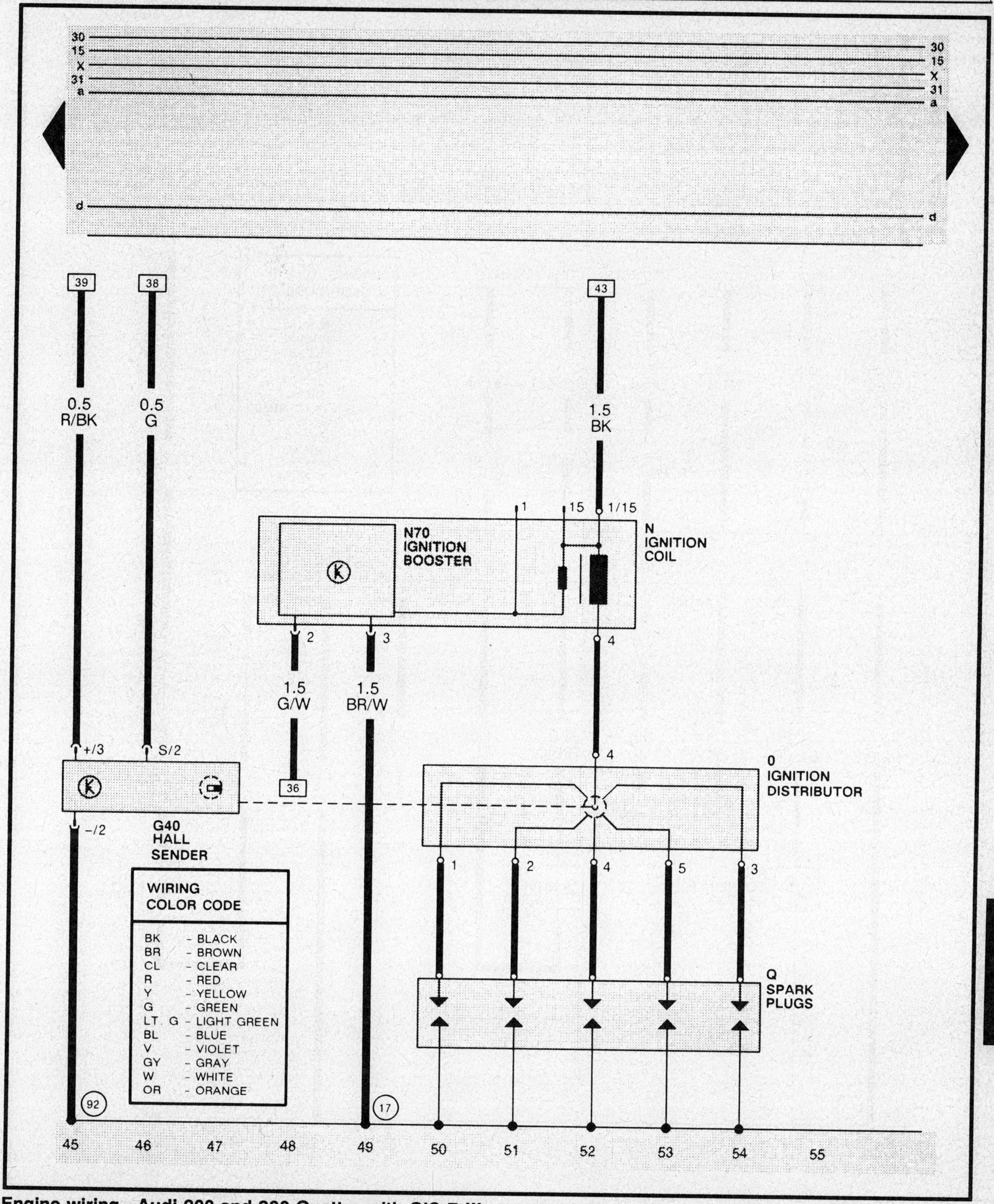

Engine wiring—Audi 200 and 200 Quattro with CIS-E III

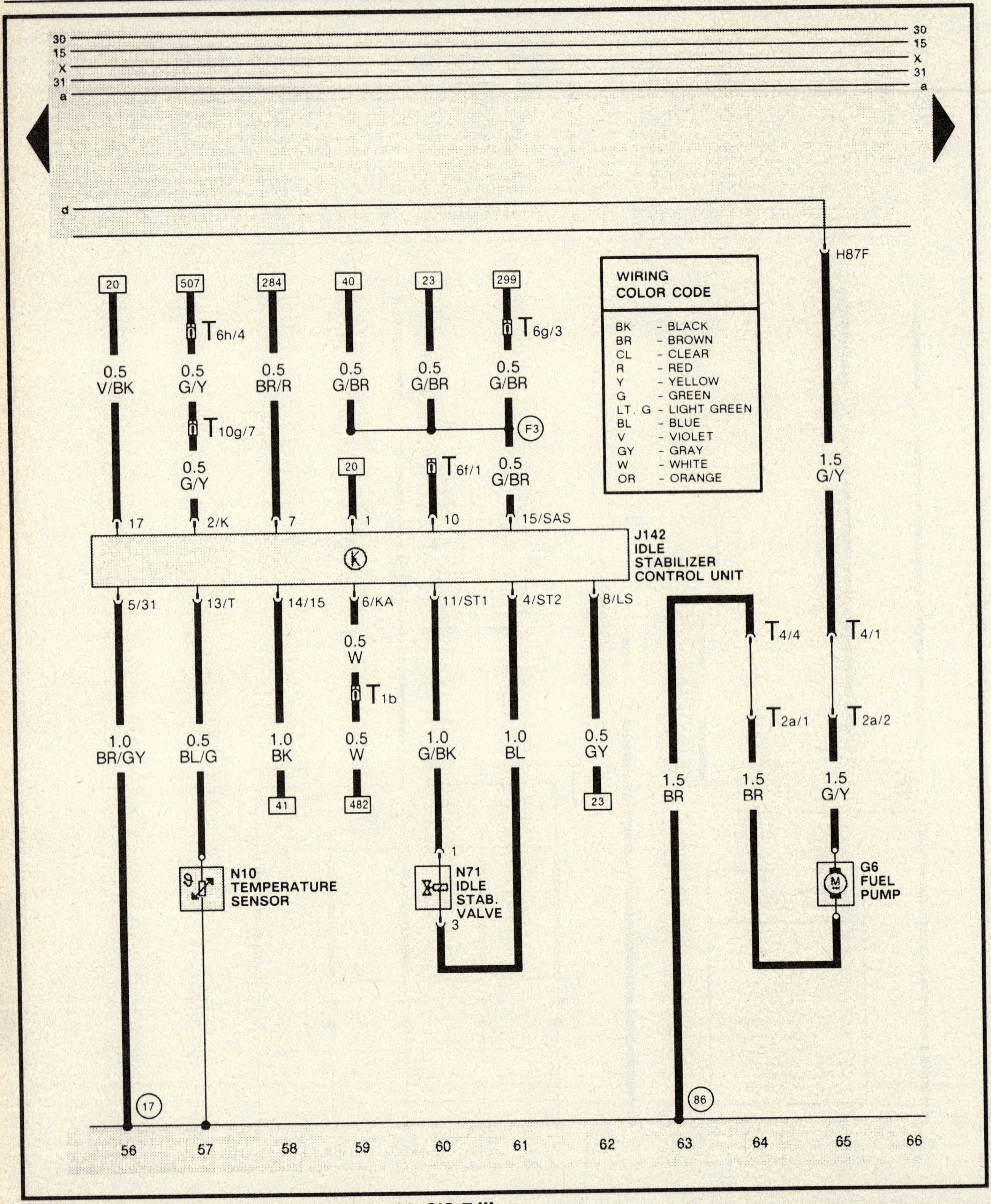

Engine wiring—Audi 200 and 200 Quattro with CIS-E III

BMW Fuel Injection

7 SECTION

BMW DIGITAL MOTOR ELECTRONICS (DME) FUEL INJECTION

General Information

All vehicles covered in this section are equipped with the Bosch Digital Motor Electronics (DME) system. The DME system operates the ignition and fuel injection systems with a single control unit. In 3 Series vehicles the unit is accessed through the glove compartment opening and is located above the glove compartment under a trim piece. In the 5, 7 and 8 Series vehicles the unit is mounted under the hood, in a covered box on the right side of the firewall. Vehicles with the 12 cylinder M70 use 2 complete DME systems and control units, mounted in the same location.

The DME control unit accepts input signals from the intake air flow or air mass sensor, crankshaft pulse sensor, cylinder identifying sensor, coolant and air temperature sensors, an exhaust oxygen sensor, a throttle position switch and if so equipped, the automatic transmission control unit. Optimum fuel quantity and spark timing requirements are calculated based on sensor inputs and internal programs and are adjusted as required for each engine revolution. Output signals operate the idle control valve, fuel pump relay, fuel injectors and the ignition system. Fuel injectors are operated sequentially and air/fuel ratio, ignition timing and idle speed are not adjustable.

Most items of the DME system can be tested with a volt/ohm meter, however an oscilloscope is required for testing some control functions. A variety of test connectors is also required but some of these can be easily made up as needed. The control unit includes a self diagnostic program that will store fault codes.

The first step in repair or service to engine management systems is to gain as much information as possible from the vehicle owner, especially regarding intermittent problems. Also check any service records which may be available. Before checking fault codes, it is absolutely essential to check for any obvious mechanical faults or failures. Remember, a trouble code only indicates which sensor or circuit is effected by the problem. Simple mechanical faults such as a vacuum leak or poor electrical connection can cause a fault code.

NOTE: Never disconnect any wiring with the ignition switch ON. Always turn the ignition switch OFF when using an ohmmeter. Any time the battery or main power terminal is disconnected, the control unit fault memory will be erased. Make sure all Codes have been retrieved before disconnecting the battery or removing the control unit connector.

DME CONTROL UNIT AND RELAY

3 Series

The DME control unit is located underneath the trim in the glove compartment. The relays are located in the engine compartment. The relays on the M20 equipped 3 series are located just ahead of the drivers side strut tower; the closest relay to the strut tower is the DME relay, the middle being the fuel pump relay and the furthest being the oxygen sensor relay. The relays on the M3 are located at the back edge of the engine compartment on the passenger side; the relay closest to the engine is the fuel pump relay, the middle being the master relay and the furthest being the tank venting relay. The relays on M42 equipped 3 series are located at the back edge of the engine compartment on the drivers side; the relay closest to the engine is the DME relay, the middle being the fuel pump relay and the furthest being the oxygen sensor relay.

DME SYSTEM COMPARISON

SYSTEM	DME 3.1	DME 1.1–1.3
Ignition	Direct, solid state	Rotary Distributor
Cylinder Identification	Camshaft sensor	Sensor on cylinder 6 ignition lead for semi-sequential fuel injection
Load sensing	Hot wire air mass metering	Volumetric air flow metering
Type of fuel injection	Fully sequential	Parallel or semi-sequential
Throttle position sensing	Throttle butterfly potentiometer; adaptive idle setting; part and full load by way of resistance value	Throttle butterfly switch, idle and full load by mechanical switch contact
Starting	Advance fuel pump run and injection; double spark (max. 250 ms); fully sequential injection	Parallel injection
Full load ignition timing	Altitude compensation; dependent on engine speed and cylinder charge	Dependent on engine speed
Idling control	Adaptive value with/without air conditioning	Single adaptive value
Model application identification	By mapped characteristic	By variant coding
Self-diagnostics	Extended	
Emergency running	Improved from DME 1.1–1.3	
Memory board	40 Kb	32 Kb
Plug pins	88	55
Electromagnetic compatibility	4X multilayer technique	Additional suppression

5 and 7 Series

M30 AND S38 ENGINES

The DME control unit is mounted in the electronic control box under the hood, on the right side of the firewall. With the cover removed, the DME control unit is closest to the firewall. The box also contains the cruise control module, the automatic transmission control module, the main DME relay, the fuel pump relay, a main power connection and with the M30 engine, the oxygen sensor heater relay. Some models have a fan below the DME control unit that runs when ever the ignition switch is on.

The main DME relay coil receives 12 volts at all times on terminal 86 and the control unit supplies ground to terminal 85 to operate the relay. When the relay is closed, voltage is supplied to the fuel injectors, idle speed actuator, carbon canister purge valve, air mass sensor, ABS control unit, ignition system and the fuel pump relay. On vehicles equipped with the factory installed anti-theft system, the control unit will not operate the relay when the system is armed.

7 and 8 Series

M70 ENGINE

The management system for this engine is two complete and separate fuel and ignition systems, from the fuel pumps all the way to the injectors. The engine is also operated with a "drive by wire" system, meaning the throttles are operated by electric motors instead of being mechanically linked to the pedal. An additional electronic control unit, called the EML, is used to synchronize and operate the throttle motors. If the battery is disconnected or if a control unit is disconnected or replaced, the 3 control units must be synchronized. Other than these features, the control strategy and DME system components are basically the same as other models. Testing the DME relay is the same as described above, be sure to use the correct wiring diagram.

DATA SENSORS

Air Flow Meter

Two types of air flow sensor are used, an electronic air mass sensor and an electro-mechanical vane type air flow meter. On the vane type, when air enters the housing and moves the air vane, the vane moves a potentiometer. The control unit sends a voltage to the potentiometer and the return signal to the control unit is calculated as air flow. The air flow meter housing also contains a temperature sensor at the inlet side. If either the flow meter or the sensor fails, the control unit will substitute preprogrammed values to allow "limp home" operation of the engine. On previous models, the CO and idle speed adjustments were on the air flow meter. These adjustments have been eliminated, CO and idle speed can no longer be adjusted. The air flow meter and temperature sensor cannot be repaired.

Air Mass Sensor

This unit is a fully electronic hot wire air mass sensor with no moving parts. It consists of a platinum resistor wire that is part of a bridge circuit that includes a measuring resistor. An amplifier mounted on the side of the housing controls and monitors this circuit. The temperature of the wire is maintained at a constant 311°F (155°C). As air flow changes, the current required to maintain the temperature changes. As the circuit adjusts the current, the measuring resistor reports the change in voltage to the control unit, which then calculates the mass of air flowing through the sensor housing. This type of air measurement system automatically compensates for changes in air temperature, humidity and altitude. When the engine is stopped and after a delay of approximately 4 seconds, the bridge circuit is altered for approximately 1 second. The hot wire is heated to a temperature of approximately 1832°F (1000°C) to burn free any deposits which could influence the resistance or measuring results. On the 12 cylinder M70 engine, the intake manifolds cross over the engine. This means the left air mass sensor and throttle body are for the right bank of cylinders (1-6) and operate with DME system 1. The air mass sensor and throttle body on the right side control the left bank of cylinders (7-12) and operate with DME system 2.

When testing or removing the air mass meter, do not touch the hot wire element or any other electronic parts inside. Protect the element from dirt and fluids. The screen in front of the element can be cleaned or replaced but there are no other serviceable parts.

Intake Air Temperature Sensor

On vehicles with a vane type air flow sensor, the air temperature sensor is inside the housing and must be tested as part of the air flow meter circuit. On 4 and 6 cylinder vehicles with an air mass sensor, the air temperature sensor is in the air filter housing or in the intake manifold. On 12 cylinder engines, the sensor is threaded into the rear of the right side intake manifold. The sensor reports to DME control unit 1 on the same circuit as the coolant temperature sensor.

Coolant Temperature Sensor

On 6 cylinder engines, 2 sensors are mounted near the thermostat housing. The blue sensor is for the DME control unit, the brown sensor is for the temperature gauge. The 12 cylinder engine uses 3 sensors. The temperature gauge and the EML control unit each have their own sensor. The DME control units share the same sensor, one terminal connected to each DME control unit with control unit 1 providing the ground connection. All 3 sensors are mounted in the coolant passage at the rear of the engine, under the fresh air housing. On the M42 engine, the coolant temperature sensors are located under the intake manifold on the engine. The forwardmost sensor is the DME sensor and the rear sensor is for the temperature gauge. The coolant temperature sensors on the M3 are located in the coolant pipe above the exhaust side of the engine. The rear sensor services the DME and the forward services the temperature gauge.

The sensors are an NTC (negative temperature coefficient) type, meaning as the temperature increases, electrical resistance of the sensor decreases. The control unit reads the voltage change in the circuit as temperature. They are the same type as the air temperature sensors and can be tested the same way using the same temperature/resistance chart. The schematic for the 4 and 6 cylinder engines appears with the air mass meter schematic.

Throttle Position Switches

The throttle position switches are used to tell the control unit when the throttle is fully closed or fully open. They are mounted in a single housing on the end of the throttle assembly. If either switch fails, the control unit will use air meter signals to calculate idle and full load control and a fault code will be set in memory. Before replacing or adjusting the switch, make sure the throttle cable is properly adjusted.

Accelerator Pedal Sensor

On vehicles equipped with the EML (throttle by wire) throttle control system, a pedal sensor is used to send pedal position and speed of movement to the EML control unit. This unit is mounted in the box with the DME control unit and has its own diagnostic program and fault codes. The sensor is attached to a bracket above the accelerator pedal and mechanically connected to the pedal. When the sensor is removed or replaced, it must be adjusted and tested for correct output signal. The EML system can only be adjusted and tested with the BMW Diagnostic Tester.

Speed/Position Reference Sensor

Engine rpm and crankshaft position are detected with a sensor mounted on the front of the engine. If this sensor fails, the engine will not run. A special wheel is mounted on the harmonic balancer on the front of the crankshaft. As the teeth in the wheel pass the sensor, a magnetic pulse is generated and sent to the control unit, which reads the pulse frequency as engine rpm. A wide gap in the teeth indicates TDC of number 1 cylinder but the gap will not be at the same place on the wheel as the 0/T timingmark stamped on the wheel used to manually set the engine to TDC. On 12 cylinder engines, 2 sensors are positioned to read the same wheel and each reports to a DME control unit.

The sensor is a magnetic pick-up mounted in a bracket that allows adjustment of the gap between the sensor and the wheel. Power to the sensor and the sensor resistance can be checked with a volt/ohm meter. Sensor function can only be checked with an oscilloscope.

Cylinder Identifying Sender

This is an inductive pulse sensor similar to the crankshaft sensor. It consists of a coil attached to an ignition wire, supplied with 5 volts from the control unit. When that spark plug is fired, the fluctuation in the return signal is used as a real time reference for sequential fuel injection and ignition control functions. If the sender fails while the engine is running, the engine will be operated in parallel fuel injection control instead of sequential.

There are 2 types of sensors. One type of sensor is attached to spark plug wire No. 6 or 12 near the distributor. The sender can be replaced by cutting the distributor end off the ignition wire and sliding the sender off the end. The other type of sensor reads off the camshaft and is similar to the speed reference sensor. It is bolted into the cam cover and is used on the M42 and M50 engines.

Oxygen Sensor

The oxygen sensor consists of a ceramic tube coated with a thin layer of platinum. The closed end of the tube is in the exhaust stream and the reference end is open to the air. When the sensor is at operating temperature, oxygen in the exhaust will cause the unit to generate a voltage. The control unit reads the signal and adjusts the fuel injection to control the amount of oxygen in the exhaust. This is called "Closed Loop" control strategy. Under normal conditions, the control unit will constantly oscillate between rich and lean to create an average ideal air/fuel ratio. This means the output of the oxygen sensor also oscillates between 0 and 1 volt.

The sensor is equipped with an internal resistance heater to bring it to operating temperature quickly. Power for the heater comes from a separate relay mounted near the control unit. If the heater fails, the engine will take longer to switch into Closed Loop operation. If the sensor fails completely, the engine will be operated in Open Loop control and slightly lean to protect the catalytic converter from overheating.

CONTROLLED OUTPUTS

Throttle Motors

Vehicles with 12 cylinder M70 engine and others with the EML system do not have a throttle cable or idle speed actuator. Each throttle is operated with a motor and controlled by the same EML control unit. Power is supplied to the control unit from the No. 2 main DME relay. Inputs are accepted from the accelerator pedal sensor, vehicle speed sensor, coolant temperature sensor, throttle motor feedback potentiometers, air conditioner relay, the clutch switch on manual transmissions and the control units for anti-lock braking, anti-skid, traction control, door locks, body and instrument control units and the automatic transmission control unit. Output signals are sent to all of the control units as well as to the throttle motors.

The motors are a single winding DC type that is driven in both directions. The motor unit includes a position feedback potentiometer to tell the control unit throttle position and speed of movement. This pot also includes the throttle position switches whose state (open or closed) is compared with the state of the switches in the pedal sensor. If both switches are open at the same time, the engine will be operated at reduced power and a warning light is displayed on the instrument panel. If this condition exists, check the pedal sensor adjustment first.

Since this system communicates with so many of the other control units in the vehicle, it cannot be tested without the BMW Diagnostic Tester or equivalent.

As this system relies strictly on computer controls, with no direct mechanical control of the throttle plates, the integrity of the system is of utmost importance. Any time any work has been performed on the EML system, a check of the safety system incorporated into the system has to be made. This will examine the safety controls of the EML system. This is called the external safety path.

CAUTION

The throttle motors are strong enough to cause injury if the throttle closes on a finger. Use caution when working around the throttles.

Idle Speed Actuator

On vehicles with a throttle cable, idle speed is controlled with a motor operated rotary valve. The motor is double wound and can be driven in both directions. Power is supplied from the main DME relay anytime the ignition switch is **ON**. The control unit selects which of 2 ground circuits to complete to open or close the valve as required to maintain the programmed idle speed. Further idle speed control is accomplished with very fine ignition timing adjustments. This allows very fast idle speed control so that turning on the air conditioner compressor at idle will produce almost no momentary drop in idle speed.

Idle speed and idle CO cannot be adjusted or changed. If the either is incorrect and the valve works properly, look for a vacuum leak, ignition timing or fuel injection problems. The schematic for this circuit appears with the fuel injector circuit.

FUEL DELIVERY SYSTEM

Fuel Pump

EXCEPT M70 ENGINE

The fuel pump is mounted in the tank and can be reached through a panel in the floor of the trunk. On the top of the pump are a pressure damper a check valve that can be replaced. Wiring for the pump and gauge unit are on the same connector. The pump can be tested separately for delivery quantity but the pressure test is for the whole fuel system.

M70 ENGINE

On vehicles with the 12 cylinder engine, two fuel pumps and a sending unit are mounted side by side in the fuel tank. They can be reached through the access panel under the right rear seat. The pump and hose connection towards the front supply fuel to cylinders 7–12. Each pump incorporates a pressure damper and check valve to maintain system pressure. These cannot be replaced separately. The 2 pin wiring connector towards the front of the vehicle is for the gauge sending unit. The rear pump and hose connection are for cylinders 1–6 and the rear electrical connection is for power to both fuel pumps. The panel under the left rear seat provides access to the left sender unit and the fuel return and vent hoses. The smaller hose is the fuel return line from the engine.

Fuel Injector

EXCEPT M42 AND M50 ENGINES

The fuel injectors work under the same basic principles as the M42 and M50 engines, but the mode of operation is semisequential. The injectors open in groups of 3 as opposed to full sequential where each injector opens individually. The injectors are provided with 12 volts and grounded through the DME module.

M42 AND M50 ENGINES

The electric fuel injectors are operated sequentially according to engine firing order. If the cylinder identification sensor fails, all injectors will be operated in parallel. Injected fuel quantity is controlled by varying the pulse width (open time) of the injector signal. When the engine is above idle and the throttle is closed, the injectors are kept closed to improve fuel economy and emissions during deceleration. At engine speeds slightly above the rpm red line, the injectors are only allowed to operate every other engine revolution to prevent overspeeding the engine.

The injectors are basically precision solenoid valves. All injectors are supplied with 12 volts whenever the ignition switch is **ON**. The injectors are grounded through the DME unit and the DME unit controls the operation of the injectors through controlling the ground paths.

If the vehicle is equipped with the factory anti-theft system and the control unit receives a signal from the N4 body computer, the main DME relay is opened and the injectors will not operate. If the engine speed is above 480 rpm when the signal is received, the injectors will be cut off for 1 second intervals to prevent stalling in traffic.

Power to each injector from the DME relay can be checked with a voltmeter but the operation of the control unit circuit can only be observed with an oscilloscope. The injector coil can be tested with an ohmmeter. The injectors are held to the fuel supply rail with clips and the rail holds the assembly to the engine. On 6 cylinder engines with 2 valves per cylinder, the upper intake manifold must be removed to access the injectors and connectors. On engines with 4 valves per cylinder, the injectors can be reached by removing the cover. Use new O-rings when installing the injectors.

Diagnosis and Testing

SERVICE PRECAUTIONS

BMW offers the following cautions that should be observed to protect the DME system and components:

- Always disconnect the battery, the DME control unit and the ignition coils when using an electric welder on the vehicle or when charging the battery or when the vehicle is placed in a paint drying oven.
- Remove the main DME relay to disable both the fuel and ignition systems when checking the compression. Never crank the engine after removing the distributor cap or disconnecting the high tension wire on the ignition coil.
- Never disconnect the battery or wires on the alternator, starter or spark plugs when the engine is running.
- Never connect a test lamp on terminal 1 of the ignition coil.
- Never connect terminal 1 of the ignition coil with ground or B+. This means if installing a burglar alarm, terminal 1 should not be used for starter interlocking.
- Whenever performing work on the EML throttle by wire system, a check of the external safety path must be made. This is to check the proper operation of the EML safeguards.
- When working with the EML throttle by wire systems do not place hands or fingers in or near the throttle plates. The throttle motors can close the throttle plates with considerable force.

USING THE DIAGNOSTIC CHARTS

Diagnostic charts are provided for use with the BMW Diagnostic Tester, Bosch Tester or equivalent. The charts provide information that will be useful when used as a reference for the above testers and will not be of great use if used as a stand alone diagnostic aid. Note that DME system operating parameters are changed by the manufacturer during the model year to obtain gains in vehicle performance or drivability and the information contained within the diagnostic charts may not reflect the most current changes. The software provided by the diagnostic tester manufacturer will contain information pertinent to the DME unit and coding for that vehicle. The 3 Series vehicles contain a number of engines (M20, M42, M50 and S14) and a chassis change (E30 and E36), there may be DME coding changes and diagnostic information changes not reflected in the defect code charts. Check the component specification charts for component testing data.

EXTERNAL SAFETY PATH (EML)

Checking

The external safety path must be checked anytime a component of the EML throttle by wire system is removed or adjusted.

1. Read all fault codes from the EML memory before checking the external safety path.
2. Turn the ignition switch to **OFF**. Unplug the EML connector. The EML connector is grouped with the DME engine connectors. On the 850i it is the connector closest to the engine. On the 750iL and 535i equipped with EML, the EML connector is next to the engine diagnostic connector, centered in the group of 3 circular connectors.

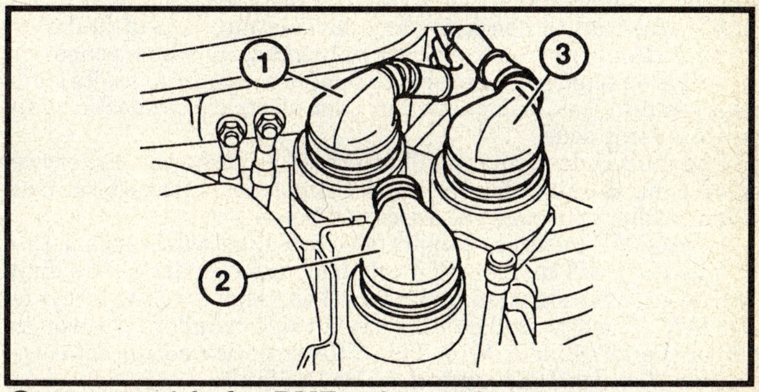

Connector 1 is for DME unit 1 and connector 2 is for DME unit 2. Connector 3 is the EML connector—BMW 850i

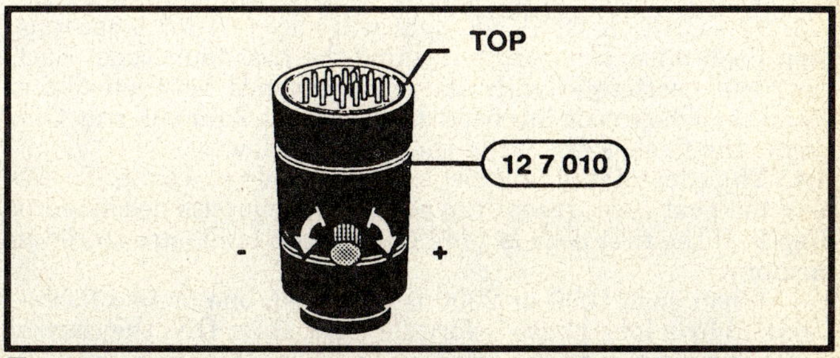

Tester 12 7 010 or equivalent, is used in checking the EML external safety path. Check the external safety path any time the EML components are replaced or adjusted—BMW with EML

3. Install the EML testing adapter 12 7 010 or equivalent to the EML connector inline. Make sure all connections are tight.
4. Turn the knob on the tester to the left position and start the engine. Adjust the engine speed to 2500 rpm with the tester knob.
5. Depress the brake pedal. The engine speed should drop immediately due to the overrun shutoff provision.
6. If the engine speed does not drop immediately, check the EML system for faults. If the speed does drop immediately, the EML external safety path is functioning.
7. Shut off the engine and remove the tester. Reconnect the EML connector and erase any fault codes from the EML control unit.

READING CODES

The DME control unit is equipped with a self diagnostic program that will detect emissions related malfunctions and turn the CHECK ENGINE light **ON** while the engine is running. Emissions related fault codes stored in the DME control unit can be read by activating the program and counting the flashes of the CHECK ENGINE light on the instrument panel. Fault codes stored in all other control units are only accessible with the BMW Service Tester that connects to the diagnostic connector in the engine compartment. The tester is menu driven and will provide operation instructions on the screen.

In the DME control unit, fault codes provide information about which circuit or component is malfunctioning at the time of the test. Intermittent malfunctions are also stored as an existing fault. As useful as these codes are, the first step in repair or service to engine management systems is to gain as much information as possible from the vehicle owner, especially regarding intermittent problems. Also check any service records which may be available. Before repairing any reported faults, it is absolutely essential to check for any obvious mechanical faults or failures. Remember, a fault code only indicates which sensor or circuit is effected by the problem. Simple mechanical faults such as a vacuum leak, bad battery or poor electrical connection can cause a fault code.

The fault code memory will be erased anytime the battery or DME control unit is disconnected. Fault codes can also be erased after reading the codes as described below.

1. All DME codes are 4 digits. The first digit will be either **1** or **2**. The digit will be 1 on all 4 and 6 cylinder engines. The digit will be **1** for the right side cylinders (1 through 6) of the 12 cylinder M70. The digit will be **2** for the left side cylinders (7 through 12) of the 12 cylinder M70. This differentiation allows determination of which DME system holds the fault.
2. With the ignition switch **ON** and the engine not running, press the accelerator pedal to the floor 5 times in less than 5 seconds. This will close the full load switch to activate the diagnostic program. If the program will not activate, the full load switch on the throttle body or accelerator pedal sensor is not working.
3. The CHECK ENGINE light will stay **ON** for 5 seconds, then flash once, then begin flashing the first fault code. Each digit will be flashed with about 2.5 seconds between digits. When the whole code has been displayed, the light will stay **ON**. Count the flashes and write the numbers down.
4. The control unit will display only 1 code at a time. To display the next code, repeat the activation sequence described in Step 2. If the first code is 1444 or 2444, no faults are stored in memory.
5. When code 1000 or 2000 is displayed, one or two flashes followed by a long pause before the light stays **ON**, the control unit has reported all fault codes.

CLEARING CODES

1. To erase the fault code memory, hold the accelerator pedal down to close the full load switch for 10 seconds with the ignition **ON** and when code 1000 or 2000 is displayed.

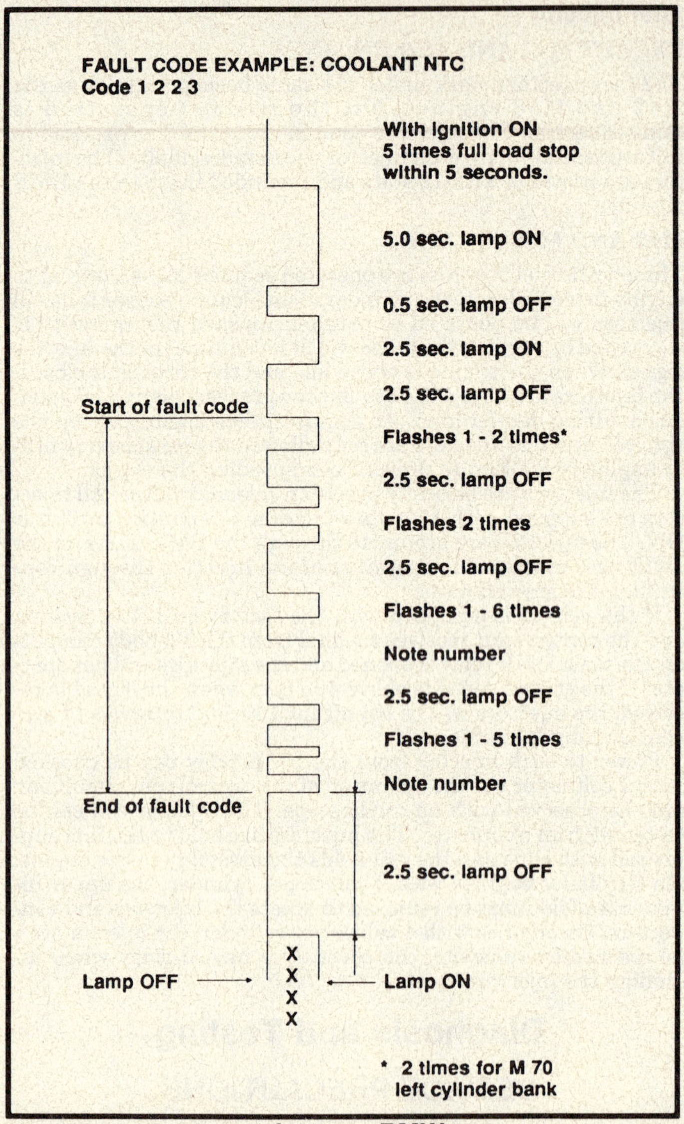

Trouble code output format—BMW

2. Alternately the DME unit can be unplugged to remove power from the memory circuits. If this method is used, make sure the ignition is **OFF**.

FUEL SYSTEM TESTING

Fuel Pump Electrical Testing

When the ignition switch is first turned to **START**, the control unit will complete the ground circuit for the pump relay coil and the relay closes to run the pump. When the control unit receives a signal from the crankshaft pulse sender, it will continue to operate the pump relay with the ignition switch in the **RUN** position. If the engine is not running, the pump will not run. The pump relay terminals can be jumpered to test the electrical system and to run the pump for fuel system pressure testing.

EXCEPT M70 ENGINE

1. Remove the pump relay and check for 12 volts at terminal 30 of the relay socket. With the ignition switch **ON**, there should be 12 volts at terminal 86 from the main DME relay.
2. Jumper terminals 30 and 87 on the relay socket. There should be power to the pump through fuse 23.

Fault Flash Codes

x111—Ignition	x262—Idle speed control
x211—Control unit (self test)	x263—Tank venting
x215—Air mass sensor	x264—Oxygen sensor heating relay
x221—Oxygen sensor	x268—Idle speed, CO potentiometer
x222—Oxygen control	x278—Ignition timing correction
x223—Coolant temperature sensor	x512—AC compressor shutoff
x224—Intake air temperature sensor	x513—ECU power input
x231—Control unit voltage monitor	x515—Converter lockup clutch
x232—Idle speed switch	x516—Speed signal
x233—Full load switch	x517—MSR signal (engine drag torque control, part of ASC)
x251—Injector group 1 (cylinders 2, 4, 6)	x518—ASC signal (Automatic Stability Control)
x252—Injector group 2 (cylinders 7, 9, 11)	x818—Final stage
x261—Electric fuel pump relay	x=1 or 2

NOTE: The x represents the numerals 1 or 2. This designates on which DME unit the fault occurred. The numeral 1 indicates the DME unit for the right side of the M70 engine (cylinders 1–6) and it will be the numeral used on all non-M70 engines. The numeral 2 indicates the DME unit for the left side of the M70 engine (cylinders 7–12).

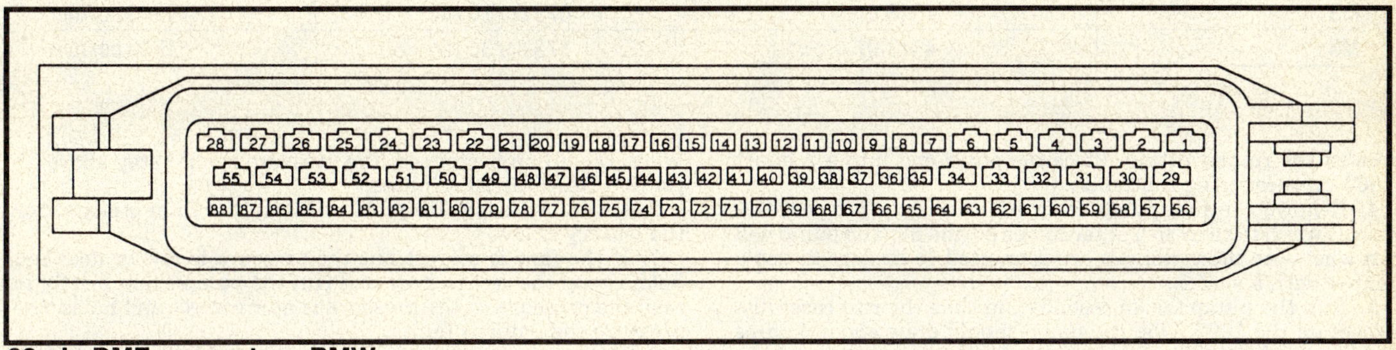

88 pin DME connector—BMW

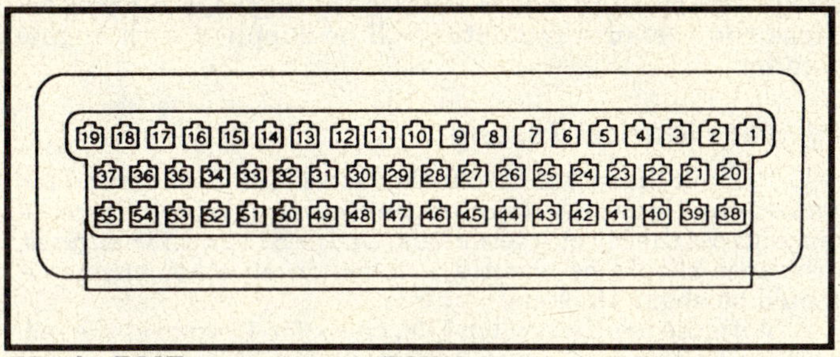

55 pin DME connector—BMW

3. Use jumper wires to supply 12 volts to terminal 86 and ground to terminal 85 on the relay itself. Do not reverse the polarity or the relay will be destroyed. The relay should close and there should be continuity between terminals 30 and 87.

4. If the pump runs when the terminals are jumpered and if the relay is good, but the pump stops when the ignition switch is in the **RUN** position, the control unit may not be receiving a signal from the crankshaft pulse sender.

M70 ENGINE

Power is supplied to the pump relay coils from the main DME relays. The ground circuit for each coil is supplied by each control unit during starting and when the control unit is receiving an rpm signal. If the engine stops running, the control units open the relays to stop the pumps. Pump No. 1 receives power through fuse No. 23, pump No. 2 receives power from fuse No. 24. The relay for pump No. 1 also supplies power to the oxygen sensor heater relay.

1. If the engine stalls or will not start, remove the pump relay and check for 12 volts at terminal 30 of the relay socket. With the ignition switch **ON**, there should be 12 volts at terminal 86 from the main DME relay.

2. Jumper terminals 30 and 87 on the relay socket. There should be power to the pump even if the ignition is **OFF**.

3. Use jumper wires to supply 12 volts to terminal 86 and ground to terminal 85 on the relay itself. Do not reverse the polarity or the relay will be destroyed. The relay should close and there should be continuity between terminals 30 and 87.

4. If the pump runs when the terminals are jumpered and if the relay is good, but the pump stops when the ignition switch is in the **RUN** position, the control unit may not be receiving a signal from the crankshaft pulse sender.

Fuel Pump Flow and Pressure Testing

CAUTION

The following procedure will produce fuel vapors. Make sure there is proper ventilation and take the appropriate fire safety precautions otherwise personal injury may occur.

EXCEPT M70 ENGINE

1. Make sure the battery is in good condition and the fuel pump wiring and ground connections are good. It is important the pump receive a full 12 volts.

2. Near the oil dipstick, disconnect the fuel supply hose and install a 13 3 60 or equivalent 100 psi pressure gauge with a tee fitting. The supply line is always the lower hose.

3. Disconnect the fuel return hose and securely install a long

FUEL PUMP TESTING SPECIFICATIONS

Model	Operating Pressure psi (BAR)	Fuel Delivery	
		System Rate	Pump Rate
318i	43 (3.0)	875 cc/30 sec	1900 cc/min
318is	43 (3.0)	875 cc/30 sec	1900 cc/min
318iC	43 (3.0)	875 cc/30 sec	1900 cc/min
325i (1990–91)	43 (3.0)	875 cc/30 sec	2100 cc/min
325i (1992)	≥ 58 (4.0)	1030 cc/30 sec	≥ 2000 cc/min
325iC	43 (3.0)	875 cc/30 sec	2100 cc/min
525i	43 (3.0)	875 cc/30 sec	1900 cc/min
535i	43 (3.0)	875 cc/30 sec	1900 cc/min
735i	43 (3.0)	875 cc/30 sec	1900 cc/min
735iL	43 (3.0)	875 cc/30 sec	1900 cc/min
750iL	43 (3.0)	875 cc/30 sec	≥ 1700 cc/min
850i	43 (3.0)	875 cc/30 sec	≥ 1700 cc/min
M3	43 (3.0)	875 cc/30 sec	1900 cc/min
M5	43 (3.0)	875 cc/30 sec	1900 cc/min

hose to the return fitting. Place the other end into a 2 quart (2000 cc) measuring container.

4. Remove the fuel pump relay and connect a jumper wire with a switch to relay socket terminals 30 and 87. The pump will run whenever the terminals are connected so be sure to use a jumper with a switch.

5. Run the pump for 30 seconds and time the run carefully. Except on the 1992 325i, the pump should draw about 5 amps and flow 29.6 oz. (875 cc) in 30 seconds. The pressure should be about 43 psi (3.0 BAR). On the 1992 325i, the pump should draw about 5 amps and flow 34.8 oz. (1030 cc) in 30 seconds. The pressure should be about 43 psi (3.0 BAR). When the pump is stopped, the pressure should stay above 38 psi (2.6 BAR) for 20 minutes.

6. If the flow is low, the pump is faulty or the filter is clogged.

7. If the flow is correct but the pressure is low or does not hold, clamp the return hose and run the pump again briefly to read the pressure. If the pressure is now correct and holds, the pressure regulator at the front of the supply rail is faulty.

8. If the pressure is still low or does not hold, there is a leak in the system, possibly an injector or the fuel pump check valve. The pressure will be lower if the engine is running because the pressure regulator will be supplied with engine vacuum.

M70 ENGINE

1. Make sure the battery is in good condition and the fuel pump wiring and ground connections are good. It is important the pump receive a full 12 volts.

2. Remove the fresh air housing and engine covers as required to gain access to the fuel supply rails. At the rear of the engine, disconnect the fuel supply hose and install a 13 3 60 or equivalent 100 psi pressure gauge with a tee fitting. The supply line is the upper hose.

3. Disconnect the fuel return hose and securely install a long hose to the return fitting. Place the other end into a 2 quart (2000 cc) measuring container.

4. Remove the fuel pump relay and connect a jumper wire with a switch to relay socket terminals 30 and 87. The pump will run whenever the terminals are connected so be sure to use a jumper with a switch.

5. Run the pump for 30 seconds and time the run carefully. The pump should draw about 5.5 amps and flow 29.6 oz. (875 cc) in 30 seconds. The pressure should be about 43 psi (3.0 BAR). When the pump is stopped, the pressure should stay above 38 psi (2.6 BAR) for 20 minutes.

6. If the flow is low, the pump is faulty, the filter is clogged or the battery is low.

7. If the flow is correct but the pressure is low or does not hold, clamp the return hose and run the pump again briefly to read the pressure. If the pressure is now correct and holds, the pressure regulator is faulty.

8. If the pressure is still low or does not hold, there is a leak in the system, possibly an injector or the fuel pump check valve. The check valve is part of the pump and cannot be replaced separately. The pressure will be lower if the engine is running because the pressure regulator will be supplied with engine vacuum.

Injector Electrical Testing

1. Remove the intake manifold to access the injectors and the connectors. Unplug the injector connectors and check the resistance across the injector terminals. On the S14 and S38 engines, resistance should be about 2.5 ohms, on all other engines it should be about 16 ohms.

2. With the ignition switch **ON**, check for 12 volts at the red/white wire terminal on the injector connectors. This power comes from the DME relay.

3. To check the signal from the control unit, connect an oscilloscope between the brown wire terminal and ground and operate the starter. If there is no signal from a single connector, the wiring is faulty. If there is no signal from any connector, the control unit may not be receiving an rpm signal or the control unit is faulty.

Injector Leak Testing

1. Remove the injector wiring harness and remove the injectors and supply rail as an assembly. Make sure all hose clamps and injector holder clips are secure.

2. Examine the nozzle tips for signs of damage or leaking and wipe them clean.

3. Remove the fuel pump relay and jumper socket terminals 30 and 87 to run the pump. If fuel drips from the tip of the nozzle faster than about 1 drip per minute at the correct system pressure, replace the injector. Be sure to use new O-rings.

INJECTOR TEST SPECIFICATIONS

Model	Engine	Flow Rate cc/min	Test Pressure psi (BAR)	Coil Resistance Ohms @ 68°F (20°C)	Plug Color	Ejection Angle
318i	M42	170	43 (3.0)	15–17.5	blue	30
318is	M42	170	43 (3.0)	15–17.5	blue	30
318iC	M42	170	43 (3.0)	15–17.5	blue	30
325i (1990–91)	M20	170	43 (3.0)	15–17.5	blue	30
325i (1992)	M50	180	43 (3.0)	15–17.5	green	20
325iC	M20	170	43 (3.0)	15–17.5	blue	30
525i (1990–91)	M20	170	43 (3.0)	15–17.5	blue	30
525i (1992)	M50	180	43 (3.0)	15–17.5	green	20
535i	M30	220	43 (3.0)	15–17.5	yellow	30
735i	M30	220	43 (3.0)	15–17.5	yellow	30
735iL	M30	220	43 (3.0)	15–17.5	yellow	30
750iL	M70	170	43 (3.0)	15–17.5	blue	30
850i	M70	170	43 (3.0)	15–17.5	blue	30
M3	S14	236	43 (3.0)	2–3	gray	30
M5	S38	270	43 (3.0)	NA	ivory	30

AIR FLOW METER

Testing

1. Unplug the sensor connector and connect a voltmeter to terminal 3 of the connector. With the ignition switch **ON**, there should be about 5 volts to the sensor.
2. Turn the ignition **OFF**. Remove the upper portion of the air cleaner housing with the flow meter attached. Connect an ohmmeter across terminals 2 and 3 of the meter.
3. Use a non-metal tool to reach into the housing and move the vane. The vane should move smoothly and the resistance should change smoothly. If there are jumps or dead spots in the vane movement or resistance changes, replace the air flow meter.
4. To check the temperature sensor, connect the ohmmeter to terminals 1 and 4 and make sure the air vane is fully closed. Measure the air temperature and check the chart to determine the correct sensor resistance. Ohmmeter readings should be within 10 percent of specification.

AIR MASS SENSOR

Testing

1. The air mass meter is mounted on or near the air cleaner housing. Unplug the connector and turn the ignition switch **ON**. There should be 12 volts at terminal 2 directly from the main DME relay. Terminal 4 goes directly to ground. If there is no power, check the relay and control unit as described earlier.
2. There should be 5 volts at terminal 3 from the control unit.
3. If the voltages are not correct, check the wiring between the air mass sensor, DME relay and the control unit. If the wiring is good and the DME relay is functioning, the control unit is faulty. If the voltages are correct but the fault code remains in memory, the air mass sensor may not be burning itself clean.
4. To test the burn-off function, connect all wiring but leave the air duct off to look into the sensor. Run the engine up to operating temperature, raise the speed above 2000 rpm, then turn the ignition switch **OFF** and watch the sensor. After a delay of about 4 seconds, the wire should glow for about 1 second to burn itself clean. If it does not, the control unit is faulty.

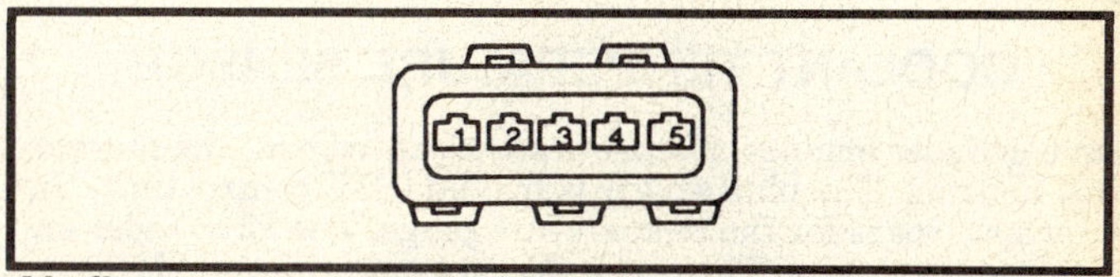

Air flow meter connector pinout—BMW except M70 engine

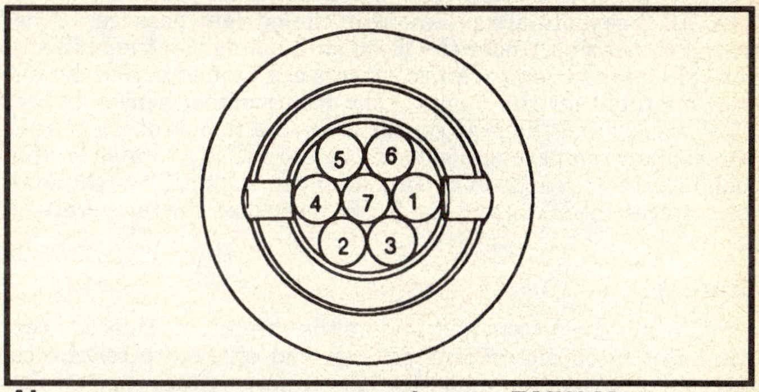

Air mass sensor connector pinout—BMW M70 engine with DME M1.7

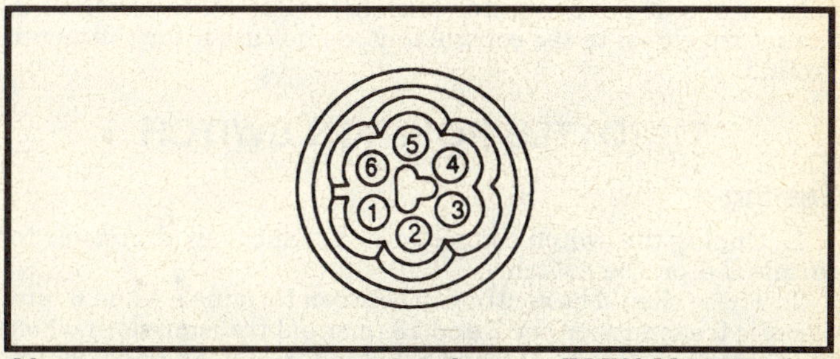

Air mass sensor connector pinout—BMW M70 engine with DME M1.2

AIR AND COOLANT TEMPERATURE SENSOR RESISTANCES

°F (°C)	kilo-ohms
14 ± 2 (−10 ± 1)	8.2–10.5
68 ± 2 (+20 ± 1)	2.2–2.7
176 ± 2 (+80 ± 1)	0.3–0.36

INTAKE AIR TEMPERATURE SENSOR

On vehicles with a vane type air flow sensor, the air temperature sensor is inside the housing and must be tested as part of the air flow meter circuit. On 4 and 6 cylinder vehicles with an air mass sensor, the air temperature sensor is in the air filter housing or in the intake manifold. On 12 cylinder engines, the sensor is threaded into the rear of the right side intake manifold. The sensor reports to DME control unit 1 on the same circuit as the coolant temperature sensor.

Testing

1. Locate the sensor and unplug the connector. Measure the air temperature and check the resistance across the sensor terminals. Use the chart to determine the correct reading.
2. With the ignition switch **ON** the sensor should receive 5 volts at one of the connector terminals. The other terminal is the return signal to the control unit, do not use it for voltmeter ground.

COOLANT TEMPERATURE SENSOR

On 6 cylinder engines, 2 sensors are mounted near the thermostat housing. The blue sensor is for the DME control unit, the brown sensor is for the temperature gauge. The 12 cylinder engine uses 3 sensors. The temperature gauge and the EML control unit each have their own sensor. The DME control units share the same sensor, one terminal connected to each DME control unit with control unit 1 providing the ground connection. All 3 sensors are mounted in the coolant passage at the rear of the engine, under the fresh air housing. On the M42 engine, the coolant temperature sensors are located under the intake manifold on the engine. The forwardmost sensor is the DME sensor and the rear sensor is for the temperature gauge. The coolant temperature sensors on the M3 are located in the coolant pipe above the exhaust side of the engine. The rear sensor services the DME and the forward services the temperature gauge.

Testing

1. Locate the sensor and unplug the connector. Measure the temperature of the coolant passage and check the resistance across the sensor terminals. On 12 cylinder engines, check resistance between terminal 1 and ground and terminal 3 and ground. Use the chart to determine the correct reading.
2. With the ignition switch **ON** the sensor should receive 5 volts at one of the connector terminals. The other terminal is the return signal to the control unit, do not use it for voltmeter ground.

THROTTLE POSITION SWITCH

Testing

1. Unplug the switch connector and connect an ohmmeter to terminal 4 on the switch.
2. There should be continuity between terminals 4 and 6 (automatic transmission) or 2 and 18 (manual transmission) when the throttle is closed (idle) and between terminals 4 and 5 (automatic transmission) or 3 and 18 (manual transmission) when the throttle is fully open.

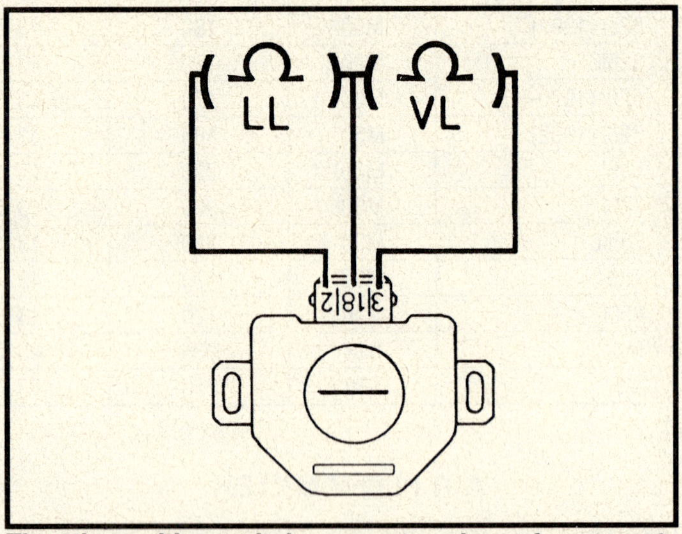

Throttle position switch connector pinout for manual transmission equipped vehicles. LL is idle position and VL is full load position—BMW

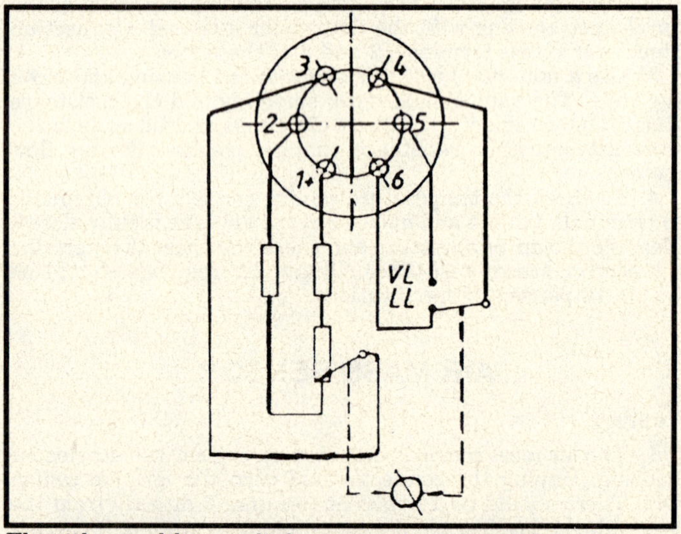

Throttle position switch connector pinout for automatic transmission equipped vehicles. LL is idle position and VL is full load position—BMW

3. Loosen the screws on the side of the housing to adjust the switch. If the switch cannot be correctly adjusted for both test results, replace the switch.

ACCELERATOR PEDAL SENSOR

NOTE: It is possible to physically adjust the accelerator pedal sensor, but to check the adjustments the BMW

Diagnostic Tester or equivalent must be utilized. Any time an EML related part has been replaced, tested or adjusted, the external safety path check must be run.

Testing

1. Install the sensor to the mounting bracket and connect the linkage. Snug the mounting screws and loosen the linkage nut on the end of the sensor.
2. Move the linkage to produce gap **A** at the idle stop and secure a 0.118 in. (3.0mm) feeler gauge in place to maintain the gap. Support the linkage and tighten the nut.
3. Connect the BMW Diagnostic Tester or equivalent, and turn the ignition switch **ON**.
4. If equipped with a manual transmission, move the pedal to produce a sensor output reading or 3.15–3.3 volts. Turn the pedal stop knurled nut to touch the pedal and tighten the locknut.
5. If equipped with an automatic transmission, move the pedal to produce a sensor output reading or 3.7–3.8 volts. Turn the pedal stop knurled nut to produce a gap of 0.236 in. (6mm) between the stop and the pedal and tighten the locknut.

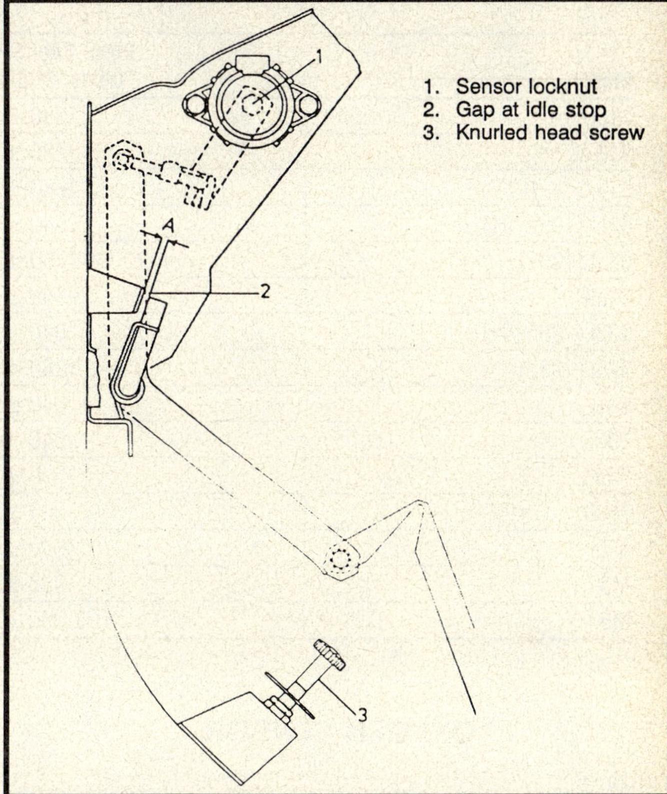

Pedal sensor adjustment—BMW with EML throttle by wire

IDLE SPEED ACTUATOR

Testing

1. Unplug the idle speed control valve connector and check for 12 volts at terminal 2 with the ignition switch **ON**. Power comes from terminal 87 of the main DME relay.
2. To check the motor resistance, connect an ohmmeter to terminals 2 and 3 of the actuator. It should read about 20 ohms. Terminals 2 and 1 should also read 20 ohms. Resistance across terminals 1 and 3 should be about 40 ohms. The terminal numbers will be imprinted on the connector body with terminal 2 in the middle position. Terminals 1 and 3 are the outer terminals and the exact positions are not critical to this test.
3. If voltage and resistance readings are correct, remove the actuator and use a tool to carefully push the rotary valve sideways to the fully open position. With the wiring connected and the ignition switch **ON**, the valve should move to about the 50 percent position and remain stable.

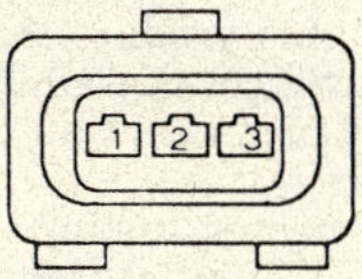

Idle speed actuator, cylinder ID sender and speed/position sensor connector pinout—BMW

SPEED/POSITION REFERENCE SENSOR

Testing

1. Locate the sensor and disconnect the wiring. With the ignition switch **ON**, there should be 5 volts across the connector terminals from the control unit. If not, check for 5 volts between terminal 1 and ground.
2. Turn the ignition switch **OFF** and connect the ohmmeter to the sensor connector terminals. Check the resistance between terminals 1 and 2. The result should be within 10% of the specified resistance. The sensor should have a resistance more than 100,000 ohms at terminal 3, the shield wire.
3. Connect an oscilloscope to the sensor terminals. While operating the starter, the sensor should produce about a 3 volt peak-to-peak sine wave with a wider wave indicating the reference mark. If there is no signal, check the gap or remove the sensor and check for damage. The gap between the sensor and the wheel should be 0.040 in. (1.0mm).

CYLINDER IDENTIFYING SENDER

Testing

1. With the ignition switch **OFF**, unplug the connector and check the resistance across terminals of the sender connector. Across terminals 1 and 2, resistance should be less than 1 ohm.
2. Terminal 3 is the shield wire, resistance between it and terminal 2 should be more than 10 mega-ohms.

CYLINDER IDENTIFICATION SENSOR RESISTANCE

Terminal Number	Resistance @ Temperature
Pins 1 and 2	less than 1 ohm @ 68°F (20°C)
Pins 2 and 3	greater than 10M ohm @ 68°F (20°C)

PULSE SENDER TEST SPECIFICATIONS

Model	Pulse Sender Resistance Ohms @ 68°F (20°C)	Pulse Sender Spacing mm from tooth wheel
318i	1280 ± 10%	1.0 ± 0.3
318is	1280 ± 10%	1.0 ± 0.3
318iC	1280 ± 10%	1.0 ± 0.3
325i (1999–91)	540 ± 10%	1.0 ± 0.3
325i (1992)	1280 ± 10%	1.0 ± 0.3
325iC	540 ± 10%	1.0 ± 0.3
525i (1990–91)	540 ± 10%	1.0 ± 0.3
525i (1992)	1280 ± 10%	1.0 ± 0.3
535i	540 ± 10%	1.0 ± 0.3
735i	540 ± 10%	1.0 ± 0.3
735iL	540 ± 10%	1.0 ± 0.3
750iL	540 ± 10%	1.0 ± 0.3
850i	540 ± 10%	1.0 ± 0.3
M3	960 ± 10%	1.0 ± 0.3
M5	960 ± 10%	1.0 ± 0.3

OXYGEN SENSOR

Testing

1. To test the sensor with a voltmeter, unplug the sensor connector and connect a digital voltmeter to the sensor terminals. With the engine at idle at operating temperature, the control unit will operate the engine in Open Loop control. The sensor signal should be about 450 millivolts and fairly steady. Rapid throttle movements should produce changes in the signal.
2. If the sensor is putting out a signal but the control unit will not operate in Closed Loop, check all other sensors before replacing the control unit.
3. To test the sensor heater, unplug the sensor connector and check for continuity to ground at terminal 4.

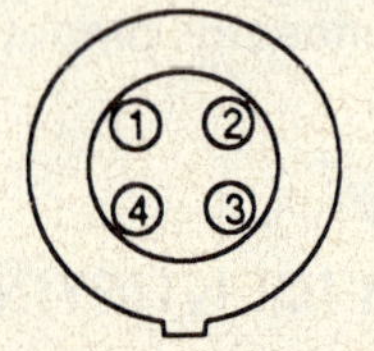

Oxygen sensor connector pinout – BMW except 8 Series

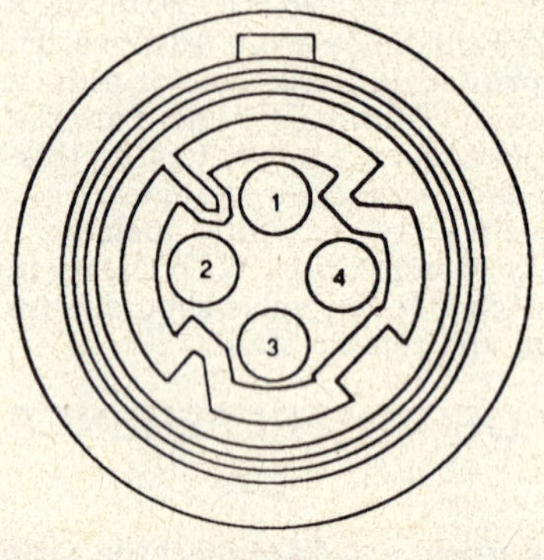

Oxygen sensor connector pinout – BMW 8 Series

4. Remove the heater relay and check for 12 volts between socket terminal 6 and ground and terminal 8 and ground with the ignition switch **ON**. Relay power comes directly from the main DME relay. The ground circuit for the heater relay coil is supplied by the control unit.

2.5L M50 ENGINE—DME M3.1 TROUBLESHOOTING—1992 325i AND 1991-92 525i

Self–Diagnosis

The task of the self–diagnosis feature is to detect and store malfunctions in DME components, the associated lines or in the memory of the control unit (ROM, RAM).

The DME control unit makes available substitute values in the event of failure of sensors (see troubleshooting – explanation). These substitute values are cancelled once again when normal operation can be resumed.

To facilitate fast and convenient troubleshooting, current values can be monitored from the DME control unit.

Notes on the BMW DIAGNOSTIC SYSTEM

Effective Range of the Self–Diagnosis Function

The self–diagnosis function is an aid to facilitate fast fault tracing in the vehicle's electrical/electronic system. The BMW DIAGNOSTIC SYSTEM (Service Tester, MoDiC) provides the necessary access for the user. The self–diagnosis function detects defects provided the necessary conditions have been fulfilled (time, temperature, load, etc.). However, defects can only be registered where there is an electrical connection to the control unit. If this is not the case, the check must be performed using conventional methods (e.g. ignition voltage, fuel pressure, mechanical functions etc.).

Control Unit Identification

If necessary, the control unit can be quickly identified with these data (works request). Automatic assignment of the diagnostic programs is also based on these control unit numbers.

2.5L M50 ENGINE—DME M3.1 TROUBLESHOOTING—1992 325i AND 1991-92 525i

Storage of Defect Codes

Defect codes are stored after the engine has been started (with the exception of defect codes 26 and 200). If several defects are detected, they appear in the order they first occurred.

The operating conditions as stored apply only to conditions under which the defect was detected for the first time. An intact voltage supply (charged battery) is the prerequisite for maintaining the memory contents.

Substitute values may be active if there are **no** malfunctions despite a defect code being stored. Troubleshooting should still, however, be carried out.

In US models, defects which influence the mixture are indicated by a lamp (engine check) in the instrument cluster. As a check, this lamp lights constantly during the period of time between "Ignition ON" and engine start.

If the defect code memory has no defect code stored although there is a defect in the vehicle, either the test conditions (e.g. temperature, time, load, etc.) for defect detection were not reached or it was technically not possible to store a defect code.

Clearing Defect Codes

Carry out this step only when the defect has been rectified or when a special instruction has been given. After being cleared, current defect codes are immediately stored in the defect code memory again. When the battery or the DME control unit are disconnected, all stored defect codes as well as currently active substitute values are lost. The adapted values for idle control (idle air) and emission control (mixture composition) are neutralized. During a new start, this may result in temporary irregular engine operation. A test run, during which an engine temperature of >80 °C is reached, is therefore recommended.

Status Monitoring

Inputs and outputs can be monitored independently of the defect code storage function. In this case, current values or operating statuses are displayed. In order to better illustrate relevant relationships, several status inquiries can be combined to form a block.

Component Activation

To facilitate fast fault finding, various components can be activated with the DIAGNOSTIC SYSTEM. The response of components, including the supply line in perfect working order can therefore be by acoustic means. Activation is only possible with the component connected so that signal testing is not possible at the detached component connector.

2.5L M50 ENGINE—DME M3.1 TROUBLESHOOTING—1992 325i AND 1991-92 525i

AG	Automatic transmission
DK	Throttle valve
DWA	Antitheft system
EGS	Electronic transmission control
EKP	Electric fuel pump
EV	Injector valve
HG	Manual transmission
KD	Kick–down
KW	Crankshaft
LL	Idle
LLS	Idle actuator
HLM	Hot–wire air mass meter
MF	Microfiche
MV	Solenoid valve
NW	Camshaft
RxD	Diagnosis initiation line
TD	Engine speed signal (ignition interval)
TE	Fuel evaporation control
ti	Injection time
TL	Part load
tL	Load signal
TR	Crankshaft reference
TxD	Diagnosis data line
SG	Control unit
U–batt	Battery voltage
U–Vers	Supply voltage
v–signal	Vehicle speed signal
VL	Full load

2.5L M50 ENGINE—DME M3.1 PIN ASSIGNMENTS—1992 325i AND 1991-92 525i

Pin Assignments

If not otherwise specified, the signal type data apply to the engine when idling

Assignments at the 88–pole DME control unit connector

Pin	Type	Description/Function	Connection	Signal type	Tester display	Measurement notes
1	A	EKP relay and crankshaft reference	EKP relay and EGS control unit	Minus with overlaid engine speed signal	ON	Can only be measured with crankshaft rotating as of start speed
2	A	Idle actuator–close winding	LLS	Minus clocked	–	See component activation
3	A	Injector valve cyl. 1	EV	Minus clocked	Injection time ti	See component activation, check pulse duration
4	A	Injector valve cyl. 3	EV	Minus clocked	Injection time ti	See component activation, check pulse duration
5	A	Injector valve cyl. 2	EV	Minus clocked	Injection time ti	See component activation, check pulse duration
6	M	Ground EV output stages	Ground terminal	Ground	–	–
7	–	–	–	–	–	–
8	A	Defect lamp (US only)	Instrument cluster	Minus	–	In case of defect, check function when ignition ON
9	–	–	–	–	–	–
10	–	–	–	–	–	–
11	A	DK potentiometer signal	EGS control unit	Pulse width modulated square wave signal	–	Oscilloscope
12	E	Throttle (DK) position	DK potentiometer	0.5...4.5 V	DK signal	Status monitoring

2.5L M50 ENGINE—DME M3.1 PIN ASSIGNMENTS—1992 325i AND 1991-92 525i

Pin	Type	Description/Function	Connection	Signal type	Tester display	Measurement notes
13	A	Free–burn	Air mass meter	Voltage increase for 1 s	–	Signal sent 4 s after shutting down engine
14	M	Ground–air mass meter	Air mass meter	Ground	–	–
15	–	–	–	–	–	–
16	E	Camshaft sensor	NW sensor (connection 2)	AC voltage	Oscillogram	Once per camshaft revolution
17	A	ti measurement signal	Fuel consumption indicator	Square–wave signal	–	Oscilloscope
18	–	–	–	–	–	–
19	–	–	–	–	–	–
20	–	–	–	–	–	–
21	–	–	–	–	–	–
22	–	–	–	–	–	–
23	A	Ignition terminal 1	Ignition coil for cyl.2	Primary signal	Oscillogram	Engine test, test step 05
24	A	Ignition terminal 1	Ignition coil for cyl.3	Primary signal	Oscillogram	Engine test, test step 05
25	A	Ignition terminal 1	Ignition coil for cyl.1	Primary signal	Oscillogram	Engine test, test step 05
26	E	B+, terminal 30	B+ terminal point	Battery voltage	–	No DME functions possible if not fitted
27	A	DME main relay	DME–Hauptrelais Kl.85	Minus	ON	No DME functions when main relay OFF, connect jumper between terminal 30 and 87
28	M	Ground for electronics and shielding of sensors	Ground terminal	Ground	–	–
29	A	Idle actuator–opening winding	LLS	Minus clocked	–	See component activation

2.5L M50 ENGINE—DME M3.1 PIN ASSIGNMENTS—1992 325i AND 1991-92 525i

Pin	Type	Description/Function	Connection	Signal type	Tester display	Measurement notes
30	–	–	–	–	–	–
31	A	Injector valve cyl. 5	EV	Minus clocked	Injection time ti	See component activation, check pulse duration
32	A	Injector valve cyl. 6	EV	Minus clocked	Injection time ti	See component activation, check pulse duration
33	A	Injector valve cyl. 4	EV	Minus clocked	Injection time ti	See component activation, check pulse duration
34	M	Ground remaining output stages (except ignition and EV)	Ground terminal	Ground	–	–
35	–	–	–	–	–	–
36	A	Fuel evaporation control valve	TE valve	Minus clocked	OPEN / CLOSE	Clock cycle takes place dependent on engine speed and load, status monitoring, component activation
37	A	Oxygen sensor heating	Sensor heating relay	Minus	ON	As of ignition ON
38	–	–	–	–	–	–
39	–	–	–	–	–	–
40	–	–	–	–	–	–
41	E	Air mass signal	Air mass meter	0...5 V	HLM signal	Dependent on air mass intake, status monitoring
42	–	–	–	–	–	–
43	M	Ground for sensors	Temperature sensor	Ground	–	–
44	E	NW sensor	NW sensor (connection 1)	AC voltage (approx. 7 Hz in LL)	Oscillogram	Once per camshaft revolution
45	–	–		–	–	–

2.5L M50 ENGINE–DME M3.1 PIN ASSIGNMENTS–1992 325i AND 1991-92 525i

Pin	Type	Description/Function	Connection	Signal type	Tester display	Measurement notes
46	–	–	–	–	–	–
47	–	–	–	–	–	–
48	A	A/C compressor cut–out (as of 9/90)	A/C compressor relay	Minus	–	Only at full load and <13 km/h for 10 seconds
49	–	–	–	–	–	–
50	A	Ignition terminal 1	Ignition coil for cyl.4	Primary signal	ignition oscillogram	Engine test, test step 05
51	A	Ignition terminal 1	Ignition coil for cyl.6	Primary signal	ignition oscillogram	Engine test, test step 05
52	A	Ignition terminal 1	Ignition coil for cyl.5	Primary signal	ignition oscillogram	Engine test, test step 05
53	–	–	–	–	–	–
54	E	U–batt from DME main relay	Main relay terminal 87	>10 V	–	No DME functions when main relay OFF, connect jumper between terminal 30 and 87
55	M	Ignition ground	Ground terminal	Ground	–	–
56	E	Ignition, terminal 15	Ignition lock	>10 V	–	–
57	–	–	–	–	–	–
58	–	–	–	–	–	–
59	A	5V component supply	DK potentiometer	5 V	–	–
60	E	Programming voltage	Diagnostic connector	Voltage value	–	Only activated during programming
61	–	–	–	–	–	–
62	–	–	–	–	–	–
63	–	–	–	–	–	–
64	E	Ignition timing intervention	EGS control unit	Ground	OFF	Only active during gearshift, status monitoring

2.5L M50 ENGINE—DME M3.1 PIN ASSIGNMENTS—1992 325i AND 1991-92 525i

Pin	Type	Description/Function	Connection	Signal type	Tester display	Measurement notes
65	E	Drive range P/N	Gear selector lever	Minus	ON	Status monitoring
66	–	–	–	–	–	–
67	E	Inductive pulse generator (terminal A)	Pulse generator for speed and crankshaft position	AC voltage (30...40 Hz in LL)	Engine speed	Status monitoring as of start speed
68	E	Inductive pulse generator (terminal B)	Pulse generator for speed and crankshaft position	AC voltage	Engine speed	Status monitoring as of start speed
69	–	–	–	–	–	–
70	E	Oxygen sensor signal	Oxygen sensor	Pulsating voltage 0.1...1.2 V	–	Status monitoring only with engine running and oxygen sensor at operating temperature
71	M	Oxygen sensor ground	Oxygen sensor	Ground	–	–
72	–	–	–	–	–	–
73	E	Vehicle speed signal	Instrument cluster	Square–wave signal	km/h	Status monitoring
74	A	Engine speed signal TD	Instrument cluster	Square–wave signal	–	One signal per ignition
75	–	–	–	–	–	–
76	–	–	–	–	–	–
77	E	Intake air temperature	Intake air temperature sensor	0...5 V	°C	Status monitoring
78	E	Engine temperature	Coolant temperature sensor	0...5 V	°C	Status monitoring
79	–	–	–	–	–	–
80	–	–	–	–	–	–
81	E	Antitheft system signal	DWA control unit	Minus	–	When DWA armed >10 V
82	–	–	–	–	–	–
83	–	–	–	–	–	–
84	–	–	–	–	–	–
85	E	A/C compressor	A/C compressor clutch	>10 V	ON	With compressor clutch engaged
86	E	A/C switch	A/C relay	>10 V	ON	With air conditioning system switched on
87	E	RxD diagnosis data line	Diagnostic connector	>10 V	–	When cover of diagnostic connector closed
88	E/A	TxD diagnosis data line	Diagnostic connector	Minus	–	When cover of diagnostic connector closed

E = Input, A = Output, M = Ground

2.5L M50 ENGINE–DME M3.1 TROUBLESHOOTING–1992 325i AND 1991-92 525i

Troubleshooting

Notes

Relevant safety codes and accident prevention regulations must be observed when carrying out all tests and work on the engine electrical and electronic systems.

Caution! Dangerous high voltage in the primary and secondary circuit of the ignition system.

Important! Disconnect control units from the wire harness **only with the ignition switched** off. After renewed operation, neutral basic values are set in the control unit for idle and emission control. Any irregularities in engine operation can be corrected by means of renewed adaptations (new learned values) during a test run, in which an engine temperature of >80 °C is reached.

Tests

The 88–pole universal adapter should be used for picking–off signals from the DME control unit and from the engine wire harness. Perform wire checks **only with the control unit or component connectors disconnected**.

Defect Location Procedure

1. Quick test with DIAGNOSTIC SYSTEM

This quick test can be used to check that other systems are operating correctly. Defects in EGS, DWA or BC can also affect the DME.

2. Defect code memory inquiry

This is recommended for all systems which were found to be defective during the quick test. If several defect codes are stored, interrelationships can often be deduced, indicating whether the defect can be attributed to central causes – e.g. a defective ground supply can lead to storage of the defect codes "temperature sensor, air mass meter, idle actuator, DK potentiometer".

3. Status monitoring and component activation

Irrespective of whether defect codes are stored or not, status monitoring and component activation should always be used for fault tracing.

4. Test instructions and schematics

The test instructions explain and supplement the self–diagnosis system. If defects occur, for which no defect code is stored, they provide the opportunity of performing troubleshooting with the aid of the subsequent tests. The schematics should always be used since they clearly show the interaction of various components.

2.5L M50 ENGINE–DME M3.1 DEFECT CODES–1992 325i AND 1991-92 525i

Defect Code Table

Malfunction: Engine Operation not Possible

♦D♦ Defect code memory – No defect code stored

Malfunctions – without stored defect codes:

Tests when engine operation not possible. For further information refer to next chapter – Component Testing.

Prerequisites:

– Starting system OK. (battery charged, ignition switch OK.,...)
– Ground connection from battery to engine or body OK
– Specified fuel in tank
– Engine mechanically OK. (timing, compression ...)

Troubleshooting:

a) Check voltage supply for DME control unit in accordance with schematic.

b) Measure input signals with ♦**D**♦ status monitoring

– Check engine speed signal from the inductive pulse generator at starting speed and
– Air mass meter signal

c) Measure output signals.

– Check fuel pump relay activation in accordance with schematic,
– Ignition output stages by means or ENGINE TEST and
– Injector valves with ♦**D**♦ component activation

d) Fuel pressure test in accordance with repair instructions

2.5L M50 ENGINE—DME M3.1 DEFECT CODES—1992 325i AND 1991-92 525i

Defect Code 0

♦D♦ Defect code memory: 0 UNDEFINED DEFECT

Explanation: This defect code is displayed when an undefined defect code has been stored in the defect code memory.

Note: If the DME program was selected manually, check again that the correct program has been selected.

If the correct program has been selected or if selection took place automatically, data has been stored in the defect code memory for some unknown reason.

The defect code can be ignored if there are no apparent effects in the system.

Defect Code 1

♦D♦ Defect code memory: 1 EKP Relay

Explanation: This defect code is stored in the case of break or short–circuit at:

– relay for the electric fuel pump
– wire from DME control unit Pin 1 to EKP relay
– output stage in the DME control unit

Troubleshooting: The EKP relay must switch on from starting speed – check with test lamp 12 V 10W at terminal 85/86.

Defect Code 2

♦D♦ Defect code memory: 2 IDLE ACTUATOR Closing Winding

Explanation: This defect code is stored in the case of a break or short–circuit at

– idle actuator
– wire from DME control unit Pin 2 to idle actuator
– output stage in the DME control unit

The defect code 29 is additionally stored in the case of a break in the voltage supply or plug–and–socket connection to the idle actuator.

Note: The defect codes 2 and 29 are also stored when engine operation is interrupted abruptly without the ignition being switched off beforehand. The defect is then detected at an engine speed <600 rpm.

Troubleshooting: Perform component activation with tester. The idle actuator is then activated at one second intervals. If there is no response, connect test adapter 61 1 491 between the wire harness and idle actuator. A voltage of approx. 12 V should be measured at terminal 2. The value between terminal 1/2 and 2/3 is >10 V. See component testing for resistance values.

Defect Code 3

♦D♦ Defect code memory: 3 INJECTOR VALVE Cyl. 1

Explanation: This defect code is stored in the case of a break or short–circuit at:

– injector valve cyl. 1
– wire from DME control unit Pin 3 to injector valve
– output stage in the DME control unit

Note: The defect code 6 can be additionally stored since a cylinder cannot be assigned when the defect occurs temporarily. The defect code 6 can be ignored in the case of an injector valve defect with cylinder assignment.

Troubleshooting: Perform component activation with tester. The activation cycle at a frequency of 10 Hz should be audible.

2.5L M50 ENGINE – DME M3.1 DEFECT CODES – 1992 325i AND 1991-92 525i

Defect Code 4

♦D♦ Defect code memory: 4 INJECTOR VALVE Cyl. 3

Explanation: Procedure as for defect code 3.

Defect Code 5

♦D♦ Defect code memory: 5 INJECTOR VALVE Cyl. 2

Explanation: Procedure as for defect code 3.

Defect Code 8

♦D♦ Defect code memory: 8 DEFECT LAMP

Explanation: Only applies to US models.

This defect code is stored if a break or short–circuit occurs at the defect lamp or its supply wire.

Troubleshooting: Connect wire (control unit Pin 8) to ground and switch on ignition. The defect lamp in the instrument cluster should light, if necessary, check lamp and voltage supply.

Defect Code 6

♦D♦ Defect code memory: 6 INJECTOR VALVE – OUTPUT STAGE without Cylinder Reference

Explanation: This defect code is stored if a break or short–circuit occurs temporarily at:

– an injector valve
– wire from DME control unit to an injector valve
– injector valve output stage in the DME control unit

Note: Defect code 6 can be ignored if an injector valve defect code with cylinder assignment is also stored.

Troubleshooting: Perform component activation with tester at each injector valve. If the defect does not occur momentarily, check wires and terminals.

Defect Code 12

♦D♦ Defect code memory: 12 DK POTENTIOMETER

Explanation: This defect code is stored in the case of:

– a break or short–circuit at the DK potentiometer or its supply wires
– mechanical defects, e.g. defective connection to throttle valve.
Substitute values are made available dependent on the current throttle (DK) position. A constant 14 % throttle opening is output at the DME control unit output Pin 11.

Troubleshooting: Perform status monitoring. Connect test adapter 61 1 491 between wire harness and potentiometer. Check 5 V supply at terminal 1/3. Check resistance at potentiometer
Check mechanical connection to throttle valve.

2.5L M50 ENGINE—DME M3.1 DEFECT CODES—1992 325i AND 1991-92 525i

Defect Code 16

♦D♦ Defect code memory: 16 CAMSHAFT PULSE GENERATOR

Troubleshooting: Check signal with oscilloscope. Connect universal adapter to engine wire harness. Set up multimeter test M13 (50 Hz) and M24 (Scope 25 ms) and measure at terminal 16. The signal is output once per camshaft revolution.

Explanation: This defect code is stored when the pulse generator signal is interrupted or defective. The ignition coils for cylinders 1/6, 5/2 or 3/4 are then activated synchronously, resulting in one ignition cycle during the exhaust stroke.

Defect Code 18

♦D♦ Defect code memory: OUTPUT STAGE PIN 18

Explanation: This defect code is stored in the case of:

– a short to plus or minus of the output stage at Pin 18

Note: PIN 18 of the DME control unit is generally not used. Clear defect code if no other defect code is stored.

Defect Code 19

♦D♦ Defect code memory: 19 OUTPUT STAGE

Explanation: This defect code is stored in the case of:

– a short to plus or minus of an output stage.

Note: The defect occurs for such a short space of time that it cannot be assigned to a certain output stage. It therefore frequently occurs in conjunction with other defects.

This defect code can be ignored if the engine is running properly.

2.5L M50 ENGINE—DME M3.1 DEFECT CODES—1992 325i AND 1991-92 525i

Defect Code 23

♦D♦ Defect code memory: 23 IGNITION CYLINDER 2

Explanation: This defect code is stored in the case of:

– a break at the ignition output stage of this cylinder

Note: No fuel is injected in this cylinder for as long as the defect applies.

Troubleshooting: Check ignition signal with the aid of the primary adapter.

Defect Code 24

♦D♦ Defect code memory: 24 IGNITION CYLINDER 3

Explanation: Procedure as for defect code 23

Defect Code 25

♦D♦ Defect Codeursache: 25 IGNITION CYLINDER 1

Explanation: Procedure as for defect code 23

Defect Code 26

♦D♦ Defect code memory: 26 CONTROL UNIT SUPPLY B+

Explanation: This defect code is stored if:

– the battery voltage is exceeded (with engine off)
– the battery is disconnected from the vehicle

Note: After the correct battery voltage has been protected, the defect code is cleared automatically after the next test cycle (new start).

2.5L M50 ENGINE—DME M3.1 DEFECT CODES—1992 325i AND 1991-92 525i

Defect Code 29

♦D♦ **Defect code memory: 29 IDLE ACTUATOR Opening Winding**

Explanation: Procedure as for defect code 2

Defect Code 31

♦D♦ **Defect code memory: 31 INJECTOR VALVE Cyl. 5**

Explanation: Procedure as for defect code 3

Defect Code 32

♦D♦ **Defect code memory: 32 INJECTOR VALVE Cyl. 6**

Explanation: Procedure as for defect code 3

Defect Code 33

♦D♦ **Defect code memory: 33 INJECTOR VALVE Cyl. 4**

Explanation: Procedure as for defect code 3

2.5L M50 ENGINE—DME M3.1 DEFECT CODES—1992 325i AND 1991-92 525i

Defect Code 36

♦D♦ Defect code memory: 36 TANK VENTILATION

Explanation: This defect code is stored when a break or short–circuit occurs at:

– fuel evaporation control valve
– wire from DME control unit to fuel evaporation control valve
– output stage in the DME control unit

Troubleshooting: Perform component activation with tester. The valve is activated at second intervals – this is audible. During normal operation, the valve is opened dependent on engine speed and load.

Defect Code 37

♦D♦ Defect code memory: 37 OXYGEN SENSOR HEATING

Explanation: This defect code is stored if a break or short–circuit occurs at:

– oxygen sensor (heater)
– sensor heating relay
– wire from DME control unit to sensor heating relay
– wires to oxygen sensor
– output stage in the DME control unit

Troubleshooting: Perform component activation with tester. The relay in the E–box is activated at second intervals which are audible. In the case of defect "connection to sensor signals", check whether the defect is at the sensor or in its supply wire by monitoring the oxygen sensor.

2.5L M50 ENGINE—DME M3.1 DEFECT CODES—1992 325i AND 1991-92 525i

Defect Code 41

♦D♦ Defect code memory: 41 AIR MASS METER (HLM)

Explanation: This defect code is stored if a break or short–circuit occurs at:

– air mass meter or its supply wires
– voltage supply to air mass meter
The DME control unit then operates with substitute values which are calculated from the position of the throttle potentiometer.

Troubleshooting: Check signal by means of status monitoring

The DME main relay supplies the HLM with 12 V at terminal 2 (+) and 4 (–).

Defect Code 46

♦D♦ Defect code memory: 46 OUTPUT STAGE

Explanation: Procedure as for defect code 19.

Defect Code 48

♦D♦ Defect code memory: 48 A/C COMPRESSOR CUT–OUT

Explanation: This defect code is stored in the case of a short to plus or minus.

In the case of a short to minus, the A/C compressor is not switched off at a speed below 8 km/h when accelerating with full load.

The A/C compressor never operates in the case of a short to plus.

Troubleshooting:

– Check wires to A/C compressor cut–out relay
– Check relay

2.5L M50 ENGINE—DME M3.1 DEFECT CODES—1992 325i AND 1991-92 525i

Defect Code 50

♦D♦ Defect code memory: 50 IGNITION CYLINDER 4

Explanation: Procedure as for defect code 23.

Defect Code 51

♦D♦ Defect code memory: 51 IGNITION CYLINDER 6

Explanation: Procedure as for defect code 23.

Defect Code 52

♦D♦ Defect code memory: 52 IGNITION CYLINDER 5

Explanation: Procedure as for defect code 23.

2.5L M50 ENGINE–DME M3.1 DEFECT CODES–1992 325i AND 1991-92 525i

Defect Code 54

♦D♦ Defect code memory: 54 CONTROL UNIT SUPPLY via Main Relay

Explanation: This defect code is stored if a break occurs in the supply voltage after start or when the voltage value is outside the permissible range between 10 V and 14 V.

If a break occurs, the supply is provided internally in the control unit via terminal 30.

Troubleshooting: If the voltage was outside the set range, check the alternator, voltage regulator and the battery.

The defect "break" refers to supply wires and ground connections to the control unit.

Defect Code 55

♦D♦ Defect code memory: 55 IGNITION OUTPUT STAGE

Explanation: This defect code is stored in the case of a break at the ignition output stage. The defect, however, could not be assigned to any specific ignition output stage.

Troubleshooting: Check ignition signals with the aid of the **primary adapter.**

Defect Code 62

♦D♦ Defect code memory: 62 SIGNAL–ELECTRONIC THROTTLE CONTROL (EML)

Explanation: This defect code is stored in the case of a short to plus.

Note: This defect code can be ignored if no EML control unit is installed.

2.5L M50 ENGINE—DME M3.1 DEFECT CODES—1992 325i AND 1991-92 525i

Defect Code 64

♦D♦ Defect code memory: 64 IGNITION TIMING INTERVENTION during Downshifts (with EGS only)

Explanation: This defect code is stored when the wire from the DME control unit Pin 64 to the EGS control unit has a short to ground (>2.5 s). In such a case, the ignition timing is adjusted in the direction RETARD only **once**, the signal is then ignored. The ignition timing assumes its original value.

Troubleshooting: Check wire between DME control unit and EGS control unit.

Defect Code 67

♦D♦ Defect code memory: 67 CRANKSHAFT PULSE GENERATOR

Explanation: This defect code is stored when the signal is not plausible. The ignition and injection cut out in the case of fault.

Troubleshooting: Check pulse generator and supply wire.

Defect Code 70

♦D♦ Defect code memory: 70 OXYGEN SENSOR

Explanation: This defect code is stored when a short or break occurs at the oxygen sensor or its supply wires. The prerequisite for this is an engine temperature of >70 °C.

The system sets lambda = 1 as a substitute function.

Troubleshooting: Check the oxygen sensor signal at operating temperature with status monitoring. The voltage value should fluctuate between 0.02 and 0.85 V. A faulty or no sensor signal is sent to the DME control unit for as long as a constant value of 0.45 V is indicated.

2.5L M50 ENGINE–DME M3.1 DEFECT CODES–1992 325i AND 1991-92 525i

Defect Code 73

♦D♦ Defect code memory: 73 VEHICLE SPEED SENSOR

Explanation: This defect code is stored when no plausible signal is detected while driving. The prerequisite for this is an engine speed >2500 rpm at high load for at least 10 s.

Troubleshooting: If the indicated speed (speedometer) is okay, check wire to the DME control unit – otherwise continue with diagnostic program – INSTRUMENT CLUSTER.

Defect Code 77

♦D♦ Defect code memory: 77 INTAKE AIR TEMPERATURE SENSOR

Explanation: This defect code is stored in the case of a short or break at the temperature sensor or its supply wires. The prerequisite for detecting a short to plus or a break is a time interval of 3 minutes after the start and idling. The system then operates with a substitute value of 30 °C.

Troubleshooting: Check temperature values by means of status monitoring.

Defect Code 78

♦D♦ Defect code memory: 78 ENGINE TEMPERATURE SENSOR Coolant

Explanation: This defect code is stored in the case of a short or break occurring at the temperature sensor or its supply wires. To detect a short to plus or a break, check the plausibility of the engine and intake air temperature. The system then operates for 3 minutes with the intake air temperature and then with a substitute value of 114 °C.

Troubleshooting: Check temperature values by means of status monitoring.

Defect Code 81

♦D♦ Defect code memory: 81 ANTITHEFT SYSTEM SIGNAL (DWA)

Explanation: This defect code is stored when the DWA is registered as "armed" after starting.

Troubleshooting: Check wires to DME control unit Pin 81 for a short.

2.5L M50 ENGINE – DME M3.1 DEFECT CODES – 1992 325i AND 1991-92 525i

Defect Code 82

♦D♦ Defect code memory: 82 INTERVENTION IN ENGINE DRAG TORQUE CONTROL (MSR)

Explanation: This defect code is stored in the case of a short to plus.

Note: This defect code can be ignored if no EML control unit is installed.

Troubleshooting: Check wire and control unit.

Defect Code 85

♦D♦ Defect code memory: 85 A/C COMPRESSOR

Explanation: This defect code is stored when the A/C compressor is detected as being switched on but no A/C switch signal is received.

Troubleshooting: Check wires to the DME control unit Pin 85 and 86 for a short and break.

Defect Code 100

♦D♦ Defect code memory: 100 OUTPUT STAGE

Explanation: Procedure as for defect code 19.

Defect Code 83

♦D♦ Defect code memory: 83 INTERVENTION IN AUTOMATIC STABILITY CONTROL (ASC)

Explanation: This defect code is stored in the case of a short to plus.

Note: This defect code can be ignored if no EML control unit is installed.

Troubleshooting: Check wire and control unit.

Defect Code 200

♦D♦ Defect code memory: 200 DME CONTROL UNIT

Explanation: This defect code is stored when a defect is detected in the RAM or in the ROM/EPROM.

Troubleshooting: After clearing the defect code memory, carry out test run. If the defect occurs once again, disconnect the DME control unit from the wire harness for approx. 5 minutes in order to trigger a reset. The control unit must be replaced if the defect is repeated.

2.5L M50 ENGINE—DME M3.1 DEFECT CODES—1992 325i AND 1991-92 525i

Defect Code 201

♦D♦ Defect code memory: 201 EMISSION CONTROL

Explanation: This defect code is stored when the emission control reaches its control limit and no other defect which affects the mixture has been detected. This may be attributed to deviations in the mixture composition (mixture too rich or too lean for longer than 5 s).

Troubleshooting: Measure CO content at exhaust manifold while observing the oxygen sensor voltage. For nominal values see chapter – Component Testing. Establish operating condition at the time the defect was detected.

Possible causes:

- Incorrect fuel pressure
- Defective or coked injector valves
- Incorrect output value from engine temperature sensor
- Secondary air in intake or exhaust tract
- Defect in tank ventilation
- Poor combustion (ignition, inlet/exhaust valves, compression etc. defective ...)

Defect Code 202

♦D♦ Defect code memory: 202 DEFECT CODE MEMORY

Explanation: This defect is indicated if illogical data are stored in the defect code memory.

Troubleshooting: A defect should be simulated after clearing the defect code memory – e.g. start engine and disconnect plug from idle actuator. The control unit must be replaced if defect code 202 is repeatedly stored.

Defect Code 203

♦D♦ Defect code memory: 203 IGNITION CIRCUIT MONITORING

Explanation: This defect code is stored if a break occurs in the primary circuit of the ignition.

Troubleshooting: Check ignition signals with the aid of the **primary adapter.**

Defect Code 204

♦D♦ Defect code memory: 204 STALL PROTECTION

Explanation: This defect code is stored when the signals are not plausible.

Note: This defect code can be ignored if no EML control unit is installed.

2.5L M50 ENGINE–DME M3.1 COMPONENT TESTING–1992 325i AND 1991-92 525i

Overview of defect codes which can be stored, monitoring and test options

Defect code/Defect path/Area	Status monitoring	Component activation	See quick test list No.
1 EKP relay	X	–	1
2 Idle actuator (close)	–	X	8,9
3 Injector valve cylinder 1	–	X	3
4 Injector valve cylinder 3	–	X	3
5 Injector valve cylinder 2	–	X	3
6 Injector valve output stage	–	–	3
8 Defect lamp	–	–	–
12 Throttle potentiometer	X	–	13
16 Camshaft pulse generator	–	–	12
18 Output stage Pin 18 (Control units from 10/90)	–	–	–
19 Output stage (Control units from 10/90)	–	–	–
23 Ignition cylinder 2 (Control units from 10/90)	–	–	–
24 Ignition cylinder 3 (Control units from 10/90)	–	–	–
25 Ignition cylinder 1 (Control units from 10/90)	–	–	–
26 Control unit supply	X	–	–
29 Idle actuator (open)	–	X	9
31 Injector valve cylinder 5	–	X	3
32 Injector valve cylinder 6	–	X	3
33 Injector valve cylinder 4	–	X	3

2.5L M50 ENGINE – DME M3.1 DEFECT PATH – 1992 325i AND 1991-92 525i

Defect code/Defect path/Area	Status monitoring	Component activation	See quick test list No.
36 Tank ventilation	X	X	10
37 Oxygen sensor heating	–	X	6
41 Air mass meter	X	–	5
48 A/C compressor cut–out (control units from 10/90)	–	–	–
50 Ignition cylinder 4 (Control units from 10/90)	–	–	–
51 Ignition cylinder 6 (Control units from 10/90)	–	–	–
52 Ignition cylinder 5 (Control units from 10/90)	–	–	–
54 Control unit supply via relay 2	–	–	–
64 Ignition timing intervention from automatic transmission	X	–	–
67 Crankshaft pulse generator	X	–	11
70 Oxygen sensor	X	–	6
73 Vehicle speed signal	X	–	–
77 Intake air temperature sensor	X	–	–
78 Engine temperature sensor	X	–	–
81 Antitheft system signal	–	–	–
82 Intervention in engine drag torque control	–	–	–
83 Intervention in automatic stability control	–	–	–
85 A/C compressor signal	X	–	–
100 Output stage (control units from 10/90)	–	–	–
200 Control unit	–	–	–
201 Emission control	X	–	7
202 Defect code memory	–	–	–
203 Ignition circuit monitoring (control units from 10/90)	–	–	–
204 Stall protection (control units from 10/90)	–	–	–

Defect code/Defect path/Area	Status monitoring	Component activation	See quick test list No.
Ignition timing	X	–	–
Load signal (from HLM and temperature)	X	–	5
Idle and full load	X	–	13
A/C switch	X	–	–
Drive range P/N	X	–	–

2.5L M50 ENGINE – DME M3.1 QUICK TEST – 1992 325i AND 1991-92 525i

Component–Quick Test List

No.	Test	Conditions	Possible cause of defect	Test aids / Remarks
1	Fuel pump relay	From starting speed	– Voltage supply for EKP relay defective – Activation (minus) via output stage in DME control unit defective – Signals from inductive pulse generator faulty	– Multimeter, schematic – Defect code memory, status monitoring – Status monitoring, oscilloscope
2	Fuel pump/fuel pressure	EKP relay ON	– Voltage supply defective – Fuel pump defective – Fuel lines/pressure regulator defective	– Multimeter, schematic – Pressure test to repair instructions
3	Injector valve	– As of starting speed – Ignition system OK on primary side otherwise catalytic converter protection function will be active (output stages do not switch), – Air mass meter OK.	– Voltage supply defective – Output stage in control unit defective – Injector valve defective	– Multimeter, schematic – Defect code memory – Component activation – Leakage test
4	Ignition coil	From starting speed	– Voltage supply defective – Signals from inductive pulse generator faulty – Wire 4a loose/broken – Output stage in control unit defective – Ignition coil defective	– Multimeter – Status monitoring "engine speed" – Engine test, oscilloscope
5	Air mass meter/load signal	From starting speed (a load signal of 1.5 ms is set for starting)	– Voltage supply via DME main relay/ground terminal defective – Heating wire dirty – HLM defective – signal not plausible – Air cleaner dirty	Attention! Substitute value can be active – Status monitoring HLM/load signal – Multimeter, schematic – Check clear–burn function – Defect code memory, status monitoring

2.5L M50 ENGINE – DME M3.1 QUICK TEST – 1992 325i AND 1991-92 525i

Component–Quick Test List

No.	Test	Conditions	Possible cause of defect	Test aids / Remarks
6	Oxygen sensor/sensor heating	Sensor signal with engine running and oxygen sensor at operating temperature, sensor heating from ignition ON	– Sensor dirty – Plug connection defective – Sensor heating defective – Oxygen sensor defective	Attention! Substitute value of 0.45 V may be active – Status monitoring – Component activation – Defect code memory
7	Emission control	Engine running and oxygen sensor at operating temperature	– Control at limit because mixture too lean/too rich – look for defect in mixture control system	– Defect code memory – Measure CO content
8	Idle actuator	From ignition ON	– Voltage supply defective – dle actuator mechanically blocked or dirty – Idle actuator electrically defective	– Component activation and visual inspection of removed component – Defect code memory
9	Idle control	Engine running, when idle signal is detected and none of the defects opposite	If a defect is detected at: – Oxygen sensor – Air mass meter – Engine temperature sensor – Intake air temperature sensor – Injector valve – Idle actuator or on the – voltage supply of the DME control unit, the last idle setting remains valid	– Defect simulation by applying additional air to the manifold
10	Fuel evaporation control valve	Emission control active, dependent on engine speed and load	– Voltage supply defective – Fuel evaporation control valve electrically/mechanically defective	– Component activation – Defect code memory – Leakage test
11	Crankshaft pulse generator	From starting speed	– Distance to increment wheel too great or generator loose/defective – Plug connection defective	– Status monitoring "engine speed" – Oscilloscope
12	Camshaft pulse generator	From starting speed	– Generator loose or defective – Plug connection defective	– Oscilloscope
13	Throttle potentiometer	From ignition ON	– Plug connection or component defective – Mechanical connection to throttle valve defective	Attention! Substitute value may be active – Status monitoring – Defect code memory

2.5L M50 ENGINE—DME M3.1 CO CORRECTION WITH MODIC PROGRAMMER—1992 325i AND 1991-92 525i

CO Correction on Models without Catalytic Converter

This correction can ONLY be carried out with the BMW MoDiC. The prerequisite for this is the MoDiC program status V 2.1 or the subsequent versions.

CO correction is only effective in the idle range and is necessary under the following conditions:

– When deviations in the CO content are found with the engine at operating temperature and there are no other defects effecting the mixture.

– After conversion of the vehicle to catalytic converter provisions (catalytic converter and oxygen sensor removed, control unit exchanged).

The CO content should be measured at the exhaust manifold before performing the correction.

The nominal CO value is 0.7 ± 0.5 vol.% at idle speed and a coolant temperature >80 °C.

After ensuring the mixture control system is operating correctly, connect the MoDiC to the diagnostic connector in order to change the CO adjustment value in the DME control unit.

The necessary steps to be taken are shown in the McDiC display.

– Switch on ignition and select DME (M50)

– Go to selection overview and CO adjustment

– Start engine and wait until a coolant temperature of >80 °C is reached (inquiry via status monitoring)

– The current setting is indicated on a bar and numerically, the setting is changed with the left/right arrow keys

– Only change setting in steps, wait for response time of CO measuring instrument (one step = a change by approx. 0.2 vol.%)

– Adjustment can be continued until the nominal value is set

– Then shut down engine and switch on ignition again– Store the newly set adjustment value by pressing the "Y" key

– A maximum of 50 storage operations can be carried out. The remaining number of possible storage operations is displayed beforehand.

– Correct storage is confirmed with the message "storage completed",
– other messages can be:

"Repeat entry"

"Memory is full"

"Program error"

The previously set CO adjustment value is not stored if these messages appear.

2.5L M50 ENGINE – DME M3.1 RESISTANCE VALUES – 1992 325i AND 1991-92 525i

Engine type	M50 B20 and B25
Injector valve	15...18 Ω
Speed/position pulse generator	540 ±50 Ω
Camshaft pulse generator	1.45 ±10% kΩ
Ignition coil – primary	0.4...0.8 Ω
Air mass meter – terminal 5/6	3...4 Ω
Air mass meter – terminal 5/1	16.3 kΩ
Throttle potentiometer – R total	4.6 kΩ
Idle actuator – per winding	20...25 Ω
Engine temperature sensor at –10±3 °C	8.26...10.56 kΩ
Engine temperature sensor at 20±3 °C	2.2...2.7 kΩ
Engine temperature sensor at 80±3 °C	290...364 Ω
Intake air temperature sensor at 20±3 °C	2.2...2.7 kΩ
Intake air temperature sensor at 50±3 °C	760...910 Ω
Fuel evaporation control valve	40...50 Ω

If not otherwise specified, the values refer to an air temperature of 23 ± 5 °C.

2.5L M50 ENGINE—DME M3.1 STATUS INQUIRY—1992 325i AND 1991-92 525i

Nominal Values for Status Inquiry

The nominal values apply at: idle speed, oil temperature >60 °C or coolant temperature >80 °C, no loads switched on.

If not otherwise specified, the values apply to both displacement variants B20 and B25 of the M50 engine.

Status inquiry	Nominal value	Signal comes from:
67 Engine speed		DME control unit
Idle speed	700 ± 40 rpm	
Cut–out speed	6500 ± 40 rpm	
78 Engine temperature	Dependent on coolant temperature, break = below –30 °C	Engine temperature sensor
77 Intake temperature	Dependent on coolant temperature, break = below –40 °C	Intake air temperature sensor
92 Ignition timing (angle)	M50 B 20: 10 ± 3 ° KW M50 B 25: 9 ± 3 ° KW	DME control unit
12 DK potentiometer	0.5 ± 0.2 V in idle position, 4.5 ± 0.2 V in full load position	Throttle potentiometer
41 HLM signal	idle value with no load 2.5 ±0.2 V, min. value 1.9 V Value increases proportionally to load, max. value 5.15 V	Hot–wire air mass meter
93 Load signal	Initial value at ignition ON = 1.5 ms when idling: M50 B20: 1.8 ±0.5 ms M50 B25: 1.7 ±0.5 ms	DME control unit

2.5L M50 ENGINE—DME M3.1 STATUS INQUIRY—1992 325i AND 1991-92 525i

Status inquiry	Nominal value	Signal comes from:
91 Emission control	Engine running and oxygen sensor at operating temperature, control = ON	DME control unit
36 Tank ventilation	Activation cycle dependent on engine speed and load, display = CLOSE/OPEN	DME control unit
70 Oxygen sensor	Engine running and oxygen sensor at operating temperature 0.02...0.85 V, faulty oxygen sensor signal = 0.45 V	Oxygen sensor
89 Idle signal	Detection by way of DK potentiometer, display ON only when idling!	DME control unit
90 Full load signal	Shortly before accelerator pedal reaches ON	DME control unit
26 Control unit supply	> 10 V	DME relay
85 A/C compressor	Compressor magnetic clutch activated = ON	Compressor relay
86 A/C switch	With A/C, defroster or air recirculation switch depressed and control wheel for air quantity not set to 0 = ON	Air conditioning system
73 Vehicle speed	If drive wheels are moved in raised condition, limit speed to max. 30 km/h	Instrument cluster
64 Ignition timing intervention by EGS	Temporarily to ON only during gearshift	EGS control unit
65 Drive range P/N	With drive range engaged ON	EGS control unit
Injection signal (no status inquiry)	M50 B20: 4,1 ±0.5 ms M50 B25: 3,9 ±0.5 ms	Measure on lines to the injector valves

2.5L M20 AND 3.5L M30 ENGINES – DME M1.3 TROUBLESHOOTING – 1990-91 325i, 1990 525i, 1990-92 535i AND 735i

Notes on the BMW DIAGNOSTIC SYSTEM

Control Unit Assignment

Initially, the identification data are displayed in order to determine whether the correct control unit is installed. The assignment can be determined by means of the version code (e.g. C01E) and parts microfilm without having to remove the control unit.

Sequence of identification output:

- DME version
- Model designation, engine type
- With or without oxygen control
- Transmission version: HG - Manual transmission
  AG - Automatic transmission, hydraulic
  EH - Automatic transmission, electronic-hydraulic
- Type of fuel: Leaded/unleaded or regular/premium
- Country version: International identification
- Control unit version code: See parts microfilm
- BMW hardware and software number
- Bosch hardware and software number
- Production code: Used only on special request by "Service Division"

Reading Defect Code Memory

The next step is to monitor the defect code memory (D 900). If one or several defect codes are stored, it is advisable to immediately print out these data (R button). Since it is often necessary during the testing procedure to cut the power supply or to disconnect the connector from the control unit, the content of the defect code memory will be unintentionally deleted as a consequence.

DME M1.1 and M1.3 model '87/'88:

The memory can store a maximum of 5 defect codes, whereby a defect code with higher priority can displace another defect code from the memory. Defect codes are indicated in the same sequence as they were detected.

The following defects have higher storage priority:

- air flow meter,
- oxygen control,
- engine temperature sensor and
- idle switch

DME M1.3 as of model '89:

The extended defect code memory stores all defect codes which are detected.

In US models, defect codes which influence the mixture are indicated by a lamp (engine check) in the instrument cluster. As a check, this lamp lights constantly in the time between "ignition ON" and engine start.

When a defect code is stored and the DME is not malfunctioning, substitute values can be active. The troubleshooting procedure should still, however, be carried out.

The substitute values are also lost if the battery or the DME control unit is disconnected. This may result in difficulties in starting the engine or it may not start at all.

2.5L M20 AND 3.5L M30 ENGINES—DME M1.3 TROUBLESHOOTING—1990-91 325i, 1990 525i, 1990-92 535i AND 735i

Clearing the Defect Code Memory

Carry out this step only when the defect has been repaired or when it has been stored by way of defect simulation (D999). When the battery or the DME control unit are disconnected, all stored defect codes as well as the substitute values are lost. Current defect codes cannot be cleared.

Monitoring the Current Values or Statuses

Further information can be called independent of defect code storage (D100 or D200). Current values or current operating statuses are displayed in this way. Refer to the chapter – „Component Testing" for the necessary comparison of set values. In the case of the DME M1.3, status inquiries can be compiled as required from the blocks 100 and 200. This selection (termed extract list) can then be displayed on the screen by pressing the ⬇ key.

Component Activation

To facilitate fast faultfinding, various components can be activated with the DIAGNOSTIC SYSTEM in order to diagnose the supply line or the component itself. The overview can be called by entering D300. Detailed information on this drive mode is displayed on the screen.

Abbreviations

AG	Automatic transmission
ASC	Automatic stability control
DK	Throttle valve
DWA	Antitheft system
EH	Electronic-hydraulic
EKP	Electric fuel pump
EML	Electronic throttle control
EV	Injector valve
HG	Manual transmission
KD	Kickdown
KW	Crankshaft
LL	Idle
LLS	Idle speed actuator
LMM	Air flow meter
MF	Microfilm
MSR	Engine drag torque control
MV	Solenoid valve
RxD	Diagnosis initiation line
TD	Engine speed signal
TE	Fuel evaporation control
ti	Injection timing
TL	Partial load
tL	Load signal
TR	Crankshaft reference (once per revolution)
TxD	Diagnostic data line
SG	Control unit
U-Batt	Battery voltage
U-Vers	Voltage supply
VL	Full load

2.5L M20 AND 3.5L M30 ENGINES—DME M1.3 PIN ASSIGNMENTS—1990-91 325i, 1990 525i, 1990-92 535i AND 735i

Pin Assignments

Assignments of the 55-pole Connector Strip on the DME Control Unit

Pin No.	Type	Designation
1	A (1)	Ignition terminal 1
2	M	Ignition ground
3	A	EKP relay and crankshaft reference
4	A	Open idle speed actuator (not for EML)
5	A	Fuel evaporation control valve
6	A	Engine speed TD
7	E	Signal, air flow meter LMM
8	E	Cylinder reference point (+)
9	–	–
10	(2)	Ground oxygen sensor
11	–	–
12	A	Supply LMM
13	E	Diagnostic line RxD
14	M	Ground, fuel injectors
15	A	Defect lamp (US only)
16	A	Injector valves (cyl. 1 + 3 + 5)
17	A	Injector valves (cyl. 2 + 4 + 6)
18	E	Terminal 30 permanent supply
19	M	Ground, control unit electronics
20	–	–
21	–	–
22	A	Close idle speed actuator LLS (not for EML)
23	–	–
24	M	Ground, remaining output stages
25	–	–
26	M	Ground, LMM
27	E	Ignition Terminal 15
28	E	Oxygen sensor signal
29	E	Speed signal from instrument cluster
30	–	–
31	(2)	Cylinder reference point (–)
32	A	ti-Signal to consumption display
33	A	Solenoid valve kickdown inhibit (for AG hydraulic only)
34	–	–
35	–	–
36	A	Control DME relay 2
37	E	Supply from DME relay 2
38	E	ASC and antitheft system (DWA)
39	E	Programming voltage
40	E	Compressor
41	E	A/C switch
42	E	Drive range P/N (automatic transmission only)
43	–	–
44	E	Temperature sensor – intake air
45	E	Temperature sensor – engine
46	–	–
47	E	Inductive pulse generator (+)
48	M	Inductive pulse generator (–)
49	–	–
50	E	Engine drag torque control
51	E	Ignition angle control (automatic transmission only)
52	E	Idle speed signal
53	E	Full load signal
54	E	Torque converter clutch unit (automatic transmission only)
55	E/A	Diagnostic line TxD

E = Input
A = Output
M = Ground
(1) = max. 150 V / max.0.2 A
(2) = potential-free ground

2.5L M20 AND 3.5L M30 ENGINES – DME M1.3 TROUBLESHOOTING – 1990-91 325i, 1990 525i, 1990-92 535i AND 735i

Troubleshooting

Notes

- Relevant safety codes and accident prevention regulations must be observed when carrying out tests and work on the engine electrical and electronic systems.

Caution!

Dangerous high voltage in the primary and secondary circuit of the ignition system.

*Wherever possible, the battery should not be disconnected otherwise the defect codes stored in **all** the control units will be deleted.*

Disconnect control units from the wire harness *only* when the ignition is switched off.

Conduct wire tests ***only*** with the control unit or component connectors disconnected.

Test Equipment

The universal adapter with the corresponding connection leads should be used for testing the DME control unit or the engine wire harness –

The universal adapter should only be connected between the control unit and wire harness when ***no*** diagnostic procedure (diagnostic connector not connected) is to be carried out since this can lead to malfunctions. Use only suitable test leads, terminals and probe tips.

Display of a Stored Error Code on the Screen.

– **Defect location:** The number of the defect code indicates the pin assignments at the control unit. The defect can be in the vicinity of the component, the corresponding wiring or in the control unit.

– **Type of defect:** Short-circuits, breaks, range transgressions or illogical functions are indicated.

– **Frequency**: Indication as to whether the defect is momentary or not (intermittent defect). Further indication is provided as to whether the defect occurred several times or only once.

As of model '89 an indication is provided as to how often the defect has already been detected. If the defect is no longer pending, the number of starts is indicated which have taken place since last detection of the defect.

– **detected at:** Additional defect information on the operating statuses during defect detection.

Defect Code Table

D 900 Defect code memory – stored defect codes:

One or several defect codes are stored → Defect codes 01...100

Malfunctions – defect codes not stored:

Engine operation not possible → Defect code 101

2.5L M20 AND 3.5L M30 ENGINES – DME M1.3 DEFECT CODES – 1990-91 325i, 1990 525i, 1990-92 535i AND 735i

Defect Code 01

D 900 Defect code memory: 01 DME CONTROL UNIT

Explanation: This defect code is stored when the internal self-test provides a negative result.

Possible causes: Program error or disturbed program sequence in the DME control unit.

Corrective measures: Carry out test run after clearing the defect code memory. If defect code is set once again, the DME control unit must be replaced.

Defect Code 03

D 900 Defect code memory: 03 ELECTRIC FUEL PUMP RELAY

Explanation: This defect code is stored when a short occurs at relay 1 (EKP) or on the control line to the DME control unit Pin 3.

Troubleshooting: As displayed on screen.

Defect Code 04

D 900 Defect code memory: 04 IDLE SPEED ACTUATOR (open)

Explanation: This defect code is stored when breaks or shorts occur at the idle speed actuator or on its supply wires (voltage supply and control lines). When the defect code 22 idle speed actuator (close) is also stored, then this is attributed to the link of both windings in the idle speed actuator.

Note: Defect codes 04 and 22 are also stored when engine operation is interrupted abruptly without the ignition being switched off beforehand. Such a defect is indicated by the operating status during defect code detection. This defect code can be ignored at an engine speed below 600 rpm.

Troubleshooting: As displayed on screen.

Defect Code 05

D 900 Defect code memory: 05 FUEL EVAPORATION VALVE

Explanation: This defect code is stored when a short occurs at the fuel evaporation valve or in the supply wire.

Troubleshooting: As displayed on screen. For test values, see chapter – Component Testing.

2.5L M20 AND 3.5L M30 ENGINES—DME M1.3 DEFECT CODES—1990-91 325i, 1990 525i, 1990-92 535i AND 735i

Defect Code 07

D 900 Defect code memory: 07 AIR FLOW METER

Explanation: This defect code is stored when shorts or breaks occur at the air flow meter or on its supply wires (voltage supply and signal line). Also detected is when the Up/Uv-value of LMM is no longer within the normal operating range of 0.02...0.95.

In this case, the DME control unit operates with fixed substitute values, load signal tL = 3.5...4.5 ms and ignition angle = 20° crank angle before TDC when the idle speed contact is open.

Troubleshooting: As displayed on screen.

Carry out air flow meter testing with status monitoring D200. The Up/Uv value indicates the ratio of the potentiometer voltage (wiper) with respect to the supply voltage (5V) in the air flow meter.

Defect Code 10

D 900 Defect code memory: 10 OXYGEN CONTROL

Explanation: This defect code is stored in the case of excessive deviations in the mixture composition (too rich or too lean mixture longer than 10 sec). A prerequisite is that the oxygen sensor must be ready for operation.

Possible causes:

- Incorrect fuel pressure
- Injector valves defective or coked
- Engine temperature sensor defective
- Secondary air in the intake duct
- Fuel evaporation control system defective

Troubleshooting: As displayed on screen.

Defect Code 15

D 900 Defect code memory: 15 DEFECT LAMP

Explanation: Applicable to US models only. This defect code is stored when shorts occur at the defect lamp or in its supply wire.

Troubleshooting: As displayed on screen.

2.5L M20 AND 3.5L M30 ENGINES – DME M1.3 DEFECT CODES – 1990-91 325i, 1990 525i, 1990-92 535i AND 735i

Defect Code 16

D 900 Defect code memory: 16 FUEL INJECTOR VALVES Cyl. 1 + 3 + 5

Explanation: This defect code is stored when shorts or breaks occur at the fuel injector valves or breaks in their supply wires.

Note: In the case of a short at one fuel injector valve or in its supply wire, the entire injector valve group is always deactivated. Three cylinders remain in operation.

Troubleshooting: As displayed on screen.

Defect Code 17

D 900 Defect code memory: 17 FUEL INJECTOR VALVES Cyl. 2+4+6

Explanation: Diagnosis as for defect code 16.

Defect Code 22

D 900 Defect code memory: 22 IDLE SPEED ACTUATOR (close)

Note: Diagnosis as for defect code 04.

Defect Code 23

D 900 Defect code memory: 23 OXYGEN SENSOR HEATING RELAY

Explanation: In the case of DME M 1.3, the defect code is stored when a short occurs at the relay or on the supply line from the DME control unit Pin 23.

Troubleshooting: As displayed on screen.

2.5L M20 AND 3.5L M30 ENGINES—DME M1.3 DEFECT CODES—1990-91 325i, 1990 525i, 1990-92 535i AND 735i

Defect Code 28

D 900 Defect code memory: 28 OXYGEN SENSOR

Explanation: This defect code is stored when a short or a break occurs at the oxygen sensor and on its supply wires, or when no oxygen sensor is installed (cat. provision). A prerequisite for this is an engine temperature of 70 °C and a load signal of > 2.0 ms for > 2.5 sec.

Troubleshooting: As displayed on screen.

Defect Code 29

D 900 Defect code memory: 29 ROAD SPEED SIGNAL

Explanation: This defect code is stored when the following conditions are fulfilled simultaneously:
Engine speed > 4500 rpm,
Load signal > 3.0 ms,
Speed signal < 3...5 km/h
This defect code can be ignored in models with EML. In this case, the road speed signal is transferred to the EML control unit and is therefore not detected by the DME.

Troubleshooting: As displayed on screen.

Defect Code 33

D 900 Defect code memory: 33 SOLENOID VALVE KICK-DOWN PREVENT

Explanation: Only applicable to models with DME M 1.3 and hydraulic automatic transmission. The defect code is stored when shorts occur at the solenoid valve or on the supply wire.

Troubleshooting: As displayed on screen.

Defect Code 37

D 900 Defect code memory: 37 CONTROL UNIT SUPPLY

Explanation: This defect code is stored when the voltage supply exceeds 16V. A drop in the voltage below 10V (M1.1) or 9V (M1.3) is detected 3 minutes after the start at the earliest.

Possible cause:
- Contact resistances at the wire connections
- Insufficient V-belt tension
- Voltage regulator or alternator defective

Troubleshooting: As displayed on screen.

2.5L M20 AND 3.5L M30 ENGINES – DME M1.3 DEFECT CODES – 1990-91 325i, 1990 525i, 1990-92 535i AND 735i

Defect Code 38

D 900 Defect code memory: 38 ASC/ANTITHEFT SYSTEM (DWA)

Explanation: Only applicable to models equipped with DME M1.3. This defect code is stored when a short to battery voltage occurs (>1 sec) on the wire to Pin 38.

Troubleshooting: As displayed on screen.

Defect Code 40

D 900 Defect code memory: 40 A/C COMPRESSOR

Explanation: This defect code is stored only in the case of DME M 1.3 when the control unit simultaneously receives the signals "compressor in operation" (Pin 40) and "A/C switch not pressed" (Pin 41). In this case, there is either a break at Pin 41 or a short to battery voltage at Pin 40.

Troubleshooting: As displayed on screen.

Defect Code 44

D 900 Defect code memory: 44 INTAKE AIR TEMPERATURE SENSOR

Explanation: This defect code is stored when shorts or breaks occur at the temperature sensor or on its supply wires. A prerequisite for detecting a break in the circuit is a 3 min interval after the start and the engine idling for a minimum of 30 sec. After the defect code is set, the system operates with a substitute value of 20 °C or 50 °C at 3.5 l with M 1.3.

Troubleshooting: As displayed on screen.

Defect Code 45

D 900 Defect code memory: 45 ENGINE TEMPERATURE SENSOR (Water)

Explanation: This defect code is stored when shorts or breaks occur at the temperature sensor or on its supply wires.

In the case of fault, the intake air temperature is used as a substitute value, however, only for the first 3 minutes after starting when the air temperature is < 11 °C. Otherwise, a substitute value of 50 °C is used.

Troubleshooting: As displayed on screen.

2.5L M20 AND 3.5L M30 ENGINES—DME M1.3 DEFECT CODES—1990-91 325i, 1990 525i, 1990-92 535i AND 735i

Defect Code 50

D 900 Defect code memory: 50 ENGINE DRAG TORQUE CONTROL

Explanation: Only applicable to models equipped with EML and DME M1.3. This defect code is stored when a short to battery voltage occurs (> 20 sec) on the wire to Pin 50.

Troubleshooting: As displayed on screen.

Defect Code 51

D 900 Defect code memory: 51 IGNITION ANGLE

Explanation: Applicable only to models equipped with electronic transmission control (AEGS) and DME M 1.3. The defect code is stored when the wire from the electronic transmission control unit Pin 24 to the DME control unit Pin 51 has a short to ground (> 2.5 sec). In such a case, the ignition angle intervention occurs only once, the signal is then ignored.

Troubleshooting: As displayed on screen.

Defect Code 52

D 900 Defect code memory: 52 IDLE SWITCH

Explanation: This defect code is stored when the throttle valve switch or the wire to the DME control unit Pin 52 has a short to ground. A prerequisite for this is that the engine is at operating temperature (> 80 °C) at a speed > 1000 rpm and a load signal > 3.5 ms.

Troubleshooting: As displayed on screen.

Defect Code 53

D 900 Defect code memory: 53 FULL LOAD SWITCH

Explanation: This defect code is stored when the throttle valve switch or the wire to the DME control unit Pin 53 has a short to ground. A prerequisite for this is a load signal > 1.8 ms and the idle contact closed at the same time. If the load signal drops below 4.5 ms, the full load signal is ignored by the DME control unit.

Troubleshooting: As displayed on screen.

2.5L M20 AND 3.5L M30 ENGINES—DME M1.3 DEFECT CODES—1990-91 325i, 1990 525i, 1990-92 535i AND 735i

Defect Code 54

D 900 Defect code memory: 54 TORQUE CONVERTER CLUTCH UNIT

Explanation: Only applicable to models equipped with electronic transmission control and DME M 1.3. This defect code is stored when the DME control unit detects a closed torque converter clutch (or short to battery voltage) and the selector lever is in position P or N.

Troubleshooting: As displayed on screen.

Defect Code 100

D 900 Defect code memory: 100 OUTPUT STAGE

Explanation: Only applicable to DME M 1.3. This defect code is stored when there is a loose contact at an output stage output. In the majority of cases, one of the defect codes 3, 4, 5, 16, 17, 22 or 23 also occurs, to which the defect code 100 can be assigned.

Defect Code 101

Malfunction: Engine operation not possible

Troubleshooting:

Prerequisites:
- Start system OK. (battery charged, ignition switch and starter OK. ...)
- Ground connection at battery, engine and body OK
- Specified type of fuel in tank
- Engine OK mechanically (timing, compression ...)

a) Check voltage supply for DME control unit in accordance with schematics.

b) Check input signals.
- Speed signal from inductive pulse generator D 200 at starting speed.
- Air flow meter signals D 200.

c) Check output signals.
- Drive of fuel pump relay in accordance with schematics
- Ignition output stage with ENGINE TEST, test step 09
- Fuel injector valves with D 300

d) Fuel pressure test

2.5L M20 AND 3.5L M30 ENGINES—DME M1.3 DEFECT CODES—1990-91 325i, 1990 525i, 1990-92 535i AND 735i

Overview of the Defect Codes and Component Tests

Component/Function	Short test after entry [D]...	Defect code after storage
DME control unit	–	1
Ignition coil	see engine test	–
Fuel pump relay (relay 1)	100	3
Idle speed actuator	300	4 / 22
Fuel evaporation valve	100, 300	5
Air flow meter	200	7
Pulse generator – cylinder reference point	100	–
Oxygen control	100	10
Defect lamp	–	15
Injector valves	300	16 / 17
Sensor heating relay	–	23
Oxygen sensor	100	28
Road speed signal	see instrument cluster/CC	29
Solenoid valve kickdown prevention	–	33
Main relay (relay 2)	100	–
Supply voltage	200	37
ASC	see ABS/ASC	38
A/C compressor	100	40
Intake air temperature sensor	200	44
Engine temperature sensor	200	45
Pulse generator – speed/position	200	–
MSR	see ABS/MSR	50
Ignition angle intervention	100	51
Throttle switch	100	52 / 53
Torque converter clutch unit	–	54

2.5L M20 AND 3.5L M30 ENGINES—DME M1.3 CHECK FUNCTION—1990-91 325i, 1990 525i, 1990-92 535i AND 735i

DME Control Unit

Check function:

Individual function blocks, e.g. ignition, injection, oxygen control etc. can be checked only when the necessary input signals are applied beforehand at the control unit

Ignition Coil

Check function:

Perform engine test, test step 09. With the engine at operating temperature, the ignition voltage is 10 ±2 kV. The maximum difference between the individual cylinders should not exceed 3 kV.

Fuel Pump Relay (DME Relay 1)

Check function:

D 100 – This status inquiry is only possible in the case of DME M1.3. The EKP relay must switch on during start or engine operation (display: ON).

Possible causes:

- Wire to the DME control unit Pin 3 defective
- Relay 1 or voltage supply not OK.
- Voltage supply for DME control unit not OK.
- No signal from engine speed/position pulse generator
- The control unit is defective if no ground potential is output at DME control unit Pin 3 during start and engine operation although the above points are OK.

Idle Speed Actuator (LLS)

Check function:

D 300 – The idle speed actuator is driven at intervals of one second – the orifice in the idle speed actuator must open and close completely.

Resistance test:

The resistance per winding can be measured directly at the component. Use special test lead

2.5L M20 AND 3.5L M30 ENGINES–DME M1.3 CHECK FUNCTION–1990-91 325i, 1990 525i, 1990-92 535i AND 735i

Fuel Evaporation Valve

Check electrical function:

[D] 300 – The fuel evaporation valve is driven at intervals of one second. The actuation cycles can be felt with the hand.

Check for leaks:

Release hoses from valve, switch on vacuum pump at service tester and connect to valve (8 mm connection) (note arrow indicating direction of flow). Set vacuum to 600 mbar and apply a voltage of 12 V to the valve.

Switch off vacuum pump, – after 20 sec, the pressure drop must not exceed 50 mbar.

When power is not applied, the fuel evaporation valve is closed by a non-return valve to a vacuum of 20 mbar.

Check drive phases:

[D] 100 – The valve is driven at idle speed (display: ON or CLOSED) and is therefore closed. As soon as the oxygen control is active and the engine speed is increased, the valve can be observed to open step-by-step (display: OFF or OPEN).

After switching off the engine, the valve is held closed for 3...5 sec.

Air Flow Meter (LMM)

Check LMM signal:

[D] 200 – The value Up/Uv (Up = potentiometer voltage, Uv = supply voltage of 5 V) indicates the voltage ratio derived from the position of the pressure flap in the air flow meter. At idle speed, this value is at 0.2...0.3, at full load it increases to 0.9. The value 0.02 is indicated if the supply voltage or signal line is interrupted.

Check voltage supply:

With the ignition ON, the air flow meter is supplied with 5 V by the DME control unit Pin 12 (+) and Pin 26 (–).

Pulse Generator – Cylinder Reference Point

Check function:

[D] 100 – This status inquiry is only possible in the case of DME M1.3. As soon as the control unit receives the cylinder reference point signals, a switchover takes place to semisequential injection.
Further test facilities with multimeter test 23/24. Connect oscilloscope connections at component connector and run engine at idle speed. An alternating voltage characteristic curve with positive and negative half waves should appear on the display screen.

Resistance test:

The resistance can be measured directly at the component connector (at the connector strip for injector valves).

2.5L M20 AND 3.5L M30 ENGINES – DME M1.3 CHECK FUNCTION – 1990-91 325i, 1990 525i, 1990-92 535i AND 735i

Oxygen Control

Check function:

D 100 – Prerequisite for this procedure is that the oxygen sensor is at operating temperature. The engine must run for 30...40 sec. before the control is implemented.

The control can compensate for wear-related deviations in the mixture composition and therefore conceal other defects. If necessary, disconnect the oxygen sensor during troubleshooting of the mixture composition.

Defect Lamp "Check Engine" (US only)

Check function:

The control unit supplies the defect lamp with ground potential as soon as an exhaust emission-related defect is detected. As a check, the lamp lights constantly after ignition ON.

Fuel Injector Valves

Check electrical function:

D 300 – In this case, the fuel injector valves are driven with a frequency of 10 Hz. The pulse period is 1 ms in order to avoid fuel being injected. Activation can be felt with the hand. The pulses can be tested with the multimeter at the fuel injector valve connector.

Resistance test:

The resistance of an injector valve must be measured directly at the component. Only the entire injector valve group (e.g. 3 valves on 6-cylinder engine) can be measured from the DME control unit connector.

Sensor Heating Relay (Engine M20 only)

Check function:

D 300 – This activation is only possible in the case of DME M1.3.
It takes place at one second intervals.

The relay remains switched on during normal operation until high speeds or load occur.

Possible causes:

- Wire to the DME control unit Pin 23 defective
- Relay or voltage supply not OK.

2.5L M20 AND 3.5L M30 ENGINES—DME M1.3 CHECK FUNCTION—1990-91 325i, 1990 525i, 1990-92 535i AND 735i

Oxygen Sensor

Check function:

D 200 – To carry out this test, the oxygen sensor must be at operating temperature. The indicated voltage signal is tapped off in steps and fluctuates between 0.02 and 0.85 V.

When the oxygen sensor is not in operation, or the wire to the DME control unit is interrupted, a constant value of 0.45 V is indicated. This value is also indicated when no oxygen sensor is installed.

Solenoid Valve Kick-Down Prevent (DME M1.3 and Automatic Transmission Only)

Check function:

With the engine running, the solenoid valve in the automatic transmission receives a voltage (> 10V). The positive signal is switched by the EKP relay and the negative signal comes from the DME control unit (pin 33). At excessively high speed, this signal is cut out thus preventing kick-down.

Effect of defect:

As of a speed of approx. 120 km/h, downshift from 4th. to 3rd. gear is no longer possible.

Possible causes:

– Wire to DME control unit Pin 33 defective
– Solenoid valve or voltage supply not OK.

Resistance test:

Connect universal adapter, measure resistance between pins 33 and 19.

Main Relay – Control (DME Relay 2)

Check function:

D 100 – This status inquiry is only possible in the case of DME M1.3. With ignition ON or during engine operation, the main relay is driven with ground potentiometer at terminal 85 (display: ON). The relay switches on and applies the control units DME, AEGS, EML and ABS with voltage.

Possible causes.

– Wire to DME control unit Pin 36 defective
– Relay 2 or voltage supply not OK.
– Voltage supply for DME control unit not OK.
– The control unit is defective if no ground potential is output at control unit Pin 36 although the above points are OK.

Control Unit Voltage Supply

Check permissible range:

D 100 – The displayed voltage is currently at the control unit. It should not be below 10V.

2.5L M20 AND 3.5L M30 ENGINES – DME M1.3 CHECK FUNCTION – 1990-91 325i, 1990 525i, 1990-92 535i AND 735i

Intake Air Temperature Sensor (in LMM)

Check function:

D 200 – With the engine at operating temperature, the intake air temperature is at 50...70 °C (ambient air temperature 23 ±5 °C).

A display below -40 °C indicates a break in the temperature sensor or in the supply wire to the DME control unit.

Resistance test:

Connect universal adapter, measure resistance across the terminals 44 and 26.

Engine Temperature Sensor (Coolant)

Check function:

D 200 – The indicated engine temperature should correspond to the coolant temperature. A value below -30 °C indicates a break in the temperature sensor or in the wire to the DME control unit.

Resistance test:

Connect universal adapter to the engine wire harness. Measure resistance across the terminals 45 and 19.

Engine Speed/Position Pulse Generator

Check function:

D 200 – The engine speed signals can be monitored at starting speed or when the engine is running.
Further tests are possible with the multimeter test 23/24. Connect terminals for the oscilloscope at the component connector and crank engine. An AC voltage characteristic curve with positive and negative half waves should appear on the display screen.

Resistance test:

The resistance can be measured directly at the component connector (at the connector strip for injector valves).

Possible cause:

If wire, plug connection or resistance are OK:
Check distance of the pulse generator to the incremental wheel on the vibration damper - 0.15...1.5 mm.
Adjustment at holder - 1.0 ±0.3 mm.

Throttle Switch

Check function:

D 100 – Check the setting of the idle speed and full load switch by displacing the throttle valve. For setting, see microfilm - Repair instructions.

The throttle switch is not installed in vehicles equipped with EML. The signals LL and VL are sent by the EML control unit to the DME control unit.

2.5L M20 AND 3.5L M30 ENGINES—DME M1.3 TECHNICAL DATA—1990-91 325i, 1990 525i, 1990-92 535i AND 735i

Technical Data

Resistance Values Measured at 23 ±5 °C Air Temperature

Engine type	M20	M30
Idle speed actuator (one winding)	20 ±2 Ω	20 ±2 Ω
Injector valve	15...17 Ω	15...17 Ω
Injector valve group (3 valves)	5.5 ±0.5 Ω	5.5 ±0.5 Ω
Pulse generator - engine speed/position	520 ±30 Ω	520 ±30 Ω
Pulse generator - cylinder reference point	0.2...1 Ω	0.2...1 Ω
Engine temperature sensor		
at coolant temperature 20 ±3 °C	2.3...2.7 kΩ	2.3...2.7 kΩ
at coolant temperature 80 ±3 °C	0.3...0.36 kΩ	0.3...0.36 kΩ
Intake temperature sensor		
at air temperature 20 ±3 °C	2.2...2.7 kΩ	2.2...2.7 kΩ
at air temperature 50 ±3 °C	0.76...0.91 kΩ	0.76...0.91 kΩ
Fuel evaporation valve	45 ±20 Ω	45 ±20 Ω
Ignition coil		
Primary winding	0.4 ±0.1 Ω	0.4 ±0.1 Ω
Secondary winding	9.0 ±1.0 kΩ	9.0 ±1.0 kΩ
Solenoid valve KD prevent	22...60 Ω	22...60 Ω

2.5L M20 AND 3.5L M30 ENGINES—DME M1.3 STATUS INQUIRIES—1990-91 325i, 1990 525i, 1990-92 535i AND 735i

Status inquiries 100	Set status
103 Fuel pump (EKP) relay control	during start or engine operation = ON 3)
105 Fuel evaporation valve	at idle speed = CLOSED or ON, until refuelling begins
110 Oxygen control	during engine operation = ON when oxygen sensor at operating temperature
136 Main relay control	during start or engine operation = ON 3)
140 Compressor control	when compressor relay is switched on = ON (not for EML)
141 A/C switch	with switch depressed and selector wheel position for air flow not set to 0 = ON
142 Drive range P/N (AG only) 3)	in range lever position P or N = ON
151 Ignition angle intervention (AG only) 3)	at idle = OFF, ON briefly only during kick-down
152 Idle switch	when idle contact closed = ON
153 Full load switch	when full load contact closed = ON
156 Semisequential injection 3)	after increasing engine speed once = ON

3) This status inquiry is only possible in the case of DME M1.3.

2.5L M20 AND 3.5L M30 ENGINES—DME M1.3 STATUS INQUIRIES—1990-91 325i, 1990 525i, 1990-92 535i AND 735i

Status inquiries 200	Set values
207 LMM Up/Uv ratio 1)	0.2...0.3, interruption = 0.02
237 Voltage supply - control unit	> 10V
243 Voltage - oxygen sensor 1)	0.02...0.85 V, in case of break or defective oxygen sensor = 0.45V
244 Intake air temperature	depending on the ambient temperature, approx. -40 °C = break
245 Engine temperature	depending on engine operation, approx. -30 °C = break

	M20 B20	M20 B25	M30 B30	M30 B35
201 Ignition angle in ° crank angle before TDC 1)	4±5	10±5	10±3	10± 3
206 Engine speed in rpm 1)				
...Idle speed	760±40	760±40	800±50	800±50
...Cut-out speed	6400±40	6400±40	6400±40	6200±40
232 Injection signal ti in ms 1)2)3)	4.9...5.3	4.4...4.8	3.5...5.0	3.5...5.0
277 Load signal tL in ms 1)	2.1...2.8	2.1...2.8	1.8...2.2	1.8...2.2
...CO volume % 1) 4)	0.7±0.5	0.7±0.5	0.7±0.5	0.7±0.5

1) Oil temperature > 60 °C or coolant temperature > 80 °C, no load switched on, at idle speed.
2) Valid for semisequential injection. Half value applicable to M30 up to 6/88 since only parallel injection is possible in diagnostic mode.
3) This status inquiry is only possible in the case of DME M1.3.
4) Measured before the catalytic converter, – Engine test step 09.

5.0L M70 ENGINE – DME M1.2 TROUBLESHOOTING – 1990 750iL

Performance Characteristic Switchover for EH Transmission

During the shift procedure, the AEGS control unit sends a signal to the DME control units which set the ignition angle towards retard. This ensures a smooth transition to the newly selected drive range.

As soon as the torque converter clutch unit is engaged, the DME control units are instructed to switch over to a different ignition angle characteristic.

Self-Diagnosis

The task of the self-diagnosis feature is to detect faults in the DME and subsequently store defect codes. Troubleshooting is assisted by the self-diagnosis feature in the form of status inquiries and component activation.

In the case of failure of the engine temperature sensor, intake air temperature sensor, air flow meter signal or oxygen sensor, the DME control unit makes available suitable substitute values. These substitute values are cancelled once again as soon as normal operation can be resumed.

Notes on the BMW DIAGNOSTIC SYSTEM

Control Unit Assignment

Initially, the identification data are displayed in order to determine whether the correct control unit is installed. The assignment can be determined by means of the version code (e.g. C01E) and parts microfilm without having to remove the control unit.

Explanation of Control Unit Identification

The designation of the DME version and the indication as to with which row of cylinders communication is currently established appear in the first line.

Further data are displayed in the following sequence
- Version: model designation, engine type
- With or without oxygen control
- Transmission version: HG – Manual transmission
  AG – Automatic transmission, hydraulic
  EH – Automatic transmission, electronic-hydraulic
- Type of fuel: Leaded/unleaded or regular/premium
- Country version: International identification
- Control unit version code: See parts microfilm
- BMW hardware and software number
- Bosch hardware and software number
- Production code: Used only on special request by „Service Division“

Monitoring Defect Code Memory – [D] 900

Defect code memory monitoring should be given first priority. If one or several defect codes are stored, it is advisable to immediately print out these data (R button). Since it is often necessary during the testing procedure to cut the power supply or to disconnect the connector from the control unit, the content of the defect code memory will be unintentionally deleted as a consequence.

Model '87/'88: The memory can store a maximum of 5 defect codes, whereby a defect code with higher priority can displace another defect code from the memory. The defect codes are displayed in the order of detection.

The following defects have higher storage priority:
- air flow meter,
- oxygen control,
- engine temperature sensor and
- idle switch

5.0L M70 ENGINE – DME M1.2 TROUBLESHOOTING – 1990 750iL

As of model '89: The extended defect code memory stores all defect codes which are detected. When a defect code is stored and the DME is not malfunctioning, substitute values can be active. The troubleshooting procedure should still, however, be carried out.

In US models, defect codes which influence the mixture are indicated by a lamp (engine check) in the instrument cluster. As a check, this lamp lights constantly in the time between „ignition ON" and engine start.

Clearing the Defect Code Memory - [D] 999

This step should be carried out when the defect has been repaired or was caused by simulation. When the battery or the DME control unit are disconnected, all stored defect codes as well as the substitute values are lost. Current defect codes cannot be cleared.

Status Monitoring [D] 100 and [D] 200

Further information can be called independent of defect code storage. Current values or current operating statuses are displayed in this way.

Component Activation - [D] 300

To facilitate fast faultfinding, various components can be activated with the DME control unit together with the service tester. This makes it possible to distinguish between faults in the control unit, the supply line or in the component itself. Detailed information on this simulation mode is displayed on the screen.

Abbreviations

ABS	Antilock brake system
AG	Automatic transmission
ASC	Automatic stability control
DK	Throttle valve
DWA	Antitheft system
EKP	Electric fuel pump
EML	Electronic throttle control
EV	Injector valve
HG	Manual transmission
HLM	Air flow meter heating filament
KD	Kickdown
KW	Crankshaft
LL	Idle
MF	Microfilm
MSR	Engine drag torque control
MV	Solenoid valve
PWG	Pedal value sensor
RxD	Diagnosis initiation line
SG	Control unit
TE	Fuel evaporation control
TD	Engine speed signal once per ignition
TR	Crankshaft reference once per revolution
ti	Injection timing
tL	Load signal
VL	Full load
TL	Partial load
TxD	Diagnostic data line
U-Batt	Battery voltage
U-Vers	Voltage supply

5.0L M70 ENGINE—DME M1.2 PIN ASSIGNMENTS—1990 750iL

Pin Assignments

Assignments of the 55-pole Connector Strip on the DME Control Units I and II

Pin No.	Type	Designation
1	A (1)	Ignition terminal 1
2	M	Ignition ground
3	A	EKP relay and crankshaft reference
4	–	–
5	A	Fuel evaporation control valve
6	A	Engine speed TD
7	E	Signal, air flow meter HLM
8	E	Cylinder reference point (cyl. 6 or 12)
9	–	–
10	M (2)	Ground oxygen sensor
11	–	–
12	–	–
13	E	Diagnostic line RxD
14	M	Ground, fuel injectors
15	A	Defect lamp (US only)
16	A	Injector valves (cyl. 1+3+5 or 7+9+11)
17	A	Injector valves (cyl. 2+4+6 or 8+10+12)
18	E	B+ permanent supply (Terminal 30)
19	M	Ground, control unit electronics
20	–	–
21	–	–
22	–	–
23	A	Sensor heating relay (control unit I only)
24	M	Ground, output stages
25	A	Burn-off, HLM
26	M	Ground, HLM
27	E	Terminal 15
28	E	Oxygen sensor signal
29	–	–
30	–	–
31	M (2)	Ground, cylinder reference point
32	A	ti-signal to consumption display
33	–	–
34	–	–
35	–	–
36	A	DME relay I drive
37	E	Power supply from relay I
38	E	ASC and antitheft system (DWA)
39	E	Programming voltage
40	E	Ground (control unit II only)
41	–	–
42	E	Drive range P/N
43	E	Idle speed CO potentiometer (in HLM)
44	E	Temperature sensor – intake air
45	E	Temperature sensor – engine
46	–	–
47	E	Inductive pulse generator (+)
48	M	Inductive pulse generator (–)
49	–	–
51	E	Ignition angle control from automatic transmission
52	E	Idle speed signal (from EML control unit)
53	E	Full load signal (from EML control unit)
54	E	Torque converter clutch unit from automatic transmission
55	E/A	Diagnostic line TxD

E = Input
A = Output
M = Ground
(1) = max. 150 V / max.0.2 A
(2) = potential-free ground

5.0L M70 ENGINE—DME M1.2 TROUBLESHOOTING—1990 750iL

Troubleshooting

Notes

– Relevant safety codes and accident prevention regulations must be observed when carrying out tests and work on the engine electrical and electronic systems.

Caution!

Dangerous high voltage in the primary and secondary circuit of the ignition system.

Important!

Wherever possible, the battery should not be disconnected otherwise the defect codes stored in ***all*** the control units will be deleted.
Disconnect control units from the wire harness ***only*** when the ignition is switched off.

Conduct wire tests ***only*** with the control unit or component connectors disconnected.

Test Equipment

The universal adapter with the corresponding 55-pole connection leads should be used for testing the DME control unit or the engine wire harness The universal adapter should only be connected between the control unit and wire harness when no diagnostic procedure (diagnostic connector not connected) is to be carried out since this can lead to malfunctions.

Use only suitable test leads, terminals and probe tips.

Display of a Stored Error Code on the Screen.

– **Defect location:** The number of the defect code indicates the pin assignments at the control unit. The defect can be in the vicinity of the component, the corresponding wiring or in the control unit.

– **Type of defect:** Short-circuits, breaks, range transgressions or illogical functions are indicated.

– **Frequency:** Indication as to whether the defect is momentary or not (intermittent defect). Further indication is provided as to whether the defect occurred several times or only once.
As of model '89 an indication is provided as to how often the defect has already been detected. If the defect is no longer pending, the number of starts is indicated which have taken place since last detection of the defect.

– **detected at:** Additional defect information on the operating statuses during defect detection.

Evaluation of Stored Defect Codes:

In order to achieve effective troubleshooting, it is advisable to first print out the contents of the defect code memory.

When carrying out repairs, it is often necessary to disconnect plug connections, thus, in many cases, causing a defect code to be stored. After defect code monitoring, under certain circumstances, this may lead to incorrect diagnosis.

All defect codes (also EML, AEGS) should therefore be cleared after the repair has been completed. Monitor the defect code memory once again after a test run.

Some defects cause a defect code to be set in several systems (EML, AEGS, ASC/MSR). Such interlinking defect codes can be determined by comparing the ambient conditions (e.g. engine speed, temperature etc.).

5.0L M70 ENGINE—DME M1.2 DEFECT CODES—1990 750iL

Defect Code Table

D 900 Defect code memory – stored defect codes:

One or several defect codes are stored	→ Defect codes 01...100

Malfunctions – defect codes not stored:

Engine operation not possible	→ Defect code 101

Defect Code 01

D 900 Defect code memory: 01 DME CONTROL UNIT

Explanation: This defect code is stored when the internal self-test provides a negative result.

Possible causes: Program error or disturbed program sequence in the DME control unit.

Corrective measures: Carry out test run after clearing the defect code memory. If defect code is set once again, the DME control unit must be replaced.

Defect Code 03

D 900 Defect code memory: 03 TR SIGNAL / ELECTRIC PUMP RELAY

Explanation: This defect code is stored when a short occurs at relay 1 (EKP) or in the TR signal line (control unit Pin 3).

Troubleshooting: As displayed on screen.

Defect Code 05

D 900 Defect code memory: 05 FUEL EVAPORATION VALVE

Explanation: This defect code is stored when a short occurs at the fuel evaporation valve or in the supply wire.

Troubleshooting: As displayed on screen.

5.0L M70 ENGINE – DME M1.2 DEFECT CODES – 1990 750iL

Defect Code 07

D 900 Defect code memory: 07 AIR FLOW METER (HLM)

Explanation: This defect code is stored when a short or a break is detected at the HLM or on the supply lines. The DME control unit then operates with fixed substitute values – Load signal tL = 3.0 ms and ignition angle = 18...20 °crank angle

– when the engine is not idling.

Troubleshooting: As displayed on screen.

Defect Code 10

D 900 Defect code memory: 10 OXYGEN CONTROL

Explanation: This defect code is stored in the case of excessive deviations in the mixture composition (too rich or too lean mixture longer than 10 sec). A prerequisite is that the oxygen sensor must be ready for operation.

Possible causes:

- No fuel delivery
- Incorrect fuel pressure
- Injector valves defective or coked
- Engine temperature sensor defective
- Secondary air in the intake duct
- Fuel evaporation control system defective

Troubleshooting: As displayed on screen.

Defect Code 15

D 900 Defect code memory: 15 DEFECT LAMP

Explanation: Applicable to US models only.
This defect code is stored when shorts occur at the defect lamp or in its supply wire.

Troubleshooting: As displayed on screen.

Defect Code 16

D 900 Defect code memory: 16 FUEL INJECTOR VALVES Cyl. 1 + 3 + 5 or 7 + 9 + 11

Explanation: This defect code is stored when shorts or breaks occur at the fuel injector valves or breaks in their supply wires.

Note: In the case of a short of one fuel injector valve or in its supply wire, three injector valves are always deactivated. The three intact valves of the cylinder group remain in operation.

Troubleshooting: As displayed on screen.

If the defect code 16/17 and 10 are stored, this is attributed to a fault in the ignition system (primary circuit or ignition coil). The injector valves are switched off as a result, thus also causing a defect code to be stored.

5.0L M70 ENGINE—DME M1.2 DEFECT CODES—1990 750iL

Defect Code 17

D 900 Defect code memory: 17 FUEL INJECTOR VALVES Cyl. 2 + 4 + 6 or 8 + 10 + 12

Explanation: Diagnosis as for defect code 16.

Defect Code 23

D 900 Defect code memory: 23 OXYGEN SENSOR HEATING RELAY

Explanation: The oxygen sensor heating relay is driven by the DME control unit I. A defect code is therefore stored only in DME control unit I. A defect code is stored when shorts or breaks occur at the relay or in the supply wire.

Troubleshooting: As displayed on screen.

Defect Code 28

D 900 Defect code memory: 38 OXYGEN SENSOR

Explanation: This defect code is stored when a short or a break occurs at the oxygen sensor and on its supply wires.

A prerequisite for this is an engine temperature of 70 °C and a load signal of > 2.3 ms for > 75 sec.

Troubleshooting: As displayed on screen.

Defect Code 37

D 900 Defect code memory: 37 CONTROL UNIT SUPPLY

Explanation: This defect code is stored when the voltage supply exceeds 16V. A drop in the voltage below 10V is detected 3 minutes after the start at the earliest.

Possible cause:
- Contact resistances at the wire connections
- Insufficient V-belt tension
- Voltage regulator or alternator defective

Troubleshooting: As displayed on screen.

5.0L M70 ENGINE—DME M1.2 DEFECT CODES—1990 750iL

Defect Code 43

D 900 Defect code memory: 43 IDLE CO-POTENTIOMETER

Explanation: Only applicable to vehicles without a catalytic converter. This defect code is stored when shorts or breaks occur at the potentiometer or in its supply wire. A substitute value of 1.25 V is used.

Troubleshooting: As displayed on screen.

Defect Code 44

D 900 Defect code memory: 44 INTAKE AIR TEMPERATURE SENSOR

Explanation: This defect code is stored when shorts or breaks occur at the temperature sensor or on its supply wires. A prerequisite for detecting a break in the circuit is a 3 min interval after the start and the engine idling for a minimum of 30 sec. The substitute value of 20 °C, as of model '89 – 50 °C is used.

Troubleshooting: As displayed on screen.

Defect Code 45

D 900 Defect code memory: 45 ENGINE TEMPERATURE SENSOR

Explanation: This defect code is stored when shorts or breaks occur at the temperature sensor or on its supply wires. A prerequisite for detecting a short to battery plus or a break is an intake air temperature of > 11 °C.

A substitute value of 50 °C is made available in the case of fault.

Troubleshooting: As displayed on screen.

Defect Code 51

D 900 Defect code memory: 51 IGNITION ANGLE

Explanation: The defect code is stored when the wire from the electronic transmission control unit Pin 24 to the DME control unit Pin 51 has a short to ground (> 2.5 sec). In such a case, the ignition angle intervention occurs only once, the signal is then ignored.

Troubleshooting: As displayed on screen.

5.0L M70 ENGINE – DME M1.2 DEFECT CODES – 1990 750iL

Defect Code 54

[D] 900 Defect code memory: 54 TORQUE CONVERTER CLUTCH UNIT

Explanation: Only applicable as of model '89. This defect code is stored when the DME control unit detects a closed torque converter clutch (or short to battery voltage) and the selector lever is in position P or N.

Troubleshooting: As displayed on screen.

Defect Code 100

[D] 900 Defect code memory: 100 OUTPUT STAGE

Explanation: Only applicable as of model '89. This defect code is stored when there is a loose contact at an output stage output. In the majority of cases, one of the defect codes 3, 5, 16, 17, or 23 also occurs, to which the defect code 100 can be assigned.

Defect Code 101

Malfunction: Engine operation not possible

Troubleshooting:

Prerequisites:
- Start system OK. (battery charged, ignition switch and starter OK. ...)
- Ground connection at battery, engine and body OK
- Specified type of fuel in tank
- Engine OK mechanically (timing, compression ...)

a) Check voltage supply for DME control units in accordance with schematics.

b) Check input signals.
- Check engine speed signals at starting speed –[D] 200.
- Check air flow meter signals [D] 200.

c) Check output signals.
- Drive of fuel pump relay in accordance with schematics
- Ignition output stage with ENGINE TEST, test step 09
- Fuel injector valves with [D] 300

d) Fuel pressure test
- See repair instructions.

e) Check EML for defects.

5.0L M70 ENGINE—DME M1.2 COMPONENT TESTING—1990 750iL

Component Testing

Notes

– Relevant safety codes and accident prevention regulations must be observed when carrying out tests and work on the engine electrical and electronic systems.

Caution!

Dangerous high voltage in the primary and secondary circuit of the ignition system.

Important!

Wherever possible, the battery should not be disconnected otherwise the defect codes stored in ***all*** the control units will be deleted.

Disconnect control units from the wire harness ***only*** when the ignition is switched off.

Conduct wire tests ***only*** with the control unit or component connectors disconnected.

Test Equipment

The universal adapter with the corresponding connection leads should be used for testing the DME control unit or the engine wire harness –

The universal adapter should only be connected between the control unit and wire harness when ***no*** diagnostic procedure (diagnostic connector not connected) is to be carried out since this can lead to malfunctions.

Use only suitable test leads, terminals and probe tips.

5.0L M70 ENGINE—DME M1.2 COMPONENT TESTING—1990 750iL

Overview of the Defect Codes and Component Tests

Component/Function	Short test after entry [D]...	Defect code after storage
DME control unit	–	1
Ignition coil	see Engine test	–
TR signal/EKP relay	–	3
Fuel evaporation valve	100, 300	5
Air flow meter	200	7
Pulse generator – cylinder reference point	[M] 23/24	–
Oxygen control	100	10
Defect lamp	–	15
Injector valves	300	16 / 17
Sensor heating relay	–	23
Oxygen sensor	100/200	28
Road speed signal	see instrument cluster/CC	29
Main relay (relay 2)	100	–
Supply voltage	200	37
ASC	see ABS/ASC	38
Intake air temperature sensor	200	44
Engine temperature sensor	200	45
Pulse generator – speed/position	200	–
MSR	see ABS/MSR	50
Ignition angle intervention	see AEGS	51
Idle/full load signal	100	–
Torque converter clutch unit	see AEGS	54

5.0L M70 ENGINE—DME M1.2 COMPONENT TESTING—1990 750iL

DME Control Unit

Check function:

Individual function blocks, e.g. ignition, injection, oxygen control etc. can be checked only when the necessary input signals are applied beforehand at the control unit

Ignition Coil

Check function:

Perform engine test, test step 09. With the engine at operating temperature, the ignition voltage is 10 ±2 kV. The maximum difference between the individual cylinders should not exceed 3 kV.

TR Signal / EKP Relay

Check function:

4-pole relay up to model '89

The EKP relay switches on as soon as the DME control unit sends a ground signal with overlaid engine speed signal (once per crankshaft revolution) at Pin 3.

7-pole relay as of model year '89

The EKP relay switches only during starting (via terminal 50) or when an ignition signal (terminal 1) is applied.

Possible causes:

- Wire to the EKP relay defective
- EKP relay or main relay defective
- Voltage supply for DME control unit, relay 1 or 2 defective
- No signals from the pulse generator – engine-speed/position
- Ignition output stage or ignition coil defective (model '89 only)

5.0L M70 ENGINE – DME M1.2 COMPONENT TESTING – 1990 750iL

Fuel Evaporation Valve

Check electrical function:

[D] 300 – The fuel evaporation valve is driven at intervals of one second. The actuation cycles can be felt with the hand.

Check for leaks:

Release hoses from valve, switch on vacuum pump at service tester and connect to valve (8 mm connection) (note arrow indicating direction of flow). Set vacuum to 600 mbar and apply a voltage of 12 V to the valve. Switch off vacuum pump, – after 20 sec, the pressure drop must not exceed 50 mbar.
When power is not applied, the fuel evaporation valve is closed by a non-return valve to a vacuum of 20 mbar.

Check drive phases:

[D] 100 – The valve is driven at idle speed (display: ON or CLOSED) and is therefore closed. As soon as the oxygen control is active and the engine speed is increased, the valve can be observed to open step-by-step (display: OFF or OPEN).
After switching off the engine, the valve is held closed for 3...5 sec.

Air Flow Meter (HLM)

Check HLM signal:

[D] 200 – The load signal provides an indication of the measured air flow rate in the HLM. The value should be 2.0 ms ±10 % when idling.

Check burn-off function:

Increase engine speed once above 2000 rpm (engine temperature > 60 °C), after stopping the engine, a voltage increase of approx. 3.5 V can be measured at the HLM between Pin 1 (+) and Pin 5 (–).

Pulse Generator – Cylinder Reference Point

Check function:

[M] 22 – Connect oscilloscope connections at component connector and run engine at idle speed. An alternating voltage characteristio curve with positive and negative half waves should appear on the display screen.

Switchover to semisequential injection takes place as soon as the reference signal is detected during normal operation. Only parallel injection is possible during diagnostic mode.

Resistance test:

The resistance can be measured directly at the component connector (at the connector strip for injector valves).

5.0L M70 ENGINE – DME M1.2 COMPONENT TESTING – 1990 750iL

Oxygen Control

Check function:

[D] 100 – Prerequisite for this procedure is that the oxygen sensor is at operating temperature. The engine must run for 30...40 sec. before the control is implemented.

The control can compensate for wear-related deviations in the mixture compensation and therefore conceal other defects. If necessary, disconnect the oxygen sensor during troubleshooting of the mixture composition.

Defect Lamp „Check Engine" (US only)

Check function:

The control unit supplies the defect lamp with ground potential as soon as an exhaust-emission-related defect is detected. As a check, the lamp lights constantly after ignition ON.

Fuel Injector Valves

Check electrical function:

[D] 300 – In this case, the fuel injector valves are driven with a frequency of 10 Hz. The pulse period is 1 ms in order to avoid fuel being injected. Activation can be felt with the hand.

Check pulse duration:

The pulses can be tested with the multimeter [M] 15 at the injector valve connector. Note that only parallel injection is possible during diagnostic mode. Therefore, when the pulse duration is measured with the multimeter (not in diagnostic mode), the pulse duration is doubled for semisequential injection.

Test with oscilloscope [M] 22 and universal adapter:

Connect terminal frequency OFF (+) to terminal 1 (black) and apply frequency 70 Hz, connect terminal D+ (blue) to Pin 16 or 17. The injection signal should be displayed on the oscilloscope. In contrast to pulse duration measurement, the injection fade-out when coasting is also displayed.

Resistance test:

The resistance of an injector valve can be measured directly at the component. Only the entire injector valve group (= 3 valves) can be measured at the universal adapter.

5.0L M70 ENGINE—DME M1.2 COMPONENT TESTING—1990 750iL

Sensor Heating Relay

Check function:

[D] 300 – This activation is only possible at cylinder row 1...6, it takes place at one second intervals.

The relay remains switched on during normal operation until high speeds or load occur.

Possible causes:

- Wire to the DME control unit Pin 23 defective
- Relay or voltage supply not OK.
- Fuse F 28 defective

Oxygen Sensor

Check function:

[D] 200 – To carry out this test, the oxygen sensor must be at operating temperature. The indicated voltage signal is tapped off in steps and fluctuates between 0.02 and 0.85 V.

When the oxygen sensor is not in operation, or the wire to the DME control unit is interrupted, a constant value of 0.45 V is indicated. This value is also indicated when no oxygen sensor is installed.

Main Relay – Control (DME Relay 2)

Check function:

[D] 100 – This status inquiry is only possible as of model '89.

With ignition ON or during engine operation, the main relay is driven with ground potential at terminal 85 (display: ON). The relay switches on and supplies the control units DME, AEGS, EML and ABS with voltage.

Possible causes.

- Wire to DME control unit Pin 36 defective
- Relay 2 or voltage supply not OK.
- Voltage supply for DME control unit not OK.
- The control unit is defective if no ground potential is output at control unit Pin 36 although the above points are OK.

Control Unit Voltage Supply

Check permissible range:

[D] 100 – The displayed voltage is currently at the control unit. It should not be below 10V.

5.0L M70 ENGINE—DME M1.2 COMPONENT TESTING—1990 750iL

CO Potentiometer (at HLM)

Check function:

D 200 – A voltage of 1.25 V should be indicated in the basic setting. Precise idle CO tuning (by adjusting the potentiometer) should only take place in conjunction with CO measurement. The typical range is between 1.3 and 1.5 V.

Intake Air Temperature Sensor (in Air Flow Meter)

Check function:

D 200 – With the engine at operating temperature, the intake air temperature is at 50...70 °C (ambient air temperature 23 ± 5 °C).

A display below – 40 °C indicates a break in the temperature sensor or in the supply wire to the DME control unit.

Resistance test:

Connect universal adapter, measure resistance across the terminals 44 and 26.

Engine Temperature Sensor (Coolant)

Check function:

D 200 – The indicated engine temperature should correspond to the coolant temperature. A value below – 30 °C indicates a break in the temperature sensor or in the wire to the DME control unit.

Resistance test:

Connect universal adapter to the engine wire harness. Measure resistance over the terminals 45 and 19.

3.6L S38 ENGINE—DME M1.2 DESCRIPTION—1990-92 M5

Functional description

The description of the DME M1.3 6-cylinder essentially applies to the DME M1.2 system on the S38 engine.

Changes and additions compared to DME M1.3:

The intake air flows via a heating element air mass meter (HLM). In vehicles without a catalytic converter, a potentiometer for CO correction when idling is fitted on the HLM.

The systems EML, EH transmission and ASC/MSR are not used. Additional functions are described below.

Resonance charging

The torque progression and the engine performance are improved with this control system. The vibrations in the air manifold are utilized to facilitate improved cylinder charging. The following components are used for this purpose:

- Control unit in the E-box
- Electrical changeover valve on the left-hand side of the engine
- Vacuum accumulator and non-return valve
- Vacuum cell on the air manifold
- Resonance flap in the air manifold

The resonance flap is open when at rest; air is admitted in the vacuum cell. As soon as the full load contact is closed and the engine speed is below 4150 rpm or above 6750 rpm, the non-return valve is driven by the control unit. The vacuum cell receives a vacuum, consequently closing the resonance flap. The non-return valve and the pressure accumulator maintain the vacuum supply even under unfavourable conditions (fast opening of the throttle valve).

Air injection

After a cold start at an engine temperature < 40 °C (for catalytic converter provisions < 30 °C), filtered air is injected by means of a pump into the exhaust manifold for approx. 2.5 min (12 min for catalytic converter provisions). The CO and HC content in the exhaust gas can be considerably reduced in this way.

The DME control unit drives the relay for the air pump (ground potential) which engages the solenoid clutch of the pump. At the same time, a changeover valve is switched which opens a shut-off valve by means of vacuum and releases the flow of air to the exhaust manifold. The air flowing via a non-return valve is directed through a collector pipe and individual branches to the exhaust pipes.

Notes on BMW DIAGNOSTIC SYSTEM

Change compared to DME M1.3

Status monitoring of the current values or statuses

If status monitoring takes place with the engine running, a speed of 2000 rpm should not be exceeded otherwise faults in data transfer can occur.

3.6L S38 ENGINE—DME M1.2 PIN ASSIGNMENTS—1990-92 M5

Pin assignments

Assignments of the 55-pole connector strip on the DME control unit

Pin No.	Type	Designation
1	A (1)	Ignition terminal 1
2	M	Ignition ground
3	A	EKP relay/TR signal
4	A	Open idle speed actuator
5	A	TE valve
6	A	Engine speed/TD signal
7	E	Signal HLM
8	E	Cylinder reference (cyl. 6)
9	-	-
10	(2)	Ground oxygen sensor
11	-	-
12	-	-
13	E	Diagnostic line RxD
14	M	Ground injector valves
15	A	Defect lamp (US only)
16	A	Injector valves (Cyl. 1,3,5)
17	A	Injector valves (Cyl. 2,4,6)
18	E	B+ permanent supply (terminal 30)
19	M	Ground electronics
20	-	-
21	-	-
22	A	Close idle speed actuator
23	A	Air pump relay
24	M	Ground, output stages
25	A	Burn-off HLM
26	M	Ground HLM
27	E	Terminal 15
28	E	Oxygen sensor signal
29	-	-
30	-	-
31	(2)	Ground cylinder reference
32	A	ti-Signal to consumption display
33	-	-
34	-	-
35	-	-
36	A	Main relay drive
37	E	Supply from main relay
38	E	Antitheft system (DWA)
39	E	Programming voltage
40	E	A/C compressor
41	E	A/C switch
42	-	-
43	E	Idle CO potentiometer (on HLM)
44	E	Intake air temperature
45	E	Engine temperature
46	-	-
47	E	Inductive pulse generator (+)
48	M	Inductive pulse generator (-)
49	-	-
50	-	-
51	-	-
52	E	Idle speed signal
53	E	Full load signal
54	E	Premium/regular switchover
55	E/A	Diagnostic line TxD

E = Input
A = Output
M = Ground
(1) = max. 150 V / max.0.2 A
(2) = potential-free ground

3.6L S38 ENGINE—DME M1.2 DEFECT CODES—1990-92 M5

Troubleshooting

Defect Code 07

D 900 Defect code memory: 7 AIR MASS METER (HLM)

Explanation: This defect code is stored when shorts or breaks are detected at the HLM or in the supply wires. The control unit then operates with substitute values: Load signal tL = 4.0 ms, ignition angle = 18° crank shaft angle.

Troubleshooting: As displayed on screen. Refer to the chapter on Component Testing for test values.

Defect Code 23

D 900 Defect code memory: 23 AIR PUMP RELAY

Explanation: This defect code is stored when a break or shorts occur at the relay or in the supply wire from the control unit.

Troubleshooting: As displayed on screen.

Defect Code 43

D 900 Defect code memory: 43 IDLE CO-POTENTIOMETER

Explanation: Only applicable to vehicles without catalytic converter. This defect code is stored when shorts or breaks occur at the potentiometer or in its supply wire. A substitute value of 1.25 V is used.

Troubleshooting: As displayed on screen.

Defect Code 44

D 900 Defect code memory: 44 INTAKE AIR TEMPERATURE SENSOR

Explanation: This defect code is stored when shorts or breaks occur at the temperature sensor or in its supply wires. The prerequisite for detecting a break is a time interval of 3 min after the start and idling for min. 30 sec. After the defect has been detected, the substitute value of 50°C is used.

Troubleshooting: As displayed on screen.

3.6L S38 ENGINE—DME M1.2 COMPONENT TESTING—1990-92 M5

Component Testing

Changes compared to DME M1.2 12-cylinder

Overview of Defect Codes and Component Tests

Component/Function	Quick test after entry D ...	Defect code after storage
DME control unit	-	1
Ignition coil	see engine test	-
TR signal/EKP relay	-	3
Open idle speed actuator	300	4
Activated charcoal valve	100,300	5
Air mass meter	200	7
Cylinder reference pulse sensor	[M] 23/24	-
Oxygen control	100	10
Defect lamp	-	15
Injector valves	300	16 / 17
Close idle speed actuator	300	22
Air pump relay	300	23
Oxygen sensor	100/200	28
Speed signal	see instrument cluster/CC	29
Voltage supply	200	37
LL CO potentiometer	200	43
Intake air temperature sensor	200	44
Engine temperature sensor	200	45
Speed/position pulse generator	200	-
Idle/full load signal	100	-

1) The resistance values are listed in the Appendix

3.6L S38 ENGINE—DME M1.2 COMPONENT TESTING—1990-92 M5

TR signal / EKP relay

Check function:

The EKP relay is activated when the DME main relay is switched on and the DME control unit (Pin 3) supplies a ground signal with overlaid speed signal (once per crankshaft revolution). The input signal of the speed pulse generator is a prerequisite for this.

Possible causes:

- Wire to the EKP relay defective
- EKP relay or main relay defective
- Voltage supply for DME control unit, EKP or main relay defective
- No signals from speed/position pulse generator

Air mass meter (HLM)

HLM signal check:

D 200 – The load signal provides information on the measured air throughput in the HLM - for idle value refer to Appendix. When air is injected in the HLM, the change in the load signal can be observed with the engine stationary.

Check voltage supply of 12V at the HLM pin 2 (+) and 4 (-). Measure hot wire resistance at pin 5 (+) and 3 (-).

Check burn-off function:

Increase engine speed once above 2000 rpm (engine temperature > – 60 °C), after stopping the engine, a voltage increase to approx. 3.5 V can be measured at the HLM between pin 1 (+) and 5 (-).

Air pump

Check function:

The solenoid clutch for the air pump is switched on at an engine temperature below 40 °C (< 30 °C for catalytic converter provisions) and an engine speed rpm ((rpm) for catalytic converter provisions. The clutch is triggered via the air pump relay in the E-box and has a duration of approx. 2.5 min (12 min for catalytic converter provisions).

Check vacuum supply for the changeover valve (approx. 200 mbar), otherwise the shut-off valve will not open.
The hose can be detached at the air filter housing (under the coolant expansion reservoir) in order to check whether the air pump is operating. The vacuum can be felt by hand when the pump is running.

Oxygen sensor heating

Note:

The voltage for the oxygen sensor heating is supplied via the EKP relay.

3.6L S38 ENGINE – DME M1.2 COMPONENT TESTING – 1990-92 M5

Intake air temperature sensor

Check function:

D 200 – The sensor for the intake air temperature is located on the air filter housing. The indicated temperature depends to a considerable extent on the ambient air temperature.
A temperature display below – 40 °C inidcates a break in the temperature sensor or in the supply wire to the DME control unit.

Throttle switch

Check function:

D 100 – The setting of the idle and full load switch can be changed by turning the switch housing. The idle contact should open as soon as the throttle valve is moved. The full load contact is closed before the full load limit stop.

3.6L S38 ENGINE – DME M1.2 TECHNICAL DATA – 1990-92 M5

Technical Data

Resistance Values Measured at an Air Temperature of 23 ± 5 °C

Engine type	S38 B36
Injector valve	15...17 Ω
Injector valve group (3 valves)	5.5 ± 0.5 Ω
Speed/position pulse generator	540 ± 50 Ω
Cylinder reference pulse generator	0.1...1Ω
Engine temperature sensor	
at coolant temperature 20 ± 3 °C	2.2...2.7 kΩ
at coolant temperature 80 ± 3 °C	300...360Ω
Intake temperature sensor	
at air temperature 20 ± 3 °C	2.2...2.7 kΩ
at air temperature 50 ± 3 °C	760...910 Ω
Activated charcoal valve	45 ± 20Ω
Ignition coil	
Primary winding (terminal 15/1)	0.5 ± 0.1 Ω
Secondary winding (terminal 1/4)	9.0 ± 1.0 kΩ
Idle speed actuator (per winding)	20 ± 5Ω
Heating wire element in HLM	3...5 Ω
Switchover valve	20...40 Ω
Solenoid clutch-air pump	5 ± 2Ω

3.6L S38 ENGINE—DME M1.2 STATUS INQUIRIES—1990-92 M5

Status inquiries 100	**Set status**
Oxygen control (cat. only)	during engine operation = ON when oxygen sensor at operating temperature
Activated charcoal valve	when idling = CLOSED until clock cycle begins at increased engine speed and load
Idle speed signal	when idling = ON
Full load signal	shortly before full load = ON

Status inquiries 200	**Set values**
Engine speed in rpm 1)	
Idling speed	890 ± 50
Cut-out speed	6900 ± 40
Load signal 2)	2.3 ms 10 %
Injection signal 2) (semisequential)	4.5 ms 10 % measurement with [M] 15
Control unit supply	> 10 V
Sensor voltage 2)	0.02...0.85 V, in the case of break or defective oxygen sensor = 0.45 V
Intake air temperature	dependent on the ambient temperature, in the case of break = approx. -40 °C
Engine temperature	dependent on engine operation, in the case of break = approx. -30 °C
Ignition angle in crankshaft angle before TDC 1)	1 ± 3 when idling
CO potentiometer (without cat.)	basic setting = 1.25 V
CO volume % (cat.) 2) 3)	0.8 ± 0.4
CO volume % (without cat.) 2)	1.0 ± 0.5
Max. permissible deviation per cylinder	± 0.2 %

1) Oil temperature > 60 °C or coolant temperature > 80 °C, no loads switched on.
2) When idling, operating temperature as under 1)
3) Measured ahead of the catalytic converter

5.0L M70 ENGINE–DME M1.7 DESCRIPTION–1991-92 318i, 750iL AND 850i

Functional Description

Note: The following description refers to the digital motor electronics DME M 1.7 on the 12–cylinder engine M70.

Two DME units operate ***independently,*** each with 6 cylinders. If not otherwise specified, the explanations, therefore, refer to only one row of cylinders.

New Features Compared to the Previous 12–Cylinder DME M 1.2

– 88–pole control unit connector
– Integrated ignition circuit monitoring
– A/C compressor cut–out

The DME control unit I supplies the cylinders 1...6 the identical control unit II the cylinder 7...12. To differentiate between them for diagnosis, Pin 85 of the control unit II is connected to ground. The associated components are mainly provided double. Exceptions are:

– Relay for oxygen sensor heating
– Engine temperature sensors

The relay is activated by the DME control unit I and supplies the heaters of both oxygen sensors. This relay will be omitted in series production. On models without a relay for oxygen sensor heating, the sensor heaters are activated directly by the EKP relay.

The engine temperature sensors for the DME control unit I and II are mounted together in one housing.

The DME M 1.7 makes it possible to precisely control fuel injection and ignition under variable operating conditions. The system also operates in conjunction with other systems such as:

– Electronic throttle control EML
– Electronic transmission control EGS
– Automatic stability control ASC
– Engine drag torque control MSR
– Antitheft system DWA
– Electronics vehicle module EKM
– On–board computer BC

which can influence the entire engine control system depending on requirements.

5.0L M70 ENGINE – DME M1.7 DESCRIPTION – 1991-92 318i, 750iL AND 850i

The DME system assumes functions which are listed briefly here and are described in more detail in the following.

- Injection control
- Ignition timing control
- Cold start control
- Engine speed control
- Acceleration enrichment
- Dynamic torque cut–out
- Catalytic converter protection function (ignition circuit monitoring)
- Adaptive emission (Lambda) control
- Adaptive fuel evaporation control
- Intervention in ignition and injection with ASC or MSR active
- Characteristics map switchover for EH transmission
- Relay control
- A/C compressor cut–out
- Self–diagnosis with defect code storage

The following components supply the DME control unit with input signals.

- Pulse generator – engine speed/position I and II
- Pulse generator – cylinder reference I and II
- Intake air temperature sensor I and II
- Engine temperature sensor I and II (in one housing)
- Air mass meter I and II
- Oxygen sensor I and I
- Battery
- EML control unit
- EGS control unit
- ABS/ASC/MSR control unit

The outputs of the DME are connected via output stages. They supply the following components with the necessary signals:

- Injector valves for cylinders 1–3–5 or 7–9–11 (group 2)
- Injector valves for cylinders 2–4–6 or 8–10–12 (group 1)
- Ignition coil I and II
- Fuel evaporation control valve I and II
- EKP relay I and II
- Main relay (control unit supply) I and II
- Oxygen sensor heater relay (only on DME control unit I)
- Oxygen sensor I and II
- Defect lamp (US only)
- Air mass meter I and I
- EML control unit
- EGS control unit
- ABS/ASC/MSR control unit

5.0L M70 ENGINE—DME M1.7 INTERCONNECTIONS—1991-92 318i, 750iL AND 850i

Interconnected System of Signals to the Drive Control Units (CU)

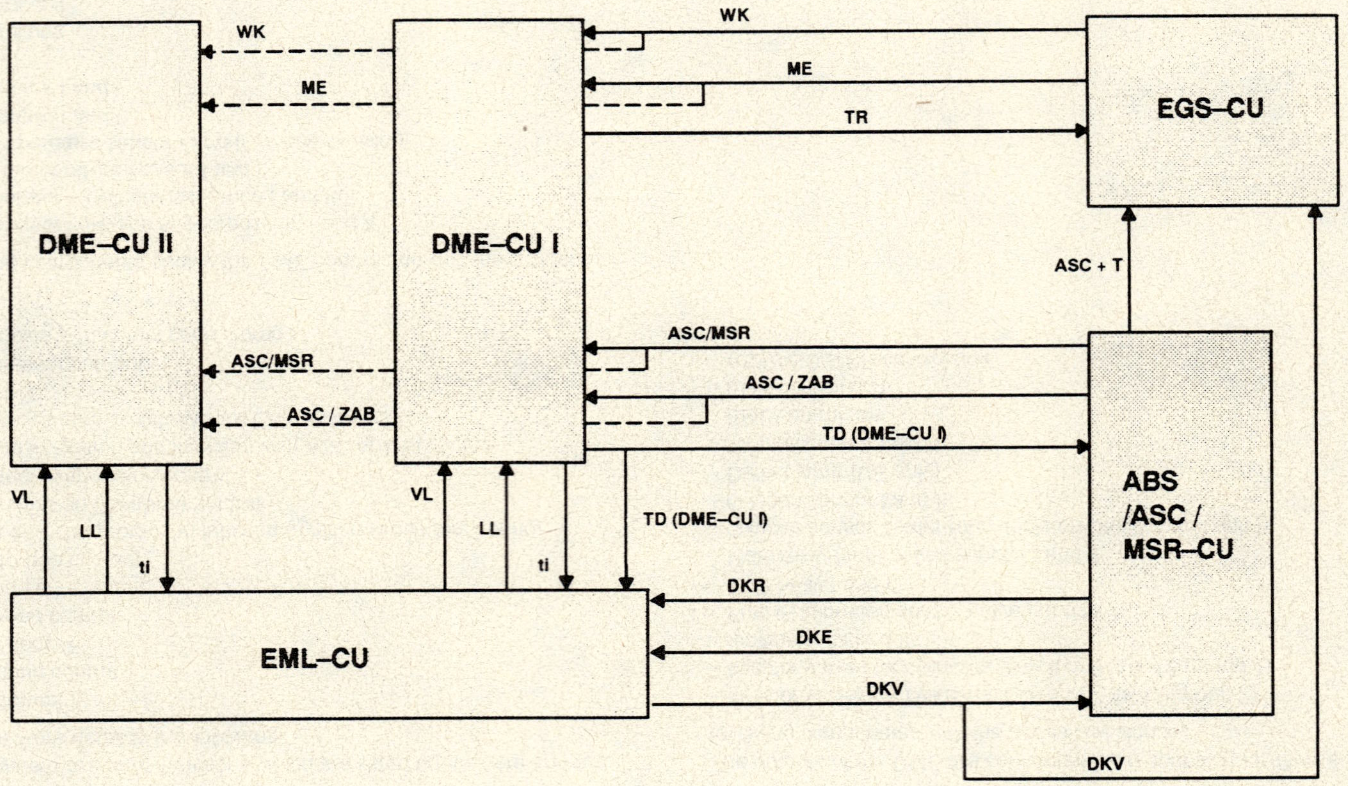

5.0L M70 ENGINE—DME M1.7 LINE DESCRIPTION—1991-92 318i, 750iL AND 850i

Abbr.	Signal Description	Type of Signal	Effect
WK	Torque converter clutch	Ground = torque converter clutch engaged (ON)	DME selects corresponding characteristics map
ME	Engine intervention (ignition angle intervention)	Ground = ON	DME reduces the torque during the gearshift procedure (ignition timing retarded).
TR	Engine speed signal (3 times per crankshaft revolution)	Square–wave voltage	EGS receives information relating to the engine speed.
ASC / MSR	Automatic stability control or engine drag torque control	U–Batt	2 functions within the DME control unit: complete torque cut–out if currently active (MSR). Otherwise, ignition timing set towards retard (ASC).
ASC / ZAB	Automatic stability control, ignition fade–out	U–Batt	DME fades out the ignition and fuel injection (not in ASC+T)
TD	Engine speed signal (once per ignition cycle)	Square–wave signal	EML undertakes idle control and engine speed limitation. ASC requires this information to ensure it does not stall the engine.
ASC + T	Automatic stability control + traction	U–Batt = ON	With ASC+T active, EGS prevents alternate gearshifts in transmission.
DKR	Throttle reduction	Square–wave voltage with fixed frequency and variable duty factor	EML reduces the throttle opening angle
DKE	Throttle increase	Square–wave voltage with fixed frequency and variable duty factor	EML increases the throttle opening angle
DKV	Set throttle value	Square–wave voltage with fixed frequency and variable duty factor	ASC/MSR receives information on the current status of the pedal position sensor (driver requirement).
LL	Idle signal	Ground = ON	DME selects corresponding characteristics map
VL	Full load signal	Ground = ON	DME selects corresponding characteristics map
ti	Injection signal	Square–wave voltage	EML synchronises the throttle valves

5.0L M70 ENGINE – DME M1.7 TROUBLESHOOTING – 1991-92 318i, 750iL AND 850i

Notes on the BMW DIAGNOSTIC SYSTEM

Initiation of Diagnostic Procedure

Communication with the control unit via the DIAGNOSTIC SYSTEM can be set up, after selecting the DME on the BMW SERVICE TESTER. If data transfer is interrupted, information appears on the display screen indicating the possible causes:

- Ignition not switched on
- Voltage supply for control unit <10 V
- DME relay not supplied with B+ (if not: bypass terminal 30 and 87)
- Control unit not connected
- Diagnostic link defective
- Data traffic inhibited by a defective control unit in the vehicle

Control Unit Assignment

Correct assignment of the control units to the engine can be determined by means of the control unit identification and the parts microfiche.

Clearing the Defect Code Memory

Carry out this step only when all defects have been rectified or when they have been stored by way of defect simulation. When the battery or the DME control unit are disconnected, ***all*** stored defect codes, the current substitute values and the adaption values are lost. After clearing, current defect codes are restored during the next start procedure.

Status Monitoring

Inputs and Outputs can be monitored independently of the defect code storage function. In this case, current values or operating statuses are displayed. Status enquiries can be compiled as required.

From the status monitoring functions, the nominal values can be displayed on the screen on special information pages.

Storing Defect Code

The first step should always be to monitor the defect code memory. If one or several defect codes are stored, it is advisable to immediately print out these data (R key). Since it is often necessary during the testing procedure to cut the power supply or to disconnect the connector from the control unit, the contents of the defect code memory will be unintentionally deleted as a consequence.

The memory stores the defect codes in the order they occurred.

With the exception of defect code 200, defect codes are stored only when the engine is running. For an overview of the monitored components and functions, refer to the overview table in the section on component testing.

Substitute values can be active when a defect code is stored and the DME is not malfunctioning. The troubleshooting procedure should still, however, be carried out.

The learned adaption values are lost if the battery or the DME control unit is disconnected. This may result in difficulties in starting and idling.

In US models, defects which influence the mixture are indicated by an "engine check" lamp in the instrument cluster. As a check, this lamp lights constantly during the period of time between "ignition ON" and engine start.

Component Activation

To facilitate fast fault finding, various components can be activated with the DIAGNOSTIC SYSTEM. The actuators can be checked by acoustic means.

Activation of the injector valves should be kept as short as possible.

5.0L M70 ENGINE – DME M1.7 TROUBLESHOOTING – 1991-92 318i, 750iL AND 850i

Abbreviations

ABS	Anti–lock brake system
AG	Automatic transmission
ASC	Automatic stability control
ASC + T	Automatic stability control plus traction control
CO	Carbon monoxide
DK	Throttle valve
DKB	Throttle–brake intervention
DKE	Throttle lift
DKR	Throttle reduction
DKV	Set throttle value
DME	Digital motor electronics
EGS	Electronic transmission control
EH	Electrohydraulic transmission
EKM	Electronic vehicle module
EKP	Electric fuel pump
EML	Electronic throttle control
EV	Injector valve(s)
HG	Manual transmission
HLM	Hot–wire air mass meter
Hz	Hertz
KD	Kick–down
KW	Crankshaft
LL	Idle
ME	Engine intervention (ignition timing)
MF	Microfiche
MSR	Engine drag torque control
MV	Solenoid valve
PWG	Pedal position sensor
RxD	Diagnosis initiation line
SG	Control unit
TE	Fuel evaporation control
TD	Engine speed signal once per ignition
TR	Engine speed signal once per revolution
ti	Injection time
tL	Load signal
TxD	Diagnostic data line
U–Batt	Battery voltage
U–Vers	Supply voltage
WK	Torque converter clutch
ZAB	Ignition fade–out

5.0L M70 ENGINE—DME M1.7 PIN ASSIGNMENTS—1991-92 318i, 750iL AND 850i

Pin Assignments

If not otherwise specified, the information on the type of signal refers to engine operation at idle speed

Assignments on the 88–Pole DME Control Unit Connectors X 6010 (DME Control Unit) and X 6020 (DME Control Unit II)

Pin	Type	Description/Function	Connection	Type of Signal	Tester Display	Measurement Notes
1	A	EKP relay and TR signal (crankshaft reference)	EKP relay and EGS control unittorque converter clutch	Ground with superimposed engine speed signal	ON	Can only be measured when crankshaft rotating (the EKP relay does not switch if the TR signal is not applied)
2	–	–	–	–	–	–
3	A	Injector valves cyl. 2–4–6 or cyl 8–10–12	Injector valve EV	Ground clocked	Injection time ti	Ground => injector valves open, see component activation
4	–	–	–	–	–	–
5	–	–	–	–	–	–
6	M	Ground EV output stages	Ground point	Ground	–	–
7	–	–	–	–	–	–
8	A	Defect lamp (US only	Instrument cluster	Ground	–	–
9	–	–	–	–	–	–
10	–	–	–	–	–	–
11	–	–	–	–	–	–
12	E	Full load signal VL	EML control unit	Ground	ON	At full load (kick–down)

5.0L M70 ENGINE—DME M1.7 PIN ASSIGNMENTS—1991-92 318i, 750iL AND 850i

Pin	Type	Description/Function	Connection	Type of Signal	Tester Display	Measurement Notes
13	A	Free combustion	Air mass meter	Voltage increase for approx. 0.5 sec	–	See chapter component testing
14	M	Ground air mass meter	Air mass meter	Ground	–	–
15	–	–	–	–	–	–
16	E	Cylinder reference sensor (ignition)	Cylinder reference sensor (connection B)	–	ON / OFF	See status monitoring
17	A	ti measuring signal (KVA signal)	EKM	Square wave signal	Multimeter function Pulse duration	–
18	–	–	–	–	–	–
19	–	–	–	–	–	–
20	–	–	–	–	–	–
21	–	–	–	–	–	–
22	–	–	–	–	–	–
23	–	–	–	–	–	–
24	–	–	–	–	–	–
25	A	Ignition terminal 1	Ignition coil	Primary signal	Oscilloscope	Engine test, test step 05
26	E	Steady pulse, terminal 30	B + terminal point	U–Batt	–	Diagnosis not possible if not fitted
27	A	DME main relay	DME main relay terminal 85 (additionally, jumper to second DME main relay)	Ground	ON	Diagnosis not possible if not fitted – connect jumper at relay terminal 30 and 87
28	M	Ground for electronics and shielding the sensors	Ground point	Ground	–	–
29	–	–	–	–	–	–
30	–	–	–	–	–	–

5.0L M70 ENGINE—DME M1.7 PIN ASSIGNMENTS—1991-92 318i, 750iL AND 850i

Pin	Type	Description/Function	Connection	Type of Signal	Tester Display	Measurement Notes
31	–	–	–	–	–	–
32	A	Injector valves cyl. 1–3–5 or cyl 7–9–11	EV and EML	Ground clocked	Injection time ti	Ground => Injector valves open,see component activation
33	–	–	–	–	–	–
34	M	Ground, remaining output stages (except for ignition and EV)	Ground point	Ground	–	–
35	–	–	–	–	–	–
36	A	Fuel evaporation control valve	TE valve	Ground clocked	CLOSED / OPEN	Perform component activation, ground => closed
37	A	Oxygen sensor heater	Sensor heater relay terminal 85	Ground	ON	Ground = > sensor heater activated
38	–	–	–	–	–	–
39	–	–	–	–	–	–
40	–	–	–	–	–	–
41	E	Signal air mass meter	Air mass meter	0...5 V	Load signal in ms	See status monitoring
42	–	–	–	–	–	–
43	M	Ground for sensors	Temperature sensor	Ground	–	–
44	E	Cylinder reference (ignition)	Cylinder reference (connection A)	–	ON/OFF	See status monitoring
45	–	–		–	–	–
46	–	–	–	–	–	–
47	–	–	–	–	–	–
48	–	–	–	–	–	–
49	–	–	–	–	–	–

5.0L M70 ENGINE–DME M1.7 PIN ASSIGNMENTS–1991-92 318i, 750iL AND 850i

Pin	Type	Description/Function	Connection	Type of Signal	Tester Display	Measurement Notes
50	–	–	–	–	–	–
51	–	–	–	–	–	–
52	–	–	–	–	–	–
53	–	–	–	–	–	–
54	E	U–Batt from DME main relay	Main relay terminal 87	U–Batt	Voltage value	Perform status monitoring. Diagnosis not possible if signal not applied – connect jumper terminal 30 and 87 on relay
55	M	Ground ignition	Ground point	Ground	–	–
56	E	Ignition terminal 15	Ignition coil	U–Batt	–	–
57	–	–	–	–	–	–
58	–	–	–	–	–	–
59	–	–	–	–	–	–
60	E	Programming voltage	Diagnostic connector	–	–	Only activated for control unit programming
61	–	–	–	–	–	–
62	E	Idle signal LL	EML	Ground	ON	See status monitoring
63	E	Torque converter clutch (WK)	EGS	Ground	ON (with engine running)	when WK engaged, see status monitoring
64	E	Ignition timing intervention	EGS	Ground	ON	Only active during gear-shift
65	E	Drive range P/N	Transmission selector lever switch	Ground	ON	When position P/N
66	–	–	–	–	–	–

5.0L M70 ENGINE—DME M1.7 PIN ASSIGNMENTS—1991-92 318i, 750iL AND 850i

Pin	Type	Description/Function	Connection	Type of Signal	Tester Display	Measurement Notes
67	E	Inductive pulse generator (connection B)	Pulse generator Engine speed/position	A/C voltage	Engine speed	Between Pins 67 and 68 (can only be measured with crankshaft rotating)
68	E	Inductive pulse generator (connection A)	Pulse generator Engine speed/position	A/C voltage	Engine speed	Between Pins 67 and 68 (can only be measured with crankshaft rotating)
69	–	–	–	–	–	–
70	E	Oxygen sensor signal	Oxygen sensor	Pulsating voltage up to approx. 1 V	Voltage value	Can only be measured with engine running at operating temperature
71	M	Ground oxygen sensor	Oxygen sensor	Ground	–	–
72	–	–	–	–	–	–
73	E	Road speed signal	EKM	Square–wave signal	km/h	–
74	A	Drehzahlsignal TD	EKM	Square–wave signal	–	one signal per ignition
75	–	–	–	–	–	–
76	E	Idle CO–potentiometer	Air mass meter	0...5 V	Voltage value	–
77	E	Intake air temperature sensor	Intake air temperature sensor	0...5 V	°C	Temperature–dependent
78	E	Engine temperature sensor	Engine temperature sensor	0...5 V	°C	Temperature–dependent
79	–	–	–	–	–	–
80	–	–	–	–	–	–
81	E	DWA	EKM	Ground	OFF	U–Batt applied when DWA armed
82	E	ASC/MSR	ASC or ASC + T	U–Batt	ON	U–Batt => signal active
83	E	ASC/ZAB	Only for ASC without T	U–Batt	ON	U–Batt => signal active

5.0L M70 ENGINE – DME M1.7 PIN ASSIGNMENTS – 1991-92 318i, 750iL AND 850i

Pin	Type	Description/Function	Connection	Type of Signal	Tester Display	Measurement Notes
84	–	–	–	–	–	–
85	E	Identifier DME control unit II	Ground point	Ground	DME cyl. 7...12	The DME control unit can only be selected with the diagnostic procedure when this link is set up.
86	E	A/C switch	A/C relay	12 V	ON / OFF	
87	E	RxD diagnostic data line	Diagnostic connector	U–Batt	–	When cover of diagnostic connector closed
88	E/A	TxD diagnostic data line	Diagnostic connector	Ground	–	When cover of diagnostic connector closed

E = Input A = Output M = Ground

Adaptors

Component	Connector Number	Description	Adaptor Number
Control unit	X 6010, X 6020	88–pole	Universal adaptor, Cartool 61 4 410 Accessories: Test lead, black, Cartool 61 4 412 Test lead, red, Cartool 61 4 413 Extensions (2), Cartool 61 4 430
Pulse generator, cylinder reference	X 6211, X 6221	3–pole, black	V–adaptor lead, Cartool 61 1 491
Pulse generator, engine speed/ position	X 6212, X 6222		
Ignition coil		1–pole	Ignition lead adaptor, Cartool 61 2 010

5.0L M70 ENGINE – DME M1.7 TROUBLESHOOTING – 1991-92 318i, 750iL AND 850i

Troubleshooting

Notes

Relevant safety codes and accident prevention regulations must be observed when carrying out tests and work on the engine electrical and electronic systems.

Caution!
Dangerous high voltage in the primary and secondary circuit of the ignition system.

Wherever possible, the battery should not be disconnected otherwise the defect codes stored in **all** the control units will be deleted.

Disconnected control units from the wire harness **only** when the ignition is switched off.

Conduct wire tests **only** with the control unit or component connectors disconnected.

Test Equipment

The 88–pole universal adaptor can be used for picking–off from the DME control unit or from the engine wire harness.

Use only suitable test leads, terminals and probe tips

Display of a stored Error Code on the Screen.

– **Defect location:** The defect can be in the vicinity of the component, the associated wiring or in the control unit. The number of the defect code indicates the pin assignments at the control unit.

– **Type of defect:** Short–circuits, wire breaks or range transgressions are indicated. Further indication is provided as to whether the defect is momentary, or whether it occurred in the past and the code is presently not stored (intermittent defect).

– **Defect frequency:** Indication of how often the defect has been detected, or the number of starts is indicated which have taken place since the defect was last detected.

– **Detected during:** Additional defect information on the operating status during defect detection.

5.0L M70 ENGINE – DME M1.7 TROUBLESHOOTING – 1991-92 850i

Evaluation of Stored Defect Codes

To facilitate effective troubleshooting, it is first advisable to print out the contents of the defect code memory.

During repairs, it is often necessary to disconnect plug connections, causing a defect code to be set in many cases. This may lead to incorrect diagnosis after monitoring the defect codes.

For this reason, all defect codes should be cleared after performing the necessary repairs (also EML, EGS, ASC/MSR). Monitor the defect code memory once again after a test run.

Some defects cause a defect code to be set in several systems (EML, EGS, ASC/MSR). Such interrelationships can be determined by comparing the ambient conditions (e.g. engine speed, temperature etc.).

Malfunction: Engine Operation Not Possible

Prerequisites for engine operation:

- Starting system OK. (Battery charge, ignition switch ...)
- Ground connection from battery to engine or body OK.
- Specified fuel in tank
- Engine mechanically OK. (timing, compression ...)

a) Check voltage supply for DME control units in accordance with schematic.

b) Measure input signals

- Check engine speed signals of the pulse generators at starting speed – ♦D♦ 2.
- Check air mass meter signals – ♦D♦ 2.

c) Measure output signals.

- Drive fuel pump relay in accordance with schematic
- Ignition output stages by means of ENGINE TEST test step 09
- Injector valves with ♦D♦ 3

d) Check fuel pressure

e) Check EML for defects

5.0L M70 ENGINE – DME M1.7 DEFECT CODES – 1991-92 318i, 750iL AND 850i

Defect Code Table

Defect Code 0

♦D♦ 1 Defect Code Memory: 0 UNDEFINED FAULT

Explanation: Data have been stored in the defect code memory for no apparent reason.

Corrective Measures: Perform test run after clearing the defect code. The defect code can be ignored if there are no noticeable effects in the system.

Defect Code 1

♦D♦ 1 Defect Code Memory: 1TR SIGNAL/EKP RELAY

Explanation: This defect code is stored when a short occurs at the EKP relay or on the control wire (from control unit Pin 1).

Effects of Defect: The fuel pump relay is not activated. If the defect occurs in the DME control unit for cylinders 1...6, the EGS takes up the emergency program.

Troubleshooting: The TR signal (engine speed signal once per revolution) is output, when the engine is running, at Pin 1 of the DME control unit for cylinders 1...6 and is routed via the EKP relay to the EGS control unit.

Possible Causes: Wires, connector or EKP relay defective.

5.0L M70 ENGINE – DME M1.7 DEFECT CODES – 1991-92 318i, 750iL AND 850i

Defect Code 3

♦D♦ 1 Defect Code Memory: 3 FUEL INJECTORS cyl. 2–4–6 or 8–10–12

Explanation: This defect code is stored when shorts or breaks occur on the fuel injector valves cyl. 2–4–6 or 8–10–12 or on their supply wires.

Effects of Defect: The entire group of injector valves is always deactivated when a short occurs at one injector valve or on its supply wire. The 3 other cylinders of the same row of cylinders remain in operation.

Troubleshooting:

- Activate injector valves (see component testing).
- Check connections and wires in accordance with schematic.
- Replace failed injector valves.

Defect Code 8

♦D♦ 1 Defect Code Memory: 8 DEFECT LAMP

Explanation: Applicable to US models only. This defect code is stored when shorts occur at the defect lamp or on its supply wire.

Effects of Defect:

- The defect lamp does not light in the case of short to plus.
- The defect lamp lights continuously in the case of a short to ground.

Troubleshooting: Normally, the defect lamp is activated by the control unit (ground signal at Pin 8) when an emission–related defect code is stored in the defect code memory.

Defect Code 16

♦D♦ 1 Defect Code Memory: 16 IGNITION

Explanation: This defect code is stored when the cylinder reference sensor on cylinder 6 or 12 no longer detects an ignition signal for a certain period of time.

Effects of Defect: The injection signal is reduced to a minimum. Fuel is no longer injected (protection of catalytic converter) however, the EML does not take up the emergency program since an injection signal is still applied.

Possible Causes:

- Cylinder reference sensor or supply wires defective
- Cylinder 6 or 12 receives no ignition voltage signal
- No spark–over at the spark plug of cylinder 6 or 12
- Ignition coil defective
- Ignition output stage of DME control unit defective

Defect Code 18

♦D♦ 1 Defect Code Memory: 18 FINAL STAGE PIN 18

Explanation: This defect code is stored when a short occurs at Pin 18. This defect code can be ignored since this pin is not connected to the wiring harness.

5.0L M70 ENGINE—DME M1.7 DEFECT CODES—1991-92 318i, 750iL AND 850i

Defect Code 32

♦D♦ 1 Defect Code Memory: 32 FUEL INJECTORS cyl. 1–3–5 or 7–9–11

Explanation: As defect code 03.

Defect Code 36

♦D♦ 1 Defect Code Memory: 36 TANK VENTING VALVE

Explanation: This defect code is stored when a short occurs at the fuel evaporation control valve or on its supply wire.

Effects of Defect:

- Short to ground: No fuel evaporation control since the solenoid valve remains closed.
- Short to plus (or break): The solenoid valve constantly remains open. As a result, the fuel vapours supplied via the fuel evaporation control valve can lead to malfunctions when idling and during acceleration.

Troubleshooting: Activate TE valve with **♦D♦ 3** component activation

Does the TE valve respond?

- YES: Check wire and connector for loose contact
- NO: Connect test lamp (12 V,10 W) and activate component.

Does test lamp flash?

- YES: Check function of TE valve, see test instructions.
- NO: Check voltage supply at TE valve. Check wires. Check output at DME control unit (pin 36).

5.0L M70 ENGINE–DME M1.7 DEFECT CODES–1991-92 318i, 750iL AND 850i

Defect Code 37

♦D♦ 1 Defect Code Memory: 37 SENSOR HEATER RELAY

Explanation: The oxygen sensor heating relay is activated by the DME control unit I. For this reason, a detected defect is only stored in the DME control unit I. The control unit detects a short or break at the relay or on its supply wire.

Effects of Defect: After a cold start, operation of the oxygen sensors is delayed.

This defect code can be ignored if no sensor relay is installed.

Troubleshooting: Drive components with ♦D♦ 3 activate the sensor heating relay. When the relay is active, ground is applied to terminal 85 of the relay (check with 12 V 3 W lamp).

Defect Code 41

♦D♦ 1 Defect Code Memory: 41 AIR MASS SENSOR (HLM)

Explanation: This defect code is stored when a short or a break occurs at the HLM or on its supply wires.

Effects of Defect: The DME control unit operates with fixed substitute values:

- When idling, the load signal TI = 2.2 ms.
- During part load and full load operation the load signal TI = 0 ms.

Troubleshooting: Observe the load signal with ♦D♦ 2 status monitoring. The nominal value is 1.8...2.2 ms when the battery is charged, air conditioning switched off and gear selector lever in position P or N. If the nominal value is not achieved:

- Check voltage supply of the HLM, wires and connector.
- Check HLM.

Defect Code 48

♦D♦ 1 Defect Code Memory: 48 A/C COMPRESSOR SHUTOFF

Explanation: This defect is stored when a short occurs at Pin 48.

Effects of Defect:

- Short to plus: Below a speed of 8 km/h, the A/C compressor is not switched off during acceleration with full load.
- Short to ground: The A/C compressor does not operate.

Troubleshooting: Check A/C cut–out relay and supply wires.

Defect Code 54

♦D♦ 1 Defect Code Memory: 54 CONTROL UNIT POWER SUPPLY

Explanation: This defect code is stored when the voltage supply exceeds 16 V. A drop in the voltage supply below 9 V is detected 3 minutes after the start at the earliest.

Troubleshooting:

- Check battery acid level
- Check charged battery under load
- Check B + wire to the DME relay
- Check charging current of alternator with consumer loads switched on
- Check voltage regulator

5.0L M70 ENGINE – DME M1.7 DEFECT CODES – 1991-92 318i, 750iL AND 850i

Defect Code 63

♦D♦ 1 Defect Code Memory: 63 CONVERTER LOCKUP CLUTCH

Explanation: The control unit receives simultaneously the signals gear selector lever in position P/N and torque converter clutch closed (ground signal). This operating status is impossible.

Effects of Defect: A permanently applied ground signal is ignored by the DME control unit. EGS may be in the emergency program.

Troubleshooting: Check the plausibility of the signals gear selector lever and torque converter clutch with ♦D♦ 2 status monitoring. The torque converter clutch inquiry is only displayed correctly when the engine is running.

Defect Code 70

♦D♦ 1 Defect Code Memory: 70 OXYGEN SENSOR

Explanation: This defect code is stored when a short or a break occurs at the oxygen sensor or on its supply wires, or when no oxygen is installed (provisions for catalytic converter). A prerequisite for defect detection is an engine temperature of 70° C and a load signal of >3.5 ms for >20 sec.

Effects of Defect: Emission control is no longer possible, the control unit operates with a substitute value.

Troubleshooting: Observe the emission control and sensor voltage with ♦D♦ 2. After the sensor has reached operating temperature, the sensor voltage between 0.02 and 0.085 V.

- If sensor voltage not OK.: Detach connector for oxygen sensor, switch on ignition and measure voltage at the connector (control unit side) (nominal value = 0.45 V). As displayed on screen.
- Check wire and oxygen sensors
- If the oxygen sensors have been interchanged (DME control unit I operates with sensor II and vice versa), this defect code will be stored in both control units. The mixture is too rich in one group of cylinders and too lean in the other.

Defect Code 64

♦D♦ 1 Defect Code Memory: 64 IGNITION TIMING TAP

Explanation: This defect code is stored when a short to ground occurs on the wire from the EGS to DME control unit Pin 64 (>2.5 sec.).

Effects of Defect: The ignition angle intervention occurs in such a case only once, the signal is then ignored. The automatic transmission is in the emergency program.

Troubleshooting:

- Check wire at connector to the EGS
- Perform defect monitoring on EGS. Monitor status once again after test run.

Defect Code 73

♦D♦ 1 Defect Code Memory: 73 ROAD SPEED SIGNAL

Explanation: This defect code is stored when, despite high load and engine speed, no road speed signal is applied.

Effects of Defect:

- For manual transmission: Irregularities at engine speeds close to the idle speed (EML controls the idle speed)
- Power loss (EML limits the throttle angle).
- FGR failure.

Troubleshooting:

- If road speed signal is in instrument cluster: check wire from DME Pin 73 to EKM.
- Otherwise check wire from instrument cluster to reed contact.

5.0L M70 ENGINE – DME M1.7 DEFECT CODES – 1991-92 318i, 750iL AND 850i

Defect Code 76

♦D♦ 1 Defect Code Memory: 76 IDLE SPEED CO POTENTIOMETER

Explanation: Only applicable to vehicles without a catalytic converter. This defect code is stored when shorts or breaks occur at the potentiometer or on its supply wire.

Effects of Defect: A substitute value of 1.25 V is set. The emission CO value may deviate from the nominal value.

Troubleshooting:

– Read out CO–potentiometer actual value with **♦D♦ 2** status monitoring.
– Check CO–potentiometer (in HLM) and supply wires.
– Measure CO content in exhaust emission.

Defect Code 77

♦D♦ 1 Defect Code Memory: 77 INTAKE AIR TEMP. SENSOR (LMM)

Explanation: This defect is stored when shorts or breaks occur at the temperature sensor or on its supply wires. A prerequisite for detecting a break is a time interval of 3 min after start and idling for at least 30 sec.

Effects of Defect: After defect detection, a substitute value o 50 °C is used.

Troubleshooting: Read out the actual value with **♦D♦ 2**.

If not plausible: Check wire to DME Pin 77 and temperature sensor.

Defect Code 78

♦D♦ 1 Defect Code Memory: 78 COOLANT TEMP. SENSOR (ENGINE)

Explanation: This defect is stored when shorts or breaks occur at the temperature sensor or on its supply wires. A prerequisite for detecting a short to plus or break is an intake air temperature above – 9 °C.

Effects of Defect: After defect detection, a substitute value of 50 °C is used.

Troubleshooting: Read out the actual value with **♦D♦ 2**.

If not plausible: Check wire to DME Pin 78 and temperature sensor.

Defect Code 82

♦D♦ 1 Defect Code Memory: 82 ENGINE DRAG TORQUE CONTROL

Explanation: This defect code is stored when the MSR signal (U–Batt at DME Pin 82) is applied constantly for more than 5 sec.

Effects of Defect: The MSR signal is subsequently ignored by the DME control unit

Troubleshooting: With **♦D♦ 2** status monitoring. With the ignition switched on or with the engine idling, the nominal value for the MSR signal = OFF. If the display shows ON, this indicates a short to plus.

5.0L M70 ENGINE–DME M1.7 DEFECT CODES–1991-92 318i, 750iL AND 850i

Defect Code 83

♦D♦ 1 Defect Code Memory: 83 ASC

Explanation: This defect code memory is stored when the ASC signal (U–Batt at DME Pin 83) is applied for more than 1 sec.

Effects of Defect: The ASC signal is subsequently ignored by the DME control unit.

Troubleshooting: With **♦D♦ 2** status monitoring. With the ignition switched on or the engine idling, the nominal value for the ASC signal = OFF. If the display shows ON, this indicates a short to plus at Pin 83 (ASC) or Pin 81 (DWA).

Defect Code 100

♦D♦ 1 Defect Code Memory: 100 OUTPUT STAGE

Explanation: This defect code is stored when the function of an output stage is not plausible and the control unit could not distinctly identify the output stage. If, at the same time, one of the following components is stored as defective, then this is the cause of defect code 100 being set and can be ignored:

- Idle speed actuator
- Injector valves
- Sensor heating relay (only DME control unit for cylinders 1...6)
- Fuel evaporation control valve
- Defect lamp (US only)
- EKP relay

Troubleshooting: Activate the corresponding components with **♦D♦ 3** component activation or check wires.

Defect Code 200

♦D♦ 1 Defect Code Memory: 200 DME CONTROL UNIT

Explanation: This defect code is stored when the internal self–test has found a defect in the control unit.

Possible Causes: Program error or disturbed program sequence in the DME control unit. This defect code is also stored when manipulations have been made in the program of the control unit. In this case, the sealing tabs on the control unit housing will be damaged.

Corrective Measures: Perform test run after clearing the defect code memory. If defect reoccurs, the DME control unit must be replaced.

Defect Code 201

♦D♦ 1 Defect Code Memory: 201 OXYGEN CONTROL

Explanation: This defect code is stored in the case of excessive deviations in the air–fuel mixture (mixture too rich or too lean for longer than 10 sec). The prerequisite is that the oxygen sensor is ready for operation.

Effects of Defect: Poor engine reaction since the DME operates with a substitute value.

Possible Causes:

- Incorrect fuel pressure
- Defective or coked injector valves
- Defective engine temperature sensor
- Secondary air (lean mixture)
- Defective fuel evaporation control system
- Defective air mass meter
- Disturbed combustion (ignition or compression not OK)

Troubleshooting: Perform emission test.

5.0L M70 ENGINE—DME M1.7 COMPONENT TESTING—1991-92 318i, 750iL AND 850i

Component Testing

Notes

Relevant safety codes and accident prevention regulations must be observed when carrying out tests and work on the engine electrical and electronic systems.

Caution!
Dangerous high voltage in the primary and secondary circuits of the ignition system.

Important!

Whenever possible, the battery should not be disconnected otherwise the defect codes stored in ***all*** the control units will be deleted. If the voltage supply to the DME is interrupted, the adaption values will be additionally forgotten (lost). This can lead to problems in starting and idling. Depending on how far the adaption values deviated from the basic values, some time will be required until the adaption values are relearned.

Disconnect control units from the wire harness ***only*** when the ignition is switched off.

Conduct wire tests ***only*** with the control unit or component connectors disconnected.

Test Equipment

The 88–pole universal adaptor can be used for picking–off from the DME control unit or from the engine wire harness.

Only use suitable test leads, terminals and probe tips.

5.0L M70 ENGINE—DME M1.7 COMPONENT TESTING—1991-92 318i, 750iL AND 850i

Overview of the Defect Codes and Component Tests

Component / Function	Quick Test after Entry ♦D♦ ...	Defect Code after Storage
DME control unit	–	200
Ignition coil	See engine test	–
TR signal/EKP relay	–	1
Fuel evaporation control valve	2, 3	36
Air mass meter	2	41
Pulse generator – cylinder reference (ignition)	2	16
Emission (lambda) control	2	201
Defect lamp	–	8
Injector valves	3	3, 32
Sensor heating relay	3	37
Oxygen sensor	2	70
Road speed signal	2, also see instrument cluster/CC	73
DME main relay	2	–
Voltage supply	2	54
ASC	2, also see ABS/ASC	83
Intake air temperature sensor	2	77
Engine temperature sensor	2	78
Pulse generator – engine speed/position	2	–
MSR	2, also see ABS/ASC	82
Ignition timing intervention	2, also see EGS	64
Idle/full load signal	2	–
Torque converter clutch	2, also see EGS	63
Drive range P/N	–	–
A/C compressor cut–out	–	48
LL–CO–potentiometer	2	76

5.0L M70 ENGINE – DME M1.7 COMPONENT TESTING – 1991-92 318i, 750iL AND 850i

DME Control Unit

Check Function:

Individual function blocks, e.g. ignition, fuel injection, emission control etc., can only be checked when the necessary input signals are applied beforehand at the control unit

TR–Signal / EKP–Relais

Check Function:

The EKP relay switches on as soon as the DME control unit outputs at Pin 1 a ground signal with superimposed engine speed signal (once per crankshaft revolution).

In addition, at ignition on, the EKP relay is briefly activated by the DME in order to build up the necessary fuel pressure before starting. The voltage increase can be measured at fuse 23.

Possible Causes:

– Wire to the EKP relay defective
– EKP relay or main relay defective
– Supply voltage for DME control unit, DME relay 1 or 2 not OK:
– No signals from engine speed/position sensor

Ignition Coil

Check Function:

Perform engine test, step 09. With the engine at operating temperature, the ignition voltage is 10 ± 2 kV. The maximum difference of the individual cylinders should not exceed 3 kV.

Fuel Evaporation Control Valve

Check Electrical Function:

♦D♦ 3 – The fuel evaporation control valve is actuated at intervals of 1 sec. The actuation cycles can be heard at ignition ON and felt with the hand.

Check Actuation Phases:

♦D♦ 2 – The valve is actuated at idle speed (display: CLOSED) and is therefore closed. As soon as the emission control system is active and the engine is increased, the valve can be seen to open step–by–step (display:OPEN).

After switching off the engine, the valve is held closed for 3...5 sec.

Checks for Leaks:

Detach hose from valve, switch on vacuum pump at the service tester and connect to the 8 mm connection (note arrow indicating direction of flow). Set vacuum to 600 mbar and apply a voltage of 12 V to the valve.

Switch off vacuum pump – the pressure drop must not exceed 50 mbar after 20 sec.

When power is not applied, the fuel evaporation control valve is closed by a non–return valve to a vacuum of 20 mbar

5.0L M70 ENGINE – DME M1.7 COMPONENT TESTING – 1991-92 318i, 750iL AND 850i

Air Mass Meter (HLM)

Check Voltage Supply HLM:

U–Batt supplied between Pins 2 and 4 on the HLM connector.

Check HLM Signal:

♦D♦ **2** – The load signal TI provides an indication of the measured air flow rate in the HLM. When idling, the value should be 2.0 ms ± 20%. In the case of a break in the voltage supply or in the signal line, the value is a constant 2.2 ms (substitute value) at idle speed and 0 ms under part load and full load.

The load signal is formed dependent upon the voltage (<5 V) applied at the hot wire of the HLM. The load signal is displayed in the BMW SERVICE TESTER only when the engine is in operation.

The voltage value of the hot wire can be measured between Pin 14 (sensor ground) and Pin 41 (signal HLM) of the DME control unit. This measurement can also be conducted with the ignition ON (nominal value approx. 1.4 V).

The load signals of the 2 rows of cylinders must scarcely deviate from each other. Check the burn–clear function if the load signal is permanently too low. A load signal which is too low tends to produce a lean mixture, thus reducing the engine output.

Check Burn–Clear Function::

Increase engine speed once to above 2000 rpm (engine temperature >6O °C). Approx. 5 sec after switching off the engine, the hot wire can be seen to briefly light up (burn–clear) at the HLM (air cleaner cover removed).

The signal can also be measured between Pin 13 and Pin 26 (U–Batt) of the 88–pole universal adaptor. A voltage of approx. 3.5 V is applied at Pin 13 to facilitate the burn–clear function (i.e. at the moment of actuation, the system voltage measured at Pin 26 reduces by this value). The signal, however, can only be measured with a fast multimeter.

5.0L M70 ENGINE – DME M1.7 COMPONENT TESTING – 1991-92 318i, 750iL AND 850i

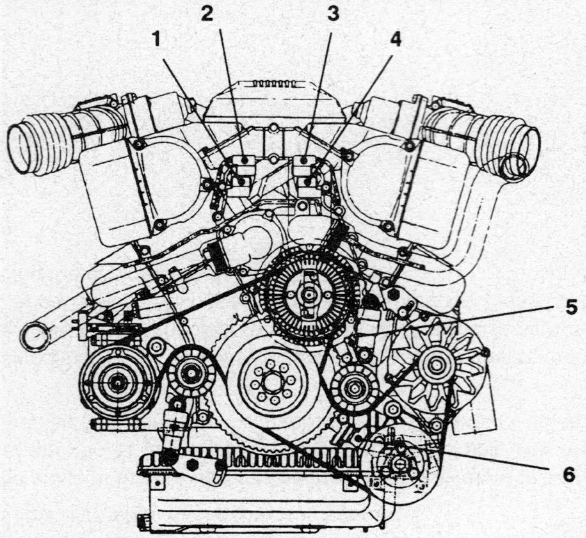

1	Connector cylinder reference	DME control unit I
2	Connector engine speed/position	DME control unit I
3	Connector cylinder reference	DME control unit II
4	Connector engine speed/position	DME control unit II
5	Pulse generator engine speed/position	DME control unit I
6	Pulse generator engine speed/position	DME control unit II

Pulse Generator – Cylinder Reference (Ignition)

Check Function:

♦M♦ 22 – Attach connections for oscilloscope at the component connector (adaptor: Cartool 61 1 491) and operate engine at idle speed. From the point the engine starts, an alternating voltage characteristic curve with positive and negative half wave should appear on the display screen. If no signal is applied, the DME control unit cuts out fuel injection.

Resistance Test:

The resistance can be measured directly at the component connector (at the connector strip for injector valves).

Pulse Generator – Engine Speed/Position

The DME does not produce an ignition and injection signal if the engine speed signal is not applied.

Check Function:

♦D♦ 2 – The engine speed signal can be monitored at starting speed or when the engine is running.

Further test options are possible with the oscilloscope **♦M♦ 23/24**. Disconnect terminals at the component connector and start engine. An alternating voltage characteristic curve with positive and negative half wave should appear on the display screen.

Possible Causes:

If wire, plug connection and resistance are OK.: check distance of pulse generator to the increment wheel on the vibration damper. Nominal value: 0.15...1.5 mm.

Defect Lamp 'Check Engine' (US only)

Check Function:

The control unit supplies the defect lamp with ground potential as soon as an exhaust emission–related defect is detected. As a check, the lamp lights constantly after ignition ON and as soon as the engine is running (if no exhaust emission–related defect code is stored).

Simulated defect: For example, detach connector from the air mass meter and start engine. The defect lamp must then light when the defect code 41 air mass meter is detected by the control unit.

– Does the control unit output the ground signal?
– Are wires at defect lamp OK.?

5.0L M70 ENGINE – DME M1.7 COMPONENT TESTING – 1991-92 318i, 750iL AND 850i

Injector Valves

Check Electrical Functions:

♦D♦ 3 – In this case, the fuel injector valves are actuated with a frequency of 10 Hz. The pulse period is 1 ms. Activation time should be kept as short as possible in order to avoid fuel being injected. Activation can be felt with the hand. The pulses can be checked with the multimeter ♦M♦ 15 at the injector valve connector or at the 88–pole universal adaptor.

Testing with oscilloscope ♦M♦ 22 and universal adaptor:

Connect terminal frequency OFF (+) to terminal 1 of the universal adaptor lead of the BMW SERVICE TESTER and apply a frequency of 70 Hz. Connect the 88–pole universal adaptor (Cartool 61 4 410) to the DME control unit and connect the terminal D+ (blue) to Pins 3 or 32. The injection signal should be displayed on the oscilloscope. In contrast to measuring the pulse duration, injection fade–out during coasting can also be seen on the oscilloscope.

Resistance Test:

The resistance of an injector valve must be measured directly at the component. Only the entire group of injector valves (3 valves) can be measured at the universal adaptor.

Sensor Heating Relay

This relay will not be fitted in series production! The voltage for the sensor heaters will then be supplied via the EKP relay.

Check Function:

♦D♦ 3 – Activation is only possible at the cylinder row 1...6 (the relay supplies both sensor heaters) and takes place at 1 sec intervals.

During normal operation, the relay remains switched on until higher speeds or loads occur.

Possible Causes:

– Wire to DME control unit Pin 23 defective
– Relay or voltage supply not OK.
– Fuse F28 defective

Oxygen Sensor

Check Function:

♦D♦ 2 – To carry out this check, the oxygen sensor must be at operating temperature. The indicated voltage signal is picked off in steps and fluctuates between 0.02 and 0.85 V.

When the oxygen sensor is not in operation, or the wire to the DME control unit is interrupted, the constant value of 0.45 V is indicated. This value is also indicated when no oxygen sensor is installed.

An oxygen sensor may be defective (e.g. due to deposits) and still indicate the required voltage signals. An indication for such cases is CO measured values deviating from the nominal value despite the emission control system being active. Test option: Disconnect DME control unit from the voltage supply for at least 5 min. Then disconnect the oxygen sensor and perform CO measurement once again. If the CO value approaches the nominal value, the oxygen sensor should be replaced and the CO measurement repeated.

Emission Control

Check Function:

♦D♦ 2 – To carry out this check, the oxygen sensor must be at operating temperature. The emission control system is dependent on the engine temperature and can have a duration of up to 5 min (at – 30 ° C). At starting temperatures > 20 ° C, the control can take up operation after approx. 30 sec.

The control system can effectively compensate wear–related deviations in the air–fuel mixture and thus conceal other defects. If necessary, disconnect the oxygen sensor when the checking the air–fuel mixture system.

5.0L M70 ENGINE – DME M1.7 COMPONENT TESTING – 1991-92 318i, 750iL AND 850i

Adaptions

The correction values stored in the control unit should **only** be used as additional information **in the case of acute engine problems**.

The correction values stored in the control unit can be of assistance in the case of acute engine problems.

If the set range is exceeded in the minus direction, the air–fuel mixture will be too rich and the DME attempts to adjust the mixture, i.e. make it leaner.

Reasons for an excessively rich mixture:

– Injector valves leaking
– Fuel pressure too high

If the set range is exceeded in the plus direction, the air–fuel mixture will be too lean and the DME attempts to adjust the mixture, i.e. make it richer.

Reasons for an excessively lean mixture:

– Leaks in intake or exhaust system (ahead of the oxygen sensor). the oxygen sensor measures the additional oxygen and signals to the DME that the mixture is too lean.
– Hot wire in the HLM defective (load signal too low). The mixture is too lean.

The adaption values are forgotten (lost) when the voltage supply to the DME is interrupted. This can lead to problems in starting and idling. Depending on how far the adaption values have deviated from the basic values, some time will be required until the adaption values are relearned.

For this reason, idle operation or test runs in order to observe the change in the adaption values are of little use.

CO–Potentiometer (without Catalytic Converter)

Check Function:

♦D♦ 2 – Measure CO content and set to nominal value with CO–potentiometer (adjusting screw on HLM). The basic setting on the potentiometer is 1.25 V.

Check Voltage Supply::

A voltage of approx. 5 V must be applied at the HLM connector between Pins 3 and 4.

Road Speed Signal

♦D♦ 2 – With the drive wheels raised, the vehicle speed should be limited to max. 30 km/h. The road speed signal is sent by the EKM to the DME..

5.0L M70 ENGINE – DME M1.7 COMPONENT TESTING – 1991-92 318i, 750iL AND 850i

DME Main Relay

Check Function:

♦D♦ **2** – With ignition ON or during engine operation, the main relay is actuated with ground potential at terminal 85 (display: ON). The relay switches on and supplies voltage to the control units DME, EGS, EML and ABS.

Possible Causes:

- Wire to the DME control unit Pin 27 defective
- Relay or voltage supply not OK.
- Voltage supply for DME control unit not OK.
- The control unit is defective if no ground potential is output at control unit Pin 27 (display:OFF) although the above points are OK.

Control Unit Supply Voltage

♦D♦ **2** – The indicated voltage is currently applied at the control unit . It should not be below 10 V.

Intake Air Temperature Sensor

Check Function:

♦D♦ **2** – With the engine at operating temperature, the intake air temperature is at 50...70 °C (ambient air temperature 23 ±5 °C).

A display below – 40 °C indicates a break in the temperature sensor or in the supply wire to the DME control unit.

Engine (Coolant) Temperature Sensor

Check Function:

♦D♦ **2** – The indicated engine temperature should correspond to the coolant temperature. A value below – 30 °C indicates a break in the temperature sensor or in the wire to the DME control unit.

Idle/Full Load Signal

Check Function:

♦D♦ **2** – The signals are sent by the EML control unit to the DME. The full load signal is sent shortly before the full load limit stop.

5.0L M70 ENGINE – DME M1.7 TECHNICAL DATA – 1991-92 318i, 750iL AND 850i

Technical Data

Resistance Values Measured at 23 ±5 °C

Engine Type	**M70 B50**
Injector valve	15...17 Ω
Injector valve group (3 valves)	5,5 ± 0,5 Ω
Pulse generator – engine speed/position	540 ± 50 Ω
Pulse generator – cylinder reference	0,1...1 Ω
Engine temperature sensor	
at coolant temperature 20 ±3 °C	2,2...2,7 kΩ
at coolant temperature 80 ±3 °C	0,3...0,36 kΩ
Intake air temperature sensor	
at air temperature 20 ±3 °C	2,2...2,7 kΩ
at air temperature 50 ±3 °C	0,8...1,2 kΩ
Fuel evaporation control valve	45 ± 20 Ω
Ignition coil	
Primary winding (terminal 15/1)	0,5 ± 0,1 Ω
Secondary winding (terminal 15/4)	–
Manufacturer: May and Christie	9,0 ± 1,0 kΩ
Primary winding (terminal 15/1)	6,0 ± 1,0 kΩ

5.0L M70 ENGINE – DME M1.7 STATUS INQUIRY – 1991-92 318i, 750iL AND 850i

The nominal values apply at: Idle speed, oil temperature > 60 °C or coolant temperature > 80 °C, no load switched on.

Status Inquiry	Nominal Values	Signal Comes From:
67 Engine speed		DME control unit
Idle speed	700 ± 50 rpm	Control by EML
Cut–out speed	6000 ± 40 rpm	Control by EML
17 Injection signal	3,8 ms ± 20 %	DME control unit
41 Load signal	2,0 ms ± 20 %	DME control unit
90 Ignition angle	15 ± 3 ° crankshaft angle (automatic transmission) 10 ± 3 ° crankshaft angle (manual transmission)	DME control unit
70 Oxygen sensor voltage	0,02...0,85 V; in the case of interruption or defective oxygen sensor = 0.45 V	Oxygen sensor
77 Intake air temperature	Dependent on ambient temperature, approx. – 50 ° C = break	Intake air temperature sensor
78 Engine temperature	Dependent on coolant temperature, approx. – 30 ° C = break	Engine temperature sensor
91 Voltage supply control unit	> 10 V	DME relay
37 Oxygen sensor heater	ON	DME control unit
92 Emission control	ON, when oxygen sensor at operating temp.	DME control unit
93 Mixture adaption II	Minus 48 to plus 32	DME control unit

5.0L M70 ENGINE – DME M1.7 STATUS INQUIRY – 1991-92 318i, 750iL AND 850i

Status Inquiry	Nominal Values	Signal Comes From:
94 Mixture adaption TL	Minus 16 to plus 16	DME control unit
16 Cylinder reference	ON	Cylinder reference sensor
36 Fuel evaporation control	As soon clock cycle active = OPEN	DME control unit
76 Co–potentiometer (without cat) CO volume (no status inquiry)	Basic setting = 1.25 V 0,7 ± 0,5 %	Co–potentiometer Measured ahead of the catalytic converter, – engine test, test step 09
62 Idle signal	ON	EML control unit
12 Full load signal	OFF, at full load = ON	EML control unit
73 Road speed	In km/h, with drive wheels raised, speed limited to max. 30 km/h	Instrument cluster
1 EKP relay	At start or during engine operation = ON	DME control unit
27 DME main relay	With ignition switched on = ON	DME control unit
63 Torque converter clutch	OFF	EGS control unit
64 Ignition angle intervention	OFF, briefly set to ON during gearshifts	EGS control unit
65 Drive range P/N	In selector lever position P or N = ON	Gear selector lever switch
81 DWA/ASC signal	OFF	DWA or ABS/ASC control unit
82 MSR signal	OFF	ABS/ASC/MSR control unit
73 Road speed	In km/h, with drive wheels raised, speed limited to max. 30 km/h	Instrument cluster
1 EKP relay	At start or during engine operation = ON	DME control unit
27 DME main relay	With ignition switched on = ON	DME control unit
63 Torque converter clutch	OFF	EGS control unit
64 Ignition angle intervention	OFF, briefly set to ON during gearshifts	EGS control unit
65 Drive range P/N	In selector lever position P or N = ON	Gear selector lever switch
81 DWA/ASC signal	OFF	DWA or ABS/ASC control unit
82 MSR signal	OFF	ABS/ASC/MSR control unit

5.0L M70 ENGINE–DME M1.7/EML DESCRIPTION–1991-92 318i, 750iL AND 850i

Functional Description

The electronic throttle control system EML replaces the previous mechanical transmission and transfer linkage of the accelerator pedal angle between the accelerator pedal and throttle valve.

The most important components involved in the EML control unit are the pedal position sensor PWG and the EML throttle valves DK.

In addition to the transmission of the accelerator pedal angle from the accelerator pedal to the throttle valve, the EML system performs several other functions.

- Idle speed control
- Engine speed limitation
- Vehicle speed limitation
- Cruise control FGR
- Throttle valve opening in accordance with various characteristic progression curves which are selected from a pedal characteristic map corresponding to the drive status.
- Synchronisation of the throttle valves
- Interface for data exchange with other control units (e.g. ASC, DME and EGS).
- Safety test of important components used by the EML system
- Maintaining vehicle operability in the event of component failure by making available substitute values
- Self–diagnosis

All functions are monitored by the control unit program. This monitoring function also includes the program run of the control unit. The program run is controlled by the hardware. When the hardware detects a defect, the program run is terminated and then restarted.

5.0L M70 ENGINE–DME M1.7/EML TROUBLESHOOTING–1991-92 318i, 750iL AND 850i

Notes on BMW DIAGNOSE–SYSTEM

Defect Code Memory

The component functions and signals listed below are processed or controlled by the DML control unit, but not stored as a defect code in the case of defect:

- LL–signal (sends the "idle" information to the DME)
- VL–signal (sends the "full load" information to the DME)
- A/C switch
- A/C compressor clutch (lock sensor)
- Range selector switch ("P" or "N" position in automatic transmission)
- Clutch switch (manual transmission)
- Brake light switch
- External safety path

The external safety path cannot be read out in the status lists of the BMW DIAGNOSTIC SYSTEM.

Defect Code Priority

The EML control unit stores a maximum of 5 defect codes. When the defect code memory is full, a new defect code with higher priority displaces a lower priority defect code already stored (see ♦D♦ # Notes).

Status Monitoring

Since all inputs and outputs and their plausibilities are not covered by the defect code memory of the EML control unit, the status monitoring functions must be used to facilitate troubleshooting.

The nominal values and actual values indicated in the status lists need not necessarily agree exactly. The EML control unit stores a defect code if the deviation between the nominal value and the actual value is outside the permissible range.

For the permissible ranges, refer to the functional description or the troubleshooting description in the BMW DIAGNOSTIC SYSTEM. If the permissible range has been exceeded and the EML control unit has detected this as a defect, the component and its line connection can be checked without complex measurements with the aid of the actual value display.

5.0L M70 ENGINE – DME M1.7/EML TROUBLESHOOTING – 1991-92 318i, 750iL AND 850i

Abbreviations

ABS	Anti–lock brake system
ASC	Automatic stability control
DME	Digital engine electronics
EGS	Electronic transmission control
EML	Electronic throttle control
MSR	Engine drag torque control
A	Output
E	Input
DK	EML throttle
DKV	Set throttle value
DKR	Throttle reduction
DKE	Throttle increase
HG	Manual transmission
ME	Engine intervention
PWG	Pedal position sensor
Poti	Potentiometer
TD	Engine speed signal (once per ignition)
TR	Engine speed signal (three times per crankshaft revolution)
TMOT	Engine temperature sensor
ti	Injection signal
LL	Idle signa
VL	Full load signal
WK	Torque converter clutch
ZAB	Ignition fade–out

5.0L M70 ENGINE—DME M1.7/EML PIN ASSIGNMENTS—1991-92 318i, 750iL AND 850i

Pin Assignments

Pin Assignments at the EML 12–cyl. Control Unit Connector X6004

Pin	Type	Description/Function	Connection	Type of signal	Test display	Measurement notes
1	A	5V (not short-circuit–proof)	PWG switch	approx. 5 V	None	
2	A	Ground (electronic ground)	Component ground PWG			
3		Not used				
4	E	DK angle reduction (DKR)	ASC control unit	Square–wave signal	Defect code is stored	♦M♦ 14 approx. 100Hz
5	A	Full load signal	DME control unit 1...6. cylinder	approx.0 V open	ON OFF	
6	A	Idle signal	DME control unit 1...6. cylinder	approx. 0 V open	ON OFF	
7	E	PWG setting	PWG potentiometer	approx. 0,4 V to 4 V	Winkel	
8	E	Road speed signal	Instrument cluster	Square–wave signal	None	♦M♦ 14
9	A	5V (not short–circuit–proof)	PWG potentiometer	approx. 5 V	None	
10	A	Ground (electronic ground)	Component ground DK – 7..12. cylinder			
11	A	5V (nicht Kurzschlußfest)	DK potentiometer DK 7..12. cylinder	approx. 5 V	None	
12	A	5V (nicht Kurzschlußfest)	DK potentiometer DK 1...6. cylinder	approx. 5 V	None	

5.0L M70 ENGINE–DME M1.7/EML PIN ASSIGNMENTS–1991-92 318i, 750iL AND 850i

Pin	Type	Description/Function	Connection	Type of signal	Test display	Measurement notes
13	E	Diagnosis	Diagnostic socket RxD			
14	E	Ground (Output ground)	Output ground DK 1...6. cylinder			
15	A	EML warning lamp	Instrument cluster	approx. 0 V	ON	*) With ignition switched off
*)	E	Voltage supply for defect code memory		approx. 12 V	None	
16	A	Drive throttle (DK) actuator	DK actuator DK – 7..12. cylinder	Square–wave signal	Defect code is stored	
17	A	Drive throttle (DK) actuator	DK actuator DK 1...6. cylinder	Square–wave signal	Defect code is stored	
18	E	Voltage supply plus	DME main relay 1...6. cylinder	On–board voltage	None	
19	E	Ground	Ground point			
20	A	Kick–down signal	EGS control unit	approx. 0 V approx. 12 V	ON OFF	
21		Not used				
22	E	DK angle increase (DKE)	ASC control unit	Square–wave signal	Defect code is stored	♦M♦ 14 approx. 100Hz
23	A	Drosselklappensignal (DKV)	ASC / EGS	Square–wave signal	Defect code is stored	♦M♦ 14 approx. 100Hz
24		Not used				
25	A	5 V (not short–circuit–proof)	DK – 1...6. cylinder (not used)	approx. 5 V		
26	A	Ground (electronic ground)	Component ground DK 1...6. cylinder			

5.0L M70 ENGINE–DME M1.7/EML PIN ASSIGNMENTS–1991-92 318i, 750iL AND 850i

Pin	Type	Description/Function	Connection	Type of signal	Test display	Measurement notes
27	E	Throttle position	DK potentiometer DK 1...6. cylinder	approx. 4,6 V to 1 V	Angle	
28	E	Throttle position	DK potentiometer DK 7..12. cylinder	approx. 4,6 V to 1 V	Angle	
29	A	5 V (not short–circuit–proof)	DK – 1...6. cylinder (not used)			
30	E	Ground (output ground)	Output ground DK – 7..12. cylinder			
31	A	Ground (electronic ground)	TMot/FGR switch			
32	A	Idle signal	DME control unit 7..12. cylinder	approx. 0 V open	ON OFF	
33	A	Full load signal	DME control unit 7..12. cylinder	approx. 0 V open	ON OFF	
34	A	Drive throttle (DK) actuator	DK actuator DK – 7..12. cylinder	Square–wave signal	Defect code is stored	
35	A	Drive throttle (DK) actuator	DK actuator DK 1...6. cylinder	Square–wave signal	Defect code is stored	
36	E	Voltage supply plus	DME main relay 7..12. cylinder	On–board voltage	None	
37	E	Ignition ON terminal 15	DME main relay 7..12. cylinder			
38	E	Throttle 17° switch	DK switch DK 1...6. cylinder	approx. 5 V open	ON OFF	
39		Not used				
40	E	Throttle 17° switch	DK switch DK – 7..12. cylinder	approx. 5 V open	ON OFF	
41	E	A/C compressor ON	Lock sensor	On–board voltage open	ON OFF	
42	E	P and N identifier (EGS)	Gear selector lever switch	approx. 12 V open	ON OFF	
43	E/A	FGR identifier	FGR switch	approx. 0 V to 5 V	Detected position	

5.0L M70 ENGINE–DME M1.7/EML PIN ASSIGNMENTS–1991-92 318i, 750iL AND 850i

Pin	Type	Description/Function	Connection	Type of signal	Test display	Measurement notes
44	E	A/C switch	A/C relay/DME	approx. 12 V open	ON OFF	
45	E/A	Engine–coolant temperature	TMot EML sensor	approx. 0 V to 5 V	°C	
46	E	PWG 9° switch	PWG switch	approx. 5 V open	ON OFF	
47	E	Engine speed signal (TD)	DME control unit 1...6. cylinder	Square–wave signal	None	
48		Not used				
49	E	Injection signal (ti)	DME control unit 1...6. cylinder	Square–wave signal	Defect code is stored	
50	E	Injection signal (ti)	DME control unit 7..12. cylinder	Square–wave signal	Defect code is stored	
51		Not used				
52	E	S–program identifier (EGS)	EGS control unit	ca. 0 V open	ON OFF	
53	E	Brake light switch	Brake light switch	approx. 12 V open	ON OFF	
54		Not used				
55	E/A	Diagnosis	Diagnostic socket TxD			

Adapters

Component	Connector No.	Description	Adapternumber	
Control unit	X 6004	55–pole 55–pole 55–pole	Connection lead, periphery Connection lead, control unit Universal adapter	HWB 81 12 9 428 907 HWB 81 12 9 428 908 HWB 81 12 9 425 091
EML safety path		Safety path testing	EML adapter	Cartool 12 7 010

Component Replacement

SERVICE PRECAUTIONS

NOTE: BMW recommends that all proper cautions be observed to protect the DME system and components and for personal safety.

FUEL PRESSURE RELIEF

CAUTION

Anytime the fuel system is serviced the generation of spilled fuel and fuel vapor expose the technician to possibility of flash fire or explosion. All work should be performed in a well ventilated area with easy access to a fire extinguisher. All spilled fuel must be removed immediately. Wear protective eyewear. Failure to observe these precautions may result in personal injury.

1. Clean the fuel rail inlet connection. Wrap a shop towel around the fuel rail inlet connection. Be prepared with a suitable container to catch fuel spilling out of the connection.
2. Slowly loosen the hose clamp on the fuel inlet connection.
3. Remove the connection and drain the fuel into the container. Cap the line and connection to prevent contaminants from entering the system.
4. On the 12 cylinder engines be sure to perform this procedure on both fuel rails. The fuel systems for either side of the engine operate independently and both sides must be treated as separate systems.

THROTTLE SYNCHRONIZATION

Control Unit Synchronization

Each control unit is equipped with an adaptive control program. All sensor input signals are compared with stored values. If the signal does not exactly match a stored value, the adaptive program allows the control unit to learn the new value and generate the most appropriate output signal value. The adaptive memory is quickly lost if the control unit loses power and the engine will not run or idle properly. To synchronize the adaptive memories of the 3 control units:

1. With all other repairs completed and the engine ready to run, install the control units or connect the power and idle the engine to reach operating temperature.
2. Drive the vehicle. With the transmission in Range 1 and the engine at 5000 rpm, allow the vehicle to coast to engine idle speed, foot off the accelerator. It should take about 10 seconds. Repeat this step 3 times.
3. With the transmission in **P** or **N**, allow the engine to idle for about 10 minutes. The 3 control units should now be synchronized.

DME CONTROL UNIT

Removal and Installation

EXCEPT 3 SERIES

1. Before removing the control units retrieve any fault codes that may be resident in the units. When power has been removed from the units all fault codes and adaptive memories will be lost.
2. Turn the ignition switch **OFF**.
3. Remove the 4 screws holding the electronics box cover. The box is located under the hood in the rear, passenger side corner.
4. On 6 cylinder equipped vehicles, the DME unit is the control unit closest to the firewall and the EML unit or cruise control unit is on the engine side of the electronics box. On the 850i the DME unit 1 is furthest from the firewall, DME unit 2 is the

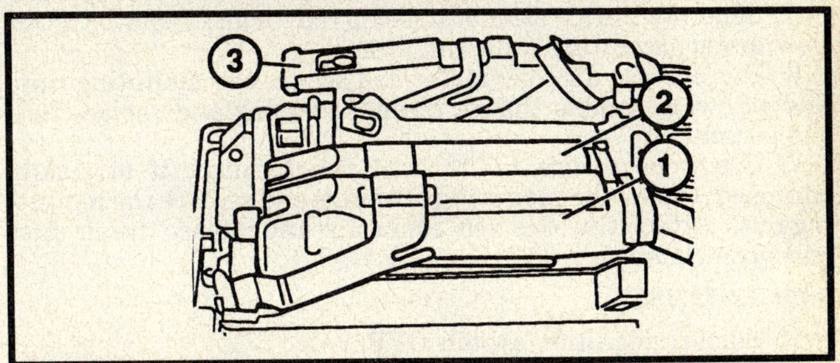

Module 1 is DME unit 1 and module 2 is DME unit 2. Module 3 is the EML control unit—BMW 850i

middle unit and the EML control unit is closest to the firewall. On the 750iL the DME unit 1 is closest to the firewall, the EML unit is next out and then DME unit 2.

5. Lift the connector lock on the control unit and pull off the connector.
6. Remove the mounting screws on the ends of the control unit or release the clamps as equipped. Pull the control unit up and out.
7. Check the code and part numbers of the replacement units and install the units into the correct locations.
8. Replace the mounting screws or attach the clamps. Carefully replace the connector on the control unit and press down the connector lock.
9. Replace the electronics box cover with the 4 screws. On EML equipped vehicles perform the external safety path check. On the M70 equipped vehicles perform the throttle synchronization procedure. Drive the vehicle to allow the DME units to relearn the adaptive settings.

3 SERIES

1. Before removing the control units retrieve any fault codes that may be resident in the units. When power has been removed from the units all fault codes and adaptive memories will be lost.
2. Turn the ignition switch **OFF**.
3. Open the glove compartment door and remove the pins holding the retaining straps. Allow the door to swing down and out of the way.
4. Remove the upper trim panel to expose the control unit.
5. Push back the control unit connector retainer and remove the connector from the control unit.
6. Remove the 4 screws retaining the control unit and remove the control unit from the glove box opening.
7. Check the code and part numbers of the replacement unit.
8. Mount the control unit and install the 4 retaining screws.
9. Attach the control unit plug and press down the retainer.
10. Install the trim panel and connect the retaining straps. Drive the vehicle to allow the DME unit to relearn the adaptive settings.

AIR FLOW METER

Removal and Installation

EXCEPT 535i AND 735i

1. Turn the ignition switch **OFF**.
2. Loosen the hose clamp securing the air duct to the body of the air flow meter. Pull off the duct.
3. Disconnect the electrical connector from the side of the air flow meter. On the M42 engine rotate the connector and remove with a screwdriver.
4. Loosen the air cleaner mounting nuts and remove the air cleaner assembly from the vehicle. There may be a wire tie holding the harness in place; cut the tie if necessary. On the 525i only the upper portion of the air cleaner housing must be removed.

5. Seperate the air cleaner housing halves and remove the air flow meter mounting nuts and screw.
6. Replace the air flow meter and secure the mounting nuts and screw. Assemble the air cleaner housing and replace into the vehicle.
7. Use a new wire tie to hold the harness, if originally equipped. Check the mounting bushings and replace the mounting nuts. Attach the electrical connector and replace the air duct and hose clamp.

535i AND 735i

1. Turn the ignition switch **OFF**.
2. Disconnect the electrical connector from the air flow meter. Loosen the hose clamps holding the air ducts to the air flow meter. Remove the ducts from the meter body.
3. Pull off the 3 fasteners holding the air flow meter body. Remove the air flow meter.
4. Check the rubber mounts and replace the air flow meter. Press the fasteners back into position and connect the air ducting and electrical connector.

AIR MASS SENSOR

Removal and Installation

EXCEPT 850i

1. Turn the ignition switch **OFF**.
2. Disconnect the air mass sensor electrical connector.
3. Loosen the air duct clamp and remove the air duct.
4. Remove the air cleaner assembly and unscrew the air mass sensor mounting bolts.
5. Remove the air mass sensor.
6. Install the air mass sensor with the web side towards the air cleaner. On the M5 make sure that the intake neck is spaced with equal clearance around its perimeter.
7. Install the air cleaner and connect the air duct and electrical connector.

850i

1. Turn the ignition switch **OFF**.
2. Disconnect the air mass sensor electrical connector.
3. Loosen the air duct clamps and remove the air ducts.
4. Lift the air mass sensor out of the vehicle.
5. Replace the air mass sensors and attach the air ducts. Connect the electrical connector.

FUEL PUMP

Removal and Installation

CAUTION

This procedure will generate gasoline fumes. Absolutely no flame or sparks should be in the area of the work being done. Gasoline fumes are heavier than air and will travel on the floor towards possible sources of ignition. Work in a well ventilated area and wipe up any gasoline spills. Failure to follow these precautions may result in personal injury.

1990-91 318i, 318iS, 325i AND M3

1. Turn the ignition switch **OFF**. Remove the rear seat bottom. Drain out some fuel from the fuel tank if the tank is full.
2. Loosen the insulating sheet and remove the screws holding the tank cover.
3. Remove the electrical connections and the fuel connections. Note which hose goes where.
4. Unscrew the fuel level sender screws and carefully withdraw the sender.
5. Turn the fuel pump holder counterclockwise and remove from the tank.
6. Remove the fuel filter and release the clamp holding the pump to the bracket. Install the new pump and filter and tighten the clamp.
7. Replace the gasket and install the new pump on the bracket into the tank. Turn the bracket clockwise until the stop is felt.
8. Install a new fuel level sender gasket and replace the fuel level sender. Tighten the mounting screws.

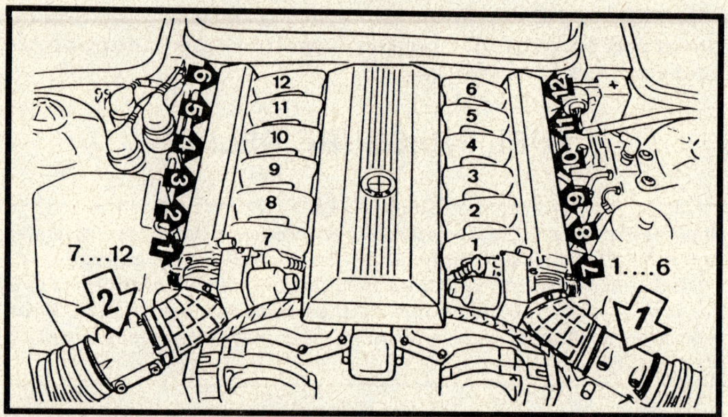

Air mass sensor locations. Sensor 1 feeds cylinders 1 through 6 and sensor 2 feeds cylinders 7 through 12—BMW 850i

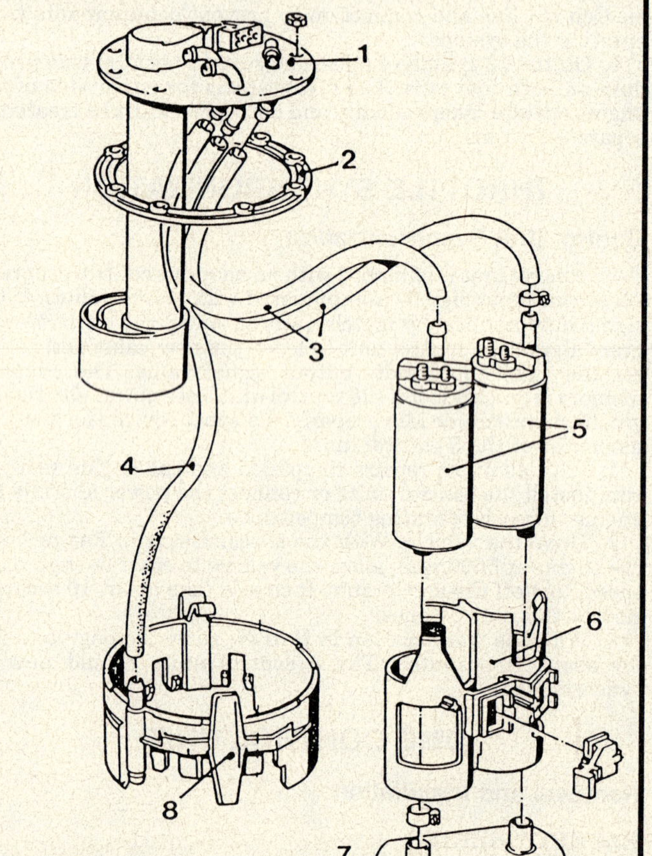

Fuel pump assembly—BMW 750iL

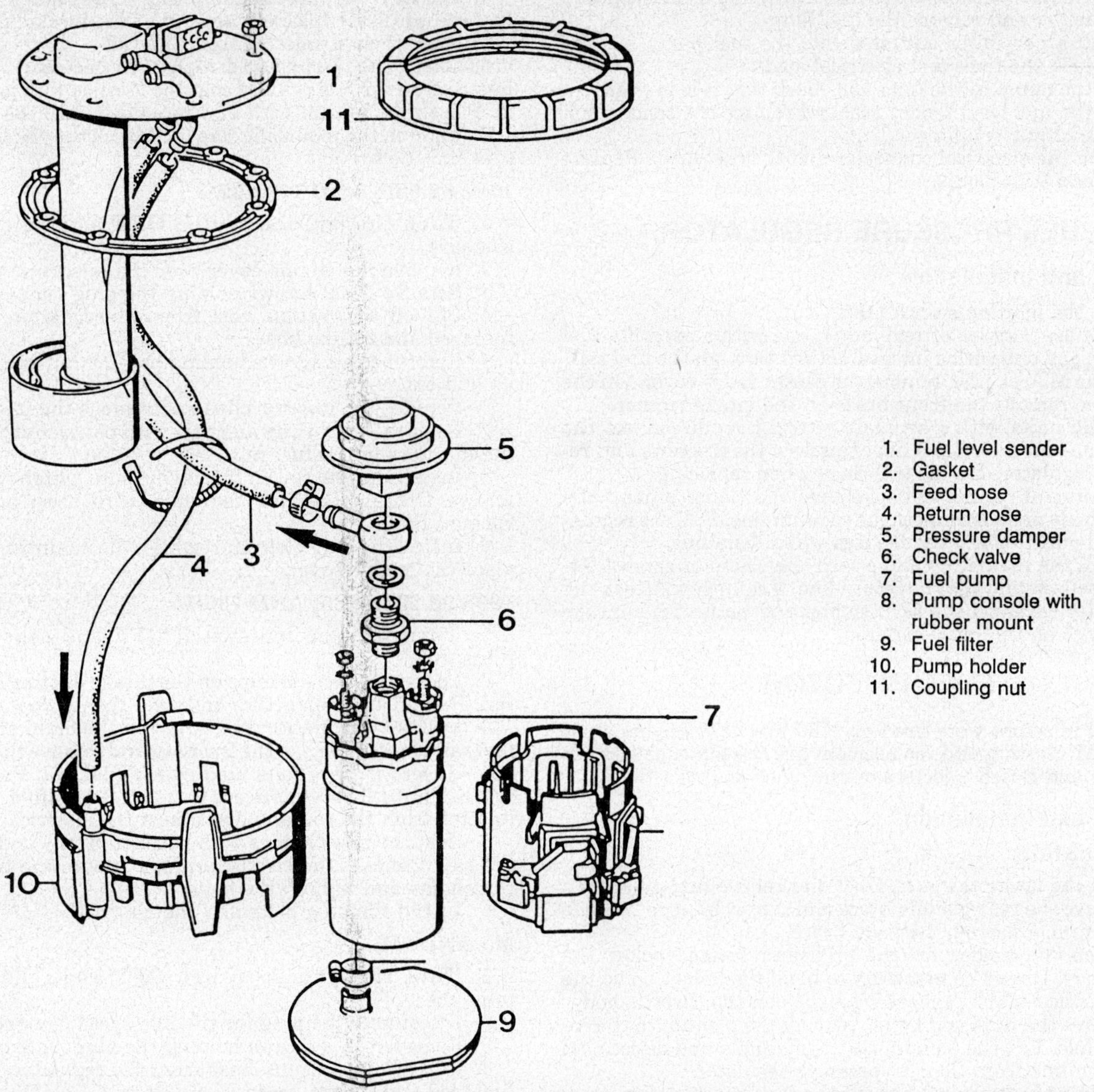

Fuel pump assembly—BMW 5 and 7 Series with 6 cylinder engines

9. Replace the fuel lines and electrical connectors. Replace the tank cover and tighten its screws. Fasten the insulator sheet and replace the rear seat bottom.

525i, 535i, 735i, 750iL AND M5

1. Turn the ignition switch **OFF**. Remove the trunk trim pieces on the right side forward corner. Drain out some fuel from the fuel tank if the tank is full.
2. Unscrew the tank cover and remove the fuel lines. Note the positions of the lines and which connection they go to.
3. Pull the electrical connector slide to the side and pull off the connector.
4. On vehicles with a plastic retaining cover over the pump bracket, use tool 16 1 020 or equivalent, to remove the coupling.
5. On vehicle with the pump bracket retained by 8 screws, remove the screws.
6. Carefully withdraw the fuel level sender from the tank and release the safety hooks holding the pump. Remove the pump from the tank.
7. Remove the lines and wires from the pump. Remove the pressure damper and the check valve. Pull the pump from the holder and remove the fuel filter.
8. Install a new filter and assemble the fuel pump into its holder. Attach the lines and electrical leads.
9. Seat the pump in the tank and check that the safety hooks are in position.
10. Replace the fuel level sender seal and replace the sender. Replace the screws or the plastic coupling, as equipped.
11. Attach the electrical connector and the fuel lines. Replace the cover and trim pieces.

850i

1. Turn the ignition switch **OFF**. Remove the right rear seat. Drain out some fuel from the fuel tank if the tank is full.
2. Unscrew the tank cover and remove the fuel lines. Note the positions of the lines and which connection they go to.
3. Pull off the electrical connector.
4. Note the position of the unit and use tool 16 1 020 or equivalent, to remove the coupling holding the pump bracket.
5. Carefully withdraw the fuel level sender and remove the pump from the tank.

6. Remove the lines and wires from the pump. Pull the pump from the holder and remove the fuel filter.
7. Install a new filter and assemble the fuel pump into its holder. Attach the lines and electrical leads.
8. Seat the pump in the tank and check that it is in position.
9.Replace the fuel level sender seal and replace the sender. Replace the plastic coupling.
10. Attach the electrical connector and the fuel lines. Replace the cover and trim pieces.

FUEL PRESSURE REGULATOR

Removal and Installation

1. Turn the ignition switch **OFF**.
2. There are 2 styles of regulators; one with a separate fuel return line and one with a integral return through the fuel rail.
3. On the M70 engine, remove the center valley cover. On the M42 engine remove the front brace on the intake runner.
4. On the units with a separate return line, disconnect the vacuum line and the return line. Unscrew the 2 screws and remove the regulator. Use new O-rings when replacing.
5. On integral return units, remove the clamp around the regulator body and disconnect the vacuum line. Pull the regulator up and out. Replace the O-rings when installing.
6. Install the regulators and attach the clamp or screws. Attach the vacuum line and the return line, if equipped. Attach the center valley cover on the M70 engine and connect the intake runner brace on the M42 engine.

FUEL INJECTOR

Lucas fuel injectors were used on M30 and M70 engine since March 1991. Testing and replacement are the same, but do not mix Lucas and Bosch injectors on the same engine.

Removal and Installation

318i AND 318iS

1. Turn the ignition switch **OFF** and relieve fuel pressure.
2. Remove the rear plenum brace and the air hose on the side of the plenum. Unscrew the front brace.
3. Loosen the holder for the preheater located below the throttle body. It may be necessary to bend the holder up so the bearing surface of the preheater touches on the throttle body.
4. Remove the nuts and bolts securing the plenum to the intake manifold. Lift the plenum off the manifold and disconnect the vacuum line from the fuel pressure regulator.
5. Pull the plug plate up and off the fuel injectors. Unscrew the line holding clamp.
6. Remove the bolts holding the fuel rail to the engine. Pull up and off the engine keeping the injectors attached to the rail.
7. Pull the injector retainers off and remove the injectors.
8. Replace the O–rings and coat with petroleum jelly. Replace the injectors in the fuel rail.
9. Install the fuel rail on the engine and tighten the mounting screws. Install the line holding clamp. Install the plug plate over the injectors.
10. Replace the plenum gasket and check the condition of the dowel sleeves. Connect the vacuum line for the fuel pressure regulator and seat the plenum on the intake manifold.
11. Tighten the plenum mounting bolts and replace the preheater holder. Install the front and rear braces and the side hose.

1990-91 325i AND 1990 525i

1. Turn the ignition switch **OFF** and relive the fuel pressure.
2. Remove the intake manifold to engine brace and disconnect the temperature sender connections. Disconnect the fuel injector electrical connector.
3. Pull up the plug plate and remove.
4. Unscrew the fuel rail mounting screws and pull up on the fuel rail until the injectors are out of the intake manifold.
5. Remove each injector from the rail.
6. Check the O–rings and replace as necessary. Replace the injectors on the fuel rail and coat the O-rings with petroleum jelly. Place the fuel rail with injectors down into the manifold.
7. Tighten the mounting screws. Attach the electrical connectors and the brace.

1991-92 525i AND 1992 325i

1. Turn the ignition switch **OFF** and relieve the fuel pressure.
2. Remove the engine cover over the injectors.
3. Remove the 2 screws holding the plug plate and remove.
4. Pull off the vacuum hose from the regulator. Remove the fuel feed and return hoses.
5. Remove the 2 screws holding the fuel rail and pull the rail up and out.
6. Remove the injector clips and remove the injectors.
7. Replace the O–rings and coat with petroleum jelly. Replace the injectors in the fuel rail.
8. Install the fuel rail on the engine and tighten the mounting screws. Connect the fuel feed and return hoses and regulator vacuum hose.
9. Install the plug plate and tighten the mounting screws. Replace the engine cover.

1990-92 535i, 735i AND 735iL

1. Turn the ignition switch **OFF** and relieve the fuel pressure.
2. Loosen the hose clamps on the intake ducting over the fuel rail. Pull off the connecting hose and the air duct.
3. Pull out the plastic caps covering the 5mm screws on the fuel rail cover. Unscrew the 2 screws and remove the plug plate.
4. Unscrew the 3 bolts holding the fuel rail. Pull up on the fuel rail until the injectors are free of the manifold. Remove the injector from the rail and disconnect the electrical plug.
5. Coat the new O-rings with petroleum jelly and install onto the rail. Connect the electrical plugs. Install the fuel rail onto the engine and tighten the bolts.
6. Install the plug plate and the air duct.

M3 AND M5

1. Turn the ignition switch **OFF** and relieve the fuel pressure.
2. Disconnect the plug for the idle speed control.
3. Remove the 2 screws holding the plug plate and remove.
4. Pull off the vacuum hose from the regulator. Remove the fuel feed and return hoses.
5. Remove the 2 screws holding the fuel rail and pull the rail up and out.
6. Remove the injector clips and remove the injectors.
7. Replace the O–rings and coat with petroleum jelly. Replace the injectors in the fuel rail.
8. Install the fuel rail on the engine and tighten the mounting screws. Connect the fuel feed and return hoses and regulator vacuum hose.
9. Install the plug plate and tighten the mounting screws. Connect the idle speed control plug.

750iL AND 850i

1. Turn the ignition switch **OFF** and relieve fuel pressure.
2. Remove the engine valley cover. On the 850i remove the fresh air housing cover at the rear of the engine compartment and remove the fresh air housing.
3. Remove the plug from the air temperature sensor. Remove the vacuum hoses from the fuel pressure regulators. Unplug the injector connectors from the injector bodies.
4. Remove the electric lead duct screws. Disconnect the cylinder ID sender plugs and unscrew the fastening hardware. Disconnect the throttle motor plugs.

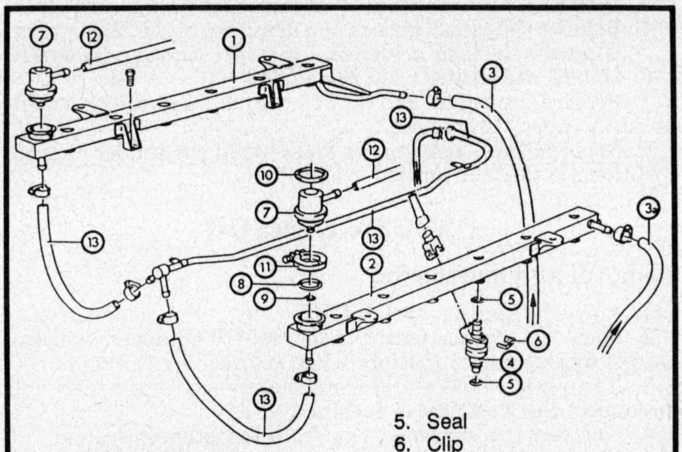

1. Fuel rail—cylinders 1 through 6
2. Fuel rail—cylinders 7 through 12
3. Fuel feed line
3a. Fuel feed line
4. Fuel injector
5. Seal
6. Clip
7. Pressure regulator
8. O-ring
9. O-ring
10. Ring
11. Clamp
12. Vacuum hose
13. Return pipe

Twelve cylinder engine fuel distribution system–BMW

5. Pull up the electric lead duct and remove. Disconnect the fuel hoses from the fuel rail and pull up and remove.

6. Pull the retaining clips from the injectors and remove the injectors.

7. Replace the O-rings and coat with petroleum jelly. Replace the injectors in the fuel rail.

8. Place the fuel rail into position and connect the fuel lines. Install the electric lead duct and connect the throttle motors, cylinder ID sensor plugs and screws, injector plugs and regulator vacuum hoses. Tighten the duct fastening hardware.

9. Connect the air temperature sensor lead and on the 850i, install the fresh air housing and cover. The bottom tab of the housing must engage the opening. Install the valley cover.

COOLANT TEMPERATURE SENDER

Removal and Installation

On 6 cylinder engines, 2 sensors are mounted near the thermostat housing. The blue sensor is for the DME control unit, the brown sensor is for the temperature gauge. The 12 cylinder engine uses 3 sensors. The temperature gauge and the EML control unit each have their own sensor. The DME control units share the same sensor, one terminal connected to each DME control unit with control unit 1 providing the ground connection. All 3 sensors are mounted in the coolant passage at the rear of the engine, under the fresh air housing. On the M42 engine, the coolant temperature sensors are located under the intake manifold on the engine. The forwardmost sensor is the DME sensor and the rear sensor is for the temperature gauge. The coolant temperature sensors on the M3 are located in the coolant pipe above the exhaust side of the engine. The rear sensor services the DME and the forward services the temperature gauge.

1. Turn the ignition switch **OFF**.

2. Unclip the sensor plug retaining spring carefully as not to lose the spring.

3. On the M42 engine it may be necessary to lift slightly on the fuel rail to gain clearance.

4. Unscrew the sender from the coolant passage.

5. Install the sender using a new crush washer. Torque to 15 ft. lbs. (12 Nm).

6. Connect the plug.

AIR TEMPERATURE SENSOR

Removal and Installation

On vehicles equipped with air flow sensors, the air temperature sensor is integral with the air flow sensor and is replaced with the air flow sensor. On air mass sensor equipped vehicles, the sensor is mounted separately. On the M70 engine the sensor is mounted on the back face of the left intake plenum covered by the fresh air housing. On other vehicles, the sensor is mounted on the air cleaner housing or on the intake manifold.

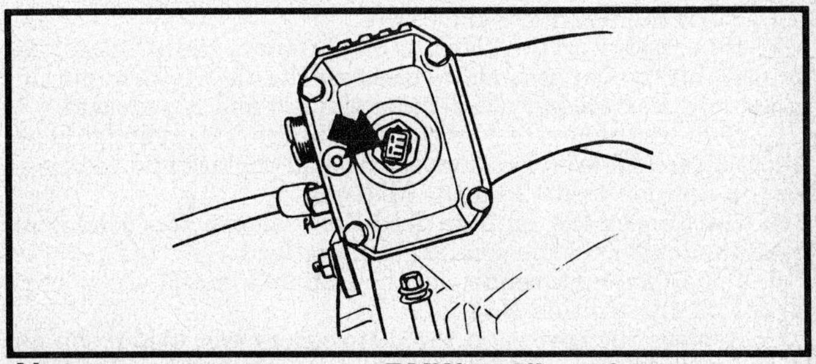

Air temperature sensor—BMW 750iL and 850i

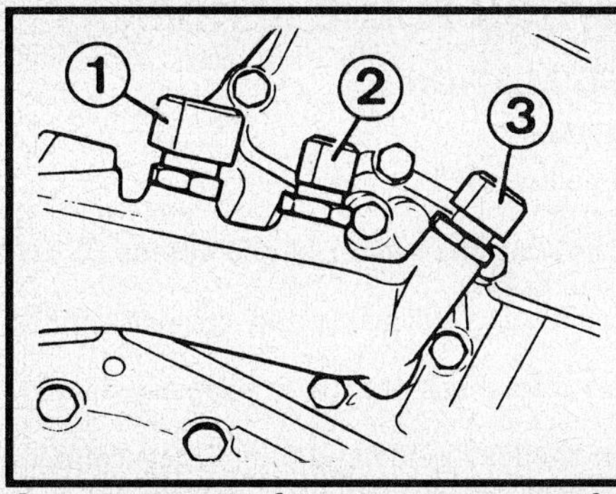

1. DME temperature sender
2. EML temperature sender
3. Gauge temperature sender

Arrangement of temperature senders—BMW 750iL and 850i

IDLE SPEED CONTROL VALVE

Removal and Installation

EXCEPT 318i AND 318iS

1. Turn the ignition switch **OFF**.

2. Disconnect the electrical connector from the body of the valve.

3. Compress the retaining strap and pull the valve free.

4. Disconnect the hoses from the valve and remove.

5. Connect the hoses and place into the retaining strap. Connect the electrical plug.

318i AND 318iS

1. Turn the ignition switch **OFF**.

2. Disconnect the electrical connector from the body of the valve.

3. Remove the nut securing the valve.

4. Pull the valve out and disconnect the hoses.

5. Replace the valve and connect the hoses. Tighten the securing nut and connect the electrical connector.

CYLINDER IDENTIFICATION SENDER

Removal and Installation

318i, 318iS, 1992 325i AND 1991-92 525i

1. Turn the ignition switch **OFF**.

2. Unplug the upper of the 2 connectors on the electric lead duct.
3. Unscrew the bolt holding the sender to the engine. Withdraw the sender from the head.
4. Unclip the lead from the lead guide and remove sender from the vehicle.
5. Replace the sender and check that the O-ring is properly seated. Tighten the bolt.
6. Replace the lead in the lead guide and plug in the connector.

EXCEPT 318i, 318iS, 1992 325i AND 1991-92 525i

1. Turn the ignition switch **OFF**.
2. If necessary on the 750iL or 850i, remove the air mass meters and air cleaner assemblies to access the distributors. On the M5 it may be necessary to remove the fan and fan shroud.
3. Remove the covers over the distributor. The cylinder identification sender is on ignition lead 6 on all engines and additionally on ignition lead 12 on the M70 engine.
4. Disconnect the ignition lead and remove the connector from the lead. Pull the sender from the lead.
5. Install the new sender on the lead and install a new connector on the ignition lead.
6. Replace the covers on the distributors and install the air cleaners.

SPEED/POSITION PULSE SENDER

REMOVAL AND INSTALLATION

EXCEPT 750iL AND 850i

1. Turn the ignition switch **OFF**.
2. Unplug the connector.
3. Unscrew the bolt holding the sender to the engine. Withdraw the sender from the holder.
4. Unclip the lead from the lead guide and remove sender from the vehicle.
5. Replace the sender and check that it is properly seated. Tighten the bolt.
6. Replace the lead in the lead guide and plug in the connector. Make sure that the lead can not rub on the drive belt.

750iL AND 850i

Left Cylinder Bank

1. Turn the ignition switch **OFF**.
2. Remove the fan from the engine.
3. Disconnect the sensor plug on the upper timing case cover.
4. Remove the splash guards to gain access to the sensor area.
5. Unscrew the lead holder on the upper timing case cover.
6. Unbolt the pulley for the power steering pump. Unbolt the cooling air duct for the alternator. Unbolt the holders for the hydraulic pipes. Unbolt the hydraulic pump holder from the engine block.
7. Loosen the power steering pump console bolts. Unbolt the pipe holder on the lower timing case cover.
8. Turn the engine over so the sender bolt is accessible through the gap on the increment wheel. Unbolt the sender and push aside the hydraulic pump to remove the sender.
9. Bolt the sender back into position. Bolt the pipe holder on the lower timing case cover. Tighten the power steering console bolts.
10. Tighten the bolts holding the hydraulic pump and the hydraulic pipes. Bolt in the alternator cooling air duct and the power steering pump pulley.
11. Install the lead holder on the upper timing case cover. Replace the splash shields. Connect the sensor plug. Install the fan.

Right Cylinder Bank

1. Turn the ignition switch **OFF**.
2. Remove the fan and belt tensioner from the engine.
3. Disconnect the sensor plug on the upper timing case cover.
4. Remove the splash guards to gain access to the sensor area.
5. Unscrew the lead holder on the upper timing case cover.
6. Unbolt the sender from the engine.
7. Replace the sender and tighten the bolt. Bolt the lead holder on the upper timing case.
8. Replace the splash guards and plug in the sender connector. Replace the belt tensioner and the fan.

OXYGEN SENSOR

Removal and Installation

1. Turn the ignition switch **OFF**.
2. Mark the oxygen sensor leads on M70 equipped vehicles. Do not mix-up the connectors when reattaching the leads.
3. Remove the heat shield from the oxygen sensor body and disconnect the electrical connector.
4. Unscrew the oxygen sensor from the exhaust system.
5. Coat the threads of the oxygen sensor with anti-seize. Do not clean or lubricate the sensor portion of the unit. Cover the unit when undercoating.
6. Check the inside threads of the oxygen sensor port. Clean any carbon build-up.
7. Thread in the oxygen sensor by hand. Torque the unit to 40 ft. lbs. (55 Nm).
8. Connect the oxygen sensor electrical connector. Make sure to connect the correct connectors on M70 equipped vehicles.

THROTTLE BODY

Removal and Installation

EXCEPT M3 AND M5

1. Turn the ignition switch **OFF**.
2. Loosen the hose clamp securing the air duct to the throttle body. Disconnect the air duct and push aside or remove.
3. On EML throttle by wire equipped vehicles, disconnect the throttle electrical connector.
4. On non-EML vehicles, disconnect the throttle cables and the throttle switch connector. On the M42 engine, remove the vacuum hoses from the throttle body.
5. Remove the 4 bolts (M42) or 5 bolts (others) securing the throttle body. Remove the throttle body.
6. The throttle body is not adjustable and must be replaced if found defective.
7. Install a new gasket and install the throttle body. Connect the vacuum lines, throttle cables and electrical connections. Install the air duct.
8. On EML equipped vehicles run the external safety path check.

M3

1. Turn the ignition switch **OFF**.
2. Disconnect the throttle cable. Loosen and remove the hose connections to the air box. Remove the capped nuts on the throttle bodies and remove the air horns/air box assembly. Cut off the hose between the engine and air box.
3. Disconnect the throttle valve switch plug and pull off the plug plate from the injectors. Remove the vacuum hose from the fuel pressure regulator. Remove the fuel rail bolts and lift out the fuel rail.
4. Remove the nut on the throttle rod joint. Remove the nuts securing the throttle bodies and remove the throttle bodies.
5. The throttle bodies are preset and marked with paint dots. Do not disturb the settings or throttle rods.
6. Install the throttle bodies and tighten the mounting nuts. Replace the self-locking nut on the throttle rod joint.
7. Replace the injectors and fuel rail. Tighten the mounting bolts and attach the vacuum hose to the fuel pressure regulator.
8. Replace the plug plate on the injectors and connect the throttle valve switch plug.

9. Check the O-rings and replace as necessary. Replace the cut hose from the air box to engine. Mount the air box assembly and tighten the mounting nuts. Connect the air ducts and hoses. Connect the throttle cable.

M5

1. Turn the ignition switch **OFF**.
2. Disconnect the throttle cable. Loosen and remove the intake hose connections to the air box. Disconnect the idle hose, the vent hose and the oil return hose.
3. Unscrew the oil trap nuts and pull off the manifold. Note the routing of the vacuum hoses.
4. Remove the plug plate from the injectors. Disconnect the fuel hoses and unbolt the fuel rail. Unplug the intermediate connector for the idle control valve and remove the fuel rail with injector. Pull off the vacuum hoses noting the routing.
5. Remove the (6) 10 mm nuts and the (9) 13 mm nuts. Unbolt the throttle rod connecting bolt and disconnect the throttle linkages. Remove the throttle bodies. The throttle bodies are preset and marked with paint dots. Do not disturb the settings or throttle rods.
6. Disconnect the connecting pipes and inspect the O–rings. Replace as necessary.
7. Replace the intake gaskets and install the throttle bodies. Make sure the connecting pipes are in place. Connect the throttle linkages and the tighten connecting bolt.
8. Tighten the 9 13 mm nuts and the 6 10 mm nuts. Route the vacuum hoses as originally placed. Install the fuel rail and injectors. Connect the intermediate plug for the idle speed valve and connect the fuel hoses. Tighten the fuel rail bolts.
9. Replace the plug plate on the fuel injectors. Check the condition of the O–rings and that the pins are in the support before mounting the manifold.
10. Install the manifold and tighten the nuts for the oil trap. Connect the hoses for the oil return, the idle control and the vent. Install the air intake hose and tighten the air box mounting screws. Connect the throttle cables.

ACCELERATOR PEDAL VALUE SENDER

Removal and Installation

1. Turn the ignition switch **OFF**.
2. Remove the lower dash trim from around the accelerator pedal location.
3. Unscrew the nut from the sender shaft and pull off the pedal lever.
4. Remove the 2 bolts from the sender and disconnect the electrical connector.
5. Install the sender, tighten the bolts and connect the electrical connector.
6. Install the pedal lever on the sensor shaft and tighten the nut.
7. Adjust the sensor. Run the external safety path check.

THROTTLE POSITION SWITCH

Removal and Installation

1. Turn the ignition switch **OFF**.
2. Unplug the connector from the switch body.
3. Remove the 2 screws from the switch body and pull the switch off the throttle body.
4. Replace the switch and tighten the 2 screws. Connect the electrical connector. Adjust the switch.

IDLE SPEED AND CO

Checking and Adjustment

The idle speed and CO are not adjustable on all BMW vehicles, except for the M3 and M5. The idle speed and CO level can be checked, but not adjusted.

M3 AND M5

1. Warm the engine to operating temperature. The air filter should be clean, the valve in good adjustment, the entire fuel injection system in good physical and operating condition and the oxygen sensor in good operating condition.
2. Insert the exhaust emission tester into the exhaust testing ports located in the exhaust system before the catalytic converter.
3. Switch off all electric accessories.
4. Remove the plug from the throttle switch and insert tool 13 4 010 or equivalent, into the wiring connector.
5. Check the idle speed. If the idle speed needs adjustment, turn the idle speed screw located just above the 'M' on the valve cover. Idle speed on the M3 should be 830–930 rpm and the idle speed on the M5 should be 930–1010 rpm.
6. With the oxygen sensor connected, measure the CO level. Disconnect the oxygen sensor and measure the CO level.
7. If the level measured is not within 0.4–1.2% CO, remove the cap on the air flow sensor with tool 13 1 011 or equivalent. Adjust the CO level with tool 13 1 100 or equivalent, to the level measured with the oxygen sensor connected.
8. Connect the oxygen sensor.
9. Check the synchronization of the throttle by measuring the vacuum at the throttle ports. Test each throttle separately and keep the caps on each port that is not being tested at the moment. Adjust each throttle bore 7 mm screw so that the vacuum at each throttle is within 0.15 in. Hg (5 mbar) of each other.
10. Check and correct the idle speed and CO level and install new anti-tamper plugs on the adjustment points.

IDLE SPEED AND CO LEVELS

Model	Idle Speed in rpm	CO Level in % volume before catalyst
318i	810–890	0.2–1.2
318is	810–890	0.2–1.2
318iC	810–890	0.2–1.2
325i (1990–91)	720–800	0.2–1.2
325i (1992)	660–740	0.2–1.2
325iC	720–800	0.2–1.2
525i (1990–91)	660–740	0.2–1.2
525i (1992)	660–740	0.2–1.2
535i	750–850	0.2–1.2
735i	750–850	0.2–1.2
735iL	750–850	0.2–1.2
750iL	750–850	0.2–1.2
850i	650–750	0.2–1.2
M3	830–920	0.4–1.2
M5	930–1010	0.4–1.2

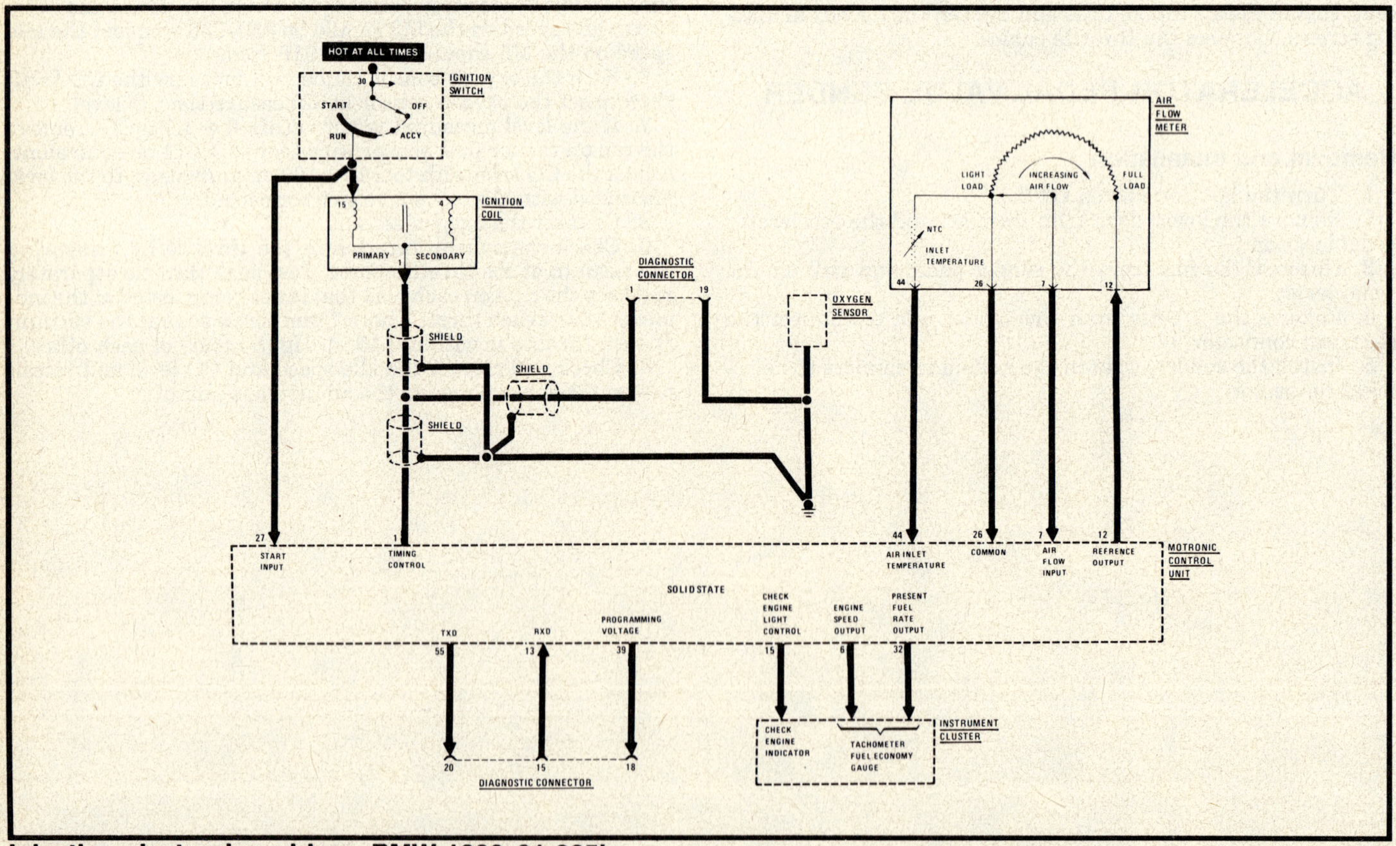

Injection electronics wiring—BMW 1990–91 325i

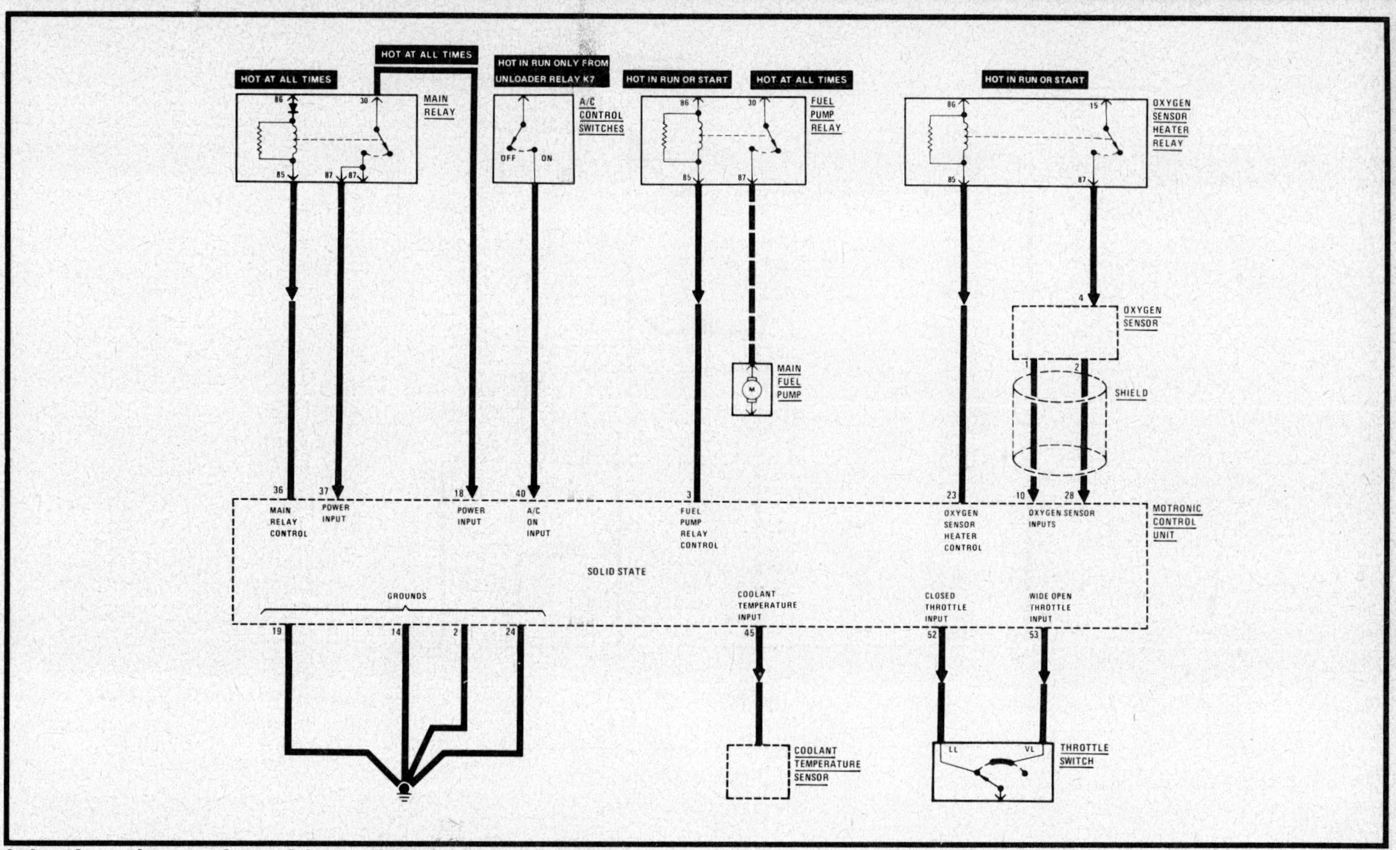

Injection electronics wiring – BMW 1990–91 325i

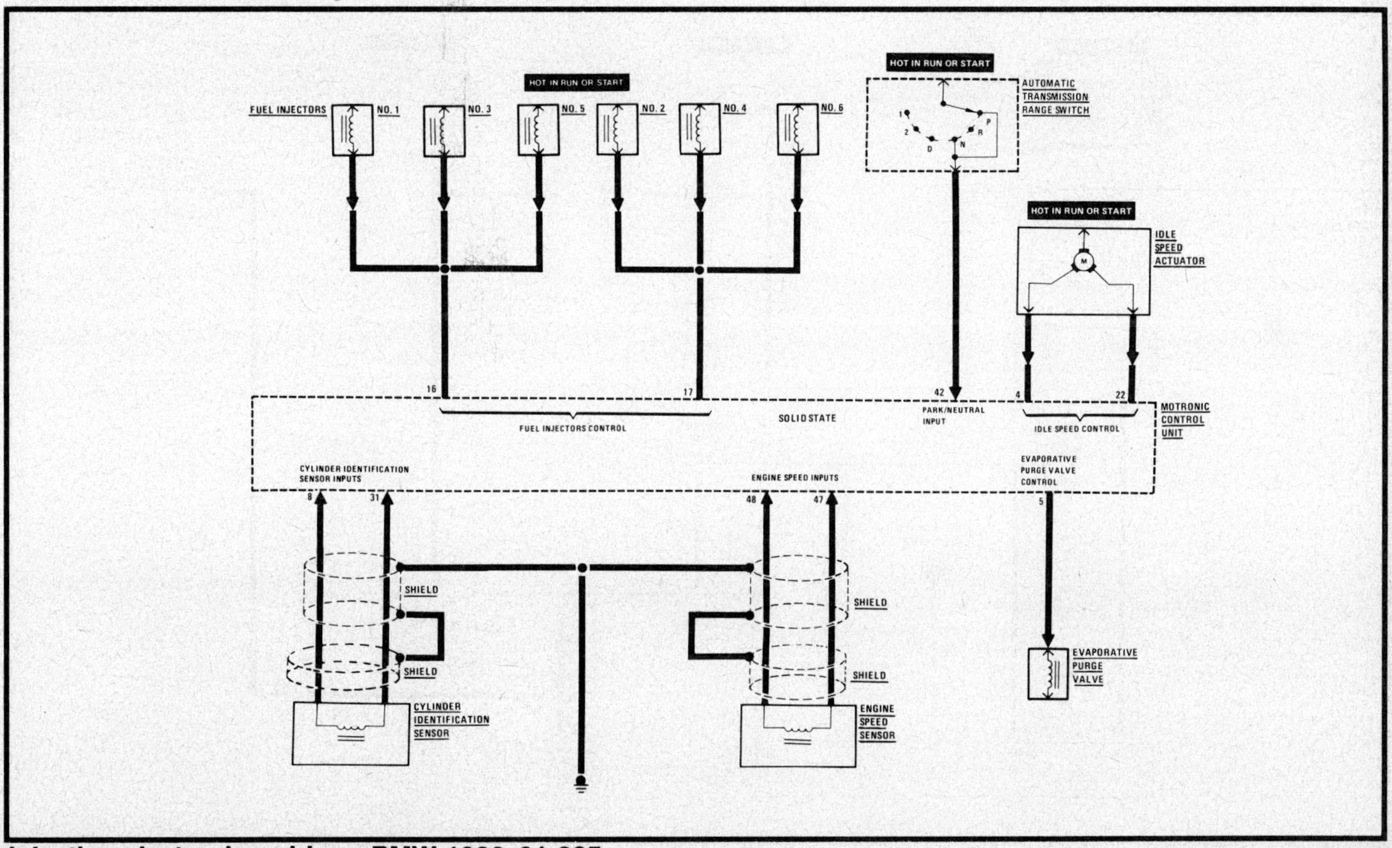

Injection electronics wiring – BMW 1990–91 325

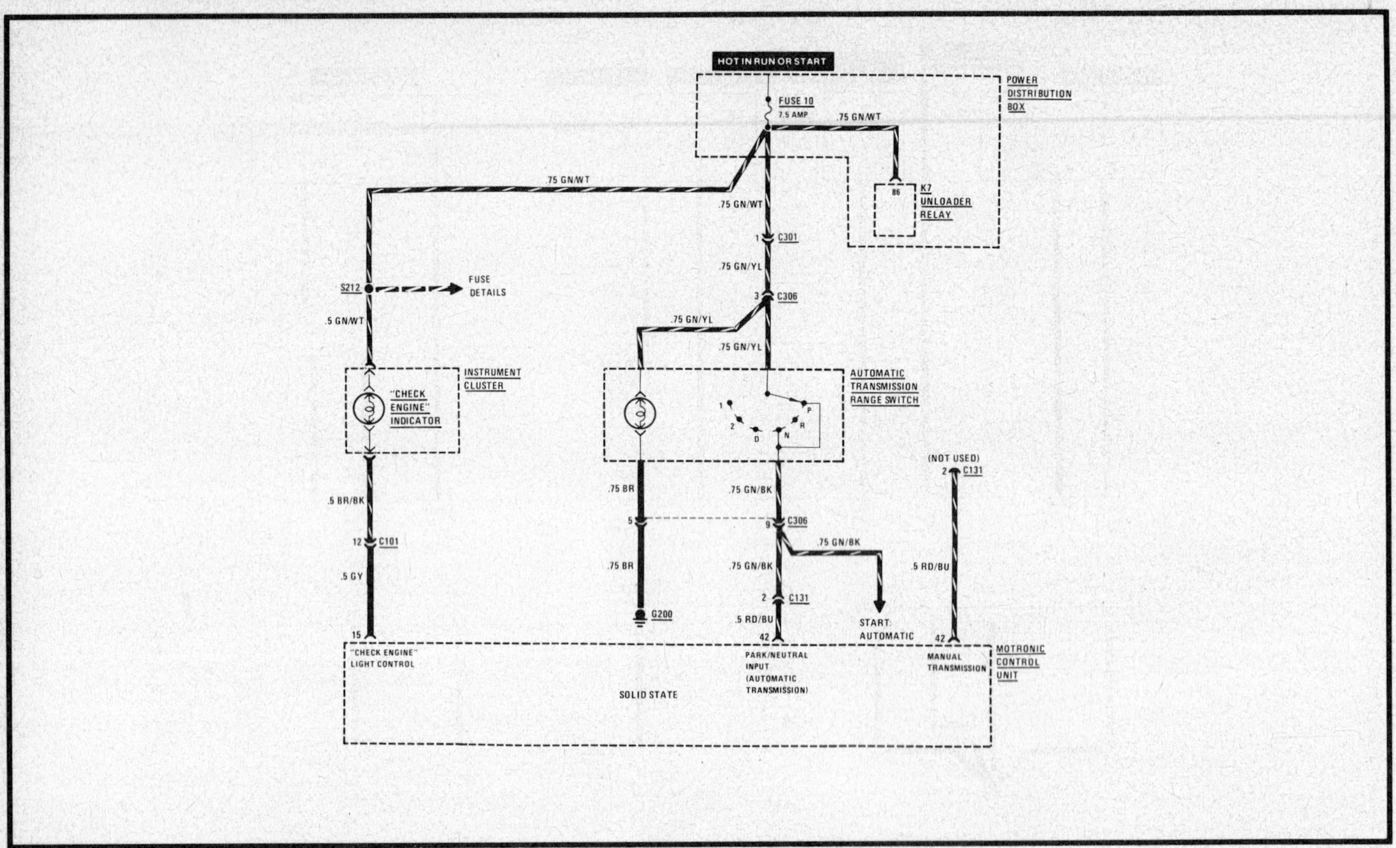

Injection electronics wiring—BMW 1990–91 325i

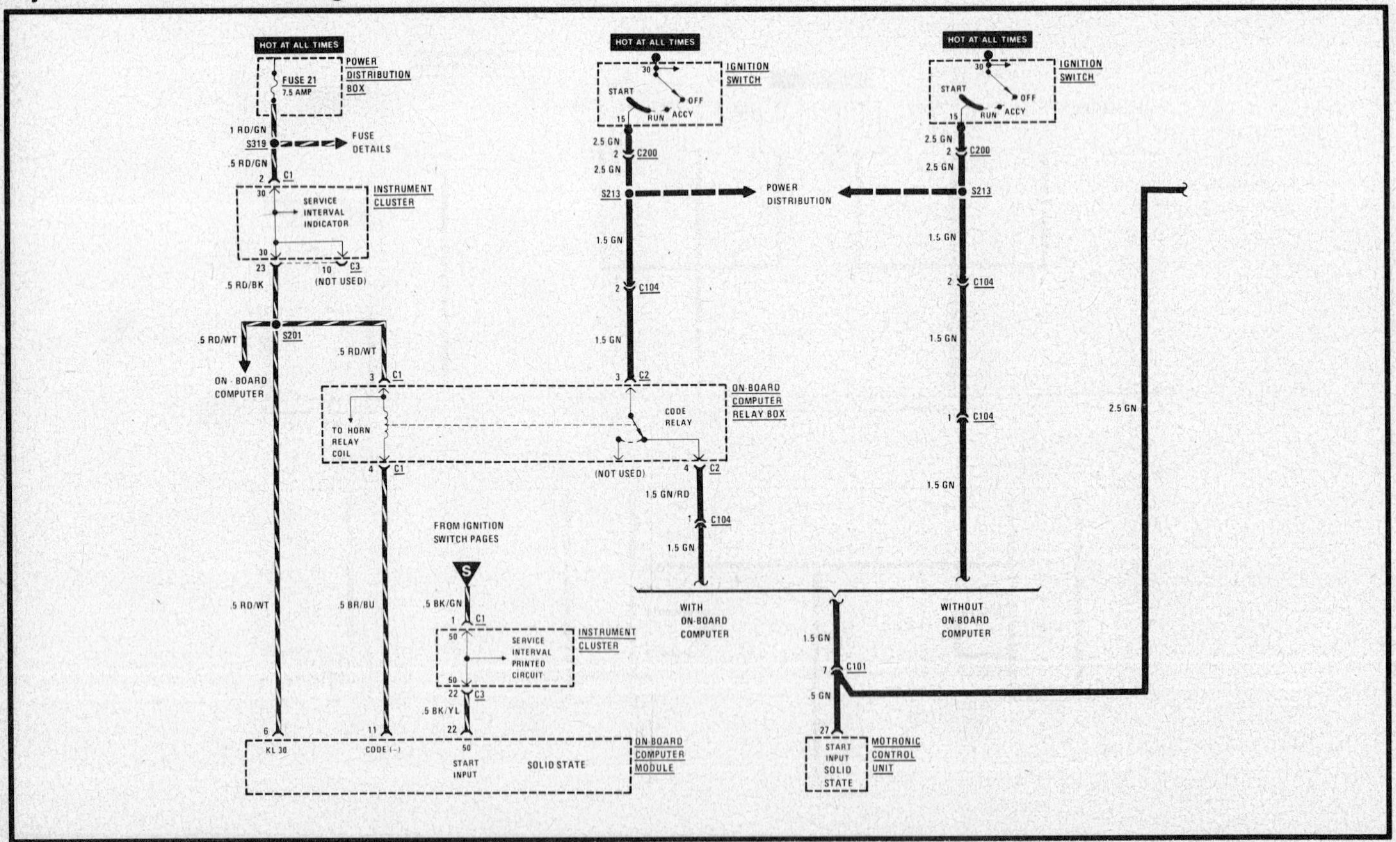

Injection electronics wiring—BMW 1990–91 325i

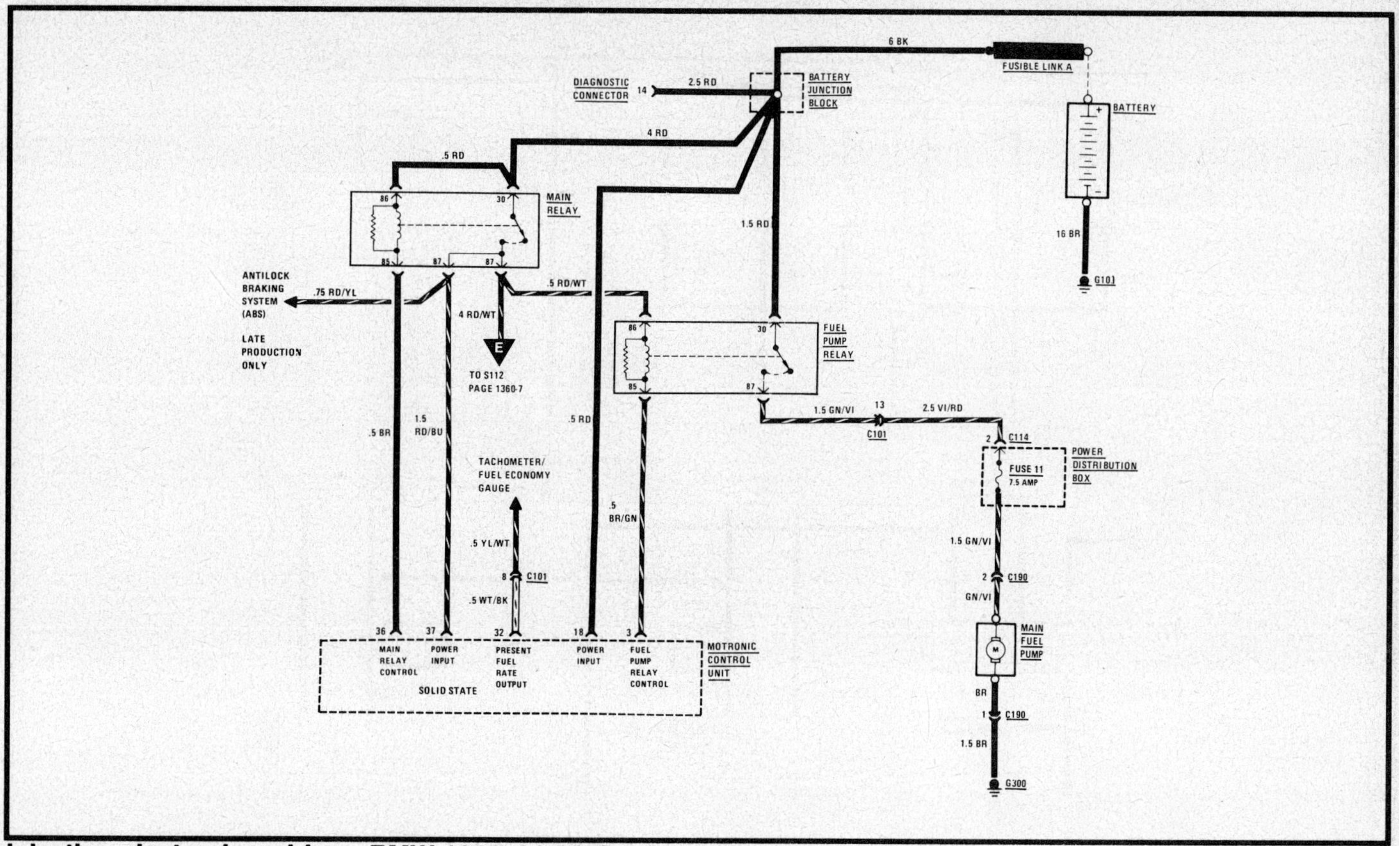

Injection electronics wiring—BMW 1990–91 325i

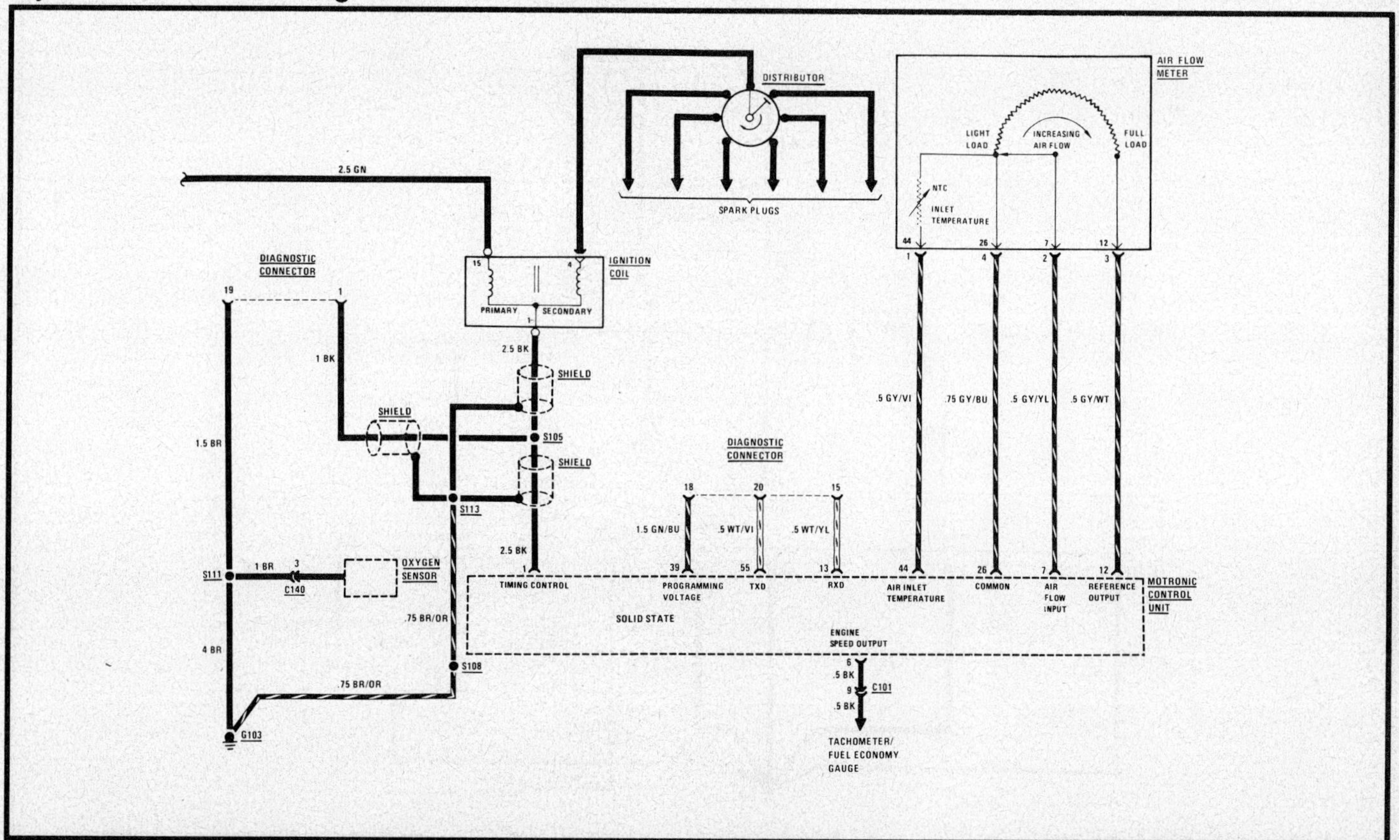

Injection electronics wiring—BMW 1990–91 325i

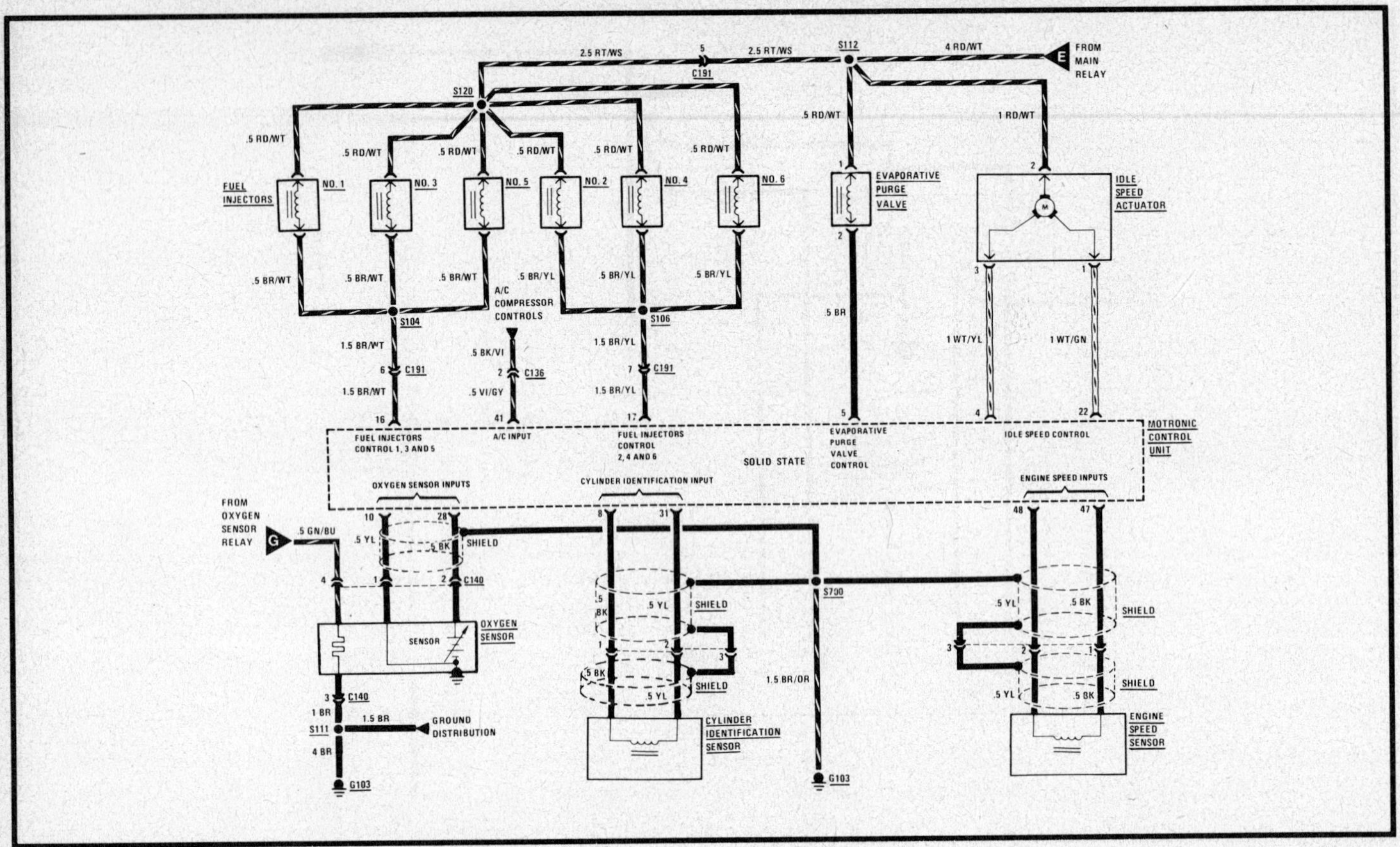

Injection electronics wiring—BMW 1990–91 325i

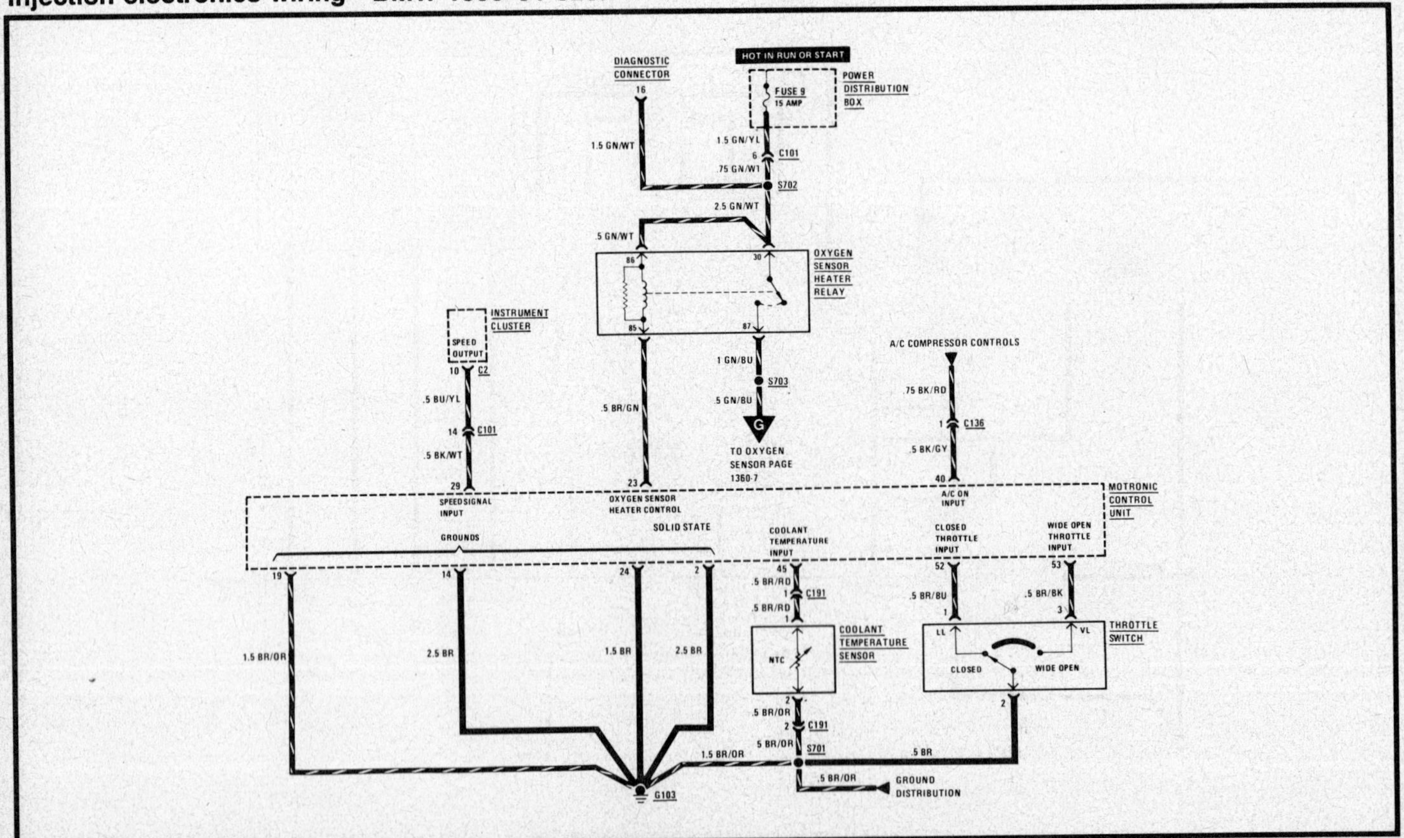

Injection electronics wiring—BMW 1990–91 325i

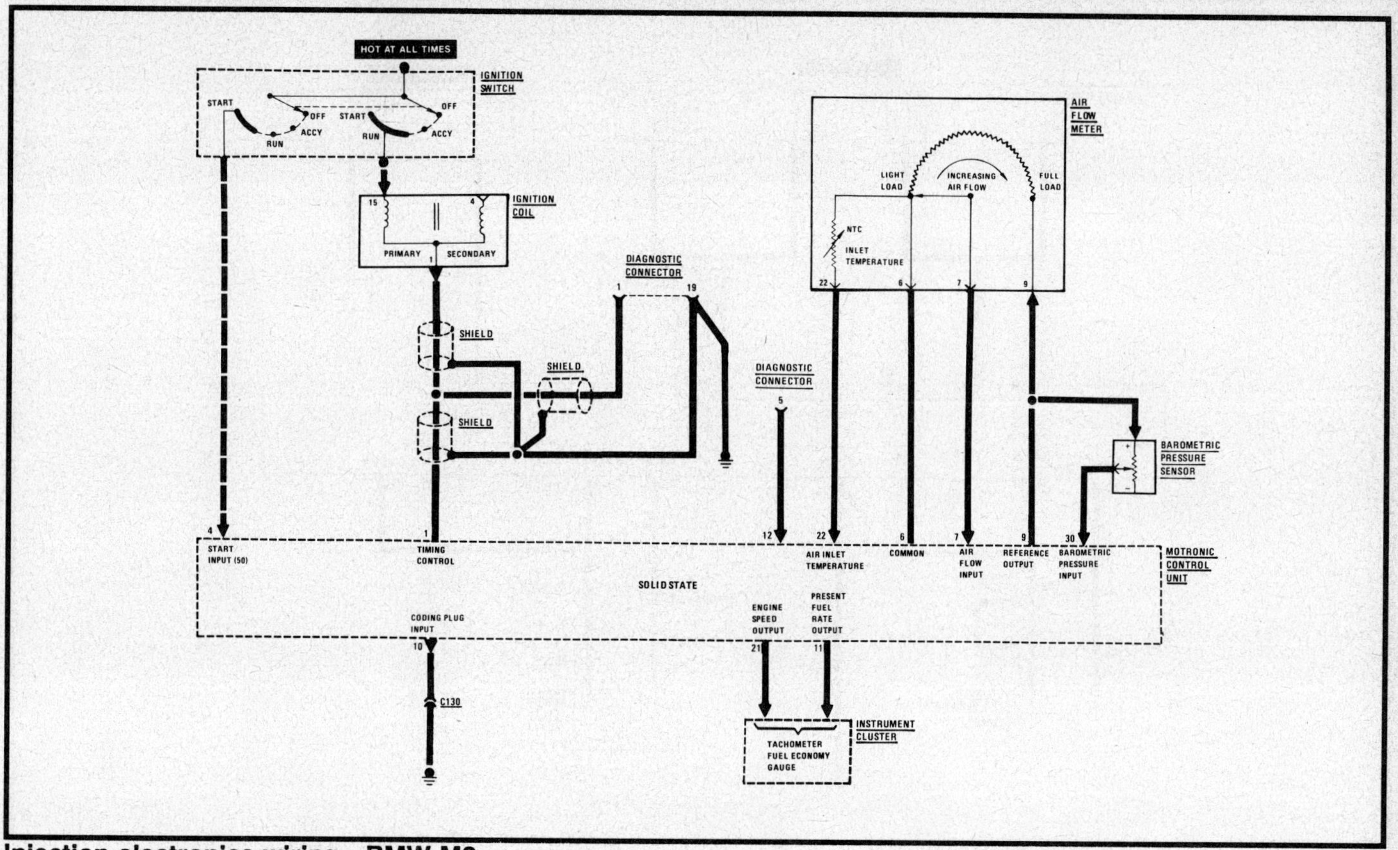

Injection electronics wiring—BMW M3

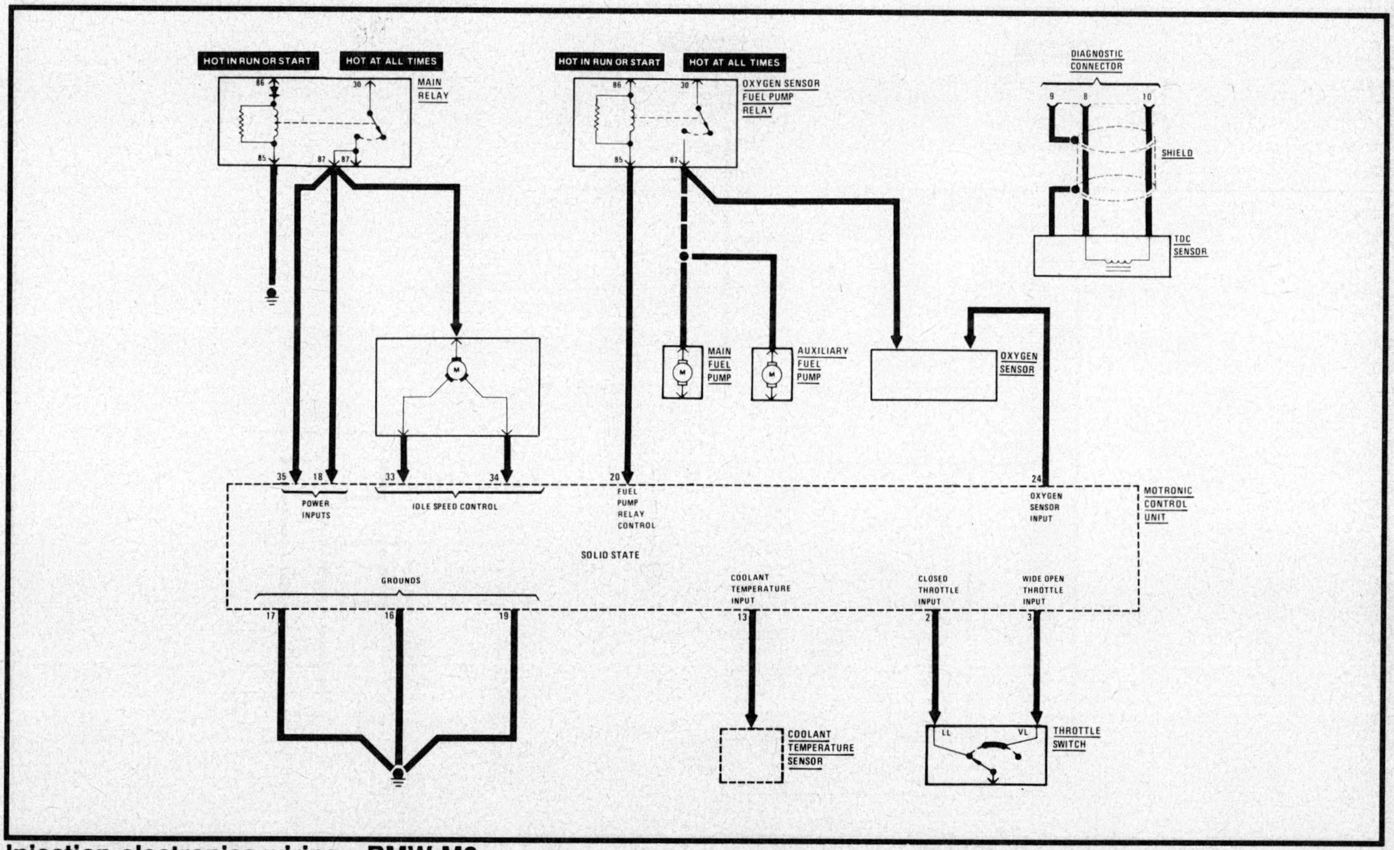

Injection electronics wiring—BMW M3

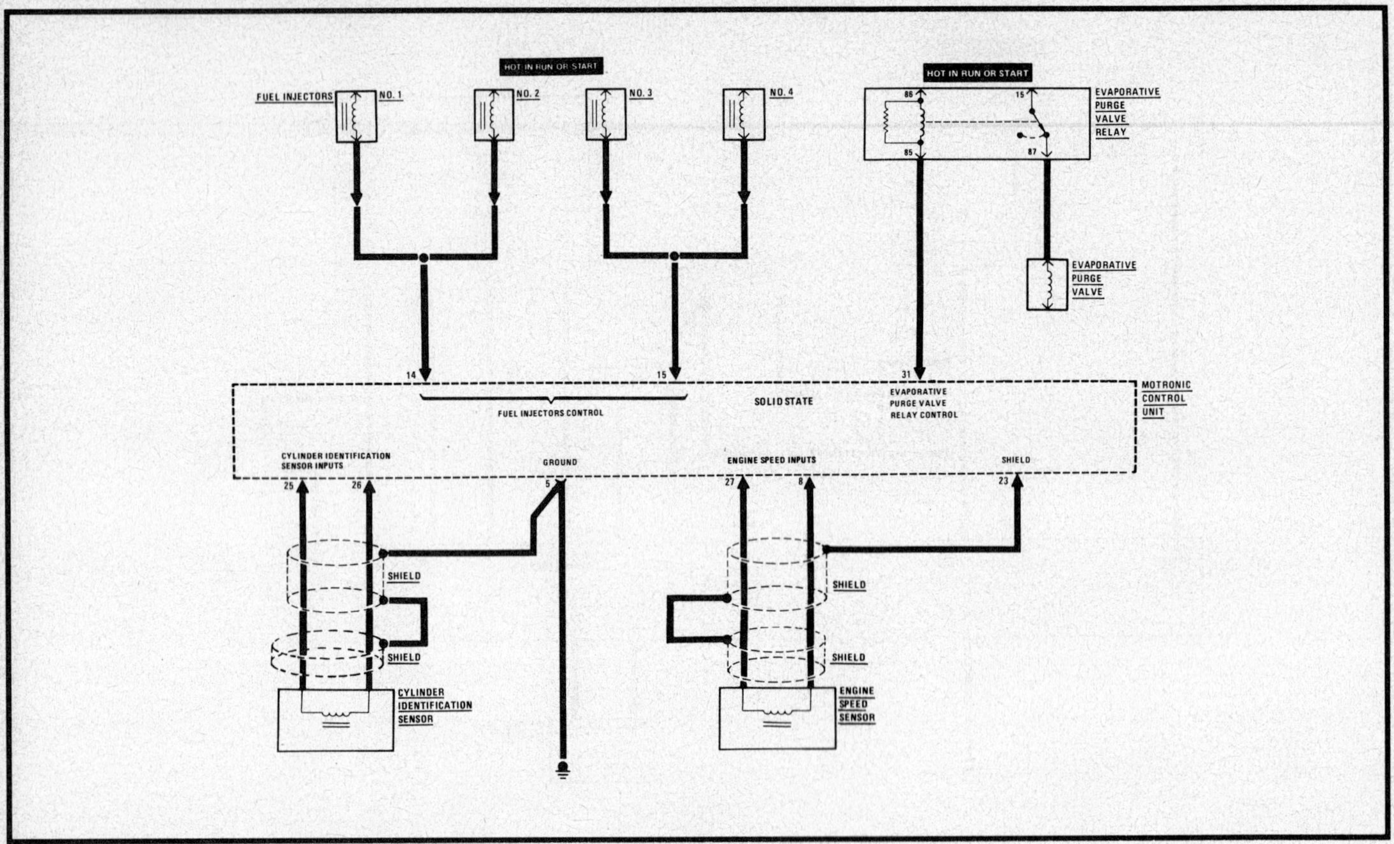

Injection electronics wiring—BMW M3

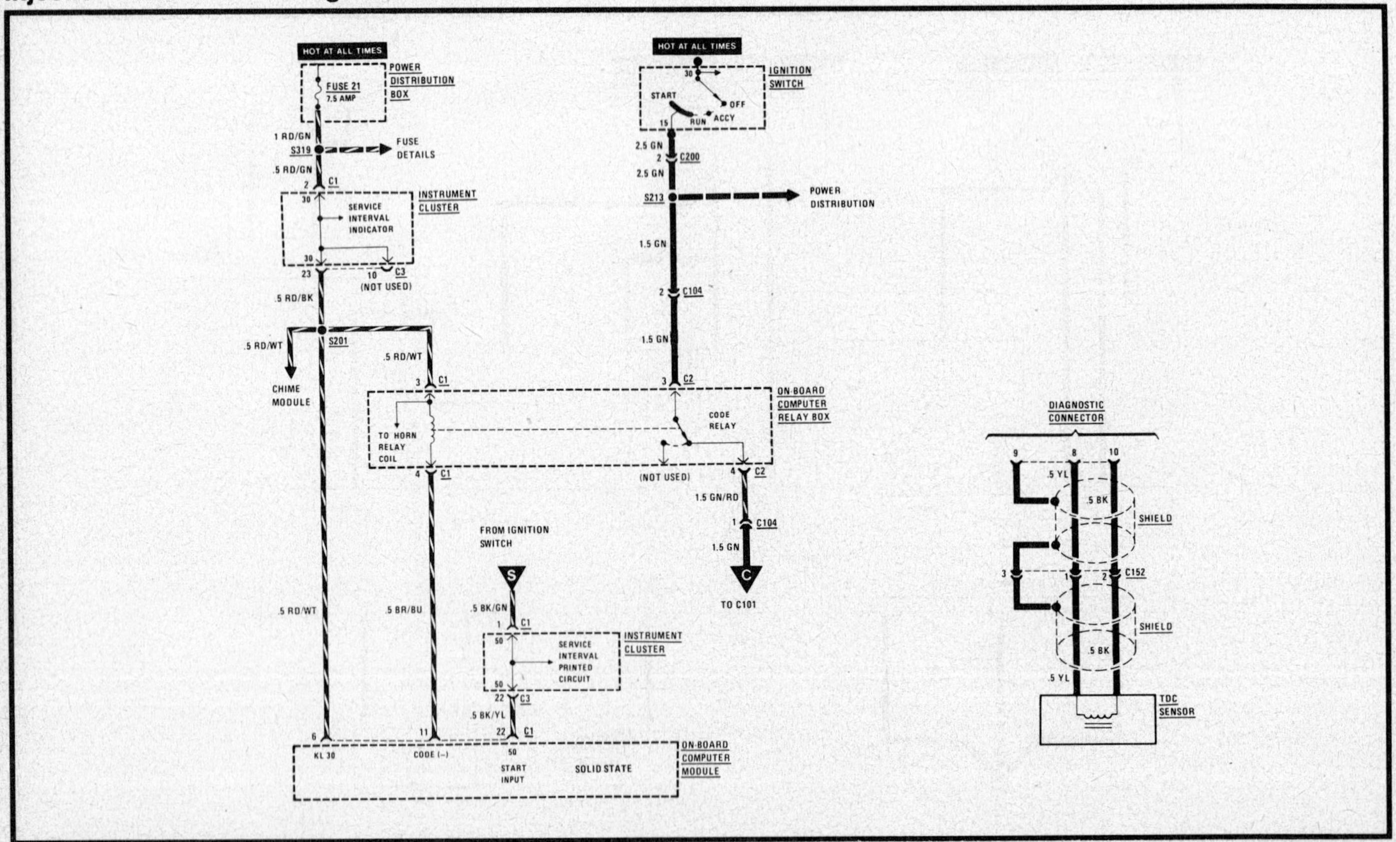

Injection electronics wiring—BMW M3

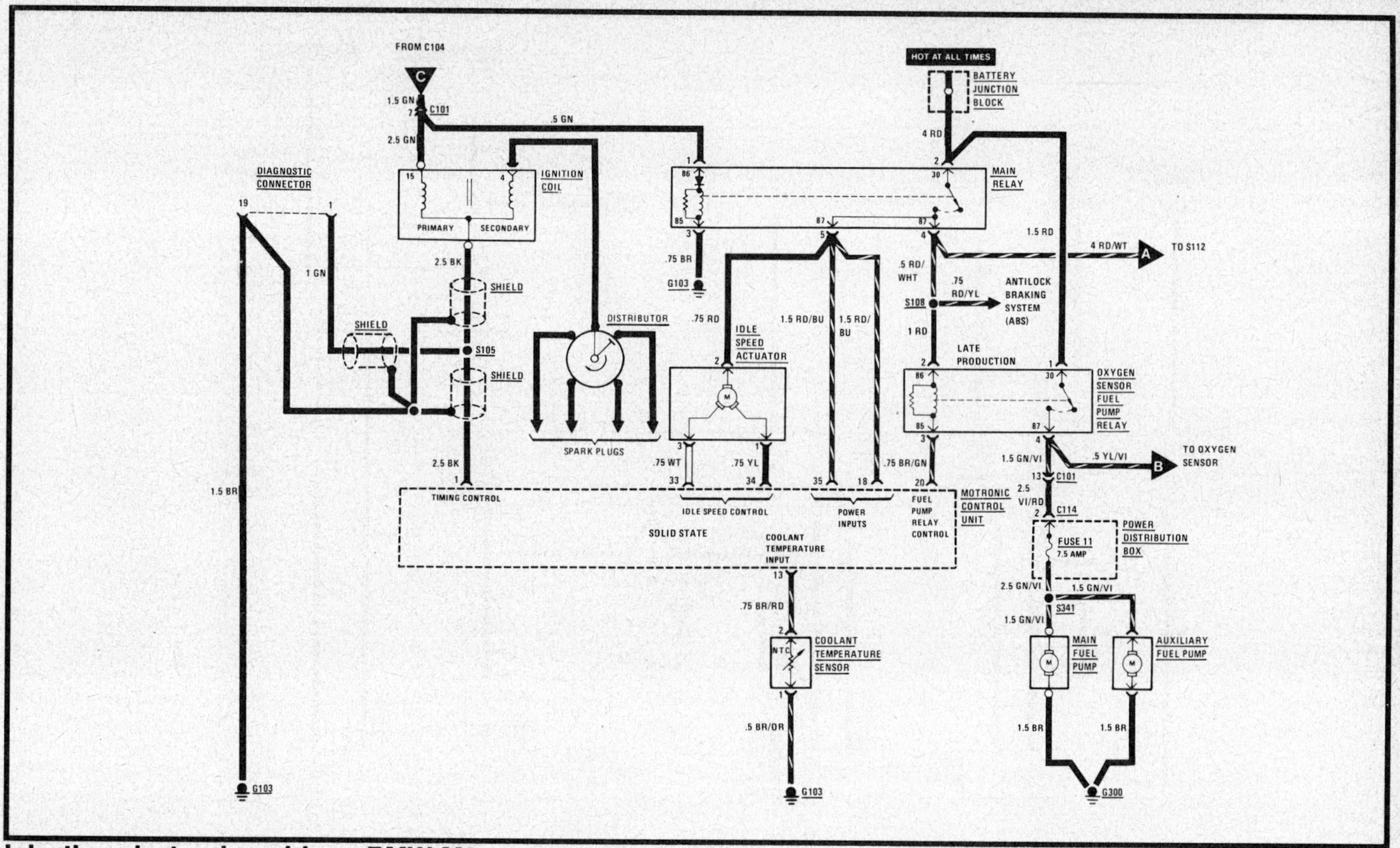

Injection electronics wiring—BMW M3

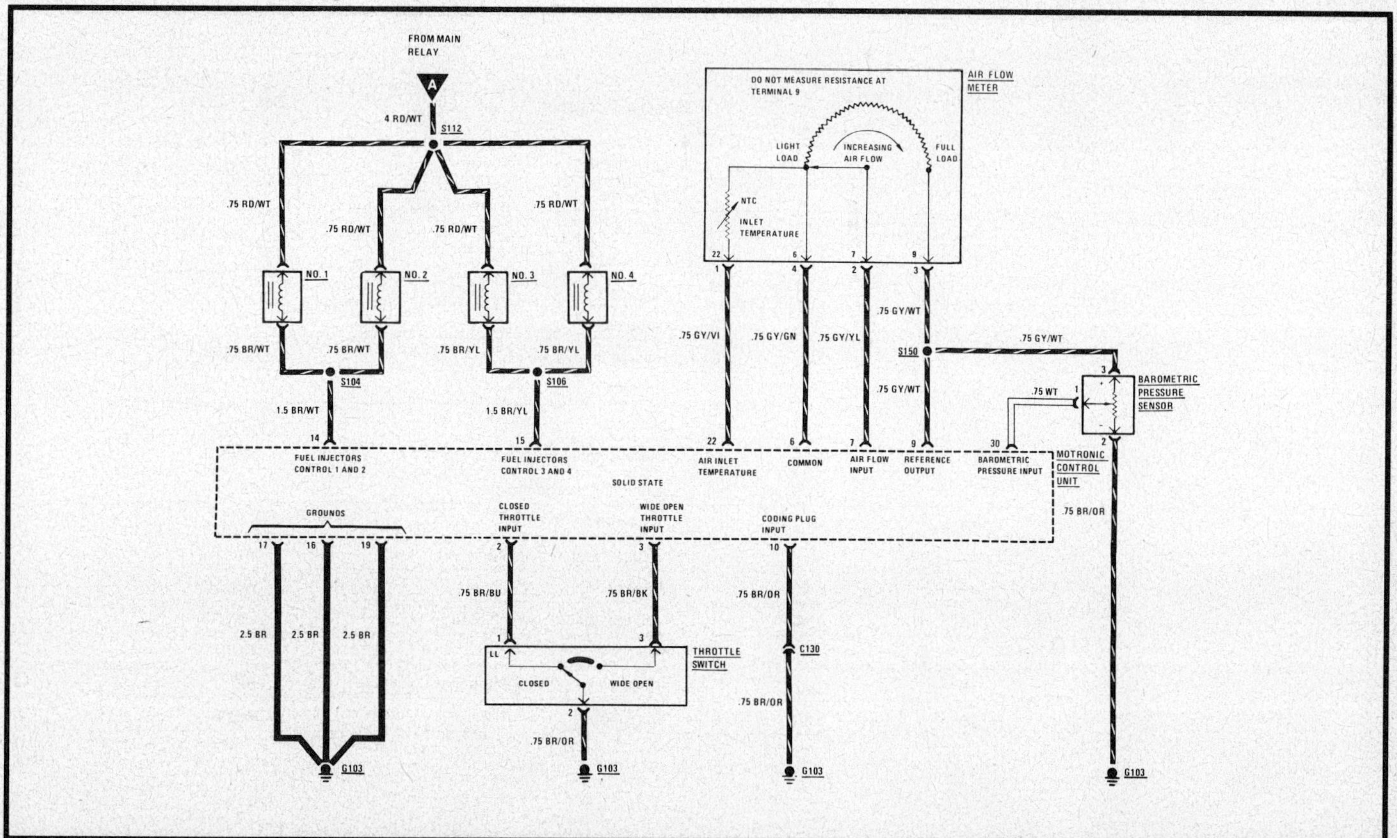

Injection electronics wiring—BMW M3

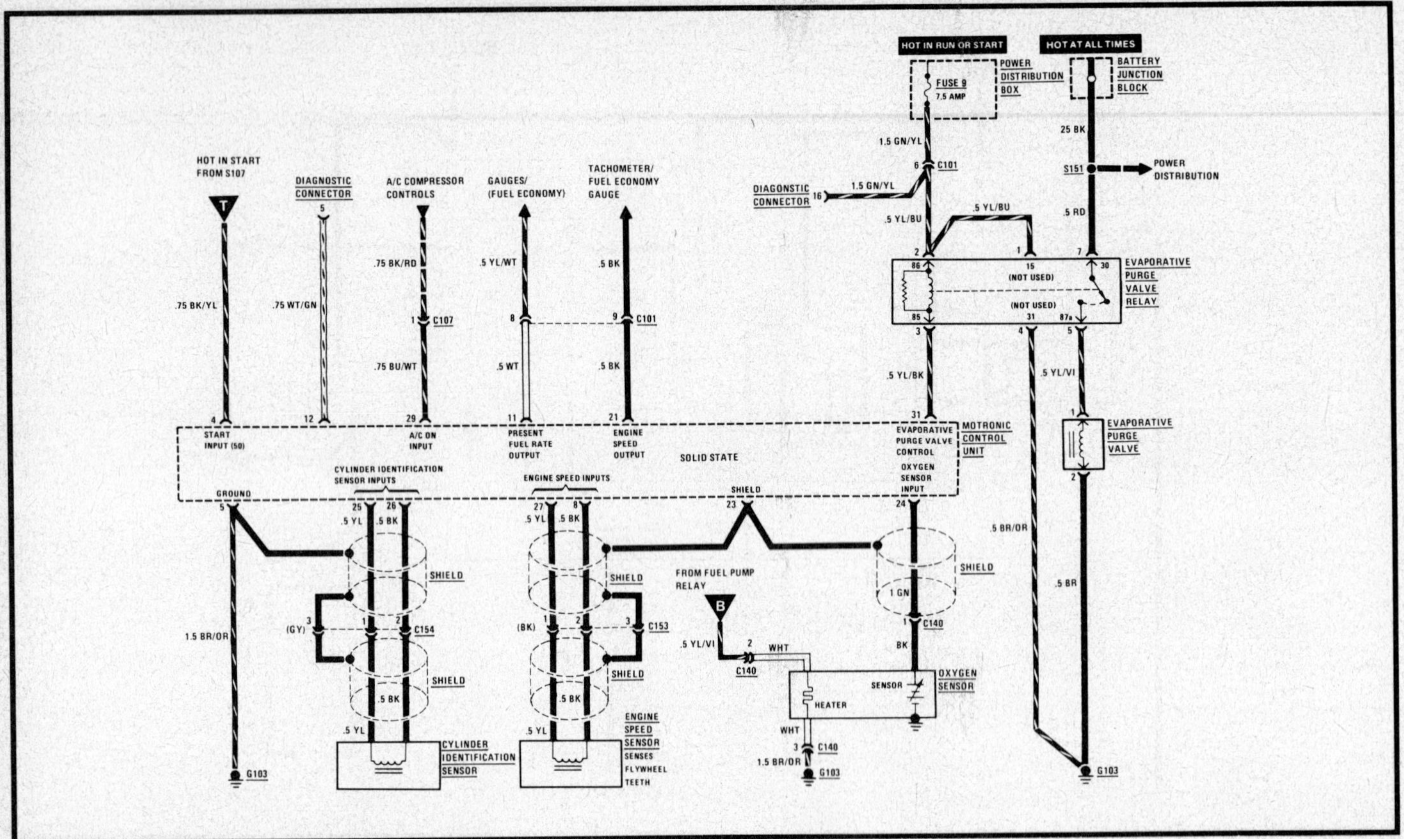

Injection electronics wiring—BMW M3

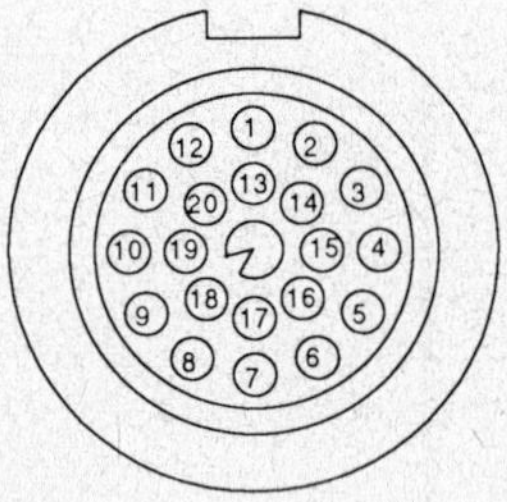

Pin number	Terminal number	Terminal description	Wire color
1	1	Ignition coil	SW
2	–	Not used	
3	–	Not used	
4	–	Not used	
5	–	Not used	
6	–	Not used	
7	SI	Service interval reset	WS/GN
8	–	Not used	
10	–	Not used	
11	50	Voltage: Ignition switch in start	SW/GE
12	61 (D+)	Alternator charge indicator	BL
13	–	Not used	
14	30	Voltage: Hot at all times	RT
15	RXD	Receive	WS/GE
16	15S	Voltage: Ignition switch in run	GN/WS
17	–	Not used	GN/BL
18	–	Program select for DME control unit	BR
20	TXD	Transmit	WS/VI

Diagnostic link pinout—BMW 5 Series

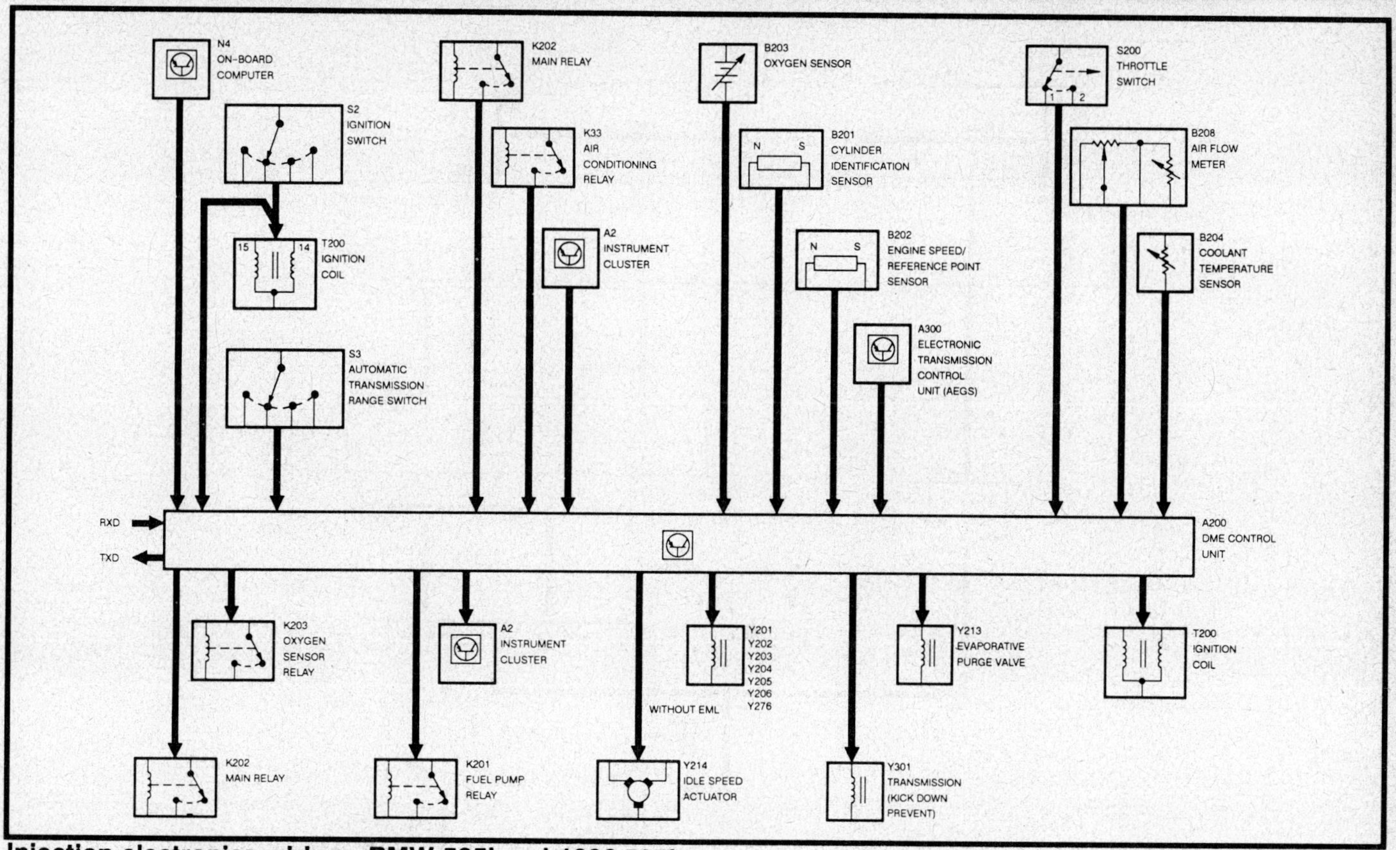

Injection electronics wiring—BMW 535i and 1990 525i

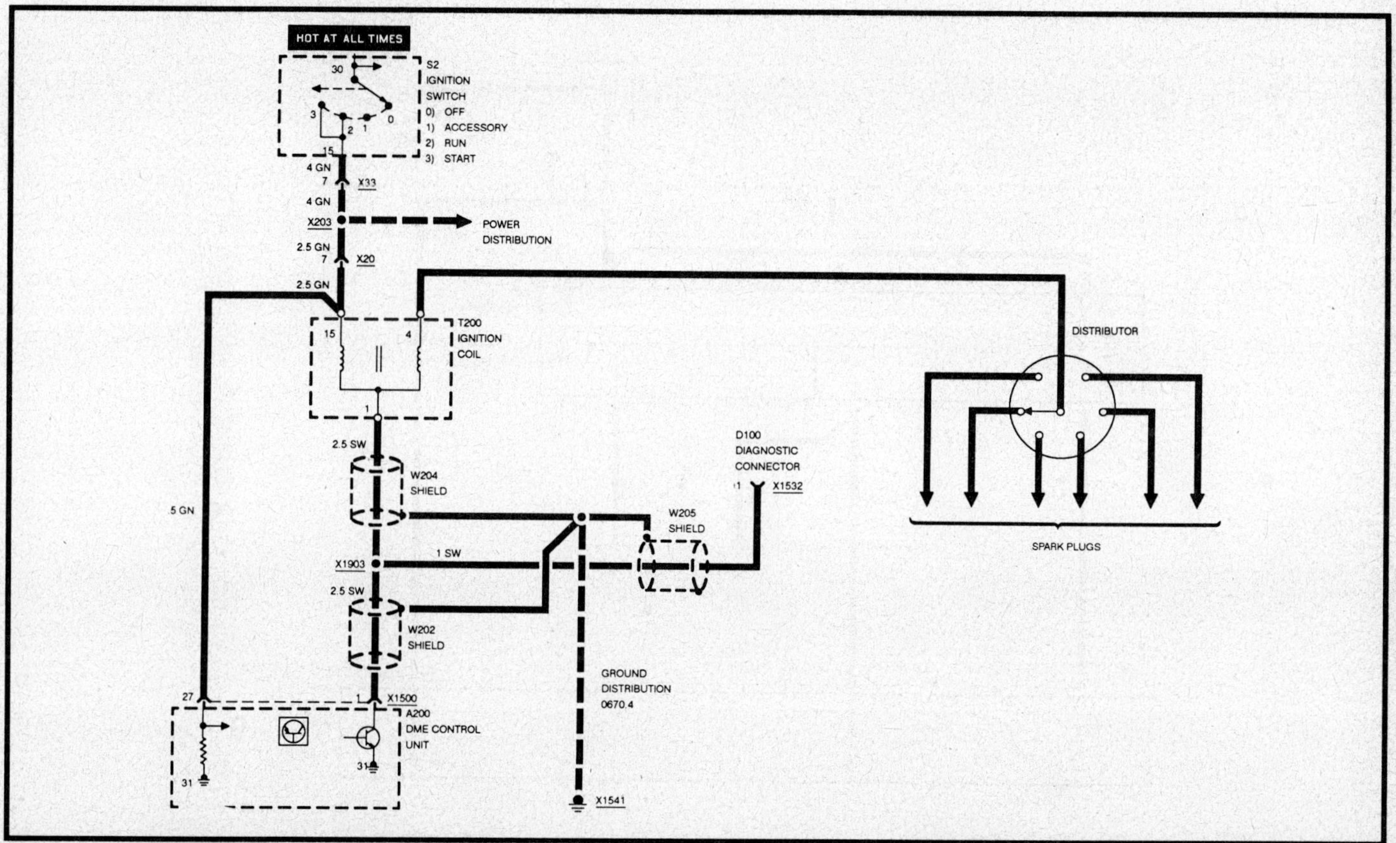

Injection electronics wiring—BMW 535i and 1990 525i

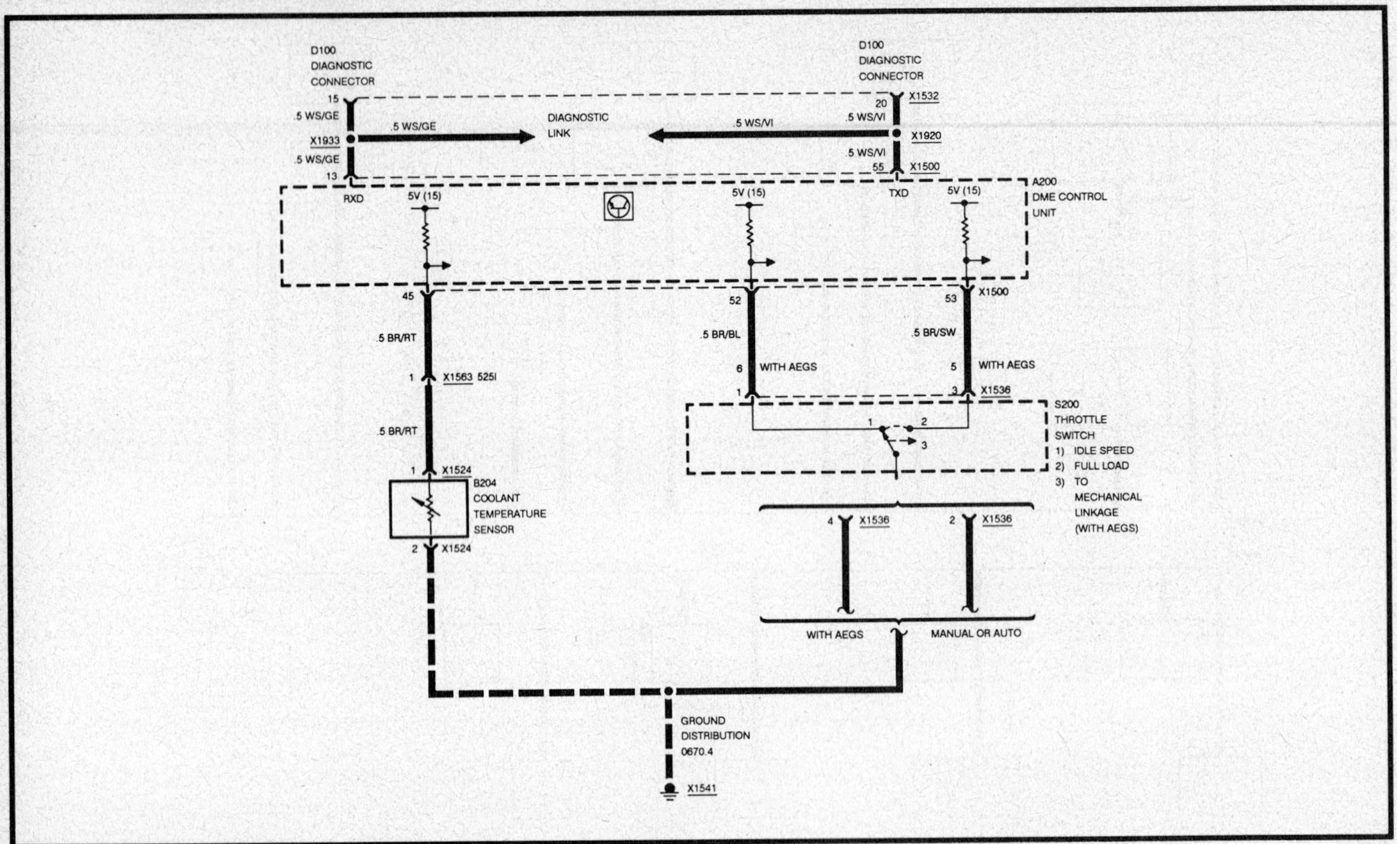

Injection electronics wiring—BMW 535i and 1990 525i

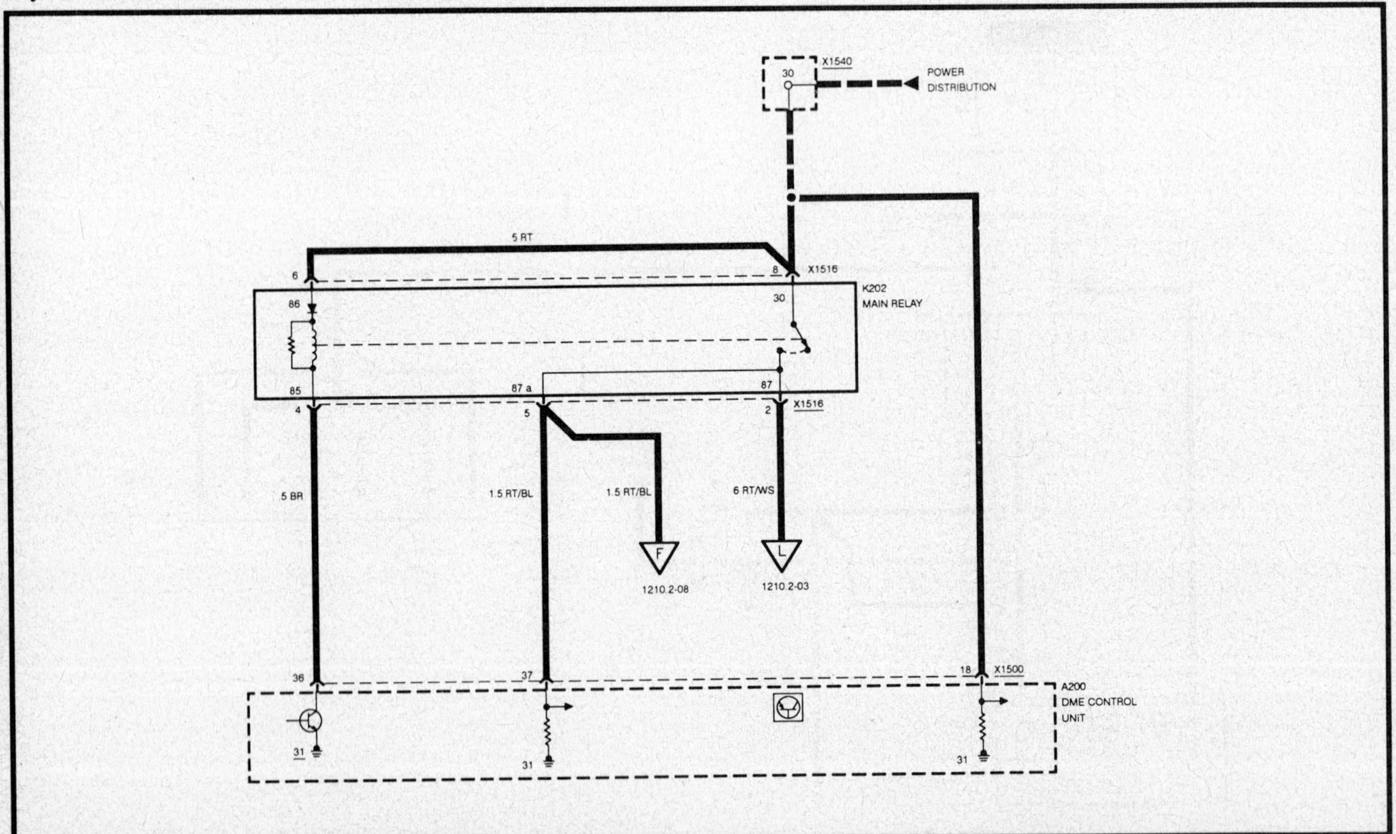

Injection electronics wiring—BMW 535i and 1990 525i

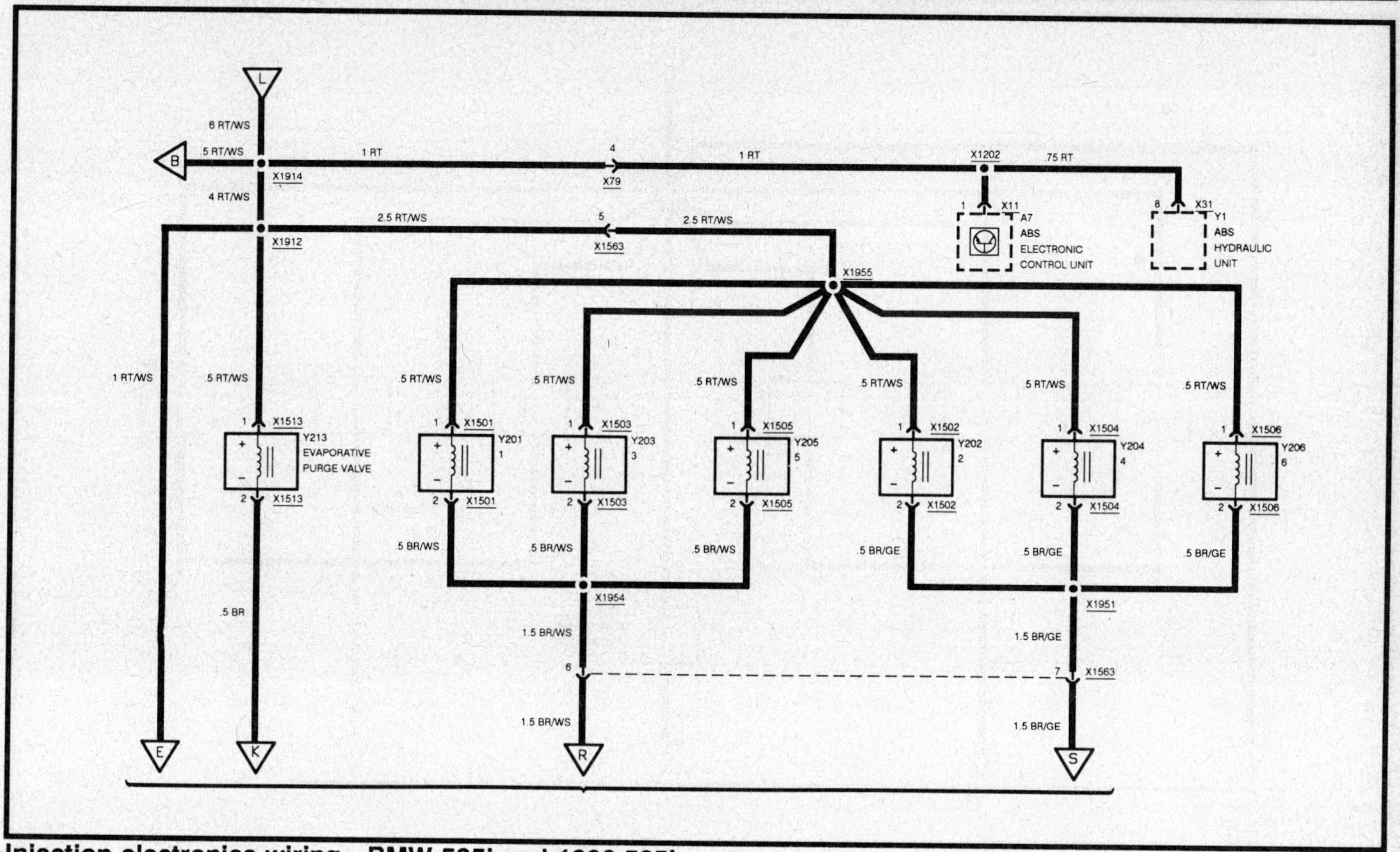

Injection electronics wiring—BMW 535i and 1990 525i

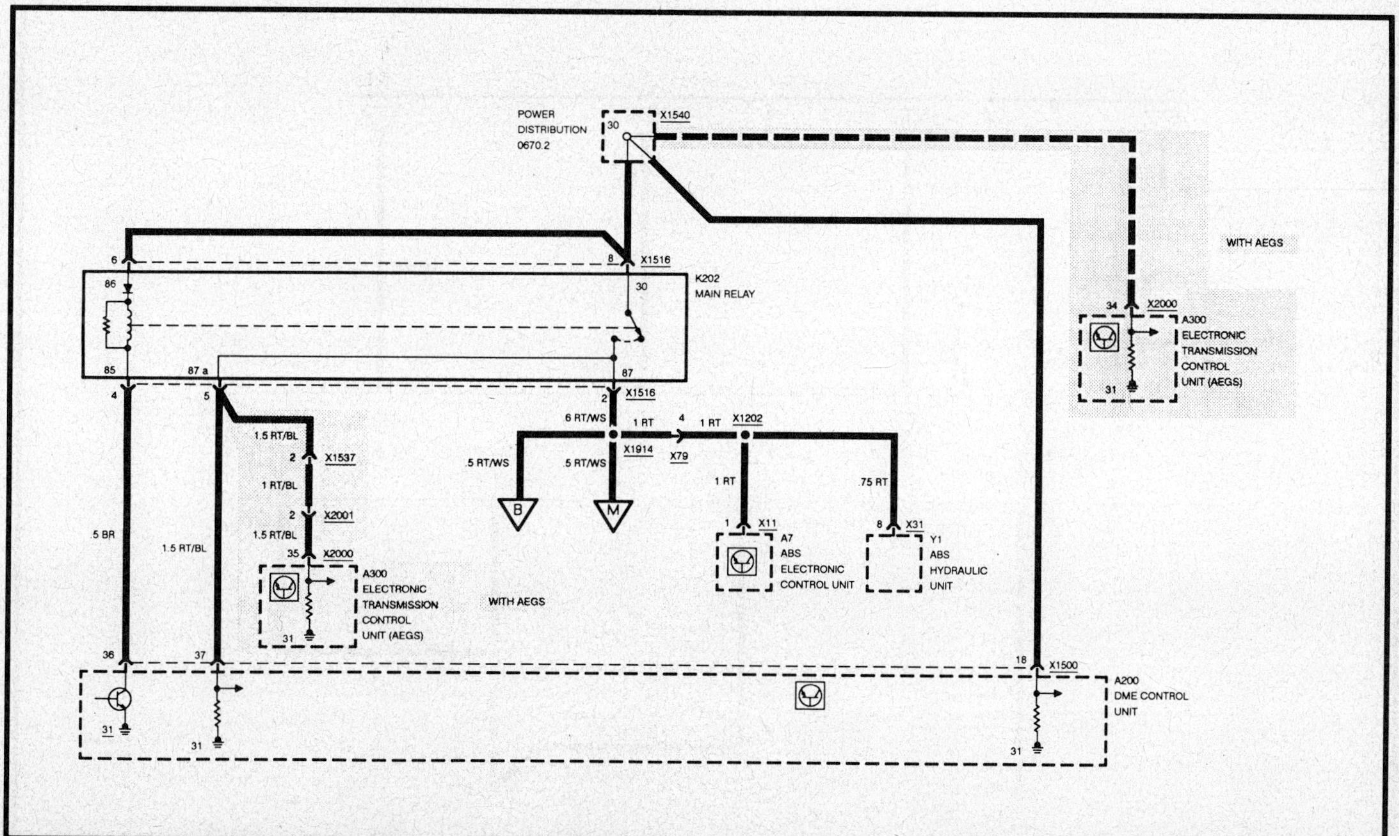

Injection electronics wiring—BMW 535i and 1990 525i

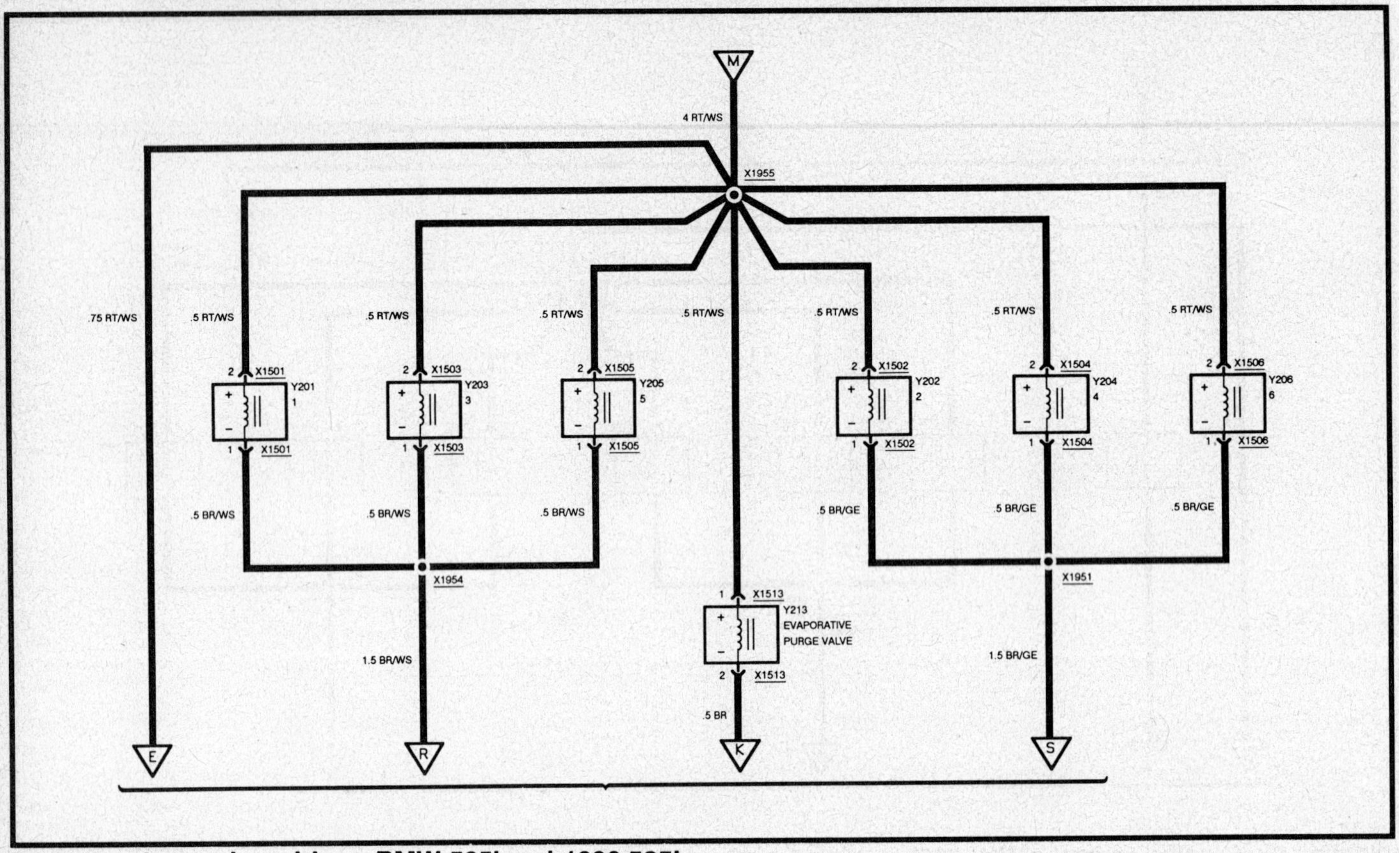

Injection electronics wiring—BMW 535i and 1990 525i

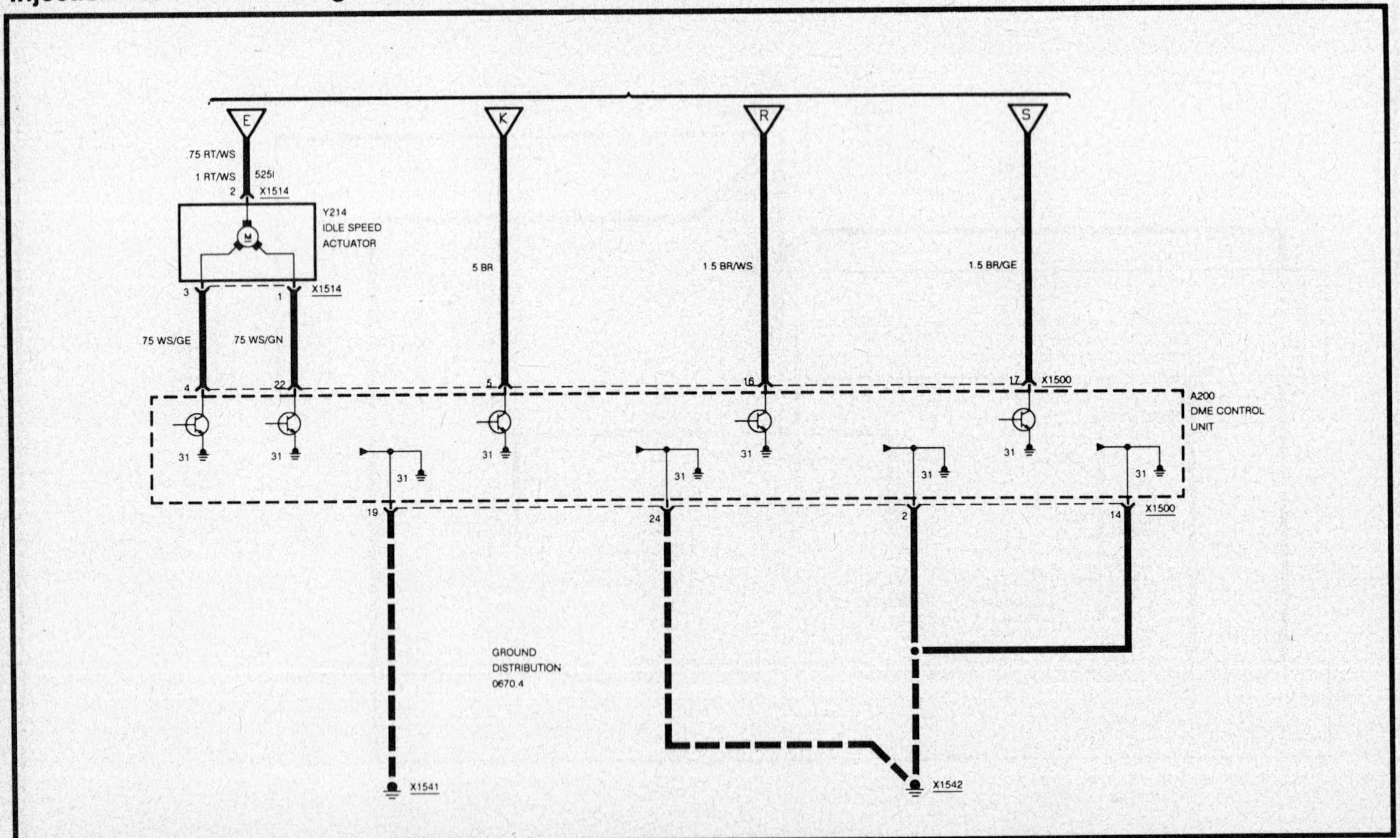

Injection electronics wiring—BMW 535i and 1990 525i

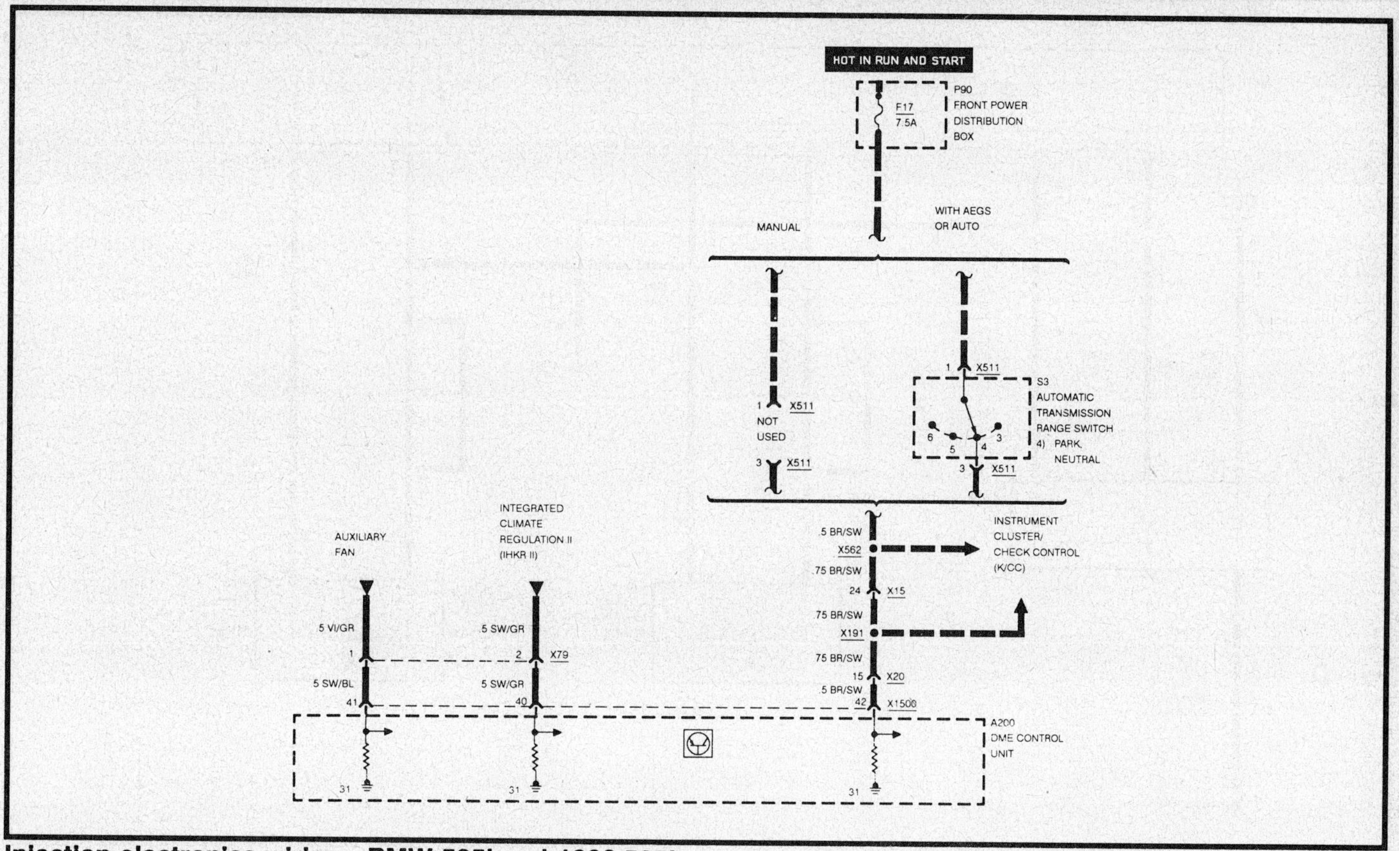

Injection electronics wiring—BMW 535i and 1990 525i

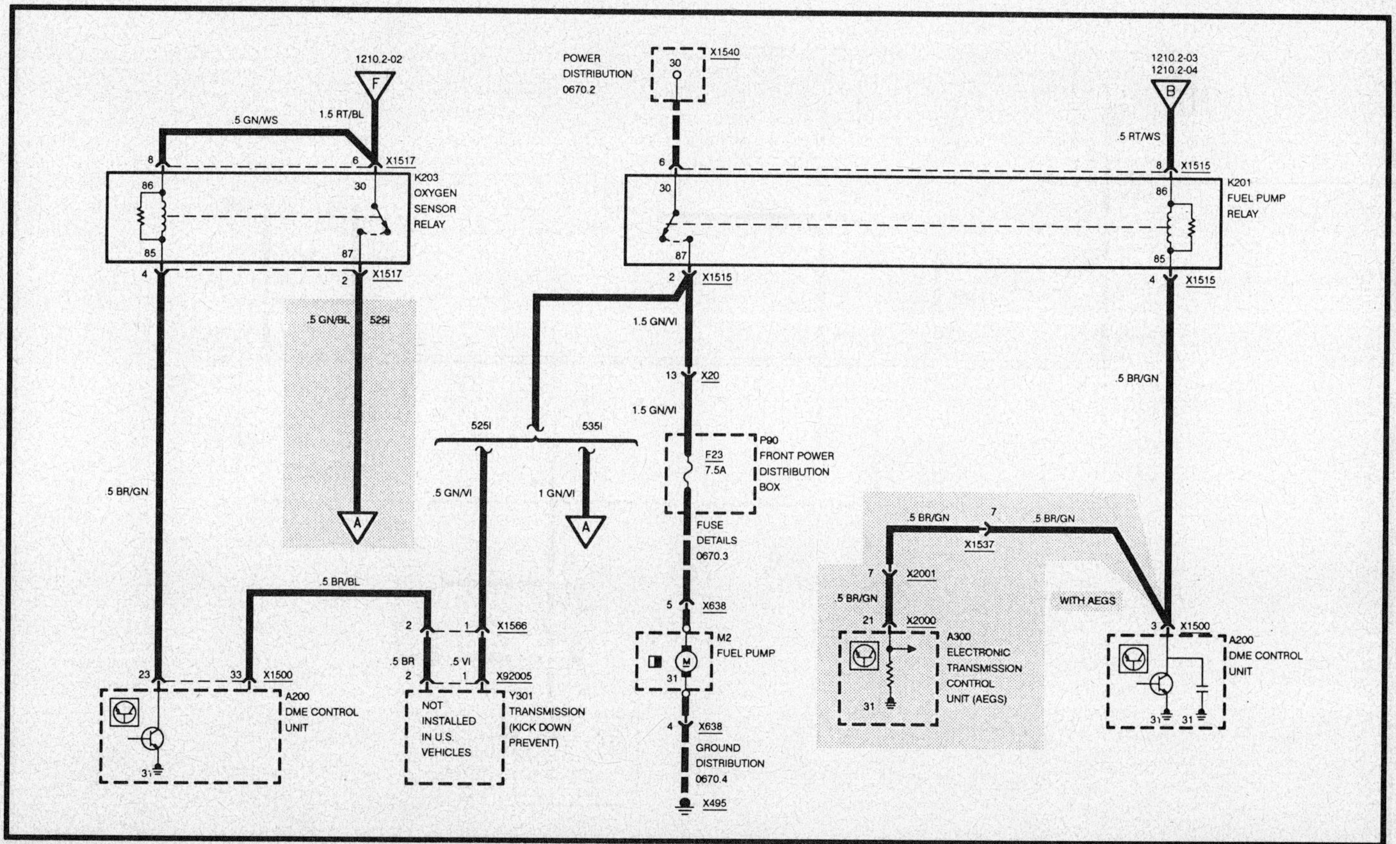

Injection electronics wiring—BMW 535i and 1990 525i

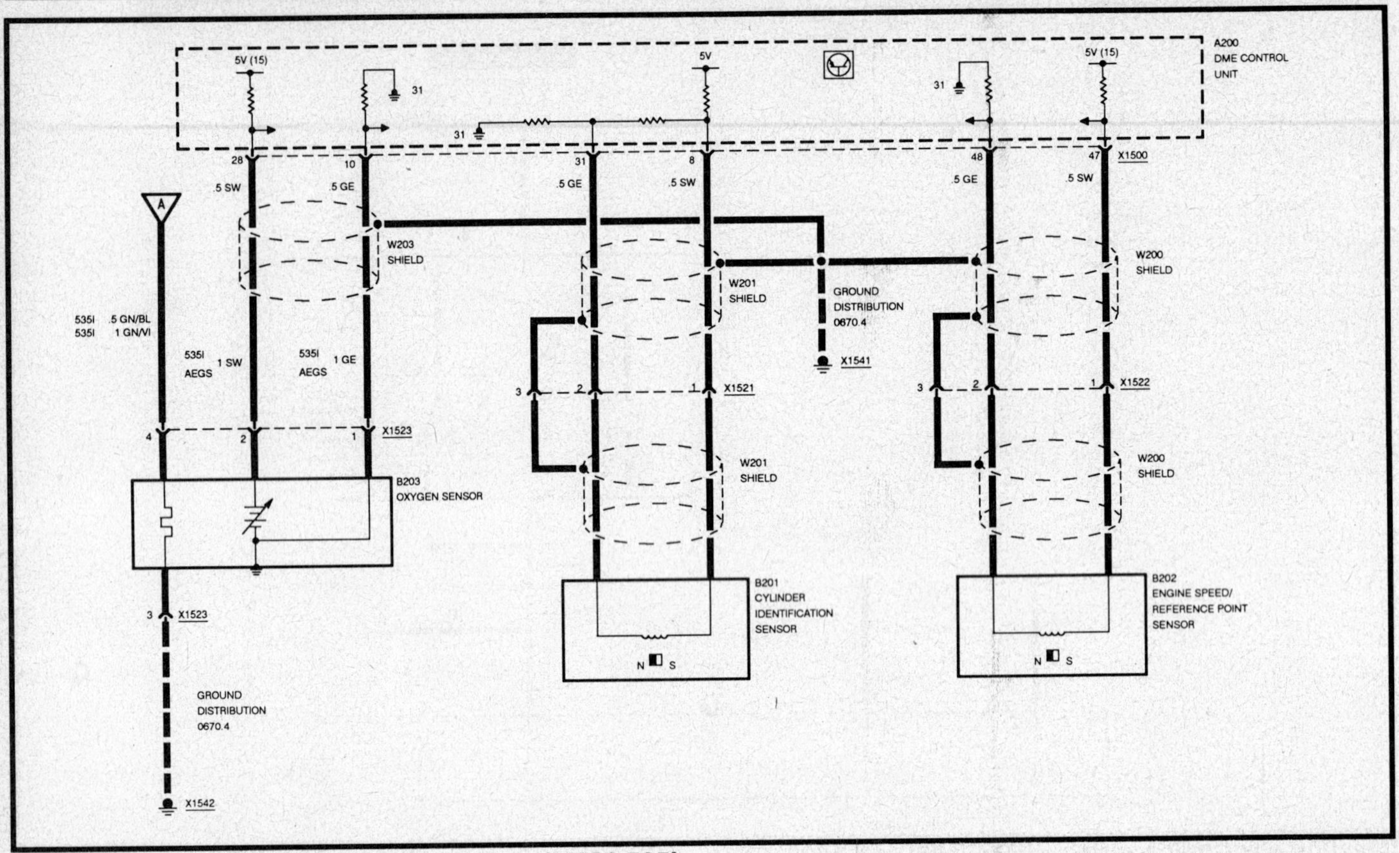

Injection electronics wiring—BMW 535i and 1990 525i

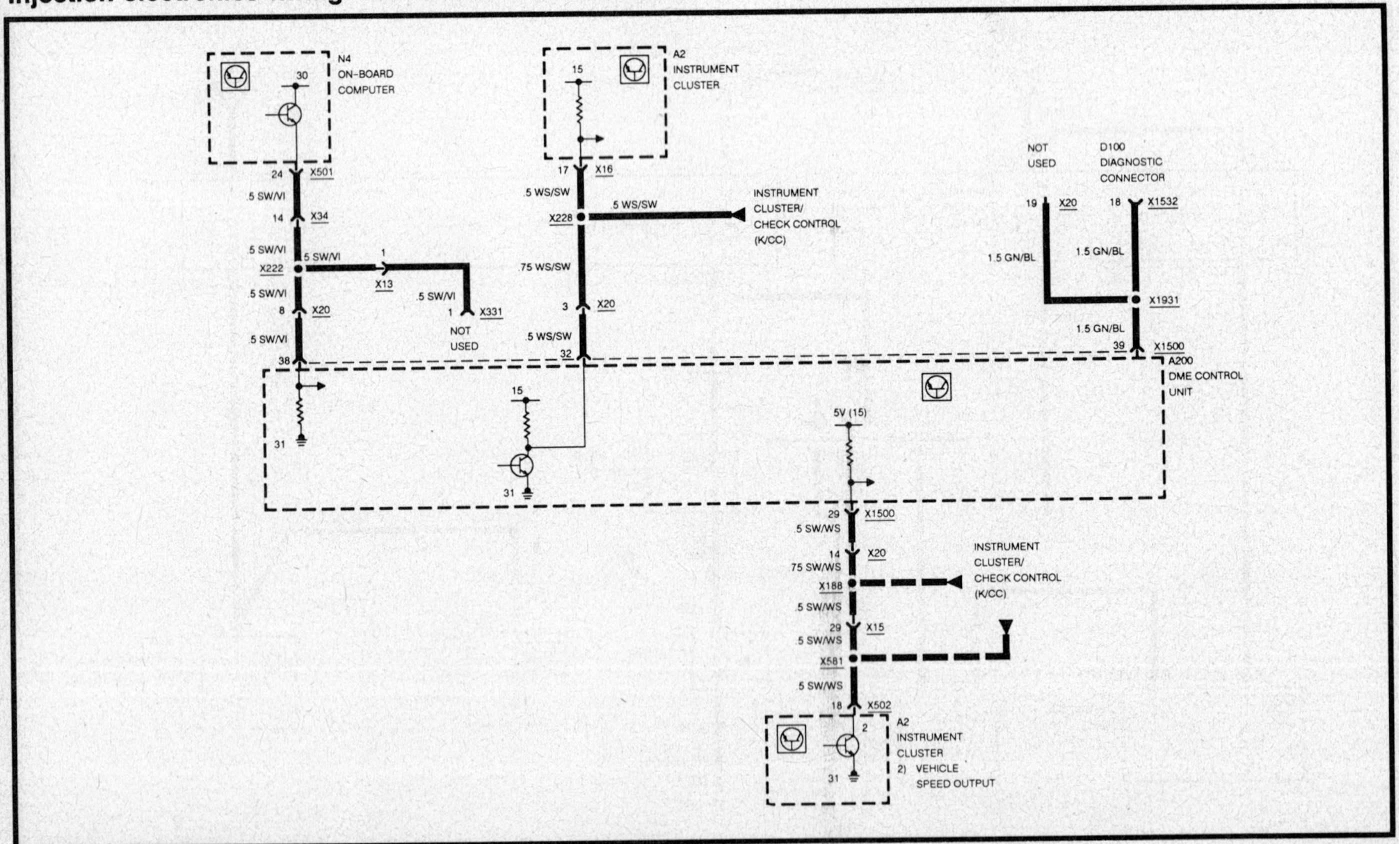

Injection electronics wiring—BMW 535i and 1990 525i

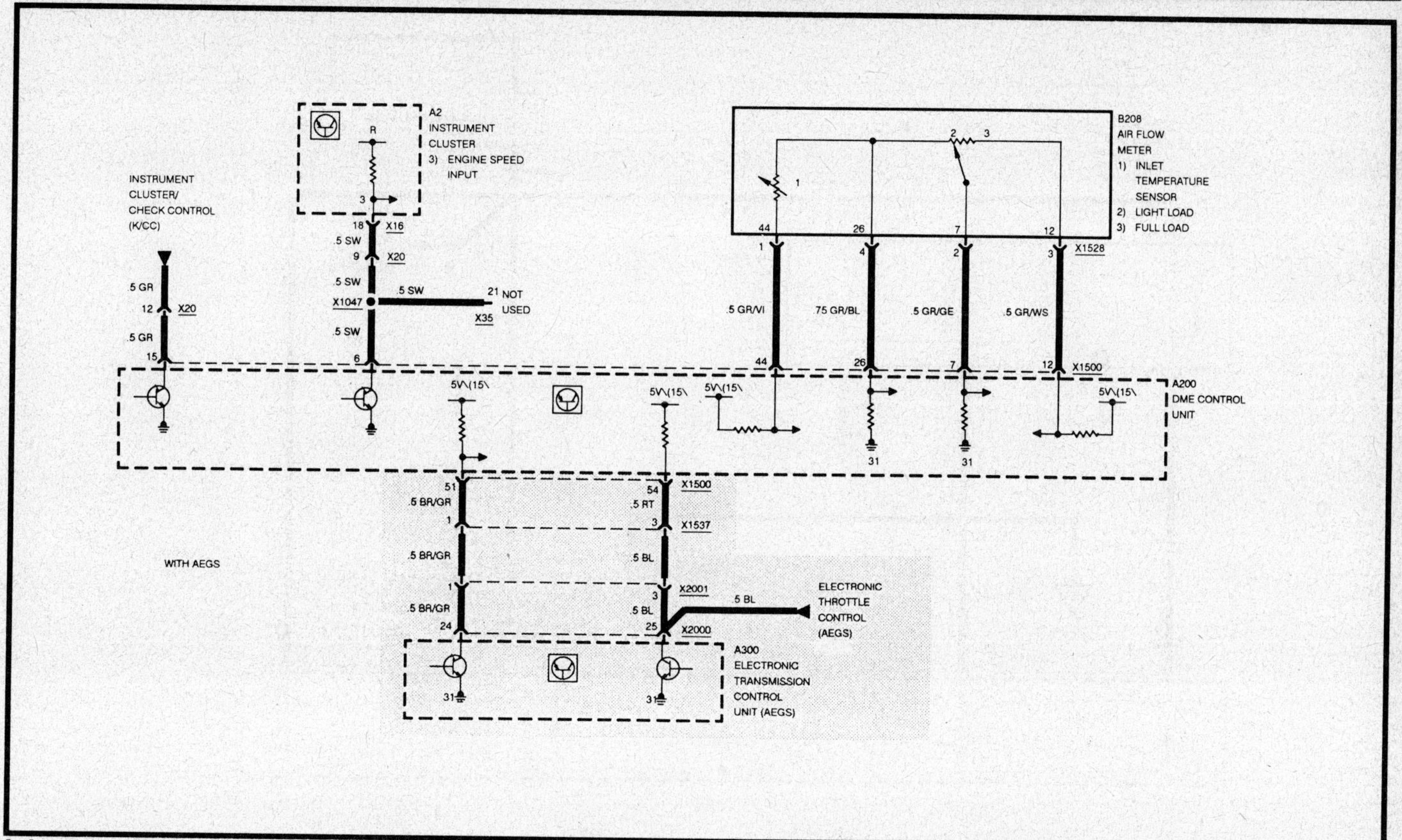

Injection electronics wiring—BMW 535i and 1990 525i

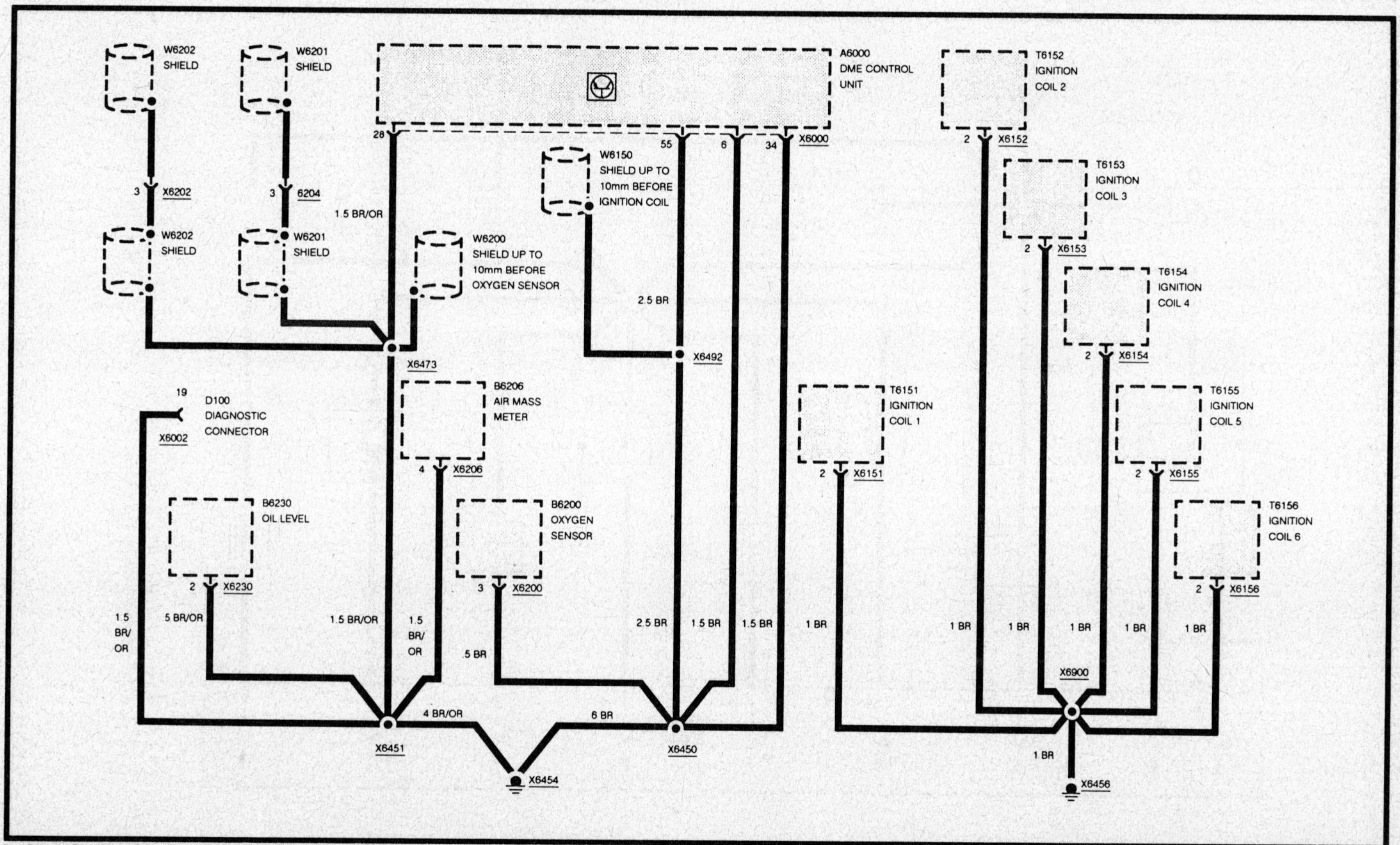

Injection electronics wiring—BMW 1991–92 525i

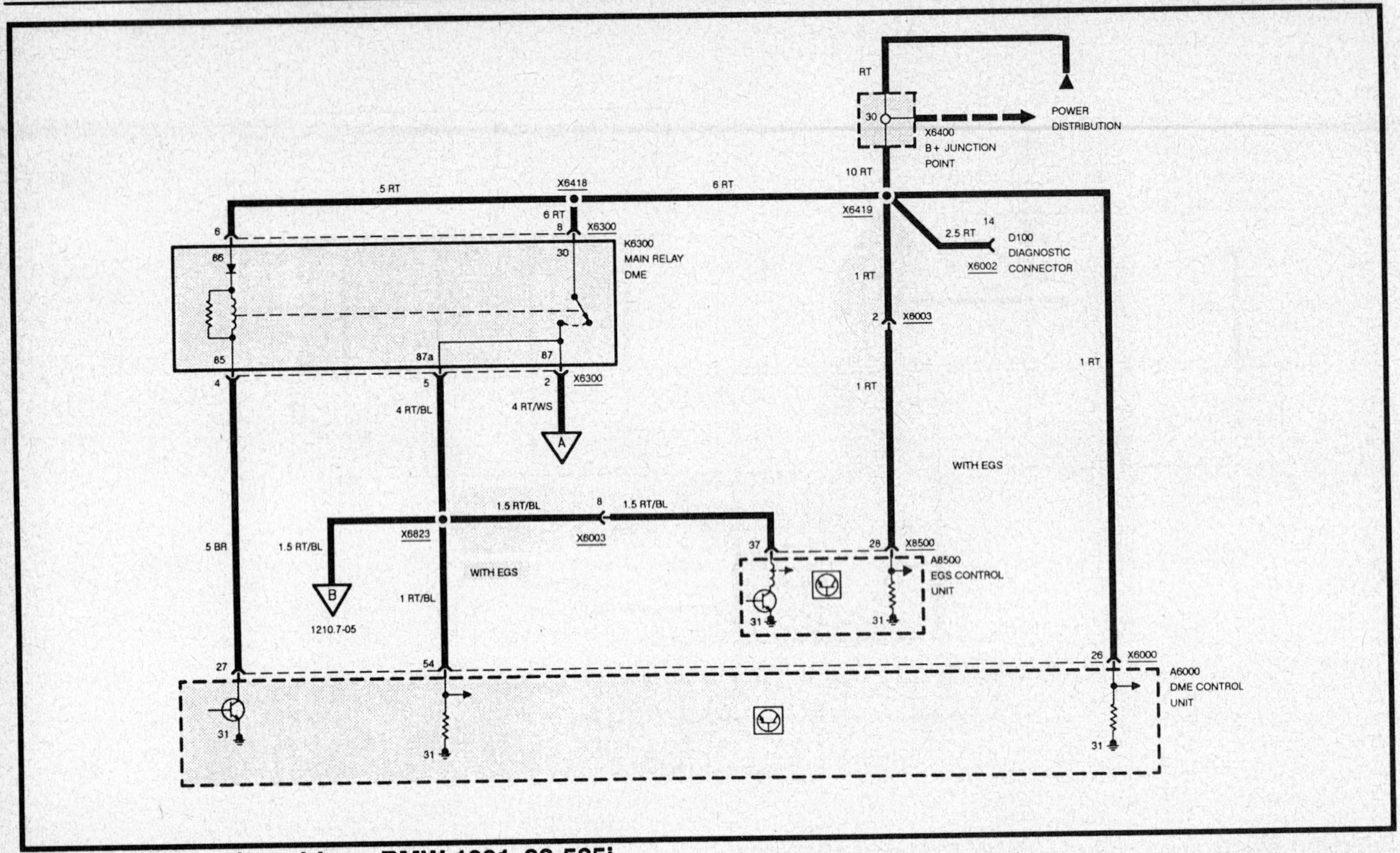

Injection electronics wiring—BMW 1991–92 525i

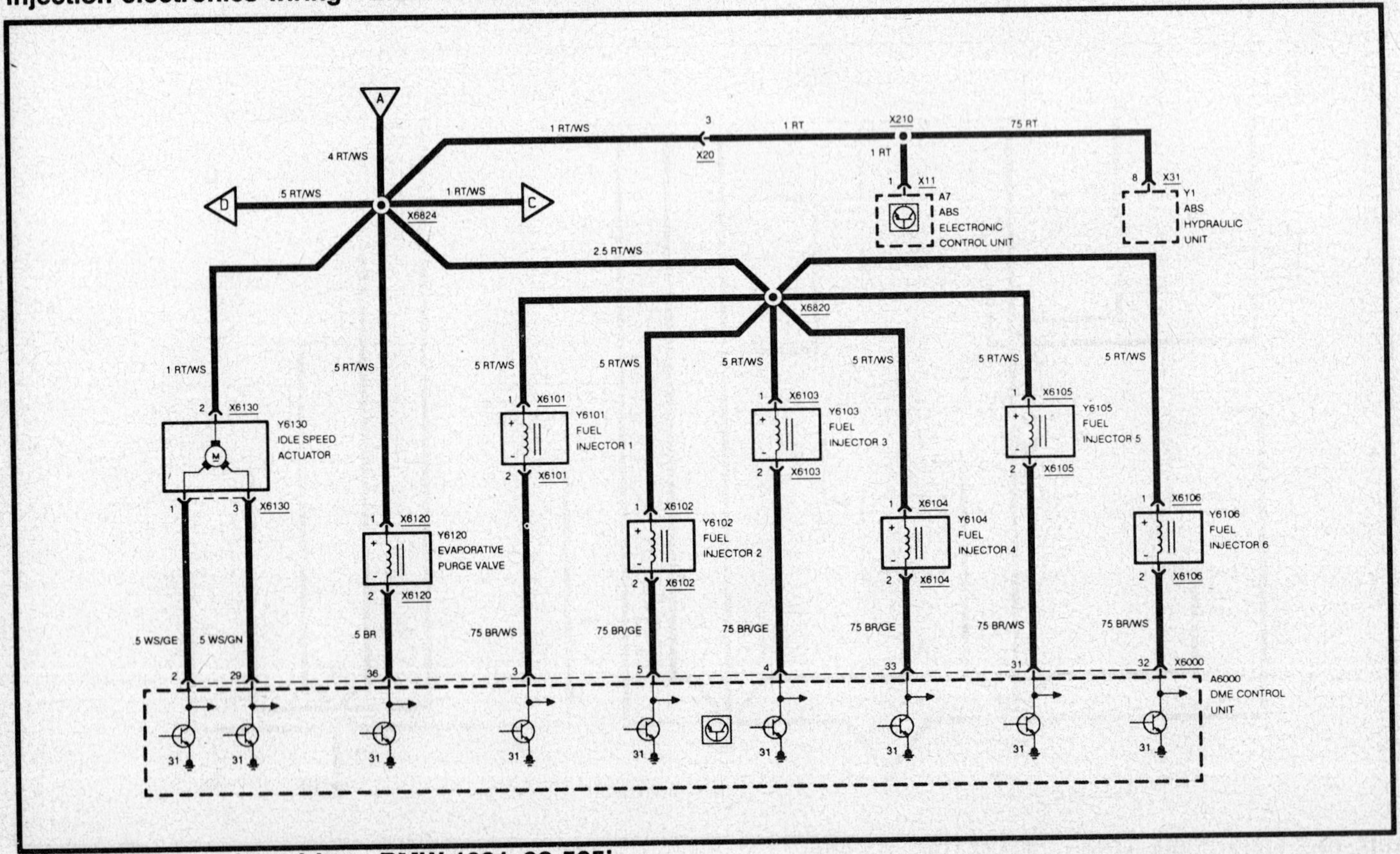

Injection electronics wiring—BMW 1991–92 525i

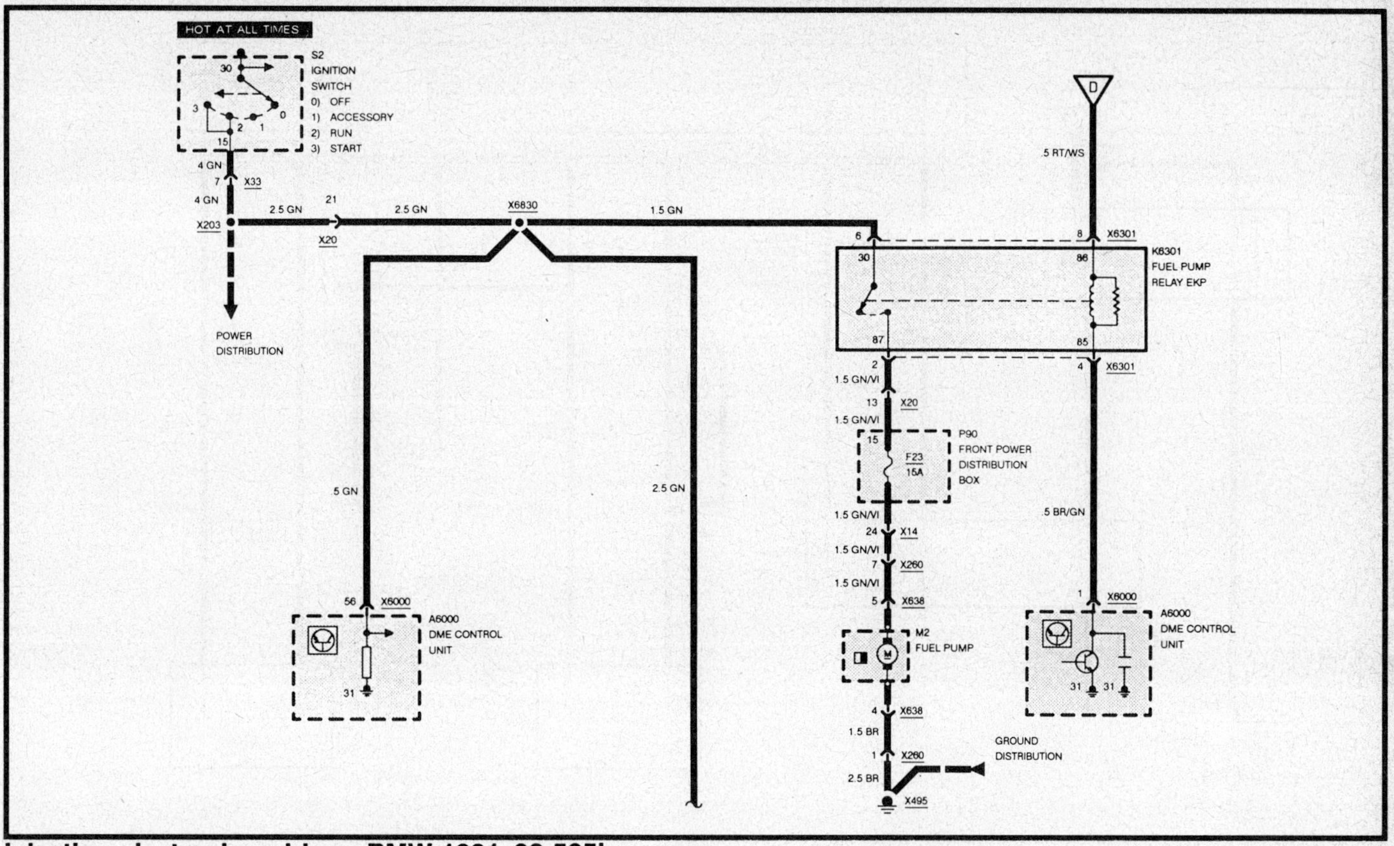

Injection electronics wiring—BMW 1991–92 525i

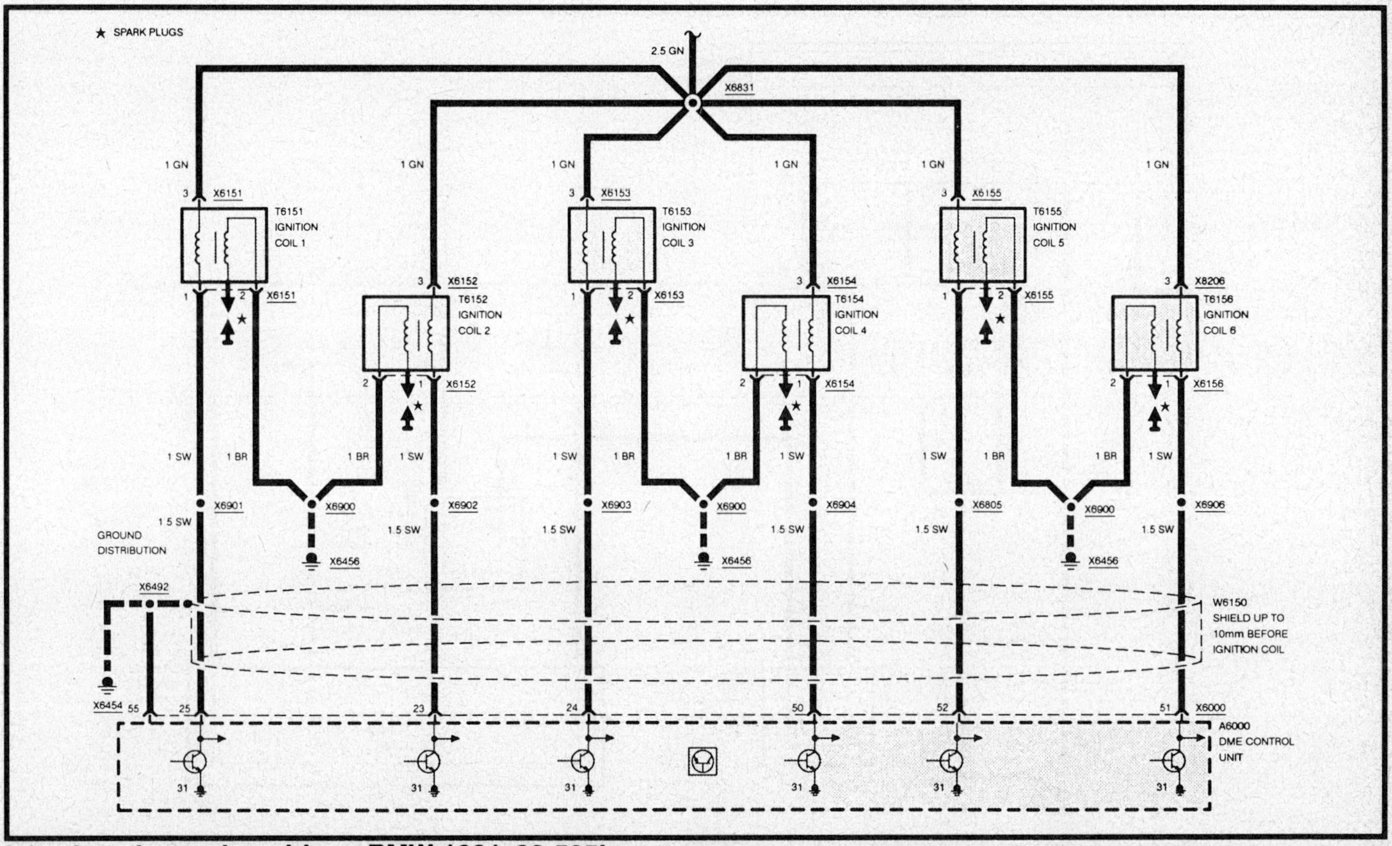

Injection electronics wiring—BMW 1991–92 525i

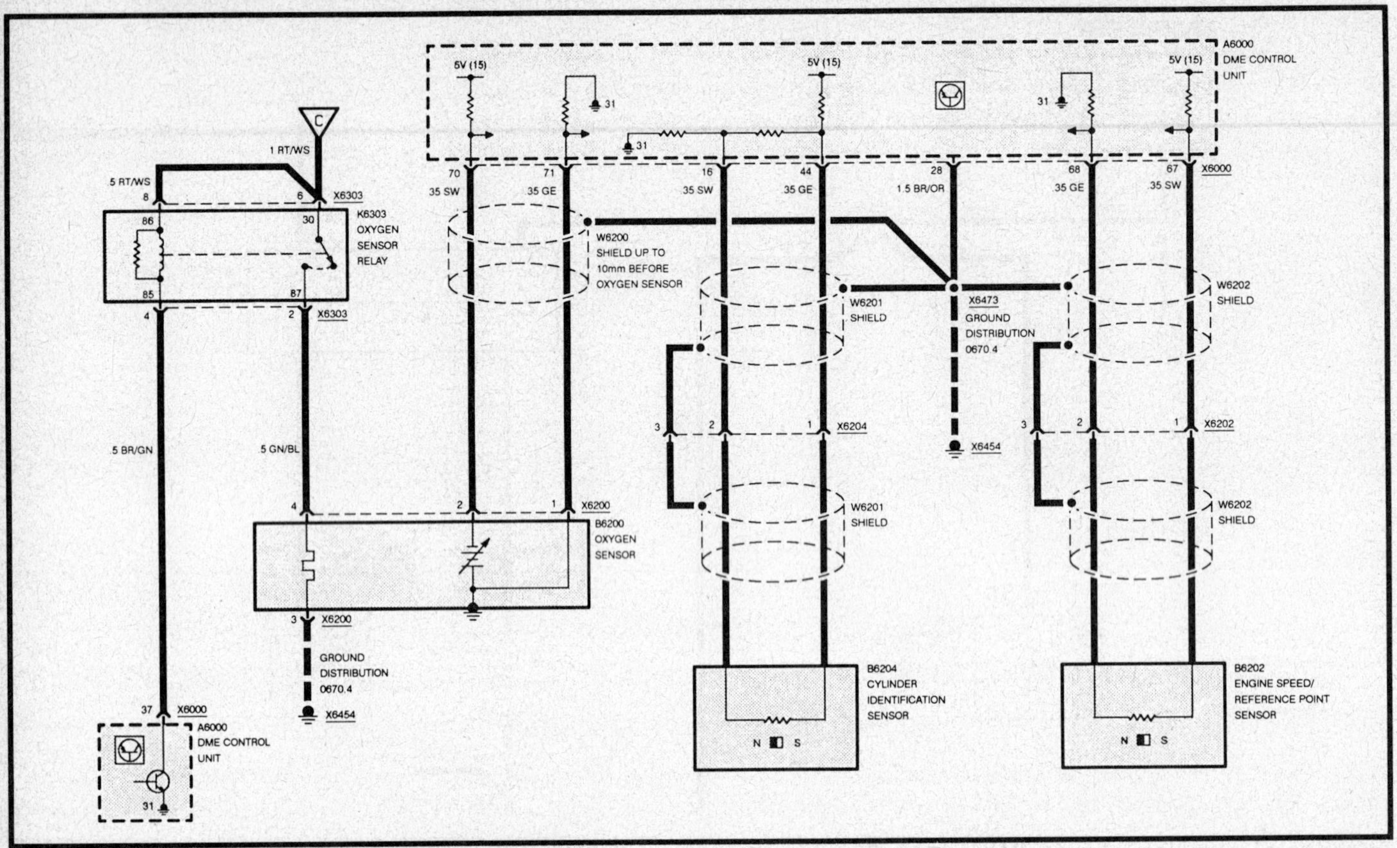

Injection electronics wiring—BMW 1991–92 525i

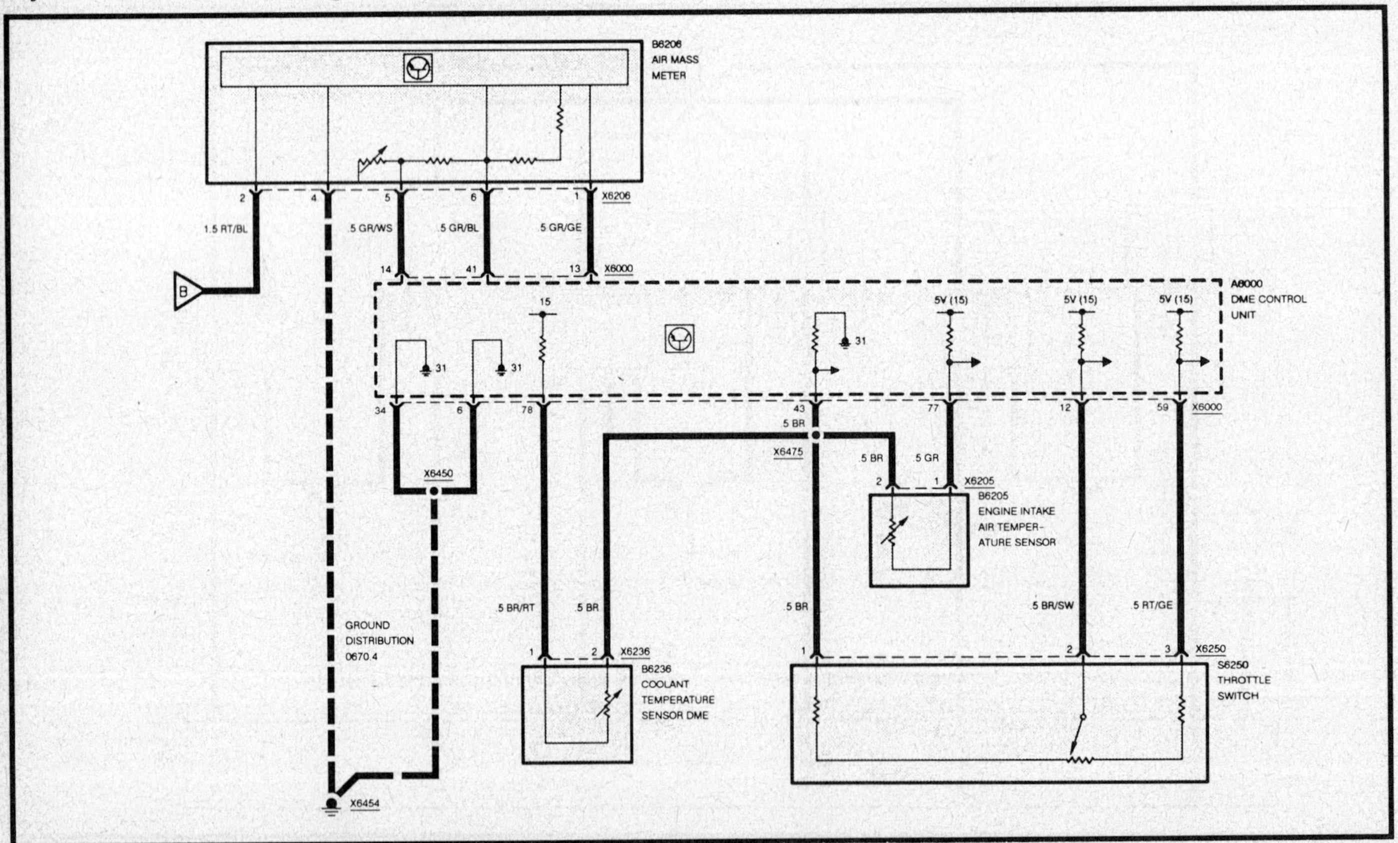

Injection electronics wiring—BMW 1991–92 525i

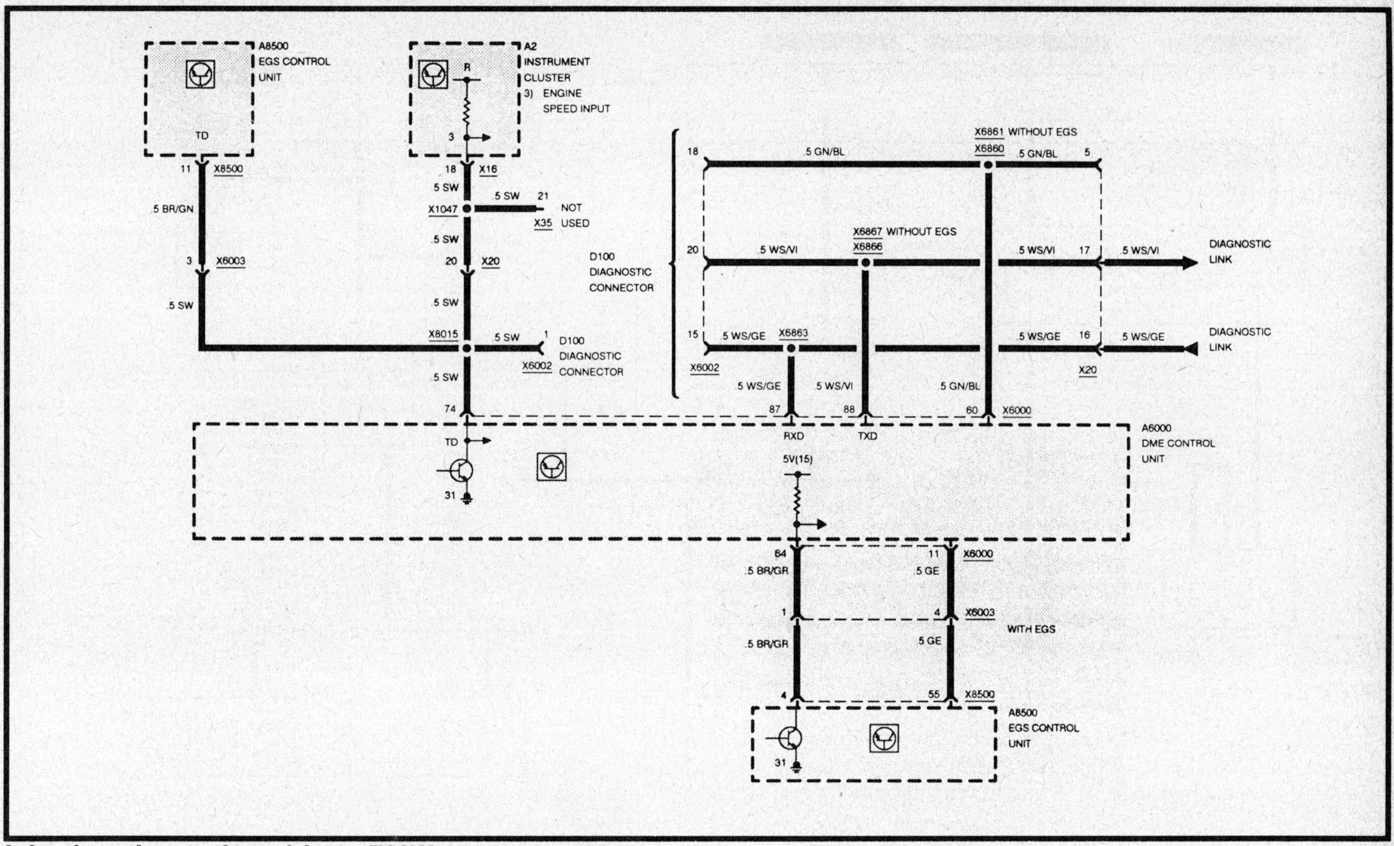

Injection electronics wiring—BMW 1991–92 525i

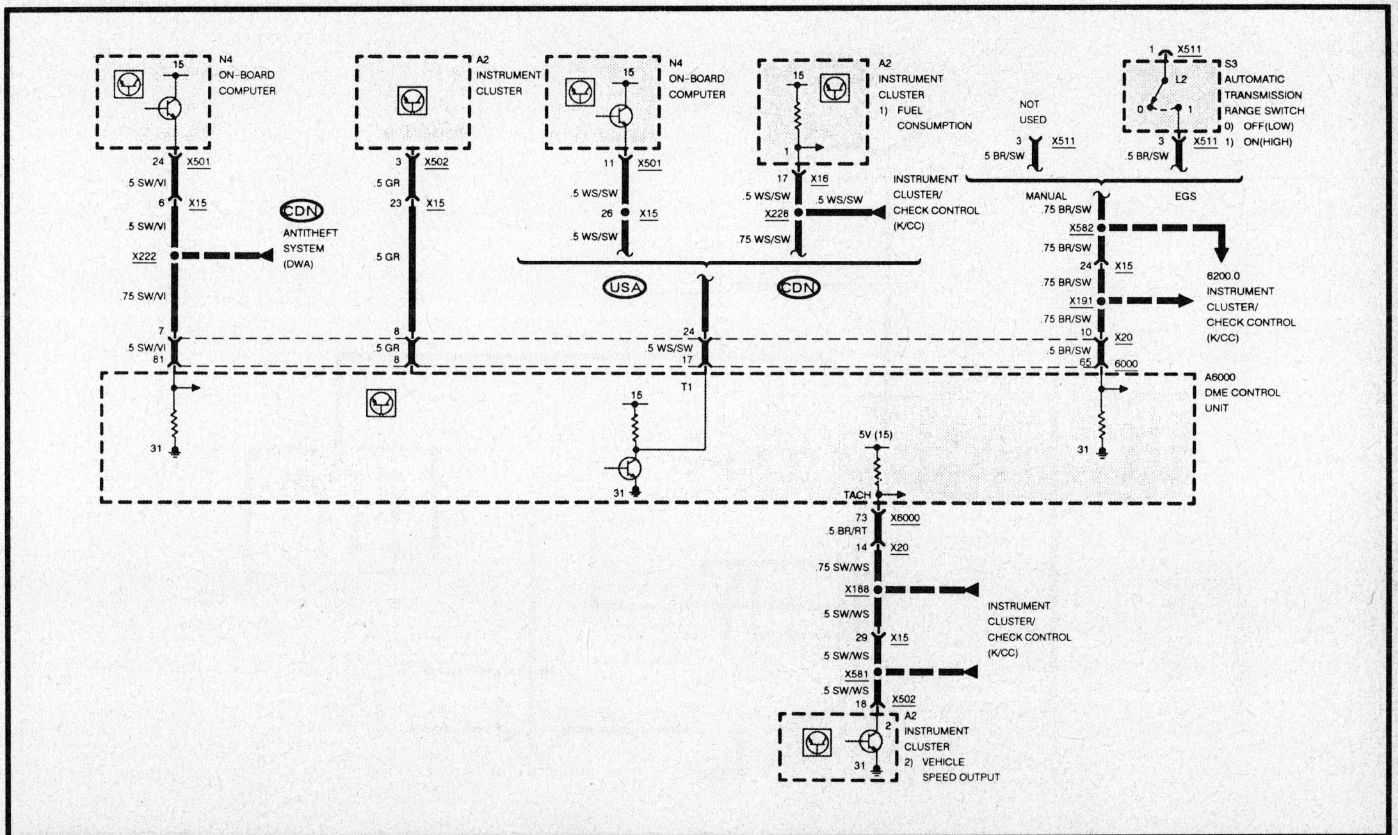

Injection electronics wiring—BMW 1991–92 525i

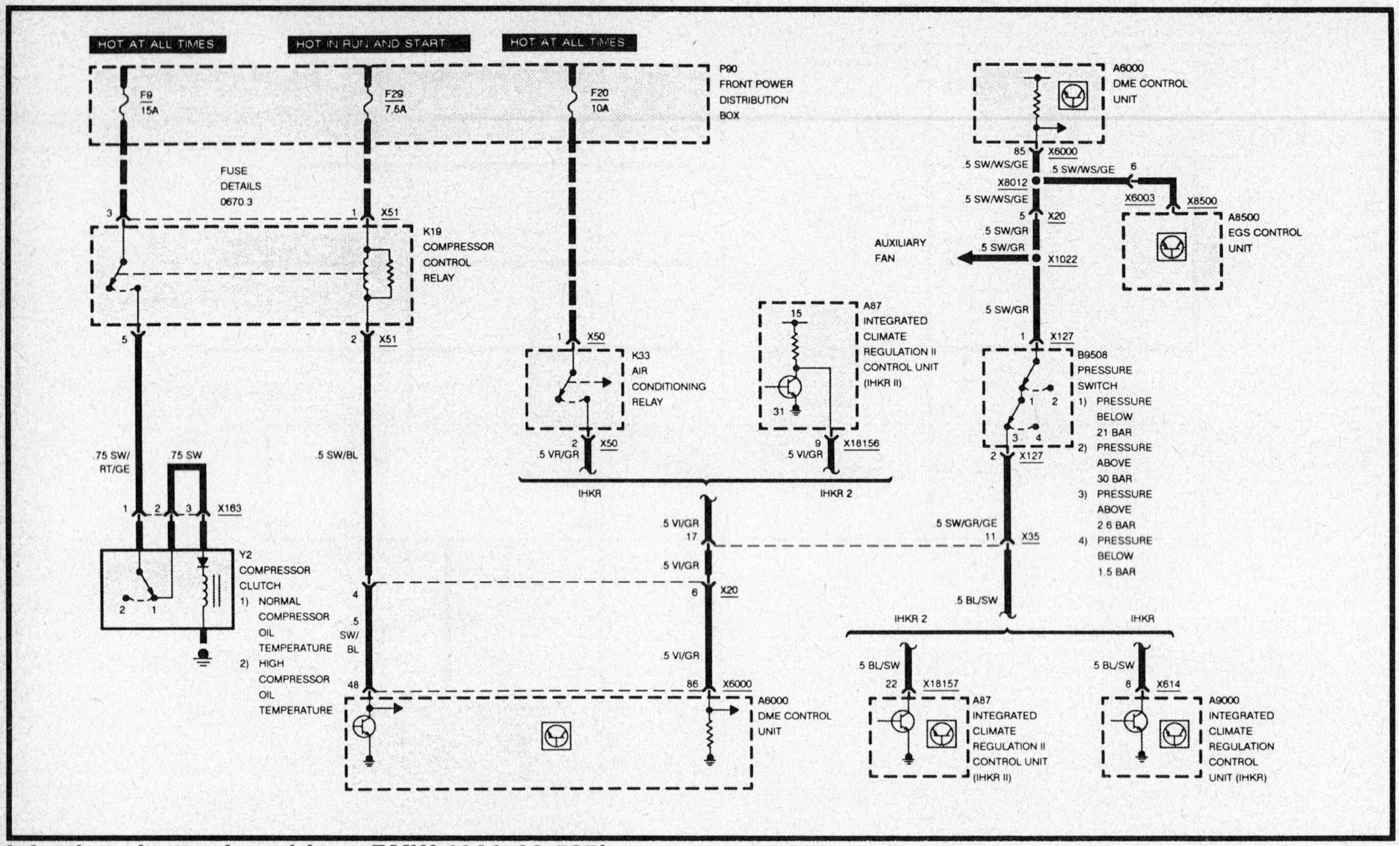

Injection electronics wiring—BMW 1991–92 525i

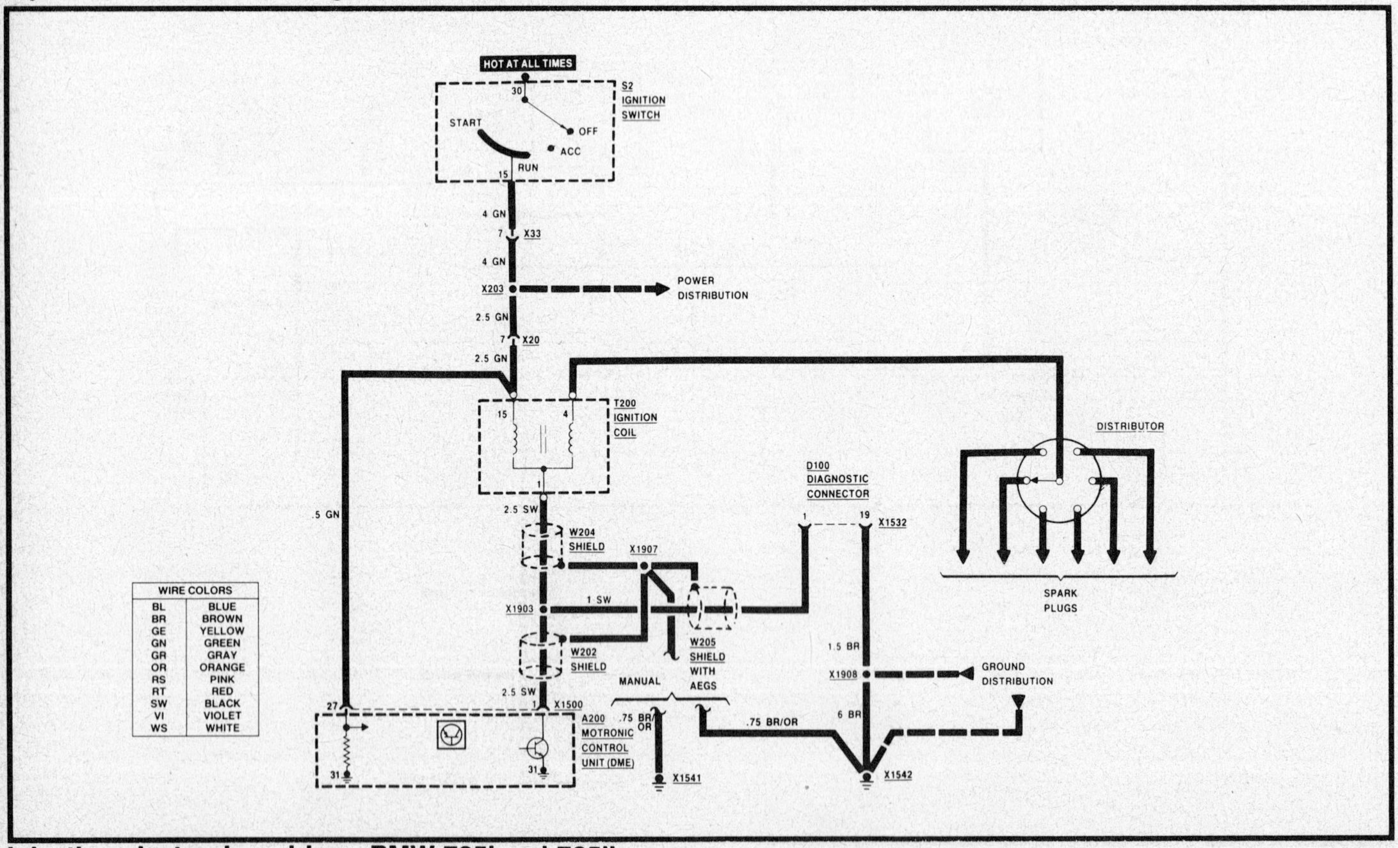

Injection electronics wiring—BMW 735i and 735iL

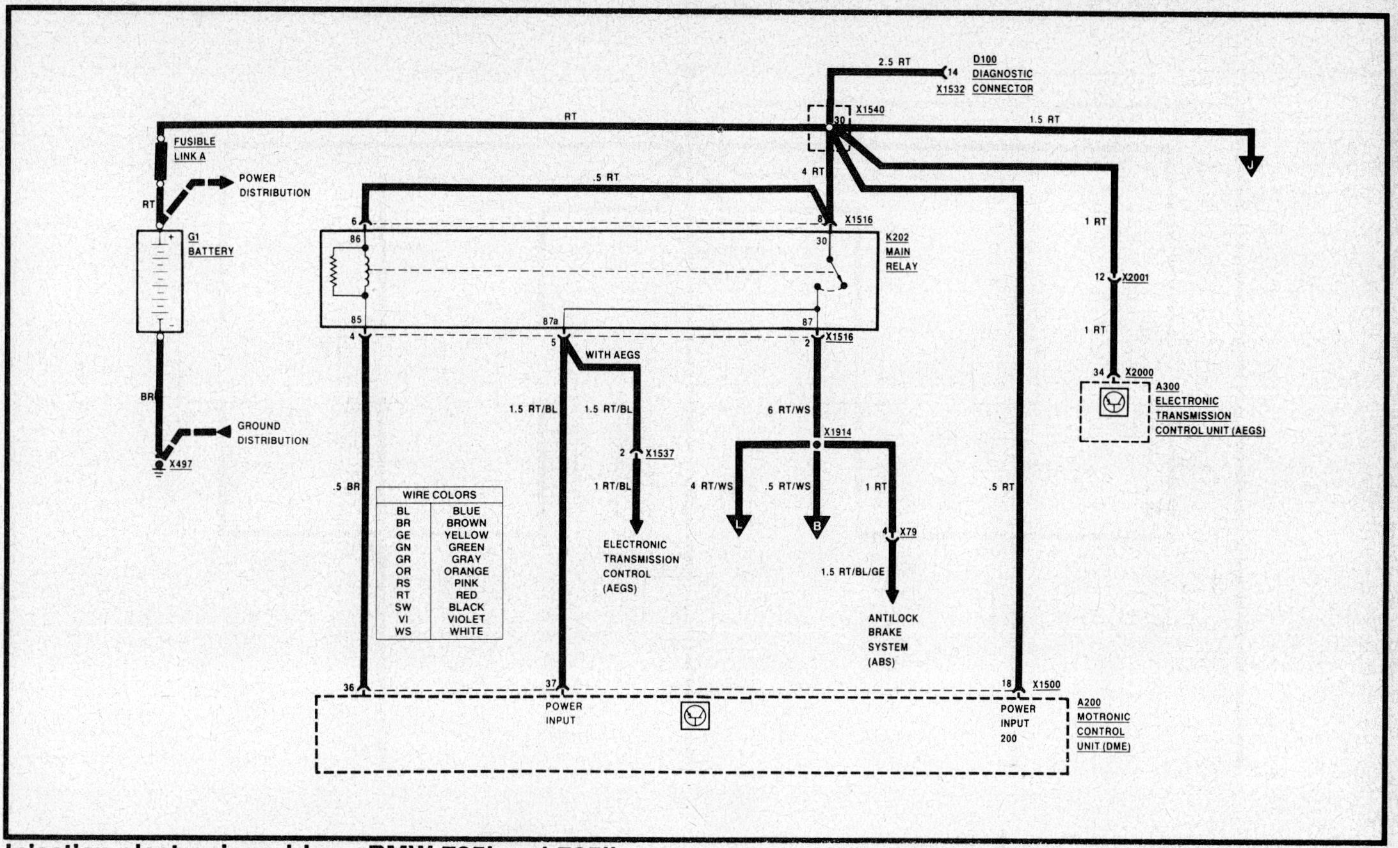

Injection electronics wiring—BMW 735i and 735iL

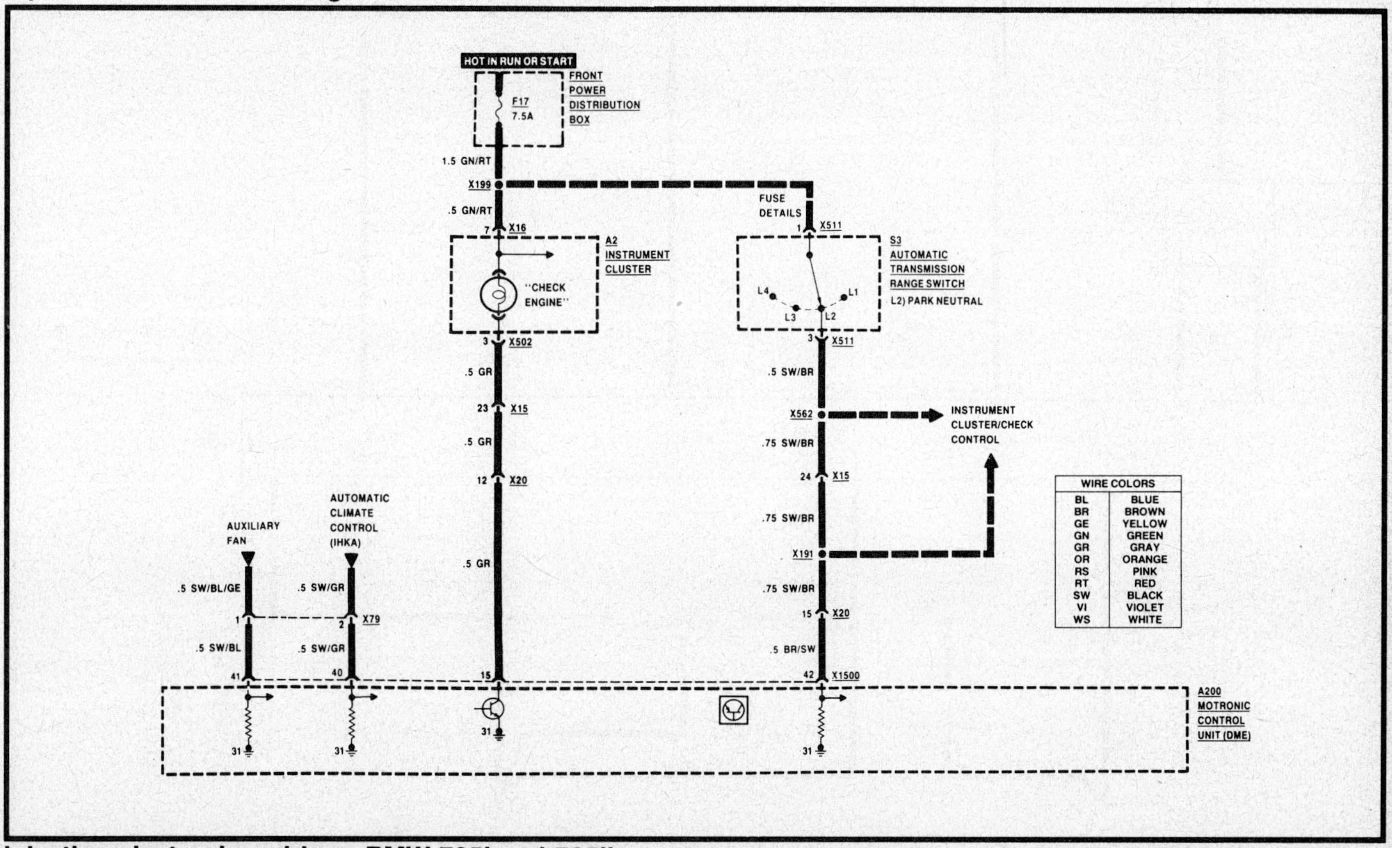

Injection electronics wiring—BMW 735i and 735iL

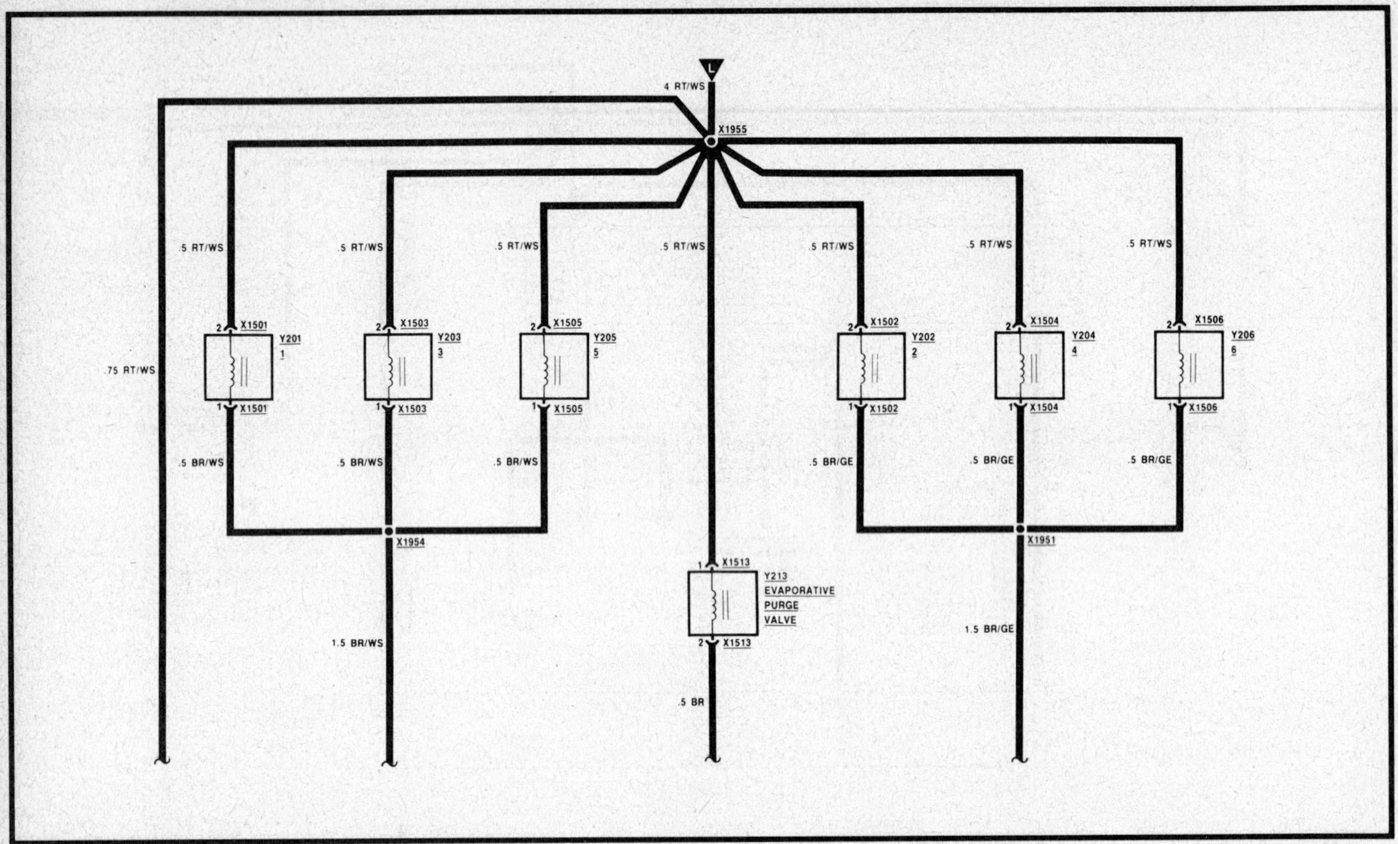

Injection electronics wiring—BMW 735i and 735iL

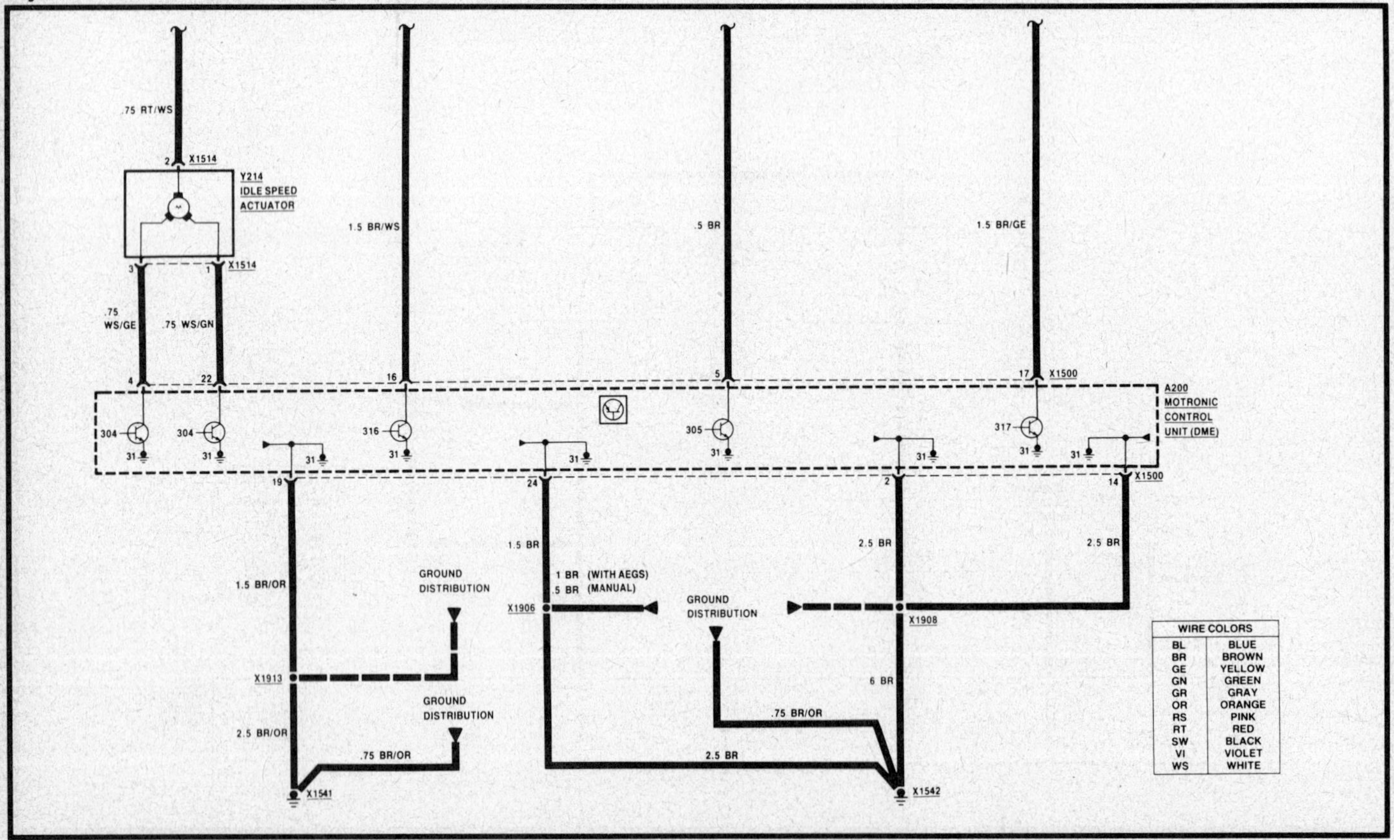

WIRE COLORS	
BL	BLUE
BR	BROWN
GE	YELLOW
GN	GREEN
GR	GRAY
OR	ORANGE
RS	PINK
RT	RED
SW	BLACK
VI	VIOLET
WS	WHITE

Injection electronics wiring—BMW 735i and 735iL

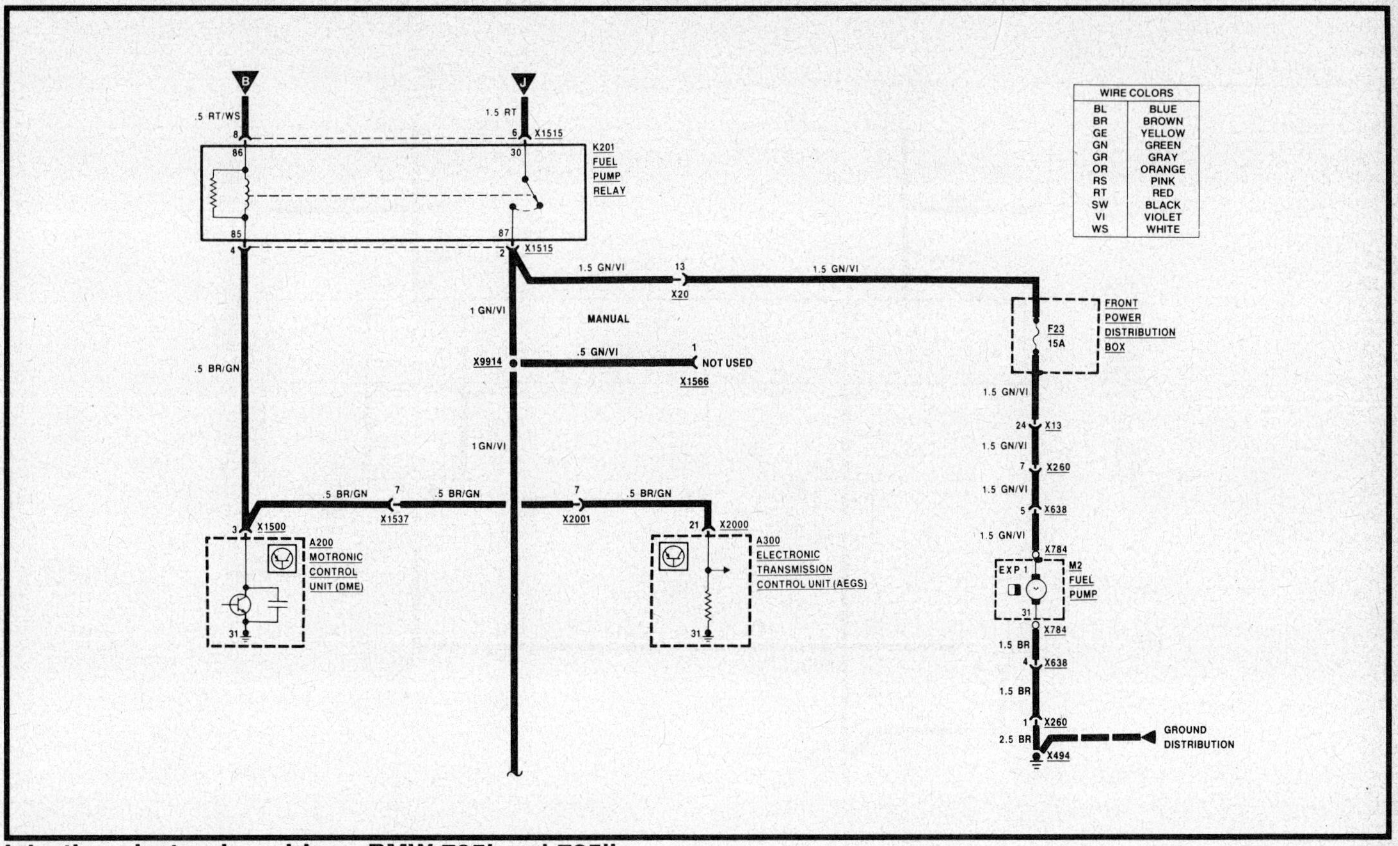

Injection electronics wiring—BMW 735i and 735iL

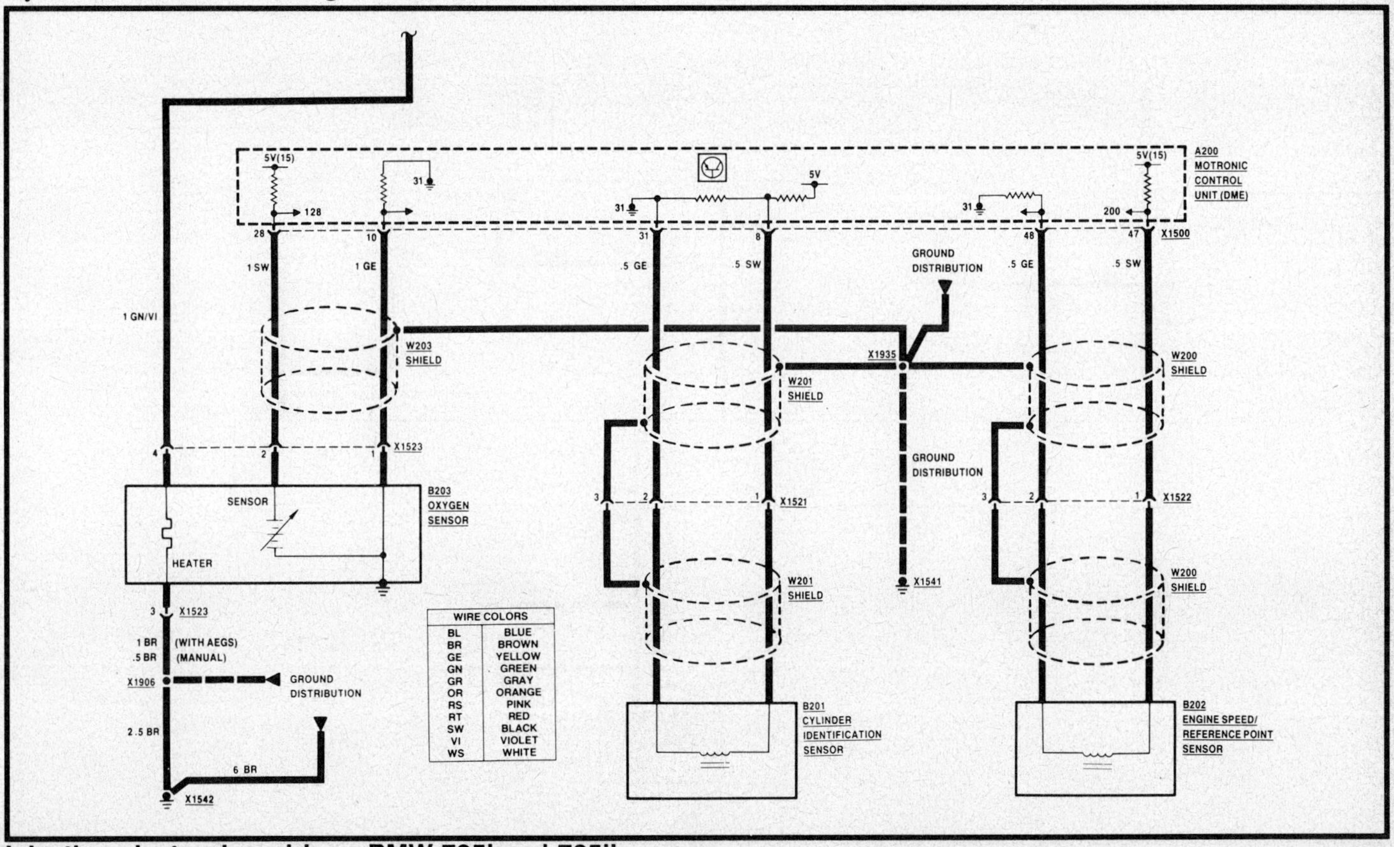

Injection electronics wiring—BMW 735i and 735iL

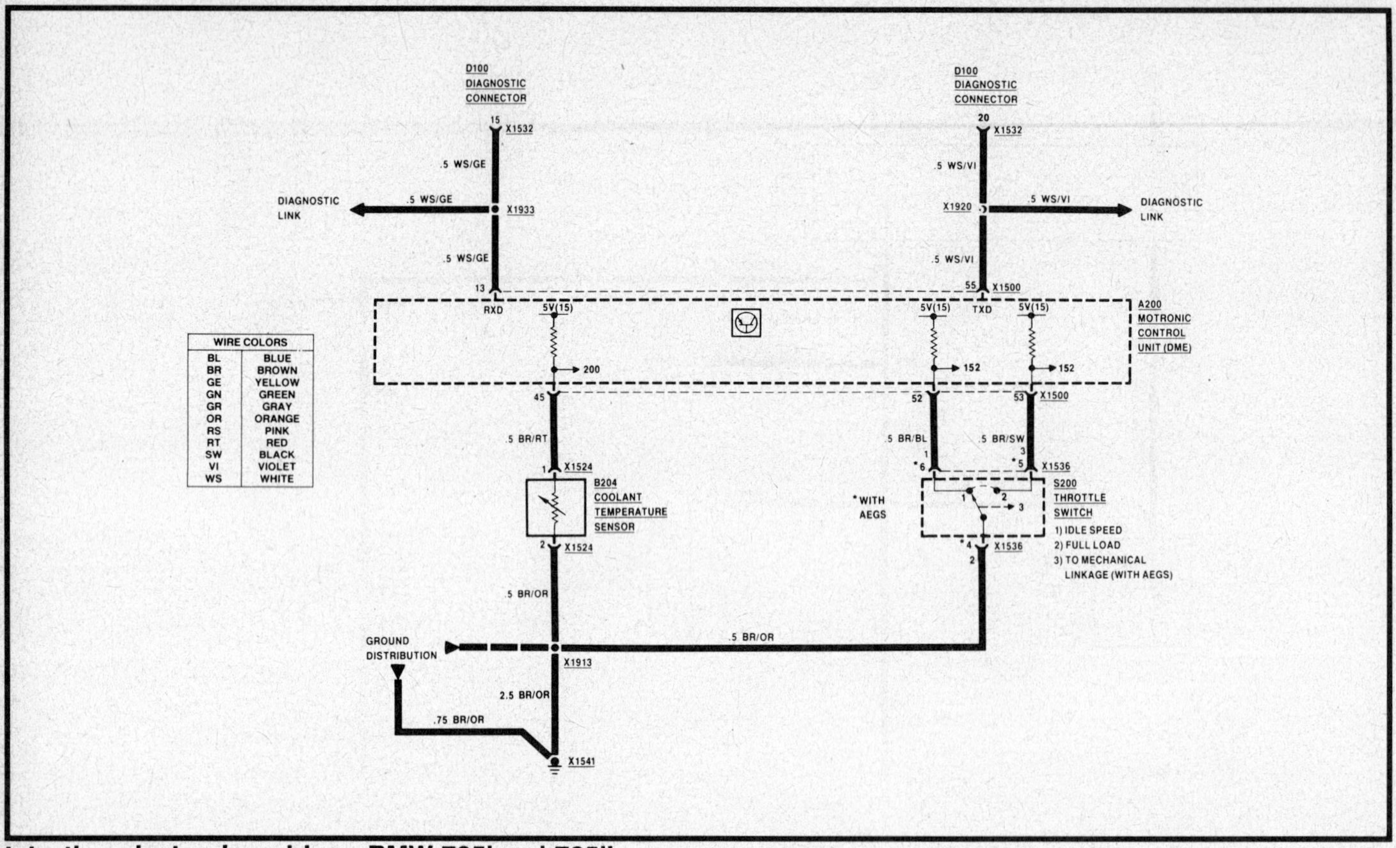

Injection electronics wiring—BMW 735i and 735iL

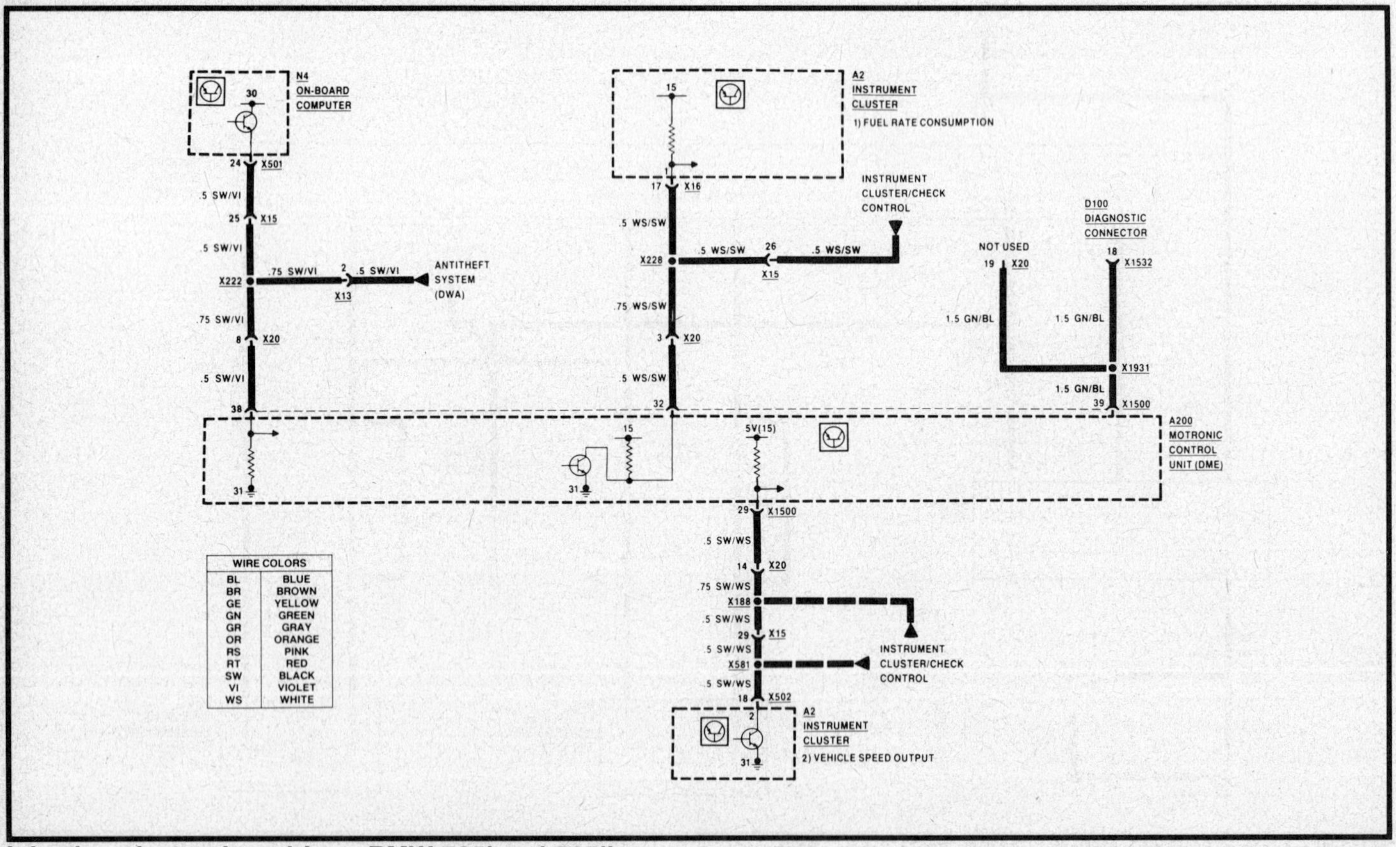

Injection electronics wiring—BMW 735i and 735iL

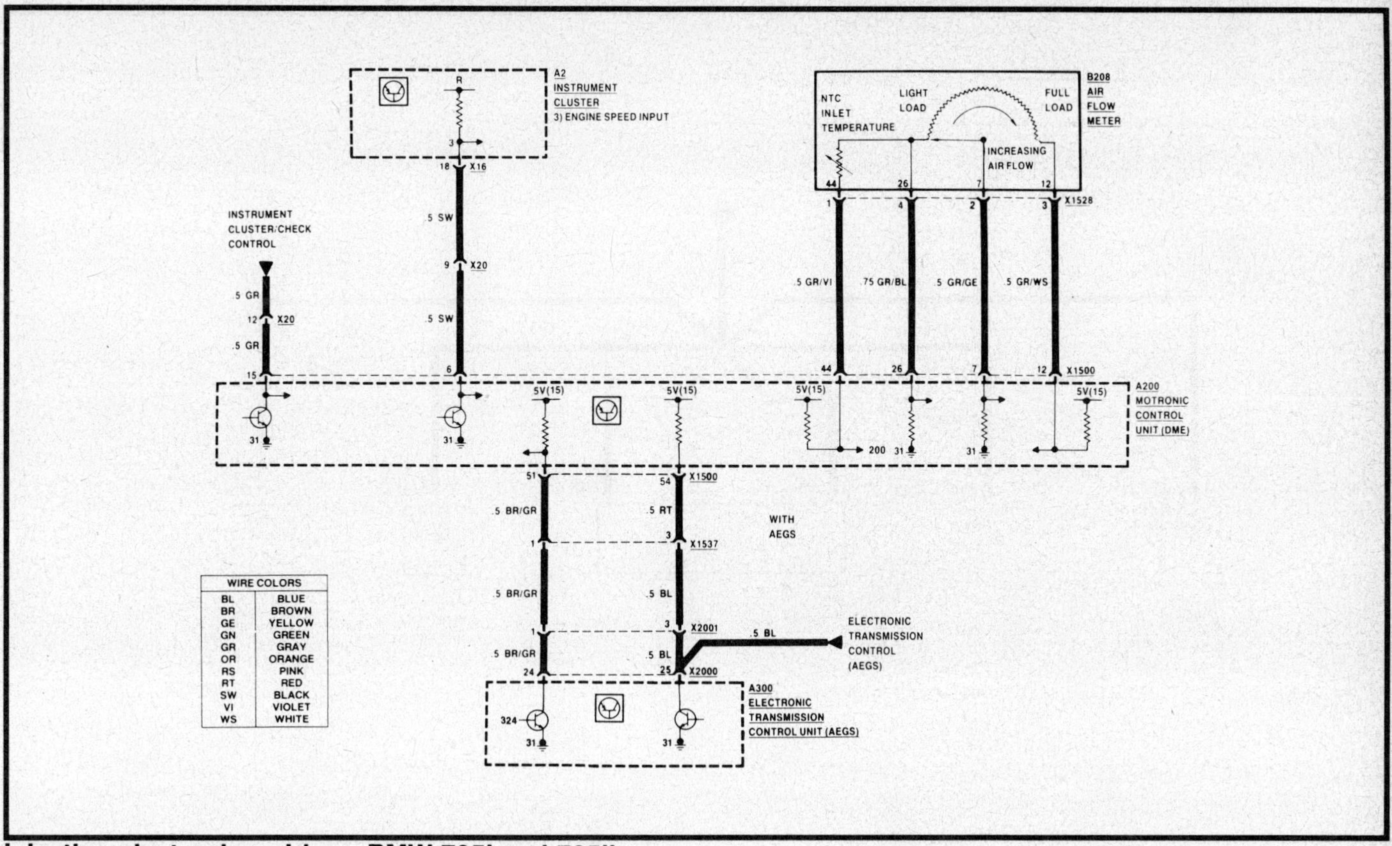

WIRE COLORS	
BL	BLUE
BR	BROWN
GE	YELLOW
GN	GREEN
GR	GRAY
OR	ORANGE
RS	PINK
RT	RED
SW	BLACK
VI	VIOLET
WS	WHITE

Injection electronics wiring—BMW 735i and 735iL

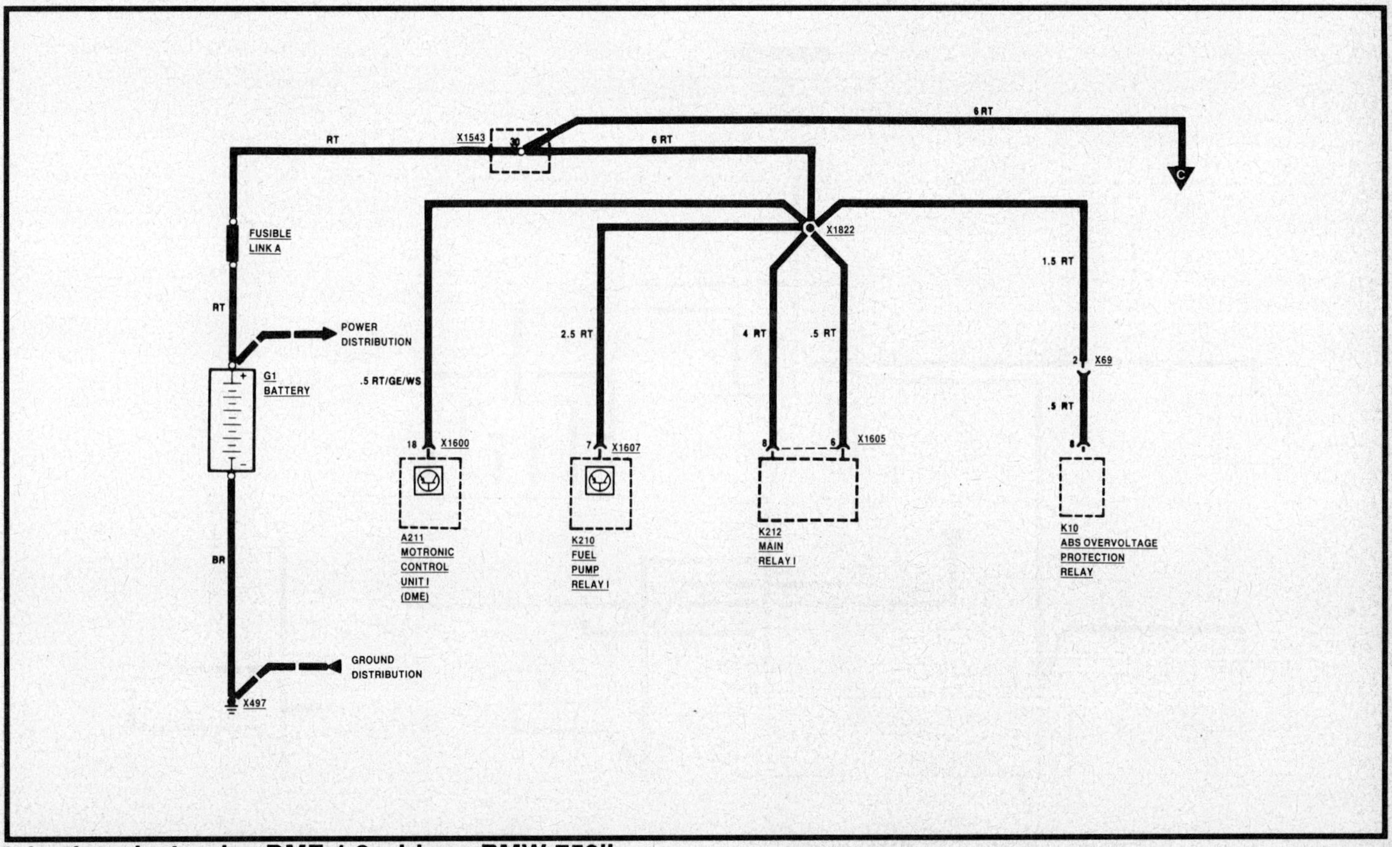

Injection electronics DME 1.2 wiring—BMW 750iL

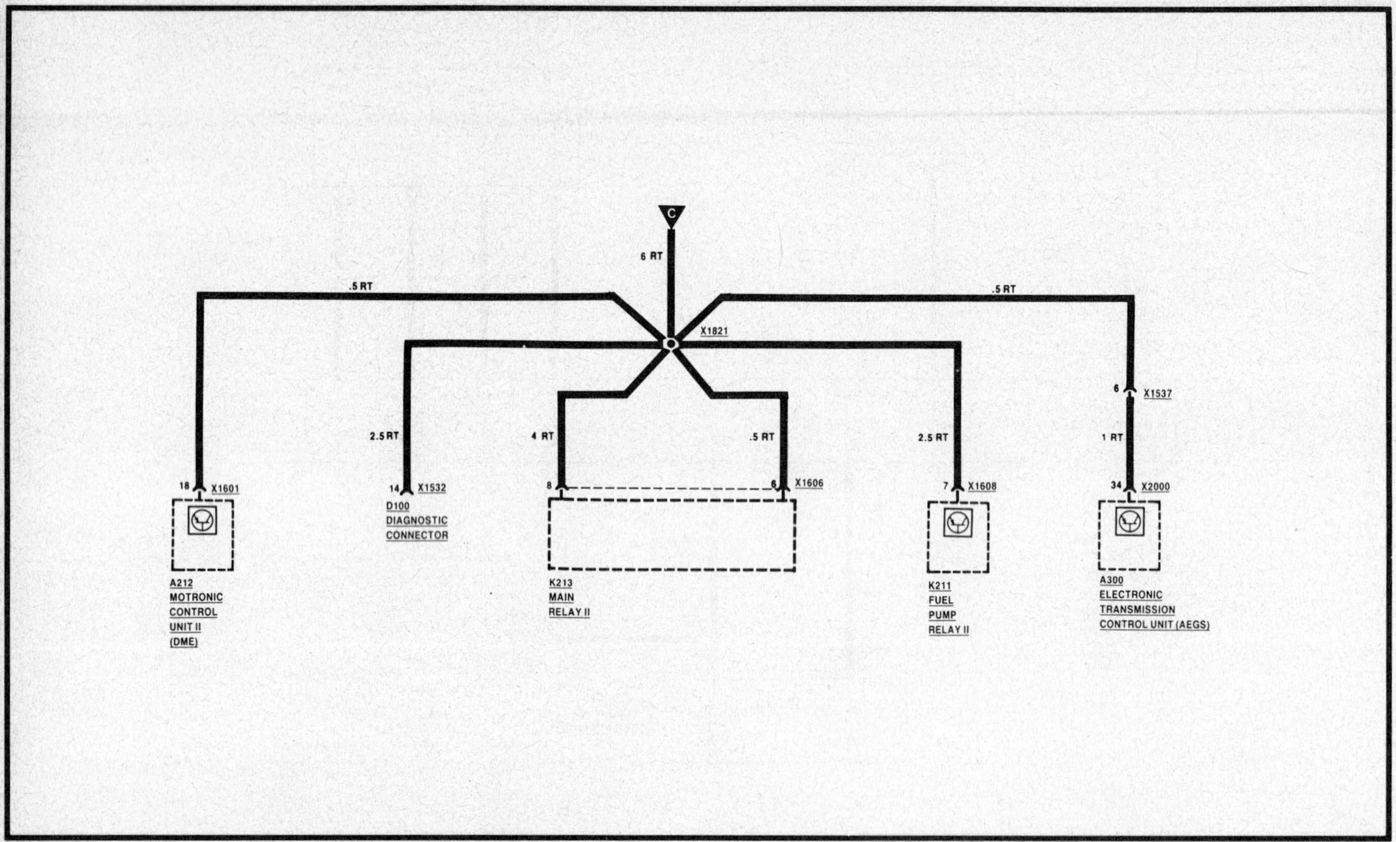

Injection electronics DME 1.2 wiring—BMW 750iL

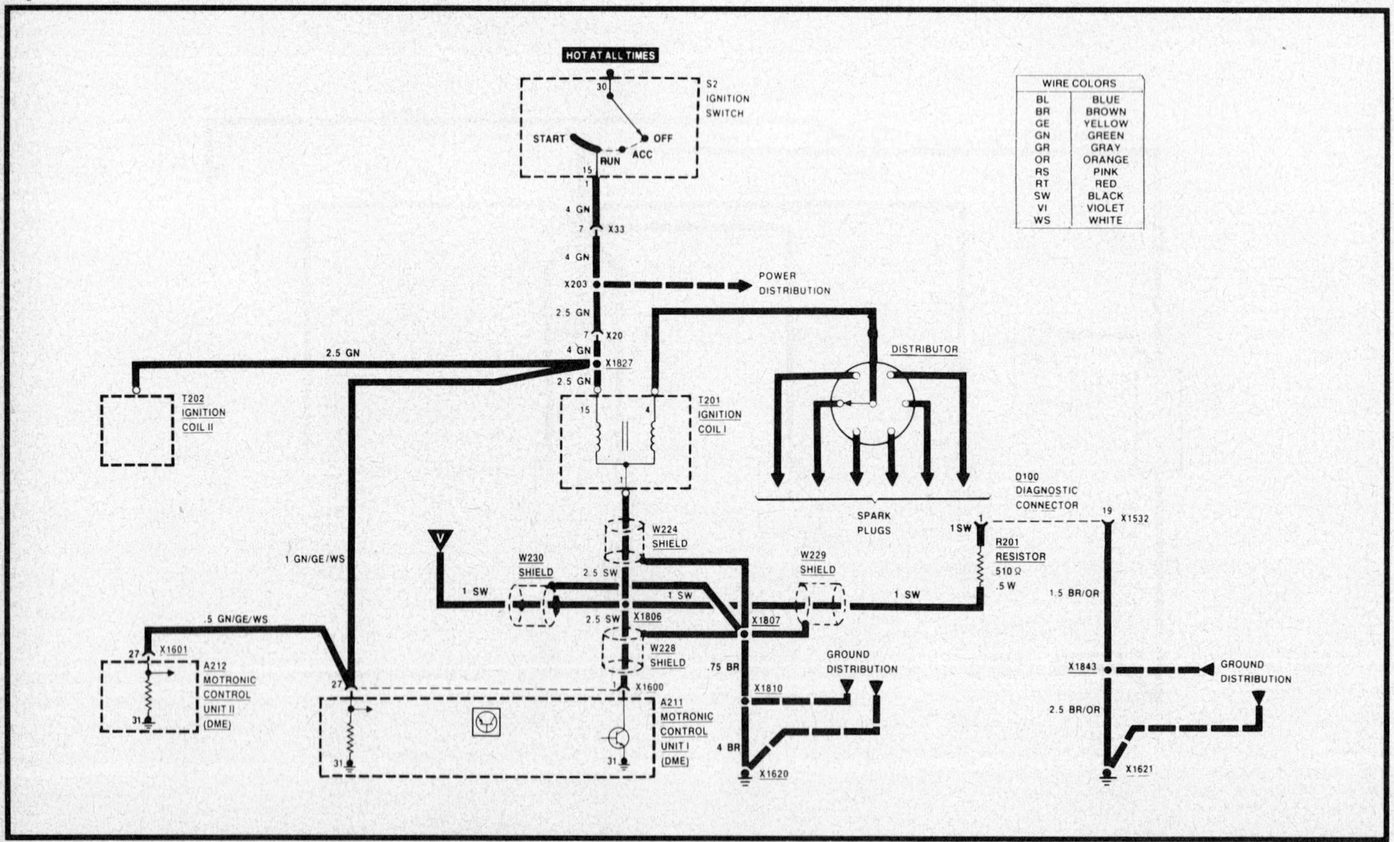

Injection electronics DME 1.2 wiring—BMW 750iL

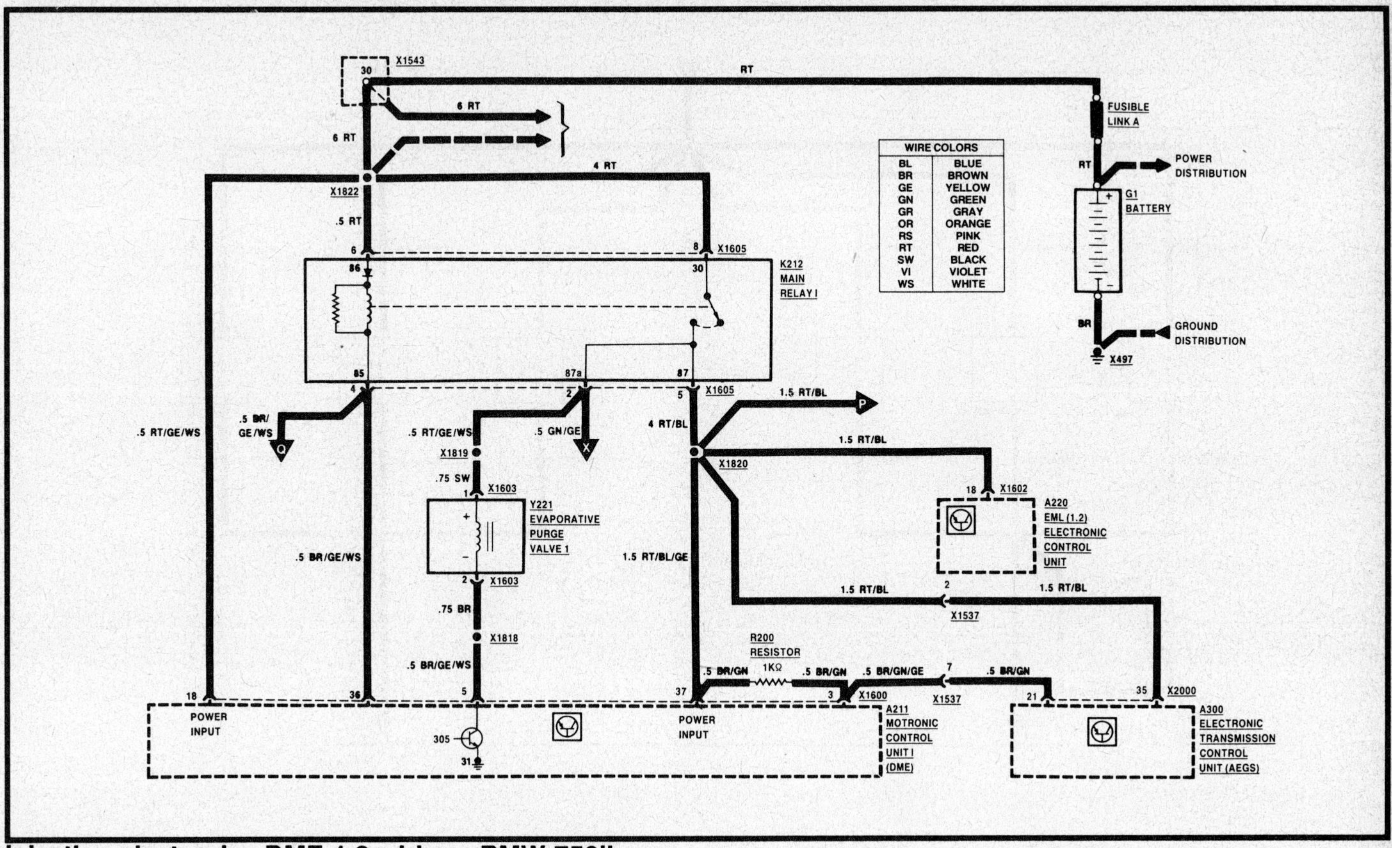

Injection electronics DME 1.2 wiring – BMW 750iL

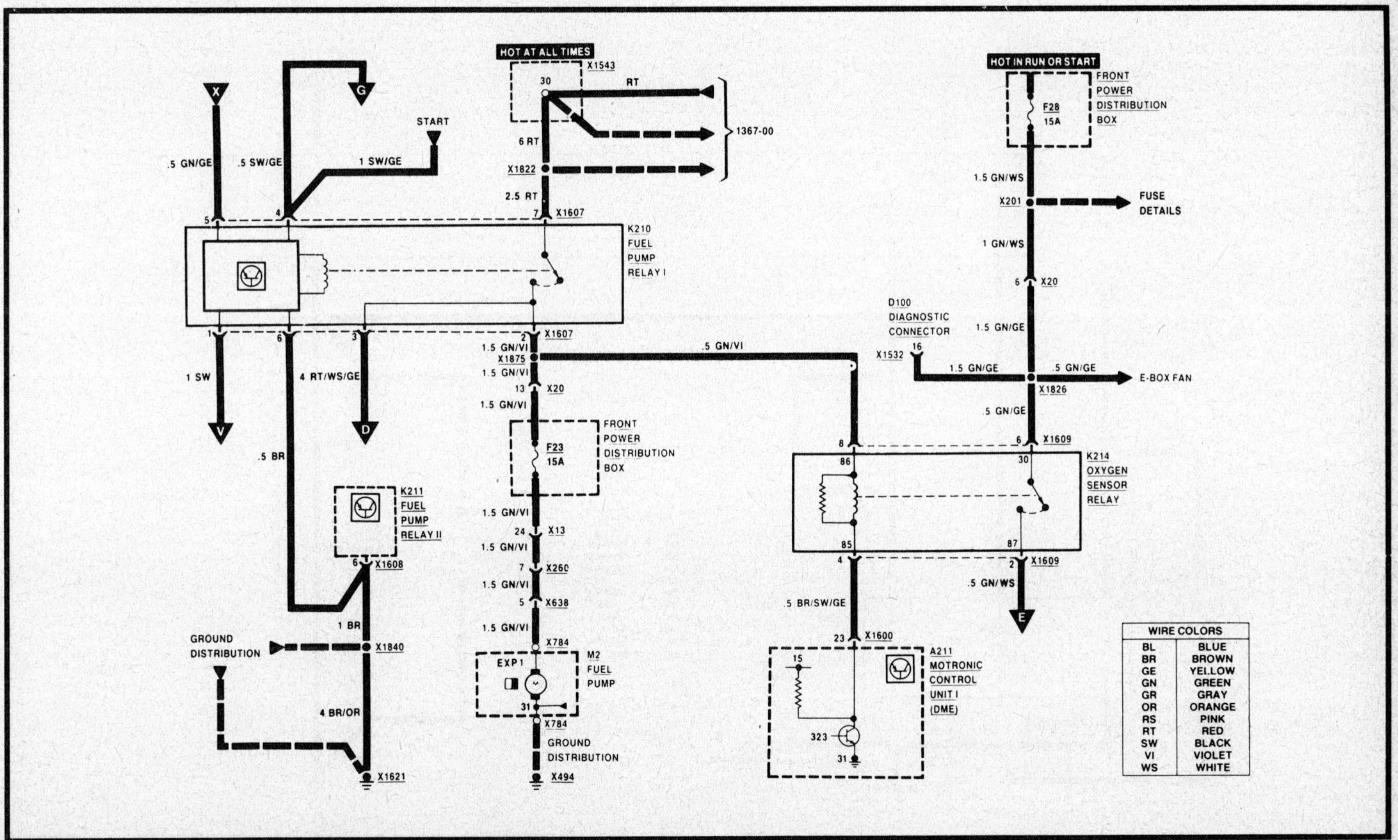

Injection electronics DME 1.2 wiring – BMW 750iL

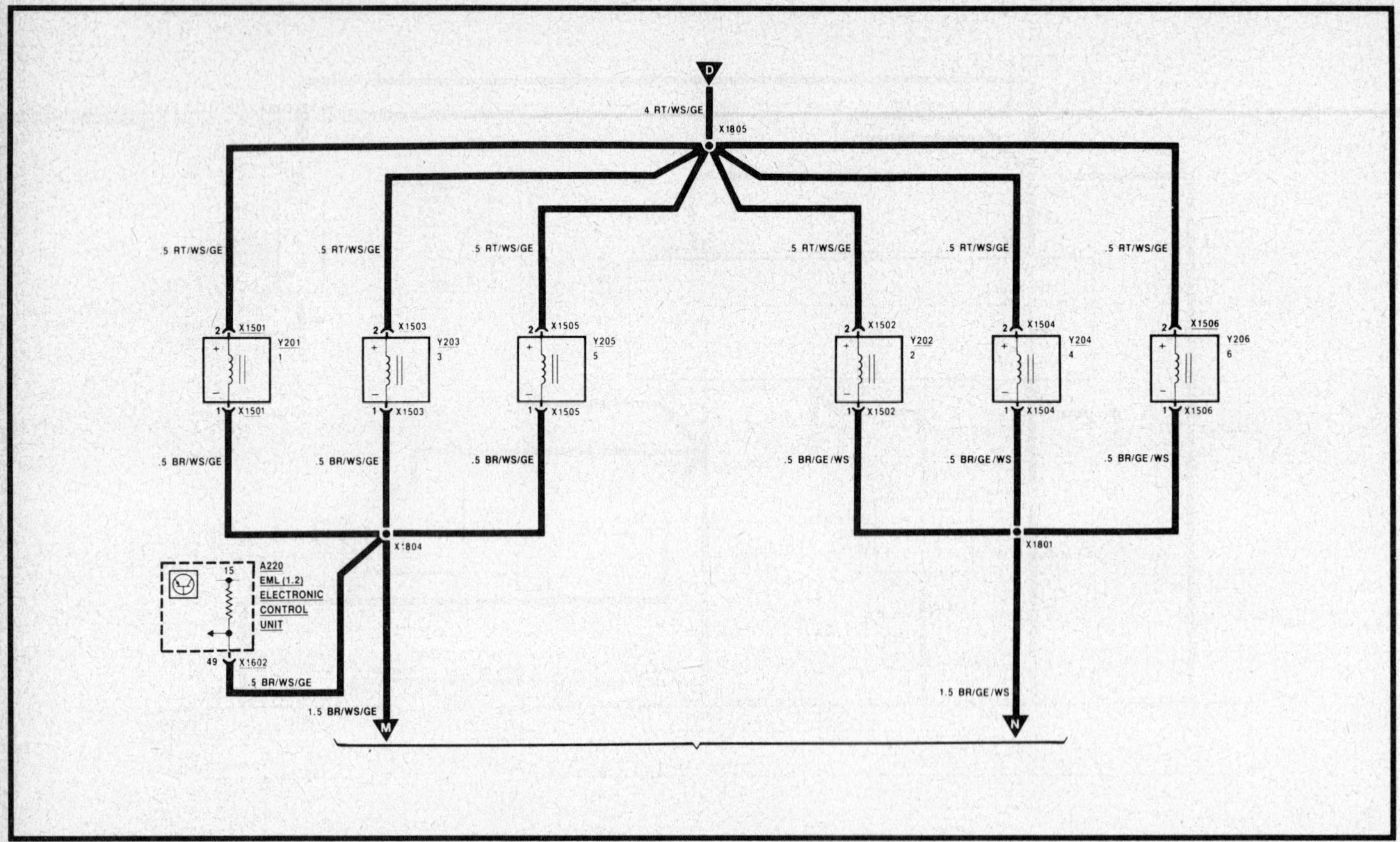

Injection electronics DME 1.2 wiring—BMW 750iL

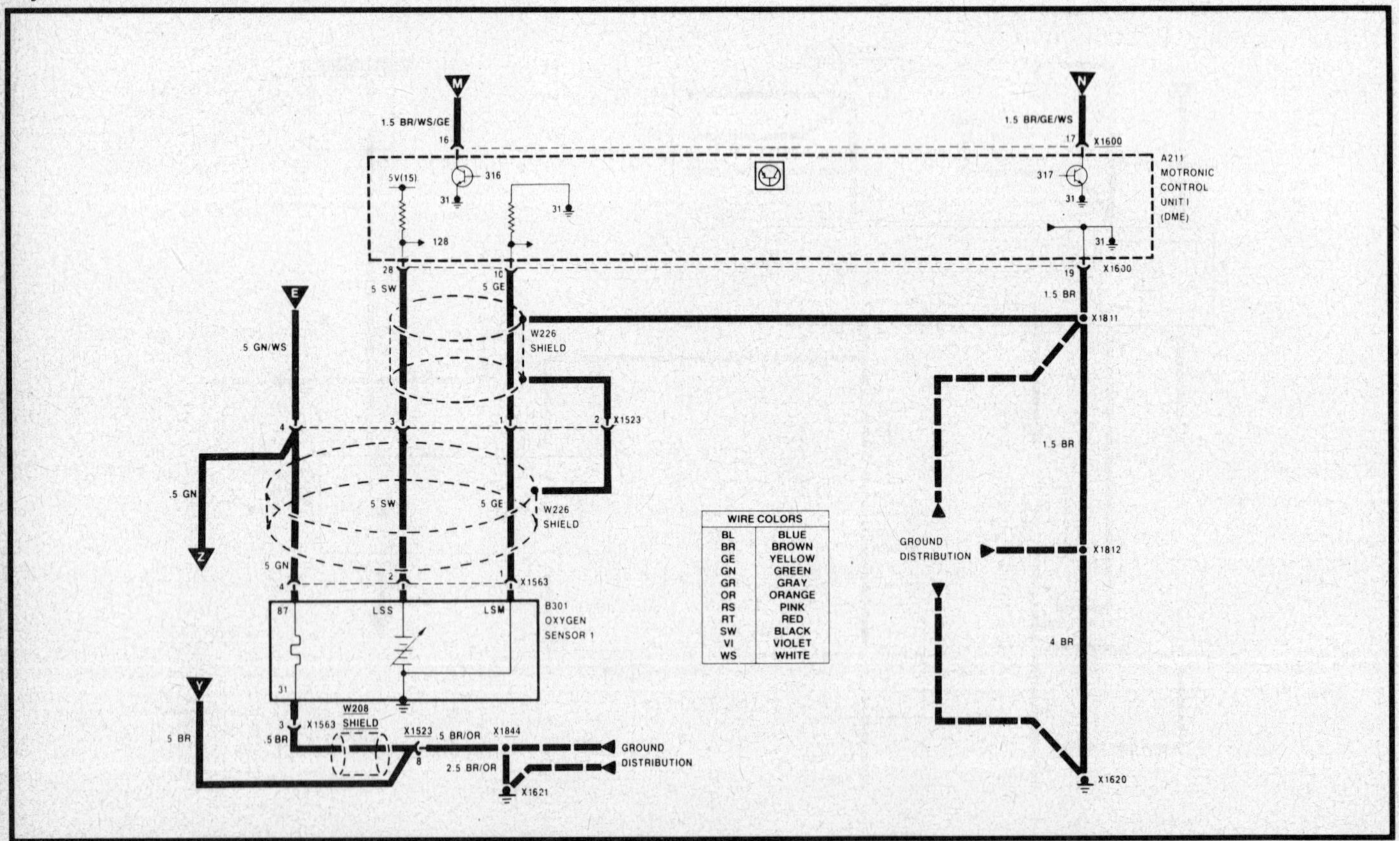

Injection electronics DME 1.2 wiring—BMW 750iL

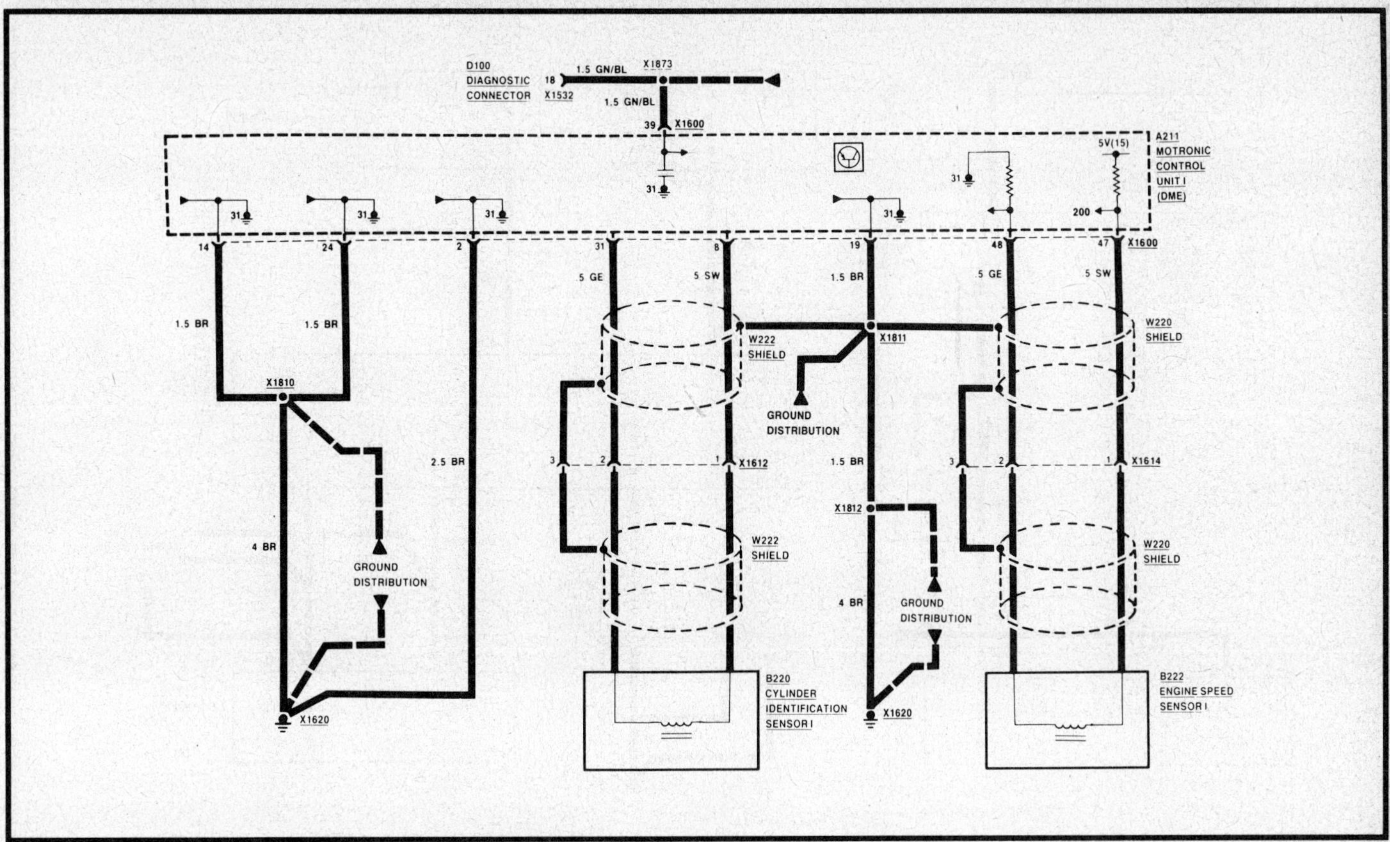

Injection electronics DME 1.2 wiring—BMW 750iL

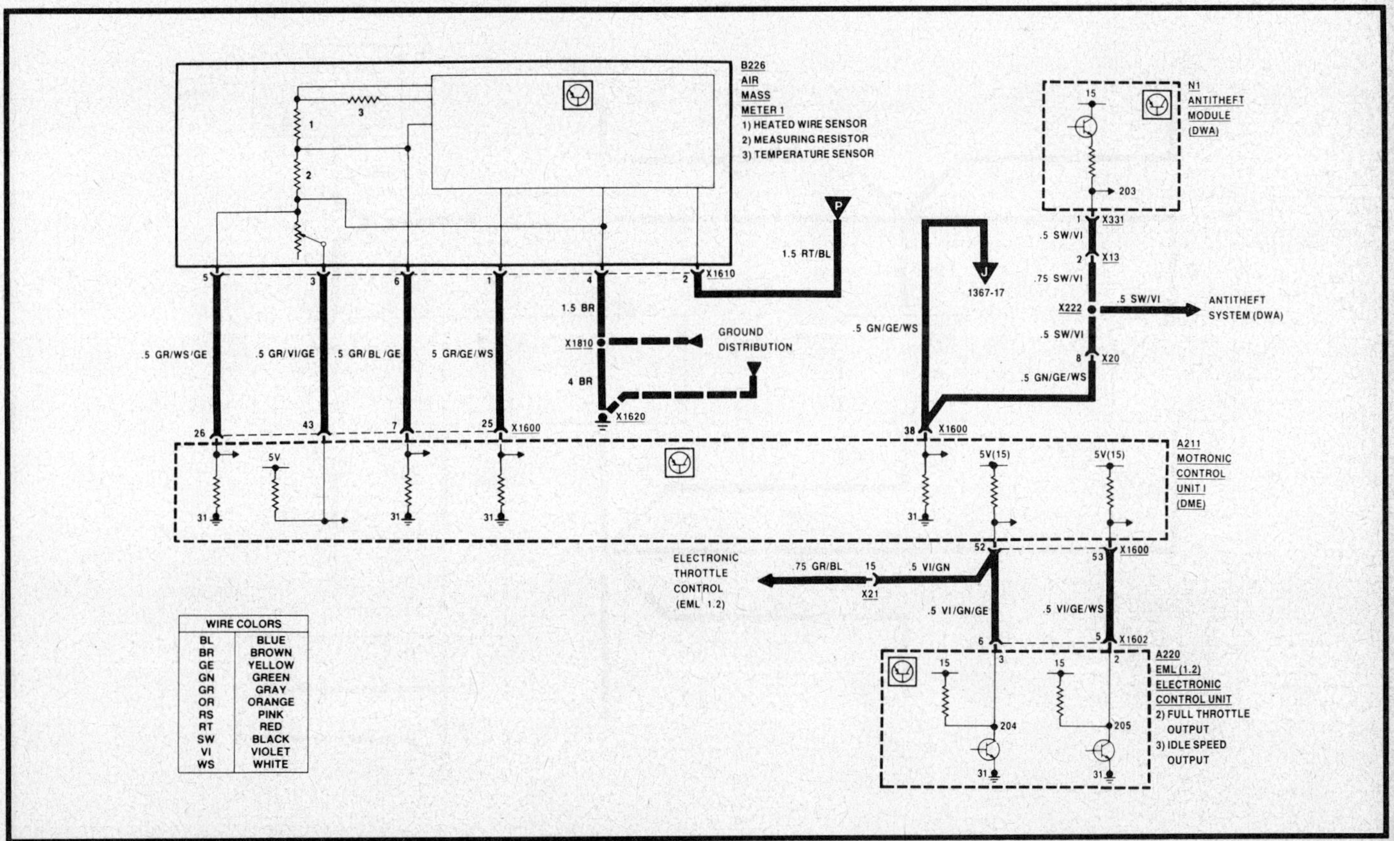

Injection electronics DME 1.2 wiring—BMW 750iL

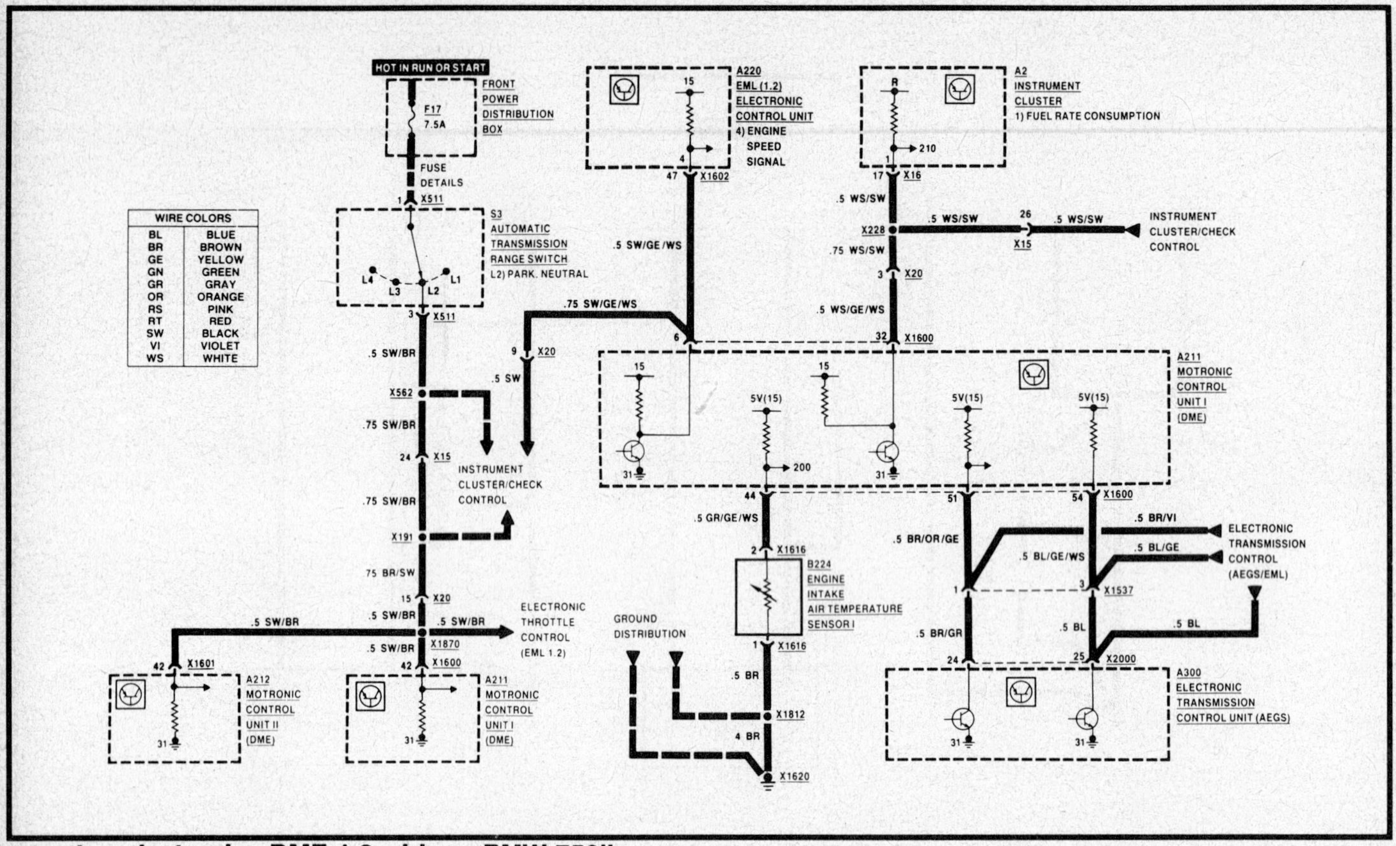

Injection electronics DME 1.2 wiring—BMW 750iL

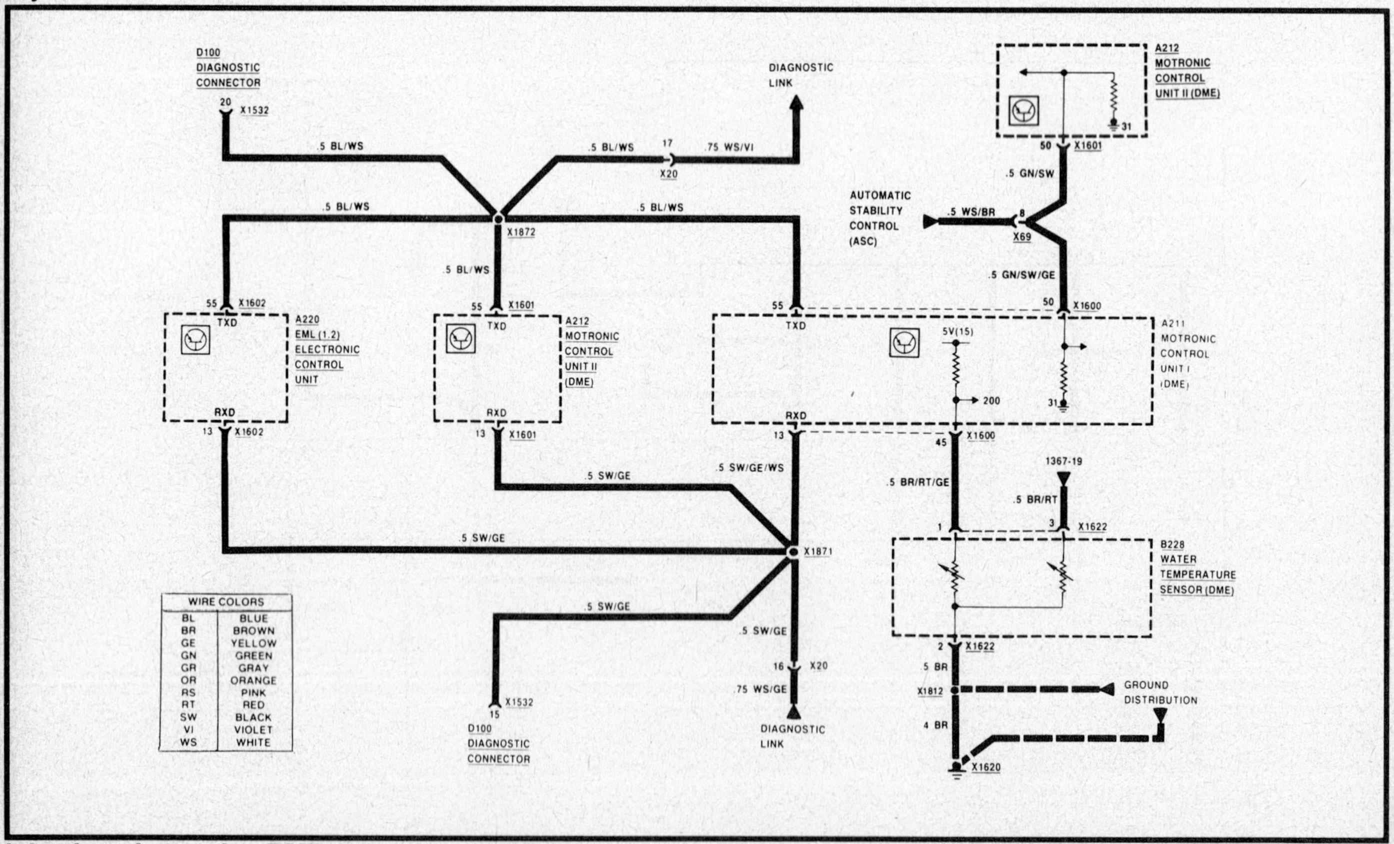

Injection electronics DME 1.2 wiring—BMW 750iL

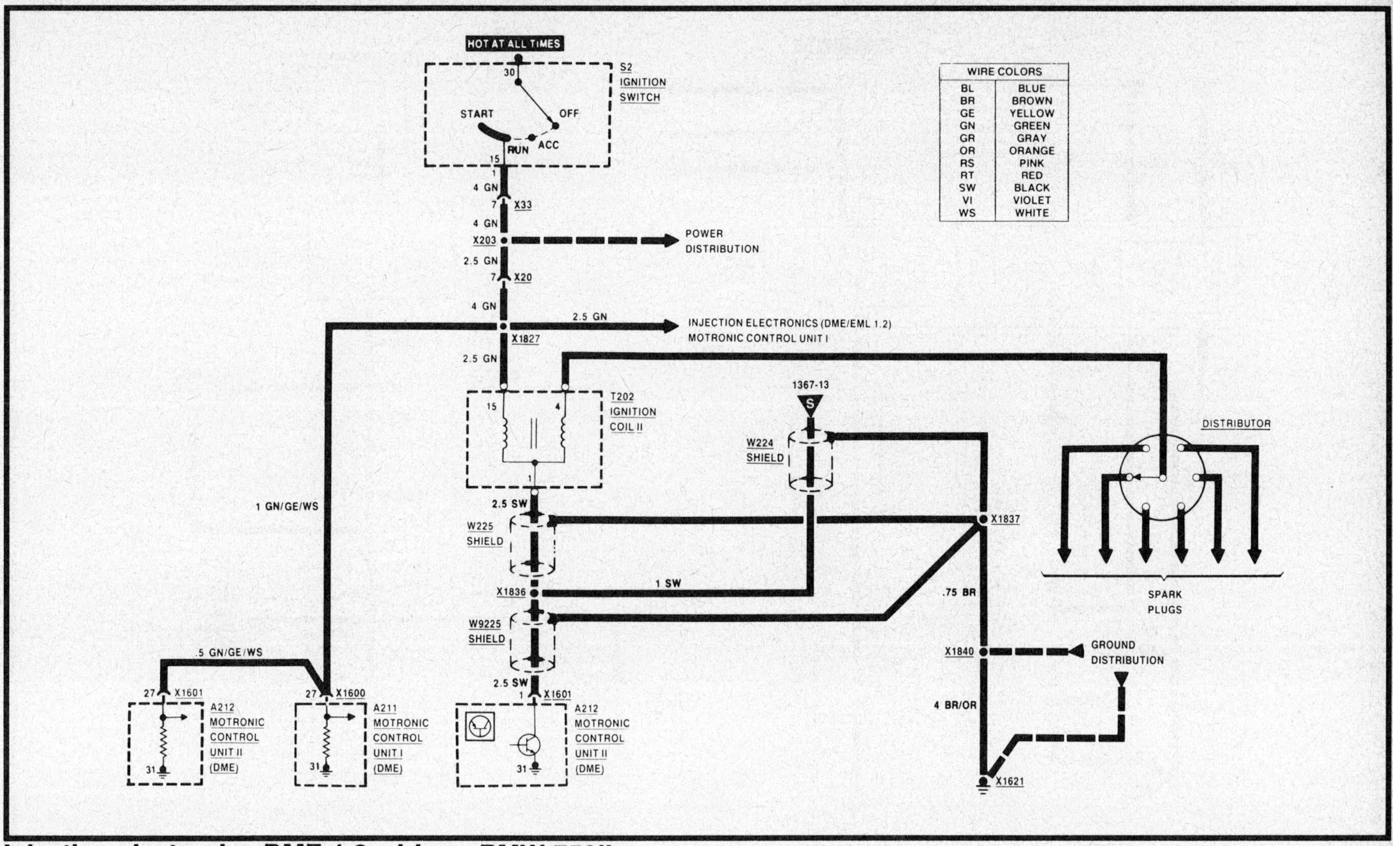

Injection electronics DME 1.2 wiring—BMW 750iL

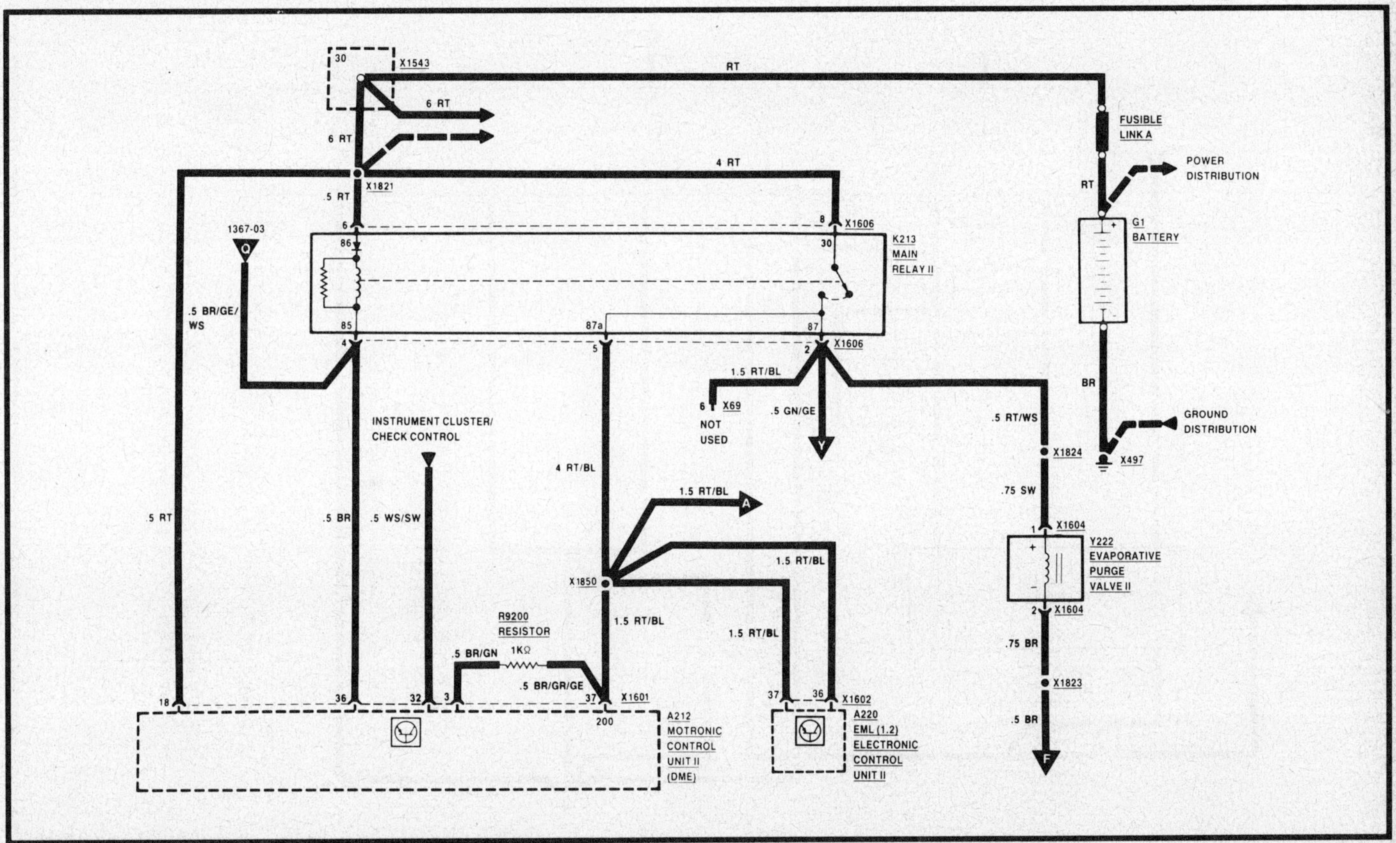

Injection electronics DME 1.2 wiring—BMW 750iL

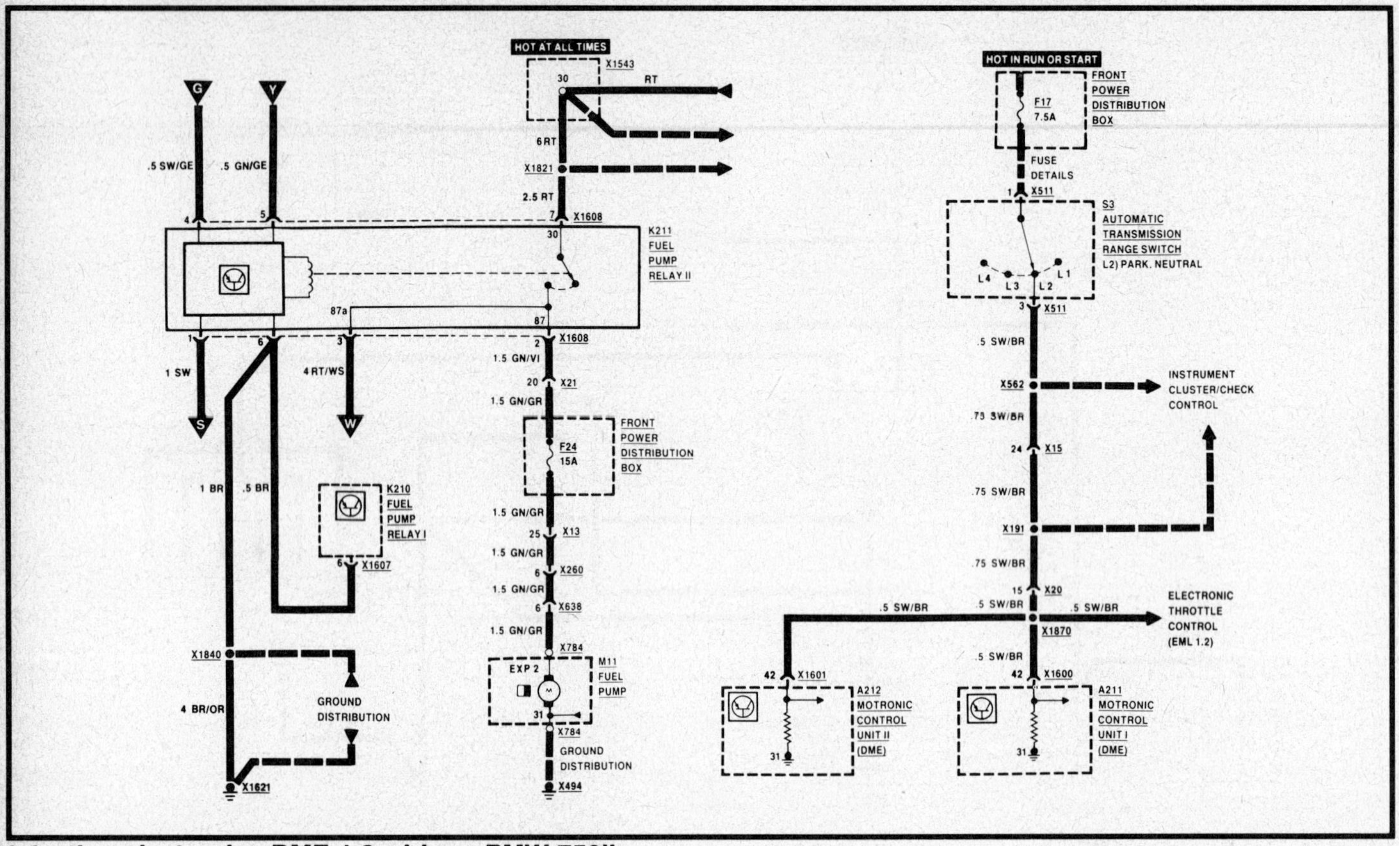

Injection electronics DME 1.2 wiring—BMW 750iL

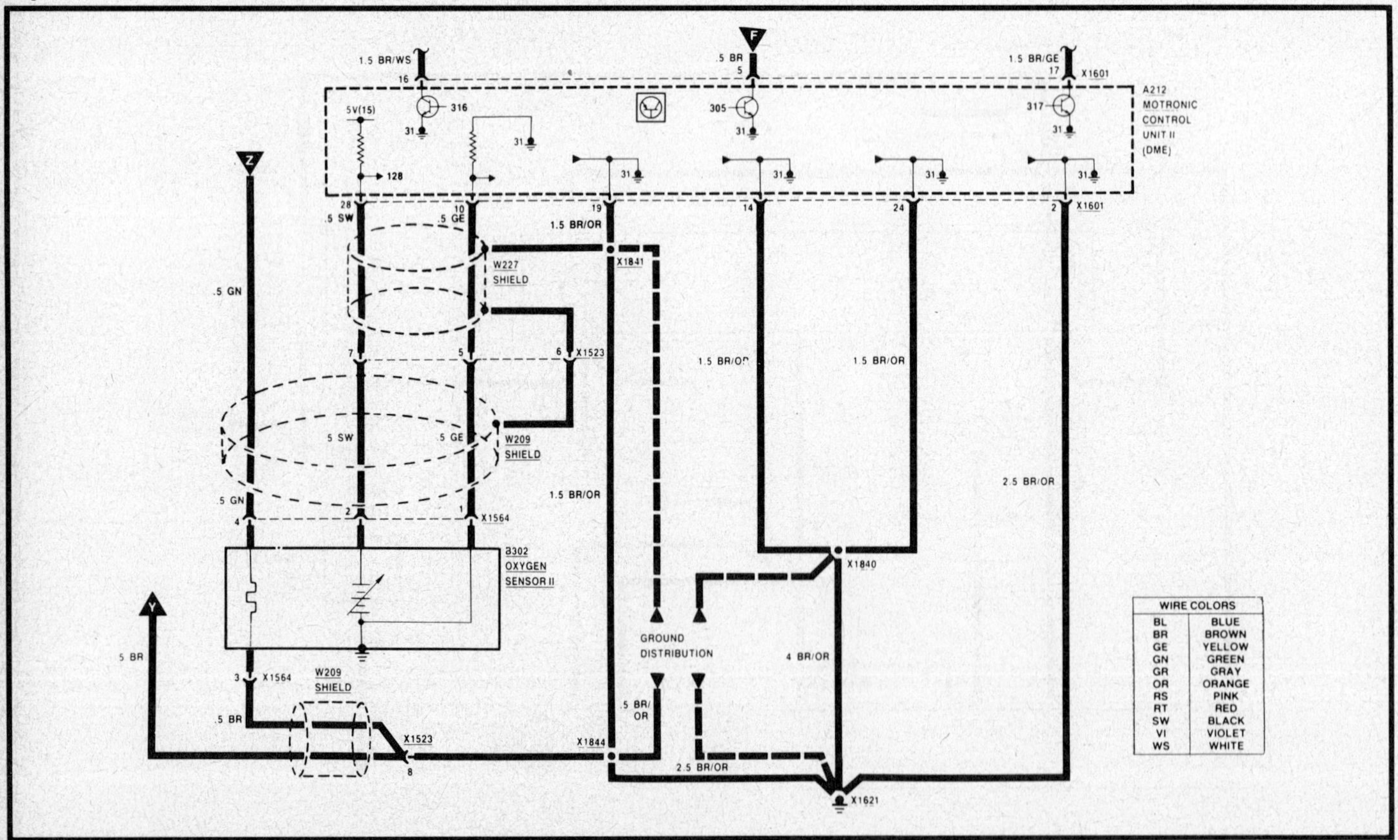

Injection electronics DME 1.2 wiring—BMW 750iL

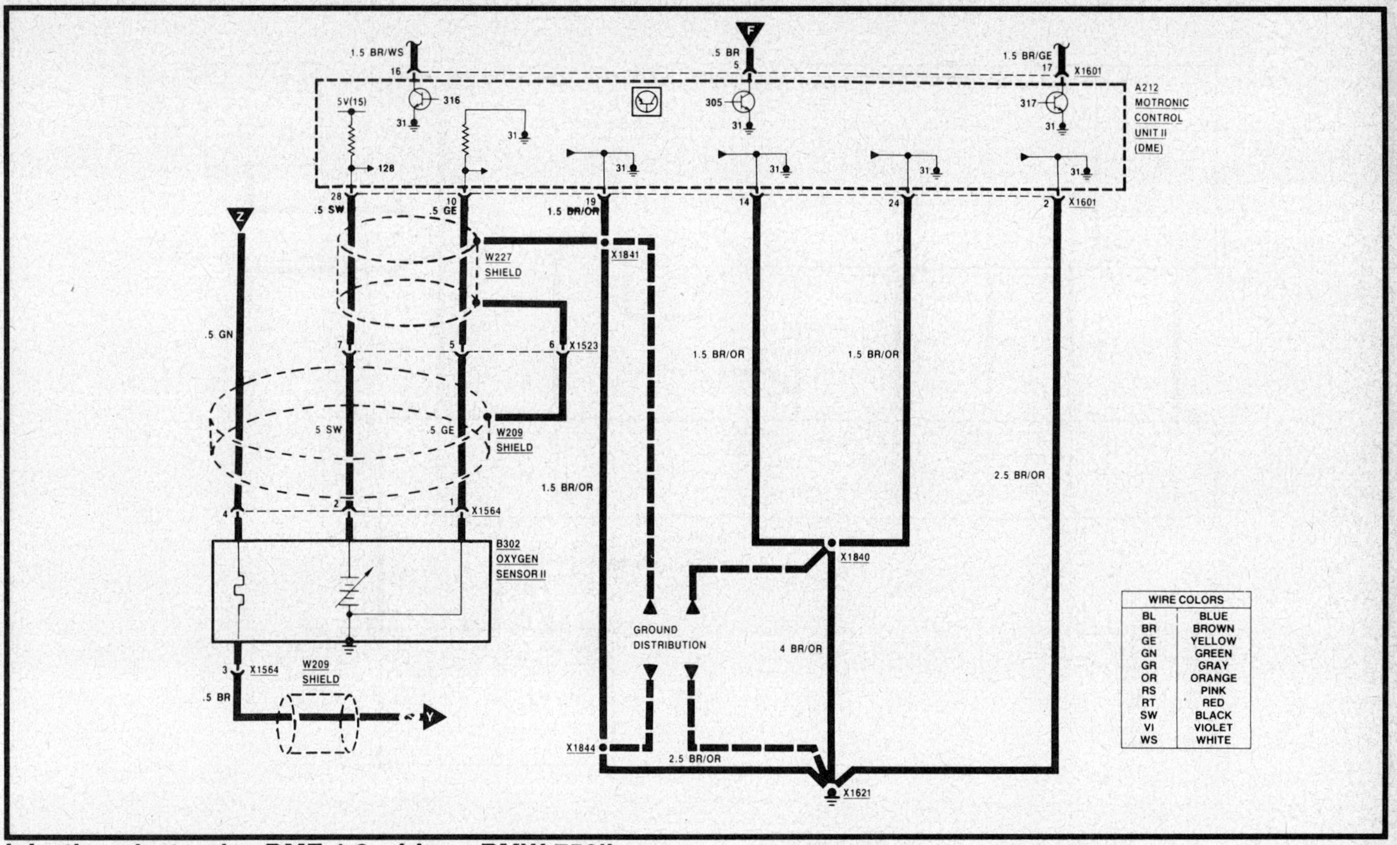

Injection electronics DME 1.2 wiring—BMW 750iL

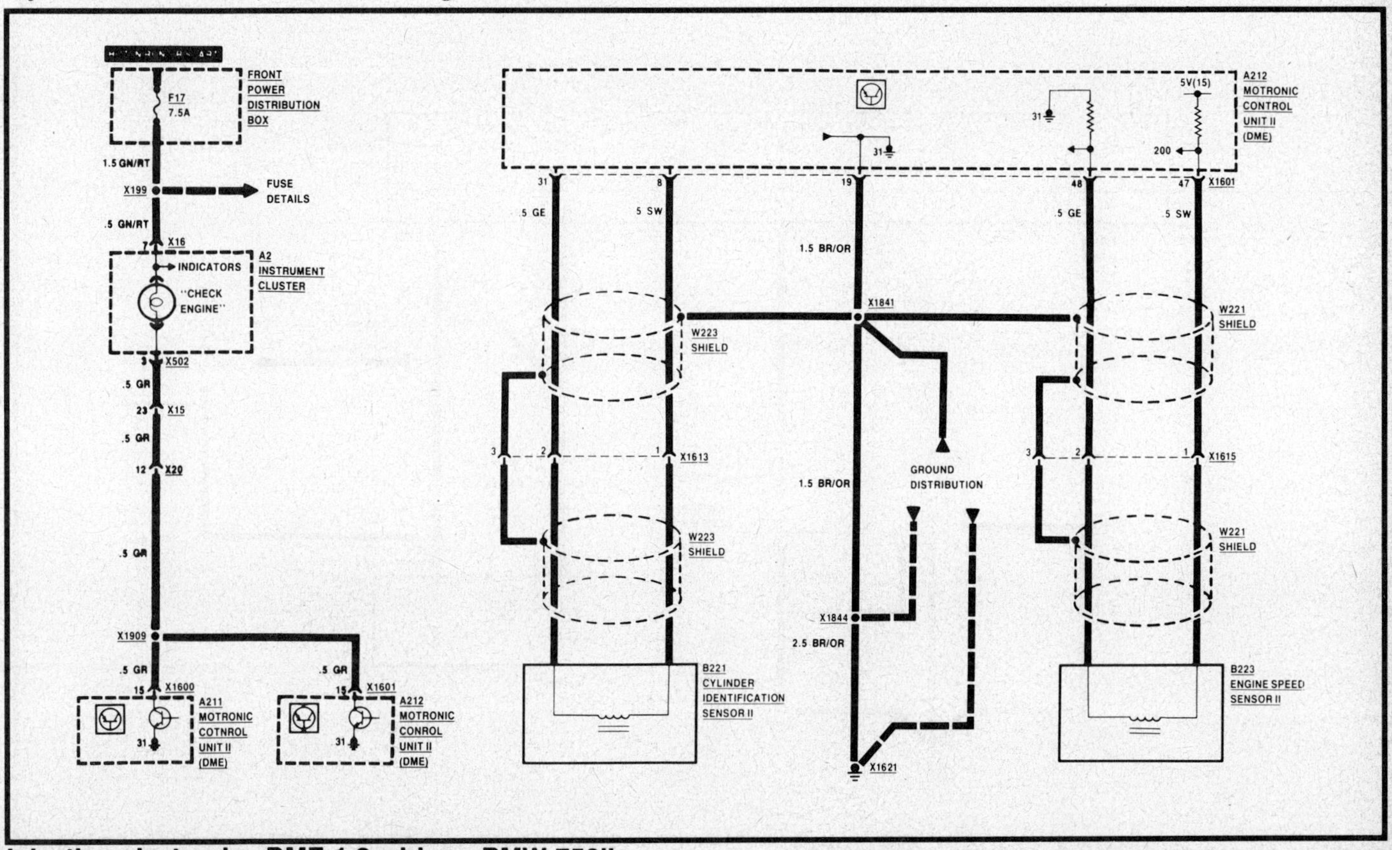

Injection electronics DME 1.2 wiring—BMW 750iL

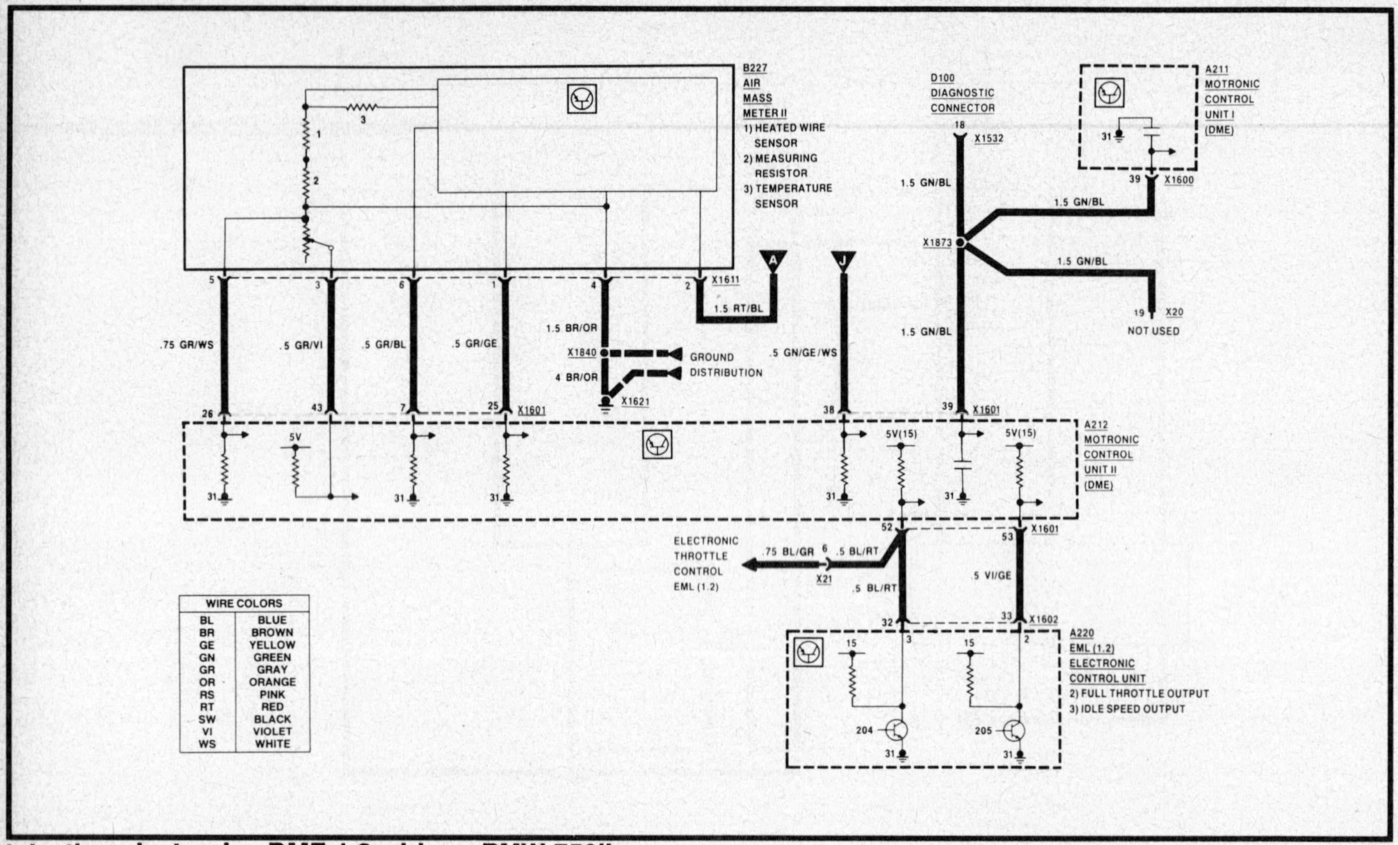

Injection electronics DME 1.2 wiring—BMW 750iL

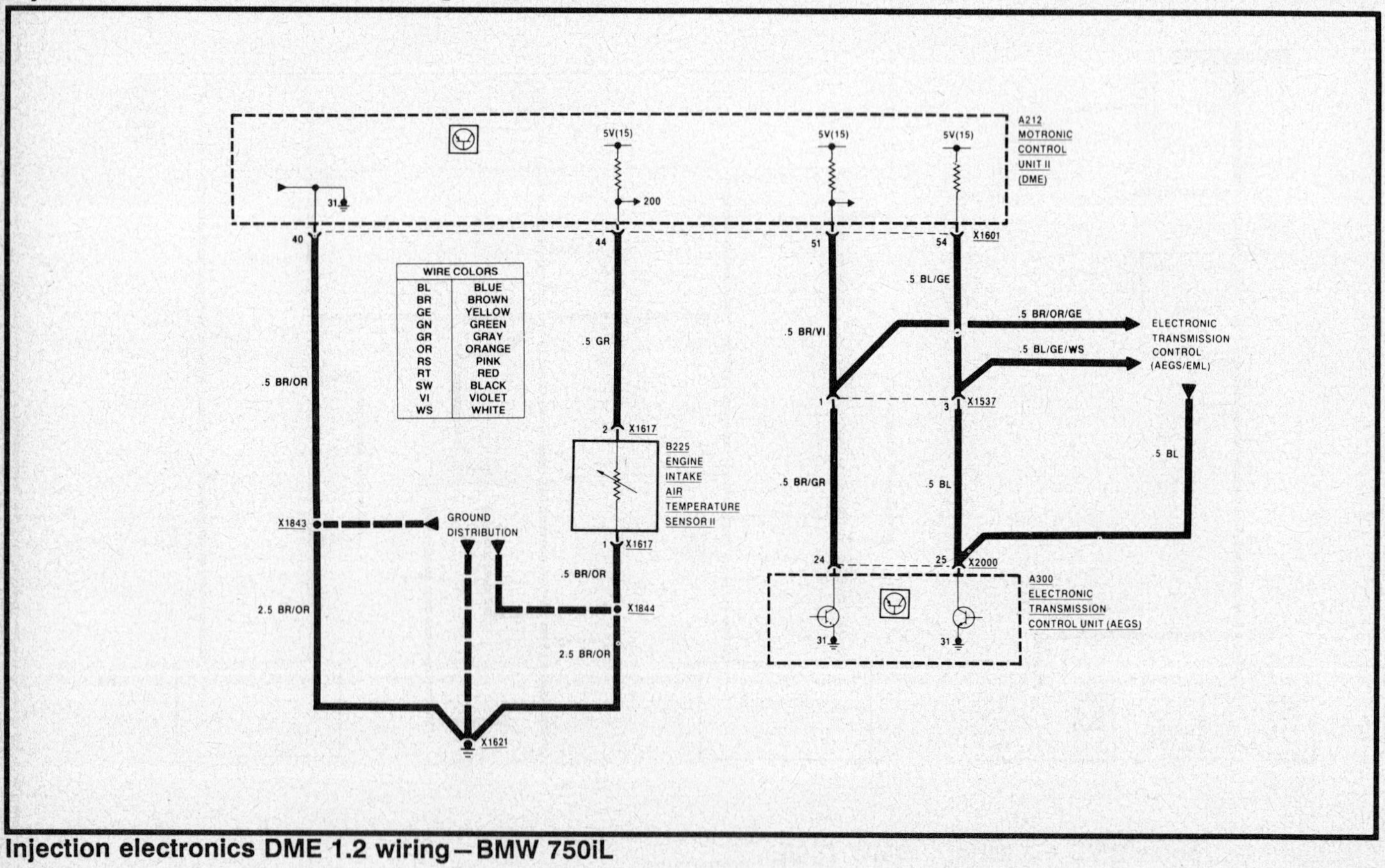

Injection electronics DME 1.2 wiring—BMW 750iL

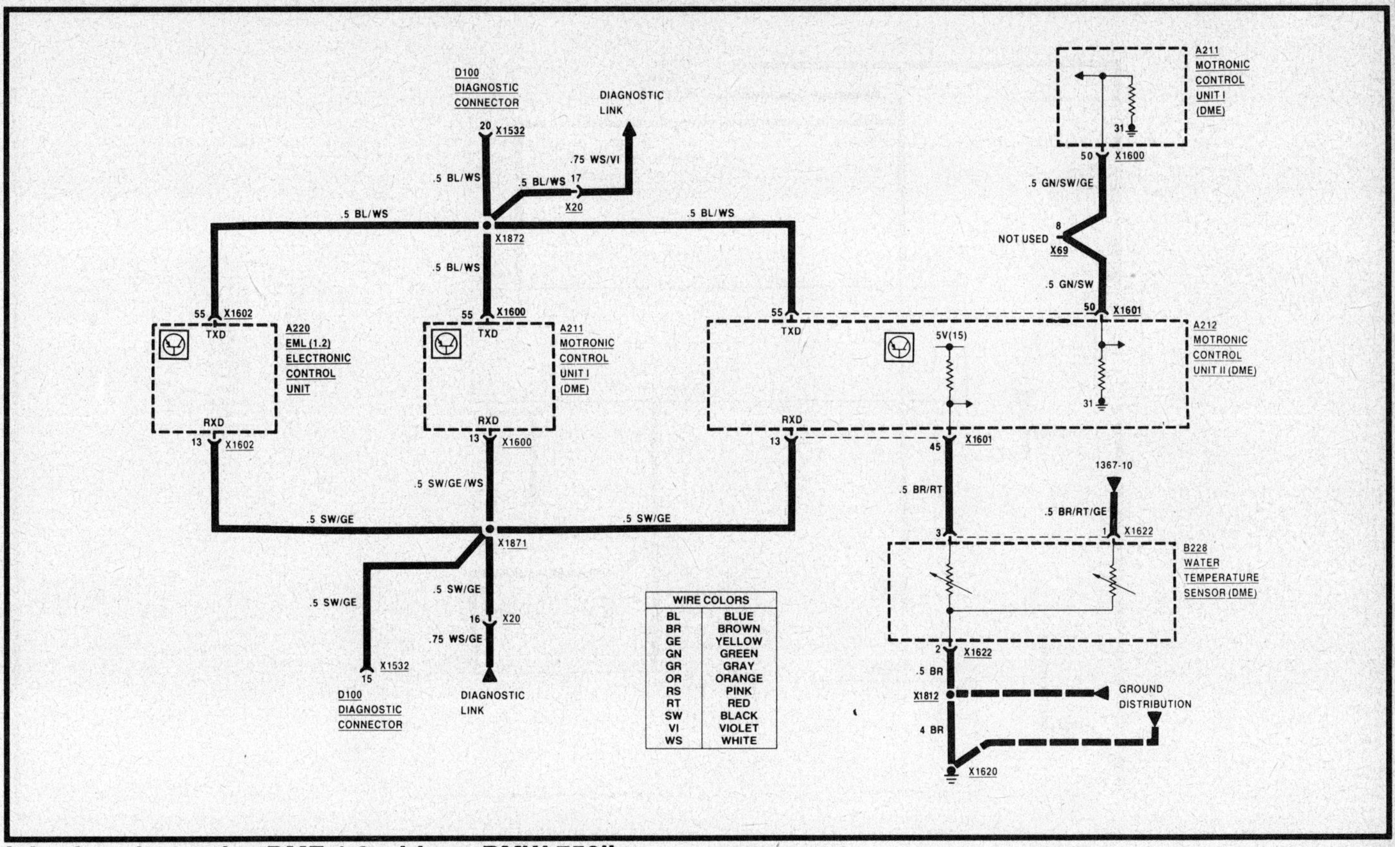

Injection electronics DME 1.2 wiring—BMW 750iL

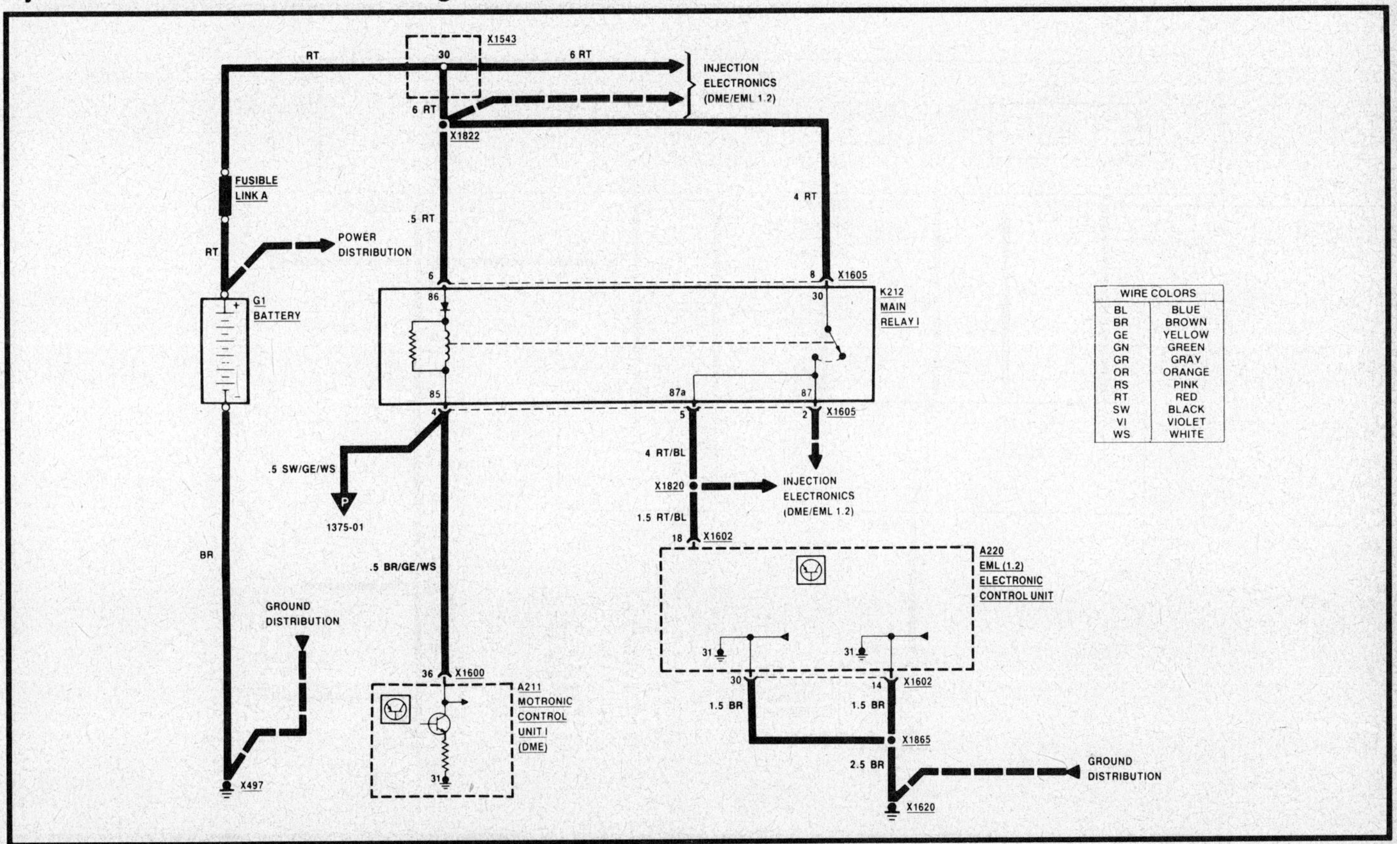

Electronic throttle control (EML 1.2) wiring—BMW 750iL

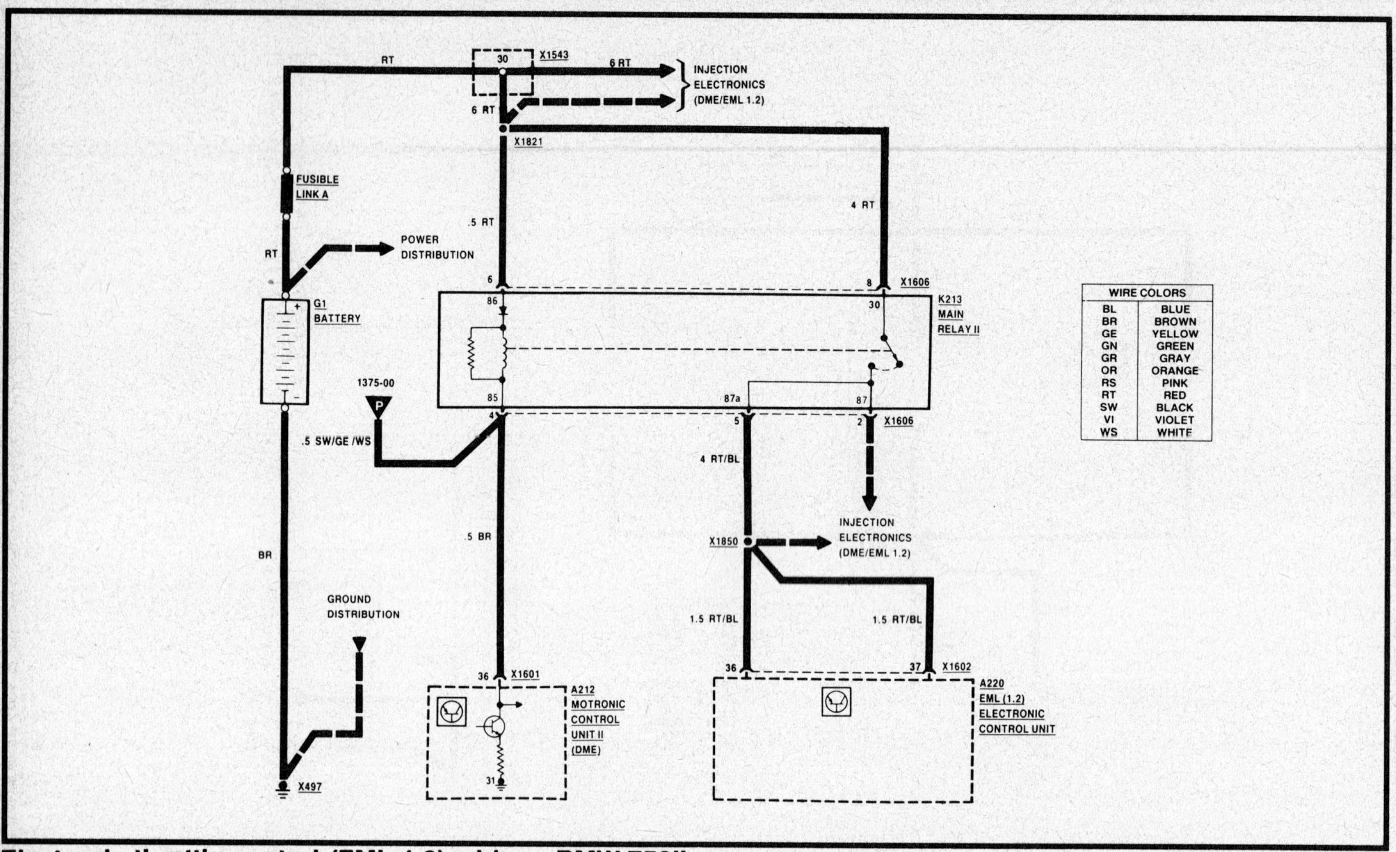

Electronic throttle control (EML 1.2) wiring—BMW 750iL

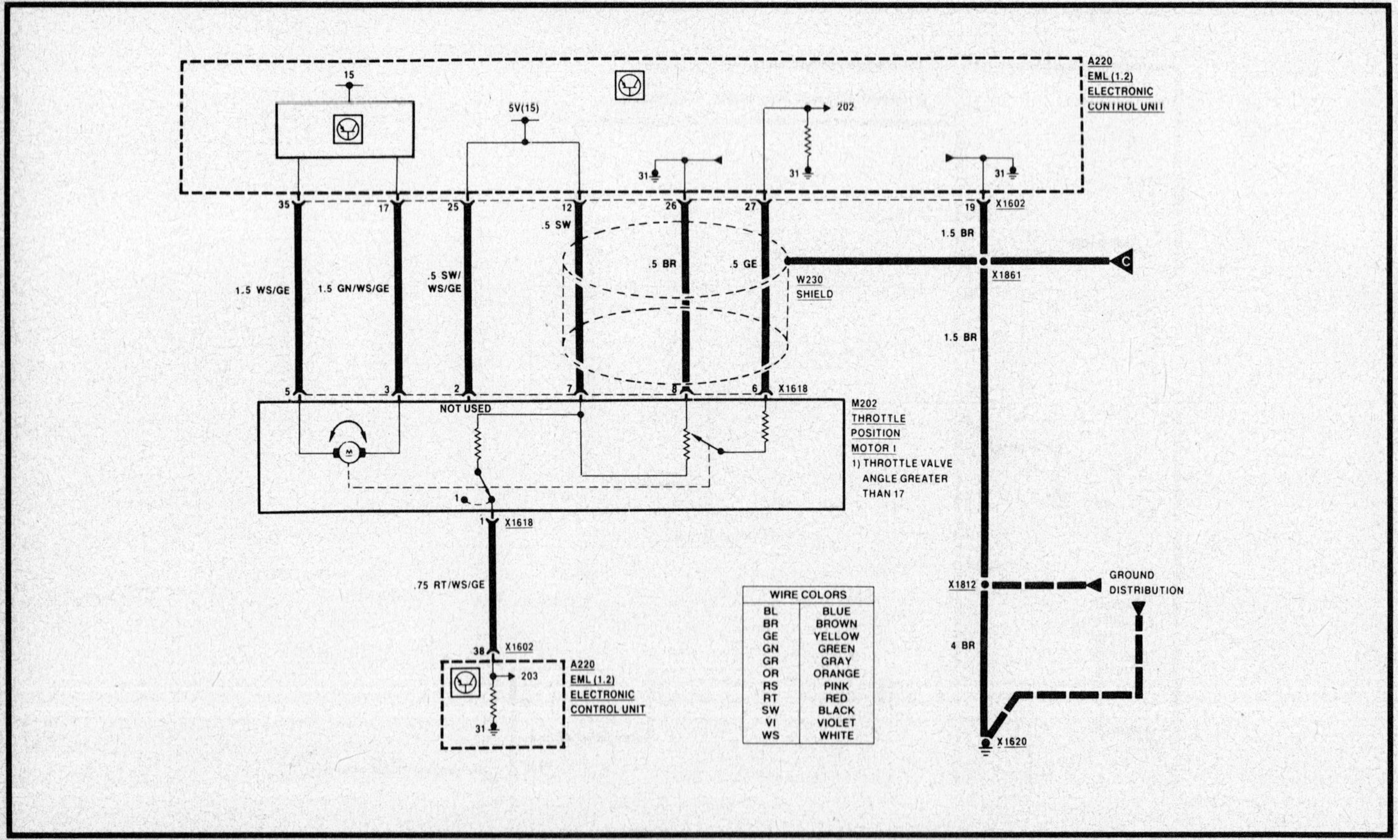

Electronic throttle control (EML 1.2) wiring—BMW 750iL

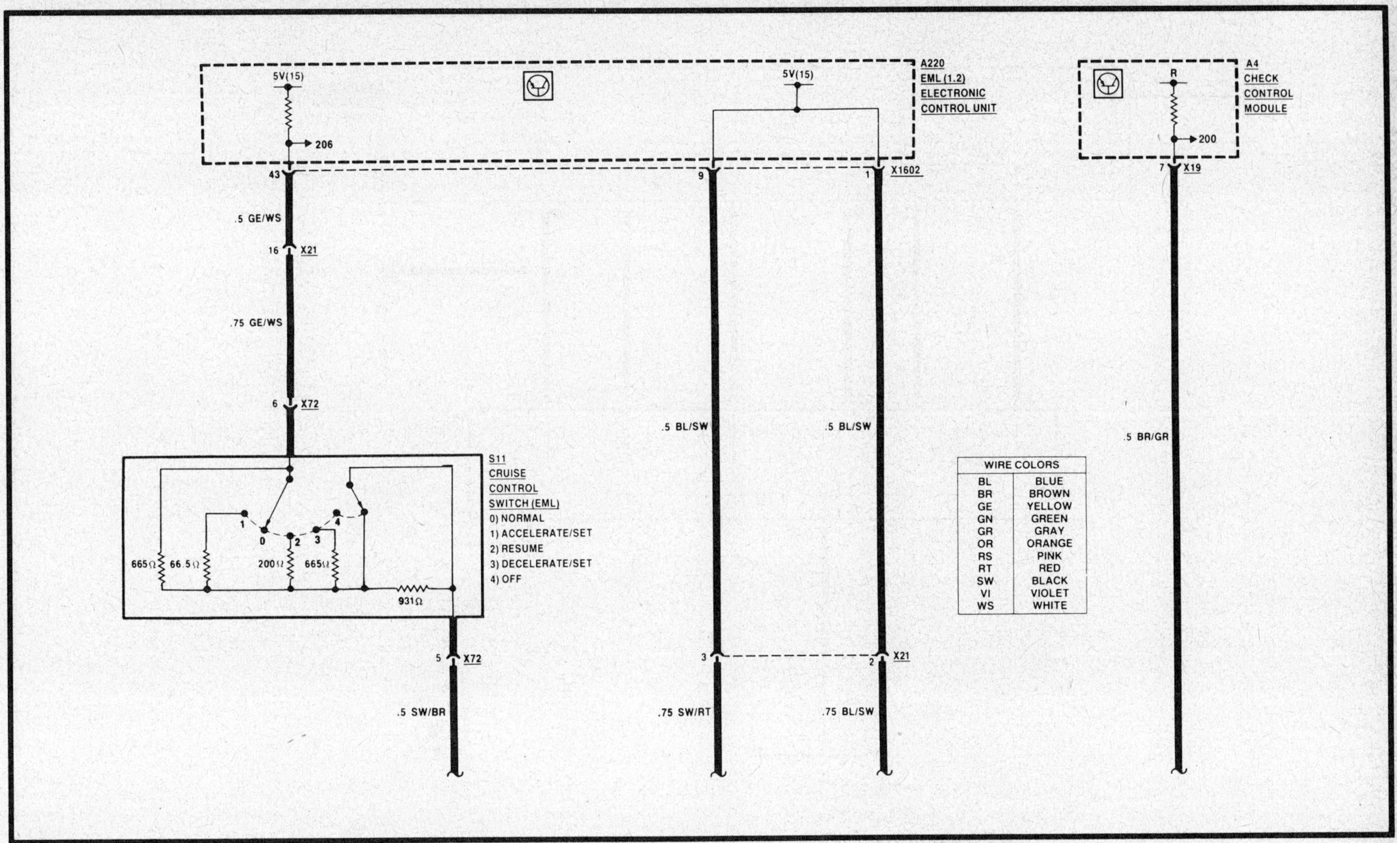

Electronic throttle control (EML 1.2) wiring—BMW 750iL

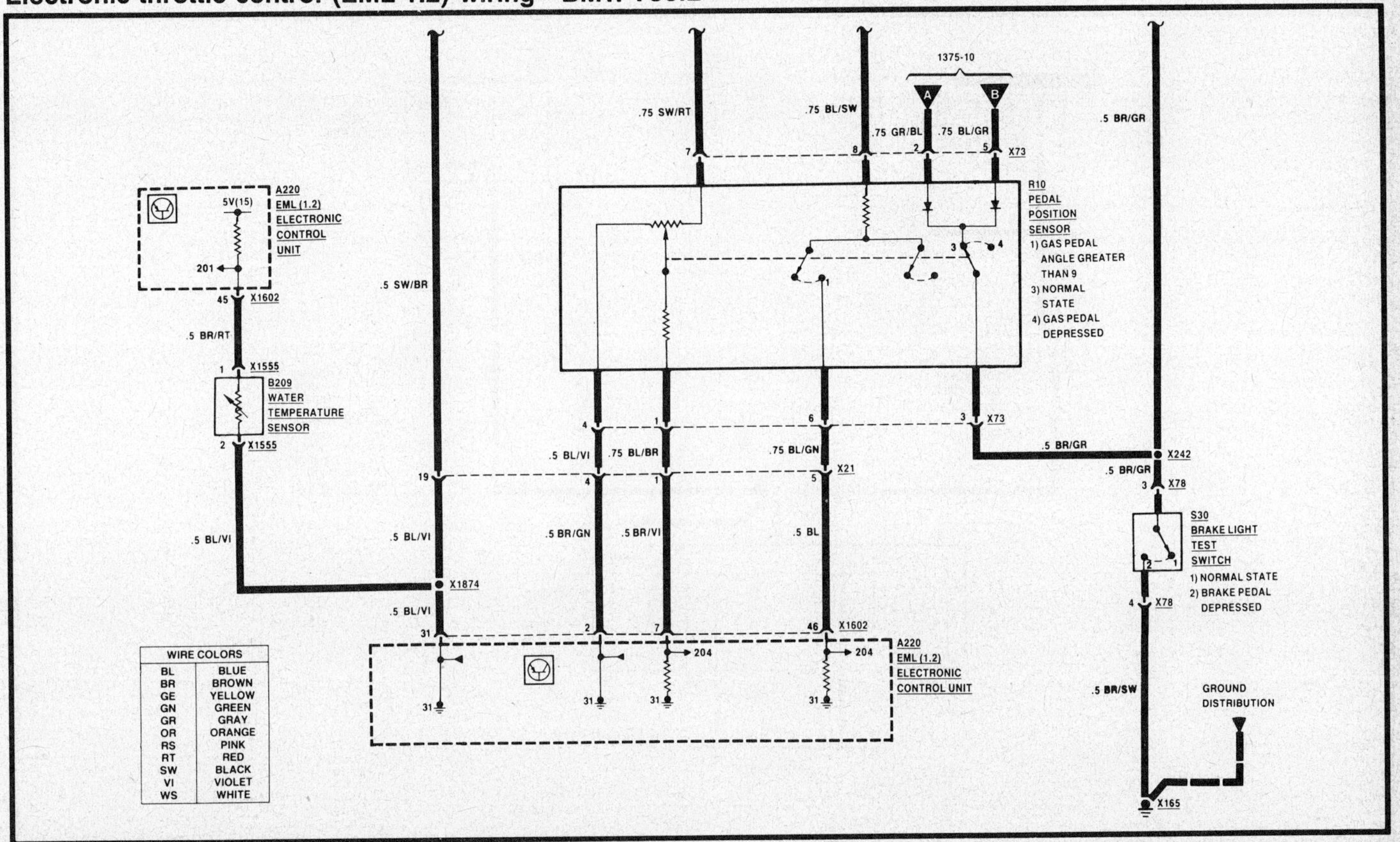

Electronic throttle control (EML 1.2) wiring—BMW 750iL

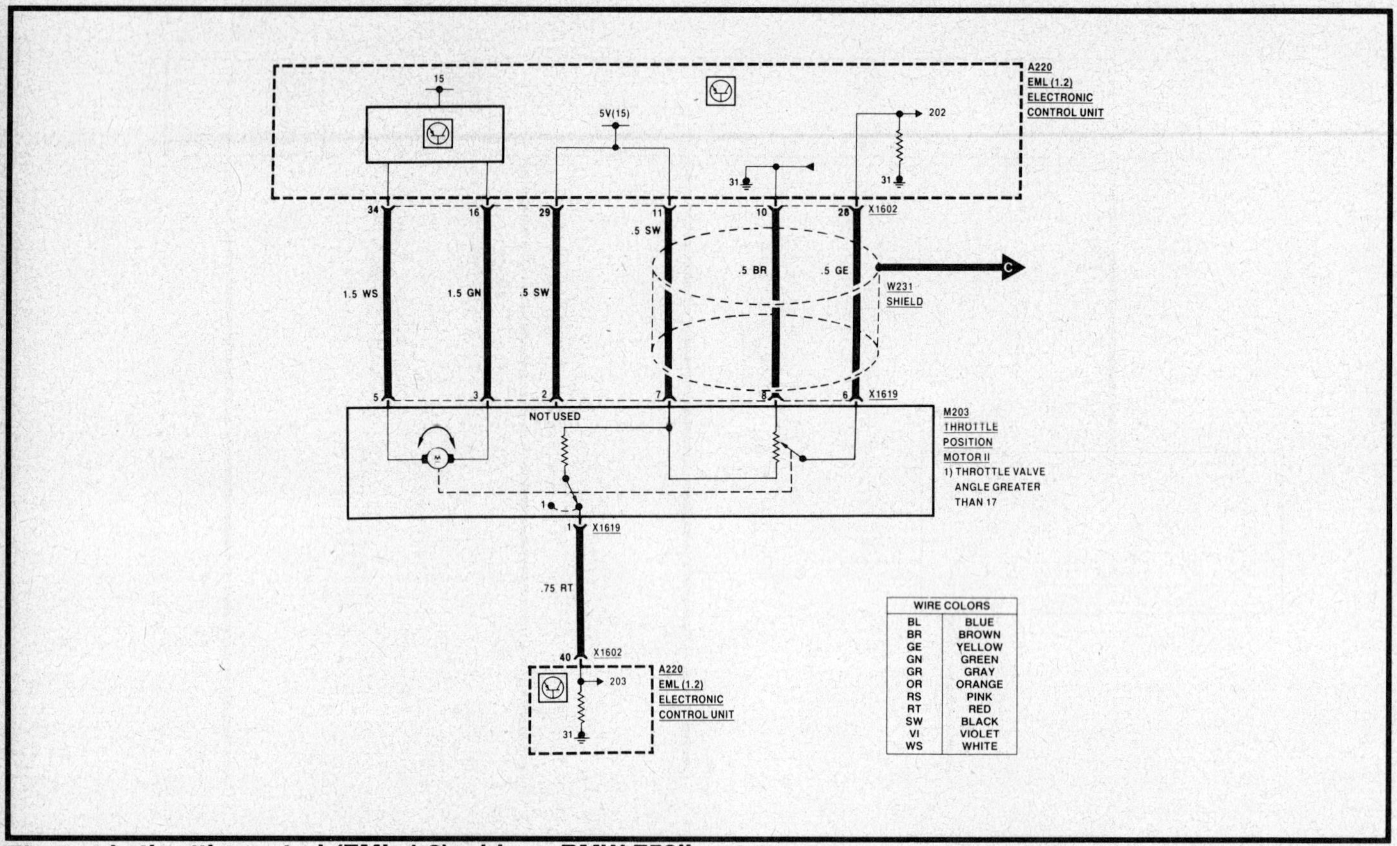

Electronic throttle control (EML 1.2) wiring – BMW 750iL

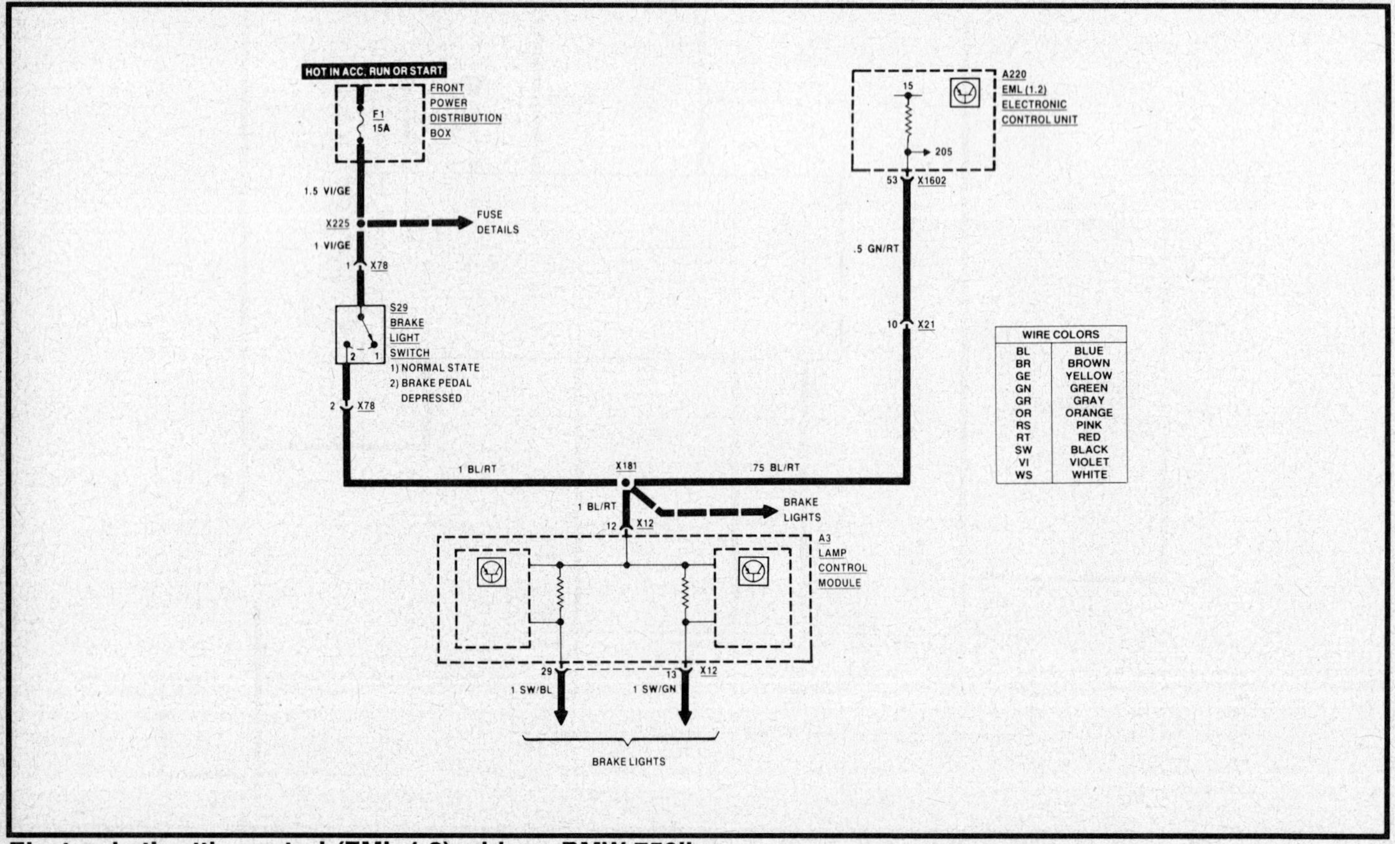

Electronic throttle control (EML 1.2) wiring – BMW 750iL

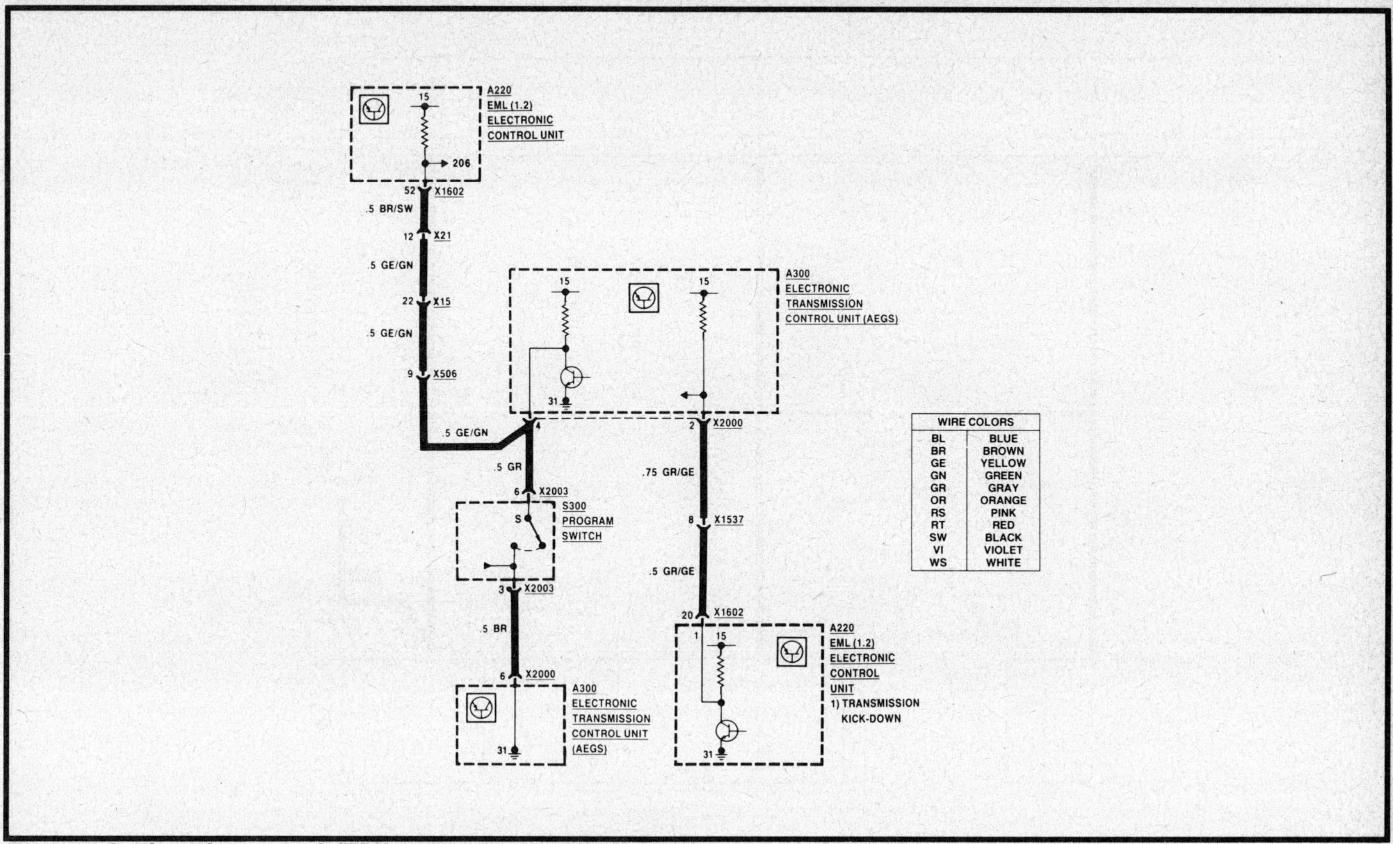

Electronic throttle control (EML 1.2) wiring–BMW 750iL

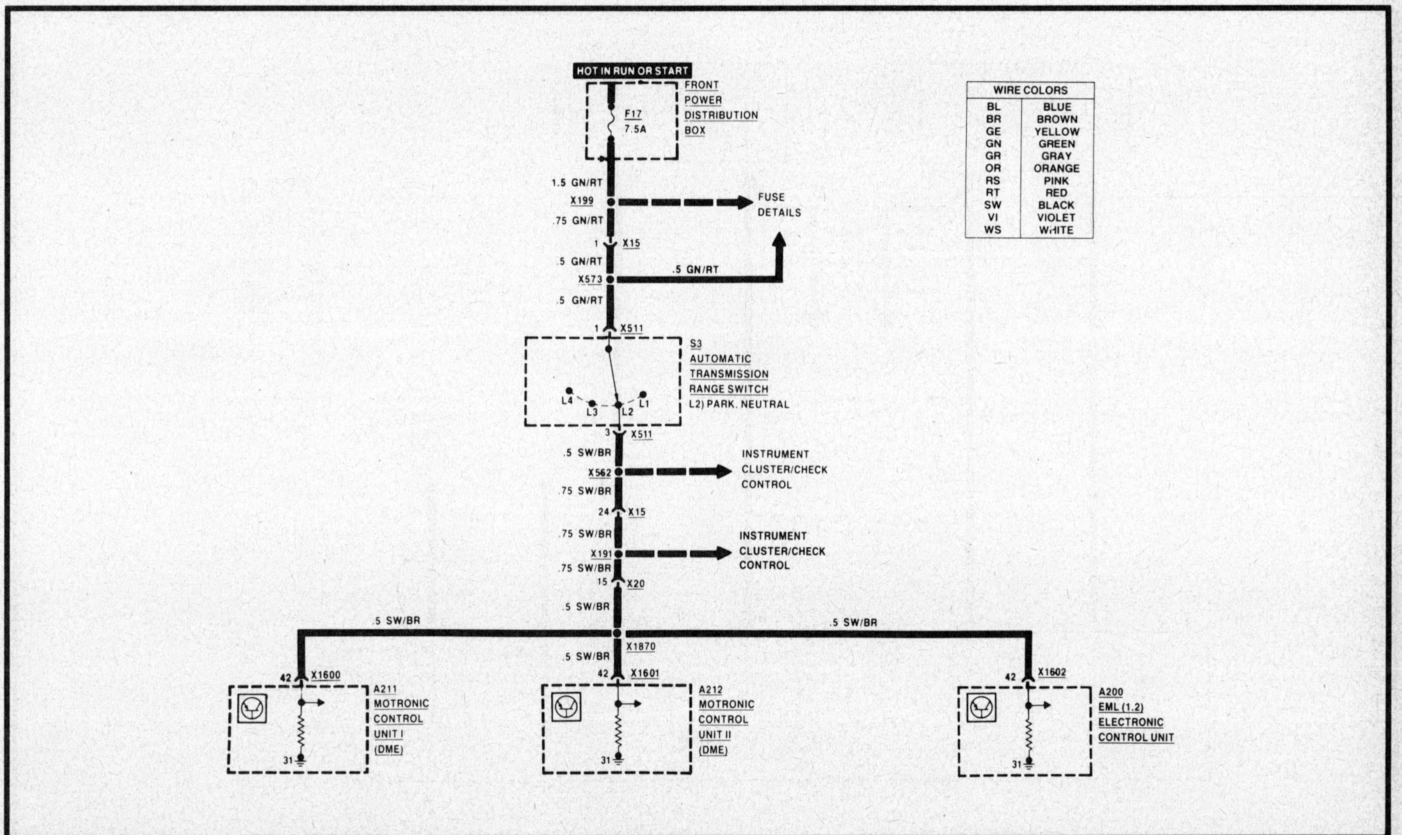

Electronic throttle control (EML 1.2) wiring–BMW 750iL

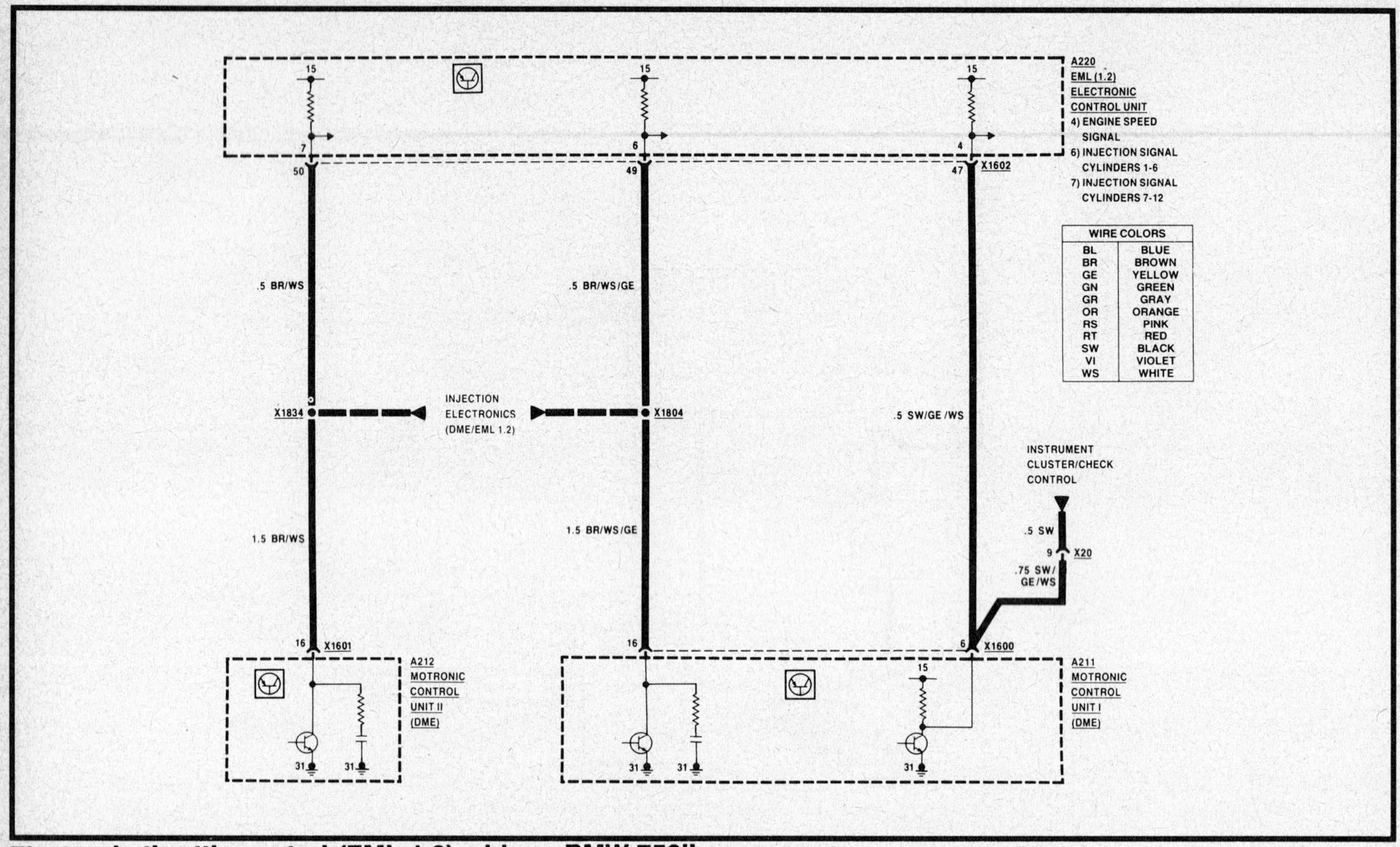

Electronic throttle control (EML 1.2) wiring—BMW 750iL

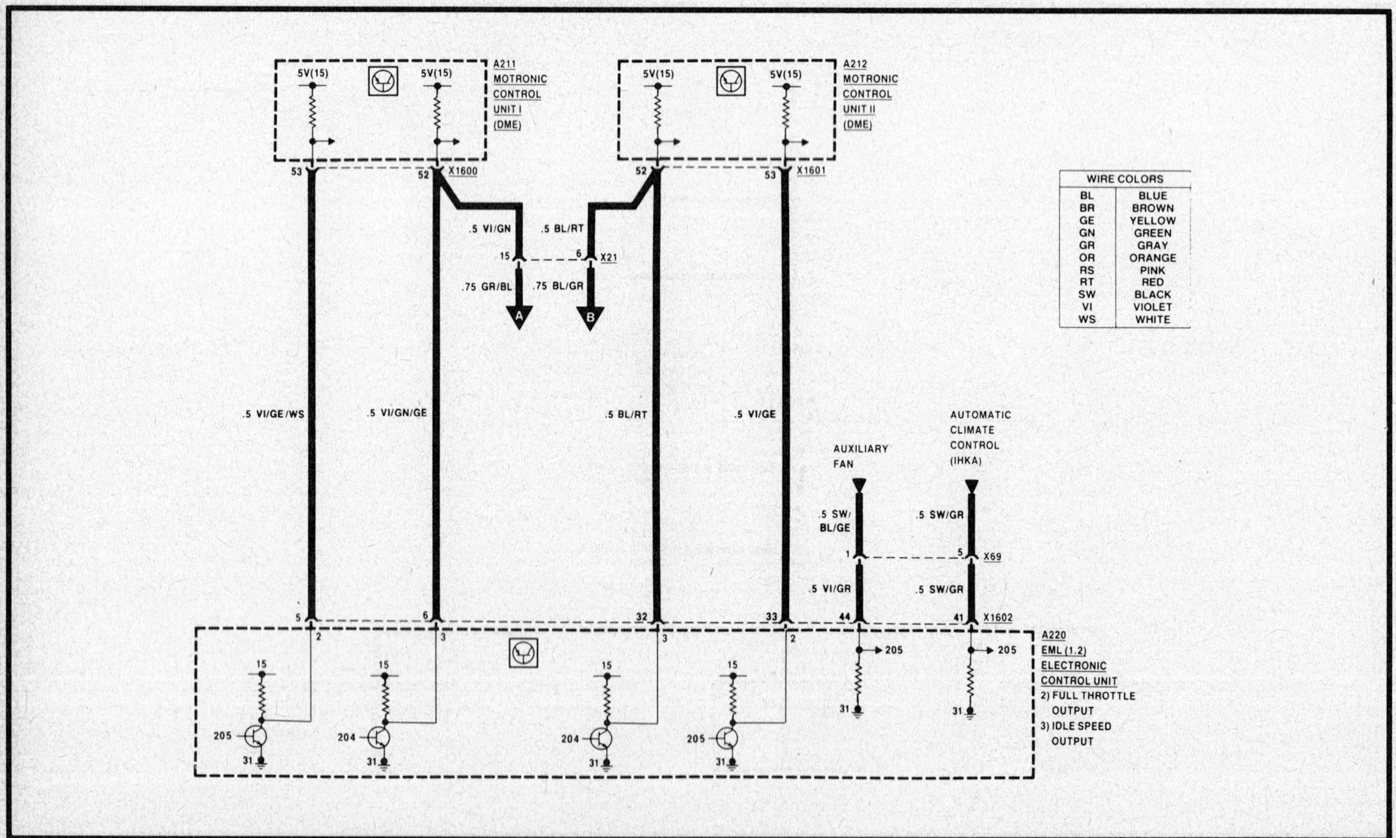

Electronic throttle control (EML 1.2) wiring—BMW 750iL

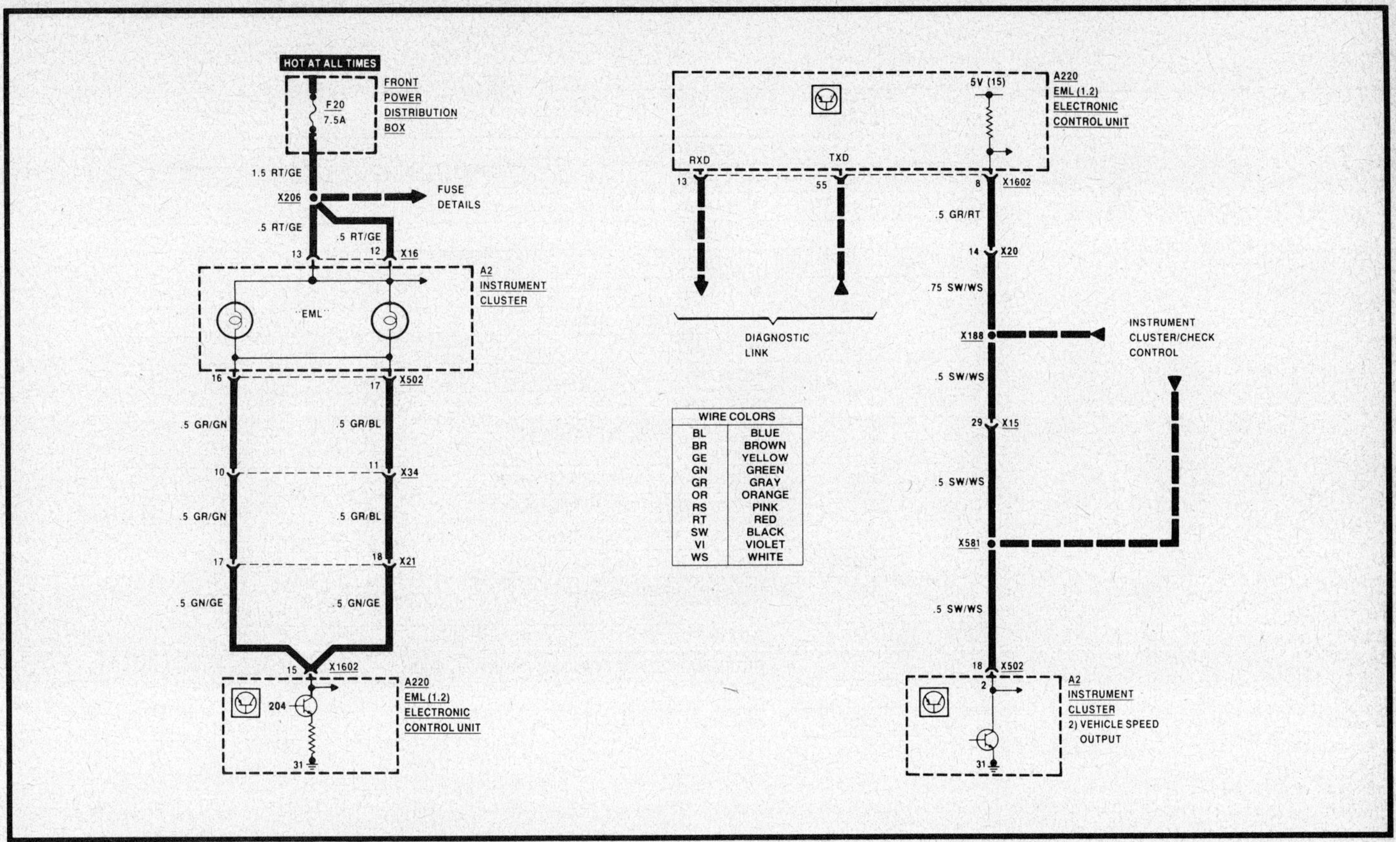

WIRE COLORS	
BL	BLUE
BR	BROWN
GE	YELLOW
GN	GREEN
GR	GRAY
OR	ORANGE
RS	PINK
RT	RED
SW	BLACK
VI	VIOLET
WS	WHITE

Electronic throttle control (EML 1.2) wiring—BMW 750iL

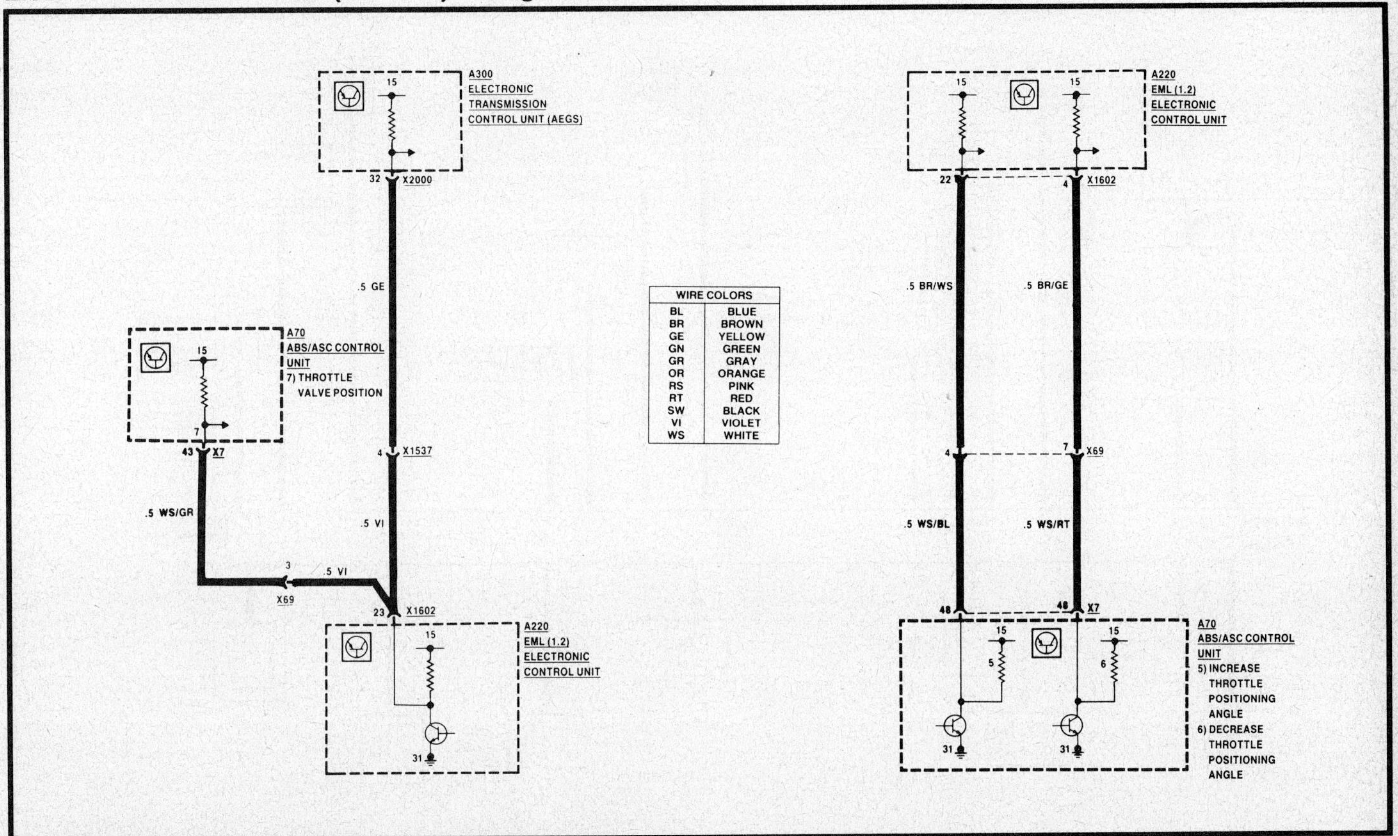

WIRE COLORS	
BL	BLUE
BR	BROWN
GE	YELLOW
GN	GREEN
GR	GRAY
OR	ORANGE
RS	PINK
RT	RED
SW	BLACK
VI	VIOLET
WS	WHITE

Electronic throttle control (EML 1.2) wiring—BMW 750iL

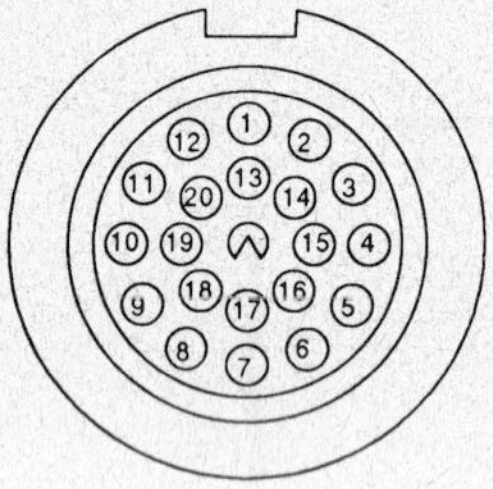

PIN NUMBER	TERMINAL NUMBER	TERMINAL DESCRIPTION	WIRE COLOR
1	1	IGNITION COIL	SW
2	–	NOT USED	
3	–	NOT USED	
4	–	NOT USED	
5	–	NOT USED	
6	–	NOT USED	
7	SI	SERVICE INTERVAL RESET	WS/GN
8	–	NOT USED	
9	–	NOT USED	
10	–	NOT USED	
11	50	VOLTAGE: IGNITION SWITCH IN START	SW/GE
12	61(D+)	ALTERNATOR CHARGE INDICATOR	BL
13	TACH	SPEEDOMETER SIGNAL-OUTPUT	SW/WS
14	30	VOLTAGE: HOT AT ALL TIMES	RT
15	RXD	DIAGNOSTIC DATALINK	WS/GE
16	15S	VOLTAGE: IGNITION SWITCH IN START	GN/WS
17	–	NOT USED	
18	–	PROGRAM SELECT FOR DME CONTROL UNIT	GN/BL
19	31	GROUND	BR
20	TXD	DIAGNOSTIC DATALINK	WS/VI

Diagnostic link pinout—BMW 850i

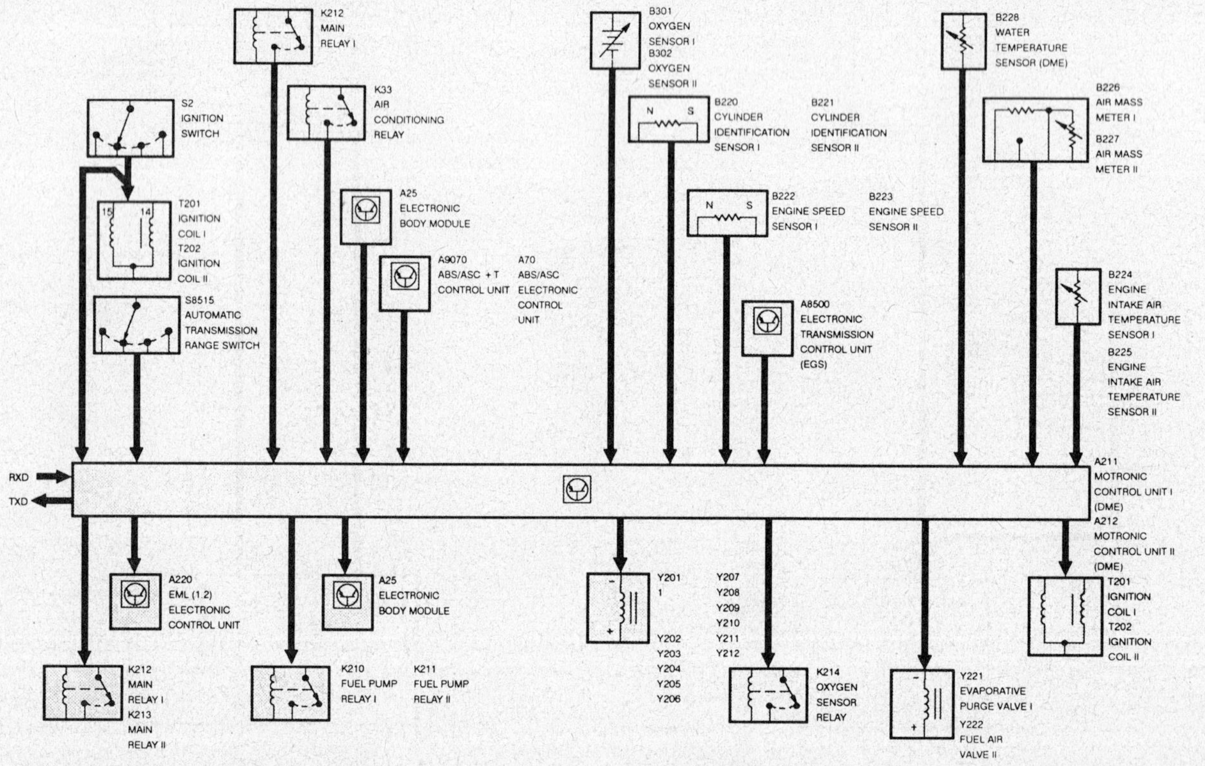

Injection electronics wiring—BMW 850i

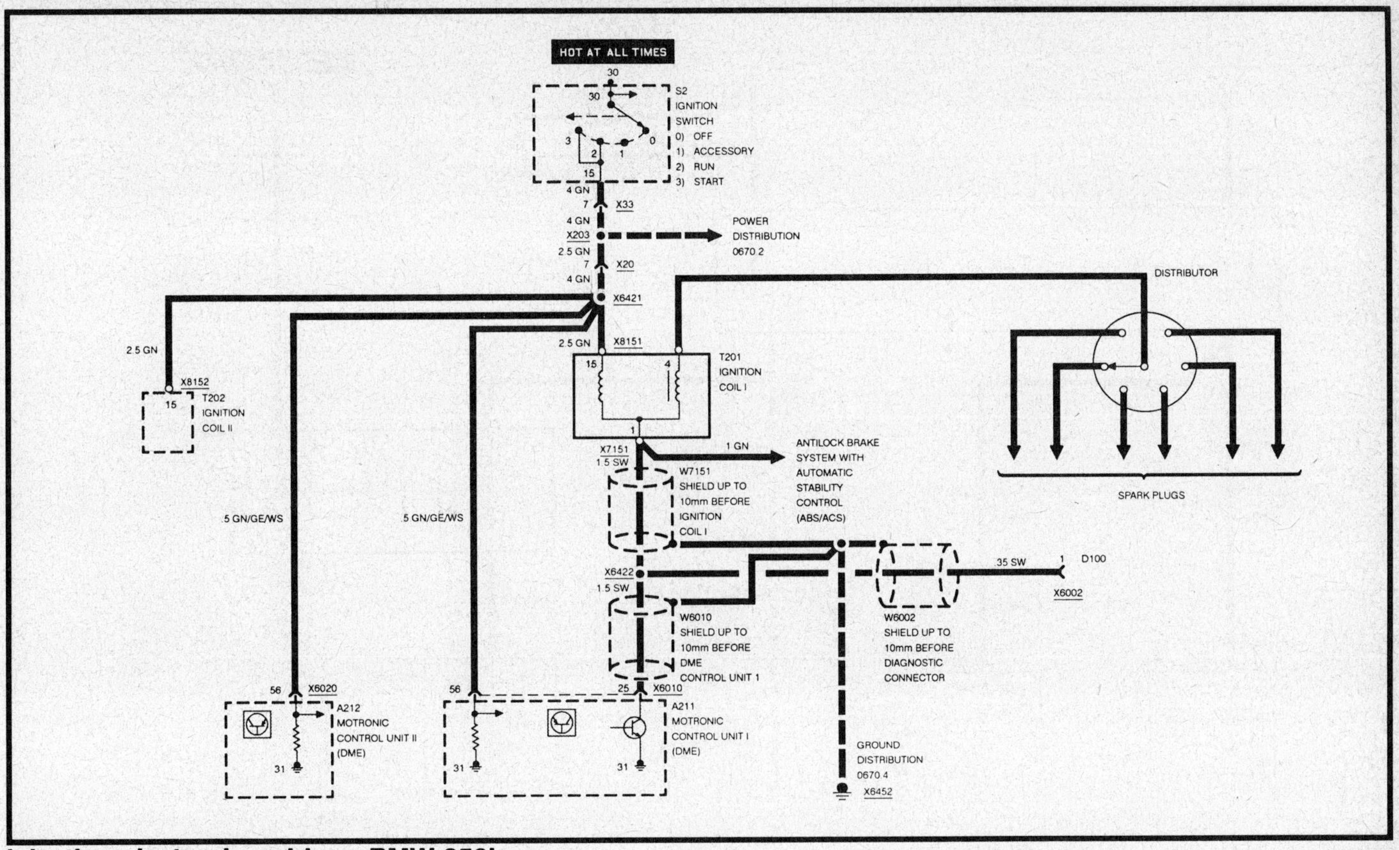

Injection electronics wiring—BMW 850i

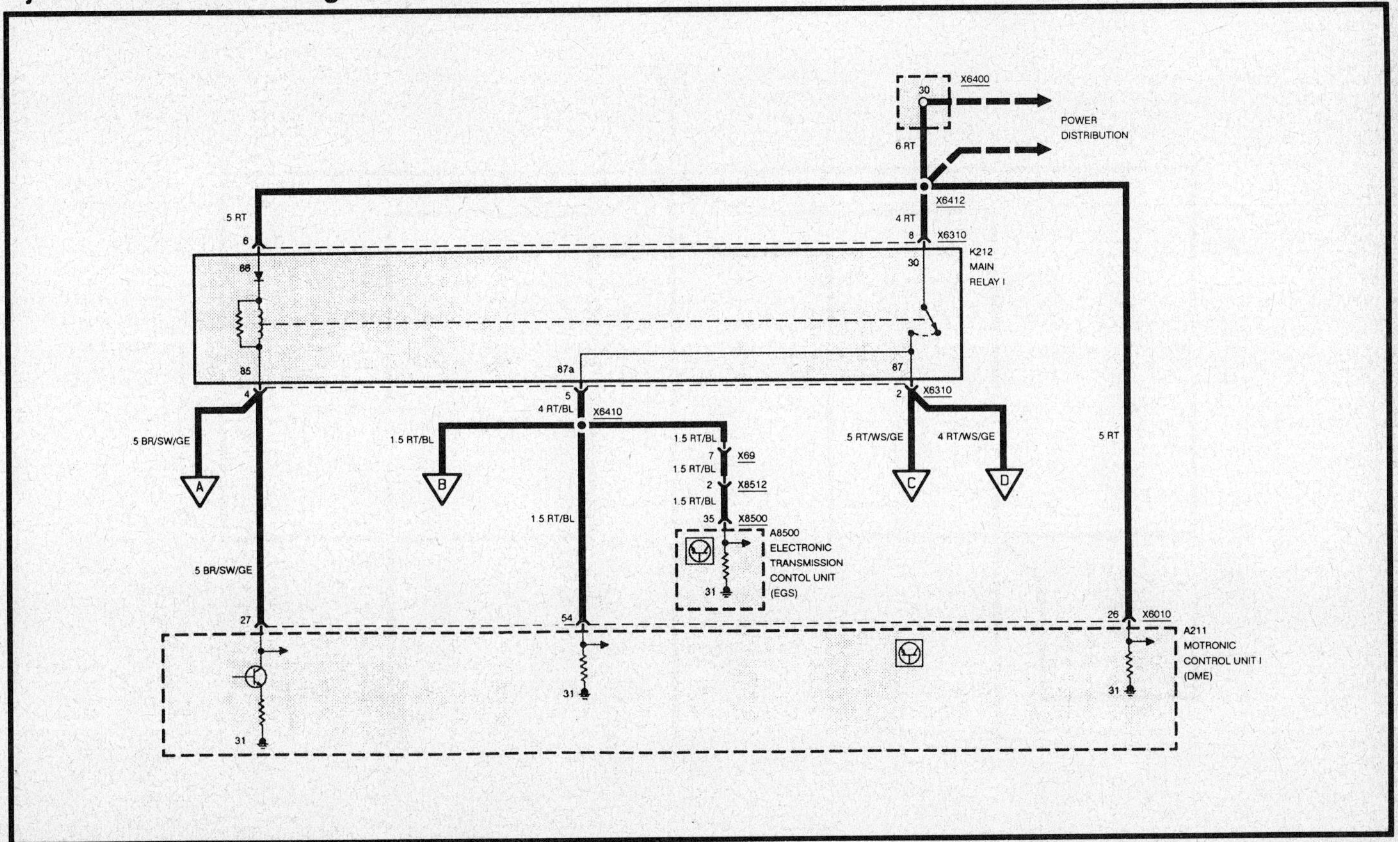

Injection electronics wiring—BMW 850i

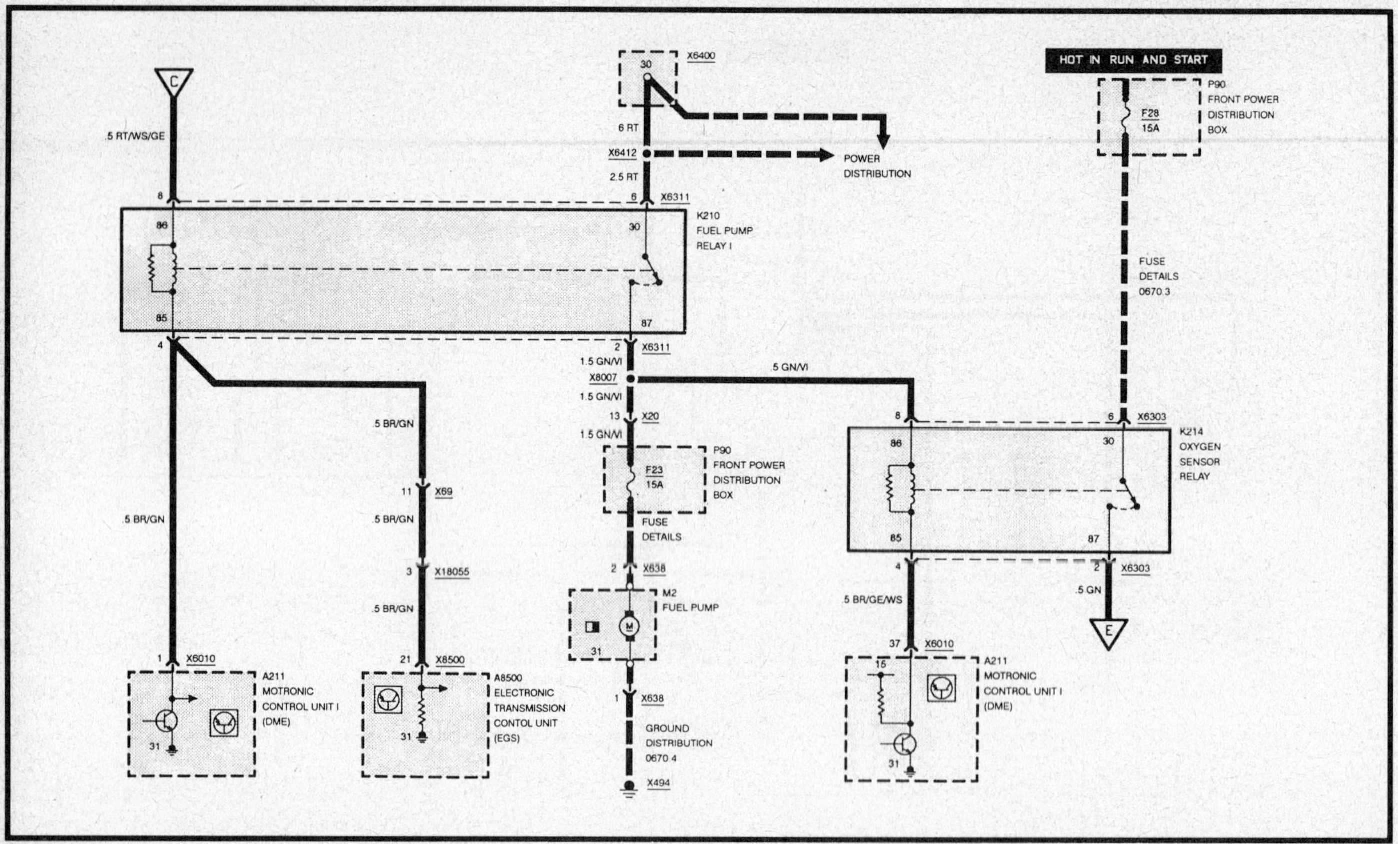

Injection electronics wiring – BMW 850i

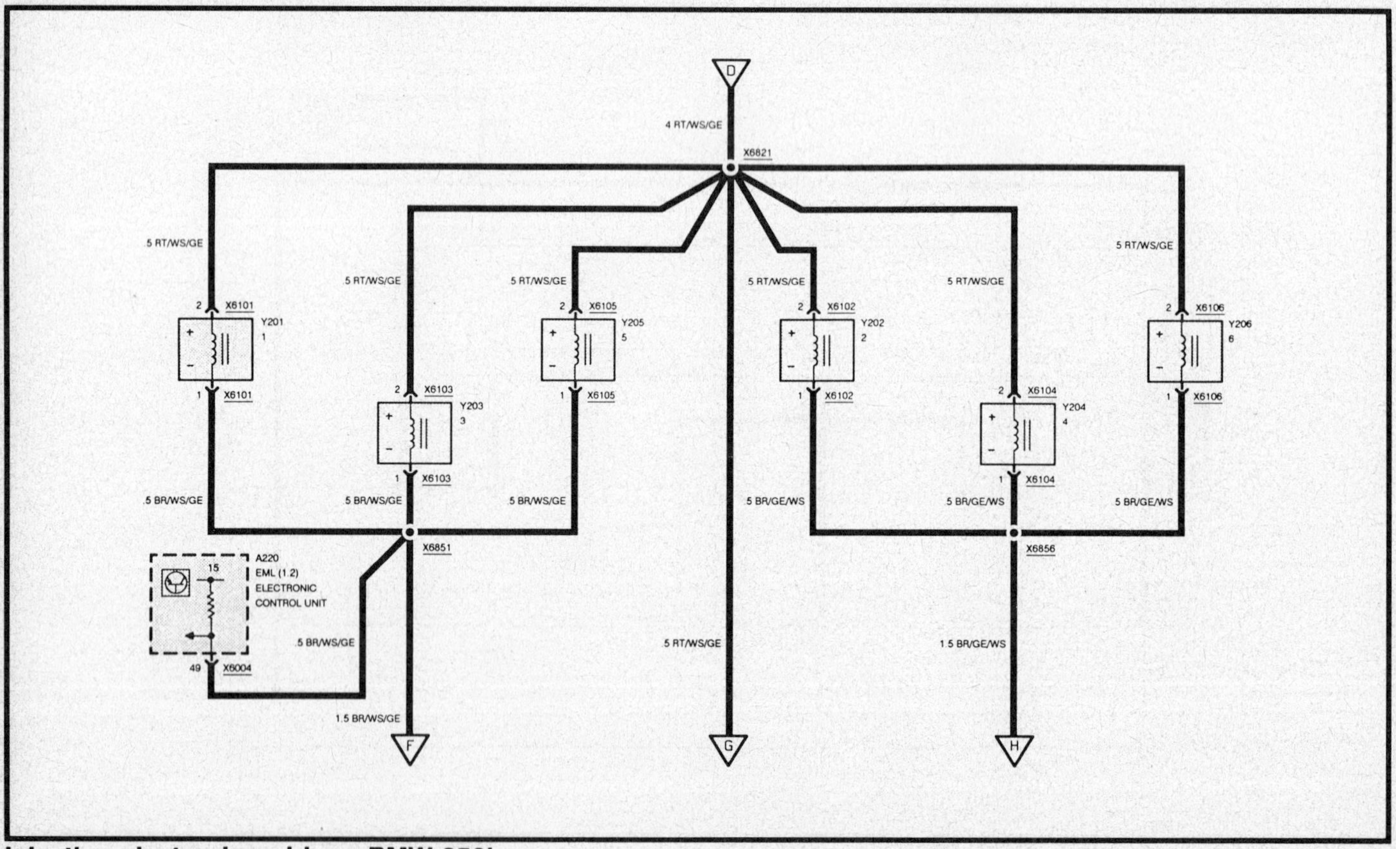

Injection electronics wiring – BMW 850i

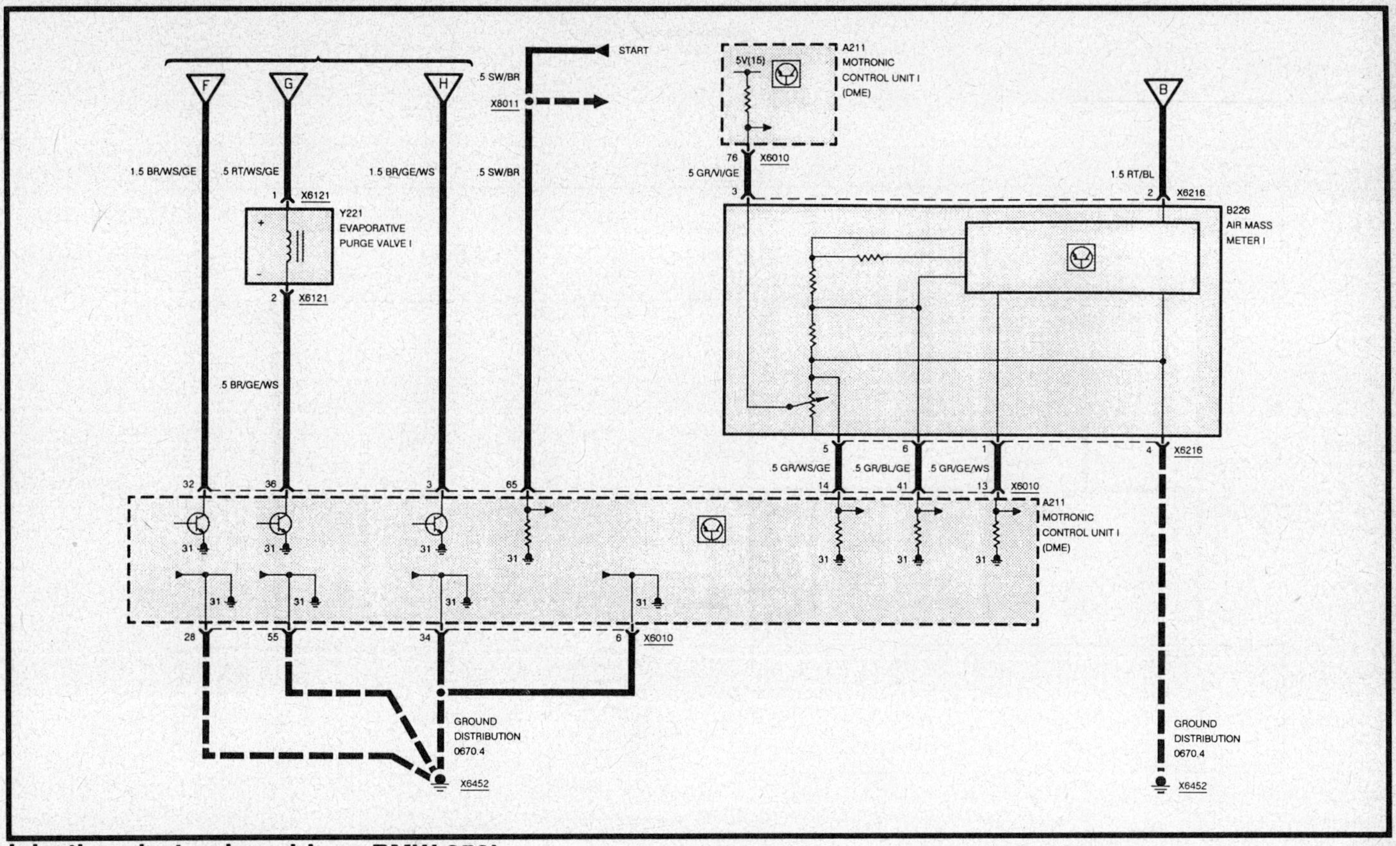

Injection electronics wiring—BMW 850i

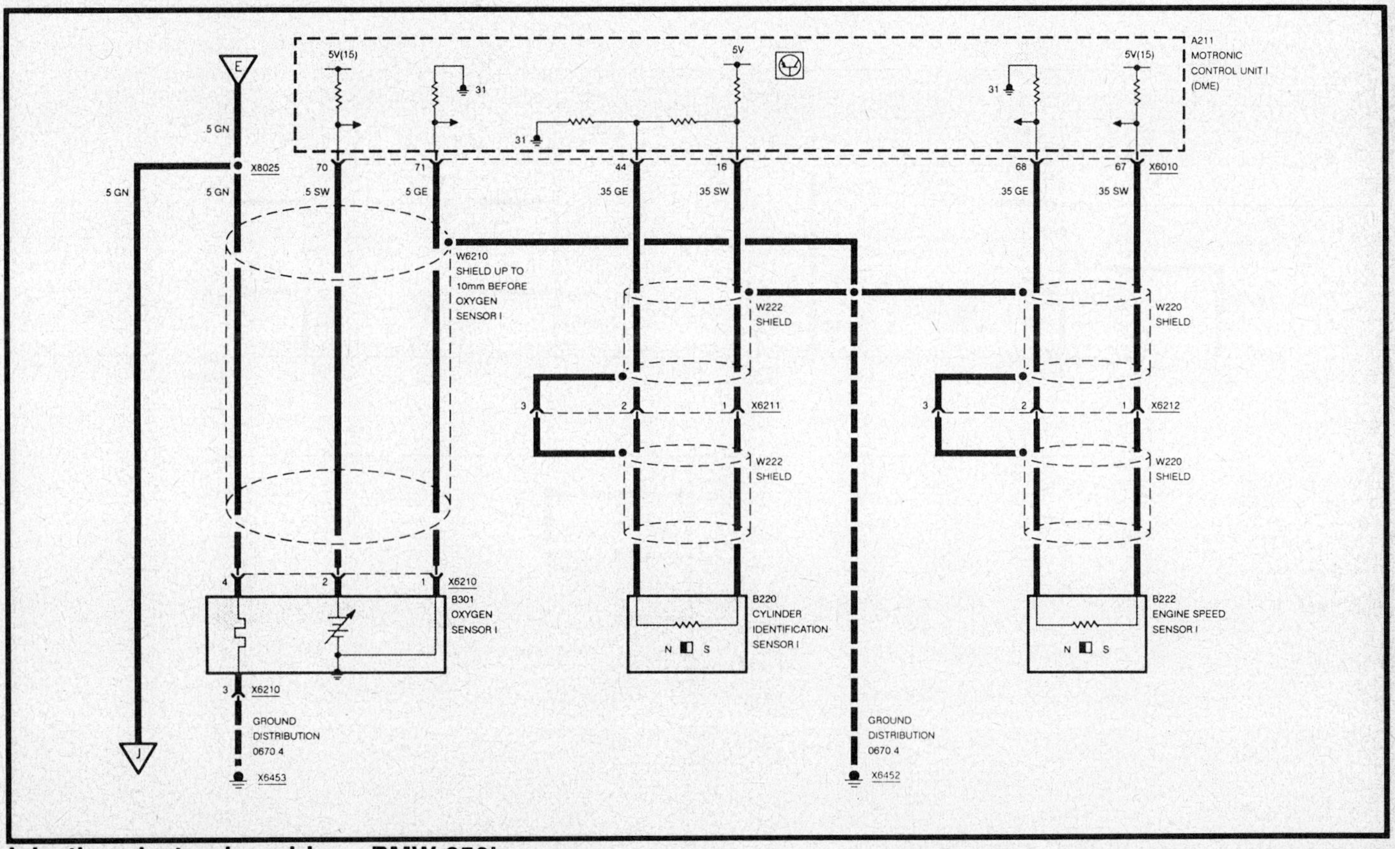

Injection electronics wiring—BMW 850i

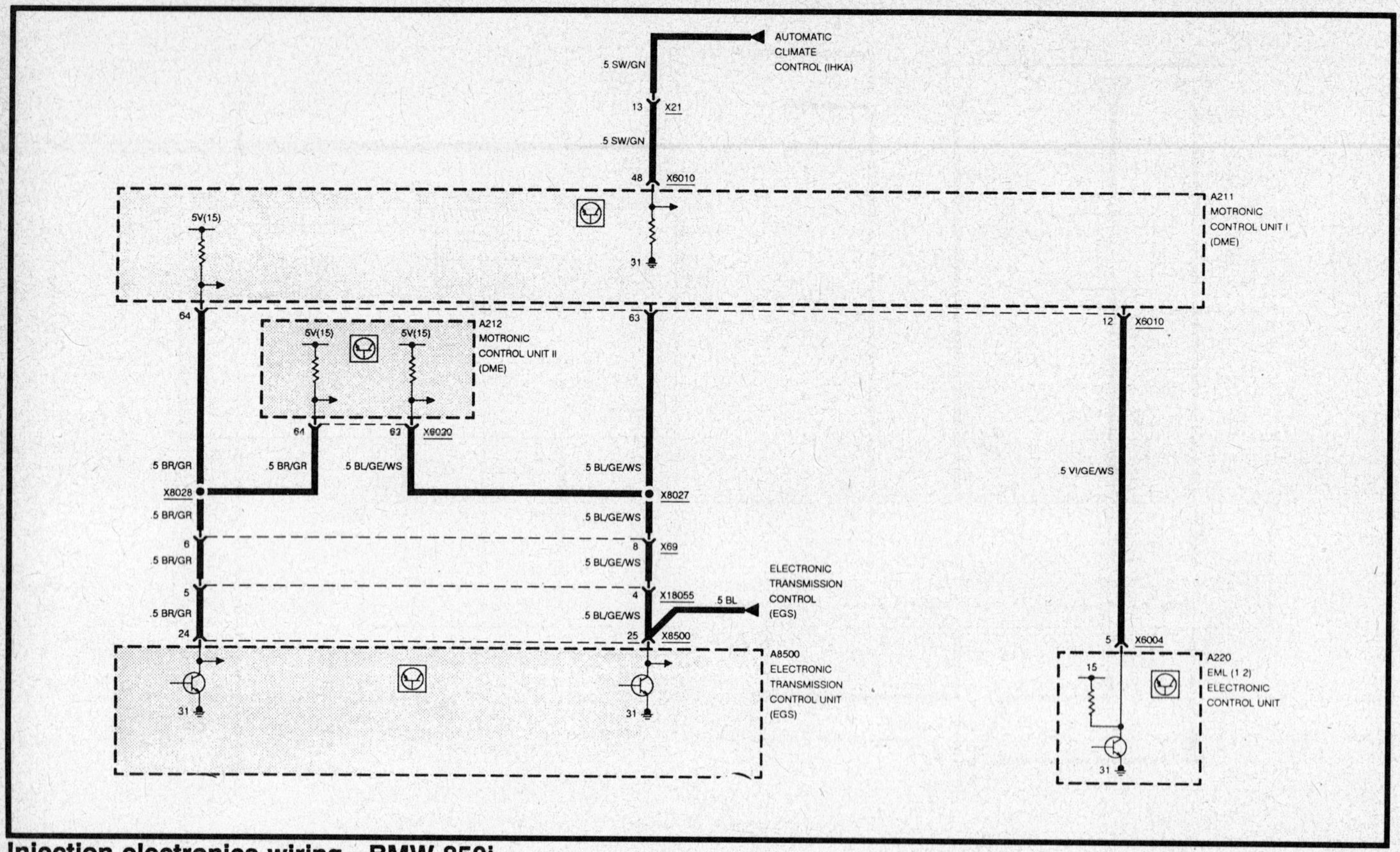

Injection electronics wiring—BMW 850i

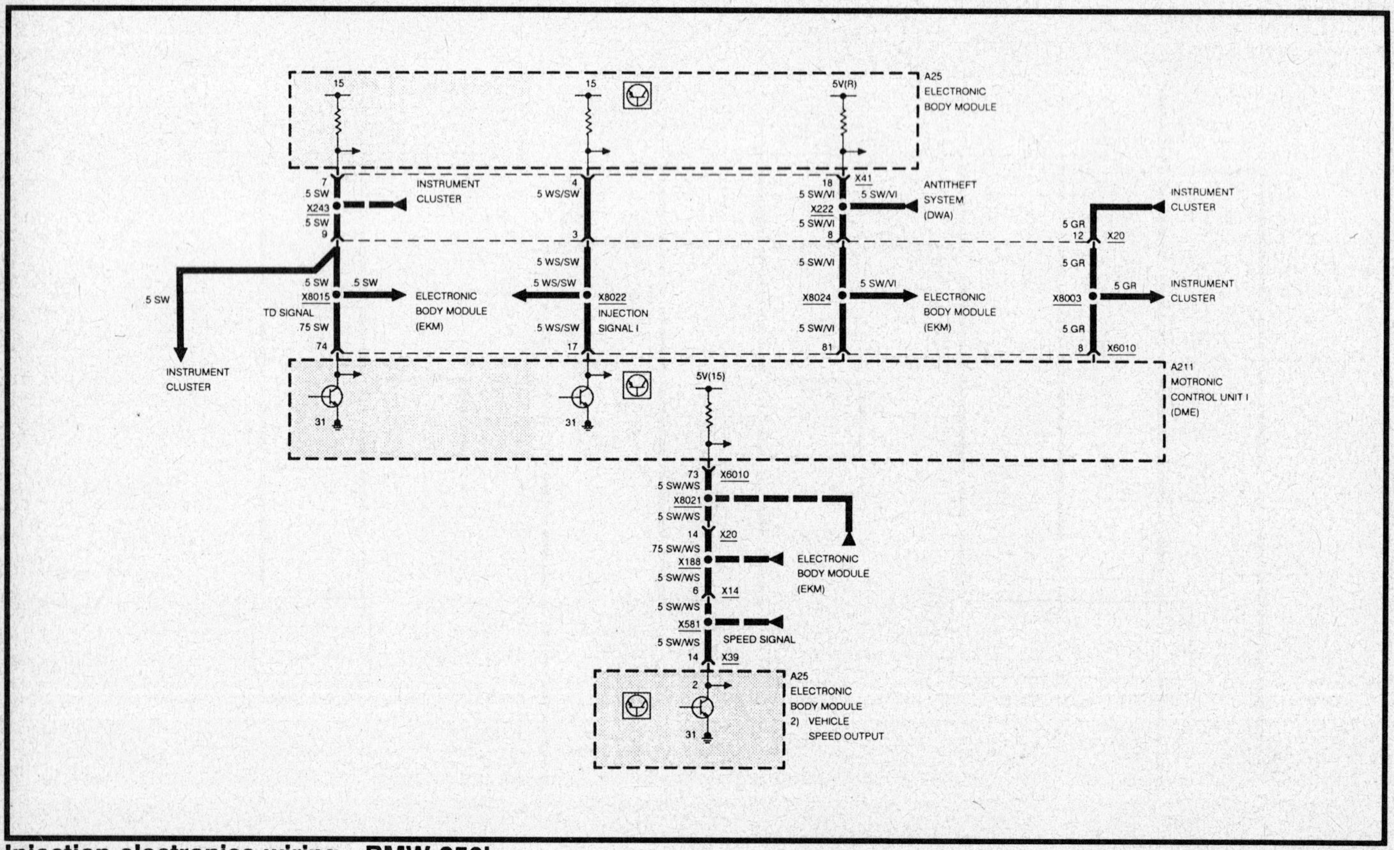

Injection electronics wiring—BMW 850i

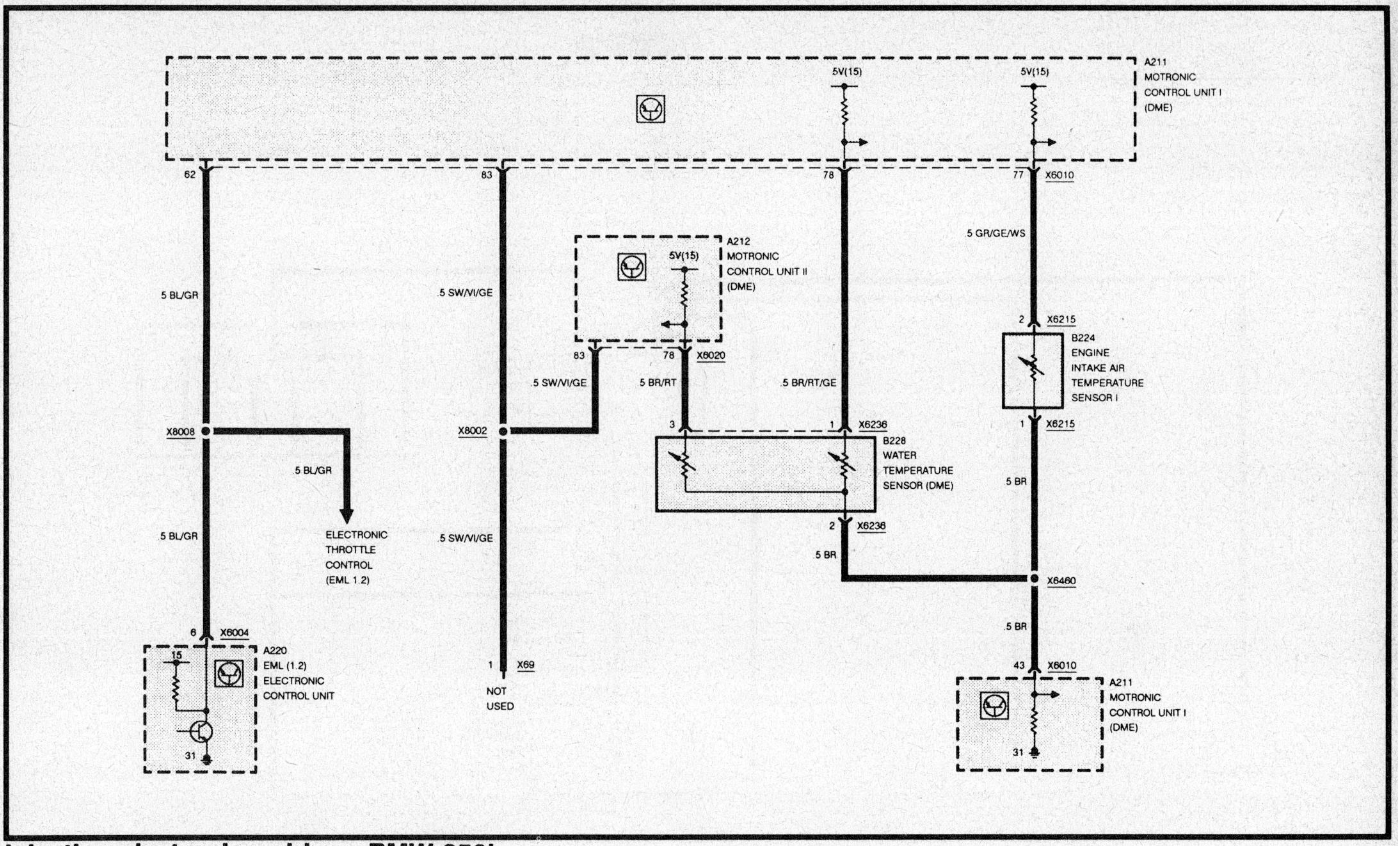

Injection electronics wiring—BMW 850i

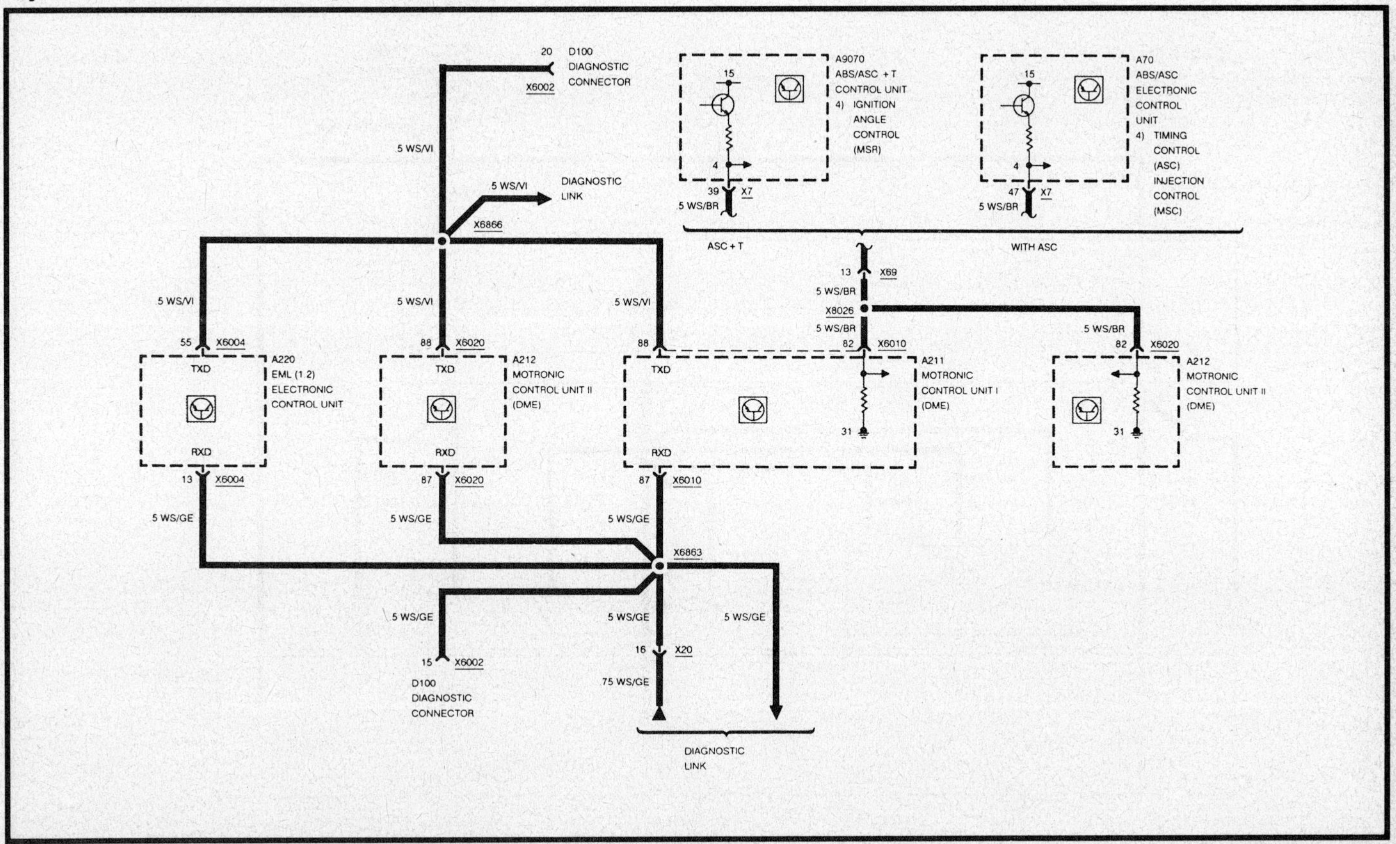

Injection electronics wiring—BMW 850i

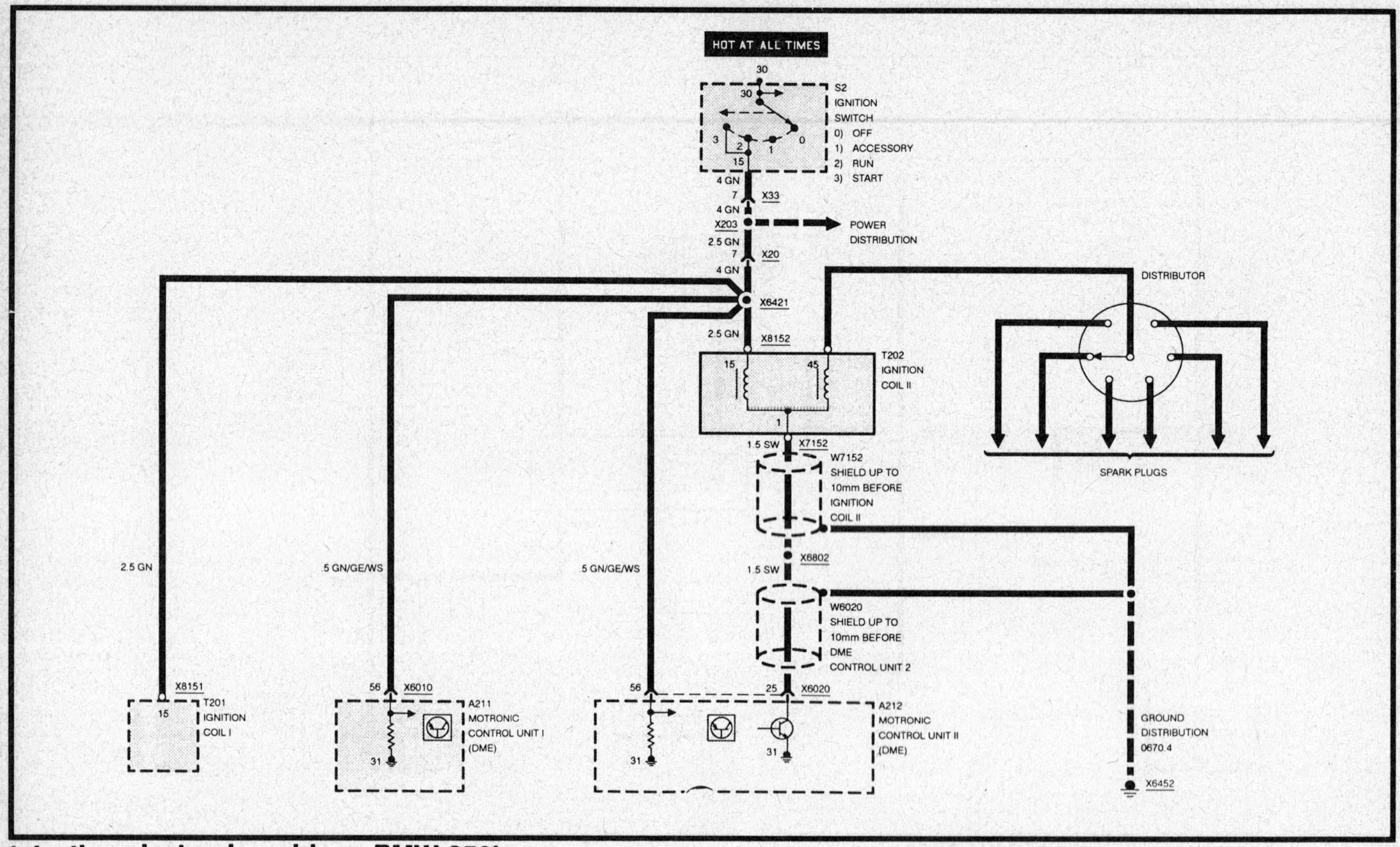

Injection electronics wiring—BMW 850i

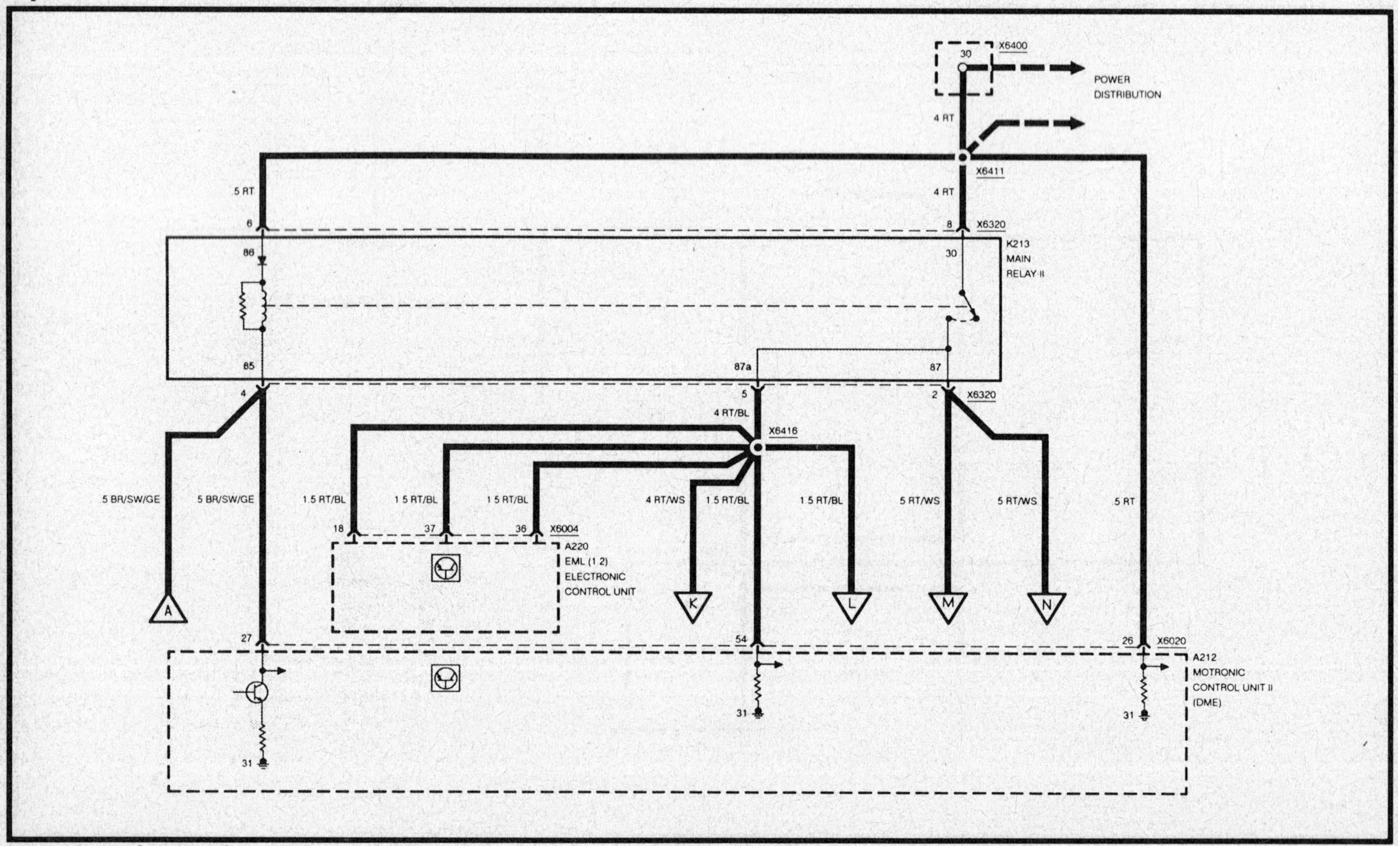

Injection electronics wiring—BMW 850i

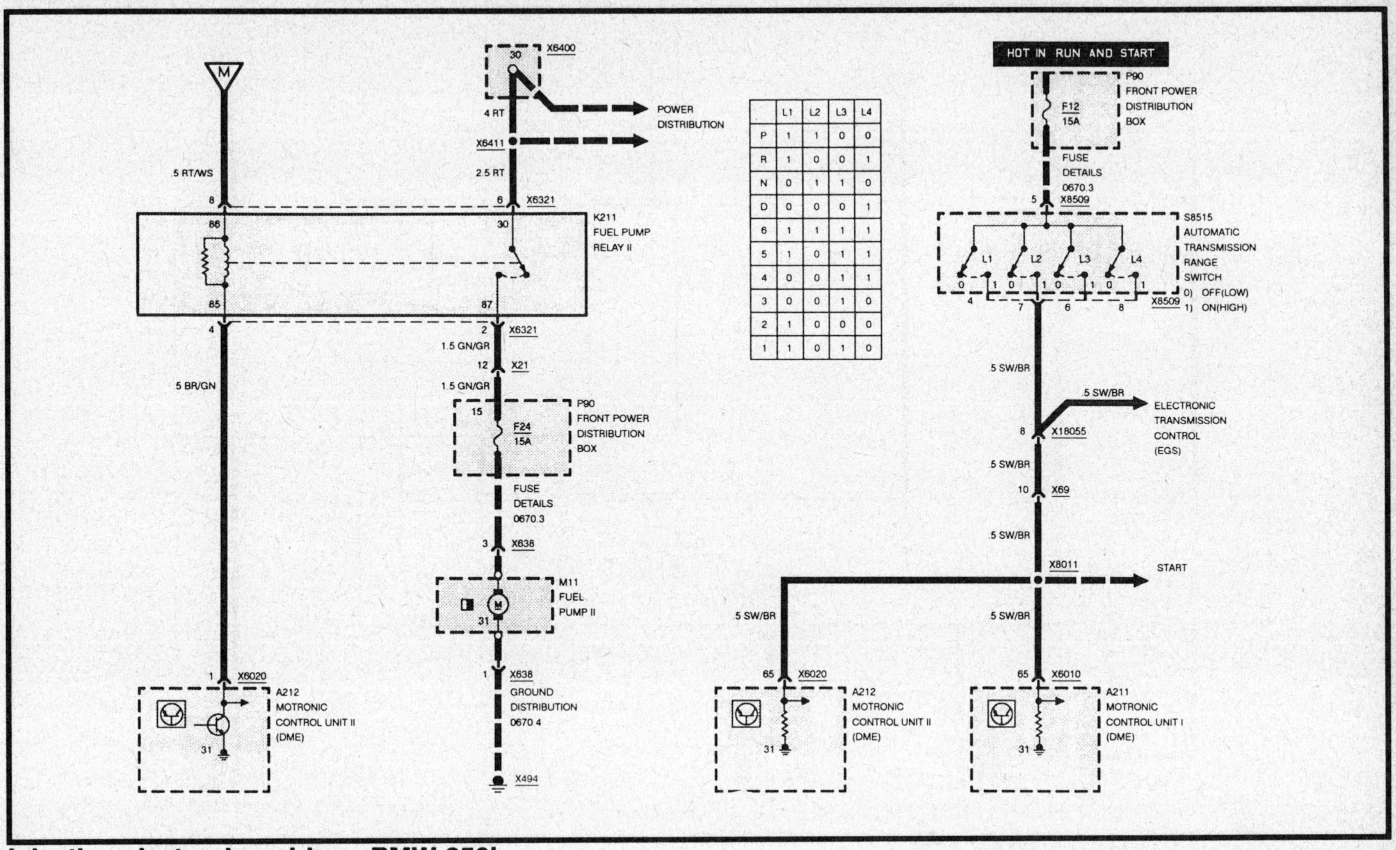

	L1	L2	L3	L4
P	1	1	0	0
R	1	0	0	1
N	0	1	1	0
D	0	0	0	1
6	1	1	1	1
5	1	0	1	1
4	0	0	1	1
3	0	0	1	0
2	1	0	0	0
1	1	0	1	0

Injection electronics wiring—BMW 850i

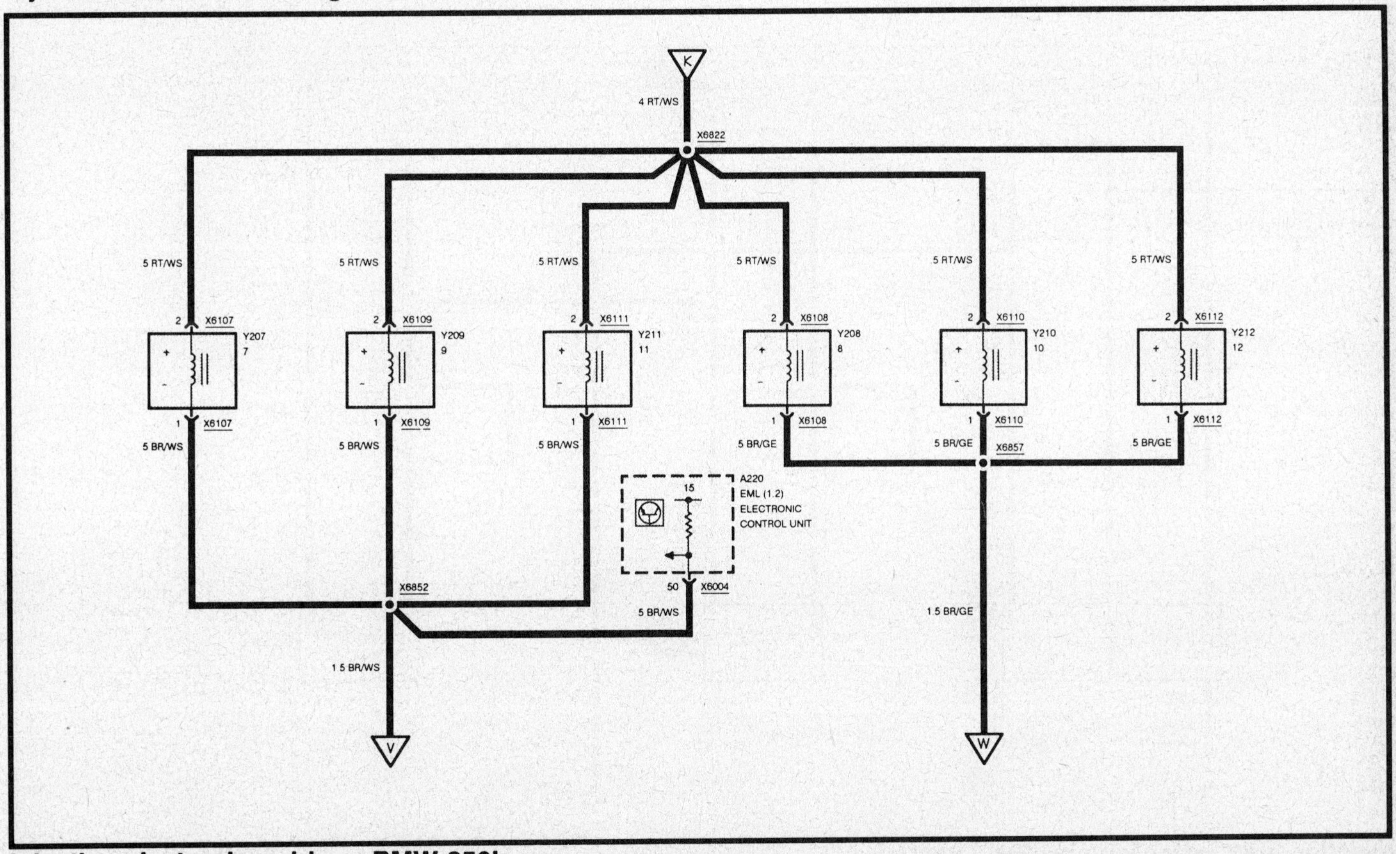

Injection electronics wiring—BMW 850i

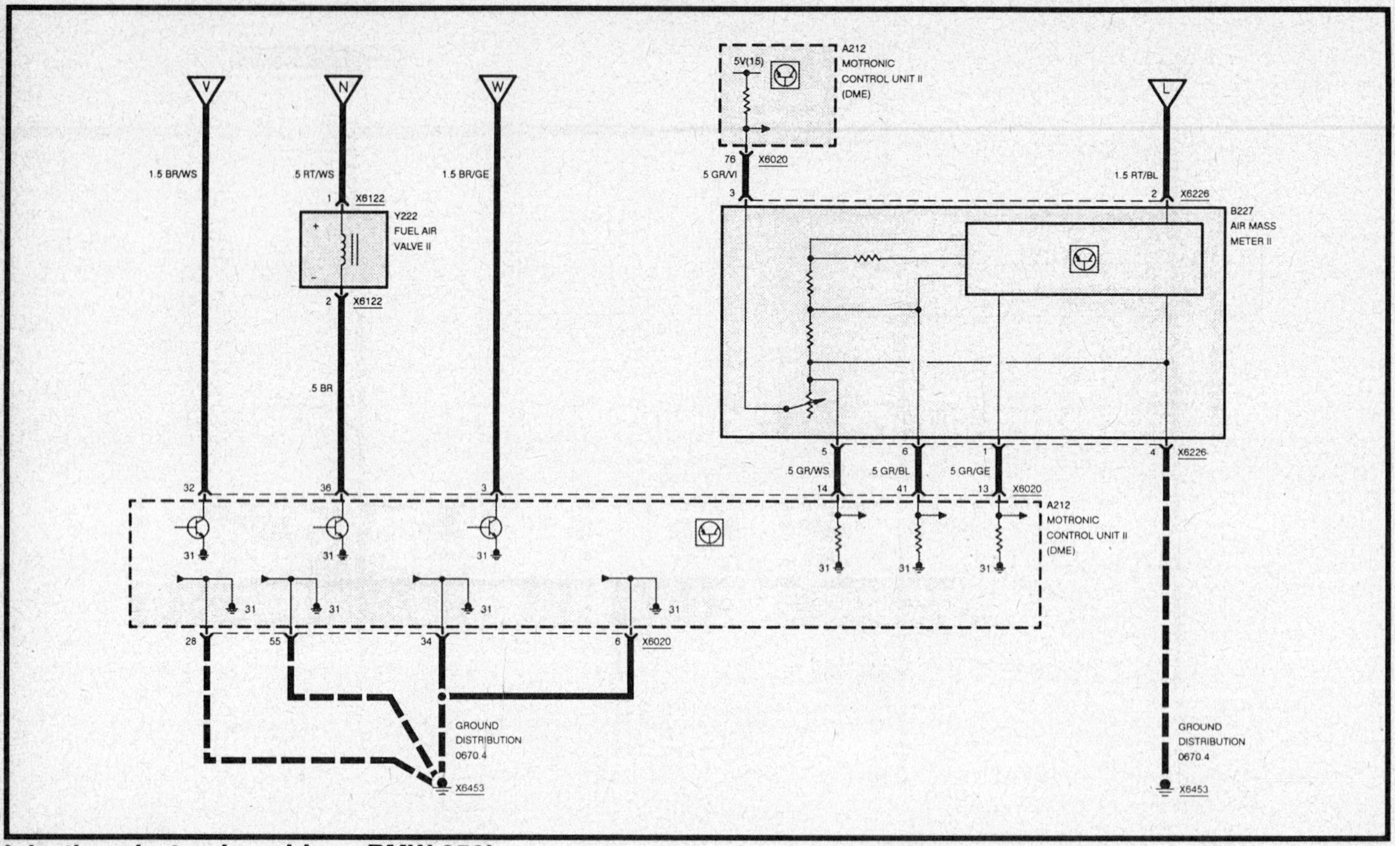

Injection electronics wiring—BMW 850i

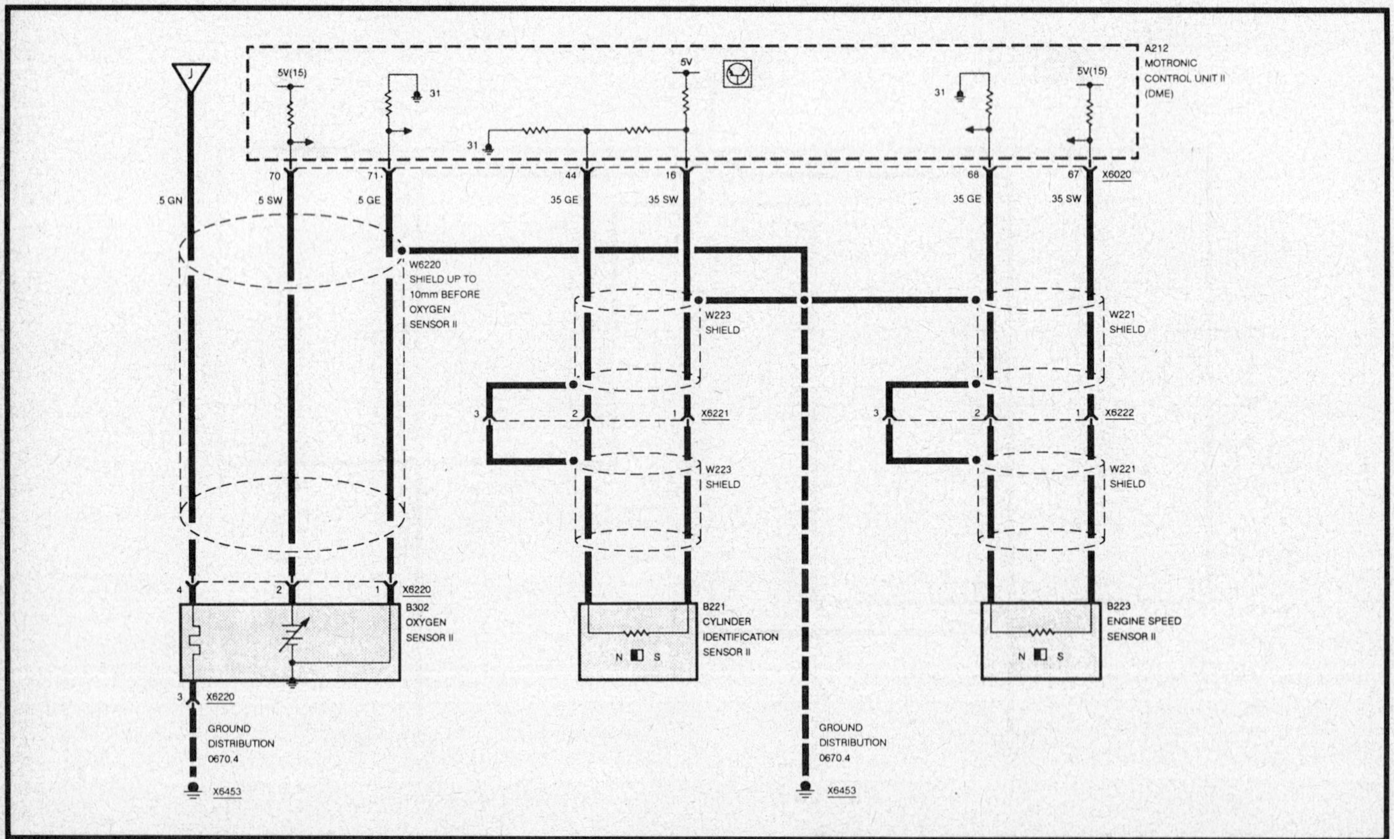

Injection electronics wiring—BMW 850i

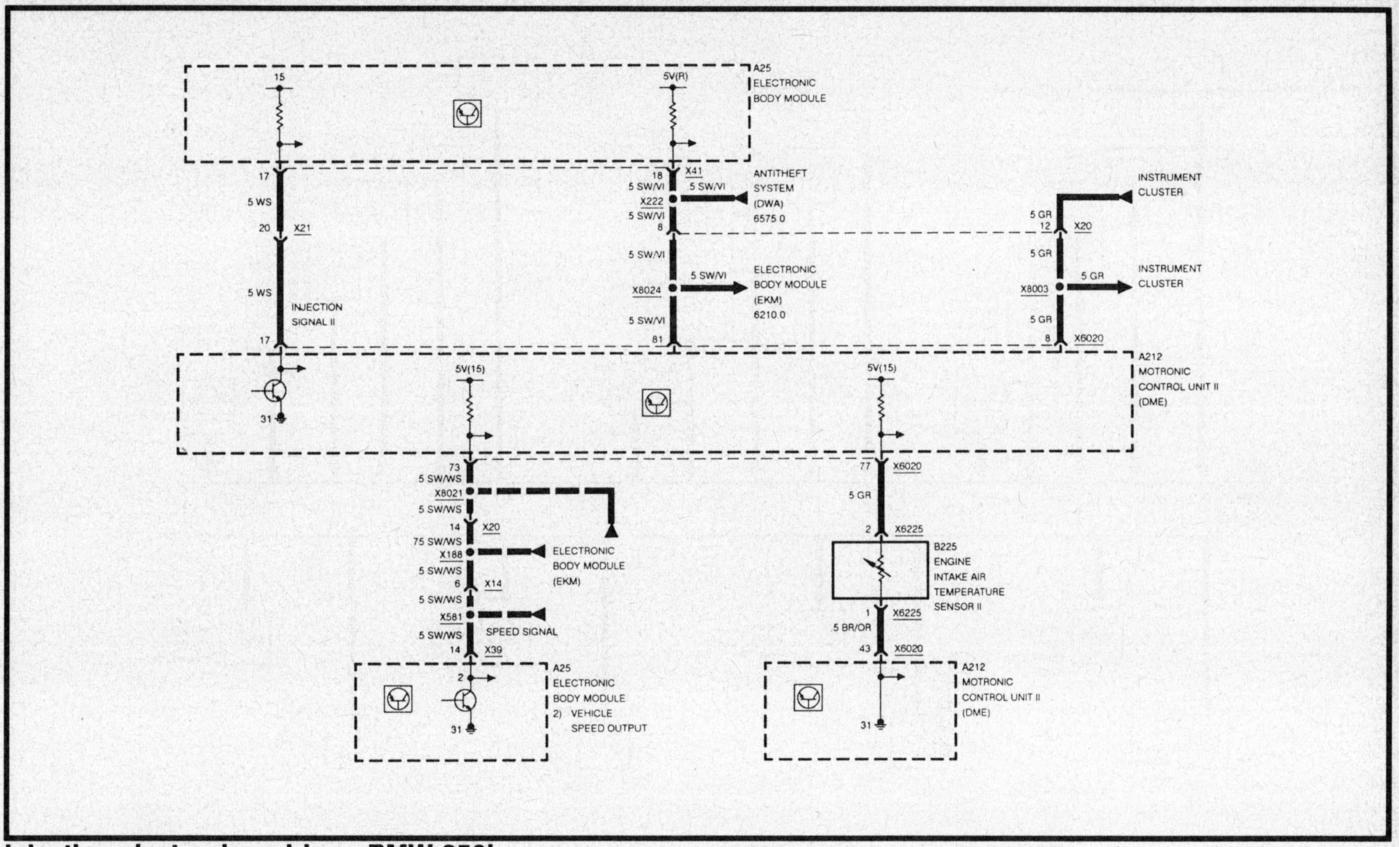

Injection electronics wiring—BMW 850i

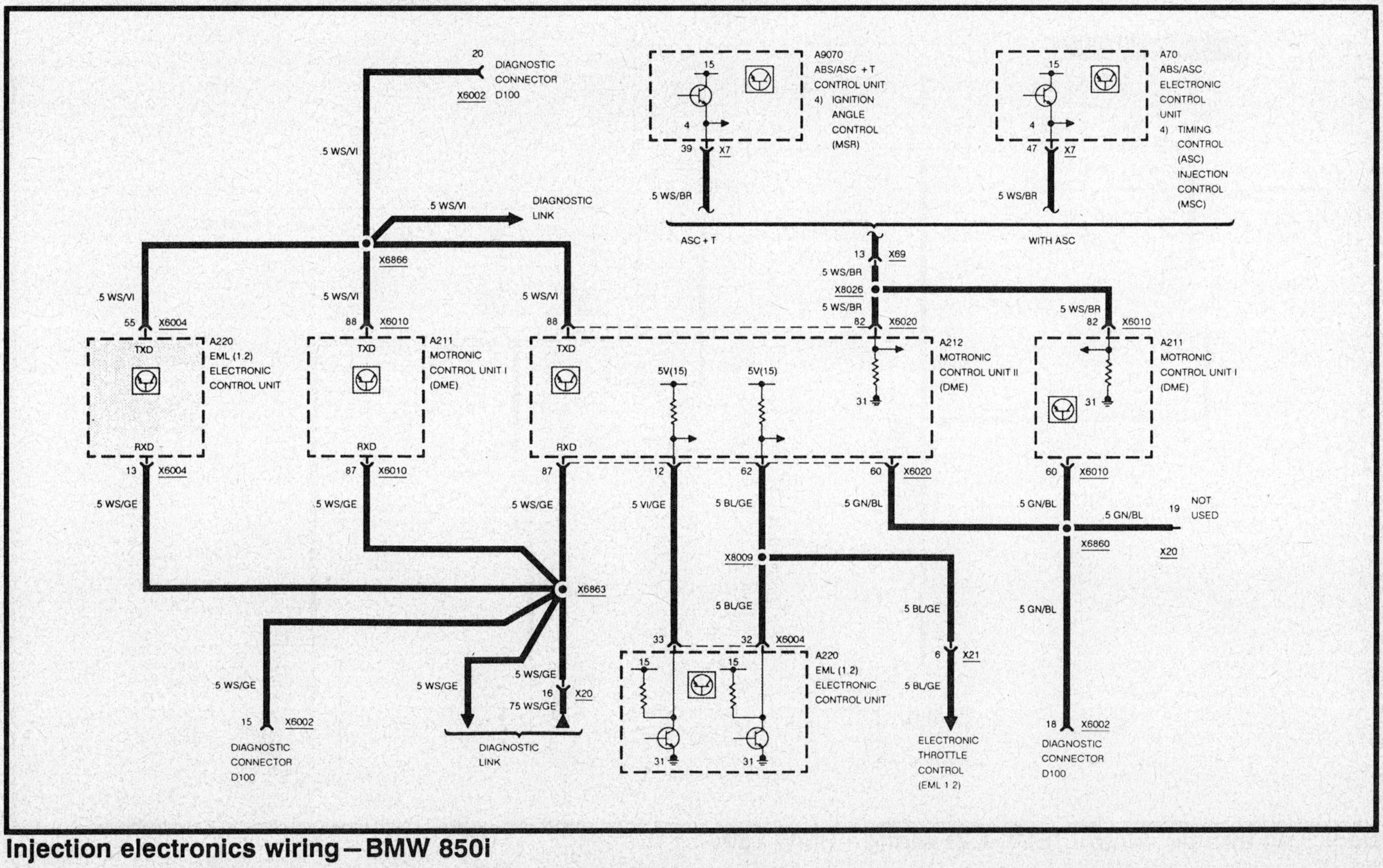

Injection electronics wiring—BMW 850i

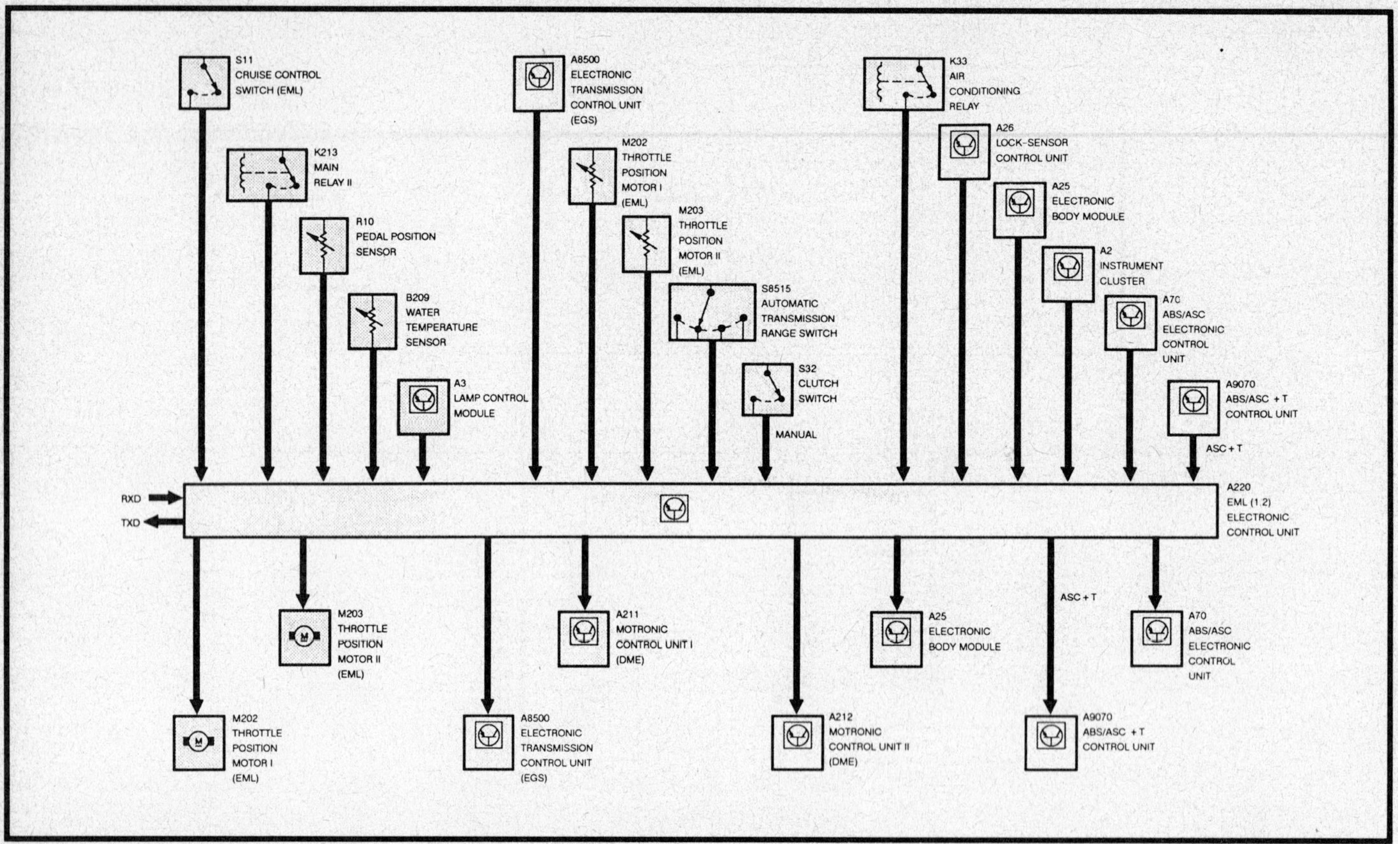

Electronic throttle control (EML 1.2) wiring—BMW 850i

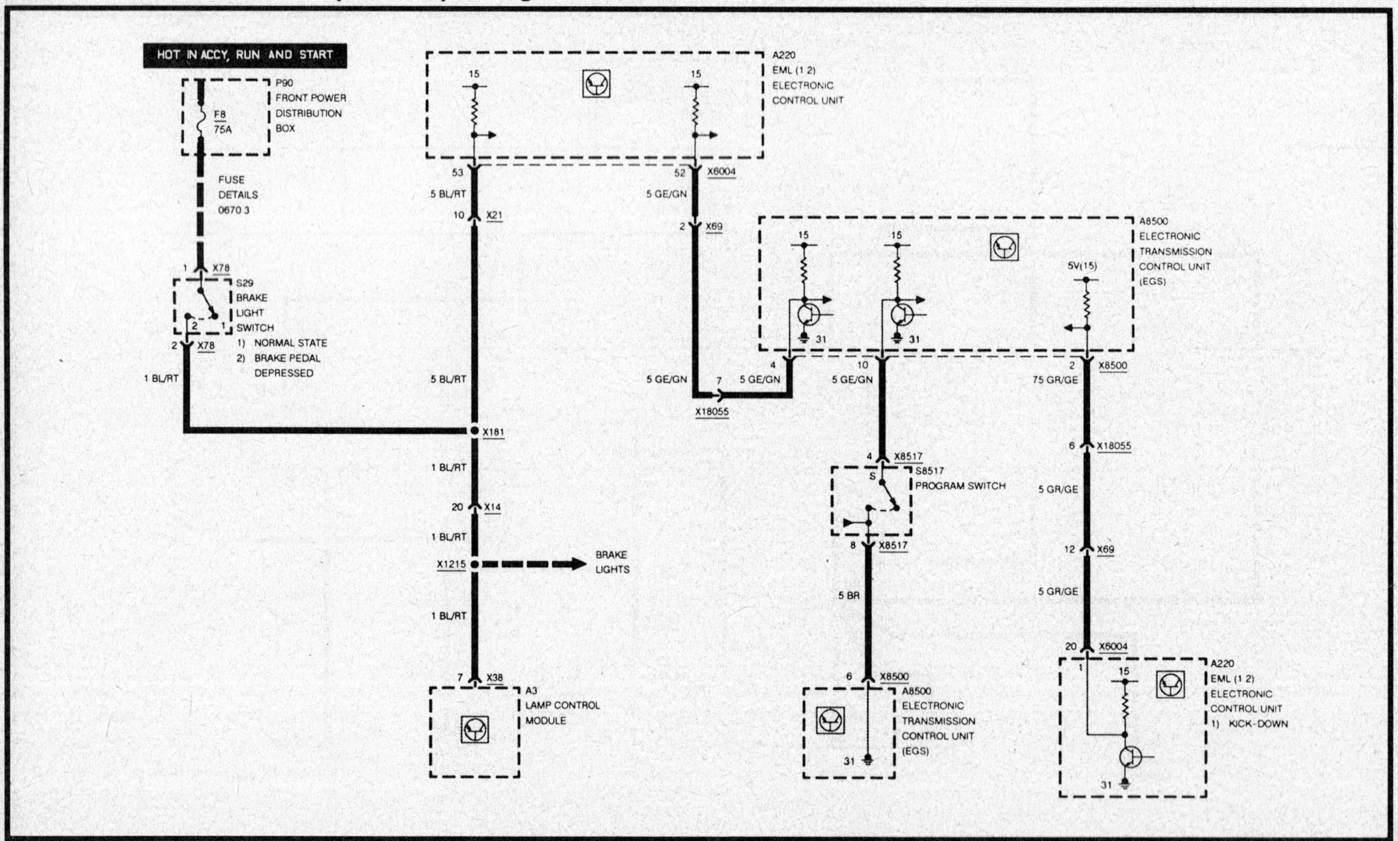

Electronic throttle control (EML 1.2) wiring—BMW 850i

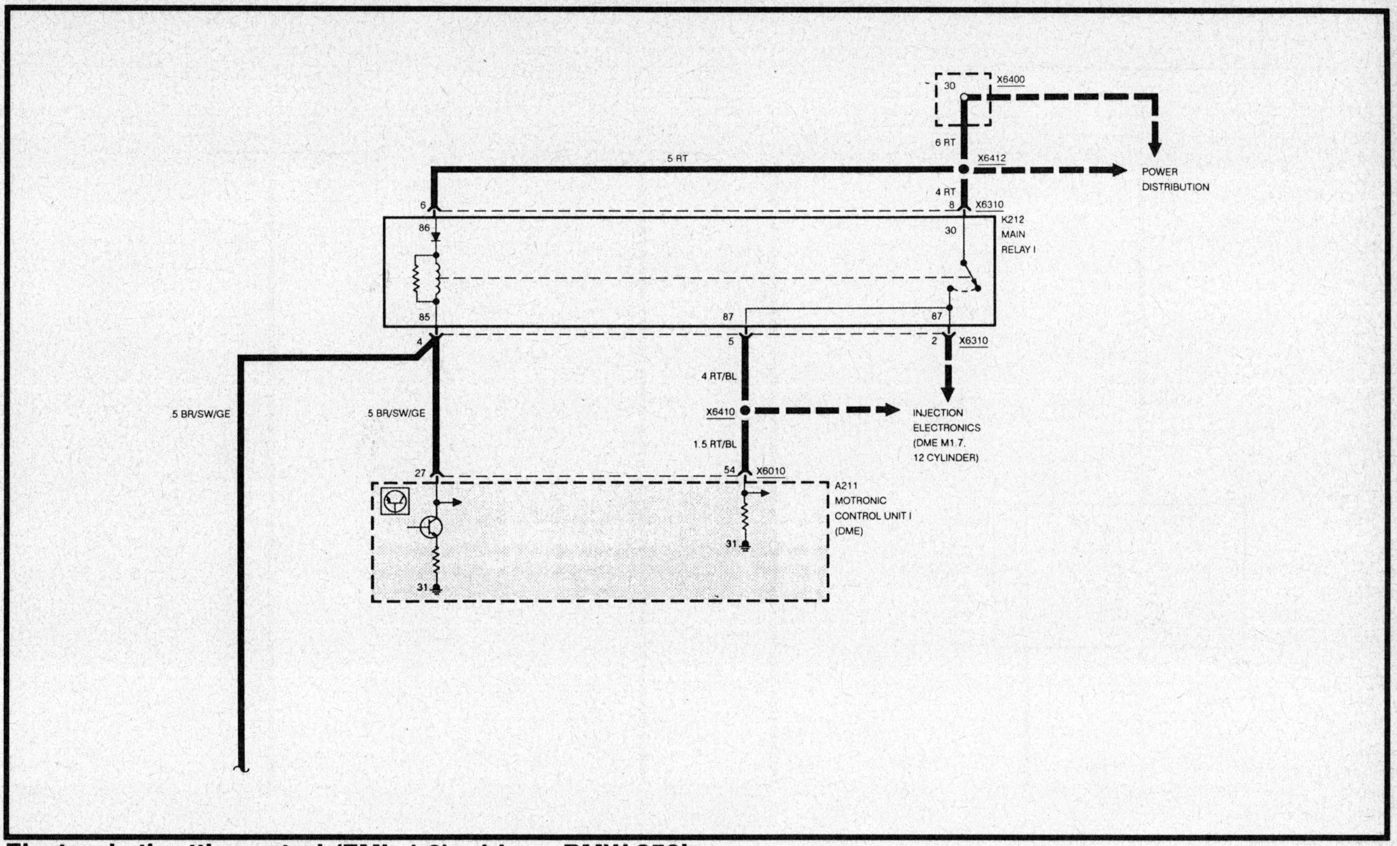

Electronic throttle control (EML 1.2) wiring—BMW 850i

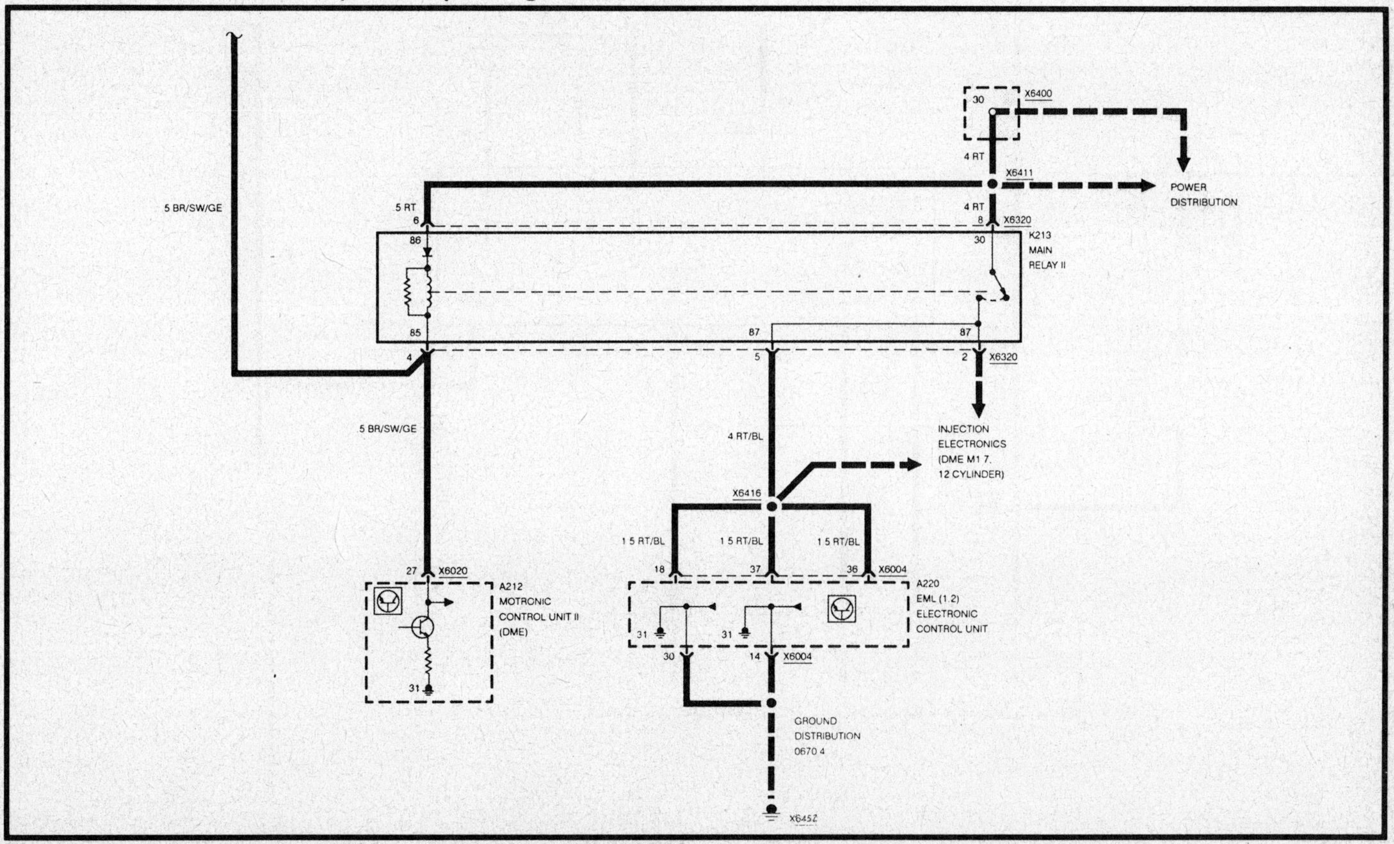

Electronic throttle control (EML 1.2) wiring—BMW 850i

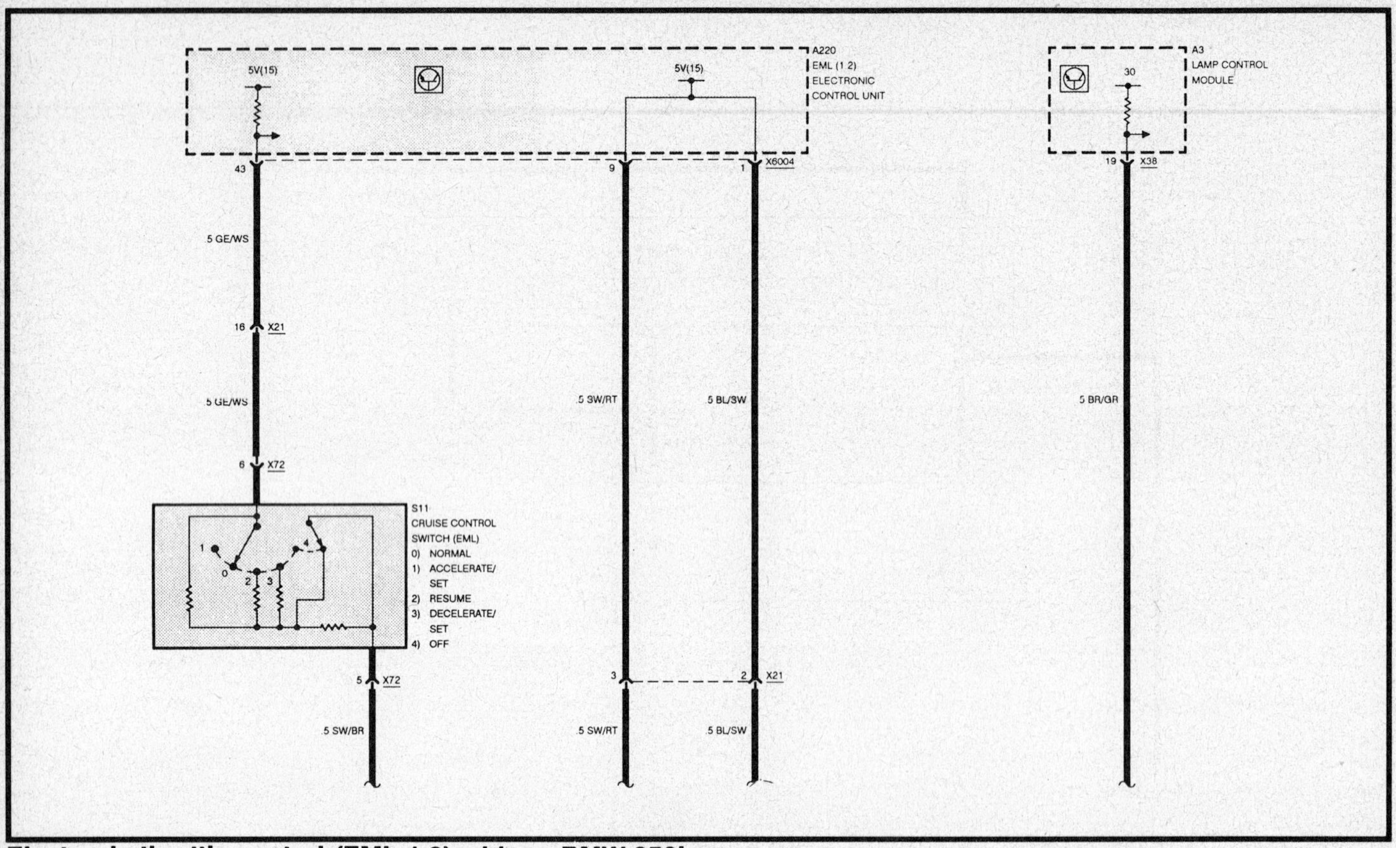

Electronic throttle control (EML 1.2) wiring—BMW 850i

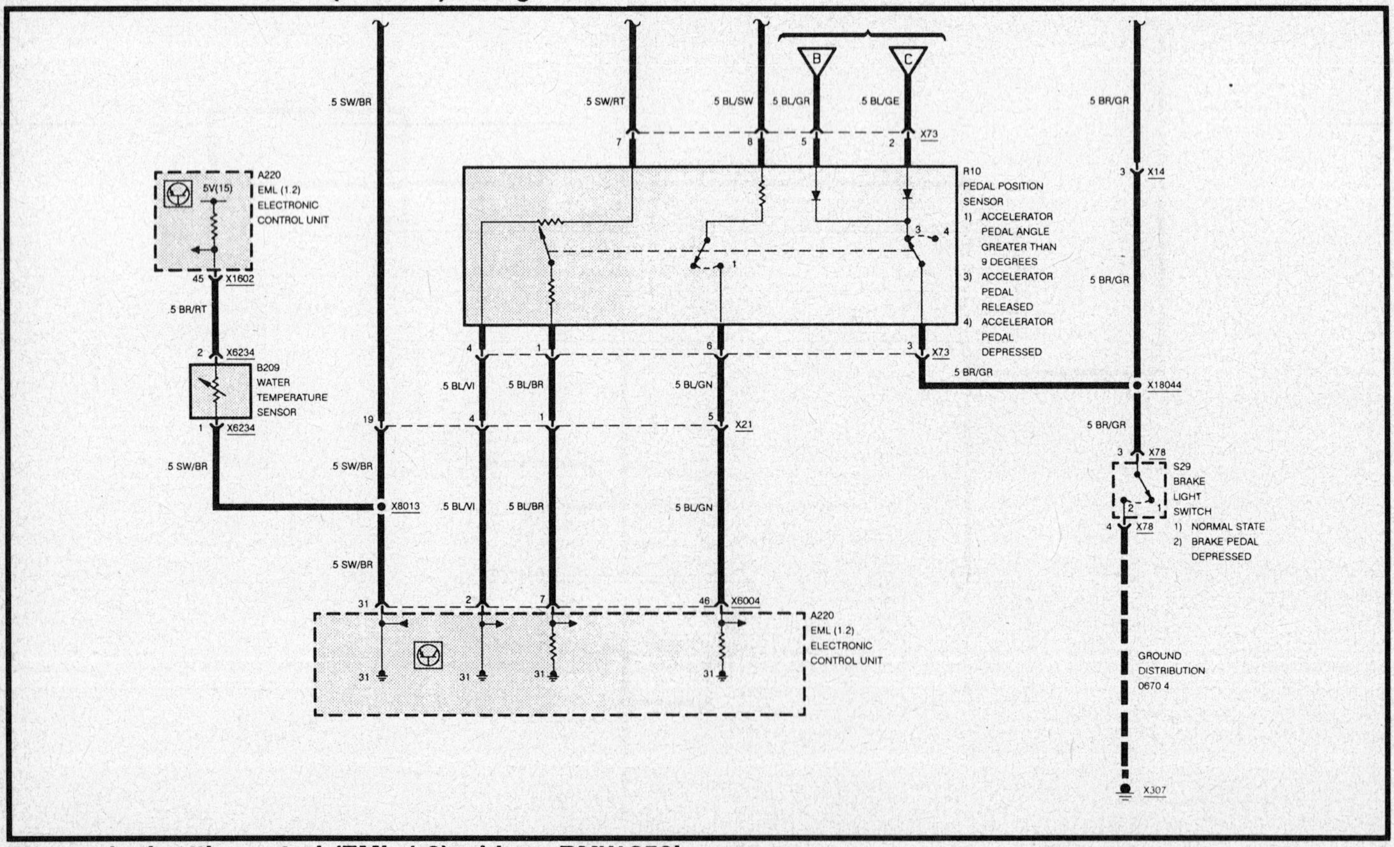

Electronic throttle control (EML 1.2) wiring—BMW 850i

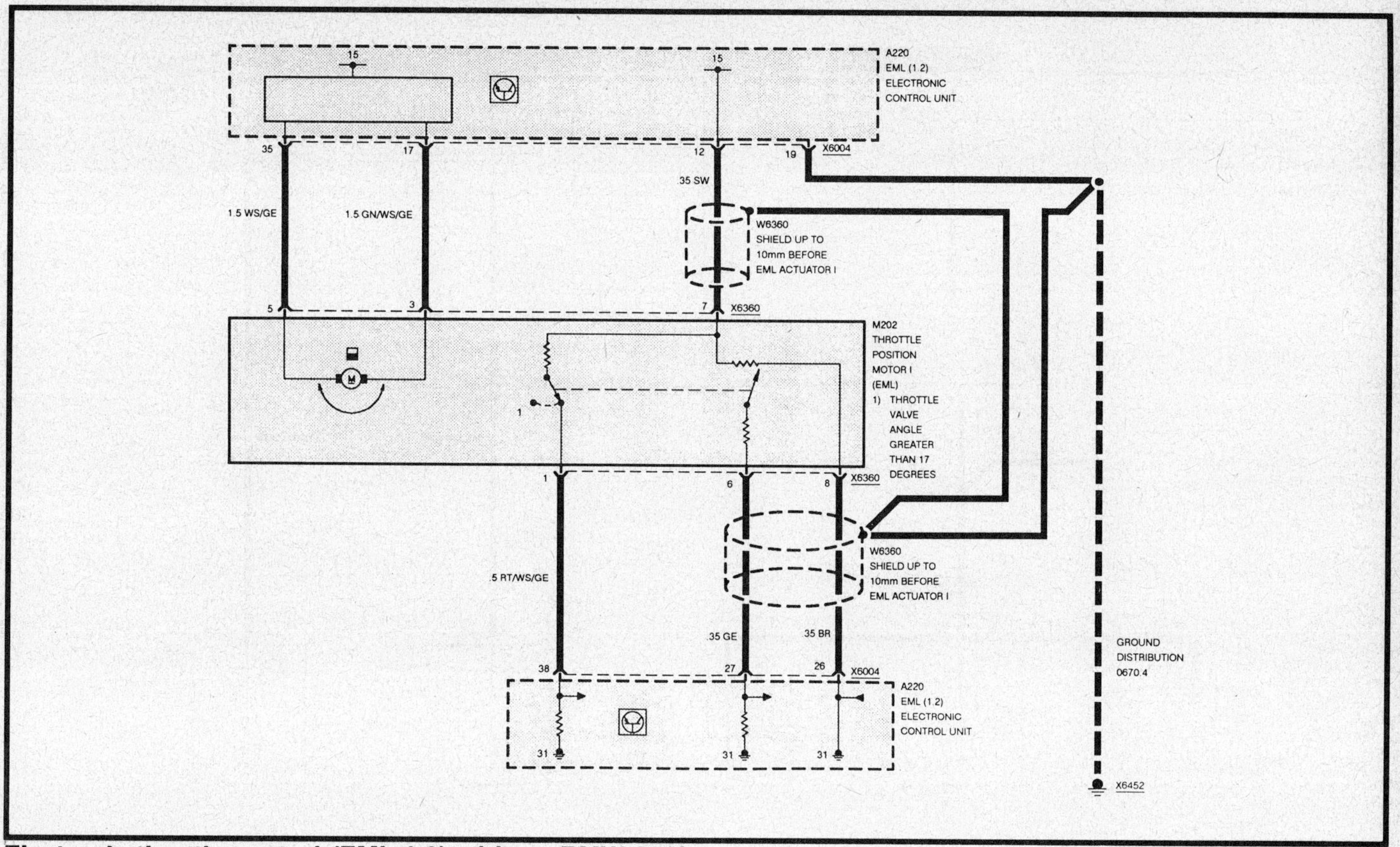

Electronic throttle control (EML 1.2) wiring – BMW 850i

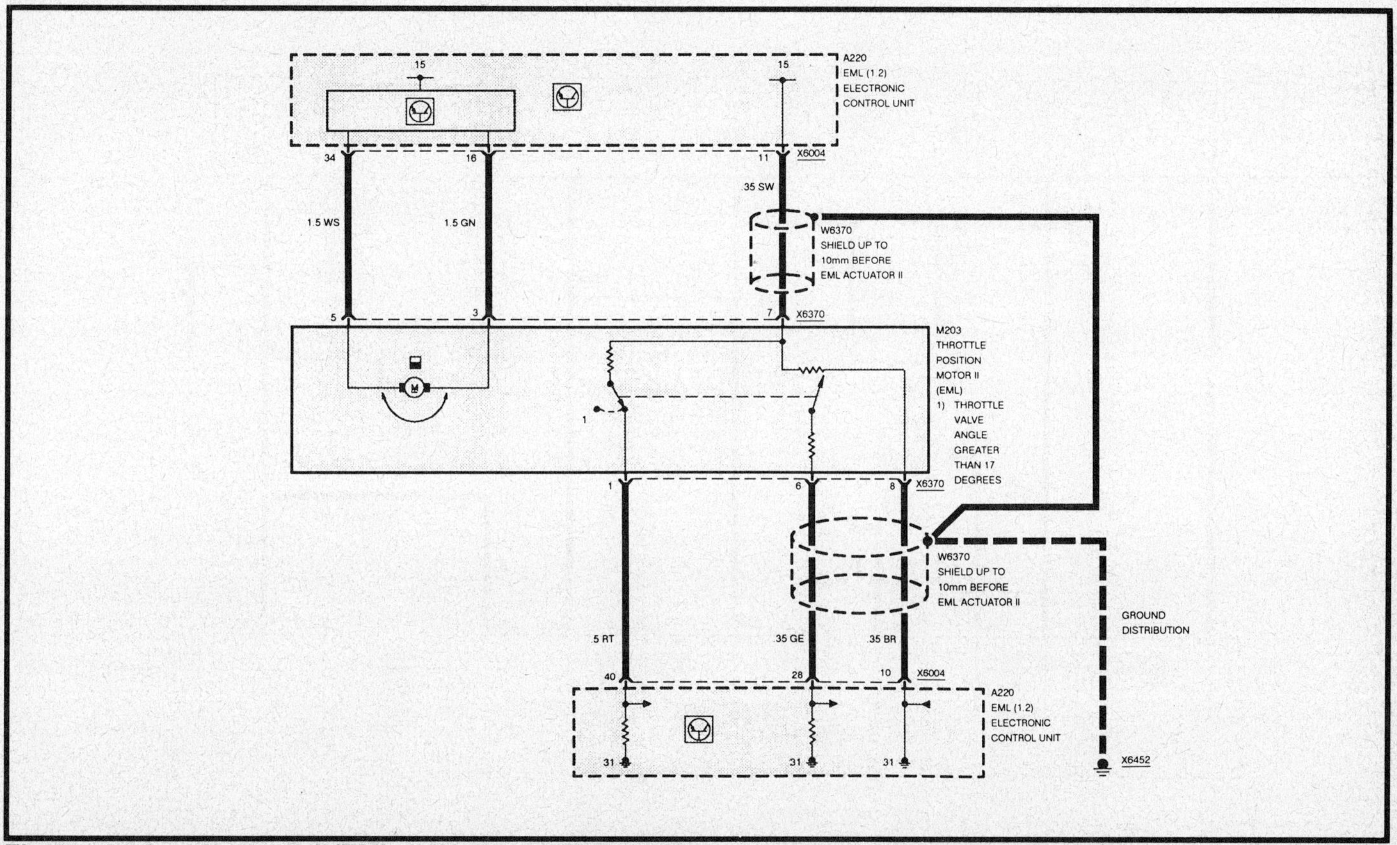

Electronic throttle control (EML 1.2) wiring – BMW 850i

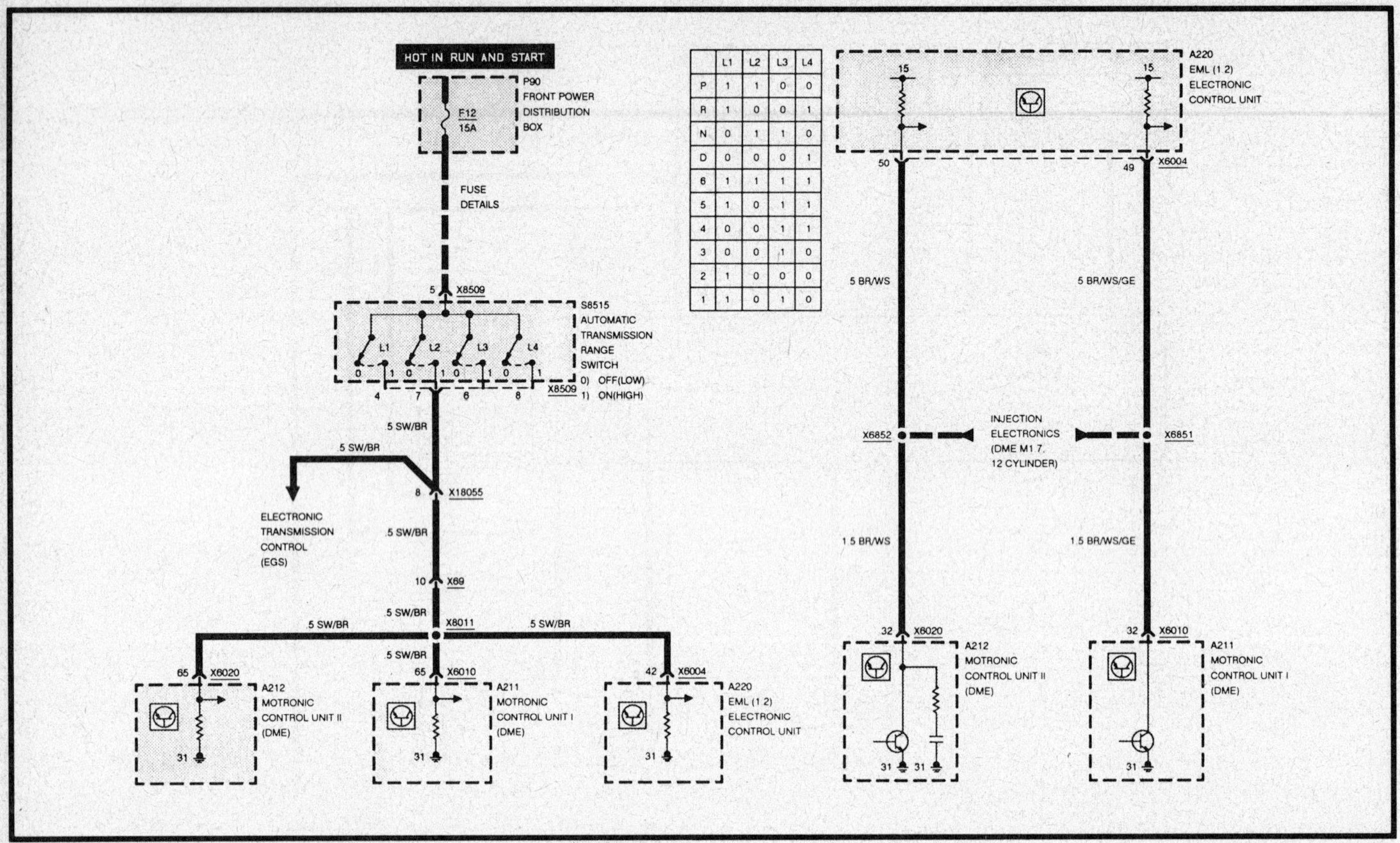

	L1	L2	L3	L4
P	1	1	0	0
R	1	0	0	1
N	0	1	1	0
D	0	0	0	1
6	1	1	1	1
5	1	0	1	1
4	0	0	1	1
3	0	0	1	0
2	1	0	0	0
1	1	0	1	0

Electronic throttle control (EML 1.2) wiring—BMW 850i

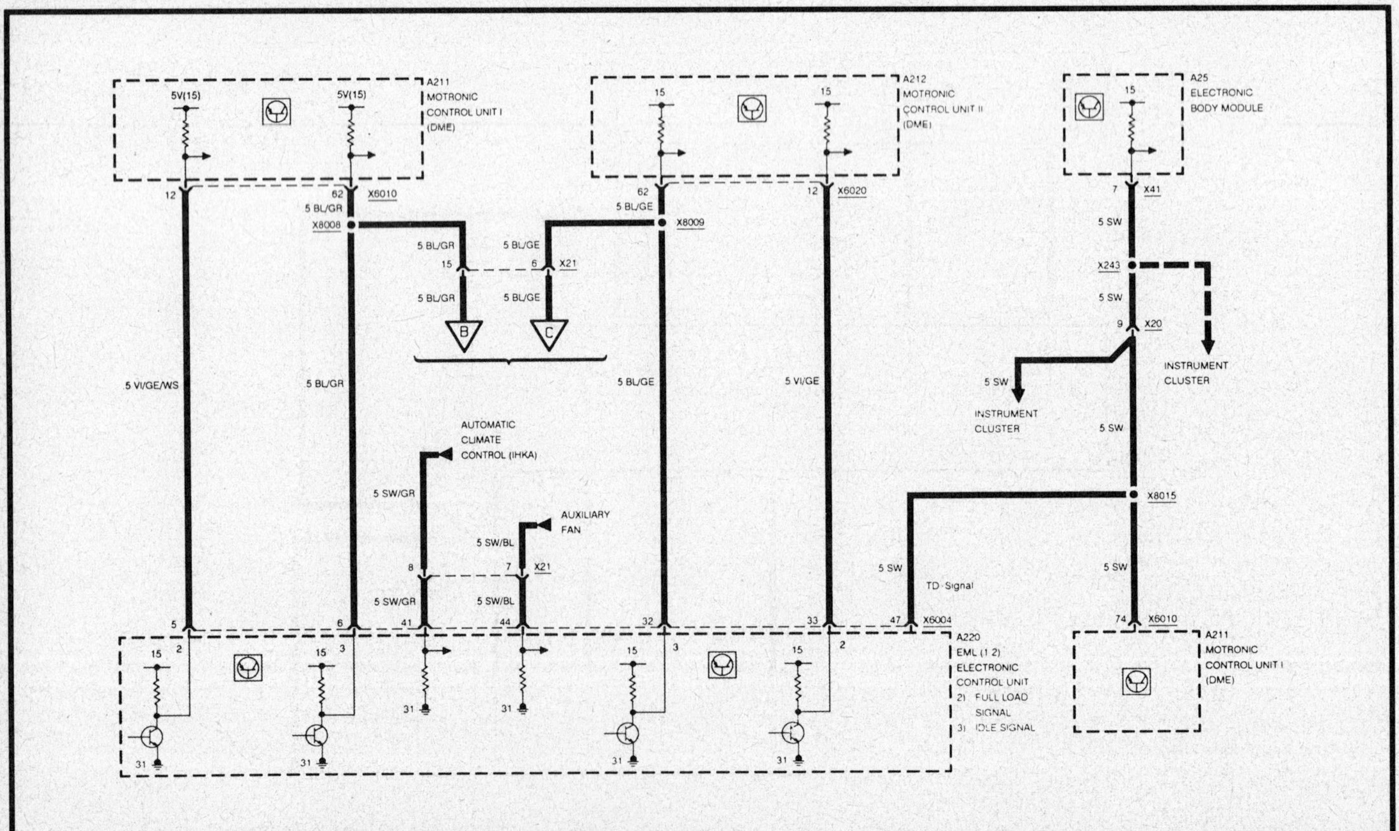

Electronic throttle control (EML 1.2) wiring—BMW 850i

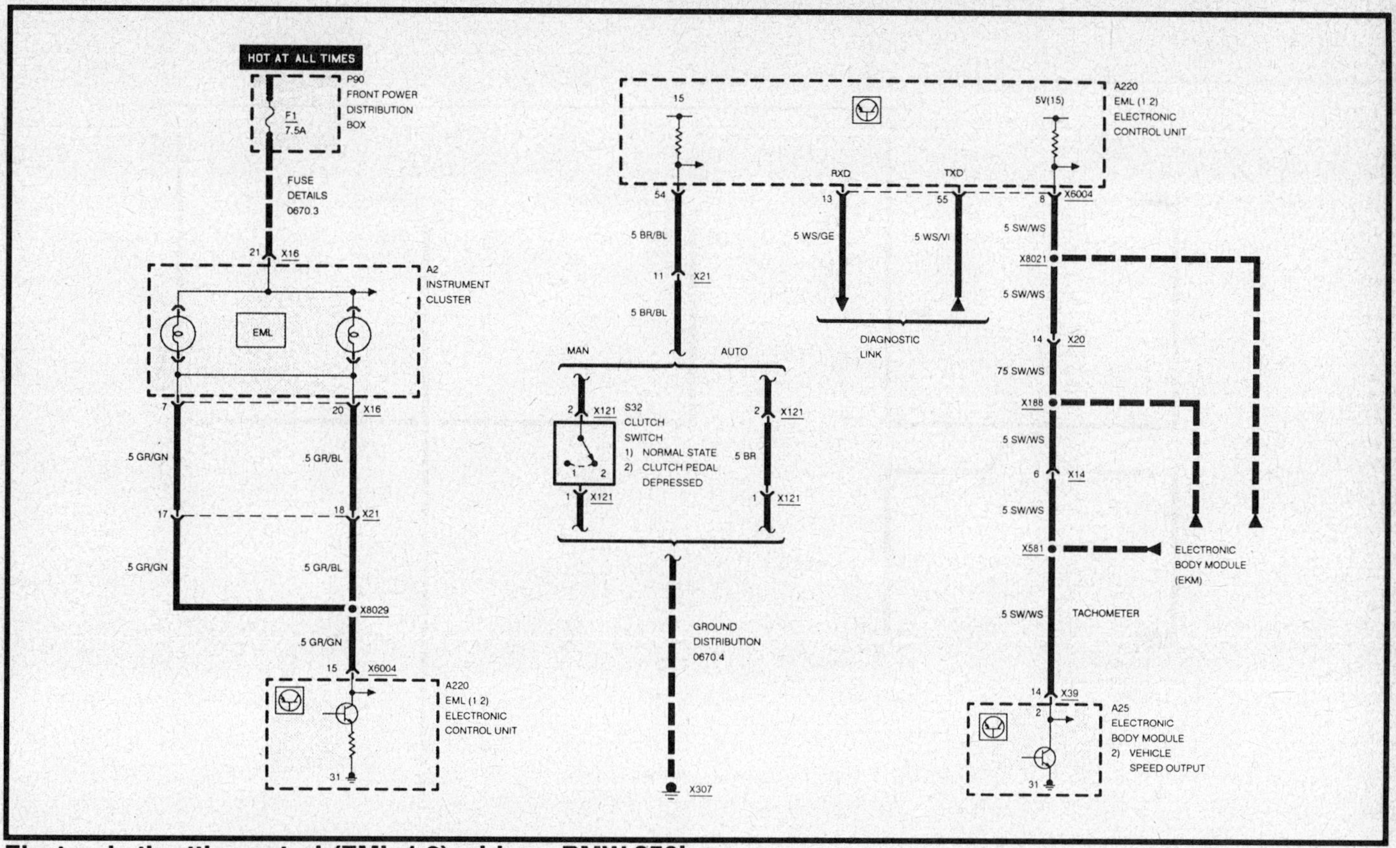

Electronic throttle control (EML 1.2) wiring—BMW 850i

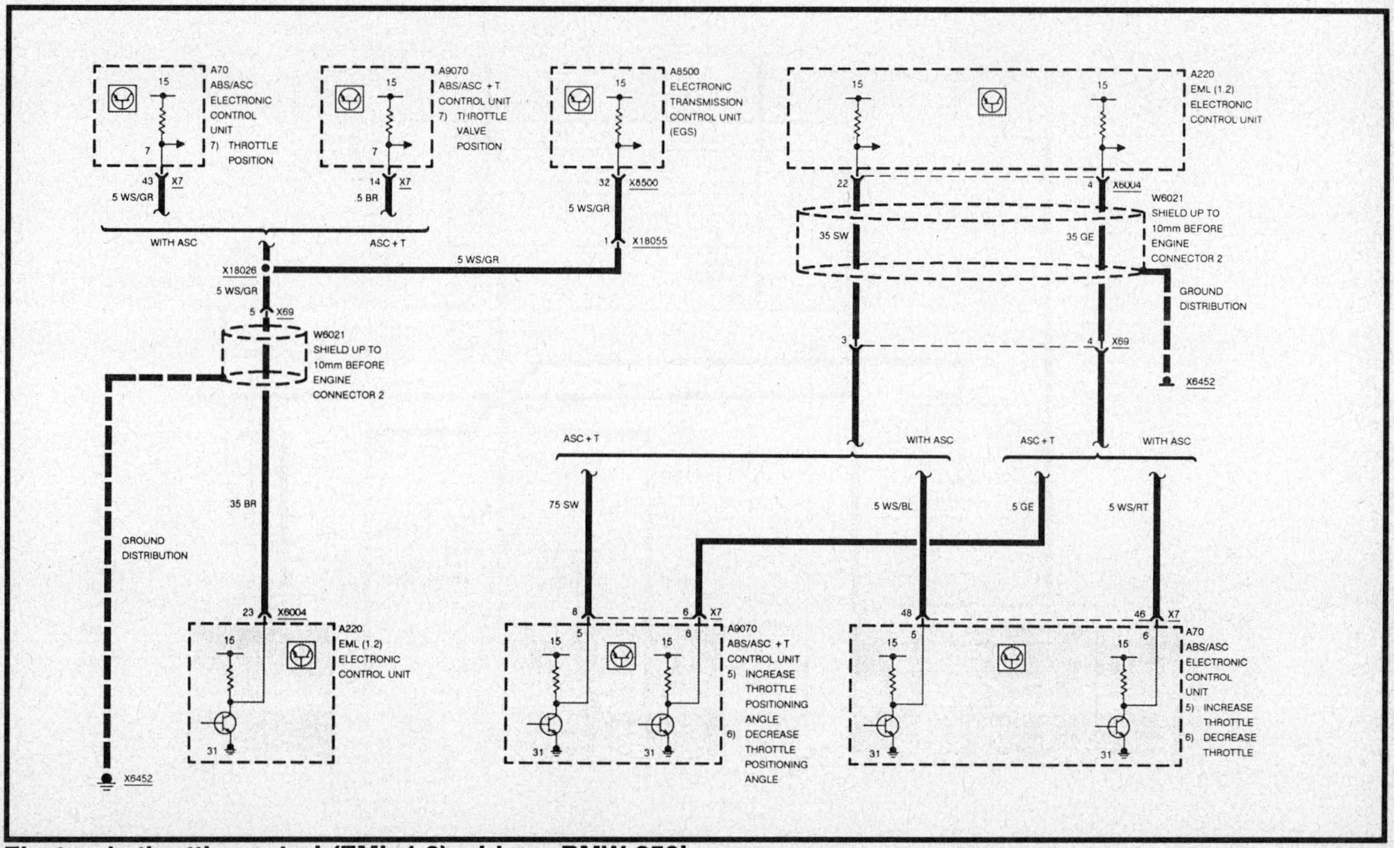

Electronic throttle control (EML 1.2) wiring—BMW 850i

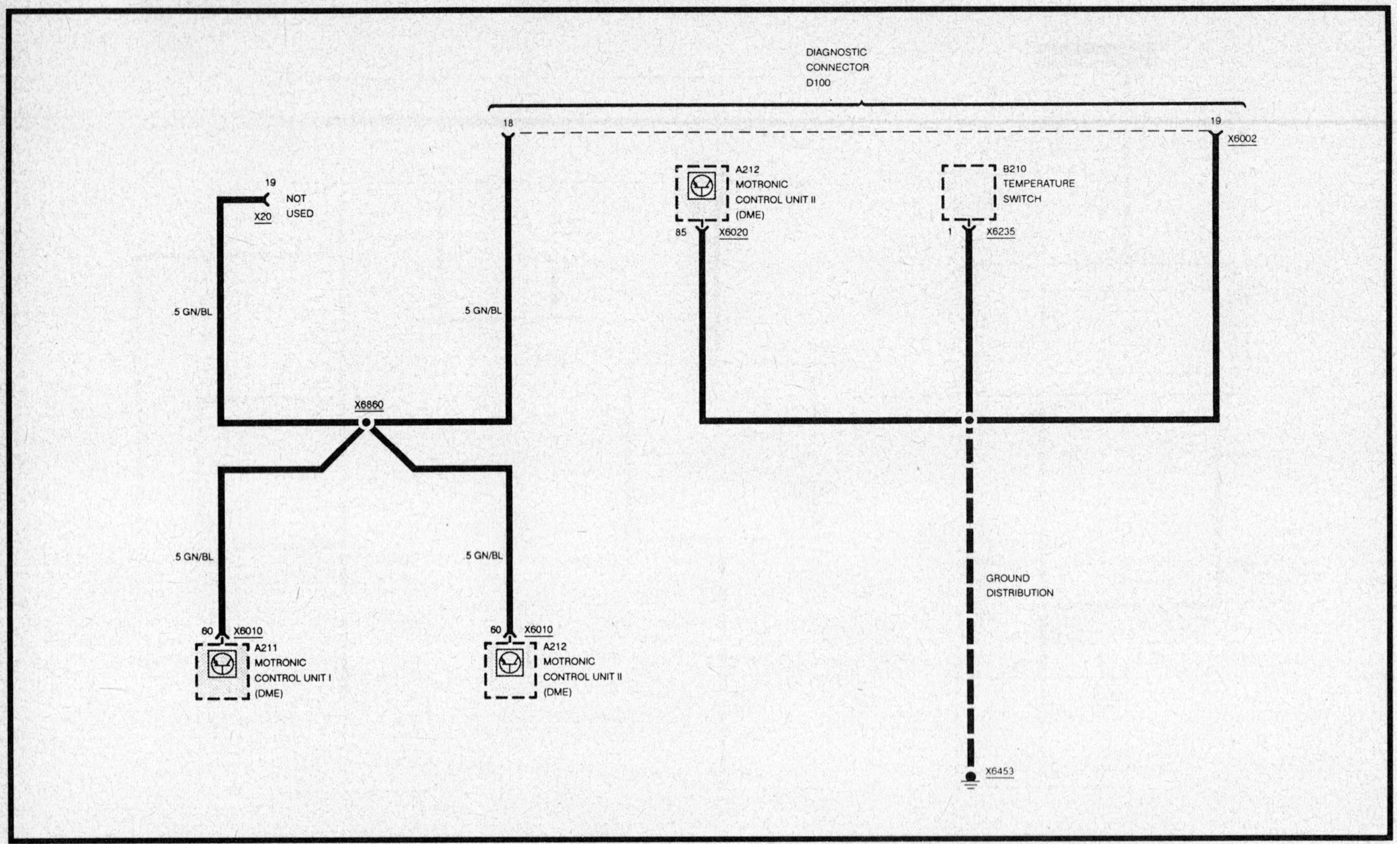

Electronic throttle control (EML 1.2) wiring—BMW 850i

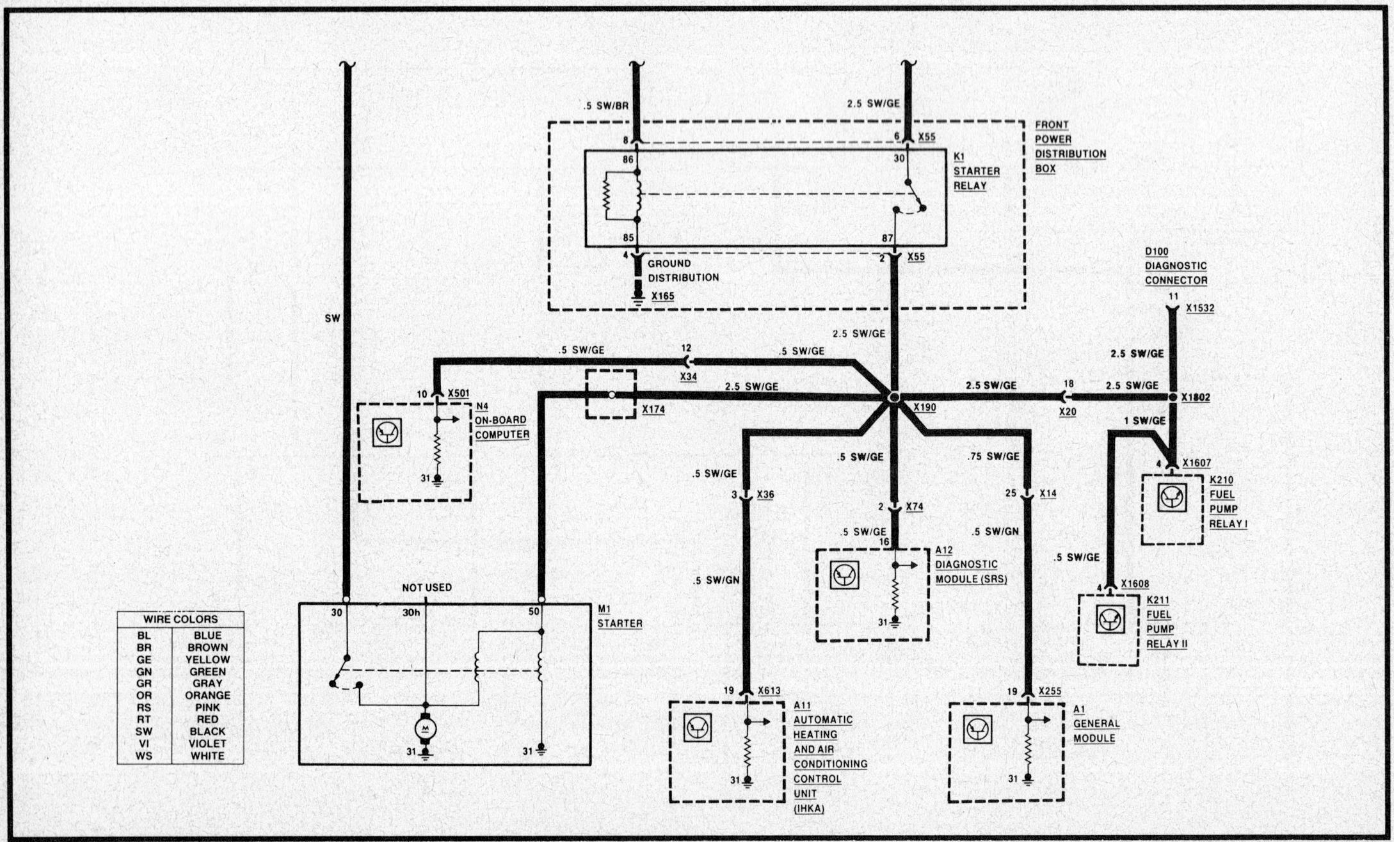

Electronic throttle control (EML 1.2) wiring—BMW 850i

Jaguar Fuel Injection

SECTION 8

JAGUAR

General Information

Fuel metering is controlled by regulating the time that the injectors are open during the engine operating cycle. Constant fuel pressure is maintained within the fuel rail; injector duration or operating time controls the volume of fuel admitted to the cylinders.

The injection system is managed by a digital Engine Control Unit (ECU). This micro-processor based unit controls the electrical signals to the injectors, triggering them for the correct time period. The ECU relies primarily on the manifold pressure and rpm signals from the engine. Once these signals, indicating engine speed and load, are received, the controller uses them to choose proper injector operating periods. This basic pulse length will be slightly modified by the signals from other engine sensors. These secondary control factors include engine coolant temperature, inlet air temperature, throttle position and battery condition.

The injectors are triggered 6 times per engine cycle; on V-12 engines, the injectors of both banks are operated 6 times per cycle.

ENGINE CONTROL UNIT (ECU)

The ECUs used on both the 6 and 12 cylinder engines are highly developed solid-state computers, capable of evaluating a constant flow of data from many sources. The data transmitted from various engine sensors is interpreted electrically and compared to a map or pre-programmed data list. Based on this built-in data, the ECU selects the fuel and ignition timing to match the engine load and rpm. Output signals are sent by the ECU to operate the controlled devices on the engine. The pre-programmed data is stored in permanent memory and cannot be altered.

In addition to overseeing normal engine operation, the engine control unit makes fuel adjustments to deal with the following conditions:

Cranking Enrichment

Based on the coolant temperature signal, the 6 cylinder ECU commands 3 injector pulses per crankshaft revolution; the V-12 controller provides 6 pulses per revolution. Enrichment is reduced as a function of time, engine speed and coolant temperature.

Warm-up Enrichment

Immediately after starting, the ECU commands extended injector ON times; enrichment decreases over a short time period at a fixed rate.

Demand and Engine Load Corrections

Idle, acceleration, deceleration and full-throttle demands are met by increasing or decreasing the injector duration or operating time. On the 4.0L engine, the throttle potentiometer provides the critical signal to the ECU. The 5.3L engine uses the signal from the manifold pressure sensor to determine engine load. The pressure sensor on the V-12 compensates for ambient atmospheric variations.

Voltage Correction

The ECU constantly monitors the state of the battery and electrical load. The time taken for an injector to open is affected by the voltage available and results in a corresponding change in fuel volume delivered. The controller compensates for any voltage change detected by adjusting the injector ON time.

Emission Control

On catalyst-equipped vehicles, the ECU responds to the signals from the oxygen (Lambda) sensor by adjusting the fuel delivery and/or ignition timing. In this fashion, the air/fuel ratio is maintained at approximately the ideal point of 14.7:1. Tailpipe pollutants are held to an absolute minimum during most operating conditions.

Anti-flooding

When the ignition is switched **ON**, the electric fuel pump will operate for 2 seconds to build fuel pressure in the fuel rail. The pump will not operate again until the ECU detects a cranking or rpm signal. At that time, the pump is re-energized by the ECU and operates normally until the signal is discontinued.

Fuel Cut-off

During certain periods of closed throttle operation, the injector function is reduced. This increases fuel economy and reduces the amount of unburned fuel introduced into the exhaust system.

Idle Speed Control

The ECU controls the idle speed and makes adjustments for such conditions as automatic transmission in gear or air conditioning compressor engaged.

The engine control unit also controls the ignition function and timing on both the 6 and 12 cylinder engines. Low voltage electric pulses are sent from the ECU through the ignition amplifier(s) to the ignition coil. The coil(s) then discharge the high tension charge to the spark plugs via the distributor. No mechanical or vacuum advance mechanisms are used; the distributor is used solely as a mechanical switch, directing the charge to the appropriate spark plug. The Lucas ignition system used on the 4.0L engine relies on signals from the airflow meter and crankshaft position sensor. The Marelli system used on the 5.3L engine uses the intake manifold vacuum and engine speed signals as the primary ignition control factors.

Back-up System

The ECU used to control the fuel injection allows the internal self-diagnostic system to evaluate the input and output signals during operation. If the ECU notes the loss or change out-of-range of a sensor signal, the computer will ignore the faulty signal and substitute a fixed value in its place. This value may not necessarily be correct for the immediate driving situation, but will allow the engine to operate. For example, many ECU decisions are based on the engine coolant temperature. Should the coolant temperature sensor signal be lost just after a cold start, the ECU will substitute the default value. This substitution value is based on an engine at normal temperature so vehicle performance will be affected until fully warmed up.

The CHECK ENGINE warning lamp on the dashboard will be lit by the ECU when a fault is detected. If the fault involved loss of a critical signal, the ECU will enter the default value for the signal. This automatic substitution is sometimes referred to as the limp-in mode of operation.

Self-Diagnostics

As the engine control computers became capable of more functions, self-diagnostic and memory circuits were added. These systems allow the controller to note a fault, assign an identity code and store the code in memory for later retrieval. The number, type and meaning of engine codes vary by year and model.

On the 5.3L V-12 engines, the ECUs used on 1990 and 1991 vehicles do not generate or store codes; stored diagnostic or fault

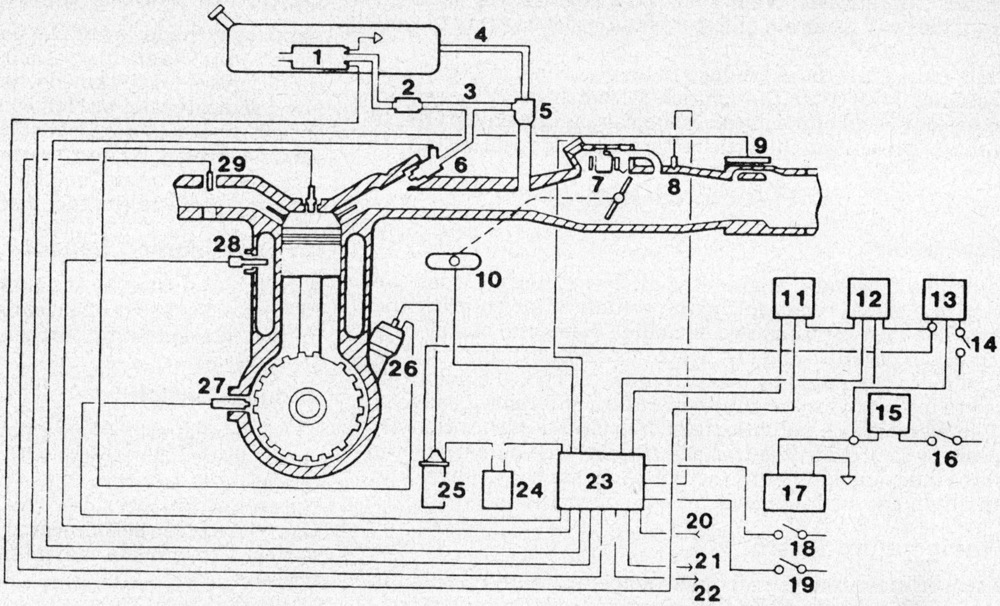

1. Pump
2. Filter
3. Fuel rail
4. Fuel return
5. Pressure regulator
6. Injector
7. Idle speed control valve
8. Air thermistor
9. Airflow meter
10. Throttle position sensor
11. Fuel pump relay
12. Main relay
13. Ignition ON relay
14. Inertia switch
15. Start relay
16. Start switch
17. Battery
18. Gearbox sensor
19. Air conditioning clutch switch
20. Idle override
21. Vehicle condition monitor VCM 1
22. Vehicle condition monitor VCM 2
23. ECU
24. Ignition amplifier
25. Ignition coil
26. High tension distributor
27. Crankshaft sensor
28. Coolant temperature sensor
29. Oxygen (Lambda) sensor

Engine management system—Jaguar 4.0L engine

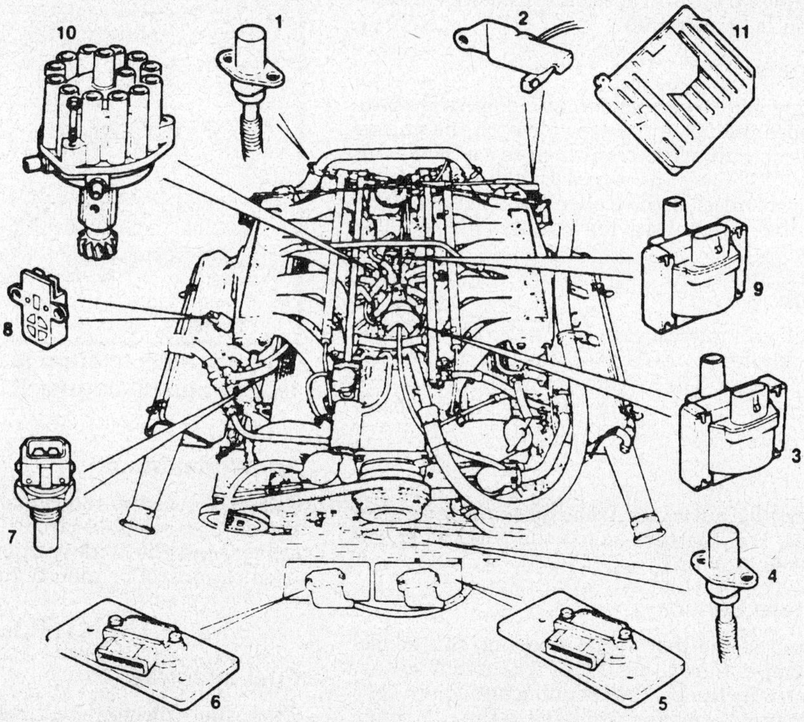

1. Flywheel sensor
2. Throttle idle switch
3. Front ignition coil
4. Crankshaft sensor
5. Left or B bank amplifier unit
6. Right or A bank amplifier unit
7. Coolant temperature sensor
8. Air temperature switch
9. Rear ignition coil
10. Distributor
11. ECU

Marelli digital ignition components—Jaguar 5.3L engine

codes may be retrieved from 1992 V-12 engines. All 1990–92 4.0L engines will generate and store fault codes relating to sensor and circuit function.

While most fault codes are held in an electronic memory and are retained even after the ignition is switched **OFF**, certain codes are only held or displayed as long as the ignition is **ON**. If the fault is present at the next restart, the code will reset.

DATA SENSORS

Airflow Meter

Used on the 6 cylinder engine, the airflow meter provides the ECU with a signal representing the volume of air entering the engine. The unit contains two signalling elements, one at normal inlet temperature and one heated by electrical current. The mass of air is computed by measuring the electrical current required to maintain a constant temperature differential between the two elements. A potentiometer is included within the airflow sensor to provide idle fuel adjustment. The potentiometer operates independently of the sensor elements but is physically mounted on the airflow meter for ease of access.

Air Temperature Sensor

The sensor measures the air inlet temperature and transmits this information to the ECU. The signal is used to alter the ignition timing under certain load/temperature conditions. The signal from this sensor is considered a secondary signal and is used to modify timing and/or fuel delivery signals.

Crankshaft Sensor

The sensor provides information on both engine speed and crankshaft position to the ECU. The sensor is mounted above a 57-tooth gear on the front of the crankshaft. The teeth passing in front of the sensor generate a small electrical voltage in the sensor; changes in the voltage signal are proportional to engine speed. The toothed wheel has 3 omitted teeth; as this space passes the sensor, the signal is interrupted and a positive indication of engine position is transmitted.

Throttle Potentiometer

The sensor is mechanically connected to the engine throttle plate pivot and provides reference voltages for both the engine ECU and the electronic automatic transmission controller, if equipped. The engine ECU uses the signal to adjust fuel delivery; the automatic transmission controller uses the signal to adapt shift points to the different driving styles as indicated by throttle input.

Throttle Idle Switch

The 5.3L engine employs a microswitch in the throttle linkage. The ON-OFF switch advises the ECU when the linkage is in the fully released or idle position. The ECU uses this signal and others to determine correct fuel and timing rates for idle or closed-throttle operation.

Gearbox Sensor

found only on vehicles with automatic transmissions, the sensor consists of 2 switches. The switches advise the ECU when the transmission selector is in **N** or **P**.

Coolant Temperature Sensor

Located on the bottom of the thermostat housing, this sensor signals the coolant temperature to the ECU. A constant voltage is provided to the sensor by the ECU; depending on coolant temperature, the sensor returns an altered voltage to the controller. Change in voltage is inversely proportional to change in temperature; resistance in the sensor decreases as coolant temperature increases.

Oxygen or Lambda Sensor

Lambda is the letter of the Greek alphabet used to denote the deviation of the air/fuel ratio from the ideal point of 14.7:1. Ideal conditions are maintained when Lambda = 1; ratios above or below 1 indicate lean or rich conditions respectively. The oxygen sensor generates a voltage indicating the oxygen content of the exhaust stream. By interpreting this voltage signal, the ECU determines that the air/fuel mixture is richer or leaner than necessary and corrects the fuel delivery accordingly.

Air Conditioning Sensor

A switch, activated by the compressor clutch, which advises the ECU of compressor engagement. The ECU commands changes in the idle speed to compensate for the added load of the compressor.

Vacuum Switch

The 5.3L engine employs a vacuum sensing switch operated by a spring loaded diaphragm within a chamber. When the engine is operating at low rpm under part throttle and at a heavy load, the switch contacts close. The ECU then enters open loop operation; the oxygen sensor signal is disregarded and fuel delivery is enriched to provide maximum power for the present throttle position.

Hot Start System

The 5.3L V-12 engine includes a hot restart system which works in conjunction with the fuel system. When fuel temperature in the right fuel rail exceeds 158°F (70°C), the electric sensor switch in the fuel rail commands an increase in fuel pressure by changing the vacuum at the fuel pressure regulator. The pressure increase is limited to 45 seconds by a timer in the circuit; vacuum control is accomplished through a vacuum solenoid valve and a vacuum delay valve.

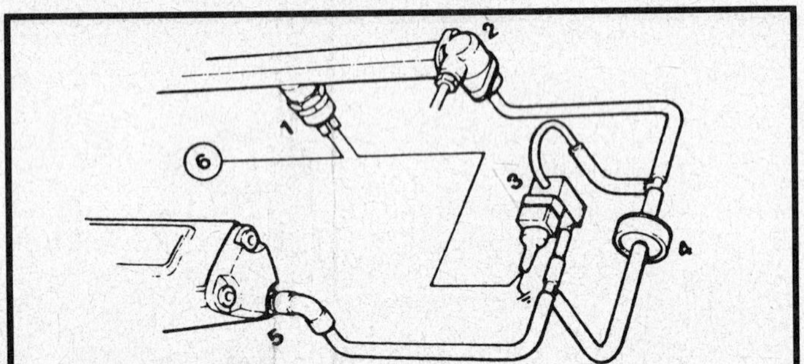

1. Fuel temperature switch
2. Fuel pressure regulator
3. Solenoid vacuum valve
4. Vacuum delay valve
5. Intake manifold connector
6. 45 second timer

The hot start system is dependent upon fuel temperature—Jaguar 5.3L engine

Additional Inputs

The ECU receives additional signals from the ignition ON relay, the starter solenoid circuit, the main relay, the vehicle speed sensor, and the V-12 ignition control circuit. These signals are used to control or modify fuel and/or ignition output signals.

CONTROLLED OUTPUTS

Fuel Injectors

Each fuel injector contains a solenoid-operated needle valve with the movable plunger rigidly attached to the needle. When closed, an internal spring holds the needle against the valve seat.

The injectors are operated in two stages. Initially triggered by a pull-in circuit, the current limiting resistors cause a change to the hold-on circuit once the injector is open. When the ECU switches the injector output circuit off, all power is discontinued and the internal spring closes the injector by forcing the needle against the seat. This 2-stage function reduces the heat build-up at the ECU output transistors and insures rapid injector response.

The 5.3L engine also includes a power resistor wired in series with each group of 3 injectors. Injectors are grouped for the right or A bank—1-3-5 and 2-4-6; the B or left bank follows the same pattern. The resistors limit output current and protect the ECU from injector faults. All 4 resistors are housed within a single unit mounted on the right side of the engine valence.

Ignition Amplifier

The ignition amplifier(s)—1 on 4.0L engine or 2 on 5.3L engine—receive the ignition output signal from the ECU, amplify it and transmit it to the ignition coil(s) on the engine. Once triggered, the coil releases the high tension energy pulse to the spark plugs via the distributor.

Idle Speed Control Actuator

Attached to the intake manifold of the 4.0L engine only, the idle speed control actuator contains an electric stepper motor connected to an air valve. The ECU positions the stepper motor to allow more or less air to bypass the throttle plate during engine idle. Correct idle speed is maintained regardless of temperature or engine load.

NOTE: The 5.3L engine accomplishes the idle control function through the use of an auxiliary air valve. The valve operates mechanically rather than electronically. The amount of air bypassing the throttle plate is controlled by the position of the valve in response to the temperature of the engine coolant.

Fuel Pump Relay

The fuel pump relay is wired in series with both the Ignition On andMain relays. The fuel pump is controlled by the ECU grounding the fuel pump relay at the appropriate time.

Trip Computer

The ECU provides signals to the trip computer on the dashboard or instrument cluster. Under normal conditions, signals regarding fuel information are supplied to the computer for computation of miles per gallon, total fuel consumed, etc.

If the ECU detects a fault and a fault code is stored, the ECU will transmit the diagnostic code through the trip computer screen when the ignition is next cycled to **ON**.

FUEL DELIVERY SYSTEM

Fuel is drawn from the rear-mounted fuel tank by the in-tank electric fuel pump. The pump delivers the fuel, through the canister filter, to the inlet port of the fuel rail.

The fuel injection rail is mounted on the engine and has a fuel pressure regulator attached at the return port. The vacuum operated fuel pressure regulator controls fuel line pressure and maintains a constant fuel pressure at the injectors. The vacuum signal is taken from the intake manifold; fuel pressure is maintained in the rail at 35–45 psi. Unused fuel is returned to the fuel tank through the return lines. Both the inlet and return ports of the fuel pump module use one-way valves for system safety.

Both the 4.0L and 5.3L engines incorporate a mechanical inertia switch into the fuel pump relay circuit. In the event of extreme deceleration or frontal impact, the switch disconnects the fuel pump relay, stopping the pump immediately and preventing fuel delivery during a collision. On XJ6 vehicles, this switch is located under the right side dash, near the door. The XJS inertia switch is located at the left front door pillar, below the dash. If tripped, the switch may be manually reset by pressing the button or plunger on top of the switch downward or inward.

NOTE: It is possible for this switch to trip without collision damage to the vehicle. Always check the inertia switch setting when diagnosing failure to start or apparent fuel pump failure.

Diagnosis and Testing

NOTE: Many procedures require disconnection of the negative battery cable. On some Jaguar models the battery is located in the trunk.

SERVICE PRECAUTIONS

- The fuel system retains pressure even when the ignition is **OFF** or the engine has not operated for a period of time. Always release the pressure using the correct procedure before any fuel line or component is loosened or disconnected.
- Observe No Smoking/No Open Flame precautions whenever the fuel system or any component is being serviced.
- Keep a charged fire extinguisher close at hand and know how to use it.
- Anticipate fuel spillage when disconnecting lines or components. Use clean rags below fuel line fittings to catch small amounts of fuel. Treat the rags as a flammable hazard and dispose of them properly.
- Large amounts of fuel, such as from the fuel tank, must be drained directly into approved, tightly-capped fuel containers. Do not use pails, plastic jugs or similar open containers. A supply of dry sand, cat litter or other absorbent material should be on hand to deal immediately with spillage.
- Keep all ECU parts and harnesses dry during service. Protect the ECU and all solid-state components from rough handling or extremes of temperature.
- Before attempting to remove any parts, turn the ignition switch **OFF** and disconnect the battery ground cable.
- Make sure all harness connectors are fastened securely. A poor connection can cause an extremely high voltage surge, resulting in damage to integrated circuits.
- Always use a 12 volt battery as a power source.
- Do not disconnect the battery cables with the engine running; never run the engine with battery cables loose or disconnected.
- Do not attempt to disassemble any ECU under any circumstances.
- If installing a 2-way or CB radio, mobile phone or other radio equipment, keep the antenna as far as possible away from the electronic control unit. Keep the antenna feeder line at least 8 in. away from the ECU harness and do not run the lines parallel for a long distance. Be sure to ground the radio to the vehicle body.
- When performing ECU input/output signal diagnosis, remove the water-proofing rubber plug, if equipped, from the connectors to make it easier to insert tester probes into the connector. Always reinstall it after testing.
- When connecting or disconnecting pin connectors from the ECU, take care not to bend or break any pin terminals. Check that there are no bends or breaks on ECU pin terminals before attempting any connections.
- When measuring supply voltage of ECU-controlled components, keep the tester probes separated from each other and from accidental grounding. If the tester probes accidentally make contact with each other during measurement, a short circuit will damage the ECU.
- Use great care when working on or around air bag systems.

Due to back-up circuitry, the system may stay armed for a period of time without battery power.

- When working on or around air bag wiring or components, always disarm the system using the correct procedures. Once disarmed, attach a flag or note to the steering wheel. Re-arm the system when repairs are completed and remove the wheel marker.

READING CODES

1990–92 4.0L and 1992 5.3L engines will provide diagnostic fault codes. The codes may be read on the trip computer display on the dashboard. Generally, codes will be held in memory and the dash warning lamp will be lit when a fault is sensed by the ECU.

NOTE: Not every stored code will cause the warning lamp to light after the first occurrence. Always check for codes during diagnosis, including cases where the dash warning lamp was not reported lit.

To read the stored codes, bring the vehicle to a complete stop. Switch the ignition **OFF** and wait at least 10 seconds. Turn the ignition to the **II** or **ON** position but do not start or crank the engine. Press the trip computer button **DISP** or **VCM**. After a short period of time, the stored codes will be displayed graphically on the trip computer panel. The display may include the designation FF, an abbreviation for FUEL system FAILURE.

The fault codes are displayed in order of priority, one at a time. The next is only displayed when the preceding one is repaired or cleared.

Engine Codes

4.0L ENGINE

Code 11—Idle potentiometer or associated wiring; idle trim pot out of normal operating range.
Code 12—Airflow meter or associated wiring; airflow meter signal out of normal operating range.
Code 14—Coolant temperature sensor (thermistor) or associated wiring; faulty signal during warm-up.
Code 16—Air intake temperature sensor (thermistor) or associated wiring; resistance out of range.
Code 17—Throttle potentiometer or associated wiring; resistance out of range.
Code 18—Throttle potentiometer or airflow meter calibration; throttle pot signal low during high air flow.
Code 19—Throttle potentiometer or airflow meter calibration; throttle pot signal high during low air flow.
Code 22—Fuel pump driver; faulty ECU output to fuel pump relay.
Code 23—Fuel supply—poor feedback control in rich direction
Code 24—Ignition driver; ECU output to ignition amplifier
Code 26—Air leak; poor feedback control in lean direction
Code 29—ECU self-check; internal function.
Code 33—Injector driver fault; ECU output to injectors
Code 34—Injector driver or associated wiring; injector dribble or staying open.
Code 44—Oxygen (Lambda) sensor or associated wiring; feedback signal out of range either rich or lean.
Code 46—Idle speed control, coil 1 driver; ECU output to idle speed control stepper motor
Code 47—Idle speed control, coil 2 driver; ECU output to idle speed control stepper motor.
Code 48—Idle speed control motor; significantly out of position with engine hot.
Code 68—Road speed sensor or associated wiring; indicated speed less than 3 mph (5 kph) at high air flow.
Code 69—Drive/neutral switch or associated wiring; shows cranking in **D** or high airflow in **N**

5.3L ENGINE

Code 11—Idle potentiometer
Code 13—ECU pressure sensor
Code 14—Coolant thermistor or associated wiring
Code 16—Air thermistor or associated wiring
Code 17—Throttle potentiometer or associated wiring
Code 18—Throttle potentiometer calibration 1
Code 19—Throttle potentiometer calibration 2
Code 23—Fuel supply
Code 34—Injector fault, A or right bank
Code 36—Injector fault, B or left bank
Code 44—Oxygen (Lambda) sensor fault, A or right bank
Code 45—Oxygen (Lambda) sensor fault, B or left bank
Code 49—Fuel injection ballast resistor or associated wiring

Jaguar Diagnostic System (JDS)

A computerized, hand-held diagnostic tool is available through Jaguar Cars Ltd. The JDS tool allows rapid testing and diagnosis of electrical circuits. The tool can be adapted to each motor through the use of specific software. The tool allows testing and limited operation of many ECU controlled components. Stored fault codes may also be erased with the JDS tool.

CLEARING CODES

Codes will not be cleared from memory until all stored codes have been displayed. Interrupting the display cycle will allow some codes to be retained and the memory will not clear.

If the JDS tool or its equivalent is not available, clear stored codes by driving the vehicle in excess of 19 mph (30 km/h); the memories will be cleared electronically. Each system will perform its self-check function and any new codes will be set. If no faults are present in a particular system, the failure memory will remain cleared.

COOLANT TEMPERATURE SENSOR

Testing

1. Allow the engine to cool sufficiently to allow removal of the either the radiator or reservoir tank cap.
2. Disconnect the negative battery cable.

Coolant Temperature °C	°F	Resistance (kilo-ohms)
−10	14	9.2
0	32	5.9
10	50	5.9
20	68	2.5
30	86	1.7
40	104	1.18
50	122	0.84
60	140	0.60
70	158	0.435
80	176	0.325
90	194	0.250
100	212	0.190

Coolant temperature sensor resistance chart—4.0L and 5.3L engines

3. Disconnect the wiring connector at the coolant temperature sensor.
4. Use an accurate thermometer to determine the temperature of the coolant.
5. Measure the resistance of the sensor by connecting an ohmmeter between the 2 terminals. The resistance will vary as the opposite of temperature; as coolant temperature increases, resistance decreases. If the resistance is not appropriate for the coolant temperature, the sensor must be replaced.
6. Use the ohmmeter to check resistance between each terminal and the body of the sensor. There should be no continuity between either terminal and the sensor body.
7. Reinstall the electrical connector.
8. Connect the negative battery cable.

FUEL PUMP RELAY

Testing

The simplest test is to listen for fuel pump operation; if the pump runs for approximately 2 seconds after switching the ignition **ON**, the relay is functioning properly. If the pump does not run under these conditions, check the system systematically.
1. Inspect the inertia switch. Make certain the reset button is pressed in. Remove the switch cover and check the security of the wiring connections.
2. Use a jumper wire to ground terminal 85 of the fuel pump relay. Switch the ignition **ON** but do not start or crank the engine.
3. Test for battery voltage at terminal 86 of the main relay. If no voltage is present, inspect the circuitry from the battery, through the ignition and inertia switches to the main relay.
4. Check for battery voltage at terminal 87 on the main relay. If no voltage is present, check the ground and connection from terminal 85 of the main relay. If voltage is present in this circuit, replace the main relay.
5. Check for battery voltage at terminal 86 of the fuel pump relay. If no voltage is present, repair the open circuit between main relay terminal 87 and fuel pump relay terminal 86.
6. Check for battery voltage at terminal 87 of the pump relay. If no voltage is present, check the ground and connection from terminal 85 of the pump relay. If voltage is present, replace the relay.
7. Check for battery voltage at the input side of fuel pump. If voltage is present in the Brown/Gray wire, either the pump is faulty or the pump ground circuit is open. If no voltage is present, repair the open circuit between terminal 87 of the fuel pump relay and the fuel pump.

AIRFLOW METER

Testing

1. With the ignition switch **OFF**, disconnect the harness connector from the airflow meter.
2. Using jumper wires with insulated connectors or clips, connect terminal 36 of the airflow meter to the ground or negative side of a 14 volt power source. Connect terminal 9 to the positive or hot side of the voltage source.
3. Cover or plug both ends of the airflow meter.
4. Connect the negative lead of a voltmeter to terminal 6. Connect the positive lead to terminal 7. With no airflow through the meter, voltage should be 0.2–0.6 volts. If 0 volts are present, the output circuit is open and the meter must be replaced. If voltage exceeds 0.7, the airflow meter must be replaced due to incorrect calibration or sensor failure.
5. Remove the plugs or covers from the ends of the airflow meter.
6. Blow one hard blast of air through the meter in the normal direction of airflow. Output voltage should rise to 1.5 volts, then fall back to 0.2–0.7 volts. If 0 volts are present, the output circuit is open and the meter must be replaced. If voltage exceeds 5.0, the airflow meter must be replaced due to incorrect calibration or sensor failure. If no change in voltage is seen, replace the airflow meter; the sensor has failed.

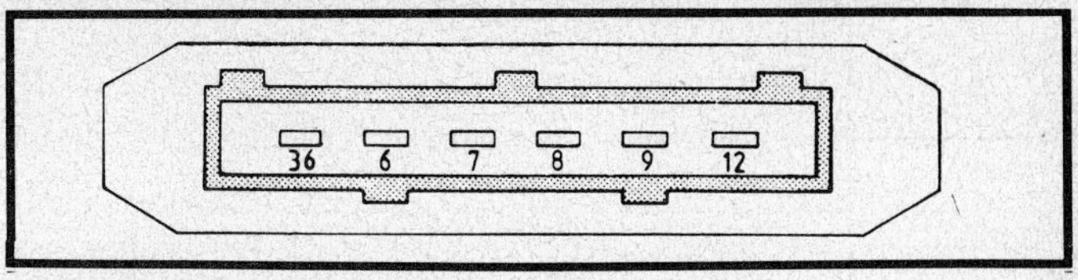

Airflow meter connector—4.0L engine

AIR TEMPERATURE SENSOR

Testing

1. With the ignition **OFF**, disconnect the negative battery cable.
2. Remove the connector from the air temperature sensor.
3. Use a reliable thermometer to determine the temperature of the air near the air intake.
4. Measure the resistance between the 2 terminals of the air sensor.
5. Resistance of the sensor will change as the inverse of air temperature; as the air becomes warmer, sensor resistance is reduced.
6. Check for continuity between each sensor terminal and the body of the sensor. No continuity should be present.
7. Reconnect the sensor connector. Connect the negative battery terminal.

Ambient Air Temperature °C	Ambient Air Temperature °F	Resistance (ohms)
−10	14	960
0	32	640
10	50	435
20	68	300
30	86	210
40	104	150
50	122	180

Air temperature sensor resistance chart—5.3L engine

AUXILIARY AIR VALVE

Testing

1. Remove the valve from the engine.
2. Close the idle speed adjustment screw completely. Note the position or number of turns to tighten the screw.
3. Carefully immerse the bulb or sensor end in boiling water. Observe the valve head through the side port; the valve should move smoothly to the closed position.
4. Quickly blow through the side port; no air should pass with the valve closed.
5. Remove the valve from the water and allow the sensor to cool. The valve head should move back to open the main air passage.
6. If the valve passes both tests, it may be reused. Reset the idle adjustment screw to its original position and reinstall the valve.

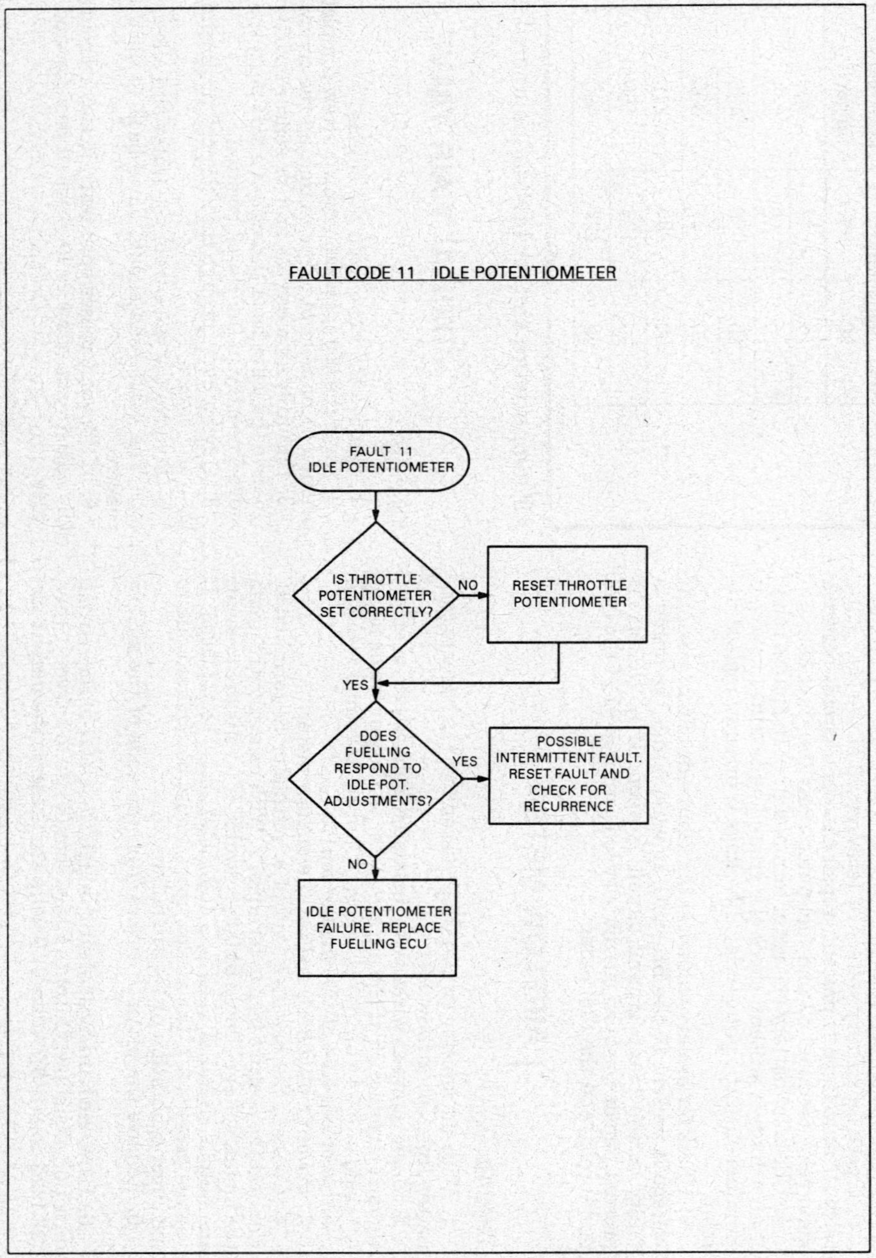
FAULT CODE 11 IDLE POTENTIOMETER
FAULT 11 IDLE POTENTIOMETER
IS THROTTLE POTENTIOMETER SET CORRECTLY?
NO
RESET THROTTLE POTENTIOMETER
YES
DOES FUELLING RESPOND TO IDLE POT. ADJUSTMENTS?
YES
POSSIBLE INTERMITTENT FAULT. RESET FAULT AND CHECK FOR RECURRENCE
NO
IDLE POTENTIOMETER FAILURE. REPLACE FUELLING ECU

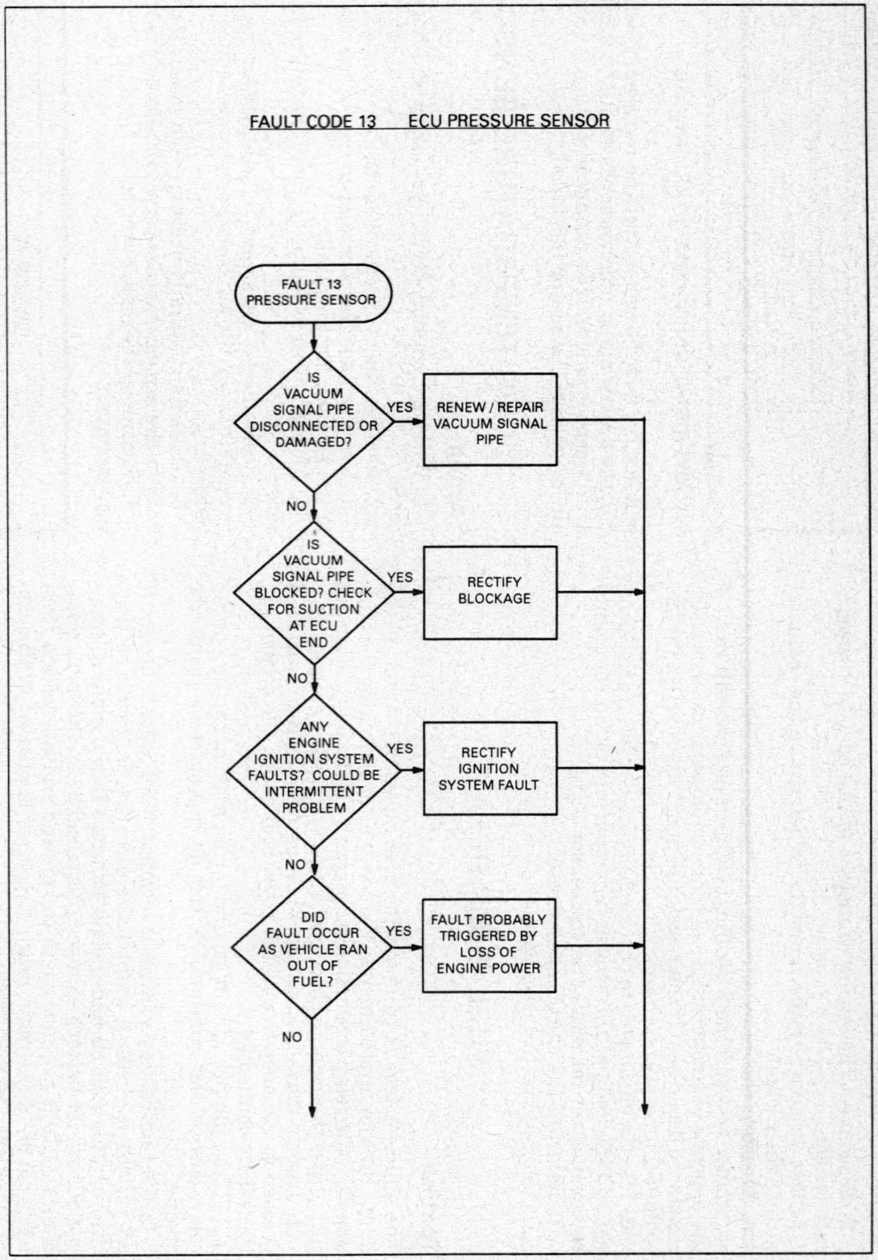
FAULT CODE 13 ECU PRESSURE SENSOR
FAULT 13 PRESSURE SENSOR
IS VACUUM SIGNAL PIPE DISCONNECTED OR DAMAGED?
YES
RENEW / REPAIR VACUUM SIGNAL PIPE
NO
IS VACUUM SIGNAL PIPE BLOCKED? CHECK FOR SUCTION AT ECU END
YES
RECTIFY BLOCKAGE
NO
ANY ENGINE IGNITION SYSTEM FAULTS? COULD BE INTERMITTENT PROBLEM
YES
RECTIFY IGNITION SYSTEM FAULT
NO
DID FAULT OCCUR AS VEHICLE RAN OUT OF FUEL?
YES
FAULT PROBABLY TRIGGERED BY LOSS OF ENGINE POWER
NO

1992 5.3L ENGINE – DIAGNOSTIC CHARTS – XJS

FAULT CODE 13 ECU PRESSURE SENSOR
(Continued)

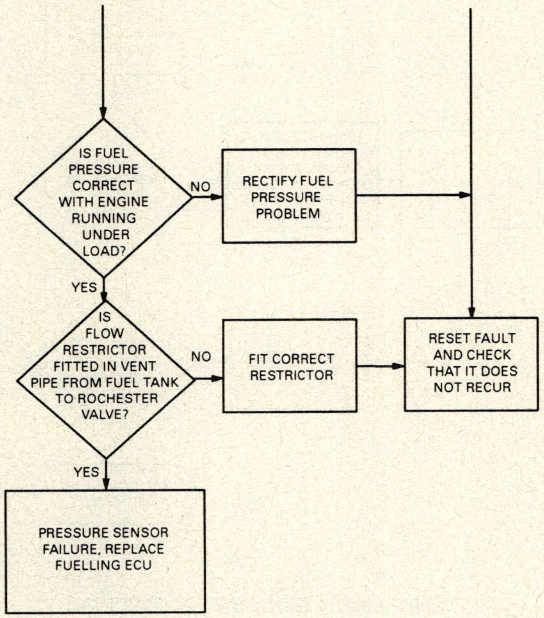

FAULT CODE 14 ENGINE COOLANT THERMISTOR

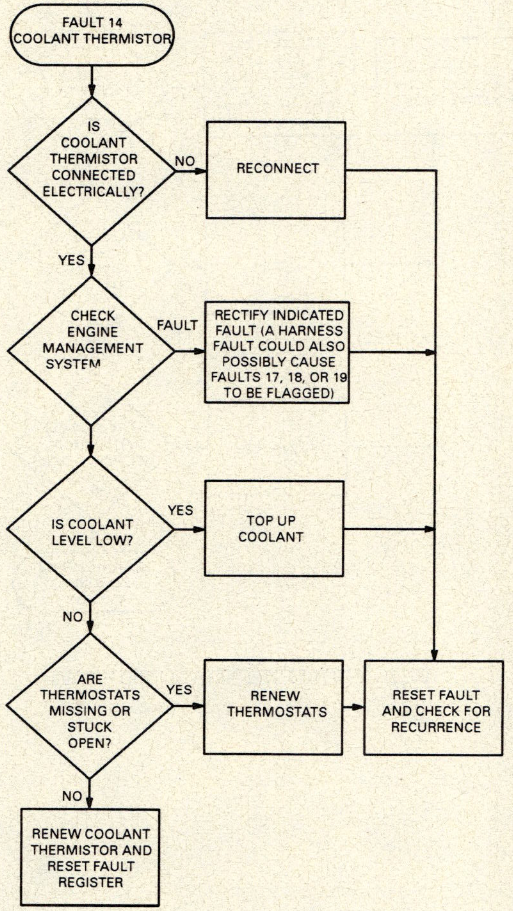

1992 5.3L ENGINE – DIAGNOSTIC CHARTS – XJS

FAULT CODE 16 AIR TEMPERATURE THERMISTOR

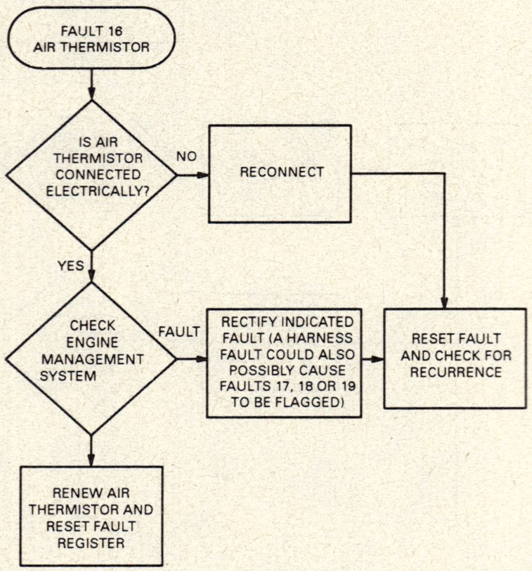

FAULT CODE 17 THROTTLE POTENTIOMETER

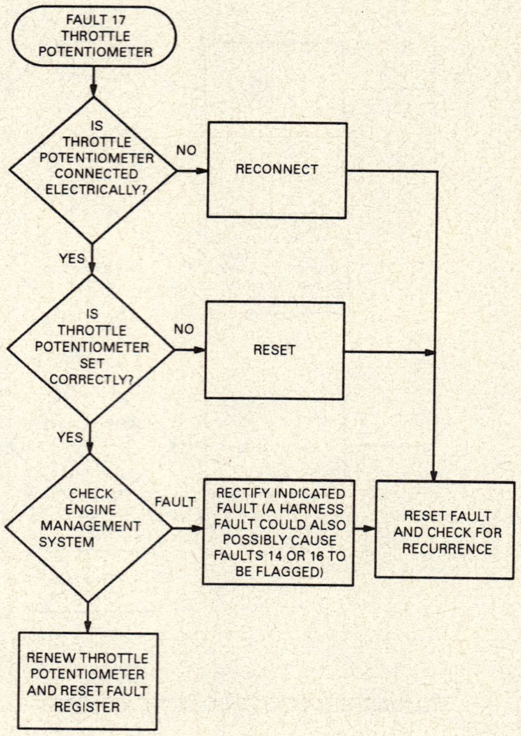

1992 5.3L ENGINE—DIAGNOSTIC CHARTS—XJS

FAULT CODES 18 AND 19 THROTTLE POTENTIOMETER CALIBRATION

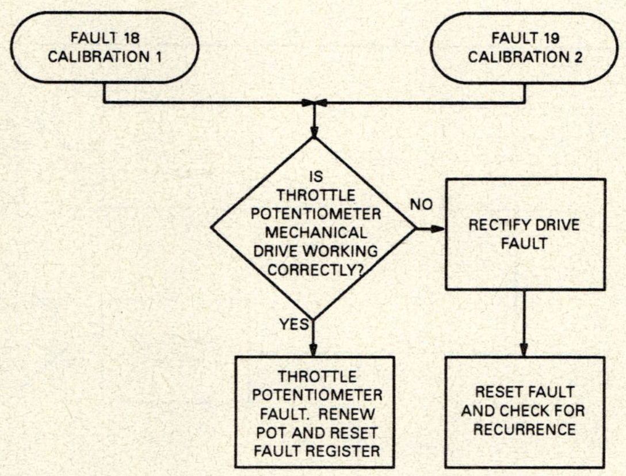

FAULT CODE 23 FUEL SUPPLY

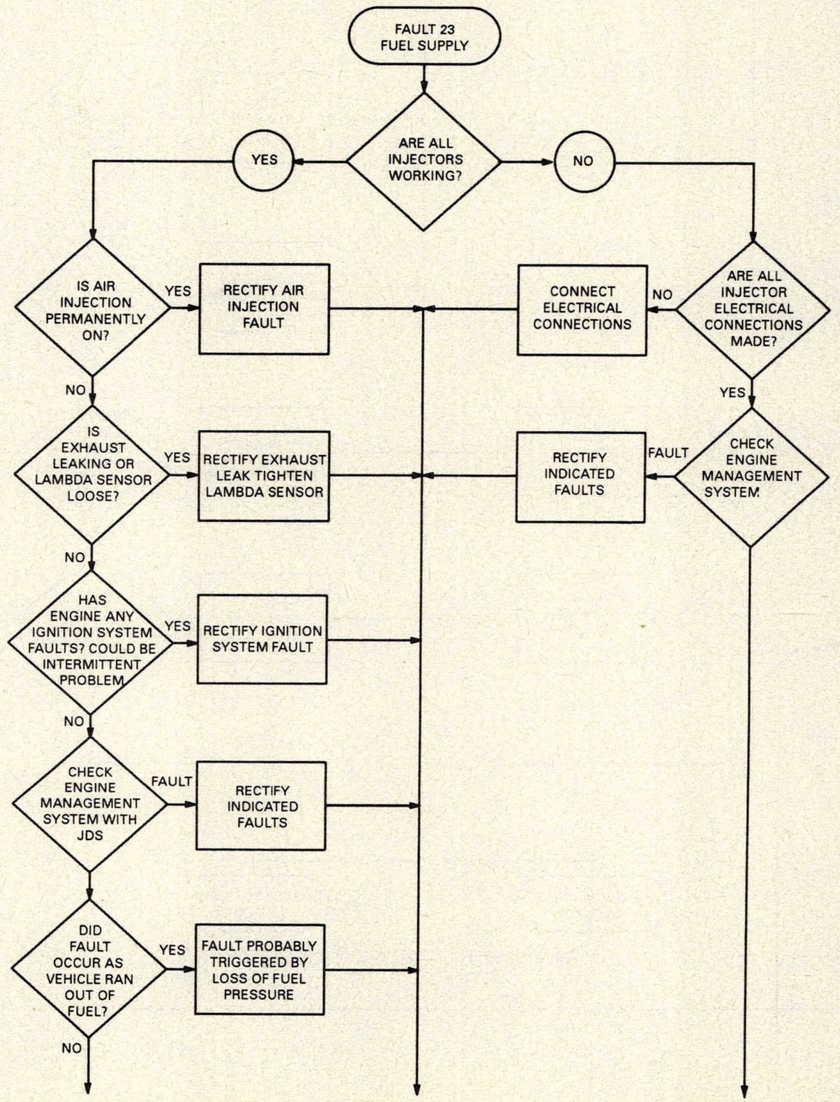

1992 5.3L ENGINE – DIAGNOSTIC CHARTS – XJS

FAULT CODE 23 FUEL SUPPLY (Continued)

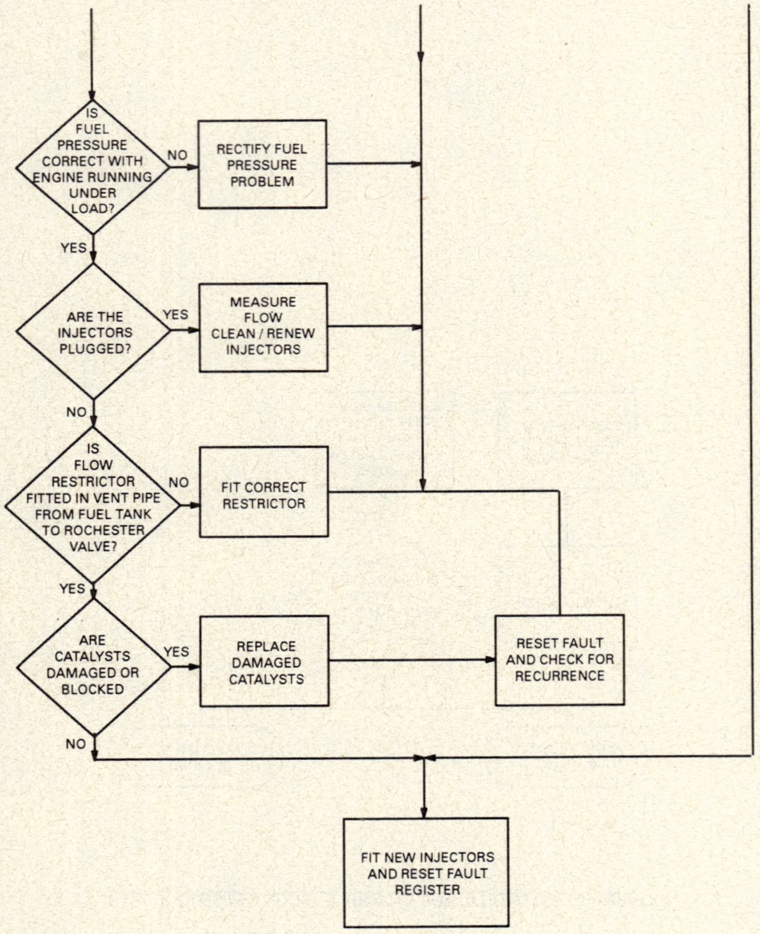

FAULT CODES 34 AND 36 FUEL INJECTORS

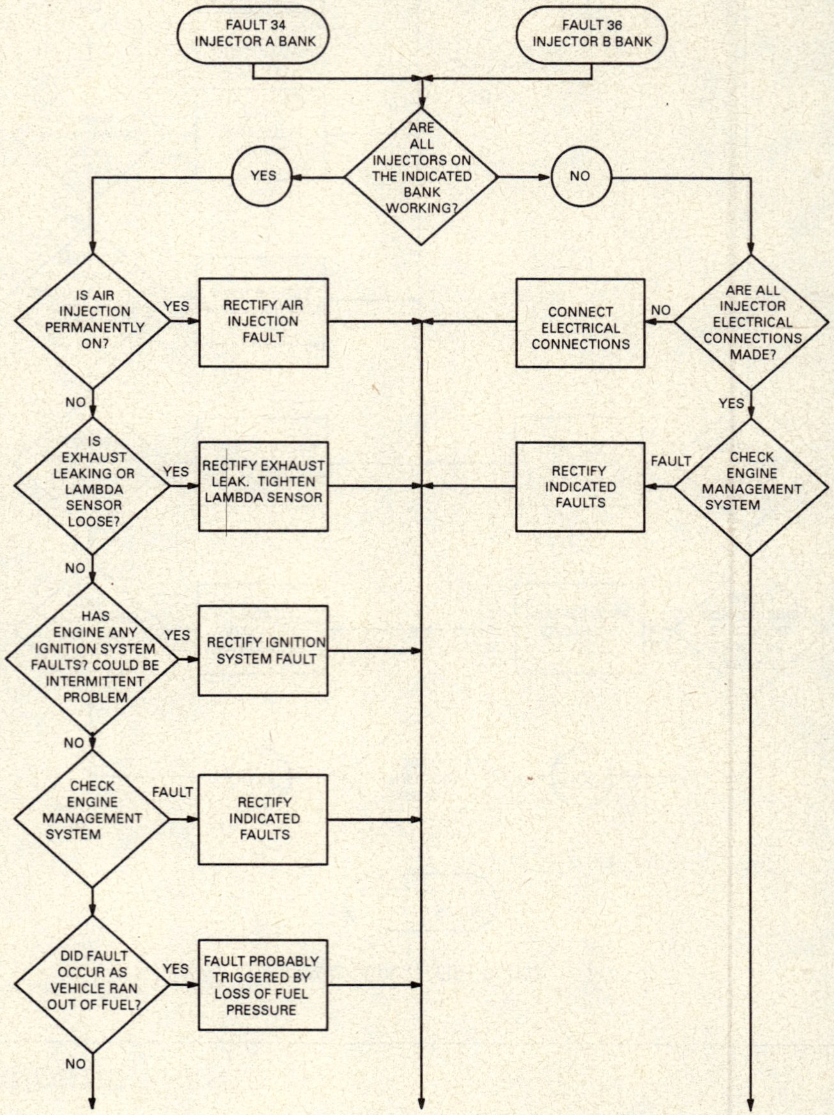

1992 5.3L ENGINE—DIAGNOSTIC CHARTS—XJS

FAULT CODES 34 AND 36 FUEL INJECTORS
(Continued)

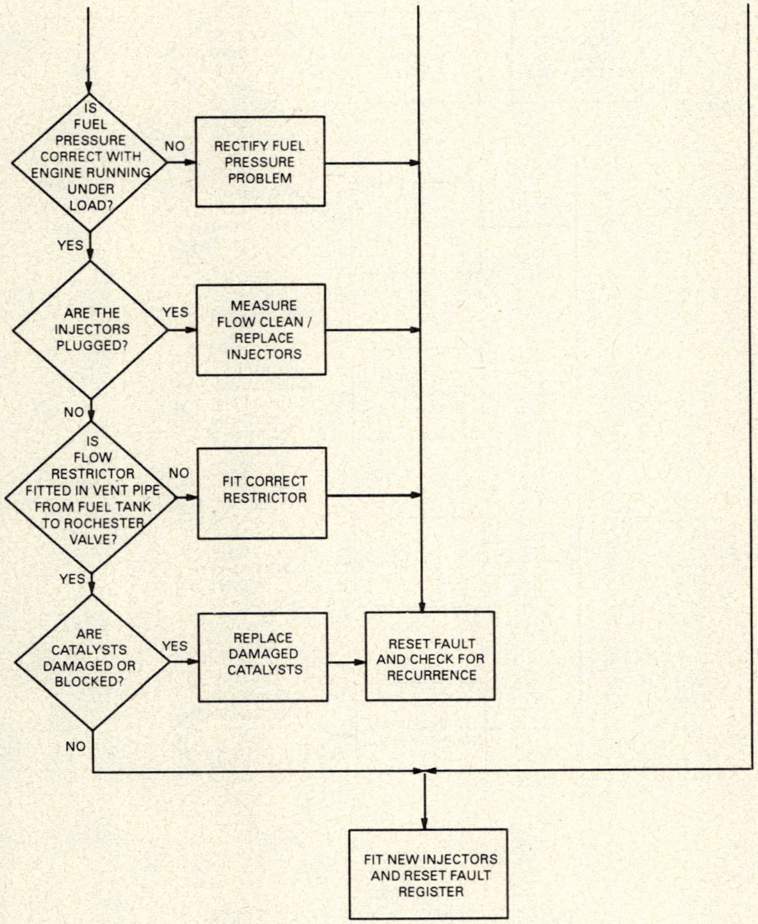

FAULT CODES 44 AND 45 LAMBDA SENSORS

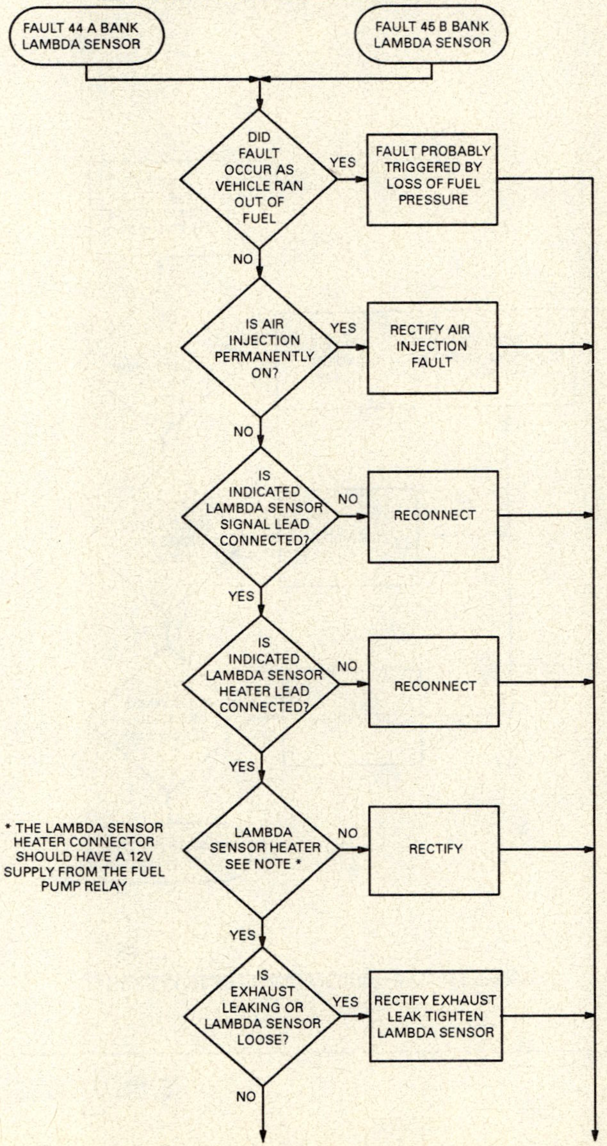

* THE LAMBDA SENSOR HEATER CONNECTOR SHOULD HAVE A 12V SUPPLY FROM THE FUEL PUMP RELAY

1992 5.3L ENGINE—DIAGNOSTIC CHARTS—XJS

FAULT CODES 44 AND 45 LAMBDA SENSORS
(Continued)

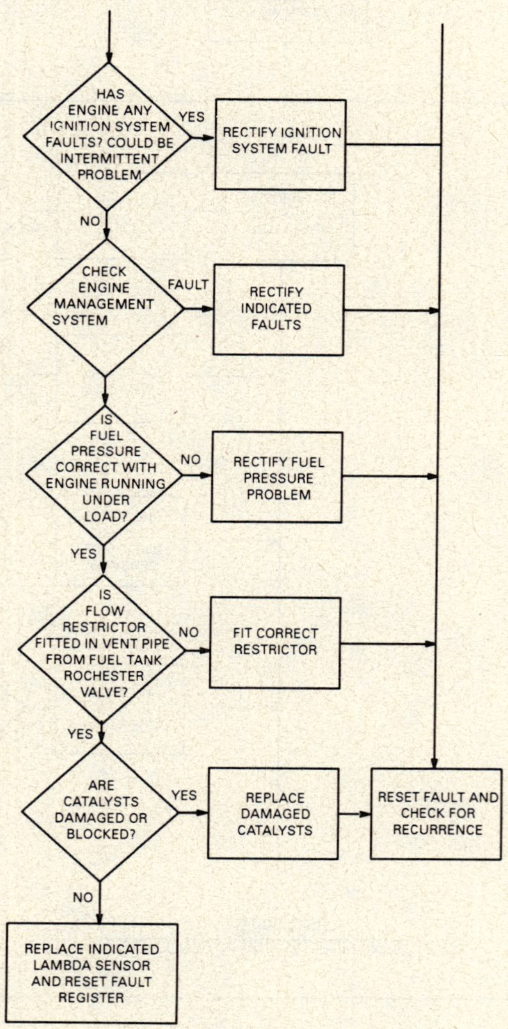

FAULT CODE 49 FUEL INJECTION BALLAST RESISTOR

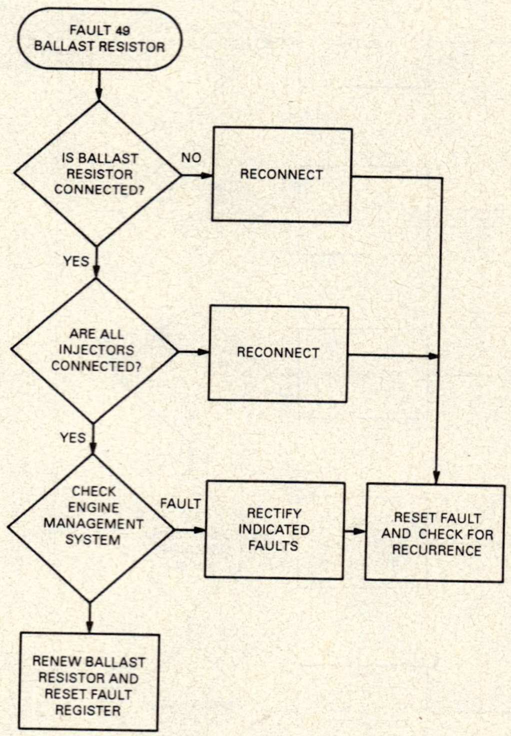

Component Replacement

FUEL PRESSURE RELIEF

CAUTION

The fuel system retains pressure even when the ignition is OFF or the engine has not operated for a period of time. Always release the pressure using the correct procedure before any fuel line or component is loosened or disconnected. Failure to observe this precaution may result in personal injury and/or property damage..

XJS

1. With the ignition **OFF**, open the trunk lid. Remove the right trim panel inside the luggage compartment.
2. Locate the relays mounted at the rear of the right panel near the ECU.
3. Loosen the relay panel from its mount. Identify the fuel pump relay; the relay is silver and is mounted on a black and yellow base.
4. Disconnect the relay from the system by removing the relay from its mount.
5. Use the ignition key to crank the engine several times. Fuel pressure will be reduced.
6. Switch the ignition **OFF**. Reinstall the relay, the mounting panel and the trim panel.
7. Fuel pressure will remain reduced until the next time the ignition switch is turned **ON** or the fuel pump is operated by other means.

XJ6

1. Disconnect the negative battery cable.
2. Open the left door; remove the lower left dash liner to expose the relay panel.
3. The fuel pump relay is the outboard or left-most relay; remove it from its mount.
4. Use the ignition key to crank the engine several times. Fuel pressure will be reduced.
6. Switch the ignition **OFF**. Reinstall the relay and the trim panel.
7. Fuel pressure will remain reduced until the next time the ignition switch is turned **ON** or the fuel pump is operated by other means.

FUEL RAIL AND INJECTORS

Removal and Installation

4.0L ENGINE

1. Safely relieve the fuel system pressure.
2. With the ignition switch **OFF**, label and disconnect the wiring connector at each injector.
3. Remove the bolts holding the fuel rail and disconnect the union nut for the fuel delivery hose.
4. Remove the nut holding the pressure regulator to the fuel rail.
5. Remove the fuel rail from the intake manifold and pressure regulator. Place the rail in a protected location, out of the immediate work area.
6. Remove the injector retaining clips and remove each injector from the fuel rail.

To install:

7. Install each injector onto the rail and secure the retaining clip.
8. Place the rail assembly onto the manifold, making sure that each injector is fully seated.
9. Connect the fuel rail to the pressure regulator and tighten the union nut.
10. Connect the fuel hose to the rail and tighten the union nut.
11. Reinstall the wiring connectors to each injector.

5.3L ENGINE

1. Safely relieve the fuel system pressure.
2. Disconnect the union joint on the fuel delivery pipe.
3. Disconnect the delivery pipe from the fuel rail.
4. Remove and discard the delivery pipe seal.
5. Disconnect the throttle breather hose from the manifold.
6. Remove the nut holding the fuel return line bracket on the camshaft cover; remove the bracket.
7. Disconnect the throttle cable from the pedestal.
8. Release the outer cable lock nut, then disconnect the throttle rods from the pedestal.
9. Disconnect the other ends of the throttle rods from the bell crank levers.
10. Disconnect the throttle return spring and remove the bracket.

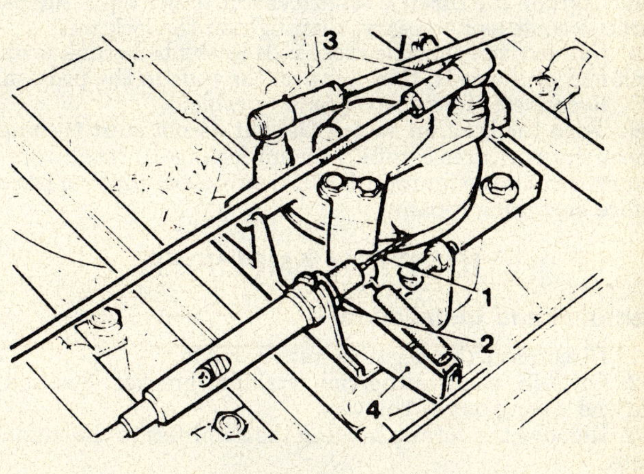

1. Throttle cable
2. Outer cable lock nut
3. Throttle cross rods
4. Throttle return spring bracket

Detail of throttle components—5.3L engine

11. Disconnect the vacuum hose from the cruise control bellows.
12. Label and disconnect the vacuum hoses at the fuel rail thermal switch.
13. Disconnect the hose from the fuel pressure regulator.
14. Disconnect the fuel return line from the fuel pressure regulator. Discard the seal or gasket. Immediately install plugs to prevent entry of dirt into the line or fuel rail.
15. Label and disconnect the wiring connector to each injector.
16. Remove the 6 fuel rail retaining bolts. Lift the fuel rail and injectors away from the engine.
17. On the workbench, remove the injector retaining clips.
18. Remove the injectors as necessary.

To install:

19. Lightly lubricate the seals of any injector before installing it to the fuel rail.
20. Remove any protective plugs from the fuel rail. Install the injectors and secure the retaining clips.
21. Lightly lubricate the manifold seals on the injectors.

22. Install the fuel rail and injector assembly to the manifold. Make certain each injector is squarely seated.
23. Install the fuel rail retaining bolts; tighten each to 10–12 Nm (8-9 ft. lbs).
24. Connect the wiring harness to each injector.
25. Connect the vacuum hoses to the thermal switch on the fuel rail. Connect the vacuum hose to the pressure regulator.
26. Install a new seal on the regulator fuel return line.
27. Install the return line to the pressure regulator; tighten the fitting to 10 Nm (8 ft. lbs).
28. Install the return line bracket to the camshaft cover; install the nut. The cam cover bolt should be tightened to 42–60 Nm (31–44 ft. lbs.). Tighten the nut to 22–30 Nm (16–22 ft. lbs).
29. Connect the vacuum hose to the cruise control; install the throttle breather hoses to the manifold.
30. Install a new seal at the fuel rail inlet. Connect the fuel delivery pipe and tighten the fitting to 10–12 Nm (8–9 ft. lbs).
31. Install the throttle return spring bracket, secure the bolt and install the spring.
32. Install the throttle cable onto the pedestal bracket. Engage the throttle cable into the throttle cam.
33. Tighten the throttle cable lock nut finger tight. Adjust the throttle cable as necessary, then tighten the lock nut.
34. Connect the throttle cross rods to the bell crank levers at each end of the rods, then connect the rods to the pedestal.
35. Reconnect the negative battery cable.
36. Turn the ignition switch **ON** but do not start the engine. Check the work area carefully for any leakage. If repairs are necessary, switch the ignition **OFF**, then relieve the fuel pressure before beginning repairs.

IGNITION AMPLIFIER

Removal and Installation

1. Disconnect the negative battery cable.
2. On XJS, remove the coil retaining bracket. Position the coil and wiring out of the way.
3. Remove the screws holding the amplifier to the mounting plate.
4. Disconnect the wiring harness from the amplifier.
5. Reinstall in reverse order, making certain the connector is firmly seated. When installing, make certain the mounting screws are secure.

COOLANT TEMPERATURE SENSOR

Removal and Installation

1. Perform this procedure on a cool or cold engine only.
2. Disconnect the negative battery cable.
3. Disconnect the wiring connector from the sensor.
4. Carefully loosen the cap on the coolant reservoir tank to relieve any pressure. Reinstall the cap tightly.
5. Coat the threads of the replacement sensor with thread sealing compound. Have the new sensor prepared and ready to install before removing the old sensor.
6. Carefully remove the sensor from its mount.
7. Install the new sensor quickly. Tighten the sensor correctly; on 4.0L engines, tighten the sensor to 8–10 Nm (6–8 ft. lbs) and on 5.3L engines, tighten it to 15–18 Nm (11–13 ft. lbs).
8. Attach the wiring connector.
9. Check the coolant level in the reservoir; add fluid if necessary.

CRANKSHAFT SENSOR

Removal and Installation

4.0L ENGINE

1. Make certain the ignition is switched **OFF**. Disconnect the negative battery cable.
2. Remove the clips holding the sensor cable.
3. Disconnect the sensor cable at the connector.
4. Remove the 2 bolts holding the sensor bracket; remove the bracket with the sensor attached.
5. Remove the sensor from the mount.
6. Reinstall in reverse order. Tighten the mounting bolts to 23–27 Nm (17–20 ft. lbs). Make certain the sensor is squarely and securely mounted in the holder. Route the wiring correctly and make certain the harness is retained by every clip.

5.3L ENGINE

1. Elevate and safely support the vehicle.
2. Remove the bolts holding the sensor. Remove the timing bracket.
3. Remove the sensor.
4. Remove the sensor cable from the lower clips.
5. From above, disconnect the sensor cable connector.
6. Tie a string to the connector end of the harness.
7. Release the harness from the upper retaining clip. From below, carefully pull the sensor and cable out, leaving the string in its place. Remove the sensor from the string.

To install:

8. Tie the drawstring to the connector of the new sensor. Pull the sensor wiring harness into position from above.
9. Install the new sensor; tighten the bolts to 22–28 Nm (16–21 ft. lbs).
10. Secure the wiring within the retaining clips.
11. Remove the drawstring. Connect the sensor connector to the wiring harness.
12. Connect the negative battery cable.

ENGINE CONTROL UNITS (ECU)

Removal and Installation

5.3L ENGINE

Fuel Control

1. Disconnect the negative battery cable.
2. Remove the right trim panel in the luggage compartment.
3. Release the harness connector lock; carefully remove the connector from the ECU.
4. Carefully release the vacuum hose from its clip.
5. Remove the mounting bolts and washers; remove the ECU. Disconnect the vacuum hose from the side of the unit.
6. Reinstall in reverse order. Take great care to insure proper installation of the wiring connector; it must be squarely and firmly mounted to the ECU.

5.3L ENGINE

Ignition Control

1. Disconnect the negative battery cable.
2. Loosen and pull back the left front carpeting.
3. Remove the cover from the inertia switch.
4. Remove the welt from the front or 'A' pillar. Remove the lower trim piece from the pillar.
5. Disconnect the ECU harness from the wiring retaining clips.
6. Remove the ECU cover. Carefully disconnect the harness connector from the ECU.
7. Disconnect the vacuum line from the ECU.
8. Remove the ECU from its mount and remove it from the vehicle.

To install:

9. Install the ECU onto the bracket and put the cover in place.
10. Connect the vacuum hose; carefully install the wiring harness connector. Make certain the connector is squarely and firmly seated.
11. Install the nuts and bolts holding the cover to the ECU.

12. Install the trim pad and welt at the A pillar. Reposition the carpet.
13. Connect the negative battery cable.

4.0L ENGINE

1. Disconnect the negative battery cable.
2. Remove the dash liner or pad from the lower left dashboard.
3. Remove the 4 bolts holding the ECU; remove the control unit from its mount.
4. Carefully disconnect the wiring harness connector from the ECU.
5. Install in reverse order. Connect the wire harness before placing the controller in the mount; make certain the connector is firmly and squarely seated.

FUEL PRESSURE REGULATOR

Removal and Installation

5.3L ENGINE

1. Safely relieve the fuel system pressure.
2. Disconnect the vacuum hose from the regulator
3. Release the fuel return line mount from the left camshaft cover.
4. Disconnect the fuel return line from the regulator. Discard the seal; install plugs in the line and regulator to prevent entry of dirt.
5. Remove the 2 bolts holding the regulator to the fuel rail. Remove the valve, discard the seal and install plugs in all openings.

To install:

6. Carefully clean the seal contact faces. Lightly lubricate new seal(s) and install them in position.
7. Install the regulator.
8. Install the fuel return line; tighten the fitting to 10–12 Nm (8–9 ft. lbs). Secure the return line in its mount at the left cam cover.
9. Connect the vacuum hose to the regulator.
10. Switch the ignition **ON** but do not start the engine. Inspect the work area closely for any sign of leakage. If repairs are required, switch the ignition **OFF**, then safely relieve fuel system pressure before beginning any work.

4.0L ENGINE

1. Safely relieve the fuel system pressure.
2. Disconnect the vacuum line from the pressure regulator.

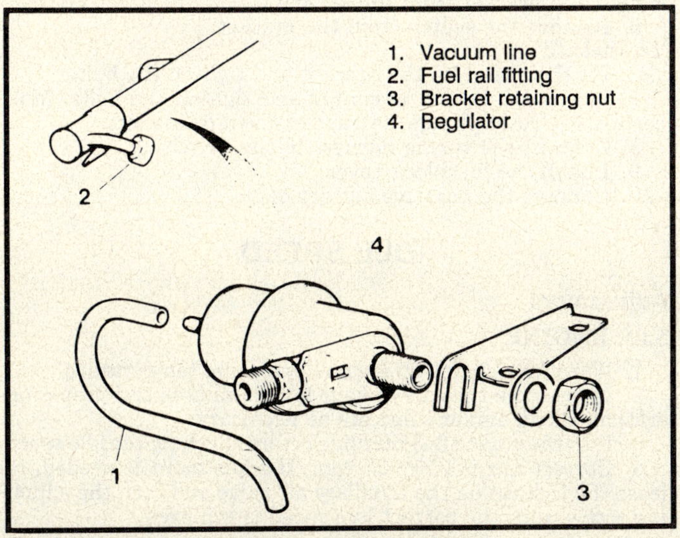

Fuel pressure regulator—4.0L engine

3. Disconnect the fuel return line from the regulator. Install plugs in the openings to prevent entry of dirt.
4. Loosen the nut holding the regulator to the fuel rail.
5. Remove the nut holding the regulator bracket; remove the regulator. Install plugs immediately to protect the system from dirt.
6. Install in reverse order. Tighten the fuel rail connector to 13–15 Nm (10–11 ft. lbs) and the bracket nut to 34–38 Nm (25–28 ft. lbs.)

THROTTLE IDLE SWITCH

Removal and Installation

1. With the ignition **OFF**, remove the bolts holding the switch to the manifold.
2. Remove the switch from its mount; remove the spacer from the manifold.
3. Disconnect the wiring connector and remove the switch.
4. Reinstall in reverse order; always adjust the switch after installation.

Adjustment

1. With the ignition **OFF**, disconnect the wiring connector from the switch.
2. Loosen the lock nut on the adjusting screw.
3. Turn the adjusting screw until the micro-switch clicks at the point of throttle closing.
4. Secure the lock nut and reconnect the wiring harness to the switch.

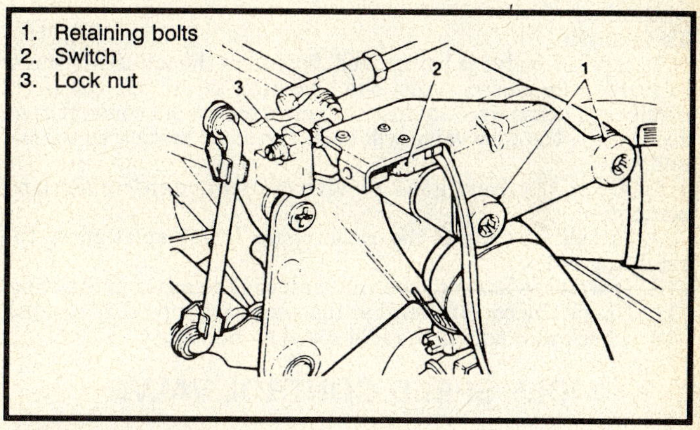

Throttle idle switch—5.3L engine

OXYGEN (LAMBDA) SENSORS

The exhaust sensor(s) are mounted in the exhaust down-pipe, forward of the catalytic converter. One sensor is used on the 4.0L engine; 2 are used for the 5.3L engine.

Removal and Installation

1. Disconnect the negative battery cable.
2. Disconnect the sensor connector from the wiring harness.
3. On 5.3L engines, remove the harness grommet from the inner fender. Elevate and safely support the vehicle, then free the sensor wiring from the wheel arch retaining clips.
4. Remove the sensor from the exhaust pipe. Use an oxygen-sensor socket of the correct size; use of other tools may damage the wiring.

NOTE: Protect the sensor tip from contact with any petroleum product, including used shop rags.

To install:

5.Coat the threads of the sensor with anti-seize compound. Use great care not to get any compound on the tip of the sensor.

6.Install the sensor into the exhaust pipe. Tighten the sensor to 50–60 Nm (37–44 ft. lbs).

7.Route the sensor wiring correctly, making sure it is held by every clip or retainer as before. On 4.0L engines, connect the sensor wiring to the harness.

8.On 5.3L engines, reinstall the grommet in the inner fender. Lower the vehicle to the ground. Connect the sensor wiring connector to the harness.

9.Connect the negative battery cable.

AUXILIARY AIR VALVE

Removal and Installation

1. Perform this procedure only on a cold or cool engine.
2. Carefully remove the pressure cap from the coolant reservoir tank, allowing any residual pressure to escape. Once pressure is released, replace and tighten the cap.
3. Loosen the clips holding the hoses running to the air valve; release the hoses from the holders.
4. Remove the 2 bolts holding the auxiliary air valve to the coolant pipe. Lift the valve away from the pipe.
5. Clean all traces of gasket or sealant from the coolant pipe. Do not damage the valve seating surface.
6. If the valve is to replaced with another unit, mark the head of the adjusting screw on the air valve. Turn the screw inward until it is fully closed; note and record the number of turns required. This step is not required if the old valve is to be reinstalled.

To install:

7. If a new valve is being used, first turn the adjusting screw in until fully closed. Turn the adjusting screw out the exact number of turns used to close the screw on the old valve. If done accurately, the new valve will be at the same setting as the old one.
8. Coat the new gasket with non-hardening sealing compound.
9. Install the valve to the coolant pipe. Install and tighten the retaining bolts.
10. Install the hoses to the valve; secure the hoses in the clips.
11. Check the coolant level at the reservoir; top off as needed.
12. Check and adjust the idle speed as necessary.

IDLE SPEED CONTROL VALVE

Removal and Installation

1. Disconnect the negative battery cable.
2. If equipped with cruise control, remove the bolts holding the actuator bracket to the intake manifold. Move the actuator out of the way, allowing access to the valve.
3. Release the clip holding the air hose to the valve; disconnect the hose from the valve.
4. Remove the bolts holding the idle speed control valve to the intake manifold. Disconnect the wiring connector and remove the valve.
5. Clean the opening in the manifold.
6. Install in reverse order. Connect the wiring before installing the valve.
7. Connect the negative battery cable.

AIRFLOW METER

Removal and Installation

1. Disconnect the negative battery cable.
2. Remove the nuts holding the air cleaner to the inner fender.
3. Disconnect the air cleaner from the airflow meter; move the air cleaner out of the way.
4. Release the clip holding the intake elbow or hose to the air flowmeter.
5. Disconnect the air flowmeter ground wire at the screw.
6. Carefully remove the wiring harness connector from the meter. Remove the air flowmeter.

To install:

7. Install the harness connector to the new air flowmeter.
8. Position the meter correctly and connect the ground wire.
9. Connect the air flowmeter to the intake elbow; tighten the hose clip. Make certain the joint is straight and tight; no air leakage is permitted.
10. Connect the air cleaner to the meter. Make sure all retaining clips are secure.
11. Attach the air cleaner to the inner fender and tighten the nuts.
12. Connect the negative battery cable.

INERTIA SWITCH

Removal and Installation

5.3L ENGINE

1. Disconnect the negative battery cable.
2. Remove the inertia switch cover, located in the left kick panel near the door.
3. Release the catch on the electrical connector; remove the connector from the switch.
4. Remove the lower left dash liner.
5. Remove the 2 mounting screws; remove the switch.

To install:

6. Install the switch and secure the mounting screws. Make certain the reset plunger on top of the switch is pushed down or inward.
7. Install the left dash trim.
8. Connect the wiring to the switch. Make certain the connector latch is engaged. Install the cover on the switch.
9. Connect the negative battery cable.

4.0L ENGINE

1. Disconnect the negative battery cable.
2. Remove the cover from the fuseblock located just inside the right windshield or 'A' pillar.
3. Disconnect the wiring connector from the inertia switch.
4. Remove the bolts holding the switch assembly. Remove the switch and bracket from the vehicle.
5. Remove the switch from the bracket.

To install:

6. Fit the switch to the bracket and tighten the bolts.
7. Install the switch assembly and tighten the bolts. Make certain the reset plunger is pushed down or inward.
8. Connect the wiring harness to the switch.
9. Install the fuseblock cover.
10. Connect the negative battery cable.

IDLE SPEED

Adjustment

5.3L ENGINE

1. Warm the engine to normal operating temperature.
2. Inspect the throttle linkage for cleanliness and proper operation. Clean and/or lubricate as necessary.
3. Run the engine 2–3 minutes before checking the idle speed.
4. Correct idle is 750 ± 25 rpm. If adjustment is needed, release the locknut on the auxiliary air valve and turn the adjusting screw until the correct idle speed is achieved.

5. If idle is too high and cannot be reduced with the adjusting screw, inspect the intake and fuel system carefully for air leaks.
6. Tighten the lock nut on the auxiliary air valve.
7. Switch the ignition **OFF**.

4.0L ENGINE

NOTE: It is recommended to set the idle using the JDS equipment. If not available, the following method may be used.

1. Operate the engine until it reaches normal operating temperature.
2. Switch the ignition **OFF**.
3. Switch the ignition **ON**, wait 5 seconds, then disconnect the connector from the idle speed controller.
4. Switch the ignition **OFF**.
5. After 15 seconds, reconnect the ISC wiring harness.
6. Repeat Steps 1–5 two more times. On the final repeat, do not reconnect the wiring harness (Step 5).
7. Start the engine with the idle controller disconnected. Base idle speed should be 575 ± 25 rpm. If the idle speed needs adjusting, turn the air bypass screw to bring the idle into specification.
8. Switch the ignition **OFF**. Reconnect the wiring to the idle speed control valve.
9. Restart the engine; idle speed should be 650–800 rpm.

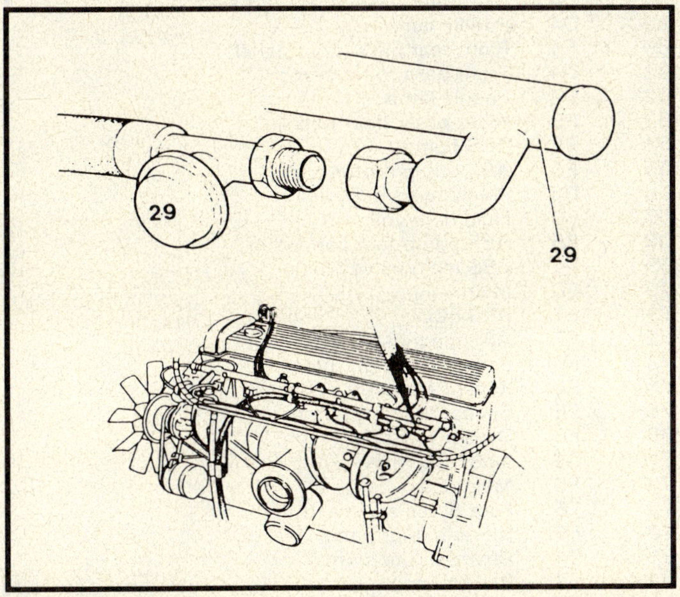

Fuel feed pipe

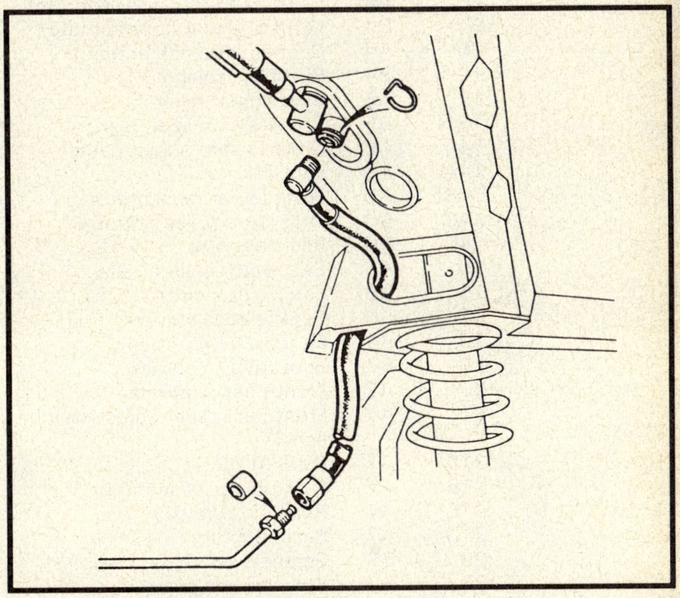

Main line return to tank hose

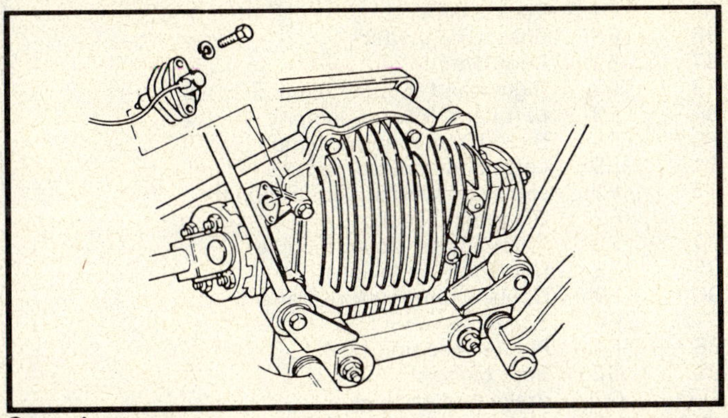

Speed sensor

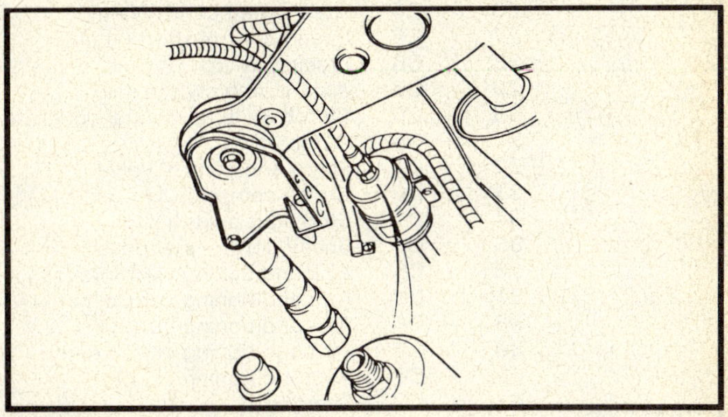

Fuel pump to filter hose

WIRE COLOR KEY

Wire colors are shown on the diagrams by a letter key. If two letters are present, the first is the main color and the second is the trace color.

N	Brown
B	Black
W	White
K	Pink
G	Green
R	Red
Y	Yellow
O	Orange
S	Slate (gray)
P	Purple
U	Blue
L	Light, as a prefix (LG: Light Green)

KEY TO CIRCUIT DIAGRAM

No	Gr./Ref	Description
1	A1	Fusebox l/h
2	A2	Fusebox r/h
3	A3	Door locks
4	B3	Infra red receiver
5	B4	Driver's front door switch and motor
6	A4	Front passenger door switch and motor
7	A4	Dealer fit security connector
8	A4	Right hand rear door actuator
9	B4	Left hand rear door actuator
10	A4	Boot lock/unlock actuator
11	B4	Fuel filler solenoid
12	A5	Seat adjust relay
13	B5	Driver's seat relay
14	A5	Driver's seat adjust panel
14A	A6	Fore/aft motor
14B	A6	Front lower/raise motor
14C	B6	Rear lower/raise motor
14D	B6	Recline motor
15	A6	Seat entry switch relay
16	A6	Entry/exit switch
17	B6	Boot release switch
18	B6	Boot lock security switch
19	B7	Boot release actuator
20	A7	Front pass.seat relay
21	A7	Front pass. seat adjust switch panel
21A	A7	Fore/aft motor
21B	A7	Front lower/raise motor
21C	A7	Rear lower/raise motor
21D	A7	Recline motor
22	B7	Screenwash relay
23	B7	Screenwash pump
24	B7	Headlamp wash relay
25	B7	Powerwash pump (headlamp)
26	C6	Auto speed control relay
27	C6	Speed control shorting plug
28	C6	Braking switch
29	C6	Auto clutch switch link
30	C6	Clutch switch (manual only)
31	C6	Dump valve
32	C7	Pump and control valve
33	D6	Cruise control ECU
34	D7	Stop light switch
35	D6	Speed control switches
36	C5	Air conditioning microprocessor
37	C5	Air conditioning control panel
38	C4	Air conditioning cut-out
39	C4	Air conditioning clutch relay
40	C4	Air conditioning clutch
41	C3	Thermostat
42	C3	Cooling fan
43	C3	Fan relay
44	C3	Right hand powerwash heater element
45	C3	Screenwash heater element
46	C3	Ambient air sensor
47	D2	Left hand powerwash heater element
48	D2	Breather pipe heater relay
49	D2	Breather heater
50	D2	Interior map light
51	E2	Pass. interior light and switch
52	E2	Driver's interior light and switch
53	D2	Bonnet lamps
54	D3	Bonnet switch
55	D3	Cubby box lamp
55A	D3	Cubby box lamp switch
56	D3	Auxiliary sidelight relay
57	D3	Lighting switch
58	D4	Right rear passenger door switch
59	D4	Puddle lamp
60	E4	Right rear pass. door panel
61	D4	Front pass. door switch
62	D4	Puddle lamp
63	E5	Front pass. door panel
64	E5	ABS pump
65	E5	ABS pump motor relay
66	D6	Pressure switch
67	E6	Fluid reservoir
68	E6	ABS pump test connector
69	E6	ABS main valve block
70	E6	Main valve
71	E7	ABS ECU
72	F7	ABS diagnostic connector
73	F6	Sensor rear right hand
74	F6	Sensor front left hand
75	F6	Sensor rear left hand
76	F6	Sensor front right hand
77	G6	ABS ECU relay
78	F6	Main relay
79	F6	Foot brake switch
80	F5	Driver's door panel
81	E4	Driver's door switch
82	E4	Puddle lamp
83	F4	Left rear pass. door panel
84	E4	Left rear pass. door switch
85	E4	Puddle lamp
86	F3	Boot lamp switch
87	F3	Boot lamp
88	F3	Right hand 'E' post lamp
89	E3	Left hand 'E' post lamp
90	E3	Right hand sunvisor lamp
91	E3	Left hand sunvisor lamp
92	F2	Lights on warning control unit
93	F2	Diagnostic socket
94	F3	Caravan and trailer module
95	F3	Coolant level sensor
96	F3	Coolant temperature transmitter
97	F4	Speed sensor
98	F4	PI interface unit (2.9L)
99	G2	Boot switch
100	G2	Brake pad sensors
101	H2	Oil pressure transmitter
102	G3	Instrument pack and trip computer
103	H4	Wash fluid level sensor
104	H4	Handbrake switch
105	H5	Accessory fusebox relay
106	H5	Fuel tank level sensor
107	G6	Trip computer keyboard
108	H6	Mirror switches

Wiring key—XJ6 with 4.0L engine

KEY TO CIRCUIT DIAGRAM

No	Gr./Ref	Description
109	I7	Passenger mirror motor (left/right)
110	I7	Passenger mirror motor (up/down)
111	I6	Driver's mirror motor (up/down)
112	I6	Driver's mirror motor (left/right)
113	I6	Battery cut-out circuit
114	H6	Battery positive terminal
115	I6	Battery post
116	I7	Lighting logic switch assembly
117	J6	High level stop lights relay
118	J5	High level stop lights
119	I5	Accessory fusebox
120	J4	Passive reverse relay (if fitted)
121	I3	Passive restraint motor (left hand,if fitted)
122	J4	Front left hand dock switch (if fitted)
123	J3	Rear left hand dock switch (if fitted)
124	J3	Seat belt prox. switch (left hand, if fitted)
125	J3	Seat belt prox. switch (right hand, if fitted)
126	J3	Rear dock switch (right hand,if fitted)
127	J3	Front dock switch (right hand,if fitted
128	J3	Passive restraint motor (right hand, if fitted)
129	I2	Passive restraint module (if fitted)
130	K2	Air pump relay
131	K2	Air pump
132	K3	Solenoid valve
133	K3	EGR check sensor
134	K3	Purge valve
135	K3	Starter motor
136	K4	Load dump module
137	K4	Alternator
138	K4	Gearbox power relay
139	K5	Decoder module
140	K6	Gear selector illum. module
141	K6	Gearbox
142	K5	Selector switch
143	L4	Starter relay
144	L4	Ignition on relay
145	L4	Diagnostic socket
146	L4	Distributor
147	L3	Ignition coil
148	L3	Fuel fail reset
149	L3	Ignition amplifier
150	L2	EGR air valve
151	L2	Fuel pump relay
152	L2	Fuel pump
153	L2	Throttle pot.
154	L2	Lambda sensor
155	M2	PI main relay
156	M2	Fuel injectors
157	M2	Idle speed control
158	M4	Air temp. sensor
159	M4	Crankshaft sensor
160	M4	Air meter
161	M4	Water temp. sensor
162	M5	Engine/gearbox connector
163	M6	Kickdown switch
164	N6	Key barrel switch
165	N6	Key barrel switch
166	N6	Gearbox control ECU
167	N4	Inertia switch
168	N4	Mode switch
169	N4	Hazard switch
170	N3	Engine management ECU
171	N2	Heated rear window relay
172	N2	Heated rear window
173	O5	Driver's mirror heater
174	O3	Passenger's mirror heater
175	P3	Centre console display board
176	P4	Microprocessor unit
177	O6	Sunshine roof control panel
178	P6	Windscreen wiper logic module
179	P6	Windscreen wiper motor
180	Q7	Sidelights on power relay
181	Q6	Dimmer and override
182	Q6	Delay wipe
183	Q6	Fast wipe
184	Q6	Slow wipe
185	Q6	Windscreen wash
186	Q6	Horn
187	Q6	Headlamp flash
188	Q6	Direction indicator (LH)
189	Q6	Direction indicator (RH)
190	Q5	Key switch
191	Q3	DI,key switch and seatbelt speaker
192	Q3	Speed audible warning module
193	R3	Radio
193A	Q2	Speaker,RH bass
193B	R2	Speaker,LH rear
193C	R2	Speaker,LH front
193D	R2	Speaker,RH rear
193E	R2	Speaker,RH front
193F	R2	Speaker LH bass
194	R3	Choke
195	R3	Radio relay
196	R4	Aerial relays
196A	R4	Aerial motor
197	R5	Horn relay
198	R5	High horn
199	R5	Low horn
200	R5	Ignition switch
201	R6	Cigar lighter relay
202	R6	Cigar lighter,rear
203	S7	Cigar lighter,front
204	S7	Squab heaters,A pass.,B driver
205	S7	Squab heaters, A pass.,B driver
206	S6	Seat heaters,A pass.,B driver
207	S6	Thermostat,A pass,B driver
208	S6	Relay and switch (pass. seat heater)
209	S6	Relay and switch (driver's seat heater
210	S3	Driver's door panel
210A	T3	Window motor UP relay
210B	T3	Window motor DOWN relay

Wiring key—XJ6 with 4.0L engine

KEY TO CIRCUIT DIAGRAM

No	Gr./Ref	Description
210C	T3	Window lift motor
211	T3	Driver's window UP/DOWN switch
212	T3	Front pass. UP/DOWN switch
213	T3	Rear pass. UP/DOWN switch
214	T3	Rear pass. UP/DOWN switch
215	S4	Window isolate
216	T5	Lamp relay and bulb failure module (RH rear)
217	U5	Fog lamp (RH rear)
218	U5	Reversing lamp (RH)
219	U5	Tail lamp (RH)
220	U5	Number plate lamp (RH rear)
221	U5	Stop lamp (RH rear)
222	U5	Direction indicator lamp(LH rear)
223	T6	Lamp relay and bulb failure module (LH rear)
224	U6	Fog lamp (LH rear)
225	U6	Reversing lamp (LH)
226	U6	Tail lamp (LH)
227	U6	Number plate lamp (LH rear)
228	U6	Stop lamp (LH rear)
229	U6	Direction indicator lamp (LH rear)
230	U6	Lamp relay and bulb failure module (LH front)
231	V6	Fog lamp (LH front)
232	V6	Dipped headlamp (LH front 35W)
233	V6	Main beam (LH front 35W)
234	V6	Main beam (LH front 50W)
235	V6	Sidelight
236	V6	Direction indicator(LH front)
237	V6	Direction indicator repeater (LH front)
238	U5	Lamp relay and bulb failure module (RH front)
239	V5	Fog lamp (RH front)
240	V5	Dipped headlamp (RH front 35W)
241	V5	Main beam (RH front 35W)
242	V5	Main beam (RH front 50W)
243	V5	Sidelight (RH front)
244	V5	Direction indicator (RH front)
245	V5	Direction indicator repeater (RH front)
246	V4	Puddle lamp relay
247	V3	Front pass. window lift relays
247A	U4	Front pass. window lift motor
248	V3	Rear pass.window lift relays (RH)
248A	V4	Rear pass. window lift motor (RH)
249	V4	Rear passenger window lift relays
249A	V3	Rear pass. window lift motor (LH)
250	T2	Centre fusebox
251	U2	Solenoid valve block
252	V2	Ride level relay
253	V2	Ride level circuit
254	V7	Driver's seatbelt switch
255	V7	Driver's seatpad switch
256	V7	Passenger's seatbelt switch
257	V7	Passenger's seatpad switch

Wiring key—XJ6 with 4.0L engine

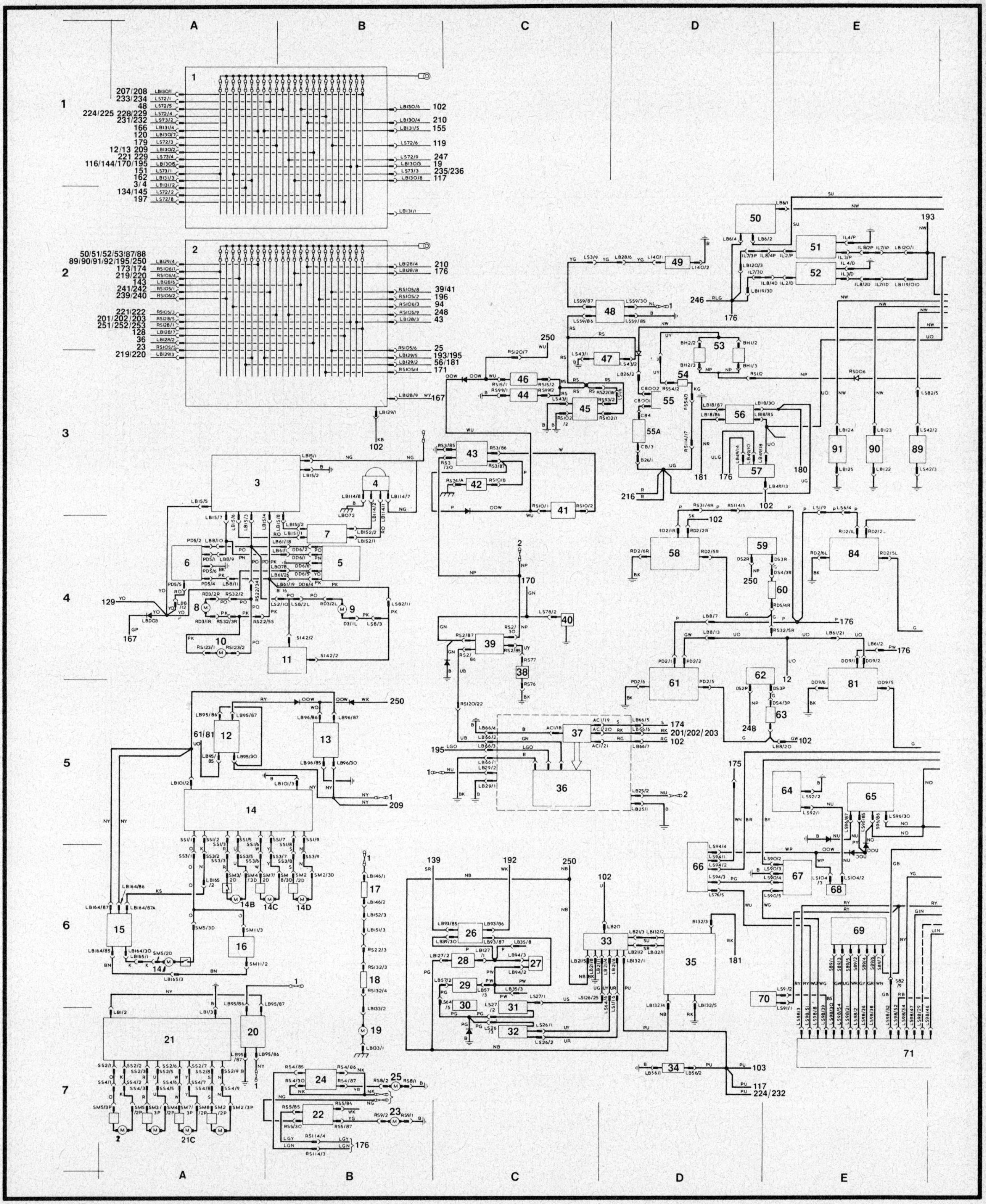

Circuit diagram—XJ6 with 4.0L engine

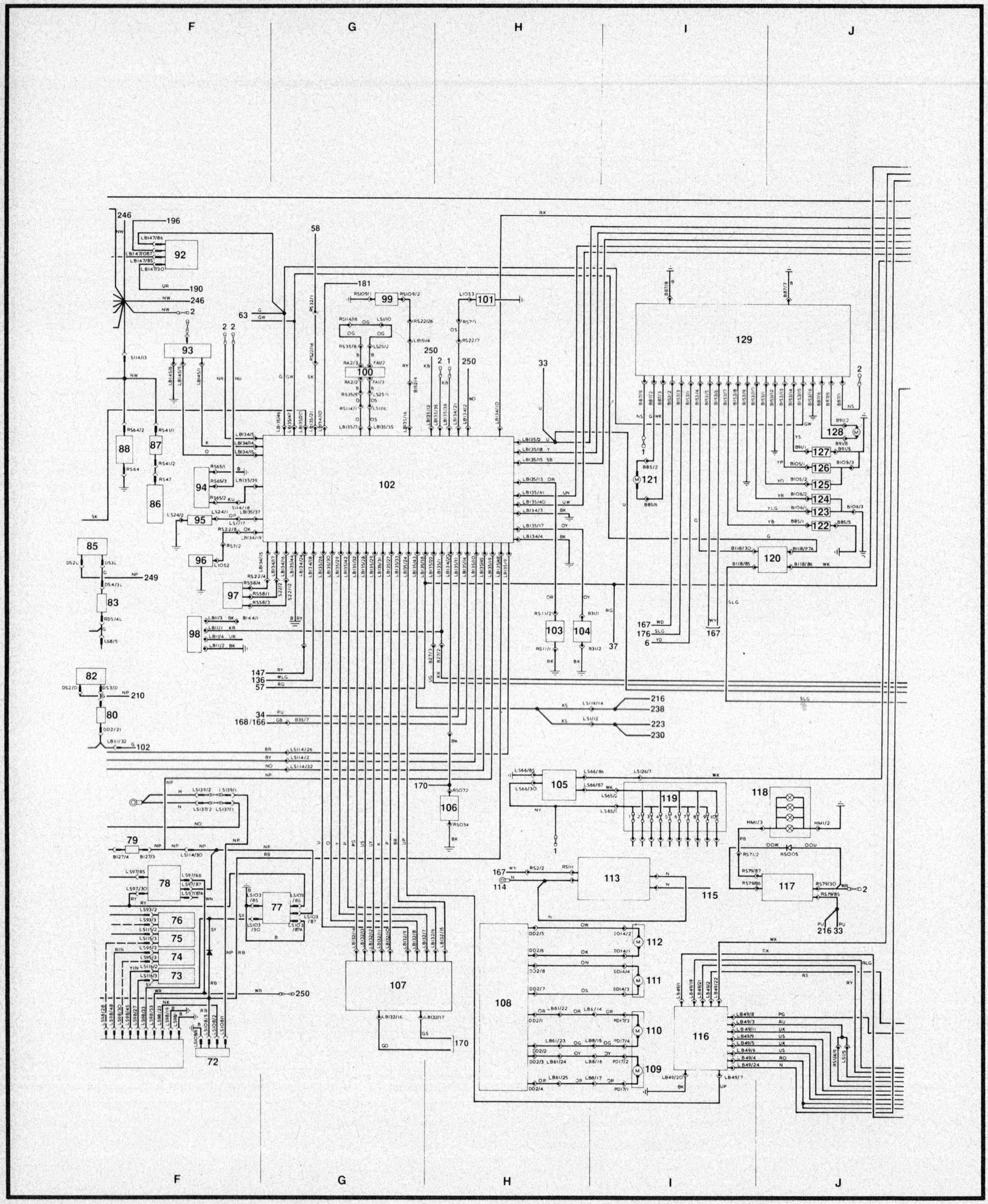

Circuit diagram—XJ6 with 4.0L engine

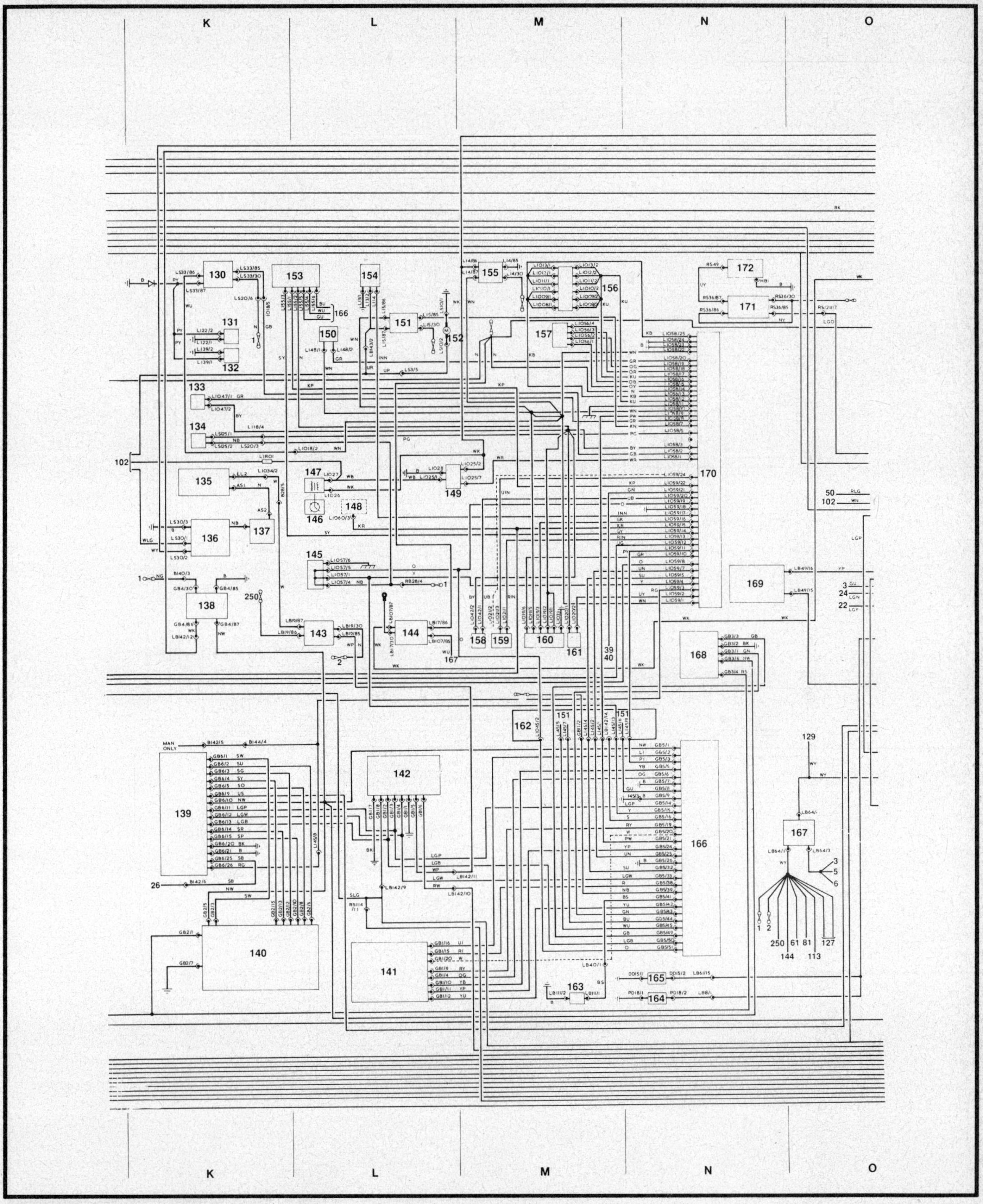

Circuit diagram—XJ6 with 4.0L engine

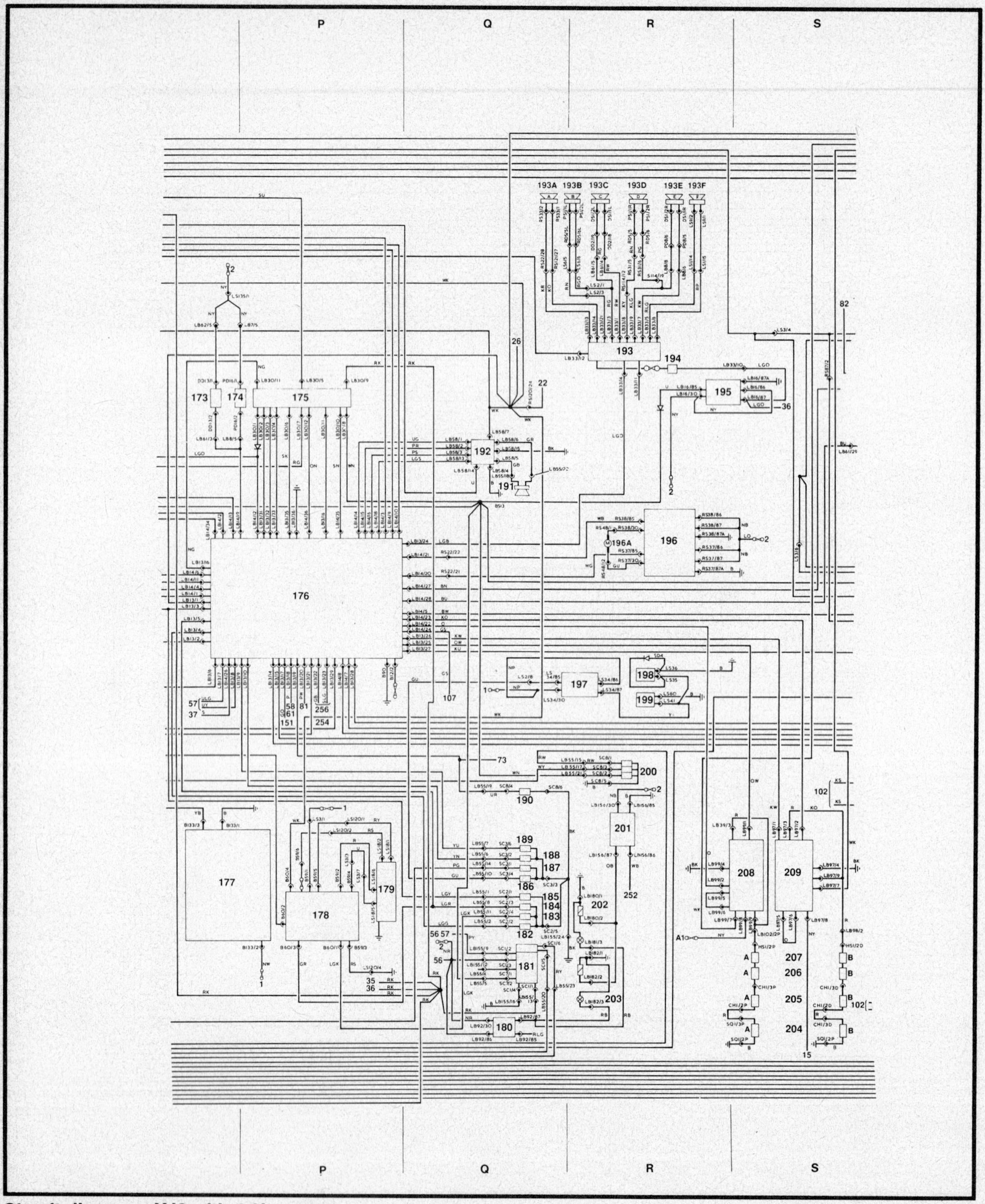

Circuit diagram—XJ6 with 4.0L engine

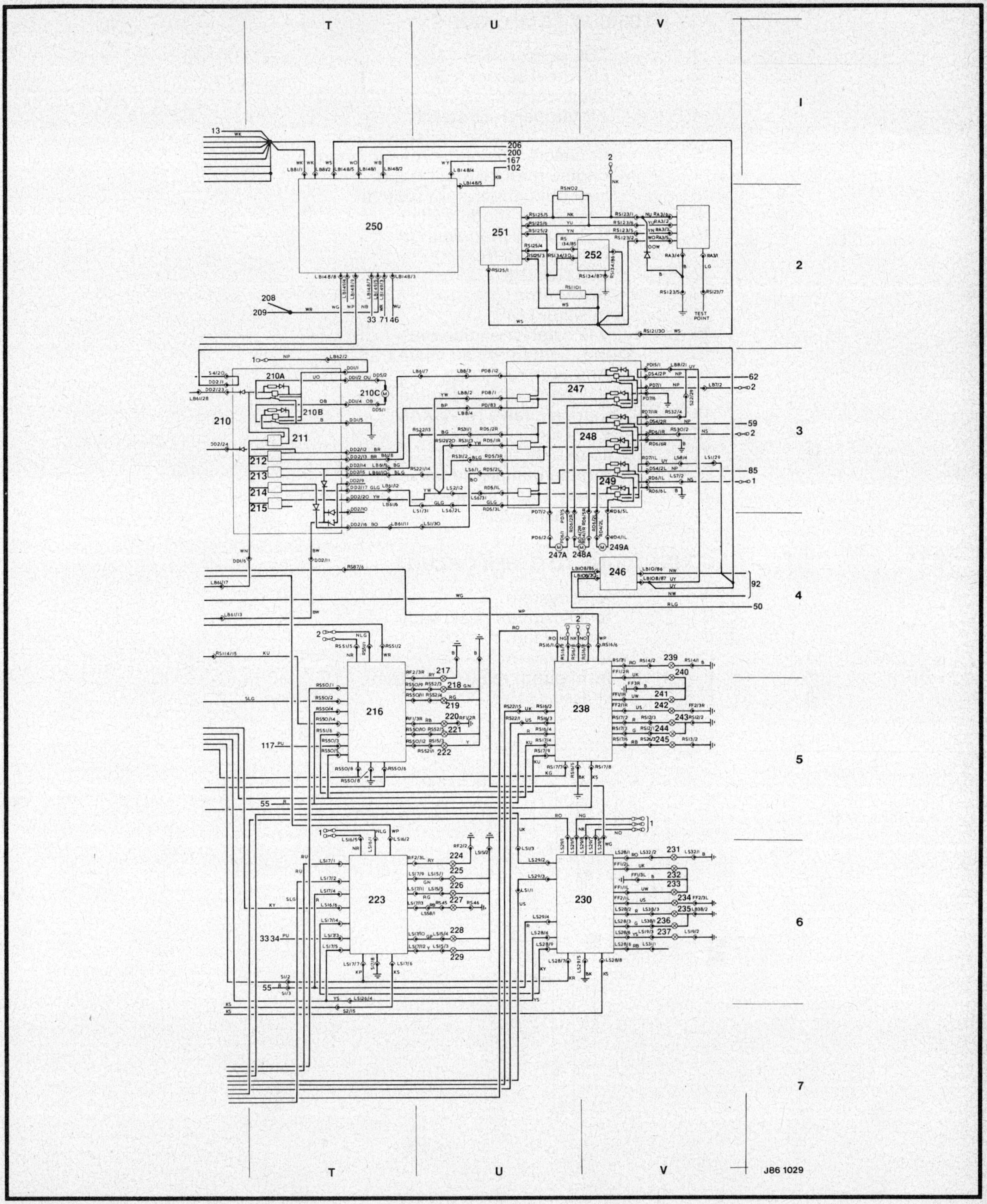

Circuit diagram—XJ6 with 4.0L engine

KEY TO CIRCUIT COMPONENTS

1	TDC sensor IH12
2	Flywheel sensor IH3
3	Idle switch
4	Air temperature switch
5	Coolant temperature sensor IH9
6	Connector to engine management E.C.U.
7	Engine management E.C.U.
8	Fuel rail temperature switch
9	L.H. power module IH13
10	R.H. power module IH14
11	Coil forward IH7
12	Coil rear IH8
13	Distributor
14	Resistor EMR1
15	V12 catalyst engine only
16	Supplementary air valve CS6
17	Anti–stall relay AS5
18	V12 non–catalyst engine only
19	Extra air valve relay XA5
20	Extra air valve relay XA4
21	15 minute timer FC1
22	Supplementary air valve XA6
23	3–way solenoid operated vac valve CL4
24	Thermal switch

CONNECTIONS TO OTHER CIRCUITS

A	to PI system
B	to Instrument pack, tacho
C	to PI system
D	to PI system
E	from pump relay (PI system)
F	chassis

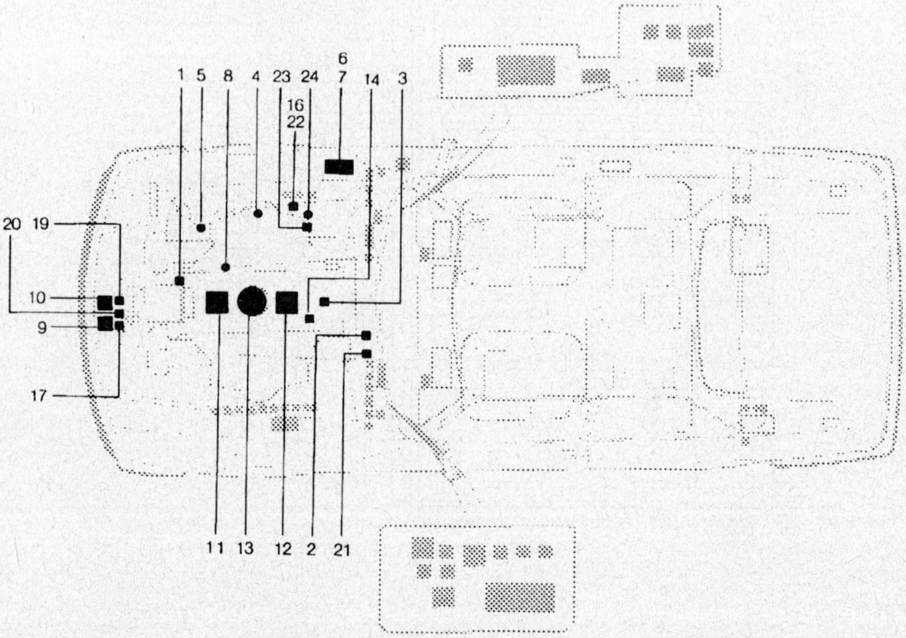

Engine management component key—XJS with 5.3L engine

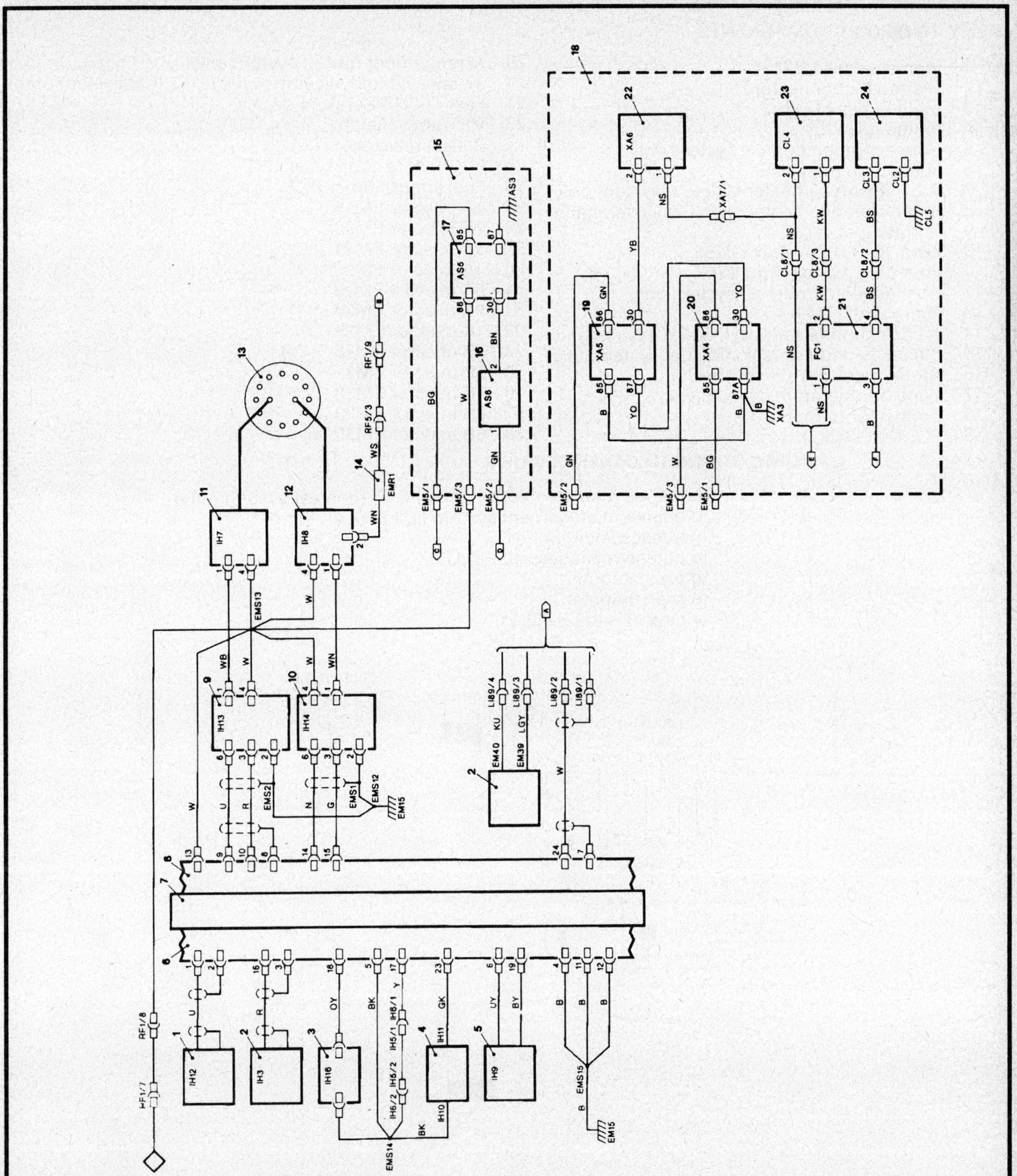

Engine management circuit diagram—XJS with 5.3L engine

KEY TO CIRCUIT COMPONENTS

1 Inertia switch LB212
2 Diagnostic connector LI90
3 Main PI relay LI47
4 In-line fule LI88
5 Connection on non-cat system only
6 Fuel pump relay LI46
7 A bank lambda heater LI76 (Cat system only)
8 B bank lambda heater LI63 (Cat system only)
9 Throttle pot LI136
10 Fuse 10, R.H.Fusebox LB268
11 Fuse 10, L.H.Fusebox LB263
12 Connection on non-cat system only
13 Fuel pump LI141
14 A bank lambda sensor LI75 (Cat system only)
15 B bank lambda sensor LI62 (Cat system only)
16 Air temperature sensor LI64
17 Water temperature sensor LI65
18 Connector to PI E.C.U.
19 PI E.C.U. LI45
20 Arrangement for Cat system only
21 3-way solenoid vacuum valve LI137 (Cat system only)
22 Fuel rail temperature switch
23 Vacuum valve (hot start) LI145
24 Ballast resistors LI73
25 Fuel failure code reset LI91
26 Fuel control timer FC2
27 1A injector EM19
28 2A injector EM20
29 3A injector EM21
30 4A injector EM22
31 5A injector EM23
32 6A injector EM24
33 1B injector EM25
34 2B injector EM26
35 3B injector EM27
36 4B injector EM28
37 5B injector EM29
38 6B injector EM30

CONNECTIONS TO OTHER CIRCUITS

A to engine management system, FC1/3 (non-cat system only)
B to engine management system FC1 & CL4 (non-cat system only)
C to engine management E.C.U.
D to trip computer
E to trip computer
F engine ok – to LB246/11

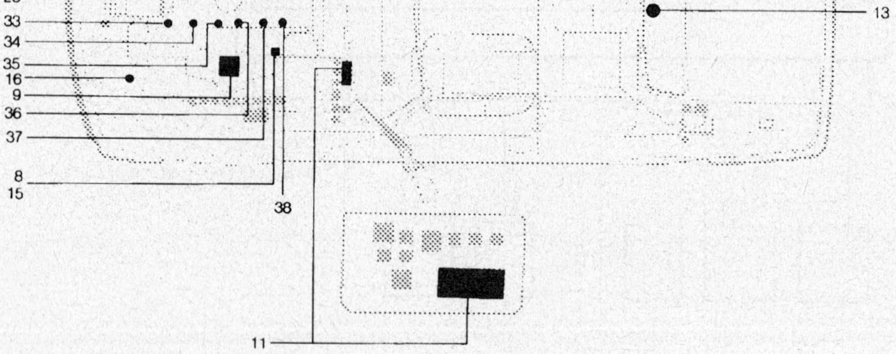

Fuel control component key—XJS with 5.3L engine

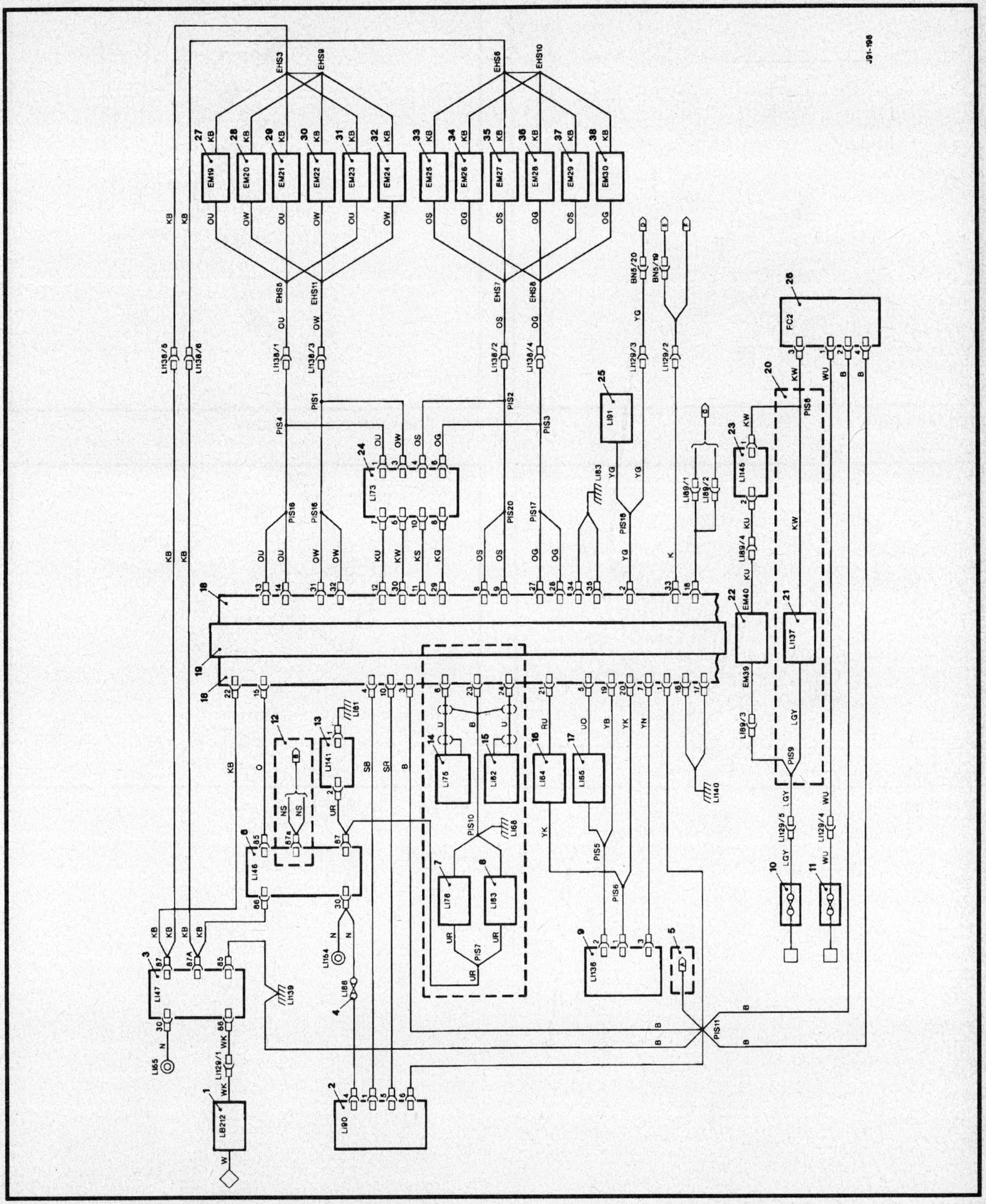

Fuel control circuit diagram—XJS with 5.3L engine

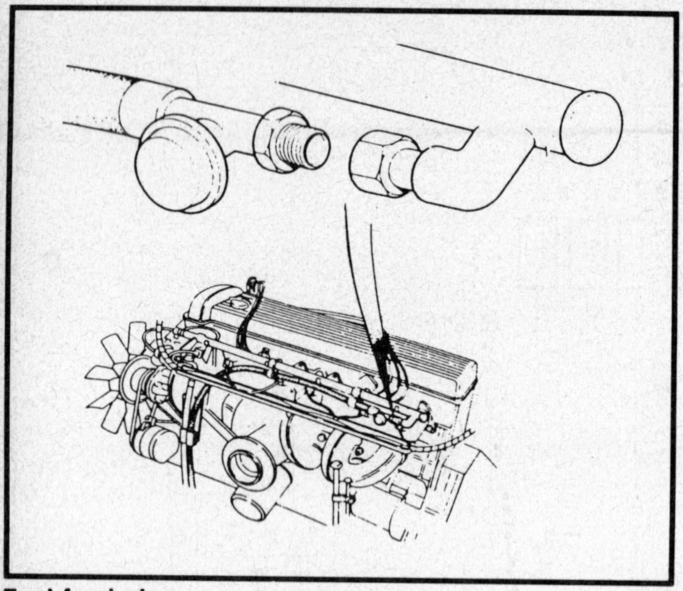

Fuel feed pipe

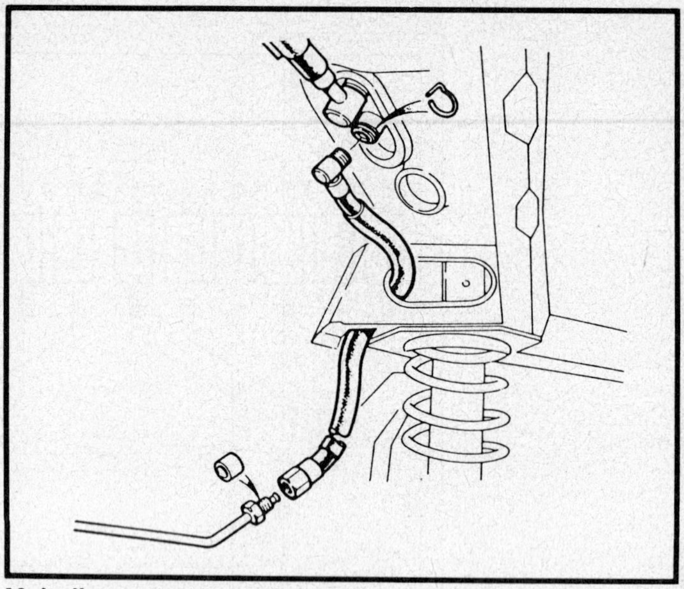

Main line return to tank hose

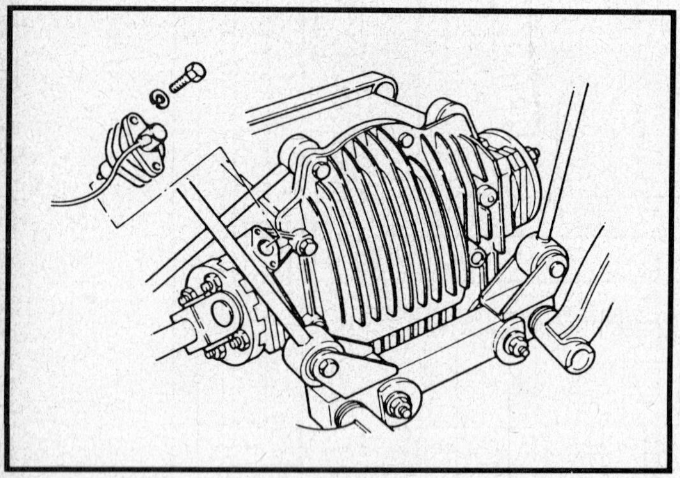

Speed sensor

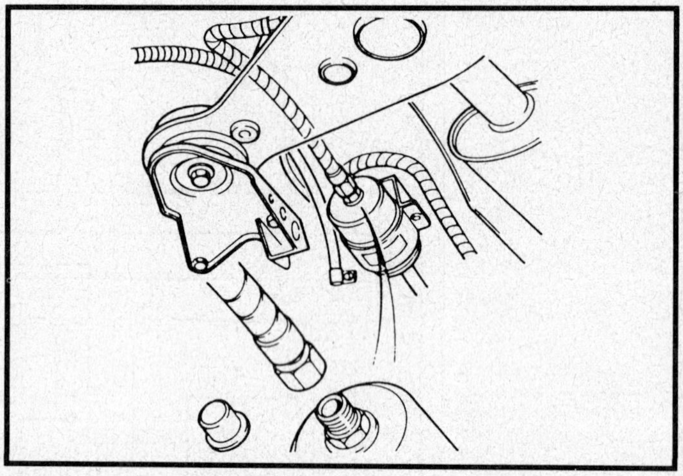

Fuel pump to filter hose

Mercedes-Benz Fuel Injection

9 SECTION

MERCEDES-BENZ CIS-E FUEL INJECTION SYSTEM

General Information

Mercedes-Benz uses an electronically controlled Bosch K-Jetronic style Continues Injection System (CIS-E). The input signals from various sensors and switches are continuously monitored by the Electronic Control Unit (ECU). In the event of an illogical signal, such as a short or open circuit, the control unit will automatically revert to a Fix Operating Mode (FOM).

If while operating at normal temperature there is a sudden change is a sensor or circuit signal, the ECU will compare this momentary change with the normal signal value in memory. The ECU will recognize the abrupt change as an open or short and will revert to FOM for continued engine operation.

The control unit does not recognize gradual changes as malfunctions, and will continue to operate as if all signals are normal.

ELECTRONIC CONTROL UNIT

The Electronic Control Unit (ECU) receives signal from the starter impulse, end start procedure, TD impulse signal, coolant temperature sensor, airflow sensor, full load switch, idle switch, oxygen sensor, shift position, air conditioner signal, altitude sensor and air temperature sensor. Based on the sensor input the ECU controls cranking enrichment, after start enrichment, warm-up enrichment, maximum engine speed, deceleration shutoff, and fuel mixture.

DATA SENSORS

Coolant Temperature Sensor

The coolant temperature is sensed by a temperature sensor. Temperature affects the resistance of the sensor and this change is detected by the electronic control unit.

Altitude Correction Capsule

Depending on the altitude, the amount of fuel will be changed based on a signal from the altitude correction capsule. This device is a barometric pressure sensor. The ECU supplies approximately a 5 volt signal to the sensor. The sensor based on atmospheric pressure returns a voltage signal the ECU. The ECU can adjust fuel mixture based on this signal.

Oxygen Sensor

The oxygen sensor sends a signal to the ECU based on oxygen content of the exhaust gas. The ECU can adjust the mixture based on this signal to emission requirements. A voltage signal from the oxygen sensor over 0.5V indicates a rich mixture while a signal of less than 0.5V indicates a lean mixture.

CONTROLLED OUTPUTS

Overvoltage Relay

The overvoltage relay is used to protect the electronic control units. There is a 10A fuse on top of this relay. When the ignition switch is turned **ON**, the relay closes and the control unit is supplied with battery voltage.

Idle Air Speed Control

The electronic air speed control is controlled by an idle speed module on 116 and 117 engines or is integrated into the CIS-E control unit on other engines. The control unit maintains idle speed based on inputs from the TD (engine rpm), coolant temperature, air temperature, idle switch, airflow, altitude, shift position, air conditioner and cruise control signals.

IDLE SPEED VALVE FUNCTION

- The electronic control unit supplies a specific voltage to the air valve which determines the respective valve opening and control engine speed.
- The nominal idle speed is controlled based on temperature. Approximately 300 rpm higher for cold engine idle.
- With the ignition switch **OFF**, the idle speed air valve is activated by a specific voltage. The port for the fixed operation mode is closed and the control port is opened.
- If the voltage supply fails, the port for the fixed operating mode is opened automatically.

Mixture Control Unit

The fuel is metered and delivered to the individual cylinders dependent on the air intake quantity. The mixture is controlled by the ECU based on the sensor input signals. The Electrohydraulic Actuator (EHA) controls the fuel delivery pressure, based on signals from the ECU. The EHA replaces the older style pressure regulator piston and and pressure compensating valve used on K-Jetronic type systems.

ELECTROHYDRAULIC ACTUATOR FUNCTION

The EHA is flanged onto the fuel distributor. It acts as a pressure regulator which operates as a plate valve. The plate valve

Altitude above sea level m	Barometric pressure mbar	Approx. voltage Volts
0	1013	4 ± 1
1000	899	3 ± 1
2000	795	2 ± 1

Attitude correction capsule table

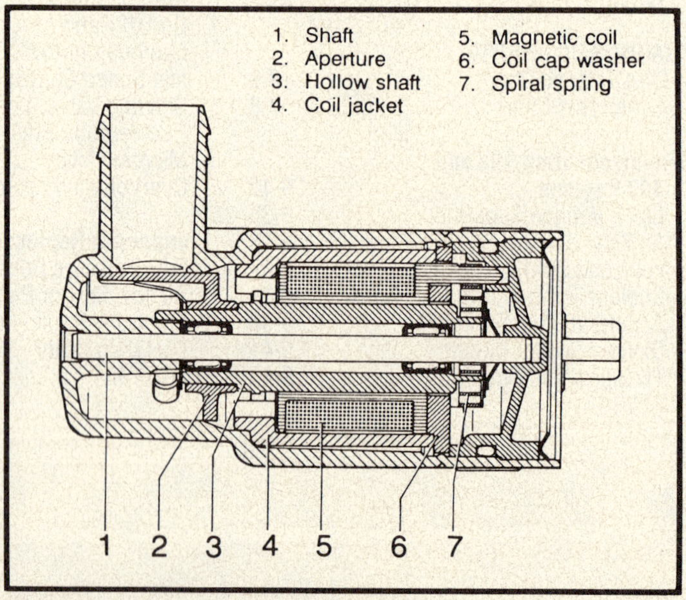

Idle speed control valve—typical

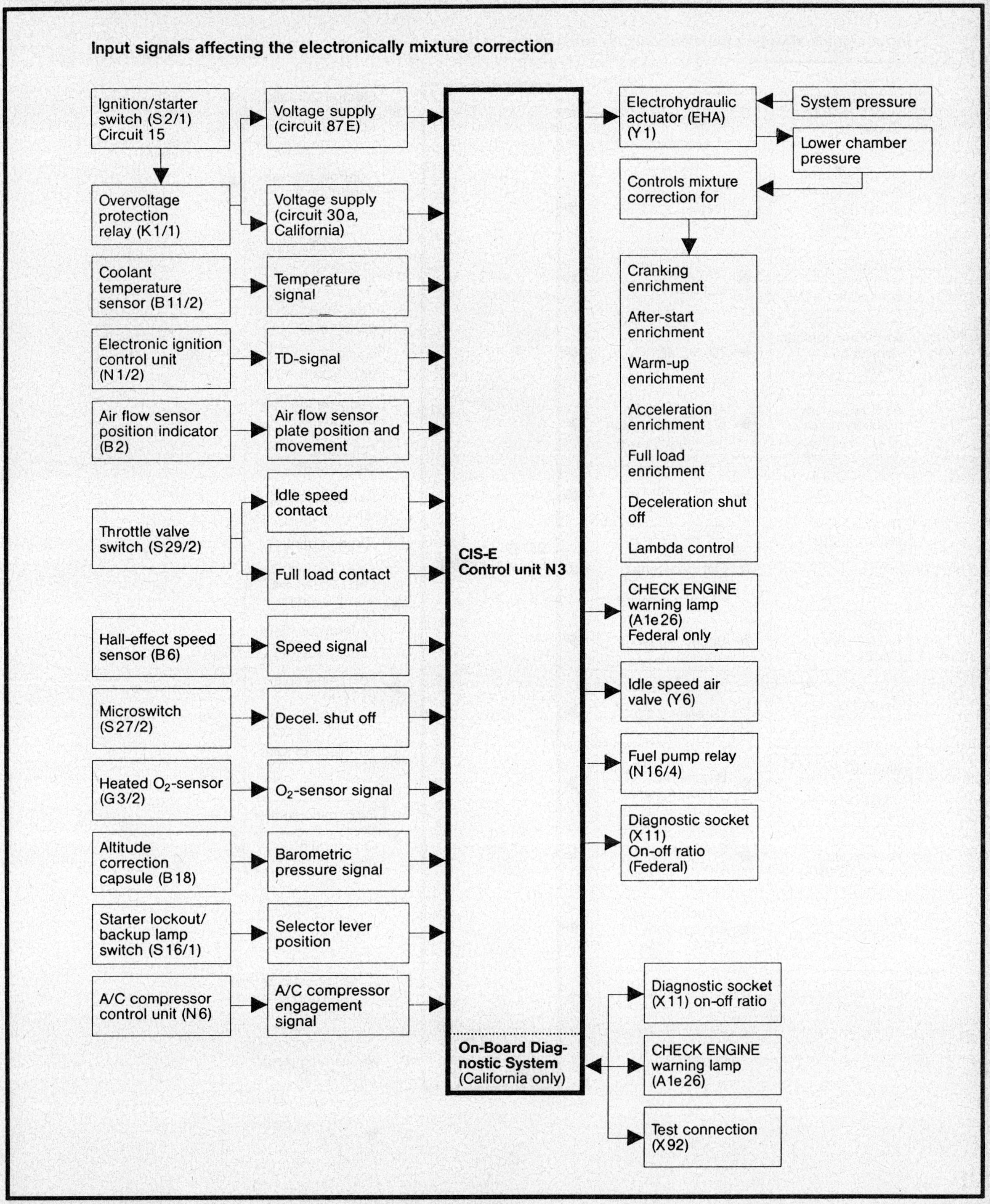

CIS-E control system—Typical 102 and 103 model engines

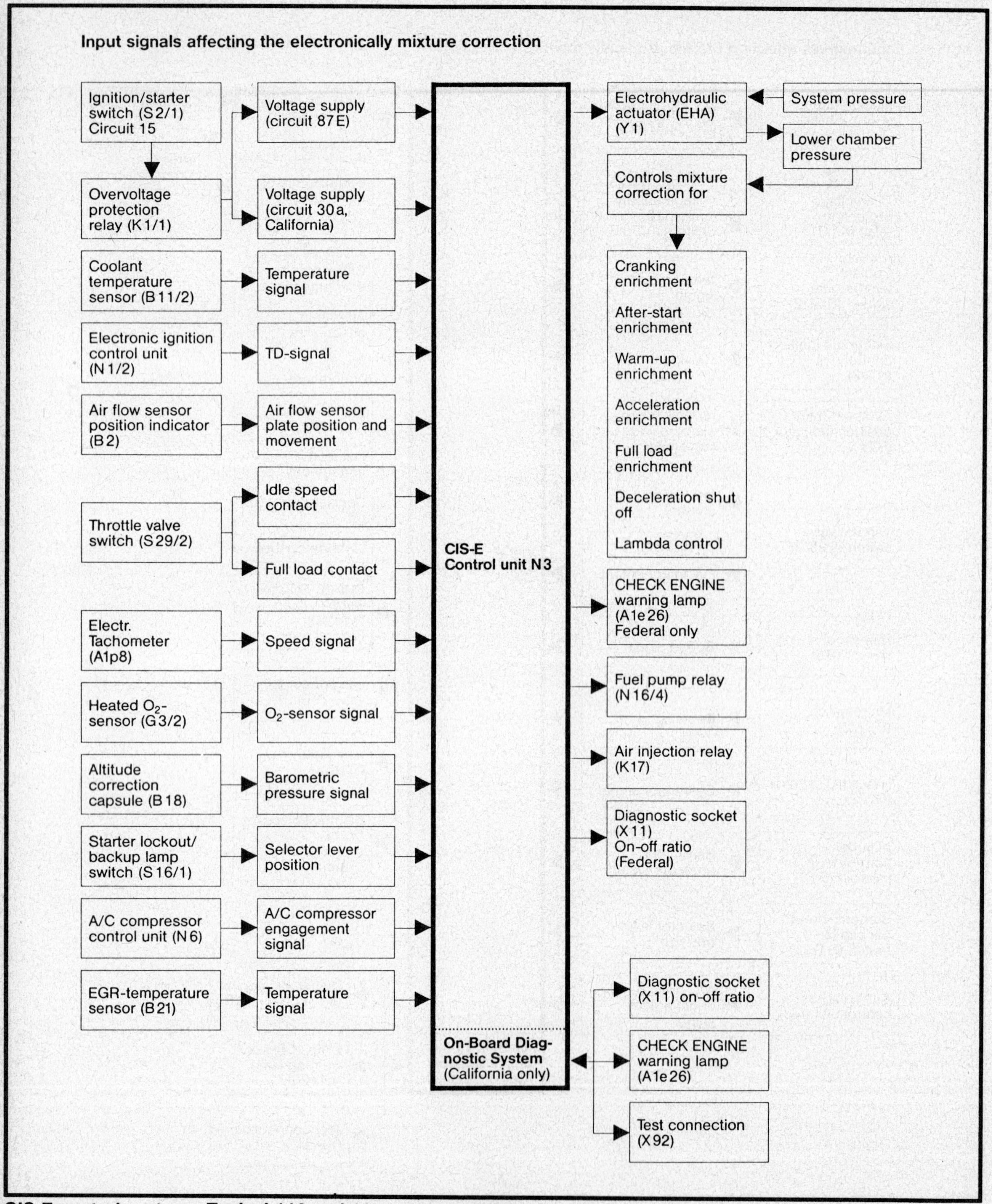

CIS-E control system—Typical 116 and 117 model engines

1. 1—Fuel distributor
2. 40—Diaphragm pressure regulator
3. 55—Fuel filter
4. 57—Fuel accumulator
5. 74—Fuel cooler
6. 75—Fuel tank
7. 76—Vent valve
8. 77—Charcoal canister
9. 78—Purge vent
10. 79—Thermo valve
11. 89—EGR valve
12. 90—Vacuum amplifier
13. 92—Injection valve
14. 125—Air pump
15. 126—Air injection shutoff valve
16. 127—Air injection check valve
17. 128—Vacuum check valve
18. 157—Primary catalyst
19. 158—Secondary catalyst
20. A-1e—Indicator lamp
21. A-1p—Electronic speedometer
22. B-2—Air flow sensor position indicator
23. B-11—Coolant temperature sensor
24. B-18—Altitude correction capsule
25. G-3—Oxygen sensor
26. L-5—Crankshaft position sensor
27. M-3—Fuel pump
28. N-1—Ignition control module (EZL)
29. N-3—CIS-E control module
30. N-8—Electronic idle speed control module

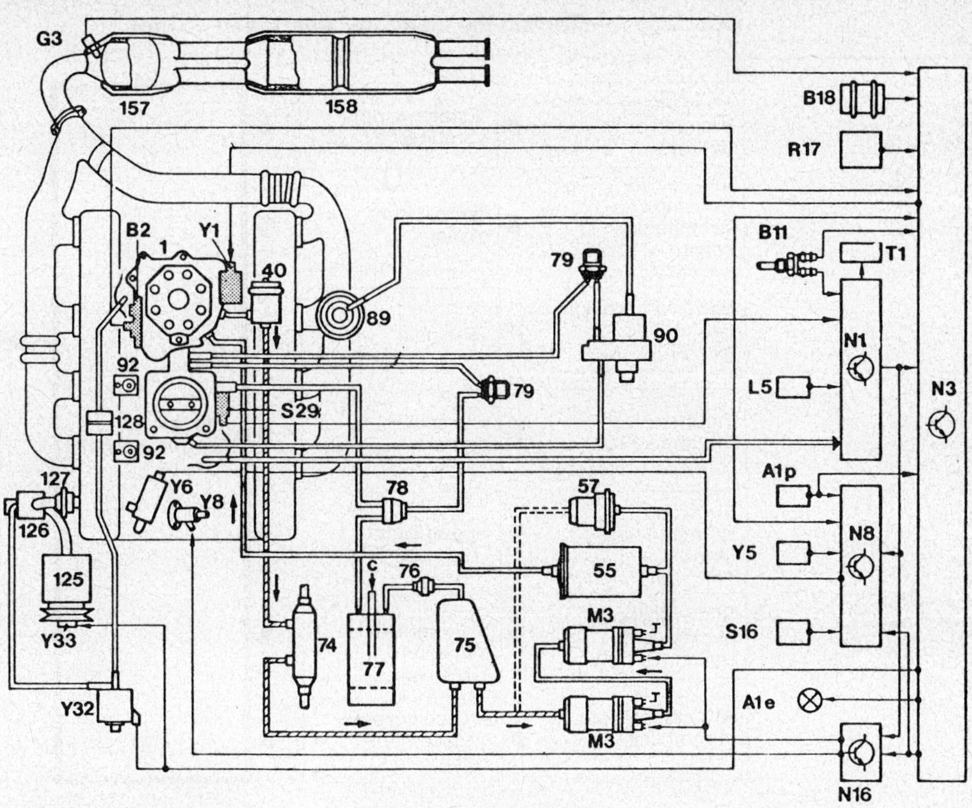

31. N-16—Fuel pump relay
32. R-17—CIS-E input resistor (if equipped)
33. S-16—Neutral safety and backup light switch
34. S 29 Throttle valve switch
35. T-1—Ignition coil
36. Y-1—Electrohydraulic actuator (EHA)
37. Y-5—A/C compressor clutch
38. Y-6—Idle speed air valve
39. Y-8—Cold start valve
40. Y-32—Air injection switchover valve
41. Y-33—Air pump clutch

CIS-E function diagram—typical

position can be varied causing a differential pressure change in the actuator and lower chamber, which provides mixture control. The ECU control voltage to the EHA is dependent on coolant temperature, engine speed and load and airflow meter plate deflection. The ECU will adjust fuel pressure through the EHA, to provide cold start, warm up, acceleration enrichment, full load enrichment, deceleration leaning and mixture control through all engine ranges.

FUEL DELIVERY SYSTEM

Fuel Pump Relay

The fuel pump relay is to activate the fuel pump when the engine is cranked and when the engine is running. It also deactivates the pump if no impulses are received from the TD of the ignition switching unit. The relay supplies power to the oxygen sensor heated and provides a kick-down fuel cutoff on automatic vehicles as well as a high engine rpm fuel cutoff, to govern maximum engine speed.

Cold Start Valve

The cold start valve depends on coolant temperature and the starter signal. Voltage is supplied to by the fuel pump relay. The length of time that the cold start injector functions depends on the coolant temperature sensor. If the engine starts before the injector timer finished its cycle, the injector power is shut off.

Injection Valves

The fuel injectors are metered relief valves. When the fuel pressure applied to them increase above approximately 60 psi (4.3 bar), they continuously inject fuel. The amount of fuel injected is dependant upon engine vacuum or load and the fuel pressure supplied to them by the fuel distributor system.

Diaphragm Pressure Regulator

The system pressure is approximately 80–93 psi (5.5–6.4 bars). The pressure regulator is set by the manufacturer and cannot be adjusted.

Diagnosis and Testing

SERVICE PRECAUTIONS

To prevent personal injury or damage to the vehicle, follow all service precautions.

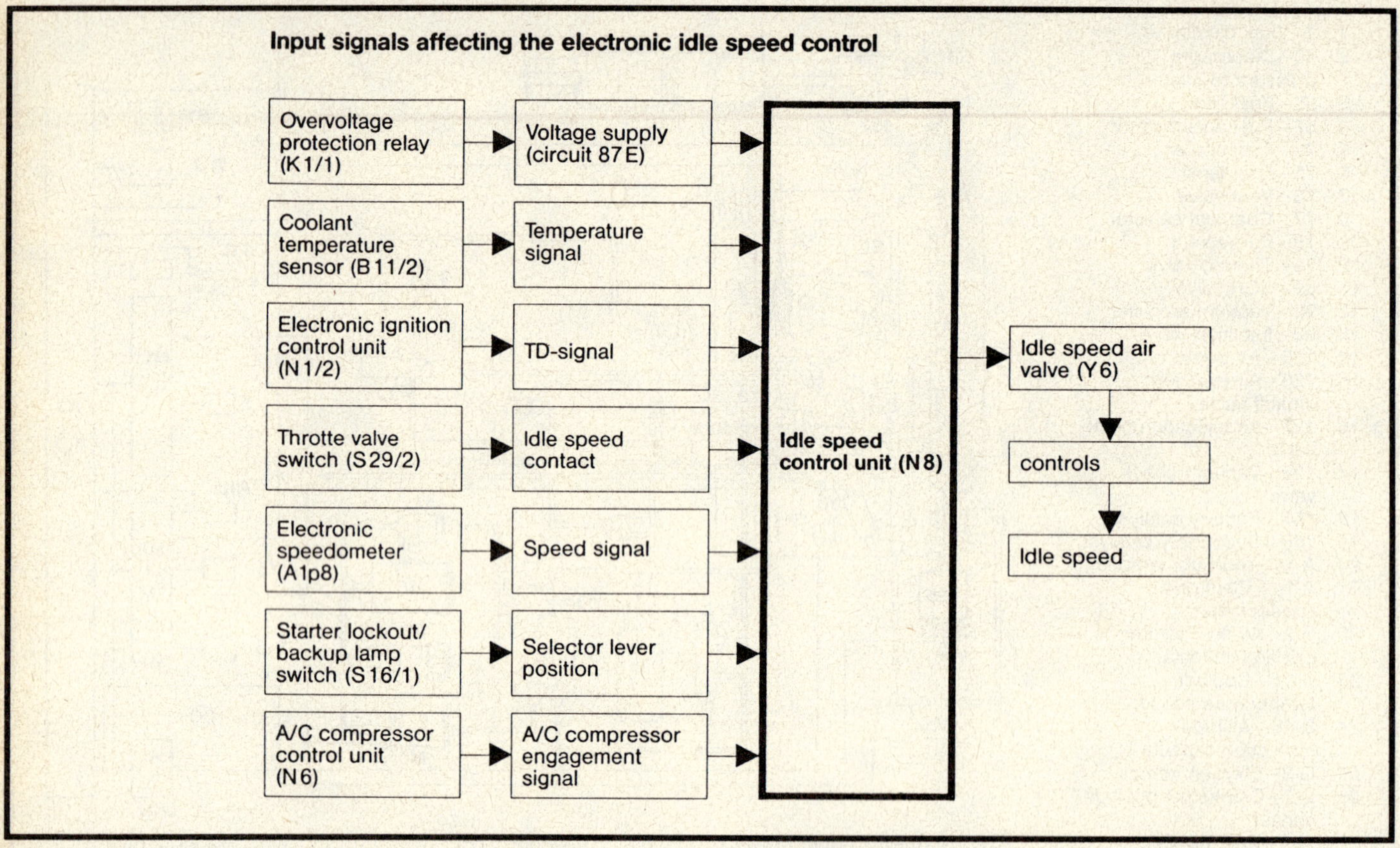

CIS-E electronic idle control system—Typical for 116 and 117 engines

- Always disconnect the negative battery terminal to prevent sparks cause by short-circuiting, circuit-breaking, etc.
- Never disconnect the battery with the engine running.
- Smoking must not be allowed when working around flammable liquids.
- Always keep a CO_2 fire extinguisher close on hand.
- Absorbant must be available to soak up any spillage.
- Ensure fuel is emptied into a suitable container.
- Always place the transmission gear shift lever in **N** (manual transmission) or **P** (automatic transmission) and set the parking brake and block drive wheels before starting the engine.
- Keep hands, tools and clothing away from engine cooling fan to help prevent personal injury.
- Thoroughly clean all unions and area near connections before disconnecting.
- Always disconnect the battery when using arc, spot or electrical welding equipment.
- When cranking engine without starting, as for a compression test, disconnect wire from distributor, and connect to ground. Do not hand hold wire.

CAUTION

To prevent personal injury, use extreme caution and always wear eye protection when opening any fuel line on a CIS fuel system. This type of system retains high fuel pressure for hours after the engine has been shut off.

READING FAULT CODES

Continuous Injection System (CIS-E)

ON-OFF RATIO TESTER (LAMBDA CONTROL)

1. Start the engine and allow it to reach operating temperature. Ensure that all accessories are **OFF**, the selector lever is in **P**, the fuse of the overvoltage protection relay is good and the battery is fully charged.
2. Connect the On-Off Ratio tester to the diagnostic connector terminal (X11) in the engine compartment.
3. Turn the ignition key **ON** and read the tester.
4. Repair or replace the component and repeat the test procedure until no faults are present.

IMPULSE TESTER (ON-BOARD DIAGNOSTICS)

NOTE: Disconnecting the CIS-E control unit, battery or overvoltage protection relay will erase all codes stored in memory.

1. Connect the impulse tester to terminals 1 and 3 of the diagnostic connector (X11/4). The connector is located on the firewall.
2. Turn the ignition **ON** with the selector lever in **P**.
3. Press the start button for no less than 2, but no more than 4 seconds and observe the indicated impulse readout.
4. Continue to press the start button to obtain additional fault codes. All codes must be read prior to servicing.
5. Repair or replace the component and repeat the test procedure to ensure no faults are present.

CLEARING CODES

Electronic Idle Speed Adjustment (ELR)

IMPULSE TESTER

Fault codes for the ELR system are not stored in memory. When the faulty component is repaired or replaced, the fault code will automatically erase.

Continuous Injection System (CIS-E)

ON-OFF RATIO TESTER (LAMBDA CONTROL)

Fault codes obtained using an On-Off ratio tester are only present during a malfunction. When the component is repaired or replaced, the fault code will automatically be erased.

IMPULSE TESTER (ON-BOARD DIAGNOSTICS)

1. With the impulse tester connected and the fault code displayed, press the start button for 6 seconds or more.
2. Each impulse code must be erased individually.
3. If there is no number displayed, then the stored fault has been erased.
4. If the display shows the number 1, then there are more system faults in memory.

Fault codes obtained using an impulse tester will be erased if the control unit, overprotection relay or negative battery terminal are disconnected. However, this is not the recommended procedure for erasing the stored codes.

ENGINE CODES

CIS-E Fuel Injection

LAMBDA CONTROL PERCENTAGE CODES

0%—CIS-E control unit receives no voltage or is defective. Circuit to diagnostic socket open or On/Off Ratio tester defective. Air/fuel mixture too rich. Oxygen sensor signal voltage is +12 volts

10%—Airflow sensor position indicator defective or polarity reversed.

20%—Wires in plug of throttle valve switch incorrectly connected connected or short circuit (full load contact closed before engine reaches full load)

30%—Short or open circuit between CIS-E control unit and coolant temperature sensor or coolant temperature sensor defective (idle speed too high)

40%—Short or open circuit to airflow sensor position indicator or airflow sensor position indicator defective.

50%—Oxygen sensor not at operating temperature, open circuit or defective

60%—Not used (4.2L and 5.6L engines)

60%—Speed signal at CIS-E control unit out of range (2.6L, 3.0L and 5.0L engines)

70%—No RPM signal (open circuit) at CIS-E control unit

80%—Short or open circuit to altitude correction capsule or capsule defective (3.0L California, 4.2L and 5.6L engines)

80%—Interrupted data exchange between EZL ignition control unit and CIS-E control unit (3.0L and 5.0L 49- state engines)

90%—Current reading out of range at electrohydraulic valve

95%—Deceleration fuel shut-off is currently in operation

100%—Overvoltage protection relay defective (idle speed too high). Air/fuel mixture too lean, oxygen sensor defective (short circuit to ground). CIS-E control unit or On/Off ratio tester defective. No power to electrohydraulic actuator (EHA).

Reading Oscillates—No malfunction within the circuit of the monitored signals. On/Off ratio reading indicated.

ON BOARD DIAGNOSTIC CODES

Code 1—No malfunctions in system
Code 2—Throttle switch full load contact
Code 3—Coolant temperature sensor
Code 4—Airflow sensor position indicator
Code 5—Oxygen sensor
Code 6—Not used
Code 7—RPM signal
Code 8—Altitude correction sensor
Code 9—Electrohydraulic actuator
Code 10—Throttle switch idle contact
Code 11—Air injection system
Code 12—EGR temperature sensor (2.3L engine, California only)
Code 12—Absolute pressure values from EZL control unit out of range (2.6L and 3.0L engines)
Code 13—Intake air temperature reading out of range
Code 14—Speed signal read by CIS-E control unit out of range
Code 15—Not used
Code 16—EGR
Code 17—Oxygen sensor line shorted
Code 18—Current to idle speed air valve out of range
Code 19—Not used
Code 20—Not used
Code 21—Not used
Code 22—Oxygen sensor heating current out of range
Code 23—Short circuit to positive in purge switchover valve circuit
Code 24—Not used
Code 25—Short circuit to positive in start valve circuit
Code 26—Short circuit to positive in shift point delay circuit
Code 27—Data exchange CIS-E control unit to EZL control unit interrupted
Code 28—Intermittent contact in coolant temperature sensor circuit
Code 29—Difference in coolant temperatures read by CIS-E control unit and EZL control unit
Code 30—Not used
Code 31—Intermittent contact in intake air temperature sensor
Code 32—Not used
Code 33—Not used
Code 34—Coolant temperature from EZL control unit out of range

ENGINE SYSTEMS (MAS) CONTROL UNIT CODES

Code 1—No system malfunctions
Code 2—Fuel pump relay not operating
Code 3—RPM signal interrupted
Code 4—Oxygen sensor heater output failure
Code 5—Air pump output failure
Code 6—Kickdown switch output failure
Code 8—Coolant temperature out of range
Code 9—Circuit 50 failure
Code 10—Start valve output failure
Code 11—A/C compressor engagement signal for idle speed regulation missing
Code 12—A/C compressor output failure
Code 13—Excessive A/C clutch slip
Code 15—Detection of a short in the fuel pump circuit

TYPICAL K-JETRONIC SYMPTOM CHART

SYMPTOM / CAUSE	Leaks in air intake system (false air)	Sensor plate and/or plunger not moving freely	Air flow sensor plate stop incorrectly set	Auxiliary air valve does not open	Auxiliary air valve does not close	Electric fuel pump not operating	Cold start system defective	Cold start valve leaking	Thermo-time switch defective	Excessive fuel delivery to control circuit	"Cold" control pressure outside tolerance	"Warm" control pressure too high	"Warm" control pressure too low	Primary system pressure out of tolerance	Fuel system leakage	Injector leaking, opening pressure low	Injected fuel quantities unequal	Basic idle setting incorrect	Throttle butterfly does not open completely	Lambda control system not functioning correctly
Engine starts but immediately dies						●				●		●	●	●						
Idle speed cannot be adjusted (too high)					●															
CO concentration at idle too low	●	●																●		●
CO concentration at idle too high		●						●					●					●		●
Driving performance unsatisfactory	●	●								●		●	●	●			●	●		●
Fuel consumption too high								●					●					●		●
Engine runs on		●	●					●								●		●		
Poor performance under load	●	●										●	●	●			●		●	●
Engine backfires		●				●						●		●		●				
Engine will not accelerate, backfires	●	●								●		●	●					●		●
Irregular idle with warm engine	●	●						●				●	●			●	●	●		●
Irregular idle during warm-up	●			●				●			●					●	●	●		●
Engine does not start/starts hard warm	●	●	●			●						●			●	●	●	●		
Engine does not start/starts hard cold		●		●		●	●		●		●					●	●	●		

Troubleshooting chart

Fixed on-off ratio %	Possible cause of failure	Tests to be performed
0 %	CIS-E control unit receives no voltage or is defective. Circuit to diagnostic socket, terminal 3 open or on-off ratio tester defective. Air/fuel mixture too rich. O_2-sensor signal voltage is +12 V.	Check voltage supply, ground connection and wires to diagnostic socket. Check on-off ratio tester. Check O_2-sensor signal.
10 %	Air flow sensor position indicator defective or polarity reversed. Increased rpm. Wires in plug of throttle valve switch incorrectly connected or short circuit (full load contact closed before engine reaches full load).	Check air flow sensor position indicator signal and routing of wires. Check throttle valve switch and wire connections according to wiring diagram. (Defective idle contact, ignition timing increased approx. 10°.)
20 %	Full load contact polarity reversed or defective. Reading at 20 % only with activated microswitch.	Check full load contact.
30 %	Short or open circuit between CIS-E control unit and coolant temperature sensor or coolant temperature sensor defective.	Check coolant temperature sensor.
40 %	Short or open circuit to air flow sensor position indicator or air flow sensor position indicator defective.	Check air flow sensor position indicator.
50 %	O_2-sensor not at operating temperature, open circuit or defective.	Check engine operating temperature. Check O_2-sensor.
60 %	Not used	–
70 %	No TD signal (open circuit) at CIS-E control unit.	Check TD signal.
80 %	Short or open circuit to altitude correction capsule or capsule defective.	Check altitude correction capsule.
90 %	Not used	–
100 %	Air/fuel mixture too lean, O_2-sensor defective (short circuit to ground). CIS-E control unit or on-off ratio tester defective. Defective overvoltage protection relay, no voltage supply.	Check adjustment of lambda control and O_2-sensor signal. Check power supply.
Reading oscillates	No malfunction within the circuit of the monitored signals. On-off ratio reading indicated.	–

Troubleshooting ratio chart

TROUBLESHOOTING CHARTS—102 AND 103 ENGINES

Test connection: Socket box, Multimeter, Battery	Test condition	Nominal value	Possible cause/remedy
Fixed on-off ratio - % Test step 1			
Ground locations			
Socket box 2 ←(V)→ Battery Plus	–	11–14 V	Ground connection (W11) loose, open circuit
Socket box 7 ←(V)→ Battery Plus	–	11–14 V	Control unit defective
Socket box 20 ←(V)→ Battery Plus	–	11–14 V	Ground connection (W10) loose, open circuit
Fixed on-off ratio - % Test step 2			
Voltage supply from N3, term 1			
Socket box 2 ←(V)→ Socket box 1	CIS-E control unit connector unplugged Ignition: ON	11–14 V	Overvoltage protection relay fuse (K1/1), relay (K1/1), defective or not connected, ground connection at K1/1, open circuit in wire to socket 1 (circuit 30), open circuit in supply to socket 3 (circuit 15), open circuit from socket 2 (circuit 87E) to socket 1 (N3)
Wiring from circuit 30			
Socket box 2 ←(V)→ X35 circuit 30	Ignition: ON	11–14 V	Loose wire, open circuit
Socket box 2 ←(V)→ K1/1 socket 1	Ignition: OFF Overvoltage protection relay K1/1 disconnected	11–14 V	Open circuit from X35 to K1/1
Wiring from circuit 15			
Socket box 2 ←(V)→ N16/ socket 9	Ignition: ON Fuel pump relay disconnected	11–14 V	Open circuit from F1 to N16/
Socket box 2 ←(V)→ K1/1 socket 3		11–14 V	Open circuit from N16/ to K1/1.
Wire circuit 87 E			
Socket box 1 ←(Ω)→ K1/1 socket 2	Ignition: OFF	$<1\,\Omega$	Open circuit from K1/1 to N3, socket 1

TROUBLESHOOTING CHARTS—102 AND 103 ENGINES

Test connection: Socket box Multimeter	Test condition	Nominal value	Possible cause/remedy
Fixed on-off ratio - % Test step 3			
Voltage supply for malfunction storage N3 (California version only)			
2 ←V→ 15	Ignition: OFF CIS-E control unit connector unplugged, overvoltage protection relay (K1/1) connected	11–14 V	Open circuit from K1/1 to N3.
Fixed on-off ratio - % Test step 4			
Test current flow at EHA Set multimeter to Ampere	Connect test cable 102 589 04 63 00 to EHA Ignition: ON	20 mA	Electrohydraulic actuator, wiring to CIS-E control unit
	Engine at idle, at operating temperature	fluctuates 0 ± 3 mA	Adjust lambda control, check O_2-sensor
Electrohydraulic actuator (EHA) 10 ←Ω→ 12	Ignition: OFF CIS-E control unit connector unplugged	$19.5 \pm 1\ \Omega$	Electrohydraulic actuator, wiring
Y1 pin 1 ←Ω→ Y1 pin 2	Unplug connector Y1	$19.5 \pm 1\ \Omega$	Replace electrohydraulic actuator
Wiring 10 ←Ω→ Y1 socket 2		$<1\ \Omega$	Open circuit
12 ←Ω→ Y1 socket 1		$<1\ \Omega$	Open circuit
Fixed on-off ratio 0 % Test step 5			
Wire from N3 to X11 23 ←Ω→ X11 socket 3	Ignition: OFF CIS-E control unit connector unplugged	$<1\ \Omega$	Open circuit On X11, open circuit at voltage supply socket 6 or ground socket 2, on-off ratio tester defective, CIS-E control unit (California version) not programmed to read on-off ratio
O_2-sensor signal	–	–	Lambda control exceeds tolerance, O_2-sensor defective (see test step 10)

TROUBLESHOOTING CHARTS—102 AND 103 ENGINES

Test connection: Socket box, Multimeter, On-off ratio tester	Test condition	Nominal value	Possible cause/remedy

Fixed on-off ratio 10 % Test step 6

Test connection	Test condition	Nominal value	Possible cause/remedy
Idle speed contact On-off ratio tester	Ignition: ON Open microswitch	–	Display 70%: idle speed contact S29/2. Display 10% or 40%: air flow sensor position indicator polarity reversed at higher idle speed. Display 20%, see test step 7, full load contact.
Socket box 2 ←Ω→ Socket box 13	Ignition: OFF CIS-E control unit connector and EZL ignition control module connector (sockets marked 1–4) unplugged	$<1\ \Omega$	–
	Activate accelerator pedal	∞	Polarity of connection S29/2x reversed, idle speed contact, wiring
S29/2x pin 1 ←Ω→ S29/2x pin 2	Unplug connector S29/2x	$<1\ \Omega$	–
	Activate accelerator pedal	∞	Adjust or replace throttle switch
Wiring Socket box 13 ←Ω→ S29/2x socket 1	Unplug connector S29/2x	$<1\ \Omega$	Open circuit
S29/2x socket 2 ←Ω→ W11 ground		$<1\ \Omega$	Open circuit
Air flow sensor position indicator Socket box 7 ←−V+→ Socket box 18	Engine at idle, at operating temperature CIS-E control unit connected	4.6 – 5.1 V	see test step 9 (40%)
Socket box 7 ←−V+→ Socket box 17		0.55 – 0.95 V	see test step 9 (40%)

TROUBLESHOOTING CHARTS—102 AND 103 ENGINES

Test connection: Socket box; Multimeter; On-off ratio tester	Test condition	Nominal value	Possible cause/remedy
Fixed on-off ratio 20 % Test step 7			
Full load contact			
2 ←Ω→ 5	Ignition: OFF CIS-E control unit connector unplugged	∞	–
	Accelerator pedal in full load position	<1 Ω	Polarity of connection S29/2x reversed, full load contact, wiring
S29/2x pin 2 ←Ω→ S29/2x pin 3	Unplug connector S29/2x	∞	–
	Accelerator pedal in full load position	<1 Ω	Adjust or replace throttle valve switch
Wiring			
5 ←Ω→ S29/2x socket 3	Unplug connector S29/2x	<1 Ω	Open circuit
W11 ←Ω→ S29/2x socket 2		<1 Ω	Open circuit
Fixed on-off ratio 30 % Test step 8			
Coolant temperature sensor			
7 ←Ω→ 21	Ignition: ON CIS-E control unit connector unplugged	see table	Coolant temperature sensor, wiring
B11/2 pin ←Ω→ B11/2 pin	Unplug temperature sensor connector	see table	Replace temperature sensor
Wiring			
21 ←Ω→ B 11/2 socket	EZL ignition control module connector (sockets marked 1-4) unplugged	<1 Ω	Open circuit
7 ←Ω→ B11/2 socket		<1 Ω	Open circuit

TROUBLESHOOTING CHARTS—102 AND 103 ENGINES

Test connection: Socket box; Multimeter; On-off ratio tester	Test condition	Nominal value	Possible cause/remedy

Fixed on-off ratio 40 % Test step 9

Test connection	Test condition	Nominal value	Possible cause/remedy
Air flow sensor position indicator			
Socket box 7 ←–(V)+→ Socket box 18	Engine at idle, at operating temperature CIS-E control unit connected	4.6 – 5.1 V	Air flow sensor position indicator, altitude correction capsule (see test step 12), CIS-E control unit
Socket box 7 ←–(V)+→ Socket box 17		0.55 – 0.95 V	Air flow sensor position indicator, CIS-E control unit
B2 pin 1 ←(Ω)→ **B2** pin 3	Ignition: OFF Unplug connector B2	3.6 – 4.4 kΩ	–
B2 pin 1 ←(Ω)→ **B2** pin 2	Slowly deflect air flow plate by hand	Ω-value increases continuously up to half of travel, then decreases again	Replace air flow sensor
Wiring			
Socket box 7 ←(Ω)→ **B2** socket 1	Ignition: OFF Measure at connector B2	<1 Ω	Open circuit
Socket box 18 ←(Ω)→ **B2** socket 3		<1 Ω	Open circuit
Socket box 17 ←(Ω)→ **B2** socket 2		<1 Ω	Open circuit

TROUBLESHOOTING CHARTS—102 AND 103 ENGINES

Test connection: Socket box Multimeter On-off ratio tester	Test condition	Nominal value	Possible cause/remedy

Fixed on-off ratio 50 % Test step 10

O_2-sensor			
7 ←(V)→ 8	Engine at idle, at operating temperature CIS-E control unit connected	fluctuates 0.1 – 0.9 V	O_2-sensor, wiring
7 ←(V)→ G3/2x2 socket	Engine at idle, connector G3/2x2 unplugged and pin grounded	> 450 mV	Replace O_2-sensor
Wiring			
7 ←(Ω)→ 8	Ignition: OFF Unplug CIS-E control unit connector	∞	Defective wire
8 ←(Ω)→ G3/2x2 pin		<1 Ω	Open circuit

Fixed on-off ratio 60 %

(not used)	–	–	–

Fixed on-off ratio 70 % Test step 11

TD-signal			
2 ←(V)→ 25	Engine at idle, at operating temperature CIS-E control unit connected	6 – 12 V	Ignition control module, wiring
Wiring			
25 ←(Ω)→ X11 socket 1	Ignition: OFF Unplug TD-signal connector on ignition control module	<1 Ω	Open circuit
N1/2 TD ←(Ω)→ X11 socket 1		<1 Ω	Open circuit

TROUBLESHOOTING CHARTS—102 AND 103 ENGINES

Test connection: Socket box Multimeter On-off ratio tester	Test condition	Nominal value	Possible cause/remedy

Fixed on-off ratio 80 % Test step 12

Altitude correction capsule 7 ←(V)→ 18	Ignition: ON CIS-E control unit connected	4.6 – 5.1 V	Altitude correction capsule, air flow sensor position indicator (see test step 9), wiring CIS-E control unit
7 ←(V)→ 11		see table	Altitude correction capsule, wiring
Wiring **B18** 7 ←(Ω)→ socket 3	Ignition: OFF Unplug connector B18	<1 Ω	Open circuit
B18 18 ←(Ω)→ socket 2		<1 Ω	Open circuit
B18 11 ←(Ω)→ socket 1		<1 Ω	Open circuit

Fixed on-off ratio 90 %

(not used)	–	–	–

Fixed on-off ratio 100 % and idle speed to high Test step 13

–	–	–	see test steps 1, 2, 3, 4, 10 and test on-off ratio

ON BOARD DIAGNOSTIC CHARTS—102 AND 103 ENGINES

Test connection: Socket box, Multimeter	Test condition	Nominal value	Possible cause/remedy

Impulse display 1

–	–	–	No system malfunction

Impulse display 2

Full load contact 2 ←Ω→ 5	Ignition: OFF CIS-E control unit connector unplugged	∞	
	Accelerator pedal in full load position	<1 Ω	Full load contact, wiring
S29/2x pin 2 ←Ω→ S29/2x pin 3	Unplug connector S29/2x	∞	
	Accelerator pedal in full load position	<1 Ω	Adjust or replace throttle valve switch
Wiring 5 ←Ω→ S29/2x socket 3	Unplug connector S29/2x	<1 Ω	Open ciruit
W11 Ground ←Ω→ S29/2x socket 2		<1 Ω	Open circuit

Impulse display 3

Coolant temperature sensor 7 ←Ω→ 21	Ignition: OFF CIS-E control unit connector unplugged	see table	Coolant temperature sensor, wiring
B11/2 pin ←Ω→ B11/2 pin	Unplug temperature sensor connector	see table	Replace temperature sensor
Wiring 21 ←Ω→ B11/2 socket	EZL ignition control module connector (sockets marked 1–4) unplugged	<1 Ω	Open circuit
7 ←Ω→ B11/2 socket		<1 Ω	Open circuit

ON BOARD DIAGNOSTIC CHARTS—102 AND 103 ENGINES

Test connection: Socket box / Multimeter	Test condition	Nominal value	Possible cause/remedy
Impulse display 4			
Air flow sensor position indicator			
7 ←(−V+)→ 18	Engine at idle, at operating temperature CIS-E control unit connected	4.6 – 5.1 V	Air flow sensor position indicator, altitude correction capsule (see impulse display 8), CIS-E control unit
7 ←(−V+)→ 17		0.55 – 0.95 V	Air flow sensor position indicator, CIS-E control unit
B2 pin 1 ←(Ω)→ B2 pin 3	Ignition: OFF Unplug connector on B2	3.6 – 4.4 kΩ	
B2 pin 1 ←(Ω)→ B2 pin 2	Slowly deflect air flow sensor plate by hand	Ω-value increases continuously up to half of travel, then decreases again	Replace air flow sensor
Wiring			
7 ←(Ω)→ B2 socket 1	Measure at connector B2	<1 Ω	Open circuit
18 ←(Ω)→ B2 socket 3		<1 Ω	Open circuit
17 ←(Ω)→ B2 socket 2		<1 Ω	Open circuit
Impulse display 5			
O_2-sensor			
7 ←(−V+)→ 8	Engine idling, at operating temperature CIS-E control unit connected	fluctuates between 0.1 – 0.9 V	O_2-sensor, wiring
7 ←(−V+)→ G3/2x2 socket	Engine idling, connector G3/2x2 unplugged and terminal grounded	> 450 mV	Replace O_2-sensor
Wiring			
7 ←(Ω)→ 8	Ignition: OFF Unplug CIS-E control unit connector	∞	Defective wire
8 ←(Ω)→ G3/2x2 pin		<1 Ω	Open circuit

ON BOARD DIAGNOSTIC CHARTS—102 AND 103 ENGINES

Test connection: Socket box, Multimeter	Test condition	Nominal value	Possible cause/remedy
Impulse display 6			
(not used)	–	–	–
Impulse display 7			
TD-signal			
2 ← V → 25	Engine idling, at operating temperature CIS-E control unit connected	6 – 12 V	Ignition control module, wiring
Wiring			
25 ← Ω → X11 socket 1	Ignition: OFF Unplug TD-signal connector on ignition control module.	<1 Ω	Open circuit
N1/2 TD ← Ω → X11 socket 1		<1 Ω	Open circuit
Impulse display 8			
Altitude correction capsule			
7 ← V → 18	Ignition: ON CIS-E control unit connected	4.6 – 5.1 V	Altitude correction capsule, air flow sensor position indicator (see impulse display 4), wiring, CIS-E control unit
7 ← V → 11		see table	Altitude correction capsule, wiring
Wiring			
7 ← Ω → B18 socket 3	Ignition: OFF Unplug connector on B2	<1 Ω	Open circuit
18 ← Ω → B18 socket 2		<1 Ω	Open circuit
11 ← Ω → B18 socket 1		<1 Ω	Open circuit

TROUBLESHOOTING CHARTS—116 AND 117 ENGINES

Test connection: Socket box, Multimeter, Battery	Test condition	Nominal value	Possible cause/remedy

Fixed on-off ratio - % Test step 3

Voltage supply for malfunction storage N3 (California version only) Socket box 2 ←(V)→ Socket box 15	Ignition: OFF CIS-E control unit connector unplugged, overvoltage protection relay (K1/1) connected	11–14 V	Open circuit from K1/1 to N3.

Fixed on-off ratio - % Test step 4

Test current flow at EHA Set multimeter to Ampere	Connect test cable 102 589 04 63 00 to EHA Ignition: ON	75 mA	Electrohydraulic actuator, wiring CIS-E control unit
	Engine idling, at operating temperature	fluctuates 0 ± 3 mA	Adjust lambda control, check O_2-sensor
Electrohydraulic actuator (EHA) Socket box 10 ←(Ω)→ Socket box 12	Ignition: OFF CIS-E control unit connector unplugged	$19.5 \pm 1\ \Omega$	Electrohydraulic actuator, wiring
Y1 pin 1 ←(Ω)→ **Y1** pin 2	Unplug connector Y1	$19.5 \pm 1\ \Omega$	Replace electrohydraulic actuator
Wiring Socket box 10 ←(Ω)→ **Y1** socket 2		$< 1\ \Omega$	Open circuit
Socket box 12 ←(Ω)→ **Y1** socket 1		$< 1\ \Omega$	Open circuit

TROUBLESHOOTING CHARTS—116 AND 117 ENGINES

Test connection: Socket box, Multimeter, Battery	Test condition	Nominal value	Possible cause/remedy

Fixed on-off ratio 0 % Test step 5

Wire from N3 to X11 Socket box X11 socket 23 ←Ω→ socket 3	Ignition: OFF CIS-E control unit connector unplugged	<1 Ω	Open circuit On X11, open circuit at voltage supply socket 6 or ground socket 2, on-off ratio tester defective, CIS-E control unit (California version) not programmed to read on-off ratio
O_2-sensor signal	–	–	Lambda control exceeds tolerance, O_2-sensor defective (see test step 10)

Fixed on-off ratio 10 % Test step 6

Air flow sensor position indicator Socket box 7 ←–V+→ Socket box 18	Engine at idle, at operating temperature CIS-E control unit connected	4.5 – 5.5 V	Polarity of air flow sensor position indicator (B2) reversed
Socket box 7 ←–V+→ Socket box 17		0.5 – 1.5 V	Polarity of air flow sensor position indicator (B2) reversed

TROUBLESHOOTING CHARTS—116 AND 117 ENGINES

Test connection: Socket box / Multimeter / Battery	Test condition	Nominal value	Possible cause/remedy
Fixed on-off ratio 20 % Test step 7			
Full load contact			
Socket box 2 ←Ω→ Socket box 5	Ignition: OFF CIS-E control unit connector and EZL ignition control module connector (sockets marked 1–4) unplugged	∞	–
	Accelerator pedal in full load position	< 1 Ω	Polarity of connection S 29/2x reversed, full load contact, wiring
S 29/2x pin 2 ←Ω→ S 29/2x pin 3	Unplug connector S 29/2x	∞	–
	Accelerator pedal in full load position	< 1 Ω	Adjust or replace throttle valve switch
Wiring			
Socket box 5 ←Ω→ **S 29/2x** socket 3	Unplug connector S 29/2x	< 1 Ω	Open circuit
W11 Ground ←Ω→ **S 29/2x** socket 2		< 1 Ω	Open circuit
Fixed on-off ratio 30 % Test step 8			
Coolant temperature sensor			
Socket box 7 ←Ω→ Socket box 21	Ignition: ON CIS-E control unit connector unplugged	see table	Coolant temperature sensor, wiring
B11/2 pin ←Ω→ **B11/2** pin	Unplug temperature sensor connector	see table	Replace temperature sensor
Wiring			
Socket box 21 ←Ω→ Socket box socket	EZL ignition control module connector (sockets marked 1–4) unplugged	< 1 Ω	Open circuit
Socket box 7 ←Ω→ **B11/2** socket		< 1 Ω	Open circuit

TROUBLESHOOTING CHARTS—116 AND 117 ENGINES

Test connection: Socket box, Multimeter, Battery	Test condition	Nominal value	Possible cause/remedy
Fixed on-off ratio 40 % Test step 9			
Air flow sensor position indicator			
7 ←(V)→ 18	Engine at idle, at operating temperature CIS-E control unit connected	4.5 – 5.5 V	Air flow sensor position indicator, altitude correction capsule (see test step 12) CIS-E control unit
7 ←(V)→ 17		0.5 – 1.5 V	Air flow sensor position indicator, CIS-E control unit
B2 pin 1 ←(Ω)→ **B2** pin 3	Ignition: OFF Unplug connector B2	3.6 – 4.4 kΩ	–
B2 pin 1 ←(Ω)→ **B2** pin 2	Slowly deflect air flow plate by hand	Ω-value increases continuously up to half of travel, then decreases again	Replace air flow sensor
Wiring			
7 ←(Ω)→ **B2** socket 1	Measure at connector B2	< 1 Ω	Open circuit
18 ←(Ω)→ **B2** socket 3		< 1 Ω	Open circuit
17 ←(Ω)→ **B2** socket 2		< 1 Ω	Open circuit

TROUBLESHOOTING CHARTS—116 AND 117 ENGINES

Test connection: Socket box, Multimeter, Battery	Test condition	Nominal value	Possible cause/remedy
Fixed on-off ratio 50 % Test step 10			
O_2-sensor			
Socket box 7 ←(V)→ Socket box 8	Engine at idle, at operating temperature CIS-E control unit connected	fluctuates 0.1 – 0.9 V	O_2-sensor, wiring
Socket box 7 ←(V)→ **G3/2x2** socket	Engine at idle, connector G3/2x2 disconnected and pin grounded	> 450 mV	Replace O_2-sensor
Wiring			
Socket box 7 ←(Ω)→ Socket box 8	Ignition: OFF Unplug CIS-E control unit connector	∞	Defective wire
Socket box 8 ←(Ω)→ **G3/2x2** pin	–	< 1 Ω	Open circuit
Fixed on-off ratio 60 %			
(not used)	–	–	–
Fixed on-off ratio 70 % Test step 11			
TD-signal			
Socket box 2 ←(V)→ Socket box 25	Engine at idle, at operating temperature CIS-E control unit connected	6 – 12 V	Ignition control module, wiring
Wiring			
Socket box 25 ←(Ω)→ **X11** socket 1	Ignition: OFF Unplug TD-signal connector on ignition control module	< 1 Ω	Open circuit
N1/2 TD ←(Ω)→ **X11** socket 1		< 1 Ω	Open circuit

TROUBLESHOOTING CHARTS—116 AND 117 ENGINES

Test connection: Socket box, Multimeter, Battery	Test condition	Nominal value	Possible cause/remedy
Fixed on-off ratio 80 % Test step 12			
Altitude correction capsule			
7 (V) 18	Ignition: ON CIS-E control unit connected	4.5 – 5.5 V	Altitude correction capsule, air flow sensor position indicator (see test step 9), wiring CIS-E control unit
7 (V) 11	–	see table	Altitude correction capsule, wiring
Wiring			
B18 7 (Ω) socket 3	Ignition: OFF Unplug connector B18	<1 Ω	Open circuit
B18 18 (Ω) socket 2		<1 Ω	Open circuit
B18 11 (Ω) socket 1		<1 Ω	Open circuit
Fixed on-off ratio 90 %			
(not used)	–	–	–
Fixed on-off ratio 100 % and idle speed too high Test step 13			
–	–	–	see test steps 1, 2, 3, 4, 10 and test on-off ratio

ON BOARD DIAGNOSTIC CHARTS—116 AND 117 ENGINES

Test connection: Socket box Multimeter	Test condition	Nominal value	Possible cause/remedy

Impulse display 1

–	–	–	No system malfunction

Impulse display 2

Full load contact 2 ←Ω→ 5	Ignition: OFF CIS-E control unit connector and EZL ignition controle module connector (terminals marked 1 – 4) unplugged	∞	
	Accelerator pedal in full load position	<1 Ω	Full load contact, wiring
S29/2x pin 2 ←Ω→ **S29/2x** pin 3	Unplug connector S29/2x	∞	
	Accelerator pedal in full load position	<1 Ω	Adjust or replace throttle valve switch
Wiring 5 ←Ω→ **S29/2x** socket 3	Unplug connector S29/2x	<1 Ω	Open circuit
W11 Ground ←Ω→ **S29/2x** socket 2		<1 Ω	Open circuit

Impulse display 3

Coolant temperature sensor 7 ←Ω→ 21	Ignition: OFF CIS-E control unit connector unplugged	see table	Coolant temperature sensor, wiring
B11/2 pin ←Ω→ **B11/2** pin	Unplug temperature sensor connector	see table	Replace temperature sensor
Wiring 21 ←Ω→ **B11/2** socket	EZL ignition control module connector (sockets marked 1-4) unplugged	<1 Ω	Open circuit
7 ←Ω→ **B11/2** socket		<1 Ω	Open circuit

ON BOARD DIAGNOSTIC CHARTS—116 AND 117 ENGINES

Test connection: Socket box, Multimeter	Test condition	Nominal value	Possible cause/remedy
Impulse display 4			
Air flow sensor position indicator			
7 ←(V)→ 18	Engine at idle, at operating temperature CIS-E control unit connected	4.5 – 5.5 V	Air flow sensor position indicator, altitude correction capsule (see impulse display 8), CIS-E control unit
7 ←(V)→ 17		0.5 – 1.5 V	Air flow sensor position indicator, CIS-E control unit
B2 pin 1 ←(Ω)→ B2 pin 3	Ignition: OFF Unplug connector on B2	3.6 – 4.4 kΩ	
B2 pin 1 ←(Ω)→ B2 pin 2	Slowly deflect air flow sensor plate by hand	Ω-value increases continuously up to half of travel, then decreases again	Replace air flow sensor
Wiring			
7 ←(Ω)→ B2 socket 1	Measure at connector B2	<1 Ω	Open circuit
18 ←(Ω)→ B2 socket 3		<1 Ω	Open circuit
17 ←(Ω)→ B2 socket 2		<1 Ω	Open circuit
Impulse display 5			
O_2-sensor			
7 ←(V)→ 8	Engine idling, at operating temperature CIS-E control unit connected	fluctuates between 0.1 – 0.9 V	O_2-sensor, wiring
7 ←(V)→ G3/2x2 socket	Engine idling, connector G3/2x2 unplugged and terminal grounded	> 450 mV	Replace O_2-sensor
Wiring			
7 ←(Ω)→ 8	Ignition: OFF Unplug CIS-E control unit connector	∞	Defective wire
8 ←(Ω)→ G3/2x2 pin		<1 Ω	Open circuit

ON BOARD DIAGNOSTIC CHARTS—116 AND 117 ENGINES

Test connection: Socket box, Multimeter	Test condition	Nominal value	Possible cause/remedy
Impulse display 6			
(not used)	–	–	–
Impulse display 7			
TD-signal 2 ←–(V)+→ 25	Engine idling, at operating temperature CIS-E control unit connected	6 – 11 V	Ignition control module, wiring
Wiring 25 ←(Ω)→ **N8** socket 10	Ignition: OFF Unplug idle speed control unit N8.	< 1 Ω	Open circuit
Impulse display 8			
Altitude correction capsule 7 ←–(V)+→ 18	Ignition: ON CIS-E control unit connected	4.6 – 5.5 V	Altitude correction capsule, air flow sensor position indicator (see impulse display 4), wiring, CIS-E control unit
7 ←–(V)+→ 11		see table	Altitude correction capsule, wiring
Wiring 7 ←(Ω)→ **B18** socket 3	Ignition: OFF Unplug connector on B18	< 1 Ω	Open circuit
18 ←(Ω)→ **B18** socket 2		< 1 Ω	Open circuit
11 ←(Ω)→ **B18** socket 1		< 1 Ω	Open circuit

ON BOARD DIAGNOSTIC CHARTS—116 AND 117 ENGINES

Test connection: Socket box / Multimeter	Test condition	Nominal value	Possible cause/remedy
Impulse display 9			
Electrohydraulic actuator			
10 ←Ω→ 12	Ignition: OFF CIS-E control unit connector unplugged	19.5 ± 1 Ω	Electrohydraulic actuator, wiring
Y1 pin 1 ←Ω→ **Y1** pin 2	Unplug connector Y1	19.5 ± 1 Ω	Replace electrohydraulic actuator
Wiring			
10 ←Ω→ **Y1** socket 2		< 1 Ω	Open circuit
12 ←Ω→ **Y1** socket 1		< 1 Ω	Open circuit
Impulse display 10			
Idle speed contact			
2 ←Ω→ 13	Ignition: OFF CIS-E control unit connector and idle speed control unit connector N8 unplugged	< 1 Ω	–
	Activate accelerator pedal	∞	Idle speed contact, wiring
S29/2x pin 1 ←Ω→ **S29/2x** pin 2	Unplug connector S29/x	< 1 Ω	
	Activate accelerator pedal	∞	Adjust or replace throttle valve switch
Wiring			
13 ←Ω→ **N8** socket 4		< 1 Ω	Open circuit
N8 socket 4 ←Ω→ **S29/2x** socket 1		< 1 Ω	Open circuit
S29/2x socket 2 ←Ω→ **W11** ground		< 1 Ω	Open circuit

ON BOARD DIAGNOSTIC CHARTS—116 AND 117 ENGINES

Test connection: Socket box, Multimeter	Test condition	Nominal value	Possible cause/remedy
Impulse display 11			
(not used)	–	–	–
Impulse display 12			
ERG temperature sensor			
7 ←(V)→ 19	Engine at operating temperature, increase engine speed to 4000 rpm for approx. 1 minute	< 4.7 V	EGR temperature sensor, EGR valve, wiring
EGR valve	–	–	EGR test
Wiring			
X80/3 19 ←(Ω)→ pin gn/vi	X80/3 disconnected	< 1 Ω	Open circuit
X80/2 7 ←(Ω)→ pin br/wt	X80/2 disconnected	< 1 Ω	Open circuit If wiring is OK, replace EGR temperature sensor.

No. of blink impulses	Malfunctioning component
1	no system malfunction
2	throttle valve switch
3	coolant temperature sensor
4	air flow sensor position indicator
5	O_2-sensor
6	(not used)
7	TD-signal
8	altitude correction capsule
9	electrohydraulic actuator (EHA)
10	throttle valve switch idle speed contact

Diagnotic codes—102 and 103 engine

No. of blink impulses	Malfunctioning component
1	no system malfunction
2	throttle valve switch
3	coolant temperature sensor
4	air flow sensor position indicator
5	O_2-sensor
6	(not used)
7	TD-signal
8	altitude correction capsule
9	electrohydraulic actuator (EHA)
10	throttle valve switch idle speed contact
11	(not used)
12	exhaust gas recirculation

Diagnotic codes—116 and 117 engine

1. A-1e26 – CHECK ENGINE lamp
2. B-2 – Air flow sensor position indicator
3. B-11/2 – Coolant temperature sensor
4. B-18 – Altitude correction capsule
5. G-3/2 – Heated oxygen sensor
6. G-3/2 × 1 – Oxygen sensor heater connector
7. G-3/2 × 2 – Oxygen sensor signal connector
8. K-1/1 – Overvoltage protection relay
9. N-3 – CIS-E control unit
10. N-16/4 – Fuel pump relay
11. S-27/2 – Deceleration shut off microswitch
12. S-29/2 – Throttle valve switch, full load/idle
13. S-29/2 × – Throttle valve switch connector
14. X-11 – Diagnostic socket
15. X-92 – Test connection for 8-pin impulse readout
16. Y-1 – Electrohydraulic actuator (EHA)
17. Y-6 – Idle speed air valve

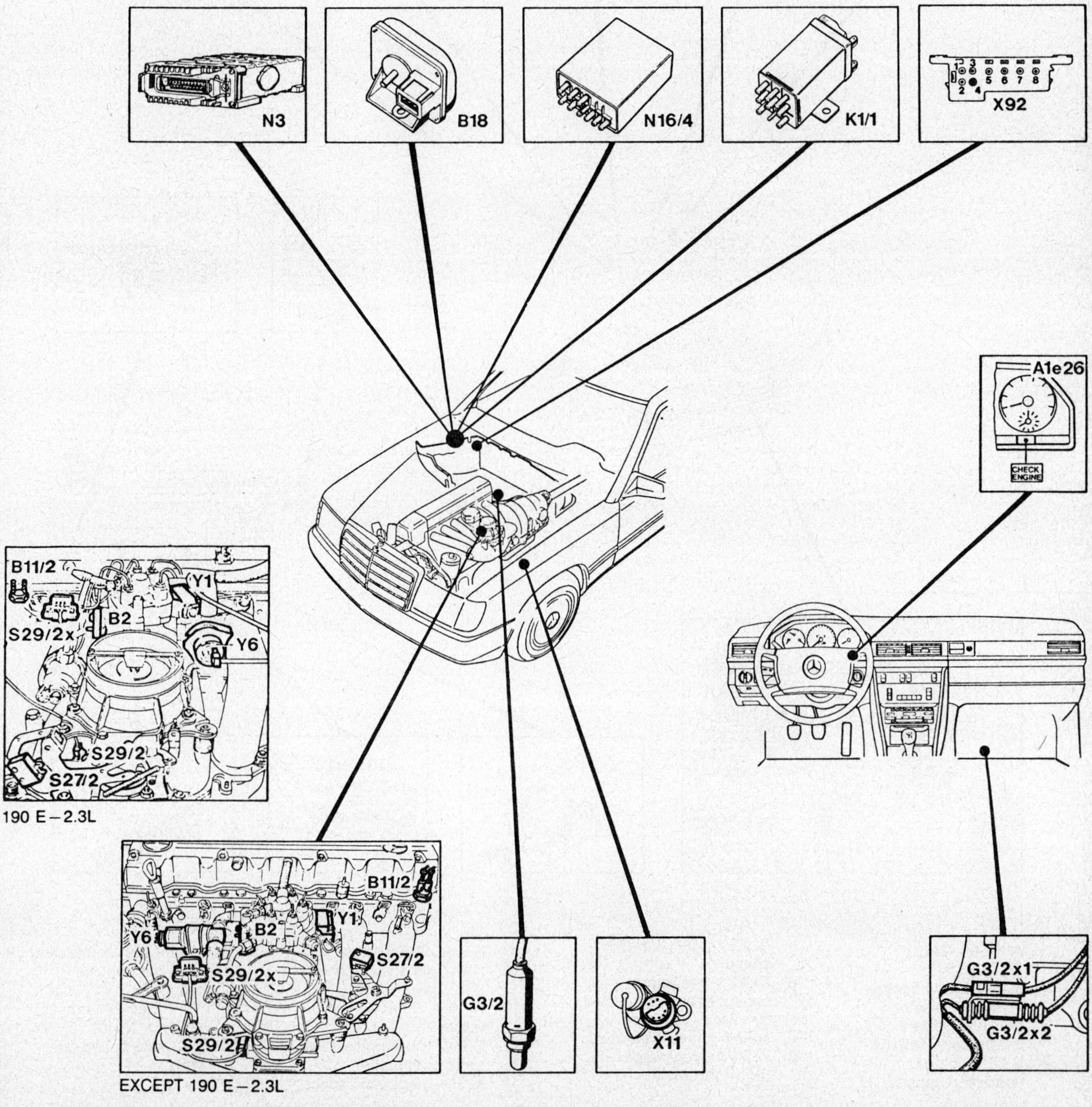

Component locations – 190E, 260E, 300E, 300CE and 300TE

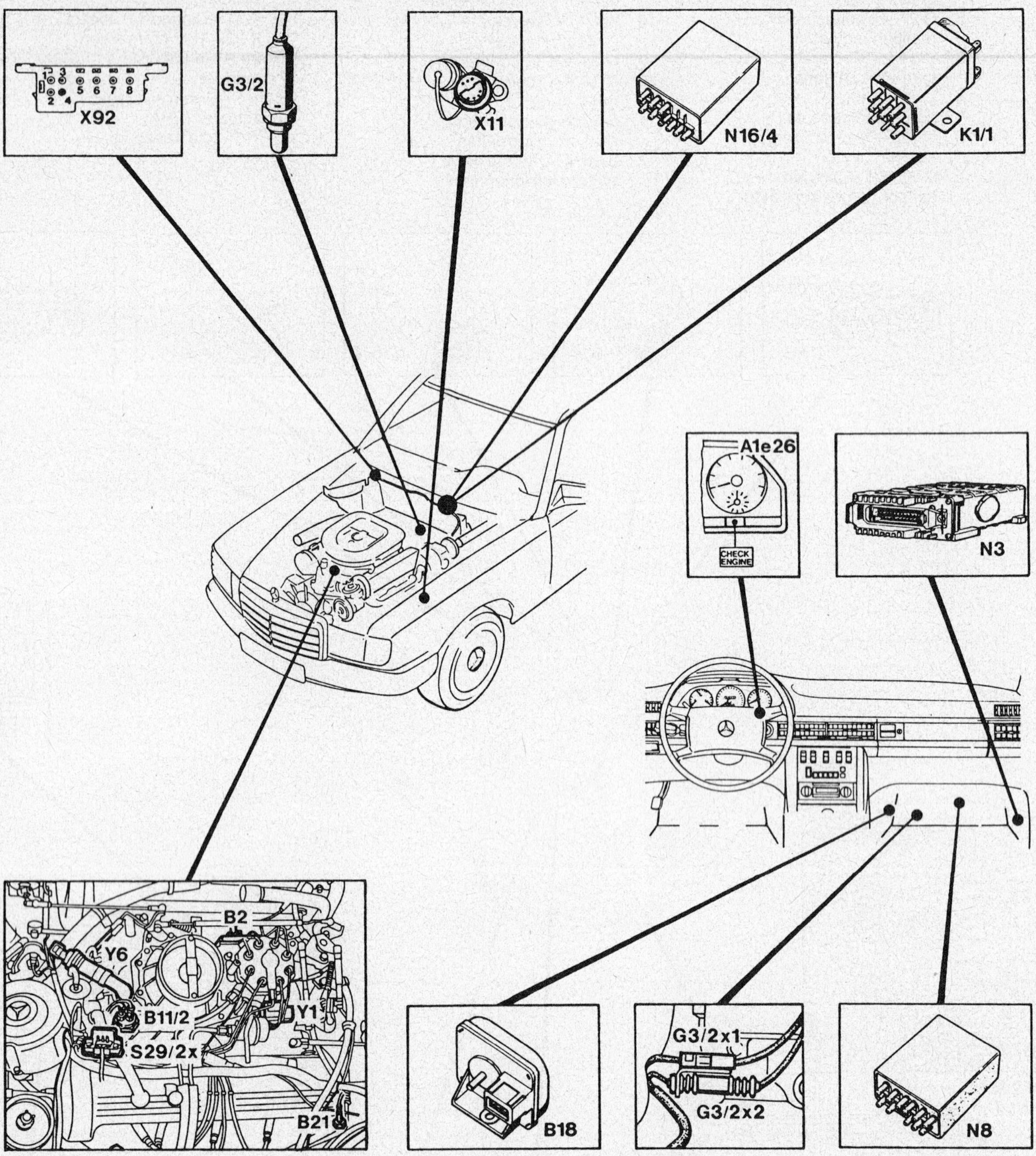

1. A-1e26 – CHECK ENGINE lamp
2. B-2 – Air flow sensor position indicator
3. B-11/2 – Coolant temperature sensor
4. B-18 – Altitude correction capsule
5. G-3/2 – Heated oxygen sensor
6. G-3/2 × 1 – Oxygen sensor heater connector
7. G-3/2 × 2 – Oxygen sensor signal connector
8. K-1/1 – Overvoltage protection relay
9. N-3 – CIS-E control unit
10. N-8 – Idle speed control unit
11. N-16/4 – Fuel pump relay
12. S-29/2 × – Throttle valve switch connector
13. X-11 – Diagnostic socket
14. X-92 – Test connection for 8-pin impulse readout
15. Y-1 – Electrohydraulic actuator (EHA)
16. Y-6 – Idle speed air valve

Component locations – 420SEL, 560SEL, 560SEC

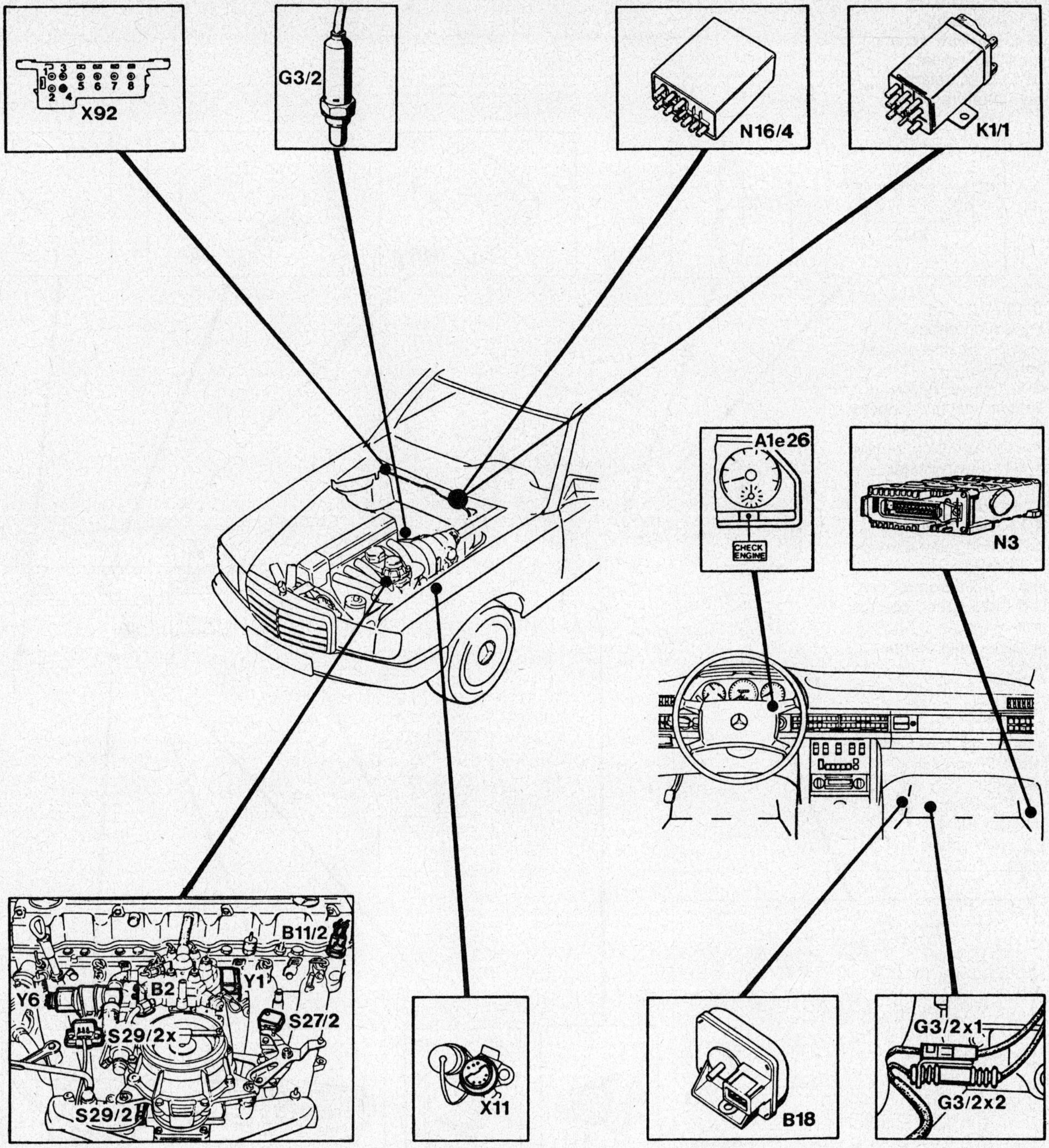

1. A-1e26 – CHECK ENGINE lamp
2. B-2 – Air flow sensor position indicator
3. B-11/2 – Coolant temperature sensor
4. B-18 – Altitude correction capsule
5. G-3/2 – Heated oxygen sensor
6. G-3/2 × 1 – Oxygen sensor heater connector
7. G-3/2 × 2 – Oxygen sensor signal connector
8. K-1/1 – Overvoltage protection relay
9. N-3 – CIS-E control unit
10. N-16/4 – Fuel pump relay
11. S-27/2 – Deceleration shut off microswitch
12. S-29/2 – Throttle valve switch, full load/idle
13. S-29/2 × – Throttle valve switch connector
14. X-11 – Diagnostic socket
15. X-92 – Test connection for 8-pin impulse readout
16. Y-1 – Electrohydraulic actuator (EHA)
17. Y-6 – Idle speed air valve

Component locations – 300SE, 300SEL

1. A-1e26 – CHECK ENGINE lamp
2. B-2 – Air flow sensor position indicator
3. B-11/2 – Coolant temperature sensor

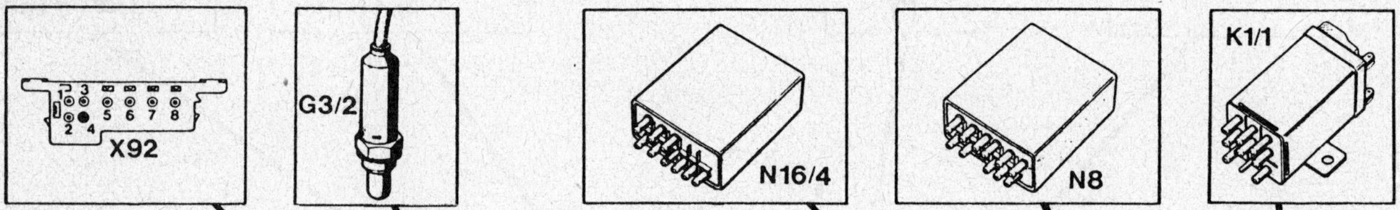

4. B-18 – Altitude correction capsule
5. G-3/2 – Heated oxygen sensor
6. G-3/2 × 1 – Oxygen sensor heater connector
7. G-3/2 × 2 – Oxygen sensor signal connector
8. K-1/1 – Overvoltage protection relay (California)
9. K-1 – Overvoltage protection relay (Except California)
10. N-3 – CIS-E control unit
11. N-8 – Idle speed control unit
12. N-16/4 – Fuel pump relay
13. S-29/2 × – Throttle valve switch connector
14. X-11 – Diagnostic socket
15. X-92 – Test connection for 8-pin impulse readout
16. Y-1 – Electrohydraulic actuator (EHA)
17. Y-6 – Idle speed air valve

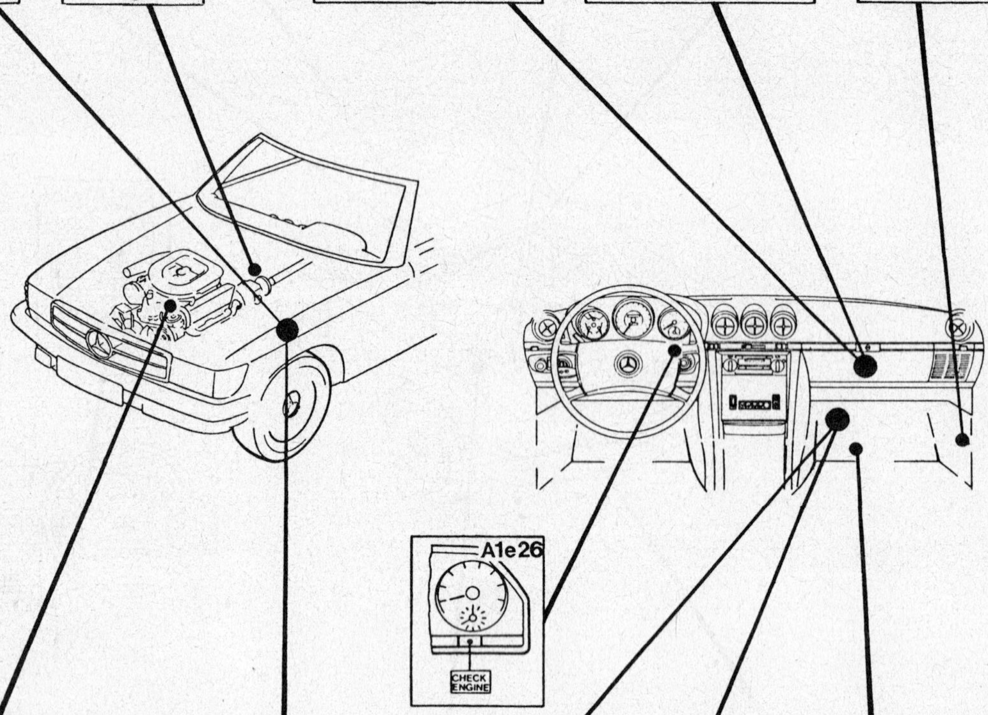

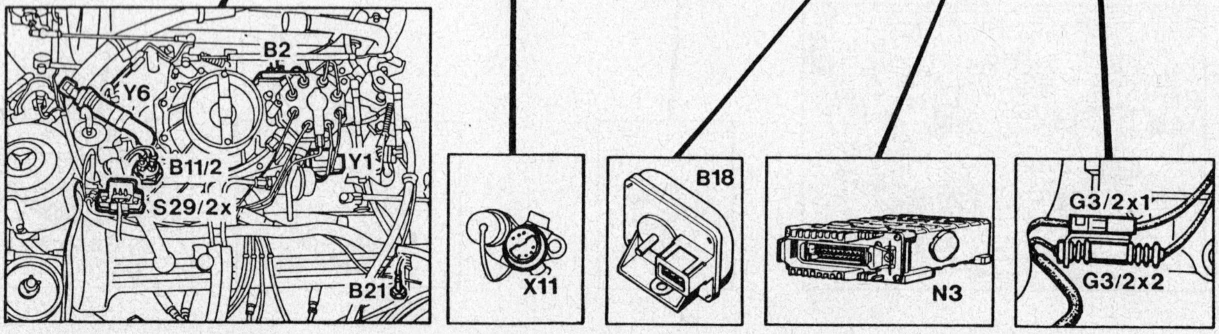

Component locations – 560SL

CIS-E VOLTAGE SUPPLY—102 AND 103 ENGINES

On-off ratio 0 %

Testing voltage supply to CIS-E control unit

Unplug CIS-E control unit. Turn on ignition and check voltage between terminal 1 (+) and terminal 2 (−) of control unit plug, with multimeter set to Volt.

Voltage approx. 12 Volts

OK	Not OK

OK: Check on-off ratio and O_2-sensor signal

Not OK:

Check voltage between terminal 1 (+) of control unit plug and ground (battery).

Voltage approx. 12 Volts

OK	Not OK

Not OK: Check fuse on overvoltage protection relay. Check wire connections and wire routing according to wiring diagram.

OK:

Check wire from terminal 2 of control unit plug to ground (W 11) for continuity, with multimeter set to Ohm.

Resistance approx. 0 Ohm

OK	Not OK

Not OK: Repair open circuit according to wiring diagram.

OK:

Check wire between diagnostic socket, terminal 3 and CIS-E control unit plug, terminal 23 for continuity.

Resistance approx. 0 Ohm

OK	Not OK

Not OK: Repair open circuit according to wiring diagram.

OK:

Disconnect O_2-sensor plug and connect male part of plug to ground. Start engine, test sensor voltage between female part of plug and ground, with multimeter set to Volt.

Sensor voltage between 450 and 1000 mV.

OK	Not OK

Not OK: Replace O_2-sensor

OK: End of test

AIR FLOW SENSOR POSITION INDICATOR—102 AND 103 ENGINES

On-off ratio 10 %

Testing air flow sensor position indicator

Loosen plug of air flow sensor position indicator so that the voltage can be measured at the terminals (do not disconnect plug). Start engine. Use multimeter set to Volt.

Terminals 1 to 3 = 4.6 to 5.1 Volts
Check wire routing.

OK | Not OK

Not OK: Repair open circuit according to wiring diagram.

OK: Measure voltage between terminal 1 and terminal 2 with engine at idle speed.

Terminal 1 to 2 = 0.57 – 0.81 Volts

OK | Not OK

Not OK: If voltage is higher or lower, replace air flow sensor.

Not OK: Voltage 0 Volts
Stop engine. Check wires between air flow sensor position indicator and CIS-E control unit plug for continuity. Check wire routing according to wiring diagram with multimeter set to Ohm.

Resistance approx. 0 Ohm

OK | Not OK

Not OK: Repair open circuit according to wiring diagram.

Stop engine. Disconnect plug (X56) of throttle valve switch. Measure resistance between terminal 1 and terminal 2.

Idle contact: Idle position = approx. 0 Ohm
Full load = ∞ Ohm
Partial load = ∞ Ohm

OK | Not OK

Not OK: Adjust or replace throttle valve switch

OK: Check continuity of wires from throttle valve switch to CIS-E control unit plug and to engine ground.

Resistance approx. 0 Ohm

OK | Not OK

Not OK: Repair open circuit according to wiring diagram.

OK: Start engine. Run at idle speed (engine oil temperature 60–80° C). Check on-off ratio at diagnostic socket.

Tester needle oscillates.

OK | Not OK

Not OK: For different fixed on-off ratio see troubleshooting chart

OK: End of test

FULL LOAD SWITCH—102 AND 103 ENGINES

On-off ratio 20 %

Testing full load contact.

Stop engine. Disconnect plug (X) of throttle valve switch. Measure resistance between terminal 3 and terminal 2, with multimeter set to Ohm.

Full load contact: Idle position = ∞ Ohm
Full load = approx. 0 Ohm
Partial load = ∞ Ohm

OK	Not OK

Not OK: Adjust or replace throttle valve switch

OK:

Check continuity of wires from throttle valve switch to CIS-E control unit plug and to engine ground.

Resistance approx. 0 Ohm

OK	Not OK

Not OK: Repair open circuit according to wiring diagram.

OK: End of test

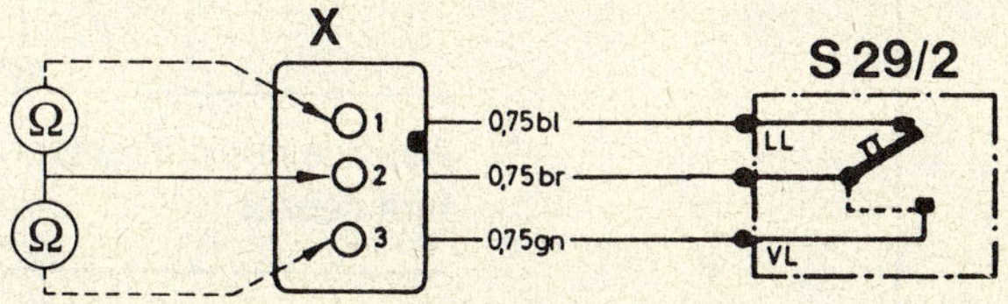

X Throttle valve switch plug (idle and full load contacts)
1 Idle contact (LL)
2 Common
3 Full-load contact (VL)

COOLANT TEMPERATURE SENSOR—102 AND 103 ENGINES

On-off ratio 30 %

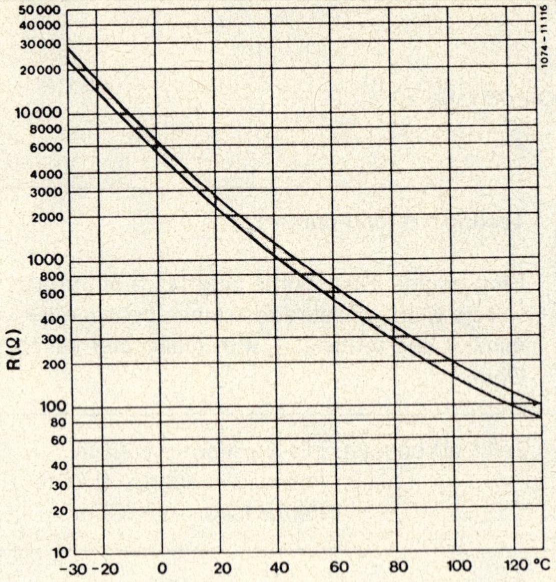

Testing coolant temperature sensor

Unplug coolant temperature sensor (green/red wire). Test resistance from sensor terminal to ground (idle speed OK with connection to ground or with open circuit), with multimeter set to Ohm.

For nominal value, refer to diagram.
Test sensor resistance at two temperatures.

Example:
+20 °C between 2.2 and 2.8 kOhm
+80 °C between 290 and 370 Ohm

OK	Not OK

Not OK → Replace coolant temperature sensor

End of test

OK →

Check wire (green/red) between terminal 21 of CIS-E control unit plug to plug of coolant temperature sensor for continuity.

Resistance approx. 0 Ohm

OK	Not OK

Not OK → Repair open circuit according to wiring diagram.

OK → End of test

ACCELERATION ENRICHMENT CIRCUIT—102 AND 103 ENGINES

On-off ratio 40 %

Testing acceleration enrichment or air flow sensor position indicator

Connect test cable (special tool) to EHA and multimeter (amperage socket). Set multimeter to A. Unplug coolant temperature sensor (green/red wire) and ground the plug with 2.5 kOhm resistance substitution unit (special tool) (simulate +20°C).

Disconnect O_2-sensor. Start engine (increased idle speed).

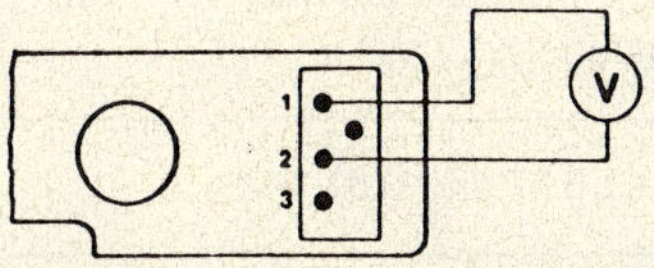

Electric current at EHA 2–6 mA (following elimination of after-start enrichment).
Revving engine abruptly should increase current.

OK	Not OK

OK:

Reconnect coolant temperature sensor and O_2-sensor.

End of test

Not OK:

Loosen plug of air flow sensor position indicator so that the voltage can be measured at the terminals (do not disconnect plug). Measure voltage between terminal 1 and 3 at engine idle speed, with multimeter set to Volt.

Voltage between 4.6 and 5.1 Volts

OK	Not OK

Not OK:

Repair open circuit according to wiring diagram.

End of test

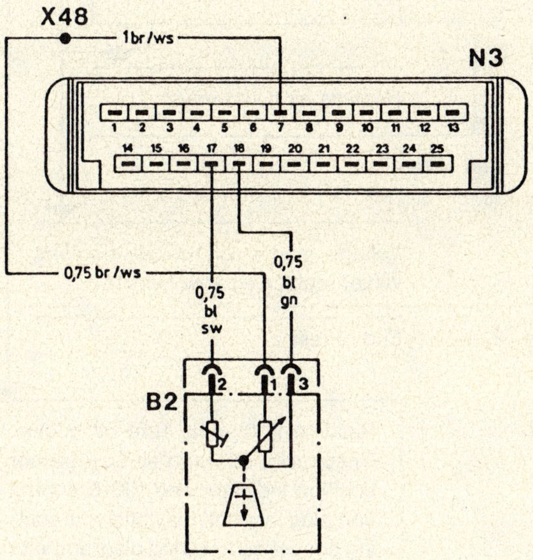

B2 Air flow sensor position indicator
N3 CIS-E control unit
X48 Connector sleeve (solder joint in harness)

ACCELERATION ENRICHMENT CIRCUIT—102 AND 103 ENGINES

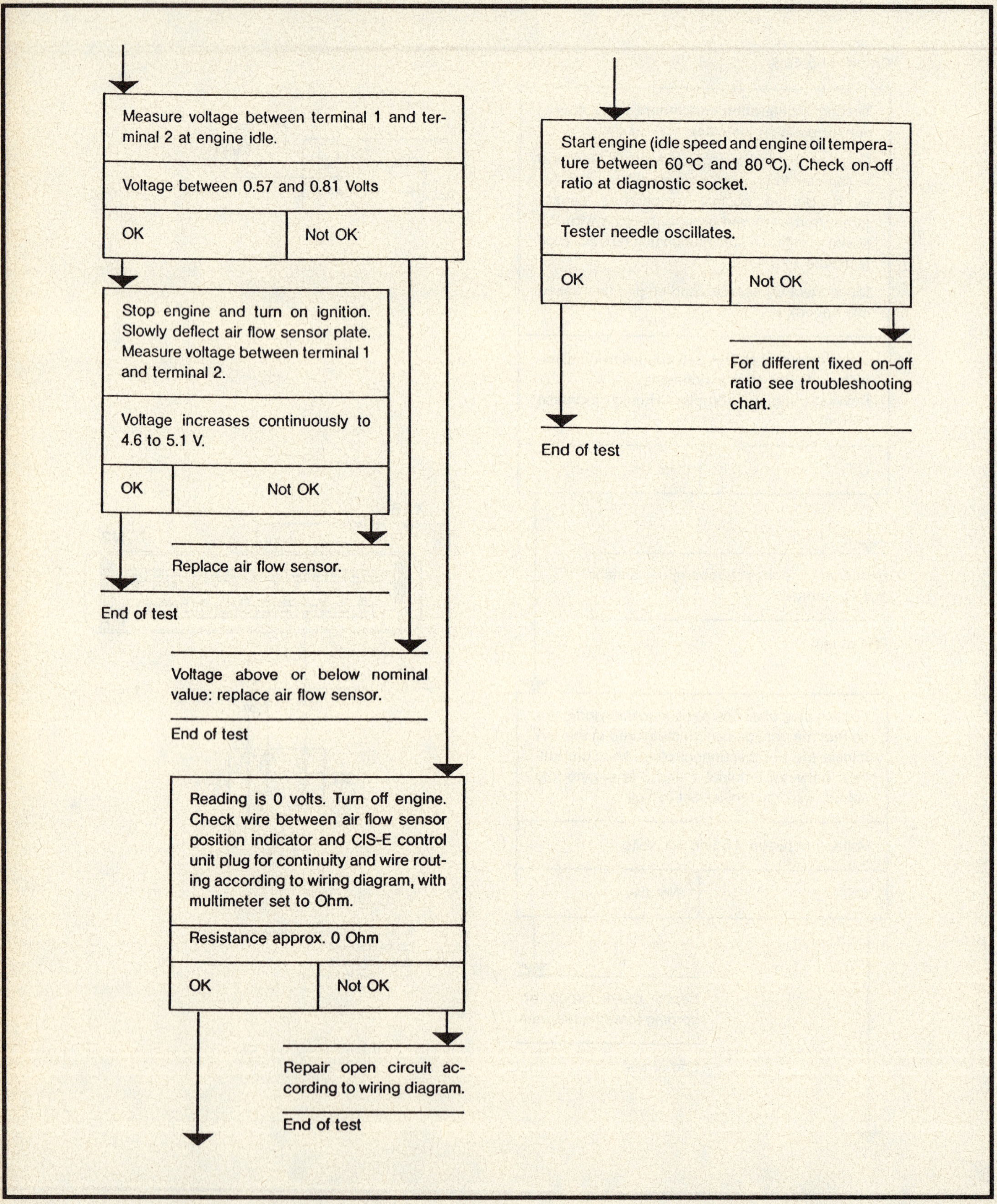

OXYGEN SENSOR CIRCUIT—102 AND 103 ENGINES

On-off ratio 50 %

Note: If the O_2-sensor signal is not present, the O_2-sensor malfunction indicator in the instrument cluster should light up within 5 minutes after starting engine.

Testing O_2-sensor

Disconnect O_2-sensor plug and connect male part of plug to ground. Start engine. Test sensor voltage between female part of plug and ground, with multimeter set to Volt.

Sensor voltage between 450 and 1000 mV

OK	Not OK
↓	Replace O_2-sensor.

Stop engine. Check wire from O_2-sensor plug to CIS-E control unit plug, terminal 8, for continuity, with multimeter set to Ohm.

Resistance approx. 0 Ohm

OK	Not OK
↓	Repair open circuit according to wiring diagram.

Disconnect plug of O_2-sensor heater. Pull off fuel pump relay and bridge terminals 7 and 8 of relay plug. Check voltage at O_2-sensor heater plug, with multimeter set to Volt.

Reading approx. 12 Volts (battery voltage).

OK	Not OK
↓	Repair open circuit according to wiring diagram.

Connect test cables 102 589 04 63 00 and 903 589 03 63 00 between heater plug, with multimeter set to A.
Measure current draw.

Nominal value between 0.5 and 1.3 A.

OK	Not OK
End of test	Replace O_2-sensor.

TD SIGNAL—102 AND 103 ENGINES

On-off ratio 70 %

Testing TD-signal

Disconnect CIS-E control unit plug. Start engine. Check voltage between terminal 25 (control unit plug) and ground, with multimeter set to Volt.

Voltage between 6 and 12 V

OK → End of test

Not OK ↓

Unplug fuel pump relay.
Check wire between terminal 10 (relay plug) and terminal 25 on CIS-E control unit plug for continuity, with multimeter set to Ohm.

Resistance approx. 0 Ohm

OK → Check wires to ignition control module for continuity.

Not OK → Repair open circuit according to wiring diagram.

End of test

ALTITUDE CORRECTION CAPSULE 102 AND 103 ENGINES

On-off ratio 80 %

Testing altitude correction capsule (B 18)

Disconnect plug on altitude correction capsule. Engine off. Ignition on. Check voltage between terminal 2 and terminal 3 of capsule plug, with multimeter set to Volt.

Voltage approx. 5 V

OK ↓

Not OK → Repair open circuit according to wiring diagram.

Turn off ignition. Check wire between altitude correction capsule plug, terminal 1, to CIS-E control unit plug, terminal 11, for continuity, with multimeter set to Ohm.

Resistance approx. 0 Ohm

OK ↓

Not OK → Repair open circuit according to wiring diagram.

Start engine and run at idle speed and at operating temperature. (60–80 °C engine oil temperature). Check on-off ratio at diagnostic socket.

Tester needle oscillates.

OK → End of test

Not OK → Replace altitude correction capsule.

End of test

ON-OFF RATIO—102 AND 103 ENGINES

On-off ratio 100 %

Note:
If on-off ratio is 100 % and idle speed is too high, check fuse on overvoltage protection relay and/or power supply and ground connection

Testing on-off ratio

On-off ratio can be adjusted, on-off ratio tester needle oscillates.

OK	Not OK

OK: End of test

Not OK:

Stop engine. Disconnect O_2-sensor plug and connect male part of plug to ground. Start engine and test sensor voltage between female part of plug and ground, with multimeter set to Volt.

Sensor voltage between 450 and 1000 mV.

OK	Not OK

Not OK: Replace O_2-sensor.

OK:

Stop engine. Check wire from O_2-sensor plug to CIS-E control unit plug, terminal 8, for continuity, with multimeter set to Ohm.

Resistance approx. 0 Ohm

OK	Not OK

Not OK: Repair open circuit according to wiring diagram.

OK:

Disconnect plug of O_2-sensor heater. Pull off fuel pump relay and bridge terminals 7 and 8 of relay plug. Check voltage at O_2-sensor heater plug, with multimeter set to Volt.

Reading approx. 12 Volts (battery voltage).

OK	Not OK

Not OK: Repair open circuit according to wiring diagram.

OK:

Connect test cables 102 589 04 63 00 and 903 589 03 63 00 between heater plug, with multimeter set to A. Measure current draw.

Nominal value between 0.5 and 1.3 A.

OK	Not OK

Not OK: Replace O_2-sensor.

OK: End of test

LAMBDA CONTROL—102 AND 103 ENGINES

C. Testing lambda control

Engine oil temperature 60°–80°C.
Pull off purge line to throttle valve housing at purge valve and plug line. Connect on-off ratio tester to diagnostic socket. Press 100% IR button.

Test on-off ratio at 2500 rpm and record the mean value. It should not exceed the idle speed mean value by more than ± 10%.

OK	Not OK

OK → End of test

Not OK →

Readout is constant. (No needle oscillation)	Needle oscillates by more than ± 10%

Readout is constant → For on-off ratio see troubleshooting chart. → End of test

Needle oscillates → Test or adjust lambda control with a lambda control tester. If no lambda control tester is available, a dwell angle tester may be used. → End of test

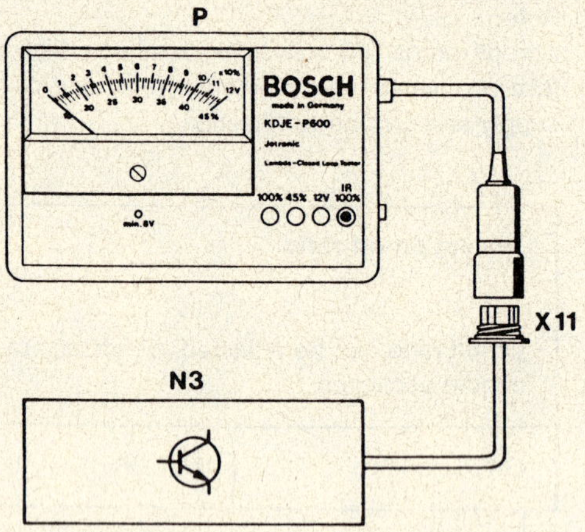

N3 CIS-E control unit
P On-off ratio tester
X11 Diagnostic socket

ELECTRONIC IDLE SPEED CONTROL—102 AND 103 ENGINES

D. Testing electronic idle speed control

Connect test cable 103 589 00 63 00 to idle speed air valve (Y6) and to on-off ratio tester 909 589 00 21 00 (Bosch KDJE-P600). Push IR 100 % button. Engine at idle (operating temperature).

Nominal value: 36 to 50 % at 700 ± 50 rpm.

OK → End of test

Not OK →

Note: Readout between 50 and 100 % = Check for air leak or Check for air restriction

Readout 0 % or 100 %.

Stop engine. Disconnect plug from idle speed air valve. Briefly apply battery voltage to idle speed air valve.

The idle speed air valve switches audibly. This can be felt by hand.

OK →

Not OK → Replace idle speed air valve.

End of test

Test voltage supply according to wiring diagram.

Switch on ignition. Voltage on terminal 2 of idle speed air valve plug (model 124 red/yellow wire, model 201 red/blue wire) approx. 12 volts. Test red/white wire for continuity between terminal 1 of idle speed air valve plug and terminal 3 of CIS-E control unit plug.

OK → Test input signals: Coolant temperature sensor, air flow sensor position indicator, idle speed contact on throttle valve switch (see troubleshooting chart or replace CIS-E control unit.

Not OK → Repair open circuit according to wiring diagram.

Engine idle speed stabilization check.
Perform tests in both columns.
Connect engine analyzer (RPM).

Automatic transmission
Engine at idle speed and operating temperature, place selector lever in a driving position. A/C compressor off.

Nominal value 550 to 650 rpm

OK → End of test

Not OK →

A/C compressor
Engine at idle speed and operating temperature. Turn on A/C compressor. Place selector lever in position "P" or "N".

Nominal value 650 to 750 rpm

OK → End of test

Not OK →

ELECTRONIC IDLE SPEED CONTROL 102 AND 103 ENGINES

Turn off engine. Test input signal on CIS-E control unit plug. Measure voltage between termin. 16 and battery (+), with multimeter set to Volt.

Position P and N: battery voltage. With driving position engaged voltage drops by approx. 1 V.

OK	Not OK

Repair open circuit according to wiring diagram.

Turn off engine. Test input signal on CIS-E control unit plug. Measure voltage between termin. 19 and ground, with multimeter set to Volt.

Engine at idle speed (idle speed rpm drop off). Compressor on: approx. 12 volts Compressor off: approx. 0 volt

OK	Not OK

Repair open circuit according to wiring diagram.

End of test

STARTING VALVE—102 AND 103 ENGINES

E. Testing starting valve activation

Ignition turned off. Pull plug off coolant temperature sensor. Connect coolant temperature sensor plug to resistance substitution unit to simulate 10 kOhm. Unplug green cable on ignition control unit. Connect multimeter, set to Volt, to plug of starting valve. Crank engine until voltage drop occurs.

Minimum voltage 10 V for approx. 5–10 sec.

OK	Not OK

Check wire from starting valve to fuel pump relay plug, terminal 4 (relay terminal 87 V) and ground wire to engine ground for continuity, with multimeter set to Ohm.

Resistance approx. 0 Ohm

OK	Not OK

Repair open circuit according to wiring diagram.

Check voltage between fuel pump relay plug terminal 12 (circuit 50) as well as terminal 2 (circuit TF) and engine ground, with multimeter set to Volt. Crank engine.

Minimum voltage at terminal 12 = 10 V. Voltage at terminal 2 = 3.0 to 5 V. (Coolant temperature sensor plug disconnected).

OK	Not OK

Replace fuel pump relay

Repair open circuit according to wiring diagram.

End of test

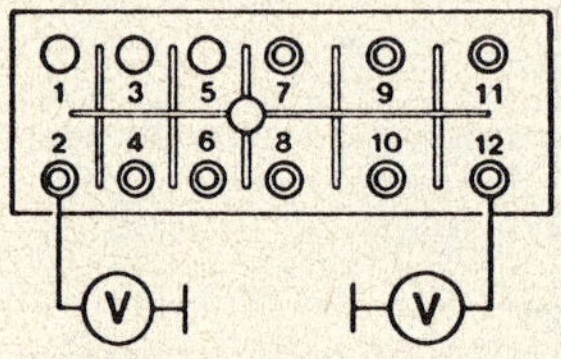

TESTING DECELERATION SHUTOFF—190E—102 AND 103 ENGINES

F. Testing deceleration shut-off

Model 201.029

Connect on-off ratio tester to diagnostic socket. Engine at idle speed, engine oil temperature between 60 °C and 80 °C.

Tester needle oscillates.

OK	Not OK

Not OK: For constant fixed on-off ratio see troubleshooting chart.

OK:

Operate vehicle on dynamometer or open road. Engine rpm above 2500. Shift selector lever in position 2 or in second gear.

When vehicle is decelerated (throttle valve closed) the on-off ratio jumps to approx. 50 % and will restart fuel injection (tester needle oscillates) below 1300 rpm.

OK	Not OK

Not OK: Check idle speed contact on throttle valve switch

Testing speed signal (speedometer). Check wire between idle speed control unit plug and speedometer for continuity according to wiring diagram, with multimeter set to Ohm.

Resistance approx. 0 Ohm

OK	Not OK

Not OK: Repair open circuit according to wiring diagram.

OK: End of test

TESTING DECELERATION SHUTOFF—260E AND 300 SERIES—102 AND 103 ENGINES

Model 124

Connect on-off ratio tester to diagnostic socket. Engine at idle speed, engine oil temperature between 60° C and 80° C.

Tester needle oscillates

OK	Not OK

Not OK: For constant fixed on-off ratio see troubleshooting chart.

OK:

Disconnect throttle valve switch and bridge terminals 1 and 2 of plug.

Engine at operating temperature. Increase engine speed to approx. 2500 rpm. Keep engine speed constant and operate microswitch (S 27/2) manually.

Engine surges. On-off ratio 50 % (constant)

Yes	No

Yes: Check adjustment of slotted lever.
Check throttle valve switch

End of test

No:

Test resistance of microswitch, with multimeter set to Ohm.

In idle speed position approx. 0 Ohm. With free travel on lever cancelled ∞ Ohm.

Yes	No

No: Replace microswitch.

Yes:

With engine at idle speed, test rpm signal (TD) on control unit plug (terminal 25, green/yellow wire). Set multimeter to Volt.

Voltage between 6 and 12 Volts

OK	Not OK

OK: Replace control unit

Not OK: Repair open circuit according to wiring diagram.

End of test

MICROSWITCH FUNCTIONS—102 AND 103 ENGINES

G. Testing microswitch functions, idle and full load contacts

Connect on-off ratio tester to diagnostic socket. Push IR 100 % button. Turn on ignition.

On-off ratio = 70 %

- OK → next step
- Not OK → Readout 10 %: Check idle speed contact of throttle valve switch

Deflect air flow sensor plate.

On-off ratio = 10 %

- OK → next step
- Not OK → Readout 70 %: Throttle valve does not close or open circuit at idle speed contact or microswitch defective.

Close air flow sensor plate (0 position).

On-off ratio = 70 %

Open throttle valve slightly while depressing microswitch.

On-off ratio = 10 %

- OK → next step
- Not OK → Microswitch defective or open circuit.

Open throttle valve fully

On-off ratio = 20 %

- OK → End of test
- Not OK → Reading 70 %: Full load contact defective

End of test

OVERVOLTAGE PROTECTION RELAY 102 AND 103 ENGINES

H. Testing overvoltage protection relay

Check fuse on overvoltage protection relay.

- OK → next step
- Not OK → Replace fuse.

Disconnect overvoltage protection relay plug.
Bridge terminals 1 and 2 of the overvoltage protection relay plug
Checkvoltagebetweenterminal1ofCIS-Econtrol unit plug and ground, with multimeter set to Volt.

Voltage approx. 12 Volts

- OK → next step
- Not OK → Check power supply according to wiring diagram.

Turn on ignition.
Check voltage on overvoltage protection relay plug.

terminals 5 and 3

Voltage approx. 12 Volts

- OK → Replace overvoltage protection relay. → End of test
- Not OK → Repair open circuit according to wiring diagram. → End of test

End of test

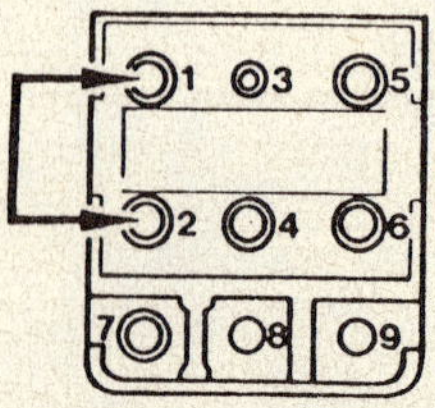

THROTTLE VALVE SWITCH 102 AND 103 ENGINES

J. Testing throttle valve switch

Disconnect plug between throttle valve switch and CIS-E control unit. Test resistance of throttle valve switch, with multimeter set to Ohm.

Idle speed contact:
- Idle speed position = 0 Ohm
- Full load position = approx. ∞ Ohm

Full load contact:
- Idle speed position = ∞ Ohm
- Full load position = approx. 0 Ohm
- Partial load position = ∞ Ohm

- OK → next step
- Not OK → Adjust or replace throttle valve switch or correct wrong wire connection.

Check continuity of wires from throttle valve switch plug to CIS-E control unit plug and to engine ground.

Resistance approx. 0 Ohm

- OK → End of test
- Not OK → Repair open circuit according to wiring diagram.

End of test

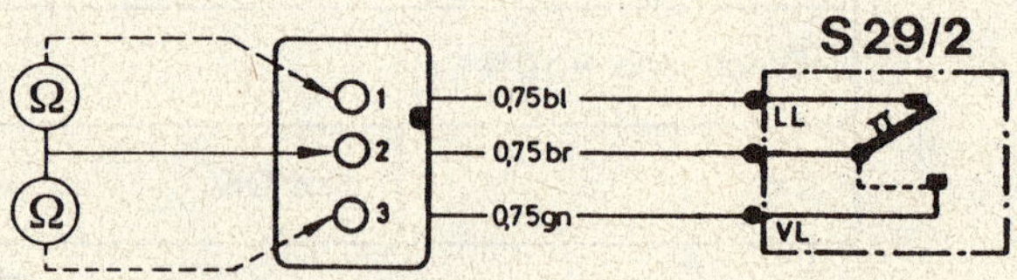

S 29/2 Throttle valve switch, full load/idle
Connector, throttle valve switch
1 Idle contact (LL)
2 Common
3 Full-load contact (VL)

HALL EFFECT SPEED SENSOR 102 AND 103 ENGINES

Test Hall-effect speed sensor

Ignition turned off. Disconnect CIS-E control unit plug. Ignition on. Measure voltage on CIS-E control unit plug between terminal 1 and 6, with multimeter set to Volt.

Move vehicle approx. 1 m (3 ft.).

Reading cycles between 0 and 12 Volts.

OK	Not OK

Not OK → Repair open circuit according to wiring diagram. Replace Hall-effect speed sensor if necessary.

OK →

Connect on-off ratio tester to diagnostic socket. Operate vehicle in second gear/selector lever position 2 on either dynamometer or open road. From 2,000 rpm accelerate at full throttle for approx. 6 seconds.

Tester needle oscillates

OK	On-off ratio is 60 % constant

On-off ratio is 60 % constant → Replace CIS-E control unit

OK → End of test

AFTER START ENRICHMENT 102 AND 103 ENGINES

Testing after-start enrichment

Connect test cable to EHA and multimeter. Set multimeter to A. Simulate +20 °C coolant temperature by using the resistance substitution unit (at 2.5 kOhm). Unplug O_2-sensor. Start engine. (increased idle speed).

Current increases from 3 to 9 mA. Value remains constant for the first 20 seconds of engine running at increased idle speed, then decreases gradually to engine warmup value between 2 and 6 mA.

OK	Not OK

OK → End of test

Not OK →

Test TD sensor signal at CIS-E control unit plug, terminal 25 (green/yellow), with multimeter set to Volt.

Voltage between 6 and 12 Volts.

OK	Not OK

Not OK → Repair open circuit according to wiring diagram.

OK →

Test wiring between CIS-E control unit plug and EHA plug for continuity, with multimeter set to Ohm (see diagram for wire colors and terminal numbers).

Resistance approx. 0 Ohm

OK	Not OK

OK → Replace CIS-E control unit

Not OK → Repair open circuit according to wiring diagram.

End of test

FUEL PRESSURE SYSTEM—102 AND 103 ENGINES

This test program is divided into the following sections:

- Testing fuel pressures and for internal leaks.
- Testing starting valve operation and for leaks.
- Testing after-start enrichment.

Testing fuel pressures and for internal leaks

Check lower chamber pressure at +20°C coolant temperature.

Nominal value 4.8 to 5.0 bar.

OK	Not OK

Not OK: Test coolant temperature sensor (see troubleshooting chart, on-off ratio 30%).

Testing starting valve operation and for leaks

Disconnect fuel line from starting valve. Remove starting valve and reconnect fuel line.

Place starting valve in a container.

Testing operation

Simulate 10 kOhm resistance at coolant temperature sensor, using the resistance substitution unit.
Disconnect green wire on electronic ignition control unit.
Crank engine while observing the starting valve eject fuel.

OK	Not OK

Not OK: Test current flow to starting valve or replace starting valve.

Testing for leaks

Wipe starting valve nozzle dry.

Observe forming of fuel drop.

Drop is formed.	No drop is formed.

Drop is formed: Replace starting valve

End of test

CIS-E VOLTAGE SUPPLY—116 AND 117 ENGINES

On-off ratio 0%

Testing voltage supply to CIS-E control unit

Unplug CIS-E control unit. Turn on ignition and check voltage between terminal 1 (+) and terminal 2 (–).

Voltage approx. 12 Volts

OK	Not OK

OK: Check on-off ratio and O_2-sensor signal (see section B) and adjust if necessary

Not OK:

Check voltage between terminal 1 (+) and ground (battery).

Voltage approx. 12 Volts

OK	Not OK

Not OK: Check fuse on overvoltage protection relay Check wire connections and wire routing according to wiring diagram.

OK:

Check wire from terminal 2 to ground (W15, model 107 or W10, model 126) for continuity.

Resistance approx. 0 Ω

OK	Not OK

Not OK: Repair interruption

OK:

Check wire between diagnostic socket, terminal 3 and CIS-E control unit plug, terminal 23 for continuity.

Resistance approx. 0 Ω

OK	Not OK

Not OK: Repair interruption

OK:

Disconnect O_2-sensor plug and connect male part of plug to ground. Start engine, test sensor voltage between female part of plug and ground

Sensor voltage above 450 mV.

OK	Not OK

Not OK: Replace O_2-sensor

OK: End of test

AIR FLOW SENSOR POSITION INDICATOR—116 AND 117 ENGINES

On-off ratio 10%

Testing air flow sensor position indicator

Loosen plug of air flow sensor position indicator so that the voltage can be measured at the pins (do not disconnect plug). Start engine.

Terminals 1 to 3 = 4.5 to 5.5 Volts
Terminal 1 to 2 = 0.5 to 1.5 Volts

OK | Not OK

Stop engine and turn on ignition. Slowly deflect air flow sensor plate. Measure voltage between terminal 1 and terminal 2.

Voltage increases continuously to 4.5 to 5.5 Volts.

OK | Not OK

Replace air flow sensor

End of test

Stop engine. Check wires between air flow sensor position indicator and CIS-E control unit plug for continuity. Check wire routing according to wiring diagram.

Resistance approx. 0 Ω

OK | Not OK

Repair interruptions

Start engine. Run at idle speed (engine oil temperature 75–85°C). Check on-off ratio at diagnostic socket.

Tester needle oscillates.

OK | Not OK

For different fixed on-off ratio see test program

End of test

FULL LOAD SWITCH—116 AND 117 ENGINES

On-Off ratio 20%

Testing full load contact and air flow sensor position indicator.

Disconnect plug (X) of throttle valve switch. Measure resistance between terminal 3 and terminal 2.

Full load contact: Idle position = approx. ∞ Ω
Full load = approx. 0 Ω
Partial load = approx. ∞ Ω

OK	Not OK

Not OK → Adjust or replace throttle valve switch

OK →

Check continuity of wires from throttle valve switch to CIS-E control unit and to engine ground.

Resistance approx. 0 Ω

OK	Not OK

Not OK → Repair interruption → End of test

OK → Check air flow sensor position indicator (see on-off ratio 40%) → End of test

End of test

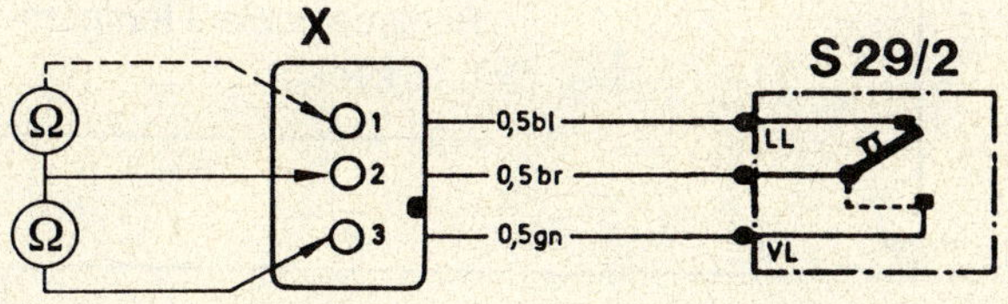

X Throttle valve switch plug (idle and full load contacts)
1 Idle contact (LL)
2 Common
3 Full-load contact (VL)

COOLANT TEMPERATURE SENSOR—116 AND 117 ENGINES

On-off ratio 30%

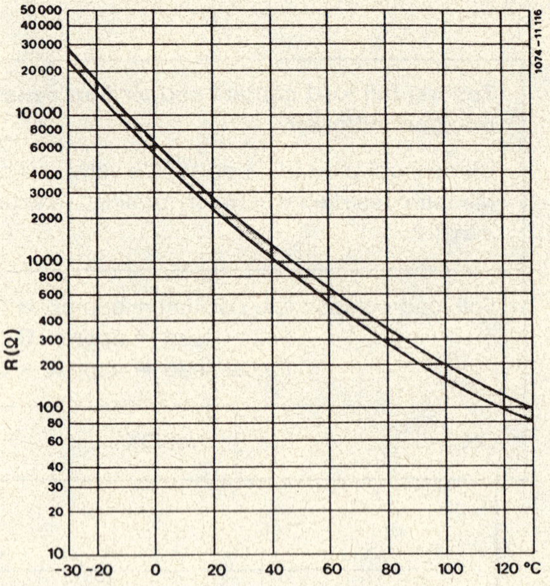

Testing coolant temperature sensor

Unplug coolant temperature sensor. Test resistance from sensor terminal to ground (idle speed OK with connection to ground or too high with interruption).

For nominal value, refer to diagram.
Test sensor resistance at two temperatures

Example:
+20 °C = 2.2–2.8 kΩ
+80 °C = 290–370 Ω

OK	Not OK

Not OK → Replace coolant temperature sensor

OK ↓

Check wire (green/red) between terminal 21 of CIS-E control unit plug to plug of coolant temperature sensor for continuity.

Resistance approx. 0 Ω

OK	Not OK

Not OK → Repair interruption

OK ↓

End of test

AIR FLOW SENSOR POSITION INDICATOR 116 AND 117 ENGINES

On-Off ratio 40%

Testing air flow sensor position indicator. Loosen plug of air flow sensor position indicator so that the voltage can be measured at the pins (do not disconnect plug). Start engine.

Terminals 1 to 3 = 4.5 to 5.5 Volts
Terminals 1 to 2 = 0.5 to 1.5 Volts

OK	Not OK

Stop engine and turn on ignition. Slowly deflect air flow sensor plate. Measure voltage between terminal 1 and terminal 2.

Voltage increases continuously to 4.5 to 5.5 V.

OK	Not OK

Replace air flow sensor

End of test

Stop engine. Check wires between air flow sensor position indicator and CIS-E control unit plug for continuity. Check wire routing according to wiring diagram.

Resistance approx. 0 Ω

OK	Not OK

Repair interruption

Start engine. Run at idle speed (engine oil temperature 75–85 °C). Check on-off ratio at diagnostic socket.

Tester needle oscillates.

OK	Not OK

For different fixed on-off ratio see Test program or replace CIS-E control unit if necessary

End of test

OXYGEN SENSOR 116 AND 117 ENGINES

On off ratio 50%

Note:
If the O_2-sensor signal is not present, the O_2-sensor malfunction indicator in the instrument cluster will light up (see test section "H").

Testing O_2-sensor

Disconnect O_2-sensor plug and connect male part of plug to ground. Start engine, test sensor voltage between female part of plug and ground.

Sensor voltage above 450 mV.

OK	Not OK

Replace O_2-sensor

Stop engine. Check wire from O_2-sensor plug to CIS-E control unit, terminal 8, for continuity.

Resistance approx. 0 Ω

Repair interruption

End of test

TD SIGNAL—116 AND 117 ENGINES

On-off ratio 70%

Testing TD-signal

Disconnect CIS-E control unit plug. Start engine. Check voltage between terminal 25 (control unit plug) and ground.

Voltage 6 to 12 V

- OK → End of test
- Not OK →

Unplug fuel pump relay.
Check wire between terminal 10 (relay plug) and terminal 25 on CIS-E control unit plug for continuity.

Resistance approx. 0 Ω

- OK → Check wires to ignition control module for continuity. → End of test
- Not OK → Repair interruption

End of test

ALTITUDE CORRECTION CAPSULE 116 AND 117 ENGINES

On-off ratio 80%

Testing altitude correction capsule (B 18).

Disconnect plug on altitude correction capsule. Turn on ignition. Check voltage between terminal 2 and terminal 3.

Voltage approx. 5 Volt

- OK →
- Not OK → Check power supply and ground according to wiring diagram.

Turn off ignition. Check wire between altitude correction capsule plug, terminal 1, to CIS-E control unit plug, terminal 11, for continuity.

Resistance approx. 0 Ω

- OK →
- Not OK → Repair interruption

Start engine and run at idle speed and at operating temperature.

On-off ratio needle oscillates

- OK → End of test
- Not OK → Replace altitude correction capsule (B 18)

End of test

ON-OFF RATIO—116 AND 117 ENGINES

On-off ratio 100%

Note:
If on-off ratio is 100% and idle speed is too high, check fuse on overvoltage protection relay and/or power supply and ground connection

Testing on-off ratio On-off ratio can be adjusted, on-off ratio tester needle oscillates.	
OK	Not OK

OK → End of test

Not OK →

Disconnect O_2-sensor plug and connect male part of plug to ground. Start engine and test sensor voltage between female part of plug and ground.	
Sensor voltage above 450 mV.	
OK	Not OK

Not OK → Replace O_2-sensor

OK →

Stop engine. Check wire from O_2-sensor plug to CIS-E control unit, terminal 8, for continuity.	
Resistance approx. 0 Ω	
OK	Not OK

Not OK → Repair interruption

OK →

Loosen plug of O_2-sensor heater so that the voltage can be measured at the pins (do not disconnect plug; Use caution – danger of short circuit). Pull off fuel pump relay and bridge terminals 7 and 8.	
Reading approx. 12 Volts (battery voltage).	
OK	Not OK

Not OK → Repair interruption

OK →

Unplug O_2-sensor heater plug. Connect test cable 102 589 04 63 00 with adapter plug 903 589 03 63 00. Measure current draw.	
Nominal valve above 0.5 A	
OK	Not OK

Not OK → Replace O_2-sensor

OK → End of test

LAMBDA CONTROL—116 AND 117 ENGINES

B) Testing and adjusting lambda control

Engine oil temperature 75°–85°C.
Pull off purge line to throttle valve housing at purge valve and plug line. Connect on-off ratio tester to diagnostic socket. Press 100% IR button.

Test on-off ratio at 2500 rpm and read mean value. Compare this value with the idle speed value. The mean value at idle speed must be between 5 and 15 % higher than the mean value measured at 2500 rpm.

OK	Not OK

OK → End of test

Not OK →

Constant reading (i.e. no needle oscillation)	Mean value is not 5 to 15 % higher

Constant reading → Test O_2-sensor or repair interruption as required. → End of test

Mean value is not 5 to 15 % higher → Adjust lambda control → End of test

ELECTRONIC IDLE SPEED CONTROL—116 AND 117 ENGINES

C) Testing electronic idle speed control.

Note:
The idle speed control is independent of the air flow sensor position indicator. Run engine at operating temperature, selecter lever in position "P" or "N", A/C compressor turned off and air intake system tight. Idle speed contact, coolant temperature sensor and EGR system is functioning properly.

[1]) At altitudes above 2000 m, the current draw can fall below 700 mA.

Connect test cable 102 589 04 63 00 to idle speed air valve. Set multimeter to mA scale. Connect tachometer. Start engine.

Current draw		
700 to 1000[1]) mA	70 to 700 mA	0 mA
Idle rpm		
650 ± 50	650 ± 50	Idle speed too high

- 700 to 1000 mA / 650 ± 50 → Perform idle speed stabilization check
- 70 to 700 mA / 650 ± 50 → Enlarge air bypass bore (engine 116 only)
- 0 mA / Idle speed too high ↓

Stop engine. Unplug idle speed control unit. Check wires between idle speed air valve and idle speed control unit for continuity.

Resistance approx. 0 Ω

OK	Not OK

- Not OK → Repair interruption
- OK ↓

Check resistance of idle speed air valve.

Resistance 3.5 to 4.5 Ω

OK	Not OK

- Not OK → Replace idle speed air valve
- OK ↓

Turn on ignition. Check voltage between terminal 9 and terminal 11 of idle speed control unit plug.

Voltage approx. 12 V

OK	Not OK

- OK → Replace idle speed control unit ↓
- Not OK → Check power supply and ground connection according to wiring diagram ↓

ELECTRONIC IDLE SPEED CONTROL—116 AND 117 ENGINES

Engine idle speed stabilization test
Perform tests in both columns

Automatic transmission in gear
Engine at idle speed and operating temperature, place selector lever in a driving position

Nominal value

Engine	rpm
116.965	500 ± 50
117.967 117.968	550 ± 50

OK → End of test

Not OK → Check input signal to idle speed control unit (between terminal 12 and battery).

Battery voltage in position "P" or "N". When engaging a driving position, voltage drops.

OK → End of test

Not OK → Repair interruption according to wiring diagram.

A/C compressor engaged
Engine at idle speed and operating temperature. Turn on A/C compressor-autom. transmission in "P" or "N".

Nominal value

Engine	rpm
116.965	650 ± 50
117.967 117.968	650 ± 50

OK → End of test

Not OK → Connect test cable to idle speed air valve. Set multimeter to mA scale. Start engine and run at idle and operating temperature. Engage and disengage A/C compressor.

When engaging the A/C compressor the current at the idle speed air valve jumps by approx. 60 mA.

OK → End of test

Not OK → Repair interruption according to wiring diagram

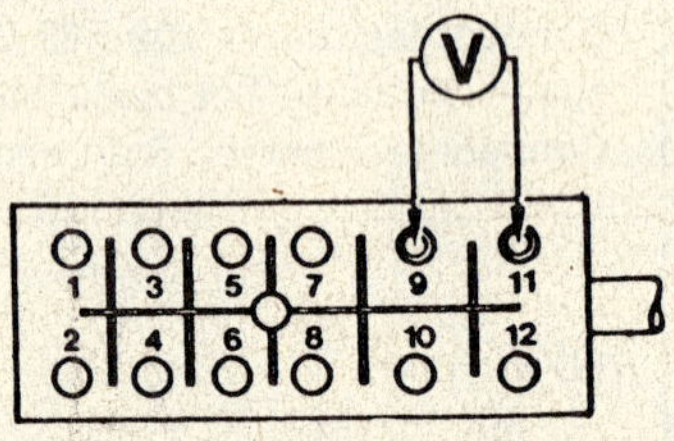

STARTING VALVE – 116 AND 117 ENGINES

D) Testing cold start valve activation.

Ignition turned off. Pull plug on coolant temperature sensor. Unplug green cable on ignition control module. Crank engine and check voltage at plug of cold start valve.

Voltage approx. 10 Volts for approx. 9 seconds.

OK	Not OK

Check wire from cold start valve to fuel pump relay, terminal 4 (circuit 87 V) and ground wire to engine ground for continuity.

Resistance approx. 0 Ω

OK	Not OK

Repair interruption

Check voltage between fuel pump relay plug, terminal 12 (circuit 50) as well as terminal 2 (circuit TF) and engine ground (coolant temperature sensor plug disconnected, Fig. 07.3/45). Crank engine.

Voltage at terminal 12 = approx. 10 Volts
Voltage at terminal 2 = 3.5 to 5.0 Volts

OK	Not OK

Repair interruption

Replace fuel pump relay

End of test

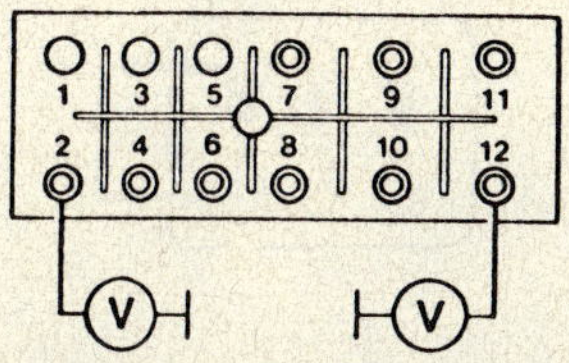

TESTING DECELERATION SHUTOFF – 190E 116 AND 117 ENGINES

E) Testing deceleration shut off

Connect on-off ratio tester to diagnostic socket. Start engine. Engine idling and at operating temperature.
On California vehicles, program the CIS-E control unit to provide the on-off ratio output.

Tester needle oscillates

OK	Not OK

For constant fixed on-off ratio see troubleshooting chart

Drive vehicle on dynometer or road with shift selector in position 2. Engine rpm ≧ 2500 rpm.

When vehicle is decelerated (throttle valve closed) the on-off ratio jumps to 95%, when engine rpm drops below 1100 rpm, the on-off ratio momentarily falls to 50% and then the control range is ended.

OK	Not OK

Check idle speed contact on throttle valve switch and TD-signal, see section "A", impulse display 7 and 10.

Test speed signal.
Check wire between CIS-E control unit connector socket 24 via idle speed control unit and connector X 26 (engine wiring harness) to electronic speedometer for continuity according to wiring diagram.

Resistance approx. 0 Ohm

OK	Not OK

Repair open circuit according to wiring diagram.

End of test

MICROSWITCH FUNCTIONS 116 AND 117 ENGINES

F) Testing idle and full load contact.

Connect on-off ratio tester to diagnostic socket and turn on ignition.

On-off ratio = 70%

↓

Deflect air flow sensor plate approx. 20 mm.

On-off ratio = 10%

OK	Not OK

Not OK → Reading 70%: Throttle valve contact does not close or interruption in wire to idle speed contact.

OK ↓

Close air flow sensor plate.

On-off ratio = 40%

↓

Open throttle valve fully.

On-off ratio = 20%

OK	Not OK

Not OK → Reading 40%: Full load contact defective

↓

End of test

OVERVOLTAGE PROTECTION RELAY (6-pole)—116 AND 117 ENGINES

G) Testing overvoltage protection relay (K 1).

Check fuse on overvoltage protection relay.

Fuse OK	Fuse not OK

Fuse not OK → Replace fuse

Fuse OK ↓

Unplug overvoltage protection relay. Bridge terminals 1 and 2 Fig. 07 Check voltage between terminal 1 of CIS-E control unit plug and ground.

Voltage approx. 12 Volts

OK	Not OK

Not OK → Check power supply according to wiring diagram.

OK ↓

Turn on ignition.
Check voltage between terminal 5 (–) and terminal 6 (+) of overvoltage protection relay plug.

Voltage approx. 12 Volts

OK	Not OK

Not OK → Repair interruption

OK → Replace overvoltage protection relay.

↓

End of test

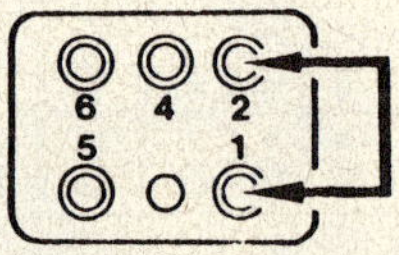

Fig. 07

OVERVOLTAGE PROTECTION RELAY (9-pole)—116 AND 117 ENGINES

Testing overvoltage protection relay

Check fuse on overvoltage protection relay.

- OK
- Not OK → Replace fuse.

Disconnect overvoltage protection relay plug. Bridge terminals 1 and 2 of the overvoltage protection relay plug
Check voltage between terminal 1 of CIS-E control unit plug and ground, with multimeter set to Volt.

Voltage approx. 12 Volts

- OK
- Not OK → Check power supply according to wiring diagram.

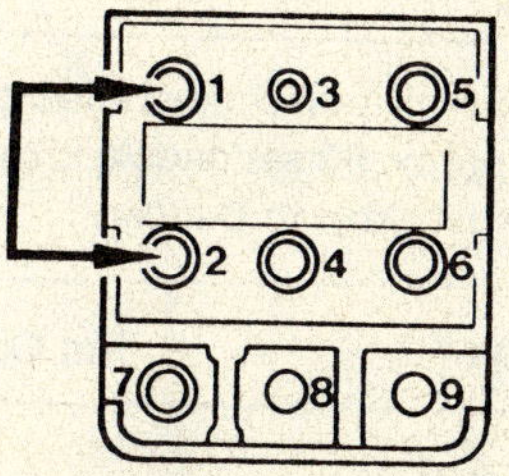

Turn on ignition.
Check voltage on overvoltage protection relay plug.

terminals 5 and 3

Voltage approx. 12 Volts

- OK → Replace overvoltage protection relay.
- Not OK → Repair open circuit according to wiring diagram.

End of test

TESTING AIR INJECTION—116 AND 117 ENGINES

H) Testing air injection

Coolant temperature above 70 °C.
Disconnect coolant temperature sensor plug and connect 2.5 k Ω test resistor to simulate +20 °C.
Disconnect vacuum line (ws/vi/bl) from side connection of switchover valve (Y 32) and connect vacuum tester with Y-fitting. Start engine.
Disconnect suction hose to air pump at air filter and close with finger.

Air pump is running and vacuum is present (approx. 600 mbar). Suction can be felt at end of hose to air pump.

OK	Not OK

The air pump must disengage after approx. 60 seconds and the vacuum must drop to 0 mbar.

OK	Not OK

End of test

Unplug air injection relay (K17)

Vacuum drops.

OK	Not OK

Check routing of vacuum lines and/or switchover valve.

Replace air injection relay (K17).

Unplug air injection relay (K17).
Check voltage between terminal 4 and 5 at plug.

Voltage approx. 12 Volts

OK	Not OK

Check power supply and ground wiring according to wiring diagram. Check air shutoff valve function and air hoses for proper routing.

Check voltage between terminal 3 and terminal 1 at plug.

Voltage approx. 12 Volts

OK	Not OK

Check power supply according to wiring diagram.

Replace air injection relay.

End of test

FUEL PRESSURE

Testing

CAUTION

To prevent personal injury, use extreme caution and always wear eye protection when opening any fuel line on a CIS fuel system. This type of system retains high fuel pressure for hours after the engine has been shut off.

1. Start engine, deactivate fuel pump until engine stalls to relieve fuel pressure. Crank engine to make certain fuel pressure is relieved.
2. Unscrew closing plug on the lower chamber of the fuel distributor.

NOTE: When loosening the closing plug, make certain that the open-end wrench is not pressing against the sensor of the airflow meter. This would damage the sensor.

3. Connect the A hose line of the pressure gauge to the lower chamber using special tool No. 102 589 06 63 000, or equivalent.
4. Unscrew the fuel line on the cold start valve on the fuel distributor.
5. Connect the B hose line of the gauge to the fuel distributor.
6. Correctly position the pressure gauge valves, according to gauge manufacturer's instructions.
7. Check system pressure with the engine **OFF**.
8. Pull off the fuel pump relay and bridge the terminals 7 and 8 to activate fuel pump.
9. Make certain pressure gauge valves are set correctly and read pressure.
10. Fuel system pressure should read:
    102 engine—77–80 psi (5.3–5.5 bar)
    103 engine—77–80 psi (5.3–5.5 bar)
    104 engine—77–80 psi (5.3–5.5 bar)
    116 engine—90–93 psi (6.2–6.4 bar)
    117 engine—90–93 psi (6.2–6.4 bar)
11. Remove jumper from fuel pump relay and open valve on pressure gauge.
12. Check lower chamber pressure as follows:
    a. Pull off the fuel pump relay and bridge terminal 7 and terminal 8.
    b. Turn ignition switch **ON**.
    c. With engine at normal operating temperature, pull the electrical connector from the electrohydraulic actuator.
    d. Make certain pressure gauge valves are set correctly.
    e. Read the lower chamber pressure, if should be approximately 6 psi (0.4 bar) below system pressure.
    f. If proper pressure is not obtained, check the coolant temperature sensor, control unit and electrohydraulic actuator.
13. Replace the fuel pump relay.
14. Check the decel shutoff as follows:
    a. With engine at normal operating temperature, run engine at idle.
    b. Increase engine speed to 2500 rpm.
    c. When the speed drops, pressure in the lower chamber should increase to system pressure.
    d. Combustion starts again at 1300 rpm.
    e. If the proper pressure is not obtained, check the microswitch(s), control current to the actuator, TD signal and the control unit.
15. Check acceleration enrichment, as follows:
    a. Make certain fuel pressure gauge valves are set properly.
    b. Simulate a cold engine by connecting a 2.5 ohm resistance in place of the coolant temperature sensor. This will simulate approximately 69°F (20°C).
    c. Start engine and suddenly increase engine speed to 2500 rpm.
    d. Pressure should read approximately 50–60 psi (3.8 bar)—102 and 103 engine or 40–45 psi (2.6–3.0 bar)—116 and 117 engine.
    e. If proper pressure is not obtained, check the airflow meter sensor, current to actuator and the control unit.

FUEL INJECTOR

Testing

1. Remove injection valve and place in appropriate container for fuel.
2. Test the opening pressure of injection valve. The minimum opening pressure should be greater than 45 psi (3.1 bar). New valve opening pressure should be 51–61 psi (3.5–4.1 bar).
3. Check spray pattern for each injection valve. Check for dripping or leaking valves.
4. Compare fuel volume for each injection valve over equal period of time. If any injector delivers unequal amount of fuel, switch the injector with a different fuel tube and retest. If the deficiency follows the injector, install a new injector. If the injection deficiency stays at the same location, the problem is in the fuel distributor unit, the unit should be replaced.

FUEL LEAKS

Testing

1. Perform fuel system pressure tests. Leave fuel pressure gauge connected to the assembly.
2. Run engine at normal operating temperature.
3. Stop engine. Pressure should drop approximately 40 psi (2.8 bar) and then hold the remaining pressure for several hours.
4. If the pressure drops immediately to 0, replace the fuel pump check valve.
5. If the pressure drops slowly, unscrew the fuel return line from the diaphragm pressure regulator. No fuel or a very slight amount of fuel should come from the line.

CAUTION

To prevent personal injury, use extreme caution and always wear eye protection when opening any fuel line on a CIS fuel system. This type of system retains high fuel pressure for hours after the engine has been shut off.

6. Pinch off the fuel line at the accumulator. If the pressure does not drop, replace the fuel accumulator.
7. If pressure drops but no leaks are found, remove and test cold start injector for leaking and proper operation. If injector tests good, check each injector leaking and for proper operation.

COOLANT TEMPERATURE SENSOR

Testing

This procedure is for the 4–pole type coolant temperature sensor. The 2–pole type sensor is tested in the same manner. The coolant temperature sensor resistance table applies to all coolant sensors. Also refer to appropriate troubleshooting and testing charts.

1. Disconnect the coolant temperature sensor.
2. Measure the resistance valve of each internal sensor, between any 2 of the diagonally opposed terminals, and compare to the each other.
3. Compare the coolant temperature sensor resistance reading to the coolant temperature sensor resistance table.
4. If either sensor circuit is not within specification through out the temperature range, replace the sensor.

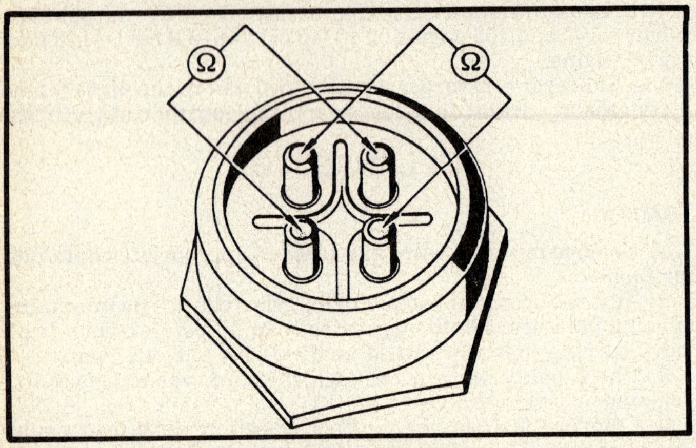

Testing coolant temperature sensor

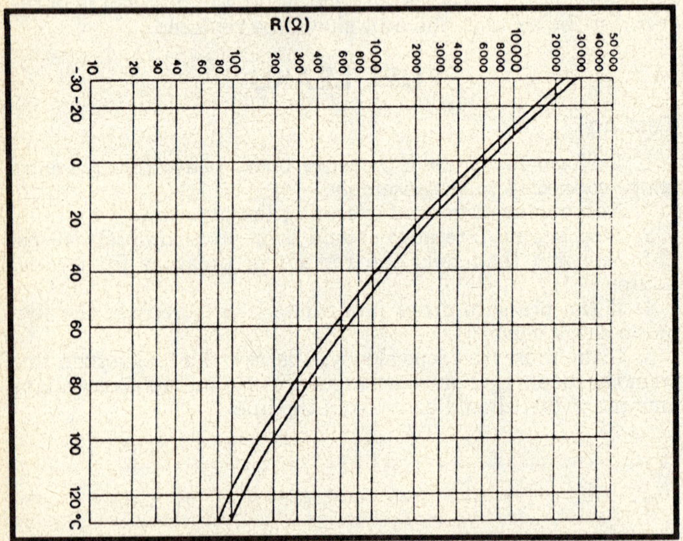

Coolant temperature sensor resistance value table

NOTE: When checking the continuity of the the wiring from the CIS-E to the connector, the terminals on the connector must be checked in clockwise order, whereby only 1 wire, respectively, should have continuity of less than 1 ohm.

EGR TEMPERATURE SENSOR

Testing

1. Disconnect the EGR temperature sensor.
2. Measure the resistance valve between the terminals.
3. Compare the EGR temperature sensor resistance reading to the EGR temperature sensor resistance table.

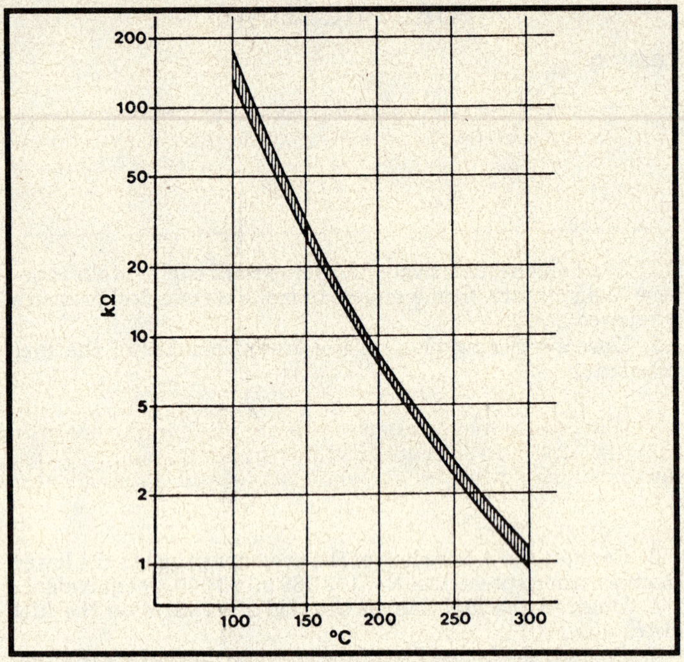

EGR temperature sensor resistance valve table

4. If the resistance is not within specification through out the temperature range, replace the sensor.

IDLE SPEED VALVE

Testing

This procedure may vary slightly, but is same basic test for idle control on any vehicle using CIS-E or K-Jetronic type systems.

1. Start the engine and allow to reach normal operating temperature.
2. Check the engine rpm. If idle control system is functioning properly it should be at the correct specifications for the vehicle being serviced.
3. Disconnect the coolant temperature sensor or simulate a cold engine.
4. The engine rpm should increase if the idle control valve and the control system is functioning properly.
5. If the engine rpm did not increase, connect a dwellmeter to the idle control valve.
6. Check the dwell and engine rpm with the coolant temperature sensor connected and at normal operating temperature.
7. Disconnect the coolant temperature sensor or simulate a cold engine.
8. The dwell and engine rpm should increase. If the dwell increases but the engine rpm stays the same, the idle speed valve or its circuit is defective. If the dwell doesn't change, the control module, sensors or their circuits must be checked.

HALL EFFECT SPEED SENSOR 190E, 300E, 300CE and 300TE

Test Hall-effect speed sensor

Ignition off. Disconnect CIS-E control unit plug. Ignition on. Measure voltage on CIS-E control unit plug between sockets 1 and 6, with multimeter set to Volt.

Push vehicle approx. 3 ft. (1 m).

Reading oscillates between 0 and 12 Volts.

OK | Not OK

Not OK → Repair open circuit according to wiring diagram. Replace Hall-effect speed sensor if necessary

OK → Connect CIS-E control unit plug. Connect on-off ratio tester to diagnostic socket. Drive vehicle in 3rd gear/selector lever position 3 on either dynometer or open road.

At approx. 2000 rpm accelerate at full throttle for approx. 6 seconds. Observe test results.

Tester needle oscillates after releasing accelerator pedal.

OK | On-off ratio is 60% constant

On-off ratio is 60% constant → Replace CIS-E control unit

OK → End of test

VEHICLE SPEED SIGNAL 300SE AND 300SEL

Test vehicle speed signal

Unplug CIS-E control unit connector. Measure voltage on CIS-E control unit plug between sockets 2 and 6, with multimeter set to Volt. Drive vehicle on either dynamometer or open road.

Nominal value: at 40 km/h (25 mph) > 1 V

OK | Not OK

Not OK → Repair open circuit according to wiring diagram. Check electronic speedometer

OK → Connect CIS-E control unit plug. Connect on-off ratio tester to diagnostic socket. Drive vehicle in 3rd gear/selector lever position 3 on either dynometer or open road.

At approx. 2000 rpm accelerate at full throttle for approx. 6 seconds. Observe test results.

Tester needle oscillates after releasing accelerator pedal.

OK | On-off ratio is 60% constant

On-off ratio is 60% constant → Replace CIS-E control unit

OK → End of test

WIRING COLOR CODES

NOTE: The German wiring color codes differ from the U.S. color codes. Take note of correct color codes before performing tests.

bl—blue
br—brown
ge—yellow
gn—green
gr—gray
rs—pink
rt—red
sw—black
vi—violet
ws—white

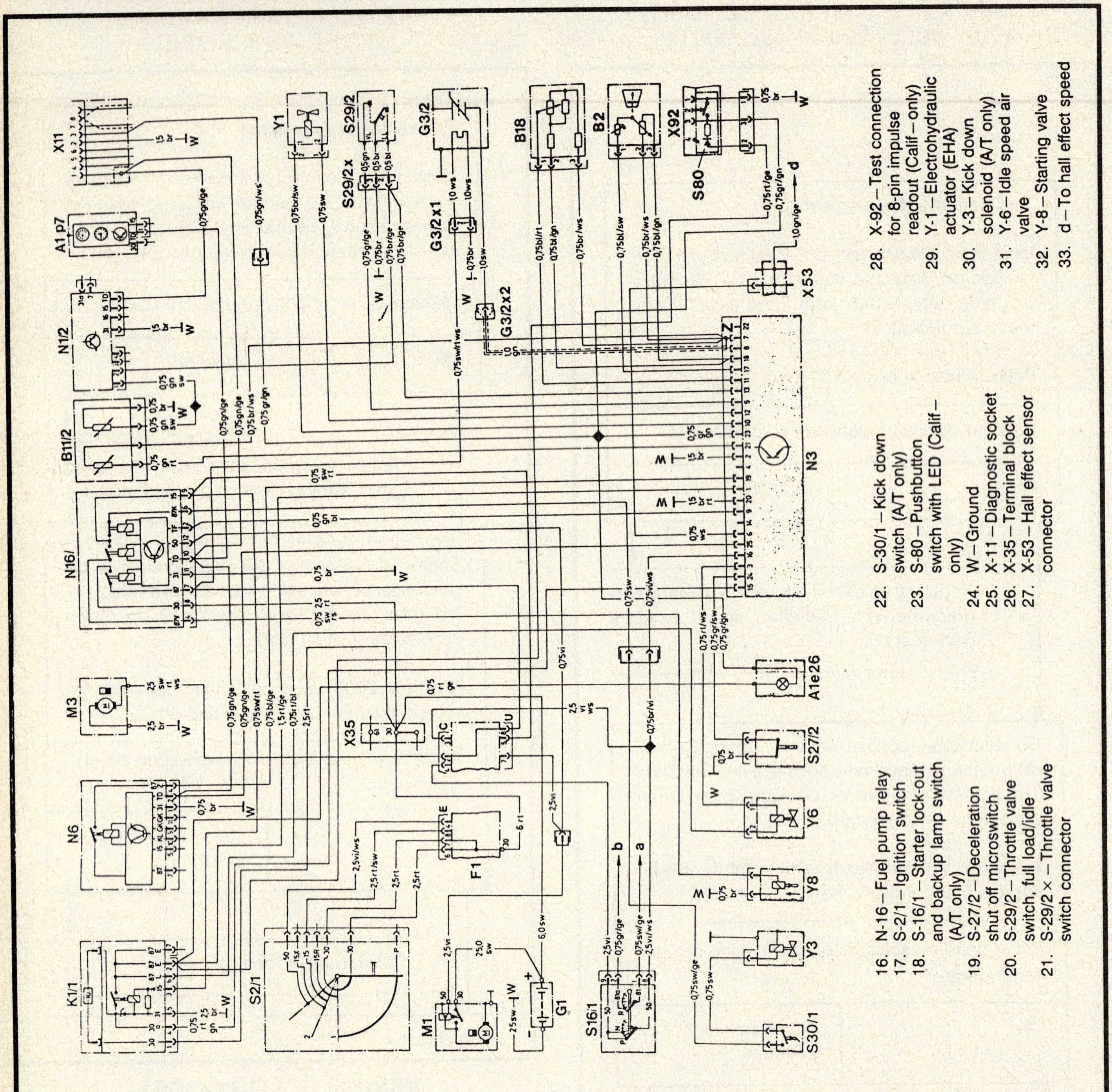

1. A-1e26 – CHECK ENGINE lamp
2. A-1p7 – Electronic clock/tachometer
3. B-2 – Air flow sensor position indicator
4. B-11/2 – Coolant temperature sensor (4-pole shown)
5. B-18 – Altitude correction capsule
6. F-1 – Electrical center
7. G-1 – Battery
8. G-3/2 – Heated oxygen sensor
9. G-3/2 × 1 – Oxygen sensor heater connector
10. G-3/2 × 2 – Oxygen sensor signal connector
11. K-1/1 – Overvoltage protection relay
12. M-1 – Starter
13. M-3 – Fuel pump
14. N-3 – CIS-E control unit
15. N-6 – A/C compressor control unit
16. N-16 – Fuel pump relay
17. S-2/1 – Ignition switch
18. S-16/1 – Starter lock-out and backup lamp switch (A/T only)
19. S-27/2 – Deceleration shut off microswitch
20. S-29/2 – Throttle valve switch, full load/idle
21. S-29/2 × – Throttle valve switch connector
22. S-30/1 – Kick down switch (A/T only)
23. S-80 – Pushbutton switch with LED (Calif–only)
24. W – Ground
25. X-11 – Diagnostic socket
26. X-35 – Terminal block
27. X-53 – Hall effect sensor connector
28. X-92 – Test connection for 8-pin impulse readout (Calif–only)
29. Y-1 – Electrohydraulic actuator (EHA)
30. Y-3 – Kick down solenoid (A/T only)
31. Y-6 – Idle speed air valve
32. Y-8 – Starting valve
33. d – To hall effect speed

CIS-E electrical circuit – 102 engine

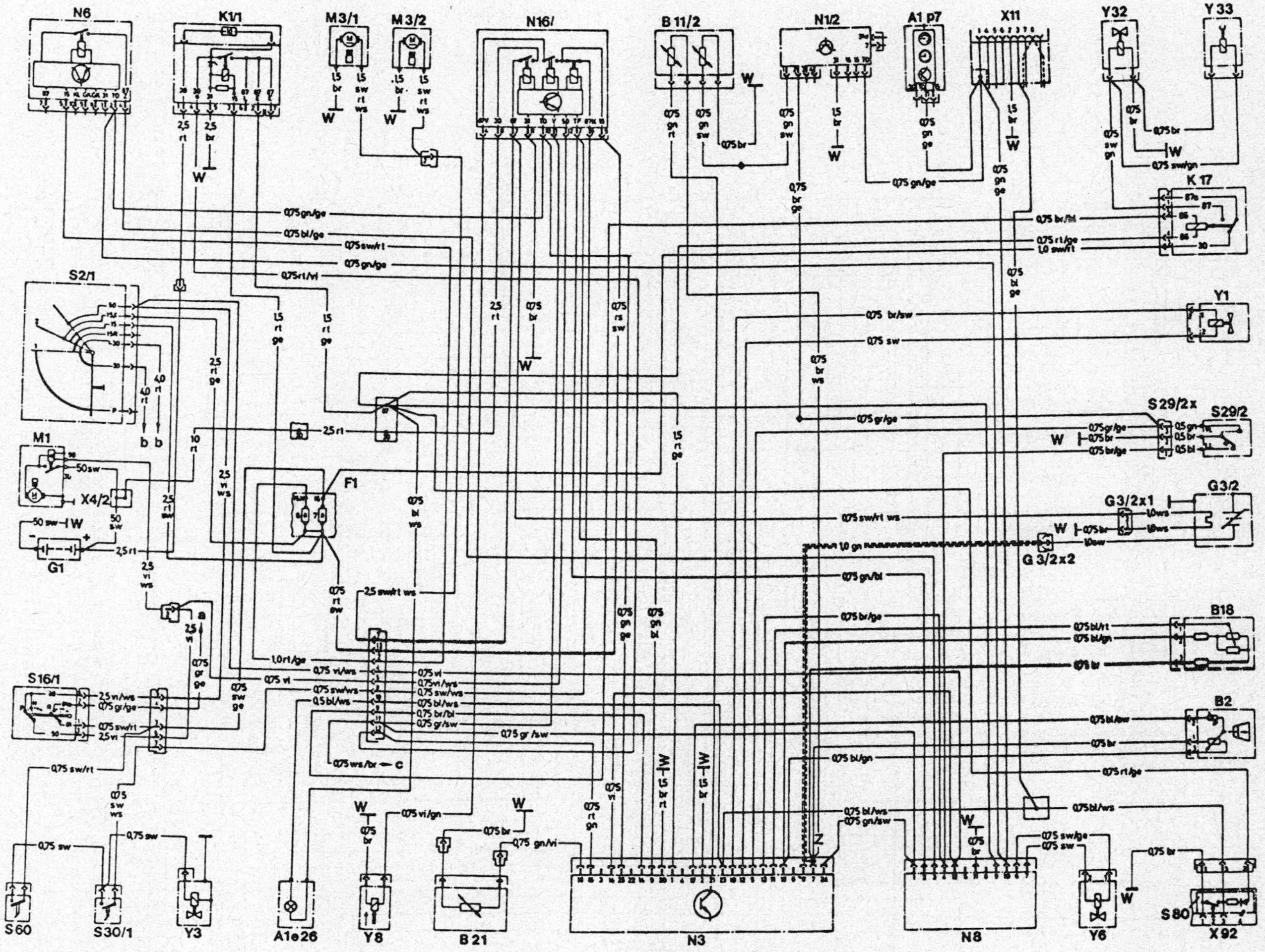

1. A-1e26 – CHECK ENGINE lamp
2. A-1p7 – Electronic clock/ tachometer
3. B-2 – Air flow sensor position indicator
4. B-11/2 – Coolant temperature sensor (4-pole shown)
5. B-18 – Altitude correction capsule
6. B-21 – EGR temperature sensor (Calif – only)
7. F-1 – Electrical center
8. G-1 – Battery
9. G-3/2 – Heated oxygen sensor
10. G-3/2 × 1 – Oxygen sensor heater connector
11. G-3/2 × 2 – Oxygen sensor signal connector
12. K-1/1 – Overvoltage protection relay
13. K-17 – Air injection relay
14. M-1 – Starter
15. M-3/1 – Fuel pump No.1
16. M-3/2 – Fuel pump No – 2
17. N-1/2 – Electronic ignition control unit (EZL)
18. N-3 – CIS-E control unit
19. N-6 – A/C compressor control unit
20. N-8 – Idle speed control unit
21. N-16 – Fuel pump relay
22. S-2/1 – Ignition switch
23. S-16/1 – Starter lock-out and backup lamp switch
24. S-29/2 – Throttle valve switch, full load/idle
25. S-29/2 × – Throttle valve switch connector
26. S-30/1 – Kick down switch
27. S-60 – B position shift lever switch
28. S-80 – Pushbutton switch with LED (Calif – only)
29. W – Ground
30. X-11 – Diagnostic socket
31. X-35 – Terminal block
32. X-92 – Test connection for 8-pin impulse readout (Calif – only)
33. Y-1 – Electrohydraulic actuator (EHA)
34. Y-3 – Kick down solenoid (A/T only)
35. Y-6 – Idle speed air valve
36. Y-8 – Starting valve
37. Y-32 – Air pump vacuum valve
38. Y-33 – Electromagnetic air pump clutch
39. a – To lamp failure monitoring unit
40. b – To cruise control

CIS-E electrical circuit – 116 and 117 engine

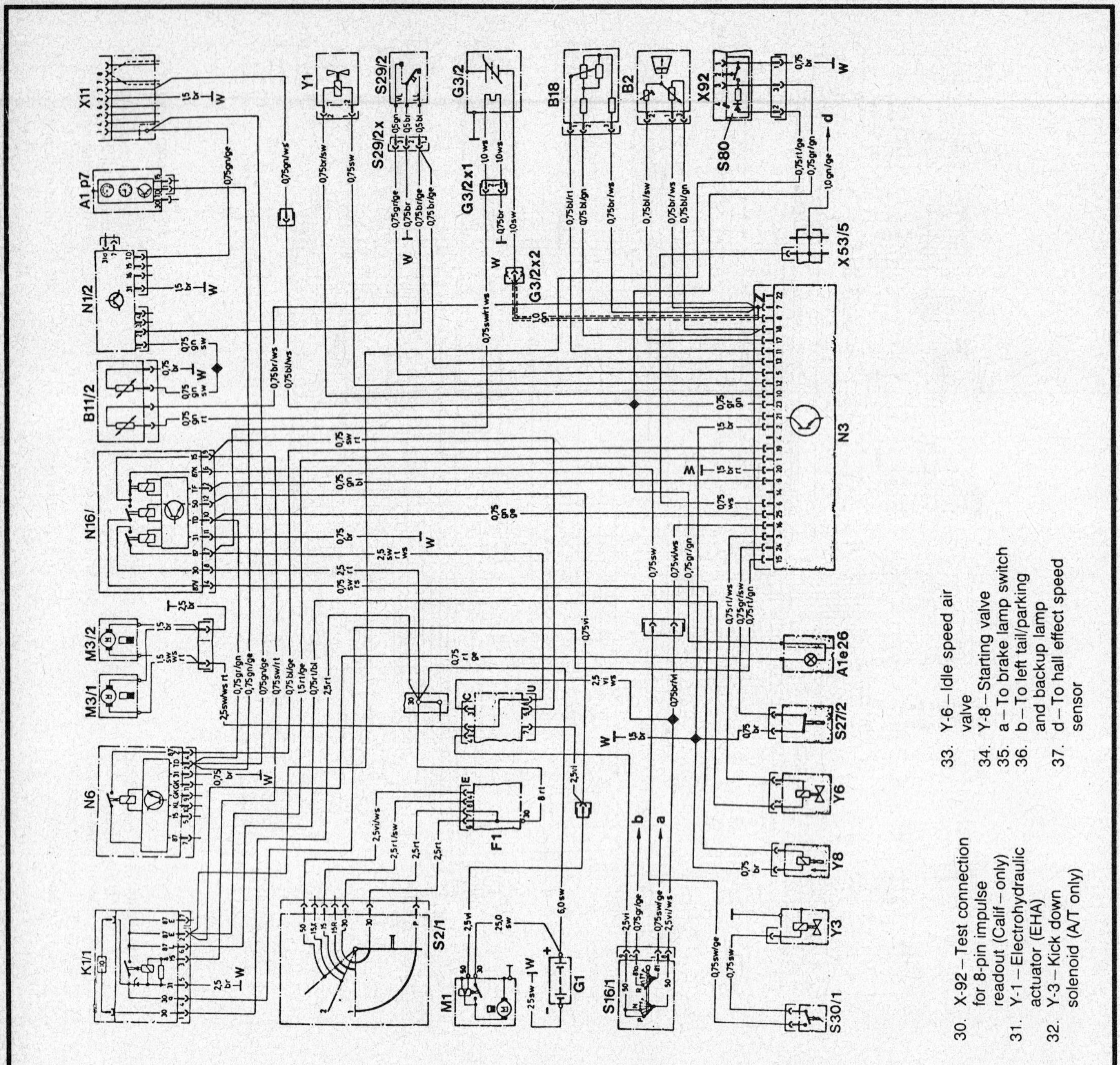

CIS-E electrical circuit—190E with 103 engine

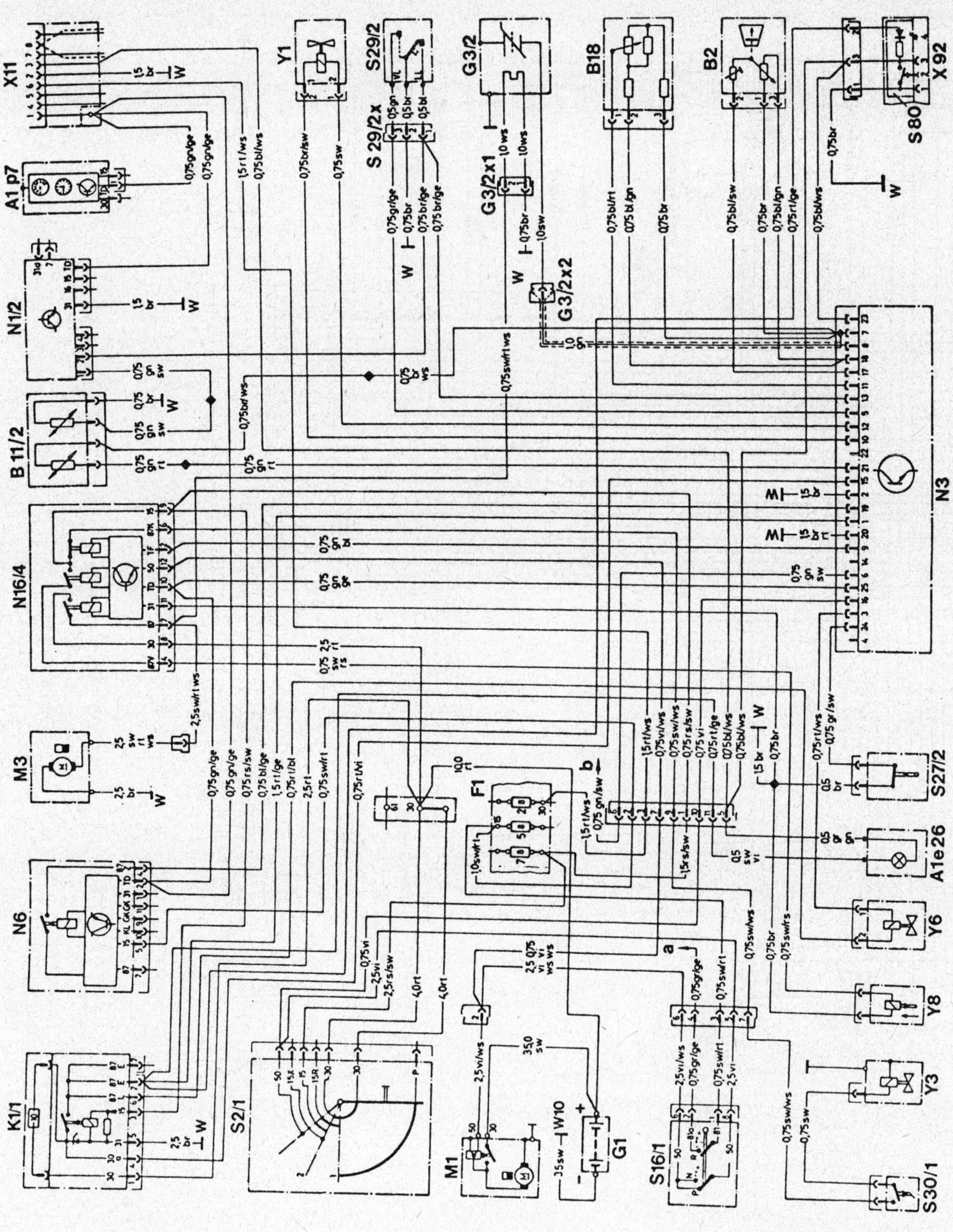

1. A-1e26 – CHECK ENGINE lamp
2. A-1p7 – Electronic clock/ tachometer
3. B-2 – Air flow sensor position indicator
4. B-11/2 – Coolant temperature sensor (4-pole shown)
5. B-18 – Altitude correction capsule
6. F-1 – Fuse/Relay box
7. G-1 – Battery
8. G-3/2 – Heated oxygen sensor
9. G-3/2 × 1 – Oxygen sensor heater connector
10. G-3/2 × 2 – Oxygen sensor signal connector
11. K-1/1 – Overvoltage protection relay
12. M-1 – Starter
13. M-3 – Fuel pump
14. N-1/2 – Electronic ignition control unit (EZL)
15. N-3 – CIS-E control unit
16. N-6 – A/C compressor control unit
17. N-16 – Fuel pump relay
18. S-2/1 – Ignition switch
19. S-16/1 – Starter lock-out and backup lamp switch
20. S-27/2 – Deceleration shutoff microswitch
21. S-29/2 – Throttle valve switch, full load/idle
22. S-29/2 × – Throttle valve switch connector
23. S-30/1 – Kick down switch (A/T only)
24. S-80 – Pushbutton switch with LED (Calif – only)
25. W – Ground
26. X-11 – Diagnostic socket
27. X-35 – Terminal block
28. X-92 – Test connection for 8-pin impulse readout (Calif – only)
29. Y-1 – Electrohydraulic actuator (EHA)
30. Y-3 – Kick down solenoid (A/T only)
31. Y-6 – Idle speed air valve
32. Y-8 – Starting valve
33. a – To lamp failure monitoring unit
34. b – To cruise control

CIS-E electrical circuit – 300SE and 300SEL with 103 engine

1. A-1e26 – CHECK ENGINE lamp
2. A-1p7 – Electronic clock/ tachometer
3. B-2 – Air flow sensor position indicator
4. B-11/2 – Coolant temperature sensor (4-pole shown)
5. B-18 – Altitude correction capsule
6. F-1 – Fuse/Relay box
7. G-1 – Battery
8. G-3/2 – Heated oxygen sensor
9. G-3/2 × 1 – Oxygen sensor heater connector
10. G-3/2 × 2 – Oxygen sensor signal connector
11. K-1/1 – Overvoltage protection relay
12. M-1 – Starter
13. M-3/1 – Fuel pump No.1
14. M-3/2 – Fuel pump No – 2
15. N-1/2 – Electronic ignition control unit (EZL)
16. N-3 – CIS-E control unit
17. N-6 – A/C compressor control unit
18. N-16 – Fuel pump relay
19. S-2/1 – Ignition switch
20. S-16/1 – Starter lock-out and backup lamp switch (A/T only)
21. S-27/2 – Deceleration shutoff microswitch
22. S-29/2 – Throttle valve switch, full load/idle
23. S-29/2 × – Throttle valve switch connector
24. S-30/1 – Kick down switch (A/T only)
25. S-80 – Pushbutton switch with LED (Calif – only)
26. W – Ground
27. X-11 – Diagnostic socket
28. X-35 – Terminal block
29. X-53/5 – Hall effect connector
30. X-92 – Test connection for 8-pin impulse readout (Calif – only)

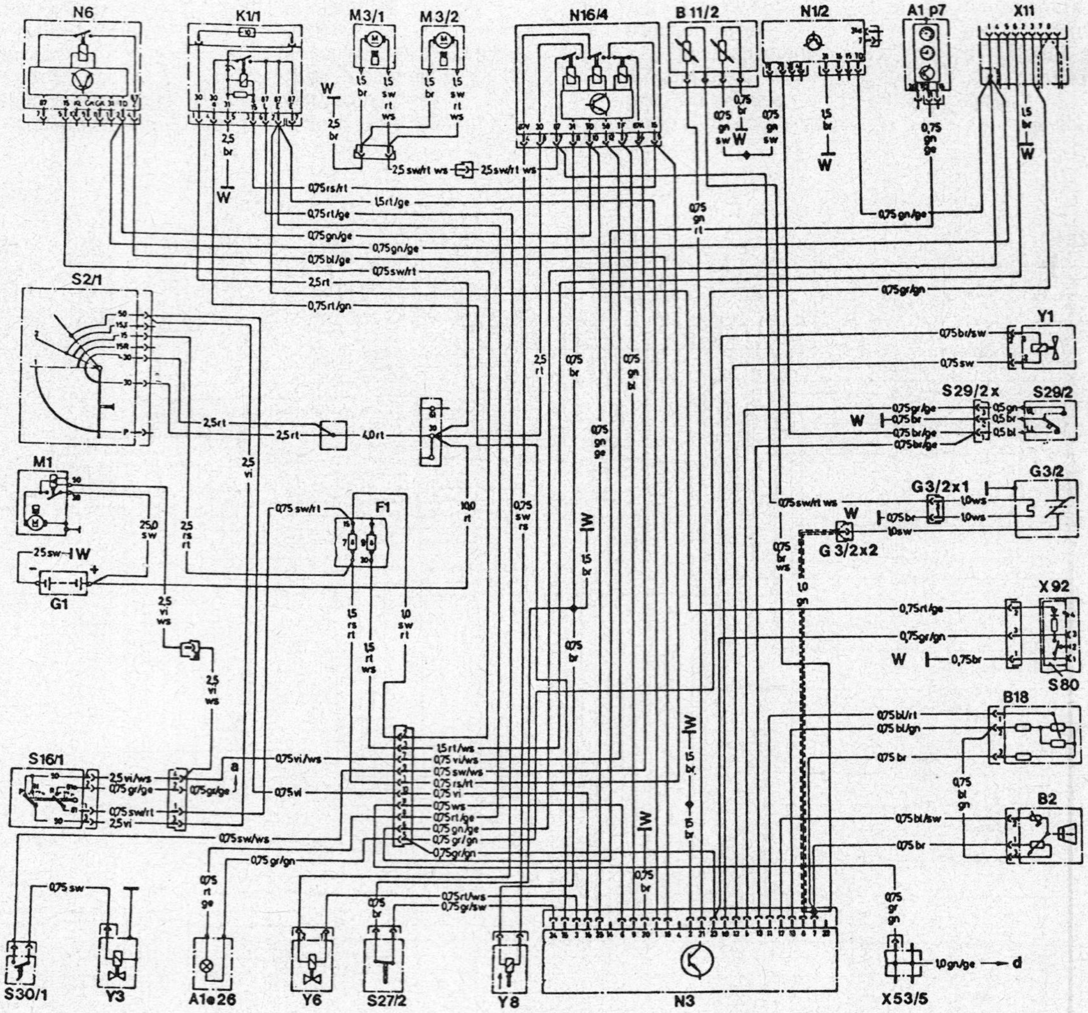

31. Y-1 – Electrohydraulic actuator (EHA)
32. Y-3 – Kick down solenoid (A/T only)
33. Y-6 – Idle speed air valve
34. Y-8 – Starting valve
35. a – To lamp failure monitoring unit
36. b – To cruise control
37. d – To hall effect speed sensor

CIS-E electrical circuit – 260E, 300E, 300CE and 300TE with 103 engine

Component Replacement

FUEL PRESSURE RELIEF

1. Follow service precautions to safely perform this procedure to prevent personal injury. Disconnect the negative battery terminal.
2. Wrap a shop rag around the feed to the fuel distributor.
3. Loosen the feed connection to relieve the system pressure. Catch any escaping fuel in the shop rag. Dispose of the rag in a safe manner.
4. Tighten the fuel distributor feed connection.

AIRFLOW SENSOR PLATE

Removal and Installation

1. Remove the air cleaner.
2. Unscrew the stop bracket.
3. Heat the fastening screw for the airflow sensor plate with a hot air blower and unscrew with care to avoid breaking.
4. Clean the threads, as needed, by using a M-6 tap.

To install:

5. Install the parts included in the repair kit.
6. Install the airflow sensor plate and washer.
7. Slightly tighten the center bolt and center the airflow sensor plate.

Adjustment

1. Jumper fuel pump relay terminals to establish pressure in fuel system.
2. Loosen the center sensor bolt slightly.
3. Make certain the airflow sensor plate is accurately centered. If the special centering tool is not available, use 4 feeler gauges of equal size every 90 degrees to ensure the plate is centered.
4. Tighten center bolt to 3.5–4.0 ft. lbs. (5.0–5.5Nm).
5. Check that the plate does not bind under light lateral pressure.
6. Check that the sensor plate does not bind through its full range of travel.
7. Release airflow plate at full travel and listen for an audible knock against the resilient stop.
8. If necessary, loosen center screw and recenter the plate.
9. Check that the plates rest position is at the narrowest part of the air funnel. A higher position of 0.008 in. (0.2mm) is permitted.
10. Adjusting the 0 position of the sensor plate requires moving the knock guide pin, by use of a mandrel. If too high, move pin down as require, take care not to move it too far. If it is too low, remove the mixture control unit and knock the guide pin from below.

NOTE: This is a press fit pin; if adjusted too often it will not hold an adjustment and will require replacement.

11. Install the fuel pump relay. Start engine and adjust CO as needed.

ELECTROHYDRAULIC ACTUATOR (EHA)

Removal and Installation

1. Remove the electrical connector.
2. Start engine, while running disconnect power from fuel pump or remove pump relay to relieve fuel system pressure.
3. Disconnect the negative battery cable.
4. Loosen fuel line to cold start injector and catch fuel in a container or rag.
5. Remove the retaining screws.

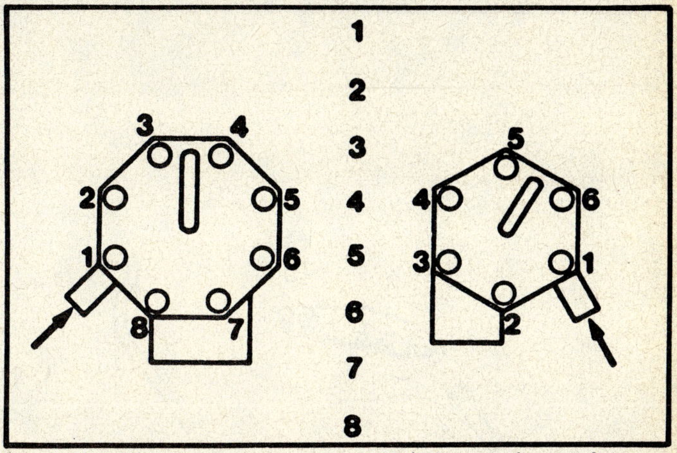

Fuel distribution sequence—116/117 engines shown

6. Carefully remove the EHA from the unit, take note of seals.

To install:

7. Clean area of EHA completely.
8. Install EHA with new seal(s).
9. Tighten bolts evenly to 22–31 inch lbs. (2.5–3.5 Nm).
10. Start engine, check for fuel leaks and proper fuel pressure.

FUEL FILTER

Removal and Installation

1. Relieve the fuel pressure.
2. Disconnect the negative battery cable.
3. Clamp the fuel suction and pressure hose with a clamp.
4. Remove protective cover and remove fuel filter.

To install:

5. When changing the filter, make certain the plastic sleeve is mounted between the fuel filter and the bracket. The sleeve should project on both sides of the bracket.

NOTE: The fuel filter should be filled with fuel before installation. The fuel pump delivers extremely high pressure. If fuel flows into the filter at this high pressure, it could dislodge part of the filter. The fuel distributor could be plug or damaged if filter material reached it.

6. Install filter hoses and bracket.
7. Remove hose clamping device.
8. Connect the negative battery cable.
9. Start engine and check for fuel leaks. Install protective cover.

FUEL PUMP

Removal and Installation

1. Relieve the fuel pressure.
2. Disconnect the negative battery cable.
3. Clamp the fuel suction and pressure hose with a clamp.
4. Remove protective cover, tag and remove fuel pump wires.
5. Disconnect the fuel pump hoses.
6. Remove fuel pump retaining screws and fuel pump.

To install:

7. Make certain the plastic sleeve is mounted between the fuel pump and the bracket. The sleeve should project on both sides of the bracket.

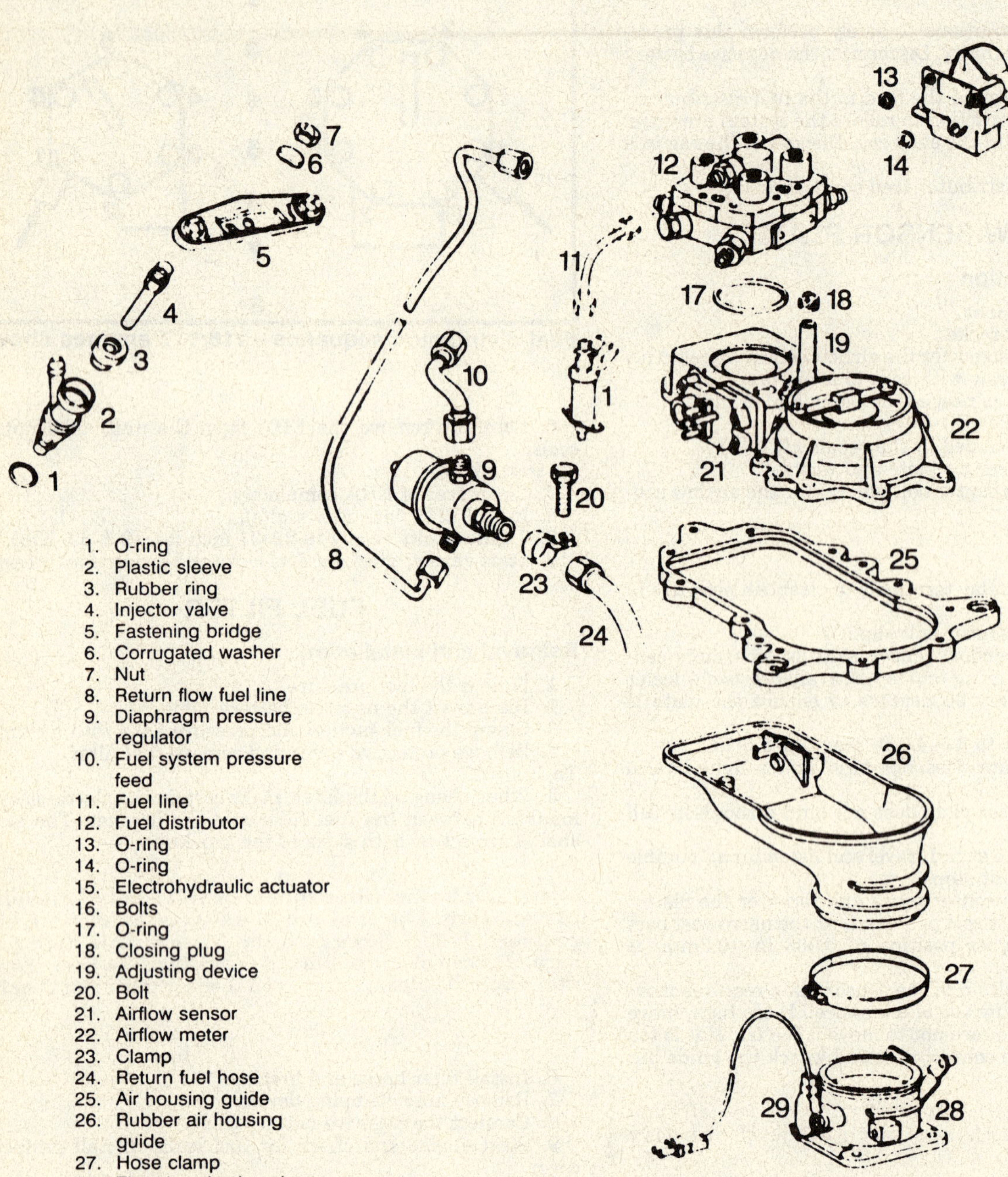

1. O-ring
2. Plastic sleeve
3. Rubber ring
4. Injector valve
5. Fastening bridge
6. Corrugated washer
7. Nut
8. Return flow fuel line
9. Diaphragm pressure regulator
10. Fuel system pressure feed
11. Fuel line
12. Fuel distributor
13. O-ring
14. O-ring
15. Electrohydraulic actuator
16. Bolts
17. O-ring
18. Closing plug
19. Adjusting device
20. Bolt
21. Airflow sensor
22. Airflow meter
23. Clamp
24. Return fuel hose
25. Air housing guide
26. Rubber air housing guide
27. Hose clamp
28. Throttle valve housing
29. Throttle valve switch

Fuel distributor assembly—Typical

NOTE: The fuel filter should be filled with fuel before installation. The fuel pump delivers extremely high pressure. If fuel flows into the filter at this high pressure, it could dislodge part of the filter. The fuel distributor could be plug or damaged if filter material reached it.

8. Install fuel hoses and bracket.
9. Make certain to connect the fuel pump wires on the correct terminals.
10. Remove hose clamping device.
11. Connect the negative battery cable.
12. Start engine and check for fuel leaks. Install protective cover.

FUEL INJECTOR

Removal and Installation

1. Relieve fuel system pressure.

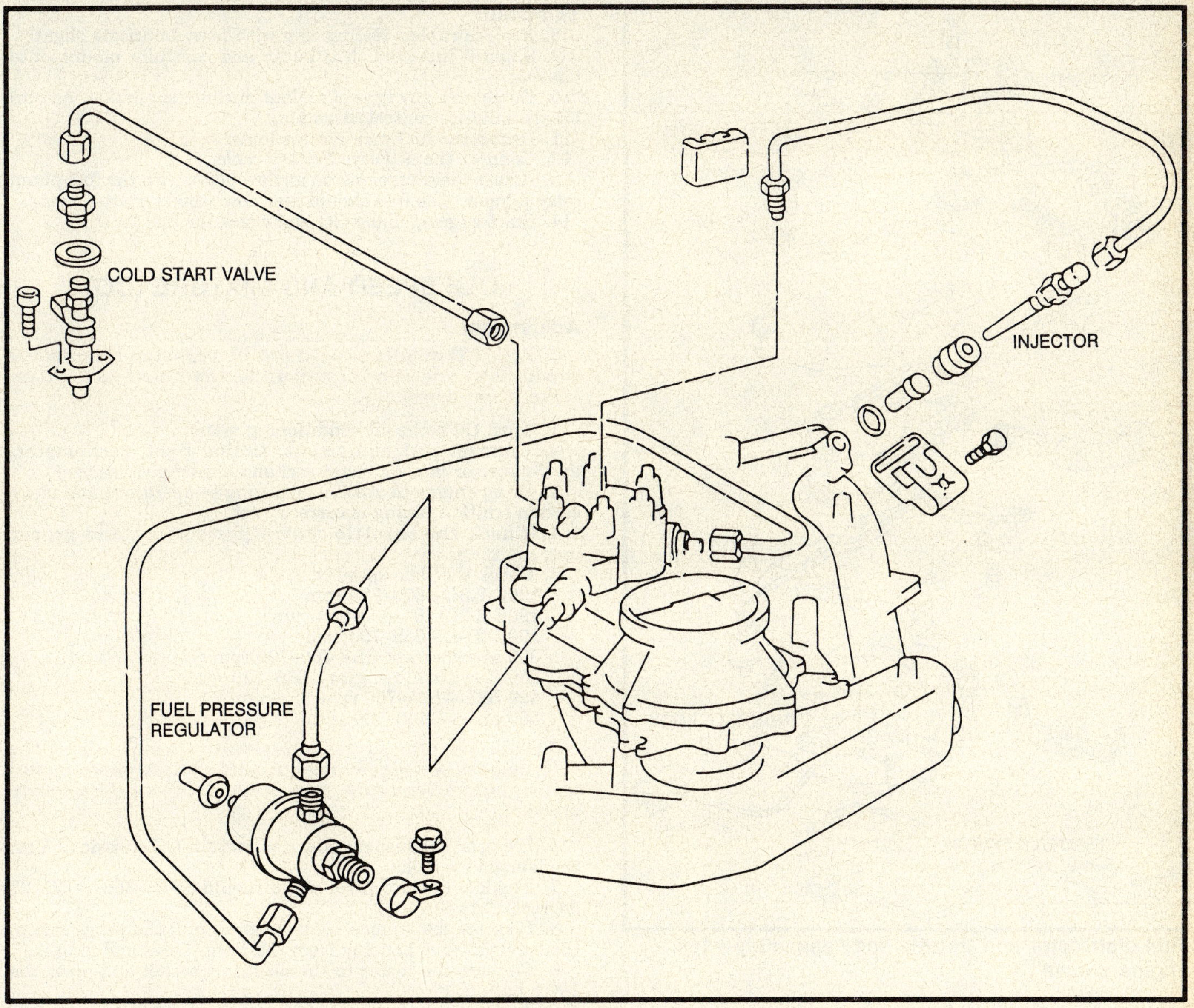

Fuel lines and fuel pressure regulator—104 engine shown

2. Disconnect the negative battery cable.
3. Carefully loosen fuel injection lines, catch fuel in container or rag.
4. Remove the fuel injection tubes from the fuel injectors.
5. Remove the injector hold-downs or fastening bridges.

NOTE: When removing the fastening bridge apply counterhold to the injection valves, so that the injection valves and insulating sleeves do not move.

6. Pull injector out from engine. If injector is stuck, a special tool may be required to assist in removal.

To install:

7. Inspect the sealing rings and insulating sleeve for damage. Make sure insulating sleeve did not pull out, if equipped with press in type.
8. Replace injector and O-ring.
9. Install injector hold-down. When installing injectors or hold-down bridge, always applied a counterhold to the injector to keep it and the insulator from moving.
10. Install fuel lines.
11. Connect negative battery cable.
12. Start engine, check for fuel leaks and proper CO adjustment.

MIXTURE CONTROL UNIT

Removal and Installation

1. Relieve fuel system pressure.
2. Disconnect the negative battery cable.
3. Carefully loosen fuel injection lines, catch fuel in container or rag.
4. Remove the fuel injection tubes from the fuel distributor.
5. Remove the fuel lines.
6. Remove the mixture control assembly together with the air guide housing.

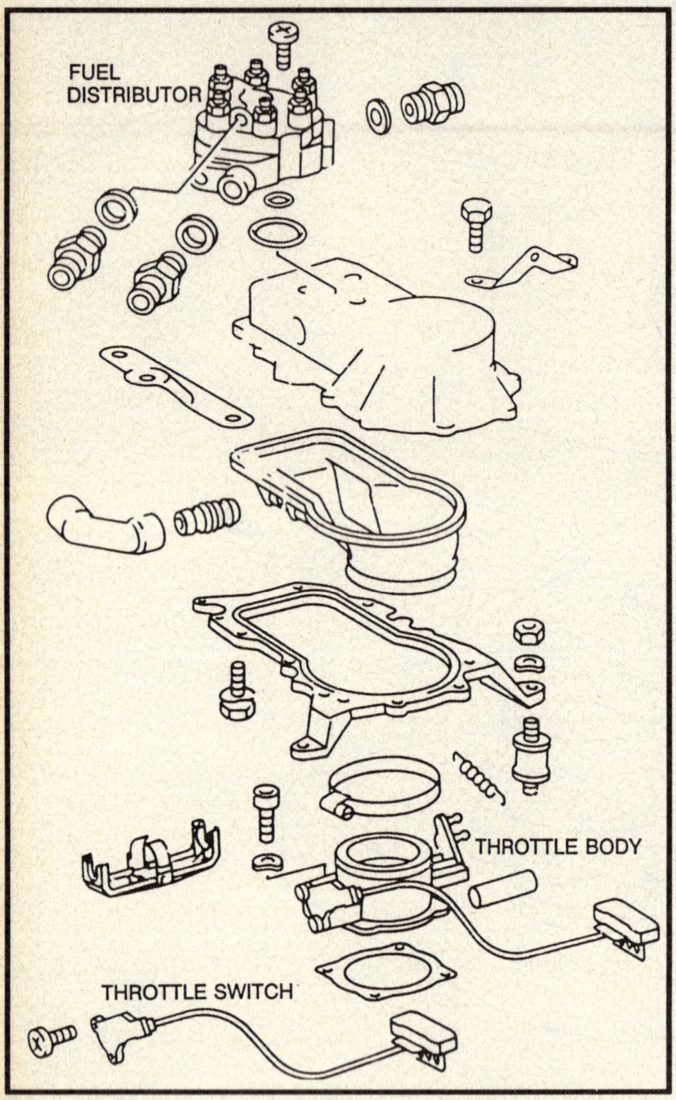

Fuel distributor and throttle body assembly—104 engine shown

7. Unscrew the mixture control unit.

To install:

8. Do not use any type of sealant during installation. Assemble the mixture control assembly.
9. Install the fuel and injector lines.
10. Connect the negative battery cable.
11. Start engine, adjust CO and check for fuel leaks.

FUEL DISTRIBUTOR

Removal and Installation

1. Relieve fuel system pressure.
2. Disconnect the negative battery cable.
3. Carefully loosen fuel injection lines, catch fuel in container or rag.
4. Remove the fuel injection tubes from the fuel distributor.
5. Remove the fuel lines.
6. Remove the fuel distributor retaining bolts.
7. Remove the fuel distributor by turning back and forth. Take care not to allow control piston to drop out of bore.

NOTE: This unit must be kept perfectly clean.

To install:

8. Replace rubber sealing ring with new. Lubricate slightly.
9. Slightly lubricate distributor and carefully mount onto housing.
10. Do not use any type of sealant during installation. Assemble the mixture control assembly.
11. Install the fuel and injector lines.
12. Connect the negative battery cable.
13. Adjust the control piston airflow plate, with the fuel pump relay jumper. The fuel should just now stop or start flowing.
14. Start engine, adjust CO and check for fuel leaks.

IDLE SPEED AND MIXTURE (CO)

Adjustment

NOTE: CO or idle should not be adjusted if engine is too hot. For specific CO setting, see emission label on vehicle being serviced.

1. Turn **OFF** the air conditioning system.
2. Connect test equipment: timing light, tachometer, dwellmeter or lambda tester and an exhaust gas analyzer.
3. Bring engine to normal operating temperature and make certain ignition timing is correctly set.
4. Check the throttle control for smooth and proper operation.
5. Check that idle speed is:
   190E 2.3L—670–770 rpm
   190E 2.3L-16—850–950 rpm
   190E 2.6L—650–750 rpm
   300 series except SL—600–700 rpm
   300SL with 104 engine—650–750 rpm
   420 SEL—625–750 rpm
   560 series—600–750 rpm

NOTE: The idle is not adjustable. Do not adjust the idle by-pass air screw, if equipped. If idle speed is not within specifications, test the electronic idle speed control.

6. Using an exhaust gas analyzer, test the CO emissions. Test the lambda control.
7. The idle speed control range should be 35–45% — 21–27 degrees of dwell.
8. Connect the lambda control tester, pull the purge line to the throttle valve housing from the purge valve and plug it.
9. Connect the tester to the diagnosis socket and press the 100% button.
10. Test the on-off ratio at 2500 rpm and read the mean valve. Compare the value with the idling speed valve, adjust as needed. The medium value at idle should not deviate the value measure at 2500 rpm by more the ±10—1988 +10—1989–90. For test, pull off the purge line and plug it.
11. To set the idle speed emission valve or lambda control, pull out the safety plug with an appropriate puller.
12. The adjusting device is provided with a protective steel plug.
13. If adjustment of the on-off ratio is necessary, the plug will require removal and an new plug kit No. 102 589 02 07, or equivalent will be required.
14. Remove the air cleaner.
15. Punchmark the breakaway safety screw in the center and drill a 2.5mm hole to a depth of 6–8mm.

NOTE: DO NOT drill completely through the screws, engine damage may be caused by the resulting chips. Thoroughly remove all metal chips with a rag.

16. Remove the breakaway safety screws with a lefthand twist extractor.
17. Install the new repair set, until the breakaway screw heads break off.

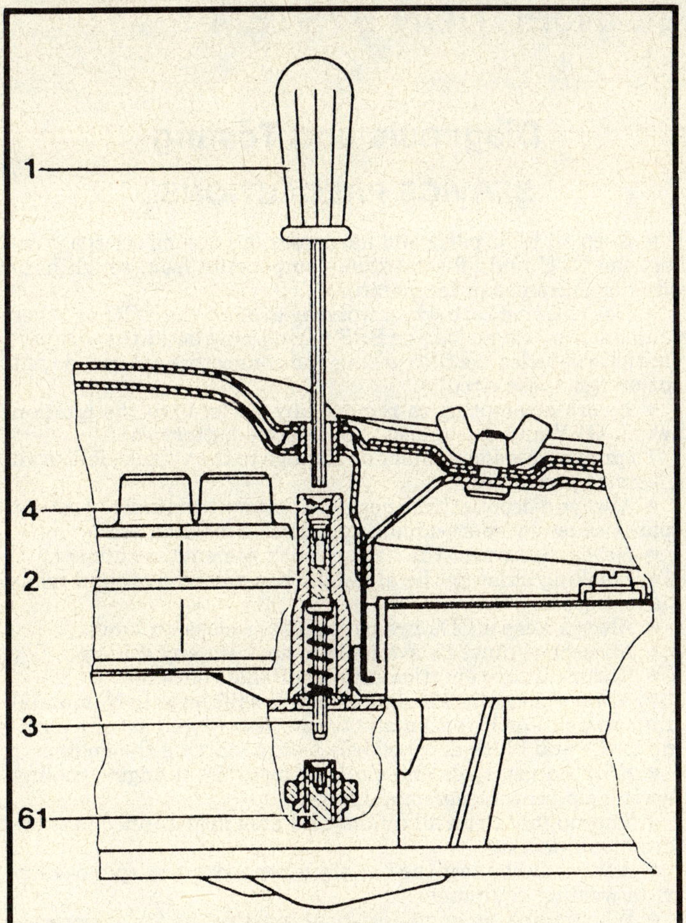

Adjusting CO or lambda control

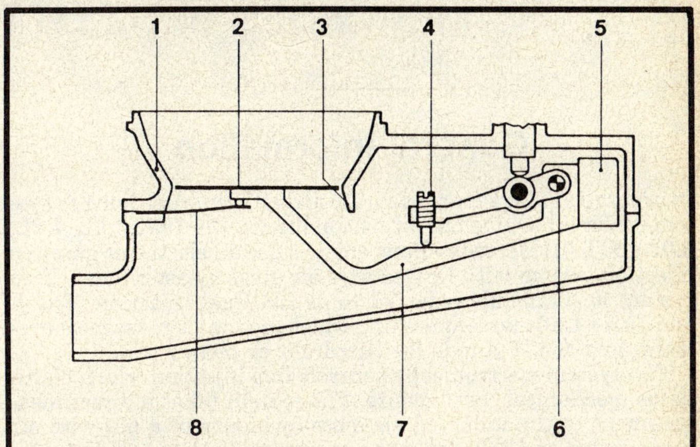

Air flow and mixture control unit

18. Install the air cleaner.
19. Push with an appropriate driver,against the adjusting device. Push this device downward by means of the appropriate driver against the spring force of the adjusting device. Turn slightly until hex head engages mixture control screw.

NOTE: Some model may require a screwdriver while others may use hex (allen) head type driver.

20. Turning the adjustment counterclockwise will make the mixture leaner, this increases the voltage. Turning the adjustment clockwise will make the mixture richer, this decreases the voltage.
21. Accelerate for a short period of time and test the mean control rod travel, readjust if necessary.
22. Remove the air cleaner.
23. Push down the new steel safety plug.
24. Install the air cleaner.
25. Reconnect the purge line.
26. Start engine, turn **ON** air conditioning, place transmission in **D** and turn wheel to full lock, the engine should not stall.

MERCEDES-BENZ LH FUEL INJECTION SYSTEM

General Information

In order to obtain a more accurate and efficient fuel delivery system under all engine operating conditions, the 1992 3.2L, 4.2L, 5.0L and 6.0L Mercedes-Benz engines use an electronic gasoline injection system with heated wire air mass measurement. This system is commonly referred to as LH Fuel Injection. The L stands for Luftmassenmessung which means air mass measurement, and the H stands for Hitzdraht or heated wire.

The system electronically controls fuel injection which eliminates mechanical components. The system takes a direct measurement of the inducted air mass by means of a hot wire air mass meter and calculates the proper amount of fuel to be injected into the cylinders. The system also uses input signal from the Lambda control (oxygen sensor) to determine injection quantity.

ELECTRONIC CONTROL MODULE

Functions performed by the LH control unit are: lambda control, oxygen sensor heater, purging of charcoal canister, air injection, raised transmission shift point, exhaust gas recirculation, camshaft timing adjustment, engine idle speed, fuel shutoff and self diagnostics.

If a malfunction should occur in the LH fuel injection air mass sensor, the system will revert to a Fixed Operating Mode (FOM). The FOM uses absolute intake manifold pressure, taken from the EZL/AKR ignition control unit, to calculate injection time.

The LH fuel injection system has several modes of operation to match current engine operating conditions. The modes are: starting control, after-start enrichment, warmup enrichment, acceleration enrichment, deceleration shutoff, maximum vehicle speed control, engine rpm limitation and fuel safety shutoff.

Inputs from various sensors allow the LH control unit to calculate current engine operating conditions. These sensor inputs are: engine speed, camshaft revolutions, coolant temperature, intake air temperature, idle speed and full load recognition, intake manifold absolute pressure, battery voltage and voltage supply.

DATA SENSORS

Hot Wire Air Mass Sensor

The air mass sensor works with a thin heated platnum resistance wire in the air intake stream. The wire is a temperature dependent resistor which increases resistance with temperature. The wire receives as much heating current as is required to heat it to 356°F (180°C) above the ambient air temperature. The temperature difference is always kept constant. The amount of current required to keep the wire at a constant temperature is used to calculate air mass by the control unit. Dirt deposited on the wire is cleaned by heating the wire to 1832°F (1000°C) after the engine has been stopped.

FUEL DELIVERY SYSTEM

Fuel is supplied by twin electric fuel pumps mounted on the frame rail. The fuel flows to a distribution manifold which ensures equal fuel pressure at all injectors. A pressure regulator, mounted at the end of the distribution manifold, regulates system pressure to 45–55 psi by returning a regulated amount of fuel to the fuel tank. The injection valves, one for each cylinder, meter a precise quantity of fuel into the intake manifold just prior to the intake valve.

Diagnosis and Testing

SERVICE PRECAUTIONS

- Keep all ECU parts and harnesses dry during service. Protect the ECU and all solid-state components from rough handling or extremes of temperature.
- Use extreme care when working around the ECU or other components; the air bag or SRS wiring may be in the vicinity. On these vehicles the SRS wiring and connectors are red; do not cut or test these circuits.
- Before attempting to remove any parts, turn the ignition switch **OFF** and disconnect the negative battery cable.

To prevent personal injury or damage to the vehicle, follow all service precautions.

- Always disconnect the negative battery terminal to prevent sparks cause by short-circuiting, circuit-breaking, etc.
- Never disconnect the battery with the engine running.
- Smoking must not be allowed when working around flammable liquids.
- Always keep a CO_2 fire extinguisher close on hand.
- Absorbant must be available to soak up any spillage.
- Ensure fuel is emptied into a suitable container.
- Always place the transmission gear shift lever in **N** (manual transmission) or **P** (automatic transmission) and set the parking brake and block drive wheels before starting the engine.
- Keep hands, tools and clothing away from engine cooling fan to help prevent personal injury.
- Thoroughly clean all unions and area near connections before disconnecting.
- Always disconnect the battery when using arc, spot or electrical welding equipment.
- When cranking engine without starting, as for a compression test, disconnect wire from distributor, and connect to ground. Do not hand hold wire.
- Do not apply battery power directly to the injectors.
- Do not disconnect the injector harness connectors with the engine running.
- When testing the ECU with a DVOM make sure that the probes of the tester never touch each other as this will result in damage to a transistor in the ECU.
- Always use a 12 volt battery as a power source.
- Do not attempt to disconnect the battery cables with the engine running or the ignition key **ON**.
- If a battery terminal is disconnected (with the ignition key **OFF**), the memory will return to the ROM value. The computer will now self-control at its initial value. The engine operation can vary slightly when the terminal is disconnected. However, this is not an indication of a problem allow the computer time to relearn before searching for a problem.
- Do not depress the accelerator pedal when starting.

READING CODES

1. Connect the impulse tester to the terminals the diagnostic connector (X11/4). The connector is located on the firewall.
2. Turn the ignition **ON** with the selector lever in **P**.
3. Press the start button for no less than 2, but no more than 4 seconds and observe the indicated impulse readout.
4. Continue to press the start button to obtain additional fault codes. All codes must be read prior to servicing.
5. Repair or replace the component and repeat the test procedure to ensure no faults are present.

CLEARING CODES

1. With the impulse tester connected and the fault code displayed, press the start button for 6 seconds or more.
2. Each impulse code must be erased individually.
3. If there is no number displayed, then the stored fault has been erased.
4. If the display shows the number 1, then there are more system faults in memory.

ENGINE CODES

Code 1—No malfunctions in system
Code 2—Oxygen sensor
Code 3—Lambda control system function
Code 4—Air injection system function
Code 5—EGR system function
Code 6—Idle speed control
Code 7—Ignition map monitoring
Code 8—Coolant temperature sensor
Code 9—Intake air temperature sensor
Code 10—Air mass sensor with hot wire
Code 11—RPM signal
Code 12—Oxygens sensor heater
Code 13—Signal preparation for camshaft position sensor
Code 14—Intake manifold pressure during starting
Code 15—Full load operation
Code 16—Idle speed operation
Code 17—Data transfer
Code 18—Camshaft position signal
Code 20—Vehicle speed
Code 21—Regeneration switchover valve
Code 22—Camshaft position sensor
Code 23—Intake manifold absolute pressure with engine running
Code 24—Flywheel segments (crankshaft position sensor)
Code 25—Knock sensors
Code 26—Delayed upshift switchover valve
Code 27—Coolant temperature sensor circuits outside tolerance range
Code 28—Coolant temperature monitoring

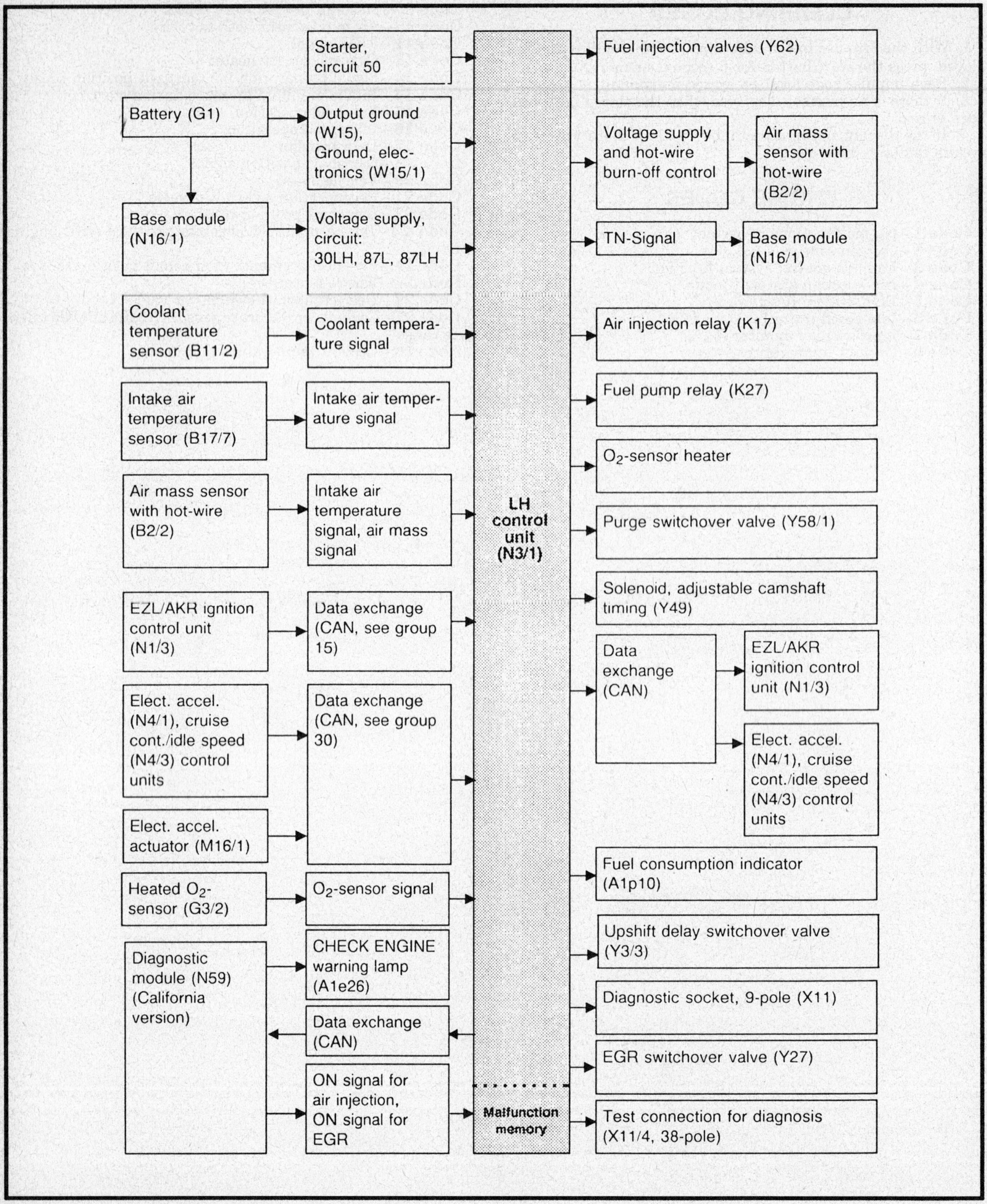

Block diagram of the LH control unit—104 engine

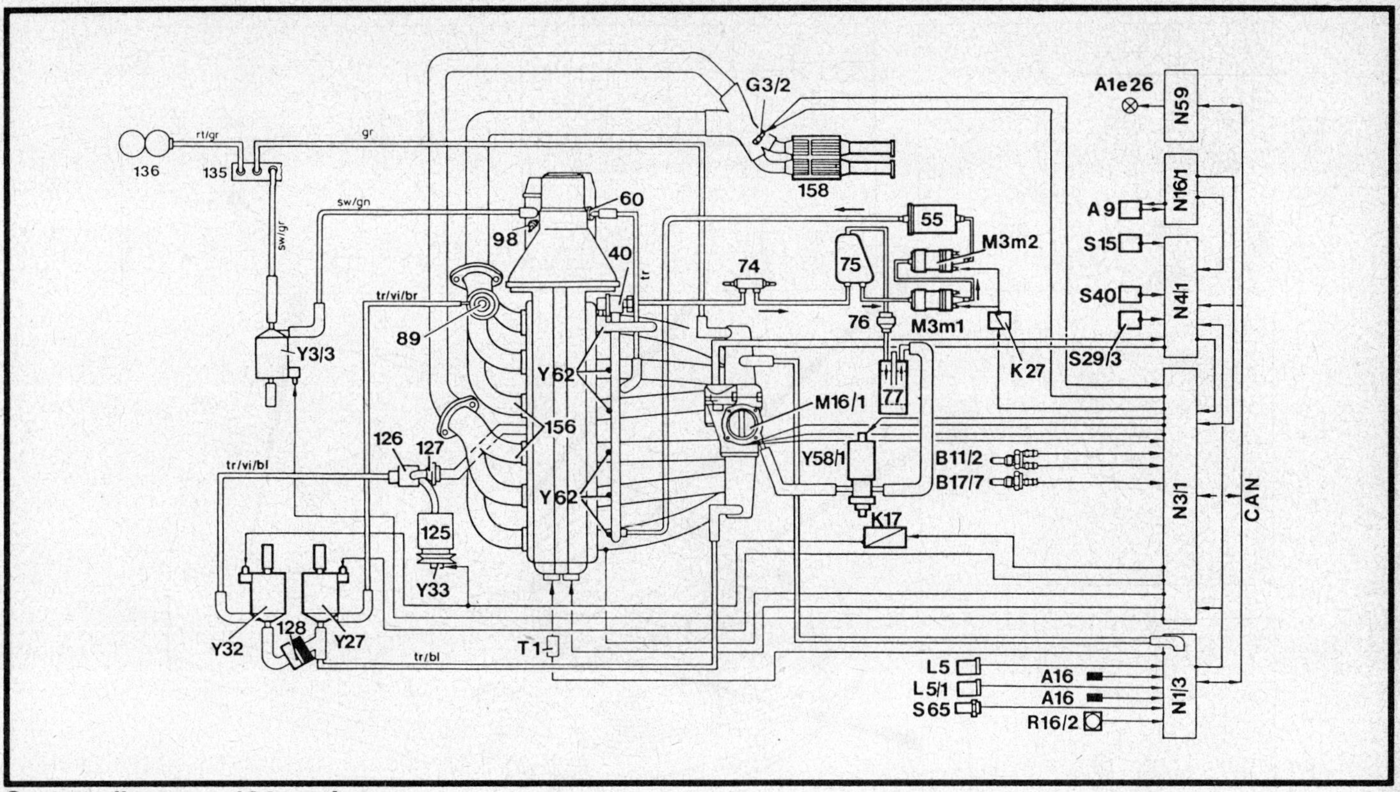

System diagram—104 engine

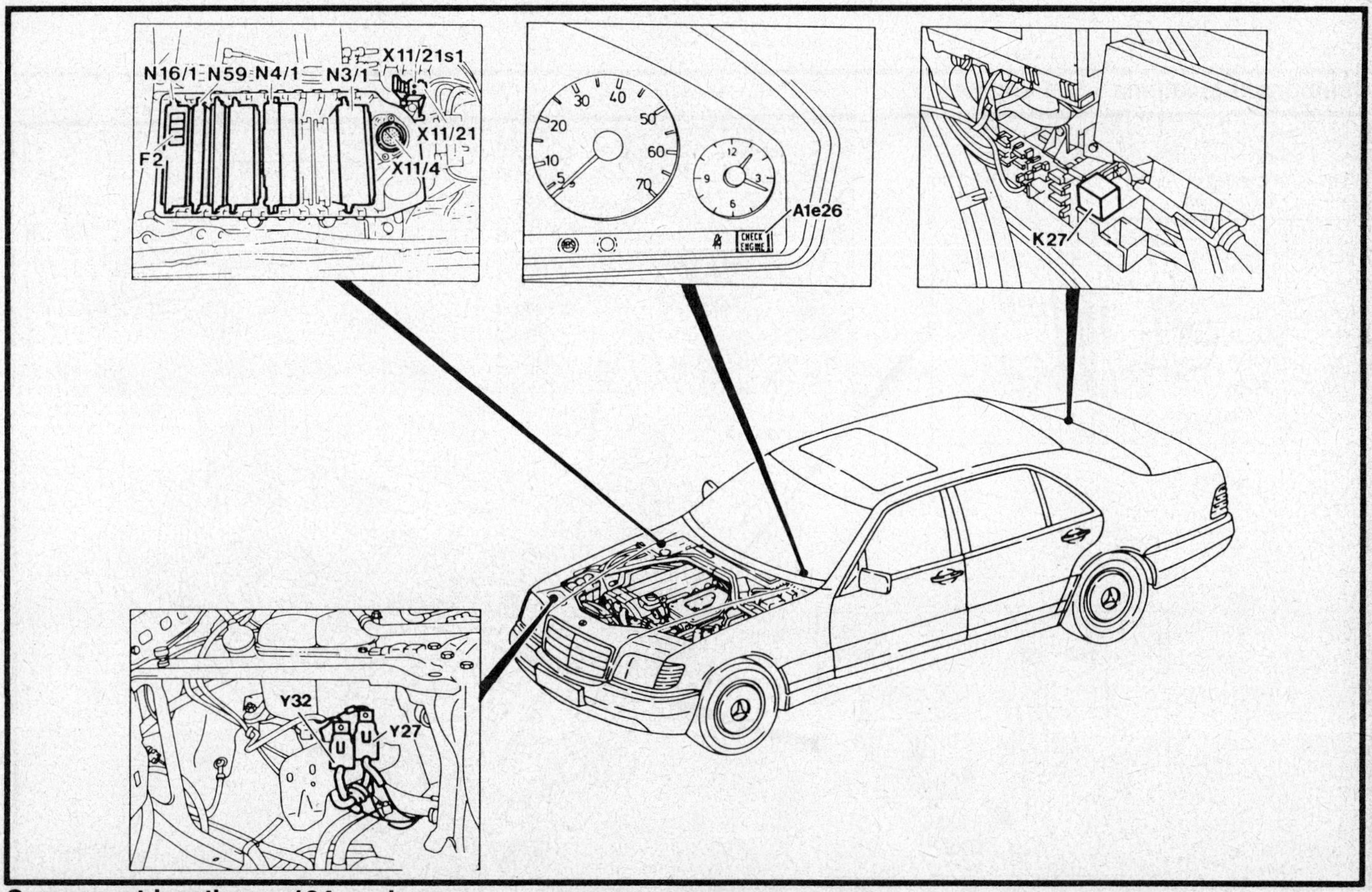

Component locations—104 engine

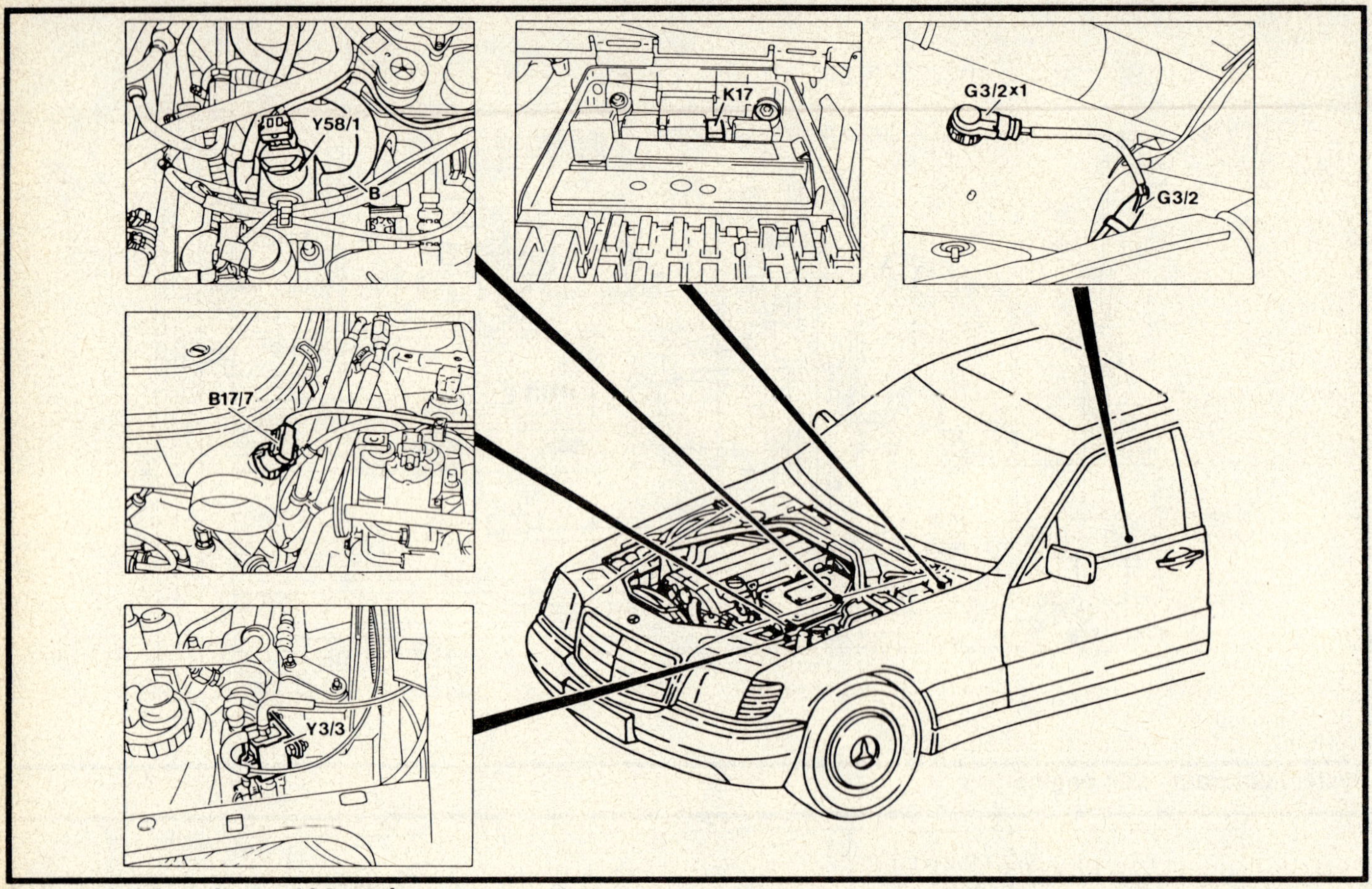

Component locations—104 engine

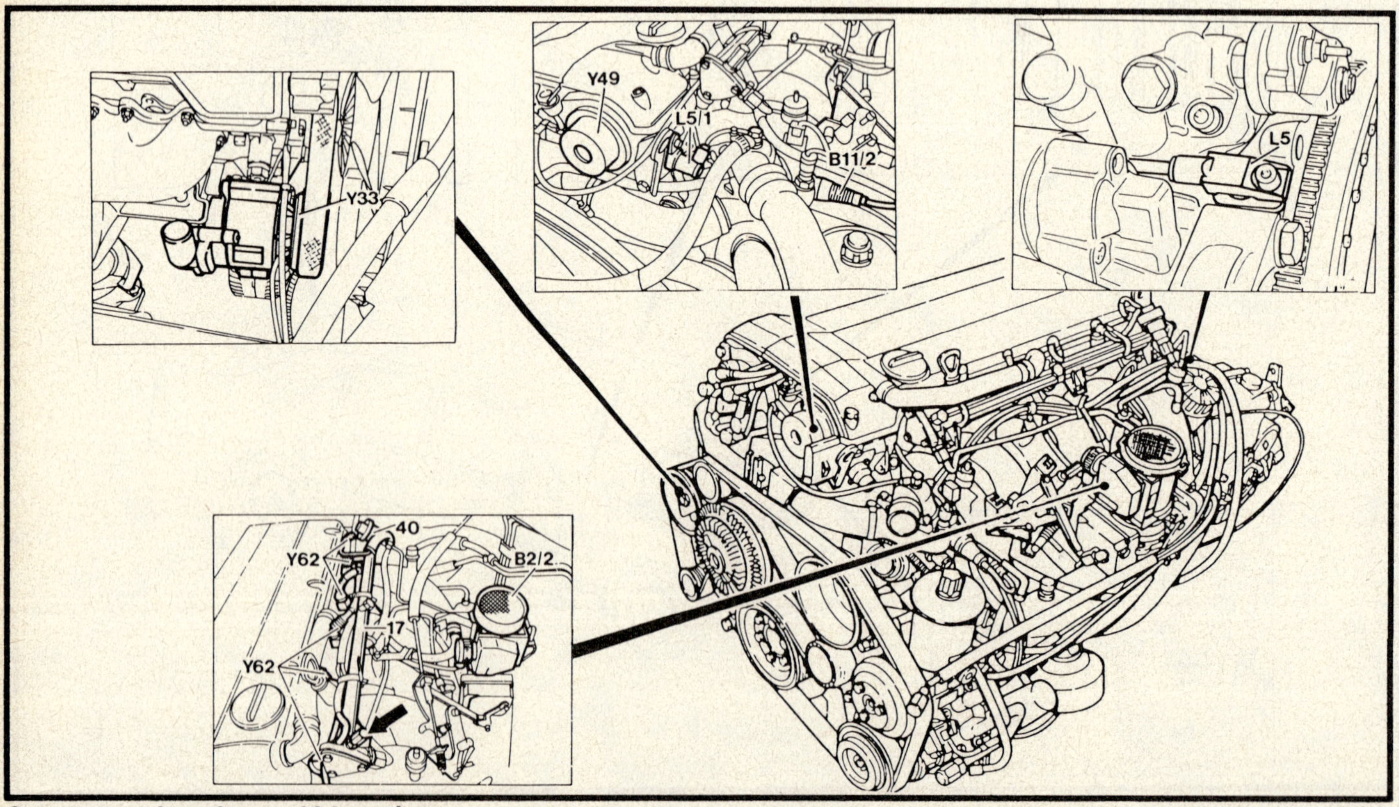

Component locations—104 engine

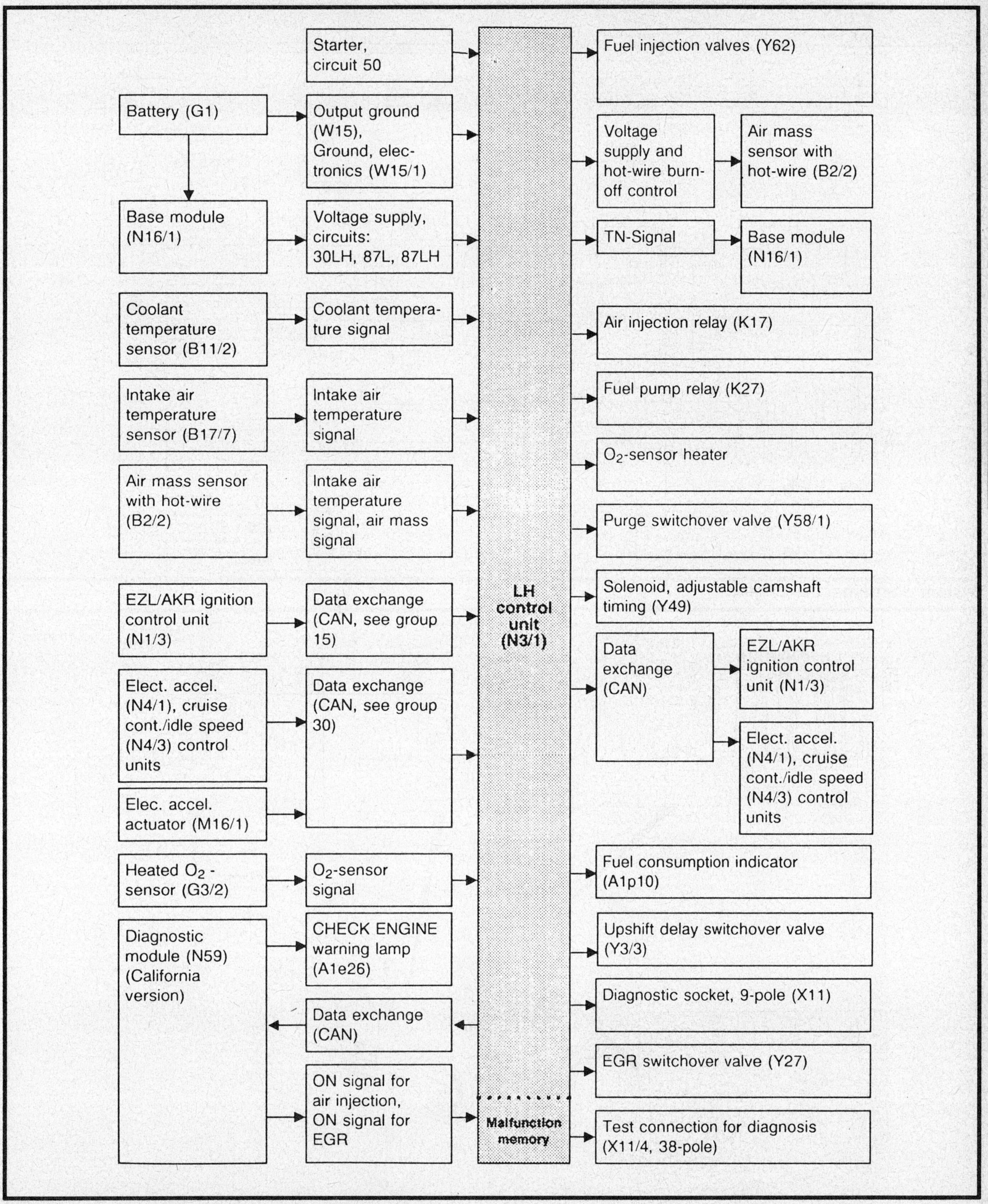

Block diagram of the LH control unit—119 engine

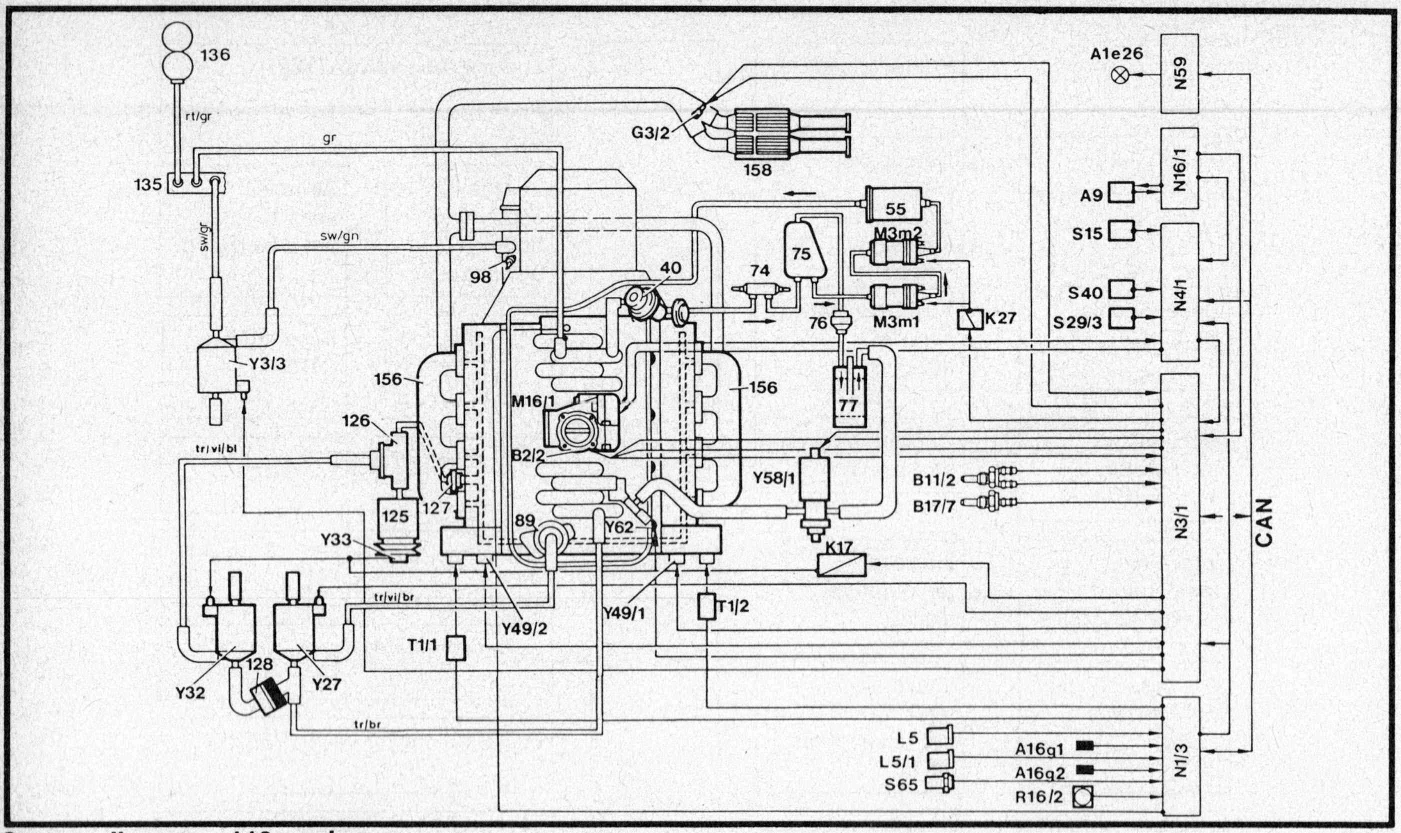

System diagram—119 engine

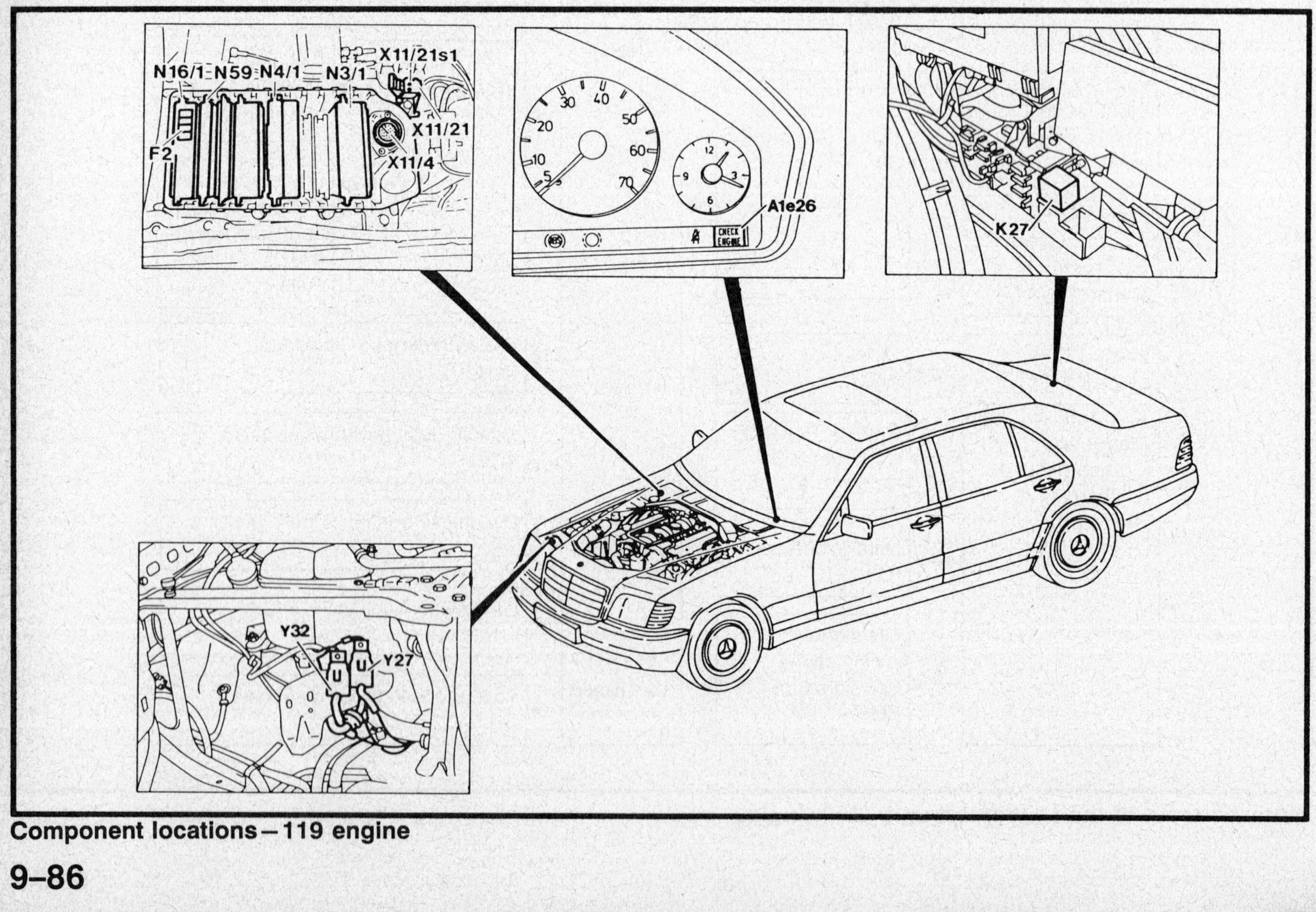

Component locations—119 engine

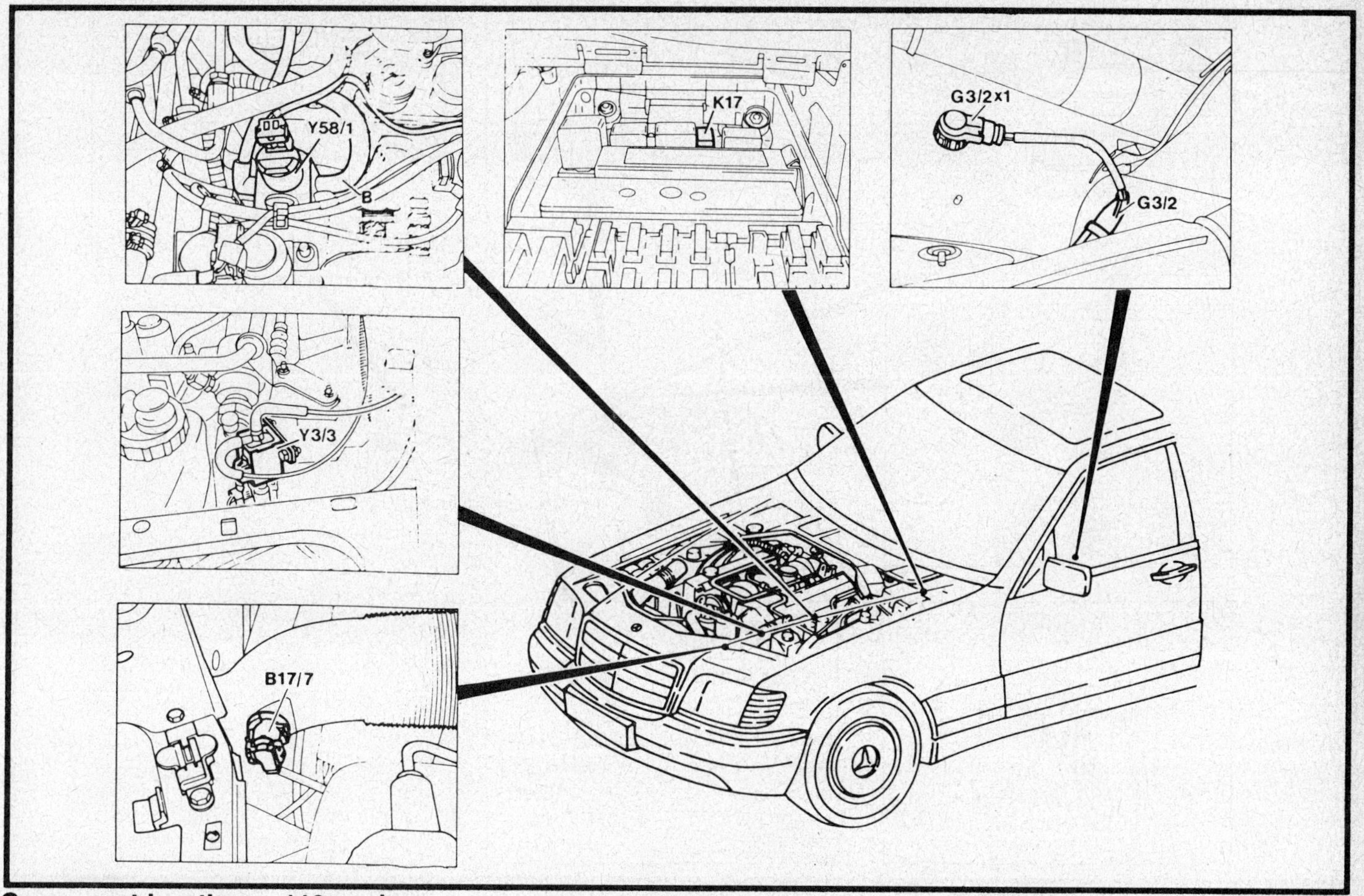

Component locations—119 engine

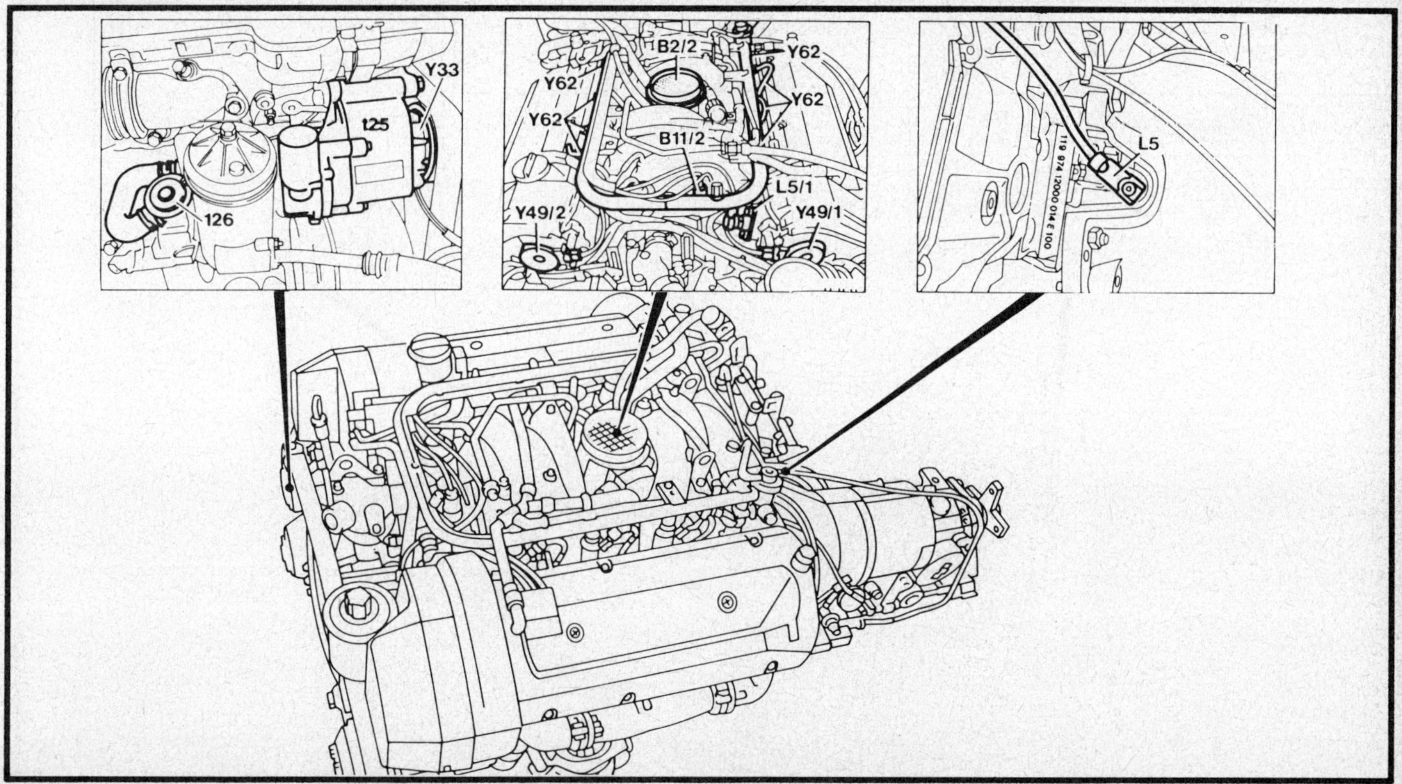

Component locations—119 engine

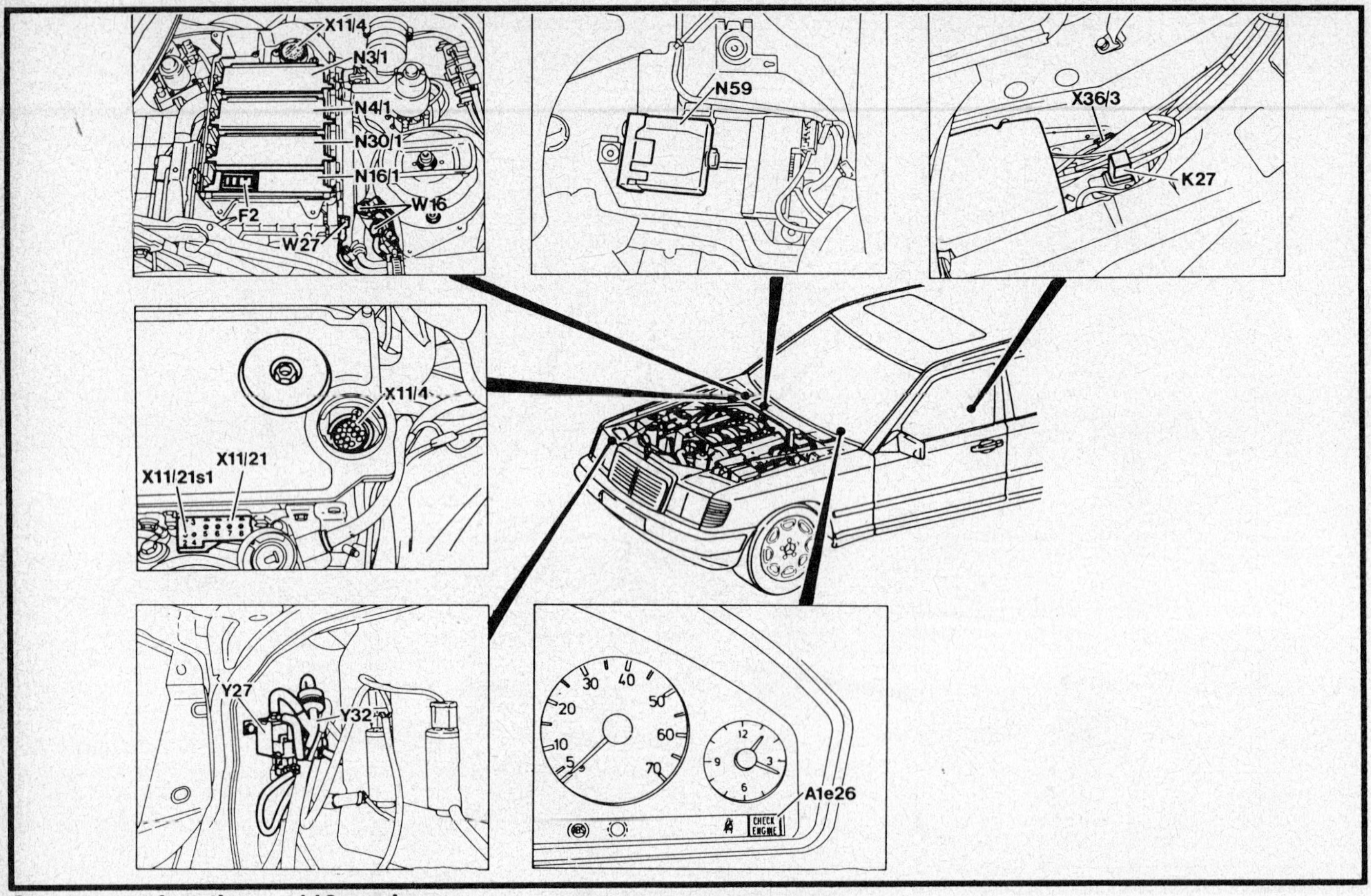

Component locations—119 engine

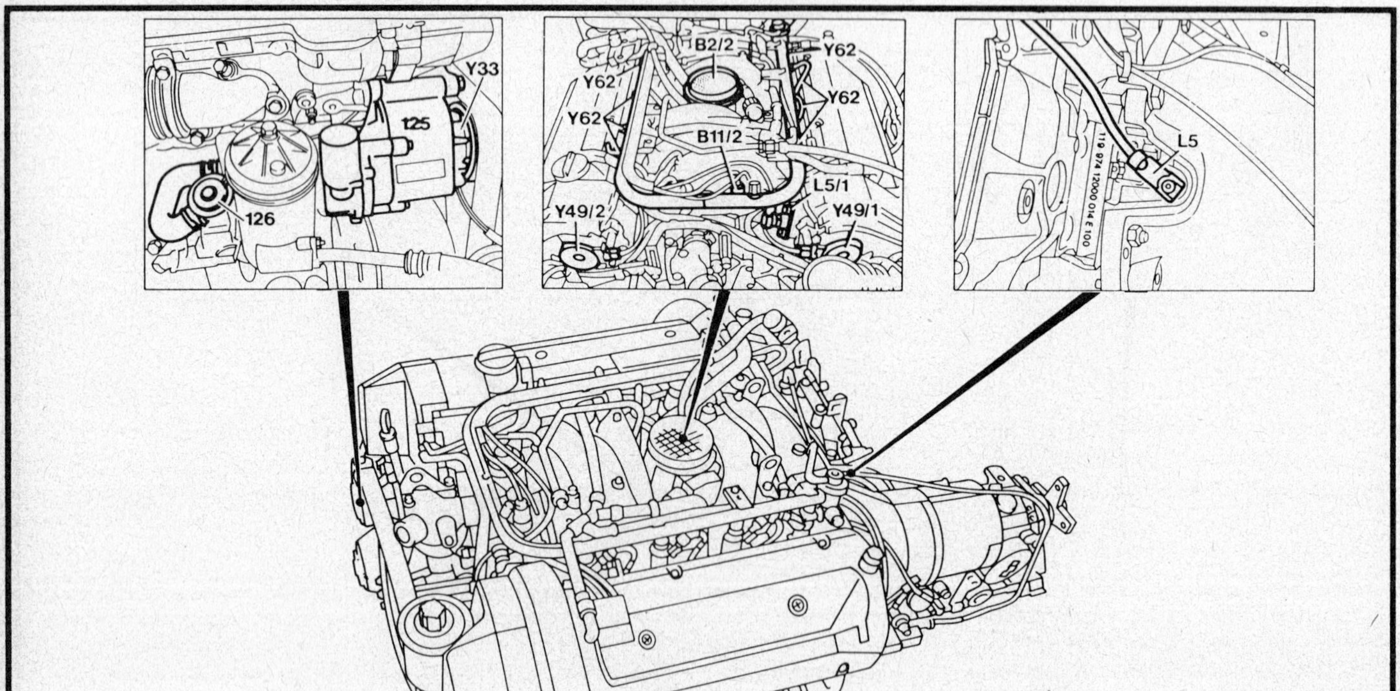

Component locations—119 engine

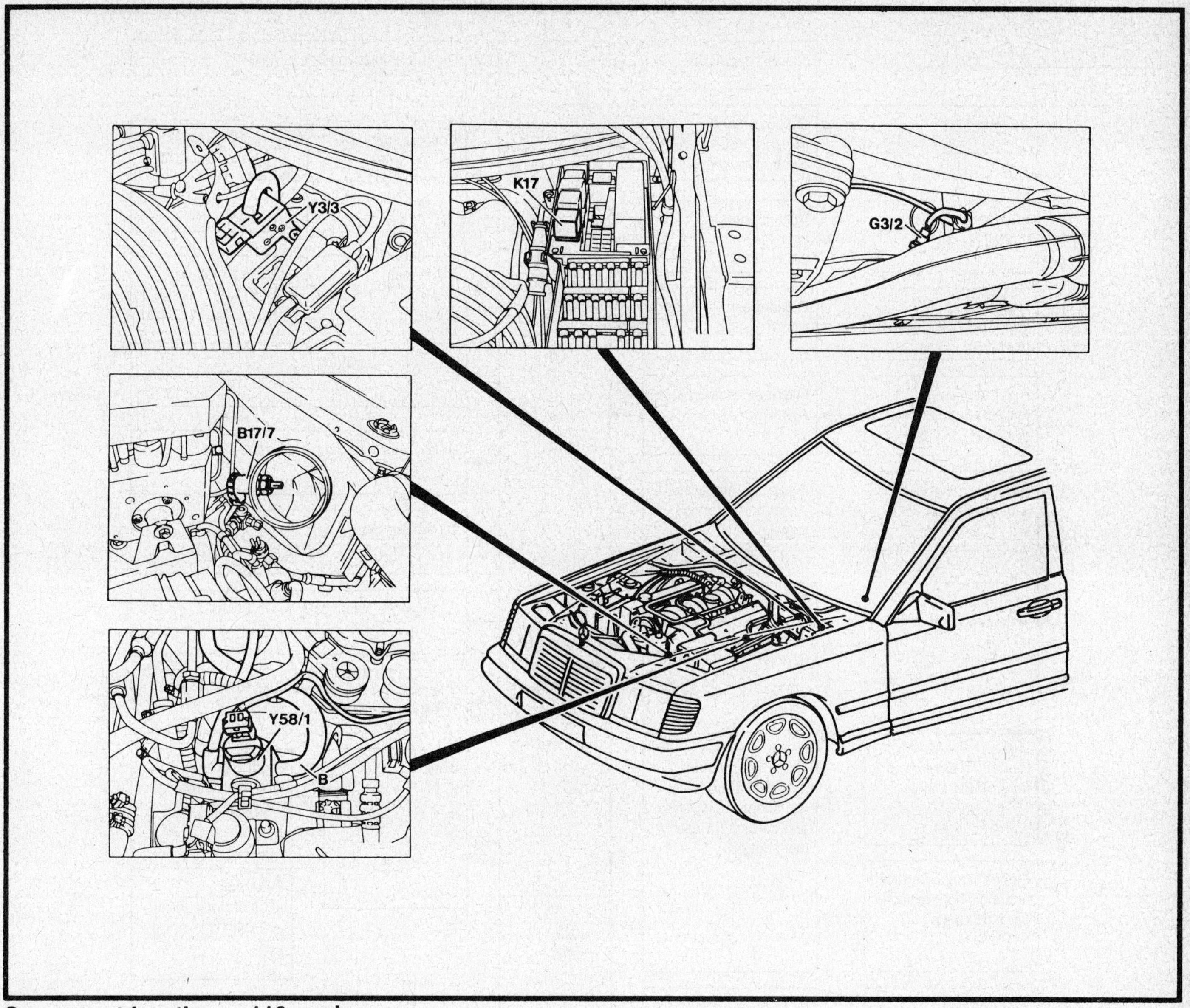

Component locations—119 engine

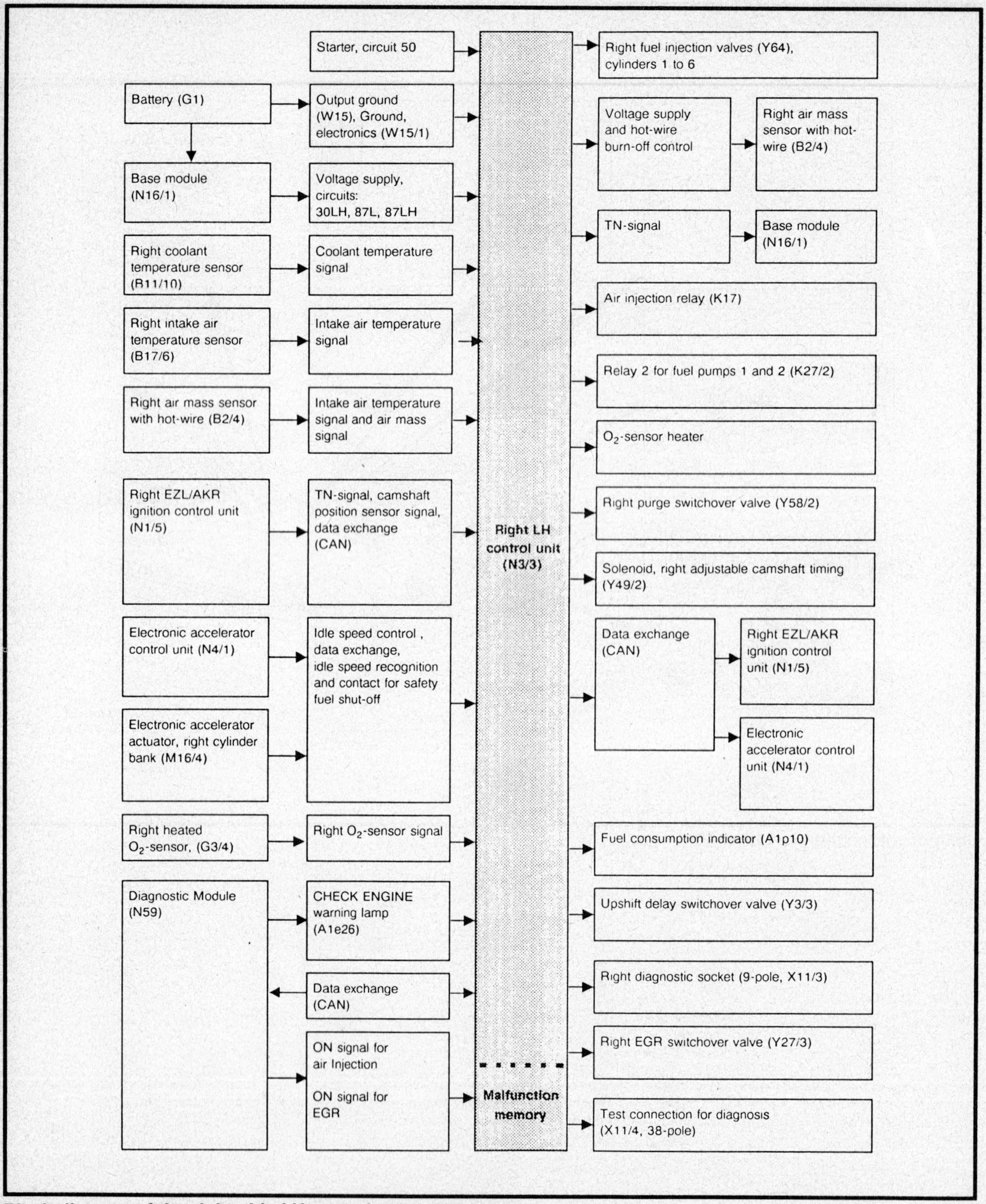

Block diagram of the right side LH control unit – 120 engine

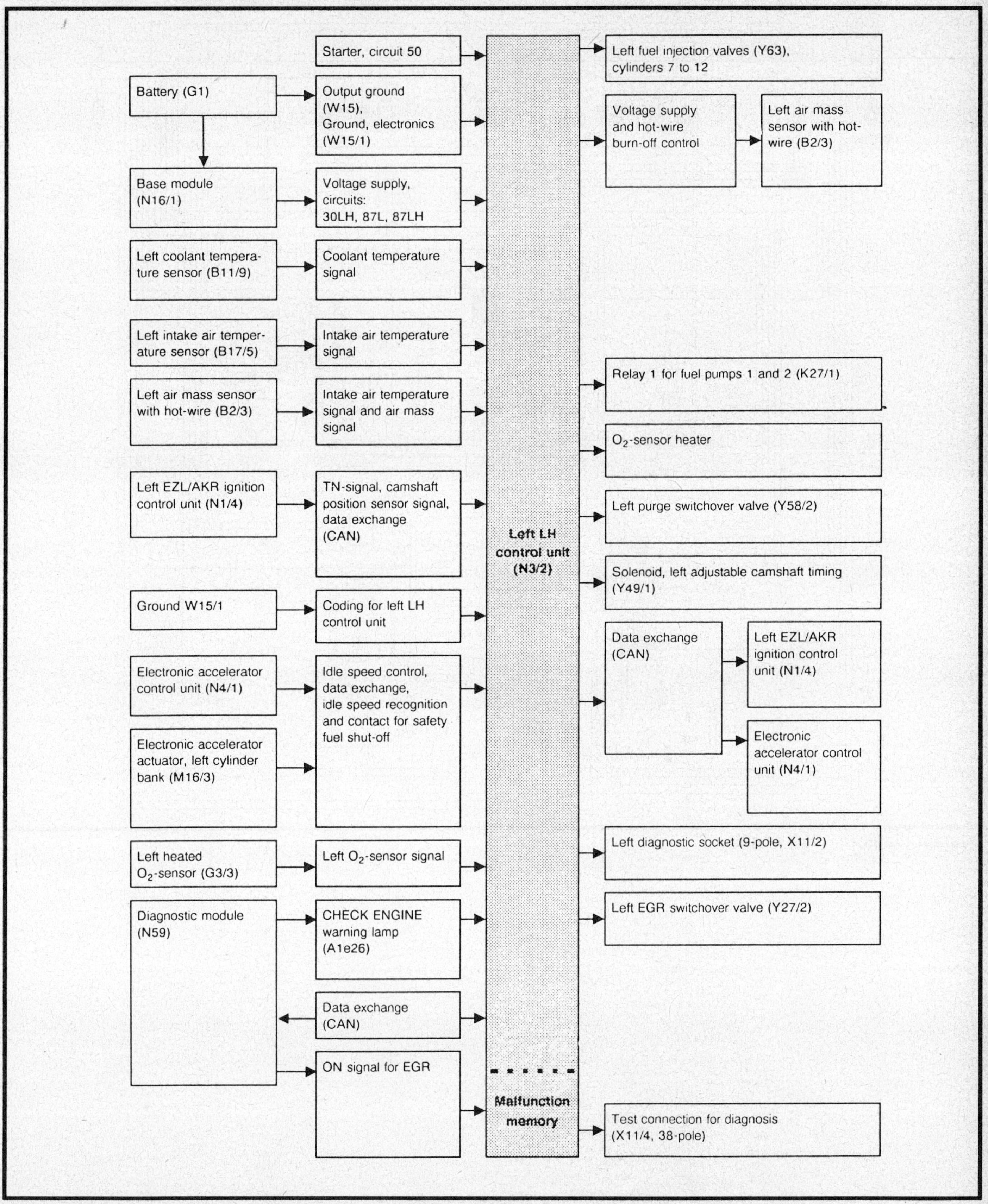

Block diagram of the left side LH control unit–120 engine

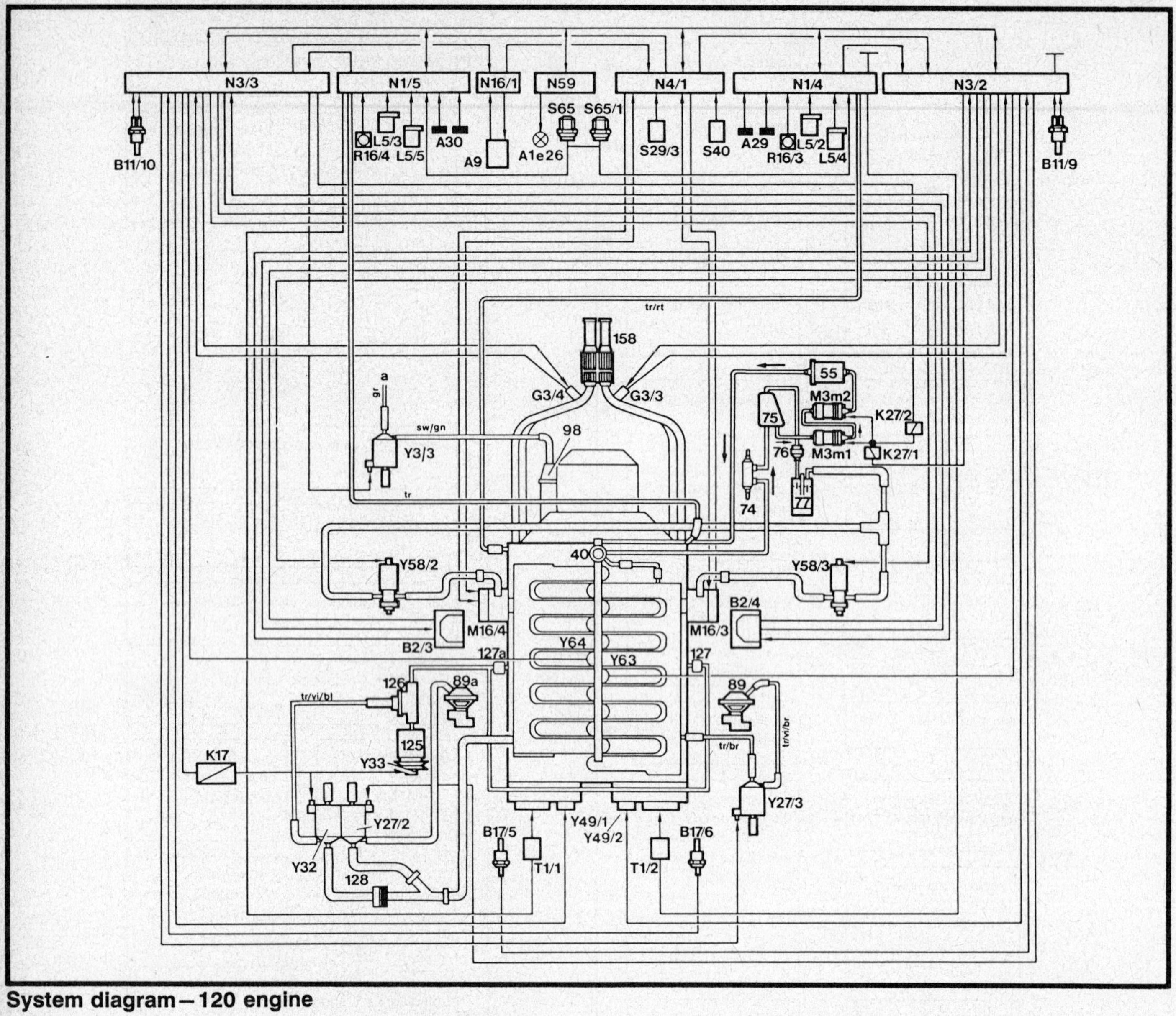

System diagram—120 engine

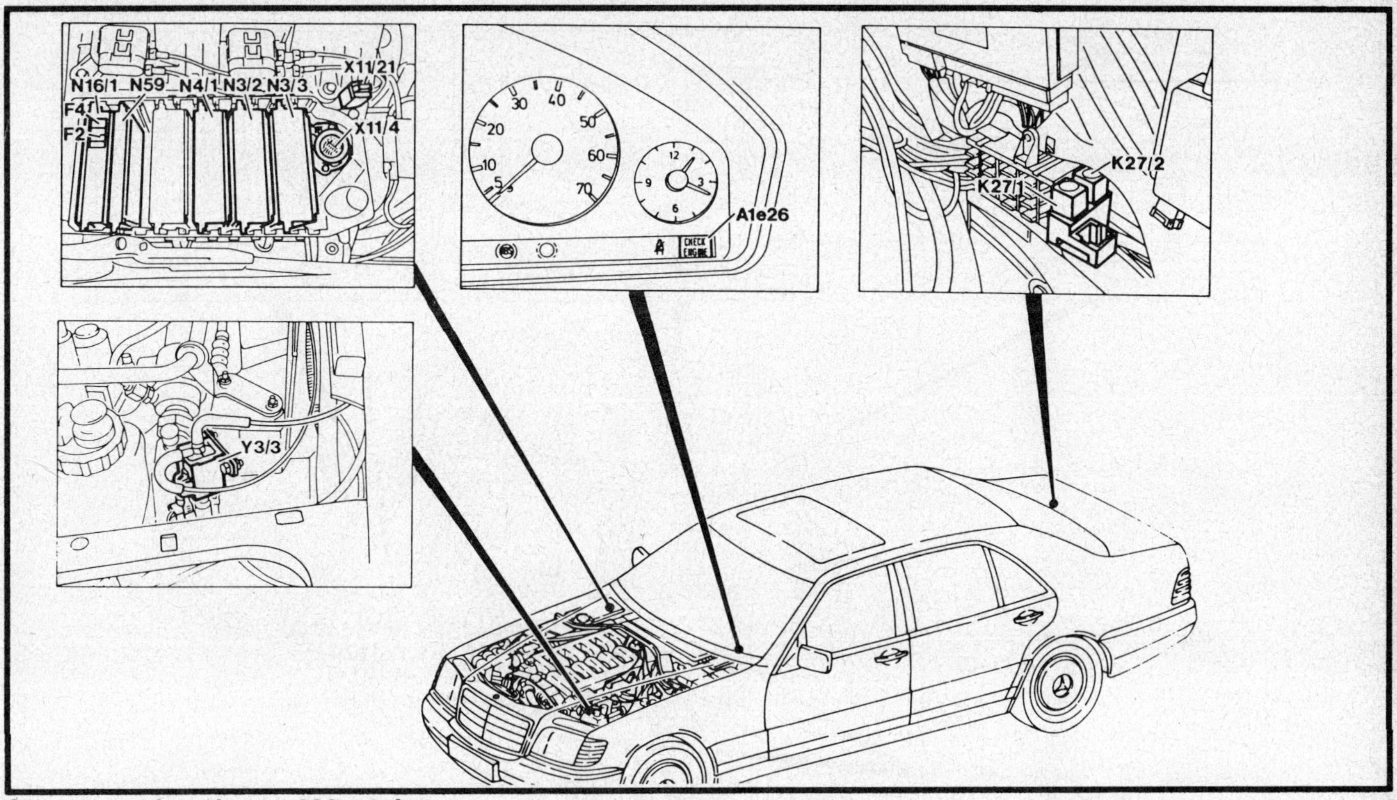

Component locations – 120 engine

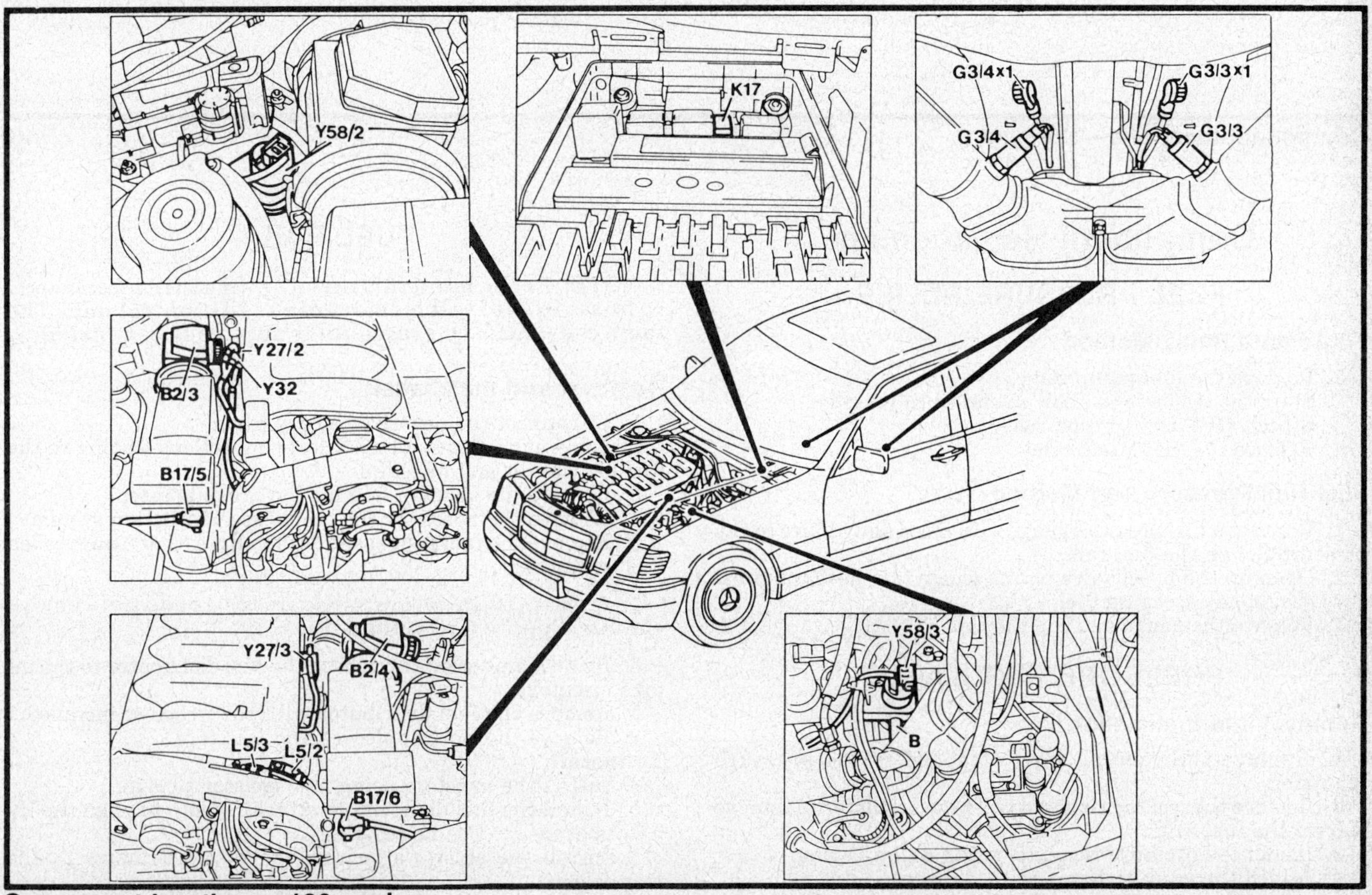

Component locations – 120 engine

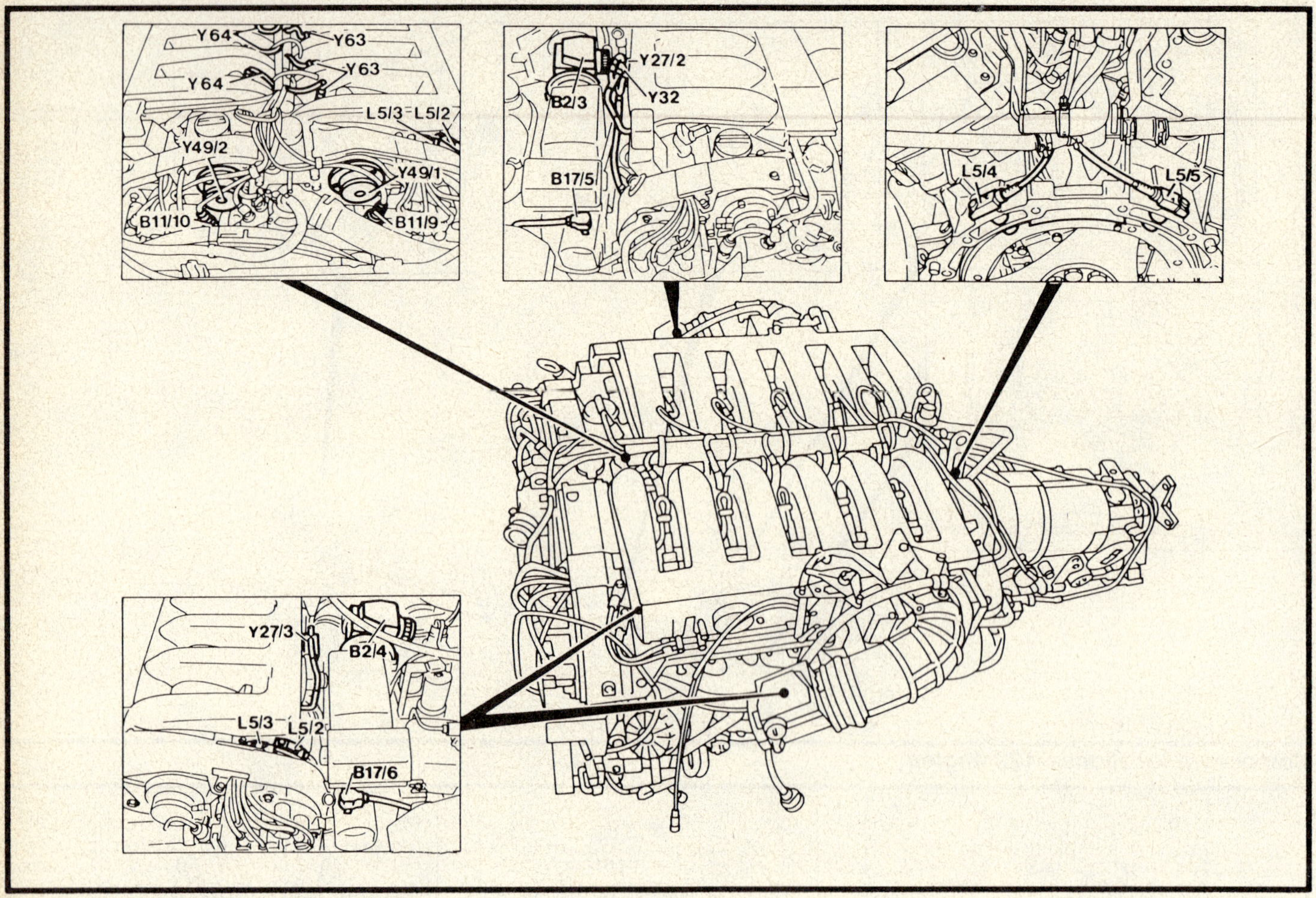

Component locations—120 engine

Component Replacement

FUEL PRESSURE RELIEF

Fuel Pump Relay Method

1. Remove the fuel pump relay.
2. Start the engine and allow to run until it stalls.
3. Switch **OFF** the ignition switch.
4. Replace the fuel pump relay.

Fuel Rail Pressure Tap Method

1. Connect a fuel pressure gauge with bleed down valve to the pressure tap on the fuel rail.
2. Open up the bleed valve on the gauge to relieve the pressure. Catch any escaping fuel.
4. Remove the gauge and cap the tap.

PRESSURE REGULATOR

Removal and Installation

1. Remove the vacuum hose attached to the pressure regulator.
2. Remove the locking ring and twist the regulator to remove it from the fuel rail.
3. Install the pressure regulator using a new O-ring.
4. Attach the vacuum hose.

FUEL RAIL

NOTE: Performing this service will cause a fault code to be set in the fault memory of the LH control unit. The fault code must be erased after service is completed.

Removal and Installation

1. Disconnect the negative battery cable.
2. Label and remove all electrical connections leading to the fuel distributor and injectors.
3. Remove the vacuum line for the brake booster.
4. Remove the hoses for the crankcase ventilation system.
5. Remove the fuel supply and return lines. Catch any spilled fuel with a rag.

NOTE: If the injectors are to be tested, do not remove the fuel supply and return lines.

6. Remove the screws attaching the fuel distributor to the intake manifold.
7. Remove the fuel distributor with the injectors installed.
8. Remove the isolator sleeves and injectors.

To install:

9. Install the injectors using new isolator sleeves.
10. Install the fuel distributor rail taking care to align the injectors properly. Use new O-rings.
11. Install the screws attaching the fuel distributor to the manifold.

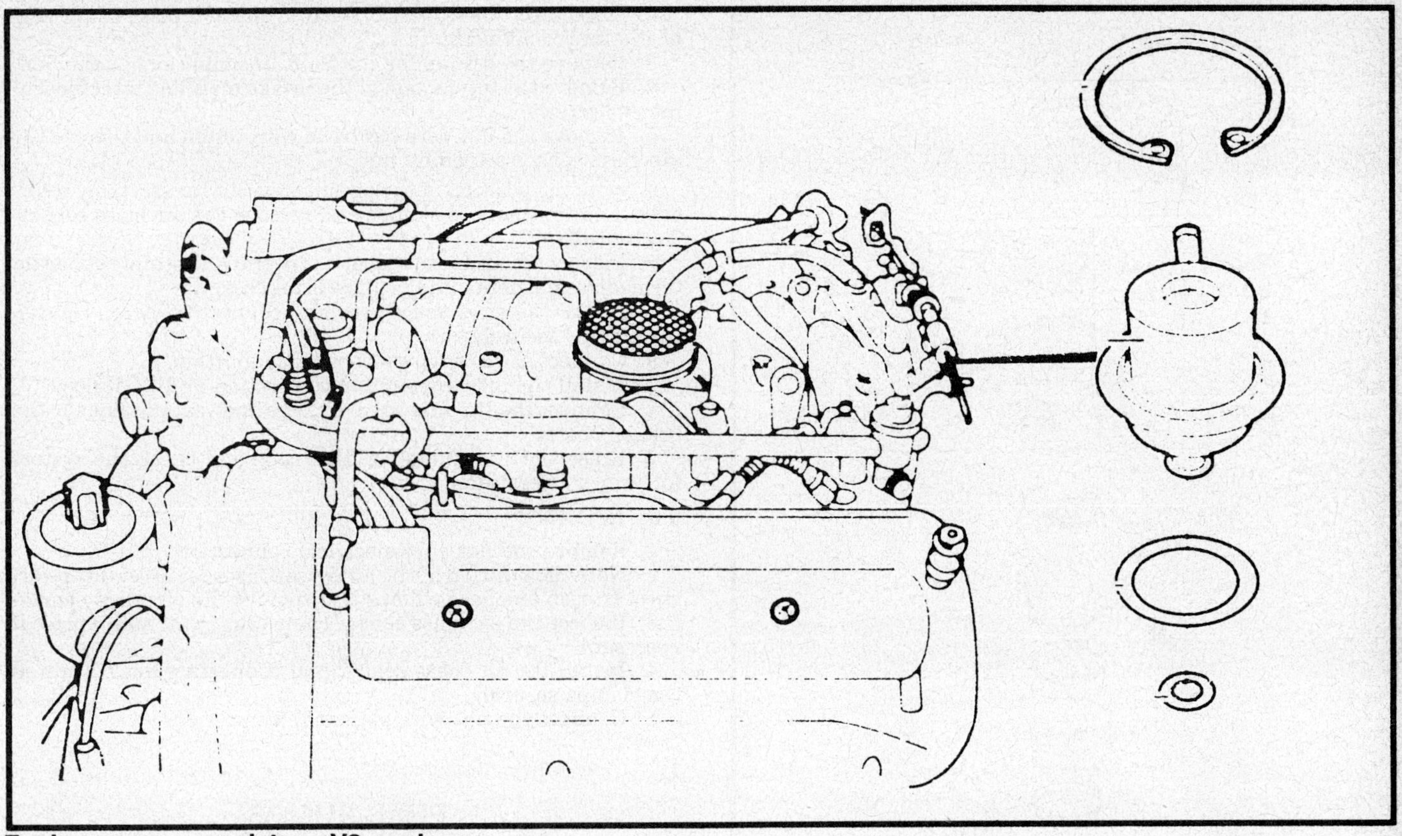

Fuel; pressure regulator—V8 engines

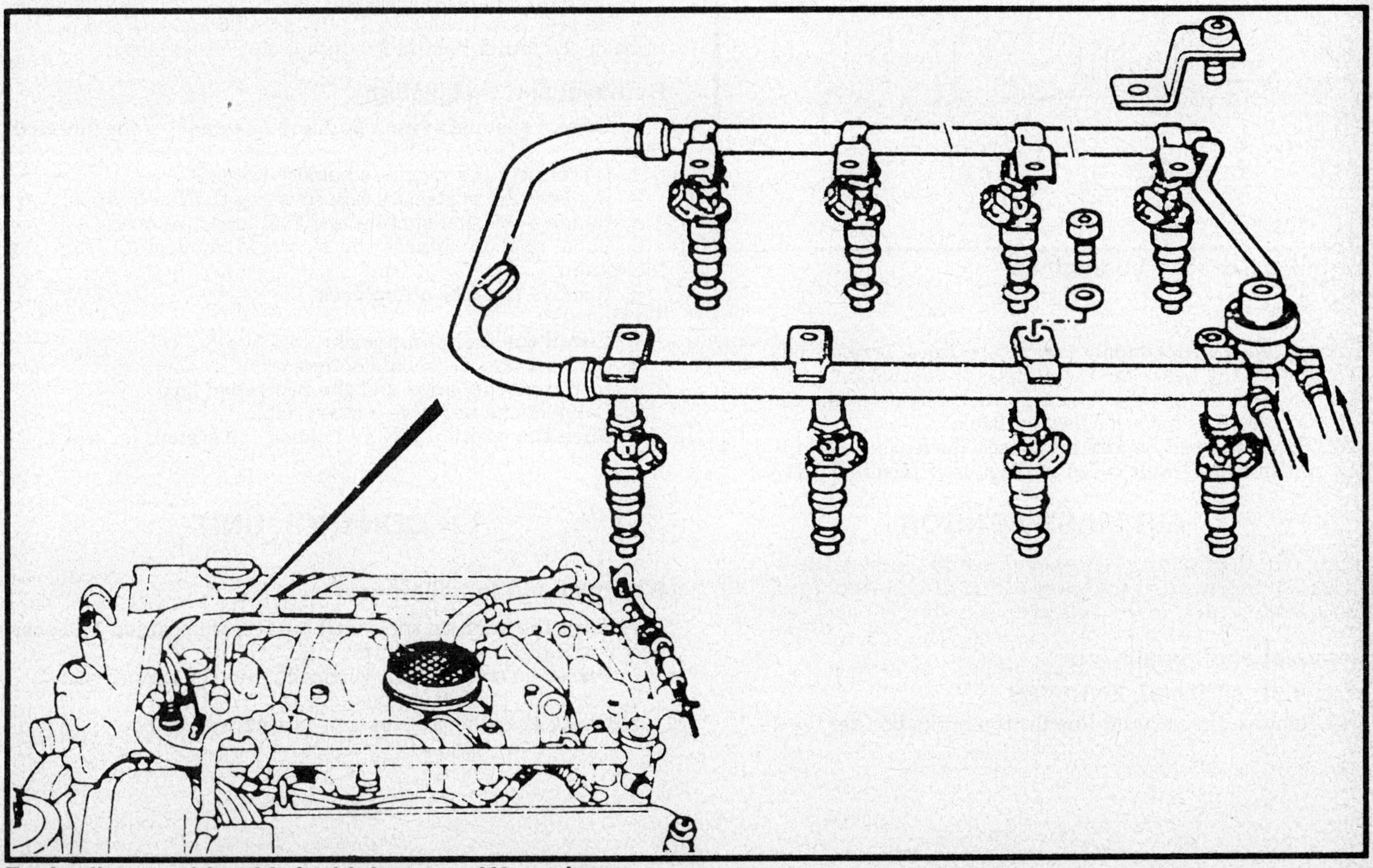

Fuel rail assembly with fuel injectors—V8 engines

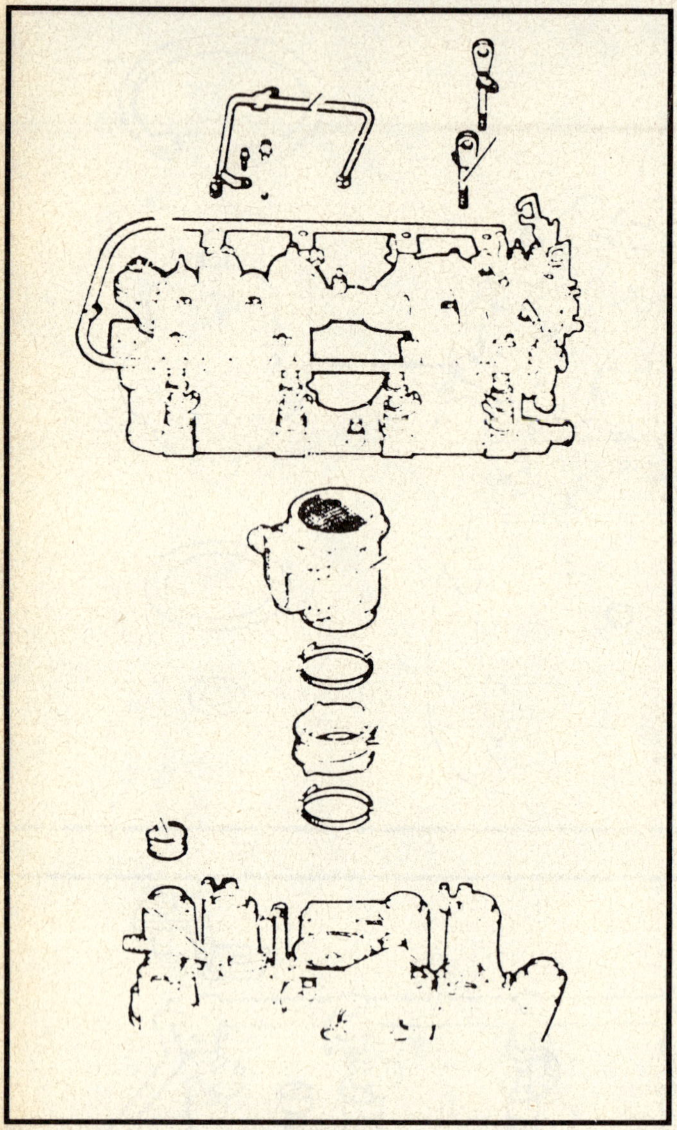

Air mass sensor—V8 engines

12. Install the fuel supply and return lines, if removed.
13. Install the hoses for the crankcase ventilation system.
14. Connect all electrical wiring previously disconnected.
15. Connect the negative battery cable.
16. Turn the ignition **ON** and check for leaks.
17. Clear stored fault codes at diagnostic terminal X11/4.

AIR MASS SENSOR

NOTE: Performing this service will cause a fault code to be set in the fault memory of the LH control unit. The fault code must be erased after service is completed.

Removal and Installation

3.2L, 4.2L AND 5.0L ENGINES

1. Remove the vacuum line for the brake booster.
2. Disconnect the valve accelerator control linkage at the rear of the intake manifold.
3. Remove the screws for the EGR line and remove the line.
4. Remove the top section of the intake manifold after loosening the screws.
5. Remove the hot wire electrical connection and remove the air mass sensor attaching nut.
6. Note the position of the air mass sensor for assembly reference. Loosen the hose clamps and remove the air mass sensor.

To install:

7. Inspect the air mass sensor connecting hose and rubber intermediate elements and replace if necessary.
8. Install the air mass sensor and connecting hose. Tighten the clamps securely.
9. Connect the hot wire electrical connection.
10. Install the intake manifold top section and EGR hose.
11. Connect the throttle valve linkage and vacuum line for the brake booster.
12. Erase the stored malfunction code and check the system for proper operation.

6.0L ENGINE

1. Remove the hot wire electrical connection.
2. Note the position of the air mass sensor for assembly reference. Loosen the hose clamps and remove the air mass sensor.
3. Inspect the air mass sensor connecting hose and replace if necessary.
4. Install the air mass sensor and connecting hose. Tighten the clamps securely.
5. Connect the hot wire electrical connection.

FUEL PUMPS

The fuel pump relay is located on the right side of the vehicle below the rear bench seat on E, CE and TE models and on the right of the trunk behind the lining on SEL models.

Removal and Installation

1. Release the fuel system pressure by removing the fuel tank cap.
2. Disconnect the negative battery cable.
3. Remove the protective box covering the fuel pumps.
4. Remove the fuel suction and fuel pressure hoses.
5. Label and disconnect the electrical connectors from the fuel pumps.
6. Remove the fuel pump pack.

To install:

7. Install the fuel pump pack.
8. Connect the electrical connectors.
9. Install all fuel lines and the protective box.
10. Connect the negative battery cable.
11. Turn the ignition **ON** and check the system for leaks.

LH CONTROL UNIT

Removal and Installation

1. Turn the knob on the control unit from position **1** to position **0** and lift the control unit out.
2. Press the control unit vertically into the connector.
3. Turn the knob from position **0** to position **1**.
4. Check that the control unit is securely seated.

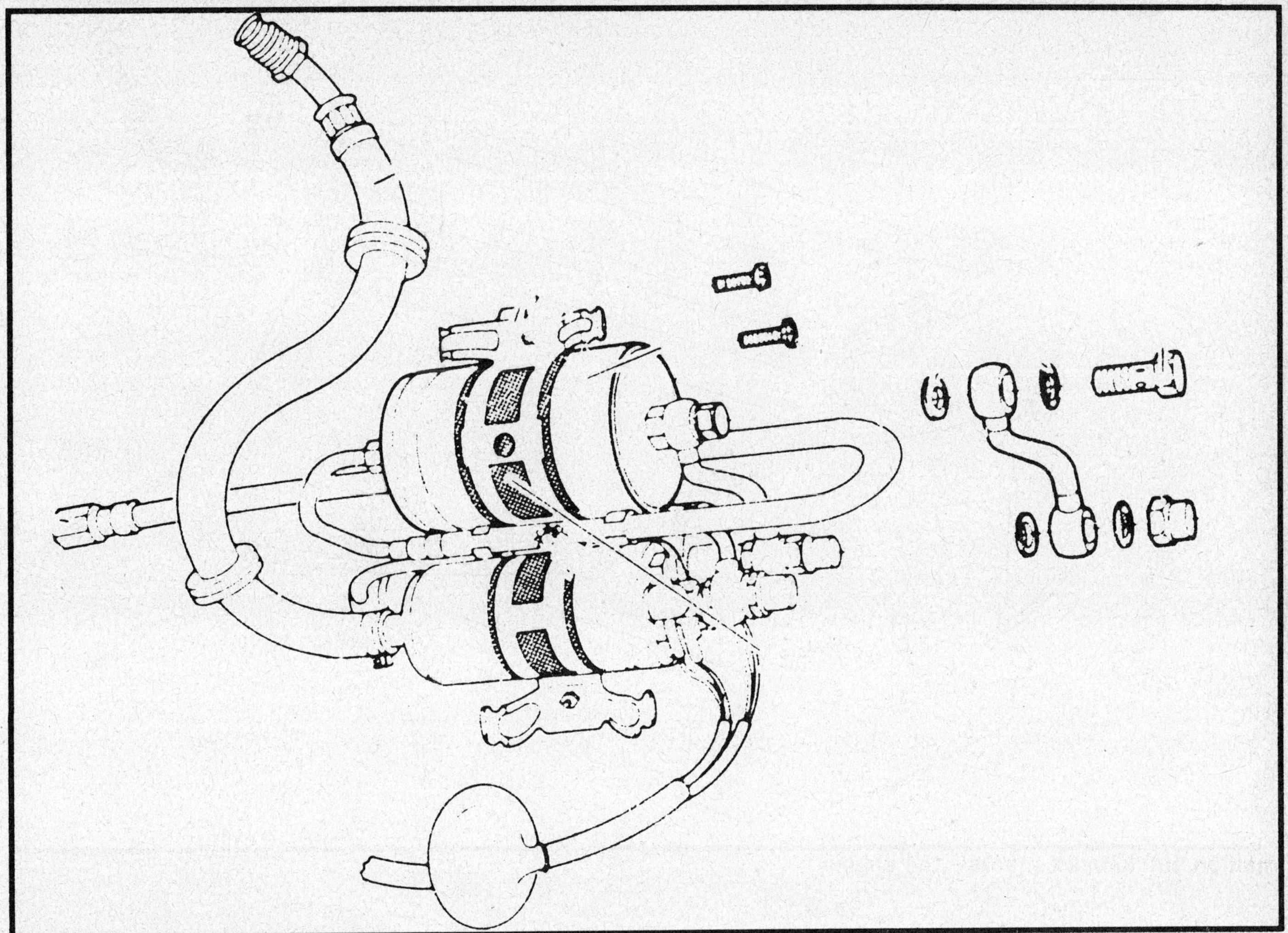

Fuel pump assembly—Mercedes

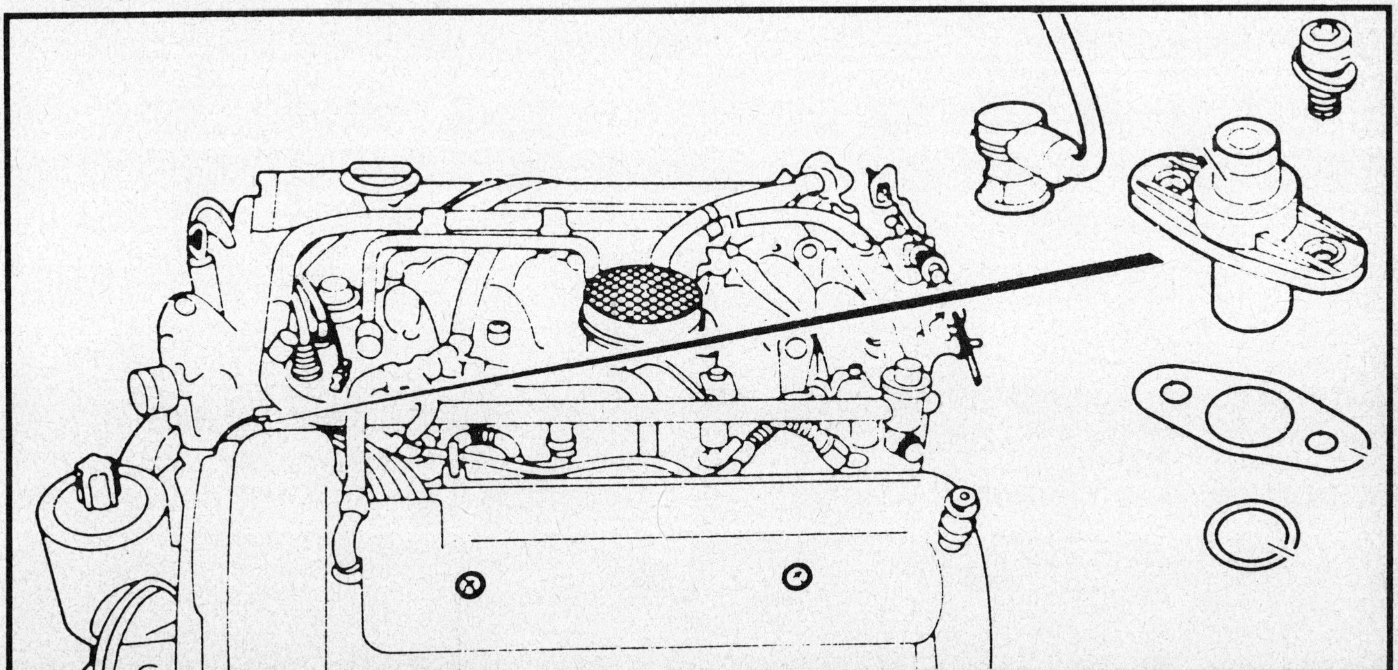

Camshaft position sensor—V8 engines

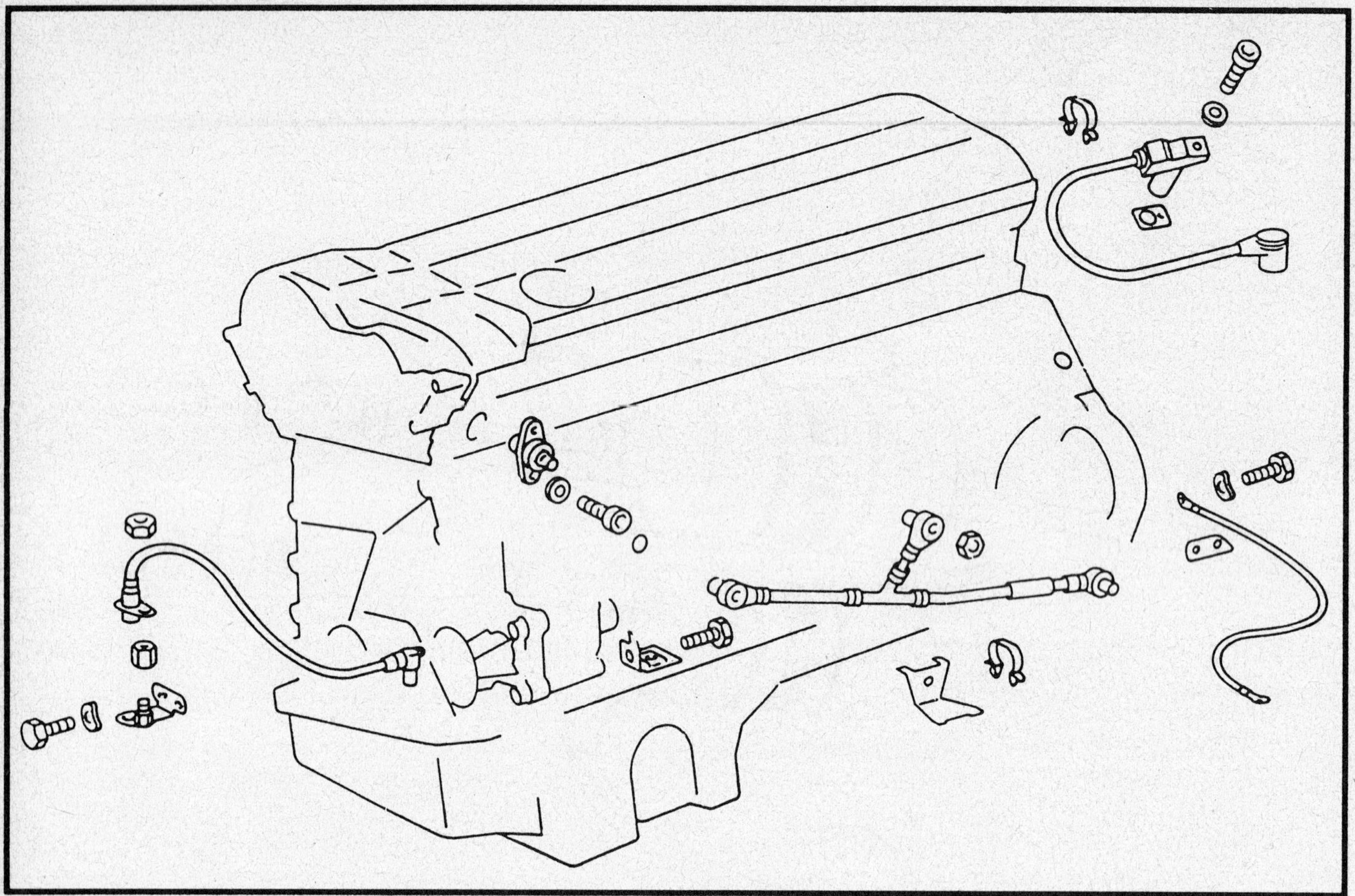

Position and knock sensors—104 engine

Merkur Fuel Injection

SECTION 10

MERKUR EEC-IV ELECTRONIC FUEL INJECTION SYSTEM

General Information

The Electronic Fuel Injection (EFI) system used in the Scorpio is classified as a multi-point, pulse time, mass air flow injection system. Fuel is metered into the intake air stream through the injectors mounted in the lower manifold near the intake ports.

An on-board vehicle Electronic Engine Control (EEC-IV) microcomputer processes inputs from various engine sensors, then outputs a prescribed fuel flow throughout the entire operating range. This allows the air/fuel ratio to be very accurately controlled. The EEC-IV engine control system automatically senses and compensates for changes in altitude, it will also to permit push starting the vehicle should it become necessary (manual transmission only).

SYSTEM OPERATION

The fuel delivery subsystem consists of an in-tank low pressure fuel pump, a high pressure chassis mounted electric fuel pump and the necessary interconnecting fuel tubing. Fuel drawn from the tank passes through a filter before entering the low pressure pump. The fuel is filtered again as it leaves the high pressure pump on its way to the pulse damper and fuel supply manifold. The pulse damper smooths the pressure variations generated by the opening and closing of the fuel injectors.

The fuel supply manifold assembly incorporates electrically actuated fuel injectors directly above each of the engines intake ports. The injectors, when energized, spray a metered quantity of fuel into the intake air stream. A constant fuel pressure drop is maintained across the injector nozzles by a pressure regulator. The regulator is mounted in series with the fuel injectors and is positioned downstream from them. Excess fuel supplied by the pump, but not required by the engine, passes through the regulator and returns to the fuel tank through a fuel return line.

The period on time the injectors are energized is referred to as injector on time or pulse width. The injector on-time or pulse width is controlled by the vehicle computer. The pulse width is calculated by input data received from the Heated Exhaust Gas Oxygen (HEGO) sensor, Engine Coolant Temperature (ECT), Engine rpm (PIP), Air Charge Temperature (ACT), Throttle Position (TP) sensor and elapsed time between engine start up.

SYSTEM COMPONENTS

FUEL INJECTORS

The fuel injector nozzles are electromechanical devices which both meter and atomize fuel delivered to the engine. The injectors are mounted in the lower intake manifold and are positioned so that their tips direct fuel just ahead of the engine intake valves. The injector bodies consist of a solenoid actuated pintle and needle valve assembly.

An electrical control signal from the Electronic Engine Control unit (EEC) activates the injector solenoid, causing the pintle to move inward off the seat, allowing fuel to flow. Since the injector flow orifice is fixed and the fuel pressure drop across the injector tip is constant, fuel flow to the engine is regulated by how long the solenoid is energized. Atomization is obtained by contouring the pintle at the point where the fuel separates.

PRESSURE REGULATOR

The fuel pressure regulator is attached to the fuel supply manifold assembly downstream of fuel injectors. It regulates the fuel pressure supplied to the injectors. The regulator is a diaphragm operated relief valve in which one side of the diaphragm senses fuel pressure and the other side is subjected to the intake manifold pressure. The nominal fuel pressure is established by a spring preload applied to the diaphragm. Balancing one side of

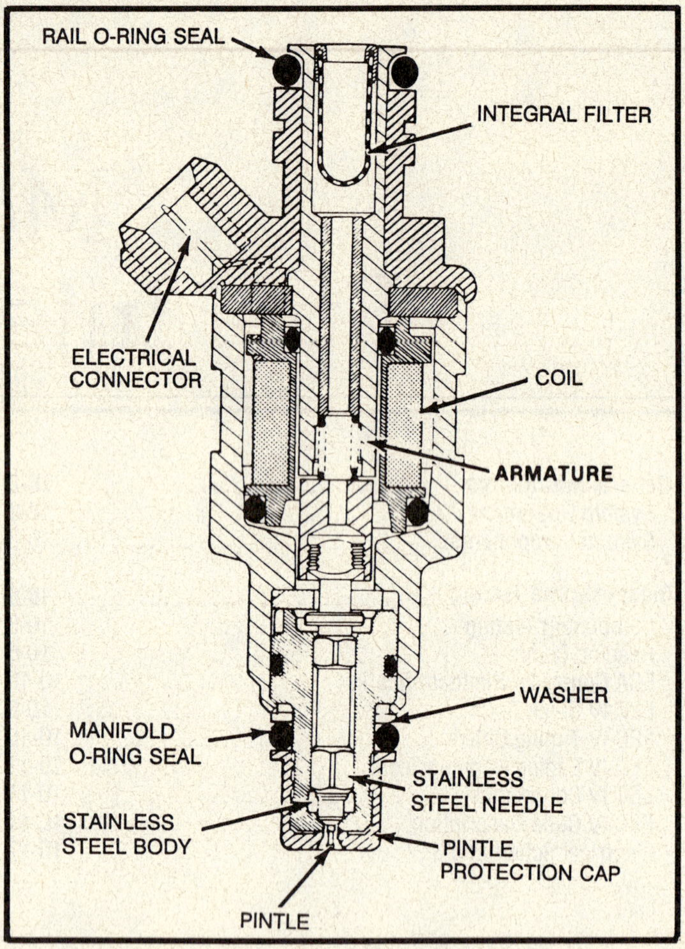

Cross section view of the fuel injector

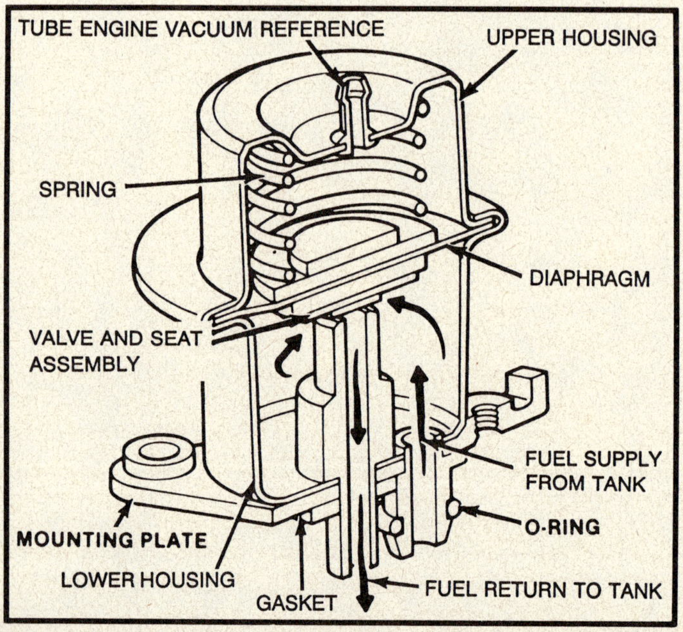

Cross section view of the fuel pressure regulator

the diaphragm with manifold pressure maintains a constant fuel pressure drop across the fuel injectors. Fuel, in excess of that used by the engine, is bypassed through the regulator and returns to the fuel tank.

FUEL SUPPLY MANIFOLD

The fuel supply manifold assembly delivers high pressure fuel from the vehicle fuel supply line to the fuel injectors. The fuel supply manifold assembly consist of a single, preformed tube or stamping with injector connectors, a mounting flange to the fuel pressure regulator and mounting attachments which locate the fuel manifold assembly and provide fuel injector retention.

AIR INTAKE MANIFOLD

The air intake manifold is a 2 piece aluminum casting. Runner lengths are tuned to optimize engine torque and power output. The upper manifold provides mounting bosses for the throttle body assembly, air bypass valve, accelerator control bracketry and vacuum taps. The lower manifold assembly includes the fuel supply manifold, fuel injectors, engine coolant temperature sensor and thermostat water outlet.

THROTTLE BODY

The throttle body controls air flow to the engine through a single butterfly type valve. The throttle position is controlled by a multiple link, progressive opening throttle linkage. The body is a single piece die casting made of aluminum. It has a single bore with an air bypass channel around the throttle plate. This bypass channel controls both cold and warm engine idle air flow as regulated by an electro-mechanical device controlled by the EEC computer. It incorporates a linear actuator which positions a variable area metering valve.

The throttle body assembly also includes, an adjustment

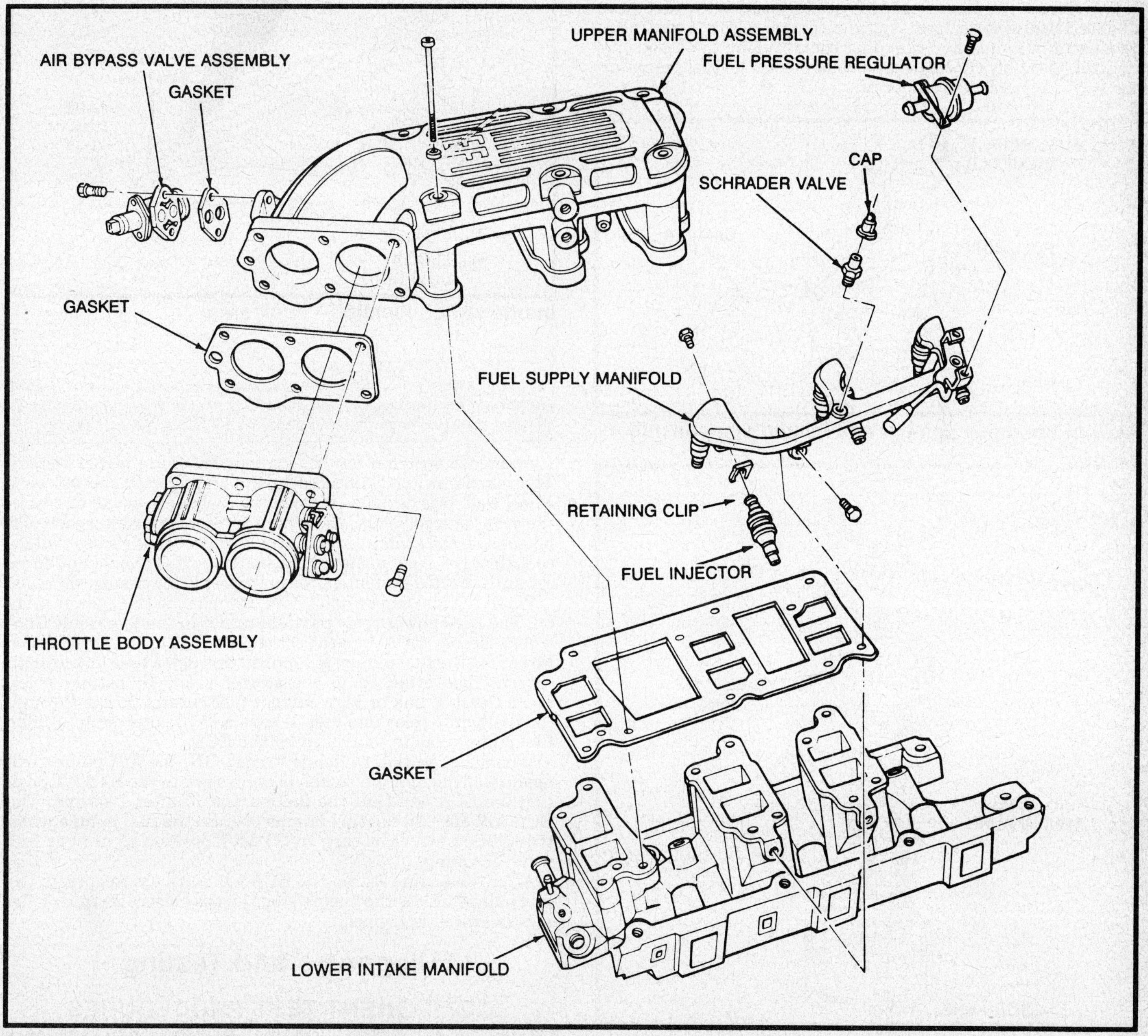

Exploded view of the fuel intake system—2.9L engine

screw to set the throttle plate at minimum idle air flow position, a preset stop to locate Wide Open Throttle (WOT), a throttle position sensor mounted to the throttle body, a canister purge port for evaporative emission control, a crankcase ventilation pressure tube and a EGR inlet port.

PULSE DAMPER

The opening and closing of the fuel injectors generates pulses within the fuel supply manifold. These pulses or pressure variations are smoothed by the pulse damper. The damper is mounted on a bracket behind the upper intake manifold between the fuel supply line and the fuel supply manifold.

FUEL PUMP

The fuel delivery system uses either a high or low-pressure inline or in-tank electric fuel pump, with some models equipped with both. It is a recirculating system that delivers fuel to a pressure regulating valve in the throttle body and returns excess fuel from the throttle body regulator back to the fuel tank. The electrical system uses 2 types of control relays, 1 controlled by a vacuum switch and the other controlled by the Electronic Control Assembly (ECA) to provide power to the fuel pump under various operating conditions.

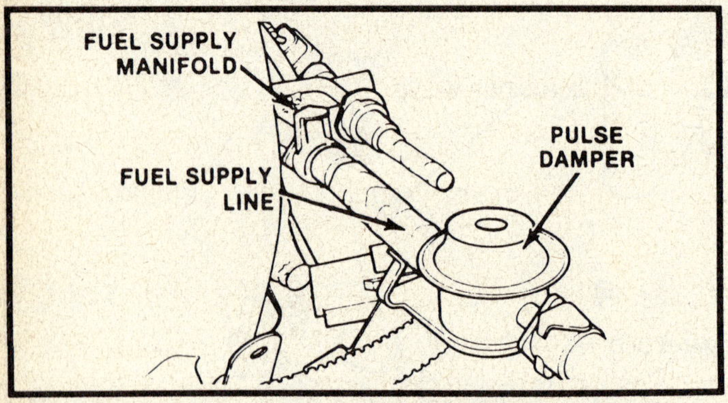

View of the pulse damper and related components

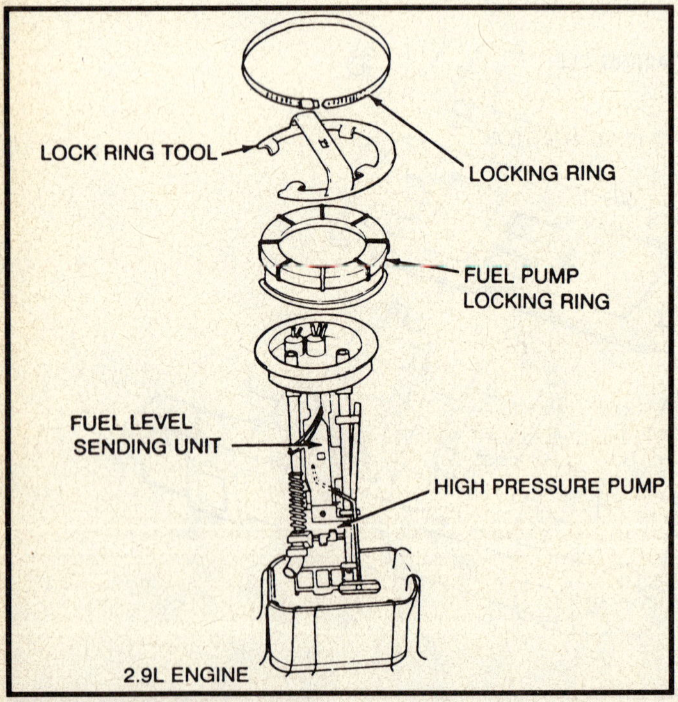

Intank electric fuel pump

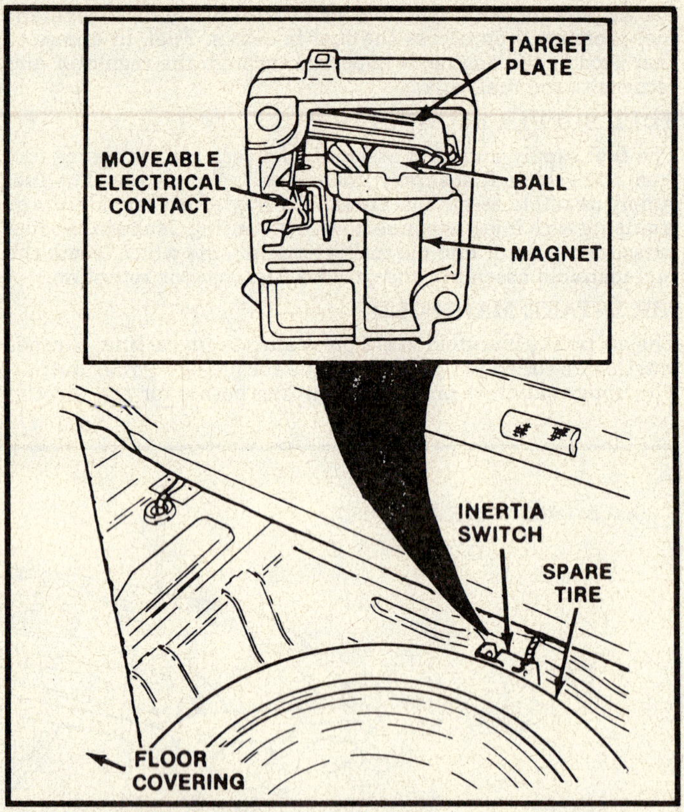

Inertia switch location—trunk area

CAUTION

Fuel supply lines on vehicles equipped with a high pressure fuel system will remain pressurized for long periods of time after engine shutdown. The fuel pressure must be relieved before servicing the fuel system.

An inertia switch is used as a safety device in the fuel system. The inertia switch is located in the trunk, near the left rear wheel well. It is designed to open the fuel pump power circuit in the event of a collision. The switch is reset by pushing each of 2 buttons on the switch simultaneously (some models use switches with only 1 reset button). The inertia switch should not be reset until the fuel system has been inspected for damage or leaks.

When the ignition switch is **ON**, it turns the EEC power relay on. The EEC power relay provides power to the Electronic Control Assembly (ECA) and the control side of the fuel pump relay. Power for the fuel pumps is supplied through a fuse link or high current fuse attached to the starter solenoid (battery side). From the fuse link or high current fuse currect flow is through the fuel pump relay and inertia switch to the fuel pump(s). The fuel pump relay is controlled by the ECA.

When the ignition switch is turned **ON**, the fuel pumps will operate. If the ignition switch is not turned to the **START** position the ECA will shut the fuel pumps off after 1 second. The ECA will operate the fuel pumps operate the fuel pumps when the ignition switch is turn to **START** position to provide fuel while cranking.

After the engine starts, the ECA will continue to operate the fuel pumps unless the engine stops, drops below 120 rpm or the inertia switch is tripped.

Diagnosis and Testing

COMPONENT TEST PROCEDURES

As in any service procedure, a routine inspection of the EEC-IV system for loose connections, broken wires or obvious damage is the best way to start. Perform the system Quick Test outlined below before going any further. Check all vacuum connections and secondary ignition wiring before assuming that the problem lies with the EEC-IV system. A self-diagnosis capability is built into the EEC-IV system to aid in troubleshooting. The primary tool necessary to read the trouble codes stored in the system is an analog voltmeter or special Self Test Automatic Readout (STAR) tester (Motorcraft No. 007-0M004, or equivalent). While the self-test is not conclusive by itself, when activated it checks the EEC-IV system by testing its memory integrity and processing capability. The self-test also verifies that all sensors and actuators are connected and working properly.

NOTE: For accessing "Quick Tests" or "Self Tests" on the EEC-IV systems, see Section 3 "Self-Diagnostic Systems".

EEV-IV System Quick Test

Correct test results for the quick test are dependent on the correct operation of related non-EEC components, such as ignition wires, battery, etc. It may be necessary to correct defects in these areas before the EEC-IV system will pass the quick test. Before connecting any test equipment to check the EEC system, make the following checks:

1. Check the air cleaner and intake ducts for leaks or restrictions. Replace the air cleaner if excessive amounts of dust or dirt are found.
2. Check all engine vacuum hoses for proper routing according to the vacuum schematic on the underhood sticker. Check for proper connections and repair any broken, cracked or pinched hoses or fittings.
3. Check the EEC system wiring harness connectors for tight fit, loose or detached terminals, corrosion, broken or frayed wires, short circuits to metal in the engine compartment or melted insulation exposing bare wire.

NOTE: It may be necessary to disconnect or disassemble the connector to check for terminal damage or corrosion and perform some of the inspections. Note the location of each pin in the connector before disassembly. When doing continuity checks to make sure there are no breaks in the wire, shake or wiggle the harness and connector during testing to check for looseness or intermittent contact.

4. Check the control module, sensors and actuators for obvious physical damage.
5. Turn **OFF** all electrical loads when testing and make sure the doors are closed whenever readings are made. Do not disconnect any electrical connector with key **ON**. Turn the key **OFF** to disconnect or reconnect the wiring harness to any sensor or the control unit.
6. Make sure the engine coolant and oil are at the proper level.
7. Check for leaks around the exhaust manifold, oxygen sensor and vacuum hoses connections with the engine idling at normal operating temperature.
8. Check the ignition and fuel systems to make sure there is fuel and spark. Check the distributor cap, rotor and internal components for damage, corrosion or signs of excessive wear.
9. Check battery for full charge and cable connections for tightness.
10. Inspect the relay connector and make sure the ECA power relay is securely attached and making a good ground connection.
11. Twist the oil filler cap on turbocharged engines to make sure it is tight. A loose oil filler cap will cause turbocharged engines to run rough at idle.
12. Only after all the above checks have been performed should the voltmeter be connected to read the trouble codes. If not, the self-diagnosis system may indicate a failed component when all that is wrong is a loose or broken connection.

QUICK TEST STEPS

1. Visually check of hoses, wiring, sensors, actuators and fuel system. Check battery for proper charge.
2. Warm engine to normal operating temperature.
3. Attach test equipment.
4. Perform—Key ON Engine OFF Self-Test.
5. Perform—Computed Timing Check.
6. Perform—Engine RUNNING Self-Test.
7. Perform—Continuous Self-Test.
8. Perform—Diagnostic by Symptom.
9. Proceed to pinpoint test only as directed by Quick-Test or Self-Test codes.
10. Replace components only when instructed in pinpoint test.

NOTE: Never measure voltage or resistance directly at the processor connector. Disconnect solenoids, switches and circuits from harness, before performing individual testing. For EEC-IV test codes and code definitions, see Section 3 "Self-Diagnostic Systems". Any ECA pin locations given are for reference only. Probing any connectors with DVOM, unless specified, will result in permanent damage to the pin connector. Always probe as directed using a breakout box.

SCORPIO DIAGNOSTIC CHART

When To Use This Diagnostic Procedure

This diagnostic procedure is only used on Electronic Engine Control (EEC-IV) equipped vehicles when the Diagnostic Routines, Section 2 instruct you to perform EEC-IV Diagnostics. It is divided into two test formats. **Quick Test**, a functional EEC-IV system test, and **Pinpoint Tests**, a number of specific component tests.

To test and service the EEC-IV subsystem, perform the Quick Test first. **If the vehicle passes all three phases of the Quick Test: Key On Engine Off, Engine Running and Continuous Testing without running any Pinpoint Tests, you will know the EEC-IV subsystem is OK and the vehicle's problem exists somewhere else other than in the EEC-IV subsystem.** However, if a step of the Quick Test fails, perform only the Pinpoint Tests specified by the failed step. Do not begin any Pinpoint Test without following the instructions at the beginning of the Section.

After all tests and services have been completed, repeat the entire Quick Test to make sure the EEC-IV subsystem works properly.

Quick Test Description

The Quick Test is a functional test of the EEC-IV system consisting of basic Test Steps (described below). These steps must be carefully followed in sequence, otherwise misdiagnosis, or the replacement of undamaged components may result.

Quick Test Steps

1. Visual Check and Vehicle Preparation
   - Checks for obvious faults.
   - Properly prepares the vehicle for testing.
2. Equipment Hookup
   - Ensures that the proper equipment for gathering test data is ready, prior to testing.
3. Key On Engine Off Self Test
   - Is a static check of processor inputs and outputs.
4. Computed Timing Check
   - Verifies the system ability to compute and maintain a fixed spark timing during Self Test.
5. Engine Running Self Test
   - Is a dynamic check with the engine in operation.
6. Continuous Self Test
   - Checks the sensor inputs for opens and shorts while the vehicle is in operation.

The Key On Engine Off and Engine Running Self Tests are intended to detect faults present at the time of testing only, not intermittent faults. Intermittent faults are detected by Continuous Self Test.

A continuous memory code is information collected and stored from an intermittent fault detected in a previous drive cycle, the fault may not be present at this time.

WARNING: Anyone who departs from the instruction provided in this publication must first establish that he compromises neither his personal safety nor the vehicle integrity by his choice of methods, tools or parts.

SCORPIO DIAGNOSTIC CHART

QUICK TEST: Visual Check And Vehicle Preparation

1.0

Correct test results for the Quick Test are dependent on the proper operation of related non-EEC components systems. It may be necessary to correct faults in these areas before EEC will pass Quick Test.

Before hooking up any equipment to diagnose the EEC system, make the following checks:

1. Verify the condition of air cleaner and ducting. These components may be removed and reinstalled as necessary for service and/or inspection.
2. Check all engine vacuum hoses for:
   - Leaks or pinched hoses.
   - Proper emission routing per Vehicle Emission Control Information (VECI) Decal.
3. Check the EEC system wiring harness electrical connections for:
   - Proper connections.
   - Loose or detached connectors, wire and terminals.
   - Corrosion.
   - Proper routing of harness.

If may be necessary to disconnect or disassemble the connector assembly to perform some of the inspections. (Note the location of each pin before disassembly.)

4. Check processor, sensors and actuators for physical damage.
5. Perform all safety steps required to start and run operational vehicle tests.
6. Apply the emergency brake. Place shift lever in Park; Neutral for manual transmission.
7. **Turn Off all electrical loads** such as the radio, lamps, air conditioner, etc. Be sure doors are closed whenever readings are made.
8. Verify engine coolant is at the specified level.
9. Start engine and idle until the upper radiator hose is hot and pressurized and the throttle is off fast idle. While the engine is operating, check for leaks around the exhaust manifold, EGO sensor, and vacuum hose connections.

   NOTE: If engine will not start, starts but stalls, idles rough or runs rough continue through Quick Test Step 3.0. If any of the above conditions are still present after a pass code 11 in the Key On, Engine Off Self-Test, go to Pinpoint Test Step [A2] for no start or diagnostics by symptom in the proper Engine Supplement Section for stalls or roughness.

10. Turn ignition Key Off.
11. Service items as required, and proceed to equipment hook-up, Step 2.0.

SCORPIO DIAGNOSTIC CHART

QUICK TEST: Visual Check And Vehicle Preparation — 1.0

Correct test results for the Quick Test are dependent on the proper operation of related non-EEC component systems. It may be necessary to correct faults in these areas before EEC will pass Quick Test. Refer to Diagnostic Routines, Section 2 for service.

Before hooking up any equipment to diagnose the EEC system, make the following checks:

1. Verify the condition of air cleaner and ducting. These components may be removed and installed as necessary for service and/or inspection.
2. Check all engine vacuum hoses for:
   - Leaks or pinched hoses.
   - Proper emission routing per Vehicle Emission Control Information (VECI) Decal.
3. Check the EEC system wiring harness electrical connections for:
   - Proper connections.
   - Loose or detached connectors, wires and terminals.
   - Corrosion.
   - Proper routing of harness.

It may be necessary to disconnect or disassemble the connector assembly to perform some of the inspections. (Note the location of each pin before disassembly.)

4. Check processor, sensors and actuators for physical damage.
5. Perform all safety steps required to start and run operational vehicle tests.
6. Apply the emergency brake. Place shift lever in PARK; NEUTRAL for manual transmission.
7. **Turn Off all electrical loads** such as the radio, lamps, etc. Doors must be closed whenever readings are made.
8. Verify engine coolant is at the specified level.
9. Start engine and idle until the upper radiator hose is hot and pressurized and the throttle is off fast idle. While the engine is operating, check for leaks around the exhaust manifold, HEGO sensor and vacuum hose connections.

   NOTE: If engine will not start, starts but stalls, idles rough or runs rough, continue through Quick Test Step 3.0. If any of the above conditions are still present after a pass code 11 in the Key On Engine Off Self Test, go to Pinpoint Test Step A2 for no start or refer to the Diagnostic By Symptom chart for stalls or rough idle.
10. Turn ignition key off.
11. Service items as required and proceed to Equipment Hookup, Step 2.0.

QUICK TEST: Equipment Hookup — 2.0

Using the STAR tester (Figure 1):

- Turn the ignition key Off.
- Connect the color-coded adapter cable leads to the STAR tester.
- Connect the adapter cable's two service connectors to the vehicle's appropriate Self-Test connectors.
- Connect the timing light.

After equipment hookup, GO to Quick Test Step 3.0.

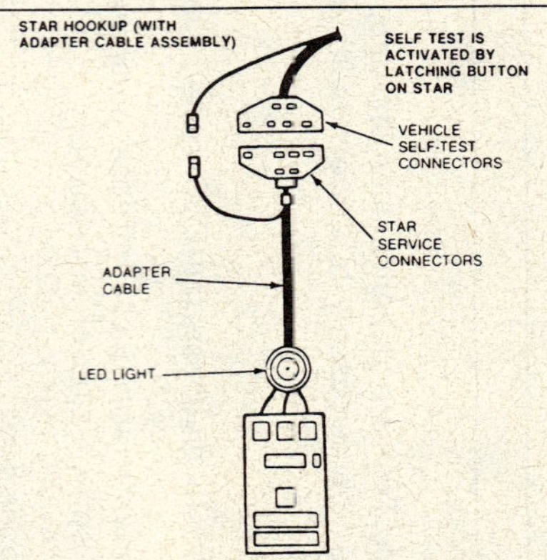

Figure 1 STAR Tester Hookup

SCORPIO DIAGNOSTIC CHART

QUICK TEST: Key On Engine Off Self Test — 3.0

A — CODE OUTPUT

Correct test results for the Quick Test are dependent on the proper operation of related non-EEC-IV components/systems. It may be necessary to correct any faults in these areas before EEC-IV will pass the Quick Test. Refer to the Diagnostic Routines, Section 2 for proper servicing.

- Verify that the vehicle has been properly prepared per Quick Test Steps 1.0 and 2.0.
- Place transmission in PARK (automatic transmission) or NEUTRAL (manual transmission). Set parking brake. Place A/C in ECONOMY position.
- Activate Self Test.
- Place key in RUN position to start Self Test.

NOTE: Do not depress throttle during Self Test.

- Observe and record all service codes indicated. One of the following outputs will occur.

NOTE: Continuous memory codes recorded in this step will be used for diagnosis later in Quick Test Step 6.0.

RESULTS — Code Format: Key On Engine Off	Separator	Continuous Memory		ACTION TO TAKE
11	1(0)	11	▶	KEY ON ENGINE OFF and CONTINUOUS MEMORY indicate a pass. Go to QUICK TEST STEP 4.0. If vehicle is a no start, Go to Pinpoint Test Step A2.
ANY CODE(S)	1(0)	11	▶	KEY ON ENGINE OFF test indicates a fault. Record codes and Go to QUICK TEST STEP 3.0B.
ANY CODE(S)	1(0)	ANY CODE(S)	▶	KEY ON ENGINE OFF and CONTINUOUS MEMORY indicate a fault. Record ALL codes. DO NOT SERVICE CONTINUOUS MEMORY CODES AT THIS TIME! KEY ON ENGINE OFF and ENGINE RUNNING codes must be serviced first. Go to QUICK TEST STEP 3.0B.
11	1(0)	ANY CODE(S) EXCEPT 15	▶	CONTINUOUS MEMORY indicates a fault. DO NOT SERVICE THESE CODES AT THIS TIME! KEY ON ENGINE OFF and ENGINE RUNNING codes must be serviced first. Go to QUICK TEST STEP 4.0.
11	1(0)	15	▶	Go to Pinpoint Test Step Q10.
NO CODES OUTPUTTED				Repeat SELF TEST and verify that no service codes are present. Go to Pinpoint Test Step Q1.

*Star tester displays a 10 in place of a 1 for the single digit separator code, therefore the separator code is expressed in the results column as 1(0).

QUICK TEST: Key On Engine Off Self Test — 3.0

B — TEST RESULTS AND ACTION TO TAKE

- Using Key On Engine Off service codes from Quick Test Step 3.0A, follow the instructions in the ACTION TO TAKE column in this step.
- When more than one service code is received, always start with the first code received.
- Whenever a service is made, REPEAT QUICK TEST!

NOTE: Before proceeding to the specified Pinpoint Test, read the instructions on how to use the Pinpoint Tests at the beginning of the Pinpoint Test section.

RESULTS		ACTION TO TAKE
Key On Engine Off Service Codes	▶	GO to Pinpoint Test Step
15 — ECA Test Failure	▶	Q12
19 — Power to Processor	▶	Q50
21 — ECT Out of Range	▶	DE1
22 — MAP Out of Range	▶	DF1
23 — TPS Out of Range	▶	DH1
24 — ACT Out of Range	▶	DB1
31 — EVR/PFE Out of Range	▶	DL1
34 — No EGR Flow in Test	▶	DL8
35 — RPM Too Low for EGR Test	▶	DL5
51 — ECT Too High	▶	DE10
53 — TPS Input Too High	▶	DH3
54 — ACT Input Too High	▶	DB10
61 — ECT Input Too Low	▶	DE20
63 — TPS Input Too Low	▶	DH10
64 — ACT Input Too Low	▶	DB20
67 — Neutral Drive Switch	▶	FA1
84 — EGR Circuit Fault	▶	DL11
85 — Canister Purge Circuit Fault	▶	KD4
87 — Fuel Pump Circuit Fault	▶	J7
No Codes/Code Not Listed	▶	Q1

SCORPIO DIAGNOSTIC CHART

QUICK TEST: Timing Check — 4.0

TEST STEP	RESULT ▶	ACTION TO TAKE
• If engine is a no start, go directly to Pinpoint Test Step A1 NO START. • If engine starts but stalls go directly to Diagnostic By Symptom	Engine stalls during timing check ▶	Go to Diagnostic By Symptom.
• Key off, wait 10 seconds. • Verify Self-Test trigger has been activated.	Timing is not 15° ± 3° ▶	Go to Pinpoint Test Step P1.
• Restart engine and check timing while in Self Test. Check Self-Test timing within two minutes after last code is displayed. **NOTE: Correct Self Test timing is 15° BTDC ± 3°. Correct base timing with spout disconnected is 10° ± 3°.**	Timing is 15° ± 3° ▶	Go To Quick Test Step 5.0.

QUICK TEST: Engine Running Self Test — 5.0

A CODE OUTPUT

NOTE: If engine is a no start go directly to Pinpoint Test Step A1. If engine stalls go directly to Diagnostic By Symptom chart.

- Deactivate Self Test.①
- Start and run engine at greater than 2000 rpm for 2 minutes. This warms up the HEGO sensor.
- Turn engine off, wait 10 seconds.
- Activate Self Test.
- Start engine. The Engine Running Self Test will progress as follows:
  - — Engine ID code.②
  - — Run test.
  - — When the dynamic response code occurs perform a brief WOT.③
  - — Engine Running service codes.
- Observe and record all service codes. One of the following outputs will occur.

NOTE: If engine stalls during Self Test, go directly to Diagnostic By Symptom chart.

RESULTS			ACTION TO TAKE
Code Format			
Engine ID	Dynamic Response	Engine Running	
3(0)	1(0) or no dynamic response pulse	11 ▶	ENGINE RUNNING SELF TEST indicates a pass. If the symptom was of an intermittent nature, Go to QUICK TEST STEP 6.0. If the drive symptom is currently present, Go to DIAGNOSTIC BY SYMPTOM. Otherwise testing is complete, EEC-IV system is OK.
3(0)	1(0) or no dynamic response pulse	ANY CODES ▶	ENGINE RUNNING SELF TEST indicates a fault. Go to QUICK TEST STEP 5.0B.
NO CODES OUTPUTTED			Repeat SELF-TEST and verify that no service codes are present, then Go To Finpoint Test Step Q1.

① Manual transmission only. If STAR Tester is connected and Self Test de-activated with engine running, after approximately 4-5 minutes from engine start-up, the LED may flash ON/OFF. This should be ignored (this will happen if ACT, ECT, MAP, TP or EDT are shorted to ground or there is an open circuit).

② Engine ID code is equal to half the number of cylinders. (A code of 3 equals a 6-cylinder engine. STAR tester adds a zero to all single digit readings, so a 30 code equals a 6-cylinder engine).

③ Star displays a 10 in place of a 1 for the single digit separator code, therefore, the dynamic response code is expressed in the results column as 1(0). Some vehicles do not require a brief WOT and therefore will not display a 1(0).

SCORPIO DIAGNOSTIC CHART

QUICK TEST: Engine Running Self Test — 5.0

B — TEST RESULTS AND ACTION TO TAKE

- Using the Engine Running service codes from Quick Test Step 5.0A, follow the instructions in the ACTION TO TAKE column in this step.
- When more than one service code is received, always start with the first code received.
- Whenever a service is made, REPEAT QUICK TEST

RESULTS ▶	ACTION TO TAKE
Engine Running Service Codes ▶	**GO to Pinpoint Test Step**
12 — RPM Out of Specification ▶	KE1
13 — RPM Out of Specification	KE10
16 — RPM Too Low ▶	KE1
21 — ECT Out of Range ▶	DE1
22 — MAP Out of Range ▶	DF7
23 — TPS Out of Range ▶	DH1
24 — ACT Out of Range ▶	DB1
31 — EVR/PFE Out of Limits ▶	DL21
32 — EGR Not Controlling ▶	DL20
33 — EVR/PFE Out of Limits ▶	DL30
34 — No EGR Flow in Test ▶	DL25
35 — RPM Too Low for EGR Test ▶	DL25
41 — System Always Lean ▶	HG10
42 — System Always Rich ▶	HG7
72 — No MAP Change In "Goose" Test ▶	DF10
73 — No TPS Change in "Goose" Test ▶	DH20
77 — Operator Did Not do "Goose" Test ▶	M1
99 — ISC Has Not Learned Yet. ▶	Rerun Quick Test Steps 3.0 through 5.0.
No Codes/Code Not Listed ▶	Q1

QUICK TEST: Continuous Self Test — 6.0

TEST STEP	RESULT ▶	ACTION TO TAKE
A — CONTINUOUS MEMORY CODES		
• To ensure proper diagnosis of Continuous Memory codes, Quick Test Steps 1.0 through 5.0 must be successfully completed. • Do the results of both Key On Engine Off and Engine Running Self Tests result in a pass code 11?	Yes ▶	GO to 6.0B.
	No ▶	RETURN to STEP 1.0 of Quick Test and service as indicated in Quick Test Steps 3.0 and 5.0 before continuing.
B — CLEAR CONTINUOUS MEMORY CODES		
• Prepare the vehicle for Key On Engine Off Quick Test per Steps 1.0, 2.0, and 3.0. • Perform Key On Engine Off Self Test. When the service codes begin, exit the Self Test program (by unlatching STAR tester). Exiting Quick Test during code output will clear all codes stored in the Continuous Memory. • Repeat Key On Engine Off Self Test. • Does Key On Engine Off Self Test output 11-10-11?	Yes ▶	GO to 6.0C.
	No ▶	CHECK STI circuit for short to ground. SERVICE as necessary. REPEAT Step 6.0B.
C — DETERMINE CODES TO BE TESTED		
• Refer to Continuous Memory Codes recorded in Quick Test Step 3.0A. • Have any of the codes been serviced either in Key On Engine Off or Engine Running Self Tests?	Yes ▶	DISREGARD codes already serviced. For remaining codes, GO to 6.0D.
	No ▶	GO to 6.0D.

SCORPIO DIAGNOSTIC CHART

QUICK TEST: Continuous Self Test — 6.0

D TEST RESULTS AND ACTION TO TAKE

- YOU SHOULD NOT BE HERE UNLESS ALL PREVIOUS QUICK TEST STEPS HAVE BEEN SUCCESSFULLY COMPLETED!
- Verify proper test equipment hookup per Quick Test Step 2.0.
- Self Test deactivated. Do not activate Self Test unless instructed.
- Using the service codes determined in Step 6.0C, follow the ACTION TO TAKE instructions. (While in the appropriate Pinpoint Test, one or both of the continuous monitor modes described below may have to be used to find the intermittent fault.)

CONTINUOUS MONITOR (WIGGLE) MODES

- Key On Engine Off (Automatic transmission only.)
  With Self Test deactivated, turn the ignition key to the ON position. You are now in the continuous monitor mode (wiggle test). A fault is indicated when the STAR LED turns off.
- Engine Running
  Activate Self Test and perform Engine Running Quick Test. After service code output has been completed do not turn engine off or deactivate Self Test. Approximately 2 minutes after code 11 has been outputted the system will enter and remain in continuous monitor mode until Self Test is deactivated or the engine is turned off. A fault is indicated the same way as Key On Engine Off.
- Refer to the appendix for a detailed description of the continuous monitor (wiggle test).

RESULTS ▶	ACTION TO TAKE
Continuous Memory Service Codes ▶	**GO to Pinpoint Test Step**
14 — Pip Was Erratic ▶	Y1
15 — ECA Test Failure ▶	Q10
18 — No Tachometer Signal ▶	N1
22 — MAP Out of Range ▶	DF90
31 — EVR/PFE Out of Limit ▶	DL90
32 — EGR Not Controlling ▶	DL94
34 — No EGR Flow in Test ▶	DL93
35 — RPM Too Low for EGR Test ▶	DL90
51 — ECT Input Too High ▶	DE91
53 — TPS Input Too High ▶	DH90
54 — ACT Input Too High ▶	DB90
61 — ECT Input Too Low ▶	DE94
63 — TPS Input Too Low ▶	DH94
64 — ACT Input Too Low ▶	DB93
No Codes/Code Not Listed ▶	Q1

APPENDIX: Self Test Description

SELF-TEST DESCRIPTION

The Self Test is divided into three specialized tests: Key On Engine Off, Engine Running and Continuous Memory. The Self Test is not a conclusive test by itself, but is used as a part of the functional Quick-Test diagnostic procedure. The processor stores the Self-Test program in its permanent memory. When activated, it checks the EEC-IV system by testing its memory integrity and processing capability, and verifies that various sensors and actuators are connected and operating properly.

The Key On Engine Off and Engine Running tests are functional tests which only detect faults present at the time of the Self Test. Continuous testing is an ongoing test that stores fault information for retrieval at a later time (during the Self Test).

Key On Engine Off

At this time, a test of the EEC-IV system is conducted with power applied and engine at rest.

For Self Test to detect errors in this test, the fault must be present at the time of testing. For intermittents, refer to Continuous Memory codes.

Separator Pulse

A single 1/2 second separator pulse is issued 6-9 seconds after the last Functional Test code (Key On Engine Off only). Then 6-9 seconds after the single 1/2 second separator pulse, the continuous codes will be issued. Refer to the example below:

NOTE: The separator code and continuous memory codes follow Key On Engine Off testing codes ONLY.

Continuous Memory Codes

Continuous Memory codes are issued as a result of information stored during continuous monitor testing, while the vehicle was in normal operation. These codes are displayed only during the Key On Engine Off testing and after the separator code. These codes should be used for diagnosis only when Key On Engine Off and Engine Running Self Tests result in code 11 and all Quick Test Steps 1.0 through 5.0 have been successfully completed.

Engine Running Test

At this time, a test of the EEC-IV system is conducted with the engine running. The sensors are checked under actual operating conditions and at normal operating temperatures. The actuators are exercised and checked for corresponding results.

Engine Identification Codes (ID Codes)

Engine ID codes are issued at the beginning of the Engine Running test and are one-digit numbers represented by the number of pulses sent out. The engine ID code is equal to 1/2 the number of engine cylinders. These codes are used to verify the proper processor is installed and that the Self Test has been entered.

Dynamic Response Check

The dynamic response check verifies the movement of the TP and MAP sensors during the brief Wide-Open Throttle (WOT) performed during the Engine Running Self Test. The signal for the operator to perform the brief WOT is a 10 code on the STAR.

SCORPIO DIAGNOSTIC CHART

APPENDIX: Diagnostic Aids

OUTPUT STATE CHECK

The output state check aids in servicing output actuators associated with the EEC-IV system. It enables the technician to energize and de-energize most of the system output actuators on command①. This mode is entered after all codes have been received from Key On Engine Off and Continuous Testing. At this time, leave Self Test activated and depress the throttle. Each time the throttle is depressed the output actuators will change state (ON to OFF or OFF to ON).

- Enter Self Test
- Code Output Ends
- Do Brief WOT
- Output Actuators Actuated
- Next Brief WOT
- Output Actuators Deactivated

①Actuators checked are as follows:

- Self Test Output
- Wide-Open Throttle Cutout Relay
- Electronic Vacuum Regulator
- Canister Purge Solenoid

APPENDIX: Continuous Self Test

The Continuous Memory service codes are separated from the Quick Test Key On Engine Off codes by a single separator pulse (Figure 1).

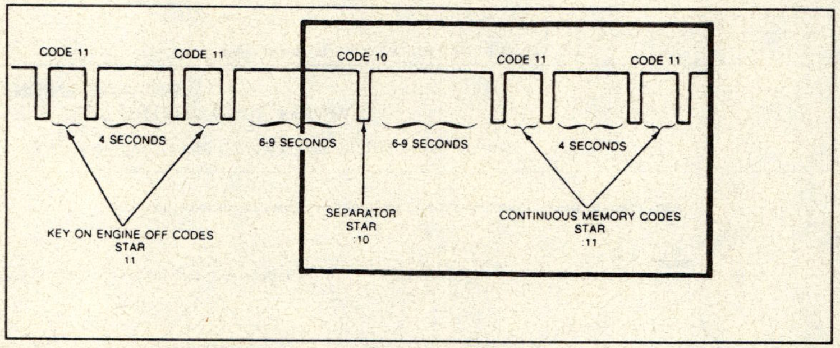

Figure 1 Continuous Memory Self-Test Output Code Format

The Continuous Memory codes should never be used for diagnosis until the Key On Engine Off and Engine Running Self Tests result in a pass code 11.

The service codes that were displayed (repeated) in both the Continuous Memory and Engine Running Self Tests are considered repaired. The unique service code(s) displayed only in Continuous Memory are considered the intermittent service codes.

During this mode of testing we continuously monitor inputs to processor for opens and shorts. The Continuous Memory codes must be retrieved within 40 engine temperature warm-up cycles. On the forty-first engine temperature cycle the service code will be automatically erased. The Continuous Memory code can also be erased by deactivating Quick Test while the service codes are being outputted.

Continuous Monitor Test

This test is intended as an aid in diagnosing intermittent failures in sensor input circuits. The Self-Test output is energized whenever the Continuous Memory mode senses a fault and de-energized when the system is OK.

- Connect STAR Tester to the Self-Test output.

SCORPIO DIAGNOSTIC CHART

APPENDIX: Continuous Self Test (Continued)

- **Key On Engine Off** Continuous Monitor Test, (Wiggle).

  To enter this mode, "verify" Self Test is not activated and then turn the key to RUN. Self Test output will be activated whenever a continuous fault is detected. If the duration is long enough, a fault code will be stored.

- **Engine Running** Continuous Monitor Test, (Wiggle).

  This mode of testing is entered approximately 2 minutes after the service codes from the Engine Running portion of Quick Test (Step 5.0) because the Engine Running test can be entered only once per ignition cycle. An alternate quick method (eliminates waiting for Self-Test to complete its cycle) is to enter the running test, exit, and re-enter (do not shut the engine off). This will put you in the Engine Running Continuous Monitor (Wiggle) Test.

  The Continuous Monitor (Wiggle) Test will allow the technician to enter this mode of Self Test and to attempt to re-create the intermittent failure (tap, move, wiggle the harness and/or the suspected sensor).

  If the STAR Tester indicates a fault (short or open) the corresponding service code will be stored. Now with the knowledge of the affected circuits, a close check of the harness and associated connectors can be made.

  NOTE: Remember to keep your eyes on the STAR tester for any change which will indicate where the intermittent is located.

EXAMPLE: How to Use the Continuous Monitor Test (Systematically)

If a service code 51 was displayed: Engine Coolant Temperature Sensor:

- Visually inspect the sensor **very closely.**
- Enter Continuous Monitor test.
- Lightly tap on the sensor.
- Push/pull on the sensor harness connector (do not disconnect, yet).
- Test and Wiggle (shake) harness vigorously working from the sensor connector toward the dash panel and from the dash panel to the processor in short sections.
- Disconnect the sensor from the harness as carefully as possible. Remove terminals from the connector. Visually inspect terminals at both ends for corrosion, bad crimps, improperly seated terminals, etc.
- Reconnect after inspection.
- Disconnect processor from harness as carefully as possible.
- Inspect terminals.
- Only remove terminals associated with the sensor being inspected.
- Reconnect the connector and erase the Continuous Monitor test service codes.
- To erase the continuous service codes, initiate Key On Engine Off Self Test.

APPENDIX: Code Output Format

SERVICE CODES

The EEC-IV system communicates service information to the outside world by way of the Self-Test service codes. These service codes are two-digit numbers representing the results of Self Test.

The service codes are transmitted on the Self-Test output (found in the Self-Test connector) in the form of timed pulses, and read by the technician on the STAR tester.

SELF-TEST OUTPUT CODE FORMAT KEY ON ENGINE OFF

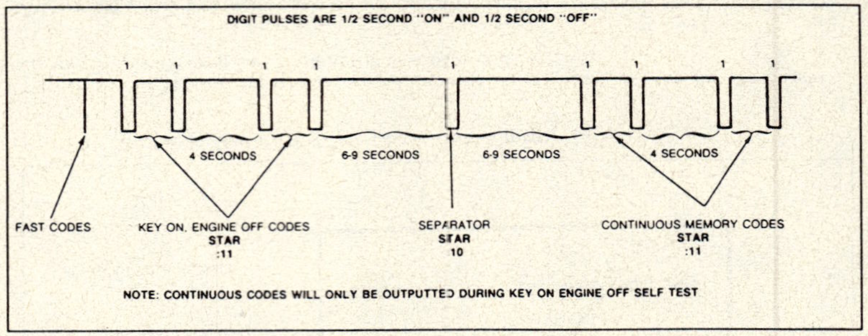

Figure 2 Key On Engine Off Self-Test Output Code Format

SELF-TEST OUTPUT CODE FORMAT ENGINE RUNNING

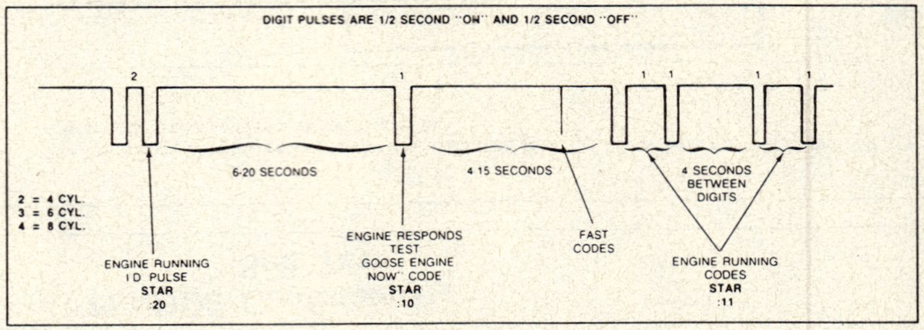

Figure 3 Engine Running Self-Test Output Code Format

Fast Codes

Fast codes are issued prior to regular service codes. These codes contain the identical information as the regular service codes but are transmitted at 100 times the normal rate. These codes can be read using the Rotunda Super Star 007-00019 or equivalent.

SCORPIO DIAGNOSTIC CHART

APPENDIX: Self Test With STAR Tester

Reading Codes — Self-Test Automatic Readout (STAR) Rotunda 007-00004 or Equivalent

After hooking up the STAR tester and turning on its power switch, the tester will run a display check and the numerals 88 will begin to flash in the display window (Figure 4). A steady 00 will then appear to signify that the STAR tester is ready to start the Self Test and receive the test's service codes.

To receive the service codes, press the push button at the front of the STAR tester. The button will latch down, and a colon will appear in the display window in front of the 00 numerals. The colon **must** be displayed to receive the service codes.

If for any reason the technician wishes to clear the display window during the Self Test, he must turn off the vehicle's engine, press the tester's push button once to unlatch it (colon will disappear), then press the button again to latch down the button (colon will appear again). Every time the STAR tester is turned off, the low battery indicator (LO BAT) should show briefly at the upper left corner of the tester's display window. If the LO BAT indicator shows steadily at any other time during the operation of the STAR tester with any service code, turn its power switch to OFF and replace the 9-volt battery in the tester.

The STAR tester will display the last service code received, even after disconnecting it from the vehicle. It will hold the service code on the display until the power is turned off or the push button is unlatched and relatched.

WARNING: Anyone who departs from the instructions provided in this publication must first establish that he compromises neither his personal safety nor the vehicle integrity by his choice of methods, tools, or parts.

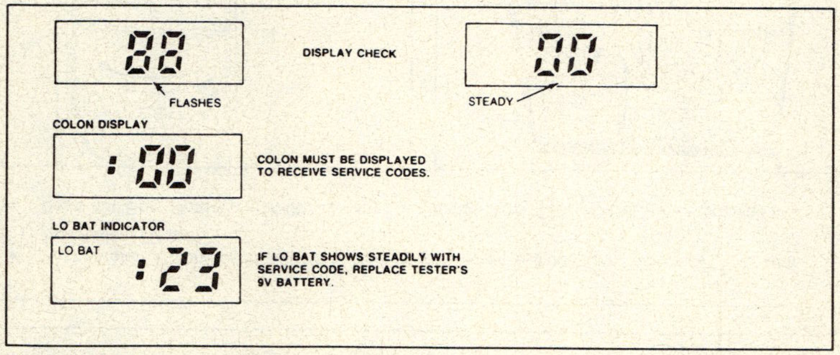

Figure 4 STAR Tester Output Code Format

APPENDIX: Test Equipment

EQUIPMENT REQUIRED

- Rotunda Self-Test Automatic Readout (STAR) 007-00004 with cable assembly 007-00010 or equivalent. Refer to STAR operation.
- Rotunda Super Star, 007-00019 or equivalent.

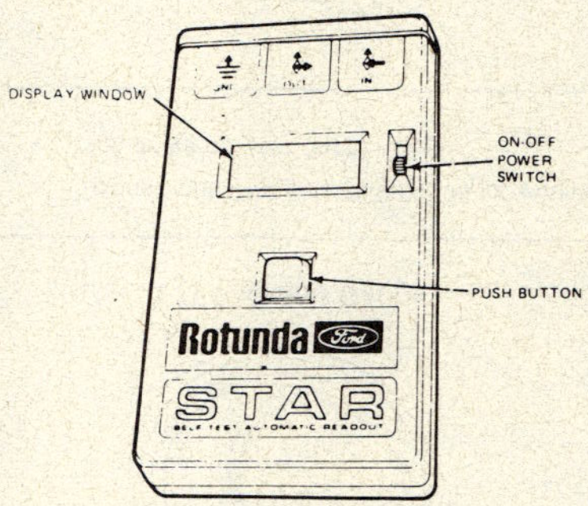

Automatic Readout (STAR) Tester

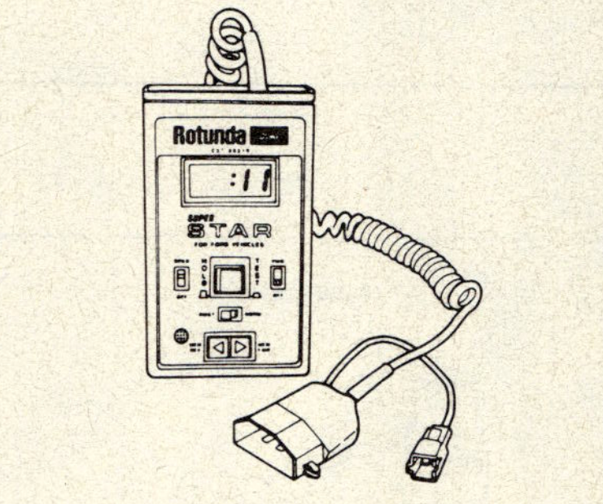

Super Star

- MAP/BP Sensor Tester, Rotunda 105-00001 or equivalent.
- TFI-IV Ignition Tester, Rotunda 105-00002 or equivalent.

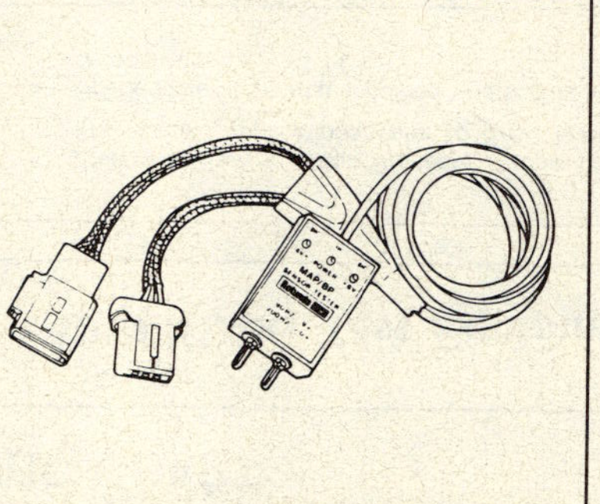

MAP/BP Sensor Tester

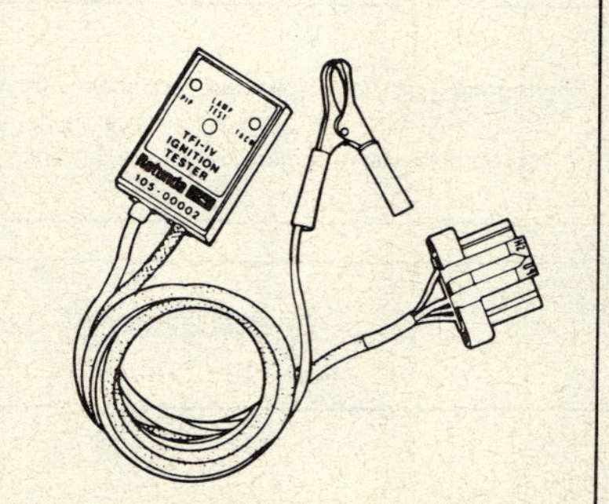

TFI-IV Ignition Tester

SCORPIO DIAGNOSTIC CHART

APPENDIX: Test Equipment (Continued)

- Tachometer, Rotunda 059-00010 or equivalent. Range 0-6,000 rpm. Accuracy ±40 rpm. Resolution 20 rpm.
- Breakout Box, Rotunda 014-00322, Special Service Tool T83L-50-EEC-IV or equivalent.

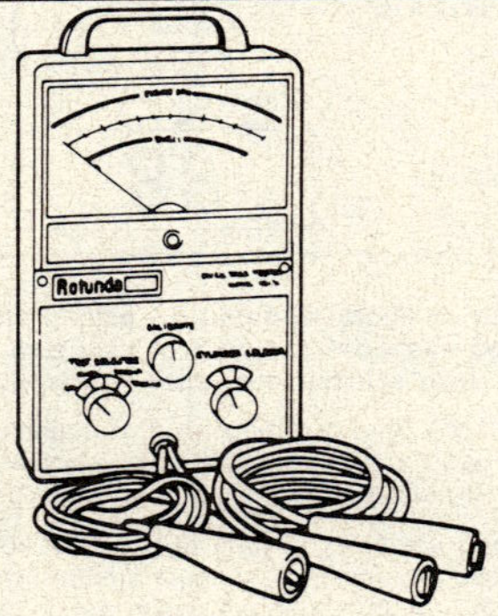

Tachometer

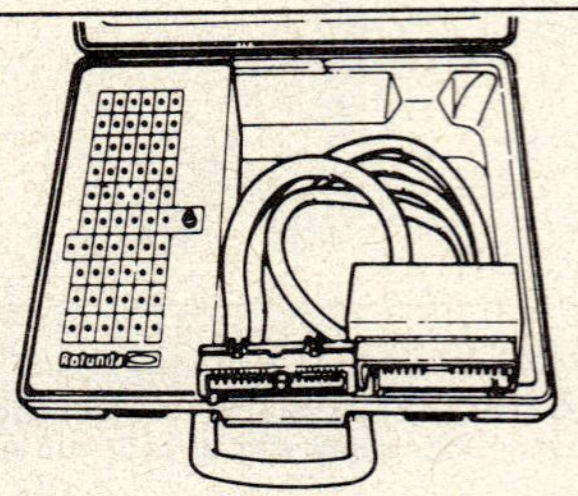
Breakout Box

- Vacuum pump, Rotunda 021-00014 or equivalent. Range 0-30 In.-Hg.
- Digital volt-ohmmeter, Rotunda 014-00407 or equivalent. Input impedance 10 Megaohm minimum.

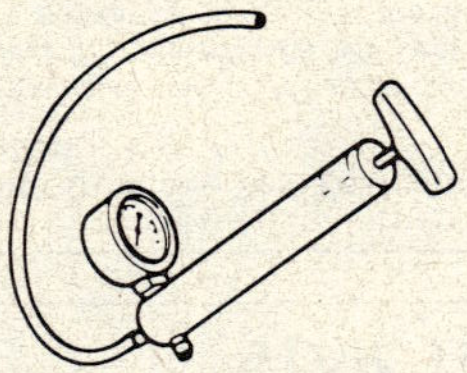
Vacuum Pump

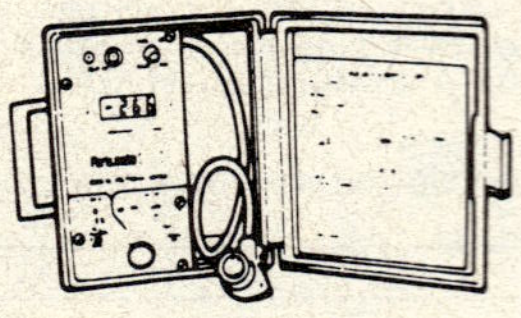
Digital Volt-Ohmmeter

APPENDIX: Test Equipment (Continued)

- Electronic Fuel Injection Pressure Gauge Tool T80L-9974-A or equivalent. (Use instructions. For specific applications, refer to Shop Manual Group 24.)
- Spark tester (optional modified spark plug with side electrode removed) Tool D81P-6666-A or equivalent.

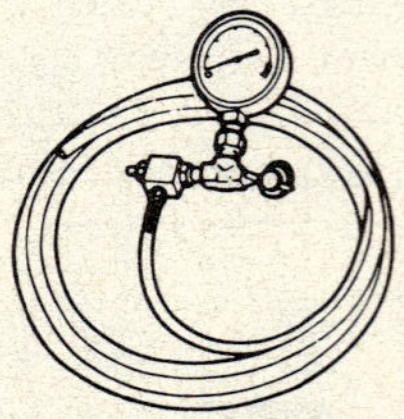
EFI Pressure Gauge

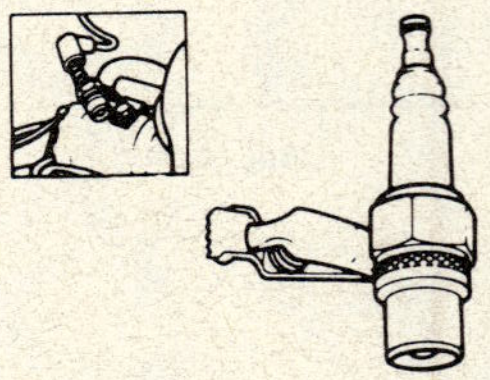
Spark Tester

- Timing light, Rotunda 059-00006 or equivalent.
- Non-powered test lamp.

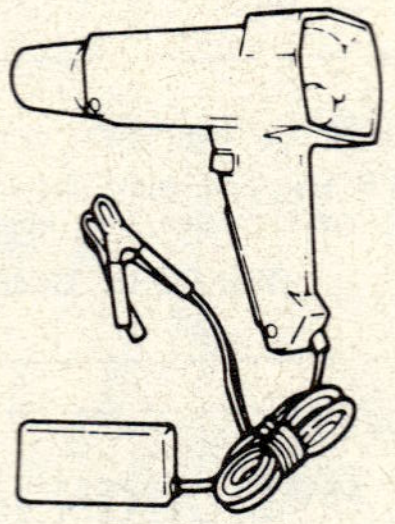
Timing Light

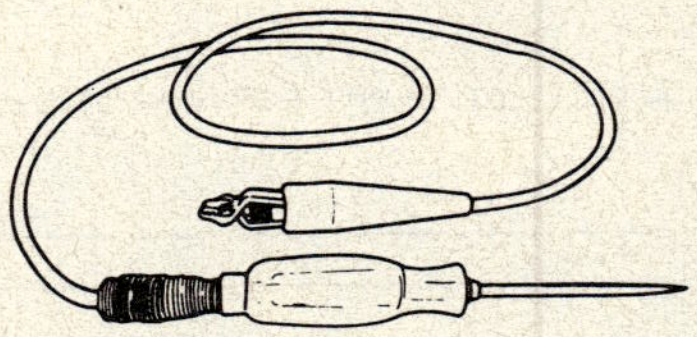
Non-Powered Test-Lamp

SCORPIO DIAGNOSTIC CHART

Diagnostic By Symptom — 2.9L EFI

SYMPTOM	RESULT	ACTION TO TAKE
• Engine runs rough or miss-fires. • Engine stalls. • Rough idle.		• GO to Pinpoint Test Step [S1]. • GO to MAP Pinpoint Test Step [DF20]. • EGR value function. • Ignition system distributor cap, rotor, wires, coil, plugs. • Base engine valves cam timing, compression etc. • Fuel delivery. • Poor power/ground connections.
• Gasoline fumes under hood.		GO to Pinpoint Test Step [KD1].
• Fuel pump always runs.		GO to Pinpoint Test Step [J15].
• Stumble after hot restart (HEGO).		GO to Pinpoint Test Step [HG21].
• No A/C.		GO to Pinpoint Test Step [X50].
• No A/C cutout at WOT.		GO to Pinpoint Test Step [X52].
• No air management (manual only).		GO to Pinpoint Test Step [Z1].

Electrical Schematic — 2.9L EFI

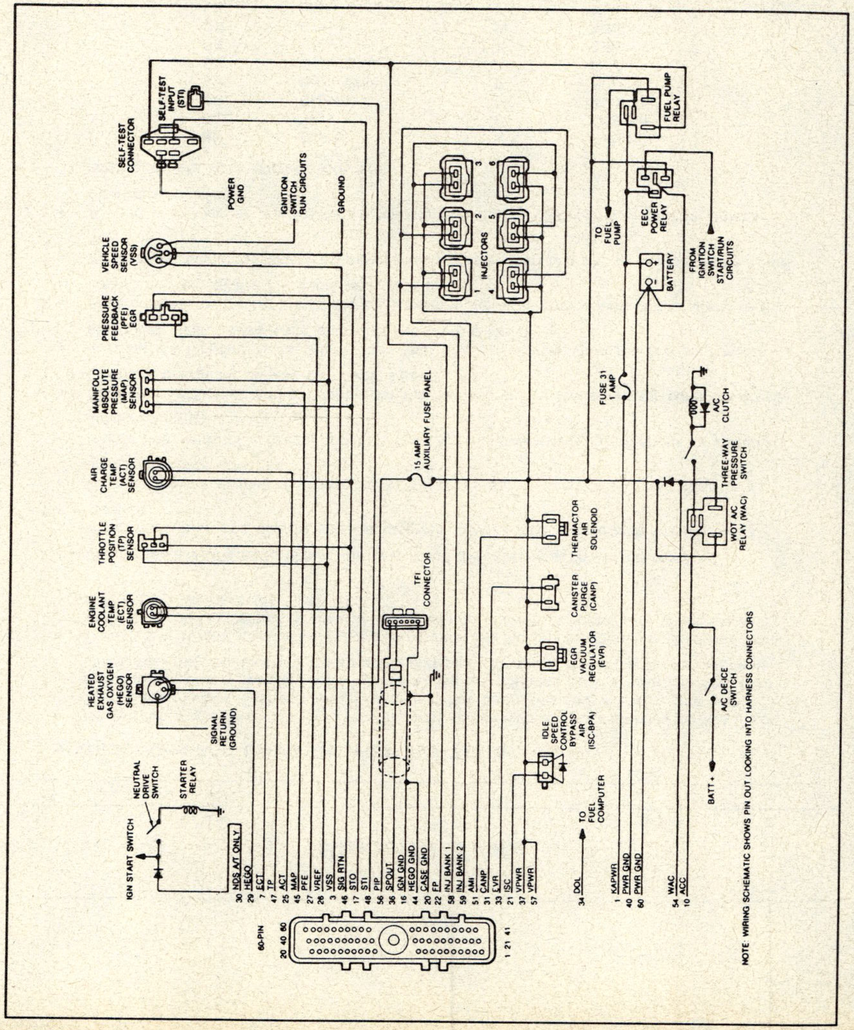

SCORPIO DIAGNOSTIC CHART

EEC-IV Module Connector Pin Usage — 2.9L EFI

Pin	Circuit	Wire Color	Application	Abbreviations
1	54-28	BK/BL	Keep Alive Power	KAPWR
3	2	Y/BR	Vehicle Speed Sensor (Hall Effect)	VSS
7	31B-75	BR/GN	Engine Coolant Temperature	ECT
10	54-37	BK/Y	A/C Clutch	ACCS
16	9-19	BL/BK	Ignition Ground	IGN GND
17	31B-40	BR/GN	Self-Test Output	STO
20	31-49	BR	Case Ground	CSE GND
21	31B-73	BR/Y	Idle Speed Control (Bypass Air)	ISC-BPA
22	9-25	BL/R	Fuel Pump	FP
25	31B-17	BR/Y	Air Charge Temperature	ACT
26	31B-20	BR/BK	Reference Voltage	VREF
27	31B-11	BR/GN	Pressure Feedback EGR	PFE
29	9-40	BL/W	Heated Exhaust Gas Oxygen (Sensor)	HEGO
30	31B-30	BR/W	Neutral Drive Switch (Auto Trans Only)	NDS
31	31B-10	BR/W	Canister Purge Solenoid	CANP
33	31B-74	BR/R	EGR Vacuum Regulator (EVR)	EVR
34	31B-3	BR/Y	Fuel Computer	FC
36	9-18	BL/R	Spark Output	SPOUT
37	15-6	BK	Vehicle Power	VPWR
40	31-29	BR	Power Ground	PWR GND
45	31B-76	BR/Y	Manifold Absolute Pressure (Sensor)	MAP
46	31-75	BR	Signal Return (Ground)	SIG RTN
47	31B-13	BR/GN	Throttle Position (Sensor)	TP
48	9-20	BL/GN	Self Test Input	STI
49	31-31	BR	Heat Exhaust Gas Oxygen Sensor Ground (Signal Return)	HEGO GND
51	31B-14	BR/R	Air Management — Thermactor Solenoid	AMI
54	31B-35	BR/GN	Wide-Open Throttle (WOT) A/C Cutoff	WAC
56	9-17	BL/Y	Profile Ignition Pickup	PIP
57	15-6	BK	Vehicle Power	VPWR
58	31-73	BR	Injector Bank 1	INJ 1
59	31-74	BR	Injector Bank 2	INJ 2
60	31-29	BR	Power Ground	PWR GND

Pin locations given for reference only. Probing 60 pin connector with DVOM probe will result in permanent damage to the pin connectors. Always probe as directed, using the Breakout Box.

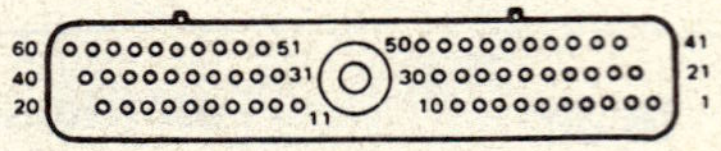

Pinpoint Tests

INSTRUCTIONS FOR USING THE PINPOINT TESTS

- Do not run any of the following Pinpoint Tests unless you are so instructed by the Quick Test. Each Pinpoint Test assumes that a fault has been detected in the system with direction to enter a specific repair routine. Doing any Pinpoint Test without direction from Quick Test may produce incorrect results and replacement of Non-Defective components.
- Correct test results for Quick Test are dependent on the proper operation of related non-EEC components/systems. It may be necessary to correct any defects in these areas before EEC will pass the Quick Test.
- Do not replace any parts unless the test result indicates they should be replaced.
- When more than one service code is received, always start service with the first code received.
- Do not measure voltage or resistance at the processor or connect any test lights to it, unless otherwise specified.
- Isolate both ends of a circuit, and turn key Off whenever checking for shorts or continuity, unless specified.
- Disconnect solenoids and switches from the harness before measuring for continuity, resistance, or energizing by way of 12-volt source.
- In using the Pinpoint Tests, follow each step in order, starting from the first step in the appropriate test. Follow each step until the fault is found.
- After completing any repairs to the EEC system, verify all components are properly reconnected and repeat the functional test (Retest).
- An open is defined as any resistance reading greater than 5 ohms unless otherwise specified.
- A short is defined as any resistance reading less than 10,000 ohms to ground, unless otherwise specified.

The standard Ford color abbreviations are:

BK	**Black**	**N**	**Natural**
BL	**Blue**	**O**	**Orange**
BR	**Brown**	**PK**	**Pink**
DB	**Dark Blue**	**P**	**Purple**
DG	**Dark Green**	**R**	**Red**
GY	**Gray**	**T**	**Tan**
GR	**Green**	**W**	**White**
LB	**Light Blue**	**Y**	**Yellow**
LG	**Light Green**		

Where two colors are shown for a wire, the first color is the basic color of the wire. The second color is the dot, hash, or stripe marking. If **D** or **H** is given, the second color is dots or hash marks. If there is no letter after the second color, the wire has a stripe.

For example:

BR/O is a brown wire with an orange stripe.

R/Y D is a red wire with yellow dots.

BK/W H is a black wire with white hash marks.

SCORPIO DIAGNOSTIC CHART

Pinpoint Test Index

PINPOINT TEST LETTER	PINPOINT TEST TITLE
A	EEC IV No Start
B	Vehicle Battery
C	Reference Voltage
	SENSOR INPUTS
DB	Air Charge Temperature (ACT) Sensor
DE	Engine Coolant Temperature (ECT) Sensor
DF	Manifold Absolute Pressure (MAP) Sensor
DH	Throttle Position (TP) Sensor
DL	Pressure Feedback EGR (PFE) EGR Valve Regulator (EVR)
	ADDITIONAL INPUTS
FA	Neutral Drive Switch A/C Input
	FUEL CONTROL SYSTEMS
HG	Fuel Control — 2.9L EFI V-6
J	Fuel Pump Circuit — (Inertia Switch)
	PROCESSOR OUTPUTS
KD	Canister Purge (CANP)
KE	Idle Speed Control (Bypass Air)
M	Dynamic Response Test
N	Ignition Diagnostic Monitor (IDM)
P	Spark Timing Check
Q	No Codes/Codes Not Listed
S	System Check
X	WOT A/C Cutoff (WAC) A/C Demand
Y	Erratic Ignition
Z	Thermactor Air (Air Management)

EEC-IV No Start — Pinpoint Test A

A

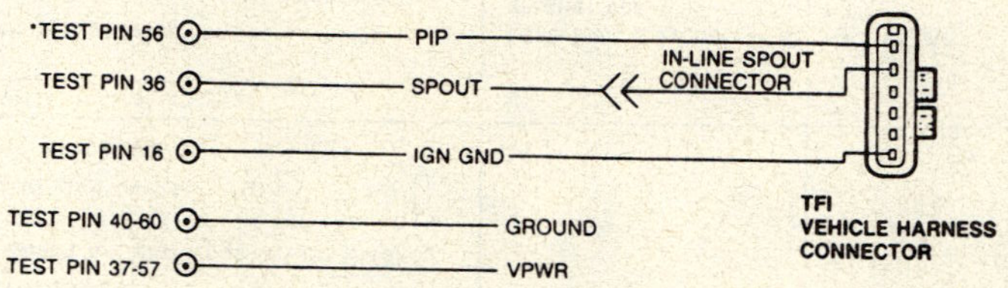

*TEST PINS LOCATED ON BREAKOUT BOX.
ALL HARNESS CONNECTORS VIEWED INTO MATING SURFACE.

NOTE: WHEN BREAKOUT BOX IS INSTALLED, ENSURE THAT TIMING SWITCH IS IN "COMPUTED" POSITION UNLESS OTHERWISE NOTED.

STOP-WARNING

You should enter this Pinpoint Test only when Steps 1.0 through 3.0 have been successfully completed and the engine is still a no start. This Pinpoint Test will not diagnose ignition system problems.

To prevent the replacement of good components, be aware that the following non-EEC areas may be at fault:

- Fuel: quantity and quality.
- Ignition: general condition, moisture, cracks, damage, etc.
- Engine: internal, valves, timing belt, camshaft.
- Starter and battery circuit.

This Pinpoint Test is intended to diagnose only the following:

- Spark (as related to EEC-IV).
- Circuits: pip, spout, ignition ground, vehicle power.

SCORPIO DIAGNOSTIC CHART

EEC-IV No Start | **Pinpoint Test** | **A**

WARNING: Stop this test at the first sign of a fuel leak and service as required.
CAUTION: No open flame — No smoking during fuel delivery checks.

TEST STEP	RESULT ▶	ACTION TO TAKE
A1 ATTEMPT TO START ENGINE	Engine cranks, but does not start, or stalls out ▶	GO to A2.
	Engine does not crank ▶	Check Starting System.
A2 CHECK FOR VREF AT THROTTLE POSITION SENSOR • Key Off, wait 10 seconds. • DVOM on 20V scale. • Disconnect TP sensor. • Key On Engine Off. • Measure voltage at the TP vehicle harness connector between VREF and signal return. **NOTE: Refer to electrical schematic in Diagnostic By Symptom chart for connector pin orientation.**	Less than 4.0V or greater than 6.0V ▶	GO to Pinpoint Test Step C1.
	4.0V to 6.0V ▶	RECONNECT TP sensor. GO to A3.

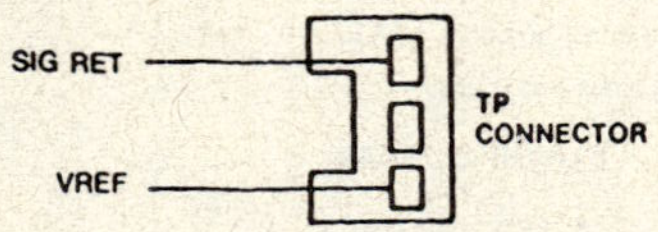

TEST STEP	RESULT ▶	ACTION TO TAKE
A3 CHECK FOR SPARK AT PLUGS • Disconnect the spark plug wire to any accessible cylinder. • Connect spark tester between spark plug wire and engine ground. • Crank engine and check for spark. • Reconnect the spark plug wire to the spark plug.	Spark ▶	GO to A13.
	No spark ▶	GO to A4.
A4 CHECK FOR SPARK AT COIL • Remove high tension coil wire from distributor and install spark tester. • Check for spark while cranking. • Reconnect high tension coil wire to distributor.	Spark ▶	CHECK TFI module using TFI-IV Ignition Tester, Rotunda 105-00002 or equivalent.
	No spark ▶	GO to A5.

EEC-IV No Start | **Pinpoint Test** | **A**

TEST STEP	RESULT ▶	ACTION TO TAKE
A5 HARNESS CHECK (IGNITION GROUND) • Key Off, wait 10 seconds. • Install Breakout box. Leave processor disconnected. • DVOM on 200 ohm scale. • Disconnect TFI. • Measure resistance between test Pin 16 at the Breakout box and TFI harness connector ignition ground.	Less than 5 ohms ▶	GO to A6.
	5 ohms or greater ▶	SERVICE harness as necessary. RERUN Quick Test.
A6 ISOLATION OF PROBLEM TO SPOUT CIRCUIT • Breakout box installed. • Connect TFI. • Connect processor. • Timing switch to "Dist" position on Breakout box. • Attempt to start vehicle. • Does the vehicle start?	Yes ▶	GO to A10.
	No ▶	GO to A7.
A7 SPOUT SIGNAL CHECK • Breakout box installed. • Timing switch to "Computed" position on Breakout box. • DVOM on 20V scale. • Measure voltage between test Pin 36 at the Breakout box and chassis ground, during crank.	Less than 3.0V or greater than 6.0V ▶	GO to A8.
	Between 3.0V and 6.0V ▶	CHECK TFI module using TFI-IV Ignition Tester, Rotunda 105-00002 or equivalent.
A8 CHECK SPOUT FOR SHORTS • Key Off, wait 10 seconds. • Breakout box installed. • Disconnect processor. • Disconnect TFI. • DVOM on 200,000 ohm scale. • Measure resistance between test Pin 36 and test Pins 16, 20, 26, 40, 60 (short to ground), 37, 57 (short to power) and 56 (short to pip) at the Breakout box.	All readings 10,000 ohms or greater ▶	GO to A9.
	Any reading less than 10,000 ohms ▶	SERVICE short in harness. RERUN Quick Test if vehicle does not start. GO to A9.

SCORPIO DIAGNOSTIC CHART

EEC-IV No Start | **Pinpoint Test** | **A**

TEST STEP	RESULT ▶	ACTION TO TAKE
A9 ISOLATE SHORT(S) IN PROCESSOR		
• Key Off, wait 10 seconds. • Breakout box installed. • Reconnect processor. • TFI disconnected. • DVOM on 200 ohm scale. • Measure resistance between test Pin 36 and test Pins 37 and 57 (short to power) also, test Pins 40 and 60 (short to ground) at the Breakout box.	All readings 5.0 ohms or greater. ▶	Connect TFI. GO to A10.
	Any reading less than 5.0 ohms ▶	REPLACE processor. RERUN Quick Test.
A10 PIP SIGNAL CHECK		
• Breakout box installed. • DVOM to 20V scale. • Measure voltage between test Pin 56 and test Pin 16 at the Breakout box. • Crank engine, record reading.	Between 3.0V and 6.0V ▶	REMOVE Breakout box. REPLACE processor. RERUN Quick Test.
	Less than 3.0V or greater than 6.0V ▶	GO to A11.
A11 CONTINUITY OF PIP CIRCUIT CHECK		
• Breakout box installed. • Key Off, wait 10 seconds. • DVOM on 200 ohm scale. • Disconnect TFI. • Disconnect processor. • Measure resistance between test Pin 56 at the Breakout box and TFI harness connector PIP circuit.	Less than 5 ohms ▶	GO to A12.
	5 ohms or greater ▶	SERVICE open PIP circuit. RERUN Quick Test.

EEC-IV No Start | **Pinpoint Test** | **A**

TEST STEP	RESULT ▶	ACTION TO TAKE
A12 CHECK PIP CIRCUIT FOR SHORTS		
• Breakout box installed. • Processor disconnected. • Key Off. • Disconnect TFI connector. • DVOM on 200,000 ohm scale. • Measure resistance between test Pin 56 and test Pins 16, 20, 26, 40, 60 (shorts to ground) and test Pins 37 and 57 (shorts to power) and test Pin 36 (short to spout) at the Breakout box.	Any resistance less than 10,000 ohms ▶	SERVICE PIP circuit. RERUN Quick Test.
	All resistance greater than 10,000 ohms ▶	CHECK TFI module using TFI-IV Ignition Tester Rotunda 105-00002 or equivalent.
A13 SPOUT SIGNAL VERIFICATION		
• Key Off, wait 10 seconds. • Disconnect processors 60 pin connector and inspect for damaged pins, corrosion, loose wires. Service as necessary. • Install Breakout box. • Processor connected. • DVOM on 20V scale. • Measure voltage between test Pin 36 at the Breakout box and chassis ground, during crank. • Ensure timing switch is in "Computed" position on Breakout box.	Between 3.0V and 6.0V ▶	GO to A21.
	Less than 3.0V or greater than 6.0V ▶	GO to A10.

SCORPIO DIAGNOSTIC CHART

EEC-IV No Start	Pinpoint Test	A

TEST STEP	RESULT ▶	ACTION TO TAKE
A21 FUEL PUMP CHECK		
• **No smoking nearby.** • Disconnect injector harness. • Connect pressure gauge. • Note initial pressure reading. • Observe pressure gauge as you pressurize fuel system. (Turn key to RUN for 1 second, then turn key to OFF. Wait 10 seconds. Repeat 5 times.) • Turn key Off. Wait 10 seconds. • Reconnect all injectors.	PRESSURE GAUGE READING: Increased ▶ Did not increase ▶	 GO to Pinpoint Test Step S2. TURN key Off, and CONTINUE to A22.
WARNING: If fuel starts leaking, turn key OFF immediately. No smoking.		
A22 INERTIA SWITCH CHECK		
• Key Off. • Fuel pressure gauge installed. • Locate fuel pump inertia switch in liftgate trim panel. • Push the button of inertia switch to reset to ON. • Check that the fuse is not open. **NOTE: If switch will not reset to ON, replace Inertia Switch and repeat Step A21. If switch button was on, GO to Pinpoint Test Step J4.** • Observe pressure gauge as you pressurize fuel system. (Turn key to RUN for 1 second, then turn key to OFF. Wait 10 seconds. Repeat 5 times.)	PRESSURE GAUGE READING: Increase ▶ No increase ▶	 RERUN Quick Test. • GO to Pinpoint Test Step J4. • All others, GO to Pinpoint Test Step J1.

Vehicle Battery	Pinpoint Test	B

B

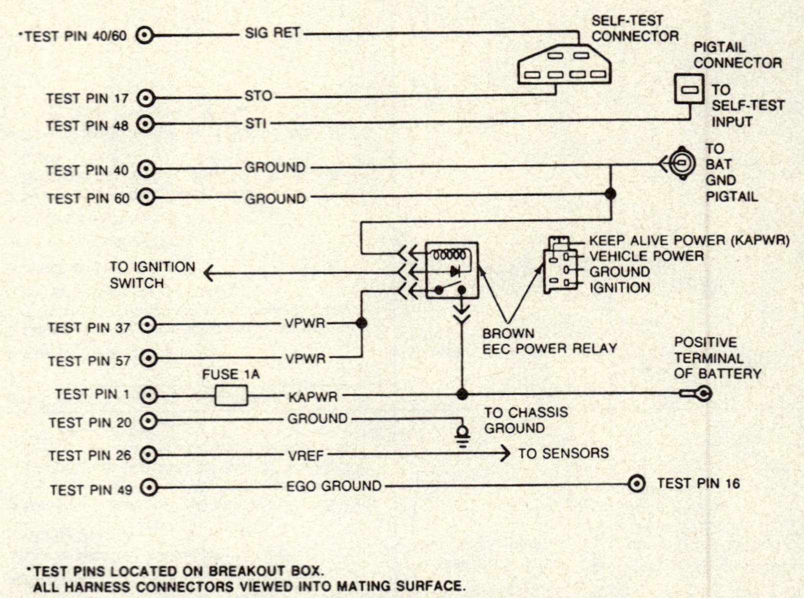

*TEST PINS LOCATED ON BREAKOUT BOX.
ALL HARNESS CONNECTORS VIEWED INTO MATING SURFACE.

STOP-WARNING

You should enter this Pinpoint Test only when directed here from Pinpoint Tests C, J or P or when a continuous memory code 78 is received in Quick Test Step 6.0C.

To prevent the replacement of good components, be aware that the following non-EEC areas may be at fault:

- Ignition switch.
- Battery Cables.
- Alternator.
- Voltage Regulator.
- Ground Straps.

This pinpoint test is intended to diagnose only the following:

- Processor.
- Harness circuits: Signal Return, STO, STI, Ground, VPWR, KAPWR, VREF, Ignition.
- Battery Voltage.
- Power Relay.

SCORPIO DIAGNOSTIC CHART

Vehicle Battery | **Pinpoint Test** | **B**

TEST STEP	RESULT ▶	ACTION TO TAKE
B1 BATTERY VOLTAGE CHECK		
• Key On Engine Off. • DVOM on 20V scale. • Measure voltage across battery terminals.	10.5V or greater ▶	GO to B2 .
	Less than 10.5V ▶	SERVICE discharged battery.
B2 BATTERY POWER GROUND CHECK		
• Key On Engine Off. • Processor connected. • DVOM on 20V scale. • Measure voltage between battery negative post and Signal Return in the Self-Test connector.	Less than 0.5V ▶	GO to B6 .
	0.5V or greater ▶	GO to B3 .
B3 GROUND FAULT ISOLATION		
• Breakout box installed. • Key On Engine Off. • Processor connected. • DVOM on 20V scale. • Measure voltage between battery negative post and test Pins 40 and 60 at the Breakout box.	Both readings less than 0.5V ▶	GO to B4 .
	One or both readings 0.5V or greater ▶	Circuit(s) with greater than 0.5V has high resistance or open. CORRECT faulty ground circuit. RERUN Quick Test.
B4 PROCESSOR GROUND FAULT ISOLATION		
• Breakout box installed. • Key Off, wait 10 seconds. • Processor connected. • DVOM on 200 ohm scale. • Measure resistance between test Pin 46 and test Pin 40 and between test Pin 46 and test Pin 60 both at the Breakout box.	Both readings less than 5 ohms ▶	GO to B5 .
	One or both readings 5 ohms or greater ▶	DISCONNECT processor connector and INSPECT for corrosion, damaged pins, etc. SERVICE as necessary and RETEST. If fault is still present, REPLACE processor. RERUN Quick Test.

Vehicle Battery | **Pinpoint Test** | **B**

TEST STEP	RESULT ▶	ACTION TO TAKE
B5 HARNESS CHECK (SIGNAL RETURN)		
• Breakout box installed. • Key Off, wait 10 seconds. • Processor connected. • DVOM on 200 ohm scale. • Measure resistance between battery negative and Signal Return in the Self-Test connector.	Less than 5 ohms ▶	System OK. RUN Quick Test.
	5 ohms or greater ▶	CORRECT cause of resistance in the harness Signal Return circuit. RERUN Quick Test.
B6 12 VOLT BATTERY POWER FAULT ISOLATION		
• Key On Engine Off. • Processor connected. • DVOM on 20V scale. • Measure voltage between the battery negative post and KAPWR circuit at EEC power relay.	10.5V or greater ▶	GO to B7 .
	Less than 10.5V ▶	CHECK KAPWR and VPWR circuits for shorts to ground and KAPWR circuit from power relay to battery positive post for opens. SERVICE as necessary. RERUN Quick Test.
B7 12 VOLT BATTERY POWER FAULT ISOLATION		
• Key On Engine Off. • Processor connected. • DVOM on 20V scale. • Measure voltage between the battery negative post and ignition circuit at EEC power relay.	10.5V or greater ▶	GO to B8 .
	Less than 10.5V ▶	CHECK for open in ignition switch circuits. SERVICE as necessary. RERUN Quick Test.
B8 12 VOLT BATTERY POWER FAULT ISOLATION		
• Key On Engine Off. • Processor connected. • DVOM on 20V scale. • Measure voltage between the battery negative post and ground circuit at EEC power relay.	Less than 0.5V ▶	If you entered this test for a code 78, GO to B10 . All others, GO to B9 .
	0.5V or greater ▶	SERVICE open or ground in Ground Circuit. RERUN Quick Test.

SCORPIO DIAGNOSTIC CHART

Vehicle Battery	Pinpoint Test	B

TEST STEP	RESULT ▶	ACTION TO TAKE
B9 12 VOLT BATTERY POWER FAULT ISOLATION		
• Key On Engine Off. • Processor connected. • DVOM on 20V scale. • Measure voltage between the battery negative post and VPWR circuit at EEC power relay.	10.5V or greater ▶	SERVICE VPWR circuit from EEC power relay to processor connector test Pins 37 and 57. RERUN Quick Test.
	Less than 10.5V ▶	REPLACE power relay. RERUN Quick Test.
B10 WIGGLE TEST VPWR CIRCUITS		
• Key On Engine Off. • STAR tester hooked up to Self-Test connector. • Self Test deactivated. • Observe STAR tester for fault indication as explained in Quick Test Step 6.0D. • Shake, bend and twist the EEC-IV harness from the EEC power relay to the processor. • Is a fault indicated or does code 78 reappear in continuous memory if Quick Test is rerun?	Yes ▶	SERVICE intermittent VPWR circuit. RERUN Quick Test.
	No ▶	INSPECT EEC-IV power relay and harness connectors for damaged pins, corrosion, etc. SERVICE as necessary. If OK, REPLACE EEC-IV power relay. RERUN Quick Test.

Reference Voltage	Pinpoint Test	C

C

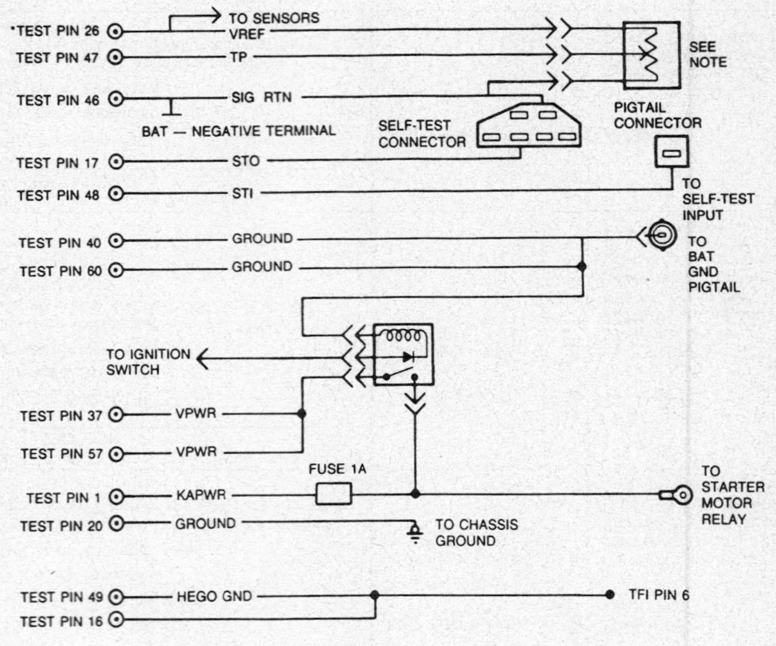

*TEST PINS LOCATED ON BREAKOUT BOX.
ALL HARNESS CONNECTORS VIEWED INTO MATING SURFACE.

NOTE: Refer to the electrical schematic in the appropriate Diagnosis by Symptoms Chart for proper connector and pin orientation.

STOP-WARNING

You should enter this Pinpoint Test only when a check for VREF has failed in the sensor Pinpoint Tests (D-Series) or Pinpoint Tests A or Q.

This pinpoint test is intended to diagnose only the following:

- Processor.
- Sensor harness circuits: Signal Return, STO, STI, Ground, VPWR, KAPWR, VREF, Ignition.

SCORPIO DIAGNOSTIC CHART

Reference Voltage — Pinpoint Test C

TEST STEP	RESULT ▶	ACTION TO TAKE
C1 VEHICLE BATTERY POWER CIRCUIT CHECK		
• Breakout box installed. • Key On Engine Off. • Processor connected. • DVOM on 20V scale. • Measure voltage between test Pin 37 at the Breakout box and Signal Return in Self-Test connector.	10.5V or greater ▶ Less than 10.5V ▶	GO to C2. CHECK battery state of charge and processor ground connection.
C2 VREF VOLTAGE CHECK		
• Breakout box installed. • Key On Engine Off. • Processor connected. • DVOM on 20V scale. • Measure voltage between test Pin 26 and test Pin 46 at the Breakout box.	6.0V or greater ▶ 4.0V or less ▶ Greater than 4.0V, less than 6.0V ▶	GO to C4. GO to C5. GO to C3.
C3 CHECK VREF AND SIGNAL RETURN FOR CONTINUITY		
• Breakout box installed. • Disconnect 60 Pin connector. Inspect for damaged pins, corrosion, loose wires, etc. Service as necessary. • Processor disconnected. • Key Off. • DVOM on 200 ohm scale. • Measure resistance from test Pin 26 at Breakout box to VREF at vehicle harness connector of the sensor that sent you here. • Measure resistance from test Pin 46 at Breakout box to signal return at vehicle harness connector of the sensor that sent you here.	Less than 5 ohms on all readings ▶ 5 ohms or greater on any reading ▶	RECONNECT sensors. Reference voltage OK. RERUN Quick Test. SERVICE open in VREF or Signal Return. RERUN Quick Test.
C4 CHECK FOR SHORT FROM VREF TO VPWR		
• Key Off, wait 10 seconds. • Breakout box installed. • Disconnect processor. • Key On Engine Off. • DVOM on 20V scale. • Measure voltage between test Pin 26 at the Breakout box and battery ground.	Less than 0.5V ▶ 0.5V or greater ▶	REPLACE processor. RERUN Quick Test. SERVICE short to battery power in EEC harness. RERUN Quick Test. If condition persists, REPLACE processor.

Reference Voltage — Pinpoint Test C

TEST STEP	RESULT ▶	ACTION TO TAKE
C5 CHECK FOR SHORTED THROTTLE POSITION SENSOR		
• Key Off, wait 10 seconds. • Breakout box installed. • Processor connected. • Disconnect Throttle Position (TP) sensor from vehicle harness. • Key On Engine Off. • DVOM on 20V scale. • Measure voltage between test Pin 26 and test Pin 46 at the Breakout box.	Less than 4.0V ▶ 4.0V or greater ▶	GO to C7. REPLACE TP sensor. RERUN Quick Test.
C6 CHECK FOR SHORTED PFE SENSOR		
• Key Off, wait 10 seconds. • Breakout box installed. • Processor connected. • Disconnect (PFE) sensor. • Key On Engine Off. • DVOM on 20V scale. • Measure voltage between test Pin 26 and test Pin 46 at the Breakout box.	Less than 4.0V ▶ 4.0V or greater ▶	GO to C7. REPLACE PFE sensor. RERUN Quick Test.
C7 CHECK FOR SHORTED MAP SENSOR		
• Key Off, wait 10 seconds. • Breakout box installed. • Processor connected. • Disconnect MAP sensor. • Key On Engine Off. • DVOM on 20V scale. • Measure voltage between test Pin 26 and test Pin 46 at the Breakout box.	Less than 4.0V ▶ 4.0V or greater ▶	GO to C8. REPLACE MAP sensor. RERUN Quick Test.

SCORPIO DIAGNOSTIC CHART

Reference Voltage — Pinpoint Test C

TEST STEP	RESULT ▶	ACTION TO TAKE
C8 SHORT TO GROUND IN VREF		
• Breakout box installed. • Processor disconnected. • Key Off, wait 10 seconds. • Disconnect TP and MAP, EVP and PFE. • DVOM on 200 ohm scale. • Measure resistance between test Pin 26 and test Pins 20, 40, 46 and 60 at the Breakout box.	Less than 5 ohms ▶	SERVICE short to ground. CONNECT all sensors. RERUN Quick Test. If original condition still exists, REPLACE processor.
	5 ohms or greater ▶	RECONNECT sensors. REPLACE processor. RERUN Quick Test.

Air Charge Temperature (ACT) Sensor — Pinpoint Test DB

DB

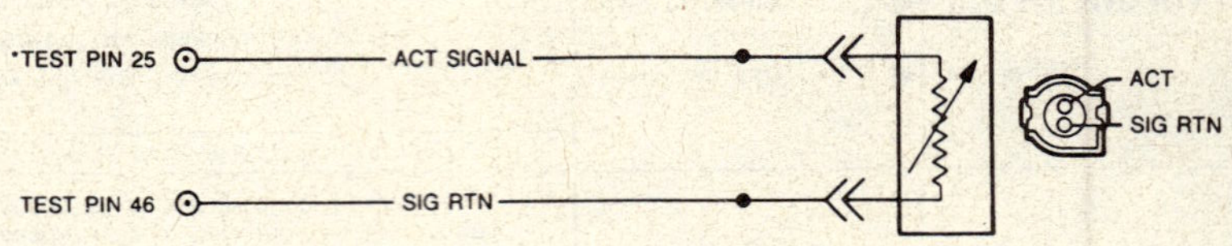

TYPICAL RESISTANCE BETWEEN TEST PINS 25 & 46	58.750 ohms	40.500 ohms	3600 ohms	1840 ohms
AT TEMPERATURE	50°F	65°F	180°F	220°F

*TEST PINS LOCATED ON BREAKOUT BOX.
ALL HARNESS CONNECTORS VIEWED INTO MATING SURFACE.

NOTE: Ambient temperature must be greater than 50°F for this test.

STOP-WARNING

You should enter this Pinpoint Test only when a service code 24, 54 or 64 is received in Quick Test Step 3.0, 5.0 or 6.0.

To prevent the replacement of good components, be aware that the following non-EEC areas may be at fault:

- Cooling system.
- Air cleaner duct problems.
- Improper engine oil level.

This pinpoint test is intended to diagnose only the following:

- ACT sensor.
- Harness circuits: ACT Signal and Signal Return.
- Processor assembly.

SCORPIO DIAGNOSTIC CHART

Air Charge Temperature (ACT) Sensor — Pinpoint Test DB

TEST STEP	RESULT ▶	ACTION TO TAKE
DB1 CHECK FOR VREF AT THROTTLE POSITION SENSOR		
• Key Off, wait 10 seconds. • DVOM on 20V scale. • Disconnect TP sensor. • Key On Engine Off. • Measure voltage at the TP vehicle harness connector between VREF and signal return.	Less than 4.0V or greater than 6.0V ▶	GO to Pinpoint Test Step [C1].
	4.0V to 6.0V ▶	RECONNECT TP sensor, GO to [DB2].
DB2 CHECK ACT SENSOR — ENGINE OFF		
• Key Off, wait 10 seconds. • Harness disconnected from ACT sensor. • DVOM on 200,000 ohm scale. • Measure resistance of ACT sensor. **NOTE: Make sure engine is warmed up prior to this test.**	Reading is between 1,100 and 58,000 ohms (approximately 50°F) ▶	GO to [DB3].
	Reading is less than 1,100 ohms or greater than 58,000 ohms ▶	REPLACE ACT sensor. RECONNECT harness to ACT sensor. RERUN Quick Test.
DB3 CHECK ACT SENSOR — ENGINE RUNNING		
• Key Off. Harness disconnected from ACT sensor. • DVOM on 200,000 ohm scale. • Run engine until warm. • Measure resistance of ACT sensor with engine running.	Reading is between 2,400 and 29,000 ohms ▶	REPLACE processor. RECONNECT harness to ACT sensor. RERUN Quick Test.
	Reading is less than 2,400 ohms or greater than 29,000 ohms ▶	REPLACE ACT sensor. RERUN Quick Test.

Air Charge Temperature (ACT) Sensor — Pinpoint Test DB

TEST STEP	RESULT ▶	ACTION TO TAKE
DB10 SERVICE CODE 54: INDUCE OPPOSITE CODE		
• Key Off, wait 10 seconds. • Disconnect vehicle harness from ACT sensor. Inspect for damaged pins, corrosion, loose wires, etc. Service as necessary. • Insert a jumper wire at the ACT vehicle harness connector between ACT Signal and Signal Return. • Run Key On Engine Off Quick Test. • Is code 64 present?	Yes ▶	REPLACE ACT sensor. REMOVE jumper wire. RECONNECT harness to ACT sensor. RERUN Quick Test.
	No ▶	REMOVE jumper wire. GO to [DB11].
DB11 CONTINUITY CHECK ACT SIGNAL AND SIGNAL RETURN		
• Key Off, wait 10 seconds. • Harness disconnected from ACT sensor. • Disconnect processor 60 pin connector. Inspect for damaged pins, corrosion, loose wires, etc. Service as necessary. • Install Breakout box. Leave processor disconnected. • DVOM on 200 ohm scale. • Measure resistance between ACT signal, at the ACT vehicle harness connector, and test Pin 25 at the Breakout box. • Measure resistance between Signal Return, at the ACT vehicle harness connector, and test Pin 46 at the Breakout box.	Both readings are less than 5 ohms ▶	REPLACE processor. REMOVE Breakout box. RECONNECT processor and ACT sensor. RERUN Quick Test.
	Either reading is 5 ohms or greater ▶	CORRECT circuit opens. REMOVE Breakout box. RECONNECT processor and ACT sensor. RERUN Quick Test.

SCORPIO DIAGNOSTIC CHART

Air Charge Temperature (ACT) Sensor — Pinpoint Test — DB

TEST STEP	RESULT ▶	ACTION TO TAKE
DB20 SERVICE CODE 64: INDUCE OPPOSITE CODE		
• Key Off, wait 10 seconds. • Disconnect vehicle harness from ACT sensor. Inspect for damaged pins, corrosion, loose wires, etc. Service as necessary. • Run Key On Engine Off Quick Test. • Is code 54 present?	Yes ▶	REPLACE ACT sensor. RECONNECT harness to ACT sensor. RERUN Quick Test.
	No ▶	GO to DB21.
DB21 CHECK FOR VREF AT THROTTLE POSITION SENSOR		
• Key Off, wait 10 seconds. • DVOM on 20V scale. • Disconnect TP sensor. • Key On Engine Off. • Measure voltage at the TP vehicle harness connector between VREF and Signal Return.	Less than 4.0V or greater than 6.0V ▶	GO to Pinpoint Test Step C1.
	4.0V to 6.0V ▶	RECONNECT TP sensor, GO to DB22.
DB22 CHECK ACT SIGNAL FOR SHORTS TO GROUND		
• Key Off, wait 10 seconds. • Harness disconnected from ACT sensor. • Disconnect processor 60 pin connector. Inspect for damaged pins, corrosion, loose wires, etc. Service as necessary. • DVOM on 200,000 ohm scale. • Measure resistance between test Pin 25 and test Pins 40, 46 and 60 at the Breakout box.	Any reading less than 10,000 ohms ▶	CORRECT circuit shorts. REMOVE Breakout box. RECONNECT processor and ACT sensor. RERUN Quick Test.
	All readings are 10,000 ohms or greater ▶	REPLACE processor. REMOVE Breakout box. RECONNECT processor and ACT sensor. RERUN Quick Test.

Air Charge Temperature (ACT) Sensor — Pinpoint Test — DB

TEST STEP	RESULT ▶	ACTION TO TAKE
DB90 SERVICE CODE 54: CONTINUOUS TEST: CHECK ACT SENSOR		
• Using continuous monitor mode, observe STAR LED for indication of a fault while performing the following: • Lightly tap on ACT sensor (simulate road shock). • Wiggle ACT connector. • Is a fault indicated?	Yes ▶	DISCONNECT and INSPECT connectors. If connector and terminals are good, REPLACE ACT sensor. RERUN Quick Test.
	No ▶	GO to DB91.
DB91 CHECK EEC-IV HARNESS		
• Observe STAR LED for a fault indication while performing the following: • Referring to the illustration in Step DB90, grasp the harness closest to the sensor connector. Wiggle, shake or bend a small section of the EEC-IV system harness while working your way to the dash panel. Also wiggle, shake or bend the EEC-IV harness from the dash panel to the processor. • Is fault indicated?	Yes ▶	ISOLATE fault and SERVICE as necessary. RERUN Quick Test.
	No ▶	GO to DB92.
DB92 CHECK PROCESSOR AND HARNESS CONNECTORS		
• Key Off, wait 10 seconds. • Disconnect processor 60 pin connector. • Inspect both connectors and connector terminals for obvious damage or faults. • Connectors and terminals are OK.	No ▶	SERVICE as necessary. RERUN Quick Test.
	Yes ▶	Unable to duplicate fault at this time. Continuous code 54 testing complete.

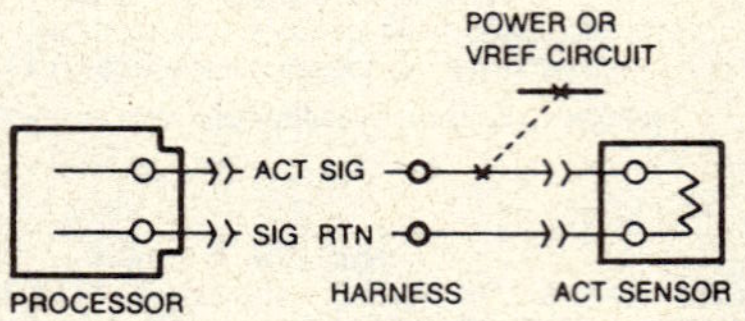

SCORPIO DIAGNOSTIC CHART

Air Charge Temperature (ACT) Sensor — Pinpoint Test — DB

TEST STEP	RESULT ▶	ACTION TO TAKE
DB93 SERVICE CODE 64: CONTINUOUS TEST: CHECK ACT SENSOR		
• Using continuous monitor mode, observe STAR LED for fault while performing the following: • Lightly tap on ACT sensor (simulate road shock). • Wiggle ACT connector. • Is fault indicated?	Yes ▶	DISCONNECT and INSPECT connectors. If connector and terminals are good, REPLACE ACT sensor. RERUN Quick Test.
	No ▶	GO to DB94 .
DB94 CHECK EEC-IV HARNESS		
• Observe STAR LED for a fault indication while performing the following: • Referring to the illustration in Step DB93, grasp the harness closest to the sensor connector. Wiggle, shake or bend a small section of the EEC-IV system harness while working your way to the dash panel. Also wiggle, shake or bend the EEC-IV harness from the dash panel to the processor. • Is fault indicated?	Yes ▶	ISOLATE fault and SERVICE as necessary. RERUN Quick Test.
	No ▶	GO to DB95 .
DB95 CHECK PROCESSOR AND HARNESS CONNECTORS		
• Key Off, wait 10 seconds. • Disconnect processor 60 pin connector. • Inspect both connectors and connector terminals for obvious damage or faults. • Connectors and terminals are OK.	No ▶	SERVICE as necessary. RERUN Quick Test.
	Yes ▶	Unable to duplicate fault at this time. Continuous code 64 testing complete.

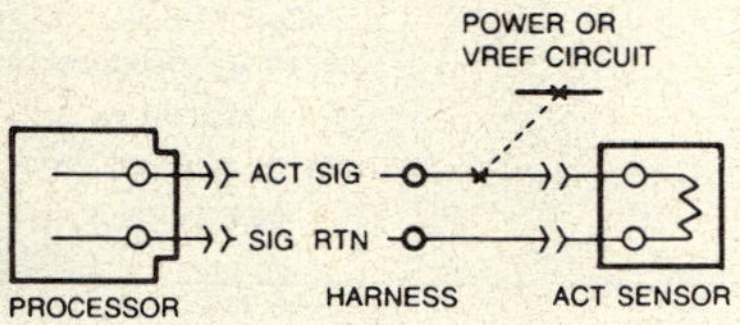

Engine Coolant Temperature (ECT) Sensor — Pinpoint Test — DE

DE

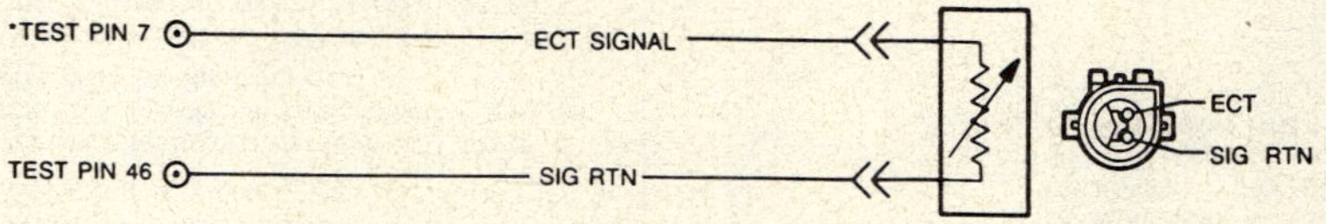

EEC COOLANT TEMPERATURE FRONT OF ENGINE

NOTE: TO PASS THIS TEST, ENGINE COOLANT TEMPERATURE MUST BE:
KEY ON ENGINE OFF (50°F TO 240°F)
ENGINE RUNNING (180°F TO 240°F)

TYPICAL RESISTANCE BETWEEN TEST PINS 7 & 46	58,750 ohms	40,500 ohms	3600 ohms	1840 ohms
AT TEMPERATURE	50°F	65°F	180°F	220°F

*TEST PINS LOCATED ON BREAKOUT BOX.
ALL HARNESS CONNECTORS VIEWED INTO MATING SURFACE.

STOP-WARNING

You should enter this Pinpoint Test only when a service code 21, 51 or 61 is received in Quick Test Step 3.0, 5.0 or 6.0.

To prevent the replacement of good components, be aware that the following non-EEC areas may be at fault:

- Coolant level.
- Oil level.
- Blocked or obstructed airflow.
- Engine not at normal operating temperature.
- Electro drive cooling fan.

This pinpoint test is intended to diagnose only the following:

- ECT sensor.
- Harness sensor circuits: ECT and Signal Return.
- Processor assembly.

SCORPIO DIAGNOSTIC CHART

Engine Coolant Temperature (ECT) Sensor — Pinpoint Test DE

TEST STEP	RESULT ▶	ACTION TO TAKE
DE1 SERVICE CODE 21: CHECK ENGINE OPERATING TEMPERATURE		
• Run engine until warm at 2,000 rpm. • Check that upper radiator hose is hot and pressurized. • Rerun Quick Test.	Vehicle stalls ▶	Do not service code 21 at this time. REFER to Diagnostic By Symptom.
	Code 21 present ▶	GO to DE2 .
	Code 21 not present ▶	SERVICE other codes as necessary.
DE2 CHECK FOR VREF AT THROTTLE POSITION SENSOR		
• Refer to illustration Q. • Key Off, wait 10 seconds. • DVOM on 20V scale. • Disconnect TP sensor. • Key On Engine Off. • Measure voltage at the TP vehicle harness connector between VREF and signal return.	Less than 4.0V or greater than 6.0V ▶	GO to Pinpoint Test Step C1 .
	4.0V to 6.0V ▶	RECONNECT TP sensor, GO to DE3 .
DE3 ECT SENSOR CHECK		
NOTE: Engine may have cooled down. Always warm engine before taking ECT resistance measurement. • Key Off, wait 10 seconds. • Harness disconnected from ECT sensor. • DVOM on 200,000 ohm scale. • Measure resistance of the ECT sensor. • Is the resistance reading: — 1300 ohms (240°F) to 7700 ohms (140°F) for engine off? — 1550 ohms (230°F) to 4550 ohms (180°F) for engine running?	Yes ▶	REPLACE processor. RECONNECT harness to ECT sensor. RERUN Quick Test.
	No ▶	REPLACE ECT sensor. RECONNECT harness to ECT sensor. RERUN Quick Test.

Engine Coolant Temperature (ECT) Sensor — Pinpoint Test DE

TEST STEP	RESULT ▶	ACTION TO TAKE
DE10 SERVICE CODE 51: INDUCE OPPOSITE CODE		
• Key Off, wait 10 seconds. • Disconnect vehicle harness from ECT sensor. Inspect for damaged pins, corrosion, loose wires, etc. Service as necessary. • Insert a jumper wire at the ECT sensor vehicle harness connector between ECT Signal and Signal Return. • Run Key On Engine Off Quick Test. • Is code 61 present?	Yes ▶	REPLACE ECT sensor. REMOVE jumper wire. RECONNECT ECT sensor. RERUN Quick Test.
	No ▶	GO to DE11 .
DE11 CHECK CONTINUITY OF ECT SIGNAL AND SIGNAL RETURN		
• Key Off, wait 10 seconds. • Harness disconnected from ECT sensor, jumper wire removed. • Disconnect processor 60 pin connector. Inspect for damaged pins, corrosion, loose wires, etc. Service as necessary. • Install Breakout box to harness leaving processor disconnected. • DVOM on 200 ohm scale. • Measure resistance between ECT signal at the ECT vehicle harness connector and test Pin 7 at the Breakout box. • Measure resistance between Signal Return at the ECT sensor vehicle harness connector, and test Pin 46 at the Breakout box.	Both readings are less than 5 ohms. ▶	REPLACE processor. REMOVE Breakout box. RECONNECT processor and ECT sensor. RERUN Quick Test.
	Either reading is 5 ohms or greater ▶	SERVICE open circuit(s). REMOVE Breakout box. RECONNECT processor and ECT sensor. RERUN Quick Test.

SCORPIO DIAGNOSTIC CHART

Engine Coolant Temperature (ECT) Sensor — Pinpoint Test DE

TEST STEP	RESULT ▶	ACTION TO TAKE
DE20 SERVICE CODE 61: INDUCE OPPOSITE CODE		
• Key Off, wait 10 seconds. • Disconnect vehicle harness from ECT sensor. Inspect for damaged pins, corrosion, loose wires, etc. Service as necessary. • Run Key On Engine Off Quick Test. • Is code 51 present?	Yes ▶	REPLACE ECT sensor. RECONNECT ECT sensor. RERUN Quick Test.
	No ▶	GO to DE21.
DE21 CHECK FOR VREF AT THROTTLE POSITION SENSOR		
• Refer to illustration Q. • Key Off, wait 10 seconds. • DVOM on 20V scale. • Disconnect TP sensor. • Key On Engine Off. • Measure voltage at the TP vehicle harness connector between VREF and signal return.	Less than 4.0V or greater than 6.0V ▶	GO to Pinpoint Test Step C1.
	4.0V to 6.0V ▶	RECONNECT TP sensor, GO to DE22.
DE22 CHECK ECT SIGNAL FOR SHORT		
• Key Off, wait 10 seconds. • Harness disconnected from ECT sensor. • Disconnect processor 60 pin connector. Inspect for damaged pins, corrosion, loose wires, etc. Service as necessary. • Install Breakout box, leave processor disconnected. • DVOM on 200,000 ohm scale. • Measure resistance between test Pin 7 and test Pins 40, 46 and 60 at the Breakout box.	Either reading is less than 10,000 ohms ▶	SERVICE circuit shorts. REMOVE Breakout box. RECONNECT processor and ECT sensor. RERUN Quick Test.
	Both readings are 10,000 ohms or greater ▶	REPLACE processor. REMOVE Breakout box. RECONNECT processor and ECT harness. RERUN Quick Test.

Engine Coolant Temperature (ECT) Sensor — Pinpoint Test DE

TEST STEP	RESULT ▶	ACTION TO TAKE
DE90 SERVICE CODE 21: CONTINUOUS TEST: TEST DRIVE VEHICLE		
• Key Off and wait 10 seconds. • Disconnect all Self-Test equipment and prepare vehicle for test drive. • Drive vehicle. Try to simulate different drive modes or mode in which drive complaint is noticed. Attempt to maintain drive complaint mode for one minute or more, if possible. • Upon completion of drive evaluation, repeat Key On Engine Off Self Test. • Is code 21 present in the continuous test results?	Yes ▶	VERIFY thermostat operating properly. If OK, REPLACE ECT sensor. RERUN Quick Test.
	No ▶	Unable to duplicate fault. Code 21 testing complete.

SCORPIO DIAGNOSTIC CHART

Engine Coolant Temperature (ECT) Sensor — Pinpoint Test DE

TEST STEP	RESULT ▶	ACTION TO TAKE
DE91 SERVICE CODE 51: CONTINUOUS TEST: CHECK ECT SENSOR		
• Using continuous monitor mode, observe STAR LED for indication of a fault while performing the following: • Lightly tap on ECT sensor (simulate road shock). • Wiggle ECT connector. • Is a fault indicated?	Yes ▶	DISCONNECT and INSPECT connectors. If connector and terminals are good, REPLACE ECT sensor. RERUN Quick Test.
	No ▶	GO to DE92.

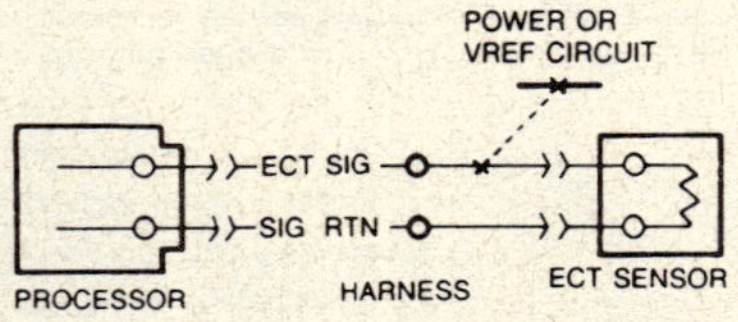

TEST STEP	RESULT ▶	ACTION TO TAKE
DE92 CHECK EEC-IV HARNESS		
• Observe STAR LED for a fault indication while performing the following: • Referring to the illustration in Step DE91, grasp the harness closest to the sensor connector. Wiggle, shake or bend a small section of the EEC-IV system harness while working your way to the dash panel. Also wiggle, shake or bend the EEC-IV harness from the dash panel to the processor. • Is fault indicated?	Yes ▶	ISOLATE fault and SERVICE as necessary. RERUN Quick Test.
	No ▶	GO to DE93.
DE93 CHECK PROCESSOR AND HARNESS CONNECTORS		
• Key Off, wait 10 seconds. • Disconnect processor 60 pin connector. • Inspect both connectors and connector terminals for obvious damage or faults. • Connectors and terminals are OK.	No ▶	SERVICE as necessary. RERUN Quick Test.
	Yes ▶	Unable to duplicate fault at this time. Continuous code 51 testing complete.

Engine Coolant Temperature (ECT) Sensor — Pinpoint Test DE

TEST STEP	RESULT ▶	ACTION TO TAKE
DE94 SERVICE CODE 61: CONTINUOUS TEST: CHECK ECT SENSOR		
• Using continuous monitor mode, observe STAR LED for indication of a fault while performing the following: • Lightly tap on ECT sensor (simulate road shock). • Wiggle ECT connector. • Is STAR LED Off?	Yes ▶	DISCONNECT and INSPECT connectors. If connector and terminals are good, REPLACE ECT sensor. RERUN Quick Test.
	No ▶	GO to DE95.

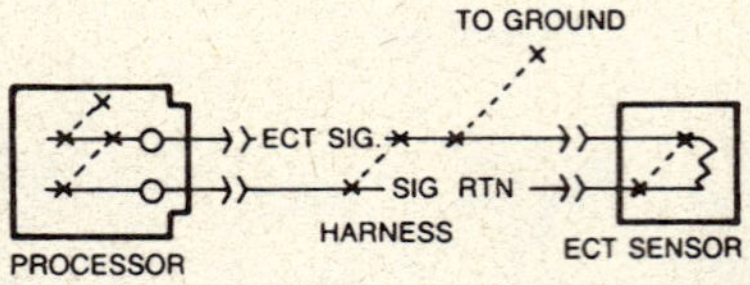

TEST STEP	RESULT ▶	ACTION TO TAKE
DE95 CHECK EEC-IV HARNESS		
• Observe STAR LED for a fault indication while performing the following: • Referring to the illustration in Step DE94, grasp the harness closest to the sensor connector. Wiggle, shake or bend a small section of the EEC-IV system harness while working your way to the dash panel. Also wiggle, shake or bend the EEC-IV harness from the dash panel to the processor. • Is fault indicated?	Yes ▶	ISOLATE fault and SERVICE as necessary. RERUN Quick Test.
	No ▶	GO to DE96.
DE96 CHECK PROCESSOR AND HARNESS CONNECTORS		
• Key Off, wait 10 seconds. • Disconnect processor 60 pin connector. • Inspect both connectors and connector terminals for obvious damage or faults. • Connectors and terminals are OK.	No ▶	SERVICE as necessary. RERUN Quick Test.
	Yes ▶	Unable to duplicate fault at this time. Continuous code 61 testing complete.

SCORPIO DIAGNOSTIC CHART

Manifold Absolute Pressure (MAP) Sensor | Pinpoint Test | DF

TEST STEP	RESULT ▶	ACTION TO TAKE
DF11 CHECK VACUUM LINES		
• Check vacuum line for proper routing. Refer to VECI decal. Check MAP sensor vacuum line for kinks or blockage. • Vacuum lines are OK.	Yes ▶	EEC-IV system OK. Check all probable subjects affecting engine vacuum.
	No ▶	SERVICE as necessary and REPEAT DF10.
DF20 CONNECTING MAP/BP TESTER		
• Key Off. • Disconnect the MAP sensor from the vehicle harness. • Connect the MAP/BP Tester between the vehicle harness and the MAP sensor. • Plug tester banana plugs into DVOM. • Set DVOM to 20V scale. • Refer to illustration DF.	Tester properly hooked up ▶	GO to DF21.
DF21 MAP/BP TESTER OUTPUT READING		
• MAP/BP Tester connected, refer to Note. • Key On. • Approximate Altitude (Ft.) / Voltage Output (± .04 Volts) 0 — 1.59 1000 — 1.56 2000 — 1.53 3000 — 1.50 4000 — 1.47 5000 — 1.44 6000 — 1.41 7000 — 1.39 **NOTE: Measure several known good MAP sensors on available vehicles. The measured voltage will be typical for your location on the day of testing.** • Is reading in range for your altitude?	Yes ▶	GO to DF22.
	No (Sensor output is out of range) ▶	REPLACE MAP sensor.

Manifold Absolute Pressure (MAP) Sensor | Pinpoint Test | DF

TEST STEP	RESULT ▶	ACTION TO TAKE
DF22 VACUUM LINE CHECK		
• Check MAP sensor vacuum line for holes, disconnections, kinks or blockage. • Are vacuum lines OK?	Yes ▶	GO to Diagnostic By Symptom chart.
	No ▶	SERVICE vacuum lines to MAP sensor. RERUN Quick Test.

SCORPIO DIAGNOSTIC CHART

Manifold Absolute Pressure (MAP) Sensor — Pinpoint Test — DF

TEST STEP	RESULT ▶	ACTION TO TAKE
DF90 SERVICE CODE 22: CONTINUOUS TEST: EXERCISE MAP SENSOR		
• Using continuous monitor mode, observe STAR LED for indication of a fault while performing the following: • Connect a vacuum pump to the MAP sensor. • Slowly apply 84 kPa (25 in.-Hg.) vacuum to the sensor. • Slowly bleed vacuum off the MAP sensor. • Lightly tap on MAP sensor (simulate road shock). • Wiggle MAP connector. • Is fault indicated?	Yes ▶	DISCONNECT and INSPECT connectors. If connector and terminals are good, REPLACE sensor. RERUN Quick Test.
	No ▶	GO to DF91.
DF91 CHECK EEC-IV HARNESS		
• Observe STAR LED for a fault indication while performing the following: • Referring to the illustration in Step DF90, grasp the harness closest to the sensor connector. Wiggle, shake or bend a small section of the EEC-IV system harness while working your way to the dash panel. Also wiggle, shake or bend the EEC-IV harness from the dash panel to the processor. • Is a fault indicated?	Yes ▶	ISOLATE fault and SERVICE as necessary. RERUN Quick Test.
	No ▶	GO to DF92.
DF92 CHECK PROCESSOR AND HARNESS CONNECTORS		
• Key Off, wait 10 seconds. • Disconnect processor 60 Pin connector. • Inspect both connectors and connector terminals for obvious damage or faults. • Are connectors and terminals OK?	No ▶	SERVICE as necessary. RERUN Quick Test.
	Yes ▶	Unable to duplicate fault at this time. Continuous code 22 testing complete.

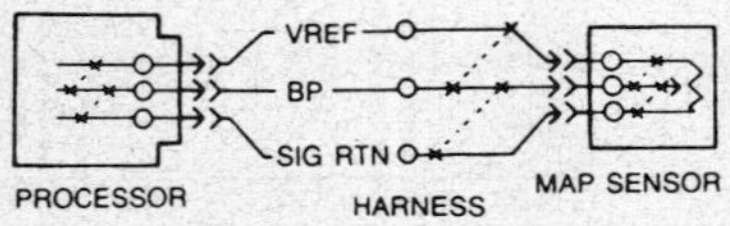

Throttle Position (TP) Sensor — Pinpoint Test — DH

DH

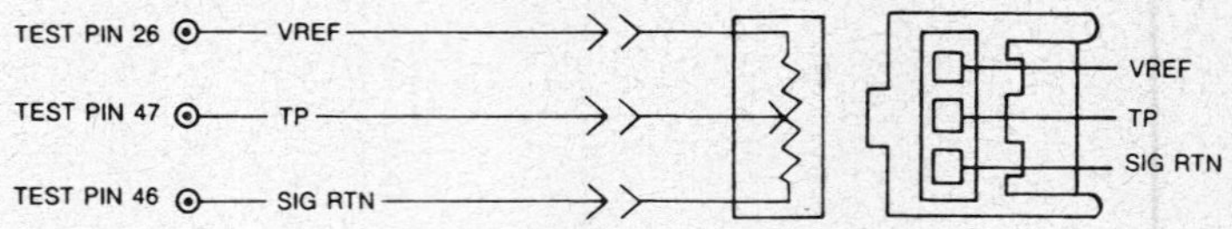

TEST PINS LOCATED ON BREAKOUT BOX.
ALL HARNESS CONNECTORS VIEWED INTO MATING SURFACE.

STOP-WARNING

You should enter this Pinpoint Test only when a service code 23, 53, 63 or 73 is received in Quick Test Step 3.0, 5.0 or 6.0.

To prevent the replacement of good components, be aware that the following non-EEC areas may be at fault:

- Idle speeds/throttle stop adjustment.
- Binding throttle shaft/linkage or speed control linkage.

This Pinpoint Test is intended to diagnose only the following:

- TP sensor.
- Sensor harness circuits: VREF, TP Signal, and Signal Return.
- Processor assembly.

SCORPIO DIAGNOSTIC CHART

Throttle Position (TP) Sensor — Pinpoint Test DH

TEST STEP	RESULT ▶	ACTION TO TAKE
DH1 SERVICE CODE 23: THE FOLLOWING CHECK MUST BE MADE BEFORE SERVICING THIS CODE		
• Check for code 31 or 41 Engine Running.	Code(s) present ▶	DISREGARD Code 23 at this time and RETURN to Quick Test Step 5.0B for codes 31 or 41. PROCEED as directed.
	No codes present ▶	GO to DH2.
DH2 CHECK FOR STUCK THROTTLE PLATE		
• Visually inspect throttle linkage for binding or sticking. • Verify the throttle linkage is at mechanical/closed throttle. Check for: binding throttle linkage, speed control linkage, vacuum line/electrical harness interference, etc.	Throttle not stuck ▶	GO to DH3.
	Throttle stuck ▶	CORRECT faults. RERUN Quick Test.
DH3 GENERATE CODE 63		
• Refer to illustration DH. • Key Off, wait 10 seconds. • Disconnect TP sensor vehicle harness connector at the throttle body. Inspect for damaged pins, corrosion, loose wires, etc. Service as necessary. • Perform Key On Engine Off Self Test and record codes. **NOTE: Ignore all other codes at this time.**	Code 63 present ▶	GO to DH4.
	Code 63 not present ▶	GO to DH5.
DH4 CHECK VOLTAGE VREF TO SIGNAL RETURN		
• Refer to illustration DH. • Key Off, wait 10 seconds. • Disconnect TP vehicle harness connector at throttle body. Inspect for damaged pins, corrosion, and pins pushed out. Service as necessary. • DVOM on 20V scale. • Key On Engine Off. • Measure voltage at the TP vehicle harness connector between VREF and Signal Return.	Voltage reading between 4V and 6V ▶	REPLACE TP sensor. RERUN Quick Test.
	Voltage reading less than 4V or greater than 6V ▶	GO to Pinpoint Test Step C1.

Throttle Position (TP) Sensor — Pinpoint Test DH

TEST STEP	RESULT ▶	ACTION TO TAKE
DH5 CHECK TP SIGNAL FOR SHORT TO POWER		
• Key Off, wait 10 seconds, TP harness disconnected. • DVOM on 200,000 ohm scale. • Disconnect processor 60 Pin connector and inspect for damaged pins, corrosion, loose wires. Service as necessary. • Install Breakout box, leave processor disconnected. • Measure resistance between test Pin 47 and test Pins 26 and 57 at the Breakout box.	Either resistance is less than 10,000 ohms ▶	SERVICE harness short. RERUN Quick Test.
	Both resistances are 10,000 ohms or greater ▶	REPLACE processor. RERUN Quick Test.
DH10 SERVICE CODE 63: GENERATE CODE 53		
• Key Off, wait 10 seconds, TP harness disconnected. • Jumper VREF to TP signal at TP vehicle harness connector. • Perform Key On Engine Off Self Test. **NOTE: If no codes are generated, immediately remove jumper and go directly to DH13.** • Check for Code 53 **NOTE: Ignore all other codes at this time.**	Code 53 present ▶	REPLACE TP sensor, RERUN Quick Test.
	Code 53 not present ▶	GO to DH11.
DH11 SERVICE CODE 63: CHECK VOLTAGE VREF TO SIGNAL RETURN		
• Refer to illustration DH. • Key Off, wait 10 seconds. • Disconnect TP vehicle harness connector at throttle body. Inspect for damaged pins, corrosion, and pins pushed out. Service as necessary. • DVOM on 20V scale. • Key On Engine Off. • Measure voltage at the TP vehicle harness connector between VREF and Signal Return.	Voltage reading between 4V and 6V ▶	GO to DH12.
	Voltage reading less than 4V or greater than 6V ▶	GO to Pinpoint Test Step C1.

SCORPIO DIAGNOSTIC CHART

Throttle Position (TP) Sensor — Pinpoint Test — DH

TEST STEP	RESULT ▶	ACTION TO TAKE
DH12 CHECK CONTINUITY OF TP CIRCUIT		
• Key Off, wait 10 seconds, TP harness disconnected. • DVOM on 200 ohm scale. • Disconnect processor 60 Pin connector and inspect for damaged pins, corrosion, loose wires. Service as necessary. • Connect Breakout box. Processor connected to Breakout box. • Measure resistance between TP Signal at the vehicle harness connector and test Pin 47 at the Breakout box.	Reading 5 ohms or greater ▶	SERVICE faulty circuit. CONNECT TP sensor. REMOVE Breakout box and RERUN Quick Test.
	Readings less than 5 ohms ▶	GO to DH13.
DH13 CHECK RESISTANCE OF TP CIRCUIT TO GROUND/SIGNAL RETURN		
• Key Off, wait 10 seconds, TP harness disconnected. • Disconnect processor 60 Pin connector and inspect for damaged pins, corrosion, loose wires. Service as necessary. • DVOM on 200,000 ohm scale. • Measure resistance between TP signal at TP vehicle harness connector and test Pin 46 at the Breakout box and between TP signal at TP vehicle harness connector and ground.	Either reading is less than 10,000 ohms ▶	SERVICE circuit short and RERUN Quick Test.
	Both readings are 10,000 ohms or greater ▶	REPLACE processor. CONNECT throttle position sensor. REMOVE Breakout box and RERUN Quick Test.
DH20 SERVICE CODE 73: TP SENSOR MOVES IN ENGINE RESPONSE CHECK		
NOTE: Code 73 indicates the TP sensor did not exceed 25 percent of its rotation in the Engine Response Check. • Key Off. • Install Breakout box. • DVOM on 20V scale. • Connect DVOM to test Pins 47 and 46 at the Breakout box. • Perform Engine Running Quick Test, Step 5.0. • Verify DVOM reading exceeds 3.5V during brief WOT at Engine Response Check.	Reading exceeds 3.5V during Engine Response Check ▶	REPLACE processor. RERUN Quick Test.
	Reading does not exceed 3.5V during Engine Response Check ▶	VERIFY TP sensor is properly attached to throttle body. If OK, REPLACE TP sensor. RERUN Quick Test.

Throttle Position (TP) Sensor — Pinpoint Test — DH

TEST STEP	RESULT ▶	ACTION TO TAKE
DH90 CONTINUOUS TEST SERVICE CODE 53: EXERCISE TP SENSOR		
• Using continuous monitor mode, observe STAR LED for indication of a fault while performing the following: • Move throttle slowly to WOT position. • Release throttle slowly to closed position and lightly tap on TP sensor (simulate road shock). • Wiggle TP harness connector. • Is a fault indicated?	Yes ▶	GO to DH91.
	No ▶	GO to DH92.

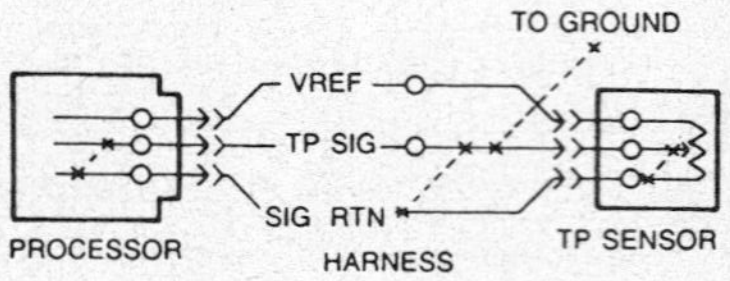

TEST STEP	RESULT ▶	ACTION TO TAKE
DH91 MEASURE TP SIGNAL VOLTAGE WHILE EXERCISING TP SENSOR		
• Key Off, wait 10 seconds. • Disconnect processor 60 Pin connector and inspect for damaged pins, corrosion, loose wires. Service as necessary. • Install Breakout box and reconnect processor. • STAR LED still connected as in previous step. • Connect a DVOM from test Pin 47 to test Pin 46. • DVOM on 20V scale. • Key On Engine Off. • While observing DVOM, repeat Step DH90. • Does the fault occur below 4.55V?	Yes ▶	DISCONNECT and INSPECT connectors. If connector and terminals are good, REPLACE TP sensor. RERUN Quick Test.
	No ▶	TP sensor overtravel may have caused the continuous code 53. Sensor service is not required. To verify harness integrity, GO to DH92.

SCORPIO DIAGNOSTIC CHART

Throttle Position (TP) Sensor — Pinpoint Test DH

TEST STEP	RESULT ▶	ACTION TO TAKE
DH92 CHECK EEC-IV HARNESS		
• Observe STAR LED for a fault indication while performing the following: • Referring to the illustration in Step DH90, grasp the harness close to the sensor connector. Wiggle, shake or bend a small section of the EEC-IV system harness while working your way to the dash panel. Also wiggle, shake or bend the EEC-IV harness from the dash panel to the processor. • Is a fault indicated?	Yes ▶	ISOLATE fault and SERVICE as necessary. REFER to appropriate figure. RERUN Quick Test.
	No ▶	GO to DH93 .
DH93 CHECK PROCESSOR AND HARNESS CONNECTORS		
• Key Off, wait 10 seconds. • Disconnect processor 60 pin connector. • Inspect both connectors and connector terminals for obvious damage or faults. • Are connectors and terminals OK?	No ▶	SERVICE as necessary. REPEAT Quick Test.
	Yes ▶	Unable to duplicate fault at this time. Continuous code 53 testing complete.
DH94 CONTINUOUS TEST SERVICE CODE 63: EXERCISE TP SENSOR		
• Using continuous monitor mode, observe STAR LED for indication of a fault while performing the following: • Move throttle slowly to WOT position. • Release throttle slowly to closed condition. • Lightly tap on TP sensor (simulate road shock). • Wiggle TP harness connector. • Is a fault indicated?	Yes ▶	DISCONNECT and INSPECT connectors. If connector and terminals are good, REPLACE TP sensor. RERUN Quick Test.
	No ▶	GO to DH95 .

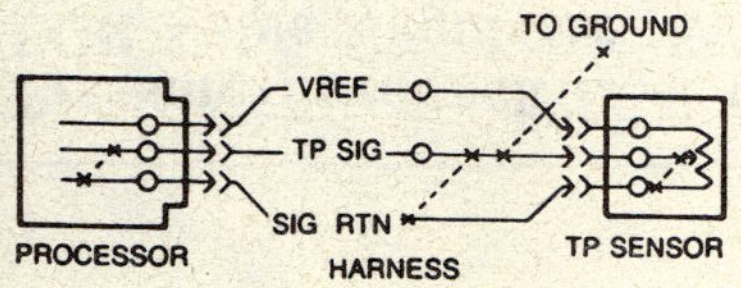

Throttle Position (TP) Sensor — Pinpoint Test DH

TEST STEP	RESULT ▶	ACTION TO TAKE
DH95 CHECK EEC-IV HARNESS		
• Observe STAR LED for a fault indication while performing the following: • Referring to the illustration in Step DH94 grasp the harness close to the sensor connector. Wiggle, shake or bend a small section of the EEC-IV system harness while working your way to the dash panel. Also wiggle, shake or bend the EEC-IV harness from the dash panel to the processor. • Is a fault indicated?	Yes ▶	ISOLATE fault and SERVICE as necessary. REFER to appropriate figure. RERUN Quick Test.
	No ▶	GO to DH96 .
DH96 CHECK PROCESSOR AND HARNESS CONNECTORS		
• Key Off, wait 10 seconds. • Disconnect processor 60 pin connector. • Inspect both connectors and connector terminals for obvious damage or faults. • Are connectors and terminals OK?	No ▶	SERVICE as necessary. RERUN Quick Test.
	Yes ▶	Unable to duplicate fault at this time. Continuous code 63 testing complete.

SCORPIO DIAGNOSTIC CHART

Pressure Feedback EGR (PFE) EGR Valve Regulator (EVR) — Pinpoint Test — DL

DL

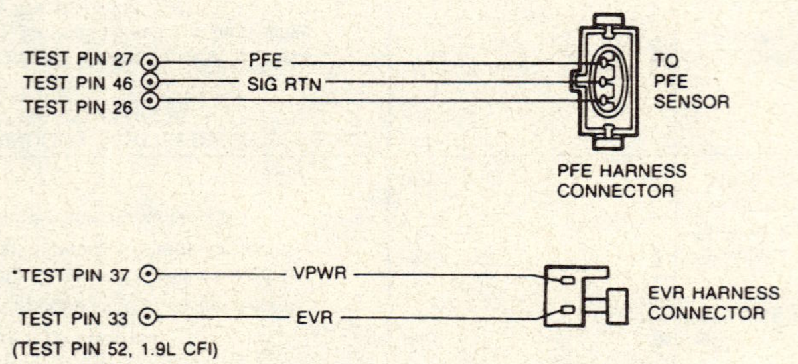

*TEST PINS LOCATED ON BREAKOUT BOX.
ALL HARNESS CONNECTORS VIEWED INTO MATING SURFACE.

STOP-WARNING

You should enter this Pinpoint Test only when a service code 31, 32, 33, 34, 35 or 84 is received in Quick Test Step 3.0, 5.0 or 6.0.

To prevent the replacement of good components, be aware that the following non-EEC area may be at fault:

- Damaged EGR valve.

This Pinpoint Test is intended to diagnose only the following:

- PFE sensor.
- Harness circuits: VREF, PFE, Signal Return, EVR, VPWR.
- EVR (EGR valve regulator).
- EGR valve assembly.
- Processor assembly.

Pressure Feedback EGR (PFE) EGR Valve Regulator (EVR) — Pinpoint Test — DL

TEST STEP	RESULT ▶	ACTION TO TAKE
SERVICE CODE 31		
DL1 ATTEMPT TO GENERATE OPPOSITE CODE (35)		
• Key Off. • Disconnect PFE vehicle harness at sensor. • Jumper VREF to PFE signal at vehicle harness sensor connector. • Perform Key On Engine Off Self Test. **NOTE: If no codes are generated, immediately remove jumper and GO directly to Step DL4.** • Check for code 35. **NOTE: Ignore all other codes at this time.**	Code 35 present ▶ Code 35 not present ▶	REMOVE Jumper. REPLACE PFE sensor. RERUN Quick Test. REMOVE jumper. GO to DL2.
DL2 MEASURE VREF TO SIGNAL RETURN VOLTAGE		
• Key Off. • PFE harness disconnected. • DVOM on 20V scale. • Key On Engine Off. • Measure voltage at PFE vehicle harness connector between VREF and Signal Return. • Refer to illustration DL.	Voltage reading between 4V and 6V ▶ Voltage reading less than 4V or greater than 6V ▶	GO to DL3. GO to Pinpoint Test Step C1.
DL3 CHECK CONTINUITY OF PFE SIGNAL		
• Key Off. • PFE harness disconnected. • DVOM on 200 ohm scale. • Disconnect processor 60 pin connector and inspect for damaged pins, corrosion, loose wires, etc. Service as necessary. • Connect Breakout box to harness. Processor connected to Breakout box. • Measure resistance between PFE signal at vehicle harness sensor connector and test Pin 27 at the Breakout box.	Reading 5 ohms or greater ▶ Reading less than 5 ohms ▶	SERVICE faulty circuit. CONNECT PFE sensor. REMOVE Breakout box. RERUN Quick Test. GO to DL4.

SCORPIO DIAGNOSTIC CHART

Pressure Feedback EGR (PFE) EGR Valve Regulator (EVR) — Pinpoint Test — DL

TEST STEP	RESULT ▶	ACTION TO TAKE
DL4 CHECK RESISTANCE OF PFE SIGNAL TO GROUND AND SIGNAL RETURN		
• Key Off. • PFE harness disconnected. • Breakout box installed. • Processor disconnected. • DVOM on 200,000 ohm scale. • Measure resistance between PFE signal at PFE vehicle harness connector and ground. • Measure resistance between PFE signal at the PFE vehicle harness connector and test Pin 46 (Signal Return) at the Breakout box.	Either reading is less than 10,000 ohms ▶	SERVICE short circuit. CONNECT PFE. REMOVE Breakout box. RERUN Quick Test.
	Both readings are 10,000 ohms or greater. ▶	REPLACE processor. CONNECT PFE sensor. REMOVE Breakout box. RERUN Quick Test.
SERVICE CODE 35		
DL5 ATTEMPT TO GENERATE OPPOSITE CODE (31)		
• Key Off. • Disconnect PFE vehicle harness at sensor. Inspect for damaged pins, corrosion, loose wires, etc. Service as necessary. • Perform Key On Engine Off Self Test, and record codes. • Check for code 31. **NOTE: Ignore all other codes at this time.**	Code 31 present ▶	GO to DL6 .
	Code 31 not present ▶	GO to DL7 .
DL6 MEASURE VREF TO SIGNAL RETURN VOLTAGE		
• Key Off. • PFE harness disconnected. • DVOM on 20V scale. • Key On Engine Off. • Measure voltage at PFE vehicle harness connector between VREF and Signal Return. • Refer to illustration DL.	Voltage reading between 4V and 6V ▶	REPLACE PFE sensor. RERUN Quick Test.
	Voltage reading less than 4V or greater than 6V ▶	GO to Pinpoint Test Step C1 .

Pressure Feedback EGR (PFE) EGR Valve Regulator (EVR) — Pinpoint Test — DL

TEST STEP	RESULT ▶	ACTION TO TAKE
DL7 CHECK PFE CIRCUIT FOR SHORT TO POWER		
• Key Off. • PFE harness disconnected. • Disconnect processor 60 pin connector and inspect for damaged pins, corrosion, loose wires, etc. Service as necessary. • Install Breakout box, leave processor disconnected. • DVOM on 200,000 ohm scale. • Measure the resistance between test Pin 27 and test Pins 26 and 57 at the Breakout box.	Either resistance is less than 10,000 ohms ▶	SERVICE harness short. REMOVE Breakout box. CONNECT PFE sensor. RERUN Quick Test.
	Both resistances are 10,000 ohms or greater ▶	REPLACE processor. REMOVE Breakout box. CONNECT PFE sensor. RERUN Quick Test.
SERVICE CODE: 34		
DL8 PFE SENSOR OUT OF RANGE		
• PFE system can sense a lack of pressure in the vehicle exhaust system. An efficient garage exhaust ventilation system, installed during Key On, Engine Off Quick Test, may deflect the PFE sensor and generate a code 34. Remove the ventilation system and retest. • Is code 34 present?	No ▶	ADDRESS any other codes in Key On Engine Off, if none CONTINUE with remaining Quick Test.
	Yes ▶	GO to DL9 .
DL9 CHECK PRESSURE LINE TO PFE SENSOR		
• Remove the pressure line from PFE sensor. • Inspect complete line, including PFE inlet for blockage. • Is blockage present?	No ▶	GO to DL10 .
	Yes ▶	SERVICE as necessary. RERUN Quick Test.

SCORPIO DIAGNOSTIC CHART

Pressure Feedback EGR (PFE) EGR Valve Regulator (EVR) — Pinpoint Test — DL

TEST STEP	RESULT ▶	ACTION TO TAKE
DL10 MEASURE VREF TO SIGNAL RETURN VOLTAGE		
• Key Off. • Disconnect PFE sensor and inspect for damaged pins, corrosion, loose wires, etc. Service as necessary. • DVOM on 20V scale. • Key On Engine Off. • Measure voltage at PFE vehicle harness connector between VREF and Signal Return. • Refer to Illustration DL.	Voltage reading between 4V and 6V ▶	REPLACE PFE sensor. RERUN Quick Test.
	Voltage reading less than 4V or greater than 6V ▶	GO to Pinpoint Test Step C1 .
SERVICE CODE 84		
DL11 MEASURE EVR SOLENOID RESISTANCE		
• Key Off. • DVOM on 200 ohm scale. • Disconnect EVR solenoid connector and measure solenoid resistance.	Resistance is between 30 and 70 ohms ▶	GO to DL12 .
	Resistance is less than 30 ohms or greater than 70 ohms ▶	REPLACE EVR solenoid assembly. RERUN Quick Test.
DL12 CHECK FOR VPWR AT EVR SOLENOID		
• EVR solenoid disconnected from harness. • DVOM on 20V scale. • Key On Engine Off. • Measure voltage between battery negative terminal and VPWR circuit at EVR solenoid vehicle harness connector.	Reading is less than 10.5V ▶	SERVICE VPWR open circuit. RERUN Quick Test.
	Reading is 10.5V or greater ▶	GO to DL13 .
DL13 CHECK CONTINUITY OF EVR CIRCUIT		
• Key Off. • EVR solenoid disconnected from harness. • Disconnect processor 60 pin connector and inspect for damaged pins, corrosion, loose wires, etc. Service as necessary. • Install Breakout box to processor harness connector. Leave processor disconnected. • DVOM on 200 ohm scale. • Measure resistance between test Pin 33 at the Breakout box and EVR signal at the EVR solenoid vehicle harness connector.	Reading less than 5 ohms ▶	GO to DL14 .
	Reading 5 ohms or greater ▶	SERVICE open circuit. REMOVE Breakout box. CONNECT process EVR solenoid. RERUN Quick Test.

Pressure Feedback EGR (PFE) EGR Valve Regulator (EVR) — Pinpoint Test — DL

TEST STEP	RESULT ▶	ACTION TO TAKE
DL14 CHECK EVR CIRCUIT FOR SHORT TO POWER AND GROUND		
• Key Off. • Breakout box installed, processor disconnected. • EVR solenoid disconnected. • DVOM on 200,000 ohm scale. • Measure resistance between test Pin 33 (EVR) and test Pins 37 and 57 (VPWR) and 40, 46 and 60 (GND) at the Breakout box.	Any reading less than 10,000 ohms ▶	SERVICE short circuit. REMOVE Breakout box. RECONNECT harness to processor and EVR solenoid. RERUN Quick Test. If code is repeated, REPLACE processor.
	All reading 10,000 ohms or greater ▶	REPLACE processor. REMOVE Breakout box. RECONNECT harness to processor and EVR solenoid. RERUN Quick Test.
SERVICE CODE 32		
DL20 VERIFY ENGINE RUNNING CODES		
The PFE system can sense a lack of pressure in the vehicle exhaust system. An efficient garage exhaust ventilation system installed during Key On Engine Run Quick Test may, on some calibrations, deflect the PFE sensor and generate a code 32. Temporarily, remove garage forced ventilation system and properly vent to atmosphere. • RERUN Engine Running Quick Test. • Is code 32 present?	Yes ▶	GO to DL21 .
	No ▶	ADDRESS any other codes in Engine Running. If none, CONTINUE with remaining Quick Test.

SCORPIO DIAGNOSTIC CHART

Pressure Feedback EGR (PFE) EGR Valve Regulator (EVR) — Pinpoint Test DL

TEST STEP	RESULT ▶	ACTION TO TAKE
DL21 ATTEMPT TO SEPARATE EVR FROM PFE		
• Key Off. • Disconnect EGR valve vacuum line at valve and plug line. • Perform Engine Running Quick Test. • Is code 31 or 32 present?	Yes ▶ No ▶	GO to DL22. GO to DL23.
DL22 CHECK PFE SENSOR SUPPLY TUBE		
• Key Off. • Check PFE sensor supply line for obstructions and/or leaks. • Are there any obstructions or leaks?	Yes ▶ No ▶	SERVICE as necessary. RECONNECT all lines and RERUN Quick Test. GO to EGR Valve Diagnosis, Section 3.
DL23 CHECK EVR FILTER		
• Key Off. • Remove and inspect EVR filter for contamination. **NOTE: Blockage of filter will cause vacuum to be applied to EGR valve prematurely.** • Is filter contaminated?	Yes ▶ No ▶	REPLACE filter. RECONNECT all lines. RERUN Quick Test. REPLACE EVR solenoid. RERUN Quick Test.
ENGINE RUNNING SERVICE CODE 34 AND 35		
DL25 CHECK FOR EXCESSIVE EXHAUST BACK PRESSURE		
• Service codes 34 and 35 in Engine Running Self Test indicate excessive exhaust back pressure; There are two possible causes: (A). The exhaust system is restricted, and (B). PFE sensor has shifted high. • Key Off. • Substitute known good PFE sensor in place of original. • Rerun Key On Engine Running Quick Test. • Is code 34 or 35 present?	No ▶ Yes ▶	Original PFE was the cause of the original 34 or 35. REPLACE PFE sensor. RERUN Quick Test. GO to Section 3, Restricted Catalytic Converter or Exhaust System Diagnosis.

Pressure Feedback EGR (PFE) EGR Valve Regulator (EVR) — Pinpoint Test DL

TEST STEP	RESULT ▶	ACTION TO TAKE
SERVICE CODE 33		
DL30 VERIFY VACUUM IS PRESENT AT VALVE		
• Key Off. • Standard vacuum gauge in.-Hg. (Mercury). • Remove vacuum line from EGR valve and attach a vacuum gauge to line. • Perform Engine Running Quick Test while observing vacuum gauge. **NOTE: Disregard code output.**	Vacuum reading 1 in.-Hg. or less ▶ Vacuum reading greater than 1 in.-Hg. ▶	GO to DL31. Check EGR system.
DL31 VACUUM SUPPLY VERIFICATION		
• Key Off. • Check vacuum line from EVR solenoid to EGR valve and source to EVR solenoid for loose or disengaged connections, cracks, etc. • Is vacuum present at EVR solenoid during run?	Yes ▶ No ▶	REPLACE EVR solenoid. RERUN Quick Test. SERVICE as necessary. RERUN Quick Test.
DL90 SERVICE CODE 31 OR 35 CONTINUOUS TEST: EXERCISE PFE SENSOR		
• Using continuous monitor mode, observe STAR LED for indication of a fault while performing the following: • Connect a vacuum pump to the PFE sensor. • Slowly apply 5 in.-Hg. vacuum to the sensor. • Slowly bleed vacuum off the PFE sensor. • Lightly tap on PFE sensor (simulate road shock). • Wiggle PFE connector. • Is fault indicated? 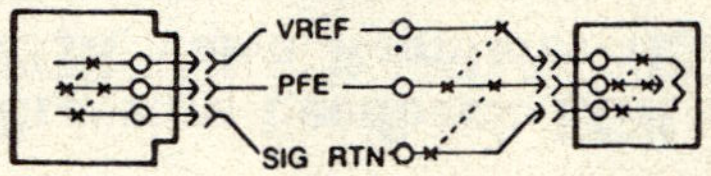	Yes ▶ No ▶	DISCONNECT and INSPECT connectors. If connector and terminals are good, REPLACE sensor. RERUN Quick Test. GO to DL91.

SCORPIO DIAGNOSTIC CHART

Pressure Feedback EGR (PFE) EGR Valve Regulator (EVR) — Pinpoint Test — DL

TEST STEP	RESULT ▶	ACTION TO TAKE
DL91 CHECK EEC-IV HARNESS		
• Observe STAR LED for a fault indication while performing the following: • Referring to the illustration in Step DL90 grasp the harness closest to the sensor connector. Wiggle, shake or bend a small section of the EEC-IV system harness while working your way to the dash panel. Also wiggle, shake or bend the EEC-IV harness from the dash panel to the processor. • Is a fault indicated?	Yes ▶	ISOLATE fault and SERVICE as necessary. RERUN Quick Test.
	No ▶	GO to DL92 .
DL92 CHECK PROCESSOR AND HARNESS CONNECTORS		
• Key Off, wait 10 seconds. • Disconnect processor 60 pin connector. • Inspect both connectors and connector terminals for obvious damage or faults. • Are connectors and terminals OK?	No ▶	SERVICE as necessary. RERUN Quick Test.
	Yes ▶	Unable to duplicate and/or identify fault at this time. Continuous code 31 or 35 testing complete.
DL93 SERVICE CODE 34: CONTINUOUS TEST — INSPECT PFE SUPPLY LINE FOR BLOCKAGE		
• Key Off. • Remove PFE sensor and inspect sensor supply inlet for liquids and/or any type of blockage. • Inspect PFE supply line to EGR valve base for liquids and/or blockage. • Is supply line free of any blockage?	No ▶	CLEAN and/or SERVICE as necessary. RERUN Quick Test.
	Yes ▶	Unable to duplicate and/or identify fault at this time. Continuous code 34 testing complete.

Pressure Feedback EGR (PFE) EGR Valve Regulator (EVR) — Pinpoint Test — DL

TEST STEP	RESULT ▶	ACTION TO TAKE
DL94 SERVICE CODE 32: CONTINUOUS TEST — INSPECT EGR VALVE FOR SMOOTH OPERATION		
• Key Off. • Connect a vacuum pump to the EGR valve. • Apply 10 in.-Hg. of vacuum to EGR valve. • While observing EGR valve, release vacuum. • Does EGR valve function in a smooth manner? **NOTE: Repeat test if necessary to ensure accurate result.**	No ▶	GO to EGR Valve Diagnosis, Section 3.
	Yes ▶	GO to DL95 .
DL95 INSPECT VACUUM LINES BETWEEN EVR SOLENOID AND EGR VALVE		
• Inspect EGR valve vacuum supply line from EVR solenoid for kinks and/or obstructions. • Is vacuum supply line to EGR valve free of any obstructions?	No ▶	SERVICE as necessary. RERUN Quick Test.
	Yes ▶	GO to DL96 .
DL96 EVR REGULATOR FILTER INSPECTION		
• Carefully check EVR filter for contamination and/or obstructions. • Is EVR filter condition acceptable?	No ▶	REPLACE EVR filter. RERUN Quick Test.
	Yes ▶	Unable to duplicate and/or identify fault at this time. Continuous code 34 testing complete.
DL97 SERVICE CODE 33: CONTINUOUS TEST — INSPECT EGR VALVE FOR FREE OPERATION		
• Key Off. • Connect a vacuum pump to the EGR valve. • While observing the EGR valve, slowly apply 10 in.-Hg. vacuum. **NOTE: EGR valve should begin to open with a very small amount of vacuum, approximately 1 to 1.5 in.-Hg. and be fully open with about 4 in.-Hg. vacuum.** • Does EGR valve move freely and smoothly?	No ▶	GO to EGR Valve Diagnosis, Section 3.
	Yes ▶	GO to DL98 .

SCORPIO DIAGNOSTIC CHART

Pressure Feedback EGR (PFE) EGR Valve Regulator (EVR) — Pinpoint Test DL

TEST STEP	RESULT ▶	ACTION TO TAKE
DL98 EVR HARNESS CHECK		
• Key Off. • Disconnect processor 60 pin connector and inspect for damaged pins, corrosion, loose wires, etc. Service as necessary. • Connect Breakout box to harness. Processor connected to Breakout box. • Enter output state check. • DVOM on 20V scale. • Connect DVOM negative test lead to test Pin 40 at the Breakout box and DVOM positive test lead to test Pin 33. • Cycle throttle if necessary to indicate 10.5V or greater. • Remain at this position. • While observing DVOM, grasp the harness closest to the EVR connector. Wiggle, shake or bend a small section of the EEC-IV system harness while working your way to the dash panel. Lightly tap EVR solenoid to simulate road vibration. • Does DVOM indicate less than 10.5V?	Yes ▶	SERVICE as necessary. RERUN Quick Test.
	No ▶	Unable to duplicate and/or identify fault at this time. Continuous code 33 testing complete.

Neutral Drive Switch A/C Input — Pinpoint Test FA

FA NEUTRAL DRIVE INPUT CHECK

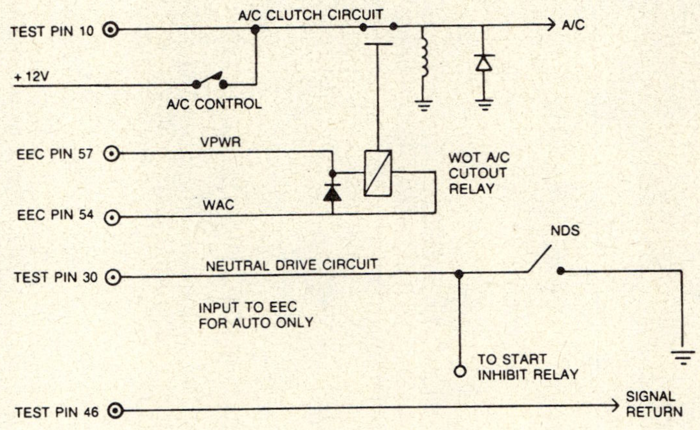

STOP-WARNING

You should enter this Pinpoint Test only when a service code 67 is received in Quick Test Step 3.0.

This Pinpoint Test is intended to diagnose only the following:

- Neutral Drive Switch
- Neutral Drive Harness Circuit
- A/C Control Module Circuit

SCORPIO DIAGNOSTIC CHART

Neutral Drive Switch A/C Input — Pinpoint Test — FA

TEST STEP	RESULT ▶	ACTION TO TAKE
FA1 CODE 67 SYSTEM IDENTIFICATION — NEUTRAL DRIVE INPUT CHECK		
• Key Off, wait 10 seconds. • Disconnect processor 60 pin connector and inspect for damaged pins, corrosion, loose wires. Service as necessary. • Install Breakout box. • Processor connected. • Key On Engine Off. • DVOM on 20V scale. • Measure voltage between test Pin 30 (neutral drive circuit) at the Breakout box and chassis ground. • Is the reading; — less than 1.0V in PARK or NEUTRAL and — 4.5-5.0V in DRIVE or REVERSE?	Yes ▶ No ▶	GO to FA2. GO to FA3.
FA2 A/C INPUT CHECK		
• Breakout box installed. • Key On Engine Running. • DVOM on 20V scale. • Measure voltage between test Pin 10 (A/C clutch circuit) at the Breakout box and chassis ground. • Using the ECON button on the electronic ATC module, cycle the A/C clutch ON and OFF. • When the compressor is cycled, are the following voltage readings observed on test Pin 10? Compressor ON — 12.0V Compressor OFF — less than 1.0V	Yes ▶ No ▶	REPLACE processor. RERUN Quick Test. SERVICE short to power in A/C clutch circuit. RERUN Quick Test.

Neutral Drive Switch A/C Input — Pinpoint Test — FA

TEST STEP	RESULT ▶	ACTION TO TAKE
FA3 NEUTRAL DRIVE SWITCH CHECK		
• Key Off, wait 10 seconds. • Breakout box installed. • DVOM on 200 ohm scale. • Disconnect vehicle harness from the neutral drive switch and measure the resistance across the switch.	Less than 5 ohms ▶ 5 ohms or greater ▶	SERVICE open in vehicle harness neutral drive circuit. RERUN Quick Test. REPLACE neutral drive switch. RERUN Quick Test.

SCORPIO DIAGNOSTIC CHART

Fuel Control — 2.9L EFI V-6 | Pinpoint Test | HG

HG

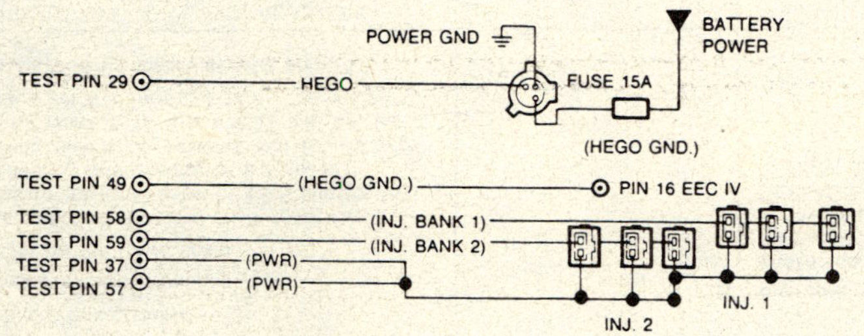

NOTE: REFER TO APPROPRIATE ENGINE SUPPLEMENT FOR PROPER INJECTOR BANK-TO-INJECTOR RELATIONSHIP.

STOP-WARNING

You should enter this Pinpoint Test only when a service code 41 or 42 is received in Quick Test Step 5.0 or when directed here from Pinpoint Test S or Diagnostic By Symptom chart.

To prevent the replacement of good components, be aware that the following non-EEC areas may be at fault:

- Ignition Coil
- Distributor Cap
- Distributor Rotor
- Fouled Spark Plugs
- Spark Plug Wires
- CANP Problems
- PCV Valves (see note below)
- EGR Valve and Gasket
- Air Filter
- Fuel Contamination, Engine Oil
- Poor Power Ground
- Fuel Pressure
- Manifold Leaks, Intake/Exhaust
- Engine Not at Normal Operating Temperatures

This Pinpoint Test is intended to diagnose only the following:

- HEGO Sensor
- Harness Circuits EGO GND, HEGO, Inj. Bank 1, Inj. Bank 2, VPWR
- HEGO Sensor Connection
- Vacuum Systems
- Fuel Injectors
- Processor Assembly

NOTE: Fuel contaminated engine oil may affect 41 and 42 Service Codes. If this is suspected, remove the PCV from the valve cover and repeat the Quick Test. If the problem is corrected, then change the engine oil and filter.

Fuel Control — 2.9L EFI V-6 | Pinpoint Test | HG

TEST STEP	RESULT ▶	ACTION TO TAKE
HG1 FUEL PRESSURE CHECK		
• Key Off, wait 10 seconds. • Install fuel pressure gauge. • Start and run engine. Fuel pressure must be 241-310 kPa (35-45 psi). • Fuel pressure must remain at 276 ± 34 kPa (40 ± 5 psi) for 60 seconds after final Key Off. **For No Starts:** • If engine will not run, cycle the key from OFF to ON several times.	Fuel pressure is within all specifications ▶	GO to HG2.
	Fuel pressure is not 241-310 kPa (35-45 psi) ▶	Check electric fuel pump and pressure regulator.
	Fuel pressure does not remain at 276 ± 34 kPa (40 ± 5 psi) for 60 seconds or longer ▶	GO to HG6.
HG2 FUEL DELIVERY TEST		
• Key Off. • Fuel pressure gauge installed. • Pressurize fuel system as per Step HG1. • Locate and disconnect the fuel pump relay. • Crank engine for 5 seconds. • Take pressure reading at the end of 5-second crank. **NOTE: Verify fuel quality; air and/or water will also pressurize and look like acceptable fuel delivery.**	Pressure Gauge Reading:	
	Pressure is approximately 69-138 kPa (10-20 psi) at the end of 5-second crank cycle. Refer to Note below ▶	The EEC system is not the fault of the No Start. Fuel and spark are present. If complaint was runs rough, misses or fuel code, GO to HG3.
	Pressure is greater or less than specified ▶	GO to HG3.

NOTE: The colder the engine, the greater the pressure drop (for example, an engine coolant temperature of 200°F equals approximately a 69 kPa [10 psi] drop in 5 seconds. 60°F equals approximately a 138 kPa [20 psi] drop in 5 seconds).

SCORPIO DIAGNOSTIC CHART

Fuel Control — 2.9L EFI V-6 | Pinpoint Test | HG

TEST STEP	RESULT ▶	ACTION TO TAKE
HG3 HARNESS INJECTOR RESISTANCE CHECK		
• Key Off, wait 10 seconds. • Disconnect processor 60 pin connector and inspect for damaged pins, corrosion, loose wires. Service as necessary. • Connect Breakout box to harness. Leave processor disconnected. • DVOM on 200 ohm scale. • Measure the resistance of injector Bank 1 between test Pin 37 and test Pin 58 at the Breakout box. • Measure the resistance of injector Bank 2 between test Pin 37 and test Pin 59 at the Breakout box. • Are both readings between 5.0 and 6.5 ohms?	Yes ▶ No ▶	GO to HG5. GO to HG4.
HG4 ISOLATE FAULTY INJECTOR CIRCUIT		
• Key Off. • Breakout box installed. • Leave processor disconnected. • DVOM on 200 ohm scale. • Disconnect all injectors on suspect Bank. Measure the resistance of each injector by connecting one injector at a time, and reading resistance between test Pin 37 and 58, for Bank 1, or test Pin 37 and 59, for Bank 2, at the Breakout box. • Are all readings between 16.0 and 18.0 ohms?	Yes ▶ No ▶	GO to HG5. SERVICE the harness and/or connectors on the suspect injector for opens or shorts. If OK, REPLACE injector. RERUN Quick Test.
HG5 INJECTOR DRIVE SIGNAL CHECK		
Requires standard non-powered 12-volt test lamp. • Key Off. • Breakout box installed. • Connect processor to Breakout box. • Connect test lamp between test Pin 37 and test Pin 58 at the Breakout box. • Crank or start engine. • Repeat above test between test Pin 37 and test Pin 59 at the Breakout box.	Dim glow at light on both tests ▶ No light on one or both tests ▶ Bright light on one or both tests ▶	GO to HG6. VERIFY 12V battery power at Pins 37 and 57. If OK, REPLACE processor. RERUN Quick Test. CHECK circuits Bank 1 and Bank 2 for shorts to ground. If OK, REPLACE processor. RERUN Quick Test.

Fuel Control — 2.9L EFI V-6 | Pinpoint Test | HG

TEST STEP	RESULT ▶	ACTION TO TAKE
HG6 INJECTOR BALANCE TEST		
• Connect tachometer to engine. Run engine at idle. • Disconnect and reconnect the injectors one at a time; note rpm drop for each injector. • Does each injector produce at least a 100 rpm momentary drop, as ISC will attempt to re-establish rpm?	Yes ▶ No ▶	Fuel delivery OK. Problem is with area common to all cylinders, i.e.: Air/vacuum leak, fuel contamination, EGR. REPLACE faulty injector. RERUN Quick Test.
HG7 FUEL CONTROL — ALWAYS RICH: CODE 42		
• Key Off, wait 10 seconds. **NOTE: Non-EEC areas could cause a code 42. Check for:** **— Fuel contaminated engine oil** **— Ignition caused misfire** **— CANP problems** • Disconnect vehicle harness at the HEGO sensor. • Jumper (ground) the HEGO circuit at the HEGO sensor vehicle harness connector to the engine block. • Repeat Engine Running Self-Test. • Service code 41 results.	Yes ▶ No ▶	REPLACE HEGO. RERUN Quick Test. If code still present GO to HG1. GO to HG8.
HG8 EGO HARNESS CHECK		
• Key Off, wait 10 seconds. • Install Breakout box. • Disconnect processor. • Disconnect HEGO sensor harness connector. • Measure resistance between test Pin 16 at the Breakout box and HEGO case. • Measure resistance between test Pin 16 and test Pin 49 at the Breakout box. • Measure resistance between test Pin 29 and HEGO harness connector. • Are all resistances less than 5 ohms?	Yes ▶ No ▶	DISCONNECT processor connector. INSPECT for damage or corrosion. If OK, REPLACE processor. RERUN Quick Test. CORRECT harness circuit with resistance greater than 5 ohms. RERUN Quick Test.

SCORPIO DIAGNOSTIC CHART

Fuel Control — 2.9L EFI V-6 | Pinpoint Test | HG

TEST STEP	RESULT ▶	ACTION TO TAKE
HG10 FUEL CONTROL — ALWAYS LEAN: CODE 41 VERIFICATION		
• Run vehicle at 2000 rpm for 2 minutes. • Key Off, wait 10 seconds. • Perform Engine Running Quick Test.	Code 41 is not present ▶ Code 41 is present ▶	GO to HG21. GO to HG11.
HG11 FUEL CONTROL — ALWAYS LEAN		
NOTE: Vacuum/air leaks in non-EEC areas could cause a code 41. Check for: **— Leaking vacuum actuator (e.g.: A/C control motor)** **— Engine sealing** **— EGR system** **— PCV system** **— Lead contaminated HEGO sensor** • Key Off. DVOM on 20V scale. Disconnect HEGO sensor from vehicle harness. Connect DVOM to HEGO sensor and engine ground. • Disconnect vacuum line to fuel pressure regulator and seal end of vacuum pipe with plug. • Start the engine and run at approximately 2000 rpm. Does the DVOM read greater than 0.5V within 1 minute?	Yes ▶ No ▶	GO to HG12. REPLACE HEGO sensor. RERUN Quick Test.
HG12 HARNESS CHECK HEGO CIRCUITS		
• Key Off, wait 10 seconds. • Install Breakout box. • Disconnect processor. • Disconnect HEGO sensor harness connector. • Measure resistance between test Pin 16 at the Breakout box and HEGO case. • Measure resistance between test Pin 16 and test Pin 49 at the Breakout box. • Measure resistance between test Pin 29 and Breakout box and HEGO harness connector. • Are all resistances less than 5 ohms?	Yes ▶ No ▶	GO to HG13. SERVICE/CORRECT as necessary circuit with greater than 5 ohms resistance. RERUN Quick Test.

Fuel Control — 2.9L EFI V-6 | Pinpoint Test | HG

TEST STEP	RESULT ▶	ACTION TO TAKE
HG13 CHECK HEGO CIRCUIT FOR SHORT TO GROUND		
• Key Off. • Breakout box installed. Processor disconnected. • DVOM to 200,000 ohms range. • Measure resistance between test Pin 29 and test Pin 40 at the Breakout box.	Reading is 10,000 ohms or greater ▶ Reading is less than 10,000 ohms ▶	GO to HG14. CORRECT cause of resistance to ground. RERUN Quick Test.
HG14 ATTEMPT TO ELIMINATE CODE 41		
• Key Off. • Reconnect HEGO sensor. • Vacuum line to pressure regulator still disconnected. • Start the engine and run at approximately 2000 rpm for 1 minute. Allow engine to return to idle. • Perform Engine Running Self Test. • Is code 41 still present? **NOTE: Disregard any other code input at this time.**	Yes ▶ No ▶	INSPECT for corrosion or damaged pins. If OK, REPLACE processor. RERUN Quick Test. RECONNECT ACT sensor. HEGO input circuit OK. GO to HG1.

SCORPIO DIAGNOSTIC CHART

Fuel Control — 2.9L EFI V-6 | Pinpoint Test | HG

TEST STEP	RESULT	ACTION TO TAKE
HG21 CHECK HEATER ELEMENT RESISTANCE ON HEGO		
• Key Off Engine Off. • Disconnect HEGO sensor. • DVOM on 200 ohm scale. • Measure resistance of heater element at HEGO connector at room temperature.	Reading between 2.5 and 5.0 ohms ▶	GO to HG22.
	Reading less than 2.5 or greater than 5.0 ohms ▶	REPLACE HEGO sensor.
HG22 CHECK FOR POWER AT HEGO HARNESS CONNECTOR		
• Key On Engine Off. • DVOM on 20V scale. • Connect positive lead to VPWR circuit and negative lead to ground at HEGO vehicle harness connector.	Reading 10.5V or greater ▶	HEGO sensor system OK, RECONNECT HEGO sensor.
	Reading less than 10.5V ▶	GO to HG23.
HG23 CHECK CONTINUITY OF GROUND TO HEGO CONNECTOR		
• Key Off, wait 10 seconds. • DVOM on 200 ohm scale. • Measure resistance of ground circuit from HEGO vehicle harness connector to battery ground.	Reading less than 5.0 ohms ▶	CHECK Fuse and/or SERVICE open in VPWR circuit.
	Reading 5.0 ohms or greater ▶	SERVICE open in ground circuit.

Fuel Pump Circuit — (Inertia Switch) | Pinpoint Test | J

J

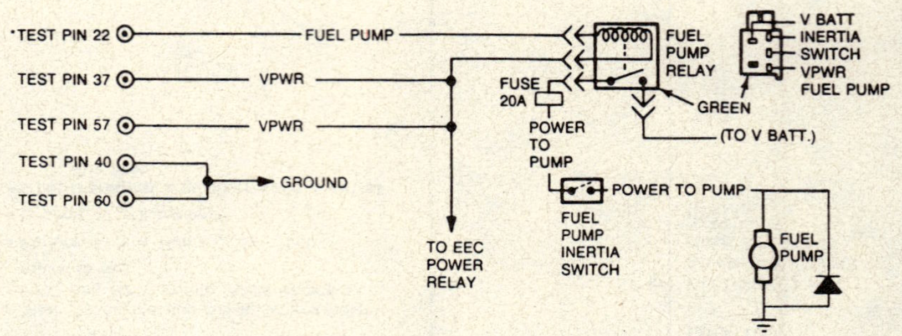

*TEST PINS LOCATED ON BREAKOUT BOX.
ALL HARNESS CONNECTORS VIEWED INTO MATING SURFACE.

STOP-WARNING

You should enter this Pinpoint Test only when a service code 87 is received in Quick Test Step 3.0 or you are directed here from Pinpoint Test Step A22 or Diagnostic By Symptom chart.

To prevent the replacement of good components, be aware that the following non-EEC areas may be at fault:

- Fuel Lines
- Fuel Filters
- Throttle Body
- Contaminated Fuel
- Fuel Pump

This Pinpoint Test is intended to diagnose only the following:

- Fuel Pump Relay
- Harness Circuits: V Batt., VPWR, FP, Ground and Power to Pump(s)
- Processor Assembly

SCORPIO DIAGNOSTIC CHART

Fuel Pump Circuit — (Inertia Switch) | Pinpoint Test | J

TEST STEP	RESULT ▶	ACTION TO TAKE
J1 NO FUEL PUMP PRESSURE		
• Fuel pressure gauge installed. • Check if fuel pump runs. • Cycle key from Off to Run, repeat several times. (Do not enter start mode.) Fuel pump should run briefly each time the key enters Run.	Yes ▶	Check electric fuel pump.
	No ▶	GO to J2.
J2 CHECK FOR VPWR TO PROCESSOR		
• Key Off, wait 10 seconds. • Disconnect processor 60 pin connector and inspect for damaged pins, corrosion, loose wires. Service as necessary. • Install Breakout box and reconnect processor. • Key On, engine Off. • DVOM on 20V scale. • Measure voltage between test Pin 37 and test Pin 40 at the Breakout box and between test Pin 57 and test Pin 60 at the Breakout box.	Either voltage reading is less than 10.5V ▶	GO to Pinpoint Test Step B1.
	Both voltage readings are 10.5V or greater ▶	GO to J3.
J3 RESISTANCE CHECK OF FUEL PUMP INERTIA SWITCH		
• Key Off, wait 10 seconds. • Leave Breakout box installed and processor connected. • Locate and disconnect fuel pump inertia switch. • DVOM on 200 ohm scale. • Measure the resistance of the fuel pump inertia switch.	Less than 5.0 ohms ▶	GO to J4.
	5.0 ohms or greater ▶	REPLACE or reset fuel pump inertia switch. RERUN Quick Test.
J4 POWER TO PUMP CHECK		
• Key On, engine Off. • Leave Breakout box installed and processor connected. • Locate fuel pump relay. • DVOM on 20V scale. • Measure voltage between chassis ground and power-to-pump circuit at fuel pump relay during crank mode.	8.0V or greater during crank ▶	GO to J5.
	Less than 8.0V during crank ▶	GO to J6.

Fuel Pump Circuit — (Inertia Switch) | Pinpoint Test | J

TEST STEP	RESULT ▶	ACTION TO TAKE
J5 POWER AT FUEL PUMP CHECK		
• Key On, engine Off. • Leave Breakout box installed and processor connected. • Locate fuel pump. • DVOM on 20V scale. • Measure voltage between chassis ground and power-to-pump(s) circuit at fuel pump during crank mode.	8.0V or greater during crank ▶	Check electric fuel pump.
	Less than 8.0V during crank ▶	SERVICE open in power to the pump circuit. RERUN Quick Test.
J6 FUEL PUMP CIRCUIT CHECK TO V BATT.		
• Key On, engine Off. • Leave Breakout box installed and processor connected. • Locate fuel pump relay. • DVOM on 20V scale. • Measure voltage between chassis ground and V Batt. at the fuel pump relay.	10.5V or greater ▶	GO to J7.
	Less than 10.5V ▶	SERVICE open in V Batt. between fuel pump relay and vehicle battery positive post. RERUN Quick Test.
J7 VPWR TO FUEL PUMP RELAY CHECK		
• Key On, engine Off. • Leave Breakout box installed and processor connected. • Locate fuel pump relay. • DVOM on 20V scale. • Measure voltage between chassis ground and VPWR circuit at the fuel pump relay.	10.5V or greater ▶	GO to J8.
	Less than 10.5V ▶	SERVICE open in VPWR circuit between the processor and the fuel pump relay. RERUN Quick Test.
J8 FUEL PUMP CIRCUIT CHECK CONTINUITY		
• Leave Breakout box installed and processor connected. • Key Off, wait 10 seconds. • DVOM on 200 ohm scale. • Measure resistance between fuel pump circuit at the pump relay and test Pin 22 at the Breakout box.	Less than 5 ohms ▶	GO to J9.
	5 ohms or greater ▶	SERVICE open in fuel pump circuit and RERUN Quick Test.

SCORPIO DIAGNOSTIC CHART

Fuel Pump Circuit — (Inertia Switch) | Pinpoint Test | J

TEST STEP	RESULT ▶	ACTION TO TAKE
J9 CHECK FOR SHORT TO GROUND		
• Key Off. • Leave Breakout box installed and processor disconnected. • Fuel pump relay disconnected. • DVOM on 200,000 ohm scale. • Measure resistance between test Pin 22 and test Pins 40 and 60 at the Breakout box.	10,000 ohms or greater ▶ Less than 10,000 ohms ▶	GO to J10 . SERVICE short in the fuel pump circuit. RERUN Quick Test.
J10 CHECK FOR SHORT TO POWER		
• Key Off, wait 10 seconds. • Leave Breakout box installed and disconnect processor. • Fuel pump relay disconnected. • DVOM on 200,000 ohm scale. • Measure resistance between test Pin 22 and test Pins 37 and 57 at the Breakout box.	10,000 ohms or greater ▶ Less than 10,000 ohms ▶	RECONNECT fuel pump relay. GO to J11 . SERVICE short to power in the fuel pump circuit. RECONNECT fuel pump relay and processor. ATTEMPT to start vehicle. If vehicle fails to start, REPLACE processor. RERUN Quick Test.
J11 FINAL SYSTEM CHECK		
• Leave Breakout box installed and processor disconnected. • Connect jumper wire from test Pin 22 to test Pin 40 or 60 at the Breakout box. • DVOM on 20V scale. • Key On, engine Off. • Measure voltage between chassis ground and power-to-pump circuit at fuel pump relay.	10.5V or greater ▶ Less than 10.5V ▶	REPLACE processor. RERUN Quick Test. REPLACE fuel pump relay. RECONNECT processor and RERUN Quick Test.

Fuel Pump Circuit — (Inertia Switch) | Pinpoint Test | J

TEST STEP	RESULT ▶	ACTION TO TAKE
J15 FUEL PUMP RELAY CHECK		
• Key Off. • Remove fuel pump relay. • Does fuel pump turn off?	Yes ▶ No ▶	REPLACE fuel pump relay. RERUN Quick Test. SERVICE short to power to pumps circuit.

SCORPIO DIAGNOSTIC CHART

Canister Purge (CANP) | Pinpoint Test | KD

KD

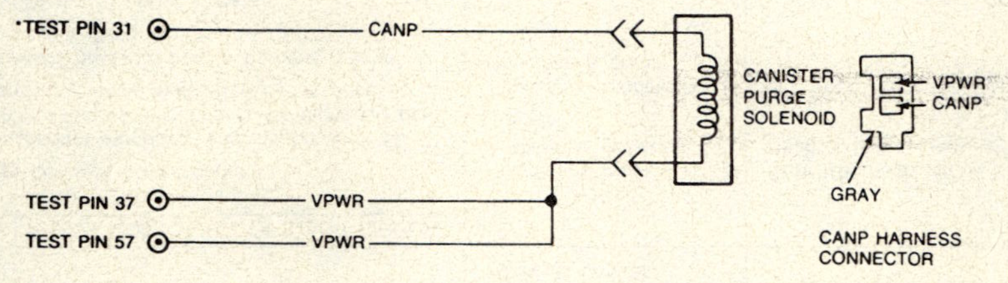

*TEST PINS LOCATED ON BREAKOUT BOX.
ALL HARNESS CONNECTORS VIEWED INTO MATING SURFACE.

STOP-WARNING

You should enter this Pinpoint Test only when a service code 85 is received in Quick Test Step 3.0 or when you are directed here from a fuel control Pinpoint Test and Diagnostic By Symptom chart.

This Pinpoint Test is intended to diagnose only the following:

- Harness Circuits: CANP and VPWR
- Processor Assembly

Canister Purge (CANP) | Pinpoint Test | KD

TEST STEP	RESULT ▶	ACTION TO TAKE
KD1 ENTER OUTPUT STATE CHECK (REFER TO APPENDIX)		
NOTE: Do not use STAR tester for this step, use DVOM. • Key Off, wait 10 seconds. • DVOM on 20V scale. • Connect DVOM negative test lead to STO circuit at Self-Test connector and positive test lead to battery positive. • Jumper STI circuit to signal return at the Self-Test connector. • Perform Key On Engine Off Self Test until the completion of the Continuous Test Codes. • DVOM will indicate zero volts when test is completed. • Depress and release the throttle. • Did DVOM reading change to a high voltage reading?	Yes ▶ No ▶	REMAIN in Output State Check. GO to KD2. DEPRESS throttle to WOT and release. If STO voltage does not go high, GO to Pinpoint Test Step Q40. Leave equipment hooked up.
KD2 CHECK CANISTER PURGE (CANP) SOLENOID ELECTRICAL OPERATION		
• DVOM on 20V scale. • Connect DVOM positive test lead to VPWR on CANP solenoid and negative test lead to CANP output circuit on the solenoid. • While observing DVOM depress and release the throttle several times (to cycle output On and Off). • CANP solenoid cycles On and Off?	Yes ▶ No ▶	GO to KD3. REMOVE jumper. GO to KD4.
KD3 CHECK CANISTER PURGE SOLENOID FOR HOLDING VACUUM		
• Key Off, wait 10 seconds. • Remove jumper from STI to Signal Return. • Disconnect vacuum hose at canister purge solenoid at PCV side. • Apply 16 in.-Hg. to CANP solenoid. • Does CANP solenoid hold vacuum for 20 seconds?	Yes ▶ No ▶	EEC-IV system OK. REPLACE CANP solenoid. RERUN Quick Test.

SCORPIO DIAGNOSTIC CHART

Canister Purge (CANP) — Pinpoint Test KD

TEST STEP	RESULT ▶	ACTION TO TAKE
KD4 MEASURE CANP SOLENOID RESISTANCE • Key Off, wait 10 seconds. • DVOM on 200 ohm scale. • Disconnect CANP solenoid connector and measure solenoid resistance.	Resistance is between 40 and 90 ohms ▶	CONNECT CANP solenoid. GO to KD5 .
	Resistance is less than 40 ohms or greater than 90 ohms ▶	REPLACE CANP solenoid. RERUN Quick Test.
KD5 CHECK VOLTAGE OF VPWR CIRCUIT • Key On, engine Off. • DVOM on 20V scale. • Measure voltage between VPWR at the CANP solenoid vehicle harness connector and battery ground.	Voltage reading is less than 10.5V ▶	SERVICE harness open circuit. RERUN Quick Test.
	Voltage reading is 10.5V or greater ▶	GO to KD6 .
KD6 CHECK CONTINUITY OF CANP CIRCUIT • Key Off, wait 10 seconds. • Disconnect processor and inspect both 60 pin connectors and inspect for damaged pins, corrosion, loose wires. Service as necessary. • Connect Breakout box to harness. Leave processor disconnected. • DVOM on 200 ohm scale. • Measure resistance between test Pin 31 at the Breakout box and CANP vehicle harness connector.	Resistance reading is 5 ohms or greater ▶	SERVICE open circuit. RERUN Quick Test.
	Resistance reading is less than 5 ohms ▶	GO to KD7 .

Canister Purge (CANP) — Pinpoint Test KD

TEST STEP	RESULT ▶	ACTION TO TAKE
KD7 CHECK FOR SHORT TO GROUND • Key Off, wait 10 seconds. • Leave Breakout box to harness. Leave processor disconnected. • Disconnect CANP solenoid. • DVOM on 200,000 ohm scale. • Measure resistance between test Pin 31 and test Pins 40, 46 and 60 at the Breakout box.	All resistance readings are 10,000 ohms or greater ▶	GO to KD8 .
	Any resistance reading less than 10,000 ohms ▶	SERVICE short to ground. RERUN Quick Test.
KD8 CHECK FOR SHORT TO POWER • Key Off, wait 10 seconds. • DVOM on 200,000 ohm scale. • Leave Breakout box installed and processor disconnected. • CANP solenoid disconnected. • Measure resistance between test Pin 31 and test Pins 37 and 57 at the Breakout box.	All resistance readings are less than 10,000 ohms ▶	SERVICE short to power. REPEAT Quick Test. If code is repeated, REPLACE processor. RERUN Quick Test.
	Any resistance reading is 10,000 ohms or greater ▶	REMOVE Breakout box. REPLACE processor. RERUN Quick Test.

SCORPIO DIAGNOSTIC CHART

Idle Speed Control (Bypass Air) — Pinpoint Test — KE

KE

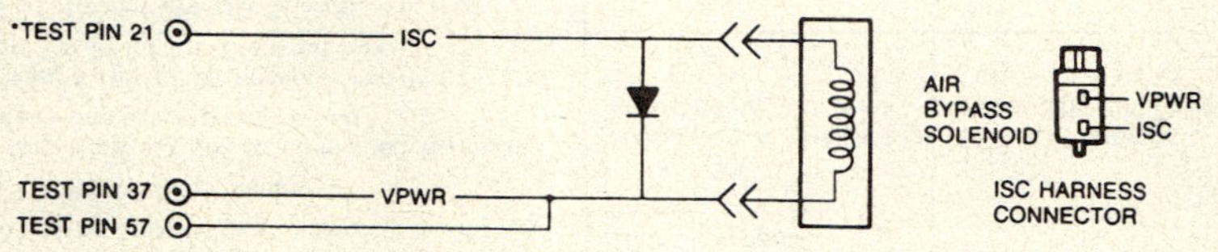

*TEST PINS LOCATED ON BREAKOUT BOX.
ALL HARNESS CONNECTORS VIEWED INTO MATING SURFACE.

STOP-WARNING

You should enter this Pinpoint Test only when a service code 12 or 13 is received in Quick Test Step 5.0 or when directed here from Diagnostic By Symptom.

CAUTION: If the engine exhibits rough running and/or idle, correct these conditions before diagnosing Idle Speed Control (ISC). Rough running/misses may be caused by:

- **Ignition System — Refer to Shop Manual Section 23-01**
- **Fuel System — Refer to Pinpoint Test Steps HG1 through HG6**
- **EGR System, Section 3**

To prevent the replacement of good components, be aware that the following non-EEC areas may be at fault:

- Engine not up to operating temperature
- Engine over operating temperature
- Improper Idle Speed Throttle Stop Adjustment
- A/C input (electrical problem)
- Throttle Speed Control Linkage

This Pinpoint Test is intended to diagnose only the following:

- RPM in Self Test only
- ISC Actuator
- Harness Circuits ISC and VPWR
- Processor Assembly

Idle Speed Control (Bypass Air) — Pinpoint Test — KE

TEST STEP	RESULT ▶	ACTION TO TAKE
KE1 RPM DROP		
• Key Off. • Connect engine tachometer. • Start engine. • Disconnect ISC harness connector. • Does rpm drop or stall?	Yes ▶	GO to KE2.
	No ▶	GO to KE3.
KE2 CHECK FOR EGR CODES		
• Are service codes 31, 32, 33 or 34 present?	Yes ▶	GO to Quick Test Step 5.0 for appropriate Pinpoint Test.
	No ▶	SERVICE next code or if code 12 or 13 only, GO to KE3.
KE3 MEASURE ISC SOLENOID RESISTANCE		
• Key Off. • DVOM on 200 ohm scale. • Disconnect ISC solenoid connector and measure solenoid resistance.	Resistance is between 7 and 13 ohms ▶	GO to KE4.
	Resistance is less than 7 ohms or greater than 13 ohms ▶	REPLACE ISC solenoid. RERUN Quick Test.
KE4 ISC SHORT TO CASE (GROUND) CHECK		
• Key Off. • DVOM on 200,000 ohm scale. • ISC solenoid disconnected. • Measure resistance from either ISC pin to ISC housing.	10,000 ohms or greater ▶	GO to KE5.
	Less than 10,000 ohms ▶	REPLACE ISC solenoid. RERUN Quick Test.
KE5 CHECK VOLTAGE OF VPWR CIRCUIT		
• Key On Engine Off. • DVOM on 20V scale. • ISC solenoid disconnected. • Measure voltage between VPWR at the ISC solenoid harness connector and battery ground.	Voltage reading is less than 10.5V ▶	SERVICE harness open circuit. RERUN Quick Test.
	Voltage reading is 10.5V or greater ▶	GO to KE6.

SCORPIO DIAGNOSTIC CHART

Idle Speed Control (Bypass Air) — Pinpoint Test — KE

TEST STEP	RESULT ▶	ACTION TO TAKE
KE6 CHECK CONTINUITY OF ISC CIRCUIT		
• Key Off, wait 10 seconds. • ISC solenoid disconnected. • Disconnect processor and inspect both 60 pin connectors for damaged pins, corrosion, loose wires. Service as necessary. • Connect Breakout box to harness. Leave processor disconnected. • DVOM on 200 ohm scale. • Measure resistance between test Pin 21 at the Breakout box and ISC circuit at vehicle harness connector.	Resistance reading is 5 ohms or greater ▶ Resistance reading is less than 5 ohms ▶	SERVICE open circuit. RERUN Quick Test. GO to **KE7**.
KE7 CHECK FOR SHORT TO GROUND		
• Key Off, wait 10 seconds. • Leave Breakout box installed and processor disconnected. • ISC solenoid disconnected. • DVOM on 200,000 ohm scale. • Measure resistance between test Pin 21 and test Pins 40, 46 and 60 at the Breakout box.	All resistance readings are less than 10,000 ohms ▶ Any resistance reading is 10,000 ohms or greater ▶	SERVICE short to ground. RERUN Quick Test. GO to **KE8**.
KE8 CHECK FOR SHORT TO POWER		
• Key Off, wait 10 seconds. • DVOM on 200,000 ohm scale. • ISC solenoid disconnected. • Leave Breakout box to connected harness. Leave processor disconnected. • Measure resistance between test Pin 21 and test Pin 37 at the Breakout box.	Resistance reading is 10,000 ohms or greater ▶ Resistance reading is less than 10,000 ohms ▶	GO to **KE9**. SERVICE short to power. RERUN Quick Test.
KE9 CHECK FOR ISC SIGNAL AT THE PROCESSOR		
• Connect processor. • Connect ISC actuator. • Vehicle prepared for Quick Test. • DVOM on 20V scale. • Connect DVOM between test Pin 21 and test Pin 40. • Start engine and observe the DVOM during Engine Running Quick Test.	Meter reading varies during Quick Test ▶ Meter reading does not vary during Quick Test ▶	REPLACE ISC actuator. RERUN Quick Test. GO to **KE10**.

Idle Speed Control (Bypass Air) — Pinpoint Test — KE

TEST STEP	RESULT ▶	ACTION TO TAKE
KE10 **CODE 13:** RPM DOES NOT DROP TO LESS THAN 1500 AT END OF TEST		
• Disconnect ISC harness connector. • Connect engine tachometer. • Repeat Engine Running Quick Test. • At end of test, record service codes for future use. • Does rpm remain below 1500 rpm during test?	Yes ▶ No ▶	REPLACE processor. RERUN Quick Test. CHECK engine vacuum hoses. REFER to VECI decal. VERIFY curb idle. CHECK that throttle plates are fully closed (against throttle stop). CHECK throttle linkage and/or speed control linkage for binding. If OK, REPLACE ISC actuator. RERUN Quick Test.
KE11 CURB IDLE CHECK		
• Key Off, wait 10 seconds. • Deactivate Self Test. • Run engine at 2000 rpm for 2 minutes or until inlet radiator hose is hot and pressurized. • Key Off, wait 10 seconds. • Activate Self Test. • Run Engine Running Self-Test. • Is code 17 still present?	No ▶ Yes ▶	SERVICE other codes as necessary. INSPECT throttle body and air inlet for contamination. SERVICE as necessary. If OK, ADJUST curb idle speed. RERUN Quick Test.

SCORPIO DIAGNOSTIC CHART

Dynamic Response Test	Pinpoint Test	M

M

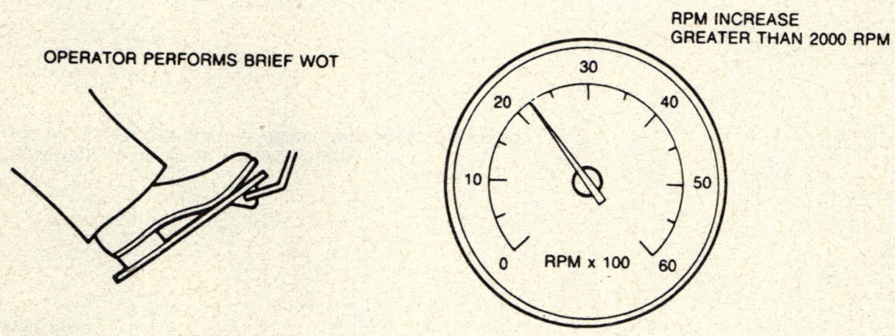

STOP-WARNING

You should enter this Pinpoint Test only when a service code 77 is received in Quick Test Step 5.0.

To prevent the replacement of good components, be aware that the following non-EEC areas may be at fault:

- Operator did not perform a brief WOT after dynamic response code.
- Mechanical engine problems; engine did not achieve greater than 2000 rpm.

This Pinpoint Test is intended to diagnose only the following:

- Throttle movement (greater than 3/4 throttle).
- RPM increase (greater than 2000 rpm).

Dynamic Response Test	Pinpoint Test	M

TEST STEP	RESULT ▶	ACTION TO TAKE
M1 CODE 77: SYSTEM FAILED TO RECOGNIZE BRIEF WOT		
NOTE: A brief snap of the throttle may not be sufficient to pass this test. Be sure to go to WOT and return.		
• Repeat Engine Running Test of Quick Test. Be sure operator is familiar with the engine running format which proceeds as follows: — With Self Test activated restart the engine. — ID Code 3 (0) start of test. — Dynamic response Code 1 (0) perform brief WOT. — Testing over. — Service code output begins.	Code 77 still present ▶	REPLACE processor. RERUN Quick Test.
	No Code 77 ▶	Dynamic Response Test passed. SERVICE any other service code(s) received as necessary.

SCORPIO DIAGNOSTIC CHART

Ignition Diagnostic Monitor (IDM)	Pinpoint Test	N

N

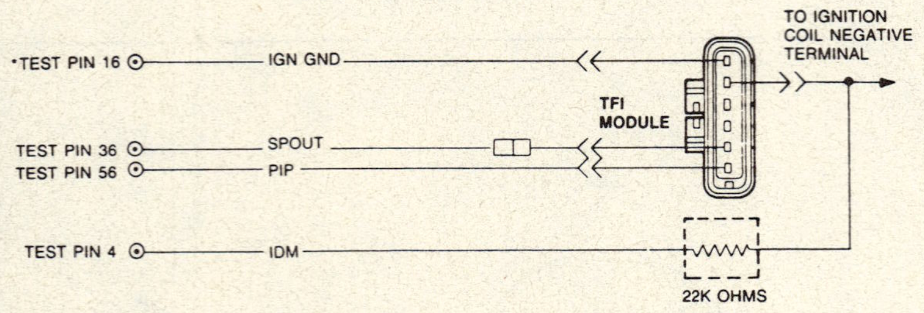

*TEST PINS LOCATED ON BREAKOUT BOX.
ALL HARNESS CONNECTORS VIEWED INTO MATING SURFACE.

STOP-WARNING

You should enter this Pinpoint Test only when a service code 18 is received in Quick Test Step 6.0.

To prevent the replacement of good components, be aware that the following non-EEC areas may be at fault:

- Ignition module.
- Ignition coil.
- Spark plugs and high tension cables.
- Distributor and PIP sensor.

This Pinpoint Test is intended to diagnose only the following:

- Harness circuits: ignition ground, spout, PIP, IDM.

Ignition Diagnostic Monitor (IDM)	Pinpoint Test	N

TEST STEP	RESULT ▶	ACTION TO TAKE
N1 CHECK CONTINUITY OF IDM CIRCUIT		
• Key Off, wait 10 seconds. • Disconnect E-core ignition connector from coil. • Disconnect processor and inspect both 60 pin connectors for damaged pins, corrosion, loose wires. Service as necessary. • Connect Breakout box to harness. Leave processor disconnected. • DVOM on 200,000 ohm scale. • Measure resistance between test Pin 4 at the Breakout box and ignition coil negative terminal.	Resistance reading is less than 20,000 ohms or greater than 24,000 ohms ▶	SERVICE open circuit. RERUN Quick Test.
	Resistance reading is less than 24,000 ohms but greater than 20,000 ohms ▶	GO to **N2**.
N2 CHECK FOR SHORT TO GROUND		
• Key Off, wait 10 seconds. • Leave Breakout box installed and processor disconnected. • DVOM on 200,000 ohm scale. • Measure resistance between test Pin 4 and test Pins 40, 46 and 60 at the Breakout box.	Any resistance reading is less than 10,000 ohms ▶	SERVICE short to ground. RERUN Quick Test.
	All resistance readings are 10,000 ohms or greater ▶	GO to **N3**.
N3 CHECK TFI MODULE		
• Key Off, wait 10 seconds. • Deactivate Self Test. • Connect STAR per Quick Test Step 2.0. • Enter Engine Running Continuous Monitor Test (as instructed in Quick Test Step 6.0D). • Observe STAR LED for indication of a fault while performing the following: • Lightly tap on TFI module (simulate road shock). • Wiggle TFI connector. • Is a fault indicated?	Yes ▶	DISCONNECT and INSPECT connectors. If connector and terminals are good, CHECK TFI module using TFI-IV Ignition Tester, Rotunda 105-00002 or equivalent.
	No ▶	GO to **N4**.

SCORPIO DIAGNOSTIC CHART

Ignition Diagnostic Monitor (IDM) — Pinpoint Test N

TEST STEP	RESULT ▶	ACTION TO TAKE
N4 CHECK EEC-IV HARNESS		
• While still in continuous monitor test from step N1 observe STAR LED for a fault indication while performing the following: • While looking for faults listed in the table below, grasp the harness close to the TFI connector. Wiggle, shake or bend a small section of the EEC-IV system harness while working your way to the dash panel. Also wiggle, shake or bend the EEC-IV harness from the dash panel to the processor. Do this test on the circuits listed one at a time if needed to locate a faulty circuit.	Yes ▶	ISOLATE fault and SERVICE as necessary. RERUN Quick Test.
	No ▶	GO to N5.

FAULT	BREAKOUT BOX NO.
PIP shorted to ground or open	Test Pin 56
Spout shorted to ground	Test Pin 36
Ign. ground open	Test Pin 16
IDM open or shorted to ground, power	Test Pin 4

• Is a fault indicated?

TEST STEP	RESULT ▶	ACTION TO TAKE
N5 CHECK PROCESSOR AND HARNESS CONNECTORS		
• Key Off, wait 10 seconds. • Disconnect processor 60 pin connector and inspect for damaged pins, corrosion, loose wires. • Are connectors and terminals OK? • Reconnect processor when this step is completed.	No ▶	SERVICE as necessary. RERUN Quick Test.
	Yes ▶	If unable to DUPLICATE an IDM fault in the EEC-IV System, testing is complete.

Spark Timing Check — Pinpoint Test P

P

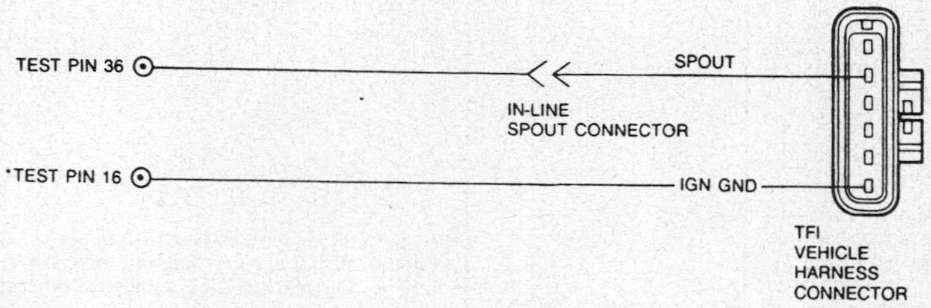

*TEST PINS LOCATED ON BREAKOUT BOX.
ALL HARNESS CONNECTORS VIEWED INTO MATING SURFACE.

STOP-WARNING

You should enter this Pinpoint Test only when directed here from Quick Test Step 4.0.

To prevent the replacement of good components, be aware that the following non-EEC areas may be at fault:

- Base Engine
- PIP Sensor
- TFI Module

This Pinpoint Test is intended to diagnose only the following:

- Harness Spout Circuit
- Base Timing
- Processor Assembly

SCORPIO DIAGNOSTIC CHART

Spark Timing Check — Pinpoint Test P

TEST STEP	RESULT ▶	ACTION TO TAKE
P1 CHECK SPARK TIMING		
• Verify that Self Test is activated during Engine Running Test. • Check timing while performing Engine Running Test. Record value. • Is timing 15 degrees ±3 degrees? **NOTE: Self Test locks the timing at 15 degrees plus base throughout this test and for two minutes after the last service code is outputted. Timing check must be made during this time period. Self-Test timing is 15 degrees (±3 degrees) BTDC. (See VECI decal for base value.)**	Yes ▶ No ▶	GO to Quick Test Step 5.0. GO to P2.
P2 CHECK SPARK OUTPUT (SPOUT) CIRCUIT TO THE TFI MODULE		
• Locate SPOUT connector and open this connection. • Start engine. • Check base timing. • Is timing 10 ± 3 degrees BTDC. **NOTE: Base timing must be checked to decal value.**	Yes ▶ No ▶	GO to P3. Adjust base timing if necessary. After timing is reset, RECONNECT SPOUT and PERFORM Quick Test Step 4.0.
P3 CHECK FOR POWER TO PROCESSOR		
• Key Off, wait 10 seconds. • Disconnect processor 60 pin connector and inspect for damaged pins, corrosion, loose wires. Service as necessary. • Install Breakout Box. • Key On Engine Off. • DVOM on 20V scale. • Measure voltage between test Pin 37 and test Pin 40 and between test Pin 57 and test Pin 60 at the Breakout box.	Either voltage reading less than 10.5V ▶ Both voltage readings 10.5V or greater ▶	GO to Pinpoint Test Step B1. GO to P4.

Spark Timing Check — Pinpoint Test P

TEST STEP	RESULT ▶	ACTION TO TAKE
P4 CHECK HARNESS FOR CONTINUITY		
• Key Off, wait 10 seconds. • DVOM on 200 ohm scale. • Disconnect TFI connector and close spout connector. • Measure resistance between test Pin 36 at the Breakout box and Pin 2 (SPOUT pin) at the TFI vehicle harness connector.	Reading 5 ohms or less ▶ Reading greater than 5 ohms ▶	GO to P5. SERVICE open circuit. RECONNECT TFI connector. CHECK timing per P1.
P5 HARNESS CHECK — IGNITION GROUND		
• Breakout box installed. • Key Off. • DVOM on 200 ohm scale. • Measure resistance between test Pin 16 at the Breakout box and TFI Pin 6 (ign. grnd.) at the TFI vehicle harness connector.	Less than 5 ohms ▶ 5 ohms or greater ▶	GO to Pinpoint Test Step A7. SERVICE harness as necessary. RECONNECT TFI harness connector. RERUN Quick Test.

SCORPIO DIAGNOSTIC CHART

No Codes/Codes Not Listed	Pinpoint Test	Q

Q

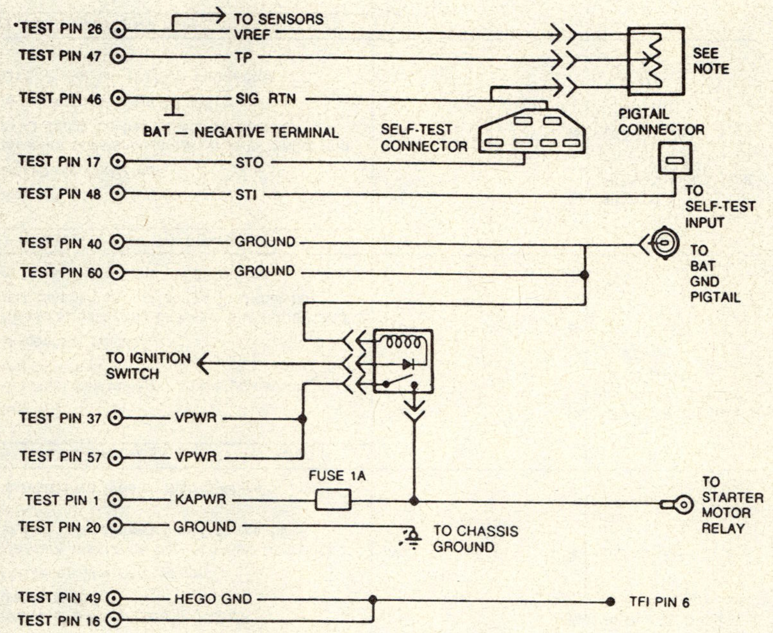

*TEST PINS LOCATED ON BREAKOUT BOX.
ALL HARNESS CONNECTORS VIEWED INTO MATING SURFACE.

NOTE: Refer to the electrical schematic in the appropriate engine supplement section for proper connector and pin orientation.

STOP-WARNING

You should enter this Pinpoint Test only when directed here from Quick Test Step 3.0, 5.0 or 6.0 or when directed here from Diagnostic By Symptom or from a Pinpoint Test Step.

This Pinpoint Test is intended to diagnose only the following:

- Processor.
- Harness Circuits: Signal Return, STO, STI, GND, VPWR, KAM, VREF, IGN, NDS, EEC Power Relay.

No Codes/Codes Not Listed	Pinpoint Test	Q

TEST STEP	RESULT ▶	ACTION TO TAKE
Q1 NO CODES, CODES NOT LISTED		
Refer to illustration Q. • Key Off, wait 10 seconds. • DVOM on 20V scale. • Disconnect TP sensor. • Key On Engine Off. • Measure voltage between VREF at the TP harness connector and Signal Return in Self-Test connector.	Less than 4.0V or greater than 6.0V ▶ 4.0V to 6.0V ▶	GO to Pinpoint Test Step C1. RECONNECT TP sensor. GO to Q2.
Q2 SELF-TEST INPUT CONTINUITY CHECK		
Refer to illustration Q. • Key Off, wait 10 seconds. • Disconnect processor 60 pin connector and inspect for damaged pins, corrosion, loose wires. Service as necessary. • Install Breakout box, leave processor disconnected. • Set DVOM to 200 ohm scale. • Measure resistance between Self-Test input at the Self-Test single pin connector and test Pin 48 at the Breakout box.	Less than 5 ohms ▶ 5 ohms or greater ▶	GO to Q3. CORRECT open in circuit.
Q3 SELF-TEST OUTPUT CIRCUIT CONTINUITY CHECK		
Refer to illustration Q. • Breakout box installed, processor disconnected. • DVOM to 200 ohm scale. • Measure resistance between Self-Test output at the Self-Test connector and test Pin 17 at the Breakout box.	5 ohms or greater ▶ Less than 5 ohms ▶	CORRECT open in circuit. GO to Q4.
Q4 EGO SENSOR GROUND CONTINUITY CHECK		
Refer to illustration Q. • Breakout box installed, processor disconnected. • Key Off. • Measure resistance between HEGO case and test Pin 16 at the Breakout box and between test Pin 49 and test Pin 16 at the Breakout box.	If both less than 5 ohms ▶ If either 5 ohms or greater ▶	GO to Q5. SERVICE open circuits and RERUN Quick Test.

SCORPIO DIAGNOSTIC CHART

No Codes/Codes Not Listed — Pinpoint Test Q

TEST STEP	RESULT ▶	ACTION TO TAKE
Q5 STO SHORT TO GROUND		
• Breakout box installed processor disconnected. • DVOM on 200 ohm scale. • Measure resistance between Self-Test output at Self-Test connector and engine block ground. • Is resistance greater than 5 ohms?	Yes ▶ No ▶	REPLACE processor. RERUN Quick Test. GO to Q6 .
Q6 HEGO HARNESS SHORT TO POWER		
• Key Off. • Breakout box installed, leave processor disconnected. • Disconnect HEGO. • Measure resistance between test Pin 29 and test Pins 37 and 57 at the Breakout box. • Is reading less than 10,000 ohms?	No ▶ Yes ▶	SERVICE harness short to power. RERUN Quick Test. GO to Q7 .
Q7 HEGO SHORT TO POWER		
• Key Off. • HEGO disconnected. • Measure resistance between RUN circuit and EGO circuit on the HEGO connector. • Is reading less than 10,000 ohms? **NOTE: Refer to HG for connector orientation.**	Yes ▶ No ▶	REPLACE HEGO. RERUN Quick Test. GO to Q8 .
Q8 INTERMITTENT NDS		
• Key Off. • Breakout box installed, processor connected. • Connect DVOM between test Pin 30 and test Pin 40 or 60 at the Breakout box. • Key On Engine Running Self Test. • Is voltage greater than 1V?	Yes ▶ No ▶	SERVICE intermittent in NDS harness, connector or switch. If OK, GO to Quick Test Step 5.0 for appropriate service codes. SERVICE intermittent in NDS harness, connector, or switch. If OK, GO to Q9 .

No Codes/Codes Not Listed — Pinpoint Test Q

TEST STEP	RESULT ▶	ACTION TO TAKE
Q9 POWER RELAY ALWAYS ON		
• Key Off. • Breakout box installed. • Connect DVOM to test Pin 37 or 57 and to test Pin 40 or 60 at the Breakout box. • Turn key On and Off. Wait 10 seconds. • Does voltage change from 10.5V or greater to zero volts?	Yes ▶ No ▶	SERVICE STO circuit for short to ground. RERUN Quick Test. REPLACE EEC Power Relay. RERUN Quick Test.

SCORPIO DIAGNOSTIC CHART

Continuous Test Code 15 — Pinpoint Test Q

TEST STEP	RESULT ▶	ACTION TO TAKE
Q10 CONDITIONS FOR CONTINUOUS CODE 15		
• Power interruption to Keep Alive Memory (KAM) Pin 1 may result in a service code being outputted.* • Clear continuous memory codes (use procedure described in Quick Test Step 6). • Repeat Quick Test Step 3.0 through Continuous memory code output. • Code 15 present on retest? *NOTE: **Anytime power is interrupted to the processor, a code 15 may be outputted the first time Self Test is run after restoration of power. Repeat Self Test to ensure correct diagnosis.**	Yes ▶ No ▶	GO to Q11. Test complete.
Q11 INSPECT ENGINE COMPARTMENT WIRING FOR PROPER ROUTING		
• Are any EEC components or EEC wiring close to ignition components or wires (High Electrical Energy Sources)? If EEC wiring close, reroute and rerun Self Test. • Is code 15 still present?	Yes ▶ No ▶	GO to Q12. Test complete.
Q12 CHECK POWER CIRCUIT TO KEEP ALIVE MEMORY		
• Key Off, wait 10 seconds. • Disconnect processor and inspect both 60 pin connectors. • Connect Breakout box to harness, leave processor disconnected. • DVOM on 20V scale. • Connect positive test lead to test Pin 1 and negative test lead to test Pin 40 or 60 at the Breakout box. • Key On. • Observe voltage reading.	Reading less than 10V ▶ Reading 10V or greater ▶	SERVICE open to KAM circuit. RERUN Quick Test. REPLACE processor. RERUN Quick Test.

Engine Stalls During Quick Test — Pinpoint Test Q

TEST STEP	RESULT ▶	ACTION TO TAKE
Q30 DIAGNOSTIC ROUTINES COMPLETE		
• Have diagnostic routines been completed?	Yes ▶ No ▶	GO to Q31. GO to Diagnostic Routines.
Q31 ATTEMPT TO BRING ENGINE TO OPERATING TEMPERATURE		
• Key Off, wait 10 seconds. • Install tachometer. • Try to maintain engine at 2,000 rpm for 2 minutes. • Can rpm be maintained?	Yes ▶ No ▶	GO to Pinpoint Test Step A13. GO to Q32.
Q32 MAINTAIN ENGINE OPERATION DURING ENGINE RUNNING QUICK TEST		
• Engine at operating temperature. • Tachometer installed. • Key Off, wait 10 seconds. • Activate Quick Test. • Perform Engine Running Self Test while maintaining 2,000 rpm.	Code 11 ▶ Any service code ▶ No codes ▶	GO to Quick Test Step 5.0. GO to Quick Test Step 5.0B. GO to Pinpoint Test Step Q1.

SCORPIO DIAGNOSTIC CHART

Output State Check Not Functioning — Pinpoint Test Q

TEST STEP	RESULT ▶	ACTION TO TAKE
Q40 CHECK FOR CODES 23, 53, 63 OR 68 • Key Off, wait 10 seconds. • Perform Key On Engine Off Self Test. • Leave key On to enter Output State Check. • Key On Engine Off. Are codes 23, 53, 63 or 68 present?	Yes ▶	GO to Quick Test Step [3.0B] and SERVICE appropriate code as instructed.
	Code 11 ▶	GO to [Q41].
	No Codes ▶	GO to [Q1].
Q41 CHECK THROTTLE LINKAGE • Check throttle and throttle linkages for sticking and binding. • Throttle OK?	No ▶	SERVICE as necessary. RERUN Quick Test.
	Yes ▶	REPLACE TP sensor. RERUN Quick Test.

Processor Power Check — Pinpoint Test Q

TEST STEP	RESULT ▶	ACTION TO TAKE
Q50 CHECK FOR POWER TO PROCESSOR • Key Off, wait 10 seconds. • Disconnect processor 60 pin connector and inspect for damaged pins, corrosion, loose wires. Service as necessary. • Install Breakout box. • Key On Engine Off. • DVOM on 20V scale. • Measure voltage between test Pin 37 and test Pin 40 at the Breakout box and between test Pin 57 and test Pin 60 at the Breakout box.	Either voltage reading less than 10.5V ▶	GO to Pinpoint Test [B1].
	Both voltage readings 10.5V or greater ▶	REPLACE processor. RERUN Quick Test.

SCORPIO DIAGNOSTIC CHART

System Check | **Pinpoint Test** | **S**

S

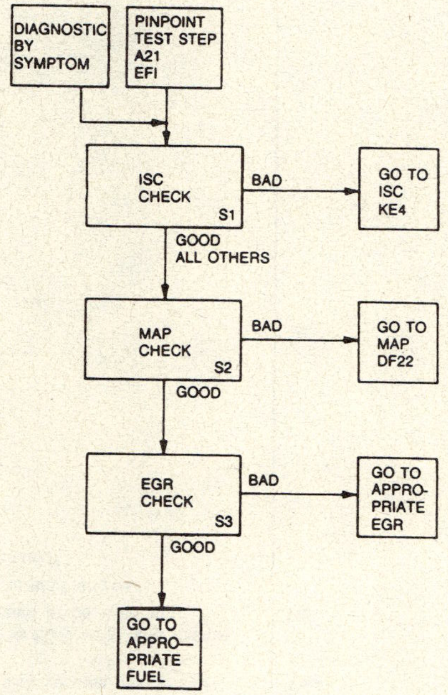

STOP-WARNING

You should enter this Pinpoint Test only after a code 11 is received in Quick Test Step 3.0, and you have been directed here from EEC-IV No-Start Pinpoint Test Step A21 or Diagnostic By Symptom.

This Pinpoint Test is intended only as a Quick Check for the basic functioning of the following:

- ISC Bypass Air System
- MAP System
- EGR System

System Check | **Pinpoint Test** | **S**

TEST STEP	RESULT ▶	ACTION TO TAKE
S1 ISC-BPA CHECK		
• If you are here for any reason other than stalls or a no start, GO to S2. • Attempt to start engine at part throttle. • Will engine run only at part throttle?	Yes, but runs rough ▶	GO to S2.
	Yes, and runs smooth ▶	GO to Pinpoint Test Step KE3.
	No ▶	GO to S2.
S2 MAP CHECK		
• Key Off. • Disconnect the MAP sensor from the vehicle harness. • Connect the MAP tester between the vehicle harness and the MAP sensor. • Plug MAP tester banana plugs into DVOM. • Set DVOM to 20V scale. • Key On. • Observe DVOM. • Approximate Altitude (Ft.) / Voltage Output (± .04V): 0 — 1.59; 1000 — 1.56; 2000 — 1.53; 3000 — 1.50; 4000 — 1.47; 5000 — 1.44; 6000 — 1.41; 7000 — 1.39 • Crank engine. • While cranking, does DVOM reading decrease from the appropriate reading for your altitude listed above?	Yes ▶	GO to S3.
	No ▶	GO to Pinpoint Test Step DF22.

Approximate Altitude (Ft.)	Voltage Output (± .04V)
0	1.59
1000	1.56
2000	1.53
3000	1.50
4000	1.47
5000	1.44
6000	1.41
7000	1.39

SCORPIO DIAGNOSTIC CHART

System Check | **Pinpoint Test** | **S**

TEST STEP	RESULT	▶ ACTION TO TAKE
S3 EGR CHECK		
• Disconnect and plug vacuum line at EGR valve. • Apply 16 in.-Hg. to EGR valve and release. Verify valve is heard to be working. • Start or attempt to start engine. • Is symptom eliminated?	Yes ▶ No ▶	Check EGR system. GO to Pinpoint Test Step HG1.

WOT A/C Cutoff (WAC) A/C Demand | **Pinpoint Test** | **X**

X

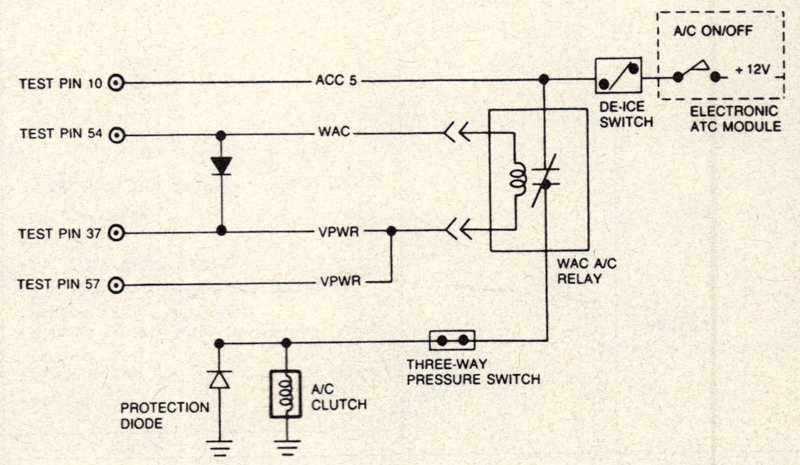

STOP-WARNING

You should enter this Pinpoint Test only when directed here from Diagnostic By Symptom chart.

This Pinpoint Test is intended to diagnose only the following:

- Harness Circuits: WAC and VPWR
- WOT A/C Cutoff Relay
- Processor Assembly
- A/C Demand Switch

SCORPIO DIAGNOSTIC CHART

WOT A/C Cutoff (WAC) A/C Demand — Pinpoint Test X

TEST STEP	RESULT ▶	ACTION TO TAKE
X50 CHECK FOR VOLTAGE AT A/C CLUTCH • Engine warm. • Key On Engine Running. • Turn the A/C On. • Wait 10 seconds. • DVOM on 20V scale. • Check voltage at A/C clutch harness connector.	10.5V or greater ▶	Check A/C System.
	Less than 10.5V ▶	GO to X51.
X51 CHECK FOR CONTINUITY FROM WAC RELAY TO A/C CLUTCH • Key Off. • Remove WAC relay. • DVOM on 200 ohm scale. • Measure resistance between Pin 87A WAC relay connector to the power circuit at the A/C clutch harness connector. • Is resistance 5 ohms or less?	Yes ▶	RECONNECT WAC relay. GO to X52.
	No ▶	CHECK A/C pressure switch is closed. If so, open in power feed to A/C. RECONNECT relay. RERUN Quick Test.
X52 ENTER OUTPUT STATE CHECK (REFER TO APPENDIX) **NOTE: Do not use STAR tester for this step, use DVOM.** • Key Off, wait 10 seconds. • DVOM on 20V scale. • Connect DVOM negative test lead to STO and positive test lead to battery positive. • Jumper STI to signal return. • Perform Key On Engine Off Self Test until the completion of the Continuous Test Codes. • DVOM will indicate zero volts. • Depress and release the throttle. • Did DVOM reading change to a high voltage reading?	Yes ▶	REMAIN in Output State Check. GO to X53.
	No ▶	DEPRESS throttle to WOT and RELEASE. If STO voltage does not go high, GO to Pinpoint Test Step Q40. LEAVE equipment hooked up.

WOT A/C Cutoff (WAC) A/C Demand — Pinpoint Test X

TEST STEP	RESULT ▶	ACTION TO TAKE
X53 CHECK WAC OUTPUT FOR PROPER ELECTRICAL OPERATION • Key Off, engine Off. • Disconnect processor 60 pin connector and inspect for damaged pins, corrosion, loose wires. Service as necessary. • Connect Breakout box to harness. Reconnect processor. • Perform Key On Engine Off. Self Test until completion of continuous test codes. • DVOM on 20V scale. • Connect DVOM positive test lead to test Pin 37 and negative test lead to test Pin 54. • While observing DVOM, depress and release the throttle several times (to cycle output On and Off). • Does voltage output cycle high and low?	Yes ▶	GO to X54.
	No ▶	GO to X57A.
X54 CHECK FOR VOLTAGE AT A/C INPUT TO PROCESSOR • Key On Engine Off. • A/C demand switch to A/C On position. • DVOM on 20V scale. • Leave Breakout box installed. • Processor and WAC relay connected. • Measure voltage between test Pin 10 and test Pin 40 at Breakout box. • Is voltage 10.5V or greater?	No ▶	GO to X56A.
	Yes ▶	GO to X55.
X55 CHECK CONTINUITY OF ACCS TO WAC RELAY • Key Off, wait 10 seconds. • Processor disconnected. • WAC relay disconnected. • DVOM on 200 ohm scale. • Measure resistance between test Pin 10 at Breakout box and test Pin 30 at WAC relay harness connector. • Is resistance 5 ohms or less?	No ▶	SERVICE open in ACCS circuit. RERUN Quick Test.
	Yes ▶	REPLACE WAC relay. RERUN Quick Test.

SCORPIO DIAGNOSTIC CHART

WOT A/C Cutoff (WAC) A/C Demand | **Pinpoint Test** | **X**

TEST STEP	RESULT ▶	ACTION TO TAKE
X56A CHECK FOR VOLTAGE AT A/C CLIMATE CONTROL MODULE SIDE OF DE-ICE SWITCH • Key On Engine Off. • A/C demanded. • DVOM on 20V scale. • Leave Breakout box installed. • Processor and WAC relay connected. • Measure voltage between Pin 4 on de-icing switch harness connector and test Pin 40 at Breakout box. • Is voltage 10.5 or greater?	Yes ▶ No ▶	GO to X56B. GO to X56C.
X56B CHECK CONTINUITY PROCESSOR TO DE-ICE SWITCH • Key Off, wait 10 seconds. • Processor and WAC relay disconnected. • De-ice switch disconnected. • DVOM on 200 ohm scale. • Measure resistance between Breakout box Pin 10 and Pin 2 on de-ice switch harness connector. • Is resistance 5 ohms or less?	Yes ▶ No ▶	GO to X56D. SERVICE open. RERUN Quick Test.
X56C CHECK CONTINUITY DE-ICE SWITCH TO CLIMATE CONTROL MODULE • Key Off, wait 10 seconds. • De-ice switch disconnected. • Climate control module disconnected. • DVOM on 200 ohm scale. • Measure resistance between de-ice switch harness connector Pin 4 and climate control module harness connector Pin 16. • Is resistance 5 ohms or less?	Yes ▶ No ▶	GO to X56E. SERVICE open. RERUN Quick Test.

WOT A/C Cutoff (WAC) A/C Demand | **Pinpoint Test** | **X**

TEST STEP	RESULT ▶	ACTION TO TAKE
X56D CHECK FOR SHORT TO GROUND • Key Off, wait 10 seconds. • Processor and WAC relay disconnected. • De-ice switch disconnected. • DVOM on 200,000 ohm scale. • Measure resistance between test Pin 10 on Breakout box and test Pins 40 and 60 at Breakout box and negative side of battery.	10,000 ohms or greater ▶ Less than 10,000 ohms ▶	REPLACE de-ice switch. SERVICE short to ground. RERUN Quick Test.
X56E CHECK FOR SHORT TO GROUND • Key Off, wait 10 seconds. • De-ice switch disconnected. • Climate control module disconnected. • DVOM on 200,000 ohm scale. • Measure resistance between de-ice switch Pin 4 and test Pins 40 and 60 at Breakout box and negative side of the battery.	10,000 ohms or greater ▶ Less than 10,000 ohms ▶	REPLACE climate control module. SERVICE short to ground. RERUN Quick Test.

SCORPIO DIAGNOSTIC CHART

WOT A/C Cutoff (WAC) A/C Demand — Pinpoint Test X

TEST STEP	RESULT ▶	ACTION TO TAKE
X57A MEASURE WAC RELAY RESISTANCE • Key Off, wait 10 seconds. • DVOM on 200 ohm scale. • Disconnect WAC relay connector and measure relay resistance.	Resistance is between 70 ohms and 120 ohms ▶	GO to X57B.
	Resistance is less than 70 ohms or greater than 120 ohms ▶	REPLACE WAC relay. RERUN Quick Test.
X57B CHECK CONTINUITY FROM WAC TO PROCESSOR • Key Off, wait 10 seconds. • Disconnect processor 60 pin connector and inspect for damaged pins, corrosion, loose wires. Service as necessary. • Connect Breakout box to harness. Leave processor disconnected. • DVOM on 200 ohm scale. • Measure resistance between test Pin 54 and Pin 85 at WAC relay harness connector. • Is resistance 5 ohms or less?	No ▶	SERVICE open in WAC circuit. RERUN Quick Test.
	Yes ▶	GO to X58.
X58 CHECK WAC CIRCUIT FOR SHORTS TO GROUND • Key Off, wait 10 seconds. • Leave Breakout box installed and processor disconnected, (8 pin). • DVOM on 200,000 ohm scale. • Measure resistance between test Pin 54 and test Pin 40 and between test Pin 54 and test Pin 46 and between test Pin 54 and battery negative post.	All resistance readings 10,000 ohms or greater ▶	GO to X59.
	Any resistance reading less than 10,000 ohms ▶	SERVICE shorts to ground in WAC circuit. RERUN Quick Test.

WOT A/C Cutoff (WAC) A/C Demand — Pinpoint Test X

TEST STEP	RESULT ▶	ACTION TO TAKE
X59 CHECK WAC CIRCUIT FOR SHORTS TO POWER • Key Off, wait 10 seconds. • Leave Breakout box installed and processor disconnected. • WAC relay disconnected. • DVOM on 200,000 ohm scale. • Measure resistance between test Pin 54 and test Pin 37 and between test Pin 54 and battery positive.	All readings 10,000 ohms or greater ▶	GO to X60.
	Any readings less than 10,000 ohms ▶	SERVICE shorts to power in WAC circuit and then GO to X60.
X60 CHECK CONTINUITY FROM WAC RELAY TO VPWR • Key Off, wait 10 seconds. • Leave Breakout box installed and processor disconnected, WAC relay disconnected. • DVOM on 200 ohm scale. • Measure resistance between test Pin 37 on Breakout box and WAC relay harness connector Pin 86. • Is resistance 50 ohms or less?	Yes ▶	GO to X61.
	No ▶	SERVICE opens in power to WAC circuit. RERUN Quick Test.
X61 CHECK WAC CIRCUIT FOR SHORTS TO GROUND • Key Off, wait 10 seconds. • Leave Breakout box installed and processor disconnected, WAC relay disconnected. • DVOM on 200 ohm scale. • Measure resistance between test Pin 37 on Breakout box and WAC relay connector harness and test Pin 40 and test Pin 60 on Breakout box and battery ground.	All resistance readings 10,000 ohms or greater ▶	GO to X62.
	Any resistance less than 10,000 ohms ▶	SERVICE shorts to ground. RERUN Quick Test.
X62 CHECK FOR VOLTAGE AT A/C CLUTCH • Key Off, wait 10 seconds. • Leave Breakout box installed. • Processor disconnected. • WAC relay connected. • A/C clutch disconnected. • A/C demanded. • Key On Engine Off. • DVOM on 20V scale. • Measure voltage at A/C clutch harness connection. • 10.5V or greater?	Yes ▶	REPLACE processor. RERUN Quick Test.
	No ▶	REPLACE WAC relay. RERUN Quick Test.

SCORPIO DIAGNOSTIC CHART

Erratic Ignition — Pinpoint Test Y

Y

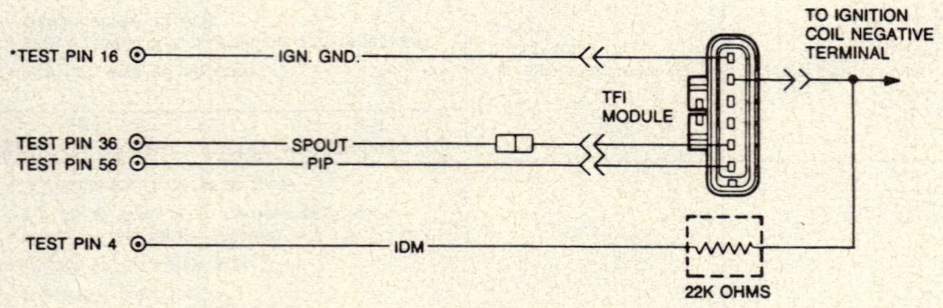

***TEST PINS LOCATED ON BREAKOUT BOX.**
ALL HARNESS CONNECTORS VIEWED INTO MATING SURFACE.

STOP-WARNING

You should enter this Pinpoint Test only when a service code 14 is received in Quick Test Step 6.0.

To prevent the replacement of good components, be aware that the following non-EEC areas may be at fault:

- TFI module.
- Arcing secondary ignition components.
  - — Ignition coil.
  - — Spark plugs and high tension cables.
- Distributor and PIP sensor.

This Pinpoint Test is intended to diagnose only the following:

- Harness circuits: IGN GND, SPOUT, PIP, IDM.

Erratic Ignition — Pinpoint Test Y

TEST STEP	RESULT ▶	ACTION TO TAKE
Y1 CODE 14: ERRATIC IGNITION		
NOTE: Code 14 indicates two successive erratic profile ignition pickup (PIP) pulses occurred, resulting in a possible engine miss or stall. • Check EEC-IV and ignition systems harnesses for: — Loose wires/connectors. — Arcing secondary ignition components (coil, cap, rotor, wires, plugs, etc.). — On-board transmitter (2-way radio).* • Are any of the above present? *Verify all 2-way radio installations. Carefully follow manufacturer's installation instructions regarding the routing of antenna and power leads.	Yes ▶ No ▶	SERVICE as necessary. RERUN Quick Test. GO to Y2.
Y2 CHECK DISTRIBUTOR		
• Key Off, wait 10 seconds. • Deactivate Self Test. • Connect STAR per Quick Test Step 2.0. • Enter Engine Running Continuous Monitor Test (as instructed in Quick Test Step 6.0). • Observe STAR LED for indication of a fault while performing the following: • Lightly tap on TFI module and distributor. • Wiggle TFI connector. • Is a fault indicated?	Yes ▶ No ▶	DISCONNECT and INSPECT connectors. If connector and terminals are good. Check Ignition system. GO to Y3.
Y3 CHECK EEC-IV HARNESS		
• While still in Continuous Monitor Test from Step Y2 observe STAR LED for a fault indication while performing the following: • **While looking for faults grasp the harness close to the TFI connector. Wiggle, shake or bend a small section of the ignition and EEC-IV systems harness while working your way to the other components and dash panel. Also wiggle, shake or bend the EEC-IV harness from the dash panel to the processor. Isolate the PIP circuit if needed for this test.** • **Is a fault indicated?**	Yes ▶ No ▶	ISOLATE fault and SERVICE as necessary. RERUN Quick Test. GO to Y4.

SCORPIO DIAGNOSTIC CHART

Erratic Ignition — Pinpoint Test Y

TEST STEP	RESULT ▶	ACTION TO TAKE
Y4 CHECK PROCESSOR AND HARNESS CONNECTORS		
• Key Off, wait 10 seconds. • Disconnect processor 60 Pin connector and inspect for damaged pins, corrosion, loose wires. • Are connectors and terminals OK? • Reconnect processor when this step is completed.	No ▶	SERVICE as necessary. RERUN Quick Test.
	Yes ▶	Unable to duplicate an erratic ignition fault in the EEC-IV System, for further diagnosis Check Ignition system.

Thermactor Air (Air Management) — Pinpoint Test Z

Z

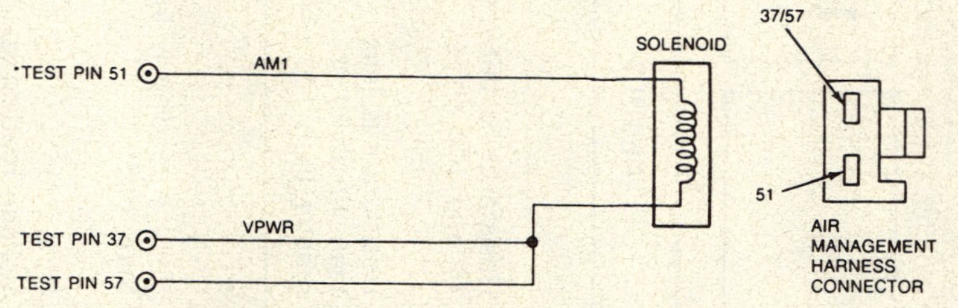

*TEST PINS LOCATED ON BREAKOUT BOX.
ALL HARNESS CONNECTORS VIEWED INTO MATING SURFACE.

STOP-WARNING

You should enter this Pinpoint Test only when you are directed here from the Diagnostic By Symptom chart in this Section.

This Pinpoint Test is intended to diagnose only the following:

- Harness Circuits: AMI and VPWR
- Processor Assembly

SCORPIO DIAGNOSTIC CHART

Thermactor Air (Air Management)	Pinpoint Test	Z

TEST STEP	RESULT ▶	ACTION TO TAKE
Z1 MEASURE AM SOLENOID RESISTANCE		
• Key Off, wait 10 seconds. • DVOM on 200 ohm scale. • Disconnect AM solenoid connector and measure solenoid resistance.	Resistance is between 40 and 100 ohms ▶	CONNECT AM solenoid. GO to Z2.
	Resistance is less than 40 ohms or greater than 100 ohms ▶	REPLACE AM solenoid. RERUN Quick Test.
Z2 CHECK VOLTAGE OF VPWR CIRCUIT		
• Key On, engine Off. • DVOM on 20V scale. • Measure voltage between VPWR at the AM solenoid vehicle harness connector and battery ground.	Voltage reading is less than 10.5V ▶	SERVICE harness open circuit. RERUN Quick Test.
	Voltage reading is 10.5V or greater ▶	GO to Z3.
Z3 CHECK CONTINUITY OF AM CIRCUIT		
• Key Off, wait 10 seconds. • Disconnect processor and inspect both 60 pin connectors and inspect for damaged pins, corrosion, loose wires. Service as necessary. • Connect Breakout box to harness. Leave processor disconnected. • DVOM on 200 ohm scale. • Measure resistance between test Pin 51 at the Breakout box and AM vehicle harness connector.	Resistance reading is 5 ohms or greater ▶	SERVICE open circuit. RERUN Quick Test.
	Resistance reading is less than 5 ohms ▶	GO to Z4.

SCORPIO DIAGNOSTIC CHART

Thermactor Air (Air Management)	Pinpoint Test	Z

TEST STEP	RESULT ▶	ACTION TO TAKE
Z4 CHECK FOR SHORT TO GROUND		
• Key Off, wait 10 seconds. • Leave Breakout box to harness. Leave processor disconnected. • Disconnect AM solenoid. • DVOM on 200,000 ohm scale. • Measure resistance between test Pin 51 and test Pins 40, 46 and 60 at the Breakout box.	All resistance readings are 10,000 ohms or greater ▶	GO to Z5 .
	Any resistance reading less than 10,000 ohms ▶	SERVICE short to ground. RERUN Quick Test.
Z5 CHECK FOR SHORT TO POWER		
• Key Off, wait 10 seconds. • DVOM on 200,000 ohm scale. • Leave Breakout box installed and processor disconnected. • AM solenoid disconnected. • Measure resistance between test Pin 51 and test Pins 37 and 57 at the Breakout box.	All resistance readings are less than 10,000 ohms ▶	SERVICE short to power. REPEAT Quick Test. If code is repeated, REPLACE processor. RERUN Quick Test.
	Any resistance reading is 10,000 ohms or greater ▶	REMOVE Breakout box. REPLACE processor. RERUN Quick Test.

Component Replacement

RELIEVING FUEL SYSTEM PRESSURE

All fuel injected engines are equipped with a pressure relief valve located on the fuel supply manifold. Remove the fuel tank cap and attach fuel pressure gauge T80L–9974–A, or equivalent, to the valve to release the fuel pressure. If a suitable pressure gauge is not available, disconnect the vacuum hose from the fuel pressure regulator and attach a hand vacuum pump. Apply about 25 in. Hg (84 kPa) of vacuum to the regulator to vent the fuel system pressure into the fuel tank through the fuel return hose. Note that this procedure will remove the fuel pressure from the lines, but not the fuel. Take precautions to avoid the risk of fire and use clean rags to soak up any spilled fuel when the lines are disconnected.

THROTTLE BODY

Removal and Installation

1. Disconnect the negative battery cable.
2. Disconnect the Throttle Position (TP) sensor.
3. Remove the air inlet ducts.
4. Disconnect the throttle link retaining clip from the ball stud.

NOTE: Firmly grasp throttle link retaining clip during removal. Failure to do so could result in loss of retaining clip.

5. Identify and note the location and color code of the vacuum hoses next the throttle position sensor. Disconnect the vacuum hoses.
6. Remove the bolts retaining the EGR tube to the throttle body and remove the tube from the throttle body.
7. Carefully separate the throttle body from the upper intake manifold.
8. Remove and discard the gasket between the throttle body and upper intake manifold.

To install:

9. Ensure that both throttle body and upper intake manifold gasket surfaces are clean.

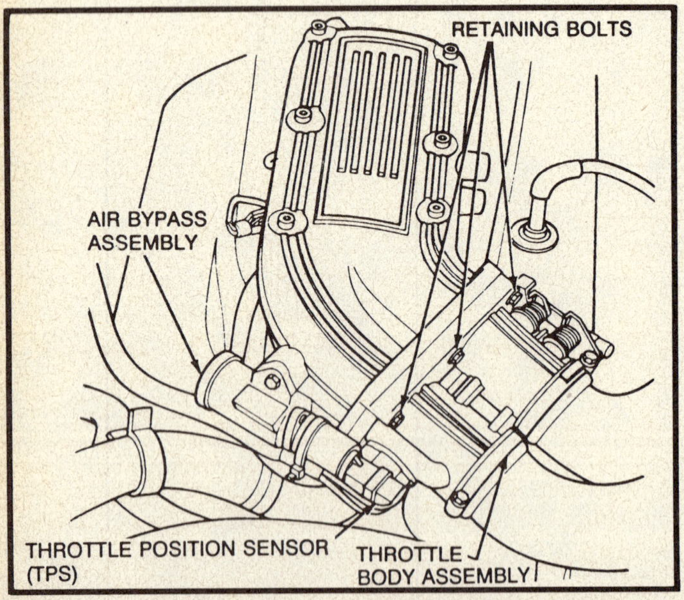

Throttle body removal and installation—2.9L engine

NOTE: If it is necessary to scrape the gasket material off of the throttle body or upper intake manifold, be careful not to damage the gasket surfaces or allow gasket material to drop into the manifold.

10. Position a new throttle body gasket to the throttle body. Place the throttle body to upper intake manifold and install the retaining bolts. Tighten the bolts to 6–8 ft. lbs. (8–11 Nm).
11. Install the vacuum hoses to the fittings next to the throttle position sensor.
12. Connect the throttle position sensor electrical connector.
13. Install the EGR tube and retaining bolts to the throttle body.
14. Install the throttle link and retaining clip.

NOTE: Firmly grasp throttle link retaining clip when installing. Failure to do so could result in loss of retaining clip.

15. Install the air inlet ducts.

THROTTLE POSITION SENSOR

Removal and Installation

1. Disconnect the electrical connector from the throttle position sensor.
2. Bend the washer retaining taps to release the retaining bolts. Remove the throttle position sensor retaining bolts.
3. Remove the throttle position sensor.

To install:

4. Position the sensor to the throttle body with the electrical connector facing forward.
5. Install the sensor and washer plate to the throttle body assembly with the retaining bolts. Tighten the retaining bolts to 10–16 inch lbs. (1.2–1.8 Nm).
6. Bend the washer plate retaining tabs upward to secure the throttle position sensor retaining bolts.

Adjustment

1. Make sure the ignition switch is **OFF**, then install an EEC-IV Breakout Box such as Rotunda T83L-50 EEC-IV, or equivalent.
2. Attach a digital volt/ohmmeter (10 megohms impedance) positive lead (+) to Pin 47 and the negative lead (–) to Pin 46. Set the DVOM on the 20 volt scale.
3. Turn the ignition key to the **ON** position (do not start the engine).
4. Loosen the mounting screws slightly and rotate the TP sensor until the DVOM reads 0.9–1.1 volts.

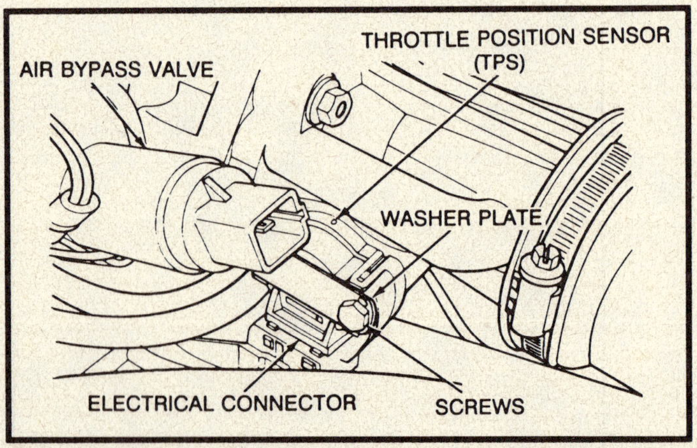

Throttle position sensor removal and installation—2.9L engine

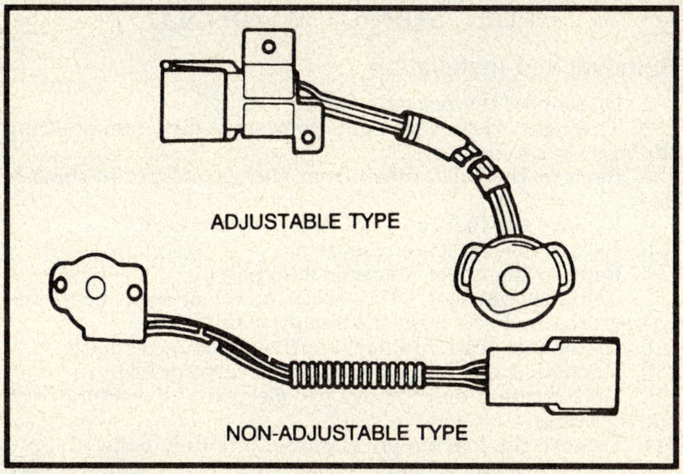

Two types of rotary throttle position sensors

5. Tighten the TP sensor mounting screws to 11–16 inch lbs. (1–1.8 Nm)
6. While watching the DVOM, move the throttle to the wide open position, then back to idle. For proper operation, the DVOM should move from about 1 volt to 4 volts and back to 1 volt during throttle movement.
7. Verify EEC-IV operation by performing the Quick Test.

UPPER INTAKE MANIFOLD

Removal and Installation

1. Disconnect the negative battery cable.
2. Remove the inlet tubes from the air cleaner-to-throttle body.
3. Remove the fuel cap to relieve tank pressure.
4. Relieve fuel system pressure.
5. Disconnect the electrical connectors at the air bypass valve, throttle position sensor and the air charge temperature sensor.
6. Disconnect the throttle link from the ball stud.

NOTE: Firmly grasp the throttle link retaining clip during removal. Failure to do so could result in loss of the retaining clip.

7. Remove the throttle cable bracket.
8. Identify and note vacuum line color codes and locations. Disconnect the upper intake manifold vacuum connectors. Disconnect the vacuum lines from the front and rear of the upper intake manifold including the EGR valve and fuel pressure regulator vacuum lines.
9. Disengage the PCV hose-to-manifold retaining clip. Disconnect the PCV hose from under the manifold.
10. Remove the EGR tube-to-throttle body attaching bolts. Disengage the EGR tube from the throttle body.
11. Remove the upper manifold retaining bolts.
12. Remove the upper intake and throttle body as and assembly from the lower intake manifold.

To install:

13. Clean and inspect the mounting faces of the lower and upper intake manifolds.

NOTE: If scraping is necessary, be careful not to damage the gasket surfaces, or allow material to drop into the manifold.

14. Positon new gaskets on the lower intake manifold and throttle body assembly to lower the manifold making sure the gaskets remain in place. Position the EGR tube to the throttle body.
15. Install the upper intake manifold retaining bolts. Tighten the retaining bolts to 11–15 ft. lbs. (16–20 Nm).
16. Install the EGR tube-to-throttle body retaining bolts.
17. Connect the PCV hose to the bottom of the upper manifold and install the retaining clip.
18. Connect vacuum lines to the vacuum tree fittings at the front and rear intake manifold.
19. Position the throttle cable bracket to the upper intake manifold and install the retaining bolts. Tighten the retaining bolts to 11–15 ft. lbs. (16–20 Nm).
20. Connect the throttle link to the throttle body.

NOTE: Firmly grasp the throttle link retaining clip when installing. Failure to do so could result in loss of the retaining clip.

21. Connect the electrical connectors at the air bypass valve, throttle position sensor and the air charge temperature sensor.
22. Install the fuel cap.
23. Connect the negative battery cable.
24. Turn the ignition **ON** and **OFF** several times without starting the engine to pressurize the fuel system and check for leaks.

NOTE: Check all connections at the fuel supply manifold, the injectors and the fuel pressure regulator. Ensure the push connect fitting is properly and securely assembled.

CAUTION

The fuel system is normally pressurized to 40 psi (270 kPa) observe care when looking for leaks.

25. Install the air inlet tubes to the throttle body.
26. Start the engine and allow it to reach operating temperature.
27. Perform a EEC-IV self test to check system functions.

LOWER INTAKE MANIFOLD

Removal and Installation

1. Disconnect the negative battery cable.
2. Remove the air inlet hoses from the throttle body and air cleaner.
3. Disconnect the throttle cable from the linkage at the throttle body.
4. Disconnect the throttle cable bracket retaining bolts. Set throttle cable and bracket aside.
5. Remove the fuel lines from the fuel supply manifold and pressure regulator.
6. Disconnect the following connectors:
   a. Air charge temperature sensor (ACT)
   b. Idle speed control valve
   c. Throttle position sensor (TP)
   d. Engine coolant temperature sensor (ECT)
7. Note location, label and remove vacuum hoses from the intake manifold plenum and throttle body.
8. Drain the radiator coolant. Remove the hose from the thermostat housing to radiator.
9. Remove the EGR tube-to-throttle body attaching bolts. Detach the EGR tube from the throttle body.
10. Remove the bolts retaining the plenum to the intake manifold. Remove the throttle body/plenum assembly.
11. Remove the distributor cap and wires as an assembly. Detach the distributor wiring connector.
12. Observe and mark the location of the distributor rotor and housing. Remove the distributor hold-down screw and clamp. Lift out the distributor.
13. Remove the rocker cover retaining bolts, the load distribution washers and the rocker covers.
14. Remove the intake manifold attaching bolts and nuts.

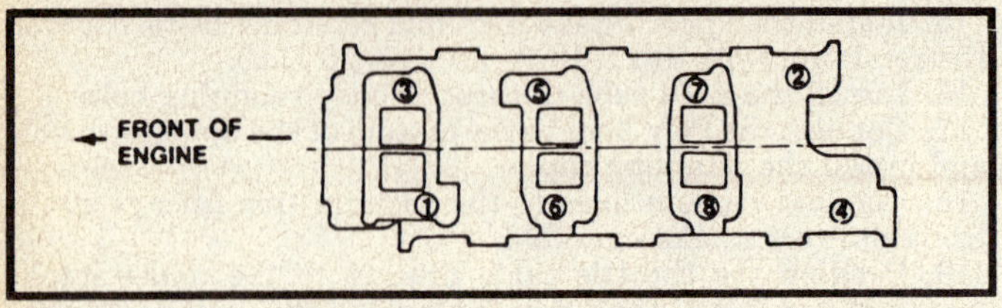

Tightening sequence for the lower intake manifold—2.9L engine

Note the length of the manifold attaching bolts to ensure installation in the original positions.

15. Tap the manifold lightly with a plastic mallet to break the seal. Remove the manifold.
16. Remove all old gasket material and sealing compound. Be careful not to drop any material into the manifold, plenum or cylinder heads.

To install:

17. Ensure that all sealing surfaces are clean.
18. Apply an appropriate sealing compound to the joining surfaces. Position the intake manifold gasket to the intake manifold. Ensure the tab on the right cylinder head gasket fits the cutout on the intake manifold gasket.
19. Position the intake manifold in place. Install and hand tighten the intake manifold retaining bolts.
20. Hand tighten the nuts onto the intake manifold retaining studs at locations 3 and 4.
21. Tighten the intake manifold bolts and nuts in sequence in 4 steps:
    a. Step 1—3–6 ft. lbs. (4–8 Nm)
    b. Step 2—6–11 ft. lbs. (8–15 Nm)
    c. Step 3—11–15 ft. lbs. (15–21 Nm)
    d. Step 4—15–18 ft. lbs. (21–25 Nm)
22. Install the rocker cover gaskets, and rocker covers, the load distribution washers and retaining bolts. Tighten the rocker cover retaining bolts to 4–5 ft. lbs. (5–8 Nm).
23. Install the distributor, clamp and attaching bolt. Install the distributor wiring connector.
24. Install the distributor cap and wires.
25. Position the throttle body/manifold and install the retaining bolts. Tighten the retaining bolts in to steps:
    Step 1, tighten to 7 ft. lbs. (10 Nm)
    Step 2, tighten to 15–18 ft. lbs. (21–25 Nm).
26. Position the EGR tube to the throttle body and install the retaining bolts. Tighten the retaining bolts to 6–9 ft. lbs. (8–12 Nm).
27. Install the fuel lines.
28. Install the coolant hose from the radiator to thermostat housing.
29. Install the vacuum hoses to the intake manifold, plenum and throttle body.
30. Connect the following electrical connectors:
    a. Air Charge Temperature (ACT) sensor
    b. Idle speed control valve
    c. Throttle Position (TP) sensor
    d. Engine Coolant Temperature (ECT) sensor
31. Position the throttle cable bracket to the manifold plenum and install the attaching bolts.
32. Install the throttle cable to the throttle linkage.
33. Install the air inlet hoses and clamps to the throttle body and air cleaner.
34. Fill and bleed the cooling system.
35. Connect the negative battery cable.
36. Start the engine and set the ignition base timing.
37. Run the engine at fast idle. Check for coolant leaks.
38. After the engine has run for awhile at operating temperature, turn it off and allow it to cool down.
39. Retighten the intake manifold bolts again in sequence.

FUEL SUPPLY MANIFOLD

Removal and Installation

1. Disconnect the negative battery cable.
2. Throughly clean the engine to prevent dirt from entering the injector cavities.
3. Remove the inlet tubes from the air cleaner-to-throttle body.
4. Remove the fuel cap to relieve tank pressure.
5. Relieve fuel system pressure.
6. Remove the upper manifold assembly.
7. Using special tool T81P–19623 or G1 or equivalent, disconnect the fuel line from the supply manifold.
8. Disconnect the fuel line from the pressure regulator.
9. Disconnect the pressure regulator vacuum line.
10. Note location and remove the electrical connectors from the injectors.
11. Remove the fuel supply manifold retaining bolts.

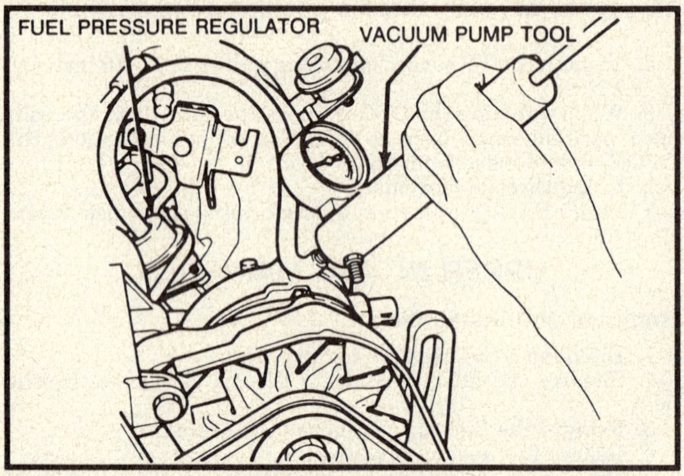

Relieving fuel system pressure using and vacuum pump tool

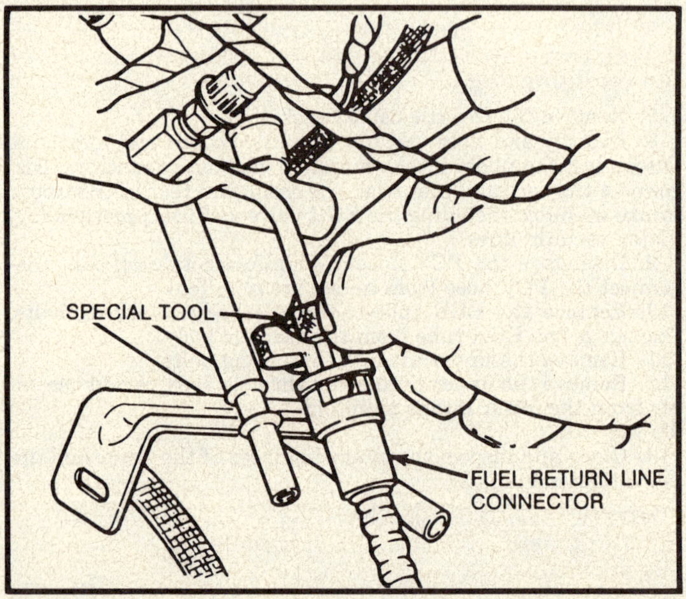

Disconnecting the fuel line coupler using a special tool

12. Remove the fuel supply manifold and injectors as an assembly.
13. Wipe the injector cups with a clean, dry cloth.

To install:

14. Ensure the injector tips are clean and free of contamination.
15. Remove and discard the fuel injector O-rings and install new O-rings.
16. Carefully wipe the injector recesses in the fuel supply manifold. The recesses must also be free of any contamination.
17. Lightly lubricate the recesses using light weight engine oil.

NOTE: Never use silicone lubricant. It will clog the injectors.

18. If the fuel injectors have been separated from the fuel supply manifold, position 3 injectors into the right side recesses of the fuel supply manifold. Locate 3 on the left side.
19. Attach the fuel injectors to the fuel supply manifold using new retaining clips.
20. Press down firmly on the fuel supply manifold and injector assembly until the retaining bolt locations align with the tapped holes in the lower manifold. Install the fuel supply manifold retaining bolts. Tighten the retaining bolts to 76–88 inch lbs. (9–10 Nm).
21. Install the pressure regulator vacuum hose.
22. Install the injector electrical connectors, ensuring they are installed into their original positions.
23. Install the fuel lines to the supply manifold and pressure regulator.
24. Install the upper manifold assembly.
25. Install the fuel cap.
26. Connect the negative battery cable.
27. Turn the ignition **ON** and **OFF** several times without starting the engine to pressurize the fuel system and check for leaks.

NOTE: Check all connections at the fuel supply manifold, the injectors and the fuel pressure regulator. Ensure the push connect fitting is properly and securely assembled.

CAUTION

The fuel system is normally pressurized to 40 psi (270 kPa); observe care when looking for leaks.

28. Install the air inlet tubes to the throttle body.
29. Start the engine and allow it to reach operating temperature.
30. Perform a EEC-IV self test to check system functions.

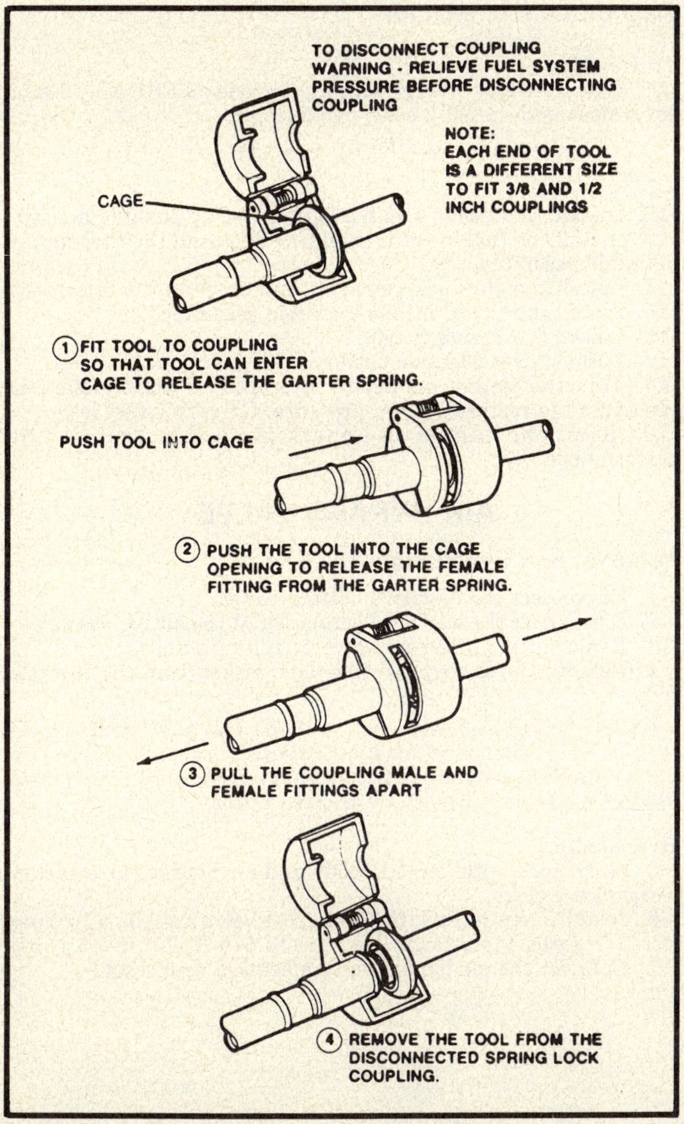

Fuel line coupling removal and installation tools

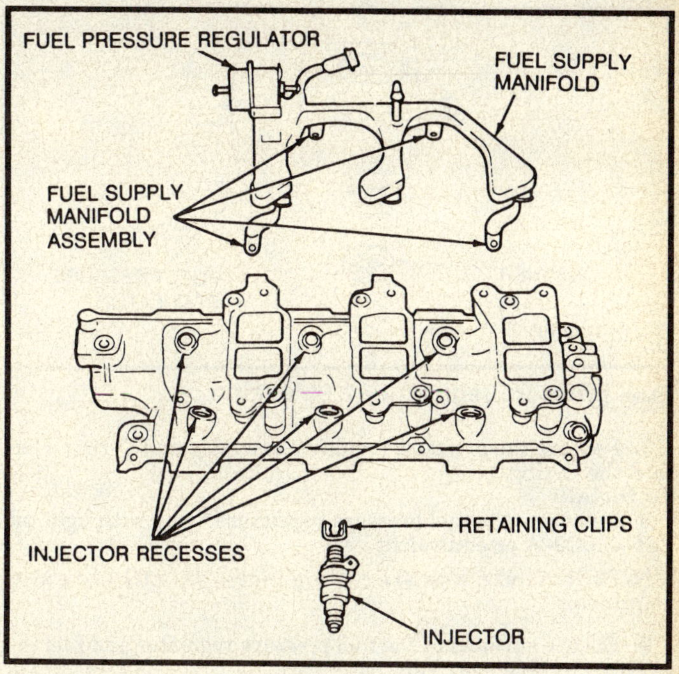

Fuel supply manifold removal and installation—2.9L engine

PRESSURE REGULATOR

Removal and Installation

1. Disconnect the negative battery cable.
2. Remove the fuel tank cap to release tank pressure.
3. Relieve fuel system pressure from the fuel lines.
4. Remove the vacuum line at the pressure regulator.
5. Remove the pressure regulator retaining bolts from the regulator.
6. Remove the fuel line from the pressure regulator.

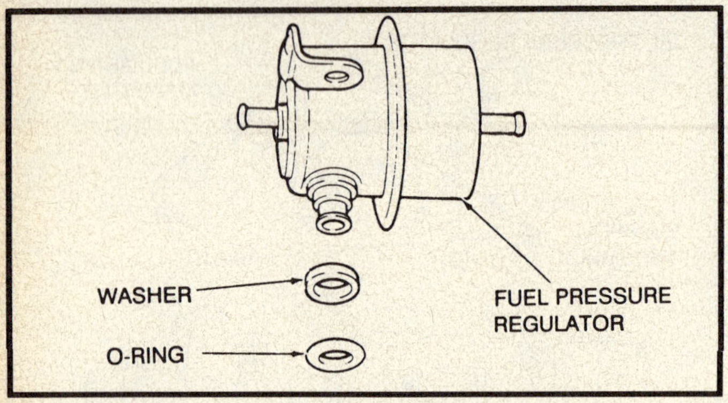

Fuel pressure regulator—2.9L engine

7. Remove the pressure regulator assembly and O-ring. Discard the O-ring.

To install:

8. Lubricate the fuel pressure regulator O-ring with light oil ESF–M6C2–A or equivalent.

NOTE: Never use silicone grease. It will clog the regulator.

9. Ensure surfaces of the fuel pressure regulator and fuel injection manifold are clean.
10. Install the new washer and O-ring on the regulator.
11. Position the fuel pressure regulator onto the fuel supply manifold and press firmly until fitting is fully engaged. Tighten the retaining screws to 71–102 inch lbs. (8–11 Nm).
12. Install the fuel line to the regulator.
13. Install the fuel tank cap.
14. Connect the negative battery cable.
15. Turn the ignition **ON** and **OFF** several times without starting the engine to restore system pressure. Check for fuel leaks.
16. Start the engine and test performance.

FUEL INJECTOR

Removal and Installation

1. Disconnect the negative battery cable.
2. Remove the fuel tank cap to release tank pressure.
3. Relieve fuel system pressure from the fuel lines.
4. Remove the upper intake manifold from the engine.
5. Note location and carefully remove the electrical harness connectors from each injector.
6. Remove the fuel supply manifold and the injector assembly from the engine.
7. Remove and discard the injector-to-fuel supply manifold retaining clip.
8. Grasping the injector body, pull outward while gently rocking the injector from side to side.
9. Remove and discard both injector O-rings. Install new O-rings.
10. Inspect the injector plastic hat covering the pintle head and washer for signs of deterioration. Replace injector if inspection reveals any signs of damage or deterioration.

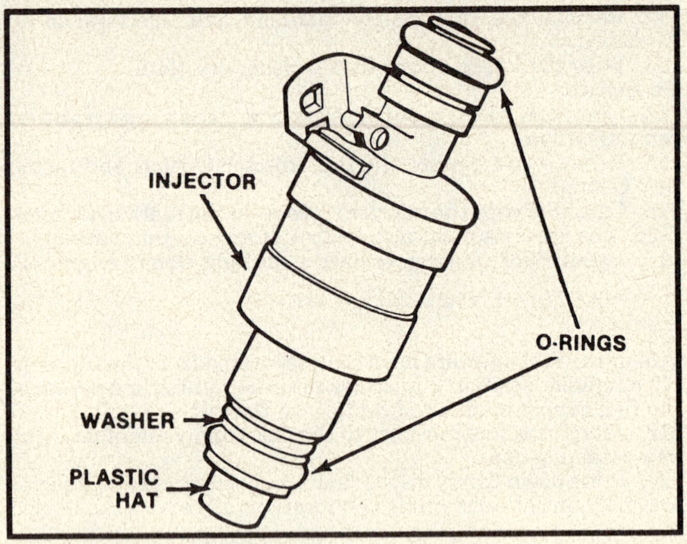

View of the fuel injector

To install:

11. Lubricate the O-rings with light grade oil ESF-M6C2-A or equivalent and install 2 seals per injector.

NOTE: Never use silicone grease. It will clog the injectors.

12. Install the injectors with a light, twisting pushing motion.
13. Install the fuel injector retaining clips and the fuel supply manifold assembly.
14. Install the electrical harness connectors to the injectors.
15. Install the upper intake manifold assembly.
16. Install the fuel tank cap.
17. Connect the negative battery cable.
18. Turn the ignition on and off several times without starting the engine to restore system pressure. Check for fuel leaks.
19. Start the engine and check for leaks. Test engine performance.

AIR BYPASS VALVE

Removal and Installation

1. Disconnect the negative battery cable.
2. Disconnect the electrical connector at the air bypass valve.
3. Remove the air bypass valve attaching screws.
4. Remove the air bypass valve and gasket from the throttle body.

NOTE: If it is necessary to scrape the gasket material off of the gasket surfaces, be careful not to damage the air bypass valve or throttle body gasket surface or drop gasket material into the throttle body.

To install:

5. Make sure both throttle body and air bypass valve gasket surfaces are clean.
6. Install a new gasket, the air bypass valve and the attaching screw. Tighten the attaching screws to 6–8 ft. lbs. (8–11 Nm).
7. Connect the air bypass valve electrical connector.

Peugeot Fuel Injection

11 SECTION

PEUGEOT

General Information

Peugeot employs 3 distinct Bosch fuel injection systems across the model line. While each system has similarities to the others, each also has significant differences. All fuel systems are computer-controlled but the amount of computer control varies by system. Each Electronic Control Unit (ECU) is unique to its system; the controllers cannot be interchanged.

Peugeot's gasoline engines include a variety of 4-cylinder models as well as a V-6. The 4-cylinder engines appear in SOHC, DOHC and turbocharged form. The V-6 engine did not appear in the 1991 models; only 4-cylinder gasoline engines were available in the US market. Peugeot withdrew from the US market at the end of 1991; no 1992 cars were imported.

LU JETRONIC

This system is found on the 2155cc 4-cylinder turbocharged engine used in the 505 T, SW8 and Turbo models. The system contains an air flow meter which provides one of the primary signals to the control unit.

Its intermittent, low pressure fuel injection system is controlled primarily by the amount of air drawn in by the engine and the rpm signal. Fuel is injected twice per engine cycle.

The primary signals are modified by signals from other sensors such as air and coolant temperature, oxygen (Lambda) sensor, and the overboost pressure switch. A fuel pressure regulator maintains a constant fuel pressure for a given intake manifold pressure. A separate ignition ECU provides engine rpm signals to the engine ECU.

The fuel system consists of 2 types of fuel pump; a primer (scavenger) pump, located in the fuel tank and an in-line (main) pump. This system provides easier start-up by avoiding vapor lock in the fuel lines between the tank and the main pump.

The LU Jetronic system is equipped with compensator systems which override the normal computations of the controller in certain circumstances.

- Fuel Injection Cut-Out On Deceleration—the injection ECU will disable fuel injector operation if the engine temperature is over 140°F (60°C) during closed-throttle deceleration at 1600 rpm or above. This avoids unburned hydrocarbons. The injection is re-established when the micro-switch on the throttle body (idle switch) is opened or at 1600 rpm.
- Full Load Enrichment—occurs when an enrichment signal input is received from the ignition ECU or from the sensor flap in the air flow sensor.
- High RPM Injection Cut-Out—at engine speeds above 6500 rpm, the injection ECU suspends the signals to the injectors. However, the ignition system continues to operate in order to burn off all the fuel vapors remaining in the cylinders.
- Altitude Correction—an altitude sensor provides the injection ECU with altitude data. The ECU will reduce the duration of the basic injection time at high altitudes.
- Fuel Injection Cut-Out—in the event of over-boost (when the intake manifold pressure is above 16±0.7 psi or 110±4.8 kPa), the overboost pressure switch signals the injection ECU. The ECU disables the injectors until the engine speed drops back to 1100 rpm.
- Engine Temperature Sensor—this device supplies the ECU with temperature information necessary to correct the quantity of air/fuel mixture during engine warm-up.
- Richness Correction—this regulation is based on a Lambda or oxygen sensor, which constantly measures the engine exhaust oxygen content.
- Cold Engine Compensators—the components in this group provide extra air and fuel when signaled into action by the ECU. The cold start injector is mounted to the auxiliary air channel. It is is controlled by the starter solenoid circuit. The cold start injector provides the engine with enrichment to compensate for the condensation of fuel against the cold combustion chamber walls and intake manifold when the engine is cold. The auxiliary air device is an auxiliary air valve located in the air by-pass circuit separate from the throttle body. Its air flow is taken upstream of the throttle plate and, therefore, accounted for by air sensor. The thermo-time switch closes and opens the cold start injector circuit in relation to engine temperature.

LU-2 AND LH-2.2 JETRONIC

The LU-2 Jetronic fuel injection system is almost identical to the LU Jetronic system; the cold start enrichment function is incorporated in the ECU eliminating the cold start injector and the thermo-time switch. During cold starting, the enrichment is calculated by the ECU based on engine coolant temperature, cranking duration and engine speed. Additionally, the engine rpm signal is provided to the ECU by the ignition coil.

The fuel injection cut-out on deceleration feature on the LU-2 and LH-2.2 Jetronic fuel injection system will discontinue fuel injection on deceleration at engine speeds above 1300 rpm.

The LH-2.2 Jetronic system uses a hot wire air mass meter rather than an air flow meter but is otherwise similar to the LU system. The air mass sensor uses a very fine platinum wire stretched across the path of the air flow. The wire is maintained electrically at a temperature 248°F (120°C) above the temperature of the air entering the intake. The intake air cools the wire.

The resistance of the wire varies with its temperature. The electronic circuit senses these variations and corrects the current flow, maintaining the wire at a constant temperature. The amount of current required to maintain the wire at a constant temperature is proportional to the amount of air drawn in. This information is sent to the ECU as a voltage signal.

On the LH-2.2 system, engine rpm signals are provided to the engine ECU by both the ignition ECU and an engine speed sensor. The LU-2 system is found on the 2.2L 4-cylinder engine used in the 505 2.2 S, DL and 1991 SW8 2.2i models.

M1.3 MOTRONIC

In order to optimize the fuel injection system and the ignition control, both functions have been regrouped into a single ECU and called the Motronic. The system operation is very similar to that of the LU Jetronic fuel injection system described previously. However, the Motronic system incorporates the following changes.

- A single ECU controls both injection and ignition. An engine sensor provides information on engine position and rpm.
- A double-throat throttle body with staggered or simultaneous openings depending on the model.
- A throttle position switch providing the ECU with idle position or full load inputs.
- Cold start enrichment incorporated in the ECU as with LU-2 system.
- An ECU-controlled idle regulation electrovalve stabilizes the idle speed regardless of engine load.
- A detonation sensor compensates for detonation, avoiding the risk of damaging the engine.
- An ECU-controlled fuel vapor recovery canister purge electrovalve recirculates fuel vapors from the fuel system.
- An on-board diagnostic system which enables the decoding of operating faults during engine operation.
- The ECU adjusts richness directly relative to the Lambda sensor input. This eliminates the need for idle mixture or speed adjustments.

The Motronic system is used on the 405 DL, S, Wagon and MI16 models. The SOHC 1.9L engines use a simplified version of the Motronic system; these versions will not have the detonation sensor or electronic idle control valve. An auxiliary air device is used to control the idle in the same manner used by the LU-2 Jetronic system.

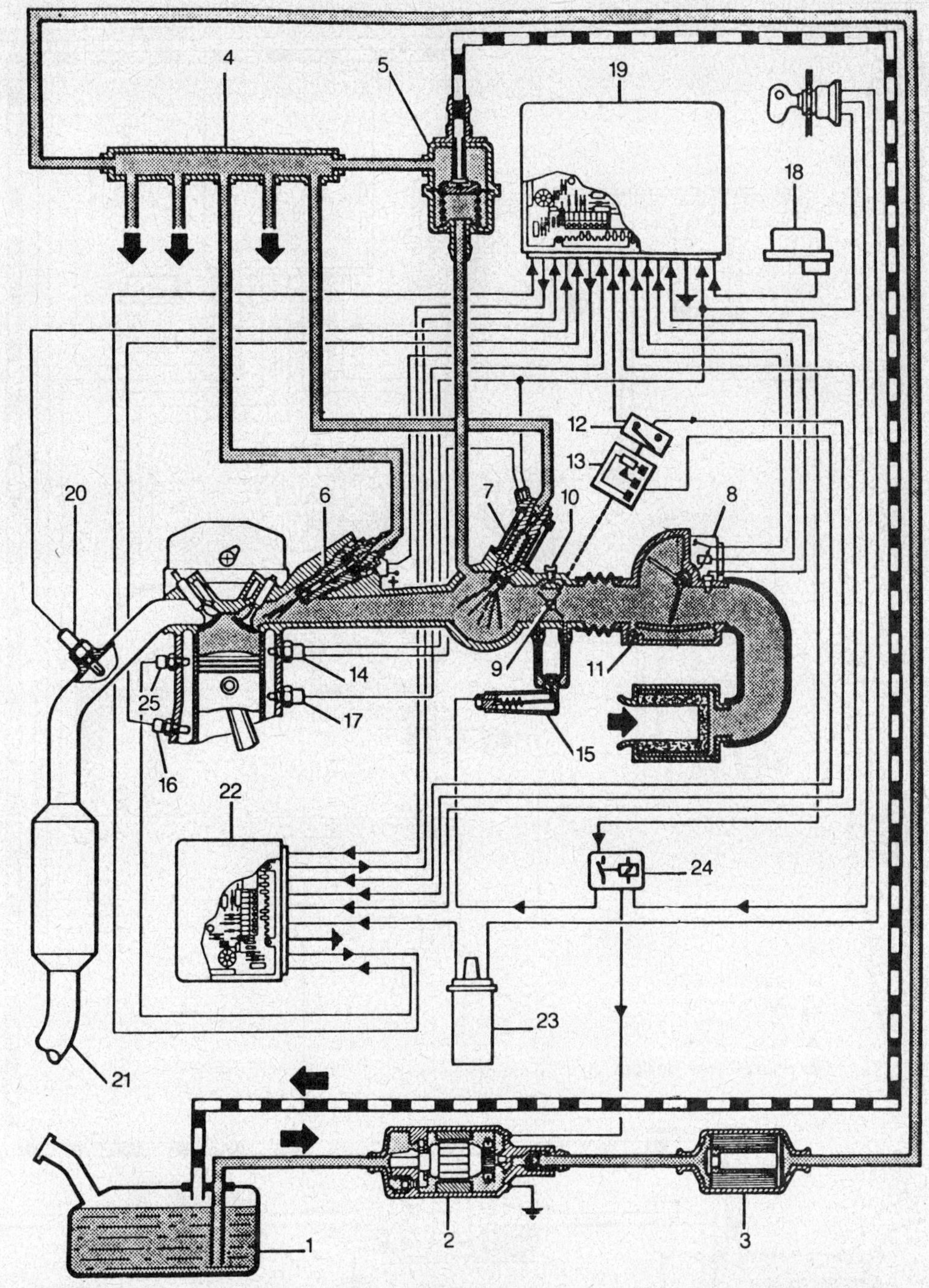

1. Fuel tank
2. Fuel pump
3. Fuel filter
4. Injector distribution pipe
5. Fuel pressure regulator
6. Injector
7. Cold start injector
8. Air flow sensor
9. Throttle body
10. Idle adjusting screw
11. Richness adjusting screw
12. Microswitch
13. Potentiometer
14. Thermo-time switch
15. Auxiliary air device
16. Temperature sensor
17. 60° thermocontact
18. Altitude sensor
19. Injection ECU
20. Lambda sensor
21. Catalytic converter
22. Ignition ECU
23. Ignition coil
24. Fuel pump relay (tachymetric relay)
25. Detonation sensor

Fuel system schematic – LU Jetronic

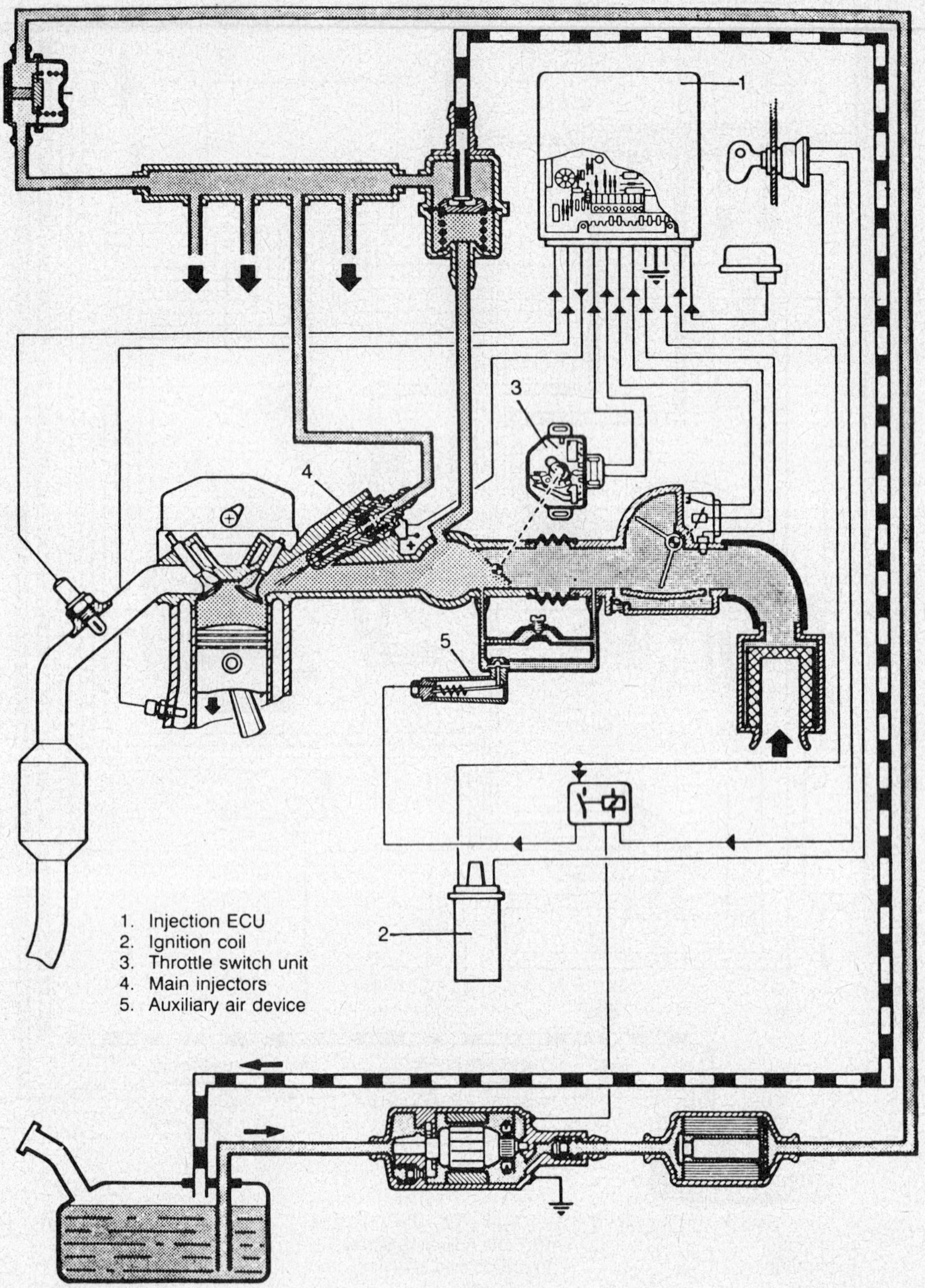

Fuel system schematic—LU-2 Jetronic

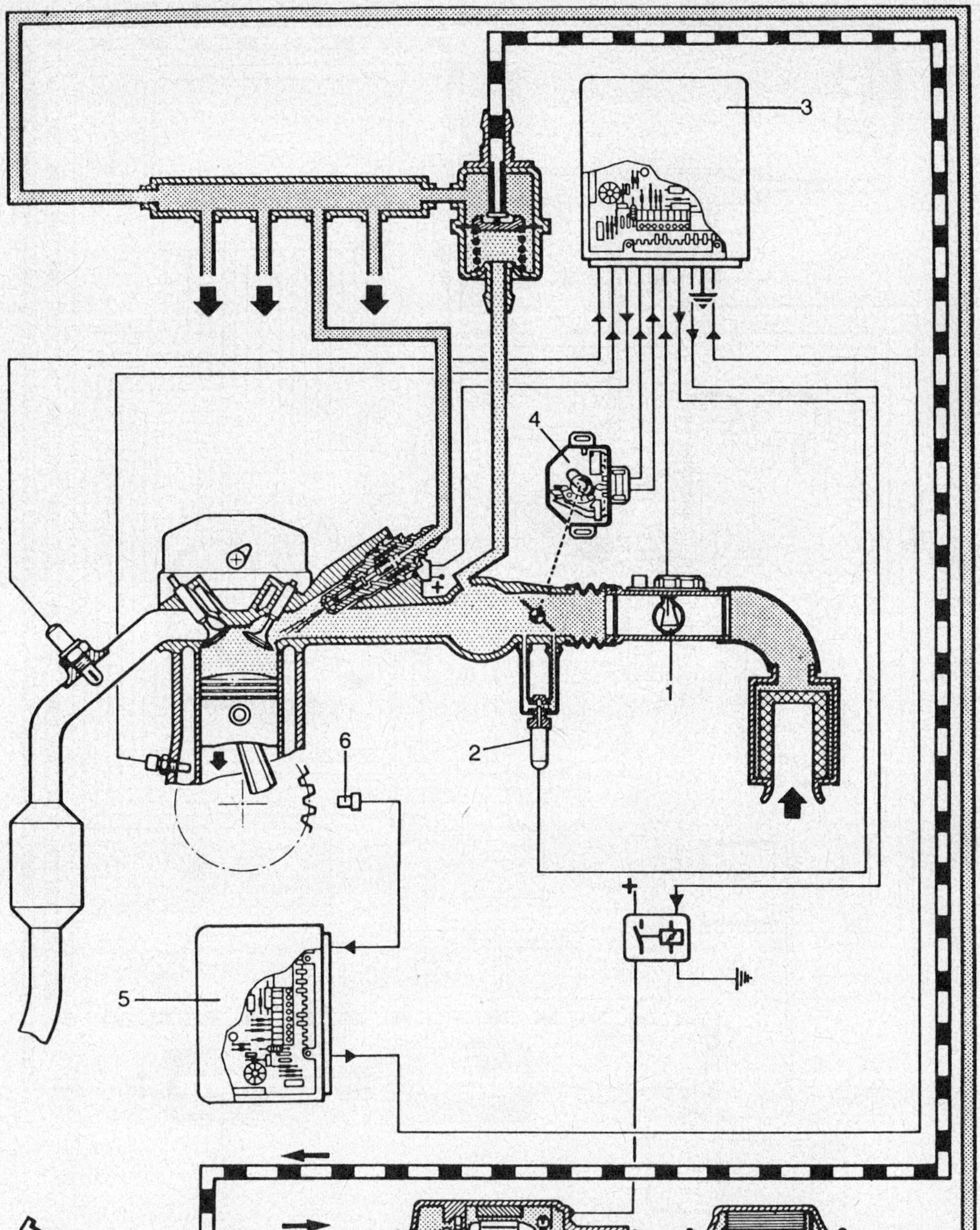

1. Hot wire air flow sensor
2. Electronic idle regulator
3. Injection ECU
4. Throttle switch unit
5. Ignition ECU
6. Engine speed sensor

Fuel system schematic—LH-2.2 Jetronic

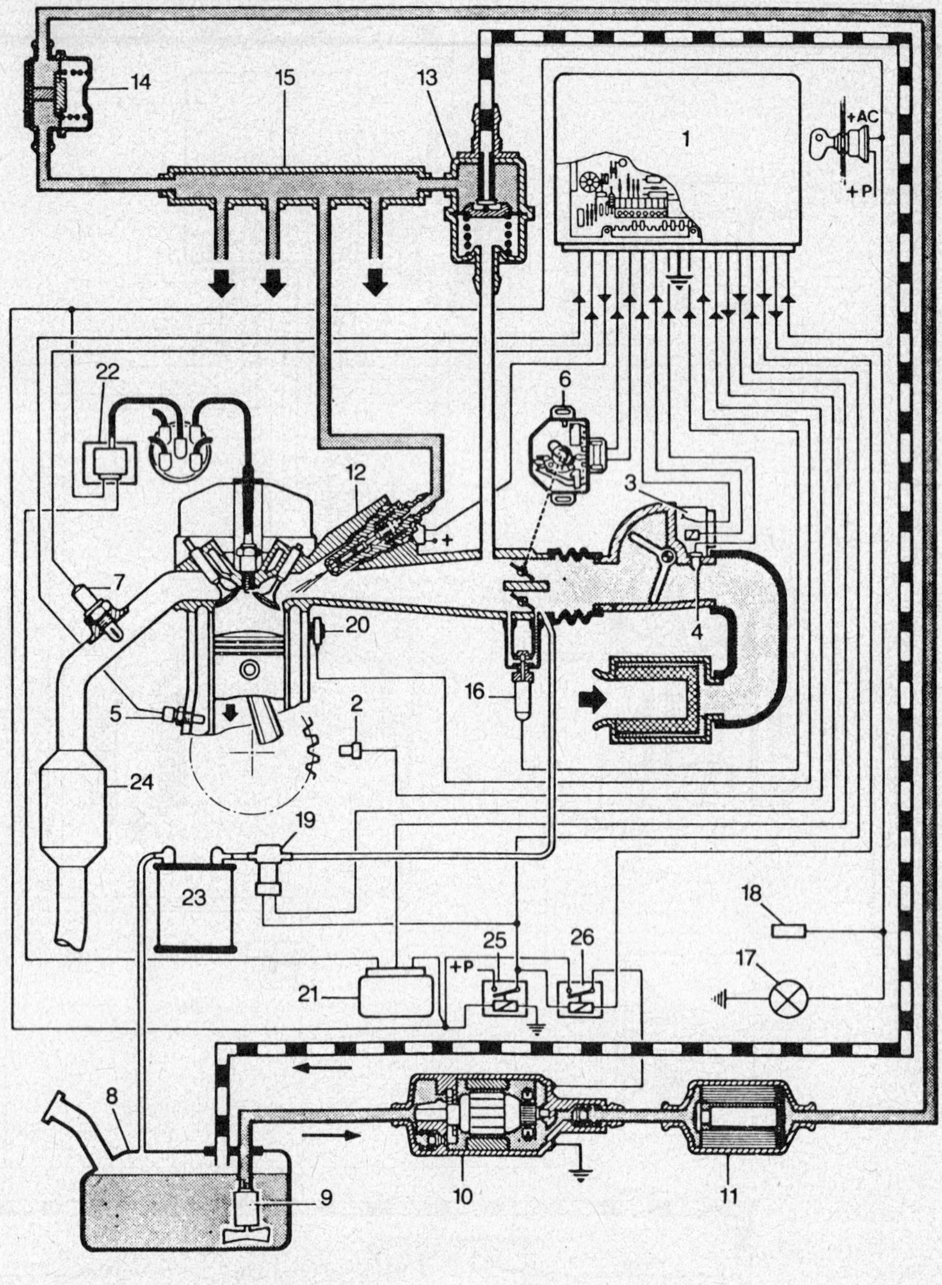

1. Electronic Control Unit (ECU)
2. TDC sensor
3. Air flow sensor
4. Intake air temperature sensor
5. Engine temperature sensor
6. Throttle switch unit (throttle flap position)
7. Lambda sensor
8. Fuel tank
9. Primer (scavenger) pump
10. Fuel pump
11. Fuel filter
12. Injector
13. Fuel pressure regulator
14. Pulsation damper
15. Injector distribution pipe
16. Idle regulation electrovalve
17. Check engine light
18. System test tap connector
19. Canister purge electrovalve
20. Detonation sensor
21. Ignition module
22. Ignition coil
23. Canister
24. Catalytic converter
25. Main relay
26. Fuel pump relay

Fuel system schematic – M1.3 Motronic

DATA SENSORS

Air Flow Sensor

Used on all systems except the LH-2.2, the air flow sensor advises the ECU of the amount of air entering the engine. Since this quantity is proportional to the engine load, the ECU can command the proper fuel delivery based on the intake air. The sensor contains a flap which is moved by the passing air column. The flap pivot is connected to a potentiometer in a sealed housing. As the flap changes position, the potentiometer changes the voltage signal being returned to the ECU. This signal is also modified by the air temperature sensor installed in the air flow sensor circuit.

The throttle flap housing contains a bypass channel, allowing a small amount of air to pass without moving the flap. The air flow in this channel can be changed by moving the richness screw on the housing.

Additionally, the housing contains a microswitch used to signal the fuel and ignition ECUs when the sensor flap is closed; this data is used for both idle and fuel cut-off control. An idle compensating electrovalve is also mounted on the housing and is used to stabilize the idle during air conditioning operation.

Hot Wire Air Mass Sensor

LH is an abbreviation of Luftmassenmes ser Hitzdraht (hot-wire air mass meter). The air mass sensor uses a very fine platinum wire stretched across the path of the air flow. The housing also contains an intake air temperature sensor. The hot wire is maintained electrically at a temperature 248°F (120°C) above the temperature of the air entering the intake. The intake air cools the wire.

The resistance of the wire varies with its temperature. The electronic circuit senses these variations and corrects the current flow, maintaining the wire at a constant temperature. The amount of current required to maintain the wire at a constant temperature is proportional to the amount of air drawn in. This information is sent to the ECU as a voltage signal.

Each time the ignition is switched **OFF**, the bridge circuit is altered for approximately 1 second; the hot wire is momentarily heated to a temperature of 1832°F (1000°C) to burn free any deposits which could influence the resistance or measuring results. The signal for this function is provided automatically by the injection ECU but will not occur if the engine did not achieve an operating temperature of at least 149°F (65°C) and a speed above 1500 rpm during its last operating cycle.

Ignition ECU

On all LU and LH systems, the fuel injection ECU receives input information from the ignition ECU regarding engine position and rpm. The output pulses to the fuel injectors are based on the input signals from the ignition ECU.

Since the Motronic system uses an integrated fuel and ignition ECU, no signal exchange is necessary in this system.

Temperature (NTC) Sensors

The ECUs rely on engine temperature signals to determine operating conditions and fuel delivery corrections. The coolant temperature sensors are sometimes referred to as NTC (Negative Temperature Coefficient) sensors; their resistance decreases as engine temperature increases.

Altitude Sensor

The atmospheric pressure changes with variations in altitude as well as with weather conditions. This sensor converts atmospheric pressure to an electrical signal; the ECU uses this signal to alter fuel delivery.

Oxygen or Lambda Sensor

Lambda is the letter of the Greek alphabet used to denote the deviation of the air/fuel ratio from the ideal point of 14.7:1. Ideal conditions are maintained when Lambda = 1; ratios above or below 1 indicate lean or rich conditions respectively. The oxygen sensor generates a voltage relative to the oxygen content of the exhaust stream. By interpreting this voltage signal, the ECU determines that the air/fuel mixture is richer or leaner than necessary and corrects the fuel delivery accordingly.

The voltage supplied by the sensor to the ECU ranges from approximately 0.1 volt when the air/fuel mixture is very lean to

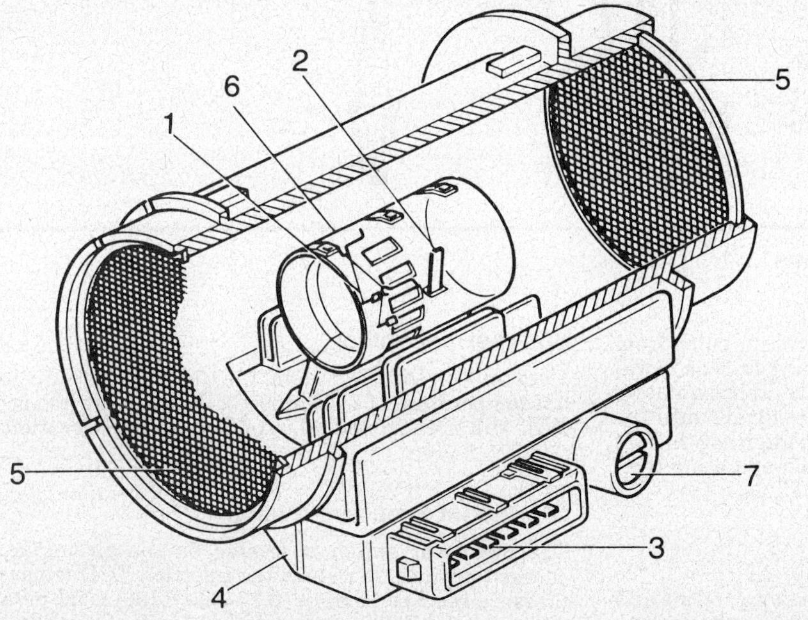

1. Hot wire(made of platinum)
2. Air temperature sensor
3. Hybrid electronic circuit
4. Housing
5. Protective grilles
6. Support
7. Idle richness adjusting screw

Hot wire air mass sensor components

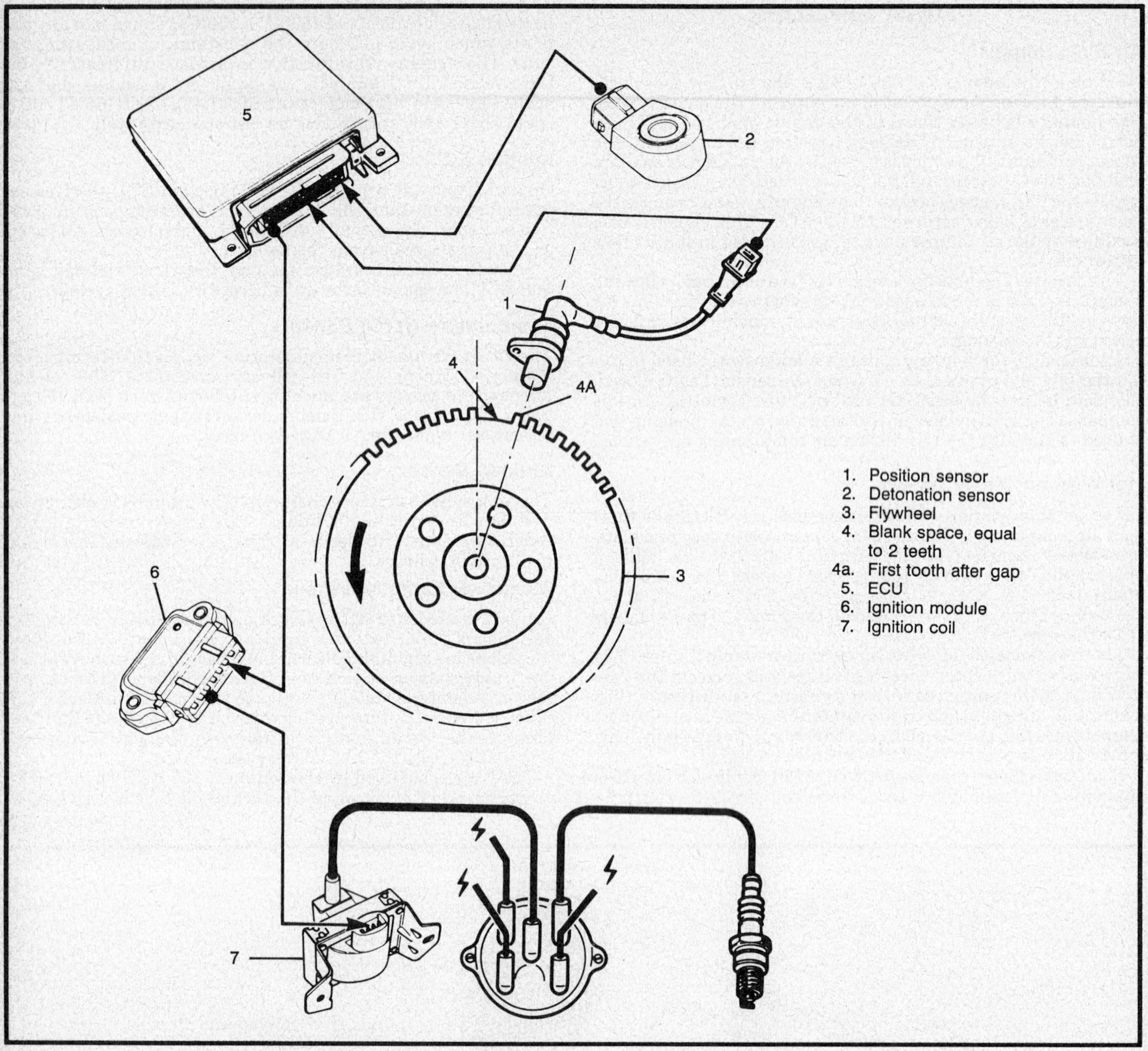

Crank and rpm sensor system – Motronic

1.0 volt if the mixture is very rich. The transition point from rich to lean occurs at 0.45–0.50 volts. The oxygen sensor will only generate accurate voltages when above approximately 635°F (300°C); the ECU will disregard sensor signals until the sensor is fully warm. Some oxygen sensors are internally heated by an electric element, allowing them to reach operating temperature earlier in the driving cycle.

Thermo-time Switch

Used only on the LU system, the thermo-time switch opens and closes the circuit to the cold start injector based on the coolant temperature. The cold start injector's operation is limited during cranking by the heating of a bi-metallic strip.

60° Thermocontact

Used only on the LU system, the thermocontact opens at a coolant temperature of 140°F (60°C). When the switch is closed, the ECU will not enable fuel cut-off during deceleration.

Overboost Pressure Switch

The overboost switch is located on the air intake of turbocharged engines. It signals the injection ECU when the intake pressure reaches 15.95 ± 0.73 psi (1100 ± 50 mbar). At this pressure, the ECU cuts injector operation, limiting the engine to 3000 rpm. Injector function is re-established when the boost pressure drops below 15.22 psi (1050 mbar).

Detonation (Knock) Sensor

This sensor, used on the Mi16 and turbocharged engines, detects detonation (knock or ping) in the engine. On LU systems, the signal is sent to the ignition ECU which immediately retards the ignition advance. On the Motronic system, the signal is sent to the combined controller with the same result.

Crank Sensor/RPM sensor

The rpm signal and crankshaft position signals for the Motronic 1.3 system originate through a sensor mounted over the flywheel. The flywheel has 60 teeth around it; a blank space equal to 2 teeth is used as a reference mark. As the first tooth after the reference mark passes, cylinder No. 1 will be 115 degrees BTDC. Engine speed is computed from the time needed for one rotation of the reference mark.

The ECU uses the engine speed and load information to compute the ignition angle based on mapped or pre-programmed values. Once the angle is calculated, the ECU determines point at which the cylinder must fire. When the appropriate number of teeth have passed the sensor, the ECU signals the ignition module which triggers the coil primary circuit.

Switch Signals

The ECU for each fuel system receives On/Off signals from various switches and circuits including the air conditioning system, the ignition switch or starter circuit. The status of these switches and circuits cause the ECU to make necessary corrections to fuel delivery and/or idle speed to compensate for additional engine load or driving conditions.

CONTROLLED OUTPUTS

Fuel Injectors

The fuel injectors, 1 per cylinder, are located at the intake ports near the intake valve. The injectors are electrically triggered by the ECU. The injector contains a solenoid armature connected to a needle valve. An internal spring forces the needle valve into the closed position when the injector is not operating. When current is passed through the solenoid winding, the needle valve lifts, opening approximately 0.1 mm. Fuel from the fuel rail is allowed to flow through the injector; both the pintle or tip of the needle and the walls of the opening in the injector are designed to promote fuel atomization.

Cold Start Injector

Used only on the LU system, the cold start injector is mounted on the intake manifold and delivers extra fuel during cold starting. The operation of the cold start injector is controlled through the thermo-time switch in proportion to engine temperature. In cases of repeated starting attempts or if actual starting takes too long, the thermo-time switch cuts off the cold start injector to prevent flooding.

Main and Fuel Pump Relays

These 2 relays work together to power both the fuel management electronics and the fuel pumps. The relays respond to the ignition switch **ON** and starter signal as well as a signal from the air flow sensor. In the event that the engine stops with the ignition **ON**, a micro-switch in the sensor disconnects the relay and shuts off the fuel pumps. This prevents fuel delivery in case of an accident.

Depending on the system and model application, the main

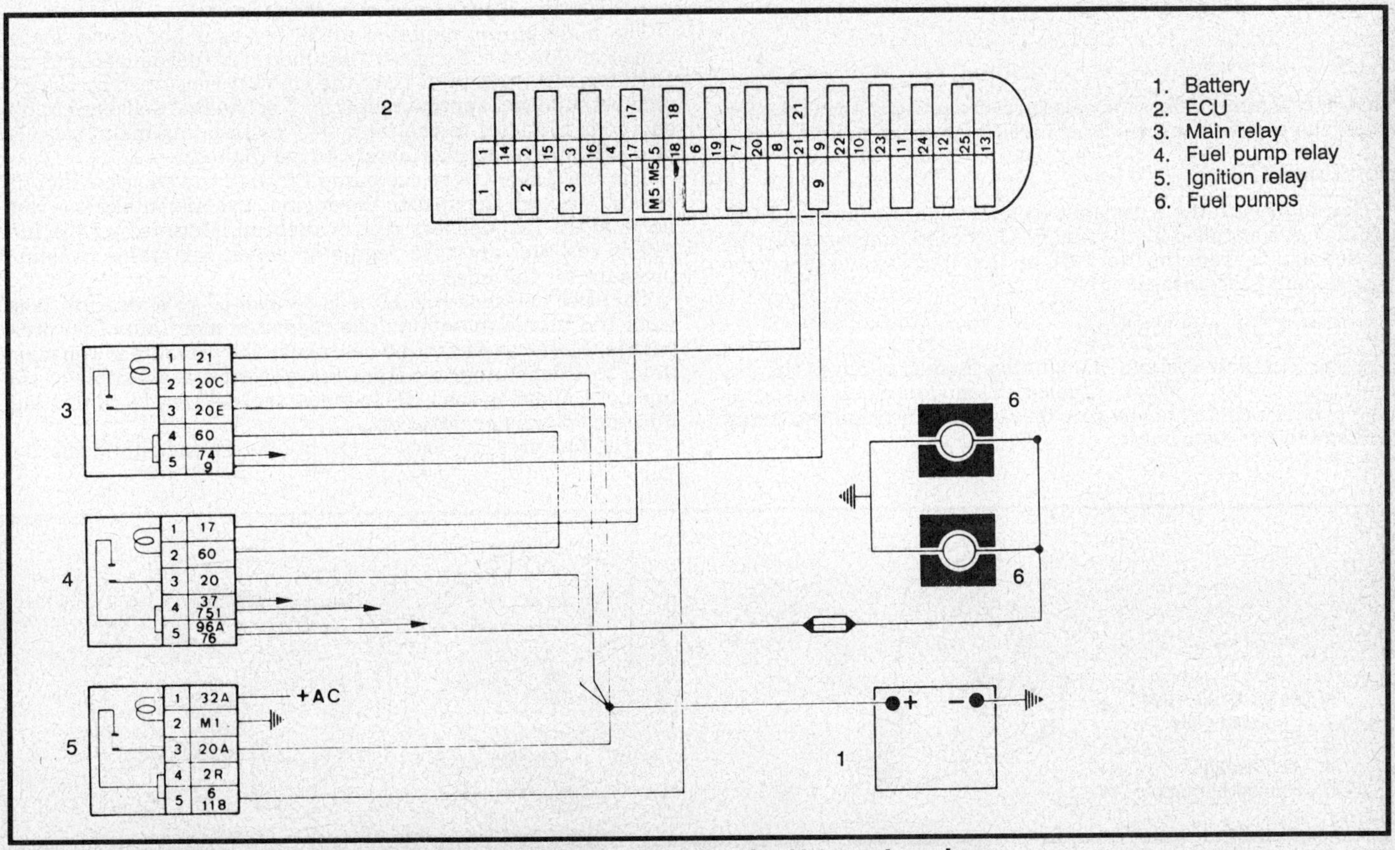

Example of main and fuel pump relay schematic. Circuitry for V-6 engine shown

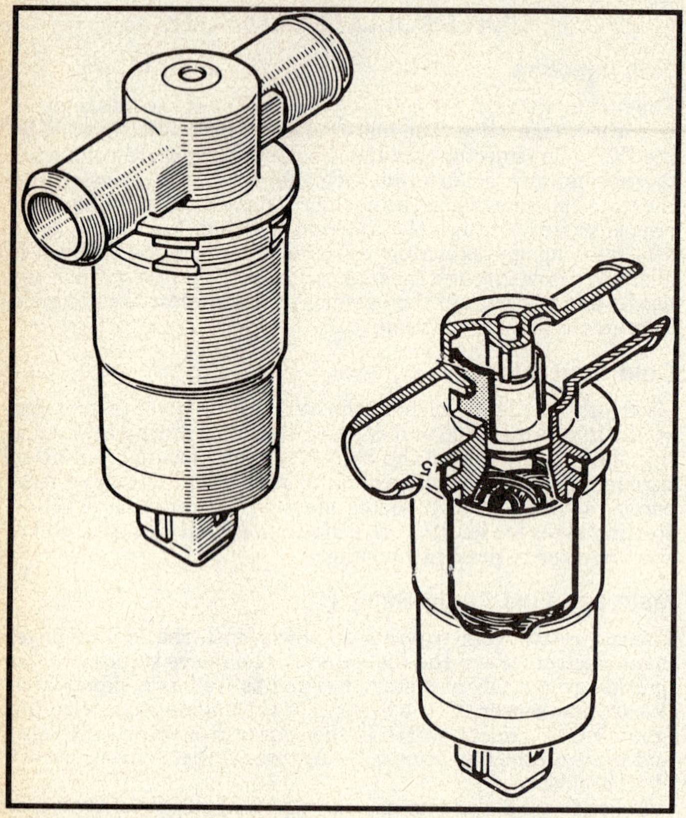

The auxiliary air valve uses an internal heater to open and close the air passage

and/or fuel pump relays may also provide power to the fuel injectors, the oxygen sensor, idle controller or other components.

Ignition ECU

All systems with the L designation use a separate ignition ECU. While not controlled by the fuel ECU, the ignition controller receives signals from the fuel ECU so that ignition and fuel delivery remain coordinated.

Ignition Coil and Module

On the Motronic system, the ignition module receives the signals from the ECU. When signaled, the module closes and opens the primary circuit in the coil, discharging the high secondary voltage to the distributor.

Auxiliary Air Device

Used only on LU systems, the auxiliary air device is energized through the fuel pump relay. It is located on the throttle body. The valve allows air to bypass the throttle plate; since it is located upstream of the air sensor, its flow is accounted for by the sensor.

The valve contains a heated bi-metallic strip. When the strip is cold, the valve allows extra air to pass; as the strip heats, a blocking plate on a shaft rotates into position, closing off the extra airflow.

Idle Actuator

The idle actuator or idle regulation valve is used only on LH and Motronic systems. It allows the ECU to maintain idle control based on signals from the coolant temperature sensor, the throttle switch, the rpm sensor and, on LH systems, the ignition ECU. The presence of this valve eliminates the need for an auxiliary air valve; a stable idle speed is maintained regardless of load factors such as air conditioning or power steering.

The valve is located in an air passage bypassing the throttle plate. The valve is controlled by the ECU, opening and closing to alter the amount of air bypassing the throttle. The valve contains a rotating slide controlled by a double-wound motor, capable of moving in both directions. As the valve changes position, the change in air is noted by the air flow sensor or air mass meter. The ECU alters the fuel delivery to match the airflow.

FUEL DELIVERY SYSTEM

Peugeot fuel systems employ 2 fuel pumps, an in-tank primer pump and an external main pump. Both pumps are electrically operated through the fuel pump relay. The primer or scavenger pump delivers fuel under light pressure to the main pump, preventing vapor lock in the fuel lines. The primer pump has its own sock-type filter attached to the pick-up.

The main pump, mounted under the rear bodywork, maintains fuel flow to the engine. The amount of fuel delivered to the engine is always greater than the amount consumed by the engine. At full load, approximately 70% of the fuel delivered to the engine is returned to the tank. The main pump maintains the line pressure at approximately 44 psi (3.0 bar).

The fuel passes from the pump through the adjacent fuel filter and through the lines to the engine. The fuel intake line connects to the fuel delivery pipe or fuel rail. Mounted on the fuel rail is the fuel pressure regulator which maintains constant pressure for the injectors.

The fuel pressure regulator is connected to a vacuum hose from the intake manifold. The regulator maintains fuel pressure in the rail at 44 psi (3.0 bar) above the pressure in the manifold. In this manner constant fuel pressure is available at the injectors, allowing the ECU to meter fuel delivery by controlling injector on-time or duration.

If the fuel pressure exceeds the pre-determined limit, the dia-

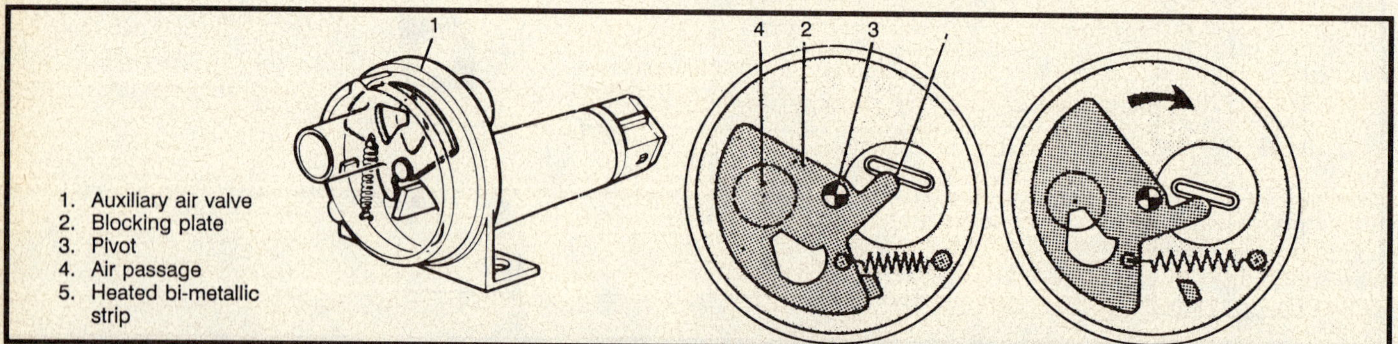

The idle speed actuator changes the airflow at the command of the ECU

phragm within the regulator lifts a valve, allowing excess fuel to return to the tank through the fuel return line.

Diagnosis and Testing

SERVICE PRECAUTIONS

- Disconnect the negative battery cable before performing any diagnostic or repair procedures. When reconnecting the cable, make certain the terminal is tight.
- Always test resistance with the harness disconnected from the ECU. Never connect the test probes to the prongs of the connector; all testing must be done from the rear or wire side of the connector body. Make certain the protective rubber boot or cover is replaced when testing is completed.
- Do not disconnect the battery while the engine is running or when the ignition switch is **ON**.
- Use only 12 volt power sources if starting assist is needed. Disconnect both battery terminals if the battery is to be recharged.
- Never disconnect the ECU when the ignition is **ON**.
- All harness connectors must be clean and tight when connected; the locking tabs must engage properly.
- Disconnect all ECUs including the radio before any welding is performed on the vehicle.
- Remove all ECUs if the vehicle is to be exposed to heat above 176°F (80°C) such as in a paint booth.
- Disconnect the main relay before performing a compression test.
- Do not use a test light to check circuit continuity.
- Never create arcing to test continuity.
- Always keep a fire extinguisher on hand; know how to use it.
- Contain and clean up any spillage immediately.
- Always place the transmission gear shift lever in **N** (manual transmission) or **P** (automatic transmission), set the parking brake and block the drive wheels before starting the engine.
- The fuel system is under pressure. In order to reduce the chance of personal injury, cover all fittings with shop towels before disconnecting.
- During checks with either a defective or by-passed ignition, remove the injection relay.
- On turbocharged vehicles, never run the engine without its air filter.

DIAGNOSTIC CODES

The Motronic M1.3 fuel system has self-diagnostic capabilities. The ECU monitors electrical signals during operation; if a signal is interrupted or goes out-of-range when compared to predetermined values, the ECU will note the fault and store an identifying code in memory. The codes may be read or retrieved during diagnosis.

Once a fault is noted, the ECU will illuminate the CHECK ENGINE warning lamp on the dashboard to advise the operator. Additionally, if the fault involves a signal critical to engine operation, the ECU will immediately substitute a fixed value for the faulty one. This pre-determined or default value may not be correct for the immediate operating conditions; engine performance may be affected when the CHECK ENGINE lamp is lit.

The LU and LH fuel systems do not provide fuel system diagnostic codes but will provide a small number of diagnostic codes generated through the ignition ECU. These codes may be helpful in diagnosing certain fuel system components.

READING CODES

Motronic M1.3

The codes may be read by connecting the Peugeot TAD 99 tester, or its equivalent, to the green 2-pin connector and to the battery. Operate the tester according to the manufacturer's instructions.

If the tester is not available, codes may be read through the flashing of the CHECK ENGINE lamp. To read codes:

1. With the ignition **OFF**, connect a switched jumper wire between Terminal 1 of the green, 2-pin diagnostic connector and a good ground.
2. Switch the ignition **ON**; do not start the engine.
3. Turn the jumper switch ON for 3 seconds; the CHECK ENGINE lamp should be lit.
4. Turn the jumper switch OFF. The CHECK ENGINE warning lamp should blink once, remain off for a short period, then blink twice. This indicates Code 12, beginning of the test. After the last 2 blinks, the lamp should remain lit.
5. Again turn the jumper switch ON for 3 seconds; when the jumper switch is turned OFF the next code, if any, will be transmitted. The first group of flashes represents groups of 10s, the second group of flashes shows the digits. For example, 5 flashes, a pause and 4 flashes represents Code 54.
6. Each time the warning lamp remains lit after a code transmission, repeat Step 5. Consecutively stored codes will be transmitted one at a time.
7. Code 11 indicates the end of the test cycle. Once this code is received, no more faults are stored. If it is necessary to repeat the cycle, the ignition must be switched **OFF**, then **ON** and the procedure repeated from Step 1.

Ignition Codes—LU and LH Jetronic

Vehicles with the EZ 115 K and EZ 200 K ignition systems provide diagnostic codes through the orange LED on the dashboard. Under normal conditions, the LED should be lit when the ignition is **ON**, then go off with the engine running. If the light stays on constantly, or never comes on, check the power supply to the lamp and the ground circuit. When a fault is detected, the ignition ECU retards the spark timing; this safety function will be noticed as a lack of power or performance in the engine.

NOTE: An occasional flash or flicker of the LED during engine operation is normal. A fault noted in the system will cause an obvious, rhythmic flashing by the LED.

When a fault is noted by the ignition ECU, the LED will immediately flash in a pattern, representing the single-digit code for the fault. During diagnosis, codes should always be read with the engine idling; on some vehicles, the speed of the flashes is relative to engine speed.

Fault Codes

When using fault codes during diagnosis, remember that the ECU checks electrical signals only; it cannot diagnose mechanical faults such as fouled injectors or vacuum leaks. The code refers to a fault in the entire circuit, not just the named component.

MOTRONIC M1.3

Code 11—End of test sequence
Code 12—Beginning of test sequence
Code 13—Air temperature (NTC) sensor
Code 14—Injection coolant temperature (NTC) sensor
Code 15—Fuel pump relay
Code 21—Throttle position switch—idle position
Code 22—Idle regulation electro-valve
Code 31—Self-correction richness regulation—air leak at intake
Code 32—Self-correction richness regulation—exhaust leak
Code 33—Airflow sensor
Code 34—Canister purge solenoid

Code 35—Throttle position switch—full throttle position
Code 42—Injector signal or circuit
Code 43—Detonation correction circuit or regulation
Code 44—Detonation sensor
Code 51—Self-correction richness regulation—oxygen (Lambda) sensor or circuit
Code 52—ECU failure
Code 53—Battery voltage
Code 54—ECU failure

NOTE: Codes 13, 31, 32 and 35 will not trigger the CHECK ENGINE warning lamp even though a code is held in memory.

IGNITION CODES

ZN3J V-6 Engine

Code 2—Coolant temperature (NTC) sensor
Code 4—Detonation sensor or ignition ECU
Code 5—Faulty load signal from injection ECU
Code 6—No. 1 cylinder sensor

N9TEA Turbocharged Engine

Code 1—Continued detonation; maximum correction reached
Code 2—Battery voltage below 10.5 volts
Code 3—Ignition ECU detonation correction circuit faulty; internal failure
Code 4—Faulty signal received from detonation sensor
Code 5—Signal from throttle position sensor (potentiometer) exceeds 4.3 volts
Code 6—Circuit fault between ignition ECU and throttle potentiometer or loss of load signal from injection ECU.

CLEARING CODES

Motronic M1.3

Codes may only be cleared after Code 11, End of Test, has been received. If using the TAD 99 tester, follow the operating instructions. If the tester is not available:

1. Install the switched jumper used for reading codes.
2. Turn the ignition switch **ON**.
3. Retrieve any stored codes. Repeat the procedure until Code 11 is received.
4. After Code 11 is received and the CHECK ENGINE light is lit, turn the jumper wire switch ON for at least 10 seconds.
5. The codes will be cleared during the 10 second period. Turn the jumper wire switch OFF; the dash indicator should remain lit.

Ignition Codes

The ignition codes are only present when the engine is running. Turning the ignition **OFF** automatically clears the codes. Any faults present at the next restart will re-set the codes and cause the dashboard LED to flash.

FUEL PRESSURE RELIEF

Pressure is maintained within the fuel lines and fuel rail even when the ignition has been **OFF** for a long period of time. This pressure must be safely relieved before any lines or fittings are loosened.

With the ignition **OFF**, wrap an absorbent cloth around the joint to be loosened. Using 2 wrenches when possible, slowly open the joint just enough to allow fuel to spray into the rag. Once the pressure is relieved, the rag may be removed and the joint disconnected. Consider the rag flammable waste; store or dispose of it properly.

While this procedure may be performed at any threaded joint, it may be easiest to relieve the pressure at the connection of the inlet hose to the fuel rail regardless of where the repair is to be made. If this is done, retighten the inlet fitting immediately after the pressure is released, then perform the repair elsewhere.

The fuel pressure will remain reduced until the next the ignition is switched **ON** or the fuel pumps are operated by other means.

FUEL PRESSURE

Testing

1. Relieve the fuel system pressure.
2. Wrap a shop towel around the fuel inlet hose to avoid fuel spraying, then remove it.
3. Install the pressure gauge set from injection kit 0141, or equivalent. Connect the gauge between the inlet hose and the fuel delivery rail.
4. Disconnect the vacuum hose from the intake manifold to the fuel pressure regulator.
5. Start the engine and check the fuel pressure. The fuel pressure should be as follows:
   XU9J2 and XU9J4 engines—between 40.6 and 46.4 psi (280 and 320 kPa)
   N9TEA engine—between 34.8 and 37.7 psi (240 and 260 kPa)
   ZN3J engine—between 33.4 and 39.2 psi (230 and 270 kPa)
   ZDJL engine—between 33.8 and 39.6 psi (235 and 275 kPa)
6. If the fuel pressure is below the specified value, check the following:
   a. Check for a plugged return hose. Replace or clean it if necessary.
   b. Check for clogging of the primer pump filter or the main fuel filter.
   c. Check the electrical feed to the fuel pump(s).
   d. Check the fuel pump(s) and if necessary, replace them.
7. If the fuel pressure is above the specified value, check the fuel pressure regulator.

FUEL OUTPUT

Testing

XU9J2 AND XU9J4 ENGINE

1. Unplug the fuel pump relay and hook up bypass harness 91.42 or equivalent, between pin 3 and pin 5. Place the bypass harness switch in the OFF position.
2. Install the pressure gauge set (from injection kit 0141, or equivalent), between the fuel inlet hose and the fuel rail.
3. Disconnect the return hose from the pressure regulator. Connect in its place a rubber hose with the opposite end placed inside a graduated container.
5. Turn the bypass harness switch ON for 15 seconds. The quantity of fuel should be greater than 540 cm^3.
6. If the fuel output quantity is incorrect, check the fuel filter and primer pump filter. Check the fuel pump electrical leads for proper connection.
7. Reconnect the return hose to the pressure regulator.

N9TEA, ZDJL and ZN3J ENGINES

1. Unplug the fuel pump relay.
2. For vehicles equipped with ZN3J or ZDJL engines, hook up bypass harness 91.42 or equivalent, to the connector. Place the bypass harness switch in the OFF position.
3. For vehicles equipped with N9TEA engines, hook up bypass harness 91.42 or equivalent, between pin 2 and pin 5. Place the bypass harness switch in the OFF position.
4. Connect the pressure gauge 0141A or equivalent between the fuel supply connector and the fuel rail.
5. Connect the back pressure pump 91.41 and pressure adapter 91.49 or their equivalent. Insert the end of the pressure adapter into the fuel pressure regulator hose.

6. Depress the bypass harness switch, turning it ON.
7. Using the pressure pump, increase the fuel pressure to 43.5 psi (300 kPa). Then, turn the bypass switch OFF.
8. Disconnect the fuel pressure regulator return hose. Connect in its place a rubber hose with the opposite end placed inside a graduated container.
9. Turn the bypass harness switch ON for 15 seconds. The quantity of fuel should be greater than 540 cm^3.
10. If the fuel output quantity is incorrect, check the fuel filter and primer pump filter. Check the fuel pump electrical leads for proper connection.
11. Reconnect the return hose on the pressure regulator; remove the test equipment.

FUEL PRESSURE REGULATOR

Testing

XU9J2 AND XU9J4 ENGINES

1. Unplug the fuel pump relay and connect bypass harness 91.43 or equivalent, between pin 3 and pin 5. Place the bypass harness switch in the OFF position.
2. Install the pressure gauge set (from injection kit 0141, or equivalent), between the fuel inlet hose and the fuel injector ramp.
3. Disconnect the vacuum hose between the intake manifold and the fuel pressure regulator. ,
4. Turn the bypass harness switch ON for 15 seconds. The fuel pressure should be 43.5 psi (300 kPa).
5. Install a vacuum pump on the fuel pressure regulator and apply 15 in. Hg of vacuum. The fuel pressure must drop by 15 psi.
6. If the fuel pressure readings are incorrect, replace the fuel pressure regulator.

N9TEA, ZDJL and ZN3J ENGINES

1. Unplug the fuel pump relay.
2. Vehicles equipped with ZN3J or ZDJL engines, hook up bypass harness 91.42 or equivalent, on the connector. Place the bypass harness switch in the OFF position.
3. Vehicles equipped with N9TEA engine, hook up bypass harness 91.42 or equivalent, between pin 2 and pin 5. Place the bypass harness switch in the OFF position.
4. Connect the pressure gauge 0141A or equivalent between the fuel supply connector and the distribution pipe.
5. Disconnect the hose at the pressure regulator. Depress the bypass switch to operate the fuel pump(s). Record the value of the fuel pressure registered.
6. Hook up a vacuum pump on the fuel pressure regulator and apply 15 in. Hg of vacuum. The fuel pressure must drop by 15 psi.
7. Hook up a pressure pump 91.41 on the pressure regulator and apply a pressure of 15 psi. The fuel pressure should increase by 15 psi.
8. If these values are not correct, replace the fuel pressure regulator.

AIR-INTAKE SYSTEM

Testing

1. Visually check that the areas around the air flow sensor or air mass sensor cylinder head, intake manifold, etc. are airtight.
2. To check the air-intake system under pressure, pinch off the PCV hose, fully open the throttle and seal off the exhaust pipe.
   a. On vehicles equipped with ZN3J engine, remove the air flow sensor.
   b. On vehicles equipped with N9TEA, ZN3J or ZDJL engines, remove the air filter cover.
3. Apply soapy water, or equivalent, to all air tubing junctions, ports or fittings, including the intake manifold and cylinder head mating faces.
4. Remove the air-intake hose from the air flow sensor.
5. Using a compressed air gun with a rubber hose on the end, blow air through the air-intake hose. Use a rag around the rubber hose to prevent leakage.
6. If there is a leak, bubbles or foam will appear at the affected area.
7. Replace the defective gaskets, seals, hoses or damaged junctions as required.

OVERBOOST PRESSURE SWITCH

Testing

Power Supply

1. Disconnect wire 96 from the switch.
2. Disconnect the harness connector.
3. Connect a voltmeter between wire 96 and ground.
4. Engage the starter. The voltmeter should read 9 volts when the starter is engaged.
5. If the reading is incorrect, check the continuity of the electrical circuit. Additionally, check the power supply to and the operation of the fuel injection relay.

Function

1. Remove the switch from the engine.
2. Connect a hand-hand operated pressure pump and gauge to the sensor port. Connect an ohmmeter to the 2 terminals.
3. The ohmmeter should show an open circuit (infinite resistance) with no pressure applied. As the pump is used to build pressure, the switch must close (showing about 1 ohm resistance) at 15.25–16.65 psi (1.05–1.15 bar).
4. All test conditions must be met. If any portion of the test is not correct, the sensor must be replaced.

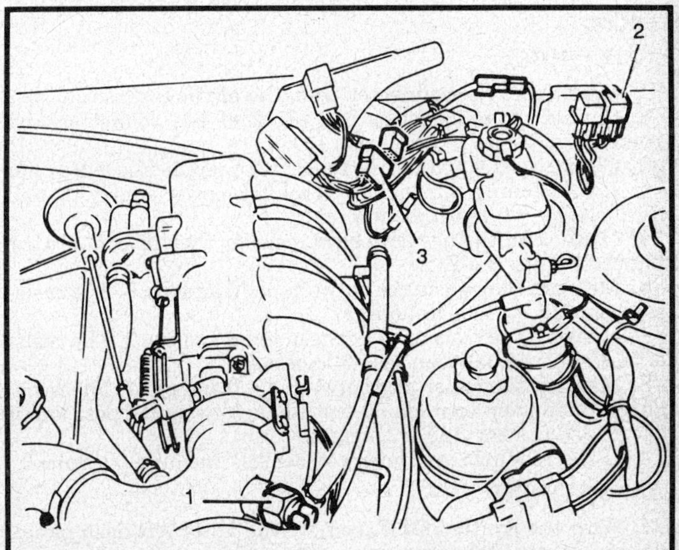

1. Overboost pressure sensor
2. Harness connector
3. Ignition relay

Overboost sensor and wiring, N9TEA engine

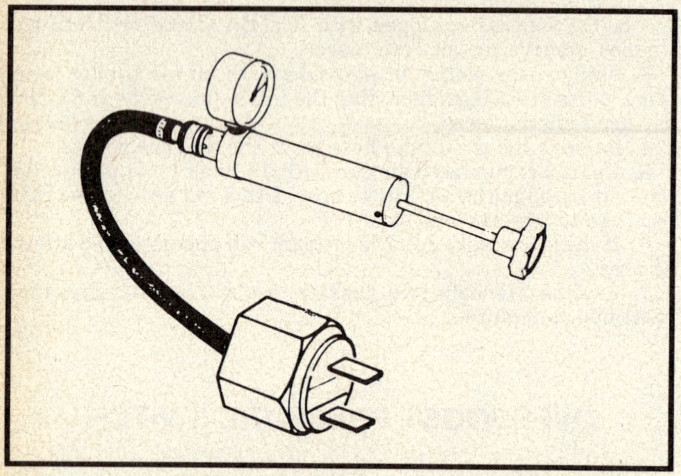

Pressure testing the overboost sensor. Connect an ohmmeter to the terminals

60° THERMOCONTACT

Testing

1. Remove the thermocontact. Dip it in a container filled with tapwater.
2. Measure the resistance across the terminals of the thermocontact.
3. Heat the water in the container with the test leads still connected to the thermocontact.
4. Check the temperature at which the resistance will go from less than 1 ohm (contact closed) to infinity (contact open). The change point should be:
   Contact open – 140 ± 5.4°F (60 ± 3°C)
   Contact closed – 122 ± 5.4°F (50 ± 3°C)

AIR FLOW SENSOR

Testing

LU JETRONIC

1. Disconnect the connector from the air flow sensor.
2. Connect a voltage between pin 2 of the connector and ground.
3. Operate the starter. The reading should be above 9 volts.
4. If the reading is incorrect, check the power supply to, and the operation of, the injection relay.
5. Disconnect the negative battery cable; then disconnect the connector from the ECU.
6. Measure the resistance between pin 5 and pin 8. The resistance should be 340–450 ohms.
7. Measure the resistance between pin 8 and pin 9. The resistance should be between 160–300 ohms.
8. Measure the resistance between pin 5 and pin 7. The resistance should vary between 60 and 1000 ohms as the position of the air flow sensor flap is changed.

10. If the readings are incorrect, replace the air flow sensor.

LU-2 JETRONIC

1. With the ignition **OFF**, remove the lower left dash panel. Disconnect the fuel pump relay connector.
2. Disconnect the negative battery cable.
3. Disconnect the auxiliary air device and the altitude sensor.
4. Measure the resistance between pin 5 and pin 8. The resistance should be 340–450 ohms.

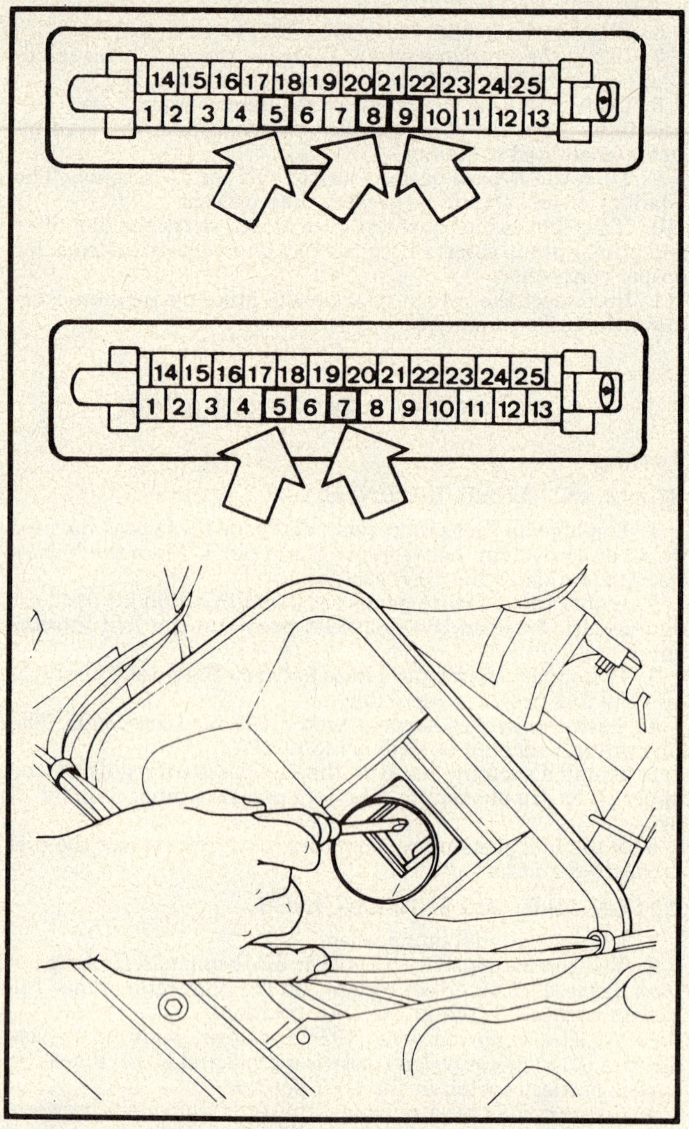

Testing the airflow sensor, LU Jetronic system

5. Measure the resistance between pin 8 and pin 9. The resistance should be between 160–300 ohms.
6. Measure the resistance between pin 5 and pin 7. The resistance should vary between 60 and 1000 ohms as the position of the air flow sensor flap is changed.
7. If the readings are incorrect, replace the air flow sensor.

M1.3 MOTRONIC

1. With the ignition **OFF**, disconnect the negative battery cable.
2. Disconnect the wiring harness from the ECU.
3. Measure the resistance between pins 12 and 26; correct resistance is 300–550 ohms.
4. Measure the resistance between pins 7 and 26 with the air flap at rest. Correct resistance is 10–200 ohms.
5. Measure the resistance between pins 7 and 12 while moving the flap in and out. Resistance should vary evenly with the motion of the flap; the circuit should never become open or discontinuous.
6. All test conditions must be satisfied. If any test is out of

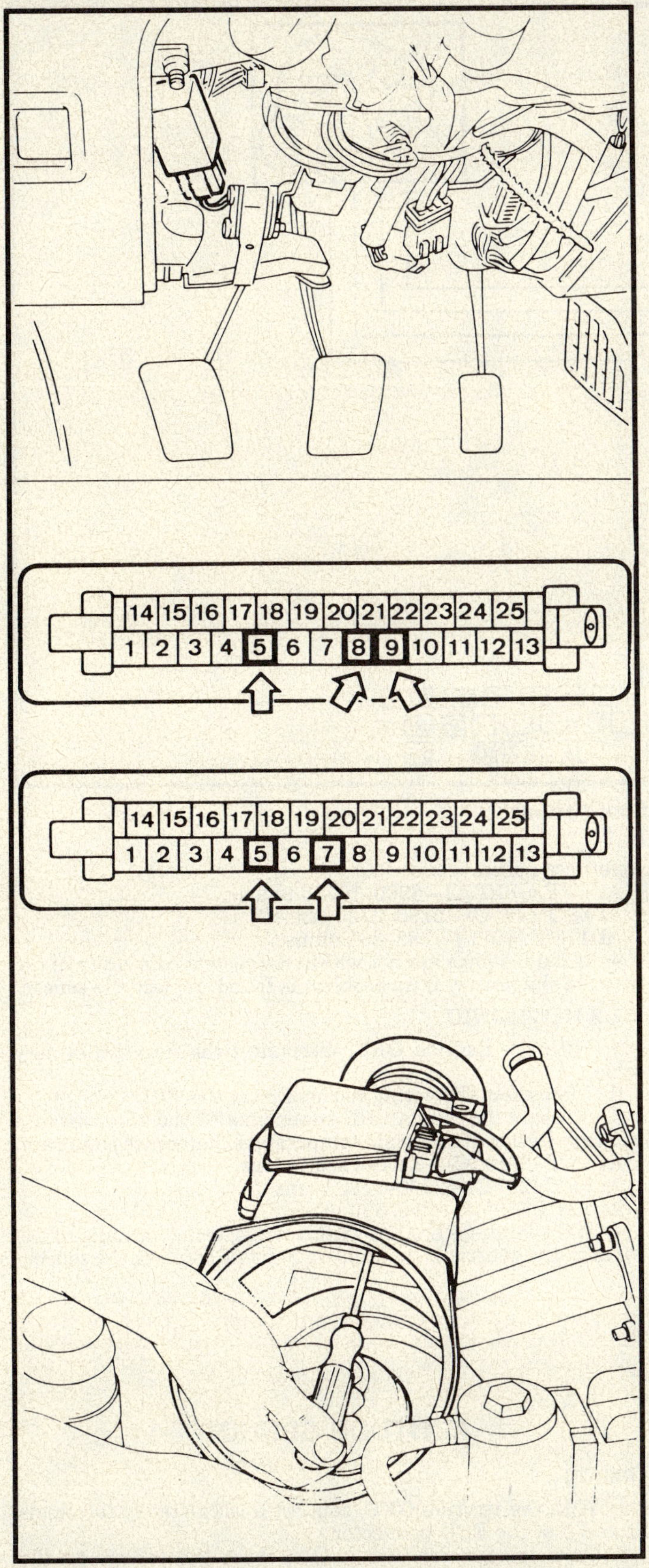

Disconnect the fuel pump relay before testing the LU-2 Jetronic airflow sensor

specification, carefully inspect the wiring between the ECU connector and the air flow sensor. If not fault is found, replace the sensor.

NOTE: This test may also be performed using the Peugeot pin-out box. If this equipment is used, perform Steps 3, 4, and 5 at pin-out terminals 9 and 19, 6 and 19, 6 and 9 respectively. The resistance values remain the same.

LH-2.2 Jetronic

1. With the ignition **OFF**, disconnect the harness connector from the ECU.
2. Reposition or remove the protective cover from the connector, allowing the terminals to be probed from the rear or wire side of the connector.
3. Reconnect the connector to the ECU.
4. Start the engine, allowing it to idle.
5. Measure the voltage between terminals 6 and 7 of the ECU by backprobing the connector. With the engine operating, voltage should be 2–5 volts.
6. If the voltage is incorrect, switch the engine off. Check continuity of each wire to the sensor. If no problem is found, replace the sensor.
6. If the voltage is correct with the engine operating, the self-cleaning function of the hot wire may be checked. The engine must be operated above 1500 rpm for a short period and must be warmed until the coolant temperature reaches at least 150°F (65°C). Pyrolysis will not occur if these 2 conditions are not met.
7. With the engine running, connect the voltmeter to terminals 8 and 11 of the ECU connector. Switch the ignition **OFF**. Five seconds after stopping the engine, the voltmeter should show 2–5 volts for 1 second.

This function may also be checked visually. Disconnect the air

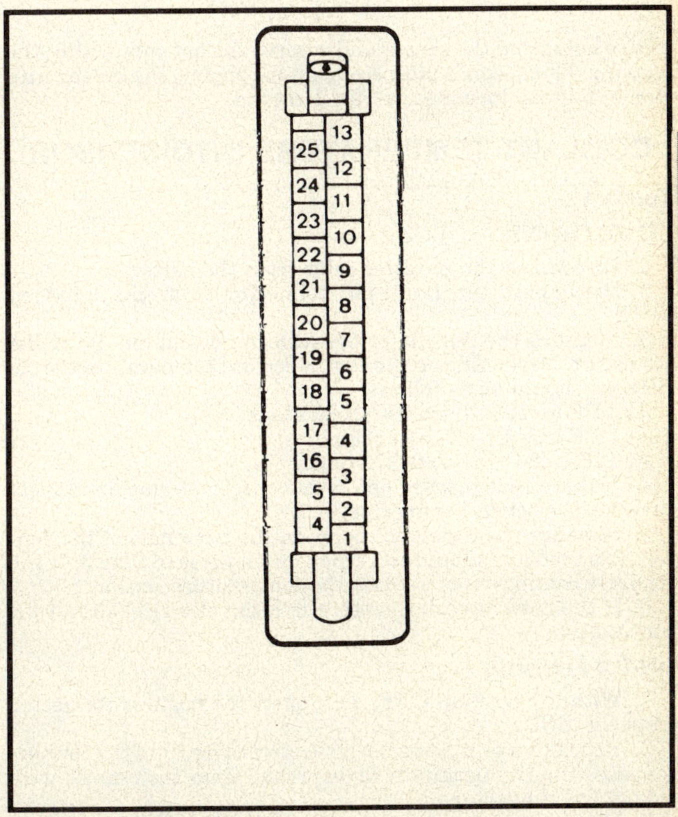

Terminal identification for LH-2.2 ECU

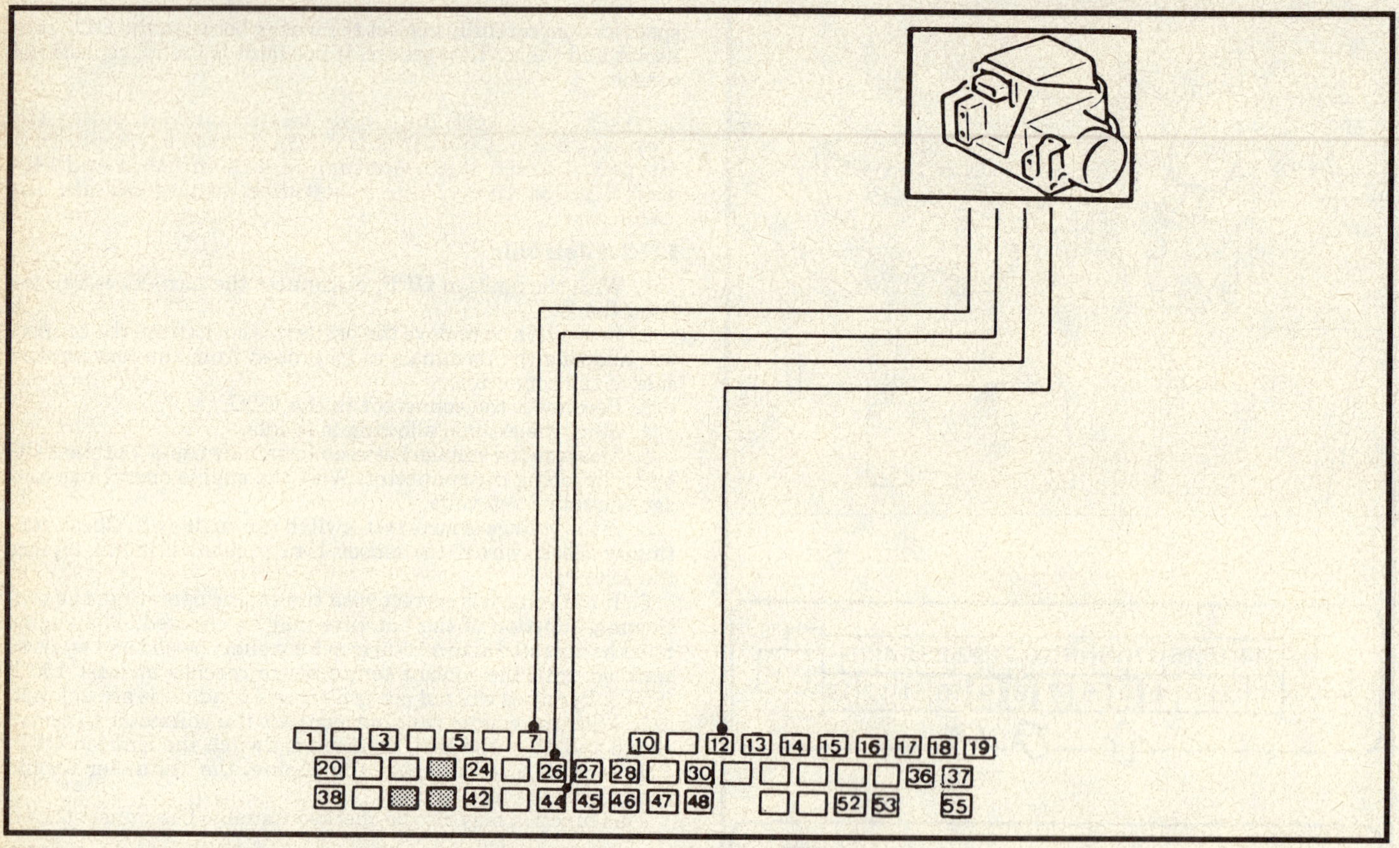

ECU terminal designations for testing M1.3 Motronic air flow sensor

intake hose from the sensor and observe the hot wire within the housing. Five seconds after stopping the engine, the sensor wire should become incandescent for 1 second.

COOLANT TEMPERATURE (NTC)SENSOR

Testing

LU JETRONIC

1. Disconnect the negative cable from the battery.
2. Disconnect the multi-pin connector from the injection ECU.
3. Measure the resistance between pin 5 and pin 10 of the connector. Depending on the engine coolant temperature the resistance should be as follows:
   At 14°F (−10° C)−8200–11000 ohms
   At 68°F (20° C)−2280–2720 ohms
   At 176°F (80° C)−290–370 ohms
4. If these readings are not within specifications, disconnect the sensor electrical connector.
5. Measure the resistance between the terminals of the sensor. The readings should correspond with those of Step 3. If the readings are incorrect, replace the temperature sensor.
6. If the readings are correct, check the electrical circuit for continuity.

LH-2.2 Jetronic

1. With the ignition **OFF**, disconnect the harness connector from the ECU.
2. Reposition or remove the protective cover from the connector, allowing the terminals to be probed from the rear or wire side of the connector.
3. Measure the resistance between pins 2 and 11 of the connector. Resistance will vary with engine coolant temperature. Reference points are:
   At 14°F (−10° C)−8200–10770 ohms
   At 68°F (20° C)−2280–2720 ohms
   At 176°F (80° C)−292–362 ohms
4. If the readings are not within range, test continuity of the wires to the sensor. If no problem is found, replace the sensor.

M1.3 MOTRONIC

1. With the ignition **OFF**, disconnect the negative battery cable.
2. Disconnect the wiring harness from the ECU.
3. Measure the resistance between pins 30 and 45; resistance will vary with engine coolant temperature. Reference points are:
   At 14°F (−10° C)−8200–11000 ohms
   At 68°F (20° C)−2280–2720 ohms
   At 176°F (80° C)−290–370 ohms
4. If the readings are not within range, test continuity of the wires to the sensor. If no problem is found, replace the sensor.

NOTE: This test may also be performed using the Peugeot pin-out box. If this equipment is used, perform Step 3 at pin-out terminals 22 and 29. The resistance values remain the same.

ALTITUDE SENSOR

Testing

1. With the ignition **OFF**, connect a voltmeter to terminals 11 and 5 of the ECU connector.
2. Turn the ignition switch **ON**; check the voltage on the meter.
3. The voltage will vary with both altitude and battery voltage. The example on the chart shows a vehicle being tested at

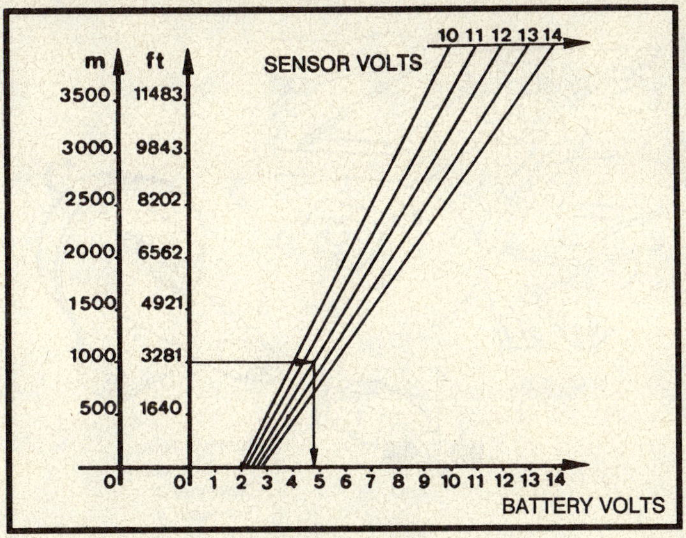

Test values for the altitude sensor

3281 ft. (1000 meters) with a battery providing 12 volts. The sensor voltage should measure 4.8 volts under these conditions.

4. If the test value is incorrect, check the wiring and connectors to and from the sensor. If no problem is found, replace the altitude sensor.

Testing

XU9J4 and XU9J2 ENGINES

1. With the sensor connected, warm the engine to normal operating temperature.
2. Pull back the rubber caps from the detonation sensor connector.
3. Connect a voltmeter – set to the A.C. scale – to the rear of the connector.
4. At idle, the voltmeter should indicate approximately 0 volts. Increase the engine speed; the voltage should increase with engine rpm, indicating around 20 milli-volts.
5. If the voltage reading is incorrect, perform the test using a known good sensor.

N9TEA ENGINE

1. Disconnect the negative battery cable.
2. Remove the ignition ECU connector.
3. Connect an ohmmeter between pin 12 and pin 13 of the ignition ECU connector. The resistance should be 270–330 kilo-ohms.
4. If out of specification, replace the detonation sensor.

INJECTORS

Testing

LU Jetronic

1. Disconnect the negative cable from the battery.
2. Disconnect the multi-pin connector from the injection ECU.
3. Measure the resistance between pin 9 and pin 12 of the connector. The resistance should be 3.5–5.0 ohms; this represents the resistance of all 4 injectors connected in parallel.
4. If the reading is incorrect, check the resistance of each injector individually. Disconnect the wiring from an injector and test resistance at the terminals; correct resistance is approximately 16 ohms.
5. Reconnect the ECU connector. Reconnect the battery.

LH-2.2 Jetronic

NOTE: This procedure requires the use of a fuel pump bypass harness, Peugeot part 91.42, or its equivalent.

1. With the ignition **OFF**, disconnect the fuel pump relay and install the bypass harness to the connector.
2. Disconnect the ECU connector; remove the protective boot to gain access to the rear of the connector.
3. Test the resistance between pin 13 of the ECU connector and wire 37 at the fuel pump relay. Resistance should be 2.5–3.0 ohms.
4. Reconnect the ECU harness; connect the fuel pump relay.

M1.3 MOTRONIC

1. With the ignition **OFF**, disconnect the negative battery cable.

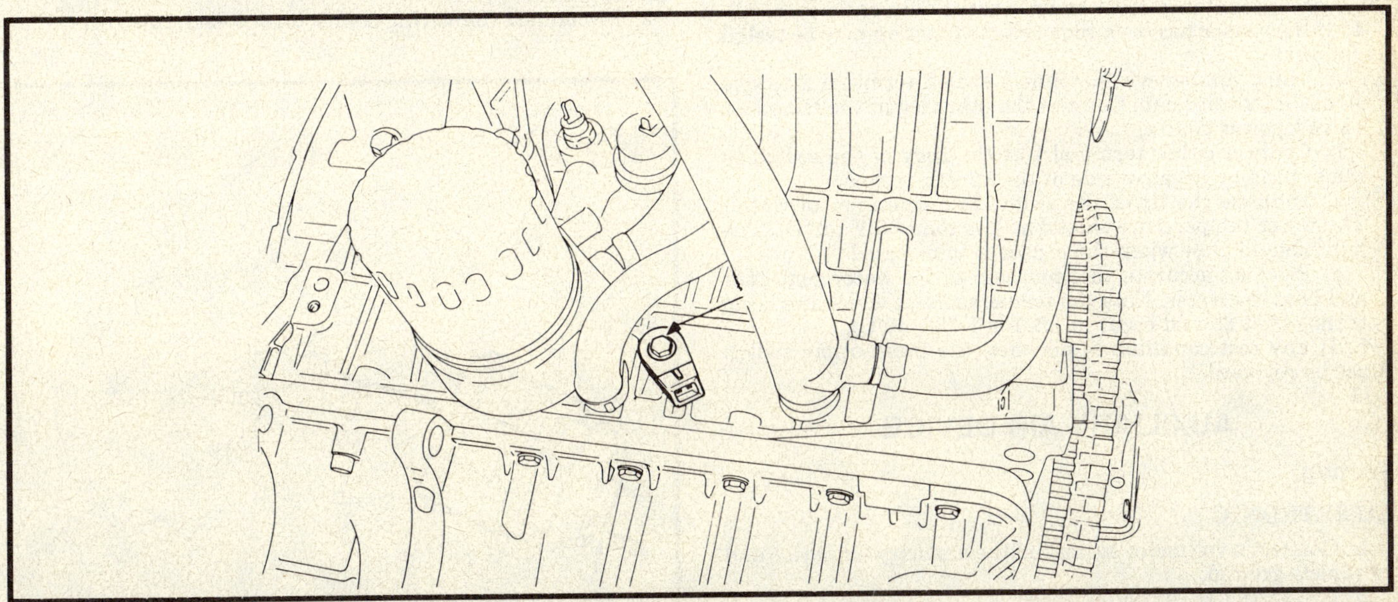

Detonation sensor for XU9J2 and XU9J4 engines. N9TEA sensor in similar location but connector points to 3 o'clock position

2. Disconnect the wiring harness from the ECU.
3. Measure the resistance between pins 16 and 37; then between pins 17 and 37. Resistance for each test should be 8 ohms.
4. If the readings are not within range, test continuity of the wires to the injectors. If no problem is found, test each injector individually. Disconnect the wiring from an injector and test resistance at the terminals; correct resistance is approximately 16 ohms.

NOTE: This test may also be performed using the Peugeot pin-out box. If this equipment is used, perform Step 3 at pin-out terminals 13 and 24, then at terminals 14 and 24. The resistance values remain the same.

COLD START INJECTOR

Testing

1. With the ignition **OFF**, disconnect the wiring harness from the cold start injector.
2. Measure the resistance across the terminals of the injector. Correct resistance is 3–5 ohms.
3. If the resistance is out of specification, replace the injector. If the sensor reads open (infinite resistance), it must be replaced.
4. Reconnect the harness to the injector.

THERMOTIME SWITCH

Testing

NOTE: Terminals 1 and 2 of the thermotime switch are labeled relative to the locating tab on the body of the switch. Facing the tab, with the connector pins at the bottom, terminal 1 is on the left. Terminal 1 is grounded through the body of the switch.

1. If the test is being performed on the vehicle, the coolant temperature must be below 86°F (30°C).
2. Disconnect the negative battery cable.
3. Disconnect the wiring from the switch. Measure the resistance at the terminals. Correct resistance is 30–40 ohms at 68°F (20°C). Resistance will vary with coolant temperature; use good judgement when interpreting test results.
3. Measure the resistance between terminal 1 and the body of the switch; resistance must be less than 1 ohm at 68°F (20°C).
4. If the sensor has been removed, its function may be tested as follows:
   a. Using jumper wires, connect switch terminal 2 to one side of a 12 volt lamp. Connect the other leg of the lamp to a 12 volt power source.
   b. Connect either terminal 1 or the body of the switch to the ground or negative side of the 12 volt source.
   c. Immerse the tip of the switch in a container of water. The water temperature should be less than 86°F (30°C). The light should glow when the sensor is immeresed.
   d. Place an accurate thermometer in the water container and heat the water. Note the temperature at which the lamp turns off; it should occur at 86–104°F (30–40°C).
5. If any test condition is not met, the thermotime switch must be replaced.

AUXILIARY AIR DEVICE

Testing

LU JETRONIC

1. Connect a voltmeter to pin 3 of the auxiliary air device and to vehicle ground.
2. Disconnect the injection relay.
3. Engage the starter. The voltmeter should show voltage at or above 9 volts, confirming power supply to the valve.

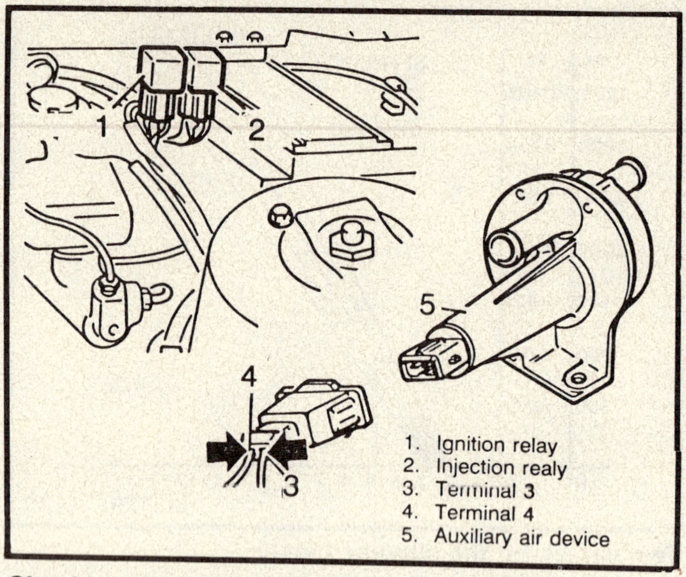

Checking the auxiliary air device on LU Jetronic system

4. Switch the ignition **OFF**. Disconnect the inlet and outlet hoses from the auxiliary air device.
5. Look into one of the hose ports on the valve. The passage in the internal diaphragm must be visible at 68°F (20°C).
6. Connect an ohmmeter across the terminals of the auxiliary air device. Correct resistance at 68°F (20°C) is 45–55 ohms. If the meter shows an open circuit (infinite resistance), replace the auxiliary air device.
7. Using jumper wires, apply 12 volts to pin 3 and ground pin 4 of the auxiliary air device. After 5 minutes, the diaphragm should completely block the air passage when viewed through the hose port as in Step 5.

LU-2 JETRONIC

1.Connect a voltmeter to pin 2 of the auxiliary air device and to vehicle ground.
2. Disconnect the wiring connectors to each injector.

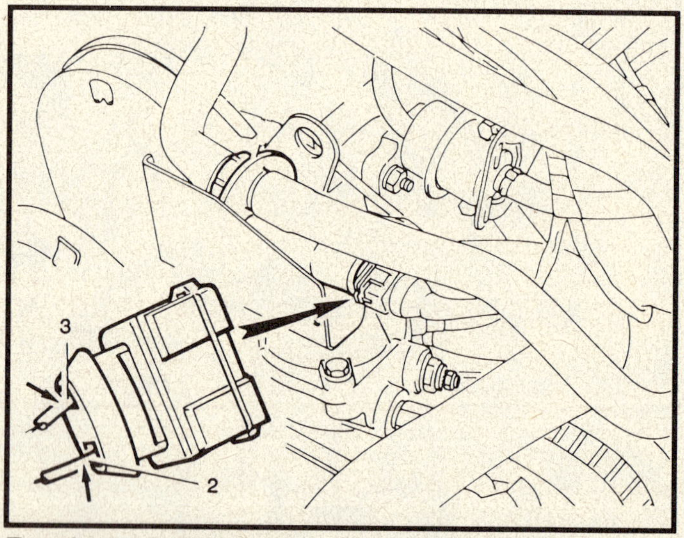

Terminal designation for auxiliary air device connector on LU-2 Jetronic system

3. Engage the starter. The voltmeter should show voltage at or above 9 volts, confirming power supply to the valve.

4. Switch the ignition **OFF**. Connect an ohmmeter between pin 3 on the auxiliary air device and vehicle ground. Correct reading must be less than 1 ohm.

5. Disconnect the inlet and outlet hoses from the auxiliary air device.

6. Look into one of the hose ports on the valve. The passage in the internal diaphragm must be visible at 68°F (20°C).

7. Connect an ohmmeter across the terminals of the auxiliary air device. Correct resistance at 68°F (20°C) is 35–45 ohms. If the meter shows an open circuit (infinite resistance), replace the auxiliary air device.

8. Using jumper wires, apply 12 volts to pin 2 and ground pin 3 of the auxiliary air device. After 5 minutes, the diaphragm should completely block the air passage when viewed through the hose port as in Step 6.

M1.3 MOTRONIC

SOHC Engine–XU9J2

1. Connect a voltmeter to pin 1 of the auxiliary air device and to vehicle ground.

2. Disconnect the wiring connectors to each injector.

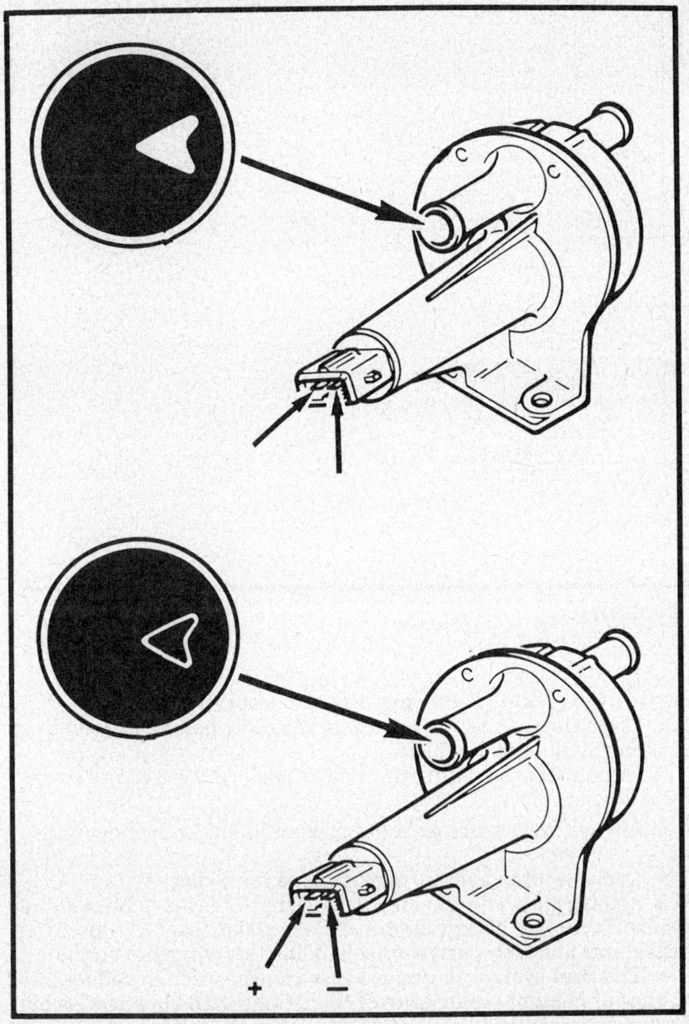

The air passage must be visible when viewed through a hose port at normal temperatures. The passage should close within 5 minutes when power and ground is applied to the AAD

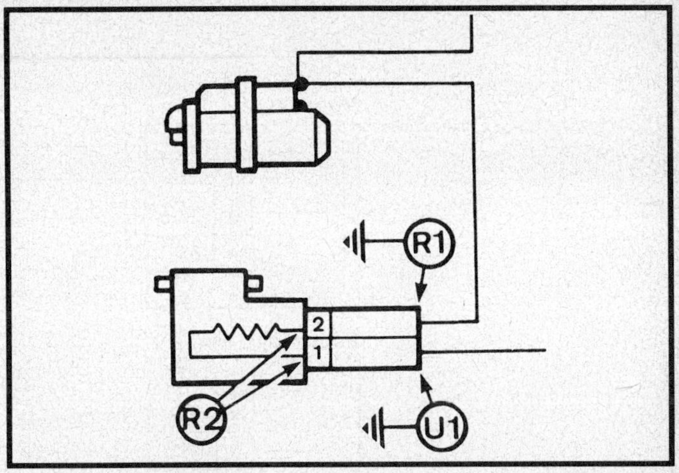

Test procedures for the auxiliary air valve on SOHC XU9J2 engines. U1 is a voltage test; R1 and R2 measure resistance

3. Engage the starter. The voltmeter should show voltage at or above 9 volts, confirming power supply to the valve.

4. Switch the ignition **OFF**. Disconnect the negative battery cable.

5. Remove the connector from the auxiliary air device. Connect an ohmmeter between pin 2 on the auxiliary air device and vehicle ground. Correct reading must be less than 1 ohm.

6. Connect an ohmmeter across the terminals of the auxiliary air device. Correct resistance at 68°F (20°C) is 46 ohms. If the meter shows an open circuit (infinite resistance), replace the auxiliary air device.

IDLE REGULATION ELECTROVALVE

Testing

LH-2.2 JETRONIC

1. With the ignition **OFF**, disconnect the harness connector from the ECU.

2. Reposition or remove the protective cover from the connector, allowing the terminals to be probed from the rear or wire side of the connector.

3. Disconnect the fuel pump relay. At pin 4 of the relay connector, install a long jumper running into the cabin of the vehicle. This jumper bridges wires No. 751 and 37 into a convenient position near the ECU connector.

4. Measure the resistance between pin 10 of the ECU connector and the jumper wire; correct resistance is 19–25 ohms.

5. Measure the resistance between pin 23 of the ECU connector and the jumper wire; correct resistance is 17–22.5 ohms.

6. If either reading is incorrect, check for wiring faults in the ECU harness. If no faults are found, replace the idle regulation electrovalve.

7. Remove the jumper wire, reconnect the relay and attach the ECU connector.

M1.3 MOTRONIC

DOHC Engine–XU9J4

1. With the ignition **OFF**, disconnect the negative battery cable.

2. Disconnect the wiring harness from the ECU.

3. Measure the resistance between pins 4 and 37; resistance for the test should be 8 ohms.

4. If the reading is not within range, test continuity of the wires to the idle valve. If no problem is found, replace the idle regulation valve.

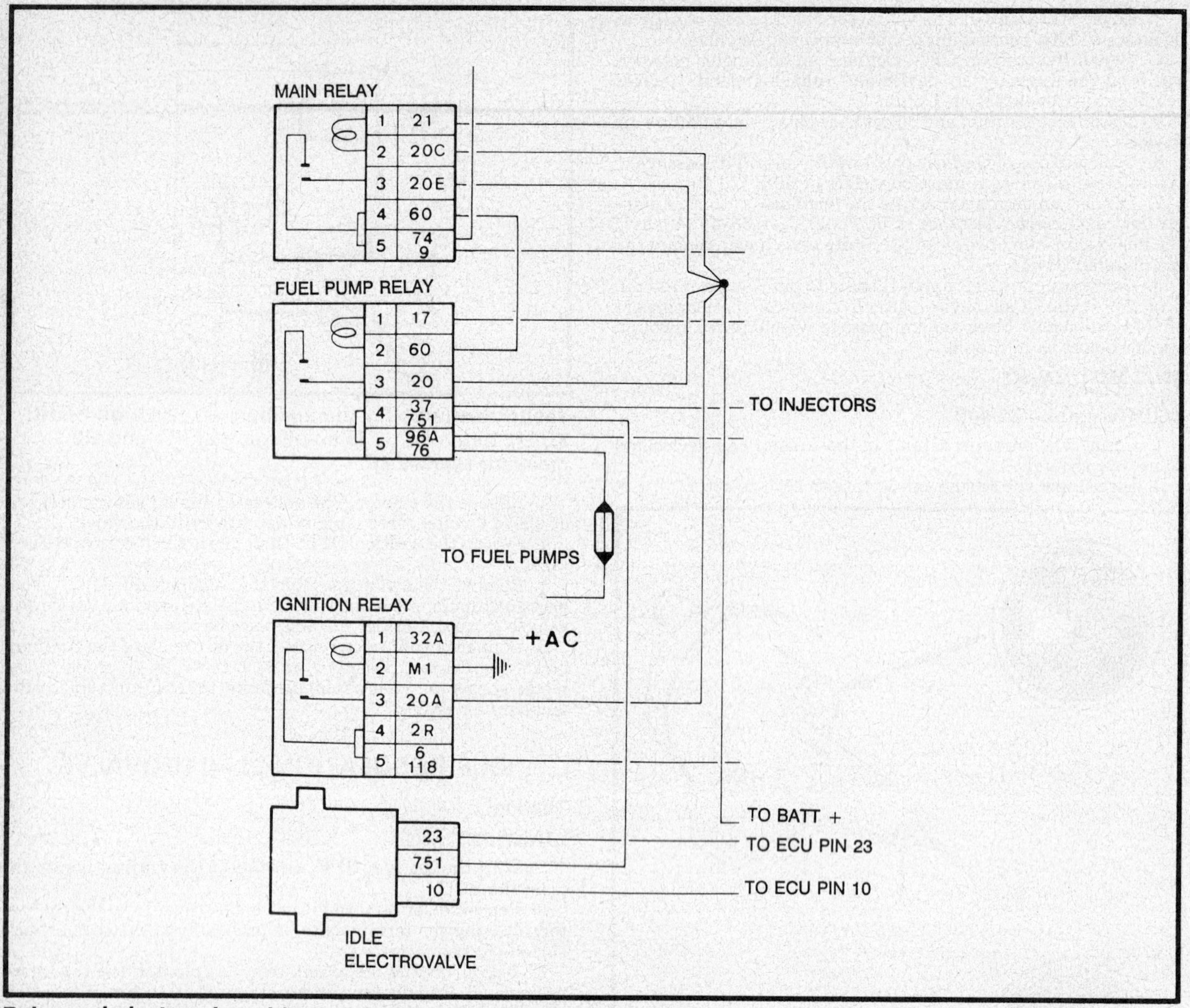

Relay and electrovalve wiring schematic, LH-2.2 Jetronic system

Component Replacement

SERVICE PRECAUTIONS

- Disconnect the negative battery cable before performing any repair procedures. When reconnecting the cable, make certain the terminal is tight.
- Never connect the test probes to the prongs of the connector; all testing must be done from the rear or wire side of the connector body. Make certain the protective rubber boot or cover is replaced when testing is completed.
- Do not disconnect the battery while the engine is running or when the ignition switch is **ON**.
- Use only 12 volt power sources if starting assist is needed. Disconnect both battery terminals if the battery is to be recharged.
- Never disconnect the ECU when the ignition is **ON**.
- All harness connectors must be clean and tight when connected; the locking tabs must engage properly.
- Disconnect all ECUs including the radio before any welding is performed on the vehicle.
- Disconnect the main relay before performing a compression test.
- Always keep a fire extinguisher on hand; know how to use it.
- Contain and clean up any spillage immediately.
- Always place the transmission gear shift lever in **N** (manual transmission) or **P** (automatic transmission), set the parking brake and block the drive wheels before starting the engine.
- The fuel system is under pressure. In order to reduce the chance of personal injury, cover all fittings with shop towels before disconnecting.
- During checks with either a defective or by-passed ignition, remove the injection relay.
- On turbocharged vehicles, never run the engine without its air filter.

FUEL PRESSURE RELIEF

Pressure is maintained within the fuel lines and fuel rail even when the ignition has been **OFF** for a long period of time. This pressure must be safely relieved before any lines or fittings are loosened.

With the ignition **OFF**, wrap an absorbent cloth around the joint to be loosened. Using 2 wrenches when possible, slowly open the joint just enough to allow fuel to spray into the rag. Once the pressure is relieved, the rag may be removed and the joint disconnected. Consider the rag flammable waste; store or dispose of it properly.

While this procedure may be performed at any threaded joint, it may be easiest to relieve the pressure at the connection of the inlet hose to the fuel rail regardless of where the repair is to be made. If this is done, retighten the inlet fitting immediately after the pressure is released, then perform the repair elsewhere.

The fuel pressure will remain reduced until the next the ignition is switched **ON** or the fuel pumps are operated by other means.

THROTTLE FLAP INITIAL POSITION

Adjustment

LU Jetronic

NOTE: This procedure is to be performed only when correct idle speed cannot be achieved using the air bypass screw (idle adjusting screw) or when the throttle flap jams or binds in the housing.

1. The following conditions must be met prior to the procedure: engine at normal operating temperature, ignition system properly adjusted, declutching fan released, and the return and compensating springs connected at the throttle linkage.
2. Remove the air intake elbow.
3. Remove the throttle flap housing.
4. Remove the tamper resistant plug. Drill the plug using a 3.5mm drill bit; thread a sheet metal screw into the drilled hole. Remove the plug.
5. Remove the throttle stop screw.
6. Clean the threads. Coat the threads with thread-locking compound.
7. Reinstall the throttle stop screw. Turn it in until the throttle flap just begins to open.

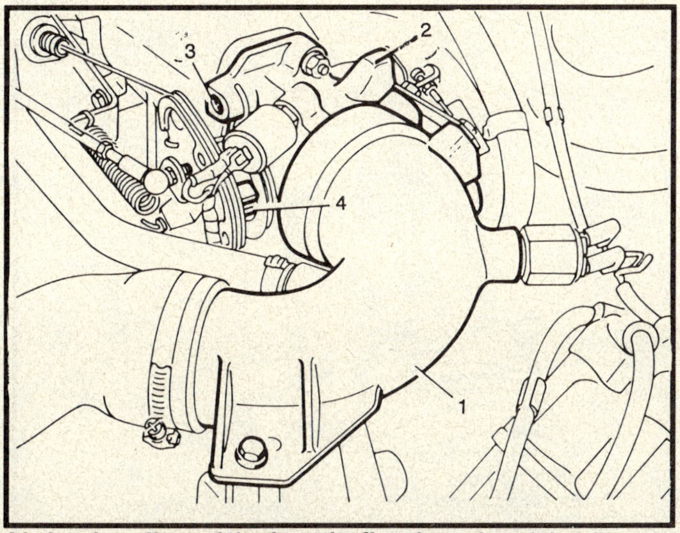

Air intake elbow (1), throttle flap housing (2), idle adjust screw or air bypass (3) and throttle stop screw (4) on LU Jetronic system

8. From the point of initial motion, turn the throttle stop screw 2 full turns inward.
9. Reinstall the throttle housing and the air intake elbow. If the vehicle is equipped with automatic transmission, adjust the kick-down cable.
10. Turn in the air-bleed screw until it bottoms; do not force it beyond the point of solid contact.
11. Start the engine. Adjust the idle speed to 650 rpm by turning the throttle stop screw.
12. Once the initial idle is set, install a new anti-tamper plug coated with thread locking compound.
13. Use the air-bleed screw to adjust the idle speed to 925 rpm for manual vehicles or 975 for automatic transmission vehicles. Allowable variation in idle speed is ±25 rpm.
14. Adjust the microswitch. Adjust the potentiometer.
15. If necessary, adjust the idle mixture.

LU-2 JETRONIC

1. Before adjusting, make certain that the ignition system is in proper operating condition and is properly adjusted. The air filter must be clean and in place. The engine must be at normal operating temperature, having run above 3000 rpm for a period of time. Inspect the throttle flap, making sure it is not binding in its housing. Additionally, it must be possible to adjust the idle at the air bypass screw if necessary.
2. Measure the vacuum available at the port on the bottom of the throttle housing; available vacuum must not exceed 2.95 in Hg. (75 mm Hg or 100 M bar).

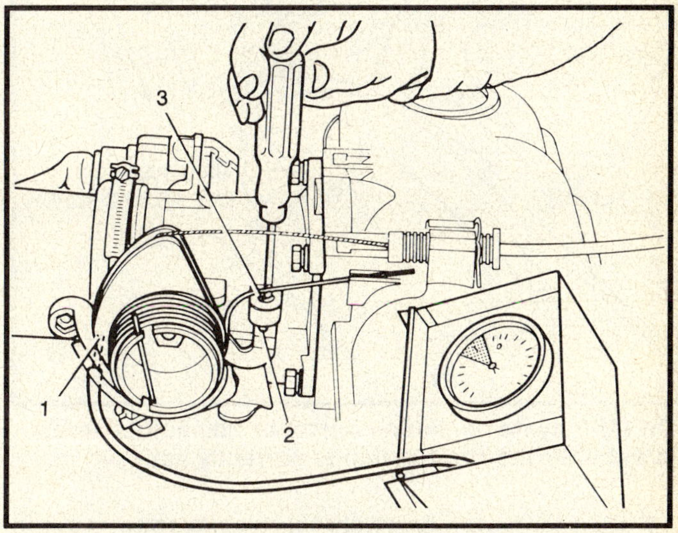

On LU-2 sytems, measure the vacuum available at the port (1) before adjusting the throttle stop screw (2). When all adjustments are complete, apply a dab of paint to the head of the screw (3)

3. Switch the ignition **OFF**.
4. Remove the anti-tamper paint, putty or wax from the head of the throttle stop screw.
5. Remove the throttle stop screw. Clean the threads of the screw and the matching threads of the housing.
6. Apply thread-locking compound to the threads of the screw. Reinstall the screw and turn it until the throttle plate just begins to move.
7. Turn the air bypass (idle adjust) screw all the way in but do not force it beyond the point of solid contact.
8. From the point of initial motion, turn the throttle flap stop screw in 6 full turns.

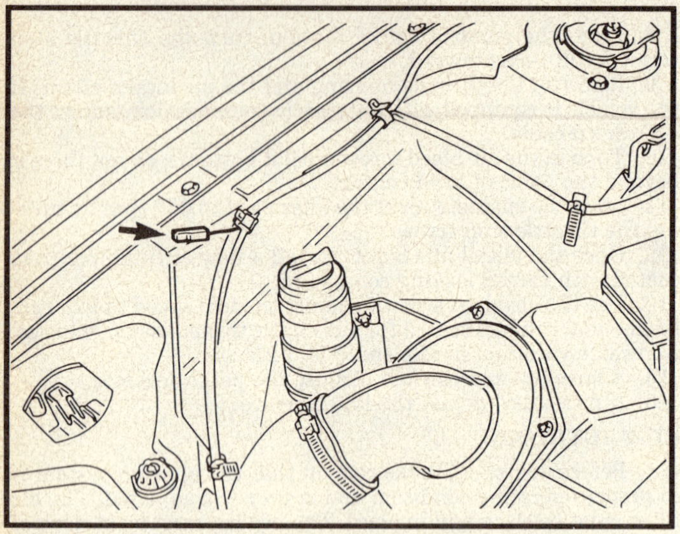

The idle richness is read as a voltage at this connector on LU-2 systems

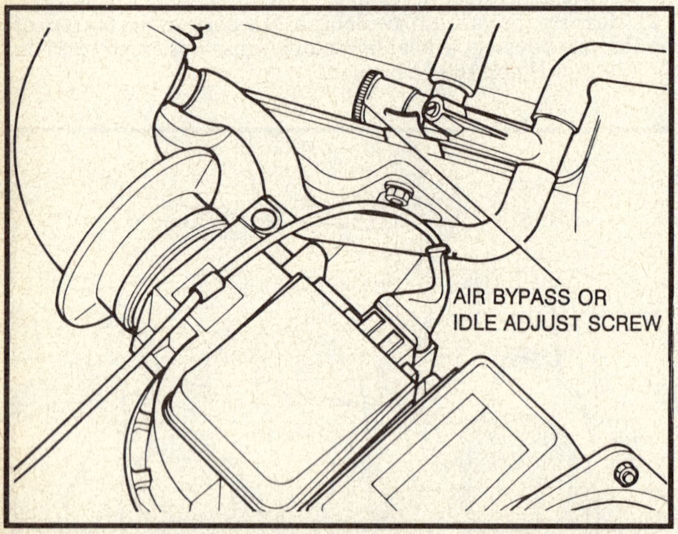

On LU-2 systems, the air screw or idle adjust screw is set after the throttle flap is correctly set

9. Start the engine. Adjust the throttle stop screw to obtain 600 rpm.
10. Connect a voltmeter to the test connector on the fender apron. Correct richness value is 6.5–7.5 volts with 7.0 volts preferred. If necessary, adjust the idle mixture.
11. Recheck the vacuum available at the throttle port. The vacuum should not exceed 2.95 in Hg. (75 mm Hg or 100 M bar). If necessary, adjust the throttle stop screw to obtain the correct vacuum.
12. When the adjustments are complete, apply a dab of paint to the head and housing of the throttle stop screw.
13. Adjust the throttle switch unit.
14. Adjust the idle speed to 900 rpm by turning the air bypass screw. The idle speed cannot be accurately adjusted if the throttle switch is out of position.

LH-2.2 Jetronic

1. Before adjusting, make certain that the ignition system is in proper operating condition and is properly adjusted. The air filter must be clean and in place. The engine must be at normal operating temperature. Inspect the throttle flap, making sure it is not binding in its housing. Make certain there is 2mm of free play between the throttle drum and its stop.
2. Start the engine. Use a jumper to ground wire No. 12, located at the right inner fender near the air filter. Grounding this wire closes the idle regulation electrovalve.
3. Screw the air bleed (idle adjust) screw all the way in. With the air bleed closed, idle speed should be 600 rpm.
4. If the engine rpm is incorrect, adjust the throttle flap as follows:
   a. Switch the ignition **OFF**.
   b. Disconnect the link rod from the throttle linkage.
   c. Loosen the locknut on the throttle stop screw.
   d. Unplug the connector from the throttle switch unit. At the back of the connector, use a small jumper to connect terminals 2 and 18.
   e. Loosen the 2 throttle switch retaining bolts.

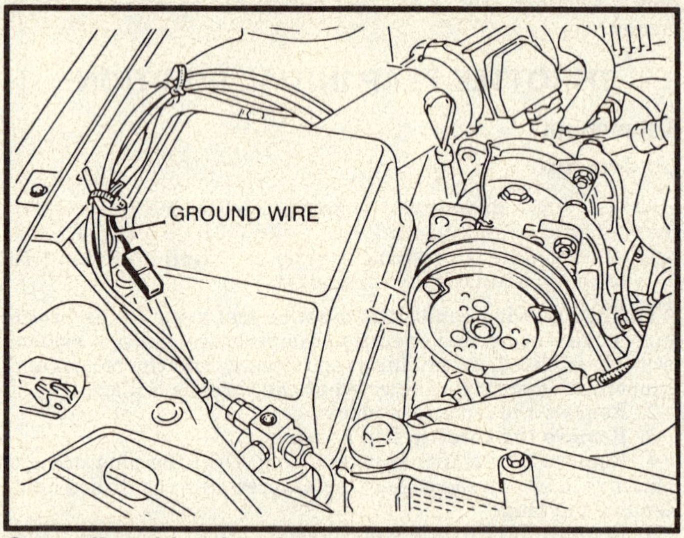

Grounding this wire closes the idle regulation valve on LH-2.2 systems

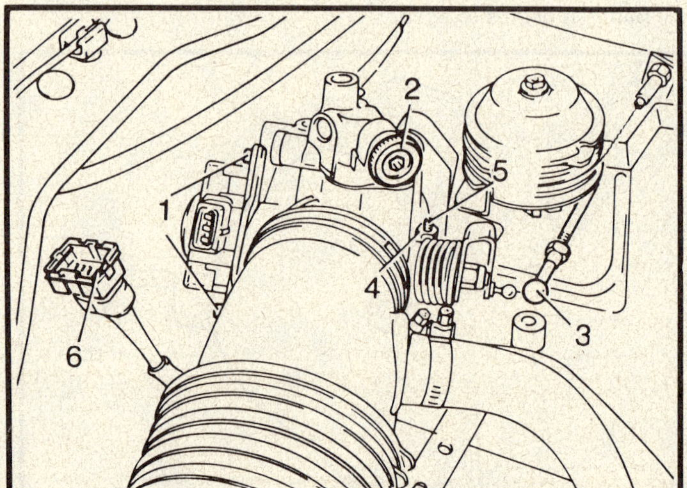

1. Throttle switch unit retaining bolts
2. Air bleed screw
3. Throttle drum rod
4. Locknut
5. Screw
6. Throttle switch unit connector

Throttle flap housing and related components—LH-2.2 system

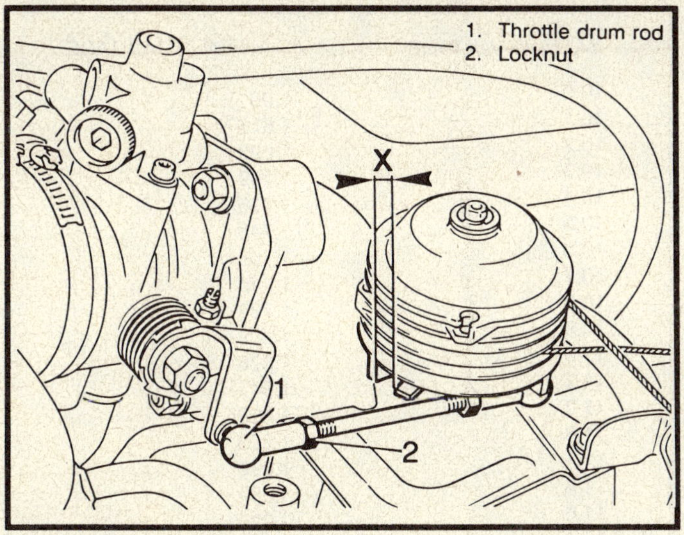

Dimension X must equal 2 mm. Loosen the locknuts and adjust the rod as needed

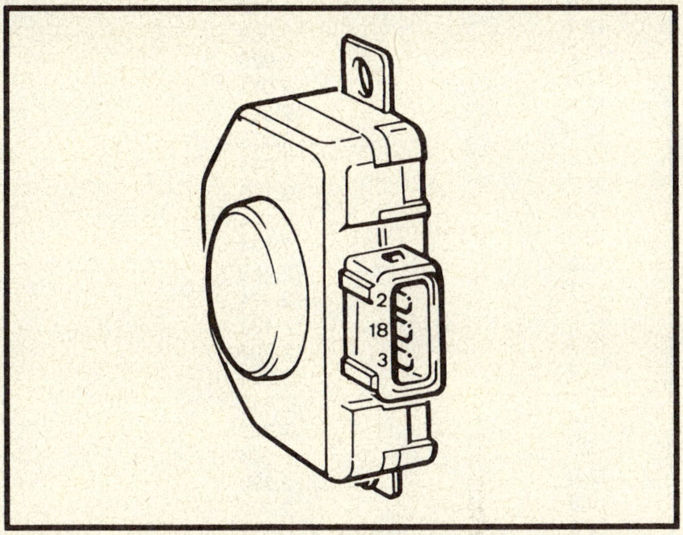

Terminal designations for throttle switch unit used on LU-2, LH-2.2 and M1.3 Motronic systems

f. Start the engine. Adjust the throttle stop screw to obtain 600 rpm.

g. Adjust the throttle switch unit and secure the retaining bolts. Connect the wiring connector to the switch unit.

h. With wire No. 12 grounded as in Step 2, adjust the air bleed screw to bring the engine speed up to 700 rpm.

i. Disconnect the ground to wire No. 12. The idle regulation electrovalve will bring the engine idle to the correct speed. Once this idle preset is complete, do not attempt to re-adjust the idle screw.

j. Reconnect the link rod to the throttle linkage. Loosen the 2 locknuts; adjust the rod to yield 2 mm of freeplay between the throttle drum and its stop. Tighten the locknuts when the freeplay is correct.

NOTE: The link rod sockets have left and right threaded fittings. With the locknuts loose, turning the rod in one direction will shorten or lengthen it.

k. If necessary, adjust the idle mixture or richness.

5. If the idle speed was correct in Step 4, perform the idle preset adjustment:

a. With wire No. 12 grounded as in Step 2, adjust the air bleed screw to bring the engine speed up to 700 rpm.

b. Disconnect the ground to wire No. 12. The idle regulation electrovalve will bring the engine idle to the correct speed. Once this idle preset is complete, do not attempt to re-adjust the idle screw.

M1.3 MOTRONIC

1. Disconnect the negative battery cable.
2. Disconnect the accelerator cable and the throttle switch connector.
3. Loosen the throttle switch unit retaining screws and rotate it counterclockwise (facing the vehicle).

First Throttle:

4. Remove the anti-tamper plug from the throttle stop or set screw.Loosen the screw until it no longer touches the lever.
5. Insert a 0.05mm feeler gauge between the lever and

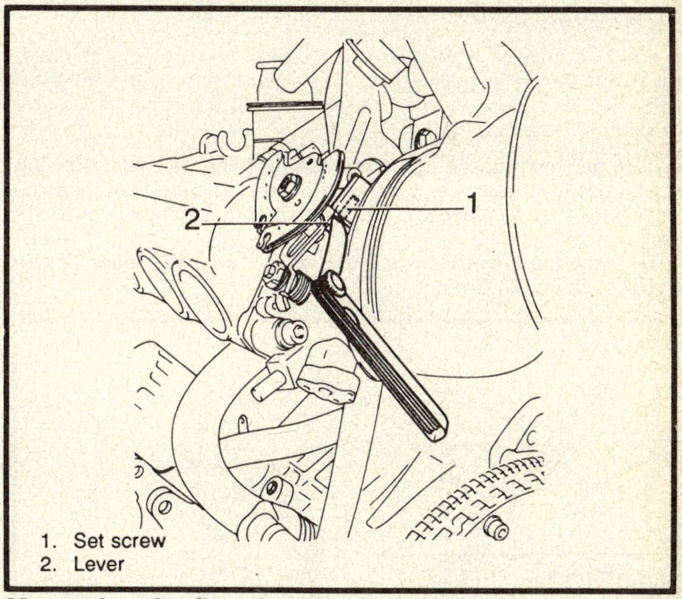

Measuring the first throttle clearance – M1.3 Motronic

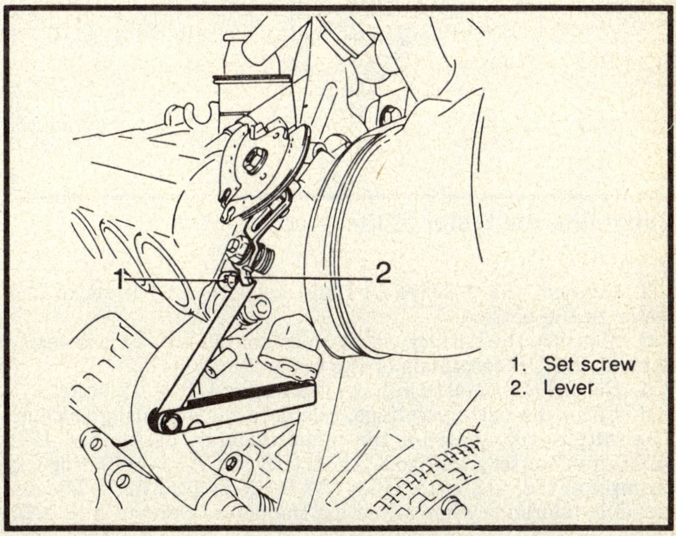

Measuring the second throttle clearance – M1.3 Motronic

screw, then re-tighten the screw until it just contacts the feeler gauge without causing the lever to move. The feeler must be free to move with no restriction.

6. Remove the feeler gauge.
7. Turn the set screw an additional ¼ turn, then reinstall a new anti-tamper plug.

Second Throttle:

8. Proceed in the same manner as the first throttle.
9. When adjustment is completed, lock the set screw with a drop of thread sealer.
10. Adjust the throttle switch and reconnect the switch unit connector.
11. Reconnect the accelerator cable and adjust it.

THROTTLE POTENTIOMETER AND MICROSWITCH

Adjustment

LU Jetronic System

NOTE: Use of Peugeot tool 91.43 is strongly recommended for this procedure. Necessary voltages may be obtained without it but the task is much easier when the test unit is used.

Before performing the adjustment, check that the throttle flap initial position is properly adjusted, the return and compensating springs are in place and the throttle flap is firmly against its stop.

1. Disconnect connectors 1 and 2. Connect the leads of the Peugeot tester in their place.

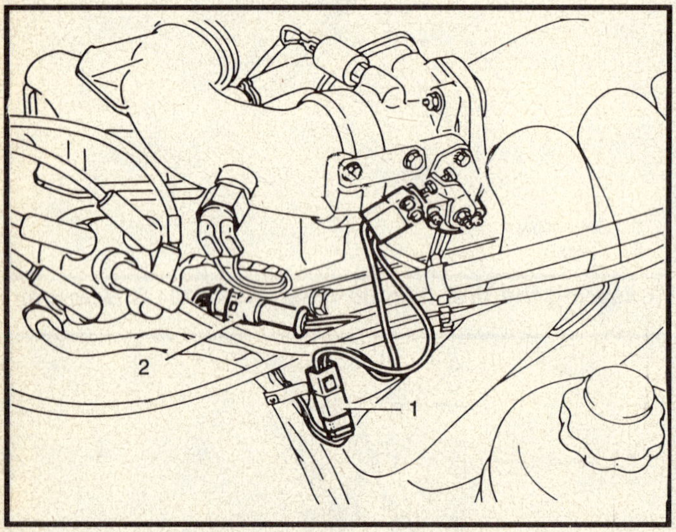

Substitute the tester at these connectors

2. Connect the positive (+) and negative (−) leads of the tester to the battery.
3. Measure the battery voltage, designated UB. This is easily done at the UB terminals of the tester.
4. Record the battery voltage (Example: UB = 12 volts).
5. From the battery voltage, calculate the adjusting voltage. The output voltage from the potentiometer, designated UR, must equal battery voltage X 0.166. (UR = UB x 0.166). For the example of UB = 12 volts, UR = 12 x 0.166 or 1.99 volts. The acceptable tolerance when adjusting the potentiometer is ± 0.10 volt.
6. Connect the voltmeter on the UR terminals of the tester.
7. Loosen the potentiometer retaining screws. Move the potentiometer to obtain the output or UR voltage calculated in Step 5.
8. Tighten the potentiometer retaining screws.
9. Connect an ohmmeter to the proper terminals on the tester.
10. Place a 0.10mm feeler gauge between the stop screw and the throttle lever.
11. Loosen the locknut on the stop screw; turn the stop screw until the ohmmeter shows a resistance of less than 1 ohm, indicating the switch has closed.
12. Coat the threads of the stop screw with thread locking compound; tighten the locknut.
13. Remove the feeler gauge. Insert a 0.20mm feeler gauge in the same location. The ohmmeter should show infinite resistance, indicating that the switch has opened.

BATTERY VOLTAGE	ADJUSTMENT VOLTAGE
10.0	1.66
10.1	1.676
10.2	1.693
10.3	1.709
10.4	1.726
10.5	1.743
10.6	1.759
10.7	1.776
10.8	1.792
10.9	1.809
11.0	1.826
11.1	1.842
11.2	1.859
11.3	1.875
11.4	1.892
11.5	1.909
11.6	1.925
11.7	1.942
11.8	1.958
11.9	1.975
12.0	1.992
12.1	2.008
12.2	2.025
12.3	2.041
12.4	2.058
12.5	2.075
12.6	2.091
12.7	2.108
12.8	2.124
12.9	2.141
13.0	2.158
13.1	2.174
13.2	2.191
13.3	2.207
13.4	2.224
13.5	2.241
13.6	2.257
13.7	2.274
13.8	2.290
13.9	2.307
14.0	2.324
14.1	2.341
14.2	2.357

Voltage calculation table—LU Jetronic potentiometer

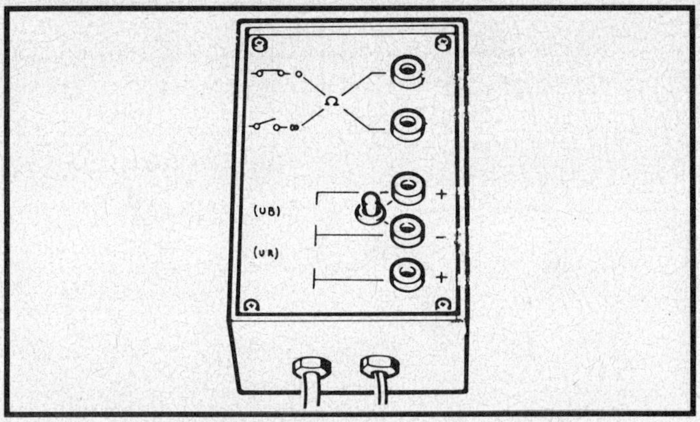

The Peugeot tester allows easy measurement of voltages UB and UR

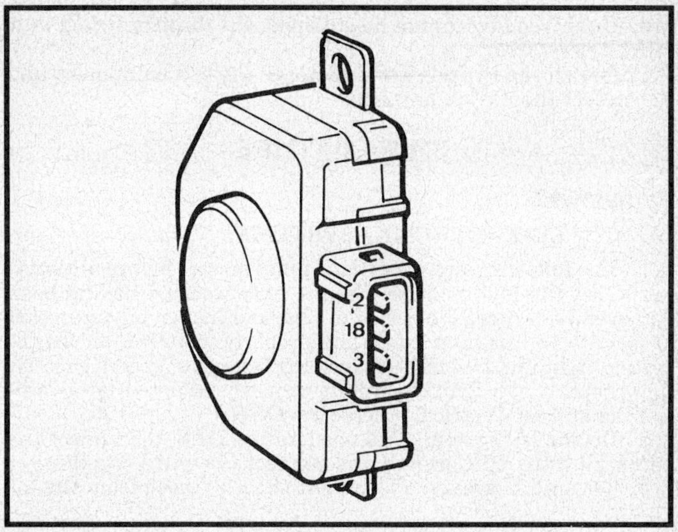

Connector pin identification for LU-2 Jetronic throttle switch

14. If the switch does not open with a 0.20mm gap, repeat the adjustment procedure at the stop screw. If the adjustment fails, the microswitch must be replaced.

THROTTLE SWITCH AND FULL LOAD SWITCH

Adjustment

LU-2 AND LH-2.2 JETRONIC

1. Before adjustment, confirm that the throttle flap position is correctly set.
2. Insert a 0.20mm feeler gauge between the stop screw and the throttle arm.
3. Loosen the 2 mounting screws holding the throttle switch unit.
4. Disconnect the wiring harness from the switch unit. Connect an ohmmeter to pins 2 and 18 of the throttle switch.
5. Turn the ignition **ON** but do not start the engine.
6. Rotate the throttle switch on its mounts until the ohmmeter reads less than 1 ohm; the contact points have closed within the switch.
7. Tighten the retaining bolts to hold the switch in this position; remove the 0.20mm feeler.
8. On LU-2, insert a 0.40 feeler gauge in the same position. On LH-2.2, insert a 0.70mm feeler in the same position. For both systems, the switch points should disengage, showing infinite resistance (open circuit) on the ohmmeter. If the circuit does not open, repeat the adjustment procedure.
9. To test the operation of the full load microswitch, connect an ohmmeter to pins 3 and 18 of the switch connector.
10. For LU-2 systems, the full load switch engages at 66 degrees of throttle travel. Open the throttle plate so that there is 4mm btween the arm and the core of the throttle housing; at this point, the ohmmeter should indicate a closed circuit or less than 1 ohm resistance. If the circuit does not close at the 4mm position, the throttle switch unit must be replaced.
11. On LH-2.2 systems, depress the throttle pedal to the floor; the ohmmeter should indicate a closed circuit or less than 1 ohm resistance.
12. If the switch does not close at full throttle, replace the throttle switch unit.
13. Turn the ignition **OFF**. Remove the test equipment and reconnect the throttle switch wiring.

M1.3 MOTRONIC

1. The position of the throttle flap must be correctly adjusted before perfoming this procedure.

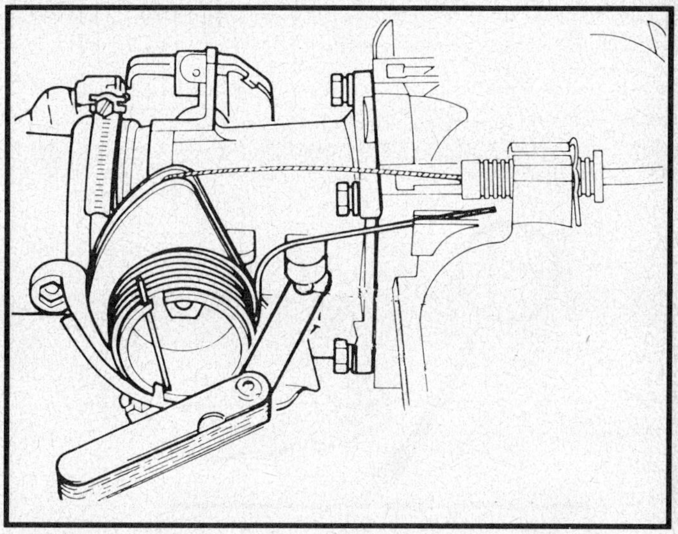

Two feeler gauge sizes must be used; the circuit must be closed at 0.20mm and open at 0.40mm

2. With the ignition **OFF**, remove the wiring connector from the throttle switch.
3. Connect an ohmmeter to pins 2 and 18 of the throttle switch.
4. Turn the ignition **ON** but do not start the engine. Make certain the throttle is fully closed; the ohmmeter should indicate a closed circuit or less than 1 ohm.
5. If adjustment is necessary, loosen the retaining screws holding the throttle switch unit.
6. Rotate the switch unit fully clockwise. Slowly return the switch counter-clockwise until an audible click occurs.
7. Tighten the screws to hold the switch in position just after the click occurs.
8. Once the idle contacts are correctly adjusted, connect the ohmmeter to pins 3 and 18 on the connector.
9. Test the full load switch by depressing the throttle pedal to the floor; the ohmmeter should show a closed circuit or less than 1 ohm when the throttle is fully open.

10. If this reading cannot be obtained, the throttle switch unit must be replaced.
11. Switch the ignition **OFF**. Remove the test equipment and re-connect the switch harness.

IDLE MIXTURE

Adjustment

LU AND LU-2 JETRONIC SYSTEMS

1. The following pre-conditions must be met before attempting to set the idle mixture: Engine fully warmed having been run over 3000 rpm for a period of time and the fan having cycled at least twice; air filter clean and properly installed; air intake system tight and free of leaks; cooling fan not engaged; throttle flap properly set; ignition system properly adjusted; air conditioning and all electrical accessories **OFF**.
2. On the LU system, with the ignition **OFF**, disconnect the wires from the 60°C switch and connect the wires together.
3. On the LU system, disconnect the PCV hose from the air intake system; plug the intake port firmly but leave the valve cover port open to the atmosphere.

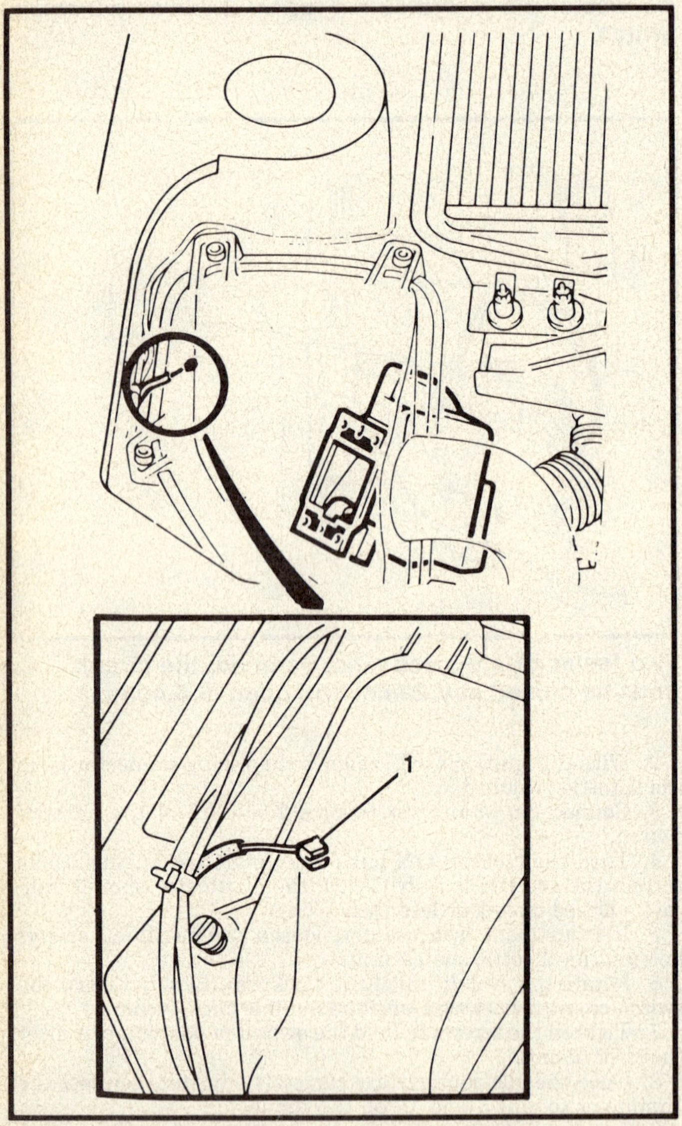

Connect the voltmeter to the test connector (1) on the right side of the engine compartment

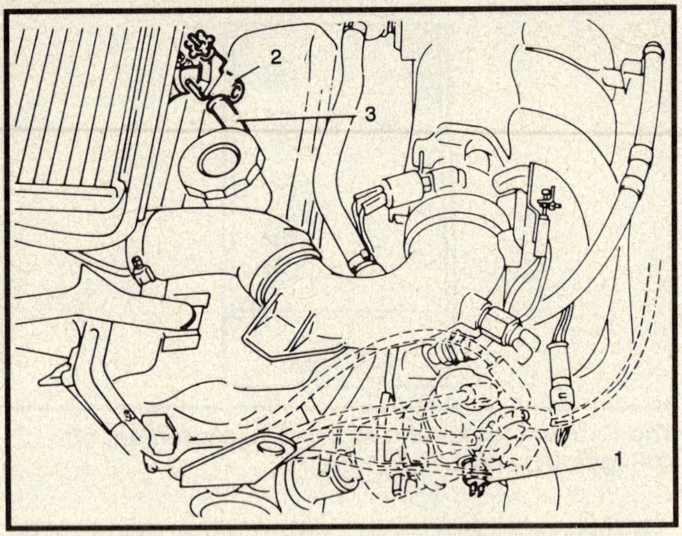

For LU controlled vehicles, disconnect the wiring to the 60°C thermoswitch (1), then remove and plug the PCV hose at the air cleaner (2). The hose or port from the valve cover (3) must remain open to the atmosphere

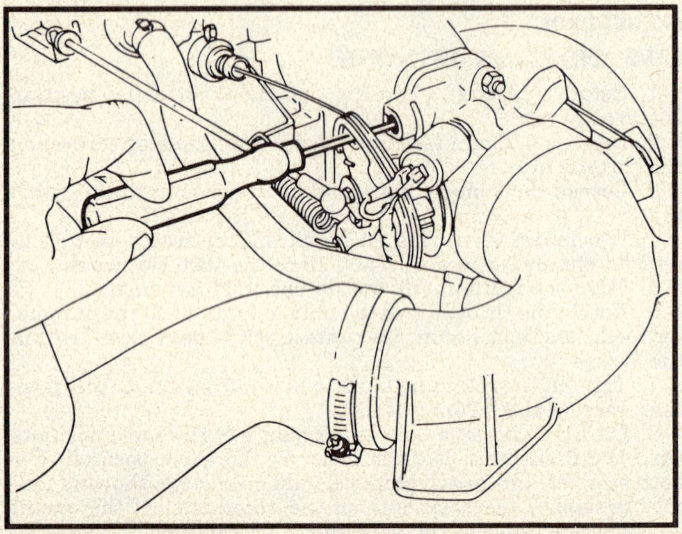

Adjust the air screw to give the correct engine rpm. LU system shown

4. Connect a voltmeter between the test wire located along the right inner fender and good ground. This wire comes from the ECU and will provide a voltage relative to the richness of the mixture.
5. Start the engine and allow it to stabilize; engine speed for LU systems should be 925 rpm for manual vehicles or 975 for automatic transmission vehicles. For LU-2 systems, correct idle speed is 925 rpm. Allowable variation in idle speed is ±25 rpm for either system.
6. Voltage at the test connector should be 7.2 ± 0.2 volts for LU systems or 7.0 volts for LU-2 systems.
7. If adjustment is necessary, drill off-center into the center of the anti-tamper plug on the air flow sensor. Use a 4.5 mm drill and do not exceed 5 mm in depth. Insert a sheet metal screw into the hole and remove the plug.

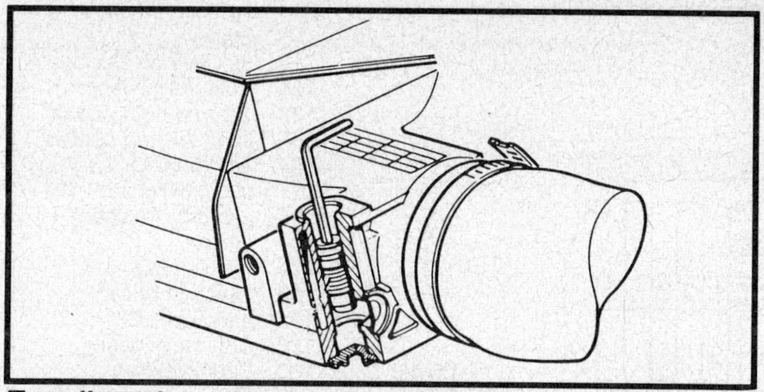

To adjust the voltage, turn the richness screw after the anti-tamper plug is removed

8. Adjust the air screw (idle adjust screw) to obtain the proper idle speed.
9. Turn the richness screw, under the anti-tamper plug, to adjust the mixture. Turn the screw in to richen the mixture and reduce the voltage; turn the screw outward to lean the mixture or increase the voltage.
10. After each adjustment is made to the richness screw, turn the idle adjust screw to reset the idle to the correct speed. If the voltage varies, repeat the Steps 9 and 10 until the voltage is correct and stable and the idle speed is correct.
11. Install a new anti-tamper plug.
12. Switch the ignition **OFF**. Remove the test equipment.
13. Reconnect the 60° switch and the PCV hose on LU-system vehicles

LH-2.2 JETRONIC SYSTEM

1. Make certain the engine is fully warmed up, the ignition system is in good order and properly adjusted and the air filter is clean and in place with no leaks in the air intake system.
2. The idle speed must be 775 ± 25 rpm.
3. Connect a digital voltmeter to the test connector located on the right inner fender, near the air filter. With the engine running, the voltmeter should display an alternating sweep from 0–13 volts and 13–0 volts. This display, when present, indicates proper adjustment of the idle richness.
4. If adjustment is necessary, switch the ignition **OFF**.

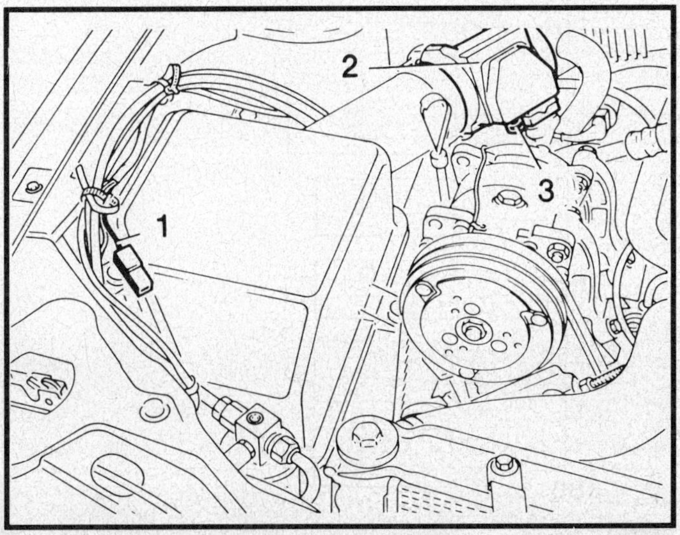

For LH-2.2 systems, connect a digital voltmeter to the test connector (1). Remove the air flow meter (2), then remove the anti-tamper plug (3) from the richness adjusting screw

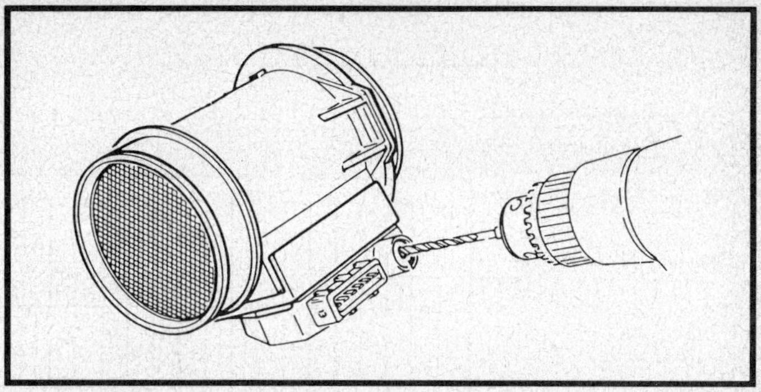

Use great care when removing the anti-tamper plug

5. Remove the air flow sensor.
6. Drill into the center of the anti-tamper plug on the air flow sensor. Use a 4.5 mm drill and do not exceed 5 mm in depth. Insert a sheet metal screw into the hole and remove the plug.
7. Reinstall the air flow sensor.
8. Start the engine, allowing the rpm to stabilize. Adjust the mixture screw to yield the correct 0–13/13–0 volt display on the voltmeter.
9. Switch the engine **OFF**. Install a new anti-tamper plug on the air flow sensor.

M1.3 MOTRONIC SYSTEM

The idle richness is not adjustable. Control of the idle mixture is performed by the ECU based on signals from the Lambda (oxygen) sensor. The ECU performs a self-check of the system during operation.

Idle speed may be adjusted if necessary. Insure that all the following conditions are met:

- Ignition system in good condition and properly set.
- Air filter clean and in place; no air intake leaks.
- Initial throttle flap position set correctly.
- Throttle switch properly adjusted.
- Engine fully warm, cooling fans OFF.
- Air conditioning **OFF**.
- Transmission in Neutral or **P**.

Turn the idle adjusting screw to alter the idle speed. Correct idle speed is 850 rpm ± 50 rpm.

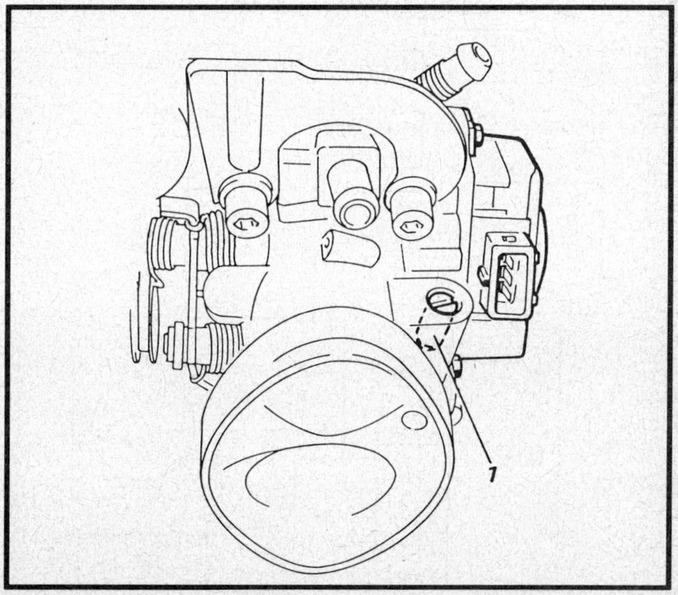

Idle speed adjust screw, M1.3 Motronic system. Idle mixture is not adjustable

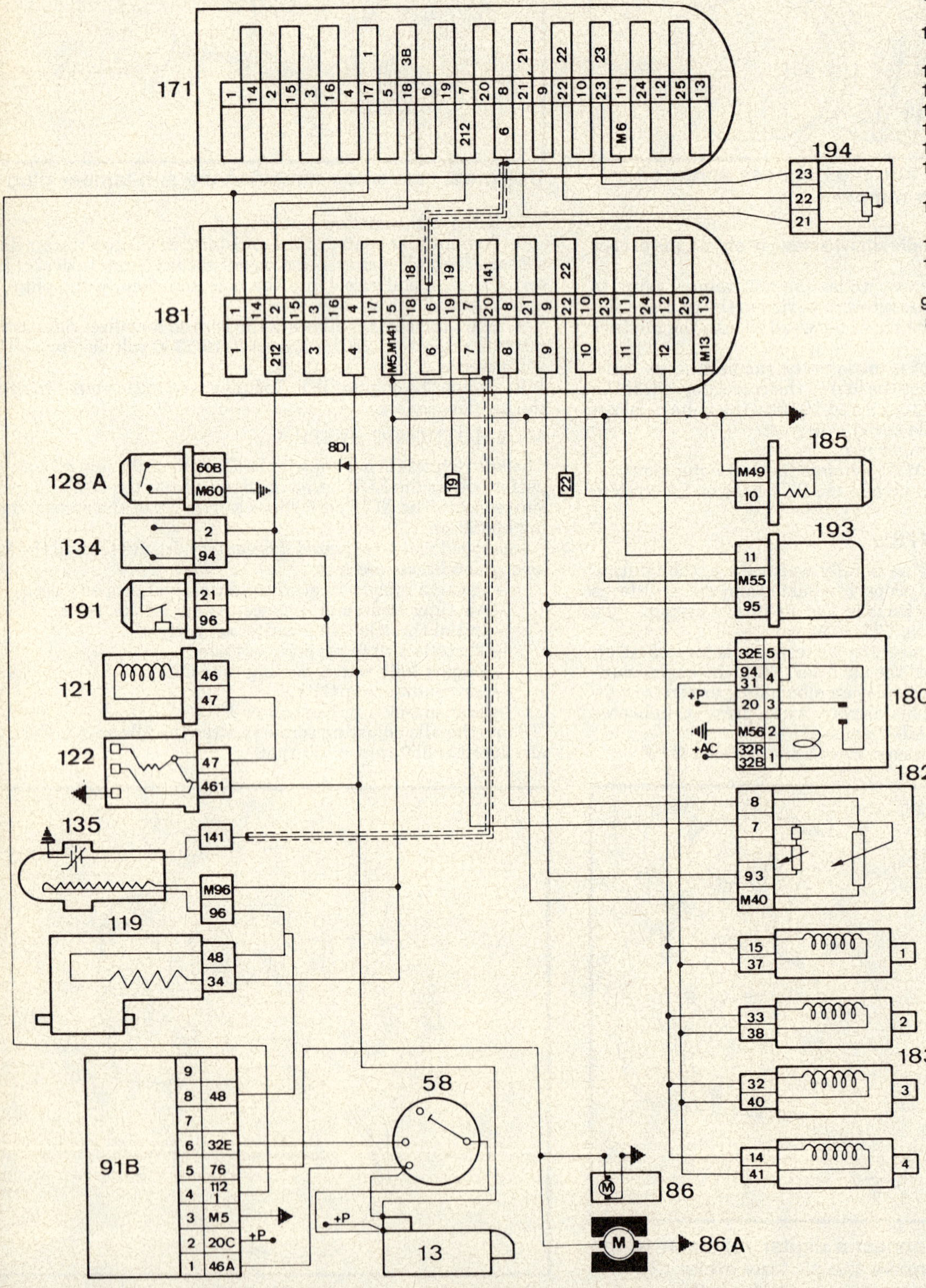

- 13. Starter
- 119. Auxiliary air device
- 121. Cold start injector
- 122. Thermo-time switch
- 128A. 60°C thermo-contact
- 134. Throttle flap microswitch
- 135. Lambda (oxygen) sensor
- 171. Ignition ECU
- 180. Injection relay
- 181. Injection ECU
- 182. Air flow sensor
- 183. Fuel injectors
- 185. Engine temperature (NTC) sensor
- 191. Overboost pressure sensor
- 193. Altitude sensor
- 194. Throttle flap potentiometer
- 91B. Tachymetric relay
- 8DI. Injection deceleration cut-out diode

Fuel injection wiring schematic—LU Jetronic

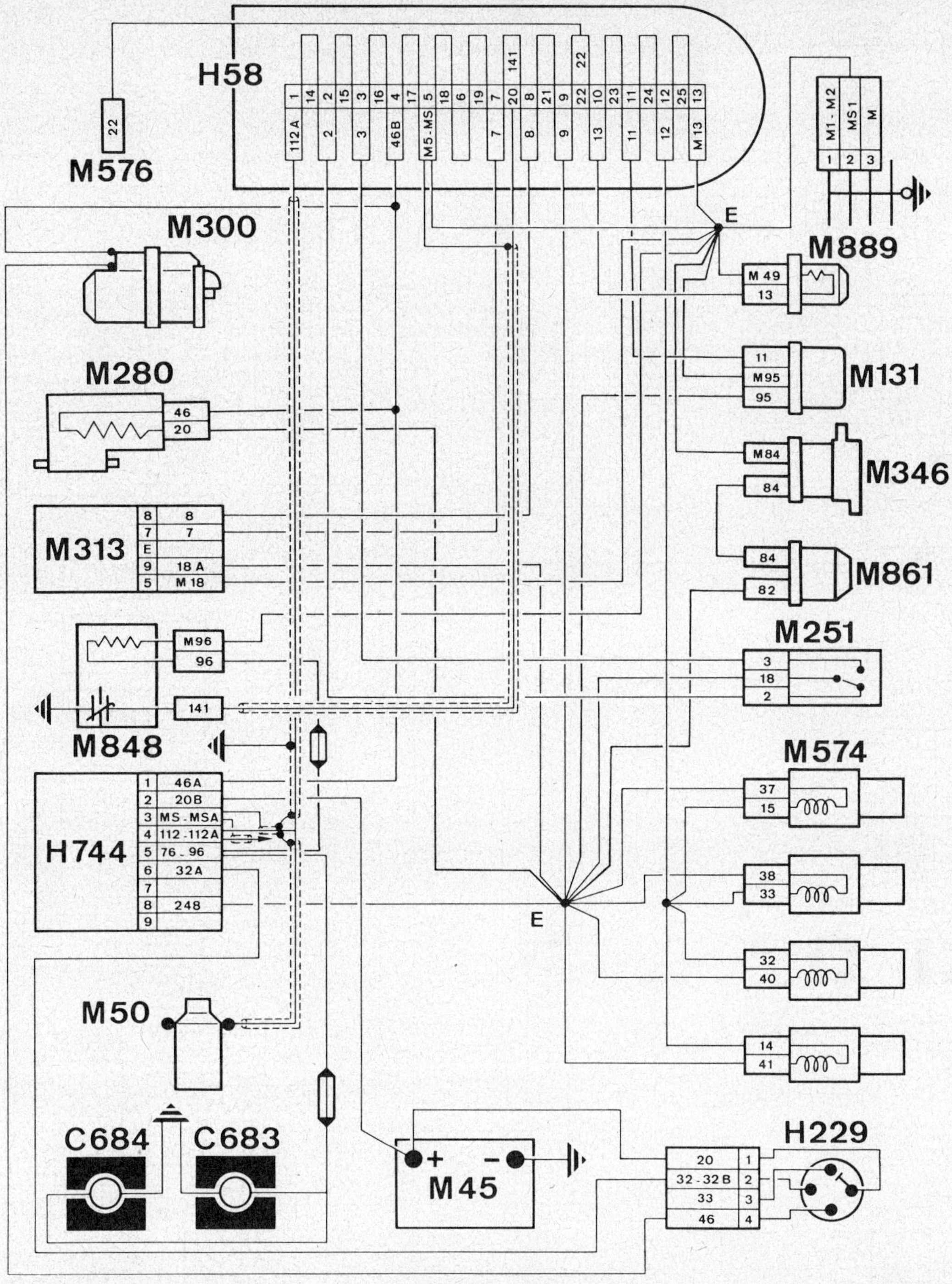

M45. Battery
M50. Ignition Coil
M131. Altitude sensor
M251. Throttle switch unit
M280. Auxiliary air device
M300. Starter motor
M313. Air flow sensor
M346. Fuel vapor recovery electrovalve
M574. Fuel injectors
M576. Test connector for idle richness
M848. Lambda (oxygen) sensor
M861. 30°C Thermo-contact
M889. Engine temperature (NTC) sensor
H58. Injection ECU
H229. Ignition switch
H744. Tachymetric relay
C683. Main fuel pump
C684. Primer fuel pump

Fuel injection wiring schematic—LU-2 Jetronic

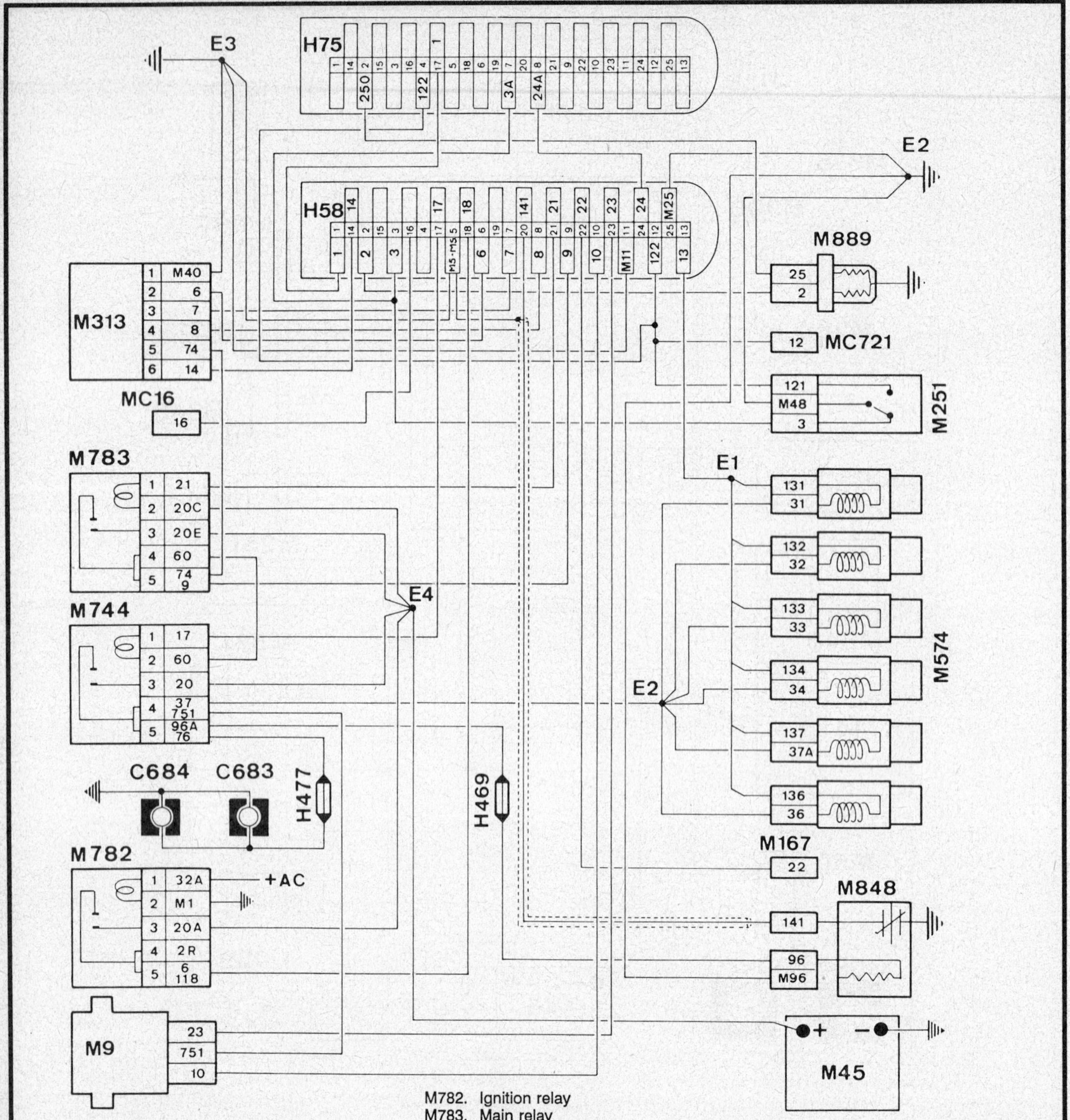

M9. Idle regulation valve
M45. Battery
M167. Test connector for idle richness
M251. Throttle switch unit
M313. Air flow sensor
M574. Fuel injectors
M744. Fuel pump relay
M782. Ignition relay
M783. Main relay
M848. Lambda (oxygen) sensor
M889. Engine temperature (NTC) sensor
MC16. A/C load signal to ECU
MC721. Test connector for idle preset
H58. Injection ECU
H477. Fuel pump fuse
H469. Lambda sensor heater fuse
C683. Main fuel pump
C684. Primer fuel pump

Fuel injection wiring schematic—LH-2.2 Jetronic

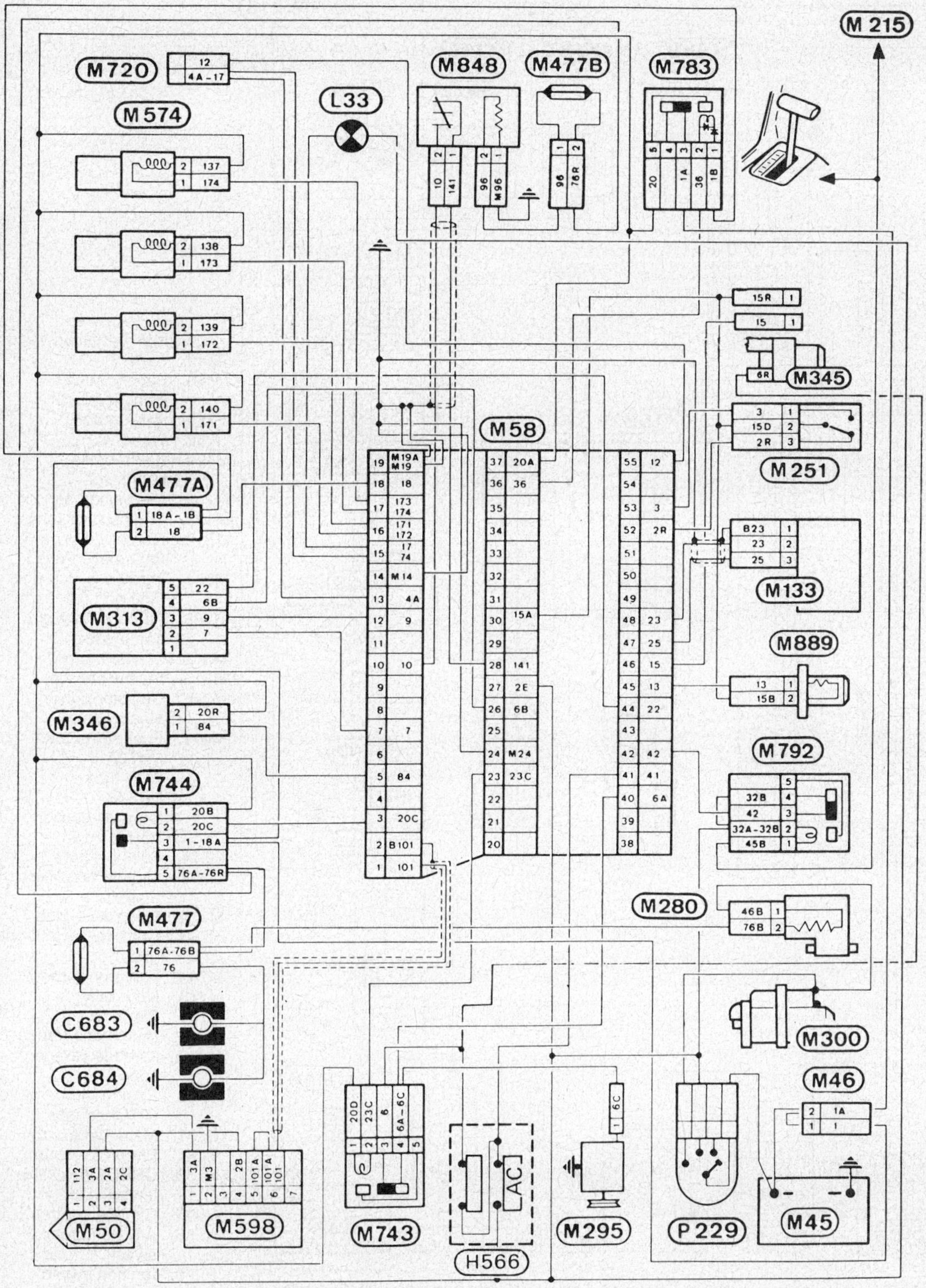

M45. Battery
M46. Battery power junction block
M50. Ignition coil
M58. Motronic ECU
M129. Radio noise supressor
M280. Auxiliary air decice
M133. Engine rpm sensor
M215. Neutral safety switch (automatic trans.)
M251. Throttle switch unit
M295. Air conditioning compressor
M300. Starter motor
M313. Airflow sensor
M345. A/C idle compensator electrovalve
M346. Canister purge electrovalve
M477. Fuel pump fuse M477A ECU fuse M477B Lambda sensor heater fuse
M574. Fuel injectors
M598. Ignition amplifier module
M720. Diagnostic plug
M743. Compressor cut-out relay
M744. Fuel pump relay
M783. Main relay
M792. Automatic transmission input relay
M848. Lambda (oxygen) sensor
M889. Engine temperature (NTC) sensor
C683. Main fuel pump
C684. Primer fuel pump
L33. CHECK ENGINE lamp
H97. Climate control ECU
H566. A/C compressor switch
P229. Ignition switch

Fuel injection wiring schematic—M1.3 Motronic, XU9J2 (SOHC) engine

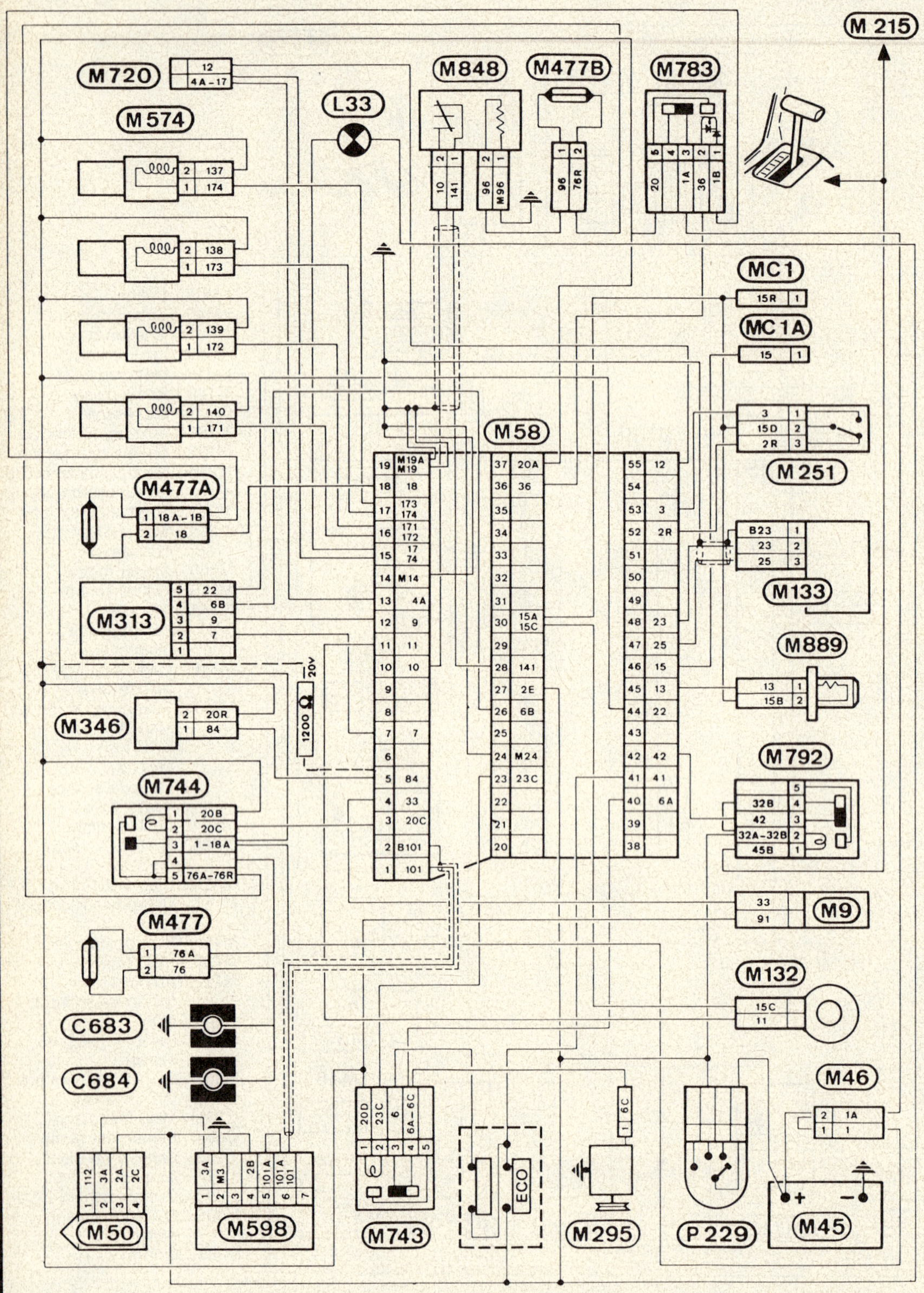

M9. Idle regulation valve
M45. Battery
M46. Battery power junction block
M50. Ignition coil
M58. Motronic ECU
M129. Radio noise supressor
M132. Detonation sensor
M133. Engine rpm sensor
M215. Neutral safety switch (automatic trans.)
M251. Throttle switch unit (idle and full load)
M295. Air conditioning compressor
M300. Starter motor
M313. Airflow sensor
M346. Canister purge electrovalve
M477. Fuel pump fuse
M477A ECU fuse
M477B Lambda sensor heater fuse
M574. Fuel injectors
M598. Ignition amplifier module
M720. Diagnostic plug
M743. Compressor cut-out relay
M744. Fuel pump relay
M783. Main relay
M793. Starter relay (Automatic trans.)
M848. Lambda (oxygen) sensor
M889. Engine temperature (NTC) sensor
C683. Main fuel pump
C684. Primer fuel pump
L33. CHECK ENGINE lamp
H97. Climate control ECU
H566. A/C compressor switch
P229. Ignition switch

Fuel injection wiring schematic—M1.3 Motronic, XU9J4 (DOHC) engine

Porsche Fuel Injection

12 SECTION

LH-JETRONIC AND DIGITAL MOTOR ELECTRONICS

911 TURBO CIS CONTINUOUS INJECTION SYSTEM

PORSCHE LH-JETRONIC AND DIGITAL MOTOR ELECTRONICS (DME) ENGINE FUEL MANAGEMENT SYSTEMS

General Information

Porsche vehicle use 2 forms of electronic fuel injection. The 928 uses LH-Jetronic fuel injection. The 911, 944 and the 968 use Digital Motor Electronics (DME) fuel injection. Both systems are advanced versions of their fuel injection systems and provide excellent control of emissions, fuel economy and performance.

LH-JETRONIC FUEL SYSTEM

LH is an abbreviation of Luftmassenmes ser Hitzdraht (hot-wire air mass meter). The most important component is the new type of air flowmeter that measures the mass flow of the air instead of its volume, as in conventional fuel injection systems. The air flow sensor consists of a compensating resistor, hot wire and measuring resistor wired together with a balancing resistor to make up a bridge circuit.

Regulation is accomplished with a variable gain amplifier, which compares the voltage at the inlet and outlet of the hot wire and makes corrections when necessary. Since the heating current flows through the measuring resistor, a current change can be taken via this resistor as a voltage drop. This voltage drop is a signal for the LH control unit, which represents the drawn in air mass.

Each time the engine is stopped and after a delay of approximately 4 seconds, the bridge circuit is altered for approximately 1 second, so that the hot wire is heated to a temperature of approximately 1832°F (1000°C) to burn free any deposits which could influence the resistance or measuring results.

A temperature sensor (with 2 negative temperature coefficient resistors) installed in the coolant circuit performs 2 operation. One of the resistors is connected with the ignition control unit and the other with the LH control unit. The resistor belonging to the LH control unit regulates the engine warming up phase, by allowing the LH control unit to extends the injection time when the engine is cold.

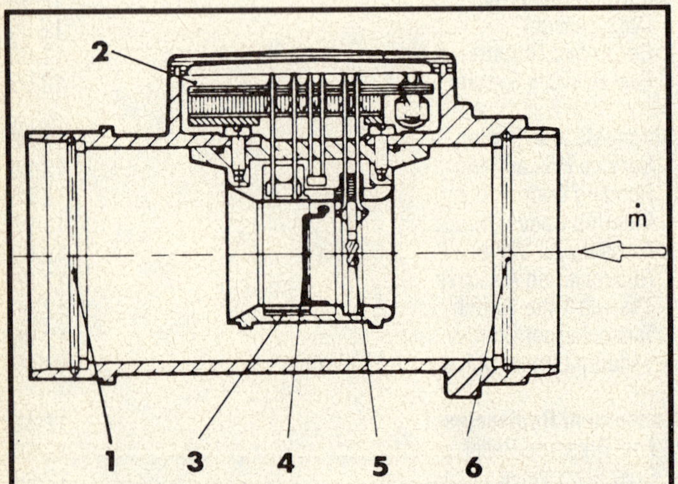

1. Protective grid
2. Regulator housing
3. Measuring resistor
4. Heated wire
5. Intake air compensation resistor
6. Protective grid

Cut away section of the air mass sensor—928 and 968

LH Control Unit And Relay

Both the LH fuel injection and the EZK electronic ignition control units are mounted next to the main fuse/relay panel behind the right side kick panel, just inside the right door. The LH unit has the black connector. Power is supplied to each unit at all times to maintain certain memory functions. The control unit completes the ground circuit of a relay coil to supply power to other components in the system. The LH relay is in position XXV on the main fuse/relay panel. The EZK relay is in position XVI.

DME FUEL SYSTEM

The DME combines the ignition, fuel and oxygen sensor systems into a single control unit. Signals from the various sensors for intake air volume, speed, crankshaft position, coolant temperature, intake air temperature, throttle position, start signal and oxygen sensor signal are sent to the control unit.

The DME control unit, a micro-computer made up of microprocessors, calculates the necessary ignition timing and optimal injection rate from the sensor signals and specific data for the engine stored in the computer according to a prescribed program. This allows both fuel economy and power to be maximized.

Earlier versions of the DME control unit consists of a 35-pin plug connection and is connected with components on the engine via a multiple-pin plug. The current version of the DME control unit consists of a 55-pin plug connection and has a fault memory for self-diagnosis, in order to be able to read out certain faults in the fuel injection and ignition.

The new DME control unit offers sequential control of the fuel injection; thereby, the entire calculated amount of fuel is distributed to each cylinder separately (once for each working stroke) according to the firing order. Injection timing is advanced before the overlapping TDC of the firing order (advanced fuel injection) for certain engine conditions.

The DME control unit is capable of self-diagnosis. Faults can be called from the memory with a specially developed diagnosis tester 9268, which can be connected on a plug connection provided on the DME wire harness. In addition to reading fault codes, the tester can activate certain functions and therefore check the function to these components.

DME Control Unit And Relay

In the 911 the DME control unit is mounted under the driver's seat. In the 944 and 968 the DME control unit is mounted under the passengers footrest. The test connector for the Diagnostic tester 9268 or 9288 is behind the right side kick panel on the 911 or under the glove compartment on the 944 and 968. The main DME relay is in the relay/fuse panel in position R41 for 911 or G5 for 944. The relay housing contains 2 relay coils and contact sets, one set for DME system power and another for the fuel pump. The power relay supplies power to the control unit, injectors and ignition system any time the ignition switch is **ON**. The control unit then completes the ground for the pump relay coil when engine speed is above about 20 rpm. If power to the control unit is less than 10 volts or more than 16 volts, a fault code will be set in memory. On vehicles equipped with the factory installed anti-theft system, the control unit will not operate the injectors when the system is armed.

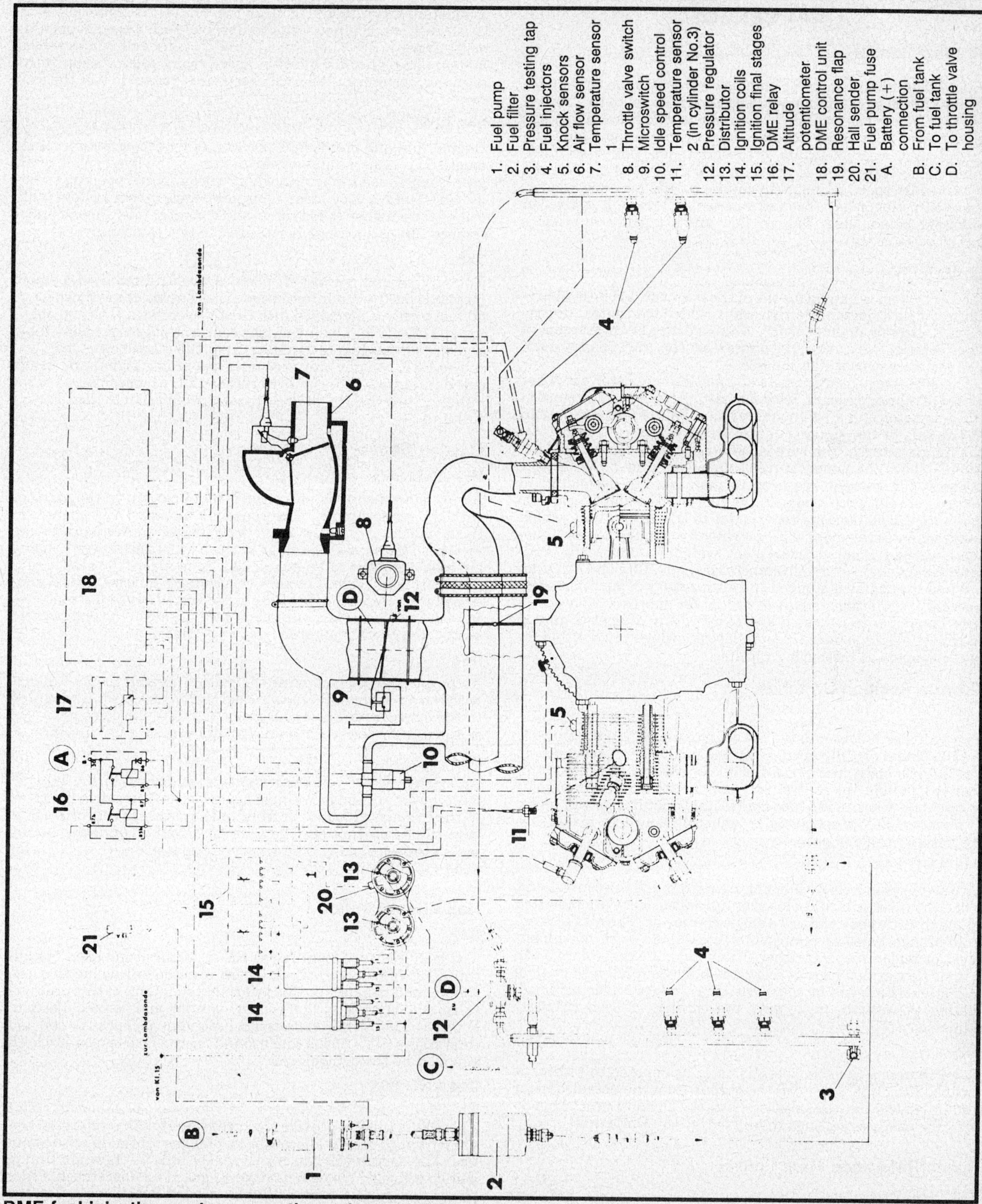

DME fuel injection system operating schematic—911

DATA SENSORS

Air Flow Sensor

When air enters the housing and moves the air vane, it changes the position of a potentiometer. The return signal to the control unit is calculated as air flow. The sensor also contains a temperature sensor at the inlet side. If either of these sensors fails, the control unit will substitute preprogrammed values to allow "limp home" operation of the engine. On previous models, the CO and idle speed adjustments were on the air flow sensor. These adjustment have been eliminated, CO and idle speed can no longer be adjusted. The air flow and temperature sensors cannot be repaired.

Air Mass Sensor

The LH system is named for the air flow sensor. LH is an abbreviation of Luftmassenmes Hitzdraht, which translates roughly to "hot wire air mass sensor". Although the DME systems of other vehicles use a vane type air flow sensor, the 968 uses a hot wire air mass sensor like the 928.

The air mass sensor consists of a platinum resistor wire that is part of a bridge circuit, with a measuring resistor, a compensation resistor and a balancing resistor. An amplifier mounted on the side of the housing controls and monitors this circuit. The temperature of the wire is maintained at a constant 212°F (100°C) above the temperature of the air around it. As air flow changes, the current required to maintain the temperature changes. As the circuit adjusts the current, the amplifier reads the change in voltage and reports this to the control unit, which then calculates the mass of air flowing through the sensor housing. This type of air measurement system automatically compensates for changes in air temperature, humidity and altitude.

When the engine is stopped and after a delay of approximately 4 seconds, the bridge circuit is altered for approximately 1 second. The hot wire is heated to a temperature of approximately 1832°F (1000°C) to burn free any deposits which could influence the resistance or measuring results.

Throttle Position Switches

911

Two switches are used to tell the control unit when the throttle is fully closed or fully open. They are mounted on each side of the throttle body, just below the idle speed control valve. If either switch fails, the control unit will use air flow sensor signals to calculate idle and full load control and a fault code will be set in memory. Before replacing or adjusting a switch, make sure the throttle cable is properly adjusted.

944 AND 928

Two switches in a single housing are used to tell the control unit when the throttle is fully closed or fully open. They are mounted to the throttle body and use a single connector. The intake manifold branch must be removed to access the switch for adjustment or replacement. If either switch fails, the control unit will use air flow sensor signals to calculate idle and full load control and a fault code will be set in memory. Before replacing or adjusting the switch, make sure the throttle cable is properly adjusted.

968

The 968 uses a throttle potentiometer as opposed to a throttle switch. The DME unit read the output from the potentiometer and infers the throttle position from the signal it gets. This allows for adaptive settings to be used by the DME unit.

Speed/Reference Mark Sensor

911

Engine rpm and crankshaft position are detected with a single magnetic pickup mounted near the top of the flywheel. A 60 tooth gear is machined into the flywheel with a special opening at 84 degrees BTDC of No. 1 cylinder. This new single sensor system is designed for a higher speed range and a greater tolerance to interference. An oscilloscope is required to test the sensor.

944 AND 968

Engine rpm and crankshaft position are detected with a single magnetic pickup mounted near the top of the flywheel. A tooth gear is machined into the flywheel with a special opening to signal BTDC of No. 1 cylinder. This new single sensor system is designed for a higher speed range and a greater tolerance to interference. An oscilloscope is required to test the sensor.

928

Engine rpm and crankshaft position are detected with a single magnetic pickup mounted on top of the engine at the flywheel. A 60 tooth gear is machined into the flywheel with a special opening at TDC of No. 1 cylinder. The sensor is not height adjustable and the sensor gear is separate from the starter ring gear.

The EZK control unit processes the sensor signal and sends speed information to the LH control unit. An oscilloscope is required to test the sensor and signal transmission to the LH control unit.

Oxygen Sensor

911 AND 968

An oxygen sensor compares the level of oxygen in the exhaust with the air around it. The sensor is mounted low enough in these vehicles that a new water proof oxygen sensor has been developed. The normal openings in the back of the sensor housing are sealed and the reference air now comes from the sensor connector. It is important to keep this connector clean, contamination of the reference side of the sensor could cause the sensor to fail. The connector is mounted on the left side of the engine, separate from the temperature and rpm sensor plugs.

944

An oxygen sensor compares the level of oxygen in the exhaust with the air around it. A single heated oxygen sensor is mounted in the exhaust pipe just in front of the catalytic converter. The sensor can be tested with a voltmeter and/or a CO meter.

928

An oxygen sensor compares the level of oxygen in the exhaust with the air around it. A single heated oxygen sensor is mounted in the exhaust pipe just in front of the catalytic converter. The wire comes up through the floor to a round connector at the main fuse/relay panel. The sensor can be tested with a voltmeter and/or a CO meter.

Hall Effect Sensor

911

The Hall sensor is mounted in the distributor and used to indicate TDC of cylinder No. 1. There is no default value that the control unit can use for this information, failure of this unit will cause a no-start condition with no ignition and no fuel injection. If only the ignition system is not functioning, check the coil and final stages first. Power and ground for the Hall sensor are both supplied by the control unit.

944 AND 968

The Hall sensor is mounted on the exhaust camshaft drive wheel. It is used to provide the control unit with a reference signal for knock sensor allocation and for sequential injection control. The engine will run if this sensor fails but ignition timing will be retarded about 6 degrees to guard against knock. Since power and ground for the Hall sensor are both supplied by the control unit, it can only be tested with an oscilloscope.

equipped with an on-board diagnostic system that reports fault codes by flashing the warning light. No special tester is needed to access the codes, although the diagnostic tester can still be used as described above. To access the codes on all 928 models and on California 911 Carrera models:

1. Turn the ignition switch **ON** without starting the engine.
2. Fully press the accelerator pedal to close the full load switch. After about 3 seconds the warning light will flash.
3. Release the accelerator to close the idle switch, then count and write down the warning light flashes. Repeat this process until Code 1000 appears.

The control unit will report all Codes stored in memory in numerical order. All Codes are 4 digit numbers. The first digit will be 1 if the fault code is stored in the DME or LH fuel injection control unit. On 928 models, if the first digit is 2, the fault code is in the EZK ignition control unit.

If the second digit is 1, the detected fault is current. If the second digit is 2, the detected fault has not occurred during the last running of the vehicle but did occur within the last 50 engine starts. The remaining 2 digits indicate which component is at fault. Consult the chart for a listing of DME, LH and EZK system fault codes.

CLEARING CODE MEMORY

The fault code memory should be cleared before returning the vehicle to service. All codes in memory and the idle control adaptation are lost when the control unit or the battery is discon-

DME ENGINE FAULT CODES—911, 944 AND 968

Fault Code	Explanation
1000	End of report
1111	Control unit supply voltage out of limit
1112	Throttle idle position switch grounded
1113	Throttle full load position switch grounded
1114	Engine temperature sensor
1115	Throttle idle position switch open
1121	Air flow sensor
1122	Idle speed control circuit
1123	Oxygen regulation out of limit
1124	Oxygen sensor signal
1125	Inlet air temperature sensor signal
1131	Knock sensor 1
1132	Knock sensor 2
1133	Knock regulation in control unit
1134	Hall sender signal missing
1141	Control unit faulty
1143	Carbon canister purge valve
1144	Resonance plate solenoid valve circuit
1145	CHECK ENGINE warning light
1151	Injector, cylinder 1 (911 only)
1152	Injector, cylinder 2 (911 only)
1153	Injector, cylinder 3 (911 only)
1154	Injector, cylinder 4 (911 only)
1155	Injector, cylinder 5 (911 only)
1156	Injector, cylinder 6 (911 only)
1500	No faults in memory

NOTE: If the second digit of the code is 2, the fault is not present at the time of testing. It may be an intermittent fault.

LH ENGINE FAULT CODES—928

Fault Code	Explanation
1000	End of report
1111	Control unit supply voltage out of limit
1112	Throttle idle position switch grounded
1113	Throttle full load position switch grounded
1114	Coolant temperature sensor
1115	Throttle idle position switch open
1121	Air flow sensor
1123	Air/fuel mixture rich
1124	Air/fuel mixture lean
1125	Oxygen sensor signal missing
1141	LH control unit faulty
1500	No faults in memory

NOTE: If the second digit of the code is 2, the fault is not present at the time of testing. It may be an intermittent fault.

nected. To avoid the loss of the learned idle program, use the instructions on the 9288 tester to clear the memory. If this tester is not available, disconnect the control unit to clear the memory. It will be necessary to drive the vehicle for at least 6 minutes and run the engine at idle for about 10 minutes so the control unit can learn idle speed, timing and mixture parameters. Make sure the engine is at operating temperature and that all accessories are **OFF**. The throttle-at-idle position switch must be closed and functioning or system adaptation will not take place.

FUEL SYSTEM TESTING

Flow Testing

CAUTION

The following procedure will produce fuel vapors. Make sure there is proper ventilation and take the appropriate fire safety precautions.

911

1. Remove the hot air blower or duct on the left side of the engine compartment.
2. At the front end of the fuel injector supply rail, disconnect the return line and install a 5 ft. (1.5m) hose onto the rail. The rail threads are M16 x 1.5. Install the line securely so it does not blow off with the fuel pressure. Place the other end of the hose into a measuring container that holds at least 32 oz. (950 cc).
3. Locate the DME relay on the fuse/relay panel in the luggage compartment, relay R41. Remove the relay and connect a jumper wire with a switch to terminals 30 and 87b. The pump will run whenever the terminals are jumpered together even without the ignition switch **ON**, so be sure to use a jumper wire with a switch.
4. Turn the switch **ON** to run the pump for 30 seconds. Time the run carefully. The pump should deliver at least 29 oz. (850 cc) in 30 seconds. If the flow is low, pressure test the system before deciding the pump is faulty.

928, 944 AND 968

1. On the 928, at the rear of the engine, the fuel rails of both sides join together and connect to a return line. On the 944 and 968 there is only 1 return line. Disconnect the return line and install a 3–5 ft. hose on the fitting. Install the line securely so it does not blow off with the fuel pressure. Securely place the other end of the hose into a measuring container that holds at least 2 quarts (2000cc).
2. Locate the fuel pump relay on the main fuse/relay panel and connect a jumper wire with a switch to terminals 30 and 87. The pump will run whenever the terminals are jumpered together even without the ignition switch **ON**, so be sure to use a jumper wire with a switch.
3. Turn the switch **ON** to run the pump for 30 seconds. Time the run carefully. On the 928 the pump should deliver at least 42 oz. (1250cc) in 30 seconds. On the 944 and 968 the pump should deliver at least 29 oz. (850cc) in 30 seconds. If the flow is not correct, pressure test the system before replacing any parts.

Pressure Testing

911

1. Remove the hot air duct on the left side of the engine compartment. If the pump has been flow tested, reconnect the fuel return line.
2. On the rear of the left fuel injector supply rail, remove the cap and ball from the test connector. Be careful not to drop the ball. Install pressure gauge P 378 or VAG 1318 or equivalent 140 psi (10 BAR) gauge to the test connection.
3. If the engine does not start, locate the DME relay on the fuse/relay panel in the luggage compartment, relay R41. Remove the relay and connect a jumper wire with a switch to terminals 30 and 87b. The pump will run whenever the terminals are jumpered together even without the ignition switch **ON**, so be sure to use a jumper wire with a switch.
4. With the pump running but with no vacuum to the pressure regulator (engine not running), pressure should be 55 psi (3.8 BAR). With the engine at idle and the pressure regulator vacuum line connected, fuel system pressure should be about 48 psi (3.3 BAR).
5. Stop the engine or pump and watch the gauge. The system should still hold at least 43 psi (3.0 BAR) after 20 minutes. If the pressure leaks off, perform the following procedure:
   a. Disconnect the return line from the pressure regulator. If fuel is coming out of the regulator, the regulator is faulty.
   b. Build up system pressure again and disconnect the fuel pump inlet hose. Clamp the hose so fuel does not run out of the tank. If fuel leaks from the pump inlet, the check valve is faulty.
   c. If the check valve and regulator are both good, leak check the injectors.
6. If pressure is low in either test but does not leak off and the pump flow is correct, the pressure regulator at the front of the left supply rail is faulty and should be replaced.
7. When installing the test connector cap nut, torque to 10 ft. lbs. (12 Nm).

928, 944 AND 968

1. At the front end of the fuel injector supply rail (right side on 928), remove the cap and sealing ball from the test connection. Be careful not to drop the ball.
2. Install pressure gauge P 378, or VAG 1318 or equivalent 140 psi (10 BAR) gauge to the test connection.
3. Remove the fuel pump relay and connect a jumper wire with a switch to terminals 30 and 87. The pump will run whenever the terminals are jumpered together even without the ignition switch **ON**, so be sure to use a jumper wire with a switch.
4. With the pump running but with no vacuum to the pressure regulator (engine not running), system pressure should be 55 psi (3.8 BAR). With the engine at idle and the pressure regulator vacuum line connected, fuel system pressure should be about 48 psi (3.3 BAR).
5. Stop the engine or pump and watch the gauge. The pressure should drop no more than 7.25 psi (0.5 BAR) after 30 minutes. Failure to hold pressure can cause hot start problems. If the pressure leaks off, perform the following procedure:
   a. Run the pump to build system pressure again. Discon-

PORSCHE FUEL SYSTEM SPECIFICATION

Model	Fuel Delivery Rate cc per 30 sec	Fuel Pressure	
		System Pressure psi (BAR)	Pump Pressure psi (BAR)
911	850	52–58 (3.6–4.0)	45–51 (3.1–3.5)
928	1250	52–58 (3.6–4.0)	45–51 (3.1–3.5)
944	850	52–58 (3.6–4.0)	45–51 (3.1–3.5)
968	850	52–58 (3.6–4.0)	45–51 (3.1–3.5)

nect the return line from the pressure regulator. If fuel is coming out of the regulator, the regulator is faulty.

b. Build up system pressure again and disconnect the fuel pump inlet hose, back at the fuel tank. Clamp the hose so fuel does not run out of the tank. If fuel leaks from the pump inlet, the check valve is faulty.

c. If the check valve and regulator are both good, leak check the injectors.

6. If pressure is low in either test but does not leak off and the pump flow is correct, the pressure regulator is faulty and should be replaced.

7. When installing the test connector cap nut, torque to 10 ft. lbs. (12 Nm).

AIR FLOW SENSOR

Testing

911 AND 944

1. Unplug the sensor connector and connect a voltmeter between terminal 3 and terminal 4 of the connector. With the ignition switch **ON**, there should be about 5 volts to the sensor.

2. Turn the ignition **OFF**. Connect a jumper wire between terminal 3 of the sensor and terminal 3 of the connector. Connect a jumper wire between terminal 4 of the sensor and terminal 4 of the connector. Connect the voltmeter between terminal 4 and terminal 2 on the sensor.

3. Remove the air outlet tube from the sensor and turn the ignition switch **ON**. Use a non-metal tool to reach into the sensor and move the vane. The vane should move smoothly and the voltage should change smoothly from 0.25 to 4.6 volts. If there are jumps or dead spots in the vane movement or voltage changes, replace the sensor.

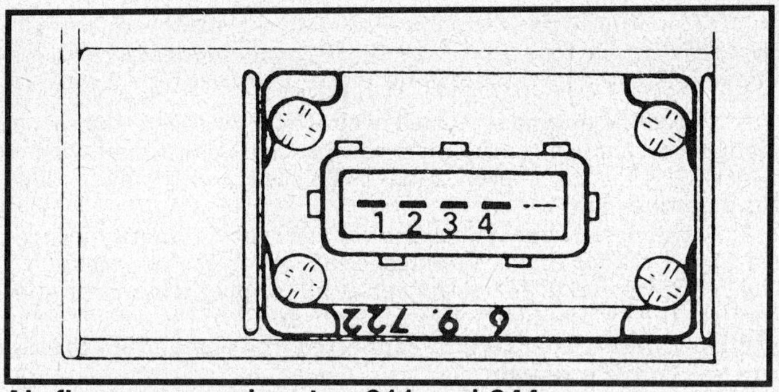

Air flow sensor pinouts—911 and 944

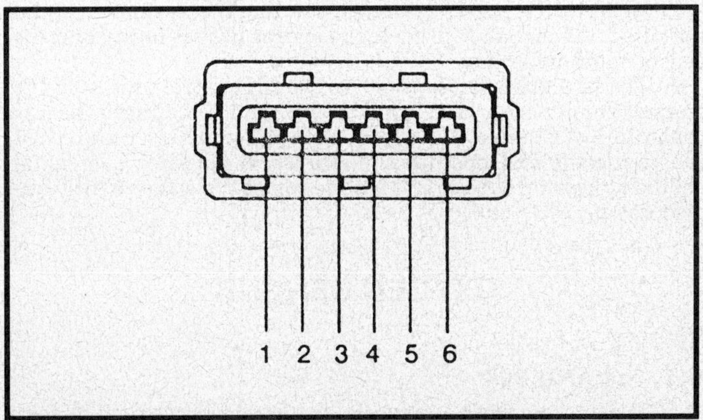

Air mass sensor pinouts—928 and 968

AIR MASS SENSOR

Testing

928 AND 968

1. To gain access to the sensor, remove the air cleaner and housing. To check for power to the sensor, unplug the control unit connector and jumper connector terminals 17 and 21 together to activate the main LH system relay. Unplug the sensor connector and check for 12 volts across terminals 2 and 4.

2. If there is no power to the connector, remove the LH relay (location XXV in the main relay panel—928) and jumper terminals 30 and 87 on the panel. If there is now power at the LH sensor connector, the relay is faulty. If there is still no power, the wiring is faulty.

3. Remove the jumper wires and check resistance across terminals 3 and 5 on the sensor housing. There should be 3.6–4.1 ohms resistance in the hot wire bridge circuit.

4. Reconnect the sensor and jumper the LH relay panel terminals 30 and 87. At the control unit connector, there should be about 1.6 volts from the sensor at terminals 6 and 7. When air is gently blown across the hot wire in the sensor, the voltage should increase. The maximum return signal is 5 volts.

5. To test the burn-off function, connect all wiring but leave the air duct off to look into the sensor. Run the engine up to operating temperature, raise the speed above 2000 rpm, then turn the ignition switch **OFF** and watch the sensor. After a delay of about 4 seconds, the wire should glow for about 1 second to burn itself clean.

THROTTLE POSITION SWITCHES

Testing

911 AND 944

1. Unplug the switch connector and connect an ohmmeter to the switch terminals. The center terminal is common, terminal 1 is the idle switch, terminal 3 is the full load switch.

2. The idle switch should be closed when the throttle is closed. Any movement of the throttle lever will open the switch just before the throttle plate actually opens. Operate the switch by hand if necessary to determine a faulty switch or faulty adjustment.

3. Connect the ohmmeter to the full load switch. The switch should close when the throttle lever is 70 degrees off of idle, about ⅔ of full stroke.

4. On the 911, if the switches are good but do not operate at the proper throttle angles, check throttle cable adjustment.

a. Make sure there is no slack in the throttle cable. The adjustment is just below the front end of the transaxle housing.

b. Make sure the throttle lever touches the idle stop on the throttle body. If not, go to Step C. Do not adjust the stop screw, this is set at the factory on a flow bench.

c. At the pedal assembly, make sure the reversing lever

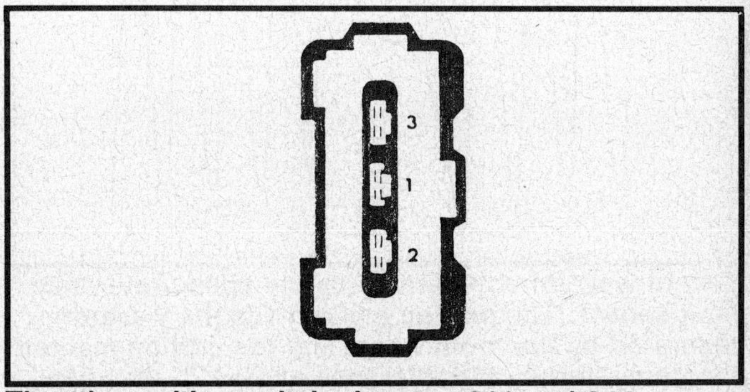

Throttle position switch pinouts—911 and 944

rests against the stop. Adjust as required with the ball socket adjuster on the reversing lever.

d. Install the floor plate and hold the throttle lever full open. Adjust the pedal push rod as required so the pedal is against the stop.

5. When adjusting the idle switch on the 911, there should be 0.035 in. (0.9mm) between the switch housing and the lever when the throttle is fully closed.

6. If the switches are good and the throttle opens and closes the switches correctly but the fault code still exists, check continuity of the wiring between the switches and the control unit. Check the control unit pin-out diagram, connect the ohmmeter to the proper terminals and move the throttle to operate the switches.

968

The throttle valve switches have been replaced by a throttle valve potentiometer on the 968. The idle position is sensed adaptively by the control unit so no adjustment is possible.

928

1. Unplug the switch connector and connect an ohmmeter to the switch terminals. The center terminal is common, terminal 2 is the idle switch and terminal 3 is the full load switch.

2. The idle switch should be closed when the throttle is closed. Any movement of the throttle lever will open the switch just before the throttle plate actually opens. Operate the lever by hand if necessary to determine a faulty switch or faulty adjustment.

3. Connect the ohmmeter to the full load switch. The switch should close when the throttle lever is about ⅔ of full stroke.

4. If the switches are good but the fault code continues to appear, check the wiring between the switches and the LH and EZK control units. Also check the throttle cable and switch housing adjustments.

SPEED/REFERENCE MARK SENSOR

Testing

911, 944 AND 968

1. Unplug the connector from the DME control unit under the driver's seat and connect an oscilloscope to terminals 47 and 48.

NOTE: When the DME control unit is disconnected, power to the unit is lost and the fault code memory will be erased. If the memory must be retained, start at Step 4.

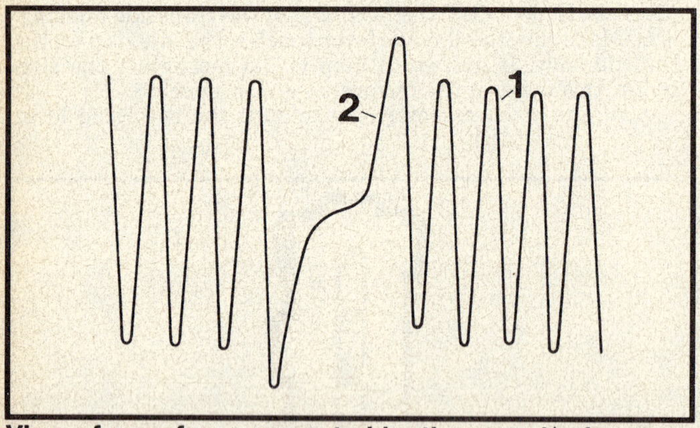

View of waveform generated by the speed/reference mark sensor. The portion marked 1 is the waveform generated by the tooth wheel and the portion marked 2 is generated by the reference gap—911, 944, 968 and 928

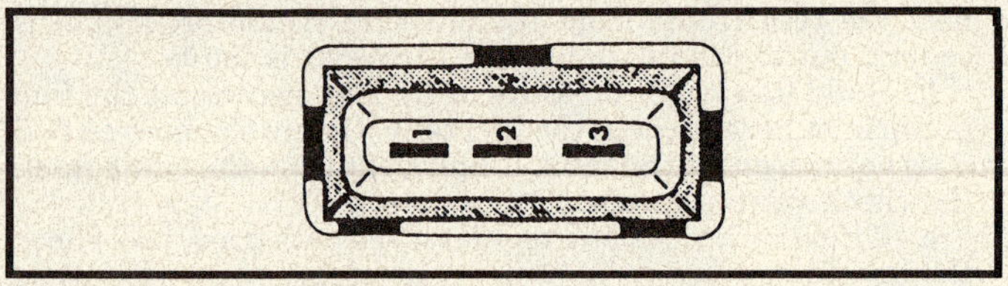

Pinout of the speed/reference mark sensor—911, 944, 968 and 928

2. When the engine is turned over at starter speed, the scope picture will show a minimum 3 volt peak-to-peak signal with a wider pulse for the reference mark. This is the only way to view the reference mark.

3. If the signal appears but the voltage is too low, adjust the distance between the sensor and the flywheel to 0.031 in. (0.8mm).

4. If the signal is still not correct or does not appear at all, locate the group of connectors at the front of the left side of the engine compartment. It may be necessary to remove the hot air blower and/or duct. The top connector is for the speed sensor.

5. Unplug the sensor connector and connect the oscilloscope to terminals 1 and 2 of the sensor. With the DME control unit connected, the engine should still run without the sensor.

6. With the engine at idle and the sensor properly adjusted, the scope should show a clean sine wave pattern. The reference mark will probably go by too fast to see.

7. Terminal 3 is a shield wire. Grounding this terminal should not affect the signal. If it does, the sensor wire is faulty and the sensor must be replaced.

928

1. Unplug the connector from the EZK control unit seat and connect an oscilloscope to terminals 6 and 23. The LH control unit must be connected.

NOTE: When the control unit is disconnected, power to the unit is lost and the fault code memory will be erased. Start at Step 4, if the memory must be retained.

2. When the engine is turned over at starter speed, the scope picture will show a minimum 3 volt peak-to-peak signal with a wider pulse for the reference mark. This is the only way to view the reference mark.

3. If the signal appears but the voltage is low, remove the sensor and check for dirt or damage, clean or replace as necessary.

4. If the signal does not appear at all, unplug the sensor and connect the oscilloscope to terminals 1 and 2 of the sensor connector. With the control unit connected, the engine should still run without the sensor.

5. With the engine at idle, the scope should show a clean sine wave pattern. The reference mark will probably go by too fast to see. Terminal 3 is a shield wire. Grounding this terminal should not affect the signal. If it does, the sensor wire is faulty and the sensor must be replaced.

6. To test signal transmission to the LH control unit, stop the engine, connect the EZK control unit and disconnect the LH control unit. Connect the scope leads to terminals 1 and 17 of the connector and operate the starter. A square wave signal should appear on the scope with the top of the wave slightly angled down.

OXYGEN SENSOR

Testing

911, 944 AND 968

1. If using the 9288 Diagnostic tester, the sensor can be tested by going to the Actual Value menu.

2. To test the sensor with a voltmeter, unplug the sensor connector and connect a digital voltmeter to the sensor terminals.

3. With the engine at idle at operating temperature, the control unit will operate the engine in open loop mode (pre-programmed air/fuel ratio) because it is not receiving an oxygen sensor signal. The sensor signal should be 150–950 millivolts and fairly steady. Rapid throttle movements should produce changes in the signal.

4. If the sensor voltage is correct but the control unit will not operate in closed loop mode, check all other sensors before replacing the control unit.

928

1. Locate and unplug the sensor connector. The single terminal is for the sensor, the 2 next to each other are for the heater. There should be 12 volts to terminal 2 any time the fuel pump relay is closed. Terminal 3 goes directly to ground.

2. To test the sensor with a voltmeter, the engine must be at operating temperature. Unplug the sensor and connect a digital voltmeter to the sensor terminal.

NOTE: When running the engine with the oxygen sensor disconnected, run the test quickly and reconnect the sensor as soon as possible. The engine will be running rich and the catalytic converter could overheat.

3. With the engine at idle, the control unit will operate the engine in open loop mode (pre-programmed air/fuel ratio) because it is not receiving an oxygen sensor signal. The sensor signal should be 0.1–1.0 volt and fairly steady. When the vacuum line is disconnected from the fuel pressure regulator, the pressure will increase and the oxygen sensor signal should also increase as the mixture becomes richer.

4. To test the control unit, install a CO meter probe into one of the test points under the hood and run the engine at idle. When the sensor is disconnected and the sensor terminal of the control unit is grounded, the CO reading should increase. Be careful not to ground the heater terminal.

5. If the sensor voltage is correct but the control unit will not respond to grounding the input wire, check all other sensors and the control unit coding before replacing the control unit.

HALL EFFECT SENSOR

Testing

911

1. Unplug the connector from the distributor and turn the ignition switch **ON**. Check for about 12 volts at terminal 3 of the connector.

2. Turn the ignition switch **OFF** and check for ground at terminal 1 of the connector.

3. The center terminal should have continuity with terminal 8 of the control unit connector. If these are all correct, the Hall effect sensor is faulty. The fault code memory in the DME control unit will be erased when the unit is disconnected.

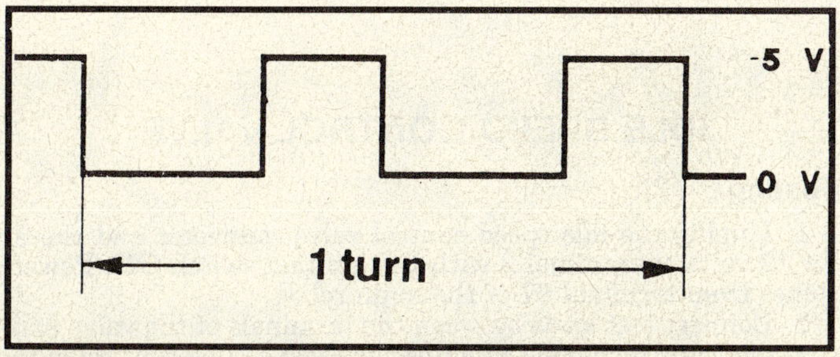

Waveform generated by the Hall effect sensor—Porsche

944 AND 968

1. Locate the sensor connector behind the camshaft drive gear. Connect test adapter lead VW 1501 or equivalent, so an oscilloscope can be connected to terminals 1 and 2 with the sensor still connected.

2. With the engine at idle, the picture on the scope should show a square wave of about 5 volts. The top of the wave form will slope down slightly.

3. If the signal is not there, the control unit will retard ignition timing about 6 degrees. Knock sensor failure will also cause the control unit to retard timing. If only the Hall sensor fault code is in control unit memory, the Hall sensor is most likely faulty.

928

1. Locate the sensor connector behind the right camshaft sprocket housing. Connect test leads so an oscilloscope can be connected to the red and green wires with the sensor still connected.

2. With the engine at idle, the picture on the scope should show a square wave of about 5 volts. The top of the wave form will slope down slightly.

3. If the signal is not there, the control unit will retard ignition timing about 6 degrees. Knock sensor failure will also cause the control unit to retard timing. If only the Hall sensor fault code is in control unit memory, the Hall sensor is most likely faulty.

INTAKE AIR TEMPERATURE SENSOR

Testing

911 AND 944

Intake air temperature is only measured on the 911 and 944. These vehicles use an air flow meter and require intake air temperature measurements for proper control unit data conversions from air flow to air mass. The 928 and 968 measure air mass directly and do not require the use of intake air temperature measurements.

1. Unplug the connector from the air flow sensor and connect an ohmmeter to terminals 1 and 4. Make sure the air flow sensor vane is not partially open. Measure the air temperature and check the chart to determine the sensor resistance. Ohmmeter readings should be within 10 percent of specification.

2. If the sensor readings are correct but the fault code still exists, check the continuity of the wiring to the control unit. The fault code memory will be erased when the control unit is disconnected.

ENGINE TEMPERATURE SENSOR

Testing

911

The 911 uses an air cooled engine and does not use an ethylene

PORSCHE TEMPERATURE SENSOR SPECIFICATIONS

Temperature °F (°C)	Resistance Ohms
32 (10)	4400–6800
60–86 (15–30)	1400–3600
104 (40)	1000–1300
175 (80)	250–390
212 (100)	160–210

These values are used for all temperature type sensors; coolant, air, engine.

glycol liquid coolant. The engine temperature sensor serves the same purpose as the coolant temperature sensor in the liquid cooled Porsche vehicles.

1. The sensor connector is on the left side of the engine, just above the fuel return line. It may be necessary to remove the hot air blower and/or duct to locate the plug.
2. Unplug the connector and connect an ohmmeter to the sensor terminals. Measure the temperature and check the chart to determine the sensor resistance. Ohmmeter readings should be within 10 percent of specification.
3. If the sensor readings are correct but the fault code still exists, check the continuity of the wiring to the control unit. The fault code memory will be erased when the control unit is disconnected.

COOLANT TEMPERATURE SENSOR

Testing

928, 944 AND 968

1. Unplug the sensor connector and check the condition of the plug and socket.
2. On the 928 dual element sensor, check the resistance between each terminal and ground. Terminal 1 connects to the EZK control unit, terminal 2 goes to the LH control unit. On other vehicle check the resistance between the 2 terminals of the sensor directly.
3. If the sensor readings are correct but the fault code still exists, check the continuity of the wiring to the control units. The fault code memory will be erased when the control unit is disconnected.

IGNITION CIRCUIT MONITORING RELAY

Testing

928

1. The relay is mounted on the bracket with the LH and EZK control units, near the main fuse/relay panel. If an LED on the Relay is **ON** when the engine is running, first check the ignition system and make sure both sides are functioning.
2. With the ignition switch **OFF**, unplug the Monitoring Relay and check the thermocouple circuit resistance across terminals E1 and E2 of the connector. The resistance for the whole circuit should be 5–10 ohms. If the circuit is open, check the thermocouples separately. They are mounted in the heads near the exhaust manifold. Thermocouples cannot be repaired.
3. Terminal 31 should have continuity to ground at all times. Terminal AL should have continuity to ground when the ignition is **OFF**. Make sure the ignition is **OFF** when checking this terminal or the system could be damaged.
4. With the relay connected and the ignition switch **ON**, there should be 12 volts at terminals 87, 15, A1 and A2. There should be about 2.7 volts at terminals E1 and E2.
5. When the engine is running, there should be 12 volts at terminals 61 and AL.

FUEL INJECTORS

Testing

911

1. Unplug the injector connectors and check the resistance across the injector terminals, it should be about 16 ohms.
2. With the ignition switch **ON**, check for 12 volts at the white wire terminal of the connector. This power comes from the DME relay R41.
3. To check the voltage from the control unit, install a test connector that allows connecting an oscilloscope to the colored wire of the injector plug when the wiring is connected. Make sure the test connector does not ground the circuit.

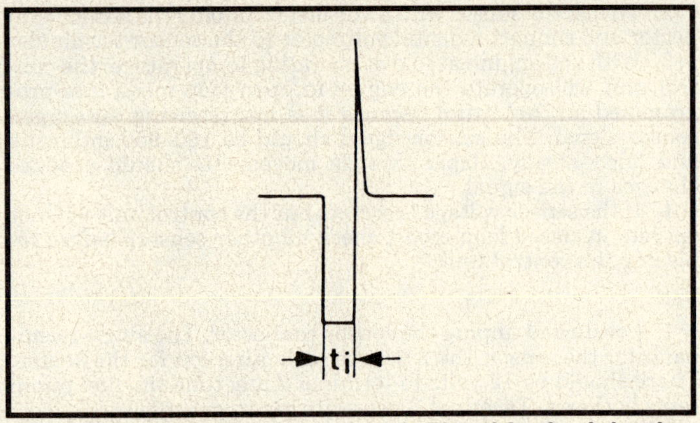

Waveform of the injector signal. The ti is the injection time—Porsche

4. With the engine at idle, the scope picture should be the same on all injectors.

944 AND 968

1. Unplug the injector connectors and check the resistance across the injector terminals, it should be about 16 ohms (2–3 ohms for 1990 944).
2. With the ignition switch **ON**, check for 12 volts at the white wire terminal of the connector.
3. To check the voltage from the control unit, install a test connector that allows connecting an oscilloscope to the colored wire of the injector plug when the wiring is connected. Make sure the test connector does not ground the circuit.
4. With the engine at idle, the scope picture should be the same on all injectors.

928

1. If the engine will run at idle, unplug each injector momentarily one at a time. The rpm should drop the same each time.
2. Stop the engine and check the resistance across both terminals of each injector. It should be about 16 ohms.
3. To check for power to the injectors, remove the LH relay from the main fuse/relay panel and jumper terminals 30 and 87. There should be 12 volts at both terminals of each connector. If there is no power to 4 injectors, check the Ignition Circuit Monitoring Relay.
4. To test the operating signal from the control unit, connect a Bosch test lead 1 684 463 093 or equivalent, between the injector and the connector. This allows connecting an oscilloscope to the injector while it is operating. Do not allow the test connections to ground or the LH control unit will be permanently damaged.
5. Crank the starter or run the engine at idle. The scope picture should look the same for all injectors. If the idle speed is low or if the engine will not start, reverse the test leads and try again.

IDLE SPEED CONTROL VALVE

Testing

1. Unplug the idle speed control valve connector and check for 12 volts at terminal 2 with the ignition switch **ON**. Power comes from terminal 87 of the main relay.
2. Connect test leads between the terminals of the valve and the connector terminals so an oscilloscope can be connected to terminal 1 to check the control unit signal. Make sure the test leads do not touch ground.

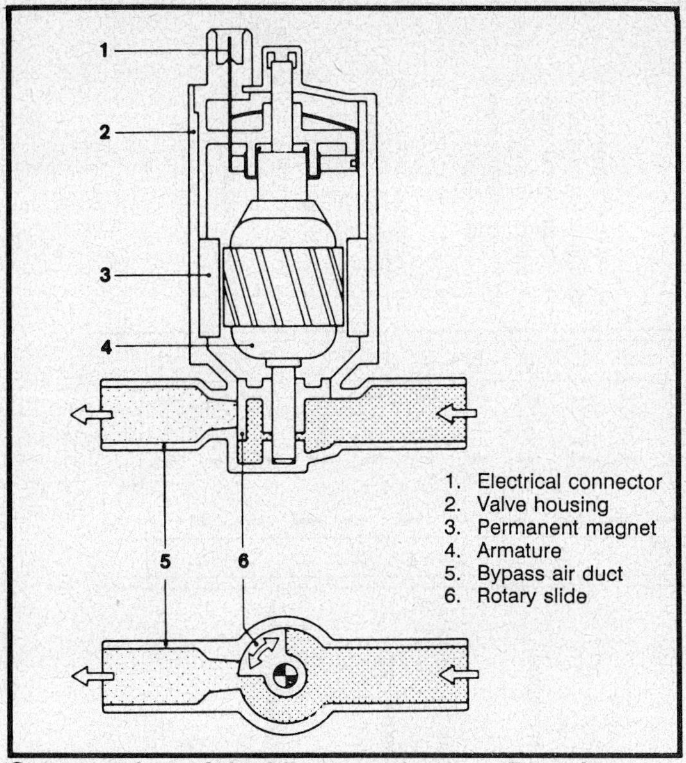

Cutaway view of the idle control valve—Porsche

3. With the engine at idle, there should be a rectangular wave form on the scope with the peaks slightly angled. Use a long tool as a stethoscope to listen for a humming noise from the valve. If the signal is there but the valve does not make any noise, the valve is faulty and should be replaced.

INTAKE RESONANCE VALVE

Testing

911

1. Have an assistant turn the ignition switch **ON** while listening at the intake manifold to hear the valve open and close.
2. If the valve does not operate, unplug the connector and connect a voltmeter across the terminals. When the ignition switch is first turned **ON**, there should be 12 volts for at least 1 second. If the voltage does not appear, the control unit or the wiring to it is faulty.
3. Operate the valve by hand to make sure it is not seized. Check the condition of the vacuum line and the reservoir, mounted near the air flow sensor.

928

1. Remove the rubber cover from the resonance valve shaft in the center of the top of the intake manifold. Mark the shaft and watch it while an assistant starts the engine. The valve should open and close while the engine is being started.
2. Run the engine above 3500 rpm and briefly snap the throttle full open. The valve should open and close again.
3. Disconnect the vacuum supply hose from the solenoid valve and attach a gauge. With the engine at idle, the vacuum in the reservoir should be about 18 in. (450mm) Hg. The resonance valve should open with about 12 in. (300mm) Hg vacuum.
4. With the engine running or the main LH relay activated, there should be 12 volts to the solenoid at terminal 2 of the connector. This power comes from the main LH relay.
5. To check the signal from the control unit, install a test lead so an oscilloscope can be connected to the solenoid valve with the wiring connected. When the test in Step 2 is repeated, the voltage signal on the scope should drop when the solenoid valve is opened.

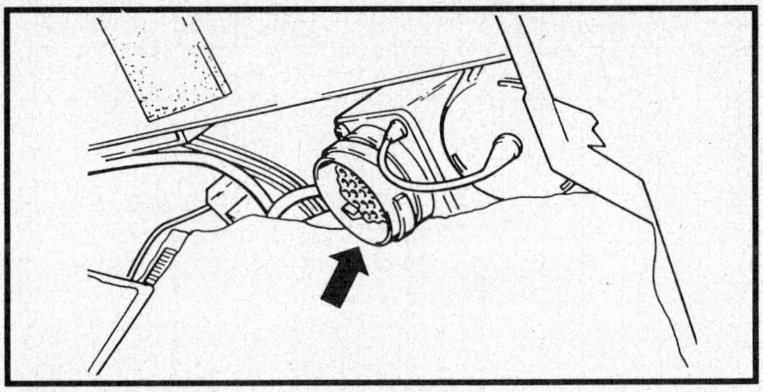

Diagnostic socket—Porsche 968

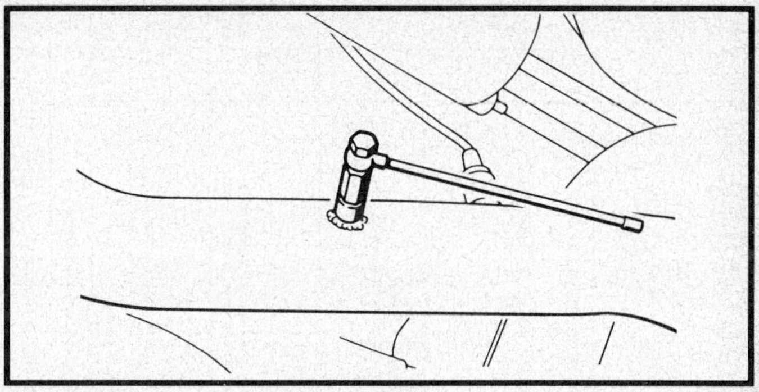

CO exhaust tap—Porsche 968

3.6L ENGINE–DME DIAGNOSIS MENU–911

Engine mechanically O.K.
Battery charged
Starter motor cranks the engine

Test point	Terms in bold letters = Fault memory display or fault path			Plug, Control unit	Engine will not start	Engine hard to start	Erratic idling	Poor pick-up	Misfiring	High fuel consumption	Engine hesitation	Poor power output	Poor hot starting	Diagnosis not possible
1	**Supply voltage**	11	V	24 → 18 24 → 37	x				x			x		x
2	**Idle speed contact ground short** **Idle speed contact break**	12 15	Ω	24 → 52		x	x			x		x		
3	**Full load contact**	13	Ω	24 → 53						x	x			
4	**Engine temperature sensor II**	14	Ω	24 → 45	x	x	x	x		x	x	x	x	
5	**Air flow sensor**	21	V	– → 3			x	x	x	x	x	x		
6	**Idle speed control activation**	22	≅		x	x	x						x	
7	**Oxygen regulation stop**	23	V				x	x	x	x	x	x	x	
8	**Oxygen sensor** (sensor signal)	24	V				x	x		x	x	x	x	
9	**Intake air temperature sensor**	25	Ω	44 → 26			x	x		x	x	x	x	
10	**Knock sensor I**	31	–							x	x			
11	**Knock sensor II**	32	–							x	x			
12	**Control unit faulty** (Knock computer)	33	–							x	x			
13	**Hall signal**	34	≅					x		x	x			
14	**Control unit faulty**	41	–		x	x	x	x	x	x	x	x	x	x
15	**DME relay**		V		x	x	x	x			x	x	x	
16	**Tank venting valve**	43	≅			x	x	x	x			x	x	
17	**Resonance plate**	44								x				
18	**Check Engine warning lamp** (Calif. as from mod. '91)	45	Ω											
19	**Injection valve - injection control unit**	51-56	≅		x	x	x	x	x	x	x	x	x	
20	Fuel pressure		⊘		x	x	x	x	x	x	x	x	x	
21	Intake system leak				x	x	x	x	x	x	x	x	x	
22	Ground and plug connections		–		x	x	x	x	x			x		x
23	Speed/reference mark sensor		≅	47→48	x	x		x	x		x	x	x	
24	Ignition system and output stage		≅		x	x	x	x	x		x	x	x	
25	Alternator/regulator		–						x			x		
26	Leads K and L													x

V = Voltmeter Ω = Ohmmeter ≅ = Oscilloscope

3.6L ENGINE–DME CONNECTOR PINOUT–911

1 Ignition pulse to output stage
2 Ground - output stage shield
3 DME relay ground (fuel pump)
4 Idle speed control
5 Tank venting valve
6 Engine speed to tachometer
7 Signal from air flow sensor
8 Signal from Hall sensor
9
10 Ground - oxygen sensor shield
11 Knock sensor I
12 Power supply, air flow sensor
13 Lead L from diagnostic plug
14 Ground
15 ti signal, cyl. 3
16 ti signal, cyl. 6
17 ti signal, cyl. 1
18 Permanent positive
19 Ground
20
21 Diagnostic plug, knocking yes + no
22 Check engine control
23 Resonance plate
24 Ground
25
26 Ground for air flow meter

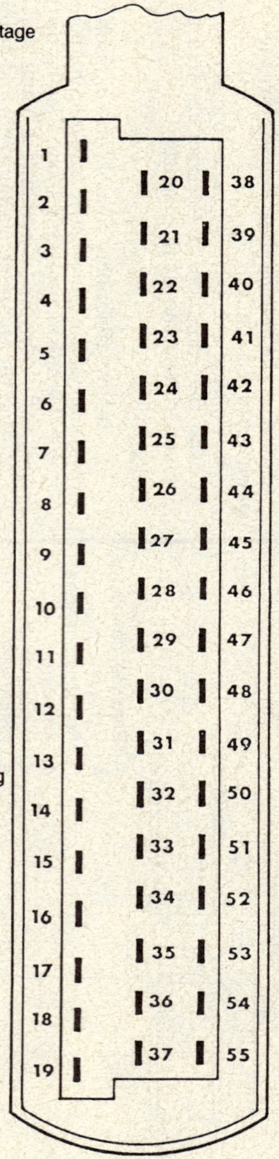

27
28 Oxygen sensor signal
29 Knock sensor II
30 Shield, for knock sensors, Hall sensor and ground for knock sensors
31 Positive to Hall sensor
32 Consumption gauge
33 ti signal, cyl. 5
34 ti signal, cyl. 4
35 ti signal, cyl. 2
36
37 Voltage from DME relay with ignition
38
39 Version coding
40 Air-conditioning compressor
41 AC switch
42
43
44 Intake air temperature NTC I
45 Engine temperature NTC II
46 Altitude potentiometer
47 Speed/reference mark positive
48 Speed/reference mark negative shield
49
50 Heating/air-conditioning regulation
51
52 Idle signal - throttle valve
53 Full load signal - throttle valve
54 Coding plug, characteristic map
55 Diagnostic plug, lead K

Fault, Fault Code	Possible Causes, Elimination, Remarks
Test point 1a Power supply for DME control unit (V)	**a) Permanent positive (B+) If there is no B+, the fault memory is cleared.** Test procedure: Connect a voltmeter to terminal 24 (–) and terminal 18 (+) of the control unit plug with the help of test leads. Display: Battery voltage No display, check current flow and ground paths in accordance with wiring diagram
Test point 1b **Power supply** for DME control unit (V) **too high/too low** Fault code 1_11	**b) Power supply via DME relay** Connect a voltmeter to terminal 24 (–) and terminal 37 (+) of the control unit plug with the help of test leads. Turn on the ignition. Display: Battery voltage No display: Check current path in accordance with wiring diagram

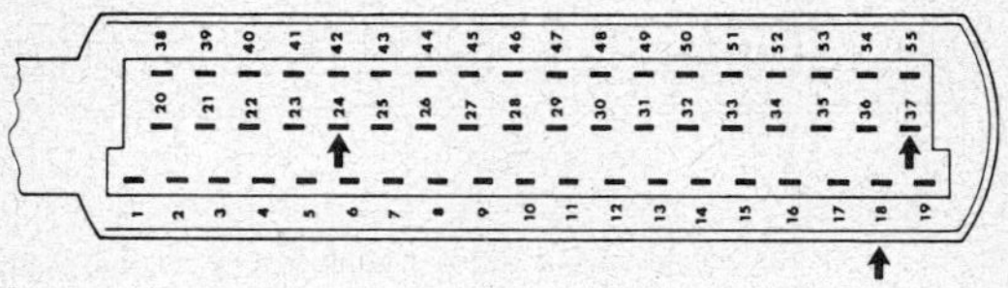

Note:

Cars with burglar alarm: Check power supply from ignition lock to terminal 86 of the DME relay via the burglar alarm control unit in accordance with the wiring diagram.

Cars without burglar alarm: Check the plug jumpers on plug section (provided for installation of a burglar alarm control unit).

Fault, Fault Code	Possible Causes, Elimination, Remarks
Test point 2: **Idle speed contact (Ω)** **Ground short** Fault code 1_12	**Check** by means of circuit input test with system tester 9288 or Tester 9268 Select test step and operate accelerator pedal. 9288 display: Idle speed contact closed Idle speed contact open. 9268 display: 1332 (idle speed contact closed) 0000 (idle speed contact open). No display: Disconnect plug from DME control unit. Connect ohmmeter between terminal 52 and terminal 24 using test leads.

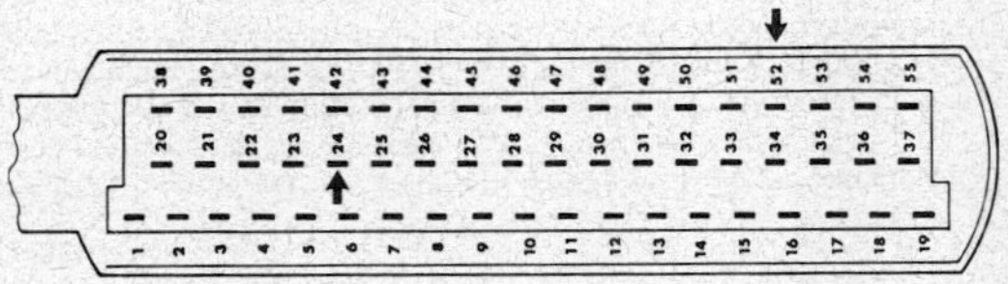

Display:
Throttle valve closed $<10\ \Omega$
Throttle valve opening angle $>1° = \infty\Omega$

If the values of the idle speed contact are not reached in this test, the test should be repeated directly at the throttle valve switch:

Disconnect the plug on the throttle valve switch. Connect an ohmmeter between terminals 1 and 2 of the throttle valve switch using test leads.

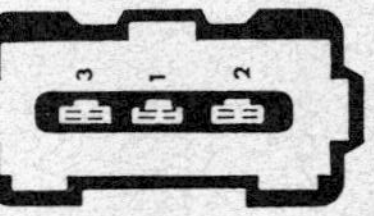

Display:
Throttle valve closed $<10\ \Omega$
Throttle valve opening angle $>1° = \infty\Omega$

Fault, Fault Code	Possible Causes, Elimination, Remarks
Idle speed contact (Ω) **Break** Fault code 1_15	Test procedure same as for ground short. (Fault will be stored in memory as from '91 models) Check whether setting is correct and adjust if necessary. Cause may be incorrectly set throttle valve switch or accelerator cable, for example.

Fault, Fault Code | **Possible Causes, Elimination, Remarks**

Test point 3

Full load contact (Ω)
Ground short
Fault code 1_13

Check by means of circuit input test with a system tester 9288 or tester 9268

Select test step and operate the accelerator pedal.
Display 9288: Full load contact open
Full load contact closed.
Display 9268: 1333 (full load contact open)
0000 (full load contact closed).

No display:

After disconnecting the DME control unit plug, connect the ohmmeter to terminal 53 and terminal 24 by means of test leads.

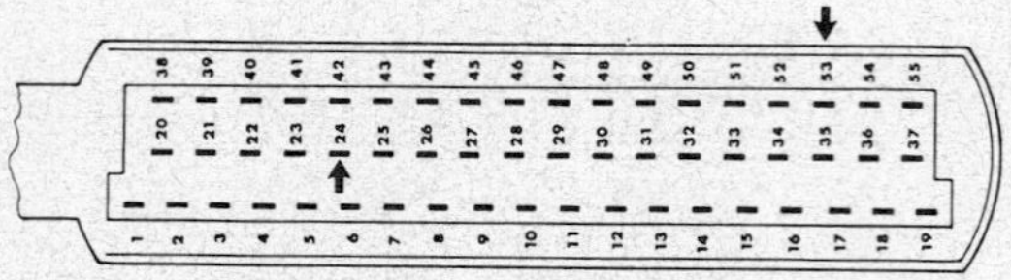

Display:

Throttle valve closed	∞ Ω
Throttle valve after approx. 2/3 opening travel	< 10 Ω

If the values of the full load contact are not reached in this test, the test should be repeated directly on the throttle valve switch:

Disconnect plug at throttle valve switch. Connect ohmmeter between terminals 3 and 2 of the throttle valve switch by means of test leads.

Display:

Throttle valve closed	∞ Ω
Throttle valve after 2/3 opening travel	< 10 Ω

Note::

If the full load contact fails, the control unit produces a load threshold via the air flow sensor. Values below this load threshold are the same as the switch being opened and values above the load threshold the same as the switch being closed.

Fault, Fault Code | **Possible Causes, Elimination, Remarks**

Test point 4

Engine temperature sensor (NTC II/Ω)
Fault code 1_14

The cylinder head temperature can be read directly in the „Actual value" menu with the system tester 9288.

If no plausible display:

After disconnecting the DME control unit plug, connect an ohmmeter with terminal 45 and terminal 30 by means of test leads.

Display at:

0°C	=	4.4 - 6.8 kΩ
15 - 30°C	=	1.4 - 3.6 kΩ
40°C	=	1-1.3 kΩ
80°C	=	250 - 390 Ω
100°C	=	160 - 210 Ω

If the values are not reached, test directly at the engine temperature sensor.

Note: The temperature sensor II informs the control unit about the cylinder head temperature. It is responsible for mixture enrichment for cold starting or warming-up.

Break (∞ Ω):

The DME control unit of the Carrera sets itself to a fixed resistance value permanently stored in the control unit which approximately corresponds to that of an engine at operating temperature. There is no enrichment by the faulty NTC II when the engine is warm (emergency running program). This results in starting problems if the engine is cold (no cold start enrichment).

Short-circuit (0 Ω):

In cold state, no engine pick-up, too lean, and engine stops.
No effect if engine is at operating temperature.

Replacement value is applicable to both types of fault!

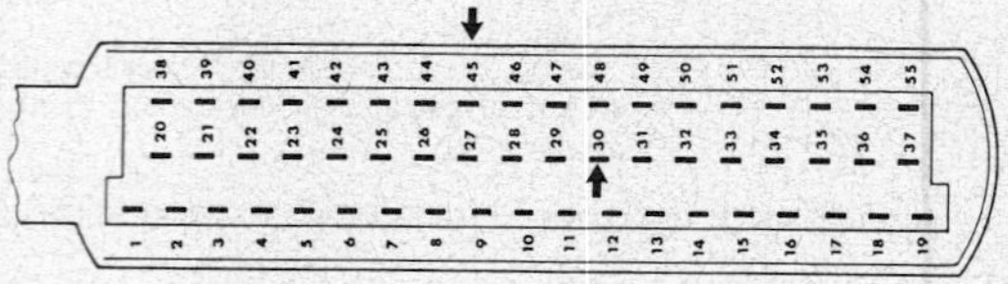

3.6L ENGINE – DME TEST POINT CHECKS – 911

Fault, Fault Code	Possible Causes, Elimination, Remarks
Test point 5a: **Air flow sensor** (V/Ω) Fault code 1_21	The ratio between Up (potentiometer voltage) and Uv (supply voltage) can be displayed in the „Actual value – menu with the system tester 9288. The ratio must be approx. 0.0 to 1 when the air flow sensor plate is opened. If not: **a) Voltage supply** Connect a voltmeter to terminal 3 and ground of the disconnected plug of the air flow sensor using test leads. Turn on ignition. Display: 5 ± 0.3 Volt
Test point 5b:	**b) Voltage drop across the air flow sensor plate potentiometer (V)** 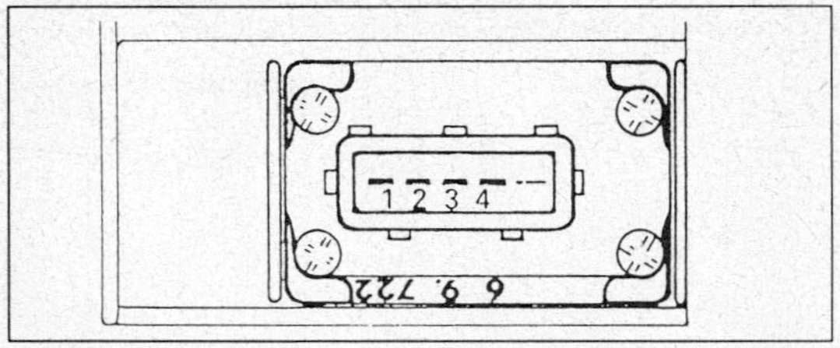 Disconnect the air filter connecting hose. Using test leads, connect pin 3 of the disconnected air flow sensor plug with air flow sensor plug terminal 3 and pin 4 with air flow sensor plug terminal 4. Connect a voltmeter to air flow sensor terminal 2 and ground (turn on ignition). Display: approx. 250 - 260 m V Using a non-metallic tool (e.g. screwdriver handle), press the air flow sensor plate to full load position. It must be possible to move the air flow sensor plate easily and without jamming. Observe the voltmeter's display at the same time. It must not be possible to observe voltage jumps or a voltage drop during the opening or closing operations of the air flow sensor plate. Air flow sensor plate at full load: Display approx. 4.60 Volt

Fault, Fault Code	Possible Causes, Elimination, Remarks
	If the voltage values are achieved, switch off the ignition and connect the plug again. **Note:** In the event of failure of the air flow sensor, substitute values are formed by the control unit (emergency running program). In idling operation, t_i signal from nominal air of the idle speed volumetric efficiency regulator. In partial and full load ranges, a constant t_i signal in each case.
Test point 6: **Idle speed control** (V) Fault code 1_22	**Drive link activation menu** If no pulse is audible, check: **Power supply** Connect the voltmeter to terminal 2 and ground of the disconnected plug of the idle speed control using test leads. Switch ignition on.  Display: Battery voltage If no display: Check power supply in accordance with wiring diagram. **Control signal test** Using the system tester 9288, it is possible to read the pulse duty factor in percent directly in the „Actual value" menu. Function test: Switch on loads during idle operation. The pulse duty factor must change, while the idling speed remains constant. If not:

Fault, Fault Code | **Possible Causes, Elimination, Remarks**

Connect a 2-pin DME test lead (Bosch No. 1 684 463 093) between the speed control and plug.

Connect and adjust engine tester in accordance with manufacturer's instructions.

Ensure that the lead connections do not have any ground contact with the vehicle. (Danger of short-circuit)

The following display must be visible when the engine is running:

Frequency approx. 100 Hz.

If no audible pulse exists in spite of the voltage and signal being present, replace the idle speed control.

Test point 7

Oxygen regulation stop
Fault code 1_23

The oxygen regulator cannot operate in its control range owing to problems with mixture preparation, e.g. too lean because of additional air or too rich because of a sticking injection valve. The oxygen regulator then moves up to the stop position.

Test 1:	Check CO level with running engine and connected CO tester (between 0.4 – 1.2 %)
too lean:	Check intake system for leaks (Test point 21)
too rich:	Test system pressure (test point 20) Test injection valves for leaks (test point 20)

Fault, Fault Code | **Possible Causes, Elimination, Remarks**

Test point 8:

Oxygen sensor (V)
(Sensor signal)
Fault code 1_24

Sensor signal test

Using the system tester 9288, the sensor voltage can be read directly in the „Actual value" menu.

If not:

Disconnect the oxygen sensor plug. Connect a digital voltmeter with the test connection (voltage signal air sensor) and ground at the sensor end.

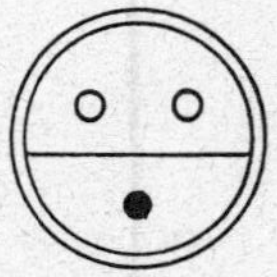

Start the engine and allow to warm up so that the oxygen sensor reaches its operating temperature. When the mixture is enriched, e.g. acceleration, a change in the voltage signal must be displayed.

Voltmeter display:
approx. 150 mV - 900 mV (depending on mixture composition).

Note:

If the control unit detects an oxygen sensor voltage signal of more than 1 V or less than 0.1 V, the control unit will switch over to operation without oxygen sensor (short-circuit or break).

If the regulator is not functioning and the sensor voltage signal is O.K., check the coding of the control unit before replacing it.

Test point 9

Air temperature sensor
NTC I (Ω)
Fault code 1_25

The intake air temperature can be read directly in the „Actual value" menu using the tester system 9288.

If not, or if no plausible value is displayed, then connect terminal 44 and terminal 26 of the disconnected DME control unit plug with an ohmmeter.

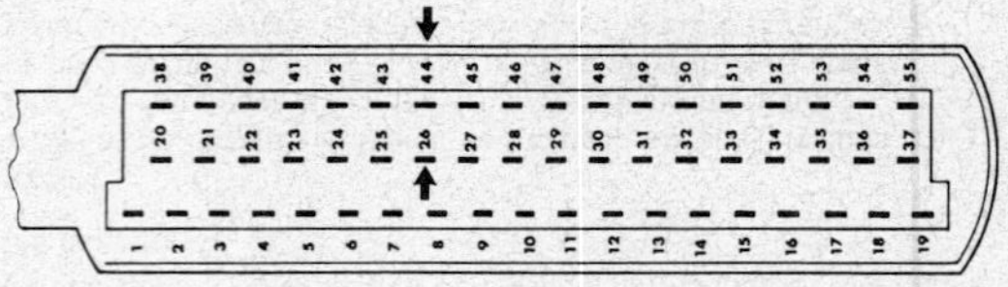

Display at:			
	0°C	=	4.4–6.8 kΩ
	15–30°C	=	1.4–3.6 kΩ
	40°C	=	1–1.3 kΩ

Fault, Fault Code	Possible Causes, Elimination, Remarks
	If there is no display, disconnect the plug at the air flow sensor. Connect an ohmmeter to air flow sensor terminals 1 and 4 using test leads. Note: In the event of a discontinuity, there will a substitute value equal to an intake air temperature of 20° C (emergency running program).

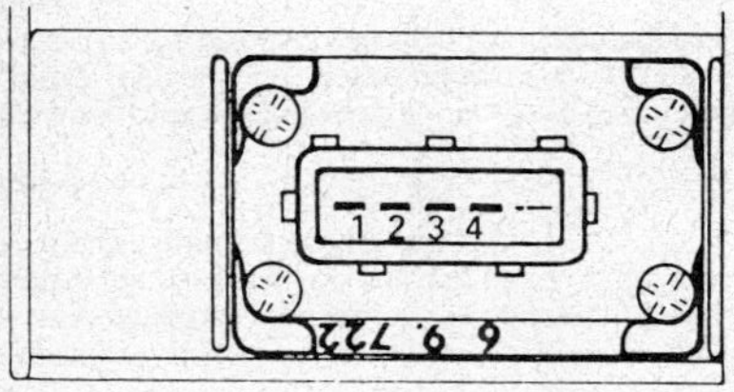

Fault, Fault Code	Possible Causes, Elimination, Remarks
Test point 10 **Knock sensor I** Fault code 1_31	Knock sensor I: Knock signal not plausible. Check: - Knock sensor mounting (check torque and type of screw) - Wiring harness and plug connections in accordance with wiring diagram reconnecting the sensors eliminates contact resistances. - Replace knock sensor In the event of a faulty knock sensor, ignition timing retarded by 6° on the crankshaft
Test point 11 **Knock sensor II** Fault code 1_32	Knock sensor II: Knock signal not plausible. Check: - Knock sensor mounting (check torque and type of screw) - Wiring harness and plug connections in accordance with wiring diagrams Reconnecting the sensors eliminates contact resistances. - Replace knock sensor In the event of a faulty knock sensor, ignition timing retarded by 6° on the crankshaft

Fault, Fault Code	Possible Causes, Elimination, Remarks
Test point 12 **Control unit:** (Knock computer) Fault code 1_33	Ignition timing is retarded by 6° if this fault occurs. Replace control unit.
Test point 13: **Hall signal** Fault code 1_34	The Hall sensor is supplied with voltage (approx. 12 V) and ground by the control unit. The signal is supplied to the control unit via the line 0. In the event of a fault display: Check voltage supply, ground supply and signal line in accordance with wiring diagram.

Plug connections Hall sensor (designation at plug)			Connections Control unit plug
(1)	-	——→	30
(2)	0	——→	8
(3)	+	——→	31

If voltage, ground and continuity exist, replace Hall sensor

Fault, Fault Code	Possible Causes, Elimination, Remarks
Test point 14: **Control unit faulty** Fault code 1_41	Replace control unit.

Fault, Fault Code	Possible Causes, Elimination, Remarks
Test point 15: **DME relay** (fuel pump) (V)	Start engine - the fuel pump must run during engine starting. If not: Switch off ignition. Disconnect control unit plug. Connect terminal 3 and terminal 24 on control unit plug with the help of a test lead. Switch on ignition - pump must run. If not: Check current-ground path, DME relay and fuel pump in accordance with wiring diagram

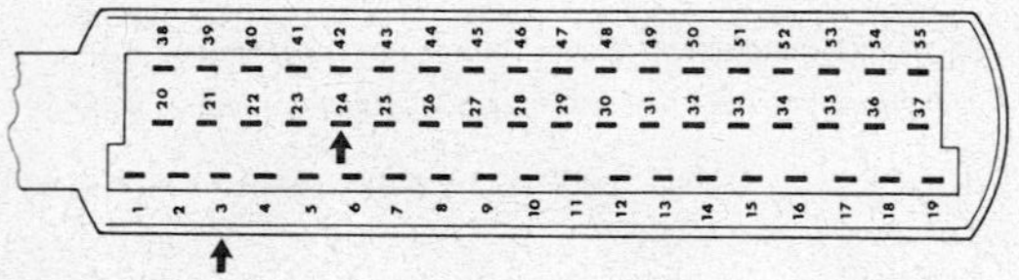

Fault, Fault Code	Possible Causes, Elimination, Remarks
Test point 16: **Tank venting valve** (V) Fault code 1_43	The electrical valve at the intake air distributor is opened for a longer or shorter period as a function of the load during engine operation at operating temperature. The opening period is determined by a ground pulse from the control unit. **a) Activation test** Activate the tank venting valve in the „Drive link activation" menu of the tester. If there is no audible pulse, check voltage supply at terminal 1 of the solenoid valve plug in accordance with the wiring diagram.

b) Control signal test

Connect DME test lead (Bosch No. 1 684 463 093) between solenoid valve and plug connection. Connect and adjust engine test in accordance with manufacturer's instructions.

Fault, Fault Code	Possible Causes, Elimination, Remarks
	Tank venting valve is not activated permanently. The test must be performed within 7 minutes after starting an engine at operating temperature. (After this, interrupt tank venting valve activation for approx. 75 s and then continue.) Start engine and accelerate. The following display must be visible on the tester for an engine which is at operating temperature.

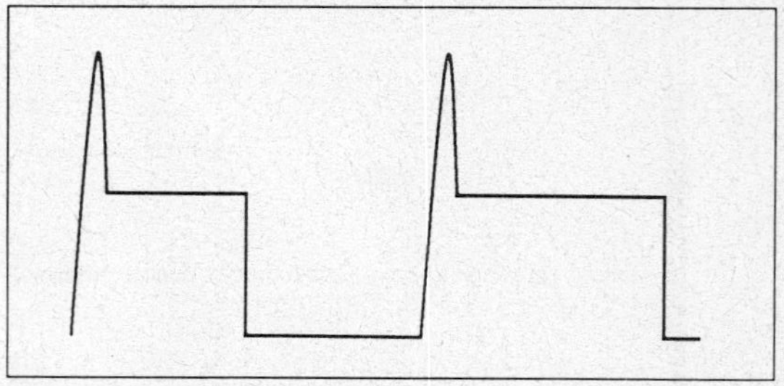

The signal becomes wider as the air throughput increases. If there is no signal, check the path in accordance with the wiring diagram. Replace control unit if necessary.

Fault, Fault Code	Possible Causes, Elimination, Remarks
Test point 17: Resonance plate Fault code 1_44	Each time the ignition is switched on, a ground pulse is supplied by the control unit to the solenoid valve of the resonance plate. If there is a vacuum in the system, the resonance plate is audibly opened and closed. The pulse can also be triggered by means of the "Drive link activation" menu. If there is no audible pulse, check: - Is there a vacuum? - Does the resonance plate move easily? - Control signal test: Disconnect the plug at the solenoid valve (sensor of intake air distributor). Connect and adjust an engine tester using test leads in accordance with the manufacturer's instructions. A signal must be visible when the ignition is switched on or for the "Drive link activation" menu. If not: - Check paths in accordance with wiring diagram. - Replace control unit if necessary.

3.6L ENGINE—DME TEST POINT CHECKS—911

Fault, Fault Code	Possible Causes, Elimination, Remarks
Test point 18 **Check Engine warning lamp** Fault code 1_45 (Calif. as from Mod. '91)	A ground signal is supplied from control unit plug 22 to the „Check Engine lamp" and switches on this lamp if a component influencing emission control fails. If the „Check Engine" warning lamp fails, the fault is stored in the fault memory. Check by supplying ground to disconnected control unit plug 22 and switching on ignition. If the display lamp is faulty, always use a specified lamp for replacement.
Test point 19: **Injection valve (V/Ω)** Fault code 1_15 for cyl. 1 to 1_56 for cyl. 6	**Drive link activation menu:** Each injection valve can be activated individually by means of the selective injection function. Opening of the injection valve can be heard only with extreme difficulty (damped). If no pulse can be heard, check: **a) Voltage supply** Disconnect valve plug, connect voltmeter with the valve plug terminal 2 and ground using test leads. Switch on ignition. **Display: Battery voltage**

If no battery voltage is displayed, check current path in accordance with wiring diagram.

Fault, Fault Code	Possible Causes, Elimination, Remarks
	b) Check coil resistance of the injection valve Disconnect valve plug. Check the coil resistance at the terminal contacts of the injection valves with an ohmmeter. **Test value: 16 Ω** **c) Injection output stage** Adjust the oscilloscope in accordance with the manufacturer's instructions. Connect a Bosch test lead (1 684 463 093) between an injection valve and plug. Connect the tester lead with the test lead in accordance with manufacturer's instructions. **Important:** Tester leads must not have any contact with ground. Start the engine. The following displays must appear if the injection output stage is working perfectly and the test connections are correct: **Starter speed**

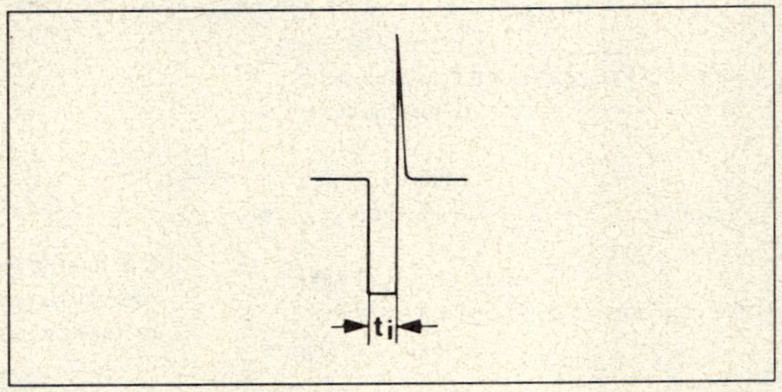

Note:

If the engine does not start or if the idle speed drops, change round tester connections to test lead and check tester adjustment.

Test point 20: Fuel pressure

The fuel collection pipe has a test connection at cyl. 1. Unscrew the cap nut of the test connection.

Important: The inserted ball seal could fall out!

Connect the pressure gauge P 378 to the test connection. Start the engine and run at idle speed with vacuum at the pressure regulator:

Test pressure 3.3 ± 0.2 bar

In the case of hot starting problems:

Perform the test with the engine stationary and with running fuel pump without a vacuum at the pressure regulator (to do this; remove the DME relay and jumper terminal 87b and terminal 30):

Test pressure 3.8 ± 0.2 bar

- Switch off the fuel pump. Read off the pressure value on the pressure gauge (nominal 3.8 ± 0.2 bar).
- Max. permissible temperature drop with warm engine: **0.5 bar in 30 min.**
- If the pressure drops below the specified value, proceed as described below:

Build up pressure again by switching on the fuel pump briefly. Remove the return pipe (fuel pump must not be switched on).

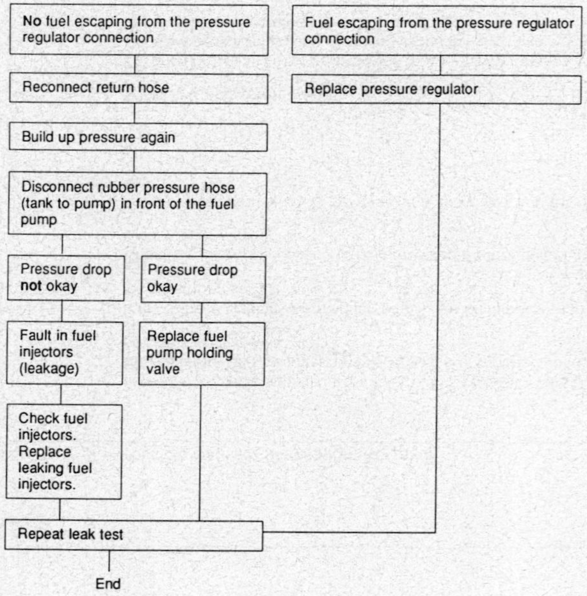

Test point 21: Intake system leaks

Check all connections after the air flow sensor for leaks. Remove the air filter bottom section with air flow sensor and lay aside. Close the hose to the activated carbon canister. Apply special tool 9264/5 to the connection hood and build up a pressure of approx. 0.5 bar in the intake system. In the event of leaks, the pressure in the intake system will fall rapidly.

Note:

Fault flashing code 1_23 (oxygen regulator at stop) appears in cars with oxygen regulation and very large leaks in the intake system (e.g. intake pipe gasket).

Test point 22: Ground connections/plug connections

- Ground point II
  fuel pump, wheelhouse front right (end tip)
- Ground point IV
  power ground under the seat, left
- Ground point V
  electronics ground under the seat, left
- Ground point VI
  central electrical ground on side member, right (under lid hinge)
- Ground point X
  battery ground
- Ground point XI
  grounding strip transmission-body (at the starter)
- Ground point XII
  ignition output stages, engine compartment, rear left (between ignition coils)

Are ground connections tight and not corroded? Loosen, clean and retighten ground connections as specified.

Note:

Never start the engine as long as the ground lead between the body and transmission is disconnected!
This will destroy the control unit immediately!

Plug connections

Are all plugs connected correctly, not loose and not corroded? Disconnect plug connections: the pins and sleeves of the plugs must not be bent or corroded.

- The 55-pin plug on the DME control unit (under the driver's seat).
- 14-pin plug (under the driver's seat).
- A 12-pin plug at the rear left in the engine compartment.
- A 14-pin plug at the rear left in the engine compartment.
- A 3-pin plug for the reference mark/speed sensor and a 3-pin plug for the oxygen sensor on the left side of the engine.
- A 3-pin plug on the throttle valve switch.
- A 2-pin plug on the solenoid valve for tank venting (center of intake air distributor).
- A 4-pin plug on the air flow sensor.
- A 2-pin plug on the NTC II on the left side of the engine. (The temperature sensor is screwed into the cylinder head 3, accessible from below.)
- Two 7-pin plugs at the ignition output stages at the rear of the engine compartment.
- 3-pin plug on the Hall sensor (at the ignition distributor).
- 3-pin plug on knock sensor I (on the left side of the engine).
- 3-pin plug on knock sensor II (center of intake air distributor).
- A 2-pin plug on the idle speed control (at the intake air distributor).
- The two-pin plug on the solenoid valve of the resonance plate (center of the intake air distributor).
- A 2-pin plug in each case at the injection valves.

Since the Carrera is supplied worldwide with only one standard control unit and the control unit is coded correspondingly depending on the country equipment, this coding must be checked in the event of a complaint.

To allow calling up of the different characteristic maps, a 2-pin or 3-pin plug connection is brought out of the DME wiring harness near the plug receptacle.

The DME control unit is supplied with power via terminal 30 (battery +) as well as via ignition lock terminal 15.

Test point 23: Speed/reference mark sensor

This test can only be performed with an oscilloscope. For this purpose, connect and adjust a workshop oscilloscope in accordance with the manufacturer's instructions.

Connect the positive tester lead via a test lead to the control unit plug terminal 47 and the negative tester lead via a test lead to control unit plug 48.

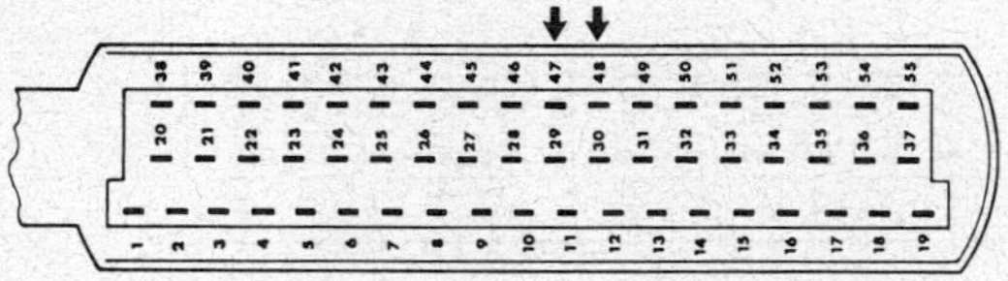

Crank the engine with the starter. Sinusoidal oscillations of at least 3 V (speed pulses) and an intermediate higher amplitude (reference mark) must be displayed on the screen.

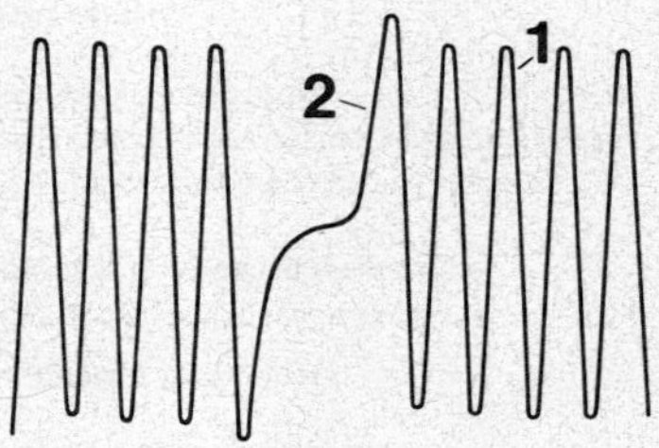

1 – Speed signal
2 – Reference mark signal

If the voltage signal is too small (< 3V), the distance between the sensor and the gear ring may be incorrectly adjusted.

Sensor distance: Between gear ring and sensor: 1 mm± 0.2 mm.

If there is no display:

Disconnect the plug of the speed sensor at the terminal strip in the engine compartment. Connect tester leads to the middle and outer plug contacts of the plug (terminals 1 and 2, pin end) on the plug section to the sensor with test leads. Start the engine. Sinusoidal oscillations must be visible on the screen.

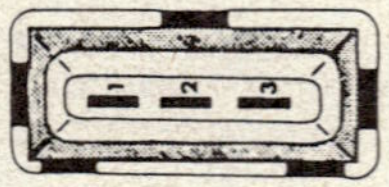

Speed sensor plug

Note:

If terminal 3 is connected to ground, the display must remain unchanged.

If this is not the case, check sensor for continuity, spacing and soiling.

Test point 24: Ignition system

In the case of dual ignition engines, each ignition circuit should be checked independently; the other ignition circuit is deactivated by disconnecting the plug of the respective ignition output stage for this. The remaining steps correspond to the sequence described in the test plan.

a) Secondary ignition display

Set the secondary ignition display on the oscilloscope. Connect the test leads according to the manufacturer's instructions.

Note:

If a fault is displayed for all cylinders, this is located in the primary or secondary circuits from the ignition coil to the distributor rotor. If a fault is displayed for one cylinder, the fault is located after the distributor rotor.

Primary resistor: Terminals 1 + 15 = 0.4-0.6 Ω
Secondary resistor: Terminals 1 + 4 = 5-7.2 kΩ

b) Spark plug connectors (Ω)

Suppression resistor: 3 kΩ
Visual inspection for damage; traces of burning

c) Distributor

Distributor rotor, distributor cap: Suppression resistor 1 k Ω each
Visual inspection for damage, soiling, incorrect installation position.

d) Ignition timing test

The ignition timing test is performed when the engine is at operating temperature.

The ignition timing and engine speed can be read directly in the „Actual value" menu with the system tester 9288.

Test values:

At 880 ± 40 rpm 0° ± 3°

e) Activation of the high-voltage end (if ignition sparks are not triggered)

The ignition trigger signal can be simulated for rapid troubleshooting at the ignition end. For this purpose, connect the oscilloscope tester lead terminal 4 to the ignition lead from the ignition coil to the high-voltage distributor (terminal 4).

Disconnect the DME control unit plug.
Switch on ignition.

If a battery voltage is now applied intermittently to terminal 1 of the control unit plug with a test lead (e.g. control unit plug terminal 18), a high-voltage peak must be displayed on the oscilloscope each time (ignition triggering).

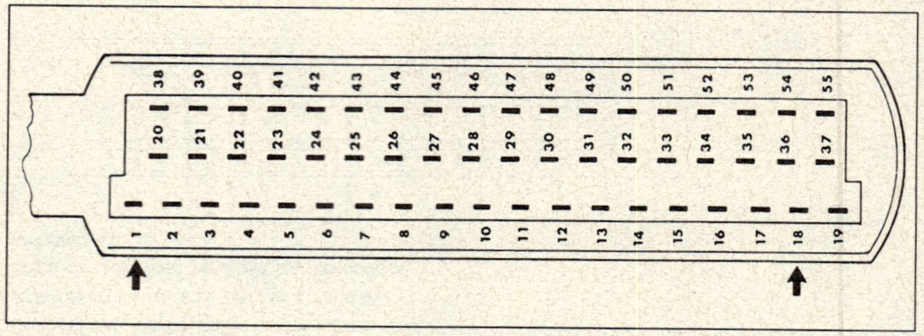

If there is no signal, check:

- Power supply to ignition output stages and ignition coils
- Ground point for ignition output stages (ground point at rear left of engine compartment).
- Plug connections to ignition output stages, ignition coils and high-voltage distributor.
- The components responsible for high voltage (e.g. ignition coil, ignition leads, spark plug connectors ...).
- Lead continuity from the control unit plug to the ignition output stage terminal 5.

3.6L ENGINE – DME TEST POINT CHECKS – 911

f) Testing the control signal for the ignition output stages

Connect the DME control unit plug.

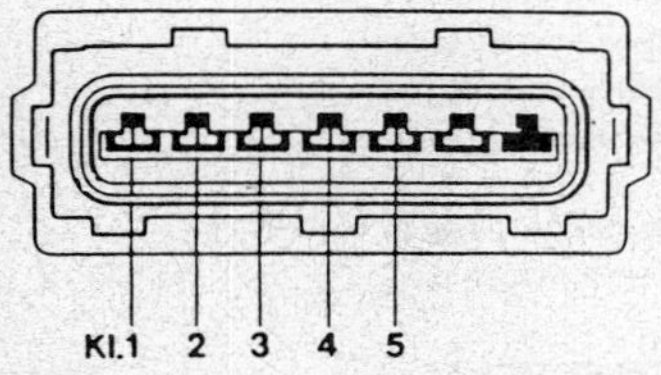

Connect the positive tester lead of the oscilloscope to terminal 5 and the negative test lead to terminal 2 of the disconnected output stage plug.

Operate the starter motor.
The oscilloscope must show the control signal of the DME control unit.

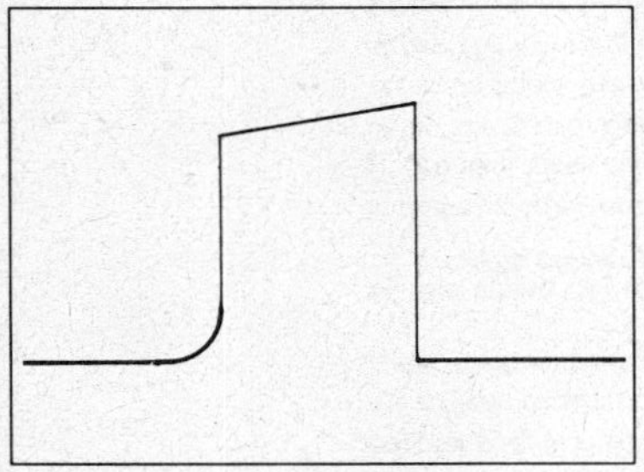

Please note:

Coat the baseplate with a heat-conducting paste prior to installation of the ignition output stage. Do not apply heat-conducting paste to painted parts.

Heat-conducting paste:
Bosch No. 5 942 860 003.

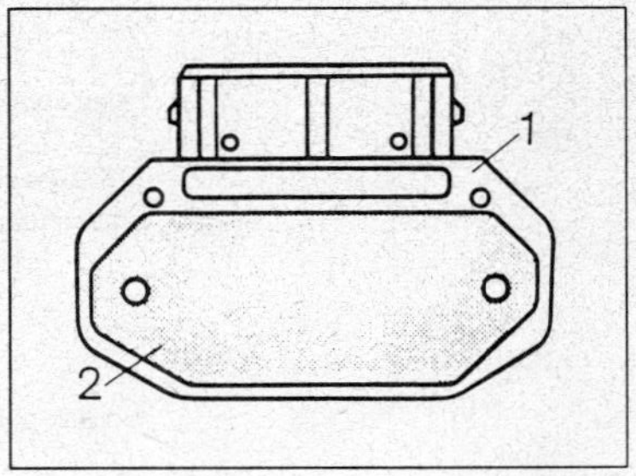

1 – Output stage
2 – Baseplate

Test point 25: Alternator, regulator

Misfiring of the engine may be caused by alternator voltage peaks.

Remove the drive belt from the alternator and start engine.

If the faults have been eliminated, check the alternator and regulator.

Test point 26: Leads K and L

The diagnosis connection between the DME control unit and testers is established by means of the two leads K and L. If no diagnosis is possible, the following points must be checked:

1. Continuity test (refer to wiring diagram)

Lead L: Pin 7 (19-pole diagnostic socket) – pin 13 (DME plug)

Lead K: Pin 8 (19-pole diagnostic socket) – pin 55 (DME plug)

2. Ground short test (refer to wiring diagram)

Lead L: The voltage at pin 7 (19-pole diagnostic socket) must be > 8 V when the ignition is turned on.

Lead K: The voltage at pin 8 (19-pole diagnostic socket) must be >8 V when the ignition is turned on.

Possible fault causes:

- Ground short or break in wiring or plug connections.
- Faulty control unit which is connected to these lines (may be any control unit capable of diagnosis, need not necessarily be the DME control unit).

Test: Disconnect plugs successively at the control units capable of diagnosis until >8 V is obtained. Replace responsible control unit.

- Tester faulty.
- No power supply at the diagnostic plug.

No ground supply at diagnostic plug.

Notes on idling speed and idle speed CO testing:

a) Idling speed

The Carrera possesses an adaptive idle speed volumetric efficiency regulator. This means that it is not necessary to adjust the idle speed of any version. In vehicles with adaptive idle speed volumetric efficiency regulation, system adaption should be performed after disconnection of the DME control unit from permanent positive and after each inspection.

Idle speed control value for engines at operating temperature:
880 ± 40 min

b) Idle speed CO of cars without catalytic converter

The CO level is still adjusted at the CO adjusting screw in the air flow sensor. The following preconditions must be observed:

- Engine at operating temperature
- Engine mechanically in order and ignition system O.K.
- All loads switched off
- CO tester at operating temperature, calibrated and adjusted correctly.

3.6L ENGINE—DME TEST POINT CHECKS—911

After exhaust gas adjustment, perform system adaption. In the case of vehicles without catalytic converter, it must be ensured that the adapter plug (911.612.422.8) is plugged onto the oxygen sensor connecting lead in order to avoid interference (see wiring diagram).

c) Idle speed CO of vehicles with catalytic converter

In these vehicles, adaption of the idle speed CO content takes place as well as adaptation of the idle speed. Consequently, it is only necessary to check the actual condition in such vehicles. The following preconditions must be observed when performing the checks:

- Engine at operating temperature
- Mechanically in order and ignition system O.K.
- All loads switched off
- CO tester at operating temperature, calibrated and adjusted correctly.
- **Oxygen sensor plug connected**

Test value: 0.4 - 1.2 % CO
at 880 ± 40 rpm and ignition timing of 0° ± 3°.

After disconnection of the control unit plug or the battery, the engine must run for at least 10 min. at idle speed before these values can be rechecked. In addition, system adaption must be performed.

5.0L ENGINE—LH DIAGNOSIS MENU—928

Engine mechanically OK
Battery charged
Starter motor cranks the engine

Test point	Terms in bold type = Fault memory display or fault path	Fault code 1_	Tester	Plug, Control unit	Engine will not start	Engine hard to start	Irregular idling	Poor pick-up	Misfiring	High fuel consumption	Low power output	Engine hesitation	Poor hot starting	Diagnosis not possible
1	**Supply voltage**	11	V	35 → 17 9 → 17	x	x			x			x		x
2	**Idle speed contact**	12 15	Ω	2 → 17						x		x		
3	**Full load contact**	13	Ω	13 → 17			x			x	x	x		
4	**Engine temperature sensor II**	14	Ω	13 → 17	x	x	x	x	x	x	x	x	x	
5	**Air mass sensor**	21					x	x	x	x	x	x		
6	**Idle speed control activation**	22	≅			x	x					x	x	
7	**Oxygen regulation stop, rich**	23	V				x	x		x	x	x		
8	**Oxygen regulation stop, lean**	24	V				x	x			x	x		
9	**Oxygen sensor, open circuit**	25	V/Ω											
10	**Injection circuit shut down**	31				x	x	x	x		x		x	
11	**Control unit faulty**	41			x									x
12	No power supply		V	35 - 17 9 - 17	x									
13	Ground and plug connections				x	x		x	x		x	x	x	
14	Fuel pump voltage		V		x	x			x		x	x	x	
15	Engine speed signal from EZK				x									
16	Injector valves / control system		≅		x	x		x	x		x	x	x	
17	Fuel pressure		(pressure gauge)		x	x	x	x	x	x	x	x	x	
18	Intake system leaks				x	x	x	x	x		x	x	x	
19	Resonance flap		≅								x			
20	Fuel tank venting		≅			x	x	x				x	x	
21	Alternator and voltage regulator		V			x			x			x		
22	Leads K and L		Ω											x

V = Voltmeter Ω = Ohmmeter ≅ = Oscilloscope

5.0L ENGINE—LH FAULT MEMORY—928

Fault memory, LH control unit, 928 S4/GT

Power supply	Oxygen sensor control - rich
Idle contact Open circuit short	Oxygen sensor control - lean
Full load contact	Oxygen sensor open circuit
Engine temperature sensor 2	Injector circuit switched off
Idle contact open circuit	Ignition circuit monitoring relay
Hot wire air mass sensor housing	Control unit faulty
Idle stabilizer control	

5.0L ENGINE—LH CONNECTOR PINOUT—928

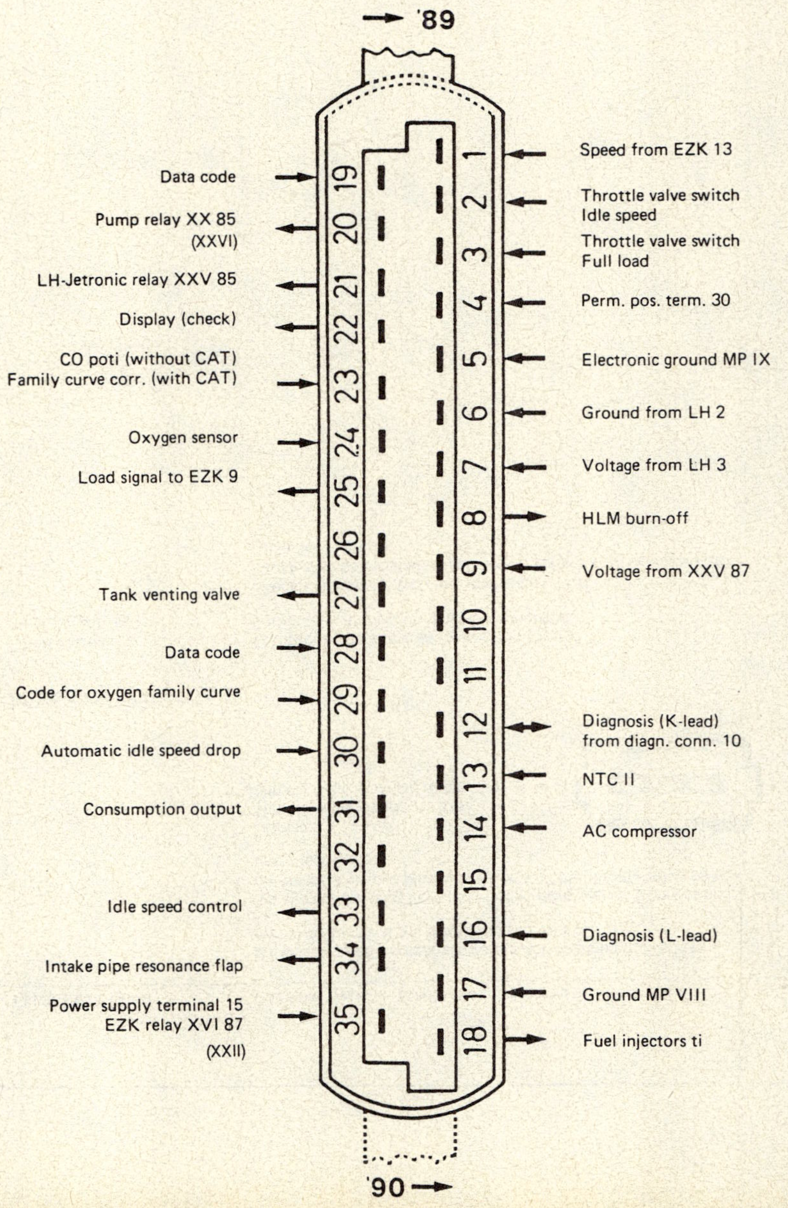

Fault, Fault Code	Possible Causes, Elimination, Remarks
Test point 1 **Power supply voltage too high / too low** **LH control unit** Fault code 1_11	If the power supply drops below 10 V or rises above q6 V, this is stored in the fault memory. Check voltage at regulator; check plug connection and terminals for firm contact and freedom from corrosion.
Test point 2 **Idle speed contact (Ω)** **Short to ground** Fault code 1_12	Check by means of circuit input test with 9288 system tester, or 9268 tester Select the test step and depress the accelerator pedal. 9288 display: Idle speed contact closed Idle speed contact open 9268 display: 1332 (idle speed contact closed) 0000 (idle speed contact open). No display: Disconnect plugs from EZK and LH control unit. Connect ohmmeter betweern terminals 2 and 5, using test leads.

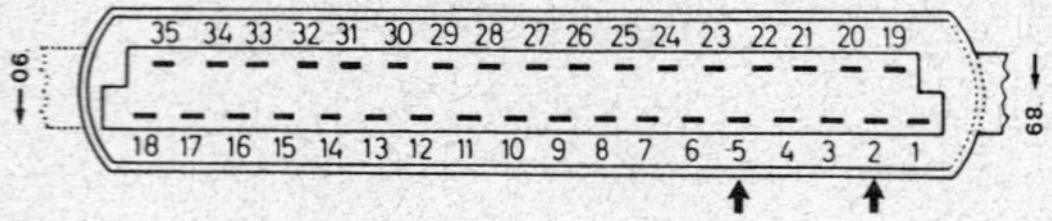

Display:
Throttle valve closed < 10 Ω
Throttle valve opening angle > 1° = ∞ Ω

The idle speed contact must open immediately.

pen circuit: Ignition timing at idle speed approx. 20° CS (crankshaft rotation).
No fuel cutoff when coasting, higher idle speed.

Short circuit: Fuel cutoff as engine runs up, once after every engine start. Ignition timing approx. 10° retarded.

Fault, Fault Code	Possible Causes, Elimination, Remarks
	If the idle speed contact values are not obtained in this test, repeat it directly at the throttle valve switch: disconnect the plug from the throttle valve switch. Connect an ohmmeter between terminals 1 and 2 of the throttle valve switch, using test leads. Display: Throttle valve closed < 10 Ω Throttle valve opening angle > 1° = ∞ Ω

Fault, Fault Code	Possible Causes, Elimination, Remarks
Idle speed contact **(Ω)** **Open circuit** Fault code 1_15	Test procedure same as for short to ground. (From 1991 models on, fault is memorized.) Check that setting is correct and adjust if necessary. Cause may be incorrectly set throttle valve switch or accelerator cable, for example.

5.0L ENGINE – LH TEST POINT CHECKS – 928

Fault, Fault Code	Possible Causes, Elimination, Remarks

Test point 3

Full load contact Ω
Fault code 1_13

Check by means of circuit input test with a 9288 system tester or 9268 tester

Select test step and operate the accelerator pedal.
9288 display: Full load contact open
Full load contact closed.
9268 display: 1333 (full load contact open)
0000 (full load contact closed).

No display:

After disconnecting the EZK and LH control unit plugs, connect the ohmmeter to terminals 3 and 5 by means of the test leads.

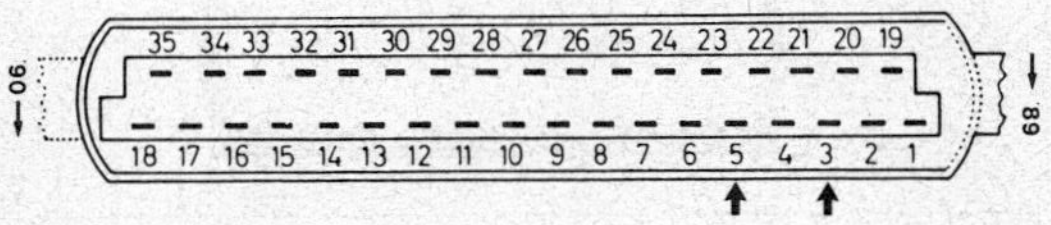

Display:
Throttle valve closed ∞ Ω
Throttle valve after opening by approx. two-thirds of its travel < 10 Ω

If the full load contact values are not reached in this test, it should be repeated directly at the throttle valve switch:

Disconnect plug at throttle valve switch. Connect ohmmeter between terminals 3 and 2 of the throttle valve switch by means of the test leads.

Display:
throttle valve closed ∞ Ω
Throttle valve open < 10 Ω

Note:

If the full load contact failsm the control uni generates a load threshold by way of the air mass sensor. Values below thisload are the same as if the switch were open, values above it the same as if the switch were closed.

Fault, Fault Code	Possible Causes, Elimination, Remarks

Test point 4

Engine temperature sensor II Ω
Fault code 1_14

Disconnect the LH control unit plug.
Connect the ohmmeter to terminals 13 and 5 or 17 of the LH control unit plug using the test leads.

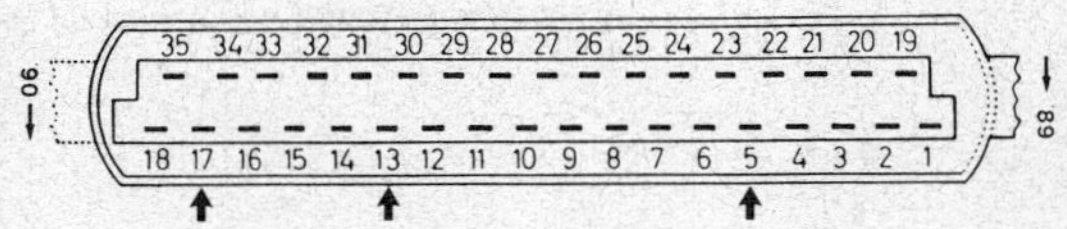

Display: resistance

Test values at:			
	0°C	=	4.4 - 6.8 kΩ
	15 - 30°C	=	1.4 - 3.6 kΩ
	40°C	=	1-1.3 kΩ
	80°C	=	250 - 290 Ω
	100°C	=	100 - 210 Ω

If the values are not reached, check directly at the engine temperature sensor. Do not connect both contact tabs to the ohmmeter, but chect each contact tab saeparately against ground (there are 2 independent temperature sensors).

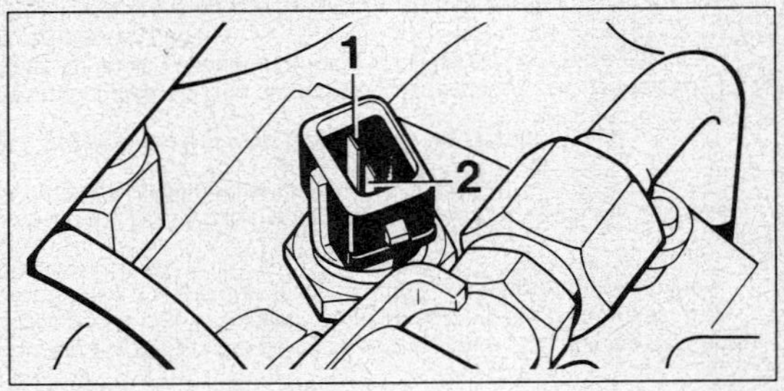

1 - EZK
2 - LH

Note: temperature sensor II supplies the cylinder head temperature to the control unit. It is responsible for mixture enrichment during cold starts and when the engine is warming up.

Fault, Fault Code	Possible Causes, Elimination, Remarks
	Break (∞ Ω): The LH control unit reverts to a fixed, previously stored program. If the engine is warm, there is no mixture enrichment. If the engine is cold, however, this causes starting problems (since there is no cold-start enrichment).
	Short circuit (0 Ω): When the engine is cold it does not respond to the throttle being opened; the mixture is too weak and the engine stalls.No effect at regular operating temperature.

Test point 5
Air mass sensor
Fault code 1_21

a) Power supply (V)

Disconnect the plug at the air mass sensor and connect the voltmeter to plug terminals 2 and 4 using the test leads.

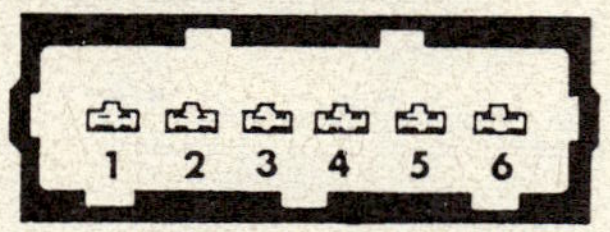

Use the test leads to connect terminals 17 and 21 on the LH control unit plug after it has been disconnected.
Display: battery voltage

No display:
Remove LH relay XXV and bridge terminals 87 and 30 at the relay base. Display on voltmeter connected to air mass sensor plug:
battery voltage.

If there is a battery voltage display, check the LH relay and its activation.
If there is no display, check leads and plug connections according to wiring diagram.

b) Checking resistance of hot wire circuit at the air mass sensor (Ω)

Disconnect the plug connection at the air mass sensor. Connect terminals 3 and 5 at the air mass sensor with the ohmmeter.
Display: 3.6 - 4.1Ω

Fault, Fault Code	Possible Causes, Elimination, Remarks

c) Checking the hot wire signal (V)

Connect the plug to the air mass sensor. Take out LH relay XXV entfernen and bridge terminals 87 and 30 at the base of the relay. Connect terminals 6 and 7 at the LH control unit plug with a voltmeter.
Display: ≈ 1.6 V

Blow on the hot wire in the air mass sensor and watch the voltmeter. The voltage reading must change. (≈ **1.6 - 5 V)**

d) Checking the hot wire burn-off circuit (visual check)

Run the engine with the air mass sensor installed and connected. When an engine temperature of > 60°C is reached, raise engine speed above 2000 min^{-1}
and then stiop the engine by switching off the ignition. After a delay of approx. 4 seconds, the hot wire must glow for approx. 1 s (burn-off function).

e) Emergency operating program

The emergency operating program stored in the LH control unit is activated if the air mass sensor fails, so that the car can be driven to the nearest authorized repair shop.

With engine warm and at engine speeds < 2000 min^{-1} ≈ 3.5 ms ti
With engine warm and at engine speeds > 2000 min^{-1} ≈ 6.3 ms ti

(ti = injection time)

Test point 6
Idle speed control (V)
Fault code 1_22

Call up actuator activation menu point

If no pulse is audible, check:

Power supply

Connect voltmeter to terminal 2 of disconnected idle speed actuator plug and ground with the test leads.
Switch on the ignition.
Display: battery voltage.
No display:

Check power supply according to wiring diagram.

5.0L ENGINE—LH TEST POINT CHECKS—928

Fault, Fault Code	Possible Causes, Elimination, Remarks
	Checking the control signal Insert 2-pin DME test lead (Bosch No. 1 684 463 093) between rotary actuator and plug connection. Connect engine tester as stated in manufacturer's instructions, and adjust. Make sure that there is no contact between the lead connectinos and ground on the vehicle (risk of short circuit). With the engine running, the following display must be seen:

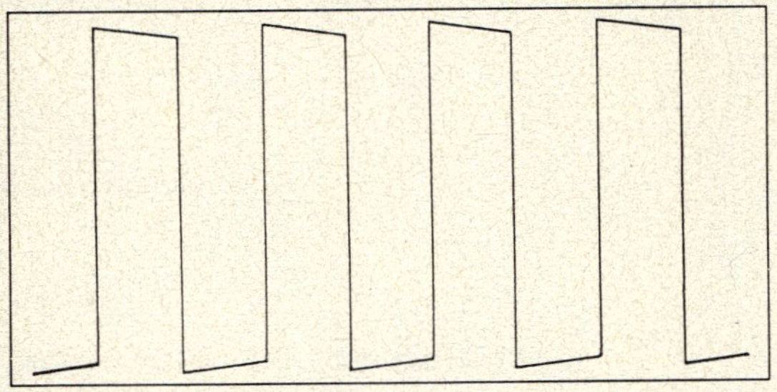

Frequency approx. 100 Hz.

If there is no audible pulse in spite of the power supply and the signal being present, renew the idle speed control.

Fault, Fault Code	Possible Causes, Elimination, Remarks
Test point 7 **Oxygen regulation Control (rich)** Fault code 1_23	Mixture preparation problems are preventing the oxygen regulator from operating within its control range. It therefore moves back to its stop. Test 1: With the engine running and a CO tester connected, note the CO content (between 0.4-1.2 %) Too lean: check intake system (test point 19)
Test point 8 **Oxygen regulator Control (lean)** Fault code 1_24	Mixture preparation problems are preventing the oxygen regulator from operating within its control range. It therefore moves back to its stop. Test 1: With the engine running and a CO tester connected, note the CO content (between 0.4-1.2 %) Too rich: Check fuel pressure (test point 17) Check injector valves for leaks
Test point 9 **Oxygen sensor Open circuit** Fault code 1_25	**a) Checking sensor signal (V)** Disconnect the oxygen sensor plug close to the central electrics. Connect voltmeter at the sensor end to the single plug pin and ground.

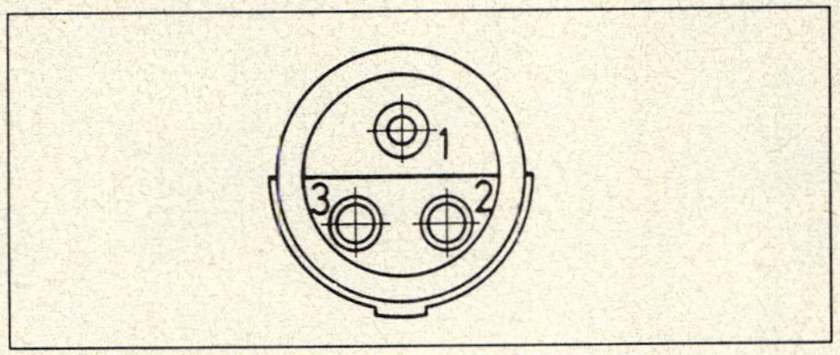

Test connection

1 = Sensor voltage
2, 3 = Sensor heating

Start the engine and warm it up so that the oxygen sensor can reach its regular operating temperature. When the mixture is richened (for example when the accelerator pedal is pressed down), the voltage signal displayed must go up.

Display on voltmeter: **0.1 - 1 V**

b) Checking the control unit

Raise engine to operating temperature. Connect the exhaust tester to the exhaust test tube. Disconnect the oxygen sensor and connect the plug sleeve to ground on the control-unit side. Watch the exhaust tester:

Fault, Fault Code	Possible Causes, Elimination, Remarks
	The CO content must go up. **Important:** Continue this test only until a change is noted in the CO content at the CO tester. **If the control unit does not process the ground signal, check its coding before replacing it.**
Test point 10 **Injection circuit shut down** Fault code 1_31	The ignition circuit monitoring relay is mounted on the retaining plate for the LH and EZK control units. If an ignition circuit fails, the power supply to the injector valves is interrupted. At the same time one of the two colored LEDs integrated into the relay lights up. Injection circuit I shut down - red LED lights up. Injection circuit I shut down - green LED lights up.
Test point 11 **Defective control unit** Fault code 1_41	Renew control unit

Fault, Fault Code	Possible Causes, Elimination, Remarks
Test point 12 **No power supply to LH control unit (V)** **a) Permanent positive supply (B+)** Connect voltmeter to terminals 17 and 4 of LH control unit plug with a test lead. **Display: battery voltage** Checking current path according to wiring diagram. **b) Power supply through LH relay XXV** At the LH control unit, connect a voltmeter to terminals 17 and 9, and run to terminal 21 (ground on body) using test leads. The LH relay must be energized. **Display: battery voltage** No display: Check relay Relais XXV and current path according to wiring diagram.	

5.0L ENGINE—LH TEST POINT CHECKS—928

Test point 13: Ground connections / Plug connections

Ground connections

- Engine ground, between engine and right side of body under engine.
- Ground for ignition output stages above the right ignition coil.
- Ground for control units under the fuel pressure regulator and the fuel pressure vaporizer.

Are the ground connections tight and free from corrosion?
Loosen the ground points, clean them and tighten them again as specified.

Note:

Never start the engine when the body-engine ground lead is not connected.
This will destroy the control unit immediately.

Plug connections

Are all plug connections correctly wired up and not loose or corroded? Separate the plug connections. Neither the pins nor the sleeves of the plug connections must be bent or corroded.

- At the central electrics, the relays for the fuel pump, EZK and LH and plug W.
- At the EZK and LH control units, the 35-pin plug in each case.
- Plug connection for activating the ignition output stages (green/white lead). Above the central electrics
- At the air mass sensor, a 6-pin plug (only 5 contacts are occupied in the plug housing).
- At temperature sensor, II a 2-pin plug.
- At the engine speed reference mark sensor, a 3-pin plug (behind the air mass sensor).
- At each of the 8 injector valves a 2-pin plug (first remove the ring pipe cover).
- Knock sensor plug connections. Front sensor: plug connection at the front left, below the fuel ring pipe.
- Rear sensor: plug connection above the right fuel ring pipe at the rear.
- Hall sensor plug connection. Behind the housing for the camshaft drive gear, cylinder bank 1-4.
- Plug connection for solenoid valve to operate resonance flap (at toothed belt cover for cylinder bank 5-8).
- Plug connection, fuel tank venting solenoid (at toothed belt cover for cylinder bank 1-4, below the water hose).
- Plug connections for the output stages (on the cover, left cross-wall).

Test point 14: Fuel pump voltage (V)

Fuel pump relay XX (XXVI) is activated by the LH control unit against ground. To check fuel pump activation, connect terminals 200 and 17 at the LH control unit plug with the test leads and switch on the ignition. The relay will be energized and the pump will run.

If the pump does not run:
Check fuse No. 42 (38). If the fuse is in good working order, remove the fuel pump relay and connect terminals 87 and 30 at the base of the relay. The fuel pump must then run.

If the pump runs, check the relay or activation circuit. If the pump does not run, examine the pipes and connections leading to the pump and the pump itself.

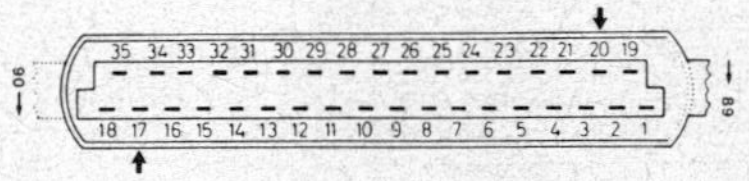

Test point 15: Engine speed signal from EZK

Important: the EZK control unit plug must be attached and the LH control unit plug detached.

At the LH control unit plug, connect terminals 17 and 1 with the engine tester leads, using the test leads.

Adjust the oscilloscope as stated by its manufacturer. Start the engine with the starter motor. A control signal must be visible on the oscilloscope.

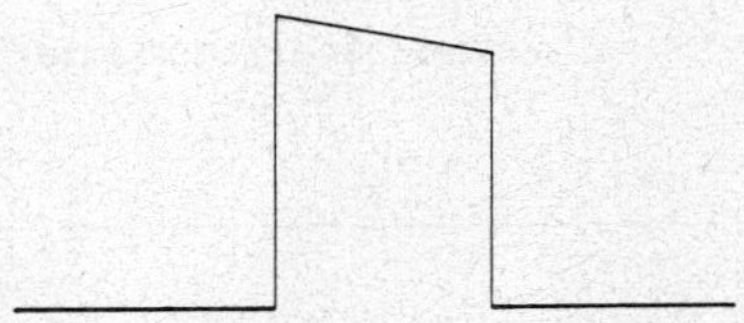

Test point 16: Injector valves Control system (V/Ω)

If the engine can be run, pull off the injector valve plugs one after the other. If the injector valve is in good working order and combustion in that cylinder is correct, engine speed will fall.
If the engine cannot be run, check:

a) Power supply
Pull off injection valve plug (to do this, first remove the
cover over the ring pipe). Remove LH relay XXV and bridge terminals 87 and 30 at the relay base. Connect voltmeter to the valve plug contacts and ground, using the test leads.
Display: battery voltage
If no battery voltage is displayed, check the current flow path according to the wiring diagram.

b)Check injection valve coil resistances

Pull off the valve plugs. Use an ohmmeter at the injection valve connection contacts to measure the coil resistance.

Test value: approx. 16 Ω

c) Injection output stage

Adjust the oscilloscope in accordance with the manufacturer's instructions. Connect a Bosch test lead (1 684 463 093) between one injection valve and plug. Connect the tester lead to the test lead in accordance with the manufacturer's instructions.

Important:

Tester leads must not have any contact with ground.

Start the engine. If the injection output stage is working correctly and the tester connections are correct, the following displays must appear:

Starter speed

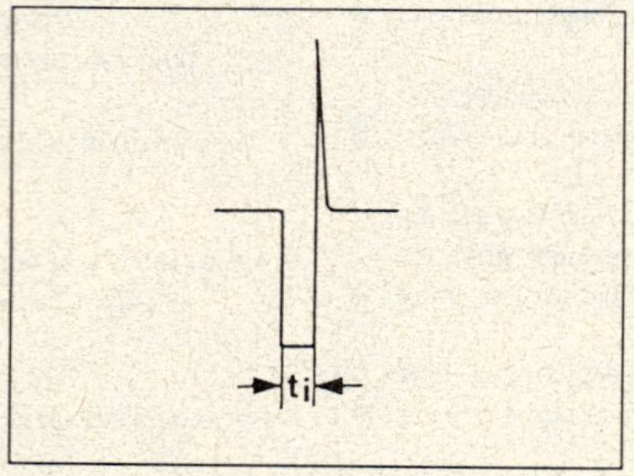

Note

If the engine does not start or if idle speed drops, interchange the tester connections at the test lead and check the tester setting.

Test point 17: Fuel pressure

The fuel collection pipe has a test connection at the front right. Unscrew the cap nut of the test connection.

Warning: the ball seal inside could fall out.

Connect a P 378 or VW 1318 pressure gauge to the test connection. Start the engine and run at idle speed with vacuum at the pressure regulator.

Test pressure 3.3 ± 0.2 bar

If there are hot starting problems:

Check with the engine stopped and the fuel pump running, without vacuum at the pressure regulator (to do this, remove the relay and bridge terminals 87 and 30):

Test pressure 3.8 ± 0.2 bar

- Switch off the fuel pump. Read off the pressure value at the test pressure gauge (nominal value 3.8 ± 0.2 bar).
- Maximum permissible pressure drop with engine warm: **0.5 bar in 30 minutes.**
- If pressure drops below the specified value, proceed as follows:

Build up pressure again by switching the fuel pump on briefly. Remove the return pipe at the pressure regulator)the fuel pump must not be switched on).

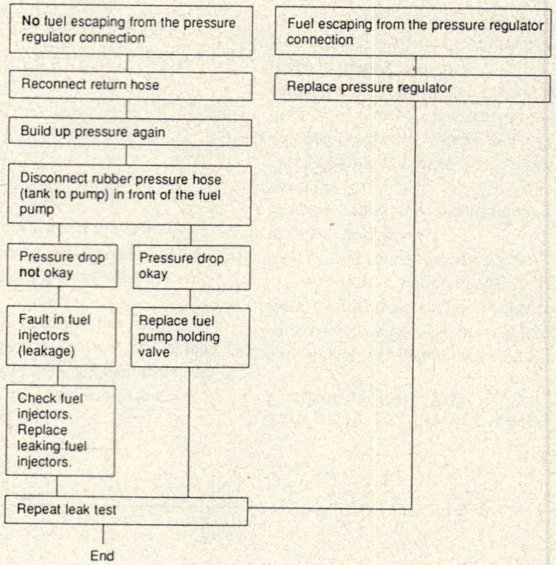

5.0L ENGINE—LH TEST POINT CHECKS—928

Test point 18: Intake system leaks

Check all connections after the air mass sensor for tightness and freedom from leaks.

Attach special tool 9264/1 to the air mass sensor with M 4 x 45 mm bolts and build up a pressure of 0.3 bar. If there are any leaks, pressure will drop in the intake system or air will be heard escaping.

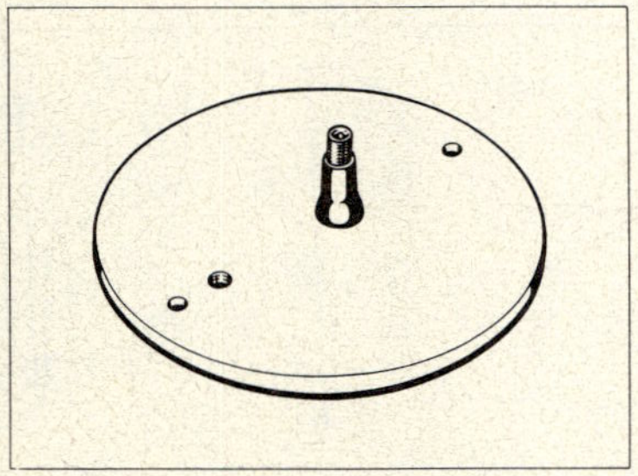

551-24

Note

On cars with oxygen regulator, flashing fault code 1_23 (oxygen regulator at rich stop) appears if there are severe leaks in the intake system (for example the intake pipe gasket).

Test point 19: Resonance flap

In certain engine speed ranges and at not less than one-third of full engine load, the resonance flap in the intake system is opened by vacuum. It is also opened whenever the engine is started (provided that vacuum is present in the reservoir).

a) Checking function of flap

Remove the resonance flap shaft cover (rubber cover) attached above the intake pipe cluster. Using a felt marker or similar, mark a central line on the surface of the flap shaft (this makes it easier to detect opening movement).

Start the engine: **the flap will open and close again as the engine is started.**

When the engine is running, depress the accelerator pedal to the full throttle position suddenly: **the flap will open and close.** (However, this does not happen unless the engine is running at more than 3500 rpm and at not less than one-third of its full load.)

b) Checking operation of resonance flap without running the engine: vacuum side

Detach the vacuum supply pipe from the vacuum reservoir at the solenoid valve. Check vacuum with a vacuum tester
≈ **0.6 bar.**

After this, build up a vacuum of approx. 0.4 bar at the pipe to the diaphragm valve (below the intake air pipe cluster = **the resonance flap will open.**

If the flap does not open:
check diaphragm valve and supply pipe.

c) Power supply (V)

Use the test leads to bridge terminals 17 and 21 at the LH control unit plug. Use other test leads to connect temrinal 34 of the control unit plug to ground = **the resonance flap will open.**

If the flap does not open:
At the LH control unit plug, connect terminals 17 and 21 with the test leads. Separate the plug connection at the solenoid valve. Connect voltmeter to plug socket (2) and ground.
Display: battery voltage

No display:
Check power supply to solenoid valve by way of LH relay XXV according to the wiring diagram. Check that there is no break in the circuit from LH control unit terminal 34 to the solenoid.

d) Checking the control signal with the oscilloscope

Attach the LH control unit plug. Disconnect the plug at the solenoid valve. Connect the engine tester with the test leads to the sleeve side of the solenoid valve plug in accordance with the manufacturer's instructions, and adjust (bei Bosch MOT 300/400, but select screen range of 20 V). While starting the engine and if the throttle is opened suddenly when it is running, the following picture must be visible on the tester:

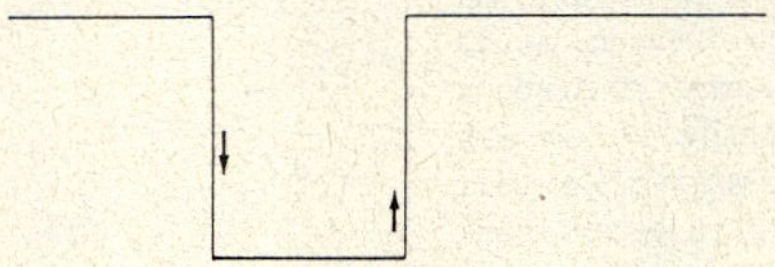

Test point 20: Fuel tank venting (V)

When the engine is at regular operating temperature and the throttle valve is opened (idle speed contact), the solenoid valve is activated by the LH control unit with an intermittent signal to ground. The greater the airflow at the air mass sensor, the longer the solenoid valve remains open.

a) Checking solenoid valve activation

Connect terminals 17 and 21 with test leads at the LH control unit plug. Use another test lead to connect terminal 27 to ground (on body): **the solenoid valve is energized.**

If the valve does not open:
Check power supply and continuity of circuit from terminal 27 to solenoid valve according to the wiring diagram.

b) Checking power supply (V)

At the control unit plug, connect terminals 17 and 21 with the test leads. Disconnect the solenoid valve plug and connect the voltmeter between the plug connection and ground with the aid of test leads.
Display: battery voltage

No display:
Check power supply through LH relay according to wiring diagram.

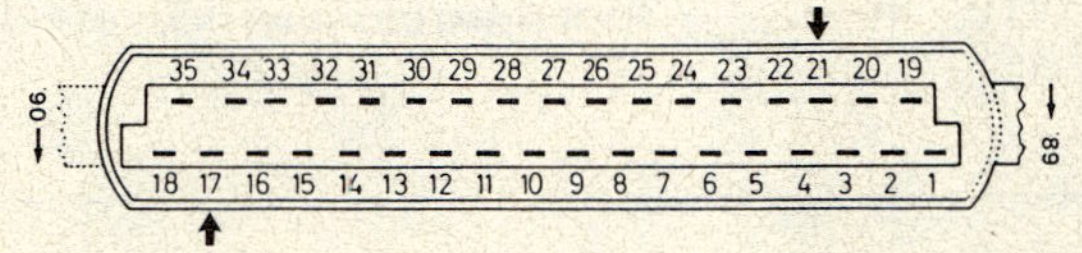

c) Checking control signal with oscilloscope

Connect DME test lead (Bosch No. 1 684 463 093) between solenoid valve and plug connection. Connect the engine tester in accordance with the manufacturer's instructions and adjust it.

Start the engine and open the throttle slightly (the idle speed contact must open, engine temperature must be > 60°C). The following display is then obtained:

Fuel tank venting

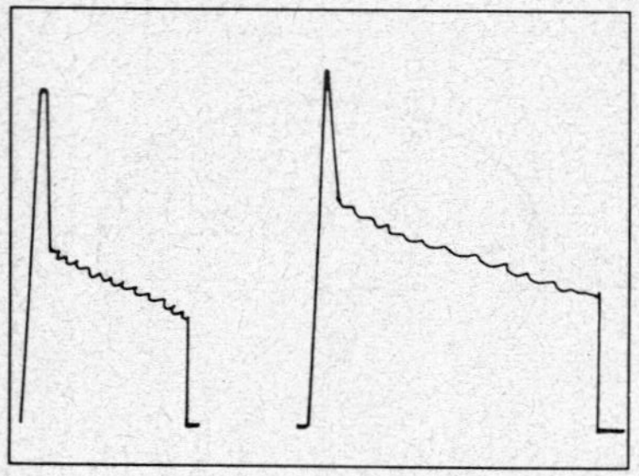

As air flow increases, the signal broadens.

Test point 21: Alternator, regulator

Engine misfiring may be caused by alternator voltage peaks.

Remove the drive belt from the alternator and start the engine.

If this eliminates the fault, check the alternator and regulator.

Test point 22: Leads K and L

The diagnostic connectino between the LH control unit and the diagnosis testers is by way of the two leads K and L. If no diagnosis is possible, the following tests must be carried out:

1. Continuity test (refer to wiring diagram)

Lead L: Pin 7 (19-pole diagnostic socket
- Pin 16 (LH plug)

Lead K: Pin 8 (19-pole diagnostic socket
- Pin 12 (LH plug)

2. Ground short test (refer to wiring diagram)

Lead L: With the ignition switched on, the voltage at Pin 7 (19-pole diagnostic socket) must be > 8 V betragen.

Lead K: With the ignition switched on, the voltage at Pin 8 (19-pole diagnostic socket) must be > 8 V betragen.

Possible fault causes:

- Ground short or break in wiring or at plug connections.

- Faulty control unit connected to these leads (this may be any control unit with diagnostic capability and must not necessarily be the LH control unit).

Checking: disconnect plugs in succession at the control units with diagnostic capability until the voltage > 8 V. Renew the control unit which is causing the fault.

- Tester faulty.

- No power supply at the diagnostic plug.

- No ground at the diagnostic plug.

Notes on idle speed and CO testing at idle speed:

a) Idle speed

The 928 S4 is equipped with an adaptive idle-speed cylinder filling control. This makes it unnecessary to adjust the idle speed of any version. On cars with adaptive idle-speed cylinder filling control, system adaptation must only be carried out after disconnecting the LH control unit from thepermanent positive supply and after each inspection.

Idle speed control value with engine at regular operating temperature: GT 775 ± 25/min
S4 675 ± 25/min

b) Idle speed CO value on cars without catalytic converter

The CO level is still adjusted at the CO adjusting screw on the potentiometer. The following preconditions must be observed:

Engine at regular operating temperature

- Engine in good working order mechanically and ignition system OK

- All electrical consumers switched off

- Exhaust emission tester at operating temperature, calibrated and correctly adjusted.

3.0L ENGINE—DME DIAGNOSIS MENU—944

Engine in perfect running condition
Battery charged
Starter motor cranks the engine

Terms in bold letters = Display Fault Memory / Fault Path

Diagnosis not practicable	Poor hot starting	Engine hesitation	Poor energie power	High fuel consumption	Misfiring	Poor pick-up	Erratic idling	Engine hard to start	Engine will not start	Test point		Fault Code 1_	Test equipment	Plug, Control unit
x		x			x				x	1	**Power Supply**	11	V	24 → 18 24 → 37
		x		x			x	x		2	**Idle Speed Contact Ground Short** **Idle Speed Contact Break**	12 15*	Ω	24 → 52
			x	x						3	**Full Load Contact**	13	Ω	24 → 53
	x	x	x	x		x	x	x	x	4	**Engine Temperature Sensor II**	14	Ω	24 → 45
		x	x	x	x	x	x			5	**Air Flow Sensor**	21	V	– → 3
	x						x	x	x	6	**Idle Speed Control Activation**	22*	⊟	
	x	x	x	x	x	x	x			7	**Oxygen Regulation Stop**	23	V	
	x	x	x	x		x	x			8	**Oxygen Sensor** (sensor signal)	24	V	
	x	x	x	x		x	x			9	**Intake Air Temperature Sensor**	25	Ω	44 → 26
			x	x						10	**Knock Sensor I**	31	–	
			x	x						11	**Knock Sensor II**	32	–	
			x	x						12	**Control Unit Faulty** (knock computer)	33	–	
			x	x						13	**Hall Signal**	34	⊟	
x	x	x	x	x	x	x	x	x	x	14	**Control Unit Faulty**	41	–	
	x	x	x			x	x	x	x	15	**DME-Relays**		V	
	x	x			x	x	x	x		16	**Tank Venting Valve**	43*	⊟	
										17	**Check Engine Warning Lamp** (USA '91 Mod. onward)	45	Ω	
	x	x	x	x	x	x	x	x	x	18	Fuel Pressure		⊘	
	x	x	x	x	x	x	x	x	x	19	Intake System Leaks			
x		x			x	x	x	x	x	20	Ground and Plug Connections		–	
	x	x	x		x	x		x	x	21	Speed/Reference Mark Sender		⊟	47→48
	x	x	x		x	x	x	x	x	22	Ignition and Final Stage		⊟	
	x	x	x	x	x	x	x	x	x	23	Fuel Injectors – Injection Control Unit		⊟	
		x			x					24	Alternator / Regulator		–	
x										25	Leads K and L			

V = voltmeter Ω = ohmmeter ⊟ = oscilloscope * not 944 S

3.0L ENGINE—DME CONNECTOR PINOUT—944

1 Ignition pulse to final stage
2 Ground - final stage shield
3 Ground - DME relay (fuel pump)
4 Idle speed control
5 Tank venting valve
6 Engine speed to tachometer
7 Signal from air flow sensor
8 Signal from hall sender
9
10 Ground from MP VIII shield for oxygen sensor
11 Knock sensor I
12 Power supply for air flow sensor
13 Lead L from diagnosis plug
14 Ground from ground point VIII
15
16
17 Signal ti
18 Permanent positive from CEL
19 Ground from ground point VIII
20
21 Knock diagnosis plug yes + no
22 Check Engine Control
23
24 Ground from MP IX
25
26 Ground for air flow sensor

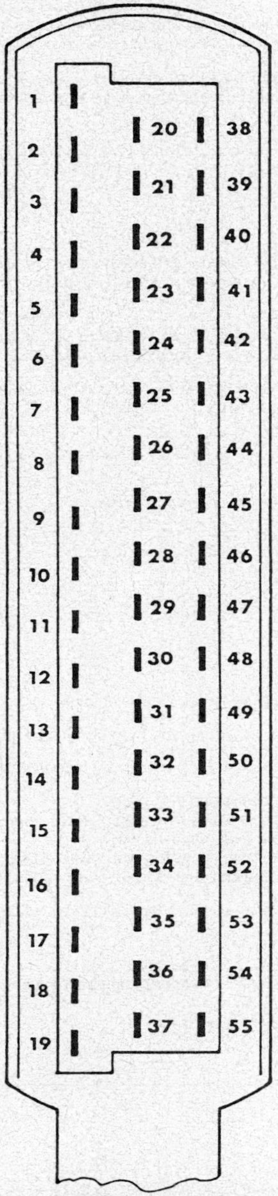

27
28 Oxygen sensor signal
29 Knock sensor II
30 Shield for knock sensors, hall sender and ground for knock sensors
31 Positive to hall sender
32 Economy control
33
34
35
36
37 Voltage from DME relay with ignition
38
39 Version code
40 AC compressor
41 Power strg. (not for 944 S2)
42 Ground from MP VIII
43
44 Intake air temperature NTC I
45 Engine temperature NTC II
46
47 Speed/ref. mark positive
48 Speed/reference mark negative shield
49
50
51
52 Idle signal - throttle valve
53 Full load signal - throttle valve
54 Coding plug curve family
55 Diagnosing plug lead K

3.0L ENGINE—DME TEST POINT CHECKS—944

Fault, Fault Code	Possible Causes, Elimination, Remarks
Test Point 1a Power supply for DME control unit (V)	**a) Perm. pos. (B+)** **Fault memory is cancelled, if there is no B +.** Testing: Connect a voltmeter on terminal 24 (–) and terminal 18 (+) of the control unit plug with help of test leads. Display: battery voltage No display: check current flow and ground paths according to the wiring diagram.
Test Point 1b **Power supply for** DME control unit (V) **too high / too low** Fault Code 1_11	**b) Power supply via DME relay** Connect a voltmeter on terminal 24 (–) and terminal 37 (+) of the control unit plug with help of test leads. Turn on ignition. Display: battery voltage No display: check current flow path according to the wiring diagram.

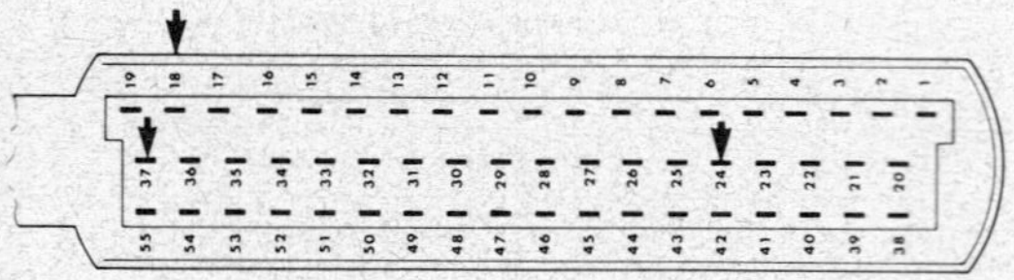

Remarks:

Cars with burglar alarm: check power supply from ignition lock via burglar alarm control unit to terminal 86 of the DME relay according to the wiring diagram.

Cars without burglar alarm: check pin bridges on plug (provided for installation of a burglar alarm control unit).

Fault, Fault Code	Possible Causes, Elimination, Remarks
Test Point 2: **Idle speed contact (Ω)** **Ground short** Fault Code 1_12	**Check** via circuit input test with 9288 System Tester or 9268 Tester. Select test step and operate accelerator pedal. 9288 Display: Idle speed contact made Idle speed contact broken 9268 Display: 1332 (idle speed contact made) 0000 (idle speed contact broken). No Display: Pull off plug on DME control unit. Connect ohmmeter between terminal 52 and terminal 24 with help of test leads.

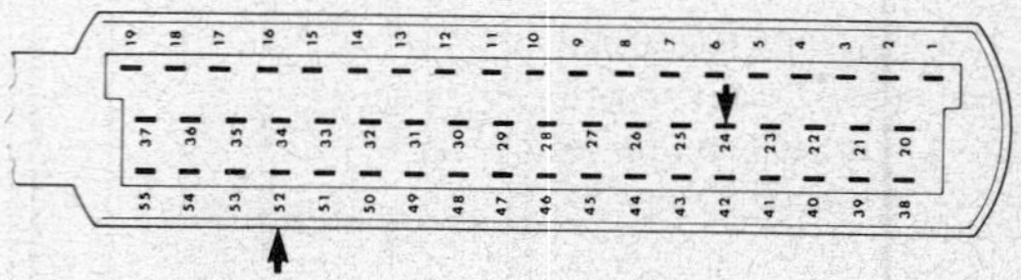

Display:
Throttle valve closed $< 10\ \Omega$
Throttle valve opened to gap of $> 1° = \infty\Omega$

If the value of the idle speed contact is not reached in this test, the test should be repeated direct on the throttle valve switch as described below.

Pull off plug on throttle valve switch. Connect ohmmeter between terminals 1 and 2 of the throttle valve switch with help of test leads.

Display:
Throttle valve closed $< 10\ \Omega$
Throttle valve opened to gap of $> 1° = \infty\Omega$

Fault, Fault Code	Possible Causes, Elimination, Remarks
Idle speed contact (Ω) **Break** Fault Code 1_15	Testing same as described for ground short (fault stored in memory as of 1991 models). Check for correct adjustment, correcting if necessary. Cause, for example, could be an incorrectly adjusted throttle valve switch or accelerator cable.

3.0L ENGINE–DME TEST POINT CHECKS–944

Fault, Fault Code	Possible Causes, Elimination, Remarks

Test Point 3

Full load contact (Ω)
Ground short
Fault Code 1_13

Check via circuit input test with a 9288 System Tester or 9268 Tester.

Select test step and operate accelerator pedal.
9288 Display: Full load contact broken
Full load contact made
9268 Display: 1333 (full load contact broken)
0000 (full load contact made)

No Display:

Connect ohmmeter on terminal 53 and terminal 24 of the disconnected DME control unit plug with help of test leads.

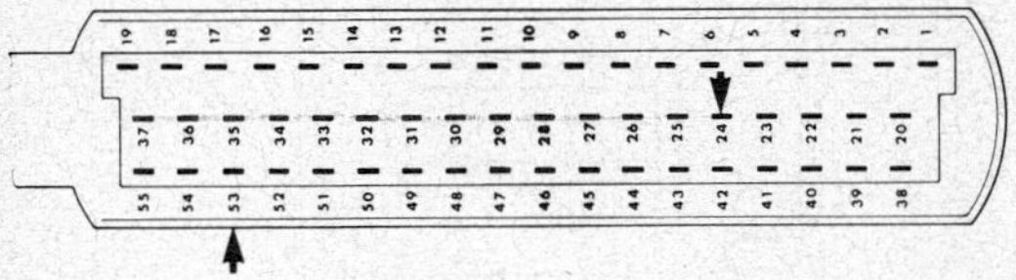

Display:

Throttle valve closed	∞ Ω
Throttle valve opened about 2/3rd of full gap	< 10 Ω

If the value of the full load contact is not reached in this test, the test should be repeated direct on the throttle valve switch as described below.

Pull off plug on throttle valve switch. Connect ohmmeter between terminal 3 and terminal 2 of the throttle valve switch with help of test leads.

Display:

Throttle valve closed	∞ Ω
Throttle valve opened about 2/3rd of full gap	< 10 Ω

Remarks:

If the full load contact fails, the control unit produces a load threshhold via the air flow sensor. Value below this load threshold same as switch broken or above load threshold same as switch made.

Fault, Fault Code	Possible Causes, Elimination, Remarks

Test Point 4

Engine temp. sensor
(NTC II/Ω)
Fault Code 1_14

Engine temperature can be read direct from the "actual value" menu with a 9288 System Tester.

No Plausible Display:

Connect ohmmeter on terminal 45 and terminal 24 of the disconnected DME control unit plug with help of test leads.

Display at:

0°C	=	4.4 - 6.8 kΩ
15 - 30°C	=	1.4 - 3.6 kΩ
40°C	=	1.0 - 1.3 kΩ
80°C	=	250 - 390 Ω
100°C	=	160 - 210 Ω

Test direct on the engine temperature sensor, if the values are not reached.

Note: Temperature sensor II sends information about the engine temperature to the control unit. It is used to provide a rich fuel/air mix-ture after a cold start or during warm-up.

Break (∞ Ω):

DME control unit of the 944 S selects a resistance value, which is stored in the control unit's memory and is approximately equal to that of an engine at operating temperature. There is no enrichment from a faulty NTC II when the engine is warm (emergency running program). This results in starting problems when the engine is cold (no cold start enrichment).

Short (0 Ω):

No engine pickup, too lean and stopping of engine in cold state. No effects in operating temperature state.

Replacement value is applicable to both types of fault!

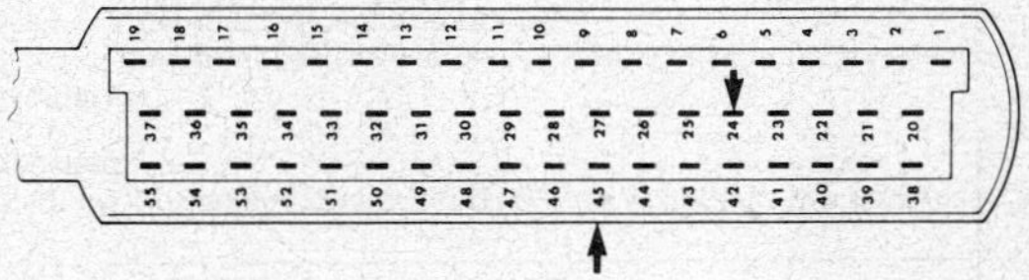

Fault, Fault Code	Possible Causes, Elimination, Remarks
Test Point 5a: **Air flow sensor** (V/Ω) Fault Code 1_21	The ratio between Up (potentiometer voltage) and Uv (power supply voltage) can be read in the "actual value" menu with a 9288 System Tester. When opening the air flow sensor plate the ratio must be about 0.0 to 1. If not: **a) Power supply voltage** Connect a voltmeter on terminal 3 and terminal 4 of the disconnected air flow sensor plug with help of test leads. Turn on ignition. Display: 5 ± 0.3 volts
Test Point 5b:	**b) Voltage drop on air flow sensor plate potentiometer (V)** 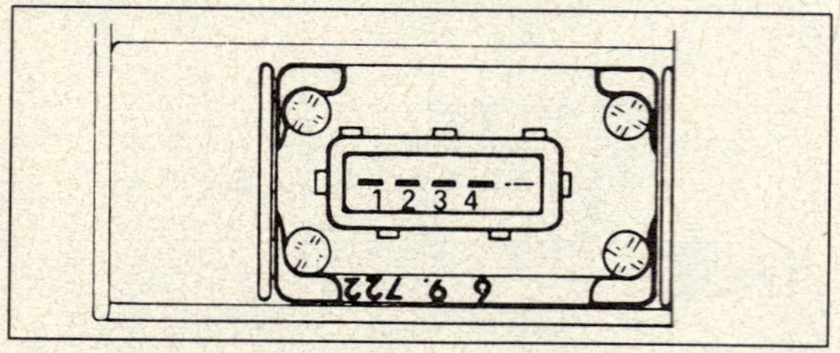 Take off air cleaner connecting hose. Connect pin 3 of the disconnected air flow sensor plug with air flow sensor plug terminal 3 and pin 4 with air flow sensor plug terminal 4 with help of test leads. Connect a voltmeter on air flow sensor terminal 2 and terminal 4 (turn on ignition). Display: approx. 250 - 260 m V It must be possible to move the air flow sensor plate easily and without hesitation. Observe the voltmeter's display at the same time. Voltage jump or voltage drop should not be observed while the air flow sensor plate is opening or closing. Display with air flow sensor plate in full load position: approx. 4.60 volts.

Fault, Fault Code	Possible Causes, Elimination, Remarks
	Turn off ignition and reconnect plug if the voltage values have been reached. **Remarks:** Control unit produces replacement values in case of air flow sensor failure (emergency running program). Injection signal t_i from nominal air of the idle speed volumetric efficiency control for idle or a constant injection signal t_i for partial load and full load ranges.
Test Point 6: **Idle speed control** (V) Fault Code 1_22 only S2	**Element Activation Menu** If pulse cannot be heard, check: **Power supply** Connect a voltmeter on terminal 2 and ground of the disconnected idle speed control plug. Turn on ignition.  Display: battery voltage No display: Check power supply according to the wiring diagram. **Control signal test** The pulse duty factor in % can be read direct in the "actual value" menu with a 9288 System Tester. Function test: Switch on electric power consumers with engine running at idle speed. The pulse duty factor must change and the idle speed must remain constant. If not:

3.0L ENGINE—DME TEST POINT CHECKS—944

Fault, Fault Code | **Possible Causes, Elimination, Remarks**

Connect two-pin DME test lead (Bosch No. 1 684 463 093) between the speed control and plug.

Connect and adjust engine tester to supplied operating instructions.

Make sure that electric lead connections do not have contact with vehicle ground (danger of short).

Displayed with the engine running:

Frequency: approx. 100 Hz.

Replace the idle speed control if the pulse cannot be heard in spite of correct power supply and received signal.

Test Point 7
Oxygen regulation on stop
Fault Code 1_23

Oxygen regulation cannot work in its regulating range in case of disturbances in air/fuel mixture; for example: too lean because of excessive air or too rich because of a sticking fuel injector. It then moves against the stop.

Test 1:	Check CO level with the engine running and CO tester connected (between 0.4 and 1.2 %).
too lean:	Check intake system for leaks (Test Point 19).
too rich:	Check system pressure (Test Point 18). Check fuel injectors for leaks (Test Point 18).

Fault, Fault Code | **Possible Causes, Elimination, Remarks**

Test Point 8:
Oxygen sensor (V)
(sensor signal)
Fault Code 1_24

Sensor signal test

Sensor voltage can be read direct in the "actual value" menu with a 9288 System Tester.

If not:

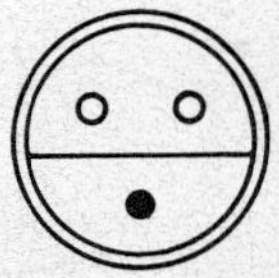

Disconnect oxygen sensor plug. Connect a digital display voltmeter on the sensor end test connection (sensor voltage signal) and ground.

Start and run engine warm so that the oxygen sensor reaches its operating temperature. A change in the voltage signal must be displayed when the mixture is richer, e.g. by accelerating.

Display on voltmeter:
approx. 150 mV - 900 mV (depending on composition of mixture)

Remarks:

If the control unit recognizes an oxygen sensor voltage signal of more than 1 V or less than 0.1 V, the control unit will switch over to operation without oxygen sensor (short or break).

If regulation does not work and the sensor voltage signal is okay, check the coding of the control unit before replacing the control unit.

Test Point 9
Air temperature sensor NTC I (Ω)
Fault Code 1_25

Intake air temperature can be read direct in the "actual value" menu with a 9288 System Tester.

If not or when a non-plausible value is displayed:

Connect an ohmmeter on terminal 44 and terminal 26 of the disconnected DME control unit plug.

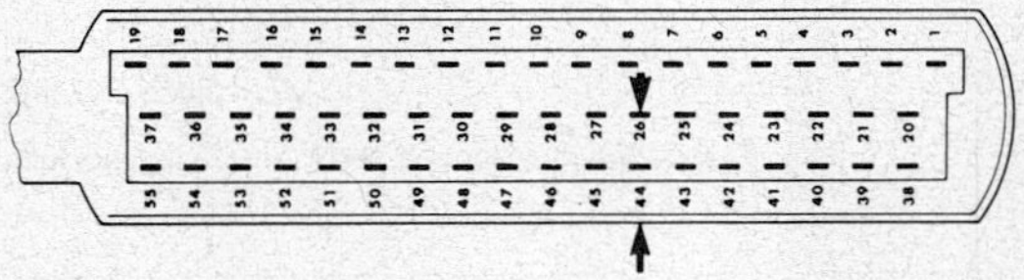

Display at:			
	0°C	=	4.4–6.8 kΩ
	15–30°C	=	1.4–3.6 kΩ
	40°C	=	1–1.3 kΩ

Fault, Fault Code	Possible Causes, Elimination, Remarks
	If no display: Disconnect plug on air flow sensor. Connect an ohmmeter on air flow sensor terminals 1 and 4 with help of test leads. Display: same as above. Note: In case of a break, there will be a replacement value equal to an intake air temperature of 20° C (emergency running program).

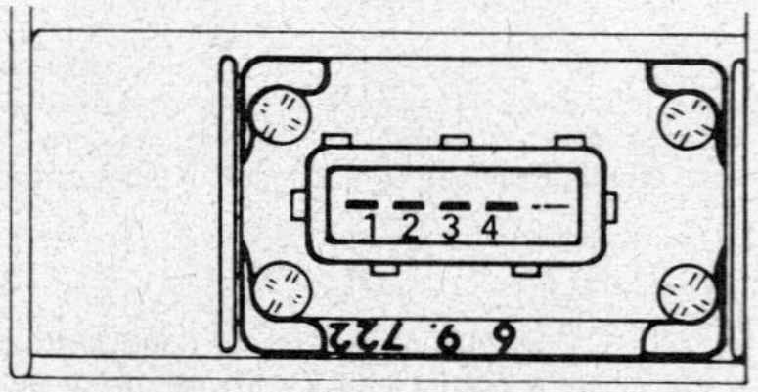

Fault, Fault Code	Possible Causes, Elimination, Remarks
Test Point 10 **Knock sensor I** Fault Code 1_31	Knock sensor I: Knock signal not plausible. Check: - Installation of knock sensor (check tightening torque and type of screw). - Wire harness and plug connections according to wiring diagram. Reconnecting sensors eliminates contact resistance. - Whether coolant or other liquids are in area of the knock sensor. - Replace knock sensor. Ignition timing retarded by 6° on crankshaft in case of a faulty knock sensor.
Test Point 11 **Knock sensor II** Fault Code 1_32	Knock sensor II: Knock signal not plausible. Check: - Installation of knock sensor (check tightening torque and type of screw). - Wire harness and plug connections according to wiring diagram. Reconnecting sensors eliminates contact resistance. - Whether coolant or other liquids are in area of the knock sensor. - Replace knock sensor. Ignition timing retarded by 6° on crankshaft in case of a faulty knock sensor.

Fault, Fault Code	Possible Causes, Elimination, Remarks
Test Point 12 **Control unit** (knock computer) Fault Code 1_33	Ignition timing is retarded 6° in case of this fault. Replace control unit.
Test Point 13: **Hall signal** Fault Code 1_34	Disconnect plug on cylinder head behind the camshaft drive gear and connect a three-pin adapter lead (VW 1501) inbetween to check the Hall signal. Connect oscilloscope positive and negative test leads on terminals 1 and 2 of the adapter lead. Start the engine. This signal should be displayed when the Hall system is okay:

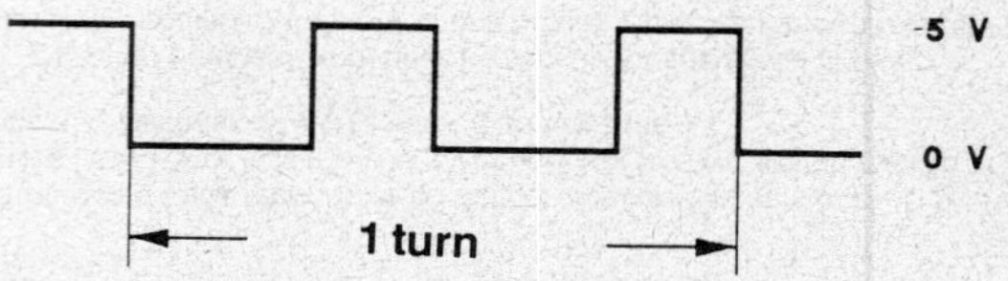

Fault, Fault Code	Possible Causes, Elimination, Remarks
	Ignition timing will be retarded approximately 6° in all speed ranges, if the DME control unit recognizes that the Hall signal is missing. Protective circuit test: Turn off ignition. Disconnect Hall plug. Start the engine. Idle speed ignition timing will adjust itself to approximately 4° before TDC (944 S). There must be a certain load signal from the air flow sensor in case of 944 S2 cars.
Test Point 14: **Control unit faulty** Fault Code 1_41	Replace control unit.

3.0L ENGINE – DME TEST POINT CHECKS – 944

Fault, Fault Code	Possible Causes, Elimination, Remarks
Test Point 15 **DME relay** (fuel pump) (V)	Start the engine – fuel pump must run during engine starting. If not: Turn off ignition. Disconnect control unit plug. Connect terminal 3 and terminal 24 on the control unit plug with help of a test lead. Turn on ignition – the pump must run! If not: Check current flow and ground paths of DME relay and fuel pump according to the wiring diagram.

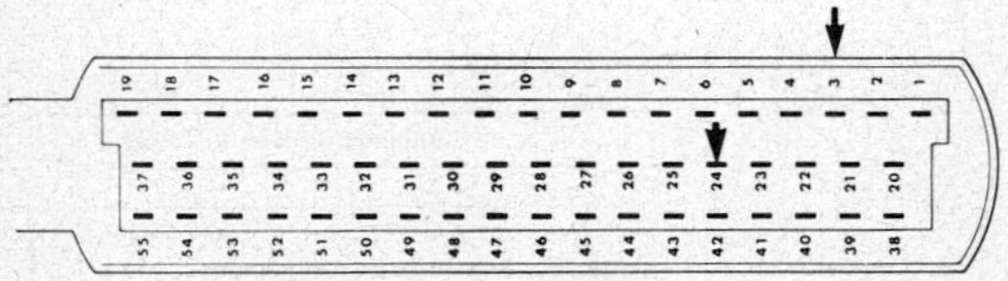

Fault, Fault Code	Possible Causes, Elimination, Remarks
Test Point 16: **Tank venting valve** (V) Fault Code 1_43 (not 944 S)	Electric valve (for brake booster) will be opened more or less long depending on the engine load when an engine having operating temperature is running. The opened time is determined by a ground pulse from the control unit. **a) Activation test** Activate the tank venting valve in the "element activation" menu of the tester. If a pulse is not heard, check power supply to terminal 1 of the electromagnetic valve plug according to the wiring diagram **b) Control signal test** Install DME test lead (Bosch No. 1 684 463 093) between the electromagnetic valve and plug connection. Connect and adjust engine tester according to the supplied operating instructions.

Fault, Fault Code	Possible Causes, Elimination, Remarks
	Tank venting valve is not activated permanently. Test must be performed within 7 minutes after starting an engine having operating temperature (interrupt tank venting valve activation approx. 75 seconds afterward and then continue). Start engine and accelerate. This must be displayed in the tester for an engine at operating temperature:

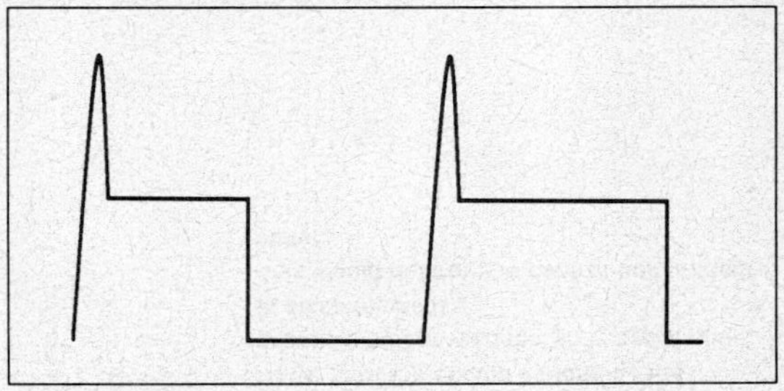

Fault, Fault Code	Possible Causes, Elimination, Remarks
	Signal will be wider as the air flow rate increases! Check path according to the wiring diagram if there is no signal. Replace control unit if necessary.
Test Point 17 **Check Engine warning lamp** Fault Code 1_45 (California as of 1991 models)	A ground signal from control unit plug terminal 22 goes to and switches on the "Check Engine Lamp", if a component influencing emission control fails. If the "Check Engine" warning lamp fails, this fault is stored in the fault memory. Check by supplying ground to control unit plug terminal 22 and turning on the ignition. Always replace a faulty display lamp with a specified lamp.

Test Point 18: Fuel Pressure

Fuel collection pipe has a test connection at left front. Unscrew cap nut of the test connection.

Important! Inserted ball seal could fall out!

Connect P 378 pressure tester on the test connection. Start the engine. Vacuum on pressure regulator at idle speed:

Test pressure: 3.3 ± 0.2 bar

Hot starting problems:

Check with the engine stopped and fuel pump running, without vacuum on the pressure regulator (this requires removing the DME relay and bridging terminals 87b and 30):

Test pressure: 3.8 ± 0.2 bar

- Stop fuel pump. Read pressure on the pressure tester (specification: 3.8 ± 0.2 bar).
- Max. permissible pressure drop on warm engine: **0.5 bar in 30 minutes**.
- If the pressure drops below the specified value, proceed as described below.

Build up pressure again by switching on the fuel pump briefly. Remove return pipe on the pressure regulator (fuel pump must not be switched on).

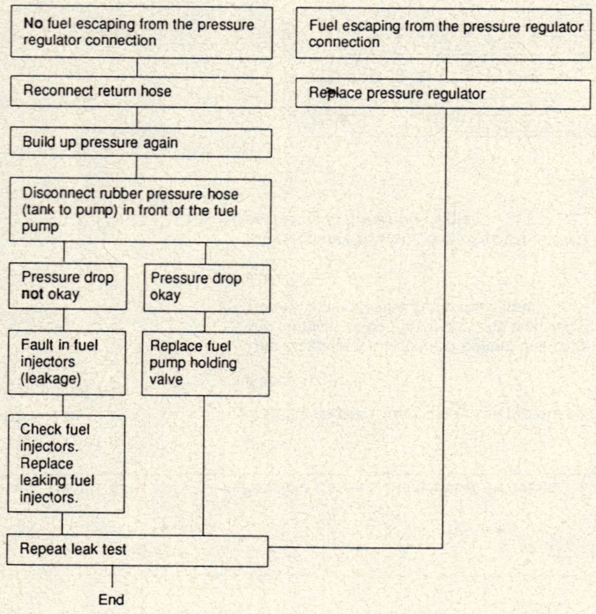

Test Point 19: Intake System Leaks

Check all connections after the air flow sensor for leaks. Remove upper air cleaner section. Apply Special Tool 9264/4 on the air flow sensor and produce approx. 0.5 bar pressure in the intake system. Leaks will cause the pressure in the system to drop quickly.

(944 S: shut the safety pressure valve for crankcase breathing.)

Remarks:

Flashing Fault Code 1_23 (oxygen regulation on stop) is displayed in cars with oxygen regulation and very large leakage in the intake system (for example: intake pipe gasket).

Test Point 20: Ground Connections (V)

- Ground lead (2 x) from wire harness to engine block.
- Battery ground lead on clutch bell housing (engine ground) and on body (body ground next to battery).
- Ignition final stage ground on side member at left front.
- Fuel pump ground on rear lock cross wall behind the spare wheel.
- Are ground connections tight and without corrosion? Loosen, clean and retighten ground connections as specified.

Caution!

Never start the engine as long as the ground lead between the body and engine is disconnected!
This would destroy the control unit immediately!.

3.0L ENGINE—DME TEST POINT CHECKS—944

Plug Connections

Are all plugs connected correctly, not loose and not corroded?
Disconnect plugs and check whether pins and terminals of plugs are bent or corroded.

- 55-pin plug on DME control unit (behind floor plate in the front passenger's foot-well).
- 14-pin plug in engine compartment next to the brake booster.
- 3-pin plug for reference mark/speed sender and 3-pin plug for oxygen sensor behind the intake pipe for cylinder no. 4 (first remove cover on the fuel ring-pipe).
- 3-pin plug on throttle valve switch.
- 2-pin plug on electromagnetic valve for tank venting (on brake booster).
- 4-pin plug on air flow sensor.
- 2-pin plug on NTC II (temperature sensor is screwed in the engine block above the left compensating shaft near cylinder no. 1).
- 7-pin plug on ignition final stage in the engine compartment at front left.
- 3-pin Hall sender plug (behind the camshaft drive gear).
- 2-pin plug for knock sensor I (between intake pipes for cylinder no. 1 and 2).
- 2-pin plug for knock sensor II (between intake pipes for cylinder no. 3 and 4).
- 2-pin plug on idle speed control (between intake pipes for cylinder no. 2 and 3).

944 S2: between cylinder no. 1 and 2.

- Plug for fuel injector ballast resistors (in engine compartment at rear right next to the plug for the brake pad wear indicator).
- 2-pin plugs on each of the four fuel injectors.

Because 944 S cars are supplied worldwide since 1987 models with only one standard control unit and said control unit is coded accordingly depending on country equipment, this coding must be checked in case of complaint.

A 2-pin or 3-pin plug leads out of the DME wire harness near the plug receptacle, in order to be able to call the different curve families.

Power is supplied to the DME control unit of a 944 S via terminal 30 (battery +) in addition to via ignition lock terminal 15.

Test Point 21: Speed/Reference Mark Sender

This test can only be performed with an oscilloscope. Connect and adjust a workshop oscilloscope according to the supplied operating instructions for this purpose.

Connect the positive tester lead via a test lead on control unit plug terminal 47 and negative tester lead via a test lead on control unit plug terminal 48.

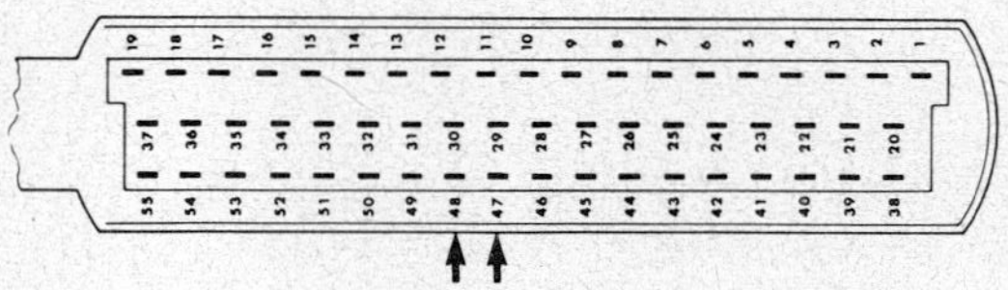

Crank engine with the starter motor. Sinusoidal oscillation of at least 3 volts (speed pulse) and an applied higher amplitude (reference mark) must be displayed on the screen.

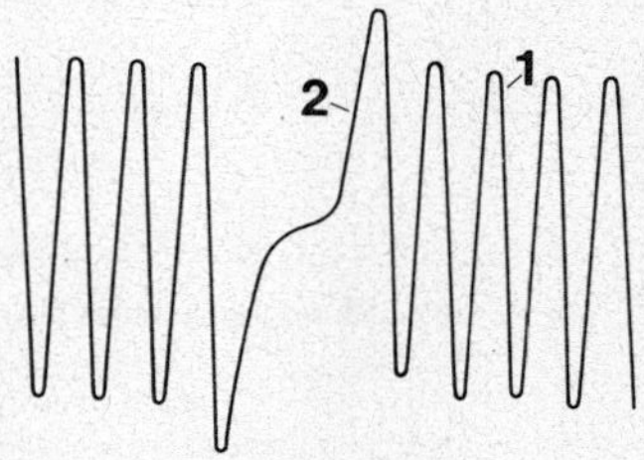

1 – Speed signal
2 – Reference mark signal

If the voltage signal is too small (< 3 V), the distance between the sender and gear ring could be adjusted incorrectly.

Sender distance: Between gear ring and sender: 0.8 ± 0.3 mm.

No display:

Disconnect plug of speed sender on plug plate in the engine compartment. Connect tester leads on middle and outside pins of the plug on the sender end of the plug with help of test leads (terminals 1 and 2, pin end). Start the engine. Sinusoidal oscillation must be displayed on the screen.

Speed sender plug

Remarks:

The display must remain unchanged after supplying ground to connection 3.

If not, check power flow, spacing and condition of the sender.

Test Point 22: Ignition

a) Secondary ignition display

Adjust secondary ignition display on the oscilloscope. Connect tester leads according to the supplied operating instructions.

Note:

If a fault is displayed for all cylinders, it is in the primary or secondary circuit from the ignition coil to the distributor rotor. If a fault is displayed for one cylinder, this fault is located after the distributor rotor.

Primary resistor: Term.1 + 15 = 0.4-0.6 Ω
Secondary resistor: Term.1 + 4 = 5-7.2 kΩ

b) Spark plug connectors (Ω)

Suppression resistor: 3 kΩ
Visual inspection for damage or traces of burning.

c) Distributor

Distributor rotor, distributor cap: suppression resistor = each 1 kΩ.
Visual inspection for damage, dirt or wrong location.

d) Ignition timing test

Ignition timing is checked with the engine at operating temperature.

Ignition timing and engine speed can be read direct in the "actual value" menu with a 9288 System Tester.

Test value: (944 S + 944 S2)

At 840 ± 40 rpm 10° ± 3°

e) Activation of high tension end (if ignition sparks are not triggered)

Ignition triggering signal can be simulated for fast troubleshooting of the ignition end. This requires connecting the oscilloscope term. 4 tester lead on the ignition lead betw. ignition coil and high tension distributor (term. 4).

Disconnect DME control unit plug.
Turn on ignition.

If battery voltage is now supplied intermittently to terminal 1 of the control unit with help of a test lead (e.g. control unit plug term. 18), a high tension peak must be displayed on the oscilloscope each time (ignition triggering).

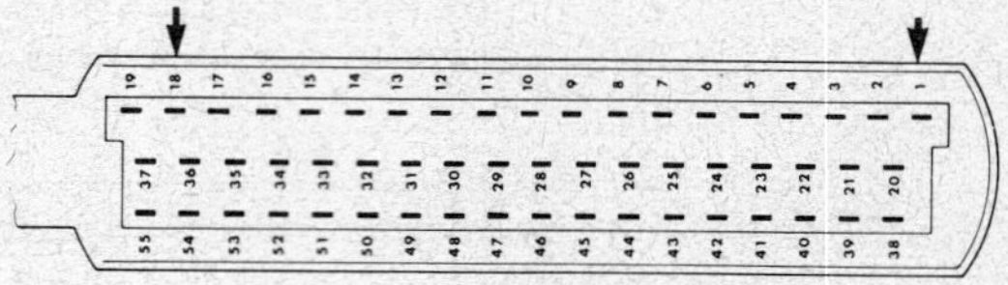

If there is no signal, check:

- Power supply to ignition final stage and ignition coil.
- Ground connection for ignition final stage (ground point on left front side member).
- Plug connections on ignition final stage, ignition coil and high tension distributor.
- Components responsible for high tension (for example: ignition coil, ignition leads, spark plug connectors, etc.).
- Power flow in lead between control unit plug and ignition final stage term. 5.

f) Ignition final stage control signal test

Connect DME control unit plug.

Connect oscilloscope's positive tester lead on terminal 5 and negative tester lead on terminal 2 of the disconnected final stage plug.

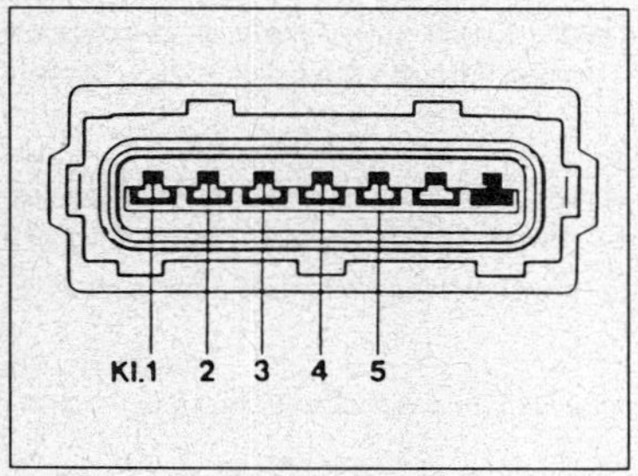

Operate starter motor.
Oscilloscope should display the control signal of the DME control unit.

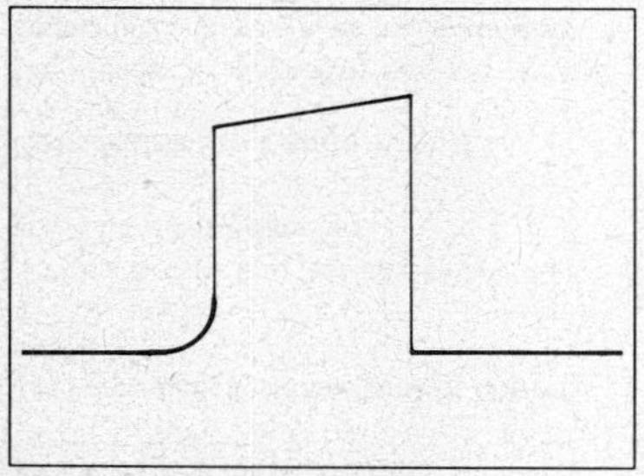

Very important:

Coat the baseplate with a heat conducting paste prior to installation of the ignition final stage. Keep heat conducting paste off of painted parts.

Heat conducting paste:
Bosch No. 5 942 860 003.

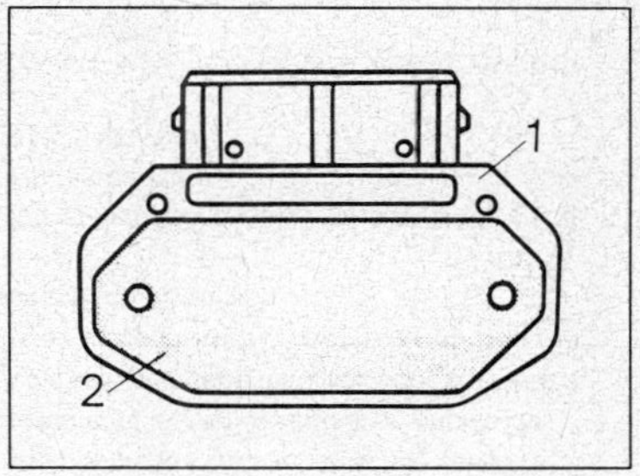

1 – Final stage
2 – Baseplate

Test Point 23: Fuel Injectors – Injection Timing (V/Ω)

Element activation menu:

If a pulse is not heard, check:

a) Power supply

Pull off injector plug. Connect voltmeter on injector plug terminal 2 and ground with help of test leads. Turn on ignition.

Display: battery voltage

If there is no display:

Disconnect 5-pin plug of fuel injector ballast resistors. Connect voltmeter between terminal 3 (terminal sleeve end) and ground.
Turn on ignition.

If battery voltage is displayed, turn off ignition and pull off all plugs on fuel injectors. Check power flow and resistance between the 5-pin plug of ballast resistors and terminal 2 of each fuel injector plug with an ohmmeter.

Resistance of each ballast resistor ≈ 6 Ω

Check current path according to the wiring diagram if battery voltage is not displayed.

b) Coil resistance test of fuel injectors

Pull off injector plugs. Check coil resistance on the contacts of the fuel injectors with an ohmmeter.

Test value: 2 – 3 Ω

c) Injection final stage

Adjust oscilloscope according to the supplied operating instructions. Connect Bosch test lead (1 684 463 093) between fuel injector and plug. Connect tester leads on Bosch lead according to the supplied operating instructions.

Caution:

Tester leads must not have any contact with ground.

Start the engine. These displays must appear in the oscilloscope if the injection final stage is working perfectly and test connections are correct.

Starter speed

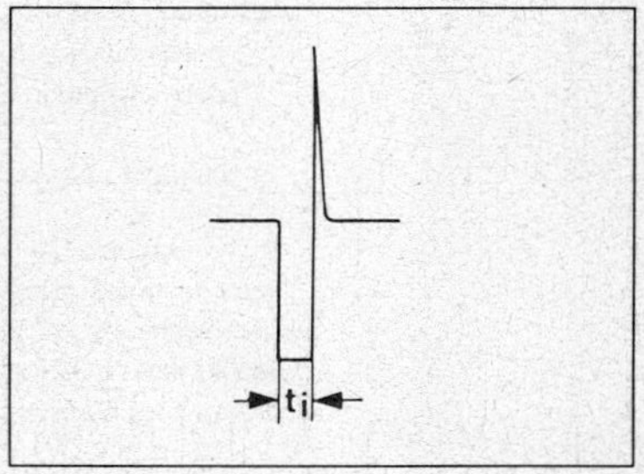

Remarks:

If the engine cannot be started or the idle speed drops, change tester connections on the Bosch lead and check tester adjustment.

Test Point 24: Alternator, Regulator

Misfiring of the engine could be caused by the peak voltage of the alternator.

Take drive belt off of the alternator. Start the engine.

Check alternator and regulator if the disturbances have been eliminated.

Test Point 25: Leads K and L

Connections for diagnosing between the DME control unit and testers are made with both leads K and L. Check the following points when diagnosing is not possible.

1. Power flow test (see wiring diagram)

Lead L: Pin 7 (19-pin diagnostics socket) – Pin 13 (DME plug)

Lead K: Pin 8 (19-pin diagnostics socket) – Pin 55 (DME plug)

2. Ground short test (see wiring diagram)

Lead L: Voltage on Pin 7 (19-pin diagnostics socket) must be > 8 V with the ignition turned on.

Lead K: Voltage on pin 8 (19-pin diagnostics socket) must be > 8 V with the ignition turned on.

Possible Causes of Fault:

- Ground short or break in wiring or plug connections.
- Faulty control unit, which is connected on these leads (could be any diagnosable control unit; does not always have to be the DME control unit).

Checking: Pull off plugs on diagnosable control units one after the other until there is > 8 V. Replace fault-causing control unit.

- Tester faulty.
- Power supply lacking on diagnostics socket.
- Ground supply lacking on diagnostics socket.

Information for idle speed and idle speed CO test:

a) Idle speed

944 S engines have adaptive idle speed volumetric efficiency regulation since 1987 models. This means that the idle speed of all versions does not have to be adjusted. The system of cars with adaptive idle speed volumetric efficiency regulation should be adapted each time the DME control unit had been disconnected on permanent positive or after each inspection.

Idle speed control value for engine at operating temperature:
840 ± 40 rpm

b) Idle speed CO of cars without catalytic converter

CO level is still adjusted on the CO adjusting screw in the air flow sensor.
Requirements:
- Engine at operating temperature.
- Engine and ignition in perfect running condition.
- All electric consumers switched off.
- CO tester at operating temperature, calibrated and adjusted correctly.

Adjusting value: 0.5 - 1.5 % CO
at 840 ± 40 rpm and ignition timing of 10° ± 3° before TDC.

Adapt the system after adjusting the CO level. An adapter plug (911.612.422.8) must be connected on the oxygen sensor connecting lead of cars without a catalytic converter, in order to avoid diffusion (see wiring diagram).

c) Idle speed CO of cars with catalytic converter

In addition to adaptation of the idle speed there is also adaptation of the idle speed CO level in these cars, so that it is only necessary to check the actual state in such cars.
Requirements:
- Engine at operating temperature.
- Engine and ignition in perfect running condition.
- All electric consumers switched off.
- CO tester at operating temperature, calibrated and adjusted correctly.
- **Oxygen sensor plug connected.**

Test value: 0.4 - 1.2 % CO
at 840 ± 40 rpm and ignition timing of 10° ± 3° before TDC.

Afterwards the engine must be run at least 10 minutes after disconnection of the control unit plug or disconnection of the battery, before it will be possible to recheck these values. In addition, the system must be adapted.

3.0L ENGINE – DME CONNECTOR PINOUT – 968

Plug Assignment

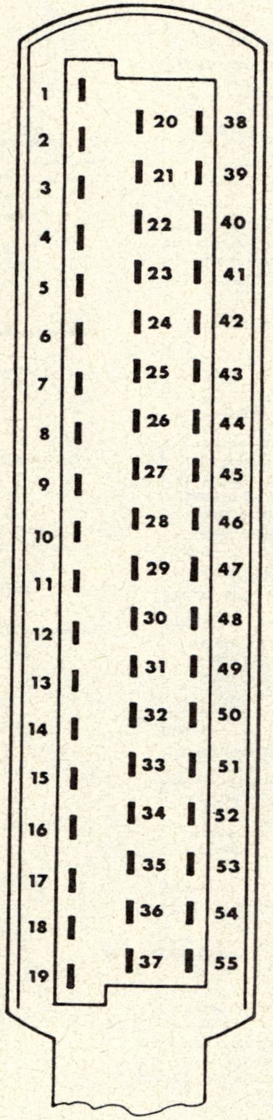

1 - Ignition coil
2 - Ground, ignition output stage (MP IX)
3 - Ground, DME relay 85b
4 - Open positioner
5 -
6 - Engine speed signal at engine speed sensor
7 - Signal from heated wire
8 - Hall-effect transmitter signal
9 - Speed signal
10 - Ground, electronics (MP VIII)
11 - Knock sensor 1
12 - Positive, Hall-effect transmitter/throttle valve
13 - Diagnosis L lead
14 - Ground, injection valves (MPIX)
15 -
16 - Injection valve, cylinder 2
17 - Injection valve, cylinder 1
18 - Terminal 30
19 - Ground, electronics
20 -
21 - Check Engine lamp
22 - Close positioner
23 - Variable camshaft (VarioCam)
24 - Ground, remaining output stages (MPIX)
25 - Burn-off, heated wire
26 - Ground, air flow sensor
27 - Terminal 15
28 - Lambda probe/idle CO potentiometer
29 - Knock sensor 2
30 - Earth NTC/screen knock sensor
31 - Diagnosis socket and knock monitoring
32 - AC relay energizing
33 - Tank breather valve
34 - Injection valve, cylinder 4
35 - Injection valve, cylinder 3
36 - DME-relay 85
37 - DME-relay 87
38 -
39 -
40 - Input from AC switch
41 - Coolant temperature switch (115°C)
42 - Position switch (selector lever)
43 -
44 - Characteristic map switch
45 - Water temperature (NTC II)
46 -
47 - Oil temperature for VarioCam (NTC I)
48 - Engine speed/reference markertransmitter, negative
49 - Engine speed/reference markertransmitter, positive
50 -
51 - Transmission control
52 - Throttle valve position on transmission control unit
53 - Signal from throttle valve potentiometer
54 - Variant encoding
55 - Diagnosis K lead

SYSTEM TESTER 9288 OPERATING INSTRUCTIONS 911, 928, 944 AND 968

1. General information

1.1 Application

The Systemtester 9288 (BOSCH KTS 301) is a microprocessor-controlled self-diagnosis tester.

All systems which have a diagnosis interface as per ISO Standard can be tested with this tester. The following tests are possible:

- Reading out the fault memory
- Testing of the actuators
- Testing the circuit inputs
- System adaptation
- Engine-knock detection
- Sensor and status checks, tire-pressure monitoring (RDK)

The Systemtester 9288 is a high-quality piece of electronic equipment. In order to prevent damage to the equipment as a result of improper use, please read the information in the operating instructions carefully and comply with it.

In addition, the instructions (specifications) of the vehicle manufacturer are also to be observed.

If the tester should fail, check the following points before sending it in for repair:

1. Has the tester been operated incorrectly?
2. Is the battery sufficiently charged?
3. Is the adapter cable OK?
   (Please note when checking the adapter cable that a highly sensitive electronic matching circuit is installed in the vicinity of the 19-pole plug).

1.2 Construction (Fig. 1)

No.	Description	Function	Remarks
1	LCD indicator	Dot matrix 5 x 8 4 lines each with 20 characters Foreign languages possible Illumination	If the Systemtester 9288 is switched on without the program module, following the self-test the tester switches off automatically and informs the user that the program module is not fitted.
2	Keyboard	Keys 1, 2, 3 = Selection key Keys < > = Previous page/next page Key H = Help menu, e.g.: Illumination, Screens stored, Control-unit overview, Setting up printer, Switching off unit Key N = Return to the next higher program level following termination of a test sequence or, during a test sequence, return to the last display Key ⇨ = Storing indication Key ⇦ = Playing back stored reading	Switching on: = Press any key Switching off: (automatic) = 180 s after last depression of a key or if no data stream flows across the serial interface. The last field in the top right-hand corner is filled completely, this means that this is a stored figure and not an actual, real-life figure.
3	Power supply	Fitted accumulator with NiCd batteries. The Systemtester 9288 must be switched off during the initial battery charging process. Charging time > 8 hours	Discharged upon delivery. Following charge: Operating time: 4 – 8 hours without scale illumination 1 – 2 hours with scale illumination
	If the voltage is not sufficient, "Charge battery" appears on the display. If this is not done, the unit switches itself off.	Connection to vehicle battery by means of vehicle-specific adapter lead (see 1.4)	Connection through ISO-interface Charging voltage supply
		Battery charger (accessory)	For test operation and for charging the NiCd batteries.
4	Connection for input and output devices	Connection facility for Printer e.g. Epson, IBM, Hewlett Packard (HP)	The Systemtester 9288 transmits data with the following configurations: 8 data bits / 1 start bit / 1 stop bit / No parity (for printer matching)
5	Connection for vehicle specific adapter lead	Reading out the data	Input for flashing-code support
6	Plug-in programme module (see also Figure 2) ⚠ C-MOS ! Do not touch plug !	Operating system LCD drive Keyboard Interface communication Computations and data conversions	Plug in module: remove rubber protector, insert module fully.

SYSTEM TESTER 9288 OPERATING INSTRUCTIONS—911, 928, 944 AND 968

1.3 Battery charger run off mains voltage (Figure 3)
– Accessory –
Item 1 Charger with connecting cable, 1.5 m long
Item 2 8-pin AMP plug

1.4 Vehicle-specific adapter cable
Porsche No. 000 721 928.81

1.5 Connecting lead (Figure 4)
– Self-fabrication –
for printer, programme load station or similar unit.
For interface-trunk assignment, see manual of corresponding unit.
Printer cable for standard D 25
BOSCH No. 1 684 465 193
Printer cable for EPSON
BOSCH No. 1 684 465 194

2. Connection

The following points must be observed:

– No gear must be engaged on the vehicle (Automatic transmission in position N-P) – Danger of Accident !
– ALL work on the vehicle must only be carried out with the ignition switched off.

After having connected the vehicle-specific adapter cable, the instructions listed under "3" are displayed on the Systemtester 9288:

2.1 Charging with the battery charger (Fig. 3)
Connecting the Systemtester 9288 to the battery charger. (Fig.1, pos.5).

2.2 Diagnosis
Connecting the Systemtester 9288 to the diagnosis plug in the vehicle by means of the vehicle-specific adapter cable.
Switch on the tester and proceed according to the instructions displayed.

3. Testing

Scope of module:
Guidance through the menu, communication with the ECU, reading out the error memory and selection of the "Help" menus, actuator diagnosis, circuit inputs and system adaptation, engine-knock detection, sensor and event check for the tire-pressure monitor (RDK).

3.1 Reading-out the error memory
Connect the Systemtester 9288 (see 2.)
Switch on the Systemtester,
(possible with every key!)
Display:

```
   PORSCHE
Eprom modul      eng
Mod. intro.   xx.xx.xx
```

If a specific instruction does not appear in a display, it is always possible to proceed by pressing the button >.

Due to the fact that the Systemtester 9288 can store error displays (see Chapter 3.7), the following display will appear if errors have been stored in the image memory:

```
Stored displays
erased ?
1 = yes
3 = no
```

Key 3
Display:

```
Print out
displays:          H

continue:          >
```

H = Help menu (see 3.6) or key 1
Display:

```
  Vehicle types
1 = 944 S
2 = 911 Carrera 4
3 = 928 S 4
```

Selection of the vehicle type with key 1, 2 or 3.
After the vehicle type has been selected, the following instruction appears:

```
Connect adapter
cable to veh. plug.
Ignition "ON".
After completion:  >
```

The following then appears:

```
Wait for
Data

Break off test:    N
```

After a short pause, the Systemtester 9288 reports all the systems that are installed in the particular vehicle. If a system is preceded by " # ", this means that at least 1 error is stored in that particular system.
Examples:

```
 Installed systems
1 = # LH
2 = # EZK
3 =   RDK
```

The particular system can be selected by means of key 1, 2 or 3.
After selection (for instance with key 1), the following display appears:

```
  LH
System:  L01 LH-JET
Ser. No.: 92861812313
RB. No.: 0280002507
```

After pressing the key >, a selection menu is displayed:

```
   Menu
1 = Fault memory
2 = Drive links
3 = Input signals   >
```

```
< Menu
1 = System adaptation
```

In the example – press key 1. There then follows the display of the number of errors which are stored (if any).

```
Number of
faults

→ 2 ←
```

Proceed with key >

```
Additional info to
every display with
key 1
continue:          >
```

Proceed with key >
Error output:

```
1: Engine
temperature sensor 2
Short to ground
not present
```

If key 1 is pressed instead of the > key, the corresponding error code display appears (the last two digits of the flashing code).

```
Fault code: – 14 –
```

Proceed with key >
Further errors are displayed (if they exist):

```
2:  Idle contact

Short to ground
present
```

If key 1 is pressed instead of key > the corresponding error-code display appears (the last two digits of the flashing code).

```
Fault code: – 12 –
```

After the last displayed error, the following instruction appears:

```
Repair fault accord-
ing to repair
instructions
Continue:          >
```

Proceed with key >

```
Fault repaired ?
1 = yes
3 = no
```

Return to display "No. of errors" with key 3.
Proceed with key 1:

```
Fault memory
1 = Erase
3 = Do not erase
```

If key 3 is pressed:
= Return to menu "error memory".
The error memory is not erased!

Proceed with key 1:

```
Fault memory
has been cleared

Return:            N
```

The test scope "Read-out error memory" is terminated at this point.

3.2 Actuator diagnosis
If an actuator is selected, this is triggered by the ECU so that it can be checked for correct functioning.
The various actuators components are gone through one after the other and are selected with the > key.
Operate the Systemtester 9288 as described under 3.1 until the following menu display appears:

```
   Menu
1 = Fault memory
2 = Drive links
3 = Input signals   >
```

After pressing key 2, the display for the first actuator appears:

```
Injector
to activate
1 = Start
Continue:          >
```

If key > is pressed, the next actuator is selected.
Pressing key 1 results in the following instruction:

```
Can injectors be
heard / felt ?
1 = yes
3 = no
```

SYSTEM TESTER 9288 OPERATING INSTRUCTIONS—911, 928, 944 AND 968

Key 1 selects the next actuator (e.g. idle actuator). Following instruction:

```
Repair fault accord-
ing to repair
instructions
Continue:          >
```

After pressing key >, the following display appears:

```
Injector
to activate
1 = Start
Continue:          >
```

Proceed with key 1

```
Can injectors be
heard / felt ?
1 = yes
3 = no
```

Proceed with key 1 to the next actuator.

```
Idle stabilizer
to activate
1 = Start
Continue:          >
```

Proceed with key 1

```
Can idle stabilizer
be heard / felt ?
1 = yes
3 = no
```

By pressing key 1, the next actuator is selected. After pressing key 3, the next instruction appears:

```
Repair fault accord-
ing to repair
instructions
Continue:          >
```

Proceed with key >

```
Idle stabilizer
to activate
1 = Start
Continue:          >
```

After pressing key 1, the following display appears:

```
Can idle stabilizer
be heard / felt ?
1 = yes
3 = no
```

By pressing key 1, the next actuator is selected. The actuators are selected one after the other and triggered until the following display appears:

```
Drive link test
completed

Return:            N
```

By pressing the key N, the operator is returned to the menu.

3.3 Circuit inputs

In addition to the actuators, the Systemtester 9288 can also check circuit inputs. To this end, operate the Systemtester 9288 in accordance with 3.1 until this menu display appears:

```
   Menu
1 = Fault memory
2 = Drive links
3 = Input signals   >
```

Press key 3

```
Idle contact
1 = Start

Continue:          >
```

By pressing key > the next circuit input is selected.

The next display appears when key 1 is pressed.

```
Activate accl. pedal
Idle contact
– closed –
Continue:          >
```

Operate the accelerator pedal, the following display appears:

```
Activate accl. pedal
Idle contact
– open –
Continue:          >
```

The next circuit input is selected by pressing key >. Repeat until this display appears:

```
Input signals
testing completed

Return:            N
```

Press key N for return to menu

3.4 System adaptation

When the function "System adaptation" is triggered, the ECU registers the basic air requirement of the engine.

To this end, operate the Systemtester 9288 as per 3.1 until the following menu display appears:

```
   Menu
1 = Fault memory
2 = Drive links
3 = Input signals   >
```

Proceed with key >

```
<  Menu
1 = System adaptation
```

Proceed with key 1

```
Prerequisite:
Eng. at oper. temp.
with all consumers
and ignition off.
```

Proceed with key >

```
System adaptation
1 = Start

Return:            N
```

If key N is pressed
= return to menu.

If key 1 is pressed:

```
Start engine !
```

Following engine start there appears:

```
System is being
adapted

Please wait !
```

After approx. 30 secs there appears:

```
System adaptation
completed

Return:            N
```

If it is impossible to carry out system adaptation (idle contact not closed, or defective), the following display appears:

```
No system adaptation
possible
Idle contact ?
Return:            N
```

After completion of the system adaptation, return to the menu with key N.

3.5 Engine-knock registration

The engine-knock registration function can only be triggered through the EZK or DME control unit.

To this end, operate the Systemtester 9288 as described in 3.1 until the following display appears:

```
Installed systems
1 = # LH
2 = # EZK
3 =   RDK
```

The particular system can be selected by means of key 1, 2 or 3. For instance with key 2 the following display appears:

```
   EZK
System:  E01EZK
Ser. No.: 92861812415
RB. No.:  0227400154
```

Proceed with key >
The following menu display appears:

```
<  Menu
1 = Fault memory
2 = Knock registration
```

Proceed with key 2

```
Condition:
Engine at operating
temperature
                   >
```

Proceed with key >

```
<  Start knock
registration before
test drive
                   >
```

Proceed with key >

```
<  A normal test
drive is a pre-
requisite
                   >
```

Proceed with key >

```
<  Stop the test
drive only if the
display with the no.
of knocks comes on.
```

Proceed with key >

```
Knock registration
1 = Start

Return:            N
```

Pressing key 1 activates the engine-knock counter:

```
Knock registration
in progress

Please wait !
```

SYSTEM TESTER 9288 OPERATING INSTRUCTIONS—911, 928, 944 AND 968

The knock counter registers 10,000 ignitions before the display with the actual number of combustion "knocks" appears.

```
Number
Knocks:        xxx
Combustion:    xxxxx
Continue:      >
```

Proceed with key >

```
Knock registration
completed

Return:            N
```

If knock registration is impossible (due to lack of engine-speed signal), the following display appears:

```
No knock regist-
ration possible.
RPM signal ?
Return:            N
```

Following completion of the knock registration test, return to the menu with key N.

3.6 Help menu

The "Help" menu can be selected from every display by pressing key H. Return to the initial display with key N.

```
    Help menu
 1 = Illumination
 2 = Display stored
 3 = Ctrl. unit chart   >
```

Proceed, for instance with key 1:

Key 1:
The scale illumination is switched on and the tester returns to the previous display.

Or with key 2:

```
Data display stored
  1 = Print
  2 = Clear
```

Proceed with key 1
Stored displays are printed out (If printer connected).

Proceed with key 2
Stored displays are erased.

With the "Help" menu, for instance

```
    Help menu
 1 = Illumination
 2 = Display stored
 3 = Crtl. unit chart   >
```

if the key > is pressed, a further section of the "Help" menu is displayed:

```
<  Help menu
1 = Printer setting
2 = Switch off equip.
3 = Baud Rate
```

Proceed for instance with key 1

```
Printer setting
1 = IBM
2 = HP Quiet Jet
3 = EPSON
```

The selection of the printer results in the tester being set up for the printer type in question.

3.7 Store measurement displays (Key ⇨)

Using key ⇨, all displays can be stored manually.

The following displays are stored automatically:

- ECU-Identity
- Installed systems
- All existing errors

When the memory limit is reached, the following instruction is displayed:

```
Data display mem.
full !

Return:            N
```

3.8 Show stored measurement displays (Key ⇦)

Using the keys < or >, the stored displays for the selected system can be shown.

The stored displays can be called up by means of the ⇦ key.

The system selection (LH - EZK - RDK) takes place with the keys 1, 2 or 3.

4. Service and wear parts (BOSCH)

Fig.	BOSCH Part No.	Designation	Comment
1/3	1 687 335 002	NC-battery	
4/1	1 684 483 152	Plug	
4/2	1 684 485 170	Socket	9pole
4/3	1 680 552 005	Screwed cap	
	1 684 465 193	Printer cable (Standard D 25)	
	1 684 465 194	Printer cable (EPSON)	

4.1 Service parts (Porsche)

Designation	Porsche Part No.	Special tool No.
Systemtester 9288	000.721.928.80	9288
Adapter cable	000.721.928.81	9288/1
Battery charger	000.721.928.82	9288/2
Module (D)	000.721.928.84	9288/4
Module (GB/USA)	000.721.928.85	9288/5
Module (F)	000.721.928.86	9288/6
Module (I)	000.721.928.87	9288/7
Module (E)	000.721.928.88	9288/8

Component Replacement

SERVICE PRECAUTIONS

- Always disconnect the battery, the control unit and the ignition coils when using an electric welder on the vehicle or when charging the battery or when the vehicle is placed in a paint drying oven.
- Remove the main relay to disable both the fuel and ignition systems when checking the compression. Never crank the engine after removing the distributor cap or disconnecting the high tension wire on the ignition coil.
- Never disconnect the battery or wires on the alternator, starter or spark plugs when the engine is running.
- Never connect a test lamp on terminal 1 of the ignition coil.
- Never connect terminal 1 of the ignition coil with ground or B+. This means if installing a burglar alarm, terminal 1 should not be used for starter interlocking.
- Keep the ignition high tension components in good working order as a high voltage flashover can destroy the control units.
- Never reverse battery polarity as it will destroy the control unit.
- Never jump start the vehicle with voltages greater than 16 volts.

FUEL PRESSURE RELIEF

Whenever working on the fuel injection system, be aware of the possibility of residual pressure existing in the fuel lines. This pressure should be relieved before disconnecting any lines when performing fuel system testing or repair.

CAUTION

This procedure will produce liquid fuel and fuel vapors. Perform this procedure with proper care and respect for the flammability of the fuel, the potential of explosion and personal injury. Have a fire extinguisher readily available. Dispose of the fuel in an enviromentally sound manner and with regard to local regulations.

1. Wrap a shop towel around a fuel line junction leading to

the fuel rail. Do not use the fuel return line from the fuel pressure regulator as this will not depressurize the system.

2. Using a backup wrench and a properly sized flare wrench gently loosen the junction.

3. Catch any escaping fuel in the shop towel and retighten the junction.

AIR FLOW SENSOR

Removal and Installation

1. Be sure the ignition switch is in the **OFF** position. The negative battery terminal can be disconnected, but the adaptive memory will be lost along with any electronic presets in the radio, alarm system or other preset type systems.

2. Disconnect the air ducting or hoses from the body of the air flow sensor.

3. Disconnect the electrical connector from the air flow sensor.

4. Remove the bolts (usually 4 or 5) from the air flow sensor mounting points.

5. Remove the air flow sensor form the vehicle. Do not subject the unit to strong shocks or other physical abuses.

6. Install the sensor and tighten the mounting bolts. Connect the electrical connector and air ducting.

AIR MASS SENSOR

Removal and Installation

1. Be sure the ignition switch is in the **OFF** position. The negative battery terminal can be disconnected, but the adaptive memory will be lost along with any electronic presets in the radio, alarm system or other preset type systems.

2. Disconnect the air ducting from the body of the air mass sensor.

3. Disconnect the electrical connector from the air flow sensor.

4. Disconnect the air mass sensor clamp attaching it to the air clearer assembly.

5. Remove the air mass sensor from the vehicle. Do not subject the unit to strong shocks or other physical abuses.

6. Install the sensor and tighten the mounting clamp. Connect the electrical connector and air ducting.

THROTTLE POSITION SWITCH

Removal and Installation

1. Be sure the ignition switch is in the **OFF** position. The negative battery terminal can be disconnected, but the adaptive memory will be lost along with any electronic presets in the radio, alarm system or other preset type systems.

2. Disconnect the electrical lead from the throttle switch.

3. Remove the switch mounting screws and remove the switch.

4. Replace the switch, its mounting screws and adjust. Connect the electrical connector.

Adjustment

911 AND 944

1. Unplug the switch connector and connect an ohmmeter to the switch terminals. The center terminal is common, terminal 1 is the idle switch, terminal 3 is the full load switch.

2. The idle switch should be closed when the throttle is closed. Any movement of the throttle lever will open the switch just before the throttle plate actually opens. Operate the switch by hand if necessary to determine a faulty switch or faulty adjustment.

3. Connect the ohmmeter to the full load switch. The switch should close when the throttle lever is 70 degrees off of idle, about ⅔ of full stroke.

4. On the 911, if the switches are good but do not operate at the proper throttle angles, check throttle cable adjustment.

a. Make sure there is no slack in the throttle cable. The adjustment is just below the front end of the transaxle housing.

b. Make sure the throttle lever touches the idle stop on the throttle body. If not, go to Step C. Do not adjust the stop screw, this is set at the factory on a flow bench.

c. At the pedal assembly, make sure the reversing lever rests against the stop. Adjust as required with the ball socket adjuster on the reversing lever.

d. Install the floor plate and hold the throttle lever full open. Adjust the pedal push rod as required so the pedal is against the stop.

5. When adjusting the idle switch on the 911, there should be 0.035 in. (0.9mm) between the switch housing and the lever when the throttle is fully closed.

6. If these parameters are not met, loosen the switch mounting screws and rotate the switch to meet the above parameters.

968

The throttle valve switches have been replaced by a throttle valve potentiometer on the 968. The idle position is sensed adaptively by the control unit so no adjustment is possible.

928

1. Unplug the switch connector and connect an ohmmeter to the switch terminals. The center terminal is common, terminal 2 is the idle switch and terminal 3 is the full load switch.

2. The idle switch should be closed when the throttle is closed. Any movement of the throttle lever will open the switch just before the throttle plate actually opens. Operate the lever by hand if necessary to determine a faulty switch or faulty adjustment.

3. Connect the ohmmeter to the full load switch. The switch should close when the throttle lever is about ⅔ of full stroke.

SPEED/REFERENCE MARK SENSOR

Removal and Installation

1. Be sure the ignition switch is in the **OFF** position. The negative battery terminal can be disconnected, but the adaptive memory will be lost along with any electronic presets in the radio, alarm system or other preset type systems.

2. The sensor is located at the back of the engine at the flywheel. On the 911, unbolt the connector bracket at the left side of the engine and rotate out of the way to access the sensor. Unplug the lead from the sensor to the harness.

3. Unbolt the sensor and pull from its mount.

4. Install the sensor and the mounting bolts. The sensor should sit 0.023–0.039 in. (0.6–1.0 mm) from the gear ring teeth.

5. Tighten the mounting bolts and connect the electrical connector.

OXYGEN SENSOR

Removal and Installation

1. Be sure the ignition switch is in the **OFF** position. The negative battery terminal can be disconnected, but the adaptive memory will be lost along with any electronic presets in the radio, alarm system or other preset type systems.

2. On the 911 the oxygen sensor plug is located on the left side of the engine compartment. On the 928, pull up the foot support on the passenger side to expose the oxygen sensor connector and the grommet that the wire goes through to the underside of the vehicle. On the 968 the connector is on the same holder as the speed/reference mark sensor connector which is bolted to the intake manifold at cylinder 4.

3. Disconnect the plug to the oxygen sensor and feed down to the underside of the vehicle to allow the sensor to come free once unscrewed from the exhaust system.
4. Unscrew the oxygen sensor and remove from the vehicle.
5. Coat the threads of the oxygen sensor with anti-seize being careful not to contaminate the sensor bulb.
6. Torque the sensor to 36–43 ft. lbs. (50–60 Nm).
7. Route the lead along its original position and connect the lead. Do not use any sealer or other substances on the connector plug as the new style oxygen sensor gets its reference air through the connector. The sensor body is sealed to prevent water or contamination from entering the sensor cavity.

ENGINE TEMPERATURE SENSOR

Removal and Installation

1. Be sure the ignition switch is in the **OFF** position. The negative battery terminal can be disconnected, but the adaptive memory will be lost along with any electronic presets in the radio, alarm system or other preset type systems.
2. The temperature sensor is located in cylinder head 3 and can be removed with a 14mm wrench and tool 9291 or equivalent.
3. Disconnect the electrical connector at the connector bracket on the left side of the engine compartment.
4. When installing use an antiseize on the threads. Tighten the sensor and connect the electrical lead.

COOLANT TEMPERATURE SENSOR

Removal and Installation

1. Be sure the ignition switch is in the **OFF** position. The negative battery terminal can be disconnected, but the adaptive memory will be lost along with any electronic presets in the radio, alarm system or other preset type systems.
2. The temperature sensor on the 928 is located on the top, front of the engine next to the fuel damper. On the 944 and 968, it is typically located in the coolant jacket, typically near cylinder 1 on the left side of the engine. If there are 2 sensors, the

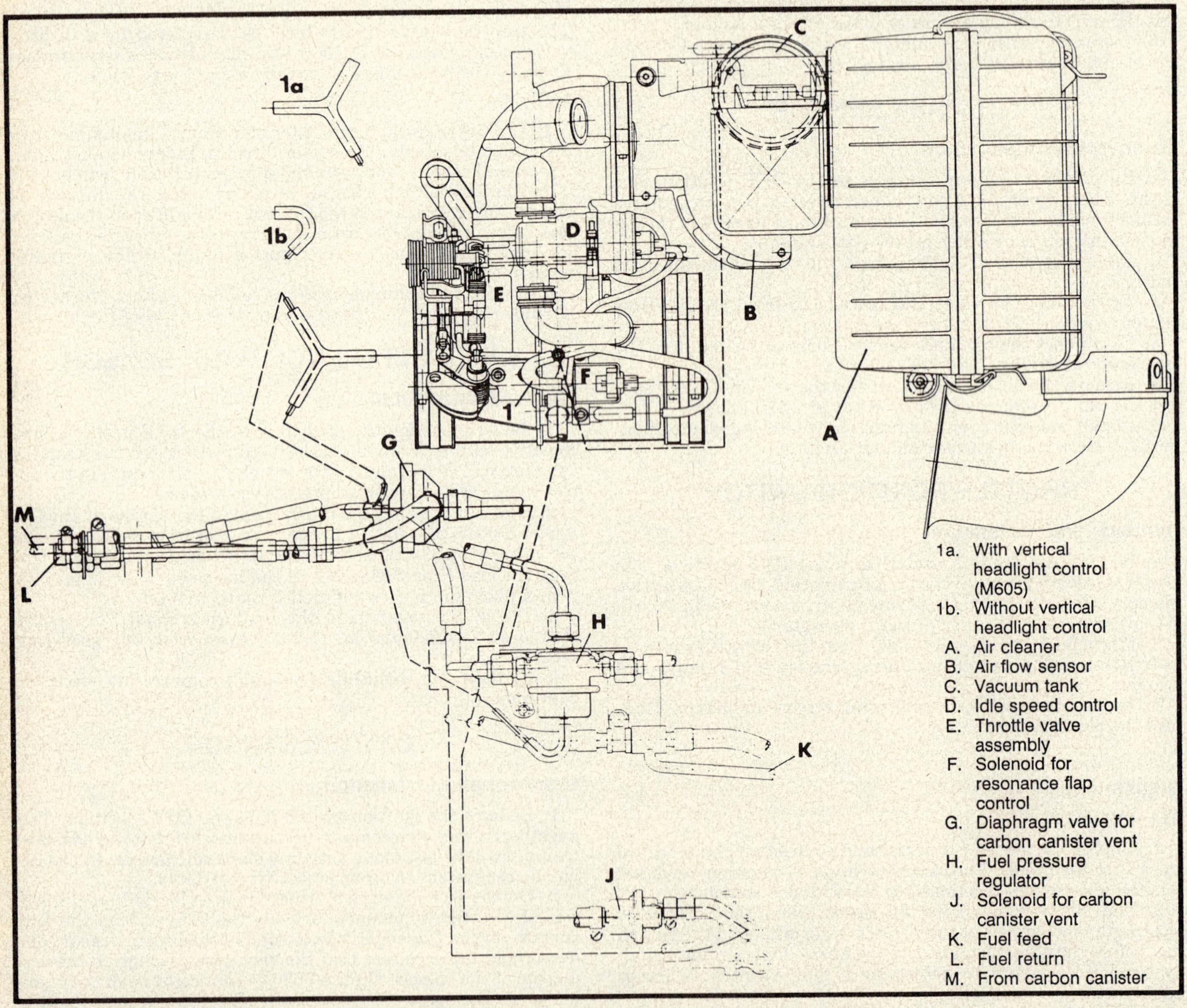

Fuel and air circuit—911

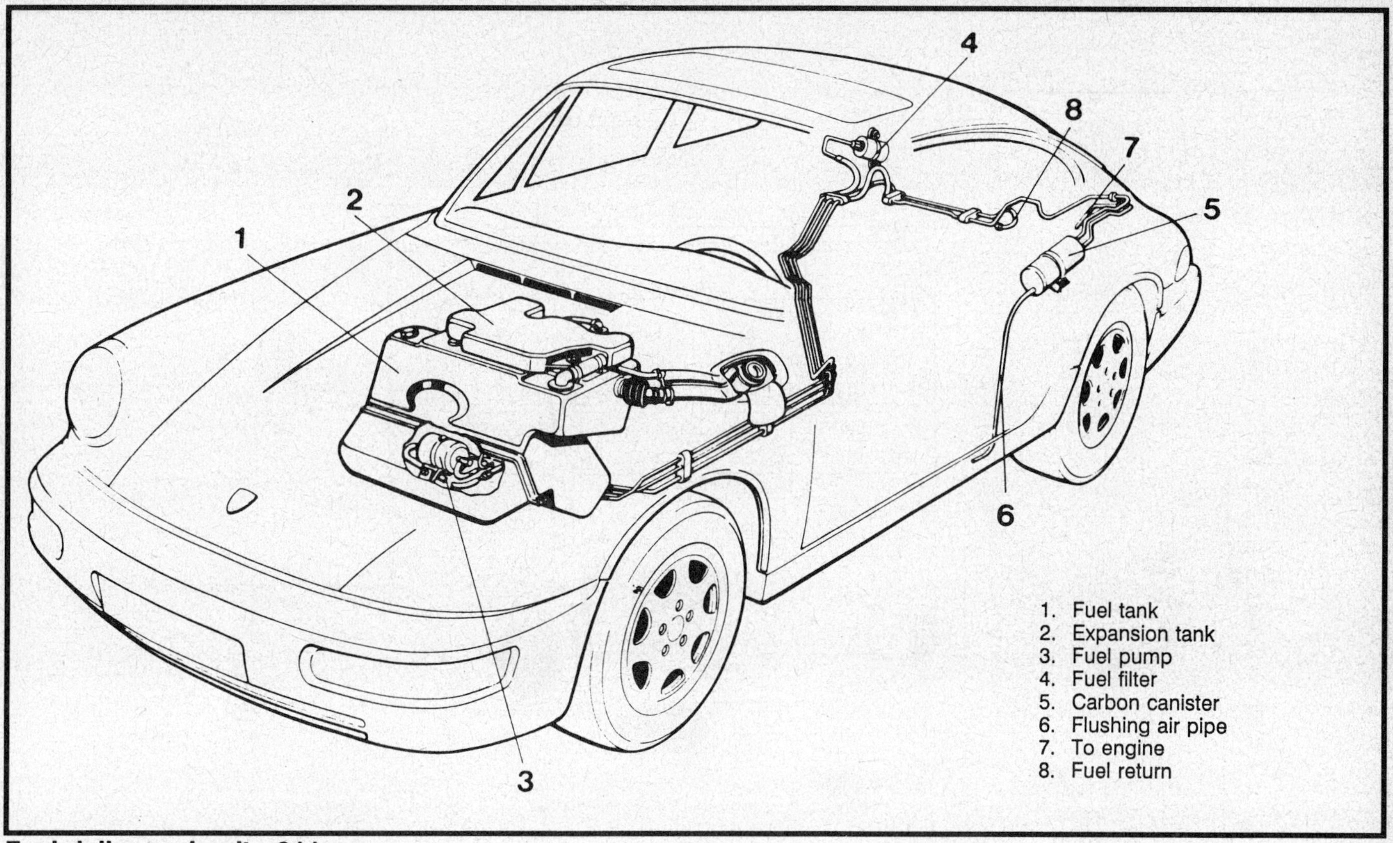

Fuel delivery circuit–911

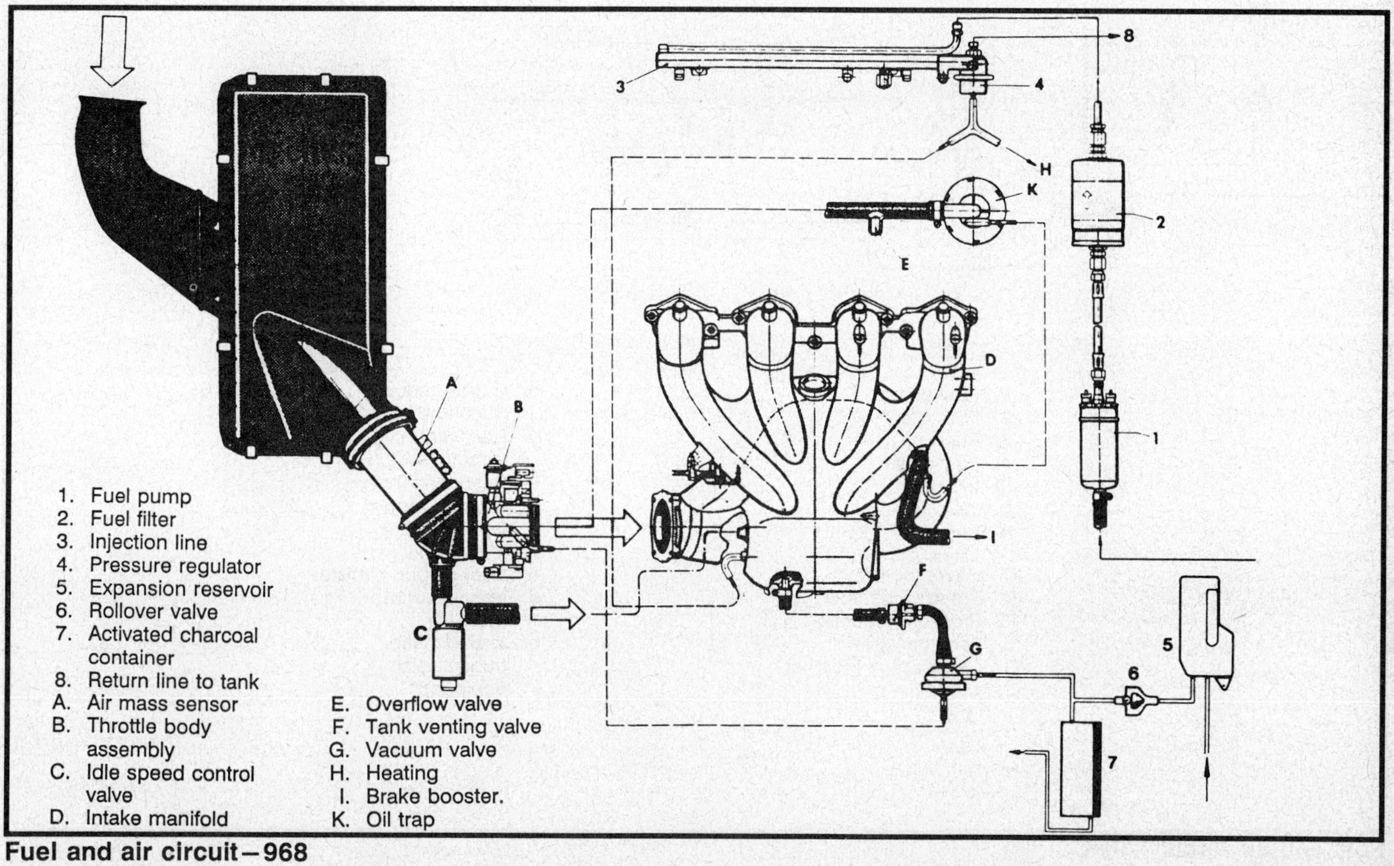

Fuel and air circuit–968

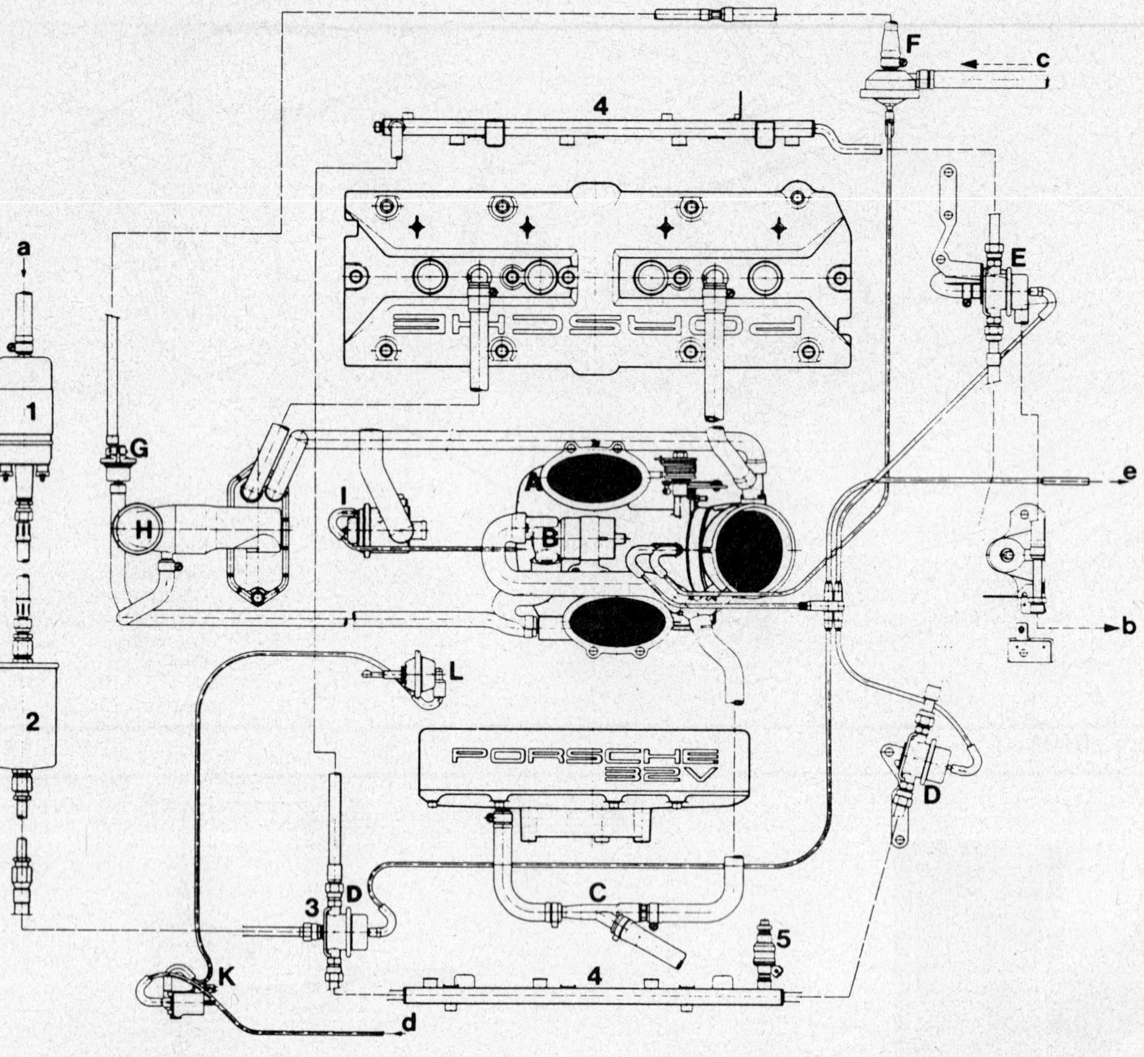

1. Fuel pump
2. Fuel filter
3. Fuel line
4. Fuel rails
5. Injectors

A. Throttle body
B. Idle speed control valve
C. Intake jet pump
D. Pressure damper
E. Pressure regulator
F. Vacuum valve
G. Electric tank breather valve
H. Oil filler neck
I. Secondary air valve
K. Resonance valve vacuum switch
L. Resonance valve vacuum motor

a. from fuel tank
b. to fuel tank
c. from carbon canister
d. from vacuum reservoir
e. to automatic transmission, if equipped

Fuel and air circuit—928

sensor with the blue plug reports to the ECU and the sensor with the black plug reports to the temperature gauge.

3. Keep in mind that if the cooling system is filled, coolant will flow out of the opening once the sensor is removed. Drain some coolant as necessary.

4. Unplug the connector and unscrew the sensor. When replacing the sensor torque to 8.7 ft. lbs. (12 Nm). Be sure to use a new crush washer, if originally equipped.

5. Connect the electrical lead and refill the cooling system.

FUEL INJECTORS

Removal and Installation

1. Be sure the ignition switch is in the **OFF** position. The negative battery terminal can be disconnected, but the adaptive memory will be lost along with any electronic presets in the radio, alarm system or other preset type systems.

2. On the 928, remove the air intake hoses and the air cleaner. On the 911 remove the hot air blower to gain access to the injector fuel rails.

3. On the 928 loosen the clamps holding the exhaust sample pipe and the intake pipe.

4. Disconnect the fuel line unions from the fuel rail. Be ready to catch any fuel that will run out of the connections.

5. Disconnect the ignition leads from the clips on the fuel rail, as equipped. Lift the rail off. Unplug the fuel injector connectors and retainers and remove the injectors from the rail.

6. When installing, coat the seals with silicone grease and replace the injectors and rails. Install the retainers and electrical connectors.

7. Connect the fuel lines, air intake hoses and the hot air blower.

IDLE SPEED CONTROL VALVE

Removal and Installation

928

1. Disconnect the negative battery terminal. Keep the vehicle supported by its own weight on its wheels. This is to maintain tension in the transverse strut rod in the engine compartment. Remove the strut rod.

2. Remove the air cleaner assembly and the air hoses as a unit. Disconnect the vacuum hose from the brake booster.

3. Remove the bracket for the accelerator linkage and the TDC sender from the intake manifold. Remove the holder for the crankcase breather and the fuel line. Remove the air mass sensor.

4. Uncover the fuel rails and disconnect the fuel lines from the rails. Catch any fuel in a suitable container. Lift the rails out.

5. Mark the vacuum lines and note their positions. Disconnect the lines. Disconnect the thermostatic switch between the switch and the intake pipe.

6. Unbolt the fuel pressure damper holder and disconnect the throttle cable. Disconnect the plug from the throttle.

7. Disconnect the electrical connector from the idle speed control valve. Be careful not to disturb the gasket in the male side of the connector. Use of tool 9266 or equivalent, will ease the removal process and can be used on other connectors of the same style. Remove the air guide dome from the throttle.

8. Unbolt the intake manifold. Lift the manifold from the cylinder heads and disconnect remaining vacuum lines and crankcase breather lines. Remove the intake manifold.

9. Remove the throttle body from the intake manifold. Remove the idle speed control valve.

10. Replace the intake manifold gaskets and install the idle speed control valve. Replace all the disconnected hoses and vacuum lines. Torque the throttle body bolts to 14 ft. lbs. (20 Nm). Torque the new self locking nuts used on the intake manifold to 11 ft. lbs. (15 Nm).

11. Attach the air guide dome, the throttle cable, throttle body plug, fuel pressure damper and holder, vacuum lines and thermostatic switch connector.

12. Connect the fuel rails and fuel lines. Install the fuel rail covers. Install the fuel line and crankcase breather holder. Install the air mass sensor.

13. Bolt in the holder for the TDC sensor and the accelerator linkage. Connect the brake booster vacuum line and install the air cleaner assembly with the air hoses. Bolt in the transverse rod. Connect the negative battery terminal.

IDLE SPEED AND CO

The idle speed and CO levels are sealed and not adjustable. The engine management system adjusts the idle speed and CO as necessary to meet operational parameters. The idle speed and CO can be checked for meeting specifications, but any deviation from the specified values would indicate a problem with the engine management system. If the values deviate, check for vacuum leaks, dripping or clogged injectors or other failures. The CO level is measured with the oxygen sensor connected and before the catalytic converter. There are exhaust gas taps built into each exhaust system.

PORSCHE IDLE SPEED AND CO

Model	Idle Speed rpm	CO % Level
911	840–920	0.4–1.2
928	650–700 S4/750–800 GT	0.4–1.2
944	800–880	0.4–1.2
968	720–800	0.2–1.2

These values are not adjustable. They are provided for checking purposes only.
① CO level in percent volume before catalyst with oxygen sensor connected.

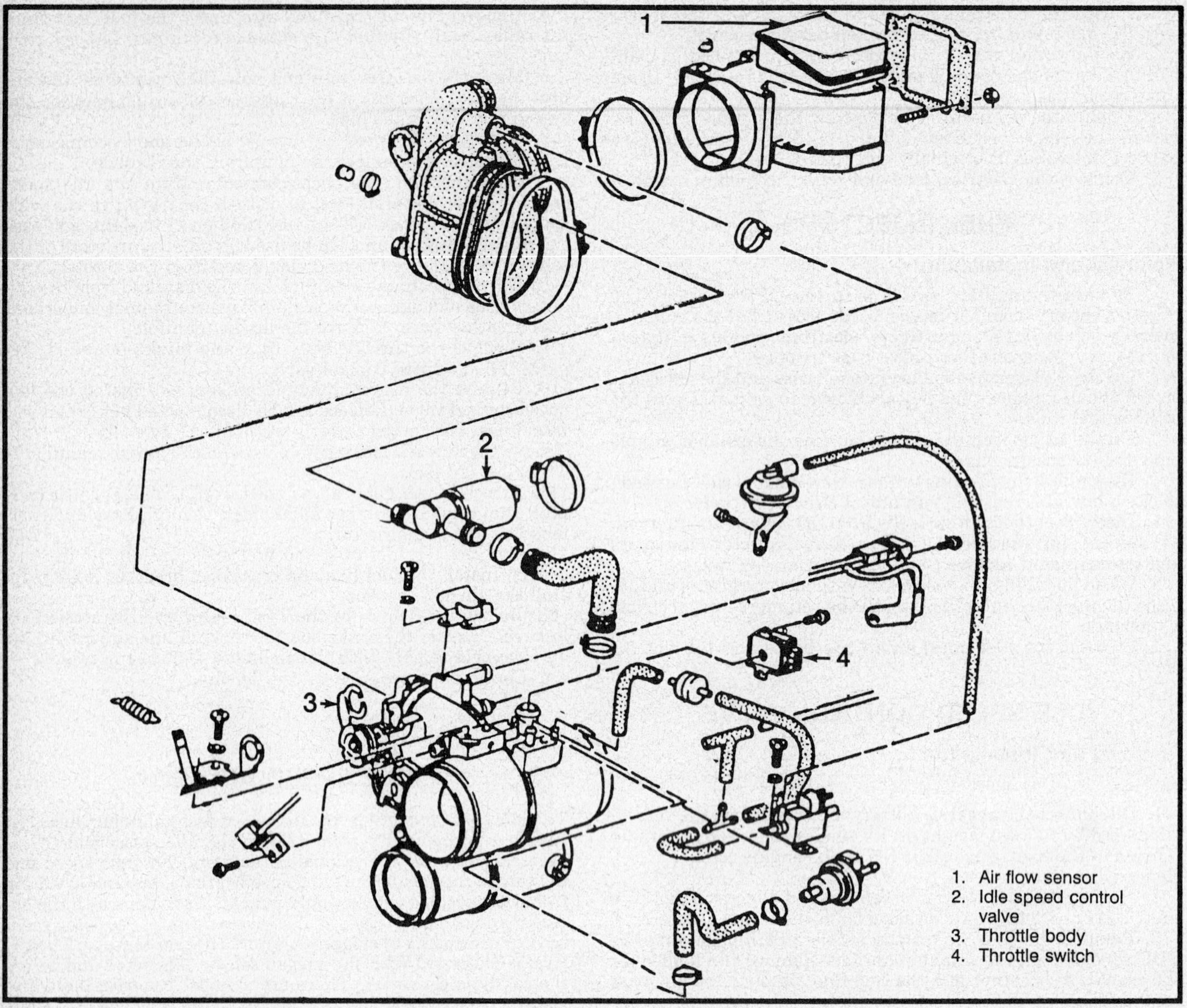

Air induction system—1990 Porsche 911

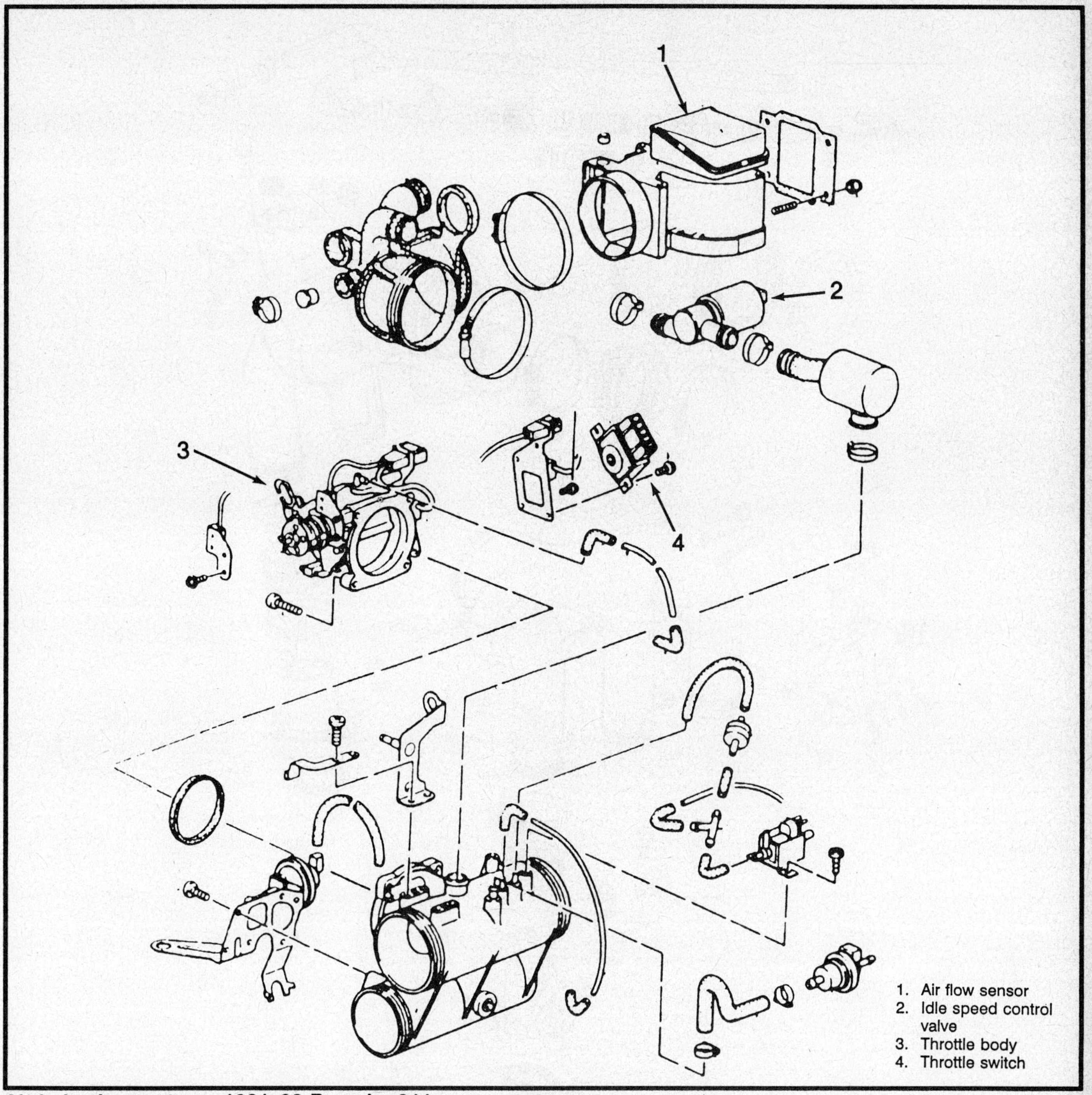

Air induction system—1991–92 Porsche 911

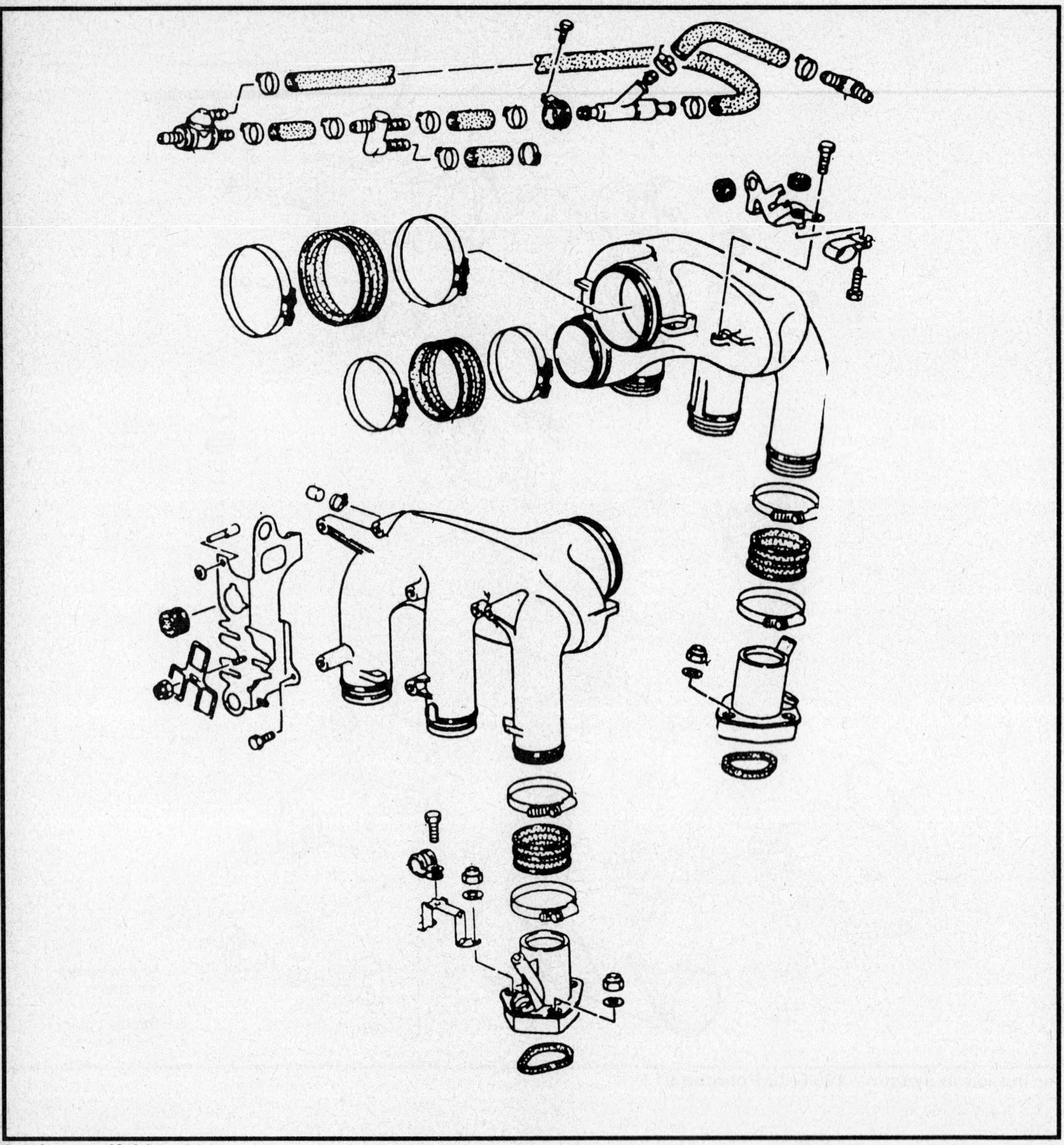

Intake manifolds—911

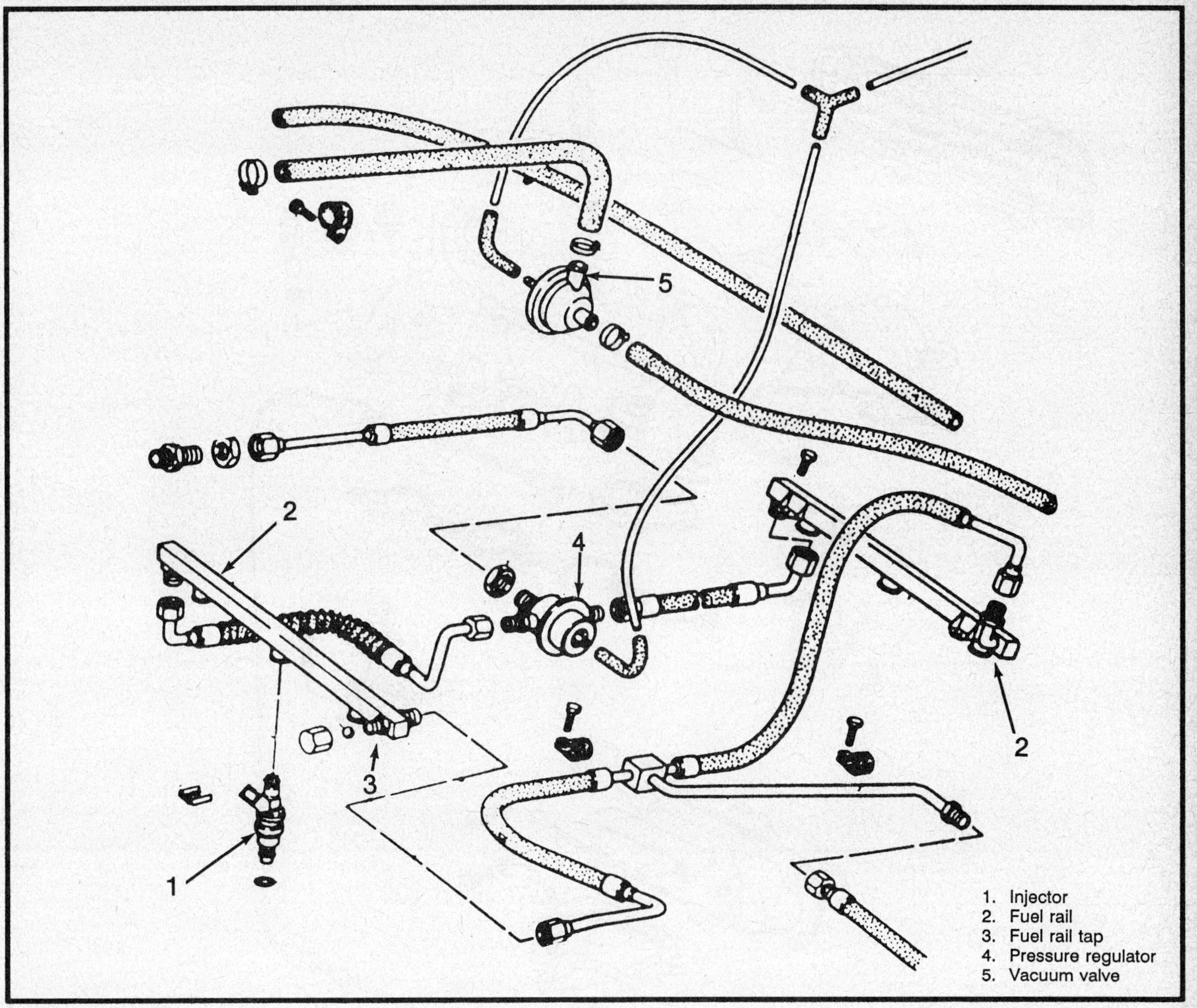

Fuel system—911

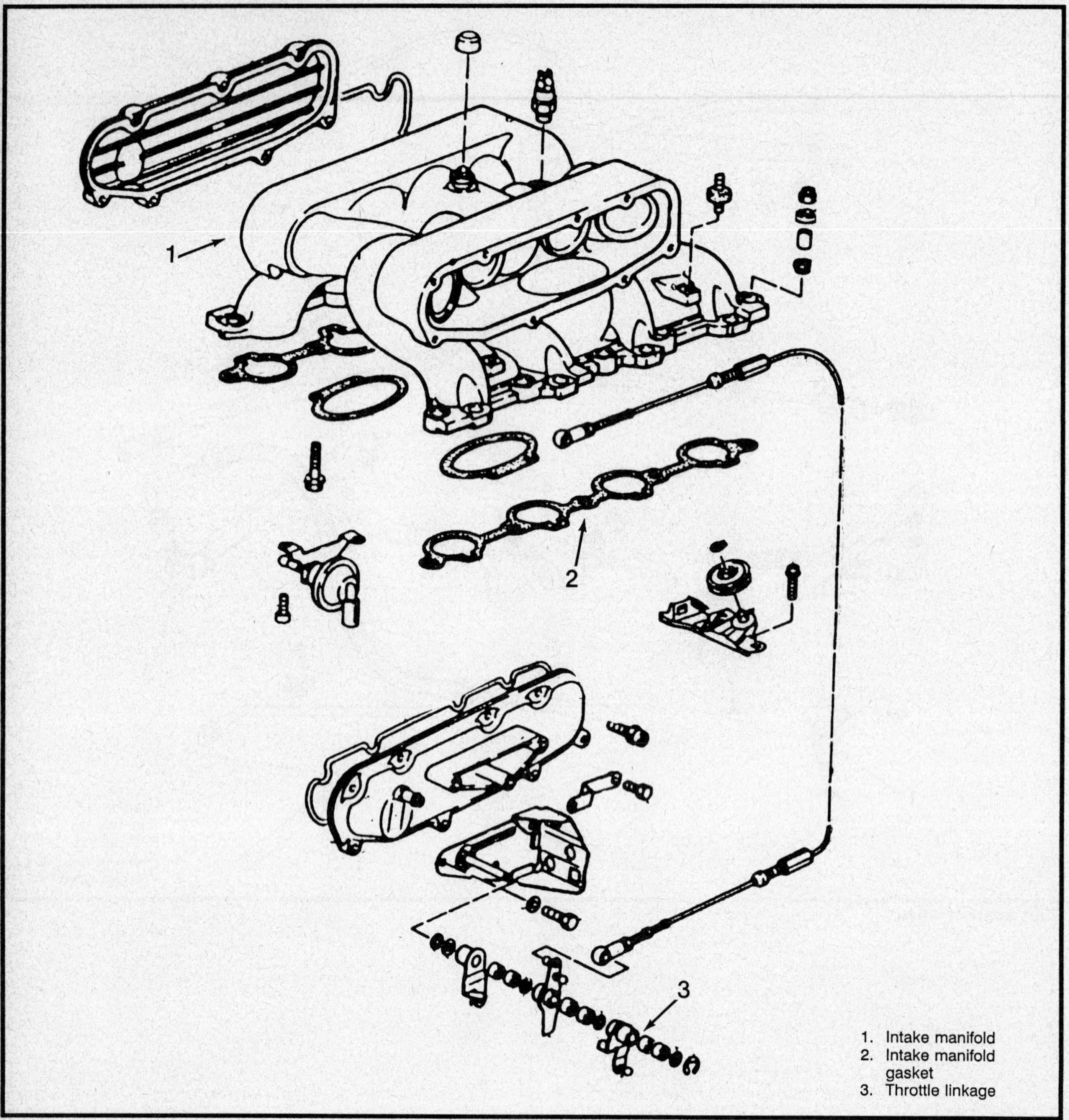

Intake manifolds—928

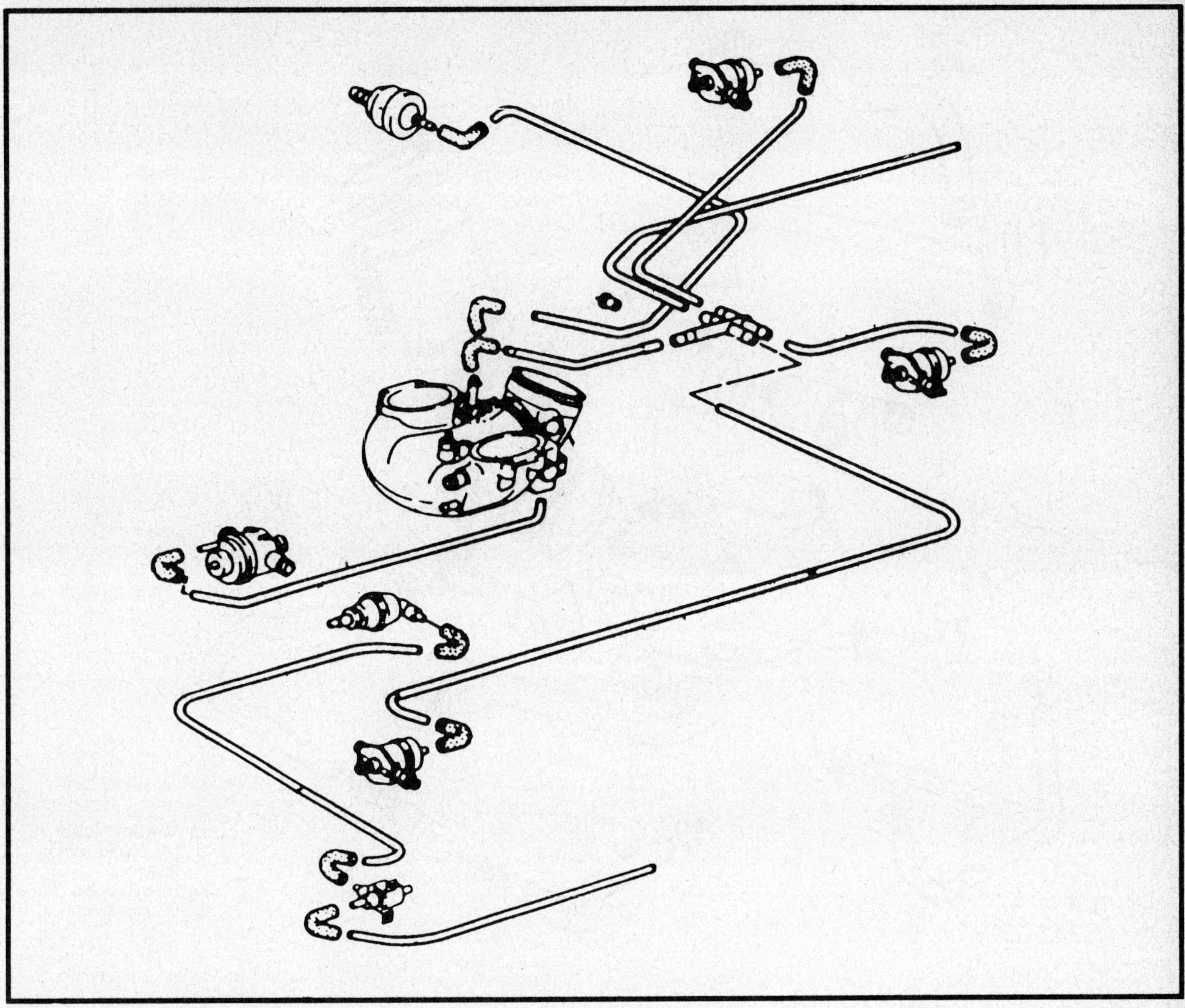

Vacuum line routing—928

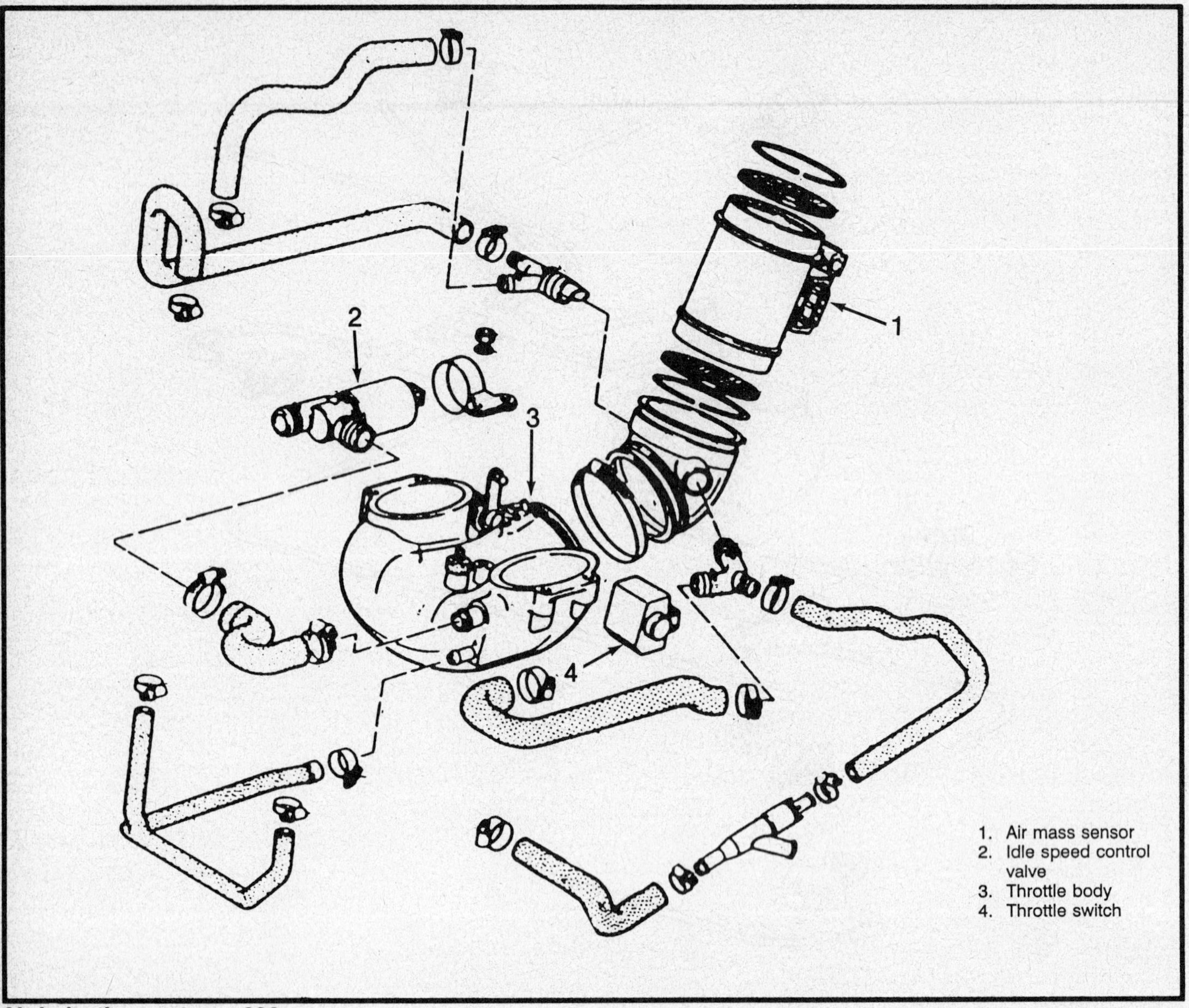

Air induction system—928

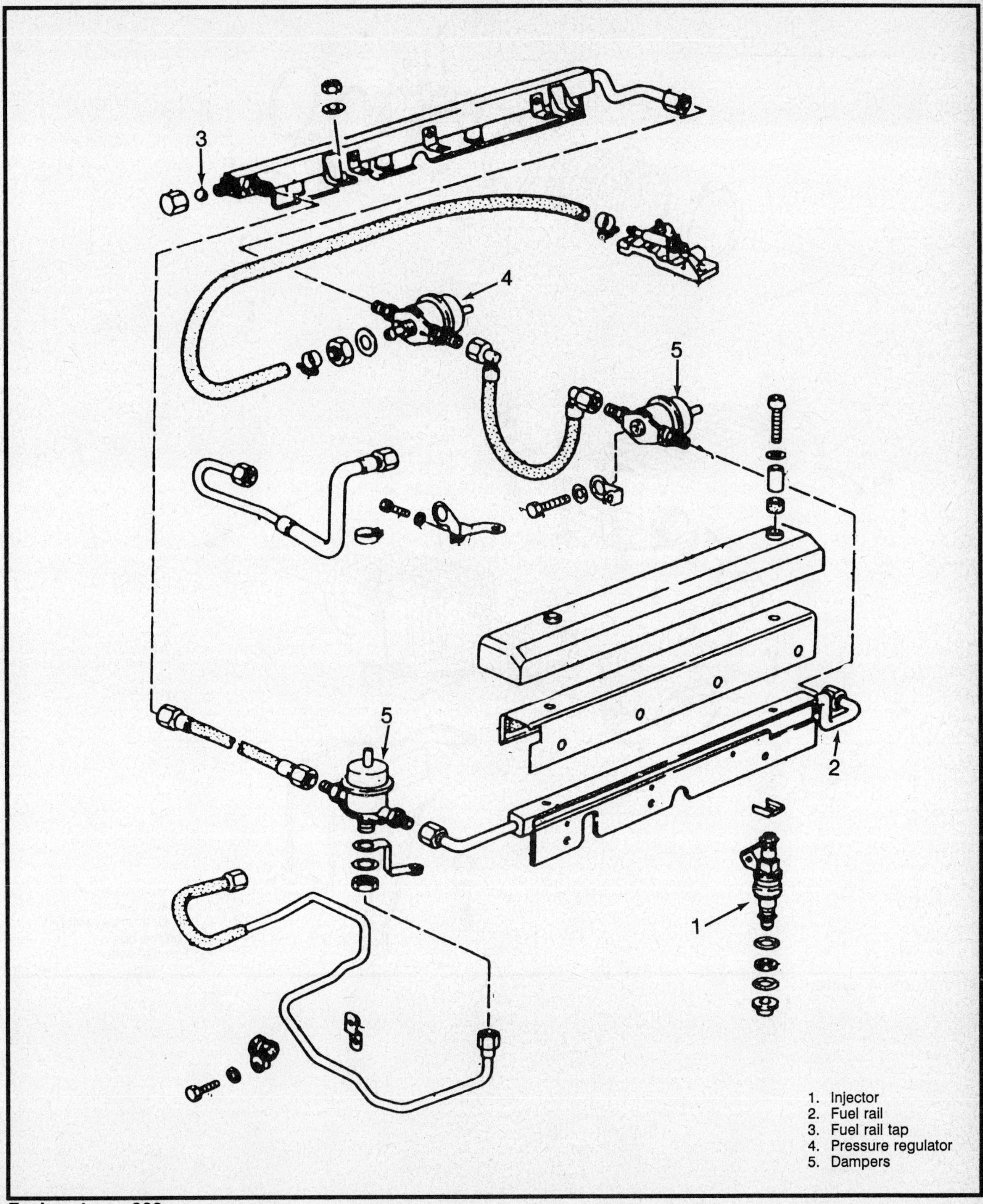

Fuel system—928

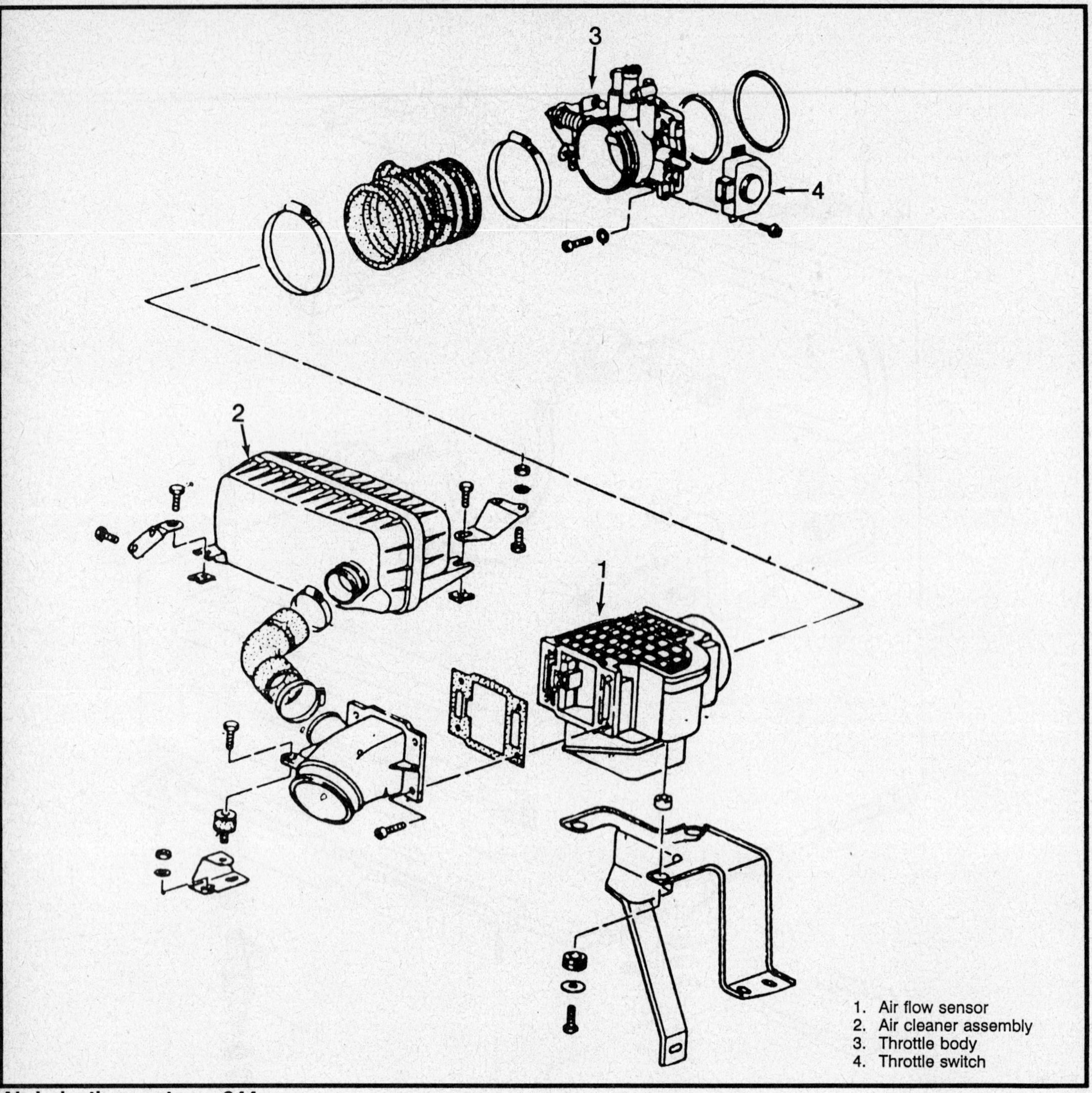

Air induction system—944

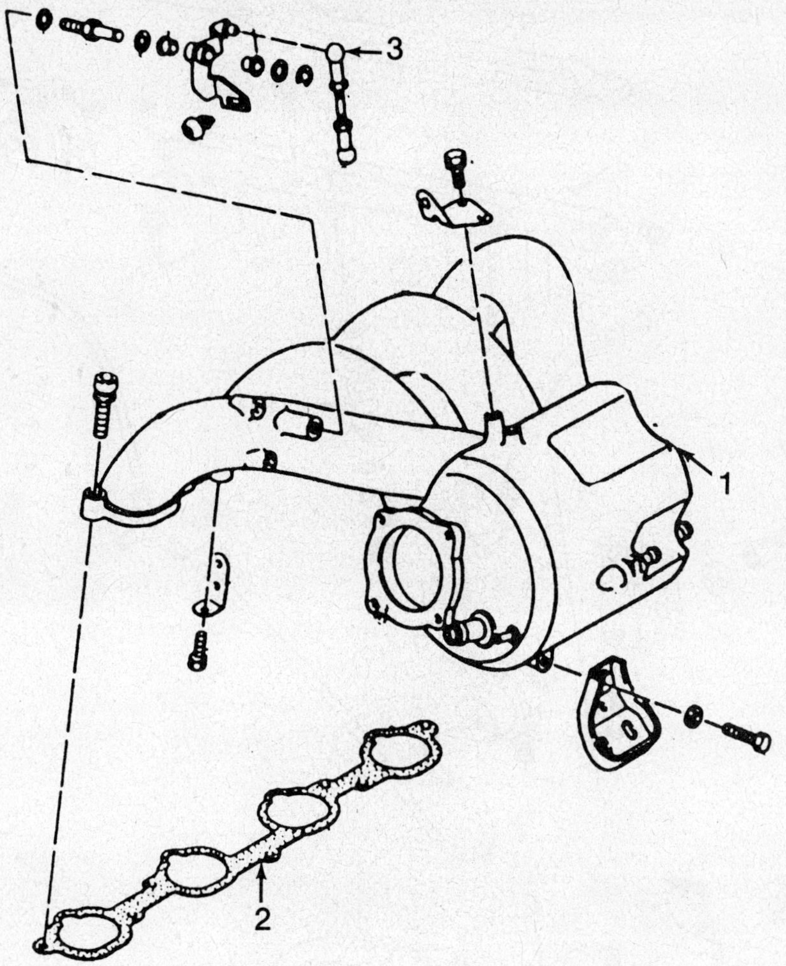

1. Intake manifold
2. Intake manifold gasket
3. Throttle linkage

Intake manifold—944

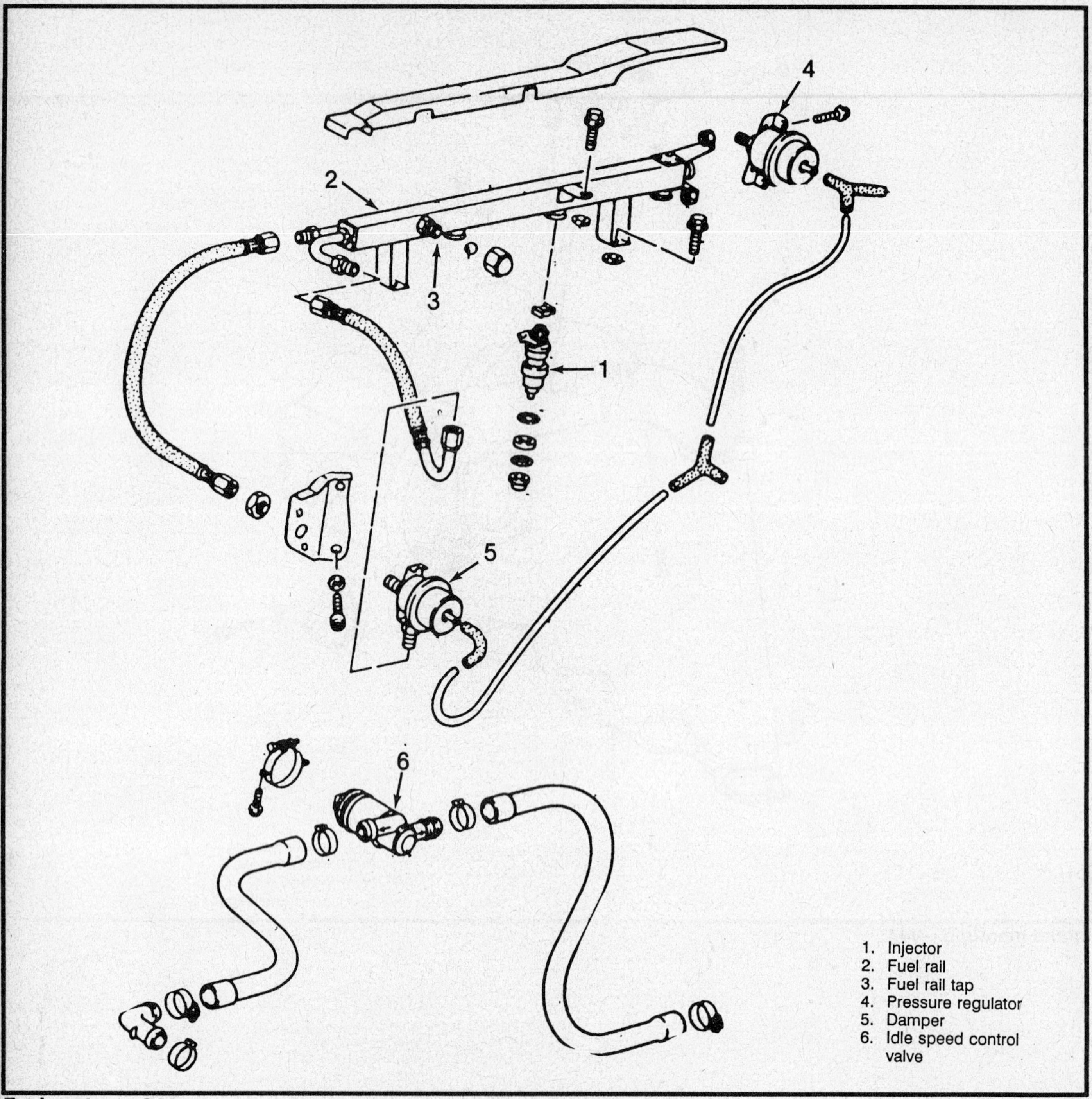

Fuel system—944

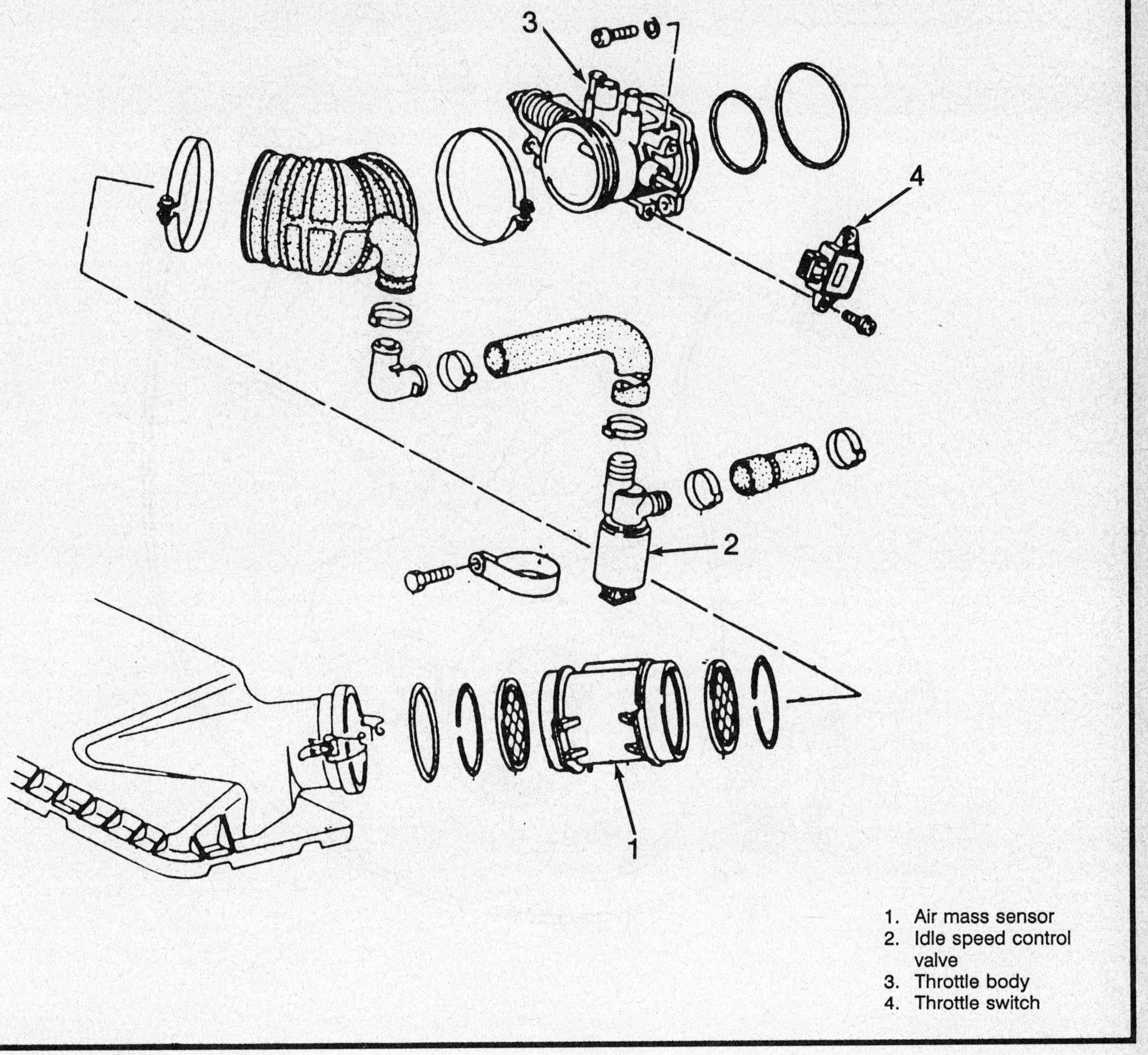

Air induction system—968

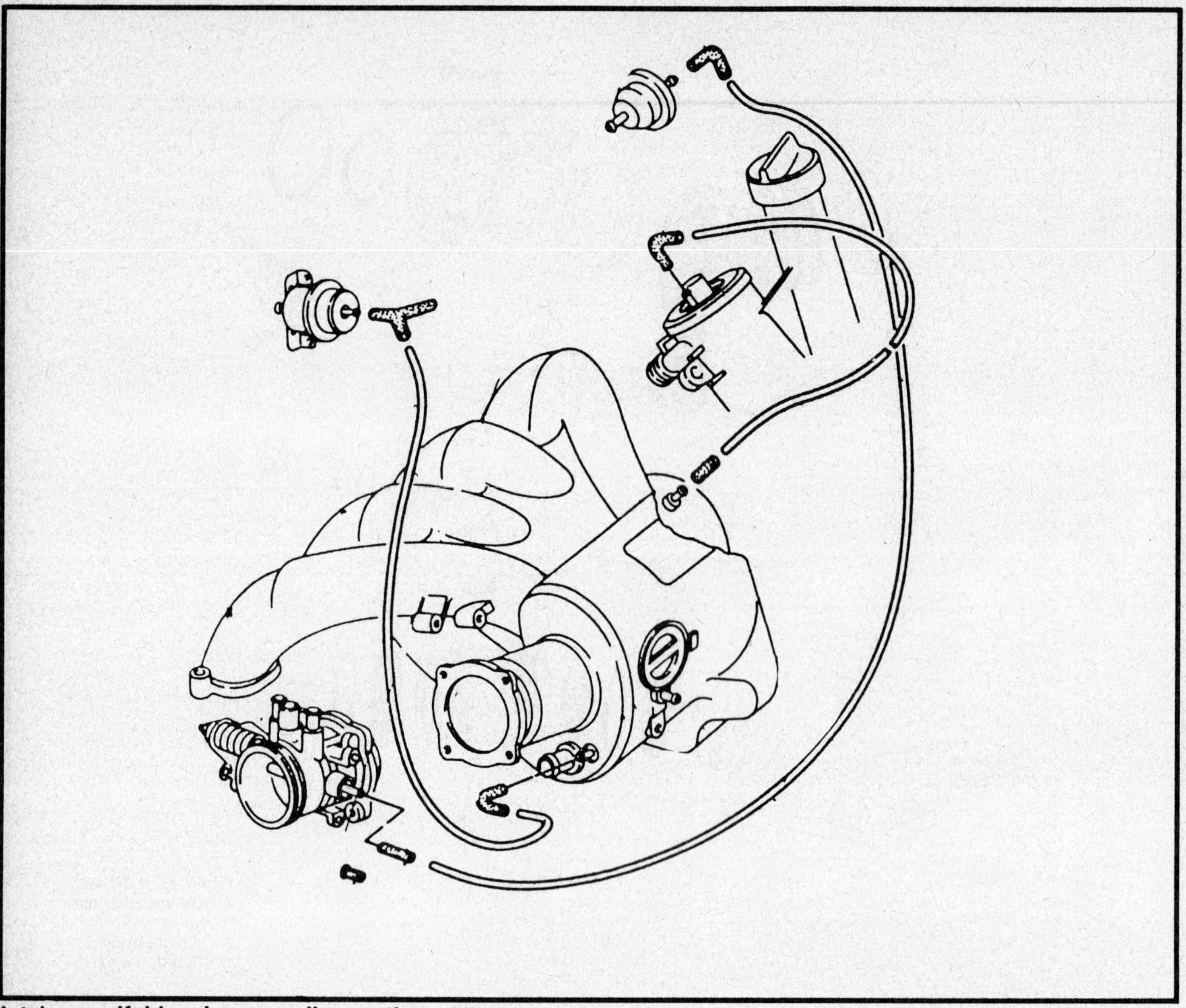

Intake manifold and vacuum line routing—968

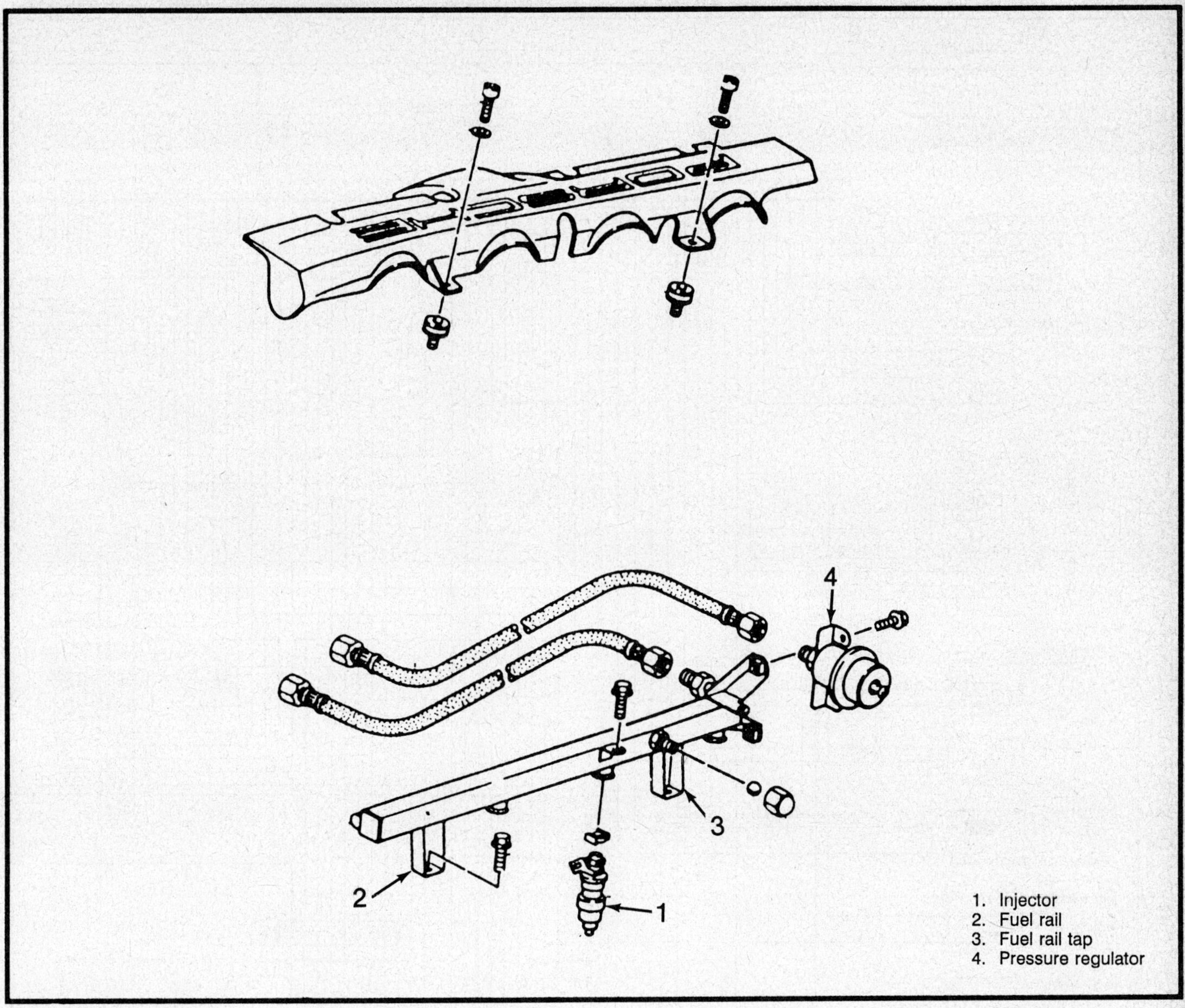

Fuel system—968

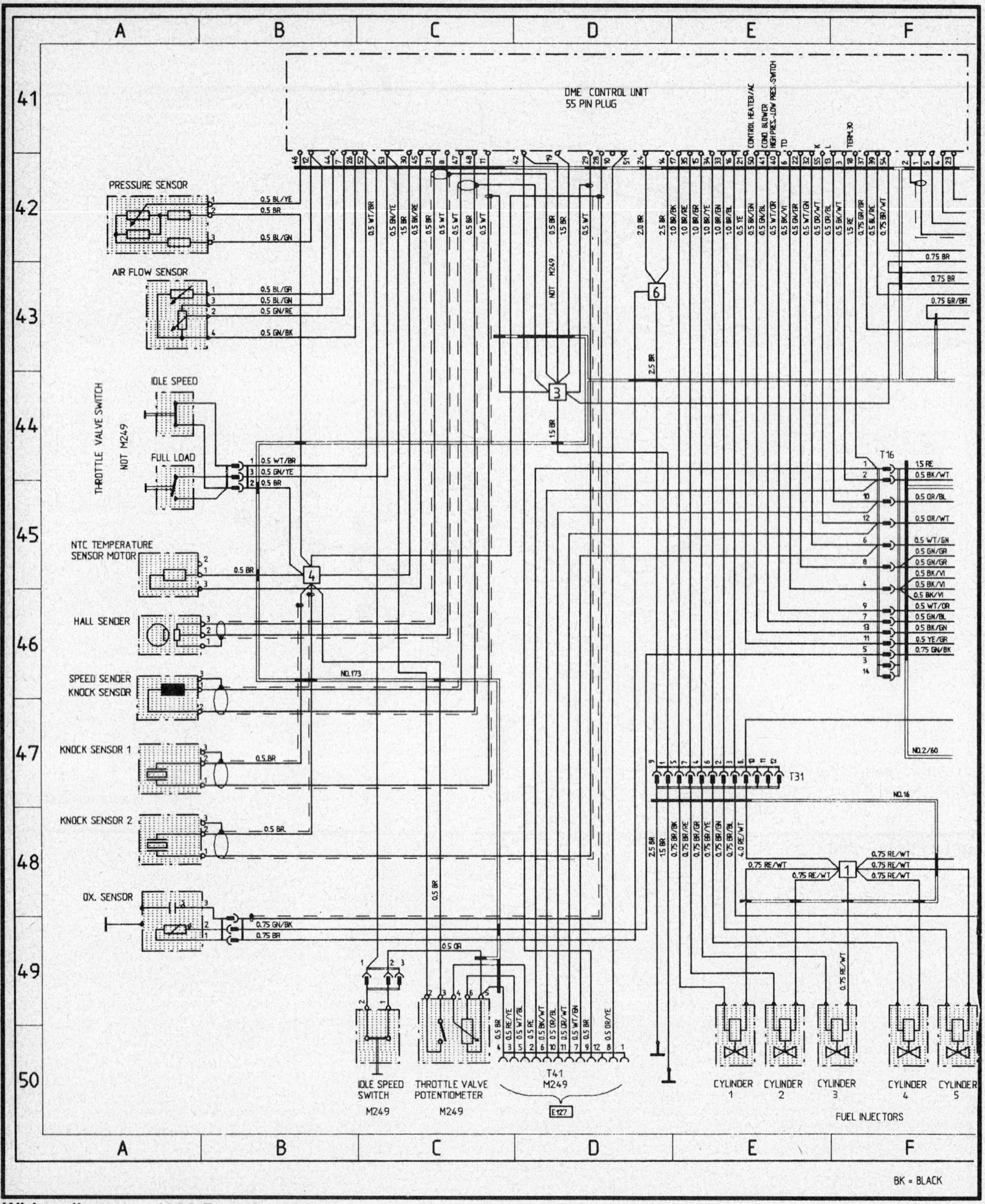

Wiring diagram—1990 Porsche 911

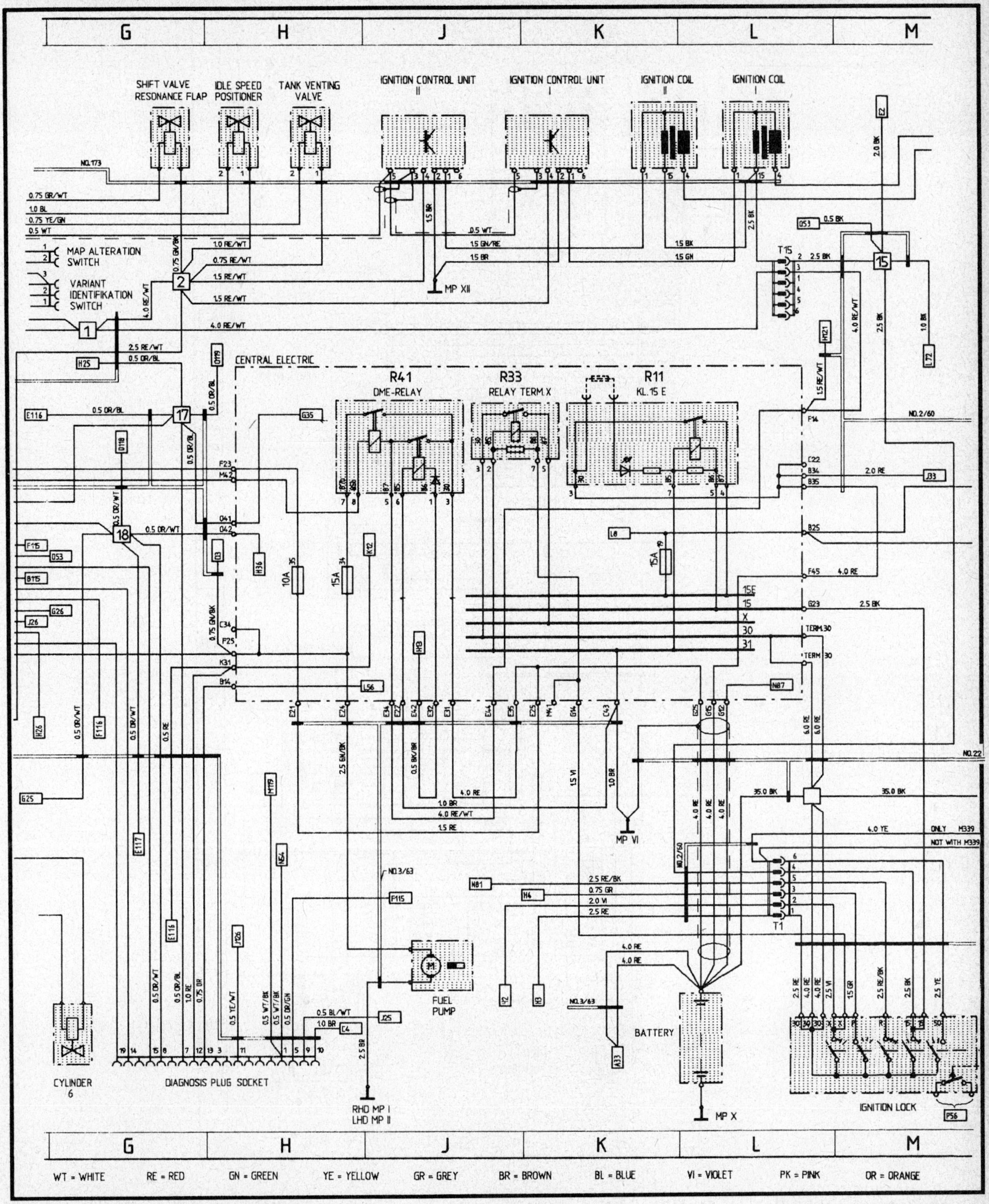

Wiring diagram—1990 Porsche 911

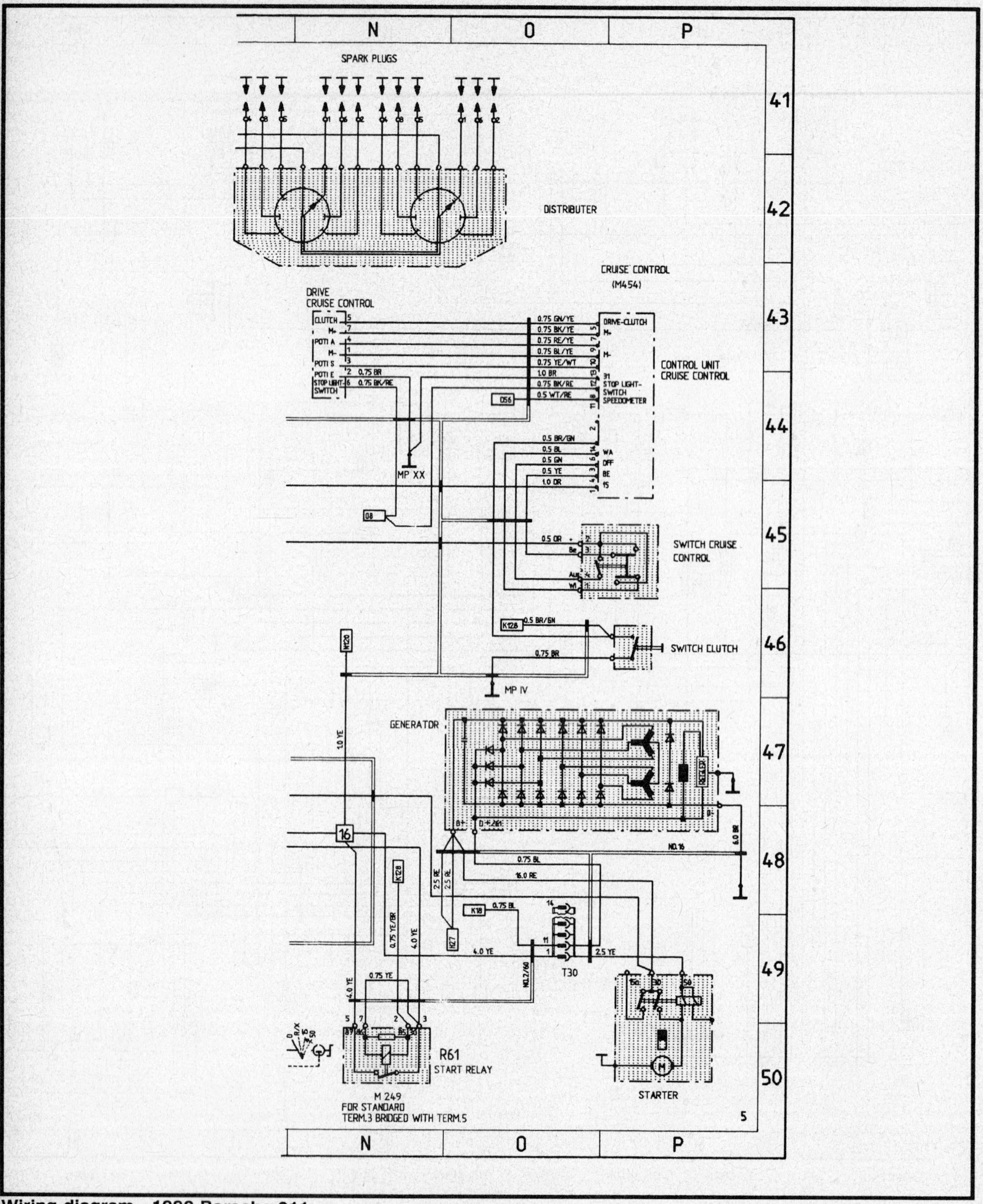

Wiring diagram—1990 Porsche 911

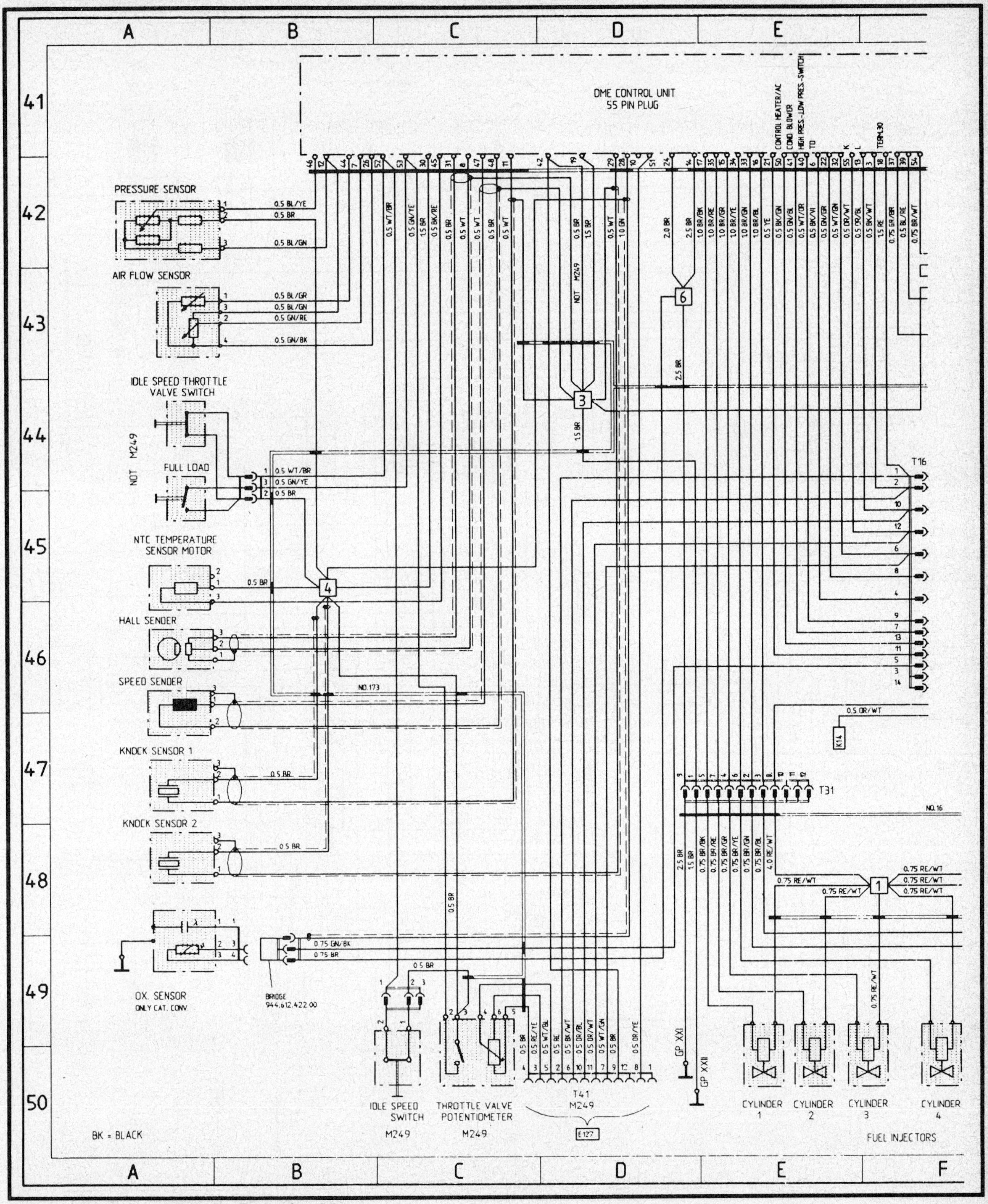

Wiring diagram—1991–92 Porsche 911

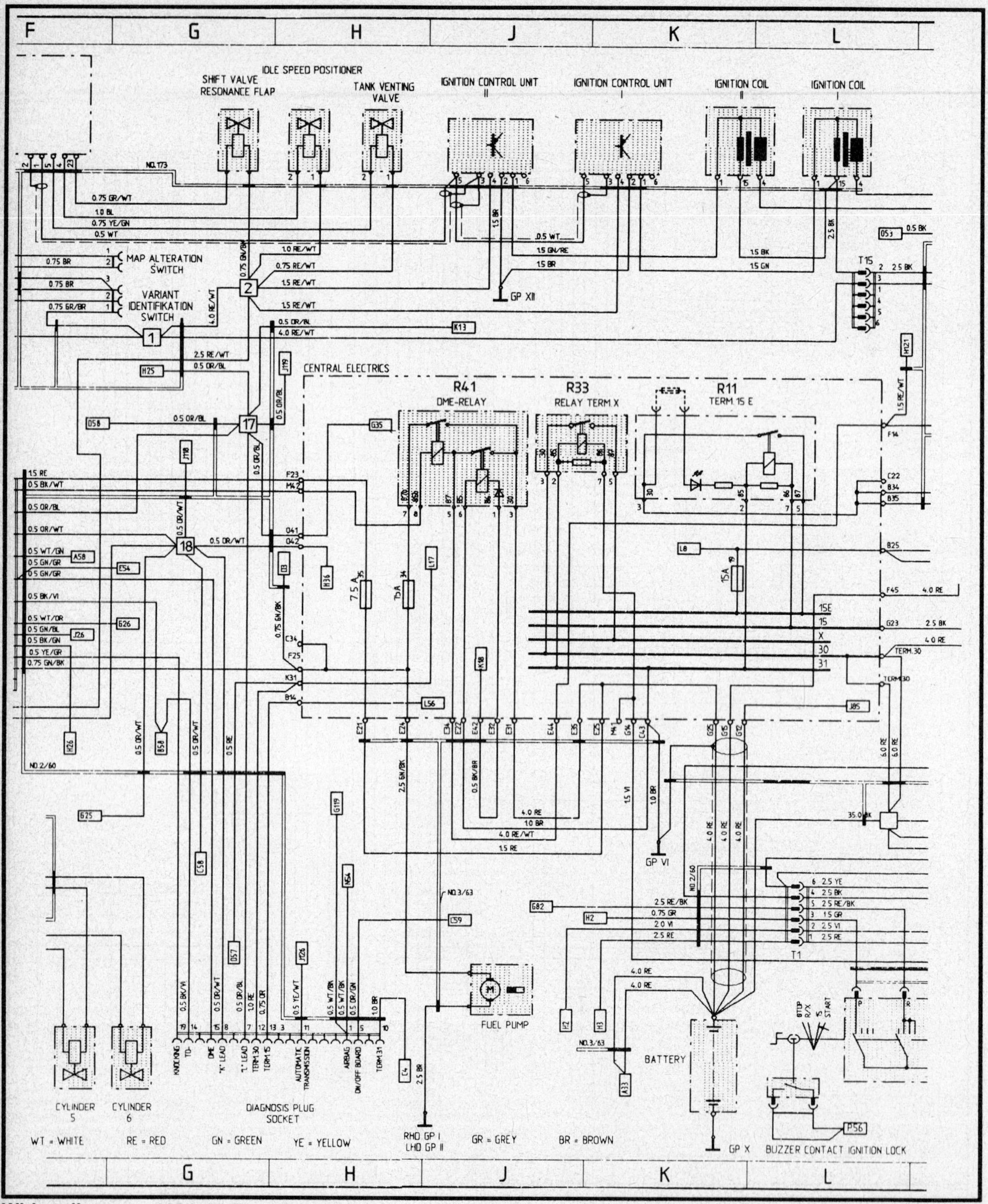

Wiring diagram—1991–92 Porsche 911

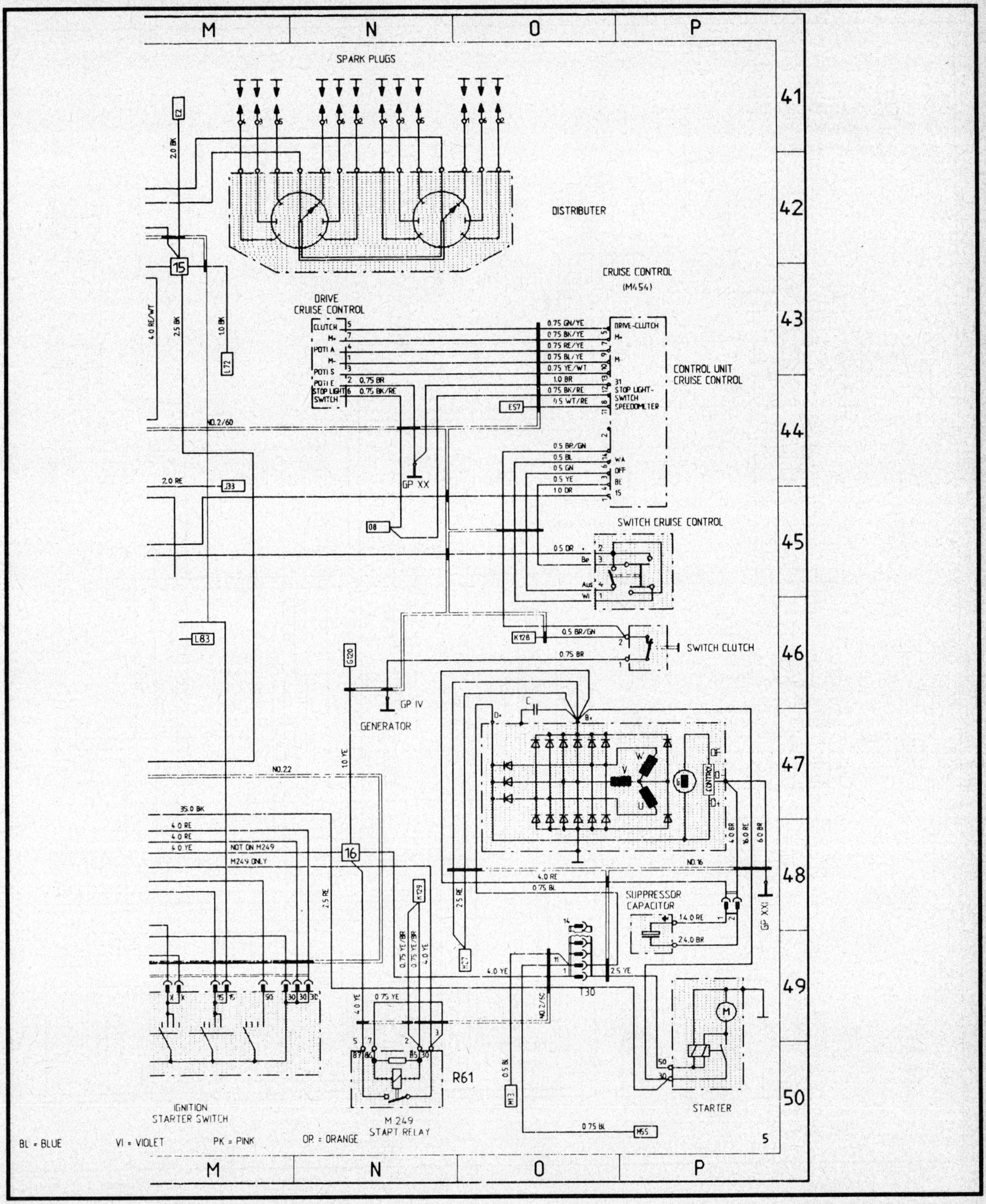

Wiring diagram—1991–92 Porsche 911

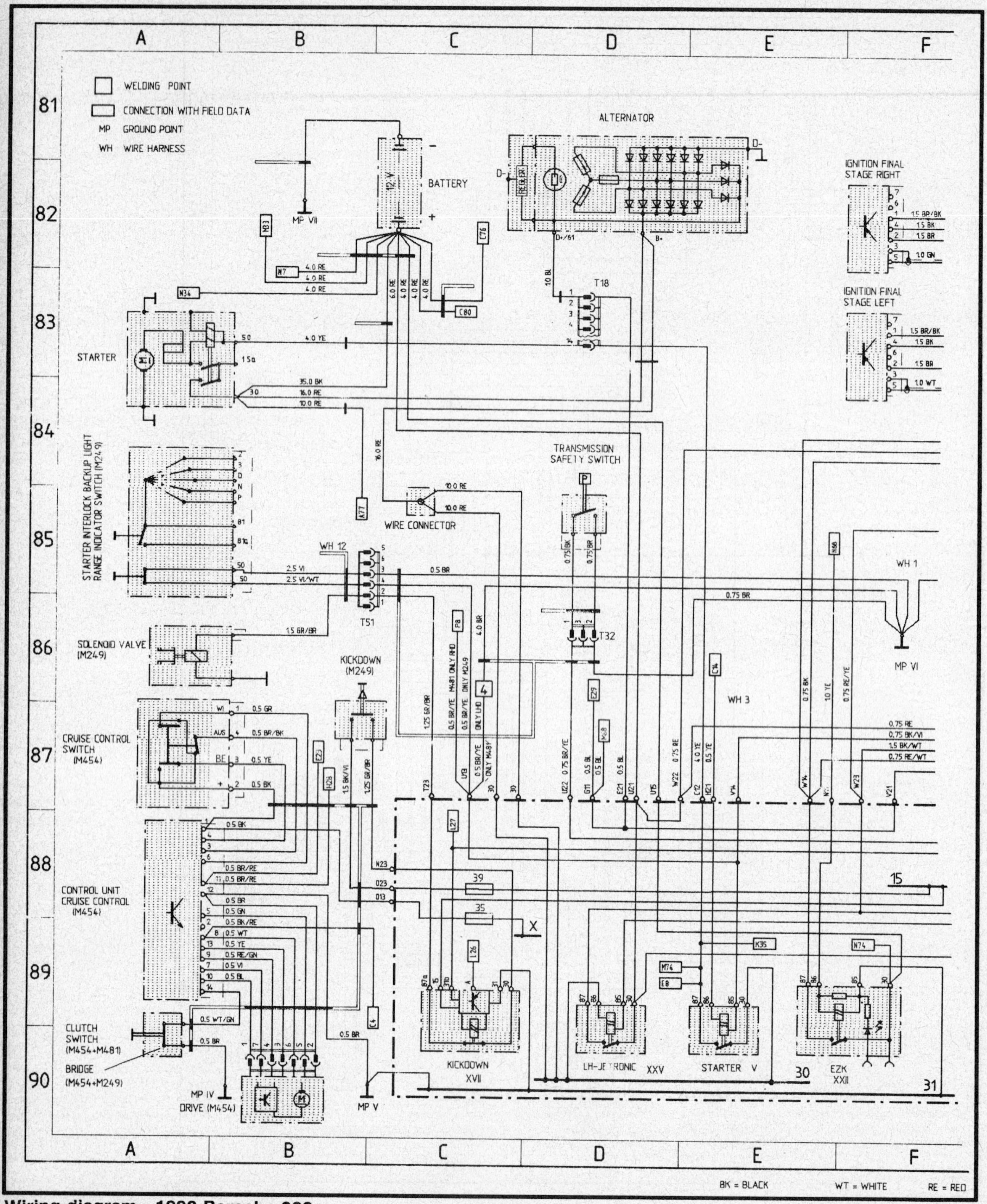

Wiring diagram—1990 Porsche 928

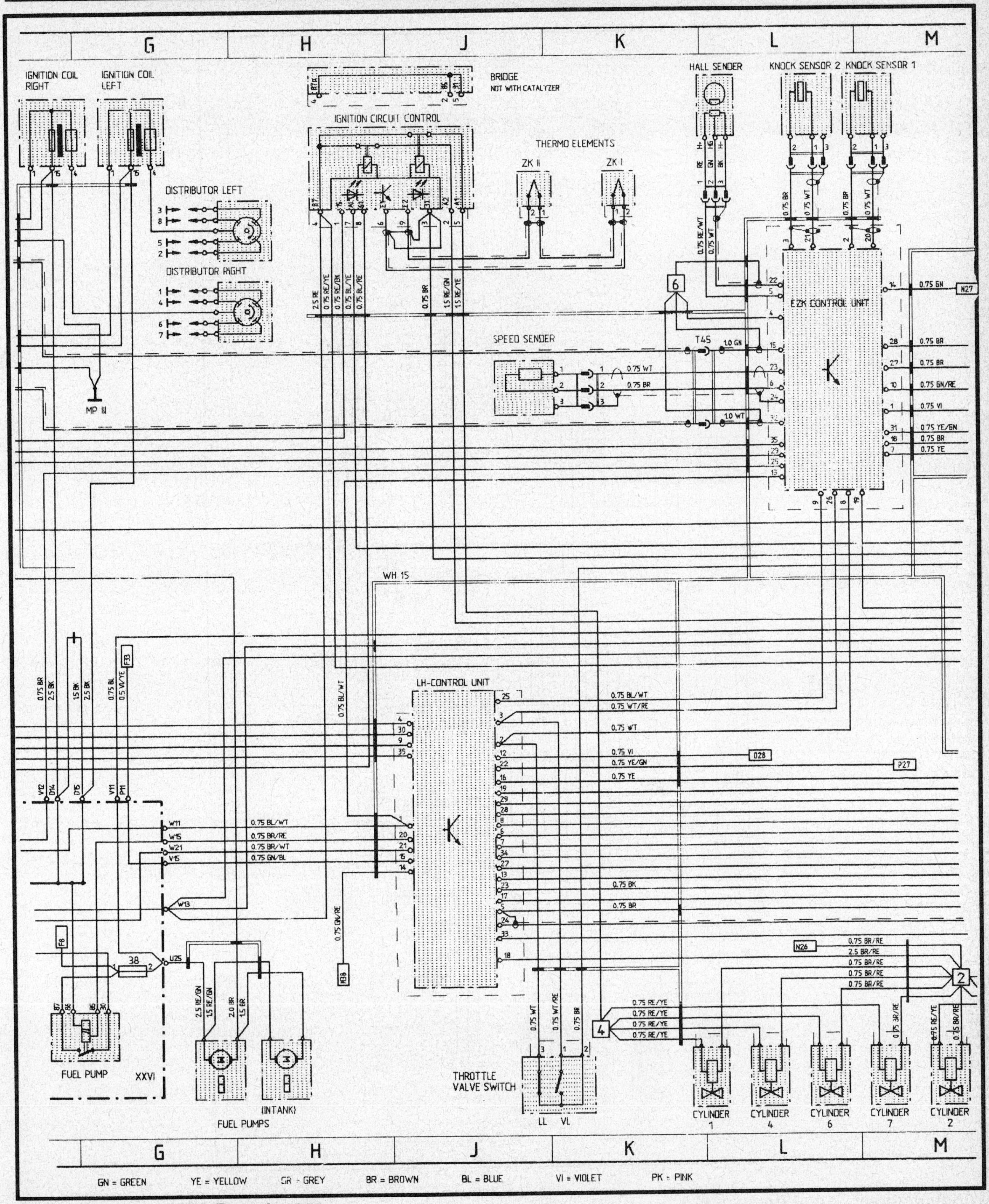

Wiring diagram—1990 Porsche 928

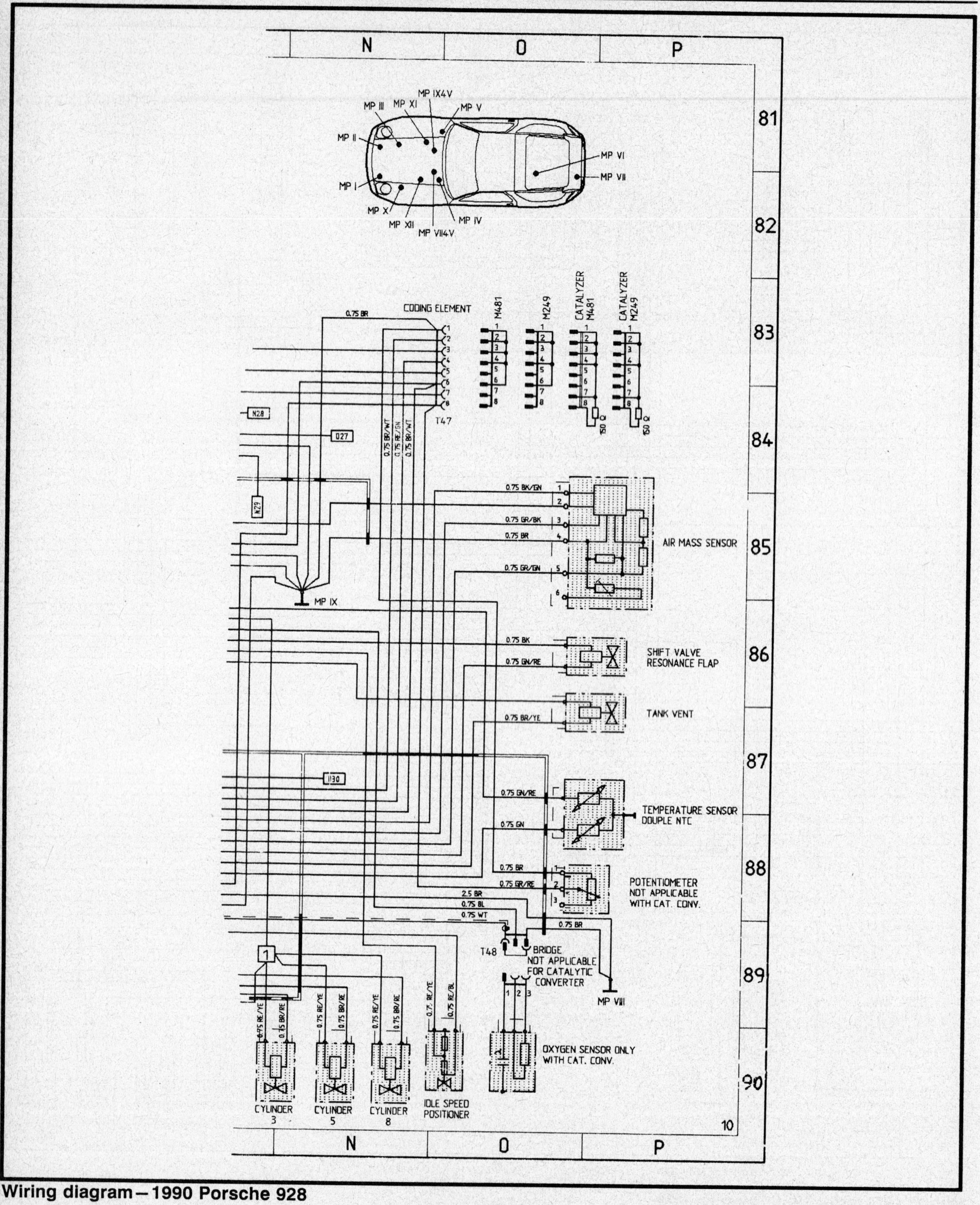

Wiring diagram—1990 Porsche 928

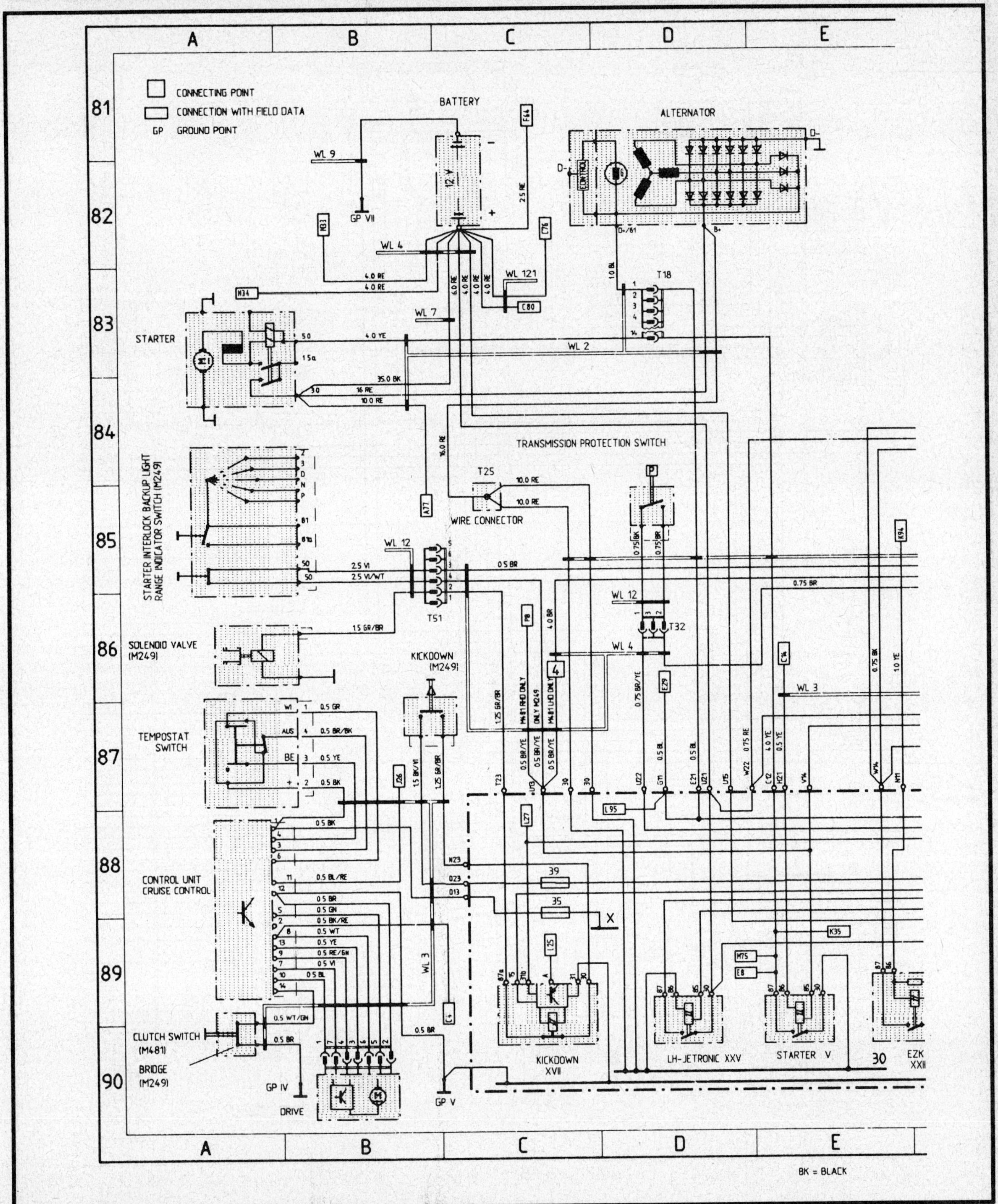

Wiring diagram—1991–92 Porsche 928

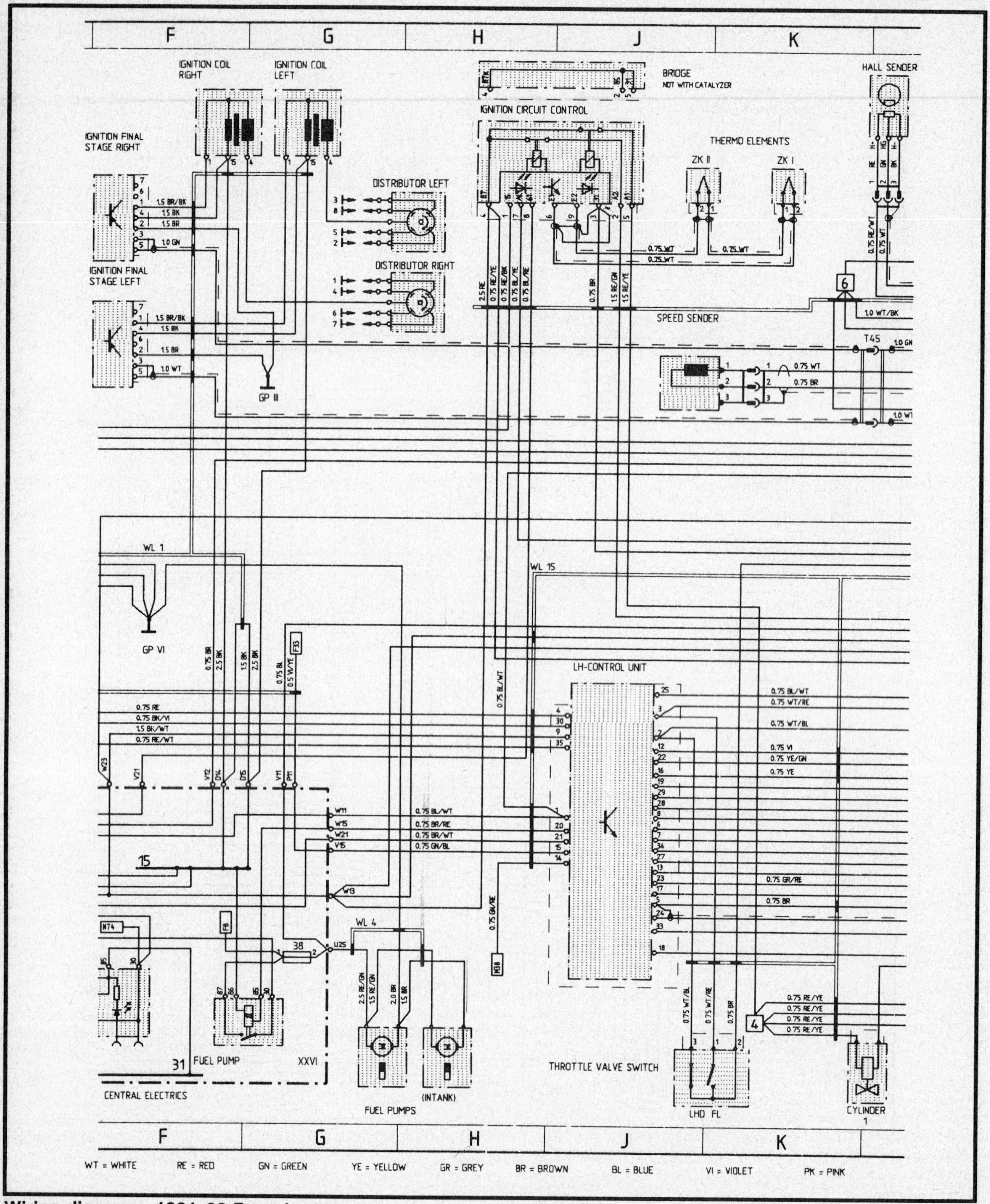

Wiring diagram—1991–92 Porsche 928

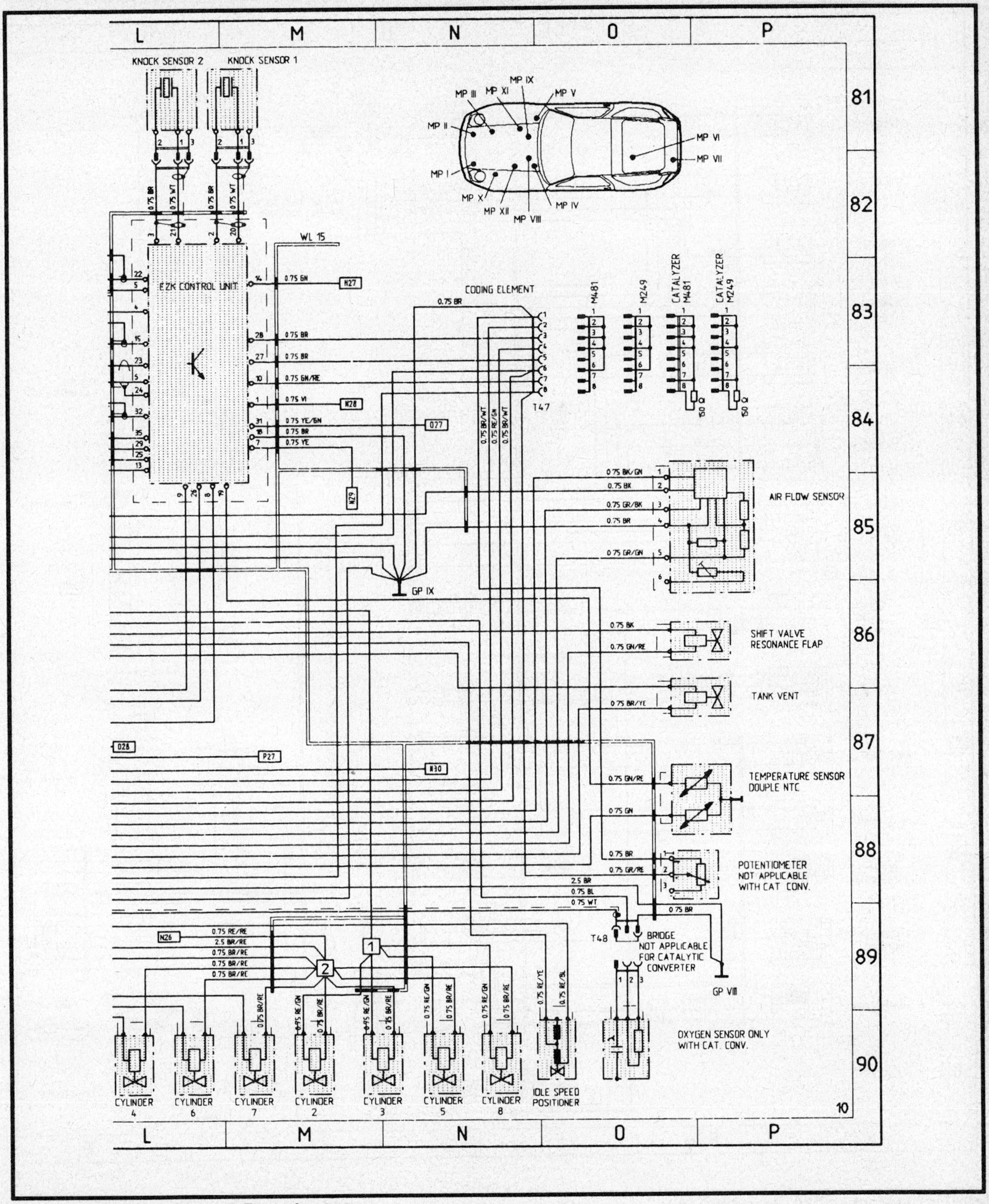

Wiring diagram—1991–92 Porsche 928

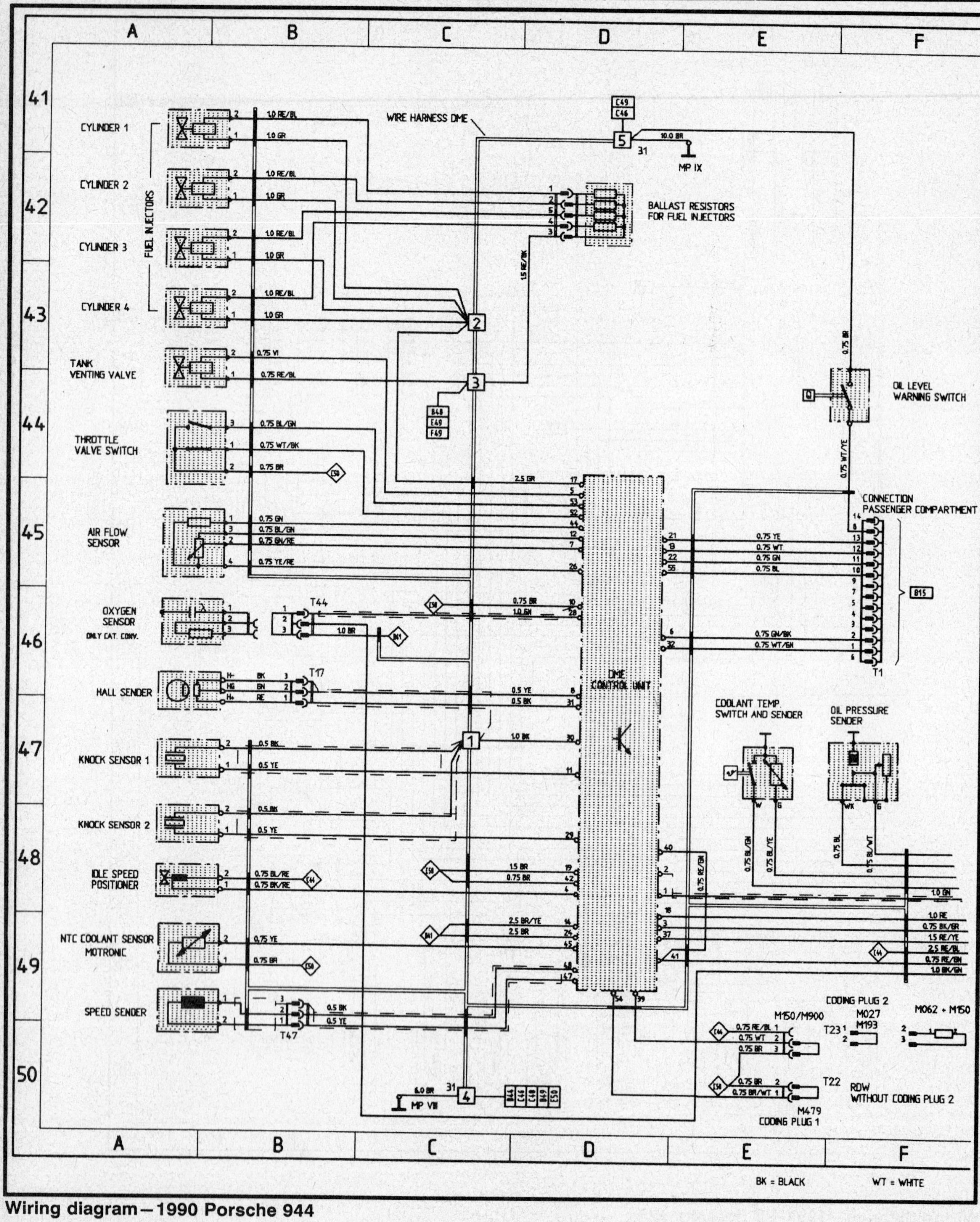

Wiring diagram—1990 Porsche 944

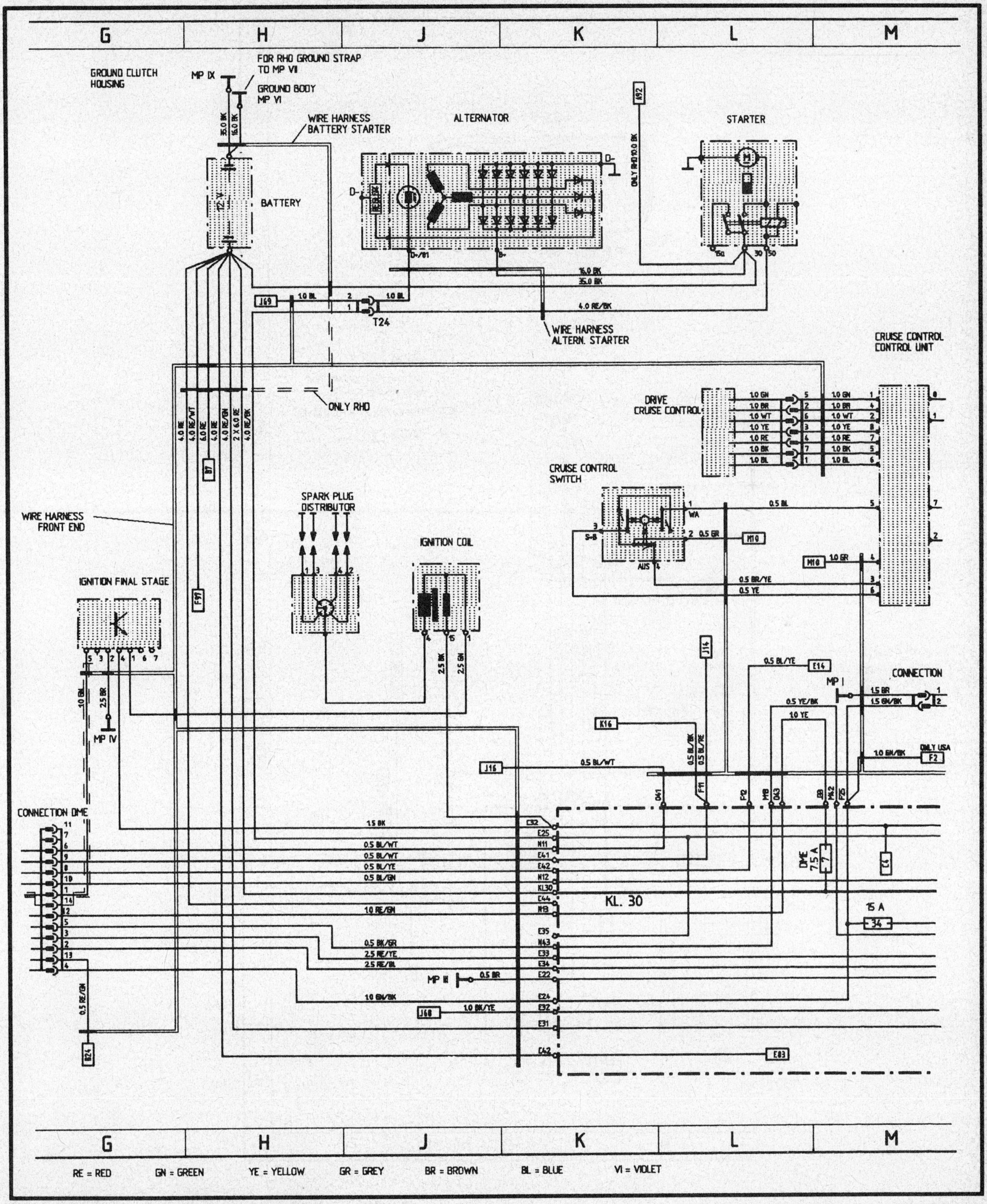

Wiring diagram—1990 Porsche 944

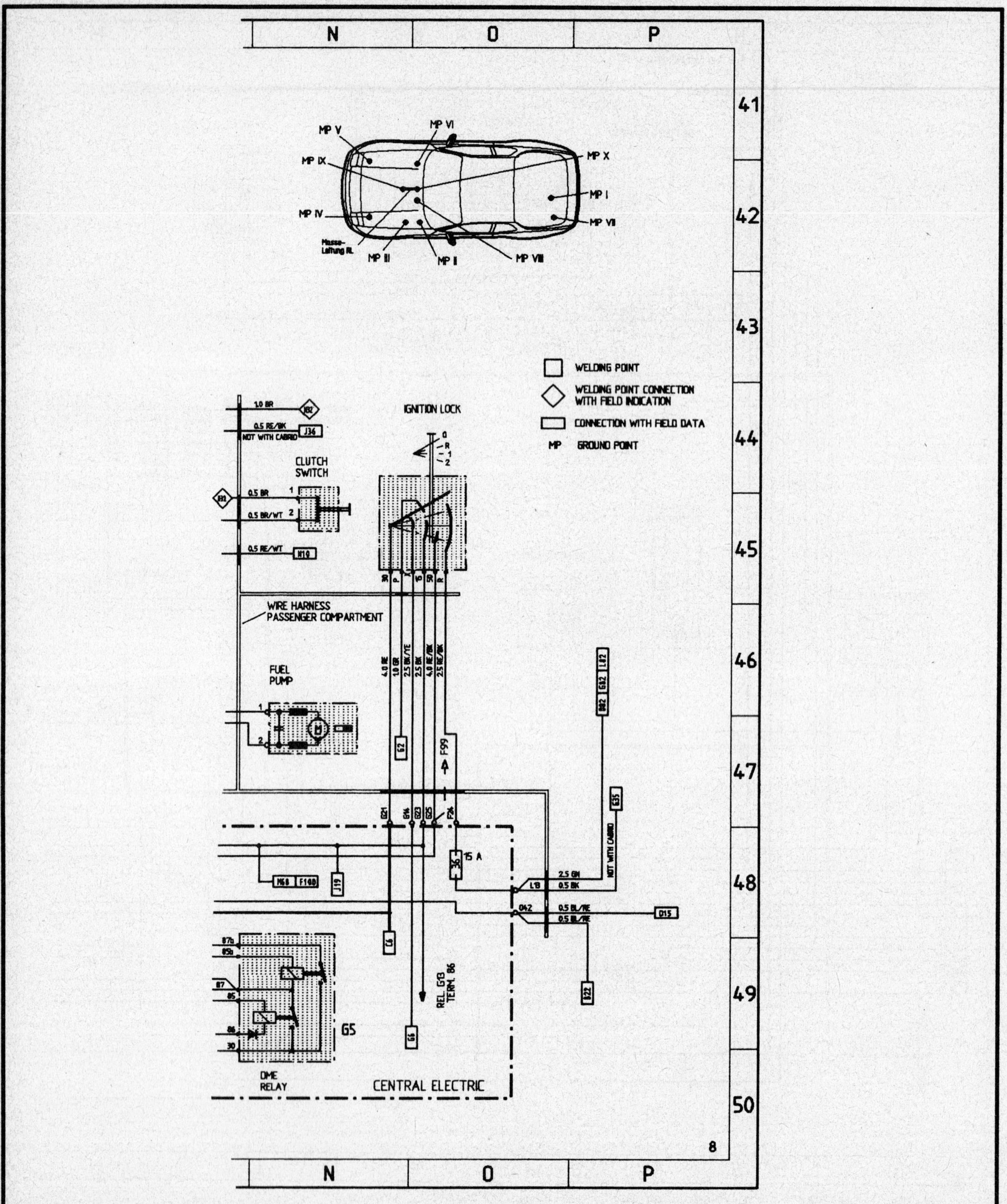

Wiring diagram—1990 Porsche 944

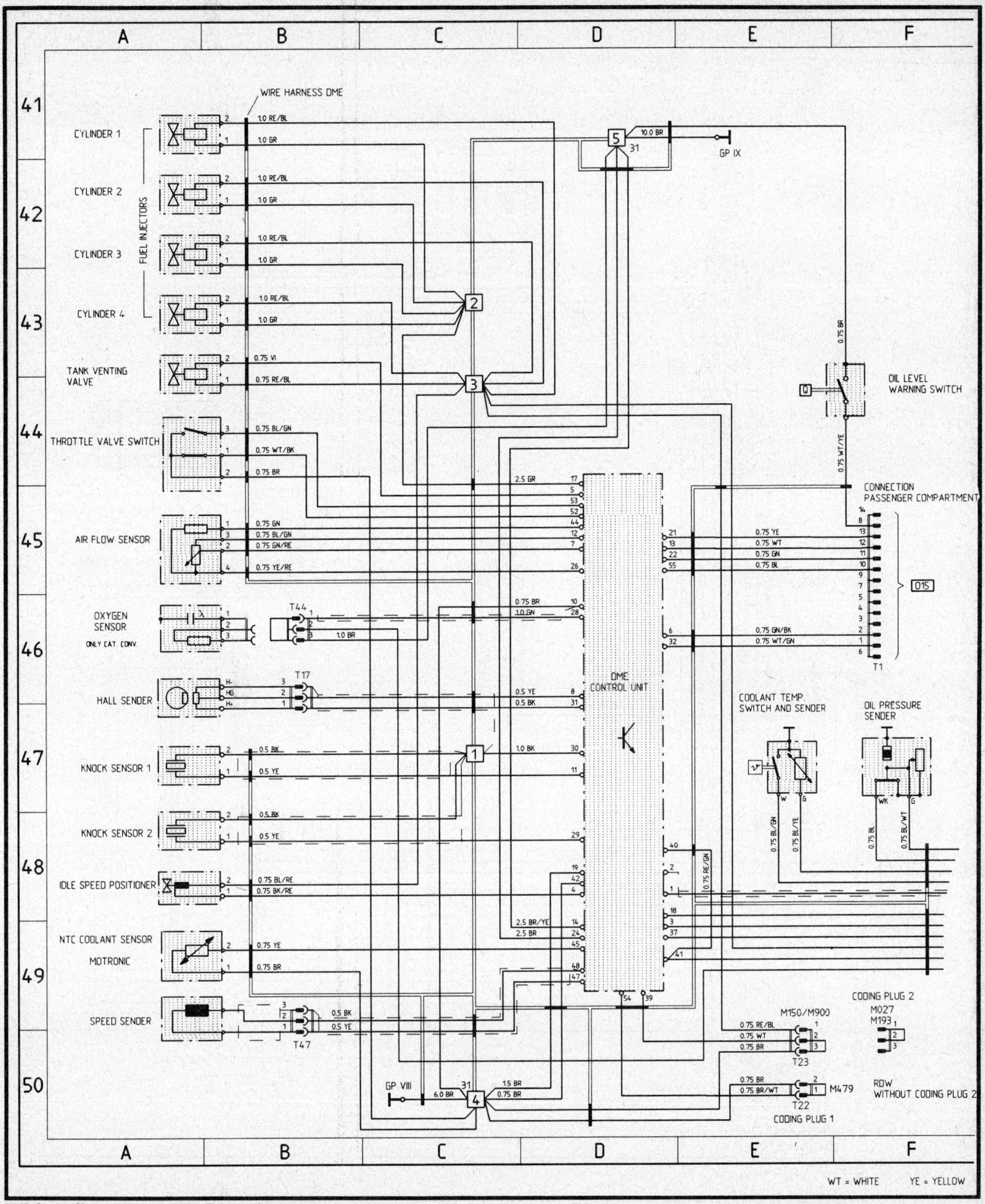

Wiring diagram—1991 Porsche 944

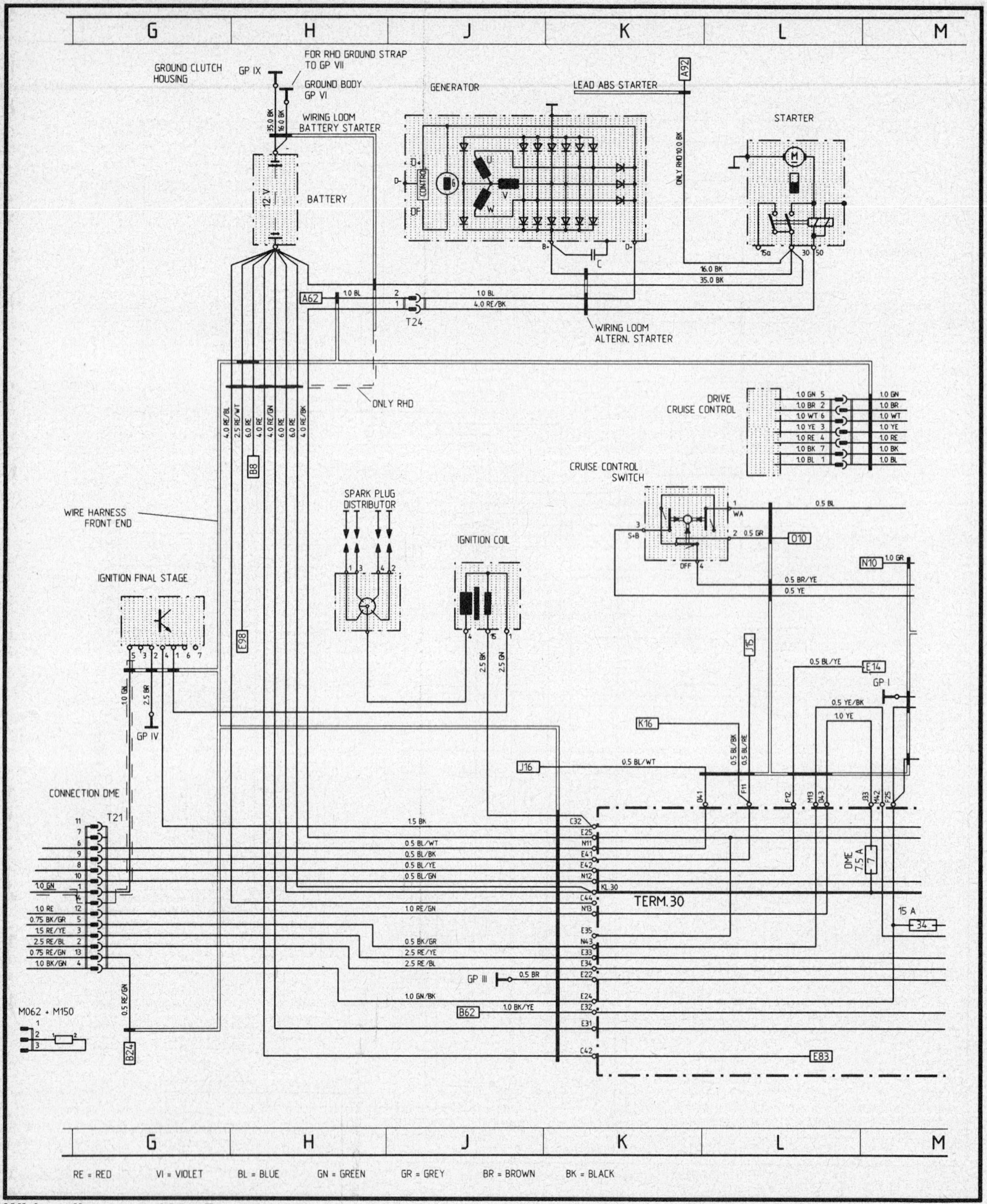

Wiring diagram—1991 Porsche 944

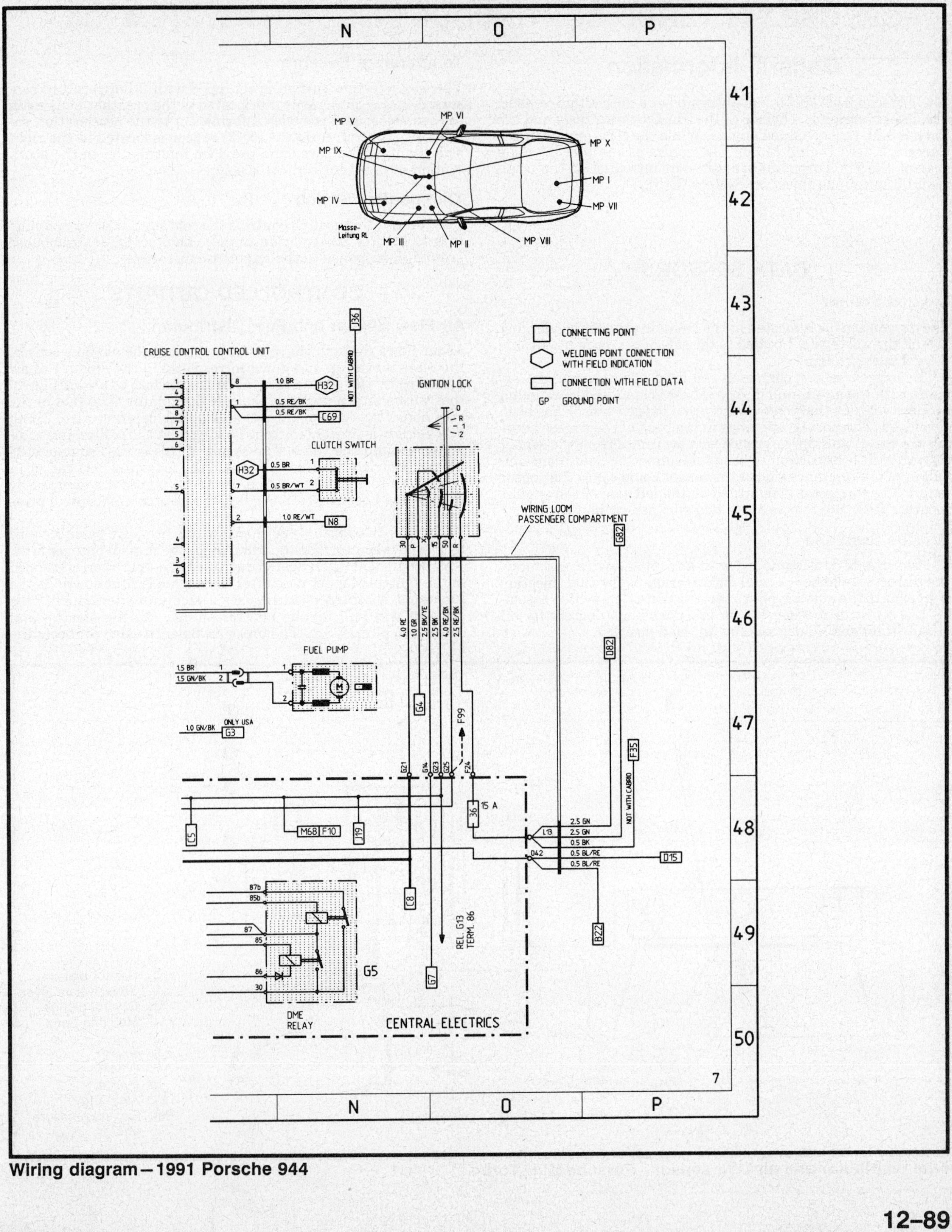

Wiring diagram—1991 Porsche 944

PORSCHE 911 TURBO CONTINUOUS INJECTION SYSTEM (CIS)

General Information

The Porsche 911 Turbo has always been a specialized vehicle. The latest version is a hybrid of the 911 Carrera 2 body and the proven 3.3L turbocharged engine. While the 3.6L engine of the Carrera 2 uses Digital Motor Electronics as the fuel injection system, the 911 Turbo has used since its introduction the traditional Continuous Injection System (CIS).

DATA SENSORS

Oxygen Sensor

The oxygen sensor is located in the left side exhaust header just behind the wastegate. Located to the side of the oxygen sensor is the exhaust gas tap.

The oxygen sensor compares the level of oxygen in the exhaust with the air around it. The sensor is mounted low enough in these vehicles that a new waterproof oxygen sensor has been developed. The normal openings in the back of the sensor housing are sealed and the reference air now comes from the sensor connector. It is important to keep this connector clean; contamination of the reference side of the sensor could cause the sensor to fail. The connector is mounted on the left side of the engine, separate from the temperature and rpm sensor plugs.

Throttle Switches

The throttle body is equipped with a throttle switch assembly. The throttle switches provide information to the fuel injection system as to the position of the throttle plate. The switch assembly has contacts at 0 degrees for idle position, 7 degrees for off idle position and 66 degrees for full load position.

Temperature Sensors

The fuel injection system is equipped with 2 temperature sensors. A 95°F. (35°C.) sensor is located in the breather housing in the crankcase and provides information to the acceleration enrichment control. A 59°F. (15°C.) sensor is located in the right side timing chain cover and provides information to the acceleration enrichment control, also.

Thermo-time Switch

The thermo-time switch controls the cold start injector opening. It is located in the left side timing chain cover. It establishes ground in response to the engine temperature.

CONTROLLED OUTPUTS

Air Flow Sensor and Fuel Distributor

As air flows through the venturi portion of the air flow sensor, the plate moves up and down in response to the amount of air drawn in by the engine. The plate is attached to a control arm that works on a plunger in the fuel distributor. The fuel in the area above the control plunger acts on the capsule valve, lifting it. The flow is prevented from flowing past the orifice and is directed around the valve. The result of this is that components are not restricted in movement and the metering port can open faster and to a greater degree. Pressure pulses are damped by the flexible hose between the fuel distributor and control pressure regulator.

The fuel distributor is equipped with a pressure balance valve to prevent vacuum from building up in the control plunger area. This occurs when fuel heated during engine operation is trapped in the control plunger area after the engine is shut down. As the fuel cools, it contracts and causes a vacuum to be formed. This vacuum will pull up the control plunger during startup and cause a rich condition. The pressure balance valve connects the

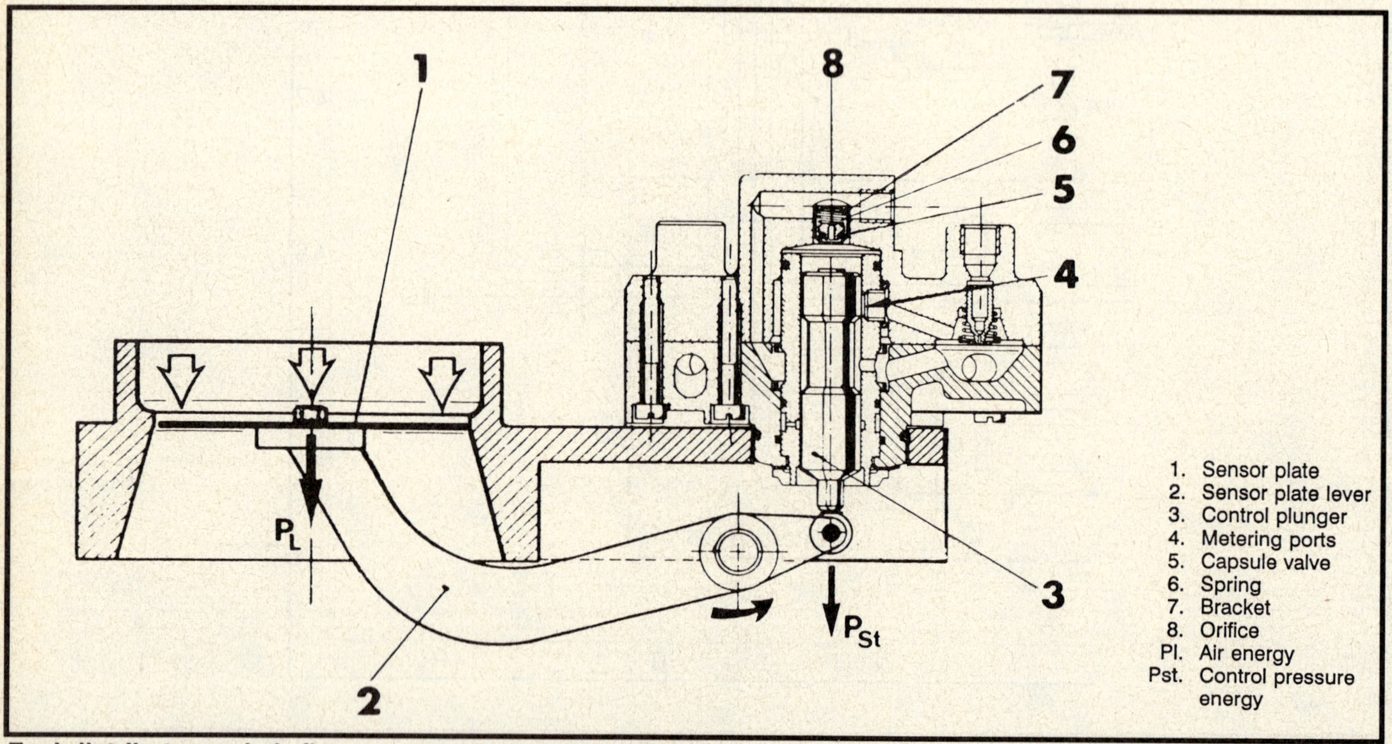

Fuel distributor and air flow sensor—Porsche 911 Turbo

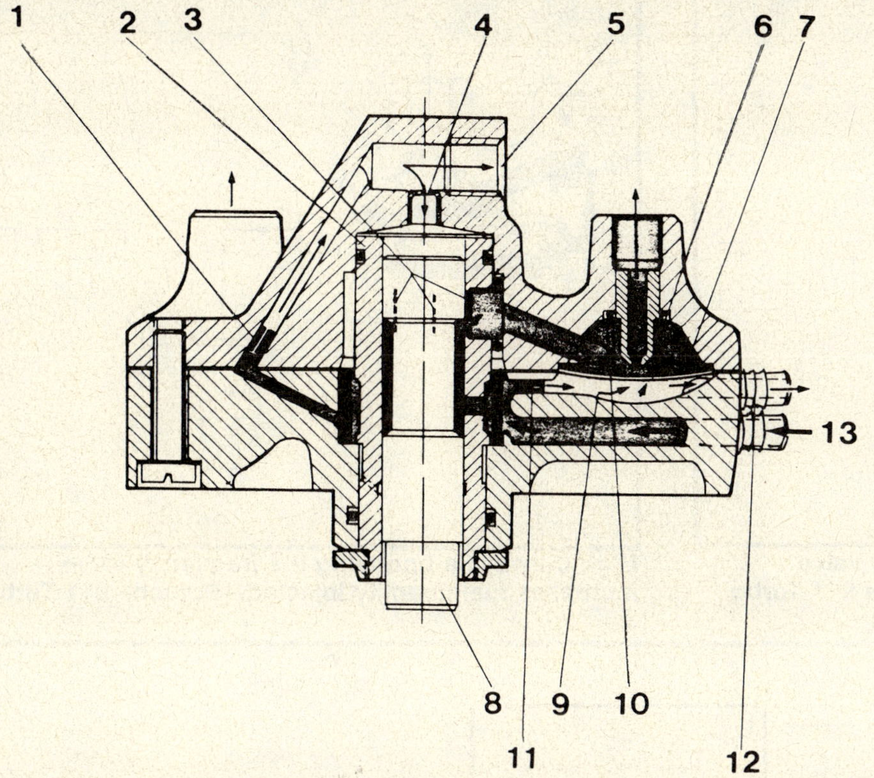

1. Orifice
2. Metering unit
3. Metering ports
4. Damping orifice
5. Control pressure regulator
6. Differential pressure valve
7. Upper chamber
8. Control plunger
9. Lower chamber
10. Diaphragm
11. Orifice
12. Return line to frequency valve
13. System pressure

Fuel distributor cross section—Porsche 911 Turbo

3.3L ENGINE— OPERATION PARAMETERS OF THE OXYGEN SENSOR CONTROL

Oil Temperature	Throttle Switch Position	Sensor Temperature	Pulse Duty Factor
<59°F (15°C)	0 Degrees	Cold	50%
	0 or 7 Degrees	Cold	75% for 0.5 seconds
	66 Degrees	Cold	50%
>59°F (15°C)	0 Degrees	Cold	50%
	0 or 7 Degrees	Cold	75% for 0.5 seconds
<95°F (35°C)	66 Degrees	Cold	50%
	0–66 Degrees	Hot	Variable
<95°F (35°C)	Independent	Cold	50%
	0–66 Degrees	Hot	Variable
	66 Degrees	Hot	50%

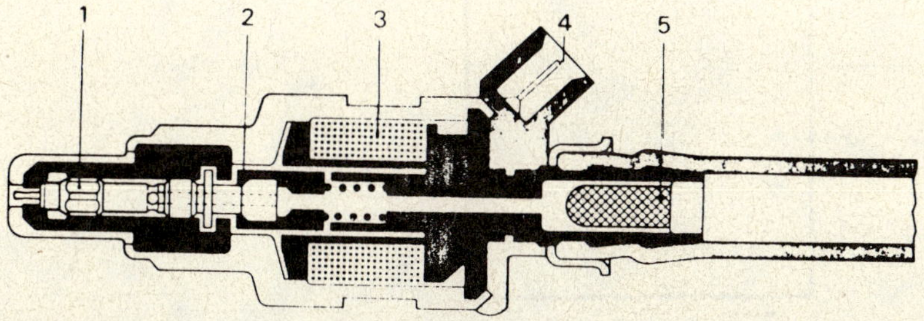

1. Valve needle
2. Solenoid armature
3. Solenoid coil—2.5 ohms
4. Electrical connector
5. Filter

Frequency valve—Porsche 911 Turbo

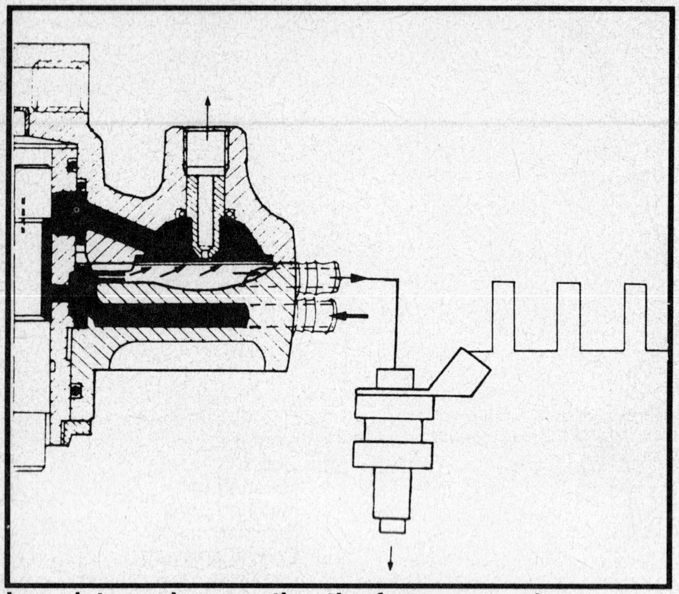

Low duty cycle operating the frequency valve reduces fuel quantity injected—Porsche 911 Turbo

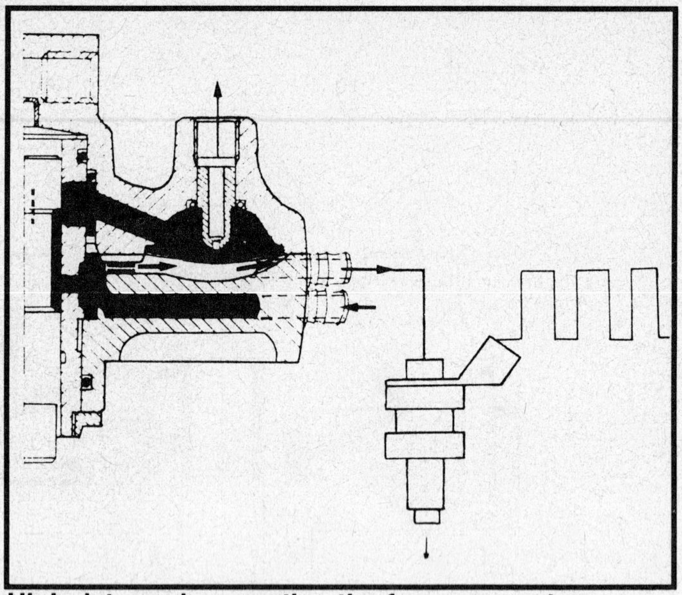

High duty cycle operating the frequency valve increases fuel quantity injected—Porsche 911 Turbo

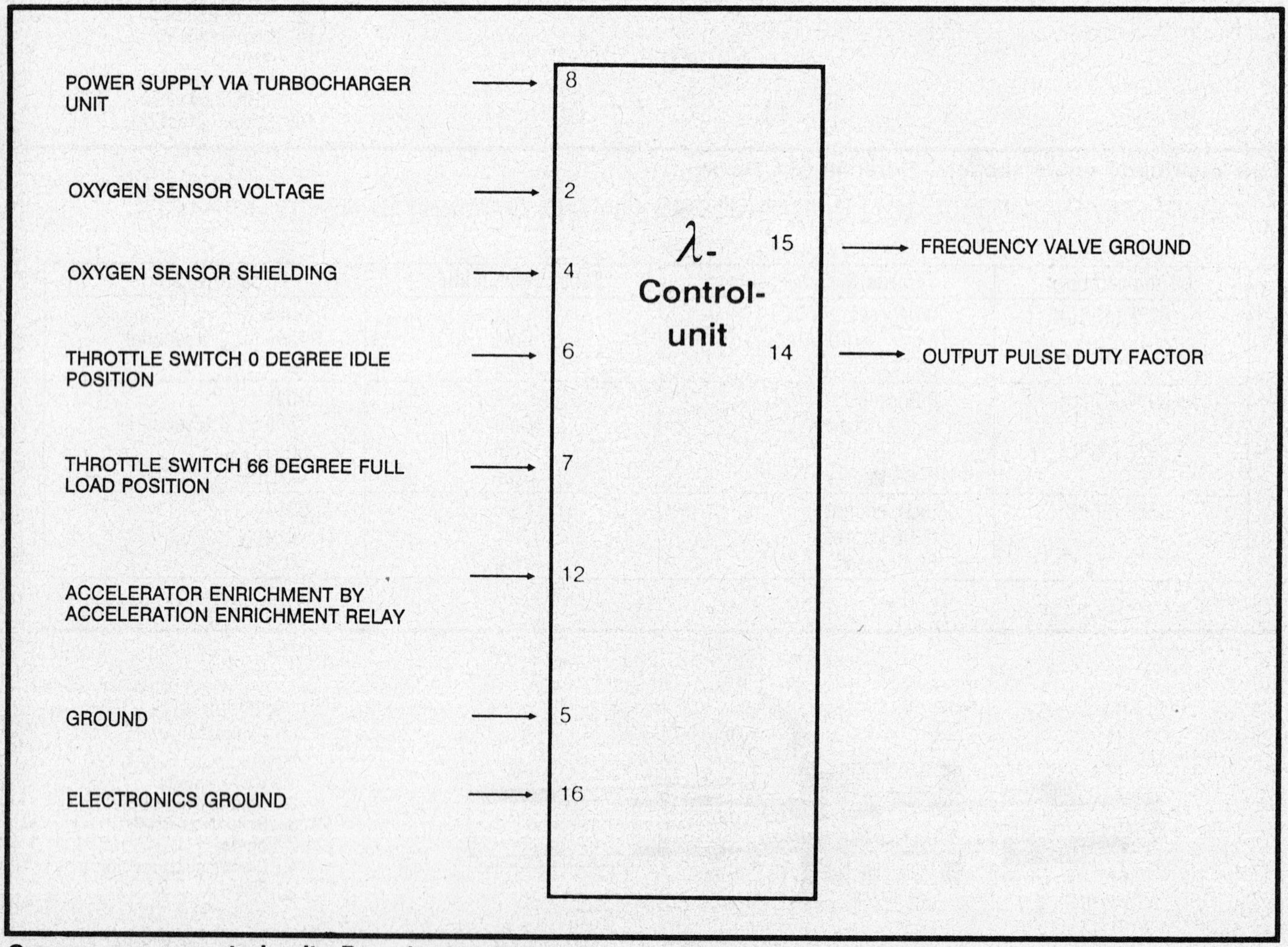

Oxygen sensor control unit—Porsche 911 Turbo

area above the control plunger and the fuel return line. When the pressure falls below 4.35 psi (0.3 bar), the valve opens to keep a vacuum from building.

Oxygen Sensor Control with Frequency Valve

To control the air/fuel mixture injected, the pressure of the lower chambers in the fuel distributor is varied. When the pressure is lowered in the chambers, the differential pressure at the metering ports is increased, thus increasing the amount of fuel injected. The 911 Turbo CIS uses lower chambers that are separated from system pressure by a fixed restriction. This is different from standard CIS systems.

Another restriction is placed between the lower chambers and the fuel return line. This restriction controls the pressure buildup in the lower chambers. The restriction is an electrically controlled valve, or frequency valve. The signal from the oxygen sensor controls the pulse duty factor of the frequency valve. This is how the mixture richness is controlled.

The frequency valve is mounted in the return line adjacent to the fuel distributor. The valve is identical to the L-Jetronic fuel injector and has an internal resistance of 2.5 ohms. The control unit provides a steady 70 Hz signal for the frequency valve. The length of the time that the valve is held open is referred to as the pulse width. The longer the valve is held open, the more the pressure in the lower chamber is reduced.

The oxygen sensor control unit is located under the drivers seat along with the acceleration enrichment control and the turbo control. The oxygen sensor control is the largest of the 3 units.

Acceleration Enrichment Control

The fuel mixture richness is increased at low engine temperatures to provide smooth running and acceleration. The acceleration enrichment control provides the enrichment information for use by the oxygen sensor control which in turn activates the frequency valve. The acceleration enrichment control unit is located under the drivers seat.

Using information provided by throttle position switches and engine temperature sensors, the control unit determines the amount of enrichment required. The enrichment only occurs at specified oil temperatures and when the system has switched to closed loop control.

When the oxygen sensor system goes closed loop and oil temperature is below 95° F. (35° C.), the pulse duty cycle of 75% is available for 0.5 seconds within 25 seconds. When the oil temperature is above 95° F. (35° C.), the pulse duty cycle of 75% is available for 0.5 seconds during a 2 minute period whenever the 0 degree and 7 degree switches are opened.

Control Pressure Regulator

The control pressure regulator on the 911 Turbo senses a number of parameters to adjust the control pressure. The control pressure regulator senses the boost pressure, the altitude, the engine temperature and an internal thermostatically controlled heating element.

Contained within the body of the control pressure regulator are 2 bimetallic elements. When the temperature is below 59° F. (15° C.), only one of the bimetallic strips is being heated. This heats the strip slowly and the control pressure builds slowly. Once the temperature exceeds 59° F. (15° C.), the second strip is heated. This speeds up the rise in the control pressure.

The body of the control pressure regulator is grounded, so it is important to keep correct polarity when connecting the unit. The negative terminal is closest to the fuel connections.

The control unit has a built-in altitude compensation capsule that adjusts the control pressure. The body of the control unit is vented and the altitude compensation capsule responds to the atmospheric pressure.

The control unit has a port connected to the intake manifold. The boost pressure signal adjusts the control pressure as a full load enrichment. When the boost reaches 3.625 psi (250 mbar), the control pressure is reduced from 63.7–66.9 psi (4.39–4.61 bar) to 44.3–48.6 psi (3.05–3.35 bar).

Cold Start Enrichment

When the engine is below 59° F. (15° C.) the bimetallic strip in the thermo-time switch is closed. This grounds the cold start injector. When the engine is cranked, voltage from terminal 50 at the starter is applied to the cold start injector. Simultaneously voltage at the thermo-time switch heats the bimetallic strip, causing it to curve away and break the cold start switch ground. At this point the injector stops flow.

The cold start injector is mounted in the intake plenum. The thermo-time switch is located in the left side timing chain cover.

Auxiliary Air Valve

The auxiliary air valve increases the amount of air allowed into the intake tract during cold starts. This is to compensate for increased engine drag during cold engine operation. The unit is temperature controlled and bypasses the air around the throttle plate.

Vacuum Limiter Valve

During coasting the engine will run excessively rich if not prevented from doing so. The vacuum limiter valve allows additional air to enter the intake tract beyond the throttle plates. This allows optimal combustion and limits how rich the engine can run. The valve is vacuum actuated.

Recirculating Air Valve

To prevent the turbocharger from providing boost against the closed throttle, a valve is fitted to return the air charge to the

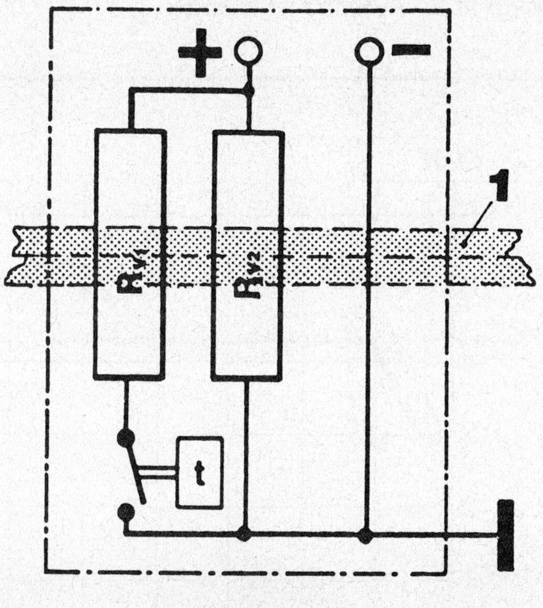

1. Bimetallic strips
Rv1. Temperature switched heater element
Rv2. Heater element
t. Temperature switch 59°F(15°C)

Dual heating elements in the control pressure regulator—Porsche 911 Turbo

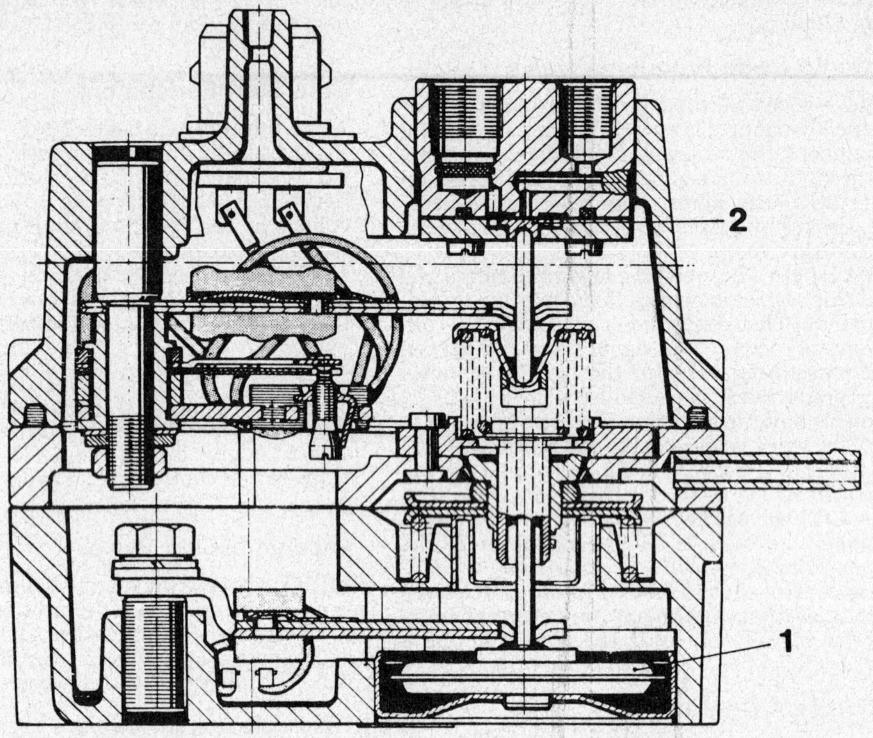

Cutaway of the control pressure regulator showing the control diaphragm at 1 and the altitude compensation capsule at 2—Porsche 911 Turbo

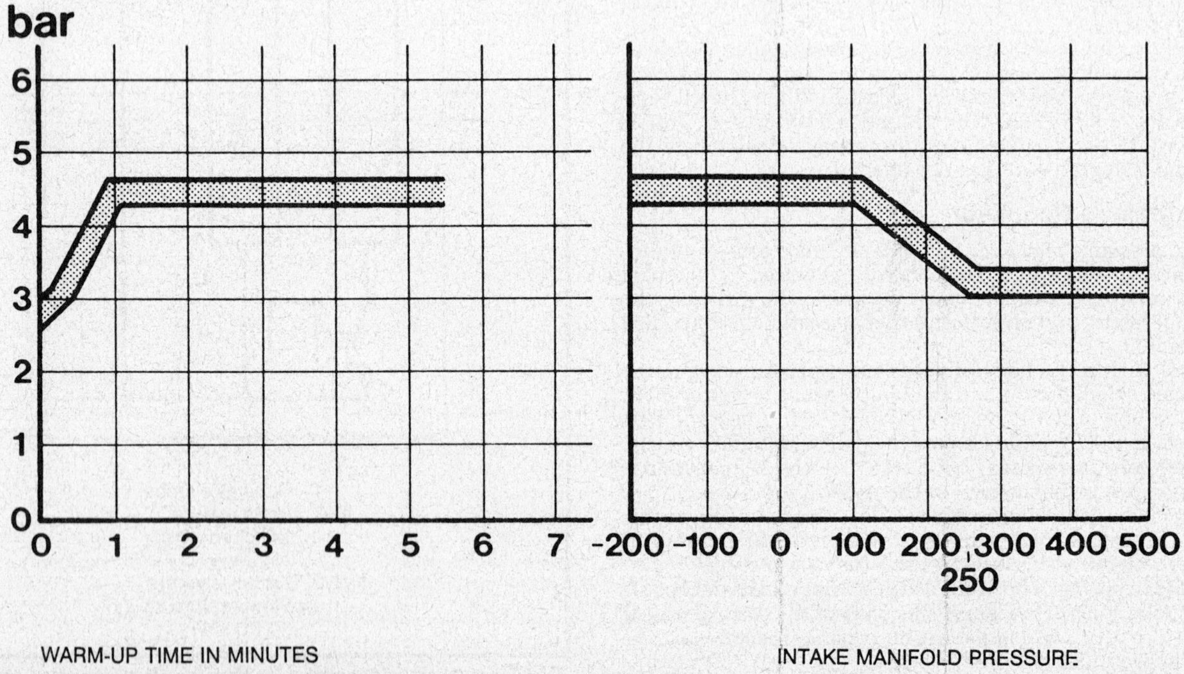

Control pressure regulator operating parameters: time and pressure—Porsche 911 Turbo

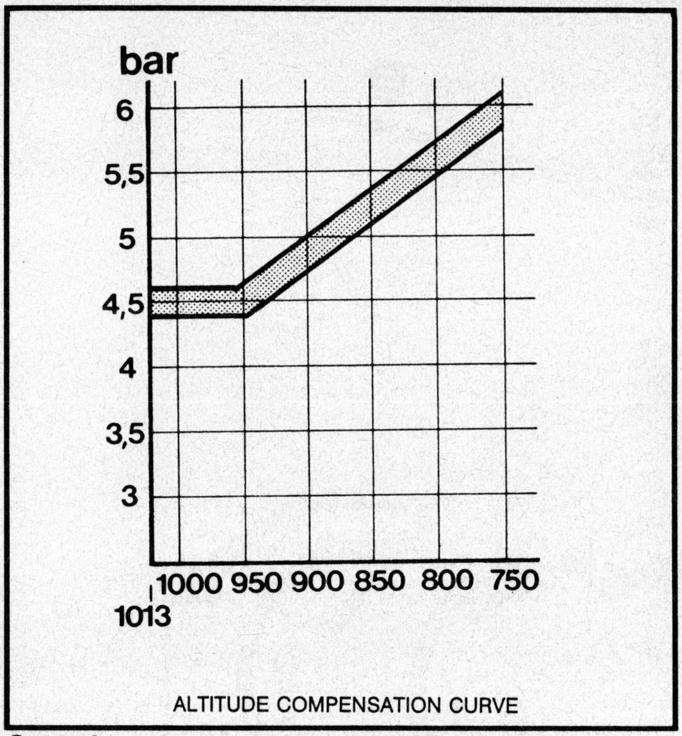

Control pressure regulator operating parameters: altitude compensation—Porsche 911 Turbo

nonpressurized side of the turbocharger. The valve is vacuum operated and is connected between the 2 sides of the turbocharger.

Turbocharger Control Unit

The turbocharger control unit performs 5 functions. The unit provides voltage to the ignition system relay, the oxygen sensor control unit, the acceleration enrichment relay, the control pressure regulator, the auxiliary air regulator, the oxygen sensor heater and the frequency valve. The unit controls the fuel pumps. The unit controls the fuel pump safety circuit, the boost pressure safety circuit and the ignition delay relay.

The ignition delay relay provides power to the ignition for 5 seconds after the ignition is turned **OFF**. This is to allow complete combustion of fuel that was injected before the engine was shut down.

The fuel pumps are connected to the fuel pump relay and are actuated when a signal from the turbocharger control unit is present. The turbocharger control unit receives an rpm signal and only when the engine is cranking or running does it allow the fuel pump relay to close. This prevents the fuel pumps from running with the engine stationary.

The turbocharger control unit will shut down the fuel pumps at engines speeds above 6800 rpm. This is to prevent engine overspeeding.

The turbocharger control unit monitors a boost switch that, in event of overboost in the 15.95–20.3 psi (1.1–1.4 bar) range, will shut down the fuelpumps. This is to prevent excessive boost situations.

FUEL DELIVERY SYSTEM

Fuel Pumps

The vehicle uses 2 fuel pumps connected in series. The first fuel pump is mounted behind the fuel tank. The second fuel pump is mounted in the right side of the luggage compartment.

Fuel Injectors

The injectors open automatically at a pressure of approximately 52.2 psi (3.6 bar). As the injector opens, it chatters at a rate of 1500 Hz, allowing for good atomization. The injector body contains a fine mesh filter to prevent tip clogging. The injector is pushed into a rubber holder on the engine.

Diagnosis and Testing

SERVICE PRECAUTIONS

- Always disconnect the negative battery terminal to prevent sparks cause by short-circuiting, circuit-breaking, etc.
- Smoking must not be allowed when working around flammable liquids.
- Always keep a fire extinguisher close on hand.
- Absorbant material must be available to soak up any spillage.
- Ensure fuel is emptied into a suitable container.
- Always place the transmission gear shift lever in **N** and set the parking brake and block drive wheels before starting the engine.
- Keep hands, tools and clothing away from engine cooling fan to help prevent personal injury.
- Thoroughly clean all unions and area near connections before disconnecting.

READING CODES

The fault codes can only be accessed using the Porsche Diagnostic Tester 9288. This hand-held unit plugs into the diagnostic connector behind the passenger side kick panel. The earlier 9268 tester includes a flashing light to report fault codes. The 9288 unit will report fault codes, sensor readings and other stored, measured or calculated values. This unit is also menu driven, meaning instructions for test procedures appear on the screen. The 9288 unit is capable of activating certain output devices for function testing. Instructions for use of the tester is provided with the unit.

CLEARING CODES

The fault code memory should be cleared before returning the vehicle to service. All codes in memory are lost when the control unit or the battery is disconnected. The memory can also be cleared using the 9288 tester.

TURBOCHARGER CONTROL UNIT

Testing

1. This test is done with the engine running. Take precautions while working with a running engine.
2. Locate the boost pressure control switch behind the intercooler on the left side.
3. With the engine running, pull off the electrical connector from the boost switch.
4. The engine must stall. The turbocharger control will cut power to the fuel pumps, thus stopping the engine. If the engine does not stop, check the wiring to the control unit and replace the control unit if necessary.

AUXILIARY AIR VALVE

Testing

1. Remove the auxiliary air valve from the engine. This will make it easier to test than if it were mounted on the engine.

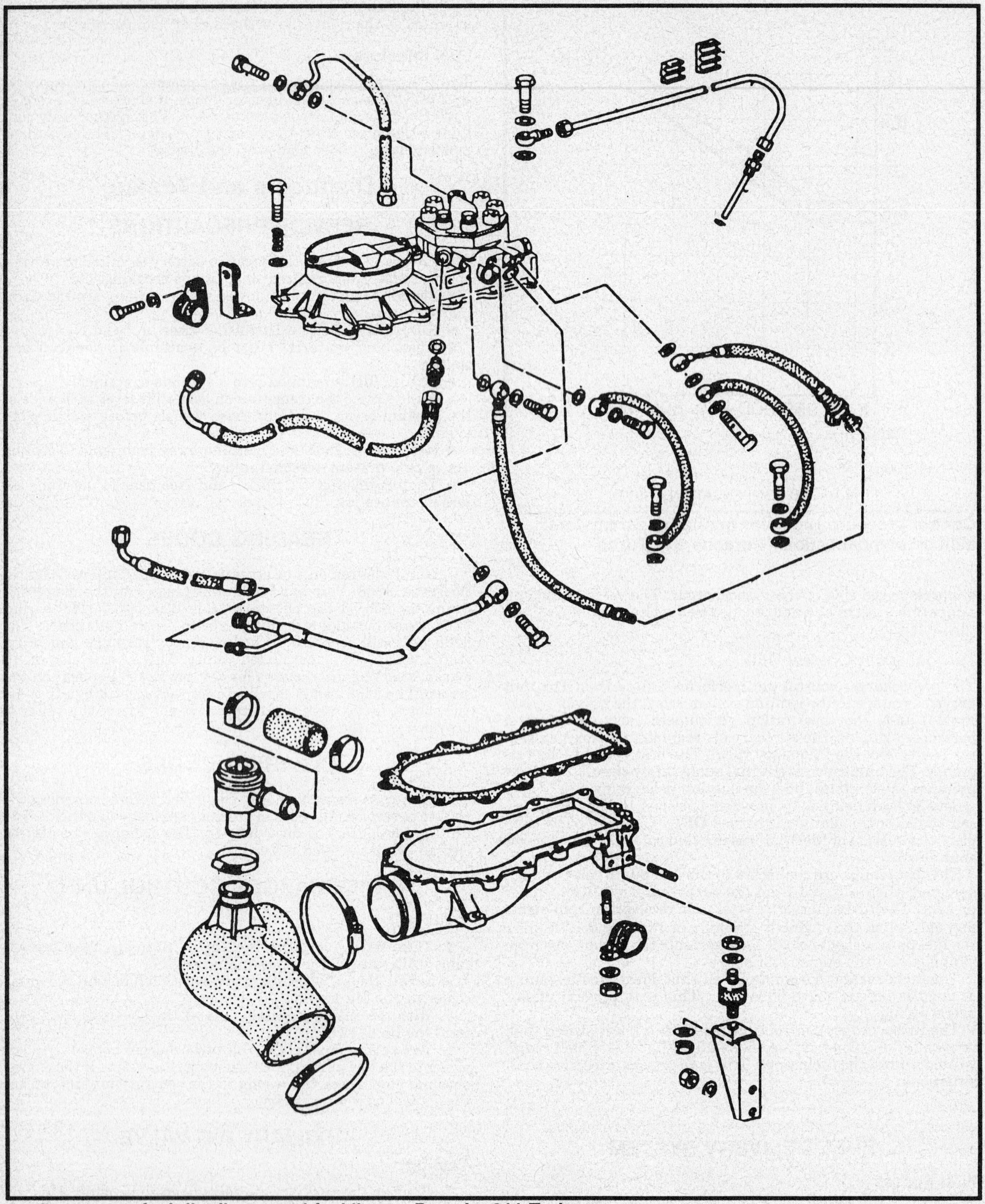

Air flow meter, fuel distributor and fuel lines – Porsche 911 Turbo

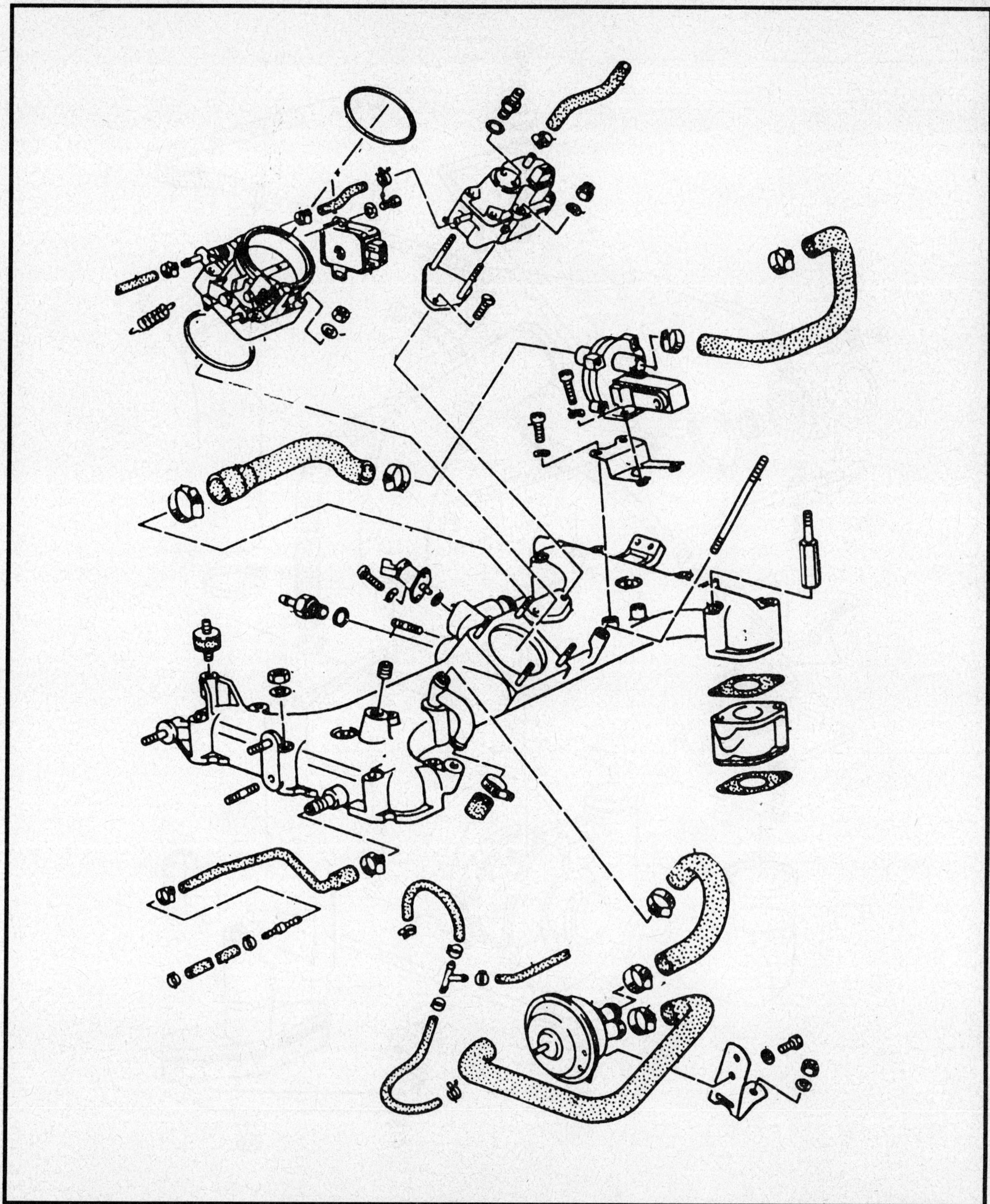

Intake manifold and throttle body—Porsche 911 Turbo

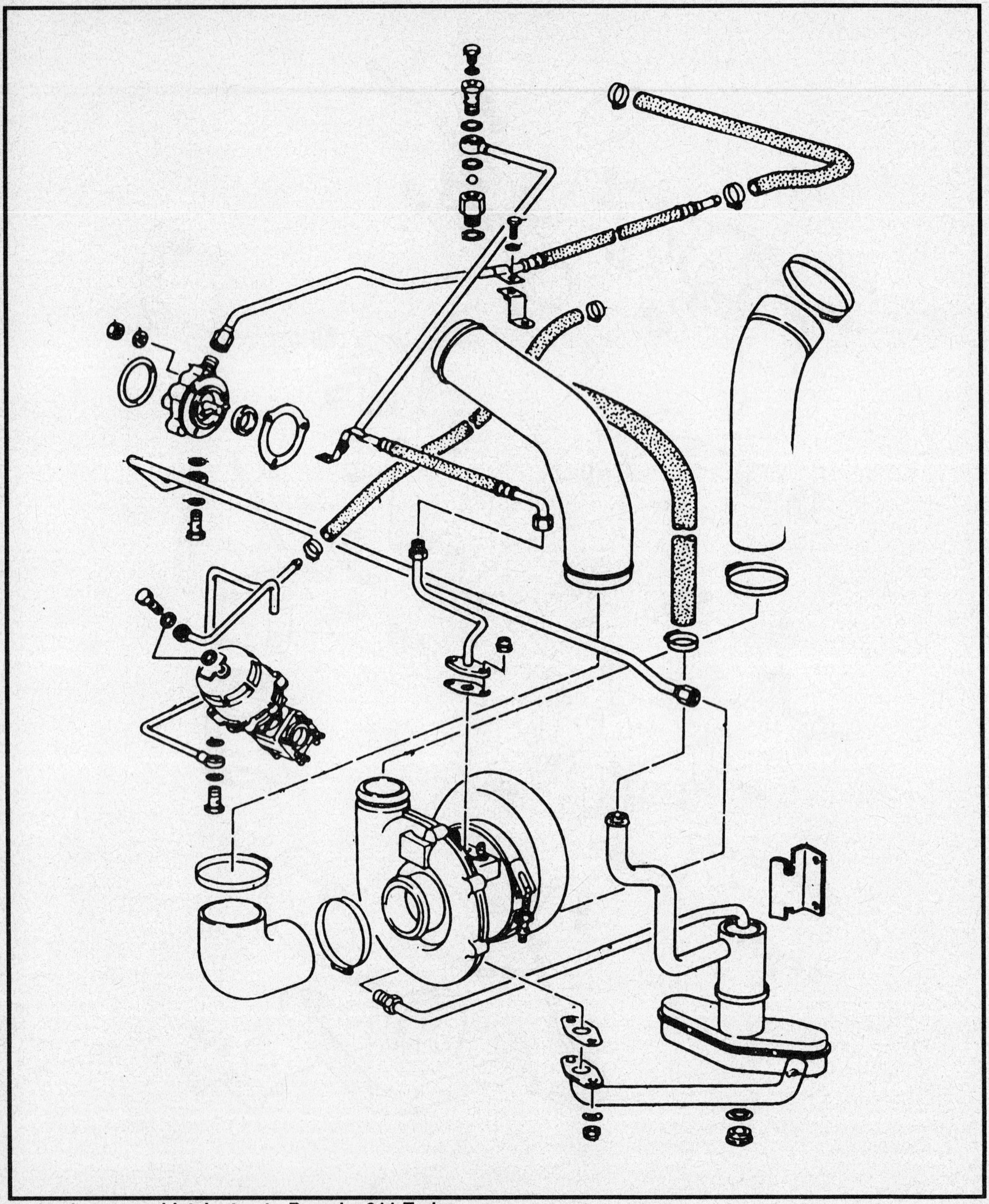

Turbocharger and intake tract—Porsche 911 Turbo

2. Apply 12 volts to the terminals of the valve.
3. The valve should open. Replace the valve if the motion of the valve is slow or sticky.

VACUUM LIMITER VALVE

Testing

1. Disconnect the hoses leading to the limiter valve at the intercooler.
2. Disconnect the vacuum line to the limiter valve and connect an external vacuum source.
3. Blow through the hose while applying vacuum to the limiter valve.
4. The valve should open with vacuum applied, and allow air to pass through. Replace the unit if it fails this test.

RECIRCULATING AIR VALVE

Testing

1. Loosen the 2 hose clamps holding the air valve to the intake duct and intercooler. Disconnect the vacuum line and remove the valve.
2. Connect an external vacuum source to the valve.
3. Blow through the valve while applying vacuum.
4. The valve should open with vacuum applied, and allow air to pass through. Replace the unit if it fails this test.

THERMO-TIME SWITCH

Testing

1. Perform this test with the engine below 59° F. (15° C.). If the temperature is not low enough, remove the switch and cool to below 59° F. (15° C.).
2. With the switch cooled enough, continuity should exist between the body of the switch and the electrical terminal W. Terminal W has the red and black wire attached to it.
3. Allow the switch to warm to above 59° F. (15° C.). Now there should be no continuity. Replace the switch if it fails these tests.

Component Replacement

SERVICE PRECAUTIONS

- Always disconnect the negative battery terminal to prevent sparks cause by short-circuiting, circuit-breaking, etc.
- Smoking must not be allowed when working around flammable liquids.
- Always keep a fire extinguisher close on hand.
- Absorbant material must be available to soak up any spillage.
- Ensure fuel is emptied into a suitable container.
- Always place the transmission gear shift lever in **N** and set the parking brake and block drive wheels before starting the engine.
- Keep hands, tools and clothing away from engine cooling fan to help prevent personal injury.
- Thoroughly clean all unions and area near connections before disconnecting.

FUEL PRESSURE RELIEF

1. Follow service precautions to safely perform this procedure. Disconnect the negative battery terminal.
2. Wrap a shop rag around the feed to the cold start injector.
3. Loosen the feed connection to relieve the system pressure. Catch any escaping fuel in the shop rag. Dispose of the rag in a safe manner.
4. Tighten the cold start injector feed connection.

CONTROL UNITS

Removal and Installation

1. Disconnect the negative battery terminal.
2. The control units are located beneath the driver's seat. Remove the drivers seat.
3. The removal of any of the control units is completed by removing the mounting screws and disconnecting the electrical plug.
4. Replace the control units and connect the electrical plug. Replace the mounting screws.
5. Replace the driver's seat and connect the negative battery terminal.

FREQUENCY VALVE

Removal and Installation

1. Disconnect the negative battery terminal. The frequency valve is located on the side of the fuel distributor secured by an Adel style clamp.
2. Relieve the fuel system pressure. Disconnect the fuel lines leading to the frequency valve.
3. Remove the holding clamp and disconnect the electrical connector. Remove the frequency valve.
4. Replace the electrical connector and fasten the mounting clamp.
5. Connect the fuel lines and negative battery terminal. Start the engine and check for fuel leaks.

OXYGEN SENSOR

Removal and Installation

1. Disconnect the negative battery terminal.
2. The oxygen sensor may be easier to reach with the car raised and safely supported. The sensor is located in the exhaust pipe before the wastegate.
3. Disconnect the oxygen sensor plug and unscrew the oxygen sensor body from the exhaust pipe. The oxygen sensor electrical connector is located on the upper left side of the engine compartment.
4. Install the new oxygen sensor and torque to 37–44 ft. lbs. (50–60 Nm).
5. Connect the electrical connector to the sensor. Connect the negative battery terminal.

COLD START VALVE

Removal and Installation

1. Disconnect the negative battery terminal. The cold start valve is located on the side of the intake plenum.
2. Relieve the fuel system pressure. Disconnect the fuel line at the cold start valve.
3. Disconnect the electrical connector and the 2 bolts holding the valve.
4. Remove the valve from the plenum.
5. Replace the valve and gasket, if used. Replace the bolts and tighten. Connect the electrical connector.
6. Connect the fuel line to the cold start valve using new crush washers. Connect the negative battery terminal.
7. Start the engine and check for fuel leaks.

CONTROL PRESSURE REGULATOR

Removal and Installation

1. Disconnect the negative battery terminal. The control pressure regulator is located in front of the fuel distributor.
2. Relieve the fuel system pressure. Disconnect the fuel lines from the regulator.
3. Disconnect the electrical connector and the 2 bolts holding the valve.
4. Disconnect the boost line. Remove the valve from the engine.
5. Install the control pressure regulator. Connect the boost line and replace the mounting bolts. Connect the electrical connector.
6. Connect the fuel lines to the control pressure regulator. Connect the negative battery terminal and start the engine. Check for fuel leaks.

IDLE SPEED AND MIXTURE

Adjusting

The 911 Turbo has adjustable idle speed and mixture. The CO adjustment will affect the entire operating range, not just the idle mixture. Make sure that engine is in proper operating condition and there is a good ventilation around the engine. It is recommended that the adjustment take place with the vehicle outside to prevent heated air from being recirculated by the engine, raising the temperature of the engine during the test. Make the tests as quickly as possible; the intake runners will heat up and affect the outcome of the tests. Between CO measurements, do not just blip the throttle; drive the vehicle to get realistic readings.

1. The air cleaner must be installed during the test. Connect the CO tester to the exhaust tap near the oxygen sensor. Disconnect the oxygen sensor at the upper air injection pump mount.
2. Disconnect and plug the line from the air injection pump to the diverter valve.
3. Run the engine to operating temperature. Oil temperature should be 185–203° F. (85–95° C.).
4. Check the CO level. It should be 0.8–1.2% for the USA and 0.6–1.0 for the rest of world at 850–950 rpm.
5. If the idle mixture is not correct, insert hex wrench 9156 or equivalent, into the mixture control screw on the control unit.

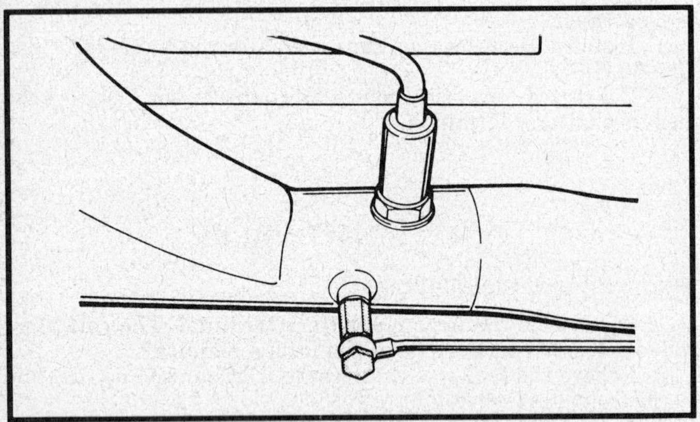
Exhaust gas tap for CO measurement—Porsche 911 Turbo

Press down to engage the screw. Turn clockwise to richen the mixture and counterclockwise to lean the mixture.

6. Connect the oxygen sensor. If the idle speed has changed, use tool P229b, P229c or equivalent, to adjust the idle speed screw at the throttle body.
7. Install the air hose and remove the exhaust tap. Coat the threads of the exhaust tap port plug with anti-sieze and torque to 11 ft. lbs. (15 Nm).

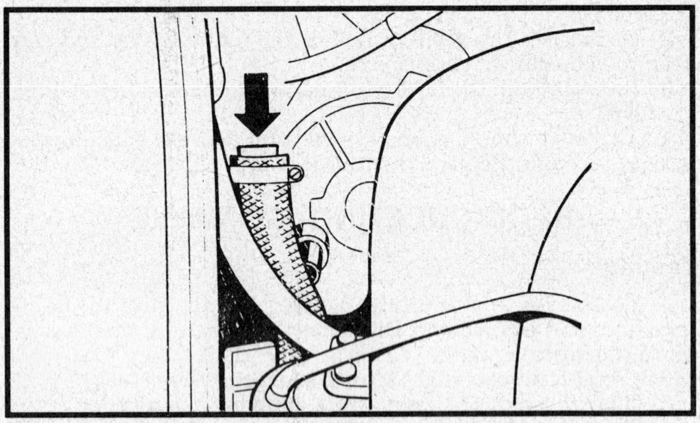
Plugging the air hose to the diverter valve—Porsche 911 Turbo

Adjusting the CO level—Porsche 911 Turbo

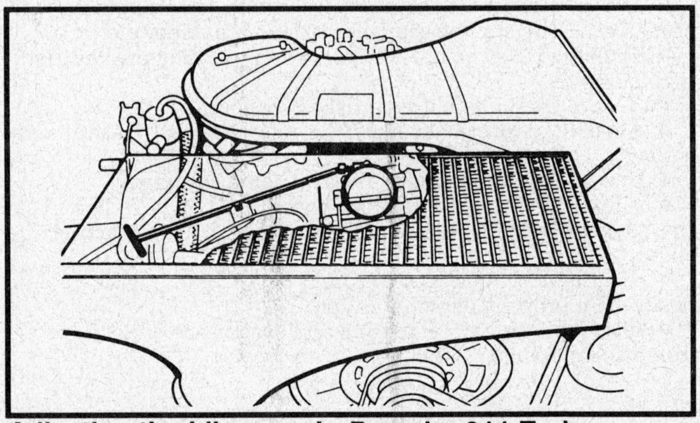
Adjusting the idle speed—Porsche 911 Turbo

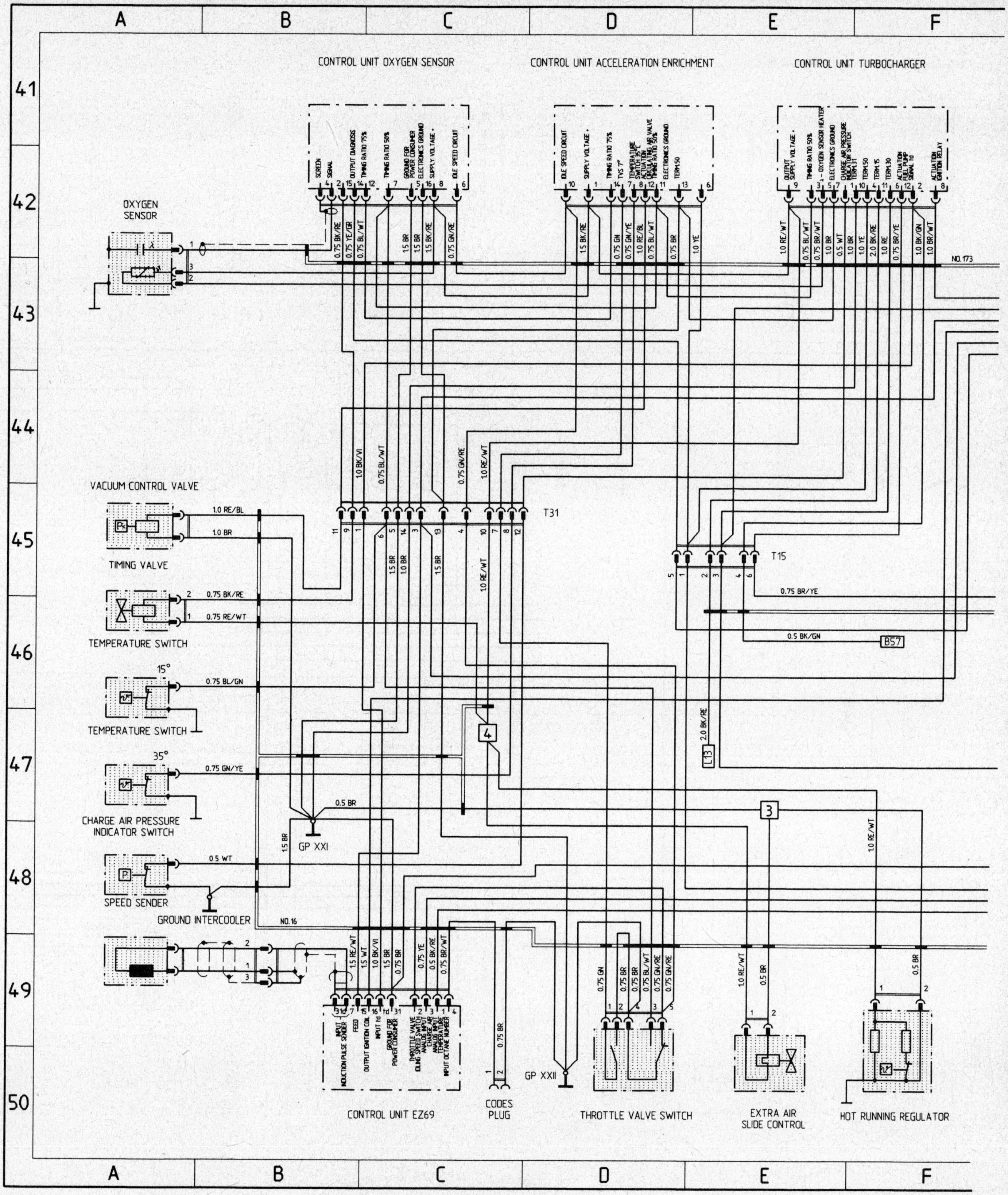

Wiring diagram—Porsche 911 Turbo

BK = BLACK

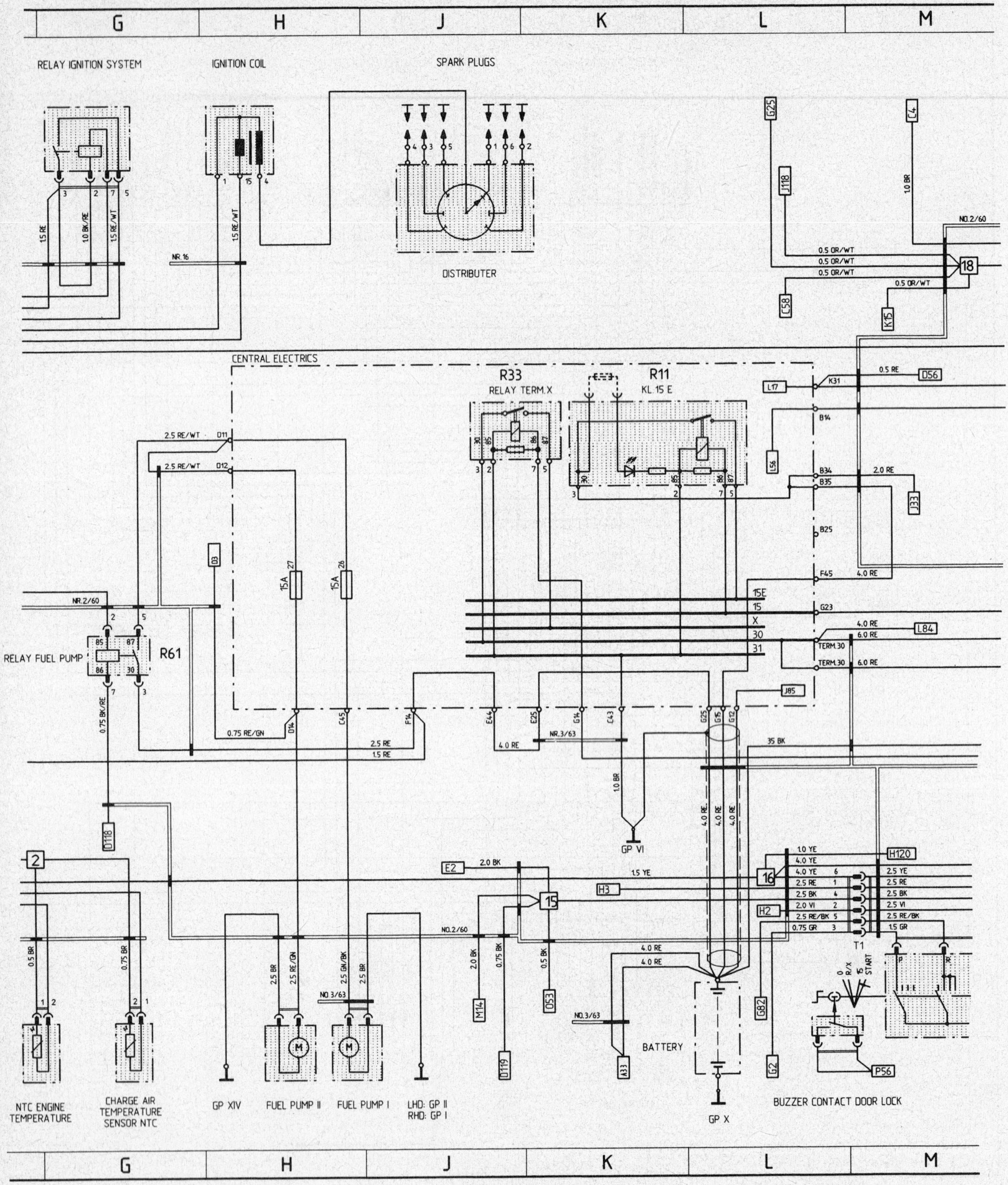

WT = WHITE RE = RED GN = GREEN YE = YELLOW GR = GREY BR = BROWN BL = BLUE VI = VIOLET PK = PINK OR = ORANGE

Wiring diagram— Porsche 911 Turbo

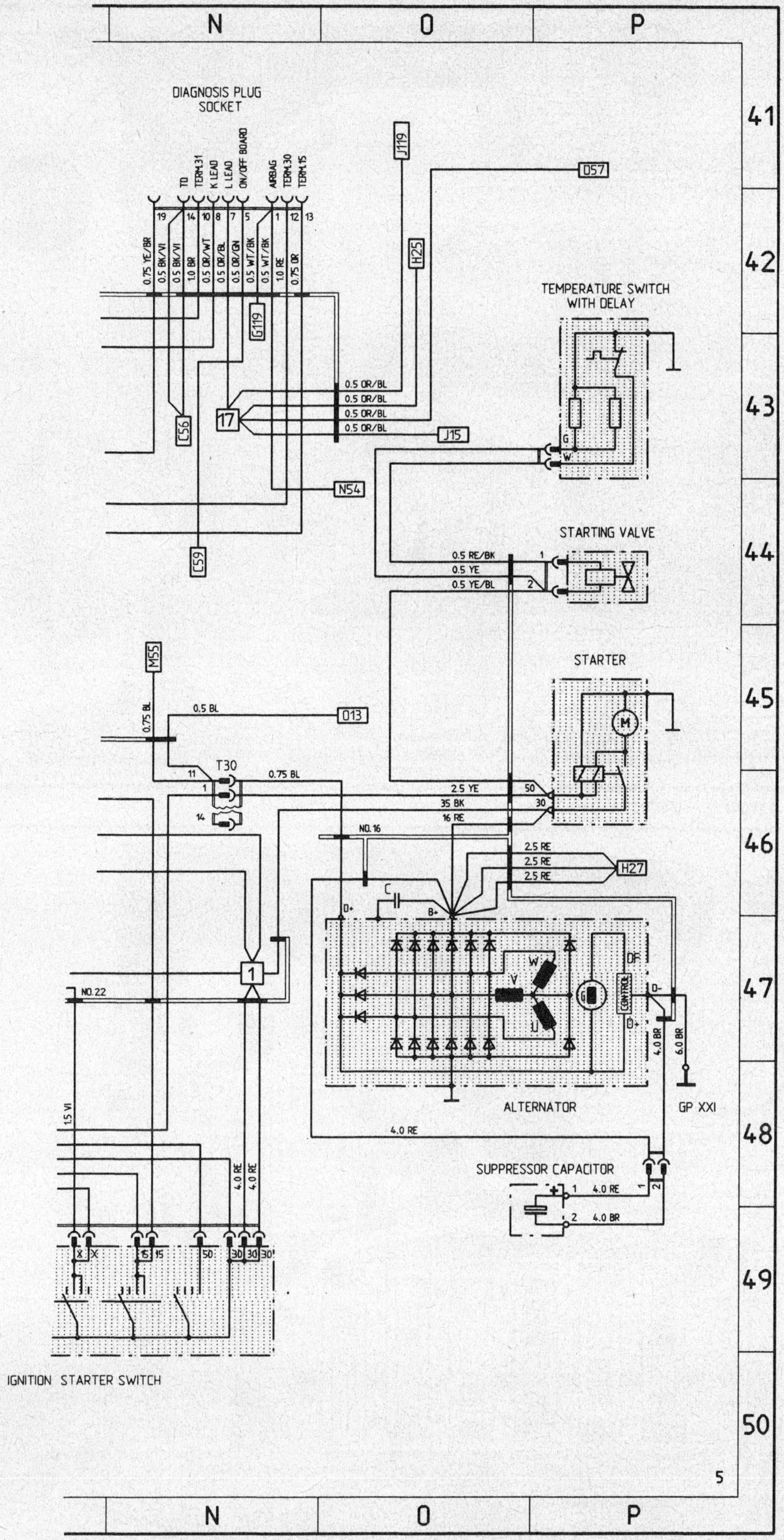

Wiring diagram— Porsche 911 Turbo

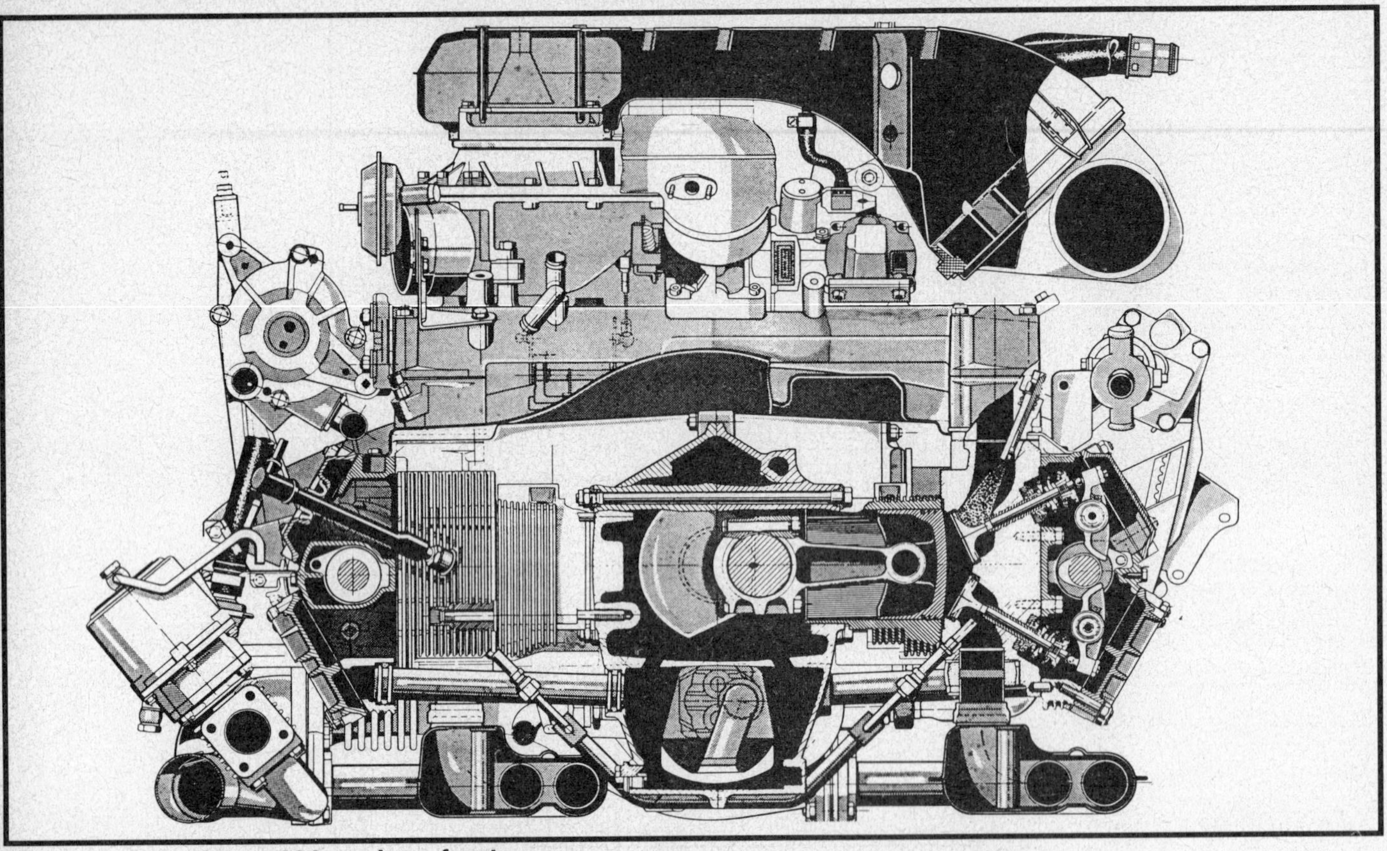

Cutaway view of type 930 engine—front

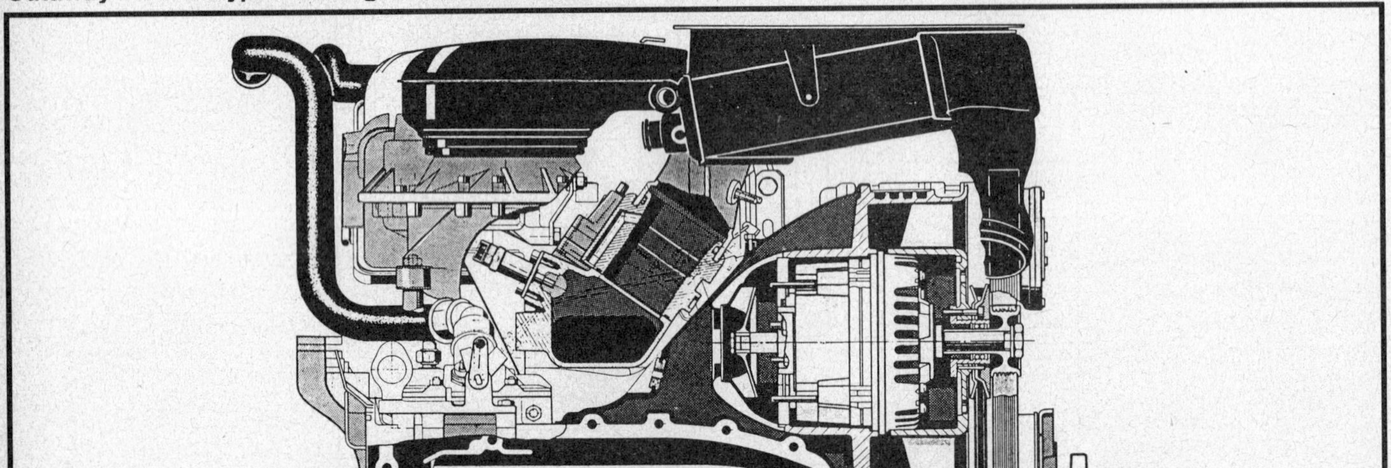

Cutaway view of type 930 engine—side

Saab Fuel Injection

SECTION 13

SAAB LH-JETRONIC FUEL INJECTION SYSTEM

General Information

The Saab 900 and 9000 are both equipped with Bosch LH-Jetronic fuel injection systems. The 1990 models and in some markets 1991–92 models are equipped with an LH 2.4 fuel system Electronic Control Unit (ECU). Most 1991–92 models are equipped with an LH 2.4.2 fuel system ECU, except the 1991 9000 with the B234 engine, which has an LH 2.4.1 fuel system ECU. Turbocharged 900 models are also equipped with an Automatic Performance Control (APC) ECU and 9000 models are equipped with an integrated Direct Ignition-Automatic Performance Control (DI/APC) control unit, which control ignition functions and turbocharger-related functions.

The LH-system ECU has the particular fuel system identification marked on it for identification. Visually, the LH 2.4 fuel system can be differentiated from the LH 2.4.2 fuel system by the pins on the Automatic Idle Control (AIC) valve. The LH 2.4 fuel system AIC valve has 2 pins and the LH 2.4.2 fuel system AIC valve has 3 pins.

The LH 2.4.1 fuel system differs from the LH 2.4 fuel system in that the cold start injector has been discontinued. The direct ignition system has taken over this function. Also, the vehicle speed sensor is used in the control of the AIC function to tell the fuel system ECU whether the car is moving or at a standstill.

The LH 2.4 (some markets) and LH 2.4.2 fuel systems with the Electronic Throttle System (ETS), used on the 1992 9000 with traction control, differ from the systems without ETS in the following ways: The electronically controlled throttle eliminates the need for automatic idling and load control throughout the load range, the throttle angle transmitter is located in the

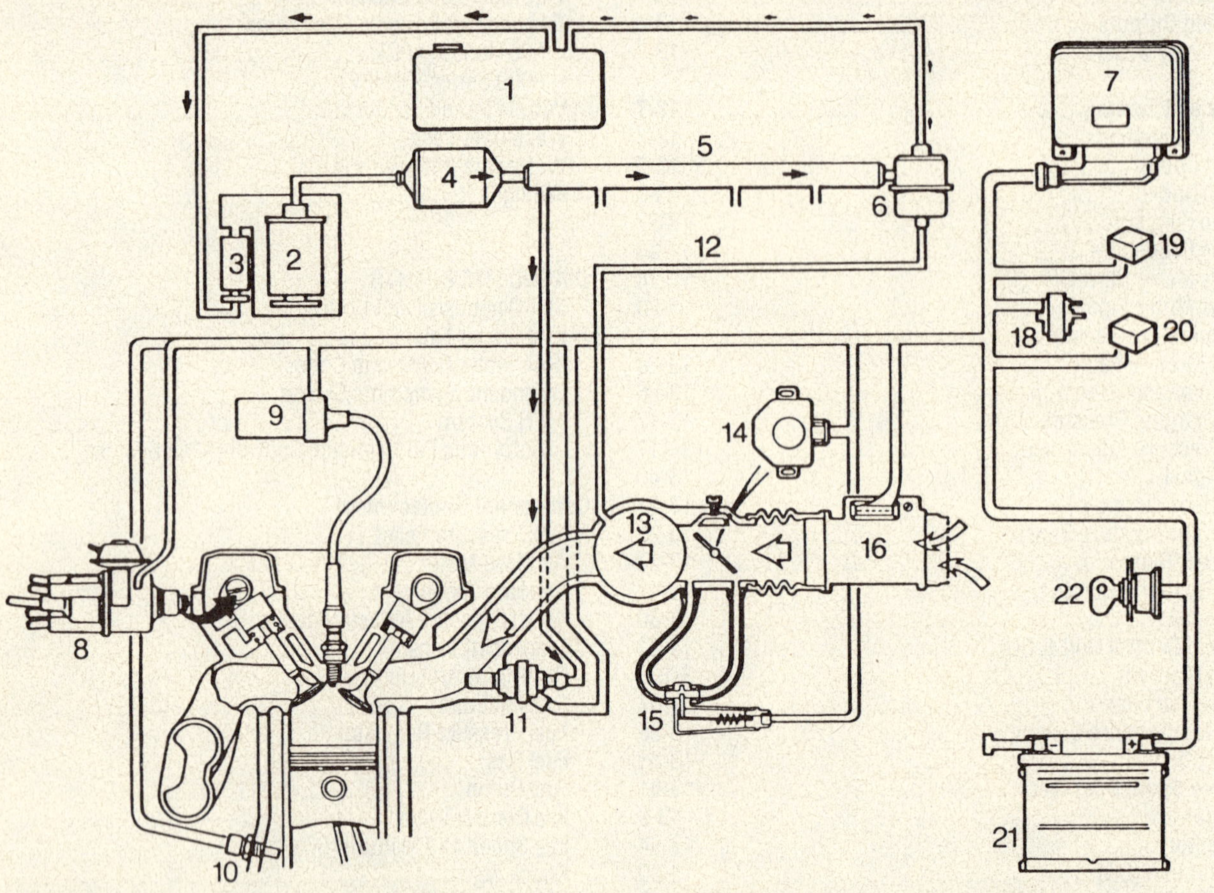

1. Fuel tank
2. Fuel pump
3. Feed pump
4. Fuel filter
5. Fuel injection rail
6. Fuel pressure regulator
7. ECU
8. Distributor
9. Ignition coil
10. Temperature sensor
11. Injector
12. Vacuum line
13. Inlet manifold
14. Throttle Position Sensor (TPS)
15. Auxiliary air valve Idling Control (AIC) valve—later vehicles
16. Air mass meter
17. Oxygen (Lambda) sensor
18. Pressure switch (turbo only)
19. System relay
20. Fuel pump relay
21. Battery
22. Ignition switch

Fuel injection system schematic—1990 Saab with LH 2.4 fuel system shown

actuator motor, which is integrated with the throttle housing and the electronically controlled throttle system carries out compensation for the air conditioning. Vehicles with ETS also have an Automatic Slip Reduction (ASR) control unit that carries out other traction control system functions.

Visually, the throttle housing on engines with ETS is larger, is vacuum operated and has an emergency cable, but in other respects this system operates in the same way as the LH 2.4 and 2.4.2 systems without ETS.

The central component of the LH-Jetronic fuel injection system is the air mass meter that measures the mass of air flow instead of the volume. The microprocessor measures how much electrical energy is used when air flow pass an electrically heated platinum wire in the air mass meter. The higher the rate of air flow, the higher the energy necessary to keep the temperature of the wire constant. At the same time, the microprocessor monitors the engine speed and temperature, calculating the exact amount of fuel needed for optimum performance. The microprocessor also incorporates an rpm limiter that ensures that no opening signals will be transmitted to the injectors at engine speeds above 6000 rpm.

The LH-Jetronic fuel injection system provides the air mass meter with a self-cleaning function. During burn-off the platinum wire in the air mass meter is quickly heated to about 1800°F (1000°C) for a 1 second duration, 4 seconds after the ignition is switched OFF. This burns away any deposits on the wire that would be detrimental to efficient operation.

The APC system on turbocharged vehicles enables the engine to achieve optimum performance and good fuel economy, regardless of the grade of fuel being used. A knock sensor, in-conjunction with the pressure transducer and ignition system information, detects knocking in the engine and sends an electrical signal to the microprocessor inside the ECU. The ECU processes these signals and sends electrical pulses to a solenoid valve that controls the charging pressure in the intake manifold. The turbocharger is designed to come into operation at fairly low engine speeds, thereby providing a high torque within the speed range of normal driving. It is water cooled and the coolant for the bearing housing is supplied by a pipe connected to the cooling system.

Charging pressure is regulated by a pressure regulator valve (known as a wastegate). The charging pressure regulator is fitted to the exhaust side of the engine and regulates the flow of exhaust gas to the compressor. The valve remains closed when the engine load is low. As the demand on the engine is increased, the wastegate opens.

The DI/APC system was updated in 1991 to include an air temperature sensor, located upstream of the throttle housing. The boost pressure is governed by the position of the throttle valve, but it is subject to temperature compensation based on information supplied by the new air temperature sensor. The separate pressure-switch function is discontinued, with the pressure-sensing function now being regulated by the DI/APC-system ECU. The load signal provided by the air mass meter is sent to the LH-system ECU, which will keep the DI/APC system ECU informed of boost status.

Starting in 1991 there is also a spark plug burn-off function that occurs when the engine is stopped. The burn-off function, which operates in all cylinders simultaneously, lasts for 5 seconds at a frequency corresponding to 6000 rpm.

Turbocharged California vehicles, except the 1991–92 9000 Turbo with the B234 engine, are equipped with an electronically controlled EGR system. A modulating valve functions as a 3-way valve as it controls the vacuum to the EGR valve. A vacuum regulator is incorporated in the modulating valve to maintain a constant vacuum to the EGR valve. A vacuum storage tank is connected via a vacuum check valve to the intake manifold. The check valve prevents the loss of vacuum in the tank during ac-

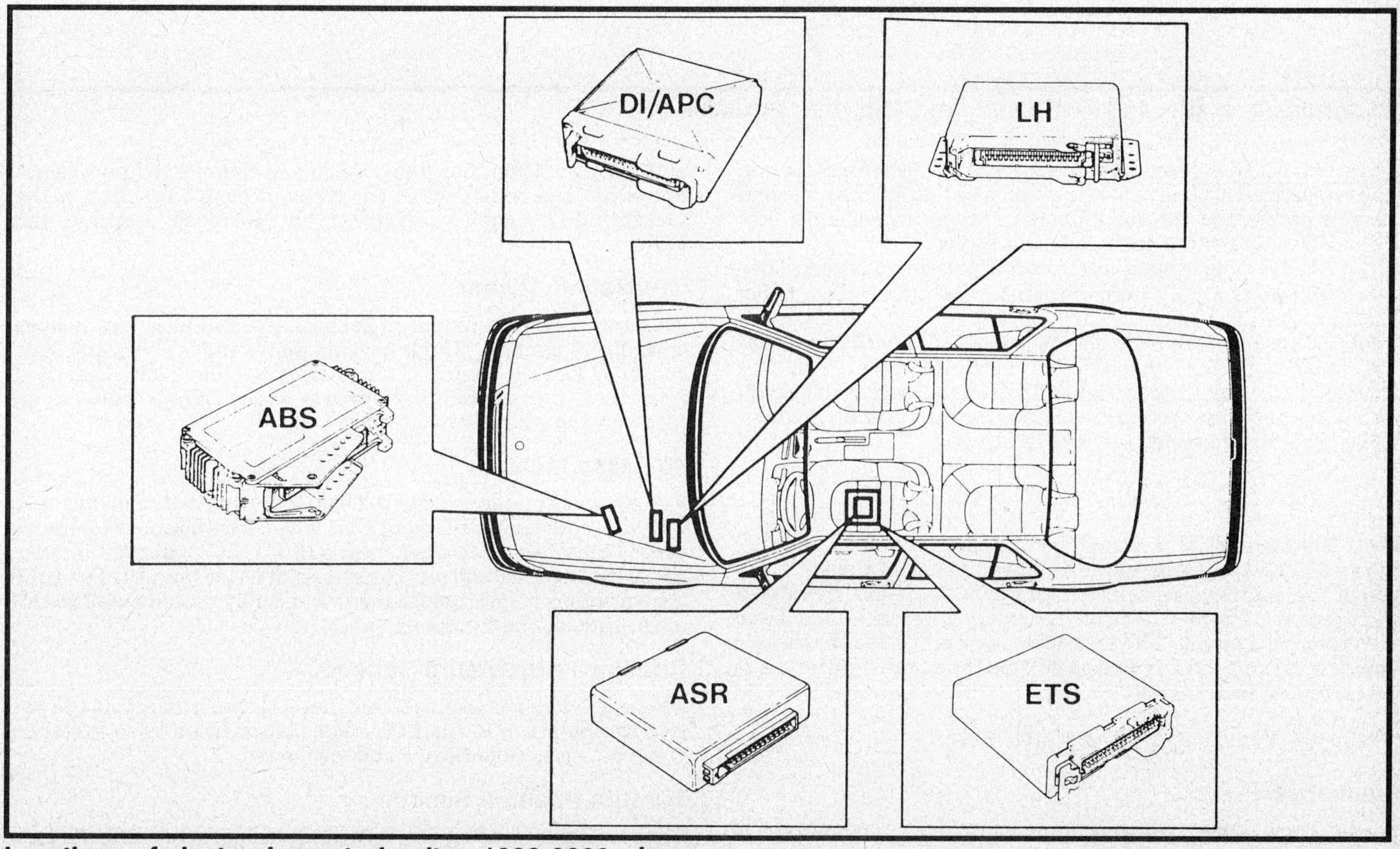

Locations of electronic control units—1992 9000 shown

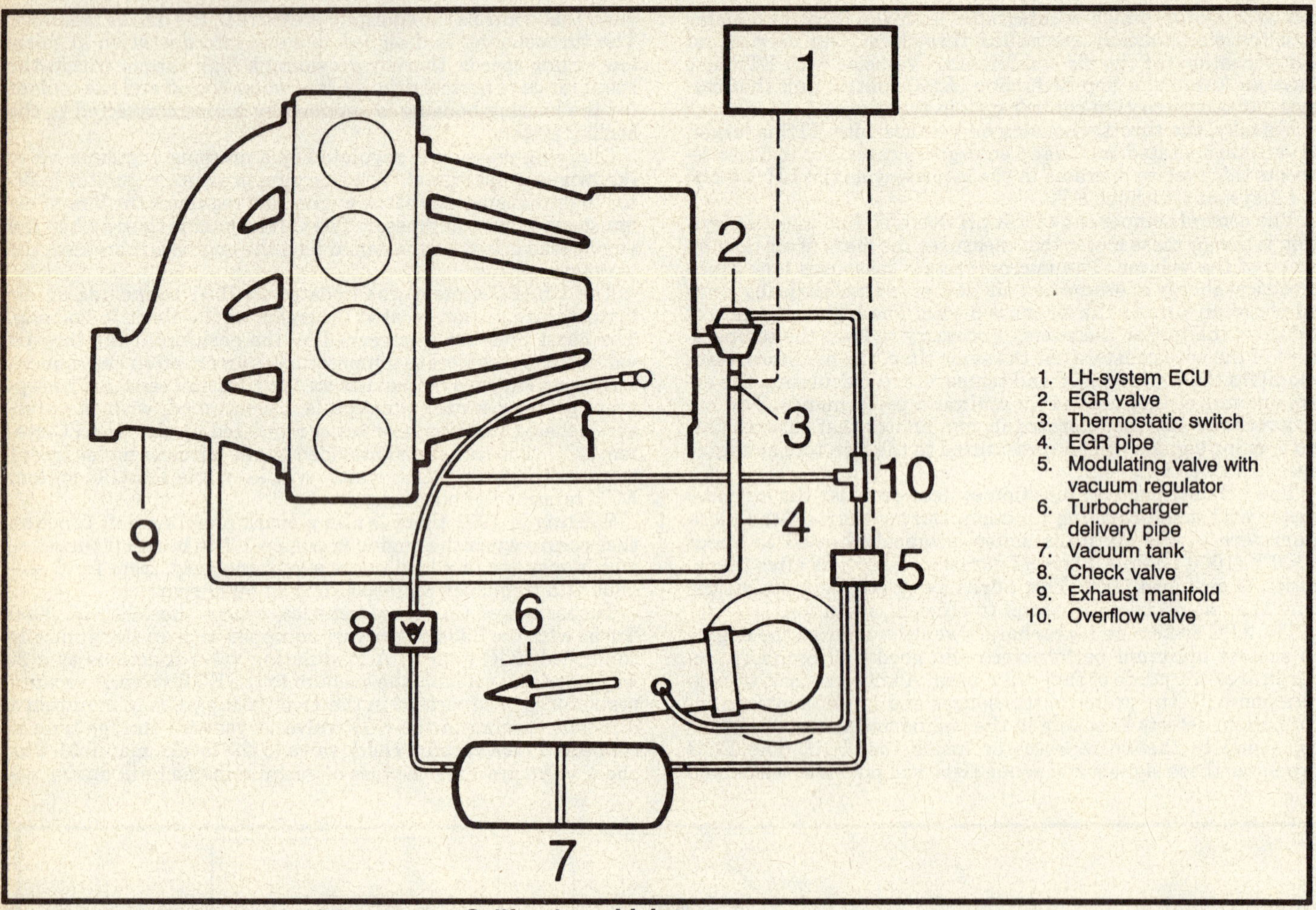

Electronically controlled EGR system—California vehicles

celeration. An EGR temperature sensor provides information to the LH-system ECU. If the temperature deviates from a normal range, a problem is indicated due to improper exhaust gas flow in the EGR pipe and a fault code will be set.

The LH-Jetronic system also incorporates an emergency system known as a limp home function. If a malfunction is detected, the limp home feature of the ECU is actuated, enabling the vehicle to continue its journey, but with somewhat diminished performance. If the vehicle is operated in this mode, the Check Engine Light (CEL) on the display panel will be illuminated. An integrated fault-storing capability enables diagnosis to be carried out efficiently.

ELECTRONIC CONTROL UNIT

The LH-system ECU is located behind the right-side kickpanel on the 900 and adjacent to the bulkhead panel on the left hand side of the engine compartment on the 9000. It incorporates a microprocessor into which all necessary reference data have been preprogrammed. The microprocessor processes the signals from the various sensors and calculates the opening duration of the injectors, for example.

DATA SENSORS

Pressure Sensor

The pressure sensor, which is found on turbocharged models, is mounted on a bracket underneath the left side instrument panel. It monitors the charging pressure so that a maximum is not exceeded. The resistance of the transducer is a function of the pressure and a signal indicating the resistance is sent to the ECU.

Crankshaft sensor

This component is found on turbocharged models and models with direct ignition, which have no distributor. The crankshaft sensor is located behind the crankshaft pulley and the slotted rotor that passes through the sensor as the engine rotates is located on the pulley itself.

Air Mass Meter

The air mass meter is a device located in the air-intake duct and contains a platinum filament (hot wire). The filament temperature is maintained at a constant 212°F (100°C) above the temperature of the intake air, regardless of the humidity. The voltage variation across the filament is directly proportional to the magnitude of the intake air mass.

Coolant Temperature Sensor

This is a device that sends a continuous signal indicating the engine temperature to the ECU. The sensor has a low resistance value at high temperature and vice versa.

Throttle Position Sensor

This device, located on the throttle housing, sends information to the ECU on the position (angle) of the throttle valve. It is a

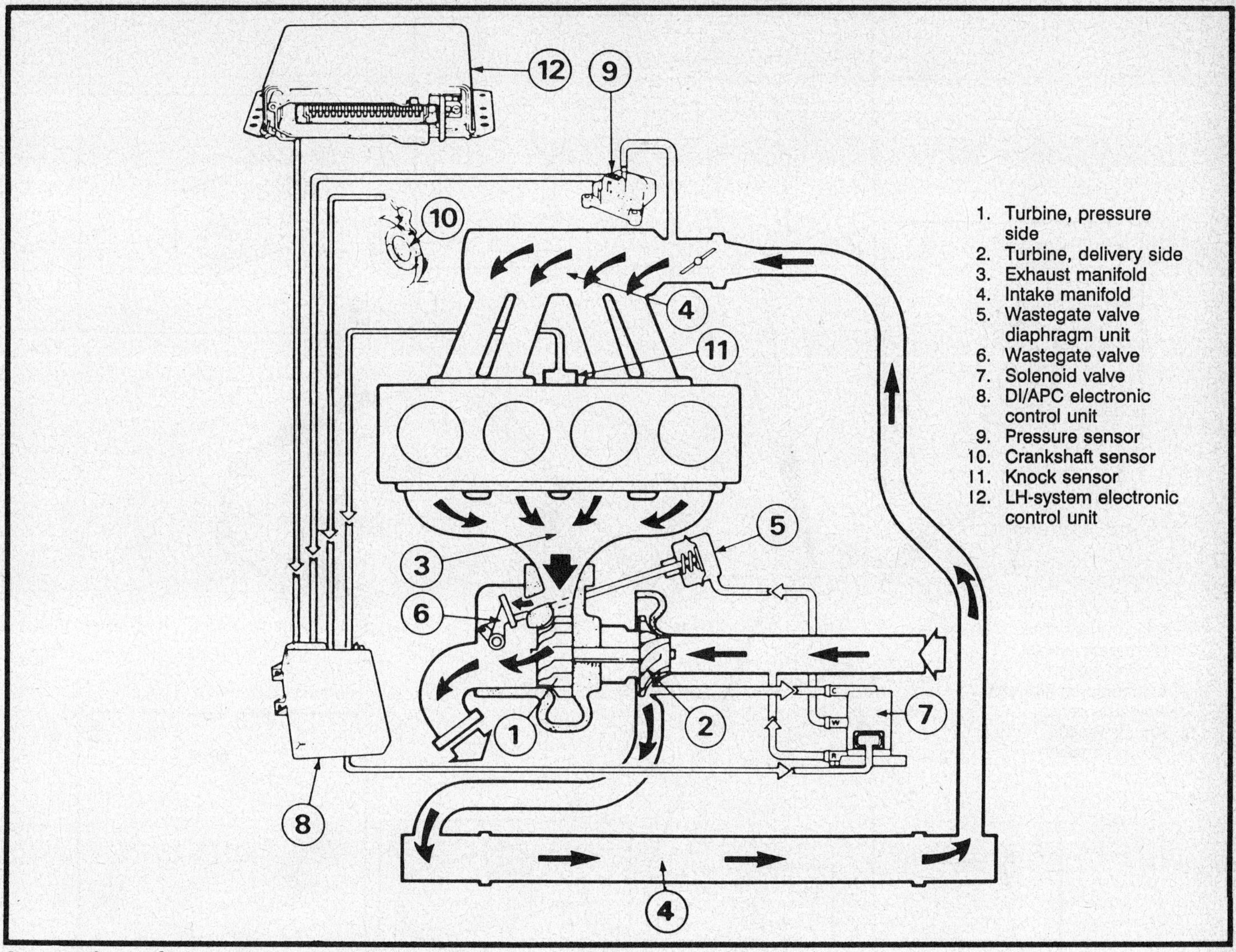

Automatic performance control system schematic— 9000 Turbo shown

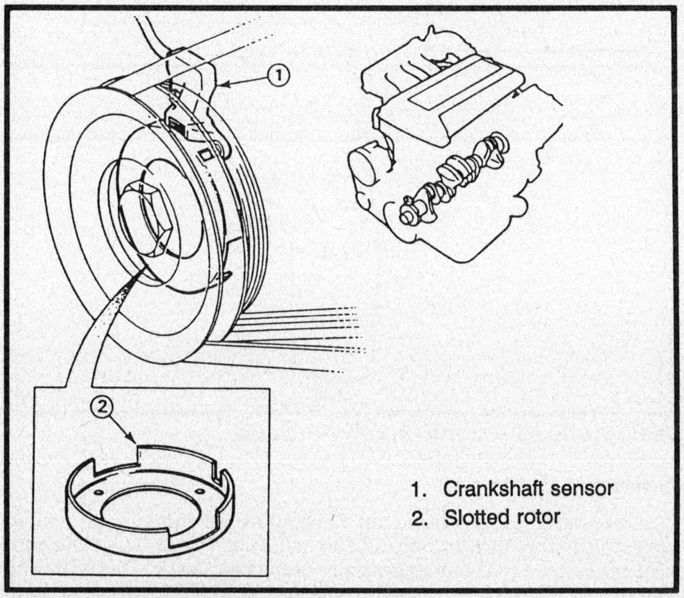

Crankshaft sensor—Saab

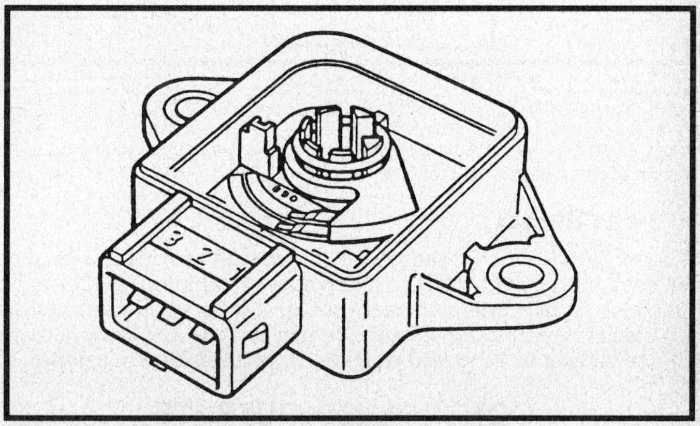

Throttle position sensor—Saab

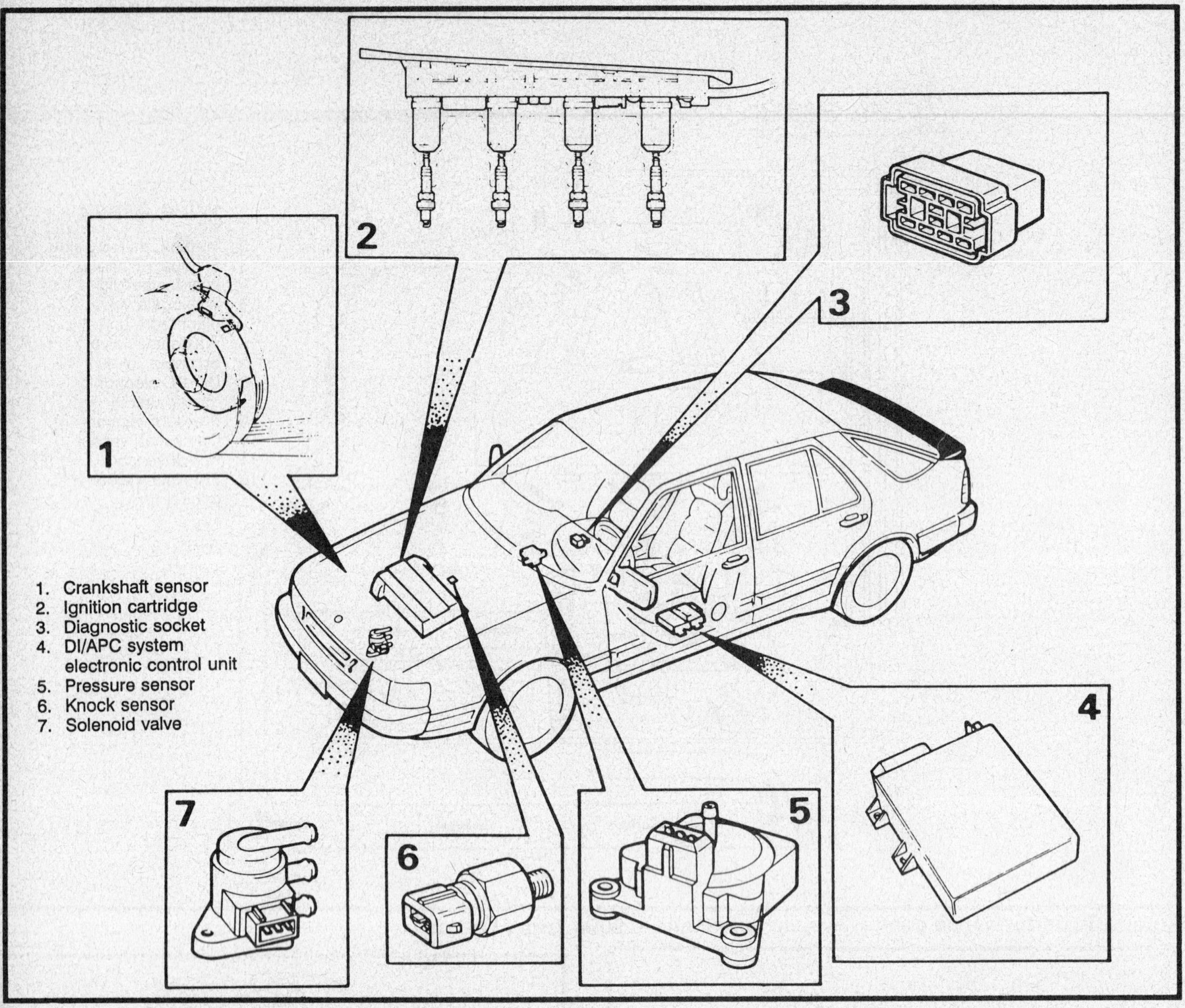

Locations of DI/APC components—1990 9000 Turbo shown

switch on the LH 2.4 fuel system and a variable resistor fuel systems other than the LH 2.4.

Oxygen Sensor

The oxygen (Lambda) sensor is a device that provides a signal to be evaluated by the ECU to determine the amount of fuel to be injected to maintain a correct fuel-air mixture. The sensor has an electric preheating element, which enables the temperature of the sensor to be raised quickly, thus lowering emissions.

CONTROLLED OUTPUTS

Automatic Idling Control Valve

The AIC valve controls the flow of air by-passing the throttle valve. The volume of air-flow is determined by the size of the opening in the actuator which is controlled by the ECU. The AIC valve is spring-loaded so that in the event of a fault, it will be put in the limp home mode to set an idling speed of about 850 rpm when there is no load on the engine.

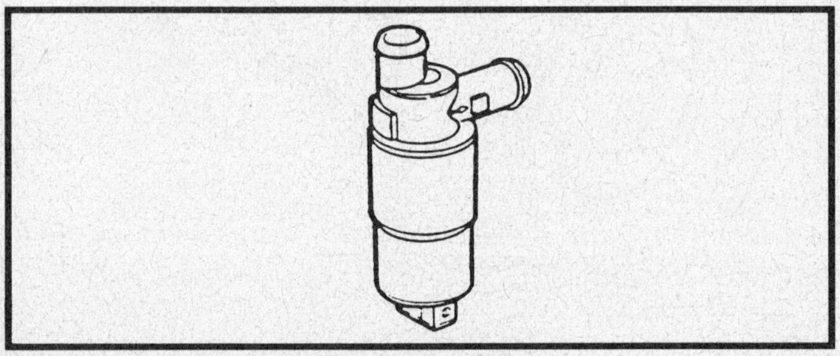

Automatic idle control valve—Saab

Solenoid Valve

The solenoid valve, found on turbocharged models, is located next to the fan shroud behind the radiator. It regulates the control pressure for the charging pressure regulator. It is inline to the pressure present in the intake manifold that is downstream of the turbocharger compressor.

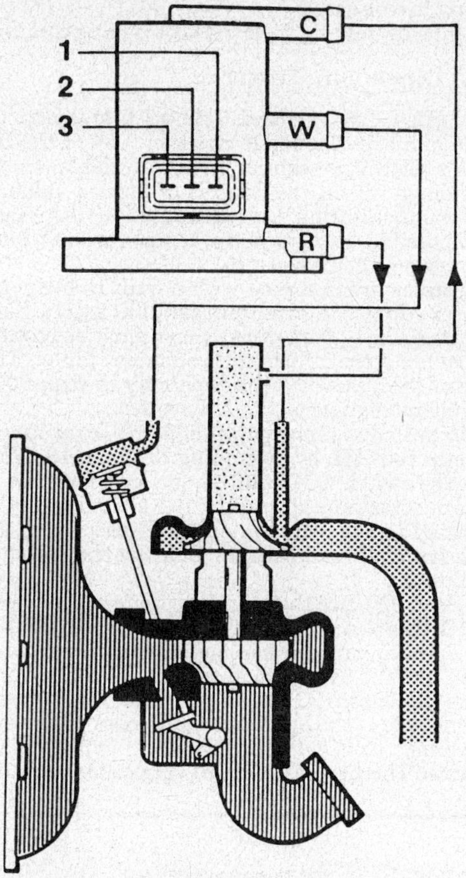

Solenoid valve connections

W = To waste-gate valve
R = To turbo inlet
C = To turbo discharge side

Solenoid valve connections—Turbocharged Saab

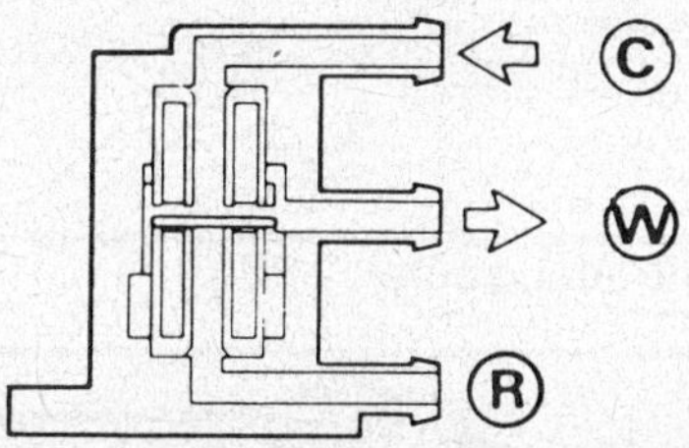

Solenoid valve vacuum passages—Turbocharged Saab

FUEL DELIVERY SYSTEM

An in-tank mounted electric fuel pump delivers fuel from the fuel tank, pressurizing the fuel system. The pressure is kept relatively constant by the fuel pressure regulator. The quantity of fuel injected is therefore governed only by the amount of time the injector is open and not by pressure variations in the intake manifold.

Fuel Injector

The fuel is delivered to the cylinders by injectors (one for each cylinder) that are located in the intake manifold adjacent to the intake valves. The injectors all open simultaneously, once for each engine revolution, except on starting, when they open twice per engine revolution. The length of time that the injectors operate is determined in part by the engine load and speed monitored by the ECU.

Fuel Pressure Regulator

The fuel pressure regulator, located at the end of the fuel injection rail, is a device that maintains the necessary pressure needed for efficient operation of the engine under all load conditions. In the turbocharged engines, the fuel pressure is maintained at a constant 36 psi (248 kPa) and in the normally aspirated engines 43 psi (296 kPa).

Fuel Pump

The fuel pump is an electrically driven rotary pump and is fitted inside the fuel tank. The pump incorporates a relief valve which comes into operation if the pressure exceeds its preset value. A check valve in the delivery connection of the pump prevents the fuel line from losing pressure immediately after the pump has stopped.

Evaporative Loss Control Device

The Evaporative Loss Control Device (ELCD) is basically a filter canister controlled by a signal from the ECU. The filter canister traps gaseous hydrocarbons from the fuel tank in its activated charcoal filter for subsequent burning in the engine on purging of the filter. The purging function is dependent on engine load and engine speed.

Diagnosis and Testing

SERVICE PRECAUTIONS

- Always disconnect the negative battery terminal to prevent sparks caused by short-circuiting, circuit-breaking, etc.
- Always keep a CO_2 fire extinguisher close on hand.
- Dry sand must be available to soak up any spillage.
- If using a battery charger, make sure that the battery leads are disconnected.
- Never disconnect the battery while the engine is running.
- Never disconnect or connect an ECU connector with the ignition switch in the **ON** position.
- Always remove the control units from the vehicle before carrying out any electrical welding.
- Before conducting a compression test, unplug the connector on the output module.
- Never subject the vehicle to temperatures exceeding 176°F (80°C), such as in a paint booth, unless the control units have first been removed from the vehicle.

READING CODES

Code retrieval is important and necessary for efficient repair of the vehicle. Diagnosis based on symptoms should be viewed as supplementary and not as the primary basis for a course of action.

There are 3 methods for reading codes on the Saab. Two are carried out with scanner-like instruments, and the third uses the CEL. These methods yield 3 different classes of codes. For example, ISAT type codes cannot be retrieved by using the CEL, nor can CEL type codes be retrieved with the ISAT.

With LH-System Tester

The Saab LH-System Tester 8394223 was developed to simplify

service and fault diagnosis work on the LH fuel injection system. The ISAT is now most widely used, but the System Tester can still be used. The tester consists of a test unit, power supply lead, test lead incorporating a 2-way 35-pin connector and a pressure sensor with magnetic base.

The tester is equipped with an automatic program for diagnosing faults, both permanent and intermittent, while the vehicle is operating. Faults detected are then stored in memory for recall after the vehicle is road-tested.

Connect the diagnostic tester, or equivalent, as follows:

1. Insert the power supply lead between the door and body where there is a break in the seal, then run it under the back of the hood on the left hand side.
2. Clean the battery terminals to ensure proper contact with lead clips.
3. Connect the power supply lead to the tester first, then connect the lead clips to the battery.
4. Remove the cover on the left side over the space behind the false bulkhead panel.
5. Remove the antilock brake system (ABS) ECU and bracket, if equipped.
6. Remove the LH-system ECU connector. Interface the test lead between the LH-system ECU and the connector. Fit a couple of wire ties around the connector and ECU to hold them tightly together.

The tester, designed to perform 3 basic functions, monitor mode, test mode and fuel mode, is now ready to start. If at any time the operation in progress must be interrupted, simultaneously press all 3 control buttons and the tester will revert to the starting point.

MONITOR MODE

The monitor mode can be selected either by switching the ignition **ON**, or if the ignition is **OFF** by pressing the start test button when MON (monitor) appears on the tester display.

When in monitor mode, the tester is used to manually control parameter and functional checks.

TEST MODE

Test mode can only be selected from the monitor mode. Press the start test button to select the test mode. TEST will now appear on the display. In the test mode, the program instructs by way of prompts in the upper part of the display.

FUEL MODE

To select the fuel mode, the ignition switch must be **OFF**. Initially, MON will be displayed on the tester for approximately 5 seconds. During this time the monitor mode can be selected. If none of the tester buttons is activated, FUEL will appear on the display for approximately 2 seconds. To activate the fuel mode, press the start button when FUEL is displayed.

When the tester is in fuel mode the following checks can be performed:

Fuel pump delivery flow.
Fuel pressure and fuel-pressure regulator.
Residual pressure.
Fuel pump delivery pressure.
Delivery flow from injectors.

With SAAB ISAT Tester

The Saab ISAT tester is also available starting in 1990 to extract hard or intermittent fault codes or to issue command codes. This makes fault tracing much easier and therefore enables diagnosis to be carried out more efficiently. ISAT Mode 1 is for fuel system testing Mode 2 is for DI/APC testing.

The diagnostic socket is a black 10-pin connector located underneath the RH front seat. It is protected by a plastic cover held in place by a screw. Read codes with the ISAT tester, or equivalent, as follows:

1. Connect the tester to the diagnostic socket.
2. Turn the ignition to the **ON** position.
3. Press No.1 on the ISAT tester to identify that the LH-system is being investigated.
4. Continue by reading fault codes or issuing command codes.

Without a Diagnostic Tester

The Saab LH2.4, LH2.4.1 and LH2.4.2 fuel injection systems are capable of a self-diagnostic check by way of the flashing of the vehicle's CEL if a scanner is not available. Using such a method of diagnosis on the Saab is of limited value, however, since external monitoring and manipulation of engine parameters is not possible. Furthermore, temporary and intermittent faults are more difficult to trace.

These systems are designed with a built in memory and the ability to store up to 3 intermittent faults at a time. Serious malfunctions are always given priority and must be rectified before the memory can store information on minor faults. The built in diagnosis function also has the capability to manually test the components and signals of the LH system.

The Saab switched jumper lead 8393886, or a simple ground wire activates the CEL by grounding ECU pin 16. When directed to turn the switch **ON**, connect the wire between the pin in the test connector and ground, if the switch is not available. When directed to turn the switch **OFF** disconnect the ground wire, if the switch is not available. Connect the switched jumper lead as follows:

For the Saab 900, use the switched jumper lead to connect the No. 3 pin (Green/Red wire) in the 3-pin test socket on the right-hand side in the engine compartment to battery ground. For the Saab 9000, use the switched jumper lead or a ground wire to connect the single-pin (Green/Red wire) socket in the test box on the left-hand side of the engine compartment to battery ground. The test connector is number 204 in the wiring diagrams for both vehicles. The process of retrieving codes can now begin.

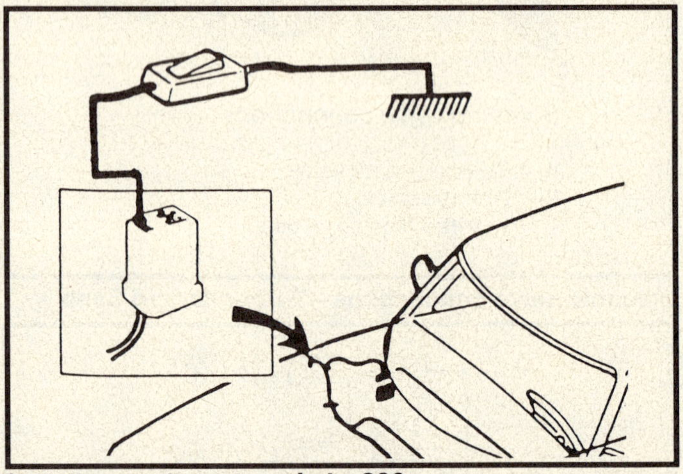

Self-diagnostic test socket—900

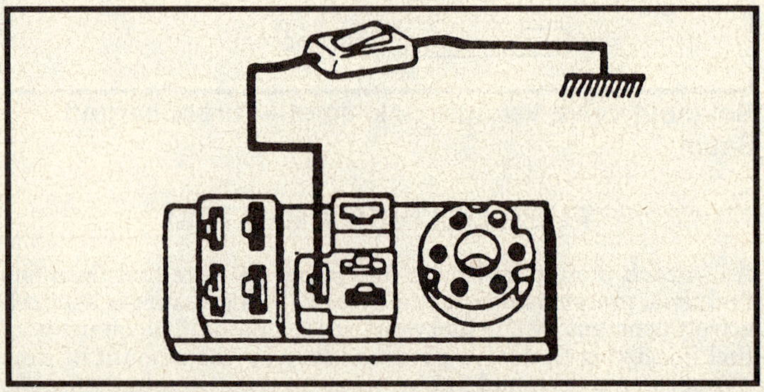

Self-diagnostic test socket—9000

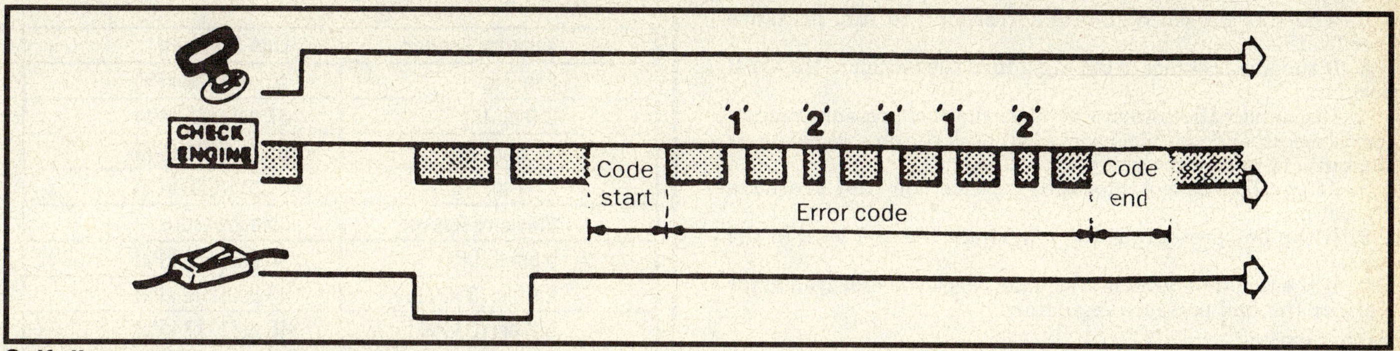

Self-diagnostic code 12112 example—Saab

1. Switch the ignition **ON**. The CEL should now illuminate.
2. Set the jumper switch to **ON** (grounding ECU pin 16). The CEL should now be extinguished.
3. Watch the CEL carefully. After about 2.5 seconds, it will flash briefly, signifying that the first error code is about to display.
4. As soon as the light has flashed, turn the jumper switch **OFF**.
5. The first of a possible 3 error codes will now be displayed by a series of short flashes. The number 1 is represented by a single flash followed by a long pause. The number 2 is represented by 2 flashes separated from each other by a short pause, but separated from the next number by a long pause. The number 3 would consist of 3 flashes separated by short pauses and followed by a long pause, and so on. The code will be displayed repeatedly until the next test step is taken.
6. To check for any additional error codes, turn the jumper switch **ON**.
7. Watch the CEL carefully. After a short flash, turn the jumper switch **OFF**.
8. If present, the next error code will now display.
9. Follow the same procedure until all faults have been identified and corrected.
10. To restart the test procedure (return to the first fault), set the jumper switch to **ON**.
11. After 2 short flashes, turn the jumper switch **OFF** or disconnect the ground wire. The fault code for the first fault should now be displayed.
12. Proceed with the test from Step 5.

TESTING COMPONENTS AND SIGNALS

1. Connect the jumper lead in the same manner as for reading fault codes.
2. Set the jumper switch to **ON**.
3. Turn the ignition switch **ON** and wait for a short flash of the CEL.
4. Immediately following the flash, turn the jumper switch **OFF**.
5. The moment the CEL begins flashing, the fuel pump should begin running
6. To move on to the next test, set the jumper switch to **ON**.
7. After a short flash, set the jumper switch to **OFF**. A test code (not a fault code) will be displayed and the corresponding component will activate.
8. Continue through the remaining items in the test sequence in the same method (set switch to **ON**, wait for a short flash, set the switch to **OFF**).

Components and signals are checked in the following order:
Fuel pump (no code displayed),
Injection valves (1.5 ms–10 Hz),
AIC valve (switches between open and closed positions),
ELCD valve (switches between closed and open)—CEL flashing stops,
Drive signal (changes when shifting from **D** to **N**)—CEL flashing stops,
Air conditioning signal (cannot be tested),
Idling signal (changes as accelerator is depressed)—CEL flashing stops,
Full-throttle signal (changes as accelerator is pressed down to the floor)—CEL flashing stops.

CLEARING CODES

With LH-System Tester

After completing repairs, reset the tester to the start mode by pressing all 3 control buttons simultaneously. If fault codes are still found, disconnect the battery for at least 10 minutes or road test vehicle for 10 minutes or until the CEL extinguishes. The fault code should erase itself after extended operation with the repaired component.

With SAAB ISAT Tester

After repairs are completed, the ISAT command Code 900 is used to clear diagnostic memory.

Without LH-System Tester

1. Set the jumper switch to **ON**.
2. After 3 short flashes, turn the jumper switch **OFF**.
3. The CEL will now either flash in a continuous series of long flashes (this represents Code 00000) or display the Code 12444, indicating that the contents of the memory have been erased.

FUEL SYSTEM TESTING

Releasing Fuel Pressure

NOTE: Soak up any escaping fuel with a shop towel and dispose of it in the proper container.

1. Disconnect the negative battery cable.
2. Loosen the banjo fitting on the fuel injection rail to release the pressure in the system.
3. Retighten the banjo fitting.

System Pressure

NOTE: The fuel tank level must be at least ¾ full before performing this test.

900 SERIES

1. Disconnect the banjo coupling at the inlet to the fuel injection manifold and connect the fuel system pressure gauge 8393852 or equivalent using the existing seals. Support the pressure gauge so that in hangs vertically.
2. Remove fuse 30. Connect jumper lead 8393886 (jumper lead with switch) or equivalent between fuse 30 and fuse 27, 28 or 29.
3. Make sure the switch is in the **OFF** position.
4. Start the pump by setting the switch to the **ON** position.

5. The line pressure should correspond to fuel pressure specifications.
6. If the line pressure is too high, turn the switch to the **OFF** position.
7. Disconnect the return line from the fuel pressure regulator. Check the line for blockage. Also check the one-way valve at the tank. Repair if necessary.
8. If the line is not blocked, replace the fuel pressure regulator.
9. If the line pressure is too low, check the fuel level in the tank.
10. If the fuel tank level is adequate, check the fuel system for leaks or the fuel pressure regulator.

9000 SERIES

1. Disconnect the banjo coupling from the bottom of the fuel filter and connect the fuel system pressure gauge 8393852 or equivalent using the existing seals. Support the pressure gauge so that in hangs vertically.
2. Remove fuses 14 and 22 from the panel inside the glove compartment and connect jumper lead 8393886 (jumper lead with switch) or equivalent.
3. Make sure the switch is in the **OFF** position.
4. Start the pump by setting the switch to the **ON** position.
5. The line pressure should correspond to fuel pressure specifications.
6. If the line pressure is too high, turn the switch to the **OFF** position.
7. Disconnect the return line from the fuel pressure regulator. Check the line for blockage. Repair if necessary.
8. If the line is not blocked, replace the fuel pressure regulator.
9. If the line pressure is too low, check the fuel level in the tank.
10. If the fuel tank level is adequate, check the fuel system for leaks or the fuel pressure regulator.

Pressure Regulator

NOTE: The fuel tank level must be at least ¾ full before performing this test.

900 SERIES

1. Remove fuse 30. Connect jumper lead 8393886 (jumper lead with switch) or equivalent between fuse 30 and fuse 27, 28 or 29.
2. Make sure the switch is in the **OFF** position.
3. Connect pressure gauge 8393514 or equivalent to the inlet and outlet on the fuel pressure regulator. Support the pressure gauge so that in hangs vertically.
4. Connect a cooling system tester and a vacuum pump to the pressure gauge.
5. Start the pump by setting the switch to the **ON** position.
6. The line pressure should correspond to fuel pressure specifications.
7. Increase the vacuum in the fuel pressure regulator using the vacuum pump. The drop in the line pressure should correspond to fuel pressure specifications.
8. Increase the pressure in the fuel pressure regulator using the cooling system tester. The rise in the line pressure should correspond to fuel pressure specifications.

Vacuum Applied	Line Pressure
0 in. Hg	40 psi (276 kPa)
5.9 in. Hg	37 psi (255 kPa)
11.8 in. Hg	35 psi (241 kPa)
17.7 in. Hg	32 psi (220 kPa)
Pressure Applied	**Line Pressure**
0 psi (0 kPa)	40 psi (276 kPa)
2.9 psi (0 kPa)	43 psi (296 kPa)
5.8 psi (0 kPa)	46 psi (317 kPa)
8.6 psi (0 kPa)	49 psi (338 kPa)

Fuel pressure specifications—Saab Turbo

Vacuum Applied	Line Pressure
0 in. Hg	43 psi (296 kPa)
5.9 in. Hg	40 psi (276 kPa)
11.8 in. Hg	37 psi (255 kPa)
17.7 in. Hg	35 psi (241 kPa)

Fuel pressure specifications—Non-turbocharged Saab

9000 SERIES

1. Remove fuses 14 and 22 from the panel inside the glove compartment and connect jumper lead 8393886 (jumper lead with switch) or equivalent.
2. Make sure the switch is in the **OFF** position.
3. Connect pressure gauge 8393514 or equivalent to the inlet and outlet on the fuel pressure regulator. Support the pressure gauge so that in hangs vertically.
4. Connect a cooling system tester and a vacuum pump to the pressure gauge.
5. Start the pump by setting the switch to the **ON** position.
6. The line pressure should correspond to fuel pressure specifications.
7. Increase the vacuum in the fuel pressure regulator using the vacuum pump. The drop in the line pressure should correspond to fuel pressure specifications.
8. Increase the pressure in the fuel pressure regulator using the cooling system tester. The rise in the line pressure should correspond to fuel pressure specifications.

SELF-DIAGNOSIS FAULT CODES—SAAB

The recommended action given in the following table should be seen as the starting point for the necessary fault-diagnosis work. The self-diagnosis function incorporated in the LH 2.4 system has two important advantages: the first is the ability to store data on intermittent faults; and the second is the ability to check the effect of action taken to rectify a fault.

Error Code	CHECK ENGINE	Malfunciton indicated	Preliminary action
12231	Off	No ignition signal. (Always appears as first error code if engine off. Run starter for 5 s: if code disappears, signal is OK, indicating that fault lies elsewhere.)	Check input signal to ECU pin 1 and wiring from pin 1 to ignition system. Check ignition system.
12221	On	No air mass meter signal. System in Limp-home mode.	Check air mass meter connectors and wiring. Try a known good air mass meter. Take care to distinguish between plastic and aluminium variants.
12214	On	Temperature sensor signal faulty (temp below -90°C or above +160°C).	Check resistance across NTC resistor and between ECU pin 13 and earth. Correct values, with connector unplugged from ECU, are 2280-2720 ohm at 20°C or 290-365 ohm at 80°C
12211	Off	Incorrect battery voltage (below 10 V or above 16 V when engine running).	Check condition of battery, charging system, earthing points, etc
12225	On*	Lambda sensor faulty or preheating defective (engine temp must be above +70°C; M88: +80°C).	Check voltage present across pins of Lambda sensor connector. Check preheater fuse (M88: also that line-fuse connector making good contact). Check that Lambda signal ranges between 0 and 1.5 V (sensor hot). Fit a new Lambda sensor. Try a known good ECU.
12223	On*	Fuel-air mixture lean.	Check all screws, hoses, hose clips, O-rings, etc., to ensure there are no leaks. Check that induction system working properly without leaks. See also error code 12225 above.
12224		Fuel-air mixture rich.	
12232	Off	Memory voltage > 1 V	Check that voltage is present in ECU pin (even when ignition off).

*)Depending on how serious the fault is.

SELF-DIAGNOSIS FAULT CODES—SAAB

Error Code	CHECK ENGINE	Malfunciton Indicated.	Preliminary action
12212	Off	Throttle position sensor: malfunction of idling contacts, shorting to earth when throttle open.	Check and adjust throttle position sensor. Check wiring between sensor and ECU for shorting to earth. Try a new sensor. Try a known good ECU.
12213	Off	Throttle-position sensor: malfunction of full-throttle contacts, shorting to earth on low engine load (idling).	Check and adjust throttle-position sensor. Check wiring between sensor and ECU for shorting to earth. Try a new sensor. Try a known good ECU
12222	Off	AIC system faulty.	Check connections and wiring to AIC valve. Fit a new valve. Try a known good ECU.
12111	Off*	Lambda adaptation fault (fuel-air mixture with throttle open).	Check system for air or fuel leaks and Lambda sensor preheater function. Check that induction system working properly without leaks. Try a known good ECU.
12112	Off*	Lambda adaptation fault (fuel-air mixture on idling).	Check system for air or fuel leaks and Lambda sensor preheater function. Check that induction system working properly without leaks. Try a known good ECU.
12113	Off	AIC adaptation fault, pulse ratio too low.	Inspect and adjust throttle butterfly and housing and check for leaks. Fit a new AIC valve. Try a known good ECU.
12114	Off	AIC adaptation fault, pulse ratio too high.	Check for sticking AIC valve. Check for other mechanical defects. Try a known good ECU.
00000	Off	No more error codes or not faults detected.	The memory cannot be deleted until this code has been displayed.

*)Depending on how serious the fault is.

When faults in the LH 2.4 system can be classed as 'adaptation faults', the car must be run for about ten minutes to prevent the same error codes recurring after the fault has been rectified. This is because the system always endeavours to compensate for any changes that can affect its performance in one way or another.

SELF-DIAGNOSIS FAULT CODES—SAAB

Error-code	CHECK	Malfunction indicated	Action
12243	Off	No signal from road speed sensor (US-West)	A With ECU connector plugged in: check the voltage across ECU pin 34 and earth. With one of the front road wheels spinning, the reading should oscillate between 1 and 11V. If not: Check the green lead between pin 132 on the speed sensor and ECU pin 34. Also check the black lead between the sensor and earthing point 3, and the green/white lead to pin 15 on the ignition switch (20). B Try a known good road speed sensor. C Try a known good LH-system ECU.
12245	On	EGR function faulty (USA-West)	A Check the yellow/white lead between ECU pin 19 and connector 394, and the white lead between the connector and the modulating valve (390). Also check the green/white lead between the modulating valve and the pin for the green/red lead in the connector to the injectors. B Try a known good modulating valve. C Check the yellow/white lead between the thermostatic switch (389) and ECU pin 23, and the black lead between the thermostatic switch and earthing point 201. D Try a known good thermostatic switch. E Try a known good LH-system ECU.

ISAT FAULT CODES—LH 2.4.2 FUEL SYSTEM

Permanent Fault	Intermittent Fault	Malfunction Indicated	Correction Test
42241	22241	Voltage high, over 16 V ②	Battery Voltage
42251	22251	ECU pin 4, signal low (below 1 V)	
42252	22252	Signal low, below 10 V ②	
42291	22291	Battery voltage below 10 V or over 16 V	
42440	22440	Mixture rich: no control by oxygen sensor	Air-fuel mixture
42441	22441	Mixture rich on idling ②	
42442	22442	Mixture rich on driving ②	
42450	22450	Mixture lean: no control from oxygen sensor	
42451	22451	Mixture lean on idling ②	
42452	22452	Mixture lean on driving ②	
42460 ①	22460 ①	Lambda-sensor signal faulty	
42491	22491	Mixture incorrect on idling	
42492	22492	Mixture incorrect on driving	
44221	24221	Speed signal absent ②	RPM signal
44261	24261	Vehicle speed sensor signal faulty ②	Vehicle speed sensor
44671	24671	PRE—IGNition signal longer than 20 s ②	Pre-ignition signal
45641 ①	25641 ①	Air mass meter signal high ②	Air mass meter signal
45651 ①	25651 ①	Air mass meter signal low ②	
45691 ①	25691 ①	Air mass meter signal faulty (too high or too low)	
45723	25723	DRIVE signal, function faulty	Drive signal
45771 ①	25771	Throttle sensor indicating idle during driving (constantly grounded)	Throttle sensor
45772 ①	25772	Throttle sensor indicating full throttle and idling simultaneously	
46221 ①	26221	Temperature sensor indicating temperature below −90°C during driving (broken circuit)	Coolant temperature sensor
46271 ①	26271	Temperature sensor indicating temperature above 160°C (constantly grounded)	
46391	26391	EGR system function faulty (temp. low)	EGR temperature sensor
58121	38121	Air mass meter burn-off function absent	Air mass meter burn-off
58321	38321	AIC valve function faulty	AIC valve
58322 ①	38322	ELCD valve function faulty	Purging valve
58371 ①	38371	Injector faulty	Injector
58372 ①	38372	ELCD valve: broken circuit or shorting to earth ②	Purging valve
58382 ①	38382	ELCD valve shorting to positive ②	
67192 ①		ROM fault (ECU)	Replace ECU

① When this fault occurs, the CHECK ENGINE light will also come on.
② 1991–92.

ISAT FAULT CODES—LH 2.4 AND LH 2.4.1 FUEL SYSTEMS

Permanent Fault	Intermittent Fault	Faulty Component/Signal	Correction Test
67192	—	ROM (microprocessor chip)	Electronic Control Unit
45771	25771	Throttle-position sensor idling contacts not breaking circuit on increase in engine idling speed and load (constantly earthed).	Throttle position sensor
45772	25772	Throttle position sensor full-load contacts constantly closed.	
46261	26261	Temperature sensor signal showing constant value.	Temperature sensor
46271	26271	Temperature sensor indicating temperature above 160°C (constantly grounded).	
46221	26221	Temperature sensor indicating temperature below −90°C (open circuit).	
42291	22291	Voltage at ECU pin 4 not within 10–16 V range.	Battery voltage
42251	22251	Voltage at ECU pin 4 too low (< 1 V).	
58121	38121	Air mass meter: no filament burn-off function.	Air mass meter burn-off
45691	25691	Air mass meter signal faulty (too high or too low).	Air mass meter signal
42491	22491	Prolonged fault in fuel-air mixture (additive idling fault).	Air-fuel mixture fault
42492	22492	Prolonged fault in fuel-air mixture (multiplicative fault in conjunction with driving).	
42450	22450	Mixture too lean.	Rich or lean Air-fuel mixture
42440	22440	Mixture too rich.	
42460	22460	Faulty signal from Lambda sensor.	Oxygen sensor signal
58321	38321	AIC valve faulty.	AIC valve
44261/ 12245	24261	Vehicle speed sensor signal faulty.	Vehicle speed sensor
45723	25723	DRIVE signal faulty.	Drive signal
46391/ 12243	26391	EGR system malfunction (temperature too high or too low).	EGR function
58371	38371	Injectors: injection pulses faulty.	Fuel injector

ISAT FAULT CODES—DI/APC SYSTEM

Permanent Fault	Intermittent Fault	Malfunction Indicated	Correction Test
429B0	229B0	Coding fault. Engine starts but only base-line charging pressure available ①	Coding fault
44261	24261	Vehicle speed sensor signal absent (Turbo) ①	Vehicle speed sensor
44360	24360	Crankshaft sensor signal faulty	Crankshaft sensor
44460	24460	Load signal faulty	Load signal
44461	24461	Adaption fault—outside limits	Adaptive function
44660	24460	Pre-ignition fault	Knock sensor signal
44661	24661	Knock-sensor fault	Knock sensor
44662	24662	Pre-ignition synchronization fault	Synchronization combustion signal
46660	26660	Pressure-sensor function faulty ②	Pressure sensor
45260	25260	Throttle position sensor signal faulty (LH 2.4.2) ①	Throttle position
45360	25360	Brake signal faulty	Brake signal
46391	26391	Signal from temp sensor upstream of throttle housing faulty ①	Air temperature sensor
60000	—	Watchdog (internal monitoring)	DI/APC system ECU internal monitoring
60001	—	ECU (ROM) fault	Read only memory
60002	—	ECU (RAM) fault	Random access memory
11111	—	Acknowledgement = OK	(No faults)

① 1991–92
② 1990

1990 ISAT COMMAND CODES—900

Code	Function	Display
550	Activates AC function (0.2 Hz) (LH 2.4.1)	8A550
200	Checks status of AC (n/a TCS)	8B100 (= activated) 8B000 (= not activated)
201	Checks position of throttle butterfly	8B101 (= idling position) 8B001 (= normal position) 8B301 (= full-load position) 8B103 (= idling + full load)
552	Activates injectors (15 Hz or 1.5 ms opening duration). Fuel pump isolated.	8A552
553	Activates AIC valve (1 Hz)	8A553
554	Activates ELCD valve (1 Hz)	8A554
202	Checks status of ELCD valve	8B002 (= open) 8B102 (= closed)
555	Activates EGR valve (1 Hz) (US-West only)	8A555
203	Checks status of DRIVE signal	8B103 (= activated) 8B000 (= not activated)
205	Checks status of Lambda sensor	8B105 (= lean) 8B305 (= rich) 8B005 (= inoperative)
900	Deletes all error codes and resets all adaptive values to basic settings	11111
930	Resets all adapted values to basic settings	11011
100	Transfers all error codes from the LH-system ECU to ISAT	
207	ECU identification code (last four digits of Bosch part no.)	(To be entered only when requested by Saab-Scania AB.)

1990 ISAT COMMAND CODES—9000

Code	Function	Display
550	Activates AC function (0.2 Hz)	8A550
200	Checks status of AC	8B100 (= activated) 8B000 (= not activated)
201	Checks position of throttle butterfly	8B101 (= idling position) 8B001 (= normal position) 8B301 (= full-load position) 8B103 (= idling + full load)
552	Activates injectors (15 Hz or 1.5 ms opening duration). Fuel pump isolated.	8A552
553	Activates AIC valve (1 Hz)	8A553
554	Activates ELCD valve (1 Hz)	8A554
202	Checks status of ELCD valve	8B002 (= open) 8B102 (= closed)
555	Activates EGR valve (1 Hz) (US-West only)	8A555
203	Checks status of DRIVE signal	8B103 (= activated) 8B000 (= not activated)
205	Checks status of Lambda sensor	8B105 (= lean) 8B305 (= rich) 8B005 (= inoperative)
900	Deletes all error codes and resets all adaptive values to basic settings	11111
930	Resets all adapted values to basic settings	11011
100	Transfers all error codes from the LH-system ECU to ISAT	
207	Ignition pulses	8B107 (= pulses present) 8B005 (= no pulses)
382	ECU identification code (last four digits of Bosch part no.)	(To be entered only when requested by Saab-Scania AB.)

1991–92 ISAT COMMAND CODES—LH 2.4 AND 2.4.2 FUEL SYSTEMS EXCEPT 9000 TURBO

Code	Function/signal	Display text (example)
22A	Air mass meter signal	803.6 = 3.6 V 804.7 = 4.7 V
22B	Battery voltage	8010.6 = 10.6 V 8007.3 = 7.3 V
23A	EGR pulse ratio	80000 = 0% (valve closed) 80012 = 12% 80027 = 27%
100	All fault codes transferred	
200	Check status of AC	8B100 = activated 8B000 = not activated
201	Throttle-butterfly position	8B101 (= idling position) 8B001 (= normal position) 8B301 (= full-throttle position) 8B103 (= idling + full throttle)
202	ELCD valve status	8B002 (= open) 8B102 (= closed)
203	DRIVE signal status	8B103 (activated) 8B003 (not activated)
204	Gear indicator	8B104 = light on 8B004 = light off
205	Lambda sensor status	8B105 (lean) 8B305 (rich) 8B005 (de-energized)
206	PRE-IGNition signal*	8B106 = fuel boost on pre-ign 8B006 = fuel boost on knocking
207	Ignition pulses	8B007 (= pulses absent) 8B107 (= pulses present)
239	AIC pulse ratio	80035 = 35% 80043 = 43%
249	RPM	801000 = 1000 rpm 805500 = 5500 rpm
250	Coolant temperature	800-30 = -30°C 80+130 = +130°C
279	Throttle-butterfly angle (degrees)	e.g. 80030 = 30°
382	ECU code	(Four last figures of Bosch part no.)
550	Activate AC function	8A550 = activated
552	Activate injectors	8A552 (opening duration of 15 Hz or 1.5 ms)

* *During normal operation, ISAT reading should alternate between 8B006 and 8B106*

1991–92 ISAT COMMAND CODES–LH 2.4 AND 2.4.2 FUEL SYSTEMS EXCEPT 9000 TURBO

553	Activate AIC (1 Hz)	8A553
554	Activate ELCD (1 Hz)	8A554 from ECU to ISAT
555	Activate EGR (1 Hz)	8A555 = activated (US–West only)
800	Terminate communications	
900	Clear all fault codes and reset all adaptive values to initial level	11111
930	Reset all adaptive values to initial setting	11011

1991–92 ISAT COMMAND CODES–9000 TURBO

Using ISAT command codes

When ISAT command codes are being used to simulate signals or functions, the resulting readings on the display are in the form of an alphanumeric code.

The code has five character positions.

Example 1:

To check the position of the timing advance, enter command code 27A on the ISAT keypad.

With the engine idling, the last five characters on the ISAT display should be 80+20, indicating that the timing is 20° BTDC.

Example 2:

Sometimes we want to know if a signal is high or low. To check the brake signal, we enter command code 200 via the ISAT keypad.

If the signal is not activated, code 8B000 will be displayed, indicating that the signal level is low. If the signal is activated, 8B100 will be displayed, indicating a high signal level.

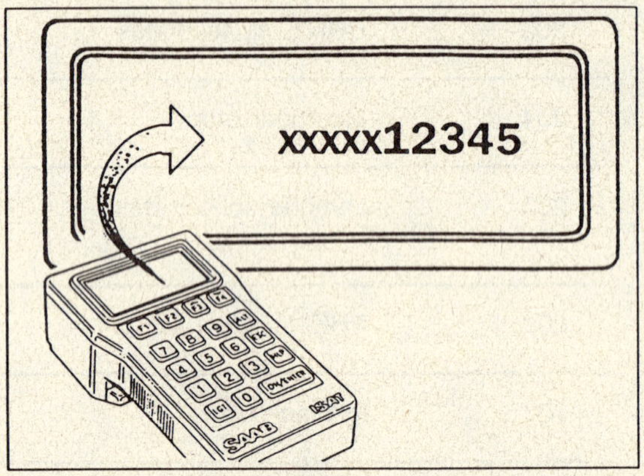

Reading on ISAT display in form of five-character alphanumeric code.

Command codes

Code	Component/signal	Code displayed; Voltage (V)
200	Brake signal	8B000 = not activated; 0 V 8B100 = activated; battery voltage
202	Crankshaft sensor	8B000 = unscreened; 0 V 8B100 = screened; 5 V
203	Ignition synchronization (all cylinders)	8B300 (combustion signal high); 5 V 8B000 (combustion signal low); ≤1 V
204	Throttle potentiometer, full throttle (LH 2.4.2)	80000 = idling (engine running); *) 80100 = full throttle (engine running); *)

*) *New reading introduced on M91*

1991–92 ISAT COMMAND CODES—9000 TURBO

Code	Component/signal	Code displayed; Voltage (V)
229	Knock-sensor signal level	80004 (e.g.) 0.4 V
27A	Current position of throttle butterfly	80+15 = 15° BTDC (M91 onwards) 80–05 = 5° ATDC
280	Engine variant: cat/non-cat	e.g. B202L.KAT-165E (M91 onwards)
282	EPROM part number	e.g. PGM.NR 912614
283	Number of times knocking detected (Turbo only)	8B0x3 = where *x* is the number of times knocking detected (M91 onwards)
285	ECU part number	e.g. ECU.NR7859721 (M91 onwards)
286	Coding—linked terminals on connector 406 (Turbo only)	8E300 = ECU pin 1 to 12 V 8E200 = ECU pin 1 to earth 8E100 = ECU pin 1 not linked 8E000 = other (invalid combination)
287	Load signal *Tq* (μs)	80015 = 15 μs pulse width
450	Activate ignition sparks	11111
451	Activate solenoid valve (Turbo only)	11111
452	Activate *CHECK ENGINE* light (blinking)	11111
453	Activate fuel-boosting function (Turbo only)	11111; 12–0 V
454	Activate RPM signal	11111 *) (crankshaft position must be 10–45° BTDC: enter 202 and move car until 8B100 changes to 8B000)

*) *Remove fuel-pump fuse before entering this code*

Note
Command codes 450–454 inclusive activate the relevant function for about one minute.

ELECTRONIC CONTROL UNIT

Testing

LH-Jetronic

If fault Code 67192 appears, indicating a ROM fault, replace the LH-system ECU with a known good and recheck for codes.

COMPONENT SYMBOLS KEY

Saab uses a standard numbering system for all of the components. Example would be and ignition coil will always have a number 5 assigned to it. So when the coil is displayed in wiring diagrams or art figures it will be designated by the number 5.

Following is the key to the numbering symbols of the components found in wiring diagrams and tests:

1—Battery.
2—Alternator.
3—Grounding point in the fascia.
4—Starter motor.
5—Ignition coil.
6—Distributor.
7—Ground point on the radiator crossmember.
20—Ignition switch.
21—Ignition switch relay.
22—Fuse box.
29—Brake light switch.
37—Radiator fan motor.
39—Temperature switch for radiator fan.
47—Instrument panel.
47P—Check engine light.
65—Ground point in back seat area.
73—Timing service instrument socket.
75—Fuse box for positive supply from battery.
76—Switch for raising idle speed, automatic transmission.
77—Inhibitor switch, automatic transmission.

79—Vacuum switch.
93—Ground point, left wheel housing.
94—Cold start valve.
95—Auxiliary air valve.
102—Fuel pump relay.
104—hot start relay.
110—Tachometer.
131—Control unit for cruise control.
132—Vehicle speed sensor.
136—Oxygen sensor.
141—Cruise control selector.
144—Pressure switch.
145—EZK test connector.
156—Time delay relay-A/C and ACC radiator fan.
159—Distribution terminal +15.
176—EZK control unit.
177—APC control unit.
178—Knock sensor, APC system.
179—Solenoid valve, APC system.
200—LH-system control unit.
201—Ground point on engine.
202—Coolant temperature sensor.
203—Throttle position sensor.
204—LH-system test connector.
205—Air mass meter.
206—Fuel injectors.
229—LH-system main relay.
230—Distribution terminal +30.
231—Distribution terminal +54.
239—Gear selector switch-automatic transmission.
272—Automatic idle control valve.
321—Valve on charcoal canister.
323—Fuel pump.
343—DI, DI/APC control unit.
344—Pressure sensor, DI/APC system.
345—Crankshaft position sensor.
346—Ignition cartridge for DI, DI/APC system.
347—10-pin diagnostic test connector.
376—ETS control unit.
377—ETS main relay.
378—ETS motor.
379—ETS accelerator pedal position transmitter.
380—ETS dump valve.
381—ETS bypass valve.
389—NTC resistor.
390—EGR modulating valve.
406—DI/APC 3 terminal connector.
407—Coolant temperature sensor-DI/APC.

CODING FAULT

This fault applies only to turbocharged vehicles. If a coding fault occurs, only base-line turbocharging pressure is available.

Testing

1. With the ignition switch in the **ON** position, enter ISAT command Code 406 to display the up-to-date coding for the DI/APC system ECU. The following display options are available: 8E000 = invalid combination, 8E100 = DI/APC system ECU pin 1 not connected, 8E200 = DI/APC system ECU pin 1 grounded and 8E300 = DI/APC system ECU pin 1 shorted to battery voltage.
2. Check the sequencing of the pins of 3-pin connector 406.
3. Check the wiring between connector 406 and the DI/APC system ECU for an open or shorted circuit or a poor connection.
4. Check that the correct DI/APC system ECU is being used.

THROTTLE POSITION SENSOR

If the signal from the throttle position sensor (TPS) is absent or incorrect, idling control will not function properly and the check engine light will illuminate. Turbocharged vehicles will run on base line turbocharging pressure only.

Testing

LH 2.4 FUEL SYSTEM

This test applies to vehicles with the LH 2.4 fuel system, except the 9000 Turbo.

1. With the ignition switch in the **ON** position, enter command Code 201 on the ISAT.
2. With the accelerator at idle, 8B101 should be displayed. With the accelerator depressed half way, 8B001 should be displayed. At full throttle, 8B301 should be displayed.
3. If 8B001 is displayed continuously on the screen, a break in the wiring to the TPS is indicated. Continue with the test.
4. If 8B103 means that the TPS is indicating idling and full throttle simultaneously. If this code is displayed continuously, unplug the connector from the TPS.
5. If 8B001 is displayed, replace the TPS with a known good.
6. If the code does not switch from 8B103 to 8B001 when the connector is unplugged, check the wiring for a short circuit.

NOTE: Before proceeding further, verify that the TPS is in adjustment. At idle there should be continuity across pins 1 and 2 of the connector.

7. With the engine idling, the TPS connector connected and the LH 2.4 system ECU connector unplugged, check that LH 2.4 system ECU pin 2 is grounded. If not, check the Gray wire from LH-system ECU pin 2 to TPS pin 1 for an open or shorted circuit.
8. With the TPS at full throttle position, with the connector connected and the LH-system ECU connector disconnected, check for a good ground at pin 3. If not, check the Green/Red wire from LH-system ECU pin 3 to the TPS pin 3 for an open or shorted circuit.
9. Check that there is a good ground at TPS pin 2.
10. If the problem still exists, replace the LH 2.4 fuel system ECU with a known good.

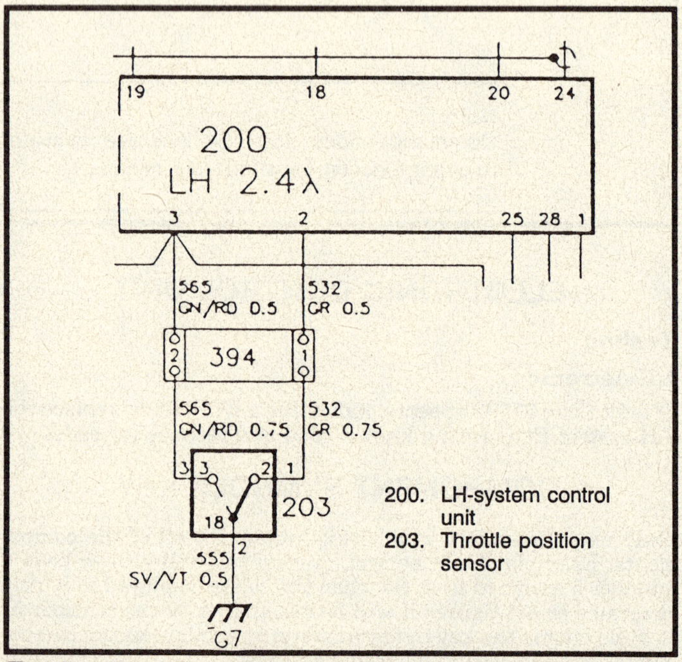

Testing throttle position sensor—Saab with LH 2.4 fuel system

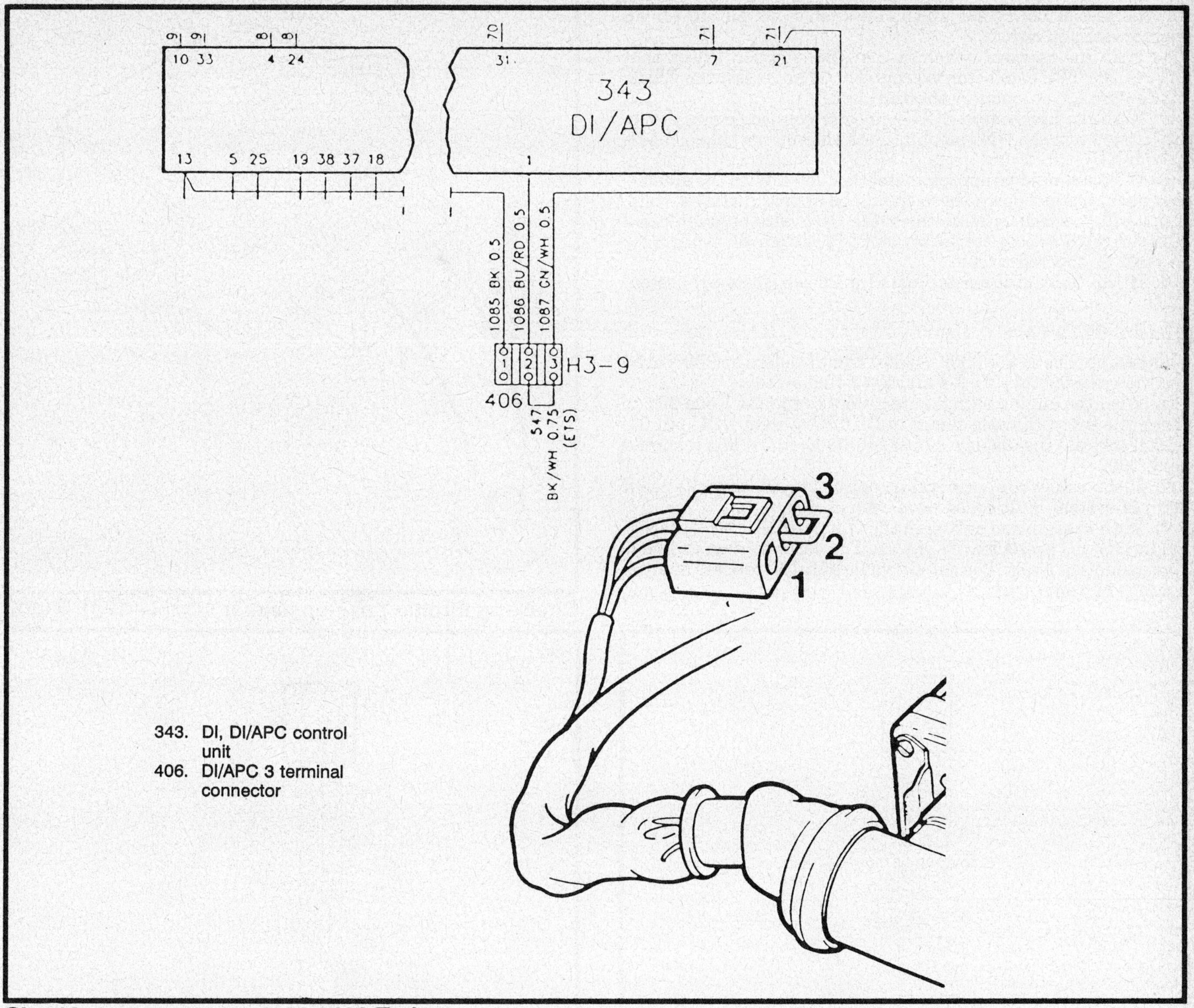

343. DI, DI/APC control unit
406. DI/APC 3 terminal connector

Checking coding fault—1991 9000 Turbo

LH 2.4.1 FUEL SYSTEM

This test applies to the non-turbocharged 1991 9000 with the B234 engine.

1. Enter command Code 201 on the ISAT.
2. Depress the accelerator slowly to the full-throttle position.
3. If 8B001 is displayed continuously on the screen, a break in the wiring to the TPS is indicated. Continue with the test.
4. If 8B103 is displayed, or the codes are shown in a different position or do not change, unplug the connector from the TPS. The display should read 8B001. If not, continue the test.
5. If 8B001 is displayed, replace the TPS with a known good.
6. If 8B101 is displayed, check the wiring between pin 2 of the LH-system ECU and pin 1 of the TPS. Check the wiring between LH-system ECU pin 2 and pin 7 of the ignition system ECU (176).
7. If 8B301 is shown, check the wiring between pin 3 of the LH-system ECU and pin 3 of the TPS. Check terminal TK on the air conditioning relay (156) and make sure that it is not grounded.
8. If 8B103 is displayed, go to Step 6.
9. If the problem still exists, replace the LH-system ECU with a known good.

LH 2.4.2 FUEL SYSTEM

This test applies to vehicles with the LH 2.4.2 fuel system, except the 9000 Turbo.

1. With the ignition switch in the **ON** position, enter ISAT command Code 279. The following throttle valve position should now be indicated on the display. With the accelerator pedal pushed down: 80085 = approximately 85° (open). With the accelerator pedal released: 80000 = approximately 0° (closed).
2. Measure the resistance across pins 1 and 3 of the throttle sensor. The resistance should be 1100–1500 ohms with the accelerator pedal pushed down and 2600–3000 ohms with the pedal released. If not within specifications, replace the throttle sensor.
3. With the ignition switch in the **ON** position, measure the voltage at pin 1 of the TPS connector. The reading should be 5 volts. If not, check the Green/Red wire between pin 1 of the

throttle sensor connector and LH-system ECU pin 10 for an open or shorted circuit.

4. With the ignition switch in the **OFF** position, check that pin 2 of the TPS connector is grounded. If not check the Black/White wire for an open or shorted circuit.

5. With the LH-system ECU connector connected and the ignition switch in the **ON** position, check the voltage between LH-system ECU pin 2 and ground. With the accelerator pedal released there should be approximately 0.2 volts. With the accelerator pedal pushed down there should be approximately 4 volts. If not within specifications, check the Grey wire between pin 3 of the throttle sensor connector and LH-system ECU pin 2 for an open or shorted circuit.

6. If the fault still exists, install a known good LH-system ECU.

DI/APC SYSTEM

This test applies to the 9000 Turbo. The tests for the 900 Turbo are the ones for the LH 2.4 and 2.4.2 fuel systems.

1. With the engine idling, enter ISAT command Code 204 to check the input throttle signal to DI/APC system ECU pin 31. At full throttle the display should be 80000 and at idle it should be 80100.

2. If the values are incorrect or not displayed, check to see if any LH-system fault codes have been stored.

3. With the ignition switch in the **ON** position, check the signal at DI/APC system ECU pin 31. The correct value at idle is approximately 1 volt. The correct value at full throttle is approximately 11 volts.

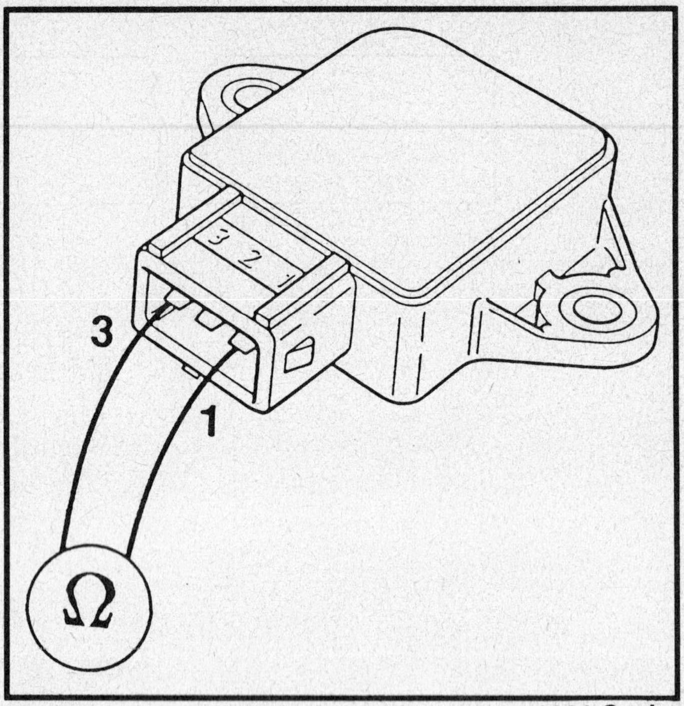

Checking throttle position sensor signal—1991 Saab

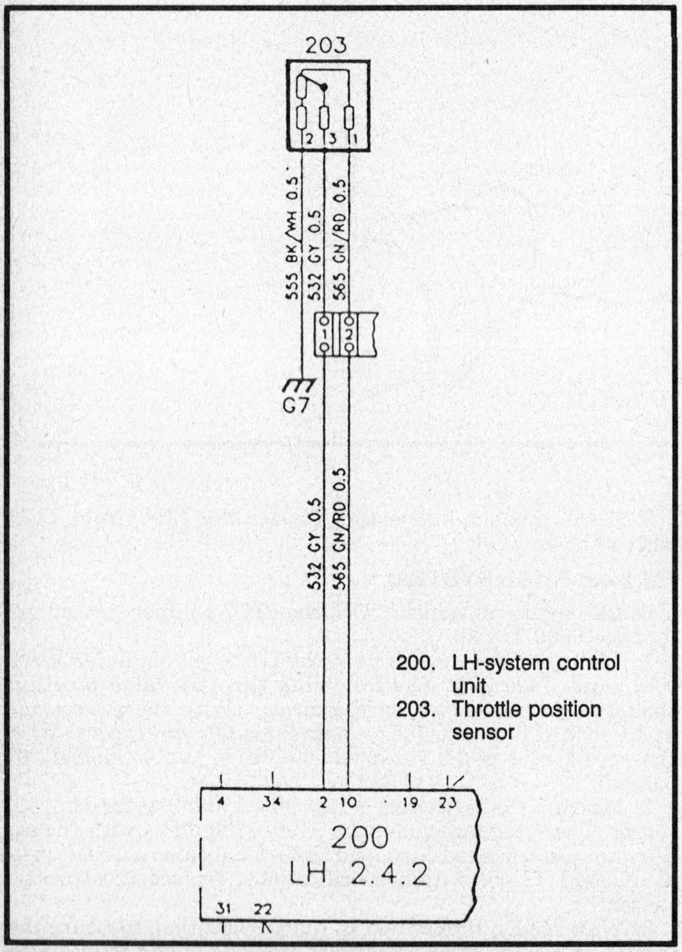

Testing throttle position sensor—Saab with LH 2.4.2 fuel system

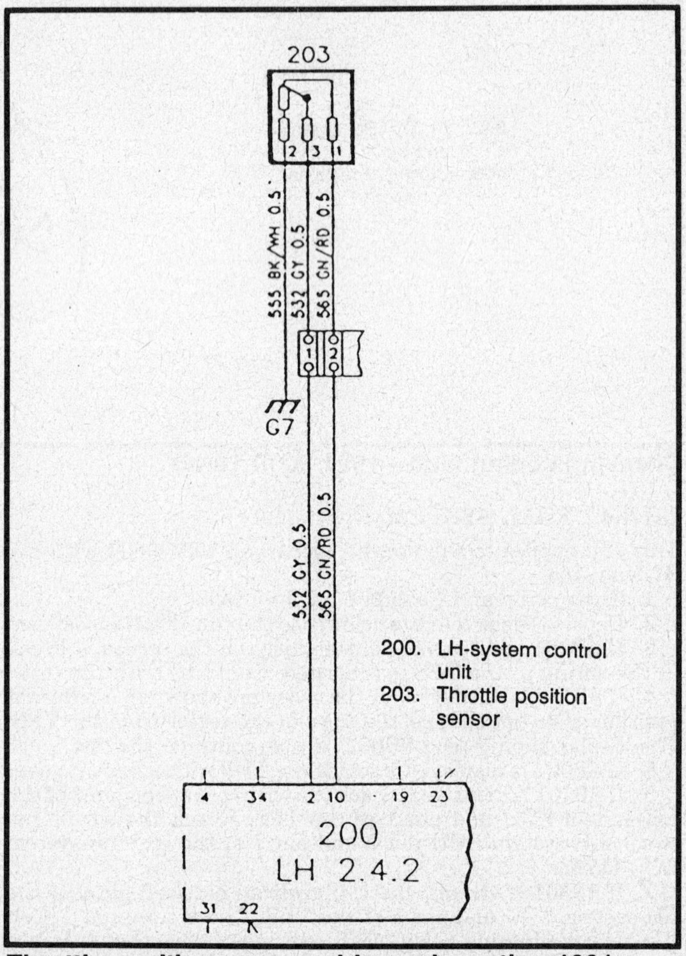

Throttle position sensor wiring schematic—1991 Saab

4. If the signal is correct, clear the fault code and test drive the vehicle. If the code is regenerated, install a known good DI/APC system ECU.
5. If there is no signal at DI/APC system ECU pin 31, check the Green/Red wire between LH-system ECU pin 3 and DI/APC system ECU pin 31.
6. If the wiring is good, replace the LH-system ECU with a known good.

COOLANT TEMPERATURE SENSOR

A fault will set if the coolant temperature sensor (CTS) shows a constant value, too high of a value or too low of a value. Poor drivability will result and the check engine light will be illuminated.

Testing

LH 2.4 AND LH 2.4.1 FUEL SYSTEMS

1. If the signal of the CTS remains constant, check the wiring between pin 13 of the LH-system ECU and pin 1 of the temperature sensor (202).
2. Check the wiring between pin 2 of the temperature sensor and ground point 201.
3. Check that ground point 201 is continuous with the battery negative terminal.
4. Check the CTS for varying resistance as it is heated in water.
5. If the problem still exists, replace the LH-system ECU with a known good.
6. If the signal is too high or too low, make sure that the ignition switch is in the **OFF** position.
7. Disconnect the LH-system ECU connector.
8. Check the resistance between pin 13 of the LH-system ECU and ground. It should be 2280–2720 ohms at 68°F (20°C) or 290–356 ohms at 176°F (80°C).
9. If out of specifications, check for an open or shorted circuit or replace the CTS.

LH 2.4.2 FUEL SYSTEM

1. Enter ISAT command Code 250. The engine temperature should be shown on the display, e.g., 80 + 080 if the temperature is 176°F (80°C).

°C	°F	Ohm (±10%)
-20	-4	14000
-10	14	9000
0	32	5800
10	50	3800
15	58	3000
20	68	2600
25	76	2000
30	86	1700
80	176	320

Coolant temperature sensor specifications—1991 Saab

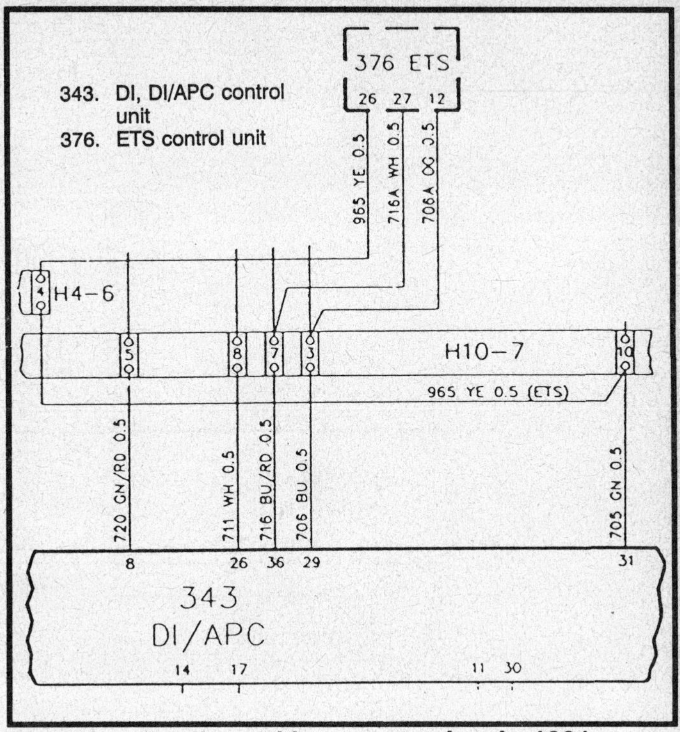

Checking throttle position sensor signal—1991 9000 Turbo

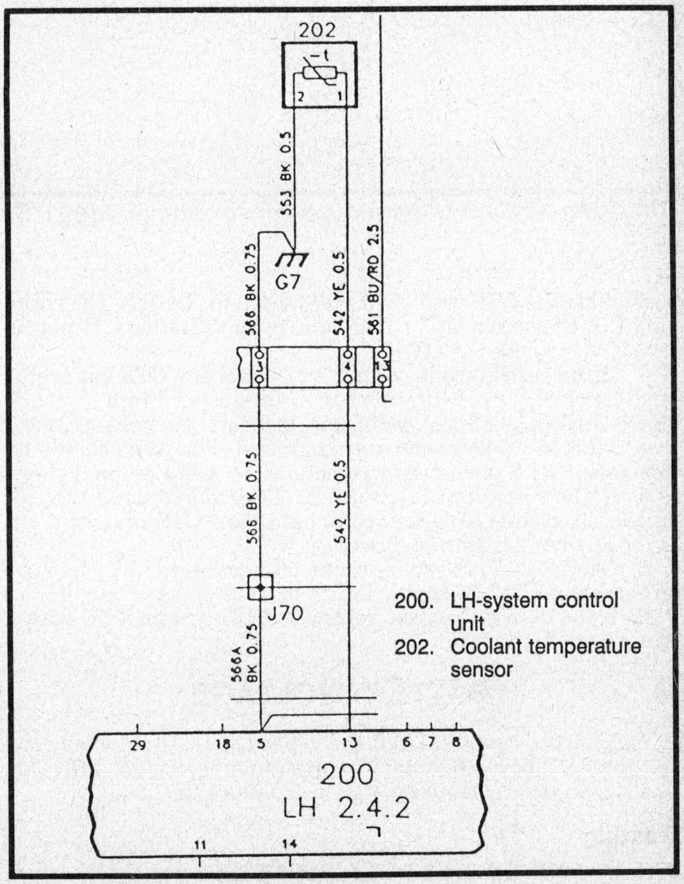

Checking coolant temperature sensor signal—1991 Saab with LH 2.4.2 fuel system

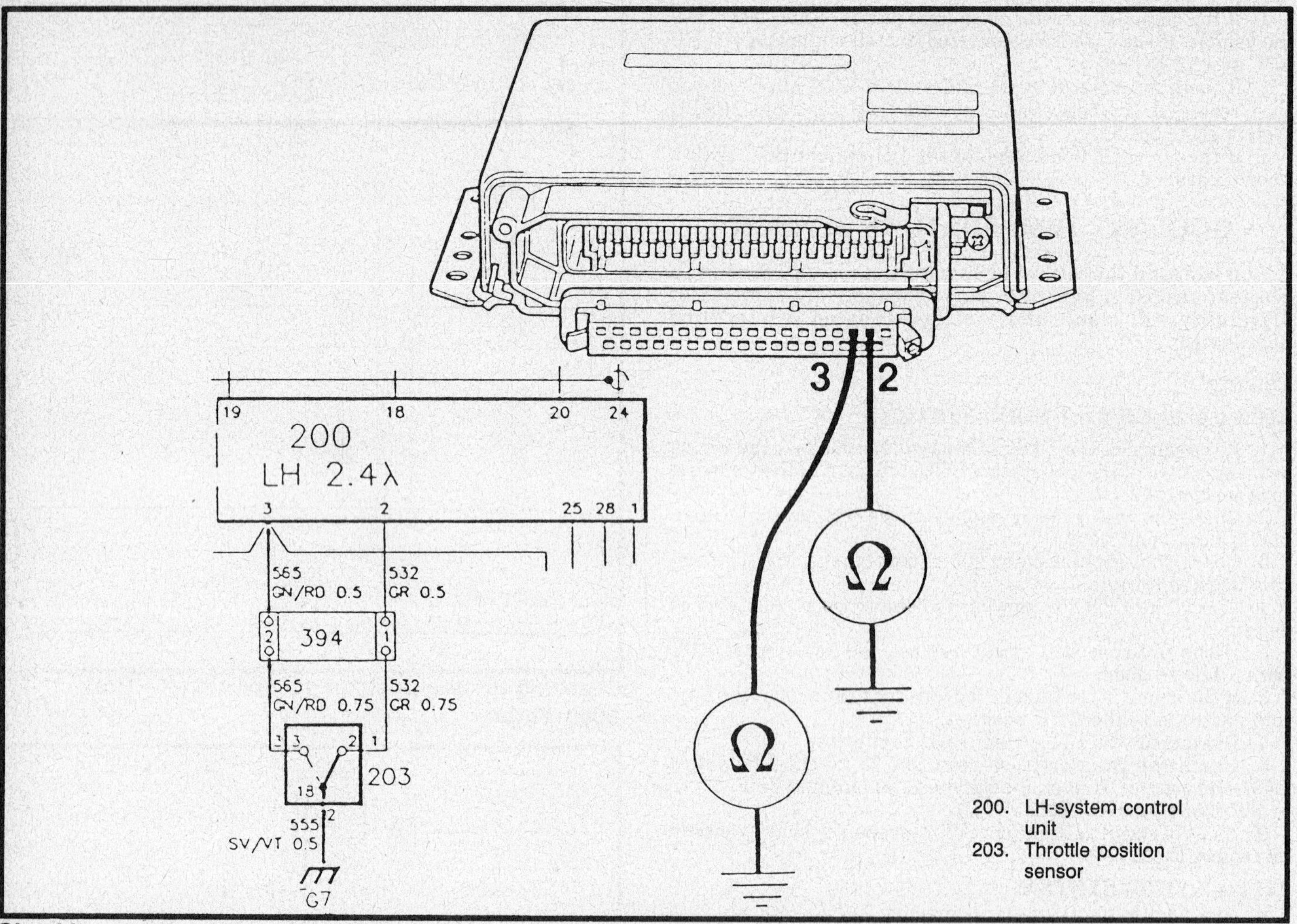

Checking coolant temperature sensor signal – 1991 Saab with LH 2.4 fuel system

2. With the CTS connector disconnected, measure the resistance of the sensor and compare it to specifications. If not as specified, replace the CTS.
3. If the resistance is correct, reconnect the CTS connector and disconnect the LH-system ECU connector. From the back of the LH-system ECU connector, measure the resistance between pin 13 (Yellow) and a good ground. The value should be consistent with previous specifications, or a few ohms higher.
4. If the resistance is incorrect, check the Yellow wire between LH-system ECU connector pin 13 and CTS connector pin 1 for an open or shorted circuit.
5. Check the Black wire between CTS connector pin 2 and the ground point on the engine for an open or shorted circuit.
6. If the fault still exists, replace the LH-system ECU with a known good.

BATTERY VOLTAGE

A malfunction is indicated if the voltage at ECU pin 4 is below 10 volts or above 16 volts. The symptom associated with this fault is an erratic running engine or failure to start.

Testing

LH 2.4 AND LH 2.4.1 FUEL SYSTEMS

Check the condition of the battery, the charging system and the grounding points.

LH 2.4.2 FUEL SYSTEM

1. Check the battery, battery cables, starter motor and charging system.
2. With the ignition key **OFF**, check that the battery voltage is available at ECU pin 4. If not, check the Red wire and Gray wire to terminal block 75 on the battery shelf.
3. Clean and check the tightness of ground points G2 behind the battery, G25 on the gearbox and G7 on the engine.

RPM SIGNAL

This fault applies to the LH 2.4.2 fuel system. Without the RPM signal from the ignition system to the ECU, the engine will fail to start.

Testing

1. Check that the voltage reading at ECU pin 1 is above 6.5 volts while the starter motor is engaged and spinning. If not, check the Blue wire between LH-system ECU pin 1 and DI/APC-system ECU pin 29 for an open or shorted circuit.
2. Check the crankshaft sensor.

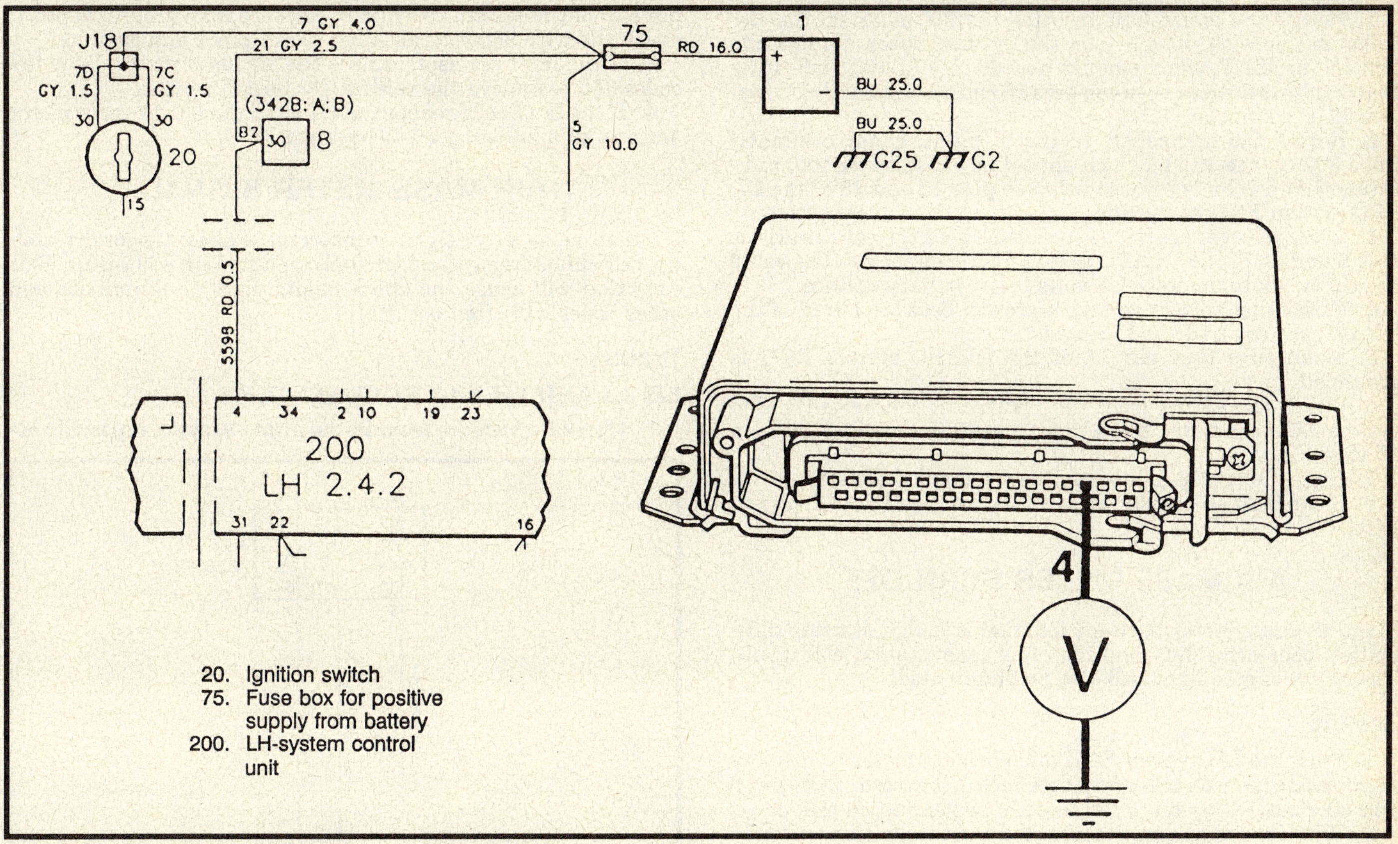

Checking voltage at ECU pin 4—1991 Saab

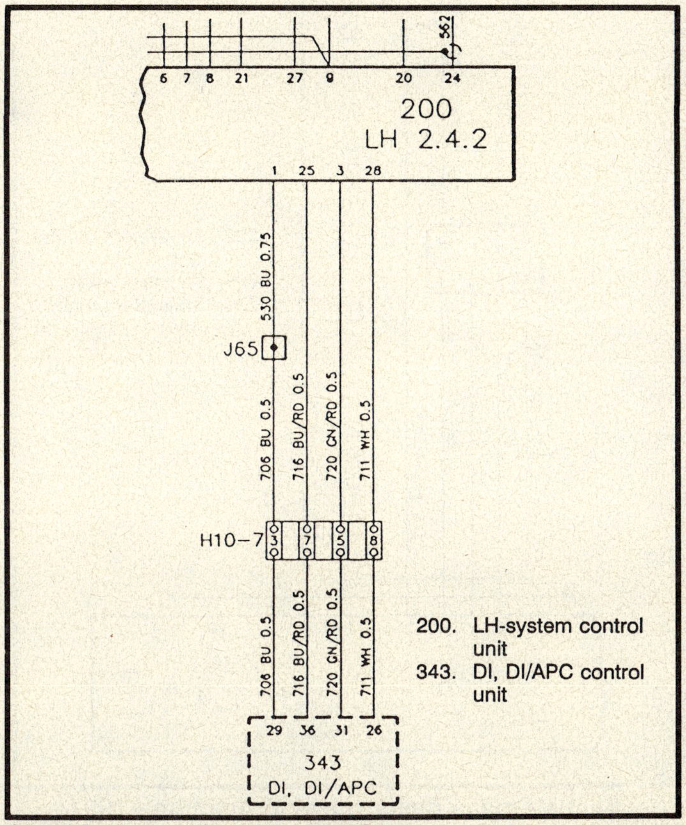

Checking rpm signal—1991 Saab

CRANKSHAFT SENSOR

Testing

The following test applies to vehicles with the DI or DI/APC ignition systems (no distributor).

CAUTION

To avoid personal injury always make sure that the ignition switch is in the OFF position before rotating the crankshaft.

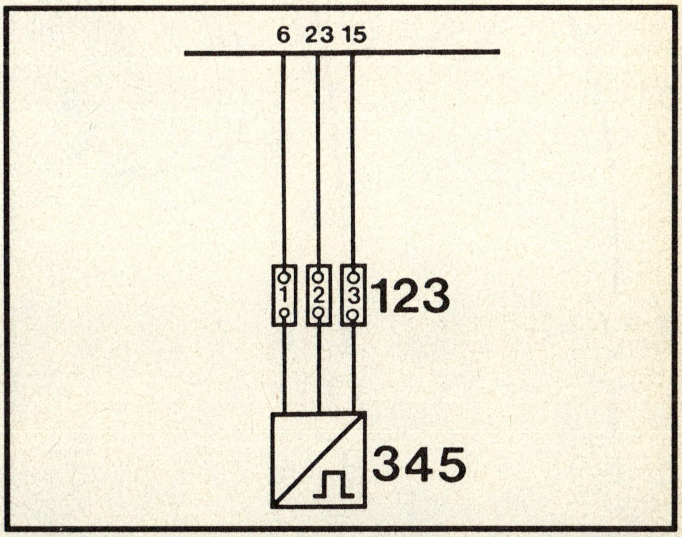

Testing crankshaft position sensor—Saab with direct ignition

1. Rotate the crankshaft to bring the 20° mark on the flywheel in line with the mark on the flywheel cover. Enter Code 302 on the ISAT, which should now display 8B000, indicating that 0 volts is present between pins 15 and 23 of the DI/APC system ECU connector.
2. Rotate the crankshaft to the 0° mark. Enter command Code 302 on the ISAT, which should now display 8B100, indicating that 5 volts is present between pins 15 and 23 of the DI/APC system ECU connector.
3. If the voltages are not as specified, check the voltage across pins 6 and 23 of the DI/APC system ECU connector. The value should be approximately 1.5 volts below battery voltage.
4. Check that battery voltage is present between pin 21 of the DI/APC system ECU and ground.
5. Make sure that pin 13 of the DI/APC system ECU is grounded.
6. Check the wiring from the crankshaft sensor to DI/APC system ECU connector pin 6 (Red), pin 23 (Green/Red) and pin 15 (Black).
7. If the fault still exists, replace the crankshaft sensor.
8. If the fault still exists, install a known good DI/APC system ECU.

AIR MASS METER BURN-OFF

If the air mass meter burn-off function is faulty, starting difficulties, poor drivability and high fuel consumption will result. The check engine light will also be illuminated.

Testing

1. With the LH-system ECU and air mass meter connectors disconnected, check the wiring between LH-system ECU pin 8 and air mass meter pin 4 for an open or shorted circuit.
2. With the connectors reconnected and the ignition switch in the **ON** position, check that battery voltage is reaching air mass meter pin 5 and that pin 1 is grounded. If no voltage is reaching pin 5, check the wires between pin 5 of the air mass meter and pin 87B of the system relay (229). If there is no ground at pin 1, check the wire between air mass meter pin 1 and ground.
3. If the fault persists, replace the air mass meter, clear the codes and test drive the vehicle. Recheck for codes.
4. If the fault code reappears, reinstall the old air mass meter and install a known good LH-system ECU.

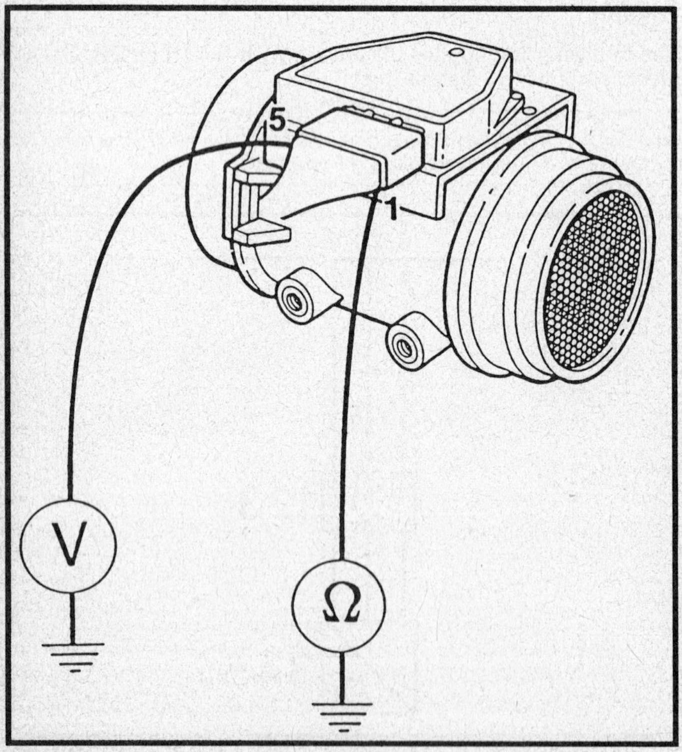

Checking air mass meter burn-off function—1991 Saab

AIR MASS METER SIGNAL

If the air mass meter signal is incorrect or absent, poor drivability, difficult starting and high fuel consumption will result. This condition will cause the check engine light to illuminate and other codes will often occur.

Testing

LH 2.4 AND LH 2.4.1 FUEL SYSTEMS

1. Check for vacuum leaks in the intake system, especially be-

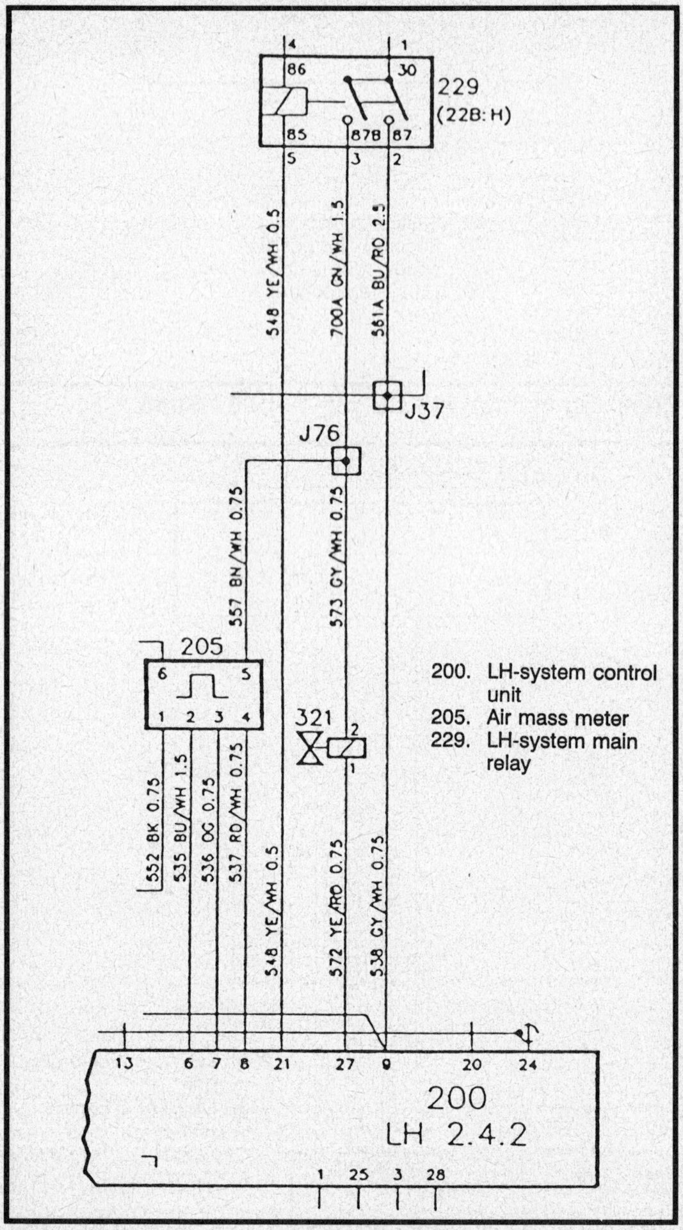

Checking air mass meter burn-off function—1991 Saab shown

tween the air mass meter and the intake manifold.

2. Check that pin 1 on the air mass meter is grounded.
3. With the ignition switch in the **ON** position, check that battery voltage is present at air mass meter pin 5.
4. If the fault code still exists, replace the air mass meter and recheck for fault codes.
5. If the fault still exists, install a known good LH-system ECU.

LH 2.4.2 FUEL SYSTEM

1. Enter ISAT command Code 22A and check that the following values are displayed: 802.0 = 2.0 volts on idling and 805.0 = 5.0 volts on full throttle.
2. Check for vacuum leaks in the intake system.
3. Check that pin 1 on the air mass meter is grounded. If not, check the Black wire between pin 1 and ground for an open or short circuit.
4. With the ignition switch in the **ON** position, check that battery voltage is present at air mass meter pin 5. If it is not, check the Brown/White wire between pin 5 of the air mass meter and pin 87 of the system relay for an open or shorted circuit.
5. Enter ISAT command Code 22A and check that the signal varies between 2–5 volts depending on the load (reading on the ISAT tester should be 802.0–805.0). If not, check the Orange wire between LH-system ECU pin 7 and air mass meter pin 3 for an open or shorted circuit.
6. Check for a good ground at air mass meter pin 2.
7. If the fault still exists, clear the fault codes and test drive the vehicle. Recheck for the same code being set.
8. If the same fault code recurs, replace the air mass meter and recheck for fault codes.
9. If the fault still exists, install a known good LH-system ECU.

AIR-FUEL MIXTURE FAULT

The symptom associated with this fault is a rough-running engine due to a rich or lean mixture. The check engine light will be illuminated.

NOTE: During the break-in period of approximately 300 miles (500 km) some of the fault codes may be generated even though no fault exists. Clear the fault code and see if the code is regenerated.

Testing

LH 2.4 AND LH 2.4.1 FUEL SYSTEMS

1. Check for leaks in the induction and fuel systems.
2. Check the operation of the oxygen sensor heater.
3. Check the operation of the air induction system.
4. If the fault still exists, replace the LH-system ECU with a known good.

LH 2.4.2 FUEL SYSTEM

1. With the engine at normal operating temperature, enter the ISAT command Code 205. The following codes should alternate on the display: 8B105 (lean), 8B305 (rich) and 8B005 (not activated). If any of these codes is displayed continuously, the oxygen sensor is faulty.
2. Check that battery voltage is available across the pins of the oxygen sensor connector. If not, check the following:
   a. Oxygen sensor fuse 28.
   b. The Blue/Red wire between pin 1 of the oxygen sensor connector and fuse 28 for an open or shorted circuit.
   c. The Black wire between pin 2 of the oxygen sensor connector and ground for an open or shorted circuit.
3. With the ignition switch **OFF**, unplug the connector for the oxygen sensor preheater and measure the resistance across the sensor White terminals. The resistance should be 4–8 ohms. If the resistance is high, replace the oxygen sensor.

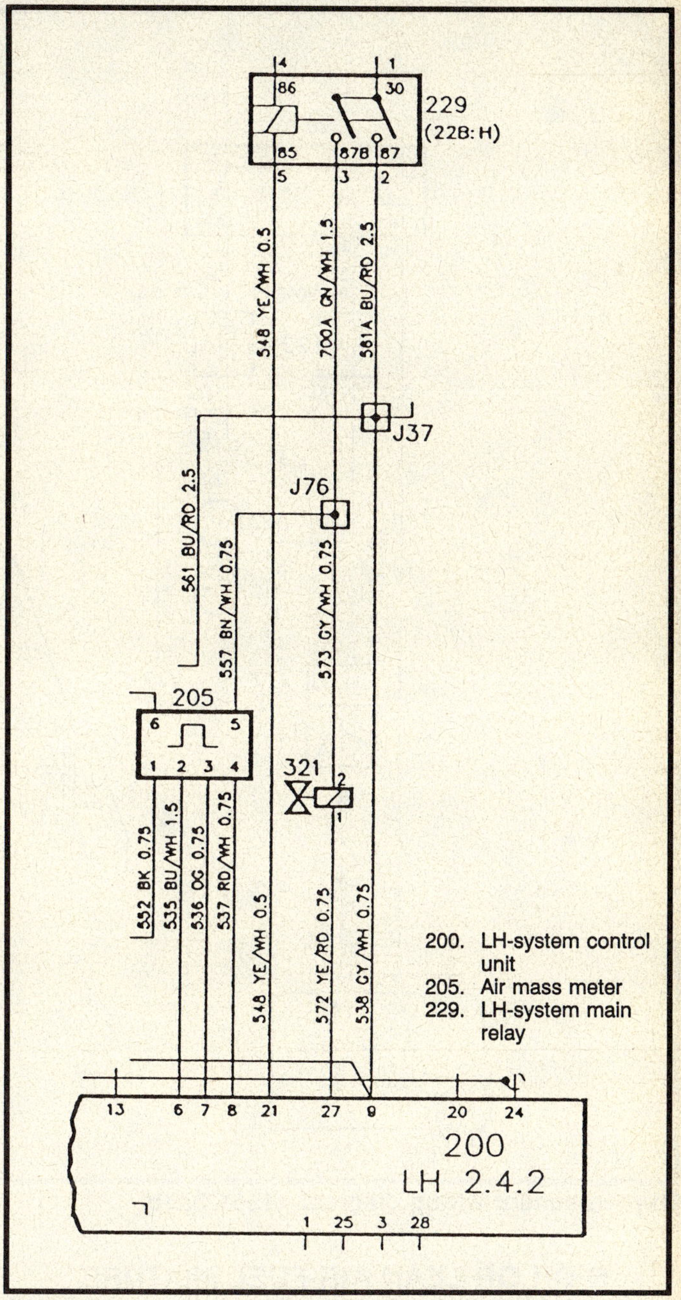

Checking air mass meter signal—1991 Saab

4. Check for vacuum leaks in the induction system.
5. Check the fuel pressure and the fuel pressure regulator.
6. With the engine idling, measure the oxygen sensor signal between ECU pin 24 and a good ground. The signal should alternate between 0–0.4 volts (lean) or 0.6–1.0 volts (rich) with the engine at normal operating temperature. If not, check the Green and the Black wires between ECU pin 24 and the oxygen sensor for an open or short circuit.
7. If the fault persists, replace the oxygen sensor and check the function again by entering ISAT command Code 205.
8. If the fault still persists, install a known good ECU.

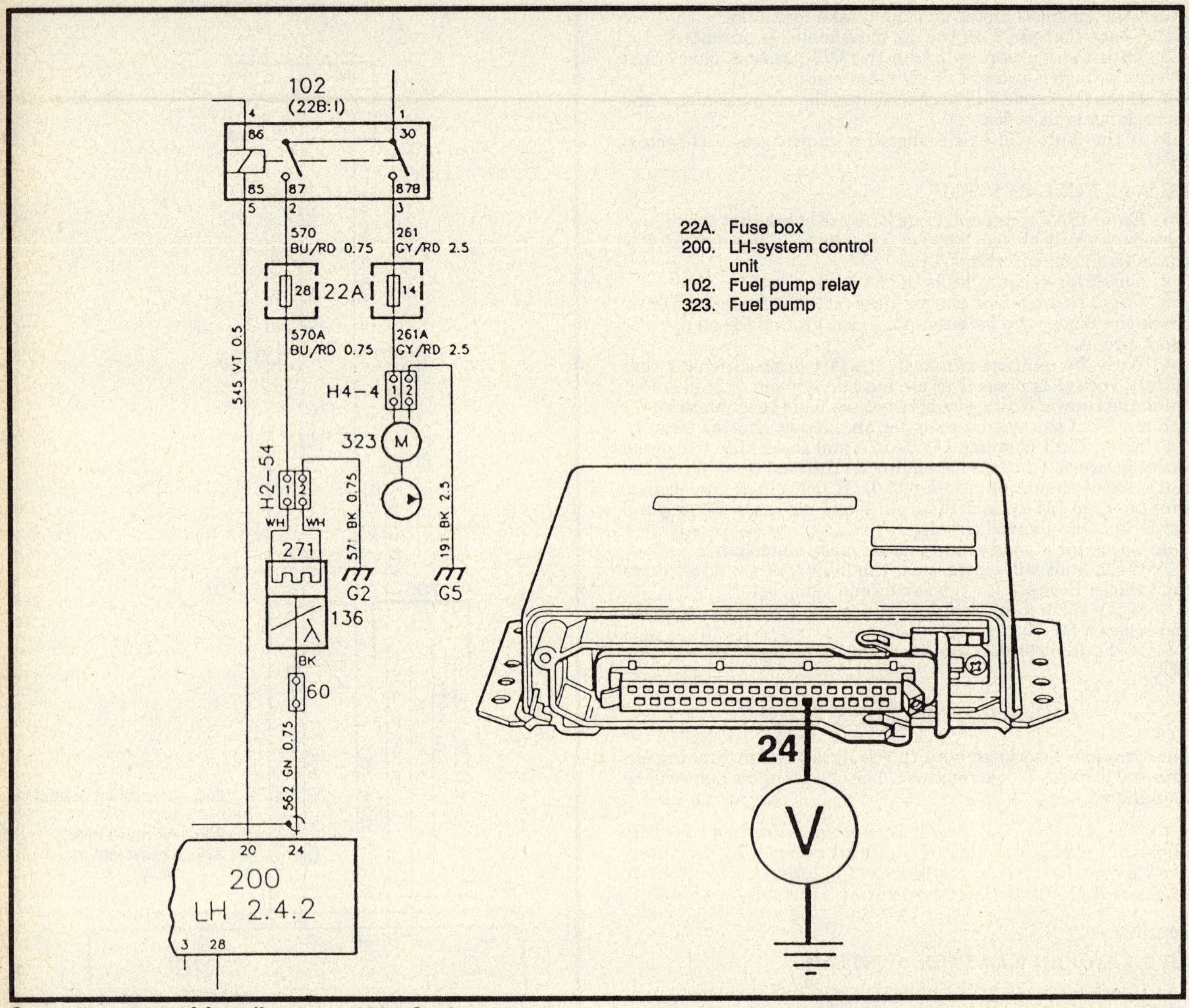

Oxygen sensor wiring diagram—1991 Saab

RICH OR LEAN AIR-FUEL MIXTURE

Testing

1. Check for loose screws or hose clamps and for leaks in hoses or around O-rings.
2. Check for leaks in the induction system and check for proper operation.
3. Check that voltage is present across the pins on the oxygen sensor connector.
4. Check the oxygen sensor heater fuse.
5. Check that the oxygen sensor signal fluctuates between 0–1.5 volts with the sensor warm.
6. If the fault still exists, replace the oxygen sensor.
7. If the fault still exists, replace the LH-system ECU with a known good.

DRIVE SIGNAL

If the drive signal is absent or incorrect, no increase in rpm will occur when a position other than **P** or **N** is selected with an automatic transmission.

Testing

1. Enter ISAT command Code 203 and observe that the following are shown on the display. With the gear selector in **P** or **N**: 8B003 (signal not activated). With the gear selector not in **P** or **N**: 8B103 (signal activated).
2. Check fuse 9.
3. Check that battery voltage is present at pin 3 of the gear selector switch. If not, check the wires between the switch and fuse 9, and the wire between fuse 9 and the +54 terminal block.
4. With the gear selector in a position other than **P** or **N**, check that battery voltage is present at pin 4 of the gear selector switch. If there is not, but pin 3 has voltage, replace the gear selector switch.
5. With the gear selector in a position other than **P** or **N**, check that battery voltage is present at LH-system ECU pin 30. If not, check the wire between LH-system ECU pin 30 and the gear selector switch for an open or shorted circuit.

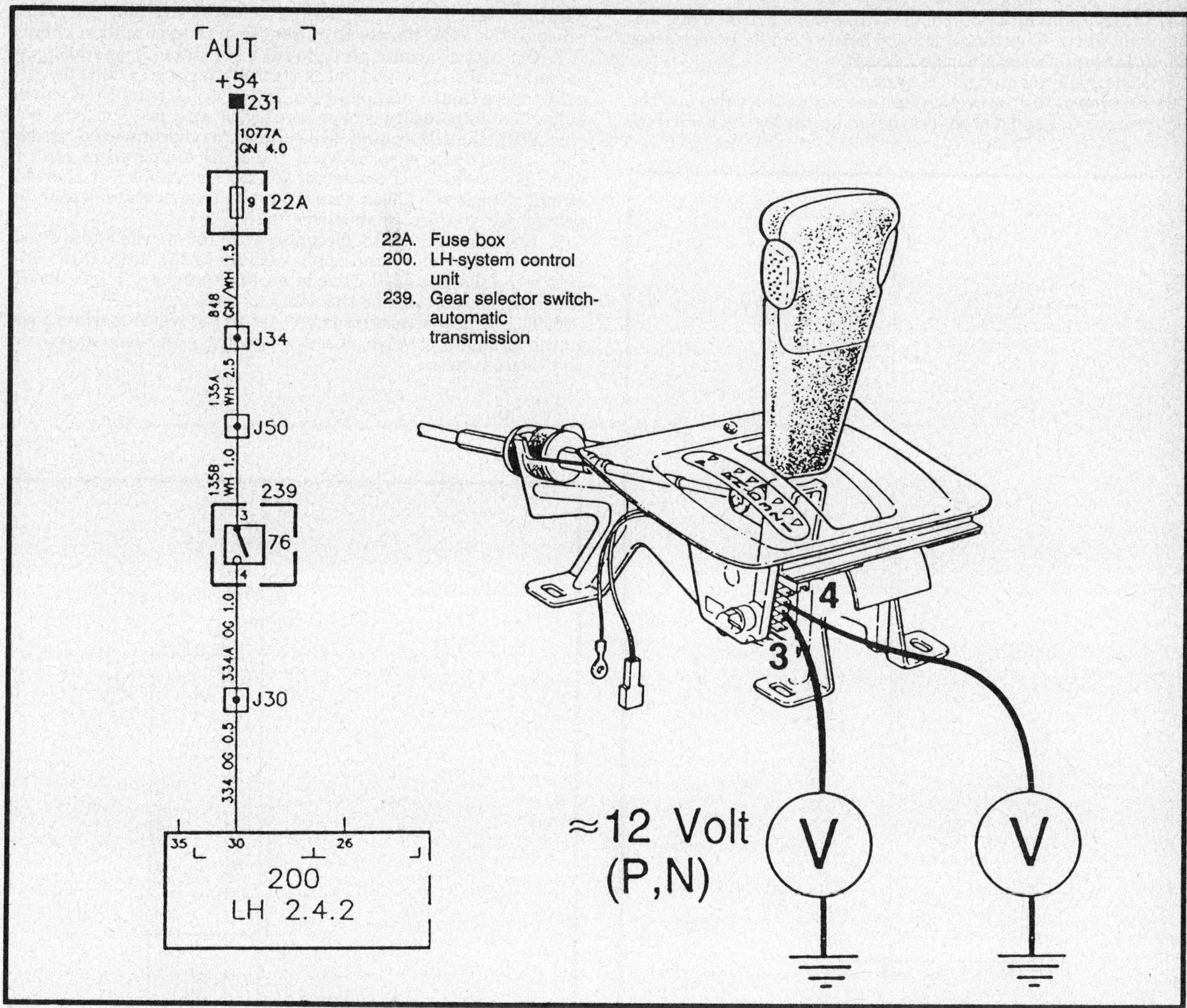

Checking drive signal—1991 Saab

EGR FUNCTION

This test applies to vehicles with California emissions. If the EGR temperature sensor signal is incorrect or absent, the check engine light will be illuminated.

Testing

LH 2.4 AND LH 2.4.1 FUEL SYSTEMS

1. With the ignition switch in the **ON** position enter command Code 555 on the ISAT.
2. Check that the modulating valve is operating. If not, check the Yellow/White wire between LH-system ECU pin 19 and connector 394, and the White wire between the connector and the modulating valve (390).
3. Check the Green/White wire between the modulating valve and the pin for the Green/Red wire in the connector to the injectors.
4. Replace the modulating valve with a known good.
5. Check the Yellow/White wire between the thermostatic switch (389) and LH-system ECU pin 23, and the Black wire between the thermostatic switch and ground point 201.
6. Replace the thermostatic switch with a known good.
7. If the problem still exists, replace the LH-system ECU with a known good.

LH 2.4.2 FUEL SYSTEM

1. With the ignition switch in the **ON** position, enter ISAT command Code 555. Listen for whether the modulating valve is working. If it is, go to Step 6.
2. If the modulating valve is not working, unplug the valve connector and measure the resistance across the pins. If the resistance is not approximately 30 ohms, replace the modulating valve.
3. With the ignition in the **ON** position, check pin 2 (Green/White) of the modulating valve for battery voltage. If there is not, check the Green/White wire and the Blue/Red wire between the modulating valve and pin 87 on the system relay for an open or shorted circuit.

4. Check the White and Yellow/White wires between pin 1 of the modulating valve connector and LH-system ECU connector pin 19 for an open or a shorted circuit.
5. Install a known good LH-system ECU.
6. Reconnect the connectors for the modulating valve and the LH-system ECU and disconnect the connector for the EGR temperature sensor. With the engine idling, check that the resistance of the EGR temperature sensor is several million ohms.
7. Connect a vacuum pump to the red-marked hose and apply vacuum to the valve and check that the resistance falls drastically to less than 1 million ohms. If not, replace the EGR valve, since the temperature sensor is integral with it.
8. With the LH-system ECU connector disconnected, check the Yellow/White wire between the EGR temperature sensor and LH-system ECU connector pin 23 for an open or shorted circuit. Check the Black wire from the temperature sensor to ground for an open or shorted circuit.
9. If the fault persists, disconnect the red-marked hose from the modulating valve and apply vacuum to the hose. If the engine rpm falls, the EGR valve is working properly. If not, check the hose, connections or the vacuum tank.
10. Check that a vacuum is present in the white-marked hose to the modulating valve. If not, check the hose, connections or the vacuum tank.

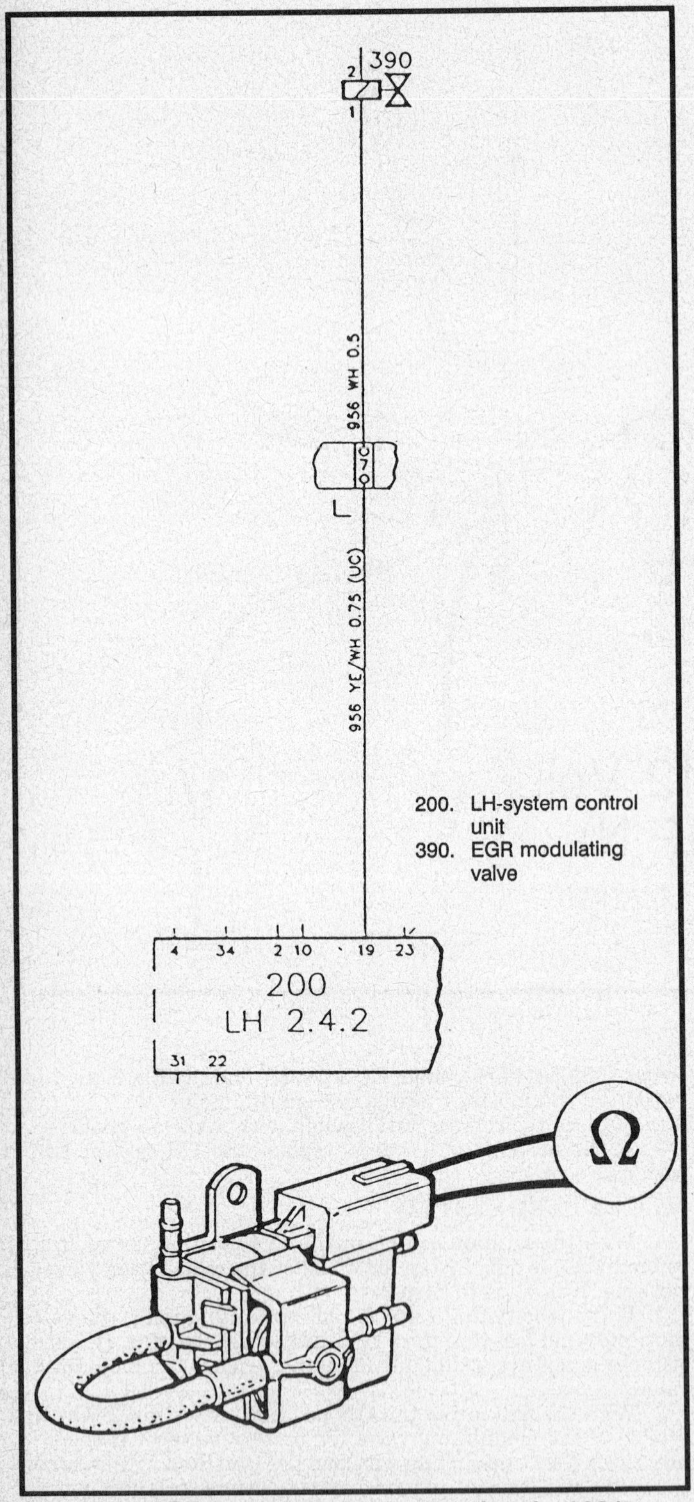

Checking EGR temperature sensor signal—1991 Saab

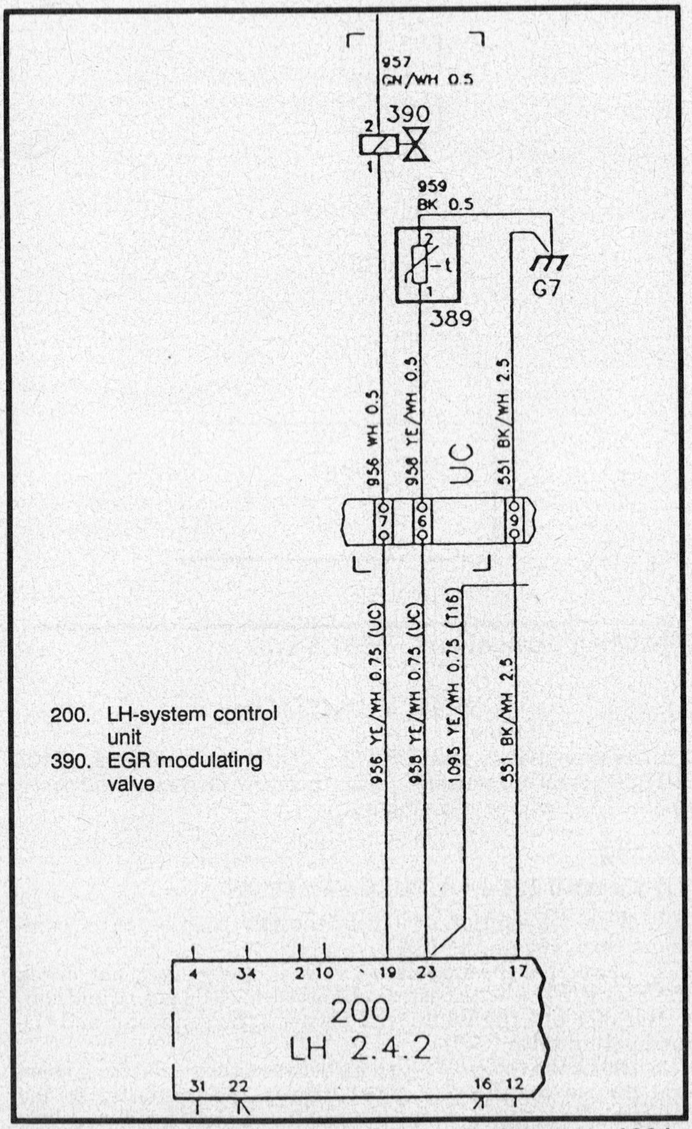

Checking EGR temperature sensor resistance—1991 Saab

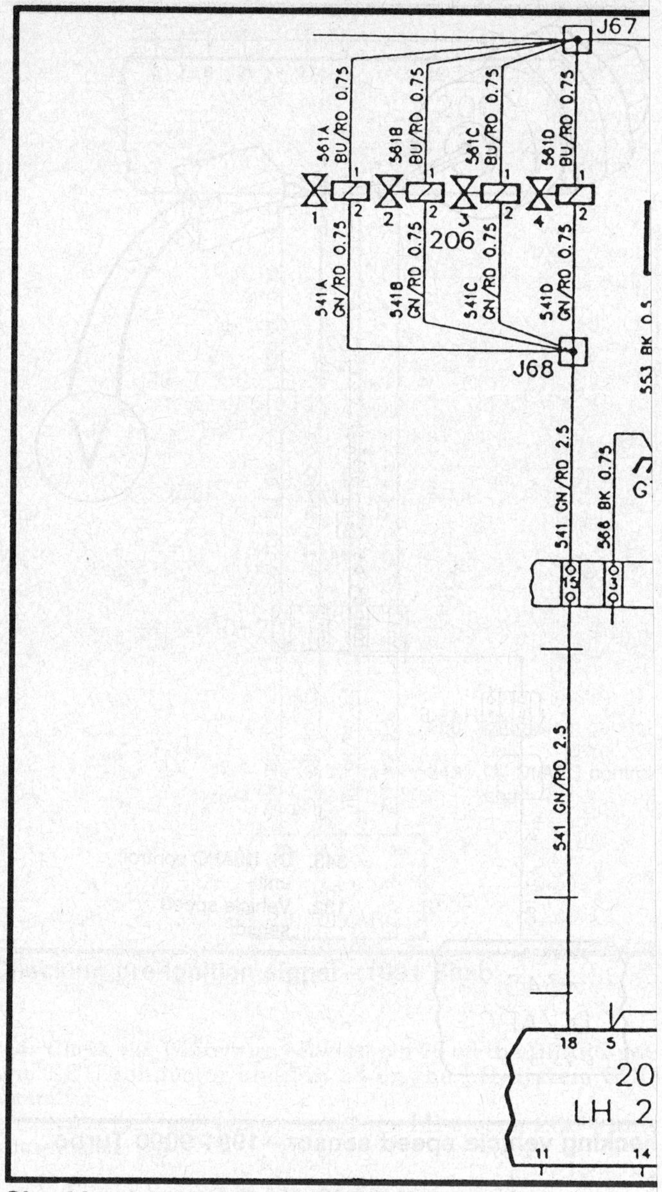

Checking injectors—1991 Saab

4. With the ignition in the **OFF** position and the LH-syster ECU connector and the purge valve connector disconnectec check the Yellow/Red wire between pin 1 on the valve connecto and pin 27 on the LH-system ECU connector for an open o shorted circuit.

5. If the fault still exists, install a known good LH-syster ECU.

VEHICLE SPEED SENSOR

The symptom related to a vehicle speed sensor (VSS) relate fault is a rough idle.

Testing

LH 2.4 AND LH 2.4.1 FUEL SYSTEMS

This test applies to vehicles with the LH 2.4 and 2.4.1 fuel sys tems, except the 9000 Turbo.

1. With the LH-system ECU connector connected, check th voltage between LH-system ECU pin 34 and ground. When ei

OXYGEN SENSOR SIGNAL

Testing

1. Check the oxygen sensor heater by checking for voltage across pins 1 and 2 of connector 59.
2. Check the wiring between pin 5 of the LH-system ECU and connector 60.
3. Check the wiring between pin 5 of the LH-system ECU and the grounding sheath around the oxygen sensor wire.
4. Start the engine and allow it to reach normal operation temperature. Disconnect the oxygen sensor at connector 60 and measure the signal voltage, which should be approximately 0.5 volts.
5. If the fault still exists, replace the oxygen sensor with a known good.
6. If the fault still exists, replace the LH-system ECU with a known good.

AUTOMATIC IDLING CONTROL VALVE

If the AIC valve becomes faulty, rough idling will result.

Testing

LH 2.4 FUEL SYSTEM

1. With the ignition switch in the **ON** position, enter command Code 553 on the ISAT. The automatic idling control (AIC) valve should begin to pulse about once every second.
2. Check the setting of the throttle position sensor by entering ISAT command Code 201.
3. With the ignition switch in the **ON** position and the LH 2.4 system ECU unplugged, check that battery voltage is available between pin 1 of the AIC valve connector and a good ground. If not, check the Blue/Red wire and the Grey/White wire between pin 2 of the AIC valve and pin 87 of the system relay for an open or shorted circuit.
4. With the ignition switch in the **OFF** position, check the Blue/White wire between pin 2 of the AIC valve and LH-system ECU pin 33 for an open or shorted circuit.
5. If the problem still exists, clear the fault codes and test drive the vehicle. Check whether the code is regenerated.
6. If the problem still exists, replace the AIC valve with a known good and repeat Step 5.
7. If the problem still exists, replace the LH 2.4 ECU with a known good.

LH 2.4.1 FUEL SYSTEM

1. With the ignition switch in the **ON** position, enter command Code 553 on the ISAT. The automatic idling control (AIC) valve should begin to pulse about once every second.
2. Check that battery voltage is available from the +30 supply terminal (230).
3. Check the wiring between the +30 supply terminal and pin 30 on the system relay (229).
4. Check the wiring between pin 87 on the system relay and pin 1 on the AIC valve.
5. Check that battery voltage is present between pin 1 on the AIC valve and ground.
6. Check the wiring between pin 33 of the LH-system ECU and pin 2 of the AIC valve.
7. Check the setting and operation of the throttle position sensor.
8. Check that the throttle valve is adjusted as near to closed as possible without binding.
9. If the problem still exists, replace the system relay with a known good.
10. If the problem still exists, replace the AIC valve with a known good.
11. If the problem still exists, replace the LH-system ECU with a known good.

LH 2.4.2 FUEL SYSTEM

1. With the ignition in the **ON** position, enter ISAT command Code 553 and listen to check that the AIC valve is working once a second.
2. With the ignition in the **ON** position, check that battery voltage is available between pin 2 of the AIC valve connector and a good ground. If not, check the Blue/Red wire and Grey/White wire between pin 2 of the AIC valve and pin 87 of the system relay for an open or shorted circuit.
3. With the ignition in the **OFF** position and the AIC valve connector unplugged, measure the resistance across pins 1 and 2, and across pins 2 and 3. The resistance for each winding should be 10–15 ohms. If not, replace the AIC valve.
4. Check the Blue/White wire between LH-system ECU pin 33 and AIC valve pin 1, and the Green/Grey wire between LH-system ECU pin 15 and AIC valve pin 3 for an open or shorted circuit.
5. Install a known good LH-system ECU.

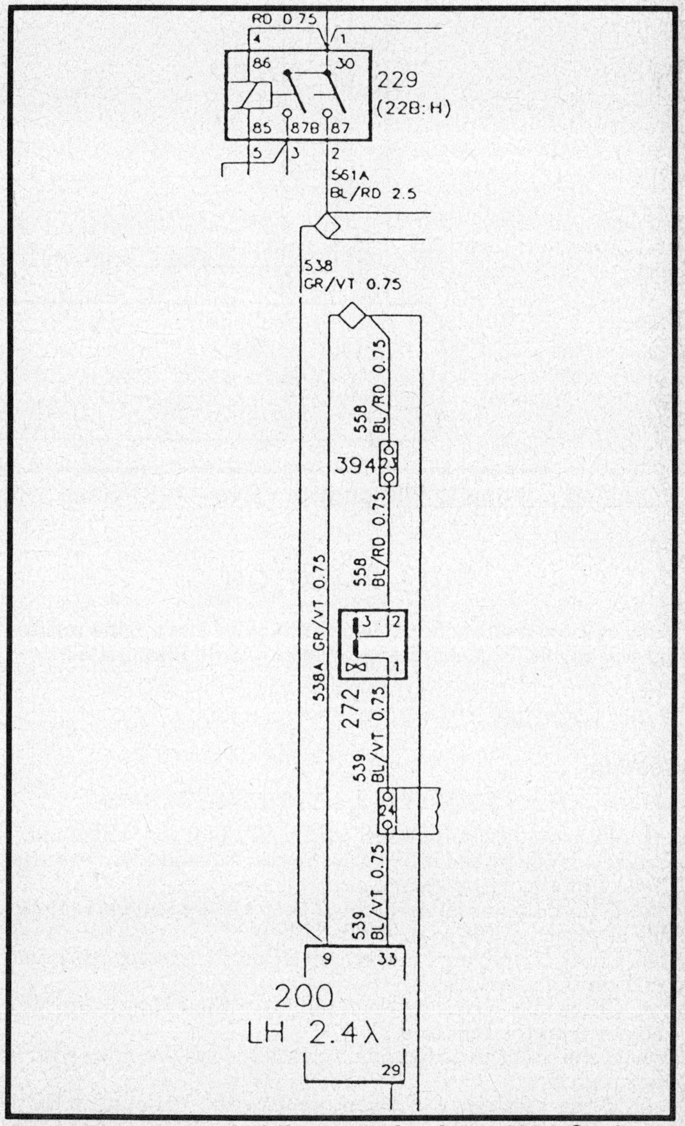

Checking automatic idle control valve—1991 Saab with LH 2.4 fuel system

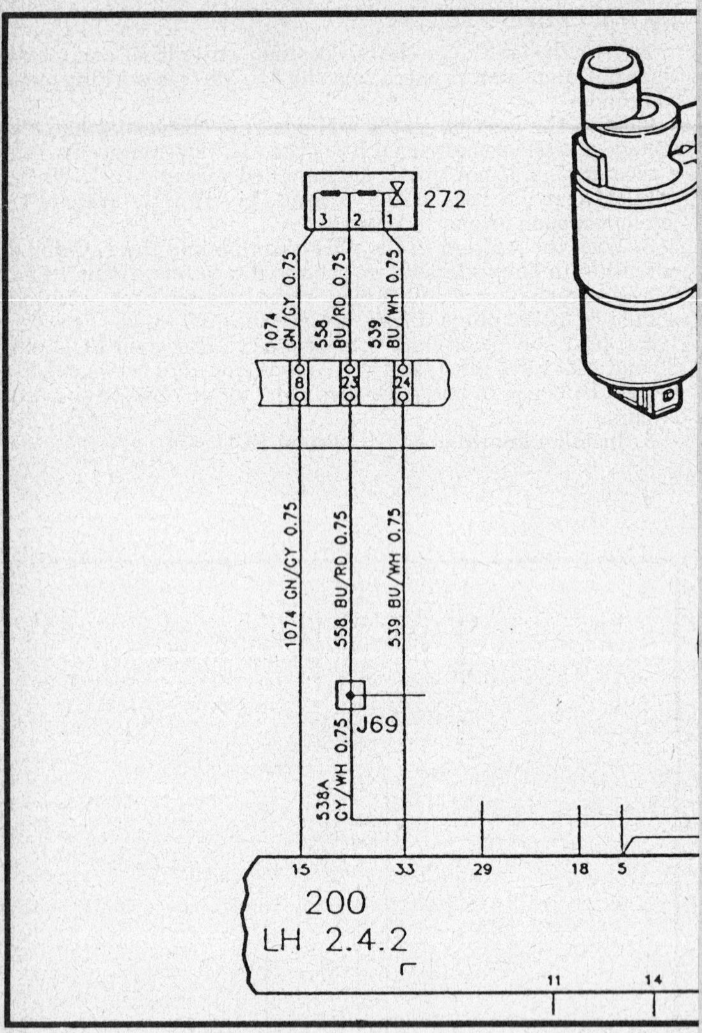

Checking automatic idle control valve—1991 Saab with

FUEL INJECTOR

If an injector malfunctions, poor drivability and engine misfiring will result. The check engine light will be illuminated.

NOTE: The fault could also be ignition related in one of the cylinders.

Testing

LH 2.4, LH 2.4.1 AND LH 2.4.2 FUEL SYSTEMS

1. With the ignition switch in the **ON** position, enter command Code 552 on the ISAT. The injectors should begin to pulse about 15 times every second.
2. If they are not pulsing, check that battery voltage is available at pin 30 of the system relay (229).
3. Check the wiring between pin 87 on the system relay and pin 1 on the injectors.
4. Check the wiring between pin 18 of the LH-system ECU and pin 2 on the injectors.
5. If the problem still exists, replace the system relay with a known good.
6. If the problem still exists, replace the LH-system ECU with a known good.
7. If some of the injectors are pulsing, disconnect the injector connectors.

4. Check the Green/Red wire between the throttle position sensor and pin 12 (LH 2.2) or pin 3 (LH 2.4) on the LH-system ECU.
5. Check that pin 2 of the throttle position sensor is grounded.
6. Enter command Code 451 on the ISAT and listen to check that the solenoid valve is operating. If not, check that voltage is present at pin 2 of the solenoid valve (Green/Red wire) and check that voltage is present between pins 1 and 2 and between pins 2 and 3 on the solenoid valve.
7. If there is not voltage, install a known good DI/APC system ECU. If there is voltage present, disconnect the connector and measure the resistance in the solenoid valve coils (pins 1 and 2, and pins 2 and 3). The resistance should be approximately 3 ohms.
8. If the resistance is not as specified, replace the solenoid valve. If the solenoid valve is operating but no control is achieved, check the condition of the hoses.
9. Check the basic charging pressure and adjust as necessary.

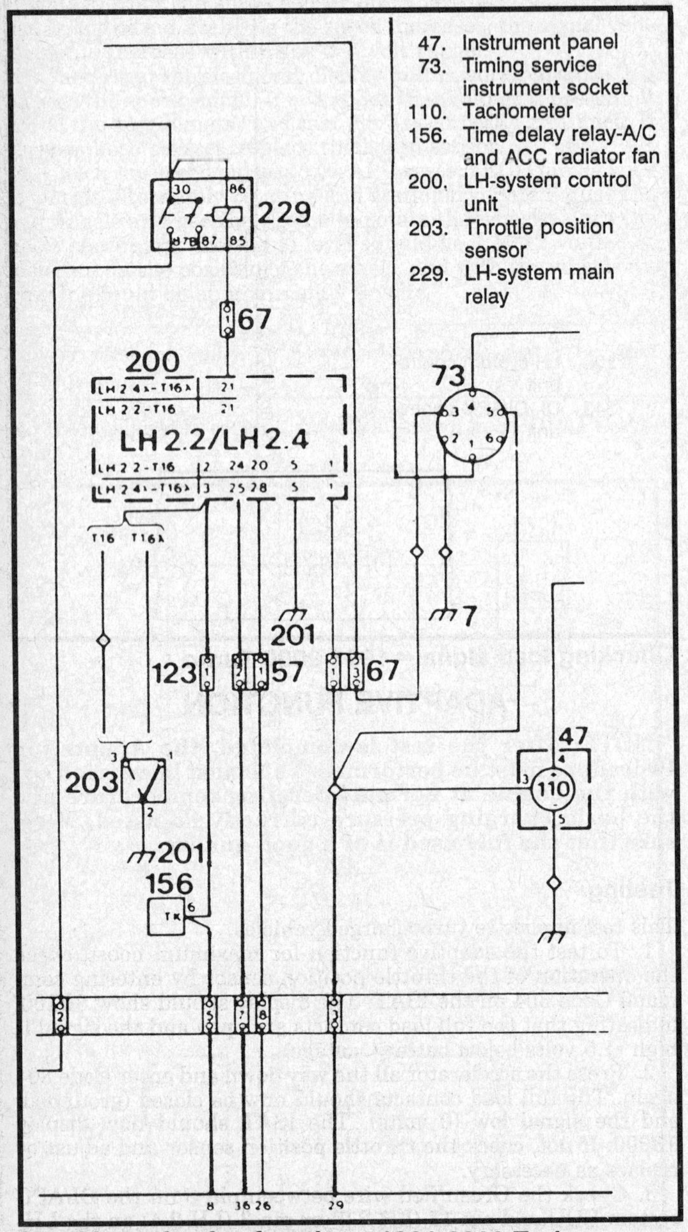

Checking adaptive function—1991 9000 Turbo

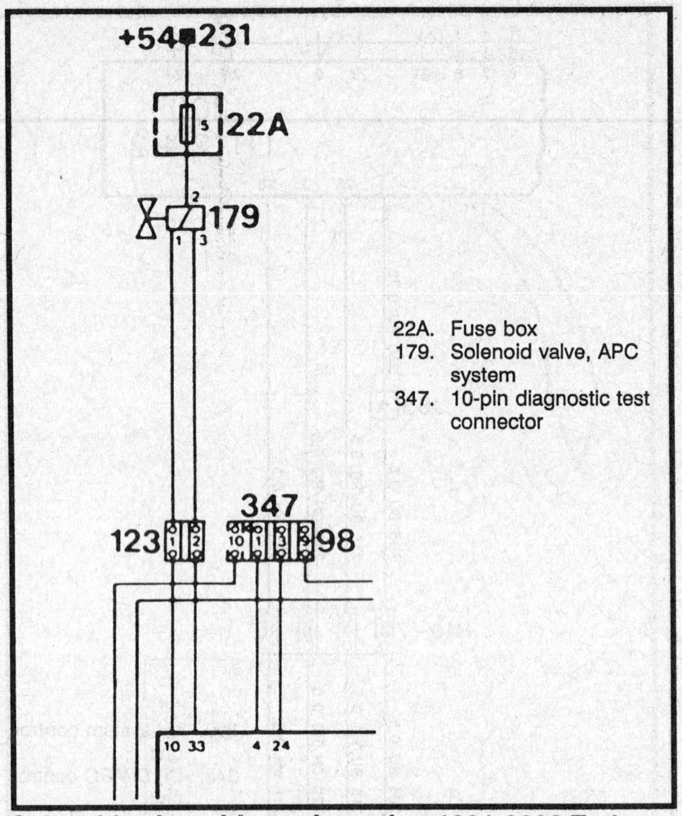

Solenoid valve wiring schematic—1991 9000 Turbo

ADAPTATION PROCEDURE

Procedure

MANUAL TRANSMISSION TRANSMISSION

1. Pull back the rubber boot on the TPS connector, and at the back of the connector connect a jumper wire between pins 1 and 2 (to similate closed contacts).
2. In the highest gear possible, accelerate at full throttle from about 2000 rpm up to 3500 rpm. Since the engine should remain within the effective speed range for adaptation (2750–3250 rpm) for over 3 seconds, ideally the vehicle should be on an uphill gradient.
3. Repeat this procedure until the maximum nominal charging pressure is obtained.
4. Remove the jumper wire.

AUTOMATIC TRANSMISSION

1. Pull back the rubber boot on the TPS connector, and at the back of the connector connect a jumper wire between pins 1 and 2 (to similate closed contacts).
2. Accelerate from approximately 2000 rpm up to 3500 rpm as hard as possible without activating the kickdown function. Since the engine should remain within the effective speed range for adaptation (2750–3250 rpm) for over 3 seconds, ideally the vehicle should be on an uphill gradient.
3. Repeat this procedure until the maximum nominal charging pressure is obtained.
4. Remove the jumper wire.

KNOCK SENSOR SIGNAL

Testing

The following test applies to turbocharged vehicles.

1. Enter command Code 453 on the ISAT and measure the

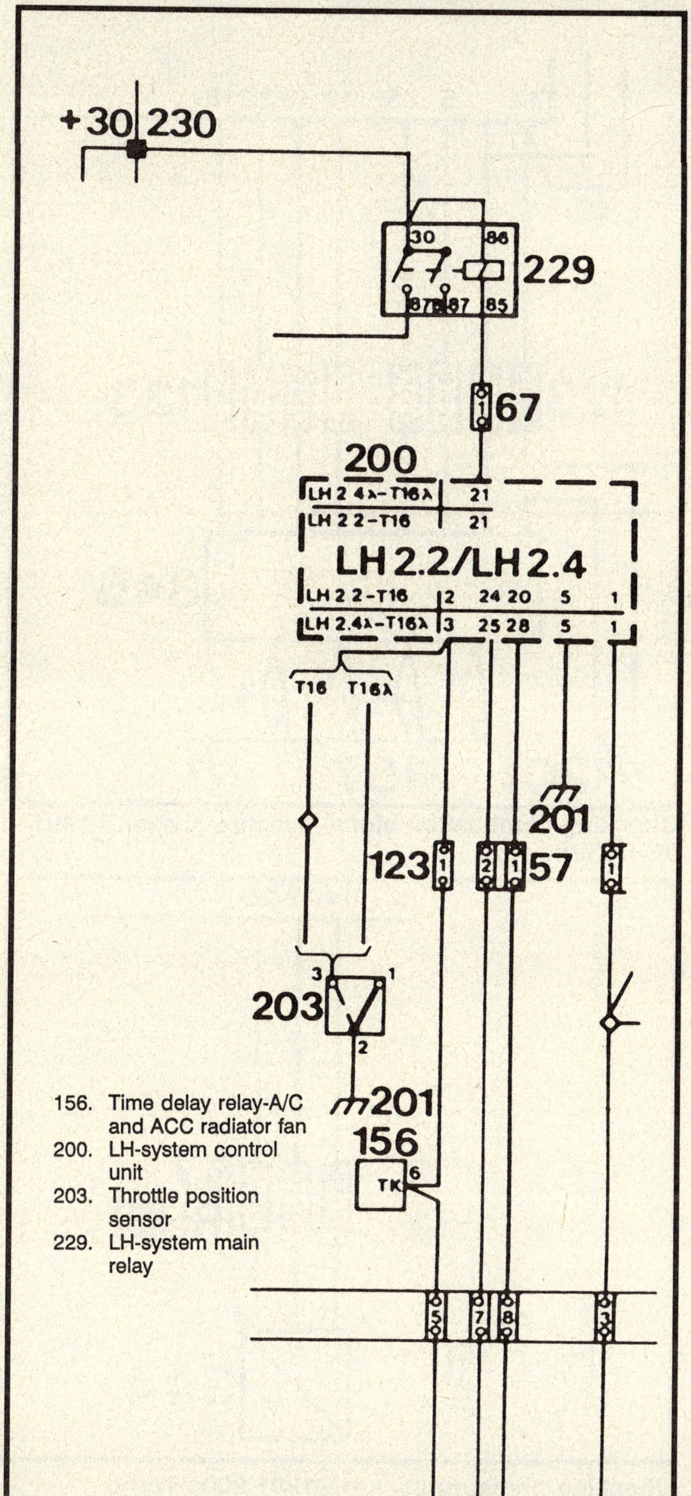

Checking pre-ignition combustion signal—1991 9000 Turbo

voltage between pins 26 and 13 of the DI/APC system ECU connector. The reading should oscillate between 0 volts and approximately 11 volts (1.5 volts below battery voltage) on vehicles with a catalytic converter and between 0 volts and approximately 5 volts on non-catalyst vehicles.

2. If not, replace the DI/APC system ECU with a known good.

3. Check the White wire between pin 26 on the DI/APC system ECU connector and pin 28 on the LH-system ECU connector.

4. If the fault still exists, replace the LH-system ECU with a known good.

KNOCK SENSOR

Testing

The following test applies to turbocharged vehicles.

1. Enter command Code 329 on the ISAT. With the engine stopped, 80000 should appear on the display, indicating the knock sensor voltage is 0.

2. With the engine idling, the ISAT should display 80001, a value corresponding to 0.1 volt, or a slightly higher value if the engine is revved. The value 80003 would correspond to 0.3 volt.

NOTE: The value of the voltage can vary slightly, owing to differences in the vibration level from one engine to another.

3. Check the screened wire between the DI/APC system ECU and the knock sensor for signs of pinching damage, or an open or shorted circuit. Replace the wire if defective or replace the knock sensor.

4. Make sure that other wires or cables are not too close to the knock sensor wire, causing electrical interference.

5. If the fault still exists, replace the DI/APC system ECU with a known good.

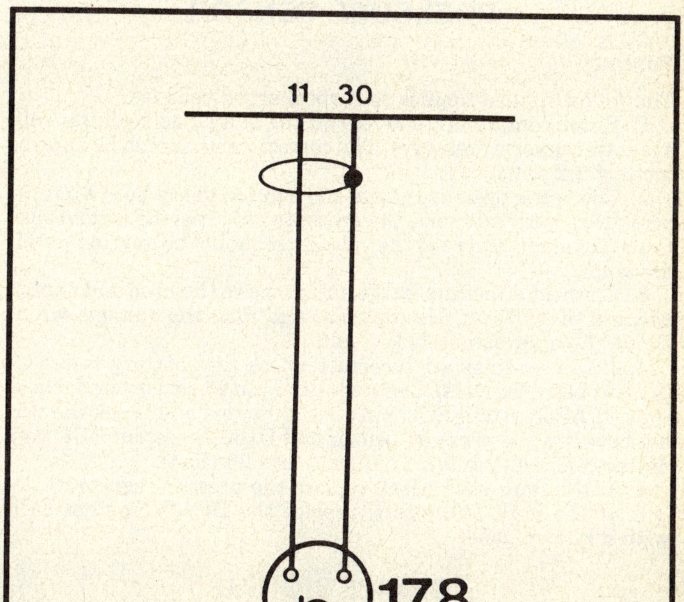

Checking knock sensor signal—1991 9000 Turbo

SYNCHRONIZATION COMBUSTION SIGNAL

Testing

The following test applies to turbocharged vehicles.

1. Repair all other fault codes first and then clear them, since this malfunction code can be caused by exceptionally high temperatures as a result of other faults.
2. Test drive the vehicle and recheck for codes.
3. Replace the spark plugs.
4. Enter command Code 303 on the ISAT. With the engine stopped, 8B000 should be displayed, indicating that the signal is low for both cylinder pairs (1 and 2, and 3 and 4).
5. Enter Code 303 again, this time with the engine running. One of these messages should display: 8B100, 8B200 or 8B300.
6. If not, disconnect the Brown and the Yellow wires from connector 123 between the DI/APC system ECU and the ignition cartridge. Measure the voltage on the ECU side of the connector between the Yellow and Black wires, and between the Brown and Black wires. The value should be approximately 5 volts.
7. Check that battery voltage is present between DI/APC system ECU pin 21 and ground on the DI/APC system ECU connector.
8. Check the Yellow and Brown wires between the ignition cartridge and DI/APC system ECU connector pins 5 and 25 for damage.
9. Check that the Black wire to the ignition cartridge and that DI/APC system ECU pin 13 are grounded.
10. If the fault still exists, replace the DI/APC system ECU with a known good.
11. Reconnect the Yellow and Brown wires to the 123 connector and measure the voltage between DI/APC system ECU connector pins 5 and 13, and 25 and 13. The readings should be less than 1 volt. If not, replace the ignition cartridge.

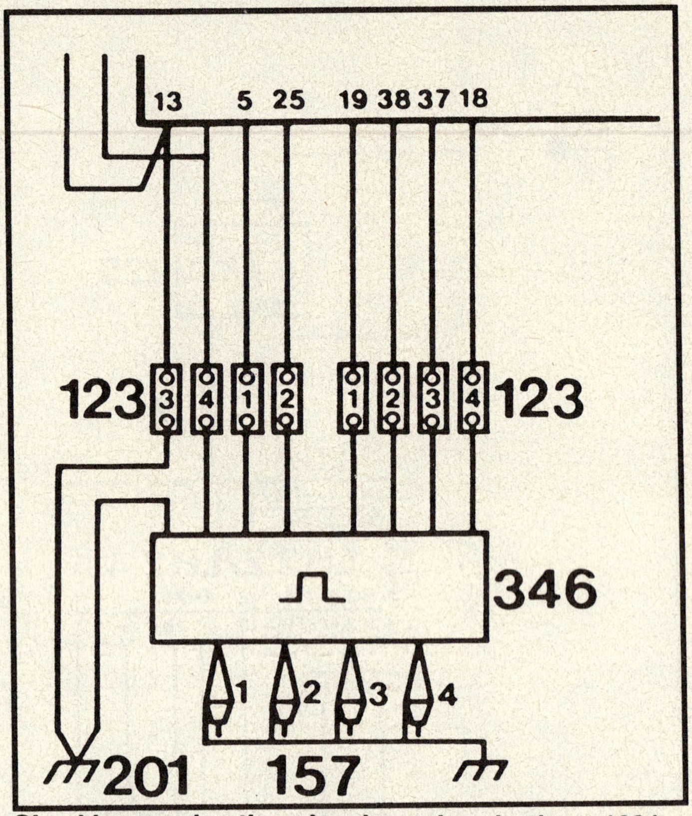

Checking combustion signal synchronization—1991 9000 Turbo

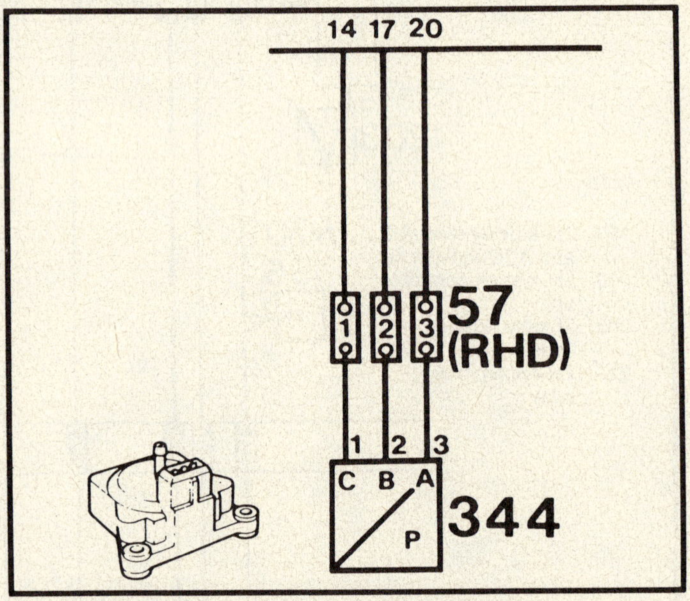

Checking pressure sensor—1991 9000 Turbo

PRESSURE SENSOR

Testing

The following test applies to turbocharged vehicles.

1. Enter command Code 320 on the ISAT and read the voltage (atmospheric pressure). The correct value should be approximately 2.2 volts.
2. Connect a pressure guage and a pump to the hose from the pressure sensor and raise the pressure to 8.7 psi (60 kPa). Enter Code 320 again and read the voltage. It should be approximately 3.6 volts.
3. Connect a vacuum guage and pump to the hose and apply a vacuum of 17.75 in. Hg (60 kPa) and read the voltage, which should be approximately 0.8 volts.
4. If the readings are incorrect, check that battery voltage is present between DI/APC system ECU pin 21 and ground, check that DI/APC system ECU pin 13 is grounded and check the wiring between the pressure sensor and DI/APC system ECU connector pins 14 (White), 17 (Black) and 20 (Red).
5. If the fault still exists, replace the pressure sensor.
6. If the fault still exists, replace the DI/APC system ECU with a known good.

BRAKE SIGNAL

Testing

The following tests apply to turbocharged vehicles.

BRAKE SIGNAL MALFUNCTION

1. Enter command Code 300 on the ISAT without applying the brake. The reading on the ISAT display should be 8B000.
2. If not, check that fuse 24 is good and that voltage is reaching it.
3. Check the functioning of the stop light switch.
4. Check the wiring between fuse 24 and DI/APC system ECU pin 32.
5. If the problem still exists, replace the DI/APC system ECU with a known good.
6. Enter Code 300 again, this time with the brake pedal depressed. The ISAT should now display 8B100.
7. If not, check that the stop light switch is operating properly. If necessary, adjust or replace the switch.

8. Check for a short circuit between the White wire from the stop light switch that runs to DI/APC system ECU connector pin 32 and another wire with voltage.

9. If the fault still exists, replace the DI/APC system ECU with a known good.

IGNITION RETARD FUNCTION

When the pressure is greater than 1.45 psi (10 kPa) and the brakes are applied, the ignition should be retarded to 2 degrees BTDC.

1. Connect a pressure guage and a pump to the hose from the pressure sensor and raise the pressure to 1.45 psi (10 kPa).
2. Connect a timing light to the Orange trigger pulse lead for cylinder number 1.
3. With the engine idling depress the brake pedal. The ignition timing should retard to 2 degrees BTDC.

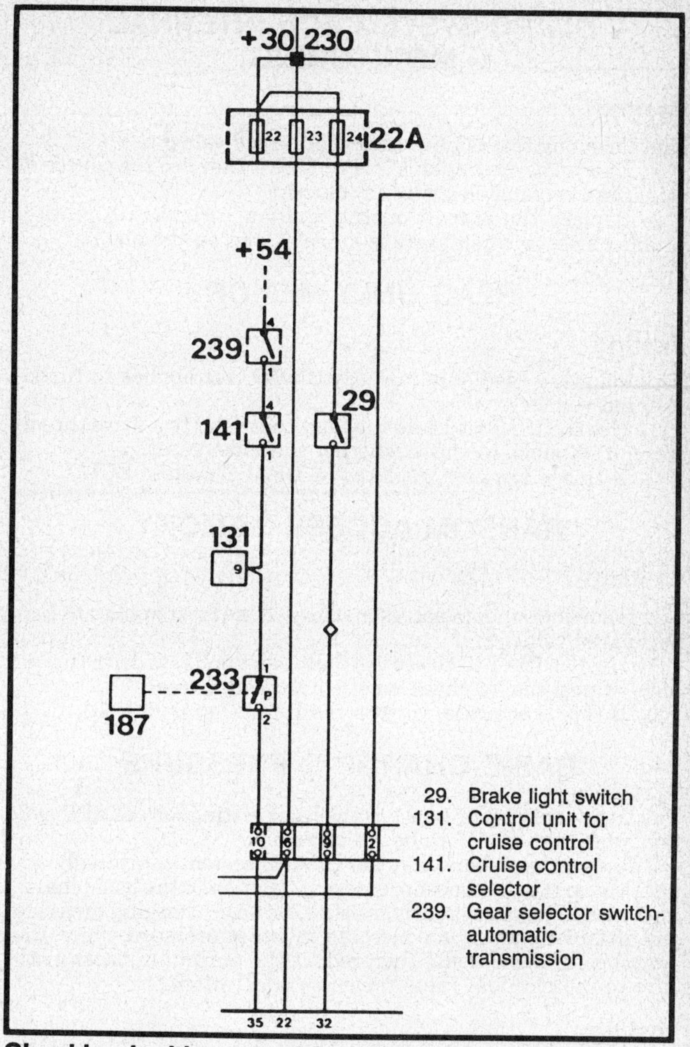

Checking ignition retard function—1991 9000 Turbo

AIR TEMPERATURE SENSOR

This test applies to turbocharged vehicles. If the air temperature signal is faulty or absent, only base line turbocharging pressure is available.

Testing

1. Measure the resistance of the air temperature sensor upstream of the throttle housing. Compare the resistance to specifications. If incorrect, replace the air temperature sensor.
2. If the resistance is correct, check the voltage across the air temperature sensor pins with the ignition switch in the **ON** position. The voltage should be approximately 5 volts.
3. If the correct value is not obtained, check the White wire between the air temperature sensor connector and DI/APC system ECU connector pin 14, and also the Black wire between the sensor connector and DI/APC system ECU pin 17 for an open or shorted circuit.
4. If the wiring is good, install a DI/APC system ECU that is known good.

Resistance values

°C	°F	Ohm (±10%)
−20	−4	14000
−10	14	9000
0	32	5800
10	50	3800
15	58	3000
20	68	2600
25	76	2000
30	86	1700
80	176	320

Air temperature sensor specifications—1991 9000 Turbo

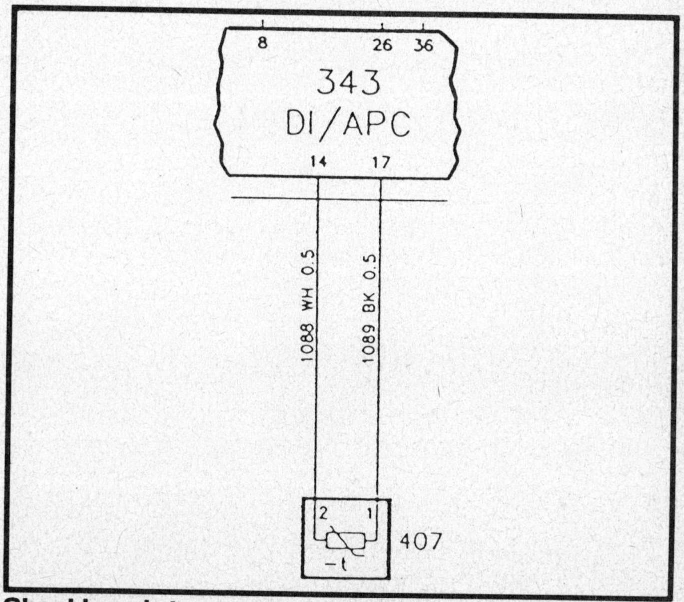

Checking air temperature sensor signal—1991 9000 Turbo

DI/APC SYSTEM ECU INTERNAL MONITORING

Testing

The following test applies to turbocharged vehicles.

1. Check the spark plugs and replace if they are the source of misfiring, causing electrical problems.
2. Replace the shroud on the ignition cartridge if ignition problems occur when there is a heavy load on the engine.

READ ONLY MEMORY

Testing

The following read only memory (ROM) test applies to turbocharged vehicles.

1. Use the ISAT to delete the fault codes and test drive the vehicle, if possible, to check whether the codes reset.
2. If this is the case, replace the DI/APC system ECU.

RANDOM ACCESS MEMORY

Testing

The following random access memory (RAM) test applies to turbocharged vehicles.

1. Use the ISAT to delete the fault codes and test drive the vehicle, if possible, to check whether the codes reset.
2. If this is the case, replace the DI/APC system ECU.

BASIC CHARGING PRESSURE

The basic charging pressure is the basic setting for the APC system and must therefore be set correctly.

The adaptive function in the DI/APC system continually adjusts the maximum boost pressure in relation to the basic charging pressure. Consequently, raising the basic charging pressure will not bring about an increase in boost pressure, since the adaptive function would then reduce the maximum boost pressure to the nominal value (negative adaptation).

Testing

1. Run the pressure gauge hose from inside the vehicle, through the opening in the front door pillar and into the engine compartment. Hang the pressure gauge 8393514 or equivalent in a vertical position to avoid faulty readings.
2. Connect the pressure gauge hose between the fitting on the intake manifold and the hose to the pressure switch.
3. Operate the engine until normal operating temperature is reached.
4. The test starting point for vehicles with manual transmissions is 3rd gear and the engine running below 1500 rpm, and for vehicles with automatic transmissions is lever position 3 and the engine running below 1500 rpm.
5. Drive the vehicle and press the accelerator hard down to accelerate at full throttle. When the engine speed approaches 3000 rpm, brake hard while keeping the accelerator fully depressed, so that full load will be put on the engine at 3000 rpm. Read the pressure on the gauge. The basic charging pressure should be approximately 5.1 ± 0.4 psi (35.2 ± 2.8 kPa).
6. If the basic charging pressure is outside specifications, adjust the basic charging pressure setting.
7. Disconnect the pressure guage.
8. Perform the Adaptation procedure of the Adaptive Function test.

WASTEGATE VALVE AND DIAPHRAGM UNIT

If the basic charging pressure fails to respond to adjusting, perform the following test.

Testing

1. Remove the exhaust bend and make a visual check of the position of the wastegate valve relative to the turbocharger housing.
2. Remove the anti-tamper seal and lock washer and disconnect the pushrod from the wastegate valve actuator.
3. Check that the actuating arm for the wastegate valve moves freely. If not, replace the wastegate valve.
4. Check that the diaphragm unit pushrod moves freely.
5. Check the diaphragm and spring by connecting low-pressure air to the diaphragm unit. Apply pressure and check that the pushrod moves freely and does not bind.
6. Release the pressure and check that the spring returns the pushrod smoothly inside the unit. If not, replace the unit.

ENGINE FAILS TO START—LH FUEL INJECTION SYSTEM

1. Engine fails to start

Symptom or component	Malfunction indicated/possible cause	Action
a. Electrical system	a. Low voltage, break in wiring, poor connection, relay faulty relay or pressure switch	a. Check that battery voltage is 11.5 V minimum. Check system and fuel-pump relay Check for breaks, poor connections or corrosion in wiring and connectors Turbo: check the pressure switch
b. Fuel pump	b. Fuel pump not working or poor delivery	b. Check fuel pump operation: delivery pressure and flow
c. Fuel system	c. Leakage, faulty fuel pressure regulator, blockage in return line.	c. Check for leakage in fuel system Check the line pressure Check fuel-pressure regulator
d. Induction system	d. Air leakage system:	d. Check for leakage in induction downstream of throt tle housing using pressure tester, and visually between air mass meter and throttle housing
e. Injectors	e. Opening duration out of spec; leakage	e. Check opening duration of injectors using a multimeter
f. Temperature sensor	f. Not functioning, break in wiring or poor connection	f. Use multimeter to check the sensor
g. ECU	g. Poor connection, no input signals or ECU faulty	g. Use multimeter to check all ECU input signals
h. Auxiliary	h. Malfunction or air leak air valve	h. Use multimeter to check the valve

ENGINE STARTS BUT STOPS IMMEDIATELY—LH FUEL INJECTION SYSTEM

2. Engine starts but stops immediately

Symptom or component	Malfunction indicated/possible cause	Action
a. Electrical system	a. Low voltage, break in wiring, poor connection, faulty relay or pressure switch	a. Check that battery voltage is 11.5 V minimum. Check system relay and fuel-pump relay Check for breaks, poor connections or corrosion in wiring and connectors Turbo: check the pressure switch
b. Temperature sensor	b. Not functioning, break in wiring or poor connection	b. Use multimeter to check the sensor
c. Induction system	c. Air leakage system:	c. Check for leakage in induction downstream of throttle housing using pressure tester, and visually between air mass meter and throttle housing
d. ECU	d. Poor connection, no input signals or ECU faulty	d. Use multimeter to check all ECU input signals
e. Auxiliary air valv	e. Malfunction or air leak	e. Use multimeter to check the valve
f. CO content too high/low	f. Incorrect setting as per spec.	f. Check idling speed and CO content Check resi stance across air mass meter
g. Throttle position sensor	g. Incorrectly adjusted/ faulty	

ERRATIC IDLING – LH FUEL INJECTION SYSTEM

3. Erratic idling

Symptom or component	Malfunction indicated/possible cause	Action
a. Temperature sensor	a. Not functioning, break in wiring or poor connection	a. Use multimeter to check the sensor
b. Air mass meter/CO	b. CO content set too high/ low, filament broken	b. Check CO setting on air mass meter and for broken filament content
c. Induction system	c. Leakage	c. Use pressure gauge to check for leaks in induction system. Depression should be between −0.4 and −0.8 bar
d. Injectors	d. Opening duration out of spec; leakage	d. Check opening duration of injectors using a multimeter
e. Fuel system	e. Pressure too low	e. Check line pressure in fuel system and operation of fuel-pressure regulator
f. ECU	f. Poor connection, no input signals or ECU faulty	f. Use multimeter to check all ECU input signals
g. Auxiliary air valve	g. Malfunction or air leak	g. Use multimeter to check the valve
h. Electrical system	h. Break in wiring or connectors. Poor on engine	h. Check wiring and connectors Check earth connections earthing and battery.
i. EGR valve	i. Valve not working. Valve sticking.	i. Check the EGR system

POOR RESPONSE ON ACCELERATION—LH FUEL INJECTION SYSTEM

4. Poor response on acceleration

Symptom or component	Malfunction indicated/possible cause	Action
a. Air mass meter/CO	a. CO content set too high/ low, filament broken content	a. Check CO setting on air mass meter and for broken filament
b. Induction system	b. Leakage	b. Use pressure gauge to check for leaks in induction system. Depression should be between −0.4 and −0.8 bar
c. Injectors	c. Opening duration out of spec; leakage	c. Check opening duration of injectors using a multimeter
d. Fuel system	d. Delivery flow/pressure	d. Check line pressure and delivery flow in fuel system
e. ECU	e. Poor connection, no or weak input signals or ECU faulty	e. Use multimeter to check all ECU input signals
f. Temperature sensor	f. Not functioning, break in wiring or poor connection	f. Use multimeter to check the sensor wiring or poor connection
g. Electrical system	g. Break in wiring or connectors. Poor on engine	g. Check wiring and connectors. Check earth connections earthing and battery.

ENGINE ERRATIC IN ALL SPEED RANGE—LH FUEL INJECTION SYSTEM

5. Engine erratic in all speed ranges

Symptom or component	Malfunction indicated/possible cause	Action
a. ECU	a. Wiring fault in connector.	a. Check ECU connector pins 9, 11, 13, 17 and 25
b. Electrical system	b. Alternator. Low voltage, break in wiring, poor earth connection	b. Check operation of alternator
c. Injectors	c. Opening duration out of spec	c. Check opening duration of injectors using a multimeter
d. Turbo/APC system	d. Malfunctionorleakage	d. Check operation of Turbo Check operation of APC system
e. Throttle position	e. Incorrectly adjusted, throttle butterfly sensor.	e. Use multimeter to check throttle-position Check sensor sticking that microswitch correctly adjusted and butterfly not sticking
f. Air mass meter	f. No burn-off function. Filament broken	f. Check that filament in air mass meter is not broken and that burn-off function is working
g. Induction	g. Air leakage	g. Check for leakage in induction systemsystem: downstream of throttle housing using pressure tester, and visually between air mass meter and throttle housing
h. Fuel system	h. No fuel pressure or pressure too low.	h. Check line pressure in fuel system Inadequate delivery flow.
i. Temperature sensor	i. Not functioning, break in wiring or poor connection	i. Use multimeter to check the sensor

EXCESSIVE FUEL CONSUMPTION—LH FUEL INJECTION SYSTEM

6. Excessive fuel consumption

Symptom or component	Malfunction indicated/possible cause	Action
a. Fuel system	a. Fuel-pressure regulator	a. Check pressure in fuel system and pump delivery
b. Injectors	b. Malfunction unplugging lead.	b. Check injectors one at a time by Engine rpm should drop noticeably
c. Temperature sensor	c. Not functioning	c. Use multimeter to check the temperature sensor
d. Throttle position sensor	d. Incorrectly adjusted	d. Check microswitch in throttle-position sensor
e. Air mass meter high	e. Filament broken	e. Check filament in air mass meter—CO and that burn-off function content too operating high
f. Electrical system	f. Break in wiring or connectors. Poor on engine earthing	f. Check wiring and connectors Check earth connections and battery.
g. ECU	g. Poor connection, no or input signals	g. Use multimeter to check all ECU weak input signals or ECU faulty

ENGINE FAILS TO REV UP (NO POWER)—LH FUEL INJECTION SYSTEM

7. Engine fails to rev up (no power)

Symptom or component	Malfunction indicated/possible cause	Action
a. Induction system	a. Leakage	a. Use pressure gauge to check for leaks in induction system.
b. Throttle butterfly.	b. Faulty adjustment	b. Check that butterfly opens fully
c. Fuel system	c. Pump performance poor/low voltage	c. Check performance of fuel pump Check power supply to pump – 11.5 V minimum
d. Injectors	d. Malfunction	d. Check injectors one at a time by unplugging lead. Engine rpm should drop noticeably
e. Air mass meter	e. Filament broken	e. Check filament in air mass meter and that burn-off function operating
f. Turbo/APC system	f. Malfunction	f. Check operation of Turbo Check operation of APC system
g. Temperature sensor	g. Not functioning, break in wiring or poor connection	g. Use multimeter to check the sensor
h. Electrical system	h. Break in wiring or connectors. Poor or earth connectors	h. Check wiring and connectors Check earth connections on engine earthing and battery.
i. ECU	i. Poor connection, no or weak input signals or ECU faulty	i. Use multimeter to check all ECU input signals

CO CONTENT TOO LOW OR TOO HIGH—LH FUEL INJECTION SYSTEM

8. CO content too low or too high

Symptom or component	Malfunction indicated/possible cause	Action
a. Induction system	a. Leakage	a. Use pressure gauge to check for leaks in induction system.
b. Fuel system	b. Leakage. Pressure regulator faulty. Blockage in return line. Pump performance	b. Check the line pressure Check fuel-pressure regulator by slowly pinching the return line Check pump per formance
c. Air mass meter	c. Filament broken. Noburn-offfunction.	c. Check filament in air mass meter and that burn-off function operating
d. Temperature sensor	d. Not functioning	d. Use multimeter to check the temperature sensor
e. Electrical system	e. Break in wiring or connectors. Poor earthing	e. Check wiring and connectors Check earth connections on engine and battery.
f. ECU	f. Poor connection, no input signals or ECU faulty	f. Use multimeter to check all ECU input signals

APC TURBOCHARGED ENGINE DIAGNOSTICS—900 SERIES

Fault	Cause	Remedy
Noise or vibration from the turbo compressor	Poor lubrication of the turbo shaft bearing	Check the oil pressure and flow to the turbo. If the fault should persist after remedial action (permanent bearing damage) exchange the turbo compressor.
	Leakage in the induction or exhaust system	Tighten leaking connections and replace defective seals and gaskets
	Unbalanced turbo shaft owing to damage	Exchange the turbo compressor
Insufficient charging pressure	Leakage between the compressor and cylinder head or between the cylinder head and turbine	Tighten leaking connections and replace defective seals and gaskets
	Incorrect setting of charging pressure	Adjust the charge pressure regulator
	Valve in charge pressure regulator sticks in open position	Overhaul the charge pressure regulator
	Partially clogged exhaust system	Clean or replace exhaust system
	Clogged air cleaner	Change cartridge
	Binding turbo shaft	Exchange turbo compressor
Excessive charging	Incorrect setting of charging pressure	Adjust charge pressure regulator
	Leakage at exhaust pressure line connections	Tighten; if necessary, replace nipples
	Clogged exhaust pressure line	Remove and clean
	Damaged diaphragm unit in charge pressure regulator	Replace diaphragm unit
	Flap valve in charge pressure regulator binds in closed position	Replace the charge pressure regulator housing
	Ice formation in exhaust pressure line. (Excessive pressure occurs 1-2 min after cold start when ambient temperature below freezing)	Avoid heavy loading of engine immediately after cold starting

APC TURBOCHARGED ENGINE DIAGNOSTICS—900 SERIES

Fault	Cause	Remedy
Engine knocking	Excessive charging pressure	Adjust charging pressure
	Unsuitable fuel (octane too low)	Change fuel
	Ignition setting too far advanced	Adjust timing
Oil leakage at turbo shaft seals (oil fumes in exhaust)	Poor return flow from turbo: - Clogged return line - Excessive crankcase pressure - Air cleaner clogged, oil coating on compressor seals	 Check return line Check crankcase ventilation Change air cleaner
	Turbo unit seals damaged	Exchange turbo compressor

APC BASIC CHARGING PRESSURE DIAGNOSTICS—900 SERIES

Symptom	Possible cause	Remedy
Not possible to adjust basic pressure as detailed.	Faulty charging pressure regulator valve.	Remove the exhaust elbow. Visually check that the valve flap is in contact with the turbocharger body. Remove the seal and circlip. Unhook the diaphragm unit push rod from the charging pressure regulator valve. Check that the valve shaft rotates freely. Replace faulty charging pressure regulator valve.
	Sticking diaphragm unit push-rod bushing	Check that the diaphragm unit push rod moves freely. Replace faulty diaphragm unit.
	Blocked orifice in the solenoid valve hose connection to the turbocompressor (connection "C").	**Note** Can be temperature-sensitive. Clean orifice.

APC COMPONENT DIAGNOSTICS—900 SERIES

Symptom	Possible cause	Remedy (refers to APC system)
Solenoid valve not functioning when APC system tested in car as per"Checking of components"	The control unit has no voltage between terminal 14 (+) and 6 ground (-). **Note.** The wiring harness connector should be left connected to the control unit during the measurement.	Check that fuse 19 has not blown. Remove any oxidation present. Check the wiring in the connectors at distributor panel (fuse box) and control unit. Check that the (+) and ground (-)cables are undamaged. Rectify as required.
	Loose connector on knock detector. Broken cable between knock detector and control unit.	Check electrical connection at the connectors (in the engine compartment and on the control unit). Rectify as required.
	Knock detector not properly tightened.	Check tightness. See under 'Removing and refitting of com-ponents, knock detector'.
	Faulty knock detector.	Test system with known good knock detector. If symptoms disappear, replace defective knock detector.
	Solenoid valve not functioning despite supply greater than 5 volts. **Note.** The solenoid electrical connector is to remain connected.	Check the solenoid valve. See under 'Checking components, checking the solenoid valve'.
	Faulty pressure transducer.	Check the pressure switch. See under 'Checking components, checking the pressure transducer'.
	Faulty control unit.	Test system with known good control unit. If symptoms disappear, replace defective control unit.

APC SYSTEM DIAGNOSTICS—900 SERIES

Symptom	Possible cause	Remedy
Maximum charging pressure too low (low engine output)	Incorrectly adjusted basic pressure.	Check the basic setting of the charging pressure regulator valve. See 'Adjusting the basic pressure of the charging pressure regulator'.
	Loose or oxidized electrical connections in the APC-System.	Check electrical connections of control unit, bulkhead connector, pressure transducer (2 connections) fuel pump relay, knock detector plug and common ground in the engine compartment.
		Also check for broken cables in the knock detector-to-control unit, pressure switch-to-control unit, or fuel pump relay to control unit, wiring. Rectify if required.
	Defective (short circuit) stop light switch.	Replace.
	Faulty or wrongly adjusted cruise control vacuum switch (T-16).	Adjust switch or replace replace if defective.
	Faulty knock detector.	Test system with known good knock detector. If symptoms disappear, knock detector was at fault. Replace.
	Faulty pressure transducer.	Check the pressure transducer. See 'Checking the pressure transducer'. Replace the pressure transducer if required.
	Solenoid valve does not open.	Check the solenoid valve. See 'Checking the solenoid valve' under 'Checking components'. Replace faulty solenoid valve.
	Faulty control unit.	Test system with known good knock control unit. If symptoms disappear, control unit was at fault. Replace.

APC SYSTEM DIAGNOSTICS—900 SERIES

Symptom	Possible cause	Remedy
	The knock detector registers abnormal engine vibrations.	Connect the test wiring harness with service box (83 93 548 or 83 94 074 as appropriate), (see 'Checking the APC system knock control'). If the LED flashes when driving under light load listen for abnormal engine noise. If the engine sounds normal try a new knock detector.
	Faulty spring or sticking rod in the diaphragm unit.	Try a new diaphragm unit.
	Faulty turbocompressor.	Remove the exhaust elbow and the inlet to the compressor. Check visually that the turbine wheel and the compressor are not damaged. Replace faulty turbocompressor.
Charging pressure too high (also covers violent changes in charging pressure during acceleration). Pressure switch cuts out.	Leak in turbocompressor - solenoid hose.	Check for leaks. Replace faulty hose.
	Leak in solenoid valve - diaphragm unit hose.	Check for leaks. Replace faulty hose.
	Leak in diaphragm unit.	Check for leaks. Replace faulty diaphragm unit.
	Ruptured diaphragm unit.	Check for leaks. Replace faulty diaphragm unit. Check the mounting of the diaphragm unit housing on the bracket. Replace faulty diaphragm unit.
	Leak in hose to pressure transducer.	Check for leaks. Replace hose.
	Faulty pressure transducer.	Check pressure transducer. See 'Checking pressure transducer'. Replace faulty pressure transducer.
	Solenoid valve does not shut due to sticking valve spool.	Check the APC system in the car. Replace faulty solenoid valve.
	Blocked orifice in the solenoid hose connection to the turbocompressor (connection 'C').	**Note.** Can be temperature-sensitive. Check and clean orifice as necessary.

APC SYSTEM DIAGNOSTICS—900 SERIES

Symptom	Possible cause	Remedy
	Incorrect basic charging pressure setting.	Check the basic setting of the charging pressure regulator valve with the car on the road. See 'Checking the basic setting of the charging pressure regulator on the road'. Adjust as necessary.
	Short circuit in the pressure transducer signal circuit.	Check the wiring between pressure transducer and the control unit. Rectify short circuit.
	Faulty control unit.	Test system with known good control unit. Replace control unit if defective.
	Shaft sticking in the pressure regulator valve.	Remove the seal an circlip. Unhook the diaphragm unit push rod from the regulator valve control arm stud. Check that the valve shaft rotates easily. Replace faulty charging pressure regulator valve.
	Sticking diaphragm unit push rod bush.	Replace diaphragm unit.
Normal charging pressure but no pressure reduction despite intensive knocking/pinking.	APC system not functioning.	See 'Checking the APC system knock control'.
	Bad contact in the knock detector wiring.	Check the connector between the knock detector and the car wiring harness.
	Knock detector not properly tightened.	Check tightness. See 'Remov- and refitting of components, knock detector'.
	Faulty knock detector.	Test system with known good knock detector. If symptoms disappear, knock detector was at fault. Replace.
	Faulty control unit.	Test system with known good control unit. If symptoms disappear, control unit was at fault. Replace.

APC SYSTEM DIAGNOSTICS—900 SERIES

Symptom	Possible cause	Remedy
	Bearing sticking in the charging pressure regulator valve.	Remove the seal and circlip. Unhook the diaphragm unit push rod from the regulator valve control arm stud. Check that the valve shaft rotates easily. Replace faulty charging pressure regulator valve.
	Diaphragm unit rod bush sticking.	Replace diaphragm unit.
Solenoid valve active at idling speed. (Irregular chattering sound.)	Idling speed too low.	Increase idling speed to 875 ± 50 r/min approx.
	Broken cable between pressure transducer and control unit.	Check the wiring. Rectify as required.
	Faulty pressure transducer.	Check the pressure transducer. See 'Checking the pressure transducer'. Replace faulty pressure transducer.
	Bad contact/broken cable between knock detector and control unit wiring.	Check the electrical connections at the connector in the engine compartment and at the control unit. Check that the wiring between the knock detector and the control unit is unbroken. Rectify as required.
	Knock detector not properly tightened.	Check tightness. See 'Removal and refitting of components knock detector'.
	Faulty knock detector.	Test system with known good knock detector. If symptoms disappear, knock detector was at fault. Replace.
	Faulty control unit.	Test system with known good control unit. If symptoms disappear, control unit was at fault. Replace.

TURBOCHARGED ENGINE TROUBLE SYMPTOMS

All symptom diagnosis, whether the vehicle be equipped with a turbocharger or not, is complementary to fault-diagnosis carried out by the ISAT tester and is therefore of limited value. Temporary or intermittent faults cannot be traced or repaired according to symptoms.

900 Testing

For symptom testing, refer to tests for the APC system. The tests for the 9000 Turbo are designed for use on the DI/APC system.

9000 Testing

The following tests apply to the 9000 Turbo with the DI/APC system.

NOTE: Follow in order the steps outlined for testing.

ENGINE FAILS TO START

1. Check the spark plugs for fouling.
2. Check for a crankshaft sensor code and check the crankshaft sensor.
3. Check the ignition cartridge.
4. Check the wiring or connections at the DI/APC system ECU or replace it with a known good.

ENGINE STARTS BUT STALLS IMMEDIATELY AFTERWARDS

1. Check the spark plugs for fouling.
2. Check for a crankshaft sensor code and check the crankshaft sensor.
3. Check the ignition coils and the shroud.
4. Check the ignition cartridge.
5. Check the wiring or connections at the DI/APC system ECU or replace it with a known good.

POOR ENGINE RESPONSE TO THROTTLE, ENGINE FAILS TO REV UP OR RUNS ERRATICALLY AT ALL SPEEDS

1. Check the spark plugs for fouling.
2. Check for a pressure sensor code or check the pressure sensor.
3. Check for a knock sensor code or check the knock sensor.
4. Check the quality of fuel being used. Refuel with a good quality fuel.
5. Check the engine for loose components or connections.
6. Check the ignition cartridge.
7. Check the wiring or connections at the DI/APC system ECU or replace it with a known good.

LOSS OF POWER OR LOW BOOST PRESSURE

With ISAT

1. With the engine running, enter ISAT command Code 283 or 229 and check for pinging while driving. The number of times knocking is detected will be displayed on the screen.
2. If pinging is occurring, use the ISAT to enter diagnostics for the LH system. Enter command Code 206 and watch the ISAT display: 8B106 (fuel boosting on knocking), 8B006 (fuel boosting on pre-ignition) and alternating between the two (when the engine is running normally without pinging).
3. If there is no fuel boosting on pinging, enter command Code 453 with the ignition switch **ON** and the engine stopped. The display should now show 11111 and the voltage at ECU connector pin 26 should alternate between 0–12 volts.
4. If not, check the 12-volt feed from LH-system ECU pin 20 to pin 26 of the DI/APC system ECU. If the circuit has no voltage, check the wire between the ECU's. If the wiring is good, install a known good LH-system ECU.

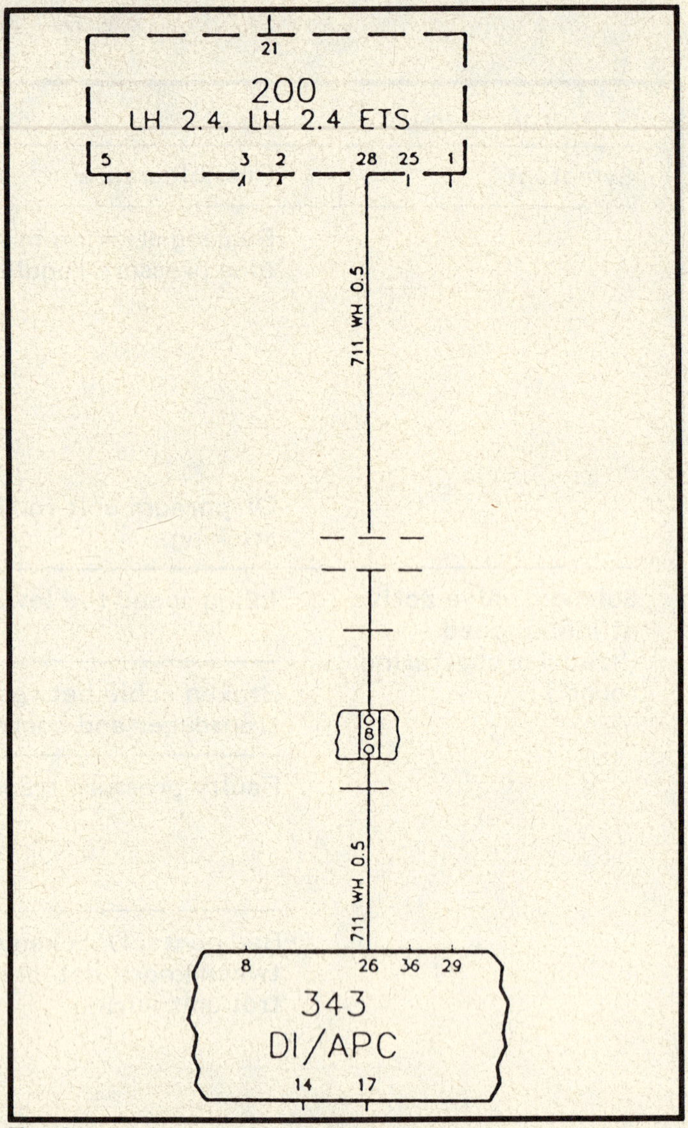

Testing boost pressure—9000 Turbo

5. If 12 volts is reaching pin 26 but the voltage is not alternating between 0–12 volts when command Code 453 has been entered, replace the DI/APC system ECU with a known good.

Without ISAT

1. Check for a Code 44461 indicating a maximum boost pressure fault.
2. Check the quality of fuel being used. Refuel with a good quality fuel.
3. Check the spark plugs for fouling.
4. Check for a knock sensor code or check the knock sensor.
5. Check for a pressure sensor code or check the pressure sensor.
6. Check for a Code 45360 indicating a brake signal fault.
7. Check for a Code 44461 indicating a throttle sensor related fault. Perform steps 1–5 of the Adaptive Function.
8. Check the 12 volt signal from the cruise control system (141) to pin 35 of the DI/APC system ECU as follows:
   a. Connect a vacuum pump to the Black outlet on the cruise control vacuum unit.
   b. Enter command Code 301 on the ISAT. The cruise control vacuum switch should now be open, and no 12 volt signal

can reach the DI/APC system ECU. The display should show 8B000.

c. Use the vacuum pump to raise a vacuum in the cruise control vacuum unit. Listen to check that the vacuum switch makes the circuit.

d. Enter command Code 301 on the ISAT again. The display should show 8B100.

e. If the results are not as described, repair the cruise control system.

9. Perform steps 6 and following of the Adaptive Function to locate a problem related to the solenoid valve.

10. Check the wastegate valve and the diaphragm unit:

a. Remove the exhaust bend and make a visual check of the position of the wastegate valve relative to the turbocharger housing.

b. Remove the anti-tamper seal and shake-proof washer (or split pin) and disconnect the pushrod from the wastegate valve actuator.

c. Check that the actuating arm for the wastegate flap valve moves freely. If defective, replace the wastegate valve.

d. Check that the diaphragm unit pushrod moves freely.

e. Check the diaphragm and spring by applying low pressure to the diaphragm unit. Check that the pushrod slides freely without any tendency to bind. Release the pressure and check that the spring returns the pushrod smoothly inside the unit. If defective, replace the diaphragm unit.

11. Replace the turbocharger unit.

12. Check the engine for loose components or connections.

13. Check the ignition cartridge.

14. Check the wiring or connections at the DI/APC system ECU or replace it with a known good.

SEVERE PINGING DURING ACCELERATION

1. Check the quality of fuel being used. Refuel with a good quality fuel.

2. Check for a pressure sensor code or check the pressure sensor.

3. Check for a knock sensor code or check the knock sensor.

4. Perform steps 6 and following of the Adaptive Function to locate a problem related to the solenoid valve.

5. Check the wastegate valve and the diaphragm unit:

a. Remove the exhaust bend and make a visual check of the position of the wastegate valve relative to the turbocharger housing.

b. Remove the anti-tamper seal and shake-proof washer (or split pin) and disconnect the pushrod from the wastegate valve actuator.

c. Check that the actuating arm for the wastegate flap valve moves freely. If defective, replace the wastegate valve.

d. Check that the diaphragm unit pushrod moves freely.

e. Check the diaphragm and spring by applying low pressure to the diaphragm unit. Check that the pushrod slides freely without any tendency to bind. Release the pressure and check that the spring returns the pushrod smoothly inside the unit. If defective, replace the diaphragm unit.

6. Replace the turbocharger unit.

7. Check for a Code 44461 indicating a throttle sensor related fault. Perform steps 1–5 of the Adaptive Function.

8. Check the wiring or connections at the DI/APC system ECU or replace it with a known good.

ENGINE MISFIRES AND FAILS TO RUN ON ALL CYLINDERS

1. Check the spark plugs for fouling.

2. Check the ignition coils and the shroud.

3. Check the ignition cartridge.

4. Check the wiring or connections at the DI/APC system ECU or replace it with a known good.

EXCESSIVE FUEL CONSUMPTION

This condition is often accompanied by a rough running engine.

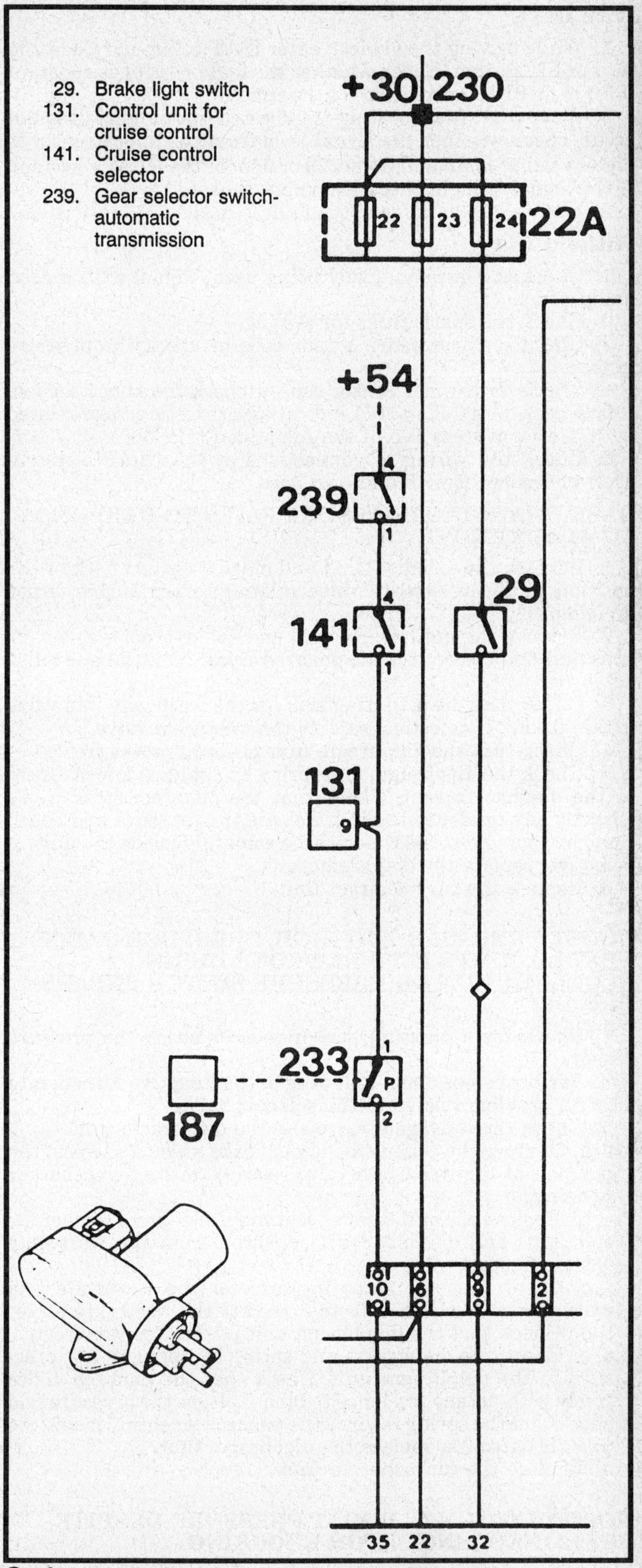

Cruise control wiring schematic—1991–92 turbocharged Saab

With ISAT

1. While driving the vehicle, enter ISAT command Code 206 for the LH system. Check whether the fuel boosting is constant (8B006 or 8B106 displayed all the time).
2. If constant fuel boosting is indicated but pinging does not occur, check whether the signal level from the knock sensor is high by using command Code 229 or 283 for the DI/APC system. If the signal is high, check the knock sensor.
3. If the signal is correct, the fault is in the LH fuel system.

Without ISAT

1. Check the quality of fuel being used. Refuel with a good quality fuel.
2. Check the spark plugs for fouling.
3. Check for a pressure sensor code or check the pressure sensor.
4. Check for a knock sensor code or check the knock sensor.
5. Check for a Code 44461 indicating a throttle sensor related fault. Perform steps 1–5 of the Adaptive Function.
6. Check the wiring or connections at the DI/APC system ECU or replace it with a known good.

BASE CHARGING PRESSURE FAILS TO RESPOND TO ADJUSTMENT

1. Remove the exhaust bend and make a visual check of the position of the wastegate valve relative to the turbocharger housing.
2. Remove the anti-tamper seal and shake-proof washer (or split pin) and disconnect the pushrod from the wastegate valve actuator.
3. Check that the actuating arm for the wastegate flap valve moves freely. If defective, replace the wastegate valve.
4. Check that the diaphragm unit pushrod moves freely.
5. Check the diaphragm and spring by applying low pressure to the diaphragm unit. Check that the pushrod slides freely without any tendency to bind. Release the pressure and check that the spring returns the pushrod smoothly inside the unit. If defective, replace the diaphragm unit.
6. Replace the turbocharger unit.

BOOST PRESSURE TOO HIGH, PRESSURE-GAUGE NEEDLE JUMPS INTO RED ZONE DURING ACCELERATION OR PRESSURE SWITCH BREAKS CIRCUIT

1. Check for a pressure sensor code or check the pressure sensor.
2. Perform steps 6 and following of the Adaptive Function to locate a problem related to the solenoid valve.
3. Check the wastegate valve and the diaphragm unit:
   a. Remove the exhaust bend and make a visual check of the position of the wastegate valve relative to the turbocharger housing.
   b. Remove the anti-tamper seal and shake-proof washer (or split pin) and disconnect the pushrod from the wastegate valve actuator.
   c. Check that the actuating arm for the wastegate flap valve moves freely. If defective, replace the wastegate valve.
   d. Check that the diaphragm unit pushrod moves freely.
   e. Check the diaphragm and spring by applying low pressure to the diaphragm unit. Check that the pushrod slides freely without any tendency to bind. Release the pressure and check that the spring returns the pushrod smoothly inside the unit. If defective, replace the diaphragm unit.
4. Replace the turbocharger unit.

NO REDUCTION IN BOOST PRESSURE DESPITE SUSTAINED PINGING OR KNOCKING

1. Check for a knock sensor code or check the knock sensor.
2. Perform steps 6 and following of the Adaptive Function to locate a problem related to the solenoid valve.

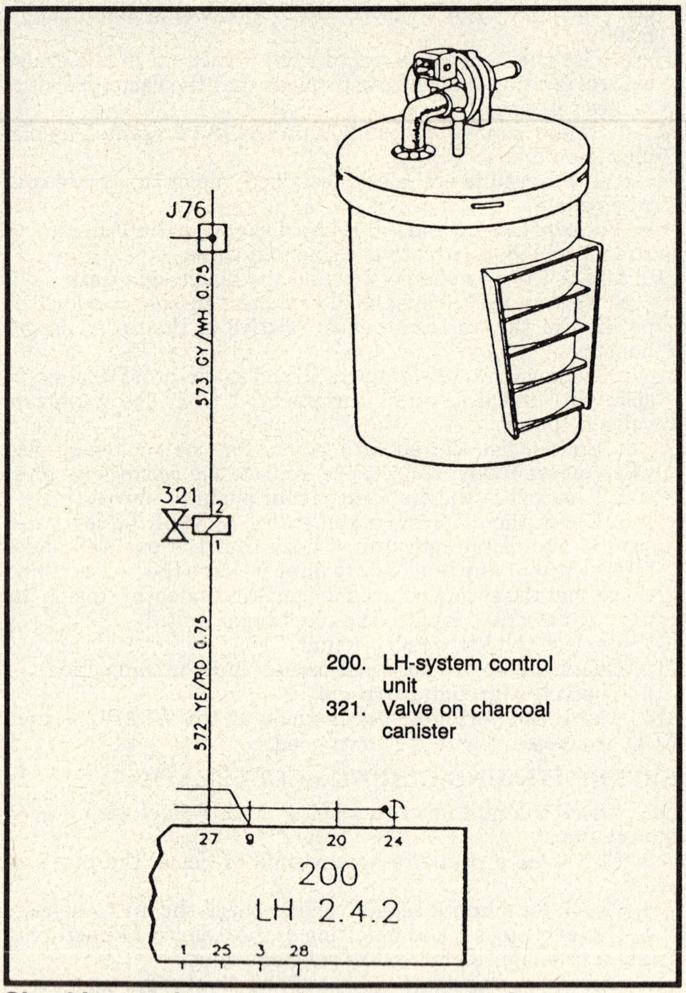

Checking canister purging valve—1991 Saab

3. Check the wastegate valve and the diaphragm unit:
   a. Remove the exhaust bend and make a visual check of the position of the wastegate valve relative to the turbocharger housing.
   b. Remove the anti-tamper seal and shake-proof washer (or split pin) and disconnect the pushrod from the wastegate valve actuator.
   c. Check that the actuating arm for the wastegate flap valve moves freely. If defective, replace the wastegate valve.
   d. Check that the diaphragm unit pushrod moves freely.
   e. Check the diaphragm and spring by applying low pressure to the diaphragm unit. Check that the pushrod slides freely without any tendency to bind. Release the pressure and check that the spring returns the pushrod smoothly inside the unit. If defective, replace the diaphragm unit.
4. Replace the turbocharger unit.

JERKY ACCELERATION AT NORMAL THROTTLE FROM IDLING UP TO 1600 RPM OR HIGHER

Check for a load signal code and repair the fault.

EXCESSIVE SPARK PLUG WEAR

1. Make sure the proper type and proper heat range plugs are installed.
2. Check for a synchronization combustion signal code and repair the fault.

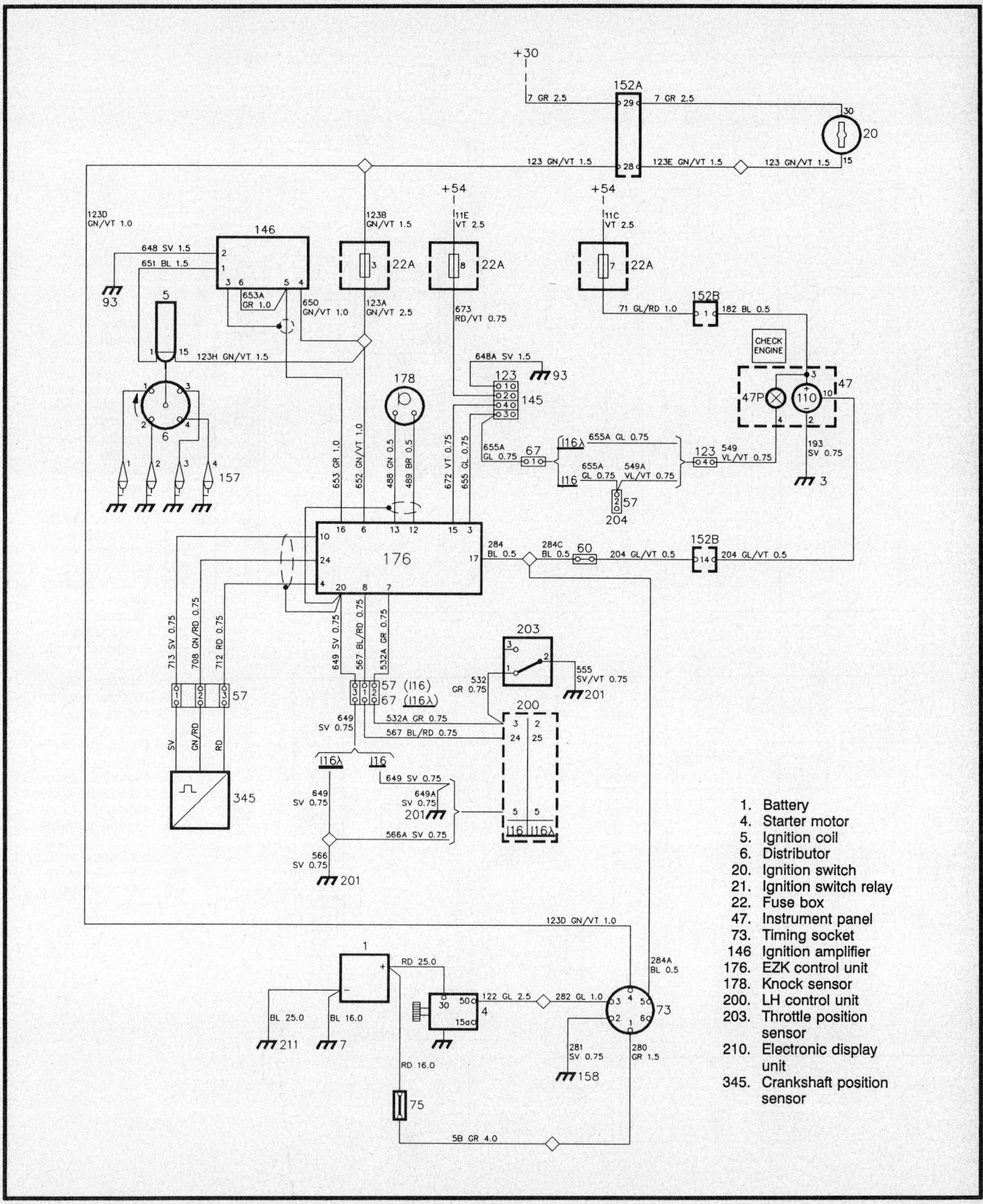

1990 EZK ignition system wiring diagram—900

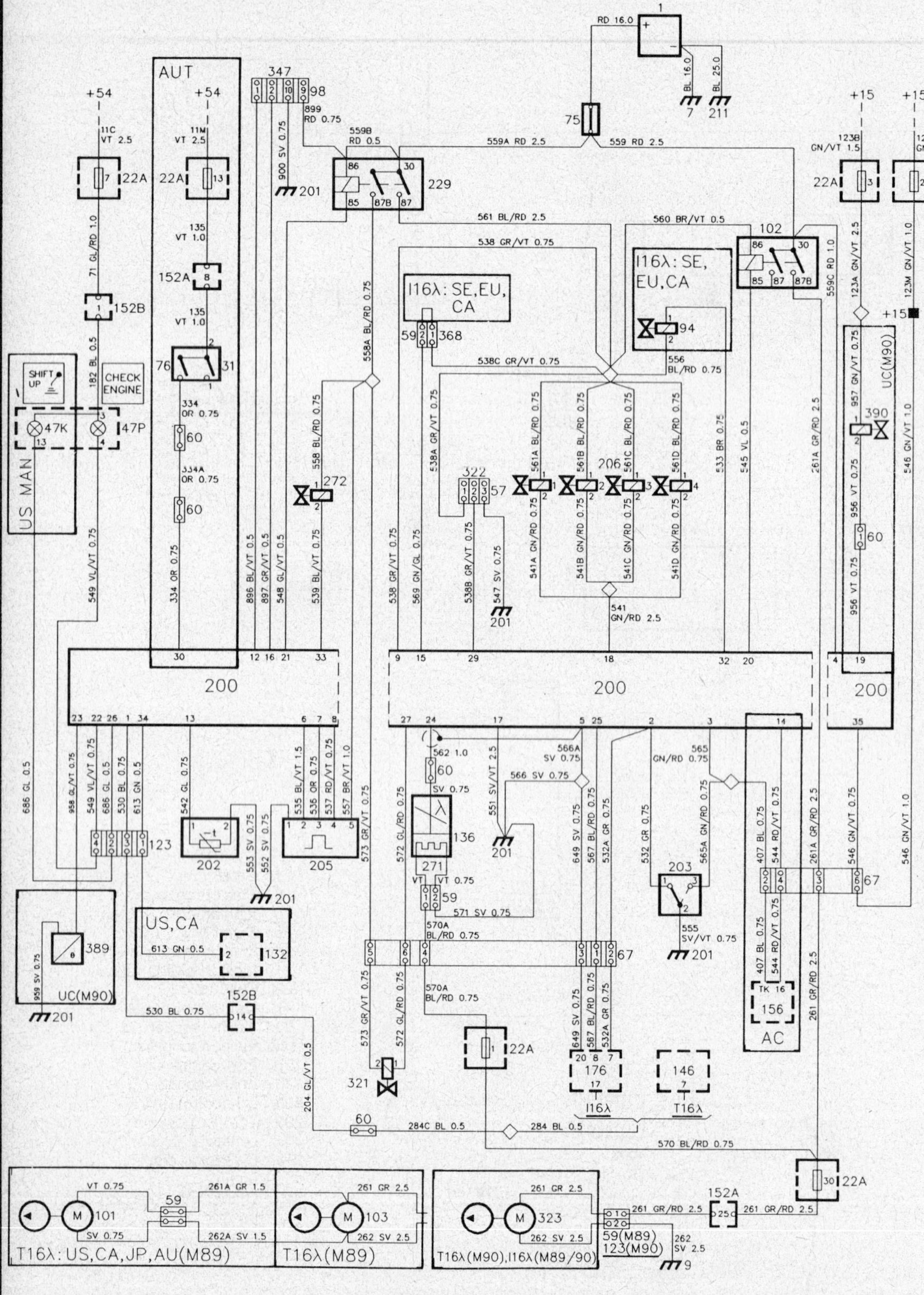

1. Battery
20. Ignition switch
22. Fuse box
47. Instrument panel
76. AT idle up switch
102. Fuel pump relay
104. Hot start relay
132. Vehicle speed sensor
136. Oxygen sensor
156. Fan time delay relay
166. A/C and ACC pressure switch
176. EZK control unit
200. LH control unit
202. CTS
203. Throttle position sensor
205. Air mass meter
206. Fuel injectors
210. Electronic display unit
229. LH main relay
239. AT Gear selector switch
272. AIC valve
343. DI, DI/APC control unit
344. Pressure sensor
345. Crankshaft position sensor
346. Ignition cartridge
347. 10-pin diagnostic connector
389. NTC resistor
390. EGR modulating valve

1990 LH 2.4 fuel system wiring diagram— Turbocharged and non-turbocharged 900

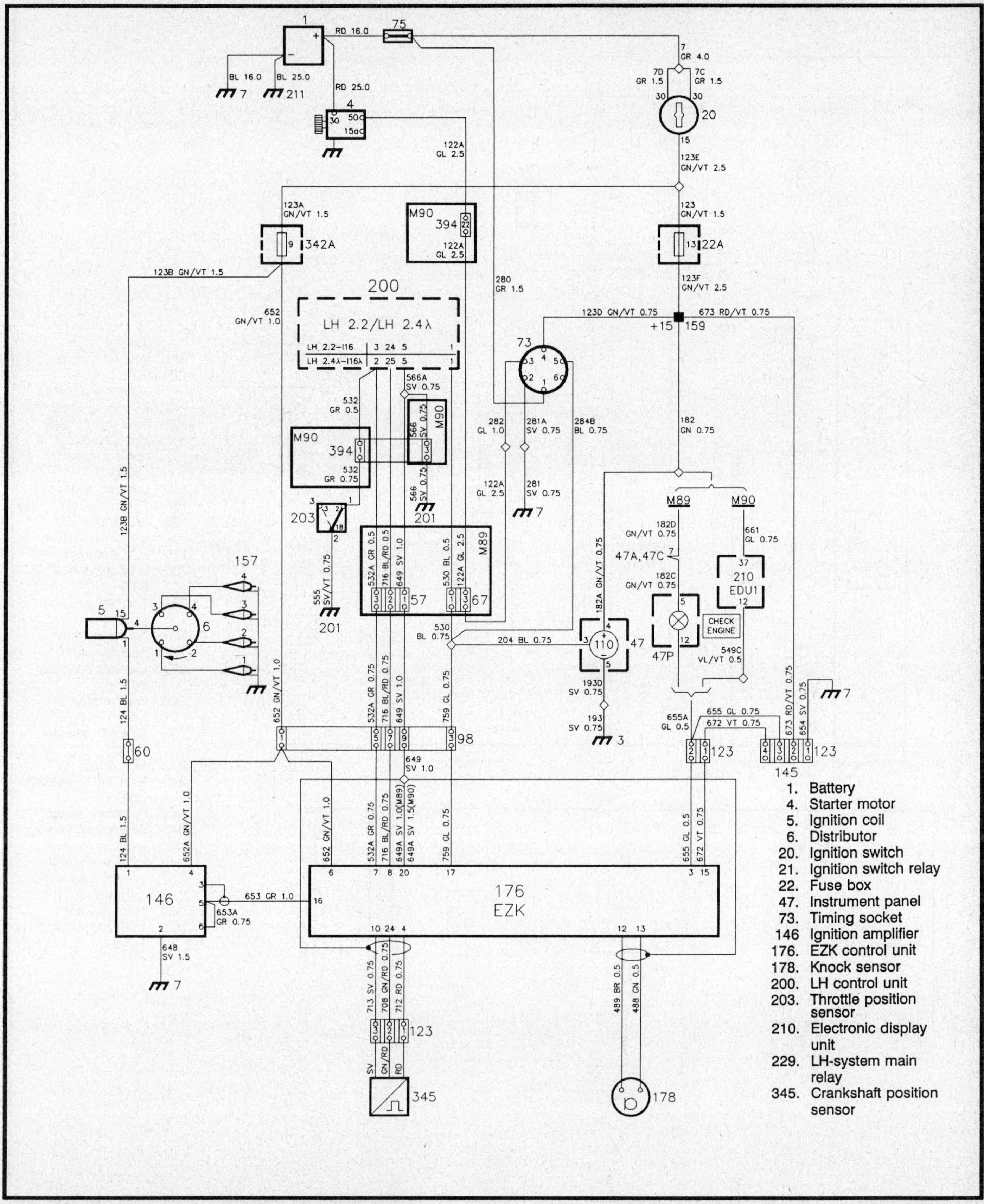

1. Battery
4. Starter motor
5. Ignition coil
6. Distributor
20. Ignition switch
21. Ignition switch relay
22. Fuse box
47. Instrument panel
73. Timing socket
146 Ignition amplifier
176. EZK control unit
178. Knock sensor
200. LH control unit
203. Throttle position sensor
210. Electronic display unit
229. LH-system main relay
345. Crankshaft position sensor

1990 EZK ignition system wiring diagram—9000

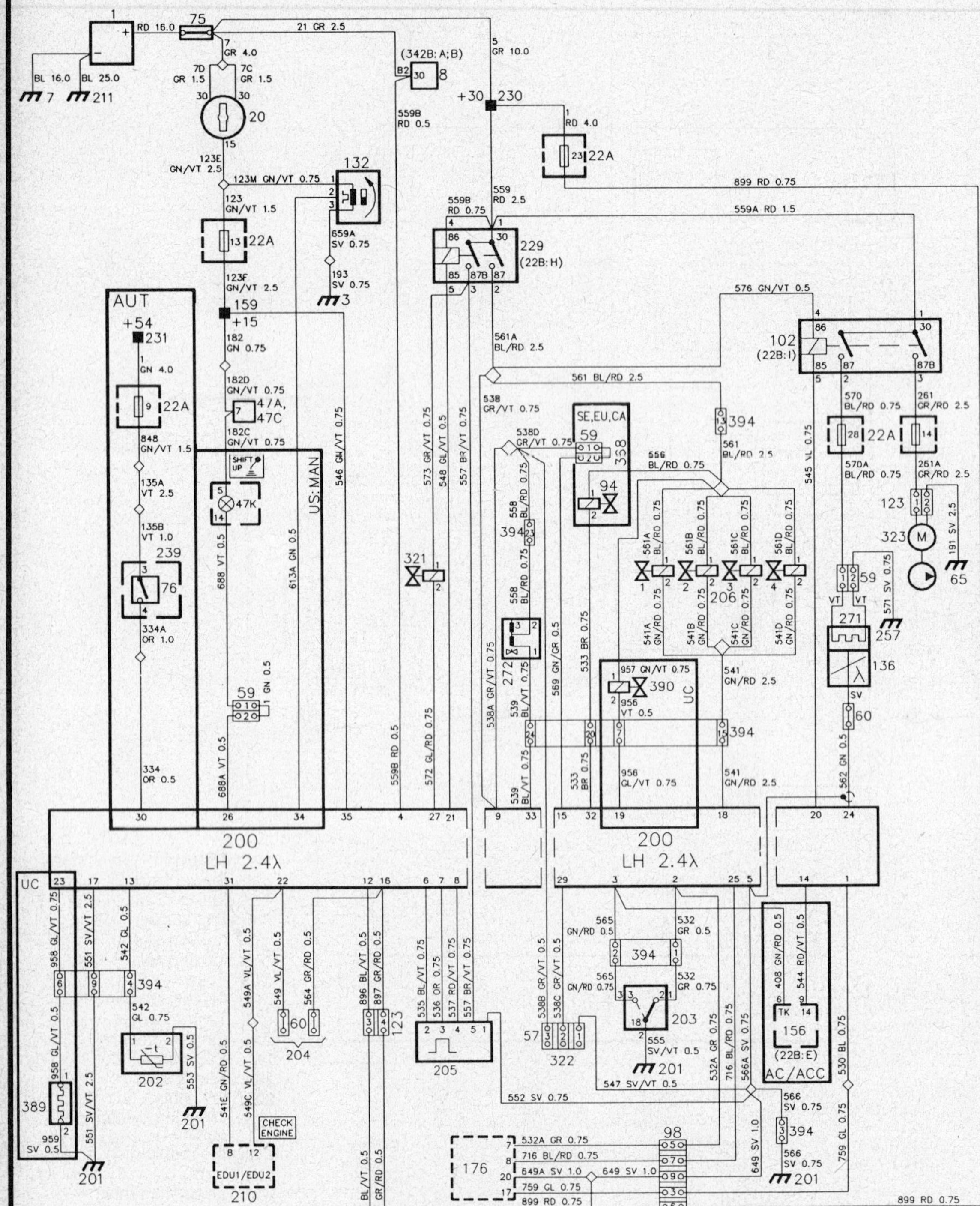

1. Battery
20. Ignition switch
22. Fuse box
47. Instrument panel
76. AT idle up switch
102. Fuel pump relay
104. Hot start relay
132. Vehicle speed sensor
136. Oxygen sensor
156. Fan time delay relay
166. A/C and ACC pressure switch
176. EZK control unit
200. LH control unit
202. CTS
203. Throttle position sensor
205. Air mass meter
206. Fuel injectors
210. Electronic display unit
229. LH-system main relay
239. AT gear selector switch
272. AIC valve
343. DI, DI/APC control unit
344. Pressure sensor
345. Crankshaft position sensor
346. Ignition cartridge
347. 10-pin diagnostic connector
389. NTC resistor
390. EGR modulating valve

1990 LH 2.4 fuel system wiring diagram—9000

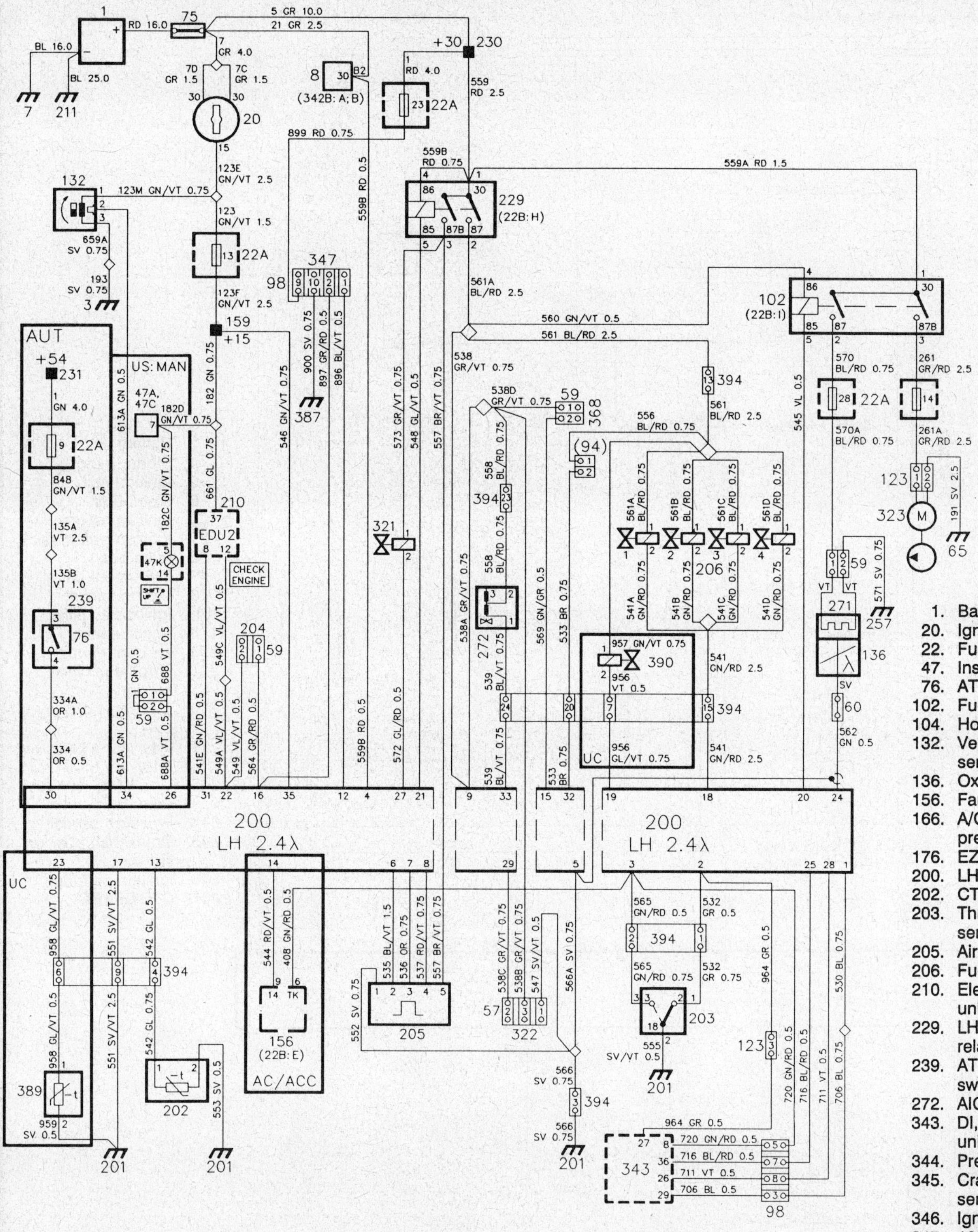

1. Battery
20. Ignition switch
22. Fuse box
47. Instrument panel
76. AT idle up switch
102. Fuel pump relay
104. Hot start relay
132. Vehicle speed sensor
136. Oxygen sensor
156. Fan time delay relay
166. A/C and ACC pressure switch
176. EZK control unit
200. LH control unit
202. CTS
203. Throttle position sensor
205. Air mass meter
206. Fuel injectors
210. Electronic display unit
229. LH-system main relay
239. AT gear selector switch
272. AIC valve
343. DI, DI/APC control unit
344. Pressure sensor
345. Crankshaft position sensor
346. Ignition cartridge
347. 10-pin diagnostic connector
389. NTC resistor
390. EGR modulating valve

1990 LH 2.4 fuel system wiring diagram—9000 Turbo

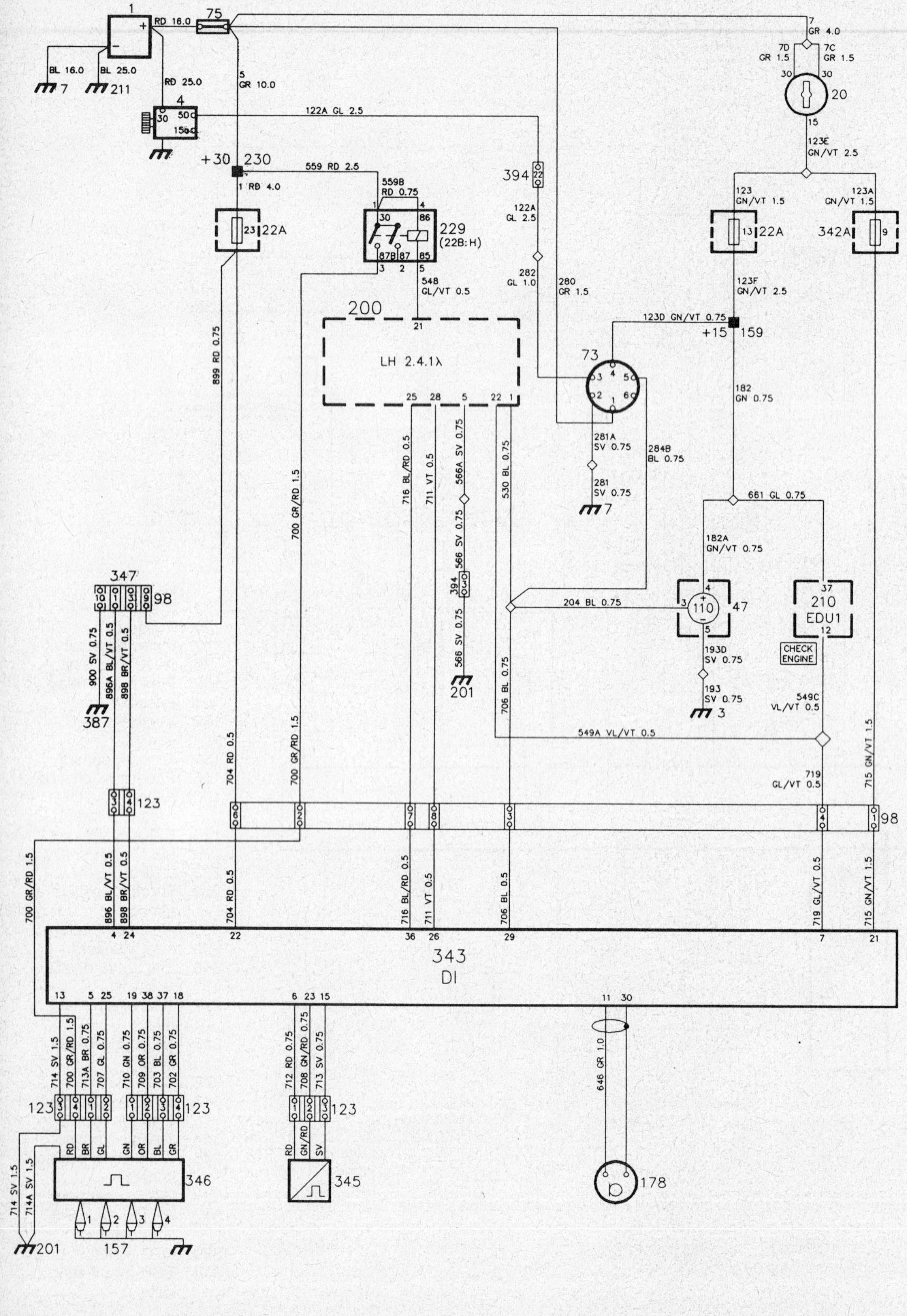

1. Battery
4. Starter motor
5. Ignition coil
6. Distributor
20. Ignition switch
21. Ignition switch relay
22. Fuse box
47. Instrument panel
73. Timing socket
146 Ignition amplifier
178. Knock sensor
200. LH control unit
203. Throttle position sensor
210. Electronic display unit
229. LH-system main relay
343. DI, DI/APC control unit
344. Pressure sensor
345. Crankshaft position sensor
346. Ignition cartridge
407. CTS-DI/APC

1990 DI system wiring diagram—9000 with LH 2.4.1 fuel system

- 1. Battery
- 20. Ignition switch
- 22. Fuse box
- 47. Instrument panel
- 76. AT idle up switch
- 102. Fuel pump relay
- 104. Hot start relay
- 132. Vehicle speed sensor
- 136. Oxygen sensor
- 156. Fan time delay relay
- 166. A/C and ACC pressure switch
- 176. EZK control unit
- 200. LH control unit
- 202. CTS
- 203. Throttle position sensor
- 205. Air mass meter
- 206. Fuel injectors
- 210. Electronic display unit
- 229. LH-system main relay
- 239. AT gear selector switch
- 272. AIC valve
- 343. DI, DI/APC control unit
- 344. Pressure sensor
- 345. Crankshaft position sensor
- 346. Ignition cartridge
- 347. 10-pin diagnostic connector
- 389. NTC resistor
- 390. EGR modulating valve

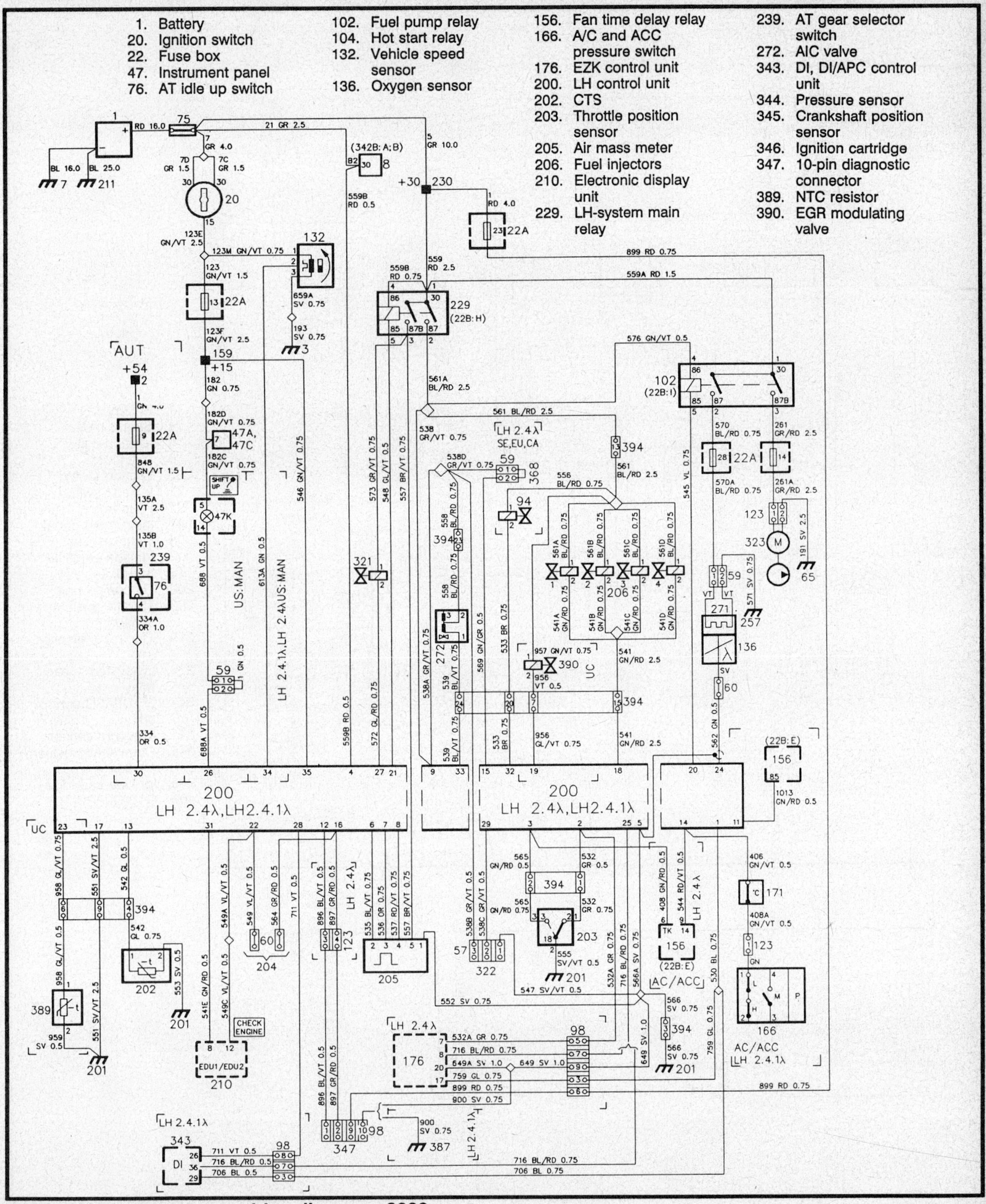

1990 LH 2.4.1 fuel system wiring diagram—9000

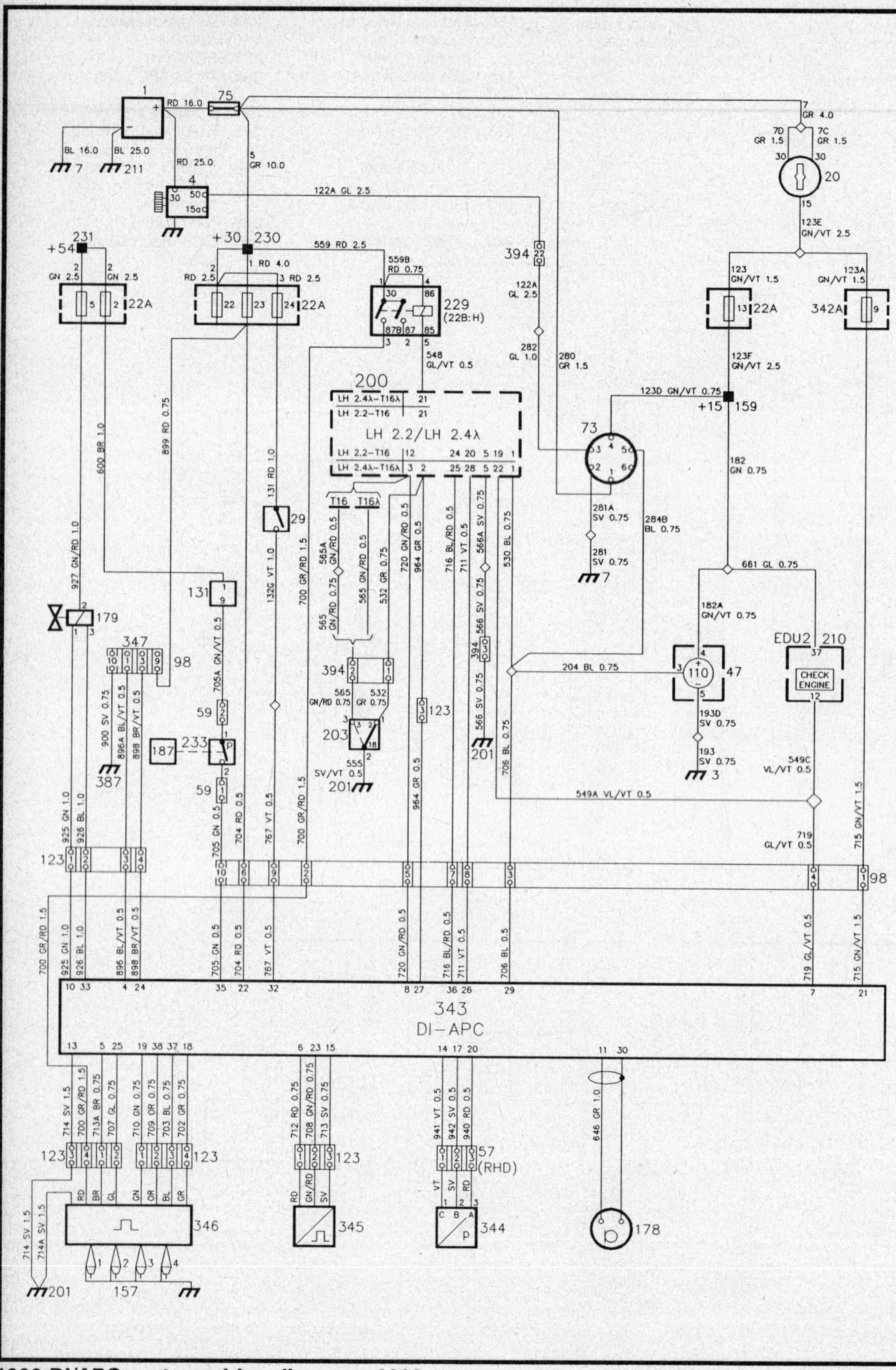

1. Battery
4. Starter motor
5. Ignition coil
6. Distributor
20. Ignition switch
21. Ignition switch relay
22. Fuse box
47. Instrument panel
73. Timing socket
146 Ignition amplifier
178. Knock sensor
200. LH control unit
203. Throttle position sensor
210. Electronic display unit
229. LH-system main relay
343. DI, DI/APC control unit
344. Pressure sensor
345. Crankshaft position sensor
346. Ignition cartridge
407. CTS-DI/APC

1990 DI/APC system wiring diagram—9000

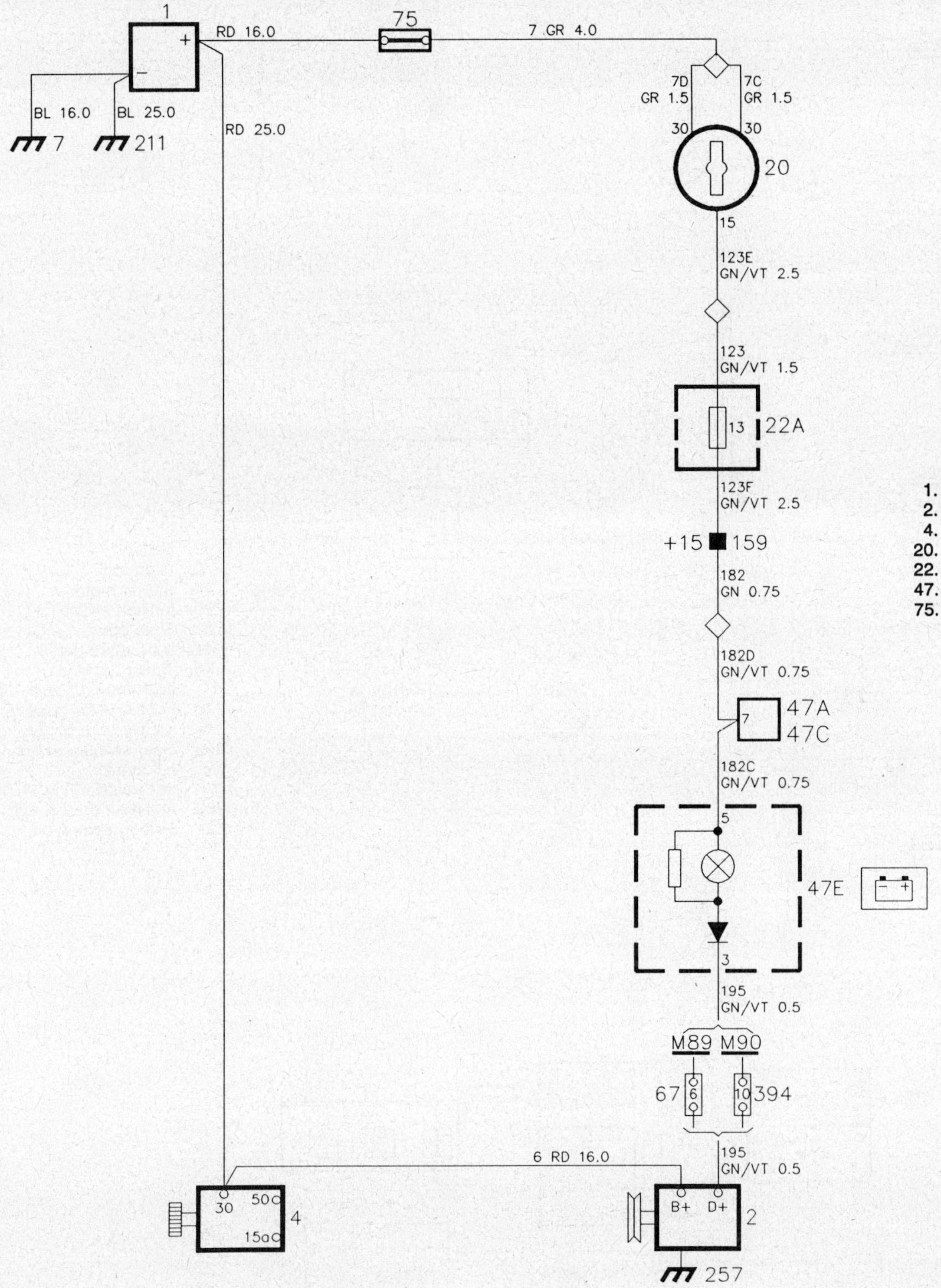

1. Battery
2. Alternator
4. Starter motor
20. Ignition switch
22. Fuse box
47. Instrument panel
75. Fuse box for positive supply from battery

1990 battery and charging system wiring diagram— 9000

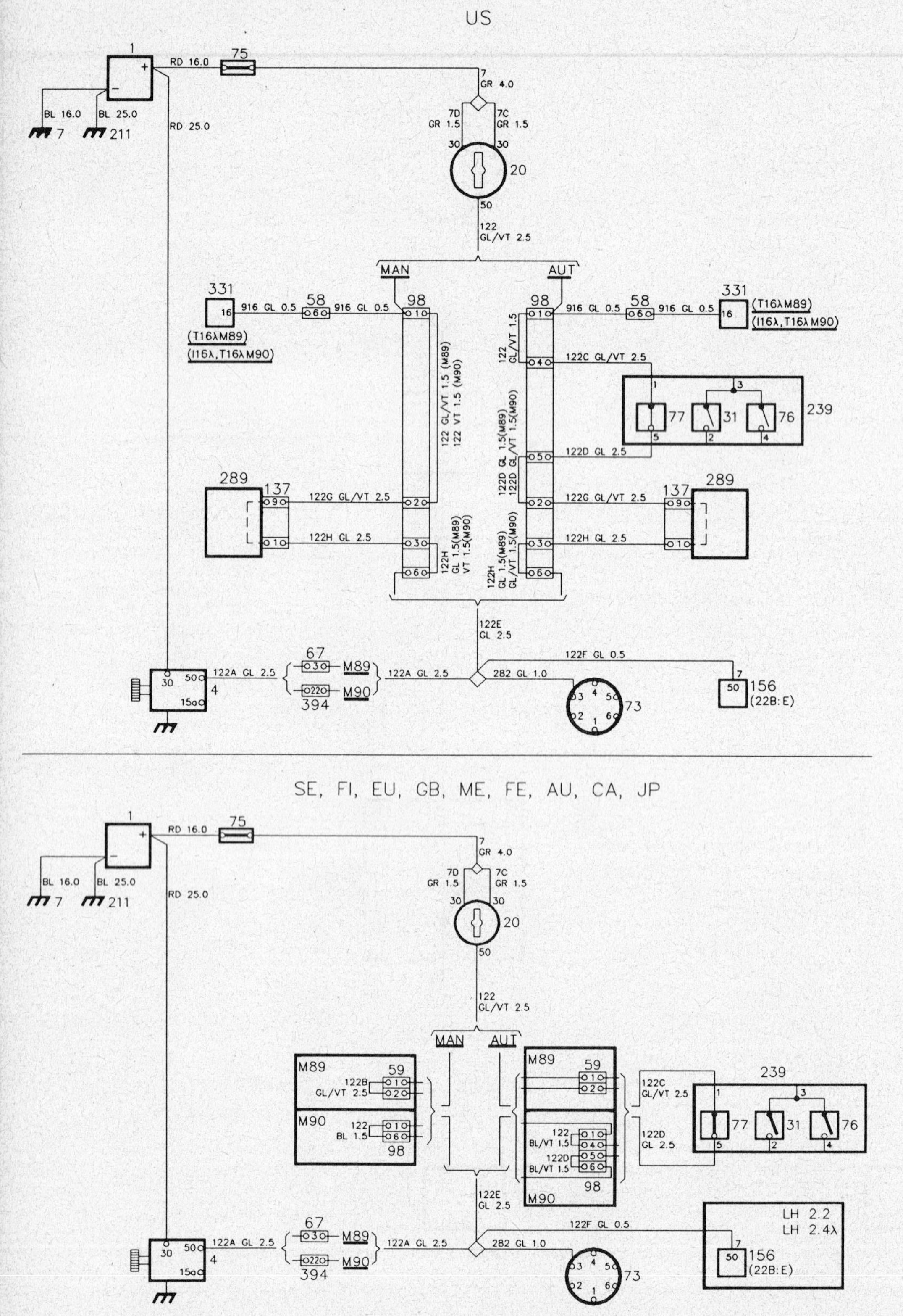

1. Battery
2. Alternator
4. Starter motor
20. Ignition switch
22. Fuse box
47. Instrument panel
73. Timing service instrument socket
75. Fuse box for positive supply from battery
239. Gear selector switch-automatic transmission
289. Anti-theft control unit
331. Air bag control unit

1990 starting system and power supply +50 wiring diagram—9000

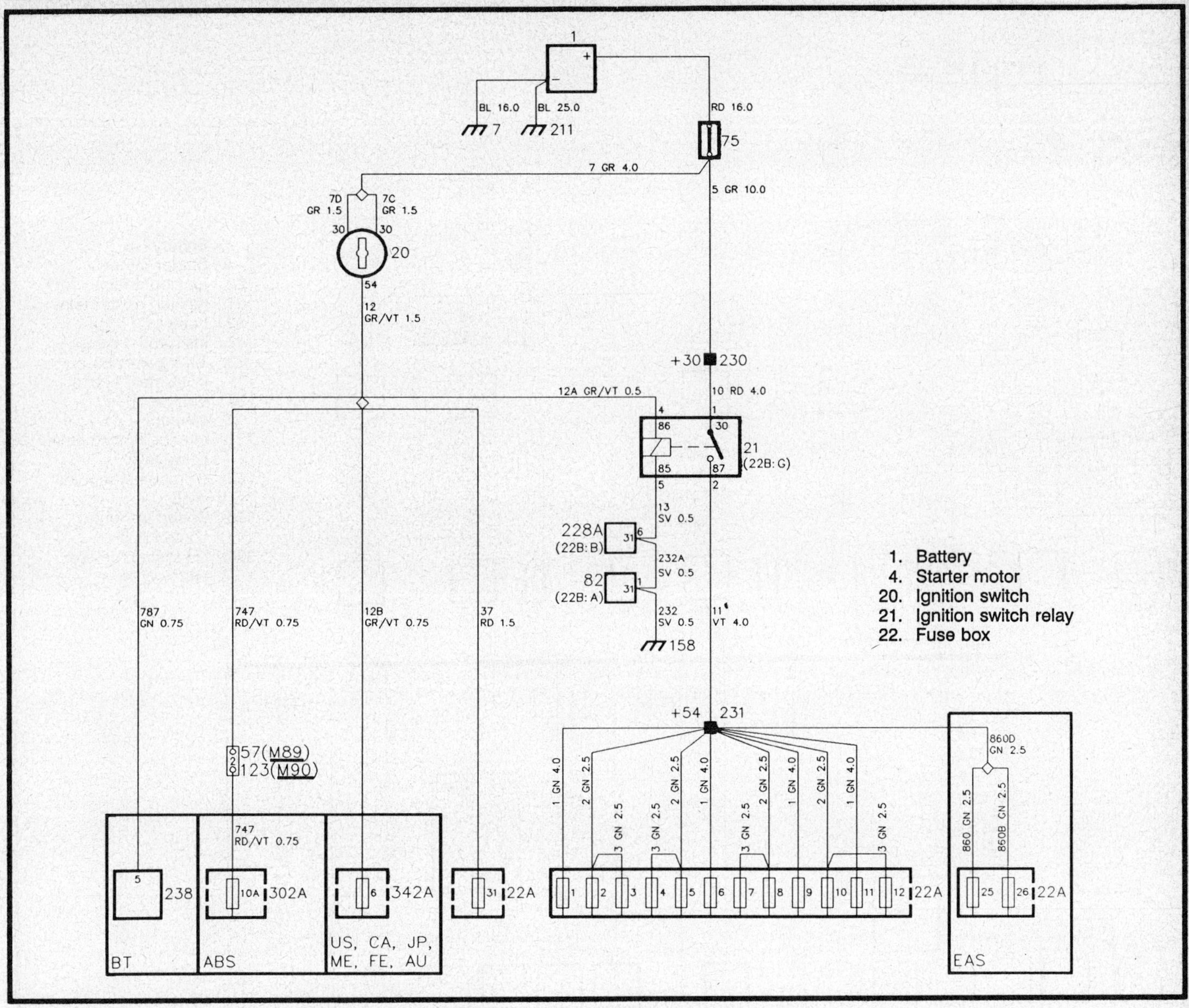

1990 power supply +54 wiring diagram—9000

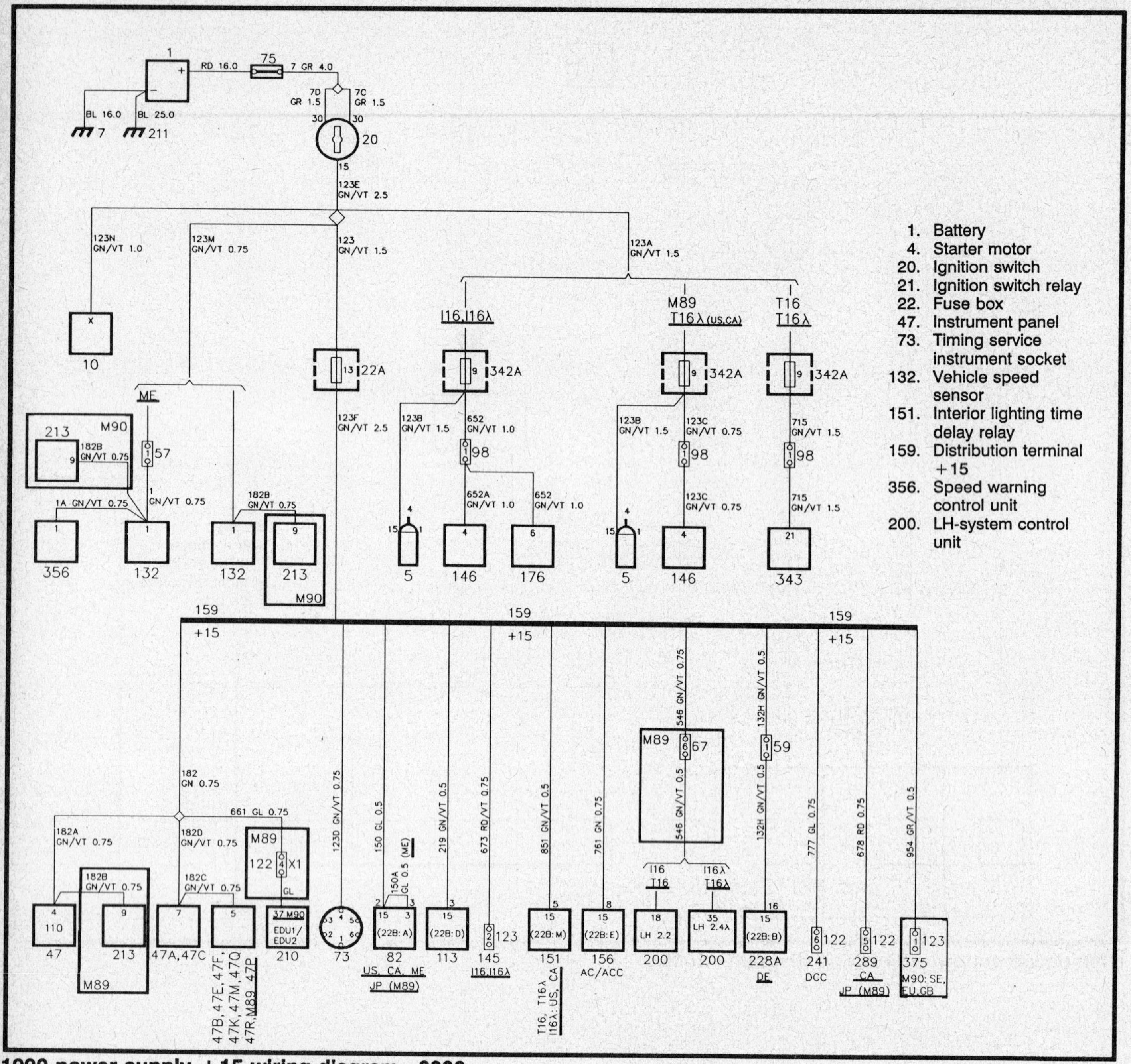

1. Battery
4. Starter motor
20. Ignition switch
21. Ignition switch relay
22. Fuse box
47. Instrument panel
73. Timing service instrument socket
132. Vehicle speed sensor
151. Interior lighting time delay relay
159. Distribution terminal +15
356. Speed warning control unit
200. LH-system control unit

1990 power supply +15 wiring diagram—9000

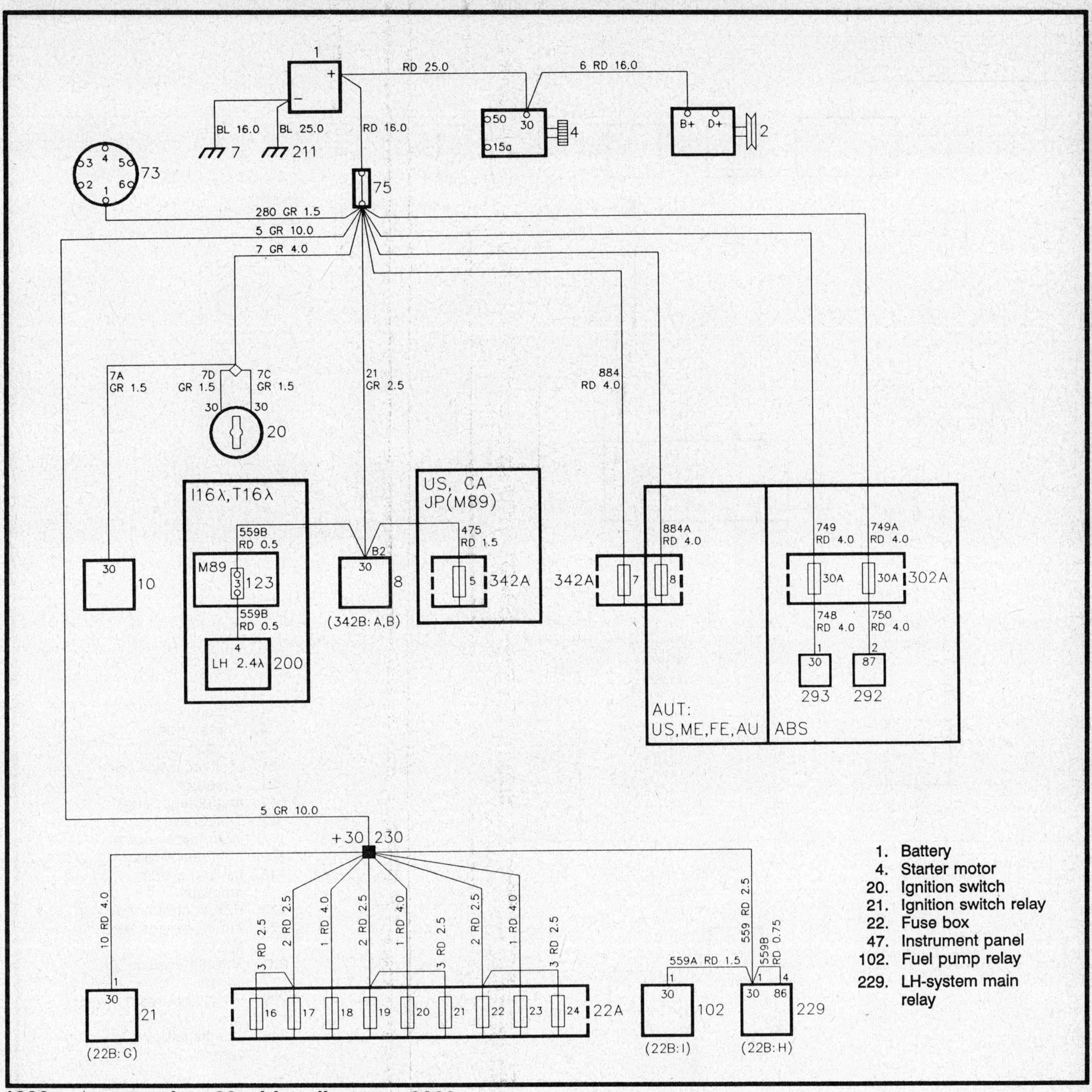

1. Battery
4. Starter motor
20. Ignition switch
21. Ignition switch relay
22. Fuse box
47. Instrument panel
102. Fuel pump relay
229. LH-system main relay

1990 power supply +30 wiring diagram—9000

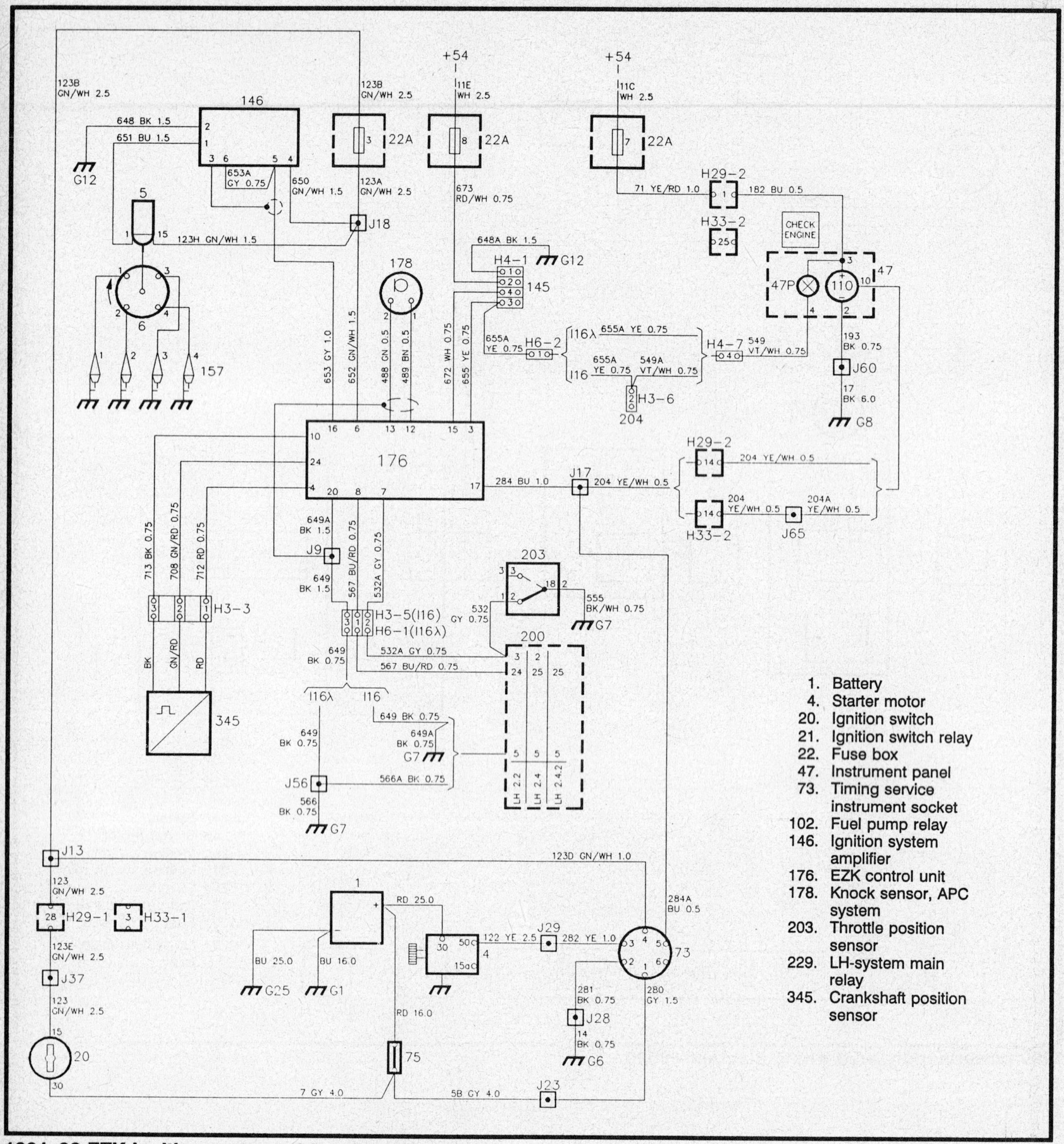

1991–92 EZK ignition system wiring diagram—900

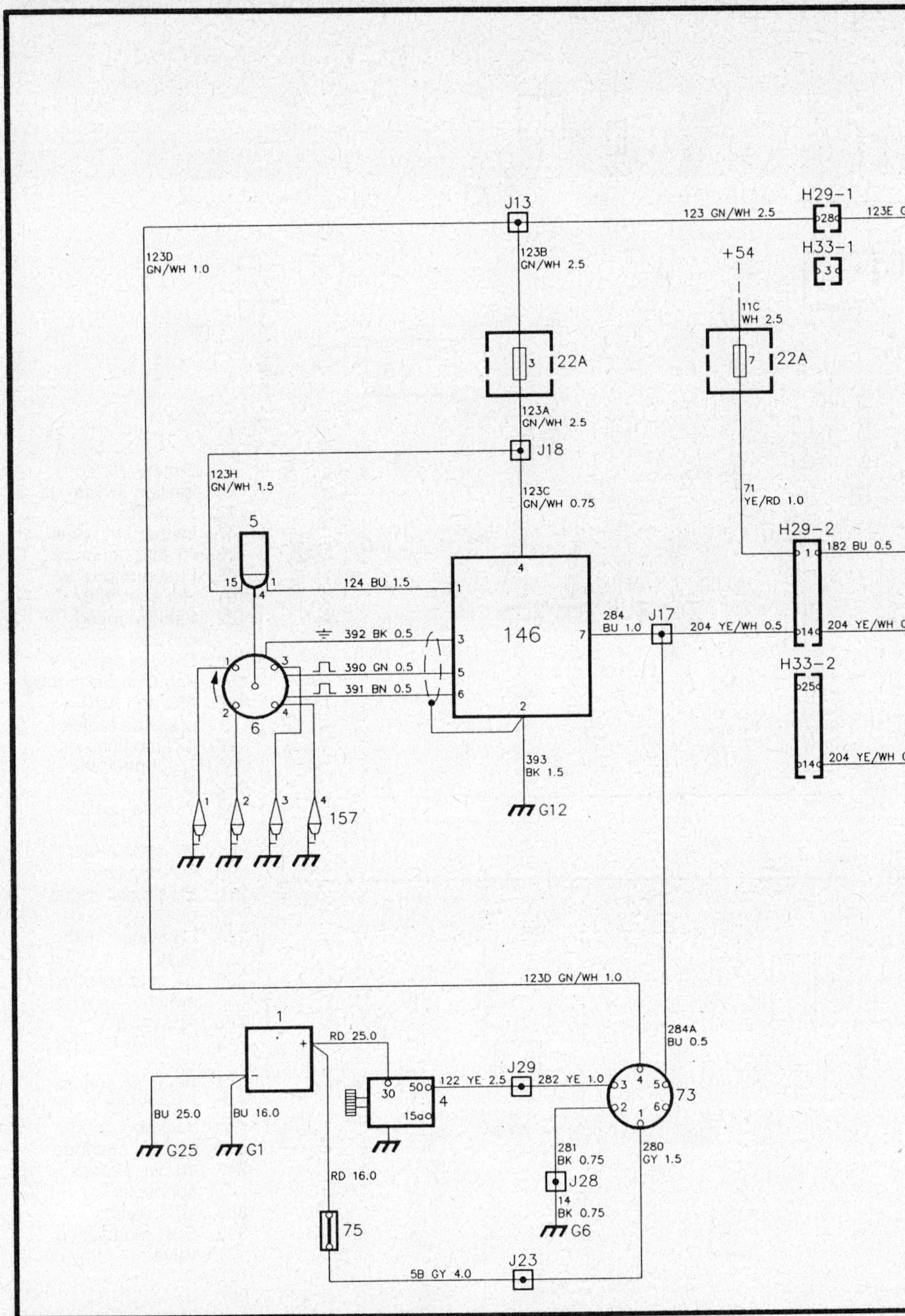

1. Battery
4. Starter motor
20. Ignition switch
21. Ignition switch relay
22. Fuse box
47. Instrument panel
73. Timing service instrument socket
102. Fuel pump relay
146. Ignition system amplifier
178. Knock sensor, APC system
203. Throttle position sensor
229. LH-system main relay

1991–92 TZ ignition system wiring diagram—900 turbo

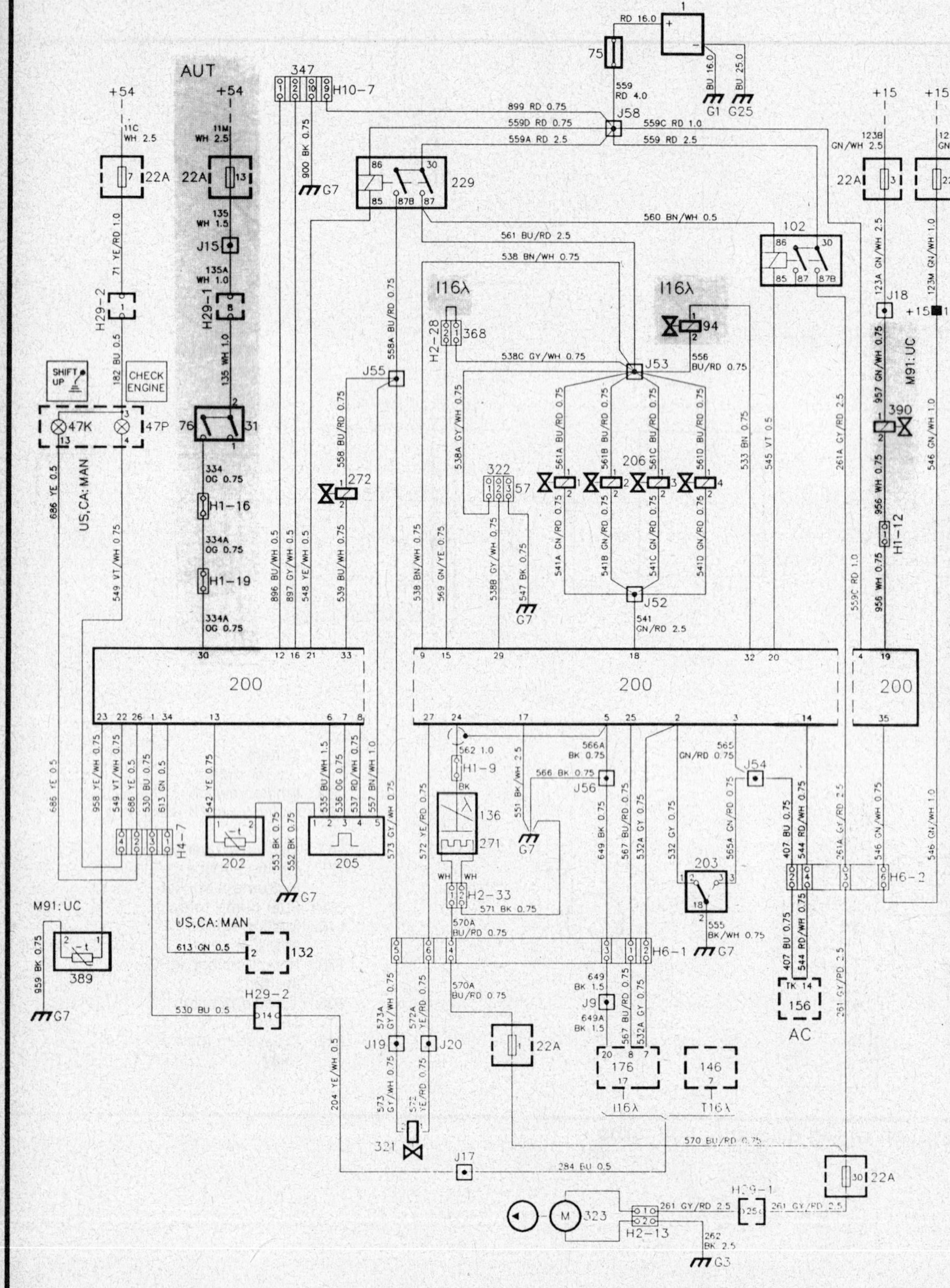

1. Battery
20. Ignition switch
22. Fuse box
47. Instrument panel
76. AT idle up switch
102. Fuel pump relay
104. Hot start relay
132. Vehicle speed sensor
136. Oxygen sensor
156. Fan time delay relay
166. A/C and ACC pressure switch
176. EZK control unit
200. LH control unit
202. CTS
203. Throttle position sensor
205. Air mass meter
206. Fuel injectors
210. Electronic display unit
229. LH-system main relay
239. AT gear selector switch
272. AIC valve
343. DI, DI/APC control unit
344. Pressure sensor
345. Crankshaft position sensor
346. Ignition cartridge
347. 10-pin diagnostic connector
389. NTC resistor
390. EGR modulating valve

1991–92 LH 2.4 fuel system wiring diagram – Turbocharged and non-turbocharged 900

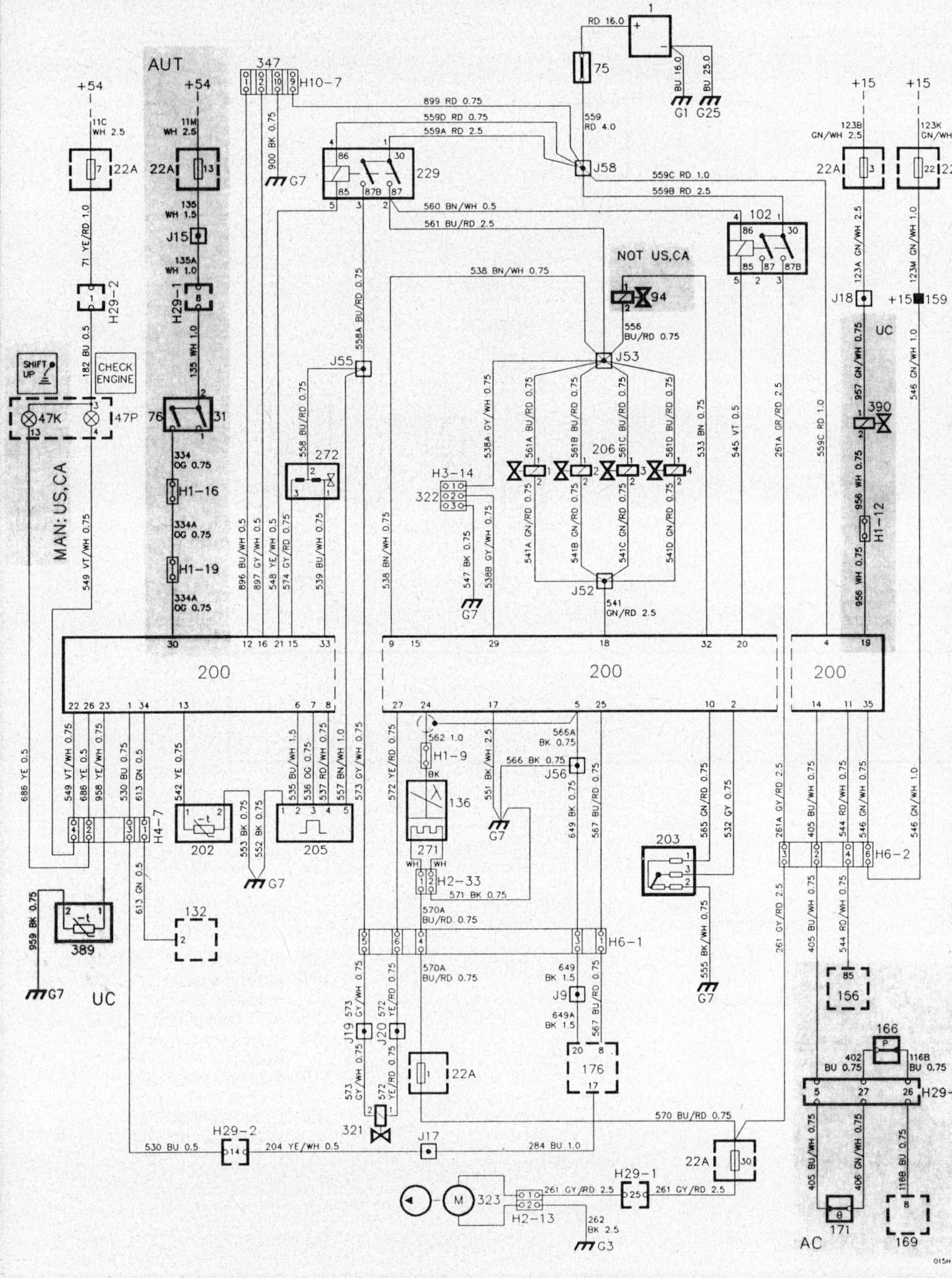

1. Battery
20. Ignition switch
22. Fuse box
47. Instrument panel
76. AT idle up switch
102. Fuel pump relay
104. Hot start relay
132. Vehicle speed sensor
136. Oxygen sensor
156. Fan time delay relay
166. A/C and ACC pressure switch
176. EZK control unit
200. LH control unit
202. CTS
203. Throttle position sensor
205. Air mass meter
206. Fuel injectors
210. Electronic display unit
229. LH-system main relay
239. AT gear selector switch
272. AIC valve
343. DI, DI/APC control unit
344. Pressure sensor
345. Crankshaft position sensor
346. Ignition cartridge
347. 10-pin diagnostic connector
389. NTC resistor
390. EGR modulating valve

1991–92 LH 2.4.2 fuel system wiring diagram—900

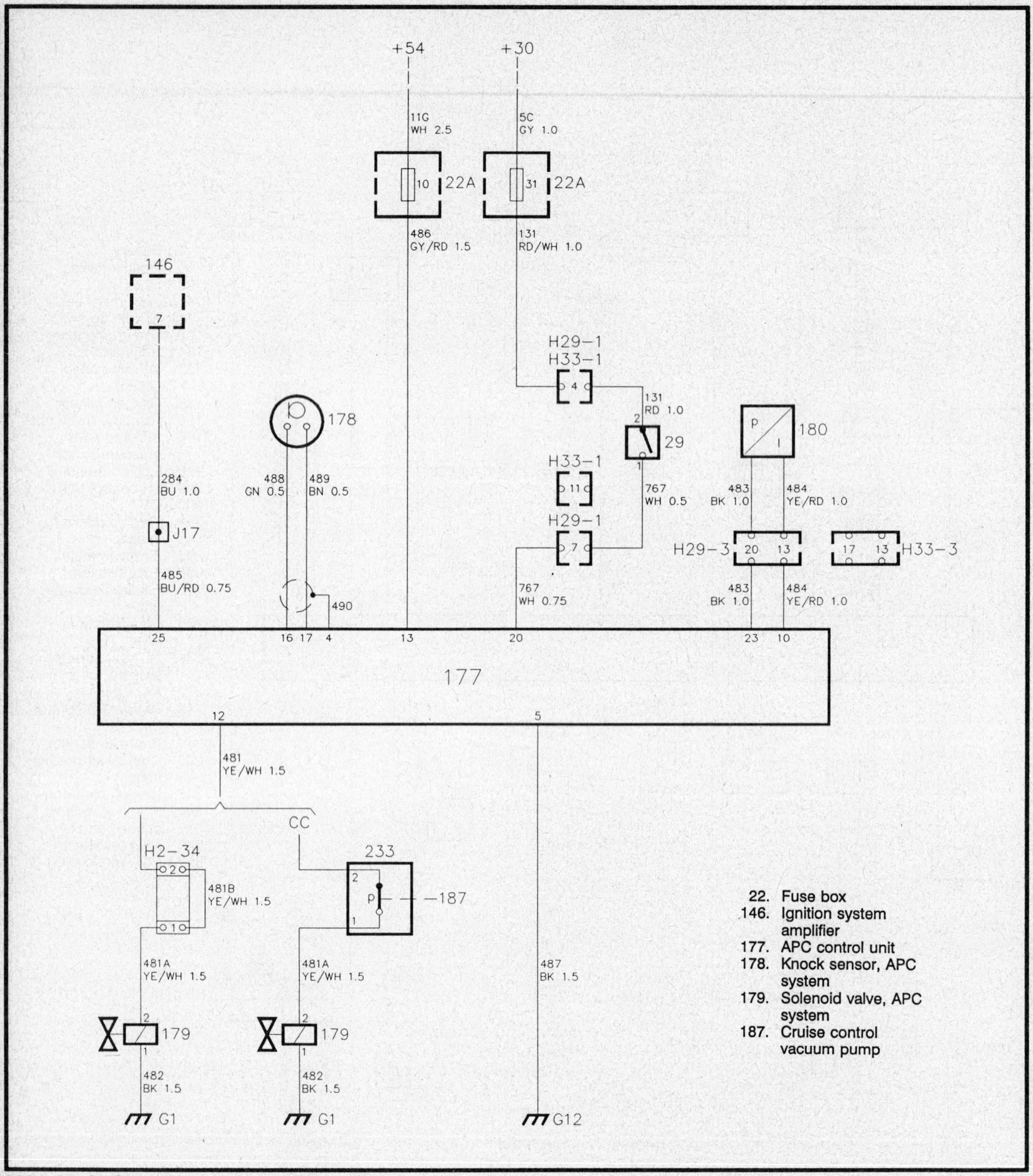

1991–92 APC system wiring diagram—900

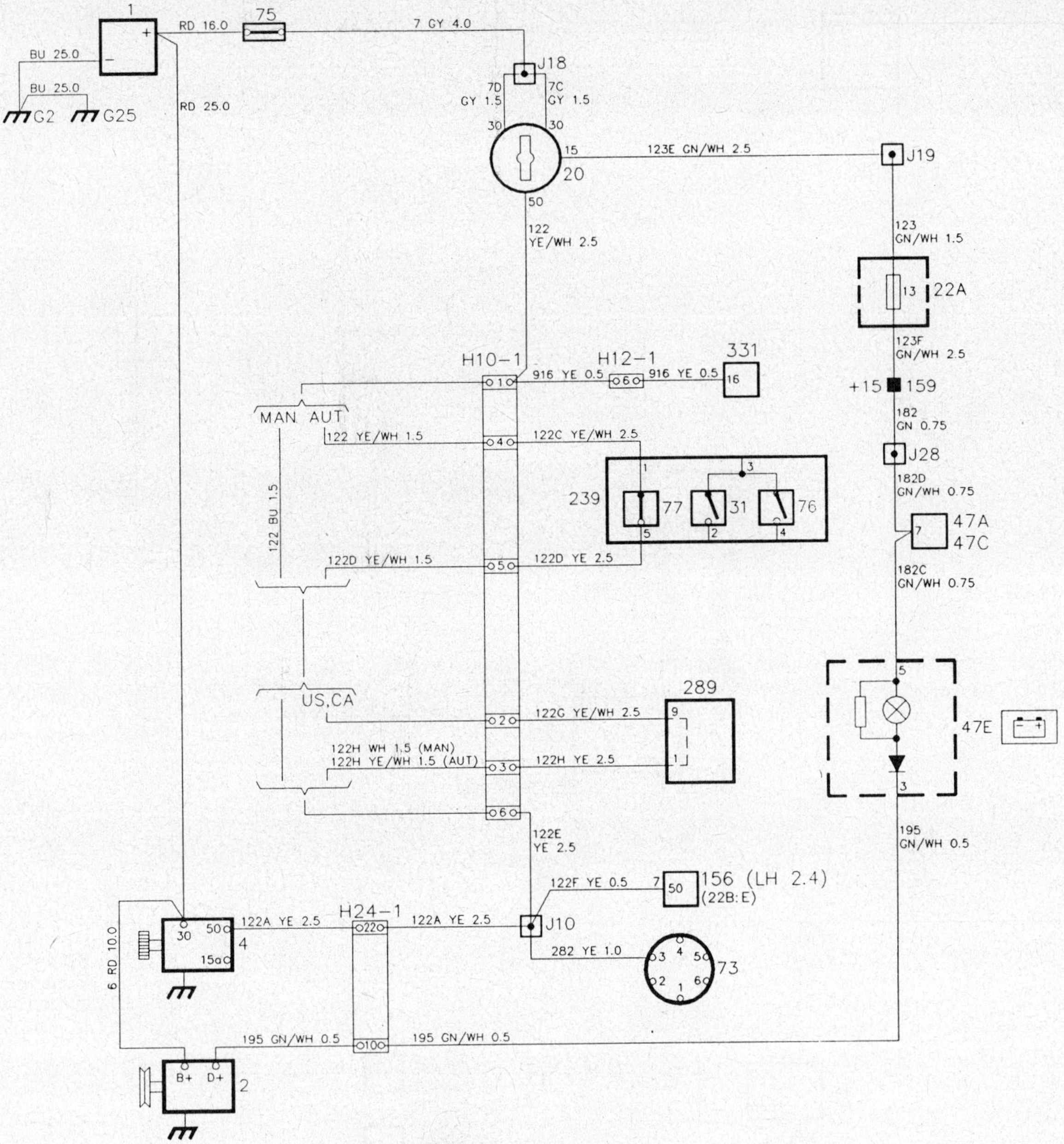

1. Battery
4. Starter motor
20. Ignition switch
21. Ignition switch relay
22. Fuse box
47. Instrument panel

1991–92 battery and charging system wiring diagram—Turbocharged and non-turbocharged 900

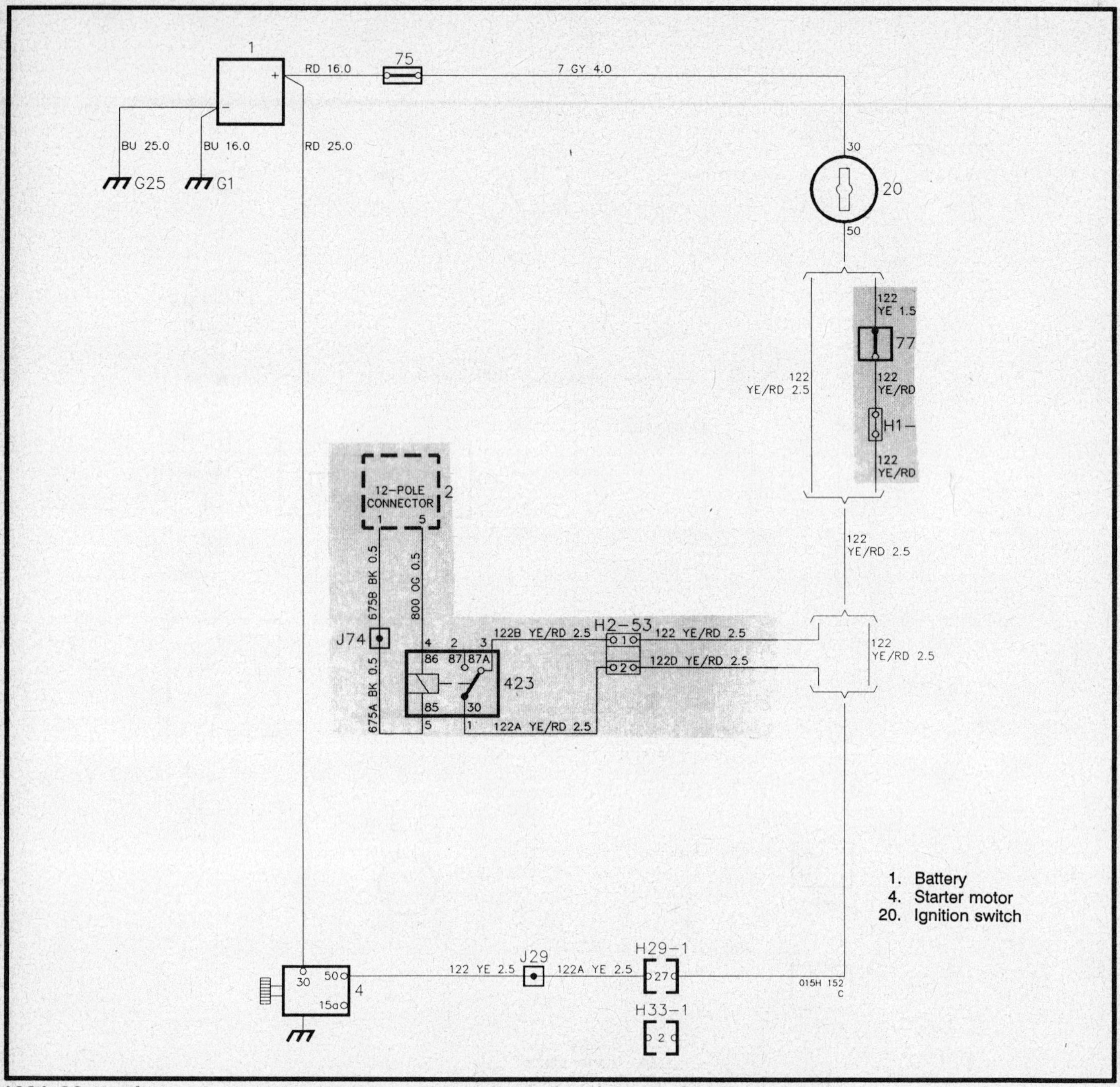

1991–92 starting system wiring diagram– Turbocharged and non-turbocharged 900

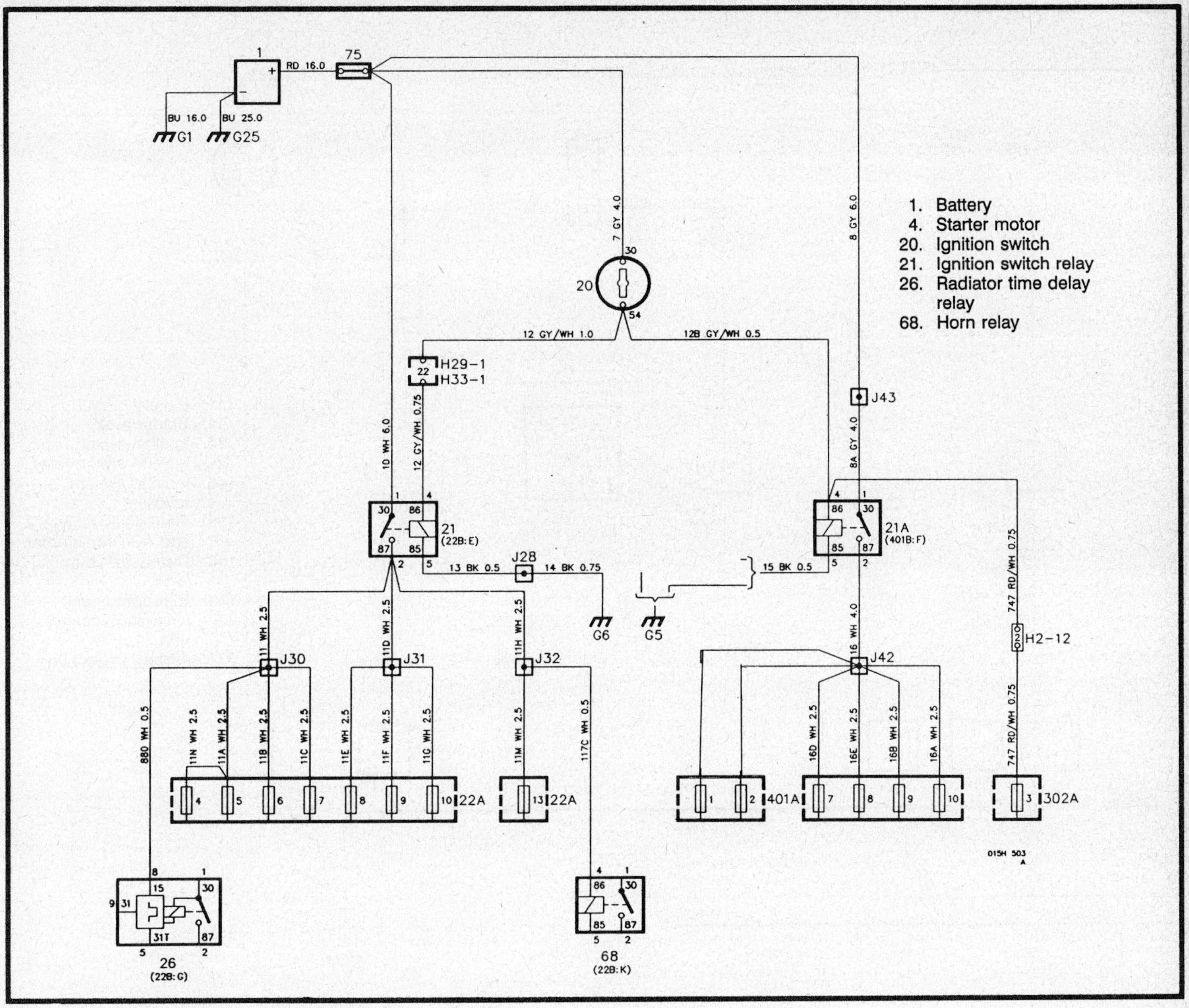

1991–92 power supply +54 wiring diagram—900

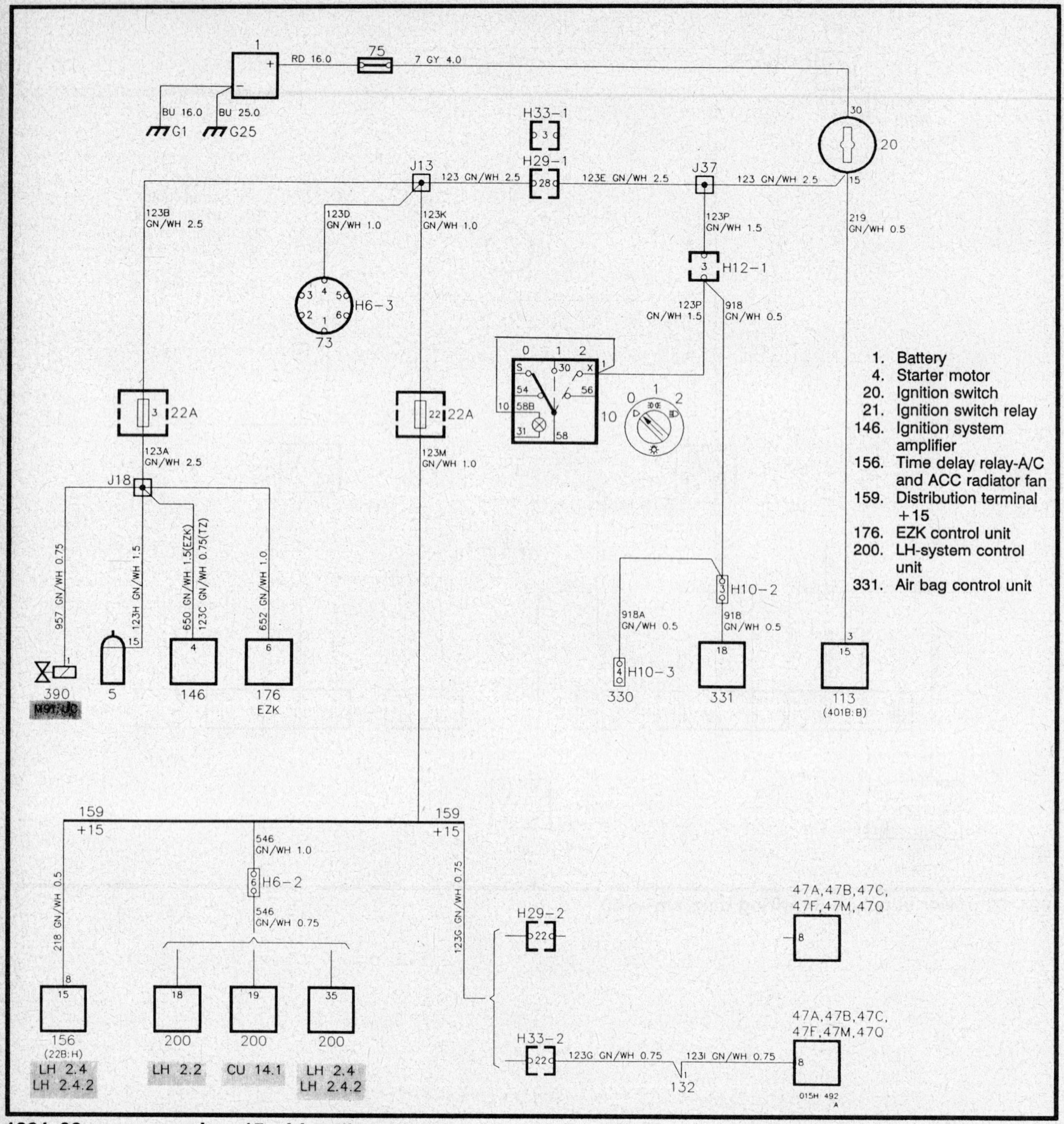

1991–92 power supply +15 wiring diagram—900

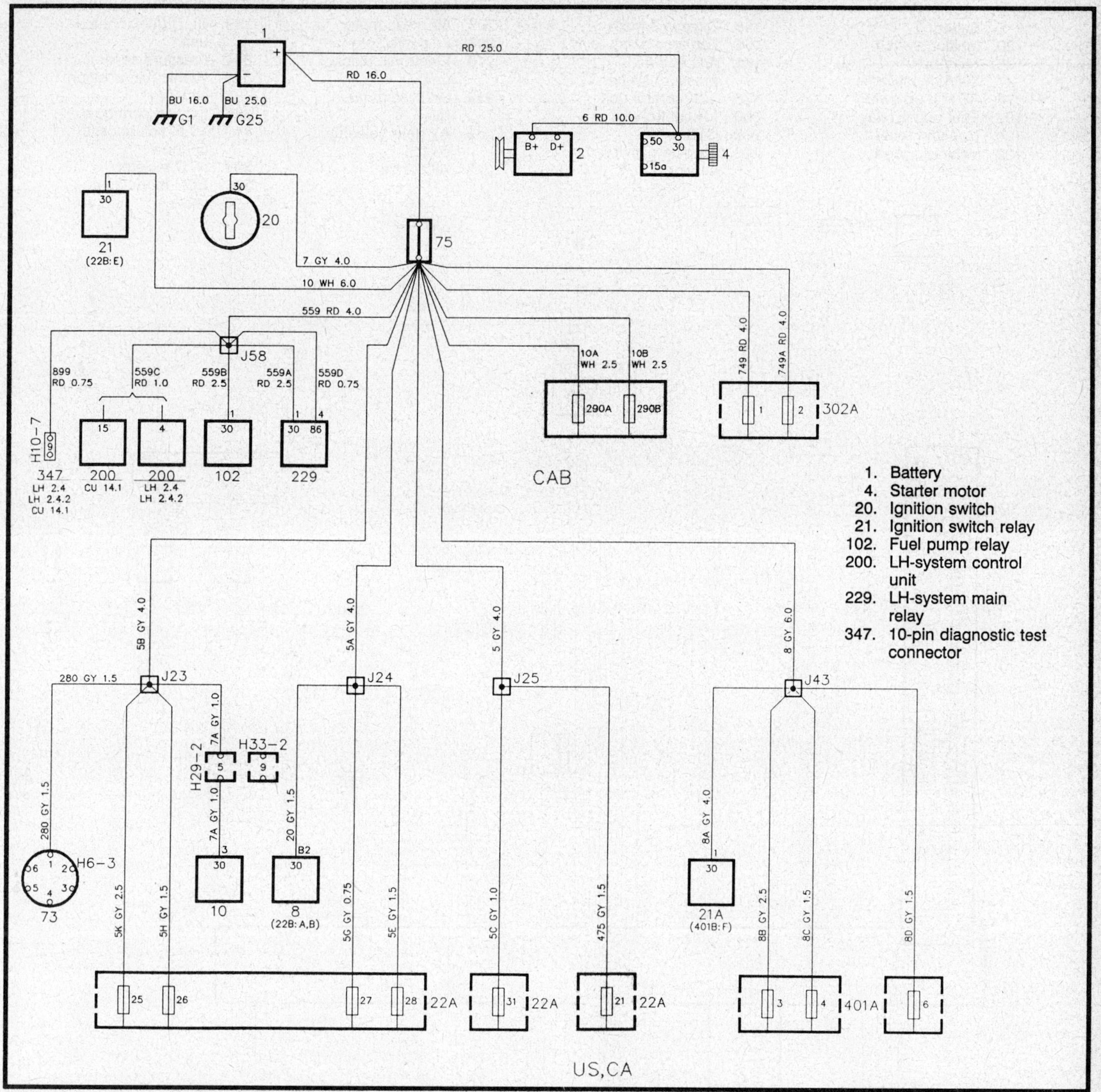

1991–92 power supply +30 wiring diagram—900

1. Battery
20. Ignition switch
22. Fuse box
47. Instrument panel
76. AT idle up switch
102. Fuel pump relay
104. Hot start relay
132. Vehicle speed sensor
136. Oxygen sensor
156. Fan time delay relay
166. A/C and ACC pressure switch
176. EZK control unit
200. LH control unit
202. CTS
203. Throttle position sensor
205. Air mass meter
206. Fuel injectors
210. Electronic display unit
229. LH-system main relay
239. AT gear selector switch
272. AIC valve
343. DI, DI/APC control unit
344. Pressure sensor
345. Crankshaft position sensor
346. Ignition cartridge
347. 10-pin diagnostic connector
389. NTC resistor
390. EGR modulating valve

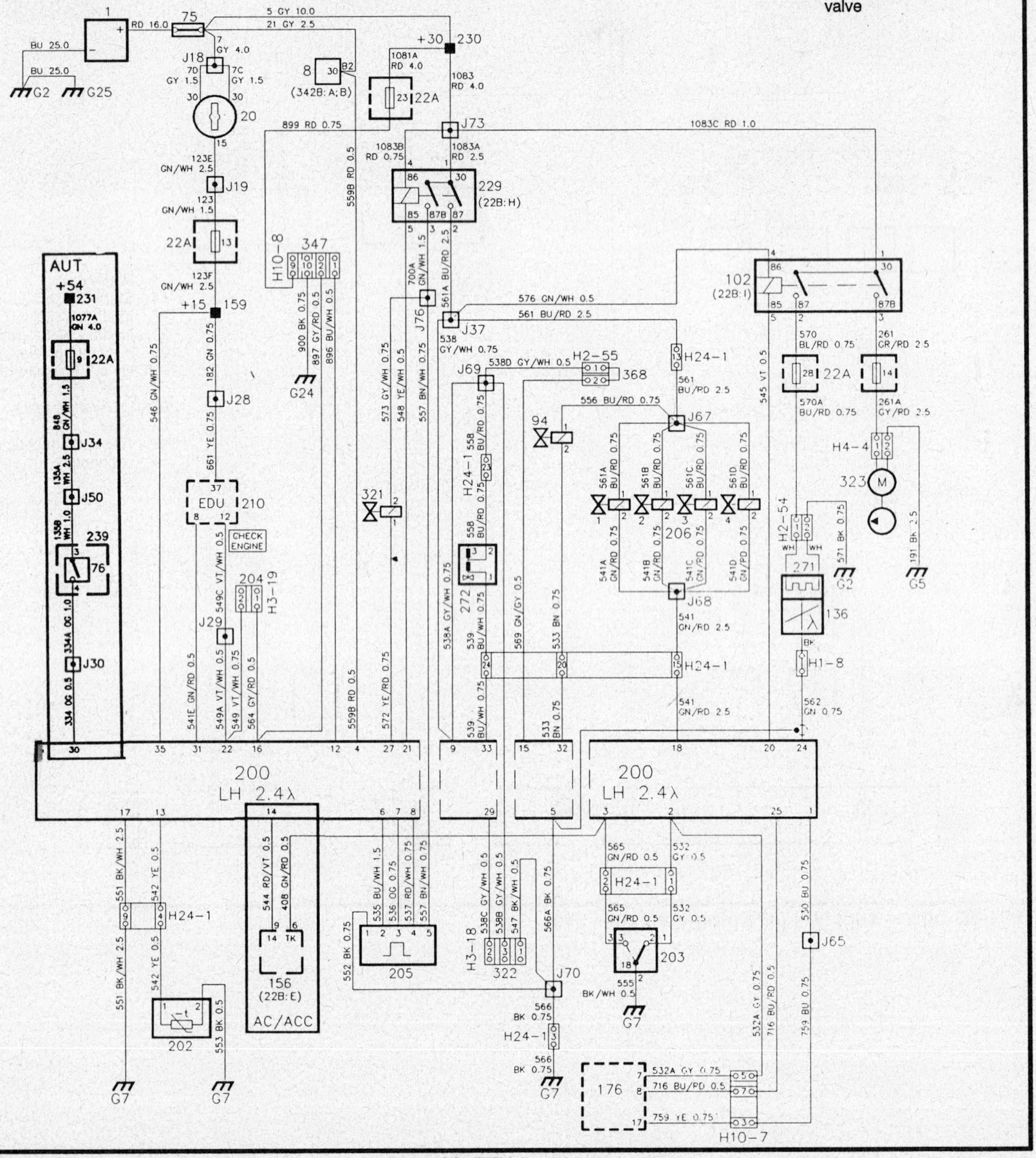

1991–92 LH 2.4 fuel system wiring diagram—9000 with EZK ignition system

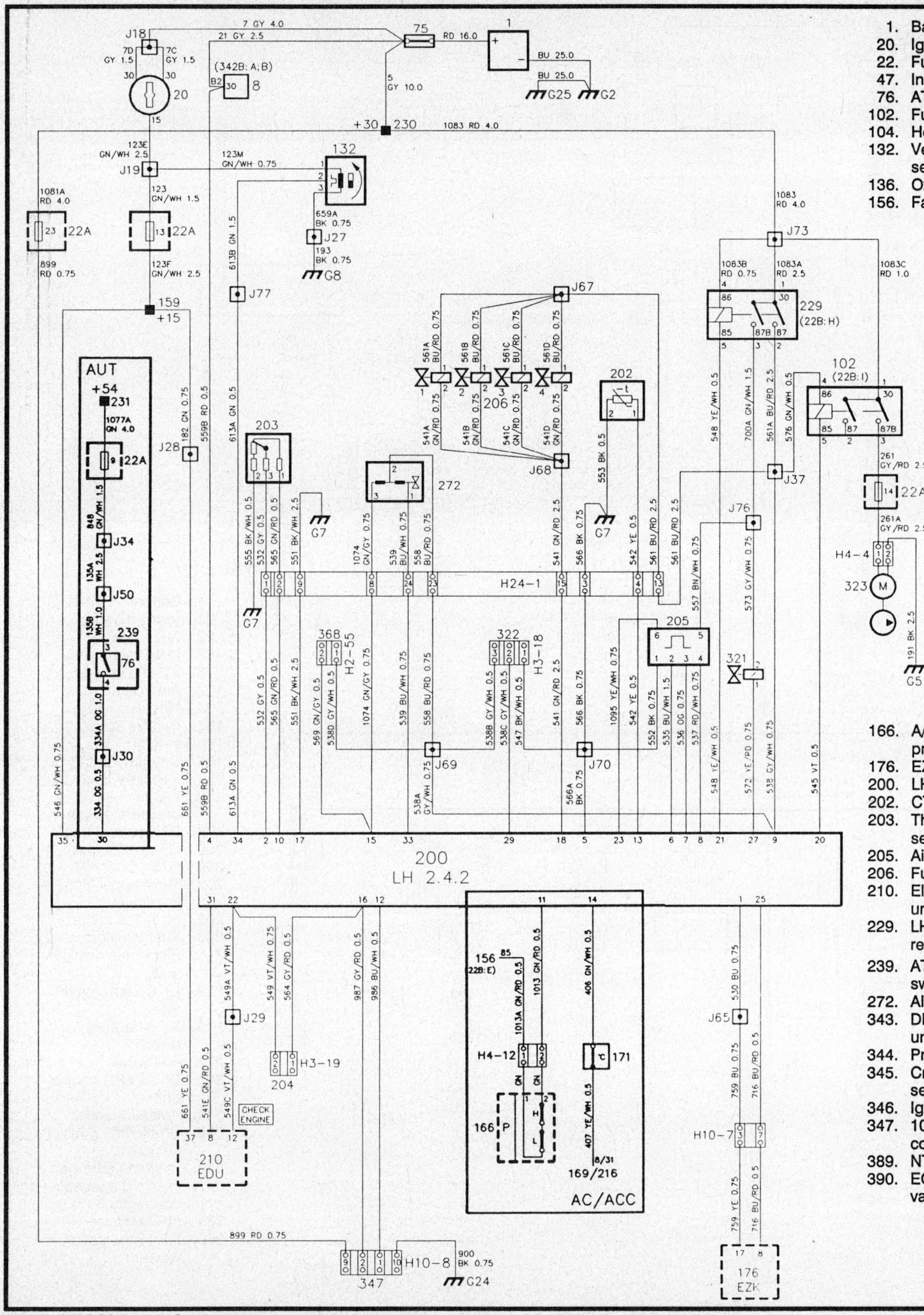

1. Battery
20. Ignition switch
22. Fuse box
47. Instrument panel
76. AT idle up switch
102. Fuel pump relay
104. Hot start relay
132. Vehicle speed sensor
136. Oxygen sensor
156. Fan time delay relay
166. A/C and ACC pressure switch
176. EZK control unit
200. LH control unit
202. CTS
203. Throttle position sensor
205. Air mass meter
206. Fuel injectors
210. Electronic display unit
229. LH-system main relay
239. AT gear selector switch
272. AIC valve
343. DI, DI/APC control unit
344. Pressure sensor
345. Crankshaft position sensor
346. Ignition cartridge
347. 10-pin diagnostic connector
389. NTC resistor
390. EGR modulating valve

1991–92 LH 2.4.2 fuel system wiring diagram—9000 with EZK ignition system

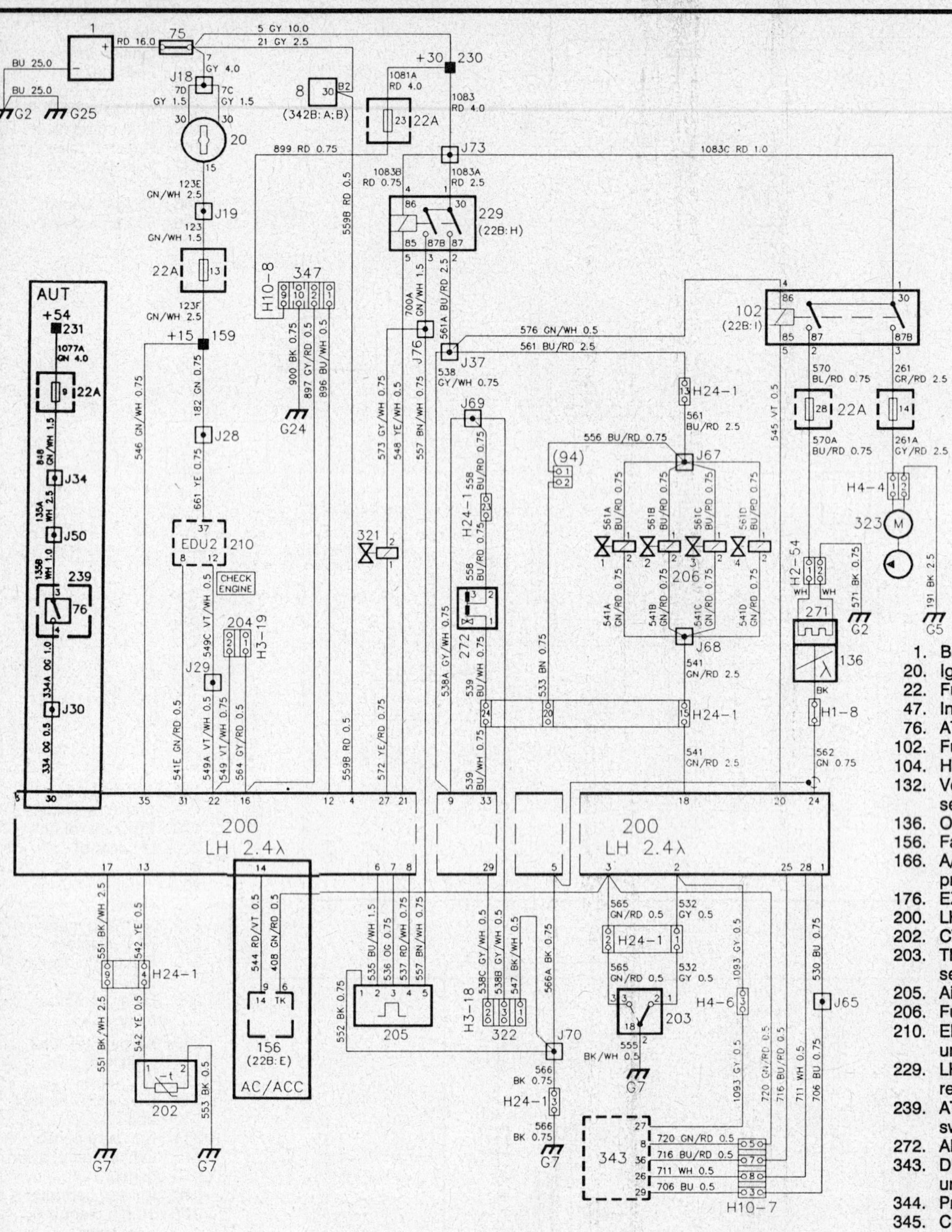

1. Battery
20. Ignition switch
22. Fuse box
47. Instrument panel
76. AT idle up switch
102. Fuel pump relay
104. Hot start relay
132. Vehicle speed sensor
136. Oxygen sensor
156. Fan time delay relay
166. A/C and ACC pressure switch
176. EZK control unit
200. LH control unit
202. CTS
203. Throttle position sensor
205. Air mass meter
206. Fuel injectors
210. Electronic display unit
229. LH-system main relay
239. AT gear selector switch
272. AIC valve
343. DI, DI/APC control unit
344. Pressure sensor
345. Crankshaft position sensor
346. Ignition cartridge
347. 10-pin diagnostic connector
389. NTC resistor
390. EGR modulating valve

1991–92 LH 2.4 fuel system wiring diagram—9000 Turbo

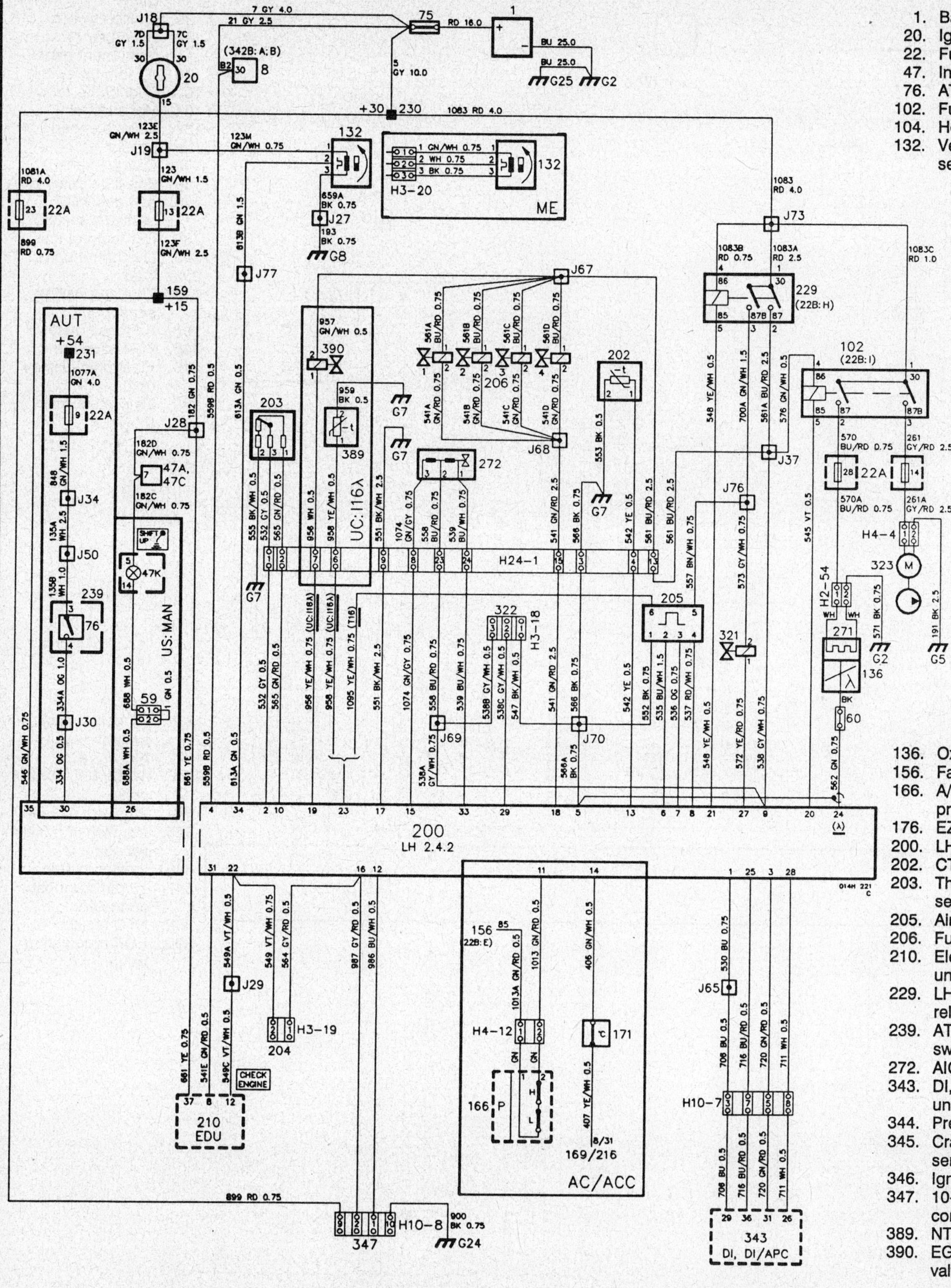

1. Battery
20. Ignition switch
22. Fuse box
47. Instrument panel
76. AT idle up switch
102. Fuel pump relay
104. Hot start relay
132. Vehicle speed sensor

136. Oxygen sensor
156. Fan time delay relay
166. A/C and ACC pressure switch
176. EZK control unit
200. LH control unit
202. CTS
203. Throttle position sensor
205. Air mass meter
206. Fuel injectors
210. Electronic display unit
229. LH-system main relay
239. AT gear selector switch
272. AIC valve
343. DI, DI/APC control unit
344. Pressure sensor
345. Crankshaft position sensor
346. Ignition cartridge
347. 10-pin diagnostic connector
389. NTC resistor
390. EGR modulating valve

1991 LH 2.4.2 fuel system wiring diagram—9000 Turbo and non-turbocharged 9000 with DI ignition system

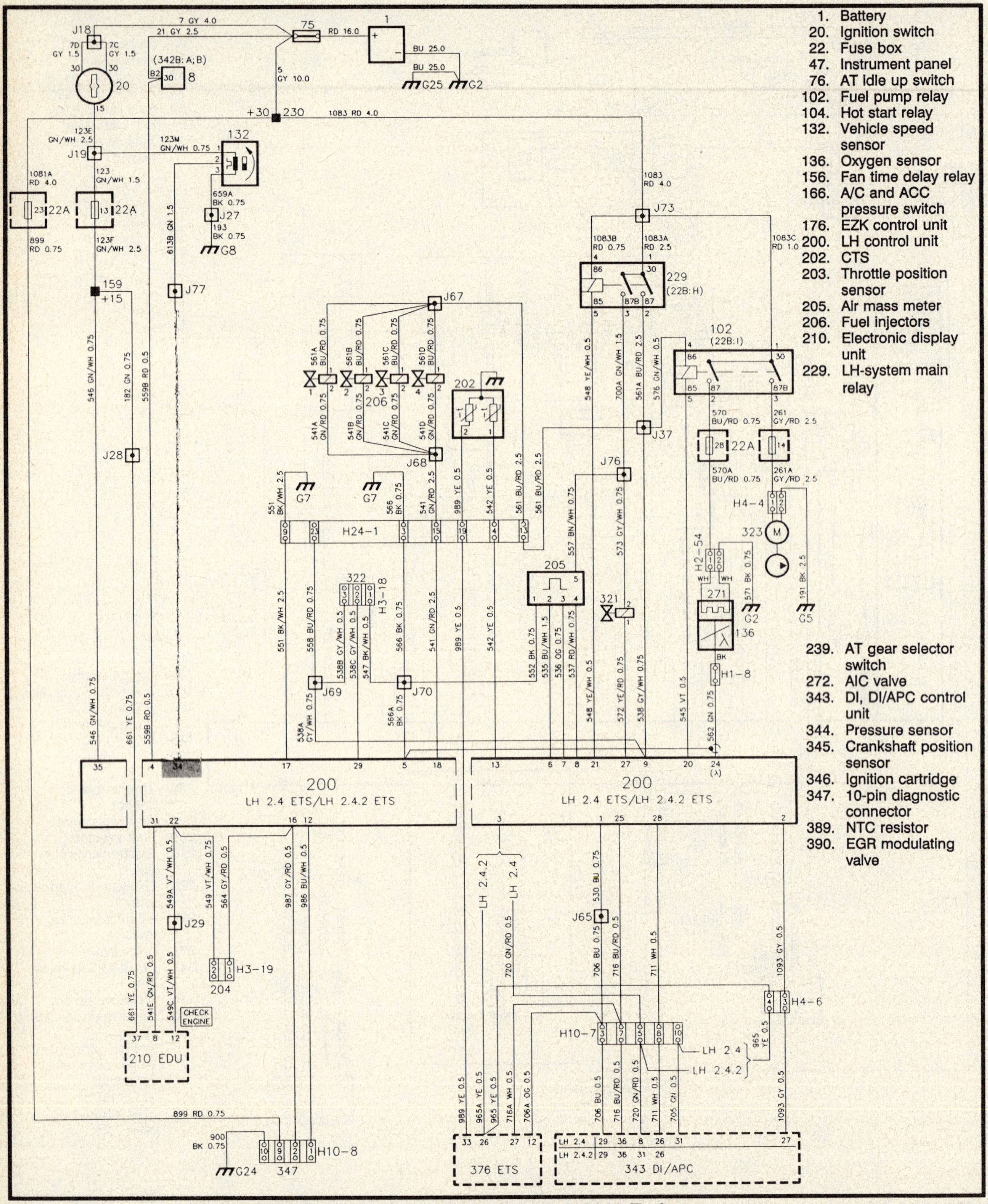

1991 LH 2.4 and LH 2.4.2 fuel systems with ETS wiring diagram—9000 Turbo

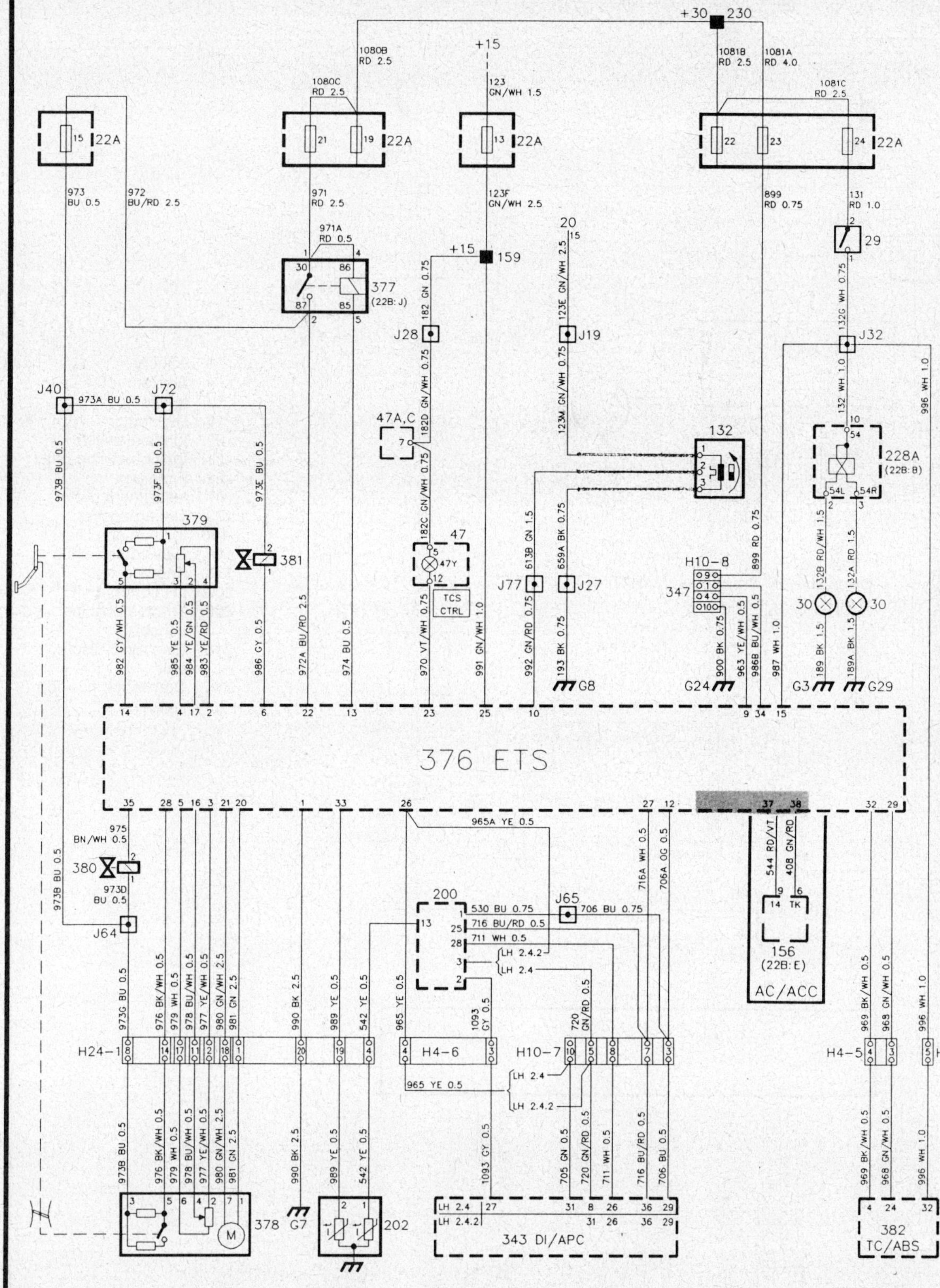

1. Battery
20. Ignition switch
22. Fuse box
47. Instrument panel
76. AT idle up switch
102. Fuel pump relay
104. Hot start relay
132. Vehicle speed sensor
136. Oxygen sensor
156. Fan time delay relay
166. A/C and ACC pressure switch
176. EZK control unit
200. LH control unit
202. CTS
203. Throttle position sensor
205. Air mass meter
206. Fuel injectors
210. Electronic display unit
229. LH-system main relay
239. AT gear selector switch
272. AIC valve
343. DI, DI/APC control unit
344. Pressure sensor
345. Crankshaft position sensor
346. Ignition cartridge
347. 10-pin diagnostic connector
389. NTC resistor
390. EGR modulating valve

1991–92 ETS wiring diagram—9000 Turbo

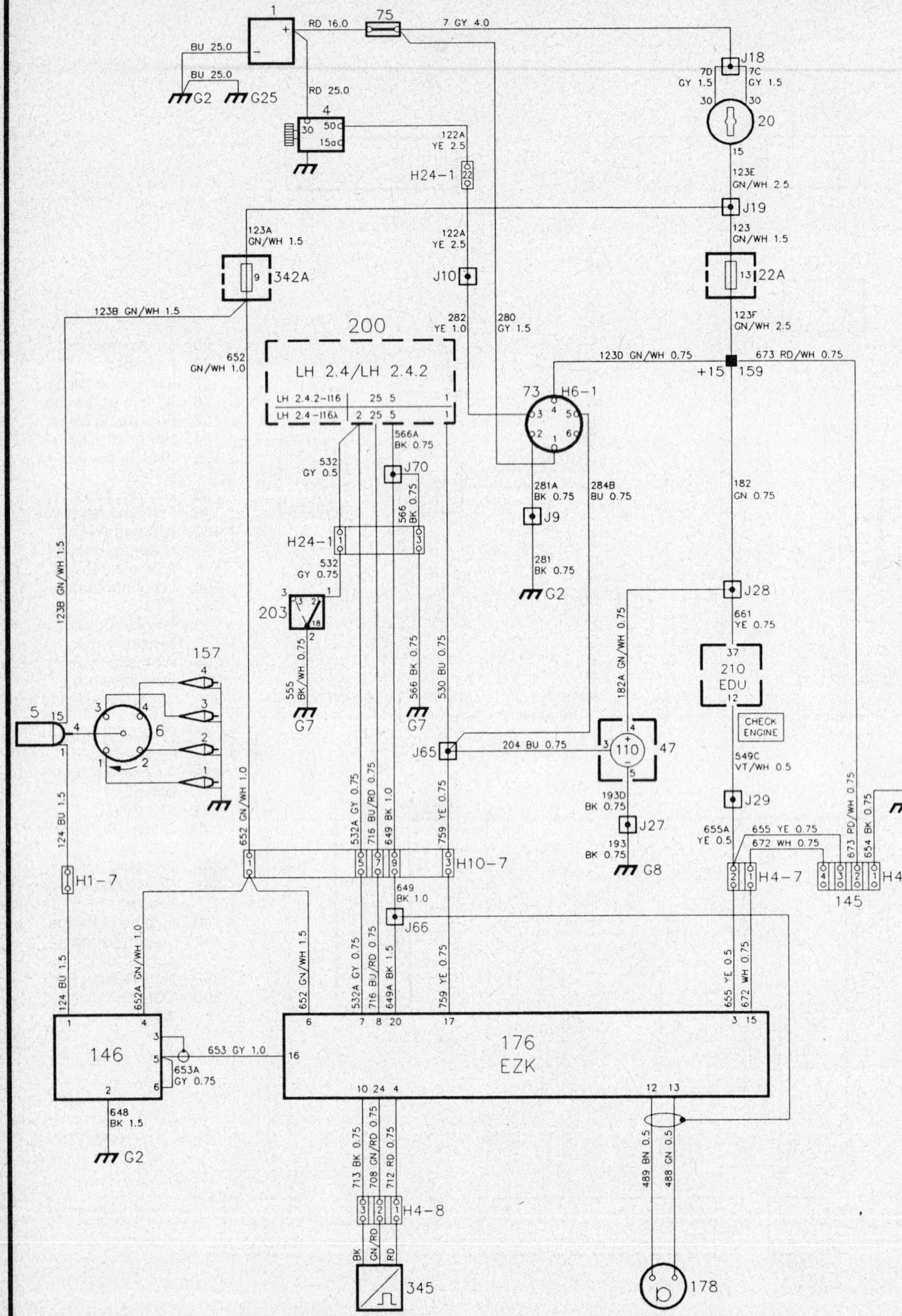

1. Battery
4. Starter motor
5. Ignition coil
6. Distributor
20. Ignition switch
21. Ignition switch relay
22. Fuse box
47. Instrument panel
73. Timing socket
146 Ignition amplifier
176. EZK control unit
178. Knock sensor
200. LH control unit
203. Throttle position sensor
210. Electronic display unit
345. Crankshaft position sensor

1991–92 EZK ignition system wiring diagram—9000

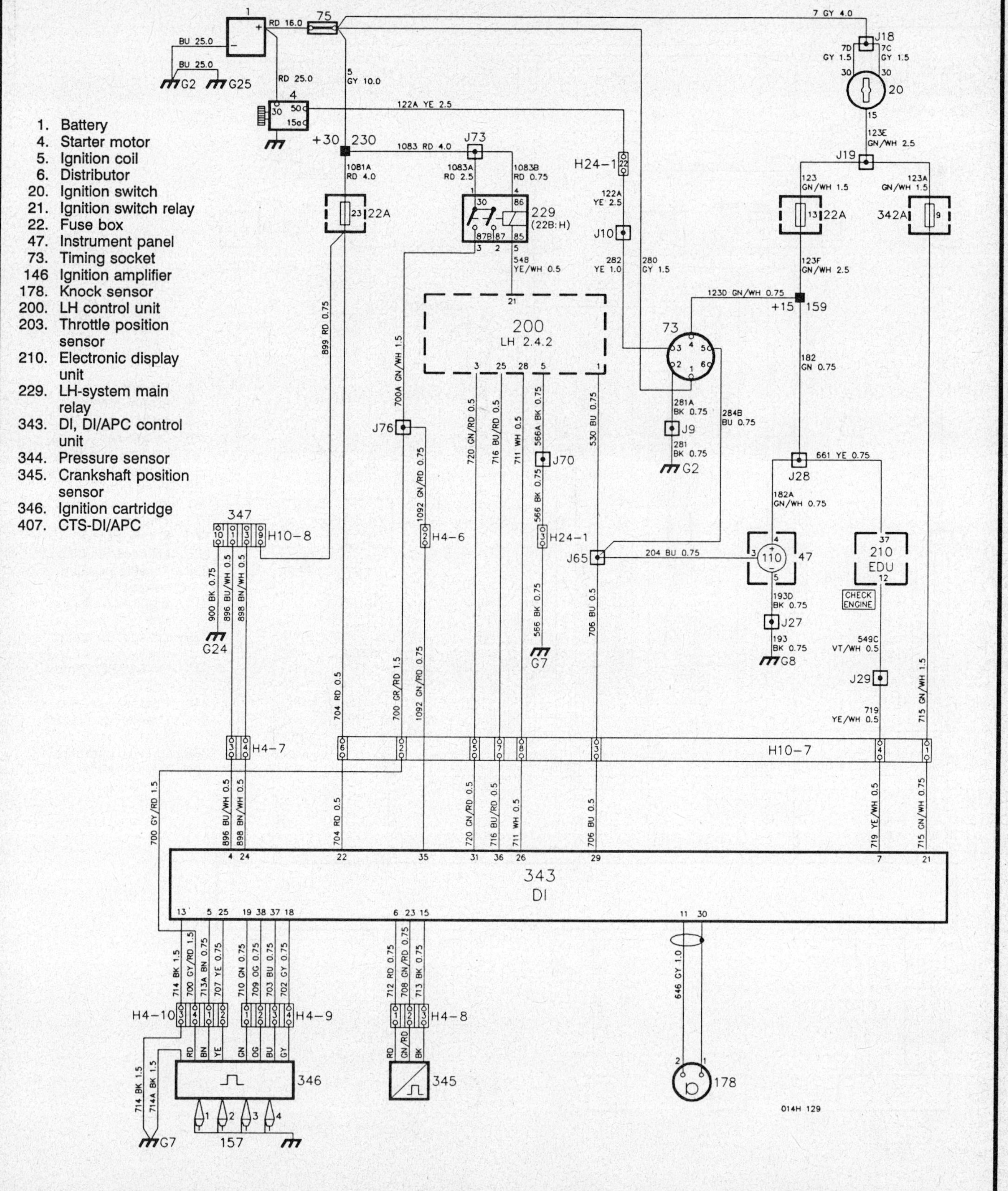

1991–92 DI ignition system with LH 2.4.2 fuel system wiring diagram—Non-turbocharged 9000

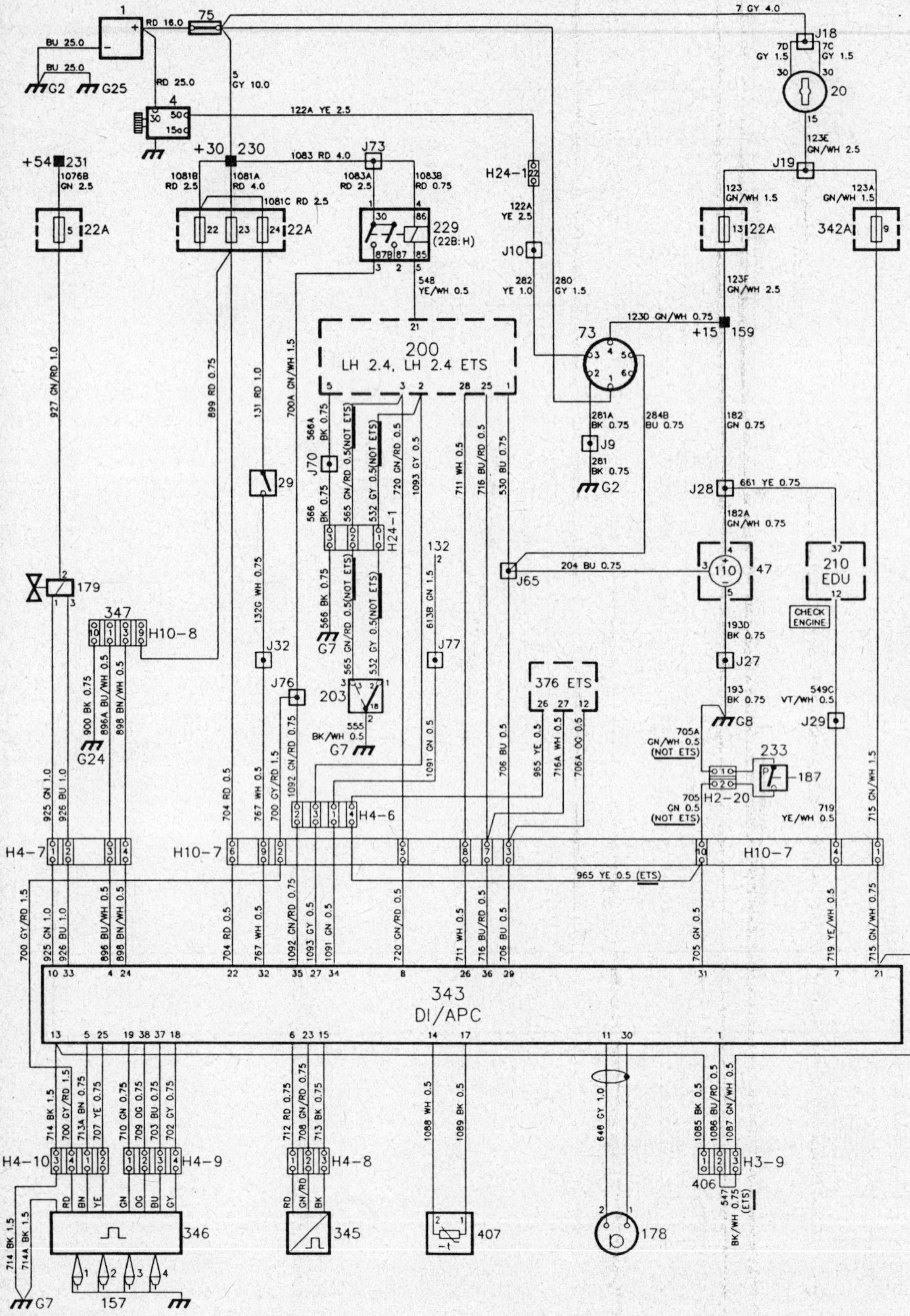

1. Battery
4. Starter motor
5. Ignition coil
6. Distributor
20. Ignition switch
21. Ignition switch relay
22. Fuse box
47. Instrument panel
73. Timing socket
178. Knock sensor
200. LH control unit
203. Throttle position sensor
210. Electronic display unit
229. LH-system main relay
343. DI, DI/APC control unit
344. Pressure sensor
345. Crankshaft position sensor
346. Ignition cartridge
407. CTS-DI/APC

1991–92 DI/APC ignition system with LH 2.4 fuel system wiring diagram—9000 Turbo

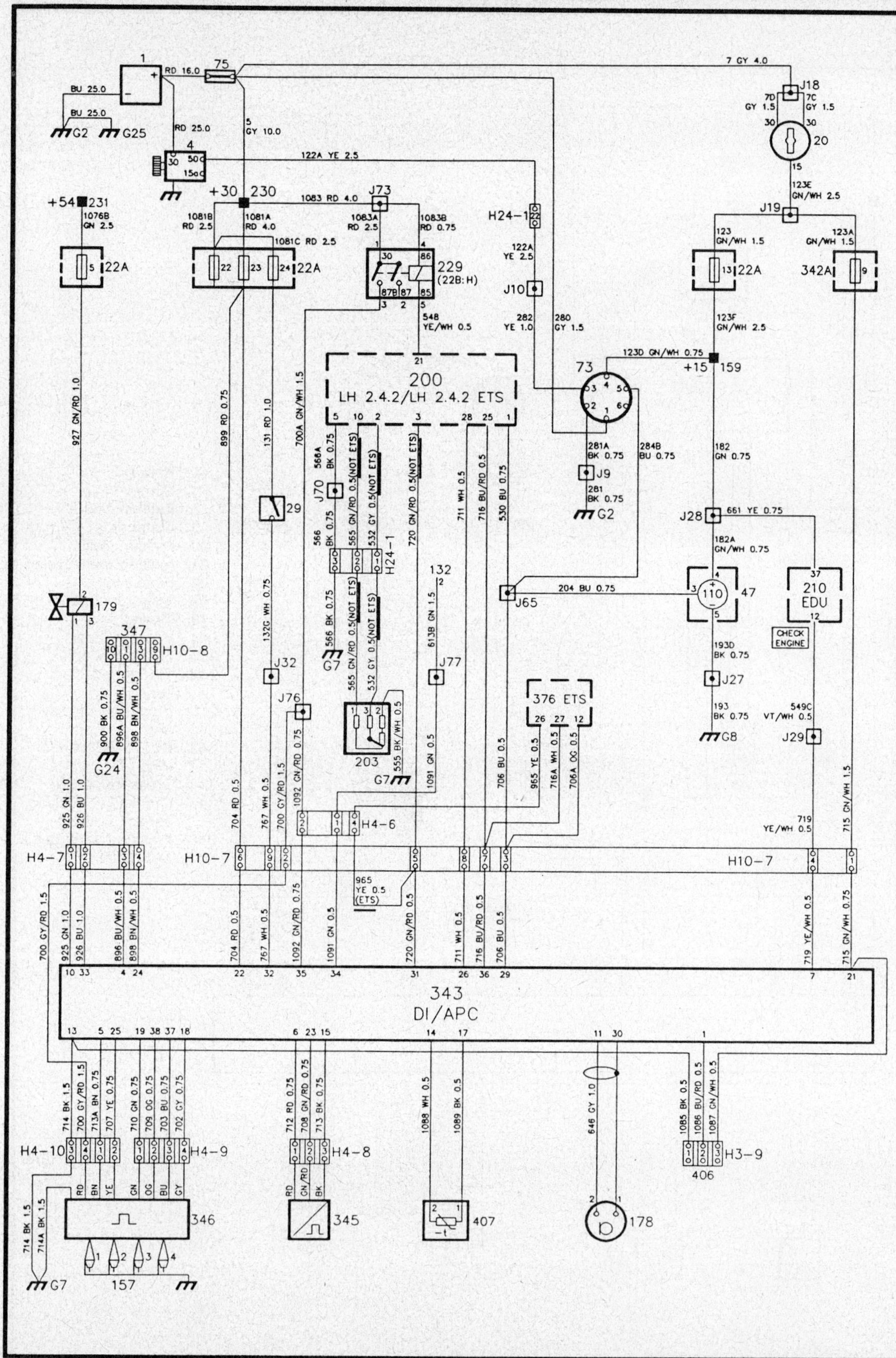

1. Battery
4. Starter motor
5. Ignition coil
6. Distributor
20. Ignition switch
21. Ignition switch relay
22. Fuse box
47. Instrument panel
73. Timing socket
178. Knock sensor
200. LH control unit
203. Throttle position sensor
210. Electronic display unit
229. LH-system main relay
343. DI, DI/APC control unit
344. Pressure sensor
345. Crankshaft position sensor
346. Ignition cartridge
407. CTS-Di/APC

1991 DI/APC ignition system with LH 2.4.2 fuel system wiring diagram—9000 Turbo

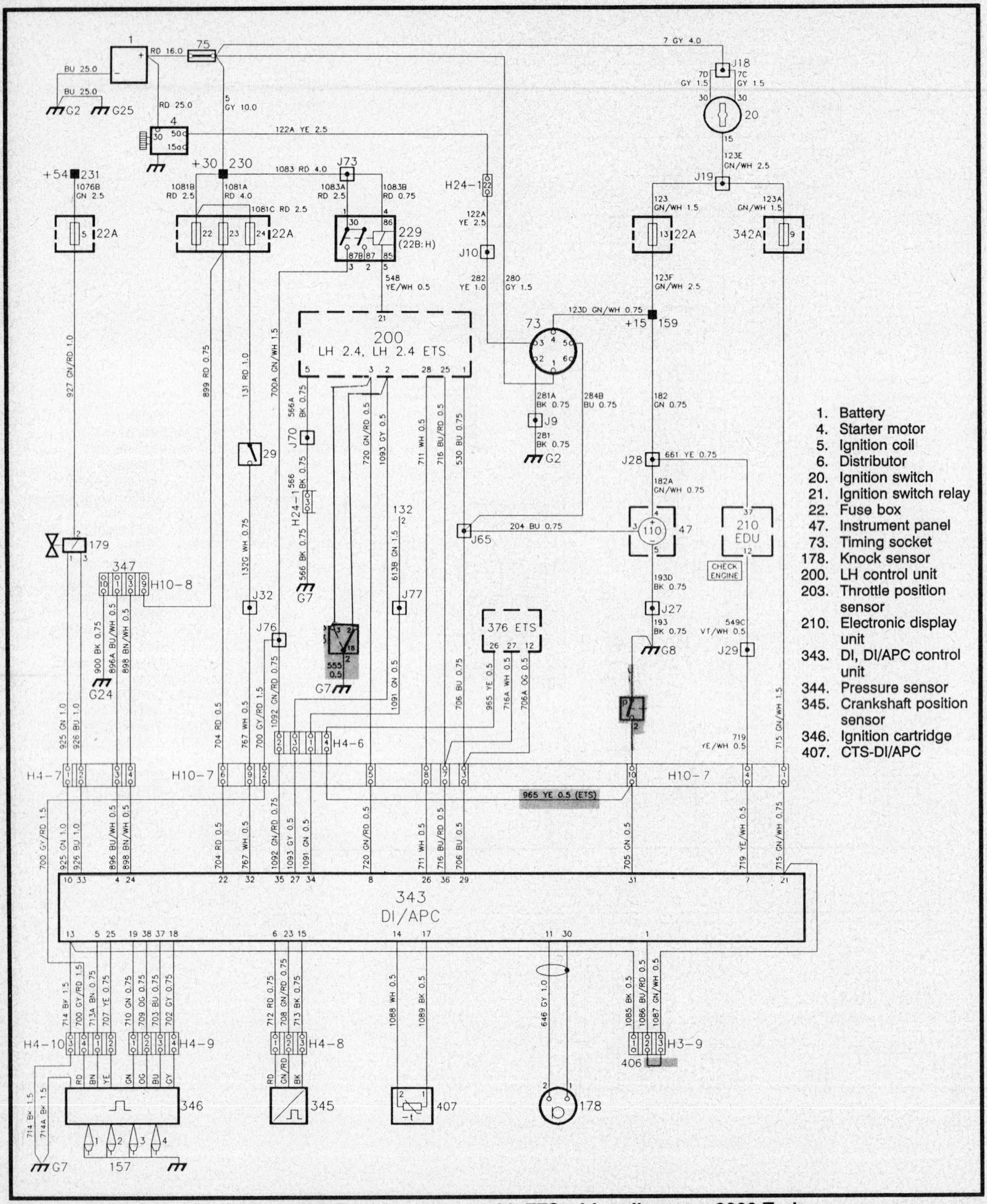

1991–92 DI/APC ignition system with LH 2.4 or LH 2.4 with ETS wiring diagram—9000 Turbo

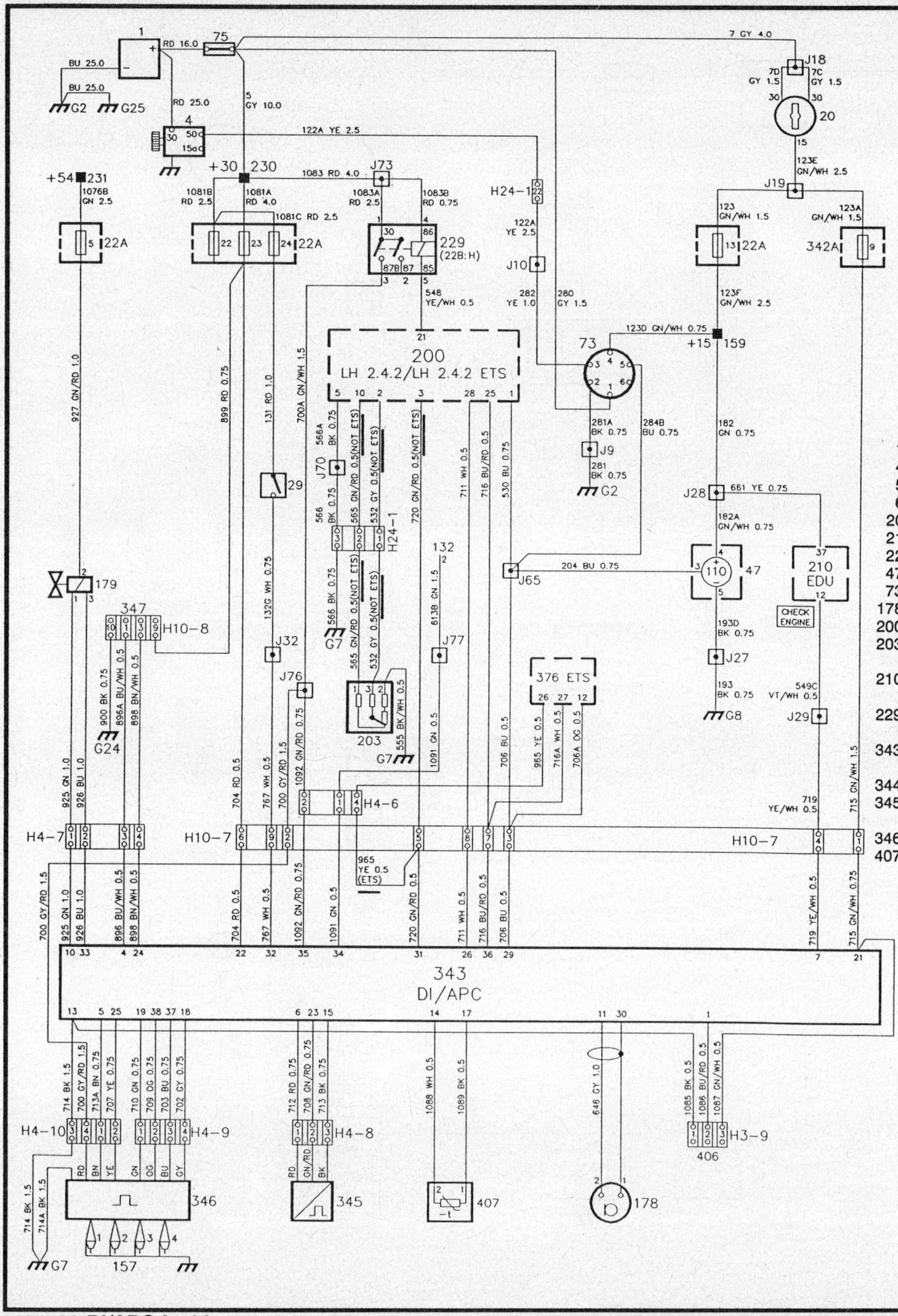

1. Battery
4. Starter motor
5. Ignition coil
6. Distributor
20. Ignition switch
21. Ignition switch relay
22. Fuse box
47. Instrument panel
73. Timing socket
178. Knock sensor
200. LH control unit
203. Throttle position sensor
210. Electronic display unit
229. LH-system main relay
343. DI, DI/APC control unit
344. Pressure sensor
345. Crankshaft position sensor
346. Ignition cartridge
407. CTS-DI/APC

1991–92 DI/APC ignition system with LH 2.4.2 or LH 2.4.2 with ETS wiring diagram—9000 Turbo

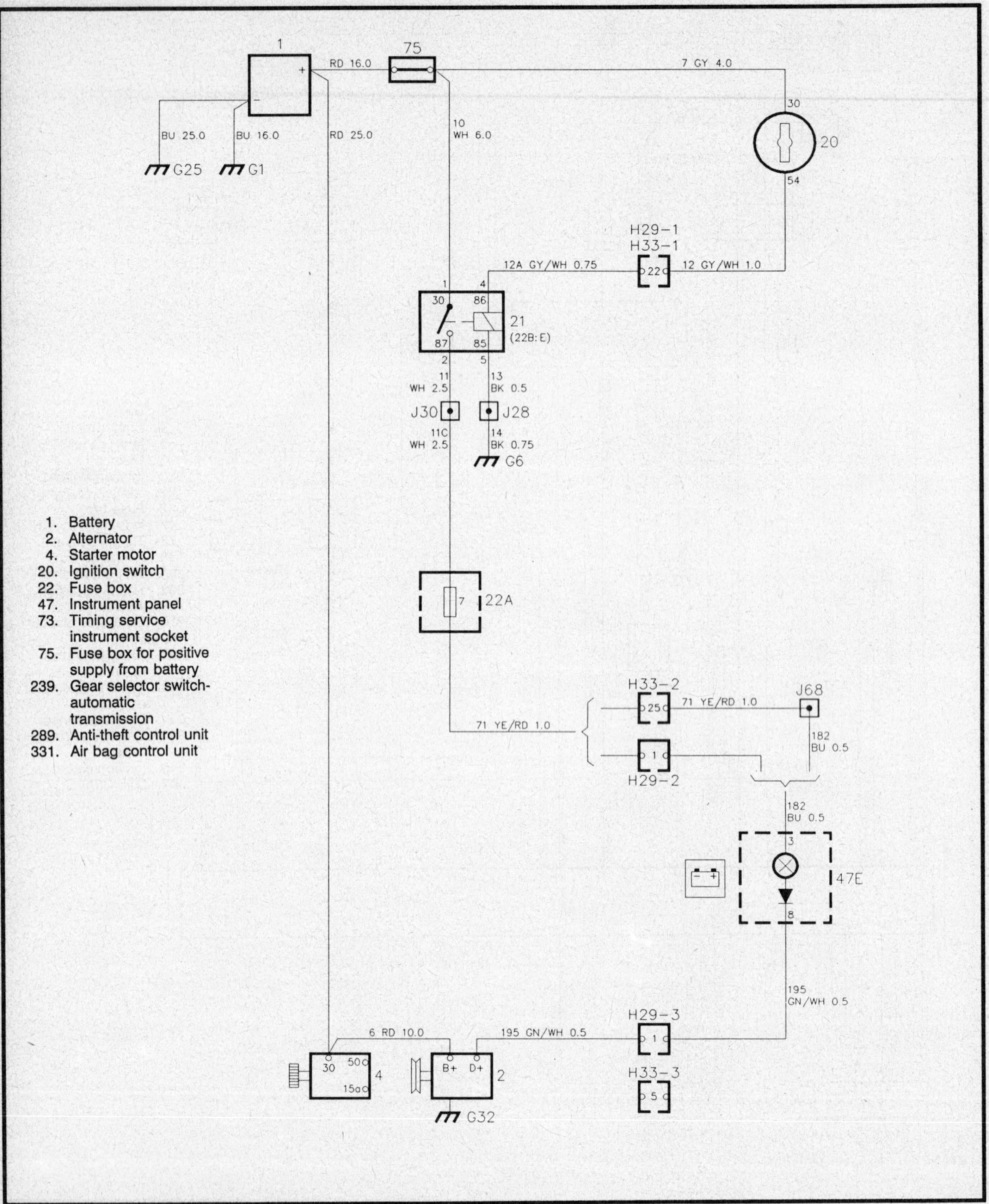

1. Battery
2. Alternator
4. Starter motor
20. Ignition switch
22. Fuse box
47. Instrument panel
73. Timing service instrument socket
75. Fuse box for positive supply from battery
239. Gear selector switch-automatic transmission
289. Anti-theft control unit
331. Air bag control unit

1991 starting and charging system wiring diagram – 9000

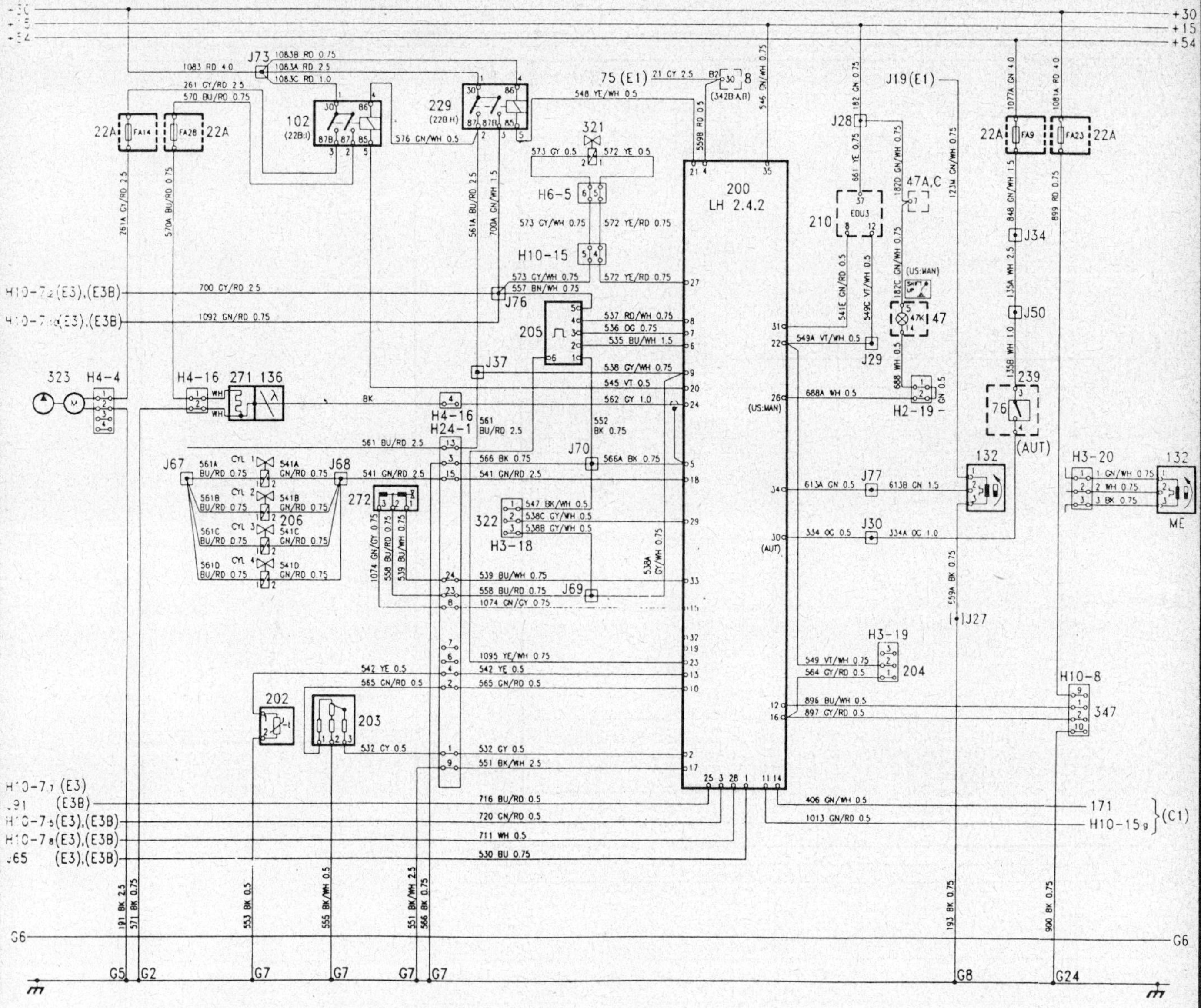

1. Battery
20. Ignition switch
22. Fuse box
47. Instrument panel
76. AT idle up switch
102. Fuel pump relay
104. Hot start relay
132. Vehicle speed sensor
136. Oxygen sensor
156. Fan time delay relay
166. A/C and ACC pressure switch
176. EZK control unit
200. LH control unit
202. CTS
203. Throttle position sensor
205. Air mass meter
206. Fuel injectors
210. Electronic display unit
229. LH-system main relay
239. AT gear selector switch
272. AIC valve
343. DI, DI/APC control unit
344. Pressure sensor
345. Crankshaft position sensor
346. Ignition cartridge
347. 10-pin diagnostic connector
389. NTC resistor
390. EGR modulating valve

1992 LH 2.4.2 fuel system with DI ignition system wiring diagram—Non-turbocharged 9000 with B234 engine

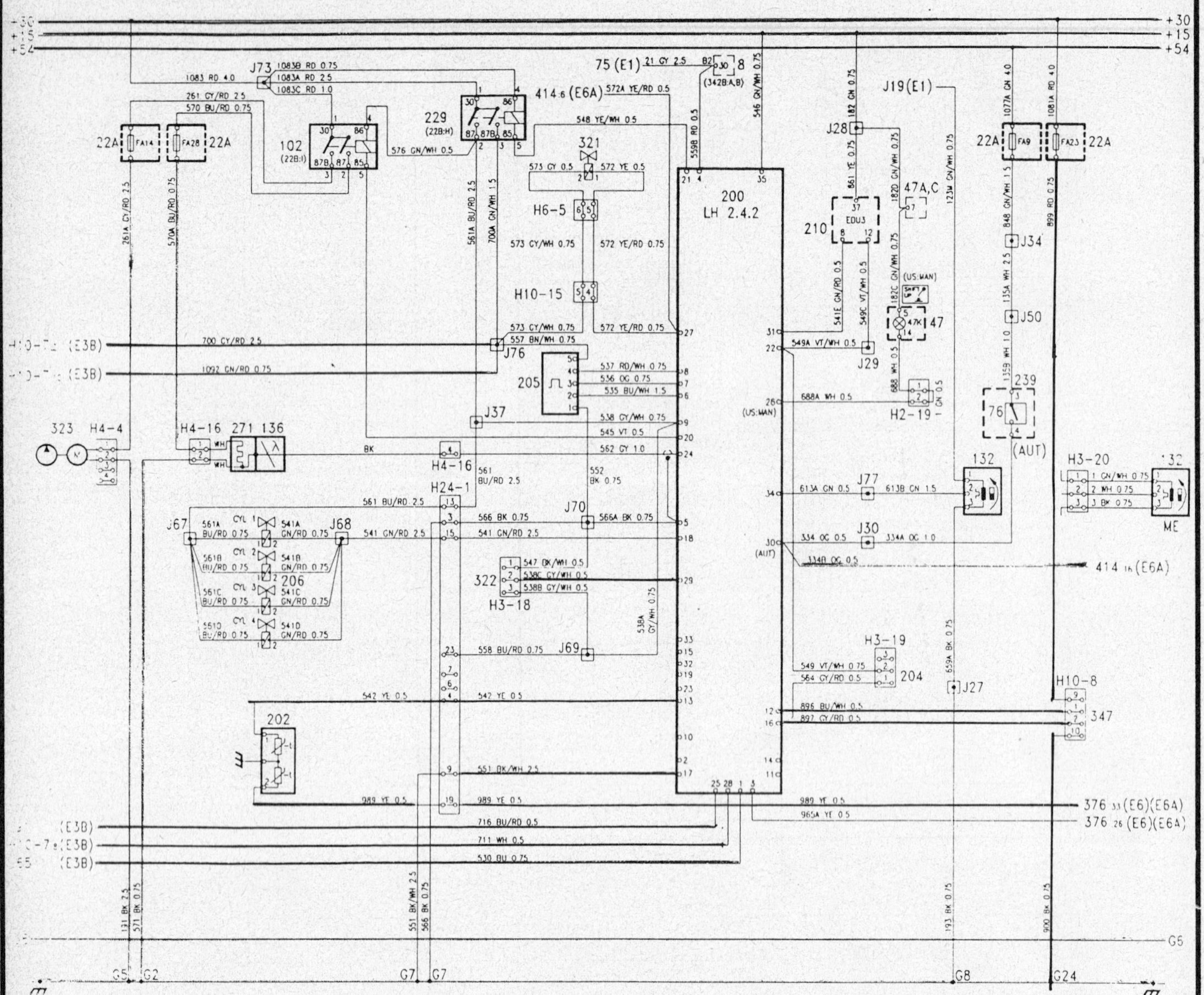

1. Battery
20. Ignition switch
22. Fuse box
47. Instrument panel
102. Fuel pump relay
104. Hot start relay
132. Vehicle speed sensor
136. Oxygen sensor
156. Fan time delay relay
166. A/C and ACC pressure switch
176. EZK control unit
200. LH control unit
202. CTS
203. Throttle position sensor
205. Air mass meter
206. Fuel injectors
210. Electronic display unit
229. LH-system main relay
239. AT gear selector switch
272. AIC valve
343. DI, DI/APC control unit
344. Pressure sensor
345. Crankshaft position sensor
346. Ignition cartridge
347. 10-pin diagnostic connector
389. NTC resistor
390. EGR modulating valve

1992 LH 2.4.2 fuel system with ETS wiring diagram—9000 Turbo

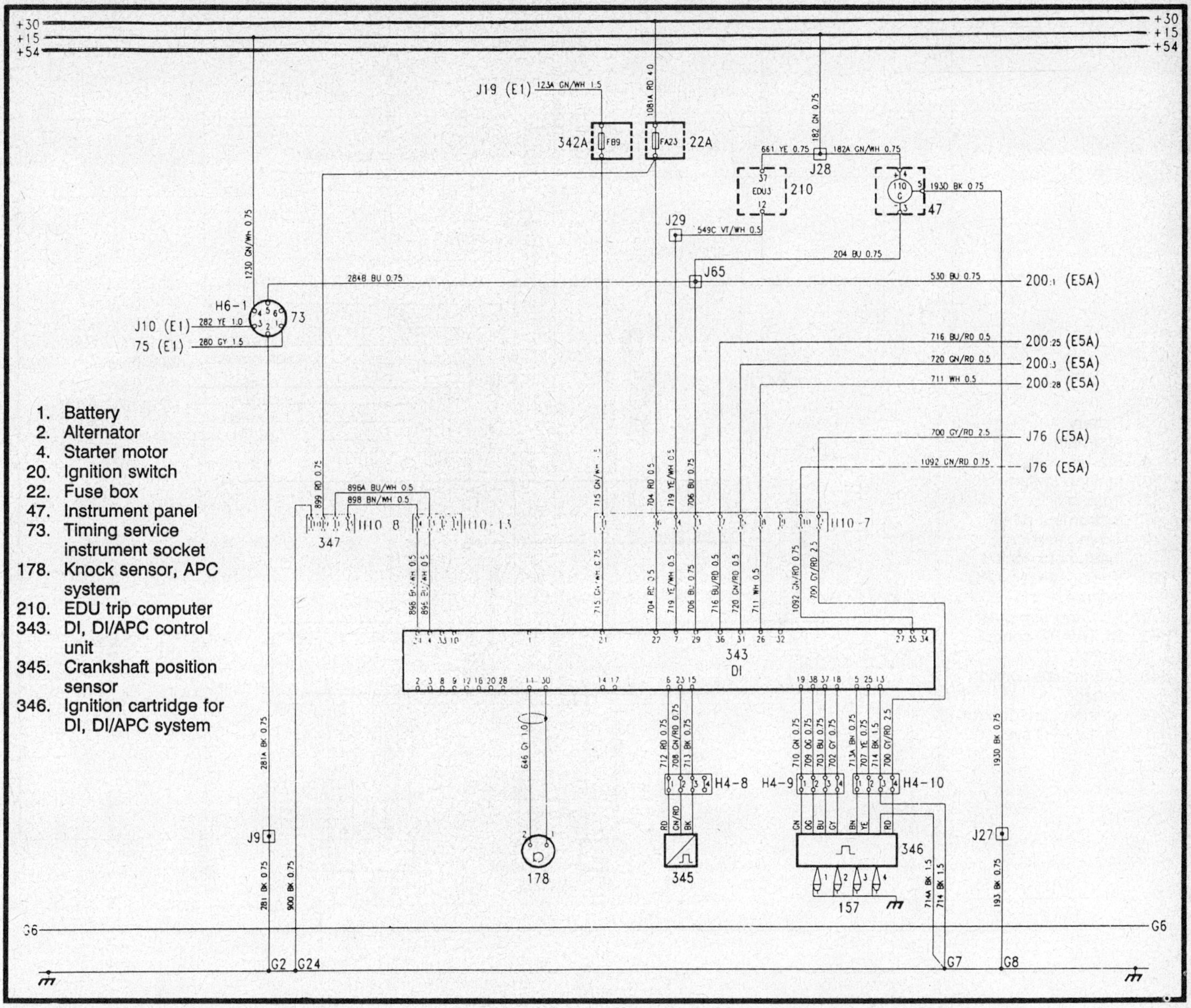

1992 DI ignition system with LH 2.4.2 fuel system wiring diagram—Non-turbocharged 9000

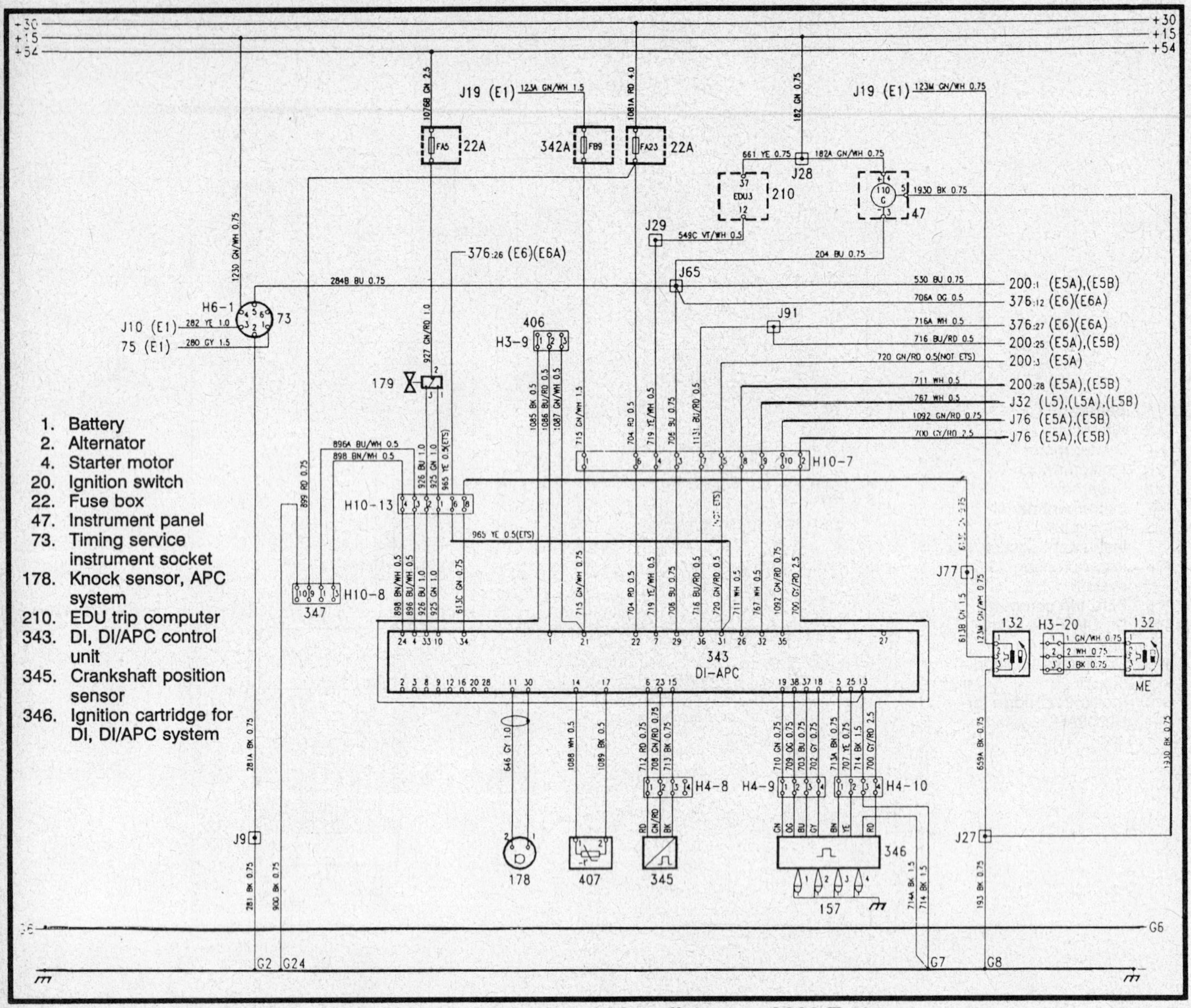

1992 DI/APC ignition system with LH 2.4.2 fuel system wiring diagram—9000 Turbo

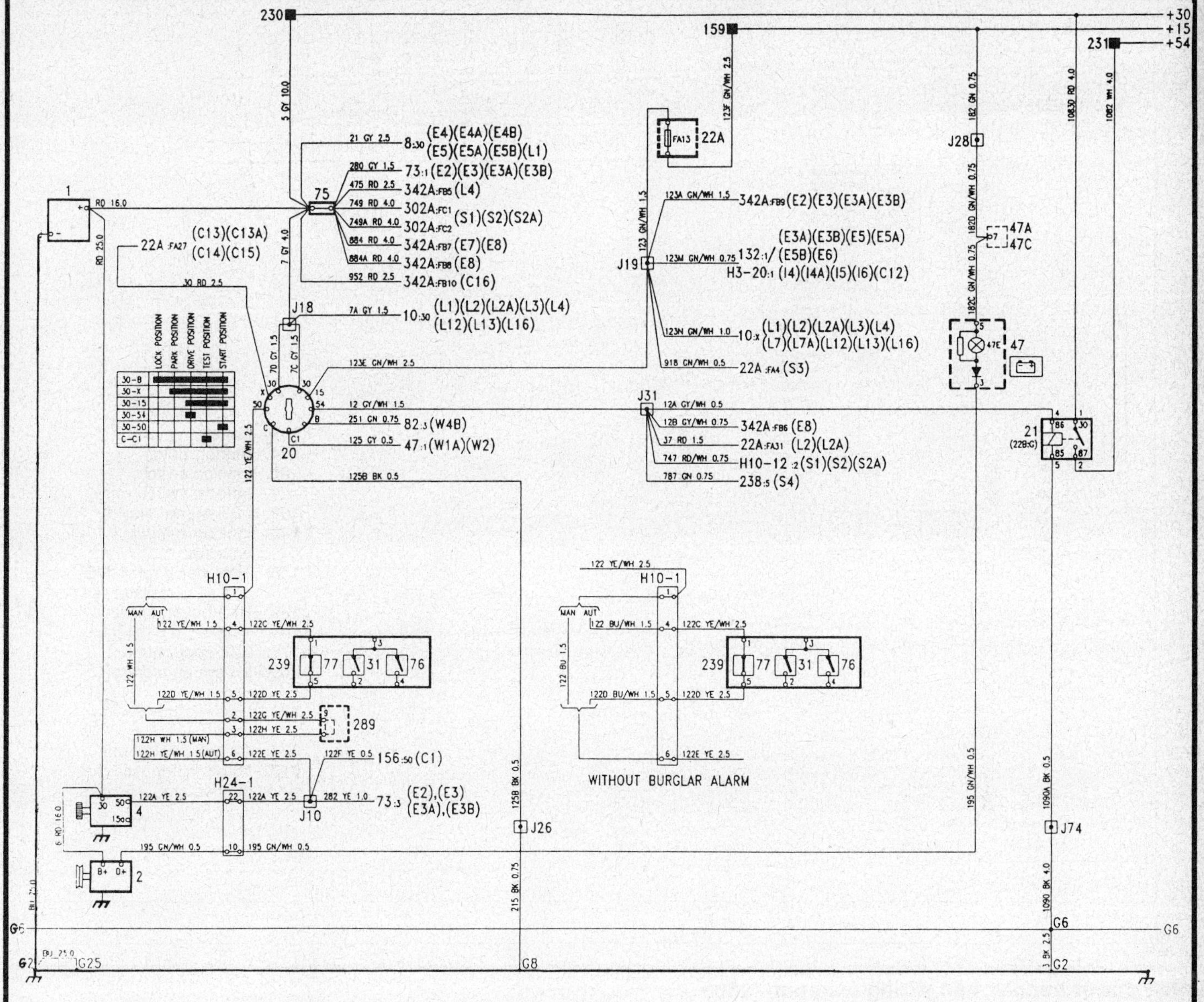

1. Battery
2. Alternator
4. Starter motor
20. Ignition switch
21. Ignition switch relay
22. Fuse box
47. Instrument panel
239. Gear selector switch-automatic transmission

1992 starting and charging system wiring diagram—9000

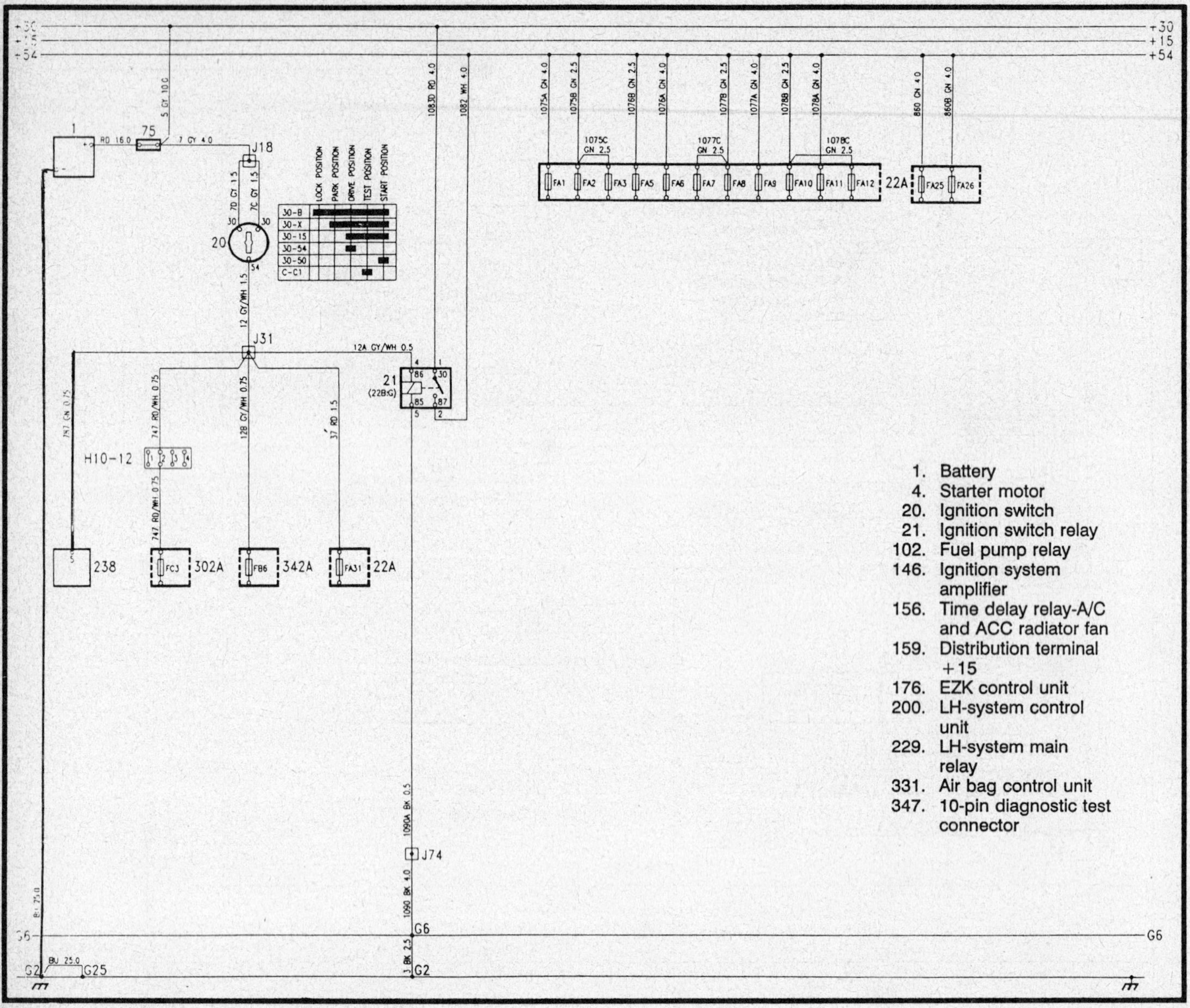

1. Battery
4. Starter motor
20. Ignition switch
21. Ignition switch relay
102. Fuel pump relay
146. Ignition system amplifier
156. Time delay relay-A/C and ACC radiator fan
159. Distribution terminal +15
176. EZK control unit
200. LH-system control unit
229. LH-system main relay
331. Air bag control unit
347. 10-pin diagnostic test connector

1992 power supply +54 wiring diagram—9000

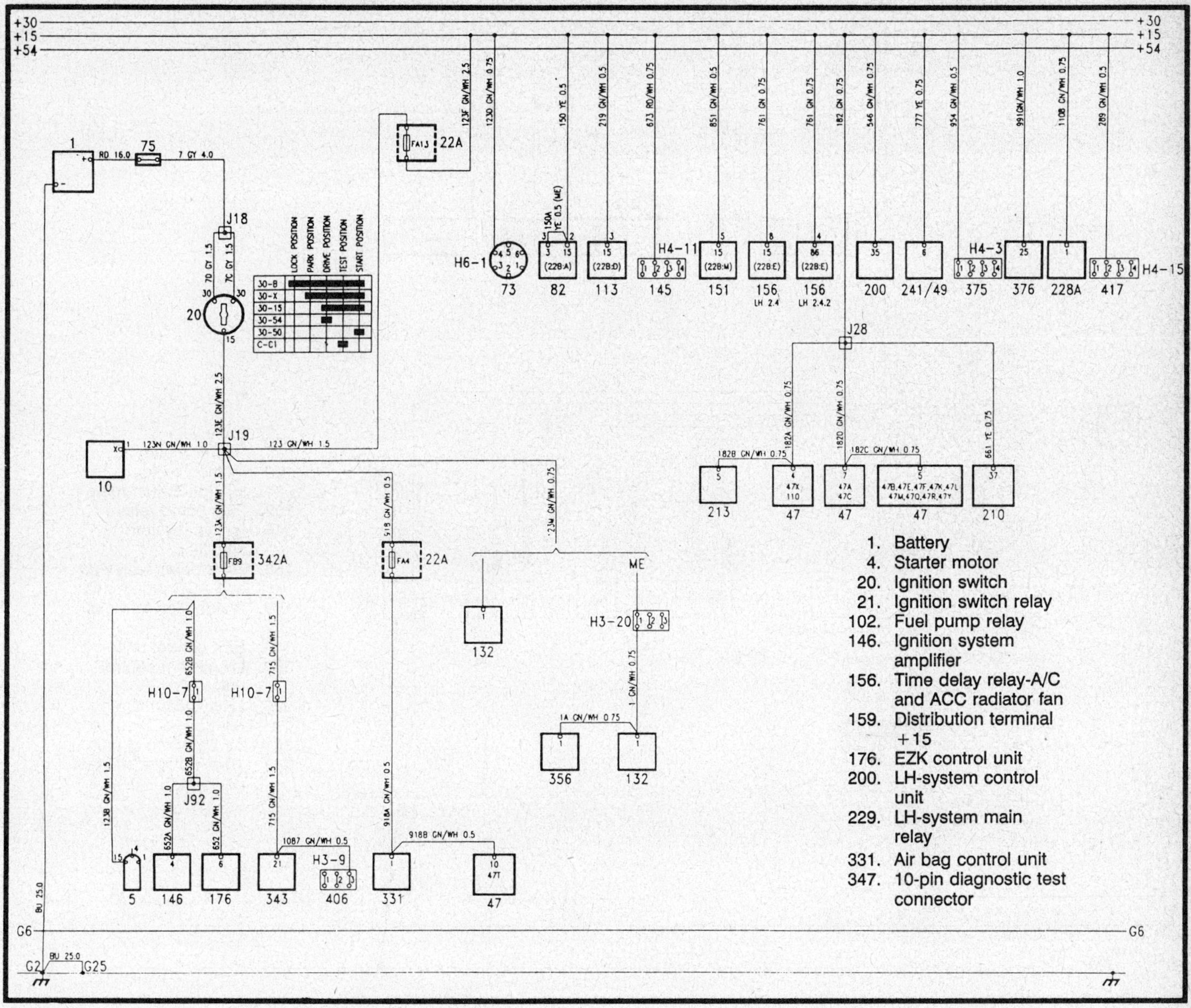

1992 power supply +15 wiring diagram—9000

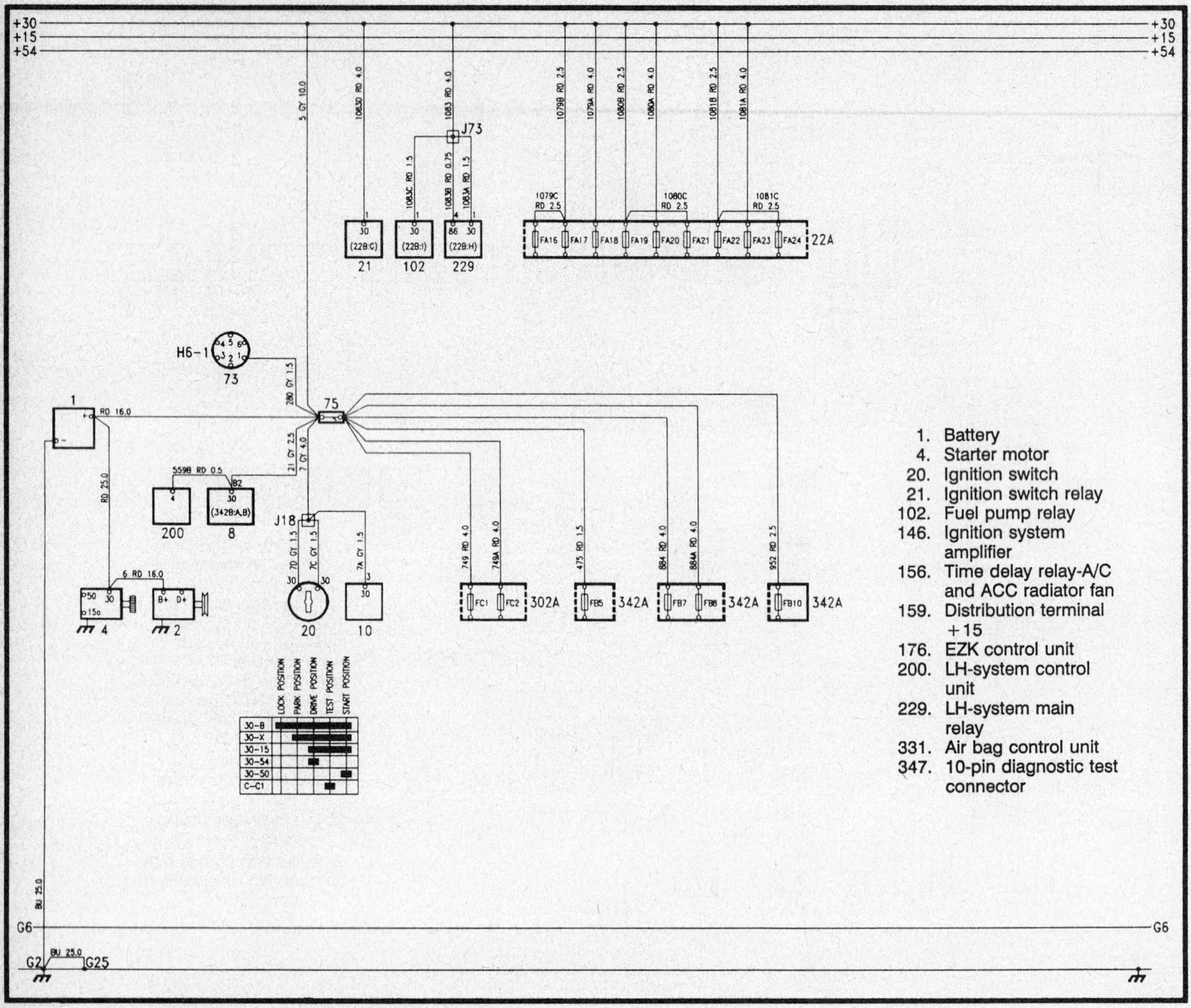

1. Battery
4. Starter motor
20. Ignition switch
21. Ignition switch relay
102. Fuel pump relay
146. Ignition system amplifier
156. Time delay relay-A/C and ACC radiator fan
159. Distribution terminal +15
176. EZK control unit
200. LH-system control unit
229. LH-system main relay
331. Air bag control unit
347. 10-pin diagnostic test connector

1992 power supply +30 wiring diagram—9000

709 OR Trigger signal from no. 1 cylinder (346)
703 BL Trigger signal from no. 3 cylinder (346)
716 BL/RD Load signal, Tq, to pin 25 on ECU (200)
705 GN Signal when cruise-control engaged (233)
926 BL Solenoid valve pin 3 (179)
967 VT Signal when brakes activated (29)
Knock-detector earth braid (176)
706 BL Ignition pulse (47, 73, 200)
711 VT Enrichment on pre-ignition pin 28 (200)
707 GL Combustion signal for nos. 1 + 2 cylinders (346)
898 BR/VT Diagnosis socket pin 3 (347)
708 GN/RD Signal from crankshaft sensor (345)
704 RD +30 voltage via fuse 23
715 GN/VT Ignition voltage via fuse 9
940 RD Live feed to pressure sensor pin 3 (344)

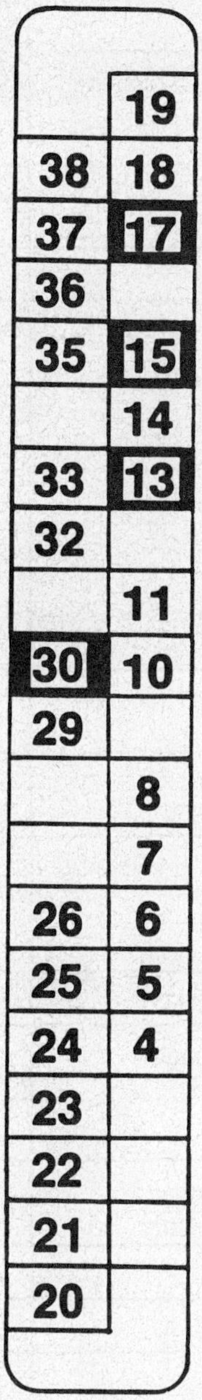

710 GN Trigger signal from no. 2 cylinder (346)
702 GR Trigger signal from no. 4 cylinder (346)
942 SV Earth for pressure sensor pin 2 (344)
713 SV Earth for crankshaft sensor
941 VT Signal from pressure sensor pin 1 (344)
714 SV Earth connection for earthing point on inlet manifold
646 GR Signal from knock detector (178)
925 GN Solenoid valve pin 1 (179)
720 GN/RD Full-load signal from throttle-position sensor (203)
719 GL/VT Earth for CHECK ENGINE light (47P)
712 RD Voltage to crankshaft sensor (344)
713 BR Combustion signal for nos. 3 + 4 cylinders (346)
896 BL/VT Diagnosis socket pin 1 (347)

896	=	Cable (lead) no.
BL	=	Blue
BR	=	Brown
GL	=	Yellow
GN	=	Green
GR	=	Grey
OR	=	Orange
RD	=	Red
SV	=	Black
VL	=	Violet
VT	=	White
(347)	=	Component no.

1990 DI/APC system ECU pin diagram—9000 Turbo

Pin no.	Cable colour	Circuit	In	Out	Voltage, V	Remarks
1	BU	RPM signal from ignition system	x		6.5 >8	When starter operating On idling
2	GY	Throttle potentiometer	x		0.2 4.0	Idling Full throttle
3						Not used
4	RD	+30 supply	x		Battery voltage	
5	BK	Signal earth			0	As distinct from chassis earth
6	BU/WH	Air mass meter earth signal	x		0	
7	OG	Air mass meter signal	x		2 5	Idling with engine running Full throttle with engine running
8	RD/WH	Air mass meter burn-off function		x	4 0	During burn-off All other times
9	BU/WH	Feed from system relay	x		Battery voltage	
10	GN/RD	Throttle potentiometer, nominal		x	5	
11	RD/WH	AC relay control circuit		x	Battery 0	AC off AC on
12	BU/WH	Diagnostics lead K (PWM)	x	x		
13	YE	Temperature sensor, engine	x		4–0.5	-20°C to +80°C
14	BU/WH	AC load signal	x		12 0	AC ON AC OFF
15	GY/RD	AIC opening control signal		x	7–11	Load-free idling; normal engine temp
16	GY/WH	Diagnostics lead L	x	x		PWM signal
17	BK/WH	Chassis earth			0	
18	GN/RD	Injectors		x		Use ISAT pulse function to check that frequency increases as throttle opens
19	WH	EGR valve		x	Battery <Battery	De-energized Energized
20	VT	Fuel pump relay control circuit		x	1 Battery	Energized Ignition on
21	YE/WH	System relay control circuit		x	1 Battery	Ignition on Ignition off

1991–92 LH 2.4 and 2.4.2 system ECU pin diagram – 900

Pin no.	Cable colour	Circuit	In	Out	Voltage, V	Remarks
22	VT/WH	*CHECK ENGINE* light		x	Battery 1	Off On
23	YE/WH	EGR temperature sensor		x	4.5 <4.5	EGR closed EGR open
24	GN	Lambda sensor signal	x		0.6–1.0 0–0.4	Rich Lean
25	BU/RD	*Tq* load signal		x		*)
26	YE	Shift-up light (US Manual)		x	Battery 0	De-energized Energized
27	YE/RD	Tank breather valve		x	Battery 1	Valve open Valve closed
28						Not used
29	GY/WH	Coding	x			Not used
30	OG	DRIVE signal	x		Battery 0	R, D, 1, 2 & 3 P, N & manual gearbox
31						Not used
32		Cold-start valve (where applicable)		x	Battery 0	De-energized Energized (<-15°C)
33	BU/WH	AIC closing control signal		x	5–11	Idling, load-free
34	GN	Signal from speed sensor	x		6 0 or 12	Wheels turning Wheels stationary
35	GN/WH	+15 feed	x		Battery	Ignition on

*) *Use ISAT pulse function to check that pulse width and frequency change with throttle setting.*

1991–92 LH 2.4 and 2.4.2 system ECU pin diagram— 900

The anomalies are valid for: • LH 2.4

ECU-pin nominal values, LH 2.4

Pin no.	Cable colour	Circuit	In	Out	Voltage, V	Remarks
2	GY	Throttle-position sensor idling contacts			0 Battery	On idling Above idling speed
3	GN/RD	Throttle-position sensor full-throttle contacts			Battery 0	On idling On full throttle
10						Not used
11						Not used
15	GN/YE	Coding for cold-start valve			12 0	Valve fitted Valve not fitted
33	BU/WH	AIC control signal			5–12	
34						Not used

1991–92 LH 2.4 system ECU pin diagram—900

Pin no.	Cable colour [1]	Circuit	In	Out	Voltage, V	Remarks
1	BU	RPM signal from ignition system	x		6.5 >8	When starter operating On idling
2	GY	Throttle potentiometer	x		0.2 4.0	Idling Full throttle
3	GN/RD	Throttle-position signal to DI/APC (PWM)		x	1.0 10	Idling Full throttle
4	RD	+30 supply	x		Battery voltage	
5	BK	Signal earth			0	As distinct from chassis earth
6	BU/WH	Air mass meter earth signal	x		0	
7	OG	Air mass meter signal	x		2 5	Idling with engine running Full throttle with engine running
8	RD/WH	Air mass meter burn-off function		x	4 0	During burn-off All other times
9	GY/WH	Feed from system relay	x		Battery voltage	
10	GN/RD	Throttle potentiometer, nominal		x	5	
11	GN/RD	AC relay control circuit		x	Battery 0	AC off AC on
12	BU/WH	Diagnostics lead K (PWM)	x	x		
13	YE	Temperature sensor, engine	x		4–0.5	-20°C to +80°C
14	GN/WH	AC/ACC load signal	x		12 0	AUTO*) ECON OFF
15	GN/GY	AIC opening control signal		x	7–11	Load-free idling; normal engine temp
16	GY/RD	Diagnostics lead L	x	x		PWM signal
17	BK/WH	Chassis earth			0	
18	GN/RD	Injectors		x		Use ISAT pulse function to check that frequency increases as throttle opens
19	YE/WH	EGR valve		x	Battery <Battery	De-energized Energized
20	VT	Fuel pump relay control circuit		x	1 Battery	Energized Ignition on
21	YE/WH	System relay control circuit		x	1 Battery	Ignition on Ignition off
22	VT/WH	*CHECK ENGINE* light		x	Battery 1	Off On
23	YE	EGR temperature sensor		x	4.5 <4.5	EGR closed EGR open
24	GN	Lambda sensor signal	x		0.6–1.0 0–0.4	Rich Lean

1991–92 LH 2.4 and 2.4.2 system ECU pin diagram – 9000

Pin no.	Cable colour [1)]	Circuit	In	Out	Voltage, V	Remarks
25	BU/RD	*Tq* load signal		x		**)
26	WH	Shift-up light (US Manual)		x	Battery 0	De-energized Energized
27	YE/RD	Tank breather valve		x	Battery 1	Valve open Valve closed
28	WH	PRE-IGNition signal	x		0 6.5 12	Fuel boosting on PRE-IGN No boosting Fuel boosting on knocking
29	GY/WH	Coding	x			Not used
30	OG	DRIVE signal	x		Battery 0	R, D, 1, 2 & 3 P, N & manual gearbox
31	GN/RD	Fuel-consumption signal to EDU3		x		***)
32		Cold-start valve (where applicable)		x	Battery 0	De-energized Energized (<-15°C)
33	BU/WH	AIC closing control signal		x	5–11	Idling, load-free
34	GN	Signal from speed sensor	x		6 0 or 12	Wheels turning Wheels stationary
35	GN/WH	+15 feed	x		Battery	Ignition on

*) *Only if anti-frost thermostat closed*

**) *Use ISAT pulse function to check that pulse width and frequency change with throttle setting*

***) *Use ISAT pulse function to check that frequency of fuel-consumption signal changes as throttle-setting changes.*

1) *New cable-colour codes have been introduced as from M91. See the table on page 24.*

1991–92 LH 2.4 and 2.4.2 system ECU pin diagram – 9000

Pin no.	Cable colour *)	Circuit	In	Out	Voltage, V	Remarks
1	RD/BU	Variant-coding lead			0 6–7 Battery	Linked to earth Not connected Linked to battery voltage
2						Not connected
3						Not connected
4	BU/WH	Diagnostics lead K (347; pin 3)				
5	BN	Combustion signal, cylinders 3 + 4			0	
6	RD	Feed to crankshaft sensor			Approx. 12	
7	YE/WH	*CHECK ENGINE* light, EDU 3			0 11–12	On Off
8	GN/RD	Throttle signal from 203 (LH 2.4 non-TCS only)			0 10–12	Full throttle Idling
9						Not connected
10	GN	Solenoid valve 179, pin 1 (Turbo only)			Battery	De-energized
11	GY	Knock-sensor signal			0	
12						Not connected
13	BK	Earth connection				
14	WH	Temperature sensor 407—sensor upstream of throttle housing (n/a DI non-Turbo)			0–5	Temperature dependent
15	BK	Earth for crankshaft sensor				
16						Not connected
17	BK	Earth for temperature sensor 407				
18	GY	Trigger signal for no. 4 cylinder			Battery	
19	GN	Trigger signal for no. 2 cylinder			Battery	
20						Not connected
21	GN/WH	+15 feed (342A)			Battery	
22	RD	+30 feed			Battery	
23	GN/RD	Signal from crankshaft sensor			0 5	Unscreened Screened
24	BN/WH	L diagnostics lead (347, pin 3)			10–12	
25	YE	Combustion signal for cylinders 1 + 2			0 5	Low High
26	WH	Enrichment on PRE-IGN/knocking (Engine running & normal operation)			12 6–7 12	Ignition on LH 2.4.2 LH 2.4
27						Not connected
28						Not connected
29	BU	RPM signal			Approx. 6.5 10–12	On starting Engine running

1991–92 DI/APC ECU pin diagram—9000 Turbo

Pin no.	Cable colour *)	Circuit	In	Out	Voltage, V	Remarks
30						Earth (screen) for knock detector 178
31	GN/RD	Throttle-position signal			1 Approx. 11	Idling Full throttle
32	WH	Brake signal (Turbo only)			0 Battery	Not activated Activated
33	BU	Solenoid valve 176, pin 3 (Turbo only)			Battery	De-energized
34	GN/GY	Speed-sensor signal (n/a DI non-Turbo)			0 or 12 0–5	Car stationary Car moving
35	GN/RD	Feed from system relay (+30)			Battery	
36	BU/RD	Load signal, *Tq*			<1	Use ISAT PULS function or command code 287
37	BU	Trigger signal for no. 3 cylinder			Battery	
38	OG	Trigger signal for no. 1 cylinder			Battery	

*) *New cable-colour codes have been introduced as from M91. See the table on page 24.*

1991–92 DI/APC ECU pin diagram—9000 Turbo

Component Replacement

SERVICE PRECAUTIONS

- Always disconnect the negative battery terminal to prevent sparks caused by short-circuiting, circuit-breaking, etc.
- Smoking must not be allowed when working around flammable liquids.
- Always keep a CO_2 fire extinguisher close on hand.
- Dry sand must be available to soak up any spillage.
- If using a battery charger, make sure that the battery leads are disconnected.
- Never disconnect the battery while the engine is running.
- Never disconnect or connect an ECU connector with the ignition switched **ON**.
- Always remove the control units from the vehicle before carrying out any electrical welding.
- Before conducting a compression test, unplug the connector on the output module.
- Never subject the vehicle to temperatures exceeding 176°F (80°C), such as in a paint booth, unless the control units have first been removed from the vehicle.

FUEL PRESSURE RELIEF

1. Disconnect the negative battery cable.
2. Loosen the banjo fitting on the fuel-injection rail to release the system pressure.

NOTE: Soak up any escaping fuel with a shop towel and dispose of it in the proper container.

3. Retighten the banjo fitting.

RELAY PANEL

Removal and Installation

1. Remove the glove compartment and power distribution panel retaining screws.
2. Unhook the retaining clip for the relay panel and lower the back of the panel.
3. Pull it forward and lift it clear of the brackets.
4. Install the panel in the reverse order of removal.

LH-JETRONIC ELECTRONIC CONTROL UNIT

Removal and Installation

1. Disconnect the negative battery cable.
2. Move the seat to the front position.
3. Remove the control unit cover and disconnect the electrical connector.
4. Remove the scuff plate and protector plate retaining screws, if required.
5. Remove the control unit.
6. Fit the control unit into position.
7. Install the scuff plate and protector plate retaining screws, if required.
8. Connect the electrical connector to the unit.
9. Install the control unit cover. Reconnect the negative battery cable.

DI/APC SYSTEM ELECTRONIC CONTROL UNIT

Removal and Installation

1. Disconnect the negative battery terminal.
2. Remove the left front seat.
3. Remove the 2 scuff plate screws and remove the plate.
4. Remove the 2 DI/APC system ECU cover plates and remove the plate.
5. Disconnect the DI/APC system ECU connectors and remove the DI/APC system ECU.
6. Installation is the reverse of removal.

AIR MASS METER

Removal and Installation

1. Disconnect the negative battery cable.
2. Unplug the air mass meter electrical leads.
3. Remove the air cleaner connecting duct and the air mass meter from the rubber elbow connector.
4. Install the new air mass meter and the air cleaner connecting duct.
5. Reconnect the air mass meter electrical leads. Reconnect the battery negative battery cable.

TEMPERATURE SENSOR

Removal and Installation

1. Disconnect the negative battery cable.
2. Remove the clips for the crankcase ventilation hose from the camshaft cover, as required.
3. Drain approximately 1 gallon (4 liters) of coolant from the cooling system.
4. Unplug the sensor electrical connector.
5. Using a ¾-inch socket and extension, remove the temperature sensor.
6. Before installing the new sensor, clean the mating surfaces and check the condition of the copper washer.
7. Install the new sensor and torque to 15 ft. lbs. (20 Nm). Reconnect the sensor electrical connector.
8. Refill the cooling system and reconnect the negative battery cable.

THROTTLE POSITION SENSOR

Removal and Installation

1. Disconnect the negative battery cable and unplug the sensor electrical connector.
2. Remove the sensor retaining screws and lift off the sensor.
3. Install the new sensor and loosely tighten the retaining screws.
4. Reconnect the sensor electrical leads and perform the TPS adjustment, if applicable.
5. Reconnect the negative battery cable.

FUEL RAIL AND INJECTORS

Removal and Installation

900 SERIES

1. Disconnect the negative battery cable.
2. Remove the fuel filler cap to relieve the fuel pressure.

NOTE: The fuel system is under residual pressure. Always wrap shop towel around the fuel lines to soak up any fuel leak.

3. Use compressed air to clean the surrounding areas before starting work.
4. Remove the crankcase ventilation hose from the camshaft cover and unplug the connectors from the injectors.
5. Remove the wiring loom clips from the injection manifold.
6. Disconnect the hose connections at either end of the injection manifold.
7. Remove the injection manifold to intake manifold retaining bolts.
8. Remove the injection manifold assembly.
9. To replace the injectors, first release the clips and remove the injectors.

To install

10. Lubricate the new injectors O-rings with vaseline and install them into position.
11. Install the injection manifold assembly. Check that the injection valves are in the correct position and pushed fully into the intake manifold. Also, check that the flange of the fuel injection manifold lines up with the groove for the clips for the injection valves.
12. Reconnect the hose connections at either end of the injection manifold.
13. Install the wiring loom clips from the injection manifold.
14. Install the crankcase ventilation hose to the camshaft cover and reconnect in the connectors to the injectors.
15. Install the fuel filler cap and reconnect the negative battery cable.

9000 SERIES

1. Disconnect the negative battery cable.
2. Remove the fuel filler cap to relieve the fuel pressure.

NOTE: The fuel system is under residual pressure. Always wrap shop towel around the fuel lines to soak up any fuel leak.

3. Use compressed air to clean the surrounding areas before starting work.
4. Remove the false bulkhead panel, if equipped. On the B234 engine, disconnect the pulsator from the fuel injection rail and position out of the way.
5. Disconnect the fuel return line from the fuel pressure regulator and the hose from the intake manifold. Disconnect the line from the fuel filter.
6. Unplug the electrical connectors from the injectors.
7. Remove the fuel injection rail retaining bolts, but leave the injectors in place.
8. Lift out the fuel injection rail complete with injectors through the space between the intake manifold and bulkhead.
9. To replace the injectors, first remove the clips and remove the injectors.

To install

10. Lubricate the new injectors O-rings with vaseline and install them into position.
11. Install the fuel rail assembly into position and secure with the retaining bolts.
12. Plug in the injector electrical connectors.
13. Connect the fuel return line and the vacuum hose to the fuel pressure regulator. Connect the line to the fuel filter. On the B234 engine, connect the fuel hose with the pulsator to the injector rail and replace the tie strap.
14. Install the false bulkhead panel and reconnect the negative battery cable.

FUEL PRESSURE REGULATOR

Removal and Installation

900 SERIES

1. Disconnect the negative battery cable.
2. Remove the fuel filler cap to relieve the fuel pressure.

NOTE: The fuel system is under residual pressure. Always wrap shop towel around the fuel lines to soak up any fuel leak.

3. Use compressed air to clean the surrounding areas before starting work.
4. Disconnect the hose from the fuel injection manifold to the pressure regulator.
5. Disconnect the hose between the intake manifold and the diaphragm side of the pressure regulator.
6. Remove the pressure regulator and bracket from the cylinder head.
7. Disconnect the return fuel line from the regulator.
8. Remove the pressure regulator from the bracket.

To Install:

9. Install the pressure regulator to the bracket and reconnect the return fuel line to the regulator.
10. Install the pressure regulator and bracket to the cylinder head.
11. Reconnect the hose between the intake manifold and the diaphragm side of the pressure regulator.
12. Reconnect the hose from the fuel injection manifold to the pressure regulator.
13. Install the fuel filler cap and reconnect the negative battery cable.

9000 SERIES

1. Disconnect the negative battery cable.
2. Remove the fuel filler cap to relieve the fuel pressure.

NOTE: The fuel system is under residual pressure. Always wrap shop towel around the fuel lines to soak up any fuel leak.

3. Use compressed air to clean the surrounding areas before starting work.
4. Remove the false bulkhead panel.
5. Disconnect the fuel return line from the fuel pressure regulator and the hose from the intake manifold. Disconnect the line from the fuel filter.
6. Unplug the electrical connectors from the injectors.
7. Remove the fuel injection rail retaining bolts and tilt up the end of the rail.
8. Wrap a rag around the hose from the pressure regulator to the fuel rail and disconnect the hose.
9. Disconnect the hose to the intake manifold from the pressure regulator.
10. Remove the clip from the return hose and unscrew the regulator from the bracket. On the B234 engine, remove the regulator retaining clip and pull out the regulator.

To Install:

11. Lubricate the O-rings with petroleum jelly and screw or push the new regulator into the bracket. Position the regulator and install the clip to the return hose.
12. Connect the hoses from the intake manifold and fuel rail to the pressure regulator.
13. Lower the fuel rail assembly into position and secure with the retaining bolts.
14. Plug in the injector electrical connectors.
15. Install the false bulkhead panel and reconnect the negative battery cable.

FUEL PUMP

Removal and Installation

NOTE: The fuel tank does not have to be removed to replace the fuel pump.

1. Disconnect the negative battery cable.
2. Release the fuel pressure by loosening the banjo fitting on

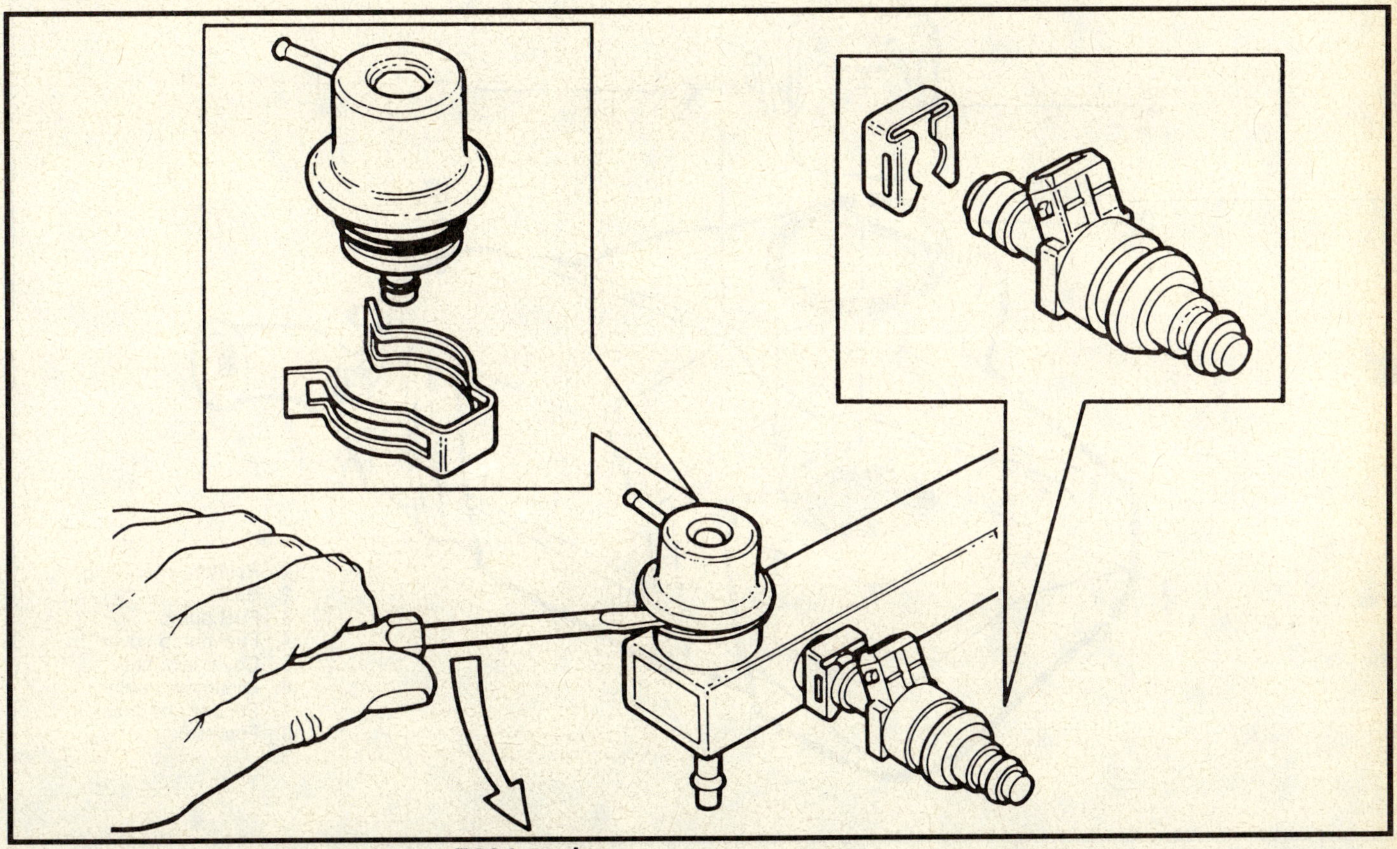

Removing fuel pressure regulator—B234 engine

the fuel injection rail. Be sure to catch any fuel with a shop towel.

3. Remove the floor panel and the fuel pump cover in the luggage compartment.
4. Release the clip and unplug the fuel pump electrical connector.
5. Disconnect the fuel lines from the pump and set them aside, using a shop towel to catch any spilled fuel.
6. Unscrew the retaining device using tool 8394462, or equivalent.
7. Take note of the alignment marks for reassembly, then carefully lift the pump from the tank, tilting the top to the right.

To Install:

8. Place a new O-ring in the tank flange, then place the fuel pump assembly inside the tank with the alignment marks in line.
9. Torque the retaining nut to 40 ft. lbs. (55 Nm) with tool 8394462, or equivalent.
10. Place new O-rings inside the fuel line fittings, then reconnect the fuel lines.
11. Reconnect the electrical connector and reinstall the clip.
12. Reconnect the battery and check the operation of the fuel pump.
13. Reinstall the panel and cover in the luggage compartment.

AUXILIARY AIR VALVE

Removal and Installation

1. Disconnect the electrical leads from the auxiliary air valve.
2. Loosen the hose clips and disconnect the hoses.
3. Unscrew the valve and remove.
4. Screw the new valve into position.
5. Connect the hoses and electrical leads to the valve.

AUTOMATIC IDLING CONTROL VALVE

Removal and Installation

1. Disconnect the electrical leads from the AIC valve.
2. Loosen the hose clips and disconnect the hoses.

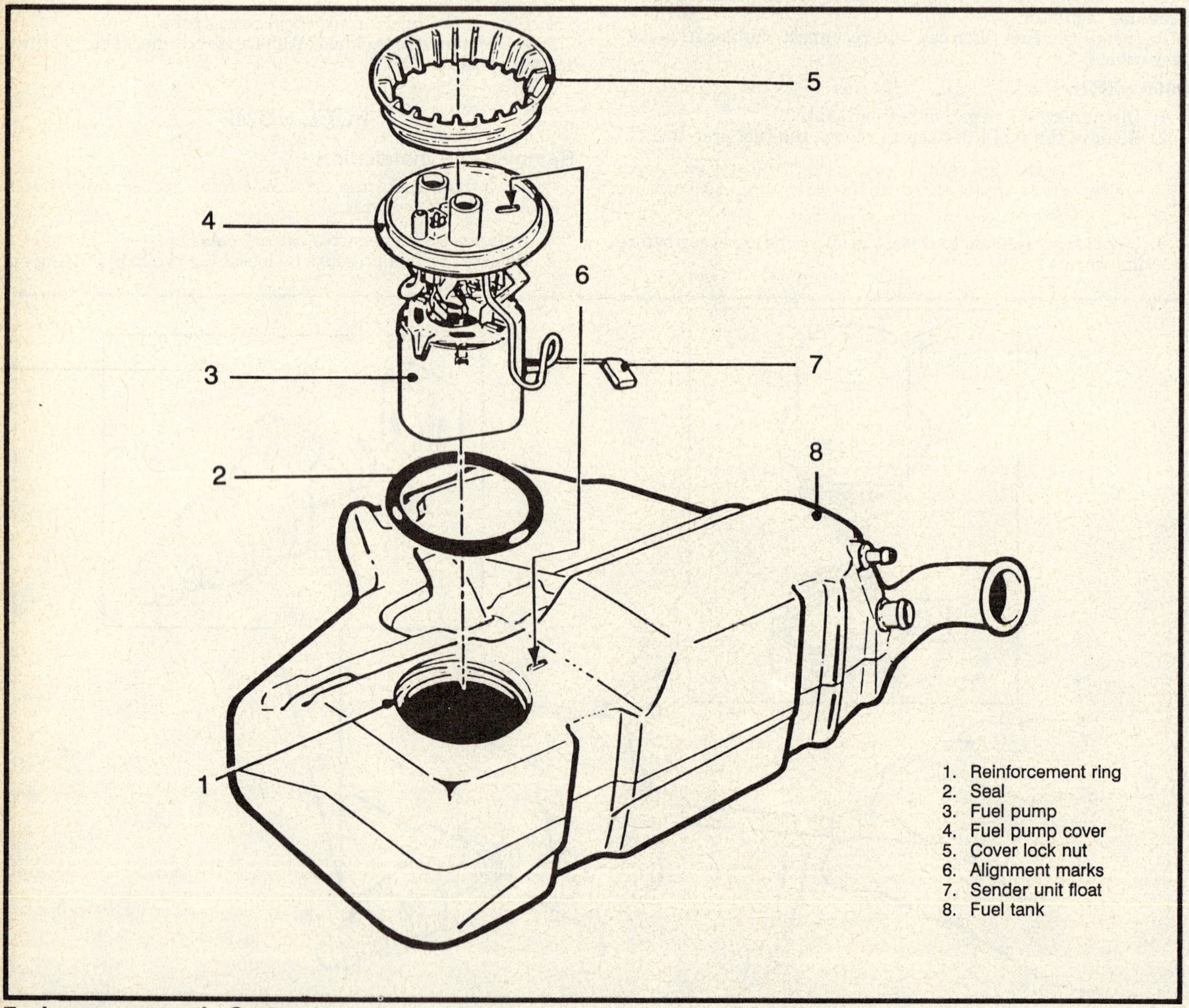

Fuel pump removal—Saab

3. Remove the valve retaining screws and remove the valve.
4. Install the new valve into position.
5. Connect the hoses and electrical leads to the valve.

PRESSURE SWITCH

Removal and Installation

1. Remove the retaining screws from the cover under the left hand side of the dash panel. Remove the mounting panel.
2. Remove the retaining clips and disconnect the hoses.
3. Remove the pressure switch.
4. Install a new pressure switch and connect the hoses.
5. Install the retaining clips.
6. Install the mounting panel.

PRESSURE SENSOR

Removal and Installation

1. Disconnect the negative battery cable.
2. Remove the acoustic insulation panel above the pedal assembly.
3. Remove the air duct above the pedal assembly.
4. Remove the screws and the pressure sensor.
5. Installation is the reverse of removal.

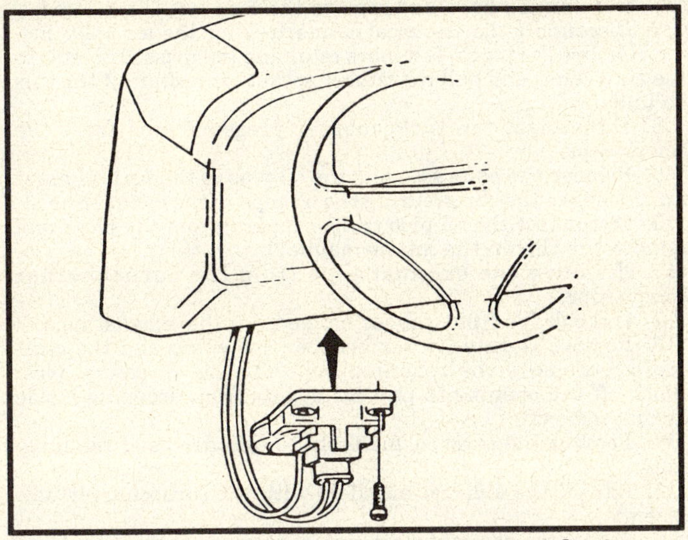

Removing pressure sensor—Turbocharged Saab

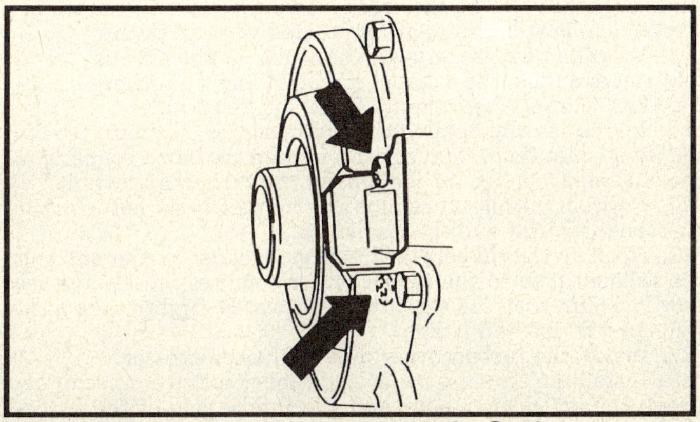

Crankshaft position sensor removal—Saab

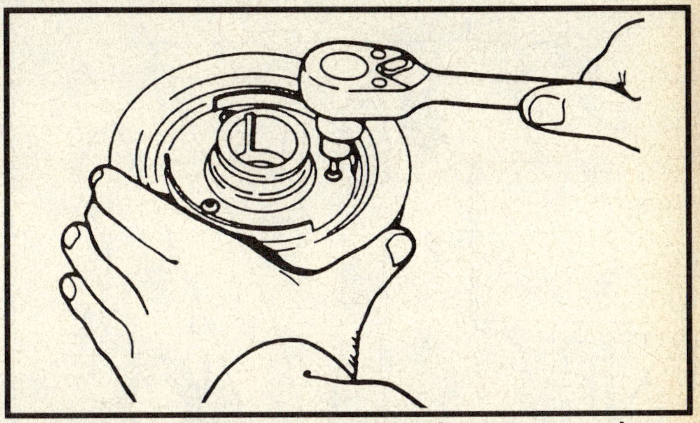

Crankshaft position sensor slotted rotor removal—Saab

CRANKSHAFT SENSOR

Removal and Installation

1. Disconnect the negative battery cable.
2. Lock the flywheel with tool 8393993 or equivalent.
3. Remove the front section of the fender liner on the right side.
4. Remove the air conditioning and alternator belts.
5. Remove the crankshaft pulley bolt and the pulley.

NOTE: Gentle heating of the pulley will make removal easier.

6. Disconnect the connector and cut the wire tie, then unbolt and remove the crankshaft sensor.
7. Installation is the reverse of removal. Use a thread-locking compound on the crank sensor bolts. Torque the crankshaft pulley bolt to 140 ft. lbs. (190 Nm).

SOLENOID VALVE

Removal and Installation

1. Disconnect the solenoid valve connector.
2. Disconnect the signal hoses from the solenoid valve.
3. Remove the self-tapping screws and remove the solenoid valve from the bracket.
4. Installation is the reverse of removal.

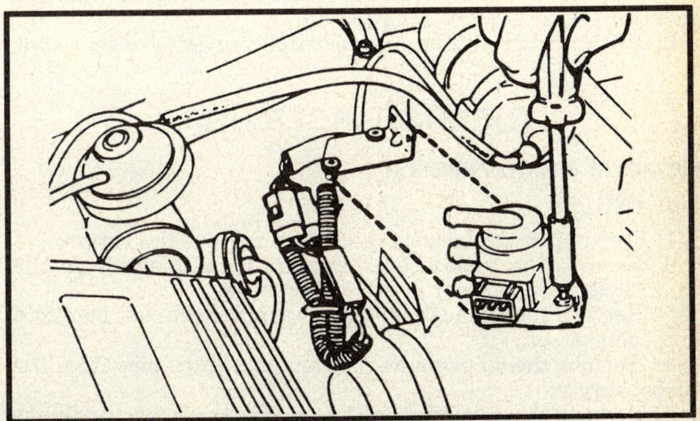

Removing solenoid valve—Turbocharged Saab

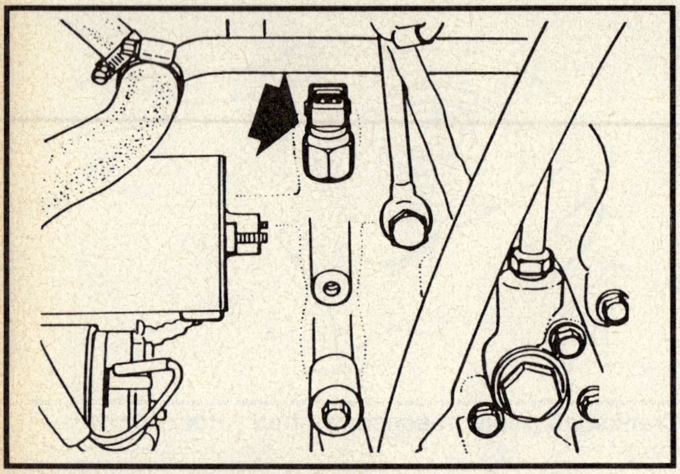
Knock sensor location—Saab

KNOCK SENSOR

Removal and Installation

1. Disconnect the negative battery cable.
2. Remove the oil filler pipe retaining bolt and position the pipe to one side.
3. Disconnect the electrical connector at the knock sensor.
4. Remove the knock sensor retaining bolt and remove the sensor from the engine.

To install:

5. Clean the thread of the sensor retaining bolt and the contact surface on the engine block.
6. Fit the sensor into position. Install the sensor retaining bolt and torque to (13 Nm).
7. Connect the sensor electrical lead. Refit the oil pipe and install the retaining bolt.
8. Reconnect the negative battery cable.

SOLENOID VALVE

Removal and Installation

1. Disconnect the negative battery cable.
2. Unplug the solenoid valve electrical connector.
3. Loosen the hose clips and disconnect the hoses from the valve.
4. Remove the solenoid-to-bracket retaining screws and remove the valve.

To install:

5. Position the solenoid valve on the bracket and install the solenoid-to-bracket retaining screws.
6. Reconnect the hoses and electrical connector to the valve.

TURBOCHARGER UNIT

Removal and Installation

900 SERIES

1. Disconnect the battery cables and remove the battery.
2. Remove the suction and pressure connections from the compressor and loosen the preheating hose.
3. Remove the exhaust elbow between the exhaust manifold and compressor.
4. Remove the oil pressure pipe and oil return pipe from the turbocharger.
5. Remove the retaining bolts from the turbocharger flange on the exhaust manifold and remove the turbocharger.

To install:

6. Install the turbocharger unit to the intake manifold using a new gasket.
7. Install the oil return pipe to the turbocharger using a new gasket.
8. Fill the oil feed channel with engine oil and install the oil pressure pipe using a new gasket.
9. Install the exhaust elbow and exhaust manifold.
10. Install the suction and pressure connections to the compressor.
11. Install the battery.
12. Disconnect the ECU. Run the engine for approximately 30 seconds on the starter motor. This will prime the turbocharger lubrication system before the engine is started.

9000 SERIES

1. Disconnect the negative battery cable.
2. Release the tension on the air conditioning compressor belt.
3. Remove the top pipe coupling on the air-cooled oil cooler and remove the clips securing the pipe to the radiator.
4. Remove air conditioning compressor mounting bolts. Insert a sheet of metal to protect the oil cooler and move the compressor towards the expansion tank.
5. Remove the solenoid valve from its mounting on the radiator and disconnect the electrical leads.
6. Disconnect the electrical leads at the radiator fan. Remove the fan retaining bolts and remove the fan.
7. Disconnect the electrical connectors for the air mass meter. Remove the toggle fasteners retaining the mass air meter to the air cleaner and pull the rubber socket connector off the turbo unit.
8. Disconnect the turbocharger pressure pipe from the compressor.
9. Remove the bolts securing the oil pipe to the turbocharger unit. Remove the clip securing the oil pipe to the cylinder head.
10. Disconnect the oil pipe banjo coupling from the block and remove the clip on the intake manifold.
11. Remove the exhaust pipe from the turbocharger compressor.
12. Unhook the front rubber hangers for the exhaust pipe.
13. Remove the steady bar between the sump and the compressor. Remove the retaining bolts for the oil return pipe. Block off the opening to prevent objects from dropping inside during removal.
14. Remove the exhaust manifold to cylinder head retaining nuts.
15. Lift off the exhaust manifold with the turbocharger unit assembly.
16. Secure the assembly in a vice and remove the nuts holding the units together.

To install:

17. Install new locknuts securing the turbocharger unit to the exhaust manifold. The tab on the locknuts must be turned inwards. The new locknuts are precoated with lubricant.
18. Install a new gasket over the studs for the exhaust manifold and assemble the exhaust manifold and turbocharger unit together. Torque the locknuts 29 ft. lbs. (40 Nm).
19. Install the clip holding the turbocharger oil supply pipe to the intake manifold. Connect and tighten the banjo coupling to the engine block. Secure the pipe to the turbocharger unit.
20. Position the oil return pipe and the steady bar between the turbocharger unit and the crankcase.
21. Hook up the rubbers for the front exhaust hanger and bolt the exhaust pipe to the turbocharger compressor. Always use new locknuts with the tab turned outwards. Tighten the locknuts 18.5 ft. lbs. (25 Nm).
22. Install the turbocharger pipe to the compressor.
23. Install the air mass meter and rubber socket connector between the air cleaner body and the intake side of the turbocharger compressor.

24. Install the fan and solenoid valve; then, plug in the electrical leads and clips. Connect the return hose to the solenoid valve.
25. Insert a sheet of metal to protect the oil cooler; then, install the air conditioning compressor.
26. Reconnect the oil pipe to the oil cooler and secure the pipe clip to the radiator.
27. Install the compressor belt and reconnect the negative battery cable.

PRESSURE TRANSDUCER

Removal and Installation

1. Disconnect the negative battery cable.
2. Remove the rubber bellows from between the front and rear center console. Remove the front center console.
3. Remove the safety padding.
4. Unscrew the pressure transducer bracket from the panel.
5. Disconnect the air hose and cables from the pressure transducer.
6. Remove the transducer to bracket retaining bolts and remove the transducer.

To install:

7. Position the transducer on the bracket and install the retaining bolts.
8. Connect the air hose and cables to the pressure transducer.
9. Screw the pressure transducer bracket to the panel.
10. Install the safety padding and front center console. Install the rubber bellows.
11. Reconnect the negative battery cable.

BASIC CHARGING PRESSURE

Adjusting

1. Loosen the locknut on the diaphragm unit.
2. Remove the lockwasher or split pin and disconnect the pushrod from the wastegate valve actuator.
3. To increase the charging pressure, rotate the threaded sleeve clockwise to shorten the length of the pushrod.

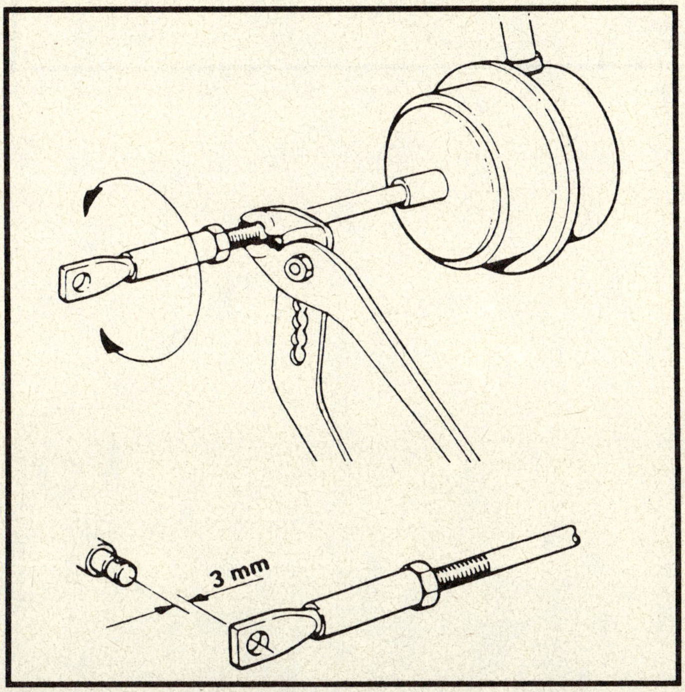

Adjusting turbocharger threaded sleeve—Saab

4. To reduce the charging pressure, rotate the threaded sleeve counterclockwise to increase the length of the pushrod.

NOTE: Do not allow burrs to be created on the pushrod while adjusting, since this can cause it to stick in the diaphragm unit, preventing proper regulation of the boost pressure. One complete turn of the threaded sleeve will result in a change in pressure of approximately 0.4 psi (2.8 kPa).

5. Reconnect the pushrod and replace the lockwasher.
6. Test drive the vehicle and check the basic charging pressure again. If the basic charging pressure fails to respond to adjusting, check the wastegate valve and diaphragm unit.

WASTEGATE VALVE BASE SETTING

The diaphragm unit of the wastegate valve is set at the factory and should only be adjusted after component replacement or if a fault is detected. During adjustment or repair, the basic charging pressure must be measured with the vehicle on the road.

Adjustment

NOTE: Always disconnect the pushrod from the valve actuator before adjusting the threaded sleeve. Do not damage the rod if holding the pushrod with pliers, since burrs can cause it to stick in the diaphragm unit, preventing proper regulation of the boost pressure.

900 SERIES

1. Hold the actuator on the wastegate valve in the closed position. Turn the threaded sleeve on the end of the pushrod to line up the hole with the pin on the actuator.
2. Screw the threaded sleeve clockwise 6 turns and tighten the locknut.
3. Pull out the lever and fit it onto the pin. Install the clip.
4. Road test the vehicle and measure the charging pressure. If necessary, adjust the charging pressure until it is within limits.

9000 SERIES

1. Hold the actuator on the wastegate valve in the closed position. Turn the threaded sleeve on the end of the pushrod to line up the hole with the pin on the actuator.

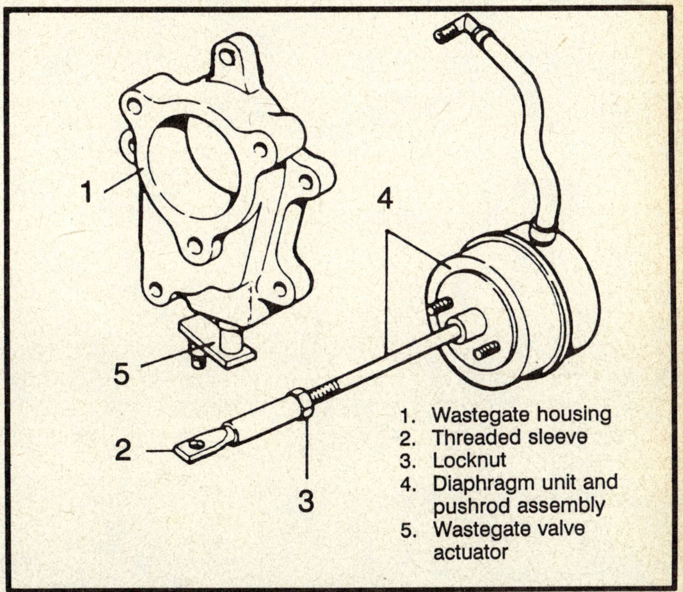

Wastegate components—Saab

2. Screw the threaded sleeve clockwise 3.5 turns on the pushrod to offset the hole by 0.118 in. (3mm) relative to the pin; then, tighten the locknut.

3. Pull the pushrod outwards and fit the end onto the pin. Fit the circlip.

4. Road test the vehicle and measure the charging pressure. If necessary, adjust the charging pressure until it is within limits.

IDLE SPEED AND MIXTURE

Adjustment

Idle speed and mixture are automatically controlled and are not adjustable.

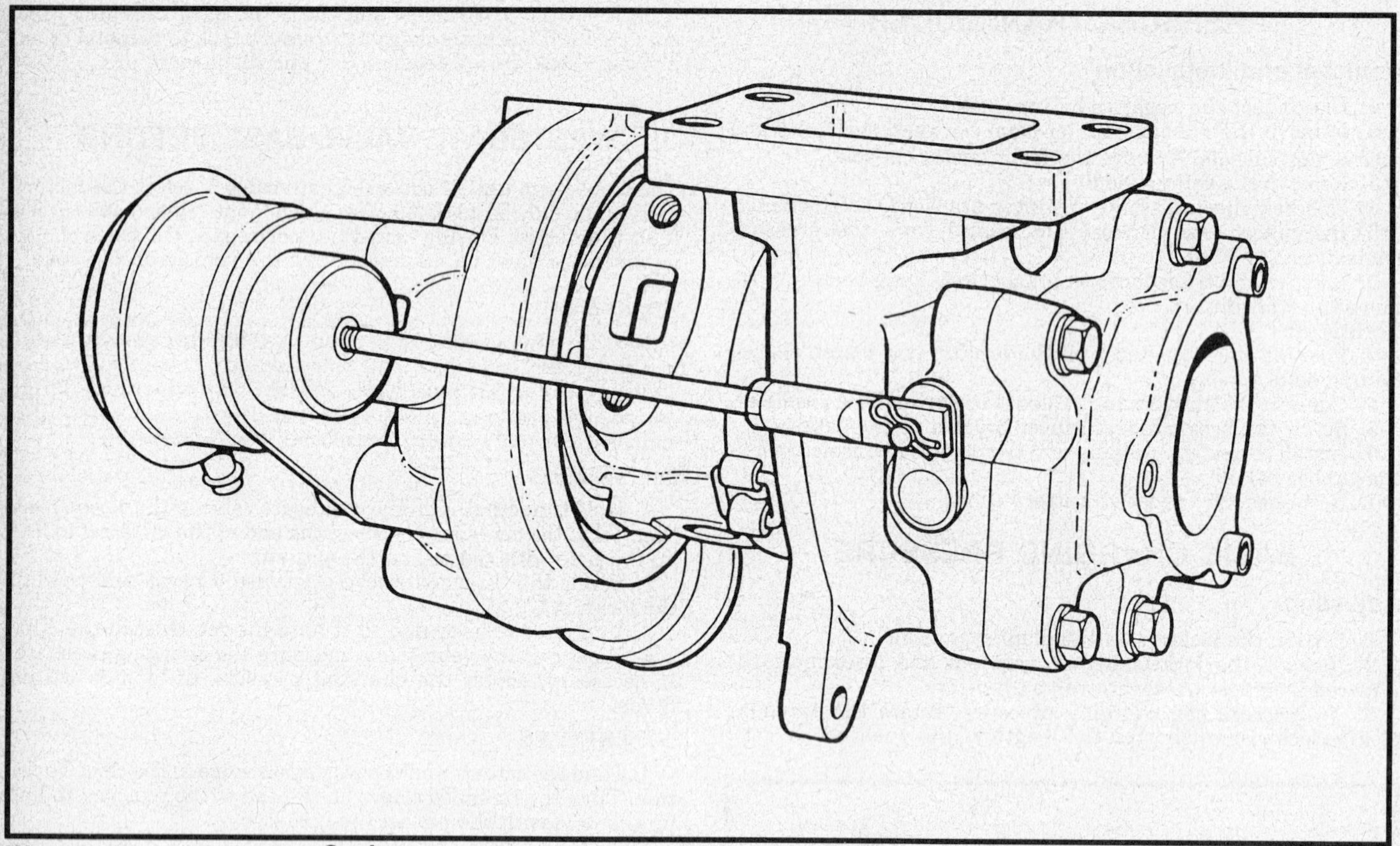

Wastegate components—Saab

Sterling Fuel Injection

14 SECTION

STERLING

General Information

The Sterling 2.7L V6 engine controls both fuel and ignition systems through the Programmed Fuel Injection (PGM-FI) system. One Electronic Control Unit (ECU) oversees operation of the fuel delivery system as well as ignition timing and advance.

The fuel injection system is sequential, employing an Electronic Idle Control Valve (EICV) and a vacuum-controlled fuel pressure regulator. The ECU contains basic injector operating times for various engine speeds and manifold pressures. The basic injector duration (open) time is modified by other sensor signals to obtain final operating time.

The control unit also contains basic timing information for the range of engine speeds and intake manifold pressures. This data is modified by signals from the throttle position sensor, the coolant temperature sensor and the intake air temperature sensor. The ECU-controlled timing signals replace the conventional centrifugal and vacuum advance mechanisms in the distributor.

ELECTRONIC CONTROL UNIT (ECU)

The ECU is a solid-state microprocessor. The controller operates by receiving signals from many engine sensors and comparing them to known combinations stored in memory. Once the correct operating condition has been identified, the ECU opens or closes output circuits to various controlled components. ECU control systems may be grouped into several sub-functions.

Starter Control

During engine start-up, the ECU will command a richer fuel mixture by extending the injector operating time.

Fuel Pump Control

When the ignition switch is turned **ON**, the ECU grounds the main relay. The fuel pump will run for 2 seconds to pressurize the system. The pump will not run again until the controller receives a cranking or rpm signal from the engine. When the engine is not running with the ignition **ON**, the ECU opens the main relay ground, disconnecting the fuel pump.

Fuel Cut-Off Control

The ECU will disable injector operation during deceleration if the throttle is closed and engine speed is above 1100 rpm. This aids in both engine braking and fuel economy. Additionally, fuel delivery is disabled if the engine speed exceeds 6700 rpm, regardless of throttle position.

Air Conditioning Relay Control

When the ECU receives a demand for cooling from the A/C system, it delays the compressor from operating until the fuel mixture has been richened; this assures a smooth transition to compressor operation.

Bypass Control Solenoid Valves A & B

Intake air flow through the dual-path manifold is controlled by the engine ECU. Butterfly valves in the system are controlled by the solenoid valves, routing the air through a long or short intake path.

Below 3200 rpm, both valves are operated, forcing the intake air through the longer intake path. This air delivery provides high torque. Between 3200 and 3800 rpm, valve A is operated; this sets the butterfly valve at mid-point, allowing the torque to be maintained. Above 3800 rpm, both solenoid valves are de-activated. This allows air to enter the engine through the shorter intake path.

Fuel Pressure Regulator Cut-off Solenoid

If coolant temperature rises above 220°F (105°C) or if the intake air temperature exceeds 176°F (80°C), the ECU will energize this solenoid. Vacuum to the fuel pressure regulator is cut, allowing higher fuel pressure and preventing vapor lock.

Automatic Transmission Solenoid

The speed signal and throttle position signal are used by the engine ECU to signal the Automatic Transmission ECU. The transmission controller uses this signal to operate the torque converter lock-up solenoid.

EGR System

The ECU receives signals indicating the amount of EGR valve lift. After comparing this signal to pre-programmed values, the ECU will reduce EGR function, if necessary, by opening the ground to the EGR control solenoid. When the solenoid is de-energized, vacuum to the recirculation valve is reduced.

Back-Up and Fail-Safe Systems

The built-in fail-safe function allows operation of the engine should a sensor signal be lost. When the ECU notes a sensor signal out-of-range, it ignores the unreasonable signal and substitutes a fixed value from its memory. Engine operation continues, although performance may be reduced due to the substituted value being inaccurate for the immediate operating condition.

The back-up system uses a complete set of fixed values. These are used in the event of an internal ECU failure; the independent back-up values allow the vehicle to operate at minimum performance rather than become stranded. This system is sometimes referred to as the limp-in or limp-home mode.

Self Diagnosis System

When the ECU notes a signal fault, the PGM-FI dashboard warning lamp will be lit. The fault is assigned a fault code and stored in memory within the controller. Should the ECU detect

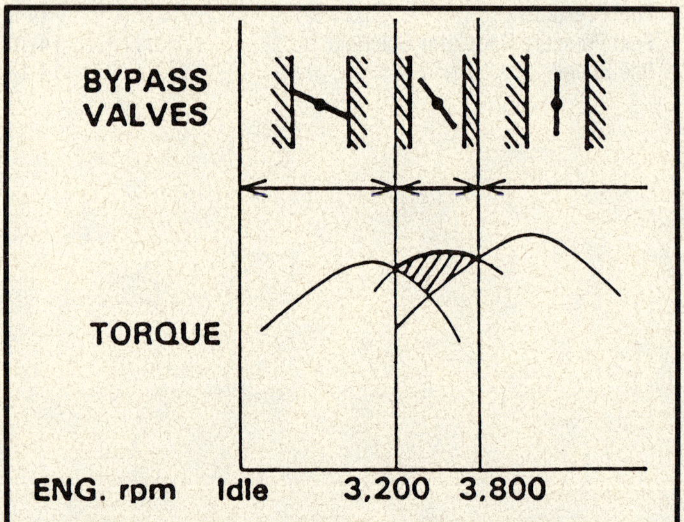

SOL. A	ON	ON	OFF
SOL. B	ON	OFF	OFF

The bypass valves improve engine torque above 3200 rpm

1. Ignition timing adjuster
2. Manifold air pressure (MAP) sensor
3. Constant vacuum control valve
4. Air suction control solenoid
5. EGR control solenoid
6. Throttle dashpot air filter
7. Air reservoir
8. Fuel pressure regulator cut-off solenoid
9. Bypass control solenoid A
10. Bypass control solenoid B
11. Engine ECU
12. Hot restart ECU
13. Injector resistor pack
14. Air conditioning clutch relay
15. Main relay
16. Oil temperature sensor
17. Inertia switch
18. Automatic transmission shift position switch
19. Absolute pressure sensor
20. Crank/cylinder sensor
21. Injectors
22. Fuel pump
23. Power steering switch
24. Oxygen sensors
25. EGR valve lift sensor
26. TDC sensor
27. Intake air temperature sensor
28. Electronic idle control valve
29. Throttle angle sensor
30. Coolant temperature sensor
31. Vehicle speed sensor
32. Resonator control solenoid

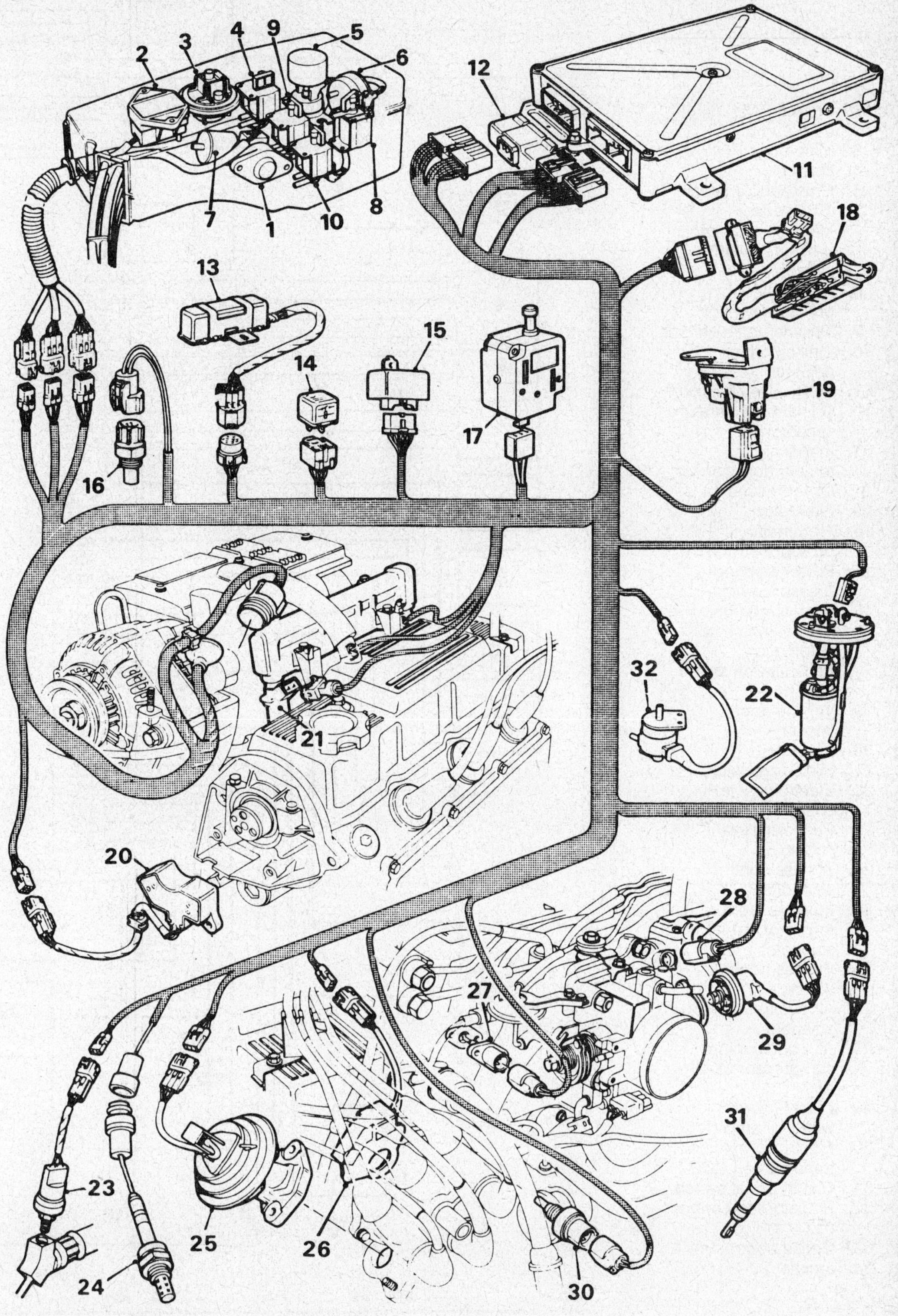

Fuel system components

1. Engine ECU
2. Fuel pump
3. Main relay
4. Starter
5. Crank angle sensor
6. TDC sensor
7. Manifold absolute pressure (MAP) sensor
8. Atmospheric pressure sensor
9. Throttle angle sensor
10. Ignition timing adjuster
11. EGR valve lift sensor
12. Coolant temperature sensor
13. Intake air temperature sensor
14. Oxygen sensors
15. Alternator
16. Air conditioning pressure switch
17. Power steering pressure switch
18. Manual gearbox neutral switch or automatic transmission shifter switch
19. Vehicle speed sensor
20. Injectors
21. Injector resistors
22. Electronic idle control valve
23. Fuel pressure solenoid
24. Bypass control solenoid B
25. Fuel pressure regulator cut-off solenoid
26. Bypass control solenoid A
27. Air suction control solenoid
28. Air conditioning compressor clutch relay
29. PGM-FI dash warning lamp
30. Electronic automatic transmission ECU
31. Clutch pedal switch
32. Radiator fan control unit
33. Cruise control unit
34. Igniter

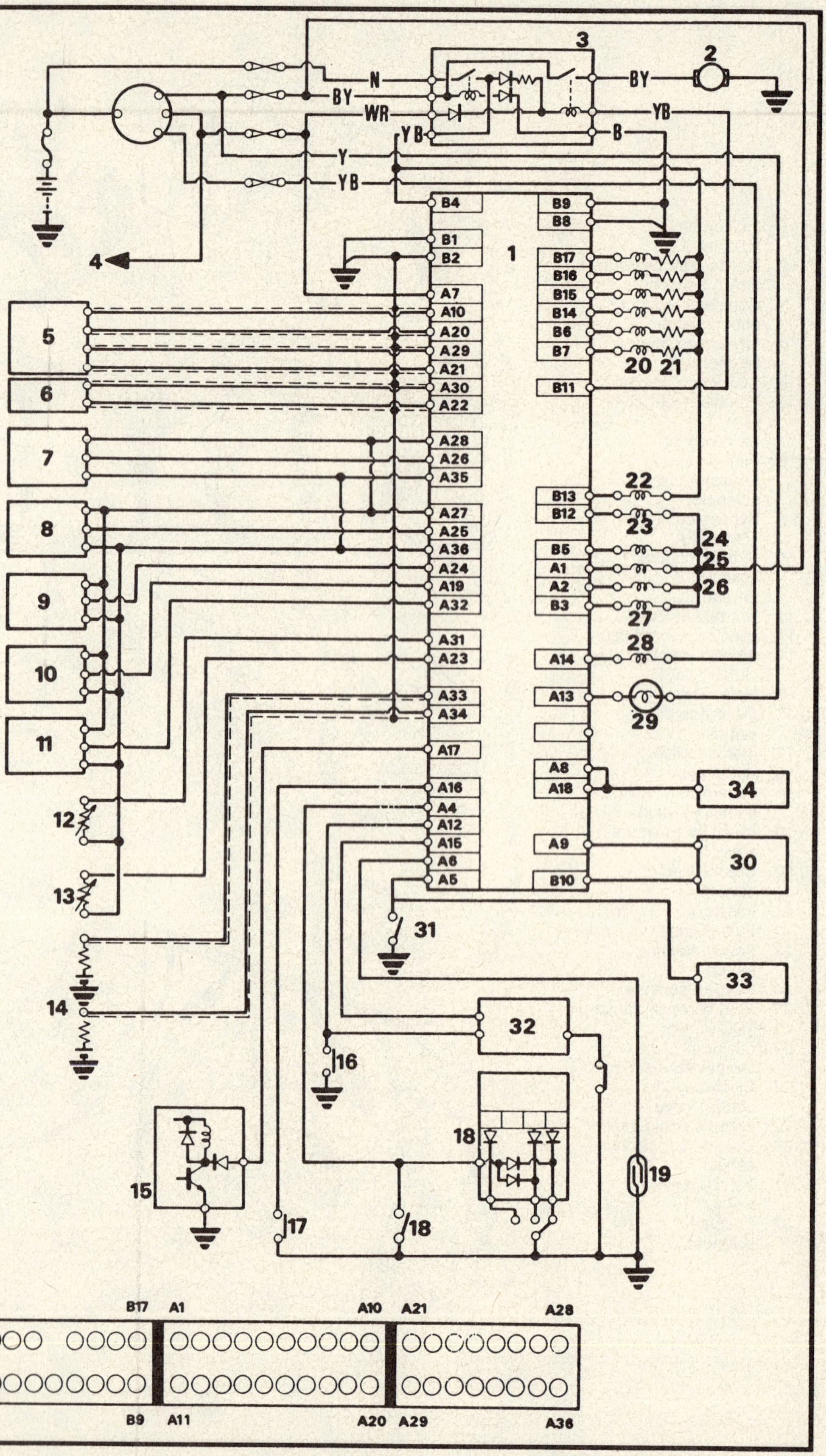

Fuel system electrical schematic

that the electronic idle control valve has exceeded its programmed limits, the yellow LED on the ECU will be lit. Depending on the limit exceeded, the yellow LED may remain on or blink as a diagnostic aid.

DATA SENSORS

Crank/Cylinder Sensor

This sensor combines a top dead center sensor and a cylinder sensor. The position of cylinder No. 1 is used as a reference for the injection sequence; the cylinder sensor determines the injection timing. The camshaft pulley contains 24 poles which pass in front of the sensor as the engine rotates. The signals from these multiple poles provide rpm and position signals to the ECU. An additional single pole on the camshaft pulley provides a TDC reference as it passes the sensor.

Manifold Absolute Pressure (MAP) Sensor

This sensor translates intake manifold pressure to an electric signal used by the ECU to determine engine load. The sensor is supplied with a constant 5 volt signal by the ECU. As manifold pressure changes, the sensor returns a different voltage signal to the ECU. The controller uses the voltage difference to control fuel delivery.

Coolant Temperature (TW) Sensor

The ECU supplies the sensor with a fixed 5 volt signal. As resistance in the sensor changes as coolant temperature changes, altering the voltage signal returned to the ECU. Ignition timing is advanced when the engine is below normal operating temperature and retarded when the engine coolant temperature exceeds normal operating range.

Intake Air Temperature (TA) Sensor

This sensor advises the ECU of changes in intake air temperature. The ECU uses this signal to modify injector opening time as well as ignition timing.

Throttle Angle Sensor

Mounted on the throttle body and attached to the throttle plate pivot, this sensor advises the ECU of the position of the throttle plate. The ECU uses the signal to alter fuel delivery to match the amount of air entering the engine.

Oxygen Sensors

The front and rear oxygen sensors react to the amount of oxygen in the exhaust stream, sending signals to the ECU. When operating in closed loop, the signals from the sensors are used by the ECU to alter the fuel delivery; based on these signals, the air/fuel ratio is maintained at 14.7:1.

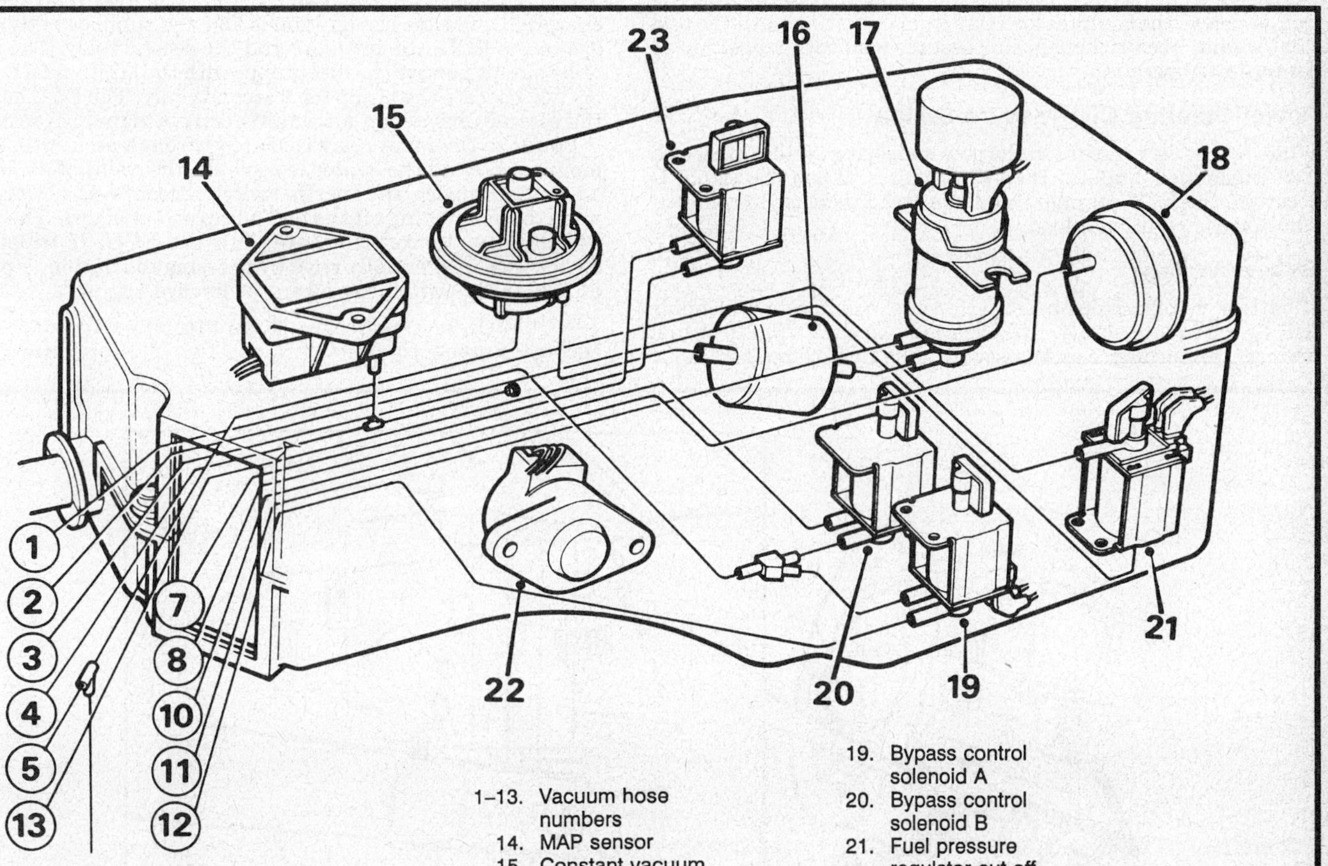

1–13. Vacuum hose numbers
14. MAP sensor
15. Constant vacuum control valve
16. Air reservoir
17. EGR control solenoid
18. Throttle dashpot air filter
19. Bypass control solenoid A
20. Bypass control solenoid B
21. Fuel pressure regulator cut-off solenoid
22. Ignition timing adjuster
23. Air suction control valve

Control box components and vacuum routing

Atmospheric Pressure (PA) Sensor

Located behind the left dashboard, this sensor allows the ECU to make fuel delivery corrections required by changes in altitude or weather.

EGR Lift Sensor

The ECU contains pre-programmed data for ideal EGR function in varying operating conditions. The lift sensor advises the controller of the amount of EGR lift or opening; the ECU compares this signal to the ideal lift for the current conditions. If necessary, the ECU will electrically disable the EGR control solenoid, stopping the flow of vacuum to the EGR valve.

Vehicle Speed Sensor

Located on the transmission case, the signal from the vehicle speed sensor is used by the engine ECU and the automatic transmission ECU. The speed signal is also used by the cruise control system.

Pedal and Shifter Position Switches

The ECU receives signals from the clutch pedal switch and manual transmission neutral switch or the automatic transmission shift position switch as well as the brake pedal switch. These signals are used to control both idle and cruise control functions.

Air Conditioning Signal

When the ECU receives a demand for cooling from the A/C system, it delays the compressor relay from operating until the fuel mixture has been richened; this assures a smooth transition to compressor operation.

Power Steering Oil Pressure Switch

When hydraulic pressure in the power steering system exceeds a pre-determined amount, the switch closes. When the signal is received, the ECU commands an idle speed increase to compensate for the additional load.

Battery Voltage

If battery voltage drops below 11.8 volts, the injector solenoids will operate more slowly. This delayed response will affect fuel delivery, performance and emissions. The ECU compensates for this voltage drop by extending injector duration in proportion to available voltage.

Starter Signal

This signal, coupled with the signal from the coolant temperature sensor, is used by the ECU to determine necessary enrichment during cold starts.

Alternator (FR) Signal

This signal to the ECU represents the electrical load on the system. With this information, the ECU controls the idle speed to compensate for the added load of headlights, rear defogger and other components.

CONTROLLED OUTPUTS

Injectors

The injectors are actuated by an internal solenoid. When current is applied via the ECU, the pintle valve is lifted and pressurized fuel is allowed to flow into the air stream near the intake valve. Because the valve lift and the fuel pressure remain constant, fuel quantity is determined solely by the amount of time the valve remains open.

Main Relay and Inertia Switch

The main relay contains two circuits. The first relay circuit is energized any time the ignition is **ON** and supplies battery voltage to the ECU, the injectors and the second relay. The second relay circuit powers the fuel pump; with the ignition **ON**, the relay will operate the pump for 2 seconds only. The ECU activates the second relay when a cranking or rpm signal is present.

Power to the main relay is routed through an inertia switch, mounted behind the center console. In the event of an accident or frontal impact, the inertia switch disconnects power to the main relay, shutting off the fuel pump and injectors. The inertia switch does not communicate with the ECU. If tripped, the switch may be manually reset by pressing the button or plunger on top of the switch downward or inward.

NOTE: It is possible for this switch to trip without collision damage to the vehicle. Always check the inertia

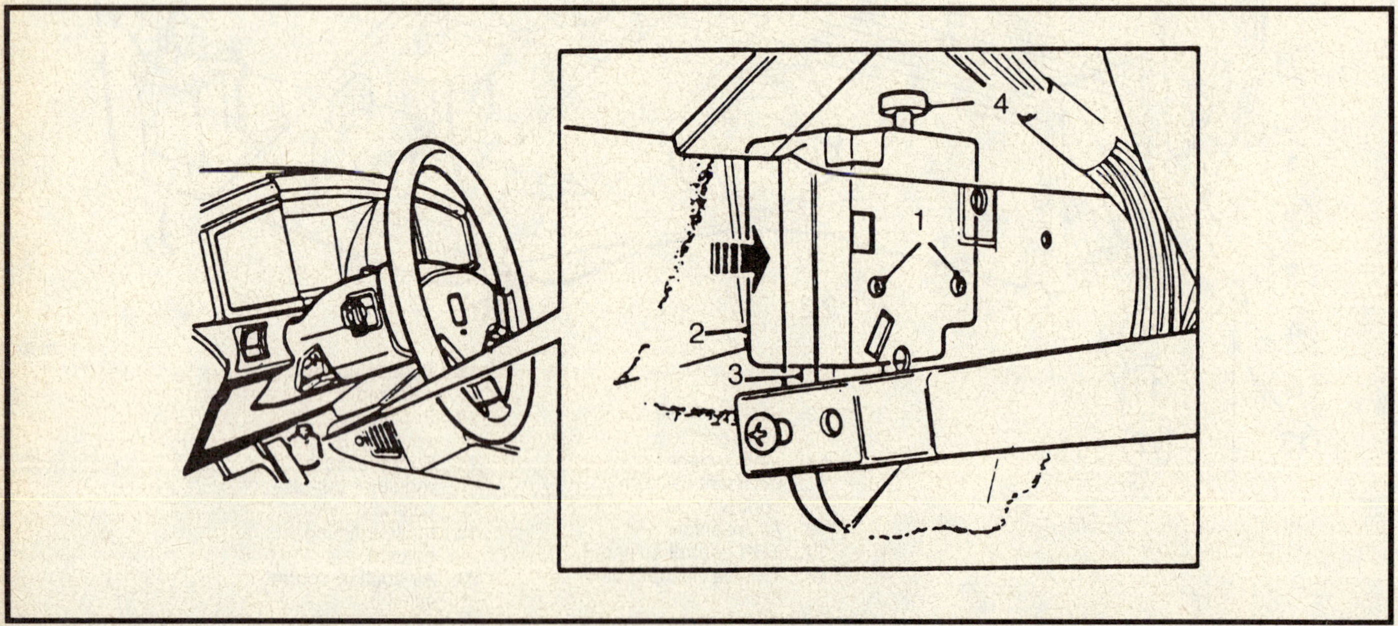

Inertia switch

switch setting when diagnosing failure to start or apparent fuel pump failure.

Dashboard Warning (PGM-FI) Lamp

The ECU will illuminate the dashboard warning lamp whenever an electrical fault is found within the system. Additionally, the lamp should light for 2 seconds each time the ignition switch is **ON** to serve as a bulb check.

Electronic Idle Control Valve (EICV)

The EICV changes the amount of air bypassing the throttle plate into the intake manifold. The valve is moved by a small electric stepper motor controlled by the ECU.

Compressor Clutch Relay

This relay allows the ECU to control engagement of the compressor clutch. Fuel delivery is altered to prevent harshness during compressor operation.

Bypass Control Valves A and B

Both valves are operated by the ECU relative to engine rpm. The position of the valves controls the path taken by the intake air stream. By changing the length of the intake path, power and torque factors are maintained throughout the engine operating range.

EGR Control Valve

The EGR control solenoid is an electrically operated valve controlling vacuum flow to the EGR valve. The valve allows the ECU to control EGR operation by switching the vacuum flow on or off. The valve is operated in response to the signal from the EGR lift sensor and other ECU inputs.

Air Suction Control Valve

This valve is part of the air injection or secondary air system. The system improves emissions by passing fresh air from the air cleaner into the exhaust manifold through the air suction control valve.

When the valve is energized by the ECU, manifold vacuum raises the diaphragm and the valve opens. The exhaust pulses draw fresh air through the reed valve into the exhaust manifold.

Fuel Pressure Regulator Cut-off Valve

When this solenoid is energized by the ECU, manifold vacuum to the fuel pressure regulator is interrupted. This allows higher fuel pressure in the rail and at the injectors, aiding in high-temperature operation and starting. The signal to the fuel pressure cut-off valve is based on input signals from both the coolant and air intake temperature sensors.

Automatic Transmission Lock-up Signal

The vehicle speed and throttle angle sensor inputs are used by the engine ECU to control an ON/OFF signal to the automatic transmission ECU. This enabling signal allows the lock-up torque converter to function only in suitable conditions.

FUEL DELIVERY SYSTEM

Fuel is delivered from the tank by the in-tank fuel pump. The pump is powered by an electric motor. When the engine is started, the main relay actuates the pump. Fuel entering the inlet port flows inside the motor from the pumping chamber and is forced through the discharge port and check valve. If the output side of the pump is blocked, a relief valve opens, rerouting the fuel to the inlet port of the pump. This prevents a excessive fuel pressure from developing in the lines.

The fuel pump is shut off by the ECU and main relay when the engine stops operating. A mechanical check valve in the line also closes by spring action, maintaining residual fuel pressure to aid in restarting the engine.

The fuel pressure regulator is mounted on the fuel delivery rail and maintains constant fuel pressure to the injectors. It is connected to the manifold vacuum through the fuel pressure control solenoid. Under normal conditions, if the difference between the fuel pressure and the manifold pressure exceeds 36.2 psi (2.55 kgf/cm^2), the regulator diaphragm is pushed upward. Excess fuel is returned to the fuel tank through the fuel return line. If coolant temperature rises above 220°F (105°C) or if the intake air temperature exceeds 176°F (80°C), the ECU will energize the regulator control solenoid. Vacuum to the fuel pressure regulator is cut, allowing higher fuel pressure and preventing vapor lock.

Diagnosis and Testing

SERVICE PRECAUTIONS

- Always disconnect the negative battery cable before replacing any fuel or electronic component. Make certain all electrical circuits, including the ignition, are **OFF** before removing the cable. If both battery cables are disconnected, remove the negative cable first and re-connect it last.
- Never reverse battery cables. Damage is instantaneous and permanent.
- Insure that the battery voltage is at least 11 volts before testing any circuit. Low battery voltage will affect system operation.
- Use great care during battery re-charging. Disconnect both battery cable before connecting the charger. Charging the battery out of the vehicle is highly recommended.
- Use extreme care when testing ignition circuits. High voltages in the secondary circuit can cause personal injury.
- Make certain all electrical connectors are in good condition before reassembly. Pins must be straight and undamaged. Insure that all connector locks and retainers are engaged after reassembly.
- Keep oil and cleaning solvents away from all electrical components and connectors.
- When working on or around ECU or solid-state components, wear a grounded wrist strap or touch automotive ground frequently. Many components can be damaged by static electricity.
- The fuel system retains pressure even when the ignition is **OFF** or the engine has not operated for a period of time. Always release the pressure using the correct procedure before any fuel line or component is loosened or disconnected.
- Observe No Smoking/No Open Flame precautions whenever the fuel system or any component is being serviced.
- Keep a charged CO_2 or dry powder fire extinguisher close at hand and know how to use it.
- Anticipate fuel spillage when disconnecting lines or components. Use clean rags below fuel line fittings to catch small amounts of fuel. Treat the rags as a flammable hazard and dispose of them properly.
- Large amounts of fuel, such as from the fuel tank, must be drained directly into approved, tightly-capped fuel containers. Do not use pails, plastic jugs or similar open containers. A supply of dry sand, cat litter or other absorbent material should be on hand to deal immediately with spillage.
- Do not disconnect the battery cables with the engine running; never run the engine with battery cables loose or disconnected.
- Do not attempt to disassemble any ECU under any circumstances.

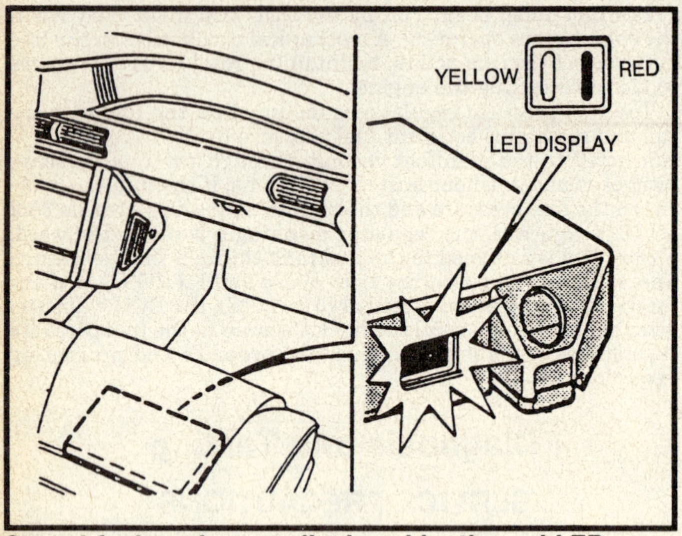

Stored fault codes are displayed by the red LED on the controller

- If installing a 2-way or CB radio, mobile phone or other radio equipment, keep the antenna as far as possible away from the electronic control unit. Keep the antenna feeder line at least 8 in. away from the ECU harness and do not run the lines parallel for a long distance. Be sure to ground the radio to the vehicle body.
- When measuring supply voltage of ECU-controlled components, keep the tester probes separated from each other and from accidental grounding. If the tester probes accidentally make contact with each other during measurement, a short circuit will damage the ECU.
- When working on or around air bag wiring or components, always disarm the system using the correct procedures. Once disarmed, attach a flag or note to the steering wheel. Re-arm the system when repairs are completed and remove the wheel marker.

READING CODES

When the ECU notes an electrical fault, the PGM-FI dashboard warning lamp will be illuminated. The fault is identified by the controller and a diagnostic code is stored in erasable memory for diagnostic use. The engine ECU, located under the right seat, contains a red and a yellow LED lamp on the case. The lamps may be viewed by looking under the seat.

If a fuel system code is stored, the code will be flashed by the red LED any time the ignition is **ON**. Single digit codes are indicated by the appropriate number of flashes; 4 flashes indicates Code 4. Two digit codes are indicated by a number of long flashes for the tens digit and a series of shorter flashes for the ones digit; 4 long flashes followed by 3 short flashes indicates Code 43. The fault will remain stored in memory until cleared; the system does not clear itself after repairs.

Note that the yellow LED is not used for fuel or emission codes. It will illuminate or flash if the ECU detects an error in the position of the EICV.

Diagnostic Codes

NOTE: If the ECU displays Codes 11, 16 or more than a total of 18 flashes (long and short), replace the ECU with a known-good unit. Test drive the vehicle and recheck for codes.

LED ON, No Blinks—Disconnected ECU ground wire
Code 1—Faulty or disconnected front oxygen sensor
Code 2—Faulty or disconnected rear oxygen sensor

Fast Checker Test Step Codes	Associated Component(s)
01-06	Fuel Injectors 1 to 6 and Fuel Injector Resisters
07	Electronic Idle Control Valve (EICV) Solenoid
08	Exhaust Gas Recirculation (EGR) Solenoid Valve
09	Fuel Pressure Cut-off Solenoid Valve
10	Bypass A Solenoid Valve and Resonator Control Solenoid
11	Bypass B Solenoid Valve
12	Air Suction Control Solenoid
13	Air Conditioning Clutch Relay
14	Check Engine Warning Indicator
15	Fuel Pump
16	Crank Signal
17	Crank/Cyl Sensor (cylinder position signal)
18	Crank/Cyl Sensor (Crank Angle Sensor)
19	Manifold Absolute Pressure (MAP) Sensor
20	Atmospheric Pressure (PA) Sensor
21	Throttle Angle Sensor
22	Ignition Timing Sensor
23	Exhaust Gas Recirculation (EGR) Lift Sensor
24	Coolant Temperature (TW) Sensor
25	Intake Air Temperature (TA) Sensor
26	Clutch Switch (manual transmission)
27	Igniter Unit
28	Alternator Load Sensor
29	Top Dead Center (TDC) Sensor
30	Over Pressure Input Switch
31	Fan Input
32	A/C Switch
33	Neutral Switch (manual transmission)
34	Speed Sensor
37	Ground
None	Rear Oxygen Sensor
None	Front Oxygen Sensor
None	Power Steering Idle Boost Switch
	Fuel Control Relay (see code 15)

Fast checker tool codes—typical Sterling

Code 3—MAP sensor circuit open/shorted
Code 4—Crank angle sensor faulty or circuit open/shorted
Code 5—Disconnected MAP sensor hose
Code 6—Coolant temperature sensor faulty or circuit open/shorted
Code 7—Throttle angle sensor faulty or circuit open/shorted
Code 8—Crank angle sensor (TDC sensor) faulty or circuit open/shorted
Code 9—Crank angle sensor (cylinder) faulty or circuit open/shorted
Code 10—Intake air temperature sensor faulty or circuit open/shorted
Code 11—Idle mixture adjuster faulty
Code 12—EGR control system faulty or control wire open/shorted
Code 13—Atmospheric pressure sensor faulty or circuit open/shorted
Code 14—EICV faulty or circuit open/shorted. Depending on nature of the failure, the yellow LED may also be lit
Code 15—Igniter unit faulty or circuit open/shorted
Code 17—Speed transducer for vehicle speed faulty or circuit open/shorted
Code 18—Ignition timing adjuster faulty or circuit open/shorted
Code 43—Ignition timing adjuster faulty or circuit open/shorted
Code 44—Ignition timing adjuster faulty or circuit open/shorted

Code 45—Ignition timing adjuster faulty or circuit open/shorted
Code 46—Ignition timing adjuster faulty or circuit open/shorted

CLEARING CODES

A code will remain stored in memory until cleared. To clear the memory, switch the ignition **OFF** and remove fuse **S** for at least 30 seconds. Codes will also be cleared when the battery is disconnected but this method is not recommended; other memories on the vehicle such as radio presets will also be removed.

ENGINE WARM-UP

Temperature will have an effect on the fuel settings at idle due to heat soak and oil drag. Warming a parked vehicle allows heat to build around the engine which will affect the temperature compensation in the fuel control system. Additionally, transmission oil will not be heated properly, resulting in erratic engine speed. If adjustments are made after an uncontrolled warm-up, engine idle speed and CO readings may be inaccurate.

Proper warming-up requires driving the vehicle at moderate speed and load for at least 4 miles. Once returned to the shop, the vehicle should be parked in an area allowing good air flow for continued cooling. Adjustments should be made before the fan cycles ON. If the fan comes on before an adjustment is complete, stabilize the engine temperature by holding 2000 rpm for 30 seconds after the fan has switched off.

If the vehicle must be warmed at idle, allow the fan to cycle on and off once. Stabilize the engine temperature by holding 2000 rpm for 30 seconds after the fan has switched off. If the fan comes on before the adjustment is completed, re-stabilize the engine at 2000 rpm after the fan shuts off.

Adjustments made to an engine warmed at idle should be rechecked after the vehicle has been driven over a moderate distance.

FUEL PRESSURE

Testing

NOTE: This procedure requires clamping fuel lines. Use only proper line clamps, never pliers or similar tools. The inner lining of the fuel hoses may be damaged by incorrect or sharp tools. Do not clamp any line more than 10 seconds; never allow the pressure to approach the limit of the gauge scale.

1. Safely relieve the fuel pressure.
2. Remove the banjo bolt from the fuel filter outlet pipe. Install Sterling adapter 18G 1500/2 or its equivalent in place of the banjo bolt. Connect Sterling fuel pressure gauge 18G 1500 or equivalent to the adapter.
3. Disconnect the vacuum hose from the fuel pressure regulator.
4. Start the engine and allow it to idle. Fuel pressure should be 36–41 psi. (2.55–2.85 bar).
5. With the engine idling, connect the vacuum hose to the pressure regulator. Fuel pressure should drop to approximately 30 psi (2.1 bar). If pressure does not drop, check the vacuum hose for blockage. Check that the pressure control solenoid is not stuck open.
6. Switch the ignition **OFF**. Observe the fuel pressure; it should not drop more than 10 psi (0.7 bar) in the first minute.
7. If the pressure drop exceeds 10 psi. in the first minute with the ignition **OFF**, clamp the return hose between the regulator and the tank. If the pressure now holds steady, the regulator is faulty and must be replaced. If the pressure continues to drop after clamping the hose, suspect a leaking injector or a failed fuel pump check valve.
   a. To test the check valve quickly, clamp off the outlet hose from the fuel pump. If the pressure now holds steady, the check valve is faulty.
8. With the fuel pump running, momentarily clamp the fuel return line to the tank. Fuel pressure on the gauge should increase to approximately 58 psi. (3.95 bar).

CRANK ANGLE SENSOR

Testing

1. With the engine **OFF**, disconnect the sensor harness.
2. Connect an ohmmeter to the Blue/Green and Blue/Yellow wire terminals at the TDC sensor. Correct resistance is 500–1200 ohms. If resistance is not correct, replace the sensor.
3. To test the CYL sensor, measure resistance at the White/Blue and Orange/Blue wires at the sensor. Correct resistance is 500–1200 ohms. If resistance is not correct, replace the sensor.

MANIFOLD ABSOLUTE PRESSURE (MAP) SENSOR

Testing

1. Operate the engine until the cooling fans have cycled on at least once. Switch the ignition **OFF**.
2. Switch the ignition **ON**.
3. Connect a voltmeter to the White wire and the Green wire at the vacuum control box harness. Do not disconnect the harness to connect the voltmeter. Voltage should be approximately 3 volts.
4. Disconnect vacuum hose No. 3. Apply vacuum to the sensor with a hand pump; the voltage on the White wire to the Blue wire at the ECU should drop.
5. If no voltage drop is seen, replace the MAP sensor.

ATMOSPHERIC PRESSURE SENSOR

Testing

1. Operate the engine until the cooling fans have cycled on at least once. Switch the ignition **OFF**.
2. Switch the ignition **ON**.
3. Connect a voltmeter to the Red wire and the Green/White wire at the vacuum control box harness. Do not disconnect the harness to connect the voltmeter. Voltage should be approximately 3 volts; the voltage will vary with air pressure.
4. If the voltage is incorrect, the sensor must be replaced.

COOLANT TEMPERATURE SENSOR

Testing

1. Operate the engine until the cooling fans have cycled on at least once. Switch the ignition **OFF**.
2. Disconnect the harness at the sensor.
3. Measure the resistance the sensor terminals. Correct resistance is 200–400 ohms.

INTAKE AIR TEMPERATURE SENSOR

1. Operate the engine until the cooling fans have cycled on at least once. Switch the ignition **OFF**.
2. Disconnect the harness at the sensor.
3. Measure the resistance the sensor terminals. Correct resistance is 1–4 ohms.

THROTTLE ANGLE SENSOR

Testing

1. Connect a voltmeter to the Red/Blue wire at the sensor. Ground the other sensor probe.

2. Turn the ignition switch **ON**.
3. With the throttle in the closed or idle position, correct voltage is 0.5 volts.
4. Move the throttle plate to the wide-open position. Voltage should increase to 4 volts
5. Both test conditions must be met. If either voltage is incorrect, the sensor must be replaced.

EGR LIFT VALVE

Testing

1. Switch the ignition **ON**.
2. Connect a voltmeter between the Yellow/White wire and the Green/White wire at the EGR lift valve. Do not disconnect the connector to install the voltmeter.
3. Voltage should be approximately 5 volts. If voltage is incorrect, look for a fault ground path or faulty power supply.
4. Measure the voltage between the White/Green and Green/White wires at the sensor. Voltage should be approximately 1 volt.
5. If voltage is incorrect, suspect a faulty EGR unit or shorted White/Green wire between the sensor and the ECU.

EGR CONTROL SOLENOID VALVE

Testing

1. Disconnect the 6-pin connector.
2. Start the engine, allowing it to idle.
3. Using jumper wires, apply battery voltage to the Light Green/Orange wire and ground the White wire.
4. Measure the vacuum at the EGR hose; approximately 8 in. Hg of vacuum should be present.
5. If the proper vacuum is not present, check carefully for leaks in the hoses or at the joints. If the vacuum supply is not leaking, replace the EGR control solenoid.

ELECTRONIC IDLE CONTROL VALVE (EICV)

Testing

1. Switch the ignition **OFF**.
2. Disconnect the 2-pin connector at the EICV.
3. Measure the resistance between the valve terminals. Resistance should be 6–20 ohms.
4. Test continuity between each valve terminal and body ground. Continuity should be present.
5. All test conditions must be met. Should any test result in readings outside of specification, the valve must be replaced.

INJECTORS

Testing

1. With the ignition **OFF**, disconnect the wiring connector from the fuel injector to be tested.
2. Measure the resistance between the injector terminals; correct resistance is 1.5–2.5 ohms.
3. If resistance is incorrect, replace the injector.

INJECTOR POWER RESISTORS

Testing

1. With the ignition switched **OFF**, disconnect the harness connector to the resistor pack.
2. Measure the resistance between each of the Black wires (resistor terminals) and the Red wire (power terminal).
3. Correct resistance for each is 5–7 ohms.

POWER STEERING OIL PRESSURE SWITCH

Testing

1. With the ignition **OFF**, disconnect the 2-pin connector at the switch.
2. Connect an ohmmeter to the Blue and Blue/Red wire terminals of the switch. Position the wires, harness and meter safely; keep them clear of heated or moving parts.
3. Start the engine, allowing it to idle.
4. With the engine running, the meter should indicate a continuous or closed circuit.
5. Turn the steering wheel from lock to lock. The meter should indicate an open or discontinuous circuit.
6. Both test conditions must be met. If either condition is failed, the switch must be replaced.

FUEL PRESSURE REGULATOR SOLENOID VALVE

Testing

1. With the ignition **OFF**, disconnect the 6-pin connector.
2. Using jumper wires, connect battery voltage to the solenoid and ground the Light Green wire. As the circuit is completed, the solenoid should be heard to click.
3. If no clicking is heard, check the integrity of the test connectors, particularly the ground circuit; retest the solenoid. If still inoperative, replace the solenoid valve.

AIR SUCTION CONTROL SOLENOID

Testing

1. With the ignition **OFF**, disconnect the 6-pin connector.
2. Use jumper wires to connect battery voltage to the solenoid. Ground the Pink wire.
3. As the circuit is completed, the solenoid should be heard to click.
4. If no clicking is heard, check the integrity of the test connectors, particularly the ground circuit; retest the solenoid. If still inoperative, replace the solenoid valve.

Component Replacement

FUEL PRESSURE RELIEF

CAUTION

The fuel system retains pressure even when the ignition is OFF or the engine has not operated for a period of time. Always release the pressure using the correct procedure before any fuel line or component is loosened or disconnected. Observe all service precautions whenever the fuel system or any component is being serviced to avoid personal injury.

1. Switch the ignition **OFF**.
2. Place a rag over the fuel filter and 6mm service bolt.
3. Counterhold the larger banjo bolt with a second wrench. Slowly loosen the 6mm service bolt 1 full turn. Pressurized fuel will be released from the system.
4. After pressure is released, remove the bolt and replace the sealing washer.

NOTE: The washer must be replaced every time the service bolt is loosened. Failure to replace the washer may cause high-pressure fuel leaks.

5. Reinstall the service bolt with the new washer; tighten the bolt to 9 ft. lbs. (12 Nm). Dispose of or store the fuel-soaked cloth properly; treat it as a flammable hazard.
6. The fuel pressure will remain reduced until the ignition is turned **ON** or the fuel pump is operated.

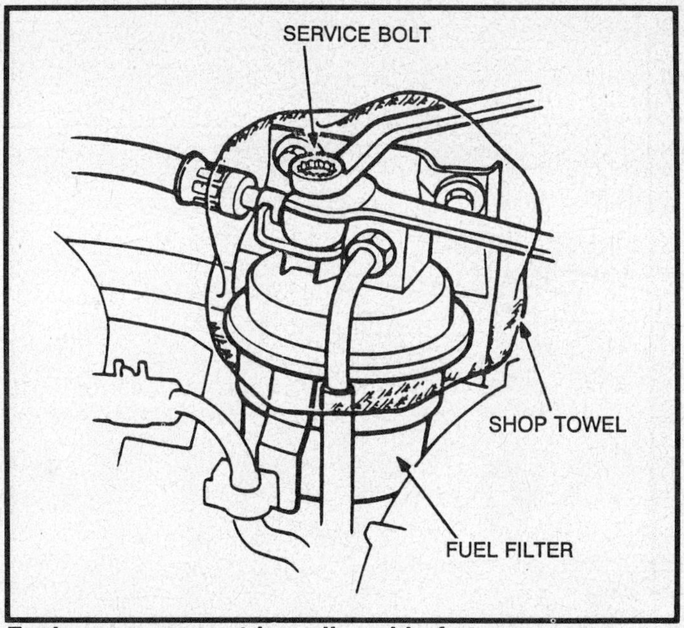

Fuel pressure must be relieved before any system repairs are begun

SERVICE PRECAUTIONS

- When inspecting terminals, check for bent or spread connector pins; fit should be firm and connectors should be free of dirt or corrosion.
- Use care when separating connector halves. Disconnect all locks or retainers before separation. At reassembly, make certain the locks are firmly engaged.
- Never disconnect the battery cable(s) with the engine running or the ignition **ON**.
- Always use a 12 volt battery as a power source, never a booster or high voltage charging unit.
- Make certain all connectors, particularly at the ECM are firmly and squarely seated. Poor connections can affect system operation and possibly damage the control unit.
- When possible, use a flashlight in place of a drop-light.
- Keep all open flame and smoking materials out of the work area.
- Always relieve the fuel system pressure before loosening a fuel-related line or component.
- Use a shop cloth or catch-pan to catch fuel when opening a fuel system line or component.
- Always use eye or full-face protection when working around fuel-bearing components.
- Keep a dry chemical (Class B-C) fire extinguisher in the work area at all times.
- Contain spillage and clean it up immediately. Treat fuel-soaked rags as a flammable hazard and dispose of them properly.
- When handling the ECU or other components, protect them from impact, extremes of temperature or exposure to moisture or water. Never disassemble the ECU.
- If the oxygen sensor is removed, protect the tip from impact, exposure to water or contact with any petroleum-based product including greasy rags.
- Always replace fuel system O-rings whenever the joint is disconnected. Always replace the washers at union bolt or banjo fittings whenever the joint is loosened or disconnected.
- After repairs are completed, inspect the work area carefully for any sign of fuel leakage.

ELECTRONIC CONTROL UNIT (ECU)

Removal and Installation

1. With the ignition **OFF**, move the right front seat rearward.
2. Remove the two bolts from the front of the ECU mounting flange, then move the right front seat forward.
3. Remove the bolt from the rear flange of the ECU; remove the cover. Don't lose the spacer(s) between the flange and the floor.
4. Carefully disconnect the multi-plugs from both the main ECU and the smaller hot restart ECU.
5. Remove the controller from the vehicle. If necessary, remove the hot restart ECU from the larger unit by removing 2 screws.

To install:

6. Assemble the hot restart ECU to the engine ECU if it was removed.
7. Connect the multi-plug connectors to the units. Make certain the plugs are squarely and firmly engaged; all connector locks must be in place.
8. Position the ECU. Install the cover and spacers.
9. Install the retaining bolts and tighten them.
10. Test drive the vehicle. Check performance and idle in as many different conditions as possible. Inspect the ECU for any fault codes after the test drive.

ATMOSPHERIC PRESSURE SENSOR

Removal and Installation

1. Disconnect the negative battery cable.
2. Remove the fuse cover from the dashboard.
3. Disconnect the wiring harness from the sensor.
4. Remove the screw holding the sensor bracket; remove the sensor and bracket.
5. Reassemble in reverse order.

MAIN RELAY

Removal and Installation

1. Disconnect the negative battery cable.
2. Remove the fuse cover.
3. Remove the bolt holding the bracket, then remove the relay from the bracket.
4. Disconnect the wiring harness from the relay.
5. Reinstall in reverse order.

ELECTRONIC IDLE CONTROL VALVE

Removal and Installation

1. Disconnect the negative battery cable.
2. Drain the engine coolant at least to a level below the throttle body.
3. Disconnect the coolant hoses from the valve manifold.
4. Disconnect the wiring harness from the control valve.
5. Remove the bolt holding the vacuum tube bracket.
6. Loosen the bolt holding the crankcase vent tubes and move the tubes out of the way.
7. Remove the two bolts holding the control valve. Remove the valve; discard the O-ring or gasket.
8. Clean all components and contact surfaces thoroughly.

To install:

9. Apply a thin coat of clean oil to a new O-ring. Install the seal and fit the control valve into place on the manifold.
10. Install the 2 retaining bolts; tighten them to 16 ft. lbs. (22 Nm).
11. Move the vent tube into position and secure the bracket bolt.

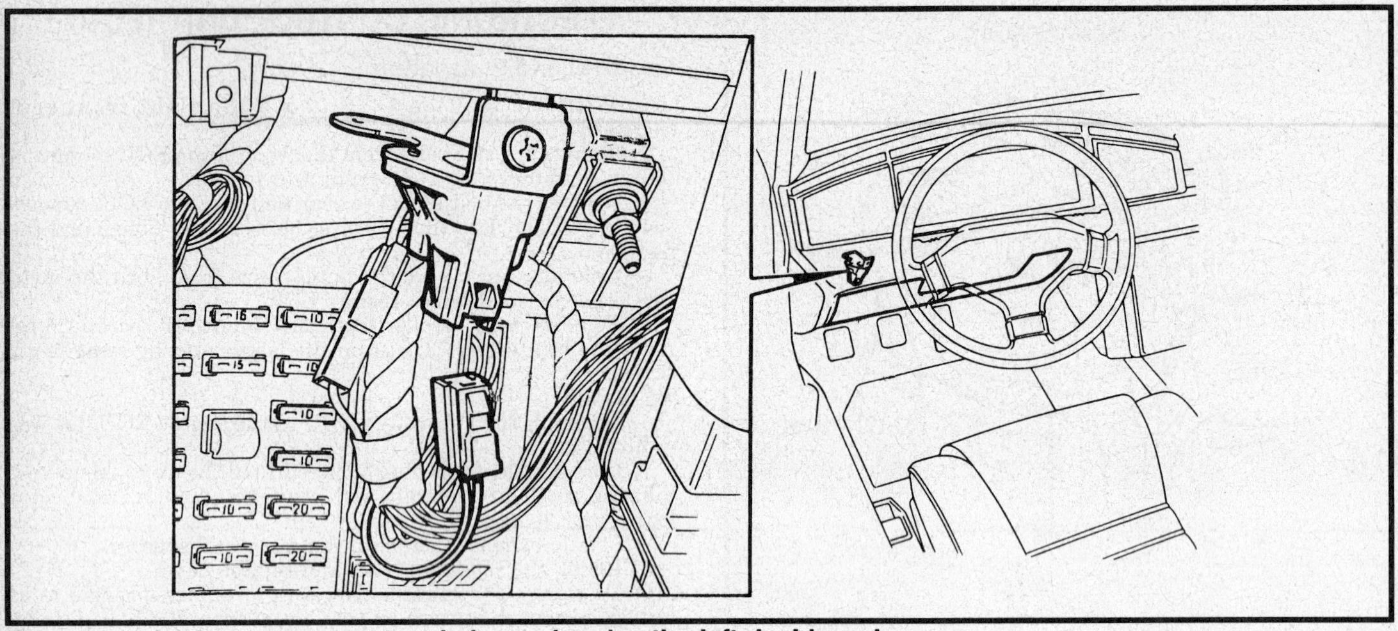

The atmospheric pressure sensor is located under the left dashboard

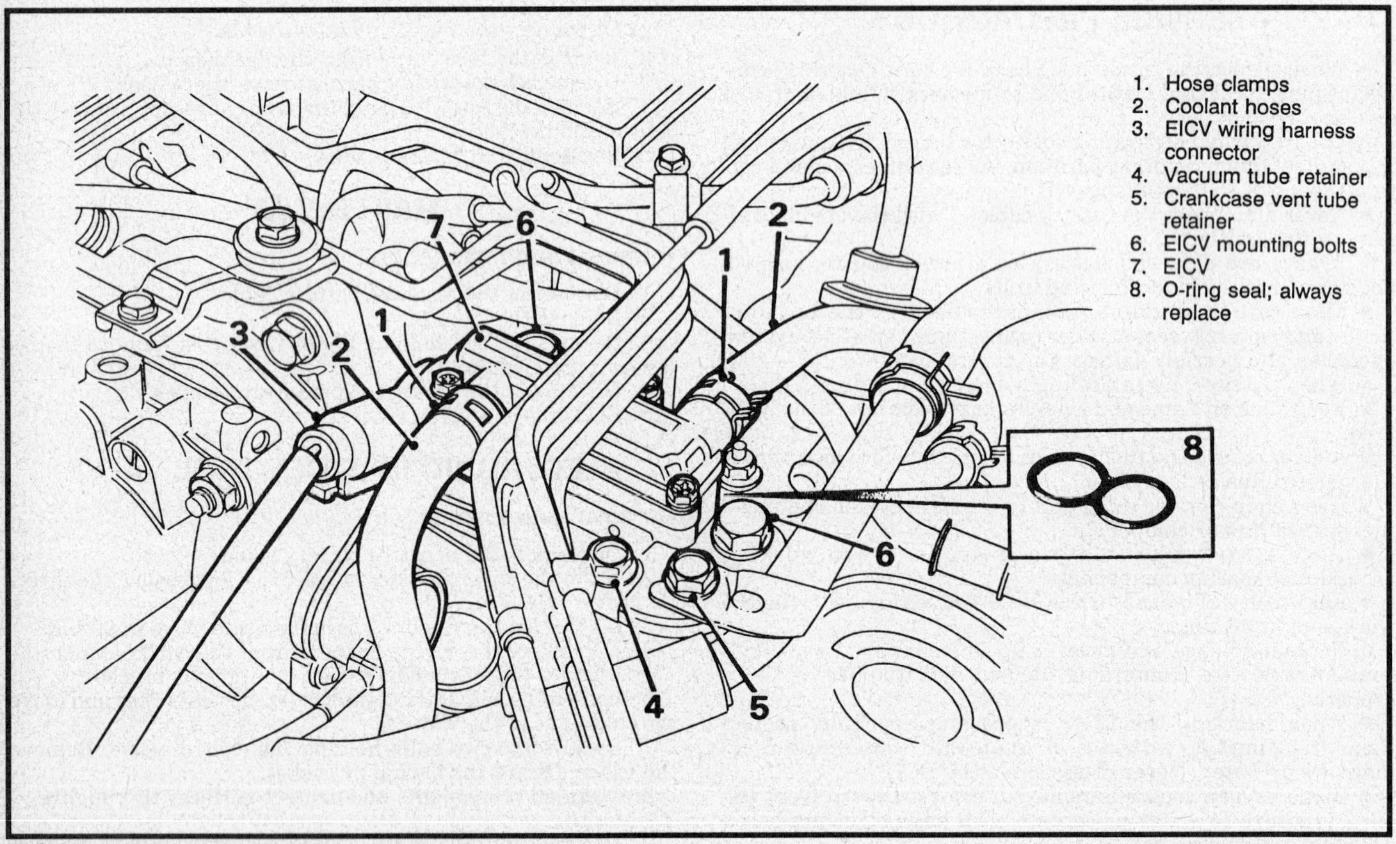

EICV location and related components

12. Install the vacuum tube bracket bolt.
13. Connect the wiring harness to the idle control valve.
14. Install the coolant hoses at the manifold.
15. Refill the coolant system.
16. Connect the negative battery cable.
17. Adjust the idle speed.

FAST IDLE VALVE

Removal and Installation

1. Disconnect the negative battery cable.
2. Drain the engine coolant at least to a level below the throttle body.
3. Disconnect the coolant hoses from the fast idle valve.
4. Remove the 3 bolts holding the valve to the manifold. Remove the valve. Recover and discard the 2 O-ring seals.
5. Thoroughly clean all components and contact surfaces.

To install:

6. Apply a light coat of clean lubricant to new O-rings.
7. Install the O-rings and position the valve on the manifold.
8. Install the retaining bolts; tighten them evenly to 9 ft. lbs. (12 Nm).
9. Connect the coolant hoses to the valve and secure the hose clamps.
10. Refill the coolant system.
11. Connect the negative battery cable.

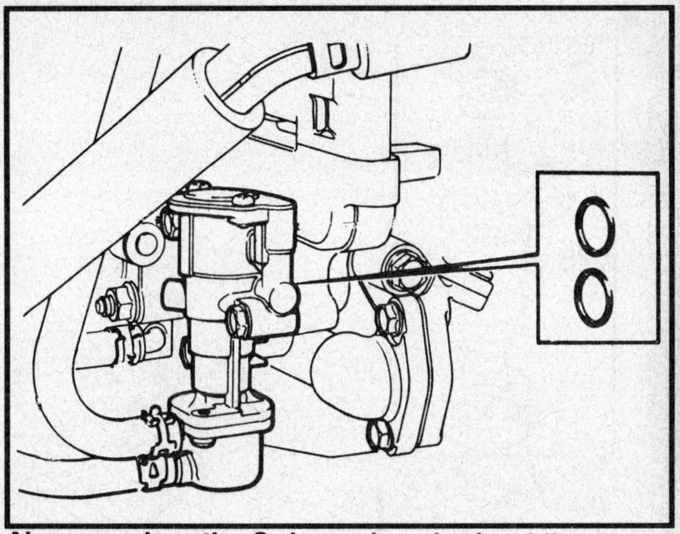

Always replace the O-rings when the fast idle valve is removed

FUEL PRESSURE REGULATOR

Removal and Installation

1. Disconnect the negative battery cable.
2. Relieve the fuel pressure.
3. Disconnect the fuel return hose from the pressure regulator.
4. Disconnect the vacuum hose from the regulator.
5. Remove the 2 bolts holding the regulator to the fuel rail. Remove the regulator. Remove and discard the O-ring from the regulator.

To install:

6. Lubricate a new O-ring with a light coat of oil.
7. Install the O-ring and position the regulator on the rail.
8. Install the retaining bolts and tighten them to 9 ft. lbs. (12 Nm).
9. Connect the vacuum hose. Install the fuel return hose.
10. Connect the negative battery cable.
11. Switch the ignition **ON**, causing the fuel pump to run for 2 seconds. Check the work area carefully for any leakage before starting the engine.

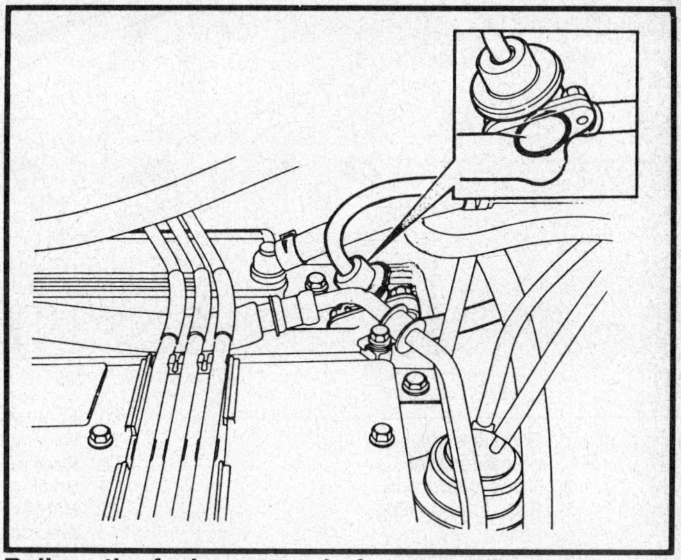

Relieve the fuel pressure before removing the pressure regulator

FUEL RAILS AND INJECTORS

Removal and Installation

1. Disconnect the negative battery cable.
2. Relieve the fuel pressure.
3. Label and disconnect the spark plug leads; move them out of the work area.
4. Remove the bolt holding the fuel line clip.
5. If the front fuel rail or injectors are to be removed:
   a. Disconnect the breather hose from the manifold cover.
   b. Disconnect the throttle cable from its clamp.
   c. Remove the cap nuts from the banjo connectors at the rail.
   d. Disconnect the fuel hoses from the rail. Immediately plug or cap the hoses and ports to prevent entry of dirt and debris. Recover and discard the 4 washers from the banjo joints. Move the fuel hoses out of the work area.
6. If the rear fuel rail or injectors are to be removed:
   a. Disconnect the vacuum hose from the pressure regulator.
   b. Disconnect the fuel return hose from the pressure regulator.
   c. Remove the cap nuts from the banjo connectors at the fuel rail.
   d. Disconnect the fuel hoses from the rail. Immediately plug or cap the hoses and ports to prevent entry of dirt and debris. Recover and discard the 4 washers from the banjo joints. Move the fuel hoses out of the work area.
7. Remove the bolts holding the wiring harness brackets on each side of the engine. Move the harnesses out of the way.
8. Disconnect the wiring connectors from the injectors on the side(s) to be removed.
9. Remove the nuts holding the fuel rail to the studs. Carefully remove the fuel rail.
10. Remove the injectors from the intake manifold. Remove and discard the O-ring and cushion from the top of the injector. Inspect the insulator and sealing ring at the bottom of the injec-

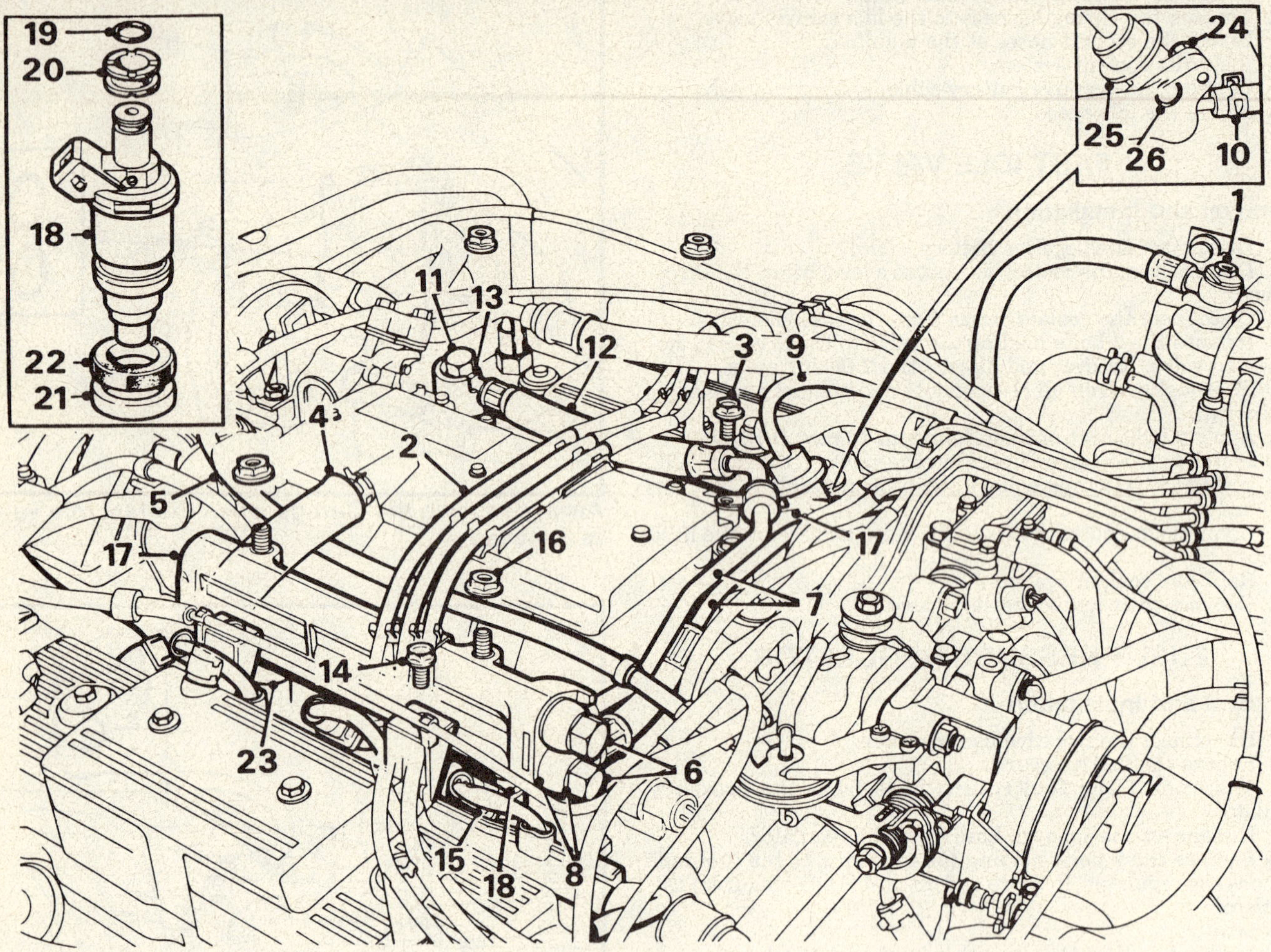

1. Service bolt/fuel pressure relief
2. Spark plug leads
3. Fuel line retaining bolt
4. Breather hose
5. Throttle cable clamp
6. Front rail cap nuts
7. Front rail fuel lines
8. Sealing washers (4)
9. Pressure regulator vacuum hose
10. Fuel return hose
11. Rear rail cap nuts
12. Rear rail fuel lines
13. Sealing washers (4)
14. Harness bracket retaining bolt
15. Injector connectors
16. Fuel rail retaining nuts
17. Fuel rails
18. Injectors
19. O-ring; always replace
20. Cushion ring; always replace
21. Insulator
22. Sealing ring
23. Spacers
24. Fuel pressure regulator
25. Regulator mounting bolts
26. Pressure regulator O-ring

Fuel rails, injectors and related components

tor. If no damage or cracking is evident, the insulator and sealing ring may be reused.

11. Remove the fuel rail spacers from the manifold to prevent loss.

12. If desired, the pressure regulator may be removed from the rear fuel rail.

To install:

13. Assemble the pressure regulator and rear fuel rail if they were separated. Always use a new O-ring when installing the regulator.

14. Assemble the injectors with new cushions and O-rings. To ease assembly, coat the new O-ring with a light coat of fuel. Fit the insulator and sealing ring into position.

15. Install the injectors in the manifold. Make certain to place the injector terminal in the correct position.

16. Install the fuel rail spacers on the manifold.

17. Install the fuel rail, making sure each injector is firmly and squarely seated. Install the retaining nuts and tighten them evenly.

18. Connect the wiring to each injector.

19. Reposition the wiring harnesses and secure the bracket bolts.

20. If the rear fuel rail was removed:

a. Install new sealing washers on the banjo fittings. Connect the fuel hose to the fuel rail and tighten the fittings to 16 ft. lbs. (22 Nm).

b. Connect the fuel return hose to the pressure regulator. Connect the vacuum hose to the regulator.

21. If the front fuel rail was removed:

a. Install new sealing washers on the banjo fittings. Connect the fuel hose to the fuel rail and tighten the fittings to 16 ft. lbs. (22 Nm).

b. Install the throttle cable in its clamp.

c. Connect the manifold breather hose.

22. Reinstall the bolt holding the fuel line retainer.
23. Connect the spark plug wires.
24. Connect the negative battery cable.
25. Turn the ignition switch **ON**, causing the fuel pump to run for 2 seconds. Carefully check the work area for any signs of fuel leakage before starting the engine.

OXYGEN SENSOR

Removal and Installation

CAUTION

Perform this operation only on a cool or cold engine. Make certain all surrounding components are cool to the touch before proceeding or personal injury may occur.

REAR

1. Disconnect the negative battery cable.
2. Raise and safely support the front of the vehicle.
3. Disconnect the sensor wiring at the connector; do not attempt to disconnect it from the sensor.
4. Remove the sensor from the manifold.
5. Once removed, protect the sensor from impact, moisture or exposure to petroleum products. Do not place the sensor on greasy or dirty rags.
6. Install the sensor in reverse order of removal. Tighten the sensor to 33 ft. lbs. (45 Nm).

FRONT

1. Disconnect the negative battery cable.
2. Raise and safely support the front of the vehicle.
3. Drain the coolant system completely.
4. Remove the radiator assembly.
5. Remove the 3 bolts holding the heat shield.

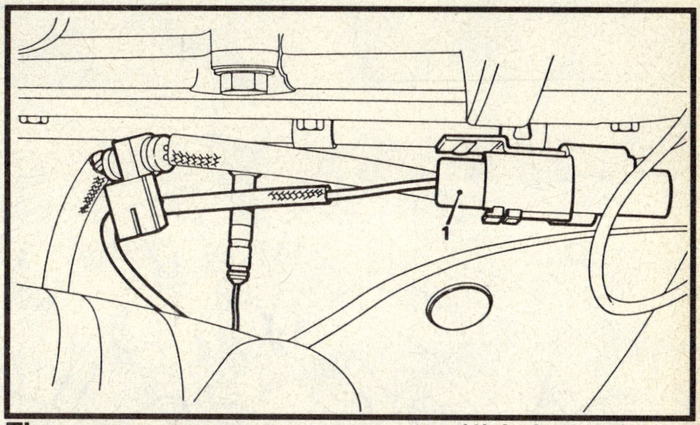

The rear oxygen sensor connector (1) is located away from the sensor

6. Disconnect the oxygen sensor wiring at the connector. Do not attempt to disconnect it at the sensor.
7. Free the sensor wiring from the retaining clips.
8. Remove the manifold heat shield.
9. Remove the sensor from the manifold.
10. Once removed, protect the sensor from impact, moisture or exposure to petroleum products. Do not place the sensor on greasy or dirty rags.

To install:

11. Install the sensor and tighten it to 33 ft. lbs. (45 Nm).
12. Install the heat shield on the manifold.
13. Place the sensor wiring in the retaining clips and fasten the harness connector.
14. Install the bolts in the heat shield.
15. Install the radiator assembly.
16. Refill the coolant system.
17. Lower the vehicle to the ground. Connect the negative battery cable.

THROTTLE BODY

Removal and Installation

1. Disconnect the negative battery cable.

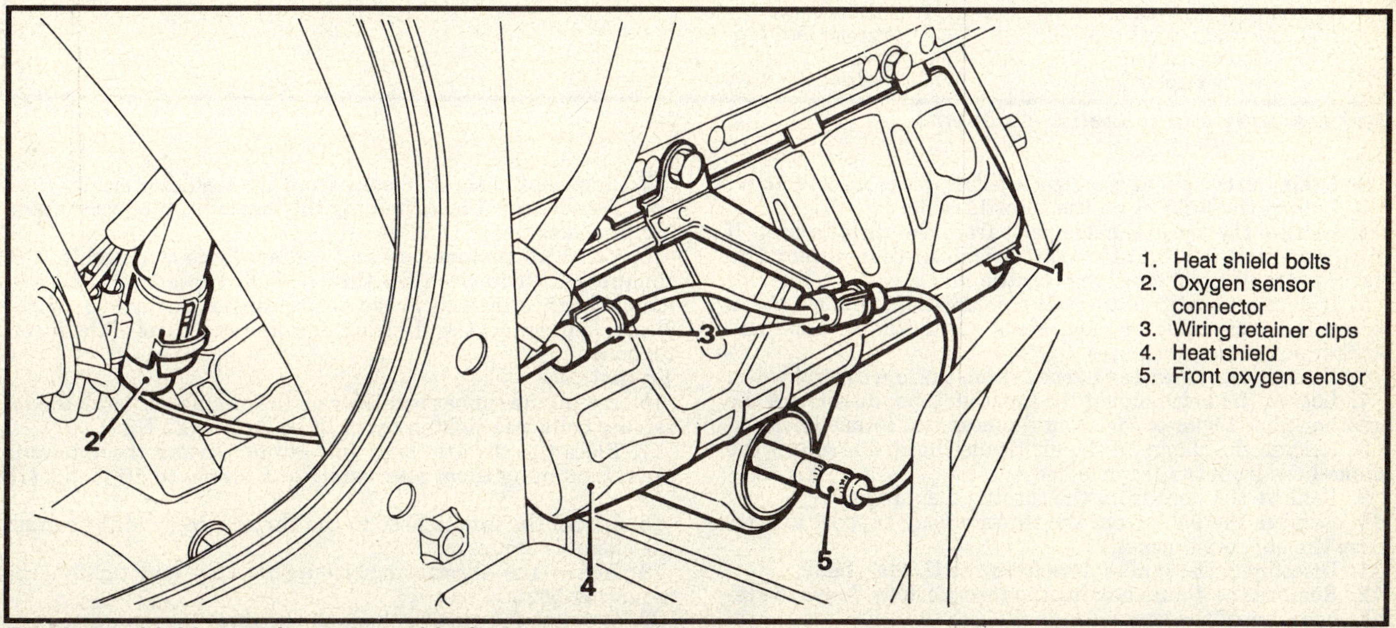

The heat shield must be removed before removing the front oxygen sensor

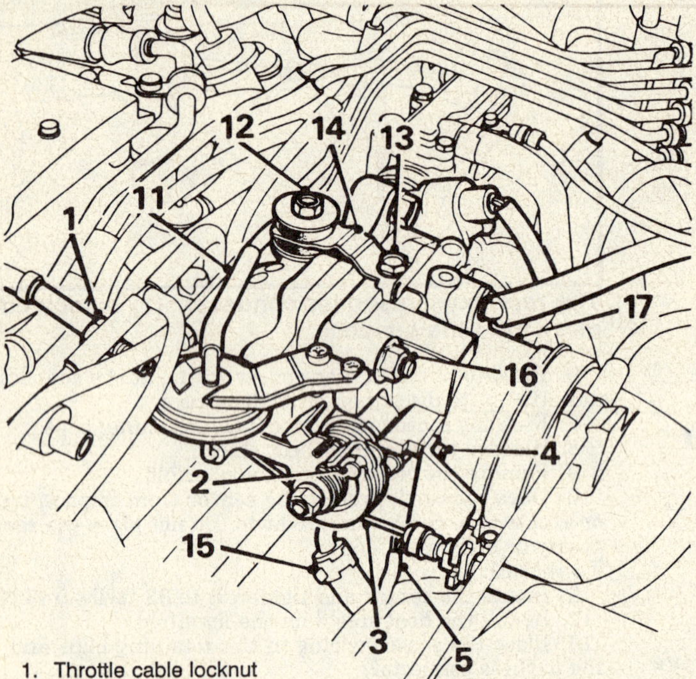

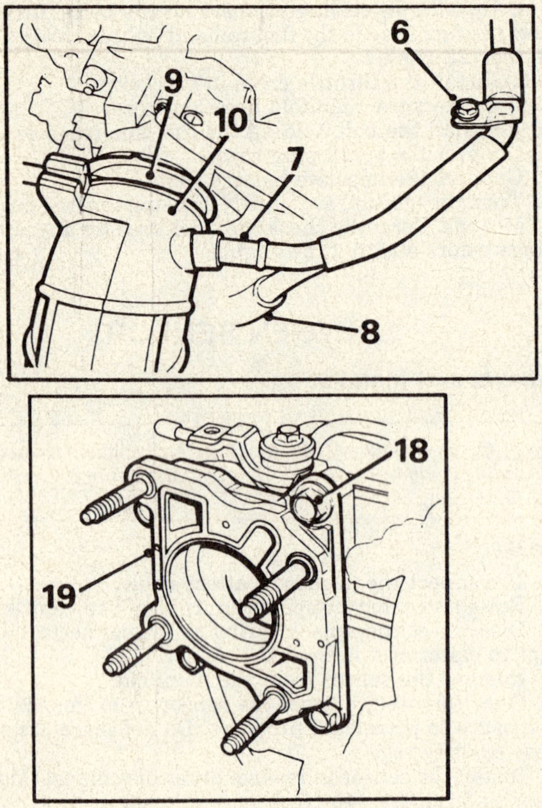

1. Throttle cable locknut
2. Throttle cable pivot
3. Kickdown cable pivot
4. Kickdown cable retaining bracket
5. Idle control valve connector
6. Breather pipe retaining bolt
7. Breather pipe joint to air intake
8. Breather pipe joint to hose
9. Intake duct clamp
10. Air intake duct
11. Vacuum hose to throttle diaphragm
12. Throttle body support mounting bolt
13. Throttle body support mounting bolt
14. Throttle body support
15. Intake hose
16. Throttle body retaining nuts
17. Throttle body
18. Insulator retaining bolts
19. Insulator

Throttle body and related components

2. Drain the coolant at least to a level below the throttle body.
3. Loosen the locknut on the throttle cable.
4. Release the throttle cable pivot from the throttle link. If equipped with automatic transmission, disconnect the kickdown cable pivot from the throttle link.
5. If equipped with automatic transmission, remove the 2 screws from the cable bracket. Move the kickdown cable and bracket out of the work area.
6. Disconnect the wiring harness from the idle control valve.
7. Loosen the bolt holding the breather pipe; disconnect the breather pipe Y connection from the hose and intake duct.
8. Release the clamp on the air cleaner hose; disconnect the intake hose from the throttle body.
9. Remove the hose from the throttle diaphragm.
10. Remove the bolts from the throttle body support and remove the support bracket.
11. Disconnect the intake hose from the throttle body.
12. Remove the 4 nuts holding the throttle body. Move the intake hose out of the way.
13. Remove the throttle body from the insulator studs.
14. Remove the 4 bolts holding the insulator; remove the rubber insulator.
15. Clean all components and contact faces thoroughly. Use compressed air to clean the throttle body passages. Do not immerse the throttle body in chemical cleaners. Do not allow spray cleaners to contact the electrical or non-metal parts of the throttle body.

To install:

16. Install the rubber insulator on the manifold. Install the retaining bolts and tighten them to 16 ft. lbs. (22 Nm).
17. Place the throttle body in position. Install the retaining nuts, tightening them alternately and evenly to 16 ft. lbs (22 Nm).
18. Install the intake hose to the throttle body. Make certain the clamp is secure.
19. Install the throttle body support stay and tighten the mounting bolts.
20. Connect the vacuum hose to the throttle diaphragm.

21. Connect the air intake duct to the throttle body and secure the clamp.
22. Connect the breather hose Y connector to the hose and intake duct. Install the bolt in the breather pipe retainer.
23. Install the connector to the idle control valve.
24. Position and install the bracket holding the kickdown cable.
25. Connect the kickdown and/or throttle cable pivots to the throttle linkage.
26. Adjust the throttle cable; tighten the locknut.
27. Refill the engine coolant.
28. Connect the negative battery cable.

THROTTLE ANGLE SENSOR

Removal and Installation

NOTE: The sensor is held to the throttle body by shear bolts. Do not remove the sensor unless replacement bolts are available.

1. Disconnect the negative battery cable.
2. Remove the throttle body.
3. Disconnect the throttle angle sensor wiring connector.
4. Remove the wiring bracket from the throttle body and free the connector from the bracket.
5. Center punch the 2 shear bolts holding the sensor for drilling.
6. Use a 3mm drill, carefully drill out the shear bolts.
7. Remove both bolts and remove the throttle angle sensor from the throttle body.
8. Remove the gasket, clamp plate and washer from the sensor.

To install:

9. Install the gasket, clamp plate and washer on the sensor.
10. Carefully fit the sensor in position; make certain it engages properly at the throttle body.
11. Install new shear bolts. Tighten them carefully until the head shears off.
12. Attach the connector bracket and mount the connector to it. Connect the wiring harness.
13. Install the throttle body.
14. Connect the negative battery cable.

IDLE SPEED

Adjustment

The yellow LED on the engine ECU will light if idle adjustment is needed. The idle speed must not be adjusted if the lamp is not lit.

Idle speed is changed, when necessary, by turning the adjusting screw on the throttle body. The adjusting screw controls the amount of air bypassing the throttle plate. This adjustment is used to set the EICV to a specified value.

During idle adjustment, if the yellow LED is blinking, turn the adjusting screw clockwise. If the LED is constantly lit but not blinking, turn the adjusting screw counter-clockwise. To adjust the idle:

1. Warm the engine to normal operating temperature. The fan must cycle on and off at least once.
2. Make certain all electrical loads are switched **OFF**.
3. Turn the adjusting screw ¼-turn (90 degrees) in the proper direction based on the LED display. Wait 30 seconds until the idle has stabilized.
4. Check the LED; if not lit, the idle is correct. If blinking or constantly lit, repeat Step 3 until the LED is off.
5. If the adjusting screw must be turned to within ¼-turn of fully closed, the EICV is not operating properly.

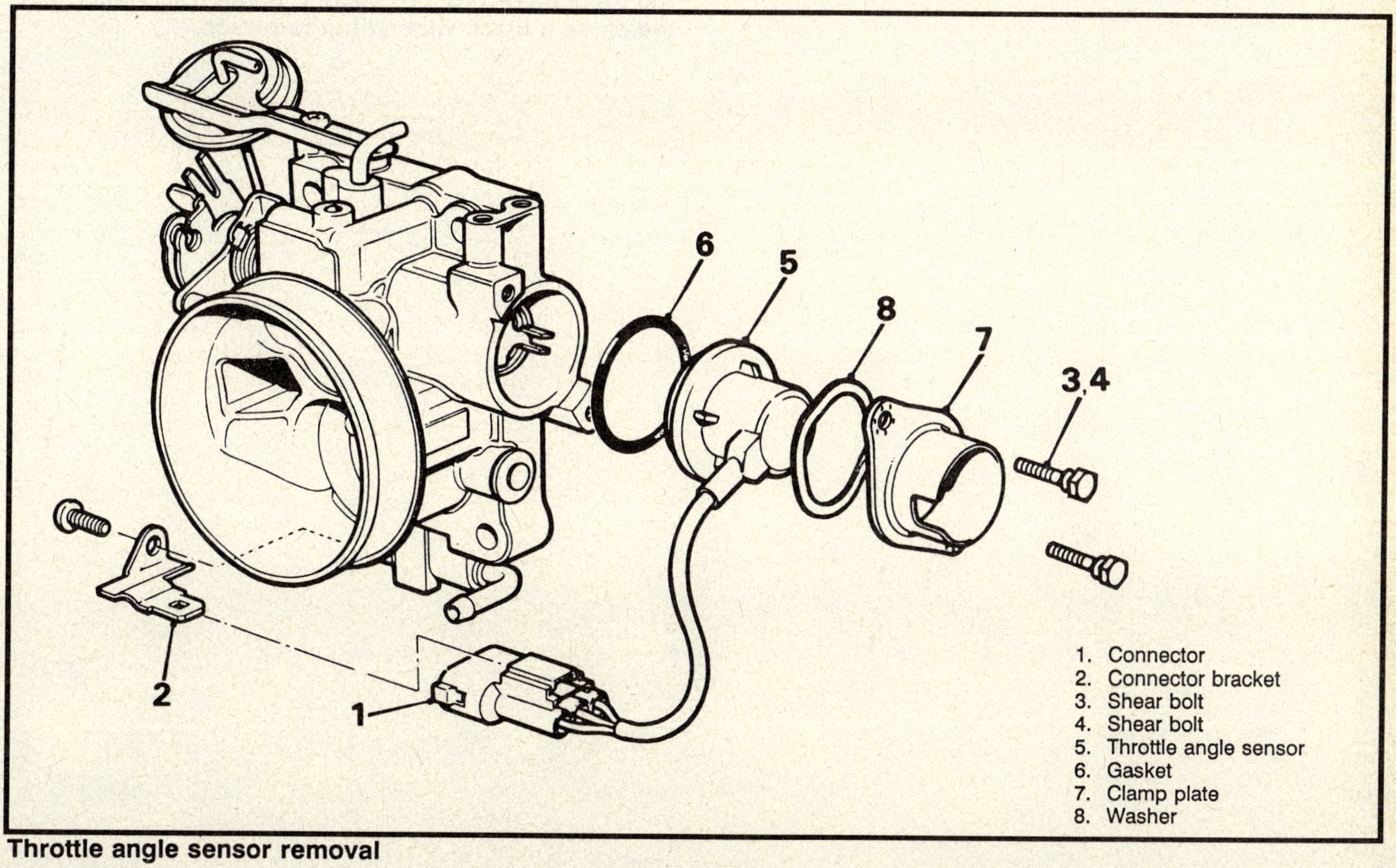

Throttle angle sensor removal

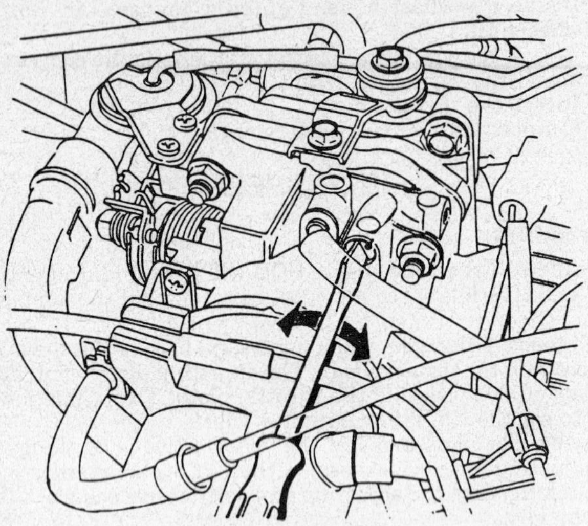

Only adjust the idle when the yellow LED is lit

Code	Color
B	Black
G	Green
K	Pink
LG	Light Green
N	Brown
O	Orange
P	Purple
R	Red
S	Slate (grey)
U	Blue
W	White
Y	Yellow

Wiring color codes. If two letters appear, the first is the main color and the second is the trace color: BR indicates a black wire with a red trace.

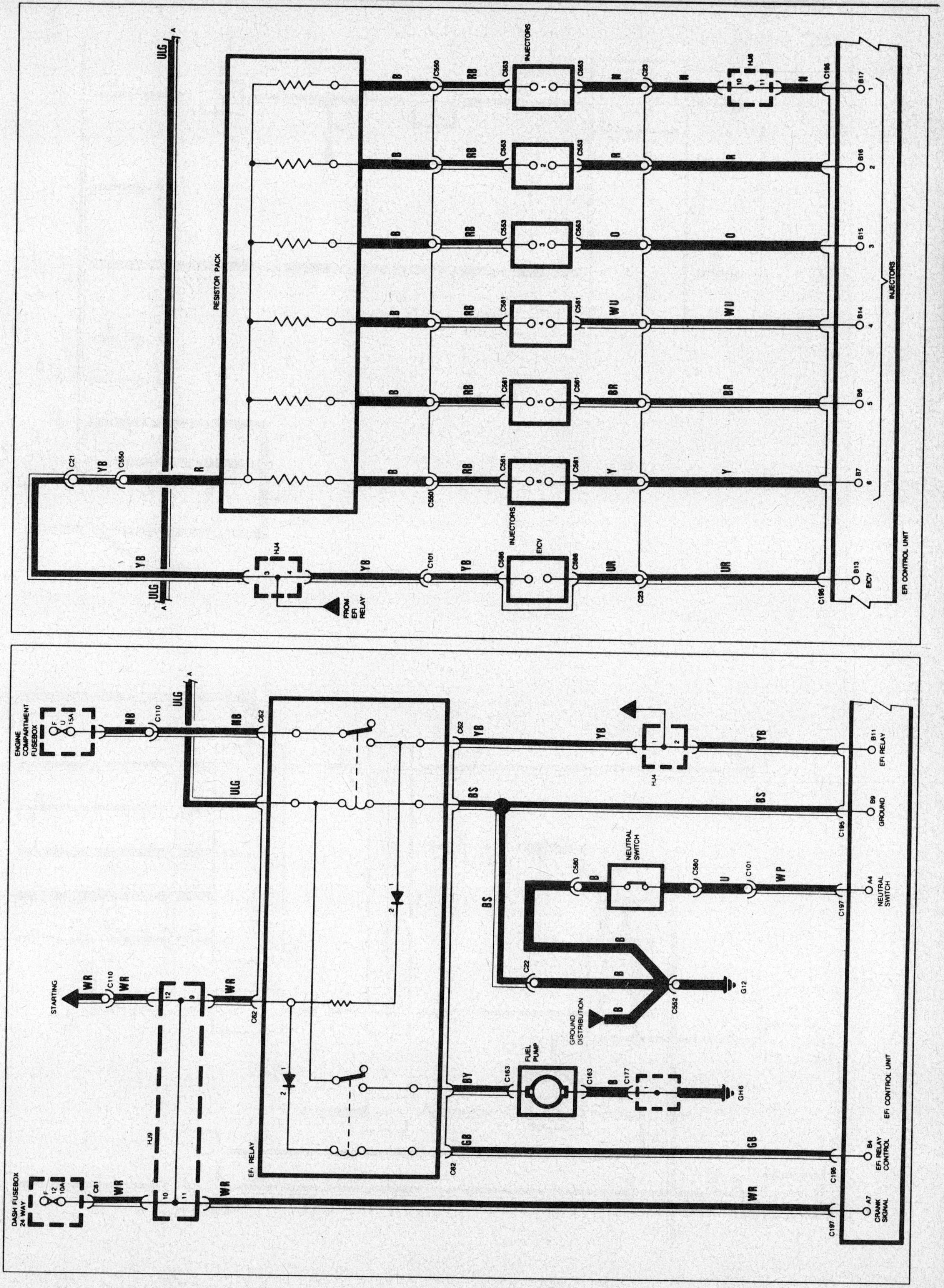

ENGINE MANAGEMENT WIRING SCHEMATIC, MANUAL TRANSMISSION—STERLING

ENGINE MANAGEMENT WIRING SCHEMATIC, MANUAL TRANSMISSION—STERLING

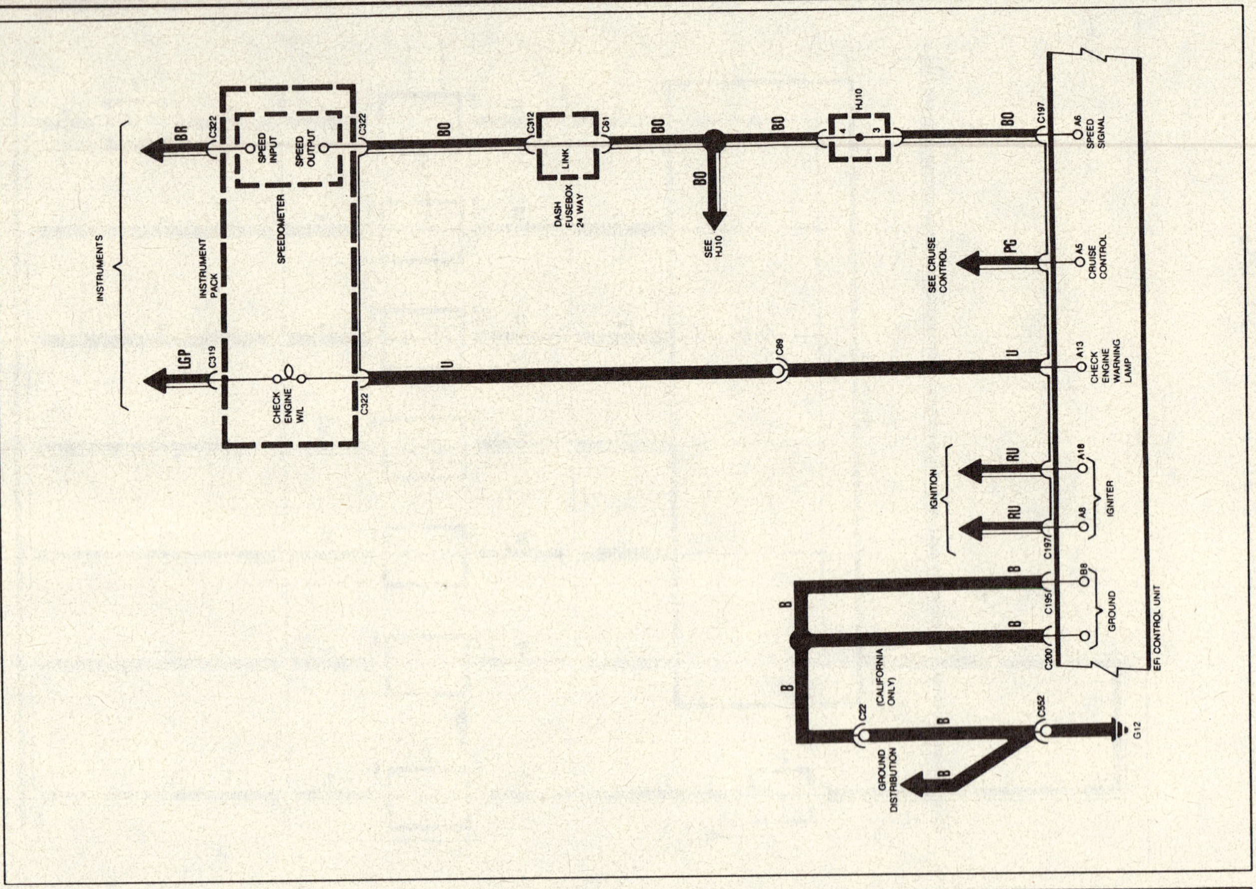

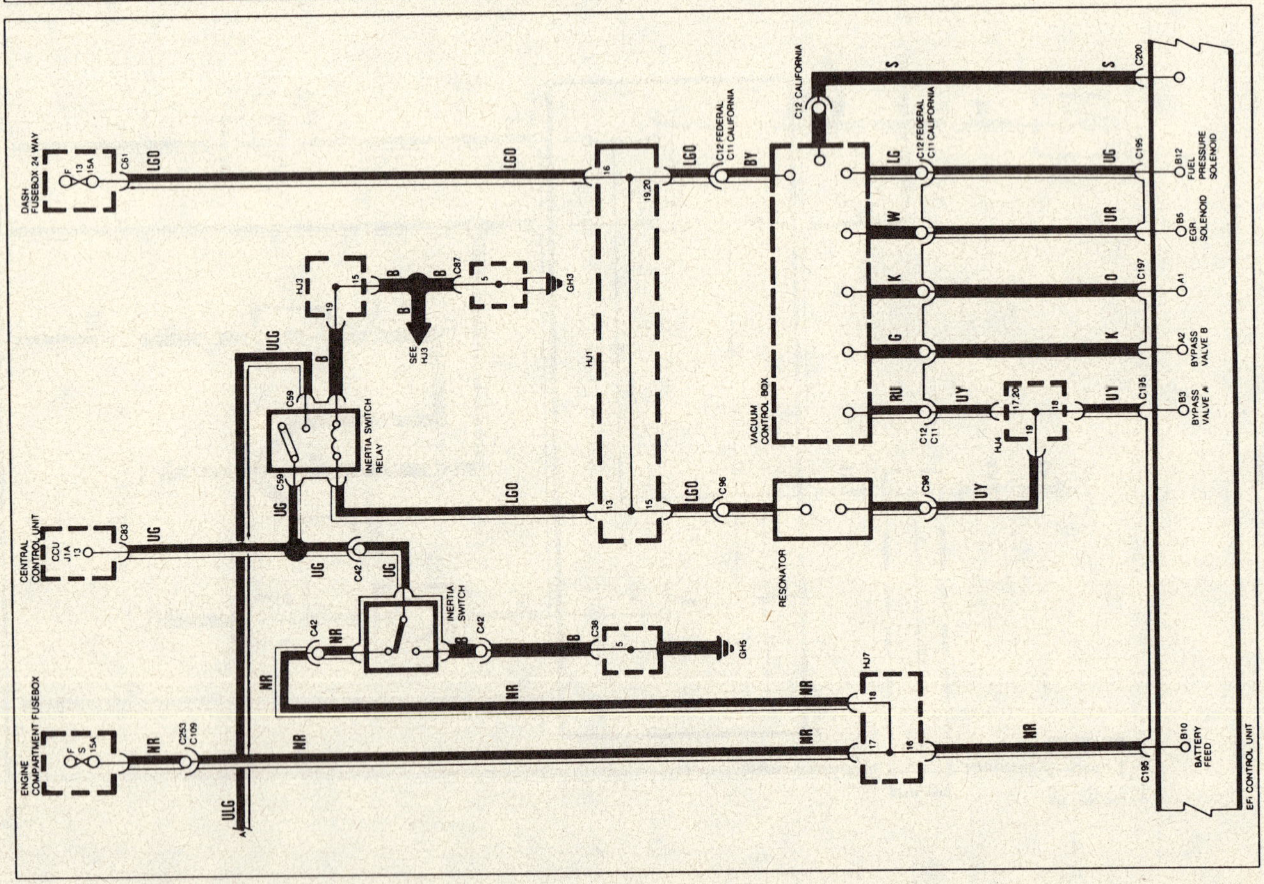

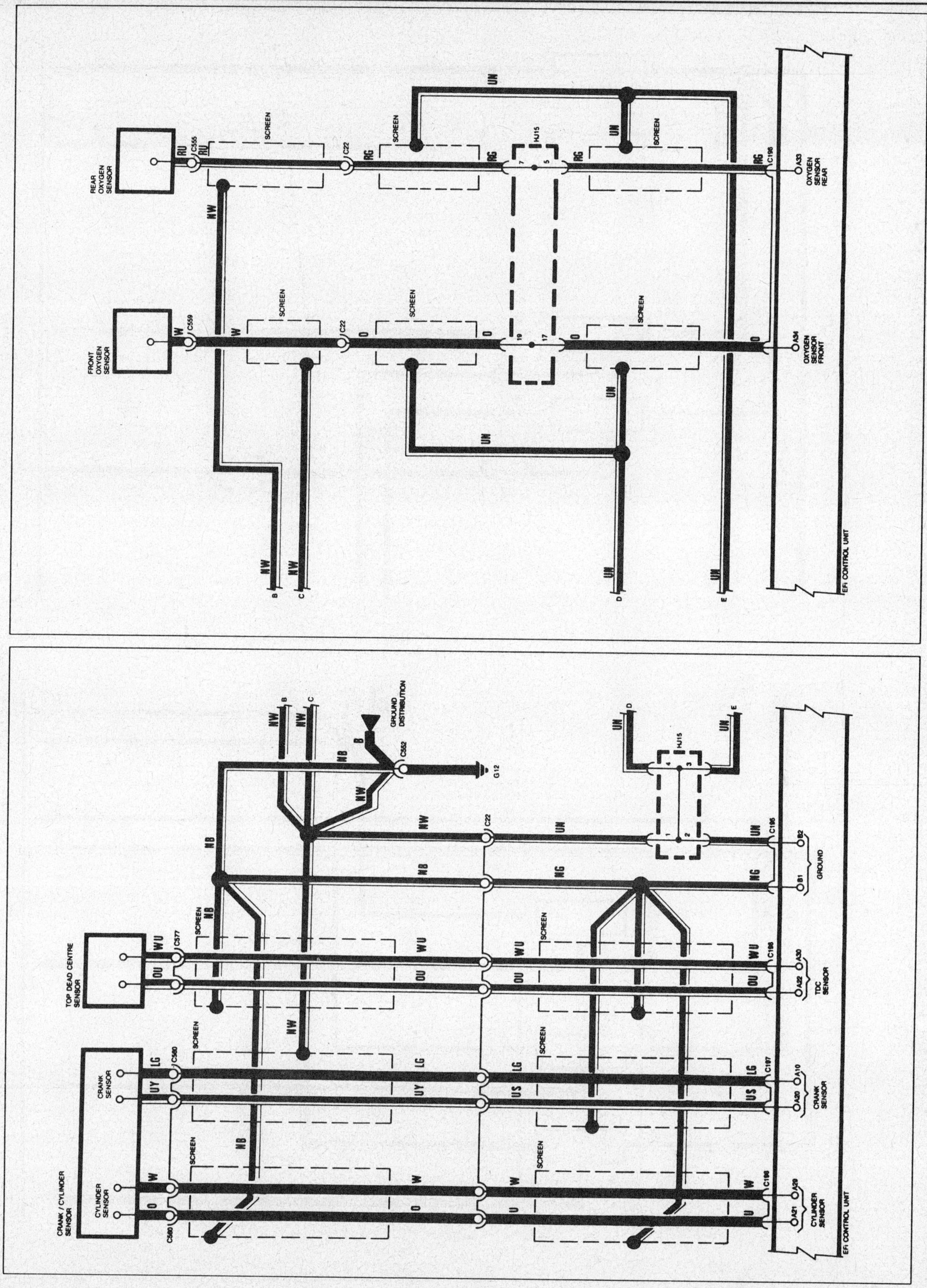

ENGINE MANAGEMENT WIRING SCHEMATIC, MANUAL TRANSMISSION—STERLING

ENGINE MANAGEMENT WIRING SCHEMATIC, MANUAL TRANSMISSION—STERLING

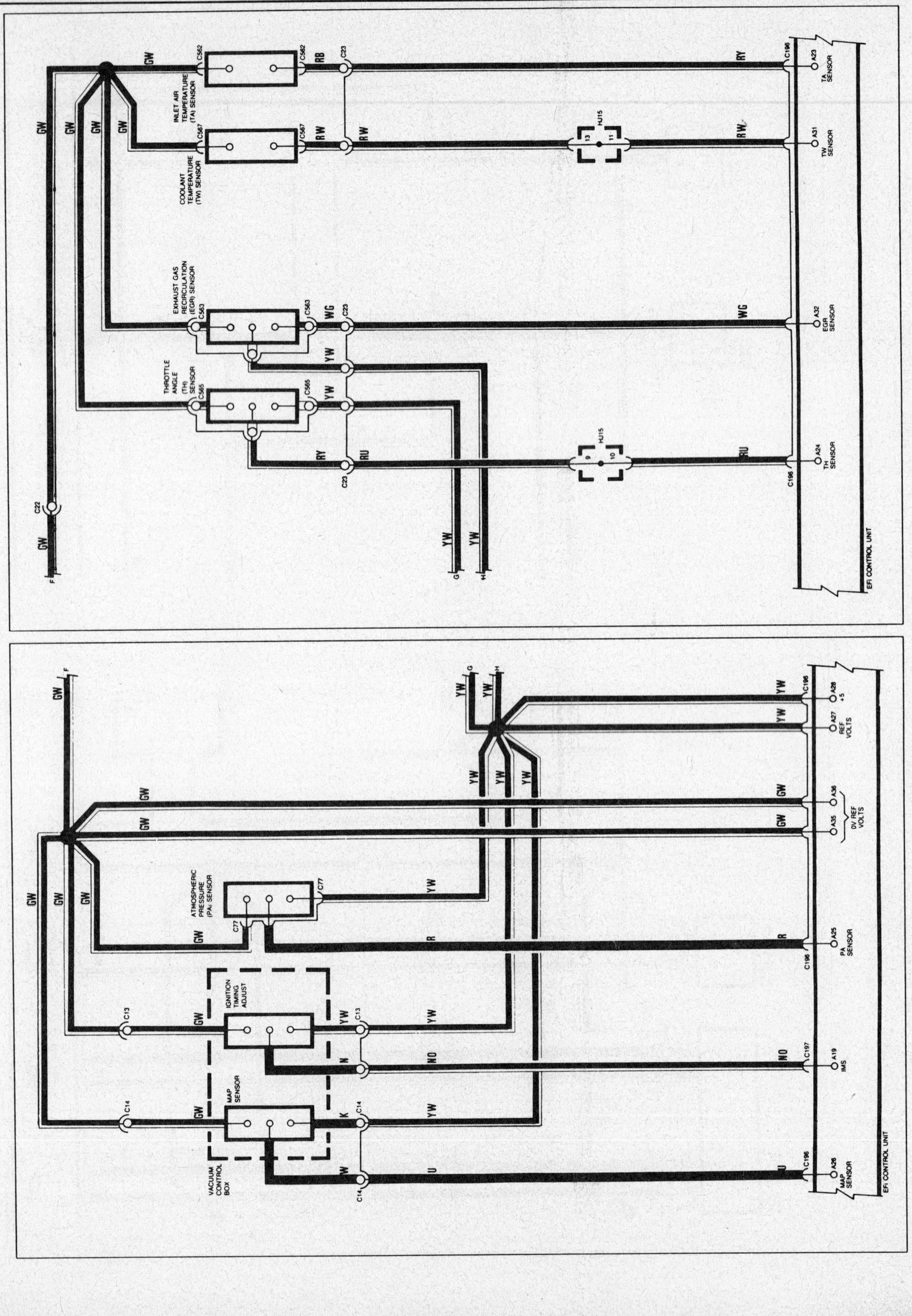

ENGINE MANAGEMENT WIRING SCHEMATIC, AUTOMATIC TRANSMISSION—STERLING

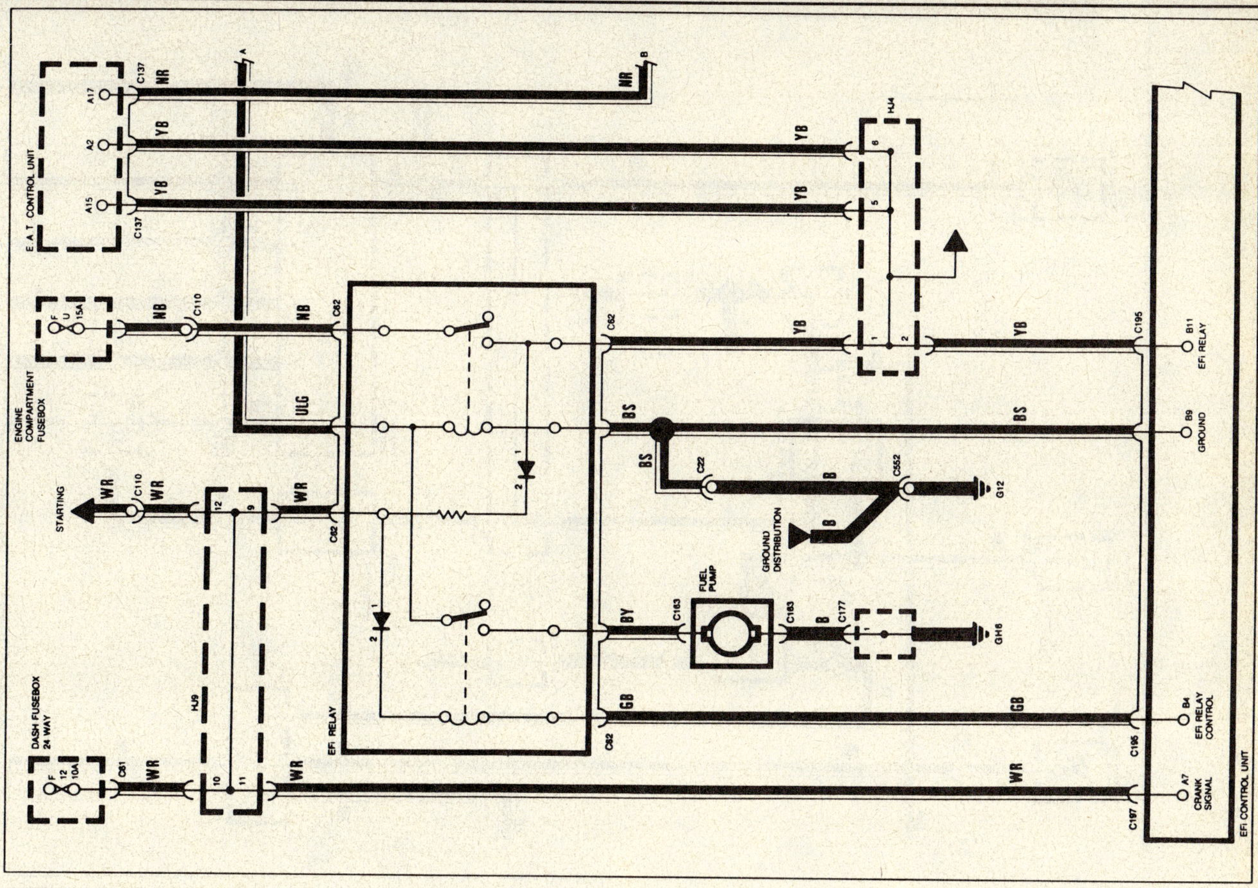

ENGINE MANAGEMENT WIRING SCHEMATIC, MANUAL TRANSMISSION—STERLING

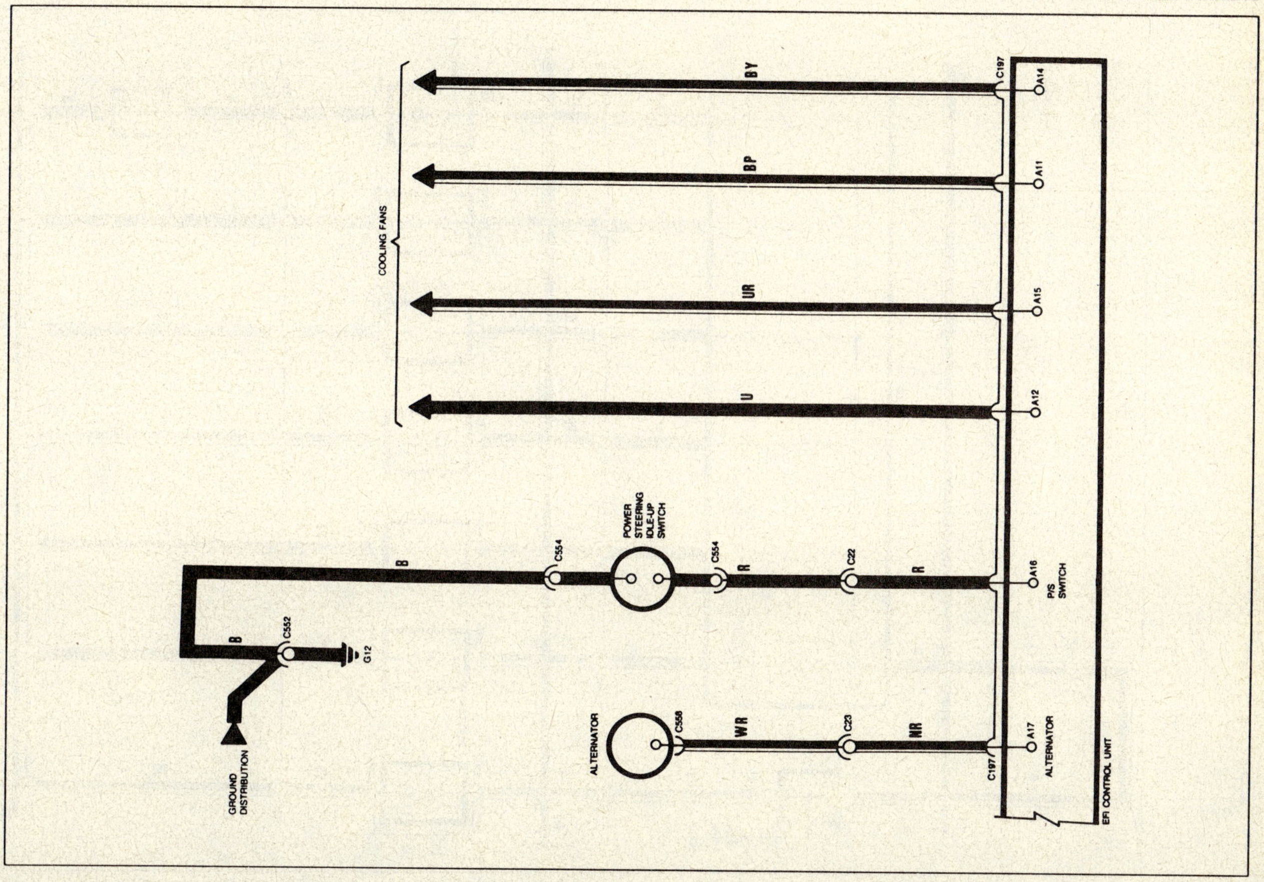

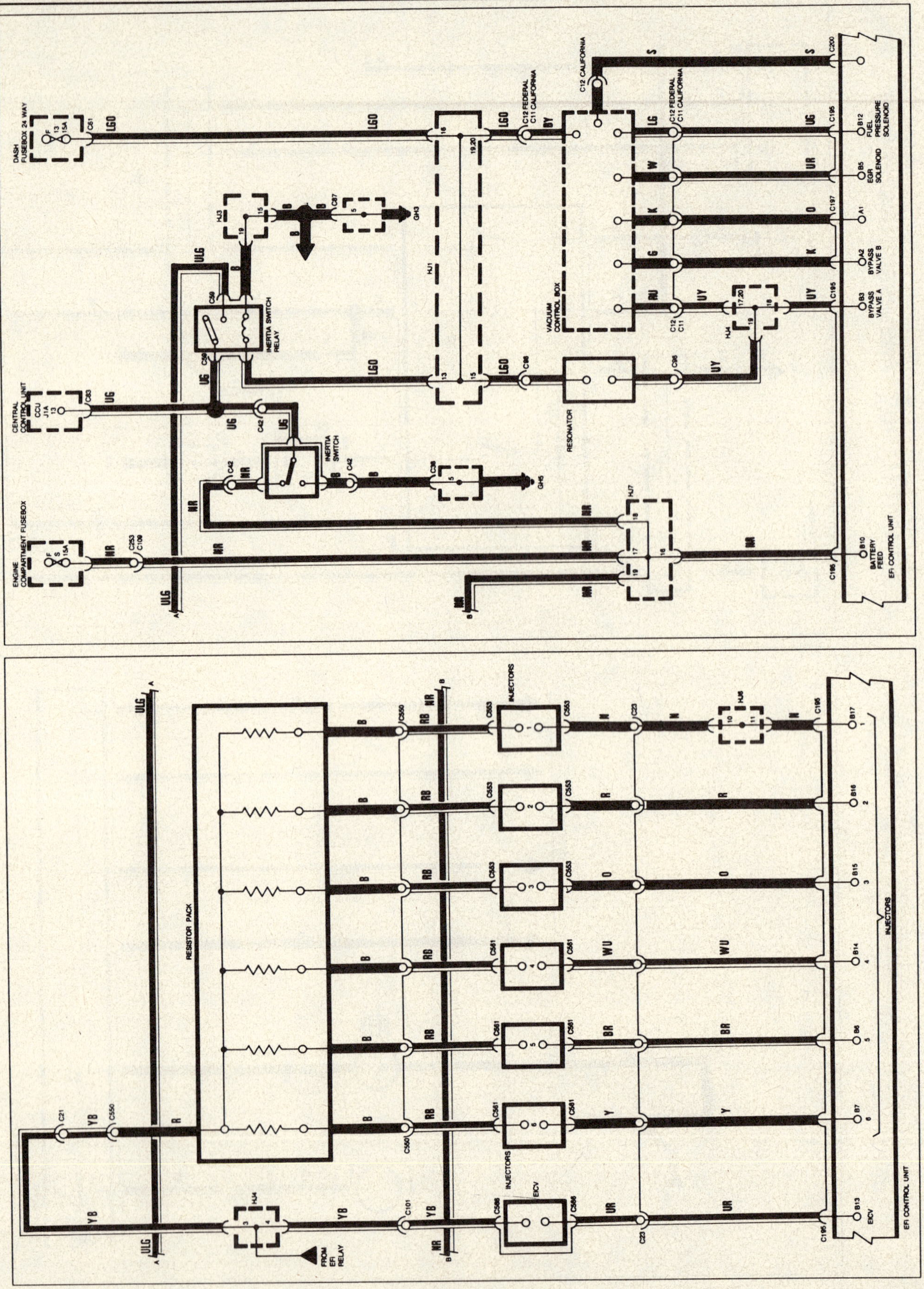

ENGINE MANAGEMENT WIRING SCHEMATIC, AUTOMATIC TRANSMISSION—STERLING

ENGINE MANAGEMENT WIRING SCHEMATIC, AUTOMATIC TRANSMISSION—STERLING

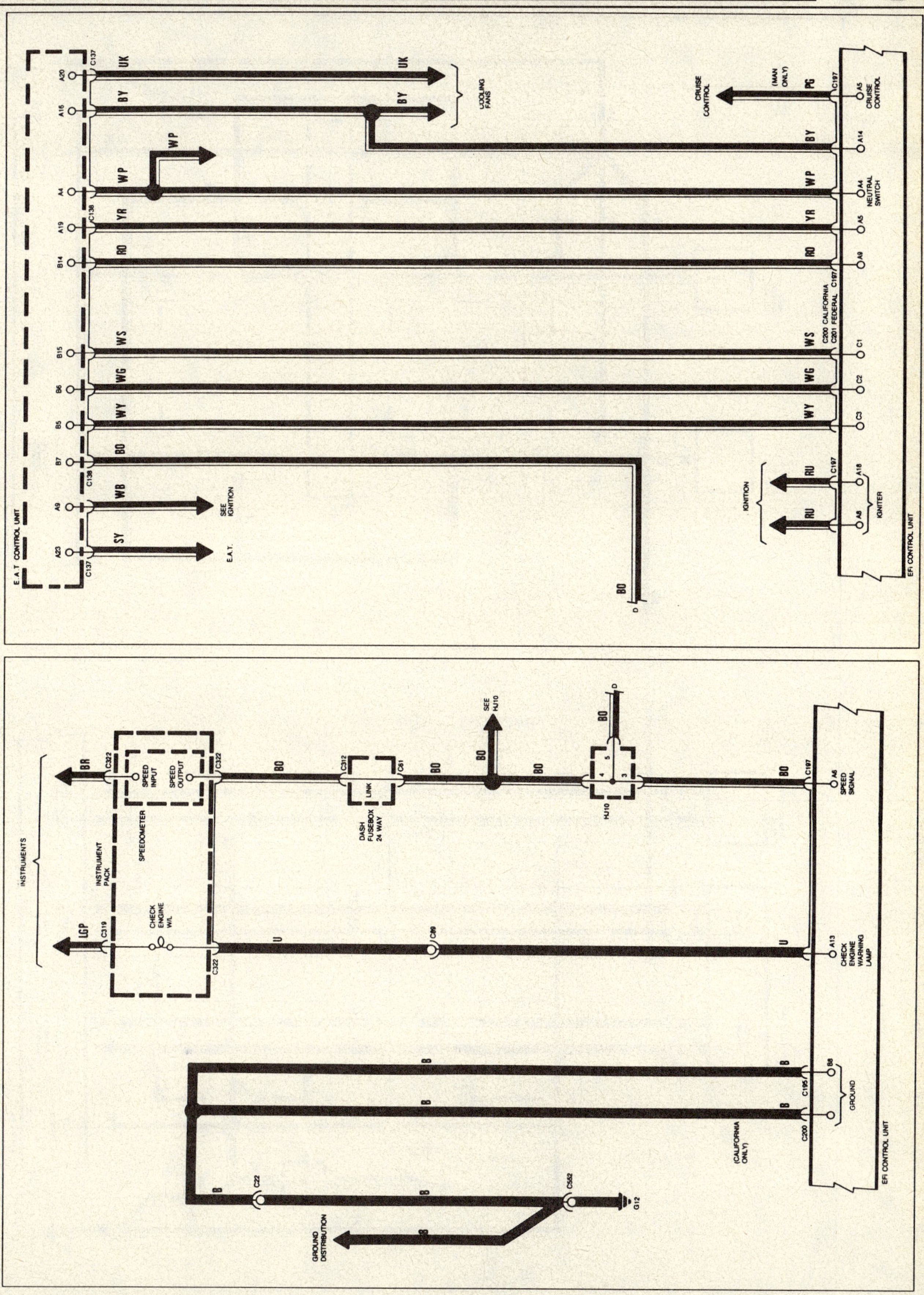

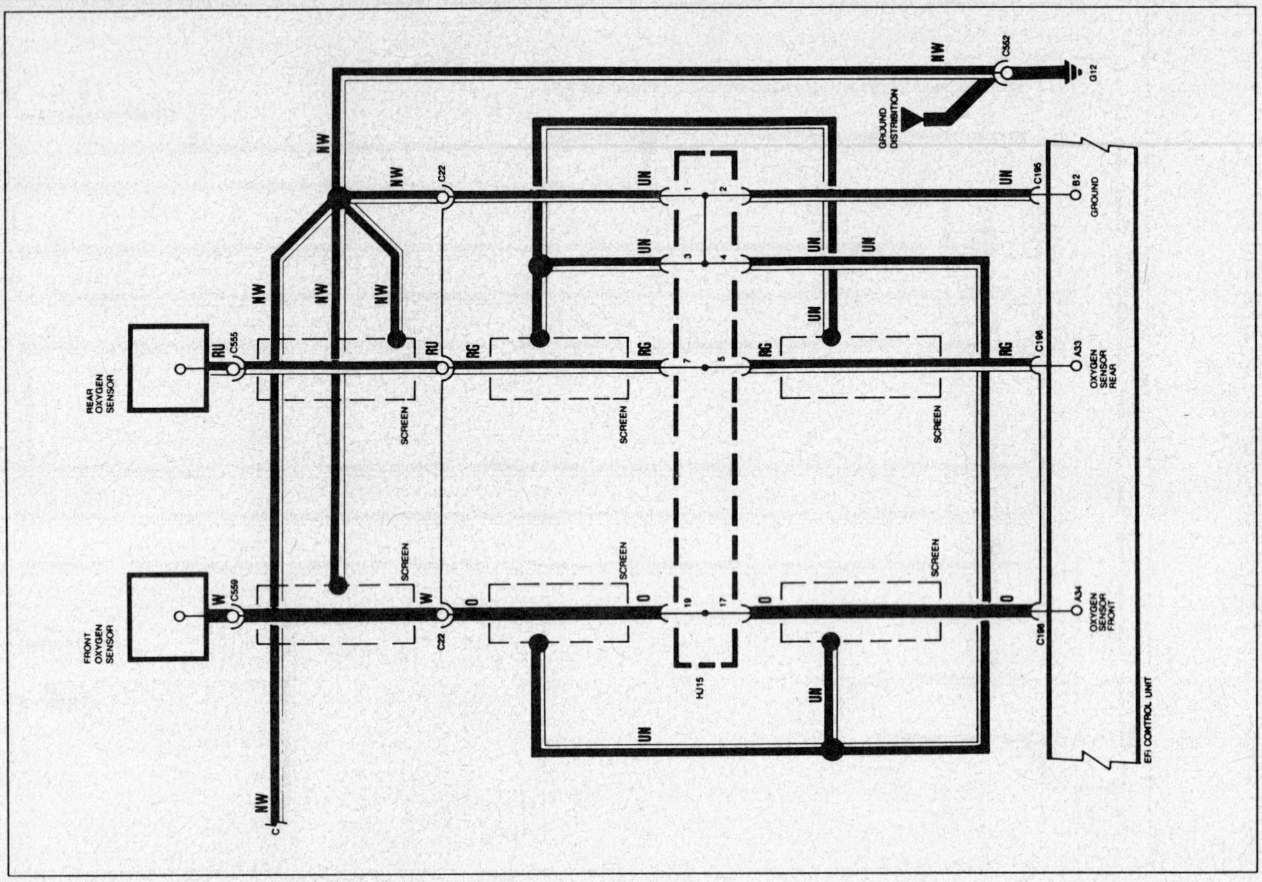

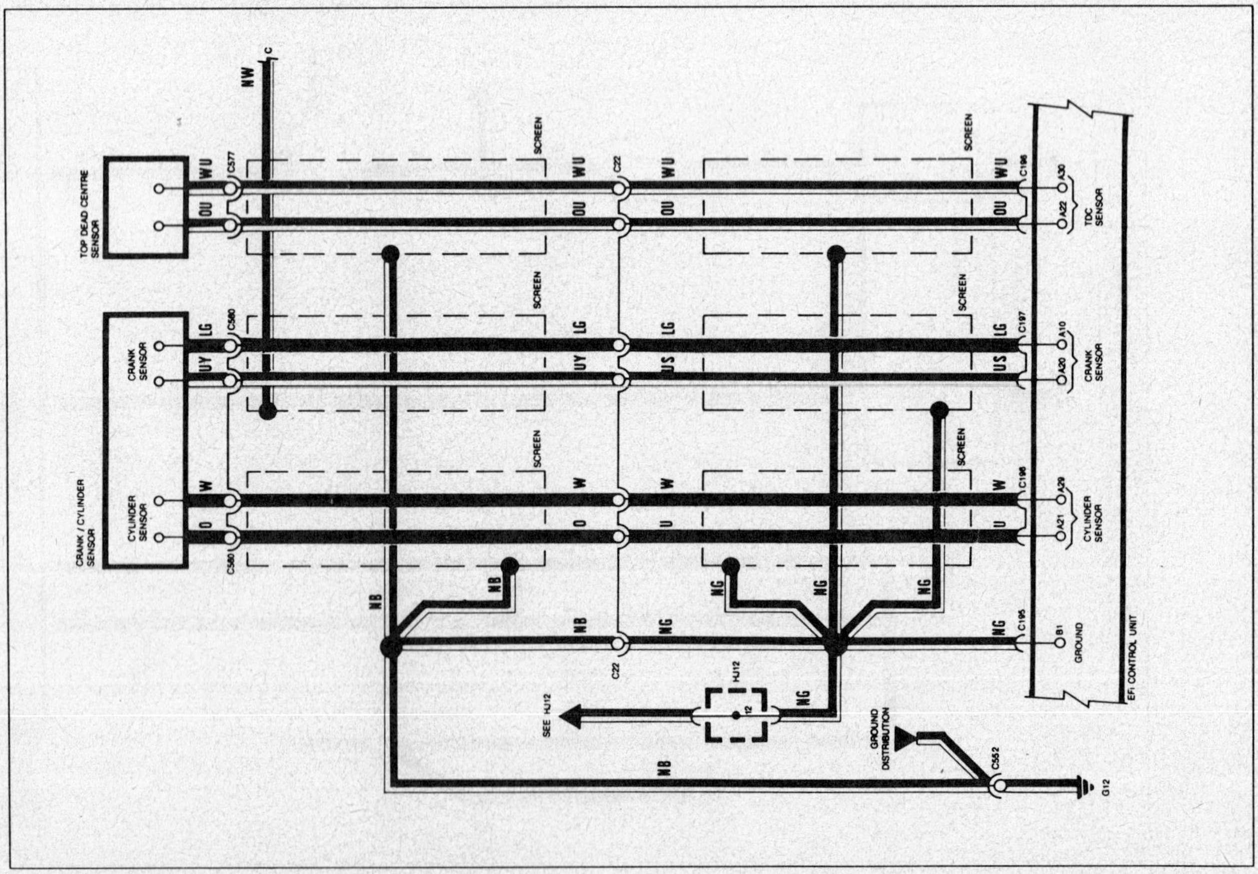

ENGINE MANAGEMENT WIRING SCHEMATIC, AUTOMATIC TRANSMISSION—STERLING

ENGINE MANAGEMENT WIRING SCHEMATIC, AUTOMATIC TRANSMISSION—STERLING

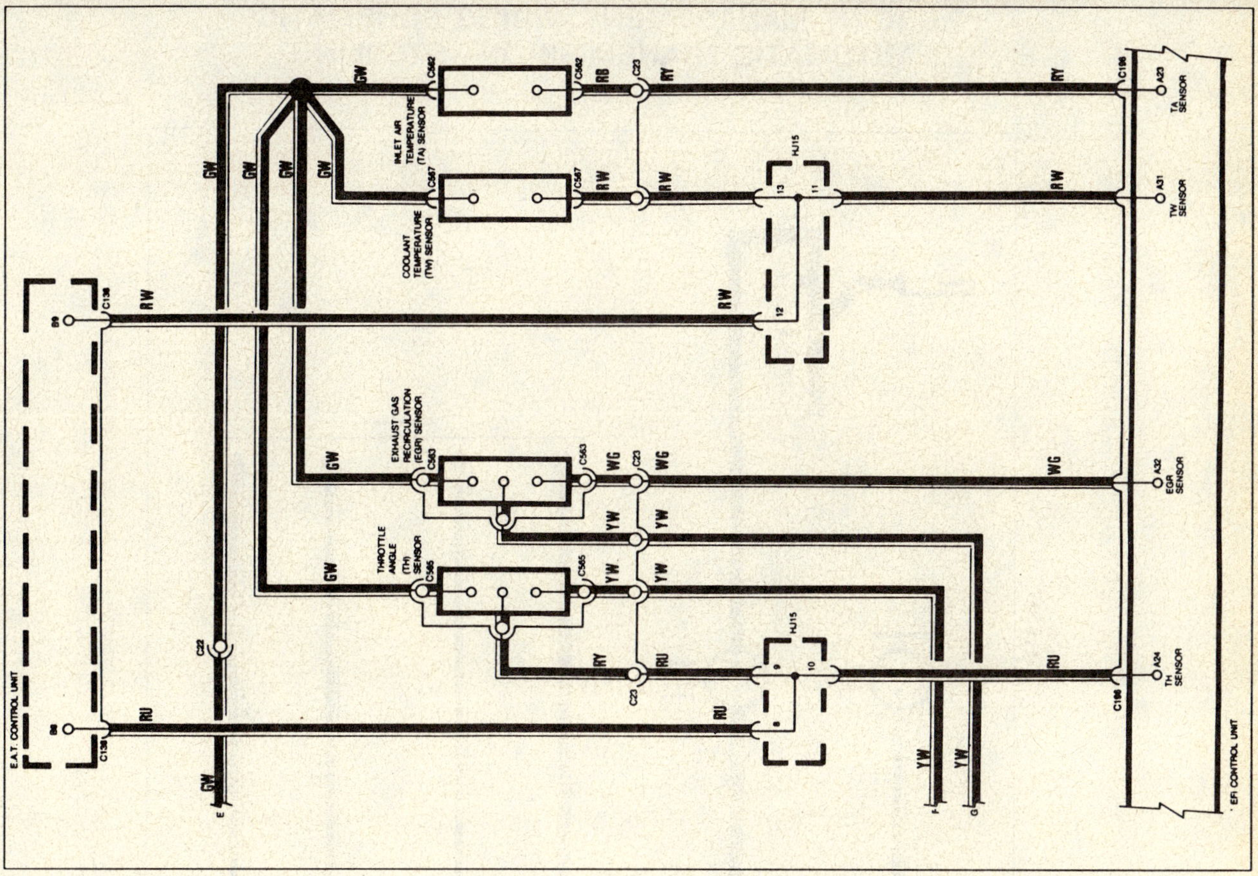

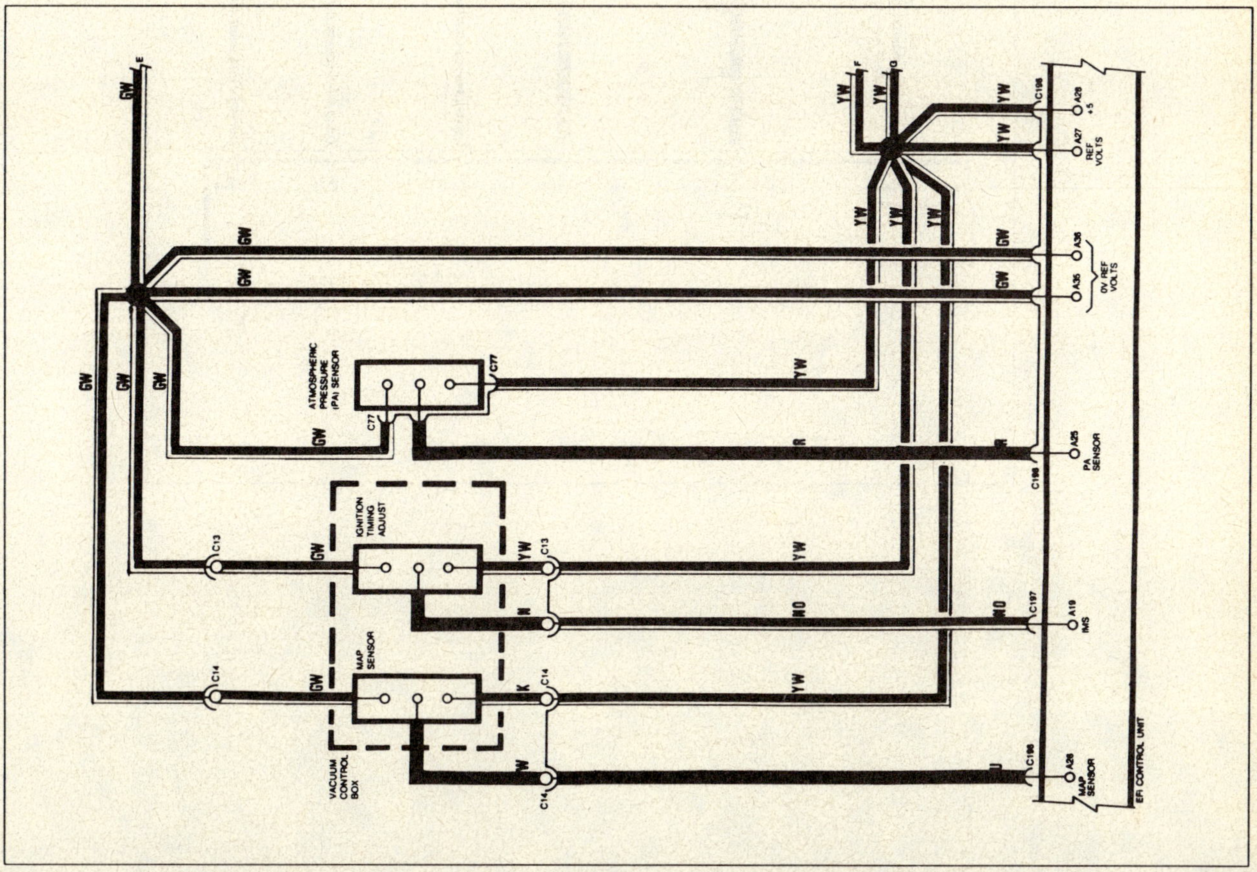

ENGINE MANAGEMENT WIRING SCHEMATIC, AUTOMATIC TRANSMISSION—STERLING

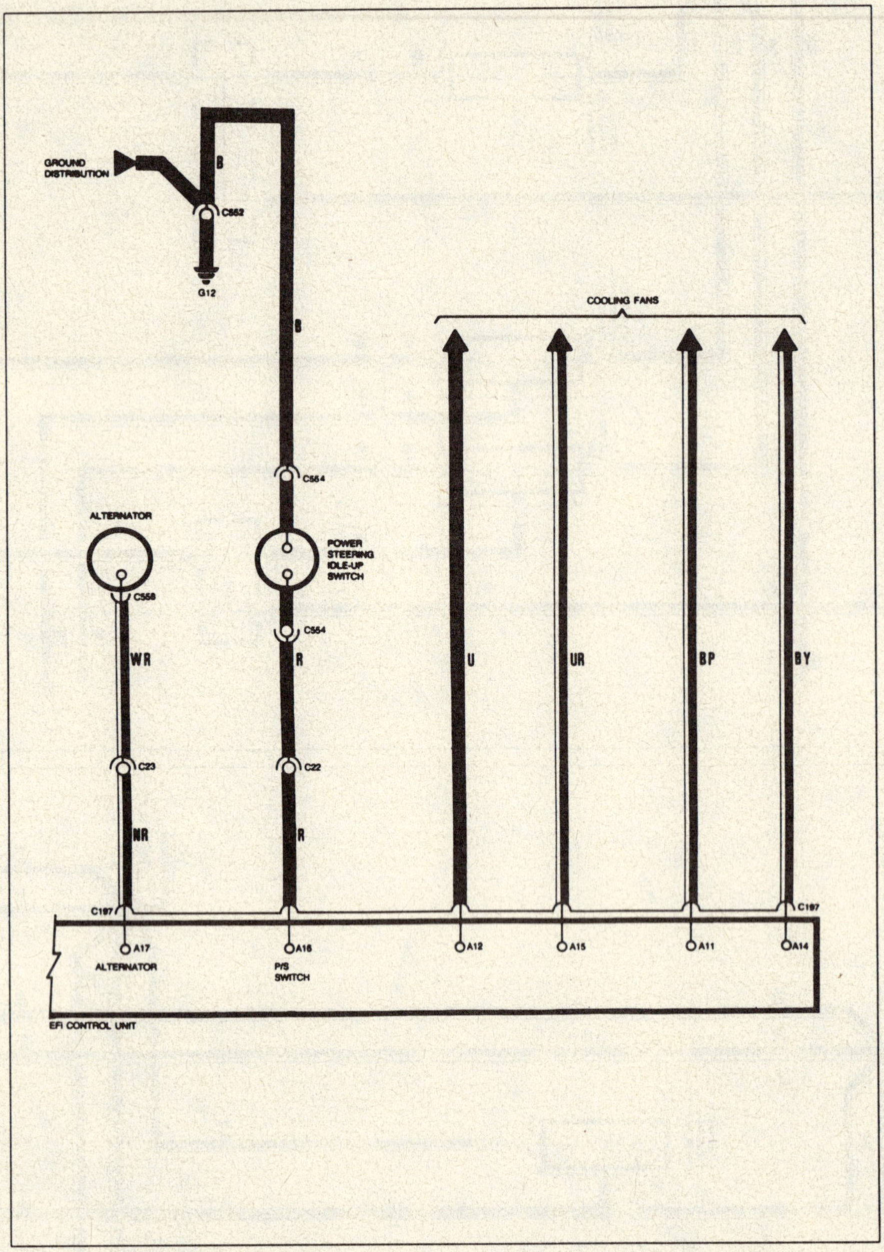

Volkswagen Fuel Injection

15 SECTION

CONTINUOUS INJECTION SYSTEM (CIS-E)

DIGIFANT AND MOTRONIC FUEL INJECTION SYSTEM

VOLKSWAGEN CIS-E FUEL INJECTION SYSTEM

General Information

The CIS-E Motronic system is the latest development of the electronically controlled mechanical Continuous Injection System (CIS). This system uses injectors, fuel pump and air flow sensor that are similar to those on earlier systems. The fuel distributor is equipped with an electronically controlled differential pressure regulator. This is operated by the ECU to control the fuel pressure in the lower chamber of the fuel distributor, which controls air/fuel mixture.

The ECU is now equipped with an adaptive learning program which allows it to learn and remember the normal operating range of the mixture control output signal. This gives the system the capability to compensate for changes in altitude, slight vacuum leaks or other changes due to things such as engine ware. Cold engine drivability and emissions are improved. The new ECU also is capable of cold start enrichment without the use of a thermo-time switch. The Fox still uses the thermo-time switch on CIS equipped vehicles.

The fuel injector pressure has been increased for better fuel atomization and residual pressure. The threads on the new injectors are different so they cannot be interchanged with older units. Some of the other components used on the CIS-E system are similar to those used on the fully electronic engine management systems. Some of the testing procedures are the same but the parts are not necessarily interchangable.

ELECTRONIC CONTROL UNIT (ECU)

This unit is located above the firewall in the engine compartment or to the right of the glove compartment. The ECU is equipped with an internal voltage correction circuit (operating

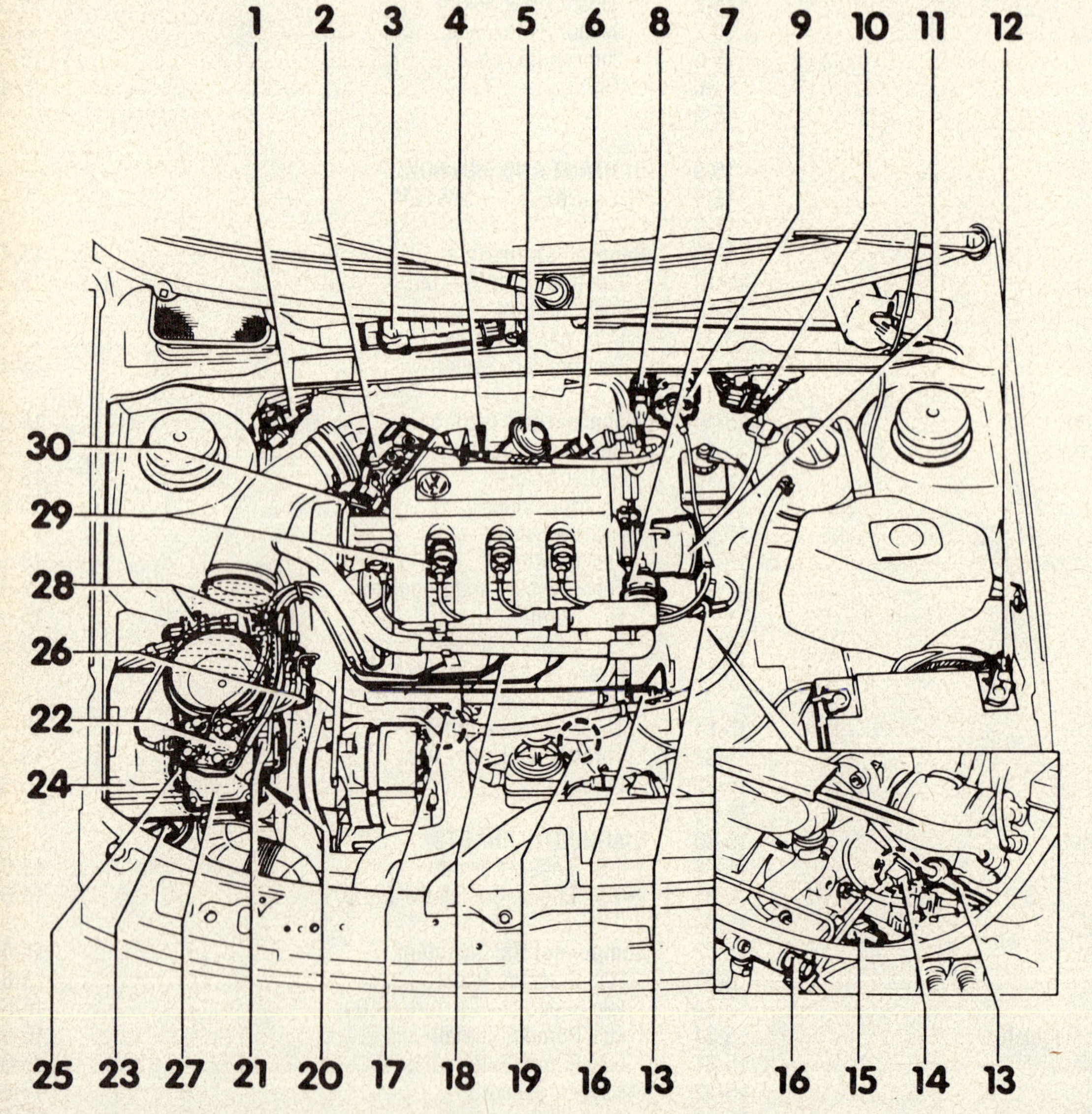

1. Oxygen sensor harness connector on right engine mount
2. Throttle body
3. Control module
4. Intake air temperature sensor—California only
5. EGR valve—California only
6. Exhaust tap
7. Idle stabilizer valve
8. Ignition coil power output stage
9. Ignition coil
10. 6-pin wiring harness connector
11. Distributor
12. EGR vacuum amplifier—California only
13. Ignition timing sensor or plug wire 4
14. EGR vacuum valve—California only
15. Coolant temperature sensor
16. Cold start valve
17. Knock sensor I
18. Fuel injector
19. Knock sensor II
20. Heated air intake control door
21. Differential pressure regulator
22. Fuel distributor
23. Charcoal canister below air cleaner
24. Air filter
25. Potentiometer
26. Fuel pressure regulator
27. Air flow sensor
28. Charcoal canister solenoid valves
29. Spark plug
30. Throttle switch harness connectors

Engine compartment layout—Volkwagon GTI 16V and Passat

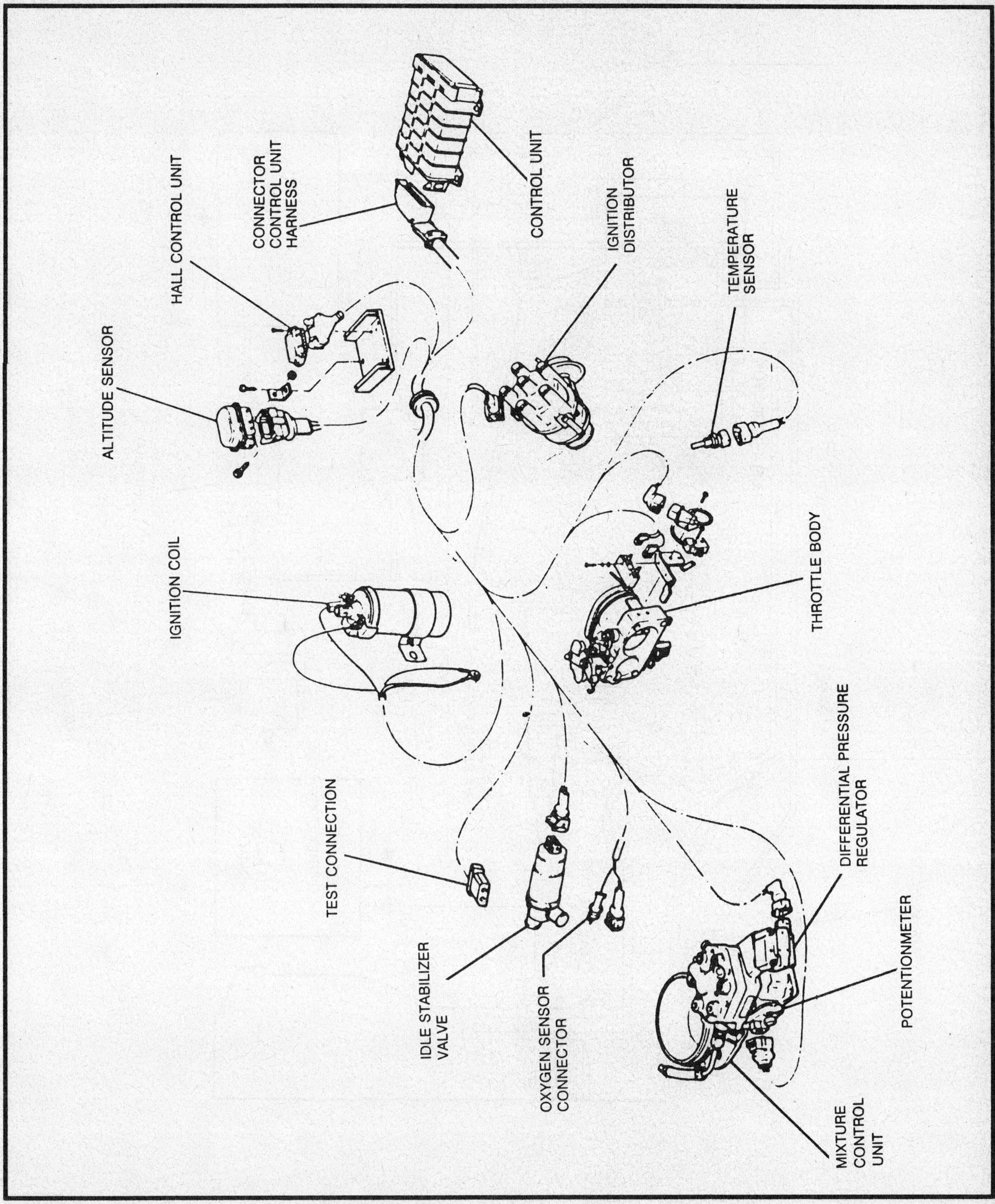

CIS-E fuel system components—Volkswagen CIS-E

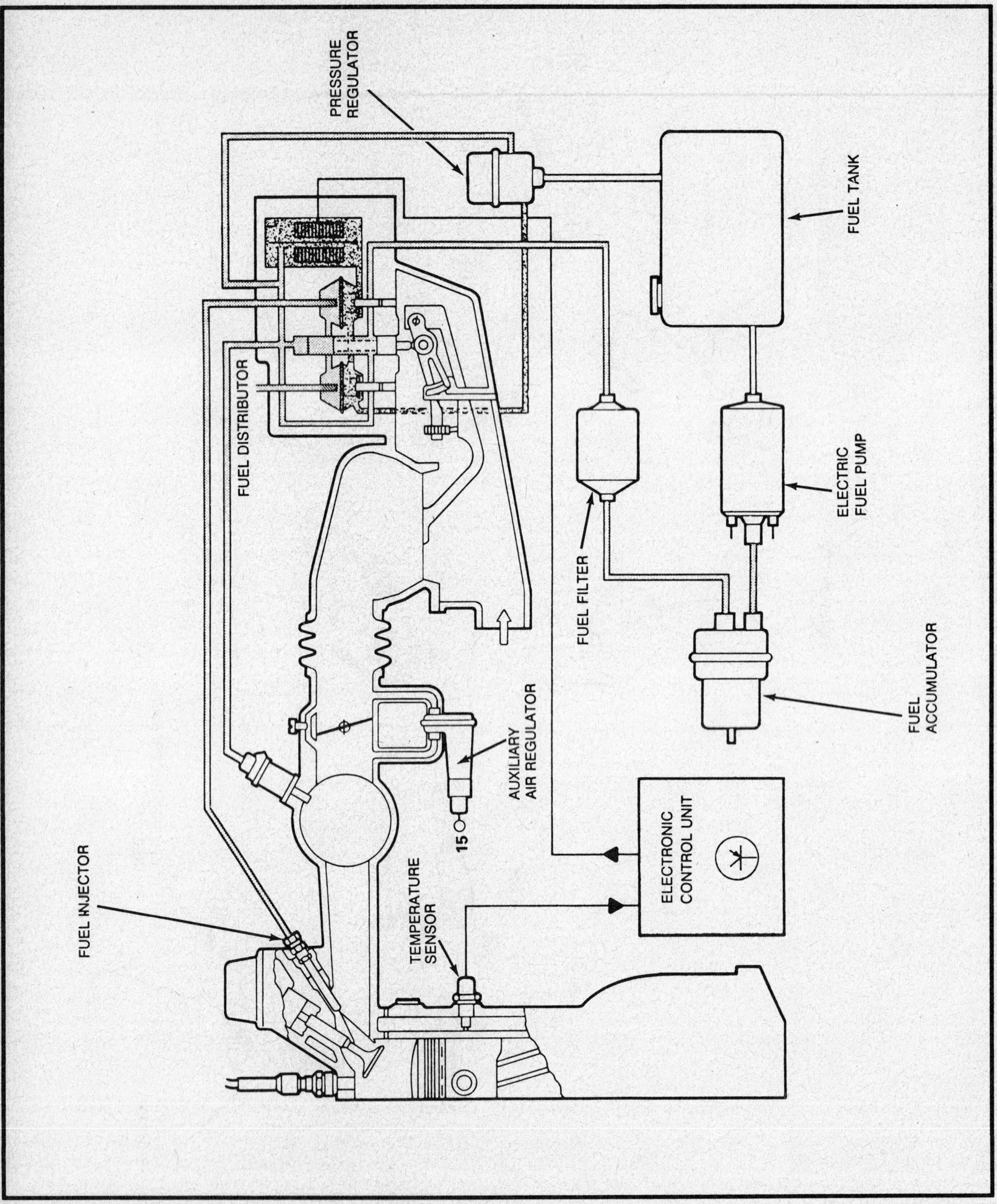

CIS-E fuel control system—Volkswagen CIS-E

voltage – 8 volts) to prevent voltage fluctuations when different vehicle components are switched on. The ECU accepts the various input signals and calculates optimum fuel injection and ignition output control signals. The main output devices are the differential pressure regulator in the fuel distributor, ignition coil and the idle air stabilizer valve. The maximum engine speed is limited by the ECU through the differential pressure regulator. The pressure in the lower chamber is increased relative to the system pressure and the fuel supply to the injectors is interrupted.

The ECU is equipped with an adaptive learning program that learns the normal operating point of the air/fuel mixture. This information comes from the oxygen sensor and is only read when the charcoal canister frequency valve is in the **OFF** cycle. As the engine ages or is driven differently or changes altitude, what the ECU considers "normal" is modified to account for deviations.

DATA SENSORS

Air Flow Sensor with Fuel Distributor

This unit is a venturi with a plate in the narrowest part of the bore. As air flow increases, the plate is pushed farther up in the bore. This lifts a piston that uncovers slots in the fuel distributor to allow more fuel to flow to the injectors. In the Motronic system, the position of the plate is measured with a potentiometer. A voltage is supplied to the position sensor and the portion of the signal returned indicates the height of the sensor plate in the bore.

The fuel distributor used in the Motronic system is very different from that used in earlier CIS systems. The unit is made of aluminum and is slightly smaller. The pressure relief valve and the frequency valve are no longer needed. Control plunger movement controls the amount of fuel supplied to the upper chamber. A differential pressure regulator mounted on the side controls the pressure in the lower chamber which ultimately controls the fuel flow to the injectors. When the engine is not running, the control plunger rests on an O-ring and there is some free-play between the plunger and air flow sensor arm. As

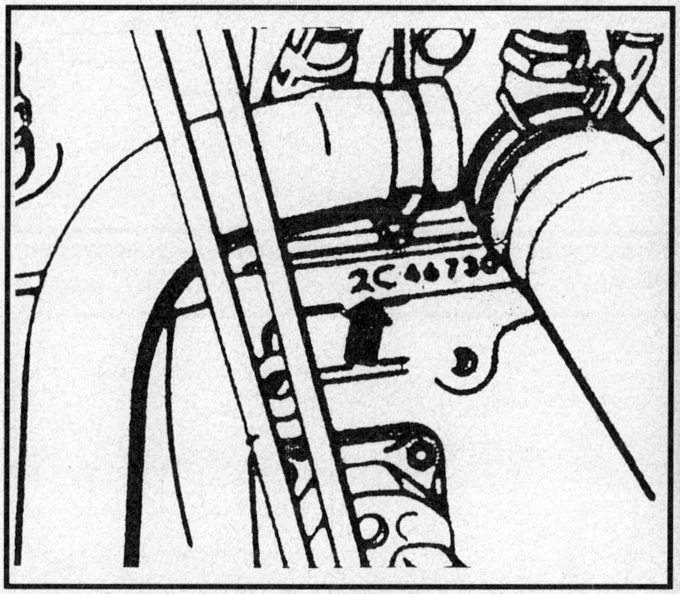

Engine identification is not provide by the VIN number, it is stamped in the cylinder block

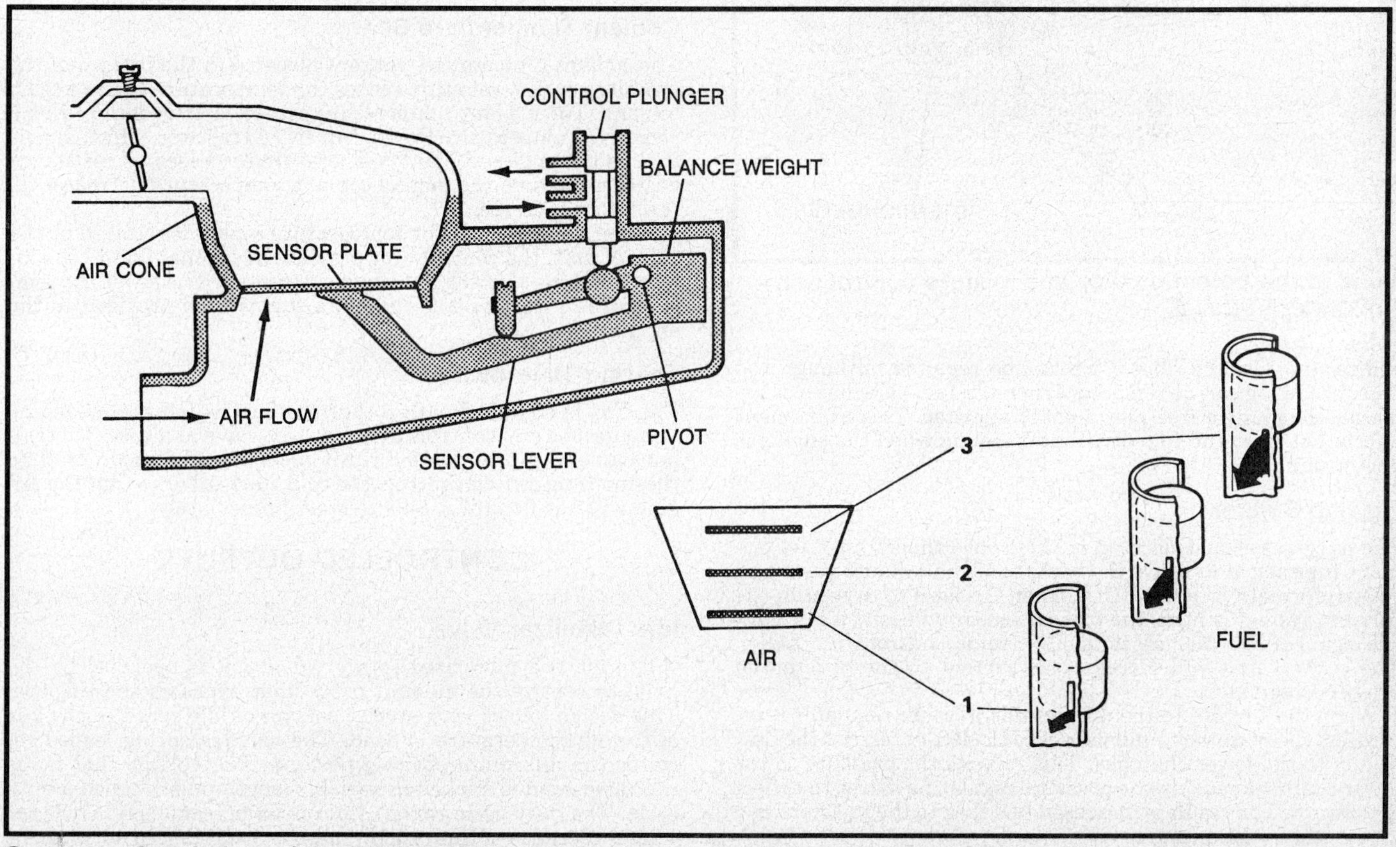

Cross sectional view of the air flow sensor and control plunger metering positions – Volkswagen CIS- E

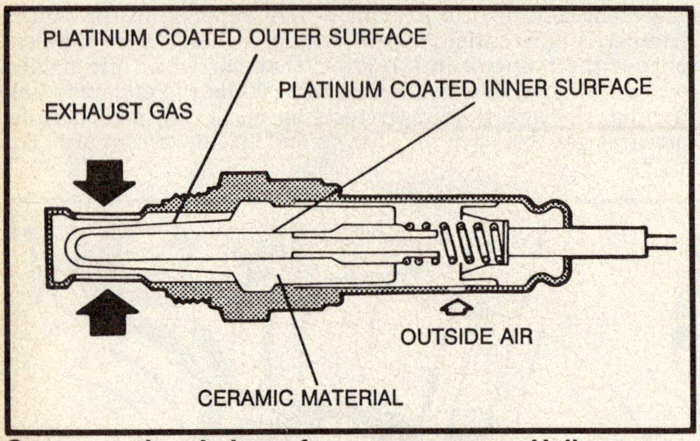

Cross sectional view of oxygen sensor—Volkswagen CIS-E

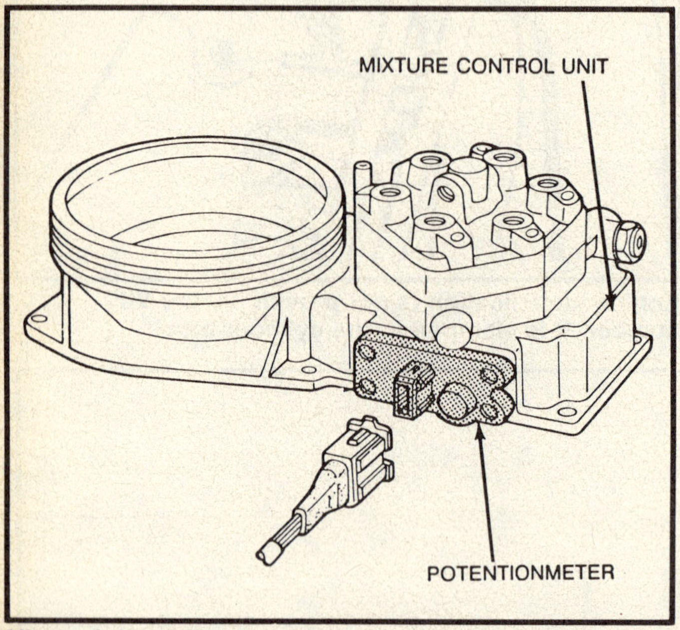

View of the potentiometer and mixture control unit—Volkswagen CIS-E

before, the fuel distributor cannot be repaired, although the plunger O-ring can be replaced. When the fuel distributor is replaced, the plunger free-play must be adjusted. This adjustment affects hot start and fuel distributor sealing when the engine is not running.

Oxygen Sensor

The oxygen sensor is located in the front exhaust pipe. Its primary function is to sense the content of the exhaust gases and relay information to the ECU. When the level of oxygen in the exhaust system is high, the oxygen sensor will send a low voltage signal of 100–500 mV to the Electronic Control Unit (ECU). The control unit will increase the current to the differential pressure regulator.

When the current to the differential pressure regulator is increased, the plate valve will deflect to the left to restrict the flow of fuel to the lower chamber. This reduces the pressure to the lower chamber causing the pressure regulating valves to deflect downward. The result is increased fuel flow to the injectors and an enriched fuel mixture.

If the amount of oxygen in the exhaust system is low, indicating a rich mixture, the sensor voltage will be high. This signal is approximately 500–900 mV, is sent to the ECU. The electronic control unit then reduces the current signal to the differential pressure regulator.

When the current to the differential pressure regulator is reduced, the plate valve will move to the right and allow more fuel into the lower chamber. The pressure in the lower chamber will then increase and defect the pressure regulating valve upward. The quantity of the fuel delivered to the injectors is then reduced and the fuel mixture is leaned.

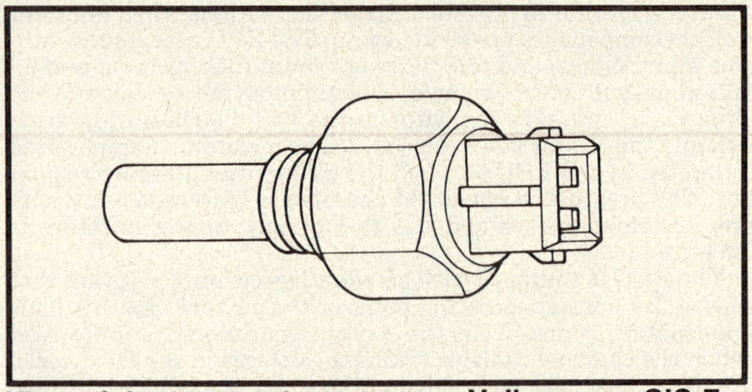

View of the temperature sensor—Volkswagen CIS-E

Potentiometer

The potentiometer is a variable resistor that sends a voltage signal to the control unit based on the sensor plate position. It consist of a wiper brush working on a carbon film. The voltage signal supplied by it is primarily for cold acceleration enrichment. It is located on the side of the air sensor housing.

Coolant Temperature Sensor

The coolant temperature sensor is located in the bottom of the cylinder coolant outlet. It senses the temperature of the engine coolant. The coolant temperature sensor is considered to be a Negative Temperature Coefficient (NTC) resistor, which means that as the engine coolant temperature increases its resistance decreases and as the engine coolant temperature decreases its resistance increases.

When a signal from the temperature sensor is received by the control unit, the control unit provides additional fuel to the injectors during starting and engine warm-up. The lower the temperature of the coolant, the more fuel will be provided to the engine.

Thermo-Time Switch

The Fox is equipped with a thermo-time switch senses engine temperature and controls the cold start valve at a specified temperature and on/off time limit measured in seconds. The thermo-time switch controls the cold start valve so that the engine will not flood if it fails to start immediately.

CONTROLLED OUTPUTS

Idle Stabilizer Valve

This unit is a motorized rotary valve that is operated by the ECU to control the amount of air that bypasses the throttle. This design allows very precise control of idle speed regardless of engine temperature or load. The valve is spring loaded towards the minimum opening position. The voltage that holds the valve open is applied in variable length pulses called a duty cycle. The duty cycle ranges from 5–95 percent "on" time per second. Because of this design, there is no idle speed adjustment on engines that use this system.

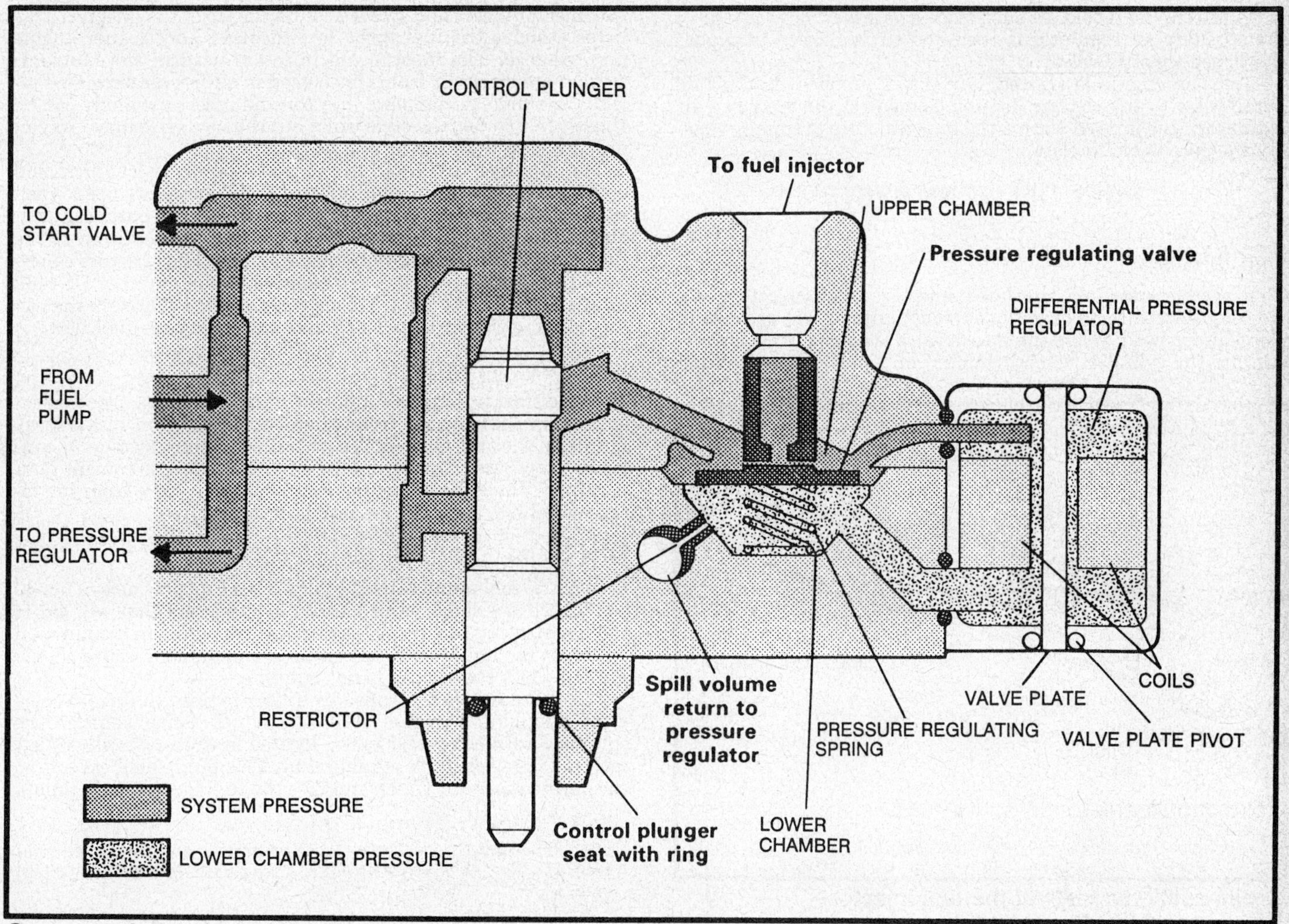

Cross sectional view of the fuel distributor — Volkswagen CIS-E

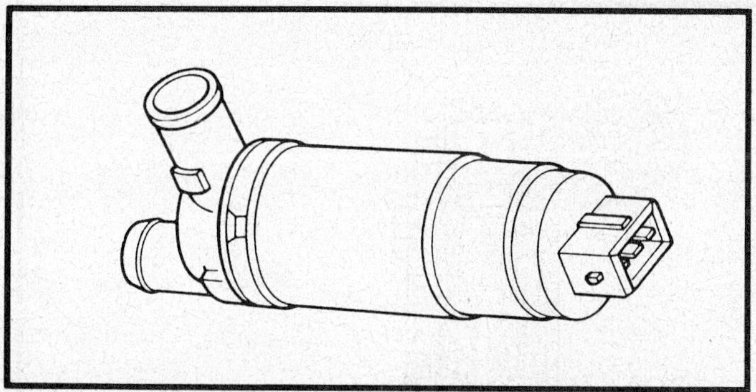

View of the idle stabilization valve — Volkswagen CIS-E

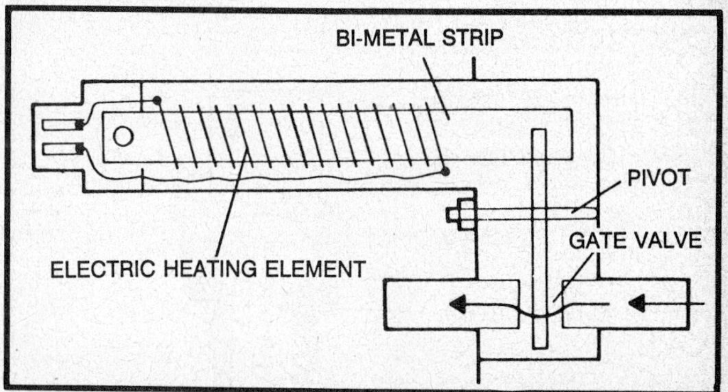

Cross sectional view of the auxiliary air regulator — Volkswagen CIS-E

Differential Pressure Regulator (DPR)

The ECU controls the air/fuel ratio by adjusting the pressure in the lower chambers of the fuel distributor. A differential pressure regulator is attached to side of the fuel distributor. It consists of a plate with an electromagnet on either side. The ECU controls the current to the magnets, moving the plate side to side, which controls the size of the opening to the lower chamber to control pressure. When the lower chamber pressure is high, the diaphragms are pushed up towards the outlets of the upper chamber, making the outlets smaller and reducing fuel flow. This provides for fine adjustment of fuel flow to the injectors. Pressure in the lower chambers ranges from 0–24 psi (0–1.6 BAR) less than system pressure.

Auxiliary Air Regulator

On the Fox, during cold engine operation, the auxiliary air regulator acts like a fast idle cam to slightly increase idle speed while the engine is warming up. The auxiliary air regulator provides additional air for idling during warm up. This is accomplished when the engine is cold. At such time, a gate way valve in side of the auxiliary air regulator is open and air is allowed to bypass the closed throttle valve.

After the engine is started, current is supplied by the fuel pump relay to the electric heating element of the auxiliary air regulator; this in turn warms the bi-metal strip gradually causing the gate valve to close.

FUEL DELIVERY SYSTEM

Fuel Injectors

This system uses air shrouded injectors. The injectors are mounted into inserts that are threaded into an air passage in the cylinder head. When the intake valve is open and air is flowing into the cylinder, a small amount of air also flows through the injector insert and swirls around the tip of the injector. This helps atomize the fuel spray for smoother and cleaner engine performance. The inserts and injectors used in the Motronic system are different from those used in earlier air shrouded injector systems. A complete injector and insert assembly can be substituted for earlier style units but the new style injectors can only be used with the new style insert.

The injectors in all CIS systems are purely mechanical and open at a pre-determined pressure. All injectors are open when the engine is running, injecting fuel to the intake ports continuously. The spray pattern is critical to good atomization of the fuel, which will affect power and emissions. The injectors can be tested for spray pattern and leakage with tool number US4480 or equivalent. This is a set of clear, graduated cylinders that allow viewing the spray pattern and measuring the fuel flow.

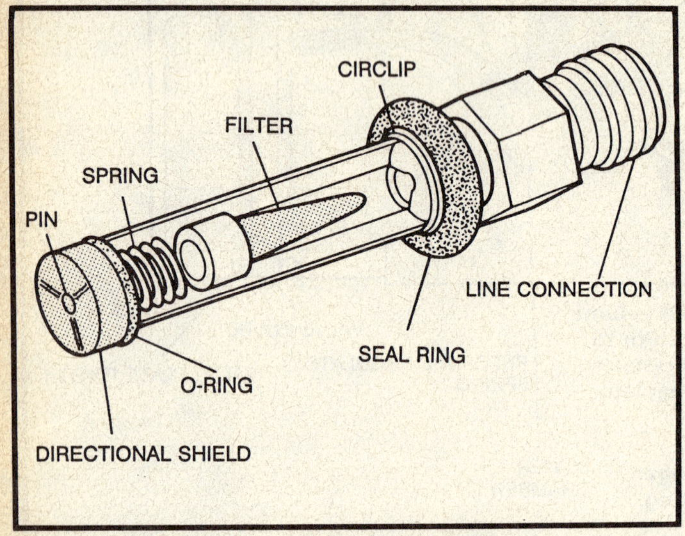

Cross sectional view of the fuel injector—Volkswagen CIS-E

Cold Start Injector

The cold start injector is supplied with system pressure whenever the pump is running. It is a simple solenoid valve with a spray nozzle on the end. On vehicles with the Motronic system, the injector is operated directly by the ECU according to coolant temperature. The injector can be tested by removing it from the intake manifold, however an assistant is required.

Fuel Pump

An electric fuel pump is used to provide a fuel pressure of 60–80 psi. There is a check valve located in the pump that will aid in the hot starting process. This check valve works in conjunction with the accumulator to maintain a constant pressure in the system when the engine is not running.

The fuel pump is controlled by a relay to prevent it from operating should the engine stall or cut off. Some vehicles are equipped with a transfer pump located in the fuel tank and attached to the fuel tank sending unit. This pump delivers fuel to a separate reservoir which supplies the fuel for the main pump.

NOTE: The fuel pump is solely depended upon the fuel flowing through it for cooling and lubrication. The fuel pump is subject to damage if it is ever allowed to run dry.

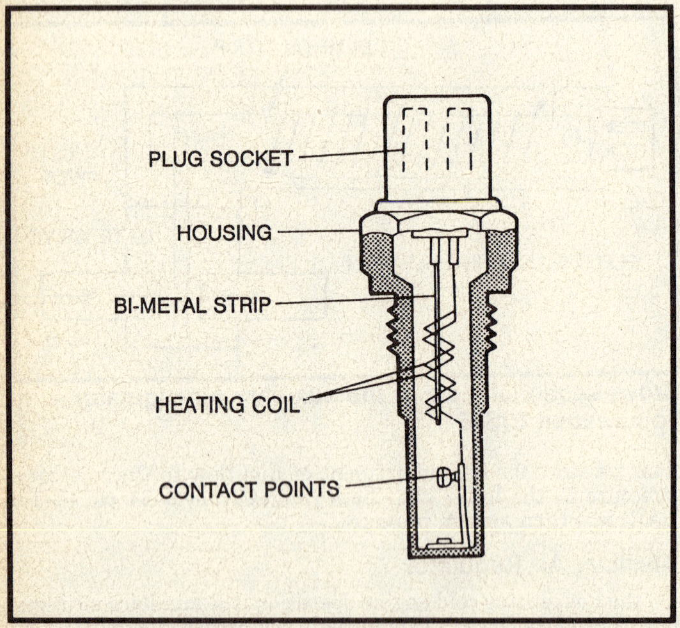

Cross sectional view of the thermo-time switch—Volkswagen CIS-E

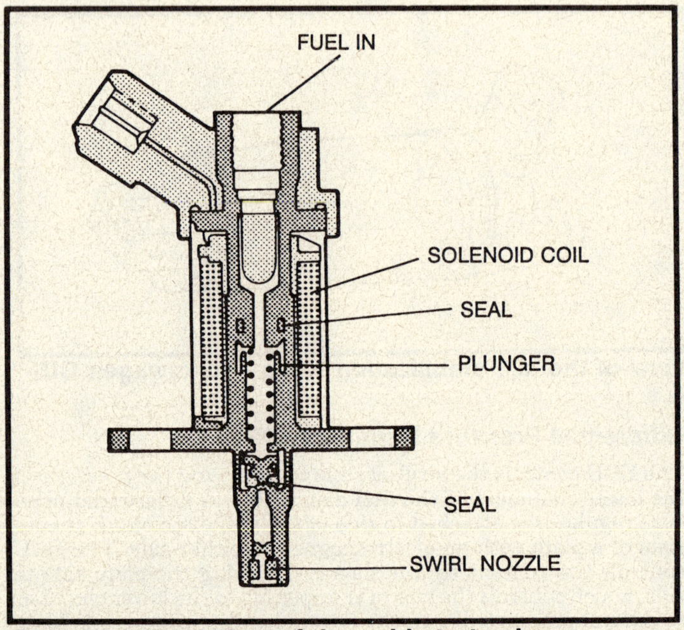

Cross sectional view of the cold start valve—Volkswagen CIS-E

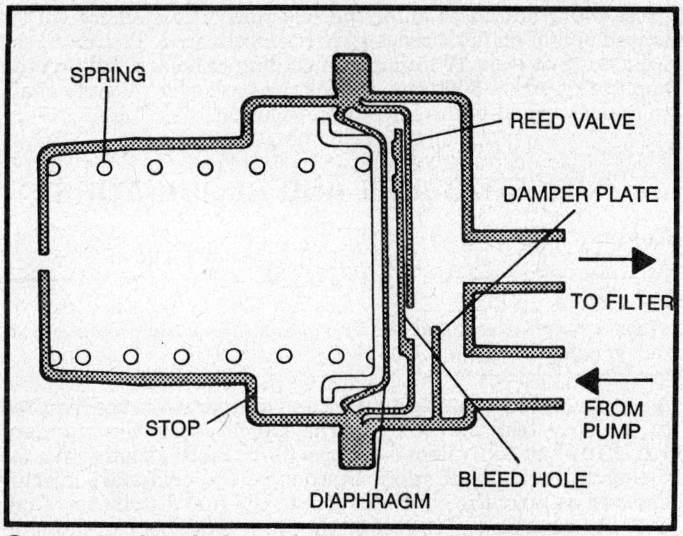

Cross sectional view of the fuel accumulator – Volkswagen CIS-E

Accumulator

The accumulator is located between the fuel pump and the fuel filter. When the fuel pump is running, a diaphragm inside of the accumulator moves to the stop against spring pressure. When the fuel pump is off, the fuel is stored under pressure in the accumulator thereby maintaining residual fuel pressure. The residual pressure is maintained by the spring slowly pushing the diaphragm to the right, forcing a small amount of fuel through an orifice in the reed valve. This in turn helps to maintain residual fuel pressure and also avoiding vapor lock.

The accumulator reduces fuel pump noise as well as protects the metal diaphragm in the fuel distributor from rapid pressure build up during pump operation.

Pressure Regulator

The control pressure regulator and pressure relief valve are not used in the CIS-E system. A diaphragm type pressure regulator is used to maintain system pressure at approximately 78.3 psi. (5.4 bar).

When the engine is started, fuel enters from the main return port of the fuel distributor and works against the regulating spring. The diaphragm assembly moves up along with the valve body which is moved by the closing pressure spring to open the sealing plate. The valve body then contacts its stop and the regulation of system pressure will begin by way of the regulating valve. Excess fuel along with the spill volume from the lower chamber in the fuel distributor is now returned back to the fuel tank by way of the opened sealing plate.

The pressure regulator also acts as a one way check valve to maintain pressure in the system once the engine is shut off.

Diagnosis and Testing

SERVICE PRECAUTIONS

To prevent personal injury or damage to the vehicle, follow these service precautions:

- Always disconnect the negative battery terminal to prevent sparks cause by short-circuiting, circuit-breaking, etc.
- Never disconnect the battery with the engine running.
- Smoking must not be allowed when working around flammable liquids.
- Always keep a CO_2 fire extinguisher close on hand.
- Absorbant must be available to soak up any spillage.
- Ensure fuel is emptied into a suitable container.
- Always place the transmission gear shift lever in **N** (manual transmission) or **P** (automatic transmission) and set the parking brake and block drive wheels before starting the engine.
- Keep hands, tools and clothing away from engine cooling fan to help prevent personal injury.
- Thoroughly clean all unions and area near connections before disconnecting.
- Always disconnect the battery when using arc, spot or electrical welding equipment.
- When cranking engine without starting, as for a compression test, disconnect wire from distributor, and connect to ground. Do not hand hold wire.

READING FAULT CODES

1. Make sure the ignition switch is **OFF** and that all fuses are good. Make sure all ground connections under the hood are good, especially those for the control unit.
2. Remove the shifter knob and boot and connect the diagnostic tester to the black and white connectors towards the left or front of the vehicle.
3. Turn the tester **ON** to make sure it is receiving power. The screen will display 2 menu options; Rapid Data Transfer and Blink Code Output. Turn the ignition switch **ON**.
4. Select Rapid Data Transfer and Address Word 01, then press the Q button to enter the selection. The tester will display a control unit part number, the system it controls and an application (country) code.

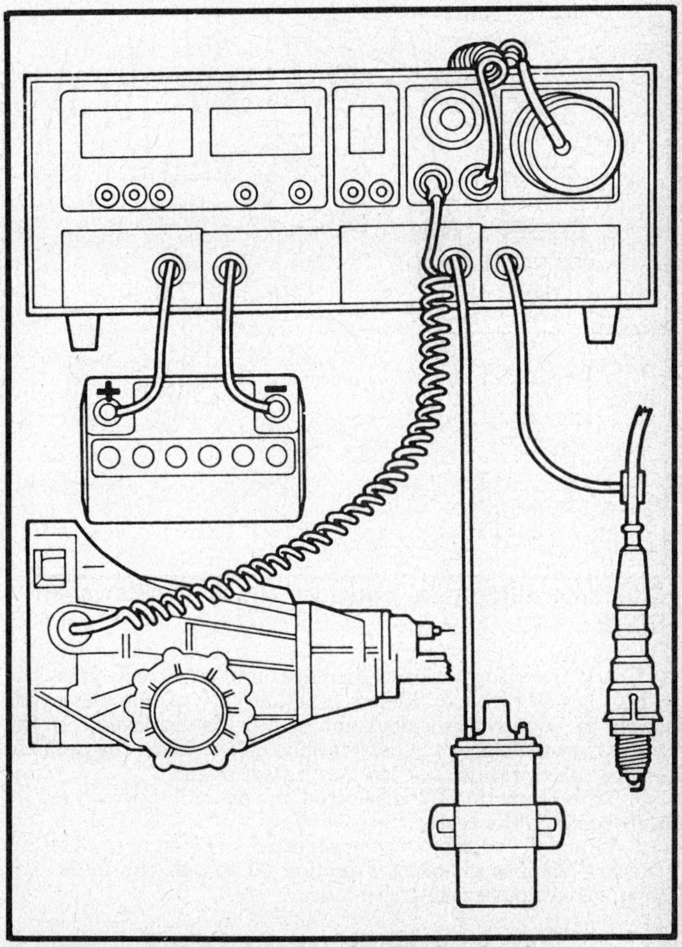

Connecting the multimeter for basic checks – Volkswagen CIS-E

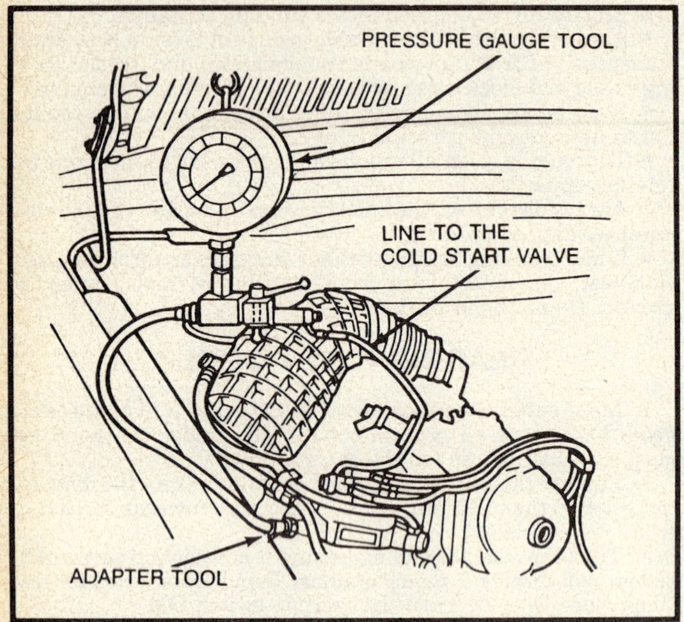

System pressure testing—Volkswagen CIS-E

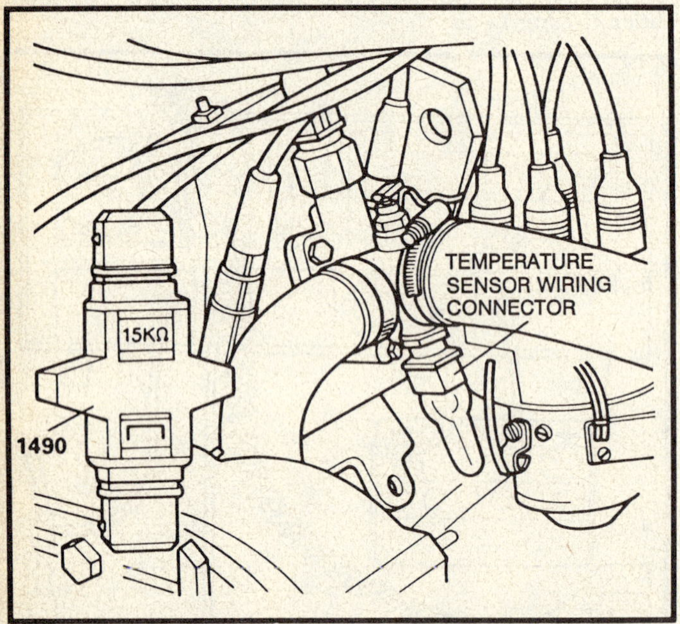

Checking differential pressure check—Volkswagen CIS-E

a. If the information is displayed and is correct, press the Run key to proceed: "Select function XX" will be displayed.

b. If "Control unit does not answer" is displayed, use the Help key to display a list of possible causes. When the problem is repaired, return to Step 1 and start again.

5. When Function 02 is selected, the control unit will report fault codes to the tester.

6. When all codes have been reported, proceed to the Output Check Diagnosis or select Function 06 to exit the fault code memory without erasing the codes.

CLEARING FAULT CODES

After all fault codes have been retrieved, select Function 05 and press the Q button to enter the selection. The memory will be erased only if all fault codes have been retrieved. Test drive the vehicle for at least 10 minutes, including at least 1 full throttle application above 3000 rpm. Check the fault code memory again to make sure all faults have been repaired.

FUEL PRESSURE AND REGULATORS

Testing

CAUTION

The following procedure will produce fuel vapors. Make sure there is proper ventilation and take the appropriate fire safety precautions to prevent personal injury.

A special fuel pressure test gauge and jumper wires are required to properly test the system. The gauge set is the standard VW1318 or an equivalent 0–140 psi (0–10 BAR) gauge with a tee fitting and a shut-off valve. Disconnect the cold start injector line and connect the shut-off valve to the fuel distributor. Connect the other side of the tee to the lower chamber test point on the fuel distributor. This means that when the valve is open, the upper and lower chambers are connected by the test gauge lines. When the valve is closed, the gauge reads only lower chamber pressure.

Locate and remove the fuel pump relay on the main fuse panel and connect a jumper wire with a switch to terminals 30 and 87 on the panel. The fuel pump will run any time the switch is **ON**.

To test the differential pressure regulator, unplug the connector and connect the VAG 1315A/1 or equivalent adapter wire. Connect a multimeter that reads milliamps inline with either wire of the connector to read current draw of the differential pressure regulator.

1. Run all tests before repairing or replacing parts. Connect the pressure gauge set and jumper wires as described above.

2. With the ignition **OFF** and the valve open, operate the fuel pump to check system pressure; it should be 88–95 psi (6.1–6.6 BAR).

3. If the pressure is correct, go to Step 5. If the pressure is too high, disconnect the fuel return line after the system pressure regulator, place it in a container and run the pump again. If the pressure is correct, there is a blockage in the return line to the tank. If the pressure is still high, the pressure regulator is faulty.

4. If the system pressure is low, the system pressure regulator or the fuel pump may be faulty. To flow test the pump, disconnect the return line after pressure regulator and hold it in a measuring container. Make sure the battery is in good condition so the pump gets a full 12 volts. Run the pump for 30 seconds and time the run carefully. The flow should be at least 12.5 oz. (420 cc). If the pressure is low but the flow is higher than about 22.5 oz (760 cc), the system pressure regulator is faulty.

5. If system pressure is correct, close the valve to read pressure in the lower chamber only and run the pump. The pressure should be about 4.5–7.5 psi (0.3–0.5 BAR) below system pressure. If it is not, disconnect the small fuel line from the system pressure regulator and hold it in a container. Run the pump for 1 minute; the flow should be 3.8–4.4 oz. (130–150 cc). If the flow is low, the differential pressure regulator may be faulty. If the flow is correct, procede.

6. With the pump running, turn the ignition switch **ON**. The multimeter should read about 100 mA at the differential pressure regulator. The pressure in the lower chamber should stabilize at 17–21 psi (1.2–1.5 BAR). If either of these is not correct and the ECU and wiring are okay, the differential pressure regulator is faulty.

7. If the engine has hot start problems, the system residual pressure may be leaking off. Open the valve and run the pump with the ignition switch **OFF**. Stop the pump and note the pressure. After 10 minutes the pressure must not be lower than 48

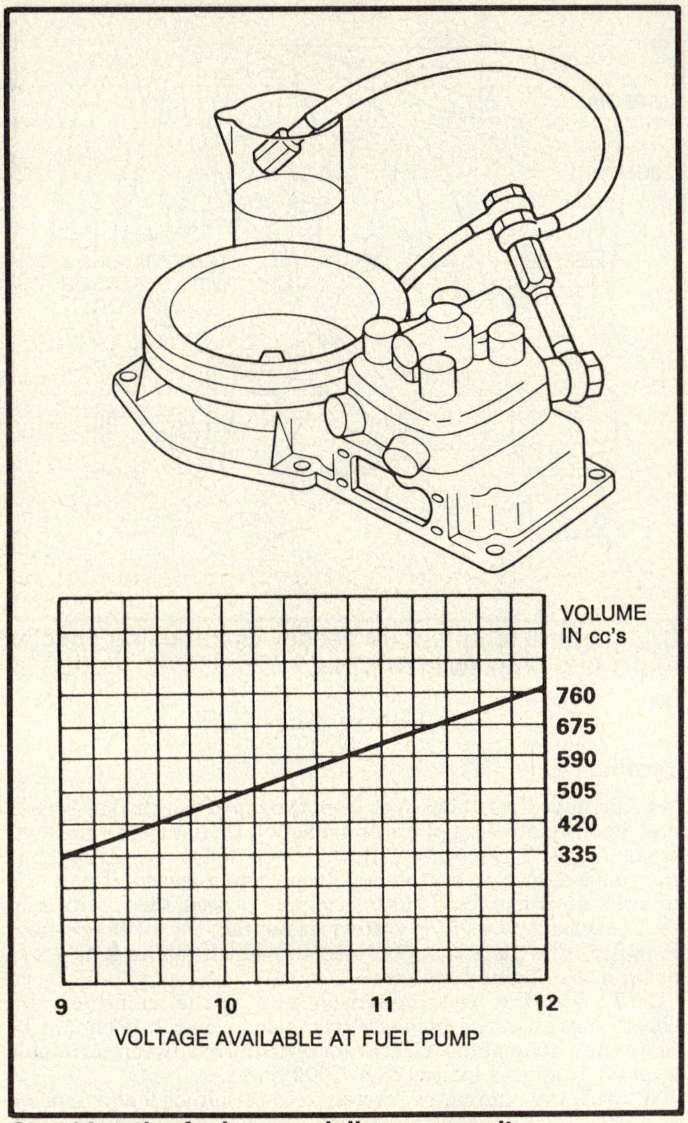

Checking the fuel pump delivery capacity—Volkswagen CIS-E

psi (3.3 BAR). After 20 minutes, the pressure must be at least 46.5 psi (3.2 BAR).

8. If residual pressure is low, look for leaks at the fuel distributor O-ring under the plunger. This will appear as fuel dripping onto the air filter. Other possibilities are the fuel pump check valve, a leaking injector, the fuel pressure regulator or improper free-play between the air flow sensor arm and the plunger in the fuel distributor. This is only likely if the fuel distributor has been replaced.

FUEL INJECTORS

Testing

CAUTION

The following procedure will produce fuel vapors. Make sure there is proper ventilation and take the appropriate fire safety precautions to prevent personal injury. Fuel will be sprayed from the injectors at high pressure. Do not point the injector at anything except the test tube.

1. Remove the injectors by prying them out of their rubber O-rings. If necessary, use a silicone spray to ease removal. Inspect the injectors for signs of leakage. They should not be wet.

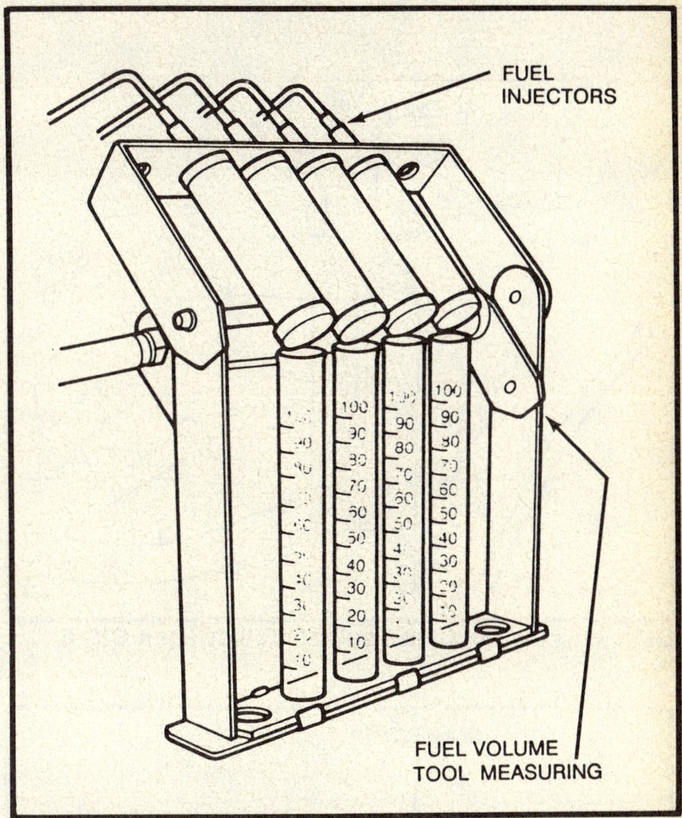

Checking the fuel injector flow capacity—Volkswagen CIS-E

2. Carefully route the fuel lines so the injectors can be inserted into the tubes on the front bumper.
3. Locate and remove the fuel pump relay on the main fuse panel and connect a jumper wire with a switch to terminals 30 and 87 on the panel. The fuel pump will run any time the switch is **ON**.
4. Remove the air flow sensor boot and insert a 0.020–0.040 in. (0.05–1.0mm) feeler gauge between the air flow sensor plate and the housing so that the plate is lifted.
5. Operate the fuel pump and observe the spray pattern. It should be a narrow conical pattern, even all the way around. Move the sensor plate up higher to see how the pattern changes. Any irregularity in the pattern indicates a partially clogged injector, raising the sensor plate all the way may clear it.
6. After one tube has about 20 cc, stop the fuel pump and observe the injectors. Note the quantity of fuel in each tube; they should all be within 2 cc of each other.
7. Repeat the test with the sensor plate all the way up to flow about 80 cc of fuel. The flows should all be with in 8 cc.
8. If an injector flows high or low, move it to another line and flow them again. If the incorrect flow follows the injector, the injector is faulty. If the incorrect flow stays with the same fuel line, the fuel distributor is faulty.
9. To leak test the injectors, remove the feeler gauge so the sensor plate is resting on its stop and run the fuel pump. There should be no fuel dripping from any injector. Any injector with a bad spray pattern, leak or incorrect flow quantity must be replaced.

COLD START INJECTOR

Testing

1. With the engine cold, remove the injector but do not disconnect the fuel line or wiring.

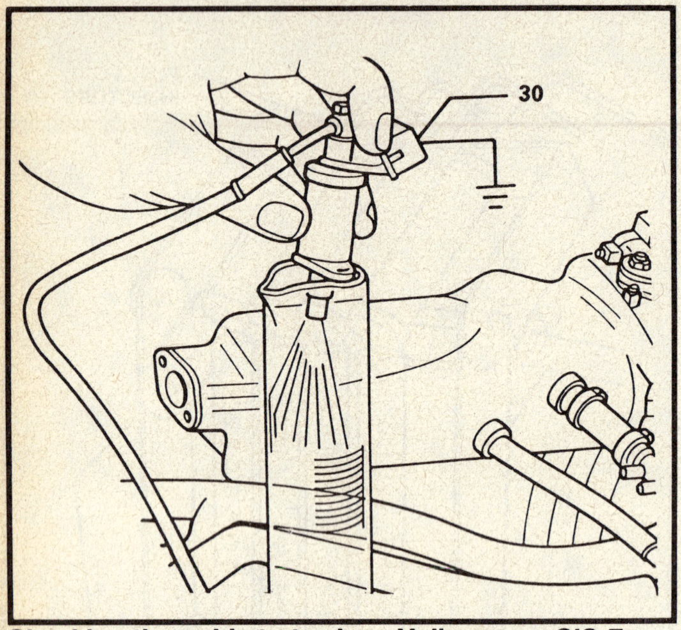

Checking the cold start valve—Volkswagen CIS-E

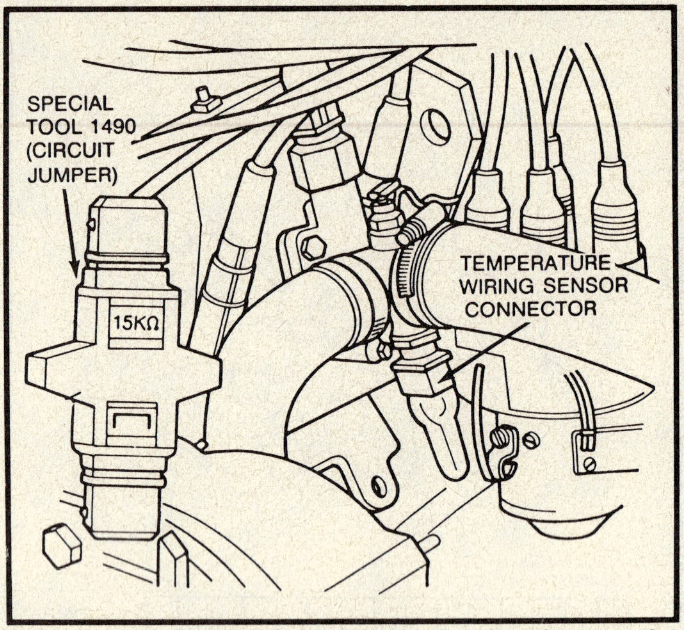

Bridging the temperature sensor circuit using special tool 1490—Volkswagen CIS-E

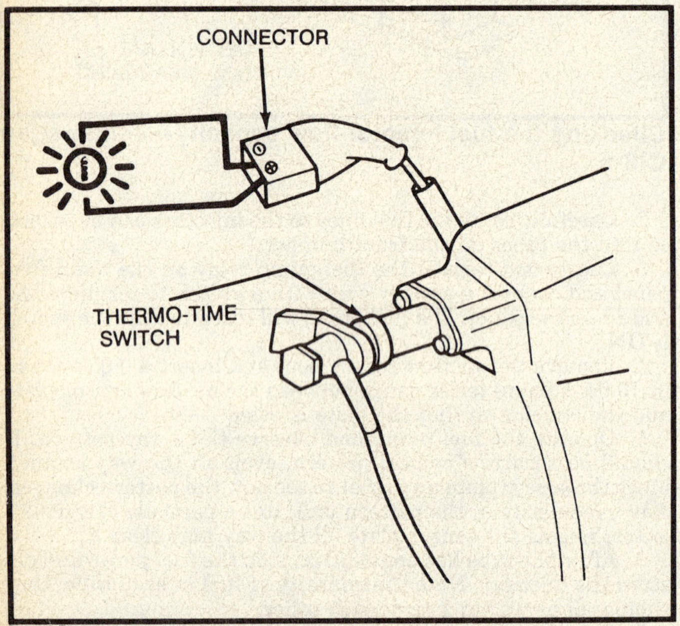

Checking the thermo-time switch circuit—Volkswagen CIS-E

2. Disconnect the power output stage wiring from the coil so the engine will not start.
3. If the engine is not cold, disconnect the blue temperature sensor connector. Connect another coolant temperature sensor or a 15,000 ohm resistor across the wiring terminals to simulate a cold engine.
4. Hold the injector over a container and have an assistant operate the starter for about 10 seconds. The injector should pulse fuel, making an even cone spray pattern, for about 8 seconds.
5. Release the ignition key and dry the end of the injector. The injector should not drip or become wet for at least 1 minute.

AIR FLOW SENSOR

Testing

1. Remove the rubber boot from the top of the air flow sensor and lift the plate with pliers or a magnet. Do not loosen the bolt. It should move smoothly with the same resistance throughout the range of motion and should drop when released. If not, it is uneven moving upward but falls when released, the fuel distributor may be faulty. If the plate does not fall, the air flow sensor is faulty. The plate rests on a spring which allows it to move down if the engine backfires.
2. To test the position sensor, unplug the connector and check the resistance between terminals 1 and 2; it should be more than 4000 ohms. Check the resistance between terminals 2 and 3; it should be less than 1000 ohms.
3. With the ohmmeter connected to terminals 2 and 3, move the plate up and down. The resistance should change smoothly to more than 4000 ohms at full travel.
4. Turn the ignition switch **ON**. Connect a voltmeter between terminals 1 and 3 of the wiring connector. There should be 4.5–5.5 volts coming from the ECU.

OXYGEN SENSOR AND CONTROL

Testing

FOX

1. Using special tool 1490, bridge the coolant temperature sensor wiring connector to the connector side opposite marked 15 ohm. The 15 ohm side remains disconnected.
2. Connect VW 1315A/1 or equivalent, inline with the differential pressure regulator. Connect a multimeter set on the mA scale to the adapter. Disconnect the oxygen sensor.
3. Turn the ignition switch **ON**. The differential pressure regulator current should be 9–11 mA.
4. Ground the male terminal (green lead) of the oxygen senor. The differential pressure regulator current should be 19–22 mA after 20 seconds.
5. If not within specifications, check the wiring from the control unit to the oxygen sensor.
6. If wiring check good, replace the control unit.
7.Connect the oxygen sensor. Run the engine to raise the oil

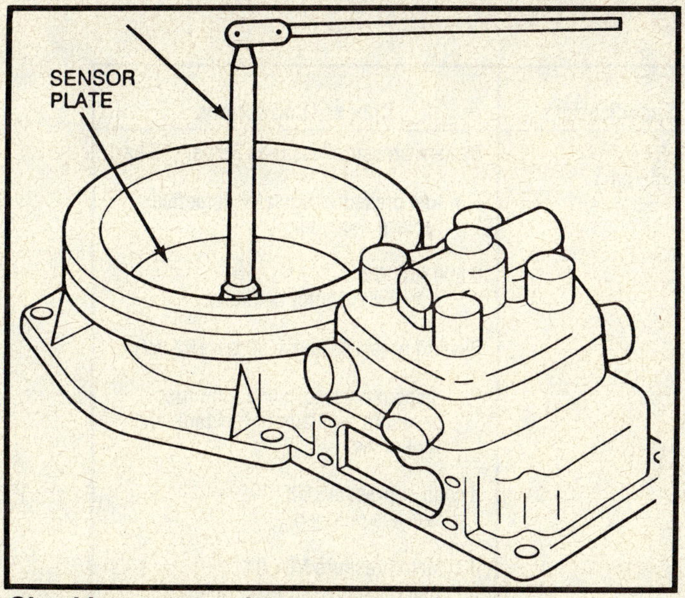

Checking sensor plate and control plunger movement—Volkswagen CIS-E

temperature to 176°F (80°C). The fan must run at least once. All accessories must not run during test.

8. Disconnect the idle boost connections and pull the breather hose off the valve cover and leave open to the air. Remove the clamp holding the hose to the T-connector with the restriction and pull off the line.

9. Idle the engine for 2 minutes to heat the oxygen sensor. Check the meter reading at idle. The reading should fluctuate. If it does not, run the engine at 3000 rpm and check again. If the read does fluctuate, the oxygen sensor is good. If the readings do not fluctuate, replace the oxygen sensor.

EXCEPT FOX

1. Connect VW 1315A/1 or equivalent, inline with the differential pressure regulator. Connect a multimeter set on the mA scale to the adapter.

3. Turn the ignition switch **ON**. The differential pressure regulator current should be 90–110 mA.

4. Run the engine to raise the oil temperature to 176°F (80°C). The fan must run at least once. All accessories must not run during test.

8. Clamp off the breather hose. The current must drop. If it does not, disconnect the oxygen sensor connector and jumper terminals 2 and 4 for 20 seconds. The connector is on the right side engine mount.

9. If the current does drop, replace the oxygen sensor. If the current does not drop, the ECU or related wiring may be defective and need replacement.

IDLE STABILIZER VALVE

Testing

1. With the valve installed, turn the ignition switch **ON** but do not start the engine. It should be possible to hear or feel the valve hum and/or vibrate.

2. Turn the ignition switch **OFF**. Unplug the connector from the idle stabilizer valve and connect an ohmmeter across the valve terminals. There should be 7–11 ohms resistance. Resistance may be higher on a warm engine.

3. Disable the ignition system so the engine will not start. Connect a voltmeter or an LED test lamp between terminal 1 of the connector and ground. When the starter is operated, there should be voltage at terminal 1.

4. Remove the valve and check for visual signs of scoring or binding on the rotating portion. Do not lubricate the valve. If it is receiving voltage but does not operate, it must be replaced.

THERMO-TIME SWITCH

Checking

NOTE: Perform the following test below 86°F (30°C) since the contact points on the thermo-time switch are designed to open at this temperature.

1. Remove the wiring connector from the cold start valve.
2. Ground the ignition coil terminal No. 4.
3. Using a test lamp, bridge the wiring connector terminals.
4. Turn the ignition switch to the **START** position.
5. Test lamp should light for approximately 1–8 seconds and then go out.

COLD START VALVE

Testing

1. Leaving the fuel line connected, remove the cold start valve from the intake manifold.
2. Remove the fuel pump relay. Using tool US 4480/3 or equivalent, energize the fuel pump.
3. Using a jumper wire, jumper 1 terminal of the cold start valve to battery positive post.
4. Hold the cold start valve in a suitable container.
5. Using a jumper wire, jumper the second terminal of the cold start valve to ground. The cold start valve should spray in a steady cone shaped pattern.
6. Disconnect the jumper wires. The cold start valve should not drip within 1 minute.

AUXILIARY AIR REGULATOR

Testing

Cold Engine

1. Make sure the engine coolant temperature is below 80°F (30°C).
2. Remove the auxiliary air regulator wiring connector.
3. Start the engine and run at idle speed.
4. Pinch either hose connected to the auxiliary air regulator. The idle speed should drop slightly.

Warm Engine

1. With the auxiliary air regulator wiring connector connected, start the engine and run it until it reaches operating temperature.
2. Pinch either hose connected to the auxiliary air regulator. The idle speed should not change.

Voltage Supply (Engine Cold/Warm)

1. Remove the auxiliary air regulator wiring connector.
2. Using a test lamp, bridge the wiring connector terminals. The test lamp should light with the engine idling. If the test lamp does not glow with the engine idling, check circuit wiring.

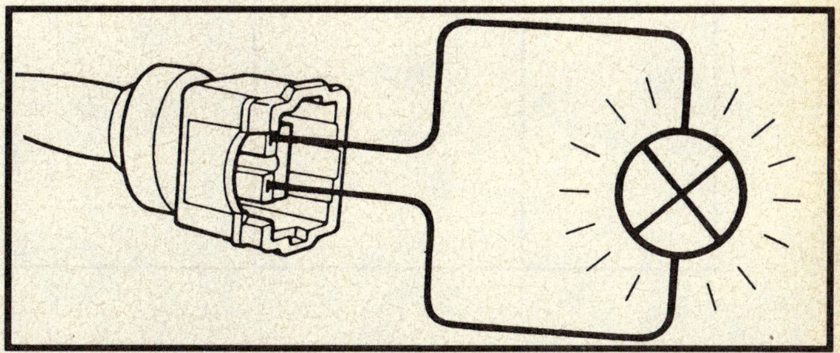
Testing supply voltage at the auxiliary air regulator—Volkswagen CIS-E

1.8L ENGINE—PRESSURE TEST—VOLKSWAGEN FOX

Test No.	Pressure Tested	Testing Conditions	Specifications	Checks/Corrections
1	System pressure ①	Pressure gauge lever in open position Remote control, US 4480/3 switched **ON** Connector of differential pressure regulator disconnected	75–82 psi (52–56 bar)	System pressure less than 75 psi (52 bar) • fuel pressure regulator defective replace regulator If still **NOT OK** • fuel distributor defective System pressure above 56 bar (82 psi) disconnect return line from fuel pressure regulator to fuel tank and repeat test If system pressure **OK** fuel line plugged If system pressure **NOT OK** fuel pressure regulator defective
2a	Differential pressure ①	Pressure gauge lever in closed position Connector of differential pressure regulator disconnected Remote control, US 4480/3 **ON**	2.9–7 psi (0.2–0.5 bar) below system pressure	Differential pressure **NOT OK** disconnect thin line from fuel pressure regulator and hold in a measuring glass seal connection on fuel pressure regulator switch remote control US 4480/3 **ON** for one minute and measure overflow quantity • specification: 130 to 150 cm^2/min If **OK** replace fuel pressure regulator If **NOT OK** replace fuel distributor
2b	Differential pressure checking is required only if there are hot start problems ②	Pressure gauge lever in closed position 15 K Ω side of resistor block VW 1490 inserted into open connector removed from temperature sensor Remote control, US 4480/3 **ON**	10–17.5 psi (0.7–1.2 bar) below system pressure Differential pressure regulator control current 59 to 70 mA	If differential pressure **NOT OK** and control current **OK** • differential pressure regulator defective If differential pressure and control current **NOT OK** remove adapter VW 1315A/1 from differential pressure regulator check electrical resistance of differential pressure regulator • specification: 17.5 to 21.5 Ω If not **OK** replace differential pressure regulator If **OK** check ground wire of temperature sensor connector If ground **OK** check power supply fuse If fuse **OK** check electrical connections on CIS-E control unit connector

1.8L ENGINE—PRESSURE TEST—VOLKSWAGEN FOX

Test No.	Pressure Tested	Testing Conditions	Specifications	Checks/Corrections
3	Residual pressure checking is required only if there are hot start problems ①	Remote control US 4480/3 ON for approx. 5 seconds	After 10 minutes minimum 38 psi (2.6 bar)	If residual pressure below 38 psi (2.6 bar) —main fuel pump check valve leak —lines and connections leaky —no clearance at sensor plate —fuel pressure regulator defective —O-rings of fuel distributor defective
4	Warm-up enrichment ②	Connector for temperature sensor disconnected Ignition **ON**	80 to 110 mA	If control current **NOT OK** —control unit defective
5	Cold acceleration enrichment ②	Connector for temperature sensor disconnected Plug in CO hole removed Sensor plate lifted fully	80 to 110 mA Briefly in excess of 110 mA	If control current **NOT OK** —potentiometer defective —wires and connections from potentiometer to control unit broken —control unit defective
6	After-start enrichment ②	Connector for temperature sensor disconnected Ground terminal 4 of ignition coil Starter operated for approx. 3 sec. (ignition remains on)	In excess of 120 mA for 20 to 40 sec., then drop in current to 80 to 110 mA	If control current **NOT OK** —no starter signal at terminal 24 of control unit —control unit defective
7	Oxygen sensor control ②	Connector from temperature sensor removed 15 K Ω side of tool VW 1490 connected to temperature sensor connector 1 Oxygen sensor disconnected Oxygen sensor male connector grounded	Control current rises from 9–11 mA to 19–22 mA after 20 seconds	If control current does not rise —oxygen sensor wire from control unit to terminal connection broken —control unit defective —wiring from control unit (terminal 8) to oxygen sensor open —control unit defective

① Test conditions
- Engine not running
- Main fuel pump delivery rate **OK**
- Pressure gauge VW 1318 connected (use new O-rings)
- High tension cable (ignition coil terminal 4) detached at distributor and connected to ground
- Multimeter US 1119 connected

② Test condition for text steps 2b, 4, 5, 6 and 7
Digital multimeter US 1119 and test lead VW 1315A/1 connected to differential pressure regulator
Ignition **ON**

1.8L ENGINE—VOLTAGE CHECKS—VOLKSWAGEN FOX

Test Terminal (ECU)	Item Tested	Testing Conditions	Specifications	Checks/Results
1+2	Voltage supply and ground connection of control unit **1** (+) from ignition switch **2** (−) from intake manifold	Ignition ON Measuring range: DCV 20	Approx. battery voltage	No voltage —fuse 23 defective —check ground connection on intake manifold —ignition switch (terminal 15) has no voltage
	Voltage supply (starter actuated)	Measuring range: DCV 20 (Starter actuated)	At least 8V	

1.8L ENGINE—RESISTANCE CHECKS—VOLKSWAGEN FOX

Test Terminal (ECU)	Item Tested	Testing Conditions	Specifications	Checks/Results
2+15	Ground connection jumper	—	0 Ω	No continuity
9+15	Ground connection jumper	—	0 Ω	—break in ground connection jumper
7+8	Electrical shielding of oxygen sensor wiring	—	α Ω	—check shield wire for short circuit
2+8	Oxygen sensor wiring	Disconnect green wiring connector from black oxygen sensor wiring and connect to ground	0 Ω	—break in green oxygen sensor wiring
		Connect oxygen sensor wiring together	α Ω	—oxygen sensor defective
10+12	Differential pressure regulator and wiring	—	17.5 to 21.5 Ω	—break in wiring 10/12 or differential pressure regulator faulty
14+17	Potentiometer and wiring	Air flow sensor plate in rest position	Less than 1K Ω	—breaks in wiring 14/17/18 or
17+18	Potentiometer and wiring • 1 to 18 2 to 17 3 to 14		more than 4K Ω	—potentiometer adjusted wrong or defective
2+21	Temperature sensor and wiring	Temperature of sensor: approx. 20°C (68°F) less than 20°C more than 20°C	Approx. 2.5K Ω less than 2.5K Ω more than 2.5K Ω	—check ground wiring to temperature sensor for break —temperature sensor defective
2+24	Wiring from starter to control unit	High tension wire (terminal 4) disconnected from distributor cap and connect to ground (Starter actuated)	Minimum 8 volts	—check for break in wiring

2.0L ENGINE—CIS-E TROUBLESHOOTING—VEHICLES MANUFACTURED BEFORE 10/90

CIS-E Motronic system, troubleshooting

- verify complaint
- satisfy tuning requirements and adjustments

If the problem still exists:

- find "symptom" in numbered top section
- scan the vertical column for ALL X's
- read across from the X to determine possible cause

Symptom

1 ENGINE DOES NOT START OR IS DIFFICULT TO START WHEN COLD
2 ENGINE DOES NOT START OR IS DIFFICULT TO START WHEN WARM (HOT-START DIFFICULTY)
3 UNEVEN IDLE OR NO IDLE IN COLD OR WARM-UP PHASE
4 ENGINE STARTS, BUT STALLS RIGHT AWAY
5 IDLE SPEED EXCEEDS SPECIFICATION
6 POOR ACCELERATION WHEN ENGINE IS COLD OR WARMING UP, TRANSITIONAL PROBLEMS

1	2	3	4	5	6	Tests and Adjustments:
X	X	X	X	X	X	First recall faults from Fault memory and eliminate displayed faults before further troubleshooting
X	X	X	X	X	X	Perform output check diagnosis
X	X	X	X	X		Perform CIS-E Motronic Electrical tests
	X					Check fan after-run circuit
X			X		X	Check fuel pump(s)
X			X			Check fuel-pump relays
	X					Check fuel pump check valve
				X		Adjust throttle cable
	X	X		X		Check carbon canister solenoid valves
X	X	X				Check fuel system for leaks
X	X	X	X	X	X	Check intake system (auxiliary air)
	X	X	X	X		Check idle speed
		X			X	Check Oxygen sensor control
X	X		X		X	Check after-start, warm-up and acceleration enrichment
X	X	X				Check cold-start valve
	X	X				Check fuel injector injection pressure
X			X		X	Check airflow sensor lever adjustment and control plunger

Troubleshooting

- verify complaint
- satisfy tuning requirements and adjustments

If the problem still exists:

- find "symptom" in numbered top section
- scan the vertical column for ALL X's
- read across from the X to determine possible cause

Symptom

7 ENGINE STOPS RUNNING WHILE DRIVING
8 DRIVING PERFORMANCE UNSATISFACTORY
9 ENGINE RUNS ON
10 FUEL CONSUMPTION TOO HIGH
11 ENGINE MAKES NOISE, KNOCKS
12 ENGINE RUNS UNEVENLY OVER ENTIRE ENGINE SPEED RANGE

7	8	9	10	11	12	Tests and Adjustments:
X	X	X	X	X	X	First recall faults from Fault memory of control unit and eliminate displayed faults before further troubleshooting
	X	X	X			Perform output check diagnosis
X	X		X			Perform CIS-E Motronic Electrical tests
X	X					Check fuel pump(s)
	X					Adjust throttle cable
		X				Check carbon canister solenoid valve II
		X	X			Check fuel system (leaks)
X	X				X	Check intake system (auxiliary air)
			X			Check Oxygen sensor control function
		X	X			Check cold-start valve
		X	X		X	Check fuel injector injection pressure
	X	X	X			Check airflow sensor lever adjustment and control plunger
	X		X			Check potentiometer
	X					Check system, differential and residual pressure
X	X		X		X	Do comparative measurement of fuel injection volume
X						Check throttle valve switch (idle switch)
X	X		X	X		Check throttle valve switch (full-throttle)

2.0L ENGINE—CIS-E TROUBLESHOOTING—VEHICLES MANUFACTURED BEFORE 10/90

Troubleshooting

- verify complaint
- satisfy tuning requirements and adjustments

If the problem still exists:

- find "symptom" in numbered top section
- scan the vertical column for ALL X's
- read across from the X to determine possible cause

Symptom

1 ENGINE DOES NOT START OR IS DIFFICULT TO START WHEN COLD
2 ENGINE DOES NOT START OR IS DIFFICULT TO START WHEN WARM (HOT-START DIFFICULTY)
3 UNEVEN IDLE OR NO IDLE IN COLD OR WARM-UP PHASE
4 ENGINE STARTS, BUT STALLS RIGHT AWAY
5 IDLE SPEED EXCEEDS SPECIFICATION
6 POOR ACCELERATION WHEN ENGINE IS COLD OR WARMING UP, TRANSITIONAL PROBLEMS

1	2	3	4	5	6	Tests and Adjustments:
X	X		X			Check sensor plate — rest position and idle path
	X					Check sealing ring under control plunger
		X		X	X	Check Potentiometer
X	X	X	X		X	Check system, differential and residual pressures
X	X	X			X	Perform comparative measurement of fuel injection volume
		X	X	X	X	Check throttle switch (idle switch)
X	X	X	X	X	X	Check temperature sensor
				X		Check throttle valve basic adjustment
		X			X	Check exhaust system
X	X	X		X	X	Check ignition timing
X		X			X	Check spark plugs
X	X					Check ignition coil
X	X				X	Check spark plug connectors, ignition cables
X					X	Check ignition distributor cap/rotor

Troubleshooting

- verify complaint
- satisfy tuning requirements and adjustments

If the problem still exists:

- find "symptom" in numbered top section
- scan the vertical column for ALL X's
- read across from the X to determine possible cause

Symptom

7 ENGINE STOPS RUNNING WHILE DRIVING
8 DRIVING PERFORMANCE UNSATISFACTORY
9 ENGINE RUNS ON
10 FUEL CONSUMPTION TOO HIGH
11 ENGINE MAKES NOISE, KNOCKS
12 ENGINE RUNS UNEVENLY OVER ENTIRE ENGINE SPEED RANGE

7	8	9	10	11	12	Tests and Adjustments:
X	X		X			Check thermostat (intake air pre-heating)
			X			Check exhaust system between cylinder head and oxygen sensor for leaks
	X					Check catalytic converter for melt/destruction
	X		X	X		Check ignition timing setting
X					X	Check spark plugs
X					X	Check spark plug connectors, ignition cables
X					X	Check ignition distributor cap/rotor

2.0L ENGINE–CIS-E ELECTRICAL TESTING–VEHICLES MANUFACTURED BEFORE 10/90

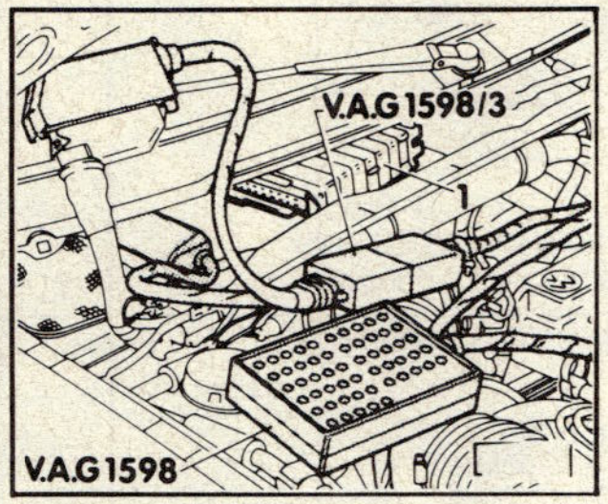

- switch **OFF** ignition
- switch multimeter **US 1119** (or equivalent) to 20 Volts DC range
- disconnect CIS-E Motronic control unit
- connect **VAG 1598** Test box to control unit harness connector using **VAG 1598/3** adaptor cable
  - CIS-E Motronic control unit **1** is **NOT** connected during the following tests

Test step	VAG 1598 terminals	Component to be tested	Test conditions and additional steps, if any	Specified value or result
1	**18 + 19**	Voltage supply	—	Battery voltage
2	**14 + 35**	Voltage supply	Ignition switched **ON**	Battery voltage
3	**17 + 35**	Idle stabilizer valve **(N 71)**	Ignition switched **ON**	Battery voltage
4	**33 + 19** bridge for short period	A/C compressor **"ON"**	Use jumper wire to bridge, do **NOT** bridge with meter	Compressor clutch must switch **ON** (audible click)
Vehicles with manual transmission:				
5	**19 + 34**	Ground to pin 34	—	Battery voltage
Vehicles with automatic transmission:				
	34 + 35	Wiring to starter lockout relay	Selector lever in Park while cranking engine	2 Volts less than battery voltage

— switch OFF ignition
— switch multimeter **US 1119** (or equivalent) to resistance range
— short measuring leads to one another
— observe resistance display value
— subtract this value from following measurements to obtain actual resistance value

Test step	VAG 1598 terminals	Component to be tested	Test conditions and additional steps, if any	Specified value or result
6	**4 + 5**	Differential pressure regulator (**N 73**)	—	15 to 20 Ohms
7	**3 + 35**	Coolant temperature sensor (**G 62**)	Check using chart 25-975 (below)	Per chart

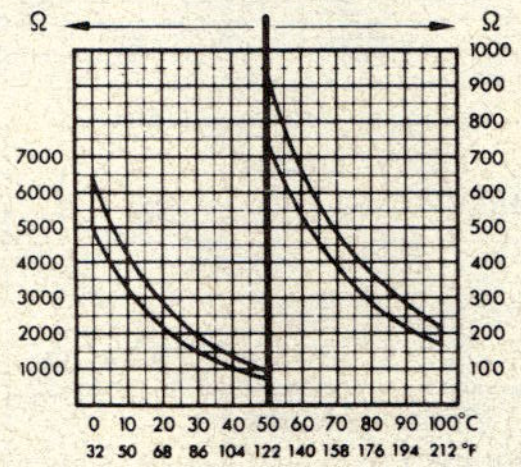

- resistance values for coolant temperature sensor (**G 62**)

NOTE
If during Self-Diagnosis a sensor fault is displayed BUT NO problem is detected during this test; there could be a temporary disconnection of the temperature signal. If so; perform test step **22**.

Test step	VAG 1598 terminals	Component to be tested	Test conditions and additional steps, if any	Specified value or result
8	**28 + 35**	Idle switch (**F 60**)	Throttle closed	Continuity, 1 Ohm maximum
			Throttle open	Open (infinite resistance)
9	**32 + 35**	Full throttle switch (**F 81**) (if installed)	Throttle closed	Open (infinite resistance)
			Throttle wide open	Continuity, 1 Ohm maximum
10	**14 + 16**	Cold start valve (**N 17**)	—	8 to 13 Ohms
11	**14 + 15**	Carbon canister solenoid valve I (**N 80**)	—	30 to 60 Ohms
12	**30 + 35**	Wires to Hall sender (**G 40**) 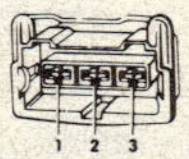 (TERMINAL NUMBERING)	Disconnect Hall sender harness connector at ignition distributor	Open, (infinite resistance)
			Bridge terminals **1** and **2** of harness connector	Continuity 1 Ohm maximum
	30 + 21		Bridge terminals **2** and **3** of harness connector	Continuity, 1 Ohm maximum
13	**11 + 14**	Wires to ignition coil power output stage (**N 70**) 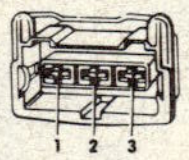 (TERMINAL NUMBERING)	Disconnect power output stage harness connector	Open, (infinite resistance)
			Bridge terminals **1** and **2** of harness connector	Continuity, 1 Ohm maximum
	11 + 18		Bridge terminals **2** and **3** of harness connector	Continuity, 1 Ohm maximum
14	**6 + 8**	Wiring to Knock sensor **II** (**G 66**) (cylinders **3** and **4**) 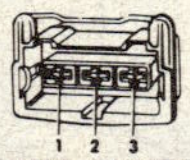 (TERMINAL NUMBERING)	Disconnect Knock sensor **II** (near ignition distributor)	Open (infinite resistance)
			Bridge terminals **1** and **2** of connector	Continuity, 1 Ohm maximum
	8 + 24	Wiring to Knock sensor **I** (**G 61**) (cylinders **1** and **2**) 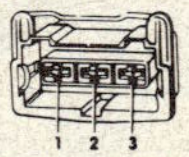 (TERMINAL NUMBERING)	Disconnect knock sensor **I** (near oil filter)	Open (infinite resistance)
			Bridge terminals **1** and **2** of connectcr	Continuity, 1 Ohm maximum

Test step	VAG 1598 terminals	Component to be tested	Test conditions and additional steps, if any	Specified value or result
15	**7 + 35**	Wiring to Oxygen sensor (**G 39**)	Disconnect Oxygen sensor harness connector and touch to ground	Continuity, 1 Ohm maximum
16	**27 + 29**	Ignition timing sensor (**G 4**)	—	2 Ohms maximum
17	**23 + 26**	Air flow sensor potentiometer (**G 19**)	Sensor plate in rest position	4500 to 5500 Ohms
	26 + 35	Air flow sensor potentiometer (**G 19**)	—	3800 to 4200 Ohms
18	**12 + 35** (bridged)	Wiring to fuel pump relay with relay connected	Ignition switched **ON**	Fuel pumps must run
19	**1 + 22**	Wiring to diagnostic connectors	Bridge terminals of white connector	Continuity, 1 Ohm maximum
	22 + 13		Connect lower terminal of white connector to upper terminal of blue connector	Continuity, 1 Ohm maximum
19A California Only	**9 + 35**	EGR temperature sensor (**G 98**)	Check using chart (below)	Per chart (below)

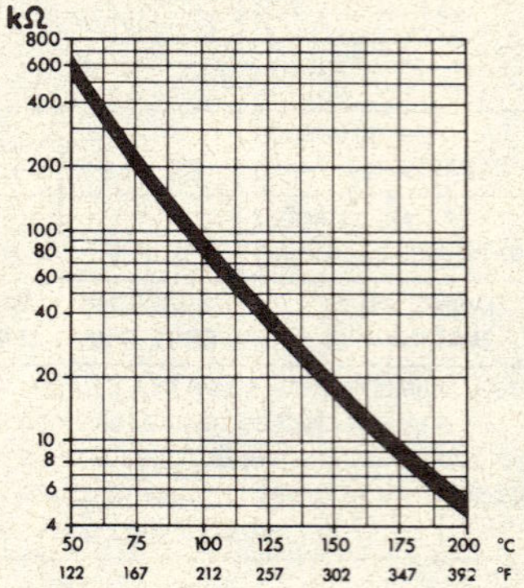

● resistance values for EGR temperature sensor (**G 98**)

2.0L ENGINE—CIS-E ELECTRICAL TESTING VEHICLES MANUFACTURED BEFORE 10/90

CAUTION

Perform the following electrical tests with great care. The CIS-E Motronic control unit could be destroyed if the following precautions are **NOT** observed.

- **ALWAYS** switch the multimeter to the desired units and range of measurements (Volts, Ohms, etc.) **BEFORE** making connections to the Test box
- Do **NOT** connect or disconnect adaptors/test instruments to the Test box when the ignition is switched **ON**

The following tests are performed with the control unit harness connector **AND** the control unit connected to the **VAG 1598/3** adaptor cable).

- switch **OFF** ignition
- connect free end of **VAG 1598/3** adaptor cable to control unit
- switch multimeter **US 1119** (or equivalent) to 20 Volt range

Test step	VAG 1598 terminals	Component to be tested	Test conditions and additional steps, if any	Specified value or result
20	**21 + 35**	Hall sender (**G 40**) voltage supply	Ignition switched **ON**	10 Volts minimum
21	**2 + 35**	Signal from vehicle Speed sender (**G 54**)	Ignition switched **ON** and L/F wheel rolled very slowly. Approximately every 10° of rotation, the voltage will switch from near 0 to roughly 10 Volts. This is an on/off signal and exact voltages are not critical.	
22	**3 + 35**	Coolant temperature sensor (**G 62**) resistance curve	Start cold engine and let warm up while observing multimeter	Voltage must uniformly become lower
For next test: — switch **OFF** ignition — disconnect multimeter and adaptors from Test box — disconnect power output stage of ignition coil — connect **US 1115 (VAG 1527B)** LED tester to Test box terminals **30 + 35** using **VW 1594** adaptor kit				
23	**30 + 35**	Signal from Hall sender (**G 40**)	Operate starter	LED tester must flicker
— switch **OFF** ignition — disconnect test equipment and adaptors — re-connect CIS-E Motronic control unit — re-connect **VAG 1551** and check for stored faults, repair as necessary, repeat until memory is clear				

2.0L ENGINE—CIS-E TROUBLESHOOTING VEHICLES MANUFACTURED FROM 10/90

CIS-E Motronic system, troubleshooting

- verify complaint
- satisfy tuning requirements and adjustments

If the problem still exists:

- find "symptom" in numbered top section
- scan the vertical column for ALL X's
- read across from the X to determine possible cause

Symptom

1 ENGINE DOES NOT START OR IS DIFFICULT TO START WHEN COLD
2 ENGINE DOES NOT START OR IS DIFFICULT TO START WHEN WARM (HOT-START DIFFICULTY)
3 UNEVEN IDLE OR NO IDLE IN COLD OR WARM-UP PHASE
4 ENGINE STARTS, BUT STALLS RIGHT AWAY
5 IDLE SPEED EXCEEDS SPECIFICATION
6 POOR ACCELERATION WHEN ENGINE IS COLD OR WARMING UP, TRANSITIONAL PROBLEMS

1	2	3	4	5	6	Tests and Adjustments:
X	X	X	X	X	X	First recall faults from Fault memory and eliminate displayed faults before further troubleshooting
X	X	X	X	X	X	Perform output check diagnosis
X	X	X	X	X		Perform CIS-E Motronic Electrical tests
	X					Check fan after-run circuit
X			X		X	Check fuel pump(s)
X			X			Check fuel-pump relays
	X					Check fuel pump check valve
				X		Adjust throttle cable
	X	X		X		Check carbon canister solenoid valves
X	X	X				Check fuel system for leaks
X	X	X	X	X	X	Check intake system (auxiliary air)
	X	X	X	X		Check idle speed
		X			X	Check Oxygen sensor control
X	X		X		X	Check after-start, warm-up and acceleration enrichment
X	X	X				Check cold-start valve
	X	X				Check fuel injector injection pressure
X			X		X	Check airflow sensor lever adjustment and control plunger

2.0L ENGINE – CIS-E TROUBLESHOOTING – VEHICLES MANUFACTURED FROM 10/90

Troubleshooting

- verify complaint
- satisfy tuning requirements and adjustments

If the problem still exists:

- find "symptom" in numbered top section
- scan the vertical column for ALL X's
- read across from the X to determine possible cause

Symptom

1 ENGINE DOES NOT START OR IS DIFFICULT TO START WHEN COLD
2 ENGINE DOES NOT START OR IS DIFFICULT TO START WHEN WARM (HOT-START DIFFICULTY)
3 UNEVEN IDLE OR NO IDLE IN COLD OR WARM-UP PHASE
4 ENGINE STARTS, BUT STALLS RIGHT AWAY
5 IDLE SPEED EXCEEDS SPECIFICATION
6 POOR ACCELERATION WHEN ENGINE IS COLD OR WARMING UP, TRANSITIONAL PROBLEMS

1	2	3	4	5	6	Tests and Adjustments:
X	X		X			Check sensor plate — rest position and idle path
	X					Check sealing ring under control plunger
		X		X	X	Check Potentiometer
X	X	X	X		X	Check system, differential and residual pressures
X	X	X			X	Perform comparative measurement of fuel injection volume
		X	X	X	X	Check throttle switch (idle switch)
X	X	X	X	X	X	Check temperature sensor
				X		Check throttle valve basic adjustment
		X			X	Check exhaust system
X	X	X		X	X	Check ignition timing
X		X			X	Check spark plugs
X	X					Check ignition coil
X	X				X	Check spark plug connectors, ignition cables
X					X	Check ignition distributor cap/rotor

Troubleshooting

- verify complaint
- satisfy tuning requirements and adjustments

If the problem still exists:

- find "symptom" in numbered top section
- scan the vertical column for ALL X's
- read across from the X to determine possible cause

Symptom

7 ENGINE STOPS RUNNING WHILE DRIVING
8 DRIVING PERFORMANCE UNSATISFACTORY
9 ENGINE RUNS ON
10 FUEL CONSUMPTION TOO HIGH
11 ENGINE MAKES NOISE, KNOCKS
12 ENGINE RUNS UNEVENLY OVER ENTIRE ENGINE SPEED RANGE

7	8	9	10	11	12	Tests and Adjustments:
X	X	X	X	X	X	First recall faults from Fault memory of control unit and eliminate displayed faults before further troubleshooting
	X	X	X			Perform output check diagnosis
X	X		X			Perform CIS-E Motronic Electrical tests
X	X					Check fuel pump(s)
	X					Adjust throttle cable
		X				Check carbon canister solenoid valve II
		X	X			Check fuel system (leaks)
X	X				X	Check intake system (auxiliary air)
			X			Check Oxygen sensor control function
		X	X			Check cold-start valve
		X	X		X	Check fuel injector injection pressure
	X	X	X			Check airflow sensor lever adjustment and control plunger
	X		X			Check potentiometer
	X					Check system, differential and residual pressure
X	X		X		X	Do comparative measurement of fuel injection volume
X						Check throttle valve switch (idle switch)
X	X		X	X		Check throttle valve switch (full-throttle)

2.0L ENGINE—CIS-E TROUBLESHOOTING VEHICLES MANUFACTURED FROM 10/90

Troubleshooting

- verify complaint
- satisfy tuning requirements and adjustments

If the problem still exists:

- find "symptom" in numbered top section
- scan the vertical column for ALL X's
- read across from the X to determine possible cause

Symptom

7 ENGINE STOPS RUNNING WHILE DRIVING
8 DRIVING PERFORMANCE UNSATISFACTORY
9 ENGINE RUNS ON
10 FUEL CONSUMPTION TOO HIGH
11 ENGINE MAKES NOISE, KNOCKS
12 ENGINE RUNS UNEVENLY OVER ENTIRE ENGINE SPEED RANGE

7	8	9	10	11	12	Tests and Adjustments:
X	X		X			Check thermostat (intake air pre-heating)
			X			Check exhaust system between cylinder head and oxygen sensor for leaks
	X					Check catalytic converter for melt/destruction
	X		X	X		Check ignition timing setting
X					X	Check spark plugs
X					X	Check spark plug connectors, ignition cables
X					X	Check ignition distributor cap/rotor

2.0L ENGINE—CIS-E VAG 1551 VALUE BLOCKS VEHICLES MANUFACTURED FROM 10/90

VAG 1551 Function 08, channel identification

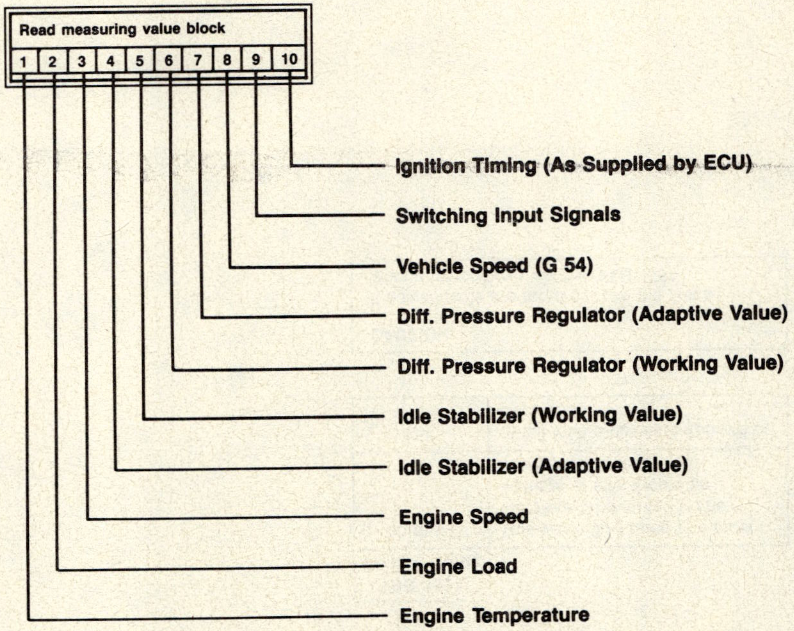

2.0L ENGINE – CIS-E VAG 1551 VALUE BLOCKS – VEHICLES MANUFACTURED FROM 10/90

VAG 1551 Function 08, channel 1 interpretation

Read measuring value block
1 2 3 4 5 6 7 8 9 10

Coolant Temperature

Interpretive Table

VAG 1551 Display	Centigrade	Fahrenheit
35	106.6	223.8
40	100.5	212.9
50	90.5	194.9
60	82.2	179.9
70	75.1	167.1
80	68.8	155.8
90	63.1	145.6
100	57.9	136.2
110	53.0	127.4
120	48.3	118.9
130	43.8	110.8
140	39.4	102.9
150	35.0	95.0
160	30.6	87.0
170	26.2	79.2
180	21.6	70.9
190	16.8	62.2
200	11.6	52.9
210	6.0	42.8
220	0.2	32.1

VAG 1551 Function 08, channel 2 interpretation

Read measuring value block
1 2 3 4 5 6 7 8 9 10

Engine Load

VAG 1551 Display value	
40 to 64	This value fluctuates and must therefore be averaged. The average at idle should be between 40 and 64
255	At full load a constant 255 will be displayed

CAUTION

If adjustments are required, use the Repair Manual procedures not the **VAG 1551**.

2.0L ENGINE—CIS-E VAG 1551 VALUE BLOCKS—VEHICLES MANUFACTURED FROM 10/90

VAG 1551 Function 08, channel 3 interpretation

Read measuring value block:
1 2 **3** 4 5 6 7 8 9 10

3 — Engine Speed

VAG 1551 Display	Engine RPM	VAG 1551 Display	Engine RPM
20	480	41	984
21	504	42	1008
22	528	43	1032
23	552	44	1056
24	576	45	1080
25	600	46	1104
26	624	47	1128
27	648	48	1152
28	672	49	1176
29	696	50	1200
30	720	64	1536
31	744	75	1800
32	768	84	2040
33	792	105	2520
34	816	125	3000
35	840	146	3504
36	864	167	4008
37	888	188	4512
38	912	209	5016
39	936	230	5520
40	960	250	6000

VAG 1551 Function 08, channel 4 interpretation

Read measuring value block
1 2 3 **4** 5 6 7 8 9 10

4 — Idle Stabilizer (Adaptive Value)

VAG 1551 Display	
0 to 24	Upper adaptive range
255 to 232	Lower adaptive range

Either range is **OK**

This number is not adjustable. If not within working range you must check all other engine adjustments, vacuum leaks, and the mechanical condition of the engine. Since this is an adaptive value it will take approximately 4 minutes or more to adapt itself after a correction is made.

VAG 1551 Function 08, channel 5 interpretation

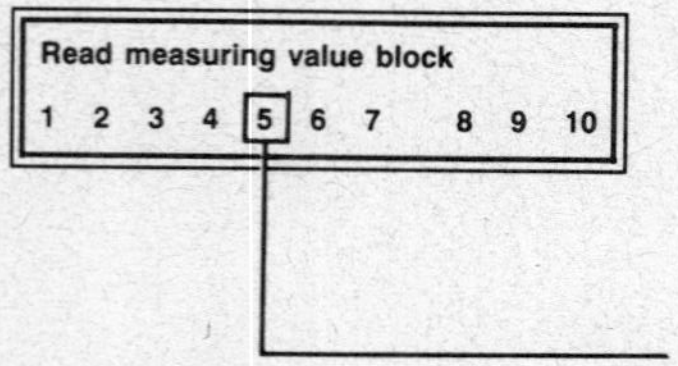

Idle Stabilizer (Working Value)

VAG 1551 Display	
0 to 48	Upper working range
255 to 206	Lower working range

Either range is **OK**

These values are not adjustable. If not within working range you must check all other engine adjustments, vacuum leaks, and the mechanical condition of the engine.

VAG 1551 Function 08, channel 6 interpretation

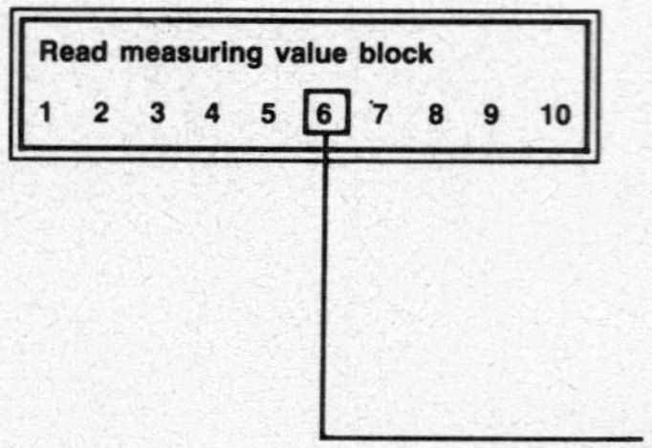

Differential Pressure Regulator (Working Value)

VAG 1551 Display	
0 to 50	Upper working range
255 to 206	Lower working range

Either range is **OK**

A Value of 50 Equals +10 Milliamps of Differential Pressure Regulator Current

A Value of 0 Equals 0 Milliamps of Differential Pressure Regulator Current

A Value of 206 Equals −10 Milliamps of Differential Pressure Regulator Current

If adjustments are required, follow the Repair Manual procedures. Adjustments cannot be made using the **VAG 1551** alone.

2.0L ENGINE—CIS-E VAG 1551 VALUE BLOCKS—VEHICLES MANUFACTURED FROM 10/90

VAG 1551 Function 08, channel 7 interpretation

Read measuring value block
1 2 3 4 5 6 [7] 8 9 10

Differential Pressure Regulator (Adaptive Value)

VAG 1551 Display	
28 to 128	Upper adaptive range
128 to 228	Lower adaptive range

Either range is **OK**

A Value of 228 Equals +10 Milliamps of Differential Pressure Regulator Current

A Value of 128 Equals 0 Milliamps of Differential Pressure Regulator Current

A Value of 228 Equals −10 Milliamps of Differential Pressure Regulator Current

VAG 1551 Function 08, channel 8 interpretation

Read measuring value block
1 2 3 4 5 6 7 [8] 9 10

Vehicle Speed Signal

VAG 1551 Display	
0	If vehicle is standing still
Fluctuates between 1 to 50	If vehicle is moving

This is a digital signal and only represents an on/off signal

VAG 1551 Function 08, channel 9 interpretation

Channel 9, interpreting

Channel **9** provides the technician with essential information that would have to be verified manually before performing the basic engine settings.

The ECU scans the system for certain conditions that must be met or for certain components that must be present or operative. It then assigns the numerical values (listed in the table below) to these conditions or components and **adds** them together to create the display value.

A value of zero will be assigned to conditions or components in the following chart that are NOT MET or NOT PRESENT.

Read measuring value block
1 2 3 4 5 6 7 8 [9] 10

Switch Input Signals to ECU

Switch input	Value
Idle switch closed	1
Full load switch closed	2
Manual transmission in gear or neutral	4
A/C compressor on	8
Fuel enrichment	64
Deceleration fuel shut off	128

Note

By analyzing the value displayed in channel **9** (using simple addition or subtraction) you can easily determine if a component is present or not.

Display Example

Idle Switch Closed	1
Manual Trans in Neutral	4
A/C Compressor On	8
Total	13

(Displayed in Channel **9**)

2.0L ENGINE—CIS-E VAG 1551 VALUE BLOCKS VEHICLES MANUFACTURED FROM 10/90

VAG 1551 Function 08, channel 10 interpretation

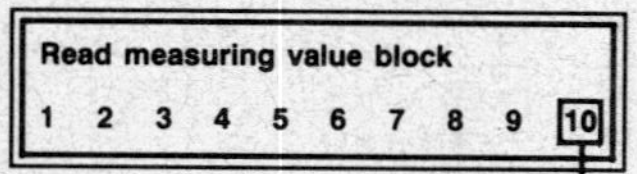

Ignition Timing

VAG 1551 Display	Engine timing	VAG 1551 Display	Engine timing
48	+43.13	164	+2.34
63	+37.85	165	+1.99
79	+32.23	166	+1.64
95	+26.60	167	+1.29
111	+20.98	168	+0.94
127	+15.35	169	+0.59
143	+9.73	170	+0.23
144	+9.38	171	−0.12
145	+9.02	172	−0.47
146	+8.67	173	−0.82
146	+8.32	174	−1.17
148	+7.97	175	−1.57
149	+7.62	176	−1.88
150	+7.27	177	−2.22
151	+6.91	178	−2.58
152	+6.56	179	−2.93
153	+6.21	180	−3.28
154	+5.86	185	−4.69
155	+5.51	190	−6.45
156	+5.16	195	−8.55
157	+4.80	200	−10.31
158	+4.45	205	−12.07
159	+4.10	210	−13.83
160	+3.75	215	−15.59
161	+3.40	220	−17.34
162	+3.05	225	−19.10
163	+2.70	230	−20.86

2.0L ENGINE—CIS-E ELECTRICAL TESTING VEHICLES MANUFACTURED FROM 10/90

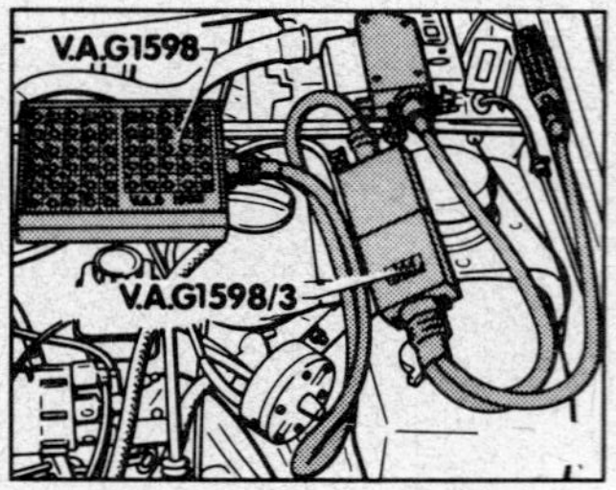

- switch **OFF** ignition
- switch multimeter **US 1119** (or equivalent) to 20 Volts DC range
- disconnect CIS-E Motronic control unit
- connect **VAG 1598** Test box to control unit harness connector using **VAG 1598/3** adaptor cable
  - CIS-E Motronic control unit is **NOT** connected during the following tests

Test step	VAG 1598 terminals	Component to be tested	Test conditions and additional steps, if any	Specified value or result
1	18 + 19	Voltage supply	—	Battery voltage
	14 + 35		Ignition switched **ON**	Battery voltage
2	17 + 35	Idle stabilizer valve (N 71)	Ignition switched **ON**	Battery voltage
3	33 + 19 bridge for short period	A/C compressor **"ON"**	Use jumper wire to bridge, do **NOT** bridge with meter	Compressor clutch must switch **ON** (audible click)
4	12 + 35 (bridged)	Wiring to fuel pump relay with relay connected	Ignition switched **ON**	Fuel pumps must run

2.0L ENGINE – CIS-E ELECTRICAL TESTING – VEHICLES MANUFACTURED FROM 10/90

— switch **OFF** ignition
— switch multimeter **US 1119** (or equivalent) to resistance range
— short measuring leads to one another
— observe resistance display value
— subtract this value from following measurements to obtain actual resistance value

Test step	VAG 1598 terminals	Component to be tested	Test conditions and additional steps, if any	Specified value or result
6	**4 + 5**	Differential pressure regulator (**N 73**)	—	15 to 20 Ohms
7	**3 + 35**	Coolant temperature sensor (**G 62**)	Check using chart (below)	Per chart

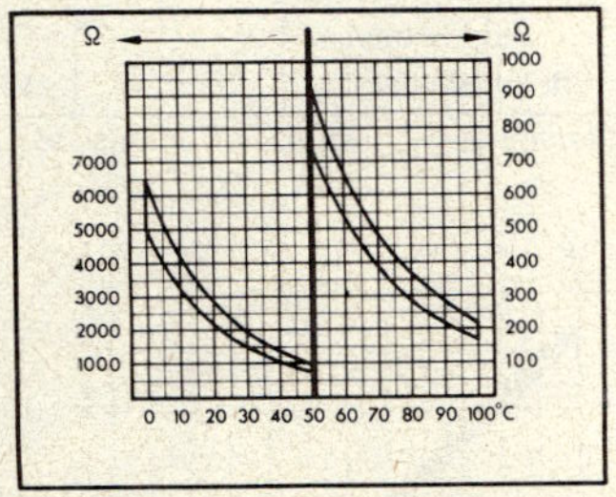

● resistance values for coolant temperature sensor (**G 62**)

Note

If during Self-Diagnosis a sensor fault is displayed BUT NO problem is detected during this test; there could be a temporary disconnection of the temperature signal. If so; perform test step **22**.

Test step	VAG 1598 terminals	Component to be tested	Test conditions and additional steps, if any	Specified value or result
8	**28 + 35**	Idle switch (**F 60**)	Throttle closed	Continuity, 1 Ohm maximum
			Throttle open	Open (infinite resistance)
9	**32 + 35**	Full throttle switch (**F 81**) (if installed)	Throttle closed	Open (infinite resistance)
			Throttle wide open	Continuity, 1 Ohm maximum
10	**14 + 16**	Cold start valve (**N 17**)	—	8 to 13 Ohms
11	**14 + 15**	Carbon canister solenoid valve I (**N 80**)	—	30 to 60 Ohms
12	**7 + 35**	Wiring to Oxygen sensor (**G 39**)	Disconnect Oxygen sensor harness connector and bridge terminals 2 + 4 of harness connector	Continuity, 1 Ohm maximum

Test step	VAG 1598 terminals	Component to be tested	Test conditions and additional steps, if any	Specified value or result
13	**30 + 35**	Wires to Hall sender (**G 40**)	Disconnect Hall sender harness connector at ignition distributor	Open, (infinite resistance)
		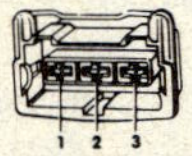 1 2 3	Bridge terminals **1** and **2** of harness connector	Continuity 1 Ohm maximum
	30 + 21	**(TERMINAL NUMBERING)**	Bridge terminals **2** and **3** of harness connector	Continuity, 1 Ohm maximum
14	**11 + 14**	Wires to ignition coil power output stage (**N 70**)	Disconnect power output stage harness connector	Open, (infinite resistance)
		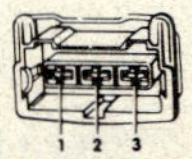 1 2 3	Bridge terminals **1** and **2** of harness connector	Continuity, 1 Ohm maximum
	11 + 18	**(TERMINAL NUMBERING)**	Bridge terminals **2** and **3** of harness connector	Continuity, 1 Ohm maximum
15	**8 + 24**	Wiring to Knock sensor I (**G 61**) (cylinders **1** and **2**)	Disconnect knock sensor I (near oil filter)	Open (infinite resistance)
		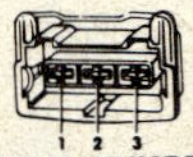 1 2 3	Bridge terminals **1** and **2** of connector	Continuity, 1 Ohm maximum
	18 + 24 **8 + 18**	**(TERMINAL NUMBERING)**	—	Infinite Ohms
16	**6 + 8**	Wiring to Knock sensor II (**G 66**) (cylinders 3 and 4)	Disconnect Knock sensor II (near ignition distributor)	Open (infinite resistance)
		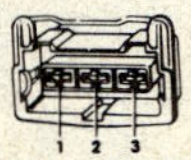 1 2 3	Bridge terminals **1** and **2** of connector	Continuity, 1 Ohm maximum
	6 + 18 **8 + 18**	**(TERMINAL NUMBERING)**	—	Infinite Ohms
17	**27 + 29**	Ignition timing sensor (**G 4**)	—	2 Ohms maximum
18	**23 + 26**	Air flow sensor potentiometer (**G 19**)	Sensor plate in rest position	4500 to 5500 Ohms
	26 + 35	Air flow sensor potentiometer (**G 19**)	—	3800 to 4200 Ohms

2.0L ENGINE—CIS-E ELECTRICAL TESTING—VEHICLES MANUFACTURED FROM 10/90

Test step	VAG 1598 terminals	Component to be tested	Test conditions and additional steps, if any	Specified value or result
19	1 + 22	Wiring to diagnostic connectors	Bridge terminals of white connector	Continuity, 1.5 Ohms maximum
20 California Only	9 + 35	EGR temperature sensor (**G 98**)	Check using chart (below)	Per chart (below)

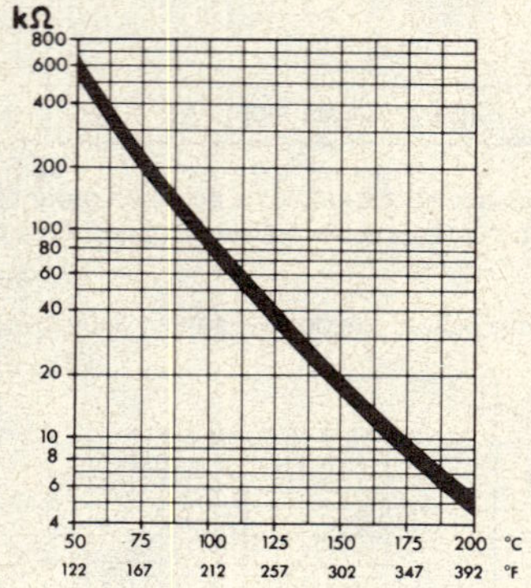

- resistance values for EGR temperature sensor (**G 98**)

Test step	VAG 1598 terminals	Component to be tested	Test conditions and additional steps, if any	Specified value or result
21	21 + 35	Hall sender (**G 40**) voltage supply	Ignition switched **ON**	10 Volts minimum
22	2 + 35	Signal from vehicle Speed sender (**G 54**)	Ignition switched **ON** and L/F wheel rolled very slowly. Approximately every 10° of rotation, the voltage will switch from near 0 to roughly 10 Volts. This is an on/off signal and exact voltages are not critical.	
23	3 + 35	Coolant temperature sensor (**G 62**) resistance curve	Start cold engine and let warm up while observing multimeter	Voltage must uniformly become lower

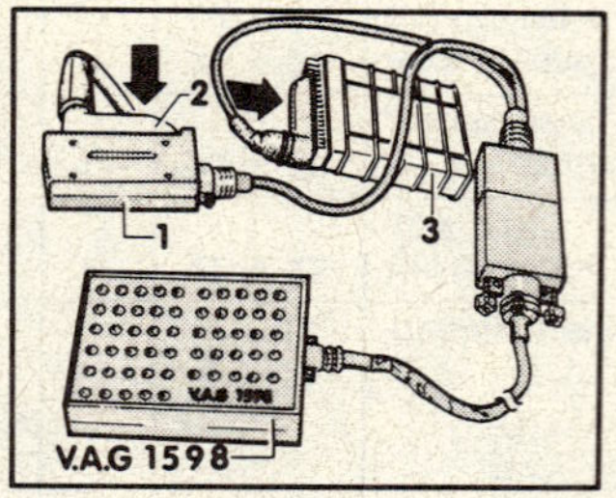

The following tests are performed with the control unit harness connector **AND** the control unit connected to the **VAG 1598/3** adaptor cable).

- switch **OFF** ignition
- connect free end of **VAG 1598/3** adaptor cable to control unit
- switch multimeter **US 1119** (or equivalent) to 20 Volt range

CAUTION

Perform the following electrical tests with great care. The CIS-E Motronic control unit could be destroyed if the following precautions are **NOT** observed.

- **ALWAYS** switch the multimeter to the desired units and range of measurements (Volts, Ohms, etc.) **BEFORE** making connections to the Test box
- do NOT connect or disconnect adaptors/test instruments to the Test box when the ignition is switched **ON**

Test step	VAG 1598 terminals	Component to be tested	Test conditions and additional steps, if any	Specified value or result
24	**35 + Intake manifold ground connection**	ECU ground*	Let engine idle Multimeter @ 200 mV range	Less than 60 mV**
For next test: — switch **OFF** ignition — disconnect multimeter and adaptors from Test box — disconnect power output stage of ignition coil — connect **US 1115 (VAG 1527B)** LED tester to Test box terminals **30 + 35** using **VW 1594** adaptor kit				
25	30 + 35	Signal from Hall sender (**G 40**)	Operate starter	LED tester must flicker
26	11 + 18	Ignition signal from (**J 204**) control unit	Briefly operate starter Note: **Test step 15 must be OK**	LED tester must flicker
— switch **OFF** ignition — disconnect test equipment and adaptors — re-connect CIS-E Motronic control unit — re-connect **VAG 1551** and check for stored faults, repair as necessary, repeat until memory is clear				

*Measure directly between actual ground point on manifold and terminal **35** of **VAG 1598** Test box.
**If greater than 60 mV check ground connection at intake manifold and clean or repair as necessary.

Component Replacement

SERVICE PRECAUTIONS

To prevent personal injury and damage to the vehicle, foolow these service precautions:

- Always disconnect the negative battery terminal to prevent sparks cause by short-circuiting, circuit-breaking, etc.
- Smoking must not be allowed when working around flammable liquids.
- Always keep a CO_2 fire extinguisher close on hand.
- Absorbant must be available to soak up any spillage.
- Ensure fuel is emptied into a suitable container.
- Always place the transmission gear shift lever in **N** (manual transmission) or **P** (automatic transmission) and set the parking brake and block drive wheels before starting the engine.
- Keep hands, tools and clothing away from engine cooling fan to help prevent personal injury.
- Thoroughly clean all unions and area near connections before disconnecting.

FUEL PRESSURE RELIEF

1. Follow service precautions to safely perform this procedure to prevent personal injury. Disconnect the negative battery terminal.
2. Wrap a shop rag around the feed to the fuel distributor.
3. Loosen the feed connection to relieve the system pressure. Catch any escaping fuel in the shop rag. Dispose of the rag in a safe manner.
4. Tighten the fuel distributor feed connection.

COLD START VALVE

Removal and Installation

1. Disconnect the negative battery cable.
2. Relieve the fuel system pressure. Wrap a cloth around the connection to catch any escaping fuel.
3. Remove the harness connector and fuel line from the cold start valve.
4. Loosen the cold start valve retaining bolt and remove the cold start valve and gasket from the intake manifold.

To install:

5. Install a new gasket and install the cold start valve to the intake manifold. Install the retaining screws and tighten to 88 inch lbs. (10 Nm).
6. Connect the fuel line and the harness connector to the cold start valve.
7. Turn the ignition switch **ON** and **OFF** several times to pressurize the system. Check for fuel leaks before starting the engine and after starting.

THERMO-TIME SWITCH

Removal and Installation

1. Disconnect the negative battery cable.

CAUTION

Allow the engine to cool sufficiently before attempting any work with the cooling system to avoid personal injury.

2. Drain the engine coolant below the level of the switch. Be careful not to damage the connectors of the switch while removing it from the engine.

To install:

3. Coat the threads of the switch with an approved thread sealant compound.
4. Install the thermo-time switch in the intake manifold. Torque the switch to 22 ft. lbs (30 Nm).
5. Connect the harness connector to the switch.
6. Refill the cooling system and connect the negative battery cable.

FUEL INJECTOR

Removal and Installation

1. Disconnect the negative battery cable.
2. Relieve the fuel system pressure. Wrap a cloth around the connection to catch any escaping fuel.
3. Clean the area around the fuel injectors. Using 2 wrenches, hold the injector with 1 wrench and loosen the fuel line fitting with the other wrench.
4. Remove the fuel line from the injector, do not allow the injector to turn.
5. Carefully pull the injector out from the insert. A injector extractor tool may be required on injectors that are difficult to remove.
6. Using a 13mm hex head socket tool, remove the injector upper insert from the cylinder head.
7. Remove the lower insert and sealing ring from the cylinder head.

To install:

8. Install a new sealing washer and the lower insert into the cylinder head.
9. Coat the upper insert threads with D–6 sealing compound or equivalent. Thread the upper insert into the cylinder head.
10. Using a 13mm hex head socket tool, tighten the the upper insert to 15 ft. lbs. (20 Nm).
11. Make sure all O-rings are in place on the injector. Lubricate the O-rings with fuel prior to installing.
12. Push the injector firmly in position into the upper insert seat.
13. Connect the fuel line to the injector. Using 2 wrenches tighten the fuel line fitting.
14. Turn the ignition switch **ON** and **OFF** several times to pressurize the system. Check for fuel leaks before starting the engine and after starting.

AUXILIARY AIR REGULATOR

The auxiliary air regulator is normally located below the throttle body, mounted to the intake manifold.

Removal and Installation

1. Disconnect the negative battery cable.
2. Remove and plug all vacuum lines to the valve.
3. Disconnect the harness connector at the valve.
4. Remove the auxiliary valve retaining bolts and remove the auxiliary air valve from the intake manifold.

To install:

5. Position the auxiliary air valve to the intake manifold and install the retaining bolts.
6. Connect all vacuum hoses to thé auxiliary air valve.
7. Connect the harness connector to the auxiliary air valve.
8. Connect the negative battery cable.

FUEL PUMP

The fuel pump is normally located under the vehicle near the fuel tank or in front of the rear axle on the right side. It is attached to a bracket which also accommodates the fuel filter and the accumulator as an assembly.

Removal and Installation

1. Disconnect the negative battery cable.
2. Relieve the fuel system pressure. Wrap a cloth around the connection to catch any escaping fuel.

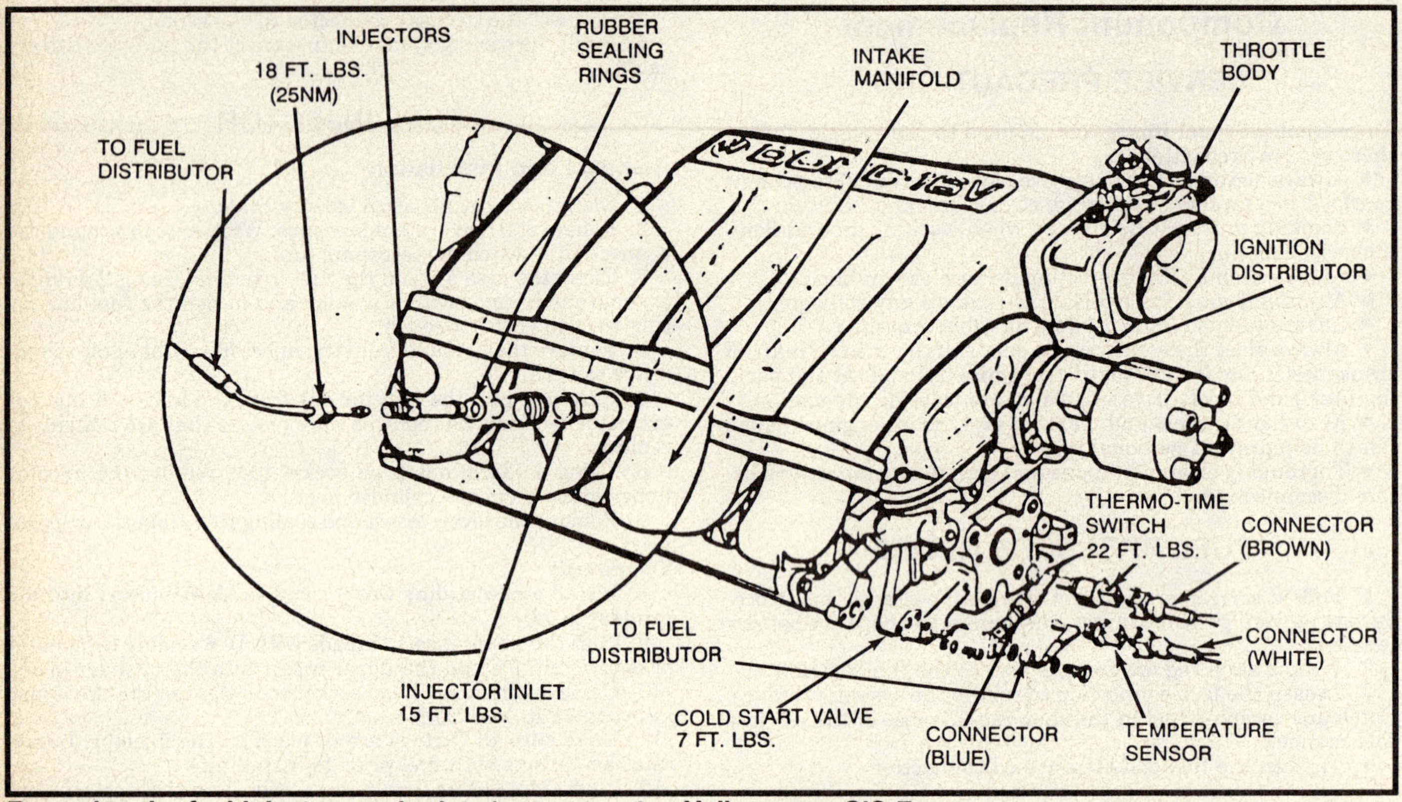

Removing the fuel injectors and related components—Volkswagen CIS-E

NOTE: When replacing the old fuel pump with new pump Part No. 191-906-091C, the new pump has a positive lock connector that differs from the earlier pumps. The electrical connectors of the fuel pump wiring harness must be changed when installing this newer pump.

3. Raise the vehicle and support it safely.
4. Disconnect the electrical connector from the fuel pump.
5. Remove both wire terminals from the connector housing.
6. Discard the old connector and protective boot.
7. Install the new protective boot over the wires and insert the wire terminals into the new connector.
8. Slide the protective boot over the connector.
9. Disconnect fuel line from the fuel pump.
10. Remove pump mounting bolts and remove the pump.

To install:

11. Position the fuel pump it place and secure it with the mounting bolts.
12. Install new O-ring seals when connecting the fuel lines to the pump.
13. Connect the electrical connector to the fuel pump.
14. Lower the vehicle.
15. Connect the negative battery cable. Turn the ignition switch **ON** and **OFF** several times to pressurize the system.
16. Raise the and support it safely, start the engine and observe for fuel leaks.
17. Shut the engine off and lower the vehicle.

TRANSFER PUMP

This pump is located inside the fuel tank and is attached to the end of the sending unit. It is not used in all applications.

Removal and Installation

1. Disconnect the negative battery cable.
2. Relieve the fuel system pressure. Wrap a cloth around the connection to catch any escaping fuel.
3. Open the rear hatch or trunk and pull back the carpet.
4. Remove the access plate retaining screws and remove the plate from the floor.
5. Disconnect the electrical connector from the sending unit.
6. Remove the fuel hoses from the sending unit and plug.
7. Unscrew the plastic cap and lift the sending unit and transfer pump assembly from the fuel tank.
8. Remove the transfer pump from the sending unit.

To install:

9. Assemble the transfer pump to the sending unit.
10. Lubricate the sealing ring with fuel prior to installing and install it.
11. Carefully lower the transfer pump and sending unit assembly into the tank.
12. Using a special tool, screw the plastic cap in place.
13. Connect the fuel hoses to the fuel pump and secure the clamps.
14. Connect the electrical connector to the sending unit.
15. Connect the negative battery cable. Turn the ignition switch **ON** and **OFF** several times to pressurize the system and check for leaks.
16. Install the access plate and secure it with the retaining screws.

FUEL FILTER

The fuel filter is located under the vehicle, in front of the fuel tank. On some vehicles it is mounted in a bracket assembly with the fuel pump. In other vehicles it is separate.

Removal and Installation

1. Disconnect the negative battery cable.

2. Relieve the fuel system pressure. Wrap a cloth around the connection to catch any escaping fuel.
3. Raise the vehicle and support it safely.
4. Throughly clean all fuel lines connected to the fuel filter and remove them.
5. Unscrew the bracket clamping bolt and remove the fuel filter from the bracket.

To install:

6. Slide the fuel filter in to its mounting bracket, making sure the arrow is pointing to flow towards the fuel distributor.
7. Connect the fuel lines to the filter and tighten the retaining bolts to 18 ft. lbs (25 Nm).

NOTE: Always install new seals and hoses clamps.

8. Connect the negative battery cable. Turn the ignition switch **ON** and **OFF** several times to pressurize the system and check for leaks.

ACCUMULATOR

The accumulator is located under the vehicle, in front of the fuel tank. On some vehicles it is mounted in a bracket assembly with the fuel pump and fuel filter. In other vehicles it is mounted separately.

Removal and Installation

1. Disconnect the negative battery cable.
2. Relieve the fuel system pressure. Wrap a cloth around the connection to catch any escaping fuel.
3. Throughly clean all fuel lines connected to accumulator and remove them.
4. Loosen the bracket clamping bolt and slide the accumulator from the bracket.

To install:

5. Slide the accumulator into its mounting bracket and secure the clamping bolt.
6. Connect the fuel lines and tighten the fittings to 15 ft. lbs. (20 Nm).
7. Connect the negative battery cable. Turn the ignition switch **ON** and **OFF** several times to pressurize the system and check for leaks.

OXYGEN SENSOR

The oxygen sensor is located in the exhaust manifold. The exhaust system should be slightly warm to make removal easier.

Removal and Installation

1. Disconnect the negative battery cable.
2. Disconnect the electrical connector at the oxygen sensor.
3. Unscrew the oxygen sensor and remove it from the exhaust manifold.

To install:

4. Before installing the oxygen sensor, coat the threads with anti-seize compound and install.
5. Connect the electrical sensor to the oxygen sensor. Connect the negative battery cable.

AIR FLOW SENSOR ASSEMBLY

Removal and Installation

1. Disconnect the negative battery cable.
2. Relieve the fuel system pressure. Wrap a cloth around the connection to catch any escaping fuel.
3. Remove rubber air duct from the mixture control unit. If equipped with a venturi air intake screen, remove it from venturi.
4. Throughly clean all fuel lines on the fuel distributor.
5. Disconnect the fuel lines at the fuel distributor. Plug the ends of the lines and the openings of the fuel distributor.
6. Disconnect the harness connectors from the differential pressure regulator and the potentiometer.
7. Remove the pressure regulator and bracket assembly and place it to the side.
8. Remove the screws retaining the air flow sensor assembly to the upper air filter housing and carefully lift the unit out.

NOTE: Prior to installing the mixture control unit, the basic adjustment of the sensor plate must be preformed whenever the fuel distributor or air flow sensor has been replaced.

To install:

9. Inspect the air filter element and replace if needed. Clean any dirt or contamination from the upper and lower air filter housings.
10. Install a new gasket on the air filter upper housing and position the mixture control unit in place.
11. Install the mixture control unit and retaining screws. Tighten the retaining screws to 31 inch lbs. (3.5 Nm) in a diagonal pattern.
12. Install the pressure regulator and bracket assembly. Install the retaining screws and torque it to 31 inch lbs. (3.5 Nm).
13. Connect the harness connectors to the differential pressure regulator and the potentiometer.
14. Connect the fuel injector lines at the top of the fuel distributor. Torque the retaining screws to 7 ft. lbs. (10 Nm).
15. Connect the fuel line from the cold start valve to the fuel distributor. Torque the retaining screw to 15 ft. lbs. (20 Nm).
16. Install the air intake screen, if equipped. Install the rubber air duct on the mixture control unit and tighten the retaining clamp.

NOTE: Improper installation or distortion of the air intake screen may result in interference with the sensor plate. Make sure the screen is installed with the word (open) on the flange facing up.

17. Check the air flow sensor plate for proper position. Adjust if needed.
18. Connect the negative battery cable.
19. Turn the ignition switch **ON** and **OFF** several times to pressurize the system. Check for fuel leaks before starting the engine and after starting.

FUEL DISTRIBUTOR

Removal and Installation

1. Disconnect the negative battery cable.
2. Relieve the fuel system pressure. Wrap a cloth around the connection to catch any escaping fuel.
3. Throughly clean all fuel lines on the fuel distributor and remove them.
4. Disconnect the harness connectors from the differential pressure regulator and potentiometer.
5. Remove the pressure regulator and bracket retaining screws and remove it as an assembly. Place to the side.
6. Remove the air flow sensor retaining screws and remove both the air flow sensor and fuel distributor as an assembly.
7. Remove the fuel distributor retaining screws and carefully lift it from the air flow sensor. Make sure that the control plunger does not fall out of the fuel distributor.

To install:

8. If the control plunger has been removed from the fuel distributor, moisten it with clean gasoline before installing it into the fuel distributor. Be sure to insert the small shoulder on the control plunger first.
9. Install a new O-ring seal before assembling the fuel distributor to the air flow sensor.
10. Assemble the fuel distributor to the air flow sensor and in-

stall the retaining screws. Tighten the retaining screws to 31 inch lbs. (3.5 Nm).

NOTE: When ever the air flow sensor or fuel distributor is replaced, the sensor plate basic adjustment must be performed.

11. Install the air flow sensor and fuel distributor as an assembly to the upper air filter housing.
12. Install the air flow sensor retaining screws. Tighten the retaining screws to 31 inch lbs. (3.5 Nm).
13. Install the pressure regulator bracket assembly and retaining screws. Tighten the retaining screws to 31 inch lbs. (3.5 Nm).
14. Connect the fuel injector lines at the top of the fuel distributor. Torque the retaining screws to 7 ft. lbs. (10 Nm).
15. Connect the fuel line from the cold start valve to the fuel distributor. Torque the retaining screw to 15 ft. lbs. (20 Nm).
16. Install the air intake screen if equipped and the air intake hose.
17. Connect the negative battery cable.
18. Turn the ignition switch **ON** and **OFF** several times to pressurize the system. Check for fuel leaks before starting the engine and after starting.

Adjustment

1. Remove the air inlet boot from the sensor plate and operate the starter for about 10 seconds to build fuel system pressure. The engine will not start.
2. Measure the distance from the top of the air flow sensor housing to the sensor plate at the point nearest the fuel distributor.
3. Use a magnet on the sensor plate bolt to lift the plate until resistance is felt. The free-play movement should be 0.019–0.118 in. (0.5–3mm).
4. To adjust, remove the fuel distributor and use a depth caliper to measure the distance from the shoulder of the nut to the stop screw. Distance should be 0.024 in. (0.6mm).
5. Turn the stop screw in or out as required. Turning the screw in ¼ turn will increase the free-play by 0.050 in. (1.3mm).
6. After adjusting plunger free-play, idle speed and CO must be adjusted.

PRESSURE REGULATOR

The pressure regulator is normally located in a bracket, mounted on or near the air flow sensor assembly.

Removal and Installation

1. Disconnect the negative battery cable.
2. Relieve the fuel system pressure. Wrap a cloth around the connection to catch any escaping fuel.
3. Throughly clean the fuel line fittings attached to the pressure regulator and remove them.
4. Unplug the vacuum hose from the pressure regulator.
5. Loosen the bracket clamping bolt and remove the pressure regulator.

To install:

6. Slide the pressure regulator into its mounting bracket and tighten the bracket clamping bolt.
7. Connect the fuel lines to the pressure regulator and plug in the vacuum hose.
8. Connect the negative battery cable.
9. Turn the ignition switch **ON** and **OFF** several times to pressurize the system. Check for fuel leaks before starting the engine and after starting.

DIFFERENTIAL PRESSURE REGULATOR

The differential pressure regulator is attached to the side of the fuel distributor.

Removal and Installation

1. Disconnect the negative battery cable.
2. Relieve the fuel system pressure. Wrap a cloth around the connection to catch any escaping fuel.
3. Disconnect the harness connector from the differential pressure regulator.
4. Before removing, thoroughly clean the exterior and surrounding area of the differential pressure regulator to prevent dirt from entering the fuel distributor or the regulator.
5. Remove the screws retaining the differential pressure regulator to the fuel distributor.
6. Carefully remove the differential pressure regulator from the fuel distributor. Do not lose the O-ring seals as the regulator is being removed.

To install:

7. Replace the O-ring seals before installing the differential pressure regulator.
8. Position the differential pressure regulator to the fuel distributor and install the retaining screws. The retaining screw is nonmagnetic and should not be replaced with a standard screw. Use the original retaining screw or an OEM replacement screw.
9. Connect the negative battery cable.
10. Turn the ignition switch **ON** and **OFF** several times to pressurize the system. Check for fuel leaks before starting the engine and after starting.

SENSOR PLATE HEIGHT

Adjustment

1. Operate the fuel pump for a few seconds to build up system pressure.
2. Remove the rubber connecting boot, from the air sensor housing.
3. Check the position of the sensor plate. The upper edge of the sensor plate must be 1.9 plus (+) 0.2 mm below the cone edge of the air venturi on the side closest to the fuel distributor.
4. Adjust the sensor plate by lifting if fully and bending the wire clip. Be careful not to scratch the venturi of the air flow sensor. Do not bend the leaf spring.

SENSOR PLATE CLEARANCE

Adjustment

1. Lift the sensor plate until even resistance is felt. Check the plated clearance. Ther should be noticable clearance at the minimum and a maximum clearance of 1.9 plus (+) 0.2mm. The sensor plate should be even with edge of the venturi cone.
2. If the sensor plate clearances check good, continue to the Sensor Plate and Control Plunger Movement checks.
3. If the sensor plate clearances do not check good, perform a sensor plate lever basic adjustment.

SENSOR PLATE LEVER

Basic Adjustment

NOTE: This adjustment must be made whenever the air flow sensor or the fuel distributor is replaced.

1. Disconnect the negative battery cable.
2. Remove the fuel distributor from the mixture control unit.
3. Check the distance between the seating pads of the removed fuel distributor on the air flow sensor and the roller on the sensor plate lever.
4. The distance should be 0.748 ± 0.004 in. (19.0 ± 0.1mm).
5. If the measurement is incorrect, adjust the CO% adjusting screw to obtain the correct measurement.
6. Reinstall the fuel distributor and recheck the sensor plate clearance.

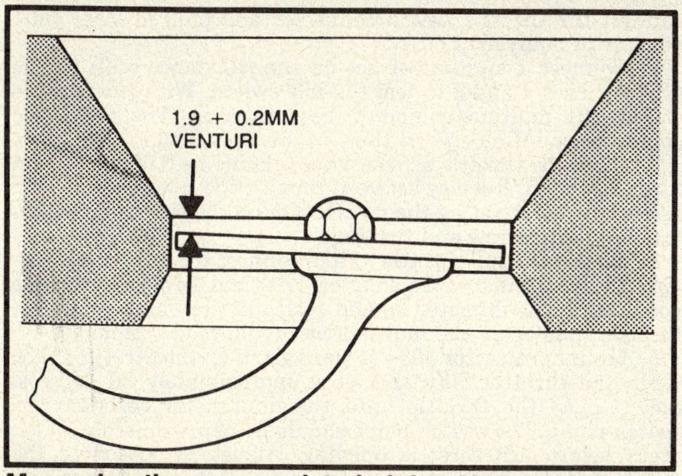

Measuring the sensor plate height—Volkswagen CIS-E

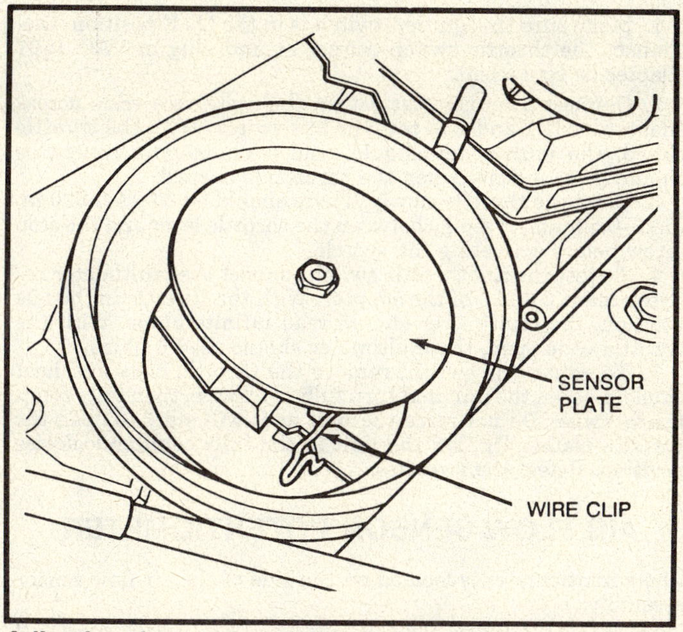

Adjusting the sensor plate height—Volkswagen CIS-E

7. If the sensor plate clearance is correct, readjust the CO%.
8. If it is not correct, remove the fuel distributor and adjust the sensor plate clearance with the control plunger stop screw.
   a. Turn clockwise to increase clearance
   b. Turn counterclockwise to decrease clearance.
   c. Each ¼ turn will change the sensor plate position approximately 0.50 in. (1.3mm).
9. Reinstall the fuel distributor and check the sensor plate height. The distance should be 0.024 in. (0.6mm) between the stop screw and flange of the hex nut.
10. Connect the negative battery cable.

NOTE: After adjusting the sensor plate position, readjust the idle speed and mA current (CO value) reading.

SENSOR PLATE

Centering Adjustment

1. Disconnect the negative battery cable.

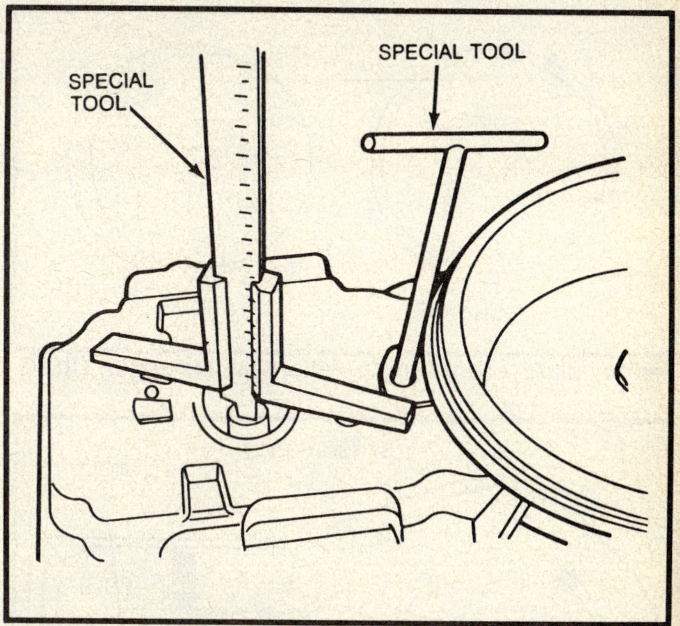

Sensor plate basic adjustment, with fuel distributor removed—measuring distance between seating pads and roller—Volkswagen CIS-E

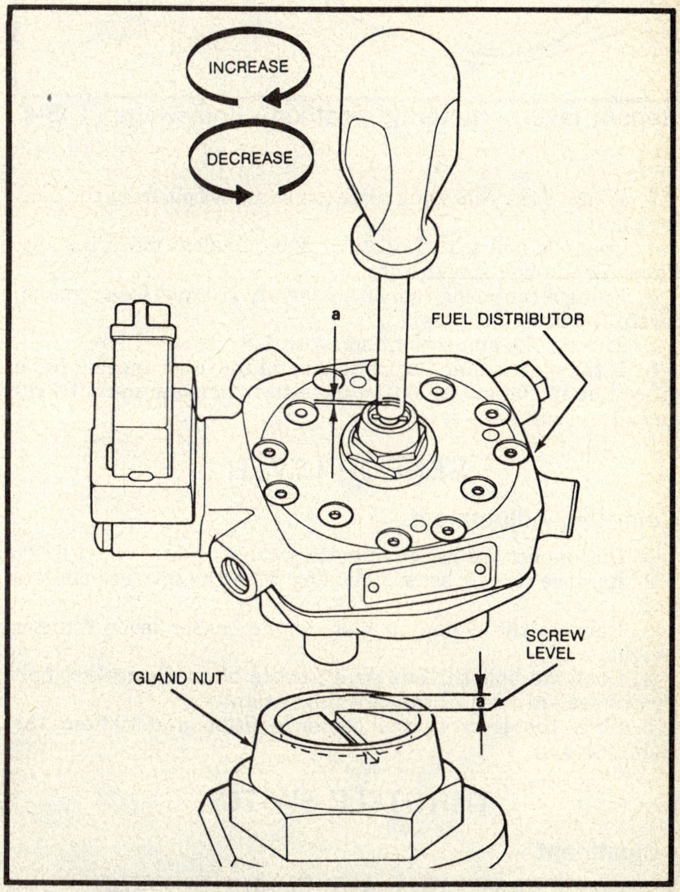

Adjusting the sensor plate with the control plunger stop screw—Volkswagen CIS-E

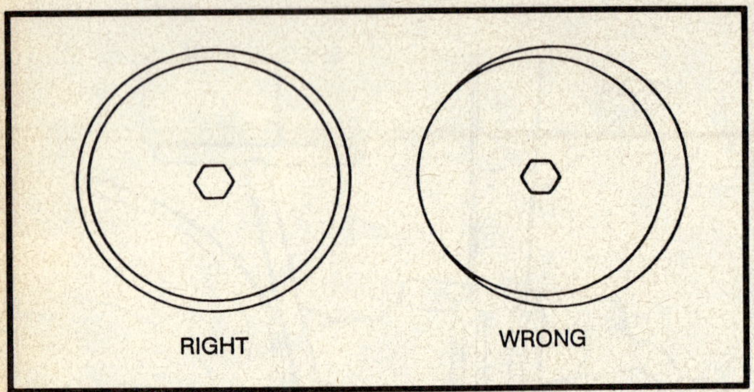

Sensor plate—centering position—Volkswagen CIS-E

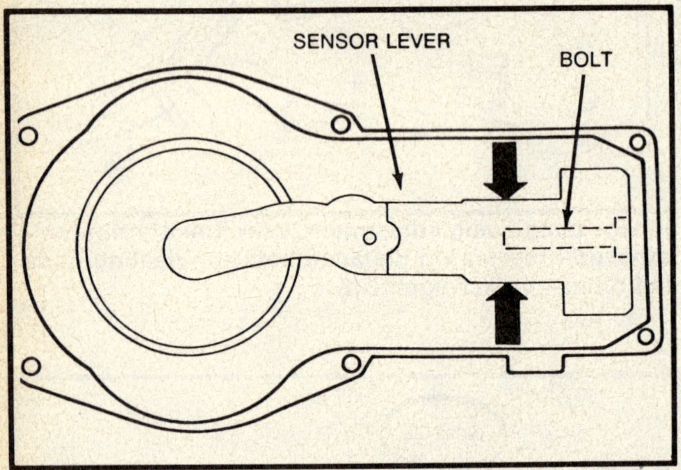

Sensor lever—centering position—Volkswagen CIS-E

2. Remove the centering plate adjustment bolt from the sensor plate.
3. Coat the bolt with Loctite or an equivalent tread sealant. Reinstall the bolt finger tight.
4. Using a centering tool or a 0.004 in. (10mm) feeler gauge, carefully center the plate.
5. Torque the adjustment screw to 4 ft. lbs. (5–7 Nm).
6. If the plate cannot be centered and the lever appears to be off-center in the air cone, the air sensor housing must be removed to center the lever.

SENSOR LEVER

Centering Adjustment

1. Disconnect the negative battery cable.
2. Remove the air sensor housing from the mixture control unit.
3. Remove the clamping bolt on the sensor lever counter weight.
4. Coat the bolt threads with Loctite or an equivalent bolt thread sealant. Install the bolt finger tight.
5. Slide the lever to the center position and tighten the clamping bolt.

THROTTLE SWITCH

Adjustment

WITH SEPARATE IDLE AND FULL THROTTLE SWITCHES

1. Make sure the ignition switch is in the **OFF** position. Disconnect the throttle switch connector and plug in VAG 1501 adapter or equivalent.
2. Connect a multimeter set on the resistance scale across adapter leads 1 and 2 to test the idle switch. With the throttle closed, the multimeter should read 0 ohms. The resistance should go to infinity when the throttle is opened.
3. Close the throttle slowly. There should be 0.006–0.020 in. (0.15–0.50 mm) freeplay between the throttle lever and the stop screw before activating the switch. Adjust the switch by loosen the mounting screw and turning.
4. To test the full throttle switch connect the multimeter between leads 2 and 3 of the adapter. With the throttle in the idle position, the multimeter should read infinite ohms. With the throttle wide open, the multimeter should read 0 ohms.
5. Mount protractor 3084 or equivalent, on the throttle shaft. Open the throttle fully and close approximately 30 degrees. Slowly open the throttle until the multimeter registers the switch point. The switch point should be approximately 10 degrees before full throttle opening. Adjust by loosening the switch and moving the unit.

WITH INTEGRATED IDLE, FULL THROTTLE SWITCH AND THROTTLE POTENTIOMETER

1. Make sure the ignition switch is in the **OFF** position. Disconnect the throttle switch connector and plug in VAG 1501 adapter or equivalent.
2. Connect a multimeter set on the resistance scale across adapter leads 1 and 2 to test the idle switch. With the throttle closed, the multimeter should read 0 ohms. The resistance should go to infinity when the throttle is opened.
3. Close the throttle slowly. There should be 0.006–0.020 in. (0.15–0.50 mm) freeplay between the throttle lever and the stop screw before activating the switch.
4. To test the full throttle switch connect the multimeter between leads 2 and 3 of the adapter. With the throttle in the idle position, the multimeter should read infinite ohms. With the throttle wide open, the multimeter should read 0 ohms.
5. To adjust the switch, remove the throttle body and heat shield. Loosen the unit and turn fully clockwise, then fully counterclockwise. Do not force the unit as it will start to open the throttle plates. Tighten the unit in the fully counterclockwise position. Retest the switches.

AIR FLOW SENSOR POTENTIOMETER

The potentiometer is located on the side of the air flow sensor assembly.

Removal and Installation

1. Disconnect the negative battery cable.
2. Remove the intake air boot from the mixture control unit.
3. Throughly clean all fuel lines on the fuel distributor and remove them.
4. Completely remove the air flow sensor from the air filter housing.
5. Disconnect the harness connector from the potentiometer.
6. Remove the retaining screws from the potentiometer and carefully remove the potentiometer from the air flow sensor body.

NOTE: After the potentiometer has been removed, do not touch the resistance strip or the contact whiskers inside of the potentiometer. This will cause damage to the potentiometer.

To install:

7. Carefully position the potentiometer to the air flow sensor and install the retaining screws.
8. Adjust the potentiometer.
9. Install the mixture control unit to the air filter housing.
10. Connect the fuel lines to the fuel distributor.

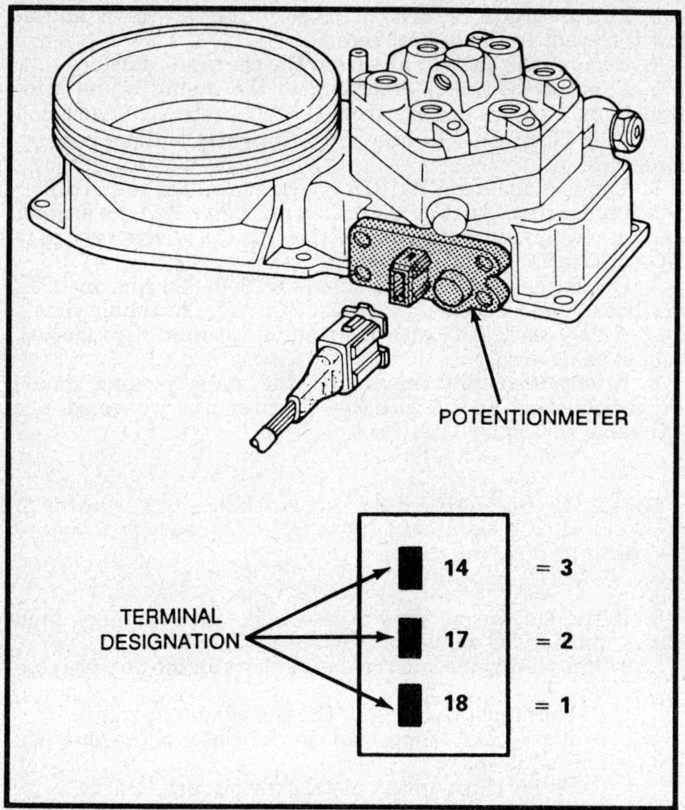

Checking the potentiometer—Volkswagen CIS-E

11. Connect the negative battery cable.
12. Position the air intake duct to the mixture control unit and tighten the retaining clamp.
13. Turn the ignition switch **ON** and **OFF** several times to pressurize the system. Check for fuel leaks before starting the engine and after starting.

Adjustment

FOX

1. Lift the sensor plate using special tool VW 1348/1, VW 377 or equivalent, up to the lower edge (on the fuel distributor side) of the air flow sensor.
2. Loosen the potentiometer screws just enough to hold it in position.
3. Connect adapter VW 1501 between potentiometer and the wiring connector.
4. Connect multimeter between terminal No. 2 and ground.
5. Set the multimeter to the 20 VDC scale and turn the ignition switch **ON**.

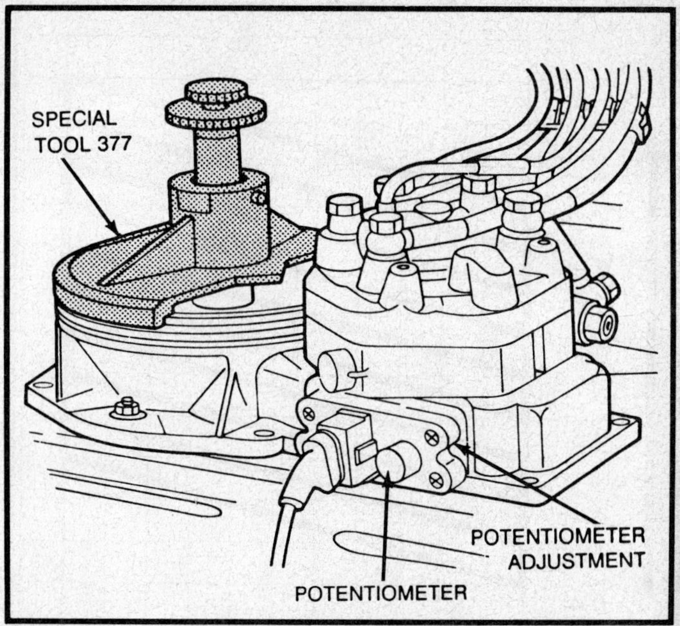

Adjusting the potentiometer—Volkswagen CIS-E

6. Adjust the potentiometer until the voltage reads 0.2–0.3 volts.
7. Lift the sensor plate up to the stop until the voltage reads 7.0 volts.
8. Turn the ignition switch **OFF**.
9. Tighten the potentiometer screws and seal with sealing compound or paint.

EXCEPT FOX

1. Connect adapter VW 1501 between potentiometer and the wiring connector.
2. Connect multimeter between terminal No. 1 and 3.
3. Set the multimeter to the 20 VDC scale and turn the ignition switch **ON**. The voltage must be between 4.35 volts and 5.35 volts.
4. Turn the ignition switch **OFF**. Connect the multimeter between terminals 1 and 3. Run the engine for a minimum of 2 minutes.
5. Measure the voltage (V1) between terminals 1 and 3. Measure the voltage (V2) between terminals 2 and 3. Compare the voltages to the chart.
6. If the voltage does not fall within the range of the chart, remove the sealing compound over the trim screw and adjust the screw until the range is met.
7. If the range can not be met, replace the air flow sensor. Cover the trim screw with silicone sealer.

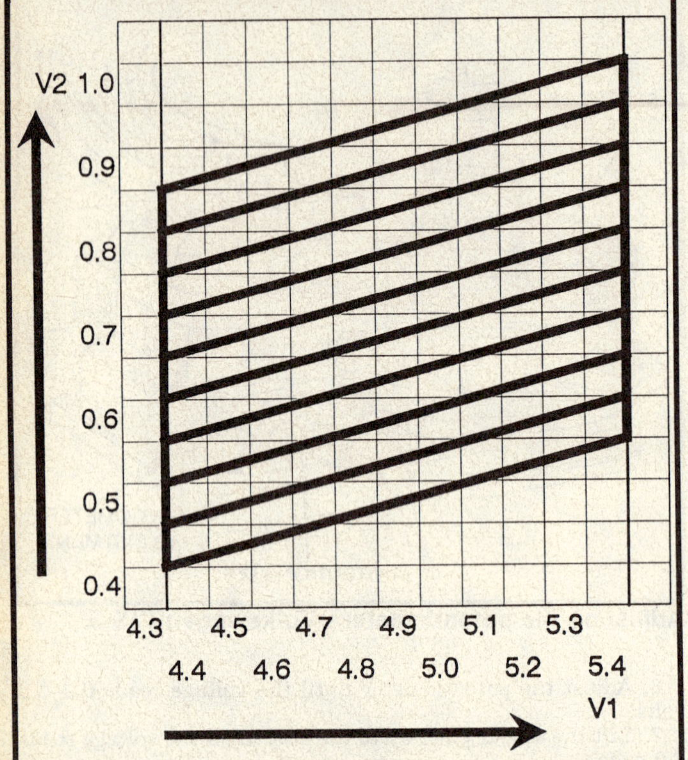

V1 and V2 values for adjusting the air flow sensor potentiometer

IDLE SPEED AND MIXTURE (CO)

Adjustments

NOTE: The following information is being published from the latest information available at the time of publication. If the information differs from the values given on the underhood emission control label, use the data on the label.

FOX

The idle speed, mixture adjustment and ignition timing must be checked together. Exhaust gas mixture must be checked and adjusted before measuring milliamps (mA), with a special meter and with the oxygen sensor connected. Refer to the underhood vacuum diagram for all hose locations.

1. Pinch close the vacuum line from the idle boost valve to the throttle intake chamber. Start the engine and allow it to reach normal operating temperature. The radiator fan must come on at least once.
2. Switch **OFF** all accessories. The radiator fan and air conditioner must not be running during the test.
3. Pull the crankcase breather hose off the valve cover and allow it to vent to the atmosphere.
4. Remove the suction line from the charcoal canister
5. Connect CO tester according to the manufacturer's instructions. Remove the cap from the CO probe receptacle and insert the CO probe. The hose must fit tightly so there is no exhaust leak.
6. Connect adapter VW 1315 A/1 or equivalent to the differential pressure regulator. Connect a multimeter to the adapter and the oxygen sensor test connection. Set the selector switch to DCA 200 mA.
7. Check the idle. The idle should be 840–880 rpm with the idle boost hose blocked off. Check the timing. The timing should be 5–7 degrees BTDC with distributor vacuum disconnected. Adjust as necessary.
8. Briefly accelerate the engine, the meter reading should range between 4–16 mA with the oxygen sensor connected. The CO value should be 0.3–1.2%.

NOTE: If the engine does not run after the adapter is connected, the connections may be improper. Reverse the plug and try again.

9. If the current reading is less than 4mA or more than 15mA, remove CO adjustment plug as follows:
   a. Stop the engine and remove the boot on the mixture control unit.
   b. Centerpunch the plug in the CO adjusting hole.
   c. Drill a $^{3}/_{32}$ in. (2.5mm) hole in the center of the plug $^{9}/_{64}$–$^{5}/_{32}$ (3.5mm) deep.
   d. Carefully clean up any metal shavings and insert a ⅛ in. sheet metal screw into the drilled hole and remove the screw and plug.
10. Start the engine and allow it to idle.
11. Adjust the current reading on the multimeter by turning the CO adjusting screw with tool P377 or equivalent to obtain 9–11mA (reading fluctuates). Turn the adjusting screw clockwise to raise the current reading (CO higher) and screw clockwise to raise the current reading (CO lower).

NOTE: Do not push adjustment tool down when making CO adjustment and do not accelerate the engine with the tool in place. Remove the tool after each adjustment and accelerate the engine briefly before reading the multimeter. Always adjust in direction from high to low.

12. Run the engine at 3000 rpm for approximately 3 minutes and recheck the meter readings. After CO adjustment is complete, recheck the idle speed and adjust if necessary. Reconnect the crankcase breather hose. If, after reconnecting the breather hose, the current reading drops below specifications, an oil change may be necessary, due to excessive gasoline dilution.
13. Turn the ignition switch **OFF** and drive in a new mixture

plug, flush with the mixture control unit. This must be done to meet federal anti-tamper regulations.

14. Remove all test equipment, attach CO probe cap and remove the device used to pinch the idle speed boost valve hose.

EXCEPT FOX

Before 9-90 production

1. Start the engine and allow it to reach normal operating temperature. The radiator fan must come on at least once.
2. Switch **OFF** all accessories. The radiator fan and air conditioner must not be running during the test.
3. Switch **OFF** the engine. Connect VAG 1367 engine analyzer or equivalent. Connect the CO tester to the CO tap tube. Connection must be gas tight or readings may be incorrect.
4. Start the engine and check the timing at idle. Timing should be 4–8 degrees BTDC. Adjust the timing if necessary. Check the idle speed. The idle speed should be 700–900 rpm. The idle speed is not adjustable.
5. Check the CO. The CO level should be 0.2–1.2%.
6. If the CO is not within range, turn **OFF** the engine and adjust as follows:
   a. Disconnect the electrical connector from the black colored charcoal canister solenoid valve.
   b. Disconnect the crankcase breather hoses from the breather housing. Be sure to leave the restriction remains in the hose. The hoses should only be able to draw in fresh air.
   c. With the ignition switch in the **OFF** position, disconnect the differential pressure regulator plug and install test adapter VAG 1315 A/1 or equivalent.
   d. Set a multimeter to the 200 mA range and connect to the test adapter. With the ignition switch in the **ON** position the control current must be a positive 90–110 mA. If the current is negative, reverse the connections.
   e. Start the engine and allow to idle. Adjust the CO screw so the current is 0.0–5.0 mA, fluctuating. The current should center around 2.5 mA. Do not press down on the adjusting wrench or accelerate the engine with the wrench inserted. Blip the throttle before each measurement with the wrench removed.
   f. If the current does not fluctuate, check the oxygen sensor and control.
7. After all testing is done, return the hoses and connection to their original states and read the fault codes. Erase the fault codes.

EXCEPT FOX

After 10-90 production

1. Start the engine and allow it to reach normal operating temperature. The radiator fan must come on at least once.
2. Switch **OFF** all accessories. The radiator fan and air conditioner must not be running during the test.
3. Switch **OFF** the engine. Connect VAG 1367 engine analyzer or equivalent. Connect VAG 1551 diagnostic tester. Connect the CO tester to the CO tap tube. Connecttion must be gas tight or readings may be incorrect.
4. Start the engine. Press button 1 on the tester to select the 'rapid data transfer' mode. Press buttons 0 and 1 to select 'engine electronics' mode. Press Q to enter.
5. Press the right arrow button and then 0 and 4 to select the 'initiate basic adjustment' mode. Press the Q button to enter, then press the 0 button twice and press the Q button again.
6. With the engin. in this mode the digital idle stabilizer is disables and the charcoal canister solenoid valve is activated.
7. Channel 9 of the display will show a value of 5. If any value but 5 is shown, check that the timing is 4–8 degrees BTDC and the idle speed is 800–100 rpm. The idle speed can not be adjusted. Check that the CO level is 0.2–1.2%.
8. If the values are within specifications, press the right arrow button, the 0 and 6 buttons and the Q button to select the 'end function' mode. If the CO level is not within specifications, adjust as follows:
   a. Disconnect the electrical connector from the black colored charcoal canister solenoid valve.
   b. Disconnect the crankcase breather hoses from the breather housing. Be sure to leave the restriction remains in the hose. The hoses should only be able to draw in fresh air.
   c. With the ignition switch in the **OFF** position, disconnect the differential pressure regulator plug and install test adapter VAG 1315 A/1 or equivalent.
   d. Set a multimeter to the 200 mA range and connect to the test adapter. With the ignition switch in the **ON** position the control current must be a positive 90–110 mA. If the current is negative, reverse the connections.
   e. Start the engine and allow to idle. Adjust the CO screw so the current is 0.0–5.0 mA, fluctuating. The current should center around 2.5 mA. Do not press down on the adjusting wrench or accelerate the engine with the wrench inserted. Blip the throttle before each measurement with the wrench removed.
   f. If the current does not fluctuate, check the oxygen sensor and control.

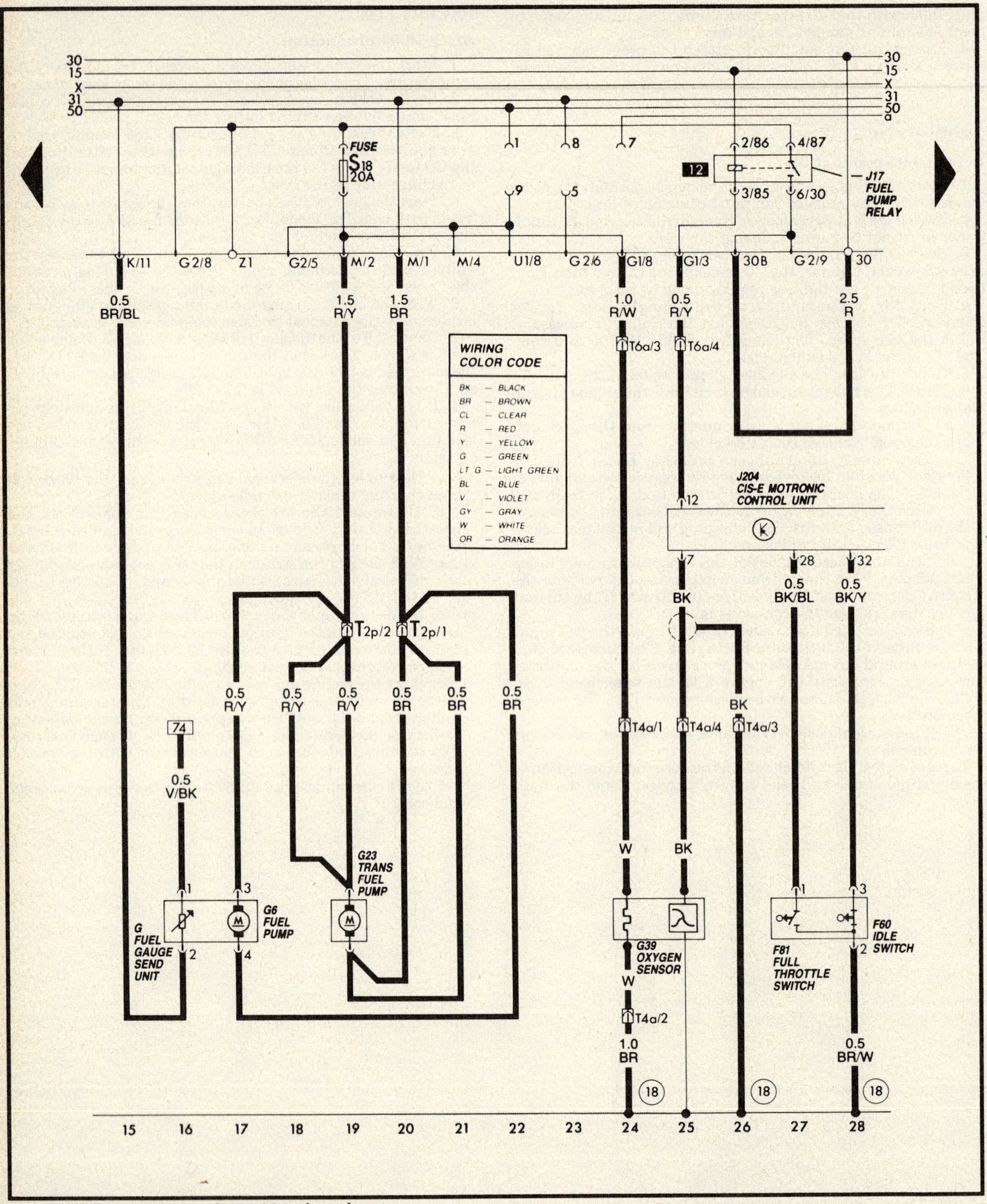

CIS-E engine system—Jetta 16 valve shown

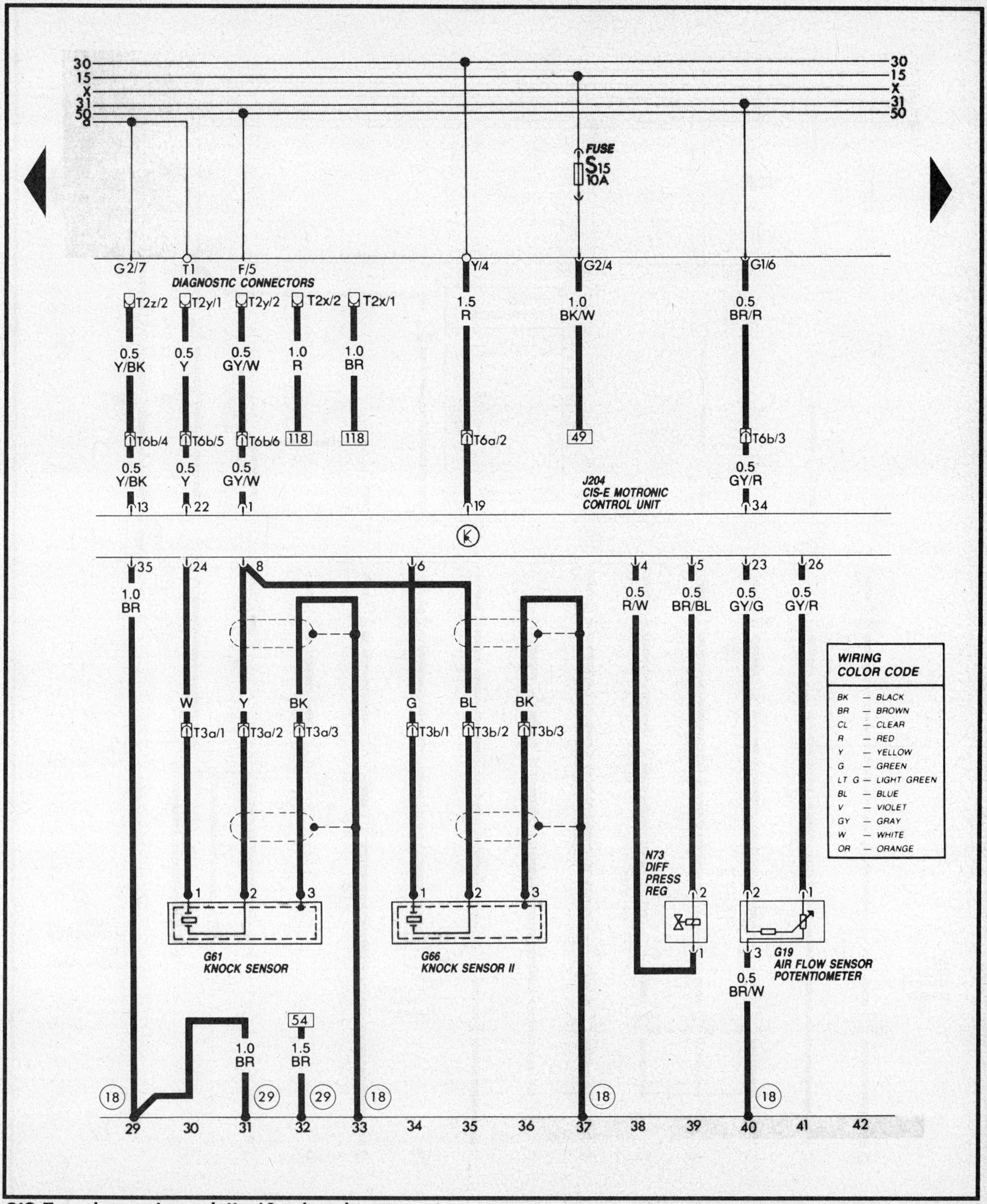

CIS-E engine system—Jetta 16 valve shown

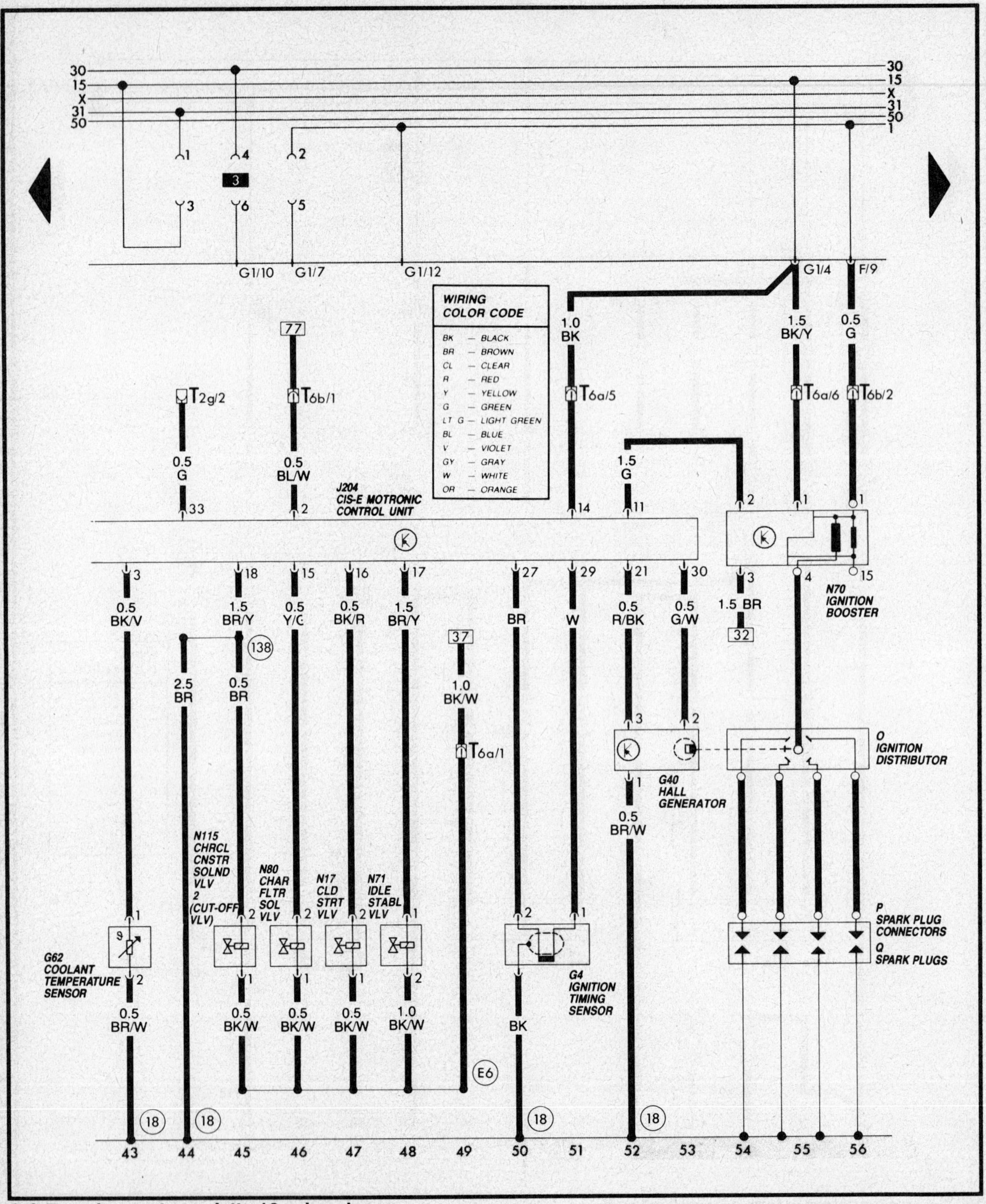

CIS-E engine system—Jetta 16 valve shown

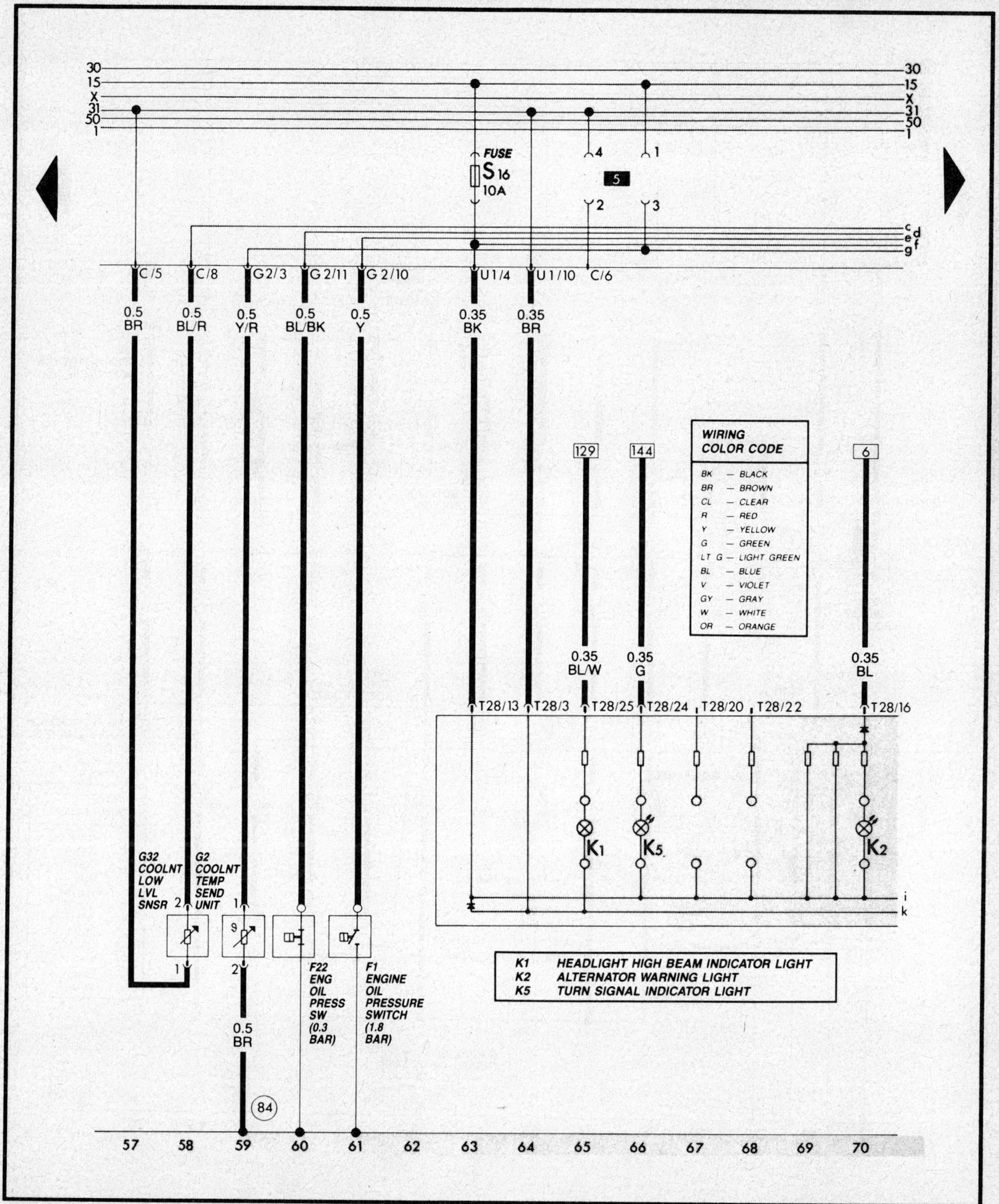

CIS-E engine system—Jetta 16 valve shown

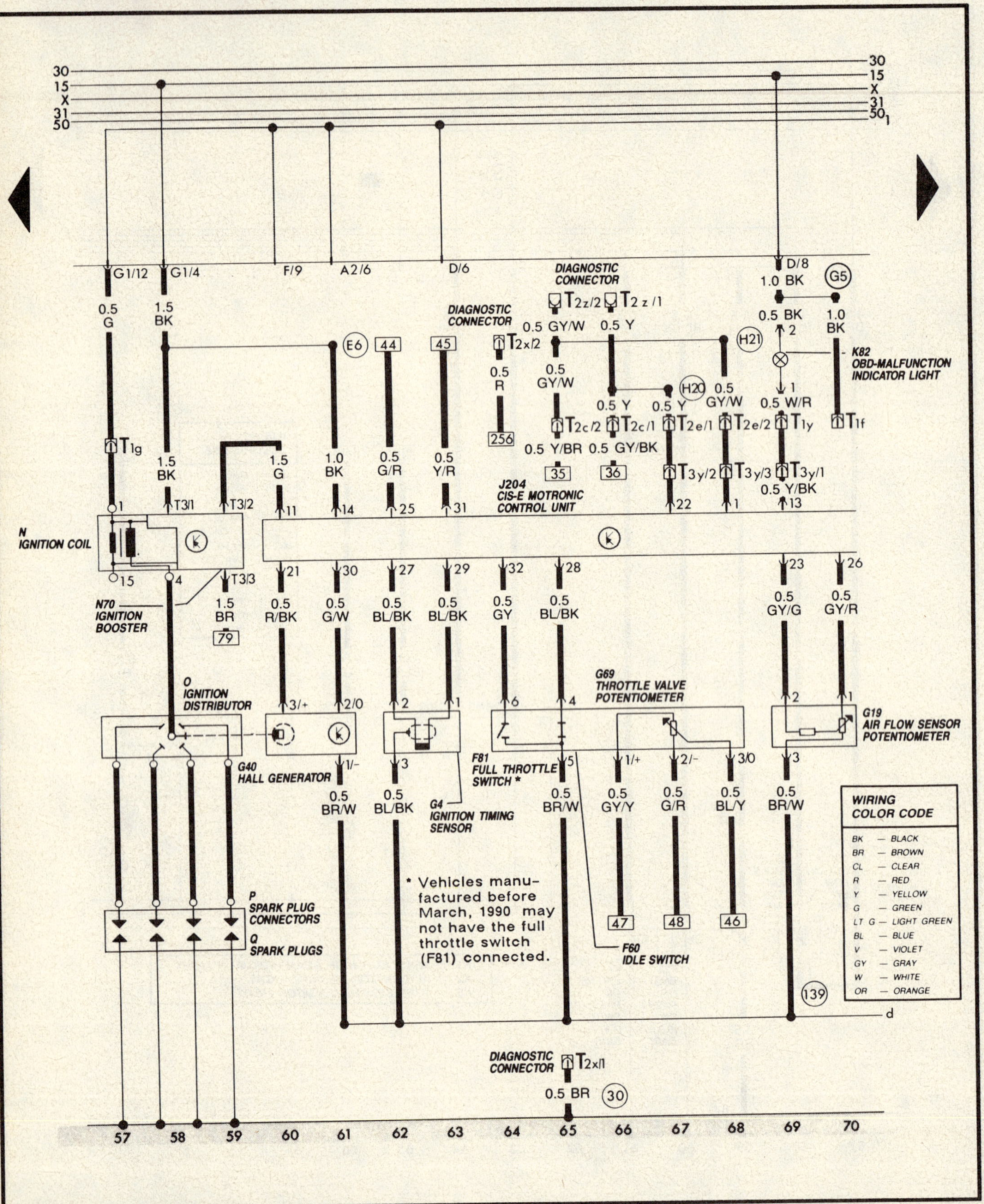

CIS-E engine system—Passat shown

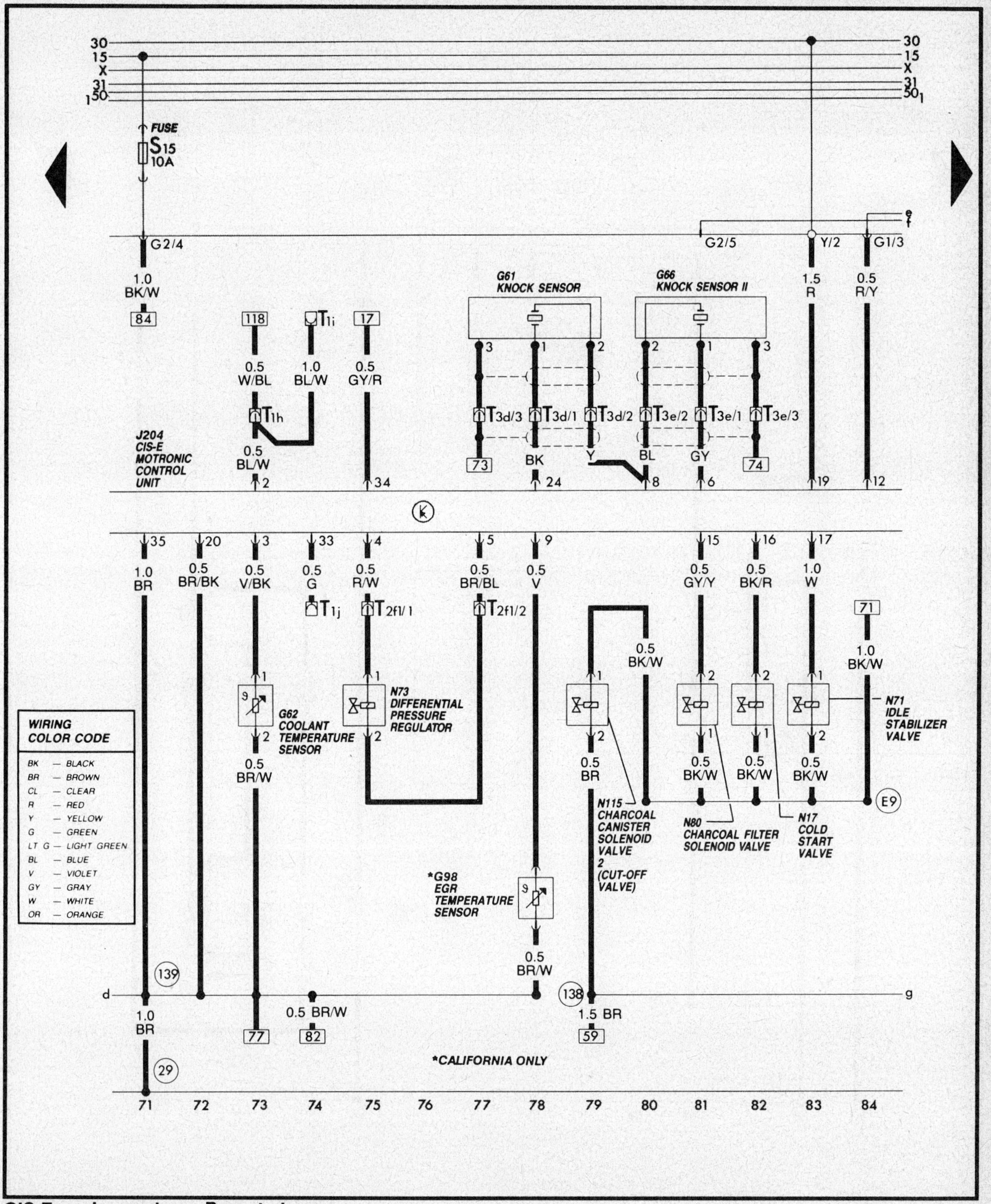

CIS-E engine system—Passat shown

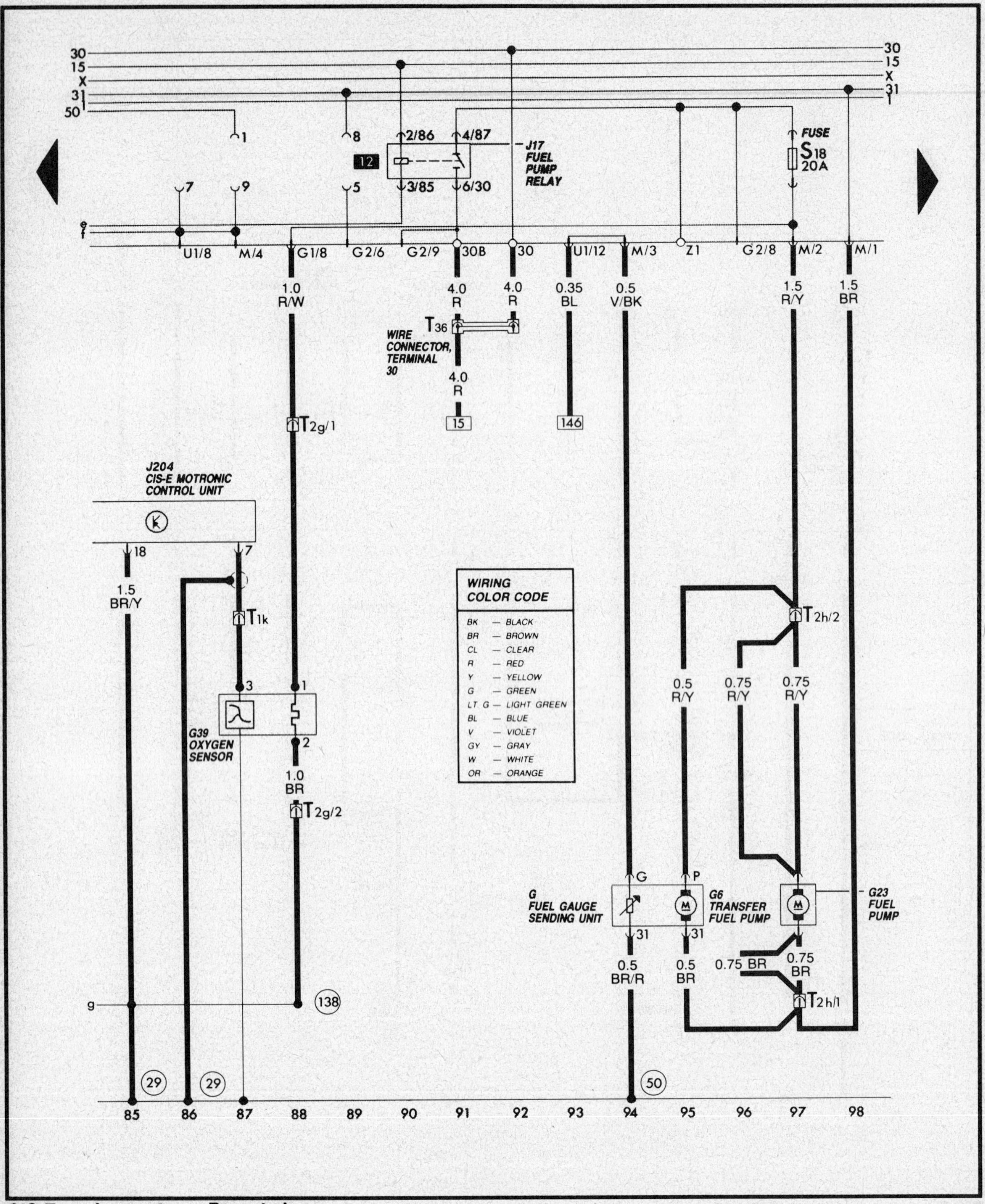

CIS-E engine system—Passat shown

VOLKSWAGEN DIGIFANT AND MOTRONIC FUEL INJECTION SYSTEMS

General Information

Volkswagen uses 2 similar all electronic fuel injection systems, Digifant and Motronic. These systems operate in similar fashions and use similar components. The names Digifant and Motronic may be used interchangebly for the purpose of this section. Digifant and Motronic systems are all electronic, using electric injectors at a lower injection pressure than the CIS type injection system. All injectors are operated by the Electronic Control Unit (ECU) so that injector opening timing and duration can be closely controlled for reduced emissions and improved fuel mileage. During deceleration above a specific rpm, the injectors are shut off to save fuel and reduce emissions. Except on Corrado, the Digifant system uses an air vane and potentiometer type air flow sensor. The Corrado uses a Manifold Absolute Pressure (MAP) sensor in the ECU and an inlet air temperature sensor to calculate air flow into the engine. All engines are equipped with an idle stabilizer valve to control idle speed. This is a motorized rotary valve that controls the amount of air allowed to bypass the throttle plate. Very precise control is maintained over a wide range of idle loads and temperatures.

The ECU on both systems has the ability to store fault codes but codes and other self diagnostic functions can only be read with the VAG 1551 tester. Procedures described here are for testing without the VAG 1551.

This system is all electronic and uses inputs from an air flow sensor with intake air temperature sensor, Hall effect sender, exhaust oxygen sensor, coolant temperature sensor, knock sen-

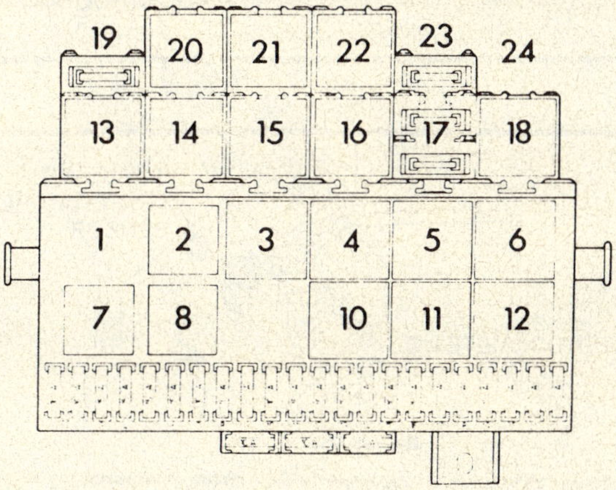

Relay locations on fuse/relay panel

1 — Digifant control unit relay
2 — Fuel pump relay
3 — Seat belt warning system relay
4 — Open
5 — A/C relay
6 — Dual horn relay
7 — Open
8 — Load reduction relay
10 — Intermittent wiper relay
11 — Open
12 — Emergency flasher relay
13 — Open
14 — Radiator cooling fan afterrun control unit
15 — Open
16 — Open
17 — A/C thermofuse
18 — Coolant low level control unit
19 — Power window fuse

Relay locations—Cabriolet shown

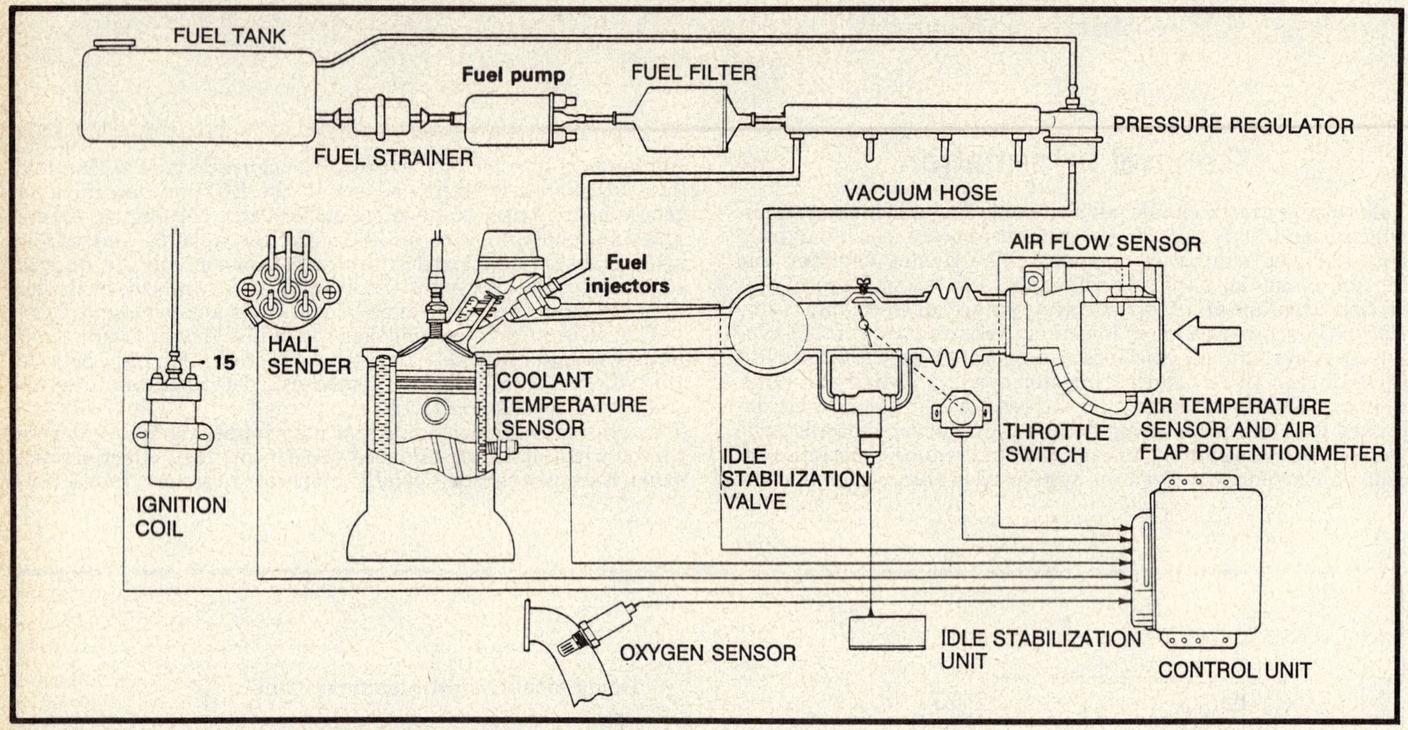

Digifant fuel and ignition control system—Volkswagen

sor and throttle position switches. The main output devices are an idle air stabilizer, electric fuel injectors and the ignition coil. The Corrado also uses an air temperature sensor in the intake tract and a manifold pressure sensor incorporated into the ECU. Most items can be tested separately.

ELECTRONIC CONTROL UNIT (ECU)

The ECU controls all fuel injection and ignition functions and also acts as a rev limiter. Above the maximum rpm rating, the ECU will shut off the fuel injectors to prevent over-revving the engine. If an input or output device fault is detected and it is serious enough to cause engine control problems, the ECU will switch into a "limp home" mode and the engine will be operated at a pre-set condition. The ECU includes a self diagnostic program that will store fault codes. These codes can be retrieved only with the special diagnostic tester VAG 1551 or equivalent. HYowever, most components can be tested individually. Tests described here are for trouble shooting without the special test equipment.

DATA SENSORS

Air Flow Sensor

This unit converts air flow to a voltage signal. Air enters and moves a spring-loaded vane which moves a potentiometer. The potentiometer modulates a voltage sent from the ECU and the return signal represents the mass of air flowing to the intake manifold. Air temperature is also measured and reported to the ECU. The air/fuel mixture adjustment is also in this unit, which is an adjusting screw that allows a certain amount of air to bypass the vane and enter the engine unmeasured. More unmeasured air means a leaner mixture.

Throttle Position Switches

On the throttle body are 2 switches that signal the ECU when the throttle is at idle or full throttle position. This information is used to calculate fuel shut-off and ignition timing during deceleration, idle stabilizer valve operation and full throttle enrichment. If there is a problem with any of these functions, check these switches first.

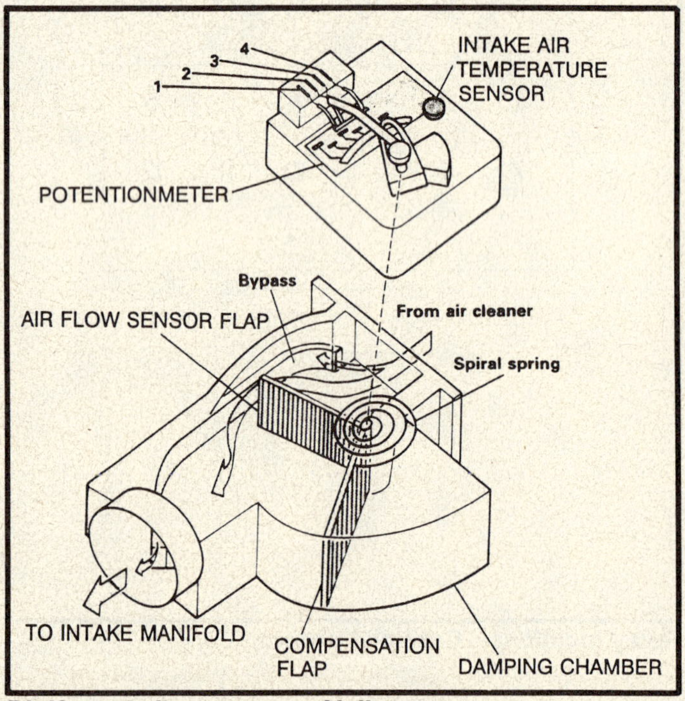

Digifant air flow sensor—Volkswagen

Throttle Potentiometer

On Corrado equipped with automatic transaxle, a throttle po-

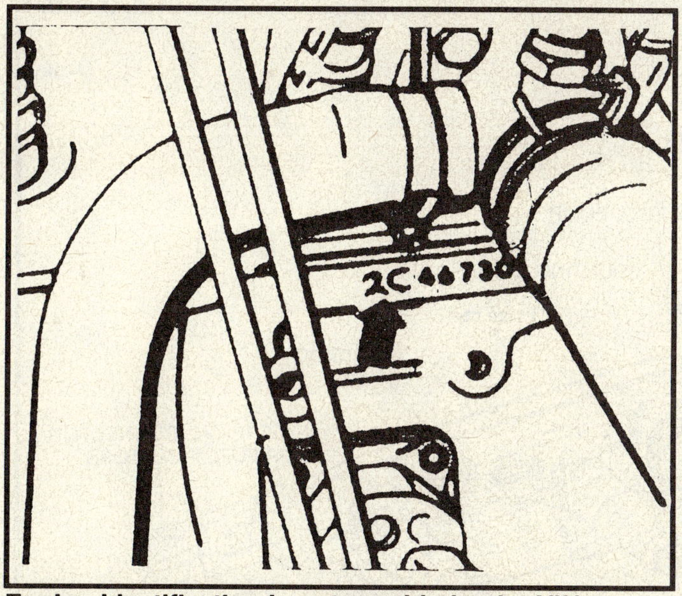

Engine identification is not provide by the VIN number, it is stamped in the cylinder block

tentiometer is mounted on the under side of the throttle body. It moves with the throttle to report to the ECU throttle position and speed of movement. This information is used to calculate fuel injection, ignition and idle stabilization control signals. This component should be checked if there is a problem with idle control or engine performance.

Coolant Temperature Sensor

Mounted in the coolant passage are 2 coolant sensors. The black sensor is for the gauge on the instrument panel. The blue sensor is for the engine management system. This sensor is a thermistor type, meaning as the temperature increases, its electrical resistance decreases. The ECU sends voltage to the sensor and the higher the temperature, the higher the return signal voltage. The sensor can be tested with an ohmmeter, using the graph and the coolant temperature gauge on the instrument panel.

Intake Air Temperature Sensor

EXCEPT CORRADO

The intake air temperature sensor is contained in the air flow sensor housing. The sensor is not servicable and can not be replaced separately.

CORRADO

Since this engine is almost constantly under positive manifold pressure, a mechanical air flow sensor could never be accurate. Air flow into the engine is calculated by the ECU using signals from manifold pressure and intake air temperature sensors. The air/fuel mixture can be adjusted with the CO potentiometer that modifies the temperature sensor signal to the ECU. The air temperature sensor and CO potentiometer are incorporated into a single assembly, mounted to the air inlet tube between the throttle body and the intercooler. The sensor is a thermistor type, like the water temperature sensor. As temperature increases, electrical resistance decreases. If the sensor is faulty, high emissions or drivability problems could result. Use the graph, a thermometer and an ohmmeter to test the sensor.

This system is extremely reliable. Before breaking the seal to adjust air/fuel mixture, make sure all other components have been tested. Something as simple as a bad connection or vacuum leak could effect the whole system.

Oxygen Sensor

This sensor is used to report the concentration of oxygen in the exhaust. It consists of a tube coated with platinum on the outside and zirconia on the inside. The tube is protected with a slotted outer shield. The platinum side is exposed only to exhaust gas. If there is any oxygen in the exhaust, a voltage is generated across the dissimilar metals that the ECU can read. The sensor operates only when it is above a certain temperature. Some vehicles use a sensor heater to warm the sensor faster and stabilize the output. The sensor is threaded into the exhaust manifold or the exhaust pipe with an anti-seize compound on the threads. When replacing it, be careful not to get anti-seize in the slots of the outer shield.

Manifold Absolute Pressure (MAP) Sensor

Since the Corrado engine is supercharged, manifold pressure can be both below or above atmospheric pressure. The ECU uses manifold pressure, engine rpm and inlet air temperature to calculate the amount of air flowing into the engine. The pressure sensor is inside the ECU and cannot be tested or replaced separately. However since it is protected, failures are unlikely. The vacuum hose to the sensor is 1 meter (39.37 in.) long. The length and wall thickness of the hose is critical. If either is incorrect, poor engine performance could result.

CONTROLLED OUTPUTS

Idle Stabilizer Valve

This unit is a linear motor solenoid valve that is operated by the ECU. The linear motion moves a plunger to control an opening in the valve which controls the amount of air that by-passes the throttle. This design allows very precise control of idle speed regardless of engine temperature or load. The voltage to the valve is constant but a variable current is fed to the solenoid to control valve opening. To test the valve in operation, a special adapter is required that allows connecting a meter that reads milliamps while the wiring is connected to the valve.

Injectors

The electric injectors are secured in place by the fuel rail, which also houses the wiring. All injectors are wired together in parallel and are operated at the same time. Power is supplied to all injectors any time the ECU is receiving an rpm signal. The ECU operates the injectors by completing the ground circuit. Injectors can be tested electrically as an assembly or tested individually.

FUEL DELIVERY SYSTEM

Fuel Pumps

Most vehicles with the Digifant system use 2 fuel pumps; a small transfer pump inside the tank and the main pressure pump under the vehicle. The transfer pump is part of the fuel gauge sending unit but it can be replaced separately. It is accessed through a panel under the rear seat. The transfer pump fills a reservoir. On Fox, the reservoir is a separate container next to the main pump. On other models, the main pump is inside the reservoir. The pump incorporates a check valve to maintain pressure in the system when the pump is off to aid hot starts. The fuel filter is mounted to the side of the reservoir and is a "lifetime" filter which only needs to be replaced if the fuel has been contaminated.

When the ignition switch is first turned **ON**, the pump will run for about 2 seconds and then stop if the ECU is not receiving an rpm signal. The pump is operated by the ECU through a relay located on the main fuse/relay panel. On Corrado, the fuel pump after-run control unit is also on the relay panel. The main

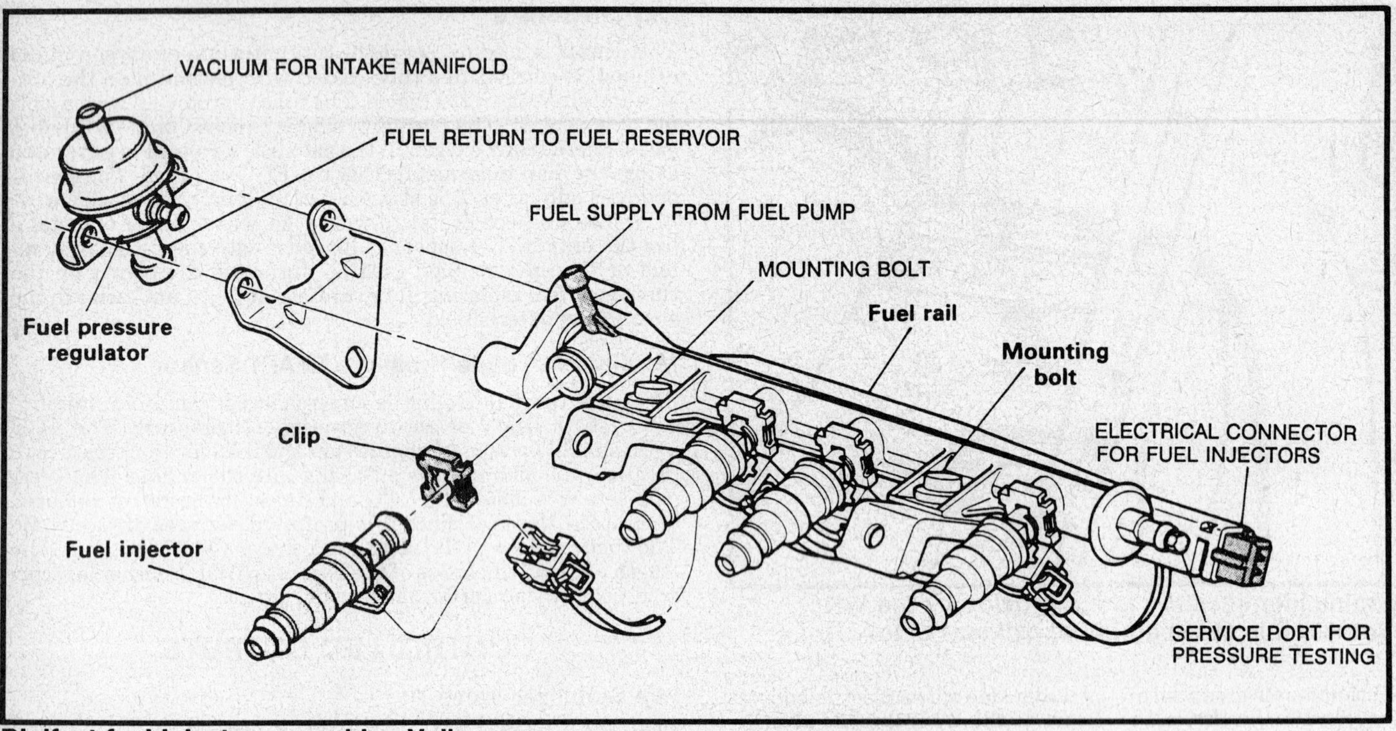

Digifant fuel injector assembly—Volkswagen

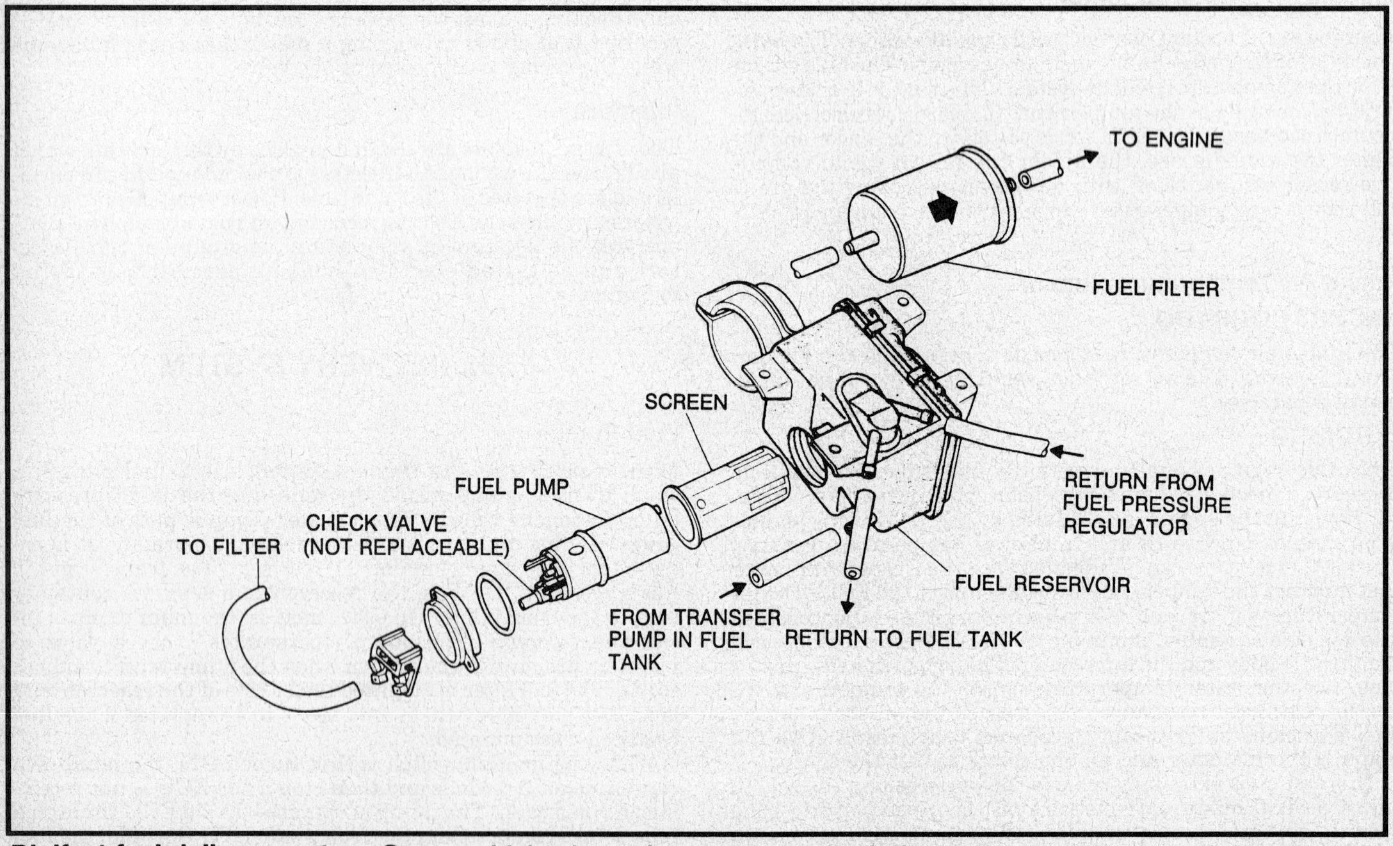

Digifant fuel delivery system. Some vehicles have the pump mounted directly in the fuel tank—Volkswagen

relay terminals can be jumpered to run the pump for testing. A pump pressure test is described in the pressure regulator test.

Fuel Pressure Regulator

This is a purely mechanical device, controlled by manifold vacuum. Fuel is supplied to the injectors through the common rail. The regulator controls the pressure in the rail by regulating the amount of fuel returned to the tank. It responds to manifold vacuum fluctuations and thereby compensates for engine load changes. When the engine is shut off, the regulator closes and seals to maintain residual fuel pressure in the injector lines for improved hot-start capability. On Corrado, a fuel pump after-run feature is provided to improve hot-start capability.

Fuel Pump After-run

The Corrado is equipped with a feature to prevent vapor-lock on a hot engine. If the vehicle has a hard start hot condition, test the after-run system. When the engine is shut off and underhood temperature is above 195°F (90°C) and system pressure is above 17 psi (1.2 BAR), the fuel pump will turn on after about 2 minutes. The pump will run for up to 8 minutes, recirculating fuel back to the tank. A pump after-run control unit is on the main fuse/relay panel, position No. 16. A fuel pump after-run switch is connected to the injector rail by a hose. An underhood temperature sensor is next to the pressure regulator. This signal is also used to operate the radiator fans when the engine is not running. The pump after-run controller also receives a signal from the tank level sending unit. If there is less than about 1.5 gallons of fuel, the after-run relay will not be activated.

Diagnosis and Testing

SERVICE PRECAUTIONS

To prevent personal injury and damage to the vehicle, follow these service precautions:

- Do not disconnect the battery or the control unit before reading the fault codes. On the Motronic system, fault code memory is erased when power is interrupted.
- Make sure the ignition switch is **OFF** before disconnecting any wiring.
- Before removing or installing a control unit, disconnect the negative battery cable. The unit receives power through the main connector at all times and will be permanently damaged if improperly powered up or down.
- Keep all parts and harnesses dry during service. Protect the control unit and all solid-state components from rough handling or extremes of temperature.

READING CODES

California Vehicles with Check Engine On-Board Diagnostic (OBD) Light

1. Make sure the ignition switch is **ON** and that all fuses are good. Make sure all ground connections under the hood are good, especially those for the control unit.
2. Remove the shifter knob and boot and connect the diagnostic jumper 357 9971 514E or equivalent, to the black and white connectors towards the left or front of the vehicle.
3. Keep the jumper connected for 5 seconds. The OBD light will start to blink and disconnect the jumper. Read the blink codes.
4. Repeat the procedure until Code 0000 or 4444 is displayed indicating the end of the sequence. The procedure can be stopped by switching **OFF** the ignition.

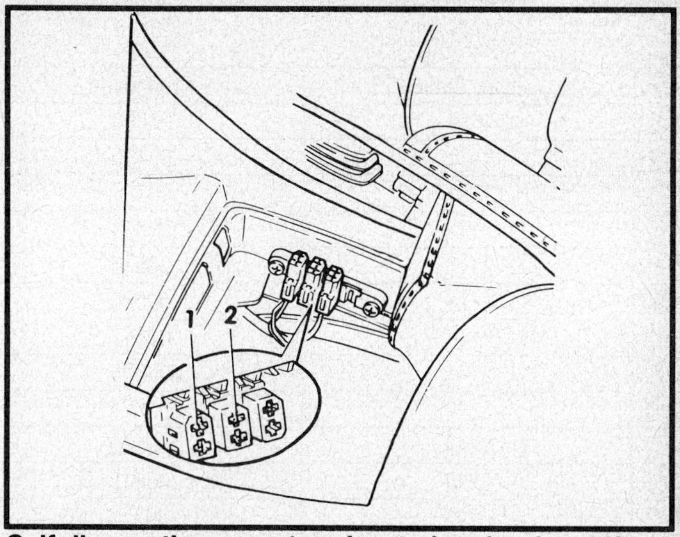

Self diagnostic connectors located under the shifter boot—Volkswagen

Self diagnostic jumper for retrieving blink codes. A jumper is provided with each vehicle—Volkswagen

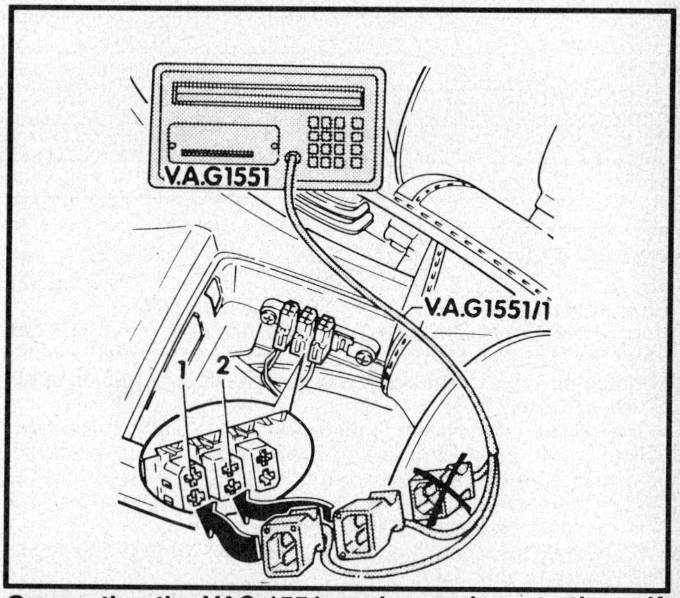

Connecting the VAG 1551 engine analyzer to the self diagnostic connectors—Volkswagen

With VAG 1551

1. Make sure the ignition switch is **OFF** and that all fuses are good. Make sure all ground connections under the hood are good, especially those for the control unit.
2. Remove the shifter knob and boot and connect the diagnostic tester to the black and white connectors towards the left or front of the vehicle.
3. Turn the tester **ON** to make sure it is receiving power. The

FAULT CODES VOLKSWAGEN DIGIFANT SYSTEM

Blink Code	Fault Code	Explanation
4444	00000	No faults in memory
1111	65535	Control unit defective
1231	00281	Vehicle speed sensor
2112	00514	Ignition reference sensor
2113	00515	Hall sender signal missing
2121	00516	Idle switch
2123	00517	Full throttle switch
2141	00535	Knock control program
2142	00524	Knock sensor 1 signal missing
2144	00540	Knock sensor 2 signal missing
2212	00518	Throttle potentiometer
2222	00519	MAP sensor
2231	00533	Idle speed regulation out of limit
2232	00520	Air flow sensor signal missing
2242	00521	CO potentiometer
2312	00522	Coolant temperature sensor
2322	00523	Intake air temperature sensor
2341	00537	Oxygen sensor signal out of limit
2342	00525	Oxygen sensor signal missing
—	00558	Adaptive limit mixture control lean
—	00559	Adaptive limit mixture control rich
—	00587	Air/fuel mixture regulation
2411	00560	EGR temperature sensor signal out of limit
4411	01249	Fuel injector
4431	01257	Idle stabilizer valve circuit
0000		End of output

screen will display 2 menu options; Rapid Data Transfer and Blink Code Output. Turn the ignition switch **ON**.

4. Select Rapid Data Transfer and Address Word 01, then press the Q button to enter the selection. The tester will display a control unit part number, the system it controls and an application (country) code.

a. If the information is displayed and is correct, press the Run key to proceed: "Select function XX" will be displayed.

b. If "Control unit does not answer" is displayed, use the Help key to display a list of possible causes. When the problem is repaired, return to Step 1 and start again.

5. When Function 02 is selected, the control unit will report fault codes to the tester.

6. When all codes have been reported, select Function 06 to exit the fault code memory without erasing the codes.

With Dashboard Switch

Some vehicles are equipped with a dashboard mounted switch that can activate the self diagnostic blink codes. This eliminates the need for jumpers or a scan tool.

1. Switch the ignition to the **ON** position.
2. Depress the switch for 4 seconds and release. The light will start blinking.
3. Press the switch again for 4 seconds to retrieve the next code.
4. The end code is 0000 or code 4444 will blink to indicate that there are no codes in memory.

CLEARING CODES

California Vehicles with Check Engine On-Board Diagnostic (OBD) Light

1. Make sure the ignition switch is **OFF** and that all fuses are good. Make sure all ground connections under the hood are good, especially those for the control unit.
2. Remove the shifter knob and boot and connect the diagnostic jumper 357 9971 514E or equivalent, to the black and white connectors towards the left or front of the vehicle.
3. Switch **ON** the ignition. Remove the jumper after 5 seconds.
4. Code 4444 is displayed indicating no faults. The procedure is completed by switching **OFF** the ignition.

With VAG 1551

1. After all fault codes have been retrieved, select Function 05 and press the Q button to enter the selection. The memory will be erased only if all fault codes have been retrieved.
2. Test drive the vehicle for at least 10 minutes, including at least 1 full throttle application above 3000 rpm.

3. Check the fault code memory again to make sure all faults have been repaired.

With Dashboard Switch

1. Turn the ignition switch **OFF**.
2. Depress the switch and turn the ignition switch **ON**.
3. Hold the switch depressed for 5 seconds or more.
4. Turn the ignition switch **OFF**.
5. Drive the vehicle for 10 minutes or more to allow the adaptive memories to undergo the relearning process.

FUEL PUMPS

Testing

ELECTRICAL

1. With the engine cold, turn the ignition switch **ON** and listen to hear the fuel pump run for about 2 seconds, then stop.
2. Use the schematics at the end of this section to locate and remove the fuel pump relay. With the ignition switch **ON**, there should 12 volts at terminal 50 on Fox, terminal 86 on all others. This is power for the relay coil that comes from the ignition switch.
3. The 2 largest terminals operate the fuel pumps. With or without the ignition switch **ON**, there should be 12 volts at terminal 48 on Fox, terminal 30 on all others.
4. When the large terminals are jumpered, the fuel pump will receive 12 even without the ignition. Use a jumper with a switch, such as a remote starter button, to check for voltage at the pump or to run the pumps for pressure and flow testing.

FLOW

CAUTION

The following procedure will produce fuel vapors. Make sure there is proper ventilation and take the appropriate fire safety precautions to prevent personal injury.

1. Locate and remove the fuel pump relay and connect a jumper wire with a switch to the 2 largest terminals in the relay socket. Make sure the switch is **OFF**.
2. Disconnect the fuel supply line at the injector rail and secure the line in a minimum 1 quart measuring container.
3. Use the switch to operate the fuel pump for 30 seconds. Time the operation carefully. With a full 12 volts available to run the pump, the flow should be at least 10 oz. (300 cc) on all except Corrado.
4. On Corrado with a single fuel pump, the flow should be at least 22 oz. (750 cc). If the vehicle is equipped with a transfer pump, the flow will be lower because of the greater power requirement. The flow should be at least 19 oz. (650 cc).

AFTER-RUN

The Corrado is equipped with the fuel pump after-run system. Other vehicle do not have this feature.

1. Make sure the battery voltage is okay and that there is at least ¼ tank of fuel.
2. Install a tee in the hose between the fuel rail service port and pressure switch and connect a 0–60 psi (0–4.1 BAR) gauge.
3. Run the engine at idle and verify a fuel system pressure of about 36 psi (2.5 BAR).
4. Turn the ignition **OFF**, disconnect the wiring to the pressure switch and check the switch with an ohmmeter to make sure the switch is closed.
5. Disconnect the wire from the ambient temperature switch near the idle stabilizer valve connector. Ground the wire with a jumper. The fans should run.
6. Reconnect the pressure switch wiring and cycle the ignition switch **ON** and **OFF**. After about 2–3 minutes, the pumps should run if the fuel pressure is above 14.5 psi (1 BAR). If the pressure drops, look for a leak in the system or a faulty pressure regulator.
7. When the ambient temperature switch wire is disconnected from ground, the pumps and fans should stop.

INJECTORS

Testing

1. Unplug the connector on the end of the fuel rail and read the resistance across the terminals in the rail. A reading of 3.5–5.0 ohms indicates all injectors are good. A reading of 5.0–6.7 indicates 1 bad injector; 7.5–10.0 ohms means 2 bad injectors; 15.0–20.0 ohms indicates 3 bad injectors.
2. If a high resistance is read, disconnect the wiring to all injectors and check each injector resistance individually; it should be 14–18 ohms each. If any injector is replaced, repeat the test in step 1 to verify the wiring is correct.
3. To check the voltage to the injectors, connect a test light or voltmeter to the connector and operate the starter. The light will flicker and a voltmeter will show some voltage but probably not a full 12 volts. If no voltage shows, try the same test with one test lead connected to engine ground. If no voltage shows at either terminal, check for power to the injectors and check the wiring between the injectors and the ECU.

FUEL PRESSURE REGULATOR

Testing

1. With the engine **OFF**, remove the test connector plug from the service port on the end of the fuel rail. On Corrado, remove the hose from the rail to the after-run switch.

CAUTION

A small amount of fuel will be released under pressure. Use a rag to shield and catch the spray and take appropriate fire safety precautions to prevent personal injury.

2. Attach a 0–60 psi (0–4.1 BAR) pressure gauge to the rail and start the engine. With the engine warm and at idle, the pressure should be about 36 psi (2.5 BAR). With the manifold vacuum hose disconnected, the pressure should increase to about 43 psi (3.0 BAR).
3. Stop the engine and note the pressure. After about 10 minutes, the pressure should be no lower than 29 psi (2 BAR).
4. If the pressure has fallen below specification with the engine off, run the engine at idle to build the pressure again. Stop the engine and immediately crimp the fuel return line. If the pressure holds, the regulator is faulty. If the pressure falls, there is a leak somewhere else in the system. Check for leaking injectors or fuel pump check valve or an external leak.

AIR FLOW SENSOR

Testing

1. With the ignition **OFF**, unplug the sensor connector and measure the resistance across the end terminals of the sensor. These are for the air temperature sensor. At 60°F (15°C), there should be about 3000 ohms. At 80°F (27°C), there should be about 1900 ohms.
2. The resistance between terminals 3 and 4 should be 500–1000 ohms when the flap is in the rest position.
3. Connect the ohmmeter between the center terminals 2 and 3 and move the vane inside the sensor. The resistance should change smoothly and linearly as the vane is moved.

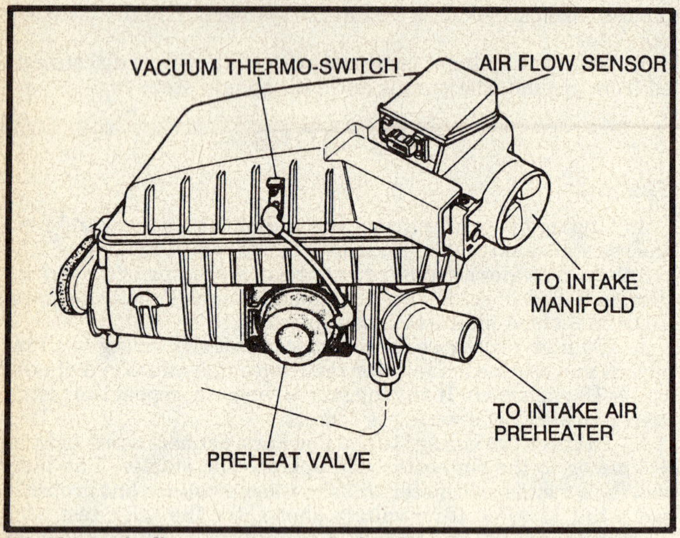

Digifant air flow meter and air cleaner assembly—Volkswagen

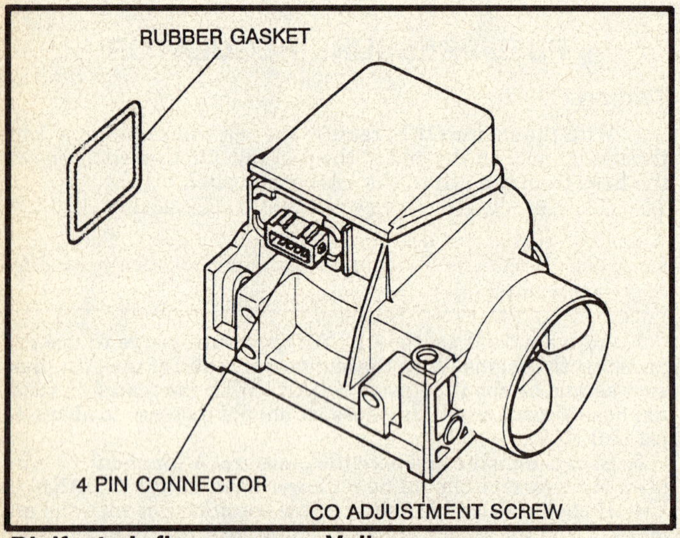

Digifant air flow sensor—Volkswagen

THROTTLE SWITCHES

Testing

1. Locate the switches on the throttle body and unplug the connector.
2. Connect an ohmmeter to the idle switch and make sure it is closed when the throttle is against the stop. Open the throttle, position a 0.024 in. (0.60mm) feeler gauge against the stop and let the throttle close on the gauge. The switch must remain open.
3. Connect the ohmmeter to the full throttle switch. Open the throttle all the way to the stop and make sure the switch closes. The switch should open when the throttle is closed to 10 degrees from the stop.
4. Turn the ignition switch **ON** and use a voltmeter to check for 5 volts at each switch connector. This signal comes directly from the ECU.

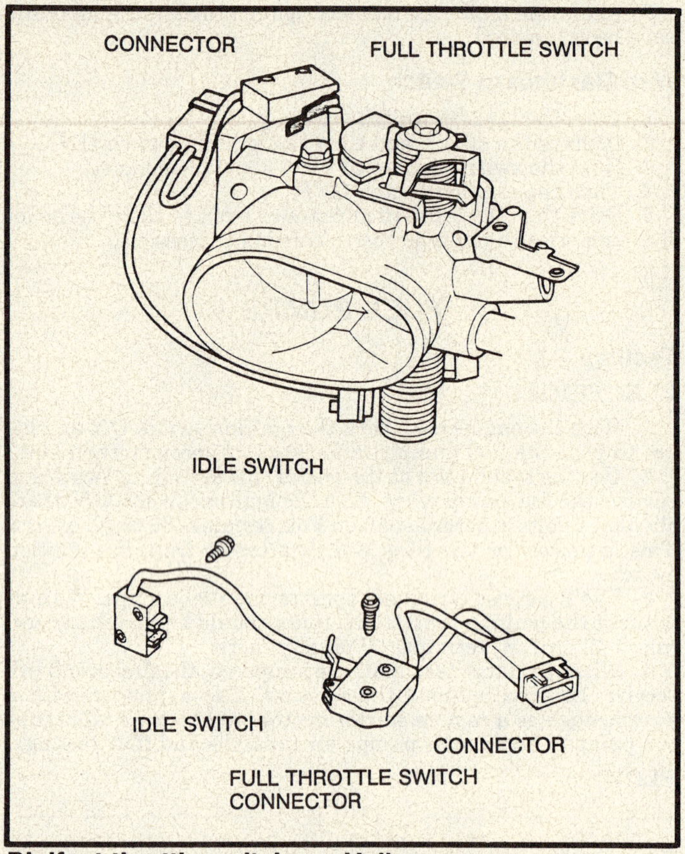

Digifant throttle switches—Volkswagen

THROTTLE POTENTIOMETER

Testing

1. Locate the potentiometer under the throttle body and unplug the connector. It may be necessary to remove the intake air tube.
2. With the ignition switch **ON**, check for 5 volts between one of the end connector terminals and ground. This is the supply voltage from the ECU to the potentiometer.
3. On the potentiometer, connect an ohmmeter to the center and one of the end terminals. When the throttle is moved, the resistance should change.
4. If voltage is supplied and the potentiometer resistance changes with throttle position, the component is okay. Check the continuity of the wiring between the component and the ECU.

IDLE STABILIZER VALVE

1. With the ignition **ON** but the engine not running, the valve should buzz and hum. If not, make sure the idle switch on the throttle body is working properly and that the throttle is fully closed.
2. If the valve makes no noise, turn the ignition **OFF**, unplug the connector and use an ohmmeter to check the resistance across the terminals on the valve. There should be continuity.
3. Connect an adapter so a multimeter can be connected. With the engine at operating temperature and idling, the current to the valve should be 460 mA and fluctuating about 60 mA. With the blue temperature sensor wiring disconnected, the current should be steady.
4. If the current is not correct, remove the valve and check for visual signs of sticking. Do not lubricate the valve. If no other

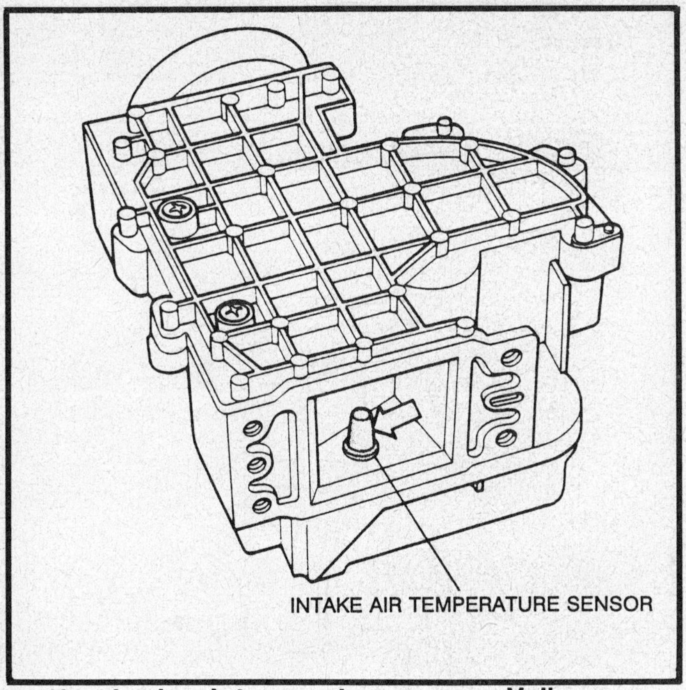

Digifant intake air temperature sensor—Volkswagen

problem is found, check the continuity of the wiring between the valve and the ECU with the ignition **OFF**.

5. If the idle stabilizer valve seems to work properly but engine idle is out of specification, check for a vacuum leak, a faulty coolant temperature sensor or some other problem with the engine control system.

INTAKE AIR TEMPERATURE SENSOR

Testing

1. Disconnect the sensor wiring and read the resistance between terminals 2 (center terminal) and 3 on the sensor. On the connector, the brown/white wire connects to terminal 3. The resistance will depend on the temperature. At 68°F (20°C) the resistance should be 2300–2700 ohms.
2. The resistance between terminals 1 (blue wire) and 3 or 2 and 3 for Vanagon, should be 0–2000 ohms, depending on the position of the CO adjusting screw. A smottoth change in resistance is more important than the actual resistance value.

OXYGEN SENSOR

Testing

1. With the ignition switch **ON**, there should be 12 volts be-

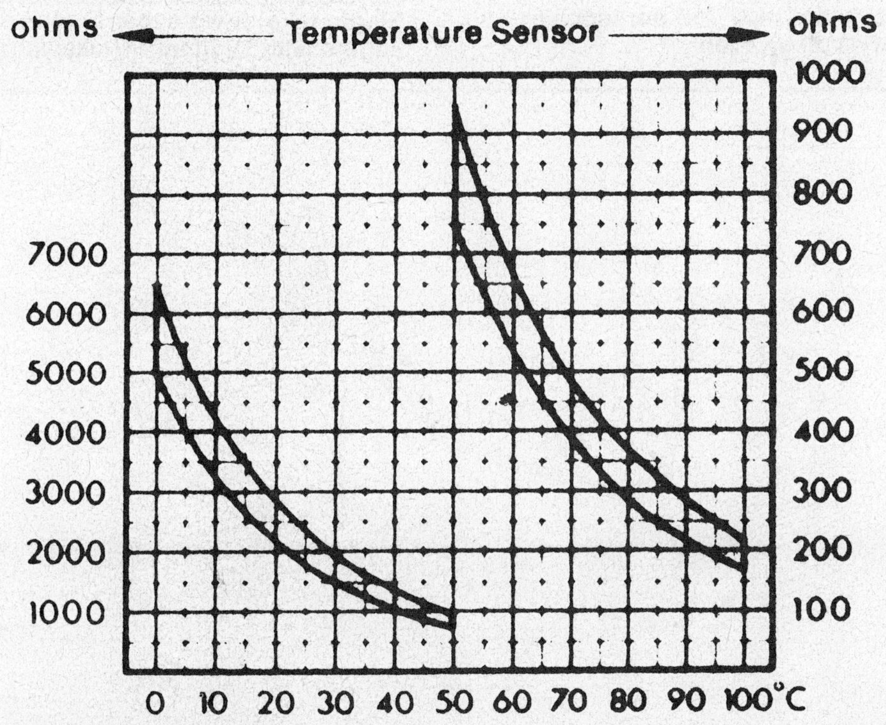

Value chart for testing the temperature sensors— Volkswagen

tween 1 of the sensor heater wires and ground, usually a red/white wire. The heater resistance should be about 3–4 ohms. The brown wire is heater ground, the green wire is sensor output to the ECU.

2. To test the sensor output, remove the blue cap from the exhaust gas sample tap and insert the probe of an exhaust gas analyzer. With the engine at operating temperature and at idle, disconnect the fuel pressure regulator vacuum hose from the intake manifold and plug the port. The CO reading should increase momentarily, then return to the original value.

3. If the CO reading does not return to normal, connect a jumper wire to the sensor wire from the ECU. Touch the wire alternately to ground and to the positive battery terminal. If the CO reading changes, the oxygen sensor is faulty. If the reading does not change, the wiring or the ECU is faulty.

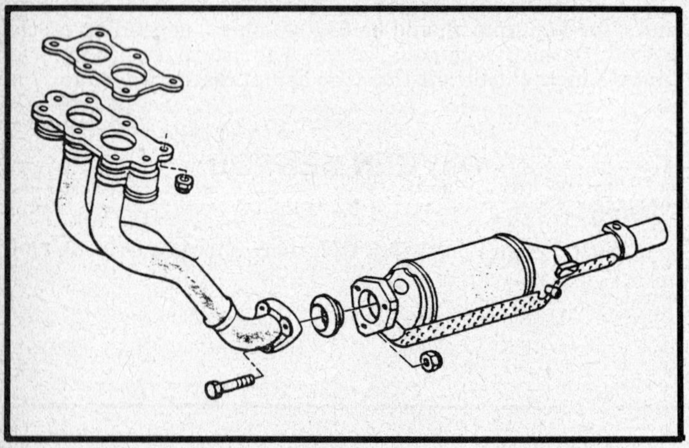

Double take down exhaust pipe 105 horsepower engine with Digifant—Volkswagen

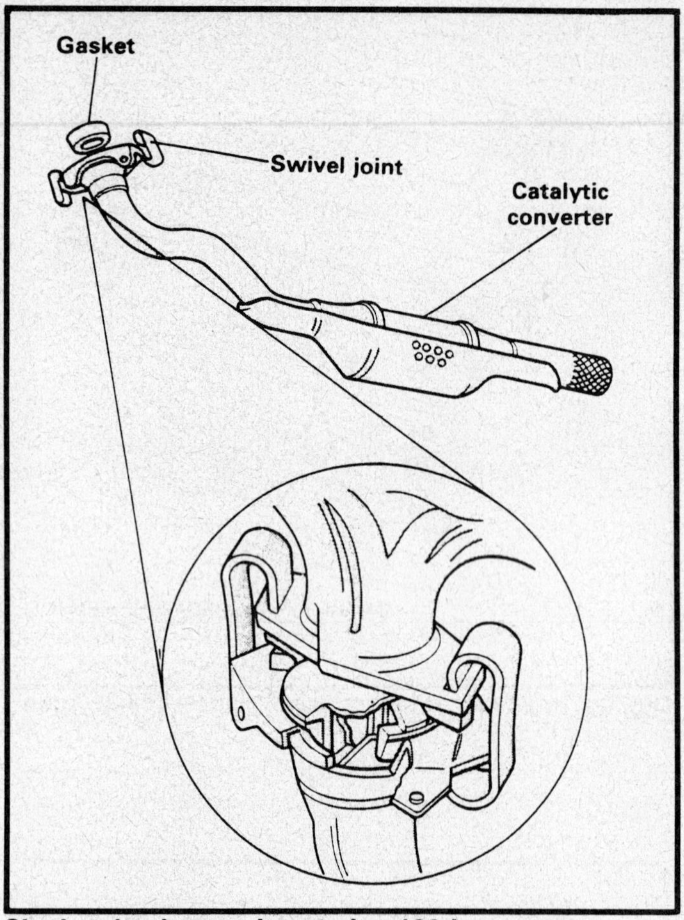

Single take down exhaust pipe 100 horsepower engine with Digifant—Volkswagen

Corrado fuel system view

1.8L ENGINE—DIGIFANT ELECTRICAL TESTING—CORRADO EXCEPT 1991–92 CALIFORNIA

Test A

Test step	VAG 1598 terminals	To be tested	Test conditions or additional steps	Specified values
1	13 + 23 (starting m.y. 1990)	Current supply relay (**J 176**) for control unit (**J 169**)	Switch **ON** ignition	Approx. battery voltage
2	13 + 14 14 + 19	Voltage supply for control unit (**J 169**)	Switch **ON** ignition	Approx. battery voltage
3	13 + 25	Wire to terminal **1** of ignition coil	Switch **ON** ignition	Approx. battery voltage
4	3 + 14 bridged	Wiring to fuel pump relay	Switch **ON** ignition	Fuel pumps must run audibly
5	12 + 13	Wiring to fuel injectors (**N 30, N 31, N 32, N 33**)	Switch **ON** ignition Remove fuse **18** Bridge terminals **3 + 13**	Approx. battery voltage
6	13 + 16	Signal from A/C	A/C system operation **OK** Switch **ON** ignition Switch **ON** A/C	Approx. battery voltage
7	For auto trans. only: 1 + 14	Wiring to relay (**J 226**) for starter lockout	Switch ignition **ON** Selector lever in **P** then **N** Switch selector lever to **1, 2, 3, D** and **R**	Approx. battery voltage No voltage
8	For manual trans only: 1 + 13	Wires from starter terminal **50**	Operate starter then switch **OFF** ignition	8 Volts minimum
■ switch **OFF** ignition ■ switch multimeter to resistance range				
9	6 + 10	Coolant temperature sensor (**G 62**)	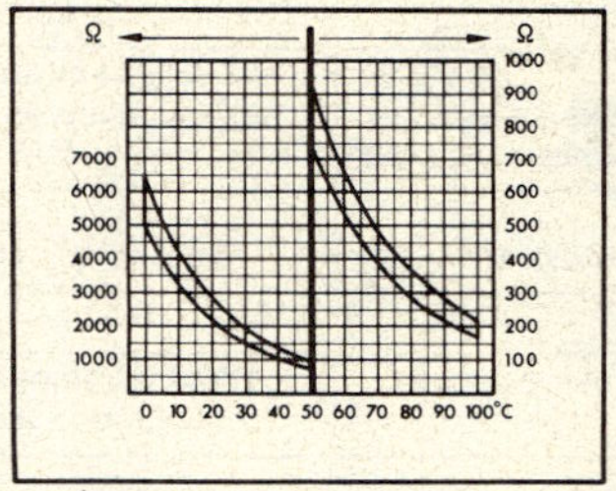	

Note

Engine malfunctions that **only** occur at certain temperatures are due to an intermittent signal from the coolant temperature sensor.

Test A continued

Test step	VAG 1598 terminals	To be tested	Test conditions or additional steps	Specified values
10	6 + 9	Intake air temp. sensor (**G 42**) in CO potentiometer (**G 74**)	—	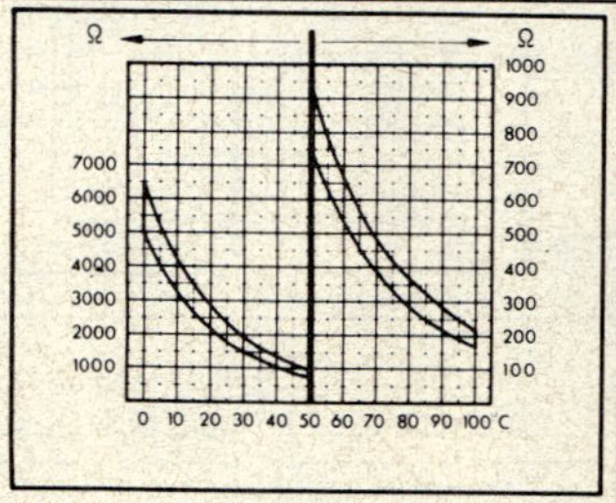
11	5 + 6	CO potentiometer (**G 74**)	—	0 to 2000 Ohms
12	14 + 22	Idle stabilizer valve (**N 71**)	—	2 to 10 Ohms
13	6 + 11	Idle switch (**F 50**)	Throttle closed Throttle open	1.5 Ohms maximum Infinite Ohms (open)
14	6 + 15	Full throttle switch (**F 81**)	Throttle closed Throttle fully open	Infinite Ohms (open) 1.5 Ohms maximum (6 times)
15	6 + 8 6 + 18	Wiring to Hall sender (**G 40**)	Disconnect Hall sender and bridge all 3 terminals of harness connector	1.5 Ohms maximum 1.5 Ohms maximum
16	4 + 7	Wiring to knock sensor (**G 61**)	Disconnect knock sensor and bridge all 3 terminals of harness connector	1.5 Ohms maximum
17	2 + 13	Wiring to Oxygen sensor	Disconnect OXS probe and connect black or lilac wire to ground Reconnect OXS probe	1.5 Ohms maximum Infinite Ohms (open)

1.8L ENGINE—DIGIFANT ELECTRICAL TESTING CORRADO EXCEPT 1991–92 CALIFORNIA

Test B

Test step	VAG 1598 terminals	To be tested	Test conditions or additional steps	Specified values
■ perform the following test step **only** if the engine will not start				
1	8 + 6	Voltage supply for Hall sender (**G 40**)	Switch **ON** ignition	10.0 Volts minimum
2	6 + 10	Resistance curve for coolant temp. sensor (**G 62**)	With engine cold, start and let warm up while observing multimeter	Indicated voltage must decrease as engine warms up
■ perform the following test step **only** if the engine will not start ■ switch **OFF** ignition ■ use **US 1115** LED tester with **VW 1594** adaptors instead of the multimeter to perform the following checks ■ disconnect harness connector from fuel rail (main fuel injector connector)				
3	6 + 18	Signal from Hall sender (**G 40**)	Operate starter	LED tester must flicker
■ perform the following test step **only** if the engine will not start				
4	3 + 25	Ignition signal function of Digifant control unit (**J 169**)	Operate starter	LED must flicker
5	19 + 24 (m.y. 1991)	RPM signal	Operate starter	LED must flicker

1.8L ENGINE—DIGIFANT ELECTRICAL TESTING CORRADO 1991–92 CALIFORNIA

Test A

Test step	VAG 1598 terminals	To be tested	Test conditions or additional steps	Specified values
1	1 + 19 19 + 35	Voltage supply for Digifant control unit (**J 169**)	Ignition switched **ON**	Approx. battery voltage
2	1 + 6	Wiring to ignition coil (**N 152**)	Ignition switched **ON**	Approx. battery voltage
3	1 + 4	Fault indicator light (**K 83**) (OBD light)	Ignition switched **ON**	Approx. battery voltage
4	1 + 23	Idle stabilizer valve (**N 71**)	Ignition switched **ON**	Approx. battery voltage
5	1 + 24 bridged	Wiring to fuel pump relay (**J 17**)	Ignition switched **ON**	Fuel pump must run audibly
6	 1 + 20 1 + 21 1 + 3 1 + 2	Wiring to injectors (**N 30** to **N 33**)* Cylinder 1 Cylinder 2 Cylinder 3 Cylinder 4	Ignition switched **ON**. Remove fuse **18**. Bridge terminals **1** and **24**.	Approx. battery voltage
*Check fuel injector resistance				
7	For Auto trans **only**: 19 + 25	Wiring to starter lockout relay (**J 226**)	Ignition switched **ON**. Selector lever in **P** or **N**. Move selector lever to **1, 2, 3, D** and **R**.	Approx. battery voltage No voltage at any of these positions
8	For manual trans **only**: 1 + 25	Wiring from starter terminal **50**	Operate starter and then switch **OFF** ignition	8 Volts minimum
■ switch **OFF** ignition ■ switch multimeter to resistance range				
9	11 + 28	Coolant temperature sensor (**G 62**)		

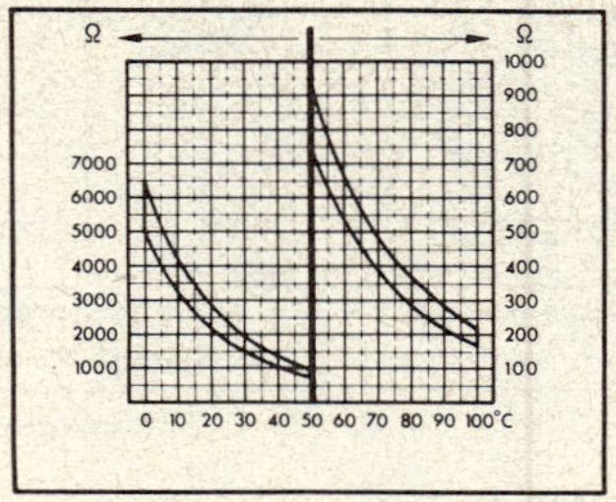

Note

Engine malfunctions that **only** occur at certain temperatures are due to an intermittent signal from the coolant temperature sensor.

1.8L ENGINE—DIGIFANT ELECTRICAL TESTING—CORRADO 1991–92 CALIFORNIA

Test A continued

Test step	VAG 1598 terminals	To be tested	Test conditions or additional steps	Specified values
10	28 + 30	Intake air temp. sensor (**G 42**) in CO potentiometer (**G 74**)	—	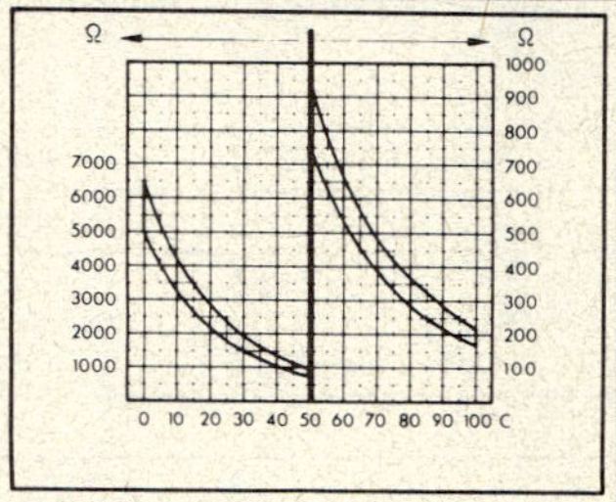
11	17 + 36	Diagnostic connector (**T 2**)	Bridge terminals in white diagnostic connector	1.5 Ohms maximum
12	10 + 28	CO potentiometer (**G 74**)	—	0 to 2 Ohms
■ switch **OFF** ignition ■ switch multimeter to resistance range ■ disconnect and perform the following check on removed harness connector for throttle potentiometer				
13	34 + terminal 1 of connector 33 + terminal 2 of connector 28 + terminal 3 of connector	Throttle potentiometer wiring (**G 69**)	Disconnect throttle potentiometer	1.5 Ohms maximum
■ disconnect and perform the following check on removed harness connector for Hall sender				
14	28 + terminal 1 of connector 37 + terminal 2 of connector 38 + terminal 3 of connector	Hall sender (**G 40**) wiring	Disconnect Hall sender	1.5 Ohms maximum

Test A continued

Test step	VAG 1598 terminals	To be tested	Test conditions or additional steps	Specified values
■ disconnect and perform the following check on removed harness connector for knock sensor				
15	29 + terminal 1 27 + terminal 2 8 + terminal 3	Wiring to knock sensor (**G 61**)	Disconnect knock sensor	1.5 Ohms maximum
■ disconnect and perform the following check on removed harness connector for Oxygen sensor				
16	1 + terminal 2 of connector 16 + terminal 3	Oxygen sensor (**G 39**) wiring	Disconnect Oxygen sensor	1.5 Ohms maximum
	9 + terminal 3			Infinite Ohms (open)
	1 + 16		Re-connect Oxygen sensor	Infinite Ohms (open)
17	Vehicles with A/C Only: 32 + 35	Wiring to A/C relay (**J 32**)	—	1 Ohm maximum

Note

To check the wiring to the automatic transmission control unit (**J 217**) terminals **14** and **18**; read fault memory.

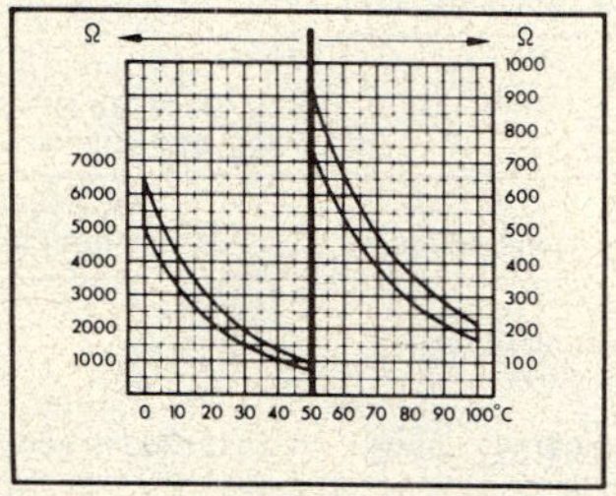

Resistance values for:
- intake air temperature sensor (**G 42**) in air flow sensor (**G 19**)
- blue coolant temperature sensor (**G 62**)

1.8L ENGINE—DIGIFANT ELECTRICAL TESTING
CORRADO 1991–92 CALIFORNIA

Test B

Test step	VAG 1598 terminals	To be tested	Test conditions or additional steps	Specified values
■ read fault memory and then erase				
1	**33 + 34**	Voltage supply for throttle potentiometer (**G 69**)	Disconnect throttle potentiometer	approx. 3.5 to 5.0 Volts
	28 + 34		Switch on ignition	4.5 . . . 5.0 Volts
■ read fault memory and then erase				
2	**33 + 28**	Throttle potentiometer (**G 69**)	Switch **ON** ignition. Re-connect throttle potentiometer Throttle setting: ● idle: ● full throttle:	 0.3-1.7 Volts 3-5 Volts
■ perform the following test step **only** if the engine will not start				
3	**28 + 38**	Voltage supply for Hall sender (**G 40**)	Ignition switched **ON**	10 Volts minimum
■ switch **OFF** ignition ■ use **US 1115** LED tester with **VW 1594** adaptors instead of the multimeter to perform the following checks ■ disconnect harness connector from fuel rail (main fuel injector connector) ■ perform the following test step **only** if the engine will not start				
4	**28 + 37**	Signal from Hall sender (**G 40**)	Operate starter	LED must flicker
■ perform the following test step **only** if the engine will not start				
5	**1 + 6**	Ignition signal function of Digifant control unit (**J 169**)	Operate starter	LED must flicker
■ perform the following test step **only** if the engine will not start				
6	**18 + 28**	RPM signal	Operate starter	LED must flicker

1.8L ENGINE—DIGIFANT ELECTRICAL TESTING
GOLF, GTI AND JETTA EXCEPT 1991–92 CALIFORNIA

Test step	VAG 1598 terminals:	To be tested	Test conditions and additional steps (if any)	Specified value or result
1	**13 + 14**	Voltage supply to ECU (**J 169**)	Switch **ON** ignition	approximate battery voltage
	14 + 19		Ignition switched **ON**	approximate battery voltage
2	**1 + 13**	Wire from starter terminal **50**	Operate starter	8 volts minimum
3	Up to 7-89 **12 + 14**	Fuel injectors (**N 30, N 31, N 32, N 33**)	—	3.7 to 5.5 Ohms (all injectors summed together)
	Starting 8-89 **12 + terminal 5** (red) of 5 pin connector		Disconnect 5 pin connector	3.7 to 5.5 Ohms (all injectors summed together)
4	**25 + terminal 6** of Hall control unit harness connector	Wire to Hall control unit (**N 41**)	Disconnect Hall control unit harness connector	0.5 Ohms maximum (continuity)
5	**6 + 10**	Coolant temperature sender (**G 62**)	Use chart to look up value 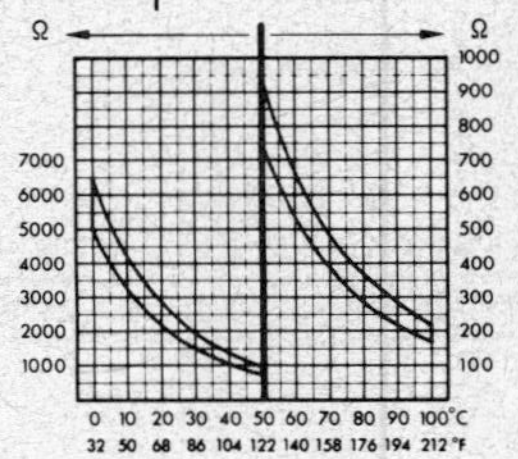 	
6	**6 + 11**	Idle switch (**F 60**) and full throttle switch (**F 81**)	Throttle valve closed	0.5 Ohms maximum
			Throttle valve opened just short of full throttle	Infinite Ohms (open)
			Throttle completely open	0.5 Ohms maximum (continuity)

1.8L ENGINE—DIGIFANT ELECTRICAL TESTING—GOLF, GTI AND JETTA EXCEPT 1991–92 CALIFORNIA

Test step	VAG 1598 terminals:	To be tested	Test conditions and additional steps (if any)	Specified value or result
7	**6 + 17**	Air flow sensor potentiometer **(G 19)**	—	500 to 1000 Ohms
	17 + 21		Move air flow sensor or operate starter	Change in resistance
8	**6 + 9**	Intake air temperature sender **(G 42)** in air flow sensor	— 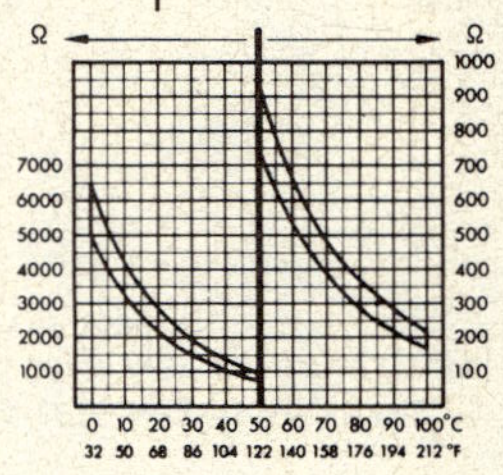	
9	**3 + 13** Bridged	Wiring to fuel pump relay **(J 17)**	Switch **ON** ignition	Fuel pumps must run audibly
10	**6 + 8**	Wiring to Hall sender **(G 40)**	Disconnect Hall sender and bridge all three terminals of harness connector	0.5 Ohms maximum (continuity)
	6 + 18			0.5 Ohms maximum (continuity)
11	**4 + 5**	Wiring to knock sensor **(G 61)**	Disconnect knock sensor and bridge all three terminals of harness connector	0.5 Ohms maximum (continuity)
	4 + 7			0.5 Ohms maximum (continuity)
12	**22 + 23**	Wiring to the idle stabilizer valve **(N 71)**	—	2 to 10 Ohms
13	**16 + 13**	Wiring to A/C compressor	Disconnect connector (on brake booster) and connect green wire to ground	0.5 Ohms maximum (continuity)

Test step	VAG 1598 terminals:	To be tested	Test conditions and additional steps (if any)	Specified value or result
14	**2 + 13**	Wiring to Oxygen sensor **(G 39)**	Disconnect oxygen sensor and connect harness wire to ground If equipped with a 3 pin terminal, bridge terminals **1 + 2** Reconnect oxygen sensor wires	0.5 Ohms maximum (continuity) Infinite Ohms (open)

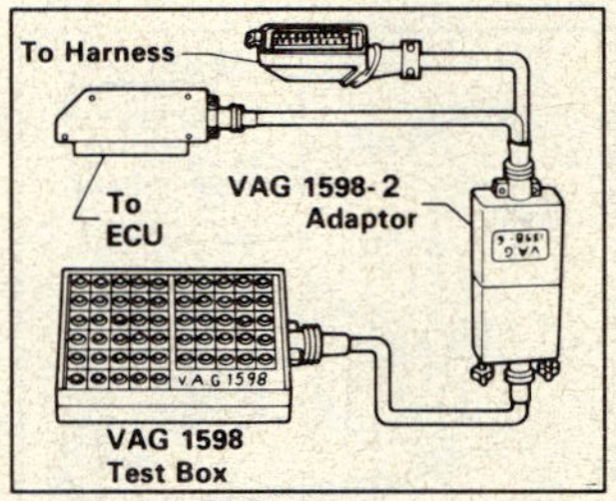

B – Electrical testing with Digifant control unit connected

Note

Perform the following tests **only** if the engine will not start.

- switch **OFF** ignition
- connect Test box **VAG 1598** to the Digifant control unit harness connector and to the Digifant control unit using adaptor cable **VAG 1598/2**

> **CAUTION**
>
> Always switch the multimeter to the desired units and range of measurement (Volts, Ohms, etc.) before making any connections to the Test box or the Digifant control unit will be permanently damaged.
>
> Do not connect or disconnect adaptors/test instruments to the **VAG 1598** Test box while the ignition is switched **ON**.

Test step	VAG 1598 terminals:	To be tested	Test conditions and additional steps (if any)	Specified value or result
1	**8 + 6**	Voltage supply to Hall sender **G 40)**	Switch **ON** ignition	10 volts minimum

1.8L ENGINE—DIGIFANT ELECTRICAL TESTING GOLF, GTI AND JETTA EXCEPT 1991–92 CALIFORNIA

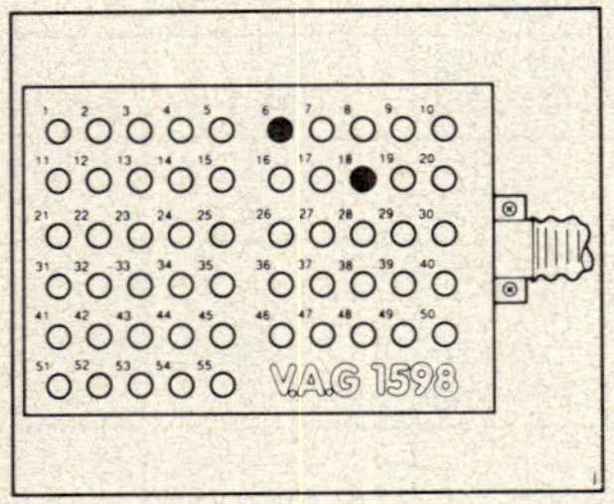

- switch **OFF** ignition
- disconnect multimeter from Test box
- connect **US 1115** LED tester (**VAG 1527B**) to terminals **6** and **18** of Test box using adaptor cables from the **VW 1594** kit
- disconnect main harness connector for fuel injectors (from fuel injector wiring guide)

Test step	VAG 1598 terminals:	To be tested	Test conditions and additional steps (if any)	Specified value or result
2	6 + 18	Signal from Hall sender (**G 40**)	Operate starter	LED tester must flicker

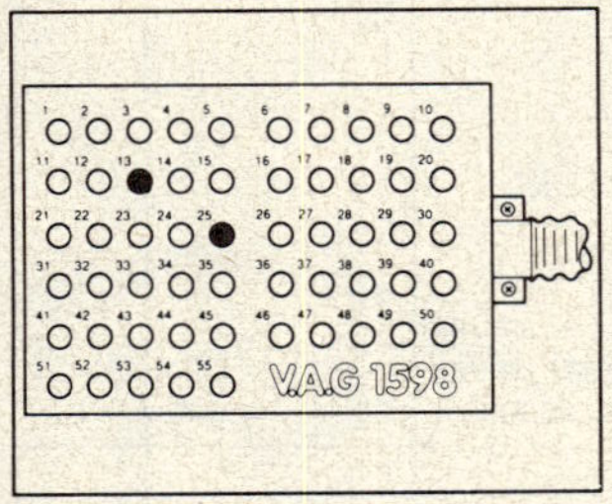

- switch **OFF** ignition
- connect **US 1115** LED tester (**VAG 1527B**) to terminals 13 and **25** of Test box using adaptor cables from the **VW 1594** kit

Test step	VAG 1598 terminals:	To be tested	Test conditions and additional steps (if any)	Specified value or result
3	13 + 25	Ignition switching function of Digifant control unit (**J 169**)	Operate starter	LED tester must flicker

1.8L ENGINE—DIGIFANT ELECTRICAL TESTING GOLF, GTI AND JETTA 1991–92 CALIFORNIA

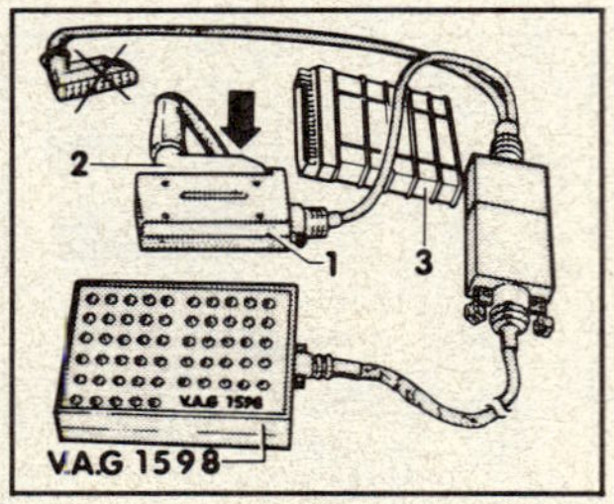

- switch **OFF** ignition
- disconnect Digifant control unit (**J 169**) harness connector
  - control unit is installed in plenum on drivers side
- connect **VAG 1598** Test box to control unit harness connector **2** using adaptor wire **VAG 1598/9**
  - control unit **3** remains disconnected
- switch multimeter to 20 Volt range
- perform checks using the following table

Test step	VAG 1598 terminals	To be tested	Test conditions or additional steps	Specified values
1	1 + 19	Voltage supply for Digifant control unit (**J 169**)	Ignition switched **ON**	Approximately battery voltage
	19 + 35			
2	1 + 6	Wiring to ignition coil (**N 152**)	Ignition switched **ON**	Approximately battery voltage
3	1 + 4	Fault indicator light (**K 83**)	Ignition switched **ON**	Approximately battery voltage
4	1 + 23	Idle stabilizer valve (**N 71**)	Ignition switched **ON**	Approximately battery voltage
5	1 + 24 bridged	Wiring to fuel pump relay (**J 17**)	Ignition switched **ON**	Fuel pump must run audibly
6	1 + 20	Wiring to injectors (**N 30** to **N 33**)* Cylinder 1	Ignition switched **ON** Remove fuse **18** Bridge terminals **1** and **24**	Approximately battery voltage
	1 + 21	Cylinder 2		
	1 + 3	Cylinder 3		
	1 + 2	Cylinder 4		
*	Check fuel injector resistance			
7	For automatic transmission **ONLY**: 19 + 25	Wiring to starter lockout relay (**J 207**)	Ignition switched **ON** Selector lever in **P** or **N**	Approximately battery voltage
			Move selector lever to 1, 2, **D** and **R**	No voltage at any of these positions

1.8L ENGINE – DIGIFANT ELECTRICAL TESTING – GOLF, GTI AND JETTA 1991–92 CALIFORNIA

Test step	VAG 1598 terminals	To be tested	Test conditions or additional steps	Specified values
8	For manual transmission **ONLY:** 1 + 25	Wiring from starter terminal **50**	Operate starter and then switch **OFF** ignition	8 Volts minimum
■ switch **OFF** ignition ■ switch multimeter to resistance range				
9	11 + 28	Coolant temperature sensor (**G 62**)	—	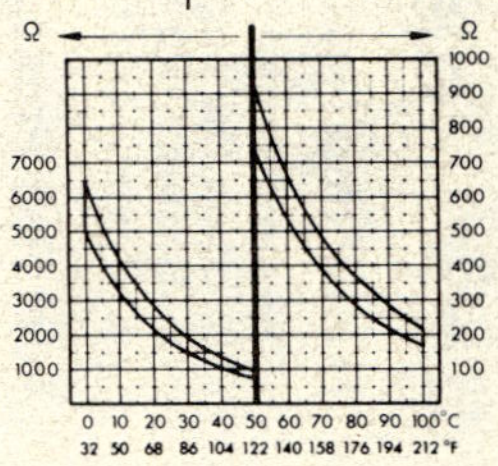
Note Engine malfunctions that **only** occur at certain temperatures are due to an intermittent signal from the coolant temperature sensor.				
10	28 + 30	Intake air temperature sensor (**G 42**) in air flow sensor (**G 19**)	—	Refer to chart (above)
11	17 + 36	Diagnostic connector (**T2**)	Bridge connectors with white diagnostic connector	1.5 Ohms maximum
12	15 + 28	Air flow sensor potentiometer (**G 19**)	—	500 to 1000 Ohms
	13 + 15		Move air flow sensor flap or operate starter	Change in resistance
■ switch **OFF** ignition ■ switch multimeter to resistance range				
13	34 + terminal **1** of connector	Throttle potentiometer wiring (**G 69**)	Disconnect 3 pin connector from throttle potentiometer	1.5 Ohms maximum
	33 + terminal **2** of connector			
	28 + terminal **3** of connector			

Test step	VAG 1598 terminals	To be tested	Test conditions or additional steps	Specified values
14	28 + terminal **1** of connector	Hall sender (**G 40**) wiring	Disconnect harness connector from Hall sender	1.5 Ohms maximum
	37 + terminal **2** of connector			
	38 + terminal **3** of connector			
15	29 + terminal **1** of connector	Knock sensor (**G 61**) wiring	Disconnect knock sensor from harness connector	1.5 Ohms maximum
	27 + terminal **2** of connector			
	8 + terminal **3** of connector			
16	1 + terminal **2** of connector	Oxygen sensor (**G 39**) wiring	Disconnect Oxygen sensor from harness connector	1.5 Ohms maximum
	16 + terminal **3** of connector			
	9 + terminal **3** of connector			Infinite Ohms (open)
	1 + 16		Re-connect harness connector to Oxygen sensor	Infinite Ohms (open)
17	Vehicles with A/C **ONLY**: 32 + green wire terminal	Wiring to A/C compressor	Disconnect single pin connector in area of coolant reservoir	1.5 Ohms maximum

1.8L ENGINE—DIGIFANT ELECTRICAL TESTING—GOLF, GTI AND JETTA 1991–92 CALIFORNIA

Test step	VAG 1598 terminals	To be tested	Test conditions or additional steps	Specified values
■ check fault memory and then erase				
1	34 + 28	Voltage supply for throttle potentiometer **(G 69)**	Disconnect throttle potentiometer Switch **ON** ignition	Approximately 4.5 to 5.5 Volts
■ check fault memory and then erase				
2	33 + 28	Throttle potentiometer **(G 69)**	Switch **ON** ignition Re-connect harness connector to throttle potentiometer Throttle setting: • idle: • full throttle:	 0.3-1.7 Volts 3-5 Volts
■ perform the following test step **only** if the engine will not start:				
3	28 + 38	Voltage supply for Hall sender **(G 40)**	Ignition switched **ON**	10 Volts minimum
■ switch **OFF** ignition ■ use **US 1115** LED tester with **VW 1594** adaptors instead of the multimeter to perform the following checks ■ disconnect harness connector from fuel rail, (main fuel injector connector) ■ perform the following test step **ONLY** if the engine will not start:				
4	37 + 38	Signal from Hall sender **(G 40)**	Operate starter	LED must flicker
■ perform the following test steps **ONLY** if the engine will not start:				
5	1 + 6	Igniton switching function of Digifant control unit **(J 169)**	Operate starter	LED must flicker
6	10 + 13	RPM signal	Operate starter	LED must flicker

Component Replacement

SERVICE PRECAUTIONS

To prevent personal injury and damage to the vehicle, follow these service precautions:

- Do not disconnect the battery or the control unit before reading the fault codes. The fault code memory is erased when power is interrupted.
- Make sure the ignition switch is **OFF** before disconnecting any wiring.
- Before removing or installing a control unit, disconnect the negative battery cable. The unit receives power through the main connector at all times and will be permanently damaged if improperly powered up or down.
- Keep all parts and harnesses dry during service. Protect the control unit and all solid-state components from rough handling or extremes of temperature.
- Smoking must not be allowed when working around flammable liquids.
- Always keep a proper fire extinguisher close on hand.
- Absorbant material must be available to soak up any spillage.
- Ensure fuel is emptied into a suitable container.
- Always place the transmission gear shift lever in **N** (manual transmission) or **P** (automatic transmission) and set the parking brake and block drive wheels before starting the engine.
- Keep hands, tools and clothing away from engine cooling fan to help prevent personal injury.
- Thoroughly clean all unions and area near connections before disconnecting.

FUEL PRESSURE RELIEF

Fuel Pump Relay Method

1. Remove the fuel pump relay.
2. Start the engine and allow to run until it stalls.
3. Switch **OFF** the ignition switch.
4. Replace the fuel pump relay.

Fuel Rail Pressure Tap Method

1. Connect a fuel pressure gauge with bleed down valve to the pressure tap on the fuel rail.
2. Open up the bleed valve on the gauge to relieve the pressure. Catch any escaping fuel.
4. Remove the gauge and cap the tap.

EXTERNAL FUEL PUMP

The fuel pump is normally located under the vehicle near the fuel tank or in front of the rear axle on the right side. It is attached to a bracket which also accommodates the fuel filter and the accumulator as an assembly.

Removal and Installation

1. Disconnect the negative battery cable.
2. Relieve the fuel system pressure. Wrap a cloth around the connection to catch any escaping fuel.
3. Raise the vehicle and support it safely.
4. Disconnect the electrical connector from the fuel pump.
5. Remove both wire terminals from the connector housing.
6. Discard the old connector and protective boot.

7. Install the new protective boot over the wires and insert the wire terminals into the new connector.
8. Slide the protective boot over the connector.
9. Disconnect fuel line from the fuel pump.
10. Remove pump mounting bolts and remove the pump.

To install:

11. Position the fuel pump it place and secure it with the mounting bolts.
12. Install new O-ring seals when connecting the fuel lines to the pump.
13. Connect the electrical connector to the fuel pump.
14. Lower the vehicle.
15. Connect the negative battery cable. Turn the ignition switch **ON** and **OFF** several times to pressurize the system.
16. Raise the and support it safely, start the engine and observe for fuel leaks.
17. Shut the engine off and lower the vehicle.

INTERNAL TANK PUMP

This pump is located inside the fuel tank and is attached to the end of the sending unit. It is not used in all applications.

Removal and Installation

1. Disconnect the negative battery cable.
2. Relieve the fuel system pressure. Wrap a cloth around the connection to catch any escaping fuel.
3. Open the rear hatch or trunk and pull back the carpet.
4. Remove the access plate retaining screws and remove the plate from the floor.
5. Disconnect the electrical connector from the sending unit.
6. Remove the fuel hoses from the sending unit and plug.
7. Unscrew the plastic cap and lift the sending unit and pump assembly from the fuel tank.
8. Remove the pump from the sending unit.

To install:

9. Assemble the pump to the sending unit.
10. Lubricate the sealing ring with fuel prior to installing and install it.
11. Carefully lower the pump and sending unit assembly into the tank.
12. Using a special tool, screw the plastic cap in place.
13. Connect the fuel hoses to the fuel pump and secure the clamps.
14. Connect the electrical connector to the sending unit.
15. Connect the negative battery cable. Turn the ignition switch **ON** and **OFF** several times to pressurize the system and check for leaks.
16. Install the access plate and secure it with the retaining screws.

OXYGEN SENSOR

The oxygen sensor is located in the exhaust manifold. The exhaust system should be slightly warm to make removal easier.

Removal and Installation

1. Disconnect the negative battery cable.
2. Disconnect the electrical connector at the oxygen sensor.
3. Unscrew the oxygen sensor and remove it from the exhaust manifold.

To install:

4. Before installing the oxygen sensor, coat the threads with anti-seize compound and install.
5. Connect the electrical sensor to the oxygen sensor. Connect the negative battery cable.

FULL THROTTLE SWITCH

Adjustment

1. Disconnect the throttle switch connector.
2. Connect an ohmmeter to the 2 male connectors at the throttle switch (No. 1).
3. Start the engine, the ohmmeter reading must be 0 ohms at idle.
4. Slightly open the throttle, the ohmmeter reading should go to infinity.
5. Slowly close the throttle to idle position. The ohmmeter reading must return to 0 ohms.

NOTE: Correct adjustment is very important. If the switch is misadjusted, the engine may surge or cut-out during steady or light throttle acceleration.

6. Stop the engine.
7. Using the accelerator pedal, open the throttle fully, the ohmmeter must read infinity.
8. If the ohmmeter readings are not correct, adjust the throttle switch.

THROTTLE IDLE SWITCH

Adjustment

1. Loosen the allen head retaining screw slightly.
2. Open the throttle and allow it to slowly return closed.
3. Measure the switch point with a feeler gauge between the the idle stop and idle stop screw. At the point where the switch contact closes, the gap must be 0.008–0.024 in. (0.2–0.6mm) or 0.002–0.004 in. (0.05–0.10mm) for Vanagon.
4. If not within specification, adjust the position of the switch using the allen head cam screw. Do not tamper with the idle stop screw.
5. Tighten the retaining screw.
6. Connect an ohmmeter to the throttle switch terminals. With the switch must be open at all points except throttle closed and throttle wide open.

THROTTLE VALVE

Adjustment

NOTE: The stop screw is set at the factory and should not be moved. If the screw position has been altered, check and adjust as required.

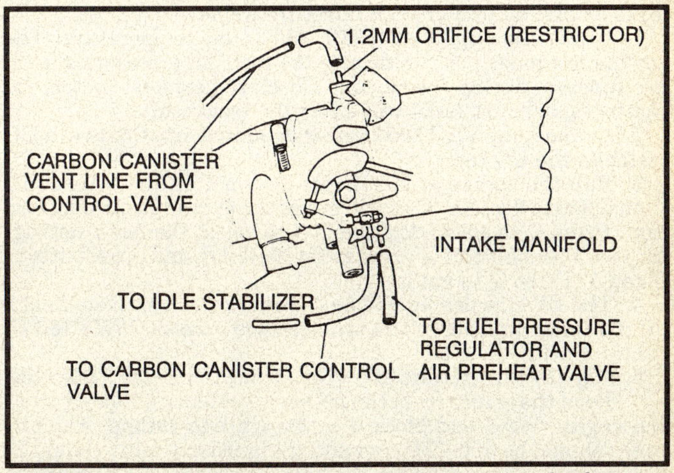

Digifant vacuum hose layout manual transmission

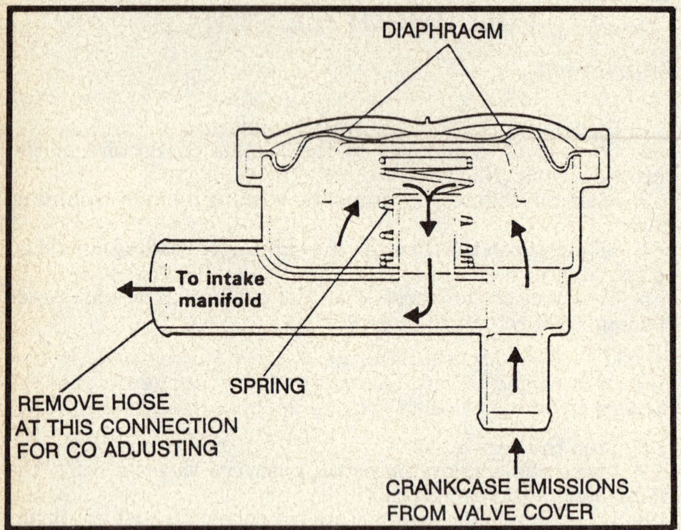

Digifant crankcase emission control valve—Volkswagen

1. Turn the adjusting screw out until a gap exists between the stop and the screw. Turn the screw in until it touches the stop.
2. In order to determine the exact point of contact with the screw stop, place a thin piece of paper between screw and the stop. Slide the paper and turn screw at same time until the screw pinches the paper.
3. Turn the screw clockwise an addition ½ turn. Check the idle speed and the CO content. Adjust if necessary.

IDLE SPEED AND MIXTURE (CO)

Adjustments

NOTE: The following information is being published from the latest information available at the time of publication. If the information differs from the values given on the underhood emission control label, use the data on the label.

CORRADO SLC AND PASSAT GLS

The idle speed, ignition timing and mixture are not adjustable on these models. The following information is provided to allow checking of values. While the idle speed, ignition timing and mixture are not adjustable, outside influences, such as vacuum leaks or fuel diluted oil, can effect the values.

1. Warm the engine to 176°F (80°C) oil temperature. The cooling fan must not run during the test. All accessories must not function during the testing. Check the exhaust system for tightness; exhaust leaks will effect the test results.
2. Connect the VAG 1551 tester or equivalent. Connect a CO tester to the CO tap.
3. Run the engine at idle. Press button 1 on the VAG 1551, then buttons 0 and 1. This will select the engine electronics section. Press Q to select, the 0 and 4 to select the basic settings section. Press button Q to enter the selection and press buttons 0 and 1. Press Q to enter.
4. The EVAP solenoid will be closed for the test. Note field 2 on the tester and when the temperature exceeds 176°F (80°C) continue with the test.
5. Blip the throttle and allow the engine to idle for 2 minutes.
6. Read the values from the information blocks. Block 1 show the engine speed and block 4 is the ignition timing. The idle speed should be 650–750 rpm and the ignition timing should be 6 degrees BTDC and be fluctuating. Check the CO tester. The value should be 0.3–1.5%.
7. Press buttons 0 and 6, then Q. This will exit the tester program.
8. If the values were not correct, check for vacuum leaks, leaking injectors and other faults with the fuel and air systems.

CORRADO G60

1. Warm the engine to 176°F (80°C) oil temperature. The cooling fan must not run during the test. All accessories must not function during the testing. Check the exhaust system for tightness; exhaust leaks will effect the test results. Check that the idle stabilizer buzzes and hums with the ignition switched **ON**.
2. Switch **OFF** the ignition and connect VAG 1367 or equivalent, and a CO tester. Disconnect the crankcase ventillation hose from the emission control valve.
3. Start and idle the engine. When the engine idle drops 100 rpm after 1–4 minutes, disconnect the blue water temperature sensor connector.
4. Raise and lower the engine speed 3 times in excess of 3000 rpm. Let the engine idle and check the test values.
5. The idle speed should be 750–850 rpm and the CO 0.3–1.1%.
6. If the values are not correct, adjust the screws on the throttle body and the intake hose. Alternate the adjustment of these 2 screws. Reconnect the blue temperature sensor connector, blip the throttle, raise the engine speed over 3000 rpm 3 times and check the CO. Repeat the procedure until specifications are met.

FOX

Except California

1. Warm the engine to 176°F (80°C) oil temperature. The cooling fan must not run during the test. All accessories must not function during the testing. Check the exhaust system for tightness; exhaust leaks will effect the test results.
2. Switch **OFF** the ignition and connect VAG 1367 engine analyzer or equivalent, and a CO tester.
3. Run the engine and raise the engine speed above 2100 rpm 4 times, allowing the throttle to close completely between pulls. Disconnect the coolant temperature sensor.
4. Check the ignition timing at 2200–2300 rpm. The timing range should be 4–8 degrees BTDC. If the timing needs to be adjusted, loosen the distributor clamp and rotate to obtain timing of 5–7 degrees BTDC. Tighten the clamp bolt to 18 ft. lbs. (25 Nm).
5. Disconnect the idle speed boost solenoid valve electrical connector. If equipped with air conditioning, disconnect the second idle speed boost valve electrical connector. Pinch the vacuum hose leading to the valves and check the idle speed. The idle speed should be 875–925 rpm. Adjust as necessary.
6. Connect the coolant temperature sensor connector. The idle should be 800–1000 rpm. Turn **OFF** the ignition and disconnect the crankcase ventilation hose from the emission control valve mounted on the valve cover and plug.
7. Start the engine and cycle the engine speed to 2100 rpm 4 times allowing the speed to drop to idle between pulls. Disconnect the coolant temperature sensor electrical connection while the engine is at idle.
8. Check the CO level. The CO should be 0.3–1.1% at sea level through 935 feet above sea level. At altitude above 935 feet add 0.2% for each 300 ft. increase.
9. If the CO needs adjustment, drill the anti-tamper plug, screw in a sheet metal screw and pull out the plug. The screw is located in the air flow meter assembly. Use a 5mm hex wrench to turn the screw.
10. Connect the coolant temperature sensor, the crankcase ventilation hose, the idle boost solenoids and unpinch the vacuum hose leading to the idle boost solenoid valves. The idle should be 875–925 rpm and the CO 0.3–1.2%, regardless of the altitude. Install a new anti-tamper plug.

11. If the idle speed is not correct, check the idle speed boost solenoid wiring. If the CO value is not correct, check the oxygen sensor system. If both the idle speed and the CO level are incorrect, check the coolant temperature sensor and wiring.

California

1. Warm the engine to 176°F (80°C) oil temperature. The cooling fan must not run during the test. All accessories must not function during the testing. Check the exhaust system for tightness; exhaust leaks will effect the test results.
2. Switch **OFF** the ignition and connect VAG 1367 engine analyzer or equivalent, and a CO tester.
3. Run the engine and raise the engine speed above 2100 rpm 4 times, allowing the throttle to close completely between pulls. Disconnect the coolant temperature sensor.
4. Check the ignition timing at 2200–2300 rpm. The timing range should be 4–8 degrees BTDC. If the timing needs to be adjusted, loosen the distributor clamp and rotate to obtain timing of 5–7 degrees BTDC. Tighten the clamp bolt to 18 ft. lbs. (25 Nm).
5. Disconnect the idle speed boost solenoid valve electrical connector. If equipped with air conditioning, disconnect the second idle speed boost valve electrical connector. Pinch the vacuum hose leading to the valves and check the idle speed. The idle speed should be 875–925 rpm. Adjust as necessary.
6. Connect the coolant temperature sensor connector. The idle should be 800–1000 rpm. Turn **OFF** the ignition and disconnect the crankcase ventilation hose from the emission control valve mounted on the valve cover. Plug the hose.
7. Start the engine and allow the cooling fan to cycle at least once. Disconnect the coolant temperature sensor electrical connection while the engine is at idle.
8. After 1 minute, connect the coolant temperature sensor. Do not connect it while the cooling fan is running.
9. Turn **OFF** the engine and connect the idle boost connections, crankcase ventilation hose and unpinch the vacuum hose. Erase the fault code memory.

CABRIOLET

Except 1991–92 California

1. Warm the engine to 176°F (80°C) oil temperature. The cooling fan must not run during the test. All accessories must not function during the testing. Check the exhaust system for tightness; exhaust leaks will effect the test results. Check that the idle stabilizer buzzes and hums with the ignition switched **ON**.
2. Switch **OFF** the ignition and connect VAG 1367 engine analyzer or equivalent, and a CO tester.
3. Run the engine and raise the engine speed above 2100 rpm 4 times, allowing the throttle to close completely between pulls. Disconnect the coolant temperature sensor.
4. Check the ignition timing at 2250–2350 rpm. The timing range should be 4–8 degrees BTDC. If the timing needs to be adjusted, loosen the distributor clamp and rotate to obtain timing of 5–7 degrees BTDC. Tighten the clamp bolt to 18 ft. lbs. (25 Nm).
5. Connect the coolant temperature sensor connector. Turn **OFF** the ignition and disconnect the crankcase ventilation hose from the emission control valve mounted on the valve cover and plug.
6. Start the engine and cycle the engine speed to 2100 rpm 4 times allowing the speed to drop to idle between pulls. Disconnect the coolant temperature sensor electrical connection while the engine is at idle. Adjust the idle speed to 800–900 rpm.
7. Check the CO level. The CO should be 0.3–1.1% at sea level through 935 feet above sea level. At altitude above 935 feet add 0.2% for each 300 ft. increase.
8. If the CO needs adjustment, drill the anti-tamper plug, screw in a sheet metal screw and pull out the plug. The screw is located in the air flow meter assembly. Use a 5mm hex wrench to turn the screw. Adjust the idle speed, if necessary.
9. Connect the coolant temperature sensor, raise the engine speed above 3000 rpm 3 times and allow to idle. The idle speed should be 750–850 rpm and the CO 0.3–1.1%, regardless of the altitude. Install a new anti-tamper plug.
11. If the idle speed is not correct, check the idle switch. If the CO value is not correct, check the oxygen sensor system. If both the idle speed and the CO level are incorrect, check the coolant temperature sensor and wiring.

1991–92 California

1. Warm the engine to 176°F (80°C) oil temperature. The cooling fan must not run during the test. All accessories must not function during the testing. Check the exhaust system for tightness; exhaust leaks will effect the test results. Check that the idle stabilizer buzzes and hums with the ignition switched **ON**.
2. Switch **OFF** the ignition and connect VAG 1367 engine analyzer or equivalent, and a CO tester.
3. Run the engine and raise the engine speed above 2100 rpm 4 times, allowing the throttle to close completely between pulls. Disconnect the coolant temperature sensor.
4. Check the ignition timing at 2200–2300 rpm. The timing range should be 4–8 degrees BTDC. If the timing needs to be adjusted, loosen the distributor clamp and rotate to obtain timing of 5–7 degrees BTDC. Tighten the clamp bolt to 18 ft. lbs. (25 Nm).
5. Connect the coolant temperature sensor connector. The idle speed should be 750–850 rpm and the CO 0.3–1.2%. Turn **OFF** the ignition and disconnect the crankcase ventilation hose from the emission control valve mounted on the valve cover. Plug the hose.
6. Start the engine and allow the cooling fan to cycle at least once. Disconnect the coolant temperature sensor electrical connection while the engine is at idle.
7. After 1 minute, connect the coolant temperature sensor. Do not connect it while the cooling fan is running.
8. Connect the crankcase breather hose and erase the fault code memory.

GOLF AND JETTA

There are 2 engines commonly used in the Golf and Jetta vehicles. Either the RV or PF engine will be used depending on the transmission, model year and other variables. The engine type is stamped in the block just below the cylinder head joint, on the middle-right side. The engine type is the first 2 letters and will be followed by the engine number.

Engine Type RV

1. Warm the engine to 176°F (80°C) oil temperature. The cooling fan must not run during the test. All accessories must not function during the testing. Check the exhaust system for tightness; exhaust leaks will effect the test results. Check that the idle stabilizer buzzes and hums with the ignition switched **ON**.
2. Switch **OFF** the ignition and connect VAG 1367 engine analyzer or equivalent, and a CO tester.
3. Run the engine and raise the engine speed above 2100 rpm 4 times, allowing the throttle to close completely between pulls. Disconnect the coolant temperature sensor.
4. Check the ignition timing at 2000–2500 rpm. The timing range should be 4–8 degrees BTDC. If the timing needs to be adjusted, loosen the distributor clamp and rotate to obtain timing of 5–7 degrees BTDC. Tighten the clamp bolt to 18 ft. lbs. (25 Nm).
5. Connect the coolant temperature sensor connector. The idle speed should be 750–850 rpm and the CO 0.3–1.2%. Turn **OFF** the ignition and disconnect the crankcase ventilation hose from the emission control valve mounted on the valve cover. Plug the hose.
6. Start the engine and allow the cooling fan to cycle at least

once. Disconnect the coolant temperature sensor electrical connection while the engine is at idle.

7. After 1 minute, connect the coolant temperature sensor. Do not connect it while the cooling fan is running.

8. Connect the crankcase breather hose and erase the fault code memory.

Engine Type PF

1. Warm the engine to 176°F (80°C) oil temperature. The cooling fan must not run during the test. All accessories must not function during the testing. Check the exhaust system for tightness; exhaust leaks will effect the test results. Check that the idle stabilizer buzzes and hums with the ignition switched **ON**.

2. Switch **OFF** the ignition and connect VAG 1367 engine analyzer or equivalent, and a CO tester.

3. Run the engine and raise the engine speed above 2100 rpm 4 times, allowing the throttle to close completely between pulls. Disconnect the coolant temperature sensor.

4. Check the ignition timing at 2250–2350 rpm. The timing range should be 4–8 degrees BTDC. If the timing needs to be adjusted, loosen the distributor clamp and rotate to obtain timing of 5–7 degrees BTDC. Tighten the clamp bolt to 18 ft. lbs. (25 Nm).

5. Raise the dipstick to allow crankcase breathing and clamp the crankcase breather hose near the emission control valve. Adjust the idle speed to 750–850 rpm.

6. Check the CO level. The CO should be 0.3–1.1% at sea level through 935 feet above sea level. At altitude above 935 feet add 0.2% for each 300 ft. increase.

7. If the CO needs adjustment, drill the anti-tamper plug, screw in a sheet metal screw and pull out the plug. The screw is located in the air flow meter assembly. Use a 5mm hex wrench to turn the screw. Fully seat the dipstick and remove the clamp on the crankcase breather hose. Adjust the idle speed to 900–1000 rpm.

8. Connect the coolant temperature sensor, raise the engine speed above 3000 rpm 3 times and allow to idle. The idle speed should be 750–850 rpm and the CO 0.3–1.1%, regardless of the altitude. Install a new anti-tamper plug.

9. If the idle speed is not correct, check the idle switch. If the CO value is not correct, check the oxygen sensor system. If both the idle speed and the CO level are incorrect, check the coolant temperature sensor and wiring.

VANAGON

1. Warm the engine to 176°F (80°C) oil temperature. The cooling fan must not run during the test. All accessories must not function during the testing. Check the exhaust system for tightness; exhaust leaks will effect the test results. Check that the idle stabilizer buzzes and hums with the ignition switched **ON**.

2. Switch **OFF** the ignition and connect VAG 1367 engine analyzer or equivalent, and a CO tester. Connect the CO tester to the exhaust tap on the left side exhaust pipe.

3. Disconnect and plug the crankcase breather hose. Disconnect the oxygen sensor plug and the idle stabilizer connector.

4. Start the engine and run at idle for 2 minutes. The idle speed should be 830–930 rpm. Adjust the idle speed with the throttle body adjusting screw.

5. The CO should be 0.3–1.1%. If the CO needs adjusting, drill the anti-tamper plug, screw in a sheet metal screw and pull out the plug. The screw is located in the air flow meter assembly. Use a 5mm hex wrench to turn the screw.

6. Alternately turn the screw in the air flow meter and on the throttle body to obtain the correct CO.

7. Connect the idle stabilizer and the oxygen sensor and allow the engine to idle for 2 minutes. Check the CO and idle speed. Repeat the procedure to meet the specifications.

8. Stop the engine and install a new anti-tamper plug.

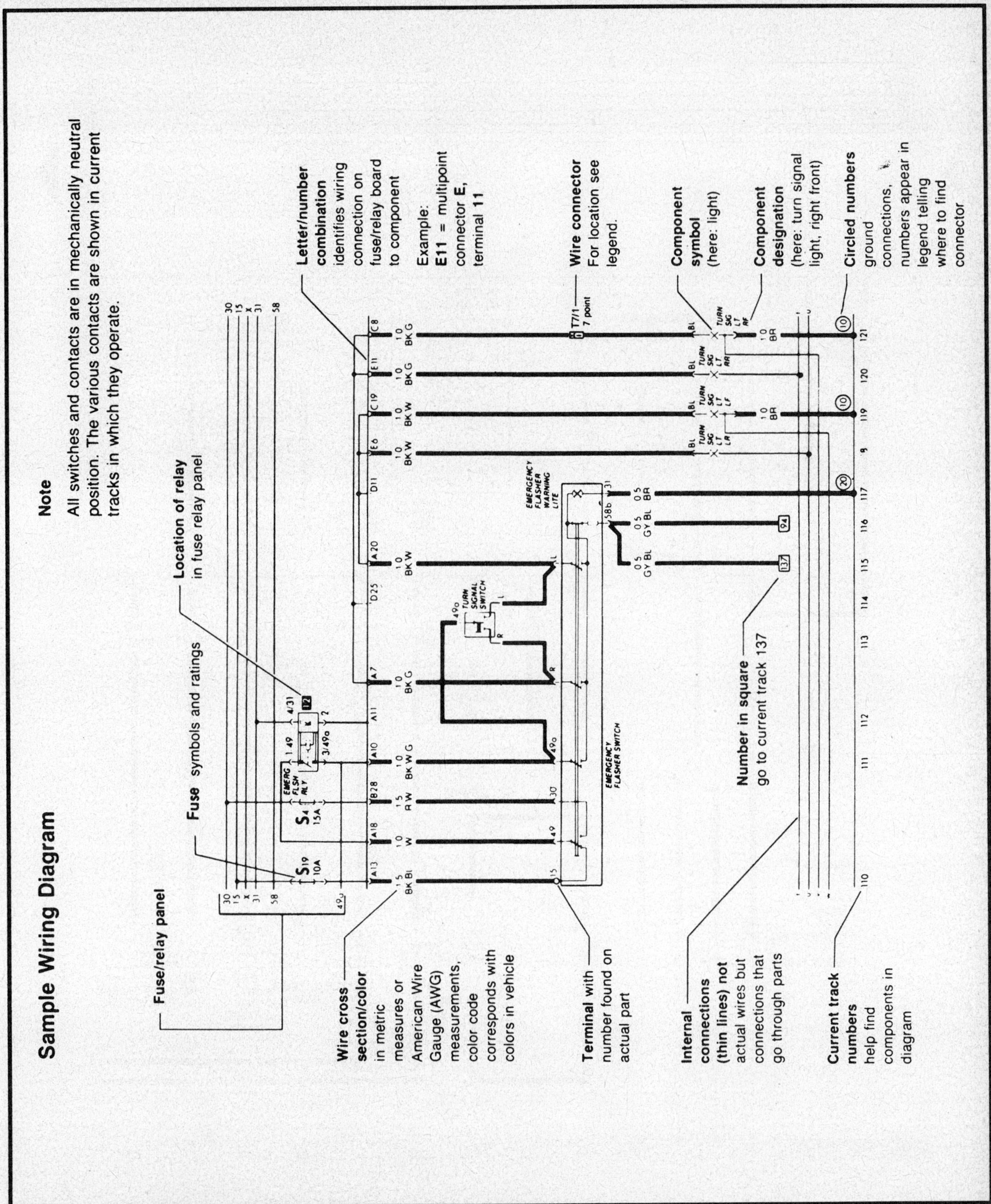

How to use Volkswagen wiring diagrams

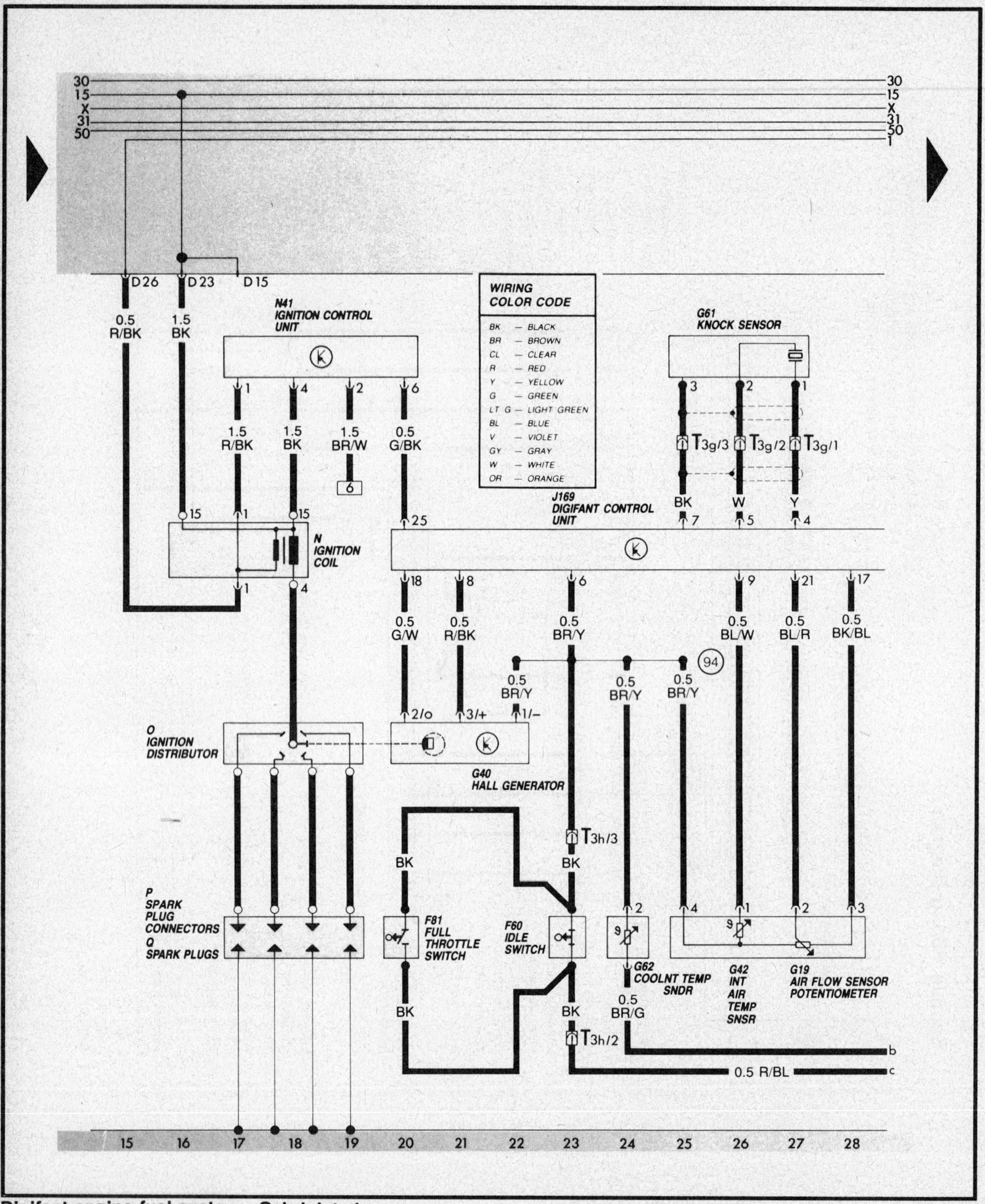

Digifant engine fuel system—Cabriolet shown

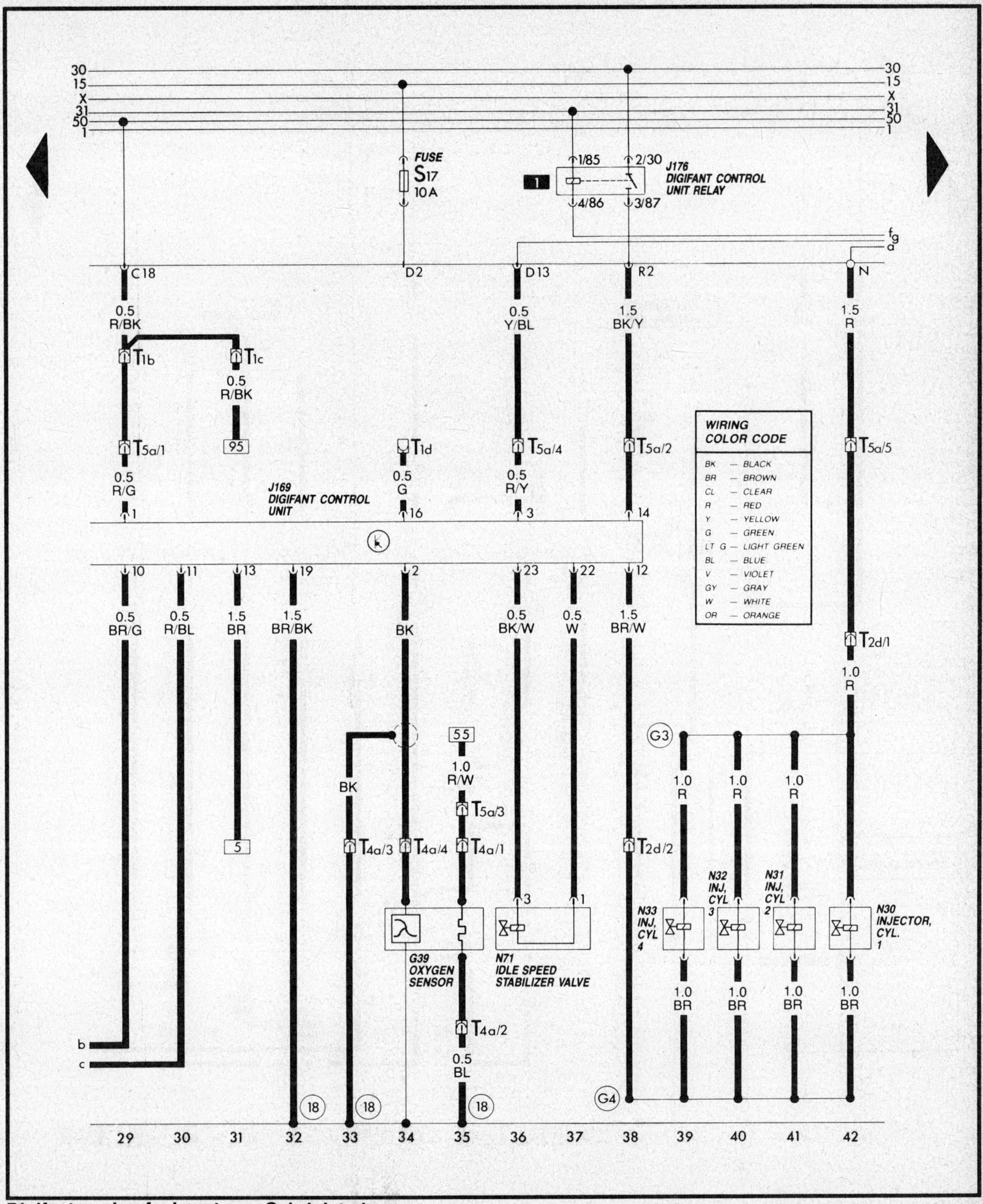

Digifant engine fuel system—Cabriolet shown

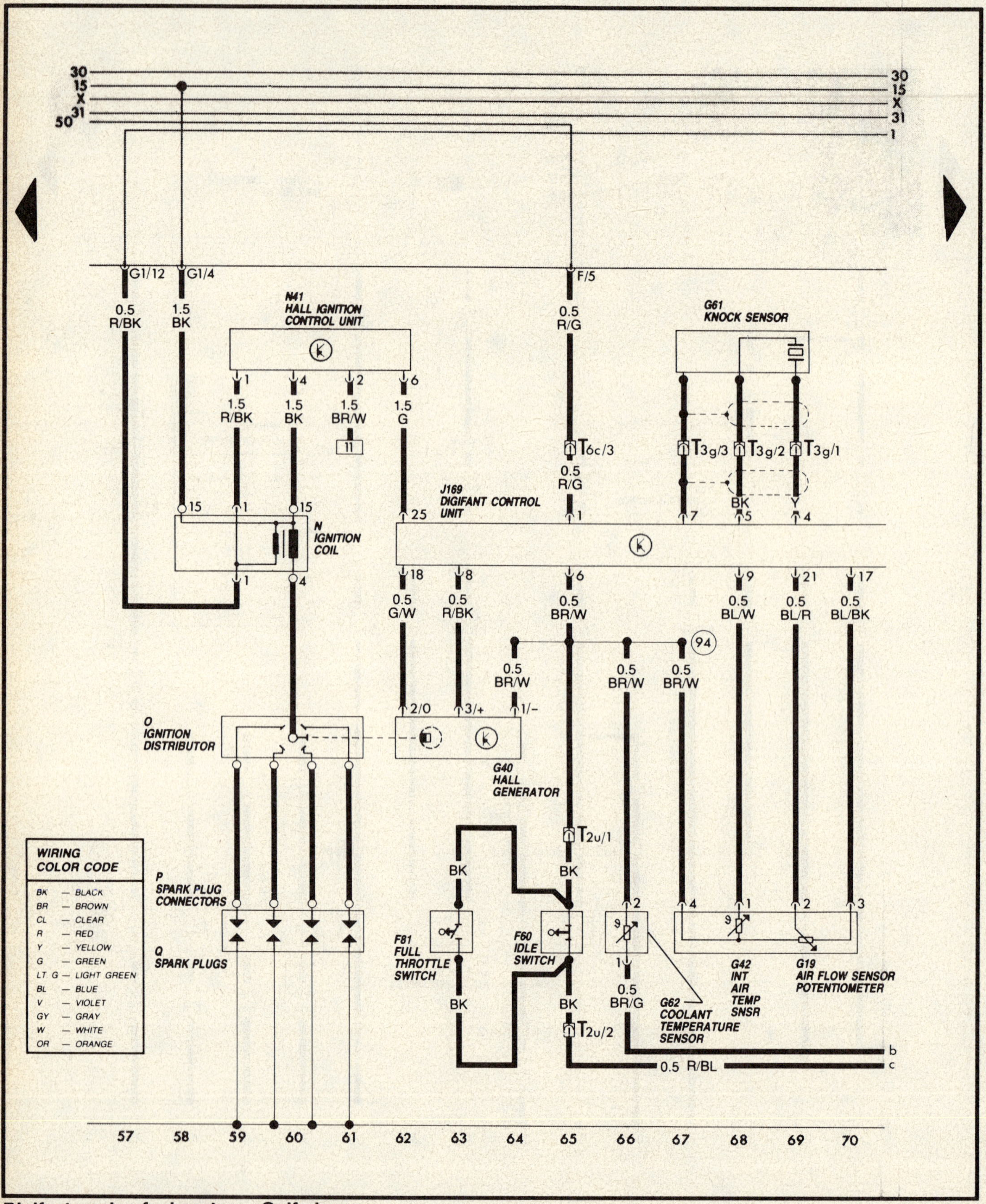

Digifant engine fuel system—Golf shown

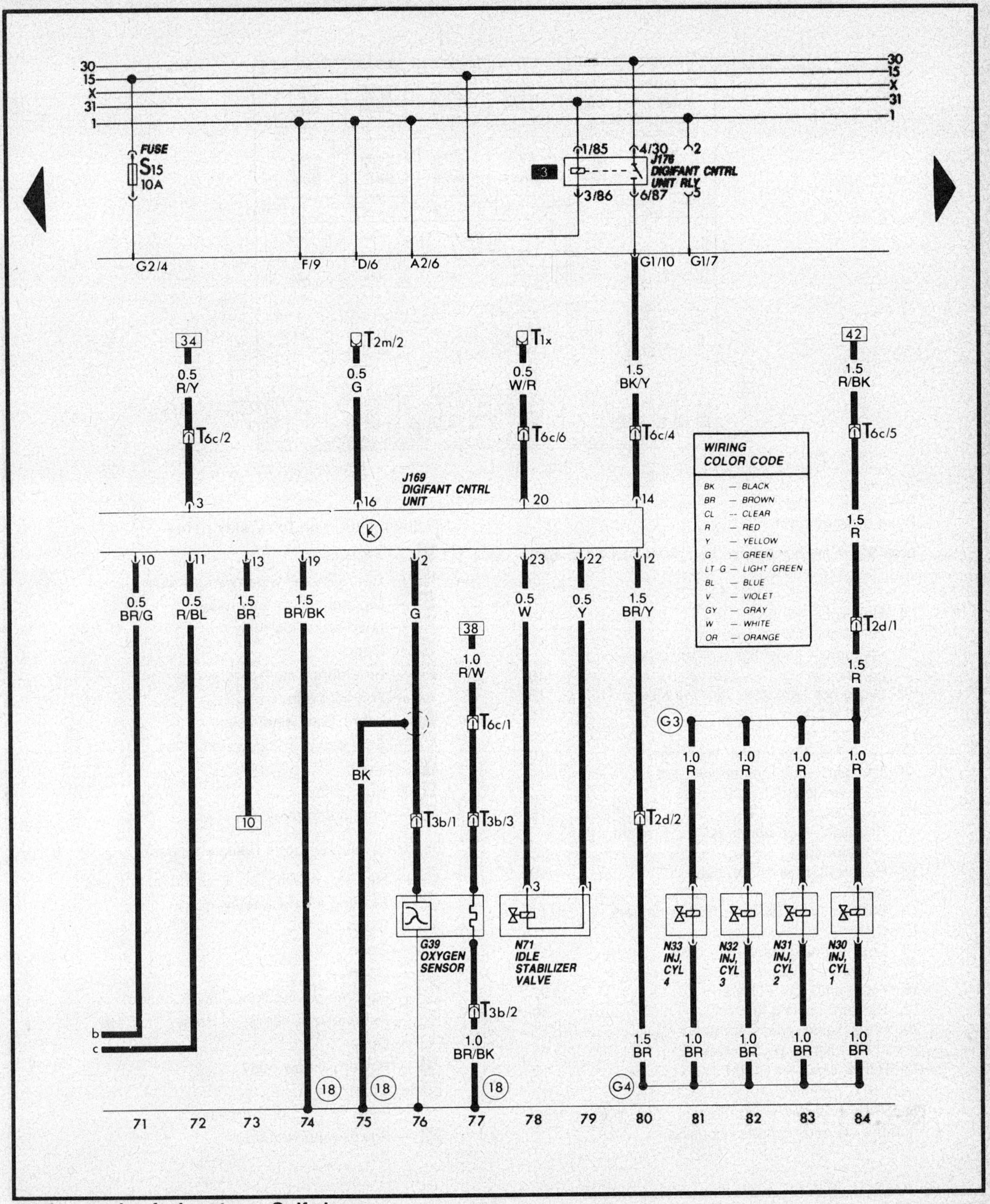

Digifant engine fuel system—Golf shown

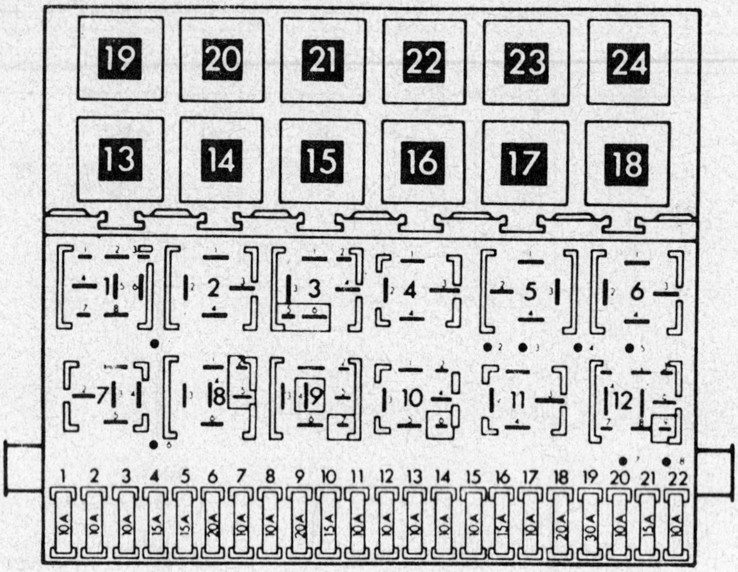

Fuse arrangement

according to the numbers on fuse panel from left to right:

		Amp.
1	Headlight, low beam, left	10
2	Headlight, low beam, right	10
3	Instrument panel lights, license plate lights	10
4	Glove box light, rear window wiper	15
5	Wiper/washer	15
6	Fresh air fan	20
7	Tail lights, side marker light, right	10
8	Tail lights, side marker light, left	10
9	Rear window defogger	20
10	Fog lights	15
11	Headlight, high beam, left, high beam indicator light	10
12	Headlight, high beam, right	10
13	Horn	10
14	Back-up lights, washer nozzle heaters	10
15	Open	
16	Dash warning lights	15
17	Emergency flasher switch, turn signals	10
18	Fuel pump, OXS heater	20
19	Radiator cooling fan, A/C	30
20	Brake lights, cruise control	10
21	Interior lights, digital clock	15
22	Radio, cigarette lighter	10

*Numbers in parentheses indicate production control number stamped on relay housing.

Relay location on fuse/relay panel

- **1** — A/C relay
- **2** — Rear window wiper/washer relay
- **3** — Digifant control unit relay
- **4** — Load reduction relay
- **5** — Open
- **6** — Emergency flasher relay
- **7** — Fog light relay
- **8** — Intermittent wiper relay
- **9** — Seat belt warning system relay
- **10** — Open
- **11** — Horn relay
- **12** — Fuel pump relay

Separate relays above fuse/relay panel

- **13** — Radiator cooling fan after run control unit
- **14** — Starting interlock relay (53)*
- **15** — Open
- **16** — Open
- **17** — Open
- **18** — Daytime driving light relay (Canada)
- **19** — Automatic transmission relay
- **20** — Open
- **21** — Power window relay
- **22** — Open
- **23** — Fuse for A/C, power seats
- **24** — Power window fuse

Digifant engine control fuse and relay locations— typical

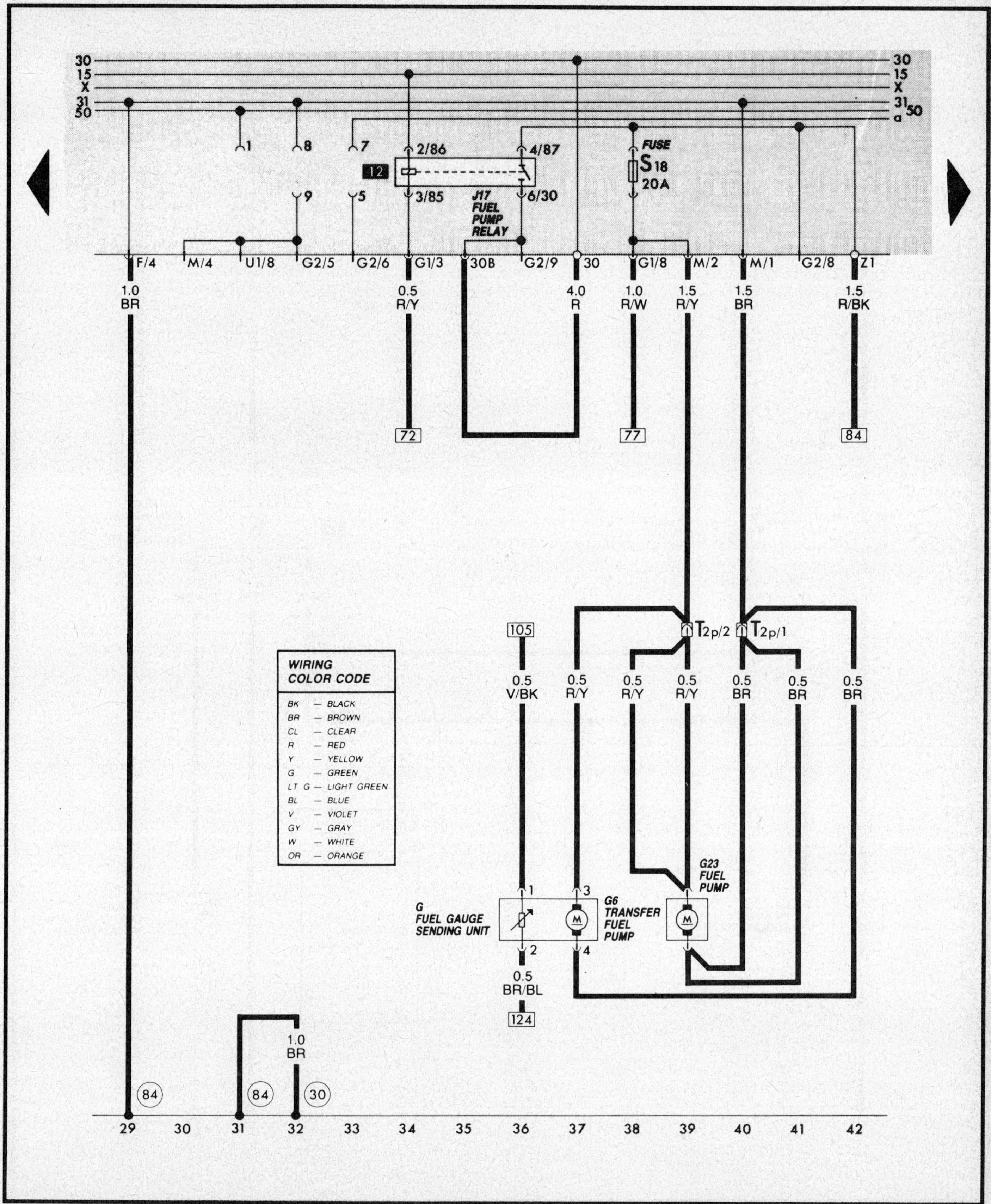

Digifant engine fuel system—Jetta shown

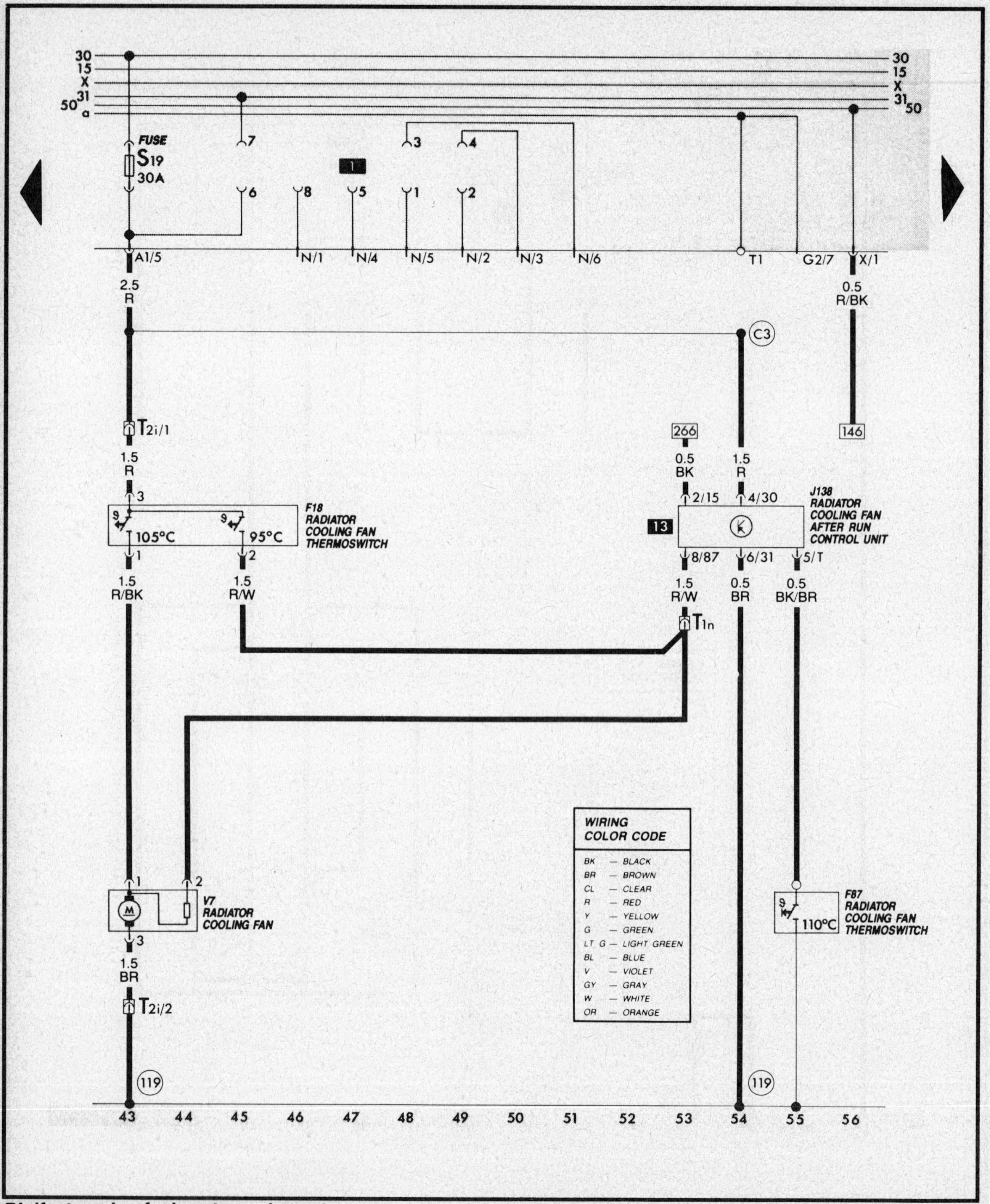

Digifant engine fuel system—Jetta shown

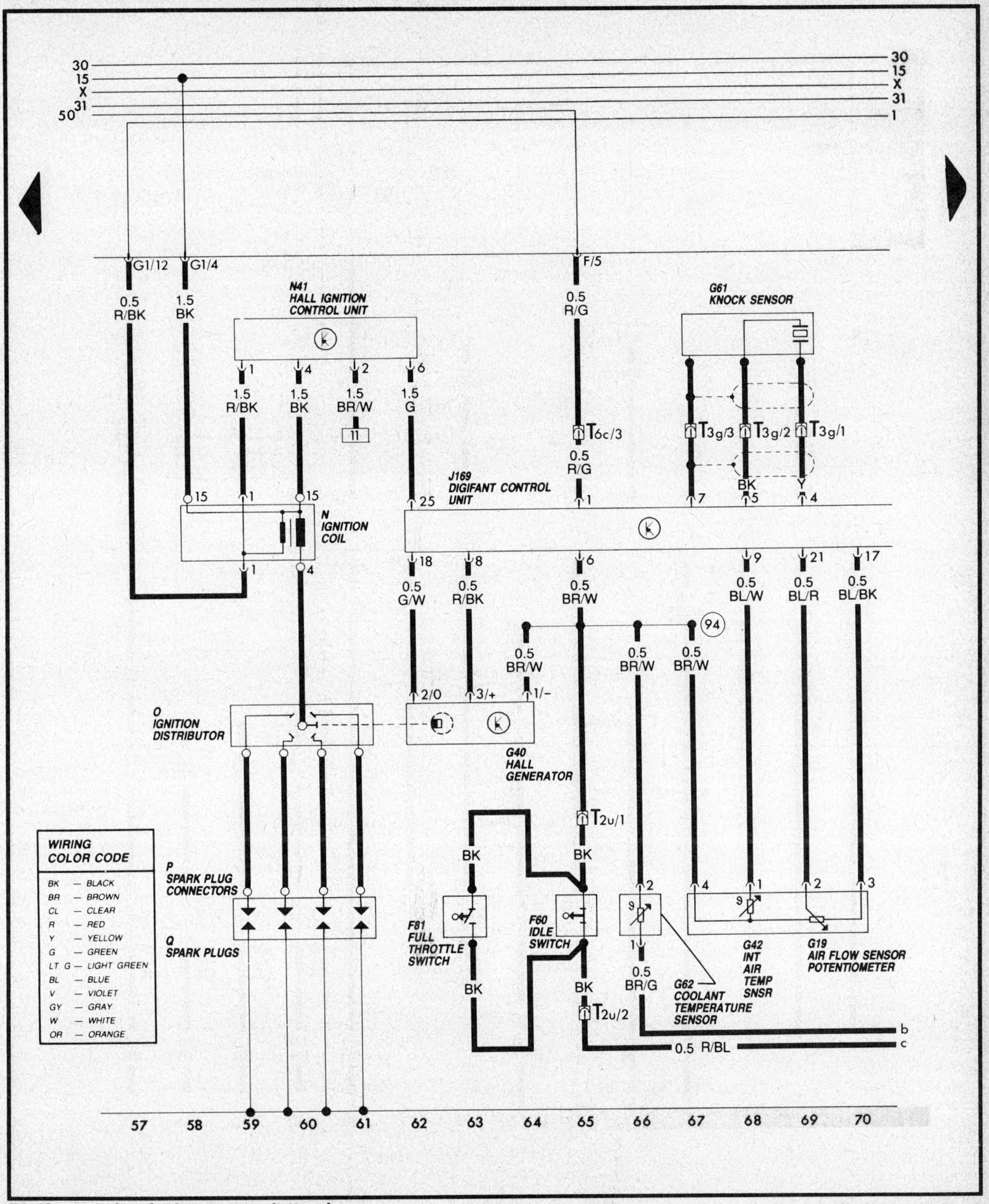

Digifant engine fuel system—Jetta shown

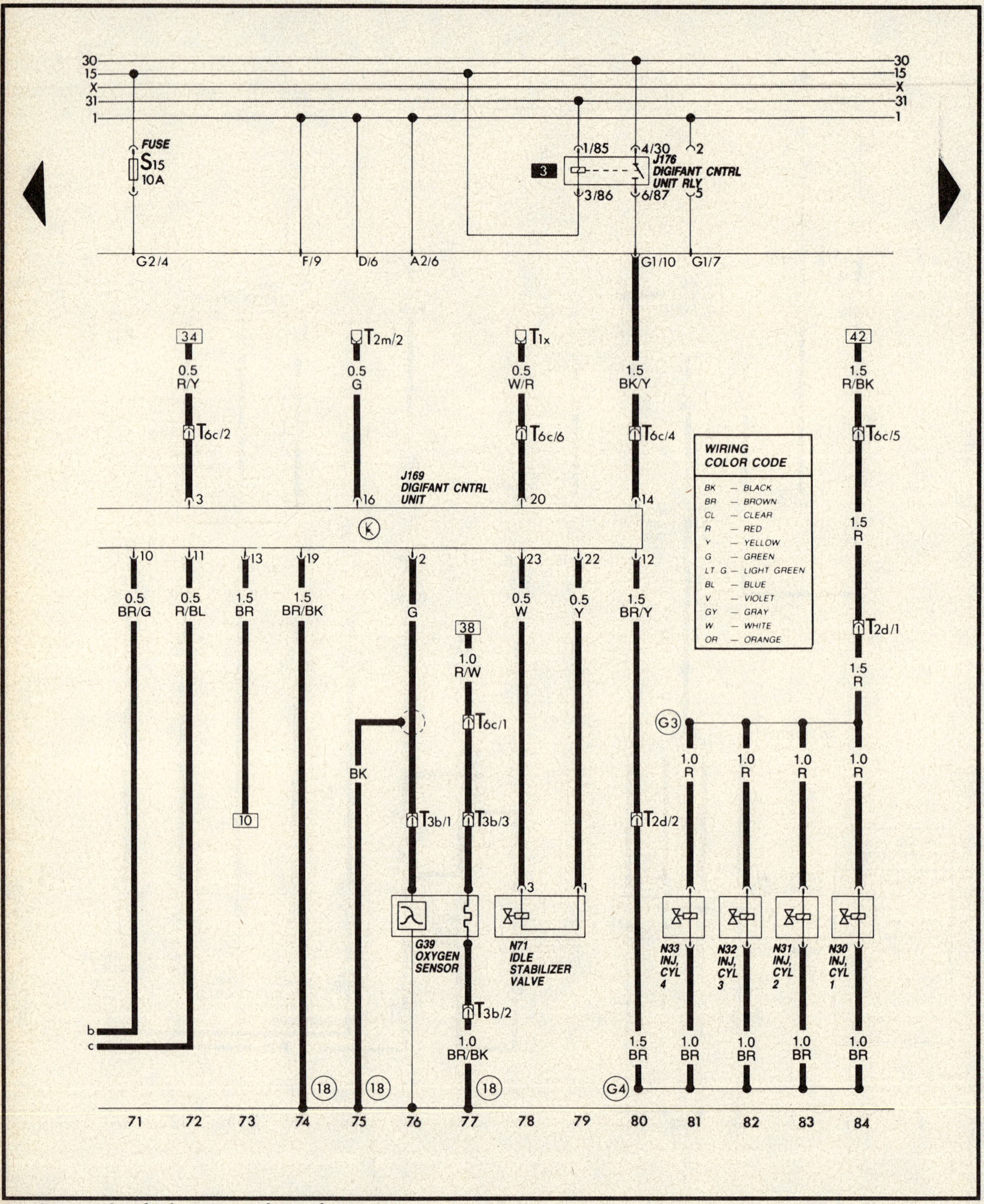

Digifant engine fuel system—Jetta shown

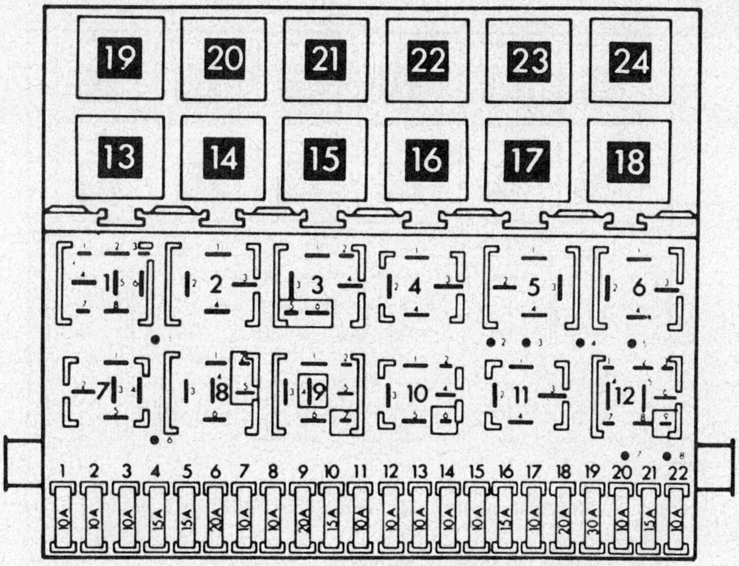

Fuse arrangement

according to the numbers on fuse panel from left to right:

		Amp.
1	Headlight, low beam, left	10
2	Headlight, low beam, right	10
3	Instrument panel lights, license plate lights	10
4	Glove box light	15
5	Wiper/washer	15
6	Fresh air fan	20
7	Tail lights, side marker light, right	10
8	Tail lights, side marker light, left	10
9	Rear window defogger	20
10	Open	
11	Headlight, high beam, left, high beam indicator light	10
12	Headlight, high beam, right	10
13	Horn	10
14	Back-up lights, washer nozzle heaters	10
15	Open	
16	Dash warning lights	15
17	Emergency flasher switch	10
18	Fuel pump, OXS heater	20
19	Radiator cooling fan, A/C	30
20	Brake lights	10
21	Interior lights, digital clock	15
22	Radio, cigarette lighter	10

*Numbers in parentheses indicate production control number stamped on relay housing.

Relay location on fuse/relay panel

- 1 — A/C relay
- 2 — Open
- 3 — Digifant control unit relay (32)*
- 4 — Load reduction relay (18)*
- 5 — Open
- 6 — Emergency flasher relay (21)*
- 7 — Open
- 8 — Intermittent washer/wiper relay (19)*
- 9 — Seat belt warning system relay (4, 29)*
- 10 — Open
- 11 — Horn relay (53)*
- 12 — Fuel pump relay (67)*

Separate relays above fuse/relay panel

- 13 — Radiator cooling fan after run control unit (31)*
- 14 — Starting interlock relay (53)* or Coolant temperature indicator relay (44)*
- 15 — Open
- 16 — Open
- 17 — Open
- 18 — Power seat fuse
- 19 — Automatic transmission relay (53)*
- 20 — Open
- 21 — Power window relay (24)*
- 22 — Open
- 23 — Fuse for A/C, power seats
- 24 — Power window fuse

Fuse and relay locations on Jetta

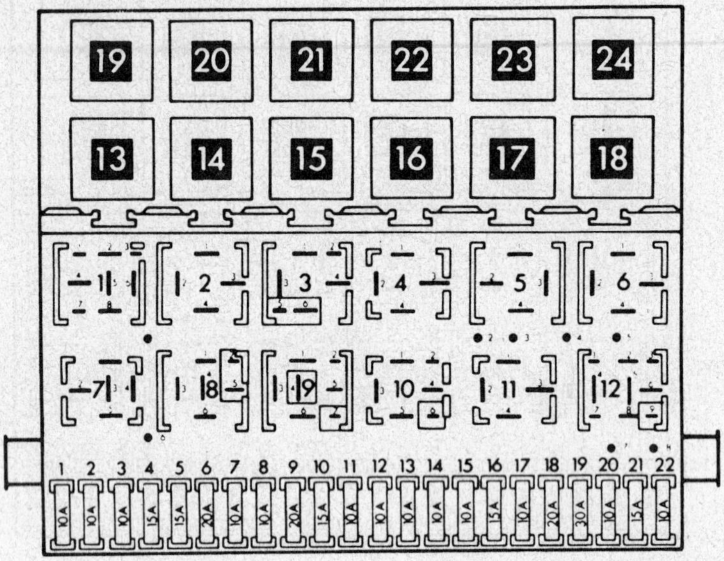

Fuse arrangement

according to the numbers on fuse panel from left to right:

1	Headlight, low beam, left	10
2	Headlight, low beam, right	10
3	Instrument panel lights, license plate lights	10
4	Glove compartment light	15
5	Windshield wiper/washer	15
6	Fresh air fan, A/C	20
7	Tail lights, side marker lights, right	10
8	Tail lights, side marker lights, left	10
9	Rear window defogger	20
10	Fog lights	15
11	Headlight, high beam, left, high beam indicator light	10
12	Headlight, high beam, right	10
13	Horn, radiator cooling fan	10
14	Back-up lights, washer nozzle heaters	10
15	Engine electronics, idle stabilizer valve, cold start valve	10
16	Dash warning lights, multi-function indicator	10
17	Emergency flashers, turn signals	10
18	Fuel pump, oxygen sensor	20
19	Radiator cooling fan, A/C	30
20	Brake lights, cruise control	10
21	Interior lights, luggage compartment lights, multi-function indicator, central locking system	15
22	Radio, cigarette lighter	10

*Numbers in parentheses indicate production control number stamped on relay housing.

Relay location on fuse/relay panel

- 1 — A/C relay, J32 (13)*
- 2 — Open
- 3 — Open
- 4 — Load reduction relay, J59 (18)*
- 5 — Open
- 6 — Emergency flasher relay, J2 (21)*
- 7 — Open
- 8 — Intermittent wiper relay, J31 (19)*
- 9 — Seat belt warning system control unit, J34 (4, 29)*
- 10 — Fog light relay, J5 (53)*
- 11 — Horn relay, J4 (53)*
- 12 — Fuel pump relay, J17 (80)*

Separate relays above fuse/relay panel

- 13 — Starter interlock relay, J207
- 14 — Radiator cooling fan after run control unit, J138 (31)*
- 15 — ABS hydraulic pump relay, J185 (78)*
- 16 — ABS relay, J102 (79)*
- 17 — Open
- 18 — Power seat circuit breaker, S45
- 19 — Open
- 20 — Open
- 21 — Power window relay (24)*/ABS hydraulic pump fuse
- 22 — ABS valves fuse
- 23 — Fuse for A/C, power seats
- 24 — Power window fuse

Fuse and relay locations on Jetta 16 Valve

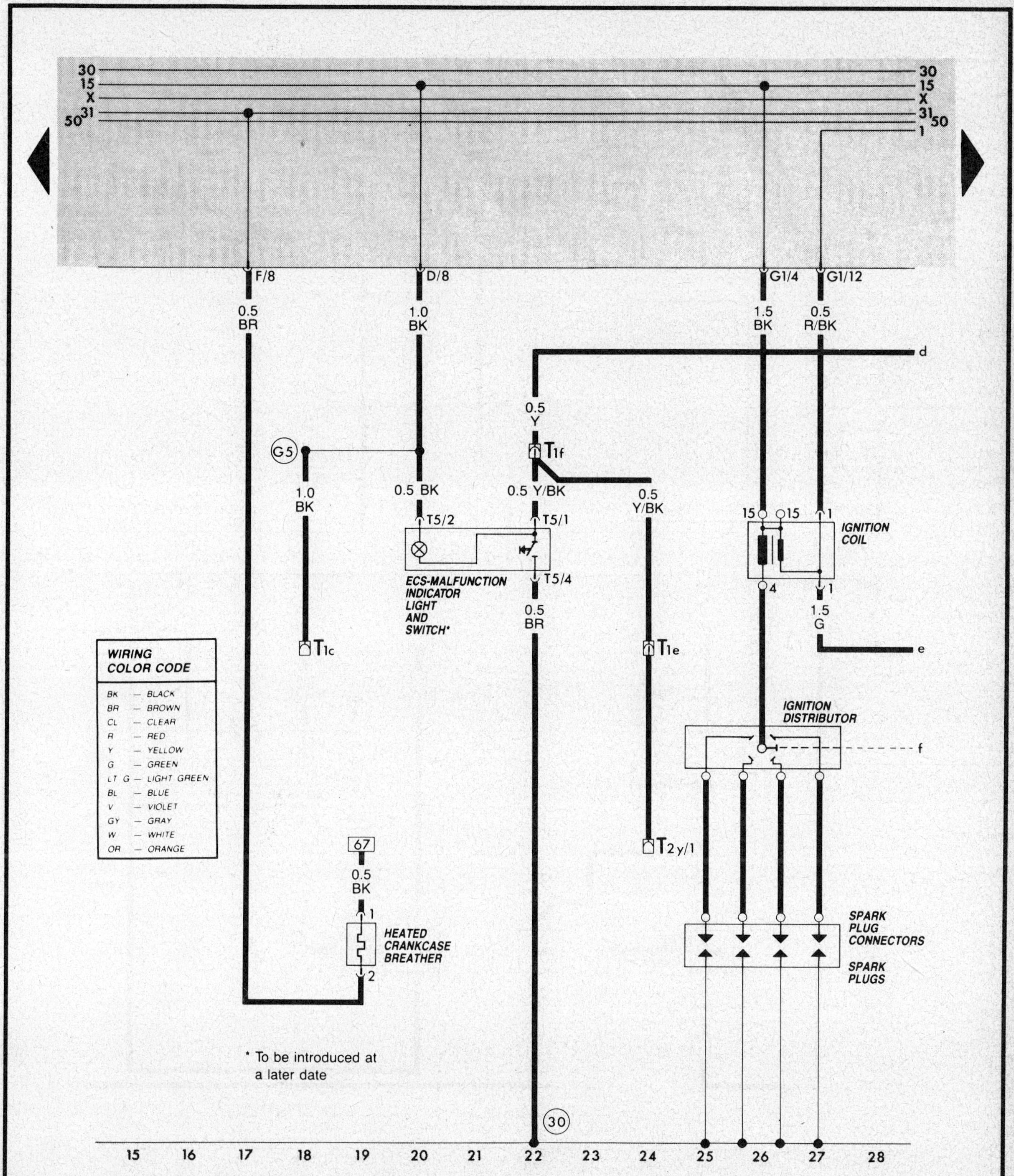

Digifant engine fuel system—Corrado shown

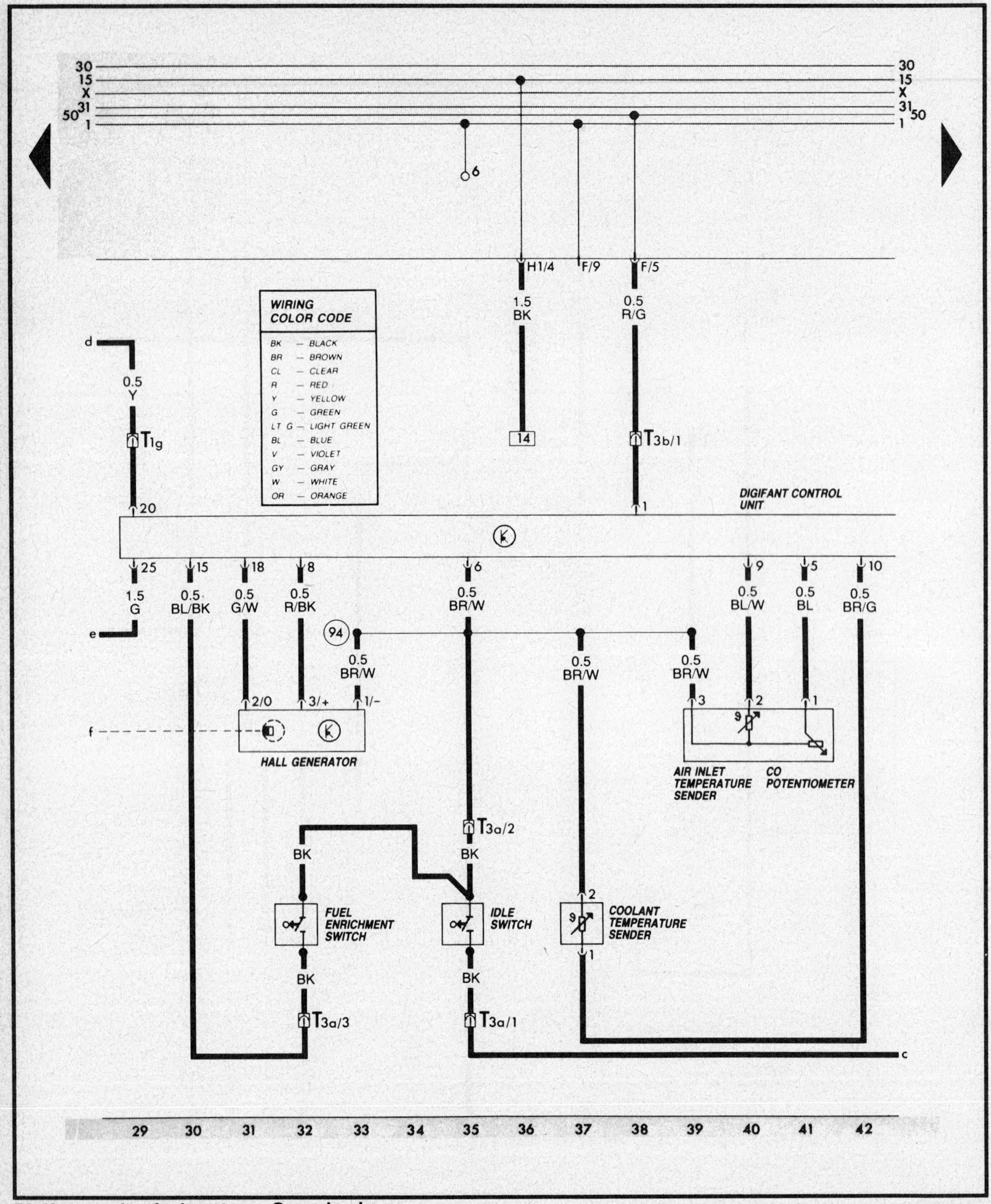

Digifant engine fuel system—Corrado shown

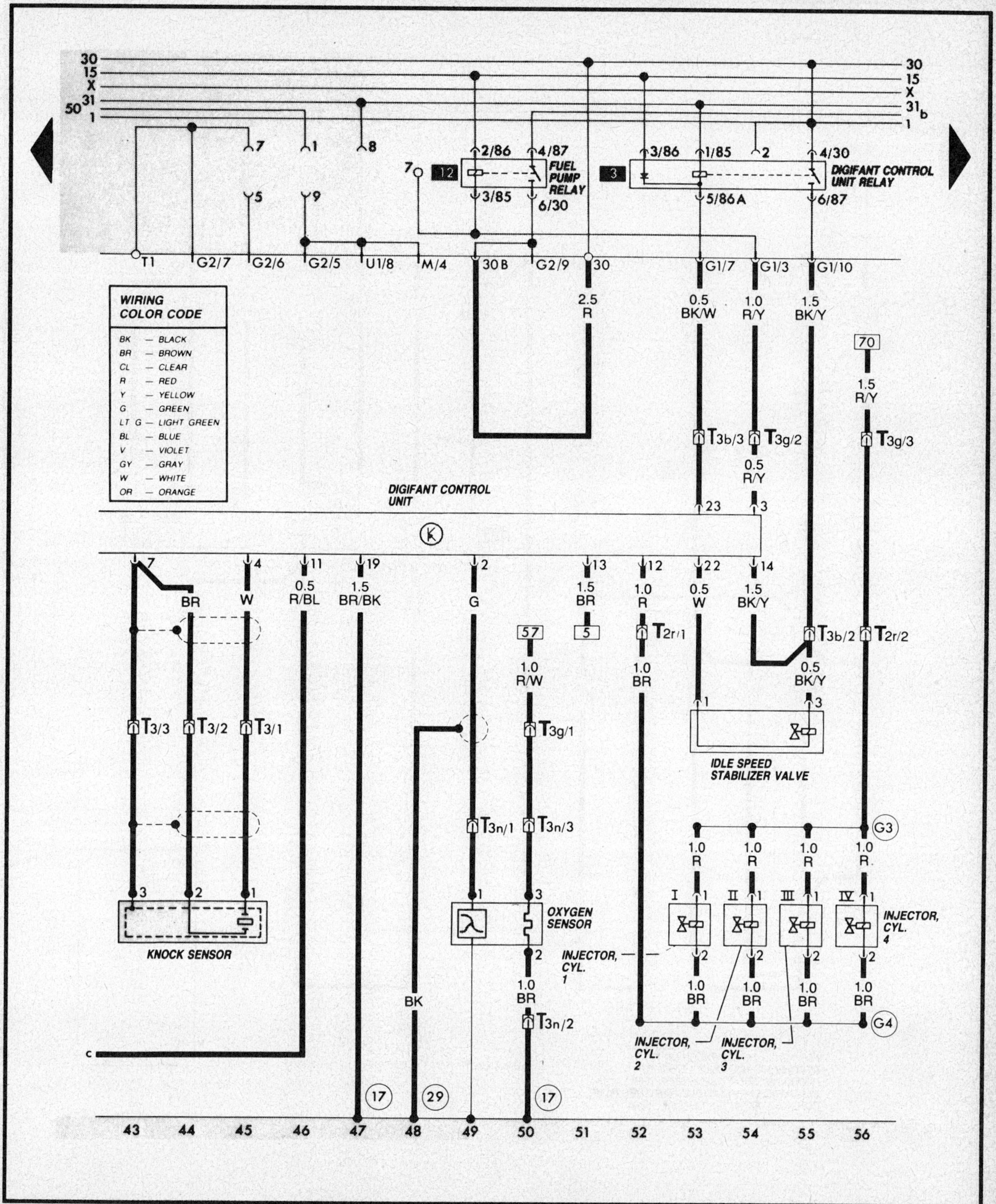

Digifant engine fuel system—Corrado shown

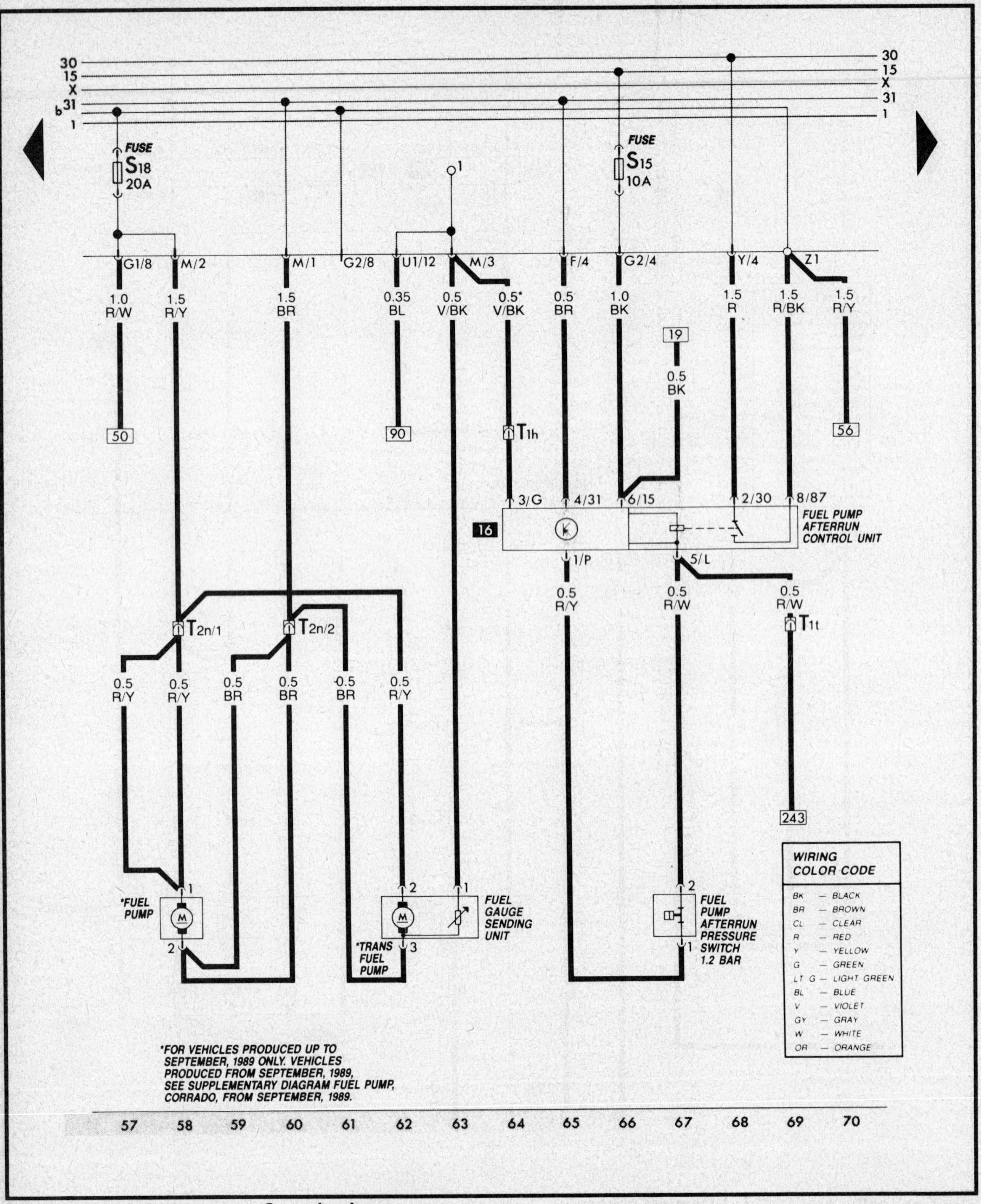

Digifant engine fuel system—Corrado shown

Volvo Fuel Injection

SECTION 16

VOLVO LH-JETRONIC FUEL INJECTION SYSTEM

General Information

The LH-Jetronic 2.2, 2.4 and 3.1 fuel injection systems are used on 1990–92 240, 700 and 900 series vehicles. The LH 2.4 and 3.1 systems are monitored by a self-diagnostic system that lights up a warning lamp on the instrument panel, but the LH-Jetronic 2.2 does not have self-diagnostic ability. Many different fault codes can be set, however only three can be stored at any one time. Fault tracing can be carried out by utilizing the diagnostic unit.

Distinguishing features of the fuel injection system are as follows: measurement of intake air through the air mass meter of the hot wire type, operation on a fuel pressure of 42 psi, use of a primary pump in the fuel tank, and an undercar fuel pump on the left side of the vehicle, use of a separate cold start valve on 1990–91 vehicles that supplies extra fuel, at or below 60°F (16°C), provision of a richer fuel mixture to counter-act knock when the fuel system's anti-knock timing control system is unsuccessful, feedback control of carbon monoxide levels and provision of a limp home setting of the idle valve. In case of a loss of current, the idle valve remains partially open to provide emergency air intake.

ELECTRONIC CONTROL UNIT

The control unit incorporates a microprocessor that receives signals from the various sensors regarding operational conditions. It evaluates them in relation to pre-programmed values and calculates the correct injector opening durations.

The control unit governs idling rpm by regulating the amount of air by-passing the throttle valve. It also controls the cold start valve (1990–91), the fuel pump and the fuel pump relay. An im-

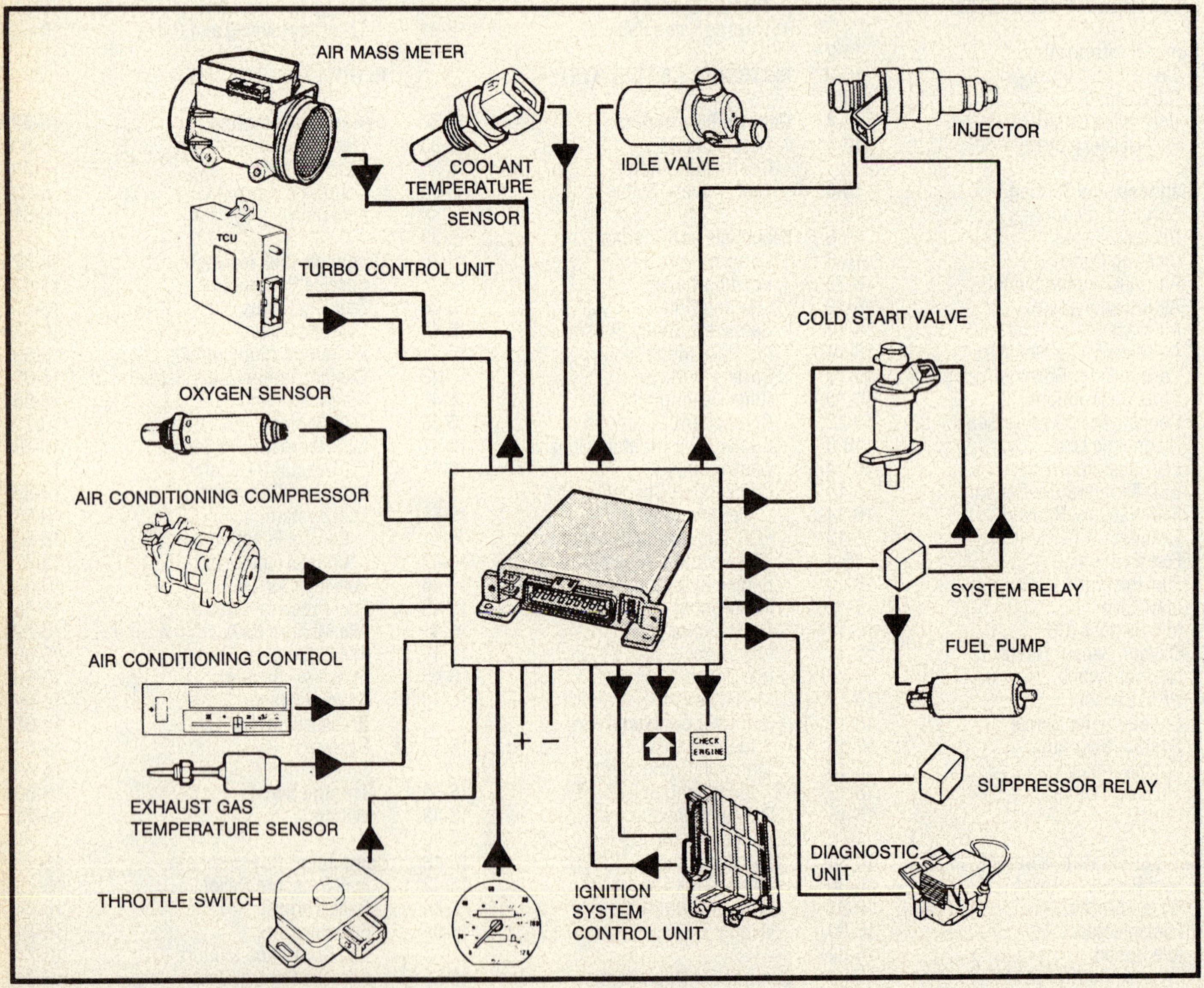

Fuel system components—LH-Jetronic

portant function of the control unit is monitoring conditions leading to faults, accessible through the diagnostic unit.

Over a period of time the throttle valve will become worn, causing less air to enter the intake system. Instead of working from a pre-programmed value, the idle valve receives a signal that is adapted to the experiences which the control unit has learned from previous driving conditions.

The limp home function allows the vehicle to be driven at a low speed in the event of an air mass meter malfunction, such as the hot wire breaking. A pre-programmed value is used for injection duration, allowing the vehicle to make its nearby destination safely.

DATA SENSORS

Several sensors feed the control unit information to precisely control fuel injection. To accomplish this, the control unit evaluates: exhaust gas oxygen content from the oxygen sensor (Lambda-sond), engine RPM and crankshaft position information from the ignition system control unit (If information is not received, the fuel system control unit will not function), engine temperature from the coolant temperature sensor, engine load information from the air mass meter, information from the throttle switch, which indicates if the throttle is closed or wide open, electrical system voltage from the battery current and signals from the A/C switch and clutch, indicating whether they are operating.

Air Mass Meter

This device measures the engine intake air mass. Factors that affect air density, such as temperature, humidity and pressure are compensated for during measurement.

The sensor inside the air mass meter consists of a wire which is maintained at 250°F (121°C) higher than the ambient air entering the engine. As the air mass passing over the wire increases, more current is required to maintain the correct temperature. The amount of current required to maintain this temperature is used to calculate the intake air mass.

When the engine is turned off, any dirt on the wire is burned off electrically by heating the wire to 1800°F (1000°C). If dirt were to remain on the wire, it would cause incorrect signals to be sent to the control unit and result in an incorrect air/fuel mixture.

Coolant Temperature Sensor

The coolant temperature sensor provides the control unit with information necessary for proper adjustment of injection duration and radiator fan control.

Throttle Switch

The throttle switch indicates to the fuel system whether the throttle valve is closed. It does not function as a variable resistor. It also indicates whether the throttle valve is fully open, except on turbocharged models.

Oxygen Sensor

The oxygen sensor is mounted in the exhaust manifold in front of the catalytic converter. It produces a measureable current by comparing the amount of oxygen in the exhaust gas with the amount in the ambient air.

Under normal conditions, the optimum air/fuel ratio is 14.7:1. This ratio is monitored by post-combustion measurement of the oxygen content in the exhaust gas using the oxygen sensor.

The oxygen sensor operating temperature ranges from 545°–1530°F (285°–832°C). It is electrically heated to enable it to reach operating temperature quickly. When ignition is turned to the **ON** position, current is sent to a resistor whose resistance increases with rising temperature. As a result the oxygen sensor quickly reaches operating temperature, even at low exhaust gas temperature.

The exhaust gases reach the outer surface of the sensor through slits in the protective sleeve. Ambient air reaches the sensors inner surface through channels. The sensor consists of a platinum covered zirconium-oxide pipe.

The signal strength is in direct proportion to the amount of oxygen in the exhaust gases. This depends on the air/fuel ratio. The Lambda value of 1, represents the theoretically perfect ratio. A rich mixture results in higher voltage and a lean mixture results in lower voltage. The current sent by the oxygen sensor to the control unit varies between 0.1–1.0 volt. The difference between high and low voltage occurs when the Lambda value is at 1. The control unit uses this information to adjust the amount of fuel injected.

EGR Temperature Sensor

This sensor, found on California vehicles, measures the temperature of the exhaust gases returned to the intake manifold. Detection of the temperature variations enables the control unit to determine whether the EGR system is operating. Its resistance increases with temperature and it is able to measure temperatures up to 930°F (500°C).

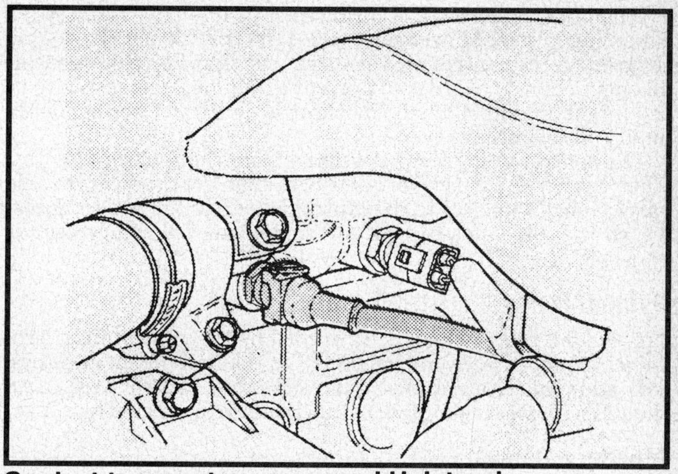

Coolant temperature sensor – LH-Jetronic

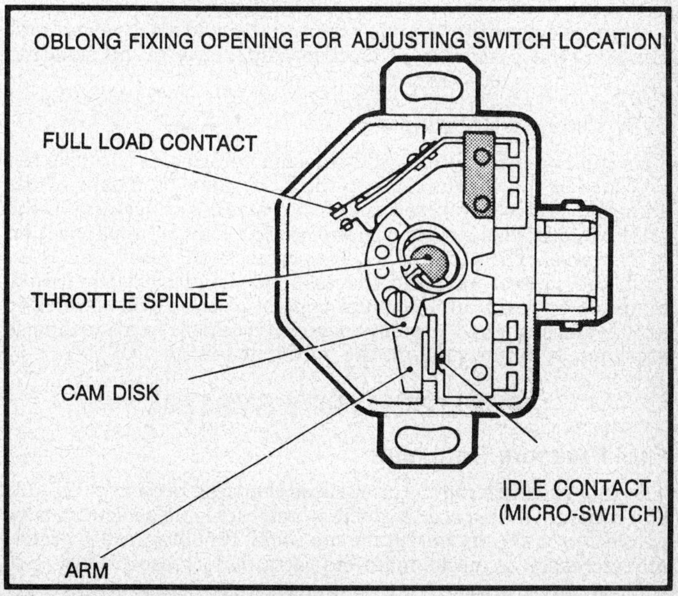

Throttle switch – LH-Jetronic

CONTROLLED OUTPUTS

In response to various operating conditions, the control unit sends commands to maintain efficient operation:

1. Provides system voltage by grounding the system relay.
2. Breaks the system relay ground if engine is stopped. This prevents the fuel pump from continuing to run and prevents the battery from being drained.
3. Grounds injectors, which regulates opening, timing and injection duration.
4. Controls air valve for constant idle speed.
5. Provides the ignition system control unit with information.
6. Protects against too high rpm by shutting off fuel injection until normal rpm is obtained.
7. Controls the check engine warning and shift indicator lamps.

Idle Valve

An idle valve is incorporated into the system to set the correct air valve opening, and constant idle speed. The control unit uses information from the air mass meter regarding the amount of air entering the engine, and from the ignition system control unit regarding rpm.

When the engine is off, a spring sets the idle valve opening for a starting idle speed between 1000–1100 rpm. When the engine is running, the control unit ensures that the idle valve opens at all engine speeds in order to prevent a high negative pressure in the intake manifold when the throttle shutter closes suddenly during deceleration.

The control unit receives a signal from the A/C control unit when the A/C is turned ON or OFF, to enable it to adjust the idle valve. Signals are sent to the control unit from the A/C compressor so the idle valve can be adjusted when the compressor is engaged.

System Relay

The system relay, which is controlled by the control unit, provides current directly, or indirectly, to the fuel pump, injectors, cold start valve, air mass meter and other control unit functions. The system relay and its functions are protected by a fuse.

Auxiliary Relay

The auxiliary relay, mounted in the engine compartment, just in front of the right shock tower, is controlled by the system relay. It supplies current to the injectors and the cold start valve. It diminishes the possibility of electrical interference, by isolating the cable harnesses from the system.

EGR Vacuum Booster

This component is found on California vehicles. It controls the pressure in the vacuum line to the EGR valve by means of the admission valve (lower section of the valve). The vacuum in the intake manifold is supplied to the reducing valve (upper section of the valve). The vacuum booster processes the electrical signal from the control unit and the vacuum signal from the intake manifold to maintain optimum control of the EGR valve. The unit is also designed to compensate for changes in atmospheric pressure. It is mounted on the left strut tower.

FUEL DELIVERY SYSTEM

Fuel Pressure Regulator

The fuel pressure regulator ensures that fuel pressure remains at a constant 42 psi at idle at the injectors. Using a vacuum tube connected to the engine intake manifold, fuel pressure is gradually increased as intake manifold vacuum decreases. Because of this, adequate pressure is maintained, regardless of throttle position. The amount of fuel injected largely depends on injection

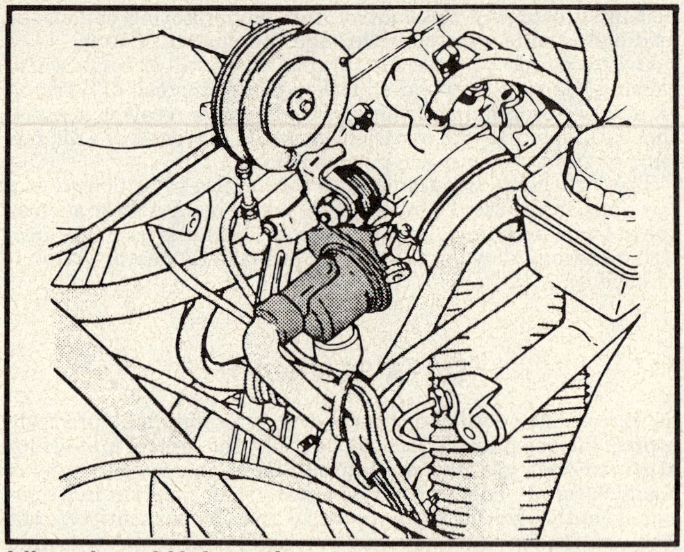

Idle valve—LH-Jetronic

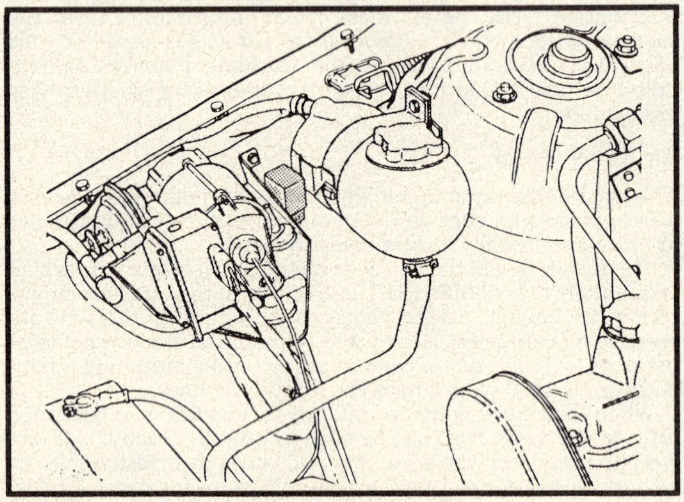

Auxiliary relay location—LH-Jetronic

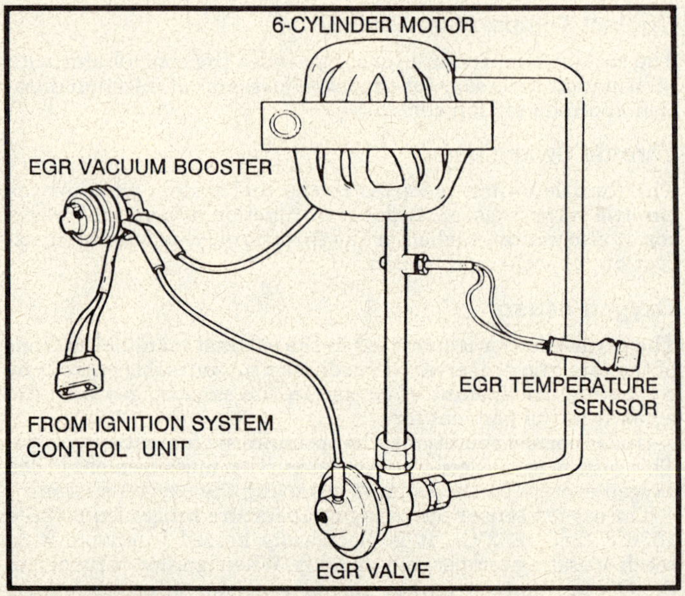

EGR system components—California

duration. Excess fuel is returned to the fuel tank through a return pipe.

Injectors

The fuel injectors as designed as electric solenoids. While the starter motor is operating, there are 2 injections per engine rotation. This is reduced to 1 during normal driving. Injection occurs in the intake manifold near the intake valves.

Cold Start Valve

The cold start valve is found on 1990–91 vehicles. It is discontinued in 1992 with the LH2.4 multi-point fuel injection. Models that are equipped with a cold start valve have it to improve engine cold starting. At cold start, fuel condenses on the cold surfaces in the form of drops. Its positioned farther away from the engine block than the injectors and delivers the fuel more as a gas than as drops. The cold start valve is controlled by the control unit, rather than the thermal switch. It operates when the temperature is approximately 5°F (−15°C), and when engine rpm is below 900 rpm. On 1992 vehicles the control unit commands extra fuel enrichment via the fuel rail injectors.

NOTE: Although the valve may continue to be physically placed in certain engines during a transition period, it will be inoperative.

Fuel Pump

The fuel pump is an electric pump, cooled by the fuel that flows through it. It incorporates a check valve and an overflow valve that opens when fuel pressure gets too high.

Both the primary pump and secondary pump operate when the starter motor or the engine is running. Should the engine stop while the ignition remains on, the control unit will cut the current to the pumps.

The in-tank fuel pump is an electric fuel pump that keeps pressure in the fuel line prior to the main fuel pump to prevent vapor lock. The pump incorporates a coarse strainer type filter and a non-return valve to maintain a constant amount of pressure in the system, even if the main pump is not operational.

Fuel Filter

The fuel filter is adjacent to the main fuel pump, which is mounted on a plate underneath the vehicle on the left side.

Diagnosis and Testing

SERVICE PRECAUTIONS

- Use caution when working around ignition system components. the ignition system operates at high power with dangerous voltages in both the low-tension and high-tension circuits, including connectors and similar fittings.
- Do not operate the fuel pumps when the fuel lines are empty.
- Do not arc weld on the vehicle unless all the control units have been removed.
- The control unit must not be exposed to temperatures above 175°F (80°C), such as in a paint booth.
- Make sure that the ignition switch is **OFF** before connecting or disconnecting the control unit, ignition coil, spark plug wiring or test instruments.
- Always use new gaskets and seals when remaking a fuel connection.
- When running a compression test, remove the electrical leads from connection 1 on the ignition coil (to prevent arcing) and remove the connectors from the injectors (to prevent flooding).
- Make sure all control unit connectors are fastened securely. A poor connection can cause an extremely high surge voltage, resulting in damage to integrated circuits.
- Keep any control unit harness at least 4 in. away from adjacent harnesses to prevent a system malfunction due to external electronic interference.
- Switch **OFF** the radio before disconnecting or connecting the battery ground lead to prevent radio damage.
- Keep all parts and harnesses dry during service.
- Before attempting to remove any parts, turn off the ignition switch and disconnect the battery ground cable.
- Always use a 12 volt battery as a power source.
- Do not attempt to disconnect the battery cables with the engine running.
- Do not attempt to disassemble a control unit under any circumstances.
- If a battery cable is disconnected, the fault code memory will be erased.
- Always disconnect the yellow SRS (air bag) connector when performing any other diagnostic, trouble shooting or service procedure that is not associated with the SRS system. Failure to do so can cause damage to the crash sensor. Also, make sure the ignition switch is **OFF** when disconnecting the connector.

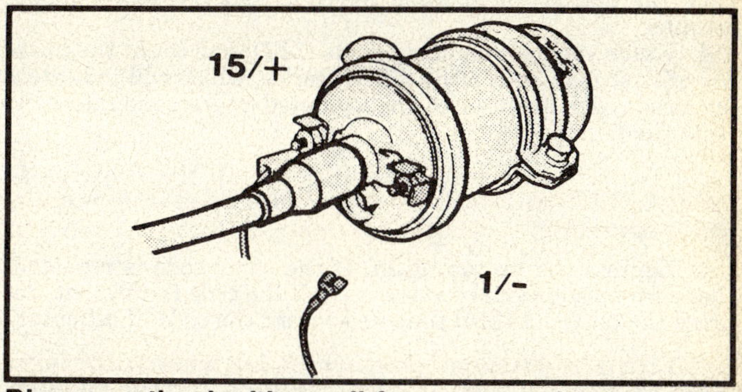

Disconnecting ignition coil for compression test—LH-Jetronic

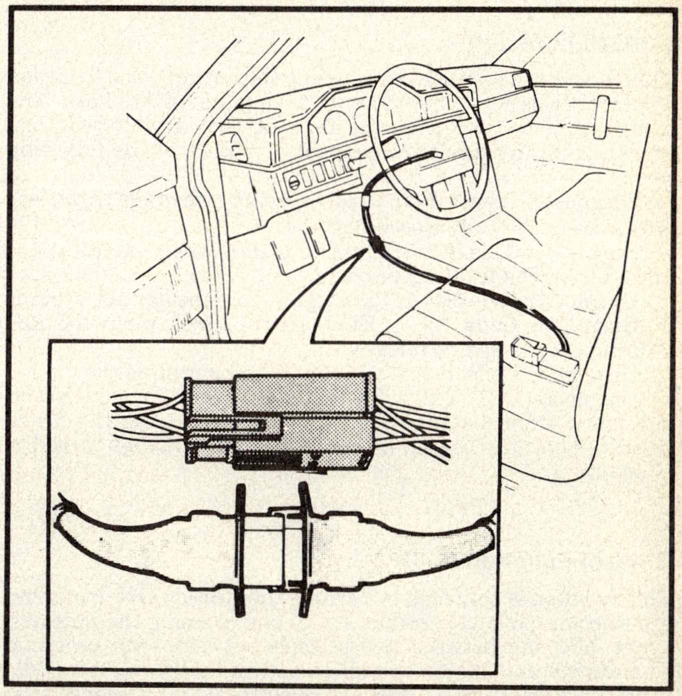
Yellow air bag connector location—LH-Jetronic

READING CODES

Control Function 1

The fuel system incorporates a built-in self-diagnostic system. It is located behind the left spring strut tower in the engine compartment. The diagnostic system uses socket No. 2 for the fuel system and socket No. 6 for the ignition system and EGR system (California).

There are many possible fault codes in the diagnostic system, and the system is capable of storing up to 3 system faults at 1 time. When the engine is running, the fuel system control unit continuously checks the control units internal functions, oxygen sensor, battery current, coolant temperature sensor, air temperature sensor, pressure sensor, throttle valve, engine rpm, speedometer and the idle speed according to the air valve position.

The following procedure is for retrieving FAULT codes.

1. Open the diagnostic socket cover in the driver's side rear of the engine compartment and install the selector cable into socket No. 2 for ignition or fuel system codes, or into socket No. 6 for EGR system codes on California vehicles.
2. Turn ignition to the **ON** position.
3. Enter Control Function 1 by pressing the button once. Hold the button for at least 1 second, but not more than 3 seconds.
4. Watch the light emitting diode (LED) and count the number of flashes in each ofthe 3 flash series, which will indicate a fault code. The flash series are separated by 3 second intervals. Note the fault codes.

NOTE: If there are no fault codes in the diagnostic unit, the LED will flash 1–1–1 and the fuel system is operating correctly.

5. Depress the button again. If the same code is repeated, there are no additional codes stored. If the code is different, depress the button a third time and record the code, if different.

NOTE: The diagnostic memory is full when it contains 3 fault codes. Until those 3 are rectified and the memory erased, the system cannot give information on any other problems.

Control Function 2

This function is activated by pressing the button twice (between 1–3 seconds between operations), causing the LED to flash. The control unit delivers a diagnostic code (not a fault code) acknowledging the receipt of a signal from each of the following items:

Diagnostic Code 3–3–2: Throttle switch, when the throttle is moved from the fully closed position

Diagnostic Code 3–3–3: Throttle switch, when the throttle is moved from the full-load position

Diagnostic Code 3–3–1: Rpm signal from the ignition system

Diagnostic Code 1–1–4: ECC control panel, when the A/C button is depressed or released

Diagnostic Code 1–3–4: Air conditioning compressor

Diagnostic Code 1–2–4: Testing idle speed compensation on automatic transmission vehicles. When depressing the brake pedal and placing the gear selector lever in **D** and then moved to N again.

Control Function 3

This function is activated by turning the ignition **ON**, installing the selector cable into socket No. 2. and pressing the button 3 times, allowing between 1–3 seconds between each pressing. The control unit then responds by activating the injectors, idle valve, cold start valve, auxiliary relay (radio interference suppression relay) and fuel pump.

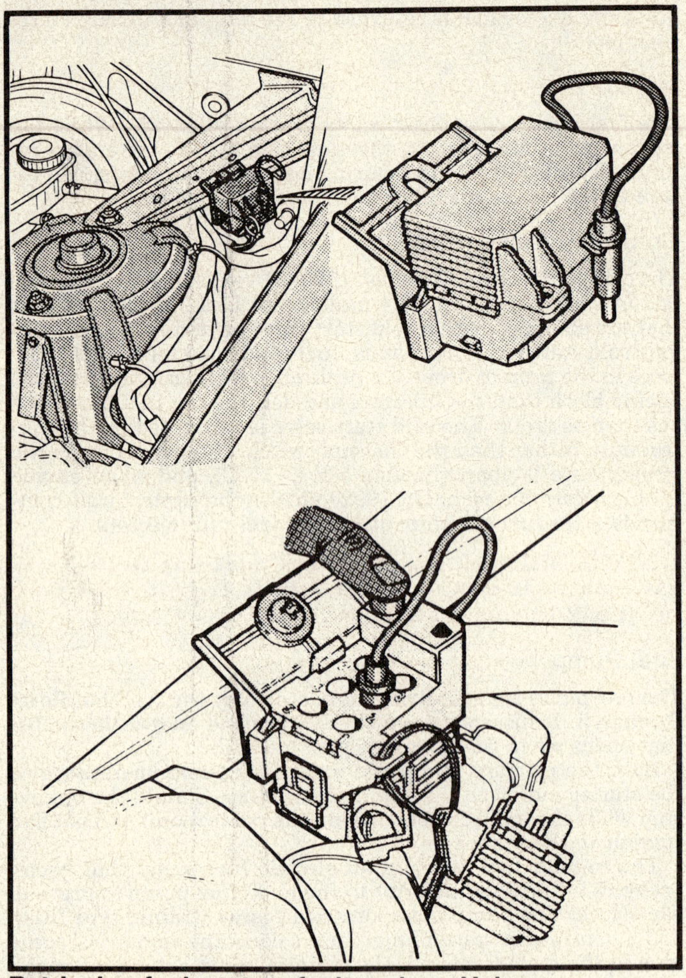

Retrieving fuel system fault codes—Volvo

CLEARING CODES

When all the fault codes have been read and corrected, the diagnostic system memory can be erased as follows:

1. Turn ignition switch to the **ON** position.
2. Read fault codes again.
3. Press and hold diagnostic socket button for more than 5 seconds. Release button. After 3 seconds, the LED should light.
4. While the LED is lit, press button again for more than 5 seconds. Release button. The LED should go out. Memory is cleared.

NOTE: To ensure that the memory has been erased, press button again for 1 second but not more than 3 seconds. The LED should flash 1–1–1, indicating that the memory is erased.

5. Start and run the engine. If it will not start, enter Control Function 2.
6. Recheck for fault codes. If Code 1–1–1 comes up, there are no additional system fault codes.

FUEL SYSTEM TESTING

Pressure Relief

1. Remove the fuel pump relay. It is located next to the control unit on the 240 and in the fuse box behind the ash tray on 700 and 900 series. The relay is furthest to the left on the second row.

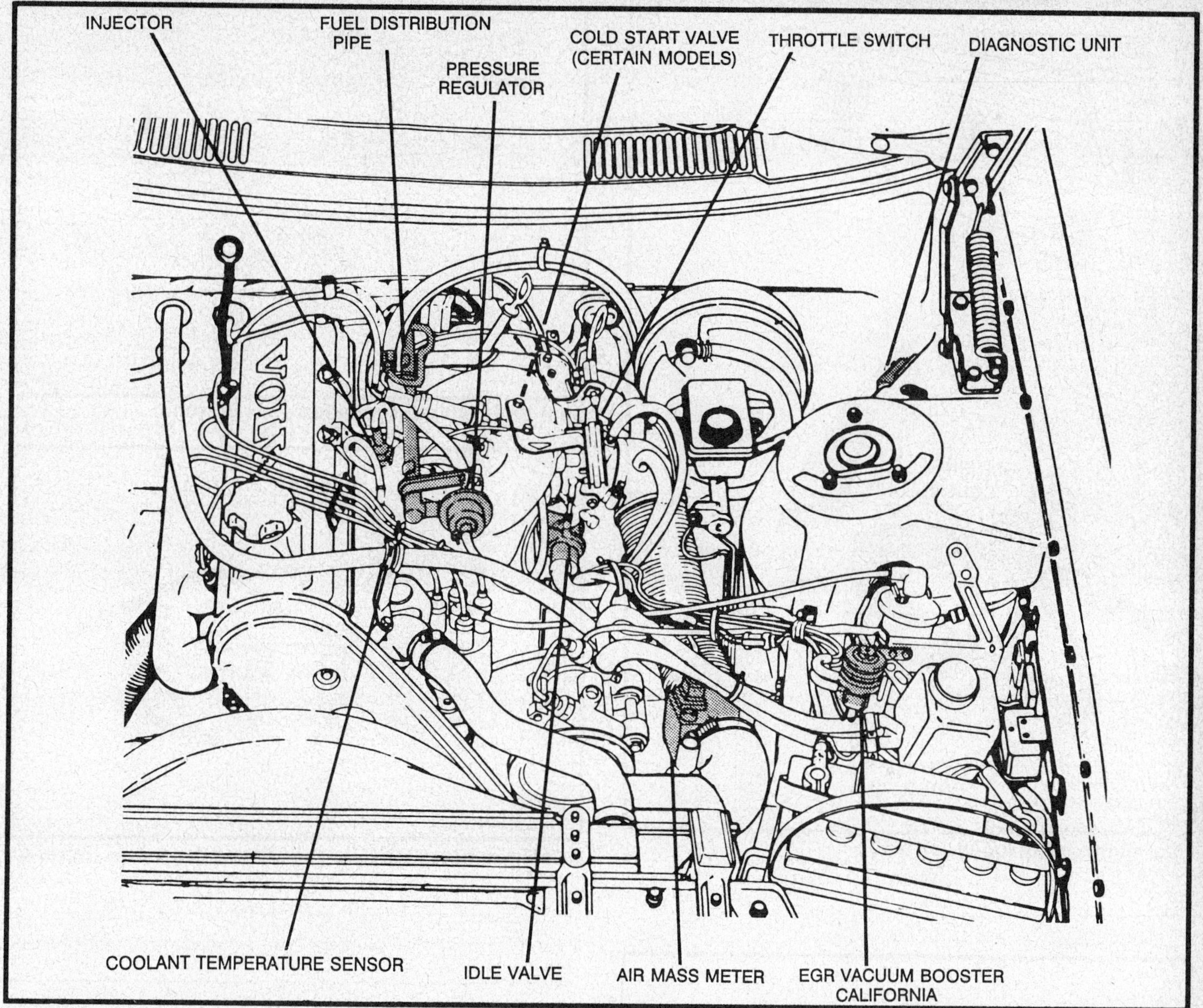

Component locations – LH-Jetronic

2. Start the engine repeatedly until the engine will not start anymore, indicating that the fuel pressure is relieved.

Pressure Check

1. Relieve the fuel pressure.
2. Connect a fuel pressure gauge. Hold a shop towel under the fuel line to catch any spilled fuel when it is disconnected.
3. With the fuel pump relay still removed, remove the seat belt reminder on 700 and 900 series, since this makes the test more easily performed. It is located in the middle of the top row in the fuse box.
4. Start the fuel pump by connecting an electrical lead between terminals 30 and 87/2 on the relay socket. Verify pump operation by removing the fuel cap and listening.
5. The fuel pressure should be 42 psi (300 kPa).
6. Remove the lead between terminals 30 and 87/2.
7. Relieve the fuel pressure and remove the pressure gauge, catching any spilled fuel with a shop towel.
8. Reinstall the relay.

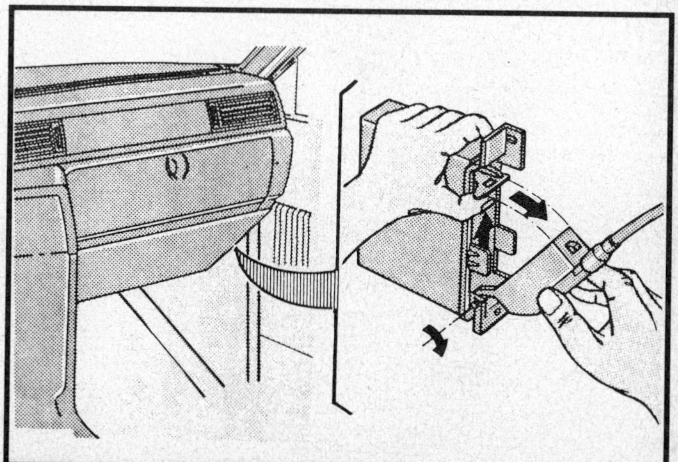

Control unit location – LH-Jetronic

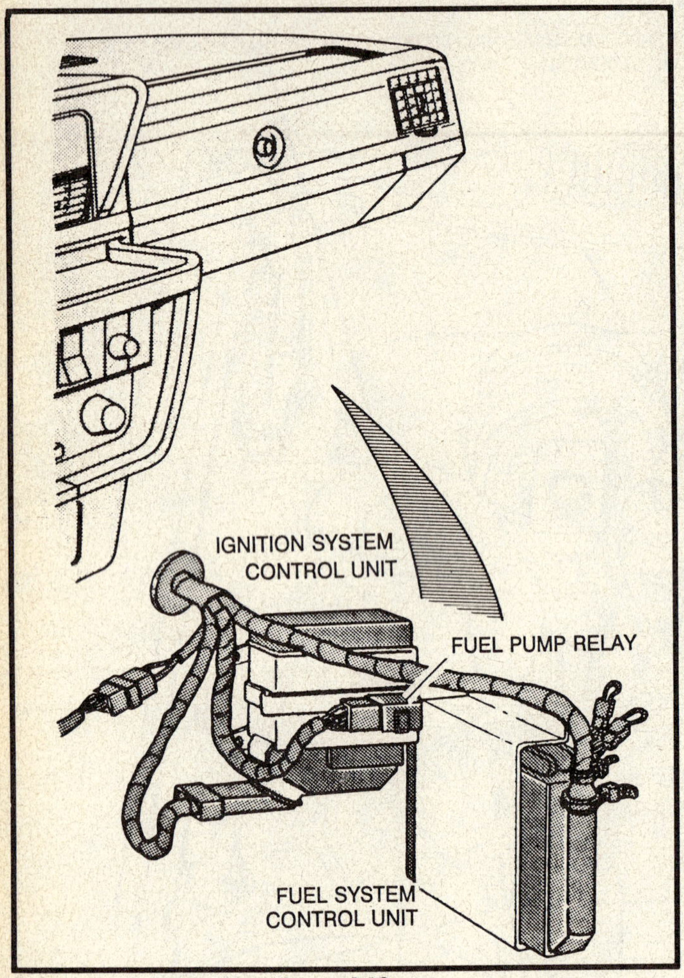

Fuel pump relay location—240

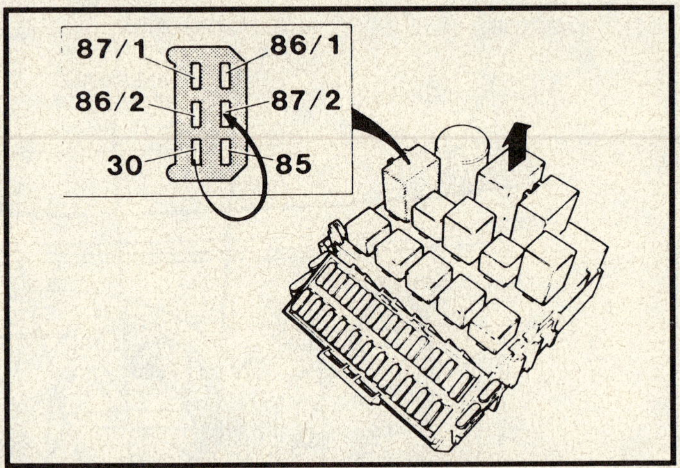

Seat belt reminder location—LH-Jetronic

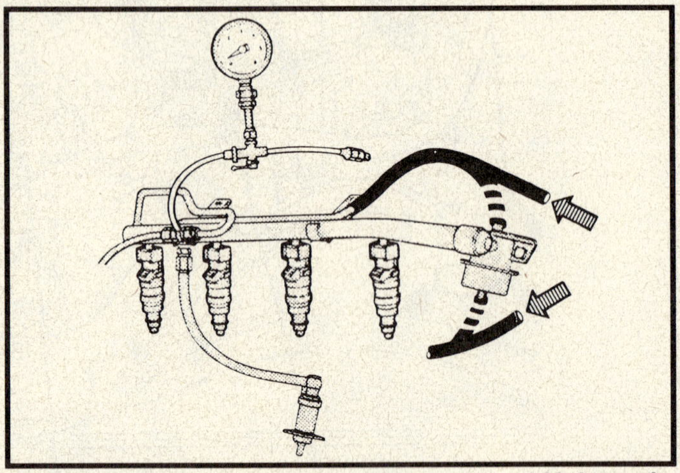
Checking fuel pressure regulator hoses—LH-Jetronic

Fuel pump relay location—Except 240s

CAUTION

To avoid fire and personal injury, replace any plastic tie bands removed from the fuel lines.

PUMP INOPERATIVE

1. If the fuel pump does not start, remove the lead between terminals 30 and 87/2.
2. Check for voltage at terminal 30. If there is none, check the lead between the relay and the battery.
3. Connect an electrical lead between terminals 30 and 87/2 on the relay socket. If the pump still does not start, check the lead between the pump and the relay. Check the lead between 87/1 and 85 for continuity with an ohmmeter.

PRESSURE TOO HIGH

1. Remove the lead between terminals 30 and 87/2 on the relay socket.
2. Remove the return hose from the pressure regulator and blow air into the hose.
3. Remove the vacuum hose from the pressure regulator and blow air into the hose.
4. If both hoses are open, the pressure regulator is faulty. Replace it and recheck the fuel pressure.

PRESSURE TOO LOW

1. Squeeze the return hose by hand, not with a tool, and check if the pressure rises.

CAUTION

To avoid personal injury and fire, do not allow the pressure to exceed 84 psi (600 kPa).

2. If the pressure rises rapidly, the pump and hoses are good. Replace the regulator and recheck the pressure.
3. If the pressure rises slowly, the fuel filter, pump strainer or the fuel lines are clogged or blocked.
4. If the pressure does not rise, the fuel pump is faulty.

Pressure Regulator

1. Connect a vacuum pump to the pressure regulator.
2. Apply vacuum to the regulator and check that the fuel pressure falls as more vacuum is applied.
3. Remove the lead between terminals 30 and 87/2.
4. Relieve the fuel pressure and remove the pressure gauge, catching any spilled fuel with a shop towel.
5. Reinstall the relay.

CAUTION

To avoid fire and personal injury, replace any plastic tie bands removed from the fuel lines.

DIAGNOSTIC UNIT TESTING

Testing

1. Remove the diagnostic unit cover and install selector plug into socket No. 2.
2. Turn ignition switch to the **ON** position.
3. Enter Control Function 1 by pressing the button once. Hold the button for at least 1 second, but not more than 3 seconds.
4. Watch the LED and count the number of flashes in the 3 flash series corresponding to a fault code. The flash series are separated by 3 second intervals. Note fault codes.

NOTE: If there are no fault codes in the diagnostic unit, the LED will flash 1-1-1. Proceed to Control Function 2.

5. If LED does not flash when button is pressed, or no code is flashed, proceed as follows:
   a. Check ground connections on the intake manifold, and the ground connection for the oxygen sensor at the right front mudguard.
   b. Check fuse 4 and the in-line fuse on the 240, fuses 1 and 11 on 740, 780 and 940 and fuses 30 and 31 on the 760 for the pump relay and the primary pump. The 240 main fuse box is located in front of the left front door pillar and the in-line fuse is in the engine compartment. On 700 and 900 series, the fuse box is located behind the ash tray in the center console.
   c. Remove glove compartment, and check control unit ground connections.
   d. Turn ignition switch to the **OFF** position. Remove control unit connector and connector protective sleeve.
   e. Connect a voltmeter between ground and pin 4 on the control unit connector. The reading should be 12 volts. If no voltage is present, check the wire between control unit connector and fuse 1 in the fuse/relay box.
   f. Turn the ignition switch to the **ON** position, and install the selector plug into the No. 2 socket. Connect a voltmeter between ground and pin 12 on the control unit connector. The reading should be 12 volts. Press the button on diagnostic unit and note voltage reading. The reading on the voltmeter should be 0 volts. If no voltage at the control unit is present, take the reading at the diagnostic unit connector. If the reading remains at 12 volts when button is pressed, check the diagnostic unit.
   g. Connect a voltmeter between ground and the Red/Black lead on the diagnostic unit connector. The reading should be 12 volts.

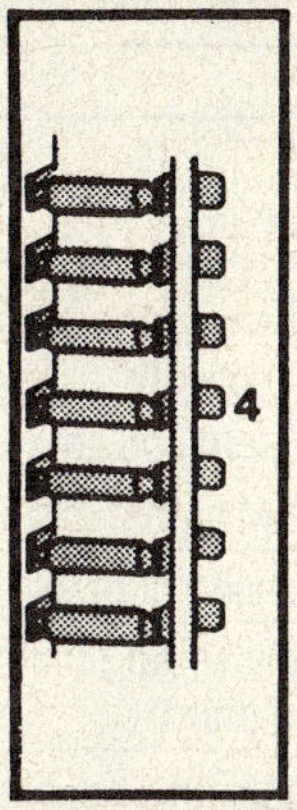

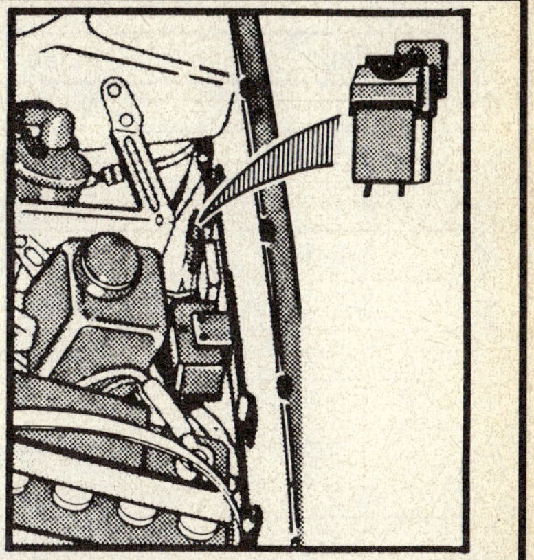

Fuse locations—240

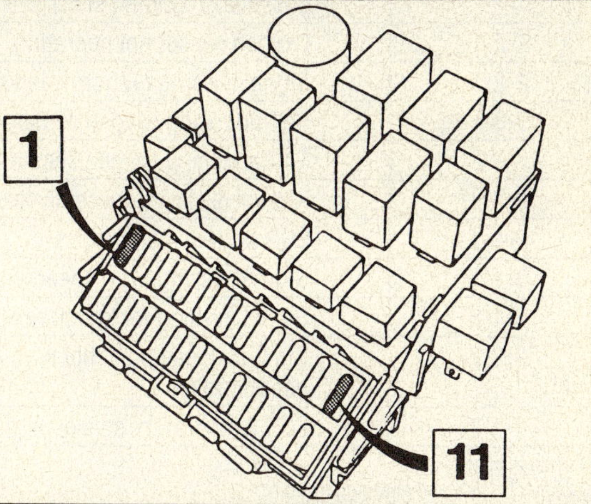

Fuse locations—740 and 780 shown

   h. Connect an ohmmeter between ground and the Brown/Black lead in the diagnostic unit connector. The reading should be 0 ohms.
   i. Turn ignition to the **OFF** position. Connect ohmmeter between diagnostic unit selector plug and the pin directly underneath the selector button. The ohmmeter should read infinity. Press the button, and note the reading. It should be 0 ohms.
   j. Connect a diode tester between the pin directly underneath the LED and the selector plug: Connect the positive test lead from the tester to the pin directly underneath the LED and the negative test lead from tester to selector cable, then reverse the leads. If current flows one way and not the other, the diode is good. If current flows both ways, replace diagnostic socket.
   k. Check the system relay by connecting a voltmeter between ground and pin 9 on the control unit connector, then connect a jumper wire between ground and pin 21 on the control unit connector. The relay should activate and the reading should be 12 volts.
6. Enter Control Function 1 again and check for any fault codes. If the LED flashes Code 1-1-1, there are no other codes in the memory.

NOTE: The diagnostic system memory is full when it contains 3 fault codes. Until those codes are corrected and the memory erased, the system cannot give information on any other problems.

7. When all fault codes have been read and corrected, erase the diagnostic system memory and recheck for codes.

Code	Fault	Correction Test
1-1-1	No faults	—
1-1-2	Fault in control unit ①	Control unit
1-1-3	Fault in injector	Injector
1-2-1	Signal to/from air mass meter faulty	Air mass meter
1-2-3	Signal missing to/from coolant temperature sensor	Coolant temperature sensor
1-3-1	Ignition System RPM signal missing	RPM sensor
1-3-2	Battery voltage too low/high ②	Charging system
1-3-3	Throttle-Switch; idle setting faulty, possible grounding short	Throttle switch
2-1-2	Oxygen sensor signal missing or is faulty	Oxygen sensor or heater
2-1-3	Throttle-Switch; full load setting faulty possible grounding short	Throttle switch
2-2-1	Oxygen sensor not operating	Oxygen sensor or heater
2-2-3	Signal missing to/from idle valve	Idle valve
2-3-1	Self-adjusting oxygen sensor not operating	Air intake system or fuel system
2-3-2	Self-adjusting oxygen sensor not operating	Air intake system or fuel system
2-3-3	Idle valve closed, possibly leaking air	Idle valve
2-4-1	EGR system malfunction ③	EGR vacuum booster
3-1-1	Signal missing from speedometer	Speedometer signal
3-1-2	Signal missing for knock related fuel enrichment	Knock enrichment signal
3-2-2	Burn-off cleaning of hot wire in air mass meter not operating	Air mass meter hot wire
4-1-3	EGR temperature sensor signal absent or faulty ③	EGR temperature signal

① Change control unit
② Check battery and charging system
③ California

Fault code descriptions—LH-Jetronic

AIR CONDITIONING SWITCH

Testing

1. Enter Control Function 2.
2. On vehicles equipped with A/C, check the on/off function of the compressor by placing A/C controls in the **ON** position. If the LED turns off, then flashes diagnostic code 1–1–4, the A/C switch is operating properly.
3. If LED continues to flash, check the lead from A/C control unit by connecting an ohmmeter between ground and pin 15 on the control unit connector. With A/C switch **OFF**, ohmmeter should read 1 kilo-ohm. With A/C switch **ON**, ohmmeter should read 10 ohms.
4. The LED will return to rapid flashes prior to A/C compressor turning on. When the compressor turns on the diode light should go off and then flash diagnostic code 1–3–4.
5. If LED continues to flash rapidly, check lead to A/C compressor by connecting an ohmmeter between ground and pin 14 connection on the control unit connector. Ohmmeter should read 0–5 ohms.
6. Turn engine **OFF**.

NEUTRAL SWITCH

Testing

1. Enter Control Function 2.
2. On vehicles equipped with automatic transmission, depress brake pedal and place gear selector lever in position **D** and then in position **N**. The LED should go off and then flash diagnostic code 1–2–4.
3. If LED continues to flash, check gear selector signal by placing gear selector in the **N** position.
4. Connect an ohmmeter between ground and pin 30 on control unit connector. Ohmmeter should read 0 ohms.
5. Move gear selector into position **D**. Ohmmeter should read infinity. On vehicles equipped with manual transmissions, ohmmeter should read 0 ohms in all gears.

CARBON FILTER SOLENOID VALVE

Testing

This test applies to vehicles with the B234F engine.

1. Enter Control Function 3.
2. While the LED continues to flash, the injectors, idle valve and carbon filter solenoid valve should all begin to operate.
3. If carbon filter solenoid valve does not operate, connect an ohmmeter between pin 9 and pin 27 on the control unit. Ohmmeter should read approximately 45 ohms.

INJECTOR

Resistance Testing

1. Test fuel injectors as follows:
   a. On 700 and 900 series, remove auxiliary relay from its

socket, then connect a jumper wire between pin 2 and pin 3 on the auxiliary relay socket.

b. Connect an ohmmeter between connections pin 9 and pin 18 on the control unit connector. Ohmmeter should read 4 ohms.

c. If reading obtained is higher, current is not going through the injectors.

d. If on non-turbocharged vehicles the ohmmeter reads 5.3 ohms, fault is in 1 injector or its leads. If ohmmeter reads 8 ohms, fault is in 2 injectors or their leads. If ohmmeter reads 16 ohms, fault is in 3 injectors or their leads.

e. If on turbocharged vehicles the ohmmeter reads 2.9 ohms, fault is in 1 injector or its leads. If ohmmeter reads 4.4 ohms, fault is in 2 injectors or their leads. If ohmmeter reads 8.5 ohms, fault is in 3 injectors or their leads.

f. If measured resistance is wrong, remove injector connectors and test them individually. Ohmmeter should read for each injector 16 ohms for non-turbocharged engines and 2 ohms for turbocharged engines.

2. On turbocharged engines disconnect the ballast resistor connector.

3. When measuring between the center pin and each of the other pins the resistance should be 5.5–6.5 ohms.

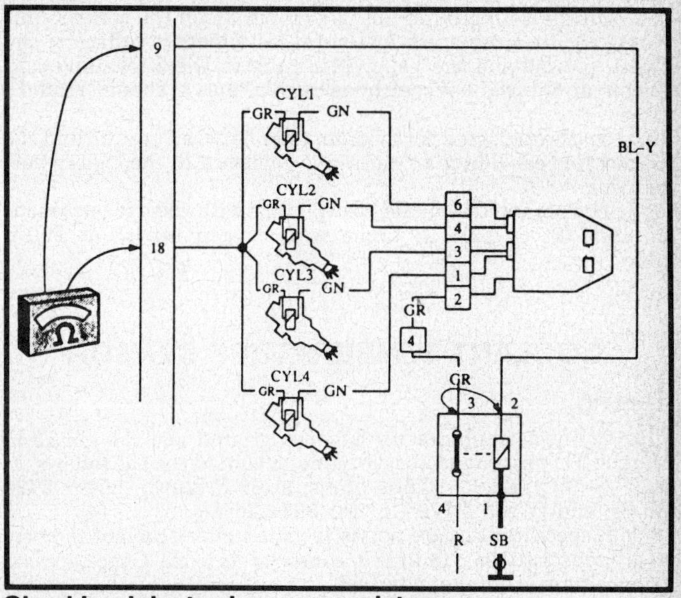

Checking injector harness resistances—Turbocharged shown

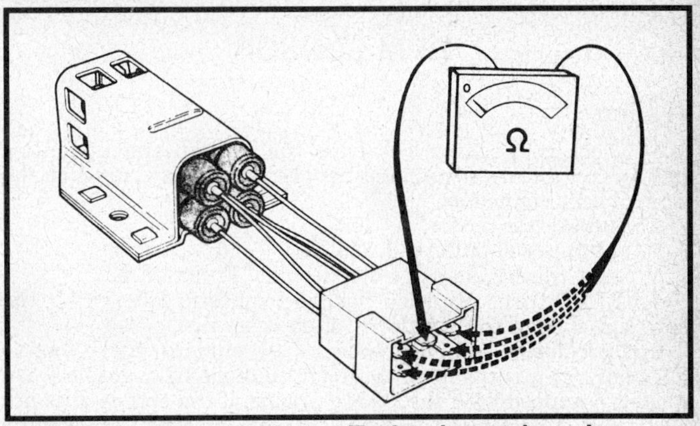

Checking ballast resistor—Turbocharged engine

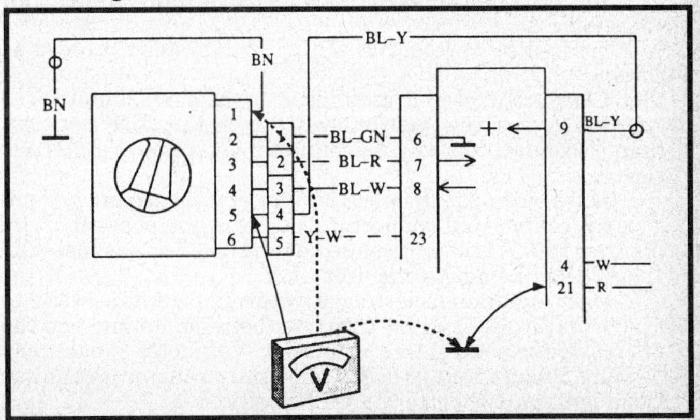

Testing air flow meter—LH-Jetronic

Operation Testing

Check fuel injectors as follows:

1. Turn ignition switch to the **ON** position, then remove diagnostic socket cover and enter Control Function 3. The injectors will begin to operate at this point.

2. The diagnostic socket LED will flash in a continuous pattern. The control function will repeat itself until interrupted by changing the control systems or turning the ignition switch to the **OFF** position.

3. Listen to and feel by hand each injector to ensure injectors are operating correctly. If an injector is not operating, change injector connector to a injector that's operating. If fault moves to the injector that was operating previously, then the fault is in the connector head.

4. If the injector is still not operating, the fault is in the injector. Check injector separately by connecting an ohmmeter between the injector pins. The ohmmeter should read 16 ohms for non-turbocharged engines and 2 ohms for turbocharged engines. If not, replace injector.

AIR MASS METER

Testing

1. Connect an ohmmeter between pin 6 and pin 7 on the control unit connector. Ohmmeter should read 2.5–4.0 ohms.

2. Slide back the rubber boot from the air mass meter connector. Connect a voltmeter between ground and the pin 5 on the air mass meter connector, then a jumper wire from ground to pin 21 on the control unit connector. Voltmeter should read 12 volts.

3. Connect voltmeter between pin 1 (ground) and pin 5 (current feed from system relay) on air mass meter connector. Voltmeter should read 12 volts.

4. Remove jumper wire from pin 21 on the control unit connector. Start and allow engine to reach normal operating temperature.

5. Connect a voltmeter between pin 1 and pin 4. Rev engine up to 2100 rpm, then turn engine off. After approximately 4 seconds, the voltmeter indicator should swing back and forth for approximately 1 second (burn-off cleaning). Remove voltmeter and slide back rubber boot.

OXYGEN SENSOR HEATER

Testing

1. On 700 and 900 series, remove panels from under instrument panel right side, then remove glove compartment. Disconnect control unit connector. Connect a voltmeter between ground and pin 8 in the 25 terminal connector.

2. On the 240, ground pin 20 and pin 21 on the control unit connector in order start fuel pumps. Connect a voltmeter between ground and the Yellow/Red cable in the 2 terminal connector at the right wheel housing. Voltmeter should read 12 volts.

3. On the 240, connect an ohmmeter between ground and the Yellow/Red cable in the 2 terminal connector at the right wheel housing.

4. The ohmmeter should read 3 ohms with the oxygen sensor at 68°F (20°C) and 13 ohms with oxygen sensor at 660°F (349°C).

COOLANT TEMPERATURE SENSOR

Testing

1. Connect an ohmmeter between ground and the pin 13 in the control unit connector. Ohmmeter should read as follows: at 14°F (−9°C), 8,260–10,560 ohms; at 68°F (20°C), 2,280–2720 ohms; and at 176°F (79°C), 290–364 ohms.

2. If specifications are not as specified, check coolant temperature sensor to see if fault is in sensor or its leads. Check ground connections at intake manifold.

RPM SENSOR

Testing

1. Check rpm sensor lead from the ignition system control unit by connecting a voltmeter between ground and pin 1 on the control unit connector.
2. Run starter motor.
3. Voltmeter should read 12 volts.
4. Start the engine and enter Control Function 2.
5. If LED turns off, then flashes diagnostic code 3–3–1, the rpm signal from the ignition system is correct.
6. On vehicles equipped with B230F engines, if the engine does not start, run starter motor until diode light goes off. On models equipped with the B234F engine, if the engine does not start, run starter motor until the diagnostic code flashes out.
7. If the LED continues to flash, check for faults in the ignition system.
8. If no faults are found in the ignition system, proceed as follows:
   a. Check control unit grounds near the control unit.
   b. Make sure the ignition switch is in the **OFF** position, then disconnect control unit connector and remove protective sleeve.
   c. Make sure that there is 12 volts between ground and pin 4 on the control unit connector. If there is not, check the wire between the control unit connector and fuse 1 in the fuse box.
   d. Check the diagnostic unit itself.
   e. Check ignition lock voltage by turning ignition switch to the **ON** position. Connect voltmeter between ground and the pin 35 on the control unit connector. Voltmeter should read 12 volts. Ensure voltage exists when starter motor is running. Turn ignition switch to the **OFF** position.
   f. Check ground connections by connecting a ohmmeter between ground and pins 5, 17, 19 and 29 on the control unit connector. Ohmmeter reading at all connections should be 0 ohms. All leads are grounded to the intake manifold.
   g. Ensure that the oxygen sensor lead is connected to the pin 5 on the control unit connector.
   h. Check the rpm sensor lead from the ignition system control unit by connecting a voltmeter between ground and the pin 1 on the control unit connector. Run starter motor and note reading. Voltmeter should read 12 volts.

THROTTLE SWITCH

Testing

1. Turn the ignition switch to the **ON** position, open the diagnostic unit cover and install selector plug into socket No. 2.
2. Turn the throttle control to the full load position.
3. Enter Control Function 2.
4. Release the throttle control. If the LED flashes diagnostic code 3–3–3, the function of the throttle switch is correct in the full load position.
5. If the LED continues to flash rapidly, proceed as follows:
   a. Check the throttle switch setting by opening the throttle slightly and listen to the switch. A click should be heard when the shutter opens (idle switch). If not, adjust throttle switch by loosening the mounting bolts, turn switch slightly clockwise, then counterclockwise until the switch clicks. Tighten mounting bolts and check setting.
   b. Turn the ignition switch to the **OFF** position.
   c. Remove control unit connector, then the connector protective sleeve.
   d. Check the throttle switch by connecting a ohmmeter between ground and pin 2 on the control unit connector. The reading should be 0 ohms (throttle switch closed).
   e. Depress gas pedal slightly. The resistance should increase to 2–3 kilo-ohms (throttle switch opens).
   f. Connect an ohmmeter between ground and pin 3 on the control unit connector. Ohmmeter should read infinity. Depress gas pedal fully. Ohmmeter should read 0 ohms. If a fault occurs, check the resistances at the throttle switch, and check the ground connections at intake manifold.
6. Turn the throttle slightly. If LED turns off, then flashes diagnostic code 3–3–2, the throttle switch is operating correctly in the idle position. If LED continues to flash, check throttle switch to see if the fault is in the switch or in its leads. Check ground connections at intake manifold.

AIR INTAKE SYSTEM

Testing

1. Check the air mass meter area for leaks.
2. Check intake manifold between the air filter and the manifold, hoses and hose connections to the intake manifold, and bolted joints and seals (throttle housing).

IDLE VALVE

Testing

1. Check idle valve by connecting an ohmmeter between connections pin 9 and pin 33 on the control unit connector.
2. Ohmmeter should read approximately 8 ohms.
3. Enter Control Function 3 for a function check.
4. If idle valve does not operate, but LED flashes, connect an ohmmeter between pin 9 and pin 33 on the control unit connector. Ohmmeter should read 8 ohms.

SPEEDOMETER SIGNAL

Testing

1. Remove access panel from under the instrument panel on the drivers side.
2. On the 240, disconnect connector from speedometer and connect an ohmmeter between the Blue cable and pin 34 on the control unit connector.
3. On 700 and 900 series, disconnect connector from speedometer and connect an ohmmeter between Violet/White cable and pin 34 on the control unit connector.
4. The ohmmeter should read 0 ohms. If Fault Code 3–1–1 has flashed and ohmmeter reads 0 ohms, replace speedometer assembly.

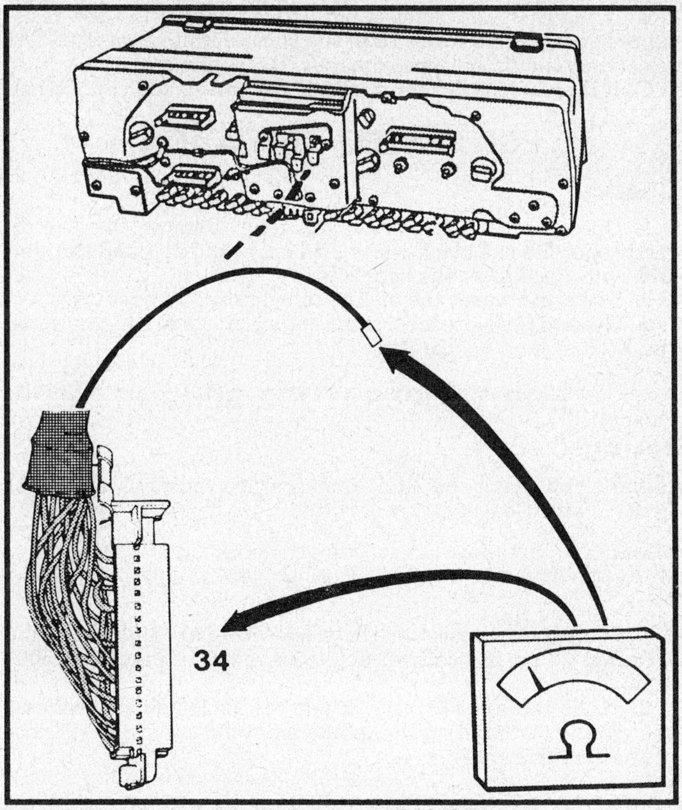

Checking speedometer signal—LH-Jetronic

SYSTEM RELAY

Testing

1. Check the system relay/primary relay by connecting a voltmeter between ground and the pin 9 on the control unit connector, then connect a jumper wire from ground to pin 21 on the control unit connector. The relay should activate and voltmeter reading should be 12 volts.
2. With pin 21 still grounded, connect voltmeter between ground and pin 9 on control unit connector, then a jumper wire from ground to pin 20 of control unit connector. The pump relay should close and fuel pumps activate.
3. On the 740 and 940, connect voltmeter between ground and pin 8 in the 25-terminal connector located at the right A-post. Voltmeter should read approximately 12 volts.

EXHAUST CARBON MONOXIDE CONTENT

Testing

1. Check carbon monoxide content by removing the tapped plug on the catalytic converter.
2. Warm the engine and check the carbon monoxide content. If carbon monoxide content is not within specification, check the oxygen sensor.
3. Check oxygen sensor by disconnecting oxygen sensor electrical connector. Ground the lead coming from the control unit. Carbon monoxide content reading should rise, indicating that the control unit and its connections are good.
4. Connect a voltmeter to the oxygen sensor electrical connector. The reading should move back and forth to indicate the functioning of the oxygen sensor. Reading on voltmeter with correct carbon monoxide content should be approximately 0.5 volts.

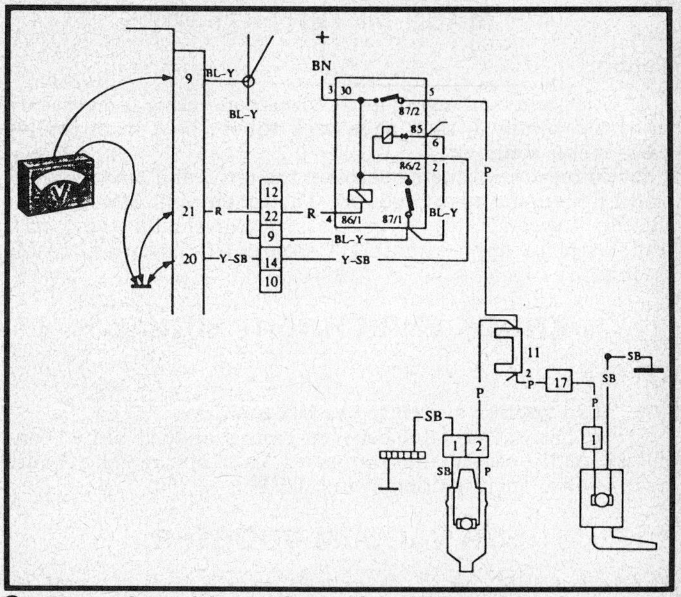

System relay wiring schematic—700 series shown

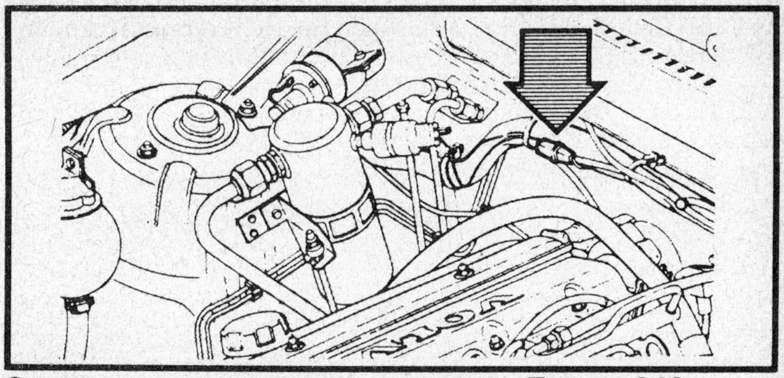

Oxygen sensor connector location—Except 240

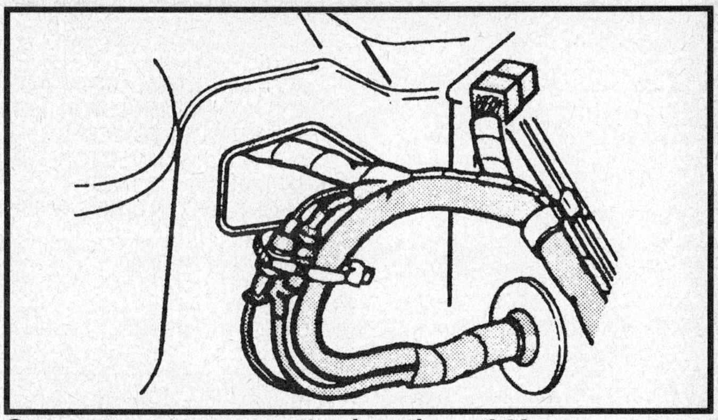

Oxygen sensor connector location—240

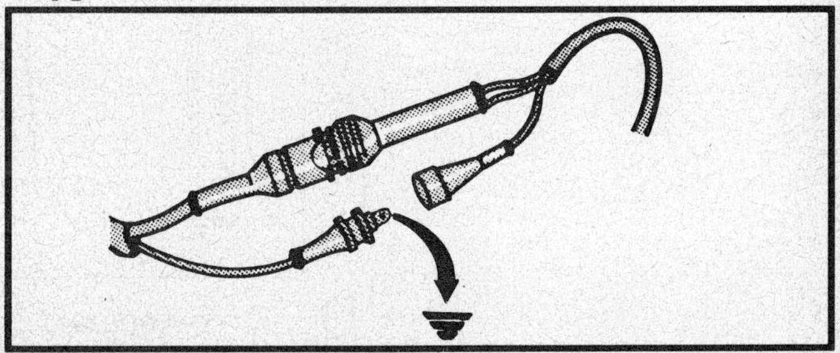

Grounding oxygen sensor lead to control unit—LH-Jetronic

AIR MASS METER HOT WIRE

Testing

1. Start and run engine until normal operating temperature has been obtained, then slide back rubber boot from the air mass meter connector.
2. Connect a voltmeter between the pin 1 and pin 4 connections. Rev engine up to 2100 rpm. Turn engine off. After approximately 4 seconds, the voltmeter indicator should swing back and forth for approximately 1 second (indicating burn-off cleaning).

KNOCK ENRICHMENT SIGNAL

Testing

1. Turn ignition switch to the **ON** position.
2. Connect a voltmeter between ground and the pin 28 connection on the control unit connector. Voltmeter reading should be 0.7 volts. Turn ignition switch **OFF**.

EGR VACUUM BOOSTER

Testing

On California vehicles an EGR system malfunction is indicated by Fault Code 2–4–1. It results when the EGR temperature sensor detects that the exhaust gas recirculation flow is too low.

1. Turn the ignition switch **ON**.
2. Enter Control Function 3. The lamp should flash according as the EGR converter is activated.
3. Turn the ignition switch **OFF**.
4. If the EGR converter is not activated, measure the resistance between pins 6 and 15 in the control unit connector. The resistance should be approximately 75–95 ohms.
5. If the value is different, check the wiring and connection.

EGR SYSTEM OPERATION

Check

1. Open the EGR converter connector. Remove the lead directly opposite the black-white lead between pin 1 and the control unit. Reconnect the connector.
2. Start and warm the engine to operating temperature.
3. Ground the loose lead. If the engine starts to run unevenly, the EGR system is operating.

EGR TEMPERATURE SENSOR

Testing

If the signal from the EGR temperature sensor is absent or faulty, Fault Code 4-3-1 will result.

NOTE: The sensor is extremely sensitive to shock and impact. When replacing, torque to 7–13 Nm (5–10 ft. lbs.).

1. Measure the resistance between terminal 14 (ground) and terminal 22 on the control unit connector. It should be 500–1000 ohms.
2. If the value is different, check the leads and connections, the ground terminal on the intake manifold and the resistance of the sensor connector.

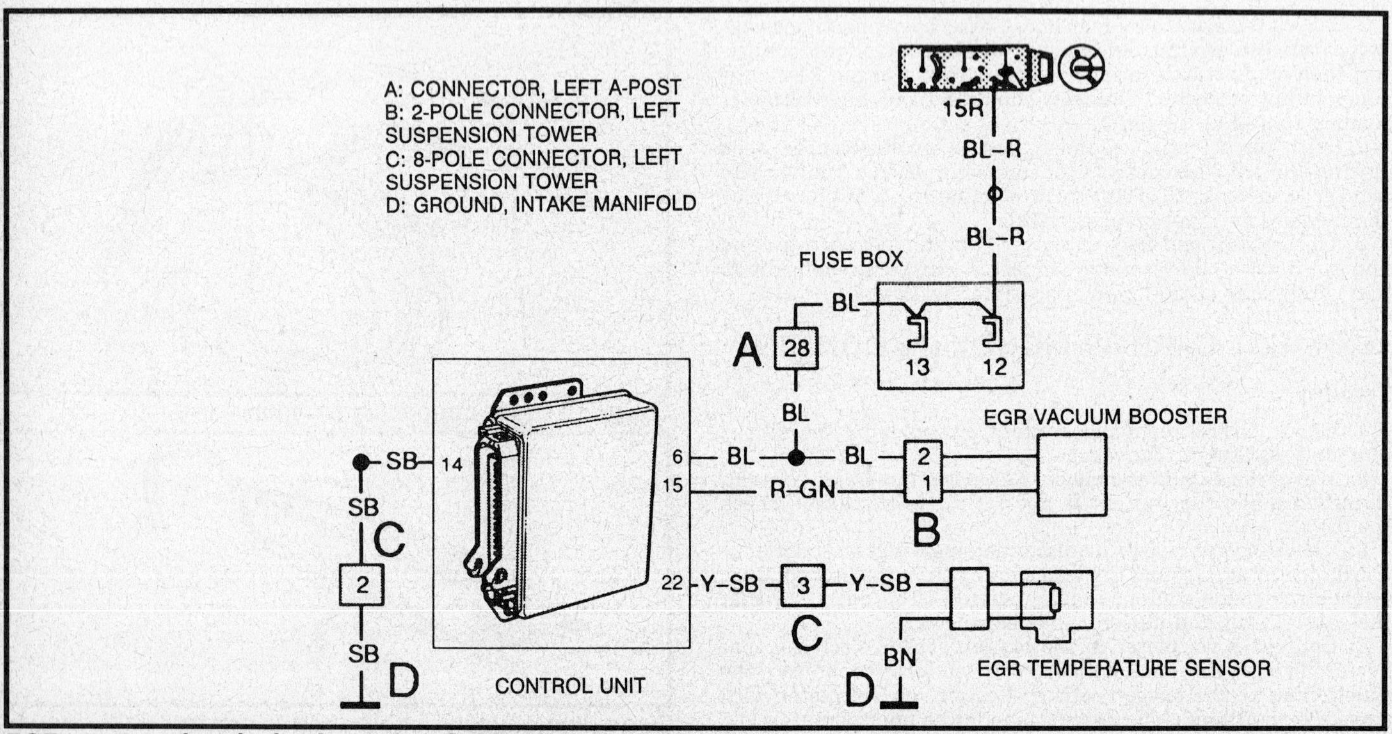

EGR system electrical schematic—California

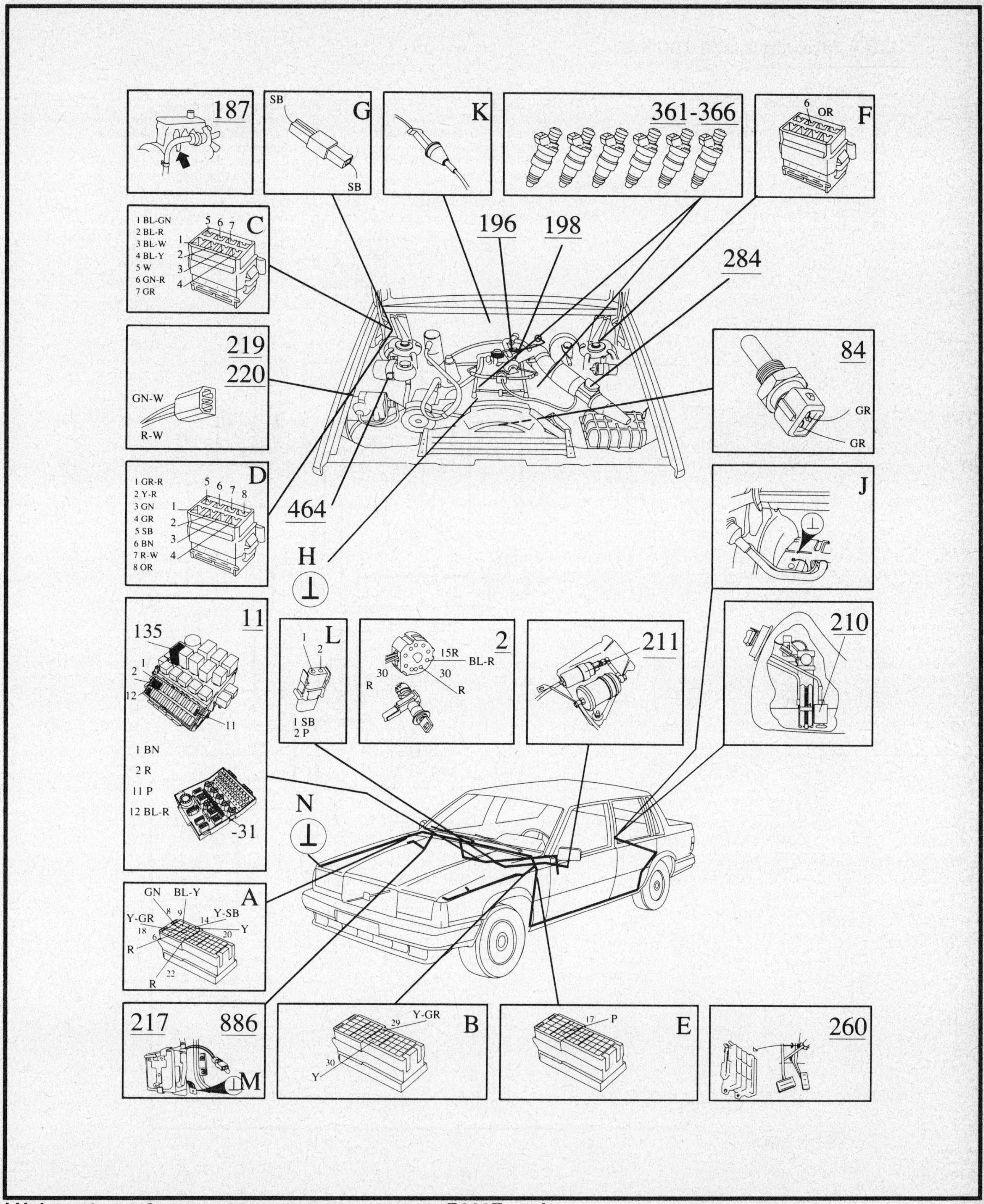

LH-Jetronic 2.2 System wiring diagram—780 with B280F engine

L.H - Jetronic 2.2 (B 280 E/F)

1. Battery
2. Ignition switch
4. Ignition coil
29. Positive terminal test point
31. Ground connection rail in central electrical unit
84. Coolant temperature sensor
135. Relay, fuel injection
187. Lambda sond
196. Air control valve
198. Throttle switch LH-Jetronic
210. Tank pump 1.6A
211. Fuel pump 6.5A
217. Control unit LH-Jetronic
219. Test point Lambda sond system
220. Test point CIS system
260. Control unit, ignition system
284. Air mass meter
361-366. Injectors
464. Radio interference relay
886. Bridge connector.

A. Connector right A post
B. Connector left A post
C. Connector right suspension tower
D. Connector right suspension tower
E. Connector left A post
F. Connector left suspension tower
G. Connector right suspension tower
H. Ground point on intake manifold
J. Ground point in trunk
K. Connector on firewall
L. Connector in central electrical unit
M. Ground point, right A post
N. Ground point right front wing

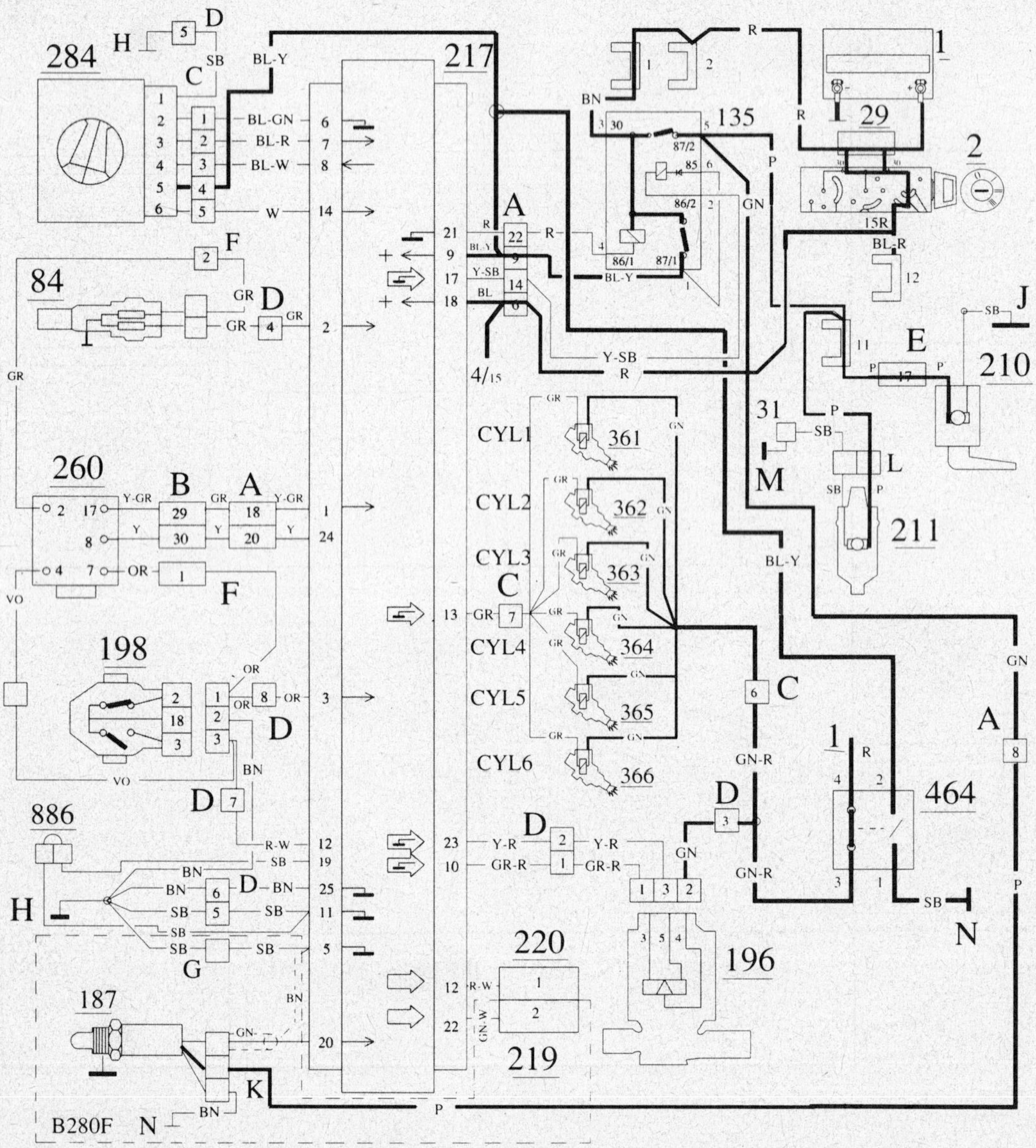

LH-Jetronic 2.2 component terminal identification— 780 with B280F engine

LH-Jetronic 2.4 (B230F)

3 Ignition switch
7 Starter motor
10 Fuse box
32 Connection at instrument
95 Cold start injector
97 Tank pump
100 Fuel pump
112 Joint
129 AC Relay
131 Diagnostic unit
132 Idle valve
162 Lambda sond
211 Electronic control unit, EZ116K
212 In—line fuse
214 Electronic control unit, LH2.4
215 Air mass meter
217 Main relay
218 Throttle switch
221 Coolant temperature sensor
223 AC pressure switch
225 Injectors

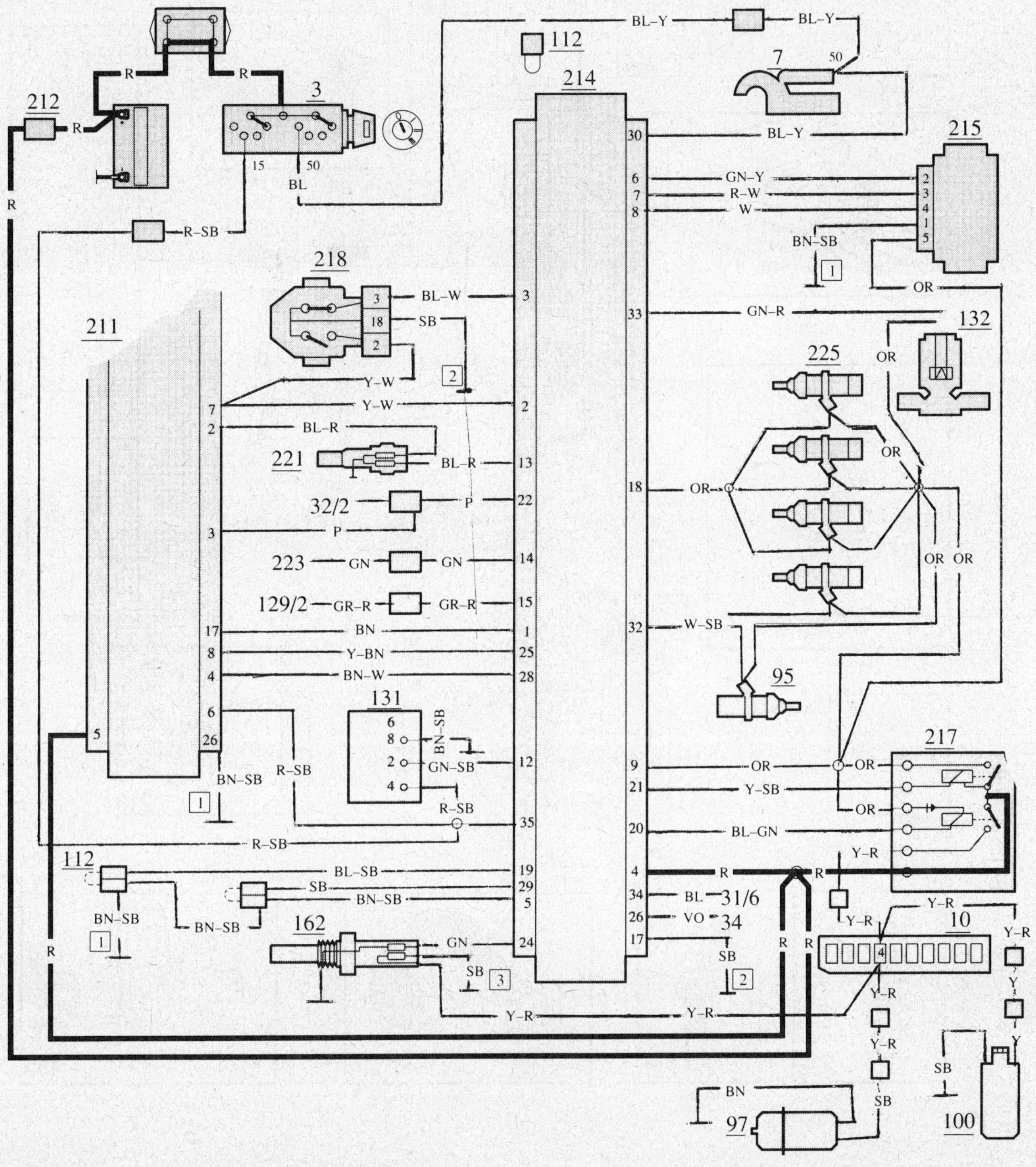

LH-Jetronic System wiring diagram—240

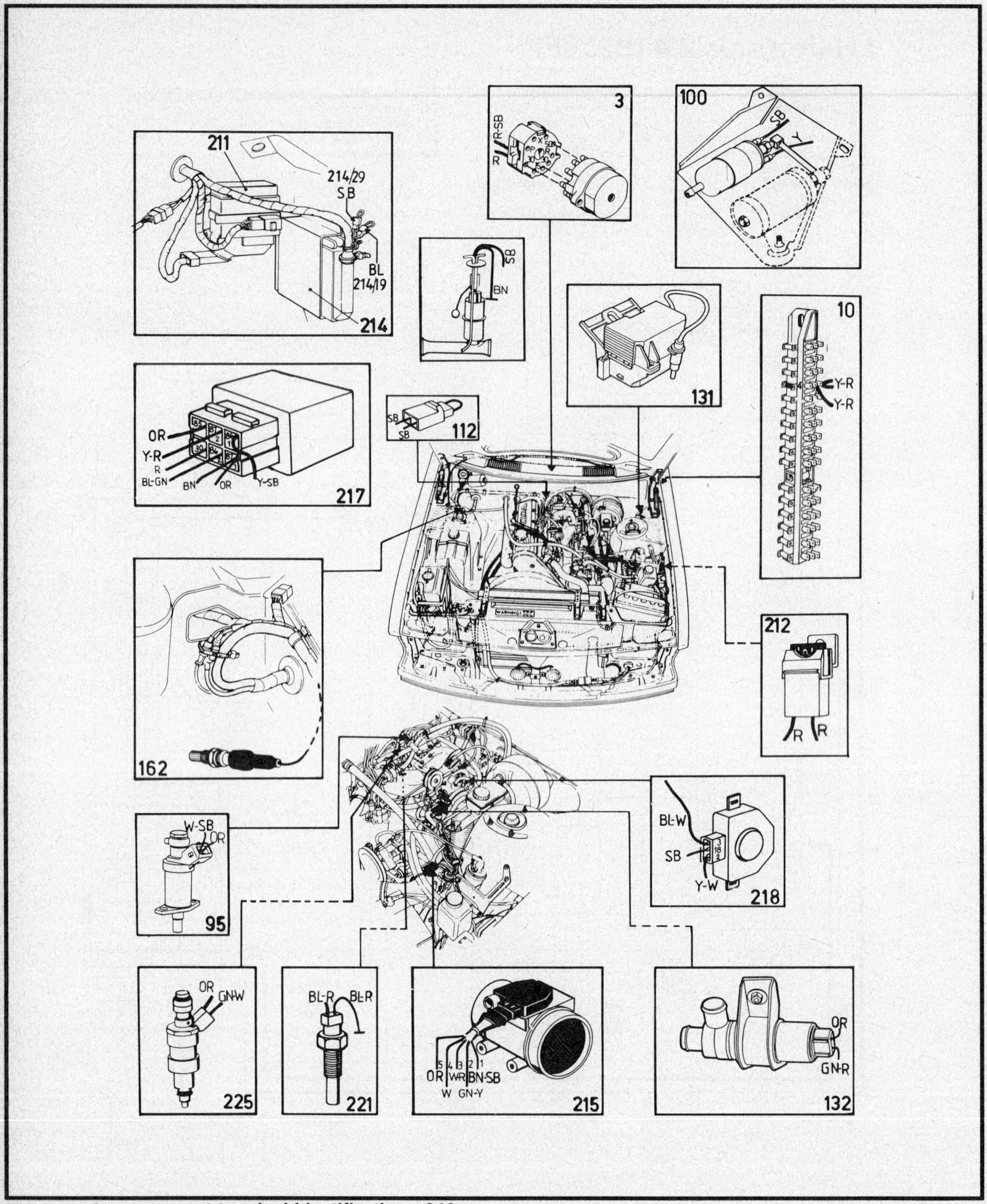

LH-Jetronic component terminal identification—240

LH-Jetronic 2.4 (B 200 F, B 230 F) 740

1 Battery
2 Ignition switch
11 Central electrical unit
12 12-pole instrument panel connector
29 Positive terminal
31 Ground terminal strip in central electrical unit
81 AC pressure sensor
84 Coolant temperature sensor
85 Speedometer
135 Fuel injection relay
187 Lambda probe (oxygen sensor)
188 Cold start valve
196 Idle valve
198 Throttle switch
200 AC compressor
202 Heater control
210 Feed pump
211 Fuel pump
212 Test point
260 Ignition system control unit
284 Air mass meter
361–364 Injectors
419 Power stage, EZK
464 Suppressor relay
472 Control unit, LH-Jetronic 2.4
482 Test output (self-diagnosis)
495 ECC control unit
886 Bridge

A Connector right A pillar
B Connector left A pillar
C Connector right wheel arch
D Connector right wheel arch
E Connector left A pillar
F Connector left wheel arch
H Ground point on intake manifold
J Ground point in boot
K Connector on firewall
L Connector on central electrical unit
M Ground point right A pillar
N Ground point right front wing
O Connector on firewall
P Connector on central electrical unit
S Connector on firewall
T Connector on left wheel arch, 3-pole
V Ground point on left front wing
Z Connector in boot, 4-pole
○ Weld point

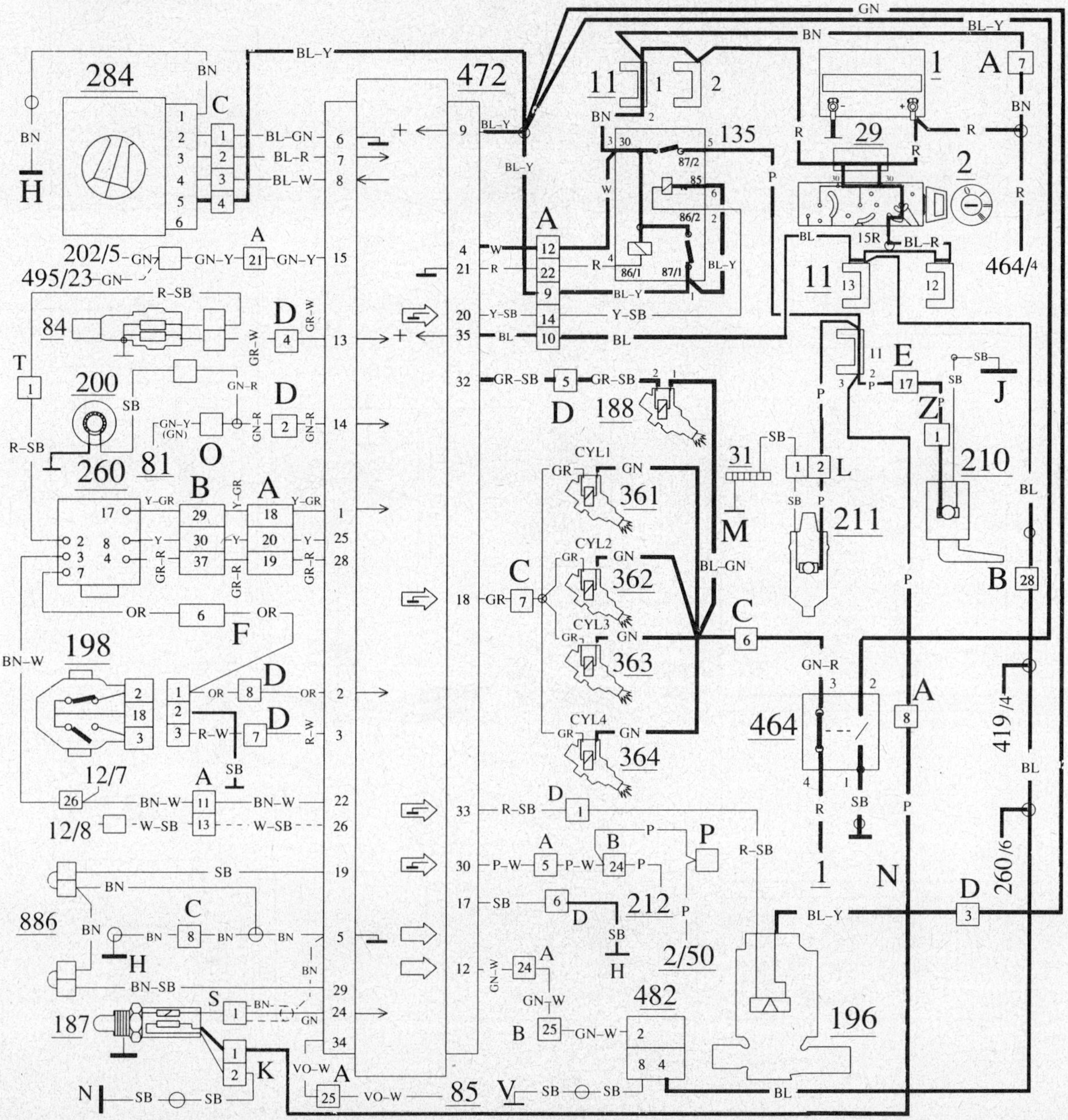

LH-Jetronic wiring diagram—740 with B230F engine

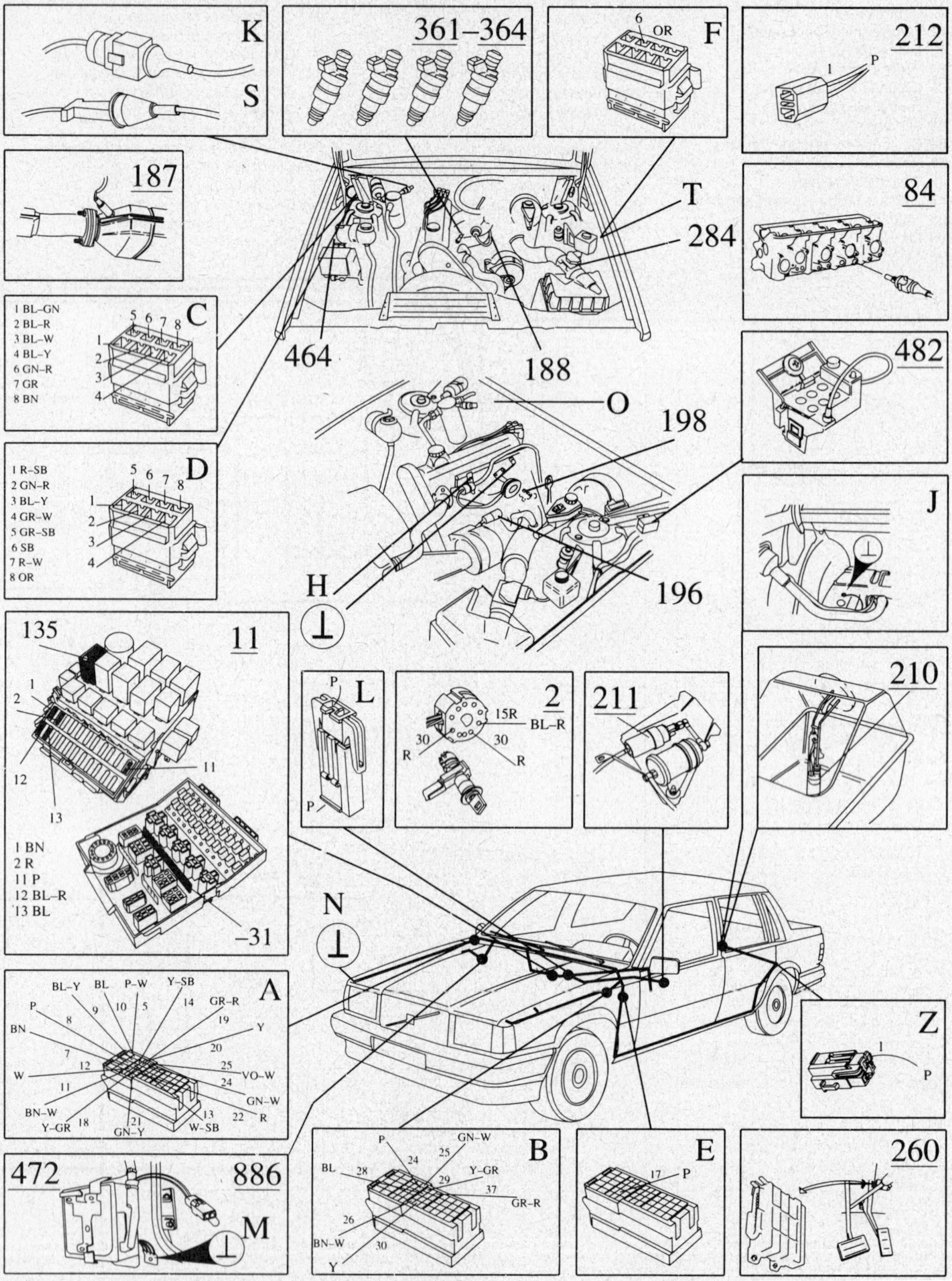

LH-Jetronic component terminal identification—740 with B230F engine

LH-Jetronic 2.4 (B 230 FT/GT) 740

1 Battery
2 Ignition switch
11 Central electrical unit
12 12-pole instrument panel connector
29 Positive terminal
31 Ground terminal strip in central electrical unit
81 AC pressure sensor
84 Coolant temperature sensor
85 Speedometer
135 Fuel injection relay
187 Lambda probe (oxygen sensor)
188 Cold start valve
196 Idle valve
198 Throttle switch
200 AC compressor
202 Heater control
210 Feed pump
211 Fuel pump
212 Test point
260 Ignition system control unit
284 Air mass meter
361–364 Injectors
376 Ballast resistor
377 PTC resistor
419 Power stage, EZK
464 Suppressor relay
472 Control unit, LH-Jetronic 2.4
482 Test output (self-diagnosis)
495 ECC control unit
886 Bridge

A Connector right A pillar
B Connector left A pillar
C Connector right wheel arch
D Connector right wheel arch
E Connector left A pillar
F Connector left wheel arch
H Ground point on intake manifold
J Ground point in boot
K Connector on firewall
L Connector on central electrical unit
M Ground point right A pillar
N Ground point right front wing
O Connector on firewall, single-pole
P Connector on central electrical unit
S Connector on firewall
T Connector on left wheel arch
U Accessory
V Ground point on left front wing
Z Connector in boot, 4-pole
○ Weld point

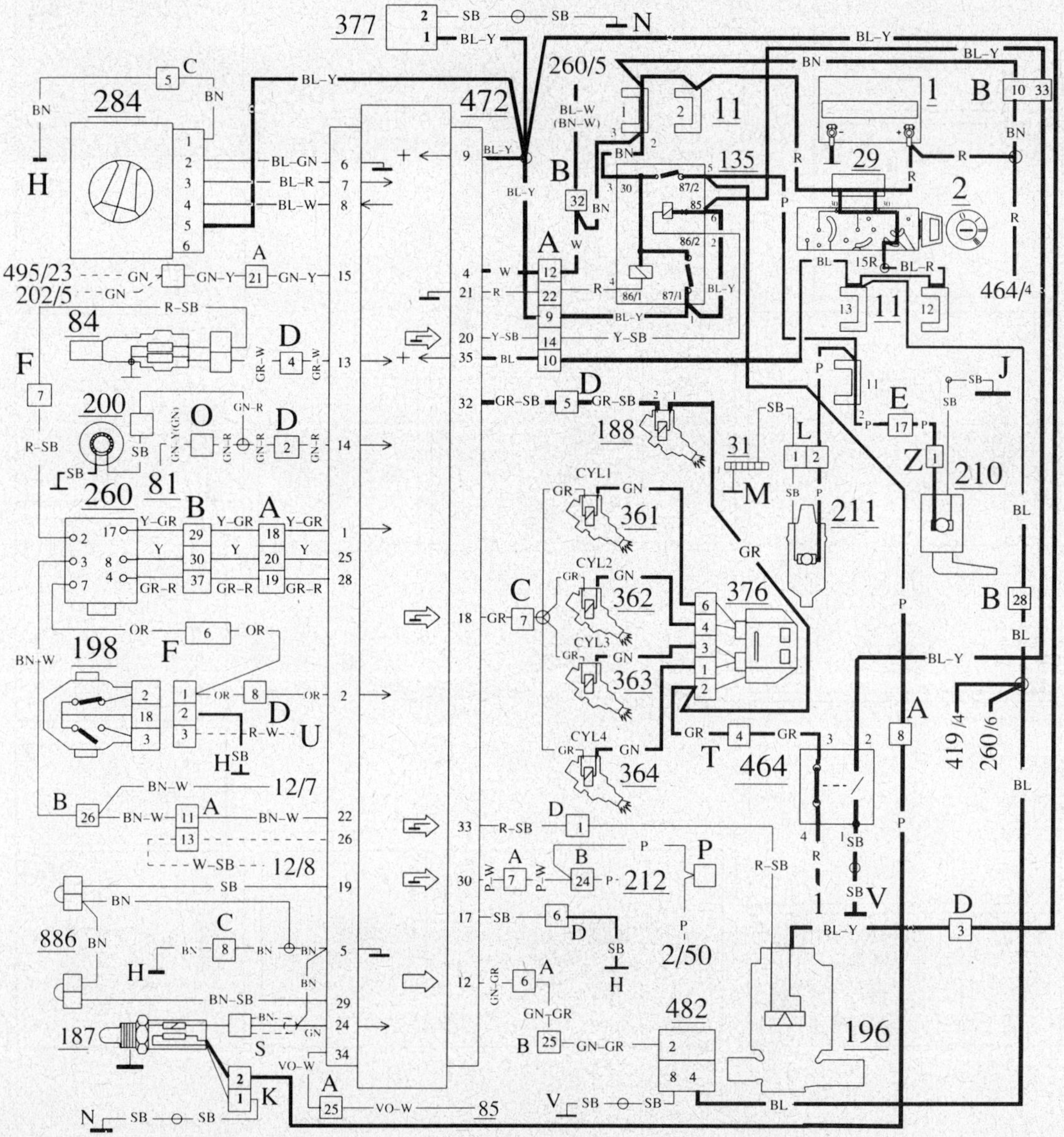

LH-Jetronic wiring diagram—740 and 940 with B230FT engine

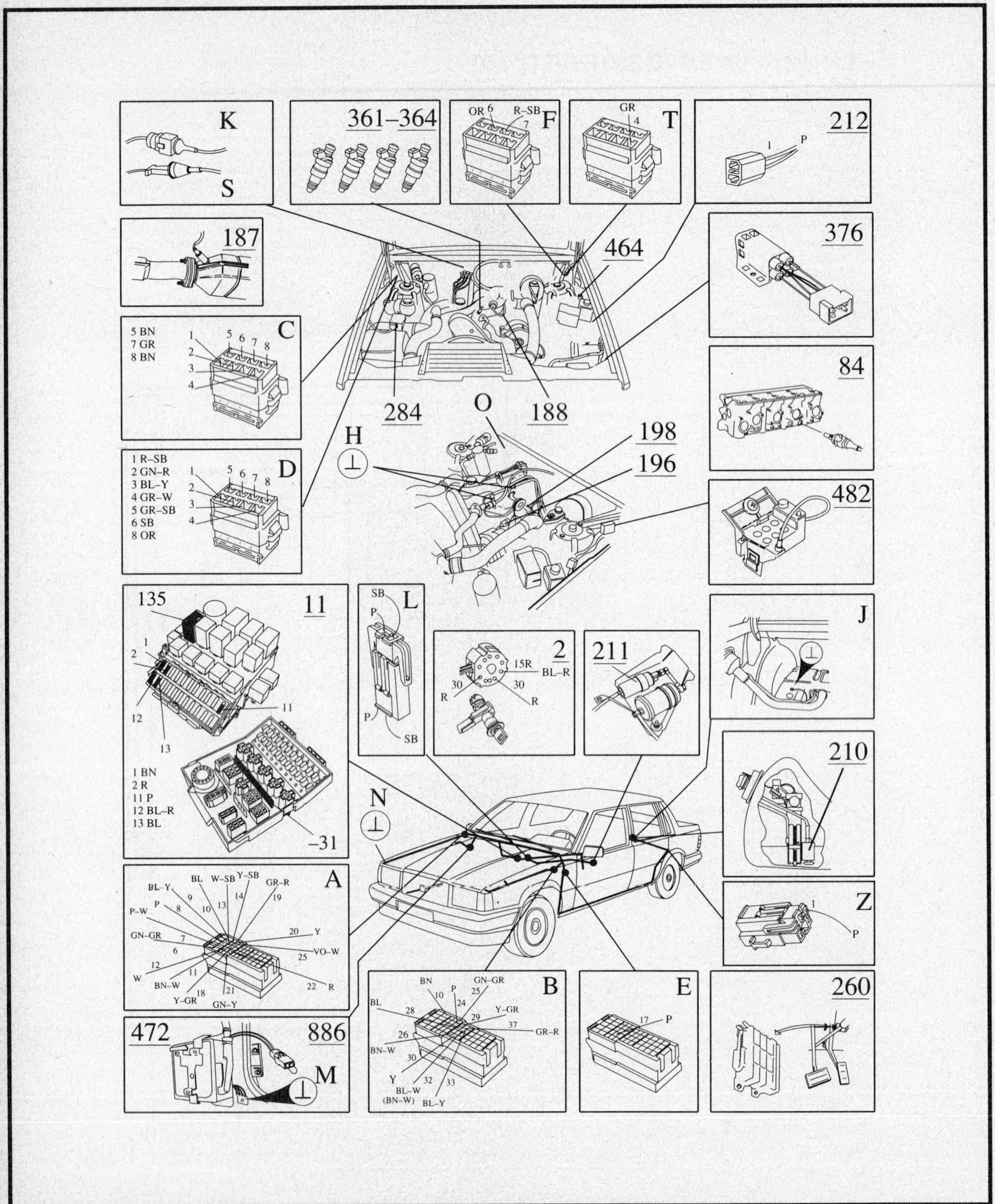

LH-Jetronic component terminal identification—740 and 940 with B230FT engine

LH-Jetronic 2.4 (B 234 F, B 204 E) 740

1 Battery
2 Ignition switch
11 Central electrical unit
12 12-pole instrument panel connector
29 Positive terminal
31 Ground terminal strip in central electrical unit
81 AC pressure sensor
84 Coolant temperature sensor
85 Speedometer
120 AC cut-out relay
135 Fuel injection relay
183 Solenoid, charcoal filter (only B 234 F)
187 Lambda probe (oxygen sensor) (only B 234 F)
196 Idle valve
198 Throttle switch
202 Heater control
210 Feed pump
211 Fuel pump
212 Test point
260 Ignition system control unit
284 Air mass meter
361–364 Injectors
419 Power stage, EZK
464 Suppressor relay
472 Control unit, LH-Jetronic 2.4
482 Test output (self-diagnosis)
495 ECC control unit
886 Bridge

A Connector right A pillar
B Connector left A pillar
C Connector right wheel arch
D Connector right wheel arch
E Connector left A pillar
F Connector left wheel arch
H Ground point on intake manifold
J Ground point in boot
K Connector on firewall
L Connector on central electrical unit
M Ground point right A pillar
N Ground point right front wing
O Connector on firewall, single-pole
P Connector on central electrical unit
S Connector on firewall
T Connector on left wheel arch, 3-pole
V Ground point on left front wing
Z Connector in boot, 4-pole
○ Weld point

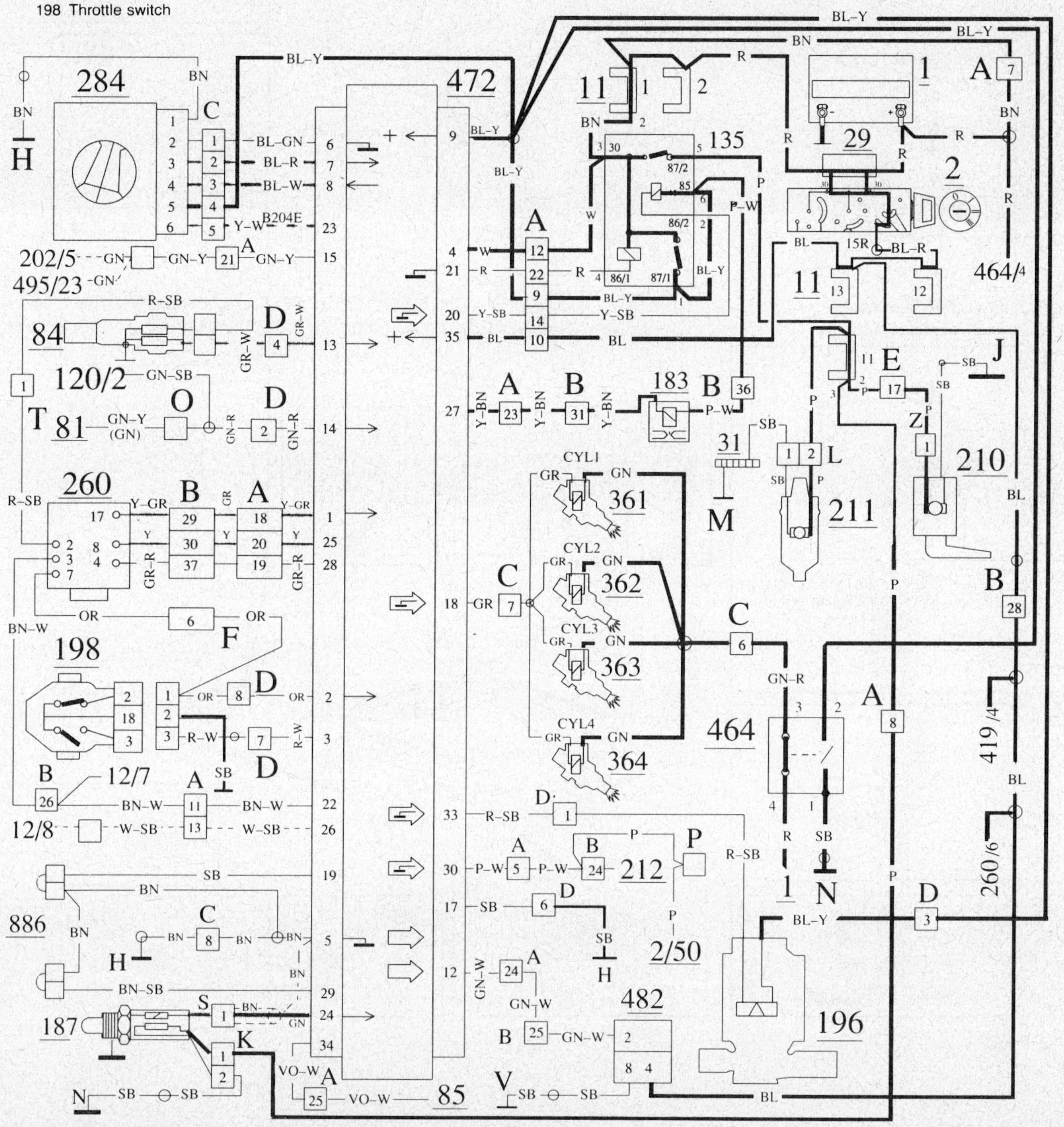

LH-Jetronic wiring diagram—740 and 940 with B234F engine

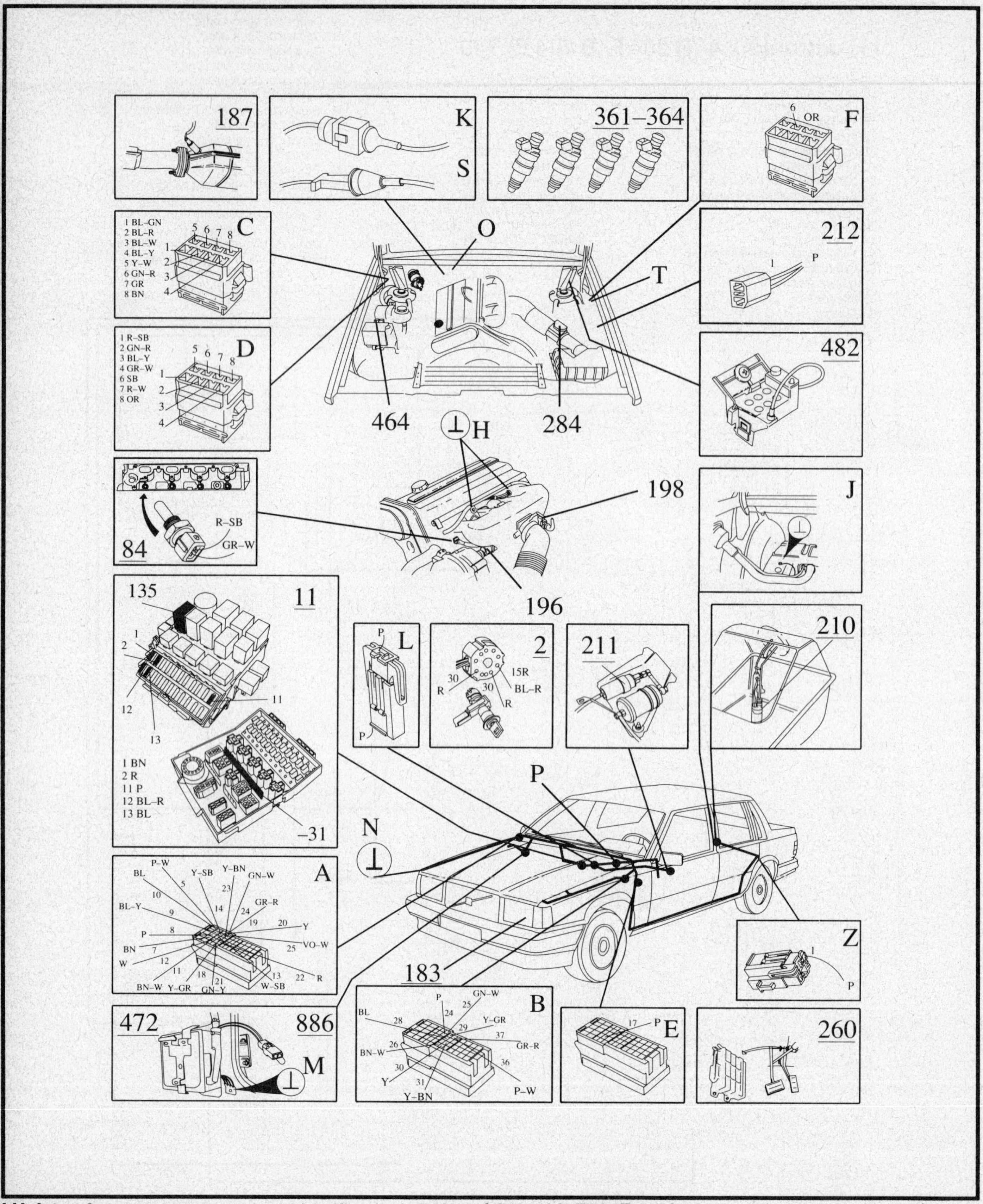

LH-Jetronic component terminal identification—740 and 940 with B234F engine

LH-Jetronic 2.4 (B 230 FT) 760

1/1 Battery (1)
2/12 Suppressor relay (464)
2/13 Fuel injection relay (135)
3/1 Ignition switch (2)
3/50 Throttle switch (198)
4/6 ECC control unit (495)
4/10 Ignition system control unit (260)
4/15 Power stage, EZK (419)
4/23 Control unit, LH-Jetronic 2.4 (472)
5/1E Speedometer (85)
5/18 18-pole connector on instrument panel (18)
6/31 Fuel pump (211)
6/32 Feed pump (210)
7/13 AC pressure sensor (81)
7/15 Lambda probe (oxygen sensor) (187)
7/15 Lambdasond (187)
7/16 Coolant temperature sensor (84)
7/17 Air mass meter (284)
7/47 PTC resistor (235)
8/3 AC compressor (200)
8/5 Idle valve (196)
8/6–8/9 Injectors (361–364)
8/12 Cold start valve (188)
11/1–35 Fuses
14/105 Bridge (886)
15/1 Positive terminal (78)
17/7 Test output (self-diagnosis) (482)
17/11 Test point (212)
20/15 Ballast resistor (376)
31/31 Ground terminal strip in central relay unit (31)

A Connector right A pillar
B Connector left A pillar
C Connector right wheel arch
D Connector right wheel arch
E Connector left A pillar
F Connector left wheel arch
G Connector, 10-pole on instrument panel
H Ground point on intake manifold
J Ground point in boot
K Connector on firewall
L Connector on central electrical unit
M Ground point in central relay unit
N Ground point right front wing
O Connector on firewall
P Connector, 3-pole, right wheel arch
Q Connector, 12-pole on instrument panel
R Connector, 10-pole on instrument panel
S Connector on firewall
T Connector on left wheel arch
V Ground point on left front wing
Z Connector in boot, 4-pole
○ Weld point

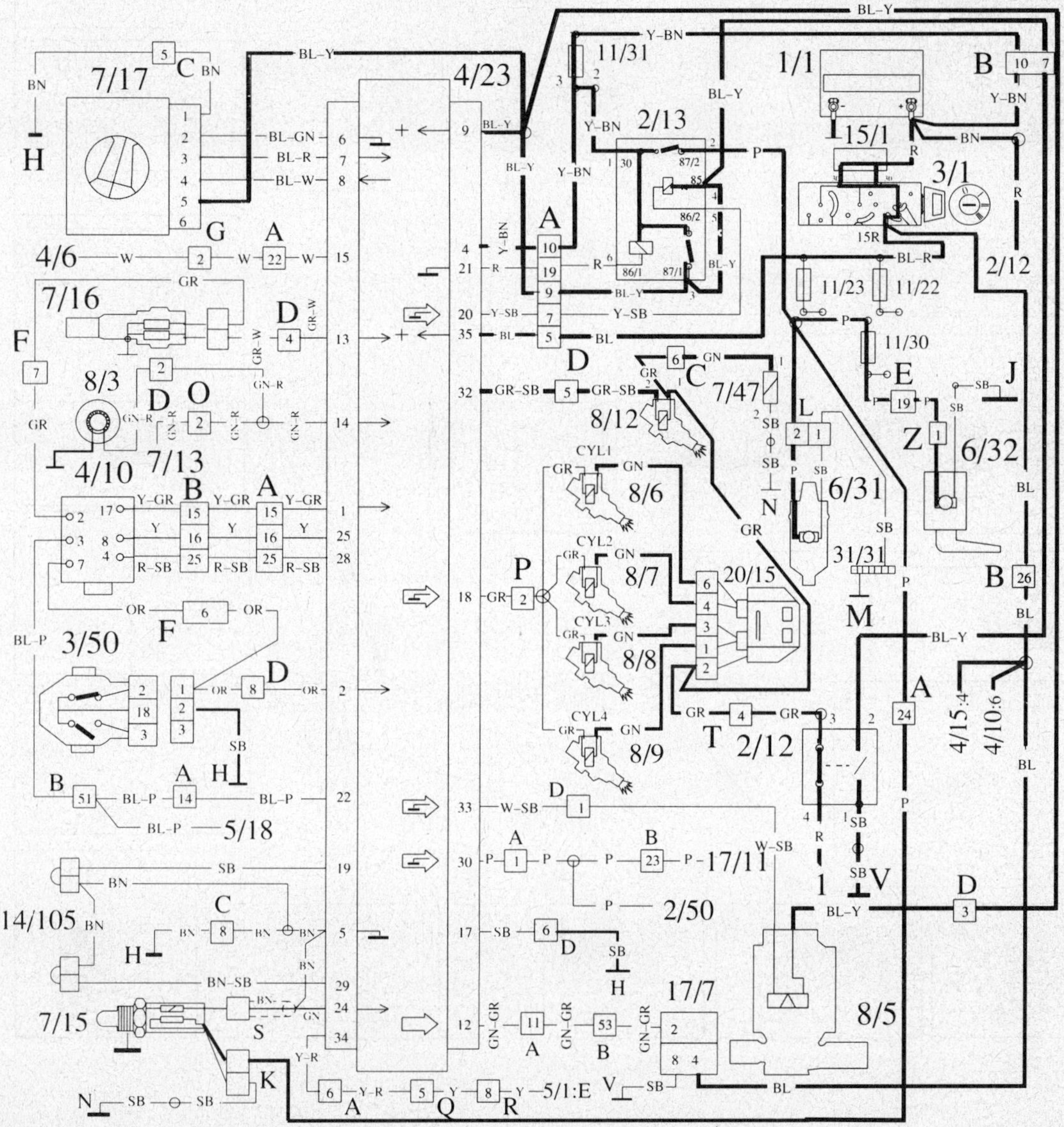

LH-Jetronic wiring diagram—760 with B230FT engine

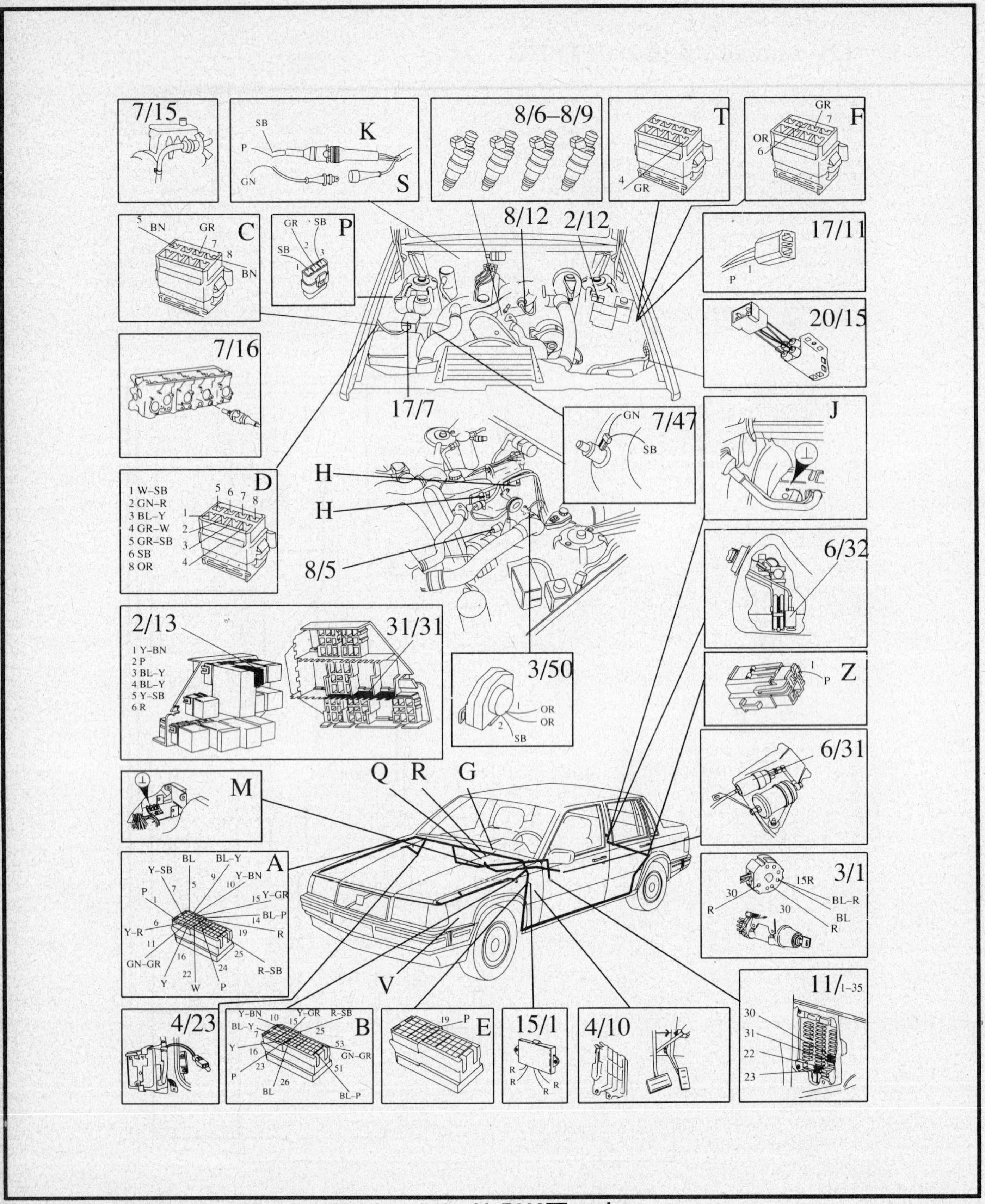

LH-Jetronic component terminal identification—760 with B230FT engine

LH-Jetronic 2.4 (B 230 FT) 780

1 Battery
2 Ignition switch
11 Central electrical unit
12 12-pole instrument panel connector
29 Positive terminal
31 Ground terminal strip in central electrical unit
81 AC pressure sensor
84 Coolant temperature sensor
85 Speedometer
135 Fuel injection relay
187 Lambda probe (oxygen sensor)
188 Cold start valve
196 Idle valve
198 Throttle switch
200 AC compressor
210 Feed pump
211 Fuel pump
212 Test point
259 High power system control unit
260 Ignition system control unit
284 Air mass meter
361–364 Injectors
376 Ballast resistor
419 Power stage, EZK
420 Throttle switch test point
464 Suppressor relay
472 Control unit, LH-Jetronic 2.4
482 Test output (self-diagnosis)
495 Control unit, ECC
886 Bridge

A Connector right A pillar
B Connector left A pillar
C Connector right wheel arch
D Connector right wheel arch
E Connector left A pillar
F Connector left wheel arch
H Ground point on intake manifold
J Ground point in boot
K Connector on firewall
L Connector on central electrical unit
M Ground point right A pillar
N Ground point right front wing
O Connector on firewall, single-pole
P Connector on central electrical unit
S Connector on firewall
T Connector left wheel arch
V Ground point on left front wing
Z Connector in boot, 4-pole
○ Weld point

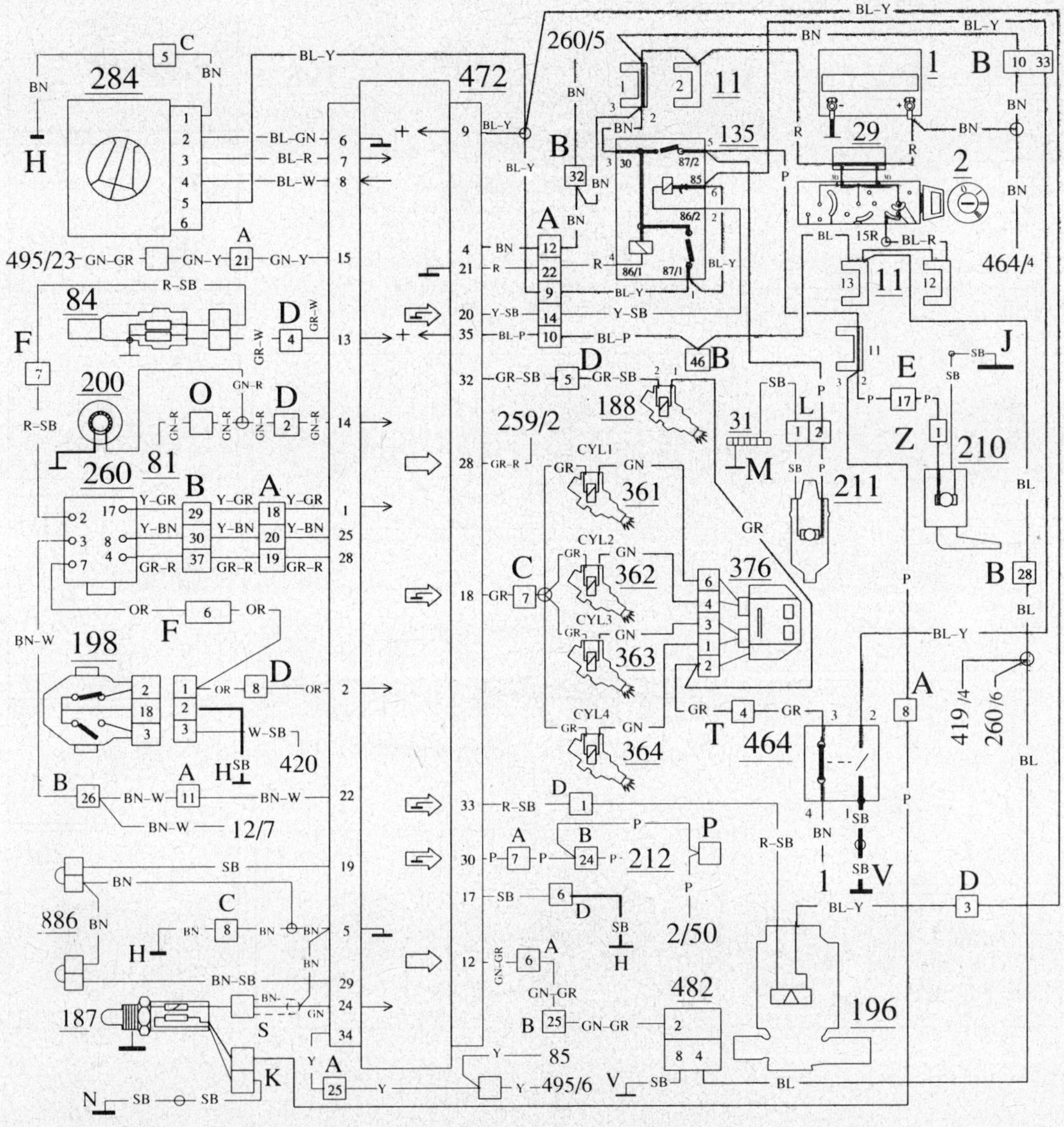

LH-Jetronic wiring diagram—780 with B230FT engine

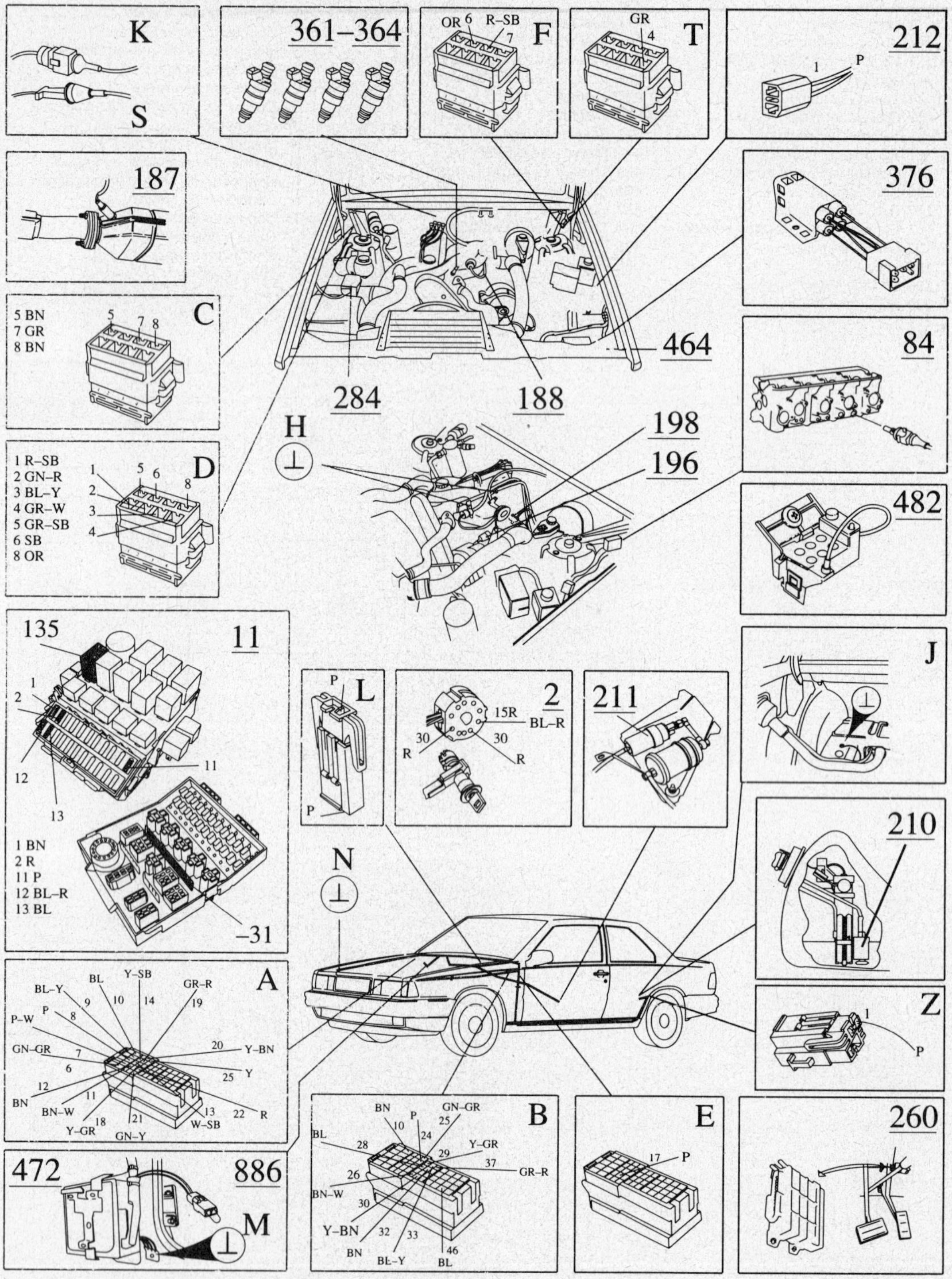

LH-Jetronic component terminal identification—780 with B230FT engine

Electric cooling fan 740, 940

1/1	Battery
2/11	Relay, electric cooling fan
4/23	Electronic Control Unit, LH-Jetronic 2.4
6/29	Motor, electric cooling fan
7/14	Termostat, electric cooling fan
7/38	Low pressure sensor
7/40	High pressure sensor
31/1	Ground point, front right fender
31/2	Ground point, front left fender
31/64	Ground point, electric cooling fan
A	3-pole connector at right wheelhouse tower

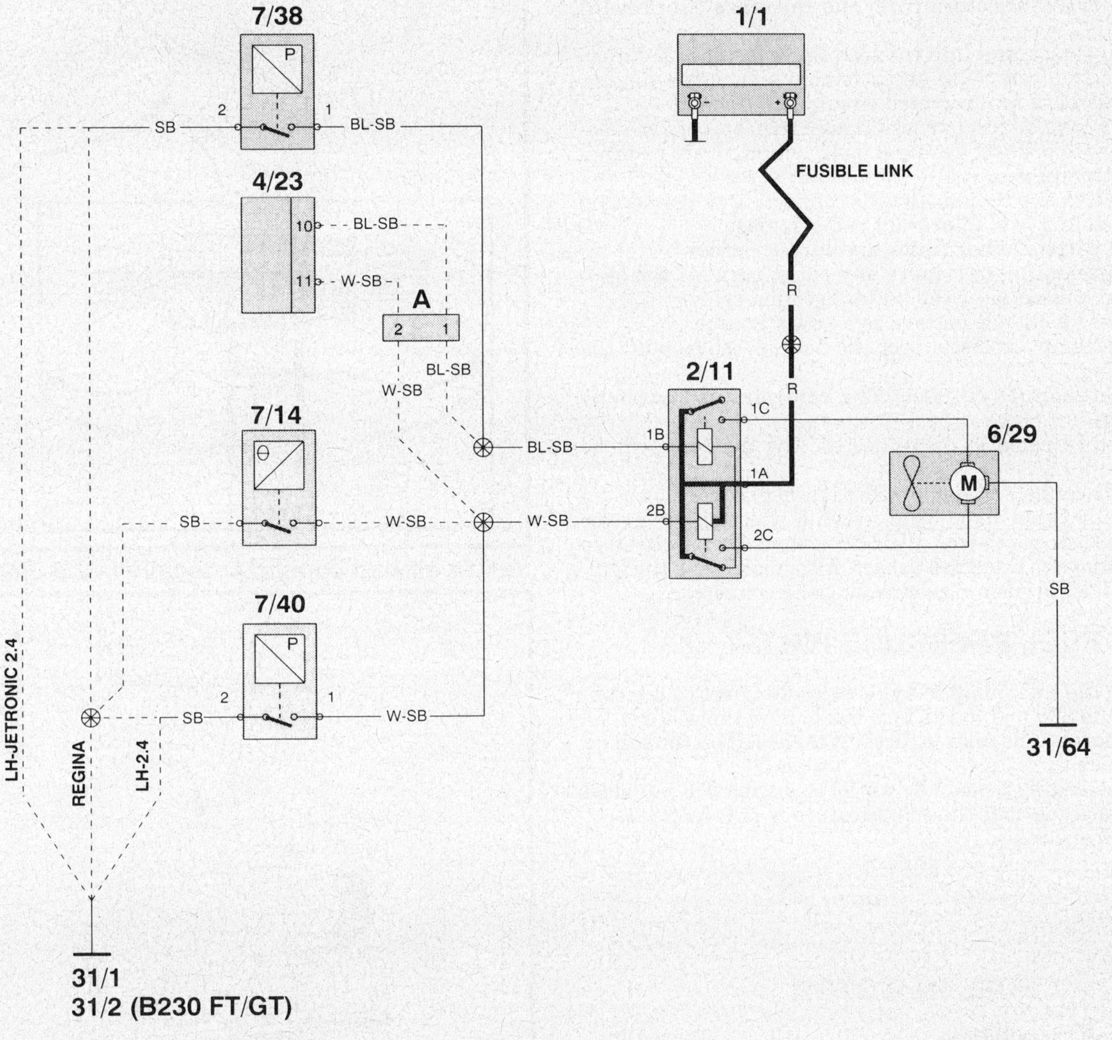

Cooling fan wiring diagram—740 and 940

Component Replacement

SERVICE PRECAUTIONS

- Use caution when working around ignition system components. the ignition system operates at high power with dangerous voltages in both the low-tension and high-tension circuits, including connectors and similar fittings.
- Do not operate the fuel pumps when the fuel lines are empty.
- Do not arc weld on the vehicle unless all the control units have been removed.
- The control unit must not be exposed to temperatures above 175°F (80°C), such as in a paint booth.
- Make sure that the ignition switch is **OFF** before connecting or disconnecting the control unit, ignition coil, spark plug wiring or test instruments.
- Always use new gaskets and seals when remaking a fuel connection.
- When running a compression test, remove the electrical leads from connection 1 on the ignition coil (to prevent arcing) and remove the connectors from the injectors (to prevent flooding).
- Make sure all control unit connectors are fastened securely. A poor connection can cause an extremely high surge voltage, resulting in damage to integrated circuits.
- Keep any control unit harness at least 4 in. away from adjacent harnesses to prevent a system malfunction due to external electronic interference.
- Switch **OFF** the radio before disconnecting or connecting the battery ground lead to prevent radio damage.
- Keep all parts and harnesses dry during service.
- Before attempting to remove any parts, turn off the ignition switch and disconnect the battery ground cable.
- Always use a 12 volt battery as a power source.
- Do not attempt to disconnect the battery cables with the engine running.
- Do not attempt to disassemble a control unit under any circumstances.
- If a battery cable is disconnected, the memory will be erased.
- Always disconnect the yellow SRS (air bag) connector when performing any other diagnostic, trouble shooting or service procedure not associated with the SRS system. Failure to do so can cause damage to the crash sensor. Also, make sure the ignition switch is **OFF** when disconnecting the connector.

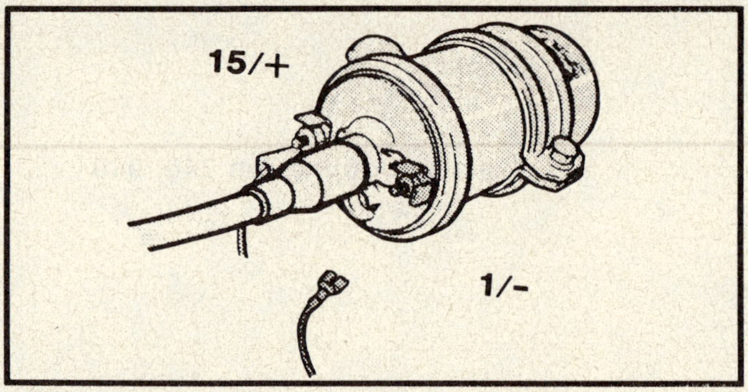

Disconnecting ignition coil for compression test—LH-Jetronic

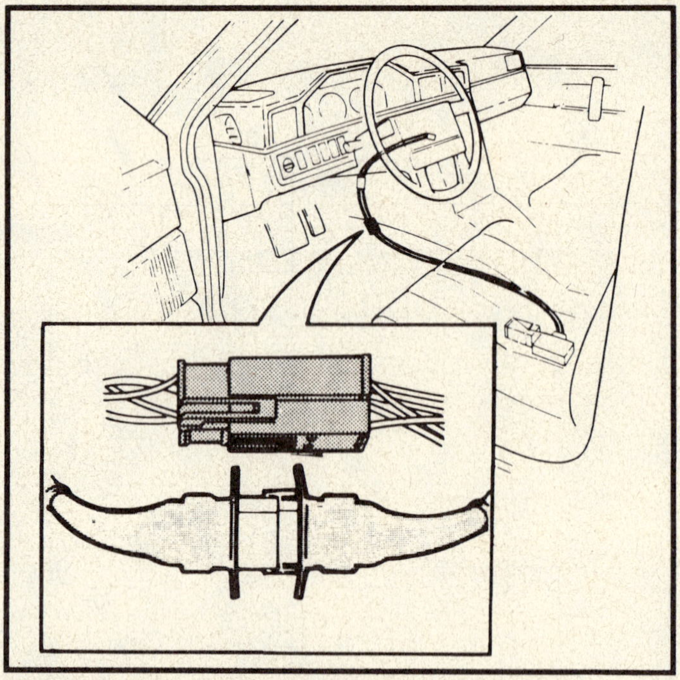

Yellow air bag connector location—LH-Jetronic

FUEL PRESSURE RELIEF

1. Remove the fuel pump relay. It is located next to the control unit on the 240 and in the fuse box behind the ash tray on 700 and 900 series. The relay is furthest to the left on the second row.
2. Start the engine repeatedly until the engine will not start anymore, indicating that the fuel pressure is relieved.
3. Reinstall the relay.

CAUTION

To avoid fire and personal injury, replace any plastic tie bands removed from the fuel lines.

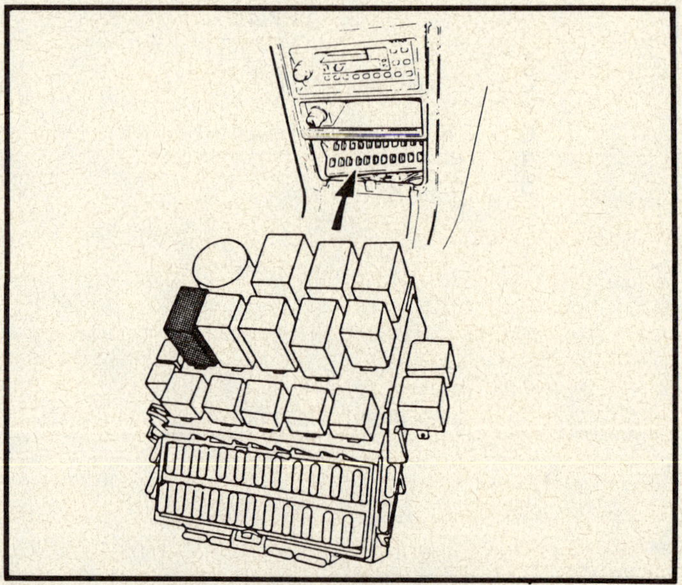

Fuel pump relay location—Except 240

FUEL INJECTOR

Removal and Installation

NOTE: Remove and install the fuel distribution pipe, the injectors and the cold start valve as one unit.

1. Unbolt the pressure regulator from the fuel rail bracket.
2. Remove the fuel rail bolts without distorting the metal of the rail.

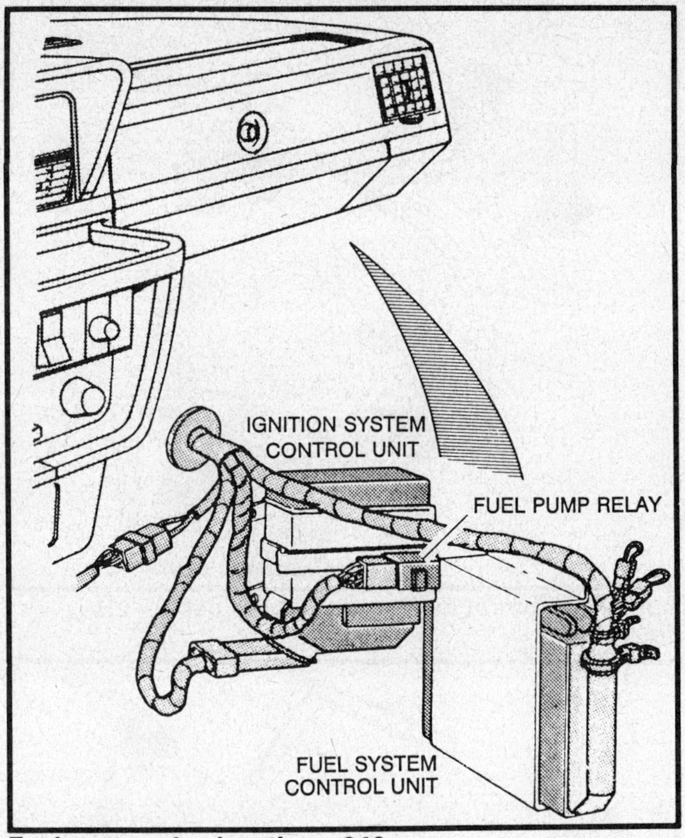

Fuel pump relay location—240

3. Use 2 line wrenches and separate the line fittings, then remove the injector assembly.
4. Installation is the reverse of removal. Use new O-rings and lubricate with petroleum jelly.

BASE THROTTLE

Adjustment

1. Loosen the throttle locknut.

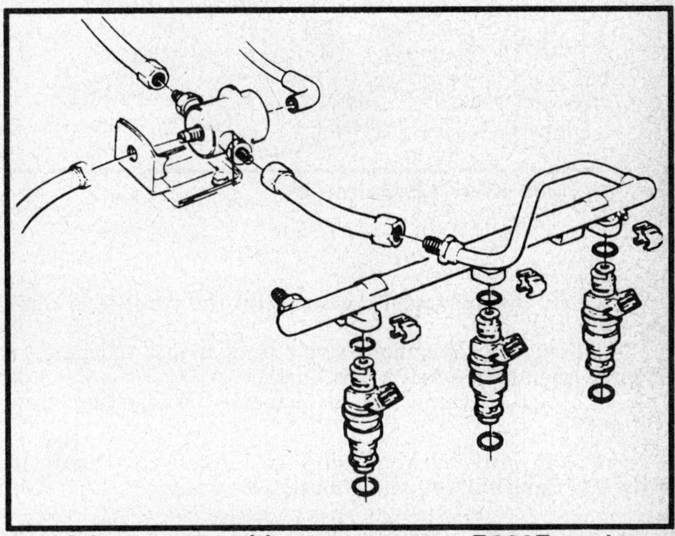
Fuel injector assembly components—B280F engine

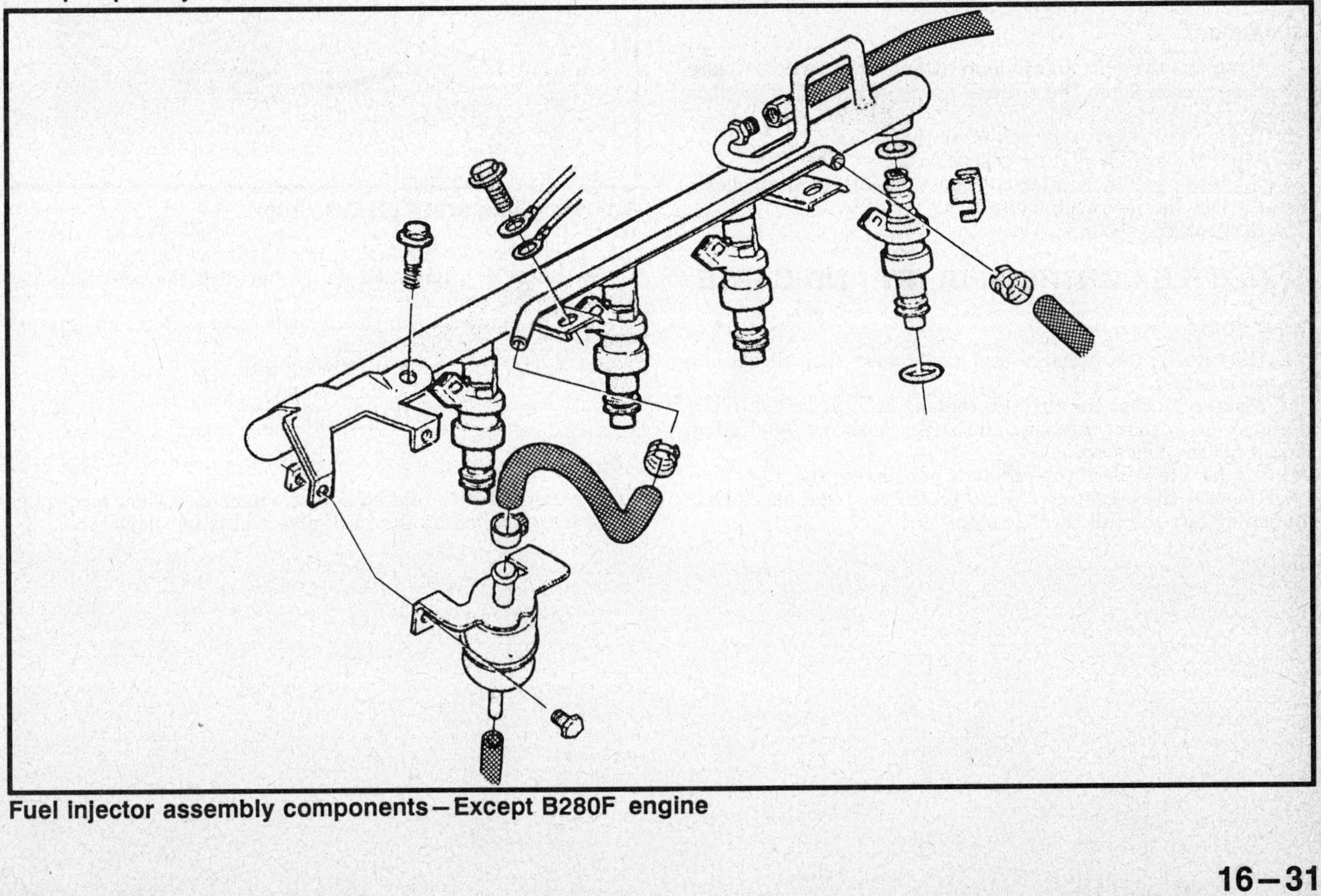
Fuel injector assembly components—Except B280F engine

Setting base idle—LH-Jetronic

2. Loosen the adjustment screw until the throttle is completely closed.
3. Tighten the adjustment screw until it just touches the linkarm, then turn a half a turn further.
4. Tighten the locknut without changing the adjusting screw position.

NOTE: It may be necessary to loosen the throttle switch before setting the throttle.

THROTTLE SWITCH

Adjustment

1. Open the throttle slightly and listen to the switch. There should be a click when the shutter opens, if the adjustment is correct.
2. Loosen the throttle switch mounting bolts.
3. Turn the switch slightly clockwise.
4. Turn the switch counter-clockwise until the switch clicks, then tighten the mounting bolts.
5. Recheck the setting.

THROTTLE CONTROL PULLEY AND CABLE

Adjustment

1. Disconnect the linkarm and make sure that the pulley moves smoothly.
2. Make sure that the throttle cable is fully extended in the idle position without affecting the pulley position. The pulley should hit the idle stop.
3. Adjust the cable if the pulley is not hitting the idle stop.
4. Depress the accelerator pedal all the way and check that the pulley hits the full throttle stop.

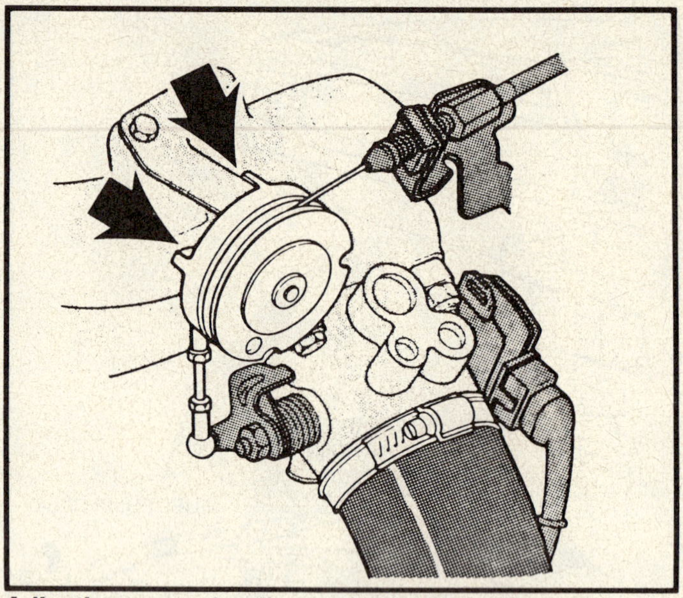

Adjusting control pulley and throttle cable—LH-Jetronic

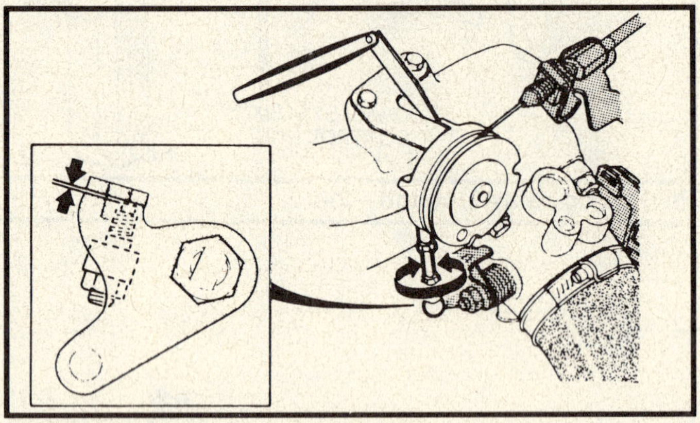

Adjusting link arm—LH-Jetronic

5. Reconnect the linkarm and adjust it as follows:
   a. Insert a 1mm feeler gauge between the control pulley and the idle stop.
   b. The play between the throttle lever and the adjustment screw should be 0.1–0.3mm.
   c. Adjust the linkarm to bring into specifications.

IDLE SPEED

Adjustment

After other throttle related adjustments have been made, idle speed is controlled by the idle valve and is not adjustable.

VOLVO MOTRONIC 1.8 FUEL INJECTION

General Information

The Motronic 1.8 fuel injection system, found on 1992 960 models, is equipped with a powerful control unit that controls the ignition and fuel injection functions by means of individual ignition coils and injectors.

In addition to controlling the ignition and fuel injection functions, Motronic also:

- Determines whether the A/C compressor may be switched in;
- Reduces the engine torque in response to a signal from the automatic transmission control unit, to insure smooth engagement of the different gears and also supplies the transmission control unit with information on engine running conditions for computing gear changes;
- Controls the operation of the radiator fan.

The control unit is provided with adaptive Lambda control and idling control functions as well as timing retardation function to eliminate knock. The service requirement is minimal, since neither the carbon monoxide level nor the idling speed require adjustment.

DATA SENSORS

Air Mass Meter

This device measures the engine intake air mass. Factors that affect air density, such as temperature, humidity and pressure are compensated for during measurement.

The sensor inside the air mass meter consists of a wire which is maintained at 250°F (121°C) higher than the ambient air entering the engine. As the air mass passing over the wire increases, more current is required to maintain the correct temperature. The amount of current required to maintain this temperature is used to calculate the intake air mass.

When the engine is turned off, any dirt on the wire is burned off electrically by heating the wire to 1800°F (1000°C). Any dirt remaining on the wire would cause it to send incorrect signals to the control unit and result in an incorrect air/fuel mixture.

Coolant Temperature Sensor

The coolant temperature sensor provides the control unit with information necessary for proper adjustment of injection duration.

Throttle Switch

The throttle switch indicates to the fuel system whether the throttle valve is closed. It does not function as a variable resistor. It also indicates whether the throttle valve is fully open.

Oxygen Sensor (Lambda-Sond)

The oxygen sensor is mounted in the exhaust manifold in front of the catalytic converter. It produces a measureable current by comparing the amount of oxygen in the exhaust gas with the amount in the ambient air.

Under normal conditions, the optimum air/fuel ratio is 14.7:1. This ratio is monitored by post-combustion measurement of the oxygen content in the exhaust gas using the oxygen sensor.

The oxygen sensor operating temperature ranges from 545°–1530°F (285°–832°C). It is electrically heated to enable it to reach operating temperature quickly. When ignition is turned to the **ON** position, current is sent to a PTC resistor (Positive Temperature Coefficient) whose resistance increases with rising temperature. Due to this system, the oxygen sensor quickly reaches operating temperature, even at low exhaust gas temperature.

The exhaust gases reach the outer surface of the sensor through slits in the protective sleeve. Ambient air reaches the sensors inner surface through channels. The sensor consists of a platinum covered zirconium-oxide pipe.

The signal strength is in direct proportion to the amount of oxygen in the exhaust gases. This depends on the air/fuel ratio. The Lambda value of 1, represents the theoretically perfect ratio. A rich mixture results in higher voltage and a lean mixture results in lower voltage. The current sent by the oxygen sensor to the control unit varies between 0.1–1.0 volt. The difference between high and low voltage occurs when the Lambda value is at 1. The control unit uses this information to adjust the amount of fuel used.

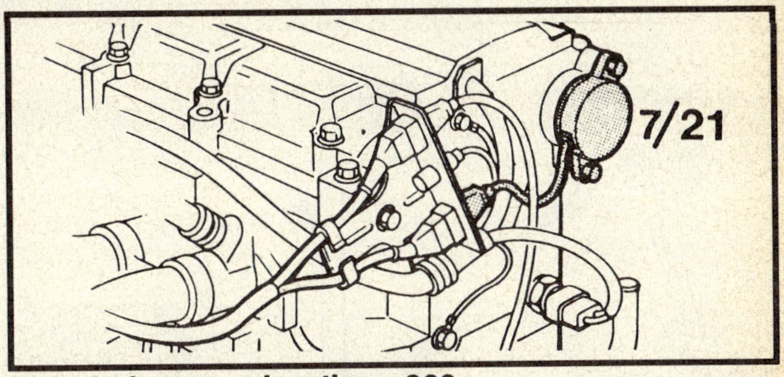

Camshaft sensor location—960

Camshaft Sensor

The camshaft sensor is mounted at the rear of the cylinder head on the exhaust side. Its signals are used by the control unit to identify which pair of pistons is simultaneously approaching top dead center. Together with the timing pick-up signals, this enables the unit to identify the stroke being performed by each of the pistons in question.

The main components of the camshaft sensor are a Hall generator, trigger rotor and cover. As the rotor rotates with the camshaft, it alternately shields and exposes the Hall element, generating high 5 volt and low 0 volt signals in turn. Since the vane and opening on the rotor are of the same width, and since the camshaft rotates at only half the crankshaft speed, the output signal changes only after a full crankshaft revolution. After 2 crankshaft revolutions, the signal from the Hall generator has reached 1 maximum and 1 minimum.

EGR Temperature Sensor

This sensor, found on California vehicles, measures the temperature of the exhaust gases returned to the intake manifold. Detection of the temperature variations enables the control unit to determine whether the EGR system is operating. It is able to measure temperatures up to 930°F (500°C) and the resistance increases with temperature.

CONTROLLED OUTPUTS

EGR Vacuum Booster

This component is found on California vehicles. It controls the pressure in the vacuum line to the EGR valve by means of the admission valve (lower section of the booster). The pressure in the intake manifold is supplied to the reducing valve (upper section of the booster). The vacuum booster processes the electrical signal from the intake manifold to maintain optimum control of the EGR valve. The unit is also designed to compensate for changes in atmospheric pressure. It is mounted on the left strut tower.

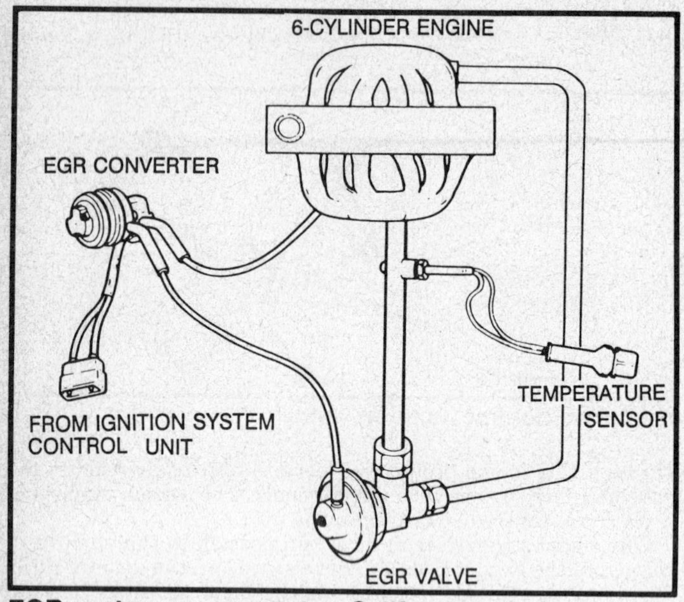

EGR system components—California vehicles

Fuel Control

The injectors are divided into 2 groups of 3. One group of injectors supplies fuel to cylinders 1, 2 and 4, while the second serves cylinders 3, 5 and 6. The 2-group arrangement has been adopted to ensure that fuel injection takes place as soon as possible after injector opening, supplying each of the cylinders with as uniform a mixture as possible.

Under normal conditions, each group of injectors is activated once every other crankshaft revolution, delivering the full fuel charge required for combustion in 1 cylinder.

To compute the fuel injection, the control unit is supplied with the following information:

1. An engine load signal by the air mass meter.
2. An engine speed signal and crankshaft position signal by the timing pick-up.
3. A piston stroke signal by the camshaft sensor, enabling the unit to determine which injector group is to be activated.
4. The exhaust gas oxygen content by the oxygen sensor (Lambdasond).
5. An engine temperature signal by the coolant temperature sensor, for functions including fuel enrichment on starting.
6. A throttle position signal by the throttle switch, for functions including acceleration and full-load enrichment.
7. A road speed signal by the speedometer.

Idle Speed Control

Factors such as whether gear position **D** or **P** is selected, and whether the air conditioning compressor is running are among those used to control the idling (CIS) valve. The necessary signals are supplied by the automatic transmission control unit and air conditioning compressor electromagnetic clutch respectively.

Radiator Fan Control

A fully-electric radiator fan is operated by a relay, either in response to the engine temperature signal supplied to the Motronic control unit or directly by the pressure switches mounted in the A/C high-pressure circuit. The fan relay is mounted on a bracket in front of the battery.

If the coolant temperature is unusually high when the engine is switched **OFF**, the Motronic control unit will keep the fan running for a further period not exceeding 5 minutes. (The fan cannot, however, be restarted until some time has elapsed after stopping the engine.)

Air Conditioning Compressor Control

The Motronic control unit controls the operation of the A/C compressor electromagnetic clutch by means of a relay in the electronic climate control (ECC) power unit. A request to start the A/C compressor is initiated from the ECC control panel.

The control unit prevents the compressor from starting under the following conditions, to reduce the load on the engine at full load and to prevent overheating:

1. When the throttle is fully open, using the signal from the throttle switch.
2. When the coolant temperature approaches 239°F (115°C), in order to remain safely within the boiling point of approximately 275°F (134°C). The signal from the coolant temperature sensor is used for this purpose.

EGR System Control (California)

The EGR electrical circuit is equipped with a temperature sensor that enables the control unit to determine if the EGR system is operating as intended. The purpose of the sensor is to detect the temperature variations that occur when the system is in operation. The sensor is connected to the control unit, which can activate 2 EGR fault codes. The sensor has no effect on system operation, only on the diagnostic function.

FUEL DELIVERY SYSTEM

Fuel Pressure Regulator

The fuel pressure regulator ensures that fuel pressure remains at approximately 43.5 psi (300 kPa) at idle at the injectors. Using a vacuum tube connected to the engine intake manifold, fuel pressure is gradually increased as intake manifold vacuum decreases. Because of this, adequate pressure is maintained, regardless of throttle position. The amount of fuel injected largely depends on injection duration. Excess fuel is returned to the fuel tank through a return pipe.

Injectors

The fuel injectors as designed as electric solenoids. While the starter motor is operating, there are 2 injections per engine rotation. This is reduced to 1 during normal driving. Injection occurs in the intake manifold near the intake valves.

Fuel Pump

The fuel pump is an electric pump, cooled by the fuel that flows through it. It incorporates a check valve and an overflow valve that opens when fuel pressure gets too high.

Both the primary pump and secondary pump operate when the starter motor or the engine is running. Should the engine stop while the ignition remains on, the control unit will cut the current to the pumps.

The in-tank fuel pump is an electric fuel pump that keeps pressure in the fuel line prior to the main fuel pump to prevent vapor lock. The pump is part of the fuel level transmitter assembly and incorporates a coarse strainer type filter and a non-return valve to maintain a constant amount of pressure in the system, even if the main pump is not operational.

Fuel Filter

The fuel filter is adjacent to the main fuel pump, which is mounted on a plate underneath the vehicle on the left side.

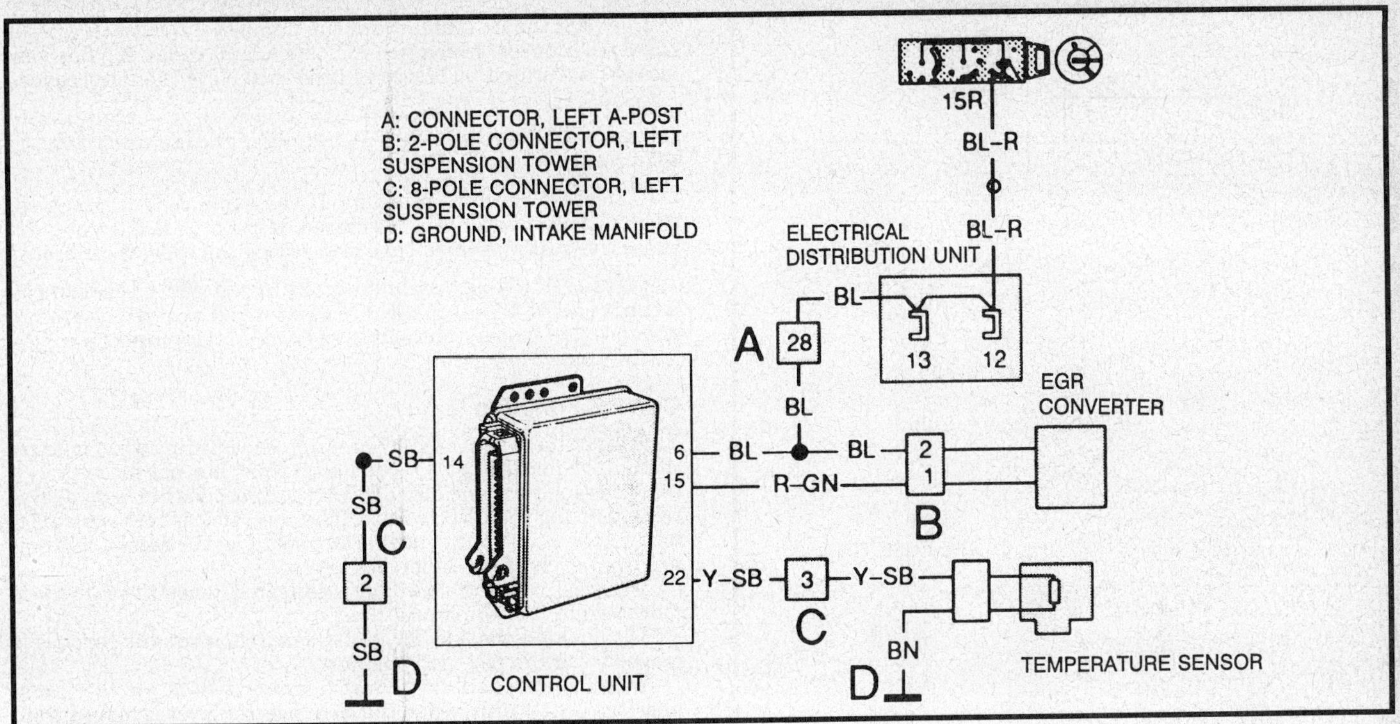

EGR electrical circuit—California vehicles

Diagnosis and Testing

SERVICE PRECAUTIONS

- Use caution when working around ignition system components. The ignition system operates at high power with dangerous voltages in both the low-tension and high-tension circuits, including connectors and similar fittings.
- Do not operate the fuel pump when the fuel lines are empty.
- Do not arc weld on the vehicle unless all the control units have been removed.
- The control unit must not be exposed to temperatures above 175°F (80°C), such as in a paint booth.
- Make sure that the ignition switch is **OFF** before connecting or disconnecting the ignition coil, spark plug wiring or test instruments.
- Always use new gaskets and seals when remaking a fuel connection. Torque the fuel filter fittings to 15–26 ft. lbs. (20–35 Nm).
- Do not test the compression on the 960 until the ignition coils are removed, the ignition coil relay is disconnected and fuse No. 31 (to disarm fuel pumps) is removed.
- Do not disconnect the control unit harness connector before fuse No. 24 has been removed.
- Make sure all control unit connectors are fastened securely. A poor connection can cause an extremely high surge voltage, resulting in damage to integrated circuits.
- Keep any control unit harness at least 4 in. away from adjacent harnesses to prevent a system malfunction due to external electronic interference.
- Keep all parts and harnesses dry during service.
- Before attempting to remove any parts, turn off the ignition switch and disconnect the battery ground cable.
- Always use a 12 volt battery as a power source.
- Do not attempt to disconnect the battery cables with the engine running.

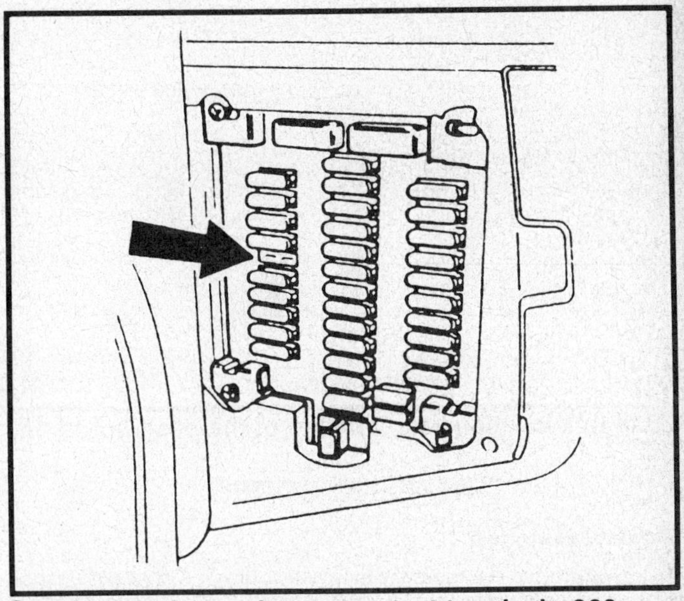

Supplemental restraint system test terminal—960

- Do not attempt to disassemble a control unit under any circumstances.
- If a battery cable is disconnected, the memory will be erased.
- Before connecting or disconnecting control unit ECU harness connectors, make sure the ignition switch is **OFF** and the negative battery cable is disconnected to avoid the possibility of damage to the control unit.
- A spare location in the fuse panel is used as a test terminal for air bag diagnostics on the 960. Never install a fuse in this position or connnect accessories to this terminal.

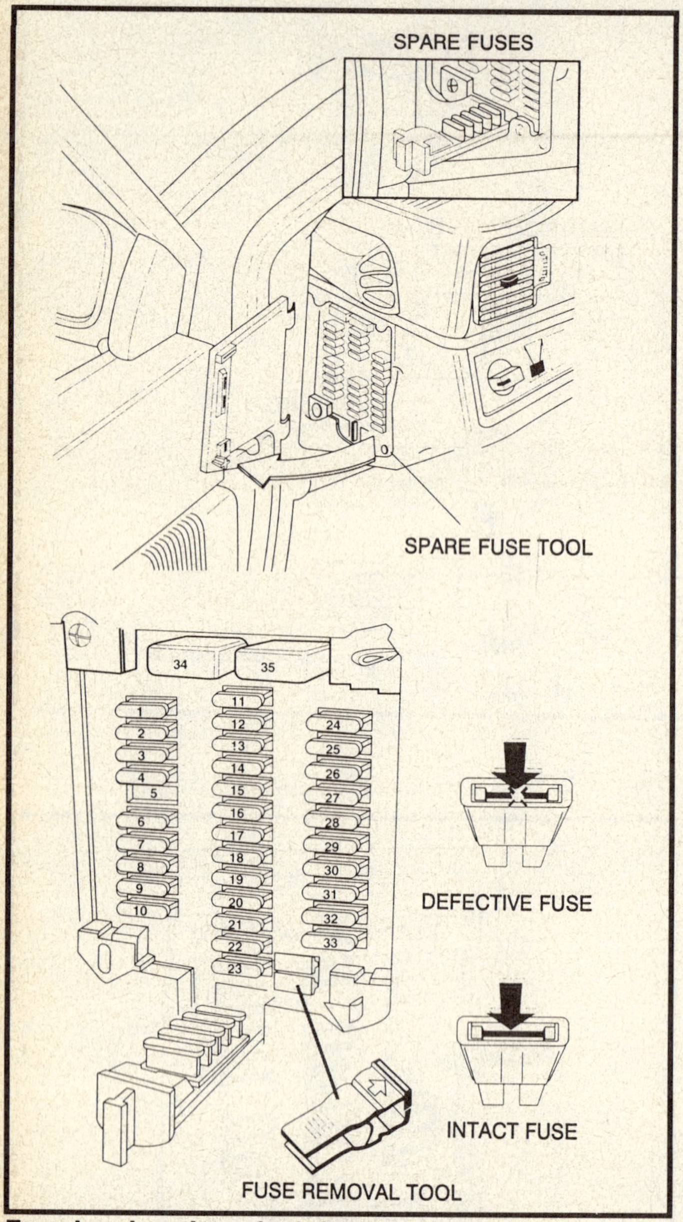

Fuse box location—far left side of the dash board on 960

READING CODES

Control Function 1

Faults stored in the memory are read through a series of flashes from the diagnostic socket LED. All fault codes have 3 digits, each capable of ranging from 1 through 9. The fault codes are read from the series of flashes delivered by the LED. Each code requires 3 series of uninterrupted flashes, with a 3 second interval between each series of flashes to make the codes easy to read.

The following procedure is for retrieving fault codes.

1. Open the diagnostic socket cover in the left side rear of the engine compartment and install the selector plug into socket No. 2 for ignition or fuel system codes, or into socket No. 6 for EGR system codes on California vehicles.
2. Turn ignition to the **ON** position.
3. Enter Control Function 1 by pressing the button once. Hold the button for at least 1 second, but not more than 3 seconds.
4. Watch the LED and count the number of flashes in each of the 3 flash series, which will indicate a fault code. The flash series are separated by 3 second intervals. Note the fault codes.

NOTE: If there are no fault codes in the diagnostic unit, the LED will flash 1-1-1 and the fuel system is operating correctly.

5. Depress the button again. If the same code is repeated, there are no additional codes stored. If the code is different, depress the button a third time and record the code, if different.

NOTE: The diagnostic memory is full when it contains 3 fault codes. Until those 3 are rectified and the memory erased, the system cannot give information on any other problems.

Control Function 2

This function is activated by pressing the button twice, holding the button for at least 1 second each time, but not more than 3 seconds each time and also allowing 1–3 seconds between operations, causing the LED to flash. The control unit delivers a diagnostic code (not a fault code) acknowledging the receipt of a signal from each of the following items:

Diagnostic Code 3-3-2: Throttle switch, when the throttle is moved from the idling position

Diagnostic Code 3-3-3: Throttle switch, when the throttle is moved from the full-load position

Diagnostic Code 1-2-4: Start inhibitor switch, via the transmission control unit, when the gear selector lever is moved from 1 of the drive positions to the **N** or **P** position

Diagnostic Code 1-4-1: Timing pick-up, when the starter motor is running

Diagnostic Code 1-1-4: Electronic Climate Control (ECC) control panel, when the A/C button is depressed or released

Diagnostic Code 1-3-4: Relay in ECC power unit and, as a result, A/C compressor electromagnetic clutch, when A/C is switched **ON**

Control Function 3

This function is activated by pressing the button 3 times, holding the button for at least 1 second each time, but not more than 3 seconds each time and also allowing 1–3 seconds between operations. The control unit then responds by activating, in order:

1. Radiator fan at half speed for 3 seconds
2. Radiator fan at full speed for 3 seconds
3. Injectors at 13 Hz
4. Idling (CIS) valve at 1 Hz
5. Relay in ECC power unit and, as a result, A/C compressor electromagnetic clutch, at 1 Hz.

This activation is repeated twice.

CLEARING CODES

When all the fault codes have been read and corrected, the diagnostic system memory can be erased as follows:

1. Turn ignition switch to the **ON** position.
2. Read fault codes again.
3. Press and hold diagnostic socket button for approximately 5 seconds. Release button. After 3 seconds, the LED should light.
4. While the LED is lit, press button again for approximately 5 seconds. Release button. The LED should go out. Memory is cleared.

NOTE: To ensure that the memory has been erased, press button again for 1 second but not more than 3 seconds. The LED should flash 1-1-1, indicating that the fuel system is operating properly.

FUEL SYSTEM TESTING

Fuel Pressure Relief

1. With the ignition **OFF**, remove fuel pump fuses 30 and 31.
2. Try starting and allowing the vehicle to run a few times until it will not stay running any more.

Fuel Pressure Testing

1. Turn the ignition switch **ON**. The pumps should run for 3 seconds.
2. Turn the ignition switch **OFF**.
3. Connect a fuel gauge between the fuel line and the distribution manifold.
4. Remove fuses 30 and 31.
5. Connect a jumper lead across the right-hand terminals in the fuse cavities. The fuel pumps should now start and be clearly audible.
6. Measure the fuel pressure, which should be 43.5 psi (300 kPa).

LINE PRESSURE TOO HIGH

1. Disconnect the jumper lead between the fuse terminals.
2. Disconnect and blow air through the pressure regulator return line.
3. Disconnect and blow through the vacuum line.
4. If both lines are clear, replace the pressure regulator and recheck the pressure.

LINE PRESSURE TOO LOW

1. Clamp the return line by hand and check whether the pressure rises.

CAUTION

Do not allow the pressure to exceed 87 psi (600 kPa) to avoid fire and personal injury.

2. If the pressure rises rapidly, the pump and lines are fault-free. Replace the regulator and recheck the line pressure.
3. If the pressure rises slowly, the fuel filter, pump strainer or lines are blocked.
4. If the pressure does not rise, replace the fuel pump.

Fuel Pressure Regulator Testing

1. Connect a vacuum pump to the regulator.
2. Check that the line pressure falls by the same amount as that in the regulator.
3. Disconnect the jumper, check and install the fuses.
4. The residual pressure should not fall below 29 psi (200 kPa) in less than 20 minutes.

CAUTION

If removed, the plastic tape securing the fuel lines must be resecured to prevent fire and personal injury.

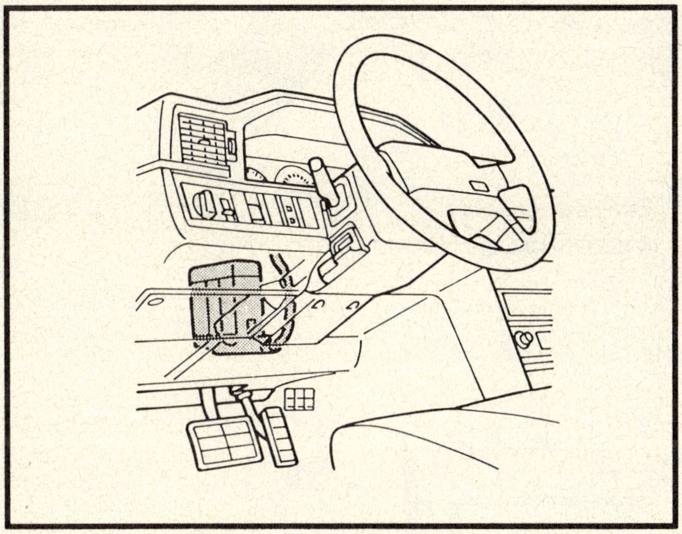

Motronic control unit location—960

Fuel pump relay location—960

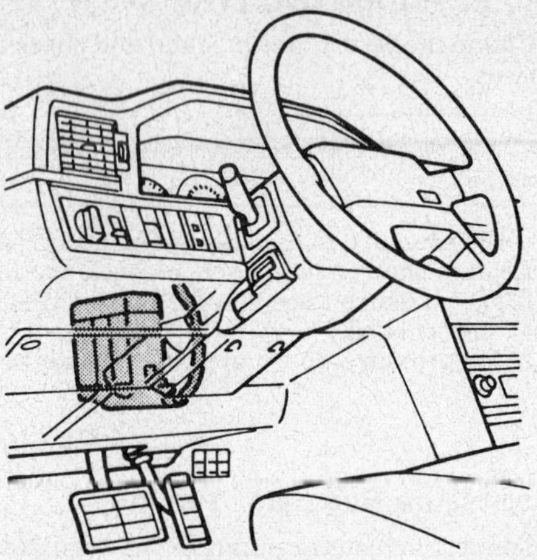

Control unit (Motronic 1.8)

INSIDE PANEL ON DRIVER'S SIDE

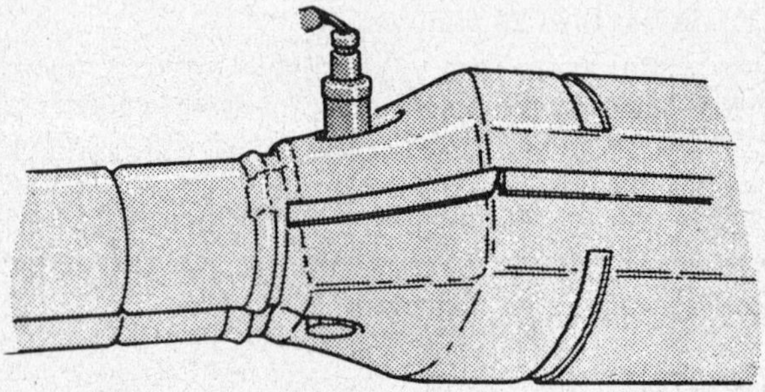

Oxygen sensor (Lambdasond)

INSTALLED IN CATALYTIC CONVERTER

Fuel pump relay

IN ELECTRICAL DISTRIBUTION UNIT

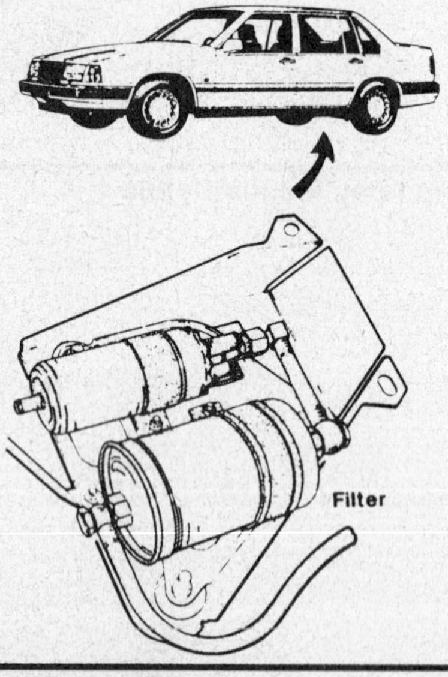

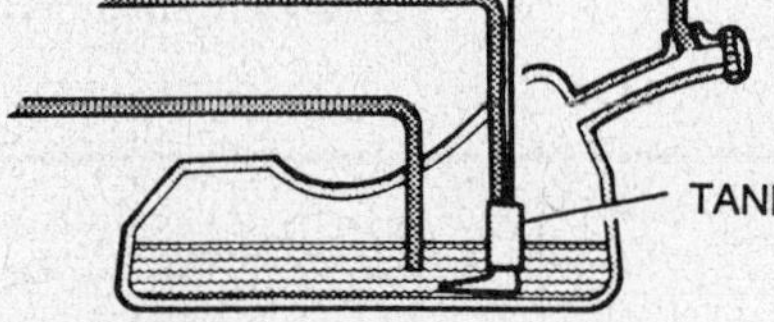

Tank pump

INSTALLED IN TANK (AS PART OF FUEL LEVEL TRANSMITTER ASSEMBLY)

Main fuel pump and fuel filter

ON SHELF UNDERNEATH CAR, UNDER REAR SEAT

Component locations—960

Engine compartment

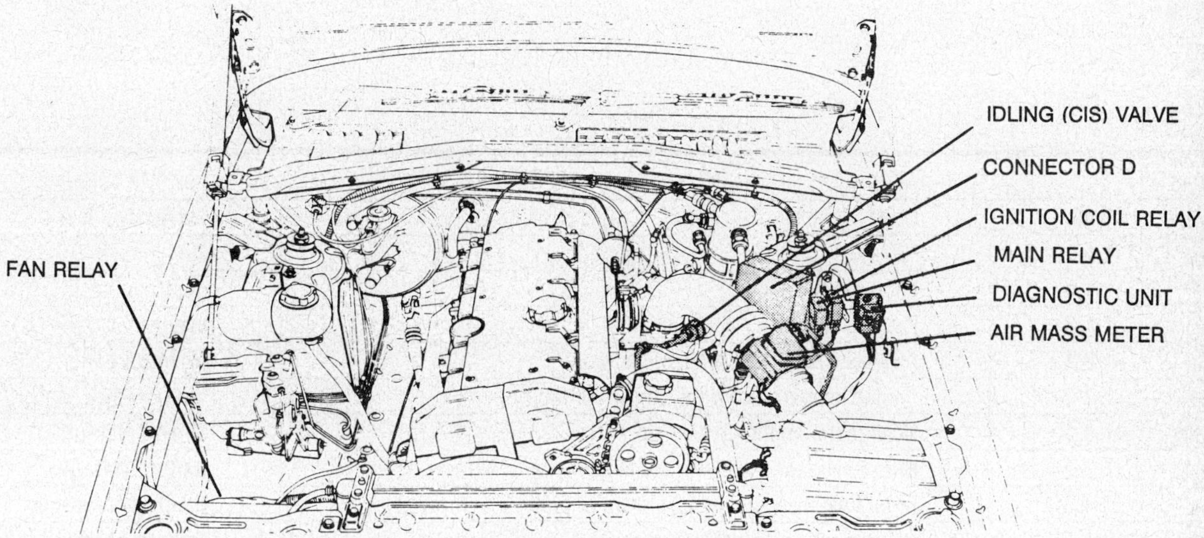

Rear of engine

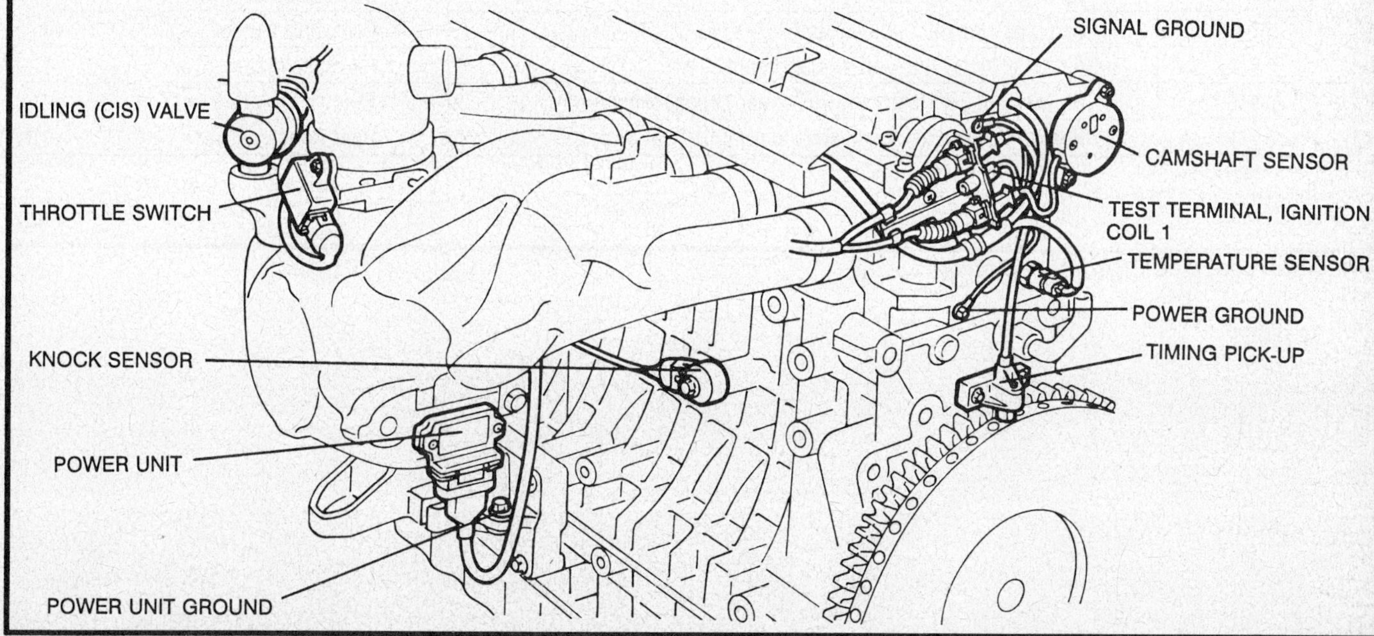

Component locations—960

FUEL PUMPS

Testing

UNDER-CAR PUMP

The under-car pump discharge capacity at a line pressure of 43.5 psi (300 kPa) and a temperature of 68°F (20°C) is 34 oz. (1 liter) in 30 seconds.

The current draw of the under-car pump at a line pressure of 43.5 psi (300 kPa), a temperature of 68°F (20°C) and at 12 volts is 3–4 amps.

TANK PUMP

The current draw of the tank pump is 6.5 amps.

Code	Fault	Correction Test
1-1-1	No faults	—
1-1-2	Fault in control unit	Control unit
1-1-3	Fault in injector	Injector
1-2-1	Air mass meter signal absent or faulty	Air mass meter
1-2-3	Engine temperature signal absent or faulty	Temperature sensor
1-3-1	No signal from timing pick-up	Timing pick-up
1-3-2	Battery voltage too low or too high ①	Battery voltage
1-4-3	Front knock sensor signal absent or faulty	Front knock sensor
2-1-2	Oxygen sensor signal absent or faulty	Oxygen sensor
2-1-4	Timing pick-up signal absent intermittently	Timing pick-up
2-2-3	Idling valve signal absent or faulty	Idle valve
2-3-1	Adaptive lambda control lean or rich in part-load range	Oxygen sensor
2-3-2	Adaptive lambda control lean or rich at idling	Oxygen sensor
2-3-3	Adaptive idling control outside control range	Idle speed
2-4-1	EGR system malfunction ②	EGR vacuum booster
2-4-3	Signal from throttle switch absent or faulty	Throttle switch
3-1-1	Speedometer signal absent	Speed sensor
3-1-4	Camshaft sensor signal absent	Camshaft sensor
3-2-2	Air mass meter burn-off signal absent	Air mass meter
4-3-1	Signal from EGR temperature sender absent or faulty ②	EGR temperature sensor
4-3-3	Rear knock sensor signal absent or faulty	Rear knock sensor

① Check charging system
② California

Correction test descriptions—Motronic

CONTROL UNIT

Testing

The Motronic control unit checks its program when the engine is switched **OFF**, whereas the knock control circuit is checked while driving. The fault code 1–1–2 is displayed if this check reveals that a fault is present.

1. Clear the fault code.
2. Start and shut off the engine a few times.
3. Check whether the fault code is redisplayed.
4. Drive the vehicle at high load with the engine speed exceeding 3000 rpm to check the knock sensor circuit.
5. Replace the control unit if the fault code recurs.

NOTE: The power to the control unit must be completely dissipated when connecting or disconnecting the connector or measuring unit, otherwise the component may be damaged. Remove fuse No. 24 when connecting or disconnecting.

INJECTOR

Testing

If the difference in current consumption between the injector groups is excessive, the control unit will interpret the condition as an injector open-circuit or short-circuit, and will store the fault code. The entire group will be isolated in the event of a short-circuit, while the injectors not affected will continue to operate in the event of an open-circuit. Lambda control will become inactive once the fault code is reported.

CONTROL FUNCTION 3 CHECK

Check the operation of the injectors in Control Function 3.

INJECTOR INPUT SIGNALS CHECK

1. Connect a voltmeter across terminals 16 and 24 of the control unit. The voltmeter should indicate battery voltage with the ignition switch **ON**.
2. Connect a voltmeter across terminals 17 and 24 of the control unit. The voltmeter should indicate battery voltage **ON**.

RESISTANCE ACROSS INJECTOR GROUPS CHECK

1. With the ignition switch **OFF**, connect an ohmmeter across terminals 16 and 37. The resistance should be approximately 5.3 ohms.
2. With the ignition switch **OFF**, connect an ohmmeter across terminals 17 and 37. The resistance should be approximately 5.3 ohms.

INDIVIDUAL INJECTOR RESISTANCE CHECK

1. With the ignition switch **OFF**, measure the resistance of each injector.
2. If the resistance of each injector is not 16 ohms at 68°F (20°C), replace the injector.

AIR MASS METER

Testing

Since the injection period is largely determined be the computed air mass, the system is dependent on this signal. If the signal is absent, and thus setting Fault Code 1–2–1, the injection period is computed on the basis of the throttle switch angle (Limphome Mode). Oxygen feedback control and idling control are disabled under these conditions.

Burn-off to clean the air mass meter element should occur if the engine speed has exceeded 1800 rpm and the coolant temperature has exceeded 140°F (60°C). Approximately 4 volts is delivered to the platinum wire in the air mass meter from control unit terminal 5, returning to terminal 7. The fault code 3–2–2 is recorded if the voltage between ground and terminal 7 does not exceed 2.5 volts.

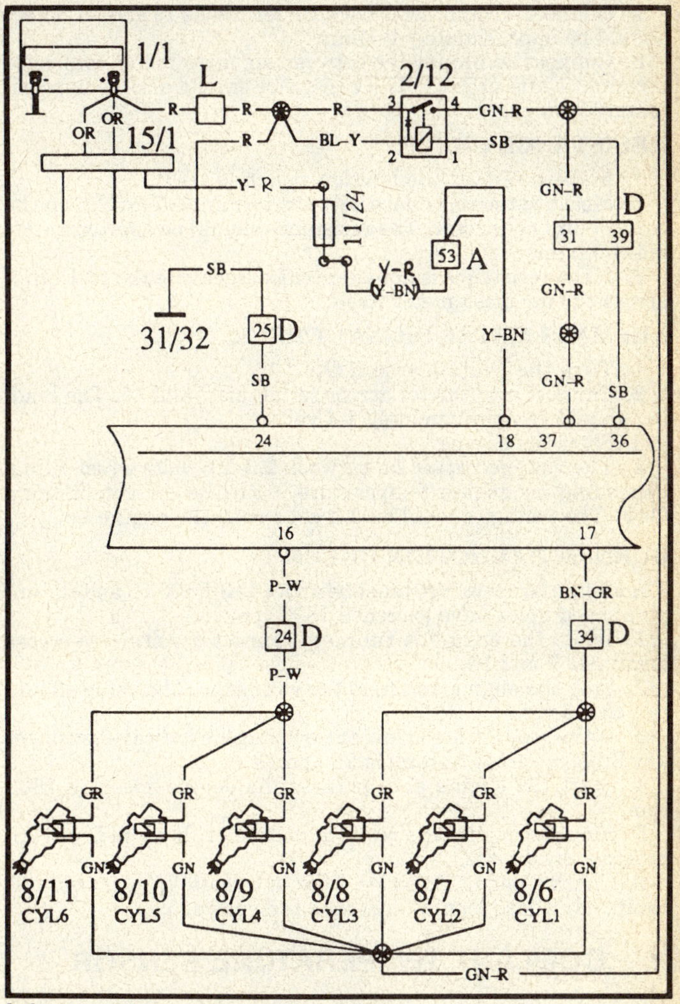

Injector wiring schematic—960

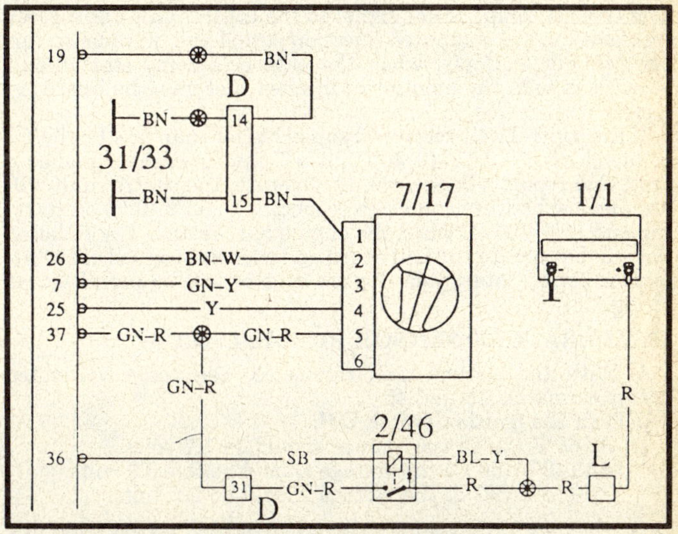

Air mass meter wiring schematic—960

AIR MASS METER GROUND CHECK

1. Make sure that the ignition switch is **OFF**.

2. Connect an ohmmeter between terminals 19 and 26. There should be approximately 0 ohms.
3. Connect an ohmmeter between air mass meter connector terminal 1 and ground. The resistance should also be approximately 0 ohms.

RESISTANCE CHECK

1. Make sure that the ignition switch is **OFF**.
2. Measure the resistance between terminals 7 and 26 on the control unit connector. The resistance should be approximately 2.5–4.0 ohms.
3. If the resistance is not as specified, measure the resistance across the air mass meter itself.

AIR MASS METER SIGNAL CHECK

1. Turn the ignition switch **ON**.
2. Connect a voltmeter across terminals 7 and 19. The reading should be approximately 1.4 volts.
3. Start the engine.
4. The voltage should be between 2.3–2.6 volts when idling, depending on engine temperature, with the air conditioning OFF. The voltage should be 2.3 volts with the engine hot.

BURN-OFF FUNCTION CHECK

The engine temperature must be above 140°F (60°C) and the engine speed must have exceeded 1800 rpm.

1. While the engine is running, connect a voltmeter across terminals 7 and 19.
2. Stop the engine and read the voltmeter. The value should rise to 2.5 volts.
3. If the reading is too low or no voltage is indicated, connect a voltmeter across terminals 25 and 19.
4. Start the engine and increase the engine speed to 1800 rpm.
5. Stop the engine and read the voltmeter. It should read approximately 4 volts.
6. Check Control Function 1 for automatic transmission (if equipped) if load signal fault code is displayed.

COOLANT TEMPERATURE SENSOR

Testing

The coolant temperature signal to the control unit has a great influence on the computed injection period and on radiator fan control. For example, when the engine is being started and when it is cold, the amount of injected fuel must be relatively large.

If the control unit receives a signal higher than 302°F (150°C) or lower than −40°F (−40°C), it will interpret the signal as a fault and report a fault code. In this case, the control unit will assume a substitute value corresponding to 32°F (0°C) on starting and 68°F (20°C) when the engine has started. The radiator fan will be operated at half speed and adaptive control of the injection period interupted. Knock control will be permitted at idling.

TEMPERATURE SENSOR SIGNAL CHECK

1. With the control unit connected, connect a voltmeter across terminals 30 and 45.
2. Turn the ignition switch **ON**.
   At 68°F (20°C) the voltage should be 1.7 volts.
   At 122°F (50°C) the voltage should be 0.8–3.9 volts.
   At 194°F (90°C) the voltage should be 2.2 volts.

NOTE: At 122°F (50°C), the control unit alters the voltage range for better signal resolution.

ABSENT OR FAULTY SIGNAL CHECK

1. Make sure that the ignition switch is **OFF**.
2. Remove fuse No. 24 before disconnecting the control unit, then disconnect it.

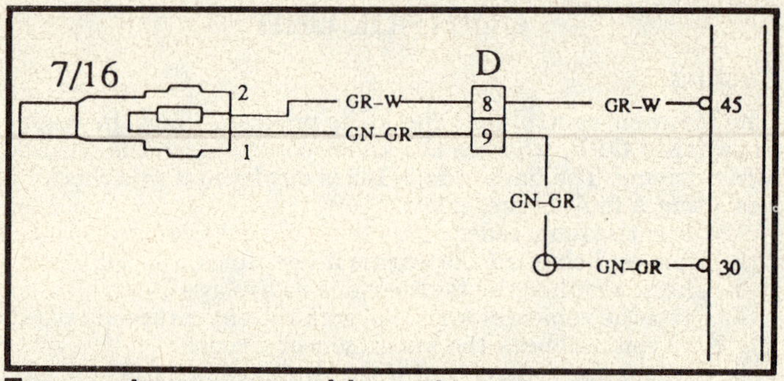

Temperature sensor wiring schematic—960

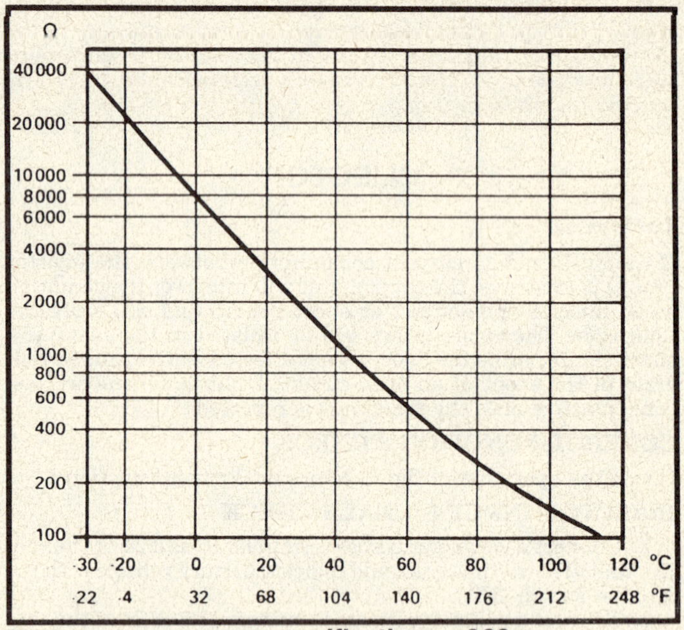

Temperature sensor specifications—960

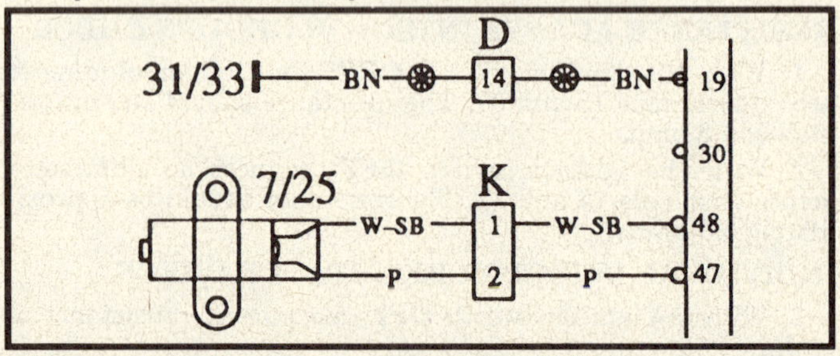

Timing pick-up wiring schematic—960

3. Measure the resistance between terminals 30 and 45. If the sensor does not match specifications at several different temperatures, replace the temperature sensor.
4. Reconnect the control unit and reinstall the fuse.

TIMING PICK-UP

Testing

Fault Code 1-3-1 is reported if no signal is received from the camshaft sensor pick-up for 10 engine revolutions. The engine will not start as a result of this.

Fault Code 2-1-4 is reported only if a fault occurs when the engine is running. The code is set if what the control unit computes and what it actually sees are different. A slight jerk may be a symptom felt in this event.

TESTING IN CONTROL FUNCTION 2

1. Enter Control Function 2 and operate the starter motor.
2. Diagnostic Code 1–4–1 should be displayed, indicating a timing pick-up signal.

TIMING PICK-UP SIGNAL CHECK

1. Connect a voltmeter across terminals 47 and 48, and set on the AC millivolt scale.
2. Operate the starter motor and check that the voltmeter displays 300–400 millivolts.

ABSENT OR FAULTY SIGNAL CHECK

1. With the ignition switch **OFF**, remove fuse No. 24 and disconnect the control unit.
2. Connect an ohmmeter between terminals 47 and 48.
3. Check that the ohmmeter reads 200–400 ohms, depending on the temperature.
4. Reconnect the control unit.
5. Measure the resistance between terminals 48 and 19. The resistance should be 0 ohms.
6. Measure the resistance between terminals 48 and 30. The resistance should be 0 ohms.
7. On vehicles with automatic transmissions, check in Control Function 1 if speed signal fault is displayed.

BATTERY VOLTAGE

Testing

Fault Code 1–3–2 will set if the battery voltage falls below 8 volts or exceeds 16 volts for 5 seconds after the engine has started.

VOLTAGE CHECK

1. Connect a voltmeter across terminals 18 and 19 of the control unit.
2. Start the engine and read the battery voltage.
3. Check the battery and charging system if the voltage is in abnormal range.

KNOCK SENSORS

Testing

Fault Codes 1–4–3 and 4–3–3 indicate that the front and rear knock sensor signals, respectively, are absent or faulty. The fault codes will be reported if the following 2 conditions are fulfilled:

1. The engine speed is at least 2970 rpm and a certain engine load has been exceeded.
2. The signal from either of the knock sensors to the control unit falls below 255 millivolts during 96 successive ignitions. The control unit will then safety-retard the timing.

WIRING CHECK

The knock sensors themselves cannot be checked; only the wiring can.

1. Disconnect the knock sensor connector and connect a jumper wire between the connector terminals.
2. Remove fuse No. 24 and disconnect the control unit.
3. Connect an ohmmeter between terminal 2 and terminal 11 or 29, depending on which knock sensor is affected. The reading should be 0 ohms.
4. Replace the knock sensor if the reading is correct. Do not replace it if the reading is incorrect. Torque the sensor to 15 ft. lbs. (20 Nm).
5. Reconnect the control unit.

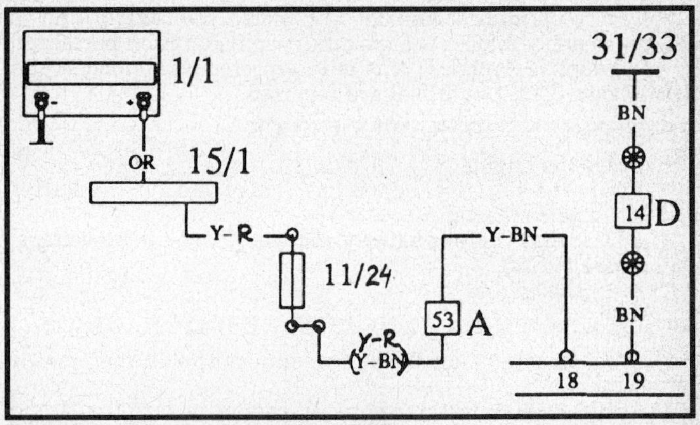

Battery wiring schematic—960

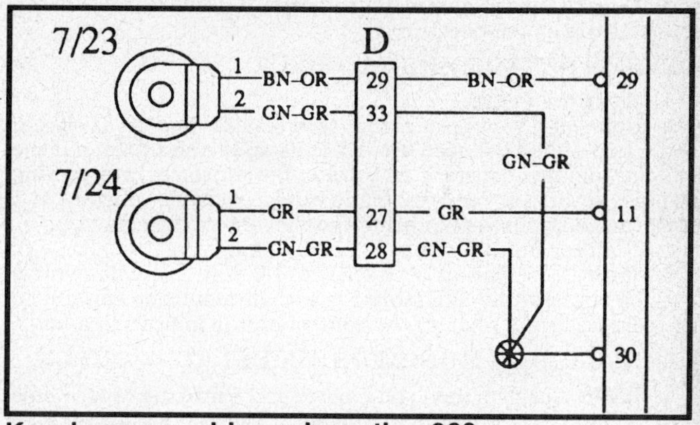

Knock sensor wiring schematic—960

Testing knock sensor circuit—960

OXYGEN SENSOR

Testing

Fault Code 2–1–2 indicates that the oxygen sensor signal is absent or faulty. The control unit cannot exercise Lambda (feedback) control if the signal is absent. The fault code will be reported if the sensor signal is constantly below 0.9 volts or above 1.1 volts, or varies from its mid-point value (0.5 volt) for 3 minutes after the engine temperature has exceeded 167°F (75°C) and a certain load has been maintained. Lambda control will become inactive under these conditions.

The control unit receives information from the oxygen sensor on whether the fuel/air mixture is too rich or too lean in the part-load and idling ranges. The control unit adaptive system compensates for this by making the mixture leaner or richer as required to maintain Lambda = 1. When the control unit is obliged to apply almost the maximum compensation permitted by the adaptive function, this is interpreted as abnormal and Fault Codes 2-3-1 or 2-3-2 are reported.

OXYGEN SENSOR SUPPLY CHECK

1. Start the engine.
2. Connect a voltmeter across terminals 1 and 2 in the starter motor 2-pole connector.
3. Check that there is battery voltage. If there is no voltage, check fuse No. 30.
4. Stop the engine.

OXYGEN SENSOR PREHEATER RESISTOR CHECK

1. With the ignition **OFF**, disconnect the starter motor connector.
2. Connect an ohmmeter between connector terminal 1 and terminal 2.
3. Check that the ohmmeter indicates 3–13 ohms, depending on the resistor temperature.

OXYGEN SENSOR SIGNAL CHECK

1. Start the engine.
2. Connect a voltmeter across terminals 28 and 30. It should begin to oscillate between 0.1–1.0 volts after a short time. If the voltage remains constant at 0 volts, the sensor is interpreting the fuel/air mixture as lean. If the voltage remains constant at 1 volt, the sensor is interpreting the fuel/air mixture as rich.
3. If either Fault Code 2-3-1 or 2-3-2 is displayed although the meter is oscillating between 0.1–1.0 volt, Lambda control may be successfully adjusting the carbon monoxide content to the correct level although the control unit is indicating a fault.

BASIC CARBON MONOXIDE ADJUSTMENT CHECK

1. Insert a carbon monoxide meter probe into the tapped hole on the catalytic converter by means of special connector tool 5410, or equivalent.

CAUTION

To avoid personal injury, be aware that the catalytic converter area may be extremely hot.

2. Run the engine up to normal operating temperature.
3. Disconnect the oxygen sensor to ensure that feedback control does not correct for the fault that has been set.
4. Reset adaptivity by displaying, recording and clearing the fault codes. The adaptive system is reset when the diagnostic system memory is erased.
5. Check the carbon monoxide content after the memory has been erased.
   a. If the meter indicates low carbon monoxide content, the engine is running lean and the system is having to enrich the mixture to compensate. Check for an air leakage problem, for example.
   b. If the meter indicates high carbon monoxide content (over 2.5%), the engine is running rich and the system is having to make the mixture leaner to compensate. Check for high fuel pressure or fuel leakage, for example.

IDLE VALVE

Testing

The role of the control unit is to maintain constant idling conditions. The control unit delivers pulsed ground signals to both idling valve coils (opening and closing). If the signal is absent, or if either of the coil circuits is open or shorted, the control unit cannot exercise idling control. In this case, no ground pulses will be delivered by the control unit and the engine speed will be controlled by the spring-loaded function 6, which maintains a small valve opening at all times.

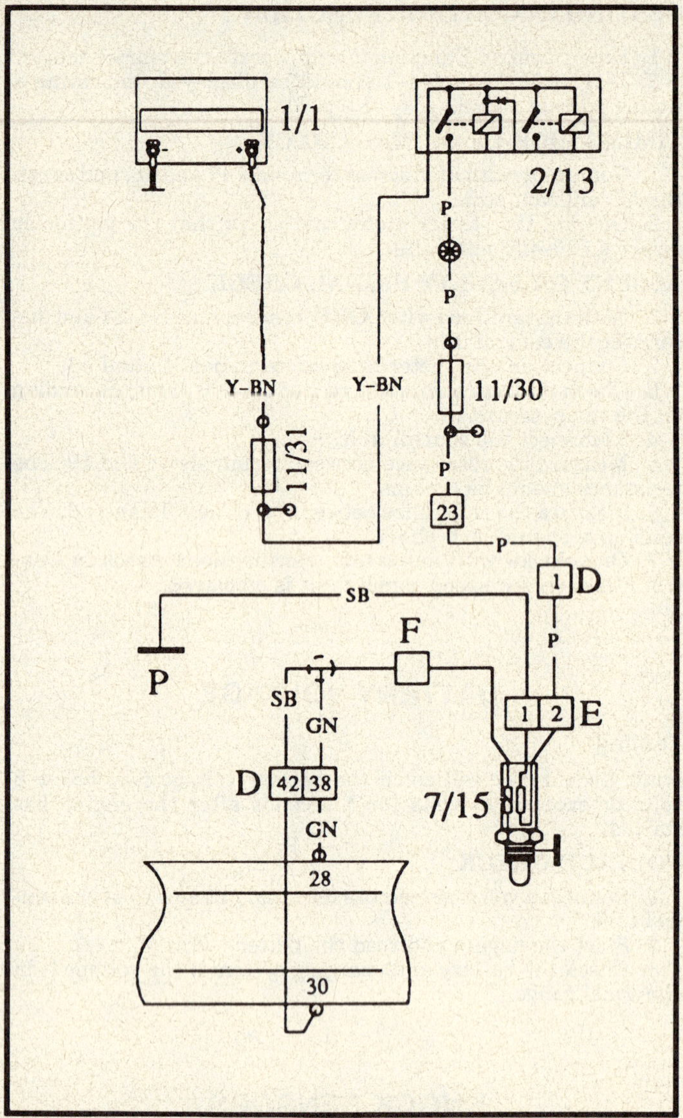

Oxygen sensor wiring schematic—960

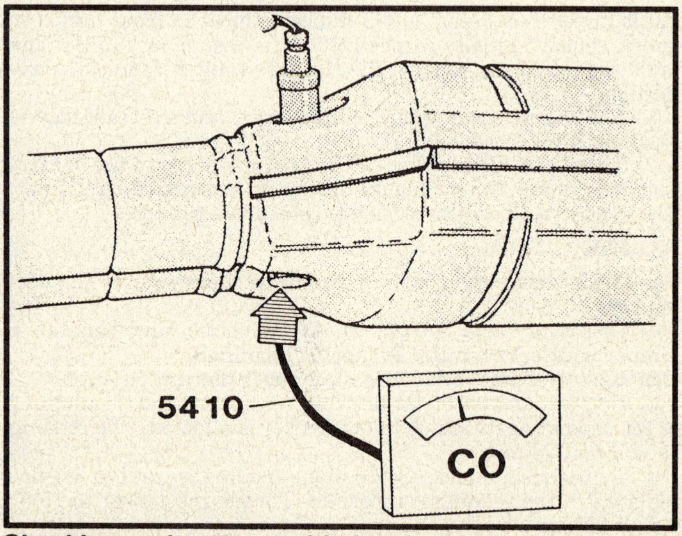

Checking carbon monoxide level—960

CONTROL FUNCTION 3 CHECK

1. In control function 3 check the operation of the idling valve. The valve should be activated by the control unit.
2. If the valve does not operate, check the wiring and the supply to and from the valve.

IDLING VALVE SIGNAL CHECK

1. Connect a voltmeter across terminals 4 and 24 of the control unit with the engine idling. There should be approximately 11 volts with the engine hot.
2. Connect a voltmeter across terminals 22 and 24 of the control unit with the engine idling. There should be approximately 7.5 volts.

ABSENT OR FAULTY SIGNAL CHECK

1. Check for voltage at terminals 4 and 22 with the ignition switch **ON**. The voltage should be approximately 8–9 volts.
2. If there is no voltage, check the supply to the valve.
3. Remove fuse 24 and disconnect the control unit.
4. With the ignition **OFF**, check the resistance across the idle valve coils at terminals 4 and 22 of the control unit. The resistance should be approximately 25 ohms.
5. Reconnect the control unit and replace the fuse.

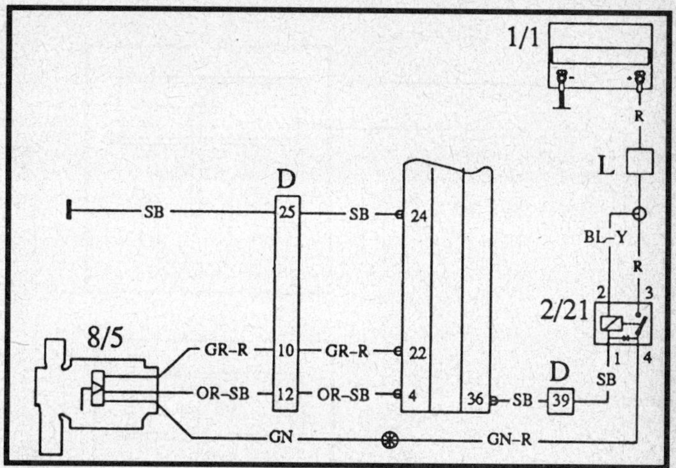

Idling valve wiring schematic—960

IDLE SPEED

Testing

If the control unit is forced to open or close the idling valve too much to maintain constant idling conditions, the control unit will interpret this as a fault and record the Fault Code 2-3-3. The most probable causes are incorrect measurement by the air mass meter or air leakage into the intake manifold downstream of the throttle housing.

1. Start and run the engine up to normal operating temperature, with the gear selector lever in **N** and the air conditioning switch **OFF**.
2. Clamping the air valve hose and observing the engine speed will indicate whether air leakage is present.
3. If the engine speed exceeds 600–700 rpm, the fault is probably due to air leakage or incorrect adjustment of the throttle disc.
4. Check for air leakage and adjust the throttle housing as required.

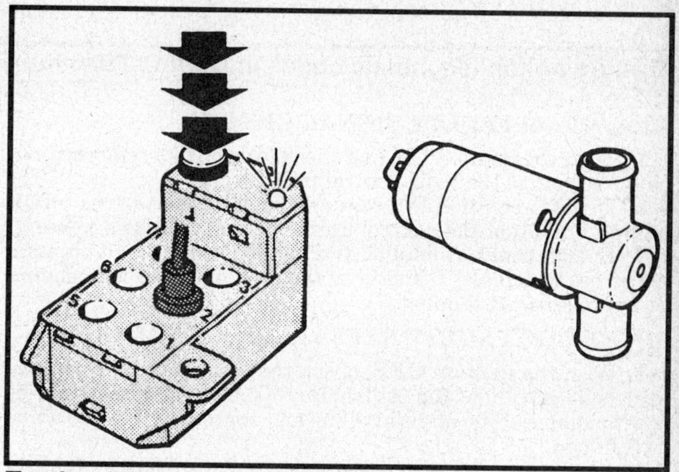

Testing idling valve in Mode 3—960

THROTTLE SWITCH

Testing

The function of the throttle switch is to detect the throttle position from fully closed to fully open. If the switch signal is constantly below 0.06 volt or above 4.9 volts, the control unit will interpret this as a fault and will report the Fault Code 2-4-3. Under these conditions, a substitute value equivalent to an opening of approximately 20 degrees will be adopted.

CONTROL FUNCTION 2 TESTING

1. Enter Control Function 2.
2. Turn the throttle slightly. If flashing Diagnostic Code 3-3-2 is displayed, the throttle function is correct in the idling position.
3. Turn the throttle to the full-load position and release. If flashing Diagnostic Code 3-3-3 is displayed, the throttle function is correct in the full-load position.

THROTTLE SWITCH SIGNAL CHECK

Turn the ignition **ON** and connect a voltmeter across terminals 30 and 53 of the control unit. Instrument should indicate approximately 0.5 volt at idling and approximately 4.5 volts at full load.

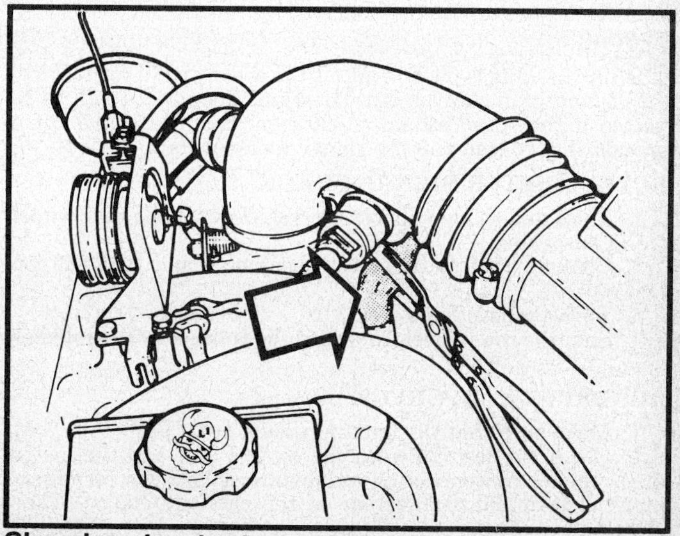

Clamping air valve hose—960

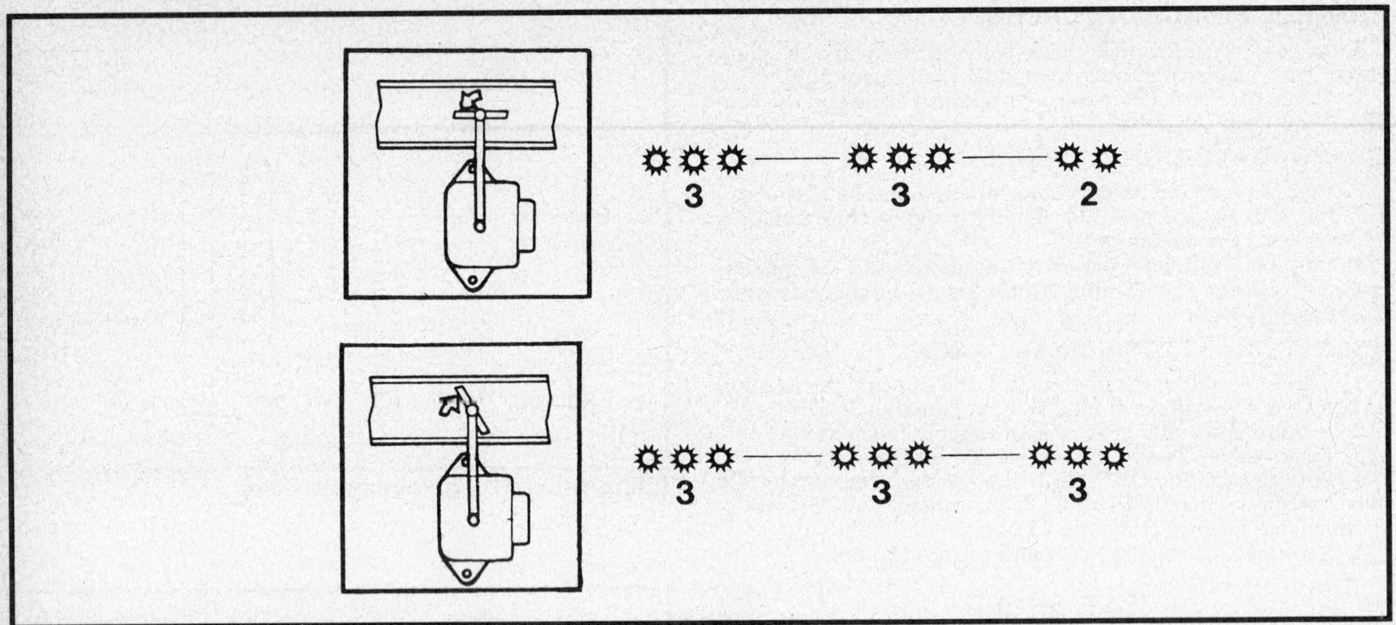

Throttle switch diagnostic codes in Control Function 2—960

ABSENT OR FAULTY SIGNAL CHECK

1. Turn the ignition **ON** and check for a 5 volts reference signal at terminal 12 of the control unit.
2. With the ignition **ON** connect the voltmeter across terminals 12 and 30 of the control unit. There should be 5 volts.
3. Ensure that terminal 30 is grounded. Connect an ohmmeter across terminals 19 and 30 of the control unit. There should be approximately 0 ohms.

ABSENT OR FAULTY REFERENCE SIGNAL CHECK

1. With the ignition **OFF**, check the resistance between terminals 53 and 30 of the control unit. The resistance should be approximately 1000 ohms at idling and approximately 700 ohms at full load.
2. Check vehicles with automatic transmissions in Control Function 1 if speed signal fault code is displayed.

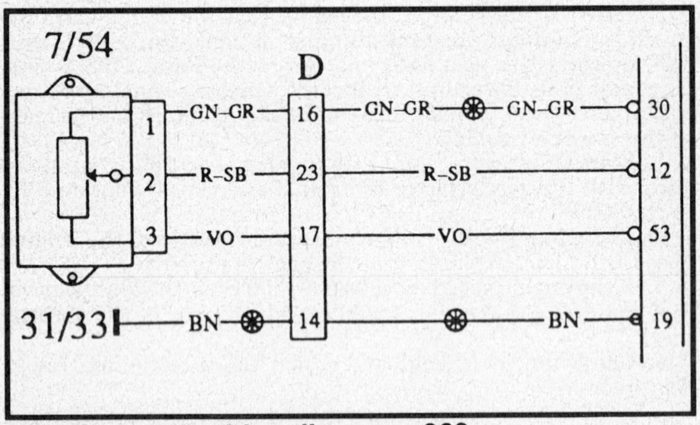

Throttle switch wiring diagram—960

SPEED SENSOR

Testing

The control unit is dependent on the speed signal for full-load enrichment, fan control, etc. The Fault Code 3–1–1 will be reported if the speed exceeds 3390 rpm, the load signal corresponds to part load and the signal is absent for 3 seconds.

SPEEDOMETER SIGNAL CHECK

1. Raise and safely support the vehicle with the rear wheels freely suspended.
2. Connect a voltmeter across terminals 9 and 19 of the control unit.
3. Turn the ignition switch **ON**.
4. Spin the rear wheels and read the voltmeter. It should oscillate between 0.1–12 volts.

ABSENT SIGNAL CHECK

1. Make sure that the ignition switch is **OFF**.
2. Check the lead between the control unit and the instrument cluster by connecting an ohmmeter between terminal 9 and the terminal at the rear of the cluster (Gn/Or). There should be 0 ohms.
3. Check the wire between the cluster and the rear axle.

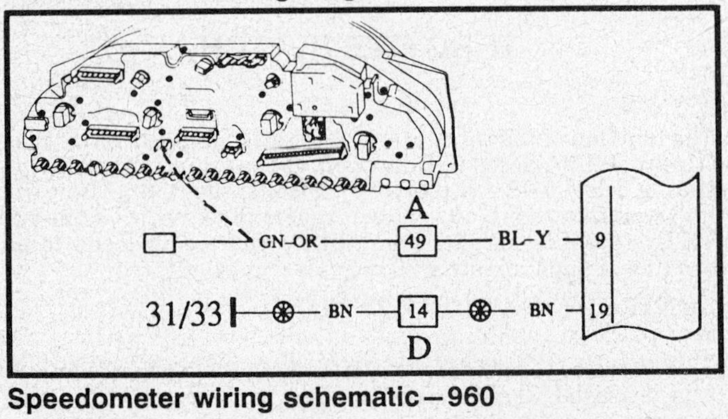

Speedometer wiring schematic—960

4. Check vehicles with automatic transmissions in Control Function 1 if speed signal fault code is displayed.

CAMSHAFT SENSOR

Testing

The camshaft sensor is mounted at the end of the camshaft and enables the control unit to determine which cylinder requires fuel and ignition. As the camshaft performs a revolution every 2 crankshaft revolutions, the sensor signal goes high on 1 crankshaft revolution and low on the next. Fault Code 3–1–4 is reported if the signal remains constantly high or low for more than 30 seconds. Under these conditions, the control unit supplies double injection and double ignition while safety-retarding the timing.

CAMSHAFT SENSOR SIGNAL CHECK

1. Remove fuse No. 31 to prevent the engine from starting.
2. Connect a voltmeter across terminals 8 and 19 of the control unit.
3. Operate the starter motor.
4. Check that the reading oscillates between 0.1–0.5 volts.

ABSENT OR FAULTY SIGNAL CHECK

1. Check that the voltage across terminals 10 and 30 of the control unit is approximately 11 volts.
2. Check that terminal 30 is grounded.
3. Connect an ohmmeter between terminals 30 and 19 of the control unit. It should read 0 ohms.

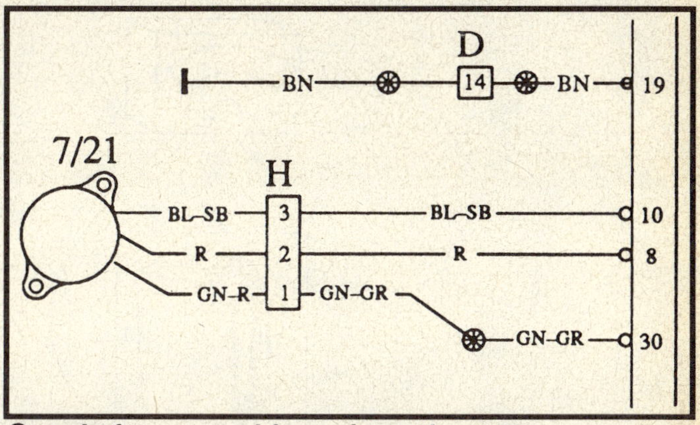

Camshaft sensor wiring schematic—960

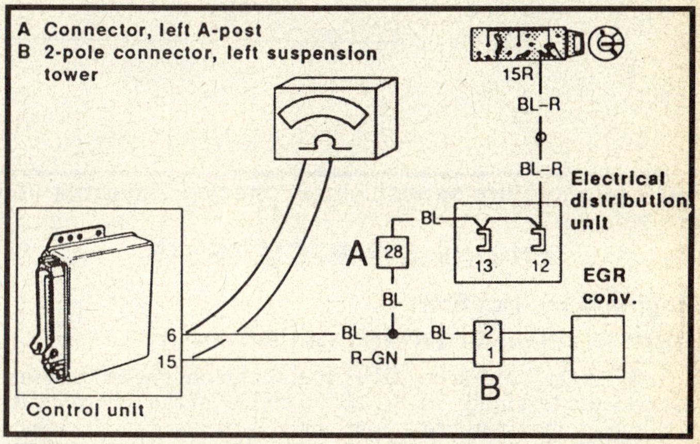

EGR converter resistance check—California vehicles

EGR VACUUM BOOSTER

Testing

An EGR system malfunction is indicated by Fault Code 2–4–1 (California vehicles). It results when the EGR temperature sensor detects that the exhaust gas recirculation flow is too low.

EGR VACUUM BOOSTER CHECK

1. Turn the ignition switch **ON**.
2. Enter Control Function 3. The lamp should flash as the EGR vacuum booster is activated.
3. Turn the ignition switch **OFF**.
4. If the EGR vacuum booster is not activated, measure the resistance between pins 6 and 15 in the control unit connector. The resistance should be approximately 75–95 ohms.
5. If the value is different, check the wiring and connection.

EGR SYSTEM OPERATION CHECK

1. Open the EGR converter connector. Remove the lead directly opposite the Black/White lead between pin 1 and the control unit. Reconnect the connector.
2. Start and warm the engine to operating temperature.
3. Ground the loose lead. If the engine starts to run unevenly, the EGR system is operating.

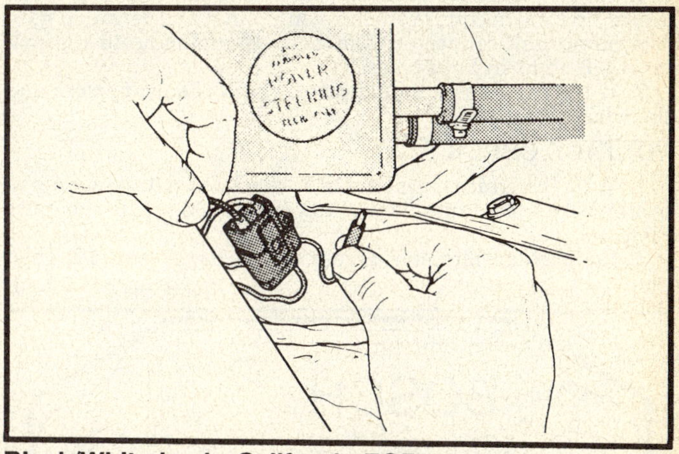
Black/White lead—California EGR system

EGR TEMPERATURE SENSOR

Testing

If the signal from the EGR temperature sensor is absent or faulty, Fault Code 4–3–1 will result (California vehicles).

NOTE: The sensor is extremely sensitive to shock and impact. When replacing, torque to 7–13 Nm (5–10 ft. lbs.).

1. Measure the resistance between terminal 14 (ground) and terminal 22 on the control unit connector. It should be 500–1000 ohms.
2. If the value is different, check the leads and connections, the ground terminal on the intake manifold and the resistance of the sensor connector.

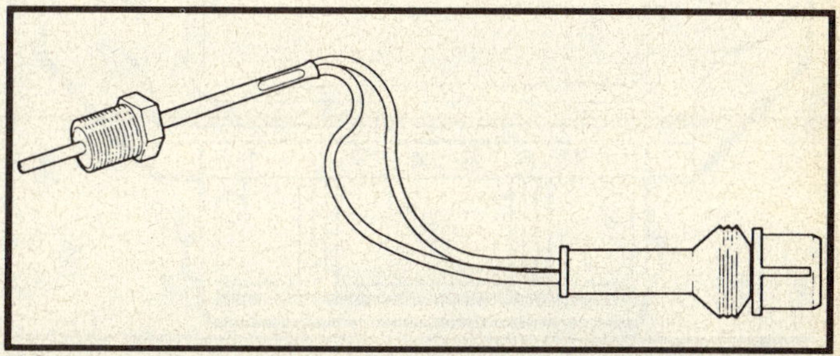
EGR temperature sensor—California vehicles

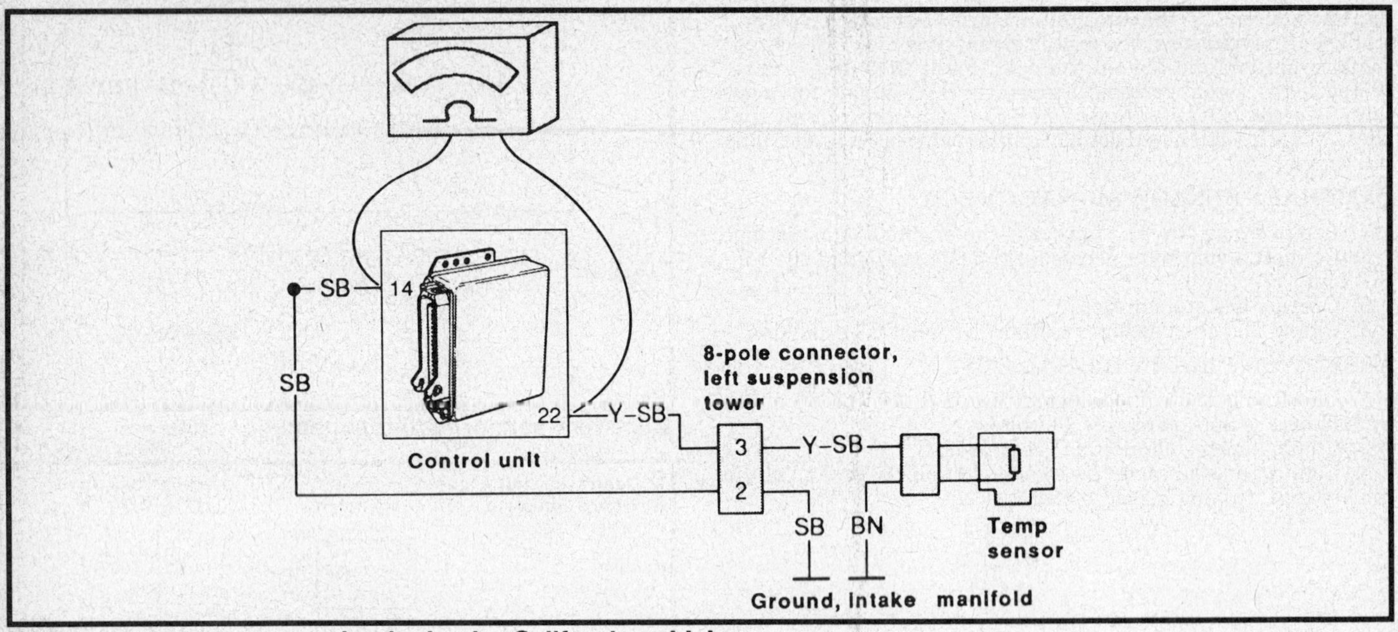

EGR temperature sensor circuit check—California vehicles

ENGINE SYMPTOM TESTS

Engine Does Not Start

GROUND CONNECTIONS CHECK

1. With the ignition key **OFF**, connect an ohmmeter between ground and terminals 19, 24, 26, 30 and 48 of the control unit. The meter should indicate approximately 0 ohms in all cases.
2. Check the power unit ground connections under each unit. Connect an ohmmeter between ground and terminal 4 on front power unit connector, then between ground and terminal 4 on rear power unit connector. The meter should indicate approximately 0 ohms in both cases.
3. If high resistance is indicated, check the connectors and the wiring.

VOLTAGE CHECK

1. With the ignition switch **OFF**, check for voltage supply to terminal 18 across fuse 24 and to terminal 36 across the main relay.
2. Connect a voltmeter across terminals 18 and 19, then

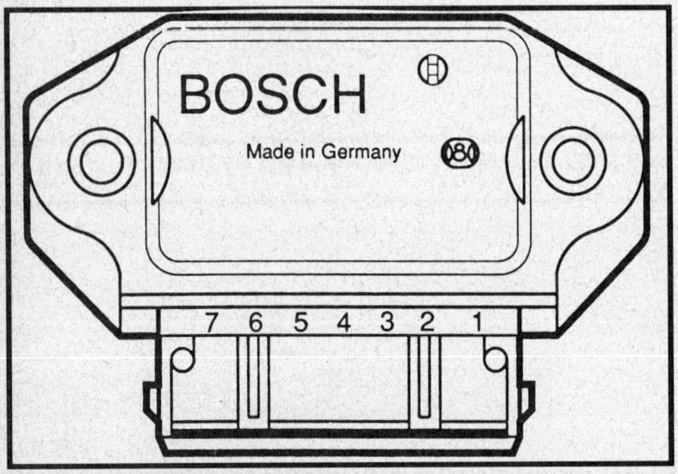

Power transistor terminals—960

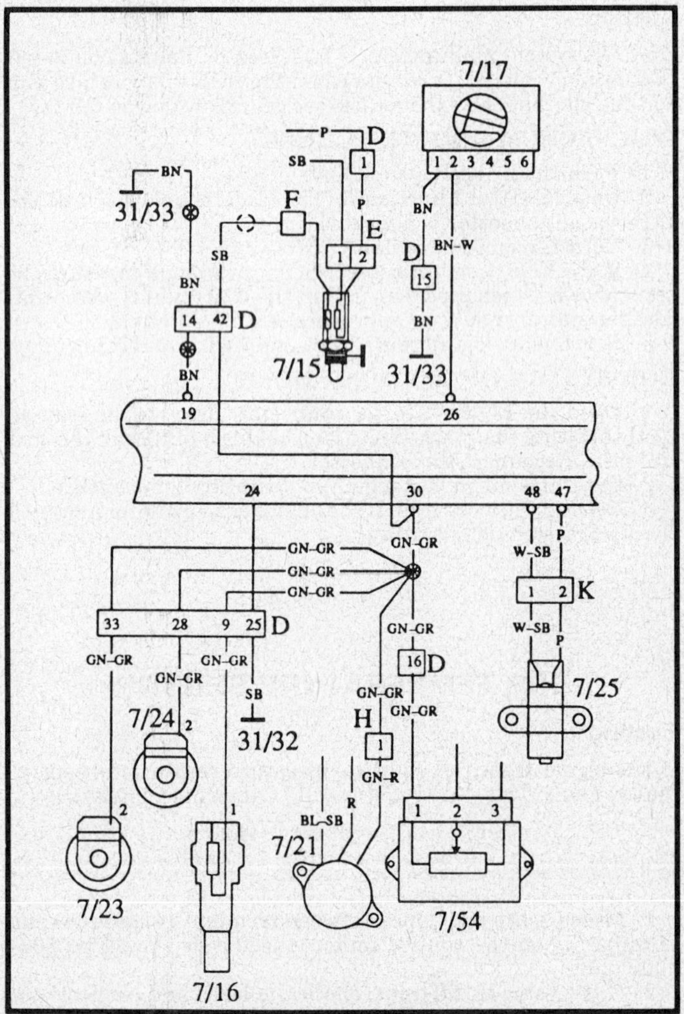

Checking ground connections—960

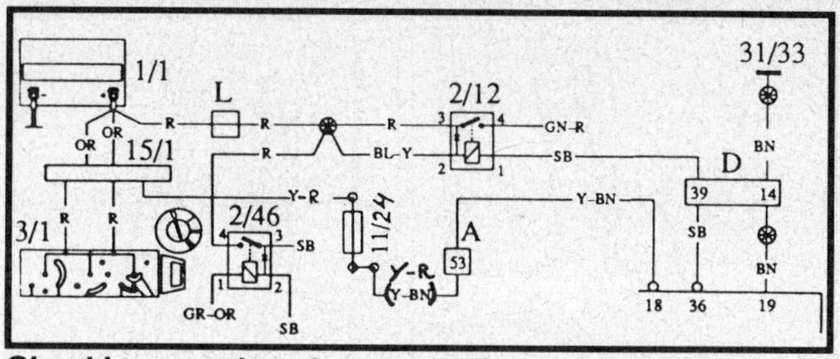

Checking supply voltages—960

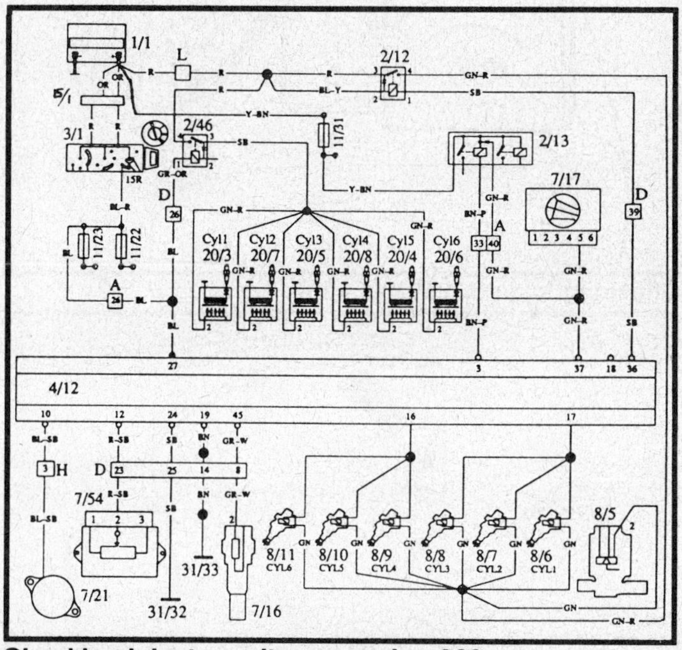

Checking injector voltage supply—960

across terminals 36 and 19 of the control unit. The voltmeter should indicate battery voltage in both cases. Check the fuse, wiring and the relay if no reading is obtained.

3. Turn the ignition switch **ON** and check that the voltage falls from battery voltage to approximately 0.1 volt at terminal 36.

FUEL PUMP OPERATION CHECK

1. Remove the fuel tank filler cap.
2. Check that the fuel pumps operate for 3 seconds when the ignition is turned **ON**, listening for noise.
3. If the pumps do not operate when the ignition is turned **ON**, connect a voltmeter across terminals 3 and 19 of the control unit. Battery voltage should be indicated.
   a. If there is no voltage, check the wiring and the supply to the pump relay.
   b. If there is voltage, check the wiring and the supply to the fuel pump.

BUSBAR NUMBER 15 SUPPLY CHECK

1. Connect a voltmeter across terminals 27 and 19, then across terminals 37 and 19 of the control unit.
2. Turn the ignition switch **ON**. The meter should indicate battery voltage in both cases.
3. Check the wiring to the control unit if no reading is obtained.

INJECTOR SUPPLY CHECK

1. Connect a voltmeter across terminals 16 and 24, then across terminals 17 and 24 of the control unit.

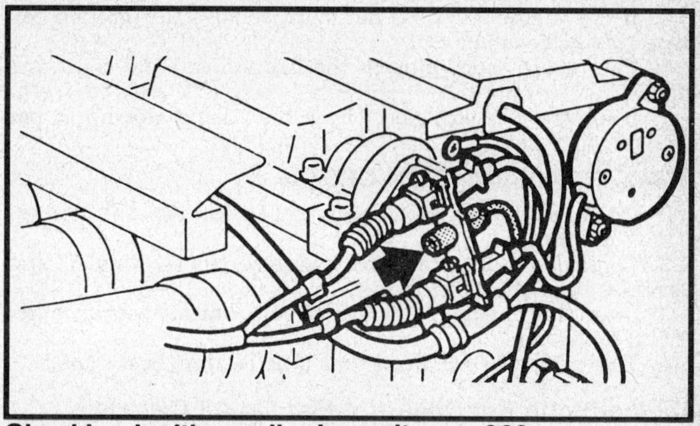

Checking ignition coil relay voltage—960

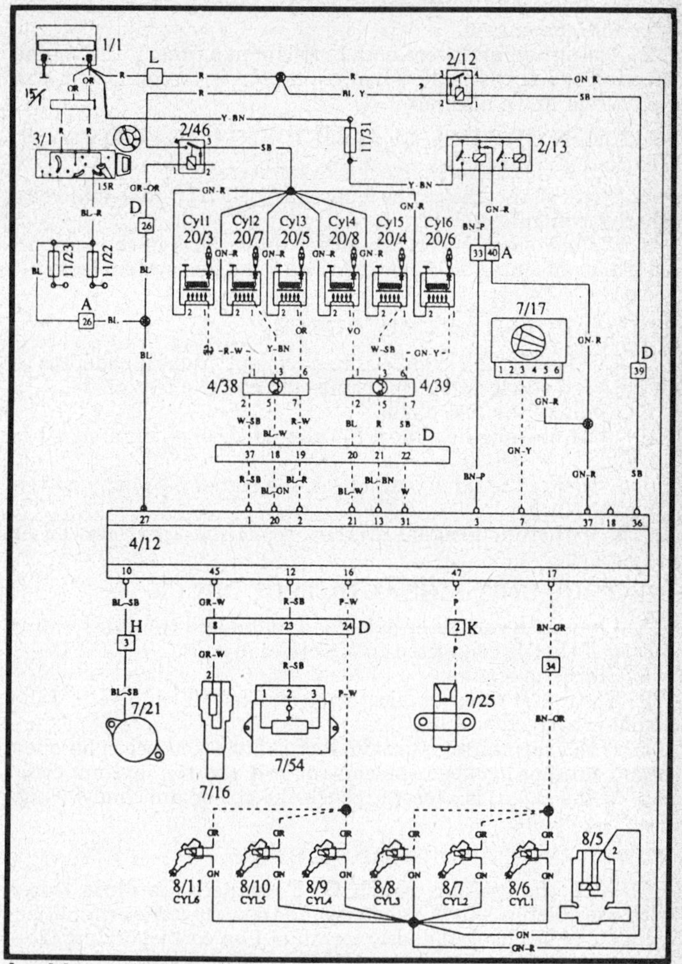

Ignition wiring circuit—960

2. Turn the ignition switch **ON**. The meter should indicate battery voltage in both cases.
3. Check the wiring and the connectors if no reading is obtained.

IGNITION COIL VOLTAGE SUPPLY CHECK

1. Turn the ignition switch **ON**.
2. Connect a voltmeter between ground and the test terminal at the rear of the engine. The reading should be battery voltage.
3. If the reading is not battery voltage, check the coil relay, wiring and the connectors up to the ignition coils.
4. Attempt to start the engine.

5. If the engine still does not start, remove the ignition coil cover and remove one coil.
6. Fit a spare spark plug in the ignition coil and crank the engine.
7. If no spark is obtained, check the timing pick-up as per Fault Code 1–3–1.

INJECTOR OPERATION CHECK

1. If the pick-up is intact, check that the control unit is operating the injectors.
2. Connect a millivoltmeter between supply terminal 37 and terminal 16 of the control unit.
3. Crank the engine. The meter should indicate approximately 500 millivolts.
4. Test with a new control unit if no reading is obtained.

Engine Starts But Does Not Fire On All Cylinders

MISSING CYLINDER IDENTIFICATION

1. Idle the engine.
2. Disconnect and reconnect 1 injector at a time. If the engine speed drops appreciably when connector is disconnected, the cylinder is firing normally.

IGNITION OR FUEL SYSTEM IDENTIFICATION

1. Use a voltmeter to check the injectors.
2. The voltage should be approximately 250–400 millivolts when the engine is hot.
3. If the voltage is correct, either the fault is in the ignition system or the injector is blocked. Check the fuel system for fault codes.

FIRING PULSE AT COIL CHECK

1. Use test diode 999–5280 or equivalent. Make an adaptor to fit between the ignition coil connector and the test diode.
2. Connect the test diode.
3. Start the engine. The LED should flash when a signal is present.
   a. If there is no signal, check the signal output from the control unit.
   b. If there is a signal present, replace the ignition coil or spark plug.

CONTROL UNIT SIGNAL OUTPUT CHECK

1. Connect a voltmeter between ground and the control unit terminal that is connected to the ignition coil.
2. Start the engine.
3. The meter should indicate between 60–100 millivolts if the signal is correct.
4. If the reading is approximately 3.5 volts, there is an open circuit present between the control unit and the ignition coil.
5. If the signal is present, check the power unit and wiring. Stop the engine.

POWER UNIT AND WIRING CHECK

1. With the ignition switch **OFF**, connect the diode tester with the positive side connected to the control unit terminal for cylinder to be checked and to terminal 1 on corresponding ignition coil connector. Also check in reverse direction.
2. Connect the diode tester with the positive side connected to ground and to terminal 1 on the ignition coil connector. Also check in the reverse direction.
3. Check the wiring if a faulty diode is found.

Idling Speed Drops When Drive Is Engaged Or When A/C Is Switched In

IDLING COMPENSATION IN DRIVE CHECK

1. Start the engine.
2. Enter Control Function 2.
3. Move the gear select lever to **D**, then to **N**, and await response.

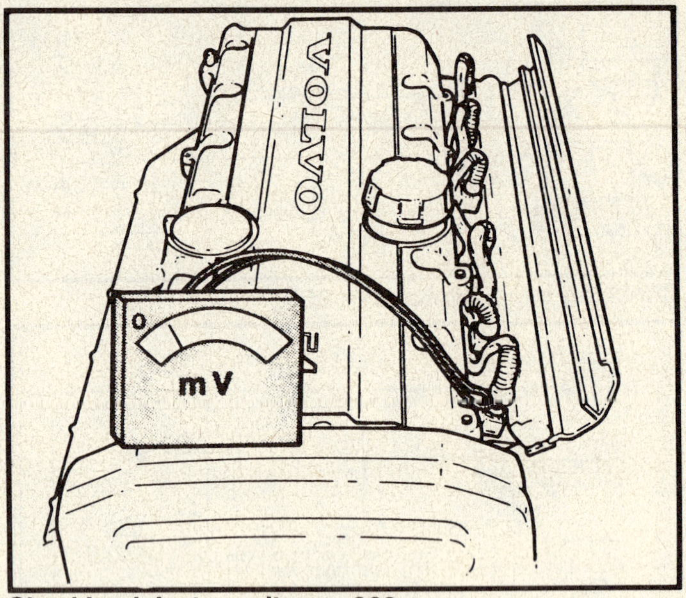

Checking injector voltage—960

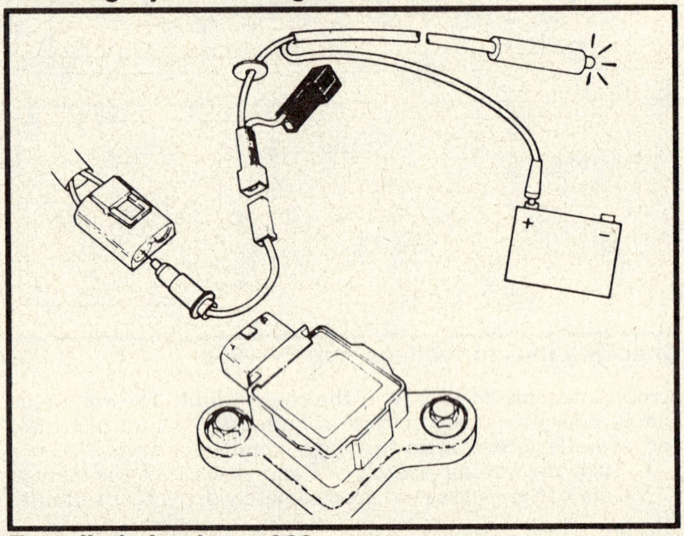
Test diode hookup—960

4. Code 1–2–4 indicates contact between the transmission control unit and the Motronic control unit.
5. If no code is displayed, check the voltage signal and the wiring.
6. Switch the ignition **OFF**.

VOLTAGE SIGNAL AND WIRING CHECK

1. Connect a voltmeter across terminals 42 and 19 of the control unit.
2. Turn the ignition switch **ON**.
3. Check the meter reading. It should be approximately 5.5 volts in **P** or **N**.
4. Check the meter reading. It should be approximately 0.5 volts in **D** or **R**.
5. If there is no signal, check the lead between the control unit and the transmission control unit.
6. If the lead is intact, check the signal from the gear selector mechanism to the transmission control unit.

AIR CONDITIONING IDLING COMPENSATION

1. Enter Control Function 2.
2. Start the engine.

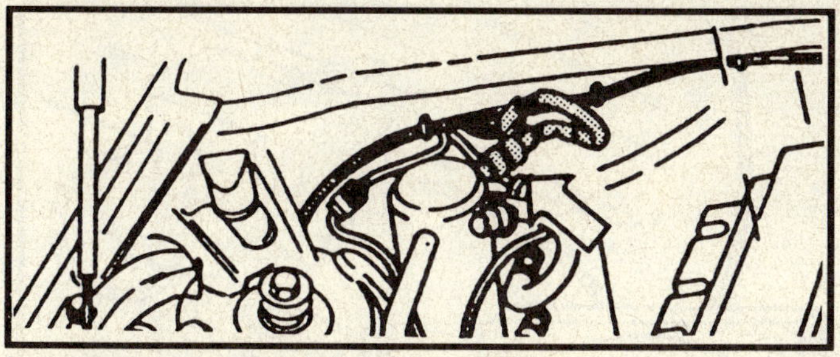
Low-pressure switch location—960

3. Disconnect the low-pressure switch connector on the dryer bottle.
4. Switch the blower motor **ON**, otherwise the A/C will not operate.
5. Depress and release the A/C button.
6. If Diagnostic Code 1–1–4 is displayed, go to step 8.
7. If no code is displayed, go to step 12.
8. Reconnect the low-pressure switch connector.
9. Diagnostic Code 1–3–4 should be displayed when the A/C compressor clutch operates.
10. If the code is not displayed or if it is displayed without the operation of the compressor, go to step 19.
11. Turn **OFF** the engine.
12. If Diagnostic Code 1–1–4 was absent, start the engine.
13. Connect a voltmeter across terminals 41 and 19.
14. Depress the A/C OFF button.
15. The instrument should indicate approximately 11 volts.
16. Release the A/C OFF button (A/C request position). The compressor should start after a brief interval and the voltage should fall drastically to a value below 1 volt.
17. If there is no voltage on A/C request, check the lead between the control unit and ECC control.
18. Turn **OFF** the engine.
19. If Diagnostic Code 1–3–4 is absent, enter Control Function 3.

NOTE: As the last operation in Control Function 3, the control unit will operate the ECC power unit relay, causing the A/C compressor clutch to engage and disengage (once per second).

20. If neither the relay nor the compressor clutch operates, check the supply to the control unit terminal 40.
21. Connect a voltmeter across terminals 23 and 19. It should indicate battery voltage.
22. If there is no voltage, check the wiring and the supply to the ECC power unit.
23. If voltage is present, connect a voltmeter across terminals 40 and 19. Ground terminal 23.
24. The voltmeter should indicate battery voltage. If not, the fault is in the A/C system.

Electric Radiator Fan Testing

1. Enter Control Function 3.

NOTE: As the first operation in Control Function 3, the control unit will operate the radiator fan relay at half-speed for 3 seconds, and then at full speed for a further 3 seconds.

2. If the fan does not operate, check the wiring and the relay.
3. Connect a voltmeter across terminals 14 and 19 of the control unit. It should indicate battery voltage.
4. Connect the voltmeter across terminals 14 and 19. It should indicate battery voltage.
5. If there is no voltage to the control unit, check the supply to the relay.
6. If voltage is present, check the wiring between the fan motor, the relay and ground.

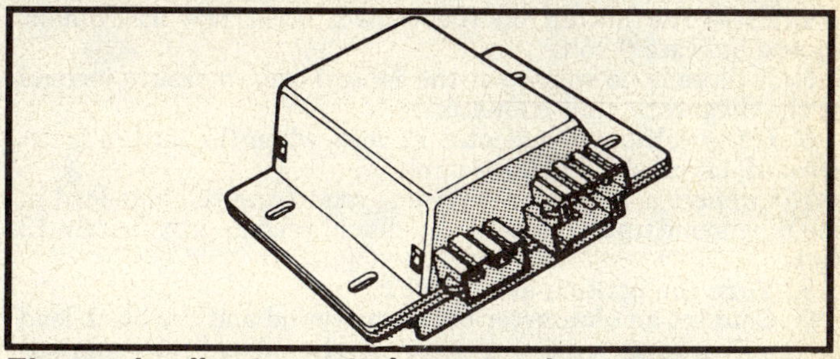
Electronic climate control power relay—960

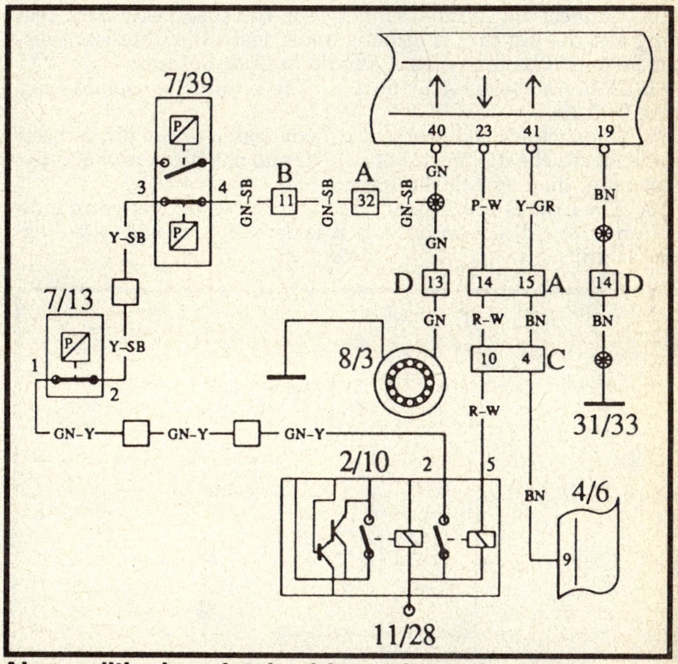

Air conditioning clutch wiring schematic—960

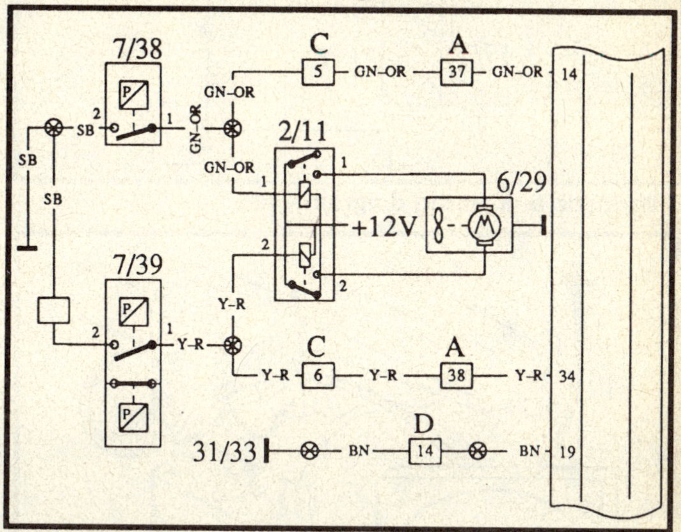

Radiator fan wiring schematic—960

Diagnostic Unit Testing

1. Turn the ignition switch **ON**.
2. Connect the test lead to socket 2 on the diagnostic unit.
3. Connect a voltmeter across terminals 55 and 19 on the control unit connector. The instrument should indicate 12 volts.

4. Press the button on the control unit. The instrument should indicate 0 volt.

5. If there is no voltage at the control unit, measure instead at the diagnostic unit connector.

6. If the voltmeter indicates 12 volts when the button is depressed, check the diagnostic unit.

7. Connect a voltmeter between ground and the blue lead in the diagnostic unit connector. It should read approximately 12 volts.

8. Turn the ignition switch **OFF**.

9. Connect an ohmmeter between ground and the black lead in the diagnostic unit connector. It should indicate approximately 0 ohms.

10. Connect an ohmmeter between the diagnostic unit test plug and the pin that is directly underneath the function selector button. The instrument should indicate infinity.

11. Depress the selector button. The ohmmeter should indicate 0 ohms.

12. Connect the red probe of a diode tester to the pin directly underneath the diagnostic unit LED and the black probe to the test plug, then switch the leads.

13. The LED is operating correctly if the tester shows an indication in one direction only. If it is defective, replace the diagnostic unit.

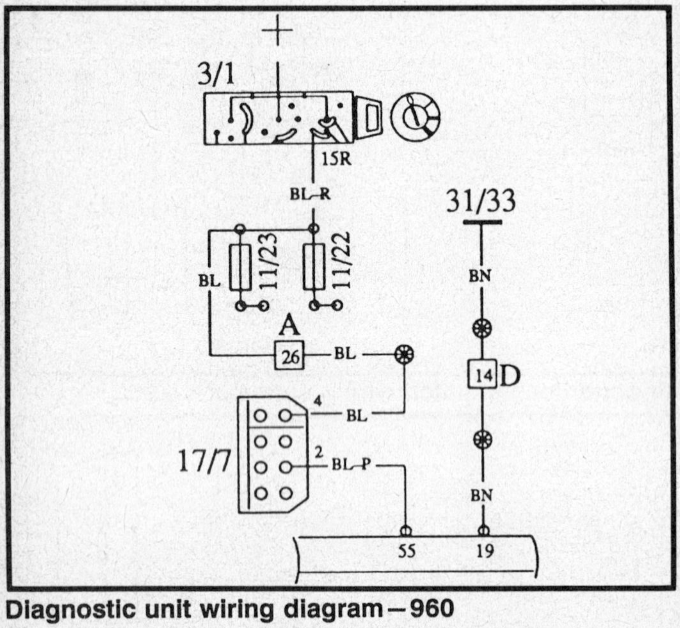

Diagnostic unit wiring diagram—960

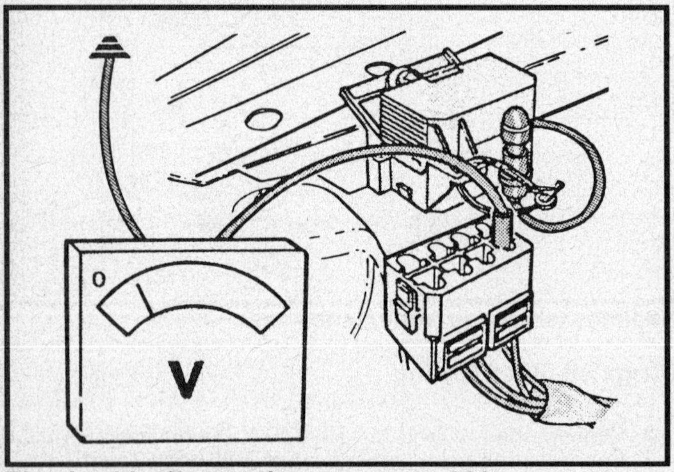

Blue lead in diagnostic connector—960

Black lead in diagnostic connector—960

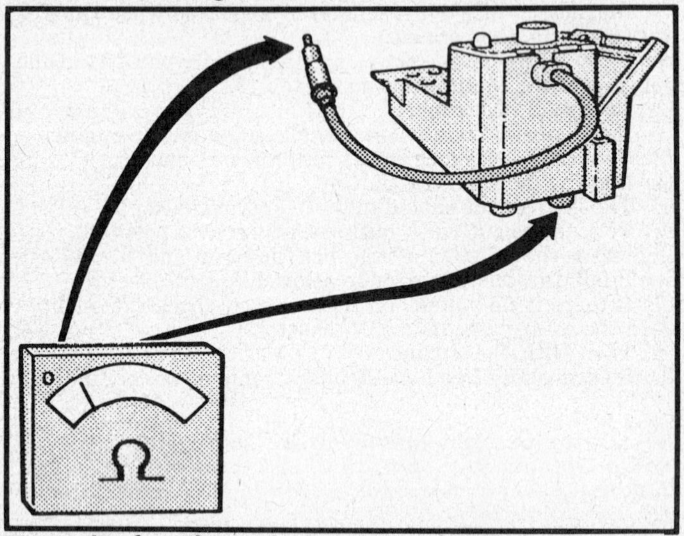

Pin under function selector button—960

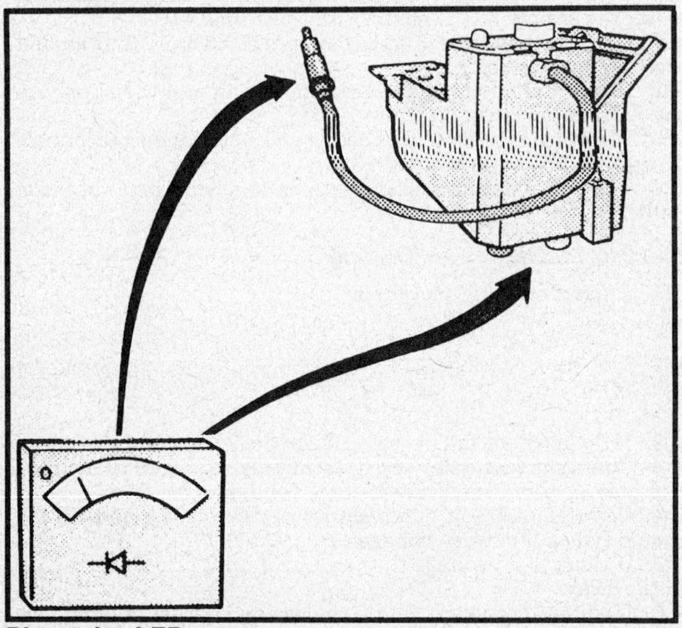

Pin under LED—960

CONNECTIONS

Conn.	Function	Type of signal
1	Ignition signal, cylinder 1	output
2	Ignition signal, cylinder 3	output
3	Fuel pump relay signal	output
4	Idling valve opening coil	output
5	Spare	
6	Speed signal	output
7	Air mass meter signal	input
8	Camshaft sensor signal	input
9	Speedometer signal	input
10	Camshaft sensor + supply	output
11	Front knock sensor signal	input
12	Throttle switch + supply	output
13	Ignition signal, cylinder 2	output
14	Radiator fan, half speed	output
15	Environmental warning lamp	output
16	Injectors, group 1	output
17	Injectors, group 2	output
18	Busbar 30 supply	supply
19	Signal ground	ground
20	Ignition signal, cylinder 5	output
21	Ignition signal, cylinder 6	output
22	Idling valve closing coil	output
23	AC on/off control signal	output
24	Power ground	ground
25	Air mass meter burn-off	output
26	Air mass meter ground	ground
27	Busbar 15 supply	supply
28	Oxygen sensor (Lambdasond) signal	input
29	Rear knock sensor signal	input
30	Sensor ground	ground
31	Ignition signal, cylinder 4	output
32	Spare	
33	Spare	
34	Radiator fan, full speed	output
35	Spare	
36	Main relay signal	output
37	Busbar 15 supply from main relay	supply
38	Spare	
39	Load signal to transmission	output
40	AC idling compensation	input
41	AC request from ECC	input
42	Drive/neutral idling compensation	input
43	Timing retardation response to transmission	output
44	Spare	
45	Temperature sensor signal	input
46	Spare	
47	Timing pick-up signal	input
48	Timing pick-up ground	ground
49	Throttle position information to transmission	output
50	Timing retardation request from transmission	input
51	Timing retardation request from transmission	input
52	Spare	
53	Throttle switch signal	input
54	Sprare	
55	Diagnostic lead	

Motronic control unit pin identification—960

1/1	Battery (1)
2/10	ECC power unit (499)
2/11	Relay, radiator fan (131)
2/12	Main relay (464)
2/13	Relay, fuel pump (135)
2/46	Relay, ignition coils (134)
4/3	Control unit, cruise control (226)
4/6	Control unit, ECC (495)
4/12	Control unit, Motronic (216)
4/28	Transmission control unit (414/415)
4/38	Power unit 1 (290)
4/39	Power unit 2 (291)
5/1	Combined instrument
6/31	Fuel pump (211)
7/15	Oxygen sensor (Lambdasond) (187)
7/16	Coolant temperature sensor (84)
7/17	Air mass meter (284)
7/21	Camshaft sensor (214)
7/23	Knock sensor 2 (218)
7/24	Knock sensor 1 (218)
7/25	Timing pick-up (412)
7/38	Low-speed pressure switch, radiator fan (82)
7/39	Safety and high-speed pressure switch, radiator fan (83)
7/54	Throttle switch
8/3	AC compressor solenoid (200)
8/5	Idling (CIS) valve (196)
8/6–8/11	Injectors (361–366)
11/1–35	Fuses
17/7	Diagnostic unit
17/8	Service terminal, coil terminal 1, No. 1 cylinder
20/3–8	Ignition coils and spark plugs 1–6
31/14	Ground terminal on relay board
31/31	Ground, busbar 31 on relay board
31/32	Ground, engine (power)
31/33	Ground, engine (instrument)
31/36	Ground, oxygen sensor
A	Connector, LH A-post
B	Connector, driver's A-post
C	Connector, 10-pole
D	Connector at LH suspension strut tower
E	Connector, 2-pole, for oxygen sensor
F	Connector, 2-pole, for oxygen sensor (screened)
G	Connector for transfer box (at control unit)
H	Connector at rear of engine
J	Connector at electrical distribution unit
K	Connector at rear of engine
L	Connector, I-pole, at LH engine mounting
M	Connector, 4-pole, luggage compartment
N	Ground, power unit 4/38
O	Ground, power unit 4/39

Motronic wiring diagram component key—960

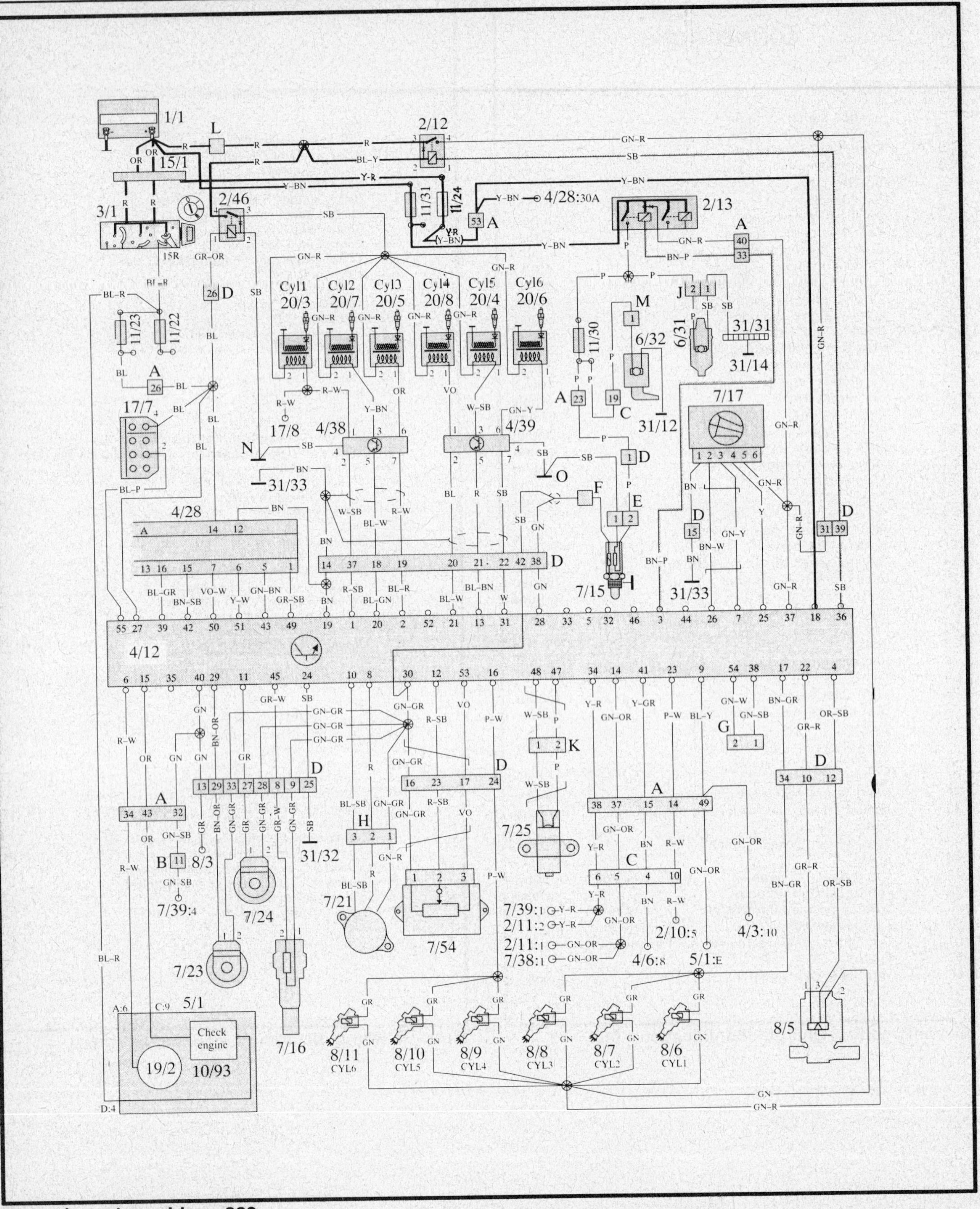

Motronic system wiring—960

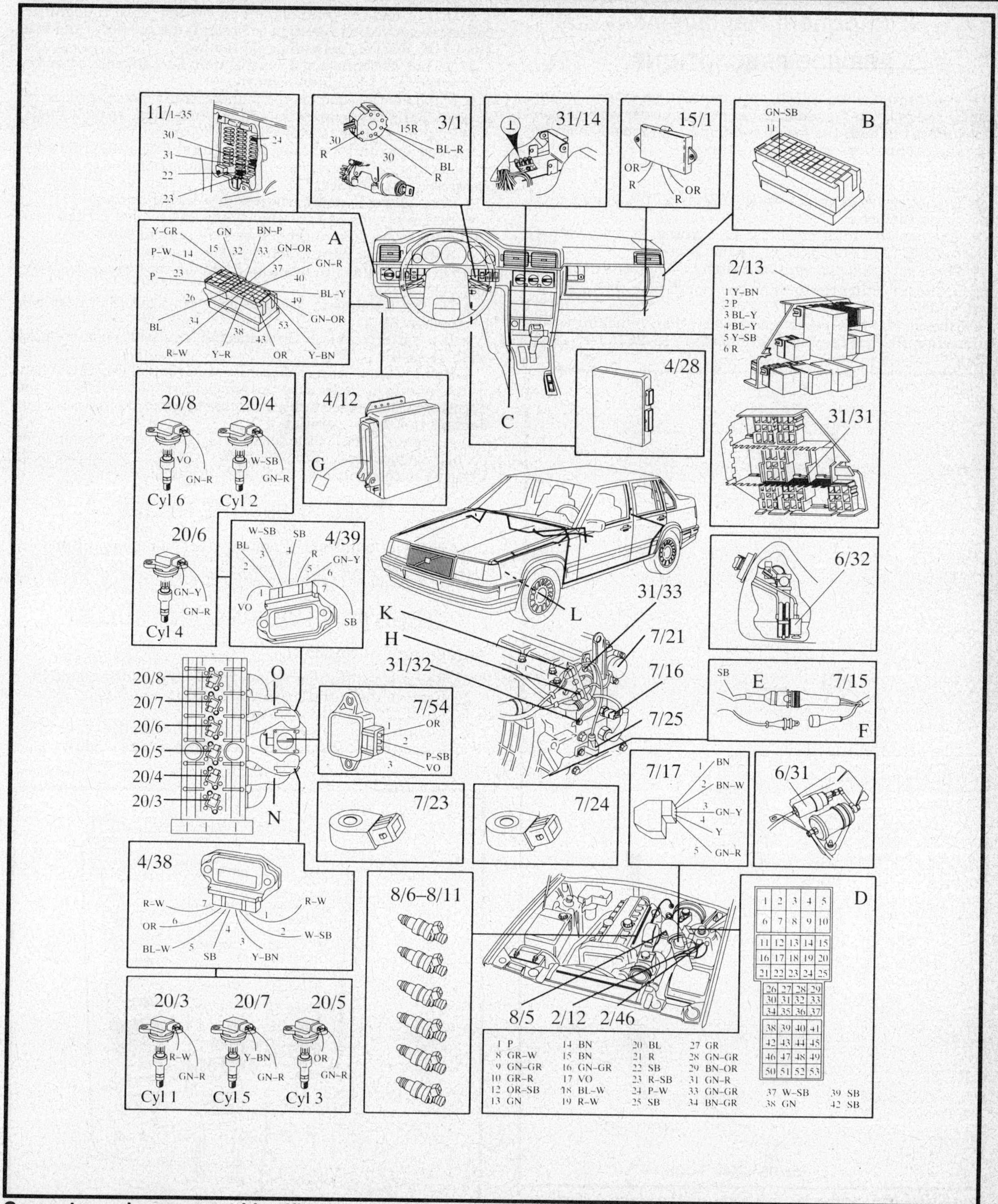

Connector and component identification—960

Component Replacement

SERVICE PRECAUTIONS

- Use caution when working around ignition system components. The ignition system operates at high power with dangerous voltages in both the low-tension and high-tension circuits, including connectors and similar fittings.
- Do not operate the fuel pump when the fuel lines are empty.
- Do not arc weld on the vehicle unless all the control units have been removed.
- The control units must not be exposed to temperatures above 175°F (80°C), such as in a paint booth.
- Make sure that the ignition switch is **OFF** before connecting or disconnecting the ignition coil, spark plug wiring or test instruments.
- Always use new gaskets and seals when remaking a fuel connection. Torque the fuel filter fittings to 15–26 ft. lbs. (20–35 Nm).

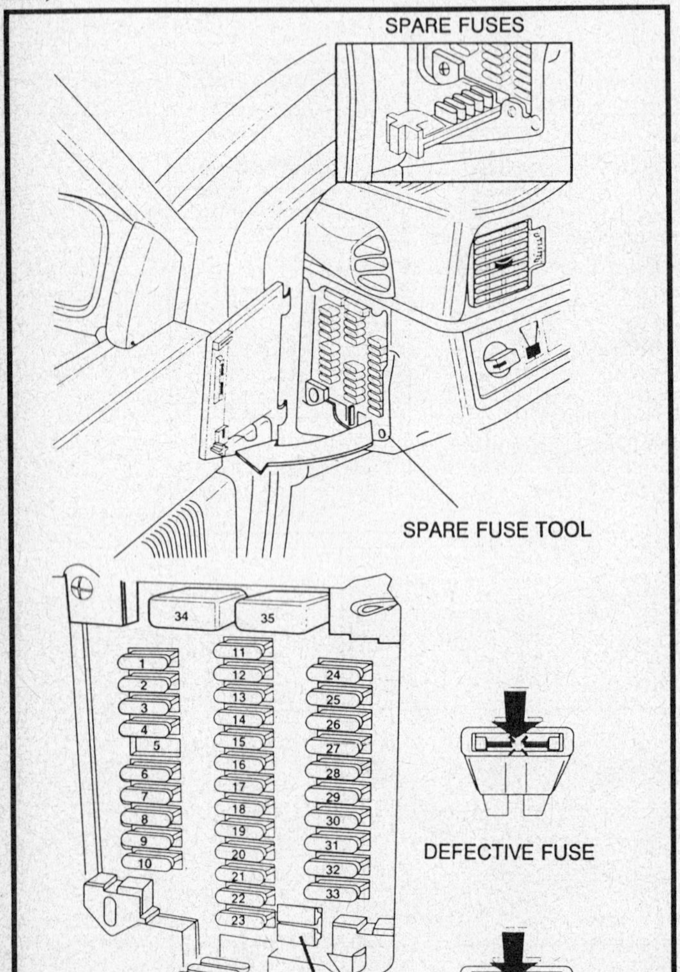

Fuse box location—far left side of the dash board on 960

- Do not test the compression on the 960 until the ignition coils are removed, the ignition coil relay is disconnected and fuse No. 31 (to disarm fuel pumps) is removed.
- Do not disconnect the control unit harness connector before fuse No. 24 has been removed.
- Make sure all control unit connectors are fastened securely. A poor connection can cause an extremely high surge voltage, resulting in damage to integrated circuits.
- Keep any control unit harness at least 4 in. away from adjacent harnesses to prevent a system malfunction due to external electronic interference.
- Keep all parts and harnesses dry during service.
- Before attempting to remove any parts, turn off the ignition switch and disconnect the battery ground cable.
- Always use a 12 volt battery as a power source.
- Do not attempt to disconnect the battery cables with the engine running.
- Do not attempt to disassemble a control unit under any circumstances.
- If a battery cable is disconnected, the fault code memory will be erased.
- Before connecting or disconnecting control unit ECU harness connectors, make sure the ignition switch is **OFF** and the negative battery cable is disconnected to avoid the possibility of damage to the control unit.
- A spare location in the fuse box is used as a test terminal for air bag diagnostics on the 960. Never install a fuse in this position or connnect accessories to this terminal.

FUEL PRESSURE RELIEF

1. With the ignition **OFF**, remove fuel pump fuses 30 and 31.
2. Try starting and allowing the vehicle to run a few times until it will not stay running any more.

THROTTLE HOUSING AND PULLEY

Adjustment of housing

1. Remove the throttle housing from the intake manifold.
2. Remove the throttle switch from the housing.
3. Carefully clean the housing.
4. Loosen the adjusting screw so that the throttle closes fully.
5. Loosen the 2 throttle disc retaining screws slightly.

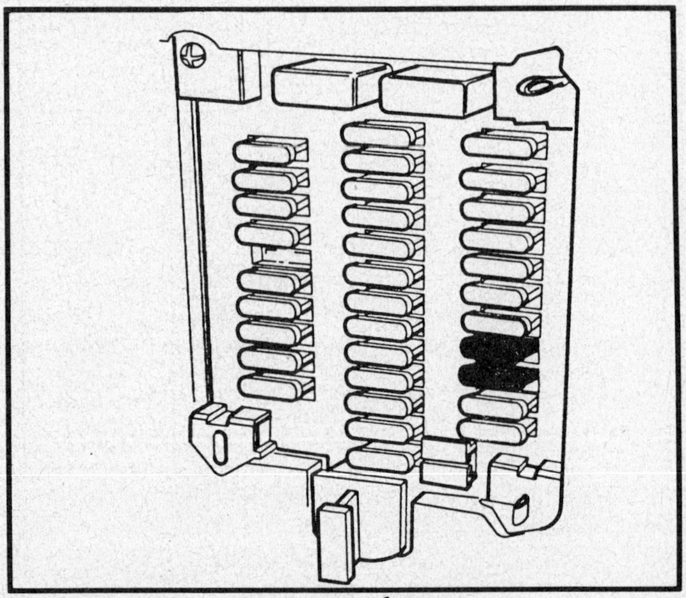

Fuses 30 and 31—Fuel pump fuses

NOTE: Check that the disc is correctly installed or facing in the right direction. Since the edges of the disc are thin, careful observation is required to detect chamfering.

6. Open the throttle approximately one-third using the lever.
7. Release the lever suddenly so that the disc centers itself in the throttle opening.
8. Check that the throttle spindle can be moved easily forward and backward.
9. Adjust the throttle spindle so that it protrudes 12.5mm outside the housing.
10. Tighten the disc retaining screws.

NOTE: It is important that the spindle does not project more than 12.5mm, otherwise it may foul the bottom of the throttle switch.

11. Screw in the adjusting screw until it just touches the arm. Tighten a further half turn, then tighten the locknut.
12. Mount the throttle switch on the spindle.
13. Install and tighten the switch retaining screws.

NOTE: The throttle switch does not require adjustment.

14. Mount the throttle housing on the intake manifold, then reconnect all hoses.

Adjustment of Throttle Housing and Pulley

1. Attach the link to the throttle housing.
2. Adjust the link exactly to fit the ball joint on pulley and press into position.
3. Tighten the adjusting nuts.
4. Check that the pulley moves freely without sticking.
5. The cable should be taut in the idling position, without al-

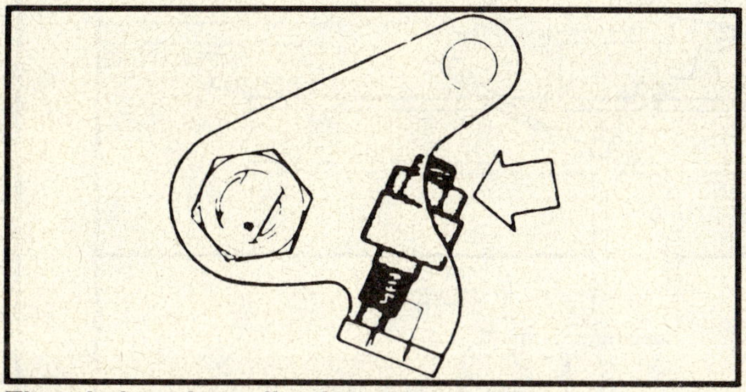

Throttle housing adjusting screw—960

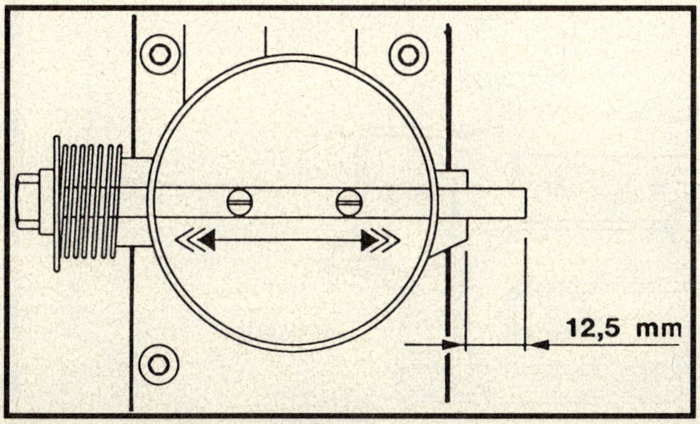

Adjusting throttle disk—960

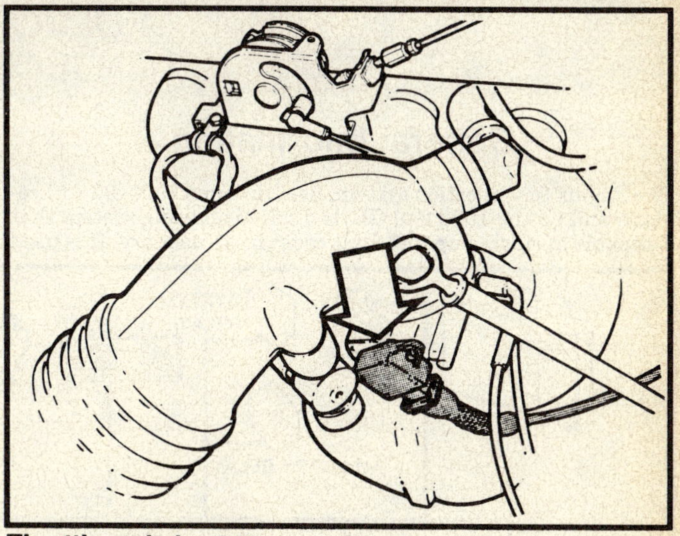

Throttle switch—960

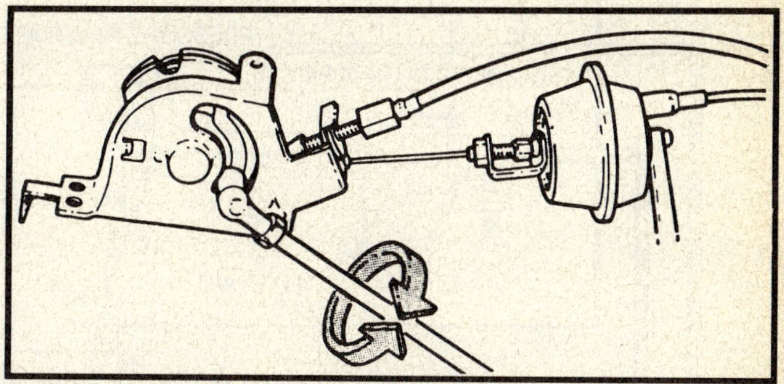

Adjusting throttle housing link—960

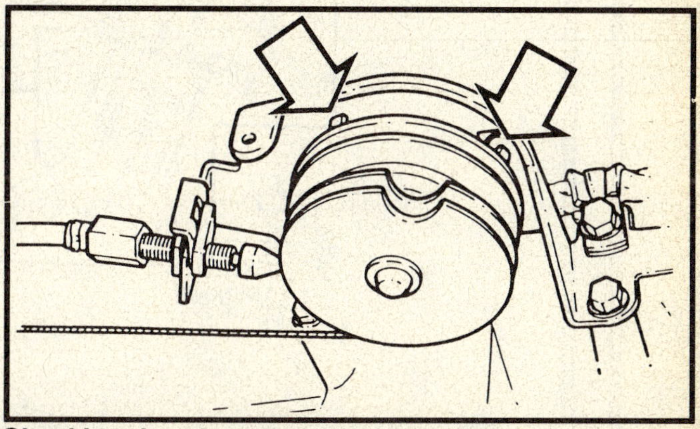

Checking throttle pulley and cable—960

tering the pulley position. The pulley should bear against the idling stop. Adjust the pulley as required.

6. Press the accelerator pedal itself to the floor and check that the pulley reaches full-load stop.

IDLE SPEED AND MIXTURE

The predetermined idle speed of 750 rpm and the mixture are not adjustable.

VOLVO REGINA FUEL INJECTION SYSTEM

General Information

The Regina fuel injection system, used on some 1990–92 740 vehicles and on the 1992 940 GL, is a self-diagnosing system that is capable of storing up to 3 fault codes in its memory. It is used in conjunction with the REX 1 ignition system. Both are adaptive systems that are capable of multiple adjustments based on previous driving. If a fault occurs, a warning lamp lights up the instrument panel. Fault tracing can be carried out using the diagnostic program.

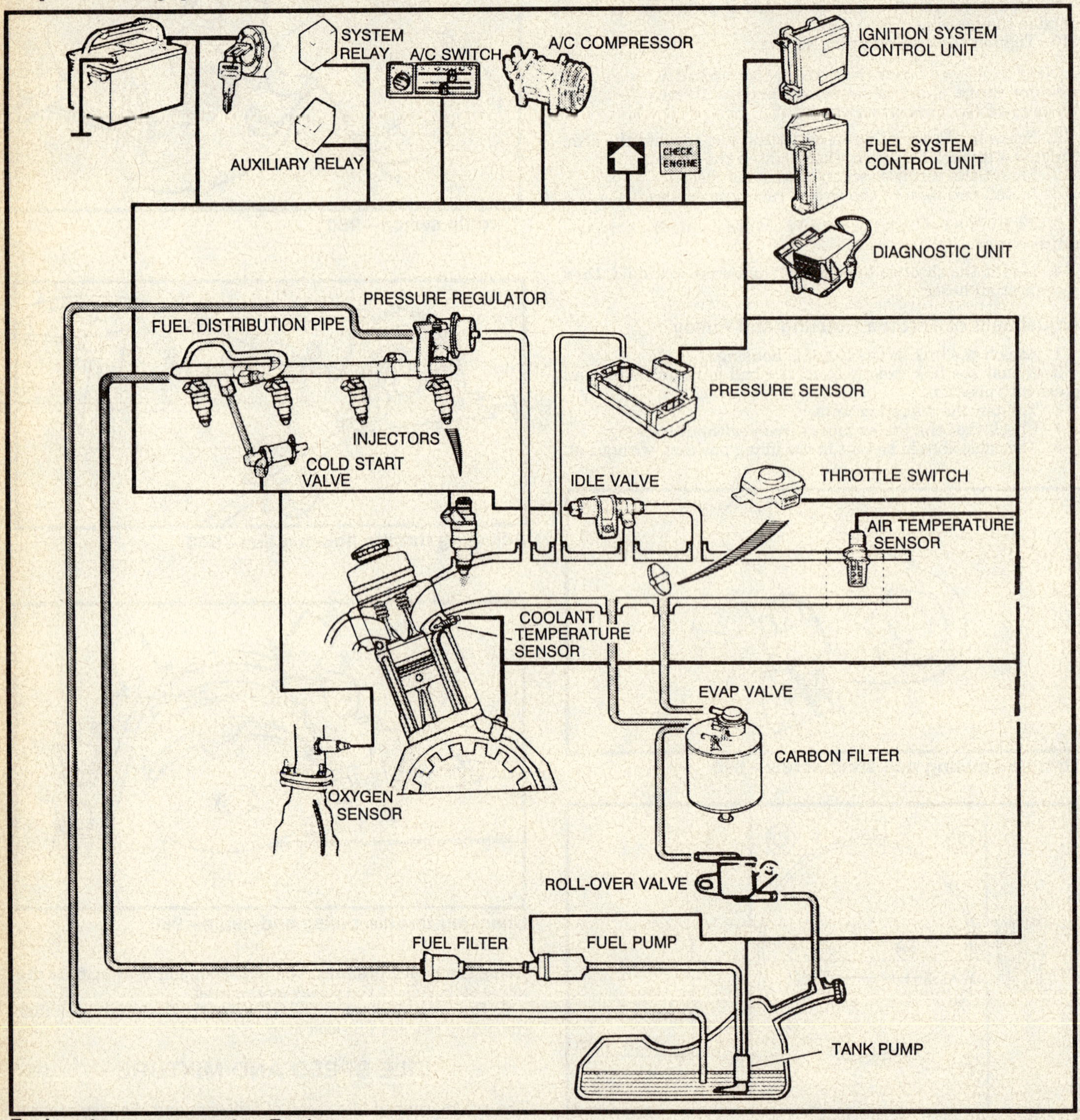

Fuel system components—Regina

Characteristics of the fuel injection system are: Utilizing a pressure sensor for measuring engine load and an air mass meter for measuring air intake volume, maintaining a fuel pressure of 42 psi, utilizing a separate cold start valve to ensure starting at low temperatures, incorporation of an automatic idle shut-off valve if power is lost, utilization of an induction sensor, which is mounted on the flywheel to indicate rpm and crankshaft position through the ignition system control unit and use of an electrically heated oxygen sensor (Lambda-sond).

CONTROL UNIT

The Regina fuel system control unit is extremely compact and uses a double-sided Printed Circuit Board (PCB). Discrete components are soldered to one side of the PCB and surface mounted components are on the other side. The multi-pin plug uses a filter to reduce interference and protect the electronics. It can operate at system voltage down to a minimum of 6 volts. The microprocessor operates at a voltage of 5 volts.

The control unit incorporates a microprocessor that receives signals from various sensors regarding operational conditions, then evaluates them in relation to pre-programmed values and calculates the correct injector opening duration.

The control unit controls idling rpm by regulating the amount of air bypassing the throttle valve. It also controls other functions, such as the cold start valve, fuel pump and the relay. The most important function is monitoring faults through the diagnostic unit.

In time, wear will affect the operation of the throttle valve, causing less air to enter the the intake system. Instead of operating from a pre-programmed value, the idle valve receives a signal that is adapted to the condition which the control unit has learned from previous driving periods.

The oxygen sensor determines if the fuel mixture is rich or lean and adjusts the control unit's Lambda regulator. The self-adjusting mechanism keeps the control unit function at mid point. Therefor, no adjustment is necessary for carbon monoxide content and automatically compensates for the effects of tolerances and wear in the injection system. Whenever the vehicle is started and driven, the control unit will use the value that has been stored from previous driving.

Incorporated is a limp home function for the pressure sensor, air temperature sensor and the coolant temperature sensor. If no load information is obtained, the control unit uses 3 pre-programmed substitute values for intake manifold pressure. It results in the injection pattern being varied according to engine revolutions, allowing the vehicle to be driven without any major problems.

If the signal from the coolant temperature sensor stops, the vehicle can still be driven without a problem. However, there can be difficulties when starting the engine since information on engine temperature is used to adjust the fuel mixture through the cold start system.

DATA SENSORS

Various input sensors feed information that is interpreted by the control unit to achieve optimum efficiency. The control unit receives signals from the pressure sensor, air intake temperature sensor and receives crankshaft position information from the ignition control unit, without which the system will not function. The coolant temperature sensor, oxygen sensor and throttle switch also send information to the control unit.

Pressure Sensor

The most important information for measuring engine fuel

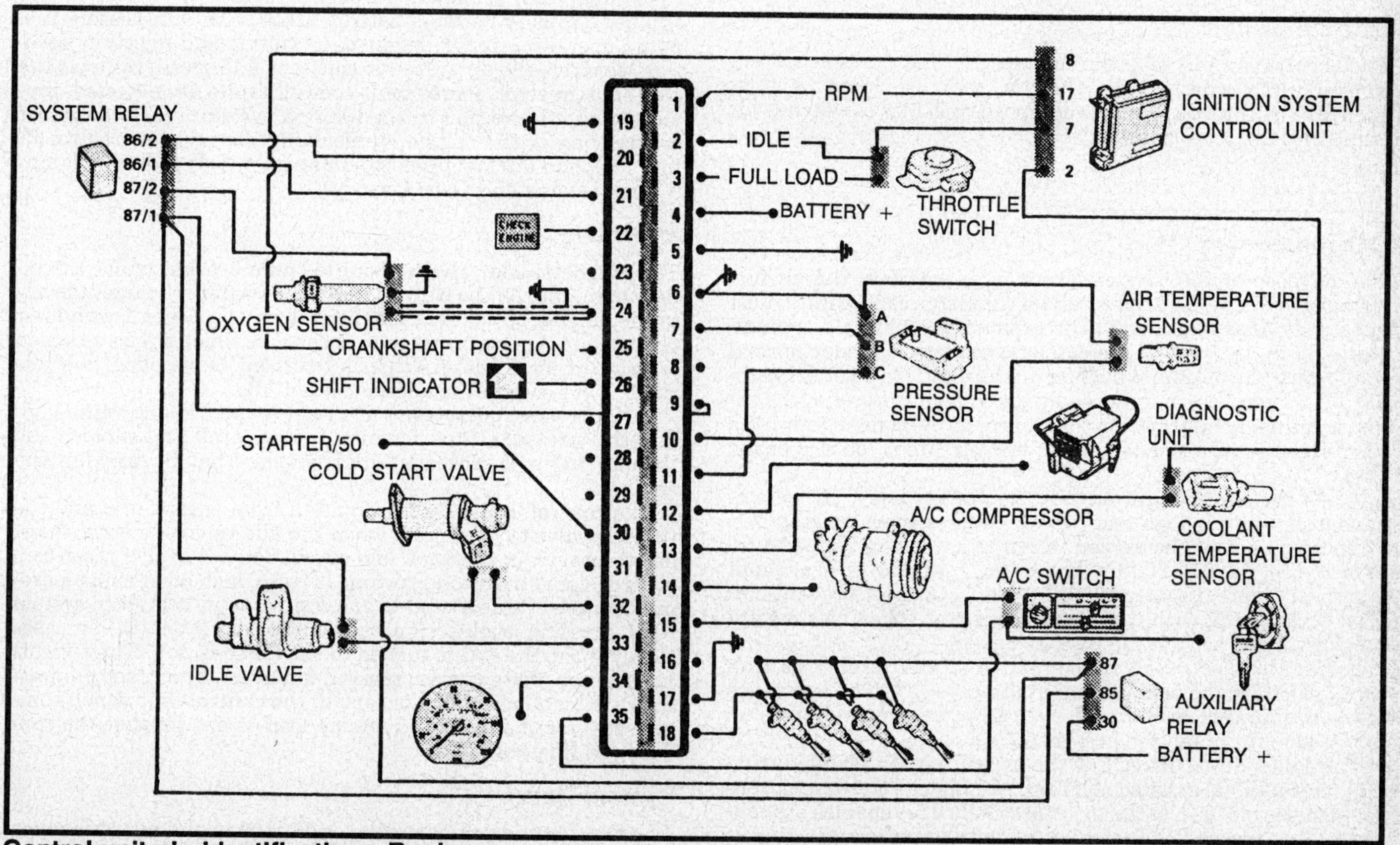

Control unit pin identification—Regina

comes from the pressure sensor. Using the pressure and temperature data, the micro-processor calculates the intake air mass. It is connected to the engine intake manifold through a hose and takes its reading there of the present pressure. A piezo-electric crystal changes a voltage input to an electrical output that reflects the pressure in the intake manifold.

Atmospheric pressure is measured when the engine is started and when driving fully loaded, then the pressure sensor information is adjusted accordingly.

The pressure sensor connector has 3 terminals: A, B and C. A is ground, B carries the output signal to the control unit and C provides the pressure sensor with 5 volts of current from the control unit. The output varies between 0.5–5 volts depending on intake manifold pressure. When the engine is not running, atmospheric pressure will register at full potential. The sensor is sensitive to electrical disturbances, therefore it is protected by a metal cover.

Air Temperature Sensor

Information from this sensor is added to the pressure sensor information to calculate the air mass being sent to the cylinders. The air mass passes through holes in the sensor and over a bulb which sends a temperature signal to the control unit. The sensor is an NTC (Negative Temperature Coefficient), meaning its resistance lowers as the temperature increases.

Coolant Temperature Sensor

While the engine is warming up, the choke mixture enrichment function is regulated by the control unit using information received by the coolant temperature sensor. This sensor is also a NTC type (Negative Temperature Coefficient), meaning resistance lowers as temperature increases. The sensor has 2 resistors. The second is connected to the ignition system's control unit and is grounded to the engine block.

Throttle Switch

The throttle switch tells the fuel system and ignition system control units whether the throttle valve is closed or fully opened. The throttle switch is equipped with micro-switches for idling and full load running.

Oxygen Sensor

The purpose of the oxygen sensor is to maintain the air/fuel mixture value so that the catalytic converter can operate most effectively. It is mounted in the exhaust gas manifold approximately 6 in. in front of the catalytic converter. Under normal conditions, the optimum air/fuel mixture ratio is 14.7:1 (Lambda = 1). The ratio is monitored by the oxygen content in the exhaust gas using information that is sent to the control unit from the oxygen sensor to determine if the mixture is too rich or too lean.

When the control unit signals the system relay, 12 volts of current is sent through resistor R1. It takes approximately 20–30 seconds to heat the oxygen sensor to a working temperature of 392°–572°F (200°–300°C). The sensor reaches a level of about 9 ohms at approximately 1292°F (700°C). Current branches off after resistor R1, with part going to resistor R2 and a constant 1.2 volts being sent to the oxygen sensor.

Exhaust gases reach the sensitive titanium-dioxide film through slits in the exhaust manifold sleeve pipe. The resistance of the titanium-dioxide film rises or drops in direct relationship to the amount of oxygen present.

The control unit disregards the oxygen sensor signal, during and after a choke assisted start-up. At full load, the oxygen sensor signal is blocked by the throttle switch. This enables the control unit to enrich the fuel mixture and prevent overheating of the engine, oxygen sensor and the catalytic converter.

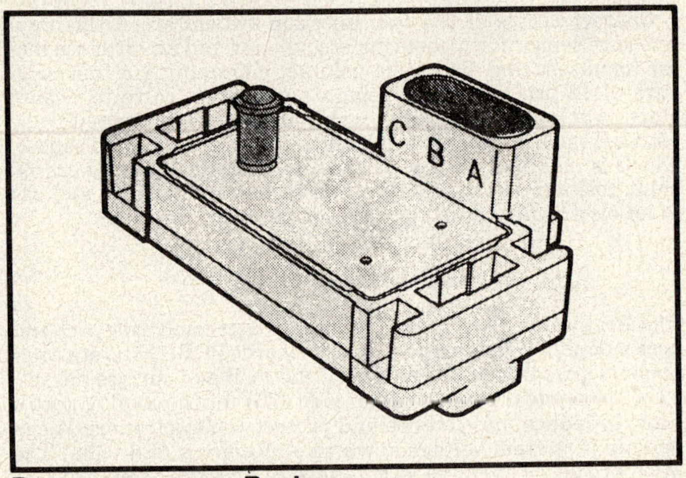

Pressure sensor—Regina

EGR Temperature Sensor

This sensor, found on California vehicles, measures the temperature of the exhaust gases returned to the intake manifold. Detection of the temperature variations enables the control unit to determine whether the EGR system is operating. It is able to measure temperatures up to 500°C (930°F). Before this limit the resistance increases with temperature.

CONTROLLED OUTPUTS

Based on information provided by the input sensors, the control unit sets system voltage by grounding the system relay, removes the system relay ground if the engine turns over too slowly (has stopped), thus preventing battery drain from unnecessary fuel pump operation, regulates opening, timing and injection duration, controls the air valve for constant idle speed, controls the cold start injector, stores fault codes if faults are detected, provides load information to the ignition system, protects against overrevving of the engine by disabling the injectors, shuts off fuel injection during deceleration and controls indicator lamps (Check Engine and Gear Change).

Idle Valve

The magnetic idle valve is mounted on a bracket under the intake manifold. As the control unit sends different current levels to the magnet actuator, a magnetized core is either drawn in or pushed out. A plate is attached to the core which allows a chosen amount of air to pass. When no current is supplied, the idle valve is closed and the engine cannot idle.

The idle valve operates at all engine rpm. This prevents high negative pressure from forming in the intake manifold if the throttle valve is closed at high engine rpm by braking the engine.

The control unit uses information from engine rpm and the throttle valve to determine when the idle valve will open, in order to provide a constant idle speed. The idle valve receives a control signal based on previous driving, instead of using a preprogrammed value stored in the control unit. A built in system for increasing engine rpm causes the idle speed to increases slowly when the A/C is turned to the **ON** position. Thus, creating an immediate rpm increasing function. When the compressor cuts in, a load signal is sent to the control unit which compensates for the increased engine load to ensure that the rpm does not fluctuate.

System Relay

The fuel injection system relay is located in the fuse/relay box, mounted in the center console underneath the radio, and is controlled by the fuel system control unit. It supplies current to the fuel pump and certain control unit functions, and controls the auxiliary relay.

Auxiliary Relay

The auxiliary relay, mounted in the engine compartment, just in front of the right shock tower, is controlled by the system relay. It supplies current to the injectors and the cold start valve. It diminishes the possibility of electrical interference, by separating the cable harnesses from the system.

EGR Vacuum Booster

This component is found on California vehicles. It controls the pressure in the vacuum line to the EGR valve by means of the admission valve (lower section). The converter processes the electrical signal from the intake manifold to maintain optimum control of the EGR valve. The unit is also designed to allow for the ambient air pressure. It is mounted on the left strut tower.

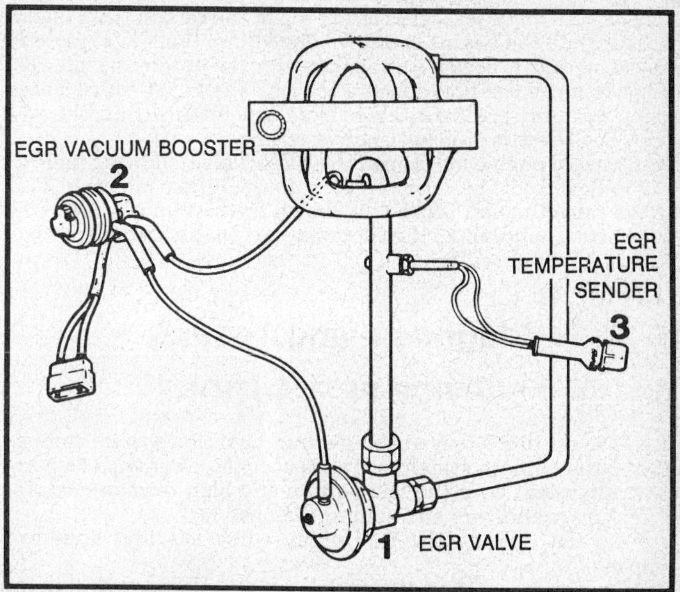

EGR system components—California

FUEL DELIVERY SYSTEM

Fuel Pressure Regulator

The fuel pressure regulator which is mounted on the distribution pipe, ensures that the fuel pressure remains constant at the injectors. Using a vacuum hose connected to the engine intake manifold, fuel pressure is kept at approximately 42 psi to ensure that the pressure over the injectors is kept constant, regardless of throttle position. The amount of fuel injected depends entirely on the injection time. Excess fuel is returned to the fuel tank through a return pipe.

Injectors

By activating the relay, the control unit sends current to the injectors. It controls injection duration by grounding the injectors. When the starter motor is operating, there are 2 injections per rotation, and 1 during normal driving. Injection occurs in the intake manifold near the intake valves.

Cold Start Valve

The cold start valve is controlled directly by the control unit, and is located under the intake manifold near the throttle valve. It is supplied with fuel through a hose connected to the distribution pipe.

At cold start, fuel condenses on the cold surfaces in the form of drops. Using a separate cold start valve improves engine cold starting. Its positioned farther away from the engine block than the injectors and delivers the fuel more in the form of vapor than in drops. It operates when the temperature is below freezing and when engine rpm is below 1050. It cuts out when engine rpm exceeds the permissible limit.

Fuel Pump

All models are equipped with a 2-stage electric fuel pump. It replaces the primary pump and the pre-pump which were used on previous models. It is located in the fuel tank.

Fuel Filter

The fuel filter is mounted on a plate underneath the vehicle on the left side.

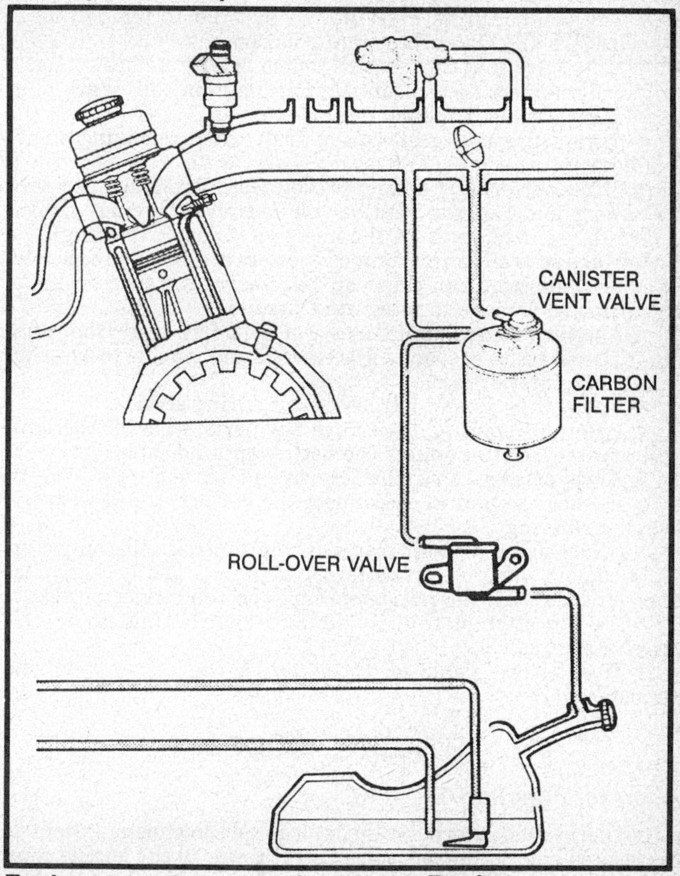

Fuel evaporation control system—Regina

Evaporative Control System (EVAP)

This system controls the gases that result from normal fuel tank evaporation, keeping them from polluting the air. Fuel vapor passes from the filter opening through a roll-over valve to a reservoir (canister, carbon filter). Fuel vapor is absorbed in the reservoir. The reservoir is equipped with EVAP valve which prevents leakage of fuel vapor when the engine is not running.

Fuel vapors from the fuel tank enter into the top of the carbon filter and are absorbed. Air is then pushed out through a channel from the bottom of the filter. Depending on temperature and other conditions, the filter can store approximately 90 grams of fuel. If for any reason, the vehicle leans sideways at more than a 45 degree angle, a roll-over valve is incorporated into the system to prevent any fuel spills.

An EVAP valve is located at the top of the carbon filter and is closed when the engine is off. During idling, the EVAP valve is closed so that it does not interfere with the automatic idle settings or make the fuel mixture to rich. The EVAP valve closes using vacuum pressure taken from the intake manifold and from the throttle shutter positive terminal.

Increased engine loads open the EVAP valve, allowing fuel vapor to flow from the carbon filter into the intake manifold, and at the same time air is drawn through the bottom channel. Under normal conditions, the filter discharges the fuel in approximately 15–20 minutes.

Diagnosis and Testing

SERVICE PRECAUTIONS

- Use caution when working around ignition system components. the ignition system operates at high power with dangerous voltages in both the low-tension and high-tension circuits, including connectors and similar fittings.
- Do not operate the fuel pump when the fuel lines are empty.
- Do not arc weld on the vehicle unless all the control units have been removed.
- The control units must not be exposed to temperatures above 175°F (80°C), such as in a paint booth.
- Make sure that the ignition switch is **OFF** before connecting or disconnecting the control unit, ignition coil, spark plug wiring or test instruments.
- Always use new gaskets and seals when remaking a fuel connection.
- When performing a compression test, disconnect the wire from terminal 1 on the ignition coil (to prevent arcing) and remove the connectors from the injectors (to avoid flooding).
- Make sure all control unit connectors are fastened securely. A poor connection can cause an extremely high surge voltage, resulting in damage to integrated circuits.
- Keep any control unit harness at least 4 in. away from adjacent harnesses to prevent a system malfunction due to external electronic interference.
- Keep all parts and harnesses dry during service.
- Before attempting to remove any parts, turn off the ignition switch and disconnect the battery ground cable.
- Always use a 12 volt battery as a power source.
- Do not attempt to disconnect the battery cables with the engine running.
- When using a tester probe on the control unit connector, first remove the protective sleeve, then always gently check connections through the numbered holes on the connector side.
- Do not attempt to disassemble a control unit under any circumstances.
- If a battery cable is disconnected, the memory will be erased.

READING CODES

Control Function 1

The fuel system incorporates a built-in self-diagnostic system. It is located behind the left spring strut tower in the engine compartment. The diagnostic system uses socket No. 2 for the fuel system and socket No. 6 for the ignition system and EGR system (California).

There are many possible fault codes in the diagnostic system, and the system is capable of storing up to 3 system faults at 1 time. When the engine is running, the fuel system control unit continuously checks the control unit internal functions, oxygen sensor, battery voltage, coolant temperature sensor, air temperature sensor, pressure sensor, throttle position, engine rpm and the idle speed through the air valve

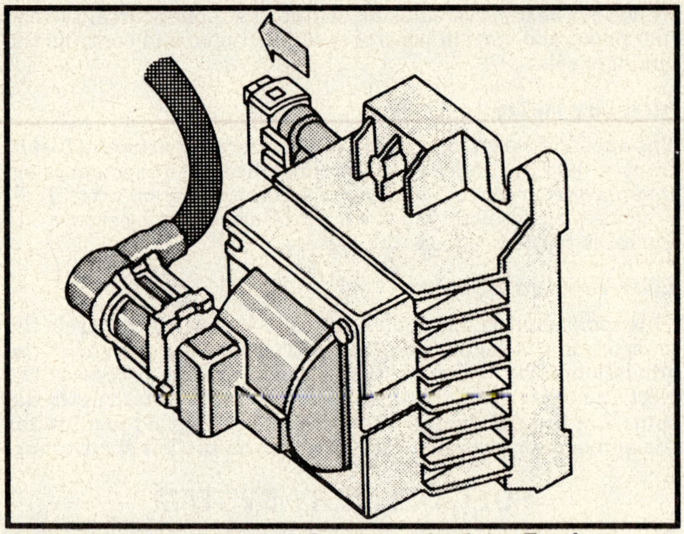

Disconnecting coil wire from terminal 1 – Regina

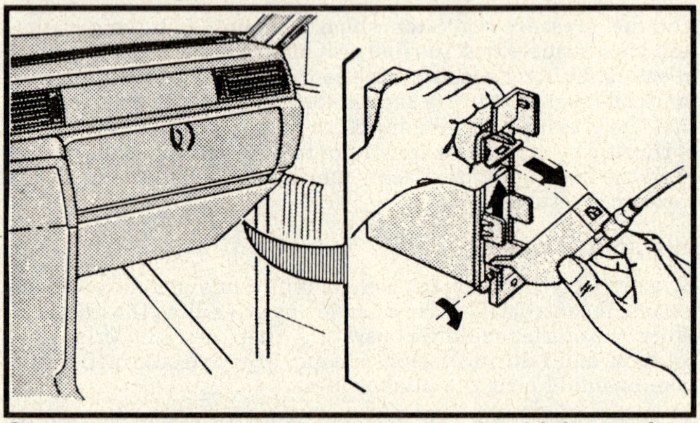

Control unit location – Behind right side kick panel

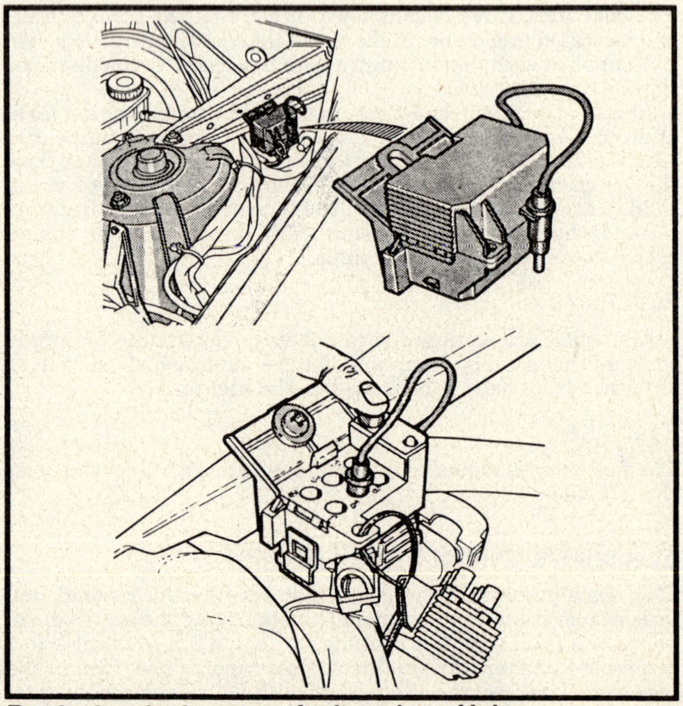

Retrieving fuel system fault codes – Volvo

The following procedure is for retrieving fault codes.

1. Open the diagnostic unit cover in the driver's side rear of the engine compartment and install the selector plug into socket No. 2 for ignition or fuel system codes, or into socket No. 6 for EGR system codes (California).
2. Turn ignition to the **ON** position.
3. Enter Control Function 1 by pressing the button once. Hold the button for at least 1 second, but not more than 3 seconds.
4. Watch the LED and count the number of flashes in each ofthe 3 flash series, which will indicate a fault code. The flash series are separated by 3 second intervals. Note the fault codes.

NOTE: If there are no fault codes in the diagnostic unit, the LED will flash 1–1–1 and the fuel system is operating correctly.

5. Depress the button again. If the same code is repeated, there are no additional codes stored. If the code is different, depress the button a third time and record the code, if different.

NOTE: The diagnostic memory is full when it contains 3 fault codes. Until those 3 are rectified and the memory erased, the system cannot give information on any other problems.

Control Function 2

This function is activated by pressing the button twice (between 1–3 seconds between operations), causing the LED to flash. The control unit delivers a diagnostic code (not a fault code) acknowledging the receipt of a signal from each of the following items:

Diagnostic Code 3–3–2: Throttle switch, when the throttle is moved from the fully-closed position

Diagnostic Code 3–3–3: Throttle switch, when the throttle is moved from the full-load position

Diagnostic Code 3–3–1: Rpm signal from the ignition system

Diagnostic Code 1–1–4: Electronic Climate Control (ECC) panel, when the A/C button is depressed or released

Diagnostic Code 1–3–4: Air conditioning compressor

Diagnostic Code 1–2–4: Testing idle speed compensation on automatic transmission vehicles. When depressing the brake pedal and placing the gear selector lever in **D** and then moved to **N** again.

Control Function 3

This function is activated by turning the ignition **ON**, installing the selector plug into socket No. 2. and pressing the button 3 times (between 1–3 seconds between operations). The control unit then responds by activating the injectors, idle valve, cold start valve, auxiliary relay (radio interference suppression relay) and fuel pump.

CLEARING CODES

When all the fault codes have been read and corrected, the diagnostic system memory can be erased as follows:

1. Turn ignition switch to the **ON** position.
2. Read fault codes again.
3. Press and hold diagnostic unit button for more than 5 seconds. Release button. After 3 seconds, the LED should light.
4. While the LED is lit, press button again for more than 5 seconds. Release button. The LED should go out. Memory is cleared.

NOTE: To ensure that the memory has been erased, press button again for 1 second but not more than 3 seconds. The LED should flash 1–1–1, indicating that the memory is erased.

5. Start and run the engine. If it will not start, see Control Function 2.

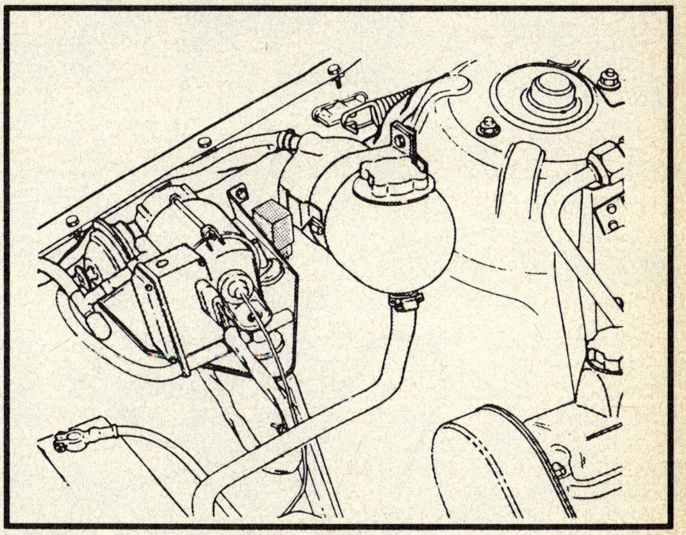

Auxiliary relay location—Regina

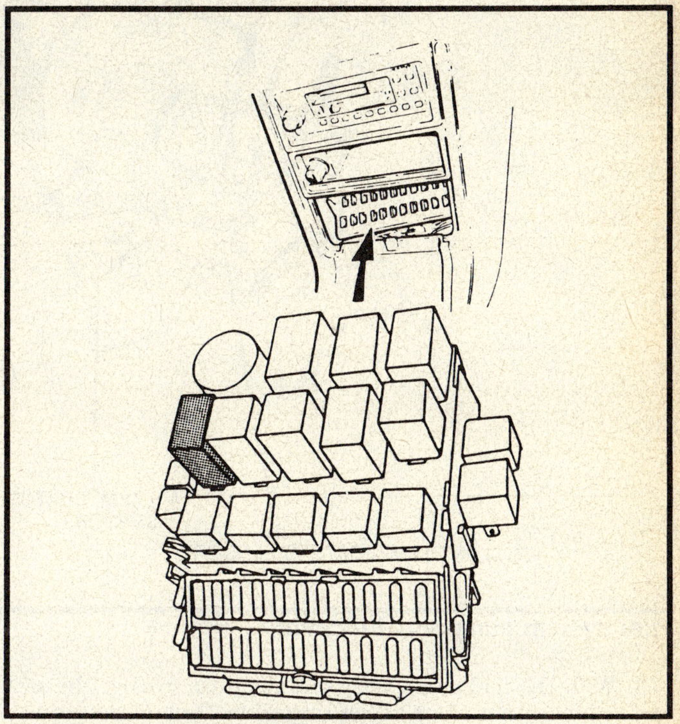

Fuel pump relay location—Regina

6. Recheck for fault codes. If Code 1–1–1 comes up, there are no additional system fault codes.

FUEL SYSTEM TESTING

Pressure Relief

1. Remove the fuel pump relay from the fuse box behind the ash tray. The relay is furthest to the left on the second row.
2. Crank the engine until the engine will not start anymore, indicating that the fuel pressure is relieved.

Pressure Check

1. Relieve the fuel pressure.
2. Connect fuel pressure gauge 5011, or equivalent, holding a shop towel under the fuel line to catch any spilled fuel.

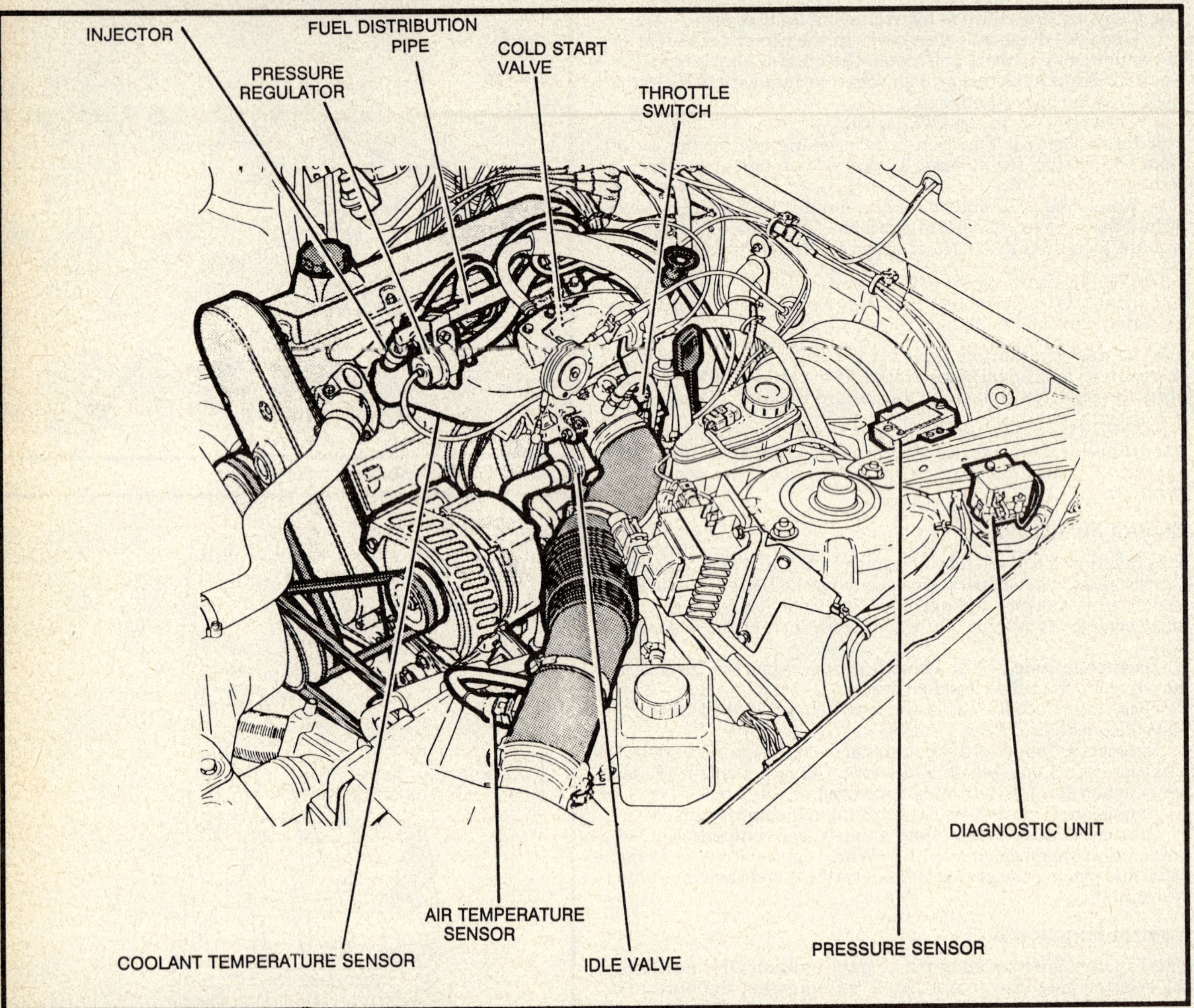

Fuel system component locations – Regina

3. With the fuel pump relay still removed, remove the seat belt reminder, since this makes the test more easily performed. It is located in the middle on the top row.
4. Start the fuel pump by connecting an electrical lead between terminals 30 and 87/2 on the relay base. Verify pump operation by removing the fuel cap and listening.
5. The fuel pressure should be 300 kPa (42 psi).
6. Remove the lead between terminals 30 and 87/2.
7. Relieve the fuel pressure and remove the pressure gauge, catching any spilled fuel with a shop towel.
8. Reinstall the relay.

CAUTION

To avoid fire and personal injury, replace any plastic tie bands removed from the fuel lines.

PUMP INOPERATIVE

1. If the fuel pump does not start, remove the lead between terminals 30 and 87/2.

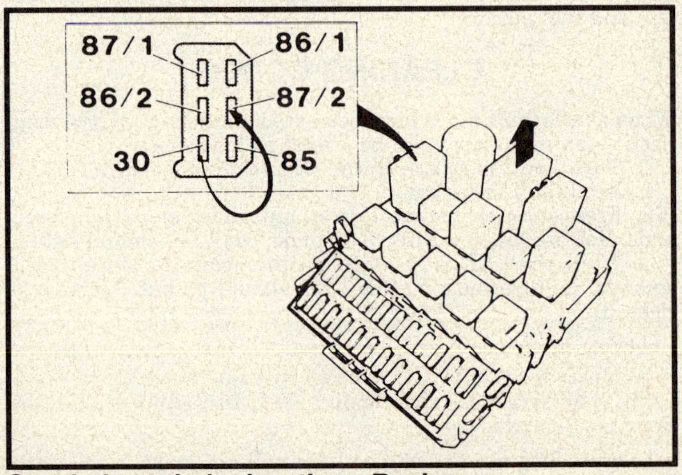

Seat belt reminder location – Regina

2. Check for voltage at terminal 30. If there is none, check the lead between the relay and the battery.
3. Connect an electrical lead between terminals 30 and 87/2 on the relay base. If the pump still does not start, check the lead between the pump and the relay. Check the lead between 87/1 and 85 for breaks with an ohmmeter.

PRESSURE TOO HIGH

1. Remove the lead between terminals 30 and 87/2 on the relay base.
2. Remove the return hose from the pressure regulator and blow into the pipe.
3. Remove the vacuum hose from the pressure regulator and blow into the pipe.
4. If both hoses are open, the pressure regulator is faulty. Replace and recheck.

PRESSURE TOO LOW

1. Squeeze the return hose by hand, not with a tool, and check if the pressure rises.

CAUTION

To avoid personal injury and fire, do not allow the pressure to exceed 600 kPa (84 psi).

2. If the pressure rises rapidly, the pump and hoses are good. Replace the regulator and recheck the pressure.
3. If the pressure rises slowly, the fuel filter, pump strainer or the fuel lines are clogged or blocked.
4. If the pressure does not rise, the fuel pump is faulty.

PRESSURE REGULATOR

1. Connect a vacuum pump to the pressure regulator.
2. Apply vacuum to the regulator and check that the fuel pressure falls as more vacuum is applied.
3. Remove the lead between terminals 30 and 87/2.
4. Relieve the fuel pressure and remove the pressure gauge, catching any spilled fuel with a shop towel.
5. Reinstall the relay.

CAUTION

To avoid fire and personal injury, replace any plastic tie bands removed from the fuel lines.

DIAGNOSTIC UNIT

Testing

1. Enter Control Function 1. If the LED does not flash when the button is pressed, or if no code is flashed, continue the diagnostic unit test.

NOTE: If there are no fault codes the LED should flash fault code 1-1-1, which indicates that no other fault codes are stored.

2. Check ground connections at the intake manifold. Poor ground connections can cause different fault symptoms.
3. Check oxygen sensor ground connection, located at the right front mud guard.
4. Check the pump relay fuse (fuse 1), located in the fuse/relay box.
5. Remove panels from under the instrument panel right side, in front of right firewall side, then remove glove compartment. Check control unit grounds.
6. Disconnect control unit electrical connector, then remove the protective sleeve.
7. Check diagnostic unit by connecting a voltmeter between ground and pin 4 on the control unit connector. Voltmeter should read 12 volts. If no voltage is present, check lead between the control unit connector and fuse 1 in the fuse/relay box.

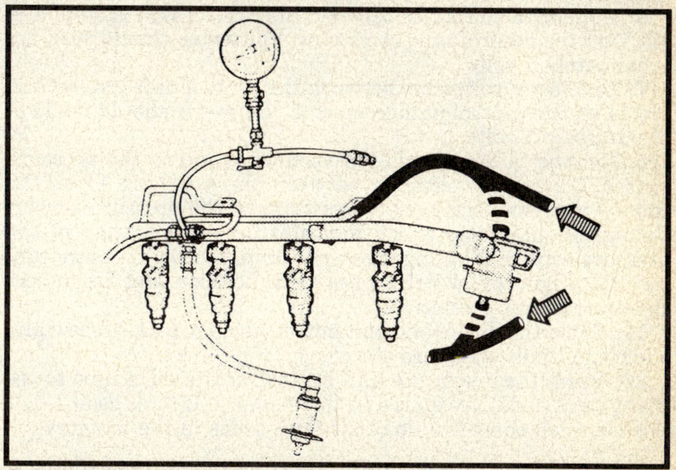

Checking fuel pressure regulator hoses – Regina

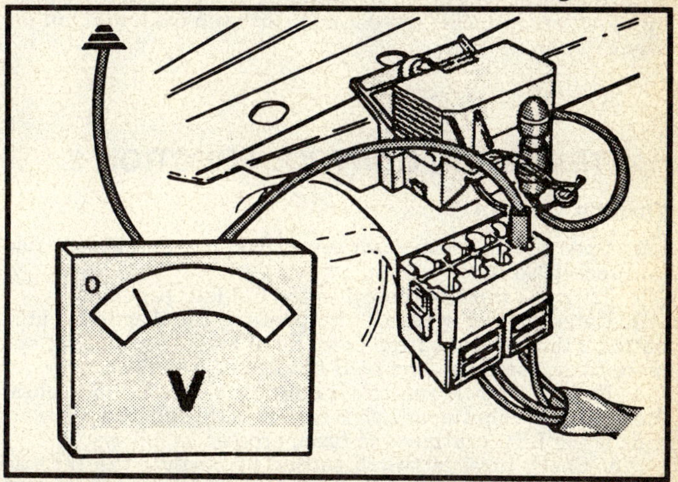

Blue lead on diagnostic connector – Volvo

8. Turn the ignition switch to the **ON** position, then install selector plug into socket No. 2 on the diagnostic unit.
9. Connect a voltmeter between ground and pin 12 on the control unit connector. Voltmeter should read 12 volts.
10. Press the button on the diagnostic unit. Voltmeter should read 0 volts. If no voltage is present at the control unit, take reading at the diagnostic unit connector. If voltmeter remains reading 12 volts when button is pressed, check the diagnostic unit.
11. Disconnect the diagnostic unit electrical connector, then connect a voltmeter between ground and the blue lead on the diagnostic unit connector. Voltmeter should read 12 volts.
12. Connect an ohmmeter between ground and the black lead in the diagnostic unit connector. Ohmmeter should read 0 ohms. Turn ignition switch to the **OFF** position.
13. Connect an ohmmeter between the diagnostic unit selector plug and the pin under the selector button (on bottom of diagnostic unit). Ohmmeter should read infinity. Press button. Ohmmeter should read 0 ohms.
14. Connect a diode tester between the selector plug and the diagnostic unit LED pin: Connect the positive test lead from the tester to the pin directly underneath the LED and the negative test lead from the tester to the selector plug, then reverse the leads. If current flows only one way, the diode is good. If current flows both ways, replace diagnostic unit.
15. Check the pressure sensor by disconnecting the pressure sensor vacuum hose, then connect the control unit connector. Turn the ignition switch to the **ON** position.

16. Connect a voltmeter between the control unit ground and pin 7 on the control unit connector. Voltmeter should read approximately 5 volts.
17. Connect a voltmeter between the control unit ground and pin 11 on the control unit connector. Voltmeter should read approximately 5 volts.
18. Connect a hand operated vacuum pump to the pressure sensor. Connect a voltmeter between pin 7 and pin 11 on the control unit connector. Apply vacuum to the pressure sensor. Voltage should drop, which indicates proper operation of the pressure sensor. If voltage does not drop, replace pressure sensor. Turn ignition switch to the **OFF** position and disconnect the control unit connector.
19. Press the diagnostic unit button again. Check to see if any additional fault codes are stored.
20. Press the diagnostic unit button for the third time to see if any fault codes are stored in the memory. If fault code 1-1-1 flashes, then there are no other fault codes in the memory.

NOTE: The diagnostic system memory is full when it contains 3 fault codes. Until those 3 are rectified and the memory erased, the system can not give information on any other problems.

TESTING IN CONTROL FUNCTION 2

Throttle Control

1. Open the diagnostic unit cover, then install selector cable into socket No. 2.
2. Turn the throttle control to the full load position.
3. Enter control system 2, by pressing the diagnostic unit button 2 times. Hold button down for at least 1 second, but not more than 3 seconds. The LED should begin to flash.
4. Release throttle control. If diagnostic code 3-3-3 flashes, function of the shutter switch is correct in the full load position.
5. If the LED continues to flash, proceed as follows:
   a. Check throttle switch setting by opening the throttle slightly and listening to the switch. A click should be heard when the shutter opens (idle switch).
   b. Turn switch slightly clockwise, then counterclockwise until the switch clicks. Tighten mounting bolts and check setting.
   c. Test the throttle switch by disconnecting the control unit electrical connector. Connect an ohmmeter between ground and pin 2 on the control unit connector. Ohmmeter reading should be 0 ohms (switch closed).
   d. Depress gas pedal slightly. Resistance should increase to 2000–3000 ohms (switch opens).
   e. Connect an ohmmeter between ground and pin 3 on the control unit connector. Ohmmeter should read infinity (full load switch open).
   f. Depress gas pedal all the way. Ohmmeter should read 0 ohms. If a fault appears, take reading at throttle switch to see if fault is in the throttle switch or its leads. Also, check the ground connections at the intake manifold.
6. Start the engine. If LED goes off, then flashes diagnostic code 3-3-1, the rpm signal from the ignition system is correct. If the engine does not start, run the starter motor until the LED goes off.
7. If the LED continues to flash rapidy, the ignition system has faults that must be repaired. If the ignition system has no faults, continue testing.
8. Check control unit ground connections.
9. Disconnect control unit electrical connector and remove protective sleeve.
10. Check diagnostic unit by connecting a voltmeter between ground and pin 4 on the control unit connector. Voltmeter should read 12 volts.
11. If no voltage is present, check lead between the control unit connector and fuse 1 in the fuse/relay box (located in center console).
12. Turn the ignition switch to the **ON** position, then install selector plug into socket No. 2 on the diagnostic unit.
13. Connect a voltmeter between ground and pin 12 on the control unit connector. Voltmeter should read 12 volts.
14. Press the button on the diagnostic unit. Voltmeter should read 0 volts. If no voltage is present at the control unit, take reading at the diagnostic unit connector. If voltmeter remains reading 12 volts when button is pressed, check the diagnostic unit.
15. Disconnect the diagnostic unit electrical connector, then connect a voltmeter between ground and the blue lead on the diagnostic unit connector. Voltmeter should read 12 volts.
16. Connect an ohmmeter between ground and the black lead in the diagnostic unit connector. Ohmmeter should read 0 ohms. Turn ignition switch to the **OFF** position.
17. Connect an ohmmeter between the diagnostic unit selector plug and the pin under the selector button (on bottom of diagnostic unit). Ohmmeter should read infinity. Press diagnostic button. Ohmmeter should read 0 ohms.
18. Connect a diode tester between the selector plug and the diagnostic unit LED pin: Connect the positive test lead from the tester to the pin directly underneath the LED and the negative test lead from the tester to the selector plug, then switch the leads. If current flows one way only, the diode is good. If current flows both ways, replace the diagnostic unit.
19. Check ignition lock voltage by turning ignition switch to the **ON** position. Connect a voltmeter between ground and pin 35 on the control unit connector. Voltmeter should read 12 volts. Ensure that voltage exists when the starter motor is operating. Turn ignition switch to the **OFF** position.
20. Check ground connections by connecting an ohmmeter between ground and pin 5, pin 17 and pin 19 on the control unit connector. Ohmmeter reading at all connections should be 0 ohms. Pin 5, pin 17 and pin 19 are ground leads at the intake manifold.
21. Ensure that the oxygen sensor wire is connected to pin 5 on the control unit connector.
22. Check the rpm sensor lead from the ignition system control unit by connecting a voltmeter between ground and pin 1 on the control unit connector. Turn ignition switch to the **ON** position. Voltmeter should read 12 volts.
23. Turn ignition switch to the **OFF** position. Check the throttle switch by connecting an ohmmeter between ground and pin 2 on the control unit connector. Ohmmeter should read 0 ohms (switch closed). Depress gas pedal slightly. Resistance should increase 2–3 kilo-ohm (switch opens).
24. Connect an ohmmeter between ground and pin 3 connection on the control unit. Ohmmeter should infinity (full load switch open). Depress gas pedal all the way. Ohmmeter should read 0 ohms. If specifications are not as specified, measure at throttle switch to see if fault is in the switch or its leads. Also, check ground connections at intake manifold..
25. Check the air temperature sensor by connecting an ohmmeter between pin 6 and pin 10 on the control unit connector. Ohmmeter reading depends on temperature; at −4°F (−20°C), reading should be 15000 ohms, at 68°F (20°C), reading should be 2500 ohms and at 212°F (100°C), reading should be 160 ohms. It is possible to to check the coolant temperature sensor resistance at various temperatures by warming the sensor using a hot air gun. If specifications are not as specified, measure the air temperature sensor to see if fault is in the sensor or its leads.
26. Check the pressure sensor by disconnecting pressure sensor hose. Ensure that the ignition switch is in the **OFF** position, then install control unit connector loosely.
27. Turn ignition switch to the **ON** position. Connect a voltmeter between ground and pin 7 on the control unit connector. Voltmeter should read approximately 5 volts. Then, connect voltmeter between ground and pin 11 connection on the control

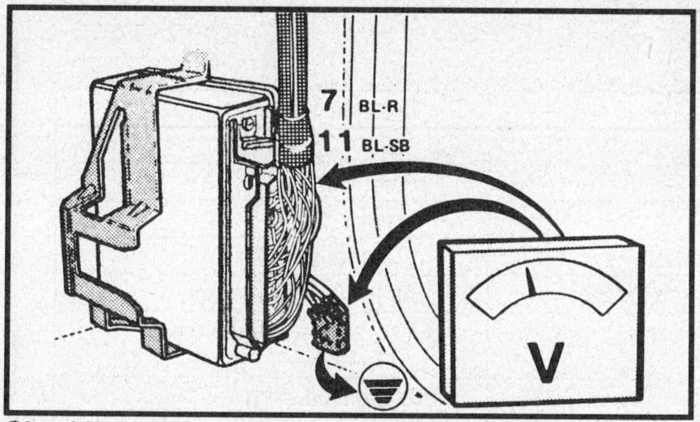

Checking pressure sensor voltage—Regina

unit connector. Voltmeter reading should be approximately 5 volts.

28. Install a hand operated vacuum pump to pressure sensor. Connect a voltmeter between pin 7 and pin 11 on the control unit connector and apply vacuum. Voltage should drop, indicating proper operation of the pressure sensor. If voltage does not drop, replace pressure sensor.
29. Turn ignition switch to the **OFF** position, then disconnect connector from control unit.
30. Test system relay/primary relay by connecting a voltmeter between ground and pin 9 on the control until connector, then a jumper wire from ground to pin 21 on the control unit connector. Relay should activate and voltmeter reading should read approximately 12 volts.
31. If fault has not been found, proceed to Control Function 3.

TESTING IN CONTROL FUNCTION 3

Air Conditioning Functions

On vehicles equipped with A/C, check the on/off function of the compressor as follows:

1. Turn ignition switch and A/C switch to the **ON** positions.
2. If LED goes off and then flashes fault code 1–1–4, A/C switch is operating correctly.
3. If the LED continues to flash rapidly, check lead from A/C control unit. Connect an ohmmeter between ground and pin 15 on the control unit connector. Note reading. Position A/C controls to A/C position. Resistance should change.
4. The LED will now return to rapid flashes, prior to the A/C compressor turning on. When the A/C compressor turns on, the LED light should go off and then flash fault code 1–3–4. If LED continues to flash rapidly, check lead from A/C compressor. Connect an ohmmeter between ground and pin 15 on the control unit connector. Ohmmeter should read 0–5 ohms. Turn ignition switch to the **OFF** position.

Automatic Transmission Idle Speed Compensation

On vehicles equipped with automatic transmission, proceed as follows:

1. Depress brake pedal and place selector lever into the **D** position and then into the **N** position. The LED should go off, then flash fault code 1–2–4.
2. If LED continues to flash rapidly, place gear selector lever into the **N** position. Connect ohmmeter between ground and pin 30 on the control unit connector. Ohmmeter should read 0 ohms.
3. Depress brake pedal and place gear selector lever into the **D** position. Ohmmeter should read infinity. On vehicles equipped with manual transmissions, ohmmeter should read 0 ohms in all gears.

Injectors

If injectors are not operating, but the LED flashes, proceed as follows:

1. Disconnect auxiliary relay.
2. Install a jumper wire between pin 2 (BL/Y) and pin 3 (GN/R) on the auxiliary relay socket.
3. Connect an ohmmeter between pin 9 and pin 18 on the control unit connector. Ohmmeter should read 4 ohms. If reading obtained is higher, current is not going through the injectors.
4. If ohmmeter reading is approximately 5.3 ohms, fault is in 1 injector or its leads; 7 ohms, fault is in 2 injectors or its leads; 14 ohms, fault is in 3 injectors or its leads.
5. If measured resistance is not as indicated, remove injector connectors and test them separately. Ohmmeter reading for individual injectors should be 14 ohms.

Idle Valve

If the idle valve is not operating, but the LED flashes, connect an ohmmeter between pin 9 and pin 33 on the control unit connector. Ohmmeter reading should be approximately 4 ohms.

Cold Start Valve

If the cold start valve is not operating, but the LED flashes, connect an ohmmeter between pin 9 and pin 32 on the control unit connector. Ohmmeter should read 10 ohms.

Auxiliary Relay

If the auxiliary relay is not operating, but the LED flashes, proceed as follows:

1. Connect a jumper wire from ground to pin 21 on the control unit.
2. Connect a voltmeter between ground and pin 18 on the control unit connector.
3. Both the system relay and auxiliary rely should operate and voltmeter reading should be approximately 12 volts. Remove jumper wire from pin 21.

Fuel Pump

If the fuel pump is not operating, but the LED flashes, proceed as follows:

1. Connect a jumper wire from ground to terminal 20 on the control unit connector. The fuel pump relay should close and the fuel pump should start to operate.

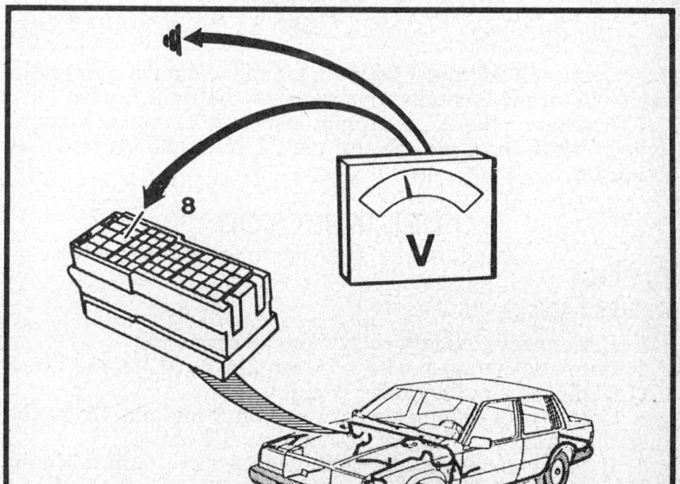

25-pin connector at right A-post—Regina

Code	Fault	Correction Test
1-1-1	No faults	—
1-1-2	Fault in control unit ①	Control unit
1-1-3	Fault in injector	Injector
1-2-1	Signal missing or faulty, to/from pressure sensor	Pressure sensor
1-2-2	Signal missing or faulty, to/from air temperature sensor	Air temperature sensor
1-2-3	Signal missing to/from coolant temperature sensor, possible grounding short	Coolant temperature sensor
1-3-2	Battery voltage too low/high ②	Charging system
1-3-3	Throttle-Switch; idle setting faulty, possible grounding short	Throttle switch
2-1-2	Oxygen sensor signal missing or faulty	Oxygen sensor
2-1-3	Throttle-Switch; full load setting faulty, possible grounding short	Throttle switch
2-2-1	Oxygen sensor not operating	Oxygen sensor or air leak
2-2-2	Fault in system relay	System relay
2-2-3	Signal missing to/from idle valve	Idle valve
2-3-1	Self-adjusting oxygen sensor not operating	Intake air or fuel system
2-3-2	Self-adjusting oxygen sensor not operating	Intake air or fuel system
2-3-3	Idle valve closed	Idle valve
2-4-1	EGR system malfunction ③	EGR vacuum booster
3-1-1	Signal missing from speedometer	Speedometer signal
3-2-1	Signal missing or faulty to/from cold start valve	Auxiliary relay
4-1-3	EGR temperature sensor circuit ③	EGR temperature sensor

① Change control unit
② Check battery and charging system
③ California

Diagnostic fault codes—Regina fuel system

2. Connect a voltmeter between ground and pin 8 connection in the 25 terminal connector, located at the right A-post.
3. Voltmeter should read approximately 12 volts. Remove jumper wires from pin 20 and pin 21 from the control unit connector.

FUEL INJECTOR

Testing

RESISTANCE

1. Disconnect auxiliary relay from its connector.
2. Connect a jumper wire between pin 2 (BL/Y) and pin 3 (GN/R) on the auxiliary relay connector.
3. Connect an ohmmeter between pin 9 and pin 18 on the control connector. Ohmmeter should read 4 ohms.
4. If ohmmeter reads approximately 5.3 ohms, fault is in 1 injector or its leads; 7 ohms, fault is in 2 injectors or its leads; 14 ohms, fault is in 3 injectors or its leads.
5. If measured resistance is not as indicated, disconnect injector connectors and test separately. Reading for individual injectors should be 14 ohms.

OPERATION

1. Turn the ignition switch to the **ON** position.
2. Open diagnostic cover and enter Control Function 3.
3. The injectors should begin to operate, followed by the idle valve and the LED should flash a continuous pattern.
4. Listen and feel by hand each injector to ensure they all are operating.
5. If an injector is not operating, change connector from an injector that's operating. If the fault moves to the injector that was operating, then the fault is in the connector lead.
6. If the injector is still not operating, then the injector is faulty.
7. Test injector separately, by connecting an ohmmeter between the injector pins. Ohmmeter should read approximately 14 ohms.

PRESSURE SENSOR

Testing

1. Disconnect the pressure sensor hose. Ensure that the ignition switch is in the **OFF** position.

2. Install the control unit connector loosely on the control unit.
3. Turn ignition switch to the **ON** position.
4. Connect a voltmeter between ground and pin 7 connection on the control unit connector. Voltmeter should read approximately 5 volts.
5. Connect voltmeter between ground and and pin 11 on the control unit connector. Voltmeter should read approximately 5 volts.
6. Connect a hand operated vacuum pump to the pressure sensor. Connect a voltmeter between pin 7 and pin 11 on the control unit connector then apply vacuum. Voltage should drop, indicating proper operation of the sensor. If voltage does not drop, replace pressure sensor.

AIR TEMPERATURE SENSOR

Testing

1. Turn ignition switch to the **OFF** position.
2. Check the air temperature sensor by connecting an ohmmeter between pin 6 and pin 10 on the control unit connector. Ohmmeter should read as follows:
   At −4°F (−20°C), approximately 15,000 ohms.
   At 68°F (20°C), approximately 2,500 ohms.
   At 212°F (100°C), approximately 160 ohms.
3. To check the sensor at various temperatures, heat sensor using a hot air gun.
4. If specifications are not as specified, test the air temperature sensor to see if fault is in the sensor or its leads.

COOLANT TEMPERATURE SENSOR

Testing

1. Turn ignition switch to the **OFF** position.
2. Check the coolant temperature sensor by connecting an ohmmeter between ground and pin 13 on the control unit connector. Ohmmeter reading various depending on temperature.
   At 14°F, reading should be between 8,260–10,560 ohms.
   At 68°F, reading should be between 2,280–2,720 ohms.
   At 176°F, reading should be between 290–364 ohms.
3. If specifications are not as specified, test the coolant temperature sensor to see if fault is in the sensor or its leads.
4. Check ground connections at the intake manifold.

THROTTLE SWITCH

Testing

1. Turn ignition switch to the **OFF** position.
2. Check the throttle switch by connecting an ohmmeter between ground and pin 2 on the control unit connector. ohmmeter should read 0 ohms (switch closed).
3. Depress gas pedal slightly, resistance should increase to 2000–3000 ohms (switch opens).
4. Connect an ohmmeter between ground and pin 3 connection on the control unit connector. Ohmmeter should read infinity. (full load switch open).
5. Depress gas pedal all the way. Ohmmeter should read 0 ohms.
6. If specifications are not as specified, check at throttle switch to see if fault is in throttle switch or its leads. Check ground connections at intake manifold.

SYSTEM RELAY

Testing

1. Turn ignition switch to the **OFF** position.
2. Check the system relay/primary relay as follows:
   a. Connect a voltmeter between ground and pin 9 connection on the control unit connector.
   b. Install a jumper wire from ground to pin 21 on the control unit connector. Relay should operate and voltmeter should read approximately 12 volts.
3. Check the system relay/pump relay as follows:
   a. Install a jumper wire from ground to pin 20 on the control unit connector. The pump relay should close and start the fuel pumps.
   b. Connect a voltmeter between ground and pin 8 in the 25 terminal connector located at the right A-post. Voltmeter should read approximately 12 volts. Remove jumper wires from pin 20 and pin 21.

EXHAUST CARBON MONOXIDE CONTENT

Testing

1. Check carbon monoxide content by inserting a carbon monoxide meter probe into the tapped hole on the catalytic converter.
2. Start and run engine. Check carbon monoxide content.
3. If carbon monoxide content is not within specifications, check the oxygen sensor.

OXYGEN SENSOR

Testing

1. Disconnect oxygen sensor electrical connector.
2. Ground the lead coming from the control unit. carbon monoxide content should rise, indicating that the control unit and its connections are operating correctly.
3. Connect a voltmeter between ground and the connector coming from the oxygen sensor. Reading at correct carbon monoxide content should be approximately 0.5 volts.

INTAKE AIR SYSTEM

Testing

Check the intake system for leaks. System air leaks could cause the fuel mixture to be too lean. Check for leaks in the following locations:

1. Intake manifold between the air filter and the manifold.
2. All hoses and hose connections to the intake manifold.
3. The intake manifold and the throttle housing.

IDLE VALVE

Testing

1. Turn ignition switch to the **OFF** position.
2. Check the idle valve by connecting an ohmmeter between pin 9 and pin 33 on the control unit connector. Ohmmeter should read approximately 4 ohms.

SPEEDOMETER SIGNAL

Testing

1. Turn ignition switch to the **OFF** position.
2. Check the speedometer signal as follows:
   a. Remove panel from under the instrument panel on the drivers side.
   b. Disconnect cable (VO/W) from the speedometer, then connect an ohmmeter between the cable and pin 34 on the control unit connector. Ohmmeter should read 0 ohms.
   c. If fault code 3-1-1 flashes and ohmmeter reads 0 ohms, the speedometer signal is missing. Replace speedometer assembly.

AUXILIARY RELAY

Testing

1. Remove auxiliary relay from its connector. Install a jumper wire between pin 2 (BL/Y) and pin 3 (GN/R) on the auxiliary relay socket.
2. Check the cold start valve by connecting an ohmmeter between pin 9 and pin 32 on the control unit connector. Ohmmeter reading should be approximately 10 ohms.

EGR VACUUM BOOSTER

Testing

On California vehicles an EGR system malfunction is indicated by Fault Code 2–4–1. It results when the EGR temperature sensor detects that the exhaust gas recirculation flow is too low.

1. Turn the ignition switch **ON**.
2. Enter Control Function 3. The lamp should flash according as the EGR converter is activated.
3. Turn the ignition switch **OFF**.
4. If the EGR converter is not activated, measure the resistance between pins 6 and 15 in the control unit connector. The resistance should be approximately 75–95 ohms.
5. If the value is different, check the wiring and connection.

EGR SYSTEM

Testing

This test applies to California vehicles.

1. Open the EGR vacuum booster connector. Remove the lead directly opposite the black-white lead between pin 1 and the control unit. Reconnect the connector.
2. Start and warm the engine to operating temperature.
3. Ground the loose lead. If the engine starts to run unevenly, the EGR system is operating.

EGR TEMPERATURE SENSOR

Testing

If the signal from the EGR temperature sensor is absent or faulty, Fault Code 4–3–1 will result.

NOTE: The sensor is extremely sensitive to shock and impact. When replacing, torque to 5–10 ft. lbs. (7–13 Nm).

1. Measure the resistance between terminal 14 (ground) and terminal 22 on the control unit connector. It should be 500–1000 ohms.
2. If the value is different, check the leads and connections, the ground terminal on the intake manifold and the resistance of the sensor connector.

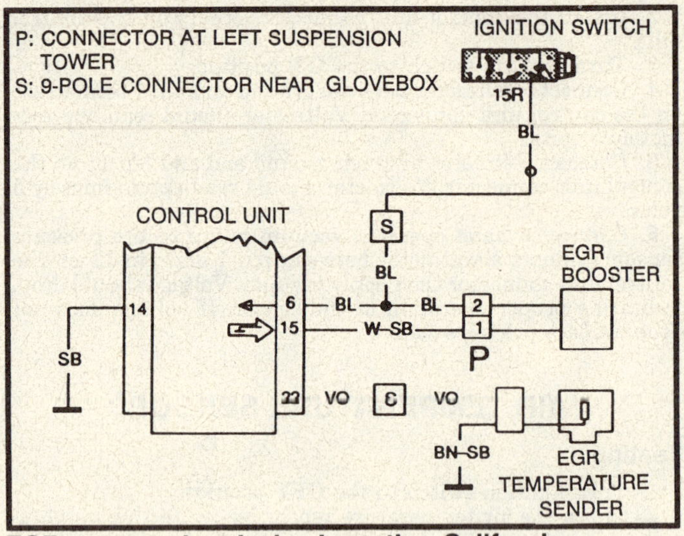

EGR system electrical schematic – California

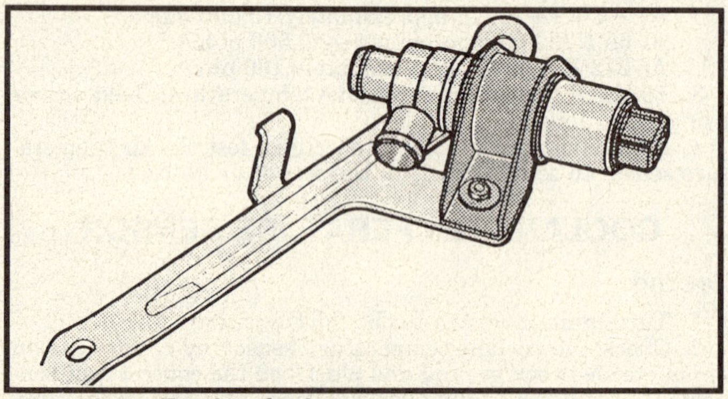

Idle valve – Regina fuel system

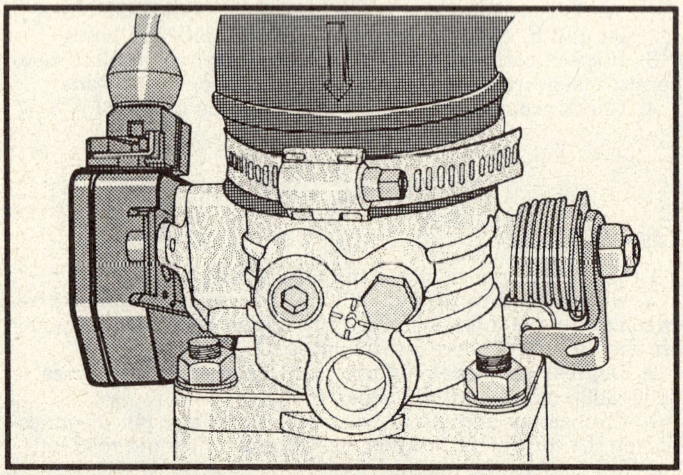

Throttle switch – Regina fuel system

Regina

1 Battery
2 Ignition switch
9 Starter motor
11 Central electrical unit
12/7 Conn. CHECK ENGINE lamp
12/8 Conn. SHIFT indicator lamp
29 Positive terminal
31 Ground point in central electrical unit
81 Pressure sensor, AC
84 Coolant temperature sensor
85 Speedometer
135 Relay fuel injection
185 Intake–air temperature sensor
187 Lambda sond
188 Cold start injector
196 Idle valve
198 Throttle switch
200 AC compressor solenoid
202 Climate unit
210 Tank pump
211 Fuel pump
212 Test point
260 Electronic control unit, ignition system
361–364 Injectors
419 Ign. system power stage
440 Pressure sensor
464 Relay, radio interference suppression
472 Electronic control unit, fuel system
482 Diagnostic unit

A Connector, right A–post
B Connector, left A–post
C Connector, right wheelhouse
D Connector, right wheelhouse
E Connector, left A–post
F Connector, left wheelhouse
H Ground point, inlet mainfold
J Ground point in trunk
K Connector on bulkhead
N Ground point, right fender
O Connector, bulkhead
P Connector, central electrical unit
Q Connector at bulkhead, 4–terminal
R Ground point at intake manifold
S Connector at bulkhead, 1–terminal
∘ Solder joint

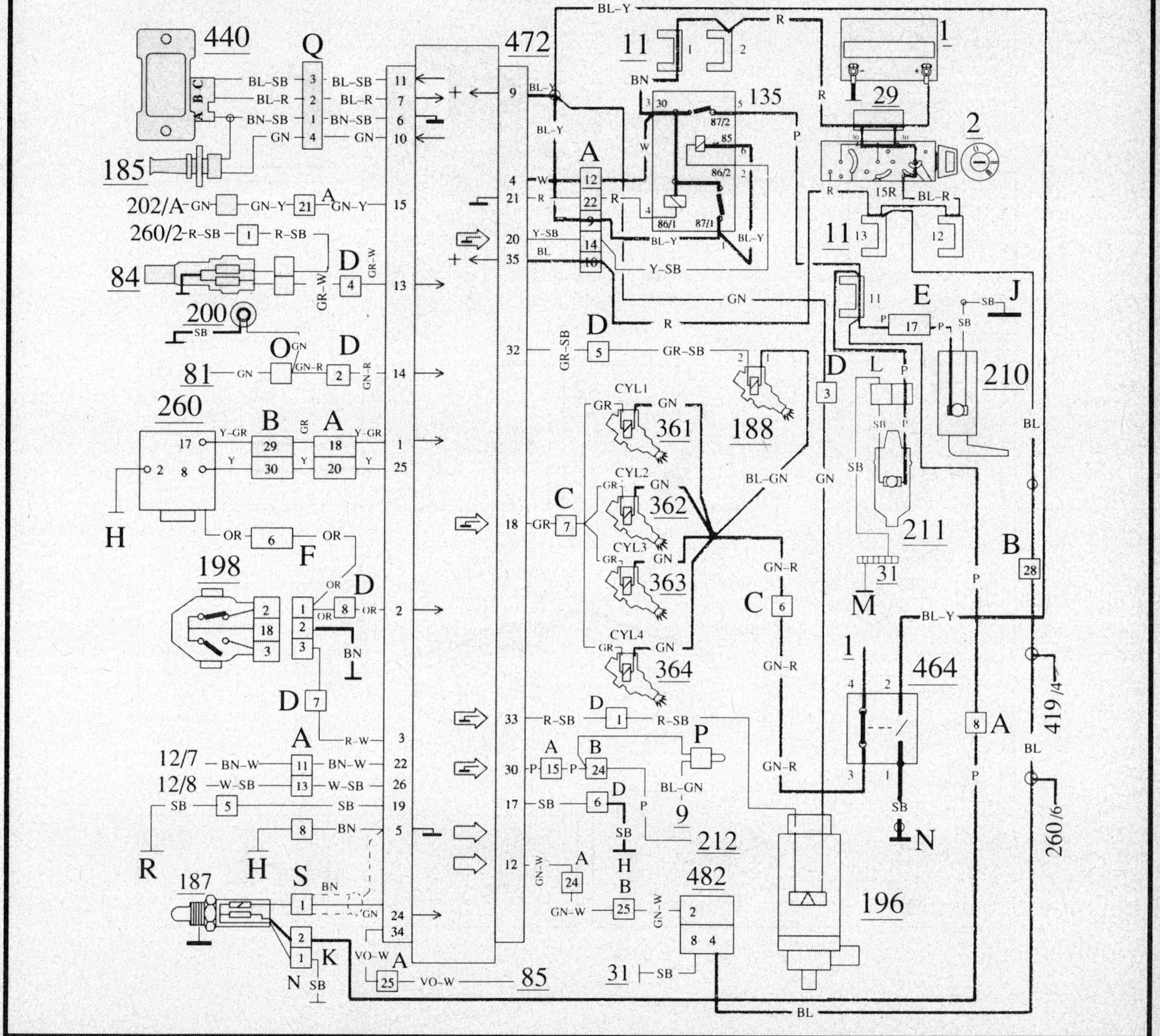

Regina wiring diagram—740 and 940

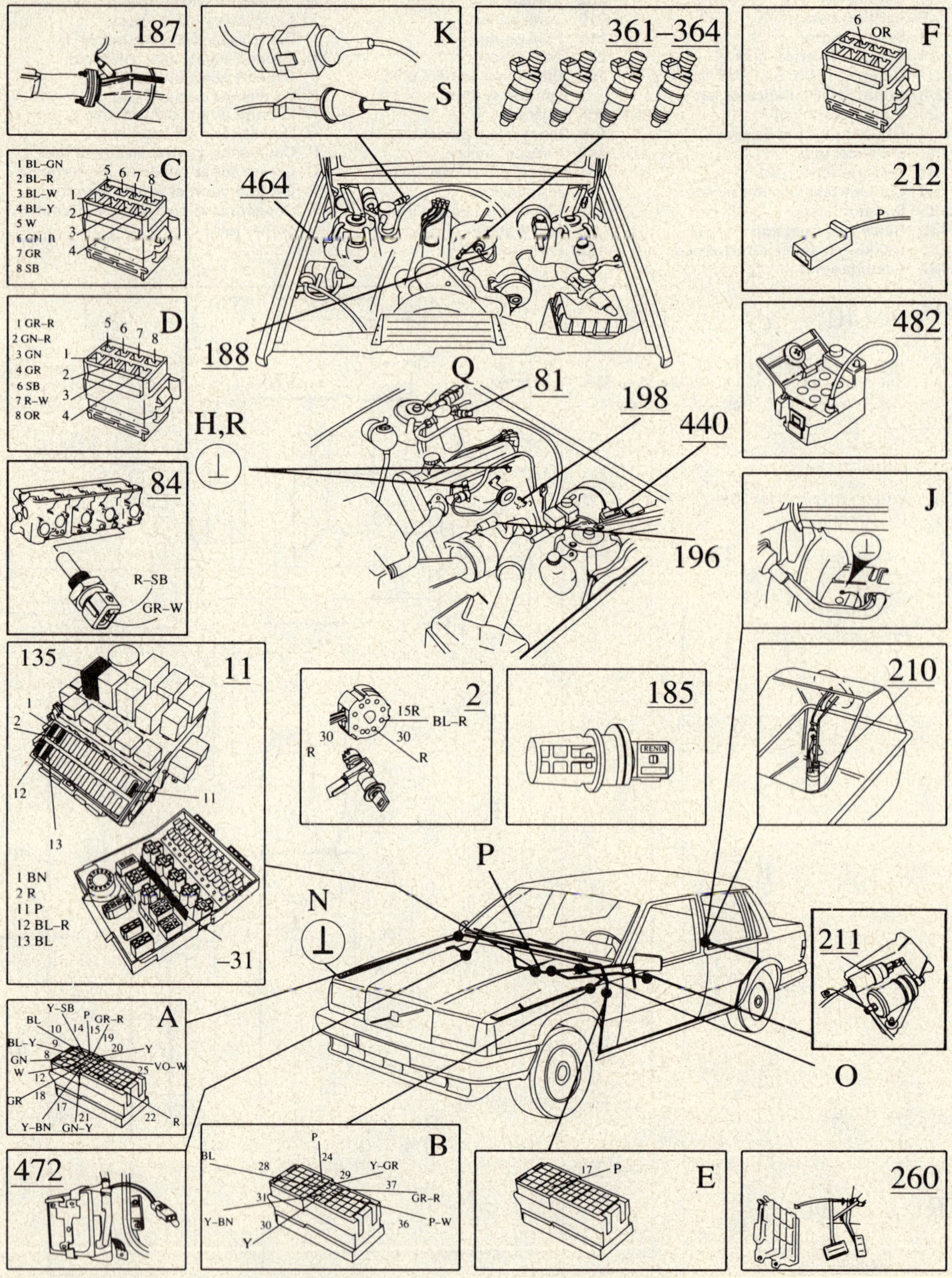

Regina component terminal identification–740 and 940

Component Replacement

SERVICE PRECAUTIONS

- Use caution when working around ignition system components. the ignition system operates at high power with dangerous voltages in both the low-tension and high-tension circuits, including connectors and similar fittings.
- Do not operate the fuel pump when the fuel lines are empty.
- Do not arc weld on the vehicle unless all the control units have been removed.
- The control units must not be exposed to temperatures above 175°F (80°C), such as in a paint booth.
- Make sure that the ignition switch is **OFF** before connecting or disconnecting the control unit, ignition coil, spark plug wiring or test instruments.
- Always use new gaskets and seals when remaking a fuel connection.
- When performing a compression test, disconnect the wire from terminal 1 on the ignition coil (to prevent arcing) and remove the connectors from the injectors (to avoid flooding).
- Make sure all control unit connectors are fastened securely. A poor connection can cause an extremely high surge voltage, resulting in damage to integrated circuits.
- Keep any control unit harness at least 4 in. away from adjacent harnesses to prevent a system malfunction due to external electronic interference.
- Keep all parts and harnesses dry during service.
- Before attempting to remove any parts, turn off the ignition switch and disconnect the battery ground cable.
- Always use a 12 volt battery as a power source.
- Do not attempt to disconnect the battery cables with the engine running.
- When using a tester probe on the control unit connector, first remove the protective sleeve, then always gently check connections through the numbered holes on the connector side.
- Do not attempt to disassemble a control unit under any circumstances.
- If a battery cable is disconnected, the memory will be erased.

FUEL PRESSURE RELIEF

1. Remove the fuel pump relay from the electrical distribution unit. The relay is furthest to the left on the second row.
2. Crank the engine until the engine will not start anymore, indicating that the fuel pressure is relieved.
3. Reinstall the relay.

CAUTION

To avoid fire and personal injury, replace any plastic tie bands removed from the fuel lines.

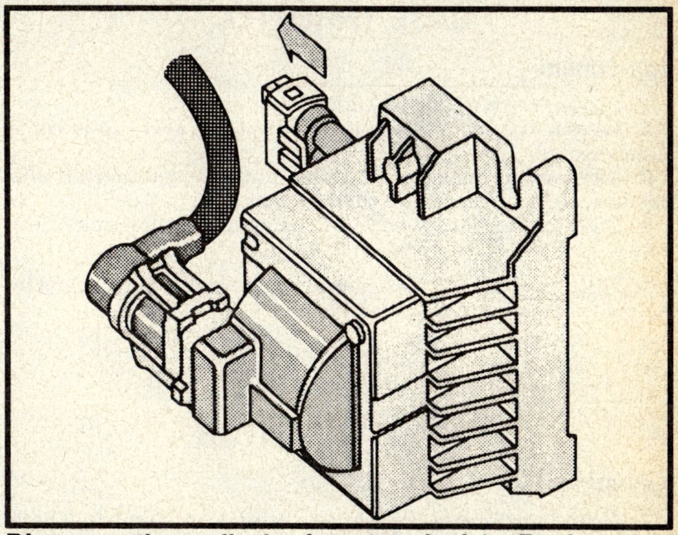

Disconnecting coil wire from terminal 1 – Regina

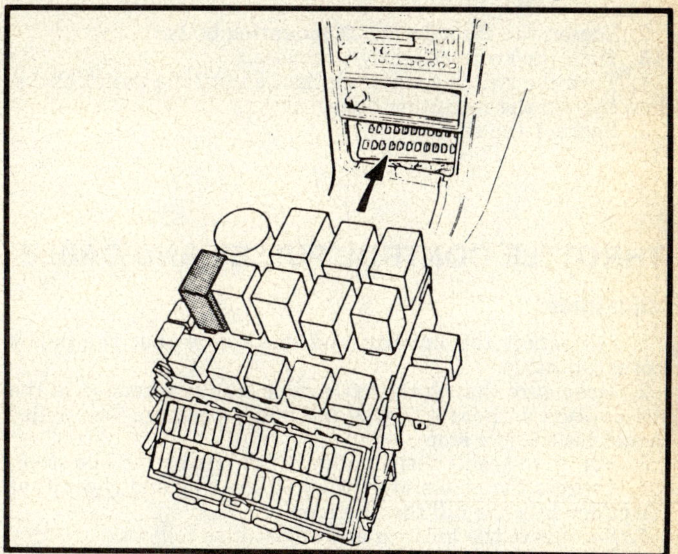

Fuel pump relay location – Regina

FUEL INJECTOR

Removal and Installation

NOTE: Remove and install the fuel distribution pipe, the injectors and the cold start valve as one unit.

1. Unbolt the pressure regulator from the fuel rail bracket.
2. Remove the fuel rail bolts without distorting the metal of the rail.
3. Use 2 line wrenches and separate the line fittings, then remove the injector assembly.
4. Installation is the reverse of removal. Use new O-rings and lubricate with petroleum jelly.

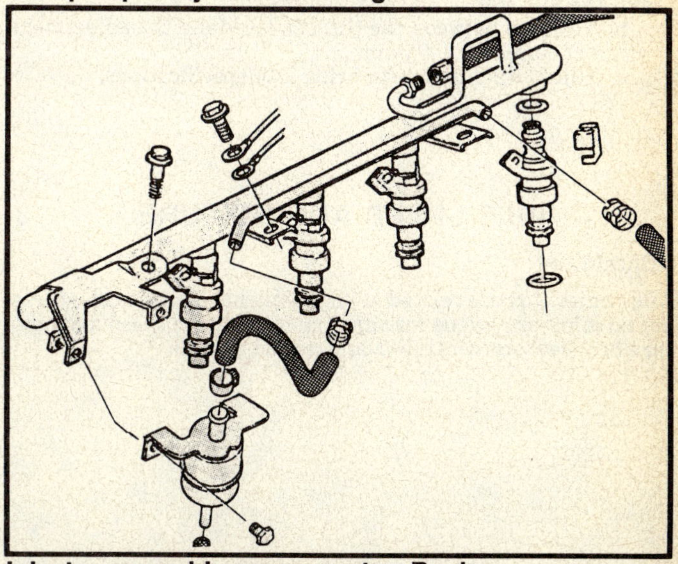

Injector assembly components – Regina

BASE THROTTLE

Adjustment

1. Loosen the throttle locknut.
2. Loosen the adjustment screw until the throttle is completely closed.
3. Tighten the adjustment screw until it just touches the linkarm, then turn a half a turn further.
4. Tighten the locknut without changing the adjusting screw position.

NOTE: It may be necessary to loosen the throttle switch before setting the throttle.

Setting base idle—Regina

THROTTLE SWITCH

Adjustment

1. Open the throttle slightly and listen to the switch. There should be a click when the shutter opens, if the adjustment is correct.
2. Loosen the throttle switch mounting bolts.
3. Turn the switch slightly clockwise.
4. Turn the switch counter-clockwise until the switch clicks, then tighten the mounting bolts.
5. Recheck the setting.

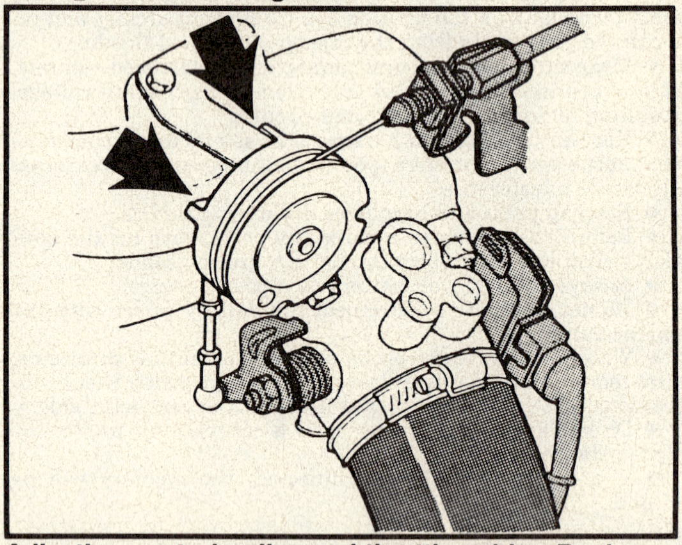

Adjusting control pulley and throttle cable—Regina

THROTTLE CONTROL PULLEY AND CABLE

Adjustment

1. Disconnect the linkarm and make sure that the pulley moves smoothly.
2. Make sure that the throttle cable is fully extended in the idle position without affecting the pulley position. The pulley should hit the idle stop.
3. Adjust the cable if the pulley is not hitting the idle stop.
4. Depress the accelerator pedal all the way and check that the pulley hits the full throttle stop.
5. Reconnect the linkarm and adjust it as follows:
   a. Insert a 1mm feeler gauge between the control pulley and the idle stop.
   b. The play between the throttle lever and the adjustment screw should be 0.1–0.3mm.
   c. Adjust the linkarm to bring into specifications.

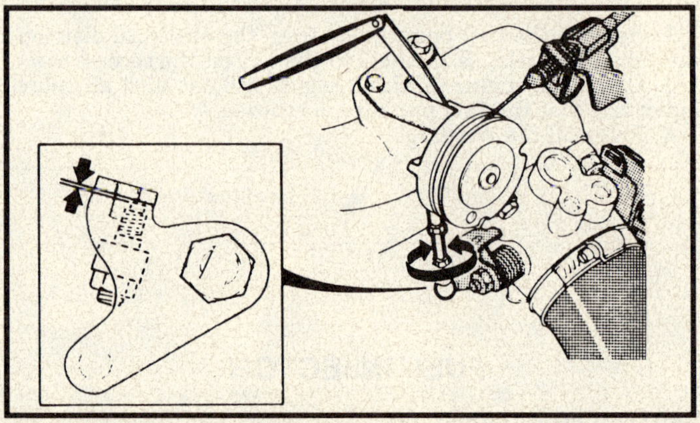

Adjusting link arm—Regina

IDLE SPEED AND MIXTURE

Adjustment

After other throttle related adjustments have been made, there are no other provisions for adjustment. The idle speed and mixture are feedback controlled and not adjustable.

Yugo Fuel Injection

SECTION 17

YUGO MOTRONIC MP 3.1 FUEL INJECTION

General Information

The Yugo has been upgraded to use the Bosch Motronic MP 3.1 fuel injection system. This is an improvement over the older design Carter/Weber carburetor. The new fuel injection system allows for improved performance, drivability, fuel economy and the ability to self-diagnose.

The Motronic unit controls both the fuel injection and the electronic ignition. This reduces the total cost of the 2 separate systems since the same sensors are shared by a common control unit.

The Motronic control unit receives data from sensors, interprets the information, compares it to the preprogrammed parameter map and outputs signals to the various output devices such as the injectors. The unit controls fuel to the engine and electronically meters it. It regulates the fuel enrichment for cold starts, warm-up and acceleration. The unit can compensate for additional engine loads, such as the air conditioner compressor.

MOTRONIC CONTROL UNIT AND RELAY

The control unit is located below the dash panel on the passenger side of the vehicle. There is a multi-pin connector leading to it and a vacuum line. Fitted to the vacuum line is a chamber that acts as a moisture condenser and pulsation damper.

The control unit is supplied with power through the main relay. When the ignition switch is turned **ON** the main relay energizes. The main relay in turn supplies power to the fuel pump relay through an inertia switch. The inertia switch is located at the left rear of the engine compartment.

The Motronic unit controls the engine operating parameters based on the signal it receives from the data sensors. It takes this information, processes it, and adjusts the air/fuel mixture to suit. The Yugo Motronic system has no direct way of measuring the air mass flowing to the engine. It calculates the air mass from information provided by the engine speed sensor, the manifold absolute pressure (MAP) sensor and the intake air temperature sensor.

The control unit controls the fuel mixture by adjusting the pulse signal to the injectors. The longer the pulse to the injectors, the longer the injector remain open and passing fuel. This increases the richness of the mixture. The reverse is true when the pulses are shortened.

During the warm-up phase of operation, the control unit modifies the injection pulses to increase the mixture richness at colder temperatures. At -22° F. (-30° C) the pulse width is increased to 140 mS; at 32° F (0° C) the width is 50 mS; at 176° F (80° C) the width is 6 mS.

Additional modification to the fuel mixture is effected by information provided by the oxygen sensor and processed by the control unit. The control unit changes the fuel mixture to maintain an optimal air/fuel ratio, stoichiometric, for best combustion and emission level control.

During deceleration fuel is cut off to the injectors. The fuel cut off prevents the mixture from going rich and emitting excessive emissions. The fuel is reinstated at 1800 rpm.

Motronic control unit pin numbering—Yugo

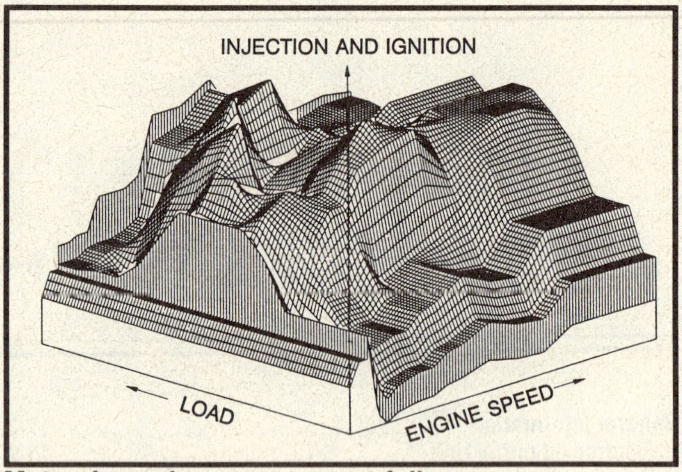

Motronic engine management follows a preprogrammed map of engine operating parameters—Yugo

Under full power conditions, the control unit modifies the injection time. Under conditions where the mixture needs to be enriched for proper full power operation, the control unit increases the injection time by 10%.

DATA SENSORS

Engine Speed and Position Sensor

A tooth wheel is mounted on the crankshaft at the front of the engine. The tooth wheel has a gap in the teeth to provide TDC information to the control unit. An inductive pickup mounted above the tooth wheel produces a pulse output that is read by the control unit. The tooth gap information and pulse speed information is separated by the control unit to provide individual reference sources for engine speed and position.

Intake Air Temperature Sensor

Located on the underside of the intake manifold, the intake air temperature sensor provides the temperature information for the control unit. The sensor is of the Negative Temperature Coefficient (NTC) type. As the temperature rises, the resistance of the sensor drops. The control unit uses the temperature information to adjust the fuel injection rate by means of calculating the air mass in conjunction with the engine speed information and MAP sensor information.

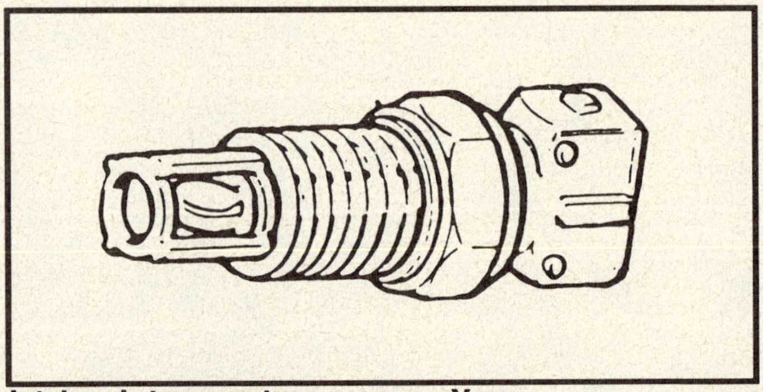

Intake air temperature sensor—Yugo

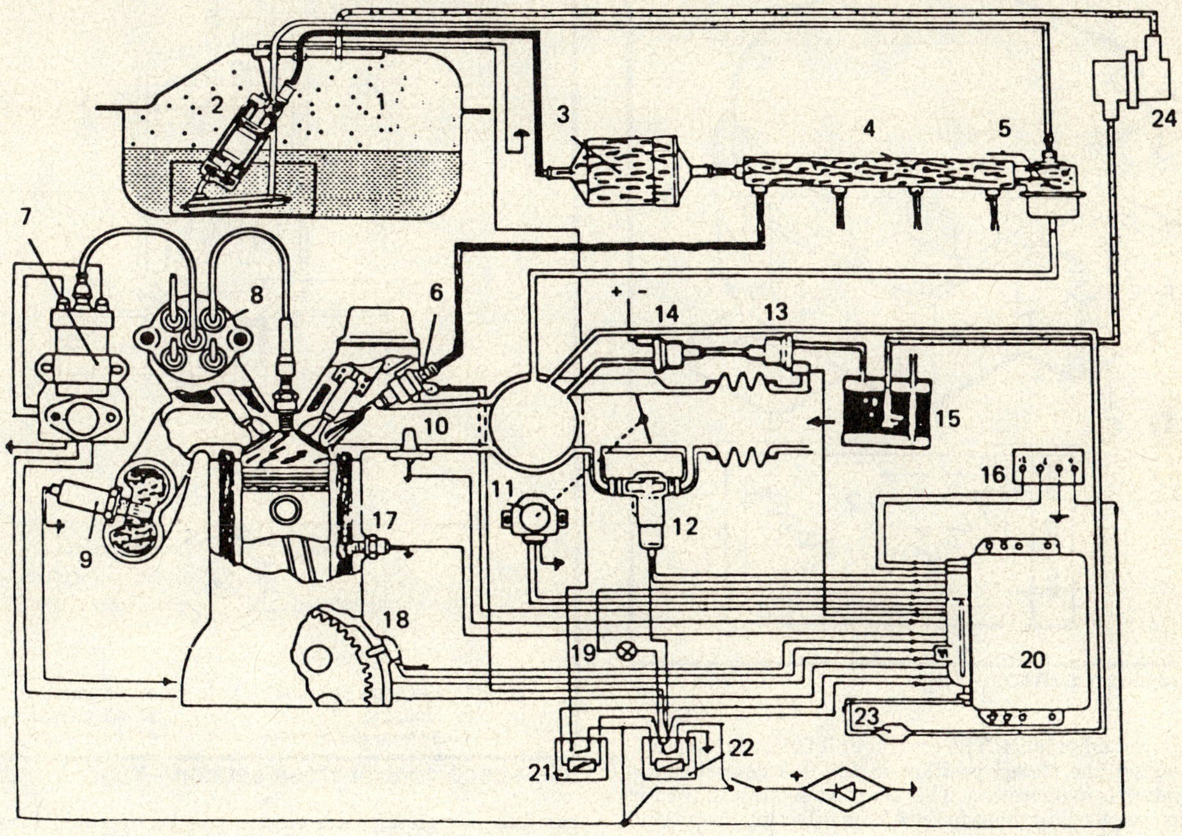

1. Fuel tank
2. Electric fuel pump
3. Fuel filter
4. Fuel rail
5. Fuel pressure regulator
6. Injector
7. Ignition coil
8. Distributor
9. Oxygen sensor
10. Intake air temperature sensor
11. Throttle position sensor
12. Idle speed control
13. Tank emission valve
14. Tank emission switch
15. Vapor canister
16. Diagnostic plug
17. Engine temperature sensor
18. Engine speed and position sensor
19. Check engine light
20. Motronic control unit
21. Fuel pump relay
22. Main relay
23. Vacuum signal damper
24. Rollover valve

Motronic system MP 3.1 diagram—Yugo

Engine Temperature Sensor

Located on the thermostat housing, the engine temperature sensor provides the temperature information for the control unit. The sensor is of the Negative Temperature Coefficient (NTC) type. As the temperature rises the resistance of the sensor drops. The control unit receives the engine temperature signal and adjusts the fuel mixture during warm-up and running conditions.

Throttle Position Sensor

The Throttle Position Sensor (TPS) is mounted to the throttle body and connected to the throttle shaft. It provides throttle opening information to the control unit by means of a varying resistance. The resistance range is 1150–2775 ohms. The control unit supplies the TPS with a nominal operating voltage of 5 volts.

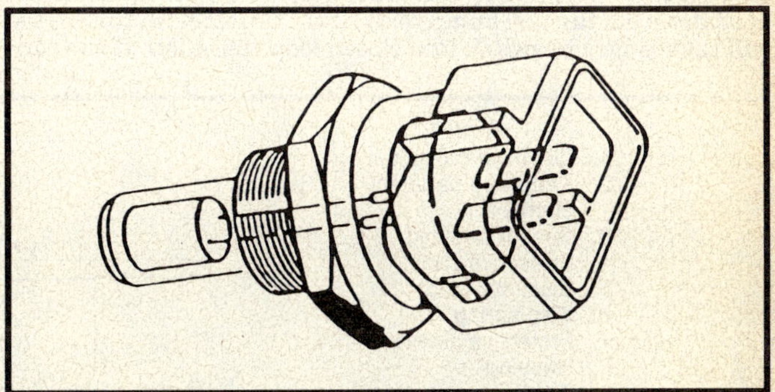

Engine temperature sensor—Yugo

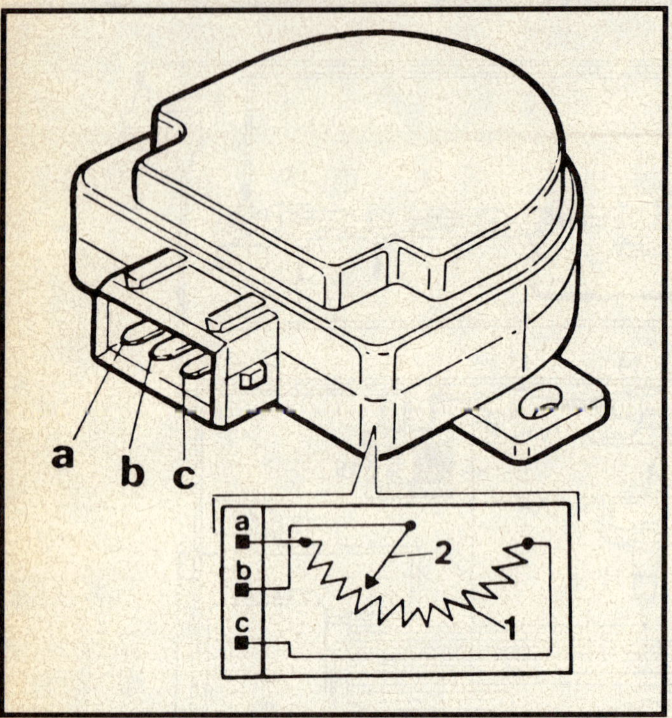

Throttle position sensor—Yugo

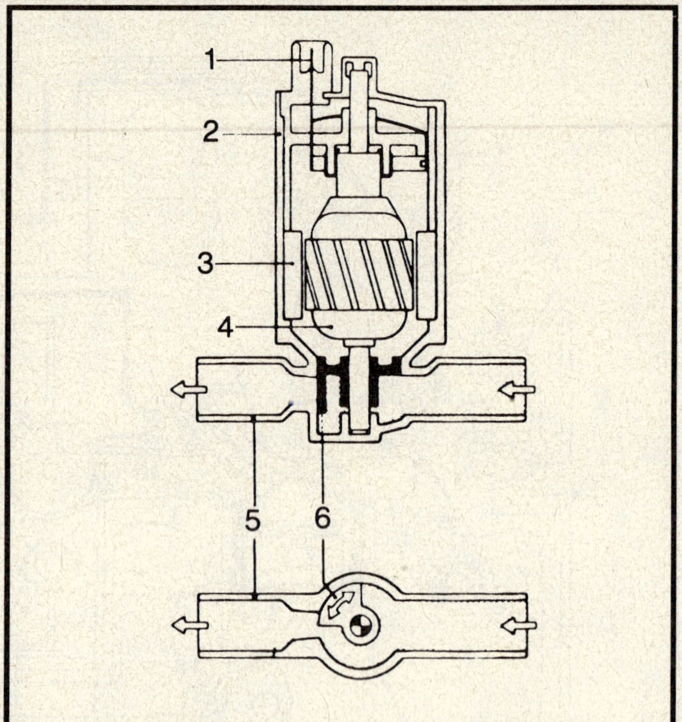

1. Electrical terminal
2. Housing
3. Permanent magnet
4. Armature
5. Air passage
6. Rotary slider

Idle speed control cross section—Yugo

The TPS is not adjustable. The control unit uses adaptive settings that adjust the closed position in the 0–3 degree range while the system is in operation. This adaptive setting allows for wear and age, preventing those factors from affecting the engine operation.

Oxygen Sensor

The oxygen sensor consists of a ceramic tube coated with a thin layer of platinum. The closed end of the tube is in the exhaust stream and the reference end is open to the air. When the sensor is at operating temperature, oxygen in the exhaust will cause the unit to generate a voltage. The control unit reads the signal and adjusts the fuel injection to control the mixture. This is called closed loop control strategy. Under normal conditions, the control unit will vary the mixture between rich and lean to create an average ideal air/fuel ratio.

Most sensors are equipped with an internal resistance heater to bring it to operating temperature quickly. Power for the heater comes from the fuel pump relay. If the heater fails, the engine will take longer to switch into closed loop operation. If the sensor fails completely, the engine will be operated in open loop control and slightly lean to protect the catalytic converter from overheating.

The sensor best functions at temperatures above 662° F. (350° C.). At this temperature level and higher the oxygen sensor will most quickly react to changing mixture conditions. At this temperature the oxygen sensor acts in the 0.2 to 0.8 volt range. At temperatures below this the system will revert to open loop.

The system will operate in open loop during cold start operation. The system will revert to open loop during abrupt acceleration, full load operation and during coast at closed throttle.

CONTROLLED OUTPUTS

Idle Speed Control

The idle speed is controlled by a rotary valve idle controller. The idle speed control is mounted on the intake manifold and by-

1. Electrical contact
2. Protective ceramic element
3. Sensor ceramic
4. Protective cap
5. Electrical lead
6. Disc spring
7. Protective sleeve
8. Housing
9. Electrode-negative
10. Electrode-positive

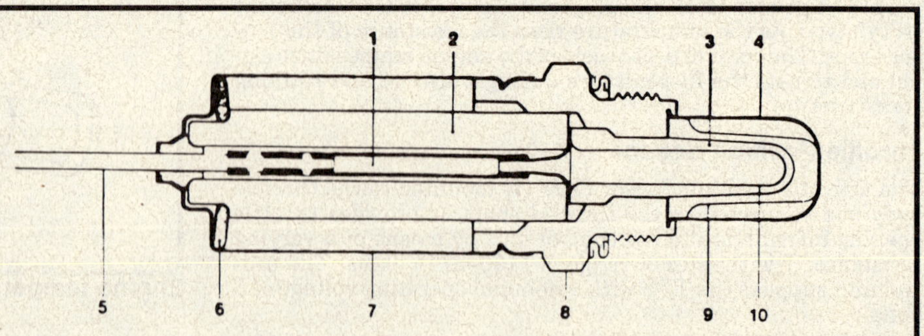

Cross section of a non-heated oxygen sensor. Heated sensor has an electrical heating element integral to the body—Yugo

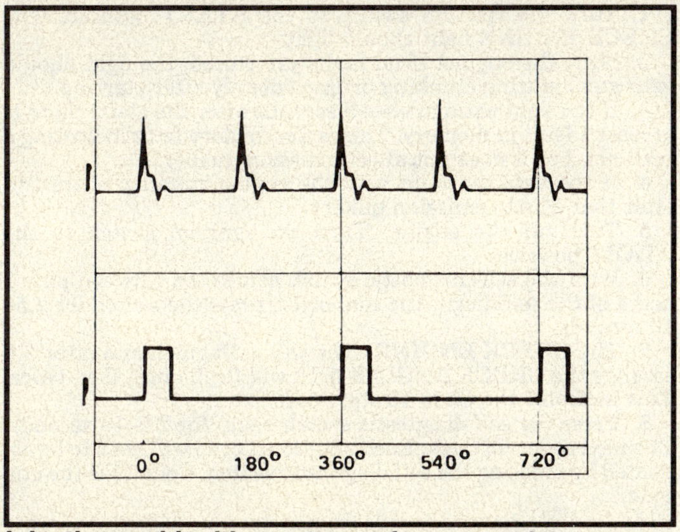

Injection and ignition pattern of warm engine operation. The upper pattern is the spark output and the lower pattern is the injector pulses. There are 2 injector pulses for each engine cycle—Yugo

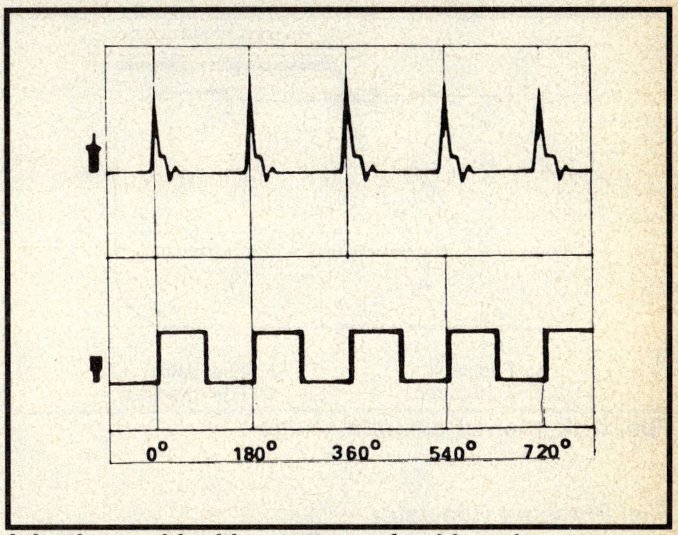

Injection and ignition pattern of cold engine operation. The upper pattern is the spark output and the lower pattern is the injector pulses. There are 4 injector pulses for each engine cycle to provide additional fuel for warm-up—Yugo

passes air around the throttle plate. The idle speed is adjusted by varying the valve opening in the controller. The Motronic control unit receives engine speed information from the engine speed sensor and compares it to the programmed speed. The control signal is developed from this comparison and the information is sent to the idle speed control.

The idle speed control contains 2 windings that are alternately supplied with voltage. The windings exert opposing forces on the valve armature. Because the armature exhibits inertia, the opposing forces move the valve varying degrees in proportion to the voltage ratio of the 2 windings.

The valve opening can be varied from 18% to 82% open. With a voltage ratio of 25% the opening assumes a small opening at warm operation and no auxiliary loads on the engine. With increased loads and the need for higher idle speeds during warm-up the ratio is increased until the fully open position of 82% is achieved.

Fuel Injectors

The fuel injectors are controlled by the Motronic unit. They are solenoid actuated with pintle-type valves. The solenoid pulls the pintle up 0.1 mm from its seat when an injection pulse is received by the injector. A coil spring returns the pintle to the closed position within 1–1.5 mS of the pulse being removed. The injectors are electrically connected in parallel. All 4 injectors spray simutaneously. The injectors spray twice each crank revolution during normal running; 4 times each crank revolution during cold start.

FUEL DELIVERY SYSTEM

Fuel Injectors

The fuel injectors are solenoid-operated valves. The injectors are mounted in rubber bushings to isolate them from heat and vibration. The injectors are positioned in the intake tract spraying towards the back of the valves.

Fuel Pump

The electric fuel pump is mounted in the fuel tank and is accessible by removing the fuel tank cover plate. The fuel pump receives power from the fuel pump relay. The Motronic control unit will only turn on the fuel pump relay at engine speeds greater than 20 rpm. There is a check valve built into the output end of the fuel pump to prevent fuel drainback and ensure system pressure at startup.

Fuel Filter

The fuel filter is located after the fuel pump and before the fuel pressure regulator. It is physically located under the vehicle in front of the fuel tank. The filter is marked as to the fuel flow direction by an arrow on the filter body.

Fuel Distribution Rail

The fuel rail acts as a fuel accumulator and as a fuel distributor to the injectors. It is mounted on the intake manifold and holds the injectors in place. The internal volume helps damp any pressure variations.

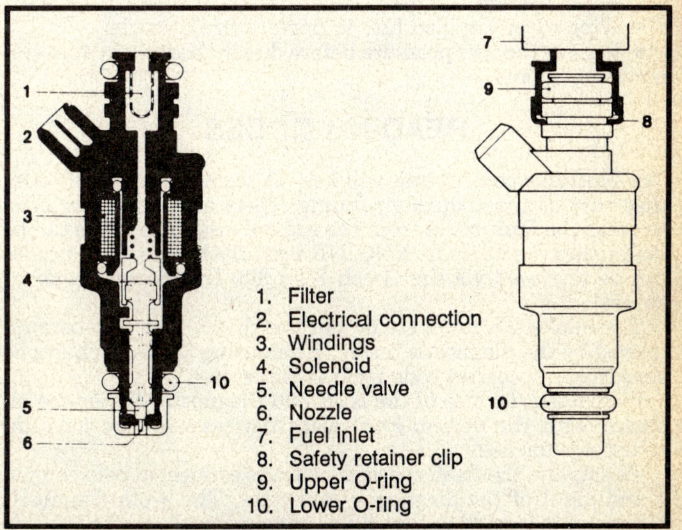

Cross section of the fuel injector—Yugo

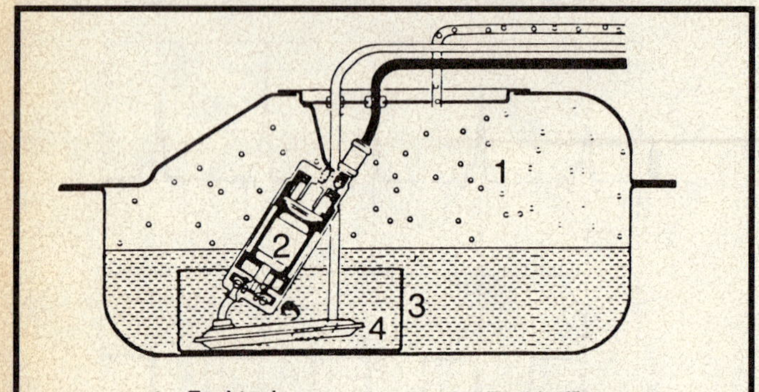

Fuel tank internal diagram—Yugo

Fuel Pressure Regulator

The fuel pressure regulator is mounted on the fuel rail. By means of a vacuum signal from the intake manifold, the regulator maintains a constant pressure differential between the intake manifold and fuel pressure. Fuel that is bypassed by the regulator is directed back to the fuel tank.

Diagnosis and Testing

SERVICE PRECAUTIONS

The following cautions that should be observed to protect the Motronic system and components:

- Always disconnect the battery, the Motronic control unit and the ignition coils when using an electric welder on the vehicle or when charging the battery or when the vehicle is placed in a paint drying oven.
- Remove the main Motronic relay to disable both the fuel and ignition systems when checking the compression. Never crank the engine after removing the distributor cap or disconnecting the high tension wire on the ignition coil.
- Never disconnect the battery or wires on the alternator, starter or spark plugs when the engine is running.
- Always work in a well ventilated area to prevent the build up of hazardous fumes.
- Always keep a fire extinguisher within reach when working of the fuel system.
- Wipe up any spilled fuel to prevent fire hazards.
- Relieve the fuel pressure before disconnecting any fuel system components.

READING CODES

The Motronic control unit will keep in memory any fault codes that have occurred during running. There are 2 methods of retrieving the fault codes from the control unit. The codes can be read using the CHECK ENGINE light flash code or the codes can be retrieved via the Bosch KTS 300 Diagnostic Tester or equivalent.

The blink code numbers do not match the code numbers retrieved by the diagnostic tester. When using the code charts be sure that the correct code form is being used.

Instruction for use of the KTS 300 Diagnostic Tester are included with the device. Equivalent devices will also have instructions for use.

To activate the flash codes, a switch is connected between pin 1 and pin 4 of the diagnostic connector. The 4-pin diagnostic connector is located in the engine compartment, on the left side near the fire wall.

1. Turn the ignition switch to the **START** position. The CHECK ENGINE light should light.
2. Start the engine. If no faults are stored, the light should extinguish during cranking or immediately after starting.
3. If the light extinguishes 3 seconds after the start, there is at least 1 fault in memory. This is a secondary fault indicating a problem, but it does not affect emission quality.
4. If the light stays on with the engine running, there is a fault that affects emission quality.
5. Turn off the engine. Turn the ignition switch to the **START** position.
6. With the self-diagnostic switch connected between pins 1 and 4 of the selfidiagnostic connector, press the switch for 2.5–10 seconds.
7. The CHECK ENGINE light will extinguish and after 3.6 seconds the CHECK ENGINE light will flash once, then twice. This indicates the Code 12 start code.
8. Press the self diagnostic switch again for 2.5–10 seconds. The next code will flash. Each additional code will need to be activated by pressing the self diagnostic switch. Code 11 is the end code.

CLEARING CODES

1. After end Code 11 is flashed, press the self diagnostic switch for more than 10 seconds. This erases the code memory.
2. The memory can also be erased by disconnecting the negative battery terminal. Any other battery dependant memories, such as radio or alarm system, will be erased, also.
3. When using the KTS 300 Tester, using the ERASE command will erase the code memory.

OUTPUT PERIPHERY TESTING

This test allows the control unit to self-actuate the outputs. Check the outputs by audible means; each output will make a distinctive sound.

1. To activate the output periphery test, a switch is connected between pin 1 and pin 4 of the diagnostic connector. The 4-pin diagnostic connector is located in the engine compartment, on the left side near the fire wall.
2. Press and hold the self diagnostic switch and turn the ignition switch to the **START** position.
3. Keep the diagnostic switch depressed for 3–5 seconds after the ignition switch is turned to the **START** position.
4. The CHECK ENGINE light will extinguish after 2.5 seconds. Code 91 will flash and the injectors will start to operate.
5. Press the self diagnostic switch again for 3–5 seconds. Code 92 will flash and the idle speed regulator will operate.
6. Press the self diagnostic switch again for 3–5 seconds. Code 94 will flash and the tank emission valve will operate.
7. Turn the ignition switch to the **OFF** position.

FUEL SYSTEM TESTING

Fuel Pump

1. If the fuel pump does not run, a check of the electrical feed should be done.
2. With the engine cranking, check for 12 volts at the fuel pump terminals.
3. If there is no voltage, check the inline fuse near the fuse box.
4. Check the inertia switch at the left rear of the engine compartment near the carbon canister. The white reset button should be in the down position and there should be continuity at the contacts on the bottom of the switch.
5. Check the leads to the fuel pump relay for continuity and check the ground connections. Check the pump relay terminal 30 for 12 volts during cranking. Terminal 20 on the control unit

must go to ground during the same period. Replace the control unit if defective.

6. Check the pump winding resistance. It should be 2.5–3.5 ohms. Replace the pump if the resistance is not within specification.

Fuel Pressure

1. Connect a fuel pressure gauge to the connection on the fuel rail at the opposite end from the fuel pressure regulator.
2. Jumper the fuel pump relay terminals 30 and 87. The fuel pressure should be 33.3–39.1 psi (2.3–2.7 bar).
3. If the fuel pressure doesn't reach this point, clamp the fuel return line to restrict slightly the fuel flow.
4. If the pressure rises to 72 psi (5 bar) replace the fuel pressure regulator. Do not exceed 87 psi (6 bar) during this test.
5. If the pressure does not rise, check the fuel line for collapses and the fuel filter for clogging.
6. If the pressure was higher than specified, check the fuel return line for pinches or collapsing. Remove the fuel return line from the fuel pressure regulator and replace with a line emptying into a suitable container.
7. If the pressure was within specifications, repair the restriction in the return line. If the pressure is still out of specification replace the fuel pressure regulator.
8. Remove the jumper from the relay socket and replace the relay. Start the engine. The fuel pressure should be 29 psi (2 bar) at idle. If not, check the vacuum line going to the fuel pressure regulator. Replace the fuel pressure regulator if the pressure is still out of specification.
9. Shut off the engine. The pressure should not drop more than 14.5 psi (1 bar) in 20 minutes.
10. If the pressure drops more than specified, check the fuel pressure regulator for a broken diaphragm and the fuel pump for a bad outlet check valve.

Injector Leakage

1. Unbolt the fuel rail from the intake manifold. Be careful not to dislodge the injectors from the rail when removing.
2. Pull the rail up enough to observe the injector nozzles.
3. Jumper the fuel pump relay socket between terminals 30 and 87. Turn the ignition switch to the **ON** position. The fuel pump should be running.
4. No fuel should drop from the injectors during a 1 minute period. Replace any injector that fails.
5. Coat the injector O-rings with silicone grease before replacing the fuel rail.

Injector Resistance

1. Before testing the injector resistance, check that the injectors are receiving voltage from main relay terminal 87 with the ignition switch in the **ON** position.
2. If the voltage is 12 volts, turn off the ignition switch and disconnect the injector connectors.
3. Check the resistance at each connector. The resistance should be 15.55–16.25 ohms. Replace any injector that doesn't meet specifications.

INTAKE AIR TEMPERATURE SENSOR

Testing

Use this test when a blink code of 13 is set or a tester code of 44 is set. This test is for a shorted or open wiring to the intake air temperature sensor.

1. The intake air temperature sensor is located on the bottom of the intake manifold. Disconnect the electrical connector from the sensor.
2. Use an ohmmeter and measure the resistance between the terminals on the sensor.
3. Compare the resistance value to the chart for the sensor

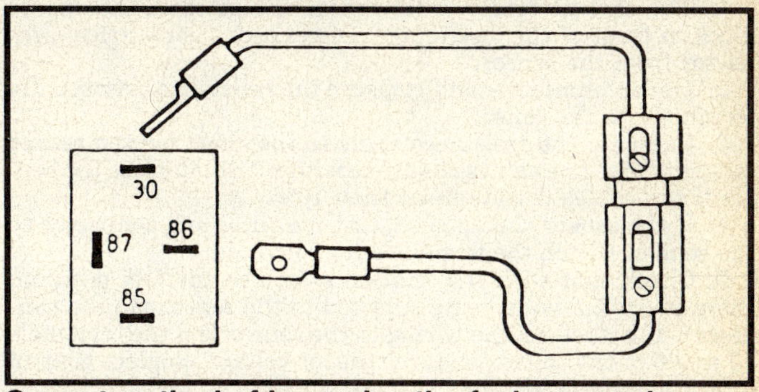

Correct method of jumpering the fuel pump relay connector terminals 30 and 87 with a fused lead—Yugo

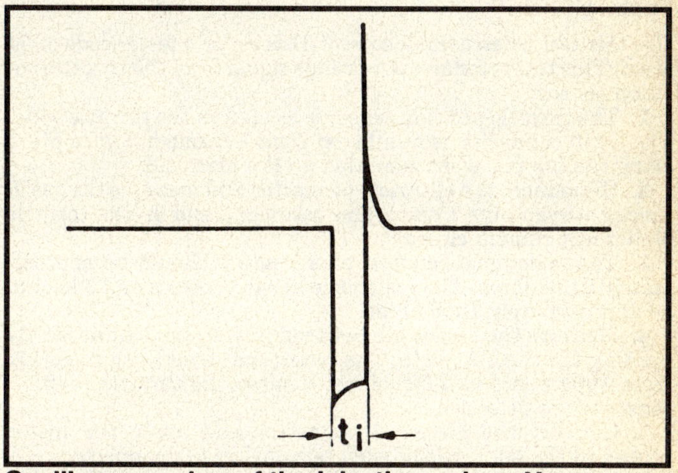

Oscilloscope view of the injection pulse—Yugo

temperature. The values should be within 10%. Replace the sensor if the values deviate more than 10%.

4. If the sensor checks good, test the wires and connector to the sensor. Check the terminals for corrosion.
5. Check that with the ignition switch in the **ON** position, there is 4.5–5.5 volts being supplied to the sensor via the connector. If not, check the wiring to the sensor and terminals 22 and 6 of the control unit for proper voltage. Replace control unit if voltage is incorrect.

ENGINE TEMPERATURE SENSOR

Testing

Use this test when a blink code of 14 is set or a tester code of 45 is set. This test is for a shorted or open wiring to the engine temperature sensor.

TEMPERATURE SENSOR VALUES

Temperature °F (°C)	Resistance Ohms
212 (100)	180
176 (80)	320
140 (60)	600
104 (40)	1100
68 (20)	2500
32 (0)	5200
−4 (−20)	14000
−40 (−30)	28000

1. The coolant temperature sensor is located in the engine block in front of the distributor. Disconnect the electrical connector from the sensor.
2. Use an ohmmeter and measure the resistance between the terminals on the sensor.
3. Compare the resistance value to the chart for the sensor temperature. The values should be within 10%. Replace the sensor if the values deviate more than 10%.
4. If the sensor checks good, test the wires and connector to the sensor. Check the terminals for corrosion.
5. Check that with the ignition switch in the **ON** position, there is 4.5–5.5 volts being supplied to the sensor via the connector. If not, check the wiring to the sensor and the terminals 13 and 6 of the control unit for proper voltage. Replace control unit if voltage is incorrect.

THROTTLE POSITION SENSOR

Testing

Use this test when a blink code of 21 is set or a tester code of 54 is set. This test is for an out of range signal from the throttle position sensor.
1. The throttle position sensor is located on the throttle body. The 3-pin connector is numbered from 1 through 3 with pin 1 being the top pin when mounted on the manifold.
2. Disconnect the electrical connector and measure the resistance between pins 1 and 2 and between 2 and 3. The throttle plate must remain closed.
3. The resistance between pins 1 and 2 should be approximately 2100 ohms. The resistance between pins 2 and 3 should be approximately 1000 ohms.
4. Measure the resistance between pins 2 and 3 while slowly opening the throttle plate. The resistance should rise smoothly from 1000 ohms to 2775 ohms. Replace the throttle position sensor if the defective.
5. Connect the electrical connector and apply the meter probes to the back side of each terminal in the connector.
6. With the ignition switch in the **ON** position, the voltage between pins 1 and 2 should be 4.5–5.5 volts. The voltage between pins 2 and 3 should increase with the opening of the throttle plate.
7. Check the wiring and connectors for faults. Check Terminals 9 and 3 on the control unit for faults. Replace the control unit if no external faults are found.

IDLE SPEED CONTROL

Testing

Use this test when a blink code of 22 is set or a tester code of 04 or 22 is set. This tests the condition of the idle speed control.
1. The idle speed control is located above the intake manifold runners.
2. Disconnect the 3-pin connector from the body of the idle speed control.
3. Measure the resistance between terminals 1 and 2. The resistance should be 19–25 ohms.
4. Measure the resistance between terminals 2 and 3. The resistance should be 17–23 ohms.
5. Replace the unit if the resistances are not as specified. Terminal 1 is connected to terminal 34 of the Motronic control unit. terminal 2 is connected to the main relay contact 87. Terminal 3 is connected to terminal 33 of the Motronic control unit.
6. With the ignition switch in the **ON** position the middle terminal, terminal 2, should have 12 volts. If not, check the main relay and related wiring for opens and shorts.
7. Pull off the hoses from the idle speed unit. Check that the internal valve rotates smoothly. If there is any resistance to rotation, the unit should be replaced.
8. If there are no faults found with the idle speed control unit or wiring, replace the Motronic control unit.

VACUUM SIGNAL

Testing

Use this test when a blink code of 33 is set or a tester code of 08 is set. This test is for an out of range condition of the vacuum signal.
1. Without the proper vacuum signal to the Motronic control unit, the engine will not run correctly.
2. Check the vacuum line from the intake manifold to the control unit damper mounted next to the control unit. The line should not have any kinks or leaks.
3. Check the line for blockages and moisture build-up. Check the damper for condensation. Blow out the line and damper if condensation is found.
4. Disconnect the vacuum line from the intake manifold. Connect a vacuum pump to the line and draw 14.7 in. Hg (0.5 bar). The vacuum must not leak out. Check the lines if it does. If the leak is traced to the control unit, the complete control unit must be replaced as the vacuum sensor is not serviceable.
5. Check for vacuum leaks in the intake tract as this will affect the vacuum signal to the control unit.
6. Clear the fault memory and run the engine. Check for fault codes. If the fault code appears again, check with a new control unit.

ENGINE SPEED AND POSITION SENSOR

Testing

Use this test when a blink code of 41 is set or a tester code of 42 is set. This test is for an out of range condition of the engine speed and position sensor.
1. The engine speed and position sensor is mounted in the front of the engine by the crank pulley. The sensor reads pulses from the tooth wheel mounted to the crank pulley. A missing tooth provides the position signal and the frequency of the pulses provide the speed signal.
2. Disconnect the sensor electrical connector. Measure the resistance between terminal 1 and terminal 2. At 68°F (20°C) the resistance should be 486–595 ohms. Replace if defective.
3. Check wiring to the control unit terminal 23 and terminal 25.
4. Check the clearance between the tooth wheel and the sensor. The clearance should be 0.020–0.040 in. (0.5–1.0 mm). Check for damage and dirt deposits.

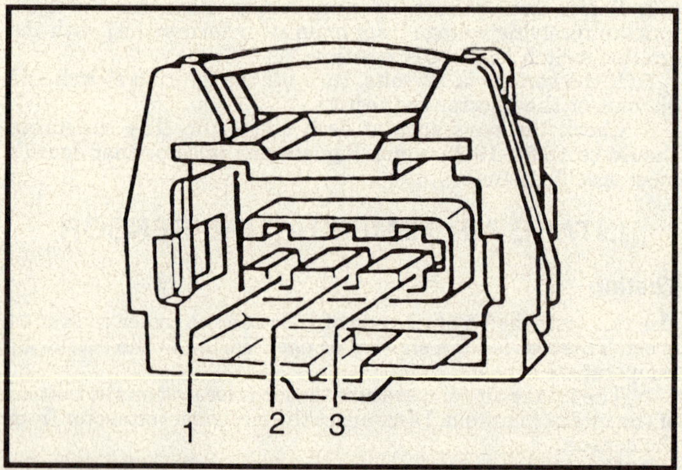

Engine speed and position sensor connector pin numbering—Yugo

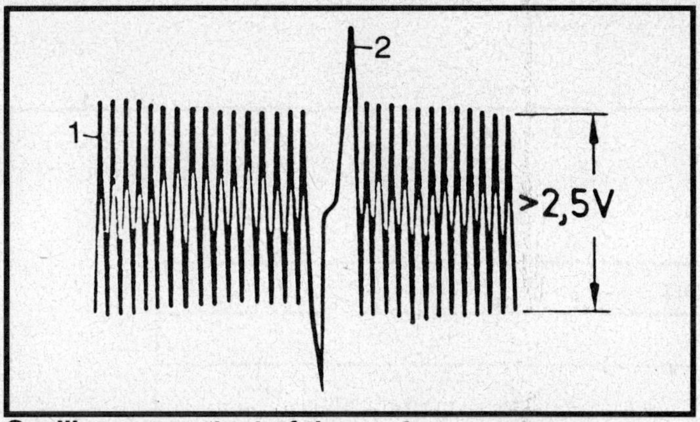

Oscilloscope output of the engine speed and position sensor. Section 1 is the speed signal and peak 2 is the position mark—Yugo

5. Connect an oscilloscope to terminal 1 and terminal 2. The signal must be regular with a greater position signal starting with a negative amplitude.
6. Replace the sensor if the operating parameters are not met.

OXYGEN SENSOR

Testing

BLINK CODE 51/TESTER CODE 28

Use this test when a blink code of 51 is set or a tester code of 28 is set. This test is for a shorted or open wiring to the oxygen sensor.

1. Check the condition of the lead from the oxygen sensor to the control unit terminal 24.
2. Check the oxygen sensor connector and the control unit connector for corrosion or looseness.
3. Check the mechanical condition of the oxygen sensor. Remove the oxygen sensor and check for deposits on the oxygen sensor bulb. Use anti-seize when installing. Be sure not to contaminate the oxygen sensor bulb with the anti-seize.
4. After completing the above checks, erase the fault memory. With the engine warm and at idle, check the voltage of the oxygen sensor at the connector. The voltage should be approximately 0.450 volts. Replace if defective.

BLINK CODE 52/TESTER CODE 10

Use this test when a blink code of 52 is set or a tester code of 10 is set. This test is for an oxygen sensor voltage out of range condition.

1. Check the CO at the tap ahead of the catalytic converter. The mixture should be 0.4–1.0%. This code would usually indicate a mixture that is very high or very low.
2. Check for high resistance connections at pins 5, 8, 16 and the grounds of the Motronic control unit. Check the resistance between the grounds of the control unit plug and the negative battery terminal. There should be less than 1 ohm resistance. Correct any high resistances.
3. Check the intake and exhaust systems for leaks. Leaks near the oxygen sensor will cause the largest change in the mixture.
4. Check the fuel pressure at the fuel rail. Check for clogged injectors and replace as necessary.
5. Check the tank emission valve for operation with the output periphery test.
6. If the pressure sensor in the Motronic control unit is defective or isn't receiving the proper vacuum signal, the mixture will not be correct.
7. Check the mechanical condition of the oxygen sensor. Remove the oxygen sensor and check for deposits on the oxygen sensor bulb. Use anti-seize when installing. Be sure not to contaminate the oxygen sensor bulb with the anti-seize.
8. After completing the above checks, erase the fault memory. With the engine warm and at idle, check the voltage of the oxygen sensor at the connector. The voltage should be approximately 0.450 volts. Replace if defective.

CONTROL UNIT POWER SUPPLY BATTERY VOLTAGE

Testing

Use this test when a blink code of 53 is set or a tester code of 37 is set. This test is for a control unit supply voltage out of range condition.

1. With the engine running check the voltage between terminals 18, positive lead and 5, negative lead. The voltage should be 10–16 volts.
2. If the voltage is below 10 volts, check the battery terminals, the engine ground and all wiring and contacts to the control unit. All connections should be clean and tight with no faults in the wiring.
3. If the voltage exceeds 16 volts, check the voltage regulator and replace as necessary.
4. Check the condition of the battery. Check the main relay for 12 volts on terminal 87 with the ignition in the **ON** position. Check that the main relay is getting 12 volts from the battery to terminal 30. Check that the main relay operates with the ignition switch.
5. Bad connections are leading causes of supply voltage problems. Clean and tighten all connections if a specific voltage fault can not be found.

FAULT CODES—YUGO

Code Output		
Flash	Tester	Fault Explanation
13	44	Intake Air Temperature Sensor—shorted or open
14	45	Engine Temperature Sensor—shorted or open
21	54	Throttle Position Sensor—signal out of range
22	04	Idle Speed Control—first winding
22	22	Idle Speed Control—second winding
33	08	Vacuum Sensor—out of range
41	42	Engine Speed and Position Sensor—out of range
51	28	Oxygen Sensor—short or open
52	10	Oxygen Sensor—voltage out of range
53	37	Battery Voltage—high or low
54	01	Motronic Control Unit—defective

TROUBLESHOOTING GUIDE

SYMPTOM

1. The starter functions but the engine will not start
2. The engine starts, but stops running
3. Poor idle quality
4. Difficult cold - starting
5. Engine power is low
6. Engine misses
7. Excessive fuel consumption
8. CO value deviates from the nominal value
9. Engine overheating
10. Difficult warm - up

1	2	3	4	5	6	7	8	9	10	CAUSE OF DEFECT
▲										No voltage for supplying the electronic control unit
▲	▲				▲					No fuel to injectors
	▲			▲		▲	▲			Insufficient (excessive) fuel pressure
▲										Inertial switch
▲	▲	▲		▲	▲					Fuel injectors
▲										R P M and T D C sensor
	▲	▲			▲					Throttle position sensor
		▲								Idle adjuster
▲					▲					Ignition coil
▲		▲	▲						▲	Intake air temperature sensor
▲		▲	▲	▲					▲	Coolant temperature sensor
		▲			▲		▲			Lambda sensor
				▲			▲	▲		Ignition timing
▲	▲	▲	▲	▲	▲	▲	▲	▲	▲	Electronic control unit

MOTRONIC CONTROL UNIT FAILURE CODE

Testing

Blink code 54 or tester code 01 indicate a faulty Motronic control unit. If these codes are set replace the control unit with a new unit. Retest the system and check for fault codes.

Component Replacement

SERVICE PRECAUTIONS

To protect the Motronic system and components:

- Always disconnect the battery, the Motronic control unit and the ignition coils when using an electric welder on the vehicle or when charging the battery or when the vehicle is placed in a paint drying oven.
- Remove the main Motronic relay to disable both the fuel and ignition systems when checking the compression. Never crank the engine after removing the distributor cap or disconnecting the high tension wire on the ignition coil.
- Never disconnect the battery or wires on the alternator, starter or spark plugs when the engine is running.
- Always work in a well ventilated area to prevent the build up of hazardous fumes.
- Always keep a fire extinguisher within reach when working of the fuel system.
- Wipe up any spilled fuel to prevent fire hazards.
- Relieve the fuel pressure before disconnecting any fuel system components.

FUEL PRESSURE RELIEF

1. Wrap a shop towel around the fuel inlet on the fuel rail.
2. Loosen the fuel line connection and allow the pressure to escape.
3. Dispose of the shop towel in a safe manner to prevent fire. Catch any escaping fuel in a suitable container. Wipe up any spilled fuel.
4. Tighten the fuel line connection after performing the required service.

MOTRONIC CONTROL UNIT

Removal and Installation

1. Turn the ignition switch to the **OFF** position.
2. The control unit is located on the passenger side under the dashboard.
3. Unscrew the retaining bolts and lower the unit.
4. Unplug the electrical connector and disconnect the vacuum hose.
5. Check for corrosion and bent pins in the connector. Check for moisture build-up or blockage in the vacuum line.
6. Connect the vacuum line and the electrical connector. Replace the unit under the dashboard and secure.

INTAKE AIR TEMPERATURE SENSOR

Removal and Installation

1. Turn the ignition switch to the **OFF** position.
2. The intake air temperature sensor is located on the bottom of the intake plenum.
3. Disconnect the electrical plug from the body of the sensor.
4. Unscrew the sensor from the intake plenum.
5. Clean the threaded hole in the intake plenum of any debris or carbon deposits.
6. Replace the sensor. Connect the electrical connector.

ENGINE TEMPERATURE SENSOR

Removal and Installation

1. Turn the ignition switch to the **OFF** position.
2. The engine temperature sensor is located in the engine block in front of the distributor.
3. Disconnect the electrical plug from the body of the sensor.
4. Unscrew the sensor from the engine.
5. Clean the threaded hole in the intake plenum of any debris or mineral deposits.
6. Replace the sensor. Connect the electrical connector.

THROTTLE POSITION SENSOR

Removal and Installation

1. Turn the ignition switch to the **OFF** position.
2. Disconnect the electrical connector from the body of the sensor.
3. Remove the 2 screws holding the sensor and remove the sensor.
4. Replace the sensor and fasten the 2 screws.
5. The Motronic control unit has adaptive capabilities for the throttle position so it is not necessary or possible to adjust the sensor. Replace the electrical connector.

IDLE SPEED CONTROL

Removal and Installation

1. Turn the ignition switch to the **OFF** position.
2. The idle speed control is located next to the intake plenum and has 2 hoses and an electrical lead connected to it.
3. Disconnect the electrical connector from the end of the unit.
4. Release the clamps holding the hoses to the body of the unit and remove the hoses.
5. Loosen and remove the clamp holding the body of the unit and remove the unit.
6. Replace the unit in the clamp and tighten the clamp. Replace the hoses and electrical connector.

ENGINE SPEED AND POSITION SENSOR

Removal and Installation

1. Turn the ignition switch to the **OFF** position.
2. The sensor is located at the crank pulley and is accessible from the underside of the vehicle. Raise and safely support the vehicle.
3. Disconnect the lead coming from the sensor.
4. Loosen and remove the bolt holding the sensor. Remove the sensor.
5. Remove the protective cover from the new sensor. Apply LONGTERM 2 anti-seize or equivalent, to the sensor and slide into place. Be careful not to damage the sensor or make contact with metallic parts.
6. Fully seat the sensor and replace the mounting bolt. The clearance between the sensor and the tooth wheel should be 0.020–0.040 in. (0.5–1.0 mm).
7. Connect the lead from the sensor.

OXYGEN SENSOR

Removal and Installation

1. Turn the ignition switch to the **OFF** position.
2. Disconnect the electrical lead from the sensor.
3. Unscrew the sensor from the exhaust pipe.

1. Air filter
2. Air cleaner housing
3. Air cleaner housing
4. Connector
5. Air intake duct
6. Connecting hose
7. Clamp
8. Bracket
9. Bushing
10. Bushing
11. Bushing
12. Clamp
13. Bolt
14. Washer
15. Gasket
16. Bracket
17. Throttle body and brackets
18. Intake air temperature sensor
19. Throttle position sensor
20. Stud
21. Heat shield

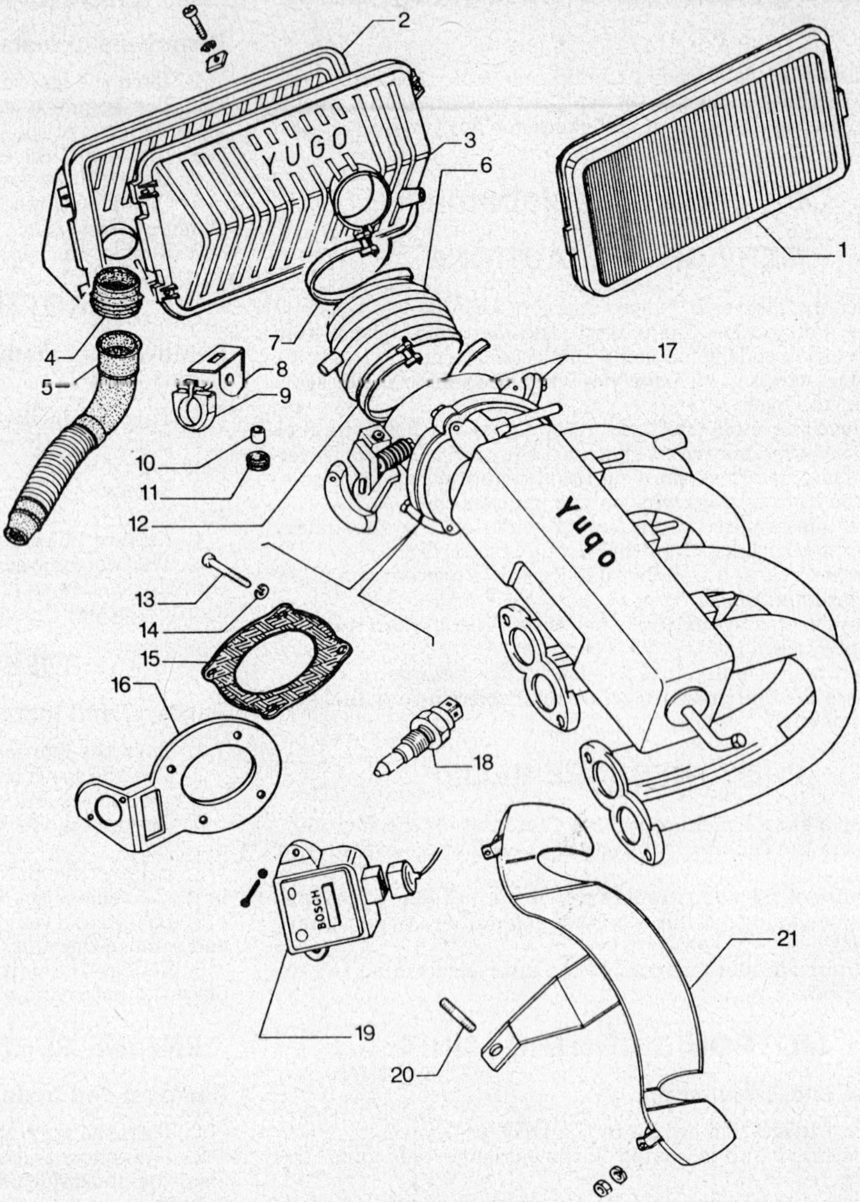

Air intake system—Yugo

4. Apply anti-sieze to the threads of the sensor. Do not contaminate the bulb of the sensor.
5. Screw the sensor into the exhaust pipe and connect the lead.

FUEL INJECTORS

Removal and Installation

1. Turn the ignition switch to the **OFF** position.
2. Relieve the fuel system pressure. Pull off the injector electrical connectors.
3. Disconnect the fuel feed to the fuel rail. Disconnect the fuel return line and the vacuum line to the fuel pressure regulator.
4. Remove the bolts holding the fuel rail. Pull the fuel rail with the injectors away from the engine.
5. Remove the injector retainer clips and remove the injectors.

To Install:

6. Inspect the injector seals for damage and replace as necessary. Apply silicon lubricant to the seals.
7. Assemble the injectors to the fuel rail. Install the fuel rail to the engine.
8. Install the mounting bolts. Install the fuel line, return line, vacuum line and the electrical connectors.
9. Turn the ignition switch **ON** and check for leaks from the connections.

FUEL PRESSURE REGULATOR

Removal and Installation

1. Turn the ignition switch to the **OFF** position.
2. Relieve the fuel system pressure.
3. Disconnect the fuel return line and the vacuum line to the fuel pressure regulator.

4. Unscrew the 2 bolts holding the fuel pressure regulator and remove the regulator.
5. Replace the fuel pressure regulator and the mounting bolts. Install the vacuum line and the fuel return line.

FUEL PUMP

Removal and Installation

1. Perform this procedure with the tank in a less than full condition. Turn the ignition switch to the **OFF** position.
2. Relieve the fuel system pressure.
3. Remove the hatch area covering to expose the access for the fuel tank cover plate.
4. Mark and disconnect the fuel lines and the electrical connections at the top of plate.
5. Remove the 10 screws holding the plate and pull the plate up and out of the tank.
6. Remove the screws holding the pump to the plate and remove the pump.
7. Install the pump and the mounting screws to the plate. Install the plate to the tank with a new gasket.
8. Install the 10 screws and the fuel lines. Connect the electrical connectors. Replace the hatch area cover.

TANK EMISSION VALVE

Removal and Installation

1. Turn the ignition switch to the **OFF** position.
2. The tank emission valve is located on the left corner of the engine and is connected by a vacuum line to the emission canister. It is the light colored valve.
3. Disconnect the vacuum lines from the body of the valve. Disconnect the electrical connector from the valve.
4. Install the vacuum lines and the electrical connector.

IDLE SPEED AND MIXTURE

Adjustment

The idle speed and mixture are controlled by the Motronic control unit. The values are not adjustable, but can be checked. If the idle speed is high check for vacuum leaks. If the idle mixture is incorrect, check the oxygen sensor system and fuel system.

1. With the engine running and at operating temperature, the idle speed should be 800–900 rpm.
2. The CO measured before the catalytic converter at the exhaust system testing tap should be 0.4–1.0%.
3. The CO measured after the catalytic converter should be 0.0–0.2%. If the CO before the catalytic converter is within specifications, but the CO after the catalytic converter is too high, replace the converter.

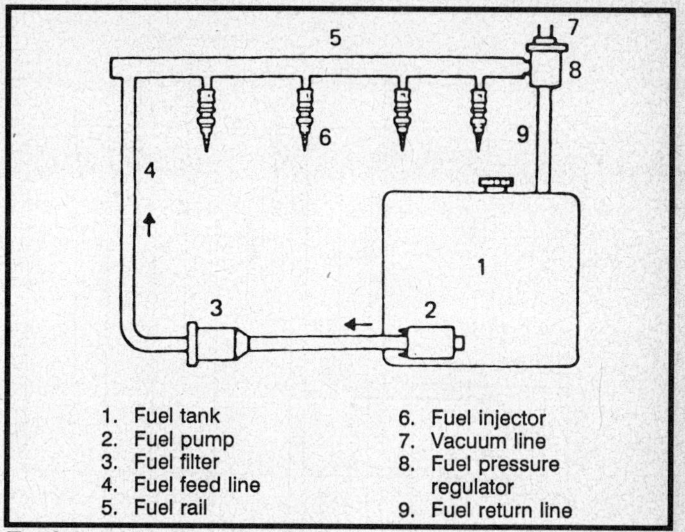

Fuel system diagram—Yugo

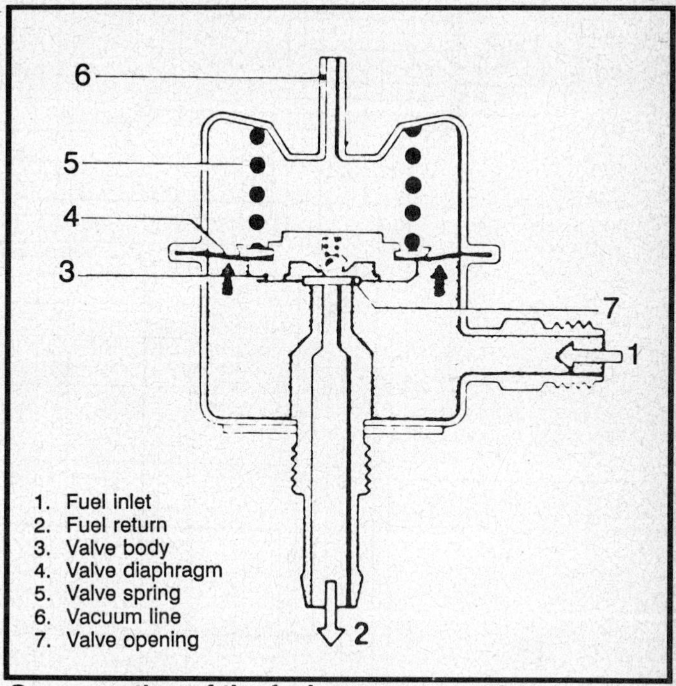

Cross section of the fuel pressure regulator—Yugo

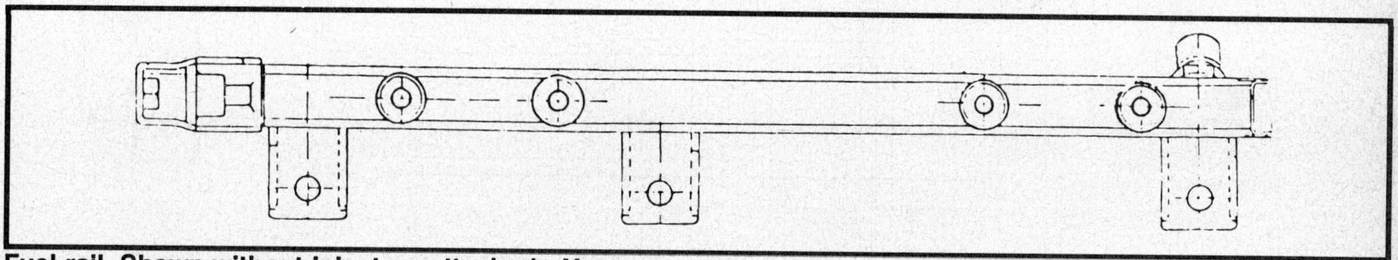

Fuel rail. Shown without injectors attached—Yugo

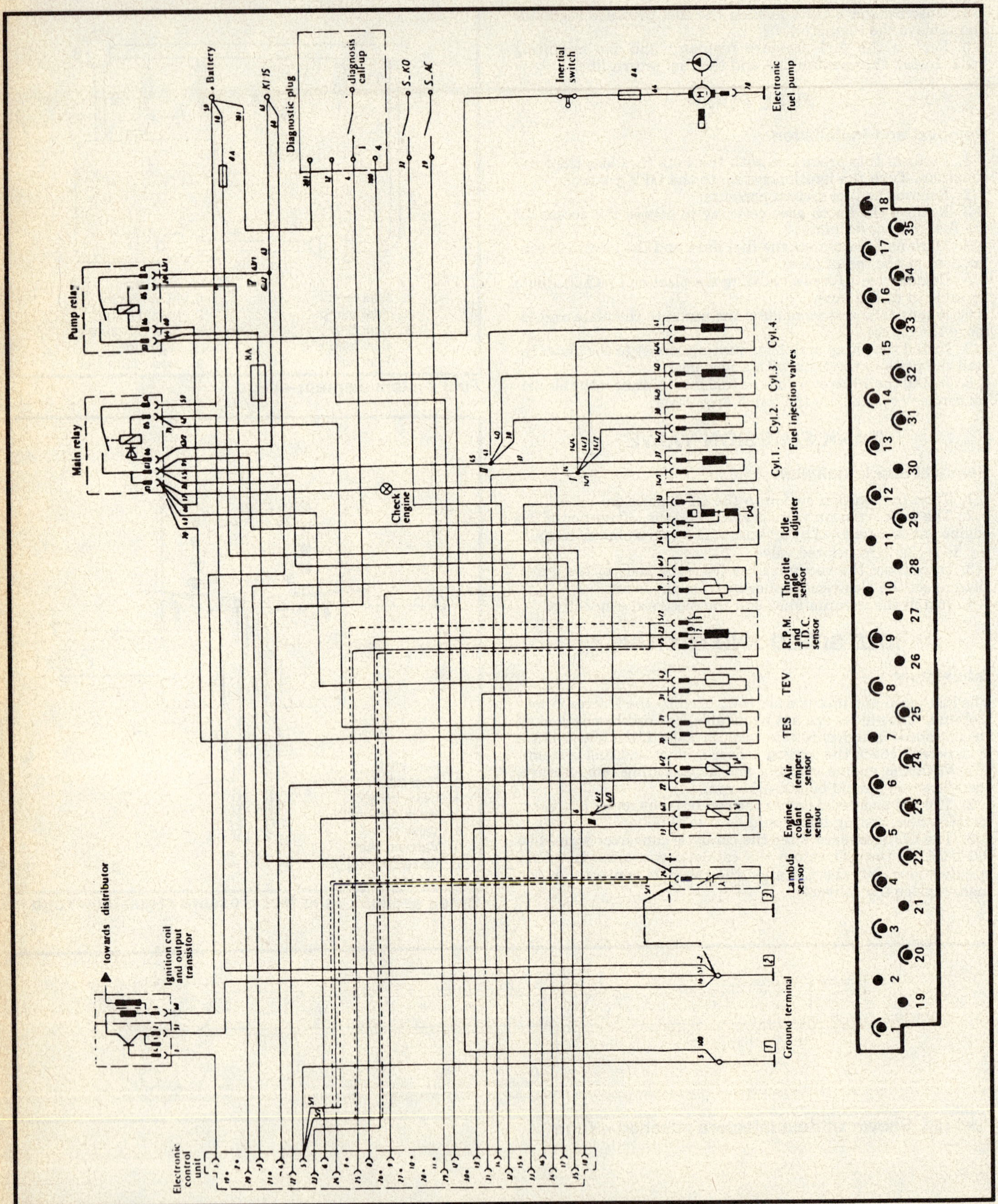

Wiring diagram—Yugo

Carburetor Systems 18 SECTION

YUGO CARBURETED FUEL SYSTEM

YUGO CARBURETED FUEL SYSTEM

General Information

The Yugo fuel system is quite basic in comparison to most fuel systems today. It utilizes a Carter/Weber 2 barrel down-draft carburetor, with a vacuum control secondary, mechanical and vacuum actuated accelerator pumps and an automatic choke. The fuel pump is of the mechanical diaphragm type, driven of the camshaft. For better atomization of the fuel during warm-up, a water heated intake manifold is used.

SERVICE PRECAUTIONS

- Always disconnect the negative battery terminal to prevent sparks cause by short-circuiting, circuit-breaking, etc.
- Smoking must not be allowed when working around flammable liquids.
- Always keep a CO_2 fire extinguisher close on hand.
- Dry sand must be available to soak up any spillage.
- Ensure fuel is emptied into a suitable container.
- Always place the transmission gear shift lever in **N** (manual transmission) or **P** (automatic transmission) and set the parking brake and block drive wheels before starting the engine.
- Keep hands, tools and clothing away from engine cooling fan to help prevent personal injury.

Diagnosis and Testing

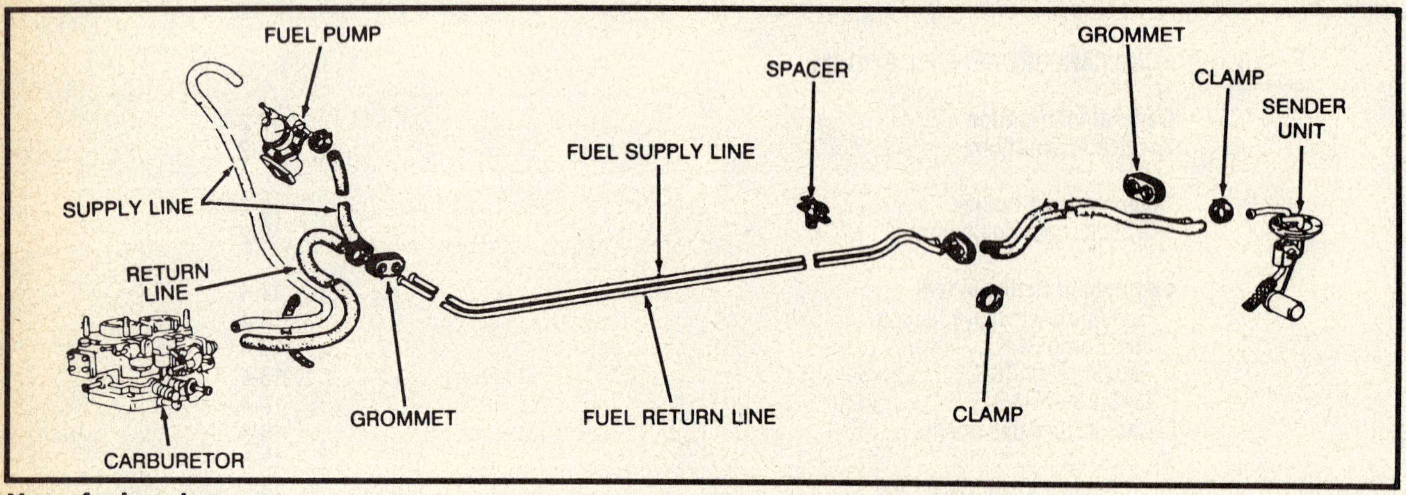

Yugo fuel system

TROUBLE SHOOTING CHART

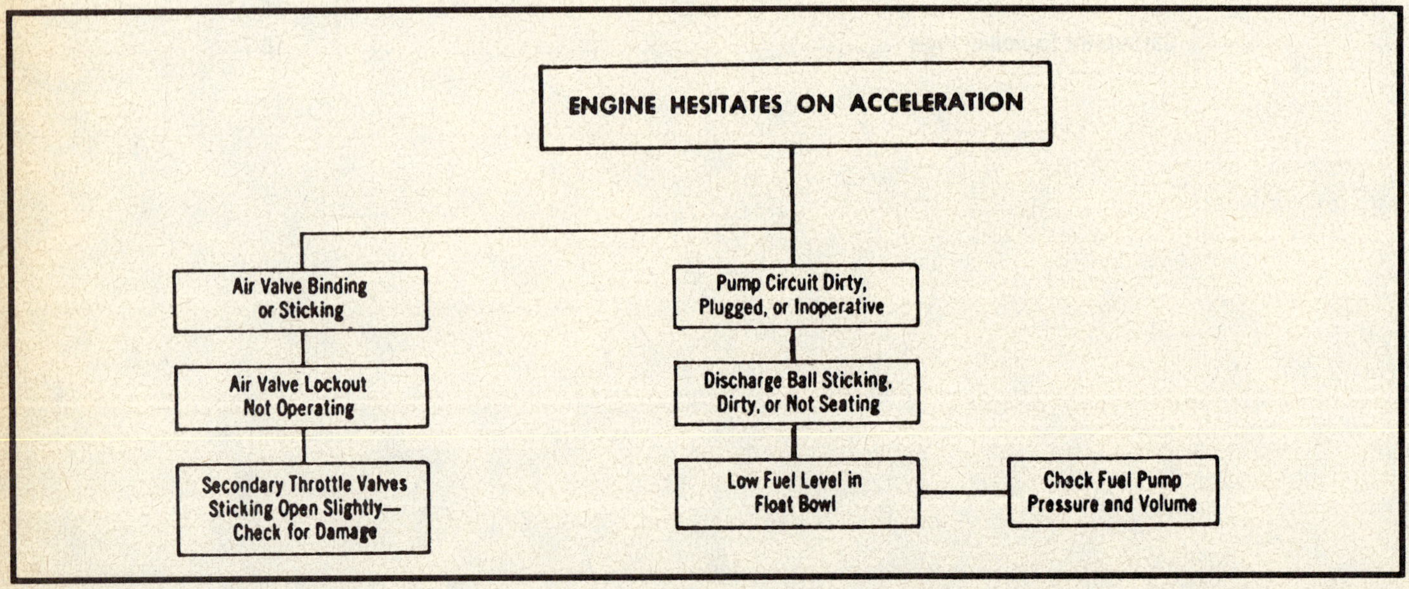

TROUBLE SHOOTING CHART

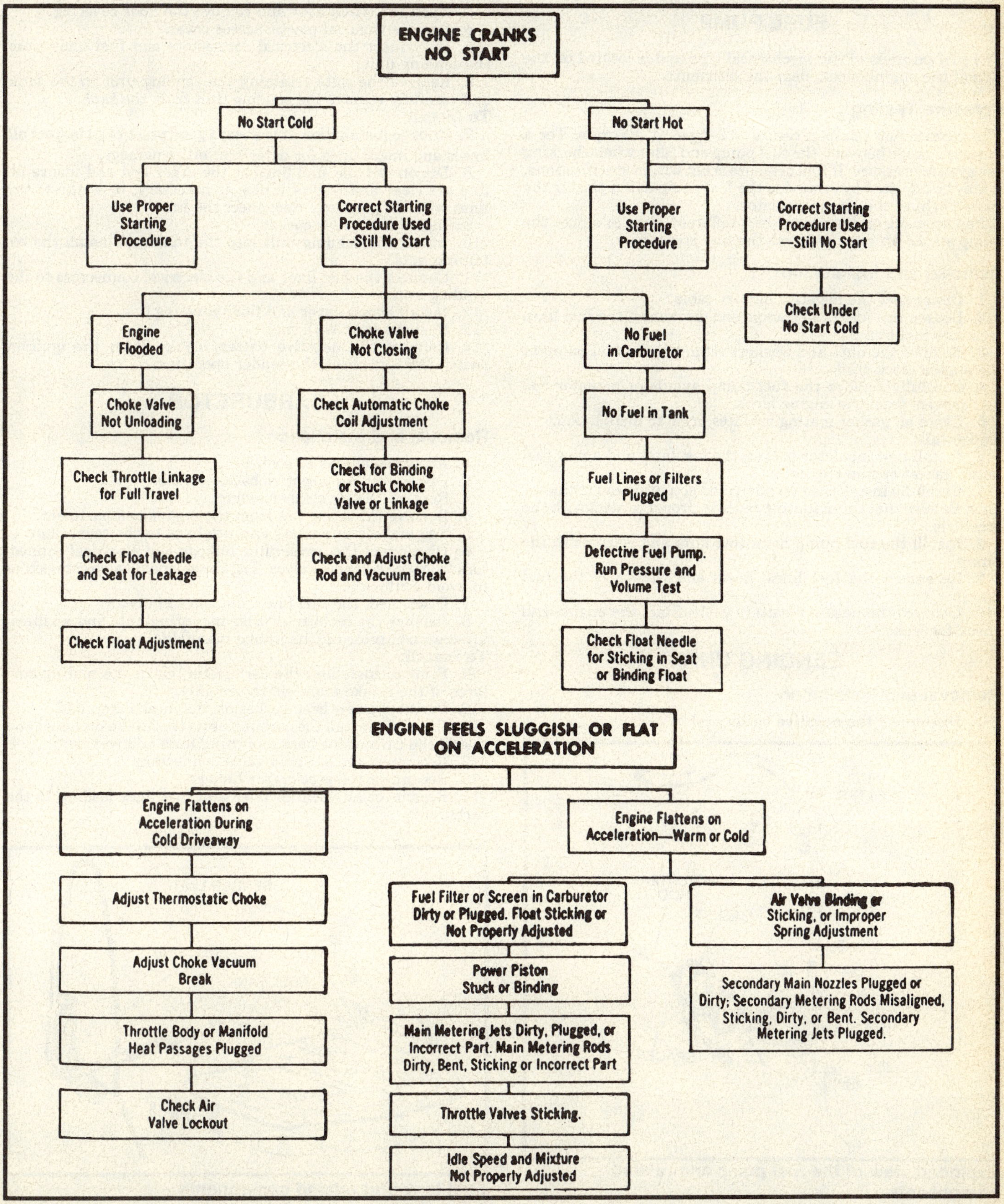

Component Replacement

FUEL PUMP

The fuel pump is of the mechanical type and is located on the side of the engine block, near the distributor.

Pressure Testing

The normal fuel pump pressure is 2–3 psi @ 400 rpm. Tee a pressure gauge between the fuel pump and filter when checking the system pressure. If fuel pressure is not within specifications, it may be adjusted by removing the fuel pump and replacing the gasket behind the pump insulator.

Replacement gaskets come in 3 different sizes to adjust the pump stroke an raise or lower the fuel pressure.

Removal and Installation

1. Disconnect the negative battery cable.
2. Loosen the fuel hose clamps and disconnect the fuel lines from the fuel pump.
3. Remove the nuts and washers retaining the fuel pump to the engine block studs.
4. Carefully remove the fuel pump, insulator actuator rod and gaskets from the engine block.
5. Clean all gasket mating surfaces prior to installation.

To install:

6. Install new insulator gasket, the insulator and a new fuel pump gasket on the engine.
7. Carefully install the fuel pump and actuating rod to the engine. Be sure that the actuating rod is in proper contact with the eccentric.
8. Install the fuel pump mounting nuts and washers to the pump.
9. Reconnect the fuel lines, hoses and clamps to the fuel pump.
10. Connect the negative battery cable. Start the engine and check for leaks.

SENDING UNIT

Removal and Installation

1. Disconnect the negative battery cable.

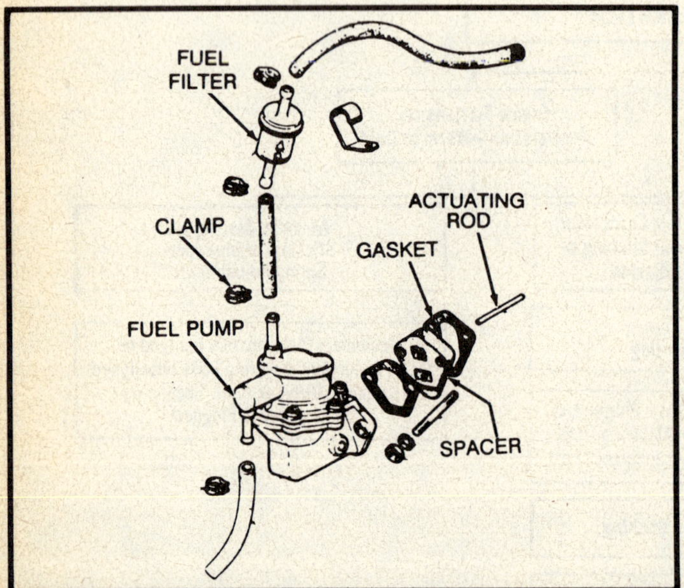

Exploded view of the fuel pump and related components

2. Remove the fuel tank cap. Using a siphon tool, siphon the fuel from the tank.
3. Open the hatch door and remove the floor covering.
4. Remove the oval plastic access cover.
5. Disconnect the electrical connectors and fuel lines from the sending unit.
6. Remove the nuts retaining the sending unit to the tank and carefully remove the sending unit from the tank.

To install:

7. Prior to installation, it is always good practice to inspect all hoses and metal lines for deterioration or damage.
8. Disconnect the fuel lines at the other end and insure all line are clear and free of kinks. If necessary, blow threw the lines with air pressure. Reconnect the lines.
9. Install the a new gasket.
10. Install the sending unit into the tank and install the retaining nuts.
11. Connect the fuel lines and the electrical connectors to the sending unit.
12. Install access cover and floor covering.
13. Fuel the tank with fuel.
14. Connect the negative battery cable. Turn the ignition switch **ON** and check the sender operation.

CARBURETOR

Removal and Installation

1. Allow the engine to cool.
2. Disconnect the negative battery cable.
3. Remove the air cleaner assembly.
4. Remove the spare tire from the engine compartment.
5. Tag and disconnect all vacuum lines to the carburetor.
6. Disconnect the accelerator linkage and electrical connectors from the solenoid valves. Tag the solenoid valve connectors to avoid confusion.
7. Disconnect the fuel line from the carburetor.
8. Remove the carburetor base mounting nuts and washers, lift the carburetor off the intake manifold.

To install:

9. Prior to installing the carburetor, clean the mating surfaces of the intake manifold thoroughly.
10. Position a new base gasket on the intake manifold.
11. Carefully install the carburetor on the intake manifold and install the carburetor base mounting nuts and washers.
12. Reconnect the solenoid valve connectors.
13. Reconnect the accelerator linkage.
14. Reconnect all vacuum lines and fuel lines leading to the carburetor.

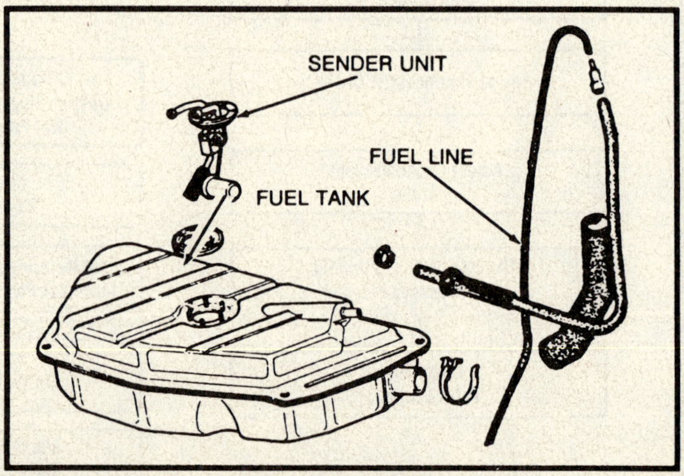

Fuel tank and related components

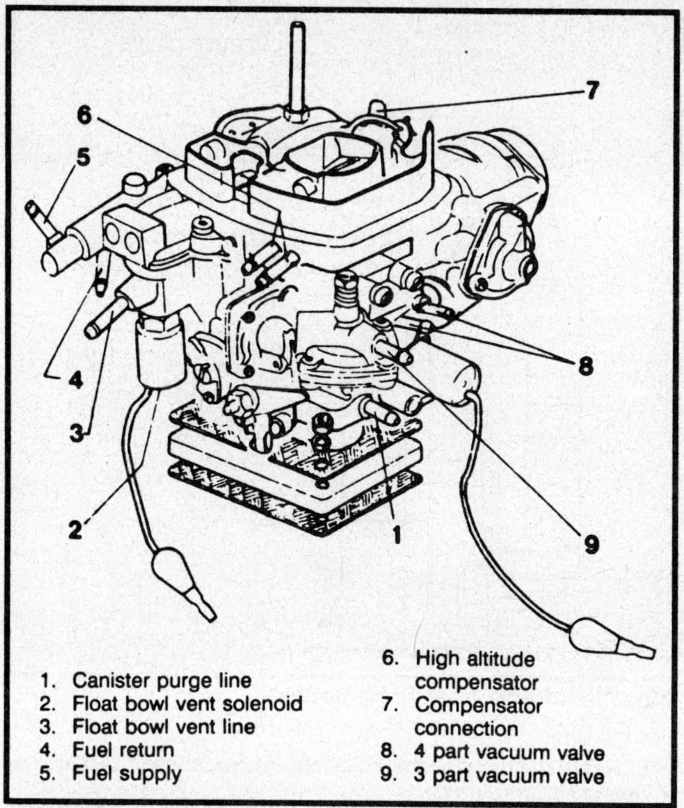

Carter/Weber 740 carburetor

15. Install the spare tire back into the engine compartment.
16. Install the air cleaner assembly.
17. Connect the negative battery cable.
18. Start the engine, check for fuel leaks and adjust as needed.

Adjustments

FAST IDLE

1. Disconnect the negative battery cable.

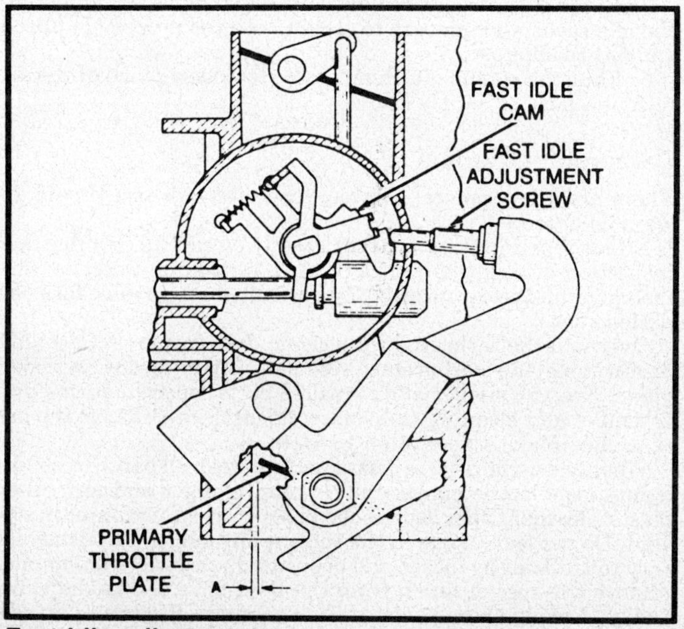

Fast idle adjustment

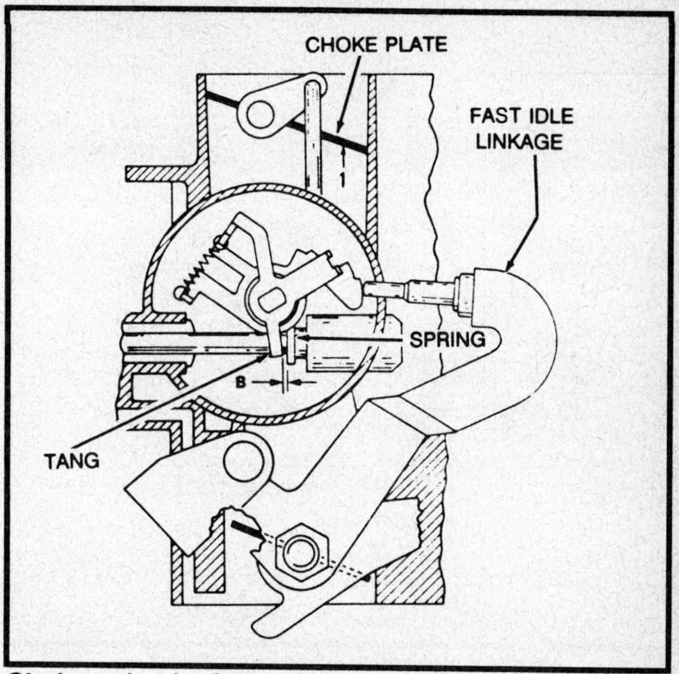

Choke unloader lever clearance adjustment

2. Remove the carburetor from the vehicle. Remove the automatic choke cover retaining screws and remove the gasket.
3. Set fast idle on 2nd step of the cam. Check that the primary throttle opening (gap A) is 0.033–0.037 in. (0.85–0.095mm). If gap A is not correct, adjust screw accordingly.

UNLOADER LEVER CLEARANCE

1. Pull the fast idle linkage back and close the choke plate.
2. Release the linkage. Measure gap B between the lever and shoulder of the bushing. A wire gauge should be used. The gap should be 0.012–0.039 in. (0.3–1.0mm). If the gap is not correct, carefully bend the tang to adjust.

FAST IDLE CAM INTERMEDIATE SETTING

1. Set the fast idle screw on the 3rd step of the cam. Make

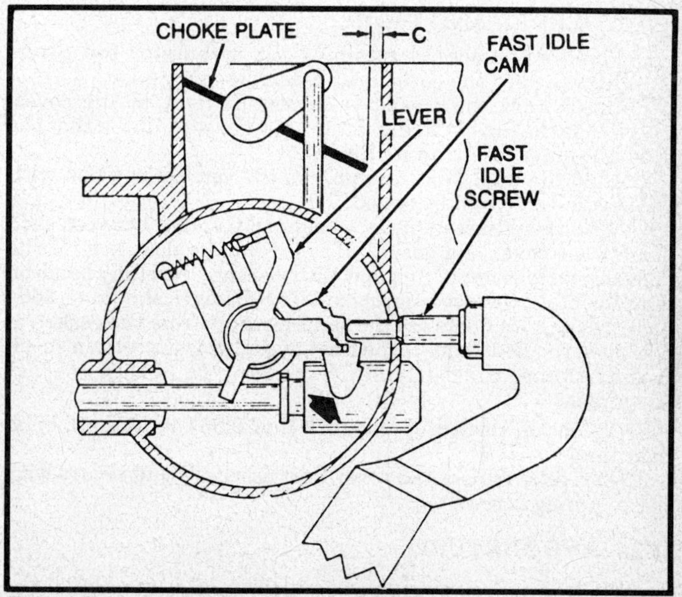

Fast idle cam intermediate setting

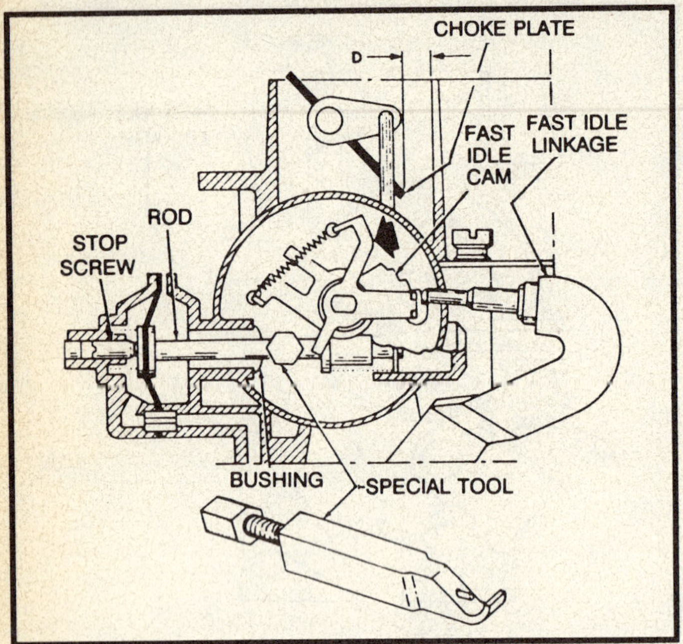

Choke unloader minimum opening setting

sure the screw is in proper contact with the 3rd step on the cam. If necessary, rotate the cam up.

2. Check that gap C is 0.098–0.118 in. (2.5–3.0mm). If gap C is not correct, bend the lever to adjust.

UNLOADER MINIMUM OPENING

1. Place the linkage on the 2nd step of the cam and position special tool No. 4460 on the rod.
2. Push the tool to the left until the rod contacts the stop screw and hold it there.

NOTE: Make sure the tool is not touching the bushing in the housing. This will restrict travel of the rod.

3. Measure gap D. The gap should be 0.128–0.147 in. (3.25–3.75mm). If the gap is not correct, adjust the travel of the rod by turning the screw. Clockwise to decrease the gap, counterclockwise to increase the gap.

FLOAT LEVEL AND DROP

1. Remove the screws retaining the carburetor top cover. Carefully remove the cover/float assembly and gasket.
2. Check that the needle valve seat is tight in the cover. Check that the float is free of dents or punctures. Check that the float can move freely on its hinges.
3. Hold the dual float assembly in the vertical position with the needle valve lightly seated.
4. With the gasket in place, measure distance A between each float to the cover and gasket. If not, bent the float arm. If the float assembly is not within specified distance, carefully bendthe tang to obtain correct adjustment. The distance should (0.266–0.66). Make sure floats are the same distance from the gasket. If not, bend the float arm. If the float assembly is not within specified distance, carefully bend the tang to obtain correct adjustment.
5. To obtain float drop distance B of 0.354 in. (9mm), bend float tang.
6. Carefully replace the cover/float assembly and secure with the retaining screws.

IDLE AND MIXTURE

NOTE: For proper specifications, always refer the EPA conformity tag in the engine compartment.

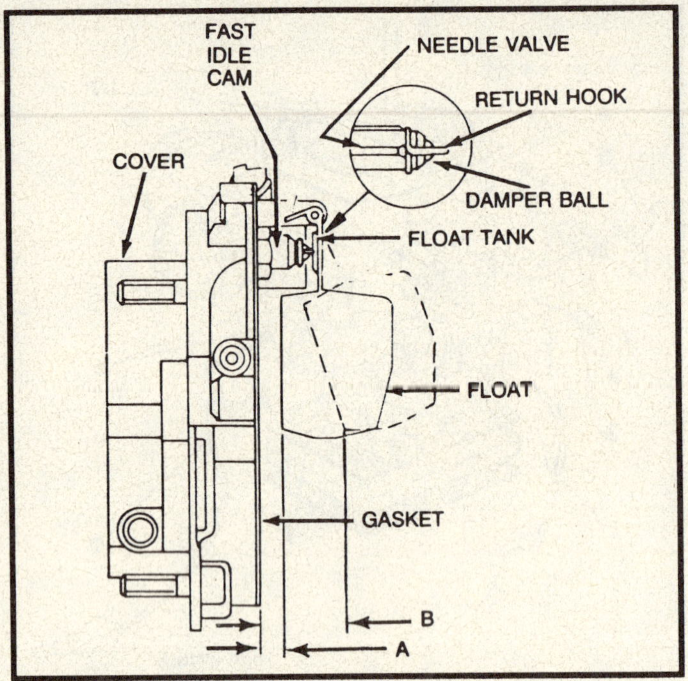

Float level and float drop setting

1. Apply the hand brake. Start the engine allow it to warm up to operating temperature.
2. Insert a CO tester probe into the tailpipe. Connect a tachometer.
3. On vehicles without an air pump, pinch off the supply line to the reed valve. On vehicles with and air pump, pinch off the air injection supply line upstream of the check valve.
4. Adjust the carburetor idle speed adjusting screw to set the engine speed to 800–900 rpm.
5. Adjust the idle mixture adjusting screw to set the CO to 2.0%.
6. Turn the screw clockwise to decrease the CO or counterclockwise to increase the CO.
7. Recheck the idle speed and, if necessary, repeat adjusting the idle speed and mixture to obtain both the proper idle speed and CO reading.
8. Shut the engine off. Remove the tool used pinch of the air injection hose or supply line.

Carburetor Overhaul

There are a few general overhaul procedures which should always be observed.

Efficient carburetion depends greatly on careful cleaning and inspection during the overhaul, since dirt, gum, water or varnish in or on the carburetor parts are often responsible for poor performance.

Overhaul the carburetor in a clean, dust free area. Carefully disassemble the carburetor, referring often to the exploded views. Keep all similar and lookalike parts separate during disassembly and cleaning to avoid accidental interchange during assembly. Make a note of all jet sizes.

When the carburetor is disassembled, wash all parts except diaphragms, electric choke units, pump plunger and any other plastic, leather, fiber or rubber parts in clean carburetor solvent. Do not leave parts in the solvent any longer than is necessary to sufficiently loosen the deposits. Excessive cleaning may remove the special finish from the float bowl and choke valve bodies, leaving these parts unfit for service. Rinse all parts in clean solvent and blow them dry with compressed air or allow

1. Carburetor cover
2. Breather valve
3. Choke plate
4. Gasket
5. Choke linkage
6. Choke plate shaft
7. Filter
8. Float assembly
9. Center venturi
10. Idle jet housing
11. Idle jet
12. Secondary throttle vacuum diaphragm assembly
13. Secondary Throttle shaft
14. Secondary throttle plate
15. Throttle actuating lever assembly
16. Choke fast idle screw
17. Progressive throttle linkage
18. Return spring
19. Stop lever/primary shaft
20. Return spring
21. Carburetor body assembly
22. Idle speed screw
23. Fast idle diaphragm assembly (A/C only)
24. Idle shut off solenoid
25. Idle mixture screw
26. Accelerator pump cover
27. Pump diaphragm
28. Diaphragm
29. Primary throttle plate
30. Primary throttle shaft
31. Accelerator pump actuating cam
32. Fast idle control lever (A/C only)
33. Power valve cover
34. Bowl vent valve rod
35. Power valve return spring
36. Power valve diaphragm
37. Accelerator pump jet
38. Main fuel jet
39. Main emulsion tube
40. Main air jet
41. Vacuum accelerator pump cover
42. Vacuum accelerator pump return spring
43. Vacuum accelerator pump diaphragm
44. Automatic choke hot water cover
45. Gasket
46. Automatic choke hot water backing plate
47. Bi-metalic spring case
48. Gasket
49. Automatic choke assembly
50. Choke unloader adjustment screw
51. Choke unloader adjustment cover
52. Choke unloader adjustment return spring
53. Choke unloader adjustment diaphragm
54. Fast idle cam return spring
55. Needle valve assembly

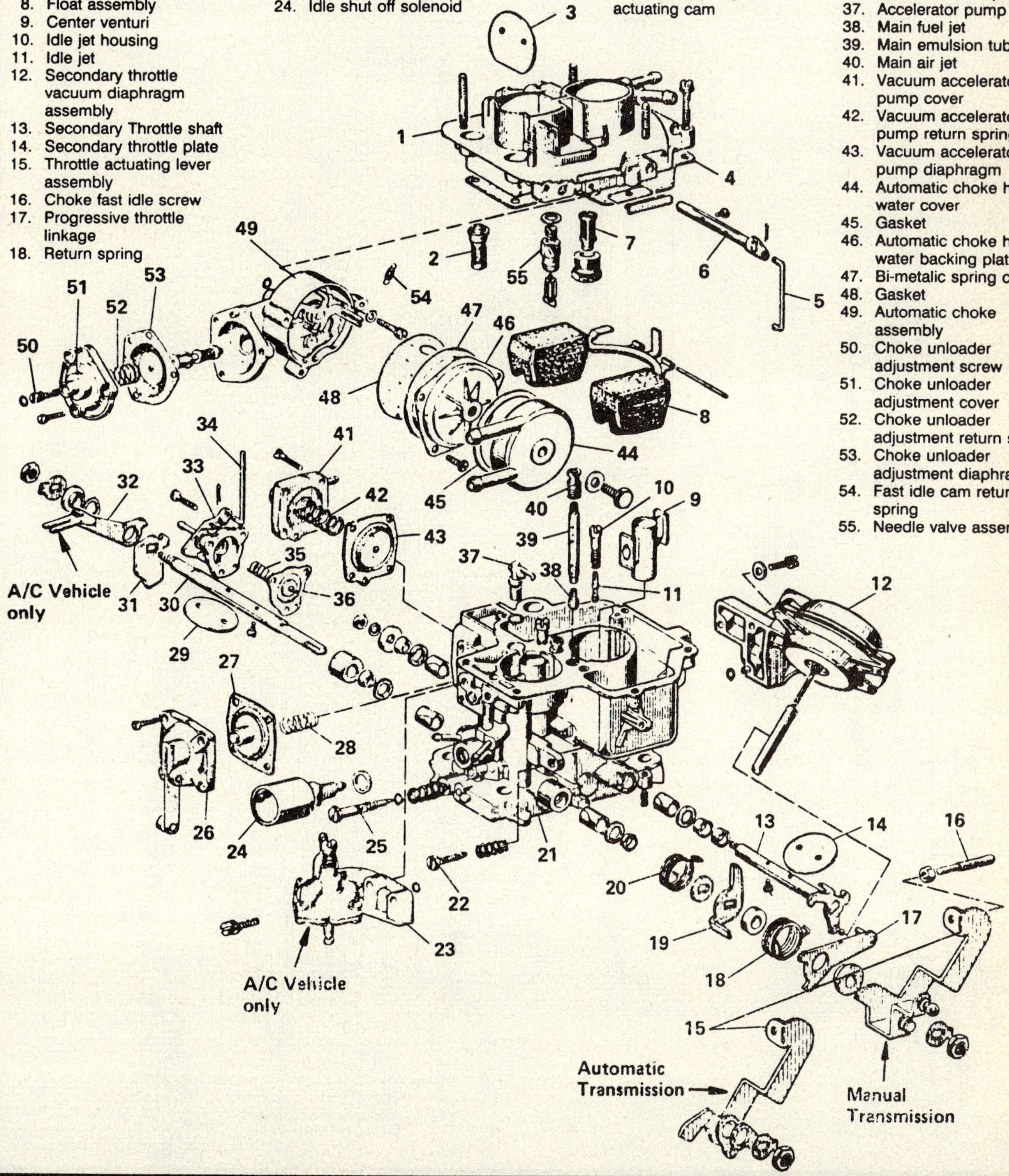

Exploded view of the Carter/Weber 740 carburetor

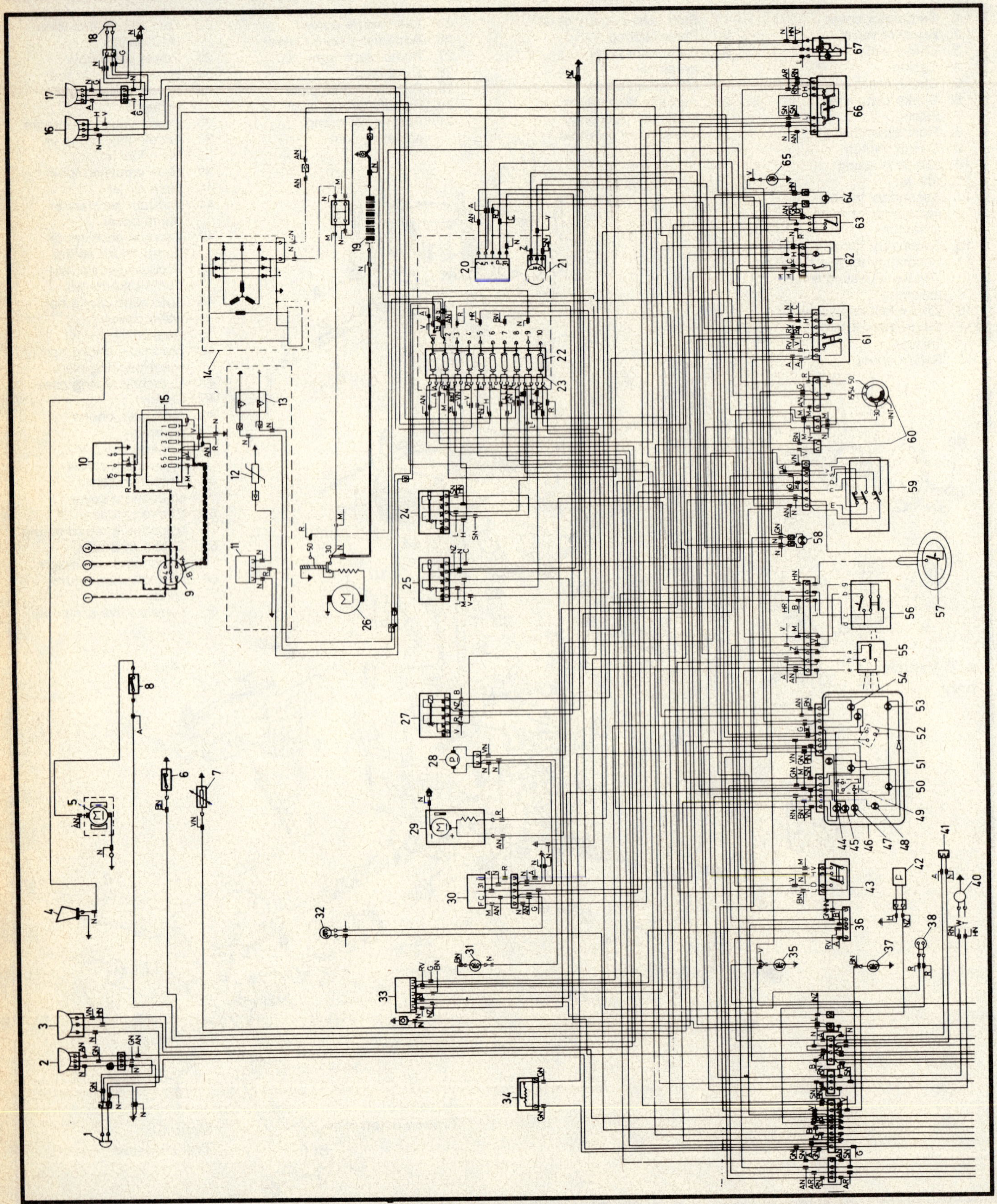

Vehicle circuits, including the defogger circuit—Yugo

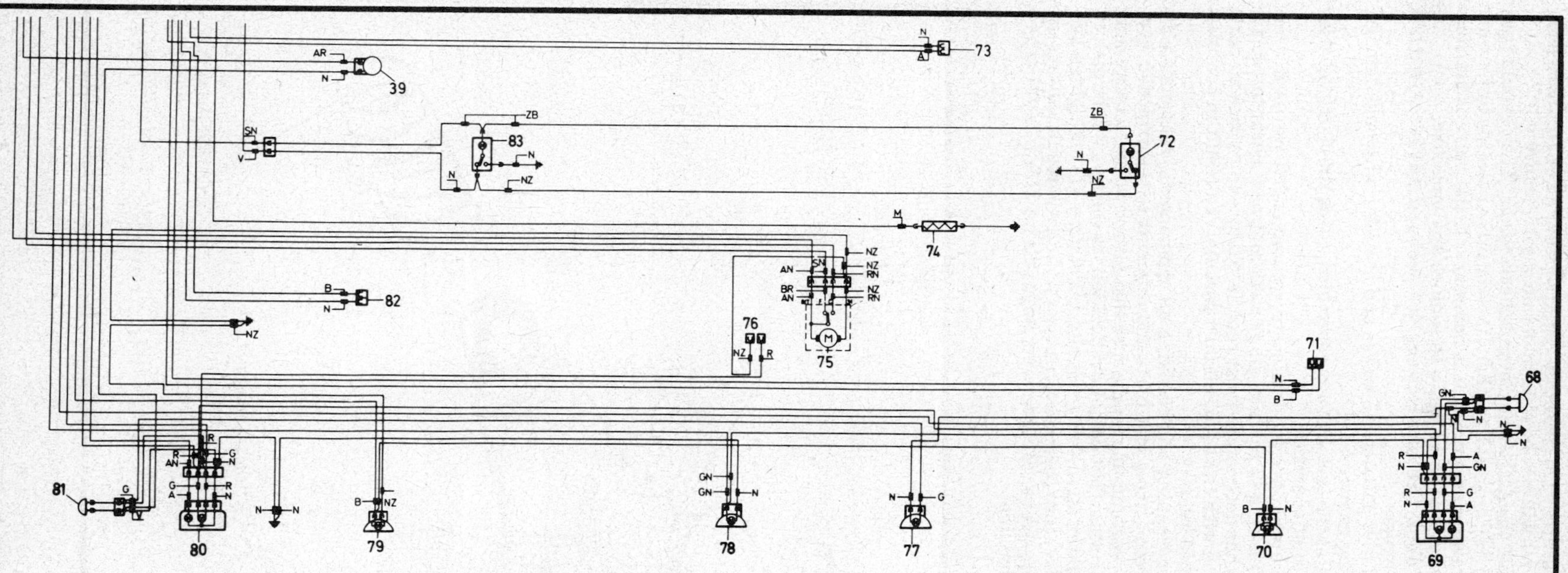

1. Left turn and parking light
2. Left front turn signal
3. Left headlamp
4. Horn
5. Cooling fan
6. Oil pressure sender
7. Coolant temperature sender
8. Cooling motor sender
9. Distributor sender
10. Ignition coil
11. Vacuum switch
12. Choke
13. Carburetor electric valve
14. Alternator
15. Electronic module
16. Right headlamp
17. Right front turn signal
18. Right turn and parking light
19. Battery
20. Hazzard lamp
21. Flasher
22. Fuses
23. Fuse box
24. Horn relay
25. Defogger relay
26. Starter
27. Highbeam relay
28. Windshield washer pump
29. Heater motor
30. Wiper motor
31. Brake failure switch
32. Backup light switch
33. Buzzer
34. Rheostat
35. Light switch
36. Radio
37. Brake failure warning light switch
38. Brake light switch
39. Rear window washer
40. Fuel level sender
41. Left door speaker
42. Seat belt switch
43. Light rocker switch
44. Hign beam indicator
45. Parking lamp warning light
46. Instrument cluster
47. Oil warning lamp
48. Seat belt warning lamp
49. Fuel level gauge
50. Fuel reserve warning lamp
51. Turn signal warning lamp
52. Temperature indicator
53. Brake failure light
54. Charge light
55. Turnsignal switch
56. High/low beam switch
57. Horn button
58. Light cables
59. Wiper switch
60. Ignition lock
61. Hazard warning switch
62. Defogger switch
63. Heater switch
64. Heater indicator lamp
65. Interior light switch
66. Rear wiper switch
67. Cigarette lighter
68. Rear turn and parking lamp
69. Rear parking, turn and brake lamp
70. Back up lamp
71. Right rear speaker
72. Dome light switch
73. Right door speaker
74. Defogger
75. Rear wiper motor
76. Additional stop lamp
77. Licence plate lamp
78. Licence plate lamp
79. Back up lamp
80. Rear parking, turn and brake lamp
81. Rear side turn and parking lamp
82. Left rear speaker
83. Dome light and switch

CABLE COLORS

A — Light blue
B — White
C — Orange
G — Yellow
H — Grey
L — Dark blue
M — Brown
N — Black
R — Red
V — Green
Z — Violet
RV — Red — green
AR — Blue — red
AN — Blue — black
BN — White — black
BR — White — red
GN — Yellow — black
HN — Grey — black
HR — Grey — red
NZ — Black — violet
RN — Red — black
SN — Pink — black
VN — Green — black
ZB — Violet — white

Vehicle circuits, including the defogger circuit—Yugo

them to air dry. Wipe clean all cork, plastic, leather and fiber parts with a clean, lint free cloth.

Blow out all passages and jets with compressed air and be sure that there are no restrictions or blockages. Never use wire or similar tools to clean jets and valves separately to avoid accidental interchange.

Check all parts for wear or damage. If wear or damage is found, replace the defective parts. Especially check the following:

- Check the float needle and seat for wear. If wear is found, replace the complete assembly.
- Check the float hinge pin for wear and the float(s) for dents or distortion. Replace the float if fuel has leaked into it.
- Check the throttle and choke shaft bores for wear or an out-of-round condition. Damage or wear to the throttle arm, shaft or shaft bore will often require replacement of the throttle body. These parts require a close tolerance of fit. Wear may allow leakage, which could affect starting and idling.

NOTE: Throttle shafts and bushings are not included in overhaul kits. They can can be purchased separately.

- Inspect the idle mixture adjusting needles for burrs or grooves. Any such condition requires replacement of the needle, since you will not be able to obtain a satisfactory idle.
- Test the accelerator pump check valves. They should pass air one way but not the other. Test for proper seating by blowing and drawing on the valve if necessary. If the valve is satisfactory, wash the valve again to remove any moisture.
- Check the bowl cover for warped surfaces with a straightedge.
- Closely inspect the valves and seats for wear and damage, replacing as necessary.
- After the carburetor is assembled, check the choke valve for freedom of operation.

Carburetor overhaul kits are recommended for each overhaul. These kits contain all gaskets and new parts to replace those that deteriorate most rapidly. Failure to replace all parts supplied with the kit (especially gaskets) can result in poor performance later.

After cleaning and checking all components, reassemble the carburetor, using new parts and referring to the exploded view. When reassembling, make sure that all screws and jets are tight in their seats, but do not overtighten, as the tips will be distorted. Tighten all screws gradually, in rotation. Do not tighten needle valves into their seats; uneven jetting will result. Always use new gaskets. Be sure to adjust the float level when reassembling.

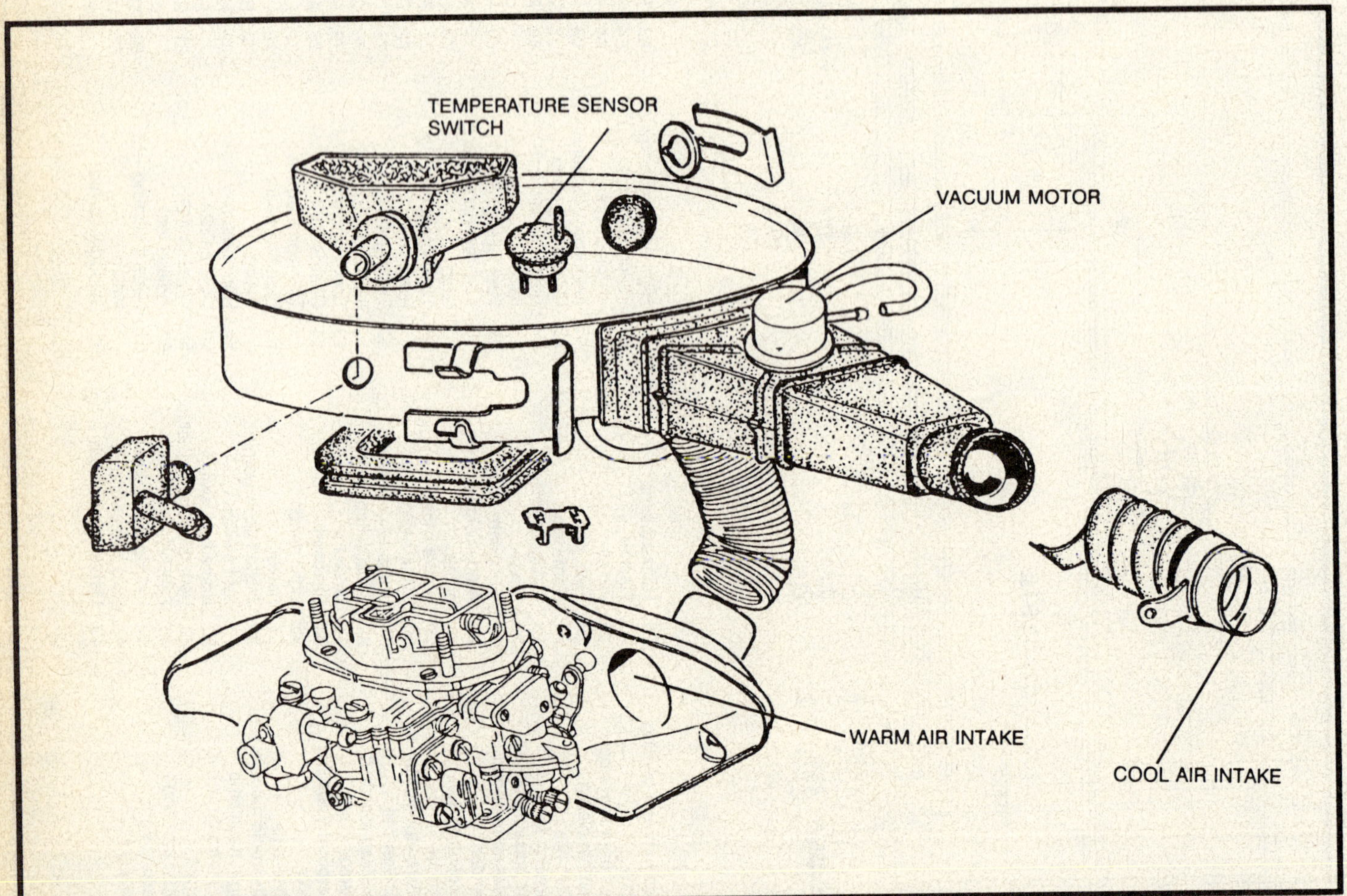

Air cleaner and related components

Glossary

SECTION 19

ABBREVIATIONS AND DEFINITIONS

A/C—Air Conditioning.

A/F—Air/Fuel Ratio.

A/T—Automatic Transmission.

AAC—Auxiliary Air Control Valve.

AAV—Anti Afterburning Valve.

ACC—Air Conditioner Clutch Compressor.

ACT—Air Charge Temperature Sensor.

ACV—Air Control Valve.

AFC—Air Flow Controlled fuel injection.

AFS—Air Flow Sensor.

AG—Automatic Transmission.

AI—Air Injection.

AIC—Automatic Idling Control Valve.

AIR GAP—The distance or space between the reluctor tooth and pick-up coil.

AIR SENSOR—An air cone with a floating plate which measures air flow and determines plunger position on K—Jetronic type systems.

AIS—Air Injection System.

AIS—Automatic Idle Speed motor.

AIV—Air Injection Valve.

AMMETER—An electrical meter used to measure current flow (amperes) in an electrial circuit. Ammeter should be connected in series and current flowing in the circuit to be checked.

AMPERE (AMP)—The unit current flow is measured in Amperage equals the voltage divided by the resistance.

APC—Automatic Performance Control.

ARMATURE—See Reluctor for definition.

ASC—Automatic stability control.

ASC+T—Automatic stability control with traction control.

ASR—Automatic Slip Reduction, part of traction control system.

ATS—Air Temperature Sensor.

AUXILIARY AIR REGULATOR—A rotary gate valve which stabilizes idle speed during engine warm-up.

AWG—American Wire Gauge system.

BAC—Bypass Air Control system.

BAR—Unit of pressure (1 bar is approximately 14.5 psi).

BID—Breakerless Inductive Discharge ignition system.

BOB—Ford Breakout Box. Device which connects in series with the computer and the EEC-IV harness and permits measurements of the processor inputs and outputs.

BP—Barometric Pressure.

BPA—ByPass Air solenoid. Used to control idle speed on Ford EFI and SEFI engines

BPA—Bypass Air Valve.

BPCSV—Bypass Control Solenoid Valve.

BREAKOUT BOX—Ford Breakout Box. Device which connects in series with the computer and the EEC-IV harness and permits measurements of the processor inputs and outputs.

BTDC—Before Top Dead Center.

CALIBRATION ASSEMBLY—A memory module that plugs into an on—board computer and contains instructions for engine operation.

CANISTER—A container in an emission control system that contains charcoat to trap fuel vapors from the fuel system.

CANP—Canister Purge solenoid that opens the fuel vapor canister line to the intake manifold when energized.

CAPACITOR—A device which stores an electrical charge.

CATALYST—Special metals such as platinum and palladium that are contained within a catalytic converter. The catalyst contacts the hot exhaust gases and promotes more complete combustion of unburned hydrocarbons and reduction of carbon monoxide.

CC—Catalytic Converter.

CCO—Converter Clutch Override.

CFC—Coasting Fuel Cut.

CHM—Cold Mixture Heater.

CIS—Continuous Injection System, Bosch K-Jetronic type system.

CIS-E—Electronic Continuous Injection System, Bosch L or LH-Jetronic type system.

CIS-III—Electronic Continuous Injection System, Bosch L or LH-Jetronic type system.

CO—Carbon monoxide.

COMPUTER TIMING—The total spark advance in degrees before top dead center. Calculated by the Ford EEC-IV processor, based on sensor input.

CONDUCTOR—Any material through which an electrical current can be transmitted easily.

CONTINUITY—Continuous or complete circuit. The type of circuit that can be check with an ohmmeter.

CONTINUOUS SELF-TEST—A continuous test of the Ford EEC-IV system conducted whenever the vehicle is in operation.

CONTROL PRESSURE—The lower chamber pressure, which is controlled by the EHA, to control mixture, warm-up and deceleration air/fuel ratio.

CP—Crankshaft Position sensor.

CPU—Central Processing Unit.

CYL SENSOR—Crankshaft Angle Sensor.

DAMPING RESTRICTION—Bore of small cross section in the fuel distributor of the K—Jetronic type system. It dampens sensor plate movement in the air flow sensor during high load and low rpm conditions.

DI—Direct Ignition system, each spark plug has its own ignition coil.

DI/APC—Direct Ignition system with Automatic Performance Control, found on turbocharged vehicles.

DIELECTRIC SILICONE COMPOUND—Non—conducting silicone grease applied to spark plug wire boots, rotors and connectors to prevent arcing and moisture from entering a connector.

DIFFERENTAIL PRESSURE—The difference in pressure between the upper and lower chambers of a CIS fuel distributor.

DIFFERENTAIL PRESSURE REGULATOR—An electronically controlled valve which regulates the fuel flow to the lower chamber of the CIS fuel distributor.

DIODE—An electrical device that will allow current to flow in one direction only.

DK—Throttle Valve.

DKB—Throttle-brake intervention.

DKE—Throttle lift.

DKR—Throttle reduction.

DKV—Set throttle valve.

DME—Digital Motor Electronics; a single control unit for both fuel injection and ignition control.

DSV—Deceleration Solenoid Valve.

DWA—Anti-theft system.

E/L—Electrical Load control unit.

EACV—Electronic Air Control Valve.

ECA—Electronic Control Assembly. Ford's engine controlling computer.

ECCS—Electronic Concentrated Control System.

ECIT—Electronic Control Ignition Timing.

ECM—Electronic Control Module.

ECS—Electronic Control Injection.

ECS—Emission Control System.

ECT—Electronic Controlled Transmission.

ECT—Engine Coolant Temperature sensor.

ECU—Electronic Control Unit.

EDU—Electronic Display Unit.

EEC—Evaporative Emission Control.

EEC-IV—Electronic Engine Control design 4. A computer controlled system of engine control used on Ford.

EFC—Electronic Fuel Control.

EFE—Early Fuel Evaporation.

EFI—Electronic Fuel Injection.

EGI—Electronic Gasoline Injection.

EGO—Exhaust Gas Oxygen sensor.

EGR—Exhaust Gas Recirculation.

EGRC—EGR Control solenoid.

EGRV—EGR Control solenoid.

EGRV—EGR Vent solenoid.

EGS—Electronic transmission control.

EH—Electro-hydraulic.

EHA—Electrohydraulic actuator, same as differentail pressure regulator.

EICV—Electronic Idle Control Valve.

EIS—Electronic Ignition System which uses a reluctor and a pick-up coil along with a module to replace the ignition points and condenser.

EIS—Electronic Injection System.

EKM—Electronic vehicle module.

EKP—Electric fuel pump.

ELCD—Evaporative Loss Control Device.

ELECTROHYDRAULIC ACTUATOR (EHA)—An electronically controlled valve which regulates the fuel flow to the lower chamber of the CIS fuel distributor.

ELECTRONIC CONTROL UNIT (ECU)—On board computer module, used to control ignition, fuel or other engine functions.

EML—Electric throttle control.

EMR—Emissions Maintenance Reminder.

EMW—Emission Maintenance Warning.

ER—Engine running, used on some Ford system tests.

ESA—Electronic Spark Advance.

ESC—Electronic Spark Control.

ESS—Engine Speed Sensor.

ETS—Electronic Throttle System, part of traction control system.

EV—Injector valve.

EVP—EGR Valve Position sensor.

EZK—Elektrische Zundkontrolle, electronically controlled ignition.

FCS—Float Chamber Ventilation system.

FCS—Fuel Control Solenoid.

FICB—Fast Idle Cam Breaker.

FOM—Fix operating Mode, limp—in mode.

FPM—Fuel Pump Monitor. A circuit in the Ford EEC system used to monitor fuel pump operation.

FUEL ACCUMULATOR—Diaphragm unit which helps maintain residual fuel pressure for hot starting on CIS type fuel systems.

FUEL DISTRIBUTOR—The component which feeds fuel to the individual engine cylinders corresponding to the air flow rate metered by the air flow sensor on CIS systems.

GND or GRD—Ground or negative (−).

HALL EFFECT PICK-UP ASSEMBLY—Used to input a signal to the electronic control unit. The system operates on the Hall Effect principle whereby a magnetic field is blocked from the pick-up by a rotating shutter assembly.

HC—Hydrocarbons. Any compound composed of hydrogen and carbon, such as petroleum products, that is considered a pollutant.

HCV—Exhaust Heat Control Valve.

HEGO—Heat Exhaust Gas Oxygen sensor.

HG—Manual transmission.

HIC—Hot Idle Compensation.

HLM—Air flow meter heating filament.

HLM—Hot wire air mass sensor.

HT—High Tension.

I/O—Input/Output, for computer data transmission.

IAS—Inlet Air Solenoid.

IC—Integrated Circuit.

IDLE AIR STABILIZATION VALVE—Electronically controlled valve used to maintain idle speed at a predetermined level.

IG—Ignition.

IGNITER—Term used by Japanese automotive and ignition manufacturers for the electronic control unit or module.

IGNITION COIL—Step up transformer consisting of a primary and a secondary winding with an iron core. As the current flows in the primary winding stops, the magnetic field collapses across the secondary winding inducing the high secondary voltage. The coil may be oil filled or of an epoxy type design.

IMA Sensor—Idle Mixture Adjuster Sensor.

IMS—Inferred Mileage Sensor.

INDUCTION—A means of transferring electrical energy in the form of a magnetic field. Principle used in the ignition coil to increase coltage.

INFINITY-An ohmmeter reading which indicates an open circuit in which no current will flow.

INJECTION VALVE—Same as injector.

INJECTOR-A solenoid or pressure-operated fuel delivery valve used for fuel injection systems.

INTERGRATED CIRCUIT (IC)—Electronic micro—circuit consisting of a semi—conductor components or elements made using thick—film or thin—film technology. Elements are located on a small chip made of a semi—conducting material, greatly reducing the size of the electronic control unit and allowing it to be incorporated within the distributor.

ISC—Idle Speed Control device.

ITS— Idle Tracking Switch. An input device that sends a signal to the control module to indicate throttle position.

JSV—Jet Mixture Solenoid Valve.

K-JETRONIC—Fuel induction by means of CIS fuel injection dependent upon the measurement of air intake by a mechanical air flow sensor.

KD—Kickdown.

KDLH—Kick-Down Low Hold.

KS—Knock Sensor. An input device that responds to spark knock caused by excessively advanced ignition timing.

KW—Crankshaft.

L-JETRONIC—a CIS-E fuel injection system dependent upon the measurement of air intake by an vane type electrical air flow meter and air temperature sensor. The L stands for air flow.

LED—Light Emitting Diode.

LH-JETRONIC—a CIS-E fuel injection system dependent upon the measurement of air intake by a hot wire type mass air flow sensor. The LH stand for air flow and hot wire sensor.

LIS—Limited Operation Strategy.

LL—Idle.

LLS—Idle speed actuator.

LMM—Airflow meter.

M/C—Mixture Control.

M/T—Manual Transmission.

MAF—Mass Airflow sensor. A device used to measure the amount of intake air entering the engine on some fuel injection systems.

MAP—Manifold Air Pressure sensor.

MAS—Mixture Adjust Screw.

MCSV—Mixture Control Solenoid Valve.

MCT—Manifold Charge Temperature sensor.

MCV—Mixture Control Valve.

ME—Engine intervention, ignition timing.

MICROPROCESSORS—A miniature computer on a silicone chip.

MIL—Malfunction Indicator Light.

MODULE—Electronic control unit, amplifier or igniter of solid state or integrated design which controls the current flow in the ignition primary circuit based on input from the pick-up coil. When the module opens the primary circuit, the high secondary voltage is induced in the coil.

MONITOR BOX—An optional Ford EEC-IV test device which connects in series with the EEC-IV processor and its harness, and permits measurements in various units of the processor inputs and outputs.

MPFI—Multi—Point Fuel Injection.

MPS—Motor Position Sensor.

MS—Millisecond.

MSR—Engine drag torque control.

MV—Solenoid valve.

NDS—Neutral/Drive Switch.

NGS—Neutral Gear Switch.

NOx—Nitrous Oxide. A compound formed during the engine combustion process when oxygen in the air combines with nitrogen to form photochemical smog.

NTC—Negative Temperature Coefficient Resistor.

OCC—Output Cycling Check.

OHM—The electrical unit of resistance to current flow.

OHMMETER—The electrical meter used to measure the resistance in ohms. Self—powered and must be connected to a voltage free circuit or damage to the ohmmeter will result.

OVCV—Outer Vent Control Valve.

OVERLAY CARD—A plastic card used for the Ford Breakout or Monitor box to identify connections for each engine tested.

OXYGEN SENSOR—Used with the feedback system to sense the presence of oxygen in the exhaust gas and signal the computer which can reference the voltage signal to an air/fuel ratio.

PA SENSOR— Atmospheric Pressure Sensor.

PCV—Positive Crankcase Ventilation.

PGM-FI—Programmed Fuel Injection system.

PGM-IG—Programmed Ignition system.

PHENOMENA—Basis of symptoms.

PICK-UP COIL—Inputs signal to the electronic control unit to open the primary circuit. Consists of a fine wire coil mounted around a permanent magnet. As the reluctor's ferrous tooth passes through the magnetic field an alternating current is produced, signaling the electronic control unit. Can operate on the principle of metal detecting, magnetic induction of Hall Effect. Is also referred to as a stator or sensor.

PIP—Profile Ignition Pickup.

POTENTIOMETER—A variable resistor used to change a voltage signal.

PRC—Pressure Regulator Control solenoid.

PRE-PUMP—In—tank fuel pump.

PRIMARY CIRCUIT—Is the low voltage side of the ignition system which consists of the ignition switch, ballast resistor or resistance wire, bypass, coil, electronic control unit and pick-up coil as well as the connecting wires and harnesses.

PSPS—Power Steering Pressure Switch.

PTC HEATER—Positive Temperature Coefficient Heater.

PULSE GENERATOR—Also called a pulse signal generator. Term used by Japanese and German automotive and ignition manufacturers to describe the pick-up and reluctor assembly. Generates an electrical pulse which triggers the electronic control unit or igniter.

PULSE WIDTH—The amount of time the control unit energizes the fuel injectors to spray fuel into the intake manifold, usually measured in milliseconds.

PVS—Ported Vacuum Switch. A temperature-activated switch that changes vacuum connections when the coolant temperature changes.

PWG—Pedal position sensor.

Poti—Potentiometer.

QUICK TEST—A functional diagnostic test for Ford EEC-IV system, consisting of test hookup, key on engine off, engine running and continuous self test modes.

RAD—Radiator Temperature Switch.

RAM—Random Access Memory.

RELAY—A switching device operated by a low—current circuit which controls the opening and closing of another circuit of higher current capacity.

RELUCTOR—Also called an armature or trigger wheel. Ferrous metal piece attached to the distributor shaft. Made up of teeth of which the number are the same as the number of engine cylinders. As the reluctor teeth pass through the pick-up magnetic field an alternating current is generated in the pick-up coil.

RESIDUAL PRESSURE—Pressure remaining in the fuel system after the engine has been shut off.

RESISTANCE—The opposition to the flow of current through a circuit or electrical device, and is measured in ohms. Resistance is equal to the voltage divided by the amperage.

ROM—Read Only Memory.

RXD—Receive data line.

RxD—Diagnosis initiation line.

SAS—Speed Adjusting Screw.

SCSV—Slow Cut Solenoid Valve.

SECONDARY—The high voltage side of the ignition system, usually above 20,000 volts. The secondary includes the ignition coil, coil wire, distributor cap and rotor, spark plug wires and spark plug.

SELF-TEST—One of the 3 subsets of the Ford EEC-IV quick test modes.

SENSOR—Also called the pick-up coil or stator. See pick-up coil for definition.

SENSOR PLATE—A round plate bolted to the air flow sensor lever which floats in the stream of intake air on the CIS type systems.

SG—Control unit.

SHUTTER—Also called the vane. Used in a Hall Effect distributor to block the magnetic field from the Hall Effect pick-up. The shutter is attached to the rotor and is grounded to the distributor shaft.

SOLENOID—A wire coild with a movable core which changes position by means of electromagnetism when current flows through the coil.

SPARK DURATION—The length of time measured in milliseconds (1/1000th second) the spark is established across the spark plug gap.

SPFI—Single Point Fuel Injection.

SPOUT—Spark Output.

STAR—Self Test Automatic Readout (Ford), used to access codes from the EEC-IV processor.

STATOR—Another name for a pick-up coil. See pick-up coil for definition.

STI—Self Test Input (Ford) circuit in the EEC system. Used to place the computer into testing mode.

STO—Self Test Output (Ford) circuit in the EEC system. Used by the computer to send testing and fault codes to tester.

SWITCHING TRANSISTOR—Used in some electronic ignition systems, it acts as a switch for high current in response to a low voltage signal applied to the base terminal.

SYSTEM PRESSURE—Fuel pressure in the CIS fuel distributor controlled by the pressure regulator.

TA SENSOR—Intake Air Temperature Sensor.

TAB—Thermactor Air Bypass solenoid.

TAD—Thermactor Air Diverter Solenoid.

TAS—Throttle Adjust Screw.

TCA—Thermostat Controlled Air Cleaner.

TCP—Temperature Compensated Accelerator Pump.

TCS—Traction Control System.

TD—Engine speed signal.

TDC—Top Dead Center.

TE—Fuel evaporation control.

TFI—Thick Film Ignition module. Controls the coil and ignition operation on most Ford vehicles.

THERMISTOR—A device that changes its resistance with temperature.

THERMO-TIME SWITCH—A switch which interupts the electrical circuit of the cold start injector based on temperature and time.

ti—Injection timing.

TK or TKS—Throttle Kicker Solenoid. An actuator moves the throttle linkage to increase idle rpm.

tL—Load signal.

TL—Partial load.

TMOT—Engine temperature sensor.

TP or TPS—Throttle Position Sensor.

TPI—Tuned Port Injection.

TR—Crankshaft reference signal.

TRANSFER PUMP—Fuel pump located in the fuel tank, usually used with a 2 pump system.

TRANSISTOR—A semi—conductor component which can be actuated by a small voltage to perform an electrical switching function.

TRIGGER WHEEL—See Reluctor for definition.

TSI—Timing Service Instrument.

TVSV—Thermostatic Vacuum Switching Valve.

TW SENSOR—Coolant Temperature Sensor.

TWC—Three-way catalyst, sometimes referred to as a dual catalytic converter. Combines 2 catalytic converters in one shell to control emissions of NOx, HC and CO.

TWSV—Three Way Solenoid Valve.

TXD—Transmitted data line.

TxD—Diagnostic data line.

U-Batt—Battery voltage.

U-Vers—Voltage supply.

VAF—Vane Airflor Meter.

VAT—Vane Air Temperature.

VB VOLTAGE—Battery Voltage.

VCM—Vehicle Condition Monitor.

VCV—Vacuum Control Valve.

VECI—Vehicle Emission Control Information label.

VF VOLTAGE—Battery Voltage.

VIS—Variable Induction System.

VL—Full load.

VLC—Vacuum Sensor and Vacuum Line Charging solenoid valve.

VM—Vane Meter.

VOLT—The unit of electrical pressure or electromotive force.

VOLTAGE DROP—The difference in voltage between one point in a circuit and another, usually across a resistance. Voltage drop is measured in parallel with current flowing in the circuit.

VOLTMETER—An electrical meter used to measure voltage in a circuit. Voltmeters must be connected in parallel across the load or circuit.

VPWR—Vehicle Power supply voltage. 10-14 voltages DC.

VREF—Reference Voltage supplied by the EEC-IV processor to some sensors, 4-6 volts DC.

VREF—The reference voltage or power supplied by the computer control unit to some sensors regulated at a specific voltage.

VSS—Vehicle Speed Sensor.

VSV—Vacuum Switching Valve.

VTV—Vacuum Transmitting Valve.

VVC—Variable Voltage Choke.

WK—Torque converter clutch.

WOT—Wide Open Throttle switch.

ZAB—Ignition fade-out.